Recent Advances in
Computational Mechanics and Simulations

VOLUME I

Recent Advances in
Computational Mechanics and Simulations

VOLUME I

**Proceedings of the
Second International Conference on
Computational Mechanics and Simulations (ICCMS-06)
held at Indian Institute of Technology, Guwahati, India,
8-10 December 2006**

Editors

Damodar Maity

S.K. Dwivedy

I.K. International Publishing House Pvt. Ltd.

NEW DELHI   •   MUMBAI   •   BANGALORE

Published by
I.K. International Publishing House Pvt. Ltd.
S-25, Green Park Extension
Uphaar Cinema Market
New Delhi 110 016 (India)
E-mail: ik_in@vsnl.net

Branch Offices:
A-6, Royal Industrial Estate, Naigaum Cross Road
Wadala, Mumbai 400 031 (India)
E-mail: ik_mumbai@vsnl.net

G-4 "Embassy Centre", 11 Crescent Road
Kumara Park East, Bangalore 560 001 (India)
E-mail: ik_bang@vsnl.net

ISBN 81-89866-16-8 (Set)
ISBN 81-89866-17-6 (Vol. I)
ISBN 81-89866-18-4 (Vol. II)

© 2007 I.K. International Publishing House Pvt. Ltd.

Published by Krishan Makhijani for I.K. International Publishing House Pvt. Ltd., S-25, Green Park Extension, Uphaar Cinema Market, New Delhi 110 016. Printed by Rekha Printers Pvt. Ltd., Okhla Industrial Area, Phase II, New Delhi 110 020.

Preface

For a country to become an economic leader – and to remain so – research in science and technology must be at the forefront of its agenda. India, being a developing country, needs to make concerted efforts towards the goal of transforming itself into a developed nation. Conferences/ symposia/workshops, where specialized minds exchange ideas and arrive at significant conclusions, which set the future directions for research, contribute immensely towards realizing this goal. Problems in mechanics of solid, fluid, solid-fluid interaction and their scientific disciplines are investigated through computational mechanics and implemented in engineering design and manufacturing. Computational mechanics has emerged as a key area of research and application.

The Indian Association for Computational Mechanics (IndACM) was founded on 1 January 2000 to bring together this community to have meaningful interaction to further the growth of computational mechanics in different disciplines. To achieve this objective the Second International Congress on Computational Mechanics and Simulation (ICCMS-06) is being organized at the Indian Institute of Technology Guwahati, Guwahati, India, during on 8-10 December 2006. The purpose of this Congress is to provide a forum for scientists, engineers and designers in universities, laboratories and industry to share their research findings to further the cause of computational mechanics.

We are overwhelmed with response to the call for contributed papers in different fields of engineering. After peer reviewing, 270 contributed full-length papers have been accepted for presentation encompassing three broad areas of Computational Solid Mechanics, Fluid Mechanics and Multi Physics. Also there will be 41-invited lectures from eminent scientists and researchers from all over the world out of which 26 full-length papers are included in this proceeding. These manuscripts are kept in two volumes of the proceedings. In Volume I, 13 invited papers and 137 contributed papers in the field of composite structures, fracture and damage mechanics, vibration and control, smart structures, electro-magnetic, nonlinear mechanics, structural analysis and new computational techniques for solid structures are given. Volume II includes 13 invited papers and 133 contributed papers in the field of fluid flow through moving boundaries, channels and pipes, turbulence modeling, hydrodynamics, multiphase flow, wave dynamics and tsunamis, fluid structure interaction, Porous media mechanics, New computational techniques in fluids, Transient dynamic problems, heat transfer, geomechanics, bridge and dam structures, biomechanics, micro and nanomechanics.

We are grateful to all the authors for sharing their technical expertiese by contributing to the proceedings. We thank all the members of the technical committee and reviewers for spending their valuable time to review the contributed papers. We thank our students C. Saran, P. Srujana, V. Dinesh Kumar, Y. Srinivas, Barun Pratiher, M. Sivasankar for their assistance in the process of finalizing the proceedings. We sincerely thank all the sponsors viz., ABAQUS Engineering India (P) Ltd, Council for Scientific and Industrial Research, Department of Science and Technology, Board of Research in Nuclear Sciences, Indian National Science Academy for funding this conference. We sincerely thank Professor Tarun Kant, Founder President of IndACM for his kind help, encouragement to make this conference. Last but not the least, we thank one and all who helped to make this congress successful in all aspects.

IIT Guwahati

October, 2006

Damodar Maity

S.K. Dwivedy

List of Reviewers

1. Arun Narasimhan, Department of Mechanical Engineering, IIT Madras
2. Ayothiraman R., Department of Civil Engineering, IIT Guwahati
3. Balakrishnan, N, Department of Aerospace Engineering, IISc Bangalore
4. Bhattacharya, Bishakh,, Department of Mechanical Engineering, IIT Kanpur
5. Bhattacharyya, R, Department of Mechanical Engineering, IIT Kharagpur
6. Bhatti, T. S., Center for Energy Studies, IIT Delhi
7. Chakraborty Debabrata, Department of Mechanical Engineering, IIT Guwahati
8. Chakraborty Goutam, Department of Mechanical Engineering, IIT Kharagpur
9. Chandiramani Naresh K., Department of Civil Engineering, IIT Bombay
10. Chandramouli, P, Department of Mechanical Engineering, IIT Madras
11. Chatterjee Dhiman, Department of Mechanical Engineering, IIT Madras
12. Dass Anoop K., Department of Mechanical Engineering, IIT Guwahati
13. Das Manab Kumar, Department of Mechanical Engineering, IIT Guwahati
14. Dash S.K. Department of Civil Engineering, IIT Guwahati
15. Date, A. W., Department of Mechanical Engineering, IIT Bombay
16. Dewan Anupam, Department of Mechanical Engineering, IIT Guwahati
17. Dixit Uday S., Department of Mechanical Engineering, IIT Guwahati
18. Dutta S., Department of Civil Engineering, IIT Guwahati
19. Dwivedy S. K., Department of Mechanical Engineering, IIT Guwahati
20. Gaitonde, U. N., Department of Mechanical Engineering, IIT Bombay
21. Ganesan N., Department of Mechanical Engineering, IIT Madras
22. Ganesan, V, Department of Mechanical Engineering, IIT Madras
23. Gokhale Sharad, , Department of Civil Engineering, IIT Guwahati
24. Gosami, P, Centre for Mathematical Modeling and Computer, Bangalore
25. Gupta Ashok, Department of Civil Engineering, IIT Delhi
26. Gupta, K, Department of Mechanical Engineering, IIT Delhi
27. Hirani Harish, Department of Mechanical Engineering, IIT Bombay
28. Jha Pradeep K., Department of Mechanical Engineering, IIT Guwahati
29. Kapuria, S, Department of Applied Mechanics, IIT Delhi
30. Kishore, N. N, Department of Mechanical Engineering, IIT Kanpur
31. Krishna Murty, K. S. R., Department of Mechanical Engineering, IIT Guwahati
32. Kurian Job, Department of Aerospace Engineering, IIT Madras
33. Lyngdoh T., Department of Civil Engineering, IIT Guwahati
34. Maity D., Department of Civil Engineering, IIT Guwahati
35. Maity Soumen, Department of Mathematics, IIT Guwahati
36. Maiti, S. K., Department of Mechanical Engineering, IIT Bombay

37. Muralidhar K, Department of Mechanical Engineering, IIT Kanpur
38. Muthukumar, P.,Department of Mechanical Engineering, IIT Guwahati7
39. Nageswara rao, Department of Civil Engineering, IIT Madras
40. Naik, N K, Department of Aerospace Engineering, IIT Bombay
41. Natesan Srnivasan, Department of Mathematics, IIT Guwahati
42. Palaninathan, R, Department of Civil Engineering, IIT Madras
43. Pandey Manmohan, Department of Mechanical Engineering, IIT Guwahati
44. Pathak Pushparaj Mani, Dept. of Mechanical and Industrial Engg, IIT Roorkey
45. Paul, P J, Department of Aerospace Engineering, IISc Bangalore
46. Ravi, M, R., Department of Mechanical Engineering, IIT Delhi
47. Saha Ujjwal K., Department of Mechanical Engineering, IIT Guwahati
48. Sahoo Niranjan, Department of Mechanical Engineering, IIT Guwahati
49. Saravana Kumar G, Department of Mechanical Engineering, IIT Guwahati
50. Saravanan C.G, Department of Mechanical Engineering, Annamali University
51. Sarma A.K., Department of Civil Engineering, IIT Guwahati
52. Senthilvelan, S, Department of Mechanical Engineering, IIT Guwahati
53. Seshu, P, Department of Mechanical Engineering, IIT Bombay
54. Singh B., Department of Civil Engineering, IIT Guwahati
55. Sing, B N, Department of Aerospace Engineering, IIT Kharagpur
56. Singh K.D., Department of Civil Engineering, IIT Guwahati
57. Singh Satinder Paul, Department of Mechanical Engineering, IIT Delhi
58. Singh Yogendra, Department of Earthquake Engineering, IIT Roorkey
59. Siva Prasad N, Department of Mechanical Engineering, IIT Madras
60. Sivakumar M S, Department of Applied Mechanics, IIT Madras
61. Sivasambu M., Department of Mechanical Engineering, IIT Kanpur
62. Sreedeep S., Department of Civil Engineering, IIT Guwahati
63. Srinivasan, K, Department of Mechanical Engineering, IISc Bangalore
64. Srithar, K, Dept. of Mechanical Engineering, Thiyagarajar College of Engineering
65. Suresh Kumar G., Department of Civil Engineering, IIT Guwahati
66. Talukdar S., Department of Civil Engineering, IIT Guwahati
67. Tarun Kant, Department of Civil Engineering, IIT Bombay
68. Tiwari Rajiv, Department of Mechanical Engineering, IIT Guwahati
69. Velmurugan, R, Composite Technology Center, IIT Madras
70. Vengatesan, S, Department of Applied Mechanics Engineering, IIT Madras
71. Venkitanarayanan P, Department of Mechanical Engineering, IIT Kanpur
72. Verma A., Department of Civil Engineering, IIT Guwahati
73. Vyas Nalinaksh S, Department of Mechanical Engineering, IIT Kanpur

Message

Indian Association for Computational Mechanics (IndACM) was founded on 1st January 2000 with its headquarters at the Indian Institute of Technology Bombay (IITB) with the objectives, viz.,

- to bring together men and women having a common interest in computational mechanics and not necessarily belonging to the same discipline,
- to liaise with other bodies having common interest (such as International Association for Computational Mechanics (IACM), Asian-Pacific Association for Computational Mechanics (APACM) and such other national/regional/international bodies),
- to have free exchange of views in the area of computational mechanics,
- to organize meetings/ seminars/ conferences/ conventions/ symposia in the area of computational mechanics, and
- to engage in other activities meant to promote development of computational mechanics as a branch of engineering science, to mention a few.

The first call through e-mail enlisted over twenty-five as founder life members. Thereafter, there has been slow but steady increase in its membership. Most of these members have joined the Association through the word of mouth. IndACM became an affiliate of the International Association for Computational Mechanics (IACM) in January 2001. Through this arrangement, IndACM members can join the international body by paying an annual membership fee of US$10 only as against US$25 by routing their application through the IndACM headquarters. This provides an international exposure to our esteemed members and these initiatives prove mutually beneficial to Indian computational mechanics as well as the international community.

It was felt by many that the state of research culture in engineering science, more particularly in the area of computational mechanics, which has established itself as an enviable distinct discipline in engineering science in the second half of the last four decades after the advent of digital computers which led to development of computational techniques for solution of real-life complex problems, in the country was not encouraging at all, seeing the investments that have been made both by the government immediately after the country became independent and later by private initiatives in technical education and research. There are presently an Indian Institute of Science (IISc) at Bangalore, seven Indian Institutes of Technology (IITs) at Kharagpur, Mumbai, Chennai, Kanpur, Delhi, Guwahati and Roorkee, 20 National Institutes of Technology (NITs), state government funded engineering colleges/ institutes and very large number of engineering colleges/ institutes set up by private initiatives throughout the country. However, both the quality and the quantity of research output is dismal when we see our neighbour China, which opened its doors to the outside world not long ago and in addition it has the disadvantage of the proficiency of its academicians/ researchers in the international language of technical communication—English. Something is needed to be done urgently if our academicians/ researchers wanted to be seen in the international arena. Quality research must be emphasized and encouraged by the managers of our science and technology temples. To meet this specific end, it was thought appropriate to organize meeting of active young and also not so young academicians/ researchers biennially at International Congress of

Computational Mechanics and Simulation (ICCMS) with a view to facilitate interaction amongst researchers at one-to-one level and present their research results at such an august gathering and finally to motivate them to publish the well-tested and thoroughly discussed results of their research in reputed peer refered journals. Thus, ICCMS was conceived just as a platform to enable our researchers to improve their research output and quality so that they perform better at international level through their quality research papers.

The association's first congress, the ICCMS-04 was held at the Indian Institute of Technology Kanpur (IITK) on 9-12 December 2004 under the joint convenorship of Professors N.G.R. Iyengar and Ashwini Kumar. There was good participation from both within and outside the country. I am happy to say that both Dr. Damodar Maity and Dr. Santosh Kumar Dwivedy of the Indian Institute of Technology Guwahati (IITG) came forward to organize the second congress, the ICCMS-06 at this Institute on 8-10 December 2006. There has been overwhelming response to this meeting and the credit goes to the convenors and I would like to congratulate them for their untiring efforts.

I am confident that IndACM will rise to great heights of achievement as it moves forward and I wish ICCMS-06 a great success.

Tarun Kant
Ph.D, FNA, FA.Sc., FNAE
Founder, IndACM

Website: indacm.civil.iitb.ac.in

Contents

VOLUME I

Part III: Fracture and Damage Mechanics   264-403

Part IV: Fracture and Damage Analysis for Composite Structures 404-445

Part VII: Smart Structures   642–691

Part VIII: Electro-Mechanics/Magnetics   692–740

Part XI: Computational Techniques in Structures   **933-1088**

Part XIV: Turbulence Modeling   1281-1338

Part XXI: Transient Dynamic Problems   1721-1800

Part XXII: Heat Transfer   1801-1883

1

Distortional Buckling in Monosymmetric I-Beams

ASHWINI KUMAR AND AVIK SAMANTA

Department of Civil Engineering, Indian Institute of Technology Kanpur-208016, India
email: ashwini@iitk.ac.in

ABSTRACT

The paper deals with distortional buckling in three types of monosymmetric I-beams: (i) simply supported, (ii) propped cantilever, and (iii) braced cantilever, under three types of load: a point load, a uniformly distributed load and a uniform moment. ABAQUS is used to obtain results. It is found that for relatively short beams the buckling is governed by distortion of the web. Moment modification factors are calculated, which account for the distortion of web, and these are compared with those based on SSRC Guidelines. It is seen that the provisions in SSRC Guidelines may seriously overestimate the critical load for relatively short span beams. For braced cantilever beams, the effect of different types of bracing and their position along the length of the beam is also investigated.

Keywords: Buckling, beams, distortion.

1. INTRODUCTION

I-section beams with loads in its stiffer principle plane may buckle in the *lateral-torsional* mode or in the *local buckling* mode. Sometimes these two buckling modes interact and beam may buckle in a new mode called the *distortional buckling* mode. In this mode, the web distorts and allows flanges to deflect laterally with comparatively little twist, and thus reduces the effective torsional resistance of the member. This is more significant in plate girders with slender, un-stiffened webs than in I-sections with comparatively stocky webs.

2. PRESENT INVESTIGATION

In this paper, distortional buckling of simply-supported, propped-cantilever and braced-cantilever I-beams are studied for three types of load; a central point load, a uniformly distributed load and a uniform moment, for different degrees of mono-symmetry of beams. ABAQUS is used in the present investigation. The following beam sections (Table 1) are used for the finite element modeling:

Table 1. Beam Sections

Sec. No.	ρ	h (mm)	B_T (mm)	T_T (mm)	B_B (mm)	T_B (mm)	t (mm)
1	0	600	0	0	210	20	12
2	0.1	600	100	20	210	20	12
3	0.3	600	150	23	210	20	12
4	0.5	600	210	20	210	20	12
5	0.7	600	210	20	150	23	12
6	0.9	600	210	20	100	20	12
7	1	600	210	20	0	0	12

In Table 1 the quantity ρ is a measure of the beam mono-symmetry and is defined as:

$$\rho = \frac{I_{YT}}{I_{YT}+I_{YB}} = \frac{I_{YT}}{I_Y} \qquad ...(1)$$

in which I_{YT}, I_{YB} and I_Y are the second moment of area about the section y-axis of the top flange, the bottom flange and the whole section, respectively. For an equal flange section $\rho = 0.5$, for a T- section $\rho = 1.0$ and for an inverted T- section $\rho = 0$.

The size of the flanges is changed to vary ρ, the degree of mono-symmetry of the cross-section. One of the flanges is fixed at 210 mm × 20 mm while the size of the other flange is varied. The web thickness is kept 12 mm. The ratio ρ varies from 0 to 1.0.

For the finite element modeling, the flanges and webs are modeled with quadratic shell elements S8R5 (*eight-node shell element with five degrees of freedom per node*). These elements are chosen because they give accurate results for thin shells in large rotations, small-strain simulations. Young's modulus E = 206.6 GPa and Poisson's ratio ν = 0.3 are used. The results have been validated with those of Ma and Hughes [1] for a symmetric I-section, i.e., for $\rho = 0.5$.

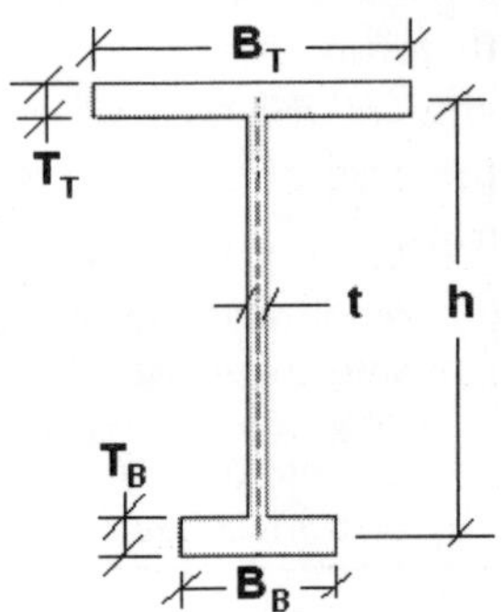

Fig. 1. Beam Cross-section.

3. MOMENT MODIFICATION FACTORS

The concept of a Moment Modification Factor, C_b, is quite common in providing solutions for buckling in beams to account for different load cases; this factor is applied to the solutions with uniform moment loading. The SSRC Guidelines [2] recommend the use of expression given by Eq. (2) for beams under single and reverse-curvature bending:

$$C_b = \frac{12.5M_{max}}{2.5M_{max}+3M_A+4M_B+3M_C}(1.4^{2\xi/h})(0.5+2\rho^2); \; 0.1 \le \rho \le 0.9 \qquad ...(2)$$

The term $(0.5 + 2\rho^2)$ in the above equation is a modifier to account for reverse-curvature bending on monosymmetric shapes; this modifier equals 1.0 for single-curvature bending. The factor $(1.4^{2\xi/h})$ in Eq. (2) accounts for the effect of load height. For results of this investigation, the C_b factor is calculated by dividing the maximum moment at buckling by the moment capacity of the beam subjected to a uniform moment; that is:

$$C_{b(ABAQUS)} = \frac{M_{cr\left(\substack{moment \\ gradient}\right)}}{M_{cr\left(\substack{uniform \\ moment}\right)}} \qquad ...(3)$$

The value of the $M_{cr\ (moment\ gradient)}$ in Eq. (3) is the maximum moment within the unbraced length; Here, $M_{cr\ (uniform\ moment)}$ is the moment at buckling in the flexural-torsional mode, while $M_{cr(moment\ gradient)}$ is for distortional buckling of the beam.

4. SIMPLY SUPPORTED BEAMS

In Figs. 2-4 results are presented in terms of non-dimensional quantities. Flexural-torsional buckling loads are obtained using the expressions suggested by Kitipornchai *et al.* [3]. $\bar{K}$ is the non dimensional beam parameter and the load is non-dimensionalized as

$$P\frac{L^2}{4\sqrt{EI_yGJ}} \quad \text{(point load)} \qquad \qquad ...(4)$$

$$M\frac{L}{\sqrt{EI_yGJ}} \quad \text{(uniform moment)} \qquad \qquad ...(5)$$

For long beams (in the present case, $0<\bar{K}<1.2$) critical loads corresponding to flexural-torsional buckling and distortional buckling are almost same. However, in the case of short beams ($\bar{K}>1.2$), distortional buckling is predominant. Trends in results for a uniformly distributed load are also similar. It is also seen that critical load increases for bottom flange loading. On the other hand, for the case of uniform moment (Fig. 4), it is observed that the critical loads remain, more or less, the same for lateral-torsional buckling as well as for distortional buckling.

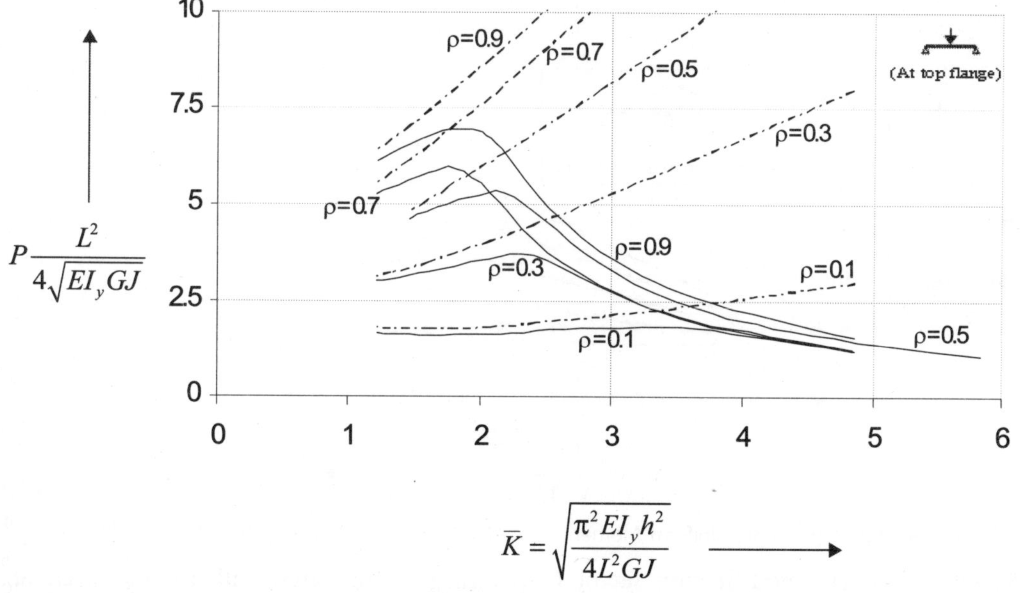

Fig. 2. Variation in Buckling Loads: Point Load at Top Flange [distortional (———) and torsional (- - - - -)].

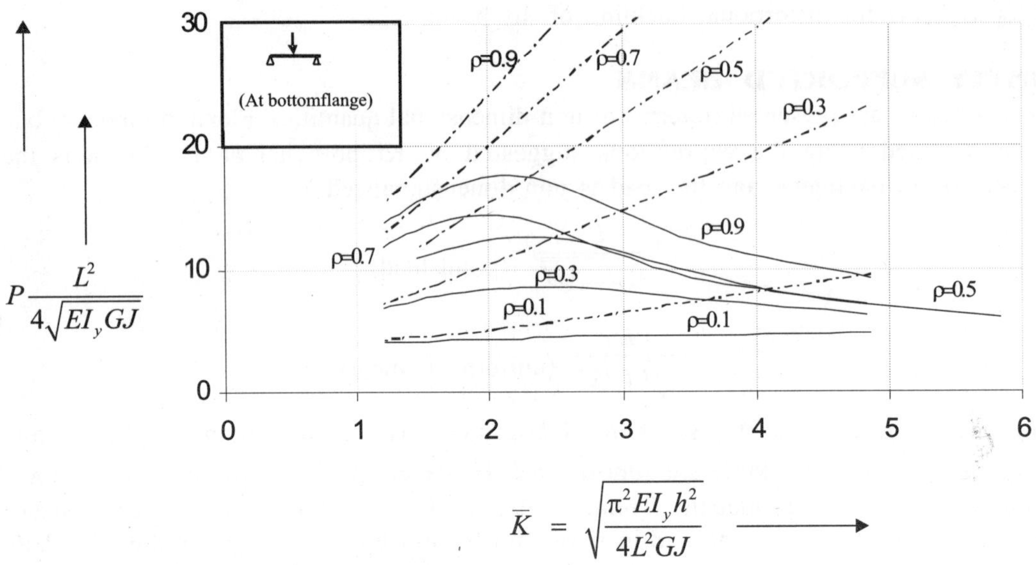

$$P\frac{L^2}{4\sqrt{EI_y GJ}}$$

$$\overline{K} = \sqrt{\frac{\pi^2 EI_y h^2}{4L^2 GJ}}$$

Fig. 3. Variation in Buckling Loads: Point Load at Bottom Flange [distortional (———) and torsional (`------`)].

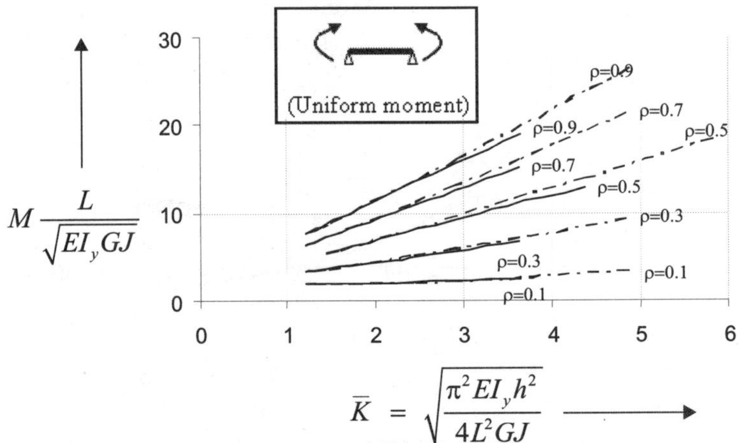

$$M\frac{L}{\sqrt{EI_y GJ}}$$

$$\overline{K} = \sqrt{\frac{\pi^2 EI_y h^2}{4L^2 GJ}}$$

Fig. 4. Variation in Buckling Loads: Uniform Moment [distortional (———) and torsional (`------`)].

In Figs. 5 and 6, moment modification factor $C_{b\ (ABAQUS)}$ is compared with the C_b based on SSRC Guidelines. Firm lines are for ABAQUS results and dashed lines correspond to the solution based on SSRC Guidelines. It is found that the effect of distortional buckling is more pronounced in short beams and when the compression (top) flange is bigger than the tension (bottom) flange ($\rho > 0.5$). Results are similar for the point and the uniformly distributed load. Details of the results can be seen in the recent work of Samanta and Kumar [4].

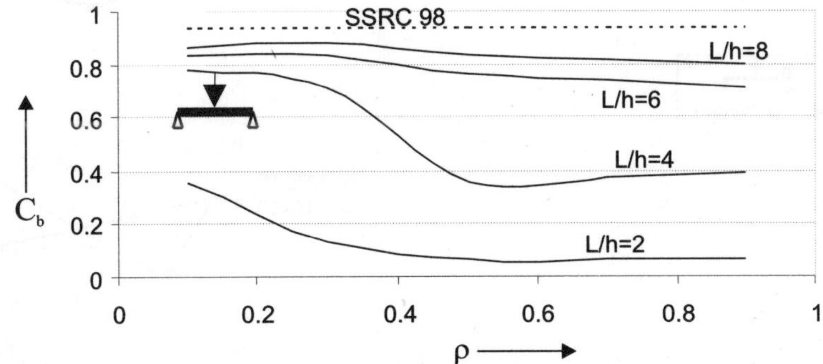

Fig. 5. Comparison of Moment Modification Factors C_b with ρ :
Point Load at Top Flange [$\rho = 0 : \perp$, $\rho = 0.5$: I, $\rho = 1$: T].

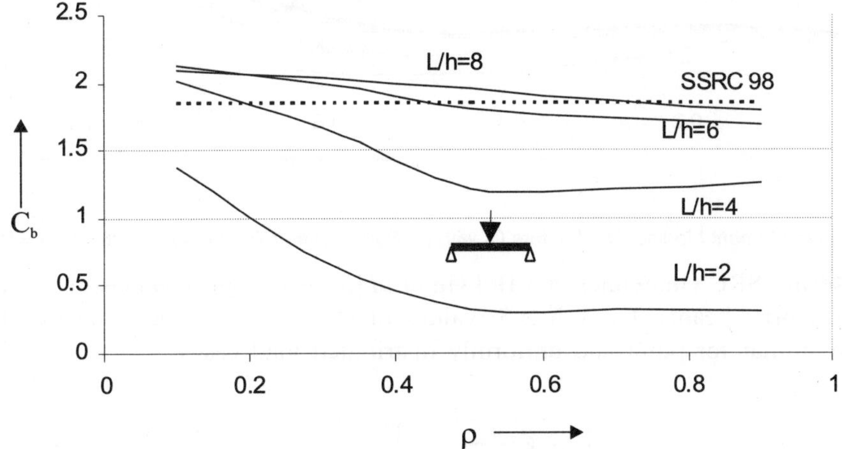

Fig. 6. Comparison of Moment Modification Factors C_b with ρ :
Point Load at Bottom Flange [$\rho = 0 : \perp$, $\rho = 0.5$: I, $\rho = 1$: T].

5. PROPPED CANTILEVER BEAMS

The buckling of beams subjected to reverse-curvature bending (in the present case, a propped cantilever beam) is more complex than the case of a single-curvature bending, because both of the flanges remain under compression at different locations along the beam length. In Fig. 7 and Fig. 8, the moment modification factor $C_{b(ABAQUS)}$, which is based on distortional buckling, is compared with the C_b based on Eq. (2) and with recommendations of Helwig *et al.* [5] which are based on lateral-torsional buckling analysis. For obtaining $M_{cr(uniform\ moment)}$, which is the moment at buckling in the flexural-torsional mode, same elements and the process as stated in section 2 are used. In addition, closely spaced transverse stiffeners are used to avoid cross-sectional distortion. Shell elements are used to model transverse stiffeners, which are spaced at approximately half the girder depth.

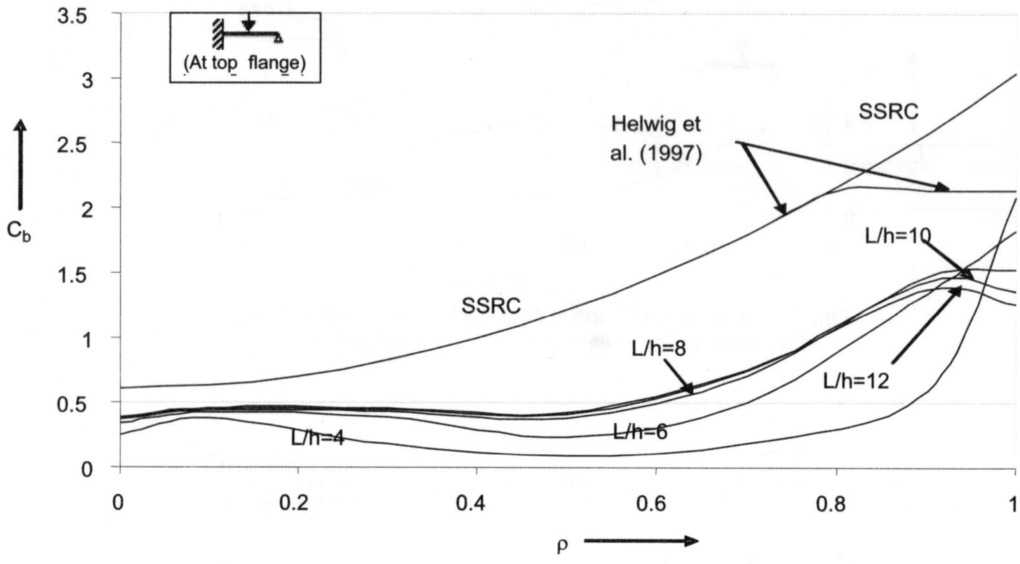

Fig. 7. Comparison of Moment Modification Factors C_b with ρ : *Point Load at Top Flange* [ρ = 0 : ⊥, ρ = 0.5: I, ρ = 1: T].

As can be seen, SSRC Guidelines and Helwig *et al.*[5]. give higher values of C_b. The deviation is too much for shorter beams. For L/h ≥ 8, values of C_b are almost the same for all load cases. Results are also similar for point and uniformly distributed load cases.

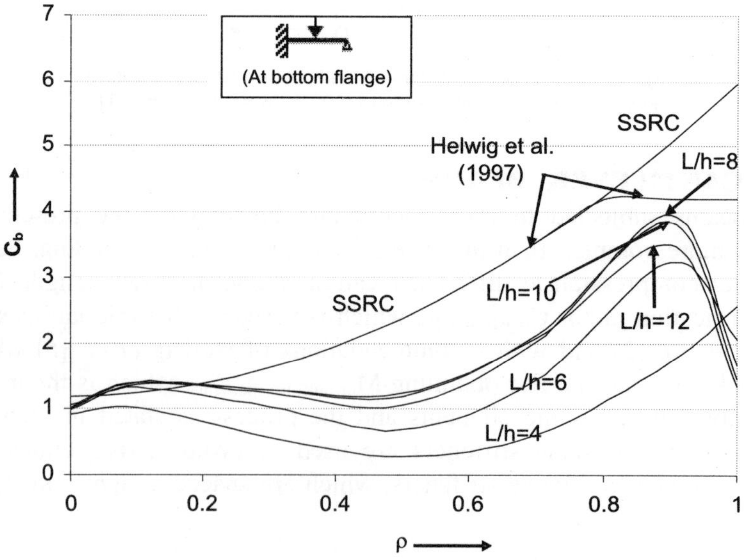

Fig. 8. Comparison of Moment Modification Factors C_b with ρ : *Point Load at Bottom Flange* [ρ = 0 : ⊥, ρ = 0.5: I, ρ = 1: T].

6. BRACED CANTILEVER BEAMS

Three types of lateral bracing are considered along the span of the cantilever beam: (a) lateral translational bracing at top flange only (T-bracing), (b) lateral translational bracing at bottom flange only (B-bracing) and (c) lateral translational bracing at top & bottom flanges (TB-bracing).

6.1 Point Load at End and Uniformly Distributed Load

6.1.1. T-bracing: Generally, for the top flange loadings, T-bracing helps most when placed at the free-end for the point load (Fig. 6) and at about 0.8L from the fixed end for the distributed load. For r = 1 (T-section), the location of the T-bracing has no effect on the buckling capacity of the beam. For bottom flange loadings (Fig. 7) the position of bracing has practically no effect on a beam of given section, except for r = 0 (inverted T-section), where the bracing is effective at about 0.8L from the fixed end.

6.1.2. B-bracing: Generally, for top flange loadings the increase in buckling load is maximum when the bracing is towards the free-end (in case of point load) (Fig. 8) and at about 0.7L from the fixed end for the distributed load. For bottom flange loadings (Fig. 9) the bracing practically does not help in increasing the buckling capacity, except for r = 0 (inverted T-section), where the bracing is effective towards the free-end.

6.1.3. TB-bracing: For top flange loadings (Fig. 10), the bracing is more effective at the tip for the point load and at about 0.8L from the fixed end for the distributed load for r = 0. In the case of bottom flange loadings (Fig. 11), the bracing is more effective for beams having larger bottom flange, like an inverted T-section beam. T-sections (r = 1.0) are most ineffective for both load cases.

6.2 End Moment

The T-bracing is more effective for beam sections with larger bottom flange. Generally (for sections with ρ = 0.3-0.7), the B-bracing is effective when it is placed near the free end of the cantilever. In general, the TB-bracing is more desirable towards the tip of the beam for ρ = 0.3-0.7.

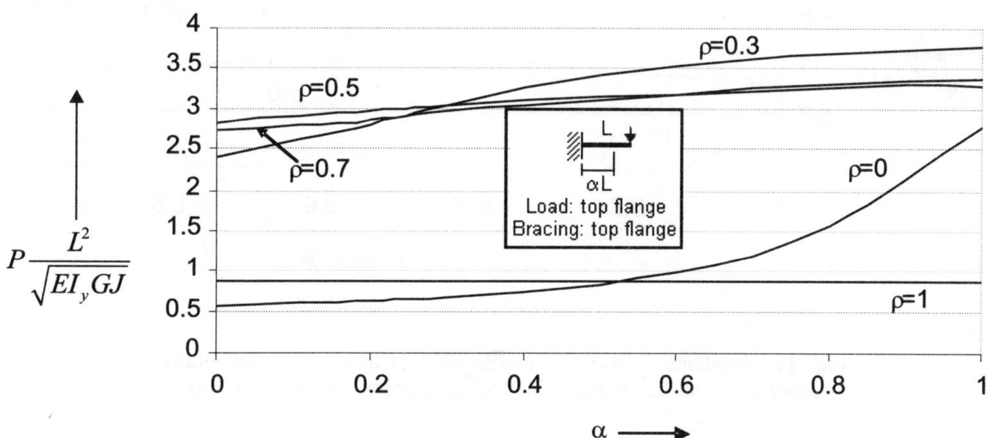

$$P\frac{L^2}{\sqrt{EI_y GJ}}$$

Fig. 9. Variations in Buckling Loads with Location of Lateral Brace:
Point Load at Top Flange, Bracing at Top Flange (L = 2m).

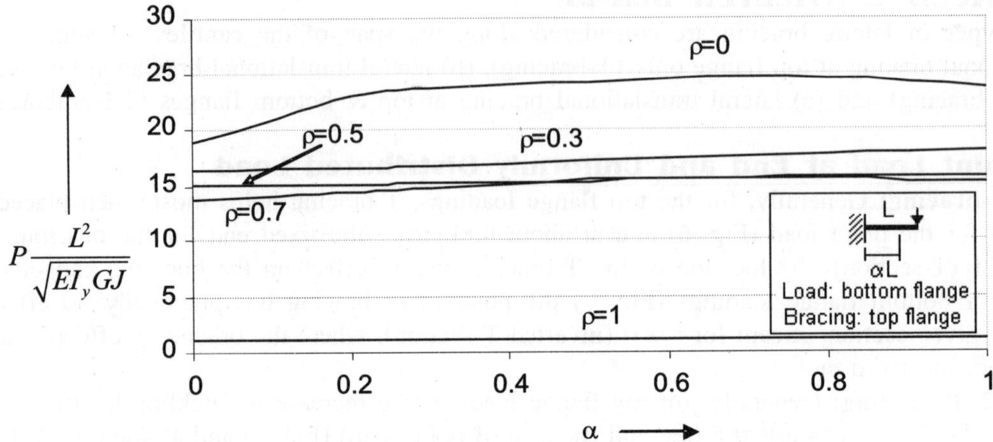

Fig. 10. Variations in Buckling Loads with Location of Lateral Brace: Point Load at Bottom Flange, Bracing at Top Flange (L = 2m).

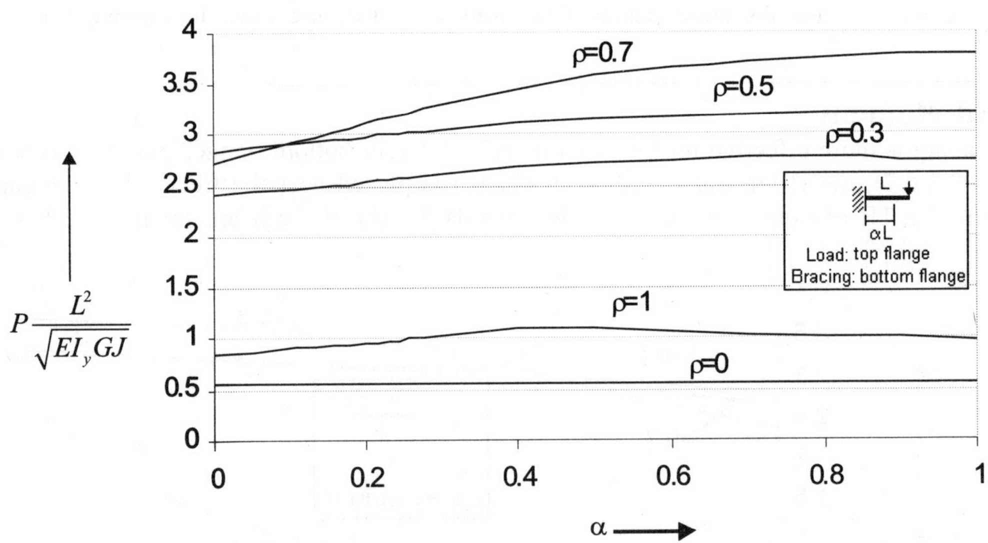

Fig. 11. Variations in Buckling Loads with Location of Lateral Brace: *Point Load at Top Flange*, Bracing at Bottom Flange (L = 2m).

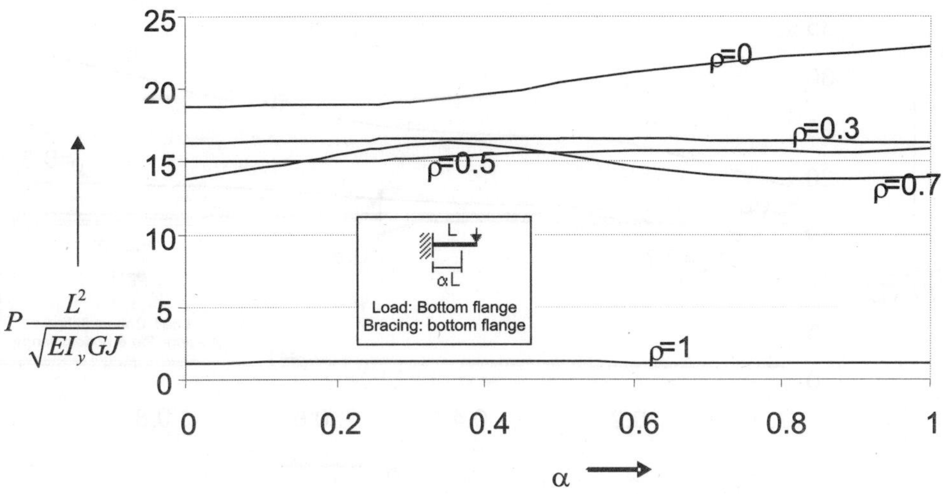

$$P\frac{L^2}{\sqrt{EI_yGJ}}$$

α

Fig. 12. Variations in Buckling Loads with Location of Lateral Brace: *Point Load at Bottom Flange*, Bracing at Bottom Flange (L=2m).

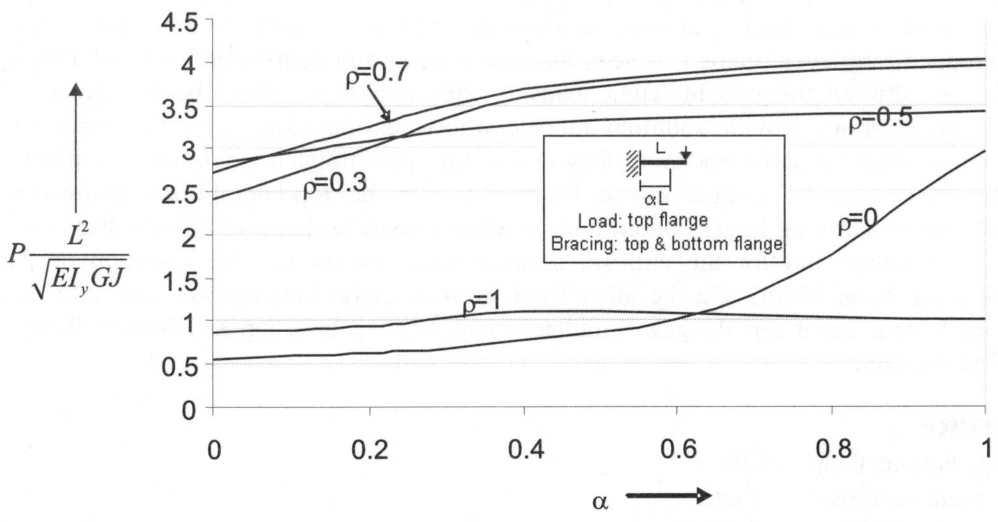

$$P\frac{L^2}{\sqrt{EI_yGJ}}$$

α

Fig. 13. Variations in Buckling Loads with Location of Lateral Brace:
Point Load at Top Flange, Bracing at Top & Bottom Flange (L = 2m).

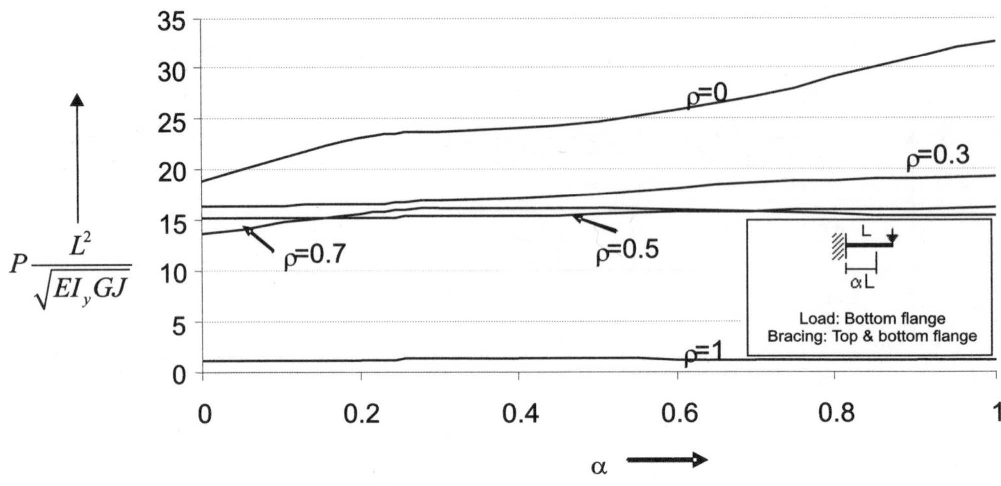

$$P\frac{L^2}{\sqrt{EI_y GJ}}$$

Fig. 14. Variations in Buckling Loads with Location of Lateral Brace: *Point Load at Bottom Flange*, Bracing at Top & Bottom Flange (L=2m).

CONCLUSION

A general observation based on the present investigation is that the distortional buckling has a significant effect on short span beams, at least for the types of beam considered here. For simply supported and propped-cantilever beams it is seen that, for a uniformly distributed load and a point load, the difference between the two buckling loads is considerable for short beams. Since all available design specifications provide solutions for lateral-torsional buckling only, these solutions are to be used with caution for short beams, as they may yield overestimated values of the buckling load for these two load cases. For long cantilever beams it is seen that top lateral bracings are very effective for beam sections having larger bottom flanges when a point load or a uniformly distributed load acts at the top flange and for the uniform moment case, except for the T-section or the inverted T-section cantilever beams. On the other hand, bottom lateral bracings are very effective for beam sections having larger top flanges. Simultaneously, bracing both top and bottom flanges is definitely more effective.

NOMENCLATURE

B_T, B_B	Top, bottom flange width
C_b	Moment modification factor
E	Young's modulus of elasticity
G	Shear modulus of elasticity
h	Distance between flange centroids
I_Y	Moment of inertia about y-axis
I_{YB}	Second moment of area about the section y-axis of the bottom flange
I_{YT}	Second moment of area about the section y-axis of the top flange

J	Torsional section constant
L	Length of the beam
M_A	Absolute values of the moments at the quarter point in the braced segment
M_B	Absolute values of the moments at the center point in the braced segment
M_C	Absolute values of the moments at the three-quarter point in the braced segment
P	Point load
t	Web thickness
T_T, T_B	Top, bottom flange thickness
α	Fractional distance of the bracing position from the fixed end of the cantilever beam
ξ	Distance of the load position from the mid-height between the flange centroids
ρ	Measure of beam mono-symmetry
ν	Poisson's ratio

REFERENCES

1. M. Ma, O. Hughes, 1996, Lateral–Distortional Buckling of Monosymmetric I-Beams under Distributed Vertical Load, Thin Walled Structures, 26(2), 123-145.
2. T. V. Galambos, 1998, Guide to Stability Design Criteria for Metal Structures, 5th Ed., John Wiley & Sons, Inc., New York.
3. S. Kitipornchai, C. M. Wang and N. S. Trahair, 1986, Buckling of Monosymmetric I-Beams under Moment Gradient, Journal of Structural Division, ASCE, 112(4), 781-799.
4. A. Samanta, A. Kumar, 2006, Distortional Buckling in Monosymmetric Beams, Thin-Walled Structures, 44(1), 51-56.
5. T. A. Helwig, K. H. Frank, J. A. Yura, 1997, Lateral-Torsional Buckling of Singly Symmetric I-Beams, Journal of Structural Engineering, ASCE, 123(9), 1172-1179.

2

Transient Response Analysis of Plates with Eccentrically Located Patch Loadings

M. Tanveer and A.V. Singh

Department of Mechanical and Materials Engineering. University of Western Ontario, London, Ontario, Canada, N6A 5B9. email: avsingh@eng.uwo.ca

ABSTRACT

Linear and geometrically non-linear transient vibrations of isotropic and orthotropic plates subject to uniform load distributed over the area and also on localized area termed as patch on the surface have been investigated in this study. The plate geometry is defined by quadrilateral region using the natural coordinates in conjunction with the Cartesian coordinates. Different sizes of the circular patch loading area are considered in this study for the simply supported and clamped boundary conditions. The equation of motion is obtained by the Hamiltonian principle. The solution procedure is developed based primarily on the Newmark beta-m method for the time integration. In addition, the Newton-Raphson method is used in each time step to get the converged results. The numerical procedure is validated successfully by comparing the obtained results with those from the literature. Numerical calculations are carried out for square and circular plates and presented in this paper. The deflection of the central point of the plates and the stress are plotted against time and discussed.

Keywords: Non-linear plate problem, patch loaching, transient vibration, p-type formulation.

1. INTRODUCTION

A study of the linear and non-linear elastic transient response of square and circular plates using a single domain p-type method [1] is reported in this paper. The plates are assumed to be subjected to uniform load on a localized area, referred to as the patch. Such problems generally have practical significance as they appear quite frequently in aerospace, nuclear, marine and other industries. Equations are based on the first order shear deformable thin plate theory, wherein rotary inertia and shear deformation are retained for both linear and geometrically non-linear analyses. The non-linearity is present only with the in-plane strain components and the transverse shear strains are kept linear [2]. Square plate having all sides simply supported are analysed first and the results are then compared successfully with those reported previously in the literature [3]. To obtain the degree of accuracy of the present method, one will require literally hundreds of quadrilateral elements in the finite element method. Transient displacement and stress results are presented in graphs and discussed. It is believed that the p-type method is numerically better suited for the non-linear analysis of plates in terms of the numerical stability and accuracy.

2. FORMULATION

The plate geometry is defined in the Cartesian coordinates, where x – y axes lie on the middle plane of the plate and the coordinate z is the distance measured in the perpendicular direction of the plate. It is assumed that u', v' and w' represent displacement components in the x, y, and z directions respectively and are written as

$$u' = u + z\beta_2; \ v' = v + z\beta_2 \ \& \ w' = w \qquad ...(1)$$

In the above, u, v, and w are the displacements at the middle plane and β_1 and β_2 are the components of rotation of the normal to the plate. The thickness (h) of the plate is assumed to be uniform and small in comparison with the in-plane dimensions. The strain-displacement relations for the geometrically non-linear analysis of plates can be described by the following.

$$\{\varepsilon'\} = ([Z][d_L] + [d_{NL}])\{\Delta\} \qquad ...(2)$$

where, $\{\varepsilon'\}^T = \{\varepsilon'_x \ \varepsilon'_y \ \gamma'_{xy} \ \gamma'_{yz} \ \gamma'_{zx}\}$ and $\{\Delta\}^T = \{u \ v \ w \ \beta_1 \ \beta_2\}$. Matrix $[Z]$ and the differential linear and nonlinear operator matrices $[d_L]$ and $[d_{NL}]$ can be easily obtained from the strain-displacement relations [2]. The strain energy under a given state of stress (or strain) for an infinitesimal volume ($dV = dx \ dy \ dz$) is given by

$$dU = (1/2) \{\varepsilon'\}^T [E]\{\varepsilon'\} \ dx \ dy \ dz \qquad ...(3)$$

Matrix $[E]$ is composed of the elastic modulus (E), the Poisson's ratio (v) and the shear correction factor $k = 5/6$. Similarly the kinetic energy of the plate is

$$dT = (1/2) \ \rho\{\dot{u}\}^T \{\dot{u}\} dx \ dy \ dz \qquad ...(4)$$

where, ρ = mass density and $\{\dot{u}\}^T = \{\dot{u}' \ \dot{v}' \ \dot{w}'\}$.

In order to evaluate the stiffness and mass parameters, the plate geometry is taken as a quadrilateral area and is defined in coordinates x and y. Subsequently, the natural coordinates ξ and η bounded by $-1 \le (\xi, \eta) \le +1$ are used to interpolate the coordinates at a point within the boundary of the plate in the following manner.

$$x = \sum_{i=1}^{n} N_i(\xi, \eta) x_i \ ; \ \ y = \sum_{i=1}^{n} N_i(\xi, \eta) y_i \qquad ...(5)$$

Here, n represents the number of points used to accurately represent the geometry, (x_i, y_i) = coordinates of the ith point, also referred to as the geometric point, and $N_i(\xi, \eta)$ = ith shape function with $i = 1, 2, 3, ..., n$. Similarly, the displacement components can also be interpolated using a different set of pre-selected points on the quadrilateral domain.

$$u = \sum_{i=1}^{p} f_i(\xi, \eta) U_i \ ; \ v = \sum_{i=1}^{p} f_i(\xi, \eta) V_i \ ; \ \ w = \sum_{i=1}^{p} f_i(\xi, \eta) W_i \ ;$$

$$\beta_1 = \sum_{i=1}^{p} f_i(\xi,\eta)\theta_i \; ; \beta_2 = \sum_{i=1}^{p} f_i(\xi,\eta)\phi_i \qquad\qquad ...(6)$$

In the above, $f_i(\xi, \eta)$ is the displacement shape function and indices U_i, V_i, W_i, θ_i, and ϕ_i correspond to u, v, w, β_1, and β_2 respectively at the ith displacement node. If ℓ and m denote the orders of the polynomials in ξ and η respectively, the number of displacement nodes required is: $p = (\ell + 1)(m + 1)$.

With the help of the above equation, it is quite straightforward to deduce the equation of motion for the plate. It is given by

$$[M]\{\ddot{\Gamma}\} + [K]\{\Gamma\} = \{P(t)\} \qquad\qquad ...(7)$$

where, mass matrix = $[M]$, stiffness matrix = $[K] = [K_1] + [K_2] + [K_3]$, displacement vector consisting of all degrees-of-freedom of all the displacement nodes = $\{\Gamma\}$ and load vector = $\{P(t)\}$, derivation method for which is given in ref. [1]. Also, $[K_1]$ = linear stiffness matrix, $[K_2]$ stiffness matrix from the product of linear and non-linear strains, and $[K_3]$ = non-linear stiffness matrix.

3. RESULTS AND DISCUSSIONS

Case 1: A comparative study of the present method is performed first with results from the paper by Akay [3], who carried out large deflection dynamic analysis of a simply supported square plate using four node isoparametric mixed quadrilateral finite elements. The calculations are done using the following parameters.

Length of the plate, a = 243.8 cm; thickness, h = 0.635 cm; suddenly applied uniform (with respect to both time and plate area) load per unit area, p = 4.883 × 10^{-4} N/cm^2; Poisson's ratio, v = 0.25; mass density, ρ = 2.547 × 10^{-6} Ns^2/cm^4; and the modulus of elasticity, E = 7.031 × 10^5 N/cm^2.

The above parameters are also used in the present calculations and the numerical results are plotted in Fig. 1 along with the results by Akay [3] which are shown by points (+, o, & ×) for three values of the load per unit area: p, $5p$ and $10p$, respectively. Excellent comparison is observed here.

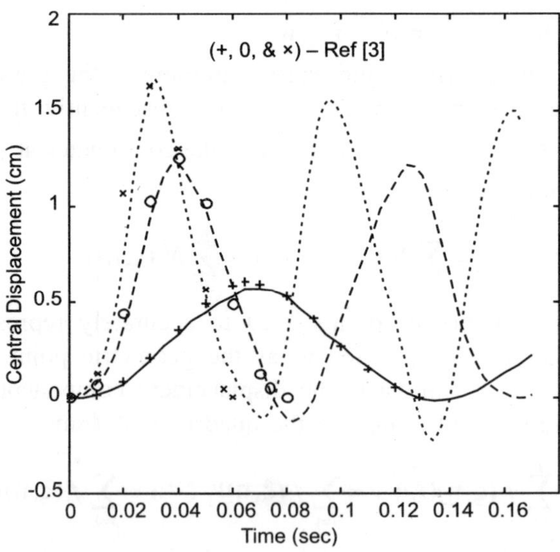

Fig. 1. Displacement versus time curves for a simply supported square plate.
Solid line – pm (present method) with load p, dashed line – pm with load $5 p$, dotted line – pm with load *10 p*.

Case 2: Simply supported square plate with $a = 1$ m, $h = 0.02$ m, $E = 1$ N/m^2, $\nu = 0.3$, $\rho = 1$ Ns^2/m^4; and has been analyzed for three loads. Firstly, the plate is subjected to a uniform load $p = 4.5 \times 10^{-6}$ N/m^2 over its entire area and remains uniform also with time. The other two loading

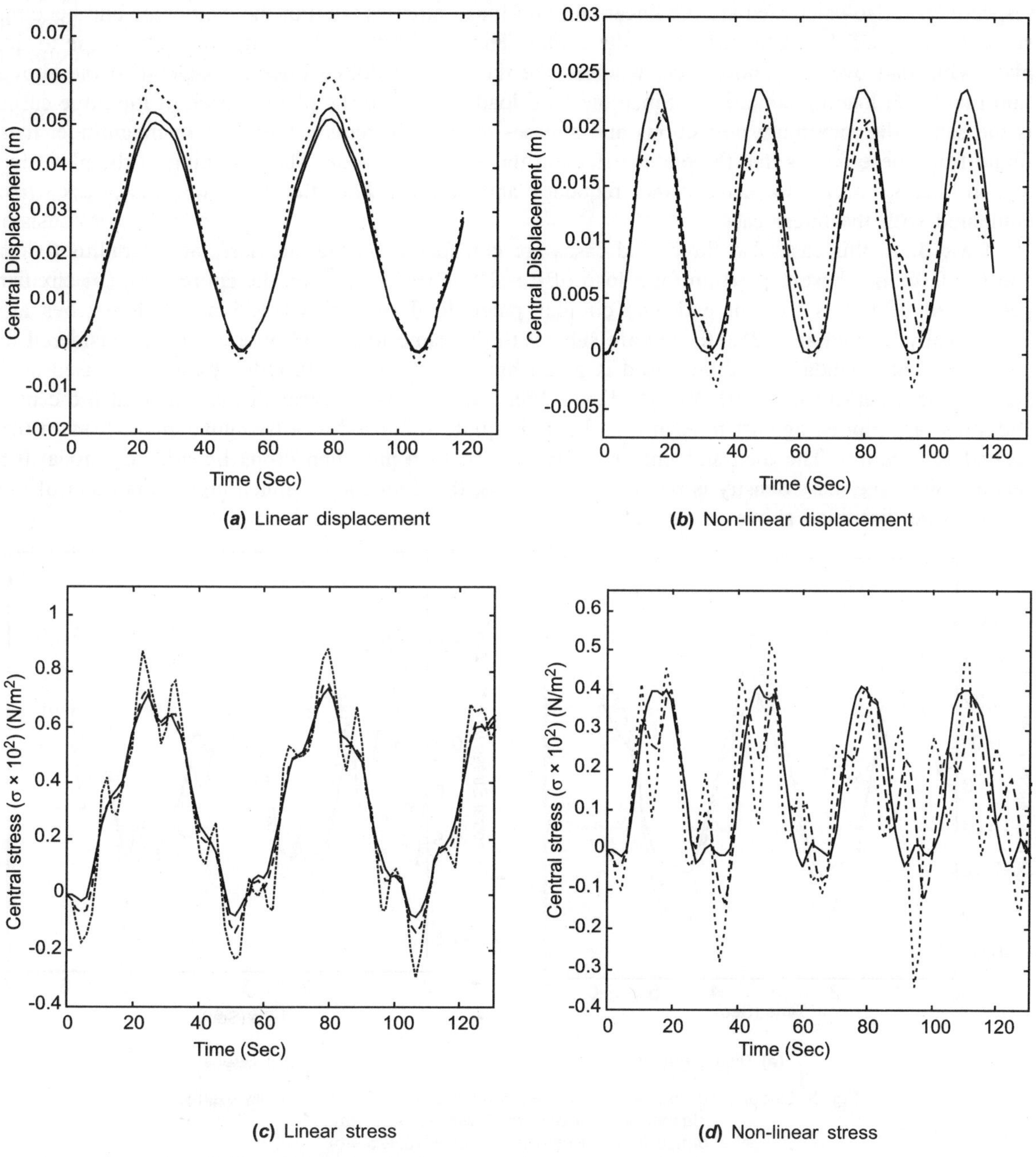

(*a*) Linear displacement   (*b*) Non-linear displacement

(*c*) Linear stress   (*d*) Non-linear stress

Fig. 2. Time history for simply supported square plate, Solid line - Fully loaded, dashed line - Patch load with $R = 0.25$ (*m*), dotted line - Patch load with $R = 0.125$ (*m*).

conditions involve the same total load on the plate, but applied uniformly to circular patches of radii 0.25 m and 0.125 m respectively, both eccentrically located at (0.25,0.25). Linear and non-linear transverse displacements at the centre of the plate are calculated and presented in Figs. 2a and 2b respectively. Also calculated are the linear and non-linear stresses (sum of the membrane and bending stresses on top of the plate surface) at the center. The solid line in these figures corresponds to the plate with load over the entire area, whereas the dashed and dotted lines correspond to the above said two patch loading conditions. Since the total load applied to the plate in each of the three cases is the same, the maximum deflections and stresses essentially remain the same in magnitude. It is found from these results that the plate stiffens in its non-linear mode. The stiffening of the plate due to non-linearity seems to increase the frequency and decrease both the deflection and stress when compared with the linear case.

Case 3: In this case also three load cases are considered on the circular plate of radius 0.5 m and $h = 0.02$ m. These are: (*i*) uniform load of $p = 2.5 \times 10^{-5}$ (N/m^2) on the entire area, (*ii*) circular patch load of radius 0.25 m and (*iii*) circular patch load of radius 0.125 m. Both patches are eccentrically located at (0.25,0.0) and are subject to the same total load on a patch as considered in (*i*) above. The boundary condition used is given by: $w = \beta_1 = \beta_2 = 0$. Other parameters used here are: the mass density $\rho = 0.01$ Ns^2/m^4, $E = 1$ N/m^2, and $v = 0.3$. Linear displacement at the centre and stress are plotted against time in Fig. 3 for the plate subjected to a triangular pulse load of 0.5 second in duration. The load suddenly rises from zero to p and then drops linearly to zero at 0.5 second. Since the mass density is reduced in this case, the frequency is much higher than that of the square plate discussed above in case 2.

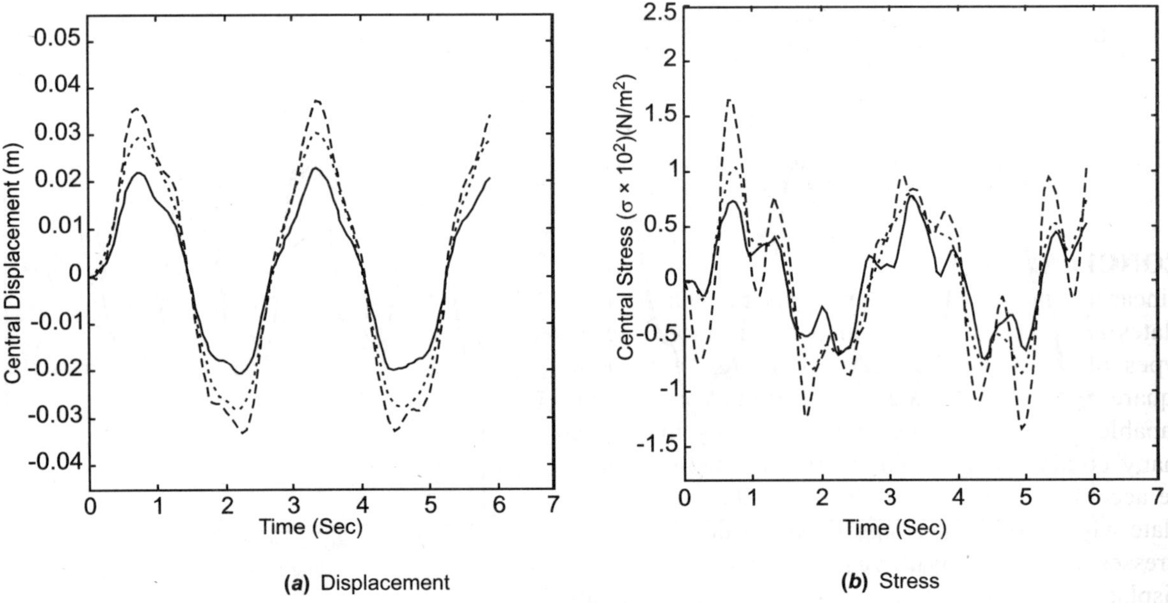

(*a*) Displacement   (*b*) Stress

Fig. 3. Linear time history for clamped circular plate, Solid line - Fully loaded.
dashed line - Patch load with R=0.25 (*m*),
dotted line - Patch load with R=0.125 (*m*).

Case 4: Linear transient analysis is carried out on an orthotropic circular plate having the same boundary condition as in case 3 and radius 0.5 m. Parameters used are: mass density $\rho = 0.01$ Ns^2/m^4, $h = 0.02$ m, $E_2 = 1\ N/m^2$, $v_{12} = 0.3$ and the moduli of rigidity $G_{12} = G_{23} = G_{13} = 0.5\ E_2$. The triangular pulse load of the same magnitude and duration as considered above is applied on a patch of radius 0.125 m located at (0.25, 0.0). Numerical results for the displacement and stress against time are presented in Fig. 4 for three different values of E_1 such that $E_1/E_2 = 1.0$, $E_1/E_2 = 10.0$ and $E_1/E_2 = 25.0$.

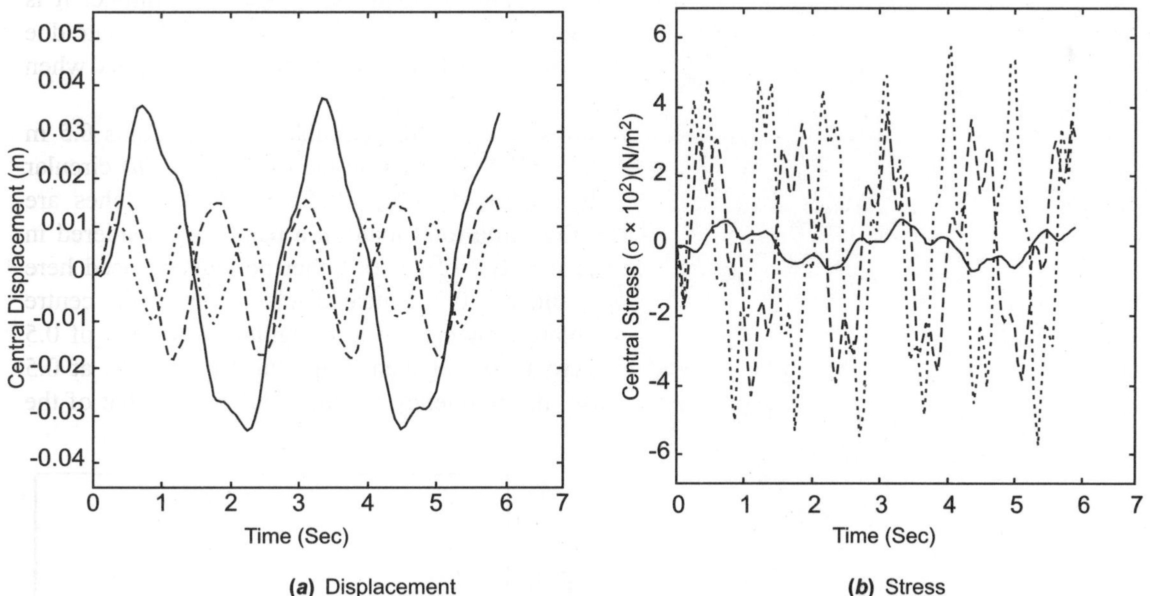

(*a*) Displacement   (*b*) Stress

Fig. 4. Linear time history for orthotropic clamped circular plate with patch load,
R = 0.125 (*m*), Solid line – (E_1/E_2 = 1), dashed line – (E_1/E_2 = 10), dotted line – (E_1/E_2 = 25).

CONCLUSION

Linear and non-linear transient response analyses of square and circular first order shear deformable plates are presented in this paper. The model consists of a single plate area upon which different types of loads and boundary conditions are applied. Though the results are presented only for square and circular plates, the formulation and the computational algorithm is quite general and capable of solving plates under various loading conditions, of different shapes, and subject to many combinations of boundary conditions. Both isotropic and composite material properties can be accommodated adequately. From the numerical results, it is found that there is stiffening of the plate when it undergoes non-linear mode of deformation. This causes the frequency to increase and stresses to reduce. With regard to the loading type, it is observed that the mean values of the displacement and stress are positive for the suddenly applied step load. But the mean responses are nearly zero for the case of the triangular pulse which ceases after 0.5 second. The present method is numerically stable and yields accurate results.

REFERENCES

1. T. Muhammad and A. V. Singh, 2004, A p-type solution for the bending of rectangular, circular, elliptic and skew plates, *Solids and Structures*, 41, 3977-3997.
2. A. V. Singh, Y. Elaghabash, 2003, On the finite displacement analysis of rectangular plates, *Non-Linear Mechanics*, 38, 1149-1162.
3. H. U. Akay, 1980, Dynamic large deflection analysis of plates using mixed finite elements, *Computers & Structure*, 11, 1-11.
4. M. G. Katona, O. C. Zienkiewicz, 1985, A unified set of single step algorithms, part 3: The beta-m method, a generalization of the Newmark scheme, *Numerical Methods in Engineering*, 21, 1345-1359.

3

Determination of the Complete Displacement Field of Hybrid Crack Element and its Coupling with XFEM

B.L. Karihaloo[1] and Q.Z. Xiao[2]

School of Engineering, Cardiff University, Cardiff CF24 3AA, UK
email: [1] karihaloob@cardiff.ac.uk [2] xiaoq@cardiff.ac.uk

ABSTRACT

The hybrid crack element (HCE) is one of the most accurate and convenient finite elements (FEs) for the direct calculation of the stress intensity factor (SIF) and coefficients of the higher order terms of the Williams expansion. This paper introduces two new developments of the element. Firstly, a least squares method (LSM) is introduced to recover the rigid body modes excluded from the HCE formulation, which create jumps between the truncated asymptotic displacements and element boundary displacements. The LSM minimises these jumps. Then an approach is introduced to combine the HCE and the extended/generalized finite element method (XFEM). The HCE is used for the crack tip region, while the XFEM is used for modelling crack faces behind the crack tip with jump functions. The coupled method retains the advantages of both HCE and XFEM. Numerical results are presented to illustrate these developments.

Keywords: Hybrid crack element, XFEM, combination of HCE with XFEM.

1. INTRODUCTION

The elastic displacement, strain and stress fields in the vicinity of the tip of a plane crack with traction-free faces subjected to arbitrary loading can be expressed in the so-called Williams expansions [1]. If the crack lies on the negative x-axis, and the polar coordinates centred at the crack tip are designated r and θ (θ is measured counterclockwise from the positive x-axis), the Williams expansions for the displacement and stress fields near the tip of the crack are (see, e.g. [2-4]):

$$u = \sum_{n=0}^{\infty} \frac{r^{\frac{n}{2}}}{2\mu}\left\{ a_n\left[\left(\kappa+\frac{n}{2}+(-1)^n\right)\cos\frac{n}{2}\theta - \frac{n}{2}\cos(\frac{n}{2}-2)\theta\right] - b_n\left[\left(\kappa+\frac{n}{2}-(-1)^n\right)\sin\frac{n}{2}\theta - \frac{n}{2}\sin(\frac{n}{2}-2)\theta\right]\right\}$$

$$...(1)$$

$$v = \sum_{n=0}^{\infty} \frac{r^{\frac{n}{2}}}{2\mu} \left\{ a_n \left[\left(\kappa - \frac{n}{2} - (-1)^n \right) \sin \frac{n}{2}\theta + \frac{n}{2}\sin(\frac{n}{2}-2)\theta \right] + b_n \left[\left(\kappa - \frac{n}{2} + (-1)^n \right) \cos \frac{n}{2}\theta + \frac{n}{2}\cos(\frac{n}{2}-2)\theta \right] \right\}$$

...(2)

$$\sigma_x = \sum_{n=1}^{\infty} \frac{n}{2} r^{\frac{n}{2}-1} \left\{ a_n \left[\left(2 + \frac{n}{2} + (-1)^n \right) \cos(\frac{n}{2}-1)\theta - (\frac{n}{2}-1)\cos(\frac{n}{2}-3)\theta \right] \right.$$

...(3)

$$\left. - b_n \left[\left(2 + \frac{n}{2} - (-1)^n \right) \sin(\frac{n}{2}-1)\theta - (\frac{n}{2}-1)\sin(\frac{n}{2}-3)\theta \right] \right\}$$

$$\sigma_y = \sum_{n=1}^{\infty} \frac{n}{2} r^{\frac{n}{2}-1} \left\{ a_n \left[\left(2 - \frac{n}{2} - (-1)^n \right) \cos(\frac{n}{2}-1)\theta + (\frac{n}{2}-1)\cos(\frac{n}{2}-3)\theta \right] \right.$$

...(4)

$$\left. - b_n \left[\left(2 - \frac{n}{2} + (-1)^n \right) \sin(\frac{n}{2}-1)\theta + (\frac{n}{2}-1)\sin(\frac{n}{2}-3)\theta \right] \right\}$$

$$\tau_{xy} = \sum_{n=1}^{\infty} \frac{n}{2} r^{\frac{n}{2}-1} \left\{ a_n \left[(\frac{n}{2}-1)\sin(\frac{n}{2}-3)\theta - \left(\frac{n}{2} + (-1)^n \right) \sin(\frac{n}{2}-1)\theta \right] \right.$$

...(5)

$$\left. + b_n \left[(\frac{n}{2}-1)\cos(\frac{n}{2}-3)\theta - \left(\frac{n}{2} - (-1)^n \right) \cos(\frac{n}{2}-1)\theta \right] \right\}$$

where, $\mu = E/(2(1 + v))$ is the shear modulus; the Kolosov constant is $\kappa = 3 - 4v$ for plane strain or $\kappa = (3 - v)/(1 + v)$ for plane stress; E and v are Young's modulus and Poisson's ratio, respectively. The displacements corresponding to $n = 0$

$$u_0 = \frac{\kappa+1}{2\mu} a_0, \quad v_0 = \frac{\kappa+1}{2\mu} b_0 \qquad \qquad ...(6)$$

are rigid body translations at the crack tip. The displacements corresponding to b_2

$$\hat{u}_2 = -\frac{\kappa+1}{2\mu} b_2 r \sin\theta = -\frac{\kappa+1}{2\mu} b_2 y \,,$$

$$\hat{v}_2 = \frac{\kappa+1}{2\mu} b_2 r \cos\theta = \frac{\kappa+1}{2\mu} b_2 x \qquad \qquad ...(7)$$

represent the rigid body rotation $\theta_0 = -(\kappa + 1)b_2/(2\mu)$ with respect to the crack tip, with $x = r \cos\theta$ and $y = r \sin\theta$. Terms involving a_0, b_0 and b_2 do not contribute to the strains or stresses. Terms involving coefficients a_n (b_n), $n \geq 1$, correspond to the mode I (II) expansion. The coefficients of the singular terms ($n = 1$) in (1) – (5), a_1 and b_1, are related to the mode I and mode II stress intensity factors (SIFs) as

$$a_1 = \frac{K_I}{\sqrt{2\pi}}, \quad b_1 = -\frac{K_{II}}{\sqrt{2\pi}} \qquad \qquad ...(8)$$

The second term in the mode I expansion in (1) – (5) corresponds to a uniform normal (elastic) *T*-stress

$$\sigma_x = T = 4a_2 \qquad \qquad ...(9)$$

acting near the crack tip in the direction parallel to the crack plane.

With the use of (6) and (7), the displacements (1) and (2) can be rewritten as

$$u = u_0 + \hat{u}_2 + u_{\text{asym}}, \quad v = v_0 + \hat{v}_2 + v_{\text{asym}} \qquad \qquad ...(10)$$

where u_{asym} and v_{asym} represent the displacements excluding terms involving coefficients a_0, b_0 and b_2.

The SIFs have been used traditionally for the determination of the initiation and propagation of cracks in brittle materials. However, more recent studies show higher order terms of the asymptotic field (1) – (5) are of great relevance to predicting the constraint of elasto-plastic crack tip fields (see, e.g. [4 – 7]) and to interpreting the size effect of quasi-brittle materials [8, 9].

The inherent singularity in the crack tip field (1) – (5) creates great difficulty in the numerical determination of the coefficients of the Williams expansions. Tong et al. [10] introduced a hybrid crack element (HCE) for direct evaluation of the SIFs in homogeneous materials (see Fig. 1). It is formulated from a simplified variational functional using truncated asymptotic crack tip displacement and stress expansions and interelement boundary displacements compatible with the surrounding regular elements.

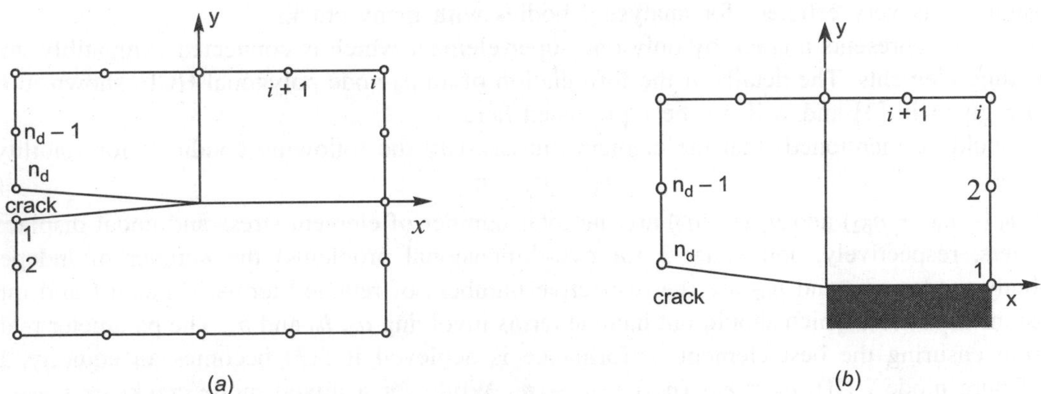

Fig. 1. A typical n_d-node polygonal HCE. (a) For mixed mode crack;
(b) For a pure mode I or mode II crack.

When the body force is ignored, the simplified variational functional for the formulation of HCE can be written as [3, 10]

$$\Pi_m^e = \int_{S_\sigma^e} (\frac{1}{2}T_i - \bar{T}_i)u_i\,ds - \int_{\partial \tilde{A}^e} T_i(\frac{1}{2}u_i - \tilde{u}_i)\,ds \qquad \qquad ...(11)$$

or in matrix form

$$\Pi_m^e = \int_{S_\sigma^e} u^T (\frac{1}{2}T - \bar{T})\,ds - \int_{\partial \tilde{A}^e} (\frac{1}{2}u^T - \tilde{u}^T)T\,ds \qquad \qquad ...(12)$$

The boundary of the element, ∂A^e, is composed of the segment S_σ^e on which the tractions $\bar{T}_i$ are prescribed and the interelement boundary, $\partial \tilde{A}^e = \partial A^e - S_\sigma^e$, common with the adjacent elements.

Displacements u_i and boundary tractions $T_i (= \sigma_{ij} n_j)$ are independent of the other elements, but boundary displacements $\tilde{u}_i$ have to be the same for the two elements over their common boundary $\partial \tilde{A}^e$. σ_{ij} denote stresses, and n_j are the direction cosines of the unit outward normal to ∂A^e. Subscripts i and j take the values 1 and 2, and the summation convention on repeated indices is used. The assumed element displacement and stress fields, u_i and σ_{ij}, should meet the equations of equilibrium and the stress-displacement relations. They should not include any rigid body or zero energy modes. Integrations on $\partial \tilde{A}^e$ are performed only along the outer boundary of the element away from the crack tip, thus standard quadrature rules can be directly adopted, avoiding errors due to the area integration of the singular integrand. Karihaloo and Xiao [3] showed that three-point Gauss integration ensures good accuracy when $\tilde{u}$ varies linearly between any two adjacent nodes. Integrations on crack faces S_σ^e may be carried out analytically; if they are traction-free, as in most cases, the integrals vanish.

The many applications of HCE to the analysis of cracks and notches as well as voids and inclusions have been reviewed in [11] with an extensive collection of references. Extensive studies have been carried out to prove that the HCE is one of the most accurate and convenient elements for the direct calculation (i.e. without the use of the energy related quantities like the J-integral, or other extra post-processing) of the SIFs and coefficients of the higher order terms of the Williams expansions. It is very efficient for analysing bodies with many cracks.

The HCE represents a crack by only one super-element which is connected compatibly with the surrounding elements. The details of the formulation of an n_d-node polygonal HCE, shown in Figure 1(a), are given in [3] and will not be reproduced here.

It should be mentioned that the element must satisfy the following condition for stability [12]

$$n_\beta \geq n_q - n_r \qquad \qquad ...(13)$$

where n_β ($= n_{\beta 1} + n_{\beta 2}$) and n_q ($= 2n_d$) are the total number of element stress and nodal displacement parameters, respectively, and n_r ($= 3$ for two-dimensional problems) the number of independent rigid body modes. $n_{\beta 1}$ and $n_{\beta 2}$ are the respective numbers of retained terms in mode I and mode II expansions in (1)-(5), which should not include terms involving a_0, b_0 and b_2. The parameter matching condition ensuring the best element performance is achieved if (13) becomes an equality. In the case of pure mode I (II), $n_\beta = n_{\beta 1}$ ($n_{\beta 2}$) $= n_q - n_r$. While for a mixed mode crack, $n_\beta = n_{\beta 1} + n_{\beta 2} = n_q - n_r$, $n_{\beta 1} = n_{\beta 2} + 1$ if n_β is odd; $n_{\beta 1} = n_{\beta 2} + 2$ otherwise [13].

For pure mode I or mode II crack problems, the half polygonal element with n_d nodes shown in Figure 1(b) may be used for constructing the HCE by exploiting the symmetric (mode I) or asymmetric (mode II) conditions along the line of extension of the crack. We need only integrate along the outer boundary and avoid integrations along the line of extension of the crack if the expansion for the relevant mode only is used in the HCE formulation [13]. The shape of the HCE can be chosen flexibly for the convenience of implementation. It is not even required to be convex.

In its implementation, the HCE is generally designed first at each crack tip and then the whole domain is meshed taking into account the boundaries of the domain as well as those of the HCE (see, e.g. Figure 2(b) below). This complicates the meshing task and hinders its incorporation into commercial FE packages. Karihaloo and Xiao [14] showed that a general FE mesh can actually be used by forming the HCE from elements surrounding the crack tip. The elements used for the formulation of the HCE will be skipped in the element analysis process, since their stiffness matrices are not required. The nodes lying inside the HCE are not actually used, and the corresponding degrees of freedom will result in zero pivots in the factorization process. This problem can be easily handled by changing these zero pivots to one and continuing the solution process, as in the HSL MA57 package [15]. The HCE can thus be included in any commercial package as conveniently as normal hybrid stress elements.

In the formulation of the HCE, only displacements u_{asym} and v_{asym} (10) are used. Since the rigid body modes corresponding to terms involving a_0, b_0 and b_2 in the truncated asymptotic displacements are excluded from its formulation, the displacements inside HCE are not complete. Along the boundary of the HCE, Xiao et al. [13] showed that, for a finite plate with an inclined edge crack under uniaxial tension (Fig. 2(a)), obvious non-constant jumps exist between element boundary displacements $\tilde{u}$, $\tilde{v}$ and displacements u_{asym}, v_{asym}, (Fig. 3). This creates doubts about the applicability of the HCE. In order to obtain the actual complete displacements with the use of HCE, it is necessary to determine a_0, b_0 and b_2. In this paper, a least squares method (LSM) proposed in [11] will be introduced to recover a_0, b_0 and b_2 by minimizing the displacement discontinuity along the HCE boundary, i.e., by minimizing the jumps between $\tilde{u}$ and u_{asym} and between $\tilde{v}$ and v_{asym}.

If the HCE only is used, the part of the crack inside the HCE need not conform to the mesh. However, crack faces away from the crack tip (outside the HCE) need to conform to the mesh. This disadvantage can be avoided by combining the HCE with the extended/generalized FEM (XFEM) [16-20]. The XFEM enriches the standard local FE approximations with a displacement discontinuity across a crack, and the asymptotic solution at the crack tip, with the use of the partition of unity (PU). In contrast with the FEM, it avoids the use of meshes conforming with the discontinuity and also adaptive remeshing as the discontinuity grows. XFEM offers great flexibility in the modelling of the fracture process. However, although the implementation of crack tip asymptotic displacements as the enrichment function enhances the accuracy [21], it also complicates the quadrature because

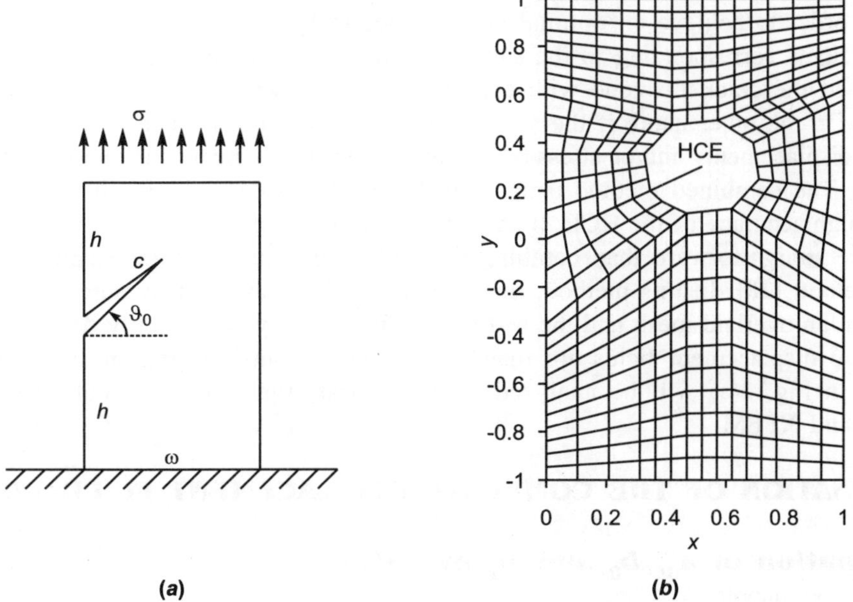

Fig. 2. (*a*) A finite plate with an inclined edge crack under uniaxial tension; (*b*) Finite element mesh for a 25-node HCE surrounded by 423 quadrilateral elements giving a total of 487 nodes [13].

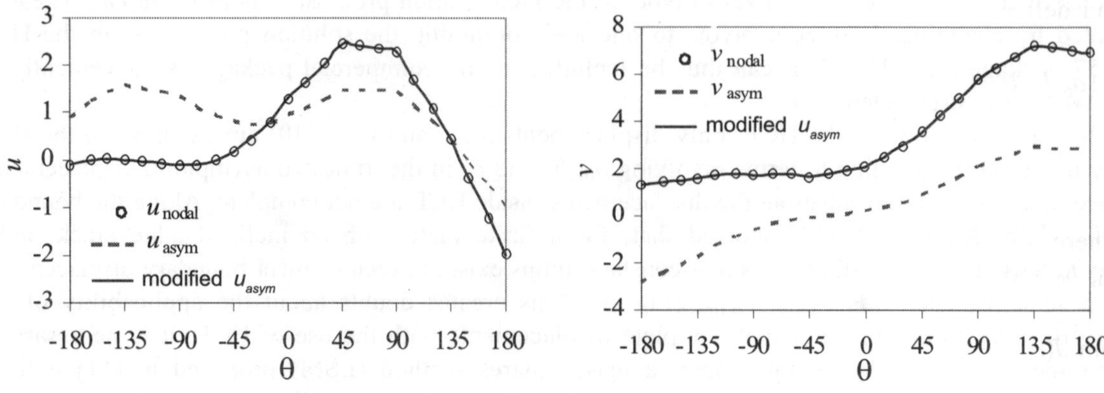

Fig. 3. Angular distribution of displacements along the boundary of the HCE (Fig. 2(b)). $u_{nodal} = \tilde{u}$ and $v_{nodal} = \tilde{v}$ are nodal displacements compatible with the surrounding elements. u_{asym} and v_{asym} are the retained asymptotic fields obtained from HCE without consideration of the terms involving a_0, b_0, or b_2. Modified u_{asym} and v_{asym} take into account a_0, b_0 and b_2 recovered from (18) — (20).

of the singular integrand [17, 21-23]. Furthermore, if a few layers of nodes surrounding the crack tip (definition of layers of elements and nodes can be found in [24]) are enriched with not only the first term but also the higher order terms of the crack tip asymptotic field, and the additional coefficients corresponding to the enrichments are forced to be equal [24, 25], thus ensuring that the displacement approximations reduce to the actual asymptotic fields adjacent to the crack tip, mixed mode SIFs can be directly evaluated, but not the coefficients of higher order terms. (Recently, Béchet *et al.* [22] and Laborde *et al.* [23] also enriched more than one layer of nodes surrounding the crack tip in order to achieve an optimal convergence rate. Laborde et al. [23] also forced the additional coefficients corresponding to the same term of the crack tip field to be equal as in [24, 25], and combined the crack tip zone with the rest via a pointwise collocation matching method similar to [26]). On the other hand, if the additional coefficients are independent at each node, the accuracy of the displacements and/or stresses in a few layers of elements surrounding the crack tip is still low [21]. The combined method using both HCE and XFEM inherits the flexibility of the XFEM and the high accuracy of the HCE. It can also avoid the implementation of blending elements [27], and is more straightforward than combing the XFEM and quarter point elements via a mesh superposition method [28]. In this method, the near-tip field is modelled by superimposed quarter point elements on an overlaid mesh and the rest of the discontinuity is implicitly described by a step function. The two displacement fields are matched through a transition region. The typical crack problem shown in Fig. 2(a) will be analyzed to demonstrate the efficiency and accuracy of the combined HCE and XFEM.

2. DETERMINATION OF THE COMPLETE DISPLACEMENT FIELD INSIDE THE HCE

2.1 Determination of a_0, b_0 and b_2 by LSM

For convenience, we denote

$$\frac{\kappa+1}{2\mu}\Delta u = \tilde{u} - u_{asym}, \quad \frac{\kappa+1}{2\mu}\Delta v = \tilde{v} - v_{asym} \qquad \qquad ...(14)$$

and define

$$J\left(a_0,b_0,b_2\right)=\frac{1}{2}\frac{4\mu^2}{\left(\kappa+1\right)^2}\sum_{k=1}^{n_d}\left[\left(u_0+\hat{u}_2-\frac{\kappa+1}{2\mu}\Delta u\right)^2_{\left(x_k,y_k\right)}+\left(v_0+\hat{v}_2-\frac{\kappa+1}{2\mu}\Delta v\right)^2_{\left(x_k,y_k\right)}\right]$$

$$=\frac{1}{2}\sum_{k=1}^{n_d}\left[\left(a_0-b_2 y_k-\Delta u\left(x_k,y_k\right)\right)^2+\left(b_0+b_2 x_k-\Delta v\left(x_k,y_k\right)\right)^2\right]\quad\ldots(15)$$

where, x_k and y_k, $k = 1, \ldots, n_d$, are the coordinates of node k on the boundary of HCE. Coefficients a_0, b_0 and b_2 are then obtained by

$$\text{min } J(a_0,\ b_0,\ b_2) \text{ or equivalently, } \frac{\partial J}{\partial a_0}=\frac{\partial J}{\partial b_0}=\frac{\partial J}{\partial b_2}=0 \qquad\ldots(16)$$

If we denote

$$\overline{x}=\frac{1}{n_d}\sum_{k=1}^{n_d}x_k\ ,\quad \overline{y}=\frac{1}{n_d}\sum_{k=1}^{n_d}y_k\ ,\quad \Delta\overline{u}=\frac{1}{n_d}\sum_{k=1}^{n_d}\Delta u\left(x_k,y_k\right),\quad \Delta\overline{v}=\frac{1}{n_d}\sum_{k=1}^{n_d}\Delta v\left(x_k,y_k\right) \qquad\ldots(17)$$

the results can be written as

$$a_0=\frac{\sum_{k=1}^{n_d}\left[b_2 y_k+\Delta u\left(x_k,y_k\right)\right]}{n_d}=\overline{y}b_2+\Delta\overline{u} \qquad\ldots(18)$$

$$b_0=-\frac{\sum_{k=1}^{n_d}\left[b_2 x_k-\Delta v\left(x_k,y_k\right)\right]}{n_d}=-\overline{x}b_2+\Delta\overline{v} \qquad\ldots(19)$$

$$b_2=\frac{\sum_{k=1}^{n_d}\left\{y_k\left[\Delta\overline{u}-\Delta u\left(x_k,y_k\right)\right]-x_k\left[\Delta\overline{v}-\Delta v\left(x_k,y_k\right)\right]\right\}}{\sum_{k=1}^{n_d}\left(-\overline{y}y_k+y_k^2-\overline{x}x_k+x_k^2\right)} \qquad\ldots(20)$$

Note that although the above results are derived in the coordinate system centred at the crack tip, they are actually applicable to any coordinate systems. It is only required that the coordinates of the nodes on the boundary of the HCE, their displacement jumps, and the coordinates used in (6) and (7) are defined in the same coordinate system.

2.2 Numerical Results

The finite plate with an inclined edge crack under uniaxial tension, Figure 2(a) that has been studied in [13], is chosen as the benchmark problem here. The geometrical parameters are: $c = 0.6$, $w = h = 1$, $\partial_0 = 30°$; and the tensile stress $\sigma = 1$. A state of plane stress is considered and the thickness is assumed to be unity. Young's modulus $E = 1$, Poisson's ratio $v = 0.25$. The units are self-consistent.

Computational details are the same as in [13]. The 4-node plane hybrid stress element PS [29, 30] is used in conjunction with the HCE, and the traction-free conditions on the crack face as well as the exterior boundary (but excluding the elements adjacent to the loading / prescribed displacement boundary and to the HCE on the crack face) are exactly satisfied using the special hybrid stress element HBE [31]. 2×2 and 3×3 Gauss quadratures are employed for the formulation of PS and HBE, respectively.

A 25-node HCE together with 423 quadrilateral elements giving a total of 487 nodes, as shown in Fig. 2(b), is used. Three-point Gauss integration is used in each segment (side) of the HCE. In the HCE formulation, 24 terms in the mode I expansion ($n_{\beta 1} = 24$) and 23 terms in the mode II expansion ($n_{\beta 2} = 23$) are used. All nodes on the bottom line ($y = -1$) are fixed in the vertical (y) direction, but just the extreme left point ($x = 0$, $y = -1$) on the line is fixed also in the horizontal (x) direction (i.e., the plate is subjected to uniaxial tension on both ends). The displacements, u_{asym} and v_{asym}, corresponding to the calculated coefficients $a_1 \sim a_{24}$, and b_1, $b_3 \sim b_{24}$ are shown as broken lines in Fig. 3 along the boundary of the HCE, and in Figure 4 along a circle with $r = 0.05$ inside the HCE (the boundary nodes of the HCE are located roughly on a circle with $r = 0.2$). Nodal displacements u_{nodal} ($= \tilde{u}$) and v_{nodal} ($= \tilde{v}$) are also shown as circles in Fig. 3. In Fig. 4, the displacements (circled) corresponding to the PS element are obtained by a mesh highly refined to the crack tip [11] but without resorting to the HCE or the HBE. Obviously, non-constant jumps exist between u_{asym}, v_{asym} and the complete displacements. From (18) – (20), we have

$$a_0 = 0.304866, \quad b_0 = 0.687008, \quad \text{and} \quad b_2 = -1.64981 \qquad \qquad ...(21)$$

Taking into account the contributions from these three terms, (6) and (7), the modified u_{asym} and v_{asym} shown as solid lines in Fig. 3 and 4 agree very well with the complete displacements.

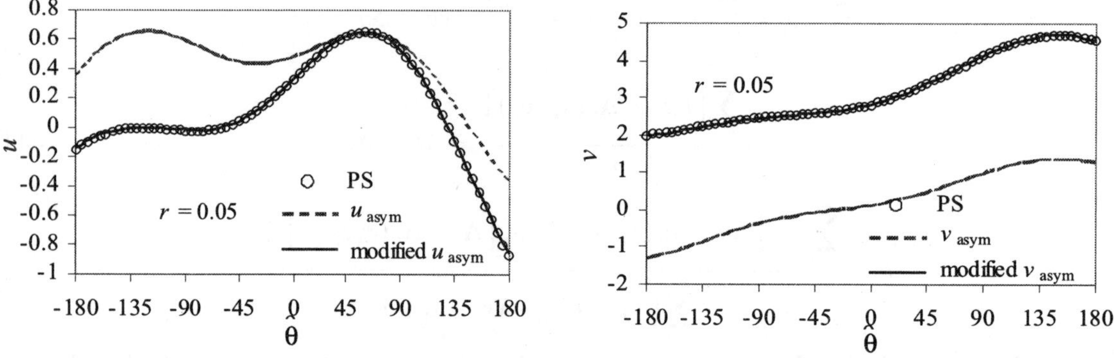

Fig. 4. Angular distribution of displacements along the circle r = 0.05 surrounding the crack tip (inside the HCE). Displacements by PS are obtained from a mesh highly refined to the crack tip but without resorting to the HCE. u_{asym} and v_{asym} are the retained asymptotic fields obtained from HCE with mesh Figure 2(b) without consideration of the terms involving a_0, b_0, or b_2. Modified u_{asym} and v_{asym} take into account a_0, b_0 and b_2 recovered from (18)-(20).

3. COUPLING OF THE HCE WITH XFEM

3.1 A Brief Introduction to the XFEM

For a FE mesh and a crack on a domain shown in Fig. 5, in XFEM a standard local displacement approximation around the crack is enriched with discontinuous Heaviside functions along the crack faces behind the crack tip (squared nodes) and the crack tip asymptotic displacement fields at nodes surrounding the crack tip (circled nodes) using the PU. The approximation of displacements for an element can be expressed in the following form

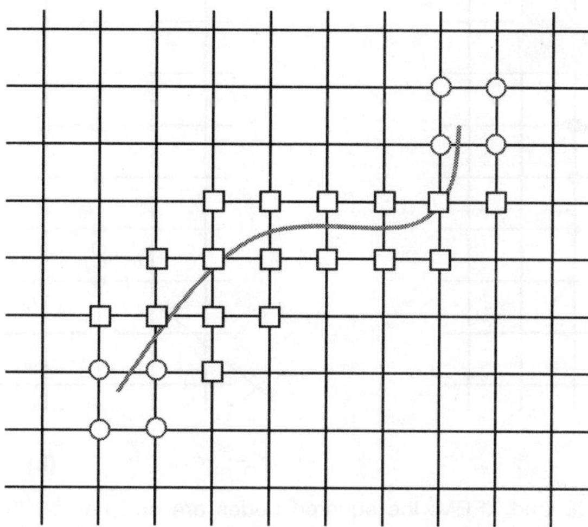

Fig. 5. Representation of a crack on a uniform mesh in XFEM: the squared nodes are enriched by the jump function whereas the circled nodes are enriched by the crack tip branch functions.

$$\begin{Bmatrix} u^h(x) \\ v^h(x) \end{Bmatrix} = \sum_{i \in I} \phi_i(x) \begin{Bmatrix} u_{0i} \\ v_{0i} \end{Bmatrix} + \sum_{j \in J \cap I} \phi_j(x) H(x) \begin{Bmatrix} b_{1j} \\ b_{2j} \end{Bmatrix} + \sum_{m \in M_k \cap I} \phi_m(x) \begin{Bmatrix} u_m^{(tip\ k)} \\ v_m^{(tip\ k)} \end{Bmatrix} \qquad ...(22)$$

where I is the set of all nodes in the element, (u_{0i}, v_{0i}) are the regular degrees of freedom at node i, ϕ_i is the FE shape function associated with node i. J is the set of nodes whose support is intersected by the crack but do not cover any crack tips, the function H(x) is the Heaviside function centred on the crack discontinuity, and (b_{1j}, b_{2j}) are the corresponding additional degrees of freedom.

M_k is the set of nodes that are enriched around the crack tip k with the asymptotic displacements $u^{(tip\ k)}$ and $v^{(tip\ k)}$. The layers of elements and nodes around the crack tip are defined in the same way as in [24]. We define the elements that include the crack tip k as the first layer elements of the crack tip with enriched nodes. The nodes in the first layer elements are called the first layer enriched nodes. The elements immediately adjacent to the first layer elements are defined as the second layer elements, and the nodes that are in the second layer elements, but not in the first layer elements, are called the second layer enriched nodes. Additional layers of elements and enriched nodes can be defined in a similar manner.

3.2 Coupling of HCE with XFEM

When a crack is modeled independent of the mesh using XFEM, a HCE can be constructed as in Figure 6(a). In order to formulate the stiffness matrix of the actual HCE shown in Figure 6(a), we construct a virtual HCE as shown in Figure 6(b), and investigate the relationship between the actual HCE with enriched nodes but without a double node on the crack face, and the virtual HCE without enriched nodes but having a double node on the crack face.

For the sake of simplicity but without loss of generality, in the actual HCE (Figure 6(a)), only nodes i_1 and i_n are enriched with jump functions, other nodes on the boundary of HCE or inside it are not enriched. Nodes i_0, i_{n+1} in the virtual HCE are a double node on the crack face. We denote

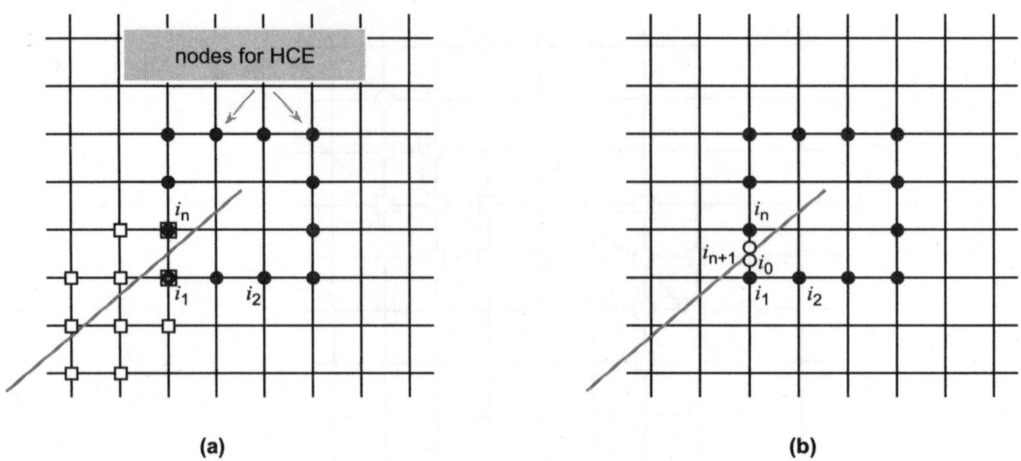

Fig. 6. (a) Combination of HCE and XFEM: the squared nodes are enriched by the jump function whereas the circled nodes are used to form the HCE. **(b)** A virtual HCE has identical size of the actual HCE in Fig. 6(a). It has a double node i_0, i_{n+1} (unfilled circles) on the crack face but without enriched nodes.

the nodal displacements of the actual and virtual HCE as u, v and $\hat{u}$, $\hat{v}$, respectively. Nodal displacements at nodes $i_2 - i_{n-1}$ in both elements are identical; the displacements $\hat{u}$, $\hat{v}$ at nodes i_0, i_1, i_n and i_{n+1} in the virtual element can be represented by the displacements and coefficients of the enrichment function at nodes i_1 and i_n of the actual element as follows with the help of (22).

$$\begin{Bmatrix} \hat{u} \\ \hat{v} \end{Bmatrix}_{i_0} = \phi_{i_1}\left(x_{i_0}\right)\begin{Bmatrix} u \\ v \end{Bmatrix}_{i_1} + \phi_{i_n}\left(x_{i_0}\right)\begin{Bmatrix} u \\ v \end{Bmatrix}_{i_n} + \phi_{i_1}\left(x_{i_0}\right)H\left(x_{i_0}\right)\begin{Bmatrix} b_1 \\ b_2 \end{Bmatrix}_{i_1} + \phi_{i_n}\left(x_{i_0}\right)H\left(x_{i_0}\right)\begin{Bmatrix} b_1 \\ b_2 \end{Bmatrix}_{i_n} \qquad \ldots(23)$$

$$\begin{Bmatrix} \hat{u} \\ \hat{v} \end{Bmatrix}_{i_1} = \begin{Bmatrix} u \\ v \end{Bmatrix}_{i_1} + H\left(x_{i_1}\right)\begin{Bmatrix} b_1 \\ b_2 \end{Bmatrix}_{i_1} \qquad \ldots(24)$$

$$\begin{Bmatrix} \hat{u} \\ \hat{v} \end{Bmatrix}_{i_n} = \begin{Bmatrix} u \\ v \end{Bmatrix}_{i_n} + H\left(x_{i_n}\right)\begin{Bmatrix} b_1 \\ b_2 \end{Bmatrix}_{i_n} \qquad \ldots(25)$$

$$\begin{Bmatrix} \hat{u} \\ \hat{v} \end{Bmatrix}_{i_{n+1}} = \phi_{i_1}\left(x_{i_{n+1}}\right)\begin{Bmatrix} u \\ v \end{Bmatrix}_{i_1} + \phi_{i_n}\left(x_{i_{n+1}}\right)\begin{Bmatrix} u \\ v \end{Bmatrix}_{i_n} + \phi_{i_1}\left(x_{i_{n+1}}\right)H\left(x_{i_{n+1}}\right)\begin{Bmatrix} b_1 \\ b_2 \end{Bmatrix}_{i_1} + \phi_{i_n}\left(x_{i_{n+1}}\right)H\left(x_{i_{n+1}}\right)\begin{Bmatrix} b_1 \\ b_2 \end{Bmatrix}_{i_n} \qquad \ldots(26)$$

The above relationship can be written in matrix form as

$$\left[\begin{Bmatrix} \hat{u} \\ \hat{v} \end{Bmatrix}_{i_0}^T \quad \begin{Bmatrix} \hat{u} \\ \hat{v} \end{Bmatrix}_{i_1}^T \quad \cdots \quad \begin{Bmatrix} \hat{u} \\ \hat{v} \end{Bmatrix}_{i_{n+1}}^T \right]^T = T \cdot \left[\begin{Bmatrix} u \\ v \\ b_1 \\ b_2 \end{Bmatrix}_{i_1}^T \quad \begin{Bmatrix} u \\ v \end{Bmatrix}_{i_2}^T \quad \cdots \quad \begin{Bmatrix} u \\ v \end{Bmatrix}_{i_{n-1}}^T \quad \begin{Bmatrix} u \\ v \\ b_1 \\ b_2 \end{Bmatrix}_{i_n}^T \right]^T \qquad \ldots(27)$$

Then the stiffness matrix $\hat{K}$ of the virtual HCE can be transformed into the stiffness matrix (K) of the actual HCE as follows

$$K = T^T \hat{K} T \qquad \ldots(28)$$

In evaluating the n_β coefficients of the truncated crack tip asymptotic field, similar transformations need to be performed.

3.3 Numerical Results

We consider again the finite plate with an inclined edge crack under uniaxial tension (Figure 2(a)). It is now divided into 20×40 square elements giving a total of 861 nodes (Figure 7). The conventional 4-node bilinear isoparametric Q4 elements are used as background elements. A 22-node HCE ($n_q = 48$ with four enrichment degrees of freedom and $n_\beta = 45$) is formulated along the third layer of nodes surrounding the crack tip. Elements cut by the crack and outside the HCE are integrated as in [21]. The elements used for the formulation of the HCE are skipped in the element analysis process as in [14], since their stiffness matrices are not required. In the HCE formulation, 23 terms in the mode I expansion ($n_{\beta 1} = 23$) and 22 terms in the mode II expansion ($n_{\beta 2} = 22$) are used. Other computational details are the same as in Section 2.2. We now have

$$a_1 = 1.3881,\ a_2 = 0.1451,\ a_3 = -1.2453,\ a_4 = 0.2615,\ a_5 = -0.5348 \qquad \ldots(29)$$
$$b_1 = -0.37751,\ b_3 = -0.21353,\ b_4 = -0.20442,\ b_5 = 0.20892$$

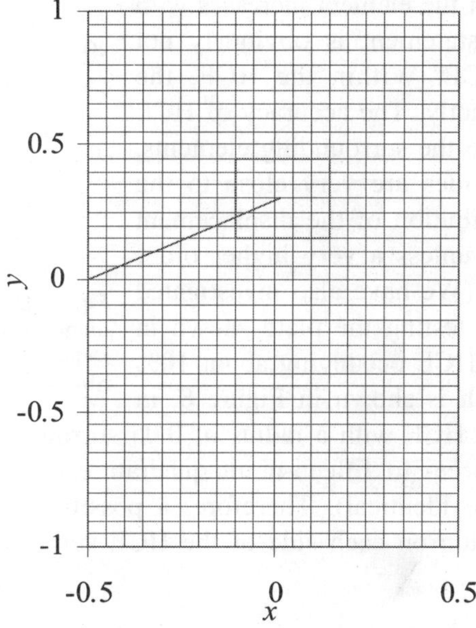

Fig. 7. Mesh with 20×40 uniform square elements and 861 nodes used to test the combination of HCE and XFEM. The HCE formed by the third layer of nodes surrounding the crack tip is shown in thicker solid line.

The above results (29) obtained by combining the HCE and XFEM using the very simple mesh shown in Figure 7 agree reasonably well with those obtained by combining the HCE with traditional FEM using a more complicated mesh shown in Figure 2(b) [13]

$$a_1 = 1.3867,\ a_2 = 0.1463,\ a_3 = -1.2419,\ a_4 = 0.24,\ a_5 = -0.5226 \qquad \ldots(30)$$
$$b_1 = -0.3762,\ b_3 = -0.2141,\ b_4 = -0.1954,\ b_5 = 0.1888$$

4. DISCUSSION AND CONCLUSION

Coefficients a_0 (18), b_0 (19) and b_2 (20) corresponding to rigid body translations and rotation which are excluded from the HCE formulation, are recovered by minimising the jumps between the element nodal displacements compatible with the surrounding elements and the truncated Williams expansions via a least squares method. Inclusion of these terms gives the complete displacements in the HCE, and eliminates the above jumps.

The combined HCE and XFEM can model a crack completely independent of the FE mesh. Moreover, it can evaluate directly and very accurately SIFs and coefficients of the higher order terms of the crack tip asymptotic field. The methodology is straightforward for modelling multiple and propagating cracks. However, in these cases, it may be necessary to resort to a HCE that is small compared to the whole crack. We have noticed that the accuracy of the HCE will reduce if its size is too small when it is combined with linear elements [3]. At a first glance, this seems in contradiction with the convergence theory of the FEM which states that the accuracy of the element increases with reducing size. However, this statement is obviously not true for the semi-analytic HCE. Within the HCE, the problem has been modelled exactly. The accuracy of HCE is controlled by the transition to the surrounding elements. If the HCE is too small, its sides are very close to the crack tip, so that a linear distribution of the displacement on each side is not sufficient, unless a very higher order HCE with many sides is used. (We have also investigated the influence of quadrature by testing the plate shown in Figure 2(a) using a very small HCE occupying about 10% of the crack. The adopted mesh is shown in Figure 8, in

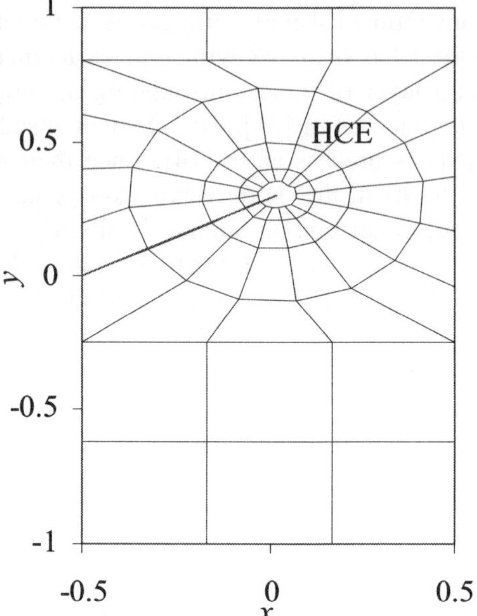

Fig. 8. Discretization of the plate in Figure 2(a) using a small circular 17-node HCE with a radius of 0.05 surrounded by 73 quadrilateral elements giving a total of 92 nodes.

which a small circular 17-node HCE with a radius of 0.05 surrounded by 73 quadrilateral elements giving a total of 92 nodes. Three- to fifteen-point quadrature rules are used for each side of the HCE. However, the results are identical). Therefore, a possible improvement is to use a higher order displacement approximation on each side of the HCE surrounded by higher order FEs.

ACKNOWLEDGEMENTS

Financial support from the European Commission KMM-Network of Excellence is gratefully acknowledged. Access to HSL packages MC30 and MA57 was graciously provided by the Numerical Analysis Group at Rutherford Appleton Laboratory.

REFERENCES

1. Williams ML (1957) On the stress distribution at the base of a stationary crack. ASME J Appl Mech 24: 109-14.
2. Owen DRJ, Fawkes AJ (1983) Engineering Fracture Mechanics: Numerical Methods and Applications. Pineridge Press, Swansea, UK.
3. Karihaloo BL, Xiao QZ (2001) Accurate determination of the coefficients of elastic crack tip asymptotic field by a hybrid crack element with p-adaptivity. Engng Fract Mech 68: 1609-30.
4. Karihaloo BL, Xiao QZ (2003) Linear and non-linear fracture mechanics. In Milne I, Ritchie RO, Karihaloo BL, editors. Comprehensive Structural Integrity, Volume 2: Fundamental Theories and Mechanisms of Failure (Volume editors, Karihaloo BL, Knauss WG), Chapter 2.03. Elsevier Pergamon, UK, p. 81-212.
5. Larsson SG, Carlsson AJ (1973) Influence of non-singular stress terms and specimen geometry on small-scale yielding at crack tips in elastic-plastic materials. J Mech Phys Solids 21: 263-77.

6.   Du ZZ, Hancock JW (1991) The effect of non-singular stresses on crack tip constraint. J Mech Phys Solids 39: 555-67.

7.   Chao YJ, Zhang X (1997) Constraint effect in brittle fracture. In Piascik R, editor. ASTM STP 1296, Fatig Fract Mech 27: 41-60.

8.   Dyskin AV (1997) Crack growth criteria incorporating non-singular stresses: size effect in apparent fracture toughness. Int J Fract 83: 191-206.

9.   Karihaloo BL, Abdalla HM, Xiao QZ (2006) Deterministic size effect in the strength of cracked concrete structures. Cem Concr Res 36: 171-188.

10. Tong P, Pian THH, Lasry SJ (1973) A hybrid element approach to crack problems in plane elasticity. Int J Numer Meth Engng 7: 297-308.

11. Xiao QZ, Karihaloo BL (2006) Determination of the complete displacement field inside a hybrid crack element. Submitted to Engng Fract Mech.

12. Wu CC, Bufler H (1991) Multivariable finite elements: consistency and optimization. Sci China (A) 34: 284-99

13. Xiao QZ, Karihaloo BL, Liu XY (2004) Direct determination of SIF and higher order terms of mixed mode cracks by a hybrid crack element. Int J Fract 125: 207-25.

14. Karihaloo BL, Xiao QZ (2004) Implementation of HCE on a general FE mesh for interacting multiple cracks. In Proc ECCOMAS, July 24 - 28, Jyväskylä, Finland (CD-ROM).

15. Duff IS, Reid JK (1983) The multifrontal solution of indefinite sparse symmetric linear systems. ACM Trans Math Softw 9: 302-25.

16. Moës N, Dolbow J, Belytschko T (1999) A finite element method for crack growth without remeshing. Int J Numer Meth Eng 46: 131-50.

17. Strouboulis T, Copps K, Babuška I (2001) The generalized finite element method. Comput Meth Appl Mech Eng 190: 4081-193.

18. Karihaloo BL, Xiao QZ (2003) Modelling of stationary and growing cracks in FE framework without remeshing: a state-of-the-art review. Comput Struct 81: 119-29.

19. Babuška I, Banerjee U, Osborn JE (2003) Survey of meshless and generalized finite element methods: A unified approach. Acta Numerica 12: 1-125.

20. Xiao QZ, Karihaloo BL (2005) Recent developments of the extended/generalized FEM and a comparison with the FEM. In Developments and Applications of Solid Mechanics (Proceedings of the Symposium on Prof M. G. Huang's 90th Birthday), Wu XP (ed). Press of University of Science and Technology of China, Hefei, China, 303-24.

21. Xiao QZ, Karihaloo BL (2006) Improving the accuracy of XFEM crack tip fields using higher order quadrature and statically admissible stress recovery. Int J Numer Meth Engng (in press).

22. Béchet E, Minnebo H, Moës N, Burgardt B (2005) Improved implementation and robustness study of the X-FEM for stress analysis around cracks. Int J Numer Meth Eng 64: 1033-56.

23. Laborde P, Pommier J, Renard Y, Salaün M (2005) High-order extended finite element method for cracked domains. Int J Numer Meth Eng 64: 354-81.

24. Liu XY, Xiao QZ, Karihaloo BL (2004) XFEM for direct evaluation of mixed mode SIFs in homogeneous and bi-materials. Int J Numer Meth Eng 59: 1103-18.

25. Xiao QZ, Karihaloo BL (2003) Direct evaluation of accurate coefficients of the linear elastic crack tip asymptotic field. Fatig Fract Engng Mater Struct 26: 719-30.

26. Xiao QZ, Dhanasekar M (2002) Coupling of FE and EFG using collocation approach. Adv Engng Softw 33: 507-15.

27. Chessa J, Wang HW, Belytschko T (2003) On the construction of blending elements for local partition of unity enriched finite elements. Int J Numer Meth Eng 57: 1015-38.

28. Lee SH, Song JH, Yoon YC, Zi GS, Belytschko T (2004) Combined extended and superimposed finite element method for cracks. Int J Numer Meth Eng 59: 1119-36.

29. Pian THH, Sumihara K (1984) Rational approach for assumed stress finite elements. Int J Numer Meth Eng 20: 1685-95.

30. Pian THH, Wu CC (1988) A rational approach for choosing stress term of hybrid finite element formulations. Int J Numer Meth Eng 26: 2331-43.

31. Xiao QZ, Karihaloo BL, Williams FW (1999) Application of penalty-equilibrium hybrid stress element method to crack problems. Engng Fract Mech 63: 1-22.

4

Modelling of Material Damage at Lower Length Scales

B.K. Dutta

*Reactor Safety Division, Bhabha Atomic Research Centre, Trombay,
Mumbai-400085, India email: bkdutta@barc.gov.in*

ABSTRACT

Modelling strategies at three length scales, namely micro, meso and nano are shown. At micro-level, the modelling of cleavage and ductile fractures has been carried out using corresponding micromechanisms. At meso scale, a new methodology is shown to model irradiated materials using the principles of dislocation dynamics. Atomistic level analysis is done to model high speed irradiation process using molecular dynamics.

Keywords: Cleavage fracture, Beremin's parameters, Ductile fracture, Charpy specimen, Dislocation dynamics, Irradiated material, Molecular dynamics.

1. INTRODUCTION

Modelling of material damage at lower length scales has drawn considerable attention of researchers in recent years. A number of models exist to address various phenomena of material behavior during damage. At micron level, the plastic or ductile rupture follows the nucleation, growth and coalescence of micro-voids with significant plastic deformation. The coupled constitutive models of Gurson and Rousselier are used for modelling ductile material. The Beremin model for cleavage fracture is based on a critical fracture stress together with the "weakest link" assumption and Weibull statistics. One needs to adopt the hybrid analytical methodology during the transition region from ductile to brittle fracture of material. The irradiation of materials by energetic particles causes significant degradation of the mechanical properties under service. This leads to an increase in yield stress and decrease in ductility limiting the lifetime of materials used in nuclear reactors. The microstructure of irradiated materials evolves over a wide range of length and time scales, making the radiation damage inherently a multi-scale phenomenon. An attempt is made here to find out the cascading effects in Aluminum crystal subjected to irradiation using molecular dynamics finite temperature modelling. In a second kind of study, dealing with higher length scale, the two-dimensional discrete dislocation dynamics (DD) modelling is done to study the irradiation effects on material stress-strain response.

2. TEMPERATURE DEPENDENT BEREMIN'S PARAMETERS FOR $20M_nM_oN_i5$-5 MATERIAL

Safety assessment of reactor pressure vessels (RPV) in light water reactors needs the analysis on possible fast fracture under normal and accidental conditions. During the reactor life, low alloy carbon steel which is generally used for the fabrication of pressure vessel material, tends to become brittle under the effect of neutron irradiation. In such cases, irradiation embrittlement is manifested by an increase in the ductile-brittle transition temperature. This kind of material embrittlement combined with a hypothetical accidental scenario such as "loss of coolant accident" (LOCA), may bring the material into ductile-brittle transition region. The reactor core flooded with emergency-core-cooling-system lowers the temperature of the pressure vessel and may fall in the transition domain. Under this scenario, micro-cracks in the component subjected to tensile stresses may lead to unstable cleavage fracture. The knowledge on fracture behavior of the material at transition region is required to quantify the inherent safety margin available under such undesirable events.

2.1 Experimental Details

In all, 60 round grooved tensile specimens (30 at −50°C and 30 at −100°C) have been tested using vaporized liquid nitrogen as cryogenic medium. The groove arc radius is 1.25 mm. All the experiments have been conducted under displacement controlled loading conditions. The load-displacement data were directly measured from the instrumentation provided in the machine. The images of the specimens at grooved region were taken at regular intervals during tests. These are used to calculate the diametral contraction as a function of loads. Image processing technique has been used for this purpose as explained below. Fig. 1 shows a typical arrangement of the experimental set-up.

The loading was applied at a strain rate of 2×10^{-4} per sec. The temperature inside the experimental chamber was controlled by maintaining proper flow rate of vaporized nitrogen into it, using semiconductors, thermocouples and attached sensors. The loading was provided through hydraulic servo-mechanism, using embedded control panel and controlling software.

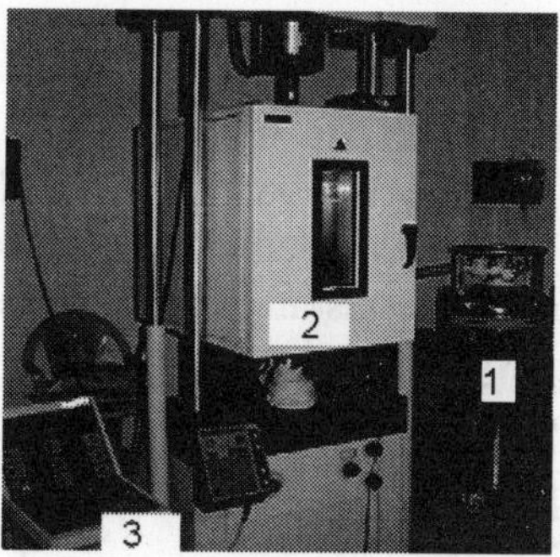

Fig. 1. Experimental set-up.

2.2 Experimental Results

The variations of load with diametral contraction are obtained using the experimental load-displacement data and the images taken. An in-house image processing code has been used for this purpose. The steps used to get the diametral contraction are as follows:

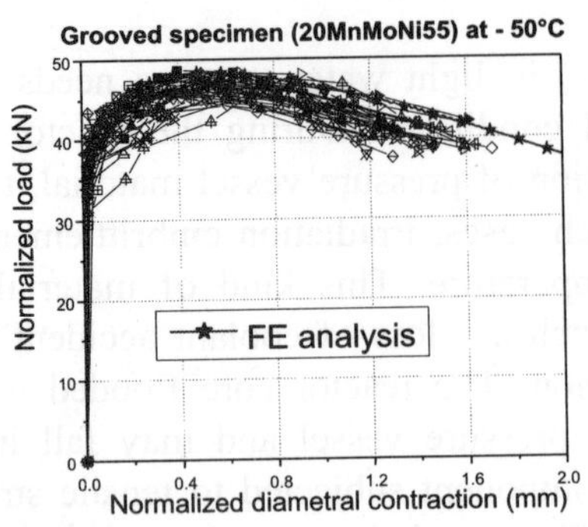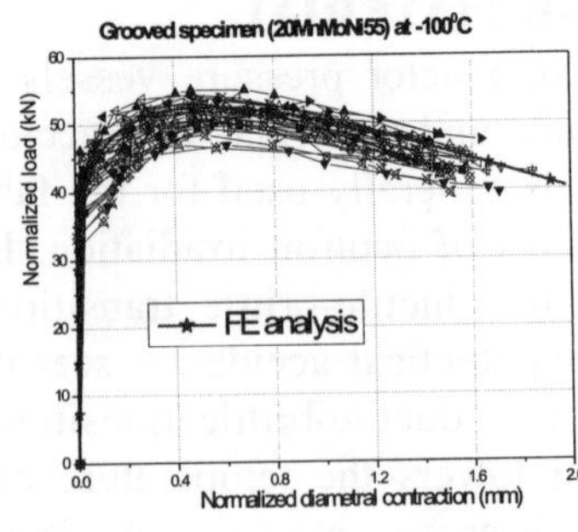

Fig. 2. Load v/s diametral contraction for tests carried out at −50°C and −100°C along with FE results.

(a) Any two vertical lines, each on the opposite diametral edges of the large-diameter-segment are marked.

(b) Corresponding pixels of these two lines are used to calculate the scale, i.e., mm/pixel. This is done with the assumption that the deformation of the large-diameter-segment is negligible during the experiment.

(c) Another vertical line, tangent to the groove is marked and the corresponding pixel number is read.

(d) The groove-diameter now is calculated using the scale of the image.

All the images have been processed using this procedure.

Load v/s diametral contraction for each specimen was calculated using this procedure. There was variation in the initial grooved diameters of the specimens due to tolerance in fabrication. Hence, the results based on actual diameters were normalized with respect to an average diameter to avoid any statistical variation due to this factor. The average diameter for −50°C specimens is 7.35 mm and for −100°C is 7.34 mm. The calculated load v/s diametral contraction curves are shown in Fig. 2.

2.3 Beremin's Model and Definition of Weibull Stresses

Based on weakest link assumption [1] and Weibull statistics, Beremin developed a model for the analysis of brittle fracture process [2]. The two Weibull parameters are determined using experimental fracture data and FE analysis. These parameters are useful to predict probability of cleavage fracture of a component. The failure probability is given as follows.

$$P_f(\sigma_w) = 1 - \exp\left[-\left(\frac{\sigma_w}{\sigma_u}\right)^m\right] \qquad \sigma_w = \left(\sum_{i=1}^{n_{pl}} (\sigma_I^{(i)})^m \frac{V_i}{V_0}\right)^{1/m}$$

Here, σ_u is the scaling parameter, which describes the mean value of distribution function on the stress axis at $P_f = 0.632$, or 63.2% failure probability and 'm' is the Weibull exponent or shape parameter which describes the slope of P_f. The second expression shows the definition of Weibull stress σ_w. Here, σ_I is the maximum principal stress, n_{pl} is the number of elements undergone plastic deformation, V_i is the volume of the ith element and V_0 is the reference volume, which is taken as 10^{-3} mm^3.

2.4 Experimental Probability of Failure as a Function of Weibull Stresses

Elastic-plastic FE analyses of the specimens have been carried out to get the stress distribution at various loads. An in-house FE code MADAM has been used for this purpose. Eight node axisymmetric isoparametric elements have been used for modelling with a minimum mesh size of 50 μm near the highly stressed region. Computed load v/s diametral contraction curves are also depicted in Fig. 2, showing appreciable matching between the experimental and the FE results. As the ductile material damage at grooved region is not considered in conventional elastic-plastic analysis, the computed results have been found to be upper bound in comparison to the experimental results. Experimental results show that a fair amount of ductility is retained in the material at both –50°C and –100°C. Hence, Beremin's parameters for the material at these two temperatures have been calculated by ranking the specimens in order of diametral contraction at failure. Experimental probabilities of failure based on diametral contraction are shown in Fig. 3.

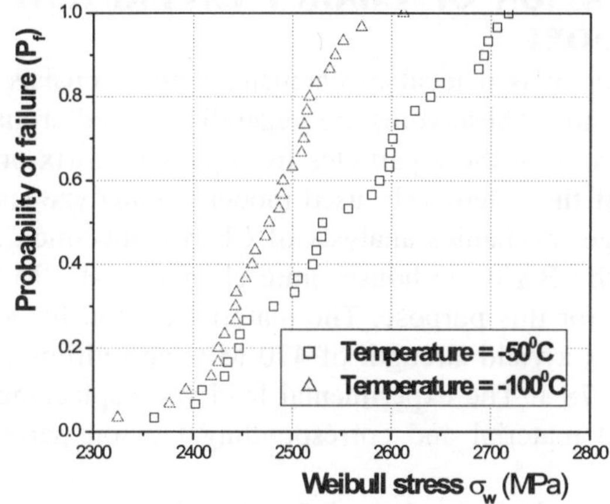

Fig. 3. Variation of P$_f$ with Weibull Stress for two different temperatures.

2.5 Maximum Likelihood Method and Determination of Beremin's Parameters

Based on [3, 4], the values of the two parameters 'm' and 'σ$_u$' are sought to have a Weibull function describing the experimental data that are most likely. The estimated set of Weibull parameters should maximize probability of occurrence of experimental results. This is equal to the probability that all 'σ$_{wi}$' values occur simultaneously, which is given by the product of all the fracture probabilities. The expressions for 'σ$_u$' and 'm' are obtained by maximizing the likelihood function. The expressions for 'σ$_u$' and 'm' derived using such procedure are shown below. These two non-linear equations are solved iteratively to arrive at the values of 'm' and 'σ$_u$'.

$$\frac{N}{m} + \sum_{i=1}^{N}\ln\left(\sigma_{w_i}\right) - N.\frac{\sum_{i=1}^{N}\left(\sigma_{w_i}\right)^{m}.\ln\left(\sigma_{w_i}\right)}{\sum_{i=1}^{N}\left(\sigma_{w_i}\right)^{m}} = 0 \qquad \sigma_u = \sqrt[m]{\frac{1}{N}\sum_{i=1}^{N}\left(\sigma_{w_i}\right)^{m}}$$

2.6 Temperature Dependency of Beremin's Parameters

The above procedure is adopted using in-house software to find out Beremin's parameters for the present material $20M_nM_oN_i5$-5 at $-50°C$ and $-100°C$. Table 1 shows the computed values of 'σ_u' and 'm'. Based on the results quoted in Table 1, two equations $\sigma_u = 2200 + 1.78T$ and $m = 100 - 0.32T$ have been derived for determining the Beremin's parameters as a function of temperature. Here, T is the material temperature in degree Kelvin.

Table 1. Beremin's parameters at $-50°C$ and $-100°C$

$-50°C$ Temperature		$-100°C$ Temperature	
σ_u	m	σ_u	m
2598	28.50	2506	44.14

3. ANALYTICAL SIMULATION OF CHARPY EXPERIMENT USING GURSON MODEL

The ductile fracture phenomenon is a local mechanism, which consists of nucleation, growth and coalescence of microscopic voids. These voids are generally formed around the inclusions or second phase particles due to de-bonding of these particles from parent matrix or cracking of these particles itself. Gurson model is one of the extensively used models to analyze such kind of ductile fracture. In the present work, a damage mechanics analysis of Charpy specimen has been carried out under high strain rate of loading. The BARC in-house finite element code 'MADAM' based on modified Gurson model has been used for this purpose. The material chosen for analysis is German standard steel StE460. The material has a yield strength of 470 MPa and ultimate strength of 633 MPa. The upper shelf Charpy energy is 78 J. The experimental load v/s displacement curve of a Charpy side-grooved specimen of StE460 material and corresponding Gurson parameters are available in the literature.

A unique feature of the analysis is to apply *node release technique* along with the Gurson model. The node release technique is generally used in conventional elastic-plastic fracture mechanics analysis along with material J_R curve to obtain crack growth. However, such efforts face computational difficulties leading to divergence. The primary reason for the divergence in such analysis is the non-consideration of stress unloading in the surrounding material at the crack tip. The unloading occurs due to the formation of voids at crack tip just before ductile crack growth, which is not considered in the material constitutive law of conventional elastic-plastic analysis. On the other hand, the Gurson material constitutive law considers the formation of voids and growth at the crack tip leading to stress unloading before fracture. Hence, one can adopt the node release technique along with the Gurson model to simulate crack growth, without facing any difficulty of divergence. In the present work, such strategy has been used to obtain crack growth as a function of time in a Charpy test.

3.1 Two-Dimensional Analysis of Side-grooved Charpy Specimen

Figure 4. shows a schematic drawing of a Charpy test arrangement. The total length of the specimen is 55 mm and the rectangular cross-section is 10×10 mm. The specimen has 2 mm deep V-shaped notch with radius 0.25 mm and the depth of side-grooves is 1 mm. The specimen is positioned upon two anvils with a span of 40 mm. For FE analysis, only half of the specimen is modelled due to symmetricity. The FE model is shown in the Fig. 5, which use 8-noded isoparametric quadrilateral elements. The number of nodes is 1537 and the number of elements is 480. The smallest element size is 0.2 × 0.2 mm, which has been used all along the ligament of crack line. The load is applied at a single node. Plane strain case has been assumed due to the presence of side-grooves in the specimens.

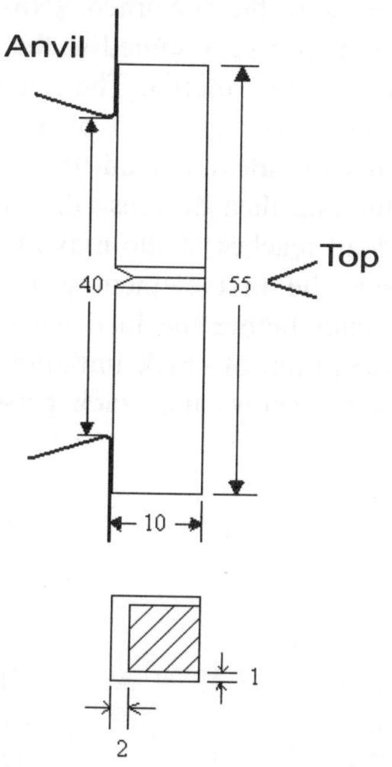

Fig. 4. Schematic arrangement of a Charpy test.

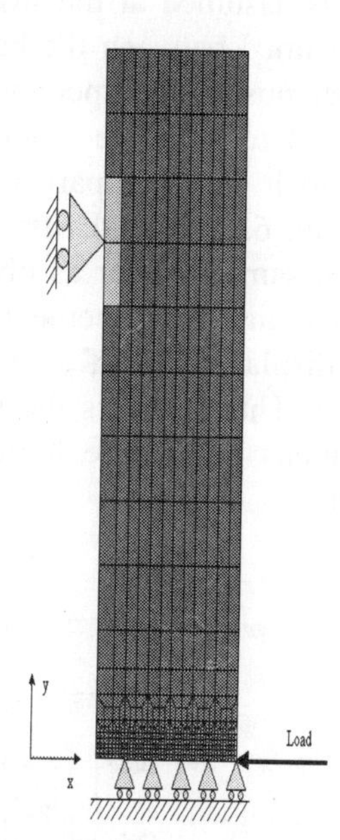

Fig. 5. Discretized model of one half of the Charpy specimen.

3.2 Material Properties and FE Code MADAM

The Gurson parameters for StE460 are shown in the following table.

Parameter	n	f_0	f_c	f_f	q_1	q_2	q_3	ε_n	s_n	f_n	f_u^*
Value	0.013	0.001	0.021	0.19	1.50	1.00	2.25	0.3	0.1	0.01	0.505

The equation $\sigma_y = \sigma_{y0}(\bar{\varepsilon}_m^{pl}, T)\left(\dfrac{\dot{\bar{\varepsilon}}_m^{pl}}{\dot{\bar{\varepsilon}}_{m0}^{pl}}\right)^n$ is used to modify the true stress-strain curve to consider the effect of high strain rate on plastic behavior. The analysis is done at room temperature. An in-house finite element code MADAM, which is based on the principles of continuum damage mechanics (Modified Gurson's model) has been used for the analysis. The code has the ability to solve both two and three-dimensional problems. The geometric non-linearity is considered by using the updated Lagrangian formulation. The large deformation and rotation in the elastic-plastic analysis is considered by transforming the computed Cauchy stress to Jaumaan stress rate using the spin rate tensor. The SPARSE solver has been used for solving linear set of simultaneous equations. The load-deformation equilibrium conditions are obtained by using Newton-Raphson algorithm. The generalized mid-point algorithm formulated by Zhang has been used to integrate pressure dependent elastic-plastic Gurson yield model.

3.3 Results

The zero time is assumed at the instant hammer comes in contact with the specimen. 'No-friction' condition is assumed between the hammer and specimen surfaces. It is also assumed in the analysis that the hammer pushes the specimen at constant velocity due to its large inertia. The node release technique is used to simulate crack growth for the elements completely damaged by the Gurson constitutive model. The comparison between the calculated and experimental load-time curves is shown in the Fig. 6. The load first increases to a maximum value and then decrease due to ductile crack growth in same manner as observed during the test. The load reaches to the maximum value of 14kN at load line displacement of about 1.35 mm. Whereas, the first element is completely damaged at a displacement of about 0.7 mm; i.e., the crack initiates before the load reaches to its maximum point. This supports the well-known experimental observation of crack initiation at side-grooved specimens before the force maximum is reached. Figure 8 shows the crack growth with time of impact.

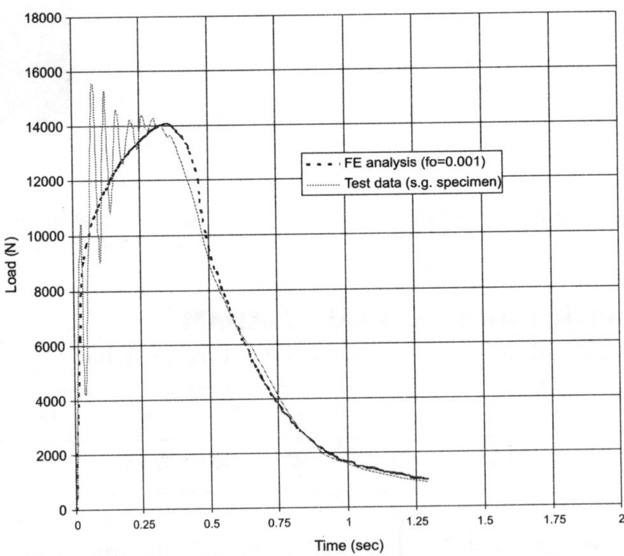

Fig. 6. Comparison of load-time curve.

3.4 Analysis of Charpy Test at Various Strain Rates

The FE model shown above has been then used to check the effect of strain rate on the load-time characteristic of the Charpy test. Fig. 7 shows the computed load-time curves for various strain rates. The increase in maximum load due to strain rate effect is visible in these curves.

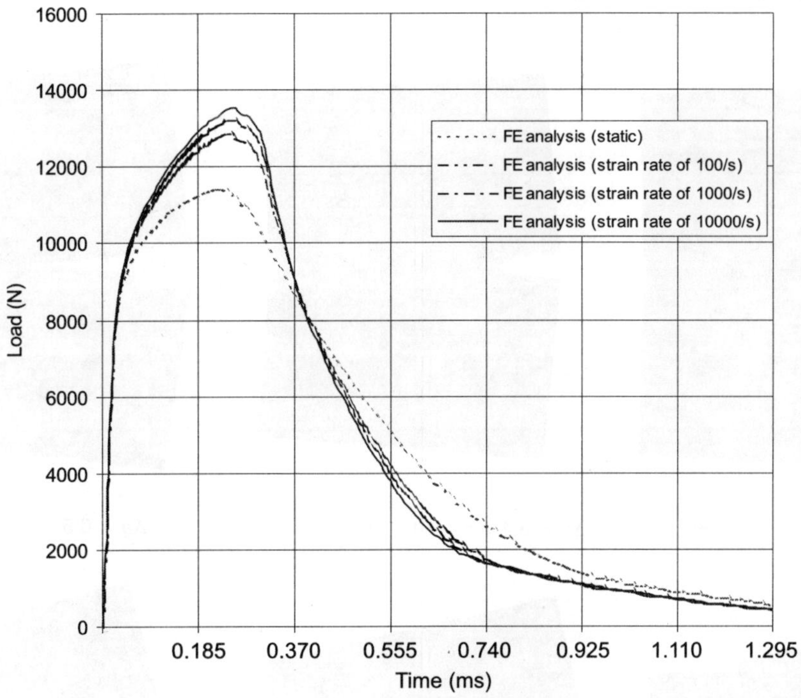

Fig. 7. Load-Time curves for different values of strain rates.

4. MODELLING OF IRRADIATED MATERIAL USING DISLOCATION DYNAMICS

4.1 Discrete Dislocation Dynamics Modelling

In a different kind of study dealing with lower length scale, the two-dimensional discrete dislocation dynamics (DD) modelling has been done to study the irradiation effects on material stress-strain response. An in-house code for DD analysis is developed based on [9]. The plastic flow is represented by collective motion of a large number of edge dislocations. The dislocation fields are specified by continuum elastic theory. Since, the elastic fields act infinite in medium, corrections for boundaries are specified by a complimentary problem which consists of solving a linear elastic boundary value problem through finite element method. The dislocation phenomenon like annihilation, generation and pinning of dislocations by obstacles are incorporated in the model through some constitutive rules. Irradiation effects are numerically modelled by locking all the dislocations with irradiation induced defects thus characterizing the fluence. These dislocations get unlocked when the stress on them exceeds a critical stress due to irradiation defects. The stress-strain response of an irradiated Copper as a function of total fluence is studied.

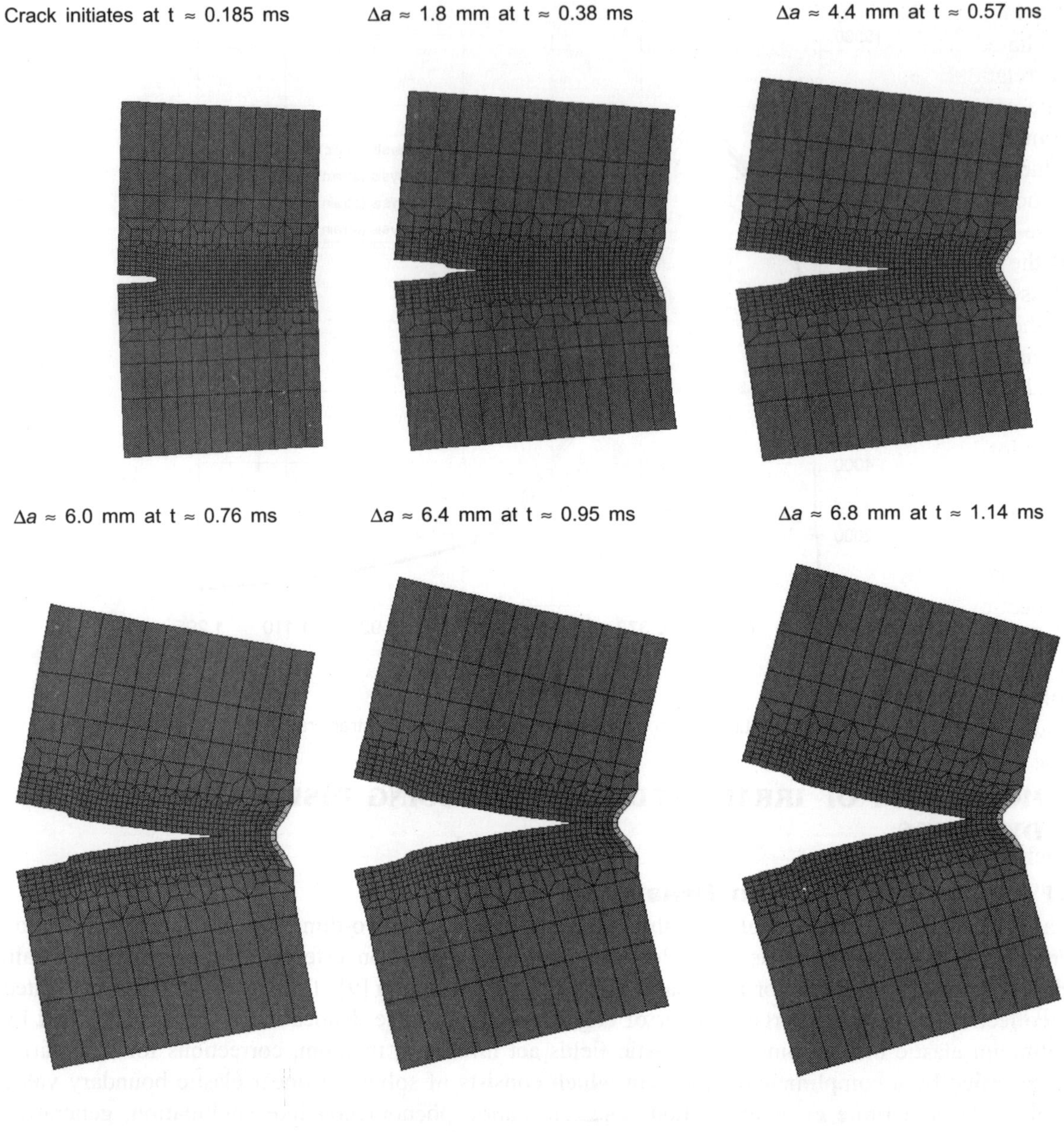

Crack initiates at t ≈ 0.185 ms   Δ*a* ≈ 1.8 mm at t ≈ 0.38 ms   Δ*a* ≈ 4.4 mm at t ≈ 0.57 ms

Δ*a* ≈ 6.0 mm at t ≈ 0.76 ms   Δ*a* ≈ 6.4 mm at t ≈ 0.95 ms   Δ*a* ≈ 6.8 mm at t ≈ 1.14 ms

Fig. 8. Computed Crack Growths during Charpy Test at various time instants.

4.2 Stress-Strain Response of Irradiated Copper

Here, we consider the problem of irradiation-induced hardening in Copper. Plastic deformation and
hardening in irradiated materials is controlled primarily by the defects due to irradiation (vacancies,
self-interstitial atoms and SFT's) and their interaction with dislocations. These defect clusters will
tend to form loops around existing dislocations, leading to their decoration and immobilization

[10, 11]. In order to understand the effect of this phenomenon on deformation behavior, we consider irradiated copper under simple shear using two-dimensional DD model. An attempt is made to find the relationship between the irradiation fluence and yield stress of Copper using DD modelling. In spite of the fact that this phenomenon is fully three-dimensional in nature, first order prediction can be made of yield stress as a function of fluence level using the two-dimensional model. The irradiation induced hardening may be understood in terms of cascade induced source hardening in which the dislocations are considered to be locked by the loops decorating them. The density and strength of these defect loops depend on the fluence level. To mimic these irradiation effects, we assume that all the dislocations are locked by the defect loops thus characterizing the fluence. When the total stress on the dislocations exceeds a critical value σ_{cr}, they get unlocked and become free to move on their glide planes. During gliding, any dislocation may get pinned down by the point obstacles or may get annihilated by an opposite dislocation. New dislocations will be generated by Frank-Read mechanism, which is mimicked here by point sources. These sources will nucleate a dislocation dipole when stress on a source exceeds a critical value. The effect of irradiation on stress-strain behavior is then obtained by subjecting the copper unit cell to simple shear. The simulated copper cell is assumed to be of dimensions 2 µm × 2 µm. The plastic flow is represented by a collection of large number of edge dislocations. We assume that the material consists of randomly distributed defect structure, such as point obstacles and Frank-Read dislocation nucleation sources. The material is assumed to have an initial dislocation density, which is then relaxed. During relaxation, the dislocations will interact with each other and try to attain equilibrium position. In the relaxed configuration, obstacles and nucleation sources are randomly generated. All the obstacles are assumed to have same strength. The strength of sources is selected randomly from a Gaussian distribution with a mean strength. Typical predicted stress-strain curves for various critical values of σ_{cr} are shown in Fig. 9. In order to eliminate any numerical fluctuations, averaging over 5-6 samples is done to obtain the observed stress-strain behavior. The yield stress σ_y increases from 6 MPa to 40 MPa by increasing σ_{cr} from 14 MPa to 100 MPa. Without irradiation, the single Copper crystal yields at 2-4 MPa [12]. When irradiation induced-defects are present, the yield point rises drastically which can be attributed to the locking of dislocations by these defects. Moreover at very high values of fluence, the experimentally observed instability can also be reproduced by DD modelling as shown in Fig. 9 for high values of σ_{cr}. At larger values of fluence, two characteristics are seen. First, the system yields at very high stress and second, a sudden instability occurs after reaching a maximum stress value. This can be attributed to a sudden unlocking of a large chunk of dislocations from the irradiation-induced defects. This is clearly seen by points a, b and c in Fig. 9 corresponding to σ_{cr}=100 MPa. The system yields at point a; there is strain hardening up to point b and then a sudden drop in stress value occurs (point c). This effect can be clearly seen again at point a, b and c in Fig. 10, where the variation of number of dislocations being locked by irradiation-induced defects is shown as a function of shear strain. As seen from point b to c, a large number of dislocations are unlocked from irradiation induced defects and hence the instability in stress-strain response occurs. These results show that such instabilities can be reproduced by numerical DD modelling. Finally, to correlate the fluence and critical stress σ_{cr}, we made use of experimental yield stress values (at 77°K) of single crystals of Copper irradiated at different fluence levels. Table 2 shows these experimental upper yield points as a function of irradiation fluence ϕ.

From the results of present DD model (see Fig. 9), the critical stress values σ_{cr} corresponding to the experimental yield stress values are obtained. In Fig. 11, these critical stress values

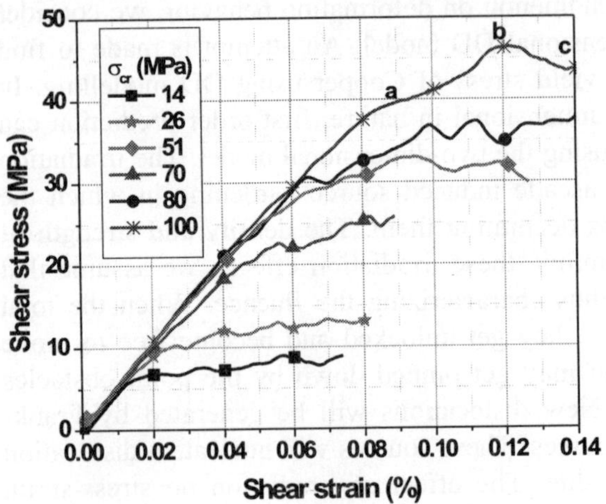

Fig. 9. Stress-strain response of irradiated Copper as a function of critical locking stress σ_{cr}.

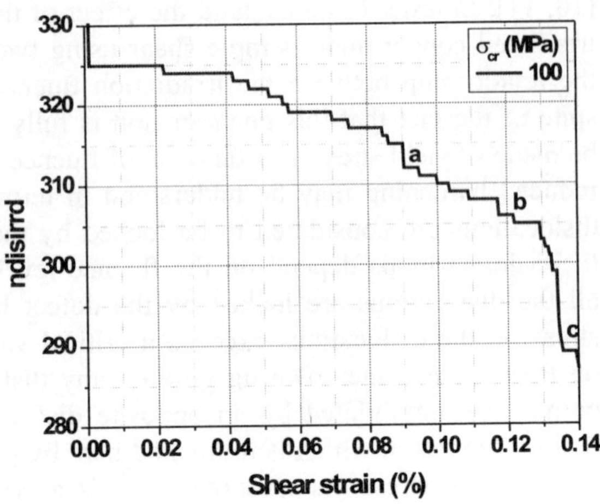

Fig. 10. Variation of locked dislocations for $\sigma_{cr} = 100$ MPa.

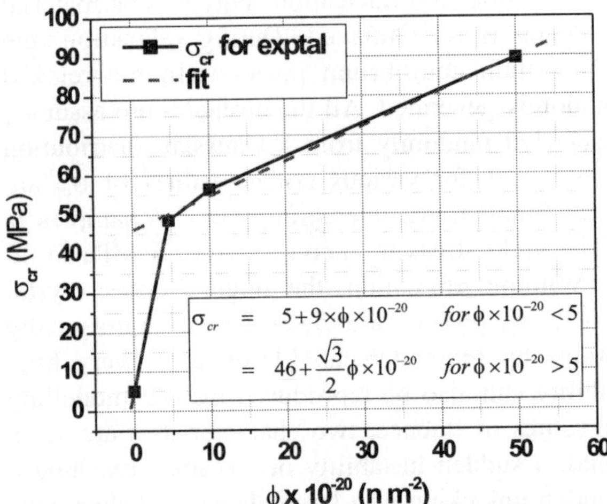

Fig. 11. Variation of critical locking stress σ_{cr} as a function of fluence ϕ.

Table 2. Experimental yield stress values of irradiated Copper single crystal at 77°K

Irradiation fluence ϕ (nm^{-2})	Upper yield stress (MPa)
0	3
4.7×10^{20}	18.75
1.0×10^{21}	21.7
5.0×10^{21}	35.38

(σ_{cr} -exptal) are plotted as function of irradiation fluence ϕ. We also tried to fit curves to correlate critical stress values in terms of parameter ϕ. These are given below:

$$\sigma_{cr} = 5+9\times\phi\times10^{-20}; \quad for\ \phi\times10^{-20} < 5$$

$$= 46+\frac{\sqrt{3}}{2}\phi\times10^{-20}; \quad for\ \phi\times10^{-20} > 5$$

Such expressions can then be used to find out critical locking stress σ_{cr} at any other fluence ϕ. To find out the yield stress σ_y at any fluence ϕ, first σ_{cr} is determined using these expressions. Then the DD model is run with this σ_{cr} to find out the yield stress of the crystal at that fluence level.

Even though current model is two-dimensional and considers only edge dislocations, the results illustrate the use of DD simulations for first order prediction of σ_y as a function of fluence.

5. ATOMISTIC MODELLING

The process of irradiation involves extremely small time and length scales, which makes experimental study of the phenomenon impractical. With the ever-increasing computational power associated with advance techniques to simulate the material at atomistic level, it is possible to simulate the process of irradiation and displacement cascading. Molecular Dynamics (MD) provides a suitable tool to simulate the process of irradiation with high degree of accuracy. A typical MD system for irradiation simulation consists of model of crystal with thousands of atoms, which interacts with each other with many body potential (MBP). Embedded Atom Method (EAM) potential is a MBP and is used to simulate the metals successfully.

One of the atoms in the domain is selected as PKA and it is provided with desired kinetic energy. The kinetic energy of PKA is dissipated in the lattice with the time and in this process displacement cascades is produced. A region of highly perturbed atoms is formed in the core of the system and in this core region temperature reaches extremely high values. The temperature of the core is gradually dissipated into the system. The simulation is continued till the system reaches uniform temperature. MD simulation is performed in NVE ensemble (constant volume and energy) and no attempt has been made to control the temperature. The output of the simulation is in the form of trajectory of atoms displaced due to cascade, number of produced SIA-vacancy pairs (N_F) and their distribution in the lattice [5].

5.1 Semi-Empirical Relations

NRT model by Norgett *et al.* [6] provides theoretical assessment of defect production in displacement cascades by deriving simple relationship between the number of SIA-vacancy pairs (N_{NRT}) created by cascade and the kinetic energy E_P of PKA.

$$N_{NRT} = 0.8\,E_{dam}\,/(2\,\bar{E}_d)$$

where N_{NRT} is the value of N_F in the Norgett, Robinson and Torrens (NRT) formulation, E_d is the value of the threshold displacement energy over all crystallographic directions and E_{dam} is the damage energy available for elastic collision, i.e., E_P with inelastic losses subtracted (since, losses have not been included in most MD simulations the replacement of E_{dam} by E_P is appropriate for comparing its prediction with N_F obtained from MD).The binary collision model on which the NRT formula is based does not accurately describe atomic interactions take place in the heated core due to cascade and is not suitable for modelling the actual configuration of defects. MD simulation on the other hand, use inter-atomic potential fitted too many of the equilibrium and defect properties of metals and do offer more realistic description of cascade process [7].

NRT model does not predict the accurate number of defects. In fact, N_F is typically 20% to 40% of N_{NRT} for a given cascade energy larger than about 1-2 Kev. Considering MD generated N_F data for several metals, it was shown that an empirical relationship between N_F and E_P, which gives a good fit to the simulation data for E_P up to 10 Kev, is given as follows [8].

$$N_F = A\left(E_p\right)^m$$

Here, *A* and *m* are constants and are weakly dependent on the material and temperature.

5.2 MD Simulation Details

The number of SIA-vacancy pairs (N_F) produced in displacement cascade using in-house MD simulation code for Nickel with different recoil energy of PKA is estimated. Cubical and parallelepiped domain has been taken for the analysis and the size of computational domain is taken proportional to the recoil energy of PKA so that the final temperature of the domain should remain in desired limits. A typical size of domain taken for 5 Kev recoil energy of PKA is 20 unit cells in each direction with 32000 atoms in the computational domain. The system is equilibrated until the temperature of the domain becomes uniform (approximately 5 pico-second). In the cascading process, a number of collisions take place and atoms come very close to each other. This reduces atomic distance and poses stringent requirement on the size of the time step. It is possible to use adaptive time step, one can speed up the MD simulation. The force of repulsion between two closely spaced atoms is exponential in nature. This prompts us to choose time step having exponential dependency on the minimum value of separation and inversely proportional to the maximum value of speed. Hence,

$$\Delta t = A \exp(-B\, r_{min})/v_{max}$$

Here, A controls the minimum value of the time step, B controls the rate of decrement of time step and v_{max} is the maximum speed in the domain. We have used this adaptive time step in our MD run to simulate the cascading process optimally. The size of time step varied from 10^{-16} to 10^{-18} seconds.

5.3 Results of Atomistic Simulations

We have performed MD simulation for three different recoil energies of PKA, these are 2 Kev, 3 Kev and 5 Kev. Domain size for 2 Kev and 3 Kev MD run is $20 \times 16 \times 16$ unit cells having 20480 atoms in the domain. Cubical domain has been taken for the 5 Kev run with 20 unit cells in each direction and the total number of atoms in the domain is 32000. Initial temperature of the domain for each run is taken 50 K. The final temperatures after the irradiation process for 2, 3 and 5 Kev simulations are found to be 464 K, 594 K and 667 K, respectively. The trajectory of the perturbed atoms in displacement cascade are depicted in Fig. 12. Figure 13 shows the distribution of SIA-vacancy defects (N_F) in the domain in typical displacement cascade. The values of N_F generated for Nickel with different recoil energy of PKA are shown in Table 3. The N_F values generated are in close agreement with the values predicted by using above analytical equations. Values of A and m are taken from as 4.37 and 0.74, respectively.

Table 3. Vacancy-interstitial pairs for different PKA energies

Recoil energy of PKA (Kev)	No. of vacancy/ interstitials pairs N_F	
	Numerical Simulation	Proposed Empirical Relation
2.0	9	7
3.0	10	10
5.0	12	14

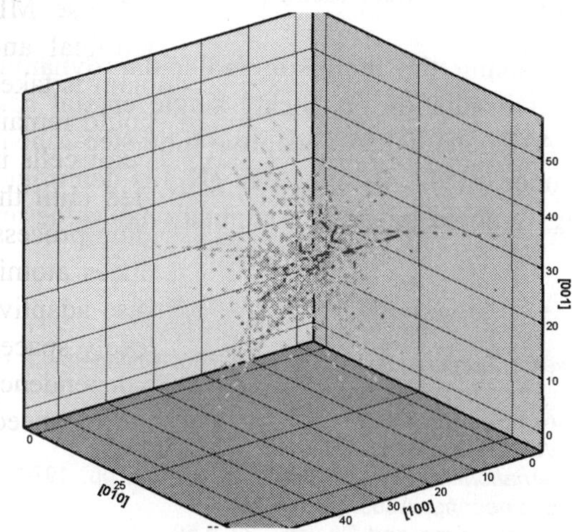

Fig. 12. Trajectory of atoms which acquire at least 1 ev KE.

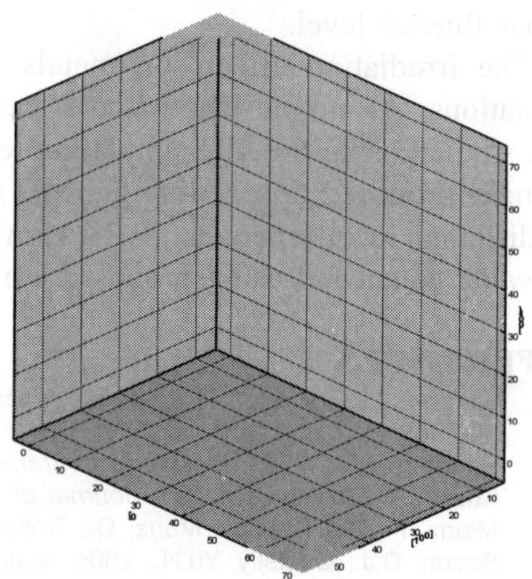

Fig. 13. Vacancy (open circles) and interstitial (filled circles) pairs generated as result of displacement cascade.

CONCLUSION

Cleavage fracture in low alloy steel usually occurs by unstable micro-cracks initiated at brittle second phase inclusions. Due to scatter in size, orientation and shape of such inclusions and the non-uniform stress field around a loaded crack tip, the event of cleavage fracture follows a statistical nature. The results shown here confirm that the probability of failure is higher at lower temperature for a particular value of Weibull stress. This behavior is expected due to better ductility at higher temperature. It is also clearly revealed that the Beremin's parameters are temperature dependent. The value of σ_u increases with the temperature in conformity with the variation of P_f with temperature. At the same time, the value of 'm' decreases with temperature showing the large spread in σ_w at failure. These Beremin's parameters are obtained as first approximations based on the results from two temperatures only. A number of tests have been planned at various temperatures for the same material to confirm and fine tune the constants derived the equations of 'σ_u' and 'm' as a function of temperature.

The Gurson model along with the node release technique may be used to simulate a Charpy test to generate load-displacement and time-crack growth curves. The comparison between computed and experimental results showed the power of Gurson model to simulate such complex material testing. The simulation is further extended to generate these curves for various strain rates.

Two-dimensional discrete dislocation modelling is used to determine irradiation induced hardening of Copper single crystal. Dislocations get pinned down due to irradiation induced defects. This is modelled by locking all the dislocations by critical stress σ_{cr}, thus characterizing the effect of fluence phenomenologically. The dislocations are required to overcome this σ_{cr} to get unlocked before they can move on their glide planes under external stress. Parametric study with respect to the critical locking stress and its effect on stress-strain response is studied. An expression for this critical locking stress in terms of irradiation fluence is proposed. The yield stress of irradiated Copper crystal can then be determined by the DD model by taking the critical locking stress corresponding to that fluence level.

The irradiation effects on metals are also studied using the atomistic molecular dynamics simulations. At atomic length scale, cascading effects of irradiation on Nickel single crystal have been studied using the MD simulation technique with EAM potential. An adaptive time step scheme has been proposed to speed-up the MD simulation. Number of SIA-vacancy pair (N_F) is determined for different recoil energies of PKA. A close matching is noted between computational value and the value calculated using proposed empirical relation.

REFERENCES

1. Landes, J.D., Shaffer, D.H., 1980, *Fracture Mechanic, Twelfth Conference,* ASTM STP 700, Philadelphia, pp. 368-382.
2. Beremin, F.M., 1983, *Metallurgical Transactions A,* Vol. 14A, pp. 2277-2287.
3. Khalili, A., Kromp, K., 1991, *Journal of Material Science,* Vol. 26, pp. 6741-6752.
4. Miami, F., Bruckner, F., Munz, D., Trolldenier, B., 1992, *International Journal of Fracture,* Vol. 54, pp. 197-210.
5. Bacon, D.J., Osetsky, Yu.N., 2004, *Material Science and Engineering,* A365, pp. 46-56.
6. Norgett, M.J., Robinson, M.T., Torren, I.M., 1975, *Nuclear Engineering and Design,* 33 p.50.
7. Bacon, D.J., Gao, F. Osetsky, Yu.N., 2000, *Journal of Nuclear Material,* 276, p. 1.
8. Bacon, D.J., Calder, A.F., Gao, F., Kapinos, V.G., Wooding , S.J., 1995, *Nuclear Instruments and Methods,* B102, p.37.
9. Giessen, E.V., Needleman, A., 1995, *Modelling and Simul. Mater. Sci. Eng.* 3, pp. 689-735.
10. Trinkaus, H., Singh, B.N., Foreman, A.J.E., 1997, *J. Nucl. Mater.* 251, p172.
11. Zbib, H.M., Rubia, T.D., Rhee, M., Hirth, J.P., 2000, *J. Nucl. Mater.* 276, pp.154-165.
12. González, H.C., Miralles, M.T., 2001, *J. Nucl. Mater.* 295, pp. 157-166.

5

A Discussion on Some Key Issues Related to Computational Analysis of Laminated Structures

C.S. UPADHYAY, P.M. MOHITE AND A. ONKAR

Department of Aerospace Engineering Indian Institute of Technology Kanpur, Kanpur-208016 India

ABSTRACT

Laminated composite components are now commonly used in engineering structures. With the increased use of composites in engineering applications, it is essential to develop suitable computational analysis tools, which can be used effectively in the design process.

Due to the layered construction of these structures, interfacial stresses are strong. Further, the directional nature of the material can lead to significant boundary-layer effects. Often, these structures are also thin. Hence, the computational analysis tool should be able to accurately take care of these effects. Several laminated plate models have been proposed in the literature. These are primarily equivalent layer theories (shear deformable theories). Recently, there has been significant interest in layer-by-layer models. These models are three-dimensional models and hence are computationally very expensive. The talk will discuss the quality of the equivalent and layer-by-layer models and will also present a new approach, called the region-by-region modeling, using which three-dimensional theories can be effectively used in regions of high stress gradients, interfaces of interest, delamination and damage while in the rest of the plate the (cheaper) equivalent model can be used. Issues related to the automatic selection of models in various regions of the domain, using a modeling error prediction algorithm, will also be discussed. It will be shown that without this approach, it will be impossible to guarantee reliability of results from a computational analysis.

The study will also discuss the use of continuum damage models to effectively study damage in laminated composite structures. The effect of existing damage on the response characteristics will also be discussed. Applications to the buckling problem will be considered as a special case.

Most of the computational analysis at the macro-level is deterministic. However, being inherently heterogeneous in nature at the micro-level, the composites are subject to uncertainties in the micro-level material constitution and distribution. The effect of these uncertainties, on the macro-level response of the structure, can be significant. Incorporating the scatter in the micro-level properties in a micromechanical analysis can give proper macro-level response variations. In this study, a simple micromechanical analysis is used to get a representation of the possible scatter in the macro-level properties. This is used to find the covariance of the buckling load and first-ply failure load for laminated plates.

PRELIMINARY RESULTS

The following table shows the influence of 5% COV in all five basic micro-level constituents, namely, E_f, v_f, E_m, and V_f, changing simultaneously on COV in the effective moduli of various fiber-reinforced composite systems.

Table 1. Effect of COV of 5% in all micro-level properties changing simultaneously on the various macro-level effective material properties for different composite systems.

Different composite system	COV in macro-level effective material properties (in %)					
	E_{ll}	E_{tt}	V_{lt}	V_{tt}	G_{lt}	G_{tt}
graphite-epoxy	6.97	3.88	3.42	3.30	11.68	2.68
boron epoxy	7.00	3.95	3.38	3.42	11.92	2.69
carbon-epoxy	6.96	3.92	3.39	3.40	11.59	2.68
glass-epoxy	6.72	3.71	3.42	3.26	10.22	2.62

These uncertainties can make the structure unstable at a load less than a specified fraction of perfect critical load. The stochastic stability of an undamaged composite laminates can be performed using linear analysis. To describe the mean and variance of buckling strength of undamaged composite plates a mean centered first-order second-moment perturbation technique is used. The following figure shows the mean buckling behavior of shear loaded square laminated plates $[-\theta / \theta]$ with and without cutout.

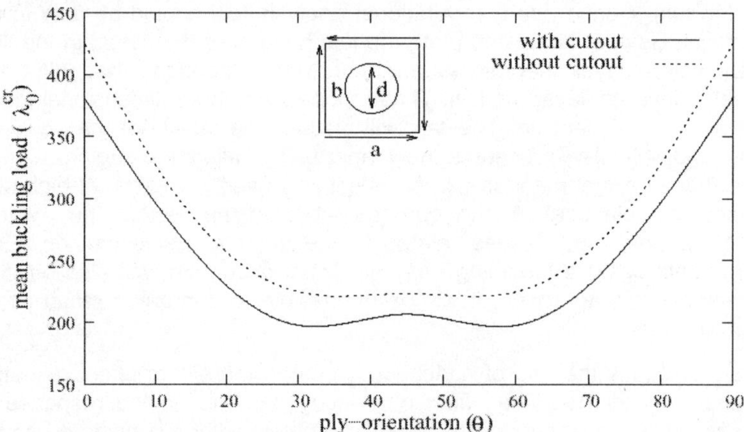

Fig. 1. The effect of ply orientation on the mean critical buckling load for $[-\theta/\theta]$ square plates with and without cutout under shear loading.

In the Fig. 1, the effect of variation in the effective material properties, obtained from micromechanics based approach, on the second order statistics of buckling load for the same above mentioned plate with and without cutout is presented. The COV obtained from micromechanics approach for E_{ll}, E_{tt}, v_{lt}, v_{tt}, G_{lt}, G_{tt} are 7%, 4%, 4%, 4%, 12% and 3%, respectively.

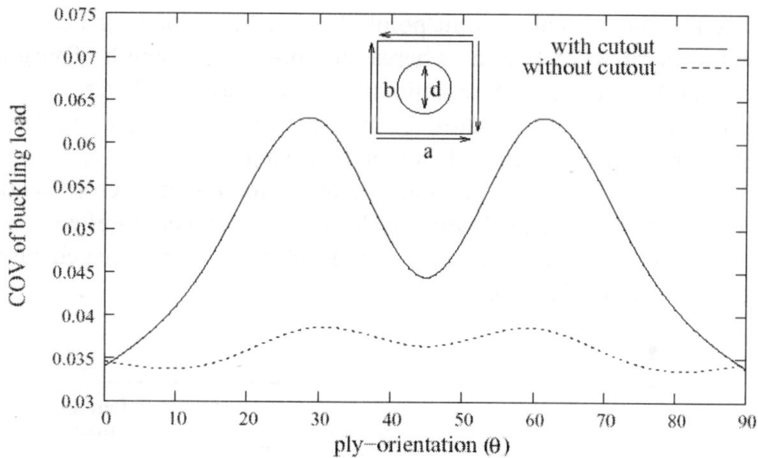

Fig. 2. Influence of dispersion in all material properties changing simultaneously on the COV of critical buckling load for [–θ/θ] square laminates under shear loading.

However, a linear buckling analysis for a damaged composite laminates cannot be done because the material non-linearity comes into effect due to the presence of damage. Hence, in order to study the buckling behavior of damaged plate up to the limit point, a non-linear finite element based on updated Lagrangian formulation is developed along with continuum damage mechanics. In this analysis a composite laminate is treated as a stacked series of elementary anisotropic layers and isotropic interfaces. The damage models are developed separately for elementary ply and interface to capture the different forms of damage at both ply and interface level.

Now, we consider a two layered antisymmetric [–30°/30°] laminates having an existing circular damage (delamination) located at the center of the interface layer. The plate has all edges simply supported and is subjected to uniform compression at the end. The size of plate is taken 12.5 × 12.5 mm and the diameter of existing damage is 2.5 mm. The value of the existing damage parameters taken is $d_{33} = d_{32} = d_{31} = 0.5$.

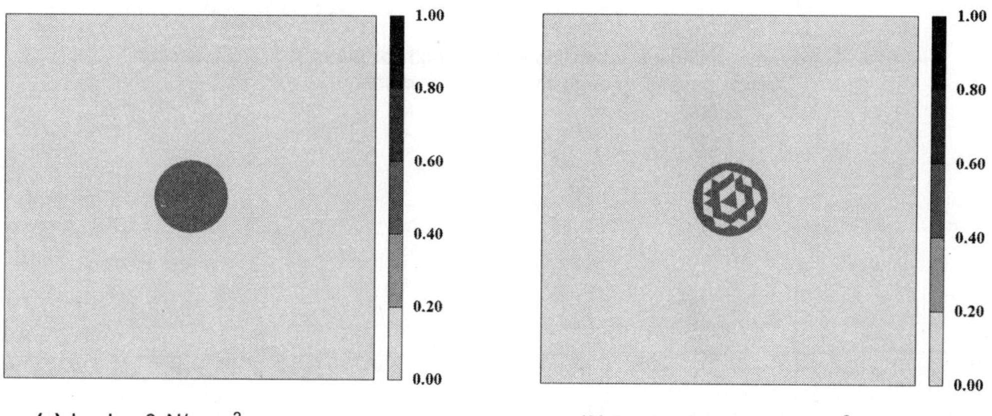

(**a**) load = 0 N/ mm².   (**b**) load = 56.8125 N/ mm².

Fig. 3. Damage state of d_{33} mode present at the interface layer of laminates under uniform compression.

Figure 3 shows the damage state of d_{33} mode at interface layer for [$-30°/30°$] laminates as the load increases and reaches the critical or limit point. The buckling load for this laminate with above damage state is found to be 56.8125 N/mm^2 whereas the buckling strength of virgin plate is 56.8164 N/ mm^2. This shows that the presence of delamination at the interface layer does not strongly affect the buckling strength of [$-30°/30°$] laminates. From the figures, the opening and closing mode of d_{33} damage parameter can also be seen at different load step.

The transverse deflection w of the laminate is also studied at a cross-section which passes through the center of the damage zone. Figure 4 shows the transverse deflection of both damaged and undamaged plate drawn at the bottom face of the laminate. For damaged plate a wavy pattern in the transverse deflection is observed around the damage zone.

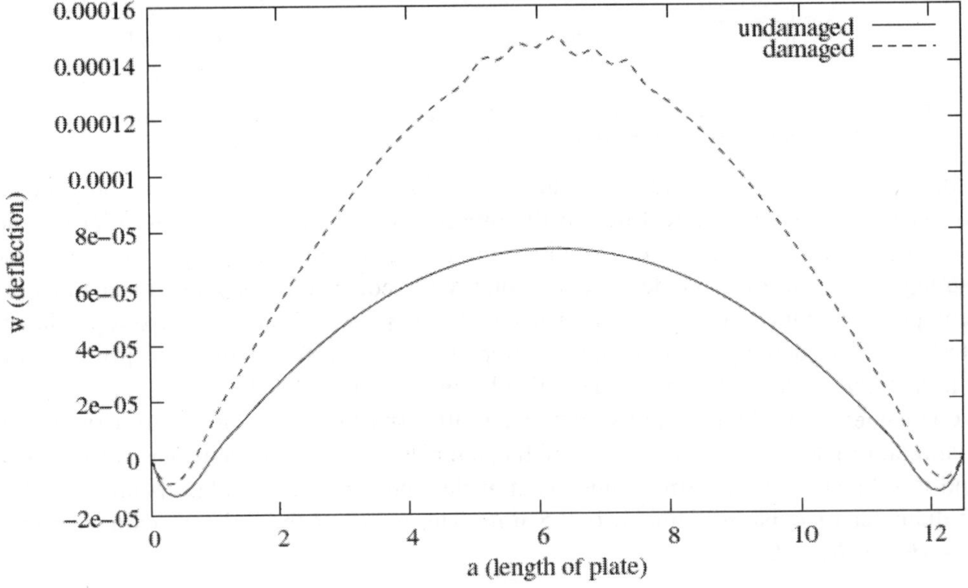

Fig. 4. Transverse deflection taken along the length of plate at a cross-section which passes through the center of damage zone.

6

Laminated Composite Stiffened Shell Elements and Elements with Embedded Stiffeners

D.N. BURAGOHAIN[1] AND P.K. RAVICHANDRAN[2]

[1]Former Director, Indian Institute of Technology Guwahati-781039 India
email : dnburagohain@yahoo.com
[2]Former Research Scholar, Department of Civil Engineering
Indian Institute of Technology, Bombay-400076, India

ABSTRACT

A stiffened shell finite element formulation is presented for the analysis of laminated composite shells with laminated eccentric stiffeners. The formulation is based on a refined higher order theory which incorporates cubic variation of the in-plane displacements across the thickness of the shell as well as of the stiffener and maintains compatibility of displacements at stiffener-shell interface. Additional degrees of freedom for the stiffener element are avoided in the proposed stiffened shell formulation by expressing the strains in the stiffener in terms of the strains in the shell element. It also has the advantage that the stiffener may be located anywhere within the shell element. The formulation is then extended to the case of a three-dimensional curved beam element with a stiffening frame embedded within it as in the case of a steel frame encased in a reinforced concrete beam. Numerical results for eccentrically stiffened isotropic and orthotropic shells establish the capabilities of the proposed formulation.

Keywords: Finite element, Stiffened shell, Eccentric stiffener, Laminated shell, Laminated stiffener, Embedded stiffener.

1. INTRODUCTION

There are mainly two approaches for the analysis of stiffened shells—the smeared and the discrete element approaches. In the smeared approach, the structure is replaced by a structurally equivalent system, while in the discrete element approach, each stiffener is considered separately. Both these approaches have been used earlier in finite element analyses of shear deformable shells [1-3]. The limitation of considering a structurally equivalent system is that the stiffeners must be of uniform size and closely and uniformly spaced so as to ensure approximate homogeneity. The discrete element approach is best suited for stiffened shells with sparsely spaced stiffeners which may be of variable size or made of different materials, but such an approach often brings in additional degrees of freedom belonging only to the stiffener elements, and requires element discretization to be such that the stiffeners lie along the element boundaries.

The stiffened shell element presented here is based on a refined higher order shear deformation theory in which warping of the cross-section is incorporated by taking cubic variation for the in-plane displacements along the thickness of the shell, and the same displacement variation is assumed for the laminated stiffener, which may lie along the element boundary or within the element, or cut across the element [4].

2. FORMULATION OF THE SHELL ELEMENT

Typical shell elements under consideration are the 7-node triangle (having nodes at the corners, mid-sides and centre) and 8- and 9-node quadrilaterals. Using the natural coordinates (x, h), the shape functions Ni (x, h) of C^0 continuity for these elements are obtained by standard methods. The global coordinates (X, Y, Z) at any point (x, h, z) on a curved element with NN nodes are obtained as,

$$\begin{Bmatrix} X \\ Y \\ Z \end{Bmatrix} = \sum_{i=1}^{NN} N_i(\xi,\eta)(\begin{Bmatrix} X_i \\ Y_i \\ Z_i \end{Bmatrix} + z\hat{v}_{3i}) \qquad \text{...(1)}$$

in which, z is the local coordinate normal to the mid-surface, and $\hat{v}_{3i}$ is the unit vector along the nodal thickness vector $\vec{T_i}$ at node i, obtained as,

$$\hat{v}_{3i} = \vec{T_i} / |\vec{T_i}| = \{l_{3i} \; m_{3i} \; n_{3i}\} \qquad \text{...(2)}$$

where, l_{3i}, m_{3i} and n_{3i} are the corresponding direction cosines.

With reference to a local Cartesian coordinate system (x, y, z) located at any point on the mid-surface of the shell element, with (x, y) tangential and z normal to the midsurface (Fig. 1), the variation of the corresponding displacements (u, v, w) along z are expressed as,

$$u(z) = u^0 + zu' + z^2u'' + z^3u''' \qquad \text{...(3a)}$$
$$v(z) = v^0 + zv' + z^2v'' + z^3v''' \qquad \text{...(3b)}$$
$$w(z) = w^0 \qquad \text{...(3c)}$$

in which, u^0, u', u'' and u''', respectively denote $u, \dfrac{\partial u}{\partial z}, \dfrac{1}{2}\dfrac{\partial^2 u}{\partial z^2}$ and $\dfrac{1}{6}\dfrac{\partial^3 u}{\partial z^3}$ at z = 0, with similar notations for v and w terms. With C^0 continuity of displacements, the nodal degrees of freedom at each node will be equal to the number of terms appearing in Eq. (3), and the element is designated as HSDT9 (higher order shear deformation theory). In a similar manner, a first order shear deformable model is obtained by limiting Eqs. (3a) and (3b) to the first two terms resulting in elements with 5 degrees of freedom per node (FSDT5).

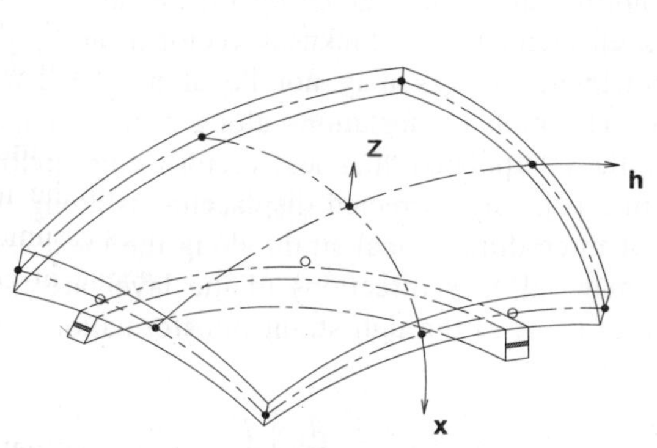

a) STIFFENED SHELL ELEMENT WITH
LAMINATED STIFFENER

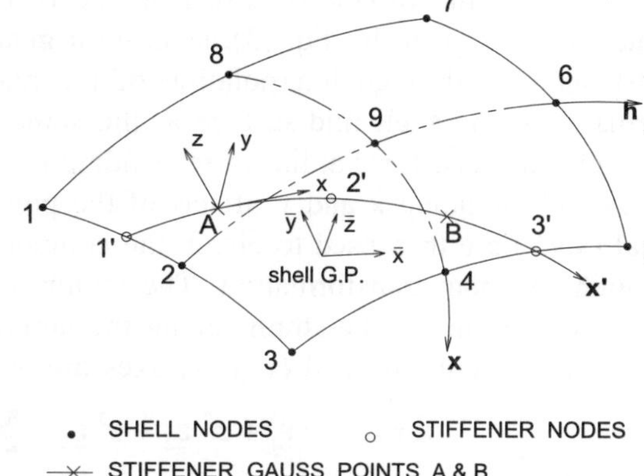

• SHELL NODES   ○ STIFFENER NODES
⟶✕ STIFFENER GAUSS POINTS A & B

b) IDEALIZATION OF SHELL AND STIFFENER

Fig. 1. Typical stiffened shell element.

Local Cartesian coordinate system (x_i, y_i, z_i) are set up uniquely at node '*i*' with zi along the thickness vector. Defining nodal displacements u_i, v_i and w_i along x_i, y_i and z_i directions respectively at node '*i*', their variations along z_i are expressed as in Eq. (3) as,

$$u_i(z_i) = u_i^0 + z_i u_i' + z_i^2 u_i'' + z_i^3 u_i'''$$

$$v_i(z_i) = v_i^0 + z_i v_i' + z_i^2 v_i'' + z_i^3 v_i''' \qquad ...(4)$$

$$w_i(z_i) = w_i^0$$

Symbolically, the nodal displacement vector d_i for the HSDT9 element is taken as,

$$\underline{d}_i = \left\{ \underline{d}_{0i}{}^T \, \underline{d}_{1i}{}^T \, \underline{d}_{2i}{}^T \, \underline{d}_{3i}{}^T \right\} \qquad ...(5)$$

in which,

$$\underline{d}_{0i}{}^T = \left\{ u_i^0 \, v_i^0 \, w_i^0 \right\}^T, \, \underline{d}_{1i}{}^T = \left\{ u_i' \, v_i' \, 0 \right\}^T,$$

$$\underline{d}_{2i}{}^T = \left\{ u_i'' v_i'' 0 \right\}^T, \, \underline{d}_{3i}{}^T = \left\{ u_i''' v_i''' \, 0 \right\}^T \qquad ...(6)$$

Defining the direction cosine matrix of the nodal local coordinates as $\underline{\theta}_i$, the translations $\bar{\underline{d}} = \left\{ U \, V \, W \right\}^T$ corresponding to global directions (X, Y, Z) at any point (x, h, z) are then obtained as,

$$\bar{\underline{d}} = \sum N_i \, \underline{\theta}_i^T \left[\underline{d}_{0i} + z_i \, \underline{d}_{1i} + z_i^2 \, \underline{d}_{2i} + z_i^3 \, \underline{d}_{3i} \right] \qquad ...(7)$$

in which, Ni are the shape functions depending on the shape and location of nodes of the element. The displacements at any point (x, h, z) within the element along the local axes (x, y, z) at the point are obtained as,

$$\underline{d} = \underline{\theta} \, \bar{\underline{d}} = \sum N_i \, \underline{\theta} \, \underline{\theta}_i^T \left[\underline{d}_{0i} + z_i \, \underline{d}_{1i} + z_i^2 \, \underline{d}_{2i} + z_i^3 \, \underline{d}_{3i} \right] \qquad ...(8)$$

in which, $\underline{\theta}$ is the direction cosine matrix of the local axes at the point.

However, the variations of displacements along normal direction z given by Eq. (8) are not the same as those given by Eq. (3), since in a general shell element, the thickness vector at any point (x,h) obtained through interpolation of the nodal thickness vectors may not lie along the actual normal z to the shell mid-surface at the same point. The global translations along z are obtained from global translations at the corresponding points on the interpolated thickness vector using Taylor's series and including x and y offsets of the point from z [4]. The corrected displacements along the local (x,y,z) are then used to obtain the components of three-dimensional strain along the local axes through Jacobian transformation. The strains in the material axes directions of the layer are then obtained in terms of the strains along the laminate axes (x, y, z) through strain transformation. The strain components in local (x, y, z) axes are arranged as $(e_z = 0)$,

$$\underline{\varepsilon} = \underline{\varepsilon}_0 + z\,\underline{\varepsilon}_1 + z^2\,\underline{\varepsilon}_2 + z^3\,\underline{\varepsilon}_3 = \sum \left[\, \underline{B}_{0i} + z\,\underline{B}_{1i} + z^2\,\underline{B}_{2i} + z^3\,\underline{B}_{3i} \right] \underline{d}_i \qquad \text{...(9)}$$

in which,

$$\underline{\varepsilon} = \{\varepsilon_x \ \varepsilon_y \ \varepsilon_z \ \gamma_{yz} \ \gamma_{zx} \ \gamma_{xy}\}^T \qquad \text{...(10)}$$

The element stiffness matrix is then obtained using standard procedure. In the present case, analytical integration is used along the shell thickness, and the determinant of the Jacobian is assumed to vary quadratically along the thickness and written in terms of the determinants of the Jacobians at the top and bottom surfaces of the total thickness of the shell.

3. STIFFENED SHELL ELEMENT

The laminated stiffener is assumed to be of rectangular cross-section and approximately normal to the mid-surface of the shell. It can have a parabolic geometry defined by the global coordinates (X, Y, Z) at the nodes located at its ends and the middle point along the shell mid-surface (Fig. 1). The lateral bending of the stiffener in the tangent plane at any point of the shell is neglected, but its contribution to in-plane shear rigidity is considered. Using the natural coordinate ξ' along the stiffener, the coordinates and displacements at any point in the stiffener are obtained in terms of their nodal values at 1′, 2′ and 3′ using parabolic shape functions $N_i'(\xi')$.

In order to avoid any additional degrees of freedom, the strains in the stiffener at the stiffener Gauss points are written in terms of the shell element strains at the corresponding points, which, in turn, are expressed in terms of the nodal degrees of freedom of the shell element using Eq. (9). For this, however, the natural coordinates (x, h) of the shell element corresponding to the stiffener Gauss points have to be obtained. This is a non-linear problem and is solved using Newton-Raphson iteration as follows:

From the known global coordinates of the stiffener nodal points, the stiffener Gauss point global coordinates are obtained as,

$$\begin{Bmatrix} X \\ Y \\ Z \end{Bmatrix} = \sum_{i=1}^{3} N_i'(\xi') \begin{Bmatrix} X_i \\ Y_i \\ Z_i \end{Bmatrix} \qquad \text{...(11)}$$

For any assumed set of values for x and h for the shell element, an approximation to the global coordinates of the stiffener Gauss point may be obtained as,

$$\begin{Bmatrix} X \\ Y \\ Z \end{Bmatrix}_G = \sum_{i=1}^{NN} N_i(\xi,\eta) \begin{Bmatrix} X_i \\ Y_i \\ Z_i \end{Bmatrix} \qquad \text{...(12)}$$

Since, both Eqs. (11) and (12) must lead to identical values for the stiffener Gauss point coordinates,

$$\begin{Bmatrix} X \\ Y \\ Z \end{Bmatrix}_G - \begin{Bmatrix} X \\ Y \\ Z \end{Bmatrix} = \{0\} \qquad \text{...(13)}$$

Taking natural coordinate z along the shell thickness direction z along with the (x, h) system, the $(n + 1)^{\text{th}}$ approximation for the natural coordinates (x, h, z) is obtained by applying Newton-Raphson scheme to Eq. (13) as,

$$\begin{Bmatrix} \xi \\ \eta \\ \zeta \end{Bmatrix}^{n+1} = \begin{Bmatrix} \xi \\ \eta \\ \zeta \end{Bmatrix}^{n} - \left[J_G^{-1} \right]^{n} \left[\begin{Bmatrix} X \\ Y \\ Z \end{Bmatrix}_G^{n} - \begin{Bmatrix} X \\ Y \\ Z \end{Bmatrix} \right] \qquad \text{...(14)}$$

in which, J_G is the coordinate Jacobian at the current (x, h) point. The first two rows of the Jacobian are given by the tangent vectors $\vec{V}_\xi$ and $\vec{V}_\eta$ respectively, and the third row given by the vector $\vec{V}_\zeta = \vec{V}_\xi \times \vec{V}_\eta$. The natural coordinates x and h are obtained by iteration from Eq. (14) such that Eq. (13) is satisfied within acceptable tolerance while z = 0.

With reference to the local coordinate system (x, y, z), with x along the tangent to the stiffener profile, as shown in Fig. 1(b), the strains in the stiffener are obtained in terms of the strains in the shell element as (refer to Eq. (3)),

$$\varepsilon_x = \frac{\partial u}{\partial x} = \frac{\partial}{\partial x}(u_0) + z\frac{\partial}{\partial x}(u') + z^2\frac{\partial}{\partial x}(u'') + z^3\frac{\partial}{\partial x}(u''')$$

$$= \varepsilon_x^0 + z\,\kappa_x^0 + z^2\,\varepsilon_x^* + z^3\,\kappa_x^* \quad \text{(symbolically)} \qquad \text{...(15)}$$

$$\gamma_{xy} = \frac{\partial v}{\partial x} + \frac{\partial u}{\partial y} = \frac{\partial}{\partial x}(v_0) + \frac{\partial}{\partial y}(u_0) + z\left\{ \frac{\partial}{\partial x}(v') + \frac{\partial}{\partial y}(u') \right\} +$$

$$z^2\left\{ \frac{\partial}{\partial x}(v'') + \frac{\partial}{\partial y}(u'') \right\} + z^3\left\{ \frac{\partial}{\partial x}(v''') + \frac{\partial}{\partial y}(u''') \right\}$$

$$= \gamma_{xy}^0 + z\,\phi_{xy}^0 + z^2\,\gamma_{xy}^* + z^3\,\phi_{xy}^* \quad \text{(symbolically)} \qquad \text{...(16)}$$

$$\gamma_{zx} = \frac{\partial w}{\partial x} + \frac{\partial u}{\partial z} = \frac{\partial}{\partial x}(w_0) + (u') + 2z\,(u'') + 3z^2\,(u''')$$

$$= \gamma_{zx}^0 + z\,\phi_{zx}^0 + z^2\,\gamma_{zx}^* \quad \text{(symbolically)} \qquad \text{...(17)}$$

The torsional strain φ_x of the cross-section is taken as,

$$\varphi_x = \frac{\partial}{\partial x}\left(\frac{\partial v}{\partial z}\right) = \frac{\partial}{\partial x}\left(v' + 2z\,v'' + 3z^2\,v'''\right)_{z=\pm t/2} \quad \text{for top/bottom stiffener}$$

$$= \varphi_x^0 \qquad \text{(symbolically)} \qquad \qquad ...(18)$$

It may be noted that the torsional strain is taken such that at any cross section, the angle of twist is equal to the slope of the warping of the shell element at the stiffener-shell interface and is constant over the entire cross-section of the stiffener.

The generalized strains $\overline{\varepsilon}_s$ in the stiffener obtained in terms of the shell element strains are now written as,

$$\overline{\varepsilon}_s = \left\{\varepsilon_x^0\kappa_x^0\varepsilon_x^*\kappa_x^* \mid \gamma_{xy}^0\phi_{xy}^0\gamma_{xy}^*\phi_{xy}^* \mid \gamma_{zx}^0\phi_{zx}^0\gamma_{zx}^* \mid \varphi_x^0\right\} \qquad ...(19a)$$

or,

$$\overline{\varepsilon}_s = \sum_{i=1}^{NN} \overline{B}_{si}\,\underline{d}_i \qquad ...(19b)$$

For stiffeners made of orthotropic layered composite materials, the constitutive matrix of each layer in the lamina principal directions is transformed to the stiffener axes (x, y, z). The stress resultants for the stiffener are defined as,

$$\overline{\sigma}_s = \left\{N_x M_x N_x^* M_x^* \mid Q_y P_y Q_y^* P_y^* \mid Q_z P_z Q_z^* \mid T_x\right\}^T \qquad ...(20)$$

in which,

$$\left[N_x Q_y Q_z\right] = \sum_{k=1}^{NLS} \int_{z_k}^{z_{k+1}} b_s\left[\sigma_x \tau_{xy} \tau_{zx}\right] dz \qquad ...(21a)$$

$$\left[M_x P_y P_z\right] = \sum_{k=1}^{NLS} \int_{z_k}^{z_{k+1}} b_s\left[\sigma_x \tau_{xy} \tau_{zx}\right] z\,dz \qquad ...(21b)$$

$$\left[N_x^* Q_y^* Q_z^*\right] = \sum_{k=1}^{NLS} \int_{z_k}^{z_{k+1}} b_s\left[\sigma_x \tau_{xy} \tau_{zx}\right] z^2 dz \qquad ...(21c)$$

$$\left[M_x^* P_y^*\right] = \sum_{k=1}^{NLS} \int_{z_k}^{z_{k+1}} b_s\left[\sigma_x \tau_{xy}\right] z^3 dz \qquad ...(21d)$$

and,

$$T_x = \left[\sum_{k=1}^{NLS} \int_{z_k}^{z_{k+1}} \mu\left(D_{44}\right)_k b_s^3\,dz\right]\varphi_x^0 \qquad ...(21e)$$

In Eq. (21), *NLS* is the number of layers in the stiffener, z_k and z_{k+1} respectively denote the bottom and top surfaces of the kth layer of the stiffener from the shell mid-surface, m is the St. Venant torsional constant for the stiffener cross-section, $(D_{44})_k$ is the shear modulus in the yz plane obtained from the transformed constitutive matrix for the kth layer, and b_s is the breadth of the stiffener cross-section. While the terms N_x, M_x, Q_y, Q_z and T_x are physically interpreted as the axial force, bending moment, shear forces in y and z directions and torsional moment, the rest of the higher order terms are defined only for convenience.

The relation between the stress resultants and the generalized strains can be written as,

$$\overline{\sigma}_s = \left[\sum_{k=1}^{NLS} \int_{z_k}^{z_{k+1}} D_{sk}^*\,dz\right]\overline{\varepsilon}_s = \overline{D}_s\,\overline{\varepsilon}_s \qquad ...(22)$$

in which, the elements of $\underline{D}^{*}_{sk}$ are obtained from the right hand side of Eq. (21) by substituting the stresses in terms of strains using the transformed constitutive matrix for the kth layer. The stiffness matrix for the stiffener is then obtained in terms of the shell nodal displacements as,

$$(\underline{K}_e)_s = \int_{-1}^{+1} \overline{B}_s^T \left[\sum_{k=1}^{NLS} \overline{D}_{sk} \left| \vec{V}_{\xi'} \right|_k \right] \overline{B}_s \, d\xi' \qquad \text{...(23)}$$

in which, the tangent vector $\vec{V}_{\xi'}$ is evaluated at the centroid of the kth layer.

4. ELEMENTS WITH EMBEDDED STIFFENERS

In many cases of reinforced concrete construction, where large spans have to be covered without intermediate supports, a regular steel frame is encased within a reinforced concrete beam. If the beam is modeled using finite elements, then each full or truncated member of the part of the frame within the element will act as an embedded stiffening member within the element (Fig. 2).

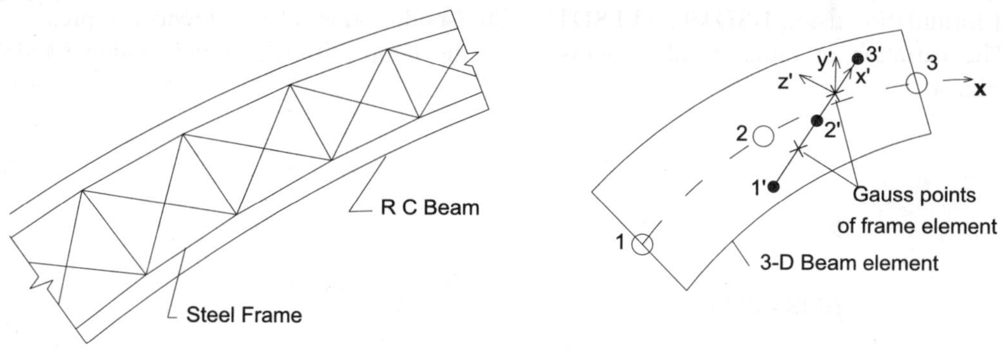

(a) Steel Frame encased in RC Beam　　　　(b) Embedded Stiffening Frame member

Fig. 2. Embedded stiffeners in a beam.

Since, the strains at any point of the embedded frame element are the same as the corresponding strains of the confining beam element at that point, the procedure adopted for the stiffened shell element can be employed to express the strains at the Gauss points of the frame element in terms of the three-dimensional beam element strains which, in turn, are expressed in terms of the nodal displacements of the beam element. The presence of the stiffening member inside the beam element thus, introduces no new degrees of freedom.

In the local coordinates (x', y', z') of the frame element as shown in Fig. 2(b), the frame element strains are written as,

$$\varepsilon_{x'} = \frac{\partial u'}{\partial x'} \quad , \quad \gamma_{x'y'} = \frac{\partial v'}{\partial x'} + \frac{\partial u'}{\partial y'} = \frac{\partial v'}{\partial x'} - \theta_{z'} \, , \quad \gamma_{z'x'} = \frac{\partial w'}{\partial x'} + \frac{\partial u'}{\partial z'} = \frac{\partial w'}{\partial x'} + \theta_{y'} \qquad \text{...(24a)}$$

$$\kappa_{y'y'} = \frac{\partial}{\partial x'}\left(\frac{\partial u'}{\partial z'}\right) = \frac{\partial \theta_{y'}}{\partial x'} \, , \quad \kappa_{z'z'} = \frac{\partial}{\partial x'}\left(\frac{\partial u'}{\partial y'}\right) = -\frac{\partial \theta_{z'}}{\partial x'} \cdot \quad \varphi_{x'} = \frac{\partial \theta_{x'}}{\partial x'} \qquad \text{...(24b)}$$

For the confining beam element, local coordinates (x, y, z) are set up at the frame element Gauss points such that x axis is tangent to the natural coordinate x, and (y, z) are located in the plane of the cross-section. The values of the beam element (x, y, z) of the frame element Gauss point are determined either by Newton-Raphson iteration, or by solving a quadratic equation in x if the (y, z) distances of the point from the centroidal axis are easily obtained. The stiffness matrix of the frame element is then obtained in terms of the nodal degrees of freedom of the beam element as in the case of the stiffener for the stiffened shell [5].

5. NUMERICAL EXAMPLES

The stiffened shell formulation described in this paper has been applied to the following three problems:

Example 1: Eccentrically stiffened cylindrical shell.

This problem was first solved by Kohnke and Schnobrich [6] and later by Venkatesh and Rao [1] and Bhimaraddi *et al.* [2]. The structure is an isotropic cantilever cylindrical shell with axial and hoop stiffeners subjected to a concentrated load at one of the corners (Fig. 3). The modulus of elasticity and Poisson's ratio of the material are, $E = 10^6$ kg/cm^2, and n = 0.3. A comparison of the radial and tangential displacements at the loaded corner obtained by the referenced methods and the stiffened shell formulation using HSDT9 and FSDT5 with 7-noded triangular elements are presented in Table 1. The variation of radial displacements along the hoop stiffener at the loaded edge is presented in Fig. 4.

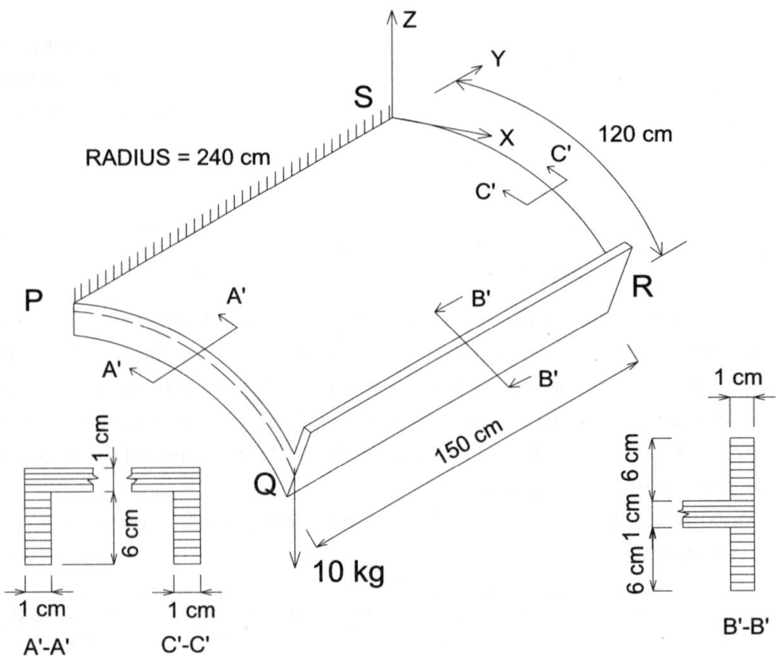

Fig. 3. Eccentrically stiffened cylindrical shell. Example 1.

Table 1 shows that the results by the stiffened shell formulation presented here are closer to those by Bhimaraddi *et al.* [3] based on higher order theory and to the flat plate solutions of

Table 1. Displacements at the loaded corner of the stiffened cylindrical shell. Example 1.

Sl. No.	Method and Mesh	Tangential displacement ($\times 10^{-3}$ cm)	Radial displacement ($\times 10^{-3}$ cm)
1.	FSDT5 (6 × 6)	−11.860	−57.560
2.	FSDT5 (7 × 7)	−12.336	−59.843
3.	HSDT9 (6 × 6)	−13.360	−64.630
4.	HSDT9 (7 × 7)	−13.632	−65.923
5.	Reference [6] Shell elements (3 × 4)	−14.985	−85.625
6.	Reference [6] flat elements (20 × 20)	−14.412	−67.394
7.	Venkatesh and Rao [1] (4 × 4)	−11.620	−55.800
8.	Bhimaraddi et al. [2] (4 × 4)	−13.020	−63.080

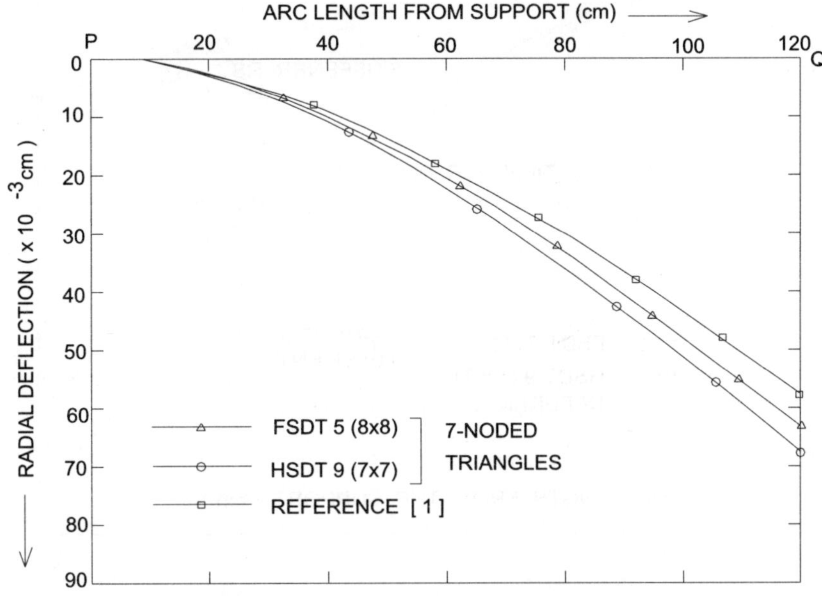

Fig. 4. Radial displacement of hoop stiffener along P-Q. Example 1.

reference [6] using a fine mesh (20 × 20). The results of Venkatesh and Rao [1] are generally found to be stiffer, while the shell element (3 × 4 mesh) results of reference [6] appear to be too flexible.

Example 2: Eccentrically stiffened laminated cylindrical shell with laminated stiffeners.

The geometry of this stiffened shell is identical to the one in Example 1. The material is chosen as glass-epoxy with the following properties:

$E_1 = 0.516 \times 10^6$ kg/cm^2; $E_2 = E_3 = 0.137 \times 10^6$ kg/cm^2; $G_{12} = 0.861 \times 10^6$ kg/cm^2; $n_{12} = 0.25$
$G_{23} = G_{13} = G_{12}$

Shell: 4 layers each of 0.25 cm thickness with (45° / − 45° / –45° / –45°/) lay up.

Hoop stiffener: 6 layers each 1 cm thick and 1 cm wide with (–45° / 45° / –45° / –45° / 45°/ –45° /) lay up.

Axial stiffener: 6 layers each 1 cm thick and 1 cm wide on either side of the shell with—
Top part: (45° / 30° / –45° / –30° / 45° / 30°) lay up.
Bottom part: (30° / 45° / –30° / –45° / 30° / 45°) lay up.

The variation of radial displacement along the hoop stiffeners obtained using the stiffened shell HSDT9 and FSDT5 formulations are shown in Fig. 5 along with the results obtained by Venkatesh and Rao [1]. As expected, the results of reference [1] are on the lower side as in example 1, and the application of the present formulation to laminated composite shells with laminated stiffeners is found to be satisfactory.

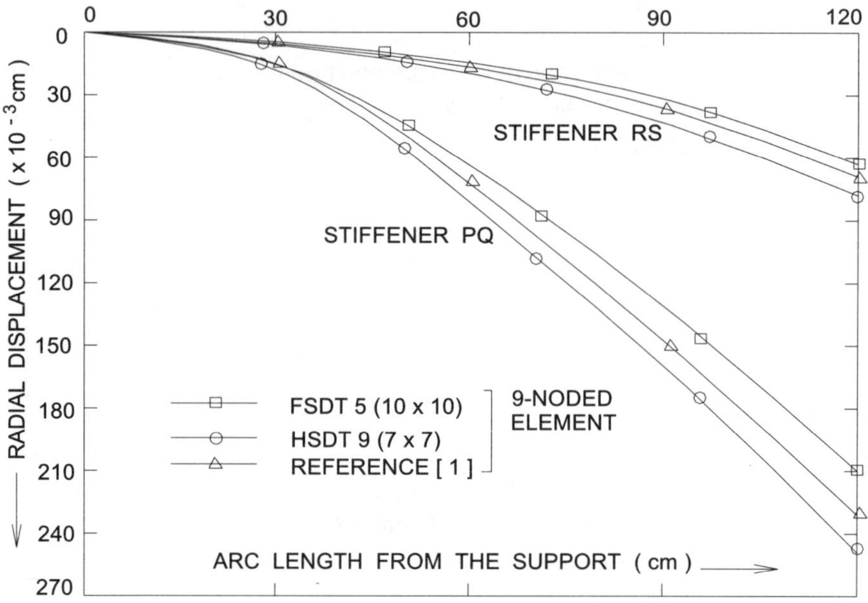

Fig. 5. Radial displacements along the hoop stiffeners. Example 2.

Example 3: Eccentrically stiffened cylindrical shell with helically oriented stiffener:

This example is devised to highlight the capability of the stiffened shell formulation presented here. The cylindrical shell in this example is stiffened by four stiffeners: two symmetric hoop stiffeners, one symmetric axial stiffener and an eccentric helical stiffener aligned between two opposite corners. The geometry of the shell is shown in Fig. 6 and it is subjected to a vertical concentrated load at one of the free corners. The material properties are the same as in example 1.

Three different schemes of discretization are used as shown in Fig. 7. The radial and tangential displacements at the free corners of the stiffened shell obtained using the three schemes of discretization are presented in Table 2. 9-noded HSDT9 quadrilateral elements are used in schemes (a) and (b), while 7-noded HSDT9 triangular elements are used in scheme (c).

It is seen that agreement between the results obtained by using different schemes of discretization and different mesh sizes is very good and this shows that the stiffened shell element presented here can model any stiffened shell with arbitrarily placed stiffeners.

Table 2. Comparison of displacements at the free corners of stiffened shell. Example 3.

Sl.No.	Scheme of discretization and mesh	Displacements at A ($\times 10^{-3}$ cm)		Displacements at B ($\times 10^{-3}$ cm)	
		Tangential	Radial	Tangential	Radial
1.	Scheme a) : (6 × 6)	−28.287	−143.55	−8.596	−47.584
2.	Scheme a) : (7 × 7)	−28.400	−144.18	−8.486	−46.954
3.	Scheme b) : (7 × 5)	−28.011	−142.21	−8.828	−48.228
4.	Scheme b) : (8 × 6)	−28.160	−143.02	−8.734	−47.646
5.	Scheme c)[1] : (6 × 6)	−28.675	−145.70	−8.706	−47.094
6.	Scheme c)[2] : (6 × 6)	−28.675	−145.71	−8.681	−47.002
7.	Scheme c)[1] : (7 × 7)	−29.127	−147.34	−9.190	−48.699
8.	Scheme c)[2] : (7 × 7)	−28.870	−146.80	−8.550	−48.280

[1]diagonal stiffener stiffness added to shell element on left
[2]diagonal stiffener stiffness added to shell element on right

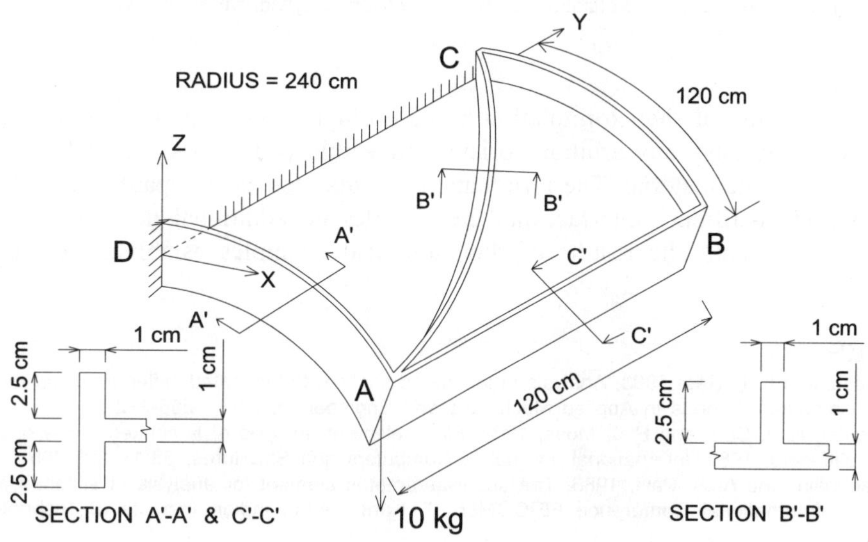

Fig. 6. Stiffened cylindrical shell with helical stiffener. Example 3.

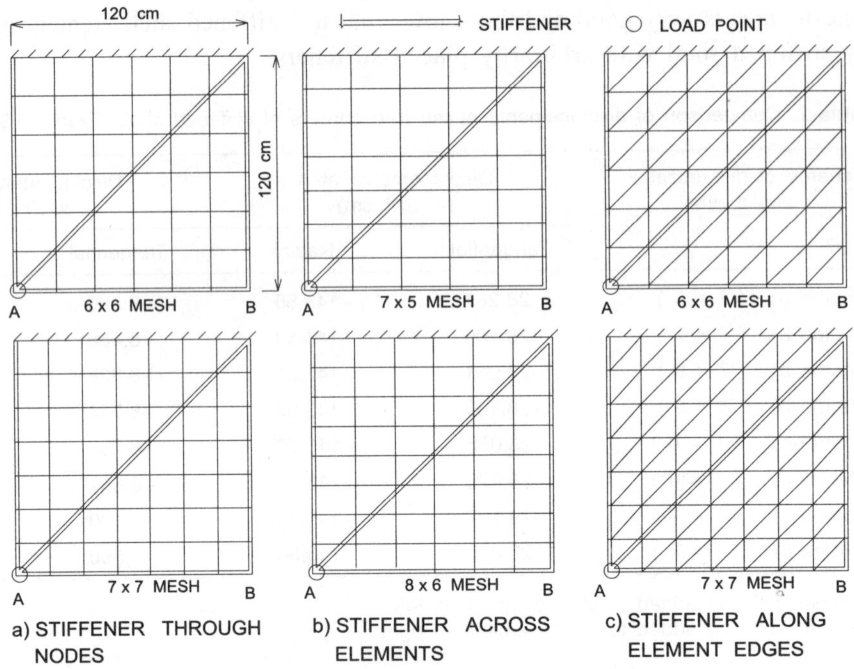

Fig. 7. Schemes of discretization for helically stiffened cylindrical shell. Example 3.

CONCLUSION

A shear deformable stiffened shell formulation based on higher order theory has been presented in which the stiffener may have any arbitrary orientation within and across the shell element and be made of layered composite material. The formulation accounts for the compatibility of displacements and strains at the shell-stiffener interface and necessitates no additional degrees of freedom over those of the shell element. The results of the numerical examples establish the validity of the formulation.

REFERENCES

1. A. Venkatesh, and K. P. Rao, 1983, Analysis of laminated shells with laminated stiffeners using rectangular finite elements, Computer Methods in Applied Mechanics and Engineering, 38(3), 255-272.
2. A. Bhimaraddi, A. J. Carr, and P. J. Moss, 1989, Finite element analysis of laminated shells of revolution with laminated stiffeners, 1989, International Journal of Computers and Structures, 33(1), 295-305.
3. D.N. Buragohain, and A. S. Patil, 1985, The super-parametric element for analysis of stiffened and composite shells, Proc. International Conference FEICOM85, T. Kant , eds., Indian Institute of Technology, Bombay, 177-186.
4. P.K. Ravichandran, 1992, Finite Element Analysis of Layered Composite and Stiffened Shells based on Refined Theories, Ph.D. thesis, Department of Civil Engineering, Indian Institute of Technology, Bombay.
5. B. Kapadia. 1992, Finite Element Analysis of Beams with Embedded Frames, B.Tech. dissertation, Department of Civil Engineering, Indian Institute of Technology, Bombay.
6. P.C. Kohnke, and W. C. Schnobrich, 1972, Analysis of eccentrically stiffened cylindrical shells and orthogonally stiffened curved panels, ASCE Journal of Structural Division, 98(ST7),1493-1510.

7

Application of Integrated Force Method To Laminated Composite Plates

H.R. DHANANJAYA[1], J. NAGABHUSHANAM[2] AND P.C. PANDEY[3]

[1]Reader, Department of Civil Engineering Manipal Institute of Technology, Manipal-576104 India
email: djaya_hr@yahoo.com, djaya.hr@manipal.edu
[2]Emeritus Professor, Department of Aerospace Engineering, Indian Institute of Science,
Bangalore-560012 India email: naga@aero.iisc.ernet.in
[3]Professor, Department of Civil Engineering, Indian Institute of Science, Bangalore,-560012 India
email: pcpandey@civil.iisc.ernet.in

ABSTRACT

A novel matrix formulation has been developed in recent years to analyze problems in structural mechanics and is termed as the Integrated Force Method (IFM). In the Integrated Force Method all independent forces, not just the redundants, are treated as unknown quantities that can be calculated by simultaneously imposing both equilibrium and compatibility conditions. This paper presents the basic theory of the IFM and its application to laminated composite plate bending problems.

Keywords: Laminated composite plate, Equilibrium matrix, Flexibility matrix, Integrated Force Method, Displacement fields, Stress-resultant fields.

1. BASIC THEORY OF IFM

In the Integrated Force Method of analysis, a structure idealized by finite elements is designated as "structure (n, m)", where n and m are force and displacement degrees of freedom of the discreet model, respectively. The 'structure $(n, m)'$ has m equilibrium equations (EE) and $r = (n-m)$ compatibility conditions (CC).

The equilibrium equation (EE) represents the vectorial summation of the internal forces $\{F\}$ to the external loads $\{P\}$ at the nodes of the finite element discretization. It can be written in symbolized matrix notation as:

Equilibrium Equations (EE):

$$[B] \{F\} = \{P\} \qquad \qquad ...(1)$$

where,

$[B]$ = global equilibrium matrix $(m \times n)$

$\{F\}$ = vector of internal forces of the structure $(n \times 1)$

$\{P\}$ = vector of external loads on the structure $(m \times 1)$

The compatibility conditions (CC) are constraints on strains, and for finite element models they are also constraints on member deformations.

In IFM, St. Venant's approach has been extended for discrete mechanics to develop the compatibility conditions. Development of CC is briefly explained below:

The Deformation-Displacement Relationship (DDR) for discrete mechanics is equivalent to the strain-displacement relationship in elasticity. The DDR for discrete analysis was obtained during the development of the variational energy formulation for the IFM.

According to work energy-conservation theorem, the internal energy (IE) stored in the structure is equal to the work done by the external load (WD), that is,

$$I\,E = W\,D$$

$$\frac{1}{2}\{F\}^{T}\{\beta\} = \frac{1}{2}\{P\}^{T}\{X\} \qquad \qquad ...(2)$$

where $\{X\}$ represents nodal displacements. Equation (2) can be rewritten by eliminating the load $\{P\}$ in favor of forces $\{F\}$, by using Eq. (1) to obtain the following relation:

$$\frac{1}{2}\{F\}^{T}[B]^{T}\{X\} = \frac{1}{2}\{F\}^{T}\{\beta\} \qquad \qquad ...(3)$$

Equation (3) can be simplified as

$$\frac{1}{2}\{F\}^{T}\left[[B]^{T}\{X\} - \{\beta\}\right] = 0 \qquad \qquad ...(4)$$

Because the n forces can be arbitrary and $\{F\}$ is not a null vector, its coefficient should be zero, which yields the DDR as

$$\{\beta\} = [B]^{T}\{X\} \qquad \qquad ...(5)$$

where, $\{\beta\}$ are member deformations.

This equation represents the Deformation-Displacement Relations (DDR) for the discrete structure. The elimination of m displacements from n deformations displacement relations given by the above equation yields $r = (n\text{-}m)$ compatibility conditions and the associated matrix [C]. It can be symbolized in matrix notations as

$$[C]\{\beta\} = 0 \qquad \qquad ...(6)$$

where, [C] is the $(r \times n)$ compatibility matrix. It is a kinematic relationship, and is independent of design parameters, material properties and external loads. This matrix is rectangular and banded. The deformation $\{\beta\}$ in the compatibility conditions (CC) given by the Eq. 6 represents the total deformation consisting of an elastic component $\{\beta_e\}$ and the initial component $\{\beta_0\}$ as

$$\{\beta\} = \{\beta_e\} + \{\beta_o\} \qquad \qquad ...(7)$$

The CC in terms of elastic deformation can be written as

$$[C]\{\beta\} = [C]\{\beta_e\} + [C]\{\beta_o\}$$

$$[C]\{\beta_e\} = \{\delta R\} \qquad \qquad ...(8)$$

where,
$$\{\delta R\} = -[C]\{\beta_o\} \qquad \qquad ...(9)$$

Using element flexibility characteristics, Eq. (6) with initial deformation can be rewritten as

$$[C][G]\{F\} = \{\delta R\} \qquad \qquad ...(10)$$

Clubbing of Eq. (1) and Eq. (10) will lead to the IFM governing equation as

$$\begin{bmatrix} [B] \\ [C][G] \end{bmatrix}\{F\} = \begin{Bmatrix} P \\ \delta R \end{Bmatrix}$$

$$[S]\{F\} = \{P*\} \qquad \qquad ...(11)$$

The solution of the Eq. (11) yields n forces $\{F\}$. The m displacements $\{X\}$ are obtained from the forces $\{F\}$ by back substitution as

$$\{X\} = [J]\{[G]\{F\} + \{\beta_o\}\} \qquad \qquad ...(12)$$

where,
$$[J] = \text{m rows of } [[S]^{-1}]^T.$$

2. FORMULATION OF ELEMENT EQUILIBRIUM AND FLEXIBILITY MATRICES

The constitutive relations required for laminated composite plates are given in the appendix A. The laminated composite plate with stress-resultants and degrees of freedom is shown in the Fig. 1. The displacement functions are assumed to be linear in the thickness direction so that the plane section remain plane. The deformations $(u_1, u_2, u_3) = (u, v, w)$ in the (x, y, z) axes, respectively are given by

$$u(x,y,z) = u^0(x,y) + z\theta_x(x,y); \quad v(x,y,z) = v^0(x,y) + z\theta_y(x,y);$$

$$v(x,y,z) = w(x,y) \qquad \qquad ...(13)$$

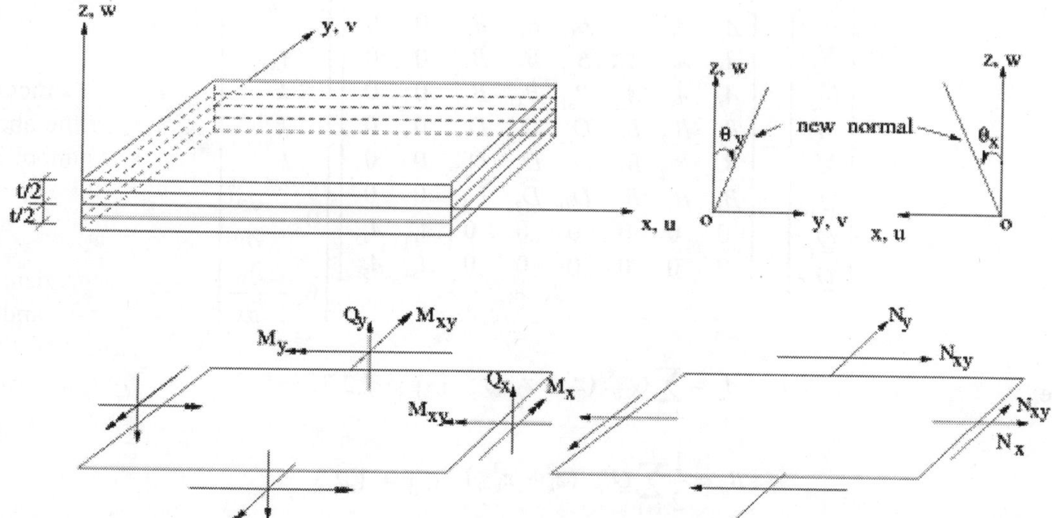

Fig. 1. A laminated composite plate with stress resultants.

The deformations u^0, v^0 represent in-plane deformations at the geometric mid-surface. The plate thickness direction is along the z-axis and the value w is referred to as transverse deformation. The functions θ_x, θ_y represent small angle of rotations of scribe line that was originally perpendicular to geometric mid-surface. The strain-displacement relations take the form

$$\varepsilon_x = \varepsilon_x^0 + zk_x \; ; \; \varepsilon_y = \varepsilon_y^0 + zk_y \; ; \; \varepsilon_z = 0$$

$$\gamma_{xy} = \gamma_{xy}^0 + zk_{xy} \; ; \; \gamma_{yz} = \theta_y + \frac{\partial w}{\partial x} \; ; \; \gamma_{xz} = \theta_x + \frac{\partial w}{\partial y} \qquad ...(14)$$

where,

$$\left\{ \begin{array}{c} \varepsilon_x^0 \\ \varepsilon_y^0 \\ \gamma_{xy}^0 \end{array} \right\} = \left\{ \begin{array}{c} \dfrac{\partial u^0}{\partial x} \\[2mm] \dfrac{\partial v^0}{\partial y} \\[2mm] \dfrac{\partial u^0}{\partial y} + \dfrac{\partial v^0}{\partial x} \end{array} \right\} \; ; \; \left\{ \begin{array}{c} k_x \\ k_y \\ k_{xy} \end{array} \right\} = \left\{ \begin{array}{c} \dfrac{\partial \theta_x}{\partial_x} \\[2mm] \dfrac{\partial \theta_y}{\partial_y} \\[2mm] \dfrac{\partial \theta_x}{\partial_y} + \dfrac{\partial \theta_y}{\partial_x} \end{array} \right\} \qquad ...(15)$$

$\varepsilon_x^0, \varepsilon_y^0$ and y_{xy}^0 represent mid-surface strains.
k_x , k_y and k_{xy} are curvatures of the plate.
The stress-resultants are calculated as

$$(N_x, N_y, N_{xy}) = \int_{-t/2}^{t/2} \left(\sigma_x, \sigma_y, \tau_{xy}\right) dz \; ; \; (M_x, M_y, M_{xy}) = \int_{-t/2}^{t/2} \left(\sigma_x, \sigma_y, \tau_{xy}\right) z dz$$

$$(Q_x, Q_y) = \int_{-t/2}^{t/2} \left(\tau_{xz}, \tau_{yz}\right) dz \qquad ...(16)$$

where, t = plate thickness. Substituting the equations (13) and (14) in the equations (16), we get moment-curvature relations and are given in equation (27).

$$\left\{ \begin{array}{c} N_x \\ N_y \\ N_{xy} \\ M_x \\ M_y \\ M_{xy} \\ Q_y \\ Q_x \end{array} \right\} = \left[\begin{array}{cccccccc} A_{11} & A_{12} & A_{13} & B_{11} & B_{12} & B_{13} & 0 & 0 \\ A_{21} & A_{22} & A_{23} & B_{21} & B_{22} & B_{23} & 0 & 0 \\ A_{31} & A_{32} & A_{33} & B_{31} & B_{32} & B_{33} & 0 & 0 \\ B_{11} & B_{12} & B_{13} & D_{11} & D_{12} & D_{13} & 0 & 0 \\ B_{21} & B_{22} & B_{23} & D_{21} & D_{22} & D_{23} & 0 & 0 \\ B_{31} & B_{32} & B_{33} & D_{31} & D_{32} & D_{33} & 0 & 0 \\ 0 & 0 & 0 & 0 & 0 & 0 & A_{44} & A_{45} \\ 0 & 0 & 0 & 0 & 0 & 0 & A_{54} & A_{55} \end{array} \right] \left\{ \begin{array}{c} \varepsilon_x^0 \\ \varepsilon_y^0 \\ \gamma_{xy} \\ k_x \\ k_y \\ k_{xy} \\ \theta_y - \dfrac{\partial w}{\partial y} \\ \theta_x - \dfrac{\partial w}{\partial x} \end{array} \right\} \qquad ...(17)$$

where,

$$A_{ij} = \sum_{k=1}^{N} Q'^{(k)}_{ij} (z_k - z_{k-1}) , \; i, j = 1,2,3 \qquad ...(18)$$

$$B_{ij} = \frac{1}{2} \sum_{k=1}^{N} Q'^{(k)}_{ij} (z_k^2 - z_{k-1}^2) , \; i, j = 1,2,3 \qquad ...(19)$$

$$D_{ij} = \frac{1}{3}\sum_{k=1}^{N} Q'^{(k)}_{ij}(z_k^3 - z_{k-1}^3), \ i, j = 1, 2, 3 \qquad \ldots(20)$$

$$A_{ij} = \sum_{k=1}^{N} K_{ij} Q'^{(k)}_{ij}(z_k - z_{k-1}), \ i, j = 4, 5 \qquad \ldots(21)$$

where, K_{ij} are shear correction coefficients.

The standard terminology for lamina locations has been employed in equations (18) to (21) and is shown in the Fig. 2. The z_k values are signed numbers varying from $z_0 = -t/2$ to $z_N = t/2$ in a laminate of N layers.

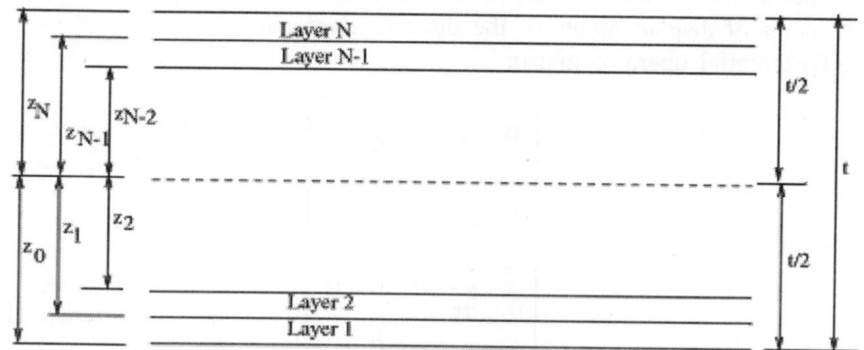

Fig. 2. A laminated composite plate with N layers.

The equation (17) can be written in matrix notation as

$$\{M\} = [C_1]\{k\} \qquad \ldots(22)$$

where, $\{k\}$ = vector of curvatures

$$= \left[\varepsilon_x^0 \ \varepsilon_y^0 \ \gamma_{xy}^0 \ k_x \ k_y \ k_{xy} \ (\theta y - \frac{\partial w}{\partial y}) \ (\theta x - \frac{\partial x}{\partial x}) \right]$$

$\{M\}$ = vector of stress-resultants

$$= \left[N_x \ N_y \ N_{xy} \ M_x \ M_y \ M_{xy} \ Q_y \ Q_x \right]^T$$

From the equation 2.10, the curvature-moment relations can be written as

$$\{k\} = [C_1]^{-1}\{M\} = [H]\{M\} \qquad \ldots(23)$$

where, $\qquad [H] = [C_1]^{-1} \qquad \ldots(24)$

$\qquad\qquad\qquad$ = matrix which relates curvatures to stress resultants

The strain for the plate is given by

$$U_p = \iint \tfrac{1}{2}\{k\}^T \{M\} dxdy \qquad \ldots(25)$$

For a discrete plate bending element, $\{M\}$ and $\{k\}$ can be written in terms of assumed stress-resultant and displacement fields, respectively. These can be expressed in the matrix notation as

$$\{M\}=[\psi]\{Fe\} \qquad \qquad ...(26)$$

$$\{k\}=\left[D_{op}\right][\phi_1]\{\alpha\}=\left[D_{op}\right][\phi]\{X_e\} \qquad \qquad ...(27)$$

where, ψ = matrix of polynomial terms for stress-resultant fields
$\{F\}$ = vector of force components of the discrete element
$[\Phi] = [\Phi_1]\,[A]^{-1}$
$\{\Phi_1\}$ = matrix of polynomial terms for displacement fields
$[A]$ = matrix formed by substituting the coordinates of the element nodes into the polynomial of displacement fields
$\{\alpha\}$ = coefficients of the displacement field polynomial
$\{W\}$ = vector of displacements of the discrete element
$[D_{op}]$ = Differential operator matrix

$$= \begin{bmatrix} \dfrac{\partial}{\partial x} & 0 & 0 & 0 & 0 \\[2mm] 0 & \dfrac{\partial}{\partial y} & 0 & 0 & 0 \\[2mm] \dfrac{\partial}{\partial y} & \dfrac{\partial}{\partial x} & 0 & 0 & 0 \\[2mm] 0 & 0 & 0 & \dfrac{\partial}{\partial x} & 0 \\[2mm] 0 & 0 & 0 & 0 & \dfrac{\partial}{\partial y} \\[2mm] 0 & 0 & 0 & \dfrac{\partial}{\partial y} & \dfrac{\partial}{\partial x} \\[2mm] 0 & 0 & \dfrac{\partial}{\partial y} & 0 & 1 \\[2mm] 0 & 0 & \dfrac{\partial}{\partial x} & 1 & 0 \end{bmatrix}$$

Substituting equations 2.14 and 2.15 into the equation 2.13, the strain energy for the discrete element can be expressed as

$$U_p = \tfrac{1}{2}\{X_e\}^T[B_e]\{F_e\} \qquad \qquad ...(28)$$

where, $[B_e]$ represents the element equilibrium matrix and is given by

$$[B_e]=\iint[\phi]^T\left[D_{op}\right]^T[\psi]\,dxdy \qquad \qquad ...(29)$$

Using the equation 2.12 the complementary strain energy for the laminated plate element is written as

$$U_c = \iint \tfrac{1}{2}\{M\}^T[H]\{M\}\,dxdy = \tfrac{1}{2}\{F_e\}^T[G_e]\{F_e\} \qquad \qquad ...(30)$$

where, $[G_e]$ represents element flexibility matrix and is given by

$$[G_e] = \iint [\psi]^T [H][\psi]\,dxdy \qquad \text{...(31)}$$

The equations 2.17 and 2.19 are used to obtain element equilibrium and flexibility matrices $[B_e]$ and $[G_e]$ respectively for laminated composite plate bending element. These element equilibrium matrix $[B_e]$ and element flexibility matrix $[G_e]$ of all elements are assembled to obtain the global equilibrium matrix $[B]$ and global flexibility matrix $[G]$ of the structure and they are used to setup the IFM governing equation to analyze the structure by IFM.

2.1 Displacement And Stress-Resultant Fields For MCP4

MCP4 is a 4-node rectangular laminated composite plate bending element developed using IFM. The Mindlin-Reissner theory is employed while formulating element matrices. Geometry of the typical 4-node plate bending element is shown in the Fig. 3.

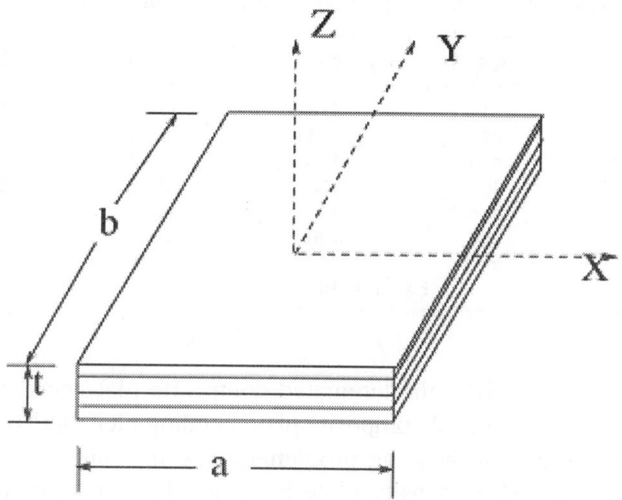

Fig. 3. A typical 4-node rectangular laminated composite plate bending element.

Five degrees of freedom namely two in-plane displacements u, v, one transverse displacement w and two rotations θ_x, θ_y are considered at each node. For this plate bending element the displacement fields for **u, v, w, θ_x,** and θy are assumed in terms of generalized coordinates as

$$u = \alpha_1 + \alpha_2\,x + \alpha_3\,y + \alpha_4\,xy; \qquad v = \alpha_5 + a_6\,x + \alpha_7\,y + \alpha_8\,xy$$

$$w = \alpha_9 + \alpha_{10}\,x + \alpha_{11}\,y + \alpha_{12}\,xy$$

$$\theta_x = \alpha_{13} + \alpha_{14}\,x + \alpha_{15}\,y + \alpha_{16}\,xy; \quad \theta_y = \alpha_{17} + a_{18}\,x + \alpha_{19}\,y + \alpha_{20}\,xy \qquad \text{...(32)}$$

The assumed stress-resultant fields in terms of independent force parameters are shown below

$$N_x = F_1 + F_2\,x + F_3\,y; \qquad N_y = F_4 + F_5\,x + F_6\,y; \qquad N_{xy} = F_7 - F_6\,x - F_2\,y$$

$$M_x = F_8 + F_9x + F_{10}y; \qquad M_y = F_{11} + F_{12}x + F_{13}y; \qquad M_{xy} = F_{14} \qquad \text{...(33)}$$

Substituting the equation 2.12 and equations 2.20 and 2.21 into the equations 2.17 and 2.19, the element equilibrium and flexibility matrices for the laminated composite plate bending can be obtained.

3. NUMERICAL RESULTS AND DISCUSSIONS

Following example problem is considered to validate the newly developed element MCP4. Although the element developed is composite type, it should estimate deflections and moments for the isotropic plates also. Therefore, the first example problem considered is isotropic (single layer) simply supported/clamped square thin plate subjected to uniform load. The parameters of the problem are $L \times B = 100 \times 100$, $t = 1$, $E = 1092000$, $v = 0.3$, $q = 1$(uniform). This problem has been analyzed for various mesh sizes using the proposed element MCP4 via the Integrated Force Method. Since geometry, boundary conditions and loading conditions of the plate are symmetrical one quadrant of the plate is considered for the analysis. Estimated central deflections for simply supported plate subjected to uniform load are summarized in Table 1. The results of MCP4 are compared with those of displacement based four node quadrilateral plate bending elements available in the literature[2]. The results are also compared with the exact solutions [3]. The proposed element yielded satisfactory results.

Table 1. Central deflection for a simply supported square plate with uniform load ($W_c(10^{-5}\ qL^4/D)$) (Example Problems 1, t/L = 0.01)).

Elements	Q4	Q4–R	DKQ	MITC4	RDKQM	MCP4
1 × 1	1.1	319.2	378.5	319.1	378.7	534.8
2 × 2	4.5	397.1	404.6	397.1	404.8	431.2
4 × 4	17.3	404.4	406.0	404.4	406.2	411.7
6 × 6	36.9	405.5	406.1	405.5	406.3	408.5
8 × 8	61.3	405.9	406.2	405.9	406.3	407.4
10 × 10	88.3	406.1	406.2	406.1	406.4	406.9

Exact = 407

CONCLUSION

The IFM has been extended to analyze the laminated composite plate bending problems. Using the Mindlin-Reissner theory, a new laminated composite plate bending element(**MCP4**) has been proposed to analyze laminated composite plate bending problems using the Integrated Force Method. **MCP4** is a 4–node rectangular laminated composite plate bending element with five degrees of freedom (two in-plane displacements **u, v,** one transverse displacement w and two rotations θ_x, and θ_y at each node. Though this element has two extra energy modes it has yielded satisfactory results.

APPENDIX A

The elastic modulii(Q_{pq}) relating the Cartesian components of stress(σ_p) and the strains(ε_q) depends on the orientation of the coordinate system. The reduced stress-strain relation of any isotropic layer under plane stress is taken as

$$\begin{Bmatrix} \sigma_x \\ \sigma_y \\ \tau_{xy} \\ \tau_{xz} \\ \tau_{yz} \end{Bmatrix} = \begin{bmatrix} Q_{11} & Q_{12} & 0 & 0 & 0 \\ Q_{21} & Q_{22} & 0 & 0 & 0 \\ 0 & 0 & Q_{33} & 0 & 0 \\ 0 & 0 & 0 & Q_{44} & 0 \\ 0 & 0 & 0 & 0 & Q_{55} \end{bmatrix} \begin{Bmatrix} \varepsilon_x \\ \varepsilon_y \\ \varepsilon_{xy} \\ \varepsilon_{xz} \\ \varepsilon_{yz} \end{Bmatrix}$$

where,
$$Q_{11} = \frac{E_1}{1 - v_{12}v_{21}} \; ; \qquad Q_{12} = \frac{v_{12}E_2}{1 - v_{12}v_{21}}$$

$$Q_{33} = G_{12} \;; \qquad Q_{44} = G_{13}\,; \qquad Q_{55} = G_{23}$$

E_1 = Young's modulus in fiber direction

E_2 = Young's modulus in the perpendicular direction to fibre orientation

$v_{12} = v_{21}$ = Poisson's ratios;

G_{12}, G_{13}, G_{23} are shear modullii

For a particular case when fibers are oriented at angle θ with the x-axis, the transformed stress-strain relations for laminae will be

$$\begin{Bmatrix} \sigma_x \\ \sigma_y \\ \tau_{xy} \\ \tau_{xz} \\ \tau_{yz} \end{Bmatrix} = \begin{bmatrix} Q'_{11} & Q'_{12} & Q'_{13} & 0 & 0 \\ Q'_{21} & Q'_{22} & Q'_{23} & 0 & 0 \\ Q'_{31} & Q'_{32} & Q'_{33} & 0 & 0 \\ 0 & 0 & 0 & Q'_{44} & Q'_{45} \\ 0 & 0 & 0 & Q'_{54} & Q'_{55} \end{bmatrix} \begin{Bmatrix} \varepsilon_x \\ \varepsilon_y \\ \varepsilon_{xy} \\ \varepsilon_{xz} \\ \varepsilon_{yz} \end{Bmatrix} \qquad \text{...(A1)}$$

i.e.,

$$\{\sigma\} = [Q']\,\{\varepsilon\}$$

where,

$$Q'_{11} = Q_{11}\cos^4\theta + 2(Q_{12} + 2Q_{33})\cos^2\theta\sin^2\theta + Q_{22}\sin^4\theta$$

$$Q'_{12} = (Q_{11} + Q_{22} - 4Q_{33})\cos^2\theta\sin^2\theta + Q_{12}(\cos^4\theta + \sin^4\theta)$$

$$Q'_{22} = Q_{11}\sin^4\theta + 2(Q_{12} + 2Q_{33})\cos^2\theta\sin^2\theta + Q_{22}\cos^4\theta$$

$$Q'_{33} = (Q_{11} + Q_{22} - 2Q_{12} - 2Q_{33})\cos^2\theta\sin^2\theta + Q_{33}(\cos^4\theta + \sin^4\theta)$$

$$Q'_{13} = (Q_{11} - 2Q_{33} - Q_{12})\cos^3\theta\sin\theta + (Q_{12} - Q_{22} + 2Q_{33})\cos\theta\sin^3\theta$$

$$Q'_{23} = (Q_{11} - 2Q_{33} - Q_{12})\cos\theta\sin^3\theta + (Q_{12} - Q_{22} + 2Q_{33})\cos^3\theta\sin\theta$$

$$Q'_{44} = Q_{44}\cos^2\theta + Q_{55}\sin^2\theta$$

$$Q'_{45} = (Q_{44} - Q_{55})\cos\theta\sin\theta$$

$$Q'_{55} = Q_{44}\sin^2\theta + Q_{55}\cos^2\theta$$

REFERENCES

1. Reddy J.N, 1985, A simple higher order theory for laminated composite plates, *Journal of Applied Mechanics*.
2. Chen Wanji, Cheun Y.K, 2000, Refined quadrilateral element based on Mindlin-Reissner plate theory, *International Journal for Numerical Methods and Engineering*; 47; 605-627.
3. Timoshenko SP, Krieger SW, 1959, *Theory of plates and shells*, Second Edition, McGraw Hill International Editions.

8

Damage Tolerant Evaluation of Structural Components Subjected to Fatigue Loading

NAGESH R. IYER, A. RAMA CHANDRA MURTHY AND G.S. PALANI

Scientists, Structural Engineering Research Centre, CSIR Campus, Taramani, Chennai-600113, India
email: nriyer@sercm.csir.res.in, nriyer@sercm.org

ABSTRACT

This paper presents methodologies for damage tolerant evaluation of structural components under constant and variable amplitude loading. An improved wheeler residual stress model for tensile overloads and combination of tensile-compressive overloads for remaining life and residual strength assessment has been presented. The remaining life prediction has been carried out by employing linear elastic fracture mechanics (LEFM) principles. Residual strength assessment has been carried out by using plastic collapse condition and fracture toughness criterion. Numerical studies on remaining life prediction and residual strength assessment of plate with a centre crack have been conducted for validating the methodologies. It is observed from the studies that the predicted remaining life using the improved Wheeler model are found to be in close agreement with the corresponding experimental values reported in the literature. It is also observed that the residual strength decreases with increase in crack length and further it is observed from the residual strength plot that most of the cracked region failure is due to fracture toughness criterion except at the end region when the crack reaches its critical length, the failure is due to plastic collapse condition.

Keywords: Stress intensity factor, fatigue loading, damage tolerant, residual strength, remaining life.

1. INTRODUCTION

Reliability and functionality are two of the most prized qualities required of engineered structures and components. Most of the structures such as nuclear containments, reactor vessels, flyovers, high-rise buildings, aerospace structures, ship hulls, bridges and offshore structures and components are required to operate under tight controllable operating conditions, while others operate under unpredictable and uncontrollable regimes. The environment may also be variable, regardless of the operating regime. Most of the above industrial structures are generally subjected to fatigue loading.

The fatigue loading may be either constant amplitude loading or random loading. Remaining life or residual strength assessment of the cracked structural components in these structures is necessary for their in-service inspection, planning, repair, retrofitting, rehabilitation, requalification and health monitoring. In view of these, it is essential to use the damage tolerant design concepts for designing the above structural components. A structural component is damage tolerant if it can sustain cracks of critical length safely until it is repaired or its economic service life has expired. Damage tolerant analysis provides information about the effect of cracks on the strength of the component/structure. This information is usually presented in the form of two diagrams, namely, the residual strength diagram and the crack growth diagram. From the residual strength diagram, it is possible to predict the maximum crack length that can be sustained safely. This data is used in the crack growth diagram to find the number of loading cycles that will be required for the crack to grow to its critical length. Fracture mechanics is a tool employed for investigation of the crack growth and fracture behaviour of structural components that are subjected to fatigue loading. From the wide literature, it has been observed that the research carried out towards achieving both the objectives of damage tolerant evaluation is limited [1-5].

This paper presents methodologies for damage tolerant evaluation of structural components under constant and variable amplitude loading. An improved wheeler residual stress model for tensile overloads and combination of tensile-compressive overloads for remaining life and residual strength assessment has been presented. The remaining life prediction has been carried out by employing linear elastic fracture mechanics (LEFM) principles. Residual strength assessment has been carried out by using plastic collapse condition and fracture toughness criterion. Numerical studies on remaining life prediction and residual strength assessment of plate with a centre crack have been conducted for validating the methodologies. It is observed from the studies that the predicted remaining life using the improved wheeler model are found to be in close agreement with the corresponding experimental values reported in the literature. It is also observed that the residual strength decreases with increase in crack length and further it is observed from the residual strength plot that most of the cracked region failure is due to fracture toughness criterion except at the end region when the crack reaches its critical length, the failure is due to plastic collapse condition.

2. WHEELER RESIDUAL STRESS MODEL

For the sake of continuity, brief description of the basic Wheeler model is presented. Wheeler employs the residual stress retardation model to quantitatively account for crack growth retardation due to cyclic interaction. The development of the Wheeler model begins with the basic crack growth rate equation [6]

$$\frac{da}{dN} = f(\Delta K) \qquad ...(1)$$

In this equation, the SIF range is a function of load range, crack length and specimen geometry. To account for crack growth retardation, a retardation parameter C_{pi} is introduced. Final crack length can be calculated by using the following expression

$$a_n = a_o + \sum C_{pi} f(\Delta K_i) \qquad ...(2)$$

The retardation parameter varies from 0.0 to 1.0, taking the maximum value immediately after the overload. Wheeler's retardation parameter is calculated as shown below:

$$C_p = \left(\frac{r_p}{(a_p - a)}\right)^{m_1} \text{ for } (a + r_p) < a_p \text{ and } C_p = 1.0 \text{ for } (a + r_p) > a_p \qquad ...(3)$$

where,

r_p = extent of current plastic zone, $(a_p - a)$ = distance from crack tip to elastic - plastic interface m_1 = shaping exponent, generally obtained through experiments, depends on material, crack size and magnitude of applied load. This is graphically illustrated in Fig. 1. It is observed that C_{pi} is minimum immediately after the application of the overload, when $(a_p - a)$ has its maximum value. As 'a' approaches a_p, C_{pi} increases.

From the Wheeler's model it can be observed that

- Retardation decreases proportionately to the penetration of the crack into the overload plastic zone.
- Retardation occurs as long as the current plastic zone is within the plastic zone created by the overload
- Retardation ceases as soon as the plastic zone touches the boundary of the plastic zone created by the overload.

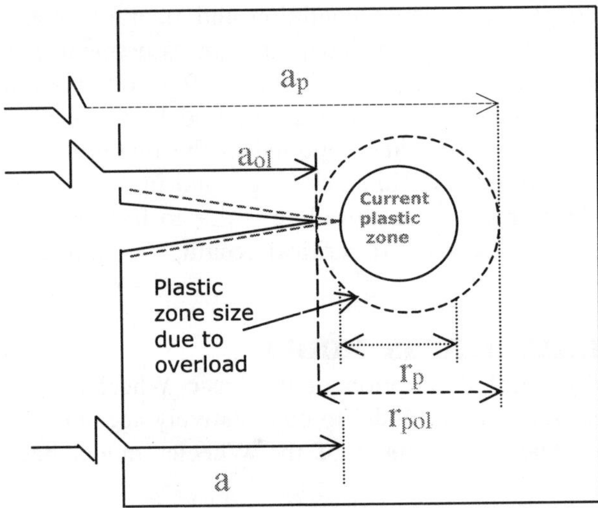

Fig. 1. Wheeler residual stress model.

2.1 Improved wheeler residual stress model for tensile overloads [1]

As mentioned earlier the shaping exponent is primarily dependent on overload ratio (OLR) and the ratio of crack length to width of plate panel. The degree of influence of OLR and shape factor (β) on shaping exponent was investigated for remaining life assessment of cracked plate panels under tensile overload. Extensive analytical investigations including regression analysis were conducted that contain different expressions for shaping exponent in terms of OLR and β to study the varying degree of influence on the remaining life prediction. Number of example problems was solved by using these expressions for different materials such as steel and aluminium alloys. Extensive studies were carried out that considered

(i)   variation of parameters β_e and β_c independently and concurrently
(ii)   variation of parameter OLR independently and concurrently
(iii)   various combinations of (i) and (ii)

Based on these studies among the various alternatives, the following shaping exponent expressions for (a) CT specimen and plate panel with (b) center crack and (c) edge crack under tensile overload were proposed in improved Wheeler model.

$$m_1 = \text{OLR} \quad \text{(for CT specimen)} \qquad \qquad ...(4)$$

$$m_1 = \text{OLR} + \beta_c^2 \quad \text{(for center crack)} \qquad \qquad ...(5)$$

$$m_1 = \frac{\text{OLR}}{2}(1 + \sqrt{\beta_e}) \quad \text{(for an edge crack)} \qquad \qquad ...(6)$$

where OLR$= \dfrac{\sigma_{max,OL}}{\sigma_{max}}$, β_e = shape factor for edge cracked panels and β_c = shape factor for center.

2.2 Improved Wheeler Model for Tensile-compressive Overloads [2]

In order to facilitate the inclusion of compressive loading effects into the Wheeler retardation model, some modifications have been proposed. It is known that Wheeler's original retardation parameter (C_p) was based on the premise that crack growth retardation should cease when the boundary of the current plastic zone has extended to or beyond the outermost boundary of any previous tensile overload. Fig. 2 summarizes the basic concepts and terms describing the Wheeler model to account for the compressive underloading effects.

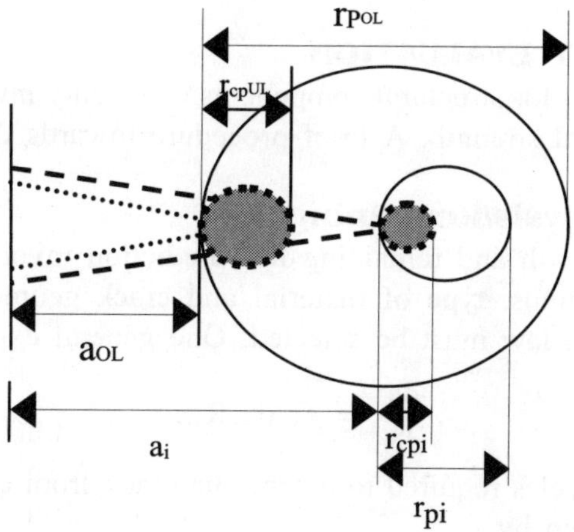

Fig. 2. Wheeler model for tensile-compressive overloads.

Effective plastic zone is given by, $\quad r_{eff} = r_p - r_{cp}$ $\qquad\qquad$ ...(7)

where, r_p = plastic zone due to tensile overload, r_{cp} = plastic zone due to compressive underload

$$r_p = \frac{1}{C_1\pi}\left[\frac{K_{max}}{\sigma_y}\right]^2 \quad \text{and} \quad r_{cp} = \frac{1}{C_1\pi}\left[\frac{\Delta K_{UL}}{2\sigma_y}\right]^2 \qquad\qquad ...(8)$$

where, C_1 $\quad$ = a factor accounting for state of stress at crack tip, K_{max} = maximum SIF,

$\qquad \sigma_y$ $\quad$ = material yield stress

$\qquad \Delta K_{UL} = K^* - K_{min,UL}$, K^* = minimum of $(K_{min,CAL}, K_{th})$

where, $K_{min,CAL}$ = SIF corresponding to min CAL and K_{th} = Threshold SIF

From Fig. 2, it can be observed that there is no retardation if

$$a_i + (r_{pi} - r_{cpi}) \geq a_{OL} + (r_{pOL} - r_{cpUL}) \qquad\qquad ...(9)$$

where, a_i $\quad$ = current crack length corresponding to the ith cycle

$\qquad r_{pi}$ $\quad$ = current plastic zone size corresponding to the ith cycle,

$\qquad r_{cpi}$ $\quad$ = current compressive plastic zone size corresponding to the ith cycle

$\qquad a_{OL}$ $\quad$ = crack length at which overload is applied,

$\qquad r_{pOL}$ = plastic zone size due to tensile overload

$\qquad r_{cpUL}$ = compressive plastic zone size due to compressive underload

Rearranging the relation given in eqn. 9 in a manner similar to Wheeler's original retardation parameter, the following relationship can be arrived at:

$$C_{pi} = \left[\frac{r_{eff,i}}{a_{OL} + r_{eff,OL} - a_i}\right]^{m_1} \quad \text{where } a_i + (r_{pi} - r_{cpi}) < a_{OL} + (r_{pOL} - r_{cpUL})$$

$$\phantom{C_{pi}} = 1 \quad\qquad\qquad \text{where } a_i + (r_{pi} - r_{cpi}) \geq a_{OL} + (r_{pOL} - r_{cpUL})$$

$\qquad\qquad\qquad\qquad\qquad\qquad\qquad\qquad\qquad\qquad\qquad$ ...(10)

where, $r_{eff,OL} = r_{pOL} - r_{cpUL}$

$\qquad m_1$ = shaping exponent, same as given in eqns. 4 to 6.

3. DAMAGE TOLERANT EVALUATION

The damage tolerant evaluation for structural components essentially involves prediction of remaining life and assessment of residual strength. A brief procedure towards this is outlined below.

3.1 Crack Growth Analysis/Remaining Life

Analysis of fatigue crack growth and remaining life prediction involves obtaining several data in relation to the loading conditions, type of material and crack geometry involved, among others. Then, a suitable crack growth law must be selected. One general expression for such a law is

$$\frac{da}{dN} = f(\Delta K, R, ...) \qquad\qquad ...(11)$$

The number of loading cycles required to extend the crack from an initial length a_0 to the final critical crack length a_f is given by

$$N = \int_{a_0}^{a_f} \frac{da}{f(\Delta K, R, ...)} \qquad\qquad ...(12)$$

Remaining life prediction using cycle-by-cycle approach is explained below:

    (a) Define the geometry, initial crack length (a_0) and crack growth increment (Δa_i)

    (b) Determine the function f(a) for calculating SIF, K

$$K_I = f(a)\sigma\sqrt{\pi a} \qquad \qquad ...(13)$$

    (c) Define the applicable crack growth law [Eqn. 11] and the retardation model.

    (d) Define the parameters for the growth model (e.g., K_c, K_{th}, the crack growth exponent, crack growth constant, etc.) and those of the retardation model, if applicable.

    (e) Obtain the stress history.

    (f) Determine K_{max} and ΔK using f(a), the crack length, Ds and s_{max}.

    (g) Apply the retardation model if needed.

    (h) Determine the no. of cycles ΔN_i observed for each incremental crack length

$$\Delta N_i = \frac{\Delta a_i}{f(\Delta K, R)}, \; a_i = a_0 + \Delta a_i \qquad \qquad ...(14)$$

    (i) Check that $a_i < a_f$, continue from (e) to resume for the cycle.

3.2 Residual Strength Evaluation

Residual strength can be computed by using (i) plastic collapse condition and (ii) fracture toughness criterion. The residual strength of a plate/panel is the least value obtained by using the above two criteria [7].

    3.2.1 Plastic collapse condition: In the plane stress condition where the stress in the entire cross-section is equal to yield strength at the time of collapse, the maximum load carrying (P_{max}) capacity of the plate with an edge crack is

$$P_{max} = t(W-a)\,\sigma_y \qquad \qquad ...(15)$$

where, t = thickness

    This failure load is called the collapse load or the limit load.

The nominal stress in full width of the component is, $\sigma = \dfrac{P_{max}}{W\,t}$ ...(16)

Hence the component fails when the nominal stress is

$$\sigma_{fc} = \frac{P_{max}}{W\,t} = \frac{t\,(W-a)\sigma_y}{W\,t} \qquad \qquad ...(17)$$

If a = W, failure will occur when the nominal stress $\sigma_{fc} = 0$.

    3.2.2 Fracture toughness criterion: The nominal stress at which fracture takes place, will be denoted as σ_{fc}

$$\sigma_{fc} = \frac{\text{Fracture toughness}}{\beta\sqrt{\pi a}} \qquad \qquad ...(18)$$

where, σ_{fc} is the residual strength or the remaining strength under the presence of cracks.

4. VALIDATION STUDIES

To demonstrate the methodologies described above, numerical studies have been conducted for remaining life prediction and residual strength assessment. One example problem on plate with a centre crack made up of 350WT Steel subjected to single tensile-compressive overload has been presented herein. This example was studied by Rushton and Taheri [8]. The data/information related to this problem is given below (Fig. 7).

Material	:	350 WT steel	Stress condition	: Plane stress
Plate dimension	:	300x100mm	Maximum stress (σ_{max}) :	114 MPa
Thickness	:	5 mm	Minimum stress (σ_{min}) :	11.4 MPa
Fracture toughness	:	50 MPa $\sqrt{m}$	Crack growth equation	: Paris
Yield Strength	:	350 MPa	C	: 1.02 e-8
Stress ratio	:	0.1	m	: 2.94
Initial crack length	:	20.0 mm		
Overload ratios	:	1.2, 1.5, 1.67, 1.83		
OL/UL	:	0.5, 1.0, 2.0		

No. of overloads are three and occurrence of overload for each specimen is at 30, 40 and 50 mm respectively. Using the above data, remaining life computation and residual strength assessment has been carried out for different OLRs and OL/UL ratios. The computed remaining life values are shown in Table 1. From Table 1, it can be observed that the predicted remaining life for different OLRs and OL/UL ratios are in good agreement with the corresponding experimental values available in literature. The predicted remaining life by considering only OL and for different OLRs is also shown in Table 1. Figure 3 shows the variation of predicted remaining life for different OLRs and OL/UL ratios. Fig. 3 also shows the plot of comparison of variation of remaining life for OLR = 1.67 considering various OL/UL ratios with the life predicted under OL alone. From Table 1 and Fig. 3, it can be observed that the predicted life decreases when OL/UL ratio decreases for a specific OLR.

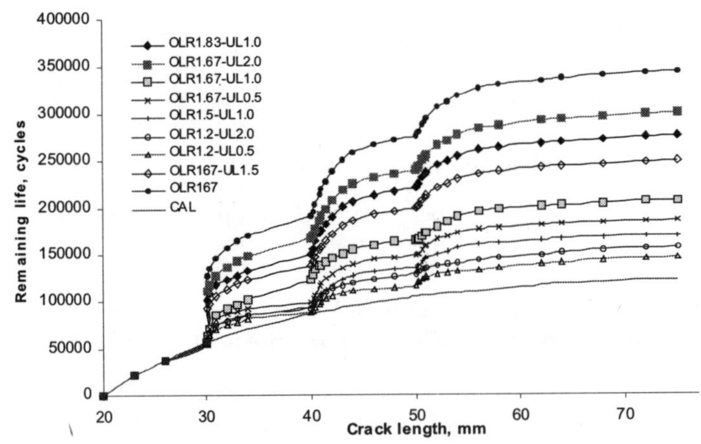

Fig. 3. Crack length vs No. of cycles.

Table 1. Remaining life values for plate with centre crack (350WT steel)

OLR	OL/UL	Remaining life		% diff.	Life predicted by considering only OL	% decrease in life compared to life with only OL
		Improved Wheeler model	Exptl. [7]			
1.2	0.5	145775	158000	7.73	174396	16.41
1.2	2.0	156444	168200	6.99		10.29
1.5	1.0	160004	174400	8.25	199271	12.48
1.67	0.5	185920	198000	6.1		47.15
1.67	1.0	206371	224100	7.91	351762	41.33
1.67	2.0	299238	328200	8.82		14.93
1.83	1.0	275299	304700	9.65	386767	28.82

*No. of cycles under CAL = 122171

Residual strength assessment has been carried out by using the plastic collapse condition as well as fracture toughness criterion. Figure 4 shows the residual strength values obtained by using the above two failure criterions. From Fig. 4, it can be observed that residual strength decreases with increase of crack length in both cases. Further, it can be observed that, most of the region, failure is due to fracture toughness criterion except in the beginning and ending regions where the failure is due to plastic collapse condition. Figure 5 shows the 3-D graphical plot of remaining life vs crack length and residual strength for CAL and OLR = 1.67. From Fig. 5, remaining life and residual strength can be obtained directly for a particular crack length. This diagram will be useful for designers, when the structures/components are designed against fatigue loading.

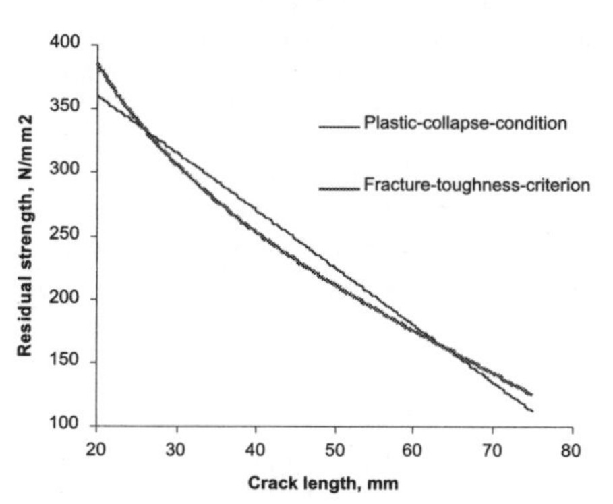

Fig. 4. Residual strength diagram.

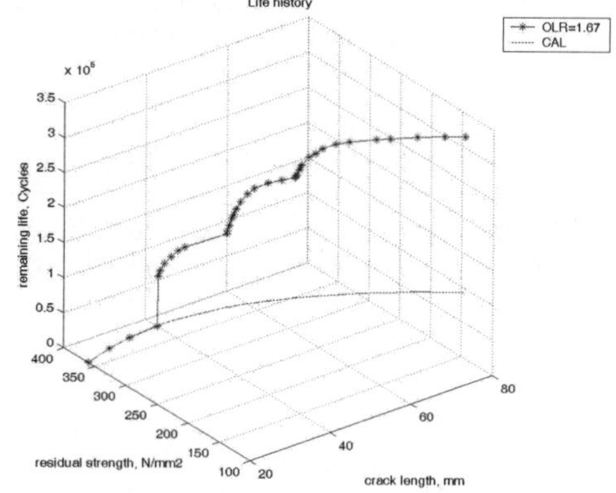

Fig. 5. Remaining life vs crack length and residual strength.

CONCLUSION

Methodologies for damage tolerant evaluation of structural components under constant and variable amplitude loading have been presented. Towards achievement of the objectives of damage tolerant, namely, residual strength and remaining life, stress intensity factor has been computed accurately for the fatigue loading. An improved wheeler residual stress model for tensile overloads and combination of tensile-compressive overloads for remaining life and residual strength assessment has been presented. The remaining life prediction has been carried out by employing linear elastic fracture mechanics (LEFM) principles. Residual strength has been carried out by using plastic collapse condition and fracture toughness criterion. Numerical studies on remaining life prediction and residual strength assessment of centre cracked plate panel subjected to tensile-compressive overloading have been carried out for validating the methodologies. It is observed from the studies that the predicted remaining life using the improved Wheeler model is found to be in close agreement with the corresponding experimental values reported in the literature. It is also observed that the residual strength decreases with increase in crack length and further it is observed that most of the cracked region, failure is due to fracture toughness criterion and when the crack reaches its critical length, the failure is due to plastic collapse condition.

ACKNOWLEDGEMENTS

The authors acknowledge with thanks the valuable suggestions provided by our colleague Mr J. Rajasankar during the course of this investigation. This paper is being published with the permission of the Director, SERC, Chennai.

REFERENCES

1. A. Rama Chandra Murthy, G.S. Palani, and Nagesh R. Iyer, 2004, An Improved Wheeler Residual Stress Model for Remaining Life Assessment of Cracked Plate Panels, Computers, Materials and Continua, .1(4), 289-300.
2. A. Rama Chandra Murthy, G.S. Palani, and Nagesh R. Iyer, 2005, An Improved Wheeler Model for Remaining Life Prediction of Cracked Plate Panels Under Tensile-Compressive Overloading, Int. Jl. Structural Integrity and durability, 1(3), 203-214.
3. T. Swift, 1984, Fracture analysis of stiffened structure. Damage Tolerance of Metallic Structures, Analysis Methods and Application, ASTM STP842, 69-107.
4. A.I. Kermanidis, P.V. Th, Petroyiannis, and SP. G. Pantelakis, 2005, Fatigue and damage tolerance behaviour of corroded 2024 T351 aircraft aluminum alloy, Theoretical and applied fracture mechanics, 43, 121-132.
5. Zerbst Uwe, Vormwald Michael, Andersch Christian, Madler Katrin and Pfuff Michael, 2005, The development of a damage tolerance concept for railway components and its demonstration for a railway axle, Engg. Fract. Mech., 72(2), 209-239.
6. O.E. Wheeler, 1972, Spectrum loading and crack growth. Transactions ASME, Jl. Basic Engg, 181-186.
7. Broek David, 1989, The practical use of Fracture Mechanics, Kluwer Academic Publishers.
8. P.A Rushton, and F. Taheri, 2003, Prediction of crack growth in 350WT Steel subjected to constant amplitude with over and underloads using a modified Wheeler approach, Marine Structures, 16, 517-539.

9

Efficient Computation of Colliding Particles in a Vertical Tumbling Sorting Machine

P. EBERHARD AND H. ALKHALDI

Institute of Engineering and Computational Mechanics, University of Stuttgart, Pfaffenwaldring 9, 70569 Stuttgart, Germany email: [eberhard, alkhaldi]@itm.uni-stuttgart.de

ABSTRACT

Particle screening is an essential and important technology in granular studies. Due to its wide variety of applications in the industrial and technological processes, it has captured great interest in recent research. The screeners which are considered as rotating sifting units while bulk materials are fed into their interior, have high importance in industrial operations. The particular problem of interest is the separation of round shape particles of different sizes using a tumbling vertical cylinder. The concept of the discrete element method that considers the motion of each single particle individually is applied in this study. Particle-particle, particle-wall and particle-mesh contacts are detected under the motion of the tumbling structure. The normal and frictional forces between particles themselves and particles and machine walls are calculated using a penalty method, which employs spring-damper models. For specific geometrical and contact parameters, particle distribution, sifting rate of the separated particles and the efficiency of the segregation process have been investigated using the uniform and stepped barrel models for both continuous screening and batch sieving processes. The rotational speed of the machine and the feeding rates of the particles flow have also a great influence on the transportation and segregation rates of the particles. In an attempt to better understand the mechanism of the particle transport between the different layers of the sifting system, different computational studies for achieving optimal operation have been performed.

Keywords: Contact mechanics; Molecular and multibody dynamics; Discrete element method; Screening and segregation phenomena.

1. INTRODUCTION

Granular studies are considered as being of great interest and are common in many engineering processes in different fields of industry [1]. To improve the performance of such processes, a good understanding of the behavior of particle motion is important and will contribute in the burgeoning of many different industrial applications of the granular technology.

Particle separation phenomena are important in granular media studies. Screeners and separators are used in a large number of industrial applications requiring separation and classification of powders or other bulk materials by particle size as well as separation of particles by density, magnetic properties or electrical characteristics. The operation of particle separation is divided into two main categories, continuous and batch operations. In continuous operation, the particles are continuously fed into the separation unit during the whole separation process. This type of particle separation is usually called 'screening'. On the other hand, batch operation is used if the particulate material is charged only once. This kind of batch separation is commonly described by the term 'sieving'.

Jansen's and Glastonbury's work in 1967 is considered being among the earliest works in studying particle screening phenomena [2]. They have tried to analyze the dynamics of screening process and to study the factors that affect the screening performance. The effect of the non-ideal aperture distribution of a sieving on the sieve residue is studied in [3], where an algorithm for deducing an effective sieve residue from the rate of powder passage through a sieve is described. The effect of some operating variables of separating sifters has been studied. Over a certain range of operating variables, the screening efficiency of two types of particles over a vibrating screen has been observed in [4]. The variables include the flow rate, deck angle, angular velocity and mesh size. The results show that the separation process is sensitive to the operating variables.

Many years ago, the discrete element method (DEM) that describes the motion of particles and models the behavior of dense solid assemblies in soil mechanics was proposed [5, 6]. The DEM was adapted in [7] to be used for the analysis of the internal dynamics of tumbling mills. An elastic-perfectly plastic contact model which only uses material parameters was developed, see [8]. Analytical works developed by [9] consider particles with or without cohesive, adhesive and frictional forces during contact using the spring-dashpot model which essentially embodies contact properties.

Classifiers or screeners are defined as sifting units which are rotated as material is fed into their interior. The finer particles should fall through the sieve opening and oversized particles are thrown out through certain outlets, see [10]. Screening round particles of different sizes using a rotary round deck separating machine is the main problem of interest in this study. To reach our goal of better understanding the processes, tumbler screening machine has been modeled. This interesting technical process of particle segregation has been investigated by our discrete element simulation program. The efficiency of the sorting process is determined by counting the number of both right-sorted and undesired-gangue particles of the whole process at the different levels of the machine. Different simulations are conducted to study the effect of some parameters on the machine performance, e.g., the machine speed and the feeding rate of particles. Continuous feeding processes with different flow rates as well as batch sieving with a limited number of particles are analyzed and compared.

2. MACHINE DESCRIPTION

The investigated tumbler screening machine (TSM) consists of several parts, see Fig. 1, these parts are: (a) *Machine foundation* which is heavy enough to hold the machine structure and contains the driving motor, (b) *Holder and adjustable plates* with angles α and β to create the tumbling motion during rotation, (c) *Rotating shafts* which are main and auxiliary shafts with a certain margin of eccentricity in between to hold the entire body of the machine and to transmit the motion to the main sifting unit, (d) *Sifting meshes* which represent the separating units and consist of a group of potentially different-sized sieves along the successive decks of the machine, (e) *Sifting vessel* which is the main cylinder in which the bulk material is charged. This vertical rotating vessel can be designed as a uniform or stepped multi level obliqued barrel with exits distributed at each sorting level and located at the outer periphery around the body of the machine, and (f) Feeding container which is used in the simulation of continuous screening through which the TSM is charged by the particle flow. The size of the exit nozzle of the container determines the particle flow rate. The machines are charged from the top and in the middle of the highest level of the sifting unit. The motion of the machine can be computed using the multibody system method, see e.g., [11].

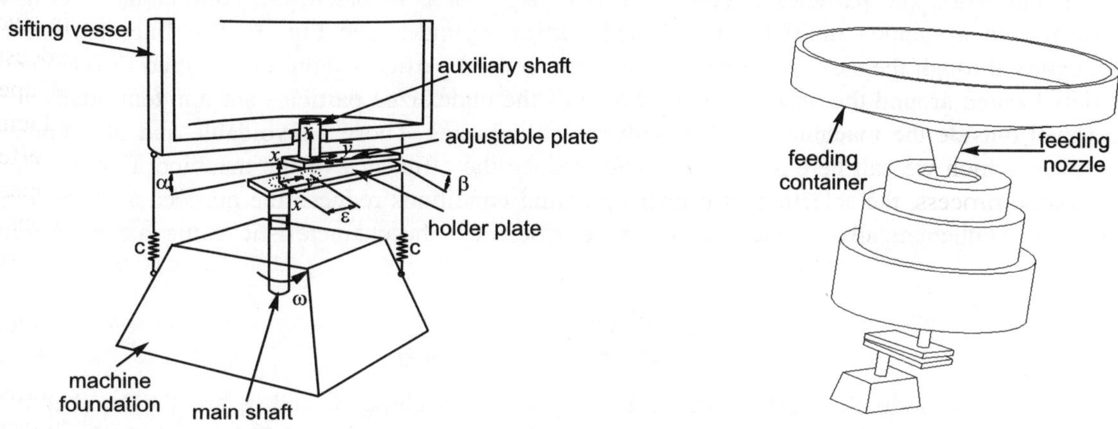

Fig. 1. Main parts of TSM (left), uniform barrel structure (middle) and stepped barrel structure (right).

3. EQUATIONS OF MOTION

For a spatial multibody system of particles, each particle has six degrees of freedom in both translation and rotation. Considering two bodies i and j, see Fig. 2, the governing Newton-Euler equations of motion can be generated for N particles as

$$m_i a_i = F_i \;\;,\;\; F_i = \sum_{j=1,\, i\neq i}^{N} F_{ij} + m_i g \;\;,\;\; F_{ij} = (k\delta_{ij} + c\dot{\delta}_{ij})n \;\;,\;\; \delta_{ij} = (r_i + r_j) - (q_i + q_j)\cdot n \;\;,\;\; n = \frac{q_i - q_j}{|q_i - q_j|}$$

$$I_i \alpha_i = M_i \;\;,\;\; M_i = \sum_{j=1,\, i\neq i}^{N} M_{ij} = \sum_{j=1,\, i\neq i}^{N} r_i \times F_{ij} \;\;,\;\; i = 1, \ldots, N \;\;,$$

...(1)

where F_i, M_i are the forces and torques acting on particle i, respectively, α_i is the vector of angular acceleration of particle i, δ_{ij} is the overlap. In order to analyze our system of particles, the DEM is used. The contact calculations are based on the soft-particle model which leads to a deterministic simulation. By using this model, we can consider a simultaneous multiple particle contact. This model assumes that the contact forces results from an overlap. Applying the penalty method which employs spring-damper models, we can determine the normal and frictional forces between particles themselves on one hand and between the particles and the machine walls on the other hand. For details see [12].

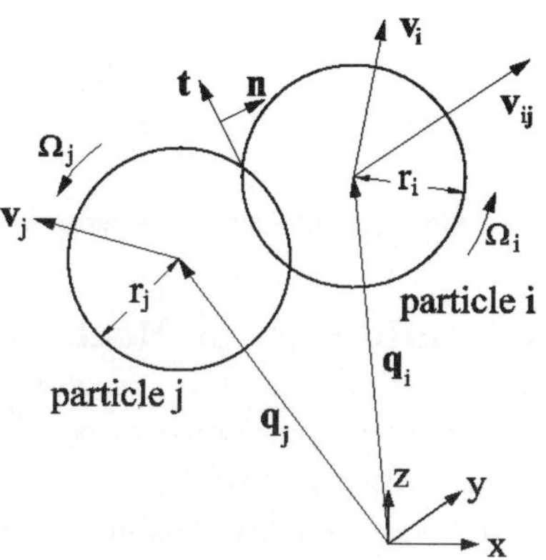

Fig. 2. Particles in overlap

4. NUMERICAL RESULTS AND DISCUSSIONS

Different simulations have been performed to investigate screening phenomena in the rotating tumbling machine. Particle distribution and sifting rates of the separated particles have been studied. Due to the complicated motions of the particles over the screen surface and the various factors that influence such motion, the understanding of the actual mechanisms involved in screening is still in its infancy. Attempts in the past to describe the performance of screening processes mathematically have adopted either a probabilistic approach or a kinetic approach, see e.g., [13].

In our work, the particles have been investigated for both batch and continuous feeding in a uniform and a stepped multi level obliqued vertical cylinder, see Fig. 3. The finer particles fall frequently through the sieve openings while the oversized particles should be ejected through certain outlets located around the machine body. Some of the undersized particles are unintentionally forced to travel outside the machine together with the sorted ones. Those unavoidable and also undesired particles, which we call here *gangue particles* reduce the efficiency of the machine. For an efficient separation process, the selection of proper operating conditions reduces the number of these gangue particles. Mathematically, to measure machine efficiency, the characteristic value c can be written as

$$c = \frac{100}{M} \sum_{i=1}^{M} c_i \, , \qquad c_i = \frac{s_i}{s_i + g_i} \qquad \qquad ...(2)$$

where, c_i is the individual efficiency of layer i of the machine, s_i and g_i are the number of the oversized sorted and the undersized gangue particles at level i of the TSM, respectively, M is the total number of barrels in the machine which has $i = 1$ as an indicator to the lowest level and $i = M$ for the highest one.

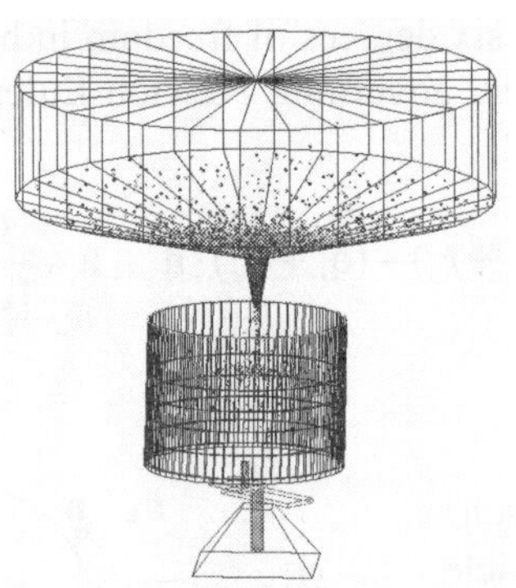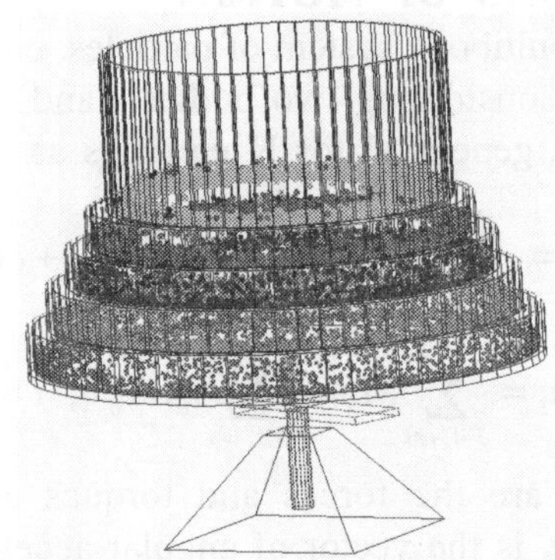

Fig. 3. Continuous screening (left) and batch sieving of 10499 bodies in stepped barrel structure (right).

4.1 Influence of Machine Velocity

The tumbling screening machine has the ability to run in different angular velocities. This angular speed which is assumed as being constant during the sorting operation process, has a great influence on the sifting rate of the mixed particles at each of the different sorting levels of the machine. Using computer simulation enables us to change the machine speed over a wide range in order to study its effect on the machine performance. The angular velocity ω should not be too high in order to avoid too fast motion of particles. Too high velocities of the particles will decrease the separation rate of the mixture and reduce the chance for undersized particles to fall down through the mesh openings. On the other side the rotational speed should not be too slow since low separation rates and a big number of undesired gangue particles can then be expected. This is due to the fact that the particles have not enough mixing due to their small velocities. This insufficient mixing reduces the chance of the undersized particles to fall through the mesh.

In order to reach the best rate of particle separation, we have to find the rotational speed of the machine which maximizes the number of the sorted particles and minimizes the gangue ones. In order to do this, we simulate our problem for different angular velocities and measure the efficiency of each set of geometrical, contact and material parameters. It is observed that the sorting under low rotational velocity leads to bad separation rates. The number of undesired gangue particles is even much higher than those of desired sorted ones in most of the machine levels, see Fig. 4a. Low efficiency of c=40% is recorded for this case. On the other hand, a relatively high velocity of 50 rpm leads to better results and higher efficiency of about c = 78%, see Fig. 4b. Going further, the very high velocity of 250 rpm will decrease the efficiency again, see Fig. 4c. From this figure, for all machine speeds the efficiency starts low and reach es its steady state in about 12 s when the machine reaches the steady state operating mode. It is observed that intermediate speed of about 50 rpm gives the best sifting rates during the continuous screening process, see Fig. 4d.

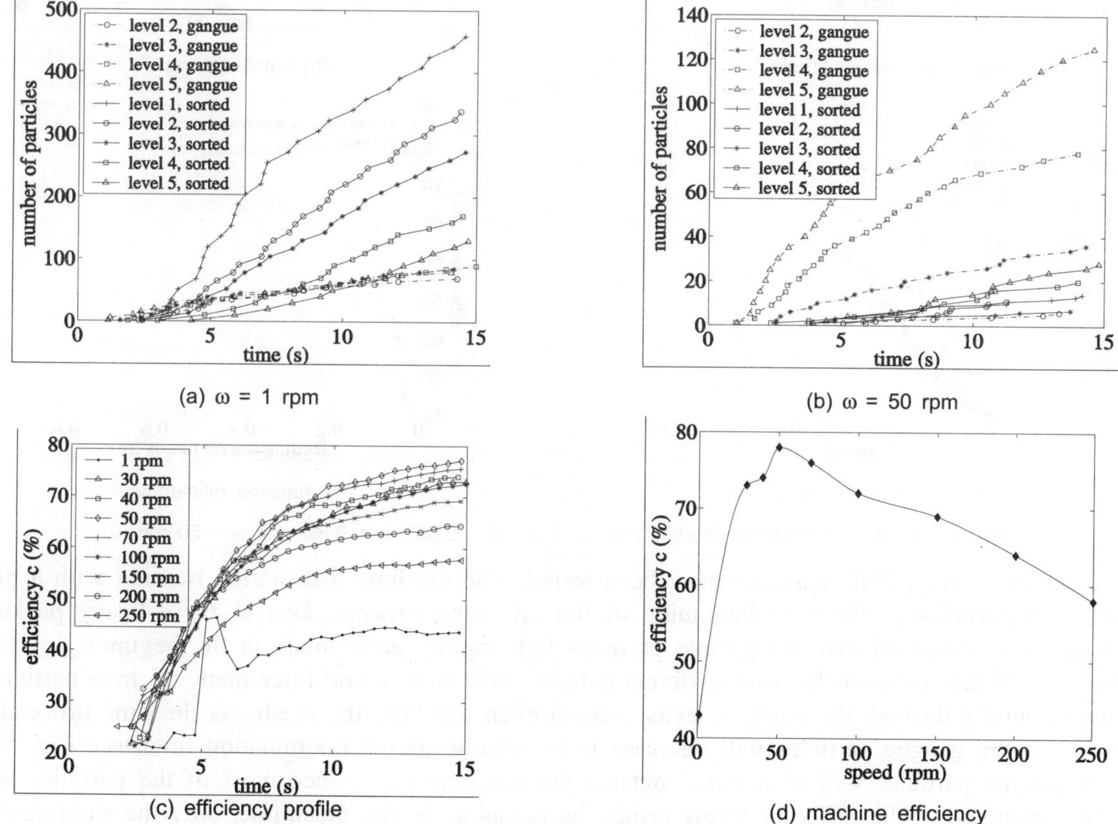

Fig. 4. Influence of rotational speed, α = 1°, β = 0.65°, b = 0.8w and feeding rate 147 particles per second.

4.2 Influence of Particles Feeding

We next study the effect of the feeding rate of the mixed particulate material on the machine efficiency. Increasing the size of the output nozzle of the feeding container will increase the feeding rate of the particle flow. In continuous screening, it is observed that low feeding rates with nozzle radius of 30 mm will lead to a bad screening performance. There, the sorted particles travel slowly to the machine exits together with some of the undesired gangue ones, see Fig. 5a. The rate of the sorted particles is observed to improve with increasing feeding rate.

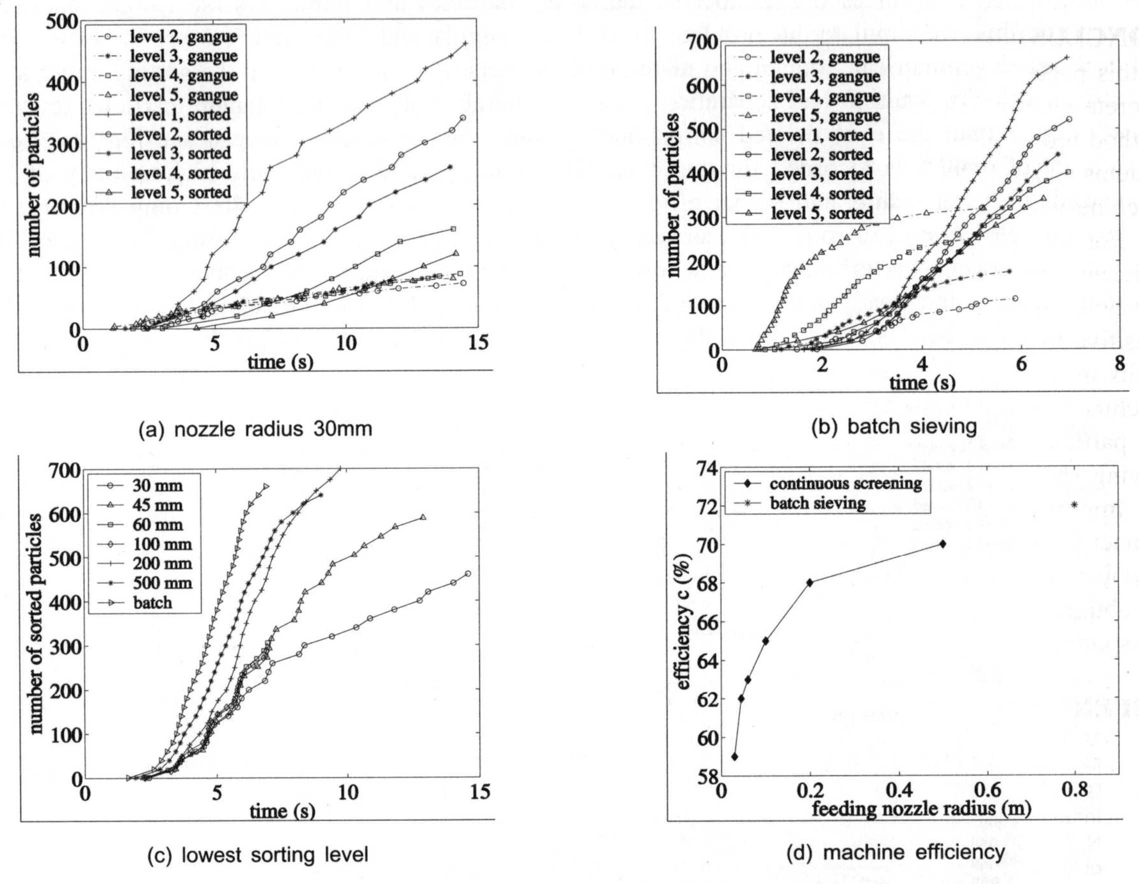

(a) nozzle radius 30mm   (b) batch sieving

(c) lowest sorting level   (d) machine efficiency

Fig. 5. Influence of particles feeding rate, $\alpha = 1°$ $\beta = 0.65°$, b = 0.8w and ω = 50 rpm.

In batch sieving, 3483 particles have been tested. The machine was started running with a big number of particles in the very beginning of the operating process. Due to these heavy packed particles, it is observed that the gangue particles will rapidly accumulate in the beginning of the simulation. Many collisions between different particles will happen and force many of these particles to travel rapidly through the machine exits without even touching the mesh. As the time proceeds, the rate of the gangue particles will decrease to be steady after 5 s simulation time, see Fig. 5b. Fewer gangue particles will then travel outside the machine exits since most of the particles are already separated at the different levels inside the machine. In the meantime, only the right-sized-particles will be sorted and directly run away through the machine exits which improve the performance of the separation process. As an indication of the influence of the feeding rates on the machine efficiency, the number of sorted particles in the lowest level of the sifting unit is recorded, see Fig. 5c. The lowest-level-particles are usually opposed to many collisions and face different obstacles during their travel. It is observed that very low feeding rates are not recommended in the case of continuous screening, see Fig. 5d. Increasing the rate of the feeding material will increase the machine efficiency. This increase in the feeding rate should be within certain limits after which the batch sieving will appear.

CONCLUSION

In this paper the particle screening phenomenon over a rotary tumbling screen is investigated. The discrete element method is applied and used as a simulation tool for the separation process. This method proves its ability to be a powerful numerical modeling tool for solving problems in particulate systems and granular media. The physical contacts between the particles themselves and the different machine walls are detected.

For specific geometrical and contact parameters particles transportation, sifting rates and machine efficiency are recorded. Particles are simulated in uniform and stepped models of tumbling cylinders. For both continuous screening and batch sieving, it was found that the segregation process is very sensitive to the rotational speed of the machine. This speed should be selected to be within certain limits to maximize the number of sorted particles and to improve the sifting rates for the different machine levels. Too high and too low speeds will lead to a bad screening performance. Furthermore, the particle feeding rates also affect the machine efficiency. For some operating conditions, batch sieving shows better results compared to continuous screening.

Improving the accuracy of the simulation requires to be more realistic in implementing the contact forces and the associated contact parameters of the dynamical system of the granular system. Physical contacts inside the TSM require some more detailed investigations. These parameters can be obtained from special experiments. For better understanding of the particle separation and transportation between the different layers of the machine, experimental studies have to be performed.

REFERENCES

1. M. Rhodes, 2005, Introduction to Particle Technology, Wiley-Interscience, Chichester.
2. M. Jansen, J. Glastonbury, 1967, The size separation of particles by screening, Powder Technology, 1, 334-343.
3. B. Kaye, N. Robb, 1979, An algorithm for deducing an effective sieve residue from the rate of powder passage through a sieve, Powder Technology, 24, 125-128.
4. N. Standish, A. Bharadwaj, G. Hariri-Akbari, 1986, A study of the effect of operating variables on the efficiency of a vibrating screen, Powder Technology, 48, 161-172.
5. P. Cundall, O. Strack, 1979, A discrete numerical model for granular assemblies, Geotechnique, 29, 47-65.
6. M. Allen, D. Tildesly, 1987, Computer Simulation of Liquids, Clarendon, Oxford, 1987.
7. B. Mishra, R. Rajamani, 1992, The discrete element method for the simulation of ball mills, Applied Mathematical Modelling, 16, 598-604.
8. B. Mishra, C. Thornton, 2002, An improved contact model for ball mill simulation by the discrete element method, Advanced Powder Technology, 13, (1), 25-41.
9. B. Muth, M. Müller, P. Eberhard, S. Luding, 2004, Contacts between many bodies, Machine Dynamics Problems, 28, (1), 101-114.
10. H. Alkhaldi, P. Eberhard, 2006, Computation of screening phenomena in a vertical tumbling cylinder, Proceedings in Applied Mathematics and Mechanics (PAMM), (submitted).
11. W. Schiehlen, P. Eberhard, 2004, Technische Dynamik (in German), Teubner, Stuttgart.
12. H. Alkhaldi, P. Eberhard, 2006, Particle screening phenomena in an oblique multi-level tumbling reservoir - A numerical study using discrete element simulation, (submitted).
13. E. Kelly, D. Spottiswood, 1999, Introduction to Mineral Processing, Wiley-Interscience, New York.

10

Assessment of Failure Strength of Fracture Prone Composite Structures with Defects

R. RAMESH KUMAR

+Engineer, Structural Design & Engineering Group, Vikram Sarabhai Space Centre, Thiruvananthapuram-695022, India.
email: r_ramesh@vssc.gov.in

ABSTRACT

Structural integrity assessment is made for compressively loaded stiffened multilayered composite flat panel and conical shell structure using the well-known MCCI approach. The flat panel has no delamination between the skin and hat stiffeners but the expected mode of failure is separation between the panel and stringer. The conical composite shell has considerable size of delaminations observed during curing process of segment joints. Strain energy release rates at the delamination interface between hat stiffener base and skin of the panel and splicer plate to basic composite skin of the conical shell are compared with interlaminar fracture toughness of the unidirectional composite material using a relationship for the critical failure load assessments. The mixed mode delamination fracture toughness of the carbon-epoxy laminate is used. In the case of stiffened panel without defect, strain energy release rate for a set of incrementally reduced delamination size are obtained and then extrapolated to zero defects. The panel specimen is designed such that its critcal buckling load is less than the compressive failure strength for a possible delamination mode of failure. For the shell structure the critical delamination fracture load obtained is estimated and found to be much higher than the given design ultimate load. Acoustic emission monitoring during the compression test also indicated the soundness of the conical composite shell structure. A good agreement between the prediction and test is achieved.

Keywords: Delamination fracture toughness, strain energy release rate, stiffened panels.

1. INTRODUCTION

Space structures made out of composites are more susceptible to damage as it is large in size with small skin thickness due its high strength and stiffness. Damages in composite are classified under interlaminar fracture which is delamination or separation between the layers and intralaminar fracture with a through the thickness crack. The well-known modified crack closure integral approach (MCCI) based on finite element method is used to assess either the critical load carrying capability of the structure with such defects [1] or fracture toughness once failure load is obtained from test [2-6].

Thus, for a given load the strain energy release rate (G) is obtained from the analysis is compared with its critical value Gc to determine the fracture load.

The interlaminar fracture toughness depends on the ratio of mode I and mode II that varies with the type loading and fibre orientations. The well-known experimental method to evaluate the delamination fracture toughness is mixed mode bending test [7].

The composite stiffened panels are normally fabricated by co-curing the skin and stiffener or by bonding the pre-cured stiffener to skin. It was also reported by author that for stringer stiffened panel, the most commonly observed failure mode under compressive loading is separation of the stiffener from the panel by delamination [1]. Once a delamination mode of failure is expected, the state-of-art is to assess G and compare it with the critical value Gc to evaluate the margin of safety. Recent study on assessment of the failure load of a closely stiffened composite panel under compression by MCCI approach showed a good agreement with the test and concluded that delamination fracture follows buckling [1]. Under such case margin on the critical buckling load drastically reduces.

In the present study, the failure load of compressively loaded carbon-epoxy conical structure with delaminations observed between splice plate and basic composite skin is assessed using MCCI approach. The delamination fracture toughness under mixed mode for the high strength and high modulus carbon-epoxy reported in literature is used [1-2].

2. STRUCTURAL CONFIGURATIONS OF PANEL AND CONICAL SHELL

2.1 Hat Stringer Stiffened Panel

Composite panel of size 400 mm X 750 mm with four hat type stringer stiffeners co-cured with the skin (panel) on one side of the panel is shown in Fig 1. The structure is made up of M55J/M18 carbon-epoxy prepreg. The material properties of unidirectional laminate are given in Table 1. The details of the lay-up sequence of the skin and stiffeners are given in Fig. 1. Each layer in a laminate is 0.1 mm thick.

2.2 Hat Stiffened Conical Structure

The structure is a closely stiffened composite shell with forty four numbers of carbon-epoxy laminate hat stiffeners, which are co-cured with the base carbon composite shell on the outer side as shown in Fig. 2. The composite shell structure is made up of in four ninety degree sectors. Each sector is made of $[0_2°, +60°, -60°]s$ carbon-epoxy laminates of 0.8 mm thick and splicers with $[+45°, -45°]s$ layers. A typical hat has four $+45°/-45°$ layers with additional four $0°$ layers fibre reinforcement in the crown and flange regions. The details of mechanical and fracture toughness properties are given in Ref. 7. Sector to sector joining is done over a $15°$ by stiffened outside splices and inside splices as shown in Fig. 2. Each sector is bonded and bolted to the end rings.

During the process of co-curing of the inner and outer splicer plates at four $90°$ locations around the circumference, debonds between the outer splicer plate and the shell segment was observed close to the segment joints below the hat stiffener base as shown in the Fig. 2. Similar debonds were observed at the four identical locations of the four bonded joints.

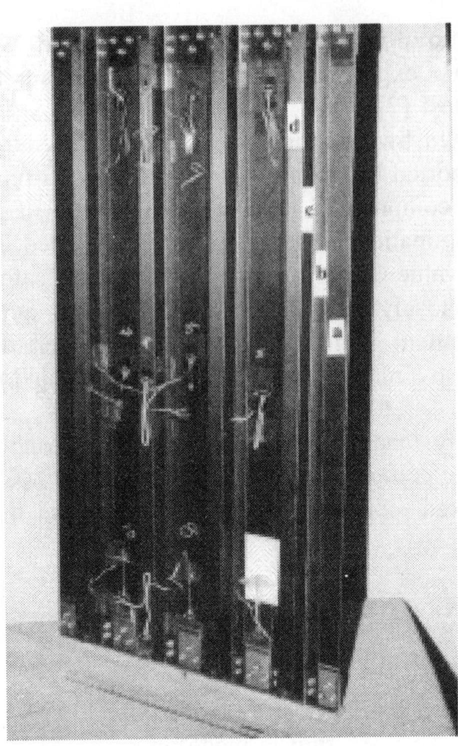

Fig. 1. Details of Stringer Stiffened (Hat Type) Composite Panel.

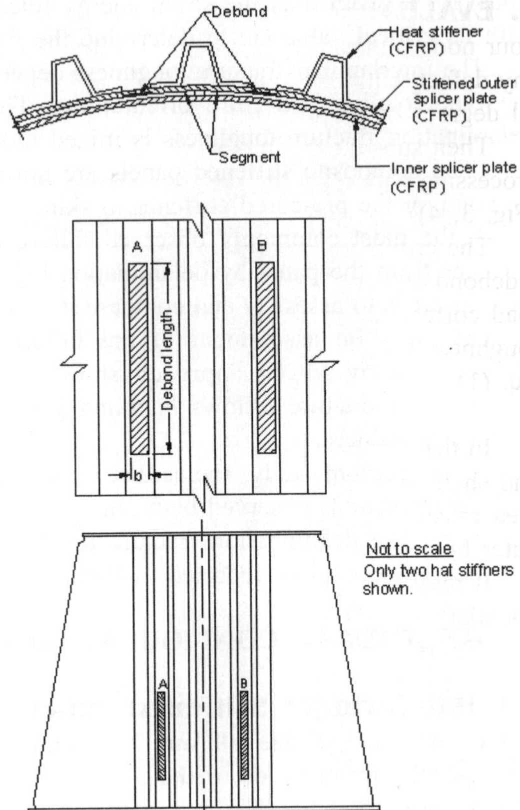

Fig. 2. Details of Debond area between the outer splicer plate and basic CFRP Shell over 15° sector.

a   -   [30°/–30°/30°/–30°/60°/-
         60°/-60°/60°/-30°/30°/-
         30°/30°].. 12 layers

Stiffener (Fig 1)
b   -   [45°/-45°/-45°/45°/45°/45°/
         45°/–45°/30°/–30°/30°/
         –30°/60°/–60°/–60°/60°/-
         30°/30°/–30°/30°]– 20 layers
c   -   [45°/–45°/–45°/45°/
         –45°/45°/45°/–45°] –8 layers
d   -   [45°/–45°/–45°/45°/45°/45°/
         45°/–45° /0°/0°]–10 layers

Table 1. Unidirectional Properties of Materials

Material	E_x (GPa)	E_y (GPa)	G_{xy} (GPa)	υ_{xy}
Carbon-epoxy [7]	300	6.0	4.5	0.35

3. EVALUATION OF G

Four node bilinear element is used to model the composite shell structure with hat stiffener without any defect. Uniform axial compressive load is applied at the small end of the shell (fore end) while all degrees of freedom of nodes at the bottom of the shell are constrained.

Then substructure model considering shell with only two hat stiffeners is extracted through post processing with retaining nodal displacement along the two side boundaries obtained from the analysis (Fig. 3, 4).

The critical load carrying capability (Pc) of the structure with defect is then estimated following a debond analysis of the substructure to obtain the strain energy release rate (G) for the compressive load corresponding to the design ultimate load (P). Then G is compared with interlaminar fracture toughness Gc of the carbon-epoxy material and Pc is obtained using the relationship given in Eq. (1).

$$Gc/G = (Pc/P)^2 \qquad \qquad ...(1)$$

In the substructure model, one of the debond areas alone is considered. The outer stiffened shell and shell segment are modelled separately with common nodes along the boundaries of the debond area as shown in Fig. 4. Elements are refined along the longitudinal direction so that at the two outer boundary elements are not refined.

It may be noted that in case element refinement is needed in the hoop direction, displacement boundary condition for the additional nodes can be given by linear interpolations.

The strain energy release rate for mode I and mode II based on MCCI approach is given by

$$G_I = \frac{1}{2ab}\begin{bmatrix} F_{yi}(u_q - u_i) + F_{yj}(u_k - u_j) + \\ F_{yi*}(u_{q*} - u_{i*}) + F_{yj*}(u_{k*} - u_{j*}) \end{bmatrix}$$

$$G_{II} = \frac{1}{2ab}\begin{bmatrix} F_{xi}(u_q - u_i) + F_{xj}(u_k - u_j) + \\ F_{xi*}(u_{q*} - u_{i*}) + F_{xj*}(u_{k*} - u_{j*}) \end{bmatrix} \qquad ...(2)$$

4. RESULTS AND DISCUSSIONS

4.1 Stringer Stiffened Composite Panel

Initially, both static and buckling analyses are carried out to assess whether the mode of failure is separation of the stringer from the panel. The compressive failure load evaluated using Tsai-Wu criterion is compared with the critical buckling load obtained from the buckling analysis. If the critical buckling load is less than the compressive failure load then the structure can fail by buckling or by delamination and hence delamination analysis is required to assess the fracture load.

Based on the analyses following Tsai-Wu criterion the compressive load is evaluated as 58.72 kN and the critical buckling load is obtained as 47.09 kN. The test showed a delamination mode of failure by which stringers separated from the panel at 46.75 kN without fibre breakage. Since, the critical buckling load is found to be less than the compressive failure load (58.72 kN), delamination analysis is carried out using strain energy release rate method.

In the finite element model with delamination is introduced between the base panel and stringer base. More details are given in Ref. 1.

The test results available for the material indicates the critical G_I and G_{II} values are 60 J/m^2 and 240 J/m^2 respectively [7]. In the actual procedure one has to evaluate the critical strain energy release rate G_c as sum of $G_I + G_{II}$ .Then compare it with as obtained from tests conducted with same mixed mode ratio as G_I/ G_c or G_{II}/ G_c. Since, in this case the value of G_I is negligibly small, only G_{II} is taken for determining the failure load due to delamination. Based on the analysis G_{IIc} is evaluated as 52.37 kN. In the finite element model the interfacial fibre orientations are +30° on the skin and + 45° on the hat stiffener. Toughness value obtained from unidirectional laminate gives a conservative estimate on the fracture load assessment.

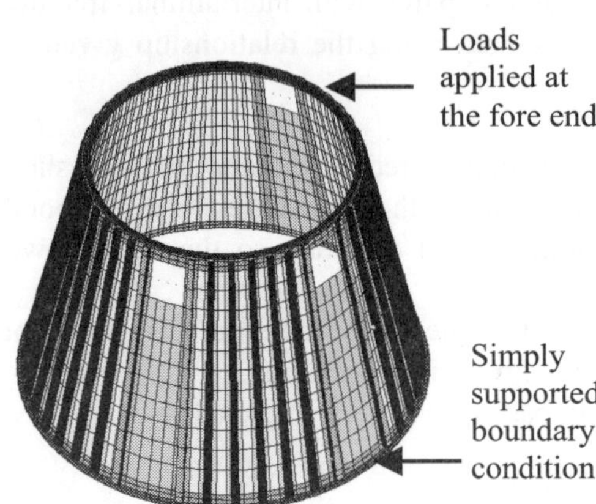

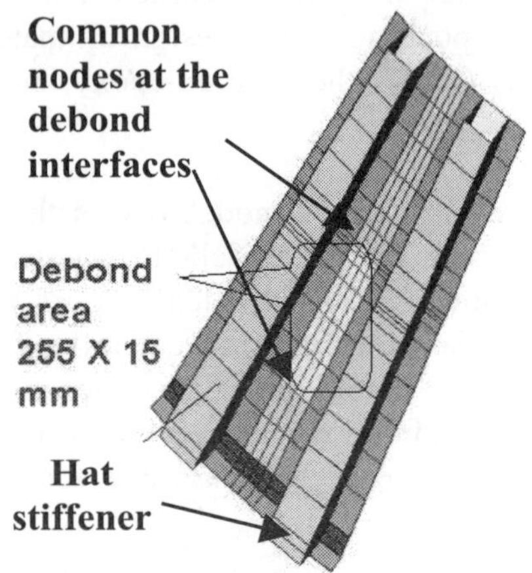

Fig. 3. Finite element model of composite structure.

Fig. 4. Finite Element model of the substructure with debond between splicer plate and segment shell.

From the finite element analysis the maximum shear stress occurring due to the given load in the panel is found to be 52 MPa as against the ultimate shear strength value of 73 MPa. The failure load evaluation based on this is 63.17 kN which could not produce a shear failure.

4.2 Composite Conical Shell

Maximum compressive displacement predicted at the fore end ring is 0.61 while using LVDT, test indicated a value of 0.58 mm at qualification load.

The measured strains obtained from the test is compared with the prediction at typical symmetric locations around 360° at the middle of the crown portion of the hat stiffener at the mid height of the structure in Table 2. It may be noted that due to the applied bending, compressive strains are reduced as expected in the diametrically (nearly) opposite locations. A good agreement is observed between the test and the prediction.

Based on the convergence study carried out, the 15 mm debond width is modelled with elements, each of size a = 4 mm and b = 9 mm at the common interface as shown in Fig. 3. The maximum value G estimated using the nodal forces and displacements for the middle element is given in Table-3. Using the Eq. (2), the G_I and G_{II} are evaluated. As anticipated the G_I is found to be negligibly small. G_{II} is obtained as 3J/m^2. The delamination fracture toughness G_c of the unidirectional carbon-epoxy laminate is reported by author in Ref. 7 as 240 J/m^2 is used to assess the critical load.

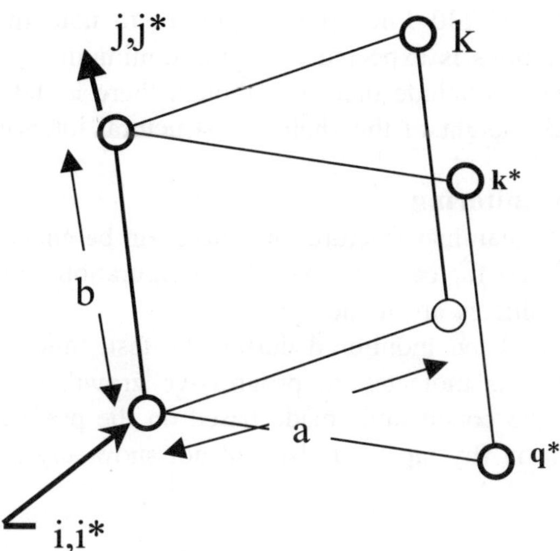

Fig. 5. Evaluation of strain energy release rate by finite element method.

i, j are the nodes for the element with nodes i-q-k-j and
i*, j* are the nodes for the element with nodes i*-q*-k*-j*

Table 2. Comparison of strains obtained from prediction and test

Outside of Crown portion at the mid height of the structure	
Predicted strains ($\mu\varepsilon$)	Test values of strains
−474	−410
−463	−441
−359	−302
−315	−144
−320	−300
−303	−339
−214	−183
−217	−104

Table 3. Typical nodal forces and displacements for the evaluation of G.

Element ID	Node ID	Nodal force F (N)	Notation
1	1	0.04176	F_{yi}
	2	0.0142	F_{yj}
2	1	−2.87	F_{yi^*}
	2	−10.0	F_{yj^*}

Node ID	Displacements (m)	Notation
1	-6.199×10^{-5}	u_i
2	-6.405×10^{-5}	u_j
3	-6.289×10^{-5}	u_k
4	-6.309×10^{-5}	u_q
5	-6.201×10^{-5}	u_{k^*}
6	-6.405×10^{-5}	u_{q^*}

It may be noted that at the qualification load, strain energy release rate is only 3.2 J /m^2 when compared to its critical value of 240 J/m^2. It is important to note that the delamination fracture toughness between 0° degree plies is expected to be the minimum.

Thus, one can theoretically conclude that even though there is delamination of 255 mm which is about one third of the slant height of the shell, the structural integrity is assessed as very safe.

4.2.1 Acoustic Emission Monitoring

Based on the present study intralaminar fracture toughness can be analytically predicted (that can be used to assess the fracture load) for certain laminate configurations once toughness values of say, [0°], [90°] and [±45°] sub laminates are made available.

The acoustic emission has been monitored during the test, upto the qualification load. Just a single signal was observed! This indicates no progressive growth or acoustic activity to cause a delamination. This confirms the conclusions made based on the prediction. The structural integrity is ensured. Post test observation (by tap test) also did not show any growth of debonds.

CONCLUSION

The structural integrity assessment of closely stiffened multilayered carbon -epoxy composite conical shell structure with most commonly observed delaminations between the basic shell and splicer plates has been made following MCCI approach and verified the soundness of the structure using acoustic emission monitoring while carrying out the load test. A good agreement for the displacement and fibre strains between the prediction and test has been obtained. When compared to the mode-II delamination fracture toughness of the carbon-epoxy unidirectional laminate value of 240 J/m^2, present substructure analysis shows a strain energy release rate of only 3.2 J/m^2 between the splicer and the basic shell for the design ultimate load applied. AE monitoring has indicated just one signal confirming the analytical prediction!

Stringer stiffened composite panel under axial compression has been analysed for the estimation of critical buckling load as 47 kN and critical compressive failure load following Tsai-Wu criterion as 58.72 kN. Using MCCI approach, maximum load carrying capacity of the panel has been assessed as 52.37 kN. Axial compression test on the panel revealed a delamination mode by failure by the separation of the hat stiffener from the panel. It has been concluded that assessment of failure load of the stringer stiffened panel can be predicted for the no defect panel even where delamination mode of failure is anticipated.

REFERENCES

1. R. R. Kumar, K. S. Praveen and G. V. Rao, 2003, Assessment of delamination fracture load of stringer stiffened composite panel, *AIAA Journal*, 41(3), 551-554.
2. S. Parhizgar, C. L. Zachary, C. T. Sun, 1982, Application of the principles of linear elastic fracture mechanics to composite materials, *Int. J Fracture*, 20, 3-15.
3. T. A. Cruse, 1973, Tensile strength of notched composites. *J Composite Materials*,1, 218-229.
4. S. Jose, R. R. Kumar, M. K. Jana and G. V. Rao, 2001, Intralaminar fracture toughness of a cross-ply laminate and its constituent sub laminates, Composite Sci. and Tech., 61, 1115-1122.
5. R. R. Kumar, P. N. Dileep, S. Renjith and G. V. Rao, 2003, A simple method for theoretical.
6. P. N. Dileep and R. R. Kumar, 2005, A simple method for the evaluation of fracture toughness of a multi layered laminate based on the failure stress of sub-laminates, Int. J Fracture, 131(1), L3-L6.
7. Jose S., Kumar R. R., G. V. Rao and P. Sriram, 2000, Studies on mixed mode interlaminar fracture toughness of M55J / M18 carbon-epoxy laminates, Adv. Composite Letters, 9 (5), 335-340. Prediction of fracture toughness of multilayered composites. Advanced Composite Letters,12,151-158.

11

The Best-Fit Paradigm of FEA—"Shadows of the Exact"

SOMENATH MUKHERJEE[1] AND GANGAN PRATHAP[2]

[1]Structures Division, National Aerospace Laboratories, Bangalore-560017, India.
email: somu@css.cmmacs.ernet.in
[2]Centre for Mathematical Modelling and Computer Simulation, Bangalore-560037, India.
email: gp@cmmacs.ernet.in

ABSTRACT

In this presentation, the mathematical and philosophical aspects of FEA as a powerful computational tool have been highlighted, using the principles of virtual work. Employing the weak forms of the differential equations, it is shown how computations in finite element method can be interpreted as orthogonal projections (or best-fits) of the analytical solutions of the strain vector in each element.

This presentation begins with a brief introduction of vector and function spaces, orthogonal projections and their relationship with FEM formulations. Simple one dimensional bar and beam elements have been used to illustrate the fundamental principles that guide finite element computations in conservative problems. The pathological problem of shear locking in Timoshenko beam elements and the rationale behind its elimination through various variational crimes, like reduced integration, have been addressed using the projection theorems.

It has been shown how the best-fit paradigm in an element level is conserved in a very subtle sense even in certain cases of prima facie violation of the best-fit rule. The message is that the FEM solutions to such systems of apparent violation respond as best fits to the stiffened analytical solutions from the discretisation errors of the nodal reactions.

Interesting geometrical patterns arising from the computational process in finite element elastodynamic problems have also been discussed. The approximate Rayleigh Quotient error is interpreted in a geometrically abstract, but elegant fashion. It has been shown how incorporation of variationally incorrect procedures (like mass lumping or reduced integration) in the FE formulations leads to the violation of the general rules of elastodynamic analysis.

Keywords: Finite element analysis, Weak forms, Virtual work, Normal equation, Orthogonal projections, Best-fits, Energy error rule, Bar element, Euler beam element, Pollution error, Timoshenko beam element, Shear Locking, Reduced integration, Variational incorrectness, Rayleigh Quotient, Frequency Error Hyperboloid.

1. INTRODUCTION

The Finite Element Method (FEM) is the mathematical tool of the engineers and scientists to determine approximate solutions, in a discretised sense, of the concerned differential equations, which are not always amenable to closed form solutions. In the conventional finite element method the following algorithm is followed. The domain of interest is first discretized into a number of elements. Equilibrium equations of each element, derived by employing the principle of minimum potential energy and using approximate interpolation functions over each element, are derived. These are then assembled in the form of matrices and solved. In essence, the method yields a discretised, matrix form $[K]\{X\} = \{F\}$ for the differential equation of the full domain. Here $[K]$ represents the global stiffness matrix. The global equations are solved for the nodal displacement vector $\{X\}$, with appropriate boundary conditions and loading $\{F\}$. Desired convergence is achieved by finer meshing. FEA is generally considered, prima facie, as a displacement method of analysis, and the general notion is that the method is driven by displacement interpolations between nodes. In the past few decades, most research papers and books [1-4], barring a few, have identified the displacement as the driving parameter for FEA, and a major section of the scientific community still sticks to the idea that in FEA computation, approximate displacement interpolations, in general, conserve the exact (analytical) nodal displacements.

This paper originated out of an attempt to present the best-fit paradigm of FEA to a greater audience and readers through a function space approach. We examine critically what happens at the element level, (which is of course of major interest to the stress analyst) and resort to a function space approach to justify that FEA computations are actually driven by the element strains, and that such computations are effectively designed to get the approximate strains in the elements as the best-fits of the analytical strains. This approach presents an alternative interpretation to that of the displacement based paradigm. It provides a logical explanation for the FEA results and for the origin of many pathological problems like locking and pollution errors and the causes of rank deficiencies in certain elements. It pivots its arguments on the rigorous mathematics of the function space algebra [5]. The function space interpretation of the finite element method in terms of projection theorems has been presented at the global level in the excellent treatise of Strang and Fix [6]. The fact that in FEA, the individual elements with approximate solutions respond in a best-fit manner to the analytical solution has been observed by Prathap [7, 8] through the orthogonality conditions that result from the Hu-Washizu's Principle of approximate computation, that apply, in general, to the classical Rayleigh-Ritz method.

Recently, extensive research in this area using the function space approach [9-15], has established the best-fit paradigm of the FEA upon rigorous mathematical foundation. The origins of shear locking and its elimination by reduced integration has been explained by this approach by Mukherjee and Prathap [9, 10] (2001, 2002). It has been shown that violation of this best-fit rule may occur when variational incorrectness is introduced through reduced integration methods [10]. Cases of prima facie violation of the best-fit rule are shown to actually conserve the best-fit paradigm in a very subtle sense by Mukherjee, Prathap and Sangeeta [11, 12]. Furthermore, it has been shown that reduced integration techniques, often adopted to eliminate locking problems, can also introduce rank deficiencies in certain elements [13]. The function space approach of the global domain, originally given by Strang and Fix [6] to elasodynamic problems has been reviewed, and new formulations and geometric insights have been presented by Mukherjee, Jafarali and Prathap [14, 15].

2. A REVIEW OF VECTORS, VECTOR SPACES, BEST-FITS

2.1 Vector and Function Spaces

Mathematically, a vector is an ordered array of numbers, and is generically denoted as $\mathbf{A}$ or $\{A\}$. A set of N vectors ($\mathbf{A_1}$, $\mathbf{A_2}$, …. AN) are linearly independent, if and only if any of their linear combination can never vanish, (except for the trivial case when all the weighting parameters α_i are equal to zero). Thus, without the trivial case, the condition for linear independence is

$$\sum_{i=1}^{N}\alpha_i A_i \neq 0 \qquad \qquad …(1)$$

A vector space is a set of vectors such that any linear combination of its members yields another vector that belongs to the space, i.e. a vector space is closed under linear combination. We note that a perfectly flat plane is two-dimensional, since only two linearly independent vectors can be chosen at a time in the plane. Any third vector in it can be expressed as a linear superposition of the earlier chosen two. In mathematical parlance, a two-dimensional space can be spanned by two linearly independent vectors, called the basis vectors. This argument can be extended to define any n-dimensional vector space which is spanned by n linearly independent vectors.

A polynomial vector space, or a polynomial function space is a set of polynomial vectors such that it is closed under any linear combination of its members. Mathematically, one can define a polynomial function space of degree n-1, as a set of vector functions of r-rows as,

$$P_n^r(\xi) = \left\{ \{p\} : \{p\} = \sum_{i=1}^{n} \{a\}_i \xi^{i-1}, \quad -1 \leq \xi \leq 1, \quad \{a\}_i \in R^r \right\} \qquad …(2)$$

An inner product in a vector space between any two of its members, $\{a\}$ and $\{b\}$, is an operation denoted by $<a, b>$. It can be defined in various forms, appropriate to the space being described. For the Euclidean space, the inner product is simply the scalar (*dot*) product, i.e., $<a, b> = a \cdot b = \{a\}^T\{b\}$. Generic forms for the various kinds of inner products are given in the following:

$$< a,b > = \{a\}^T[D]\{b\} \quad \text{for discrete vector spaces,} \qquad …(3a)$$

$$< a,b > = \int_{-1}^{1} \{a\}^T\{b\}d\xi \quad \text{for the ordinary polynomial vector space,} \qquad …(3b)$$

$$< a,b > = \int_{-1}^{1} \{a\}^T[D]\{b\}(\det J)d\xi \quad \text{for the polynomial vector space used in FEA.} \qquad …(3c)$$

Here the $r \times r$ matrix $[D]$ is positive definite and symmetric. In particular, for the Euclidean form, $[D]$ in definition 3(a) is the identity matrix $[I]$. The *Jacobian*, J is used in the definition (3c) to account for any mapping of the dimensional co-ordinate (x, say) to the non-dimensional co-ordinate ξ so that one can use the appropriate transformation rule ($dx = [\det J] \cdot d\xi$) in the integral defining the inner product. The generic norm (or absolute value) of a vector $\{a\}$ in a vector space is defined from the inner product as

$$\|a\| = \sqrt{< a,a >} \qquad \qquad …(4)$$

2.2 The Gram-Schmidt Process

From a given basis set B (of vectors in V), one can determine an *orthogonal* basis set spanning V using a method called the Gram-Schmidt (GS) Orthgonalization. The following algorithm illustrates this method.

Given: A basis set B (of cardinal number n) that spans a space V

$$B = \left(\{b\}_i : \sum_{i=1}^{n} \alpha_i \{b\}_i \neq 0 \quad \{b\}_i \in V, \quad i = 1, 2, \ldots n \right) \qquad \ldots(5)$$

Required: To find an orthogonal basis set spanning V

Let $\qquad\qquad\qquad\qquad \{v\}_1 = \{b\}_1 \qquad\qquad\qquad\qquad \ldots(6a)$

The other orthogonal basis vectors can be obtained from

$$\{v\}_k = \{b\}_k - \sum_{j=1}^{k-1} \frac{<b_k, v_j>}{<v_j, v_j>} \{v\}_j \qquad \ldots(6b)$$

The non-zero orthogonal basis vectors $\{v\}_k$ ($k = 1, 2, \ldots m$) spanning the m-dimensional V can be arbitrarily scaled. They satisfy the orthogonality condition

$$<v_i, v_j> = \{v\}_i^{T} \{v\}_j = 0 \qquad i \neq j \qquad \ldots(7)$$

Example 1: A single-row cubic polynomial p in the four-dimensional function space $P_4(\xi)$ can be expressed as

$$p = \sum_{i=1}^{4} a_i \xi^{i-1} = a_1 + a_2 \xi + a_3 \xi^2 + a_4 \xi^3, \quad -1 \leq \xi \leq 1, \quad a_i \in R$$

The four linearly independent basis vectors spanning this four-dimensional space are

$$b_1 = 1, \; b_2 = \xi, \; b_3 = \xi^2, \; b_4 = \xi^3$$

One can use the Gram-Schmidt method to identify the orthogonal basis set (scaled if necessary) spanning this same space. The orthogonal basis vectors spanning $P_4(\xi)$ are given as

$$P_1 = b_1 = 1, \; P_2 = \xi, \; P_3 = (3\xi^2 - 1), \; P_4 = (5\xi^3 - 3\xi) \qquad \ldots(8)$$

These basis vectors, spanning the polynomial space are known as the *Legendre Polynomials*. They are, as expected, orthogonal to each other,

$$<P_i, P_j> = \int_{-1}^{1} P_i P_j \, d\xi = 0, \quad i \neq j \qquad \ldots(9)$$

2.3 Orthogonal Projection (or Best-Fit) of a Vector onto a Vector Space

Consider a plane P, a vector space of two dimensions, spanned by two orthogonal vectors **u** and **v**. The orthogonal projection (or best-fit) $\overline{A}$ on plane P of any vector **A**, that does not lie in the plane P, can be expressed as a sum of the orthogonal projections of **A** along the basis vectors **u** and **v**

$$\overline{A} = \left(\frac{A \bullet u}{u \bullet u} \right) u + \left(\frac{A \bullet v}{v \bullet v} \right) v \quad ; \quad u \bullet v = 0 \quad ; \quad u, v \in P$$

Fig. 1 shows the orthogonal projection $\overline{A}$ of a vector **A** on a two-dimensional subspace (plane P).

We note from Fig. 1 the condition for the best-fit. *The error is orthogonal to the best-fit*, i.e.,

$$\left(\mathbf{A}-\overline{\mathbf{A}}\right)\bullet\overline{\mathbf{A}}=0 \quad i.e., \quad \left|\mathbf{A}-\overline{\mathbf{A}}\right|^2 = \left|\mathbf{A}\right|^2 - \left|\overline{\mathbf{A}}\right|^2 \qquad\qquad ...(10)$$

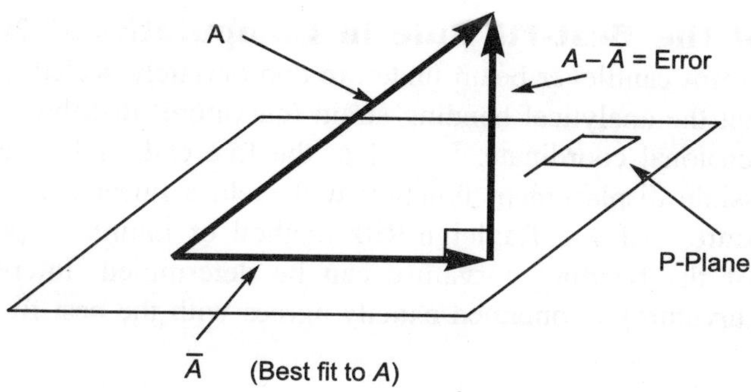

Fig. 1. Projection of the vector **A** onto a two-dimensional plane *P*.

When we seek a best-fit solution $\{\overline{y}\}=[B]\{\alpha\}$ to an original polynomial function $\{y\}$, we actually mean that we seek an orthogonal projection of $\{y\}$ upon a function space B in which the column vectors of the matrix $[B]$ lies. Mathematically, the orthogonal projection $\{\overline{y}\}$ of the original vector $\{y\}$ onto any m-dimensional vector space B is given as

$$\{\overline{y}\}=\sum_{j=1}^{m}\frac{<y,v_j>}{<v_j,v_j>}\{v_j\}, \quad <v_i,v_j>=0 \quad for \quad i \neq j \quad v_i,v_j \in B \qquad ...(11)$$

The orthogonality properties are now translated into the following forms with the generic inner product and norm as

$$<\overline{y},y-\overline{y}>=0 \quad i.e., \quad \left\|y-\overline{y}\right\|^2 = \left\|y\right\|^2 - \left\|\overline{y}\right\|^2 \qquad ...(12a,b)$$

Equation (12a) represents the *normal equation* for the best-fit computation.

Example 2: Consider a quadratic polynomial in the 3-dimensional space $P_3(\xi)$.

$$y = 1 + 2\xi + \xi^2 \quad -1 \leq \xi \leq 1$$

It can be expressed as a linear combination of the Legendre Polynomials,

$$y = 1 + 2\xi + \xi^2 = \frac{4}{3}P_1 + 2P_2 + \frac{1}{3}P_3 \qquad -1 \leq \xi \leq 1$$

The linear best-fit polynomial to $\{y\}$ can be expressed as

$$\overline{y} = \alpha_1 + \alpha_2\xi \qquad -1 \leq \xi \leq 1$$

Thus $\{\overline{y}\}$ can be determined from the generic projection formula for a determination of the orthogonal projection of a vector $\{y\}$ onto an m-dimensional function space V, given by equation (5). Thus for this case, the orthogonal projection of $\{y\}$, or best-fit to it, is given by (11),

$$\overline{y} = \sum_{i=1}^{2}\frac{<y,P_i>}{<P_i,P_i>}P_i = \frac{\displaystyle\int_{-1}^{1}y.1.d\xi}{\displaystyle\int_{-1}^{1}d\xi} + \frac{\displaystyle\int_{-1}^{1}y.\xi.d\xi}{\displaystyle\int_{-1}^{1}\xi^2 d\xi} = \frac{4}{3} + 2\xi = \frac{4}{3} + 2P_1$$

The inner products involved here are taken according to definition (3b) given earlier. The original quadratic function $\{y\}$ and its linear best fit $\{\bar{y}\}$ (orthogonal projection on a linear polynomial space) is shown in Fig 2 (i).

2.4 An indication of the Best-Fit Rule in Computational Mechanics

Suppose we consider a uniform cantilever beam under an appropriately scaled uniformly distributed load q (Fig. 2 (ii)), such that the analytical bending strain (curvature) distribution is given by $\varepsilon = 1 + 2\xi + \xi^2$; (ξ is non-dimensional coordinate, $\xi = -1$ at the free end, and $+1$ at the fixed end). A cubic polynomial as admissible displacement function will yield a linear approximation $\varepsilon^h = \alpha_1 + \alpha_2\xi$ for the bending curvature. Using a Rayleigh-Ritz method of minimum potential energy, this approximate distribution of the bending curvature can be determined. Interestingly, this linear approximation for strain (curvature) so obtained exactly agrees with the best-fit linear solution to ε, i.e., $\varepsilon^h = \bar{\varepsilon}$. (Fig 2(ii)).

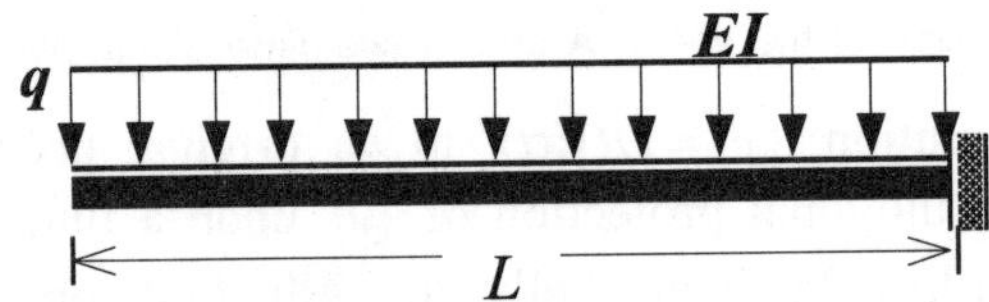

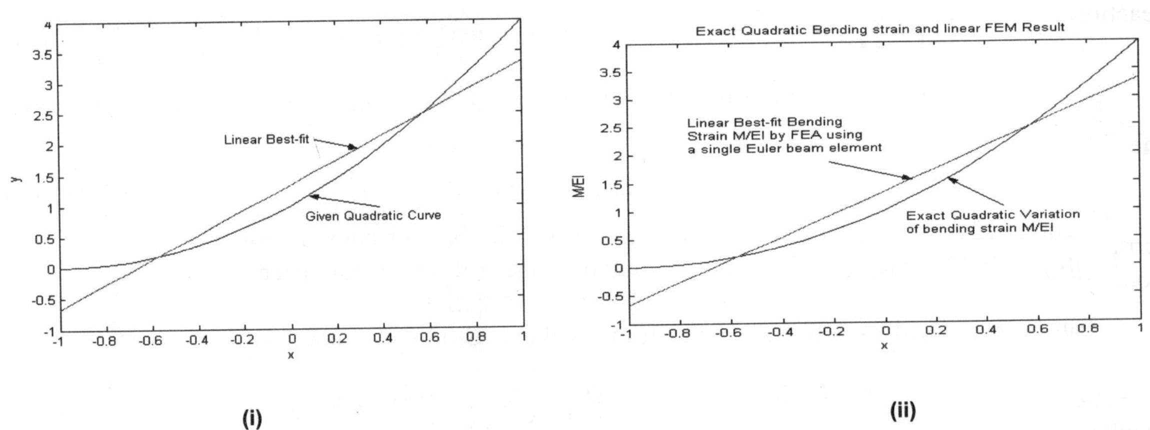

Fig. 2. (i) A linear best-fit (orthogonal projection) to a quadratic curve. (ii) The approximate bending strain of a cantilever beam is the linear best-fit to its analytical quadratic strain.

3. FINITE ELEMENT ANALYSIS AS AN ORTHOGONAL PROJECTION

3.1 A Critical Review of the Weak Form

Let us consider a conservative problem of a continuum, described by the following generic differential equation with a self-adjoint differential operator L, upon the analytical solution u,

$$L(u) - f = 0 \qquad\qquad ...(13)$$

Using the virtual work principle with an approximate function u^h, upon the differential equation with exact u, one can immediately get the weak form of (13) at an element level by

$$\int_e u^h (Lu - f)dx = 0 \qquad \qquad ...(14)$$

Through integration by parts, one arrives at the weak form

$$a(u,u^h)^e = (u^h,f) + \left[u^h, R^e\right]_B \qquad \qquad ...(15)$$

where, $a(u, u^h)^e$ is the *bilinear symmetric form* [2, 6] at the element level. Here the *analytical* nodal reaction vector acting on the element through element connectivity to the adjacent element are given $\{R^e\}$, and subscript B represents the element boundary (nodes). When the analytical function u is used in the differential equation (14), these nodal reactions are analytical, and are equal to the corresponding *internal stress resultants* across inter-element nodes. However, forms (14) or (15), with both analytical and approximate displacement solutions (*viz.* u and u^h), are not conventionally used in FEA computations. Instead, one uses only approximate solution u^h as the weighting solution and also as the operand in the differential equation. At the element level, when $u^h \neq u$, we have

$$\int_e u^h (Lu^h - f)dx = a(u^h,u^h)^e - (u^h,f) - \left[u^h, Q^{he}\right]_B \neq 0 \ \text{(in general)} \qquad ...(16)$$

The parameters Q^{he} at the element boundary now represents the *approximate internal stress resultants* from approximate function u^h, and they could be discontinuous across element nodes. Note that the condition that the weak form of (16) need not necessarily vanish. In an assembly of elements, each element is in *external equilibrium*, but *not in internal equilibrium*. This implies that with the replacement of the approximate stress resultant vector Q^{he} by an FEA computed nodal reaction $\{R^{he}\}$, the condition for external equilibrium of the element can be obtained as

$$a(u^h,u^h)^e = (u^h,f) + \left[u^h, R^{he}\right] \qquad \qquad ...(17)$$

Here the FEA computed element nodal reaction vector $\{R^{he}\}$ can be obtained explicitly from the following equilibrium equation of the element, after solution of the displacements are known,

$$\{R^{he}\} = [K^e]\{\delta^e\} - \{F^e\} \qquad \qquad ...(18)$$

The generalized forces at nodes $\{F^e\}$ for an element are defined according to the equivalence in terms of the virtual work done by the applied load f, and that done by the generalized force over nodal displacements $\{\delta^e\}$

$$(u^h,f) = \int_e u^h f.dx = \{\delta^e\}^T \{F^e\} \qquad \qquad ...(19)$$

Internal equilibrium in an element is violated since with an approximate function u^h, the stress resultants do not match the FEA computed element nodal reactions, i.e., $\{R^{he}\} \neq \{Q^{he}\}$. *This is the penalty for using an approximate function.* For instance, a two-noded uniform bar element with linear interpolation function u^h is exact for nodal loads only. But for other arbitrary loads, u^h is only an approximation to the exact solution u, hence we recognize that

(a) with exact solution u: $\{R^e\} = \left[-EA(du/dx)_1 \quad EA(du/dx)_2\right]^T = \{Q^e\}$

(b) with approximate solution u^h: $\{R^{he}\} \neq \{Q^{he}\}$; $\quad \{Q^{he}\} = \left[-EA(du^h/dx)_1 \quad EA(du^h/dx)_2\right]^T$

3.2 The Best-Fit Paradigm of FEA

Subtracting (15) from (17), we have the following equation, always valid for any element,

$$a(u^h, u^h)^e - a(u, u^h)^e = \left[u^h, R^{he} - R^e \right]_B$$

$$= \{\delta^e\}^T \{R^{he} - R^e\} \qquad \text{...(20)}$$

It can easily be established that the symmetric bilinear forms can be expressed as the inner products of the strains, i.e.,

$$a(u, u^h)^e = \int_e \{\varepsilon^h\}^T [D]\{\varepsilon\} dx = <\varepsilon^h, \varepsilon> \qquad \text{...(21a)}$$

$$a(u^h, u^h)^e = \int_e \{\varepsilon^h\}^T [D]\{\varepsilon^h\} dx = <\varepsilon^h, \varepsilon^h> = \|\varepsilon^h\|^2 \qquad \text{...(21b)}$$

where, [D] is the elastic rigidity matrix of the element. Two cases are now considered.

Case A: When the FE computed element nodal reaction vector agrees with the exact nodal reaction vector; i.e., when $\{R^{he}\} = \{R^e\}$,

$$a(u^h, u^h)^e - a(u, u^h)^e = 0 \qquad for \quad \{R^{he}\} = \{R^e\}$$

$$\Rightarrow <\varepsilon^h, \varepsilon^h> = <\varepsilon^h, \varepsilon> \quad \Rightarrow \quad <\varepsilon^h, \varepsilon - \varepsilon^h> = 0 \qquad \text{...(22)}$$

Equation (22), representing the *normal equation*, shows clearly that the *finite element strain error is orthogonal to the finite element strain ,that lies on the B subspace, since $\{\varepsilon^h\} = [B] \{\delta^e\}$. In other words, the finite element strain $\{\varepsilon^h\}$ is the orthogonal projection (best-fit) of the analytical strain $\{\varepsilon\}$ onto the B subspace, in which the column vectors of the strain-displacement matrix $[B]$ lies.*

Mathematically, the FE computed element strain is the best-fit of the exact strain vector onto an m-dimensional vector space B. Using the projection formula (equation (11),

$$\{\varepsilon^h\} = \{\overline{\varepsilon}\} = \sum_{j=1}^m \frac{<\varepsilon, v_j>}{<v_j, v_j>} \{v_j\}, \quad <v_i, v_j> = 0 \quad for \quad i \neq j \quad v_i, v_j \in B \qquad \text{...(23)}$$

An important consequence of the best-fit paradigm of FEA for each element follows immediately,

$$\|\varepsilon - \varepsilon^h\|^2 = \|\varepsilon\|^2 - \|\varepsilon^h\|^2 \qquad \text{...(24)}$$

i.e.,   *The Energy of the Error = Error of the Energy*

Thus one can now easily explain how the approximate bending strain of the cantilever beam (discussed in section 2.4, Fig 2(ii)) computed through the Rayleigh-Ritz Method, matched the linear best-fit solution of the analytical quadratic bending strain. This can also be shown to be true for simple bar elements.

Case B: In certain indeterminate problems, the FE computed element nodal reaction vector does not agree with the exact nodal reaction vector; i.e., $\{R^{he}\} \neq \{R^e\}$.

In such cases, equation (20) is still valid, but (22) is not valid anymore. Therefore, there seems to be an *apparent* violation of the best-fit paradigm. Actually this is just a *prima facie* violation, and in a deeper sense, the best-fit paradigm is still conserved. Under this circumstance, the discrete finite element model is actually constrained by a *spurious* force equal to the error of the element reactions, $\{R^{he}\} - \{R^e\}$. If one considers the original system further stiffened by this spurious force, then the FE computed strain of the original un-stiffened system solution is also identical to the FE computed strain and best-fit strain of the modified, stiffened analytical system! Equation (20) can be cast with respect to the stiffened analytical strain solution u^*,

$$a(u^h, u^h)^e - a(u^*, u^h)^e = 0 \qquad \text{...(25)}$$

Thus, the projection formula for the strain (23) is still valid for such cases, if we replace the original un-stiffened analytical strain by the stiffened analytical strain ε^*. In indeterminate structures, due to excess of constraints over the minimum required for stability, the analytical nodal reactions are not conserved by FEA with approximate solutions. Interestingly, the errors in the nodal reactions of the full domain, as well as the elements, are so adjusted by the FE computation, that the best-fit paradigm is still conserved in a subtle sense! A case of such a situation can be considered from an example of the fixed-fixed tapered bar, given by Babuska *et al.* [16], and recently studied by Mukherjee and Prathap [11]. Unlike the usual problems in determinate structures, the nodal displacements for this case are not exactly interpolated by the approximate displacement functions. This deviation of the nodal displacement from the exact has been termed as the "pollution error", though no physical reason for this behaviour has been given in terms of the displacement paradigm in reference [6]. The origin of this problem has been identified [11]. The argument has been extended by Sangeeta *et al.* [12] to explain pollution problems in the analysis of the spherically symmetric Laplace equation with Dirichlet constraints at the boundaries and the analysis of the simple bar on elastic foundation.

3.3 Shear Locking in Timoshenko Beam Elements

Locking is a pathological problem of convergence in certain types of elements, despite the fact that the continuity and completeness conditions are satisfied. It arises from *field-inconsistency* in the strain descriptions from the displacement interpolations in elements [7].

The two-noded Timoshenko beam element, designed to incorporate the shear strain deformation from bending, is prone to this kind of problem. For thin beams, the convergence is very slow, due to spurious constraints emerging from the requirement that the shear strains should vanish in the limit of the Euler beam. These spurious constraints introduce spurious bending stiffness, resulting in very stiff solutions, and spurious shear strain oscillations (Figs. 3 and 4). Elimination of this problem is usually made through reduced integration (using a one-point Gauss rule instead of the two-point necessary for accurate integration) for the shear components in the element stiffness matrix. This way, the spurious terms are dropped off from the computation, and the problem of locking is eliminated.

The function space approach has been used by Mukherjee and Prathap to study the locking problems in Timoshenko beam elements [9, 10]. In essence, elements with multi-component strain vectors suffer locking if the strain projection is performed on a field-inconsistent B space, (emerging out of a field-inconsistent strain-displacement matrix $[B]$) that cannot be spanned by a set of standard orthogonal basis vectors the elements (rows) of which are decoupled. The basis vectors spanning the field-inconsistent two-dimensional B subspace are $\{v_1\} = [0, 1]^T$ $\{v_2\} = [2/L, \xi]^T$, where L is the element length. A reduced integration scheme essentially drops the highest Legendre Polynomial (and eliminates the spurious constraint) from the $[B]$ matrix and generates a new $[B^*]$ matrix and its two-dimensional B^* space, that can be spanned by the standard orthogonal basis vectors $\{v_1^*\} = [0, 1]^T \{v_2^*\} = [1, 0]^T$. An orthogonal projection onto the B^* subspace is therefore lockfree. The basis vectors required for orthogonal projection computation can be derived by the Gram-Schmidt process upon the column vectors of the $[B]$ or $[B^*]$ matrix.

Interestingly, locked solutions, as a rule, are variationally correct, since no 'variational crimes' are introduced. The best-fit rule is always satisfied over the field-inconsistent subspace B. On the other hand. reduced integrated solutions are generally variationally incorrect, and are variationally correct only under certain condition [10] given here.

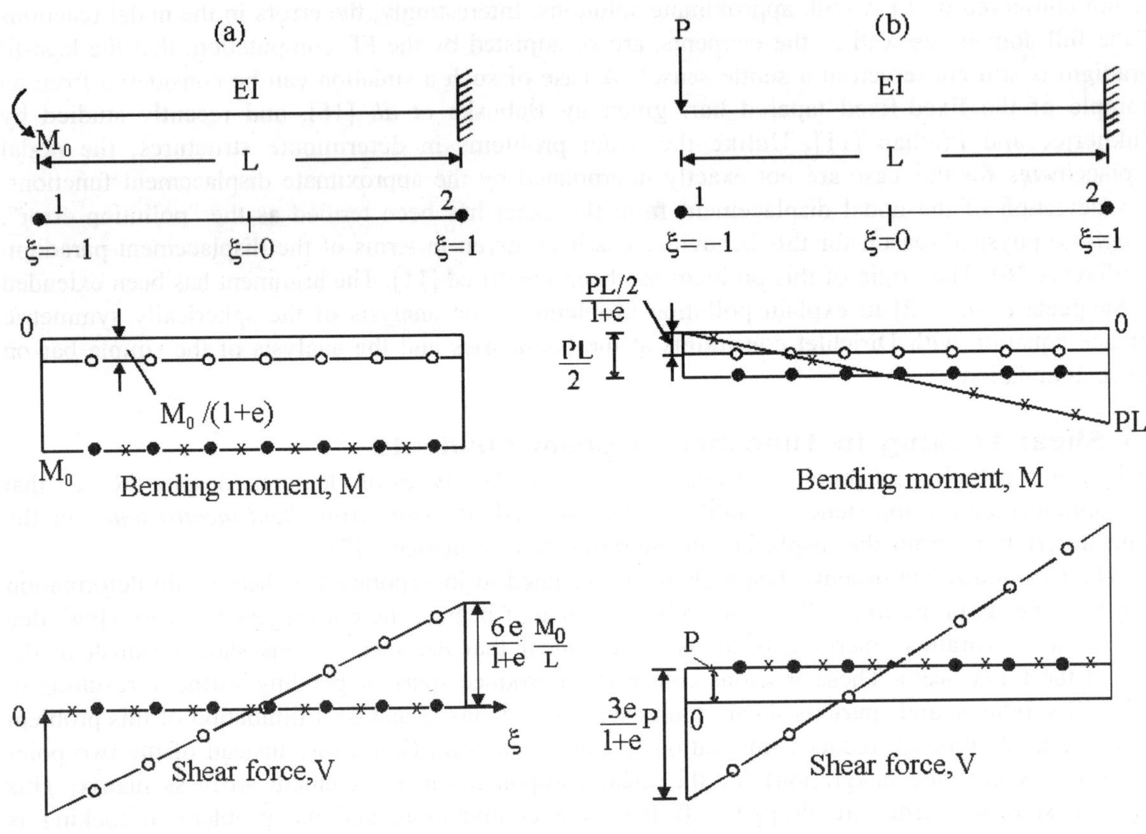

Fig. 3. Shear Locking and Lockfree Reduced integrated results in the Timoshenko beam with point loading. Analysis with one element. Exact ______ x ______ , FEM (FI) Locked ______ O ______ FEM (FC) Lockfree ______ ● ______ (variational correct), Best-fit (FC) matches FEM (FC).

If $\int_e \left[[B]-[B*]\right]^T [D]\{\varepsilon\}dx = 0$ then the reduced integrated solutions are variationally correct, and the best-fit strain on the $B*$ space agrees with the FE strain. If this condition is violated, then the best-fit strain suffers an additional strain over the FE strain due to extraneous force vector $\{F^e_E\} = -\int_e \left[[B]-[B*]\right]^T [D]\{\varepsilon\}dx$. A cantilever beam analysis using linear two-noded elements with reduced integration is variationally correct for point loads, but variationally incorrect with distributed loading. (Figs. 3 and 4).

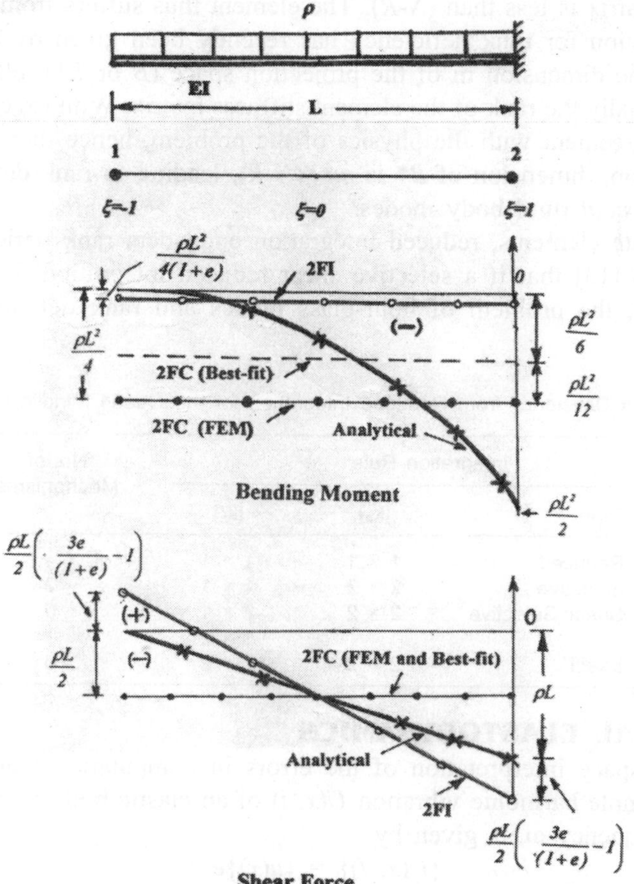

Fig. 4. Shear Locking and Lockfree Reduced integrated results in the Timoshenko beam with uniform distributed loading ρ. (One element analysis). Note that lockfree best-fit (FC) deviates from *variationally incorrect* lockfree FEM (FC) for bending because of extraneous forces $\{F_E^e\}$.

Furthermore, an explanation why superconvergent an-isoparamteric elements, (with different degrees of interpolation functions for displacements) do not lock can now be provided [9]. These elements are such that the projection space B for the strain is field-consistent, and they can be shown to be spanned by the standard, basis vectors, using the Gram-Schmidt orthgonalization.

Origins of "parasitic locking" in plane stress elements and "membrane locking" in curved beam elements can be traced to similar field-inconsistent sources, and can be eliminated by reduced integration.

4. RANK DEFICIENCY FROM REDUCED INTEGRATION

If N is the number of degrees of freedom of the element, and R the number of proper physical rigid body motions of the element, then the proper rank of the element stiffness matrix is actually N-R. If no variational crime like reduced integration is performed, this relationship is always true. With reduced integration, certain spurious zero-energy (hour-glass) modes are incorporated so that the rank of the stiffness matrix is less than (N-R). The element thus suffers from rank deficiency [4]. A function space explanation for rank deficiency has recently been given by Sangeeta *et al.* [13]. It has been shown that the dimension m of the projection space (B or B^*), obtained from the Gram-Schmidt process, is actually the rank of the element stiffness matrix. With exact integration, dimension of B is $m = N$-R in agreement with the physics of the problem, hence there is no rank deficiency. With reduced integration, dimension of B^* is $m^* < (N$-$R)$, leading to rank deficiency because of the appearance of *non-physical* rigid body modes.

For the Quad4 plate elements, reduced integration engenders rank deficiency [4]. However, it has been demonstrated [13] that if a selective shear reduced integration, instead of a full reduced integration, is adopted, the problem of hour-glass modes and rank deficiency can be eliminated (Table 1).

Table 1. Rank Deficiency from Reduced Integration for the Quad4 Mindlin Plate Element.

Element Type	Integration Rule			No. of Mechanisms	Dimension of B or B^* space
	Type	$[k_b]$	$[k_s]$		
	Reduced	1 × 1	1 × 1	4	5
Four noded	Selective	2 × 2	1 × 1	2	7
(12 d.o.f)	Shear Selective	2 × 2	2 × 1	0	9
			1 × 2		
N = 12, R = 3	Exact	2 × 2	2 × 2	0	9

5. COMPUTATIONAL ELASTODYNAMICS

Recently, a function space interpretation of the errors in computational elastodynamics has been presented [14, 15]. Simple harmonic vibration $U(x, t)$ of an elastic body in one its modes, $u(x)$, and its circular natural frequency ω, is given by

$$\{U(x,\ t)\} = \{u(x)\}e^{i\omega t} \qquad \ldots(26)$$

Using energy conservation principle, one can define the exact and approximate Rayleigh Quotients respectively by,

$$\omega^2 = \|\varepsilon\|^2 / \|u\|^2 \qquad (\omega^h)^2 = \|\varepsilon^h\|^2 / \|u^h\|^2 \qquad \ldots(27a,\ b)$$

However, using the virtual work principle, it can be shown that with a *variationally correct*, but approximate solution (u^h, ε^h), the exact Rayleigh Quotient corresponds to the following expression,

$$\omega^2 = <\varepsilon^h,\ \varepsilon>/ (u^h,\ u) \qquad \ldots(28)$$

In equations (27) and (28), the *stiffness and inertia inner products* (in a *global* sense, with summation of elements) and the corresponding *norms* are given respectively by

$$<\varepsilon^h,\varepsilon> = \sum_{ele=1}^{N^e} \int_{ele} \{\varepsilon^h\}^T [D]\{\varepsilon\}dx, \quad \|\varepsilon\|^2 = <\varepsilon,\varepsilon>, \quad \|\varepsilon^h\|^2 = <\varepsilon^h,\varepsilon^h>$$

$$(u^h,u) = \sum_{ele=1}^{N^e} \int_{ele} \{u^h\}^T [\rho]\{u\}dx, \quad |u|^2 = (u,u), \quad |u^h|^2 = (u^h,u^h) \qquad \ldots(29a,\ b)$$

where, $[D]$ is the elastic rigidity and $[\rho]$ is the inertia density of the element. From (27a, b), one gets the energy error equation for any FE solution as

$$\|\varepsilon\|^2 - \|\varepsilon^h\|^2 = \omega^2 |u|^2 - (\omega^h)^2 |u^h|^2 \qquad \ldots(30)$$

This equation shows that with normalization of the normal mode displacements for both the exact and the approximate solutions ($|u| = 1$, $|u^h| = 1$) the *error of the frequency is governed by the error of the strain norm* as

$$(\omega^h)^2 - \omega^2 = \|\varepsilon^h\|^2 - \|\varepsilon\|^2 \qquad \ldots(31)$$

From (27b) and (28) one gets the following equation for a variationally correct approximation only,

$$<\varepsilon^h, \varepsilon - \varepsilon^h> = (u^h, \omega^2 u - \omega^{h2} u^h) \qquad \ldots(32)$$

For elastostatics, the orthogonality rule follows immediately, $<\varepsilon^h, \varepsilon - \varepsilon^h> = 0$. For a *variationally correct* solution (with *conforming* elements, *without mass lumping or reduced integration*) it can be shown that [14]

$$\frac{|u - u^h|^2}{|u^h|^2} + \frac{\omega^{h2}}{\omega^2} - \frac{\|\varepsilon - \varepsilon^h\|^2}{\omega^2 |u^h|^2} = 1 \qquad \ldots(33)$$

i.e., the values of the Rayleigh Quotient ratio (ω^h/ω), the norms of the modal displacement error, $|u - u^h|$ and the norms of the modal strain error, $\|\varepsilon - \varepsilon^h\|$ of *any* variationally correct but approximate solution generates a portion of the *surface of the first octant of a hyperboloid of one sheet, with* (ω^h / ω) ≥ 1 (Fig. 5). Furthermore, it has been justified [14] that the in variationally correct formulations, the approximate Rayleigh Quotient exceeds the exact value [(ω^h / ω) ≥ 1] because the *deviation* of the approximate modal strain vector from the exact is more than that of the approximate modal displacement vector.

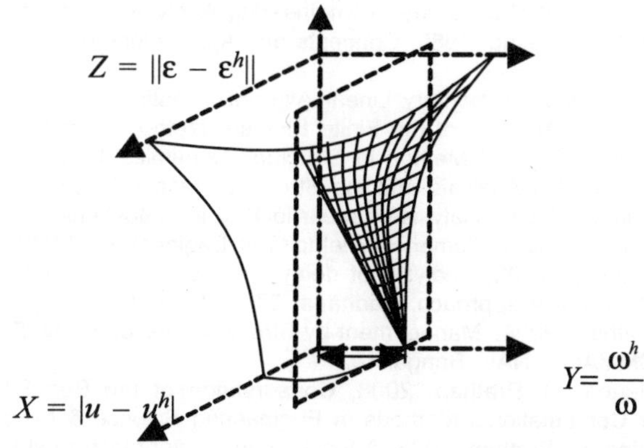

Fig. 5. Geometric interpretation of eigenvalue analysis of the variationally correct formulation using Frequency-Error Hyperboloid. Approximate eigenvalues obtained from a variationally correct formulation lie in the shaded portion of the Hyperboloid.

Free vibration analysis of a fixed-fixed bar, performed using consistent mass and lumped mass matrices and using only two linear elements, has been studied [14] with varying position of the middle node. Variationally correct formulation (with consistent mass matrix) always yields an approximate value for the Rayleigh Quotient higher than the exact one for all positions of the middle node, but variationally incorrect formulations (with lumped masses) can yield this as lower than, equal to, or greater than the exact value, according the position of the middle node. Recently, it has been demonstrated by Jafarali *et al.* [15] that in the vibration analysis of Timoshenko beams, use of reduced integration, as a variationally incorrect step, cannot guarantee that the approximate natural frequency will always be above the exact one. Locked solutions, (with no variational crimes), actually satisfy all the conditions of variationally correct results (*viz.* equations (28), (32) and (33)). Equations (27b), (30) and (31) are satisfied by *all* finite element solutions.

CONCLUSION

This paper presents a brief outline of the recent research work done on the function space interpretation of computational process that underlies all finite element computation. The projection theorems of linear algebra have been suitably adopted to show that FEA computations of elastostatics, in principle, follow the laws of orthogonal projections of the exact element strains onto function subspaces. The exact is portrayed as a shadow upon the computational subspace. This is illustrated through simple bar and beam elements. The origins of pathological problem like shear locking, and origin of pollution error are explained. The secret behind the elimination of locking problems through reduced integration, and the resulting variational incorrectness, have been shown explicitly. The cause of rank deficiency in elements from reduced integrations is identified. It has been shown clearly how only variationally correct approximations in elastodynamic problems follow certain definite geometrical patterns.

Using the function space method, one can estimate, or predict, a priori, the quality of a new element being formulated.

REFERENCES

1. O.C. Zienkiewicz , R.L. Taylor, 1991, The Finite Element Method. McGraw Hill.
2. J.N. Reddy, 1993, An introduction to the Finite Element Method II ed,, McGraw Hill.
3. K.J. Bathe, 1982, Finite Element Procedures in Engineering Analysis, Prentice Hall.
4. R.D. Cook, D.S. Malkus, M.E. Plesha, 1989, Concepts and Applications of Finite Element Analysis, III ed, John Wiley, New York.
5. L.H. Edwards, P.E. Penny, 1988, Elementary Linear Algebra, Prentice Hall.
6. G. Strang, G.J. Fix, 1973, An Analysis of the Finite Element Method, Prentice Hall.
7. G. Prathap, 1993, The Finite Element Method in Structural Dynamics, Kluwer Academic.
8. G. Prathap, 1996, Finite Element Analysis and the Stress Correspondence Paradigm, Sadhana; 21:, 525-546.
9. S. Mukherjee and G. Prathap, 2001, Analysis of shear locking in Timoshenko beam elements using the function space approach, Communications in Numerical Methods in Engineering; 17:385-393.
10. S. Mukherjee and G. Prathap, 2002, Analysis of delayed convergence in the three-noded Timoshenko beam element using the function space approach, Sadhana; 27(5): 507-526.
11. S.Mukherjee and G. Prathap, 2004, Management-by-Stress Model of Finite Element Computation, Research Report No CM 0408, CMMACS, NAL, Bangalore.
12. K. Sangeeta, S. Mukherjee, G. Prathap, 2006, Conservation of the Best-Fit Paradigm at Element Level, International Journal for Computational Methods in Engineering Science & Mechanics; 7: 1-12.
13. K. Sangeeta, S. Mukherjee, G. Prathap, 2005, A function space approach to study rank deficiency and spurious modes in finite elements, Structural Engineering and Mechanics; 21(5):539-551.
14. S. Mukherjee, P. Jafarali and G. Prathap 2005 A variational basis for error analysis in finite element elastodynamics, Journal of Sound and Vibration; 285(3): 615-635.
15. P. Jafarali, S, Mukherjee, M. Ameen and G. Prathap, (in press), Variational Correctness in Timoshenko beam finite element elastodynamics, Journal of Sound and Vibration.
16. I. Babuska, T. Stroubolis, A. Mathur, CS Upadhyay, 1994, Pollution error in the h-version of the finite element method and the local quality of a-posteriori error estimators, Finite Elements in Analysis and Design; 17: 273-321.

12

Free Vibration of Laminated Folded Plates Using a Shear Deformable Finite Element Model

Ajay Kumar Garg[1], Rakesh Kumar Khare[2] and Tarun Kant[3]

[1]Department of Civil Engineering, Government Engineering College Bilaspur-495009 India
email: akgcivil@yahoo.com
[2]Department of Civil Engineering, Shri G.S. Institute of Technology and Science Indore-452003 India
email: rakeshkhare@hotmail.com
[3]Department of Civil Engineering, Indian Institute of Technology Bombay, Mumbai-400076 India
email: tkant@civil.iitb.ac.in

ABSTRACT

A simple $C°$ isoparametric finite element model, based on a higher-order shear deformation theory is presented for the free vibration analysis of isotropic, orthotropic and layered anisotropic composite and sandwich folded plates. This theory incorporates a realistic non-linear variation of displacements through the laminate thickness, and eliminates the use of shear correction coefficients. The accuracy of the present model is demonstrated by comparing with alternative solutions available in the literature. New results are presented for natural frequencies of one- and two-fold cantilevered sandwich folded plates with various crank angles.

Keywords: Composite, higher-order theory, finite element, folded plate, free vibration, sandwich, shear deformation.

1. INTRODUCTION

A plate structure formed by assembling or folding thin elastic plates is one of the most useful components of architectural or mechanical structures. Free vibrational analysis for predicting their eigenfrequencies and eigenmodes is of great importance in structural design. The free vibration characteristics of such plates are of interest to the designer in consideration of dynamic response. Goldberg and Leve [1] presented the exact static analysis of folded plate structures. Iffland [2] reviewed the comprehensive literature dealing with thin-walled folded plate structures together with the analysis methods for them.

Irie *et al.* [3] presented an analysis of the free vibration of a thin-walled cantilever folded plate. The kinetic and strain energies of the plate are evaluated analytically, and the frequency equations are derived by the Ritz method. Subsequently, Irie *et al.* [4] obtained the natural frequencies for thin-walled one-fold folded plates with simply supported edges in the axial direction and free lateral edges. Tanaka *et al.* [5, 6] used the boundary element method for the free vibration analysis of thin-walled plate structures. Liu and Huang [7] used the finite element- transfer matrix method to study the natural frequencies of folded plate structures. They obtained frequency parameters for cantilever one-fold and two-fold folded plates with various crank angles, and many-fold cylindrical shell panels. Niyogi *et al.* [8] used a 9-noded Lagrangian plate bending finite element that incorporates first-order transverse shear deformation and rotary inertia to predict the bending, free- and forced vibration response of laminated composite and sandwich folded plate structures.

It is observed that free vibration analysis of folded plate using higher-order shear deformation theory is not attempted so far. Kant et al. [9] are the first to present finite element formulation of a higher order flexure theory. This theory considers three-dimensional Hooke's law and incorporates the effect of transverse normal strain in addition to transverse shear deformations. Reddy [10] later proposed a higher order shear deformation theory utilizing a displacement field with cubic variations with respect to the thickness direction. Kant and Mallikarjuna [11] streamlined the higher order shear deformation theory by allowing the displacement in the thickness direction to be quadratic with respect to the thickness co-ordinate, and developed a simple C^o finite element formulation and presented solutions for the free vibration analysis of general laminated composite and sandwich plate problems.

In the present work, a C^o continuous shear deformable finite element formulation based on a higher-order displacement model is presented, which does not require the use of a shear correction coefficient. The formulation is used for the free vibration analysis of one- and two-fold cantilevered composite and sandwich folded plates with various crank angles. Accuracy of the present formulation is verified against the literature values for isotropic and composite folded plates. New results are presented for the laminated sandwich folded plates using the standard material properties available in the literature.

2. THEORY AND FORMULATION

The higher-order shear deformation theory considered for investigation in the present work is based on the assumption of the displacement field in the following form.

$$u(x,y,z,t) = u_o(x,y,t) + z\ \theta_y(x,y,t) + z^2\ u_o^*(x,y,t) + z^3\ \theta_y^*\ (x,y,t),$$

$$v(x,y,z,t) = v_o(x,y,t) - z\ \theta_x(x,y,t) + z^2\ v_o^*(x,y,t) - z^3\ \theta_x^*\ (x,y,t), \qquad \ldots(1)$$

$$w(x,y,z,t) = w_o(x,y,t)$$

where, t is the time, u, v and w are the displacements of a general point (x, y, z) in the plate space in the x, y and z directions, respectively. The parameters u_o, v_o are the in-plane displacements and w_o is the transverse displacement of a point (x, y) on the laminate middle plane (Fig. 1). The functions, θ_x and θ_y are the rotations of the normal to the laminate middle plane about x- and y-axes, respectively. The parameters u_o^*, v_o^*, θ_x^*, θ_y^* are the higher order terms in the Taylor's series expansion and are also defined at mid-surface. The continuum displacement vector at the mid-plane can thus be defined as:

(1, 2, 3)—Lamina reference axes.

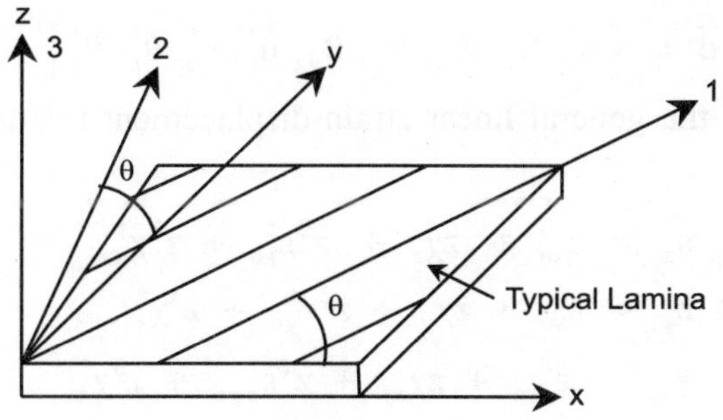

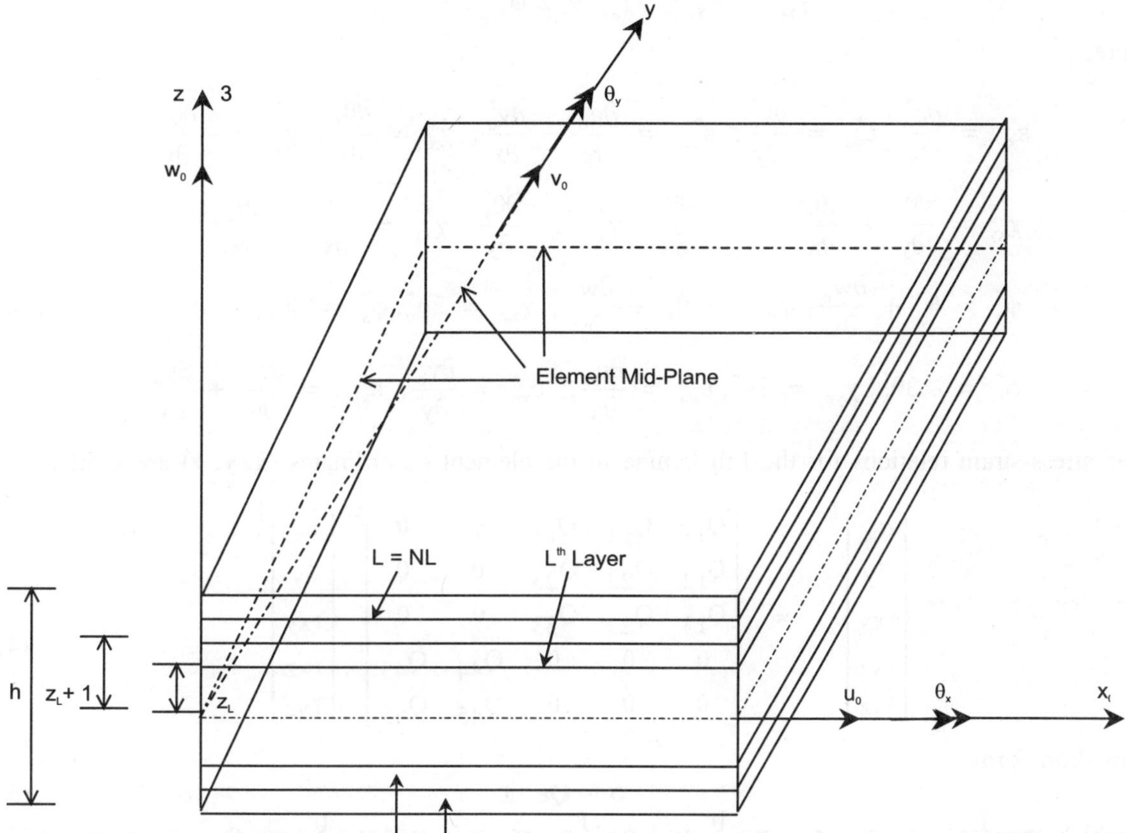

Fig. 1. Element laminate geometry with positive set of lamina/laminate reference axes, displacement components and fibre orientation.

$$\overline{\mathbf{d}} = \left\{ u_o, v_o, w_o, \theta_x, \theta_y, u_o^*, v_o^*, \theta_x^*, \theta_y^* \right\}^T \qquad \text{...(2)}$$

Substituting Eq. (1) into the general linear strain-displacement relations, the following relations are obtained.

$$\begin{aligned}
\varepsilon_x &= \varepsilon_{xo} + z\chi_x + z^2\varepsilon_{xo}^* + z^3\chi_x^* \\
\varepsilon_y &= \varepsilon_{yo} + z\chi_y + z^2\varepsilon_{yo}^* + z^3\chi_y^* \\
\gamma_{xy} &= \varepsilon_{xyo} + z\chi_{xy} + z^2\varepsilon_{xyo}^* + z^3\chi_{xy}^* \\
\gamma_{xz} &= \varphi_x + z\chi_{xz} + z^2\varphi_x^* \\
\gamma_{yz} &= \varphi_y + z\chi_{yz} + z^2\varphi_y^*
\end{aligned} \qquad \text{...(3a)}$$

where,

$$\varepsilon_{xo}^* = \frac{\partial u_o^*}{\partial x}, \ \varepsilon_{yo}^* = \frac{\partial v_o^*}{\partial y}, \ \varepsilon_{xyo}^* = \frac{\partial u_o^*}{\partial y} + \frac{\partial v_o^*}{\partial x}, \ \chi_x = \frac{\partial \theta_y}{\partial x}, \ \chi_y = -\frac{\partial \theta_x}{\partial y},$$

$$\chi_{xy} = \frac{\partial \theta_y}{\partial y} - \frac{\partial \theta_x}{\partial x}, \ \chi_x^* = \frac{\partial \theta_y^*}{\partial x}, \ \chi_y^* = -\frac{\partial \theta_x^*}{\partial y}, \ \chi_{xy}^* = \frac{\partial \theta_y^*}{\partial y} - \frac{\partial \theta_x^*}{\partial x},$$

$$\varphi_x = \theta_y + \frac{\partial w_o}{\partial x}, \ \varphi_y = -\theta_x + \frac{\partial w_o}{\partial y}, \ \chi_{xz} = 2u_o^*, \varphi_x^* = 3\theta_y, \qquad \text{...(3b)}$$

$$\varphi_y^* = -3\theta_x^*, \ \chi_{yz} = 2v_o^*, \varepsilon_{xo} = \frac{\partial u_o}{\partial x}, \ \varepsilon_{yo} = \frac{\partial v_o}{\partial y}, \ \varepsilon_{xyo} = \frac{\partial u_o}{\partial y} + \frac{\partial v_o}{\partial x}.$$

The stress-strain relations for the Lth lamina in the element co-ordinates (x, y, z) are written as

$$\begin{Bmatrix} \sigma_x \\ \sigma_y \\ \tau_{xy} \\ \tau_{xz} \\ \tau_{yz} \end{Bmatrix}^L = \begin{bmatrix} Q_{11} & Q_{12} & Q_{13} & 0 & 0 \\ Q_{12} & Q_{22} & Q_{23} & 0 & 0 \\ Q_{13} & Q_{23} & Q_{33} & 0 & 0 \\ 0 & 0 & 0 & Q_{44} & Q_{45} \\ 0 & 0 & 0 & Q_{45} & Q_{55} \end{bmatrix}^L \begin{Bmatrix} \varepsilon_x \\ \varepsilon_y \\ \gamma_{xy} \\ \gamma_{xz} \\ \gamma_{yz} \end{Bmatrix}^L \qquad \text{...(4)}$$

or in short form

$$\sigma = \mathbf{Q}\varepsilon \qquad \text{...(5)}$$

in which $\boldsymbol{\sigma} = \left\{ \sigma_x, \sigma_y, \tau_{xy}, \tau_{xz}, \tau_{yz} \right\}^T$ and $\boldsymbol{\varepsilon} = \left\{ \varepsilon_x, \varepsilon_y, \gamma_{xy}, \gamma_{xz}, \gamma_{yz} \right\}^T$ are the stress and strain vectors, respectively. Q_{ij}'s are the plane stress reduced stiffness coefficients. The transformation rule of stresses/strains between the lamina and laminate co-ordinate systems follows the usual stress tensor transformation rule.

The governing differential equations of motion can be derived using Hamilton's principle

$$\delta \int_{t_1}^{t_2} (\Pi - E)\,dt = 0 \qquad \text{...(6)}$$

where t is the time, E is the total kinetic energy of the system and Π is the potential energy of the system, including both strain energy and potential of conservative external forces. For the ideal case in which the system has no damping and no external forcing function, the mathematical statement of Hamilton's principle can be written as

$$\delta \int_{t_1}^{t_2} \left[\frac{1}{2} \int_V \bar{\varepsilon}^T \, \bar{\sigma} \, dv - \frac{1}{2} \int_V \dot{u}^T \rho \, \dot{u} \, dv \right] dt \; = \; 0 \qquad \text{...(7)}$$

where, ρ is the mass density of the material, $\dot{u}$ defines the particle velocity vector and

$$\bar{\sigma} = \left\{ N^T, M^T, Q^T \right\}^T . \qquad \text{...(8)}$$

The stress resultants in Eq. (8) for the laminate with NL number of layers are defined as follows:

$$\begin{bmatrix} N_x & N_x^* & M_x & M_x^* \\ N_y & N_y^* & M_y & M_y^* \\ N_{xy} & N_{xy}^* & M_{xy} & M_{xy}^* \end{bmatrix} = \sum_{L=1}^{NL} \int_{z_L}^{z_{L+1}} \begin{Bmatrix} \sigma_x \\ \sigma_y \\ \tau_{xy} \end{Bmatrix} (1, z^2, z, z^3) \, dz \qquad \text{...(9)}$$

$$\begin{bmatrix} Q_x & Q_x^* & S_x \\ Q_y & Q_y^* & S_y \end{bmatrix} = \sum_{L=1}^{NL} \int_{z_L}^{z_{L+1}} \begin{Bmatrix} \tau_{xz} \\ \tau_{yz} \end{Bmatrix} (1, z^2, z) \, dz \qquad \text{...(10)}$$

After integration, these stress resultants can be written in matrix form as

$$\bar{\sigma} = D\bar{\varepsilon} \qquad \text{...(11)}$$

where,

$$D = \begin{bmatrix} D_m & D_c & 0 \\ D_c^T & D_b & 0 \\ 0 & 0 & D_m \end{bmatrix} \qquad \text{...(12)}$$

in which D_m, D_b, D_c and D_s are the membrane, bending, membrane-bending coupling and shear rigidity matrices, respectively and are defined in Appendix A.

3. FINITE ELEMENT FORMULATION

Let the region of the folded plate be divided into finite number of quadrilateral elements. The continuum displacement vector within an element is discretized such that

$$D = \sum_{i=1}^{NN} N_i \, D_i \qquad \text{...(13)}$$

where, d_i is the displacement vector corresponding to node i, N_i is the interpolating or shape function associated with node i, and NN is the total number of nodes per element.

The generalized mid-surface strains at any point given by Eq. (3) can be expressed in terms of nodal displacements in matrix form as follows:

$$\overline{\varepsilon} \; = \; \sum_{i=1}^{NN} \; \mathbf{B}_i \, \mathbf{d}_i \qquad\qquad ...(14)$$

where, $\mathbf{B}_i$ is a differential operator matrix of shape functions. Substituting for $\overline{\varepsilon}$, $\overline{\sigma}$ and velocity vector $\dot{\mathbf{u}}$ in Eq. (7), we get

$$\int_{t_1}^{t_2} \sum_{e=1}^{NE} \delta \mathbf{d}_e^t \, [\mathbf{K}^e \, \mathbf{d}_e \, + \, \mathbf{M}^e \, \ddot{\mathbf{d}}_e] \, dt \; = \; 0 \qquad\qquad ...(15)$$

in which $\mathbf{K}^e$ is the stiffness matrix for an element 'e' which includes membrane, flexure and the transverse shear effects and $\mathbf{M}^e$ is the element mass matrix, which are given by

$$K^e = \int_A B^t \, DB \, dA \quad \text{and} \quad M^e = \int_A N^t m \, N \, dA \qquad\qquad ...(16)$$

The element stiffness matrix and mass matrix can be obtained by using the standard relation

$$\mathbf{K}_{ij}^e \; = \; \int_{-1}^{1} \int_{-1}^{1} \mathbf{B}_i^T \, \mathbf{D} \, \mathbf{B}_j \, |\mathbf{J}| \, d\xi \, d\eta \qquad\qquad ...(17)$$

and,

$$\mathbf{M}_{ij}^e \; = \; \int_{-1}^{1} \int_{-1}^{1} \mathbf{N}_i^T \, \mathbf{m} \, \mathbf{N}_j \, |\mathbf{J}| \, d\xi \, d\eta \qquad\qquad ...(18)$$

where $|\mathbf{J}|$ is the determinant of the standard Jacobian matrix, $\mathbf{D}$ is the rigidity matrix and $\mathbf{m}$ is the inertia matrix, which is defined in Appendix B.

The global discrete equation for free vibration in matrix form can be written as

$$\mathbf{Kd} \, + \, \mathbf{M\ddot{d}} \; = \; 0 \qquad\qquad ...(19)$$

where, $\mathbf{K}$ and $\mathbf{M}$ are the assembled stiffness and mass matrices, respectively for the structure, $\mathbf{d}$ is the nodal displacement vector and $\ddot{\mathbf{d}}$ is the second derivative of the displacements of the structure with respect to time.

To find natural modes and frequencies, we assume that the field variables can be expressed as

$$\mathbf{d} \; = \; \overline{\mathbf{d}}_a \, e^{i\omega t} \qquad\qquad ...(20)$$

where, $\overline{\mathbf{d}}_a$ is the vector of unknown amplitudes at time $t = 0$ at the nodes, and ω is the circular natural frequency of the system. Substituting Eq. (20) in Eq. (19), we get

$$(\mathbf{K} - \omega^2 \mathbf{M}) \, \overline{\mathbf{d}}_a \; = \; 0 \qquad\qquad ...(21)$$

The subspace iteration method [12] is used here to obtain the numerical solution of the eigenvalue problems.

4. NUMERICAL RESULTS AND DISCUSSIONS

A computer program has been developed, based on the foregoing finite element model to solve a number of numerical examples on free vibration of composite and sandwich folded plates. A 4×8 mesh of 9-noded elements for each plates of the folded plate is used in computations. This scheme is arrived at on the basis of a convergence study in which the fundamental natural frequency converges monotonically from a higher value. The details of the convergence study are not presented for the sake of brevity. The full integration scheme, namely 3×3 Guass-Legendre for the membrane, flexure, membrane-flexure and shear energy is used. A parallel computer code was also developed based on the Reissner-Mindlin's first-order shear deformation theory (FOST) in order to compare its results with those of higher-order shear deformation theory (HOST). A shear correction factor of 5/6 is used with this theory. The accuracy of the present finite element formulations is evaluated first for isotropic and composite folded plates with the available literature results. Subsequently, some new results are presented for cantilevered laminated sandwich folded plates.

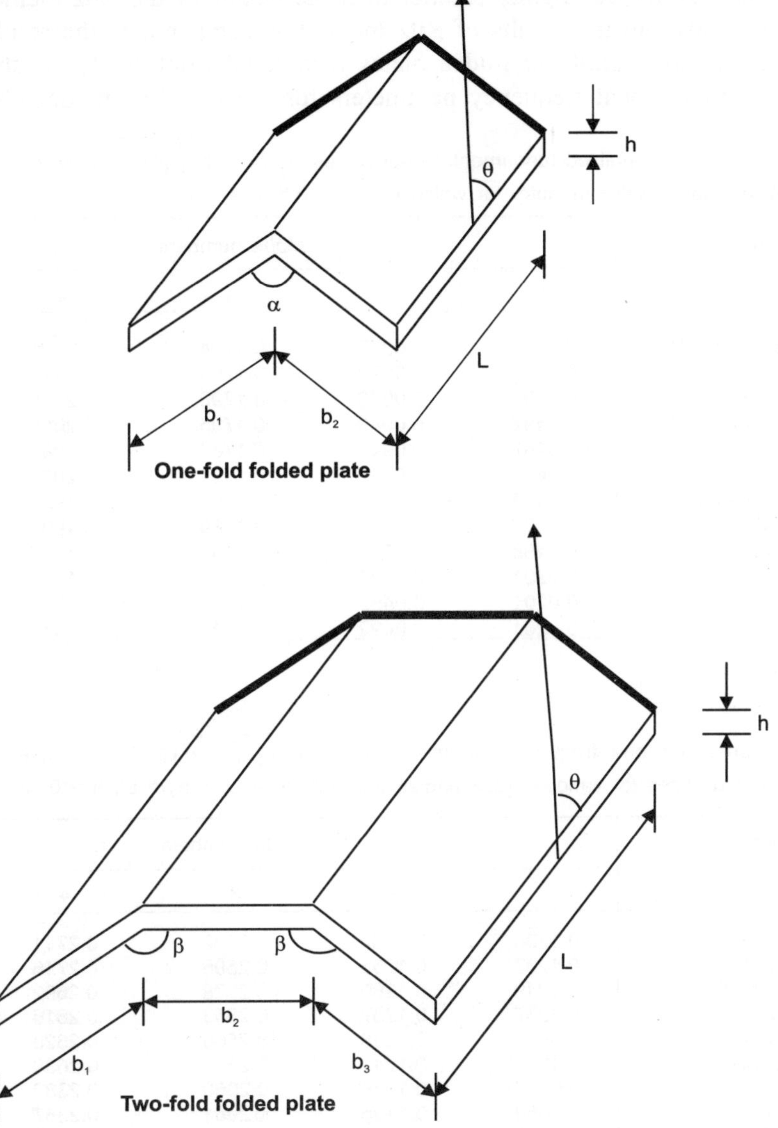

Fig. 2. Geometry of one-fold and two-fold folded plates.

4.1 Isotropic Folded Plates

One-fold isotropic cantilever folded plates as shown in Fig. 2 having, $b_1 = b_2 = 0.5L$ and various crank angles are considered. The non-dimensional frequency parameter $\bar{\omega}$ is defined as $\bar{\omega}_i = \omega_i\, L\, \sqrt{\rho(1-v^2)/E}$, where ω_i is the ith natural frequency. Numerical results obtained by the FOST and HOST models as given in Table 1, are compared with those given by Irie *et al.* [3] using Ritz method. It is observed that the frequency parameter $\bar{\omega}_i$ decreases as crank angle increases. The present finite element results are slightly smaller than the results of the Ritz method [3]. However, results of HOST are closer to the results of Ritz method as compared to the results of FOST.

Next, two-fold isotropic cantilever folded plates with equal crank angles as shown in Fig. 2 is considered and non-dimensional frequency parameters for the first five modes obtained by FOST

Table 1. Non-dimensionalized fundamental frequencies $\bar{\omega}_i = \omega_i L \sqrt{\rho(1 - v^2)/E}$ of cantilever folded plates with symmetry for which $b_1 = b_2 = 0.5$, $h = 0.02$, $n = 0.3$, $L = 1.0$.

Cranked angle	Source	Mode numbers				
		1	2	3	4	5
90°	Ritz method[a]	0.0492	0.0977	0.1794	0.2101	0.3573
	FOST	0.0490	0.0979	0.1795	0.2113	0.3480
	HOST	0.0490	0.0979	0.1796	0.2115	0.3482
120°	Ritz method[a]	0.0492	0.0949	0.1795	0.2082	0.2984
	FOST	0.0487	0.0946	0.1788	0.2081	0.2907
	HOST	0.0488	0.0947	0.1791	0.2084	0.2909
150°	Ritz method[a]	0.0492	0.0816	0.1795	0.1927	0.2227
	FOST	0.0487	0.0805	0.1789	0.1905	0.2193
	HOST	0.0488	0.0806	0.1791	0.1907	0.2196
180°	Ritz method[a]	0.0201	0.0493	0.1234	0.1577	0.1796
	FOST	0.0201	0.0492	0.1243	0.1576	0.1800
	HOST	0.0201	0.0492	0.1244	0.1577	0.1801

[a] Irie *et al.* [3]

Table 2. Non-dimensional frequency parameter $\bar{\omega}_i = \omega_i L \sqrt{\rho(1 - v^2)/E}$ for cantilever two-fold folded plates for various crank angles and with $b_1 = b_2 = b_3 = L/$, $h = 0.02$.

Crank angle	Source	Mode numbers				
		1	2	3	4	5
90°	FOST	0.1236	0.1252	0.2592	0.2711	0.3191
	HOST	0.1237	0.1253	0.2596	0.2716	0.3193
	FETMM[a]	0.1249	0.1260	0.2579	0.2692	0.3286
120°	FOST	0.0967	0.1237	0.2553	0.2616	0.2898
	HOST	0.0968	0.1238	0.2560	0.2623	0.2899
	FETMM	0.1000	0.1241	0.2571	0.2630	0.2986
150°	FOST	0.0665	0.1134	0.2060	0.2382	0.2548
	HOST	0.0666	0.1135	0.2061	0.2387	0.2555
	FETMM	0.0687	0.1145	0.2109	0.2415	0.2571
180°	FOST	0.0201	0.0491	0.1243	0.1572	0.1798
	HOST	0.0201	0.0491	0.1243	0.1572	0.1799

[a]Finite element-transfer matrix method results, Liu and Huang [7].

and HOST models are tabulated in Table 2 along with the results given by Liu and Huang [7] using the finite element-transfer matrix method. The non-dimensional frequency parameters are obtained for various crank angles and with $b_1 = b_2 = b_3 = L/3$. It is observed that as the crank angle β is increased, the frequency parameter $\bar{\omega}_i$ decreases. The results of present finite element formulations are matching well with those given by Liu and Huang [7].

4.2 Composite Folded Plates

The non-dimensional frequencies of the one-fold and two-fold cantilevered composite folded plates with fibre-oriented at 0° is presented in Table 3 along with the finite element results given by Niyogi *et al.* [8] based on the first-order shear deformation theory. E-glass-epoxy composite has been used with $E_1 = 60.7$ GPa, $E_2 = E_3 = 24.8$ GPa, $G_{12} = G_{13} = G_{23} = 12.0 \times 10^9$ GPa, $v_{12} = v_{13} = v_{23} = 0.23$ and $\rho = 1300$ Kg/m². The results of present FOST and HOST are nearly identical and in good agreement with the finite element results of Niyogi *et al.* [8].

Table 3. Non-dimensional frequency parameter $\bar{\omega}_i = \omega_i L \sqrt{\rho(1 - n_{12})^2/E_1}$ for composite cantilever folded plates for various crank angles. L/h = 50.

Description of the folded plate	Crank angle	Mode number	FOST	HOST	Niyogi et al. [8]
One-fold folded plate	90°	1	0.0390	0.0390	0.0390
		2	0.0679	0.0680	0.0675
		3	0.1558	0.1559	0.1556
		4	0.1734	0.1735	-
		5	0.2424	0.2425	-
	120°	1	0.0388	0.0389	0.0391
		2	0.0667	0.0667	0.0664
		3	0.1555	0.1557	0.1557
		4	0.1720	0.1722	-
		5	0.2448	0.2450	-
	150°	1	0.0389	0.0389	0.0391
		2	0.0613	0.0613	0.0609
		3	0.1555	0.1557	0.1557
		4	0.1611	0.1613	-
		5	0.1758	0.1760	-
Two-fold folded plate	90°	1	0.0855	0.0856	0.0844
		2	0.0944	0.0945	0.0937
		3	0.2047	0.2050	0.2049
		4	0.2134	0.2138	-
		5	0.2744	0.2746	-
	120°	1	0.0800	0.0801	0.0796
		2	0.0847	0.0848	0.0851
		3	0.2026	0.2031	0.2047
		4	0.2076	0.2081	-
		5	0.2609	0.2610	-
	150°	1	0.0568	0.0569	0.0567
		2	0.0814	0.0815	0.0817
		3	0.1793	0.1794	0.1765
		4	0.1956	0.1959	-
		5	0.2021	0.2025	-

4.3 Sandwich Folded Plates

The natural frequencies of a 1 m long cantilevered sandwich folded plates for the various crank angles are presented in Table 4. The sandwich folded plate has two identical aluminum faceplates and an aluminum honeycomb core. The physical properties for the faceplates are

$E = 68.948$ GPa, $\qquad G = 25.924$ GPa, $\qquad \nu = 0.33$, $\qquad \rho = 2768.0$ kg/m^3

and for the core

$G_{23} = 0.05171$GPa, $\qquad G_{13} = 0.13445$ GPa, $\qquad \rho = 121.83$ kg/m^3

It is assumed that core in-plane, direct and shear stiffnesses are zero. The length to thickness ratio (L/h) and thickness of the core to thickness of the flange ratio (t_c/t_f) are taken equal to 50 and 10, respectively. It is observed that natural frequency of cantilever folded plates, decreases as the crank angle increases. The results of first-order shear deformation theory are much higher as compared to the results of higher-order shear deformation theory and the difference increases for the higher modes of vibration.

Table 4. Natural frequency (Hz) of cantilever one-fold and two-fold sandwich folded plates for various crank angles with L/h=50 and t_c / t_f = 10.

Crank angle	Mode number	One-fold folded plate		Two-fold folded plate	
		FOST	HOST	FOST	HOST
90°	1	57.5667	53.7823	114.6943	111.3753
	2	113.5546	106.6179	149.1869	141.4031
	3	211.1188	187.8319	244.3621	241.7338
	4	244.8302	213.9498	303.6196	267.0252
	5	280.3772	275.1157	304.25092	267.4168
120°	1	56.8744	53.1690	87.6413	84.3225
	2	104.0760	98.8595	143.9689	136.1851
	3	209.7662	186.7110	224.4743	221.8460
	4	228.8545	206.9530	288.5343	251.9399
	5	256.9839	242.5965	296.1710	257.3369
150°	1	56.9176	53.2026	66.1119	62.9890
	2	75.7969	73.9466	110.8923	108.1772
	3	194.7592	182.2008	191.65128	184.0092
	4	209.8985	186.7983	252.0484	225.9411
	5	237.6279	214.3865	287.5893	251.9882
180°	1	24.2081	23.7668	24.2081	23.7668
	2	58.2428	54.3255	58.2428	54.3255
	3	148.0268	135.6779	148.0268	135.6779
	4	188.8084	176.3040	188.8084	176.3040
	5	213.1012	189.1577	213.1012	189.1577

CONCLUSION

A simple C° isoparametric finite element formulation based on the higher-order shear deformation theory is presented for the free vibration analysis of fibre reinforced laminated composite and sandwich folded plates. The accuracy of the present formulation is evaluated by obtaining the solutions to a wide range of problems and comparing them with the available results in the literature along with the solutions obtained using first-order shear deformation theory. Numerical results are presented for isotropic, anisotropic and sandwich folded plates. The natural frequencies of folded plates decreases as crank angle increases. It is observed that in case of composite laminates the difference in prediction of results by first-order and higher-order shear deformation theories are small. However, for sandwich laminates, in comparison to higher-order shear deformation theory, first-order shear deformation theory overestimates the natural frequency with significant margin and the margin increases for the higher modes. It is believed that the present results on laminated sandwich folded plates will serve as benchmark solutions for other researchers to validate their numerical techniques.

REFERENCES

1. J. E. Goldberg, H. L. Leve, 1957, Theory of prismatic folded plate structures. IABSE, 17.
2. J. S. B. Iffland, 1979, Folded plate structures. Journal of Structural Division. 105, 111-123.
3. T. Irie, G. Yamada, Y. Kobayashi, 1984, Free vibration of a cantilever folded plate, Journal of Acoustic Society of America. 76(6), 1743-1748.
4. T. Irie, G. Yamada, K. Tanaka, 1984, Natural frequencies of folded plates, Journal of Sound and Vibration. 95(1), 131-135.
5. M. Tanaka, T. Matsumoto, A. Shiozaki, 1998, Application of boundary-domain element method to the free vibration problem of plate structures, Computers and Structures. 66(6), 725-735.
6. M. Tanaka, Yamagiwa, K. Miyazaki, T. Ueda, 1988, Free vibration analysis of elastic plate structures by boundary element method, Engineering Analysis and Boundary Elements. 5(4), 182-188.
7. W. H. Liu, C. C. Huang, 1992, Vibration analysis of folded plates, Journal of Sound and Vibration. 157(1), 123-137.
8. A. G. Niyogi, M. K. Laha, P. K. Sinha, 2000, Finite element analysis of cantilever laminated composite and sandwich plate structures, Proc. Structural Engineering Convention-2000, IIT-Bombay, 5-8 January, 77-84.
9. T. Kant, D. R. J. Owen, O. C. Zienkiewicz, 1982, A refined higher order C^o plate bending element, Computers and Structures. 15, 177-183.
10. J. N. Reddy, 1984, A simple higher-order theory for laminated composites, Journaql of Applied Mechanics. 51, 745-752.
11. T. Kant, Mallikarjuna, 1989, Vibrations of unsymmetrically laminated plates analyzed by using a higher order theory with a C^o finite element formulation, Journal of Sound and Vibration. 134, 1-16.
12. K. J. Bathe, 1996, Finite element procedures, Englewood Cliffs Prentice-Hall.

APPENDIX A: RIGIDITY MATRICES

Assuming $H_i = (z^i_{L+1} - z^i_L)/i$, where i takes an integer value from one to seven, the elements of the submatrices of the rigidity matrix can be readily obtained in the following forms:

$$D_m = \sum_{L=1}^{NL} \begin{bmatrix} Q_{11}H_1 & Q_{12}H_1 & Q_{13}H_1 & Q_{11}H_3 & Q_{12}H_3 & Q_{13}H_3 \\ & Q_{22}H_1 & Q_{23}H_1 & Q_{12}H_3 & Q_{22}H_3 & Q_{23}H_3 \\ & & Q_{33}H_1 & Q_{13}H_3 & Q_{23}H_3 & Q_{33}H_3 \\ & & & Q_{11}H_5 & Q_{12}H_5 & Q_{13}H_5 \\ & \text{Symm.} & & & Q_{22}H_5 & Q_{23}H_5 \\ & & & & & Q_{33}H_5 \end{bmatrix}$$

$$\mathbf{D_s} = \sum_{L=1}^{NL} \begin{bmatrix} Q_{44}H_1 & Q_{45}H_1 & Q_{44}H_3 & Q_{45}H_3 & Q_{44}H_2 & Q_{45}H_2 \\ & Q_{55}H_1 & Q_{45}H_3 & Q_{55}H_3 & Q_{45}H_2 & Q_{55}H_2 \\ & & Q_{44}H_5 & Q_{45}H_5 & Q_{44}H_4 & Q_{45}H_4 \\ & & & Q_{55}H_5 & Q_{45}H_4 & Q_{55}H_4 \\ & \text{Symm.} & & & Q_{44}H_3 & Q_{45}H_3 \\ & & & & & Q_{55}H_3 \end{bmatrix}$$

The elements of the $\mathbf{D}_c$ and $\mathbf{D}_b$ matrices are obtained by replacing (H_1, H_3 and H_5) by (H_2, H_4 and H_6) and (H_3, H_5 and H_7), respectively in the $\mathbf{D}_m$ matrix.

APPENDIX B: INERTIA MATRIX

The inertia matrix $\mathbf{m}$ for the present higher order theory is given by

$$
\mathbf{m} = \begin{bmatrix}
I_1 & 0 & 0 & 0 & I_2 & I_3 & 0 & 0 & I_4 \\
0 & I_1 & 0 & -I_2 & 0 & 0 & I_3 & -I_4 & 0 \\
0 & 0 & I_1 & 0 & 0 & 0 & 0 & 0 & 0 \\
0 & -I_2 & 0 & I_3 & 0 & 0 & -I_4 & I_5 & 0 \\
I_2 & 0 & 0 & 0 & I_3 & I_4 & 0 & 0 & I_5 \\
I_3 & 0 & 0 & 0 & I_4 & I_5 & 0 & 0 & I_6 \\
0 & I_3 & 0 & -I_4 & 0 & 0 & I_5 & -I_6 & 0 \\
0 & -I_4 & 0 & I_5 & 0 & 0 & -I_6 & I_7 & 0 \\
I_4 & 0 & 0 & 0 & I_5 & I_6 & 0 & 0 & I_7
\end{bmatrix}
$$

The parameters I_1, I_2 and (I_5, I_7) are linear inertia, rotary inertia and higher-order inertia terms, respectively. The parameters I_2, I_4 and I_6 are the coupling inertia terms. They are defined as follows:

$$
(I_1, I_2, I_3, I_4, I_5, I_6, I_7) = \sum_{L=1}^{NL} \int_{Z_L}^{Z_{L+1}} (1, z, z^2, z^3, z^4, z^5, z^6)\, \rho_L\, dz
$$

where, ρ_L is the material density of the Lth layer.

13

Nonlinear Vibrations and Stability of Thermally Induced Postbuckled Laminated Composite Cylindrical Shells

S. Singh, B.P. Patel and Y. Nath

Department of Applied Mechanics, Indian Institute of Technology Delhi-110 016 India
email: nathyogendra@hotmail.com

ABSTRACT

The stability of postbuckled equilibrium configurations and the nonlinear dynamic characteristics of cross-ply laminated heated cylindrical shells are investigated employing semi-analytical shell finite element. The frequencies of small oscillations about equilibrium configuration are obtained by solving the eigenvalue problem formulated using tangent stiffness matrix of the converged equilibrium configuration and mass matrix. The present study reveals that the longitudinally symmetric postbuckled equilibrium configuration of two-layered cylindrical shells is stable and the longitudinally anti-symmetric one is unstable whereas for the eight-layered shell the symmetric postbuckled configuration is unstable and antisymmetric one is stable. The nonlinear dynamic response of eight-layered shell shows that the shell with longitudinally symmetric disturbance jumps from symmetric mode to antisymmetric mode and the predicted equilibrium configuration is of antisymmetric nature irrespective of the type of initial disturbance. The nonlinear forced dynamic response of the shell heated in the postbuckling region reveals distinct features compared to that of the prebuckling region.

Keywords: Dynamics; Stability; Cylindrical Shell; Cross-ply; Thermal; Postbuckling; Semi-analytical finite element.

1. INTRODUCTION

The structural components such as laminated composite cylindrical shells may often be subjected to mechanical loading coupled with the continuous and/or sudden exposure to elevated temperature. For the purpose of optimal utilization, the structures may be permitted to undergo elastic buckling and operate in postbuckling region. The structures in their buckled state may be expected to operate under dynamic thermo-mechanical loading conditions. Under these operating conditions, the thin shells may experience transverse deflections of the order of shell thickness or even higher necessitating the incorporation of geometric nonlinearity for the adequate analysis. The induced thermal state of stress in the prebuckling and postbuckled configurations may significantly affect the vibration and stability characteristics. Further, the frequency characteristics of small amplitude vibrations about the postbuckled configuration may reveal the stability/instability of postbuckling equilibrium path. Therefore, the study of vibration characteristics of thermally stressed composite laminated structures in the pre- and post-buckling regions gains importance. The studies on the vibration characteristics of laminated plates/shell panels subjected to thermal loading in the pre- and post-buckling configuration have been carried out [1-2]. The studies on the vibration characteristics of laminated heated cylindrical shells in the post-buckled regions and the investigation on the stability of post-buckling equilibrium configurations through dynamic analysis are scarce in the literature. Furthermore, the behavior of the circumferentially closed shells especially undergoing moderately large deformation is significantly different from the panels/plates.

In the present investigation, the stability of postbuckled equilibrium configurations and nonlinear dynamic characteristics of composite laminated heated cylindrical shells are studied using semi-analytical shell finite element approach [3]. The frequencies of small oscillations about buckled configuration are obtained. The nonlinear dynamic analysis of the shell heated in pre- and post-buckling regions is carried out incorporating a small amount of proportional damping in order to facilitate the convergence to a new stable state after jump.

2. FORMULATIONS

A laminated composite cylindrical shell is considered with the coordinates s, θ and z along the meridional, circumferential and radial/thickness directions, respectively. Using first order shear deformation theory the displacements u, v, w at a point (s, θ, z) from the median surface are expressed as functions of middle surface displacements u_0, v_0 and w_0, and rotations β_s and β_θ of the meridional and hoop sections, respectively, as

$$u(s, \theta, z) = u_0(s, \theta) + z\, \beta_s(s, \theta); \quad v(s, \theta, z) = v_0(s, \theta) + z\, \beta_\theta(s, \theta); \quad w(s, \theta, z) = w_o(s, \theta) \quad ...(1)$$

Using the semi-analytical approach, u_o, v_o, w_o, β_s and β_θ are represented by a Fourier series in the circumferential angle θ. For the nth harmonic, these can be written as

$$u_o(s,\theta,t) = u_o^o(s,t) + \sum_{i=1}^{M_1}\left[u_o^{c_i}(s,t)\cos(in\theta) + u_o^{s_i}(s,t)\sin(in\theta)\right]$$

$$v_o(s,\theta,t) = v_o^o(s,t) + \sum_{i=1}^{M_1}\left[v_o^{c_i}(s,t)\cos(in\theta) + v_o^{s_i}(s,t)\sin(in\theta)\right]$$

$$w_o(s,\theta,t) = w_o^o(s,t) + \sum_{i=1}^{M_2}\left[w_o^{c_i}(s,t)\cos(in\theta) + w_o^{s_i}(s,t)\sin(in\theta)\right]$$

$$\beta_s(s,\theta,t) = \beta_s^o(s,t) + \sum_{i=1}^{M_2}\left[\beta_s^{c_i}(s,t)\cos(in\theta) + \beta_s^{s_i}(s,t)\sin(in\theta)\right]$$

$$\beta_\theta(s,\theta,t) = \beta_\theta^o(s,t) + \sum_{i=1}^{M_2}\left[\beta_\theta^{c_i}(s,t)\cos(in\theta) + \beta_\theta^{s_i}(s,t)\sin(in\theta)\right] \quad ...(2)$$

where superscript o refers to the axisymmetric component of displacement field variables, and c_i and s_i refer to the asymmetric components of the field variables having circumferential variation proportional to $\cos(in\theta)$ and $\sin(in\theta)$, respectively. The number of terms M_1 in the approximation of variables (u_o, v_o) is, in general, twice compared to M_2 in $(w_o, \beta_s, \beta_\theta)$ for the converged solution.

Using von Karman's type nonlinearity for moderately large deformation, Green's strains can be written in terms of middle surface deformations as,

$$\{\varepsilon\} = \begin{Bmatrix} \varepsilon_p^L \\ 0 \end{Bmatrix} + \begin{Bmatrix} z\varepsilon_b \\ \varepsilon_s \end{Bmatrix} + \begin{Bmatrix} \varepsilon_p^{NL} \\ 0 \end{Bmatrix} \qquad \text{...(3)}$$

where, the membrane strains $\{\varepsilon_p^L\}$, bending strains $\{\varepsilon_b\}$, shear strains $\{\varepsilon_s\}$ and nonlinear in-plane strains $\{\varepsilon_p^{NL}\}$ and can be written as

$$\{\varepsilon_p^L\} = \begin{Bmatrix} \dfrac{\partial u_o}{\partial s} \\[2mm] \dfrac{\partial v_o}{r\partial \theta} + \dfrac{w_o}{r} \\[2mm] \dfrac{\partial u_o}{r\partial \theta} + \dfrac{\partial v_o}{\partial s} \end{Bmatrix}; \{\varepsilon_b\} = \begin{Bmatrix} \dfrac{\partial \beta_s}{\partial s} \\[2mm] \dfrac{\partial \beta_\theta}{r\partial \theta} \\[2mm] \dfrac{\partial \beta_s}{r\partial \theta} + \dfrac{\partial \beta_\theta}{\partial s} \end{Bmatrix}; \{\varepsilon_s\} = \begin{Bmatrix} \beta_s + \dfrac{\partial w_o}{\partial s} \\[2mm] \beta_\theta + \dfrac{\partial w_o}{r\partial \theta} - \dfrac{v_o}{r} \end{Bmatrix}; \{\varepsilon_p^{NL}\} = \begin{Bmatrix} \dfrac{1}{2}\left(\dfrac{\partial w_o}{\partial s}\right)^2 \\[2mm] \dfrac{1}{2}\left(\dfrac{\partial w_o}{r\partial \theta}\right)^2 \\[2mm] \dfrac{\partial w_o}{\partial s}\dfrac{\partial w_o}{r\partial \theta} \end{Bmatrix} \qquad \text{...(4)}$$

where, r is the radius of the parallel circle.

The stress resultant $\{\mathbf{N}\}$ and the moment resultant $\{\mathbf{M}\}$ are related to membrane strains $\{\varepsilon_p\}\left(=\{\varepsilon_p^L\}+\{\varepsilon_p^{NL}\}\right)$ and bending strains $\{\varepsilon_b\}$ through the constitutive relations as

$$\begin{Bmatrix} \{\mathbf{N}\} \\ \{\mathbf{M}\} \end{Bmatrix} = \begin{bmatrix} [A] & [B] \\ [B] & [D] \end{bmatrix} \begin{Bmatrix} \{\varepsilon_p\} \\ \{\varepsilon_b\} \end{Bmatrix} - \begin{Bmatrix} \{\bar{\mathbf{N}}\} \\ \{\bar{\mathbf{M}}\} \end{Bmatrix} \qquad \text{...(5)}$$

where $[A]$, $[D]$ and $[B]$ are extensional, bending and bending-extensional coupling stiffness coefficients matrices. $\{\bar{\mathbf{N}}\}$ and $\{\bar{\mathbf{M}}\}$ are the thermal stress and moment resultants, respectively.

The transverse shear stress resultant $\{Q\}$ is related to the transverse shear strains $\{\varepsilon_s\}$ as

$$\{Q\}=[E]\{\varepsilon_s\} \qquad \text{...(6)}$$

where, $[E]$ is the transverse shear stiffness coefficients matrix.

The potential energy functional $U_1(\delta)$ (due to strain energy and transverse load) is given by,

$$U_1(\delta)=\frac{1}{2}\int_A \left[\begin{Bmatrix} \varepsilon_p \\ \varepsilon_b \end{Bmatrix}^T \begin{bmatrix} A & B \\ B & D \end{bmatrix} \begin{Bmatrix} \varepsilon_p \\ \varepsilon_b \end{Bmatrix} + \{\varepsilon_s\}^T [E]\{\varepsilon_s\} - \begin{Bmatrix} \varepsilon_p \\ \varepsilon_b \end{Bmatrix}^T \begin{Bmatrix} \bar{N} \\ \bar{M} \end{Bmatrix} \right] dA - \int_A q\, w_o\, dA \qquad \text{...(7)}$$

where, δ is the vector of degrees of freedom associated to the displacement field in a finite element discretisation and q is the applied pressure load.

The potential energy $U_2(\delta)$ due to initial state of in-plane stress resultants $\{N^0\}= \left\{N_{ss}^0\ N_{\theta\theta}^0\ N_{s\theta}^0\right\}^T$ is written as

$$U_2(\delta)=\int_A \left\{\varepsilon_p^{NL}\right\}^T \left\{N^0\right\} dA \qquad \text{...(8)}$$

Following the procedure of Rajasekaran and Murray [4], the total potential energy functional $U(\delta)$ [$= U_1(\delta) + U_2(\delta)$] can be expressed as

$$U(\delta) = \{\delta\}^{\mathrm{T}} [(1/2)[[K] - [K_{\Delta_T}] + [K_G]] + (1/6)[K_1(\delta)] + (1/12)[K_2(\delta)]]\{\delta\} - \{\delta\}^{\mathrm{T}}\{F_M\} - \{\delta\}^{\mathrm{T}}\{F_T\} \qquad ...(9)$$

where, $[K]$ is the linear stiffness matrix, $[K_1]$ and $[K_2]$ are nonlinear stiffness matrices linearly and quadratically dependent on the field variables, respectively. $[K_T]$ and $[K_G]$ are the geometric stiffness matrices due to thermal and initial stress resultants. $\{F_M\}$ and $\{F_T\}$ are mechanical and thermal load vectors, respectively.

The kinetic energy of the shell is given by

$$T(\dot{\delta}) = \frac{1}{2}\int_A \left[p\left(\dot{u}_o^2 + \dot{v}_o^2 + \dot{w}_o^2\right) + I\left(\dot{\beta}_s^2 + \dot{\beta}_\theta^2\right) \right] dA \qquad ...(10)$$

where, $p = \sum_{i=1}^{N} \int_{z_i}^{z_{i+1}} \rho^i \, dz$, $I = \sum_{i=1}^{N} \int_{z_i}^{z_{i+1}} \rho^i z^2 \, dz$ and ρ^i is the mass density of the ith layer. z_i and z_{i+1} are

the z coordinate of the inner and outer surfaces of the ith layer. The dot over the variable denotes the partial derivative with respect to time.

Using Eqs. (9) and (10) the Lagrange's equation yields the governing equations of motion for the shell incorporating damping and are,

$$[M]\{\ddot{\delta}\} + [C]\{\dot{\delta}\} + \left[[K] - [K_{\Delta T}] + [K_G] + \frac{1}{2}\left[K_1(\delta)\right] + \frac{1}{3}\left[K_2(\delta)\right]\right]\{\delta\} = \{F_M\} + \{F_T\} \qquad ...(11)$$

where, $[M]$ and $[C]$ are mass and damping matrices, respectively.

The governing Eq. (11) can be employed to study the linear/nonlinear static/dynamic response and eigenvalue analyses by neglecting the appropriate terms as:

Linear Static Analysis:

$$[K] \{\delta\} = \{F_M\} + \{F_T\} \qquad ...(12)$$

Nonlinear Static Analysis:

$$[[K] - [K_{\Delta T}] + [K_G] + (1/2)[K_1(\delta)] + (1/3)[K_2(\delta)]]\{\delta\} = \{F_M\} + \{F_T\} \qquad ...(13)$$

Nonlinear Dynamic Analysis:

$$[M]\{\ddot{\delta}\} + [C]\{\dot{\delta}\} + \left[[K] - [K_{\Delta T}] + [K_G] + \frac{1}{2}\left[K_1(\delta)\right] + \frac{1}{3}\left[K_2(\delta)\right]\right]\{\delta\} = \{F_M\} + \{F_T\} \qquad ...(14)$$

Free Vibration Analysis about Deformed Equilibrium Configuration:

$$[K_t]\{\overline{\delta}\} = \omega^2 [M]\{\overline{\delta}\} \qquad ...(15)$$

where, $[K_t] = [K] - [K_{\Delta T}] + [K_G] + [K_1(\delta)] + [K_2(\delta)]$ is the tangent stiffness matrix, $\{\overline{\delta}\}$ is the vector representing the vibration mode shape measured from the deformed equilibrium configuration, and ω is the frequency.

Eigenvalue Buckling Analysis:

$$[K] \{\delta\} = \Delta T [K_G^*] \{\delta\} \qquad ...(16)$$

where $[K_G{}^*]$ is the geometric stiffness due to initial state of stress developed because of unit uniform temperature rise and ΔT is the temperature rise.

It may be noted here that for the purpose of evaluating $[K_G{}^*]$, firstly the static analysis of the shell using Eq. (12) for unit uniform temperature rise is carried out. The resulting deformation field is used to calculate the initial state of stress resultants using Eq. (5) and in turn, for evaluating the $[K_G{}^*]$ matrix.

The nonlinear static equilibrium path is being traced by solving Eq. (13) using Newton-Raphson iteration procedure coupled with the adaptive displacement control method [3]. The degree of freedom having the highest increment in the previous step is selected as a control parameter except the first step wherein the temperature increment is specified. The Eq. (14) for nonlinear dynamic response is solved using the Newmark's numerical integration scheme coupled with the Newton-Raphson technique.

3. RESULTS AND DISCUSSIONS

The nonlinear thermoelastic static response, vibration characteristics and dynamic response of laminated circular cylindrical shells subjected to uniform temperature rise, which may typically be encountered during the cruise flight of a space vehicle, are investigated using the semi-analytical finite element formulation. The presence of small magnitude initial disturbance in the form of a load spatially proportional to the linear buckling modeshape having maximum transverse displacement parameter ($\bar{w}_{0\max}/h$) equal to 0.001, unless otherwise is specified, is assumed. The nonlinear prebuckling/postbuckling characteristics is presented as relationship between maximum outward normal displacement parameter (w_{max}/h) versus temperature rise parameter λ_T ($= 10^{-6}\,\Delta Tr/h$). The stability of the equilibrium configurations is investigated from the variation of frequency parameter Ω^2 ($= \omega^2\,r^2\,\rho\,/E_T$) with temperature rise parameter (λ_T). The positive and negative values of the frequency parameter indicate the stable and unstable equilibrium configurations, respectively.

The material properties considered are: $E_L = 181$ GPa, $E_T = 10.3$ GPa, $G_{LT} = G_{TT} = 7.17$ GPa, $\nu_{LT} = \nu_{TT} = 0.28$, $\alpha_L = 0.02 \times 10^{-6}$ /°C, $\alpha_T = 22.5 \times 10^{-6}$ /⁰C, $\rho = 1600$ Kg/m³.

where, E, G, ν and α are Young's modulus, shear modulus, Poisson's ratio and coefficient of thermal expansion, respectively.

All the layers are of equal thickness and the ply-angle is measured with respect to the meridional axis (s-axis). The first layer is the innermost layer of the shell.

The simply supported immovable boundary conditions of the shells considered are:

$$u_o^o = u_o^{c1} = u_o^{s1} = u_o^{c2} = u_o^{s2} = u_o^{c3} = u_o^{s3} = u_o^{c4} = u_o^{s4} = v_o^o = v_o^{c1} = v_o^{s1} = v_o^{c2} = v_o^{s2} = v_o^{c3} = v_o^{s3} = v_o^{c4} = v_o^{s4} = 0$$

$$w_o^o = w_o^{c1} = w_o^{s1} = w_o^{c2} = w_o^{s2} = \beta_\theta^o = \beta_\theta^{c1} = \beta_\theta^{s1} = \beta_\theta^{c2} = \beta_\theta^{s2} = 0 \qquad \text{at } s = 0,\, L.$$

Based on progressive mesh refinement, 48 elements idealization is found to be adequate to model the complete length of the shells.

The nonlinear thermoelastic response and linear vibration characteristics about the deformed configuration of two-layered cross-ply laminated ($0°/90°$) simply supported cylindrical shell ($L/r_1 = 1$, $r_1/h = 200$, $n = 13$) are shown in Figure 1. The stable part of the equilibrium path is depicted in Figure 1(a) as solid lines and unstable one with the dashed lines. The prebuckling path

(Equilibrium path I) is unstable after the bifurcation point. The equilibrium configuration of the shell after bifurcation point follows the path II or III depending upon whether the small magnitude radial load (assumed for initiation of bifurcation in the nonlinear static analysis) is spatially proportional to longitudinally antisymmetric or symmetric linear buckling mode shape. The frequency parameter variation with temperature presented in Figs. 1(b) and 1(c) reveals that the decrease in the frequency corresponding to different modes in the prebuckling region is independent of the type of disturbance (antisymmetric or symmetric) as expected. However, the variation of the frequency parameters with temperature significantly depends on the nature (longitudinally symmetric or antisymmetric) of the postbuckling equilibrium configuration. The frequency parameter of one mode for the longitudinally antisymmetric postbuckling configuration is negative indicating that the longitudinally antisymmetric postbuckled configuration is unstable for the entire temperature range considered here. The results shown in Fig. 2 reveal that the antisymmetric postbuckled configuration is stable and symmetric one is unstable for eight-layered $(0°/90°)_4$ shell.

The stability of symmetric and antisymmetric postbuckled deformed configurations, the dynamic analysis of the shells is carried out incorporating the proportional damping of the form $[C] = \alpha$ $([K]+[M])$. The dynamic response of the heated eight-layered $(0°/90°)_4$ shell $(\lambda_T = 0.2)$ in the presence of longitudinally symmetric disturbance is shown in Fig. 3 for $\alpha = 0.5 \times 10^{-4}$. It can be observed from Fig. 3 that the shell response grows in the longitudinally symmetric deformation shape and exhibits the oscillations about the mean configuration. The amplitude of these oscillations reduces with the time due to the presence of damping, and when the amplitude of the oscillations is reduced to a certain level, the shell response jumps to oscillation about longitudinally antisymmetric deformation shape. It may be noted here that in the nonlinear static analysis for eight-layered shell the longitudinally symmetric disturbance leads to the deformation in the same mode that is predicted as unstable through the eigenvalue analysis. However, in the dynamic case, the response of the shell jumps to stable configuration in spite of the initial disturbance being longitudinally symmetric.

The forced nonlinear dynamic response analysis of the eight-layered $(0°/90°)_4$ heated shell subjected to external radial load $q = q_0$ Sin $(\pi \, s/L)$ Cos $(n \, \theta)$ Sin $(w_F t)$ is carried out. The load amplitude q_0 corresponds to a maximum transverse displacement parameter $(\overline{w}_{0\max}/h)$ of linear static response of the shell subjected to only radial load. The steady state response amplitude (w_o^{c1}/h) (at $s = L/2$ versus forcing frequency curves, extracted from the time response curves corresponding to different forcing frequencies, are shown in Figure 4. It can be inferred from this figure that the shell heated in prebuckling region reveals jump phenomena and the forcing frequency corresponding to the jump in the response decreases with the increase in force magnitude revealing the softening type of nonlinear effect. The response of the shell heated in the postbuckling region does not reveal distinct jump. The typical time response $[(w_o^{c1}/h)$ at $s = L/2$ versus time] curves is shown in Figures 5 and 6 for $\lambda_T = 0.2$. The change in the nature of the response pattern is associated with the change of mean equilibrium configuration from longitudinally symmetric to antisymmetric one.

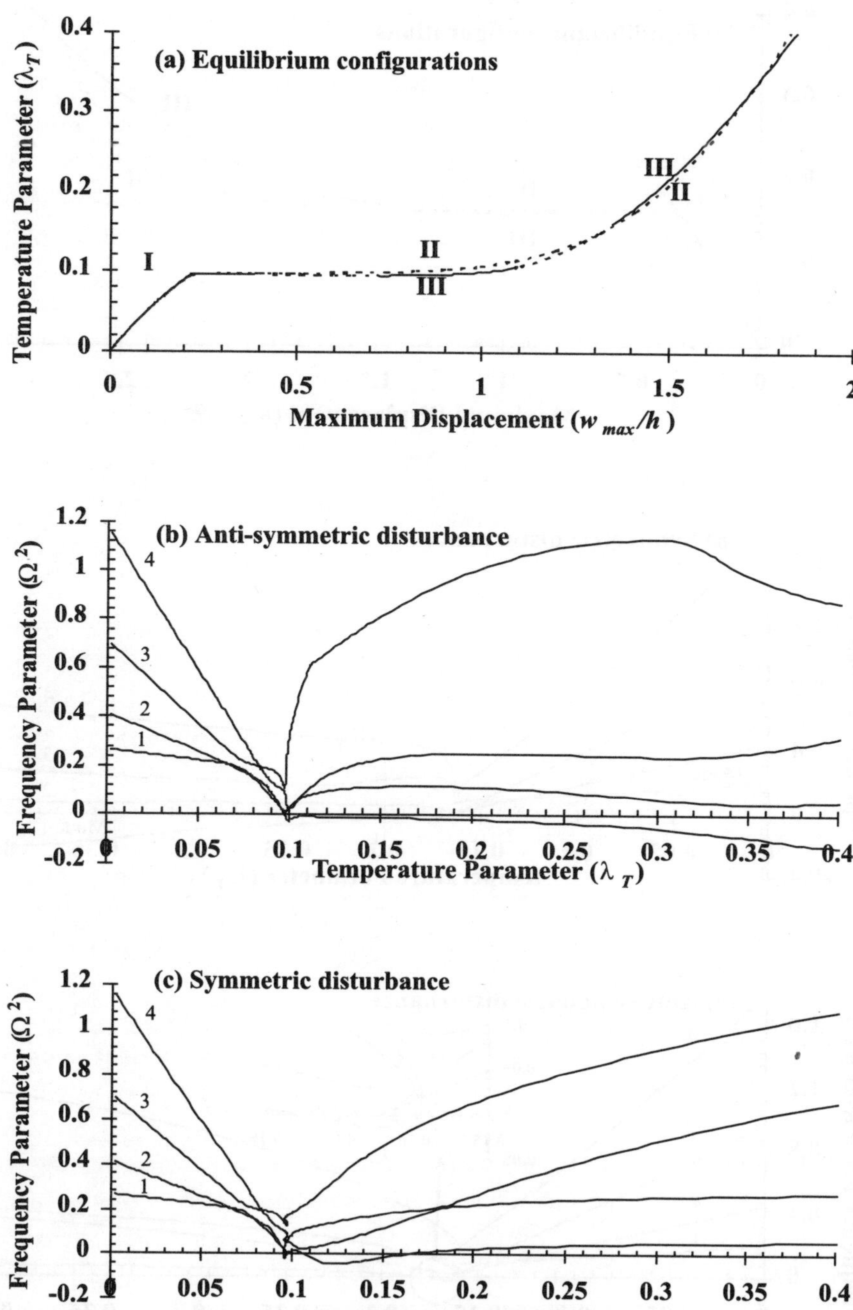

Fig. 1. Equilibrium configurations and free vibration frequencies of two-layered cross-ply (0°/90°) laminated heated cylindrical shell (L/r = 1, r/h = 200, n = 13) with longitudinally anti-symmetric and symmetric disturbance.

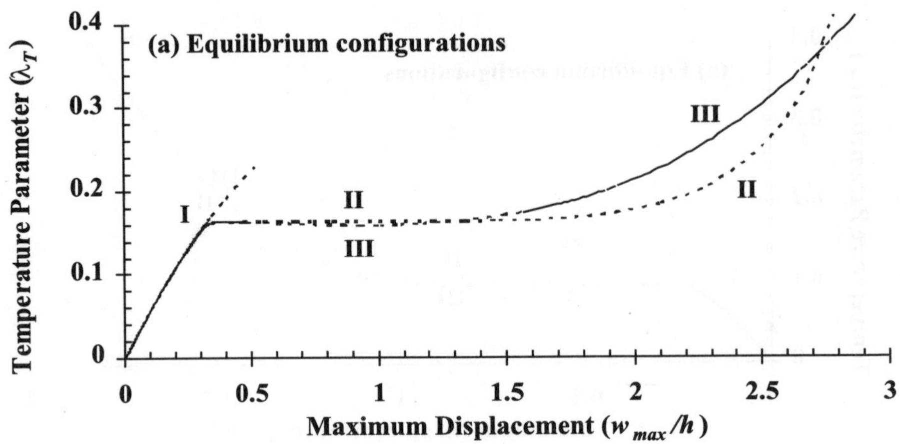

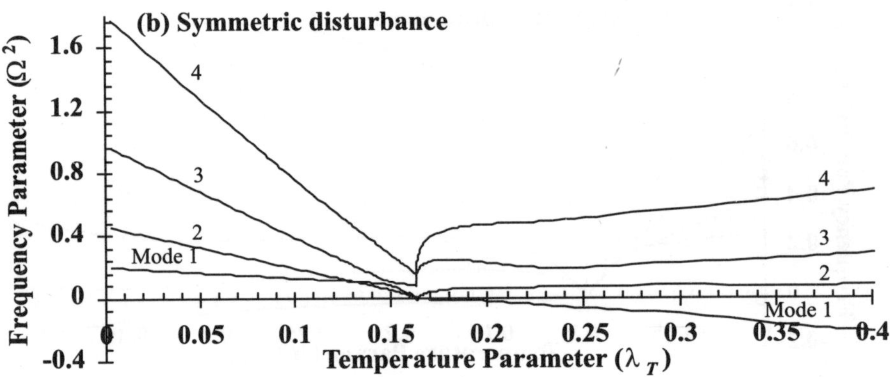

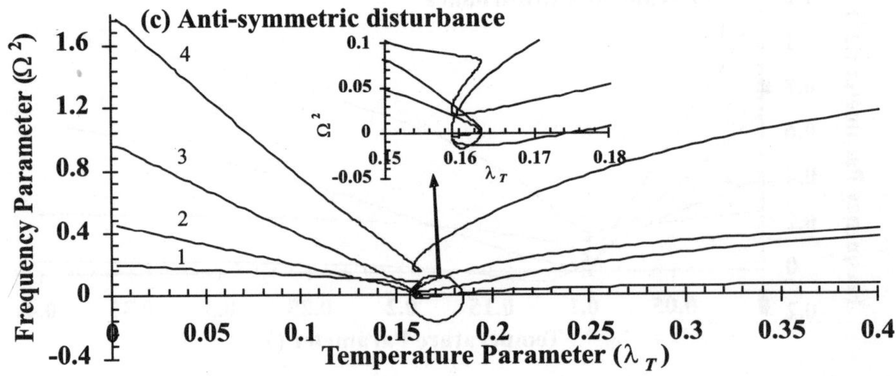

Fig. 2. Equilibrium configurations and free vibration frequencies of eight-layered cross-ply $(0^0/90^0)_4$ laminated heated cylindrical shell (L/r = 1, r/h = 200, n = 9) with longitudinally symmetric and anti-symmetric disturbance.

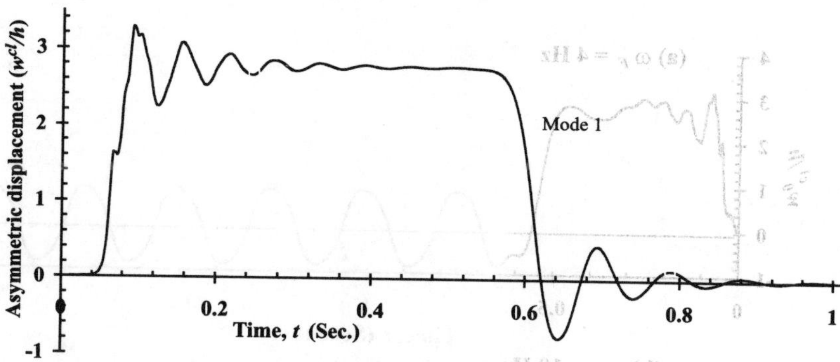

Fig. 3. Dynamic response of the heated shell ($\lambda_T = 0.2$) at $s = L/2$ with damping parameter $\alpha = 0.5 \times 10^{-4}$ in the presence of longitudinally symmetric disturbance for eight-layered shell.

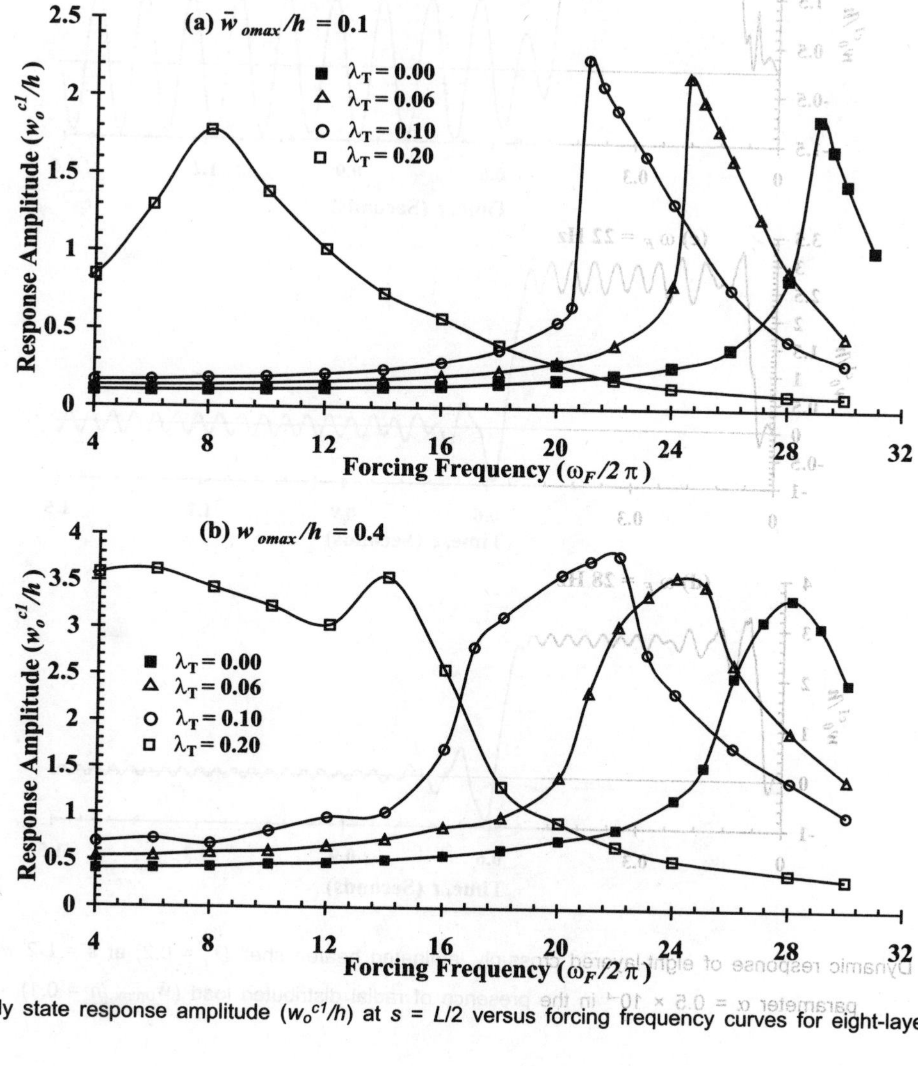

Fig. 4. Steady state response amplitude (w_o^{c1}/h) at $s = L/2$ versus forcing frequency curves for eight-layered shell.

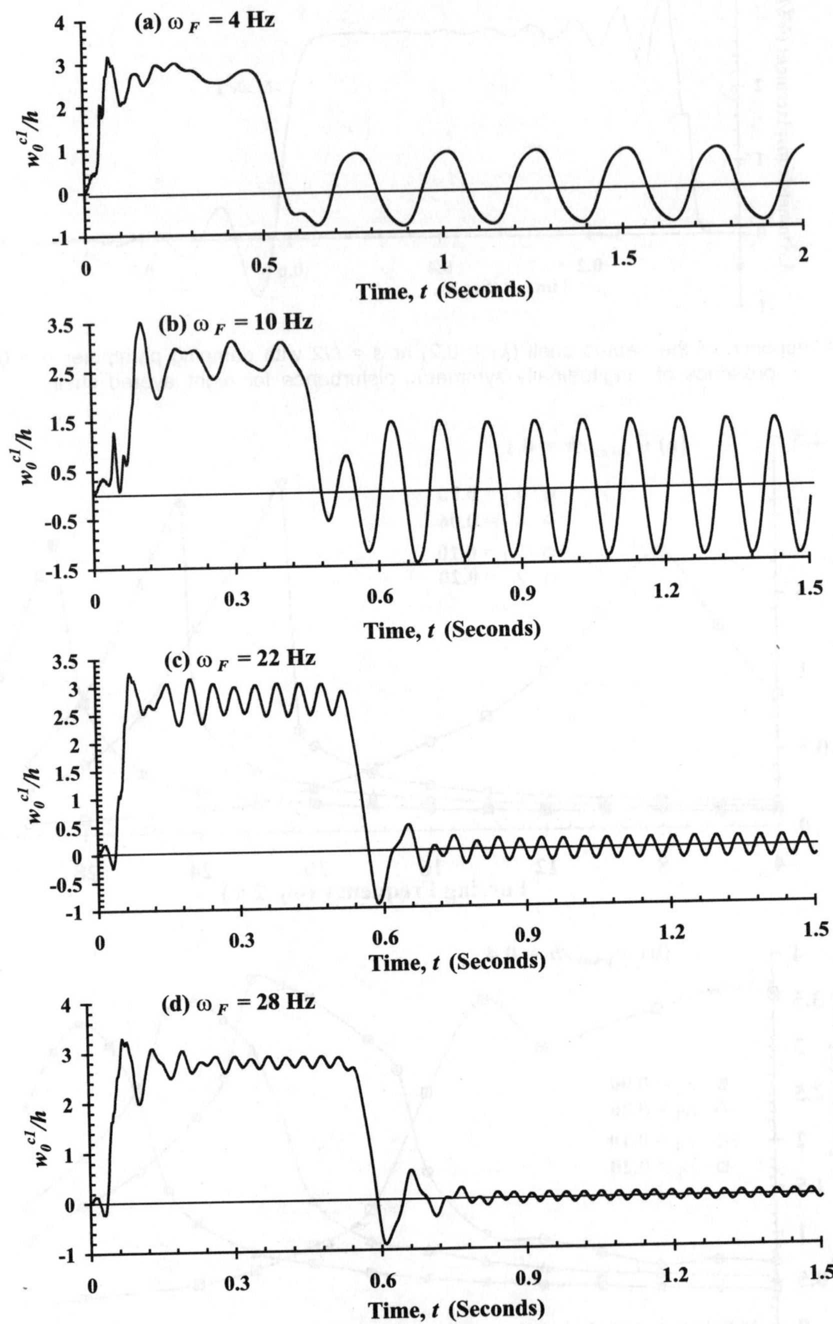

Fig. 5. Dynamic response of eight-layered cross-ply laminated heated shell ($\lambda_T = 0.2$) at $s = L/2$ with damping parameter $\alpha = 0.5 \times 10^{-4}$ in the presence of radial distributed load ($\bar{w}_{0max}/h = 0.1$).

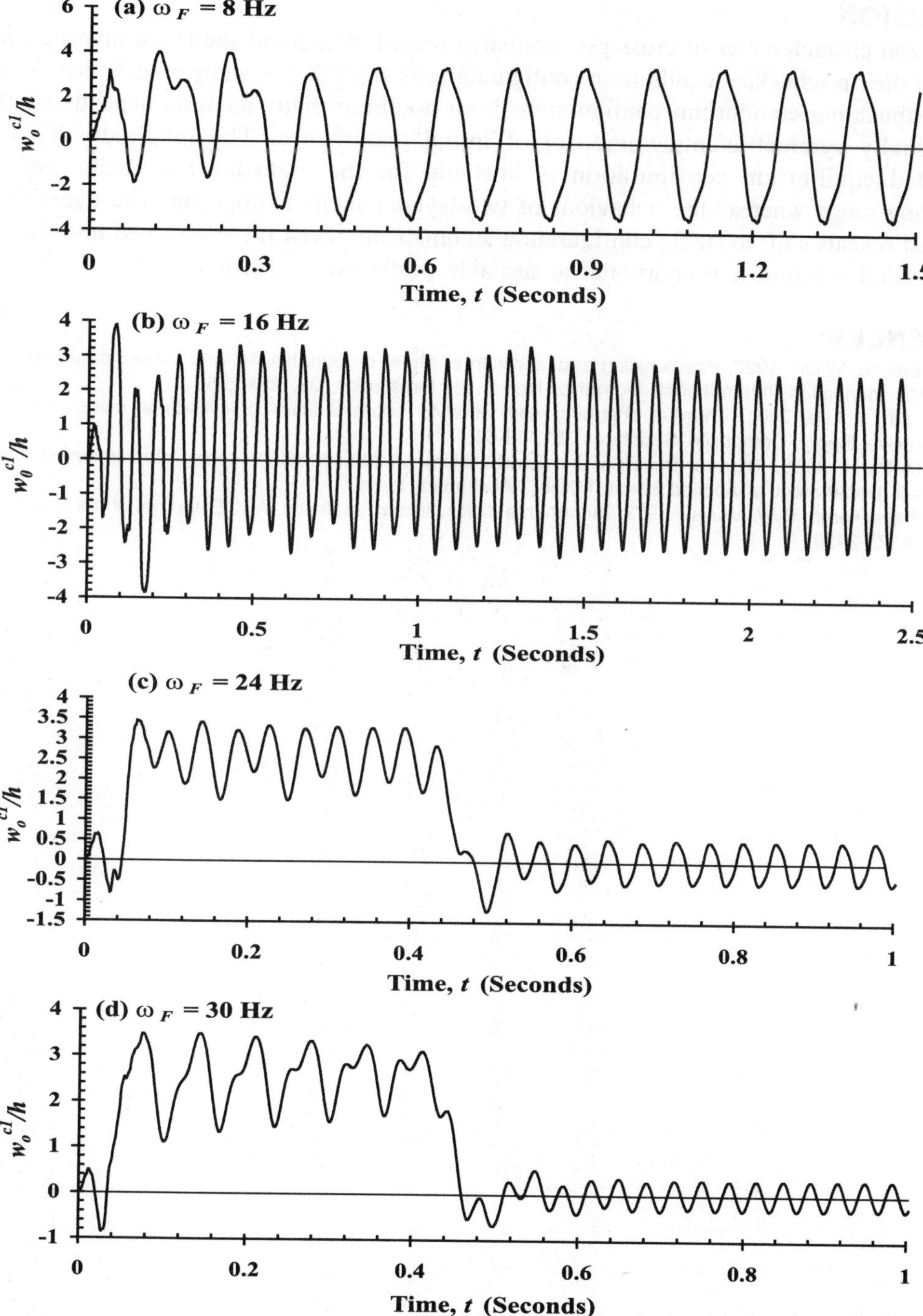

Fig. 6. Dynamic response of eight-layered cross-ply laminated heated shell ($\lambda_T = 0.2$) at $s = L/2$ with damping parameter $\alpha = 0.5 \times 10^{-4}$ in the presence of radial distributed load ($\bar{w}_{0max}/h = 0.4$).

CONCLUSION

The vibration characteristics of cross-ply laminated heated cylindrical shells are investigated and the stability of their postbuckled equilibrium configurations is studied. The study reveals that the prediction of the postbuckling equilibrium configuration from nonlinear static analysis depends on the nature (longitudinally symmetric/antisymmetric) of initial disturbance. The longitudinally symmetric postbuckled equilibrium configuration is unstable for the eight-layered shells for the entire postbuckling range whereas the behaviour of two-layered shells is opposite. The dynamic response of the shell reveals shift to stable configuration automatically even if the assumed initial disturbance/ dynamic radial pressure is proportional to unstable postbuckling configuration.

REFERENCES

1. L.Librescu, W.Lin, 1997, Postbuckling and vibration of shear deformable flat and curved panels on a non-linear elastic foundation, International Journal of Non-linear Mechanics, 32, 211-225.
2. I.K. Oh, I. Lee, 2001, Thermal snapping and vibration characteristics of cylindrical composite panels using layerwise theory, Composite Structures, 51, 49-61.
3. B. P. Patel, 2005, Thermal buckling and postbuckling characteristics of composite laminated shells. Ph. D. thesis: Department of Applied Mech, MNNIT, Allahabad, India.
4. S. Rajasekaran, D.W. Murray, 1973, Incremental finite element matrices, ASCE Journal of the Structural Division, 99, 2423-2438.

PART – II

COMPOSITE STRUCTURES

14

Failure Analysis of a Pin Loaded Glass Vinylester Composite Plates Using FEM

K. SRIDEVI[1] AND A. SATYADEVI[2]

[1]Department of Mechanical Engineering, G.V.P. College of Engineering Visakhapatnam-530041, Andhra Pradesh, India email: sridevi079@yahoo.com
[2]Department of Mechanical Engineering, College of Engineering; Andhra University, Visakhapatnam-530003, Andhra Pradesh, India email: satyadevia@yahoo.co.in

ABSTRACT

Composites are becoming an essential part of today's materials because of their advantages like low weight, high strength, etc. They are used for aircraft structures to golf clubs, electronic packaging to medical equipment, and space vehicles to home buildings due to their good mechanical properties. In the practical use of composite structures certain discontinuities like holes and cutouts arise while joining by rivets or bolts. This paper deals with the study of failure modes and failure loads of a glass vinylester composite plate with a circular hole subjected to a traction force by a rigid pin. These are investigated for two variables; the ratio of distance from the free edge of the plate to the diameter of the hole and the ratio of width of the plate to the diameter of the hole. The study is performed using a finite element analysis package ANSYS. It is observed that the numerical results are in good agreement with the experimental results from the existing literature.

Keywords: Composite material; pin loading; failure mode.

1. INTRODUCTION

Composite materials are becoming more popular and are used in almost all fields due to their high strength to weight ratio, good fatigue resistance, corrosion resistance, etc. compared to metals. They are used extensively in aircraft structures where these are essential requirements. However, in building complex structures several parts must be joined together. Mechanical joining using bolt or a rivet is one method for joining parts. The difficulty with this method is that the presence of a hole in a laminated plate subjected to external loading introduces a disturbance in the stress field. Stress concentrations are generated in the vicinity of the hole making the joint a weak one.

The knowledge of failure strength of a joint helps in selecting the appropriate joint size in a given application. Many investigators have studied the strength of mechanically fastened joints in composite structures. Chang [1] has developed a computer code, which can be used to calculate the maximum load. Kretsis and Matthews [2] showed that as the width of the specimen decreases, there is a point where the mode of failure changes from bearing to tension. Icten [3] has investigated the mechanical behavior and damage development of pin loaded woven glass fiber-epoxy composites. Vyasraj and Ukadgaoenker [4] have obtained analytical solutions for an irregular shaped hole in an orthotropic laminate. Karakuzu [5] has carried out experiments on glass vinylester composite plates to find the failure modes and loads of different specimens with pin loading. However, as the finite element analysis has become the most versatile method for the analysis of mechanical components, in the present work the failure modes and failure loads of different pin loaded glass vinylester composite plates are studied using the finite element package ANSYS and the results are validated with the experimental results in [5].

In the present work a composite rectangular plate of length L + E, width W and thickness T [5] is considered. A hole of diameter D is present at a distance E from one edge of the plate. A rigid pin is located at the centre of the hole. A load P is applied to the plate along the longitudinal axis. The plate is symmetric with respect to the longitudinal axis. The composite plate consists of 12 layers. The geometry of the specimen is as shown in Fig.1. Different models are obtained by varying E/D and W/D but keeping the parameters D and T as constant. These models are analyzed using Finite Element package ANSYS and the results are compared with the experimental results obtained from [5].

diameter of the hole, $D = 5$ mm,
length of the plate, $L+E = 200$ mm
thickness of the plate, $T = 2.8$ mm.

The material properties considered for the plate are

Maximum permissible tensile stress, $\sigma_t = 395$ MPa
Maximum permissible compressive stress, $\sigma_c = 260$ MPa
Maximum permissible shear stress, $\tau_{xy} = 217$ MPa.

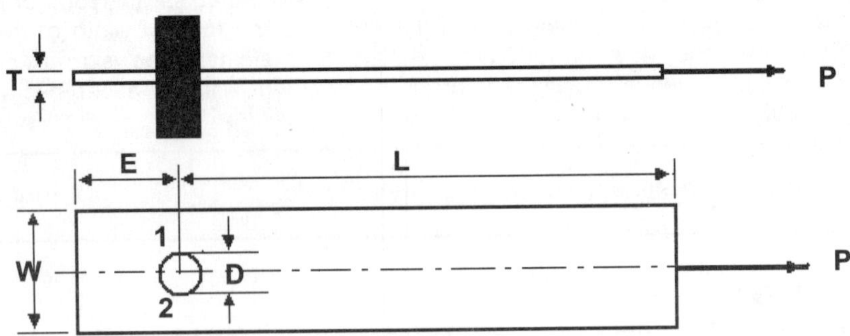

Fig. 1. Geometry of the Specimen

2. FAILURE MODES IN MECHANICAL JOINTS

Failure in mechanical joints can be classified into three types

1. Normal or tearing mode
2. Shearing mode
3. Bearing or crushing mode

These three modes of failure are shown in Fig. 2. A joint fails in one of these three modes or a combination of these. The strength of the joint is the least of normal, shearing and bearing strengths. The mode of failure depends on the type of strength, which is the least. In general, failure of a joint means either the failure of the plate or the failure of the pin / joint. The normal mode of failure occurs for plate while shearing and bearing modes of failure occur either for plate or pin depending on which one is weaker. In the present work, the pin is considered to be rigid and therefore the failure is considered only in the plate.

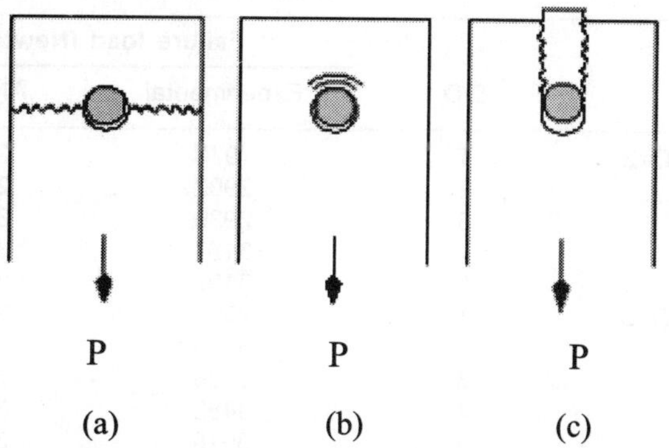

Fig. 2. Modes of failure (a) normal (b) bearing (c) shear

3. FINITE ELEMENT ANALYSIS

A 3D model of the specimen is built in ANSYS. The specimen has symmetry about the longitudinal axis. Using this only one half of the model is created. A symmetry condition is applied on the specimen model. The pin located at the centre of the hole tries to compress one half of the hole (leftside portion of 1-2 in Fig. 1) because of the load acting in the other direction. It is shown as a radial constraint on the hole. The constraints applied on the finite element model are as shown in Fig.3. The model is meshed using SOLID46 from the ANSYS element library. The maximum permissible stresses in the three modes are defined as the failure criteria.

The load on the specimen is varied using bisection method till failure occurs. Failure of the specimen takes place in one of the three modes—tension, compression, and shear. To check whether failure has occured, the failure criterion option available in ANSYS is used. The specimen is considered to fail once the failure criterion reaches a value of 1. At this point the maximum induced stress in the specimen reaches one of the permissible stresses viz. tensile, compressive and shear stresses. The failure mode is decided depending on which stress has reached its permissible limit first. This procedure is continued with different specimens. Different specimens are obtained by varying the geometric proportions of the model. The geometry is varied by taking W/D as 2, 3 and 4 and for each W/D, E/D varying from 1 to 5. These specimens are analyzed for their failure modes and failure loads and compared with the available experimental results [5].

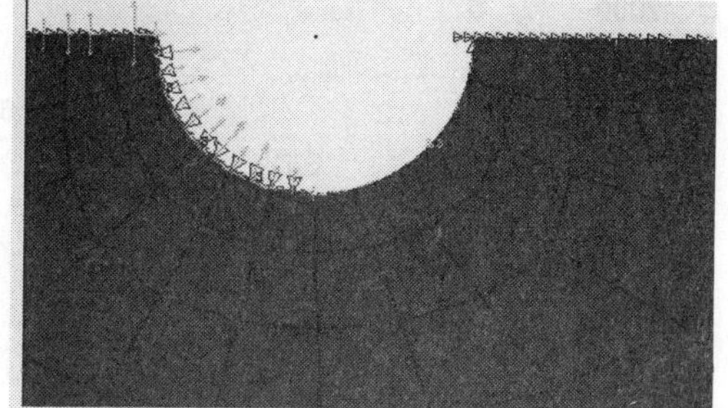

Fig. 3. Constraints applied on the ANSYS model

A comparison of failure strengths obtained from experimental results [5] and ANSYS results for different E/D and W/D ratios are shown in Table 1. The graphs in Fig. 4 show that the results obtained from ANSYS are in good agreement with the experimental results.

Table 1. Comparison of experimental and ANSYS results

	E/D	Failure load (Newton)		Modes of failure*	
		Experimental	ANSYS	Experimental	ANSYS
W/D=2	1	1970	2096	N	N
	2	2902	2996	N	N
	3	2930	3024	N	N
	4	3157	3052	N	N
	5	3192	3108	S-N	N
W/D=3	1	2535	2814	S	S-N
	2	3158	3444	B	B
	3	3330	3461	B	B
	4	3482	3528	B	B
	5	3916	3654	B	B
W/D=4	1	2550	2632	S	N
	2	3160	3360	B	B
	3	3556	3472	B	B
	4	3698	3528	B	B
	5	3897	3740	B	B

* N- normal   S- shear   B- bearing

4. RESULT ANALYSIS

It is observed from the graphs in Fig. 4 and data presented in Table 1 that:

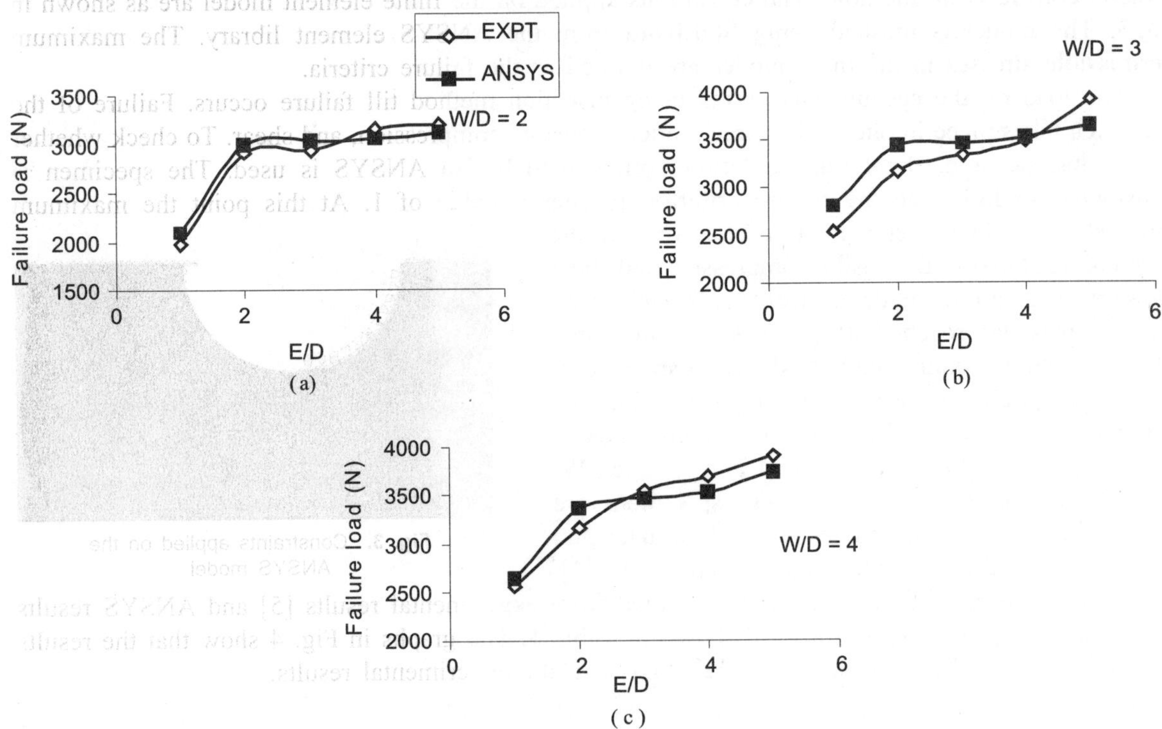

Fig. 4. Comparison of Experimental and ANSYS results for different W/D ratios

1. For a constant W/D, the failure strength of the specimen increases with increase in E/D. This is because keeping the diameter of hole constant, when E/D increases, the distance of the hole from one edge of the plate increases. This increases the shear strength of the specimen.
2. E/D has a greater influence on the failure load but the mode of failure depends on W/D.
3. For W/D = 2, the specimen fails in normal mode. As W/D increases the mode of failure changes from normal to shear or bearing. This is because as W/D increases the width of the specimen increases and hence the load bearing cross-sectional area increases. So the specimen becomes stronger in normal mode. Similarly as E/D increases the specimen becomes stronger in shear mode. Hence, the specimen fails in bearing mode only. Bearing failure is the preferred mode of failure as the specimen can sustain high amounts of load before it fails. This makes the joint a stronger one.

It can also be observed from Fig. 4 that the failure strengths obtained from ANSYS are in close agreement with the experimental values. In addition, Table 1 shows that the failure modes found from ANSYS are also in good agreement with the experimental results in most of the cases.

CONCLUSION

1. The failure load of the composite plate with hole increases as E/D increases.
2. The strength of the plate is found to increase as a result of increase in the width of the plate.
3. W/D has a greater effect on the mode of failure. The material is found to be weak and fails in tension or shear mode when W/D = 2.
4. As W/D increases, the specimen is found to fail in the bearing mode, which is preferred.

REFERENCES

1. Chang FK, Scott RA, Springer GS, 1982, Strength of mechanically fastened composite joints, Journal of Composite Materials; 16:470-494.
2. Kretsis G and Matthews FL, 1985, The strength of bolted joints in glass fiber/epoxy laminates, Journal of Composite Materials; 16; 92-102.
3. Icten BM and Sayman O, 2003, Failure analysis of pin loaded aluminum-glass-epoxy sandwich composite plates, Composite Science Technology; 63:727-37.
4. Vyasaraj and Kakhandki, 2005, Stress analysis of an orthotropic plate with an irregular shaped hole for different in-plane loading conditions-Part 1, Composite Structures, 70; 255-274.
5. Ramazan Karakuzu, Tayfun Gulem and Bulent Murat Icten, 2006, Failure analysis of woven laminated glass-vinylester composites with pin loaded hole, Composite Structures, 72 (1), 27-32.

15

Sensitivity Analysis for Predicting Parameters for Ann for Bending Moment in Continuous Composite Beams Considering Concrete Cracking

UMESH PENDHARKAR[1], SANDEEP CHAUDHARY[2] AND A.K. NAGPAL[3]

[1]Research scholar, Department of Civil Engineering, IIT Delhi-110016 India
[2]Research scholar, Department of Civil Engineering, IIT Delhi-110016 India
[3]Professor, Department of Civil Engineering, IIT Delhi-110016 India
email: aknagpal@civil.iitd.ernet.in

ABSTRACT

Sensitivity studies are reported in order to identify structural parameters, which govern the redistribution of elastic moments (M^i) in steel-concrete composite continuous beams arising from concrete cracking. The parameters can be subsequently used as input parameters in a neural network model that may be developed to predict the inelastic moments (M^i) from elastic moments (M^e).

The propagation of the effect of cracking at a support, along the beam length, is studied and based on this study significant structural parameters, which determine the redistribution in elastic moments are identified. These parameters can be used for the development of the neural network model.

1. INTRODUCTION

The cracking of concrete in a continuous composite beam [Fig. 1] in the hogging moment region at the supports [Fig. 2] results in moment redistribution and thus in the development of inelastic moments. Methods are available in the literature for analysis of beams [1], which take into account this moment redistribution. These methods are based either on incremental or iterative approach. Both the approaches require a computational effort, which is many times more than that required for the elastic analysis (neglecting cracking). The neural network model would help in reducing the computational effort in predicting inelastic moments. The methodology can be easily extended to tall composite building frames where huge saving in computational effort would result.

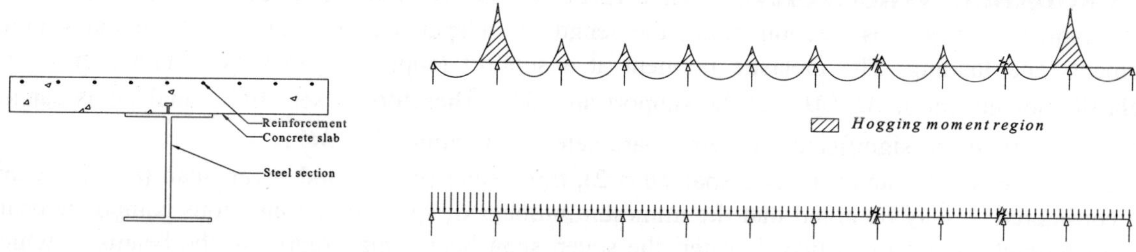

Fig. 1. Composite cross-section.　　　　**Fig. 2.** A typical multi-span continuous composite beam.

Principles and applications of neural networks in civil engineering have been summarized in the works by Flood and Kartam [2, 3] and used in many areas of structural engineering e.g., in predicting time effects in RC frames Maru, *et al.* [4].

In this paper, sensitivity analysis for identifying the important input parameters and training data sets required for a neural network model to estimate the inelastic moments, M^i (considering the cracking of concrete) from the elastic moments, M^e (neglecting the cracking of concrete) has been presented. The elastic moments can be obtained from any of the readily available softwares. This identification would lead towards development of an efficient neural network model.

2. PROPAGATION OF CRACKING EFFECT

For this purpose, first, a preliminary numerical study is carried out to estimate the significant extent of propagation of the effect of cracking at a support. Two cases are considered; case-I, cracking occurs over a penultimate support and case- II, cracking occurs over an interior support [Fig. 2]. The normalized variation η, in an elastic moment at a support, $[\eta = (M^e - M^i)/M^e]$, is chosen as a measure of the significant extent of propagation of the effect of cracking. A smaller value of η at a support would indicate that the effect of cracking is minimal and vice versa. Cracking at a support j would occur when the elastic moment M^e_j is greater than the cracking moment M^{er}. From the studies (not reported), typically, the variation of η at different supports with M^{cr}/M^e_2 for case-I and case-II, is shown in Fig. 3. It is seen from the figure that the significant effect of cracking is confined to the cracked support and the adjacent support only. This observation is made use of in identifying the significant structural parameters for the neural network model.

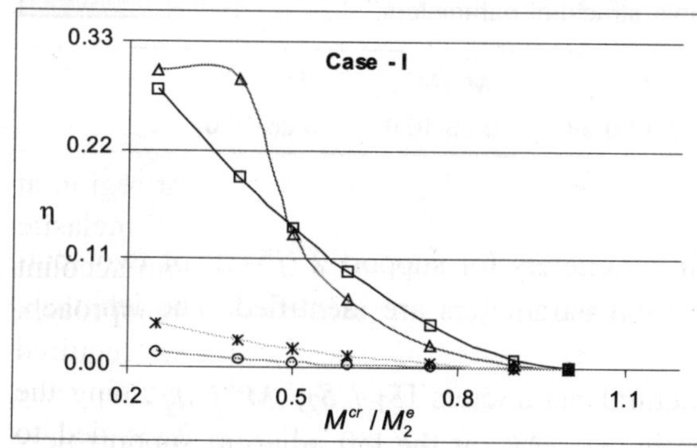

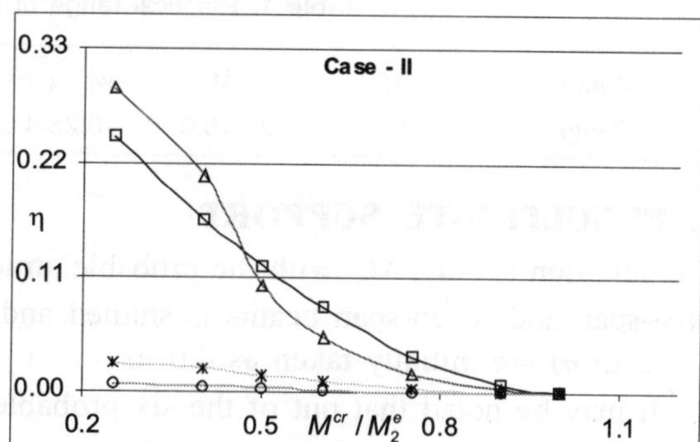

Fig. 3. Variation of η with M^{cr}/M^e_2 for Case I and Case II.

3. PROBABLE STRUCTURAL PARAMETERS AND SENSITIVITY ANALYSIS

The moment at any cross-section along the length of a span can be obtained from the support moments and loading. The inelastic moment at a support (support j) may be obtained from the inelastic moment ratio, M_j^i/M_j^e, at the support and M_j^e. Therefore, a sensitivity analysis is carried out to determine the significant structural parameters governing M_j^i/M_j^e.

For the sensitivity analysis, two span ($n = 2$), three span ($n = 3$) and seven span ($n = 7$) beams are considered. It may be noted that the minimum number of spans in a continuous composite beam in which cracking occurs is two. Further, the seven span beams may represent the beams in which number of spans is much greater than three. Therefore, these sets of beams may be considered to represent continuous composite beams with any number of spans.

As has been observed in the previous section, redistribution of moment at a support (support j) is significantly affected by the cracking at the support and adjacent supports (support $j - 1$ and support $j + 1$). Accordingly, the probable structural parameters, which may influence M_j^i/M_j^e at a support j of a beam, with the same cross-section throughout the length, are listed below:

1. Stiffness ratio of adjacent spans, S_{j-1}/S_j,
2. Cracking moment ratio at the support, M^{cr}/M_j^e.
3. Load ratio of the adjacent spans, w_{j-1} / w_j.
4. Composite inertia ratio, I^S / I^{CO}, where I^S = transformed moment of inertia of steel section.
5. Cracking moment ratio at left adjacent support, M^{cr}/M_{j-1}^e.
6. Cracking moment ratio at right adjacent support, M^{cr}/M_{j+1}^e.

The practical range of probable structural parameters is given in Table 1. Sensitive studies are reported for the left penultimate support (support 2) of two span, three span and seven span beams and for a typical interior support (support 5) of a seven span beam. In these studies, only one parameter is varied at a time keeping the other parameters constant (either the lowest value or the highest value in the practical range). It may be noted that the highest value (10.0) of M^{cr} / M^e at supports $j, j - 1, j + 1$ in the practical range denotes the case where the moment at a support is very small and has been incorporated to simulate an end support (zero moment). It may be further noted that since the cross-section of beam is the same throughout the length, therefore the required stiffness ratios S_{j-1}/S_j for the studies are achieved by varying the length of spans.

Table 1. Practical range of probable structural parameters.

Parameter	S_{j-1}/S_j	M^{cr}/M_j^e	w_{j-1}/w_j	I^S / I^{co}	M^{cr}/M_{j-1}^e	M^{cr}/M_{j+1}^e
Range	0.25-4.0	0.25-10.0	0.25-4.0	0.18-0.28	0.25-10.0	0.25-10.0

4. PENULTIMATE SUPPORT

The variation of M_j^i/M_j^e with the probable structural parameters for support 2 ($j = 2$) of two-span, three-span and seven-span beams is studied and structural parameters are identified. The lengths l_j ($j = 2$ to n) are initially taken as 4.0 m.

It may be noted that out of the six probable structural parameters [S_1 / S_2, M^{cr}/M_2^e, w_1 / w_2, I^S / I^{co}, M^{cr}/M_1^e, M^{cr}/M_3^e], the cracking moment ratio M^{cr}/M_1^e at the left adjacent support need not be considered for two-span, three-span and seven-span beams as the moment is zero at support 1. Similarly, cracking moment ratio M^{cr}/M_3^e at the right adjacent support need not be considered for two span beams, as the moment at support 3 is zero.

Effect of Stiffness Ratio (S_1 / S_2): The redistribution of elastic moment M_2^e at the support (support 2) may vary with the relative stiffness of the spans; hence stiffness ratio S_1 / S_2 has been considered a probable structural parameter. Two sets of the extreme values (Maximum and Minimum) of the other structural parameters have been considered. It may be noted that $M^{cr}/M_2^e = 0.25$ in represents the maximum cracking at the support, whereas $M^{cr}/M_2^e = 10.0$ represents no cracking at the support. The variations for the above sets are shown in Fig. 5. As expected, there is no variation of M_2^i/M_2^e with S_1 / S_2 for the set having $M^{cr}/M_2^e = 10.0$ (no cracking), however, there is significant variation for other set. Therefore, S_1 / S_2 is considered as a parameter. Since, the nature of variations of M_2^i/M_2^e with S_1 / S_2 for two span, three span and seven span beams is the same, therefore, the variations for all the beams can be represented fairly accurately by considering five values (0.25, 0.5, 1.0, 2.0 and 4.0) of $S1$ / S_2, designated as sampling points.

Effect of cracking moment ratio $\left(M^{cr}/M_2^e\right)$: Since redistribution of M^e at a support may vary with ratio of M^{cr} to M^e at the support, M^{cr}/M_2^e is selected as a probable structural parameter. The variations of M_2^i/M_2^e with M^{cr}/M_2^e for the two sets is shown in Fig. 5. The variation is significant and five sampling points (M^{cr}/M_2^e = 0.25, 0.75, 0.85, 1.0, 10.0) are considered to be sufficient to represent fairly accurately the variations for all the beams.

Effect of load ratio (w_1/ w_2): Ratio of loading on the adjacent spans of a support may affect the redistribution of M^e, therefore, w_1 / w_2 is also considered as a probable structural parameter. As the variation for the three beams for the two sets is mild, as shown in Fig. 6, two sampling points (w_1/ w_2 = 0.25, 4.0) are sufficient to represent the variations for all the beams.

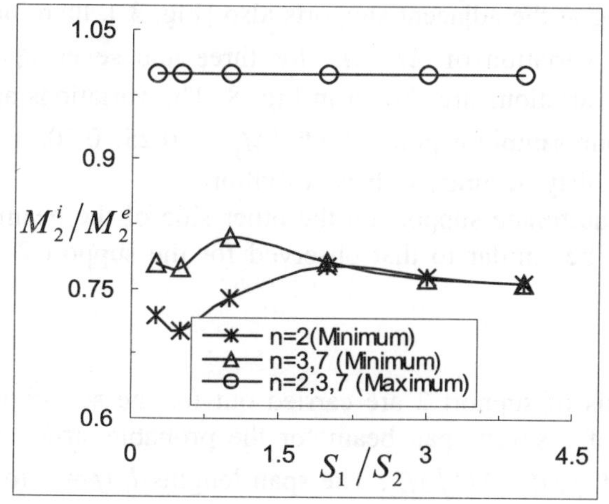

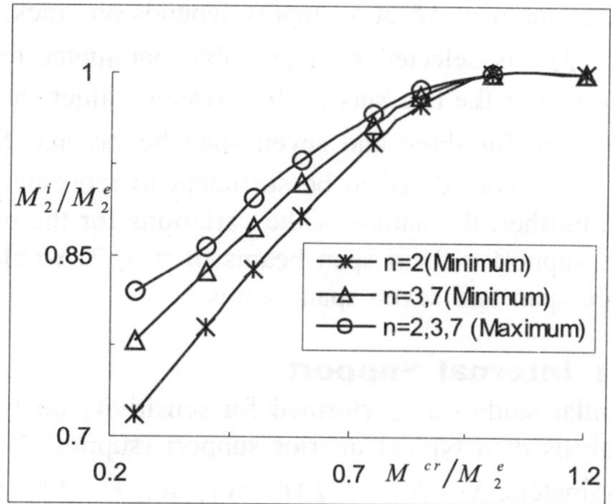

Fig. 4. Variation of M_2^i/M_2^e with S_1 / S_2 for 2, 3 and 7 span beams.

Fig. 5. Variation of M_2^i/M_2^e with M^{cr}/M_2^e for 2, 3 and 7 span beams.

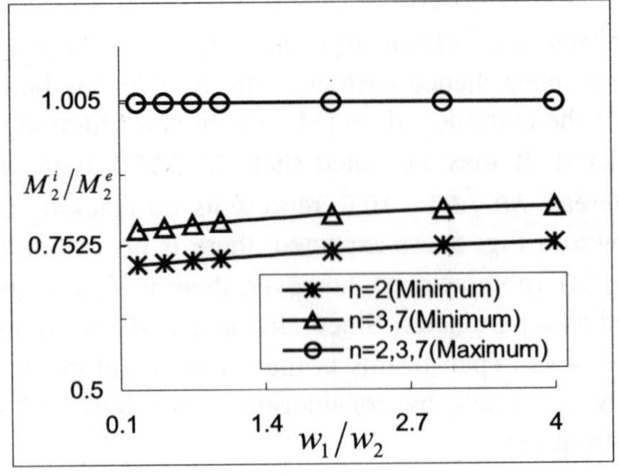

Fig. 6. Variation of M_2^i/M_2^e with w_1 / w_2 for 2, 3 and 7 span beams.

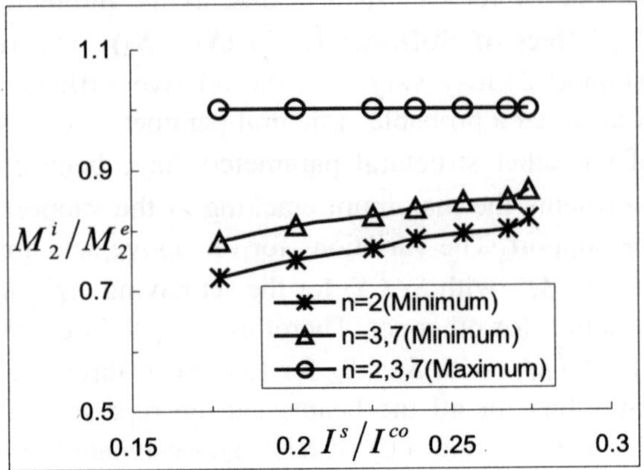

Fig. 7. Variation of M_2^i/M_2^e with I^S / I^{co} for 2, 3 and 7 span beams.

Effect of composite inertia ratio (I^S / I^{co}): The stiffness of a composite section, I^{co}, reduces to that of steel section, I^S on cracking, which leads to redistribution of forces. Therefore, I^S / I^{co} is considered as a probable structural parameter. The variations are shown in Fig. 7. Again, no variations are observed for one set due to the absence of the cracking whereas significant variations are observed for other set (Minimum). The variations for all the beams can be represented fairly accurately by four sampling points (I^S / I^{co} = 0.18, 0.21, 0.25, 0.28).

Effect of cracking moment ratio at right adjacent support $\left(M^{cr}/M_3^e\right)$: The redistribution of elastic moment M^e at a support depends on cracking at the adjacent supports also [Fig. 3], therefore M^{cr}/M_3^e is selected as a probable parameter for variation of M_2^i/M_2^e for three and seven span beams. For the two sets of the extreme values the variations are shown in Fig. 8. The variations are the same for three and seven span beams and four sampling points (M^{cr}/M_3^e = 0.25, 0.50, 1.0, 10.0) are considered to be sufficient to represent fairly accurately these variations.

Further, the nature of the variations for the penultimate support on the other side of the beams, i.e., support n for n span beams (n = 3, 7) would be similar to that observed for the support 2 of three span and seven span beams.

4.1 Internal Support

Similar studies as performed for sensitivity analysis of support 2 are carried out for the sensitivity analysis of a typical interior support (support 4) of a seven span beam for the probable structural parameters, S_3 / S_4, M^{cr}/M_4^e, w_3 / w_4, I^S / I^{co}, M^{cr}/M_3^e, M^{cr}/M_5^e. The span lengths l_j (j = 1 to 3 and 5 to 6) are initially taken equal to 4.0 m. The nature of corresponding variations for internal support is found to be similar as shown in Fig. 4-8 for penultimate support and hence same sampling points can be taken for internal support. It may be further noted that the variation of M_4^i/M_4^e with M^{cr}/M_5^e is the same as the variation of M_4^i/M_4^e with M^{cr}/M_3^e.

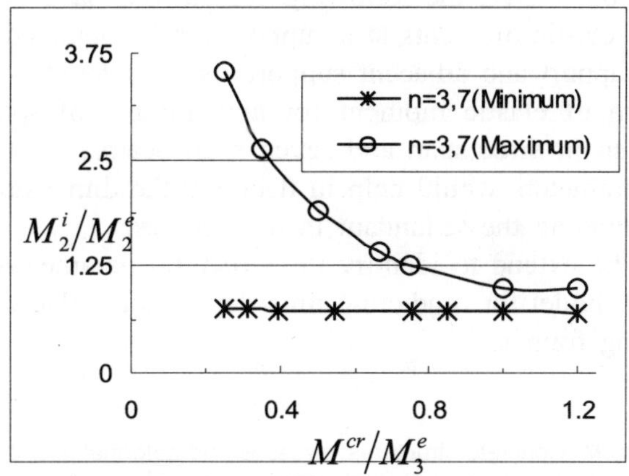

Fig. 8. Variation of M_2^i/M_2^e with M^{cr}/M_3^e for 3 and 7 span beams.

Similar studies for each of the above five parameters for both the supports, are carried out for l_j (j = 2 to n) = 8.0 m and 12.0 m also. Exactly the same variations as observed for l_j (j = 2 to n) = 4.0 are obtained. Therefore, the absolute span length is not a parameter governing the redistribution of moments at penultimate support.

4.2 Sensitivity Analysis Summary

It is observed from the sensitivity studies carried out for a penultimate support and an internal support that all the variations for three span and seven span beams overlap each other. It therefore may be inferred that a seven span beam may represent all the beams having more than two spans. It has also been inferred above that M_j^i/M_j^e for a support is independent of the absolute span lengths.

Further, it is observed that the nature of variations for internal support, is similar to the corresponding variations for penultimate support, shown in Figs. 5-9, though the numerical values are different. Therefore, the same structural parameters and sampling points are applicable for internal supports and penultimate supports.

Considering the above findings, the structural parameters (along with the sampling points) governing the variation of M_j^i/M_j^e at support j (penultimate or internal) for a beam of any number of spans, may be summarized as: S_{j-1} / S_j {0.25, 0.5, 1.0, 2.0, 4.0}, M^{cr}/M_j^e {0.25, 0.75, 0.85, 1.0, 10.0}, w_{j-1} / w_j {0.25, 4.0}, I^S / I^{co} {0.18, 0.21, 0.25, 0.28}, M^{cr}/M_{j-1}^e {0.25, 0.5, 1.0, 10.0}, M^{cr}/M_{j+1}^e {0.25, 0.5, 1.0, 10.0}.

Artificial Neural networks are being developed using the structural parameters identified above as input parameters and inelastic moment ratio as output parameter. The sampling points would be used in deciding and generating the training data sets to be used for training and testing of Artificial Neural Network Models.

CONCLUSION

Using the analysis carried out above, the following conclusions have been drawn:

1. Redistribution of elastic moments at a support can be obtained with sufficient accuracy if cracking at the support and adjacent supports is considered.
2. The redistribution of elastic moment for any number of spans can be represented by considering two span, three span and seven span beams.
3. The structural parameters would help in deciding the dimensionality of the networks and will help in eliminating the redundant in the networks.

The methodology can be extend to identify the structural parameters for continuous composite beams to develop an ANN model for predicting time dependent inelastic moment in the beams and subsequently in the building frames.

REFERENCES

1. Ghali A, Favre R, Elbadry M. Concrete structures: stresses and deformations. 3rd ed. London:Spon Press; 2002.
2. Flood I, Kartam N. Neural Networks in Civil Engineering I: Principles and understanding. Journal of Computing in Civil Engineering, ASCE 1994;8(2):131-148.
3. Flood I, Kartam N. Neural Networks in Civil Engineering II: Systems and application. Journal of Computing in Civil Engineering, ASCE 1994;8(2):149-162.
4. Maru S, Nagpal AK. Neural network for creep and shrinkage deflections in reinforced concrete frames. Journal of Computing in Civil Engineering, ASCE 2004;18(4):350-359.

16

Modeling Metal Matrix Composites for Multiple Inclusions Using Eshelby Approach

R.T. Durai Prabhakaran[1], R. Prasad[2] AND P. Mahajan[3]

[1]Department of Mechanical Engineering, Birla Institute of Technology & Science-Pilani,
Goa Campus, Zuari Nagar, Goa-403726 India email: durai_p@yahoo.com
[2,3]Department of Applied Mechanics, Indian Institute of technology, Hauz Khas, New Delhi-110001 India
email: rajesh.prasad.iitd@gmail.com, mahajan@am.iitd.ernet.in

ABSTRACT

The present paper introduces the inclusion problem and modeling of metal matrix composites for isotropic, inhomogeneous and multiple inclusions using Eshelby's approach. The article also includes different models for estimating the partitioning of loads between the constituents of composites subjected to external loads. The various models like Tucker and Liang, Dilute Eshelby, Mori-Tanaka, Chow's, Lielens, Self-consistent, Clyne and Withers, and Weng's bounding models used for calculation of composite stiffness tensor, elastic constants, anisotropic ratio. For all these theoretical models, we showed the variation of modulus of elasticity for Al-SiC whiskers as a function of volume fraction with fixed aspect ratio. The highlights of this present article are identification of some mistake in the standard expressions given in the available literature.

Keywords: Metal matrix composites, Eshelby's approach, Isotropic inclusions.

1. INTRODUCTION

Metal matrix composites gaining its importance in various engineering application like aircraft structures, aerospace structures, automotive components etc. Modeling a metal matrix composite for determining various mechanical properties like internal stresses in loaded composites, internal stresses with Eigenstrain due to external load, internal stresses due to thermal misfit, calculation of stiffness tensor, elastic constants and anisotropic ratio using micromechanics concept is mostly required for analyzing a composite structures under given loading conditions. The present paper introduces the inclusion problem and the Eshelby's method [1-2] for isotropic inclusions as well as multiple inclusions. The inclusion problem is a well-known problem of elasticity having wide ranging applications in various fields dealing with micromechanics of solids such as solid-solid phase transformations, dislocations, fracture mechanics, composite materials, etc. As inclusion is a region in elastic medium which disturbs the uniformity of the medium and thereby the elastic field, because it has either changed its form or has elastic constants different from those of the remainder. The inclusion problem involves finding the elastic field inside and outside the inclusion. Many researchers extended Eshelby's approach for analyzing microelongated elastic fields [3] and anisotropic damage mechanics [4] etc. In this present work we used Eshelby method for isotropic [5] and multiple inclusions. The method is illustrated by simple calculations of stress and strain fields both inside and around a single ellipsoidal inclusion in an infinite isotropic matrix.

The mechanical behavior of composites depends on its mechanical properties. The stresses developed inside the inclusions as well as matrix are not uniform. Most of the predictions we made on the basis of Eshelby's approach. Other than this we used models like Weng's bounding model, Dilute Eshelby model, Mori-Tanaka model, Lielens model, and Chow's model to predict the stiffness and strain concentration tensors.

2. THE ESHELBY THEORY

Eshelby first posed and solved a homogeneous inclusion by considering an infinite solid body with stiffness and initially stress-free condition. The concept of Eigenstrain introduced by Eshelby, where strain might be acquired through a phase transformation, or by a combination of a temperature change and a different thermal expansion coefficient in the inclusion. With in a matrix the stress σ^m is simply the stiffness times the strain.

$$\sigma^m(x) = C^m \, \varepsilon^C(x) \qquad \qquad ...(1)$$

but within the inclusion the transformation strain does not contribute to the stress, so the inclusion stress is

$$\sigma = C^m \, (\varepsilon^C - \varepsilon^t) \qquad \qquad ...(2)$$

Within an ellipsoidal inclusion the strain ε^C is uniform and related to Eigenstrain by

$$\varepsilon^C = S \, \varepsilon^t \qquad \qquad ...(3)$$

Where S denotes Eshelby's tensor of fourth rank tensor enables its reduction to a 6×6 matrix.

2.1 Eshelby Approach for Inhomogeneous Inclusions

If the elastic modulus of both matrix and inclusion are different then the problem is called inhomogenity problem. Eshelby developed a method called equivalent inclusion method to solve inhomogenity problem. In this method there is a hypothetical inclusion made up of matrix material, also called equivalent inclusion, which replaces the inhomogeneous inclusion with proper equivalent transformation strain, ε^t to get the constrained strain in the inclusion and in the matrix material as in the actual inhomogeneous situation. As shown above the elastically equivalent homogeneous problem has the solution. So by calculating the equivalent stress-free strain ε^t required to imitate the constrained stress state in the inhomogeneity, we have also solved the inhomogeneous problem. The stress in the inhomogeneous inclusion is given in terms of the elastic strain by

$$\sigma = C_I \, (\varepsilon^C - \varepsilon^{t*}) \qquad \qquad ...(4)$$

The modulus of homogeneous inclusion is equal to the modulus of the matrix. The constrained strain for the elastically homogeneous problem is given in terms of the equivalent stress-free transformation strain (ε^t) by equation 3. Since the stress in the equivalent homogeneous inclusion should be equal to the stress in the actual inhomogeneous inclusion we have,

$$C_I (\varepsilon^c - \varepsilon^{t*}) = C_M (\varepsilon^c - \varepsilon^t) \quad \text{...(5)}$$

Substituting equation 3 in equation 5 and solving for ε^t, we get

$$\varepsilon^t = [(C_I - C_M) S + C_M]^{-1} C_I \varepsilon^{t*} \quad \text{...(6)}$$

The equivalent homogeneous transformation strain ε^t can therefore be expressed in terms of the stress-free transformation strain of the inhomogeneous inclusion ε^{t*} now the stress in the inclusion can then be calculated by substituting ε^t value i.e equation 6 in equation 2 we get

$$\sigma = C_M (S - I) [(C_I - C_M) S + C_M]^{-1} C_I \varepsilon^{t*} \quad \text{...(7)}$$

Thus by calculating the stress-free strain of the equivalent inclusion, the stress and strains of the inhomogeneous inclusion have been calculated. In this way the equivalent inclusion concept extends the scope of the Eshelby method from the elastically homogeneous case to "Composites" involving phases of different stiffness. The model further extended to

- Calculate inclusion stress in a loaded composite as

$$\sigma_I = C_M (S-I)[(C_M - C_I)S - C_M]^{-1} (C_I - C_M)\varepsilon^A + C_M\varepsilon^A \quad \text{...(8)}$$

- Calculate internal stresses with Eigenstrain due to external load

$$\varepsilon^t = [C_M (S - I) - C_I S]^{-1}[(C_I - C_M)\varepsilon^A - C_I\varepsilon^{t*}] \quad \text{...(9)}$$

- Calculate internal stresses due to thermal misfit, the residual stresses within a single reinforcing inclusion is given by

$$\sigma_{residual} = C_M (S - I) [(C_I - C_M) S + C_M]^{-1} C_I (\alpha_I - \alpha_M)\Delta T \quad \text{...(10)}$$

2.2 Eshelby Approach for Multiple Inclusions

In the previous section we considered Eshelby's approach for single inclusion, the extended Eshelby approach for multiple inclusions is as follows

Let the volume fraction of inclusions be f and that of matrix be $(1-f)$. Then as shown by Mura (1987)

$$(1 - f) <\sigma>_M + f <\sigma>_I = 0 \quad \text{...(11)}$$

where $<\sigma>_M$ is average stresses in matrix minus applied stress,

$<\sigma>_I$ is average stresses in inclusion minus applied stress.

The above equation relates the volume-averaged internal matrix stress to the inclusion stress which can be calculated without requiring details of the form of the matrix stress field itself. The average stresses in the matrix $(\bar{\sigma}_M)$ and inclusions $(\bar{\sigma}_I)$ are $(\sigma^A + <\sigma>_M)$ and $(\sigma^A + <\sigma>_I)$ respectively. The interrelation between the inclusion and the mean matrix stress is straight forward, the mean matrix stress further strains the equivalent homogeneous inclusions, so that it is strained by $\varepsilon^c + \varepsilon^A + <\varepsilon>_M$. The general equation for the stress with in the inhomogeneous inclusion is given by

$$\sigma^A + <\sigma> I = C_I (\varepsilon^c + \varepsilon^A + <\varepsilon>_M - \varepsilon^{t*}) \text{ real inclusion} \quad \text{...(12)}$$
$$= C_M (\varepsilon^c + \varepsilon^A + <\varepsilon>_M - \varepsilon^t) \text{ equivalent inclusion} \quad \text{...(13)}$$

where $\varepsilon^c = S\varepsilon^t$; $\qquad \sigma^A = C_M\varepsilon^A$; $<\sigma>_M = C_M <\varepsilon>_M$;
$$<\varepsilon>_M = - f(\varepsilon^c - \varepsilon^t) = -f(S - I) \varepsilon^t \quad \text{...(14)}$$

After solving for Eigenstrain expression, by equating eqns. (12) and (13), and substituting ε^t in eqns. (14), we get

$$\varepsilon^t = [(C_I - C_M) [S - f(S-I)] + C_M]^{-1} [(C_M - C_I) \varepsilon^A + C_I \varepsilon^{t*}] \quad \text{...(15)}$$

3. DIFFERENT MODELS FOR PREDICTION OF MECHANICAL PROPERTIES TO METAL MATRIX COMPOSITES

This section introduces different models to predict stiffness tensor, elastic constants and anisotropic ratio.

The general expression to calculate composite stiffness tensor [7]

$$C_C = C_M + v_f (C_I - C_M)A \qquad \ldots(16)$$

- According to Tucker and Liang [7], the strain concentration tensor is defined as follows

$$A = \hat{A} [(1-v_f) I + v_f \ \hat{A}]^{-1} \qquad \ldots(17)$$

- According to Dilute Eshelby model, the strain concentration tensor is defined as follows

$$A^{Eshelby} = [I + S (C_M^{-1}) (C_I - C_M)]^{-1} \qquad \ldots(18)$$

- According to Mori – Tanaka model, the strain concentration tensor is defined as follows

$$A^{MT} = A^{Eshelby}[(1-v_f) I + v_f A^{Eshelby}]^{-1} \qquad \ldots(19)$$

- According to Chow's model, the strain concentration tensor is defined as follows

$$A^{Chow} = [I + (1 - v_f) S (C_M^{-1})(C_I - C_M)]^{-1} \qquad \ldots(20)$$

- According to weng's bounding models and Lielens model

When the matrix is chosen as the reference material, then the lower bound to strain concentration tensor is defined as follows

$$\hat{A}^{Lower} = [I + S_M (C_M^{-1})(C_I - C_M)]^{-1} \qquad \ldots(21)$$

The other bound, with the fiber as the reference material, has a strain concentration tensor is defined as

$$\hat{A}^{Upper} = [I + S_I (C_I^{-1})(C_M - C_I)] \qquad \ldots(22)$$

- Lielens and co-workers propose a model that interpolates between the upper and lower bounds, such that the lower bound dominates at low volume fractions and the upper bound dominates at high volume fractions. The expression to calculate the inverse of the strain-concentration tensor $\hat{A}$, is

$$\hat{A}^{Lielens} = \{(1-f)[\hat{A}^{Lower}]^{-1} + f [\hat{A}^{Upper}]^{-1}\}^{-1} \qquad \ldots(23)$$

The interpolating factor depends on fiber volume fraction, and they propose

$$f = \frac{\left(v_f + v_f^2\right)}{2} \qquad \ldots(24)$$

- According to self-consistent model, the strain-concentration tensor is defined as

$$A^{sc} = [I + S (C^{-1})(C_I - C)]^{-1} \qquad \ldots(25)$$

The properties C and C-1 of the embedding 'matrix' are initially unknown.

- According to Clyne and Withers model, the composite stiffness tensor is defined as follows

$$C_c = [\{C_M^{-1}\sigma^A - f[(C_I - C_M) [S - f(S-I)] + C_M]^{-1}(C_M - C_I) C_M^{-1}\sigma^A\}]^{-1} \qquad \ldots(26)$$

The engineering constants of the composite can be derived from this tensor.

The axial young's modulus is given by

$$E_{3c} = \frac{1}{C_{3c}} = \frac{\sigma\sigma^A}{(\varepsilon_3^A + f\varepsilon_3^T)} \qquad ...(27)$$

After finding composite stiffness tensor, to calculate elastic constants we use the following relationship

$$\text{Axial young's modulus } E_{3c} = \frac{1}{C_{3c}^{-1}} \qquad ...(28)$$

$$\text{Transverse young's modulus } E^{1c} = \frac{1}{C_{1c}^{-1}} \qquad ...(29)$$

$$\text{Let us define as anisotropic ratio } r = \frac{E_{1c}}{E_{3c}} = \frac{C_{3c}^{-1}}{C_{1c}^{-1}} \qquad ...(30)$$

4. RESULTS AND DISCUSSIONS

In the present work, we calculated the stresses and strain fields inside a single as well as for multiple inclusions in an infinite isotropic matrix, for both homogeneous and in-homogeneous problems. We identified some mistakes given in Clyne and Withers [8] like $S_{1111} = S_{222}$ component in basic Eshelby tensor and sign mistake in two expressions, i.e. the expressions given for the calculation of G components in Kinoshita and Mura [9] to calculate Eshelby tensor [10].

We developed software coding using "MATLAB" and "C" programming languages to implement the Eshelby method and also for the extended Eshelby approach. These programs have been used to do calculations of elastic fields in situations such as inside a homogeneous or inhomogeneous inclusion for a given stress-free strain, inside an inhomogeneous inclusion with an applied load. The variation of Eaxial and $E_{transverse}$ of Al-SiC with aspect ratio 5 as a function of volume fraction of SiC for Clyne and Withers model as shown in Figure 1. These programs also include the calculation of strain concentrator tensor from that composite stiffness tensor to the Dilute Eshelby model, Mori-Tanaka model, Weng's bounding models, Lielens model and Chow's model. The Figure 2 gives the E_{axial} and $E_{transverse}$ of Al-SiC with aspect ratio 5 as a function of volume fraction of SiC for the models considered, and observed that Mori-Tanaka, Weng's lower bound, Chow's model are satisfying the same result. The extensions of Eshelby's method to systems containing many inclusions have also been used to compute elastic properties of composite materials according to Clyne and Withers. We developed another program to calculate a composite stiffness tensor for self-consistent model.

The anisotropic ratio defined as given in Eq. (30), is used to check the variation of fiber volume fraction versus anisotropic ratio of Al-SiC using different model like Dilute Eshelby model, Mori-Tanaka model, Weng's bounding model, Lielens model as shown in Figure 3. The Figure 4 shows the variation of volume fraction at which maximum anisotropic ratio versus aspect ratio for Al-15vol%SiC composite.

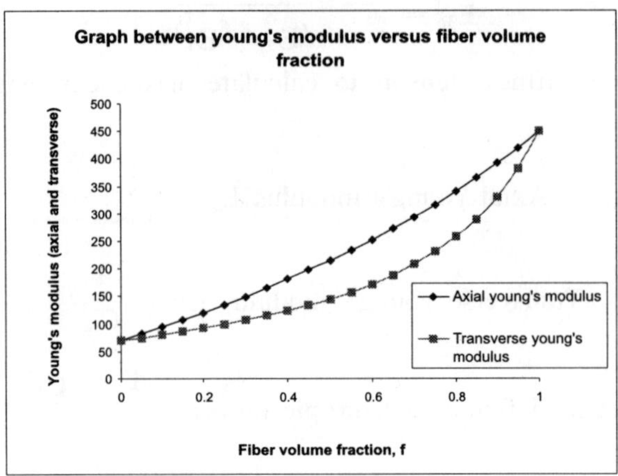

Fig. 1. Axial and transverse young's modulus of Al-SiC with aspect ratio 5 as a function of volume fraction of SiC according to Clyne and Withers.

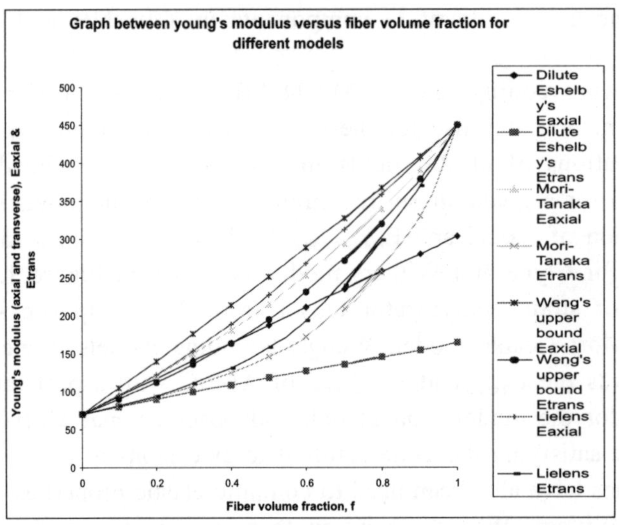

Fig. 2. The variation of Eaxial and Etransverse of Al-SiC with aspect ratio 5 as a function of volume fraction of SiC for 1. Dilute Eshelby model 2. Mori-Tanaka model 3. Weng's lower bound model 4. Weng's upper bound model 5. Lielens model 6. Chow's model. (But we observed Mori-Tanaka model, Weng's lower bound model, Chow's model are same).

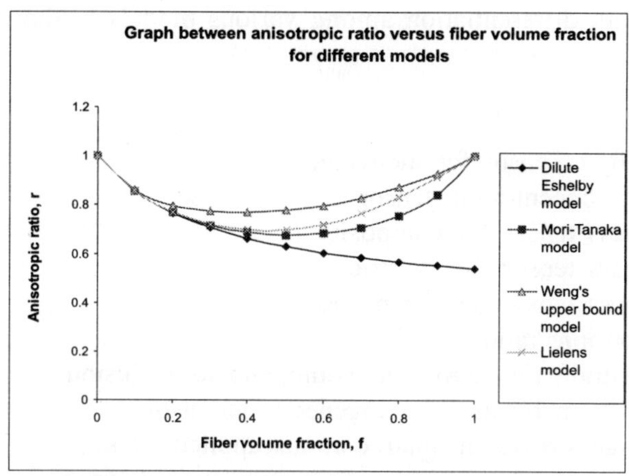

Fig. 3. The variation of anisotropic ratio versus fiber volume fraction for different models like Dilute Eshelby model, Mori-Tanaka model, Weng's bounding models, Lielens model

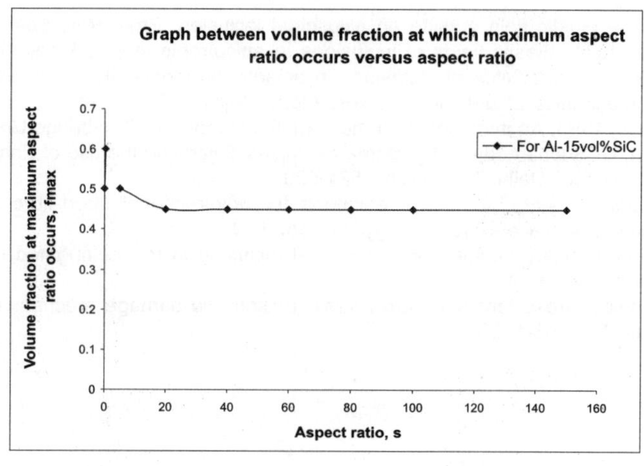

Fig. 4. Variation of volume fraction at which maximum anisotropic ratio versus aspect ratio for Al-SiC.

CONCLUSION

A mathematical model for analyzing metal matrix composites under loading conditions has been presented using Eshelby's approach. Using the tensor notation, the mean fibre and matrix stresses, which dominate an MMC's response to load or heating, are expressed concisely as a linear function of applied load and fibre/matrix misfit. In this format, the model has been successfully applied to the modeling of composite young's modulus, residual stress and the transfer of load upon elastic and plastic deformation. The numerical results discussed for E_{axial} and $E_{transverse}$ by varying fibre volume fraction of SiC whiskers with constant aspect ratio is in good agreement with the available results in the literature. A clear cut differentiation among various models is shown for Al-SiC composite system.

NOMENCLATURE

[S]	Eshelby's Tensor for inclusion
[A]	Strain concentration tensor
C_I	Stiffness tensor for composite
C_M	Stiffness tensor for inclusion
C_c	Stiffness tensor for composite
r	Anisotropic ratio
ε^t	Eigenstrain for equivalent homogeneous inclusion
ε^{t*}	Eigenstrain for an inhomogenous inclusion
$<\sigma>_M$	Average stresses in matrix minus applied stress
$<\sigma>_I$	Average stresses in inclusion minus applied stress
f	Fibre volume fraction

REFERENCES

1. J.D. Eshelby, 1957, The determination of the elastic field of an ellipsoidal inclusion and related problems, Proc. Roy. Soc. 241, 376-396.
2. J.D. Eshelby, 1959, The elastic field outside an ellipsoidal inclusion, Proc. Roy. Soc. A252, 561-569.
3. N. Kinoshita, T. Mura, 1971, Elastic fields of inclusions in anisotropic media, Phys. Stat. Sol (a). 5, 759-768.
4. S.C. Lin, T. Mura, 1973, Elastic fields of inclusions in anisotropic media (II), Phys. Stat. Sol (a). 15, 281-285.
5. T. Mura, 1987, Micromechanics of defects in solids, Martin Nijhoff: Dordrecht.
6. T.W. Clyne, P.J. Withers, 1993, An introduction to metal matrix composite, Cambridge University Press: Cambridge.
7. R. Prasad, P. Mahajan, G. Subbareddy, 1999, Some examples illustrating the use of Eshelby method, Proceedings of 11th ISME conference, IIT Delhi, New Delhi, 524-529.
8. C.L. Tucker III, E. Liang, 1999, Stiffness Predictions for unidirectional short-fibre composites: Review and Evaluation, Composites Science and Technology. 59, 655-671.
9. A. Kivis, E. Inan, 2005, Eshelby tensors for a spherical inclusion in microelongated elastic fields, International Journal of Engineering Science. 43, 49-58.
10. M. Brunig, 2004, Eshelby stress tensor in large strain anisotropic damage mechanics, International Journal of Mechanical Sciences. 46, 1763-1782.

17

Analysis of Laminates Under Cylindrical Bending Using Numerical Integration

Tarun Kant, Yogesh Desai and Sandeep Pendhari

Department of Civil Engineering Indian Institute of Technology Bombay Powai, Mumbai-400076.
email: tkant@iitb.ac.in

ABSTRACT

An effort has been made in this paper to develop a novel methodology for static analysis of laminates under cylindrical bending by using numerical integration. Present methodology is based on defining a boundary value problem (BVP) through the thickness of a laminate governed by a set of linear first-order ordinary differential equations (ODEs). Fourth-order Runga-Kutta-Gill routine is used for numerical integration. Numerical investigations are performed on various laminates with different lay-ups and results obtained through present approach are found to be in close agreement with elasticity solution, which shows the efficiency and effectiveness of the present method.

Keywords: Laminates, Boundary value problem, Composites, Cylindrical bending.

1. INTRODUCTION

The increased uses of composite materials in almost all engineering fields are seen due to their high stiffness and strength-to-weight ratios and other favorable engineering properties. Vibration, buckling and bending analyses of composites have received widespread attention in last two decades.

A number of two dimensional (2D) laminate theories, namely, classical laminate plate theory (CLPT), first and higher order shear deformation theories (FOST and HOST), are used to represent the kinematics of deformation [1, 2] and are grouped into equivalent single layer (ESL) theories. In these theories only continuity of displacement components are maintained through the thickness of laminae interface whereas transverse stresses remain discontinuous. This situation necessitated the need for layer-wise theories for analysis of composites.

In this paper, a simple and efficient analytical methodology is presented for analysis of laminates by using numerical integration. A laminate under cylindrical bending is formulated as a two-point BVP governed by a set of linear first-order ODEs,

$$\frac{d}{dz}\,\mathbf{y}(z) = \mathbf{A}(z)\,\mathbf{y}(z) + \mathbf{p}(z) \qquad \text{...(1)}$$

in the interval $-h/2 \leq z \leq h/2$, $\mathbf{y}(z)$ is an N dimensional vector of dependent variables, $A(z)$ is an $N \times N$ coefficient matrix and $\mathbf{p}(z)$ is an N dimensional vector of non-homogenous (loading) terms [3]. The boundary conditions at the two levels along the z may be written as: at $z = -h/2$ any $N/2$ elements $\mathbf{y}(z)$ of are specified and at $z = h/2$ any $N/2$ elements of $\mathbf{y}(z)$ are specified. Thus the formulation indicates a method for solving problems with mixed non-homogenous boundary conditions.

2. FORMULATION

A laminate of plan dimension $a \times b$ with thickness, h and simply supported along the edges at x = 0 and a is considered. The thickness h is composed of homogenous isotropic/orthotropic layers of uniform thicknesses. Top surface is loaded with transversely distributed load which can be expanded in the form of a Fourier series as,

$$p(x) = \sum p_{0m}\,\sin\frac{m\pi x}{a} \qquad \text{...(2)}$$

where p_{om} = peak intersity of distributed loading corresponding to m^{th} harmonic

The load distribution remains constant in y direction.

Under such condition, a laminate is in a 2D state of plane-strain in x-z plane (Fig. 1).

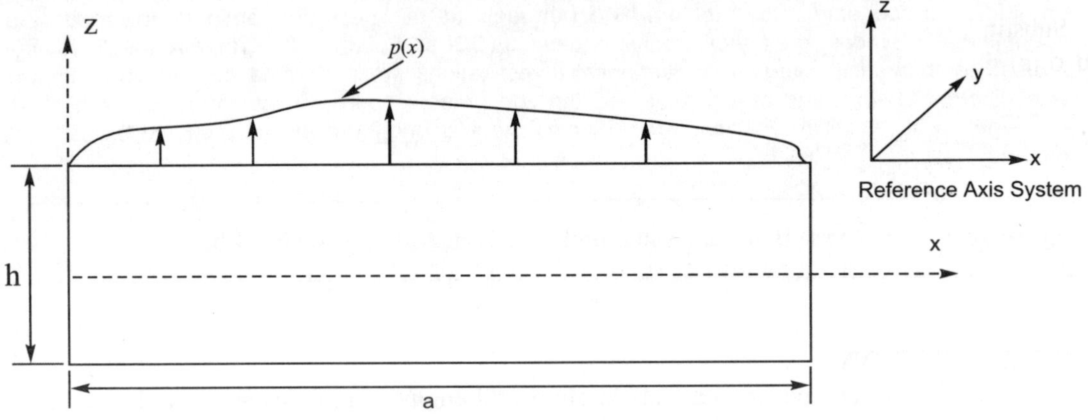

Fig. 1. Laminate in a state of plane-strain.

The 2D differential equations of equilibrium are,

$$\frac{\partial \sigma_x}{\partial x} + \frac{\partial \tau_{xz}}{\partial z} + B_x = 0$$

$$\frac{\partial \tau_{zx}}{\partial x} + \frac{\partial \sigma_z}{\partial z} + B_z = 0 \qquad \text{...(3)}$$

The material constitute relations can be written as,

$$\begin{Bmatrix} \sigma_x \\ \sigma_z \\ \tau_{xz} \end{Bmatrix} = \begin{bmatrix} C_{11} & C_{12} & 0 \\ C_{21} & C_{22} & 0 \\ 0 & 0 & C_{33} \end{bmatrix} \begin{Bmatrix} \varepsilon_x \\ \varepsilon_z \\ \gamma_{xz} \end{Bmatrix} \qquad \text{...(4)}$$

and the general linear strain-displacement relations in 2D can be written as,

$$\varepsilon_x = \frac{\partial u}{\partial x}, \qquad \varepsilon_z = \frac{\partial w}{\partial z} \quad \text{and} \quad \tau_{xz} = \frac{\partial u}{\partial z} + \frac{\partial w}{\partial x} \qquad \ldots(5)$$

The equations (3) – (5) have eight unknowns, u, w, ε_x, ε_z, γ_{xz}, σ_x, σ_z and τ_{xz}.. After simple algebraic manipulation, a system of partial differential equations (PDEs) involving only four particular dependent variables u, w, τ_{xz} and σ_z called as 'primary variables' are obtained.

$$\frac{\partial u}{\partial z} = \frac{\tau_{xz}}{C_{33}} - \frac{\partial w}{\partial x} \qquad\qquad \frac{\partial w}{\partial z} = \frac{1}{C_{22}}\left[\sigma_z - C_{21}\frac{\partial u}{\partial x}\right]$$

$$\frac{\partial \tau_{xz}}{\partial z} = \left[-C_{11} + \left(\frac{C_{12}C_{21}}{C_{22}}\right)\right]\frac{\partial^2 u}{\partial x^2} - \frac{C_{12}}{C_{22}}\frac{\partial \sigma_z}{\partial x} - B_x \qquad\qquad \frac{\partial \sigma_z}{\partial z} = -\frac{\partial \tau_{xz}}{\partial x} - B_z \qquad \ldots(6)$$

The above PDEs (Eq. 6) can be reduced to ODEs by using Fourier series expansion for primary variables. The variation along the x direction is assumed in trigonometric form such that it satisfies the simple support boundary condition at the both the ends exactly can be written as,

$$u(x,z) = \sum_m u_m(z)\cos\frac{m\pi x}{a} \qquad\qquad \tau_{xz}(x,z) = \sum_m \tau_{xzm}(z)\cos\frac{m\pi x}{a}$$

$$w(x,z) = \sum_m w_m(z)\sin\frac{m\pi x}{a} \qquad\qquad \sigma_z(x,z) = \sum_m \sigma_{zm}(z)\sin\frac{m\pi x}{a} \qquad \ldots(7)$$

Substituting equation (7) into equation (6), a set of linear first-order ODEs in $u(z)$, $w(z)$, $\tau_{xz}(z)$ and $\sigma_z(z)$ only is obtained as,

$$\frac{du_m(z)}{dz} = -w_m(z)\frac{m\pi}{a} + \frac{1}{C_{33}}\tau_{xzm}(z)$$

$$\frac{dw_m(z)}{dz} = \frac{C_{21}}{C_{22}}u_m(z)\frac{m\pi}{a} + \frac{1}{C_{22}}\sigma_{zm}(z)$$

$$\frac{d\tau_{xzm}(z)}{dz} = \left[-C_{11} + \left(\frac{C_{12}C_{21}}{C_{22}}\right)\right]\frac{m^2\pi^2}{a^2}u_m(z) - \frac{C_{12}}{C_{22}}\frac{m\pi}{a}\sigma_{zm}(z) - B_x \qquad \ldots(8)$$

$$\frac{d\sigma_{zm}(z)}{dz} = \frac{m\pi}{a}\tau_{xzm}(z) - B_z$$

Equation (8) represents a BVP with stress components known at top and bottom faces of a laminate. The availability of efficient, accurate ODE numerical integrators for initial value problem (IVP) helps in obtaining the primary variables through the thickness and then secondary variable is simply computed by substitution of the values of the primary variables into the strain-displacement and elasticity relations.

$$\sigma_x = \left[\left(\frac{C_{12}C_{21}}{C_{22}} - C_{11}\right)\sum_m u_m(z)\left(\frac{m\pi}{a}\right) + \left(\frac{C_{12}}{C_{22}}\right)\sum_m \sigma_{zm}(z)\right]\sin\frac{m\pi x}{a} \qquad \ldots(9)$$

3. NUMERICAL INVESTIGATIONS

Numerical investigations have been performed on various layered symmetric/ unsymmetric composite and sandwich laminates for validation of the presented analytical approach. In this paper, results of homogenous orthotropic and three layered symmetric composite laminates and subjected to a sinusoidal load (Equation 2) on their top surface have been presented for sake of brevity. Unidirectional graphite/epoxy composite material properties are used in all examples.

$$E_L = 25 \times 10^6 \, psi \quad ; \quad E_T = 10^6 \, psi \quad ; \quad G_{LT} = 0.5 \times 10^6 \, psi$$
$$G_{TT} = 0.2 \times 10^6 \, psi \quad ; \quad v_{LT} = v_{TT} = 0.25 \qquad \text{...(10)}$$

The dependent quantities are non-dimensionalzed in the following manner.

$$\bar{z} = \frac{z}{h} \; ; \qquad \bar{u} = \frac{E_2 u(0,z)}{h p_0} ; \qquad \bar{w} = \frac{100 E_2 h^3 w(a/2,z)}{p_0 a^4}$$
$$\bar{\sigma}_x = \frac{\sigma_x(a/2,z)}{p_0} ; \qquad \bar{\sigma}_z = \frac{\sigma_z(a/2,z)}{p_0} ; \qquad \bar{\tau}_{xz} = \frac{\tau_{xz}(0,z)}{p_0} \qquad \text{...(11)}$$

The elastic solution given by Pagano [4] is used for comparison of the results obtained through present analysis. Convergence study is performed for all examples at the initial stage to fix the number of steps required for numerical integration through the thickness. Fig. 2 shows the plot of converged solution for normalized transverse displacement and transverse shear stress with respect to number of steps for an orthotropic laminate. Based on convergence results, nearly 16 to 20 steps are used in all examples for numerical integration.

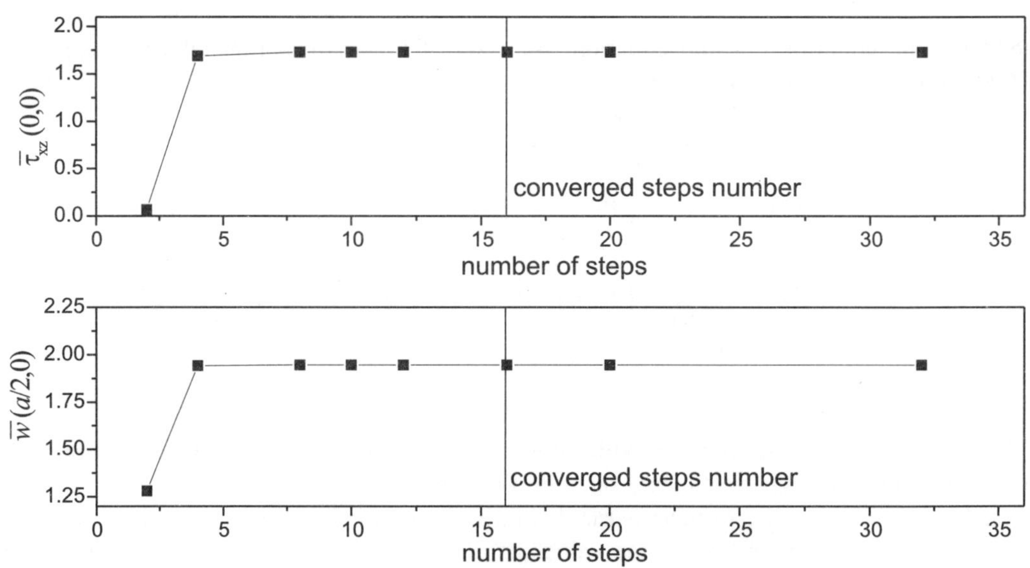

Fig. 2. Convergence study of simply supported orthotropic homogenous beam under sinusoidal loading under cylindrical bending.

The comparisons of normalized transverse displacement and stresses have been presented in Table 1 for orthotropic and layered (0°/90°/0°) beams for aspect ratios, 4 and 10. Variation of normalized inplane normal stress and transverse shear stress have been presented in Figs. 3(a) and 3(b), respectively for orthotropic beam with aspect ratio is 4, and variation of inplane displacement and transverse normal stress have been presented in Figs. 4(a) and 4(b), respectively for layered (0°/90°/0°) beam with an aspect ratio of 4.

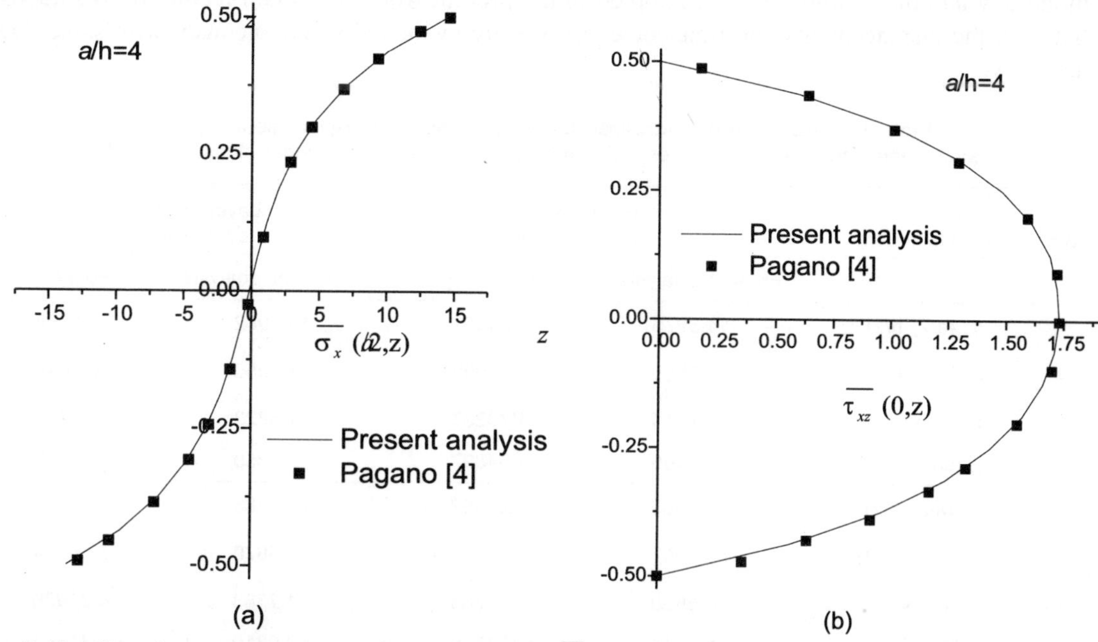

(a)　　　　　　　　　　　　　　(b)

Fig. 3. Variation of normalized (a) in-plane normal stress ($\overline{\sigma_x}$) and (b) transverse shear stress ($\overline{\tau_{xz}}$) through the thickness of simply supported orthotropic homogenous beam under sinusoidal loading under cylindrical bending.

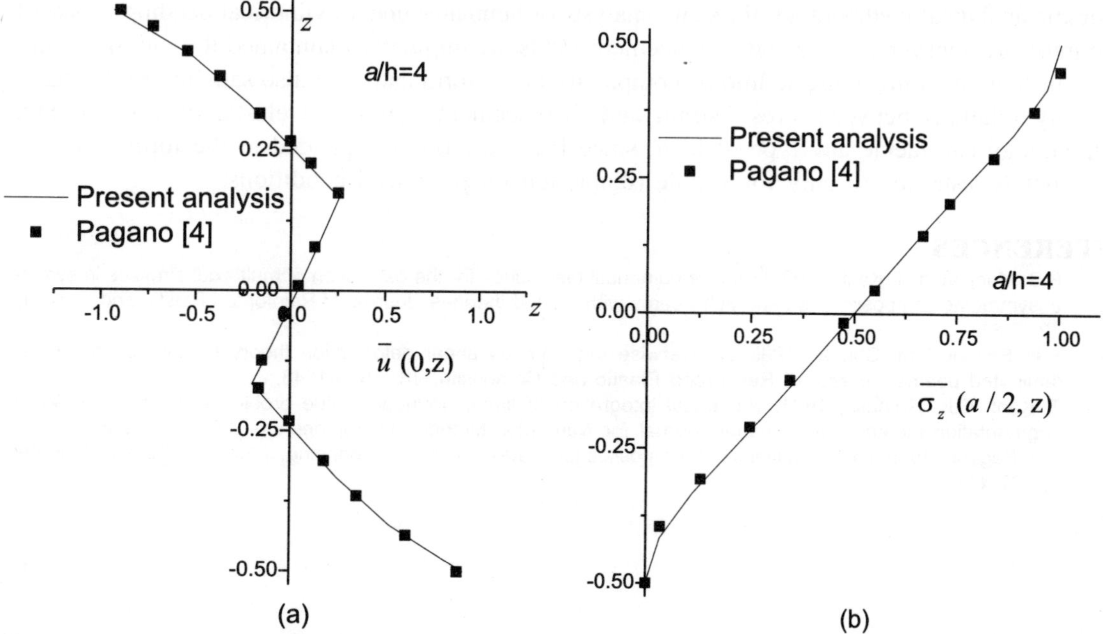

(a)　　　　　　　　　　　　　　(b)

Fig. 4. Variation of normalized (a) in-plane displacement ($\overline{u}$) and (b) transverse normal stress ($\overline{\sigma_z}$) through the thickness of simply supported layered composite beam (00/900/00) under sinusoidal loading under cylindrical bending.

From the Table and Figures, it is observed that results obtained by the present formulation show excellent agreement with the elasticity solution given by Pagano [4]. This validates the methodology and the solution scheme adopted in the present work. The main feature of this approach is that both the displacements and transverse stresses are evaluated simultaneously with same degree of accuracy.

Table 1. Comparison of normalized maximum transverse displacement, inplane stress and transverse shear stress of simply supported laminates under sinusoidal load

a/h	Stresses /displacement	Homogenous orthotropic bema		Layered (0°/90°/0°) composite beam	
		Present analysis	Pagano [4]	Present analysis	Pagan [4]
4	$\overline{\sigma}_x$ (a/2, h/2)	14.3975	14.4091	18.7962	18.808
	$\overline{\sigma}_x$ (a/2, h/2)	−13.5585	−13.5693	−18.0950	−18.1044
	$\overline{\tau}_{xz}$ (max)	1.7314	1.73237	1.5823	1.58271
	$\overline{w}$ (a/2, 0)	1.9465	1.94897	2.8850	2.88723
10	$\overline{\sigma}_x$ (a/2, h/2)	65.6859	65.692	73.6286	73.6704
	$\overline{\sigma}_x$ (a/2, -h/2)	−65.6067	−65.513	−73.5870	−73.6304
	$\overline{\tau}_{xz}$ (max)	4.6830	4.68283	4.2383	4.23928
	$\overline{w}$ (a/2, 0)	0.7314	0.731876	0.9310	0.931643

CONCLUSION

A simple analytical methodology for static analysis of laminates under cylindrical bending is described. Continuity of transverse stress and displacement fields are implicitly maintained through the thickness of laminate without involving additional complexity in the formulation. It also satisfies the fundamental elasticity relations between stress, strain and displacement within the elastic continuum. Present methodology can tackle any type of load, since loading term is expanded in the form of a Fourier series but is restricted to only for simple (diaphragm) support end conditions.

REFERENCES

1. B.S. Manjnatha, T. Kant, 1993, Different numerical techniques for the estimation of multiaxial stresses in symmetric/ unsymmetric composite and sandwich beams with refined theories, Journal of Reinforced Plastic and Composites. 12, 2-37.
2. R.P. Shimpi, Y.M. Ghugal, 1999, A layerwise trigonometry shear deformation theory for two layered cross-ply laminated beams, Journal of Reinforced Plastic and Composite. 18, 1516-1543.
3. T. Kant, C.K. Ramesh, 1981, Numerical integration of linear boundary value problems in solid mechanics by segmentation method, International Journal for Numerical Methods in Engineering, 17. 1233-1256.
4. N.J. Pagano, 1969, Exact solutions for composites laminates in cylindrical bending, Journal of Composite Materials. 3, 397-411.

18

Finite Element Bending Analysis of Composite Conoids

Hari Sadhan Das and Dipankar Chakravorty

Department of Civil Engineering, Jadavpur University, Kolkata-700 032. email: dchakravorty@vsnl.net

ABSTRACT

The conoidal shells offer a number of parallel advantages and are suitable as roofing units in many industrial applications. The advent of the laminated composites as an advanced structural material of high specific strength and stiffness has provided new impetus to the research about conoidal shells. A number of authors reported findings on conoidal shell research which are mostly on vibration aspects and on very few areas of static analysis. The present study aims at exploring the bending characteristics of composite conoids under uniformly distributed load for different laminations and practical boundary conditions. The numerical results which are obtained by employing an eight noded curved quadratic isoparametric element are studied meticulously to extract conclusions of engineering significance.

Keywords: Composites, conoids, Finite element, Force/moment resultants.

1. INTRODUCTION

Singly ruled, anticlastic, non-developable conoidal shell configurations are aesthetically appealing, structurally stiff and may be used for covering large column free spaces. Naturally, these forms received importance from the engineers and research on conoidal shells dates back to seventh decade of the twentieth centry when Hadid [1] studied conoidals both analytically and experimentally. Static aspects of simply supported compoite conoidal shells received attention from Dey *et al.* [2]. Side by side the free vibrations characteristics were looked into by Chakravorty and his colleagues [3, 4, 5] and also by Nayak and Bandopadhyay [6,7,8]. Apart from the work of Dey *et al.* [2], static bending of composite conoids has not been reported. Hence, in the present paper, the authors attempt to carry out such study for different laminations and arrangement of boundary conditions.

2. FORMULATIONS

An eight-noded curved quadratic isoparametric finite element is used to model the conoidal shell.

Five degrees of freedom taken into consideration at each node are u, v, w, α and β. The strain-displacement relations on the basis of improved first order approximation theory for thin shell are used as

$$\left\{\begin{array}{c} \varepsilon_x \\ \varepsilon_y \\ \gamma_{xy} \\ \gamma_{xz} \\ \gamma_{yz} \end{array}\right\} = \left\{\begin{array}{c} \partial u/\partial x \\ \partial v/\partial y - w/R_{yy} \\ \partial u/\partial y + \partial v/\partial x - 2w/R_{xy} \\ \alpha + \partial w/\partial x \\ \beta + \partial w/\partial y \end{array}\right\} + z\left\{\begin{array}{c} \partial \alpha/\partial x \\ \partial \beta/\partial y \\ \partial \alpha/\partial y + \partial \beta/\partial x \\ 0 \\ 0 \end{array}\right\} \qquad ...(1)$$

A laminated composite conoidal shell (Fig. 1) of uniform thickness h and radii of curvature R_{yy} and R_{xy} is considered. A given shell thickness may consist of any number of thin laminae each of which may be arbitrarily oriented at an angle q with reference to the X-axis of the coordinate system. The constitutive equations for the shell are given by

$$\{F\} = [D]\{\varepsilon\} \qquad ...(2)$$

where,
$$\{F\} = \{N_x, N_y, N_{xy}, M_x, M_y, M_{xy}, Q_x, Q_y\}^T$$

and
$$[D] = \begin{bmatrix} [A] & [B] & [0] \\ [B] & [D] & [0] \\ [0] & [0] & [S] \end{bmatrix}, \qquad ...(3)$$

$$\{\varepsilon\} = \left\{\varepsilon_x^0 \quad \varepsilon_y^0 \quad \gamma_{xy}^0 \quad k_x \quad k_y \quad k_{xy} \quad \gamma_{xz}^0 \quad \gamma_{yz}^0\right\}^T$$

The coefficients of the elasticity matrix are defined as

$$A_{ij} = \sum_{k=1}^{np}(Q_{ij})_k(z_k - z_{k-1}); \quad B_{ij} = \frac{1}{2}\sum_{k=1}^{np}(Q_{ij})_k(z_k^2 - z_{k-1}^2)$$

$$D_{ij} = \frac{1}{3}\sum_{k=1}^{np}(Q_{ij})_k(z_k^3 - z_{k-1}^3) \quad i, j = 1, 2, 6; \quad S_{ij} = \sum_{k=1}^{np}F_iF_j(G_{ij})_k(z_k - z_{k-1}) \quad i, j = 1, 2$$

where Q_{ij} are elements of the off-axis elastic constant matrix which are derived from appropriate transformation of the on-axis matrix. F_i and F_j are shear correction factors presently taken as unity. The terms of the on-axis matrix depend on the elastic moduli and Poisson's ratio of the material. The element stiffness matrix and the load vector are derived through the routine steps of finite element formulation employing numerical integration.

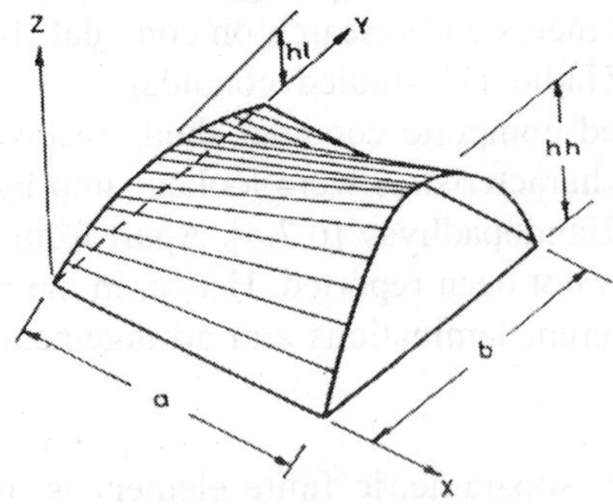

Fig. 1. Conoidal shell.

3. NUMERICAL EXAMPLE

The different stacking sequences taken are anti symmetric cross ply(0°/90°–ASCP), symmetric cross ply(0°/90°/0°–SYCP), antisymmetric angle ply(45°/–45°–ASAP) and symmetric angle ply(45°/–45°/45°–SYAP). The different boundary conditions are as follows: CCCC—all four edges clamped, SSSS—all four edges simply supported, CSCS—clamped along X = 0 and X = a and simply supported along Y = 0 and Y = b and CCSS—clamped along X = 0 and Y = 0 and simply supported along X = a and Y = b.The results of transverse deflection and different force and moment resultants are obtained but for the sake of brevity only deflection and some moment resultants are presented in the form of tables using the non-dimensional parameters $\overline{-w}$, $\overline{M_x}$, $\overline{-M_x}$, $\overline{M_y}$ and $-\overline{M}_y$. In all the cases only the converged results are presented in Tables 1 to 5. A particular shell action is taken to have converged for a particular finite element grid, if further refinement of the grid does not improve the result by more than one percent. With this criterion an 8 × 8 mesh is found to be appropriate for all the problems taken up here.

4. RESULTS AND DISCUSSIONS

An isotropic truncated conoidal shell clamped along all four edges is solved as a special case of this laminated composite formulation and this problem was earlier solved by Hadid [1]. Comparison of deflection is presented in Fig. 2.

Excellent agreement is observed.

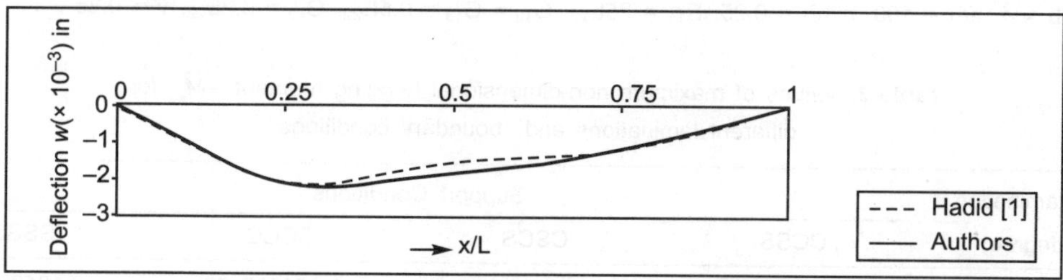

a = 95 in, b = 95 in, hh = 18.0 in, hl = 9.0 in, h = 05 in, E = 5620000 psi, υ = 0.15, q = 60 psf

Fig. 2. Deflection of isotropic conoid under uniformly distributed load along $\bar{y}$ = 0.5.

4.1 Effect of Edge Conditions on Static Deflection

Bending characteristics of shells under uniformly distributed load and SSSS, CCSS, CSCS and CCCC edge conditions are studied for four different laminates. As expected, SSSS and CCCC shells exhibit maximum and minimum static deflections respectively for all laminates. CCSS shells are stiffer than CSCS shells for SYCP(0°/90°/0°) and ASAP(45°/–45°) laminates but for ASCP(0°/90°) and SYAP(45°/–45°/45°) laminates, performance of CCSS shells are just reverse. CCSS conoidal shells are either stiffer than CSCS shells or their performances in terms of static deflections are comparable. Hence, it may be concluded that CCSS conoidal shells are better in terms of static deflection for all present combinations of laminates.

Table 1. Values of maximum non-dimensional downward deflection $\overline{w}$ for different laminations and boundary conditions.

Lamination (degree)	Support Conditions			
	CCSS	CSCS	CCCC	SSSS
0/90	−0.32777E-03	−0.32074E-03	−0.31934E-04	−0.56293E-03
0/90/0	−0.25662E-03	−0.35009E-03	−0.29839E-04	−0.47020E-03
45/−45	−0.10446E-03	−0.11184E-03	−0.72229E-04	−0.31212E-03
45/−45/45	−0.11745E-03	−0.10043E-03	−0.62884E-04	−0.24137E-03

a/b = 1, a/h = 100, hl/hh = 0.25, E11 = 25E22, G_{12} = G_{13} = 0.5E22, G_{23} = 0.2E22, ν = 0.25

Table 2. Values of maximum non-dimensional sagging moment $\overline{M_x}$ for different laminations and boundary conditions.

Lamination (degree)	Support Conditions			
	CCSS	CSCS	CCCC	SSSS
0/90	0.16749E-01	0.15814E-01	0.30043E-02	0.41572E-02
0/90/0	0.75440E-02	0.81543E-02	0.14921E-02	0.50497E-03
45/−45	0.56443E-02	0.10404E-01	0.29539E-02	0.15880E-01
45/−45/45	0.12438E-01	0.11696E-01	0.51480E-02	0.22553E-01

a/b = 1, a/h = 100, hl/hh = 0.25, E_{11} = 25E_{22}, G_{12} = G_{13} = 0.5E_{22}, G_{23} = 0.2E_{22}, ν = 0.25

Table 3. Values of maximum non-dimensional hogging moment $\overline{-M_x}$ for different laminations and boundary conditions.

Lamination (degree)	Support Conditions			
	CCSS	CSCS	CCCC	SSSS
0/90	−0.76884E-02	−0.76186E-02	−0.15110E-02	−0.10103E-01
0/90/0	−0.33187E-02	−0.35346E-02	−0.88498E-03	−0.35712E-02
45/−45	−0.64957E-02	−0.71808E-02	−0.18809E-02	−0.13948E-01
45/−45/45	−0.52448E-02	−0.66473E-02	−0.20973E-02	−0.15724E-01

a/b = 1, a/h = 100, hl/hh = 0.25, E_{11} = 25E_{22}, G_{12} = G_{13} = 0.5E_{22}, G_{23} = 0.2E_{22}, ν = 0.25

Table 4. Values of maximum non-dimensional sagging moment $\overline{-M_y}$ for different laminations and boundary conditions.

Lamination (degree)	Support Conditions			
	CCSS	CSCS	CCCC	SSSS
0/90	0.24253E-02	0.38367E-02	0.39074E-03	0.98091E-03
0/90/0	0.46238E-02	0.24200E-02	0.37459E-02	0.66384E-02
45/−45	0.43473E-02	0.77366E-02	0.24669E-02	0.15138E-01
45/−45/45	0.10137E-01	0.95336E-02	0.43128E-02	0.21583E-01

a/b = 1, a/h = 100, hl/hh = 0.25, E_{11} = 25E_{22}, G_{12} = G_{13} = 0.5E_{22}, G_{23} = 0.2E_{22}, ν = 0.25

Table 5. Values of maximum non-dimensional hogging moment $\overline{-M_y}$ for different laminations and boundary conditions.

Lamination	Support Conditions			
(degree)	CCSS	CSCS	CCCC	SSSS
0/90	−0.88843E-02	−0.86412E-02	−0.37728E-02	−0.15852E-01
0/90/0	−0.14969E-01	−0.14521E-01	−0.92396E-03	−0.24866E-01
45/-45	−0.61783E-02	−0.68288E-02	−0.18781E-02	−0.13902E-01
45/-45/45	−0.46261E-02	−0.65975E-02	−0.20299E-02	−0.15537E-01

a/b = 1, a/h = 100, hl/hh = 0.25, $E_{11} = 25E_{22}$, $G_{12} = G_{13} = 0.5E_{22}$, $G_{23} = 0.2E_{22}$, $\nu = 0.25$

4.2 Performances of Different Boundary Conditions with Respect to Different Shell Actions

Considering two and three layered cross and angle ply laminates only, corresponding to each shell action (deflection and bending moments presented in Tables 1 to 5) the best four combinations of lamination and edge condition are selected. These combinations obtained are furnished in Table 6 in ascending order of magnitude. In the present analysis upward deflection and hogging moments are considered positive. However, the values of positive (upward) deflections are left out from the tables because those are found to be negligible compared to downward deflections. This rank wise

Table 6. Shell options arranged according to ascending order of magnitudes of shell actions.

Non-dimensional shell actions	Non-dimensional location of shell actions	Materials in ascending order
$\overline{-w}$	(0.125,0.5)	CCCC/SYCP
	(0.25,0.1875)	CSCS/SYAP
	(0.25,0.1875)	CCSS/ASAP
	(0.25,0.1875)	SSSS/SYAP
$\overline{M_x}$	(0.875,0.875)	SSSS/SYCP
	(0,0.5)	CCCC/SYCP
	(1,0)	CCSS/ASAP
	(0,0.75)	CSCS/SYCP
$\overline{-M_x}$	(0.125,0.625)	CCCC/SYCP
	(0.125,0.25)	CCSS/SYCP
	(0.125,0.25)	CSCS/SYCP
	(0.125,0.8125)	SSSS/SYCP
$\overline{M_y}$	(0,0.625)	CCCC/ASCP
	(0.875,0.875)	SSSS/ASCP
	(0.875,0.875)	CSCS/SYCP
	(0,0.25)	CCSS/ASCP
$\overline{-M_y}$	(0.25,0.5)	CCCC/SYCP
	(0.125,0.125)	CCSS/SYAP
	(0.125,0.875)	CSCS/SYAP
	(0.125,0.875)	SSSS/SYAP

arrangement of the shells corresponding to the different shell actions will be helpful to a practicing engineer because, if he knows that which shell action is critical for a particular situation, he can make a choice among a number of options. It is interesting to note that the superiority of a particular combination of lamination and boundary condition in terms of deflection over another combination cannot form the basis of predicting their relative performances in terms of other shell actions.

CONCLUSION

1. The finite element model proposed here can successfully analyse bending problems of conoidal shells which is reflected by close agreement of present results with benchmark one.
2. An increase in the number of support constraints reduces the deflection but may increase other shell actions. Hence for a complete knowledge about all shell actions, a detail study is needed.
3. For all boundary conditions except CCCC, angle ply laminates show much less deflections than cross ply laminates.
4. For any stacking sequence, SSSS and CCCC shells exhibit highest and lowest value of static deflection respectively.
5. In comparison to other boundary conditions, CCCC conoidal shells offer very insignificant moments for all stacking sequences.
6. For CCSS and CSCS conoidal shells, sagging $\overline{M}_x$ values are much higher than hogging $\overline{M}_x$ values for all stacking sequences and highest values are observed for ASCP and SYCP laminates.
7. SYAP laminates and SYCP laminates offer highest value of sagging $\overline{M}_y$ and hogging $\overline{M}_y$ respectively for CCSS and CSCS shells.

Nomenclature

a, b	length and width of shell in plan
E_{11}, E_{22}	elastic moduli
G_{12}, G_{13}, G_{23}	shear moduli of a lamina with respect to 1, 2 and 3 axes of fiber
h	shell thickness
hl	lower height of truncated conoid
hh	higher height of truncated conoid
M_x, M_y	moment resultants
np	number of plies in a laminate
q	intensity of transverse uniformly distributed load
R_{xy}	radii of cross curvature of shell
x, y, z	local co-ordinate axes
z_k	distance of bottom of the kth ply from mid-surface of a laminate
u, v, w	translation degrees of freedom
α, β	rotational degrees of freedom
n	Poisson's ratio
$\overline{w}$	non-dimensional fundamental frequency ($= wE_{22}h^3/(qa^4)$)
$\overline{M}_x$	non-dimensional moment resultant ($= M_x/(qa^2)$)
$\overline{M}_y$	non-dimensional moment resultant ($= M_y/(qa^2)$)

REFERENCES

1. H.A. Hadid, 1964, An analytical and experimental investigation into the bending theory of elastic conoidal shells. Ph. D. dissertation, University of Southampton.

2. A. Dey, J.N. Bandyopadhyay and P.K.Sinha, 1992, Finite element analysis of laminated composite conoidal shell structures. Computers and structures, 43 (3), 469-476.

3. D. Chakravorty, J.N.Bandyopadhyay and P.K. Sinha, 1995, Freevibrationanalysis of point-supported laminated composite doubly curved shells - A finite element approach., Computers & Structures, 54(2), 191-198.

4. D. Chakravorty, J.N.Bandyopadhyay and P.K. Sinha, 1996, Finite element free vibration analysis of doubly curved laminated composite shells., Journal of Sound and Vibration, 191 (4), 491-504.

5. D. Chakravorty, J.N.Bandyopadhyay and P.K. Sinha, 1998, Application of FEM on free and forced vibration of laminated shells., ACSE, Journal of Engineering Mechanics, 124(1), 1-8.

6. A.N. Nayak and J.N. Bandopadhyay, 2002, Free vibration and design aids of stiffened conoidal shells., Journal of Engineering Mechanics, 128(4), 419-427.

7. A.N. Nayak and J.N. Bandopadhyay, 2002a, On the free vibration of stiffened shallow shells., Journal of Sound and Vibration, 255 (2), 357-382

8. A.N. Nayak and J.N. Bandopadhyay, 2005, Free vibration analysis of laminated stiffened shells., Journal of Engineering Mechanics, 131(1), 100-105.

19

Finite Element Analysis of Smart Multidirectional Composites in Hygro-Thermal Load

Dipak K. Maiti and P.K. Sinha

Department of Aerospace Engineering Indian Institute of Technology, Kharagpur–721 306, India
email: dkmaiti@aero.iitkgp.ernet.in

ABSTRACT

The composite materials may be exposed to moisture and/or temperature during their service life. The structure may undergo swelling or expansion due to hygrothermal environment. In the present study, a finite element modeling technique including superelement concept is developed to assess the performance of multidirectional composite structures with active fiber lamina to serve as sensors or actuators in hygrothermal environments. The FE model is validated for composite structures in hygrothermal environment. The numerical results show that the smart fiber-reinforced layer is effective to control the undesirable response due to hygrothermal effect.

Keywords: Finite element; Superelement; Multidirectional; Smart Composite; Hygro-thermal.

1. INTRODUCTION

The composite materials may be exposed to moisture and/or temperature during their service life. When exposed to extreme moist environment, they absorb moisture. Heat is conducted into the composite materials when they are exposed to hot environment. In general, when a material absorbs moisture or there is a rise in temperature, it swells (expands). Moisture and thermal expansion coefficients provide a measure of swelling or expansion that a material may undergo. If a heated/moist body is not permitted to expand freely in all the directions, some stresses will be developed inside the body, which influence the design of structural components. In a laminate, the laminae cannot swell or expand freely and as a result, stresses known as residual stresses are introduced. The strength and stiffness of the laminate are reduced in hygrothermal environment. Laminates may also undergo interlaminar failure due to bending or buckling in such environment.

The concept of distributed sensing and actuation has led to the development of new structures called smart/intelligent structures. The conventional structures can be integrated with distributed sensors/actuators to achieve self-controlling, self-monitoring capabilities. Hence, a structural element integrated with distributed sensors and actuators may be used in the design of lightweight structure [1]. The distributed sensors and actuators may be used to detect damage location or to suppress hygrothermally induced deflection.

Piezoelectric materials produce mechanical strain when subjected to an electric field or alternately, generate an electric charge when subjected to a mechanical strain. This property gives piezoelectric materials the ability to act as actuators or sensors. The use of piezoelectric actuators and sensors to control undesirable response in structures has received a lot of attention in the recent years [2-5].

The exiting monolithic piezoelectric materials being used in smart structures pose several draw backs i.e., (i) low control authority due to their low stress/strain constants, (ii) brittle natures of piezoelectric crystals and (iii) relatively higher density compared to conventional aerospace structural materials, etc. The piezoelectric fibers are currently available and they may be used to manufacture synthetic piezo-fiber reinforced composites to overcome the above short falls.

In the present study, it is aimed to utilize the active fibers in a lamina of composite structure to serve as sensors or actuators. These active fibers can be laid in one plane or multidirectional plane to achieve higher sensing and actuating capabilities. A finite element modeling technique including super element concept [6] is developed to assess the performance of multidirectional composite [6-8] structures with active fiber lamina to serve as sensors or actuators under hygrothermal environment.

2. FINITE ELEMENT FORMULATION

An N-layered rectangular plate with or without surface bonded/embedded piezoelectric sensor and actuator layers are shown in Figure 1. A three dimensional (20 noded isoparametric) multi-layered composite finite element is developed with the displacement fields along three mutually perpendicular directions, u(x, y, z), v(x, y, z) and w(x, y, z) at each node of the element. The element has the capability to model structure as passive material or passive with active piezo-composite layer.

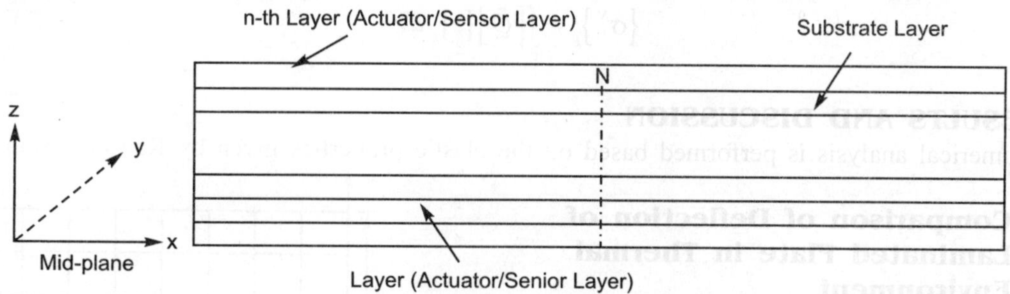

Fig. 1. Layer arrangement.

The strain-displacement field for the element is as follows
$$\{\varepsilon\} = [B_s]\ \{d_e\} \qquad \qquad ...(1)$$
The stress-strain relation for usual composite and piezo-composite can be expressed as follows
$$\{D\} = [e]^T\ \{\varepsilon\} + [\kappa]\{E\}$$
$$\{\sigma\} = [C]\{\varepsilon\} - [e]\{E\} \qquad \qquad ...(2)$$
where $\{\sigma\}$ is stress vector, [C] is compliance matrix, $\{\varepsilon\}$ is strain vector, $\{D\}$ is electric displacement, [e] is piezoelectric stress coefficient matrix, $[\kappa]$ is dielectric constants and $\{E\}$ is electric field vector.

The electric field-potential relation can be expressed as follows:

$$E_x = -\frac{\partial \phi}{\partial x}, \quad E_y = -\frac{\partial \phi}{\partial y}, \quad E_z = -\frac{\partial \phi}{\partial z} \qquad \text{...(3)}$$

The minimization of the total potential energy with respect to nodal variables leads to three sets of equilibrium equations for an element, as given below:

$$\left[K_{dd}^e\right]\{d_e\} - \left[K_{da}^e\right]\{\phi_0^{en}\} - \left[K_{ds}^e\right]\{\phi_0^{el}\} = \{F_1^e\}$$

$$\left[K_{ad}^e\right]\{d_e\} + \left[K_{aa}^e\right]\{\phi_0^{en}\} = \{F_2^e\} \qquad \text{...(4)}$$

$$\left[K_{sd}^e\right]\{d_e\} + \left[K_{ss}^e\right]\{\phi_0^{el}\} = \{0\}$$

The global equilibrium equations are obtained after assembling Eqn. (4) with respect to the global axis.

3. ELEMENT LOAD VECTOR DUE TO NON-MECHANICAL LOAD

The potential energy due to hygrothermal forces is give by

$$V_{he} = \int_V \{\varepsilon\}^T \{\sigma^N\} dV \qquad \text{...(5)}$$

Substituting Equation (1), in Equation (5), one obtains,

$$V_{he} = \int_V \{d_e\}^T [B_s]^T \{\sigma^N\} dV \qquad \text{...(6)}$$

where $\{\sigma^N\}$ is the non-mechanical force vector. The element strains $\{e\}_k$ are due to temperature and moisture along the laminate axes

$$\{e\}_k = \Delta T \{\overline{\alpha}\}_k + \Delta C \{\overline{\beta}\}_k \qquad \text{...(7)}$$

The hygrothermal force vector for the kth lamina is defined as

$$\{\sigma^N\}_k = \int_V [\overline{Q}] \{e\}_k \, dV \qquad \text{...(8)}$$

4. RESULTS AND DISCUSSION

The numerical analysis is performed based on the elastic properties given by Rao and Sinha [6].

4.1 Comparison of Deflection of Laminated Plate in Thermal Environment

To test the accuracy of the super element formulation for the linear bending of laminated plates in thermal environment the present analysis results are compared with those available in the open literature. A square simply supported antisymmetric cross ply laminate [0/90/0/90] subjected to uniform temperature at T = 400K is analysed as a validation problem. A full plate finite element model is considered and shown in Figure 2. The mid-plane deflections are computed along the x-axis at (y = b/2, z = h/2) at different points, A, B, C and D as shown in

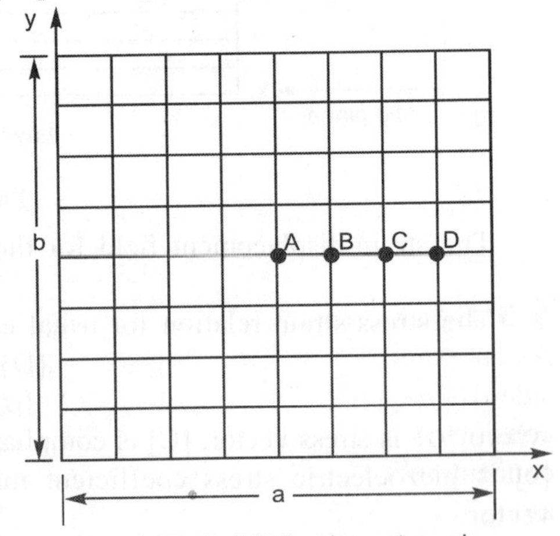

Fig. 2. Typical tinite element mesh.

Figure 2 and are listed in Table 1. The present results are compared with the closed form solution [9] and numerical results generated by developed composite plate analysis program and are found to be in good agreement.

Table 1. Deflection of simply supported [0/90/0/90] laminate subjected to uniform temperature at T = 400K.

Source	Mid-plane Deflection (mm) at Points			
	A	B	C	D
S = 100				
Ref. [9]	0.0000	0.0085	0.0267	0.0337
Present 3D FEM	0.0007	0.0095	0.0282	0.0355
2D FEM [10]	0.0000	0.0085	0.0265	0.0336
S = 50				
Present 3D FEM	0.0000	0.0042	0.0136	0.0176
2D FEM [10]*	0.0000	0.0043	0.0132	0.0167

*Hygrothermal analysis capability is added to the earlier program [2D FEM].

4.2 Comparison of Results for Smart Composite Laminate

A test problem of a simply supported square plate (a = 0.45 × b = 0.45 m) [10] made of graphite/ epoxy composite with distributed actuator and sensor on top and bottom face of the substrate respectively, is analyzed for validation purpose and shown in Table 2. The comparison is good for thin plates, but for moderately thick plates a difference is observed.

Table 2. Comparison of maximum values of non-dimensional displacement and stresses.

S	Volt	Source	$\bar{u}$ (0,b/2, ± H/2)	$\bar{w}$ (a/2, b/2, 0)	$\bar{\sigma}_{xx}$ (a/2, b/2, ± H/2)	$\bar{\tau}_{xy}$ (0, 0, ± H/2)
50	0	Present FEM	$\mp$ 0.0067	0.4318	± 0.5384	± 0.0217
		Ref. [10]	$\mp$ 0.0067	0.4409	± 0.5462	$\mp$ 0.0218
	100	Present FEM	− 0.0050 +0.0057	0.3453	+ 0.4009 − 0.4577	− 0.0157 + 0.0189
		Ref. [10]	− 0.0050 +0.0057	0.3528	+ 0.4076 −0.4635	− 0.0158 +0.0189
10	0	Present FEM	−0.0064 +0.0065	0.6289	±0.5290	−0.0250 +0.0252
		Ref. [10]	$\mp$0.0064	0.6691	±0.5216	$\mp$0.0255
	100	Present FEM	+0.0364 −0.0161	−1.8996	−2.9385 +1.3079	+0.1648 −0.0473
		Ref. [10]	+0.0343 −0.0172	−1.9145	−2.7916 +1.3933	+0.1483 −0.0686

4.3 Bending of Antisymmetric Plates in Hygrothermal Environment

As an example problem, a square simply supported antisymmetric cross ply laminate [0/90/0/90] of side 100mm and a/h=50, subjected to uniform temperature and uniform moisture distribution is considered. In the present analysis, the material properties at elevated temperature and moisture concentrations are used as given by Rao and Sinha [6]. The mid-plane linear center deflections along the x-axis at (y = b/2, z = h/2) for uniform temperature rise ($\Delta T = 50K$, $100K$) and uniform moisture concentration rise (ΔC = 0.5%, 1.0%) are shown in Figure 3. The numerical results indicate that the deflections along the diagonals are zero and similar trends were also observed by Wu and Tauchert [9].

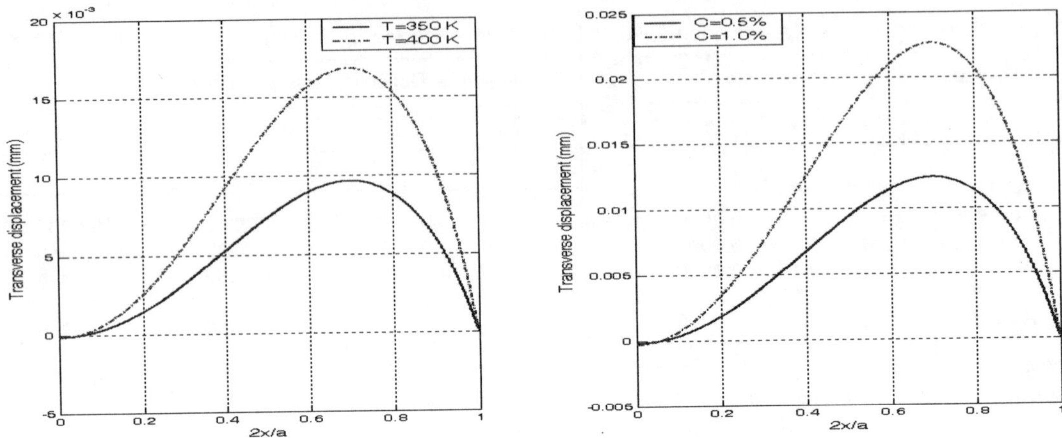

Fig. 3. Deflections of [0/90/0/90] plate along x-axis at ($y = b/2$, $z = h/2$) for different uniform temperature and moisture levels.

4.4 Analysis of Cantilever Plate with Piezo-composite Layer

The numerical results are generated for piezo-composite plates based on the present numerical modeling technique. The piezo-composite properties are calculated based on the formulation given in reference [7]. The piezo-fiber and matrix properties considered for the calculation of composite properties are listed in Table 3. The substrate structure is made of graphite-epoxy composite with 0.5 mm ply thickness and lamination sequence of 0°/90°/0°/90° with planform dimension 0.10 × 0.10 m. The following composite properties are used for the numerical data generation:

E_{11} = 130.00 GPa, E_{22} = 7.00 GPa, E_{33} = 7.00 GPa, G_{12} = G_{13} = 4.75 GPa & G_{23} = 2.375 GPa

ν_{12} = 0.30, ν_{13} = 0.30 & ν_{23} = 0.30

Table 3. Material properties of Piezo-fiber and matrix

Fiber/Matrix	C_{11} (GPa)	C_{12} (GPa)	C_{13} (GPa)	C_{33} (GPa)	e_{31} (C/m2)	e_{32} (C/m2)	κ_{33} (C/Vm)	Volume Fraction
PZT-5H	151	98	96	124	-5.1	27	13.27 × 10⁻⁹	0.7
Expoxy	3.86	2.57	2.57	3.86	0	0	0.079 × 10⁻⁹	0.3

In the present analysis, two piezo-composite layers of thickness 0.25 mm are considered to be at the top and bottom of the the substrate structure. It is considered that the laminate is exposed to 400 K temperature. Due to the temperature effect the composite plate will bend as shown in Figure 4. The voltage is applied first to top smart composite layer and then to both top and bottom layers. The results are plotted in Figure 4. It is clear from the plot that the smart composite layer is effective to control undesirable response.

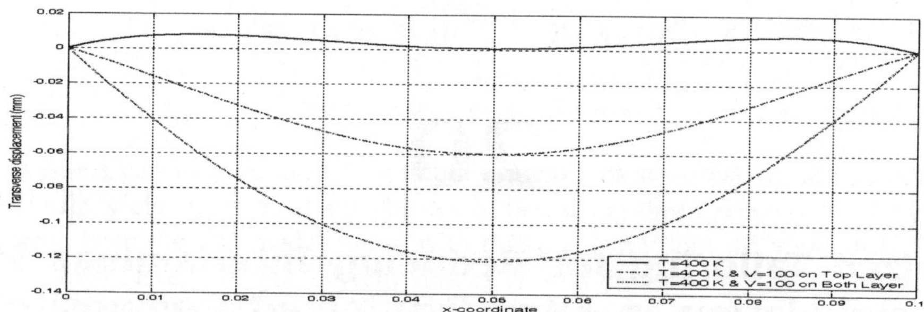

Fig. 4. Deflections of composite plate along x-axis at (y = b/2, z = h/2)
for uniform temperature and Voltage applied to Smart Layer/Layers.

CONCLUSION

In the present study, a multidirectional 3D FE model employing super element concept is developed to perform analysis of smart piezo-fiber composites under hygrothermal environment. The piezo-composite layer can be treated as actuator and/or sensor. The FE model is validated for the bending of a composite plate in thermal environment. The FE program is also validated for composite structure with piezo-patch. The numerical results show a good agreement with the published data. The analysis is performed for a square smart laminated plate in hygrothermal environment. It is observed that the smart fiber-reinforced composite laminate is capable of controlling undesirable response in hygrothermal environment.

REFERENCES

1. L. Meirovitch and M.A. Norris, 1984, Vibration Control, Proceedings of Inter-Noise, 477-482.
2. H.S. Tzou and C. I. Tseng, 1990, Distributed Piezoelectric Sensor/Actuator Design for Dynamic Measurement/Control of Distributed Parameter Systems: A Piezoelectric Finite Element Approach, Journal Sound Vibration, 138(1), 17-34.
3. W. Hwang and H.C. Park, 1993, Finite Element Modeling of Piezoelectric Sensors and Actuators, AIAA Journal, 31, 930-937.
4. S.K. Ha, C. Keiler and F.K. Chang, 1992, Finite Element Analysis of Composite Structures Containing Distributed Piezoelectric Sensors and Actuators, AIAA Journal, 30, 772-780.
5. V.V. Varadan, Y-H Lim and V.K. Varadan, 1996, Closed Loop Finite-Element Modeling of Active/Passive Damping in Structural Vibration Control, Smart Materials and Structures, 5, 685-694.
6. V.V.S. Rao and P.K. Sinha, 2004, Bending Characteristics of Thick Multidirectional Composite Plates under Hygrothermal Environment, Journal Reinforced Plastics and Composites, 23(14), 1481-1495.
7. A.A. Bent, 1997, Active Fiber Composites for Structural Actuation, PhD Thesis, MIT, USA.
8. N. Mukherjee and P.K. Sinha, 1994, Three Dimensional Thermostructural Analysis of Multidirectional Fibrous Composite Plates, Composite Structures, 28(3), 333-346.
9. C.H. Wu and T.R. Tauchert, 1980, Thermoelastic Analysis of Laminated Plates 2: Antisymmetric Cross Ply and Angle Ply Laminate, Journal of Thermal Stresses, 3, 365-378.
10. Dipak K. Maiti, E. Hemalatha, J.V. Kamesh and A.R. Upadhya, 2002, Finite Element Analysis of Structures with Surface Bonded Piezoelectric Patches, Proceedings of SPIE, Smart Materials, Structures, and Systems, 5062(1), 334-341.

20

Optimized Tailoring for Reducing Interlaminar Stress Accumulation in Fibre-Reinforced Composites

U. ICARDI AND L. FERRERO

Dipartimento di Ingegneria Aeronautica e Spaziale, Politecnico di Torino, Italy
email: ugo.icardi@polito.it, laura.ferrero@polito.it

ABSTRACT

Papers dealing with variable stiffness composites and constituent materials with properties which vary with time have been recently presented with the purpose to improve stiffness, strength or the dissipation properties. Unfortunately these techniques are unable to conjugate these contrasting properties. On the contrary, present study tries to conjugate an improvement of stiffness and delamination damage resistance seeking for an optimal distribution of plate stiffness variable from point to point which is able to tune the energy absorbed in the bending, in-plane and out-of-plane shear energy modes, so to not involve weak properties such as interlaminar strengths. In this way, the delamination damage of laminates subjected to impact and blast pulse loads can be consistently reduced maintaining a high stiffness, as shown by the numerical applications presented.

Keywords: Design of Composites, Impact Induced Damage, Delamination Analysis, Survivability.

1. INTRODUCTION

Characteristic feature, laminated and sandwich composites absorb a large amount of the incoming energy through a local damage accumulation mechanism involving matrix and fibres failure and delamination. Helpful from the standpoint of energy absorption, this damage accumulation can have detrimental effects on the overall behaviour, load carrying capacity, service life and strength. Several technological skills have been devised to limit these effects, but unfortunately represent technological complications. Sophisticated computational models, the so-called layerwie models, have been recently developed to accurately describe the evolving local damage accumulation and its effects on the overall response, in a form suitable for general design purposes. The widespread equivalent single layer models are of limited validity in the region where damage rises, since they disregard the transverse normal stress which is important in this region, while the interlaminar shear stresses require a more accurate description of that featured by these models for capturing their effects on the overall behaviour. The most accurate are the discrete-layer models which subdivide the thickness into computational layers, each viewed as an equivalent single layer with its specific displacement and stress fields, then impose the displacement and stress contact conditions as constraint conditions. Their counterparts, the zig-zag models, fulfil a priori the interfacial stress contact conditions through an appropriate C° continuous displacement field. They are less computationally intensive, but also less accurate., The sublaminate zig-zag models allow the level of accuracy to be refined adding more subdivisions, likewise the discrete-layer models . In the present paper, both these types of models are used in different contexts, to take advantage of their complementary features.

The purpose of present paper is to improve stiffness and delamination strength varying the spatial distribution of the laminate stiffness properties over the plane of the plate. This new idea arises from the former studies on variable stiffness composites, in which the orientation of reinforcement fibres minimizes the stresses (see, e.g. Pedersen [3] and Setoodeh et al. [4]) and from the studies of the variation of the laminate properties with the orientation of reinforcement fibres (see, e.g., Zinoviev and Ermakov [5] and Georgi [6]). These studies allow us to argue that fiber orientations can exist which are a good compromise among stiffness, strength and the capability to absorb or dissipate a large amount of the incoming energy. Recent studies try to comply stiffness and energy dissipation by combining different materials with different absorption and stiffness properties (see, e.g. Jung [8]), instead of incorporating viscoelastic layers which make quite compliant the laminates and reduce their strength (see, e.g. Suzuky et al. [7]). Other recent studies use the concept of structural hierarchy to obtain the desired stiffness and damping (see, e.g. Lakes [9]). At the best of the authors' knowledge no studies were published in which the potential advantages of spatially varying the laminate properties, here referred to as the tailoring capability, were exploited for improving the impact resistance or suppressing vibrations. Since the intricate expression of the first variation of strain energy under variation of the plate stiffness coefficients preclude use of discrete-layer models, the zig-zag model of Ref. [1], which is quite accurate although mathematically simple, is used for finding the extremal conditions for the bending and shear energy contributions.The mixed solid element of Ref. [2], which has the three displacements and the three interlaminar stresses as nodal d.o.f. is used for the local stress analysis of the optimal solutions. Since these computations are performed only several times, they does not heavily affect the overall computational costs. In the present paper, the material is tailored by an optimization process, so to obtain wanted energy absorption properties which oppose interlaminar stress concentrations. The key feature is finding the variable spatial distributions of plate stiffness coefficients which makes extremal the bending, in-plane and out-of-plane shear energy contributions, with the purpose to limit the energy amount absorbed involving weak properties, namely the interlaminar shear strength. This appears as an energy transfer from unwanted to wanted modes In this study, the orientation of the reinforcement fibres and/or the constituent materials are varied over the plane of the laminate according to current technologies.

2. THE STRUCTURAL MODEL

The elasticity theory requires the interlaminar stresses and the transverse normal stress gradient to be continuous at the layer interfaces and, in absence of bonding damage, the displacements to be continuous at the interfaces. These conditions, which plays a primary role for predicting realistic stress fields, require the derivatives of displacements to be discontinuous at the interfaces. This is the feature which marks layerwise models (i.e., discrete-layer and zig-zag models). The zig-zag models are partial layerwise models which account for the discrete-layer effects of transverse shears assuming an *ad hoc* displacement field based on the five functional d.o.f. of classical plate models. They need a post-processing technique for computing the interlaminar stresses. Although refined versions of these models are available which fulfills the whole set of contact conditions, in this paper, a classical zig-zag model [1] is used for the optimization process, because it is mathematically easy to handle for deriving and solving the Euler variational equations involved. A computationally more intensive mixed solid element [31] which enables the accurate layerwise description required for damage assessment is used to predict the local effects and to validate the optimized solutions,. The possibility of refinement offered by staking computational layers makes it able to capture the interlaminar stresses directly from constitutive equations even when the material properties abruptly change across the thickness, but it also makes it computationally extensive.

A piecewise cubic displacement field is featured across the thickness by the zig-zag model, while the transverse displacement is assumed constant across the thickness. The following expression for the in-plane displacement in the x direction is reported

$$U(x,y,z) = u^{(0)}(x,y) + z\left[\gamma_x^{(0)}(x,y) - w^{(0)}(x,y)_{,x}\right] - \frac{1}{2h}z^2\sum_{k=1}^{N-1}{}^{(k)}\phi_x(x,y)H_k +$$

$$-\frac{4}{3h^2}z^3\left[\gamma_x^{(0)}(x,y) + \frac{1}{2}\sum_{k=1}^{N-1}{}^{(k)}\phi_x(x,y)H_k\right] + \sum_{k=1}^{S-1}{}^{(k)}\phi_x(x,y)\left(z - {}^{(k)}Z^+\right)H_k$$

The interlaminar shear stresses are made continuous at the layer interfaces by appropriate expressions of the continuity functions Φ_x and Φ_y, which are obtained enforcing the continuity at the interfaces of these stresses.

The three interlaminar stresses and the three elastic displacements are chosen as the nodal d.o.f. in the mixed element, so to fulfil the prescribed interfacial stress and displacements continuity requirements stacking elements. The slave fields are the strains by the stress-strain and strain-displacement relations, respectively. An examination of the integrals involved within the Hellinger-Reissner multi-field governing principle reveals that a C° approximation is sufficient both for the displacement and stress fields. Accordingly, we can choose C° standard serendipity polynomials as interpolation functions for every d.o.f. and fulfill the intra-element equilibria in an approximate integral form, instead of being met in a point-wise sense. Thank to this representation, the computational effort is not larger than for displacement-based counterpart solid elements, while accuracy and convergence are dramatically improved (see [2] and the references therein quoted for the discussion of the details here omitted). The behavior of mixed elements is governed by rather complex mathematical relations which are beyond the purpose of present paper to examine into detail. It is only reminded that certain choices of the shape functions could not yield meaningful results. Present element passed successfully all the prescribed tests and appeared fast convergent, robust and stable.

3. THE OPTIMIZATION PROCESS

Assume the stiffness coefficients to be functions of position over the plane of the plate and the orthotropic relations to hold locally, with the three principal material directions with a different orientation point to point. The optmization of the energy storage starts writing the membrane, bending, in-plane and out-of-plane first variations by the zig-zag model under variation of displacements, which represents the equilibrium condition. Its first variation is then written under in-plane variation of the stiffness coefficients, obtaining the Euler-Lagrange equations which make extremal each of the desired strain energy contributions, in the form of partial differential equations in terms of the derivatives of these stiffness properties. They constitute a so intricate system of coupled, third-order, partial differential equations which cannot be report here. Let us consider arbitrary the number of layers in the laminate together their material properties and orientation. An approximate solution to this set of Euler-Lagrange equations is a second-order polynomial approximation for the transformed reduced stiffness coefficients for each constituent layer. This approximation is of technical interest since it is compatible with current technologies of construction and allows to enforce a number of desired conditions which range from imposition of a mean value (e.g., coinciding with that of standard layers with constant stiffness that are substituted within the laminate), to the choice of convexity or concavity, to the values at the bounds, to the thermodynamic constraints. The object here pursued is the obtainment of a wanted, specific structural behaviour by tuning the energy absorption, in order to improve the strength at the onset of delamination of plates subjected to impact and blast pulse loads, consequent to an achieved reduction of the interlaminar stress accumulation. Two kind of layers to incorporate in a pre-existent lay-up are studied with this aim. The first one reduces the bending of a lamina without consistently increasing shear stresses, the second one reduces these stresses without consistently increasing bending. The incorporation of these layers (with the same mean stiffness properties of the layers they substitute) will be shown to substantially change the energy amount absorbed in the various modes, appearing as a transfer of the incoming energy from unwanted to wanted modes.

4. APPLICATIONS

The first application in which optimized tailoring has been used is a multilayered rectangular plate, simply supported and loaded by an impact force in the middle. The main purpose is to investigate the advantage of substituting some classical plies (constant stiffness) with the optimal layers, over a range of thickness (from 0.5 to 2; length 10×20). Starting from a classical stacking sequence (symmetric and balanced) [0/45/90/ −45/0/0/ −45/90/45/0], let we substitute the I and the X ply firstly with the optimal layer type a: we will call this plate var1a; afterwards the layer type b has been introduced in the same positions: this plate will be indicated as var1b. In both cases symmetry and balance are satisfied. Layer type a has been designed with the same mean stiffness than classical plies, but optimized to minimizing bending energy and maximizing membrane contributions. Its effect on the plate var1a is shown in Figure 1, where two graphs compare the transversal deflection in classical plate and in optimized one; it may be observed that a reduction of 95% occurs in thin plates, whereas the gain decreases to 80% in thicker laminates. Meanwhile, the shear can decrease of 75% its initial value, in the middle region; on the contrary, near boundaries, it approximately keeps both shape and magnitude; Figure 2 reports transversal x-direction shear across the thickness under the load application point.

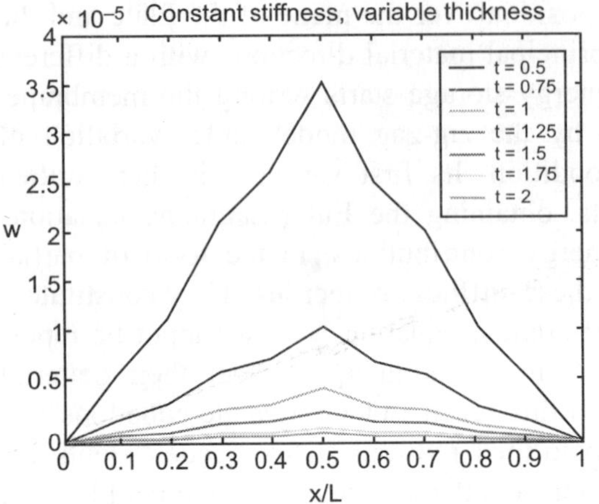
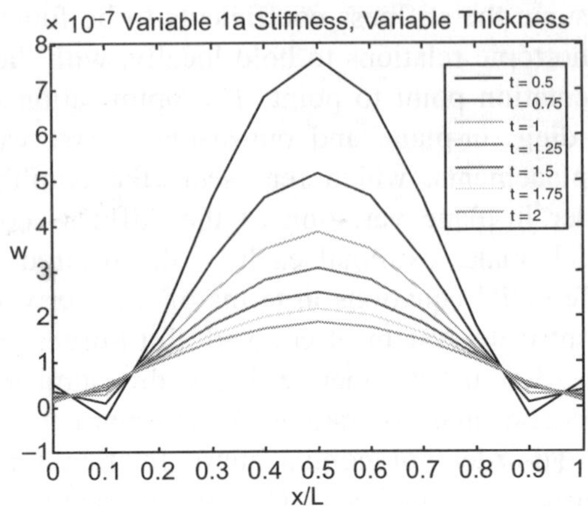

Fig. 1. Transversal deflection of classical plate (left) and optimized plate var1a (right); thicknesses considered from 0.5 to 2, with principal dimensions 20 × 10.

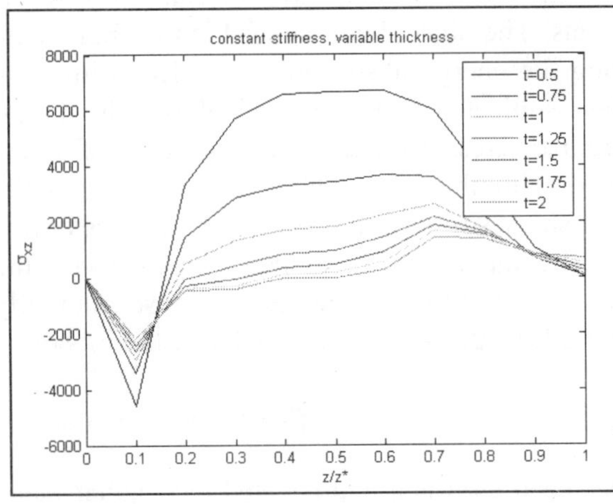
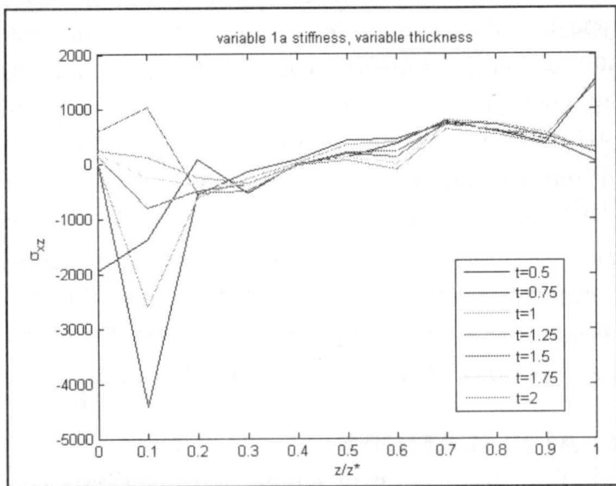

Fig. 2. Transversal shear σ_{xz} of classical plate (left) and optimized plate var1a (right) along the thickness; thicknesses considered from 0.5 to 2, with principal dimensions 20 × 10.

The introduction of plate b allows a reduction in transversal tension at approximately the same shear and deflection, while strain energy is tuned on membrane mode. Figure 3 reports transversal tension on the upper surface, where the effect of impact load is more evident. As underlined in the previous example, the effect of optimized layers is evident in the middle region, where the peak value is 75% lower than in classical plate; instead, close to boundary we may observe the same behavior for both the plates; this aspect is fundamentally due to the stiffness law of variation, minimum near the constraint and maximum in the most critical region (middle).

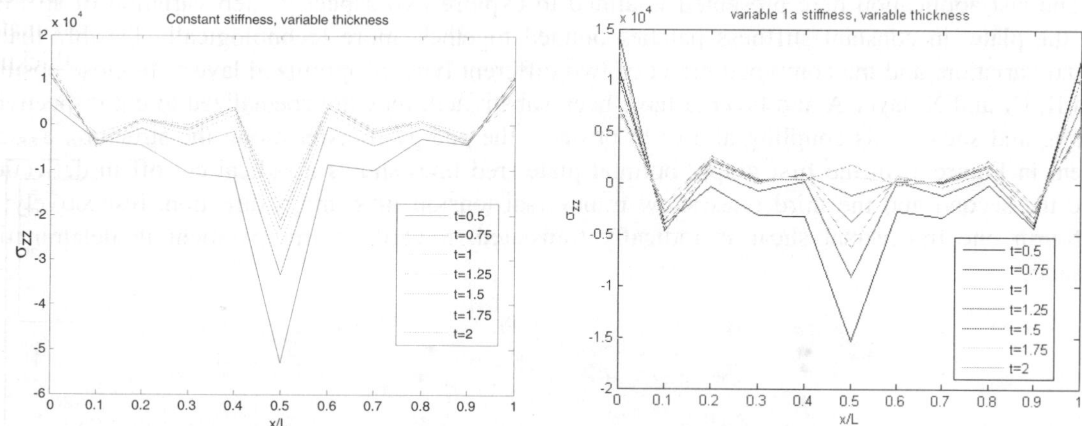

Fig. 3. Transversal tension $\acute{o}_{zz}$ of classical plate (left) and optimized plate var1a (right) on the upper surface; thicknesses considered from 0.5 to 2, with principal dimensions 20 × 10.

The optimization is aimed to improve the impact resistance of multilayered plate, whose post-impact behavior is affected by delamination occurring; therefore delamination damage can be estimated by applying Hou-Petrinic-Rouitz criterium, at each interface. Figure 4 shows the H-P-R index, for the classical plate, previous described, and for three optimized lay-ups: in plate 2a and 3a, optimized layer a fills respectively positions II and IX, III and VIII, while in plate 1b layers I and X are of type b. The index is calculated on outer layer, the opposite side of impact load application, where damage is normally more evident (see Figure 4, left side) and on the second layer (see figure 4, right side). It can be observe a deep improvement of impact resistance in optimized structures; in fact, delamination normally occurs near boundaries, due to free edge effects and in these regions in optimal plate index reaches just 50% of the value shown in classical plate; at the same time, H-P-R index increases of a negligible quantity in the middle. The improvement is more significant in the more critical layer.

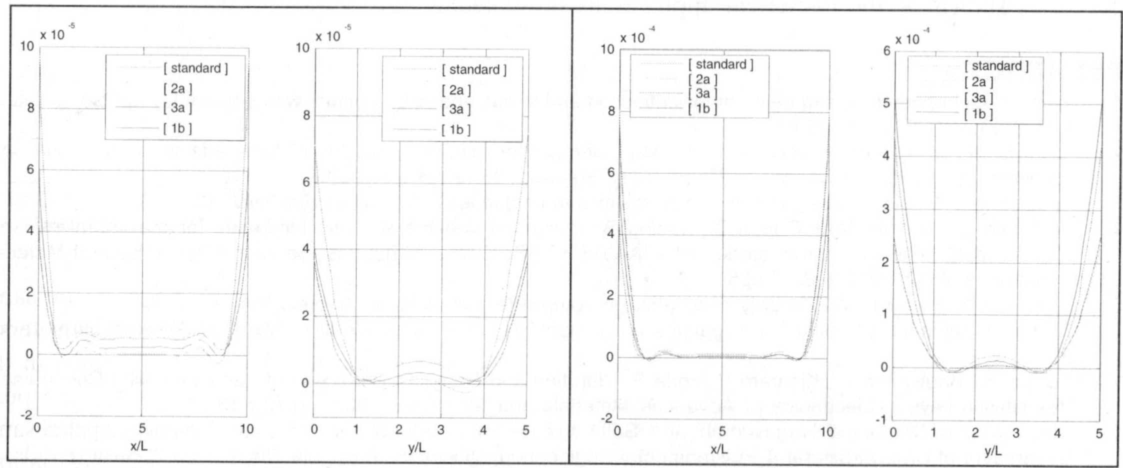

Fig. 4. Hou-Petrinic-Rouitz index on outer face (left) and on second interface for a standard plate and three variations, including optimized layers in symmetric positions.

The last application here presented is aimed to explore two aspect: a step variation of stiffness over the plate, in constant stiffness patches bonded together, more technologically feasible than a gradual variation, and the combined effect of two different type of optimized layers. In close position I and II, IX and X, layer A and layer B have been substituted; they are specialized to cut respectively bending and shear. This coupling allows to obstacle the energy transfer along the structure, as self-evident in Figure 5. In the first graph, optimal plate (red line) shows a radical cut off in deflection, while the second and the third ones show transversal tension in x and z direction, respectively; in the fourth one transversal shear is radically translated, toward an improvement in delamination resistance.

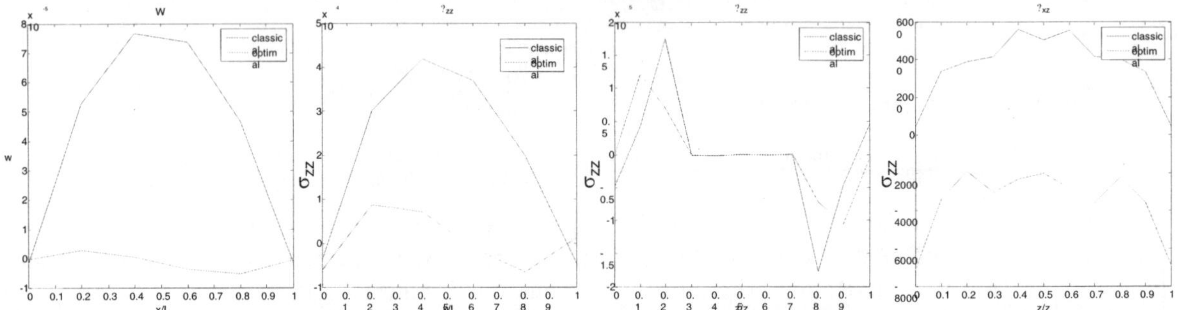

Fig. 5. Deflection, transverse tension in spanwise direction and across the thickness, transverse shear across the thichness, for classical plate (blue line) and optimal plate with layer A in I and X plies, and layer B in II and IX plies (red line).

CONCLUSION

In this theoretical study has the absorption of the incoming energy was tuned using variable stiffness properties which are defied by an optimization process in which the membrane, bending and transverse shear components of strain energy are made extremal. With this technique the unwanted stress concentrations and the delamination damage in laminates undergoing impact loads was consistently reduced, as shown by the numerical applications presented.

REFERENCES

1. Icardi U. Eight-noded zig-zag element for deflection and stress analysis of plates with general lay-up. Composites-Part B: Eng 1998; 29b: 435-41.

2. Icardi U. Atzori A. Simple efficient mixed solid element for accurate analysis of local effects in laminated and sandwich composites. Advances in Engineering Software 35 (2004), 843-839.

3. Pedersen P., A note on design of fiber-nets for maximum stiffness. Journal of Elasticity 2003; 73 (1-3), 127-145.

4. Setoodeh S., Abdalla M.M.,Gurdal Z., Tatting B. Design of variable-stiffness laminates for maximum in-plane stifness using lamination parameters. 46th AIAA/ASME/ASCE/ASC Structures, Structural Dynamics and Material Conference. AIAA 2005-2083, 2005.

5. Zinoviev PA, Ermakov YN. Energy dissipation in composite materials. Technomic Pub. Co., 94

6. Georgi H. Dynamic damping investigations on composites. Proc. 48th Meeting AGARD, Williamsbourg, 1979: 9.1-9.20.

7. Suzuky K., Kageyama K., Kimpara I., Hotta S. Vibration and Damping Prediction of Laminates with Constrained Viscoelastic Layers. Mechanics of Advanced Materials and Structures 2003; 10 (2): 43-73.

8. Jung W.Y., A Combined Honeycomb and Solid Viscoelastic Material for Structural Damping Applications. Department of Civil, Structural & Environmental Engineering, University at Bu_alo.Thrust Area 2: Seismic Retrofit of Acute Care Facilities, 2001, 41-43.

9. Lakes R.S., High Damping Composite Materials: Effect of Structural Hierarchy, Journal of Composite Materials, vol. 36 (3), 2002, 287-297.

21

Dynamic Analysis of Composite Panels Subjected to Non-Contact Explosion

K. Ramji[1], B. Satyanarayana[2] and B. Ramgopal Reddy[3]

[1]Dept. of Mech. Engg., Andhra University, Vizag, India email: ramjidme@yahoo.co.in
[2]Dept. of Mech. Engg., Andhra University, Vizag, India email: snbeela@yahoo.com
[3]Dept. of I&P. Engg., R.V.R. & J.C. College of Engg.,Guntur, India email: brgreddy_b@yahoo.com

ABSTRACT

The response of structural panels subjected to non-contact explosion is of vital importance in the design of air crafts, marine vehicles such as ships, submarines and steel off-shore towers. The prediction of modes of failure associated with non-contact explosion and possible means of protecting the composite structures against such failures is still a concern to the designers world wide. In the present work, experimental results for composite flat plates from literature [3, 4] are compared with MSC/DYTRAN results for sinusoidal and blast loading. It is found that there is a good agreement between the two results which means that DYTRAN software can be used for solving the present problem. Three types of composite hull panels such as flat, concave and convex are studied for the above loading conditions and their response is estimated using MSC/DYTRAN. The effect of stiffeners and boundary conditions on the response of composite panels is studied numerically.

Keywords: Composite Panels; DYTRAN; Geometric non-linearity; Dynamic response.

1. INTRODUCTION

Fibre-reinforced composite materials are considered to have a great potential in air craft, space craft, marine and submarine structures. With the increased use of composite materials in structural applications, the subject of dynamic response of composite structures subjected to different loading conditions has begun to receive wide spread attention in the recent years. The response prediction of composite hull panel is a complex phenomenon involving fluid structure interaction, orthotropic material modeling and geometric non-linearity.

The dynamic response of composite panels subjected to explosive loading has been studied by different authors. Ramajeyathilagam [6] has presented theoretical and experimental non-linear transient dynamic response of isotropic ship hull panels subjected to under water shock loading. An attempt has been made by the same author to predict the response and failure modes of three types of isotropic hull panels (flat, concave and convex).

Turkmen and Macitoglu [3] presented the geometrically non-linear response of laminated composite plates subjected to air blast loading. Hunpark and Lee [4] developed a methodology for determination of dynamic response of composite structures under high explosive blast wave loading taking into effect of progressive material damage and failure. In the present work, blast analysis for square plates, similar to that presented in the literature [3, 4], is carried out using DYTRAN software.

2. FINITE ELEMENT ANALYSIS

Fiber-reinforced composite plates and shells are finding increasing applications in engineering industries due to their lightweight, high strength and high stiffness. Consequently, efficient and robust computational tools are required for the analysis of such structural models. One can find a large number of finite element (FE) models for laminated plate that have been developed and incorporated in most commercial codes for structural analysis.

Finite element modeling is central to the ability to perform an engineering analysis of a model using a computer. In the present problem the modeling of composite panels such as flat, concave and convex has been done using MSC/PATRAN. One of the strengths of PATRAN is its ability to create a finite element model, either from an existing geometry model or through direct finite element operations. Geometric non-linear dynamic analysis for flat, concave and convex composite hull panels is performed using MSC/DYTRAN. DYTRAN uses explicit time integration technique to solve the problems, which uses smaller time steps. It is most suitable for short time events such as explosions and high speed impact.

3. COMPARISON OF DYTRAN RESULTS FROM LITERATURE

To verify the applicability of software for explosive loading, we first compared our DYTRAN results with experimental results available in the literature [3, 4]. Figures 1(a) and 1(b) show the comparison between DYTRAN results and experimental results under sinusoidal loading [4] and air blast load [3] respectively. The plate is clamped at all edges. The material properties, geometry and laminated sequence for the above two problems are given below.

3.1 Sinusoidal Load

Geometrically non-linear dynamic analysis is carried out for unidirectional E-glass composite plate with a side length of 0.6 m and a thickness of 0.343 m is considered for sinusoidal loading with following material properties: E_1 = 23.6 GPa, E_2 = 10 GPa, G_{12} = 1GPa , γ_{12} = 0.23, ρ = 2028 kg/m^3. Five layers are chosen and lamination sequence followed is [0/45/90/-45/0]. The model consists of 36 elements and 49 nodes with active degrees of freedom 150. The time variation pressure for sinusoidal loading is given as

$$P(t) = P_m Sin\left[2\pi \frac{t}{t_0} \right] \qquad \qquad ...(1)$$

where $P(t)$ is the pressure at any instant of time t, P_m the peak pressure and t_0 the duration of the pulse. The present problem is solved for two different load cases (*a*) P_m = 0.1 MPa, t_0 = 2 msec (b) P_m = 0.2 MPa , t_0 = 4 msec. The predicted displacement time history results by DYTRAN code are compared with numerical results in Fig. 1 and it is observed that the curves shows a close correlation between the predictions and numerical results. The results for maximum displacement are within 3% deviation.

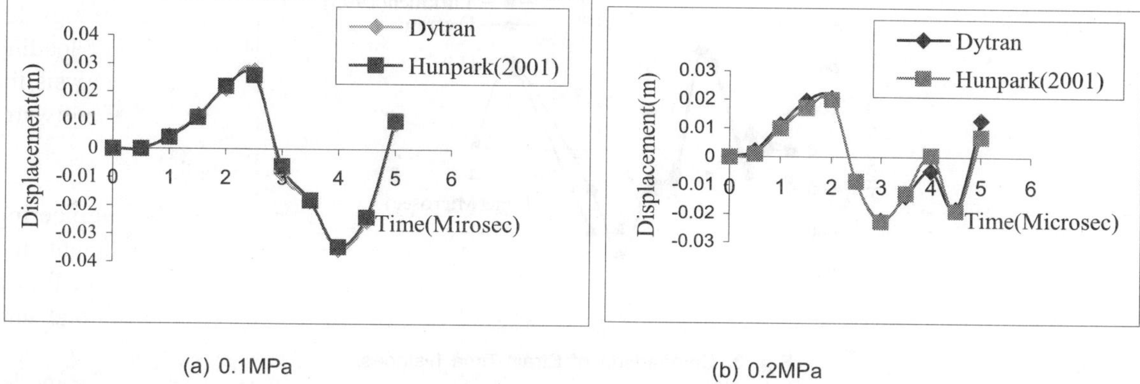

(a) 0.1MPa   (b) 0.2MPa

Fig. 1. Comparison of Displacement - Time histories for sinusoidal loading.

3.2 Blast Load

GFRP square plate with a side length of 0.22 m, consists seven plies with 0.00196 m thickness is considered. The lay-up sequence followed is $[90/0]_s$ and the material properties are: $E_1 = E_2$ = 24.014 GPa, G_{12} = 3.79 GPa, γ_{12} = 0.11, ρ = 1800 kg/m^3. The finite element model consists of 25 elements and 36 nodes with 96 active degrees of freedom. In this analysis, the pressure acting on the plate is approximated by a well-known Friedlander decay function given by

$$P(t) = P_m \left(1 - \frac{t}{t_p} \right) e^{-\alpha t / t_p} \qquad ...(2)$$

where $P(t)$ is the pressure at any instant of time t, P_m the peak pressure, t_p the pulse duration and a the wave parameter. For the present problem, a peak pressure of 28906 N/ m^2, pulse duration of 0.0018 sec and wave parameter of 0.35 are used. The displacement result predicted by DYTRAN software is compared with the experimental results [3] in Fig. 2. The curve shows a close correlation between the predictions and experimental values. The DYTRAN code predicted a maximum value of 0.00359m compared to the experimental value of .0033 m. The predicted values are showing a deviation of about 8.8%.

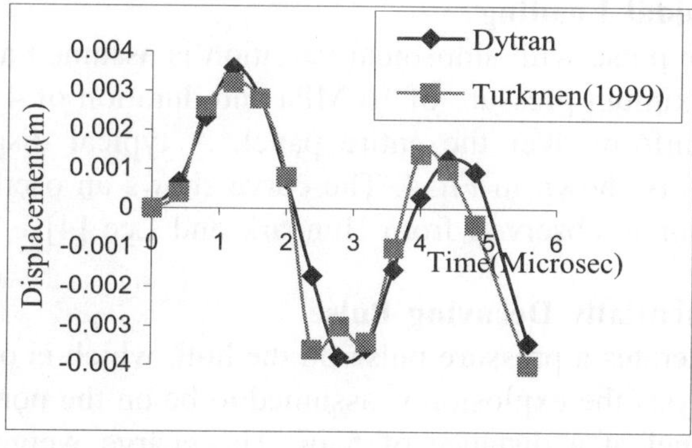

Fig. 2. Comparison of Displacement-Time histories.

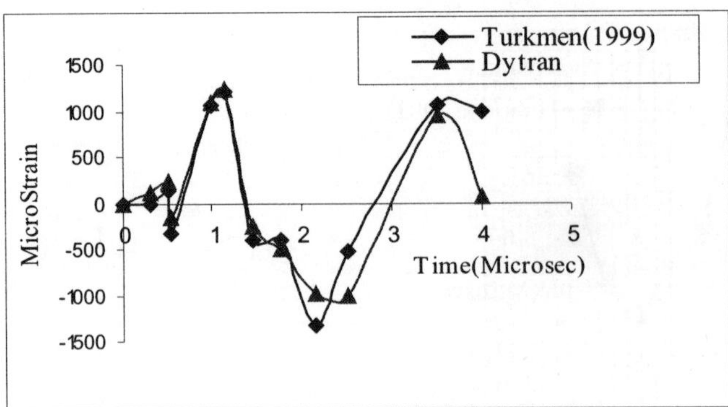

Fig. 3. Comparison of Strain-Time histories.

The strain-time history predicted by DYTRAN software is compared with experimental strain-time history [3] in Fig. 3. The DYTRAN predicted maximum strain shows a variation of 2.8% compared to experimental strain [3]. It shows slight variation in the later time steps.

4. RESULTS AND DISCUSSION

The failure of composite structures under non contact explosion generally initiates in the hull panel. In the present work three types of composite hull panels such as flat, concave and convex are studied using DYTRAN software. The shock pressure loading on the panels is estimated based on Taylor's plate theory, which is widely used for solving shock related problems.

4.1 Flat Panel

To understand the failure of a hull panel subjected to shock loading, a typical hull panel of size 1.5 m × 0.6 m × 0.02 m (thickness) is used. The panel is assumed to have 40 layers of 0.005 m thick. Lamination sequence of $[0/90]_s$ is assumed for the panel. The Finite element model consists of 50 elements, 66 nodes and active degrees of freedom 204. For the analysis, both simply supported and clamped boundary conditions at the edges are assumed. The material properties used in the analysis are: $E_1 = E_2 = 23.6$ GPa, $G_{12} = 1.0$ GPa, $\gamma_{12} = 0.23$, $\rho = 2028$ kg/m^3, $X_t = Y_t = 735$MPa, $X_c = Y_c = 600$MPa.

4.1.1 Response to Sinusoidal Loading

A time dependent pressure pulse with sinusoidal variation is assumed as the first load case. For the present problem, a maximum pressure of 16 MPa and duration of 4 msec is considered. The load is assumed to be uniform over the entire panel. A typical displacement—time history obtained from the analysis is shown in Fig.4. The curve shows an oscillating behavior over the mean line. Similar behavior is observed from Hunpark and Lee [4].

4.1.2 Response to Exponentially Decaying Pulse

Underwater explosion generates a pressure pulse on the hull, which is of exponentially decaying nature. In the present analysis the explosion is assumed to be on the normal line passing through the centre of the hull panel at a distance of 5 m. The charge weight has been varied from

0.1 kg to 1 kg. A typical displacement time history at the centre of the panel is shown in Fig. 5. The curve monotonically increases up to the maximum and shows oscillations. The permanent deformation of the panel can be obtained by subtracting the elastic deformation from the maximum displacement. The failure index (*FI*) has been estimated based on Maximum Stress failure theory to predict the sequence of failures, and shown in Table 1. From the results it can be concluded that the sequence of failure of composite panel are matrix tension, fiber tension, matrix shear, fiber compression and matrix compression.

4.1.3 Effect of Boundary Conditions

The influence of boundary conditions such as clamped-clamped and simply supported on the response of flat plate under shock loading is examined. A typical displacement-time history corresponding to

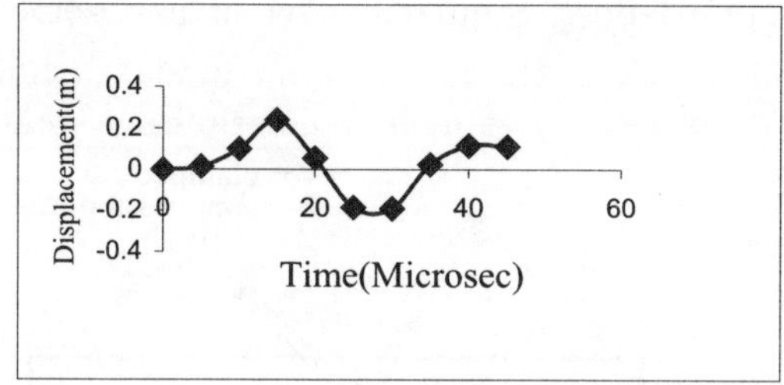

Fig. 4. Response of Flat Plate to Sinusoidal Excitation.

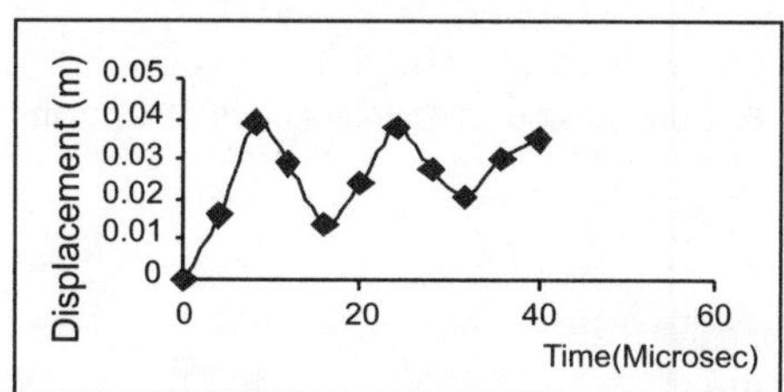

Fig. 5. Typical Displacement – Time history for flat panel under shock load.

Table 1. Failure Indices for different sequence of failures

Charge Weight (W kg)	Fiber Tension (FI)	Fiber Compression (FI)	Matrix Tension (FI)	Matrix Compression (FI)	Matrix Shear (FI)
0.1	0.148	0.181	0.321	0.253	0.0706
0.25	0.255	0.225	0.648	0.321	0.086
0.5	0.368	0.306	0.827	0.368	0.143
0.75	0.472	0.371	0.991	0.425	0.162
1.0	0.556	0.425	1.228	0.416	0.222
2.0	0.933	0.443	1.483	0.601	0.369
5.0	1.523	0.768	2.313	0.623	0.833
10.0	2.789	0.903	3.809	0.63	1.388

a charge weight of 1 kg is compared in Fig. 6. It can be seen that response of simply supported condition is slightly high compared to clamped conditions. The shock factor Vs displacement has been compared in Fig. 7, it shows that the simply supported condition is more severe. This is because the clamped end condition arrests both transverse and rotary displacement which causes lesser deformation when compared to simply supported end condition.

Small deviation in the response for clamped and simply supported end conditions is observed in the range of shock factor 0.02 to 0.06, when compared to initial and final range from Fig. 7. A typical stress – time history for a charge weight of 1 kg is compared in Fig. 8, it is observed that, the stress response for clamped boundary condition exhibits much higher values than those for simply supported conditions. This is due to the fact that the clamped end condition will be subjected to more bending moment. The shock factor vs. stress is shown in Fig. 9. For all the cases considered the stresses are more for clamped conditions. This implies that the simply supported condition can undergo more deformation before failure. Similar behaviour is observed from Gong and Lam [2].

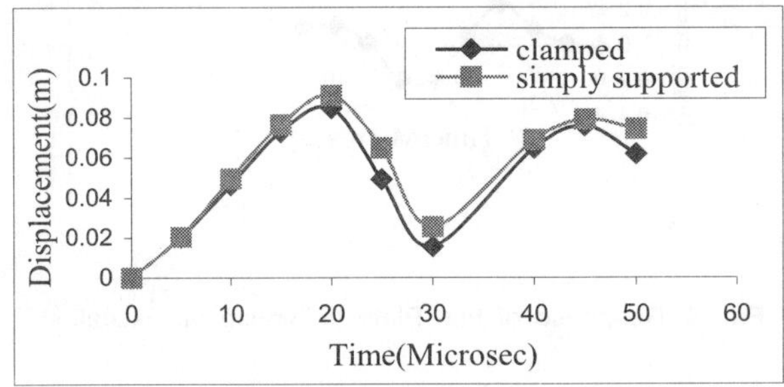

Fig. 6. Comparision of Displacement-Time histories.

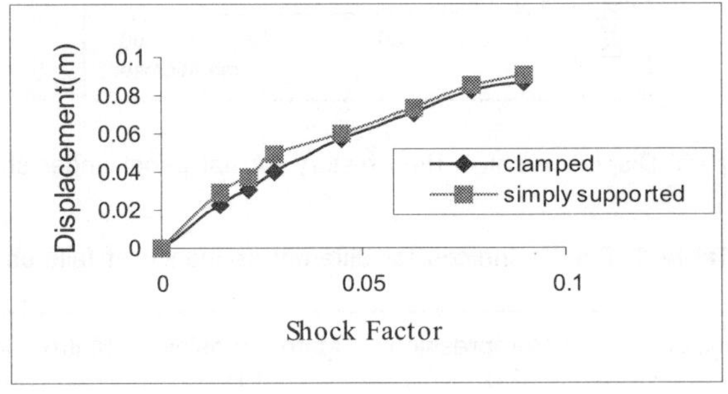

Fig. 7. Comparision of Shock factor vs. Displacement.

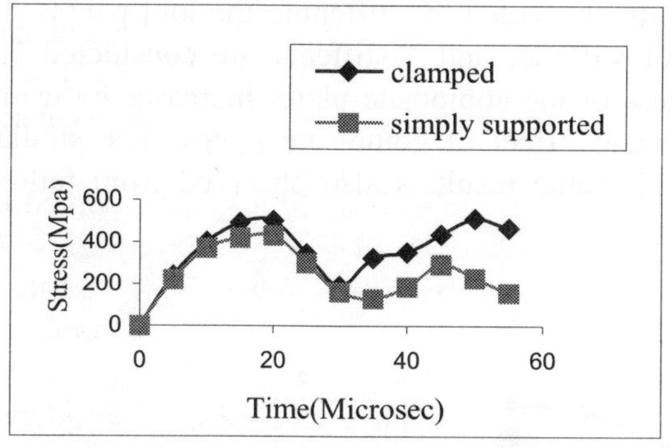

Fig. 8. Comparison of Stress vs. Time.

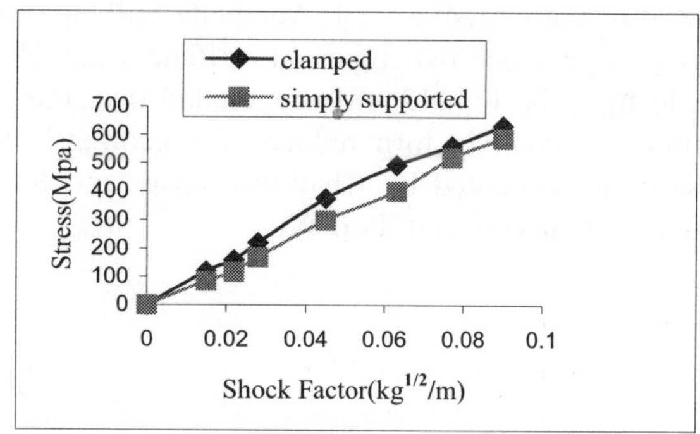

Fig. 9. Comparision of Shock factor vs. Stress.

4.2 Concave and Convex Panel

Concave and convex panels are generally used for hull structures, subjected to shock loading. Shell rise ratio 0.05 with projected dimensions 1.5 m × 0.6 m × 0.02 m (thickness) is used for both the panels. Lamination scheme of $[0/90]_s$ with 40 layers of 0.005m thick is assumed. The finite element model consists of 32 elements and nodes 45 with active degrees of freedom 114. The analysis is carried out using clamped end conditions. The material properties used in the analysis are: $E_1 = E_2 = 23.6$ Gpa, $G_{12} = 1.0$ GPa, $\gamma_{12} = 0.23$ and $\rho = 2028$ kg/m^3. The shock factor Vs displacement for all the three panels are compared in Fig. 10. From the figure, the displacement is found to increase with increase shock factor for all the three panels. It is observed from the response that concave panels offering better resistance to shock loads compared to flat and convex panels.

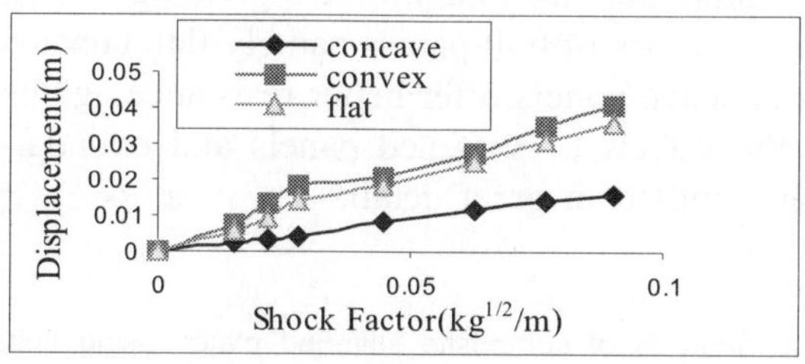

Fig. 10. Comparison of Displacement vs Shock Factor

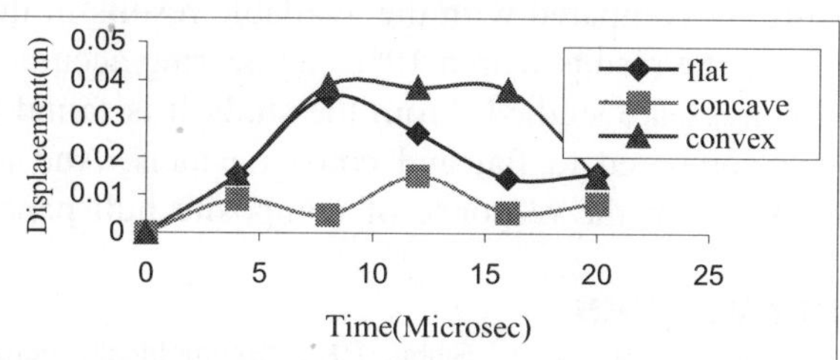

Fig. 11. Comparison of Displacement vs Time histories

The displacement-time history showing oscillatory motion for three panels is compared in Fig. 11. Up to 7 milliseconds, the flat and convex panels exhibiting same response. After that the convex panel suffers higher deformation when compared to flat panel. It is also observed that the concave cylindrical shell panels suffer lesser deformation than the flat panel because of membrane resistance offered by the initial curvature of concave panel. The convex cylindrical panels experience more deformation than that of flat panels because of snap-through-behavior.

4.3 Effect of Stiffeners

Stiffeners are the structural members generally used in ship hull panels to reduce the unsupported span of shell or deck panels and in eliminating the possibility of lateral-torsional instability under lateral or compressive load. Adequate hull rigidity is also provided by stiffening the hull panels. In this present work two types of stiffeners namely HAT-stiffener and T-stiffener are considered for study from the Fig. 12, it is concluded that, the stiffness of the composite plates increases by using stiffeners which in turn reduces the geometric non-linear effect of composite plates. The similar results are observed by Chattopadhayay and Sinha [1]. Same result is also observed from failure theories Tsai-Hill and Tsai-Wu.

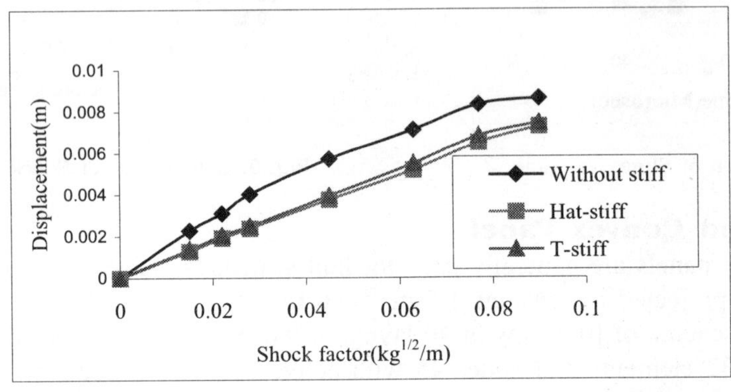

Fig. 12. Comparison of Shock Load Vs Displacement for stiffened and unstiffened rectangular composite plates.

CONCLUSION

An attempt is made for the dynamic analysis of composite panels using DYTRAN software and the results are compared with the available results in the literature and are found to be a good agreement between the results within 10% engineering accuracy. Three types of hull panels namely flat, concave and convex are studied. From the study it is found that concave panels offer better resistance against shock compared to flat and convex panels. And also the effects of stiffened panels and boundary conditions on the response of composite hull panels are studied in great detail.

REFERENCES

1. Chattopadhyay. B., Sinha. P.K., "Geometrically non-linear analysis of composite stiffened plates using finite elements", *Journal of composite structures,* vol-31, pp 107-118 (1995).
2. Gong. S.W., and Lam.K.Y., "Transient response of floating composite ship section subjected to underwater shock", *Journal of Composite Structures,* vol-46, pp 65-71 (1999).
3. Halit S.Turkmen and Zahit Mecitoglu., "Non-Linear structural response of laminated composite plates subjected to blast loading", *AIAA Journal,* vol-37, pp 1639-1647 (1999).
4. Hunpark and Sung W.Lee., "Dynamic analysis of geometrically non-linear composite structures under time-dependent pressure loading", *Annual Technical Conference American Society for Composites* (2001).
5. MSC/DYTRAN Theory Manual, MSC Software Corporation (2000).
6. Ramajeyathilagam. K., "Underwater explosion damage of ship hull panels", *Defence Science Journal,* vol-53, No-4, pp 393-402 (2003).

22

Finite Element Analysis for Effect of Fibre Orientation on Stress Concentration Factor in a Laminated Composite Plate with Central Hole Under In-plane Static Loading

N.D. MITTAL AND N.K. JAIN

Department of Applied Mechanics, MANIT Bhopal, India

ABSTRACT

A number of analytical, numerical and experimental techniques are available for the study of stress concentration around holes. The stress distribution in a rectangular laminated composite plate with central hole has been studied using finite element method. The aim of author is to analyse the effect of different fibre orientation upon stress concentration around the hole in a finite width plate under in-plane static loading. The results are obtained for laminates of three different types of orthotropic composite materials. Studies were carried out for three cases of different hole diameter to plate width ratio. A finite element study is made for whole analysis of finite width plate with a central hole under in plane static loading. The finite element formulation was carried out in the analysis section of the package, known as ANSYS.

Keywords: Finite Element Method, Stress Concentration Factor, Composite, Laminates, Fibre Orientation.

1. INTRODUCTION

A laminated composite plate with a central hole have found widespread applications in various fields of engineering such as aerospace, marine, automobile and mechanical. High stress due to discontinuity or abrupt change in geometry is known as stress concentration and always found at the edges of discontinuity. For the design of plate with a hole, accurate knowledge of stresses and stress concentration factor (SCF) at the edge of hole under in plane or transverse loading are required. Analytical solutions are available for stress concentration factor in the literature with different types of abrupt changes in shape for isotropic material and some specific orthotropic material.

Shastry and Raj [1] has analyzed the effect of fibre orientation for a unidirectional composite laminate with finite element method by using plane stress triangular elements, but results are not much accurate due to element type, limited number of elements and nodes. Also analysis has done only for one composite material type. Ukadgaonker and Rao [2] proposed a general solution for stresses around holes in symmetric laminates by introducing a general form of mapping function and an arbitrary biaxial loading condition in to the boundary conditions. In which basic formulation is extended for multilayered plates. Paul and Rao [3, 4] presented a theory for evaluation of stress concentration factor of thick and FRP laminated plate with the help of Lo-Christensen-Wu higher order bending theory under transverse loading. Meguid [5] studied the reduction of stress concentration factor by introducing defence hole system in a uni-axially loaded plate with two coaxial holes by finite element method. Peterson [13] has developed good theory and charts on the basis of mathematical analysis and presented excellent mythology in graphical form for evaluation of stress concentration factors for finite width isotropic plate as D/A ratio greater then one, but no techniques are presented for orthotropic composite plate and laminates.

In this article a study of rectangular laminated plate with central hole upon the effect of fibre orientation on stress concentration factor under in-plane static loading is made. The analytical treatment for such type of problem is very difficult and hence the finite element method adopt for whole analysis. The purpose of this research work is to investigate the effect of fibre orientations on SCF in a single layer laminate with central circular hole. Three types of different composite materials of different material properties are used for analysis to find out the sensitivity of SCF with respect to elastic constants. The work also illustrates, the variation of SCF versus hole diameter to plate width ratio in a lamina at different fibre orientation.

2. DESCRIPTION OF PROBLEM

To study the influence of fibre orientation upon the stress concentration around the hole, a laminated plate of dimension 200 mm × 100 mm × 10 mm with a circular hole at centre under uniform distributed static loading of σ N/m² is analysed by finite element method. The analysis is carried out for three different D/A ratio where A is plate width and D is diameter of hole. Figure 1 shows the basic model of the problem.

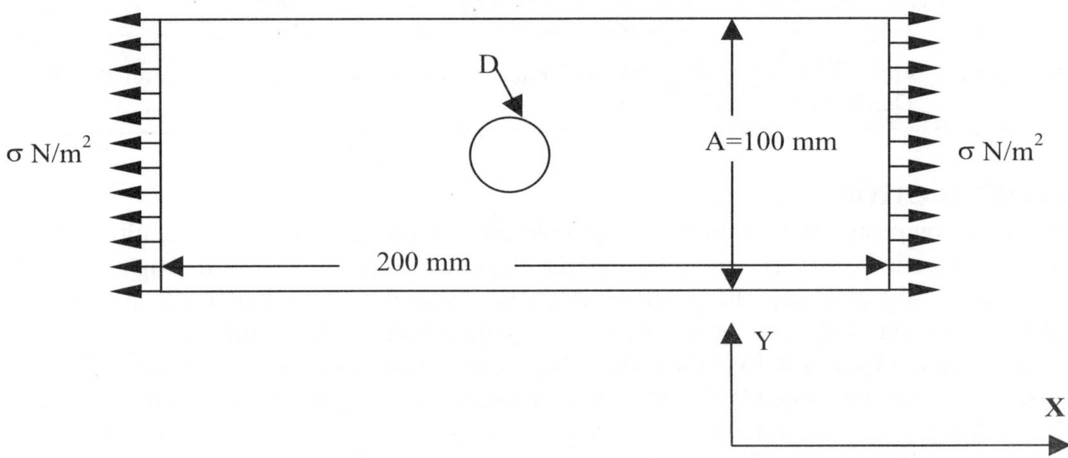

Fig. 1. Basic model (A plate with central hole and under in plane static loading).

3. FINITE ELEMENT ANALYSIS

An eight nodded Linear Layered Structural 3-D Shell Element with six degrees of freedom at each node (specified as Shell99 in ANSYS package) was selected based on convergence test and used through out the study. Each node has six degrees of freedom, making a total 48 degrees of freedom per element. Figure 2 provides the detail and geometry of element type. In order to construct the graphical image of the geometries of the three different models of different *D/A* ratio of laminated plate examined using the ANSYS (Advanced Engineering Simulation), it was necessary to input the basic geometric elements such as points, lines and arcs. Due to the un-symmetric nature of different system investigated, it was necessary to discretize the full laminated plate of each system for finite element analysis. Main task in finite element analysis is selection of suitable element type. Numbers of checks and convergence test are made for selection of suitable element type from different available elements and to decide the element length. Results were then displayed by using post processor of ANSYS programme. Figure 3 provides the example of the discretized models for D/A = 0.2 used in study.

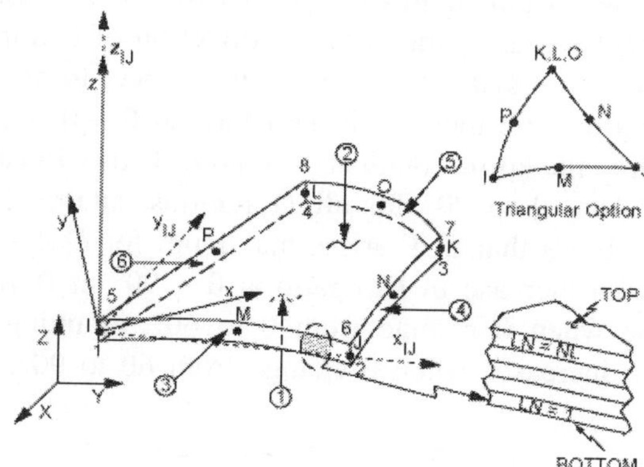

Fig. 2. Details of element type.

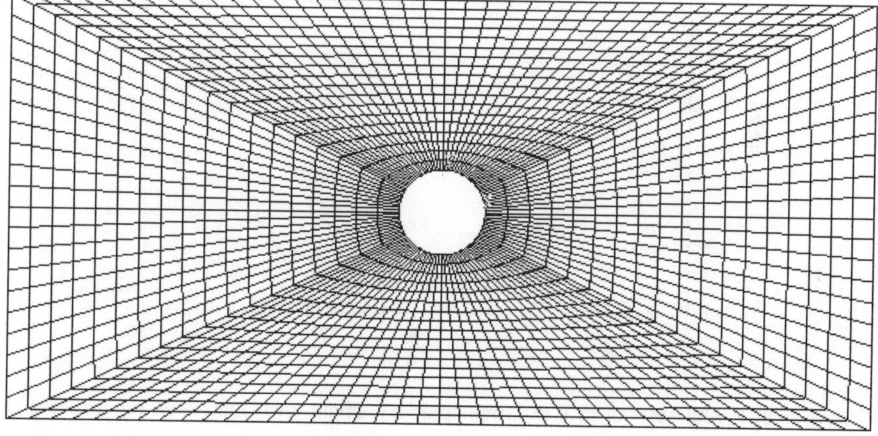

Fig. 3. Typical example of finite element model.

4. RESULTS AND DISCUSSION

Numerical results are presented for three different *D/A* ratio as 0.1, 0.2 and 0.5 i.e. for three different hole diameter 10, 20 and 50 *mm* in a finite width laminated plate. Three different orthotropic composite materials are used for analysis. The material properties are shown in following Table.

SN	Material	E_1(GPa)	E_2(GPa)	G_{12}(GPa)	μ
1	Woven glass/epoxy	294	6.4	4.9	0.23
2	Graphite\epoxy	235	137	47	0.30
3	Boron\aluminium	29.7	29.7	5.3	0.17

Where E_1, E_2, G_{12} are longitudinal, transverse, shear modulus respectively and μ is poisson's ratio.

Figures 4, 5 and 6 show the variation of SCF with respect to fibre orientation (θ) for Graphite\epoxy, Boron/Al and Woven glass/epoxy laminates respectively for three different *D/A* ratio as 0.1, 0.2 and 0.5. It is clear from figures 4 to 6 that SCF is at a maximum when the $\theta = 0°$ or $180°$ i.e. when the fibre directions are parallel to the direction of loading for all *D/A* ratios. It is also observed that the SCF follows a symmetric trend with respect to $90°$ orientations in all cases. Figure 4 shows that SCF increases with increase in *D/A* ratio at $\theta = 0°$ and decreases with increase in *D/A* ratio at $\theta = 90°$ in Graphite/epoxy laminates. Figure 4 also illustrates that SCF decreases continuously when θ change from 0 to $90°$ for all *D/A* ratios, attaining a minimum value when orientation is at $90°$. Figure 5 shows that SCF arises maximum for *D/A* = 0.2, minimum for *D/A* = 0.5 at $\theta = 0°$ and decreases with increase in *D/A* ratio at $\theta = 90°$ in Boron/Al laminates. Figure 5 also shows that SCF decrease when θ change from 0 to $60°$, attaining a minimum value when orientation is at $60°$ and again increases when θ change from 60 to $90°$.

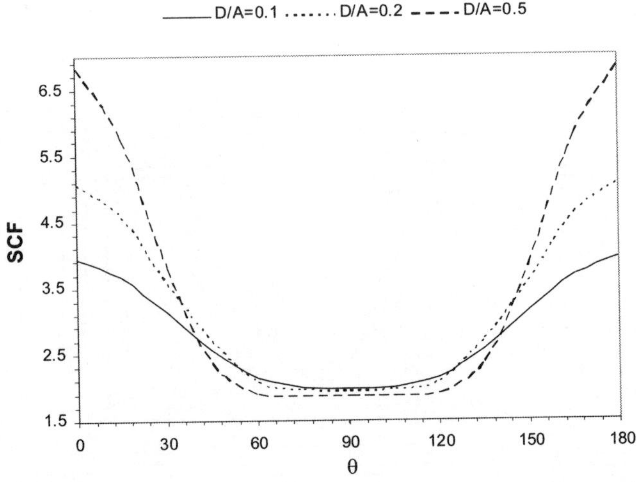

Fig. 4. Variation of SCF with respect to θ in Graphite\epoxy single layer laminates for different *D/A* ratios.

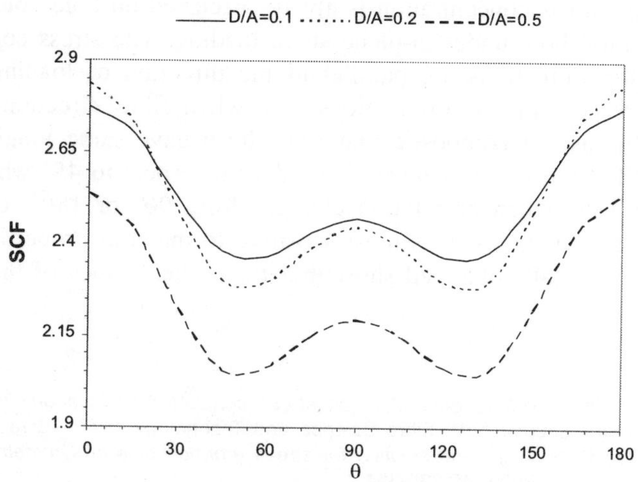

Fig. 5. Variation of SCF with respect to θ in Boron/Al single layer laminates for different *D/A* ratios.

Figure 6 shows that SCF arises maximum for *D/A* = 0.2 and minimum for *D/A* = 0.5 at θ = 0° or 90° in Woven glass/epoxy laminates. Figure 6 also illustrates that SCF decrease when θ change from 0 to 45°, attaining a minimum value when orientation is at 45° and again increases when θ change from 45 to 90°. Results obtained for woven glass/epoxy laminate show that the SCF follows a symmetric trend with respect to 45° when orientation change from 0° or 90° and to 135° when orientation change from 90° or 180° due to same value of E_1 and E_2.

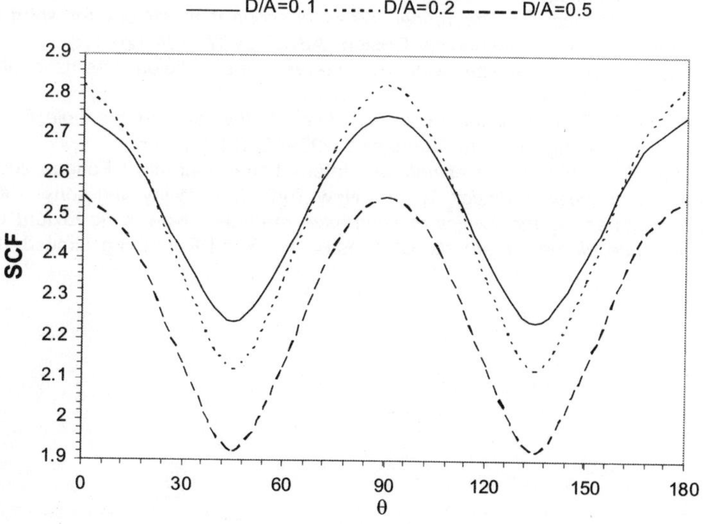

Fig. 6. Variation of SCF with respect to θ in Woven glass/epoxy single layer laminates for different *D/A* ratios.

CONCLUSION

In general, the maximum stress concentration is always occurred on hole boundary in a finite width laminated plate with central hole under in-plane static loading. The stress concentration factor is at maximum, when the fibre directions are parallel to the direction of loading. The SCF follows a symmetric trend with respect to 90° orientations (i.e. when fibre directions are perpendicular to loading) in all cases. In case of composite materials those have same longitudinal and transverse modulus of elasticity; SCF follows a symmetric trend with respect to 45° when orientation changes from 0° or 90° and to 135° when orientation changes from 90° or 180°. On the basis of results obtained, it has been seen that the SCF is most sensitive to material properties and directly depend on E_1/E_2 and E_1/G_{12}. The results obtained show that for higher values of these ratios, SCF is also higher.

REFERENCES

1.   Shastry BP, Raj GV. *Effect of fibre orientation on stress concentration in a unidirectional tensile laminate of finite width with a central circular hole*. Fibre Science and Technology 1977; 10:151-154

2.   Ukadgaonker VG, Rao DKN. *A general solution for stress around holes in symmetric laminates under in-plane loading*. Composite structure 2000; 49:339-354

3.   Paul TK, Rao KM. *Finite element evaluation of stress concentration factor of thick laminated plates under transverse loading*. Computers and Structures 1993; 48(2):311-317

4.   Paul TK, Rao KM. *Stress analysis in circular holes in FRP laminates under transverse load*. Computers and structures 1989; 33(4):929-935

5.   Meguid A. *Finite element analysis of defence hole system for the reduction of stress concentration factor in a uni-axially loaded plate with two coaxial holes*. Engineering fracture mechanics 1986; 25(4):403-413

6.   Ting K, Chen KT, Yang WS. *Stress analysis of the multiple circular holes with the rhombic array using alternating method*. International journal of pressure vessels and piping 1999; 76:503-514

7.   Chaudhuri RA. *Stress concentration around a part through hole weakening laminated plate*. Computers and structures 1987; 27(5):601-609

8.   Sinclair GB. On the effect on *stress concentration of rounding the edge of a hole through plate*. International Journal of Mechanical Sciences 1980; 22(12):731-734

9.   Troyani N, Gomes C, *Sterlacci G. Theoretical stress concentration factors for short rectangular plates with centred circular holes*. Journal of Mechanical Design, ASME 2002; 124:126-128

10.   Iwaki T. *Stress concentrations in a plate with two unequal circular holes*. International Journal of engineering Sciences 1980; 18(8):1077-1090

11.   Giare GS, Shabahang R. *The reduction of stress concentration around the hole in an isotropic plate using composite material*. Engineering Fracture Mechanics 1989; 32(5):757-766

12.   Fillipini M. *Stress Gradient calculations at notches*. International Journal of Fatigue 2000; 22(5):397-409

13.   Peterson RE. *Stress concentration design factors*. New York: John Wiley and sons, 1966

14.   Daniel IM, Ishai O. *Engineering mechanics of composite materials*. New York: Oxford University Press, 1994

15.   Ross CTF. *Advance finite element methods*. Chichester: Horwood Publishing Limited, 1998.

23

Effect of Bidirectional Fiber Orientation on the Bending Strength of Hybrid Polyester Composites

K.G. SATISH[1] AND B. SIDDESWARAPPA[2]

[1]Research Scholar, Mechanical Engineering Department, University B.D.T. College of Engineering, Davangere, Karnataka, India email: kgsati@gmail.com
[2]Professor & Chairman, Dept. of Industrial & Production Engg., University B.D.T. College of Engineering, Davangere, India, email: dr_bsiddeswar@yahoo.co.in

ABSTRACT

The composite materials are being developed to replace conventional materials for competitive reasons such as high specific strength, higher fracture toughness, good resistance to heat, cold and moisture, ease of fabrication, etc. The determination of material properties is of importance for optimum design, quality control and damage detection. An experimental investigation was conducted to study the effect of bidirectional orientation and volume fraction on steel reinforced polyester hybrid composite specimen subjected to bending test. The hybrid composite laminated specimens were fabricated from stainless steel and nylon bidirectional mesh as reinforcements and polyester as the binder according to ASTM standards. The results indicated that the bending strength varies significantly with the change in fiber orientation.

Keywords: Steel-Polyester Hybrid, Bending Strength, Bidirectional Orientation, Volume Fraction.

1. INTRODUCTION

Fiber reinforced composites are certainly one of the oldest and most widely used composite materials. Their study and development have largely carried out due to their vast structural potential and also the concept and technology of fiber reinforced polymer composites have undergone a sea change with better understanding of the basics like the bonding mechanism between the matrix and fiber reinforcement, fiber orientation, fiber reinforcement size and distribution, morphological features etc.

Fiber reinforced polymer composites have steadily gained applications in the fields ranging from aerospace to piping to sports equipment. This appeal is due primarily to their high strength-to-weight ratio, tailorable mechanical properties and fatigue resistance. In order to improve the properties of existing composites, a newer trend is towards the development of Hybrid Composites. Generally Hybrid applies to advanced composites and refers to use of various combinations of fibers or particulate in either thermoset or thermoplastic matrices. Hybrids have unique feature that can be used to meet the diverse and competing design requirements in a more cost effective way than either advanced or conventional composites.

There has been various rehabilitation techniques proposed for civil infrastructure to overcome problems associated with the aging process, increased traffic, change in use, and deterioration. Among these techniques, external strengthening provides a practical and cost effective solution when compared to other traditional repair methods. The first generation of external strengthening methods utilized steel plates bonded to the tension surface of the structure. The strengthening effectiveness was acceptable; however several problems, including durability, heavy weight, handling, and shoring, had to be resolved; thus the need for alternative materials aroused. The introduction of advanced composite materials, particularly fiber reinforced polymers (FRP), in structural engineering industries, as a second generation of externally bonded retrofit materials, has offered numerous benefits (i.e. corrosion-free, excellent weight to strength ratio, good fatigue resistance, flexibility to conform to any shape, broad applications, and easy manipulations). Retrofit of structures using glass-FRP (GFRP) and carbon-FRP (CFRP) has been studied extensively over the past decade (Meier 1995, Neale 2000, Bakis *et al.*, 2002 among many others). Although the applications of FRPs are becoming wider and popular, the cost of material is still relatively high. Recently, a new composite material has been developed to overcome this shortcoming. The steel reinforced polymer (SRP) consists of high-carbon steel unidirectional Hardwire fabrics embedded in polymer matrix [1].

The Coir/Glass Polyester hybrid composite was studied for the enhancement of properties by C. Pavithran *et al.* [2]. Mohan and Kishore [3, 4] have reported that in jute-glass hybrid composites, jute can be used as a reasonable core material. They evaluated flextural properties [3] of jute-glass reinforced epoxy laminates fabricated from filament winding technique. Four different hybrid combinations were studied for different percentage of glass and the results compared with the jute reinforced plastic. They found substantial increase in flextural properties due to hybridization. Mishra *et al.* [5] Studied the effect of glass fiber addition on tensile and flextural strength and izod impact strength of pineapple leaf fiber along with sisal fiber reinforced polyester composites. K. John *et al.* have studied on sisal-glass polyester hybrid composites with 5% and 8% volume fraction and found a considerable enhancement in flextural [6], impact, tension [7], compression properties.

The present study involves the effect of hybridization of steel/nylon/ polyester composite on the bending strength for various orientations and volume fractions in accordance with the ASTM standards.

2. MATERIALS

The different materials selected for this study are Isopthalic Polyester Resin, stainless steel and nylon. Polyester Resin is used for a wide variety of industrial and consumer applications. Typical Polyester reinforced applications are boats, cars, shower stalls, building panels, and corrosion-resistant tanks and pipes. Polyester resin composites are cost effective because they require minimal setup costs and the physical properties can be tailored to specific applications. Another advantage of polyester resin composites is that they can be cured in a variety of ways without altering the physical properties of the finished part. To make the composite material reduce in its weight, light weight reinforcement like nylon mesh which is freely available and stainless steel mesh to improve the stiffness and strength were selected. In the present investigation the materials selected for the preparation of the hybrid composite specimen are the resin system consisting of Isopthalic Polyester with MEKP catalyst and Cobalt Napthenate accelerator, stainless steel and nylon mesh as reinforcements for hybridization.

3. SPECIMEN PREPARATION

The symmetric bidirectional ply hybrid composite specimen was fabricated at room temperature under constant pressure in rectangular plate shape using hand lay-up method. The preparation of specimen includes selection of different volume fraction for ingredients in which the polyester matrix was 40% for all specimens, the percentage of stainless steel was varied from 15-45% in steps 15% and the corresponding percentage of nylon was estimated. The different bidirectional fiber orientations (0°/90°, 30°/60° and +45°/−45°) were considered for the preparation of specimen in this study. The hybrid composite specimens were evaluated by bending tests according to the ASTM D790 standards.

4. EXPERIMENTATION

Fig. 1. Experimental Set-up of 3-Point Bending Test

The bending test specimen prepared in accordance with ASTM standards were loaded on computer controlled Universal Testing Machine (Model-UTM/E-40). The specimen of each laminate family i.e., bidirectional fiber orientations 0°/90°, 30°/60° and +45°/−45° for different volume fraction were subjected to three-point bending test conducted at room temperature. The specimens were tested for different bidirectional fiber orientations and volume fractions under freezing temperature (below 0°C) and room temperature (32°C).

5. RESULTS AND DISCUSSION

The influence of bidirectional fiber orientation on steel reinforced polyester hybrid composite specimen under 3-Point Bending Test can be studied by categorizing the analysis into the following sections:

- Relation between Bending Strength and Bidirectional Fiber Orientation at Freezing and Room Temperature.
- Relation between Bending Strength and Volume Fraction of Steel at Freezing and Room Temperature.
- Effect of Working Temperature on Bending Strength for different Volume Fraction of Steel.

5.1 Relation between Bending Strength and Bidirectional Fiber Orientation at Freezing and Room Temperature

The effect of bidirectional fiber orientation at different working temperature can be analyzed using Figures 2 and 3. The specimen with higher steel percentage has shown superior bending strength at freezing and room temperatures. The bending strength of composite laminates was higher when the bidirectional fiber orientation is 0°/90° at both the temperatures compared to 30°/60° and +45°/–45° bidirectional fiber orientations. This increased in bending strength is due to the fact that the loads when applied on the specimen distribute equally and along the axis of the fibers. But in 30°/60° and +45°/–45° bidirectional fiber orientations the load distribution is unequal

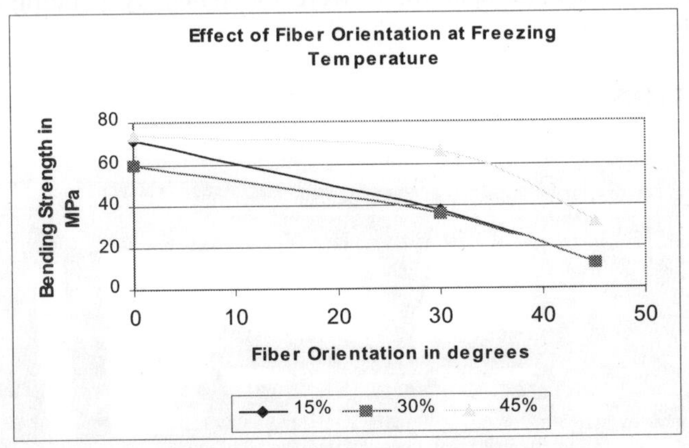

Fig. 2. Effect of fiber orientation at Freezing Temperature For various volume fractions of steel under bending load.

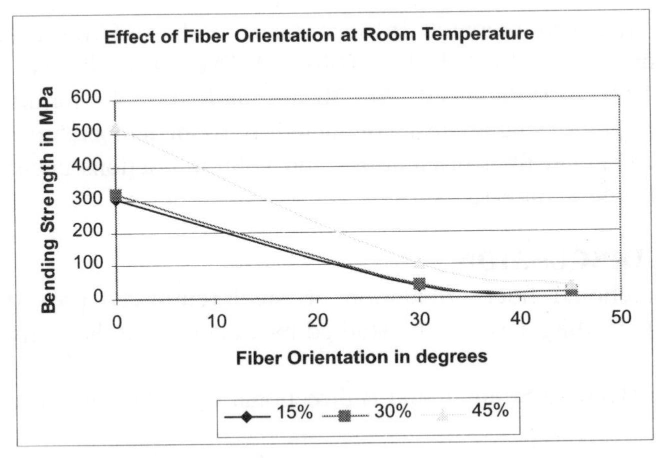

Fig. 3. Effect of fiber orientation at Room Temperature For various volume fractions of steel under bending load.

and not along the axis of the fibers and hence the load bearing capacity is less in these specimens. On the other hand, if the comparison is with respect to the service temperatures, the specimens under room temperature yielded superior results. This behavior is because of the reason that the specimens in room temperature have sufficient bonding, no internal stresses and, hence, load distribution is equal and uniform. Whereas the specimens in freezing temperature will encounter the problem of debonding, and, hence, delamination and microcracks develop resulting in earlier fracture of the specimens.

5.2 Relation between Bending Strength and Volume Fraction of Steel at Freezing and Room Temperature

It can be seen from the Figures 4 and 5 that, the variation in bending strength is not very significant with respect to the percentage of steel, but it is highly significant with respect to the bidirectional fiber orientation. The variation in bending strength is marginal as steel percentage varies from 15%

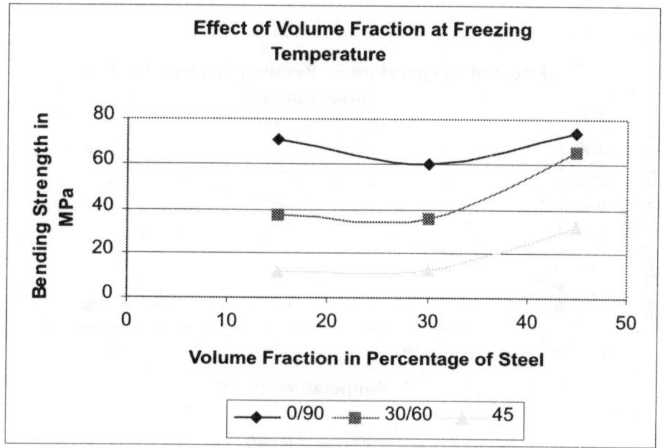

Fig. 4. Effect of Volume Fraction on Bending Strength at Freezing Temperature.

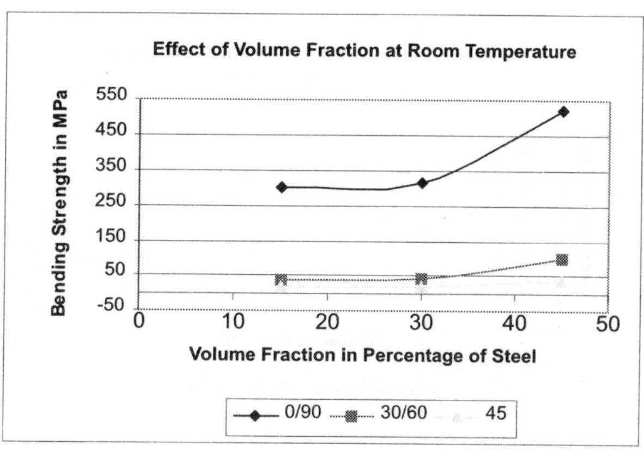

Fig. 5.   Effect of Volume Fraction on Bending Strength at Room Temperature.

to 45% at freezing temperature due to debonding. But the variation of bending strength is significantly increasing for 0°/90° bidirectional fiber orientation at room temperature when the percentage of steel is varied from 30% to 45% than with 15% to 30% steel content. The plots also confirms that the bending strength is superior in 0°/90° bidirectional fiber orientation specimen in both the temperatures. It can also be noted the bending strength is superior at room temperature than in freezing temperature.

5.3 Effect of Working Temperature on Bending Strength for different Volume Fraction of Steel

To analyze the effect of working temperature, the Figures 6, 7 and 8 are plotted. From the graphs, it can be clearly stated that the specimens in room temperature behave excellently under bending as they have greater bonding and load bearing capacity. The behavior is similar in all the three plots to prove that the steel fiber content has less significance in bending strength. The pecularity in the plots is that the 0°/90° *bidirectional fiber orientation* specimen in room temperature has

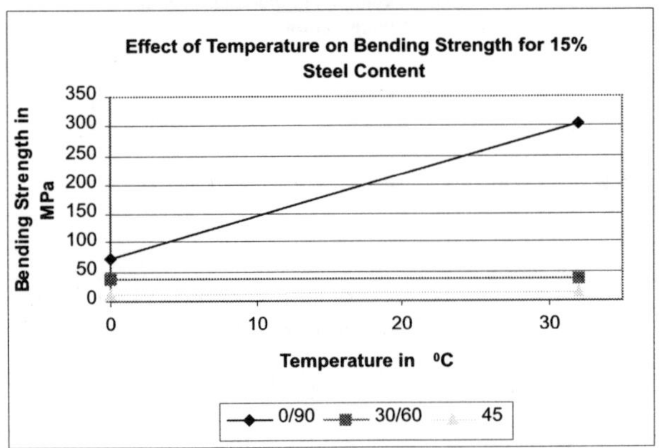

Fig. 6. Effect of Temperature on Bending Strength For 15% Steel Content.

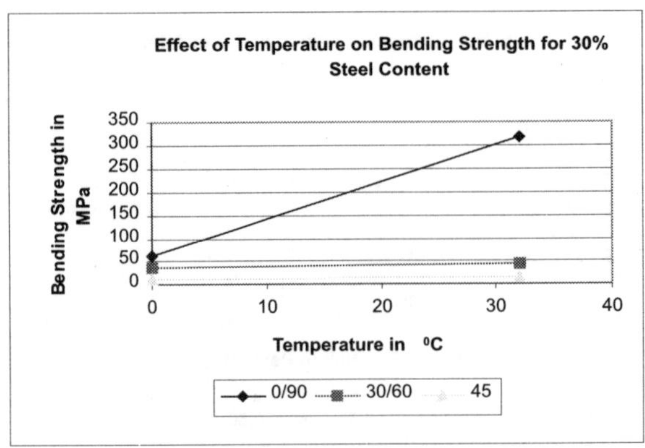

Fig. 7. Effect of Temperature on Bending Strength For 30% Steel Content.

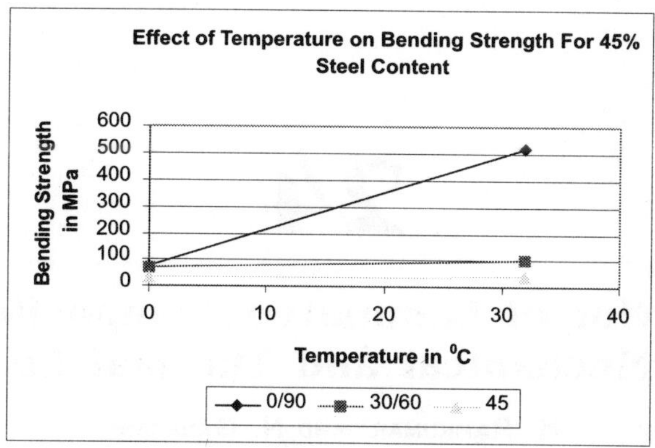

Fig. 8. Effect of Temperature on Bending Strength For 45% Steel Content.

higher bending strength compared to all other specimens. This is due to the fact that in case of specimens in freezing temperature tend to develop internal stresses as the steel fibers start shrinking at lower temperatures. Also it can be seen that the specimens in room temperature with 30°/60° and +45°/–45° bidirectional fiber orientations have lesser bending strength due to unequal load distribution.

CONCLUSION

The results of the investigation have revealed that the bending strength is superior in 0°/90° bidirectional fiber orientation specimen than in 30°/60° and +45°/–45° orientation specimens due uniform distribution of load. The bending strength is increasing significantly at room temperature for 0°/90° orientation when the steel percentage is varied from 30% to 45%. It can also be said that the bending strength is always superior at room temperature than at freezing temperature.

REFERANCES

1. Yail J. Kim, Amir Fam, Andrew Kong, and Mark F. Green, "Flexural Strengthening of RC Beams using Steel Reinforced Polymer (SRP) Composites,"
2. Pavithran C, Mukherjee P S, Brahkumar M and Damodaran A D, "Impact Properties of Sisal-Glass Hybrid Laminates", Journal of Material Science, 1991, Vol.26, pp.452-459.
3. Rangarajan Mohan and Kishore, "Jute-Glass Sandwich Composites", Journal of reinforced Plastics and Composites, 1985,Vol.4, pp.186-194.
4. Mohan R and Kishore, "Compressive Strength of Jute-Glass hybrid fiber composites", Journal of Material Science Letters, 1983, Vol.2,pp.99-102.
5. Mishra S, Mohanty A K, Drzal L T, Mishra M, Parija S and Tripathy S S, " Studies on Mechanical Performance of Biofiber/Glass Reinforced Polyester Hybrid Composites", Composites Science and Technology, 2003, Vol.63, pp.1377-1385.
6. John K and Venkata Naidu S, " Effect of Fiber Content and Fiber Treatment of Flextural Properties of Sisal Fiber/Glass Fiber Hybrid Composites", Journal of Reinforced Plastics and Composites, 2004, Vol.23, No.15, pp.1601-1605.
7. John K and Venkata Naidu S, "Tensile Properties of Unsaturated Polyester Based Sisal Fiber-Glass Fiber Hybrid Composites", Journal of Reinforced Plastics and Composites, 2004, Vol.23, No.17, pp1815-1819.
8. George Lubin, "Hand book of composites" Van Nostrand Reinhold Company, New York.
9. Robert M Jones., "Mechanics of composite materials", McGraw-Hill publications, New Delhi.

24

Buckling Behavior of Laminated Composite Box Column under Mechanical and Thermal Loadings

K. RAMKUMAR[1] AND **N. GANESAN**[2]

Machine Design Section, Department of Mechanical Engineering,
Indian Institute of Technology, Madras-600036 India.
email: [1]ramkumar2yk@yahoo.com and [2]nganesan@iitm.ac.in

ABSTRACT

In the present work, finite element method using four noded thin rectangular plate elements with five degrees of freedom per node is used for buckling analysis of laminated composite box columns. The formulation is based on classical lamination plate theory. In contrast to the analysis available in the literature, where the study was carried out under mechanical and thermal environment, here the effects of fiber angle, stacking sequence and boundary conditions on buckling behavior are taken in to account and analyzed.

Keywords: Buckling; Finite element; Composite; Box column.

1. INTRODUCTION

Fiber-reinforced laminated composite structures are finding increasing applications in engineering industries due to their lightweight and tailorable properties. In practical situations these laminated structures are subjected to various types of in plane loadings in the form of mechanical and thermal. The thickness of the majority of such structural components being very low, they are prone to buckling. It is well known that the plate structures are capable of carrying a much increased load beyond buckling without failure. In order to fully exploit their strength, an accurate prediction of their buckling load carrying capacity is essential. The importance of the field has attracted and the numerous researchers focusing on the investigation of buckling characteristics of laminated composite rectangular plates subjected to mechanical [1-3] and thermal loading [4-6].

Several studies have been already carried out on buckling behavior of plates subjected to mechanical and thermal loading and reported in the literature, mostly of plates made of isotropic and orthotropic materials. But the limited investigations have been carried out on the buckling of box type of laminated column under mechanical and thermal loading [7-8].

To author's knowledge, there is no much work reported on the buckling characteristics of the composite box column considering the effects of thermal and mechanical loading. Hence the present paper, aims to determine the critical buckling loads of laminated composite box column under mechanical and thermal loading and its corresponding buckling mode shapes using the finite element method.

2. FORMULATIONS

A plate theory proposed by Dawe and Wang [1] for thin laminated plate is discussed and extended to laminated composite square box column. The displacement field is expressed as

$$u(x,y,z) = u_o(x,y) - z\frac{\partial w_o}{\partial x}$$

$$v(x,y,z) = v_o(x,y) - z\frac{\partial w_o}{\partial y} \qquad \qquad ...(1)$$

$$w(x,y,z) = w_o(x,y)$$

where u, v and w are the displacements in x, y and z directions, respectively. u_o, v_o, w_o are the mid-plane translation, and $\dfrac{\partial w_o}{\partial x}$, $\dfrac{\partial w_o}{\partial y}$ are the independent normal rotations in the x and y direction. The strain-displacement relations are given by

$$\varepsilon_x = \frac{\partial u}{\partial x}, \; \varepsilon_y = \frac{\partial v}{\partial y}, \; \gamma_{xy} = \frac{\partial u}{\partial y} + \frac{\partial v}{\partial x} \qquad \qquad ...(2)$$

Thus, for the derived displacements u and v in Eqn. (1) and (2), the strains are

$$\varepsilon_x = \frac{\partial u_o}{\partial x} - z\frac{\partial^2 w_o}{\partial x^2}$$

$$\varepsilon_y = \frac{\partial v_o}{\partial y} - z\frac{\partial^2 w_o}{\partial y^2} \qquad \qquad ...(3)$$

$$\gamma_{xy} = \frac{\partial u_o}{\partial y} + \frac{\partial v_o}{\partial x} - 2z\frac{\partial^2 w_o}{\partial x \partial y}$$

The resultant forces $[N_x]$, $[N_y]$ and $[N_{xy}]$ and moments $[M_x]$, $[M_y]$ and $[N_{xy}]$ per unit length of the plate are given as

$$\begin{Bmatrix} N_x \\ N_y \\ N_{xy} \end{Bmatrix} = \int_{-h/2}^{+h/2} \begin{Bmatrix} \sigma_x \\ \sigma_y \\ \tau_{xy} \end{Bmatrix} dz, \quad \begin{Bmatrix} M_x \\ M_y \\ M_{xy} \end{Bmatrix} = \int_{-h/2}^{+h/2} \begin{Bmatrix} \sigma_x \\ \sigma_y \\ \tau_{xy} \end{Bmatrix} zdz \qquad \qquad ...(4)$$

2.1. Mechanical Buckling

Using eqn. (1) in the Green's expressions for in-plane non-linear strains and neglecting lower order terms in a manner consistent with the usual von Karman assumptions gives the following expressions for strains at a general point.

$$\begin{Bmatrix} \varepsilon_x \\ \varepsilon_y \\ \gamma_{xy} \end{Bmatrix} = \begin{Bmatrix} \dfrac{\partial u_o}{\partial x} - z\dfrac{\partial^2 w_o}{\partial x^2} + \dfrac{1}{2}\left(\dfrac{\partial w}{\partial x}\right)^2 \\[2ex] \dfrac{\partial v_o}{\partial y} - z\dfrac{\partial^2 w_o}{\partial y^2} + \dfrac{1}{2}\left(\dfrac{\partial w}{\partial y}\right)^2 \\[2ex] \dfrac{\partial u_o}{\partial y} + \dfrac{\partial v_o}{\partial x} - 2z\dfrac{\partial^2 w_o}{\partial x \partial y} + \left(\dfrac{\partial w}{\partial x}\right)\left(\dfrac{\partial w}{\partial y}\right) \end{Bmatrix} \qquad \ldots(5)$$

The laminated orthotropic construction of the plate is consisted of N layers. Each layer is of thickness t_k, so that $h = \sum_{k=1}^{N} t_k$ is the total thickness of the laminate. The linear stress–strain relation for each layer is expressed with x, y-axes and has the form

$$\begin{Bmatrix} \sigma_x \\ \sigma_y \\ \tau_{xy} \end{Bmatrix}_k = \begin{bmatrix} \bar{Q}_{11} & \bar{Q}_{12} & \bar{Q}_{16} \\ \bar{Q}_{12} & \bar{Q}_{22} & \bar{Q}_{26} \\ \bar{Q}_{16} & \bar{Q}_{26} & \bar{Q}_{66} \end{bmatrix}_k \begin{Bmatrix} \varepsilon_x \\ \varepsilon_y \\ \gamma_{xy} \end{Bmatrix}_k \qquad \ldots(6)$$

Eqn. (6) can be written by the use of eqn. (5) as,

$$\begin{Bmatrix} \sigma_x \\ \sigma_y \\ \tau_{xy} \end{Bmatrix}_k = \begin{bmatrix} \bar{Q}_{11} & \bar{Q}_{12} & \bar{Q}_{16} \\ \bar{Q}_{12} & \bar{Q}_{22} & \bar{Q}_{26} \\ \bar{Q}_{16} & \bar{Q}_{26} & \bar{Q}_{66} \end{bmatrix}_k \begin{Bmatrix} \dfrac{\partial u_o}{\partial x} - z\dfrac{\partial^2 w_o}{\partial x^2} + \dfrac{1}{2}\left(\dfrac{\partial w}{\partial x}\right)^2 \\[2ex] \dfrac{\partial v_o}{\partial y} - z\dfrac{\partial^2 w_o}{\partial y^2} + \dfrac{1}{2}\left(\dfrac{\partial w}{\partial y}\right)^2 \\[2ex] \dfrac{\partial u_o}{\partial y} + \dfrac{\partial v_o}{\partial x} - 2z\dfrac{\partial^2 w_o}{\partial x \partial y} + \left(\dfrac{\partial w}{\partial x}\right)\left(\dfrac{\partial w}{\partial y}\right) \end{Bmatrix} \qquad \ldots(7)$$

To obtain the resultants and moments acting on the laminate cross section in terms of A_{ij}, B_{ij}, and D_{ij} one has to substitute eqn. (7) into (4), from which:

$$\begin{Bmatrix} N_x \\ N_y \\ N_{xy} \end{Bmatrix} = \begin{bmatrix} A_{11} & A_{12} & A_{16} \\ A_{12} & A_{22} & A_{26} \\ A_{16} & A_{26} & A_{66} \end{bmatrix} \begin{Bmatrix} \varepsilon^o_x \\ \varepsilon^o_y \\ \gamma^o_{xy} \end{Bmatrix} + \begin{bmatrix} B_{11} & B_{12} & B_{16} \\ B_{12} & B_{22} & B_{26} \\ B_{16} & B_{26} & B_{66} \end{bmatrix} \begin{Bmatrix} k_x \\ k_y \\ k_{xy} \end{Bmatrix}, \quad \begin{Bmatrix} M_x \\ M_y \\ M_{xy} \end{Bmatrix} = \begin{bmatrix} B_{11} & B_{12} & B_{16} \\ B_{12} & B_{22} & B_{26} \\ B_{16} & B_{26} & B_{66} \end{bmatrix} \begin{Bmatrix} \varepsilon^o_x \\ \varepsilon^o_y \\ \gamma^o_{xy} \end{Bmatrix} + \begin{bmatrix} D_{11} & D_{12} & D_{16} \\ D_{12} & D_{22} & D_{26} \\ D_{16} & D_{26} & D_{66} \end{bmatrix} \begin{Bmatrix} k_x \\ k_y \\ k_{xy} \end{Bmatrix}$$

$$\ldots(8)$$

where $\quad \left(A_{ij}, B_{ij}, D_{ij}\right) = \int_{-h/2}^{+h/2} \bar{Q}_{ij}\left(1, z, z^2\right) dz \quad (i, j = 1, 2, 6)$ $\qquad \ldots(9)$

A_{ij} is the extensional stiffness matrix and it is associated with the plate in-plane behavior. D_{ij} is the bending stiffness matrix and B_{ij} refers to the coupling between the laminate bending and extension.

2.2. Thermal Buckling

For analyzing buckling behaviour of box columns under thermal environment, the stress-strain relation presented in eqn. (6) can be modified and expressed as

$$\begin{Bmatrix} \sigma_x \\ \sigma_y \\ \tau_{xy} \end{Bmatrix}_k = \begin{bmatrix} \bar{Q}_{11} & \bar{Q}_{12} & \bar{Q}_{16} \\ \bar{Q}_{12} & \bar{Q}_{22} & \bar{Q}_{26} \\ \bar{Q}_{16} & \bar{Q}_{26} & \bar{Q}_{66} \end{bmatrix}_k \begin{Bmatrix} \varepsilon_x - \alpha_x \Delta T \\ \varepsilon_y - \alpha_y \Delta T \\ \gamma_{xy} - \alpha_{xy} \Delta T \end{Bmatrix} \qquad ...(10)$$

where ΔT is temperature increase, α_x and α_y are the coefficients of thermal expansion in directions of x and y axes, respectively. For thermal buckling, the stress resultants and moment resultants can be obtained by integration of stresses over thickness of the plate with eqn. (10)

$$\begin{Bmatrix} N_x \\ N_y \\ N_{xy} \\ M_x \\ M_y \\ M_{xy} \end{Bmatrix} = \begin{bmatrix} A_{11} & A_{12} & A_{16} & B_{11} & B_{12} & B_{16} \\ A_{12} & A_{22} & A_{26} & B_{12} & B_{22} & B_{26} \\ A_{16} & A_{26} & A_{66} & B_{16} & B_{26} & B_{66} \\ B_{11} & B_{12} & B_{16} & D_{11} & D_{12} & D_{16} \\ B_{12} & B_{22} & B_{26} & D_{12} & D_{22} & D_{26} \\ B_{16} & B_{26} & B_{66} & D_{16} & D_{26} & D_{66} \end{bmatrix} \begin{Bmatrix} \varepsilon^o_x \\ \varepsilon^o_y \\ \gamma^o_{xy} \\ k_x \\ k_y \\ k_{xy} \end{Bmatrix} - \begin{Bmatrix} N_{tx} \\ N_{ty} \\ N_{txy} \\ M_{tx} \\ M_{ty} \\ M_{txy} \end{Bmatrix} \qquad ...(11)$$

where $[N_t]$ and $[M_t]$ are the thermal stress resultants & moment resultants, respectively and which are defined as

$$\{N_t\} = \begin{bmatrix} N_{tx} \\ N_{ty} \\ N_{txy} \end{bmatrix} = \sum_{k=1}^{n} \int_{h_{k-1}}^{h_k} \left[\bar{Q}_{ij} \right]_k \begin{Bmatrix} \alpha_x \\ \alpha_y \\ \alpha_{xy} \end{Bmatrix}_k \Delta T dz \;, \quad \{M_t\} = \begin{bmatrix} M_{tx} \\ M_{ty} \\ M_{txy} \end{bmatrix} = \sum_{k=1}^{n} \int_{h_{k-1}}^{h_k} \left[\bar{Q}_{ij} \right]_k \begin{Bmatrix} \alpha_x \\ \alpha_y \\ \alpha_{xy} \end{Bmatrix}_k \Delta T z dz$$

$$...(12)$$

The total potential energy Π of a box column under mechanical and thermal loading is equal to $\Pi = U + V$. Where U is the strain energy of bending. U represents the potential energy of in-plane loadings due to temperature change in the case of thermal buckling and due to compressive mechanical load in the case of mechanical buckling.

$$U = \frac{1}{2} \iint \left[N_x \varepsilon_x + N_y \varepsilon_y + N_{xy} \gamma_{xy} \right] dx dy \qquad ...(13)$$

$$V = \frac{1}{2} \iint \left[N_x \left(\frac{\partial w_0}{\partial x} \right)^2 + 2 N_{xy} \left(\frac{\partial^2 w_0}{\partial x \partial y} \right) + N_y \left(\frac{\partial w_0}{\partial y} \right)^2 \right] dx dy \qquad ...(14)$$

For the equilibrium, the potential energy Π must be stationary. The equilibrium equations of the box column subjected to thermal buckling and also mechanical buckling can be derived from the variational principle through use of stress–strain and strain–displacement relations. One may obtain these equations by using $\delta \Pi = 0$.

2.3. Finite Element Model and Formulations

In order to study the buckling of the box column (assemblages of plates), four-noded rectangular elements with 20 degrees of freedom system per element is used in this study. A typical mesh, generated over the laminated box column is shown in Fig. 1.

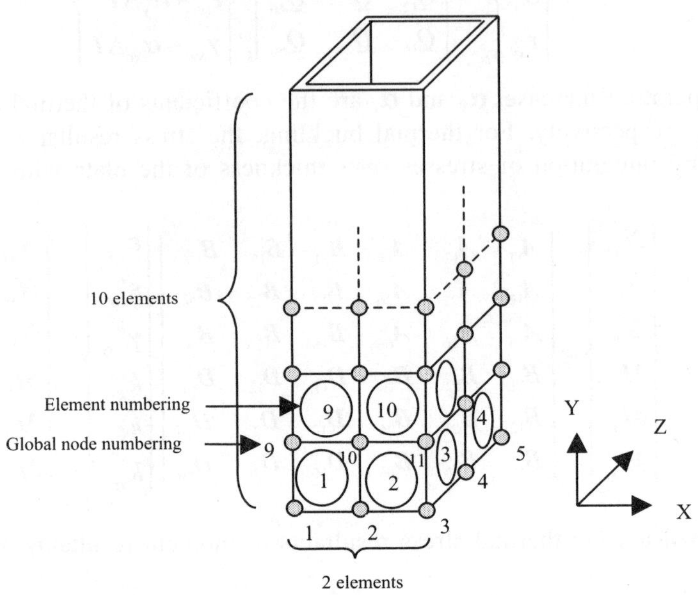

Fig. 1. Typical meshing of square box column with 80 rectangular elements.

The node (both local and global) and element numbering scheme is also shown in this figure. In what follows the finite element formulation is presented. The displacement components are approximated by the product of shape function matrix $[N_i]$ and the nodal displacement vector

$$\{d^e_i\} = \{u_{0i}, v_{0i}, w_{0i}, \partial w_{0i}/\partial x, \partial w_{0i}/\partial y,\}^T$$

i.e.
$$\{d\} = \sum_{i=1}^{4} [N_i]\{d^e_i\}$$

The stiffness matrix of the plate is obtained by using the minimum potential energy principle. Elemental stiffness $[K^e]$ and geometric stiffness $\left[K_g^e\right]$ can be expressed as

$$[K^e] = \iint_e [B_b]^T [D][B_b]dxdy, \quad [K_g^e] = \iint_e [B_g]^T \begin{bmatrix} N_x & N_{xy} \\ N_{xy} & N_y \end{bmatrix} [B_g]dxdy \qquad ...(15)$$

The total potential energy principle for the column satisfies the assembly of the element equations. In order to obtain the matrices $[K]$ and $[K_g]$ in global form, element stiffness and the geometric stiffness matrices are assembled and transformed by the following expression.

$$[K] = [T_r]^T [K^e][T_r], \quad [K_g] = [T_r]^T [K_g^e][T_r] \qquad ...(16)$$

where $[T_r]$ is the transformation matrix. The corresponding eigenvalues problem can be solved using any standard eigenvalues extraction procedures

$$\left[[K]+\lambda\left[K_g\right]\right]\begin{Bmatrix} u_i \\ v_i \\ w_i \end{Bmatrix} = \{0\} \qquad \qquad ...(17)$$

The buckling eigen values and buckling mode shapes are computed using the simultaneous iteration technique. The product of λ and the initial guessed value T is the critical buckling temperature (T_{cr}) Whereas the critical strength $(P_{cr.})$ under mechanical load can be obtained by multiplying λ and the in-plane mechanical load P.

3. NUMERICAL RESULTS AND DISCUSSIONS

The materials used in this study are Glass/epoxy (M1) and Kevlar/epoxy (M2) composite, whose properties are reported in Table 1. E, v, G, α denote, respectively, Young's modulus, Poisson's ratio, shear modulus and coefficient of thermal expansion; subscript L represents fiber direction, the subscript T the direction orthogonal to them. The symmetric composite box column with geometry of L = 0.750 m, b = 0.150 m, varied in thickness (h) and number of layers = 5 is taken for investigating the buckling behaviour. The boundary conditions for which the results have been obtained are clamped-free and clamped-clamped end conditions.

*C-F: *all dofs are arrested at the bottom end and remaining are free to translate and rotate*
*C-C1: *all dofs are arrested at the both two ends (in the case of mechanical buckling, load is applied next to the constrained nodes at the top end)*
*C-C2: *all dofs are arrested at the bottom end and at the top end fixed against all degrees of freedom except for the displacement at the loaded end in the direction of the applied load.*

Table 1. Material properties

Material	E_1(Gpa)	E_2(Gpa)	G_{12}(Gpa)	G_{12}(Gpa)	$\alpha_1(10^{-6}/°C)$	$\alpha_2(10^{-}/°C)$
M1[6]	53.4	17.9	0.25	8.6	6.3	20.5
M2[6]	76.0	5.5	0.34	2.3	-4.0	79.0

3.1. Convergence Study and Validation

A convergence study was carried out for the mechanical buckling analysis with C-F boundary condition and 320-element idealization is found to yield converged solutions. In order to minimize computational time in modeling the box column, 80-element idealization is considered in this study. Hence, critical load and buckling temperatures presented in the following section are obtained using 80-element idealization. In order to validate the code developed for evaluation of linear critical mechanical buckling load, a box column made of glass/epoxy is considered. The material properties used are given in the Table 1. Column geometry and the boundary conditions considered are given in Table 2. Table 2 shows a comparison between current finite element formulation and ANSYS-8 package. Table 2 shows an identical correlation.

Table 2. Validation of linear mechanical buckling

Boundary condition	Fiber angle	P_{cr} (KN), L = 0.750m, b = 0.150m, h = 0.004m, glass/epoxy [M1]			
		80 elements		320 elements	
		ANSYS	Present code	ANSYS	Present code
C-F	0°	50.0	49.70	61.24	63.54
	45°	60.4	60.12	71.12	70.88
	90°	52.01	49.2	61.92	63.61

3.2. Linear Mechanical and Thermal Buckling of Box Column

3.2.1.Mechanical buckling: The composite column subjected to in-plane loading due to mechanical load is considered. The variation of the critical buckling load of box column made of M1 and M2 material with the fiber angle for two different boundary conditions like clamped-free (C-F) and clamped-clamped (C-C1 and C-C2) is shown in Fig. 2 (a)-(c). It is observed that columns with clamped-clamped boundary condition exhibit a higher critical buckling load compared to those with clamped-free edge condition. It is also interesting to note that the critical buckling load decreases with an increase in length-to-thickness ratio (L/h) for both the end conditions. In addition, the critical buckling load of symmetric laminated box column is increases up to 45^0 fiber angle and then reduces for C-F and C-C2 and not much for C-C1.

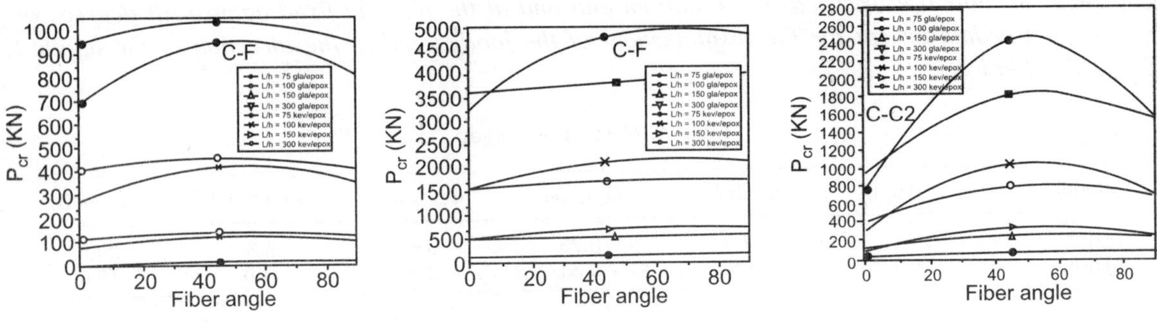

Fig. 2. Variation of critical buckling load with fiber angle for different boundary conditions.

3.2.2. Thermal buckling: The effect of fiber angle and length-to-thickness ratio on the critical buckling temperature of composite box column is shown in Fig. 3 (a)-(b). In this study, the column with clamped-free (C-F) and clamped-clamped (C-C1) end condition, under uniform temperature rise (T = 100°C) is considered. Coefficient of thermal expansion and mechanical properties used in the study for the material M1 and M2 are given in Table 1. The study indicates that the critical buckling temperature decreases with the increase in length-to-thickness ratio. Thermal buckling strength of thick plates is higher than that of thin plates in the small (but finite) deflection regime. It is also observed that columns with clamped-free boundary condition exhibit a higher critical buckling temperature compared to those with clamped-clamped edge condition. In addition, the critical buckling temperature of symmetric laminated box column is continuously decreases as fiber angle increases for C-F and for C-C1 increases.

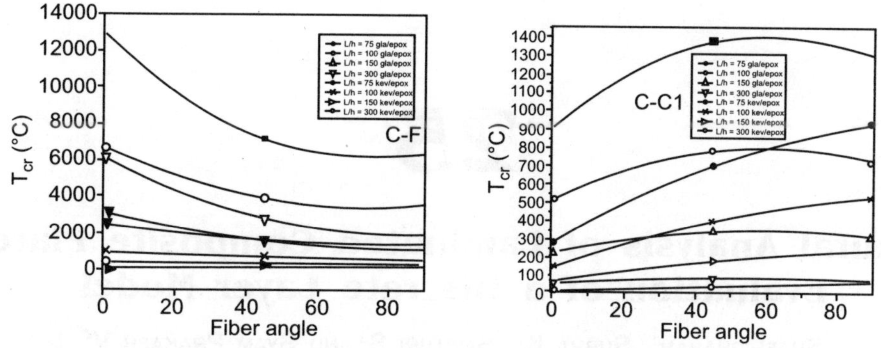

Fig. 3. Variation of critical buckling temperature with fiber angle for different boundary conditions.

CONCLUSION

Linear buckling behavior of laminated composite box column has been investigated by using finite element method under mechanical and thermal loading. The numerical results discussed for a wide range of geometry, loading and boundary conditions are in good agreement with the theory and FEM package ANSYS8. Also the effects of various parameters such as plate length-to-thickness ratio, fiber orientation angle and number of layers on buckling behaviour are studied in greater detail. Based on these observations, the said technique can be recommended for the buckling analysis of laminated composite box column under mechanical and thermal loadings to predict their behaviour with adequate accuracy.

REFERENCES

1. D.J.Dawe and S.Wang, 1998, Postbuckling analysis of thin rectangular laminated plates by spline FSM, Thin-Walled Structures. 30, pp.159-179.
2. K.K.Shukla, Y.Nath, E.Kreuzer and K.V.Sateesh kumar, 2005, Buckling of laminated composite rectangular plates, Journal of Aerospace Engineering. 18, 215- 223.
3. N.G.R. Iyengar and A. Chakraborty, 2004, Study of interaction curves for composite laminate subjected to in-plane uniaxial and shear loadings. Composite Structures. 64, 307-315.
4. Thangaratnam KR, Ramachandran J, 1989, Thermal buckling of composite laminated plates. Computers & Structures, 32, pp.1117–1124.
5. Prabhu MR, Dhanaraj R., 1994, Thermal buckling of laminated composite plates, Computers & Structures. 53, pp.1193–204.
6. Robert M. Jones, 2005, Thermal buckling of uniformly heated unidirectional and symmetric cross-ply laminated fiber-reinforced composite uniaxial in-plane restrained simply supported rectangular plates, composites part A: applied sciences and manufacturing.36, pp. 1355-1367.
7. K. Bhaskar and L. Librescu, 1995,Buckling under axial compression of thin-walled composite beams exhibiting extension-twist coupling, Composite Structures.3, pp. 203-212.
8. R. Suresh and S. K. Malhotra, 1997, Some studies on buckling of laminated composite thin walled box beams, Composite Structures, 40, pp. 267-275.

25

Flexural Analysis of Laminated Composite Plates: Evaluation of a Discrete Layer Model

SHASHIDHARAN[1], SUBHA K[2], SAVITHRI S[3] AND SYAM PRAKASH V[4]

[1]Department. of Civil Engineering, NSS College of Engineering, Kerala-678008 India
email: karkeer@sancharnet.in
[2]Department of Civil Engineering, NSS College of Engineering, Kerala-678008 India
email: shashisubha@gmail.com
[3]Regional Research Laboratory, CSIR, Trivandrum, Kerala-695019 India
email: sivakumarsavi@rediffmail.com
[4]Department of Civil Engg., Govt.Engineering College Trivandrum, Kerala-695016 India
email: syam@asia.com

ABSTRACT

The behaviour of laminated composite plate under static loading is studied by taking into account of the effects of shear deformation based on a higher order layer wise theory. The displacement model incorporates piecewise cubic variation of in plane displacements through the thickness and is layer dependent and able to maintain in plane displacements and shear stress continuity at the interface. Zero transverse shear stress boundary condition at the top and bottom of the plate are also satisfied. The numerical results indicate that the model predicts very accurate results for displacements and stresses for symmetric and anti-symmetric cross ply laminates. The results are compared with equivalent single layer theory and 3D elasticity solutions.

Keywords: Zig Zag model, Composite Laminated Plates, Higher order Shear deformation theory, Heavyside Function, Symmetric Cross Ply.

1. INTRODUCTION

The use of Composite materials has increased steadily during the past two decades, particularly in aerospace, under water, and automotive structures because of their high specific strength (failure stress/unit weight) and specific stiffness (stiffness/unit weight). The mechanical behaviour of laminated composite plates are strongly dependent on the degree of orthotropy of individual layers, the low ratio of transverse shear modulus to the in plane modulus and the stacking sequence of the laminates. The low modulus of shear necessitates higher order plate theories for analysis of composite plates.

The classical lamination theory based on Kirchhoffs' hypothesis is inaccurate for modelling moderately thick plates neglecting the transverse shear strains and transverse normal strain in the laminate. In first order shear deformation theories (Mindlin *et al.* 1956, Cheung and Zhon, 2000) the transverse shear strains are assumed to be constant in the thickness direction, shear correction factors have to be incorporated. Hence, the accuracy of the solution will be dependent on predicting better estimation for shear correction factors. In order to obtain accurate results of plate response characteristics a number of contributions based on three dimensional elasticity theory have been made to analyse laminated composite plates by Pagano [1], Pagano and Hatfield [2] etc. However, these solutions are often computationally expensive. Several refined Higher order theories have been proposed by Noor and Burton [3], Reddy [4], and Leissa [5] etc. In most of these theories the in plane displacements are assumed to be a cubic expression of the thickness coordinate. In these models the displacements and slope with respect to thickness is continuous and leads to discontinuous transverse shear stress and is not able to model the Zig zag nature of the displacement field. These are also known as Equivalent Single Layer Theories (ESL).

In the later developments, the displacement models presented are layer dependent and are able to maintain the shear stress continuity at the interface. The layer wise theories can represent the zig zag behaviour of in plane displacements through the thickness. Several Layer wise Model for Laminated Plates have been presented by Mau [6], Di Sciuva [7], Ren [8] etc. This paper presents a Layer wise theory for flexural analysis of composite plates. The displacement model accounts for piecewise linear distribution across the thickness of the in plane displacements and allows the contact condition on the transverse shearing stresses to be satisfied. The number of variables in the model is independent of layers. Based on this model equilibrium equations are derived using Hamiltons' Principle. Results are compared with Reddys' [9] ESL model of similar class and closed form solutions of the first order shear deformation theory [10] and 3 D elasticity solutions [10].

2. PROBLEM FORMULATION

A laminate composed of N laminae oriented such that their material deflection corresponds to the coordinate axes is considered. The total plate thickness is h. Taking x, y to be a set of coordinates locating the middle plane of the plate and z axis pointing downwards, and the plate subjected to an arbitrary transverse load on the top surface and the material of plate is assumed to be linearly elastic. The assumed displacement field is [11]

$$u = u_0(x,y) - z\frac{\partial w_0}{\partial x} + J_{01}(z)u_1$$

$$v = v_0(x,y) - z\frac{\partial w_0}{\partial y} + J_{02}(z)v_1$$

$$w = w_0(x,y) \qquad \qquad ...(1)$$

where

$$J_{01} = p(z) + \sum_{i=1}^{N-1}[p(z) - p(z_1)]r_iH(z - z_i)$$

$$J_{02} = p(z) + \sum_{i=1}^{N-1}[p(z) - p(z_1)]t_iH(z - z_i) \qquad ...(2)$$

where $p(z) = z - \dfrac{4z^3}{3h^2}$ and $H(z)$ is the Heaviside unit function defined as

$$H(z - z_i) = \begin{cases} 0 \text{ if } z \le z_i \\ 1 \text{ if } z > z_i \end{cases}$$

the summation is extended over N-1 interfaces and Z_i is the Z coordinate of the ith interface, and r_i and t_i are function of the material properties of the laminate under consideration.

The normal and shear strain components can be expressed as

$$\varepsilon_x = \frac{\partial u_0}{\partial x} - z\frac{\partial^2 w_0}{\partial x^2} + J_{01}(z)\frac{\partial u_1}{\partial x}$$

$$\varepsilon_y = \frac{\partial v_0}{\partial y} - z\frac{\partial^2 w_0}{\partial y^2} + J_{02}(z)\frac{\partial v_1}{\partial x}$$

$$\gamma_{xy} = \left(\frac{\partial u_0}{\partial y} + \frac{\partial v_0}{\partial x}\right) - 2z\frac{\partial^2 w_0}{\partial x \partial y} + J_{01}(z)\frac{\partial u_1}{\partial y} + J_{02}(z)\frac{\partial v_1}{\partial x} \qquad ...(3)$$

$$\gamma_{xz} = u_1 J_{11}(z) \; ; \; \gamma_{yz} = v_1 J_{12}(z) \text{ where } J_{11}(z) = \frac{dJ_{01}}{dz} \text{ and } J_{12}(z) = \frac{dJ_{02}}{dz}$$

The stress strain relationship for any layer can be written as

$$\begin{Bmatrix} \sigma_x \\ \sigma_y \\ \tau_{xy} \end{Bmatrix} = \begin{bmatrix} Q_{11} & Q_{12} & 0 \\ Q_{12} & Q_{22} & 0 \\ 0 & 0 & Q_{66} \end{bmatrix} \begin{Bmatrix} \varepsilon_x \\ \varepsilon_y \\ \gamma_{xy} \end{Bmatrix} \text{ and } \begin{Bmatrix} \tau_{yz} \\ \tau_{xz} \end{Bmatrix} = \begin{bmatrix} Q_{44} & 0 \\ 0 & Q_{55} \end{bmatrix} \begin{Bmatrix} \gamma_{yz} \\ \gamma_{xz} \end{Bmatrix} \qquad ...(4)$$

Q_{ij} are plane stress reduced elastic constants in the material axes of the layer for the layer under consideration.

3. EQUILIBRIUM EQUATIONS

The equilibrium equations obtained using the principle of variational energy and virtual work is

$$\frac{\partial N_x}{\partial x} + \frac{\partial N_{xy}}{\partial y} = 0 \; ; \; \frac{\partial N_{xy}}{\partial x} + \frac{\partial N_y}{\partial y} = 0 \; ; \; \frac{\partial^2 M_x}{\partial x^2} + 2\frac{\partial^2 M_{xy}}{\partial x \partial y} + \frac{\partial^2 M_y}{\partial y^2} = -q \; ;$$

$$\frac{\partial P_x}{\partial x} + \frac{\partial P_{xy}}{\partial y} - V_{xz} = 0 \; ; \; \frac{\partial P_{yx}}{\partial x} + \frac{\partial P_y}{\partial y} - V_{yz} = 0 \qquad ...(5)$$

where the stress resultants are defined as

$$(N_x, N_y, N_{xy}) = \sum_{k=1}^{N} \int_{z_k}^{z_{k+1}} (\sigma_x, \sigma_y, \tau_{xy}) dz$$

$$(M_x, M_y, M_{xy}) = \sum_{k=1}^{N} \int_{z_k}^{z_{k+1}} (\sigma_x, \sigma_y, \tau_{xy}) z\, dz$$

$$(P_x, P_{xy}) = \sum_{k=1}^{N} \int_{z_k}^{z_{k+1}} (\sigma_x, \tau_{xy}) J_{01}(z)\, dz$$

$$(P_y, P_{yx}) = \sum_{k=1}^{N} \int_{z_k}^{z_{k+1}} (\sigma_y, \tau_{xy}) J_{02}(z)\, dz$$

$$(V_{xz}, V_{yz}) = \sum_{k=1}^{N} \int_{z_k}^{z_{k+1}} (\tau_{xz} J_{11}(z), \tau_{yz} J_{12}(z))\, dz \qquad \ldots(6)$$

In order to obtain the Navier's solutions cross ply laminates are considered. The Navier method admits SS1 type boundary conditions for cross ply laminates. The generalised displacements u_0, v_0, w_0, u_1, v_1 can be written by assuming the following variations

$$u_0 = \sum_{m=1}^{\infty} \sum_{n=1}^{\infty} U_{mn} \cos \alpha x \sin \beta y$$

$$v_0 = \sum_{m=1}^{\infty} \sum_{n=1}^{\infty} V_{mn} \sin \alpha x \cos \beta y$$

$$w_0 = \sum_{m=1}^{\infty} \sum_{n=1}^{\infty} W_{mn} \sin \alpha x \sin \beta y$$

$$u_1 = \sum_{m=1}^{\infty} \sum_{n=1}^{\infty} X_{mn} \cos \alpha x \sin \beta y$$

$$v_1 = \sum_{m=1}^{\infty} \sum_{n=1}^{\infty} Y_{mn} \sin \alpha x \cos \beta y \qquad \ldots(7)$$

where $\alpha = \frac{m\pi}{a}$ and $\beta = \frac{n\pi}{b}$

The transverse load q is also expanded in double Fourier series as

$$q(x, y) = \sum_{m=1}^{\infty} \sum_{n=1}^{\infty} Q_{mn} \sin \alpha x \sin \beta y \qquad \ldots(8)$$

4. NUMERICAL RESULTS AND DISCUSSION

Flexural analysis of Simply Supported Five Layer (0/90/0/90/0) square plates under Sinusoidal Loading ($h_1 = h_3 = h_5 = h/6$, $h_2 = h_4 = h/4$, $E_1 = 25\, E_2$, $G_{12} = G_{13} = 0.5\, E_2$, $G_{23} = 0.2 E_2$, $\nu_{12} = 0.25$) is carried out and results are compared with the 3D elasticity solutions, Higher order Shear Deformation Theory (Reddy's Model) and Closed form solutions of First order Shear deformation Theory [10]. The solutions using Higher order Shear Deformation Theory are calculated using

Reddy's model as the results for the present problem is not available in the literature. Table 1 contains the non-dimensionalised Maximum deflection and Stresses for a symmetric cross ply plate predicted with the layer wise theory derived here. The table also contains the results obtained by 3 D Elasticity solution [1], the closed form solutions of First order Shear Deformation Theory [10] and the results calculated using Reddy's ESL model of similar class. It may be observed that the results obtained by the present model are in excellent agreement with the elasticity solution.

Table 1. Comparison of Maximum Deflection and Normal Stresses of a five Ply (0/90/0/90/0) Laminate ($h_1 = h_3 = h_5 = h/6$, $h_2 = h_4 = h/4$)

a/h	Source	$\overline{w} \times 10^2$ (a/2, b/2,0)	σ_x (a/2, b/2, h/2)	τ_{xy} (0, 0, − h/2)	τ_{xz} (0, b/2,0)	τ_{yz} (a/2, 0, 0)
	ELS	5.2949	1.3320	0.0634	0.2270	0.1860
	Present	5.2513	1.2392	0.0685	0.2125*	0.1716*
2	HSDT(Reddy)	4.7912	1.0917	0.0628	0.1580*	0.2409*
	CFS	4.8346	0.4021	0.0243	0.2112*	0.3068*
	ELS	1.8505	0.6850	0.0384	0,2380	0.2290
4	Present	1.8152	0.6911	0.0357	0.2298	0.2406
	HSDT(Reddy)	1.5949	0.6366	0.0332	0.2157	0.2579
	CFS	1.5623	0.4369	0.0235	0.2267	0.2809
	ELS	0.6771	0.5450	0.0247	0.2580	0.2230
10	Present	0.6689	0.5478	0.0241	0.2557	0.2278
	HSDT(Reddy)	0.6269	0.5397	0.0235	0.2546	0.2275
	CFS	0.6213	0.5021	0.0221	0.2559	0.2324
	ELS	0.4938	0.5390	0.0222	0,2680	0.2120
20	Present	0.4919	0.5388	0.0220	0.2670	0.2129
	HSDT(Reddy)	0.4810	0.5374	0.0219	0.2670	0.2122
	CFS	0.4796	0.5276	0.0215	0.2673	0.2135
	ELS	0.4412	0.5390	0.0214	0.2710	0.2060
50	Present	0.4410	0.5386	0.0214	0.2713	0.2066
	HSDT(Reddy)	0.4393	0.5384	0.0214	0.2714	0.2064
	CFS	0.4390	0.5368	0.0213	0.2714	0.2066
	ELS	0.4338	0.5390	0.0213	0.2720	0.2050
100	Present	0.4332	0.5387	0.0213	0.2720	0.2056
	HSDT(Reddy)	0.4333	0.5386	0.0213	0.2720	0.2055
	CFS	0.4332	0.5382	0.0213	0.2720	0.2056
	CLPT	0.4313	0.5387	0.0213	0.2722	0.2052

*** stresses computed from 3D Equilibrium equations**
ELS: 3D Elastic Solutions, **HSDT**: Higher order Shear deformation Theory, **CFS**: Closed form solutions of the first order shear deformation Theory with K = 5/6, **CLPT** : Classical Laminated Plate Theory

The following non-dimensionalized quantities are reported in the tables.

$$\overline{w} = w_0(a/2, b/2)\left(\frac{E_2 h^3}{a^4 q}\right) \times 100 \, ; \quad \sigma_x = \sigma_x(a/2, b/2, h/2)\left(\frac{h^2}{b^2 q}\right);$$

$$\tau_{xy} = \tau_{xy}(0, 0, h/2)\left(\frac{h^2}{b^2 q}\right); \quad \tau_{yz} = \tau_{yz}(a/2, 0, 0)\left(\frac{h}{bq}\right); \quad \tau_{xz} = \tau_{xz}(0, b/2, 0)\left(\frac{h}{bq}\right)$$

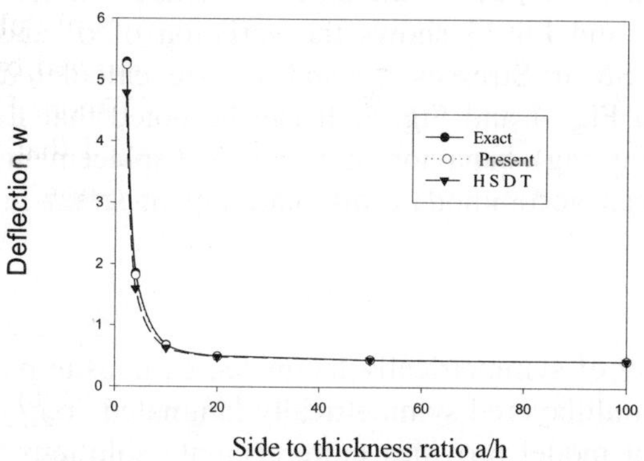

Fig. 1. Maximum Central deflection as a function of side to thickness ratio.

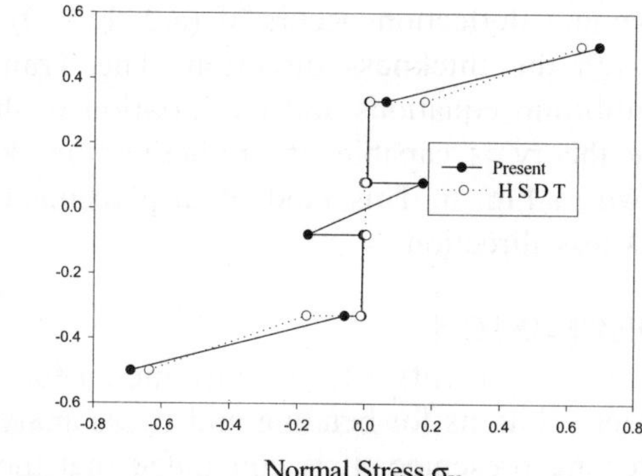

Fig. 2. Variation of Normal Stress σ_x through thickness of a 5 ply Laminated plate under Sinusoidal loading (a = b = 1) a/h = 4.

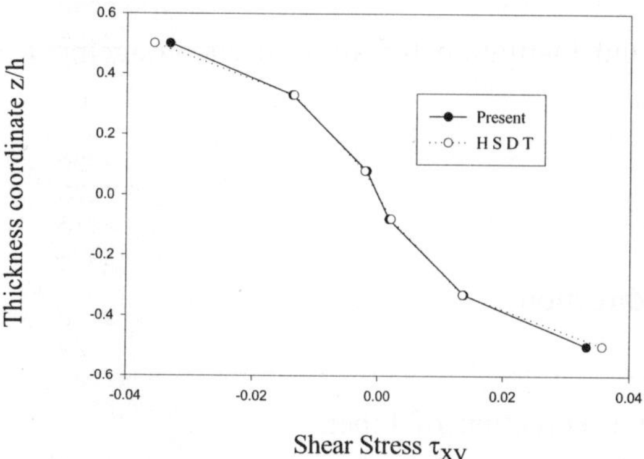

Fig. 3. Variation of Transverse Shear Stress τ_{xy} through thickness of a 5 ply Laminated plate under Sinusoidal loading (a = b = 1) a/h = 4.

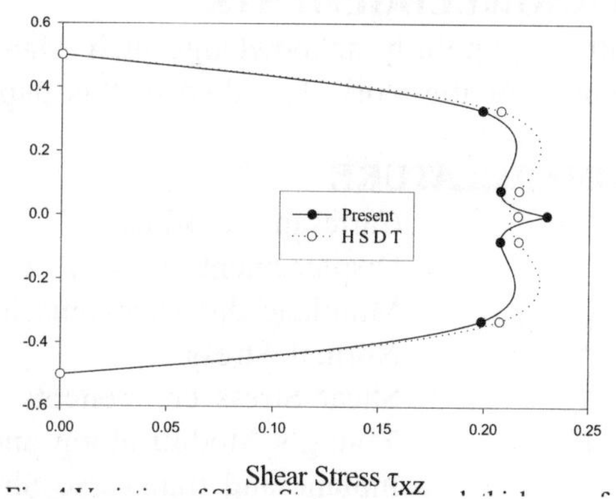

Fig. 4. Variation of Shear Stress τ_{xz} through thickness of a 5 ply Laminated plate under Sinusoidal loading (a = b = 1) a/h = 4.

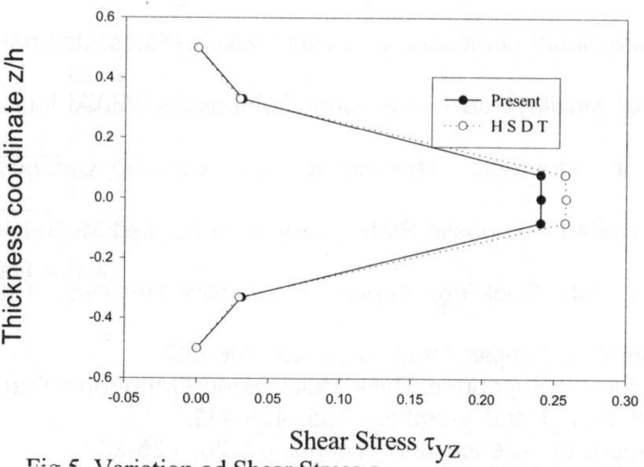

Fig. 5. Variation of Shear Stress τ_{yz} through thickness of a 5 ply Laminated plate under Sinusoidal loading (a = b = 1) a/h = 4.

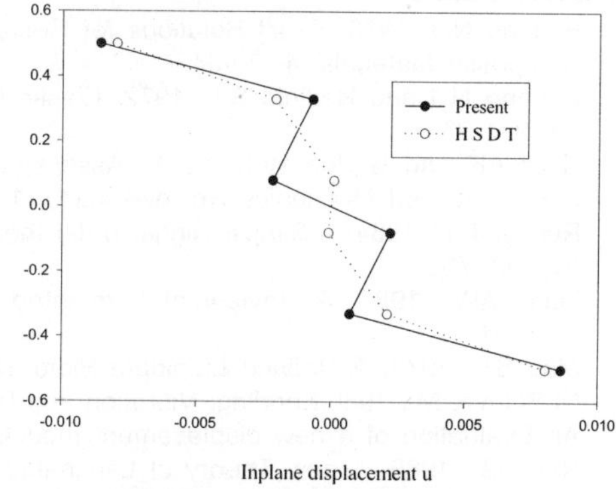

Fig. 6. Variation of inplane displacement u through thickness of a 5 ply Laminated plate under Sinusoidal loading (a = b = 1) a/h = 4.

Fig 1. shows the maximum central deflection with respect to the side to thickness ratio. The maximum deflection occurs at (a/2, b/2, 0). Fig. 2 and Fig. 3 shows the variation of σ_x and τ_{xy} through the thickness direction. The Transverse Shear Stresses τ_{xz} and τ_{yz} are calculated by Equilibrium equations and its variation is shown in Fig. 4 and Fig. 5. It can be noted that Layer wise theory is capable of predicting the 'kink' (zig zag) behavior in in-plane displacements as shown in Fig. 6. This kind of displacement field can accommodate discontinuity in strain in the thickness direction.

CONCLUSION

This paper presents a Layer wise model for the study of symmetrically laminated composite plates. Navier solutions for bending and stress analysis of multilayered symmetrically laminated composite plates are presented. It is concluded that the present model provides more accurate solutions than that of HSDT with similar level of analytical complexity when compared with 3D elasticity exact solutions.

ACKNOWLEDGEMENTS

Authors gratefully acknowledge their Management and Institution for all kind encouragement and support for the work described in this paper.

NOMENCLATURE

x, y, z	Cartesan Coordinates.
u, v, w	Displacements in x, y, z directions.
u_o, $v_0 w_0$	Midplane displacements in x, y, z directions.
σ_x, σ_y	Normal Stress
τ_{xy}, τ_{xz}, τ_{yz}	Shear Stress components
E_1, E_2	Young's Moduli along and transverse Direction of Fiber.
G_{12}, G_{13}, G_{23}	Inplane and transverse Shear Moduli.
q	Intensity of loading

REFERENCES

1. Pagano N.J, 1970, Exact Solutions for Rectangular Bidirectional composite and Sand Witch Plates, Journal of Composite Materials, 4, 20-34.
2. Pagano N.J and Hatfield SJ., 1972, Elastic Behaviour of Multilayered Bidirectional Composites, AIAA Journal, 10, 931-933.
3. Noor AK and Burton WS., 1989, Assessment of Shear Deformation Theories for Multi Layered Composite Plates, Applied Mechanics Reviews,42(1), 1-13.
4. Reddy, J. N.,1984, A Simple Higher order theory for Laminated Composite Plates, Journal of Applied Mechanics, 51, 745-752.
5. Leiss AW., 1987, A Review of Laminated Composite Plate Buckling, Applied Mechanics Reviews, 40(5), 575-591
6. Mau ST., 1973, A Refined Laminate Plate Theory, Journal of Applied Mechanics, 40, 606-607.
7. Di Sciuva, M., 1986,Bending, Vibration and Buckling of Simply Supported Thick Multilayered Orthotropic Plates: An Evaluation of a new displacement model, Journal of Sound and Vibration, 105, 425-442.
8. Ren JG., 1986, A new Theory of Laminated Plate, Composite Science and Technology, 26, 225-239.
9. Reddy, J.N., 1990, A Review of Refined theories of Laminated Composite Plates, Shock and Vibration Digest 22(7) 3-17.
10. Reddy, J.N., 1998, Mechanics of Laminated Composite Plates–Theory and Analysis CRC press Newyork.
11. Savithri, S.,1991, Linear and Non linear analysis of Thick Homogeneous and Laminated Plates, PhD Thesis, Department of Mathematics, IIT Madras.

26

Experimental Charaterization of Glass Fibres (Woven-Mat)/Polyester Laminates

PARTAP SINGH[1] AND S.C. DHAWAN[2]

[1]Department of Civil Engineering, DR. B.R. Ambedkar National Institute of Technology
(Deemed University), Jalandhar, Punjab, India email: singhp@nitj.ac.in
[2]Former Professor and Head, Department of Civil Engineering,
Punjab Engineering College, Chandigarh, India

ABSTRACT

In the present work, experimental characterization of laminates fabricated by Hand lay-up technique using glass fibres (woven-mat) of density 3.6 N/m^2 and general purpose polyester resin has been carried out. The material properties i.e. Young's moduli (E_1 – in direction 1, E_2 – in direction 2), Poisson's ratios(v_{12} – load in direction 1, v_{21} – load in direction 2), Inplane shear modulus (G_{12}) and transverse shear moduli (G_{13} and G_{23}) (Refer Fig.1) have been determined through tests on suitably designed tension and flexural test specimens. Test specimens are cut from sheet of 10 ply laminates of size 1000 mm $\times$ 1000 mm 4 mm. Tension specimens are cut in directions parallel to direction 1 for E_1 and V_{12} and direction parallel to direction 2 for E_2 and v_{21}. Flexural specimens are cut: (i) in a direction inclined at 45° to the direction 1 for G_{12} (ii) in directions parallel to direction 1 and direction 2 for G_{13} and G_{23}.

Keywords: Characterization; laminates; glass fibres (woven mat)/polyester; moduli.

1. INTRODUCTION

Fiber reinforced plastic composites (FRPC) are anisotropic and therefore, in contrast to isotropic materials, their elastic properties can not be defined by only two independent constants i.e. the elastic modulus (E) and the poison's ratio. However, they can be classified as orthotropic materials; therefore, to define their elastic characteristics one needs to identify nine independent properties. These properties are defined according to the material principal axes. Figure 1 shows the definition of these axes for a typical fiber reinforced lamina. As shown in Fig. 1, axis 1 is along the fibre length and represents the longitudinal direction of the lamina; axes 2 and 3 represent the transverse in-plane and through-the-thickness directions respectively.

Several test methods have been developed for evaluating the mechanical properties along these axes. Some methods, however, have gained more popularity because they are easy to conduct and/or they use specimens with simple geometry. Several test methods have been developed and are available for the evaluation of the longitudinal or shear properties of FRPC. Some of these methods are ASTM approved and some are considered as guidelines only.

Xie and Adams [1] studied three-and four-point shear testing of unidirectional composite materials. Cheng [2] presented test method for evaluation of shear modulus and modulus of elasticity of laminated anisotropic composite materials. Zhou and Davies [3] carried out characterization of thick glass woven-roving/polyester laminates. Jalali and Taheri [4] introduced a new test method for measuring the longitudinal and shear moduli of fiber reinforced composites. Knight [5] measured three-dimensional elastic moduli of graphite/epoxy composites from testing simple tension, compression and shear specimens. Li [6] presented two experimental methods to determine the shear moduli G_{12}, G_{13}, G_{23} of orthotropic laminates. Tsai and Daniel [7] developed an experimental method for determining in-plane and out – of plane shear moduli of composite materials. Stijnman [8] used an ultrasonic immersion technique to determine elastic constants of some composites. Rosen [9] defined a simple procedure for experimental determination of the longitudinal shear modulus of unidirectional composites. Mujika et al. [10] proposed off-axis three-point flexure test for unidirectional composites for obtaining in-plane shear modulus and in-plane shear strength.

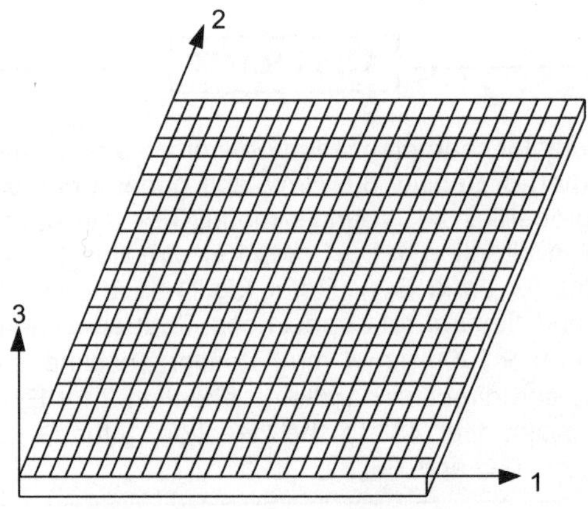

Fig. 1. (1, 2, 3) Lamina Reference Axes.

2. FABRICATION OF LAMINATED COMPOSITE PLATE

Hand lay-up technique used to fabricate laminated composite plate is described below:

Hand lay-up technique is the major process for the manufacture of GRP products. It is the only method which takes the full advantage of the two most important characteristics of unsaturated polyester resin i.e. it can be set without heat and pressure. In this technique, an operator deposits resin and reinforcement in or on a mould by using hand tools.

The steps to be taken for manufacturing by the Hand lay-up process are:

2.1 Preparation of Mould

For Hand lay-up technique, only one mould is necessary. The pattern of mould can be made either of wood (seasoned teak and rose wood preferred) or plaster of Paris, though the former is most commonly used.

2.2 Application of Release Agent

The mould surface must be thoroughly cleaned and must be free from dirt, cracks before the application of release agent. The mould surface is then coated with a silicon free wax using a smooth cloth. Then a film of Polyvinyl alcohol (PVA) is applied over the wax surface using sponge. PVA is a water soluble material and 15% solution in water is used. When water evaporates, a thin film of PVA is formed on the mould surface. PVA film should dry completely before the application of resin coat. This is very important as the surface of final article will be marred with a partly dried PVA film otherwise release will not be smooth.

2.3 Preparation of Matrix Material

The matrix material is prepared using general purpose (GP) Polyester resin. Cobalt Octate (0.35% by volume of resin) is added to act as Accelerator. Methyl ethyl ketone peroxide (MEKP) (1% by volume) is added to act as catalyst. Resin, accelerator and catalyst are thoroughly mixed. The use of accelerator is necessary because without accelerator resin does not cure properly. After adding the accelerator and catalyst to the polyester resin, it should be left for some time so that bubbles formed during stirring may die out. The amount of added accelerator and catalyst should not be high because a high percentage reduces gel time of polyester resin and may adversely affect impregnation.

2.4 Preparation of Reinforcement

Glass fiber mats (woven-mat) are cut in 10 pieces of required size (1000 mm $\times$ 1000 mm).

2.5 Preparation of Laminate

The first layer of mat is laid and resin is spread uniformly over the mat by means of a brush. The second layer of mat is laid and resin is spread uniformly over the mat by means of a brush. After second layer, to enhance wetting and impregnation, a teethed steel roller is used to roll over the fabric before applying resin. Also resin is tapped and dabbed with spatula before spreading resin over fabric layer. This process is repeated till all the ten fabric layers are placed. No external pressure should be applied while casting or curing because uncured matrix material squeezes out under high pressure. This results in surface waviness (non-uniform thickness) in the model material. A symmetry should be maintained in stacking the fibre layers. In non-symmetric laminate, a bending-stretching coupling causes an undesirable warping of the composite plate. The casting is cured at room temperature for 24 hours and finally removed from the mould to get a fine finished composite plate.

2.6 Preparation of Test Specimens

Test specimens are cut from sheet of 10 ply laminate of size 1000 mm $\times$ 1000 mm $\times$ 4 mm by using a diamond impregnated wheel, cooled by running water. All the test specimens are finished by abrading the edges on a fine corborundum paper or simple wood sand paper.

3. YOUNG'S MODULI E_1 AND E_2 AND POISSON'S RATIOS (ν_{12} AND ν_{21})

Tension specimens (Refer Fig. 2) are made with adhesively bonded and bevelled tabs for load application and to prevent edge crushing of the specimens at the ends. Tabs used are made of chopped strand mat/polyester resin. Tabs used are 100 mm long, 25 mm wide and 14 mm thick.

Two strain gauges are mounted on each of the specimen using cyno acrylate cement (instant setting) at its mid-length, one in the longitudinal direction and other in the transverse direction. The strain gauges used are 350 ohm

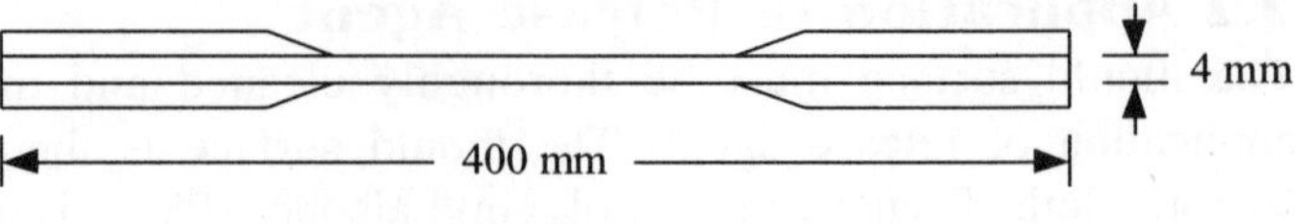

Fig. 2. Tension Specimen.

resistance with gauge factor 2.0 and of gage length 10 mm. The specimens are loaded on Universal Testing Machine and strain gage signals are measured using Data-logger (HP 3852 Data Acquisition and Control Unit). The stress-strain curves are shown in Figs. 3 to 5.

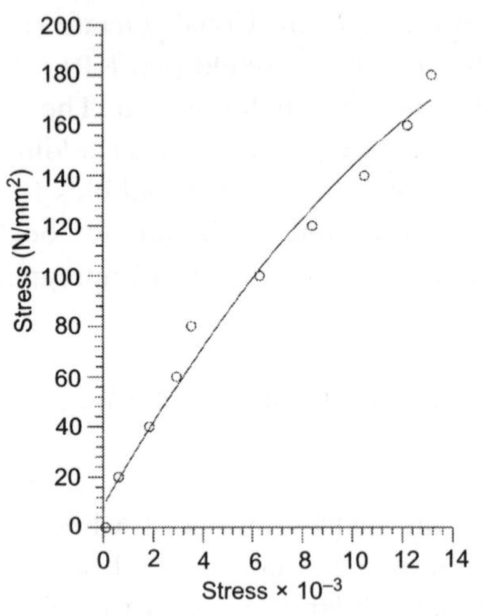

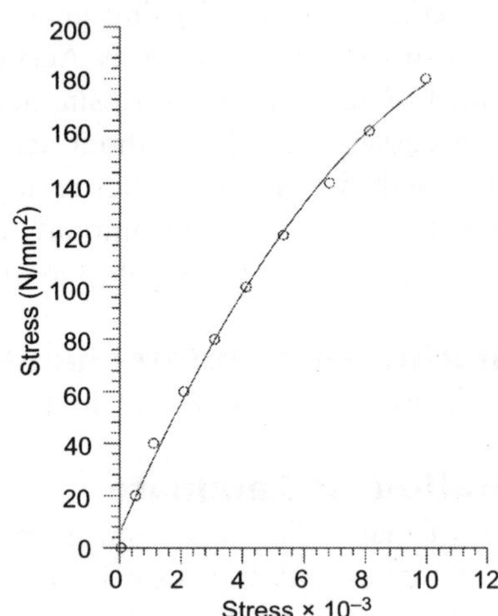

Fig. 3. Tensile Stress-Strain Curve of Specimen 1 for E_2.   **Fig. 4.** Tensile Stress-Strain Curve of Specimen 2 for E_2.

The Young's moduli and Poisson's ratios from the strain in the transverse direction and axial direction of the specimens have been determined from the stress-strain curves where tensile modulus is defined by the initial linear slope of the stress-strain curve at 0.25% axial strain [3].The Young's moduli E_1 and E_2 have been determined as 28000 MPa and 25200 MPa and Poisson's ratios (v_{12} and v_{21}) have been determined as 0.033 and 0.020.

4. INPLANE SHEAR MODULUS

Pure bending shear test of 45° off-axis laminates reported by LI [6] has been used to determine inplane shear modulus G_{12} of a laminated specimen by the relation

$$G_{12} = \frac{3Pb}{h^2 d(\epsilon_{n_1} - \epsilon_{t_1})}, \text{ where } P, b, h \text{ and } d \text{ are shown in}$$

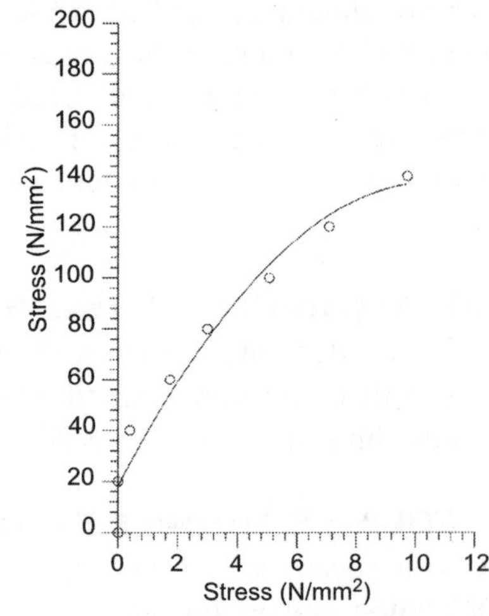

Fig. 5. Tensile Stress-Strain Curve of Specimen E_1.

Fig. 6. $\in_{n_1}$ and $\in_{t_1}$ are curvatures in n and t directions measured by strain gages stuck as shown in Fig. 6.

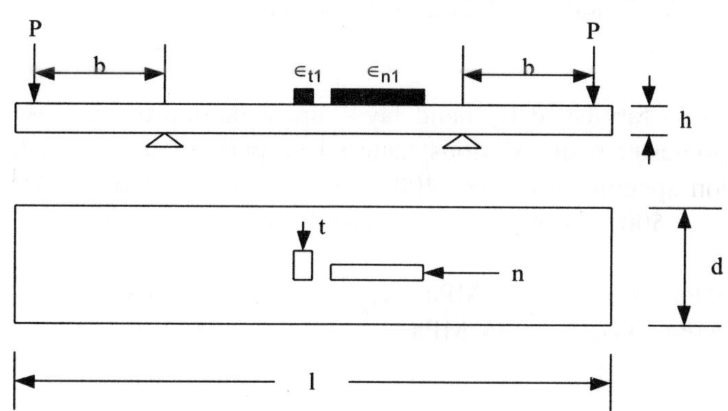

Fig. 6. Experimental set-up for Inplane Shear Modulus G_{12}.

Two specimens with two strain gages mounted on each of them at its mid-length (one in longitudinal direction and other in the transverse direction) have been tested. The specimens are loaded as per experimental set-up (Refer Fig. 6) and strain gage signals are measured using Data-logger. The width of specimens 1 and 2 are measured with vernier calliper as 37 mm, 37 mm and thickness of specimens 1 and 2 are measured with micrometer as 4.2425 mm and 4.202 mm. The inplane shear modulus is determined as 5640.593 N/mm^2.

5. TRANSVERSE SHEAR MODULI (G$_{13}$ AND G$_{23}$)

Three points loading test reported by LI [6] has been used to determine transverse shear moduli (G_{13} and G_{23}) by the relation G_{13} or $G_{23} = \dfrac{P\alpha}{2A\delta_3}$, where P, α, δ_3 are shown in Fig. 7 and A is the cross-sectional area of the specimen.

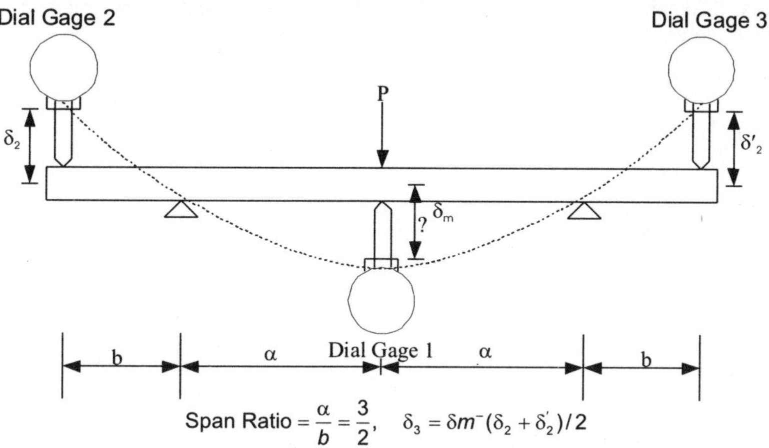

Fig. 7. Experimental Set-up for Transverse Shear Moduli G_{13} and G_{23}.

Flexural specimens cut in direction parallel to direction 1 for G_{13} and in direction parallel to 2 for G_{23} are tested as per experimental programme (Fig. 7). The transverse shear moduli G_{13} and G_{23} are obtained as 32.68 N/mm^2 and 25.74 N/mm^2 respectively.

CONCLUSION

Test specimens have been fabricated by hand lay – up technique using glass fibre (woven – mat) and general purpose polyester resin. Various material properties have been determined. For tensile properties, three tension specimens of size 400 mm × 25 mm × 4 mm and for shear moduli, six flexural specimens of size 500 × 40 mm × 4 mm have been tested and material properties determined are:

$$E_1 = 28000 \text{ MPa}, \ E_2 = 25200 \text{ MPa}, \ v_{12} = 0.033, \ v_{21} = 0.02, \ G_{12} = 5640.593 \text{ MPa},$$
$$G_{13} = 32.68 \text{ MPa}, \ G_{23} = 25.74 \text{ MPa}.$$

REFERENCES

1. Ming Xie, and Donald F. Adams,1995, Study of three-and four-point shear testing of unidirectional composite materials, Composites, 26(9), 653-659.
2. Shun Cheng, 1985, Test method for evaluation of shear modulus and modulus of elasticity of laminated anisotropic composite materials, Journal of Testing and Evaluation, 13(5), 387-389.
3. G. Zhou, and G.A.O. Davies, 1995, Characterization of thick glass woven –roving/polyester laminates: 1. Tension, compression and shear, Composites, 26, 579-586.
4. S. Javad Jalali, Farid Taheri, 1999, A new test method for measuring the longitudinal and shear moduli of fiber reinforced composites, Journal of Composite Materials, 33(23), 2134-2160.
5. Marvin Knight, 1982, Three-dimensional elastic moduli of graphite/epoxy composites, Journal of Composite Materials, 16, 153-159.
6. Gia-Ju LI, Test Methods for Inplane Shear Modulus G_{12} and Transverse Shear Moduli G_{13} and G_{23}, Advances in Composite Materials, Vol. I edited By Bunsell, Bathias, 3rd International Conference on Composite Materials, ICCM, 914-925.
7. C.L.Tsai, I.M.Daniel, 1999, Determination of In-plane and Out-of-plane shear moduli of composite materials, Experimental Mechanics, 295-299.
8. P.W.A. Stijnman, 1995, Determination of the elastic constants of some composites by using ultrasonic velocity measurements, Composites, 26(8), 597-604.
9. B.Walter Rosen, 1972, A simple procedure for experimental determination of the longitudinal shear modulus of unidirectional composites, Journal of Composite Materials, 552-554.
10. Mujika, A.Valea, P.Ganan, and I. Mondragon, 2005, Off-axis flexure test: A new method for obtaining in-plane shear properties, Journal of Composite Materials, 39(11), 953-980.

27

Laminated Composites for Structural Engineering-Perspective Applications and Challenges

Upendra K. Mallela[1], Rajeev Chandak[1] and Akhil Upadhyay[2*]

[1]Research Scholar [2]Assistant Professor, Department of Civil Engineering, Indian Institute of Technology, Roorkee, India. email: akhilfce@iitr.ernet.in

ABSTRACT

Laminated composites are a special form of FRP which belongs to the new generation of energy efficient materials, almost dominating over the metallic materials. The potential of laminated composites offer several possibilities but on the other hand the mechanical characterization of a composite structure is more complex than that of metal structures. This paper presents an overview of the material, advantages over traditional systems, perspective applications in civil engineering structures with due emphasis on highway bridge decks. Optimum design studies and 3-D FE numerical studies are carried out on laminated composite stiffened panels and box beams. It is observed from FEA that on one side laminated composites can be used in highway bridge decks with great advantage of weight savings but on the other side it is a challenge to handle some unwanted adverse effects like shear lag.

Keywords: Laminated composites, composite structures, stiffened panels, box beams, shear lag.

1. INTRODUCTION

Fiber reinforced laminated composites are composed of fibers embedded in matrix material. New technology has provided a variety of reinforcing fibers and matrices those can be combined to form laminated composites having a wide range of exceptional properties. Laminated composites are a special form of fiber reinforced plastics (FRP) which belongs to the new generation of energy efficient materials, almost dominating over the metallic materials in view of their high specific strength and modulus, fatigue and corrosion resistance and dielectric properties. In addition to this the real advantage of the use of laminated composites lies in the fact that they provide flexibility to tailor different properties of the structural elements to achieve strength and stiffness requirements. The tailoring results in large mass savings.

Continuing efforts in material development and research activities make fiber reinforced laminated composites a possible alternative to conventional type of material, offering many potential.

This paper presents an overview of the material, advantages over traditional systems, perspective applications in civil engineering structures with due emphasis on highway bridge decks. Optimum design studies and 3-D FE numerical studies are carried out on laminated composite stiffened panels and box beams. It is observed from the studies that on one side laminated composites can be used in highway bridge decks with great advantage of weight savings but on the other side it is a challenge to handle some unwanted adverse effects like shear lag.

2. LAMINATED FIBROUS COMPOSITES

Laminated fibrous composites are made by stacking the unidirectional (or woven fabric) laminas (layers) at different fiber orientations in the matrix. The unidirectional lamina may be composed of one or more layers of material in a matrix, but all fibers are in the same direction. The stiffness and strength in the fiber direction are typically much greater than in the transverse directions, depending on the matrix material and the quality of the fiber/matrix bond. The properties of a unidirectional lamina are orthotropic, with different properties in the material principal directions (parallel and perpendicular to the fibers). Thus the effective properties of the laminate vary with the orientation, thickness, and stacking sequence of the individual layers. The specific stiffness of a fiber can be more than 13 times that of structural metals, and the specific strength can be more than 16 times that of structural metals. Thus the potential for large weight savings is present through the use of advanced composites.

3. ADVANTAGES OF LAMINATED FIBROUS COMPOSITES COMPARED TO CONVENTIONAL MATERIALS

The actual advantages of laminated fibrous composites to conventional materials should be estimated from both technical and economic points of view.

From technical point of view, composite materials have significant advantages.

- Laminated composites offer significant weight saving over existing metals. Composites can provide structures that are 25-45 % lighter than the conventional structures designed to meet the same functional requirements (Fig. 1a).
- Unidirectional fiber composites have specific tensile strength (ratio of material strength to density) about 4 to 6 times greater than that of steel and aluminium (Fig. 1b).
- Unidirectional fiber composites have specific modulus (ratio of material stiffness to density) about 3 to 5 times greater than that of steel and aluminium (Fig. 1c).
- Fatigue endurance limit of composites may approach 60% of their ultimate tensile strength. For steel and aluminium, this value is considerably lower.
- Corrosion resistance of fiber composites leads to reduced life cycle cost.

From an economic point of view, the main factors contributing to their competitiveness with respect to conventional materials are time saving, flexibility, low labour costs, low tooling and machinery costs on the construction site because of the light weight and manageability of tools, possibility of restoring a structure without interrupting its utilization by users and durability.

In spite of many advantages of laminated composites over traditional materials complex mechanics involved in laminated fibrous composites poses new challenges for the construction industry.

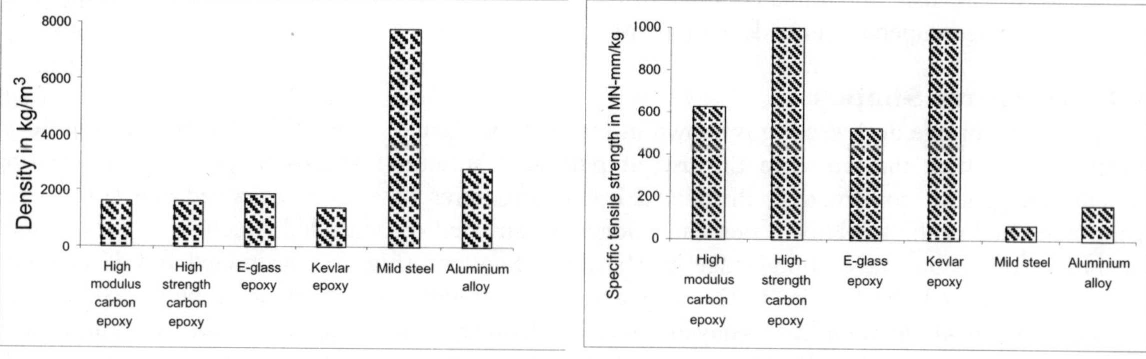

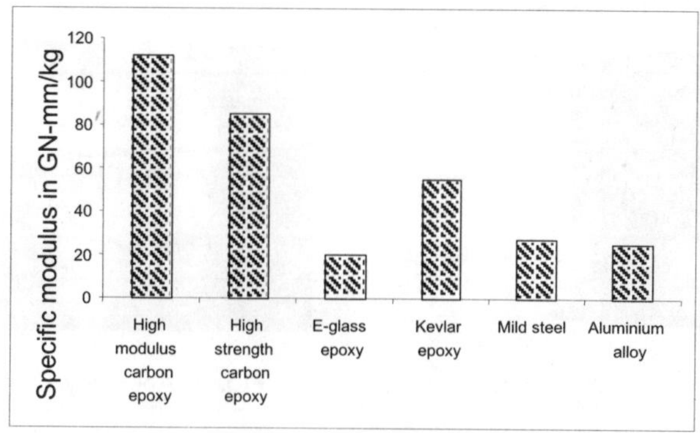

Fig. 1. Comparison of different properties of composite materials with traditional materials.

4. APPLICATION OF LAMINATED COMPOSITES IN STRUCTURAL ENGINEERING

Structural safety is always a crucial aspect, especially in seismic areas where social and economic concerns are very high. More resources are being devoted to the retrofit and upgrade of existing structures. The widespread use of the FRP system is in the restoration and seismic strengthening of historical masonry buildings. The use of laminated composites is thus becoming more widespread as an optimal innovative system able to reduce the seismic vulnerability of reinforced concrete structures. It ensures an effective confinement of concrete.

The use of laminated composites is also very effective in case of urgency, for safety and temporary preservation of structures damaged during special events.

Structures having strategic interest and identified as sensitive objectives (in case of explosion risk, terrorist or attempted attacks, etc), the adoption of laminated composites can help to limit damages to persons and structures.

Laminated composites can be efficiently used where a fast installation is a crucial factor, such as during military/army operations etc.

The use of laminated composites in the aerospace and automobile industries is now well-established and gaining momentum in structural applications such as buildings, bridges and marine applications. The use of laminated composite in structural applications is discussed along with significant bridge superstructure developments.

4.1 Structural Shapes

A typical FRP bridge deck section is shown in Fig. 2. From last 15 years, there has been a significant increase throughout the world in the use of pultruded structural shapes in primary load-bearing systems for general construction. Prominent bridge structures have been designed and constructed using pultruded profiles. Major pedestrian bridges constructed of pultruded structural shapes include the 114 m long cable-stayed Footbridge in Aberfeldy, Scotland (Fig. 3) constructed and designed by Maunsell Structural Plastics in 1992 using a proprietary interlocking modular pultruded decking section and the 40 m long cable-stayed Fiberline Bridge in Kolding, Denmark constructed and designed by the Danish firm RAMBOLL in 1997, using standard pultruded shapes in both the cable tower and the decking system.

Fig. 2. FRP girders for use in vehicular bridges.

Fig. 3. Aberfeldy Footbridge, Scotland.

4.2 Highway Bridge Decks

Within the field of highway structures, FRP bridge decks have received the greatest amount of attention in the past few years, due to their inherent advantages in strength and stiffness per unit weight as compared to traditional steel reinforced concrete decks. A laminated composite highway bridge is recently constructed at Ohio, USA–1997, shown in Figs 4(a) and 4(b) [1]. Also reducing the weight of replacement decks in rehabilitation projects presents the opportunity for rapid replacement and reduction in dead load, thus raising the live load rating of the structure.

(a)

(b)

Fig. 4. Installation of laminated composite highway bridge at Ohio, USA.

5. NUMERICAL STUDIES

Sincere efforts are required to study the behaviour of laminated fibrous composite members or shapes for the wide-spread applications of this material in structural engineering. In the present work results of a few optimum design studies carried out on hat stiffened laminated composite panels subjected to in-plane share are reported and compared with the aluminium stiffened panels along with this 3-D FE numerical studies are carried out on laminated composite box beams to see the shear lag effects. Stiffened panels and box sections are chosen for study as these are the most commonly used structural sections. The material properties used for different structural elements are given in Table 1.

Table 1. Lamina material properties.

Structural Element	E_1(MPa)	E_2(MPa)	G_{12}(MPa)	$\rho(Kg/m^3)$	υ_{12}	υ_{21}
Stiffened panel	131000	13000	6410	1522×10^5	0.380	0.0378
Box beam	145000	16500	4480	1522×10^5	0.314	0.037

5.1 Stiffened Panels

Stiffened panels are widely used as structural components in various civil, aerospace and marine applications. Stiffeners are commonly attached to the plates in the required direction so as to achieve higher stiffness/weight and strength/weight ratios. The stiffeners are of very less weight but at the same time enhance the in-plane critical load carrying capacity when compared with unstiffened panels. Laminated composite stiffened panels are being used in civil engineering applications such as bridge decks, ship deck hulls, offshore oil platforms etc.

The structural efficiency of corrugated, hat and blade stiffened aluminium panels and composite panels subjected to shear loading are tabulated in Table 2. From Table 2 it can be observed that aluminium panels are substantially heavier that composite blade stiffened panels for $N_{xy}/L = 500$ and 1000 kPa respectively. Also the difference of mass index between different composite panels is less.

Table 2. Comparison of optimal mass indices.

$\dfrac{N_{xy}}{L}$ kPa	Mass index kg/m³			
	Corrugated aluminium	Composite panel		
	panel (Ref. [2])	Corrugated (Ref.[2])	Hat (Ref.[2])	Hat (Ref.[3])
500	6.9	4.6	5.0	3.87
1000	10.3	6.2	7.2	6.02

5.2 Effect of Shear Lag in Box Beams

Only limited studies are reported on the analysis of shear lag effects in laminated composite box beams under bending loads. Pavlovic *et al.* [4], Nagaraj and Ganga Rao [5], Lopez-Anido and Ganga Rao [6],Wu *et al.* [7] and Upadhyay and Kalyanaraman [8] studied shear lag effect and reported importance of shear lag in laminated composite plate or box beam. It is particularly emphasized that shear lag effect can not be neglected in case of wide beams of laminated composites.

More investigations are required to study the shear lag behaviour in laminated composite box beams.

According to the elementary beam theory, when a box beam is under vertical load, the longitudinal normal stresses induced in the flanges are assumed to be uniformly distributed across the flange width. However, in most cases, particularly in a wide flange, these stresses are non-uniformly distributed due to the shear deformations of the flange plates. This phenomenon, characterized by the fact that the longitudinal stress on a flange near the web is much larger than that far from the web, is known as positive shear lag. In the converse case, it is referred to as negative shear lag, in which the stress near the web is much smaller than that far from the web.

The exact solution of shear lag problems is quite complicated. An approximate method of dealing with shear lag is to use an effective width concept, in which the actual width b of a flange is replaced by a reduced width b_e given by

$b_e = b * $ (Nominal bending stress/Maximum bending stress)

The concept of effective width, b_e, was first proposed by Von Karman in order to measure the effect of shear lag in thin-walled structures and has been widely adopted. In the present work, the stresses obtained in different layers of the laminated composites are integrated across the thickness of the laminate to obtain the corresponding forces. These forces are finally used to calculate the b_e/b ratio which is a useful conventional measure of the shear lag effect.

Efforts are made to study the shear lag effect in wide flanged composite box beams. The finite element method is utilized since it is proven to be one of the most reliable tools. Upadhyay and Kalyanaraman [7] studied the shear lag phenomenon in a simply supported composite box beam (Fig. 5b) subjected to central point load using MSC-NASTRAN. They reported an effective breadth ratio $\psi = 0.33$ and $\psi = 0.7$ for the beam having fiber orientations of $0°$ and $90°$ respectively, in all the elements. To validate the approach an identical box beam is analyzed using ANSYS 7.1. The comparison of results shown in Fig. 5a validates the modeling aspects used in shear lag studies.

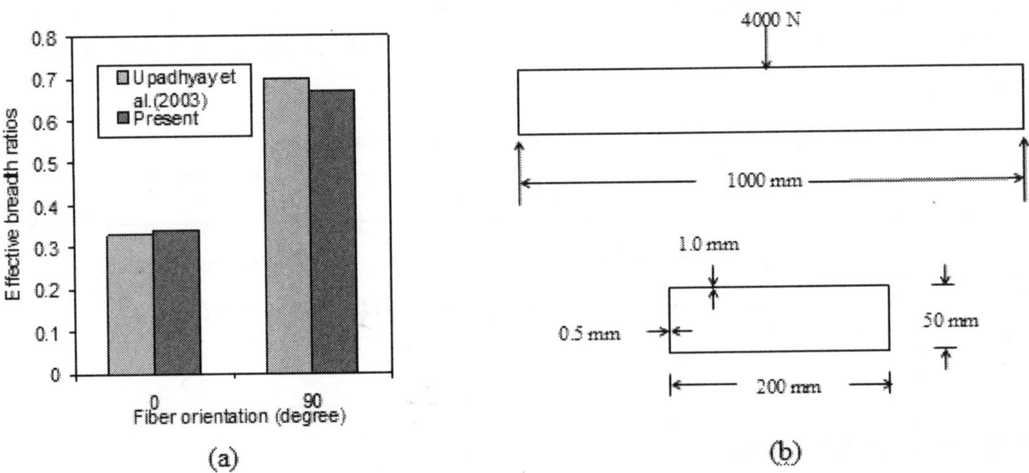

Fig. 5. Shear-lag in composite box beam: (a) beam details; (b) effect of orthotropy on shear-lag.

5.2.1 Finite Element Analysis of Shear Lag

Symmetric graphite epoxy box beams are considered with $L/b = 3$ to 7, $b/D = 3$, $t_F/t_W = 1$, $n = 6$; where b = width of the flange, L = length of the box beam, D = depth of box beam, t_F = thickness of the flange, t_W = thickness of the web, and n = number of plies in flange or web elements. All elements of the box beam are made of balanced symmetric laminates. Fiber orientations of 0° and 90° have been used in panels of box beam in order to avoid various couplings like extension-bending etc. Box beam model is discretized with 8-node isoparametric laminated shell element (SHELL 99) in ANSYS 7.1 as shown in Fig. 6.

Fig. 6. Discretized Finite element model of box beam.

5.2.2 Simply Supported Composite Box Beam under Point Load at Mid Span

A simply supported box beam subjected to an identical pair of concentrated loads at the mid span is considered. The load P = 1000 N is applied at the bottom of each web so as to produce no torsional effect. The, ply normal stress in all layers of the top flange and effective breadth ratios are evaluated in the analysis of this paper. In Fig. 7, the results are given when fibers are oriented at 0°. Variation of b_e/b (shear lag) with fiber orientations in top flange is shown in Fig. 8, for different L/b ratios.

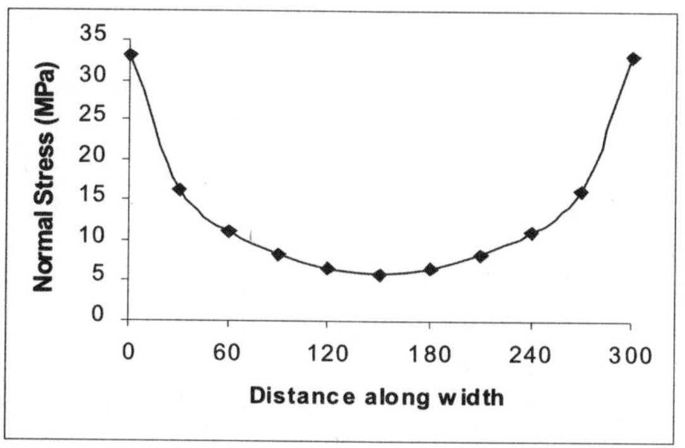

Fig. 7. Ply normal stress along width of the top flange in 1ˢᵗ layer (0°) (measured at midspan).

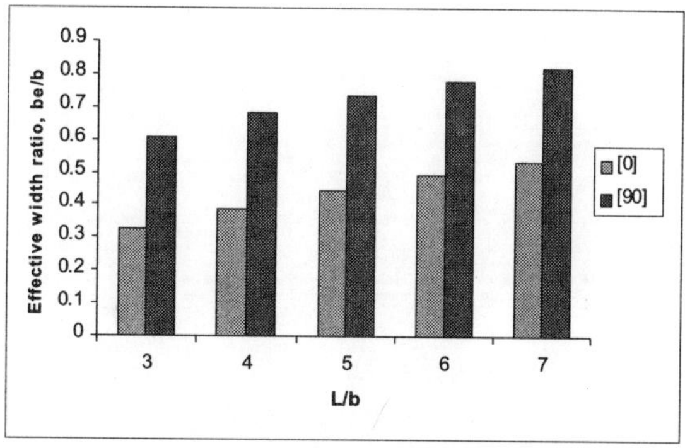

Fig. 8. Variation of shear lag in laminated composite box beam.

CONCLUSION

The outlook for laminated composites is very promising in structural engineering. Laminated composites offer a number of advantages over traditional materials and as such can be a viable alternate solution in civil engineering structures. The studies carried out in present work clearly demonstrate its weight saving capabilities which may be utilized with great advantage in bridge decks. However, the FEM studies warn the presence of substantial shear lag in box beam made of laminated composites. So, in the design of the laminated composite bridge, designers' should attempt to exploit the advantages without getting trapped into the disadvantages.

REFERENCES

1. Foster D.C., Richards D., Boqner B.R. "Design and installation of fiber-reinforced polymer composite bridge. Journal of Composites for Construction, ASCE 2000; 4(1): 33-7.
2. Stroud W.J., Agranoff N. "Minimum-mass design of filamentary composite panels under combined loads: Design procedure based on simplified equations". TN D-8257, NASA, 1976.
3. Upadhyay A. Optimum design of FRP box girder bridges. Ph.D. thesis, Indian Institute of Technology Madras, 1997.
4. Pavlovic, M.N., Tahan, N., Kotsovos, M.D. (1998). Shear lag and effective breadth in rectangular plates with material orthotropy. Part 1: Analytical formulation. J. Thin-Walled Struct., Elsevier, 30(1-4), 199-213.
5. Nagaraj, V., Ganga Rao, Hota V.S. (1997). Static behaviour of pultruded GFRP beams. J. Composites for Construction, ASCE, 1(3), 120-29.
6. Roberto L-A, Ganga Rao H.V.S. (1996). Warping solution for shear lag in thin-walled orthotropic composite beams. J. Eng. Mech., ASCE, 122(5), 449-57.
7. Wu Yaping *et al.* (2002). Analysis of shear lag and shear deformation effects in laminated composite box beams under bending loads. J. Composite Struct., Elsevier, 55, 147-56.
8. Upadhyay A., Kalyanaraman V. (2003). Simplified analysis of FRP box-girders. J. Composite Struct., Elsevier, 59, 217-25.

28

Thermal Buckling of Laminated Composite Plate with Random Material Properties

S. Bose[1], B.N. Singh[2] and Anuj Jain[3]

[1] Graduate Student, Department of Applied Mechanics, MNNIT Allahabad-211 004, India
[2] Assistant Professor, Department of Aerospace Engineering, IIT Kharagpur-721 302, India
email: bnsingh@areo.iitkgp.ernet.in
[3] Professor, Department of Applied Mechanics, MNNIT Allahabad-211 004, India
email: au_jain123@yahoo.com (corresponding author)

ABSTRACT

In the present work, the effect of variation in the material properties of laminated composite plate on its thermal buckling has been obtained by modeling material properties as random variables. Based on incremental potential energy approach, the governing equations for laminated composite plate subjected to uniform temperature rise have been formulated incorporating transverse shear effects. A Monte Carlo Simulation (MCS) approach in conjunction with C0 finite element method (FEM) is employed to obtain the mean and variance of the random thermal buckling temperature. Typical numerical results are presented for square, thin ($a/h=100$) plate, with stacking sequence $[75^0/-75^0/75^0/-75^0/75^0]$ and having all edges simply supported (S1). The results reveal that variation in elastic modulus (E_{11}, E_{22}) has about 100 fold greater influence on the buckling temperature as compared to that due to variation in shear modulus (G_{12}, G_{13}).

1. INTRODUCTION

Laminated composite plates are being used extensively in many aerospace, marine and automobile structures due to their inherent advantages. The structural members used in these applications are susceptible to buckling when subjected to thermal loads. This necessitates the development of accurate theoretical models to predict buckling responses in such loading conditions. High elastic modulus to shear modulus ratio renders classical theories inadequate for the analysis of advanced composites. In common deterministic analysis, the variations in structural properties are ignored. However, some variations in material properties are inevitable even for the composite laminates manufactured carefully in a laboratory. Composite structures have more uncertainties and variability in their structural properties compared to conventional isotropic structures as a large number of parameters are associated with complex manufacturing and fabrication processes. Composite structures are therefore more appropriately modeled with random material properties having uncertainties where an accurate analysis is required.

Sufficient literature is available on theoretical analysis of in-plane mechanical buckling and thermal buckling of laminated composite plates [1-7]. Rudolph [1] showed that simple energy criterion makes it possible to circumvent with ease the difficulties associated with singularities of the stiffness matrices. Ghosh and Dey [5] concluded that thin plate theory over predicts the buckling load due to neglect of transverse shear. Robert *et al.* [7] presented experimental, numerical and analytical result for bending and buckling of rectangular orthotropic plate.

A very limited literature is available for composite plate with random material properties [8-10]. Nakagiri *et al.* [8] studied simply supported graphite/epoxy plates with stochastic finite element method taking fiber orientation, layer thickness and number of layers as random variables. They found that the overall stiffness of the fiber reinforced laminated plate largely depends on fiber orientation. Singh *et al.* [9,10] obtained buckling response of composite structures with random material properties for composite plate and cylindrical panel using Monte Carlo Simulation (MCS). Monte Carlo method is a statistical simulation method that utilizes random numbers to perform the simulation. Monte Carlo method has been used for centuries but in the past few decades, the technique has gained the status of a full-fledge numerical method capable of addressing the many complex applications.

The objective of the present work is to investigate the effect of random variation in material properties on the critical buckling temperature of laminated composite plate. The higher order shear deformation theory along with incremental potential energy method is used to obtain the final governing equations. FEM in conjunction with MCS is employed to obtain the mean and variance of thermal buckling response corresponding to the deviation in material properties.

2. MATHEMATICAL FORMULATIONS

Consider a composite laminated plate consisting of layers of unidirectional fiber reinforced laminas subject to the in-plane uniform load $\tilde{N}_x, \tilde{N}_y, \tilde{N}_{xy}$ and temperature variation from T_0 to T. T_0 is the reference temperature corresponding to the thermal stress-free state, and T is the ultimate temperature. Following Reissner-Mindlin type approach to account for shear deformation, the higher order displacement components for the plate are taken to obtain the total potential energy which is a sum of bending strain energy, shear strain energy, and potential energy due to external work [2] as follows:

$$\bar{\Pi} = \frac{1}{2}\int_R \left\{ \left(<\varepsilon^0><\kappa><\eta> \right)\left[D_b^1 \right]\left(<\varepsilon^0><\kappa><\eta> \right)^T - 2\left(<\varepsilon^0><\kappa><\eta> \right)\left[D_b^2 \right] \right\} dR$$

$$+ \frac{1}{2}\int_V \{\alpha \Delta T\}^T \left[\bar{Q} \right]_{3\times3}^T \{\alpha \Delta T\} dV + \frac{1}{2}\int_R \left(<\gamma^0><\zeta> \right)\left[D_s \right]\left(<\gamma^0><\zeta> \right)^T dR$$

$$- \int_R \left\{ \bar{N}_X \frac{\partial u_0}{\partial x} + \bar{N}_Y \frac{\partial v_0}{\partial y} + \bar{N}_{XY} \left(\frac{\partial u_0}{\partial y} + \frac{\partial v_0}{\partial x} \right) \right\} dR \qquad \ldots(1)$$

The governing equations of buckling and thermal buckling can be derived from the incremental total potential energy by imposing an infinitesimal perturbation on the bifurcated point after the onset of buckling or thermal buckling along the bifurcated path [2]. The total potential energy can then be written as

$$\Pi = \bar{\Pi} + \Delta\Pi \qquad \qquad ...(2)$$

and Eq. (1) further used to calculate the change in value of incremental total potential energy as

$$\Delta\Pi = \frac{1}{2}\int_R \left(<\varepsilon^0><\kappa><\eta>\right)\left[D_b^1\right]\begin{Bmatrix}\{\varepsilon^0\}\\\{\kappa\}\\\{\eta\}\end{Bmatrix}dR + \frac{1}{2}\int_R \left(<\gamma^0><\zeta>\right)[D_s]\begin{Bmatrix}\{\gamma^0\}\\\{\zeta\}\end{Bmatrix}dR$$

$$-\frac{1}{2}\int_R \left(<\frac{\partial w_0}{\partial x}\frac{\partial w_0}{\partial y}>[\bar{N}]\begin{Bmatrix}\frac{\partial w_0}{\partial x}\\\frac{\partial w_0}{\partial y}\end{Bmatrix}\right)dR \qquad \qquad ...(3)$$

Where R is the area of the mid-plane of the laminate,

$$<\varepsilon^0> = <\varepsilon_x^0 \varepsilon_y^0 \gamma_{xy}^0>$$

$$<\kappa> = <\kappa_x \kappa_y 2\kappa_{xy}>$$

$$<\eta> = <\eta_x \eta_y \eta_{xy}>$$

$$<\gamma^0> = <\gamma_{xz}^0 \gamma_{yz}^0>$$

$$<\zeta> = <\zeta_{xz} \zeta_{yz}> \qquad \qquad ...(4)$$

strain fields in vector form using transformed thermal coefficients ($\alpha_x\ \alpha_y\ \alpha_{xy}$) in laminate axes,

$$\{\alpha\Delta T\} = \int_{T_0}^{T_1} < \alpha_x \alpha_y 2\alpha_{xy} >^T \ dT \qquad \qquad ...(5)$$

$$\left[D_b^1\right] = \begin{bmatrix} A_{11} & A_{12} & A_{16} & B_{11} & B_{12} & B_{16} & E_{11} & E_{12} & E_{16} \\ A_{12} & A_{22} & A_{26} & B_{12} & B_{22} & B_{26} & E_{12} & E_{22} & E_{26} \\ A_{16} & A_{26} & A_{66} & B_{16} & B_{26} & B_{66} & E_{16} & E_{26} & E_{66} \\ B_{11} & B_{12} & B_{16} & D_{11} & D_{12} & D_{16} & F_{11} & F_{12} & F_{16} \\ B_{12} & B_{22} & B_{26} & D_{12} & D_{22} & D_{26} & F_{12} & F_{22} & F_{26} \\ B_{16} & B_{26} & B_{66} & D_{16} & D_{26} & D_{66} & F_{16} & F_{26} & F_{66} \\ E_{11} & E_{12} & E_{16} & F_{11} & F_{12} & F_{16} & H_{11} & H_{12} & H_{16} \\ E_{12} & E_{22} & E_{26} & F_{12} & F_{22} & F_{26} & H_{12} & H_{22} & H_{26} \\ E_{16} & E_{26} & E_{66} & F_{16} & F_{26} & F_{66} & H_{16} & H_{26} & H_{66} \end{bmatrix} \qquad \qquad ...(6)$$

$$[D_5] = \int_{-k/2}^{k/2} \begin{bmatrix} [\overline{Q}]^T & z^2[\overline{Q}]^T \\ z^2[\overline{Q}]^T & z^4[\overline{Q}]^T \end{bmatrix} dz = \begin{bmatrix} A_{55} & A_{45} & D_{55} & D_{45} \\ A_{45} & A_{44} & D_{45} & D_{44} \\ D_{55} & D_{45} & F_{55} & F_{45} \\ D_{45} & D_{44} & F_{45} & F_{44} \end{bmatrix} \qquad ...(7)$$

where

$$\left(A_{ij} B_{ij} D_{ij} E_{ij} F_{ij} H_{ij} \right) = \sum_{i=1}^{NL} \int_{-h/2}^{h/2} \overline{Q}_{ij} \left(1, z, z^2, z^3, z^4, z^6 \right) dz \qquad ...(8)$$

and

$$\left[\overline{Q}_{ij} \right] = \begin{bmatrix} \overline{Q}_{11} & \overline{Q}_{12} & \overline{Q}_{16} & 0 & 0 \\ \overline{Q}_{12} & \overline{Q}_{22} & \overline{Q}_{26} & 0 & 0 \\ \overline{Q}_{16} & \overline{Q}_{26} & \overline{Q}_{66} & 0 & 0 \\ 0 & 0 & 0 & \overline{Q}_{44} & \overline{Q}_{45} \\ 0 & 0 & 0 & \overline{Q}_{45} & \overline{Q}_{55} \end{bmatrix}_k \qquad ...(9)$$

where subscript k denotes the kth layer.

$$\left[D_b^2 \right] = \int_{-h/2}^{h/2} \left[[\overline{Q}]^T \ z[\overline{Q}]^T \ z^3[\overline{Q}]^T \right]^T \int_{T_0}^{T_1} <\alpha_x \alpha_y \ 2\alpha_{xy} >^T dTdz \qquad ...(10)$$

where

$$[\overline{Q}]_{3\times3}^T = \begin{bmatrix} \overline{Q}_{11} & \overline{Q}_{12} & \overline{Q}_{16} \\ \overline{Q}_{12} & \overline{Q}_{22} & \overline{Q}_{26} \\ \overline{Q}_{16} & \overline{Q}_{26} & \overline{Q}_{66} \end{bmatrix} \qquad ...(11)$$

and

$$[\overline{N}] = \begin{bmatrix} \overline{N}_x & \overline{N}_{xy} \\ \overline{N}_{xy} & \overline{N}_y \end{bmatrix} \qquad ...(12)$$

where

$$\left(\overline{N}_x, \overline{N}_y, \overline{N}_{xy} \right) = \sum_{i=1}^{NL} \int_{-h/2}^{h/2} \left(\overline{\sigma}_x \overline{\sigma}_y \overline{\tau}_{xy} \right) dz \qquad ...(13)$$

It should be noted that in order to derive Eq. (3), von Karman strains should be used first since buckling and thermal buckling are basically non-linear phenomena, and then linearization can be performed due to the small perturbation.

3. NUMERICAL RESULTS AND DISCUSSIONS

The solution in the form of non-dimensional critical buckling temperature ($\alpha_0*Tcr*10^4$) is obtained by the C_0 finite element method for the uniform temperature distribution considering T_0 as zero without loss of generality. A nine noded Lagrangian isoparametric element, which results in 63 degrees of freedoms (DOFs) for HSDT model is used for discretizing the laminate. All the results reported in this paper have been obtained by employing the full (3×3) integration rule. The following laminate material properties and lamina thermal coefficients have been used for computation of critical buckling temperature $E_{22} = 10^5$ Pa, $E_{11} = 15*E_{22}$, $G_{12} = G_{13} = 0.6*E_{22}$, $G_{23} = 0.5*E_{22}$, $v_{12} = v_{21} = 0.25$, $\alpha_0 = 10^{-6}/°C$, $\alpha_1 = 0.02*\alpha_0$ and $\alpha_2 = 22.5*\alpha_0$ [4].

The convergence study has shown that change in critical buckling temperature is 0.061% when the mesh density is increased from 5×5 to 6×6. Therefore a 5×5 mesh has been selected to obtain further numerical results.

To validate the solutions with Prabhu and Dhanraj [4], the critical buckling temperature for five layered symmetric cross ply [90°/–90°/90°/–90°/90°] square ($a = b = 200$) plate are obtained with varying a/h having all edges simply supported (S1) boundary condition, i.e., $u = v = w = \Phi_y = 0$ at $x = 0$, a and $u = v = w = \Phi_x = 0$ at $y = 0$, b as shown in Fig. 1.

From the Fig.1, it can be observed that the present results are in good agreement with that of Prabhu and Dhanraj [4] as the deviation is within 6%.

MATLAB in-built command based on Monte Carlo Simulation (MCS) is used to generate random numbers corresponding to mean values and selected standard deviation (SD) of the material property having normal probability distribution, and corresponding changes in the SD/Mean of non-dimensional critical buckling temperatures are obtained.

The convergence study of the non-dimensional critical buckling temperatures with respect to random number for a standard deviation in E_{22} as 0.1 is shown in Fig. 2. The figure shows that the SD/Mean of non-dimensional buckling temperature becomes almost asymptotic when random numbers are above 6400. However, the sample size to fit the desired mean and SD is taken as 10000 in the present study.

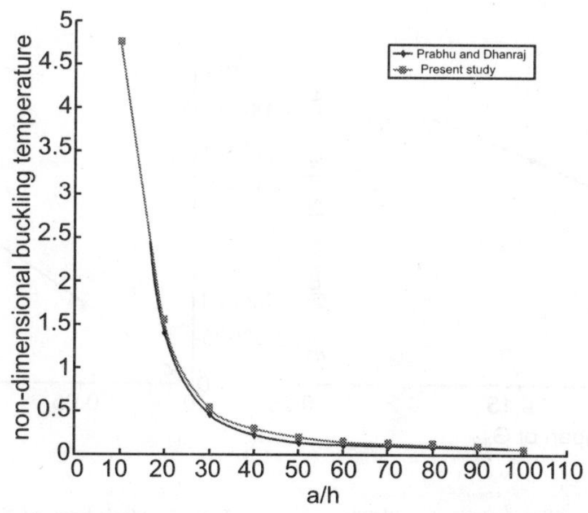

Fig. 1. Comparison of results of the present study with that of Prabhu & Dhanraj [4].

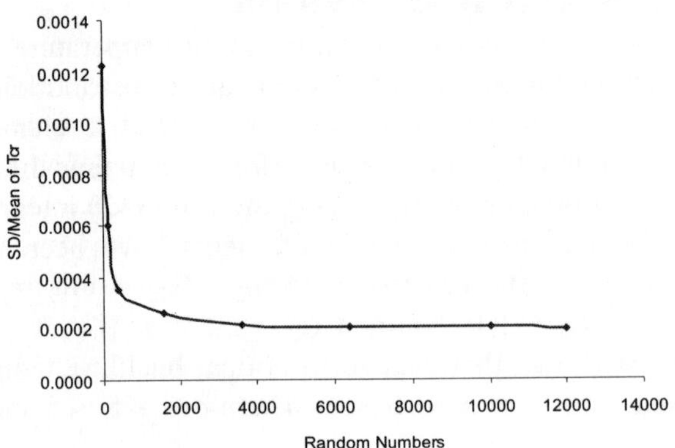

Fig. 2 Effect of increase in random numbers on SD/Mean of nondimensional *Tcr* for 10% deviation in E_{22}.

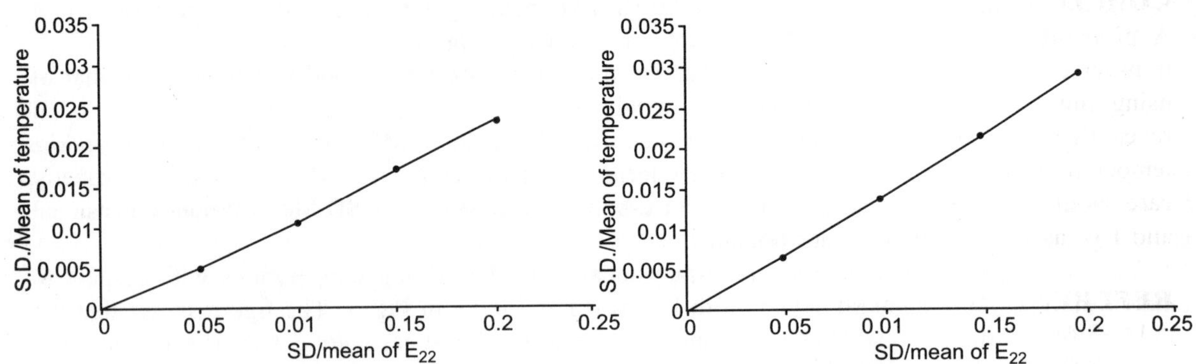

Fig. 3. Variation of SD/Mean of non-dimensional Critical buckling temperature with SD/mean of E_{11}.

Fig. 4. Variation of SD/Mean of nondimensional critical buckling temperature with SD/mean of E_{22}.

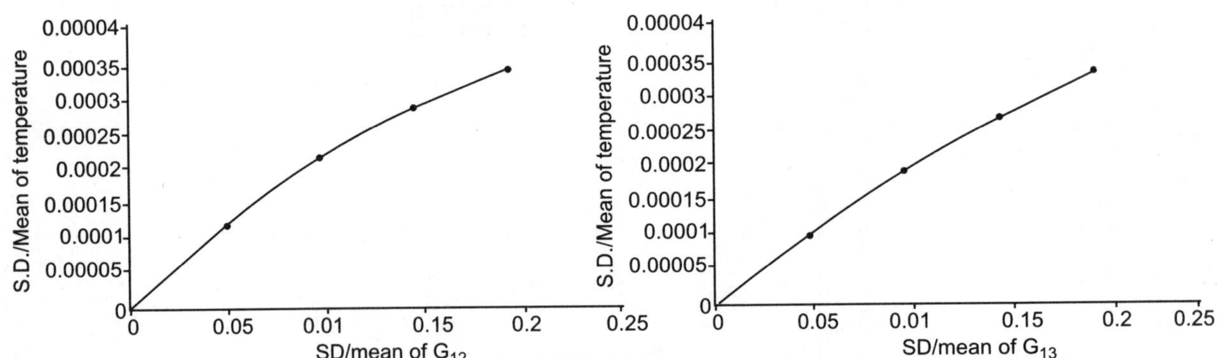

Fig. 5. Variation of SD/Mean of nondimensional critical buckling temperature with SD/mean of G_{12}.

Fig. 6. Variation of SD/Mean of non-dimensional critical buckling temperature with SD/mean of G_{13}.

Further, the typical results are presented here for five layered symmetric angle ply [75°/–75°/ 75°/–75°/75°] square ($a = b = 200$) laminate with $a/h = 100$ having all edges simply supported (S1) boundary condition. To study the sensitivity of the critical buckling temperature due to variation in the individual material properties, the variation of SD/Mean of non-dimensional critical buckling temperature with SD/mean of each material property (E_{11}, E_{22}, G_{12} and G_{13}) is obtained and shown in Figs. 3 to 6, respectively. These figures depict very little non-linearity up to the 20% deviation in the material properties carried out in the present study. Hence it can be inferred that the scattering in buckling temperature also increases with the increase in deviation of material properties almost linearly. The slope of the lines for G_{12} is nearly equal to that of G_{13}. Also, the variation in slope of E_{11} and that of E_{22} is also not large. However the slope of shear modulus (G_{12} and G_{13}) is smaller by an order of 100 as compared to that of elastic modulus (E_{11} and E_{22}). This implies that variation in elastic modulus (E_{11} and E_{22}) is having a 100-fold effect on the scattering in buckling temperature as compared to that of shear modulus (G_{12} and G_{13}).

CONCLUSION

A probabilistic study of thermal buckling behaviour of laminated composite plate incorporating transverse shear effect with all edges simply supported (S_1) boundary condition has been carried out using finite element method in conjunction with Monte Carlo simulation technique. The results reveal that variation in elastic modulus (E_{11}, E_{22}) has about 100 fold greater influence on the buckling temperature as compared to that due to variation in shear modulus (G_{12}, G_{13}). This means that the care should be taken during fabrication process so as to allow less variation in elastic modulus, E_{11} and E_{22}, as compared to that in shear modulus, G_{12} and G_{13}.

REFERENCES

1. Szilard Rudolph, 1985, Critical load and post buckling analysis by FEM using energy balance technique, J. Computers & Structures, 20, 227-286.
2. Jeng-Shian Chang, 1990, FEM analysis of buckling and thermal buckling of antisymmetric angle-ply laminates according to transverse shear and normal deformable higher order displacement theory, J. Computers and Structures, 37, 925-946.
3. Shu Xiaopingt and Sun Liangxint, 1994, Thermo-mechanical buckling of laminated composite plates with higher-order transverse shear deformation, J. Computers & Structures, 53, 1-7.
4. M.R. Prabhu and R. Dhanraj, 1994, Thermal buckling of laminated composite plates. J. Computers & Structures, 53, 1193-1204.
5. Ashok Kumar Ghosh and Shanty Shekhar Dey, 1994, Buckling of laminated plate-A simple finite element based on higher order theory, J. Finite Element in Analysis and Design, 15, 289-302.
6. Hiroyuki Matsunaga, 2005, Thermal buckling of cross-ply laminated composite and sandwich plates according to a global higher-order deformation theory, J. Composite Structures, 68, 439-454.
7. J.C. Robert, G. Bao and J.G. White, 1999, Experimental, numerical and analytical result for bending and buckling of rectangular orthotropic plate, J. Composite Structures, 43, 289-299.
8. S. Nakagiri, H. Tatabate and S. Tani, 1997, Uncertain eigen value analysis of composite laminate plates by SFEM,. J. Composite Structures, 37, 385-391.
9. B.N. Singh, N.G.R. Iyenger, D. Yadav, 2001, Initial buckling of composite cylindrical panels with random material properties. J. Composite Structures, 53, 55-64.
10. B.N. Singh, N.G.R. Iyenger, D. Yadav, 2002, A °C finite element investigation for buckling of shear deformable laminated composite plates with random material properties, Int. J. Structural Engineering and Mechanics, 13, 53-74.

29

Dynamic Buckling of Imperfect Rectangular Composite Plates

R. DARIPA AND **M.K. SINGHA**

*Department of Applied Mechanics, Indian Institute of Technology, Delhi, New Delhi-110016, India.
email: maloy@am.iitd.ernet.in*

ABSTRACT

Here, the dynamic stability characteristics of imperfect composite plates subjected to a periodic in-plane load are investigated using finite element approach. A four-nodded shallow shell element is developed for this purpose. The formulation includes the effects of transverse shear deformation, in-plane and rotary inertia. The boundaries of the instability regions are obtained using the Bolotin's method and are represented in the non-dimensional load amplitude-excitation frequency plane. Attempt is also made to study the effect of imperfection and load amplitude on the dynamic stability characteristics of composite plates from its dynamic response.

Keywords: Imperfect composite plate, dynamic stability, finite element method.

1. INTRODUCTION

Fiber reinforced composites are increasingly used in thin walled structural components of aircrafts, submarines, automobiles and industrial structures. They are sometimes subjected to dynamic in-plane load and become unstable for certain combinations of excitation frequency and corresponding load amplitude. A general theory of dynamic stability of isotropic structures is available in Ref [1], whereas, studies pertaining to composite laminates have been attempted by many investigators [2-7]. From the review of the earlier literature, it is observed that the dynamic stability characteristics of perfect composite plates have received considerable attention of the researchers, whereas, limited work has been focused on the imperfect composite plates.

Structural elements, such as composite plates may have some unavoidable geometric imperfection. Honig and Stronge [8] and Petry and Fahlbusch [9] investigated the dynamic stability characteristics of imperfect isotropic plates subjected to in-plane impact. Papazoglou and Tsouvalis [10] presented an analytical solution for the dynamic buckling of imperfect composite plates under in-plane linear and pulse load. However, a general study on the dynamic stability characteristics of imperfect composite plates appears to be scarce in the literature.

The objective of the present paper is to investigate the dynamic stability characteristics of imperfect composite plates subjected to a periodic in-plane load. A four-noded high precision shear flexible shallow shell element is developed for this purpose. The formulation includes in-plane and rotary inertia effects. The element is free from locking syndrome and has good convergence properties. The boundaries of the instability regions are obtained using Bolotin's method. The finite element equations are also solved using Newmark's time integration method to get the dynamic response. Attempt is made to study the effect of imperfection and load amplitude on the dynamic stability characteristics of composite plates from its dynamic response.

2. FORMULATION

The first order shear deformation theory, used in the analysis, assumes a linear variation of in-plane displacements, u and v through the depth of the laminate. The transverse displacement w (x, y) is assumed to be constant throughout the thickness of the plate. The displacement components at an arbitrary point (x, y, z) of a laminate can be expressed as [11]

$$u\ (x,\ y,\ z) = u_0\ (x,\ y) + z\ \{w_{,x} + \gamma_x\ (x,\ y)\}$$
$$v\ (x,\ y,\ z) = v_0\ (x,\ y) + z\ \{w_{,y} + \gamma_y\ (x,\ y)\} \qquad ...(1)$$
$$w\ (x,\ y,\ z) = w_0\ (x,\ y)$$

Here, u_0, v_0, and w are the mid-surface displacements; γ_x and γ_y are the rotations due to shear; $()_{,x}$ and $()_{,y}$ represent the partial differentiation with respect to x and y; $\theta = w_{,x} + g_x\ (x,\ y)$ and $\theta_y = w_{,y} + \gamma_y\ (x,\ y)$ are the nodal rotations.

From the Marguerre's shallow shell theory [12], the membrane strains $\{\varepsilon\}^D$, curvatures $\{k\}$ and shear strains $\{\gamma\}$ can be written as

$$\{\varepsilon\}^0 = \begin{Bmatrix} u_{0,x} \\ v_{0,y} \\ v_{0,x} + u_{0,y} \end{Bmatrix} + \begin{Bmatrix} \overline{w}_{,x}\, w_{,x} \\ \overline{w}_{,y}\, w_{,y} \\ \overline{w}_{,y}\, w_{,x} + \overline{w}_{,x}\, w_{,y} \end{Bmatrix}; \quad \{\kappa\} = \begin{Bmatrix} w_{,xx} + \gamma_{x,x} \\ w_{,yy} + \gamma_{y,y} \\ 2w_{,xy} + \gamma_{x,y} + \gamma_{y,x} \end{Bmatrix} \text{ and } \{\gamma\} = \begin{Bmatrix} \gamma_{xz} \\ \gamma_{yz} \end{Bmatrix} = \begin{Bmatrix} -\gamma_x \\ -\gamma_y \end{Bmatrix} \quad ...(2)$$

The internal strain energy (U) of the plate can be written as

$$U(\delta) = \frac{1}{2} \int_A \left[\{\varepsilon^0\}^T [A]\{\varepsilon^0\} + \{\varepsilon^0\}^T [B]\{\kappa\} + \{\kappa\}^T [B]^T \{\varepsilon^0\} + \{\kappa\}^T [D]\{\kappa\} + \{\gamma\}^T [S]\{\gamma\} \right] dA \quad ...(3)$$

where $[A]$, $[B]$, $[D]$, and $[S]$ are extensional, bending-extensional, bending, and shear stiffness coefficients respectively. For a composite laminate of thickness h, comprising of N layers with stacking angles θ_i (i = 1,2, ..., N) and layer thicknesses h_i (i = 1,2, ..., N), the necessary expression to compute the stiffness coefficients are available in the literature [13].

The potential energy due to external in-plane mechanical force (N_{xx}) in the x direction is written as

$$W(\delta) = \int_A \left[\frac{1}{2} N_{xx} \left(\frac{\partial w}{\partial x} \right)^2 + \frac{h^2}{24} N_{xx} \left\{ \left(\frac{\partial \theta_x}{\partial x} \right)^2 + \left(\frac{\partial \theta_y}{\partial y} \right)^2 \right\} \right] dA \qquad ...(4)$$

The kinetic energy of the shell is given by

$$T(\delta) = (1/2) \int_A \left[p(\dot{u}_0^2 + \dot{v}_0^2 + \dot{w}^2) + I(\dot{\theta}_x^2 + \dot{\theta}_y^2) \right] dA \quad p = \int_{-h/2}^{h/2} \rho \, dz, \quad I = \int_{-h/2}^{h/2} \rho z^2 \, dz \qquad ...(5)$$

where ρ the mass density. A dot represents the partial derivative with respect to time.

Following standard procedure, the finite element equations for the laminate under uniaxial compressive force $N_{xx}(t) = N_0 + N_1 \cos \theta t$ are derived as

$$[M]\{\ddot{\delta}\} + [K + N_{xx} K_G]\{\delta\} = 0 \qquad ...(6)$$

where, **M**, **K** and **K**$_G$ are mass, stiffness, and geometric stiffness matrices respectively and δ is the vector of degrees of freedom.

2.1 Element Description

Here, a four-noded quadrilateral shallow shell element with fourteen degrees of freedom per node, namely u_0, $u_{0,x}$, $u_{0,y}$, v_0, $v_{0,x}$, $v_{0,y}$, w, $w_{,x}$, $w_{,y}$, $w_{,xx}$, $w_{,xy}$, $w_{,yy}$, γ_x and γ_y are used. The cubic polynomial shape functions are employed to describe the field variables corresponding to in-plane displacements (u_0, v_0), quintic polynomial function is considered for the lateral displacement (w), whereas, linear polynomial shape functions are used for the rotations due to shear of the middle surfaces (γ_x, γ_y), and are expressed as follows:

$$u_0 = [1, x, y, x^2, xy, y^2, x^3, x^2y, xy^2, y^3, x^3y, xy^3] \{c_i\}, \ i = 1, 12$$

$$v_0 = [1, x, y, x^2, xy, y^2, x^3, x^2y, xy^2, y^3, x^3y, xy^3] \{c_i\}, \ i = 13, 24$$

$$w = [1, x, y, x^2, xy, y^2, x^3, x^2y, xy^2, y^3, x^4, x^3y, x^2y^2, xy^3, y^4, x^5, x^4y, x^3y^2, x^2y^3, xy^4, y^5, x^5y,$$
$$x^3y^3, xy^5]\{c_i\}, \ i = 25, 48$$

$$\gamma_x = [1, x, y, xy] \{c_i\},$$
$$i = 49, 52 \qquad ...(7)$$

$$\gamma_y = [1, x, y, xy] \{c_i\},$$
$$i = 53, 56$$

where c_k are constants and are expressed in terms of nodal displacements in the finite element discretization. The shear correction factor is taken as 5/6. The full integration scheme with 6×6 Gaussian integration rule is adopted for computing the element mass matrix **M**, whereas 4×4 Gaussian integration rule is used to calculate the stiffness matrices **K** and **K**$_G$.

2.2 Dynamic Stability Analysis

Equation (6) represents the dynamic stability problem of a laminate subjected to a periodic in-plane force $N_{xx}(t)$. The dynamic instability boundary is determined using the method suggested in Ref [1]. The present study is focused on the determination of primary instability region that occurs in the vicinity of $2\omega_n$ (ω_n –the lowest natural frequency) of composite panels. This is the most dangerous zone and has practical importance. To obtain points on the boundaries of the instability regions, the

component δ are written in the Fourier series as $\delta = \displaystyle\sum_{i=1,3,5,..} \left\{ \mathbf{a}_i \sin \dfrac{i\theta t}{2} + \mathbf{b}_i \cos \dfrac{i\theta t}{2} \right\}$ with period

$2T$ ($T = 2\pi/\theta$). Here, $\mathbf{a}_i$ and $\mathbf{b}_i$ are unknown coefficient vectors and θ represents the forcing frequency. Substituting the solutions into equation (6) and by grouping the coefficients of sine and cosine terms, two sets of linear algebraic equation in $\mathbf{a}_i$ and $\mathbf{b}_i$ are obtained. To obtain non-trivial coefficient vectors $\mathbf{a}_i$ and $\mathbf{b}_i$, the determinant of the coefficient matrix must be zero for each set. Further, only one term solution is employed here since it furnished accurate results for low values of load amplitude (N_1/N_{cr}) (N_{cr} is the critical buckling load) and significantly reduces the computational time. Hence the dynamic instability of the composite panel reduces to the following eigenvalue problem

$$\mathbf{K} + N_0 \mathbf{K}_G \pm 0.5 N_1 \mathbf{K}_G - 0.25 \mathbf{M}\,\theta_1^2 = 0 \qquad \qquad ...(8)$$

Solving equation (8) for a given value of N_0, the variation of θ with respect to N_1 can be found. Such a plot shows the boundaries of the dynamic instability regions for the given plate subjected to harmonically excited load [2-7].

Equation (8) is also solved using Newmark's time integration technique to get the dynamic response of the panel and finally an attempt is made to study the dynamic stability characteristics of imperfect composite plates through its dynamic response characteristics.

3. NUMERICAL RESULTS AND DISCUSSION

In this section, we use the above formulation to investigate the dynamic stability characteristics of composite plates. The material properties, unless specified otherwise, used in the present analysis are $E_L / E_T = 40.0$, $G_{LT} / E_T = 0.6$, $G_{TT} / E_T = 0.5$, $\mu_{LT} = 0.25$, $\rho = 1.0$ where E, G, μ and ρ are Young's modulus, Shear modulus, Poisson's ratio and density. Subscripts L and T represent the longitudinal and transverse directions respectively with respect to the fibers. All the layers are of equal thickness. The boundary conditions considered here are *simply supported on all sides* ($u_0 = v_0 = w = 0$ along the boundary nodes).

Before proceeding for the detailed analysis, the formulation developed herein is validated against the linear free vibration of laminated composite plates and cylindrical panels. The non-dimensional natural frequencies ($\Omega = \omega a^2 / \pi^2 h\sqrt{\rho / E_T}$; a & h are length and thickness of the plate) obtained for thin ($a/h = 1000$) simply supported cross-ply [0°/90°/0°/90°/0°] composite plates are presented in Table 1 along with the analytical solutions of Wang [14] and they match very well with the available results. It is also observed from Table 1 that the element developed here has a good convergence property and thus, an 8 × 8 mesh is found to be adequate to model the full plate. Further, the efficacy of the present formulation is tested by comparing the free vibration frequencies of cylindrical cross-ply panels with the exact results of Reddy [15] in Table 2 and these results are found to be in good agreement for shallow shells.

Table 1. Convergence study of non-dimensional linear natural frequency ($\Omega = \omega a^2 / \pi^2 h\sqrt{\rho / E_T}$)
of 5-layered [0°/90°/0°/90°/0°] simply supported square plate ($a/h = 1000$)

	Mesh size	Modes					
		1	2	3	4	5	6
Present study	4 × 4	1.9007	3.9304	6.2905	7.0695	8.0924	9.8203
	6 × 6	1.9128	3.9689	6.6116	7.5837	8.1247	10.4874
	8 × 8	1.9137	3.9730	6.6448	7.6381	8.1434	10.5878
	10 × 10	1.9139	3.9738	6.6519	7.6494	8.1478	10.6107
	12 × 12	1.9140	3.9741	6.6541	7.6528	8.1492	10.6180
Wang. [14]		1.9141	3.9745	6.6567	7.6564	8.1511	10.6249

Table 2. Comparison of non-dimensional fundamental frequencies $\Omega = \omega a^2 \sqrt{\rho/E_T}/h$ of a simply supported cylindrical panel ($a = b$, $E_L = 25\,E_T$, $G_{LT} = 0.5\,E_T$, $G_{TT} = 0.2\,E_T$, $n_{LT} = 0.25$).

R/a	[0°/90°/90°/0°]				[0°/90°]			
	$a/h = 100.0$		$a/h = 10.0$		$a/h = 100.0$		$a/h = 10.0$	
	Present	Reddy[15]	Present	Reddy[15]	Present	Reddy[15]	Present	Reddy[15]
4	23.181	22.749	12.338	12.289	19.935	19.509	9.068	8.930
5	20.654	20.361	12.298	12.267	16.952	16.668	9.011	8.909
10	16.716	16.634	12.244	12.236	12.015	11.831	8.934	8.887
Plate	15.183	15.184	12.225	12.226	9.703	9.687	8.908	8.899

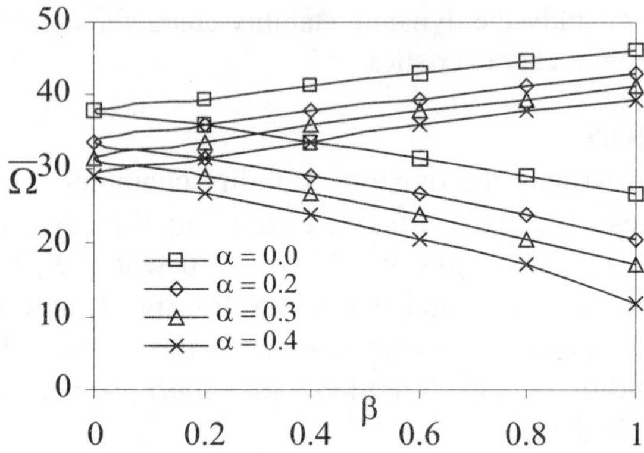

Fig. 1. Effect of static in-plane load parameter α on the dynamic instability region of perfect composite plate, ($a/h = 100$).

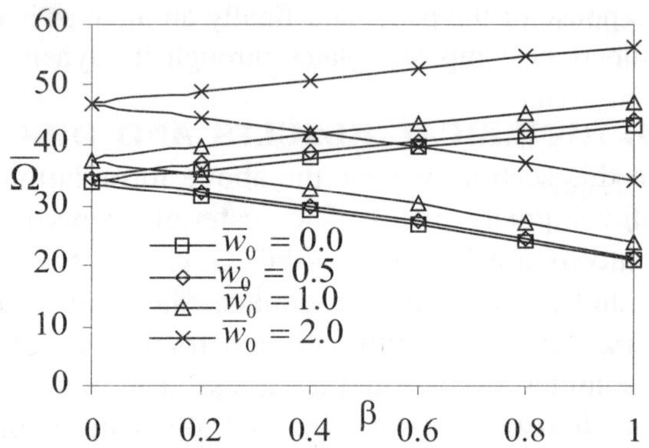

Fig. 2. Effect of imperfection on the primary instability region of a composite plate ($a/b = 1.0$, $a = 0.0$, $a/h = 100$).

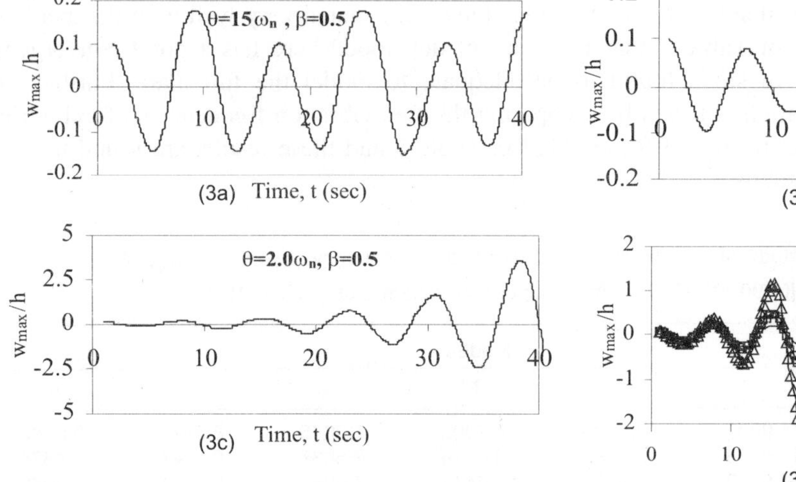

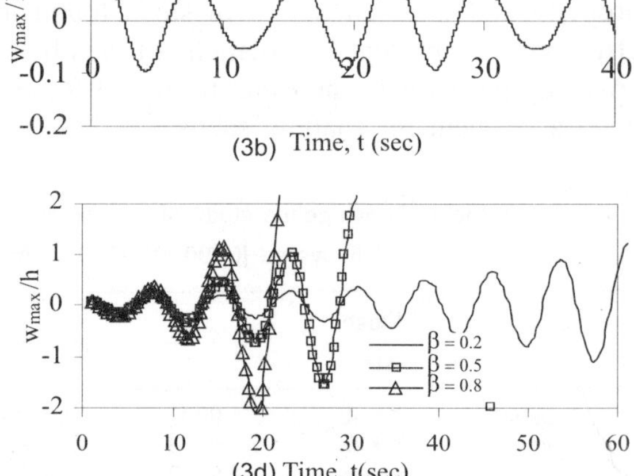

Fig. 3. Dynamic responses of perfect cross ply [0°/90°/0°/90°/0°] plate for different values of in-plane forcing frequency θ ($N_0 = 0.0$, $\beta = N_0/N_{cr}$).

Next, a detailed study is carried out considering 5-layered symmetric [0°/90°/0°/90°/0°] cross-ply laminates Dynamic instability region of a perfect square plate is presented in Fig. 1, for various values of static in plane load α (= N_0/N_{cr}). Instability regions are plotted in the plane having non-dimensional load amplitude β (N_1/N_{cr}) as abscissa and non-dimensional load frequency $\bar{\Omega}$ ($\bar{\Omega} = \theta a^2 / \pi^2 h \sqrt{\rho / E_T}$) as ordinate. Because of one term solution methodology, the calculation in the present work is restricted in the region of β = 0.0 – 0.1. It is observed from the figure, that the increase in static in-plane load (N_0) load shifts the dynamic instability area to lower frequency region and increases the width of the instability zone. Effect of the imperfection amplitude on the dynamic instability region is studied in Fig. 2, where the imperfection shape is assumed to be sinusoidal ($\bar{w} = \bar{w}_0 \sin\dfrac{\pi x}{a} \sin\dfrac{\pi y}{b}$). Figure 2 shows that with the increase in imperfection amplitude $\bar{w}_0$ dynamic instability occurs at higher disturbing frequencies.

Dynamic response of a perfect square plate is evaluated in Fig. 3 for various values of excitation frequencies (θ) and load parameter β. Material and geometric properties considered here are E_L = 172.7 GPa, E_T = 7.2 GPa, G_{LT} = G_{TT} = 3.7 GPa, μ_{LT} = 0.3, ρ = 1566 Kg/m^3, a/h = 625. A disturbance in the first mode shape with w/h = 0.1 is assumed to be the initial condition. It is observed from Fig. 3 that, the plate becomes dynamically unstable with the out-of-plane displacement increases without bound at a disturbing frequency of $2\omega_n$. Further with the increase in dynamic load amplitude (β = N_1/N_{cr}) the amplitude of vibration (w/h) increases rapidly. Dynamic response analysis of imperfect cross-ply plate is depicted in Fig 4 for various imperfection shape and amplitude.

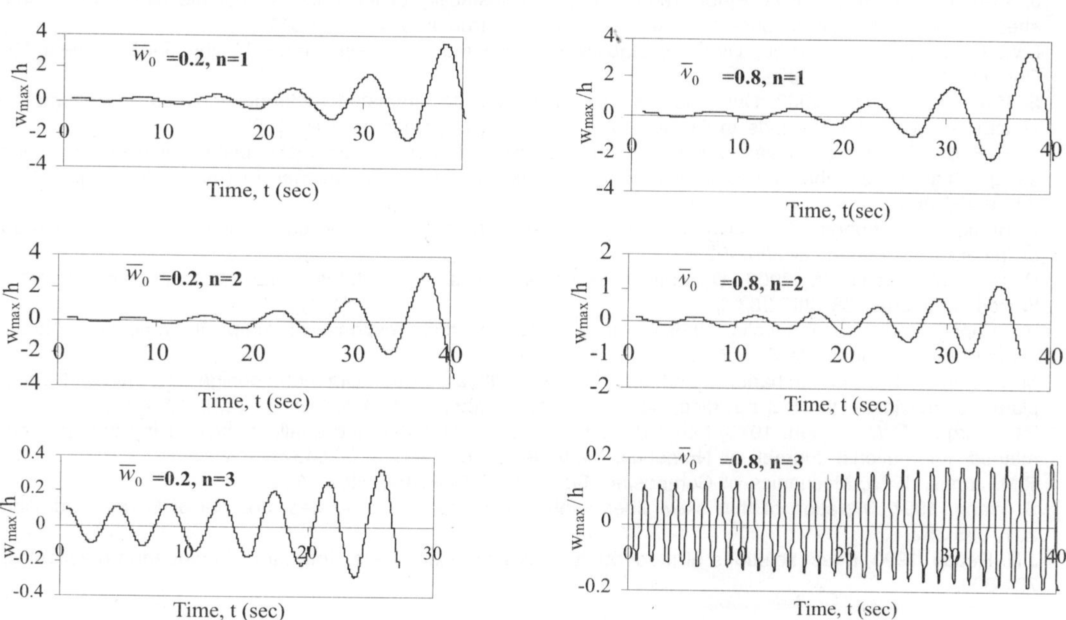

Fig. 4. Dynamic responses of imperfect cross ply plate for different values of imperfect amplitude.

(α = 0.0, β = 0.5). sinusoidal imperfection $\bar{w} = \bar{w}_0 \sin\dfrac{n\pi x}{a} \sin\dfrac{n\pi y}{b}$.

It is interesting to note from Fig. 4 that, for a given dynamic load amplitude β $(N_1/N_{cr}) = 0.5$, with the increase in imperfection amplitude, the amplitude of vibration (w/h) decreases. Furthermore, it can be noticed from Fig. 4 that the amplitude of vibrations of the plate increases rapidly when imperfection is of first order compared to higher order imperfection mode.

CONCLUSION

Dynamic stability characteristics of imperfect composite plates under periodic in-plane load have been investigated using finite element method. A four nodded high precision shallow shell element is developed for this purpose. The element is found to be free from locking and have good convergence properties. The present study leads to the following observations.

(1) The width of the instability region increases with increase in both static (N_0) and dynamic (N_1) load amplitude.

(2) Dynamic instability region shifts towards lower disturbing frequency with the increase of static in-plane load (N_0).

(3) Higher imperfection amplitude shifts the dynamic instability region towards higher disturbing frequencies, whereas, higher mode of imperfection $(n > 1)$ decreases the instability phenomenon.

REFERENCES

1. V.V. Bolotin, 1964, Dynamic Stability of Elastic Systems, Holden-day: San Francisco.
2. V. Birman, 1985, Dynamic stability of unsymmetrically laminated rectangular plates, Mechanical Research Communication. 12, 81-86.
3. C.W. Bert, V. Birman, 1987, Dynamic instability of shear deformable antisymmetric angle-ply plates, International Journal of Solids and Structure. 23, 1053-1061.
4. J. Moorthy, J.N. Reddy, R.H. Plaut, 1990, Parametric instability of laminated composite plates with transverse shear deformation. International Journal of Solids and Structures. 26, 801-811.
5. L.W. Chen, J.Y. Yang, 1990, Dynamic stability of laminated composite plates by the finite element Method, Comput. Srrucr. 36, 845-851.
6. S. Wang, D.J. Dawe, 2002, Dynamic instability of composite laminated rectangular plates and prismatic plate structures, Computer Methods in Applied Mechanics and Engineering. 191, 1791-1826.
7. A.G. Radu, A. Chattopadhyay, 2002, Dynamic stability analysis of composite plates including delaminations using a higher order theory and transformation matrix approach, International Journal of Solids and Structure. 39, 1949-1965.
8. A. Honig, W.J. Stronge, 2000, Dynamic buckling of an imperfect elastic, visco-plastic plate, International Journal of Impact Engineering. 24, 907-923.
9. D. Petry, G. Fahlbusch, 2000, Dynamic buckling of thin isotropic plates subjected to in-plane impact, Thin-Walled Structures. 38, 267-283.
10. V.J. Papazoglou, N.G. Tsouvalis, 1995, Large deflection dynamic response of composite laminated plates under in-plane loads, Composite Structures. 33, 231-252.
11. M.K. Singha, L.S. Ramachandra, and J.N. Bandyopadhyay, 2000, Optimum design of laminated composite plates for maximum thermal buckling loads, Journal of Composite Materials. 34 (23), 1982-1997.
12. P.G. Bargan, D.W. Clough, 1973, Large deflection analysis of plates and shallow shells using the finite element method, International Journal for Numerical Methods in Engineering. 5, 543-556.
13. R.M. Jones, 1975, Mechanics of Composite Materials, McGraw-Hill: New York,.
14. S. Wang, 1997, Vibration of thin skew fiber reinforced composite laminates, Journal of Sound and Vibration. 201, 335-352.
15. J.N. Reddy, 1990, Exact solutions of moderately thick laminated shells, Journal of Engineering Mechanics, 110, 794-809.

30

Buckling of Laminated Joined Conical-Cylindrical Shells Subjected to Thermo-mechanical Loads

S. SINGH, B.P. PATEL AND Y. NATH

Department of Applied Mechanics, Indian Institute of Technology Delhi-110 016, India
email: badripatel@hotmail.com, bppatel@am.iitd.ernet.in

ABSTRACT

In the present work the buckling of laminated joined conical-cylindrical shells subjected to torsion, external pressure, axial compression and uniform temperature loading are investigated using the semi-analytical finite element approach. The formulation is based on the first order shear deformation theory and field consistency principle. The variation in the stiffness coefficients along meridian direction due to the changes in ply-angle and ply-thickness of filament wound angle-ply laminated conical shells are accounted in the finite element formulation. The distribution of prebuckling stress resultant and the bifurcation loads of cross-ply/angle-ply laminated joined shell systems is investigated.

Keywords: Joined Conical-Cylindrical shell; Buckling; Semi-Analytical Finite element; Critical Load; Angle-ply.

1. INTRODUCTION

The thin-walled structures composed of two or more shells having one common axis of revolution are widely used in mechanical, marine, aeronautical, chemical, civil and power engineering. Such shells subjected to thermo-mechanical loads may be prone to buckling. The buckling analysis of isotropic joined conical-cylindrical shells subjected to external pressure has received the attention of few researchers (Flores and Godoy [1]; Anwen [2]). It is brought out that the bifurcation loads of the complex shells are lower than those of the individual components. The buckling behaviour of continuous fibre wound angle-ply laminated conical shells with variation of layer ply angle and thickness along the meridional direction is carried out by Goldfeld and Arbocz [3] and Goldfeld *et al*. [4]. The advances in composite technology have lead to the application of laminated composite structural elements like cylindrical and conical shells joined together as load bearing members in the design of more and more sophisticated futuristic structures. In order to optimally exploit the strength and load carrying capacity of laminated composite joined conical-cylindrical shells, accurate prediction and understanding of their thermo-mechanical buckling behaviour is important and undertaken here.

The critical buckling load of laminated joined shells system subjected to the torsion, external pressure, axial compression and uniform temperature change are investigated using the semi-analytical finite element approach [5]. The critical load is evaluated employing the eigenvalue buckling analysis. The study is carried out for the distribution of prebuckling stress resultant of cross-ply laminated conical, cylindrical and joined conical-cylindrical shells and the bifurcation loads of cross-ply/angle-ply laminated joined shell system subjected to thermo-mechanical loads.

2. FORMULATIONS

A laminated composite joined conical-cylindrical shell is considered with the coordinates s, θ and z along the meridional, circumferential and radial/thickness directions, respectively. Using first order shear deformation theory the displacements u, v, w at a point (s, θ, z) from the median surface are expressed as functions of middle surface displacements u_0, v_0 and w_0, and rotations β_s and β_θ of the meridional and hoop sections, respectively, as

$$u(s, \theta, z) = u_0(s, \theta) + z\,\beta_s(s, \theta)$$
$$v(s, \theta, z) = v_0(s, \theta) + z\,\beta_\theta(s, \theta)$$
$$w(s, \theta, z) = w_0(s, \theta) \qquad\qquad ...(1)$$

Using the semi-analytical approach, u_0, v_0, w_0, β_s and β_θ are represented by a Fourier series in the circumferential coordinate θ. For the nth harmonic, these can be written as

$$u_o(s,\theta)=u_o^0(s)+u_o^c(s)\cos(n\theta)+u_o^s(s)\sin(n\theta)\,;\ v_o(s,\theta)=v_o^0(s)+v_o^c(s)\cos(n\theta)+v_o^s(s)\sin(n\theta)$$

$$w_o(s,\theta)=w_o^0(s)+w_o^c(s)\cos(n\theta)+w_o^s(s)\sin(n\theta)\,;\ \beta_s(s,\theta)=\beta_s^0(s)+\beta_s^c(s)\cos(n\theta)+\beta_s^s(s)\sin(n\theta)$$

$$\beta_\theta(s,\theta)=\beta_\theta^0(s)+\beta_\theta^c(s)\cos(n\theta)+\beta_\theta^s(s)\sin(n\theta) \qquad\qquad ...(2)$$

where superscript 0, c, and s refers to the axisymmetric, cosine and sine asymmetric components of displacement field variables.

The Green's strains can be written in terms of the mid-plane deformations as,

$$\{\varepsilon\}=\begin{Bmatrix}\varepsilon_p^L\\ 0\end{Bmatrix}+\begin{Bmatrix}z\varepsilon_b\\ \varepsilon_s\end{Bmatrix} \qquad\qquad ...(3)$$

where, the membrane strains $\{\varepsilon_p^L\}$, bending strains $\{\varepsilon_b\}$, shear strains $\{\varepsilon_s\}$ in the equation (3) are written as [6]

$$\{\varepsilon_p^L\}=\begin{Bmatrix}\dfrac{\partial u_o}{\partial s}\\[2mm] \dfrac{u_o\sin\phi}{r}+\dfrac{\partial v_o}{r\partial\theta}+\dfrac{w_o\cos\phi}{r}\\[2mm] \dfrac{\partial u_o}{r\partial\theta}-\dfrac{v_o\sin\phi}{r}+\dfrac{\partial v_o}{\partial s}\end{Bmatrix}\,;\ \{\varepsilon_b\}=\begin{Bmatrix}\dfrac{\partial\beta_s}{\partial s}\\[2mm] \dfrac{\beta_s\sin\phi}{r}+\dfrac{\partial\beta_\theta}{r\partial\theta}\\[2mm] \dfrac{\partial\beta_s}{r\partial\theta}+\dfrac{\partial\beta_\theta}{\partial s}-\dfrac{\beta_\theta\sin\phi}{r}\end{Bmatrix}\,;$$

$$\{\varepsilon_s\} = \left\{ \begin{array}{c} \beta_s + \dfrac{\partial w_o}{\partial s} \\[2ex] \beta_\theta + \dfrac{\partial w_o}{r\partial\theta} - \dfrac{v_o\cos\phi}{r} \end{array} \right\} ; \qquad \qquad ...(4)$$

where r and ϕ are the radius of the parallel circle and the semi-cone angle of the conical shell.

If $\{N\}$ represents the stress resultants (N_{ss}, $N_{\theta\theta}$, $N_{s\theta}$) and $\{\mathbf{M}\}$ the moment resultants (M_{ss}, $M_{\theta\theta}$, $M_{s\theta}$), one can relate these to the membrane strains $\{\varepsilon_p\}=\{\varepsilon_o^L\}+\{\varepsilon_o^{NL}\}$ and the bending strains $\{\varepsilon_b\}$ through the constitutive relations as

$$\left\{ \begin{array}{c} \{N\} \\ \{\mathbf{M}\} \end{array} \right\} = \begin{bmatrix} [\mathbf{A}] & [\mathbf{B}] \\ [\mathbf{B}] & [\mathbf{D}] \end{bmatrix} \left\{ \begin{array}{c} \{\varepsilon_p\} \\ \{\varepsilon_b\} \end{array} \right\} - \left\{ \begin{array}{c} \{\overline{\mathbf{N}}\} \\ \{\overline{\mathbf{M}}\} \end{array} \right\} \qquad \qquad ...(5)$$

where $[A]$, $[D]$ and $[B]$ are the extensional, the bending and the bending-extensional coupling stiffness coefficient matrices of the composite laminat. $\{\overline{N}\}$ and $\{\overline{M}\}$ are the thermal stress and the moment resultants, respectively.

Similarly, the transverse shear force $\{Q\}$ representing the quantities (Q_{sz}, $Q_{\theta z}$,) are related to the transverse shear strains $\{\varepsilon_s\}$ through the constitutive relation as

$$\{Q\} = [E] \{\varepsilon_s\} \qquad \qquad ...(6)$$

where $[E]$ is the transverse shear stiffness coefficient matrix of the laminate.

For a laminated shell consisting of N layers with the stacking angles θ_i ($i = 1, ..., N$) and the layer thicknesses h_i ($i = 1, ..., N$), the necessary expressions to compute the stiffness coefficients and the thermal stress/moment resultants, available in the literature [7] are used here.

The ply-angle θ_i and the layer thickness h_i for the filament wound conical shells are expressed as

$$\theta_i = \arcsin\left(\frac{r_1}{r}\sin\theta_i^1 \right) \qquad \qquad ...(7a)$$

$$h_i = h_i^1 \frac{r_1}{r} \frac{\cos\theta_i^1}{\cos\theta_i} \qquad \qquad ...(7b)$$

where r_1, θ_i^1 and h_i^1 are the radius of the parallel circle, ply-angle and layer thickness at the left end of the shell.

The potential energy functional $U_l(\delta)$ (consisting of strain energy and potential of external loads) is given by,

$$U_1(\delta) = \frac{1}{2}\int_A \left[\left\{ \begin{array}{c} \varepsilon_p \\ \varepsilon_b \end{array} \right\}^T \begin{bmatrix} \mathbf{A} & \mathbf{B} \\ \mathbf{B} & \mathbf{D} \end{bmatrix} \left\{ \begin{array}{c} \varepsilon_p \\ \varepsilon_b \end{array} \right\} + \{\varepsilon_s\}^T [\mathbf{E}]\{\varepsilon_s\} - \left\{ \begin{array}{c} \varepsilon_p \\ \varepsilon_b \end{array} \right\}^T \left\{ \begin{array}{c} \overline{\mathbf{N}} \\ \overline{\mathbf{M}} \end{array} \right\} \right] dA - \int_A q\, w_o\, dA - v_o^o(0)T - u_o^o(0)P \qquad ...(8)$$

where δ is the vector of degrees of freedom associated to the displacement field in a finite element discretisation. q is the applied external radial pressure. T and P are the applied external torsional and axial loads at the small end ($s = 0$) of the shell, respectively.

The potential energy $U_2(\delta)$ due to initial state of in-plane stress resultants $\{N^0\}= \left\{N^0_{ss}\ N^0_{\theta\theta}\ N^0_{s\theta}\right\}^T$ is written as

$$U_2(\delta)=\int_A \{\varepsilon_{NL}\}^T \{N^0\}\,dA \qquad \text{...(9)}$$

where

$$\{\varepsilon_{NL}\}^T = \left\{\frac{1}{2}\left(\frac{\partial w_0}{\partial s}\right)^2 \quad \frac{1}{2}\left(\frac{\partial w_0}{r\partial\theta}\right)^2 \quad \frac{\partial w_0}{\partial s}\frac{\partial w_0}{r\partial\theta}\right\}$$

The total potential energy functional $U(\delta)$ $[= U_1(\delta) + U_2(\delta)]$ can be expressed as

$$U(\delta) = \{\delta\}^T [(1/2)[[K]+ [K_G]]] \{\delta\}- \{\delta\}^T \{F_M\}- \{\delta\}^T \{F_T\} \qquad \text{...(10)}$$

where $[K]$ is the linear stiffness matrix, $[K_G]$ is the geometric stiffness matrix due to the initial stress resultants. $\{F_M\}$ and $\{F_T\}$ are the mechanical and the thermal load vectors.

The condition for the extremum of total potential $U(\delta)$ given in equation (10) with respect to the vector of degrees of freedom δ leads to the governing equation for the deformation of the shell as

$$[[K] + [K_G]] \{\delta\} = \{F_M\} + \{F_T\} \qquad \text{...(11)}$$

The governing equation (11) can be employed to study the linear static and the eigenvalue buckling analyses by neglecting the appropriate terms as:

Linear Static Analysis:

$$[K] \{\delta\} = \{F_M\} + \{F_T\} \qquad \text{...(12)}$$

Eigenvalue Buckling Analysis:

$$[K] \{\delta\} = \lambda [K_G{}^*] \{\delta\} \qquad \text{...(13)}$$

where $[K_G{}^*]$ is the geometric stiffness due to the initial state of stress developed because of the unit load and λ is the load multiplier factor.

It may be noted here that for the purpose of evaluating $[K_G{}^*]$, firstly the static analysis of the shell using equation (12) for the unit load is carried out. The resulting deformation field is used to calculate the initial state of stress resultants using Eq. (5) and in turn, for evaluating the $[K_G{}^*]$ matrix.

3. RESULTS AND DISCUSSIONS

The critical buckling load of cross-ply $(0°/90°)_4$ and two layered angle-ply $(\theta/-\theta)$ laminated cone-cylinder joined shells subjected to torsional load, external pressure, axial compression and uniform temperature change is evaluated. The material properties and boundary conditions used in the study are given as following:

3.1 Material Properties Used

Cross-ply laminated shell: $E_L = 181$ GPa, $E_T = 10.3$ GPa, $G_{LT} = G_{TT} = 7.17$ GPa, $\nu_{LT} = 0.28$, $\alpha_L = 0.02 \times 10^{-6}$ /°C, $\alpha_T = 22.5 \times 10^{-6}$ /°C

Angle-ply laminated shell: $E_L = 172.25$ GPa, $E_T = 6.89$ GPa, $G_{LT} = 3.445$ GPa, $G_{TT} = 1.378$ GPa, $\nu_{LT} = \nu_{TT} = 0.25$, $\alpha_L = 6.3 \times 10^{-6}$ /°C, $\alpha_T = 18.9 \times 10^{-6}$ /°C,

where E, G, ν and α are Young's modulus, the shear modulus, the poission's ratio and the coefficient of thermal expansion.

3.2 Simply Supported Boundary Conditions

Right end (s = L): $u_o^o = u_o^c = u_o^s = v_o^o = v_o^c = v_o^s = w_o^o = w_o^c = w_o^s = \beta_\theta^o = \beta_\theta^c = \beta_\theta^s = 0$

   Left end (s = 0)

   Torsional Loading: $u_o^o = u_o^c = u_o^s = v_o^c = v_o^s = w_o^o = w_o^c = w_o^s = \beta_\theta^o = \beta_\theta^c = \beta_\theta^s = 0$

   External Radial /Thermal Loading: $u_o^o = u_o^c = u_o^s = v_o^o = v_o^c = v_o^s = w_o^o = w_o^c = w_o^s = \beta_\theta^o = \beta_\theta^c = \beta_\theta^s = 0$

   Axial Loading: $u_o^c = u_o^s = v_o^o = v_o^c = v_o^s = w_o^o = w_o^c = w_o^s = \beta_\theta^o = \beta_\theta^c = \beta_\theta^s = 0$

The bifurcation loads of eight layered cross-ply laminated conical, cylindrical and joined conical-cylindrical shells ($r_2/h = 200$, $L_1/r_2 = 1$; $L_1/L_2 = 1$, $\phi = 30°$) are listed in Table 1.

Table 1. Critical buckling pressure (p_{cr}), axial load (P_{cr}) and temperature (ΔT_{cr}) of cross-ply $(0°/90°)_4$ laminated conical, cylindrical and joined conical-cylindrical shells.

Shell	Critical external pressure, p_{cr} (N/m$_2$)	Critical axial load, P_{cr} (N)	Critical temperature rise, ΔT_{cr}(°C)
Conical	166202.8875 (8)*	10.6908×10^7 (8)	883.3395 (8)
Cylindrical	129391.9690 (9)	12.1571×10^7 (11)	733.1390 (11)
Conical-Cylindrical	118514.0477 (9)	9.1248×10^7 (8)	2084.3512 (9)

*Number in brackets denotes the circumferential wave number

The study of prebuckling stress resultants of cross-ply $(0°/90°)_4$ laminated conical, cylindrical, joined conical-cylindrical shell ($r_2/h = 200$, $L_1/r_2 = 1$; $L_1/L_2 = 1$, $\phi = 30°$) is shown in Figure 1. The results reveal that the predominant prebuckling stress resultant in joined shell system subjected to external pressure (Figure 1(a)) or axial load (Figure 1(b)) is close to that in the individual components (except near to the joint) whereas joined shells subjected to thermal loading reveal significant reduction in the prebuckling meridional stress resultant (Figure 1(c)) compared to that in the individual components. Further, the stiffness of the joined shell system decreases due to the increase in the effective length compared to that of the individual components. This reduction of stiffness is responsible for the decrease in the critical bifurcation load of joined shells (Table 1) subjected to external pressure or axial load. However, for the thermally loaded joined shells, the more relative reduction in the meridional thermal stress resultant compared to the reduction in the stiffness leads to the higher bifurcation temperature compared to the individual components.

The study of the critical buckling loads of two layered angle-ply ((15°/–15°) and (30°/–30°)) and semi-cone angles ($\phi = 15°$, –15°, 30°, –30°) for joined shell system ($r_2/h = 100$, $L_1/r_2 = 1$; $L_1/L_2 = 1$, $L_2/L_3 = 1$, $h = 0.003$ m) subjected to torque at left end, external pressure, axial compression and uniform temperature change are listed in Tables 2 to 5. The circumferential wave number (n) taken is the value corresponding to lowest critical load. The study reveals that the bifurcation loads of angle-ply laminated joined shells are lower than that of individual shells except the thermally loaded joined shells for all cases and the external pressure loaded joined shells for semi-cone angle (ϕ) = –15° and –30°. The critical buckling load decreases as the semi-cone angle increases, whereas it increases for the negative semi-cone angle (diverging cone sections) except the external pressure and thermally loaded joined shells. This is because of the changes in the ply thickness with semi-cone angle. It is also observed that the joined shell system of ply-angle (30°/–30°) is stiffer than that of the ply-angle (15°/–15°) for the semi-cone angle (ϕ = 15°, –15°) whereas it is less stiff for semi-cone angle (ϕ = 30°) and more stiff for semi-cone angle (ϕ = –30°) except the thermally loaded joined shells.

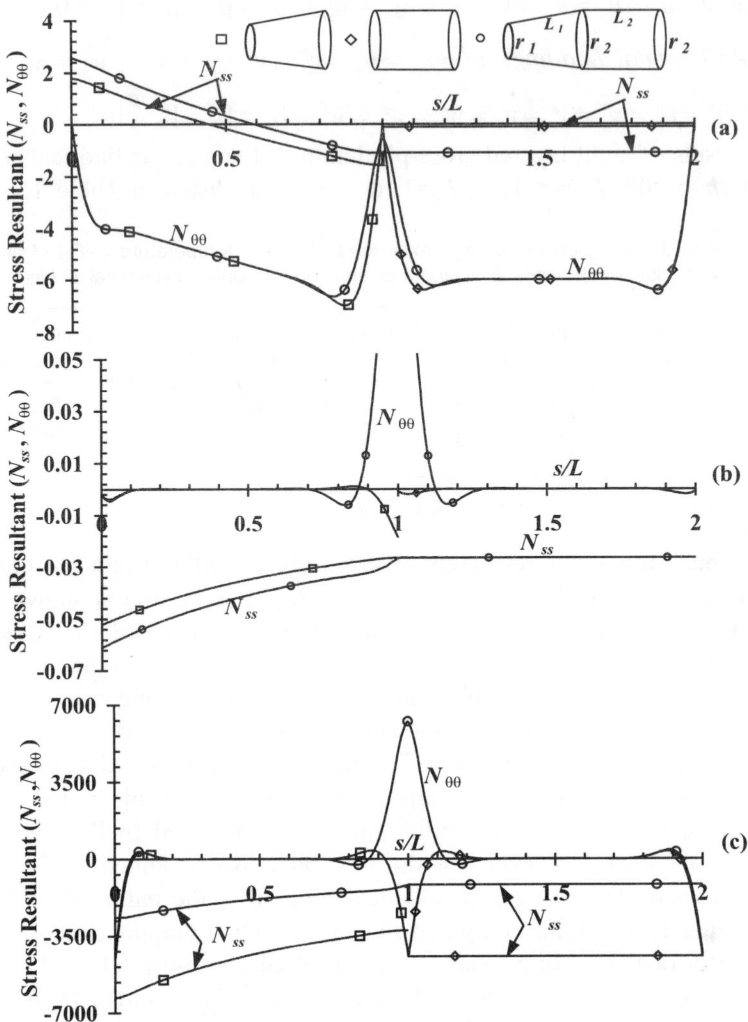

Fig. 1. Distribution of meridional, N_{ss} (N/m) and hoop, $N_{\theta\theta}$ (N/m) prebuckling stress resultants along the length (s/L) of cross-ply laminated (0°/90°)$_4$ simply supported conical, cylindrical and joined conical-cylindrical shells subjected to: (a) unit external pressure, p (N/m2) (b) unit axial load P (N) and (c) unit uniform temperature rise ΔT (°C).

Table 2. Critical buckling pressure (p_{cr}), axial load (P_{cr}), torsion (T_{cr}) and temperature (ΔT_{cr}) of angle ply (15°/–15°) laminated shells.

Shell	Critical Torsional load, T_{cr} (N–m)		Critical external pressure, p_{cr} (N/m^2)		Critical axial load, P_{cr} (N)		Critical temperature rise, ΔT_{cr} (°C)	
	$\phi = 15°$	$\phi = -15°$	$\phi = 15°$	$\phi = -15°$	$\phi = 15°$	$\phi = -15°$	$\phi = 15°$	$\phi = -15°$
Single Section Shell	207028.09 (12)	361411.36 (14)	101409.47 (11)	152416.71 (12)	322774.33 (9)	672828.38 (8)	103.01 (8)	94.63 (10)
Two Section Shell	182984.05 (14)	377975.71 (13)	59451.12 (12)	173686.66 (13)	269762.08 (9)	649229.54 (8)	605.72 (9)	130.54 (10)
Three Section Shell	191393.87 (14)	374352.05 (13)	59816.44 (12)	176105.41 (13)	272906.70(10)	640811.64 (8)	63.79 (12)	335.21 (8)

Table 3. Critical buckling pressure (p_{cr}), axial load (P_{cr}), torsion (T_{cr}) and temperature (ΔT_{cr}) of angle ply (15°/–15°) laminated shells.

Shell	Critical Torsional load, T_{cr} (N–m)		Critical external pressure, p_{cr} (N/m^2)		Critical axial load P_{cr} (N)		Critical temperature rise, ΔT_{cr} (°C)	
	$\phi = 30°$	$\phi = -30°$	$\phi = 30°$	$\phi = -30°$	$\phi = 30°$	$\phi = -30°$	$\phi = 30°$	$\phi = -30°$
Single Section Shell	128045.80 (12)	405799.14(14)	48517.13 (12)	136687.55 (12)	142284.27 (9)	710838.48 (7)	65.38 (9)	63.23 (11)
Two Section Shell	95934.49 (15)	452824.88 (14)	20769.14 (14)	159347.55 (13)	118824.50(12)	686598.38 (7)	-121.61(9)	85.36 (12)
Three Section Shell	99913.18 (15)	45389.76 (14)	21100.61 (14)	160410.59 (13)	114646.84(13)	678708.99 (7)	13.37 (13)	-237.14(10)

Table 4. Critical buckling pressure (p_{cr}), axial load (P_{cr}), torsion (T_{cr}) and temperature (ΔT_{cr}) of angle ply (30°/–30°) laminated shells.

Shell	Critical Torsional load, T_{cr} (N–m)		Critical external pressure, p_{cr} (N/m^2)		Critical axial load P_{cr} (N)		Critical temperature rise, ΔT_{cr} (°C)	
	$\phi = 15°$	$\phi = -15°$	$\phi = 15°$	$\phi = -15°$	$\phi = 15°$	$\phi = -15°$	$\phi = 15°$	$\phi = -15°$
Single Section Shell	246700.89 (13)	552896.37 (13)	130189.02 (12)	276438.69 (12)	329173.32 (4)	648182.15 (0)	-220.93(13)	130.97 (13)
Two Section Shell	192686.07 (15)	552003.82 (12)	69586.93 (13)	284031.88 (12)	279659.20 (6)	643173.67 (0)	-176.25(11)	131.63 (13)
Three Section Shell	187054.63 (14)	528965.23 (9)	61459.70 (12)	284606.77 (12)	282526.76 (5)	616783.12 (4)	30.78 (12)	–185.56(10)

Table 5. Critical buckling pressure (p_{cr}), axial load (P_{cr}), torsion (T_{cr}) and temperature (ΔT_{cr}) of angle ply (30°/–30°) laminated shells.

Shell	Critical Torsional load, T_{cr} (N–m)		Critical external pressure, p_{cr} (N/m^2)		Critical axial load P_{cr} (N)		Critical temperature rise, ΔT_{cr} (°C)	
	$\phi = 30°$	$\phi = -30°$	$\phi = 30°$	$\phi = -30°$	$\phi = 30°$	$\phi = -30°$	$\phi = 30°$	$\phi = -30°$
Single Section Shell	132272.08 (13)	685642.38 (13)	47654.17 (14)	268970.01 (13)	126902.96 (9)	658669.49 (0)	120.41 (8)	55.78 (14)
Two Section Shell	91780.73 (17)	708624.54 (13)	20382.69 (15)	290029.34 (13)	111546.04(13)	645907.06 (0)	-90.78(11)	55.79 (14)
Three Section Shell	91892.63 (17)	713225.39 (13)	19138.33 (15)	289627.97 (13)	113541.87(14)	634720.24 (0)	6.21 (14)	-214.73(12)

CONCLUSION

The bifurcation load of angle-ply laminated conical-cylindrical joined shells subjected to torsion, external pressure, axial compression and uniform thermal loading are determined using the semi-analytical finite element approach. The study shows that the critical buckling load decreases with the increase in the semi-cone angle whereas it increases for negative semi-cone angle. The study also reveals that the prebuckling stress resultant in joined shell system subjected to external pressure or axial load is close to the individual components except near the joint whereas a significant reduction in prebuckling meridional stress resultant is observed for thermally loaded joined shell throughout.

REFERENCES

1.   W. Anwen, 1998, Stresses and stability for the cone-cylinder shells with toroidal transition, International Journal of Pressure Vessels and Piping, 75, 49-56.

2.   F.G Flores, L.A Godoy, 1991, Post-buckling of elastic cone-cylinder and sphere-cylinder complex shells, International Journal of Pressure Vessels and Piping, 45, 237-258.

3.   Y. Goldfeld, J. Arbocz, 2004, Buckling of laminated conical shells given the variations of the stiffness coefficients, AIAA Journal, 42(3), 642-649.

4.   Y. Goldfeld, J. Arbocz, A. Rothwell, 2005, Design and optimization of laminated conical shells for buckling, Thin Walled Structures, 43, 107-133.

5.   B.P. Patel, 2005, Thermal buckling and postbuckling characteristics of composite laminated shells. Ph.D. thesis: Department of Applied Mechanics, Motilal Nehru National Institute of Technology, Allahabad, India.

6.   H. Kraus, 1976, Thin Elastic Shells, New York: John Wiley.

7.   R.M. Jones, 1999, Mechanics of Composite Materials, Philadelphia: Taylor and Francis.

31

Modeling of Steel Fibre Reinforced Concrete with High Fibre Volume Fractions

Y.M. Ghugal and A.G. Dahake

Department of Applied Mechanics, Govt. Engineering College, Station Road, Aurangabad-431005 (M.S.), India. email: ghugal@rediffmail.com

ABSTRACT

This paper presents the results of the experimental investigation of various strengths of steel fibre reinforced concrete (SFRC). Variables considered in the research work are various strengths and fibre volume fractions. Various strengths considered for investigation are compressive strength, flexural strength, split tensile strength, bond strength and shear strength. Concrete mix of M25 grade and crimped steel fibres with aspect ratio 50 are used. The fibre volume fraction is varied from 0.5% to 4.5% at an interval of 0.5% by weight of cement. Standard test specimens for compressive strengths, split tensile strength, flexural strength and push-off specimens for shear strength were cast and water cured for 7 and 28 days. All the test specimens were tested according to relevant Indian Standards and standard test procedures available in the literature wherever applicable. All the strengths are found to be increased continuously with increase in fibre volume fraction. The experimental results obtained for various strengths are modeled in terms of the material properties of matrix, fibre and compressive strength. The mathematical expressions developed for various strengths are presented. The inclusion of steel fibre in to the normal concrete showed the excellent strength performance in this investigation compared to the normal concrete. The results predicted by mathematically modeled expressions are in excellent agreement with experimental results.

Keywords: Composites, SFRC, aspect ratio, fibre volume fraction, strengths, mathematical modeling.

1. INTRODUCTION

Plain cement concrete is the most widely used material for construction of various structures. However, it suffers from numerous drawbacks such as, low tensile strength, brittleness, unstable crack propagation and low fracture resistance etc. Addition of steel fibres to plain cement concrete results in improved structural properties, such as better resistance against cracking, impact, thermal shocks, wear, fatigue, spalling and improved compressive, flexural, tensile, shear, bond strengths, ductility and toughness. Hence, steel fibre reinforced concrete (SFRC) has been proved as a reliable and promising composite construction material having superior performance characteristics compared to conventional concrete.

A compressive review of literature related to fibre reinforced concrete is presented by Balaguru and Shah [1] and Beaudoin [2] including guidelines for design, mixing, placing and finishing steel fibre reinforced concrete. Hannant [3] has presented the basic theoretical and simplified principles of FRC subjected to various states of stress. The improvement in various strengths of SFRC has been studied by various research workers with low fiber volume fractions of fibers [4-12]. Crimped fibres, surfaces deformed fibres and fibres with end anchorage have shown to produce more strength than that is given by smooth fibres for the same volume content. Recently Ghugal [13] has studied the structural behavior of glass fiber reinforced concrete using mathematical models.

In this paper effect of crimped steel fibres on properties of hardened concrete with high volume fractions are studied. The results of an investigation carried out on various strengths such as compressive, split tensile, flexural, bond and shear strengths of hardened concrete are presented.

2. EXPERIMENTAL PROGRAM

Ordinary Portland cement of 53 grade confirming to IS 12269 and natural sand and coarse aggregates (10, 20 mm) confirming to IS 8112 and IS 383 and crimped steel fibres were used. The cement and aggregates were tested to fulfil the IS requirements. The fineness moduli of fine and coarse aggregates were 3.2 and 7.7 respectively. The crimped steel fibers having tensile strength 825 MPa, modulus of elasticity 200 GPa and aspect ratio 50 were used. The M-25 grade of concrete having mix proportions was used throughout the experimental investigation. The concrete ingredients, namely, cement, fine, coarse aggregates and fibres were first mixed in the dry state and water was added last. Cubes of 150 mm size for compressive and bond strengths, cylinders of size 150 mm dia $\times$ 300 mm long for split tensile strength, beams of size 150 mm $\times$ 150 mm $\times$ 700 mm for flexural strength and push-off specimens for shear strength were cast incorporating 0.0% to 4.50% steel fibres at the interval of 0.5% by weight of cement. For each test, three specimens were cast with and without fibres. Compaction of all the specimens was done using table vibrator to avoid balling of fibres. All the specimens were water cured for 28 days at room temperature and were tested in surface dry condition on 1000 kN Universal Testing Machine. In all 120 specimens were cast and tested to evaluate the strengths performance. Each value of the results presented in this is the average of three test samples.

2.1 Discussion of Test Results

The tests on hardened concrete are carried out according to the relevant standards wherever applicable. New expressions for flexural strength, flexural shear; split tensile strength, and bond strength of SFRC are proposed in this investigation. Results obtained using experiments, theory and proposed expressions are presented and discussed in comparison with those of normal concrete.

2.1.1. Compressive Strength

This strength was determined by carrying out cube compressive test on 150 mm size cubes, using UTM. Compressive strength results of SFRC compared to that of normal concrete (V_f = 0.0 %) are shown in Table 1.

Table 1. Compressive strength of SFRC, MPa.

Fibre Content (V_f) %	Compressive Strength (MPa)		% increase in Compressive Strength	
	7 days	28 days	7 days	28 days
0.0	29.48	33.92	--	--
0.5	32.26	35.11	8.62	3.39
1.0	33.10	36.00	10.94	5.78
1.5	34.12	36.44	13.60	6.92
2.0	34.26	36.82	13.95	7.88
2.5	34.37	37.03	14.23	8.40
3.0	34.37	38.00	14.23	10.74
3.5	35.20	38.22	16.25	11.25
4.0	36.78	38.66	19.85	12.26
4.5	36.96	39.40	20.24	13.91

The compressive strength increased continuously with the increase in the fiber content. The increase in the strength is directly proportional to the fiber content. The maximum increase in compressive strength at 4.5% of fiber content is 13.91 % at 28 days.

2.1.2 Flexural Strength

For determining this strength each specimen of size 100 mm × 100 mm × 500 mm was supported over a span of 400 mm and a two-point load was applied at the middle third of the span. All the beams were loaded up to failure. The flexural strength using strength of materials theory was calculated by the formula:

$$f_{cr} = \frac{PL}{bh^2} \qquad \qquad ...(1)$$

The simple formula is proposed for calculation of flexural strength of SFRC in terms of elastic moduli of matrix and fibres, volume fraction of fibres and compressive strength for respective fibre volume fraction as expressed by Equation (2).

$$f_{cr} = \left[0.85 + \left(\frac{E_m}{E_f} \right) V_f^{0.011} \right] \sqrt{f_{cu}} \qquad \qquad ...(2)$$

where $E_m = 5\sqrt{f_{ck}}$ is the modulus of elasticity of the concrete, E_f is the modulus of elasticity of the fibre in GPa and f_{ck} is the 28 days compressive strength of concrete in MPa. Results obtained using expressions (1 and 2) are presented in Table 2.

2.1.3 Flexural Shear Strength

Maximum shear strength of the flexural member can be computed using theory of strength of materials by the following equation

$$\tau_s = \frac{3F_s}{2bh} \qquad \qquad ...(3)$$

where F_s = P/2 is the maximum shear force on the section of the beam, b is the width of beam and h is the depth of beam. The results obtained using this equation for various fiber volume fractions are presented in Table 2. The equation for maximum shear strength is proposed in terms of volume fraction of fiber (V_f), elastic modulus of matrix (concrete) (E_m), elastic modulus of fiber (E_f) and flexural strength (f_{cr}) obtained from four point bending test. The equations proposed is as follows

$$\tau_s = 0.19\left[1 - V_f\left(\frac{E_m}{E_f}\right)\right]f_{cr} \qquad \text{...(4)}$$

Expression for flexural shear strength of SFRC in terms of f_{cu} is given as below:

$$\tau_S = 0.9421 + 0.0052\,f_{cu} + 0.0044\,f_{cu}^2 \qquad \text{...(5)}$$

Direct shear stress is obtained using push-off type specimens subjected to axial compression. This stress is obtained using following formula:

$$\tau_s = \frac{P_s}{A_s} \qquad \text{...(6)}$$

where P_s is the shear load at failure and A_s is the area of shear plane. Results of flexural shear strength and direct shear strength obtained using above equations are given in Table 2.

Table 2. Flexural and flexural shear strengths of SFRC

Fibre Content (V_f) %	Flexural Strength (f_{cr}), MPa			Flexural Shear (τ_s) MPa			Direct Shear (τ_s) MPa
	using "Eq. (1)"	using IS 456	using "Eq. (2)"	using "Eq. (3)"	using "Eq. (4)"	using "Eq. (5)"	using "Eq. (6)"
0.0	4.98	4.08	4.95	0.95	0.93	0.94	4.27
0.5	5.08	4.15	5.86	0.97	0.95	0.95	4.44
1.0	5.26	4.20	5.96	1.00	0.99	0.95	4.59
1.5	5.55	4.23	6.00	1.05	1.04	0.96	4.80
2.0	5.69	4.25	6.04	1.08	1.07	0.97	4.98
2.5	6.08	4.26	6.06	1.15	1.14	0.98	5.01
3.0	6.24	4.32	6.15	1.18	1.17	1.00	5.16
3.5	6.51	4.33	6.18	1.23	1.22	1.01	5.33
4.0	7.11	4.35	6.22	1.34	1.33	1.03	5.51
4.5	7.68	4.39	6.29	1.45	1.44	1.05	6.22

It is observed that the flexural strength continues to increase with increase in fiber content. Results predicted using "Eq. (2)" are in close agreement with those obtained from "Eq. (1)". However, IS 456 formula underestimates the results of flexural strength as can be seen from the Table 2. Results of flexural shear strength are computed using "Eq. (3)" of strength of material theory and presented in Table 2. Results predicted from the proposed Eq. (4) and those obtained from the strength of material theory are in excellent agreement with each other. Shear strength is found to vary linearly with the flexural strength according to the law governed by the "Eq. (4)". However, when it is expressed in terms of compressive strength, it varies according to second degree polynomial, "Eq. (5)". Shear strengths predicted by this equation are slightly on lower side. It can be observed from Table 2 that the direct shear strength is nearly 4.5 times the flexural shear.

2.1.4 Split Tensile Strength

The cylinder splitting test was used to determine the tensile strength of concrete. In this test, compressive line loads are applied along a vertical symmetrical plane setting up tensile stresses normal to the plane, which causes the splitting of specimen. The formula derived using theory of elasticity has been used to calculate the tensile strength at the time of splitting. The indirect splitting tensile strength has been computed from the following expression

$$f_{cys} = \frac{2P_L}{\pi D L_c} \qquad \qquad ...(7)$$

The split tensile strength of SFRC is expressed in terms of E_m, E_f, V_f and compressive strength of concrete (f_{cu}) as given below:

$$f_{cys} = \left[1.011 + V_f\left(\frac{E_m}{E_f}\right)\right] \log_e f_{cu} \quad \text{for} \quad 0\% \leq V_f \leq 4.5 \text{ \%} \qquad ...(8)$$

The results of split tensile strength obtained using "Eqs. (7 and 8)" are presented in Table 3.

2.1.5 Bond Strength

The bond strength test was carried out according to ASTM standard C234–91a [14] on cubes of 150 mm size with 16 mm diameter tor steel bar embedded in each test specimen up to a depth of 150 mm at the centre. The bar is pulled out with the help of Universal Testing Machine. The bond strength has been computed from the following expression:

$$\tau_{bd} = \frac{P_F}{\pi d L_e} \qquad \qquad ...(9)$$

The bond strength of SFRC is expressed in terms of compressive strength of concrete (f_{cu}) as given below

$$\tau_{bd} = 13.735 \log_e fcu - 40.55 \quad \text{for } 0.0 \text{ \%} \leq V_f \leq 2.5 \text{ \%} \qquad ...(10)$$

$$\tau_{bd} = -17.5 \log_e f_{cu} + 74.18 \quad \text{for } 3.0 \text{ \%} \leq V_f \leq 4.5 \text{ \%} \qquad ...(11)$$

The results of the bond strength of SFRC obtained using "Eqs. (9 through 11)" are shown in Table 3.

Results from Table 3 show that cylinder split tensile strength of concrete increases considerably with increase in fiber content. The continuous increased in strength is observed. The results obtained

Table 3. Split tensile and bond strengths of SFRC

Fibre Content (V_f) %	Split Tensile Strength (f_{cys}), MPa		Bond Strength (ζbd.) MPa	
	using "Eq. (7)"	using "Eq. (8)"	using "Eq. (9)"	using "Eq. (10, 11)"
0.0	3.49	3.56	8.03	7.82
0.5	3.52	3.60	8.59	8.29
1.0	3.57	3.63	8.91	8.64
1.5	3.64	3.64	9.35	8.81
2.0	3.64	3.66	9.84	8.95
2.5	3.67	3.67	10.16	9.03
3.0	3.73	3.69	10.42	10.56
3.5	3.75	3.70	10.87	10.46
4.0	3.76	3.72	10.61	10.26
4.5	3.85	3.74	10.93	9.93

by "Eq. (7)" based on theory of elasticity and those predicted by the "Eq. (8)" are in close agreement with each other. This strength is found to vary with the natural logarithm of compressive strength of SFRC.

Results of bond strength of SFRC using "Eq. (9)" are shown in Table 3. Bond strength increased with the addition of fibers in the concrete and the maximum increase up to 36.11 % is observed at 4.5 % of fiber content over the normal concrete. It is found to increase with increase in the fiber content. This increase in bond strength with increase in fiber content may be due to effect of confinement of concrete and enhanced bond strength between the fibers and matrix. The trend of variation of bond strength is modelled mathematically in terms of compressive strength (f_{cu}) of SFRC and is given by the "Eq. (10)" and "Eq. (11)". Results predicted using these equations and those obtained by "Eq. (9)" are in close agreement with each other. This strength is found to be a function of natural logarithm of compressive strength.

CONCLUSION

The following conclusions are drawn from the test results and discussion of this investigation.

1. The percentage increase in cube compressive strength, flexural strength, split tensile strength, bond strength and direct shear strength achieved at 28 days are 13.91, 54.22, 10.31, 36.11 and 45.67 respectively, at 4.5 % of fiber volume fraction, when compared to that of normal concrete.

2. Mathematically modelled expressions have been established to predict the value of flexural strength, flexural shear strength, cylinder split tensile strength, and bond strength of SFRC in terms of cube compressive strength and a mathematical expression has been suggested for flexural shear strength in terms of flexural strength. Results predicted from these expressions are in close agreement with the experimental results in this investigation.

3. In general, the significant improvement in various strengths is observed with the inclusion of steel fibres in the plain concrete with high volume fractions.

REFERENCES

1. P.N. Balaguru, and S.P. Shah, 1992, Fiber Reinforced Cement Composites, McGraw-Hill: New York.
2. J.J. Beaudoin, 1990, Handbook of Fibre Reinforced Concrete—Principles, Developments and Applications, Noyes Publications: New Jersey, USA.
3. D.J. Hannant, 1978, Fibre Cements and Fibre Concretes, John Wiley and Sons: New York, USA.
4. G.R. Williamson, 1974, The effect of steel fibres on compressive strength of concrete, ACI Special Publication. SP-44, 195-207.
5. S.P. Shah and V.B. Rangan, 1971, Fibre reinforced concrete properties, ACI Journal. 68, 126-135.
6. C.D Johnston and R.A. Coleman, 1974, Strength and deformation of steel fibers reinforced mortars in uniaxial tension, ACI SP-44, 177-193.
7. C.D. Johnston, 1984, Steel fibre reinforced mortar and concrete: A review of mechanical properties, ACI SP-44, 127-142.
8. R.N. Swamy, *et al.*, 1974, The mechanics of fibre reinforcement in cement matrices: Fibre-reinforced concrete, ACI SP-44, 1-28.
9. R.N. Swami and S.A. Al-Taan, 1981, Deformation and ultimate strength in flexure of reinforced concrete beam made with steel fibre concrete, ACI Journal, 78, 395-405.
10. M.A. Mansur, *et al.*, 1986, Shear strength of fibrous concrete beams without stirrups, ASCE Journal of Structural Engineering. 112, 2068-2079.
11. CI Committee 544 Report, 1988, Design of steel fibre reinforced concrete, ACI Structural Journal, 85, 563-580.
12. Y.M. Ghugal, 2003, Effects of steel fibers on various strengths of concrete, Indian Concrete Institute Journal. 4, 23-29.
13. Y.M. Ghugal, 2006, Performance of alkali-resistant glass fibre reinforced concrete, Journal of Reinforced Plastics and Composites. 25, 617-630, SAGE Publications, USA.
14. ASTM Standard: C234-91a, Test for comparing concretes on the basis of the bond developed with reinforcing steel. ASTM Standards, Philadelphia, USA.

32

Postbuckling Response of Laminated Composite Rectangular Plates

RAMESH PANDEY, K.K. SHUKLA AND ANUJ JAIN

Department of Applied Mechanics Motilal Nehru National Institute of Technology, Allahabad-211004, India email: kkshukla@mnnit.ac.in

ABSTRACT

In the present work analytical-numerical type solutions of the post-buckling response of moderately thick laminated composite plates subjected to thermal and mechanical loadings are obtained. The mathematical formulation is based on higher order shear deformation theory (HSDT) and von-Karman's non-linear kinematics. The quadratic extrapolation technique and fast converging finite double Chebyshev series are used for linearization and spatial discretization of the governing non-linear equations of equilibrium, respectively.

Keywords: Analytical-Numerical, Post–buckling, HSDT, Non-linear, Chebyshev Series.

1. INTRODUCTION

The increased applications of advanced composite and other high performance materials in structural components especially in aerospace industry have stimulated interest of scientific community in the accurate prediction of the response characteristics of multiple layers fibre-reinforced laminated composite plates/panels. Post-buckling analysis of laminated composite plates subjected to in-plane mechanical loading and thermal loadings have been the interest of many researchers [1-2]. A considerable amount of literature exists on mechanical, thermal and thermo-mechanical buckling of laminated composite plates based on linear kinematics of the plates. But severity of the loadings demands consideration of higher-order effects i.e. non-linearity and it must be incorporated for an adequate buckling analysis. The post-buckling responses of laminated composite plates subjected to mechanical edge compressive and/or shear loadings, thermal and thermo-mechanical loadings have been investigated by many researchers [3-4]. It is evident from the literature that most of the analysis is based on numerical techniques and relatively fewer attempts are made to obtain analytical or analytical-numerical type solutions incorporating higher order shear deformation theory.

2. PROBLEM FORMULATION

Based on HSDT with cubic variation of in-plane displacements through the thickness and constant transverse displacement, the displacement field at a point in the laminated plate is expressed as;

$$U(x,y,z) = u_0(x,y) + z\psi_x(x,y) + z^2 u_1(x,y) + z^3 \phi_x(x,y) \qquad ...(1)$$

$$V(x,y,z) = v_0(x,y) + z\psi_y(x,y) + z^2 v_1(x,y) + z^3 \phi_y(x,y) \qquad ...(2)$$

$$W(x,y,z) = w_0(x,y) \qquad ...(3)$$

Where, the parameters u_0, v_0 and w_0 are mid plane transverse displacements. The functions $\psi_x, \psi_y, u_1, v_1, \phi_x$ and ϕ_y are the higher order terms in the Taylor's series expansion, representing higher-order transverse cross-sectional deformation modes. Employing von-Karman non-linear kinematics, and considering plane stress condition, the in-plane stress and moment resultants of the laminated composite plate consisting of *n* layers are expressed as:

$$\begin{Bmatrix} N \\ M \\ N^* \\ M^* \end{Bmatrix} = \begin{bmatrix} A & B & D & E \\ B & D & E & F \\ D & E & F & H \\ E & F & H & J \end{bmatrix} \begin{Bmatrix} \varepsilon^0 \\ \kappa \\ \varepsilon^1 \\ \eta \end{Bmatrix} - \begin{Bmatrix} N_b \\ 0 \\ 0 \\ 0 \end{Bmatrix} - \begin{Bmatrix} N_T \\ M_T \\ N_T^* \\ M_T^* \end{Bmatrix} \qquad ...(4)$$

Transverse shear stress resultants are written as:

$$\begin{Bmatrix} Q_y \\ Q_x \\ S_y \\ S_x \\ Q_y^* \\ Q_x^* \end{Bmatrix} = \begin{bmatrix} A & B & D \\ B & D & E \\ D & E & F \end{bmatrix} \begin{Bmatrix} \psi_y + w_{0,y} \\ \psi_x + w_{0,x} \\ 2v_1 \\ 2u_1 \\ 3\phi_y \\ 3\phi_x \end{Bmatrix} \qquad ...(5)$$

Where, $\mathbf{A, B, D, E, F, H, J}$ are plate stiffness matrices and the coefficients of these are defined as:

$$(A_{ij}, B_{ij}, D_{ij}, E_{ij}, F_{ij}, H_{ij}, J_{ij}) = \sum_{k=1}^{n} \int_{z_{k-1}}^{z_k} \overline{Q}_{ij}^{(k)}(1, z, z^2, z^3, z^4, z^5, z^6) dz; \quad (i, j = 1,2,4,56) \qquad ...(6)$$

and N_b represents the applied in-plane mechanical loading and N_T, M_T, N_T^*, M_T^* are stress and moment resultants due to thermal loading. The governing equations of equilibrium are expressed in non-dimensional form as:

$$(L_a + L_b + L_c) \, d + L_d F_T + Q = 0 \qquad ...(7)$$

where,

$$L_a = L_{a1}\frac{\partial^2}{\partial x^2} + L_{a2}\frac{\partial^2}{\partial y^2} + L_{a3}\frac{\partial^2}{\partial x \partial y} + L_{a4}\frac{\partial}{\partial x} + L_{a5}\frac{\partial}{\partial y} + L_{a6}$$

$$L_b = L_{b1}\frac{\partial^2}{\partial x^2} + L_{b2}\frac{\partial^2}{\partial y^2} + L_{b3}\frac{\partial^2}{\partial x \partial y}, \quad L_c = L_{c1}\frac{\partial^2}{\partial x^2} + L_{c2}\frac{\partial^2}{\partial y^2} + L_{c3}\frac{\partial^2}{\partial x \partial y}$$

$$L_d = L_{d1}\frac{\partial}{\partial x} + L_{d2}\frac{\partial}{\partial y} + L_{d3}, \quad \mathbf{d} = \begin{bmatrix} \overline{u} & \overline{v} & \overline{w} & \overline{\psi}_x & \overline{\psi}_y & \overline{u}_1 & \overline{v}_1 & \overline{\phi}_x & \overline{\phi}_y \end{bmatrix}^T,$$

$$\mathbf{F_T} = \begin{bmatrix} N_x^T & N_y^T & N_{xy}^T & M_x^T & M_y^T & M_{xy}^T & N_{xT}^* & N_{yT}^* & N_{xyT}^* & M_{xT}^* & M_{yT}^* & M_{xyT}^* \end{bmatrix}^T$$

The associated admissible boundary conditions are of the form:

Clamped edge **(C):** $\quad \overline{u}_0 = \overline{v}_0 = \overline{w}_0 = \overline{\psi}_x = \overline{\psi}_y = \overline{u}_1 = \overline{v}_1 = \overline{\phi}_x = \overline{\phi}_y = 0$

Simply supported edge **(S):** $\quad \overline{u}_0 = \overline{v}_0 = \overline{w}_0 = \overline{\psi}_y = \overline{u}_1 = \overline{v}_1 = \overline{\phi}_y = \overline{M}_x = \overline{M}_x^* = 0 \quad$ at $\quad$ x $\quad$ edge

$$\overline{u}_0 = \overline{v}_0 = \overline{w}_0 = \overline{\psi}_x = \overline{u}_1 = \overline{v}_1 = \overline{\phi}_x = \overline{M}_y = \overline{M}_y^* = 0 \quad \text{at} \quad y \quad \text{edge}$$

where a **bar** sign over the variable represents the Non-dimensional parameter.

3. SOLUTION METHODOLOGY

The governing non-linear equations of equilibrium along with appropriate boundary conditions are solved using an analytical technique. The coupled non-linear equations are linearized by a total linearization scheme based on quadratic extrapolation technique and the fast converging, orthogonal, double Chebyshev polynomial in range of $-1 \leq X \leq 1$ and $-1 \leq Y \leq 1$, is used for spatial discretization of the differential equations. The displacement functions and the loadings are approximated in space domain by finite degree Chebyshev polynomial [5] and expressed as:

$$(\overline{u}_0, \overline{v}_0, \overline{w}_0, \overline{\psi}_x, \overline{\psi}_y, \overline{u}_1, \overline{v}_1, \overline{\phi}_x, \overline{\phi}_x, \overline{Q})$$

$$= \sum_{i=0}^{M}\sum_{j=0}^{N} \delta_{ij} (\overline{u}_{ij}, \overline{v}_{ij}, \overline{w}_{ij}, \overline{\psi}_{xij}, \overline{\psi}_{yij}, \overline{u}_{1ij}, \overline{v}_{1ij}, \overline{\phi}_{xij}, \overline{\phi}_{yij}, \overline{Q}_{ij}) T_i(x) T_j(y); \quad -1 \leq X, Y \leq 1 \qquad ...(8)$$

where, M and N are the number of terms in finite degree double Chebyshev series based on convergence. The non-linear terms are linearized using the quadratice extrpolation technique. The displacement functions appearing in the non-linear terms in the governing equations are predicted at each step of the marching variable. The loads (marching variables) are increased in small steps and non-linear terms are evaluated at each step of the marching variables and transferred to right hand side [4]. The governing equations finally reduces to a set of linear simultaneous equations and expressed as:

$$\sum_{i=0}^{M-2}\sum_{j=0}^{N-2} F_k \left(\overline{u}_{ij}, \overline{v}_{ij}, \overline{w}_{ij}, \overline{\psi}_{xij}, \overline{\psi}_{yij}, \overline{u}_{1ij}, \overline{v}_{1ij}, \overline{\phi}_{xij}, \overline{\phi}_{yij}, \overline{Q}_{ij} \right) T_i(x) T_j(y) = 0; \quad k = 1 \text{ to } 9 \quad ...(9)$$

Similarly, the appropriate sets of boundary conditions are also discretized and expressed in form of linear simultaneous equations. The number of equations obtained from the set of generating equations (9) and boundary conditions are more than the total number of unknowns. In order to have a unique and compatible solution, the multiple regression analysis based on least-square error norms is used and finally the set of linear equations are expressed in the matrix form as;

$$a = BQ \qquad \qquad ...(10)$$

The values of the displacement vector a obtained from Eq. (10) is put into the Eq. (8) to evaluate the displacement at the desired location on mid-plane of the plate.

4. RESULTS AND DISCUSSIONS

Following material properties are used in the present work unless and otherwise stated: $E_1 =$ 181.0 GPa, $E_2 = 10.3$ GPa, $G_{12} = 7.17$ GPa, $G_{23} = 2.39$ GPa, $G_{13} = G_{12}$ for the analysis of angle-ply, moderately thick (a/h=20), square and clamped laminated plate. The convergence study of the laminated plate subjected to uni-axial in plane edge compressive loading is carried out and depicted in Fig. 1. It is clear that 9 to 10 terms expansion of each variable gives quite good convergence for critical load λ_{xcr} and postbuckling response. In the present work 9 terms expansion of the variables are considered. In order to validate the present solution methodology, the thermal buckling load $\lambda_{Tcr}(=\alpha_0 T_{cr} \times 10^3)$ of an anti-symmetric angle-ply $[(\pm45)_3]$, simply supported (SSSS), square laminated plate subjected to a uniform temperature rise are obtained and compared with the results due to Shen [3]. The comparison is shown in Fig. 2, which depicts that the results are in good agreement. The effect of number of layers on critical/limiting load and post-buckling strength of laminated plate subjected to uni-axial compression is depicted in Fig. 3. The increase in number of layers increases the stability of the plate against buckling. It may be attributed to the fact that coupling effect decreases due to increase

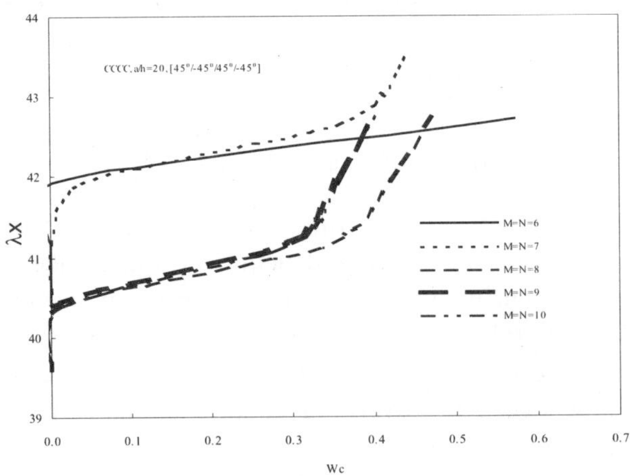

Fig. 1.

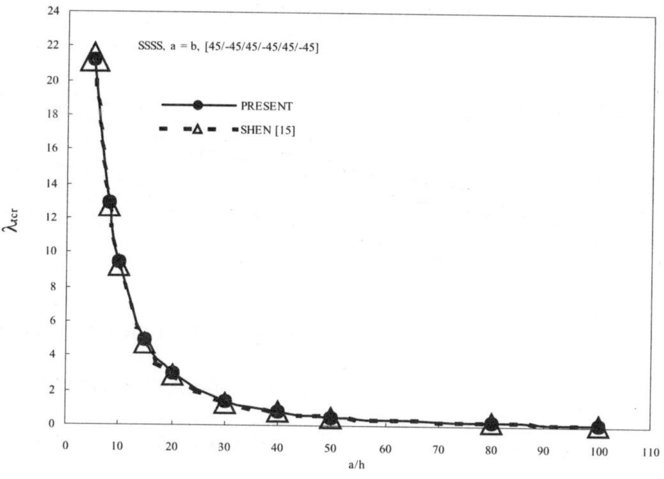

Fig. 2.

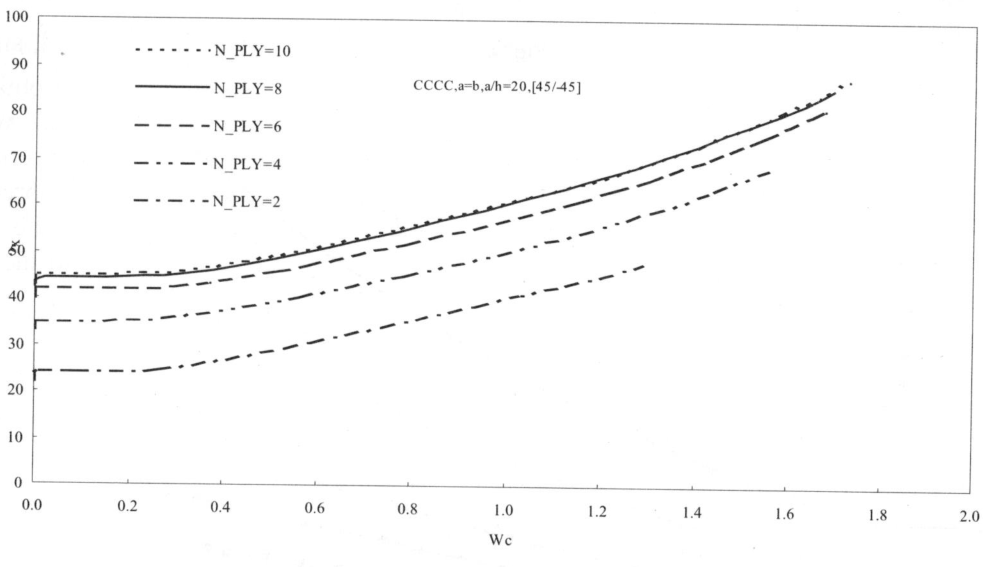

Fig. 3.

in number of layers. In general, laminates with large number of layers can be treated as having specially orthotropic properties and such laminates are free from coupling effects. The effect of stacking sequence on stability of the laminated plate subjected to uni-axial compression is shown in Fig. 4. It can be observed that the anti-symmetric angle-ply laminate is more stable than symmetric angle-ply laminate. Fig. 5 depicts the effect of aspect ratio (b/a) on critical buckling load and post-buckling strength of laminated plate subjected to uni-axial compression.

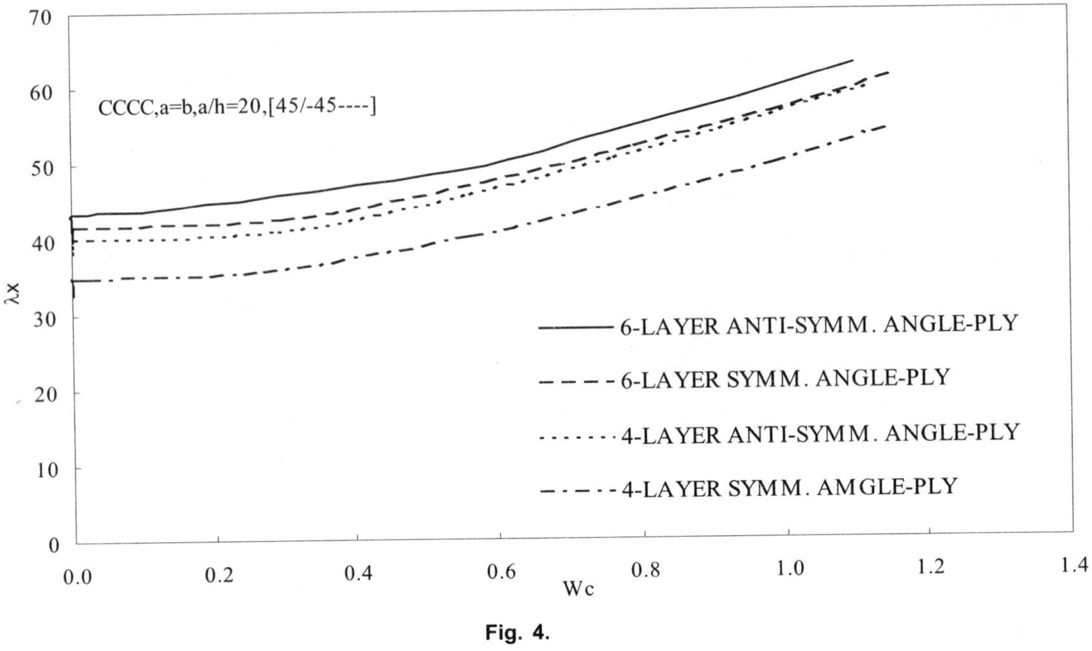

Fig. 4.

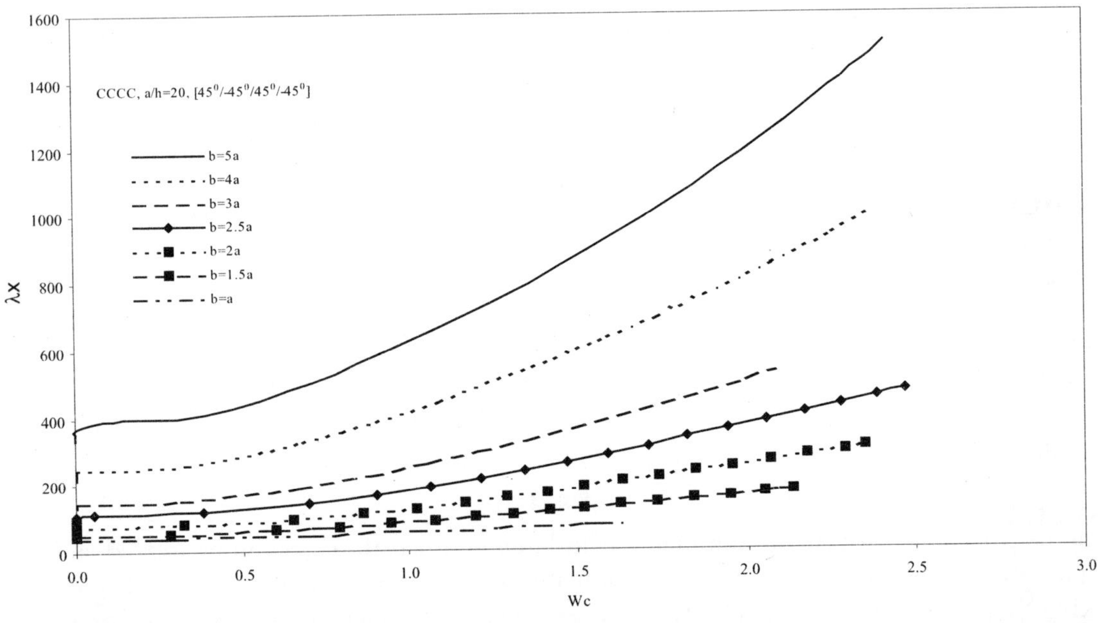

Fig. 5.

Increase in plate aspect ratio (b/a), increases the non-dimensional critical buckling load ($N_x b^2/ E_2 h^3$) and post-buckling strength. The effect of uni-axial loading, biaxial loading, shear loading and their combinations on the buckling and post-buckling strength of laminated plate are studied and depicted in Fig. 6. The buckling load and postbuckling strength of plate subjected to bi-axial loading (compression-tension) without shear loading is maximum and that of plate subjected to bi-axial loading (compression-compression) with shear loading is minimum.

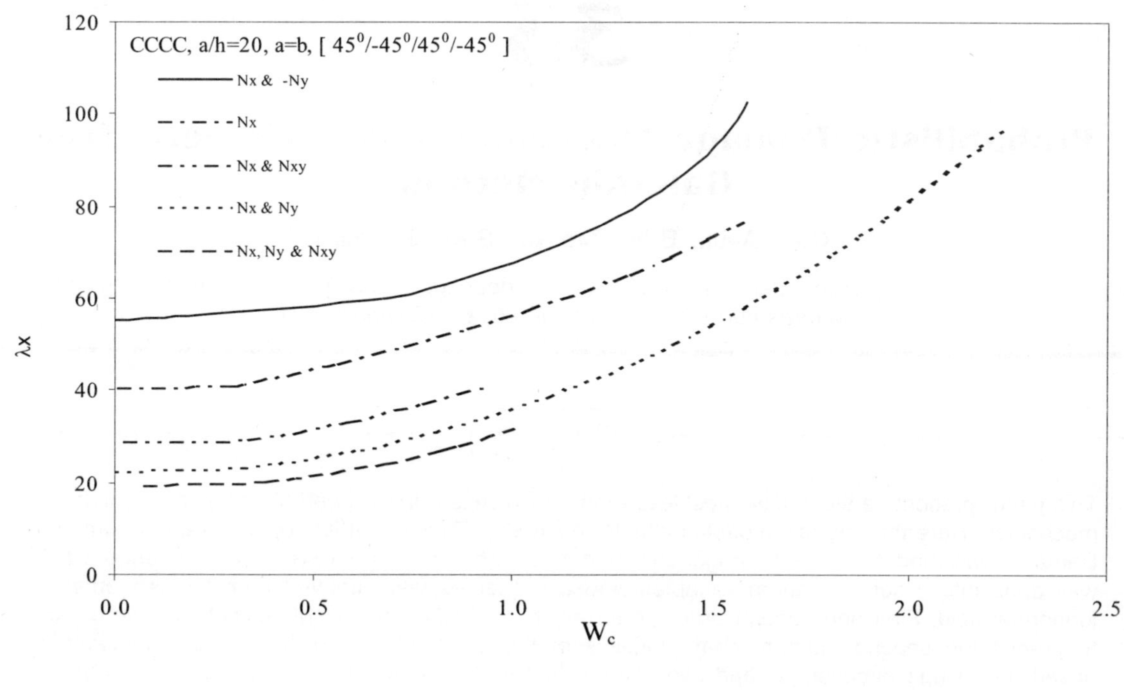

Fig. 6.

CONCLUSION

Buckling and post-buckling response of moderately thick laminated composite rectangular plates subjected to in-plane mechanical loading and uniform temperature has been presented. The buckling/limiting load and post-buckling strength increases with increase in number of layers but it becomes insignificant if the number of layers is more than 8 to10. The buckling load and post-buckling strength of an anti-symmetric angle-ply laminate is more than symmetric angle-ply laminate. The present solution methodology can be extended to other boundary value problems also.

REFERENCES

1. T.R. Tauchert, 1991, Thermally induced flexure, buckling and vibration of plates, ASME Appl. Mech. Rev. 44(8), 347-360.
2. G.J. Turvey, I.H. Marshall, 1995, Buckling and postbuckling of composite plates. London. Chapman & Hall.
3. H.S. Shen, 1998, Thermomechanical postbuckling analysis of imperfect laminated plates using a higher order shear deformation theory, Computers and Structures. 66(4), 395-409.
4. K.K. Shukla, Y. Nath, 2001, Analytical solution for the buckling and postbuckling of angle-ply laminated plates under thermomechanical loading, Int. J. Non-linear Mechanics 2001. 36,1097-1108.
5. L. Fox, I.B. Parker, 1966, Chebyshev polynomials in numerical analysis. Oxford University Press, London.

33

Probabilistic Damage Mechanics Using Element Free Galerkin Method

C.O. ARUN, B.N. RAO AND S.M. SIVAKUMAR

Structural Engineering Division Department of Civil Engineering, Indian Institute of Technology Madras, Madras-600 036 India email: co_arun@yahoo.co.in

ABSTRACT

This paper presents a stochastic meshless method which is applied in problems involving damage mechanics. Here the Young's modulus was formulated as function of isotropic damage parameter. Damage was modeled as a homogenous lognormal random field. Discretization of random field was done into a set of random variables whose properties were derived from the properties of lognormal field. First and second order perturbation methods with meshfree methods were used to predict the second moment characteristics of the structural response. Numerical examples based on an one dimensional and two dimensional elastic problems were studied.

Keywords: Moving Least Square, isotropic damage, random field, perturbation, Monte-Carlo.

1. INTRODUCTION

Many problems in mechanics include uncertainties in material, geometry or loads. This may leads to the formulation of stochastic differential equations. Stochastic meshfree methods are the current development in this area, considering the randomness in the problems. Stochastic meshfree method based on element free Galerkin approach for solving damage mechanics problems is presented. Damage is the deterioration which occurs in materials prior to failure and generally you see and you touch nothing [7, 8]. Damage appears to be highly random in nature and which makes the material properties also random and so the response. The present study was based on the concept that the actual modulus of elasticity is a function of damage, which is random in nature. Only problems modeled with isotropic damage was studied in the present study.

2. MOVING LEAST SQUARE APPROXIMATION

Consider a function, $u(x)$ over a domain, $\Omega \subseteq \Re^K$, where $K = 1$, 2, or 3. Let $\Omega_x \subseteq \Omega$ denote a sub-domain describing the neighborhood of a point, $x \in \Re^K$ located in Ω. According to moving least-squares (MLS) [1,4,6], the approximation, $u^h(x)$ of u(x) is

$$u^h(x) = \sum_{i=1}^{m} p_i(x)a_i(x) = p^{\mathrm{T}}(x)a(x) \qquad \qquad ...(1)$$

Where

$p^T(x) = \{p_1(x) \ p_2(x) \ p_3(x)....p_m(x)\}$ is a complete basis function of order m and

$a^T(x) = \{a_0(x) \ a_1(x) \ a_2(x).....a_m(x)\}$ is a vector of unknown parameters that depend on x In

equation (1), $a(x)$ is determined by the minimizing $J(x) = \sum_{I=1}^{n} w_I(x)[p^T(x_I)a(x) - d_I]^2$ where x_I is

the coordinate of node I, d_I is the nodal parameter with $w_I(x)$ denoting weight function associated with node I. n is the number of nodes in the neighborhood of x for which the weight function $w_I(x) > 0$. The stationarity of $J(x)$ gives $a(x)$ which can be substituted in equation (1) to yield

$$u^h(x) = \sum_{i=1}^{n} \Phi_I(x)d_I = \Phi^T(x)d \qquad \qquad ...(2)$$

where $d^T = [d_1, d_2....d_n]$, $\Phi(x) = \{\Phi_1(x), \Phi_2(x),...\Phi_n(x)\}^T$ and the shape function for node I,

$\Phi_I(x) = \sum_{j=1}^{m} p_j(x)(A^{-1}(x)B(x))_{jI}$, with $A(x) = \sum_{I=1}^{n} \overset{\nabla}{w_I}(x)p(x_I)p^T(x_I) = P^TWP$ and $B = [w_1(x)p(x_1),$

$w_2(x)p(x_2)....w_n(x)p(x_n)] = P^TW$. The weight function used in the current study was student's t-distribution [10].

3. VARIATIONAL FORMULATION AND DISCRETIZATION

For small displacement in two-dimensional, isotropic and linear-elastic solids, the equilibrium equations and boundary conditions are $\Delta.\sigma + b = 0$ in Ω, $\sigma.n = t$ on Γ_t (natural boundary condition) and $u = \bar{u}$ on Γ_u (essential boundary condition) respectively. Where $\sigma = C\varepsilon$ is the stress vector, C is the material property matrix, $\varepsilon = \nabla s_u$ is the strain vector, u is the displacement vector, b is the body force vector, $\bar{t}$ and $\bar{u}$ are the vectors of prescribed surface tractions and displacements respectively, n is a unit normal to domain, Ω. The variational form of above equations

$$\int_\Omega \sigma^T \delta\varepsilon d\Omega - \int_\Omega b^T \delta u d\Omega - \int_{\Gamma_t} \bar{t}^T \delta u d\Gamma - \delta W_u = 0 \qquad \qquad ...(3)$$

where δ denotes the variation operator and δW_u represents a term to enforce the essential boundary conditions. Its explicit form depends on the method by which the boundary conditions are imposed. Using MLS approximation the discrete form of equation becomes [10], $Kd = F$, here K is the stiffness matrix and F is the force vector. To construct K and F a background mesh is needed for numerical integration. However, the back ground mesh is independent of the meshless nodes. To impose the essential boundary condition, full transformation method [2, 10] was used.

4. RANDOM FIELD DISCRITIZATION

In this study the Young's modulus of elasticity, E of the material was modeled as a function of damage parameter D as, $E = E_0(1-D)$, [3,8] where E_0 is the young's modulus of virgin material. The spatial variability of damage, $D(x)$, is modeled as a homogeneous lognormal random field as. $D(x) = c_\alpha \exp[\alpha(x)]$, where $\alpha(x)$ is a zero mass, scalar, homogeneous Gaussian random field [11]. Now in stochastic meshless method, it is necessary to discretize the continuous parametered random field into a vector of random variables. Consider a discretization of zero-mean random field, $\alpha(x)$, by M random variables associated with M discrete material points in structural domain. Let $Y = \{Y_1, Y_2, Y_3...Y_m\}^T$ denotes an M-dimensional random vector comprising these random variables. Its mean vector, $\mu = \varepsilon(Y) = 0$ and the covariance matrix, $\gamma = \varepsilon((y-\mu)(y-\mu)^T)$, can be defined from the knowledge of the mean and correlation function of $\alpha(x)$.

5. PERTURBATION METHOD

The discrete equilibrium equation of the stochastic mesh less system is $K(Y)d(Y) = F(Y)$, where Y is the uncertainty in material properties, K stiffness matrix, d generalized displacement vector and F is the load vector. For the present problem F is not depended on Y. But for the sake of generality F is assumed to be a function of Y. Let $\mu_d = \varepsilon(d)$ and $\gamma_d = \varepsilon\{(d-\mu_d)(d-\mu_d)^T\}$ denote the mean vector and covariance matrix of the response vector d. Using Taylor series at $Y=0$, the first order perturbation will give

$$\mu_d = d_0 \quad and \quad \gamma_d = \sum_{i,j=1}^{M} d_{,i}\, d_{,j}^T\, \gamma_{ij} \qquad ...(4)$$

Also the second order perturbation solutions on assumption that Y follows Gaussian are

$$\mu_d = d_0 + \frac{1}{2}\sum_{i,j=1}^{M} d_{,ij}\, \gamma_{ij} \quad and$$

$$\gamma_d = \sum_{i,j=1}^{M} d_{,i}\, d_{,j}^T\, \gamma_{ij} + \frac{1}{4}\sum_{i,j,k,l=1}^{M} d_{,ij} d_{,kl}^T\,(\gamma_{il}\gamma_{jk} + \gamma_{ik}\gamma_{jl}) \qquad ...(5)$$

In order to reduce the computational effort required to evaluate the second-moment characteristics of response, modal decomposition method is used. [10].

6. NUMERICAL EXAMPLES BASED ON ABOVE METHODOLOGY

6.1. Bar with Linear Body Force (One Dimensional Problem)

A one-dimensional bar AB with constant cross-sectional area, $A = 1$ units, and with linear body force distribution $p(x) = x$ as in the Fig. 1 in the domain $0 \leq x \leq 1$ was considered. The point A of the bar is fixed and the point B is free. Damage was assumed to follow a homogeneous log-normal random field with mean (μ_D), 0.12 and standard deviation (σ_D), 0.018 and is modeled as

$D(x) = c_\alpha \exp[\alpha(x)]$ [5, 9, 11]. An exponential auto-correlation function used for $\alpha(x)$ was

$$\Gamma_{\alpha\alpha}(x_1 - x_2) = \sigma_\alpha^2 \exp\{-(\|\xi\|)/c)\} \qquad \text{...(6)}$$

Correlation length, c = 1 was assumed ξ is the separation vector between two points $x \in \mathbb{R}^k$ and $x + \xi \in \mathbb{R}^k$ both located in $\Omega \in \mathbb{R}^k$. Standard deviation σ_α is given by the expression $\sigma_\alpha = \sqrt{\ln(1 + \sigma_D^2/\mu_D^2)}$. For numerical calculations, correlation length, c = 1 was assumed. A meshless discretization involving 21 uniformly spaced nodes and background mesh with its nodes coincident with the meshless nodes were used.

The second moment characteristics of axial displacement of the bar were determined using the method discussed. The numerical integration involved one-point Gauss quadrature. Figure 2 and 3 shows, respectively, the plots of mean, $\mu_d(x)$, and standard deviation, $\sigma_d(x)$, of the axial response as a function of x. Both first and second order perturbation techniques were used. Results were validated using Monte-Carlo simulation

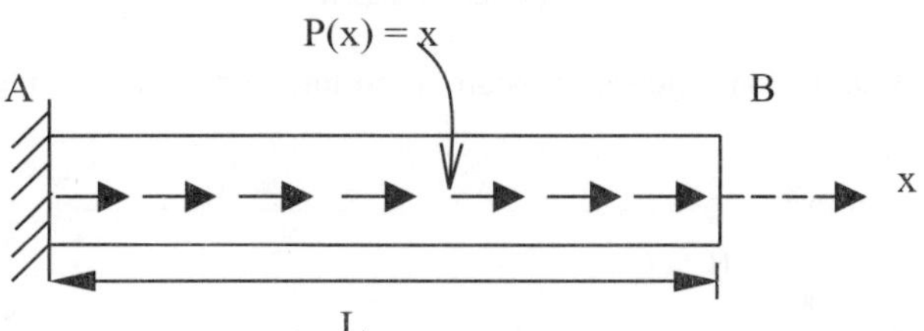

Fig. 1. A bar subjected to linear body force.

with 1.0×10^6 samples. Perturbation results agree well with that of the simulation for $\sigma_d = 0.018$. Also the convergence properties of the perturbation method were studied by changing the number of nodes, (N) from 6 to 41. Figure 4 and 5 show the response characteristics, $\mu_d(L)$ and $\sigma_d(L)$, at the free end of the bar as a function of N. The results of both first and second order perturbation of meshless methods generate convergent solution of mean and standard deviation. Figure 6 and 7 show the results of $\mu_d(L)$ and $\sigma_d(L)$ when standard deviation of $D(x)$ is varied from zero to 0.024. The results of both perturbation and simulation are presented. When σ_D is large, the perturbation technique unpredicts the response statistics by simulation. This trend is expected since the fundamental assumption of perturbation method is that uncertainties are small. Therefore for larger σ_D, perturbation methods should be made with care.

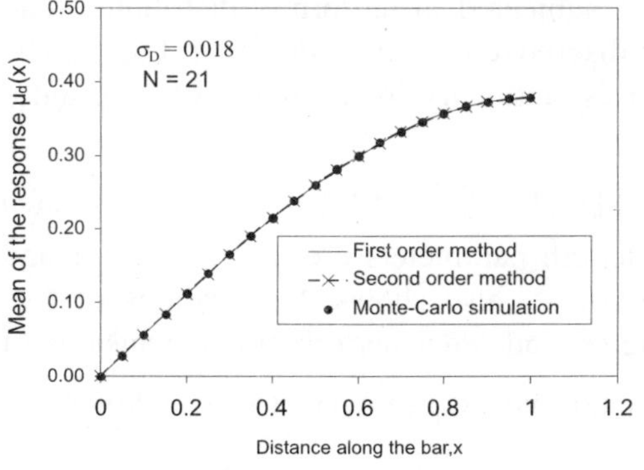

Fig. 2. Mean at Distance along the bar.

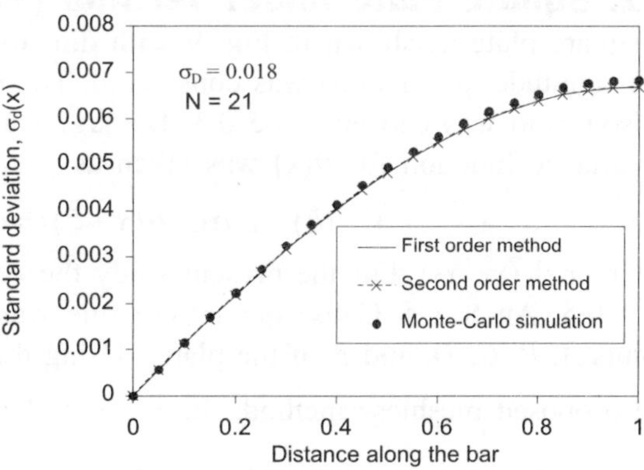

Fig. 3. Standard deviation at distance along the bar.

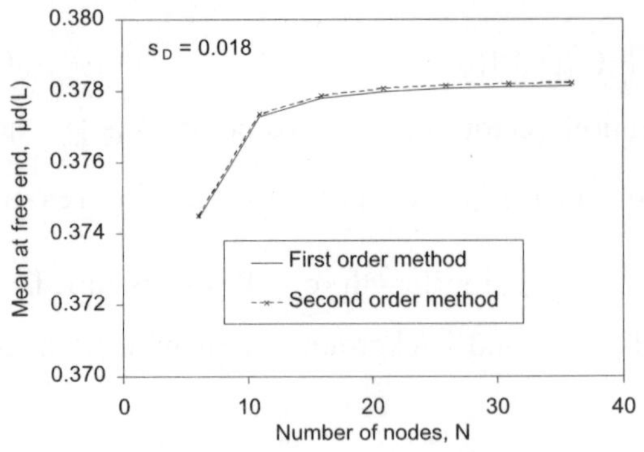

Fig. 4. Convergence of mean at the free end.

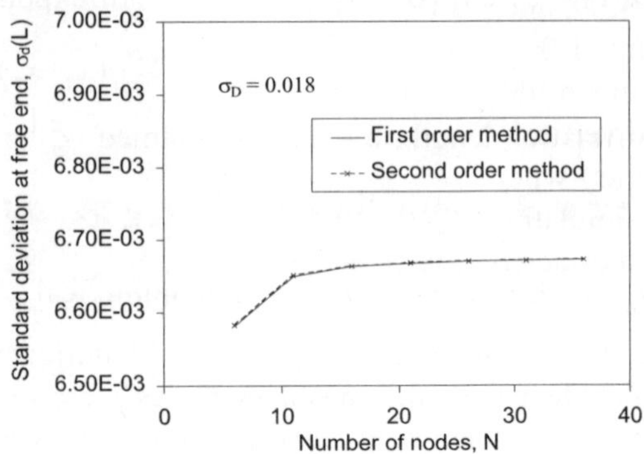

Fig. 5. Convergence of standard deviation at the free end.

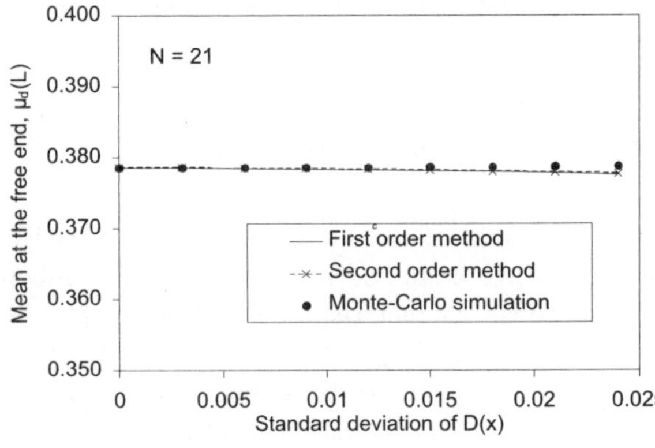

Fig. 6. Mean at the free end of the bar
for various σ_D.

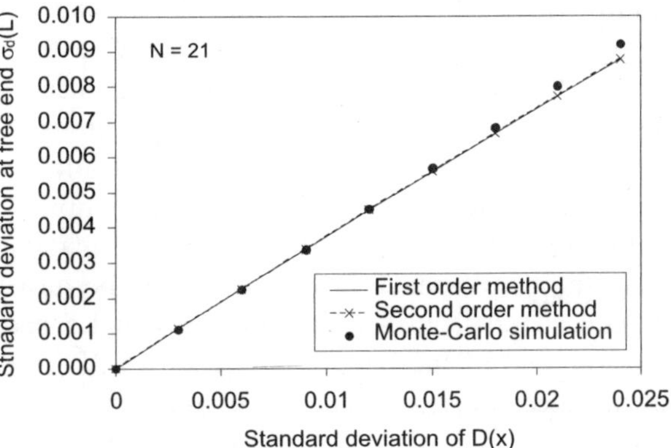

Fig. 7. Standard deviation at the free end
for various σ_D.

6.2. Square Plate under Tension (Two-Dimensional Problem)

A square plate as shown in Fig. 9 with dimensions $L = 1$ subjected to uniformly distributed load of magnitude, p = 1 units was considered. The domain is discretized by 49 nodes as in Fig. 9. The poison ratio was chosen to be 0.3. Damage was modeled as stationary lognormal field. The auto-covariance function for $\alpha(x)$ was taken as

$$\Gamma_{\alpha\alpha}(\xi) = \varepsilon[\alpha(x)\alpha(x+\xi)] = \sigma_\alpha^2 \exp\{-(\|\xi_1\|/c_1 L + \|\xi_2\|/c_2 L)\}\} \qquad \ldots(7)$$

where $x = (x_1, x_2)$. For the present study the correlation length parameters are taken as $c_1 = 1$ and $c_2 = 0.5$. An 8×8 Gauss quadrature rule was used. Figure 9 also shows the locations of five points A, B, C, D, and E of the plate. Using the first- and second-order perturbation expansions of the proposed meshless method, the mean and variance of normal $[\sigma_{11}(x), \sigma_{22}(x)]$ and/or shear $[\sigma_{12}(x)]$ stresses at A, B and C the horizontal $[u_1{}^h(x)]$ and/or vertical $[u_2{}^h(x)]$ displacements at C, D, and E were calculated and are shown in Table 1. The results of the Monte Carlo simulation using 1.0×10^6 samples are also given in Table 1. A good agreement is observed between the results of the perturbation methods and simulation. Figures 9 and 10 show the variances of horizontal and vertical displacements, respectively, at point D by the second-order perturbation method (closed points) when the correlation distances c_1 and c_2 are varied from 1 to 4. The corresponding results of simulation (open points), also shown in the same figures, indicate that the proposed method can predict accurate statistics of random response regardless of the correlation distances.

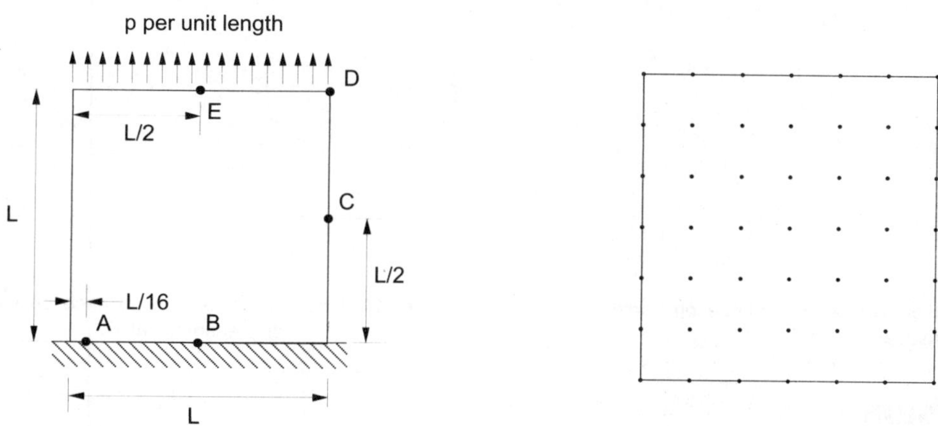

Fig. 8. Square plate subjected to uniformly distributed tension and its discretization (49 nodes).

Table 1. Mean and variances of displacements and/or stresses by various methods.

Location	Response[a]	First-Order Perturbation		Second-Order Perturbation		Monte Carlo Simulation	
		Mean	Variance	Mean	Variance	Mean	Variance
A	σ_{11}	7.400×10^{-2}	8.215×10^{-7}	7.398×10^{-2}	1.506×10^{-6}	7.398×10^{-2}	8.714×10^{-7}
	σ_{22}	$1.107 \times 10^{+0}$	1.125×10^{-4}	1.107×10^{0}	2.107×10^{-4}	1.107×10^{0}	1.183×10^{-4}
	σ_{12}	1.415×10^{-1}	2.171×10^{-6}	1.415×10^{-1}	4.821×10^{-6}	1.415×10^{-1}	2.289×10^{-6}
B	σ_{11}	2.867×10^{-01}	5.726×10^{-6}	2.867×10^{-1}	1.126×10^{-05}	2.866×10^{-1}	6.055×10^{-6}
	σ_{22}	9.764×10^{-1}	6.739×10^{-5}	9.764×10^{-1}	1.321×10^{-4}	9.763×10^{-1}	7.127×10^{-5}
C	u_1	-1.656×10^{-1}	2.707×10^{-5}	1.656×10^{-1}	2.878×10^{-5}	1.659×10^{-1}	2.863×10^{-5}
	u_2	5.668×10^{-1}	1.271×10^{-4}	5.672×10^{-1}	1.319×10^{-4}	5.678×10^{-1}	1.335×10^{-4}
	σ_{22}	9.515×10^{-1}	6.611×10^{-5}	9.515×10^{-1}	1.128×10^{-4}	9.514×10^{-1}	6.941×10^{-5}
D	u_1	-1.769×10^{-1}	1.915×10^{-4}	1.766×10^{-1}	2.026×10^{-4}	1.772×10^{-1}	2.036×10^{-4}
	u_2	$1.123 \times 10^{+0}$	4.507×10^{-4}	1.124×10^{0}	4.626×10^{-4}	1.125×10^{0}	4.723×10^{-4}
E	u_2	$1.114 \times 10^{+0}$	3.350×10^{-4}	1.115×10^{0}	3.401×10^{-4}	1.11610^{+0}	3.499×10^{-4}

1. u_1 and u_2 represent horizontal and vertical displacements, respectively. σ_{11} and σ_{22} represent normal stresses in x_1 and x_2 directions, respectively; and σ_{12} represents shear stress.

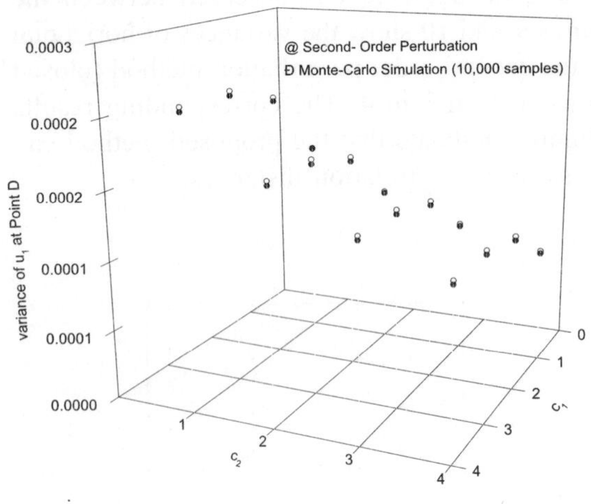

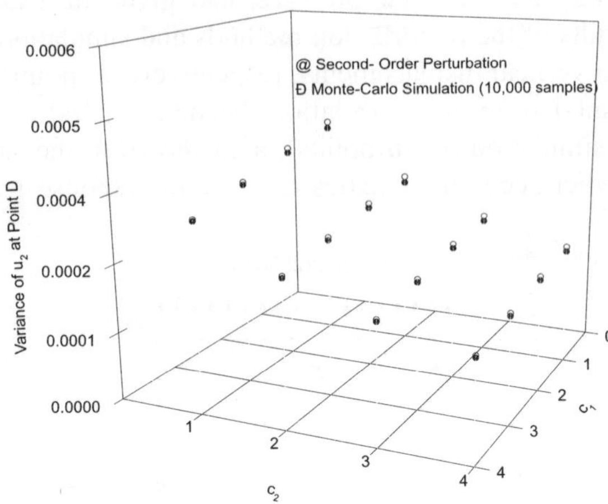

Fig. 9. Effect of correlation distance on horizontal displacement at D.

Fig. 10. Effect of correlation distance on vertical displacement at D.

CONCLUSION

A stochastic meshfree method based on first and second order perturbation techniques was discussed for damage mechanics problems. Isotropic damage parameter, which was used in this study, was considered to follow a homogenous lognormal random field distribution type. Numerical example based on one dimensional and two dimensional problems were solved to find the second moment characteristics of displacements and stresses. Validations of perturbation techniques were done by comparing the results with those obtained by Monte Carlo simulation. A good agreement is found between the results obtained by both the methods.

REFERENCES

1. T. Belytschko, Y.Y. Lu, and L.Gu, 1994, Element Free Galerkin Methods, International Journal for Numerical Methods in Engineering, 37, 229-256.
2. J.S. Chen, and H.P. Wang, 2000, New Boundary Condition Treatments in Meshfree Computation of Contact Problems, Computer Methods in Applied Mechanics and Engineering, 187, 441-468.
3. P.I Kattan, and G.Z.Voyiadjis, 2002, Damage Mechanics with Finite Elements., Springer-verlag Berlin Heidelberg.
4. J. Dolbow, and T. Belytschko, 1998, An Introduction to Programming the Meshless Element Free Galerkin Method, Achieves of Computational Mechanics, Vol. 15, No. 3, 207-241.
5. A.D. Kiureghian, and Y. Zhang, 1998, Space-variant Finite Element Reliability Analysis, Computer Methods in Applied Mechanics and Engineering, 168; 173-183.
6. P. Lancaster, and K. Salkauskas, 1981, Surfaces Generated by Moving Least Sqaures Methods, Mathematics of Computation, 37, 141-158.
7. J. Lemaitre 1992, A course on Damage Mechanics., Springer-Verlag Berlin Heidelberg.
8. J. Lemaitre, and R. Desmorat, 2005, Engineering Damage Mechanics, Springer-verlag Berlin Heidelberg.
9. P.L. Liu, and A.D. Kiureghian, 1986, Multivariate Distribution Models with Prescribed Marginals and Covariances, Probabilistic Engineering Mechanics, 1. No. 2. 105-113
10. S. Rahman, and B.N. Rao, 2001, A Perturbation Method For Stochastic Meshless Analysis In Elastostatics, International Journal for Numerical Methods in Engineering, 50; 1969-1991.
11. H. Xu, and S. Rahman, 2005, Decomposition Method for Structural Reliability Analysis, Probabilistic Engineering Mechanics, Vol. 20, 239-250.

34

Effect of Blow-Holes on the Reliability of a Cast Component

B.N. Rao[1] and Rajesh Ashokkumar[2]

[1]Asst. Professor, Department of Civil Engineering, Indian Institute of Technology, Madras,
Chennai-600 036, India email: bnrao@iitm.ac.in

[2]Post Graduate, Department of Civil Engineering, Indian Institute of Technology, Madras,
Chennai-600 036, India email: mail2raj_5@yahoo.com

ABSTRACT

Fracture mechanics requires a complete knowledge of the defect in the specimen to be analyzed. Since the size and location of defects are quite random, deterministic analysis provides an incomplete picture of the safety of a component. Moreover the randomness in external loads and geometry also influence the reliability of a component. Also in case of cast component the effect of micro defects like that of blow-holes are neglected in the calculation of reliability of the component. Probabilistic fracture mechanics, which combines fracture mechanics with stochastic methods, provides a useful tool to address these problems. Given the complexity of failure mechanism, the combination of finite element method with the theory of statistics and reliability has become a powerful method for safety and reliability analysis. In the present study, a generic cast component with blow-holes and a crack has been taken as the starting point. The reliability is calculated by a novel technique called lagrangian based univariate technique, which is validated by the Monte Carlo Simulation (MCS) technique.

Keywords: Blow Holes; Clustering of Defects; First Order Reliability Method (FORM); Second Order Reliability Method (SORM); Monte Carlo Simulation (MCS).

1. INTRODUCTION

Fracture mechanics requires a complete knowledge of the defect, or crack in the specimen to be analyzed. Since the size and location of defects are quite random, deterministic analysis provides an incomplete picture of the safety of a component. Moreover the randomness in external loads and geometry also influence the reliability of a component. From the literature, it is very evident that in the analysis of cast components, the effect of micro-defects like blow-holes on the stress intensity factors was neglected. Probabilistic fracture mechanics, which combines fracture mechanics with stochastic methods, provides a useful tool to address these problems. Given the complexity of failure mechanism, the combination of finite element method with the theory of statistics and reliability has become a powerful method for safety and reliability analysis.

The fundamental problem in time-invariant component reliability analysis entails calculation of a multi-fold integral

$$P_F \equiv P[g(X) < 0] = \int_{g(x)<0} f_x(x)dx \qquad ...(1)$$

where $X = \{X_1,........, X_N\}^T \in R^N$ is a real-valued, N-dimensional random vector defined on a probability space (Ω, F, P) comprising the sample space Ω, the σ-field F, and the probability measure P; $g(x)$ is the performance function, such that $g(x) < 0$ represents the failure domain; P_F is the probability of failure; and $f_X(x)$ is the joint probability density function of X, which typically represents loads, material properties, and geometry. For most practical problems, the exact evaluation of this integral, either analytically or numerically, is not possible because N is large, $f_X(x)$ is generally non-Gaussian, and $g(x)$ is highly non-linear function of x.

The most common approach to compute the failure probability in Eq. 1 involves the first- and second-order reliability methods (FORM/SORM), which are based on linear (FORM) or quadratic approximation (SORM) of the limit-state surface at a most probable point (MPP). The results obtained are then compared with the direct Monte-Carlo simulation (MCS) technique. Here the whole process is simulated efficiently and since the results are obtained from ABAQUS, they are found to be accurate.

2. STOCHASTIC MECHANICS

In probabilistic mechanics, both random variables and random fields/processes may be required for stochastic representation of uncertainties. In most probabilistic analysis, it is typically assumed that the basic structural parameters are discrete and represented as a single-valued random variable. This assumption is valid when the random quantity lacks variability with respect to time or space. Uncertainties associated with this type of parameter can be modeled as random variables. Hence, a more accurate representation of uncertainty requires modeling these stochastic parameters as random fields. The random field modeling allows spatial variability, which is more realistic, but also greatly increases the size and complexity of structural reliability analysis. [1]

3. RELIABILITY METHODS

3.1. Simulation Methods

Simulation methods involve sampling and estimation and are well known in the statistics and reliability literature. The direct MCS is the most widely used simulation method. In its pure mathematical form, the Monte Carlo method consists of finding the definite integral of a function by choosing a large number of independent-variable samples at random from within an interval or region, averaging the resulting dependent-variable values, and then dividing by the span of the interval or the size of the region over which the random samples were chosen. There is always some error involved with this scheme, but the larger the number of random samples taken, the more accurate the result. [2]

3.2 First and Second Order Reliability Methods

The FORM/SORM are general computational methods of structural reliability. They are based on linear (first-order) and quadratic (second-order) approximations, respectively, of the limit state surface at a point, known as the most probable point (MPP), which is closest to the origin of the standard Gaussian image of the original space of random variables. The determination of the MPP involves solving a non-linear constrained optimization problem. Based on the location of the MPP, a FORM approximation of the failure probability can be made. If the FORM estimate is not adequate, it can be improved by SORM analysis, which involves calculating the principal curvatures of the performance function at the MPP.

4. DECOMPOSITION METHODS

4.1 Multivariate Function Decomposition at Most Probable Point (MPP)

Consider a continuous, differentiable, real-valued performance function $g(x)$ that depends on $x = \{x_1,\ldots,x_n\}^T \in R^N$. The transformed limit states $h(u) = 0$ and $y(v) = 0$ are the maps of the original limit state $g(x) = 0$ in the standard Gaussian space (u space) and the rotated Gaussian space (v space), respectively, as shown in Fig. 1 for N = 2. The closest point on the limit-state surface to the origin, denoted by the MPP (u^* or v^*) or beta point, has a distance β_{HL} which is commonly referred to as the Hasofer–Lind reliability index. The determination of MPP and β_{HL} involves standard non-linear constrained optimization and is usually performed in the standard Gaussian space. [3,4]. Figure. 1 depicts FORM and SORM approximations of the limit-state surface at MPP.

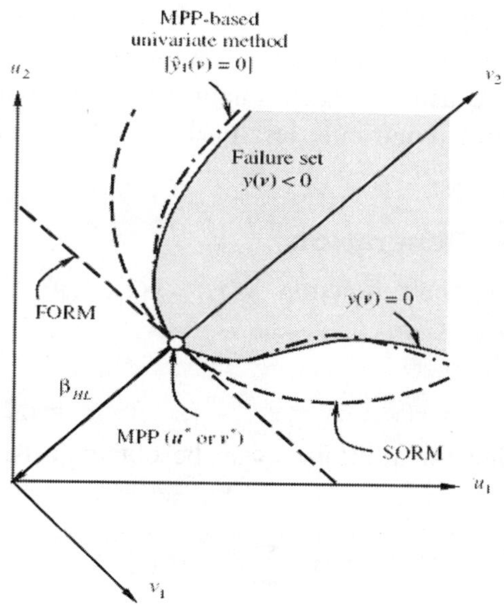

Fig. 1. Performance Function Approximation by various methods.

Suppose that $y(v)$ has a convergent Taylor series expansion at MPP $v^* = \{v_1^*,........,v_N^*\}$ and can be expressed by

$$y(v) = y(v^*) + \sum_{j=1}^{\infty} \frac{1}{j!} \sum_{i=1}^{N} \frac{\partial^j y}{\partial v_i^j}(v^*)(v_i - v_i^*)^j + R_2 \qquad ...(2)$$

or

$$y(v) = y(v^*) + \sum_{j=1}^{\infty} \frac{1}{j!} \sum_{i=1}^{N} \frac{\partial^j y}{\partial v_i^j}(v^*)(v_i - v_i^*)^j + \sum_{j1,j2>0} \frac{1}{j1!j2!} \sum_{i1<i2} \frac{\partial^{j1+j2} y}{\partial v_{i1}^{j1} \partial v_{i2}^{j2}}(v^*)(v_{i1} - v_{i1}^*)^{j1}(v_{i2} - v_{i2}^*)^{j2} + R_3 \qquad ...(3)$$

where the remainder R_2 denotes all terms with dimension two and higher and the remainder R_3 denotes all terms with dimension three and higher.

4.1.1 Univariate Approximation

Consider a univariate approximation of $y(v)$, denoted by

$$\hat{y}_1(v) \equiv \hat{y}(v_1,...., v_N) = \sum_{i=1}^{N} y(v_1^*,..., v_{i-1}^*, v_i, v_{i+1}^*,...., v_N^*) - (N-1)y(v^*) \qquad ...(4)$$

where each term in the summation is a function of only one variable and can be subsequently expanded in a Taylor series at $v = v^*$, yielding

$$\hat{y}_1(v) = y(v^*) + \sum_{j=1}^{\infty} \frac{1}{j!} \sum_{i=1}^{N} \frac{\partial^j y}{\partial v_i^j}(v^*)(v_i - v_i^*)^j \qquad ...(5)$$

Comparisons of Eqs. 2 and 5 indicates that the univariate approximation leads to the residual error $y(v) - \hat{y}_1(v) = R_2$, which includes contributions from terms of dimension two and higher. For sufficiently smooth $y(v)$ with convergent Taylor series, the coefficients associated with higher-dimensional terms are usually much smaller than that with one-dimensional terms. As such, higher-dimensional terms contribute less to the function, and therefore, can be neglected. [3,4].

4.2 Response Surface Generation

Consider the univariate component function $y_i(v_i) = y(v_1^*,..., v_{i-1}^*, v_i, v_{i+1}^*,...., v_N^*)$ in   Eq. 4. If for $v_i = v_i^{(j)}$, n function values

$$y_i(v_i^j) = y(v_1^*,..., v_{i-1}^*, v_i^{(j)}, v_{i+1}^*,...., v_N^*); j = 1, 2,......, n \qquad ...(6)$$

are given, the function value for arbitrary v_i can be obtained using the Lagrange interpolation as;

$$y_i(v_i) = \sum_{j=1}^{n} \phi_j(v_i) y_i(v_i^{(j)}), \qquad ...(7)$$

where the shape function $\phi_j(v_i)$ is defined as

$$\phi_j(v_i) = \frac{\prod_{k=1,k\neq j}^{n}(v_i - v_i^{(k)})}{\prod_{k=1,k\neq j}^{n}(v_i^{(j)} - v_i^{(k)})} \quad ...(8)$$

By using Eqs. 7 and 8, arbitrarily many values of $y_i(v_i)$ can be generated if n values of that component function are given. The same procedure is repeated for all univariate component functions, i.e. for all, $y_i(v_i)$, $i = 1,..., N$. Therefore, the total cost for the univariate approximation in Eq. 4 in addition to that required for locating MPP, entails a maximum of $nN + 1$ function evaluations. [3, 4].

More accurate bivariate or multivariate approximations can be developed in a similar way. However the cost involved in multivariate approximations are higher and therefore not discussed in the present study.

5. EXAMPLE PROBLEM

5.1 Methodology

In the present study, a generic component of the dimension $20 \times 20 \times 5$ mm^3 with a through hole of diameter 5 mm at the centre has been taken as the starting point. It is found that these blow-holes occur as a cluster. This cluster of blowhole is modeled as a single defect [5]. In this component, the crack is found to be initiated at the interface of the cut hole and the surface of the component. The component is modeled and meshed using PATRAN. The blow-hole mesh is moved relative to the crack and its critical position with respect to the crack is found out. The uncertainty in the crack geometry, blow-hole dimension, material properties and the external loads are characterized by a set of independent random variables. Statistical properties of various random parameters are shown in Table 1. A response surface function, based on the above five random parameters, is obtained by a novel technique called the lagrangian based univariate method. This response surface function is given as an input to the FORM/SORM code and thereby the reliability of the component is estimated. The results thus obtained are verified using the direct MCS technique. Here an innovative method is incorporated, where-in a FORTRAN code is written which calls ABAQUS for each iteration, reads in the required result (Stress intensity factors and crack propagation angle in this case), deletes the resulting files from each run, and proceeds with the next run. The MCS is done for 100000 samples with different distributions for the random variables. From the results obtained from ABAQUS, the effective stress intensity factor is found out using the formula,

$$K_{eff} = \left[K_1(X)\cos^2\frac{\theta(X)}{2} - \frac{3}{2}K_{II}(X)\sin\theta(X) \right]\cos\frac{\theta(X)}{2} \quad ...(9)$$

Table 1. Statistical Properties of various Random Parameters

RANDOM PARAMETERS	MEAN	STANDARD DEVIATION	PROBABILITY DISTRIBUTION
Crack Length (a)(in mm)	0.5495	0.0288	Uniform Distribution
Blow-hole diameter (d) (in mm)	0.9497	0.0288	Uniform Distribution
External Load (σ)(in N/mm^2)	175.0234	5.458	Normal Distribution
Youngs Modulus (E) (in N/mm^2)	207024.71	30.81	Log-normal Distribution
Critical Stress Intensity Factor (K_{IC}) $\left(MPa\sqrt{mm}\right)$	1161.16	1.673	Log-normal Distribution

a $^{0.5\text{-}0.6}$ d $^{0.9\text{-}1.0}$

6. RESULTS

6.1 POF Using Reliability Methods

In the present study, the response surface function is calculated by lagrangian interpolation technique. Here, 9 sets of values are taken for each of the random parameters (5 in the present case). By varying one of the parameters and keeping all the other parameters at its mean value, the response is calculated. By doing so we obtained 45 responses. The response surface function is calculated by the response thus obtained. From the response surface function, the reliability of the component is estimated by means of FORM/SORM and the importance sampling technique. Then, the direct MCS is carried out for 100000 sets of inputs, got by varying the random parameters like blow-hole diameter, crack length, elastic modulus, applied load and the fracture toughness values. The POF values thus obtained are shown in the Table 2.

Table 2. Probability of Failure by Monte-Carlo Simulation and the Lagrangian based Univariate Method.

SIMULATION TYPE	PROBABILITY OF FAILURE (%)
Direct Monte-Carlo Simulation (100000 Runs)	0.00378
Lagrangian Based Univariate Method	
First Order Reliability Method (FORM)	0.00383
Second Order Reliability Method (SORM)	0.00301
Importance Sampling	0.00283

6.2 Sensitivity Analysis

This involves computing the change in failure probability when one of the random variables is assumed to be deterministic at the mean, while the other parameters are varied randomly. The importance measure of a random variable is then expressed as

$$\Gamma_i = (P_f - P_{fi})^2 / \sum_{j=1}^{n} (P_f - P_{fi})^2 \qquad ...(10)$$

The results are shown in Table. 3.

Table 3. Sensitivity Index of the various Random Parameters

RANDOM PARAMETERS	SENSITIVITY INDEX
Crack-Length	0.1328
Blow-hole Diameter	0.5701
External Load	0.1735
Youngs Modulus	0
Fracture Toughness	0.1235

CONCLUSION

Micro-defects as that of blow-holes were neglected in estimation of the failure probability of cast component. But it is found that these defects have a significant effect on the SIF values and therefore should be considered in estimation of reliability of the component. Also, the lagrangian based univariate method for calculation of response surface function is found to be an efficient one. This response surface function, when given to the FORM/SORM code is found to give good results in estimating the reliability of the component. The probability of failure obtained by the above method is validated by the direct MCS technique. Here an interface between FORTRAN and ABAQUS has been developed and since the results are obtained from ABAQUS, they are found to be fairly accurate.

REFERENCES

1. IASSAR Subcommittee on Computational stochastic Structural Mechanics, 1997, A State-of-the-Art Report on Computational Stochastic Mechanics, Probabilistic Engineering Mechanics, edited by G.I. Schueller, 12, 197-321.
2. W.K. Liu, Y. Chen, and T. Belytschko., 1996, Three reliability methods for fatigue crack growth, Engineering fracture mechanics, 53, 733-752.
3. H. Xu and S. Rahman., 2005, Decomposition methods for structural reliability analysis, Probabilistic engineering mechanics, 20, 239-250.
4. S. Rahman and D. Wei., 2006, A univariate approximation at most probable point for higher-order reliability analysis, International journal of solids and structures, 43, 2820-2839.
5. Elena Kabo., 2002, Material defects in rolling contact fatigue — influence of overloads and defect clusters, International Journal of Fatigue, 24, 887-894.

35

Application of the FRF Curvature Energy Damage Index Detection Method to Plate-like Structures

D. Mallikarjuna Reddy[*] and S. Swarnamani

[1]Department of Mechanical Engineering, Indian Institute of Technology Madras, Chennai-600 036, India
email: dmreddy@iitm.ac.in
[2]Department of Mechanical Engineering, Indian Institute of Technology Madras, Chennai-600 036, India
email: mani46@iitm.ac.in

ABSTRACT

The objective of current work is to show the effectiveness of using frequency response function (FRF) curvature energy damage index and to establish its capability to detect and localize damage. The dimension of aluminium plate under consideration is 0.25 m x 0.25 m x 0.003 m with fixed–fixed support condition Damage is simulated by reducing the thickness of one element for different damage cases. The mode shape and FRF energy data of the square plate with damage of different sizes are obtained by free vibration analysis using MATLAB 7.0. Testing the frequency intervals, it is found that the FRF curvature energy damage index method defined in the range of frequencies include the eigen frequencies. The damage index is found to be a function of the frequency bandwidth and variation of FRF curvature energy damage index versus frequency range (band width) seems to provide further information in choice of optimum frequency range response analysis. This typical analysis presented wider frequency ranges, including several higher modes, the difference of curvature energies of the damaged and undamaged model becomes less significant. The influence of noise in the FRF data seems to be quite small. It also found that influence of excitation location is not important for the damage detection. It is observed that by using FRF curvature energy damage index data, damage location can be identified provided the reduction in elemental thickness is more than 10%.

Keywords: Structural health monitoring; Modal analysis; Damage detection; FRF curvature energy damage index method.

1. INTRODUCTION

More recently resonant methods based on modal data have been used both to identify that damage exists, and to locate it. The basic approach for not only detecting but also locating a structural fault presented a method to determine mass, stiffness and damping properties from measured frequencies response functions and showed how changes in those parameters could be used to locate the fault [1].

A method based on the decreases in modal strain energy between two structural degrees of freedom as defined by the curvature of the measured mode shapes. This method successful applied to data from a damaged bridge [2]. Absolute changes in mode shape curvature can be a good indicator of damage [3] for the FEM beam structure they considered. The displacement functions converted into curvature function, which are further processed to yield a damage index. A method [4] in place of using the displacement mode shapes, strain or curvature shapes (surface strain in a beam is proportional to curvature) are more effective at identifying the location of damage. Later presented a method [5] which requires that the mode shapes before and after damage is known, but the modes do not need to be mass normalized making it very advantageous when using ambient excitation. In reference [6] discussed a approach for locating structural damage using experimental vibration data. This method uses measured frequency response functions to obtain displacement as function of frequency.

2. FORMULATIONS

2.1 The Frequency Response Function (FRF) Curvature Method

This method presented by Pandey *et al.* [3] have found that in place of using a displacement mode shape, strain or curvature shapes (surface strain in a beam is proportional to curvature) are more effective at identifying the location of damage. This paper is extended to plate like structures by using FRF curvature data rather than mode shape data. The FRF- curvature for any frequency Ω is defined by

$$H''(\Omega)_{i,j} = \frac{-H(\Omega)_{i-2,j} + 16H(\Omega)_{i-1,j} - 30H(\Omega)_{i,j} + 16H(\Omega)_{i+1,j} - H(\Omega)_{i+2,j}}{2h^2} \qquad ...(1)$$

where $H''(\Omega)_{i,j}$: FRF curvature measured at location *i* due to a force input at position j.

'*h*': The distance between two consecutive measurement points

2.2 The FRF Curvature Energy Damage Index

In this section a new damage index based on the concept of FRF curvature energy is proposed. The damage indices are based on the variation of the FRF curvature energy at the element of the structure for a given excitation frequency. In a plate structure the FRF curvature energy can be defined as

$$\eta(\Omega) = \int_0^L \left[H''(x;\Omega) \right]^2 dx \qquad ...(2)$$

where L is Span of the plate and $H''(x;\Omega)$ is FRF curvature for a frequency Ω.

The structure system is assumed to be divided into *n* elements ($j = 1, 2, 3...n$). For the jth element the FRF curvature energy can be written as

$$\eta(\Omega) = \int_{x_k}^{x_{k+1}} \left[H''(x;\Omega) \right]^2 dx \qquad ...(3)$$

where x_k and x_{k+1} : coordinates of the modes of the elements *j*.

For applied force at point *j*, the absolute difference between the FRF curvature energy of damaged and undamaged structure at a location *i*, in a predetermined frequency range is defined as

$$\Delta\eta = \sum_{\Omega}\left|\eta^{*}(\Omega)-\eta(\Omega)\right| \qquad \ldots(4)$$

3. SIMULATION RESULTS AND DISCUSSIONS

The dimension of plate under consideration is 0.25m × 0.25m × 0.003 m with Fixed – Fixed support condition as shown in Fig 1. The material properties used in modeling the plate are E = 68.9 GPa, v = 0.33 and ρ = 2710 kg/m³ which refer to Young's Modulus, Poisson's ratio and Mass Density respectively. The Length and width of plate is equally divided into 15 elements so the simulated model has total of 225 elements. Damage is simulated by reducing the thickness of one element (100th element) by 83.33, 63.33, 50, and 10% at location shown in Fig 1. The mode shape and Frequency response function {FRF} energy data of the square plate with damage of different sizes are obtained by using MATLAB 7.0. The first five natural frequencies resulting from modal analysis for undamaged and all damaged cases are tabulated in Table 1.

Table 1. Represents different damage cases with element number

Damage cases	% of thickness reduction	Damaged element number
1	83.33	100
2	66.66	100
3	50.00	100
4	10.00	100

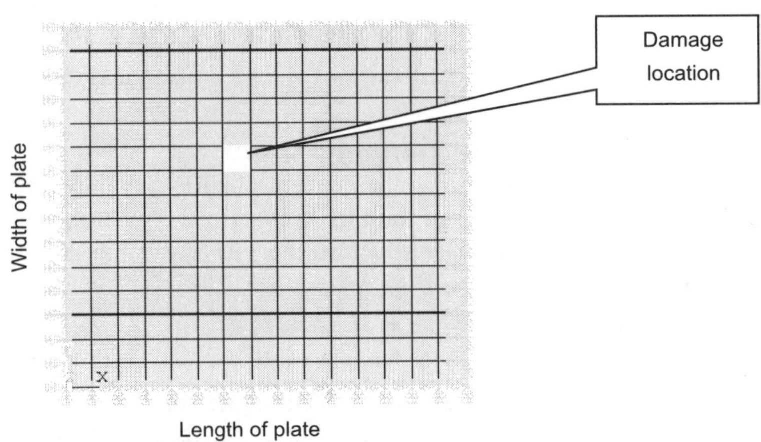

Fig. 1. Finite element mesh of a fixed—fixed plate with an area of reduced thickness.

Simultaneously harmonic analysis is done with frequency range 0-2000 Hz with load applied on structure 100N and the Frequency Response Function (FRF's) data is collected from same location. It is observed that there is no considerable shift in natural frequencies due to damage as observed from Table 2 and thus damage identification becomes difficult.

Table 2. Comparison of first five natural frequencies (Hz) for undamaged and damaged cases.

Damage cases	Frequencies (Hz) for the mode number				
	1	2	3	4	5
Undamaged	422.56	860.56	860.56	1263.6	1541.1
Case 1	422.45	859.04	860.06	1263.1	1539.6
Case 2	422.37	857.81	859.67	1262.7	1538.4
Case 3	422.41	857.32	859.41	1262.5	1537.9
Case 4	423.17	859.12	860.12	1263.3	1540.7

3.1 Influence of Frequency Range

While testing frequency intervals, it is found that the FRF curvature energy damage index method worked better for a range before the first anti-resonance or resonance. In fact for wider frequency ranges, including several modes the difference of curvature energies of the damaged and undamaged model becomes less significant when compared with the amplitude difference arising from the resonant frequencies shift. Fig. 2 shows the damage index variation for different damage cases in frequency range 0 to 500 Hz which shows great potential in detecting and estimating the damage index. Further Fig. 3 shows the FRF curvature energy damage index amplitude increases in frequency

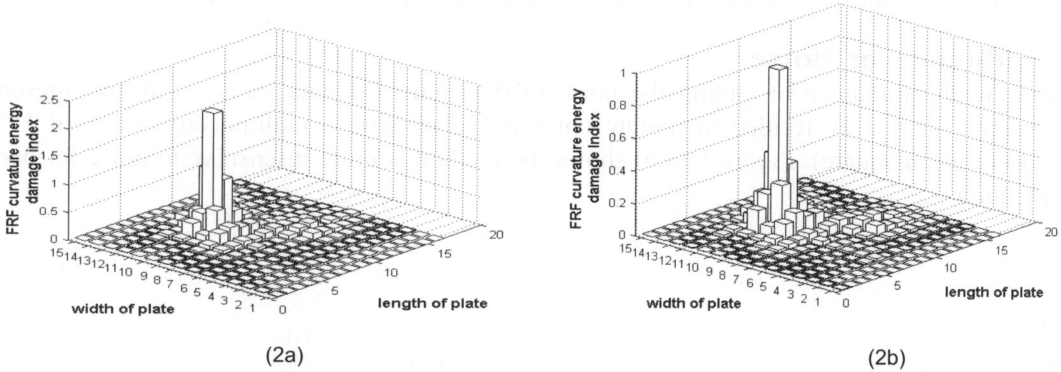

(2a)   (2b)

Fig. 2. FRF curvature energy damage index for a frequency range of 0-500 Hz. (a) Case 1 (b) Case 2.

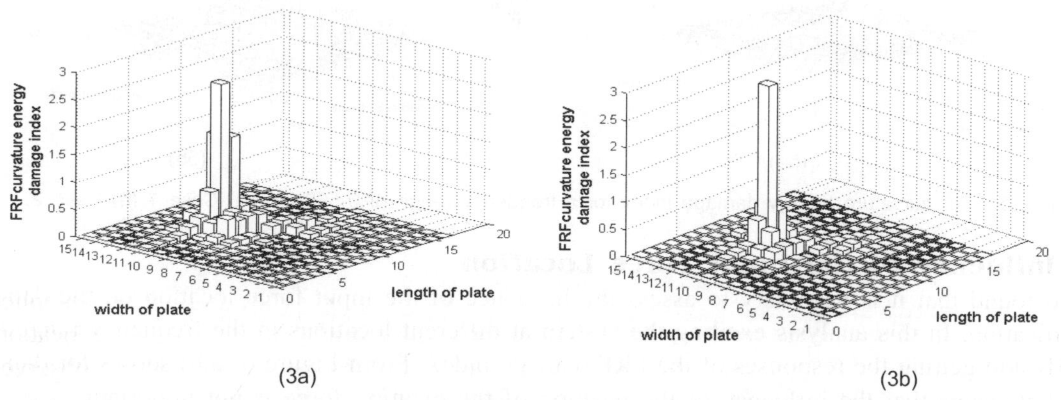

(3a)   (3b)

Fig. 3. FRF curvature energy damage index for a frequency range of 0-900 Hz. (a) Case 1 (b) Case 2.

range 0 to 900Hz for different damage cases but adjacent elements of damaged one also having amplitude which affects the quantification of damage. Further increases the frequency range from 0 to 1280 Hz there is no localization for damage as shown in Fig. 4 for different damage cases.

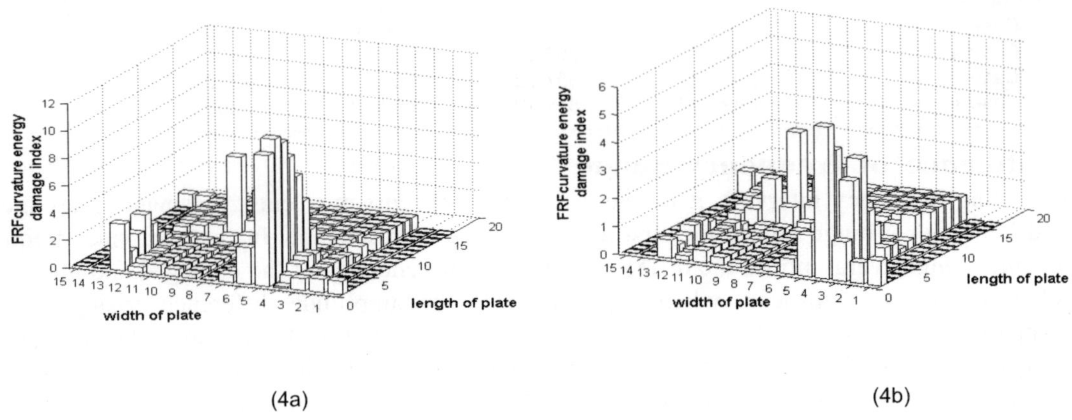

(4a)   (4b)

Fig. 4. FRF curvature energy damage index for a frequency range of 0-1280 Hz. (a) Case 1 (b) Case 2.

3.2. Influence of Noise

To find out the influence of adding the noise to the numerical data (noise is always present on experimental data) it is decided to pollute our modal data with a multiplicative error of 15% of RMS (Root mean square). Figure 5 (a,b) shows the method is quite insensitive to noise for damage identification.

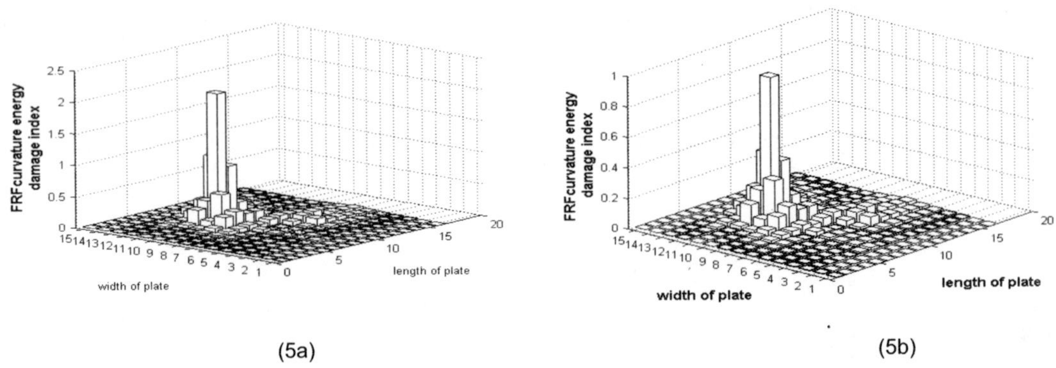

(5a)   (5b)

Fig. 5. FRF curvature energy damage index for a frequency range of 0-500 Hz. (a) Case 1 (b) Case 2.

3.3 Influence of the Input Force Location

It also found that it is necessary to assess the influence of the input force location on the damage identification. In this analysis exciting the system at different locations in the frequency range 0 to 500 Hz and getting the responses of the FRF damage index. From Figure 6 (a,b) shows for damage case 1 it seems that the influence of the position of the exciting force is not important.

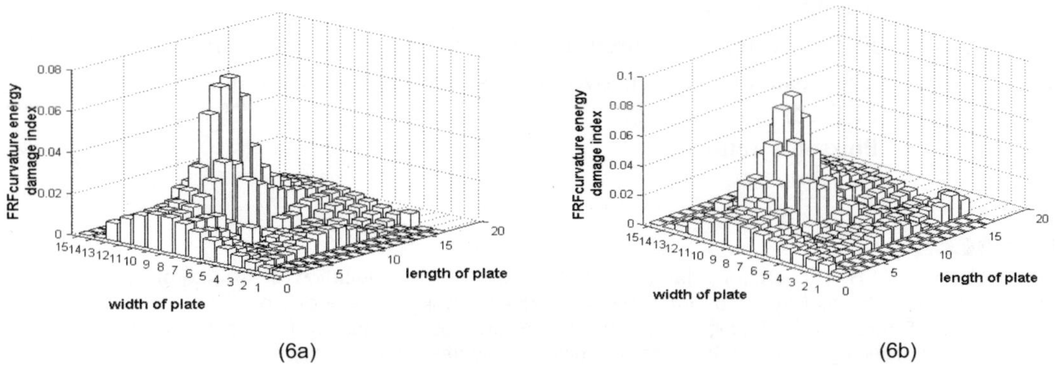

(6a)   (6b)

Fig. 6. FRF curvature energy damage index for a frequency range of 0-500 Hz. Case 1.

For the quantification of damage and also to ascertain the sensitivity of particular frequency range, the variation of maximum value of FRF curvature energy damage index versus percentage of damage for different frequency ranges is plotted as shown in Fig. 7. It is observed that the damage index value increases with damage severity for all the three range of frequencies. By comparing the sensitivity of different range of frequencies it is found that the 0 to 500 Hz range and 0 to 900 Hz is much more sensitive than the range 0 to 1280 Hz.

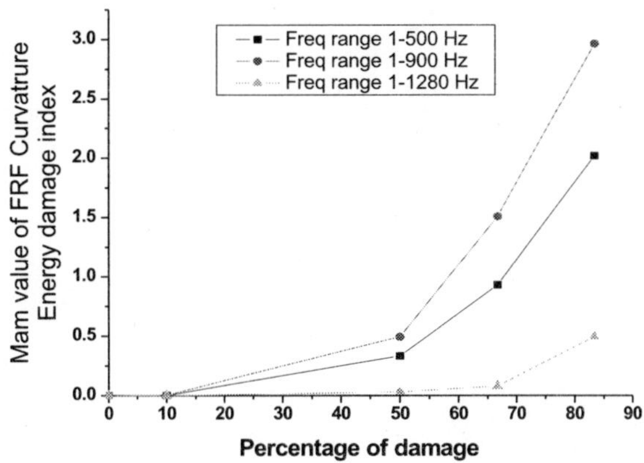

Fig. 7. Variation of percentage of damage versus maximum value of FRF curvature energy damage index for different frequency ranges.

CONCLUSION

The results show that the FRF curvature energy damage index method performed well in detecting, locating and quantifying damage. Its main advantage is its simplicity and no need of performing a modal analysis for the identification of mode shapes or resonant. It's shown that even the system having noise will not effect the localization of damage. Also shown influence of the position of the exciting force is not important for damage identification.

NOMENCLATURE

E	Young's modulus,
$H''(\Omega)_{i,j}$	FRF curvature measured at location i due to a force input at position j.
'h':	The distance between two consecutive measurement points.
$\eta\ (\Omega)$	FRF curvature energy
ν	Poisson's ratio
ρ	Density

REFERENCES

1. M.A. Mannan, M.H. Richardson, January-1990, "Detection and location of structural cracks using FRF measurements", Proceedings of 8th International Modal Analysis Conference", pp: 1-6.
2. N. Stubbs, J.T. Kim, and K. Topole, 1992, "An efficient and robust algorithm for damage localization in offshore platforms." Proceedings of the ASCE Tenth structures congress, pp: 543-546.
3. A.K. Pandey, M. Biswas, and M.M. Samman, 1994, "Damage detection from changes in curvature mode shapes", Journal of Sound and Vibration, 154, pp: 321-332.
4. M.M.N. Maia, J.M.M. Silva, and R.P.C. Sampaio, 1997, "Localization of damage using curvature of the frequency response functions," Proceedings of the 15th International Modal Analysis Conference, Society of Experimental Mechanics.1, pp: 942-946.
5. P. Cornwell, S.W. Doebling, and C.R. Farrar, 1999, "Application of the strain energy damage detection method to plate like structures". Journal of Sound and Vibration, 224(2), pp: 359-374.
6. C.P. Ratcliffe, July-2000, "A frequency and curvature based experimental method for locating damage in structures'. Transactions of the ASME, Vol-122, pp: 324-329.

36

Application of Discrete Dislocation Dynamics Modelling to Predict Stress-Strain Behavior of Irradiated Material

P.V. DURGAPRASAD[1] AND B.K. DUTTA[2]

Reactor Safety Division, Bhabha Atomic Research Centre, Trombay, Mumbai-400 085, India.
[1] email: pvdp@barc.gov.in [2] email: bkdutta@barc.gov.in

ABSTRACT

Effect of irradiation on material stress-strain behavior is analyzed using two-dimensional discrete dislocation dynamics modelling. The plastic flow is represented by collective motion of a large number of edge dislocations. To mimic these irradiation effects, we assume that all the dislocations are locked by the defect loops thus characterizing the fluence. When the total stress on the dislocations exceeds a critical value, they get unlocked and become free to move on their glide planes. The other phenomenons like pinning of dislocations by obstacles, nucleation of new dislocations are also incorporated. The stress-strain response of an irradiated Copper material as a function of critical locking stress is obtained. An attempt is made to find an expression for this critical locking stress in terms of irradiation fluence.

Keywords: Dislocation dynamics; Finite element; Irradiation.

1. INTRODUCTION

The microstructure of irradiated materials evolves over a wide range of length and time scales, making radiation damage inherently multi-scale phenomenon. The primary source of damage in the irradiated material is the displacement cascades generated by the primary knock-on atom recoiled under the collision with the energetic particle. Most of the displaced atoms regain some equilibrium position in the lattice but some of them fail to return on equilibrium position and thus form self-interstitial atom (SIA)/vacancy pairs. For irradiated materials, interaction of dislocations with these irradiation-induced defects entirely controls the plastic yield. The main notable features of irradiation-induced mechanical behavior are: an increased yield strength with irradiation dose, and an instability that results in plastic flow localization within dislocation channels leading to loss of ductility and premature failure. Because of present computer simulation techniques, it has become possible to model these irradiation phenomena and their corresponding effect on macroscopic deformation behavior.

Here, we used the two dimensional discrete dislocation dynamics (DD) modelling to study the irradiation effects on material stress-strain response. An in-house code for DD analysis is developed based on formulation given in [1]. The plastic flow is represented by collective motion of a large number of edge dislocations. The dislocation fields are specified by continuum elastic theory. Since the elastic fields act infinite in medium, corrections for boundaries are specified by a complimentary problem which consists of solving a linear elastic boundary value problem through finite element method. The dislocation phenomenon like annihilation, generation and pinning of dislocations by obstacles are incorporated in the model through some constitutive rules. Irradiation effects are numerically modeled by locking all the dislocations with irradiation induced defects thus characterizing the fluence. These dislocations get unlocked when the stress on them exceeds a critical stress due to irradiation defects. The stress-strain response of an irradiated Copper material as a function of total fluence is studied.

2. FORMULATION FOR 2D-DD ANALYSIS

The problem is formulated as follows: Consider a linear elastic body of volume V which contains a distribution of dislocations. The dislocations are treated as line defects in the elastic continuum. Each dislocation is characterized by its Burger's vector b_i and its slip plane. The body is subjected to time dependent traction and displacement boundary conditions $T = T_0(t)$ on S_f and $u = u_0(t)$ on S_u. The deformation process will lead to the motion of dislocations, mutual annihilation and generation of new dislocations and their pinning at point obstacles. The obstacles may be second phase particles, defects generated due to irradiation etc. The analysis of deformation process is performed in an incremental manner in time, where the incremental step at any instant t involves three main computational stages: (i) determining the current stress and strain state of for the current dislocation arrangement; (ii) determination of the so-called Peach-Koehler force, i.e., the driving force for changes in dislocation structure; and (iii) determination of the instantaneous rate of change of dislocation structure on the basis of a set of constitutive equations of motion, annihilation and generation of dislocations. All the three stages of computation are described below.

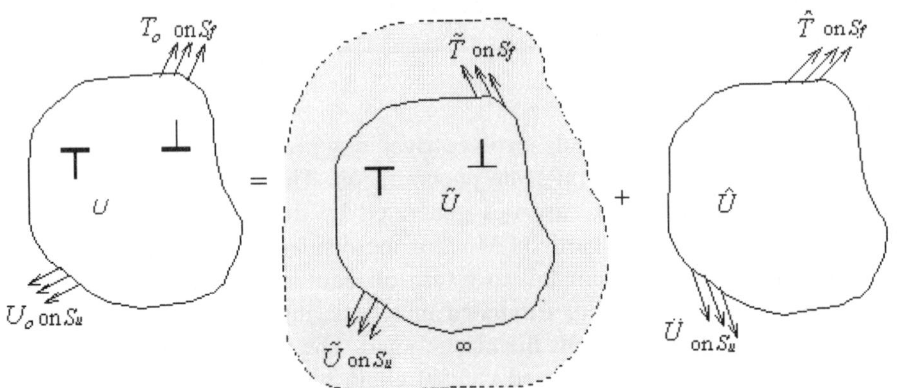

Fig. 1. Decomposition of problem into problem of dislocations in infinite solid (~fields) and complementary problem without dislocation fields (^ fields).

The current state of body in terms of the displacement, strain and stress fields is written as the superposition of two fields,

$$u = \tilde{u} + \hat{u} \quad \varepsilon = \tilde{\varepsilon} + \hat{\varepsilon} \quad \sigma = \tilde{\sigma} + \hat{\sigma} \ \ in \ V \qquad \qquad ...(1)$$

respectively, as illustrated in Fig. 1. The ($\sim$) fields are associated with the n dislocations in the current configuration but in infinitely large medium of material. These fields are obtained by superposition of the fields associated with each individual dislocation,

$$\tilde{u} = \sum_{i=1}^{n} u^i \quad \tilde{\varepsilon} = \sum_{i=1}^{n} \varepsilon^i \quad \tilde{\sigma} = \sum_{i=1}^{n} \sigma^i . \qquad \qquad ...(2)$$

The corresponding displacement ($\tilde{u}$) and stress ($\tilde{\sigma}$) fields of a dislocation i are given in [1]. As the solution for ($\sim$) fields is facilitated by virtue of absence of boundaries, the ($^$) fields are added to correct the actual boundary conditions on S. This leads to a linear elastic complementary boundary value problem, the governing equations of which are given by,

$$\left. \begin{aligned} \nabla . \hat{\sigma} &= 0 \\ \hat{\varepsilon} &= \nabla \hat{u} \end{aligned} \right\} in \ V$$

$$\left. \begin{aligned} \hat{T} &= T_o - \tilde{T} \ \ on \ S_f \\ \hat{u} &= u_o - \tilde{u} \ \ on \ S_u \end{aligned} \right\} b.c's.$$

Solution to this complementary boundary value problem is obtained using finite element method. The motion of the dislocations is governed by constitutive equations according to liner drag relation given by,

$$\tau^i b^i = B v^i \qquad \qquad ...(3)$$

where $\tau^i b^i = B v^i$, is the Peach-Koehler force, B is the drag coefficient and v is the velocity of a dislocation. The result of the above formulation is a set of non-linear first order differential equations governing the motion of the dislocations, which are solved using Euler forward time integration method. The motion of dislocations along a slip plane can be hindered in real crystals by obstacles such as dislocations on intersecting slip planes, small precipitates etc. We model this by means of point obstacles at which moving dislocations get pinned down. Such pinned dislocations will be released when the resolved shear strength on them exceeds the obstacle's strength (τ_{obs}). Two edge dislocations with opposite Burger's vector will annihilate each other when they are brought closer together within a critical annihilation distance L_e. New dislocations are being generated through the operation of Frank-Read sources. We assume that sources are point sources on the slip plane, which generate a dislocation dipole when the magnitude of the shear stress exceeds the critical stress (τ_{nuc}) during a period of time t_{nuc}. The distance L_{nuc} (see [1]) between the two dislocations is determined by the critical stress according to:

$$L_{nuc} = \frac{\mu}{2\pi(1-v)} \frac{b}{\tau_{nuc}} .$$

3. STRESS-STRAIN RESPONSE FOR IRRADIATED COPPER UNDER SIMPLE SHEAR USING DD

Here, we consider the problem of irradiation-induced hardening in Copper. Plastic deformation and hardening in irradiated materials is controlled primarily by the defects due to irradiation (vacancies, self-interstitial atoms and SFT's) and their interaction with dislocations. These defect clusters will tend to form loops around existing dislocations, leading to their decoration and immobilization. In order to understand the effect of this phenomenon on deformation behavior, we consider irradiated copper under simple shear using two-dimensional DD model described above. An attempt is made to find the relationship between the irradiation fluence and yield stress of Copper using DD modelling. In spite of the fact that this phenomenon is fully three dimensional in nature, first order prediction can be made of yield stress as a function of fluence level using the two-dimensional model.

The irradiation induced hardening may be understood in terms of cascade induced source hardening in which the dislocations are considered to be locked by the loops decorating them [2,3]. The density and strength of these defect loops depend on the fluence level. To mimic these irradiation effects, we assume that all the dislocations are locked by the defect loops thus characterizing the fluence. When the total stress on the dislocations exceeds a critical value σ_{cr}, they get unlocked and become free to move on their glide planes. The motion of the dislocations is governed by Eq.(3). During gliding, any dislocation may get pinned down by the point obstacles or may get annihilated by an opposite dislocation. New dislocations will be generated by Frank-Read mechanism, which is mimicked here by point sources. These sources will nucleate a dislocation dipole when stress on a source exceeds a critical value. The effect of irradiation on stress-strain behavior is then obtained by subjecting the copper unit cell to simple shear.

The simulated copper cell is assumed to be of dimensions $2\mu m \times 2\mu m$. The shear modulus $\mu = 55$ GPa and the Poisson's ratio is $\nu = 0.3$. The drag coefficient B in Eq. (3) is taken as 10^{-4} Pa.s. The plastic flow is represented by a collection of large number of edge dislocations. We assume that the material consists of randomly distributed defect structure, such as point obstacles, Frank-read dislocation nucleation sources. The material is assumed to have an initial dislocation density of $\rho_{disinit}= 200/hw$, which is then relaxed. During relaxation, the dislocations will interact with each other and try to attain equilibrium position. In the relaxed configuration, obstacles and nucleation sources are randomly generated. All the obstacles are assumed to have same strength $\tau_{obs}=5.7 \times 10^{-3}\mu$. The strength of sources is selected randomly from a Gaussian distribution with a mean strength of $\bar{\tau}_{nuc}=1.9 \times 10^{-3}\ \mu$ corresponding to a mean nucleation distance of $L_{nuc}= 125b$ and the nucleation time is taken as $t_{nuc}= 2.6 \times 10^6\ B/\mu$. The strength distribution is assumed to have a standard deviation of $0.2\ \bar{\tau}_{nuc}$. The critical annihilation length is taken as $L_e= 6b$. The unit cell is subjected to simple shear along top and bottom edges in time incremental manner, with a strain rate of $\dot{\Gamma}$,

$$\left.\begin{array}{rcl} u_1(t) & = & \pm h\dot{\Gamma}t \\ u_2(t) & = & 0 \end{array}\right\} \quad \text{along } x_2 = \pm h.$$

Typical predicted stress-strain curves for various critical values of σ_{cr} are shown in Fig. 2 which reveals the effect of dislocation loops on increased yield stress. In order to eliminate any numerical fluctuations, averaging over 5-6 samples is done to obtain the observed stress-strain behavior. Any fluctuation thus seen in Fig. 2 is because of the physics of dislocation motion rather than numerical computation.

When σ_{cr} is increased from 14 MPa to 100 MPa, the corresponding yield stress σ_y increases from 6 MPa to 40 MPa. Without irradiation, the single Copper crystal yields at 2-4 MPa [4]. When irradiation induced-defects are present, the yield point rises drastically which can be attributed to the locking of dislocations by these defects. Moreover at very high values of fluence, the experimentally observed instability can also be reproduced by DD modelling as shown in Fig. 2 for high values of σ_{cr}. At larger values of fluence, two characteristics are seen. First, the system yields at very high stress and second, a sudden instability occurs after reaching a maximum stress value. This can be attributed to a sudden unlocking of a large chunk of dislocations from the irradiation-induced defects. This is clearly

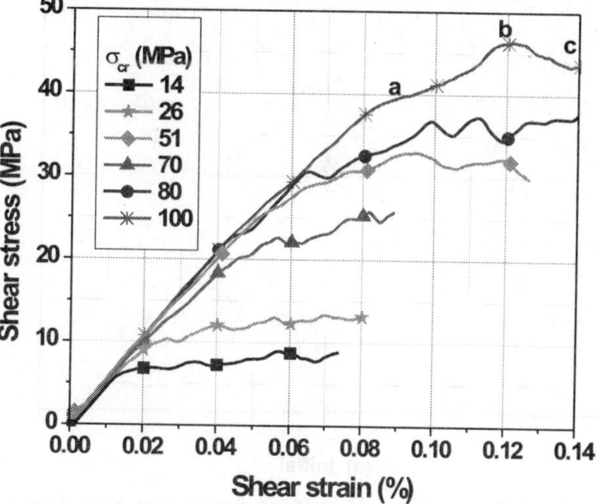

Fig. 2. Stress-strain response of irradiated Copper under simple shear as a function of critical locking stress σ_{cr}.

seen by points a,b and c in Fig. 2 corresponding to σ_{cr} = 100 MPa. The system yields at point a; there is strain hardening up to point *b* and then *a* sudden drop in stress value occurs (point *c*). This effect can be clearly seen again at point *a*, *b* and *c* in Fig. 3, where the variation of number of dislocations being locked by irradiation-induced defects is shown as a function of shear strain. As seen from point *b* to *c*, a large number of dislocations are unlocked from irradiation induced defects and hence the instability in stress-strain response occurs. These results show that such instabilities can be reproduced by numerical DD modelling. Figure 4 shows the dislocation configurations corresponding to σ_{cr} = 80 MPa at: (a) zero strain and (b) after shearing up to 0.14%.

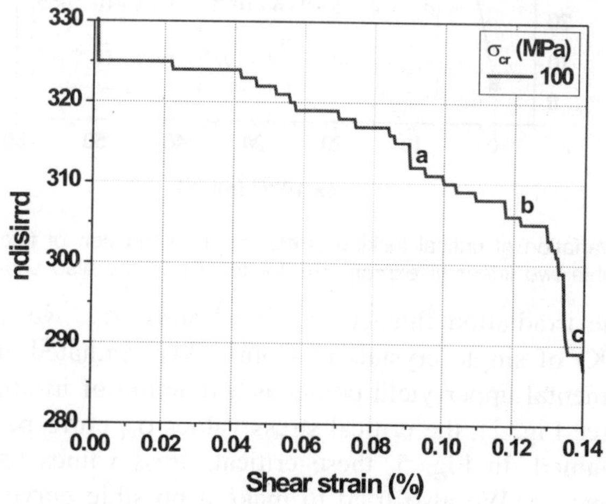

Fig. 3. Variation of no. of dislocation locked by irradiation defects (ndisirrd) Vs. shear strain for σ_{cr} = 100 MPa.

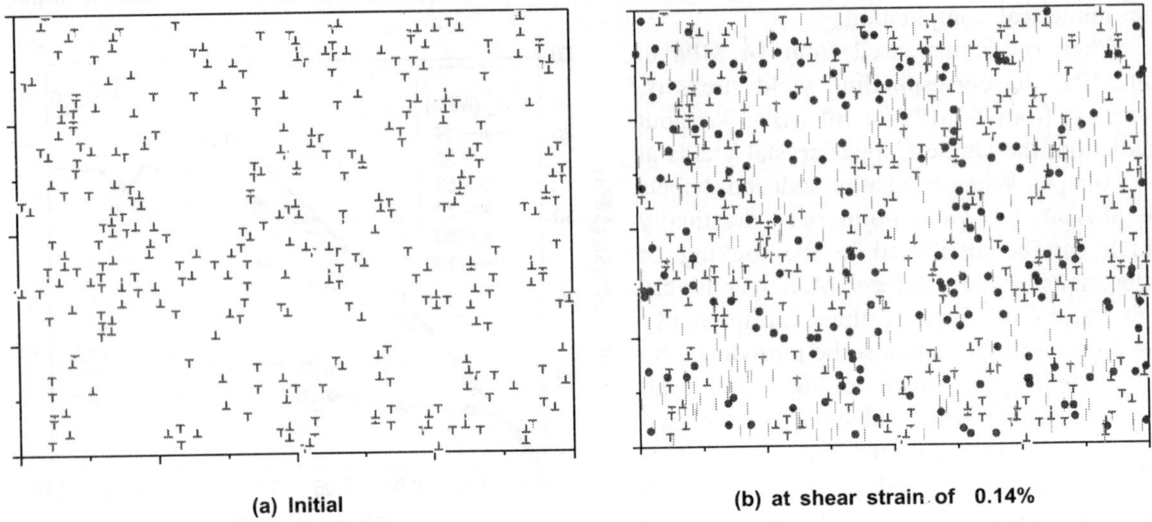

(a) Initial   (b) at shear strain of 0.14%

Fig. 4. Dislocation configurations before and after deformation in the cell for σ_{cr} = 80 MPa;. $\perp$ indicates a +ve dislocation, **T** indicates a –ve dislocation. Frank-Read sources ($\bullet$) and obstacles (|) are also shown.

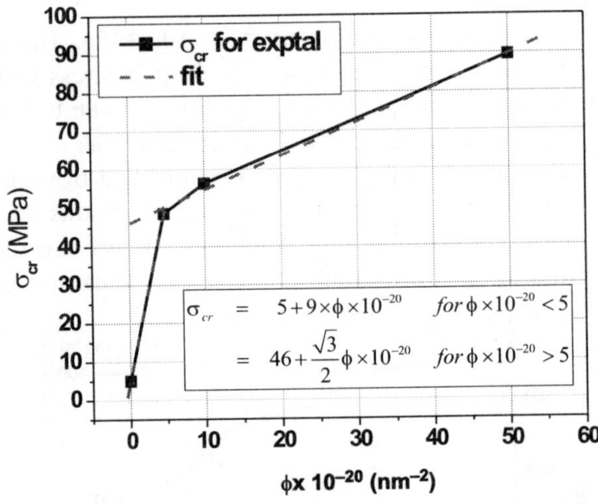

$$\sigma_{cr} = 5+9\times\phi\times10^{-20} \quad for\,\phi\times10^{-20}<5$$
$$= 46+\frac{\sqrt{3}}{2}\phi\times10^{-20} \quad for\,\phi\times10^{-20}>5$$

Fig. 5. Variation of critical locking stress σ_{cr} as a function of fluence ϕ.
Possible two linear fit expressions for the same are also shown.

Finally, to correlate the irradiation fluence to critical stress σ_{cr}, we made use of experimental yield stress values (at 77°K) of single crystals of Copper [4] irradiated at different fluence levels. Table 1 gives these experimental upper yield points as a function of irradiation fluence ϕ. From the results of our DD model (*see* Fig. 2), the critical stress values σ_{cr} corresponding to the experimental yield stress values are obtained. In Fig. 5, these critical stress values (σ_{cr} -exptal) are plotted as function of irradiation fluence ϕ. We also tried to make a possible curve fit of these experimental critical stress values in terms of ϕ as given below:

$$\sigma_{cr} = 5 + 9 \times \phi \times 10^{-20}; \quad for\ \phi \times 10^{-20} < 5$$

$$= 46 + \frac{\sqrt{3}}{2} \phi \times 10^{-20}; \quad for\ \phi \times 10^{-20} > 5$$

Such expressions can then be used to find out critical locking stress σ_{cr} at any other fluence ϕ. Now, to find out the yield stress σ_y at any fluence ϕ, first using these expressions, the corresponding σ_{cr} is determined. Then the DD model is run with this σ_{cr} to find out the yield stress of the crystal at that fluence level. Even though current model is two-dimensional and considers only edge dislocations, the results illustrate the use of DD simulations for first order prediction of σ_y as a function of fluence.

Table 1. Experimental yield stress values of irradiated Copper single crystal at 77°K [4]

Irradiation fluence ϕ (n m^{-2})	Upper yield stress (MPa)
0	3
4.7 x 10^{20}	18.75
1.0 x 10^{21}	21.7
5.0 x 10^{21}	35.38

CONCLUSION

Two dimensional discrete dislocation modelling is used to determine irradiation induced hardening of Copper single crystal. Dislocations get pinned down due to irradiation induced defects. This is modeled by locking all the dislocations by critical stress σ_{cr}, thus characterizing the irradiation fluence phenomenologically. The dislocations are required to overcome this σ_{cr} to get unlocked before they can move on their glide planes under external stress. Parametric study with respect to the critical locking stress and its effect on stress-strain response is studied. An expression for this critical locking stress in terms of irradiation fluence is proposed. The yield stress of irradiated copper crystal can then be determined by the DD model by taking the critical locking stress corresponding to that fluence level.

REFERENCES

1. Giessen, E.V., Needleman, A., 1995, *Modelling and Simul. Mater. Sci.Eng.* 3, pp. 689-735.
2. Trinkaus, H., Singh, B.N., Foreman, A.J.E., 1997, *J. Nucl. Mater.* 251, pp. 172.
3. Zbib, H.M., Rubia, T.D., Rhee, M., Hirth, J.P., 2000, *J. Nucl. Mater.* 276, pp.154-165.
4. González, H.C., Miralles, M.T., 2001, *J. Nucl. Mater.* 295, pp.157-166.

37

Energy Based Equivalence Between Damage And Fracture In Concrete Under Fatigue

T. SAIN[1] AND J.M. CHANDRA KISHEN[2]

[1]Department of Civil Engineering, Indian Institute of Science, Bangalore-560 012, India
email: trin@civil.iisc.ernet.in
[2]Department of Civil Engineering, Indian Institute of Science, Bangalore-560 012, India
email: chandrak@civil.iisc.ernet.in.

ABSTRACT

In this study, a method is proposed to correlate fracture and damage mechanics through energy equivalence method and to predict the damage scenario in concrete under fatigue loading. The objective is to study whether an already available fracture analysis can give an equivalent distribution of damage. The analytical method developed here has been exemplified with some already available data in literature. It is concluded, that through energy approach the discrete crack can be modeled as an equivalent damage zone, wherein both the cases correspond to the same energy loss, and it is also shown that by knowing the critical damage zone dimensions, the critical fracture properties could be obtained.

Keywords: Fatigue, fracture, damage index, energy equivalence.

1. INTRODUCTION

There are two main categories of models that describe the failure process in concrete member: one that uses fracture mechanics concepts and the other which uses continuum damage mechanics concepts. Fracture mechanics is well suited to describe the separation due to decohesion of two parts of the continuum in the form of a discrete crack [1]. On the other hand, damage mechanics, which includes smeared (or distributed) crack models, describes the local effects of micro-cracking, that is the evolution of the mechanical properties such as elastic stiffness degradation, inelastic strains etc. of the continuum as micro-cracking develops.

In the present work, an approach to correlate the state of diffused microcracking with an equivalent discrete crack is developed through energy based equivalence concept. The equivalent damage zone dimension is obtained as a function of increasing crack length upto failure. A finite element analysis is performed to assess the strength and stiffness degradation through modeling of the damage zone with reduced elastic moduli.

2. LEFM BASED FATIGUE LAW FOR CONCRETE

Structures such as airport/highway pavements and bridge decks are subjected to repetitive loads of high stress amplitude due to moving vehicles. The stress state in such structures is often simulated with three-point bending tests. Therefore, flexural fatigue is a common phenomenon in case of concrete members. Plain concrete subjected to flexural loading fails due to crack propagation. Repeated loading results in a steady decrease in the stiffness of the structure, eventually leading to failure. It is of interest to characterize the material behavior subjected to such loading and study the crack propagation resulting from such loading.

Based on linear elastic fracture mechanics principle, Slowik *et al.* [2] have suggested for modifications of the well known Paris law [3], to describe fatigue cracking in concrete members. The proposed law includes all the important parameters such as fracture toughness, loading history, specimen size etc. except the frequency of externally applied load. Mathematically the law is expressed as,

$$\frac{da}{dN} = C\frac{K_{\mathrm{I\,max}}^{m}\Delta K_{I}^{n}}{\left(K_{Ic} - K_{I\,\mathrm{sup}}\right)^{p}} + F\left(a, \Delta\sigma\right) \qquad ...(1)$$

where K_{Isup} is the maximum stress intensity factor ever reached by the structure in its past loading history, K_{IC} the fracture toughness, K_{Imax} is the maximum stress intensity factor in a cycle, N is the number of load cycles, a is the crack length, ΔK is the stress intensity factor range, and m, n, p, are constants. In Equation (1), C represents the crack propagation rate per fatigue load cycle. Slowik *et al.* [2] proposed a linear relationship between C and the ratio of ligament length L to characteristic length L_{ch}. In an earlier work, the authors [4] have proposed a modified empirical formula to estimate the constant C, which includes the effect of loading frequency along with the above mentioned parameters. This is done through a regression analysis, using experimental results of Slowik *et al.* [2] and Bazant and Xu [5]. The resulting expression represents a quadratic polynomial given by,

$$Cf = -0.0193\left(\frac{L}{l_{ch}}\right)^{2} + 0.0809\left(\frac{L}{l_{ch}}\right) + 0.0209 \; mm/\sec \qquad ...(2)$$

where f is the external loading frequency. From this equation one can estimate the value of parameter C for any load frequency, grade of concrete and size.

The proposed fatigue law given by Equation (1) is validated using Bazant and Xu's experimental results for small, medium and large specimens. In Fig. 1, the proposed model is compared with the experimental results of Bazant and Xu [5] on three point bend specimens. It is seen in this figure that initially, the crack growth rate is moderate, and as the stress intensity factor approaches the fracture toughness, crack growth becomes faster finally leading to failure, which is represented by the asymptotic nature of the crack propagation curve. For all the three specimens, a good agreement is obtained between the proposed model and experimental results.

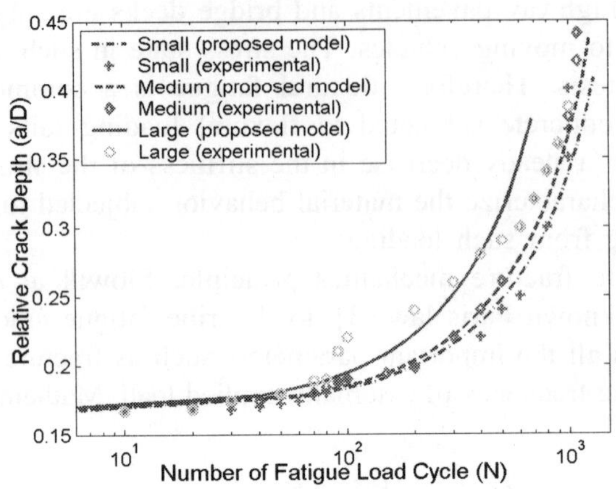

Fig. 1. Fatigue Crack propagation curve based on LEFM.

3. ENERGY BASED EQUIVALENCE BETWEEN DAMAGE AND FRACTURE

According to fracture mechanics theory, energy is required for an existing crack to propagate by an amount δa. This energy is commonly expressed as strain energy release rate per unit crack extension and denoted by 'G'. Similarly, in case of damage based analysis, the strain energy loss per unit volume of the material due to increase in damage by an amount dD is referred to as damage strain energy release rate. The idea of energetic equivalence is based on equating the energy loss due to damage, with the energy required for equivalent crack propagation within the member. In a more explanatory sense, energetic equivalence correlates two structures having the same geometry and loading condition, but different damage definitions. In a global sense they behave in the same manner, when the energy dissipation corresponding to two different damage conditions become equal for the two structures [6].

The energy release rate δU per unit crack extension δa (which in turn is equal to the potential energy lost ($\delta\Pi$) by the applied load) is related to the stress intensity factor K_I by [7],

$$\frac{1}{B}\frac{\delta U}{\delta a} = -\frac{1}{B}\frac{\delta \pi}{\delta a} = \frac{K_I^2}{E} \qquad \qquad ...(3)$$

Considering a three point bend specimen with a crack at the bottom of midspan, the stress intensity factor is given by,

$$K_I = \frac{6Y(\alpha)M_{max}\sqrt{a}}{BD^2}$$

where M_{max} is the maximum bending moment at mid-span region ($= P_{max}L/4$); 'a' is the crack length; E is te modulus of elasticity and $Y(\alpha)$ is the geometry factor; 'α' being the relative crack depth ($= a/D$).

Substituting K_I into Equation (3), and integrating over 'a' from the instant when there is no crack in the beam to the current crack length 'a', we get,

$$U(\alpha) = \frac{9}{4} \frac{P^2 L^2}{BD^2 E} F(\alpha)$$

where,

$$F(\alpha) = \int_0^{\alpha} \alpha Y^2(\alpha) d\alpha \qquad \text{...(4)}$$

Using Equation (4), the strain energy as a function of relative crack depth 'α' can be evaluated.

In order to compute the energy dissipation due to increase in the degree of damage by an amount dD, the energy dissipated in the elemental volume δV is required and this quantity can be obtained using,

$$U_D = \int_D (-Y)\, dD dV \qquad \text{...(5)}$$

where, D is the scalar damage parameter and Y is the damage strain energy release rate. For the three dimensional stress state, Y can be expressed as,

$$Y = \frac{\sigma_{eq}^2}{2E(1-D)^2} \left[\frac{2}{3}(1+v) + 3(1-2v)\left(\frac{\sigma_H}{\sigma_{eq}}\right)^2 \right] \qquad \text{...(6)}$$

where, σ_{eq} is Von Mises equivalent stress and σ_H is the hydrostatic stress at a point. As it has been mentioned earlier, that in the present analysis, the progressive fracture phenomenon has been replaced by an equivalent continuously growing damage process. As per the well known convention of the damage mechanics, $D = 0$ corresponds to no damage (zero crack opening) and $D = 1$ corresponds to full crack opening results in a zero stress transfer along the crack surface. The discrete crack is replaced with an equivalent damage zone of size $l_c \times L_D \times B$, where l_c represents the localization limiter.

Hence, the energy dissipated due to progressive damage in the zone $l_c \times L_D \times B$, results in the gradual change of damage variable D from $0 \rightarrow 1$ and is given by,

$$\Delta U_D = Bl_c \int_0^{L_D} \int_0^{\left(\frac{-y}{L_D}+1\right)} (-Y)\, dD dy \qquad \text{...(7)}$$

Equating the two energy terms in Equations (4) and (7), we can solve for the unknown length of damage zone, L_D through a trial and error procedure, after defining l_c, the internal length of the continuum as mentioned earlier.

4. CASE STUDIES

To implement the aforementioned theory of energetic equivalence, three point bend specimens under constant amplitude fatigue loading similar to the ones used by Bazant and Xu [5] in their experimental studies, have been considered. In these case studies, the geometry of the damage zone is predicted as a function of increasing crack length for all the three sizes of specimens considered. Further, the stiffness and strength reduction factors as functions of crack and damage zone length are obtained. In addition, it is also shown that the fracture parameter such as the fracture energy can be estimated by knowing the critical damage zone size.

In this case study, the length (L_D) and width (l_c) of an equivalent damage zone is determined for the three point bend specimen. The width (l_c) of the damage zone is related to the internal length parameter and depends on the characteristics of the micro structure of the material. Following Bazant's approach [8], in the present study, the dimension of l_c has been fixed by relating it to the maximum size of the aggregates used. A parametric study is done by considering three different widths of the damage zone: $l_c = 3d_a$; $4d_a$; $5d_a$. The maximum size of the aggregate used in this study is 15 mm, same as the one used by Bazant and Xu [5] in their experimental study. Finally, considering $l_c = 4d_a$ the length of the damage zone L_D is computed for small, medium and large specimens as a function of increasing crack length, since it gives closest match. Figure 2 shows the equivalent damage zone length as a function of the relative crack depth (a/D) for all the three specimens. Thus, by knowing the discrete crack length, an equivalent damage zone size could be obtained by using the energy equivalence concept.

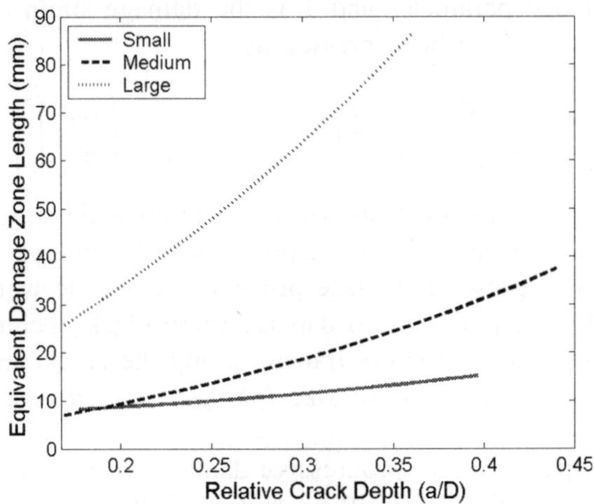

Fig. 2. Equivalent damage zone length as a function of relative crack depth.

4.1 Stiffness and Strength Reduction Factors

It is a well known fact that concrete members subjected to cyclic loading undergo both stiffness and strength degradation that accompanies the damage process. In this study, the strength and stiffness reduction in concrete beams are determined using both, fracture mechanics and damage mechanics approach. The stiffness reduction factor, is defined by Lybas and Sozen [9] as,

$$D_R = \frac{K_0}{K_r} \qquad \qquad ...(8)$$

where K_0 is the initial stiffness corresponding to no damage and K_r is the reduced secant stiffness due to damage.

Young *et al.* [10] have defined a strength reduction index as,

$$S_D = \left(\frac{\varphi - \varphi_y}{\varphi_f - \varphi_y} \right)^w \qquad \qquad ...(9)$$

where φ_y is the curvature at the peak load (monotonic) or yield level corresponding to undamaged situation; φ_f is the final curvature at failure and φ is the curvature at peak load corresponding to a particular damage level. The authors, through calibration studies have suggested a value of 1.5 for the parameter w.

The same three-point bend small specimen that was used in the previous case study is modeled using the finite element program FRANC [11]. In the fracture mechanics based analysis, a rosette of singular crack tip quarter point elements are used to model the crack tip. The damage zone, in the damage mechanics based analysis is modeled using reduced value of modulus of elasticity. According to the theoretical definition of damage, the modulus of elasticity of the elements in the damage zone should approach zero. In this finite element study, two trial values, (1/100 of undamaged E) and (1/1000 of undamaged E) are considered in order to avoid the numerical difficulties arising from using a zero value. Considering the width of the damage zone as 60 mm, maximum displacement (δ_r) under monotonic peak load is computed for different discrete crack lengths and damage zone lengths. The reduced secant stiffness K_r can be obtained using $K_r = P/\delta_r$. Finally, the stiffness drop index is computed using Equation (8) and it is shown in Fig. 3 (a). As expected, lower the value of modulus of elasticity of the elements in the damaged zone, closer is the match between finite element and damage mechanics theory.

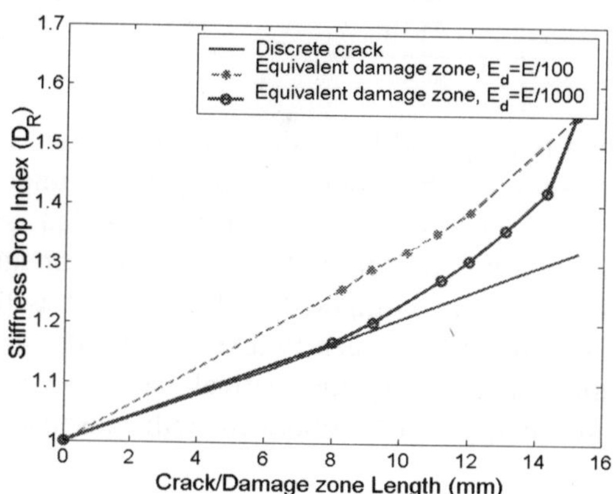

Fig. 3(a). Stiffness degradation factor.

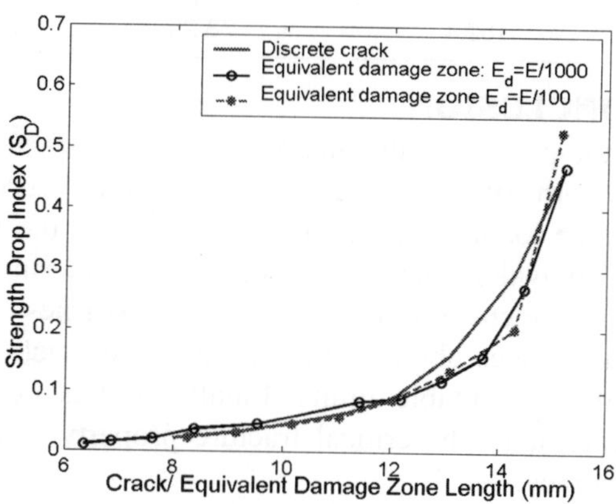

Fig. 3(b). Strength drop index as a function of crack depth/damage zone length.

Similarly the curvature (φ) values corresponding to different damage zone dimensions as well as for different discrete crack lengths are obtained from the computed strains. The strength drop index has been computed using Equation (9), and plotted in Fig. 3(b). It is seen that there is not much difference between the strength reduction factors computed for the two values of reduced modulus considered in the study. Further, a close match is obtained for the strength reduction factor between the damage theory and fracture theory. Hence, from this analysis it can be concluded, that knowing the equivalent damage zone dimension or the discrete crack length, one can easily find out the reduced secant stiffness and the residual strength of the member, without much difference in the structural response.

4.2 Determination of Fracture Parameter

Based on the energy equivalence principle, the critical fracture parameter such as the fracture energy can be obtained by knowing the critical damage parameter from experiments. In the previous section, the procedure to obtain the geometry of equivalent damage zone has been explained. If the critical damage zone dimension $L_{Dc} \times L_c \times B$ corresponding to failure of the specimen is known, then the critical damage energy release (U_{Dc}) can be computed from Equation (7). The critical fracture energy can be written as $G_c \delta a$; where G_c is the critical fracture energy release rate. Hence, based on energy equivalence, we can write,

$$G_c \delta a = U_{DC}$$

The critical fracture energy is computed for the three-point bending beams used earlier for all the three sizes using this procedure. Table 1 shows the computed values of G_C together with those reported by Bazant and Xu [5] from their experimental studies. It is seen that there is a very good match between the computed values and the experimental ones.

Table 1. Determination of G_{Ic}

Specimen Depth	Critical L_D	Predicted G_{Ic}	Reported G_{Ic}	% Error
38.1	15.02	98.01	101.8	3.7
76.2	37.425	89.5	84.01	6.5
152.4	78.25	71.5	73.19	2.3

CONCLUSION

In the present study, an energy based equivalence approach is proposed to model a discrete crack in the form of a distributed damage zone. Through case studies on three-point bending beam under fatigue loading, it has been shown, that the progressive cracking phenomenon can be modeled, as an equivalent damage zone, without altering its global structural response. Knowing the damage zone dimensions, the strength and stiffness drop index have been computed and compared with those values obtained through discrete crack analysis. The variation between two series of values is within acceptable limits. Finally, it has been shown that by knowing the critical damage zone dimensions, the critical fracture property, such as fracture energy can be obtained with reasonable accuracy.

REFERENCES

1. M. Kaplan, 1961, Crack propagation and the fracture of concrete, ACI Journal, 58, 591-610.
2. V. Slowik, G. Plizzari, V. Saouma, 1996, Fracture of concrete under variable amplitude loading, ACI Materials Journal, 93, 3, 272-283.
3. P. Paris, F. Erdogan, 1963, A critical analysis of crack propagation laws, J. of Basic Engineering, ASME 85 (3).
4. T. Sain, J. Chandra Kishen, 2003. Damage and residual life assessment using fracture mechanics and inverse method, in: Proc. 16th Engineering Mechanics Conference, EM2003, ASCE.
5. Z.P. Bazant, Xu. Kangming, 1991, Size effect in fatigue fracture of concrete, ACI Materials Journal, 88, 4, 427-437.
6. J. Mazars, G. Cabot, 1996, From damage to fracture mechanics and conversely: a combined approach, Int. Jl. of Solids and Struc., 33, 3327-3342.
7. B. Karihaloo, 1995, Fracture Mechanics and Structural Concrete, Longman Scientific and Technical, London.
8. Z.P. Bazant, B. Oh, 1983, Crack band theory for fracture of concrete, Materials and Structures 16, 155-177.
9. J. Lybas, M. Sozen, Effect of beam strength and stiffness on dynamic behavior of reinforced concrete coupled walls, Civil Engg. Studies, Structural Research Series (44), University of Illinois, Urbana.
10. S. Young, C. Meyer, M. Shinozuka, 1989, Modelling of concrete damage, ACI Structural Journal, 86, 3, 259-271.
11. Cornell Fracture Group, FRANC-2D, 1997, A Two Dimensional Fracture Mechanics Analysis Code, Cornell University.

38

Numerical Evaluation of Bi-material Fracture Parameters

J.M. Chandra Kishen[1] and Bharat Singh Joiya

Department of Civil Engineering, Indian Institute of Science, Bangalore-560 012, India
[1]email: chandrak@civil.iisc.ernet.in.

ABSTRACT

A crack at the interface between two different materials is often subjected to mixed-mode type of loading even when the geometry is symmetric and the far field applied loading is of pure mode I. This is due to the elastic mismatch between the two materials on either side of the interface which is characterized by a mismatch parameter. Thus, an interface crack has a tendency to deviate away from the interface and kink into any of the adjoining materials. The angle of kink depends on the relative proportions of mode I and mode II stress intensity factors defined by the mode mixity parameter. In this work, an analytical closed form expression between the crack kinking angle and the mode mixity is developed through finite element simulations on an integrated mixed-mode bi-material specimen.

Keywords: Stress intensity factors, bimaterial interface, mode mixity, crack kinking angle.

1. INTRODUCTION

The application of interfacial fracture mechanics has increased due to the increased use of adhesive joints, composite laminates, multi-layered electronic devices, wear-resistant coatings etc. In the civil engineering sector, fracture mechanics of interfaces could be used to study stability of dam-foundation system, behavior of patch repaired systems etc. These diversified application areas has catapulted an upsurge among the researchers to study interfacial fracture mechanics.

Fracture at a bi-material interface is essentially mixed-mode, even when the geometry is symmetric with respect to a crack and loading is pure mode I. This is due to the differences in the elastic properties across an interface which would disrupt the symmetry [1]. Consequently, both tensile and shear stresses act on the interface ahead of the crack and opening and sliding displacements of the crack flanks occur behind the crack tip. The linear elastic solutions of the crack tip stress and displacement fields show that the stresses ahead of the crack front and displacements behind the crack front behave in an oscillatory manner. Due to this oscillatory behavior, the definition of the stress intensity factors needs special consideration, and in addition crack face contact may occur at some short distance behind the crack tip. The mode I and mode II stress intensity factors cannot be decoupled to represent tension and shear stress fields as seen in the case of homogeneous materials.

Cracks at the interface of dissimilar bi-materials have a tendency to depart from the interface and kink into the adjoining material depending on the stress field at the tip of the crack. Hence, it becomes important to determine the angle of crack kinking. He and Hutchinson [2] provided the stress intensity factors and energy release rate (G) of the kinked crack in terms of the corresponding quantities for the interface crack prior to kinking. They suggested that the interface crack should kink out of the interface in the direction at which the energy release of the branched crack is maximum. Mukai *et al.* [3] studied the problem of crack branching off the interface between two bonded dissimilar isotropic materials. Results were presented in terms of the ratio between energy release rate of a branched interface crack to the energy release of a straight interface crack. They deduced that the ratio reaches a maximum when the interface crack branches into the softer material. Yuuki and Xu [4] proposed fracture criteria of an interface crack between two dissimilar isotropic elastic solids based on the maximum stress which can be characterized by stress intensity factors for an interface crack before kinking. Geubelle and Knauss [5] emphasized that a unique kink angle cannot be obtained in the bimaterial case if one tends to use the maximum circumferential stress criteria, unless an additional characteristic length is introduced and suggested that it should be some material parameter that constitutes a kind of "retrofit" to the linearized theory. Chandra Kishen and Singh [6] proposed fracture criteria of kinking and branching of interface cracks and also proposed closed form expressions to compute the kinking angle.

In this work, the fracture behavior of bi-material interface subjected to pure mode I, pure mode II and mixed-mode conditions is studied. Finite element analysis is performed on an integrated mixed-mode bi-material specimen to determine the energy release rates using the virtual crack closing technique for different loading conditions. The nominal mode mixity and crack kinking angles for different combinations of elastic constants of materials on either side of the interface are determined. A closed form relationship between crack kinking angle and nominal mode mixity is developed.

2. ELASTIC STRESS AND DISPLACEMENT FIELD SOLUTIONS FOR BI-MATERIAL INTERFACE CRACKS

Linear elastic fracture mechanics (LEFM) assumes that the material comprising the body of structure to be perfectly linear with no yield stress, allowing an infinite stress to be reached at the crack tip. Theoretical investigations of the interface crack problem date since the late fifties. It was the geology-prompted discovery by Williams [7] of the general differences between the stress field at the tip of a crack in a homogeneous, linearly elastic solid and that at an interface between two elastic solids. He performed an asymptotic analysis of the elastic fields at the tip of an open interface crack and found that the stresses and displacements behaved in an oscillatory manner as [1]

$$\sigma_{22} + i\sigma_{12} = \frac{(K_1 + iK_2)r^{i\varepsilon}}{\sqrt{2\pi r}} \qquad \text{...(1)}$$

$$r^{i\varepsilon} = \cos(\varepsilon \ln r) + i\sin(\varepsilon \ln r) \qquad \text{...(2)}$$

$$\varepsilon = \frac{1}{2\pi} \ln\left[\frac{(3 - 4\nu_1)/\mu_1 + 1/\mu_2}{(3 - 4\nu_2)/\mu_2 + 1/\mu_1}\right] \qquad \text{...(3)}$$

in which μ, v and a are the shear modulus, Poisson's ratio and crack length, respectively, and subscripts 1 and 2 refer to the materials above and below the interface, respectively. ε is known as the ocillatory index and vanishes for identical materials. K_1 and K_2 are components of the complex stress intensity factor K. The crack flank displacements, for plane strain are given by [1]

$$\delta_2 + i\delta_1 = \left(\frac{1}{\overline{E}_1} + \frac{1}{\overline{E}_2} \right) \frac{Kr^{i\varepsilon}}{2(1 + 2i\varepsilon)\cosh(\pi\varepsilon)} \sqrt{\frac{2r}{\pi}} \qquad ...(4)$$

where δ_2 and δ_1 are the opening and sliding displacements of two initially coincident points on the crack surfaces behind the crack tip. It may be noted from equation 4 that crack face interpenetration is implied for non zero ε. The energy release rate (ERR) for extension of the crack along the interface for plane strain is given by [1]

$$G = \frac{\left(1/\overline{E}_1 + 1/\overline{E}_2\right)\left(K_1^2 + K_2^2\right)}{2\cosh^2(\pi\varepsilon)} \qquad ...(5)$$

From Equation 1, it is seen that the usual definition of stress intensity factors will not work and it produces logarithmically infinite factors. Furthermore, any attempt to define the stress intensity factors without reference to a characteristic length will produce dimensionally wrong factors. However, the strain energy release rate G will still be dimensionally valid.

3. COMPUTATION OF ERR BASED ON VIRTUAL CRACK CLOSURE TECHNIQUE

The virtual crack closure technique (VCCT) has been widely applied to cracks for computation of the energy release rate. It is a technique that is relatively insensitive to the mesh pattern and density in a finite element analysis. VCCT was introduced by Rybicki and Kanninen [8] for line cracks and extended by Shivakumar *et al.* [9] for planar cracks. An overview of the VCCT for interfacial cracks with a fixed crack front has been provided by Krueger [10].

The formulations developed for VCCT of a slant crack cannot be applied to a kinking crack since the former relies on self similar crack growth while in the latter, the crack path changes direction. However, by adopting a different formulation in conjunction with the VCCT, the ERR can be computed to evaluate the trajectory of kinking cracks. In this work, the method proposed by Xie *et al.* [11] is followed, wherein the strain energy release rate (SERR) is partitioned thus providing an efficient means to compute values of the mode I and II ERR at the tip of kinking cracks. The solution procedure is computationally efficient and involves only the nodal forces and displacements near the crack tip.

Figure 1 (left) shows the the coordinate system for the parent interfacial crack and its child kinking crack. The parent crack has a length a and the child crack kinks into material 1 at an angle of ω with length Δa. The stress intensity factors K_I and K_{II} are associated with the parent crack while K_1 and K_2 correspond to the child crack. The energy release rate for a kinking crack may be written as

$$G(\omega) = G_I(\omega) + G_{II}(\omega) = \frac{F_{\theta 12}^{(1)}\Delta u_{\theta 12}^{(2)} + F_{\theta 34}^{(1)}\Delta u_{\theta 34}^{(2)}}{2\Delta aB} + \frac{F_{r12}^{(1)}\Delta u_{r12}^{(2)} + F_{r34}^{(1)}\Delta u_{r34}^{(2)}}{2\Delta aB} \qquad ...(6)$$

where $G_I(\omega)$ and $G_{II}(\omega)$ are the mode I and mode II ERR for kinking crack respectively. Following the work done by Xie *et al.* [11], the above ERR's may be obtained from the displacements and forces obtained at nodes close to the crack tip using,

$$G_I(\omega)=\frac{A}{a\Delta aB}\left[F_{\theta 12}^{(1)}\left(C_{11}\Delta v_{56}^{(1)}+C_{12}\Delta u_{56}^{(1)}\right)+F_{\theta 34}^{(1)}\left(C_{11}\Delta v_{78}^{(1)}+C_{12}\Delta u_{78}^{(1)}\right)\right] \qquad ...(7)$$

$$G_{II}(\omega)=\frac{A}{a\Delta aB}\left[F_{r12}^{(1)}\left(C_{21}\Delta v_{56}^{(1)}+C_{22}\Delta u_{56}^{(1)}\right)+F_{r34}^{(1)}\left(C_{21}\Delta v_{78}^{(1)}+C_{22}\Delta u_{78}^{(1)}\right)\right] \qquad ...(8)$$

where the coefficients C_{ii} are defined by

$$C_{11}(\omega)=1/4\left[3\cos(\omega/2)+\cos(3\omega/2)\right];C_{12}(\omega)=-1/4\left[3\sin(\omega/2)+\sin(3\omega/2)\right] \qquad ...(9)$$

$$C_{21}(\omega)=1/4\left[3\sin(\omega/2)+\sin(3\omega/2)\right];C_{22}(\omega)=1/4\left[3\cos(\omega/2)+3\cos(3\omega/2)\right] \qquad ...(10)$$

In the above equations u_{ij} and v_{ij} are the opening and sliding displacement components at nodal points i and j respectively, r and θ are the polar coordinates.

The relative proportions of mode I to mode II ERR are defined by the nominal mode mixity parameter defined as

$$\psi_0=\tan^{-1}\left(G_{II}(0)/G_I(0)\right) \qquad ...(11)$$

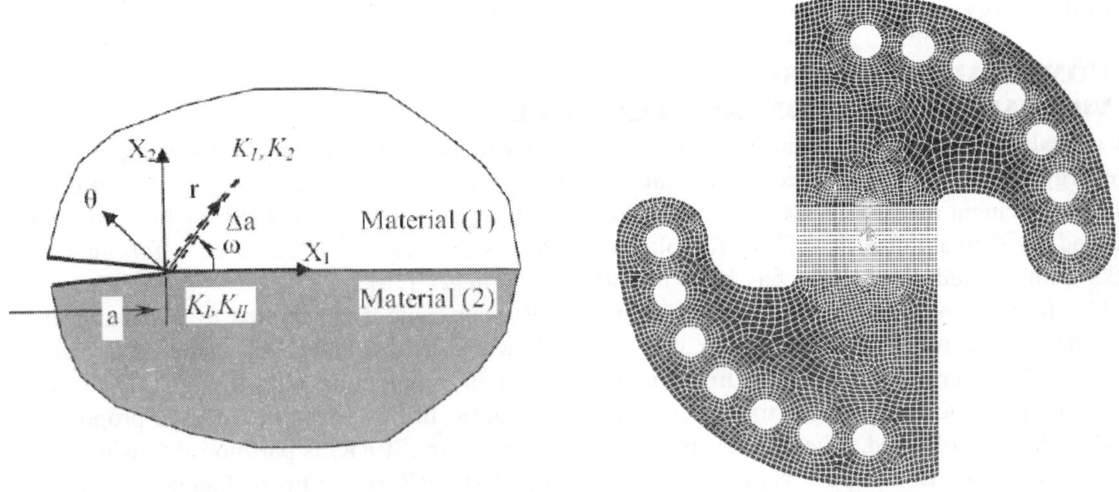

Fig. 1. Co-ordinate system for the interfacial crack (Left); Integrated mixed-mode specimen (Right).

4. FINITE ELEMENT ANALYSIS OF AN INTEGRATED MIXED-MODE SPECIMEN

Guided by the loading device and the specimen developed by Richard and Benitz [12], a new integrated mixed mode specimen was proposed by Arora *et al.* [13] wherein the loading device were merged along with the specimen as shown in Fig. 1 (Right). Finite element analysis is performed on this specimen using the commercial code Ansys. In the analysis, the modulus of elasticity of the upper half is kept constant at unity while that of the lower half is varied as 1, 2, 3, 5, 10, 50 and 100. The thickness of the specimen is considered to be unity. Mixed mode loading conditions are introduced by changing the loading angles with respect to the crack and interface. The loading angles considered were 0°, 15°, 30°, 45°, 60°, 75° and 90° with 0° implying pure tension and 90° pure shear. A radial mesh is used near the crack tip as shown in Fig. 1. The parent crack of size 27 mm is placed at the interface between the two elastic materials. Assuming that the crack kinks into the relatively less stiffer material, the parent crack kinks into material 1 at an angle ω to produce a child crack of length 0.27 mm. A load of 4 N is applied at a circular hole depending on the load angle and the hole opposite to it is hinged. Displacement and nodal forces obtained from the code Ansys are used in the computation of the energy release rates G_I and G_{II} using the VCCT as described in the previous section.

Figure 2 shows the mode I, mode II and the total ERR for loading angle 0^0 and 45^0 for same moduli ratios of the two materials. It is seen from these plots that the maximum value of mode I ERR and the total ERR occur at the same kinking angle. Hence, either of the two criteria, i.e., the maximum mode I ERR or the maximum total ERR can be used for determining the crack kinking angle. Table 1 shows the crack kinking angle obtained for different loading angles and different material combinations. It is seen that the angle of crack kinking increases as the loading angle increases. In other words, the increase in the shear component of the load (increasing load

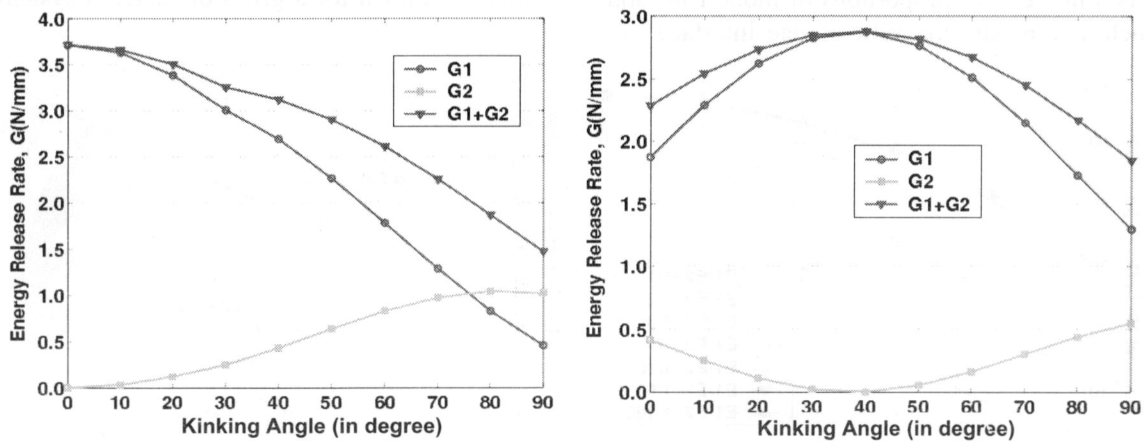

Fig. 2. ERR versus kinking angle for E_1/E_2 = 1/1 and loading angle 0^0 (left) and 45^0 (right).

Table 1. Crack kinking angles in degrees for different material combinations and loading angles.

Loading Angle → E_1 / E_2　　↓	0°	15°	30°	45°	60°	75°
1 /1	0	10	30	40	50	60
1 / 2	20	30	40	50	60	60
1 / 3	30	40	50	50	60	70
1 / 5	40	50	60	60	60	70
1 / 10	40	50	60	60	60	70
1 / 50	50	50	60	60	70	70
1 / 100	50	50	60	60	70	70

angle) increases the crack kinking angle. This has a great bearing in gravity dams wherein at the interface between concrete dam and rock foundation, the presence of shear makes the interface crack to kink into the adjoining material rather than propagate along the interface in a self similar manner. Further, it is seen from Table 1 that when the elastic mismatch between the two materials increases beyond a certain ratio ($E_1/E_2 = 1/5$), there is not much difference in the crack kinking angle.

The nominal load mixity which is defined in Equation 11 is computed for different material combinations and loading ratios and plotted in Fig. 3. It is seen that the nominal mode mixity increases with loading angle due to the increase in the mode II component of the ERR. Further, with increasing relative stiffness of the second material, the nominal mode mixity increases indicating the increase in mode II component for a given loading angle.

Parametric study is also conducted by changing elastic moduli, loading angle and crack kinking angle. From the results, a linear regression analysis is done to obtain a relationship between the crack kinking angle and the nominal mode mixity as shown in Fig. 3 (right) for one case. Table 2 shows the linear relationships between the crack kinking angle and the nominal mode mixity for varying moduli ratios. These relationships would be useful in determining the cracking behavior when the relative proportions of mode I to mode II loading is known for a given bi–material system such as a patch repaired concrete interface system.

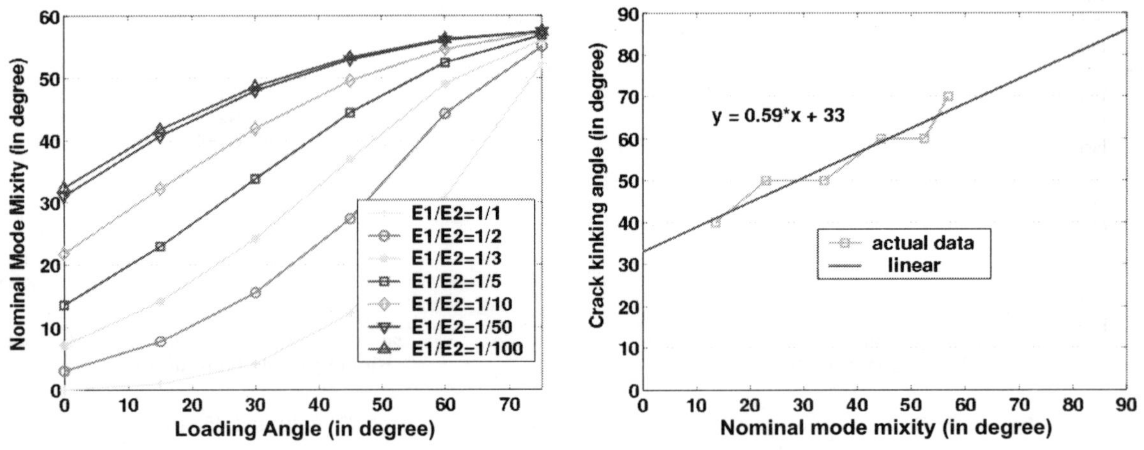

Fig. 3. Nominal mode mixity versus loading angle (Left); Linear regression plot (Right).

Table 2. Relationship between kinking angle and nominal mode mixity.

Moduli Ratio E_1 / E_2	Bi-materialconstant (ε)	Relationship between kinking angle (ω) and nominal mode mixity (ψ_0)
1 / 1	0	$\omega = 0.98\ \psi_0 + 15$
1 / 2	0.018078	$\omega = 0.89\ \psi_0 + 22$
1 / 3	0.027313	$\omega = 0.73\ \psi_0 + 29$
1 / 5	0.036796	$\omega = 0.59\ \psi_0 + 33$
1 / 10	0.045719	$\omega = 0.71\ \psi_0 + 26$
1 / 50	0.054467	$\omega = 0.81\ \psi_0 + 22$
1 / 100	0.055689	$\omega = 0.85\ \psi_0 + 19$

CONCLUSION

In this study, the mode I and mode II energy release rate for bi-material interface cracks are evaluated using the virtual crack closure technique. Relationship between the crack kinking angle and the nominal mode mixity is obtained for different combination of materials on either side of the bi-material interface through a finite element analysis of an integrated mixed mode bi-material specimen. This relationship would help in characterizing an interface cracks in a bi-material system.

REFERENCES

1. L.A. Carlsson, S. Prasad, 1993, Interfacial fracture of sandwich beams, Engineering Fracture Mechanics 44 (1993) 581-590.
2. M.Y. He and J.W. Hutchinson, 1989, Kinking of a crack out of an interface, Journal of Applied Mechanics, 56, 270-278.
3. Mukai, D.J., Ballarini, R. and Miller, G.R., (1990), Analysis of branched interface cracks, Journal of Applied Mechanics, 57, 887–893.
4. R. Yuuki, J. Xu, (1992), Stress based criterion for an interface crack kinking out of the interface in dissimilar materials, Engineering Fracture Mechanics 41, 5, 635-644.
5. P.H. Geubelle, W.G. Knauss, (1994), Crack propagation at and near bimaterial interfaces: Linear analysis, Journal of Applied Mechanics, 61, 560-566.
6. J.M. Chandra Kishen and K.D. Singh, (2001), Stress intensity factors based fracture criteria for kinking and branching of interface crack: Application to dams, Engineering Fracture Mechanics, 68, 201–219.
7. M.L. Williams, (1959), The stresses around a fault or crack in dissimilar media, Bulletin of the Seismological Society of America, 49 (2), 199-204.
8. E.F. Rybicki and M.F. Kanninen, (19770, A finite element calculation of stress intensity factors by modified crack closure integral, Engineering Fracture Mechanics, 9, 931-938.
9. K.N. Shivakumar, P.W. Tan and J. C. Newman (1988), A virtual crack closure technique for calculating stress intensity factors for cracked three-dimensional bodies, Int. Journal of Fracture, 36, R43–R50.
10. R. Krueger, (2002), The virtual crack closure technique: History, approach and applications, ICASE Report No. 2003-10-NASA/CR-20020211628, NASA Langford research center.
11. D. Xie, A.M. Waas, H.W. Shahwan and J.A. Schroeder and R.G. Boeman, (2004), Computation of ERR for kinking cracks based on VCCT, Computer Modelling in Engineering and Science, 6, 515–524.
12. H. Richard, K. Benitz, (1983), A loading device for the creation of mixed mode in fracture mechanics, Int. Journal of Fracture 22, R55-R58.
13. P. Arora, K. Yogendra Simha, S. Kurahatti, (1995), Validation of integrated mixed mode fracture specimen: A photoelastic study, Tech. Rep. ARDB-SP-TR-1995–01, Dept. of Mechanical Engg., Indian Institute of Science, Bangalore, India.

39

Fatigue Crack Growth Life Prediction of Welded T-Joints Using Different Stress Intensity Factor Solutions In Paris Equation

Gautam J. Sawaisarje[1] and K.B. Mulchandani[2]

[1]Hindustan Aeronautics Limited Nasik, Maharashtra-422 207 India email: gjsawaisarje@gmail.com
[2]Mechanical and Industrial Engineering Department, Indian Institute of Technology, Roorkee-247 667, Uttaranchal, India. email: kbm22fme@iitr.ernet.in

ABSTRACT

In this paper, Stress Intensity Factor (SIF) solutions for weld toe crack in T-joints determined by different researchers (Brennan et al., Bowness et al. and Maddox et al.) have been assessed for their accuracy and applicability for practical problems. Fatigue Crack Growth (FCG) life for welded T-joints, subjected to constant amplitude (CA) and variable amplitude (VA) bending load, has been predicted using Paris equation. Comparison with experimental FCG results shows that Brennan equation gives better FCG life prediction than Bowness and Maddox SIF solutions. Bowness equation gave over prediction for higher weld angle. Maddox solution do not consider weld angle and aspect ratio terms, therefore it cannot be used to study the effect of weld angle and aspect ratio on FCG life.

Keywords: Fatigue crack growth (FCG); Stress Intensity Factor (SIF); Stress Intensity Correction Factor for welded joint (Y_w); M_k factor (Stress Intensity Magnification factors).

1. INTRODUCTION

Welding technology has a significant impact on industrial development. Among the merits of welded structures, are high joint efficiency, water and air tightness and low fabrication cost. The types of welded joints can be classified into five basic categories as butt, fillet (T and cruciform), corner, lap and edge joints. T-joint fillet welds (Fig. 1A) are widely employed in ships, bridge structures and supporting frames for pressure vessels and piping.

It has been recognized that the large proportion of the fatigue life of welded structure can be spent on propagation of crack initiated at existing welding defects or stress concentration sites such as weld toe. Thus prediction of the fatigue life of welded joint has to be based on the accurate analysis of fatigue crack growth. This consequently requires development of fracture mechanics approaches to calculate correct stress intensity factors [1]. A pre-requisite for defect assessment using fracture mechanics approach, is the availability of reliable solutions for Stress Intensity Factors, SIF (K).

Different researchers have developed SIF solutions for welded T-joints. The aim of the present study is to assess the different SIF solutions available, for their accuracy and applicability to predict the Fatigue Crack Growth (FCG) life of welded T-joints.

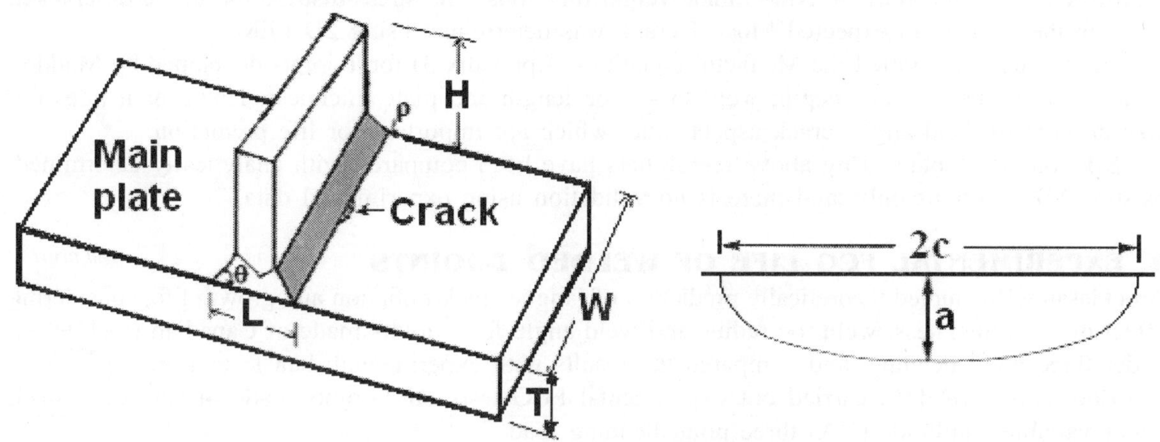

Fig. 1A. Welded T-joint geometry.   **Fig. 1B.** Crack geometry.

2. PREDICTION OF FATIGUE CRACK GROWTH (FCG) LIFE IN WELDED JOINTS

Paris [2] proposed Fatigue Crack Growth Rate (FCGR) 'da/dN' equation in terms of Stress Intensity Factor (SIF) range (ΔK), to determine the FCG life as,

$$da/dN = C\,(DK)^m \qquad \qquad ...(1)$$

where 'C', 'm' are material constants. ΔK= Stress Intensity Factor (SIF) range,

For semi-elliptical surface crack at weld toe, with depth 'a' and surface length '2c', (Fig.1B), the SIF range (ΔK) for mode I, is given by [5,6],

$$\Delta K_I = Y_w \Delta\sigma \sqrt{\pi a} = (M_k Y)\Delta\sigma \sqrt{\pi a} \qquad \qquad ...(2)$$

$Y_w = M_k Y$ = Stress intensity correction factor for welded joint, $\Delta\sigma$ = Nominal stress range,

M_k = Stress intensity magnification factor. Y = Shape factor for flat plate of thickness T, and Y can be calculated using Newman-Raju [3] equation.

3. STRESS INTENSITY FACTOR SOLUTIONS FOR WELD TOE CRACK IN T-JOINTS (Y_W /M_K)

Bowness et al. [4] derived a set of equations (Appendix 1) for estimating weld toe magnification factor (M_k) for semi-elliptical surface cracks in welded T-joints, subjected to tension and bending load, using 3D FEM. This equation is presented in terms of crack depth (a), aspect ratio (a/c), weld toe-to-toe length (L) and weld angle (θ). The element type used for the FEM analysis was the reduced integration 20-noded brick element. The virtual crack extension technique was used to provide estimates of the J-integral. In the elastic regime, SIF 'K' was calculated from the J-integral value as $K = \sqrt{JE}$; E = Young's modulus and M_k can be calculated as: $M_k = K/[Y\Delta\sigma\sqrt{\pi a}]$.

Brennan et al. [5] derived the parametric equations (Appendix 2) for stress intensity correction factors (Y_w) for weld toe crack in T-joints, as a function of plate thickness, weld toe-to-toe length, weld angle, weld toe radius (ρ), crack depth and crack aspect ratio for tension and bending. SIF solutions are derived from the Niu-Glinka weight functions. The stress distribution in the un-cracked plane in the path of an expected Mode I crack was determined using 2D FEM.

Hobbacher [6] reported the M_k factor equations (Appendix 3) for T-joints developed by Maddox *et al.* [7] in terms of crack depth; weld toe-to-toe length and plate thickness. However it does not take account of weld angle, crack aspect ratio, which are important for life prediction.

SIF solutions obtained by above researchers have been compared with analytically determined, existing SIF solutions only, and there is no validation using experimental data.

4. EXPERIMENTAL FCG LIFE OF WELDED T-JOINTS

Arockiasamy [8] studied theoretically prediction of fatigue crack initiation and growth life, considering effect of plate thickness, weld toe radius and weld angle for T-joints made of Canadian steel, tested under three point bending, and compared the results with experimental data from Ref. [9].

Bouchard *et al.* [10] carried out experimental FCG tests on T-joints made of Canadian steel, under variable amplitude (VA) three point bending load.

Otegui *et al.* [11] studied experimentally the propagation of fatigue cracks from weld defects at the toes of manual and automatic, non-load carrying welded T-joints made of structural steel, in as welded condition, subjected to constant amplitude three point bending load.

Dimensions of welded joint, loading details, initial crack depth and experimental FCG life are given in Table 1. For all specimens the final or critical crack depth was taken, $a_f = T/2$.

Table 1. Welded T-Joint Geometry & Loading Parameters For FCG Analysis. [8,10,11]

Sp No.	T	L	θ	ρ	$\Delta\sigma$	R	a_i	Sp No	T	L	θ	ρ	$\Delta\sigma$	R	a_i
1.	16	36	40	1.3	150	0.05	0.5	14.	18.5	37*	60	0.3	174	0.1	0.30
2.	26	58	40	1.4	150	0.05	0.5	15.	18.5	37*	60	0.3	176	0.1	0.60
3.	52	108	40	1.2	150	0.05	0.5	16.	18.5	37*	60	0.3	175	0.1	0.20
4.	78	154	45	1.5	150	0.05	0.5	17.	18.5	37*	45	0.2	174	0.1	0.35
5.	108	213	45	1.6	150	0.05	0.5	18.	18.5	37*	45	0.2	172	0.1	0.30
6.	18.5	37*	40	0.5	300	0.2	0.30	19.	26	56	38	1.44	106	0.05	0.5
7.	18.5	37*	40	0.5	305	0.1	0.50	20.	26	58	40	1.04	106	0.05	0.5
8.	18.5	37*	40	0.5	305	0.1	0.30	21.	26	58	33	1.04	77	0.05	0.5
9.	18.5	37*	40	0.5	250	0.1	0.25	22.	26	60	40	0.81	90	0.05	0.5
10.	18.5	37*	40	0.5	165	0.1	0.50	23.	78	156	33	1.19	105	0.05	0.5
11.	18.5	37*	40	0.5	400	0.1	0.20	24.	78	170	27	0.99	105	0.05	0.5
12.	18.5	37*	60	0.3	250	0.1	0.35	25.	78	158	37	0.8	76	0.05	0.5
13.	18.5	37*	60	0.3	252	0.1	0.25	26.	78	158	33	1.28	65	0.05	0.5

Sp. No.: Specimen No.1-5, from Ref. [8], Specimen No.6-18 [11] & Specimens 19- 26, [10]
T = main plate thickness in mm, **L** = weld toe-to-toe length in mm, θ = weld angle in Degrees
P = weld toe radius in mm, a_i = initial crack depth in mm. *– Assumed value of L.
R = stress ratio, **Δs** = stress range MPa, $\Delta\sigma_{RMS}$ = RMS value of stress range for specimens 19-26.
Experimental FCG life of specimens 1–26: 1000 × [1388, 676, 326, 314, 198, 760, 340, 520, 850, 2950, 199, 560, 570, 1350, 480, 2600, 1520, 460, 1690, 2030, 3000, 4190, 865, 905, 1795, 3680]; Respectively.

5. PREDICTION OF FCG LIFE ANALYTICALLY USING DIFFERENT SIF SOLUTIONS

1. Input data–Welded T-joint specimen dimensions, crack details-initial crack depth and load variables-$\Delta\sigma$ and stress ratio (R) have been taken from Ref. [8,10,11].
2. Material constants in Paris equation for Canadian steel (specimens 1–5,19–26) was taken as, $m = 3$ & $C = 5.36 \times 10^{-13}$ (ΔK in MP_a–$mm^{1/2}$) and for structural steel (specimens 6 – 18), $m = 3.2$ & $C = 0.15 \times 10^{-11}$ [11]. (ΔK in MPa–$m^{1/2}$).
3. Calculate 'Y' using Newman-Raju equation from Ref. [3] and 'M_k' factors using Bowness and Maddox equation. 'Y_w' using Bowness/Maddox equation would be, $Y_w = M_k Y$. Also, Calculate 'Y_w' using Brennan equations for welded T-joint at deepest point for bending load.
4. Paris equation was used to calculate FCG life for propagating the crack from 'a_i' to 'a_f', using different SIF solutions.

6. RESULTS AND DISCUSSION

6.1. Comparison of Stress Intensity Correction Factor (Y_w)

(Fig.2). It has been observed that, for Brennan equation, as weld angle increases the Y_w also increases up to a/T = 0.3. For a/T > 0.3, Y_w is approximately same for all weld angles for given (a/c) ratio. Bowness equation gave opposite trend. For a/T > 0.15, Y_w decreases as weld angle increases, and very little difference in the value of Y_w was observed for weld angle 30° and 45° for (a/T) >0.15.

6.2. Comparison of FCG Life

Figure 3 (A & B) shows the FCG life calculated using SIF solutions by Brennan *et al.* and Bowness *et al.* for crack aspect ratio of 0.40 and 0.45 and its comparison with experimental results. Results show that, as (a/c) increases, the calculated FCG life also increases.

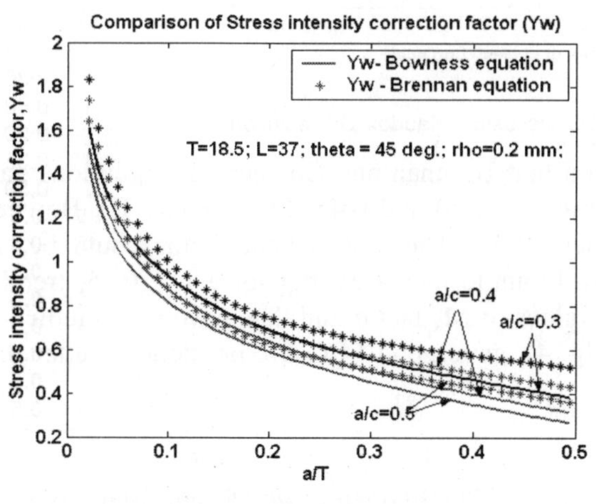

(A) For different (a/c) ratios.

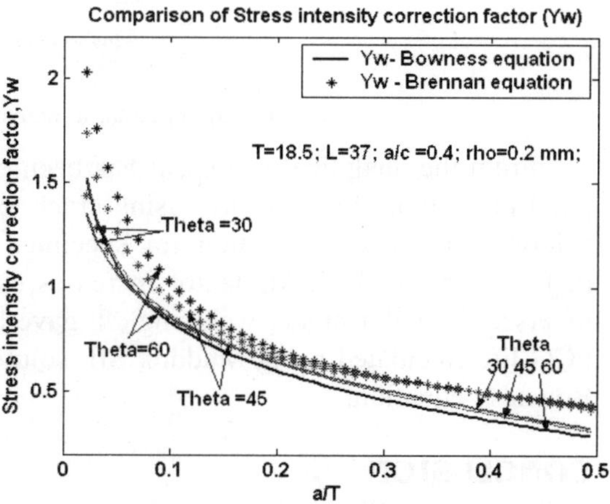

(B) for different weld angles.

Fig. 2. Comparison of SIF solution by Bowness and Brennan.

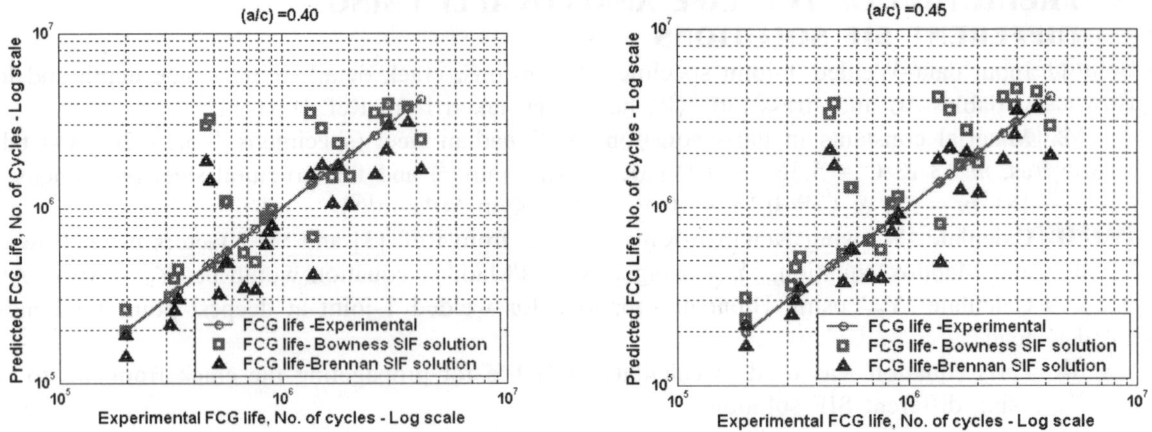

Fig. 3. (A & B) Predicted and Experimental FCG life of T-joints, using Bowness and Brennan SIF solutions.

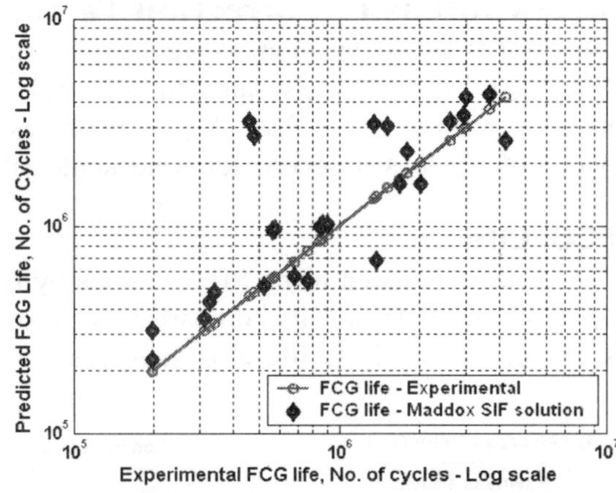

Fig. 4. Experimental & predicted FCG life using Maddox SIF solution.

From the analytical results, it has been observed that Brennan and Bowness SIF solutions gave good predictions for FCG life using crack aspect ratio of 0.40 and 0.45, Fig. 3 (A & B). Bowness solution gave over prediction for specimen number 12-15. The reason behind this could be, for $(a/T) = 0.005 - 0.15$, M_k factor increases as the weld angle increases, but for $a/T > 0.15$, trend is reversed. Thus for higher weld angle it gave lower values of M_k factor and thus gave over prediction. FCG life calculated using Maddox SIF solution (Fig. 4), gave good FCG life prediction, better than Bowness equation.

CONCLUSION

In this paper, SIF solutions developed by Bowness *et al*. [4], Brennan *et al*. [5] and Maddox *et al*. [7] has been assessed for their accuracy to predict Fatigue Crack Growth (FCG) life of welded T-joints, subjected to constant amplitude (CA) and variable amplitude (VA) bending load. FCG life has been calculated using the above-mentioned SIF solutions in the Paris equation.

Comparison of predicted FCG life with experimental data shows that Brennan SIF solution gives better FCG life prediction than Bowness and Maddox solutions. Bowness and Brennan SIF Solutions gave good FCG life predictions by taking crack aspect ratio of 0.4 or 0.45. However, Bowness equations gave over-prediction for higher weld angles (60°).

Maddox SIF Solutions is simple to use and gave good FCG predictions for CA and VA loading. However, it cannot be used to study the effect of weld angle and crack aspect ratio on FCG life.

NOMENCLATURE

a	Crack depth
C, m	Material constants in Paris equation
J	J-Integral
M_k	Magnification factor
Y_w	Stress intensity correction factor for welded joint
2c	Surface crack length
E	Young's modulus
ΔK	SIF rang
Y	Shape factor

REFERENCES

1. X. Niu and G. Glinka, 1989, Stress intensity factors for semi-elliptical surface cracks in welded joints. International Journal of Fracture. 40. 255-270.
2. P.C. Paris and F. Erdogan, 1963, A critical analysis of crack propagation laws. Trans. ASME, Journal of Basic Engineering. D85, 528-534.
3. J.C. Newman and I.S. Raju, 1981, an empirical SIF equation for surface crack, Engineering Fracture Mechanics. 15 (1/2). 185-192.
4. D. Bowness and M.M.K. Lee, 2000, Prediction of weld toe magnification factors for semi-elliptical cracks in T-butt joints. International Journal of Fatigue, 22. 369-387.
5. F.P. Brennan, W.D Dover, R.F Kare and A.K Hellier, 1999, parametric equations for T-butt weld toe stress intensity factors. Int. Journal of Fatigue. 21. 1051-1062.
6. A. Hobbacher, 1996, Recommendations for fatigue design of welded joints and components. Int. institute of welding IIW/IIS./Document XIII-1539-96/XV-8465.
7. S.J. Maddox, J.P. Lechocki and R.M. Andrews, 1980, Fatigue analysis for the revision of BS: PD 6493: 1980. Report 3873/1/86, The welding Institute, Cambridge, UK.
8. M. Arockiosamy, J.S Jani and D.V. Reddy, 1991, Weld profile and thickness effect and fatigue of welded joints applicable to offshore structures. Proceedings of International Symposium on Fatigue and Fracture, fatigue and fracture in steel and concrete structures, at SERC Madras. Dec. 19-21, Vol.2. (993-1014).
9. O. Vosikovsky and A. Rivard, 1985, Effect of thickness on fatigue life of welded plate T joints. PMRL report 85-65(TR) CANMET, Energy, Mines and resources, Canada.
10. R. Bouchard, O. Vosikovsky and A-Rivard, 1991, Fatigue life of welded plate T. joints under variable amplitude loading. Int. journal of Fatigue. 13 (1). 7-15.
11. J.L Otegui, U.H. Mohaupt and D.J. Burns, 1991, Effect of weld process on early growth of fatigue cracks in steel T. joints. Int. J. of Fatigue. 13(1). 45-58.

Appendix 1. Bowness M_k equations for welded T-joints - deepest point, bending load [4]

$M_k = f_1 + f_2 + f_3$. For $0.005 \leq (a/T) \leq 0.5$;	$M_k = 1$. For $0.5 < (a/T) \leq 0.9$;
$f_1 = 0.065916(a/T)^{(A_1 + (A_2(a/T))^{A_3})}$ $\qquad + 0.52086 \exp(a/T)^{-0.10364} + A_4$ $A_1 = -0.014992(a/c)^2 - 0.021401(a/c) - 0.23851$; $A_2 = 0.61775(a/c)^{-1.0278}$ $A_3 = 0.00013242(a/c) - 1.4744$; $A_4 = -0.28783(a/c)^3 + 0.58706(a/c)^2$ $\qquad - 0.37198(a/c) - 0.89887$;	$f_2 = A_5[1 - (a/T)]^{A_6} + A_7(a/T)^{A_8}$; $A_5 = 0.11052\theta^2 - 0.19007\theta + 0.059156$; $A_6 = -15.124\theta^2 + 15.459\,\theta - 0.0036148$; $A_7 = -0.047620\theta^2 + 0.16780\,\theta - 0.081012$; $A_8 = -17.195(a/T)^2 + 12.468(a/T) - 0.51662$;

$$f_3 = A_9(a/T)^{(A_{10}\theta^2 + A_{11}\theta + A_{12})} + A_{13}(a/T)^{A_{14}} + A_{15}(a/T)^2 + A_{16}(a/T) + A_{17}$$

$A_9 = 0.75722\,\theta^2 - 1.8264\,\theta + 1.2008$;

$A_{10} = -0.013885(L/T)^3 - 0.014872(L/T)^2 + 0.55052(L/T) - 0.072404$; $\quad$ $A_{14} = 0.43912\,\theta^2 - 1.3345\theta + 0.57647$;

$A_{11} = -0.065232(L/T)^3 + 0.54052(L/T)^2 - 1.8188(L/T) - 0.0022170$; $\quad$ $A_{15} = -0.35848(L/T)^2 + 1.3975(L/T) - 1.7535$;

$A_{12} = -0.034436(L/T)^2 + 0.28669(L/T) + 0.36546$; $\qquad\qquad\qquad$ $A_{16} = 0.31288(L/T)^2 - 1.3599(L/T) + 1.6611$;

$A_{13} = -0.61998\,\theta^2 + 1.4489\theta - 0.90380$; $\qquad\qquad\qquad\qquad\qquad$ $A_{17} = -0.0014701(L/T)^2 - 0.0025074(L/T) - 0.0089846$;

Appendix 2. Brennan equation for Y_w for bending load, deepest point on the crack [5]

$$Y_w = 0.96\,A \log(a/T) + C_0 + C_1(a/T) + C_2(a/T)^2 + c_3 + c_4 + c_5$$

For $(a/c) < 0.1$;

$\qquad A = -0.388 - 0.958\sqrt{(a/c)} + 1.111(a/c) + M_A$;

$\qquad C_0 = 0.544 - 4.125\sqrt{(a/c)} + 4.018(a/c) + M_0$;

$\qquad C_1 = -2.664 + 22.408\sqrt{(a/c)} - 22.264(a/c) + M_1$;

$\qquad C_2 = 8.758 - 41.156\sqrt{(a/c)} + 29.768(a/c) + M_2$;

For $(a/c) \geq 0.1$

$\qquad A = -0.686 + 0.310\sqrt{(a/c)} + 0.0622(a/c) + M_A$;

$\qquad C_0 = -0.645 + 1.111\sqrt{(a/c)} - 0.648(a/c) + M_0$;

$\qquad C_1 = 3.860 - 6.128\sqrt{(a/c)} + 2.876(a/c) + M_1$;

$\qquad C_2 = -1.648 + 0.926\sqrt{(a/c)} + 0.00393(a/c) + M_2$;

$$M_A = 0.597 - 0.649\,\theta - 0.0028\,(T/\rho); \quad M_0 = 1.2822 - 1.325\,\theta - 0.0077\,(T/\rho)$$
$$M_1 = -2.222 + 2.154 + 0.0170\,(T/\rho); \quad M_2 = 0.789 - 0.621\,\theta - 0.0097\,(T/\rho)$$

For $(a/T) < 0.25$;

$C_3 = -0.25\{0.25 - (a/T)\}^{0.5} \{1.1 - (a/c)\}^{0.16} \theta^2 (\rho/T)^{-0.16} (L/T)^{-0.37} [1 - \exp - \{0.25 - (a/T)/0.15\}^2]$;

For $a/T \geq 0.25$; $C_3 = 0$;

NOTE: 'θ' in radians

For $a/T < 0.05$ & $L/T < 0.455$;

$C_4 = -4\{0.05 - (a/T)\}^{0.565} \{1.1 - (a/c)\}^{0.3} \theta^{1.35} (\rho/T)^{-0.3} \{0.455 - (L/T)\}^{0.204} [1.0 - \exp - \{(0.05 - a/T)/S_4\}^2]$;

$\qquad S_4 = 0.06$; If $\theta > 0.6109$ & $\rho/T < 0.04$ & $L/T < 0.35$;   Else, $S_4 = 0.05$;

For $a/T < 0.05$ & $L/T \geq 0.455$;

$C_4 = 0.5\{0.05 - (a/T)\}^{1.1} \{1.1 - (a/c)\}^{-0.486} \theta^{-2.66} (\rho/T)^{0.11} (L/T - 0.455)^{-0.0384} [1.0 - \exp - \{(0.05 - a/T)/0.015\}^2]$;

$C_4 = 0$;   For $(a/T) \geq 0.05$;

For $(a/T) > 0.35$; $C_5 = -014\{(a/T) - 0.35\}^{0.098} \{1.1 - (a/c)\}^{0.862} \theta^{0.675} (\rho/T)^{-0.077} (L/T)^{0.148}$

$$[1.0 - \exp - \{(a/T - 0.35)/0.2\}^2];$$

For $(a/T) \leq 0.35$; $C_5 = 0$;

Appendix 3. M_k factor for welded T-Joint by Maddox et al. [6,7]

$M_k = v\,(a/T)^w$ For $L/T \leq 1$; $\qquad\qquad\qquad$ $M_k = v\,(a/T)^w$ For If $(a/T)^w$ For $(L/T) > 1$;

$v = 0.45(L/T)^{0.21}$ & $w = -0.31$; If $(a/T) \leq 0.03(L/T)^{0.55}$; $v = 0.45$; & $w = -0.31$;   If $(a/T) \leq 0.03$;

$v = 0.68$; & $w = -0.19(L/T)^{0.21}$;   If $(a/T) > 0.03(L/T)^{0.55}$; $v = 0.68$; & $w = -0.19$;   If $(a/T) > 0.03$;

40

An Efficient Quarter Point Element for Fracture Analysis of Cracks

SAYANTAN PAUL[1] AND B.N. RAO[2]

[1]Post Graduate, Department of Civil Engineering, Indian Institute of Technology, Madras, Chennai-600 036, India. email: mrpaul_in@yahoo.com
[2]Asst. Professor, Department of Civil Engineering, Indian Institute of Technology, Madras, Chennai-600 036, India. email: bnrao@iitm.ac.in

ABSTRACT

The use of modified quarter point elements in finite element method for the computation of stress intensity factors in cracked plates subjected to mixed mode loading is being presented here. The standard quarter point singular element is modified such that the near-tip crack opening displacement satisfies a known constraint; the coefficient of the term in William's expression which is the linear in the distance to the tip must vanish but keeping in mind the singularity around the crack-tip. Stress intensity factors calculated with the new modified quarter point element in conjunction with the Displacement Correlation Technique gives better results as compared to the ones given by standard quarter point element. The technique basically involves the same guidelines as that of standard quarter point element where 6- noded quarter point elements are used to mesh the region around the crack-tip and 8-noded elements are used to mesh the non-singular zone outside. Based on the new shape functions crack-tip opening displacement in the coordinate system with the crack-tip under consideration is found out taking into account the nodal displacement values of the elements around the crack-tip. Several edge-cracked, inter-cracked and off-centre problems have been solved to justify the effectiveness of the proposition.

1. INTRODUCTION

In the numerical modeling of fracture, a correct representation of the local stresses and displacement fields in crack-tip region is essential for accurate evaluation of stress intensity factors (SIFs). For a cracked geometry, Williams result for the displacement $\boldsymbol{u} = u_k$, $k = 1, 2$ in the neighborhood of the tip is

$$u_k(r,\theta) = a_k + b_k(\theta)r^{1/2} + c_k(\theta)r + d_k(\theta)r^{3/2} + \ldots\ldots \qquad \ldots(1)$$

where r, θ are the distance to, and the direction emanating from, the tip, respectively.

Thus the crack opening displacement (COD) $\Delta\boldsymbol{u} = \Delta u_k$, $k = 1, 2$, is

$$\Delta u_k(r,\theta) = b_k(\theta)r^{1/2} + c_k(\theta)r + d_k(\theta)r^{3/2} + \ldots \qquad \ldots(2)$$

In both finite element and boundary integral modeling of discrete cracks, the standard approach is used to incorporate the critical r by means of Quarter Point (QP) element developed by Shaw and Barosum. Use of singular elements has significantly improved the performance and over the years the element has been modified to enhance the performance.

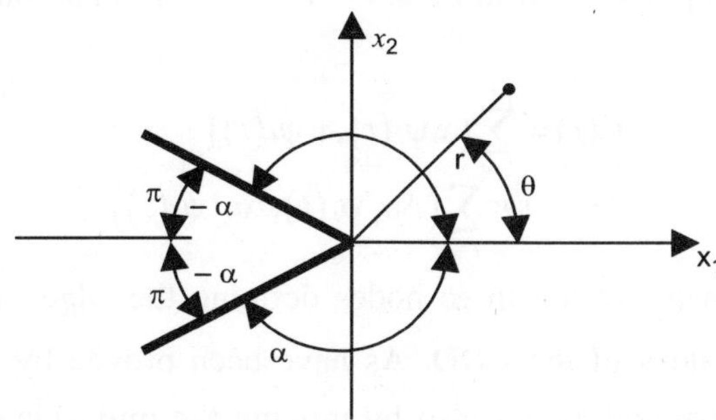

Fig. 1. Definition of the coordinate system (x_1, x_2) and (r, θ) for a notch or crack geometry.

Recently it has been proved by Gray and Paulino that, irrespective of the problem geometry or boundary conditions, the series expansion in Eqn. (2) must have $c_k = 0$ for Δu_k on the crack surface. The general QP element fails to satisfy this constraint. Therefore by forcing the linear term in Δu to be zero, we expect to have a more accurate analysis at the crack tip region. The coefficients of the linear terms in William's expansion 1, 2 are related by

$$c_k(\pi) = c_k(-\pi) \qquad \text{...(3)}$$

So the linear term vanishes from the expression of crack-tip opening displacement,

$$\Delta u(r) = u(r,\pi) - u(r,-\pi) \qquad \text{...(4)}$$

The result also follows directly from William's eigen function expansion, but the eigen function analysis is restricted to a traction-free flat crack in an infinite plate, whereas the boundary integral derivation and the finite element method make it clear for an arbitrary crack geometry and any two dimensional problem for which a boundary integral can be constructed. Once we consider the crack-edge the problem turns out to be similar as a boundary element formulation.

2. MODIFIED QUARTER POINT ELEMENT

The two-dimensional QP element is based upon the six nodded quadratic element.

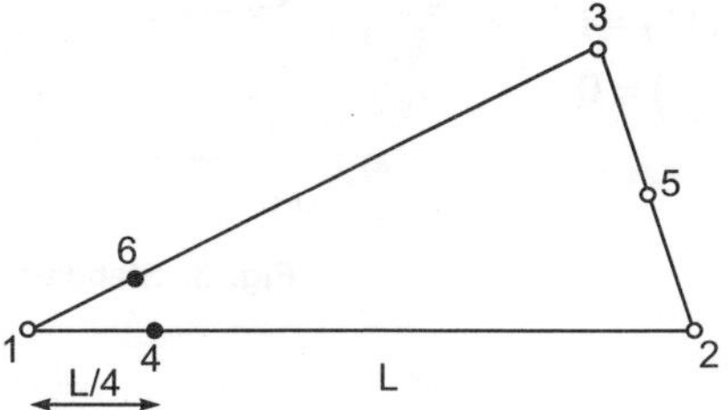

Fig. 2. QP Element.

For $t \in [0, 1]$, the shape functions for this element along the crack edge (corresponding to nodes 1, 2 and 4) are given by

$$\psi_1(t)=(1-t)(1-2t)$$
$$\psi_2(t)=t(2t-1)$$
$$\psi_4(t)=4t(1-t) \qquad \qquad \text{...(5)}$$

As $\Delta u = 0$ at the crack tip, assumed to be at $t = 0$, the representations of the crack-tip geometry and COD are

$$\Gamma(t)=\sum_{j=1}^{3}\left(x_j\psi_j(t),\,y_j\psi_j(t)\right)$$
$$\Delta u_k(t)=\sum_{j=2}^{3}\left(\Delta u_1^j\psi_j(t),\,\Delta u_2^j\psi_j(t)\right) \qquad \text{...(6)}$$

Here $(x_j,\,y_j)$ are the coordinates of the three nodes defining the edge of the element bordering the crack and Δu_k^j the nodal values of the COD. As have been proved by Barosum and Henshell and Shaw and have been demonstrated earlier also by moving the mid–side coordinates to three fourths of the way towards the tip, the parameter t becomes $(r/L)^{1/2}$, with L being the length of an element as mentioned in Anderson. As a consequence, the leading order term in Δu_k^j at $t = 0$, which is t is the correct square root of distance. However the next term which is t^2, is r/L. This term should vanish to give a more accurate representation of SIFs. So the new shape functions are assumed for nodes 1, 2 and 4 keeping the leading term $t=\sqrt{r}$ intact by adding another cubic term.

$$\hat{\psi}_2(t)=t(2t-1)+2t(1-t)(1-2t)/3=\frac{1}{3}\left(4t^3-t\right)$$
$$\hat{\psi}_4(t)=4t(1-t)-4t(1-t)(1-2t)/3=-\frac{8}{3}\left(t^3-t\right) \qquad \text{...(7)}$$

This additional contribution accomplishes the cancellation of $t^2 = r$ term without disturbing the interpolation. The alteration doesn't change the COD shape functions radically as is shown in Fig. 3.

$$\hat{\psi}_2(0)=0, \quad \hat{\psi}_2(1/2)=0, \quad \hat{\psi}_2(1)=1$$
$$\hat{\psi}_4(0)=0, \quad \hat{\psi}_4(1/2)=1, \quad \hat{\psi}_4(1)=0$$

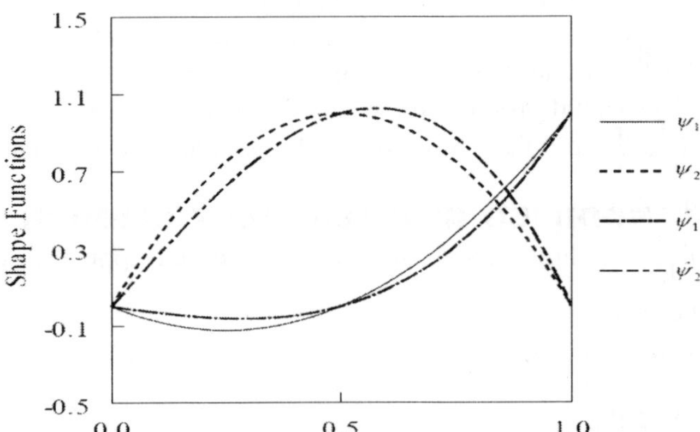

Fig. 3. Standard (ψ_2,ψ_4) and modified $(\hat{\psi}_2,\hat{\psi}_4)$ shape functions.

3. NUMERICAL IMPLEMENTATIONS

SIFs provided by the QP element (both modified and standard method) will be calculated by means of very simple DCT. However the point to be noted here is to evaluate the quality of modified QP element by means of very simple DCT. The general expression of SIFs by means of DCT are given by

$$K_{\mathrm{I}} = \frac{G}{\kappa+1} \lim_{r\to 0} \sqrt{\frac{2\pi}{r}} \Delta u_2$$

$$K_{\mathrm{II}} = \frac{G}{\kappa+1} \lim_{r\to 0} \sqrt{\frac{2\pi}{r}} \Delta u_1 \qquad ...(8)$$

where Δu_k is the COD in the coordinate system associated with the crack tip under consideration, G is the shear modulus, and v is the Poisson's ratio,

$$\kappa = 3 - 4v \text{ (Plane strain) and } \kappa = \frac{3-v}{1+v} \text{ (Plane stress)}$$

By using the modified QP shape functions in Eqn. (7), we obtain (the crack-tip is assumed to be at node 1, see the Fig. (1)

$$\Delta u_k = \Delta u_k^4 \hat{\psi}_4(t) + \Delta u_k^2 \hat{\psi}_2(t) =$$
$$\frac{1}{3}\left(8\Delta u_k^4 - \Delta u_k^2\right)t + \frac{4}{3}\left(\Delta u_k^2 - 2\Delta u_k^4\right)t^3 \qquad ...(9)$$

Use of Eqn. (9) in Eqn. (8) with $t = \sqrt{r/L}$ yields

$$K_{\mathrm{I}} = \frac{G}{3(\kappa+1)} \lim_{r\to 0} \sqrt{\frac{2\pi}{L}} \left(8\Delta u_2^4 - \Delta u_2^2\right)$$

$$K_{\mathrm{II}} = \frac{G}{3(\kappa+1)} \lim_{r\to 0} \sqrt{\frac{2\pi}{r}} \left(8\Delta u_1^4 - \Delta u_1^2\right) \qquad ...(10)$$

Thus SIFs are given directly in terms of the nodal values of the COD at the crack-tip element.

4. NUMERICAL EXAMPLES AND DISCUSSIONS

The values of SIFs obtained by the use of modified shape functions for the QP element using the DCT are compared with those obtained from standard QP elements using the same DCT. The solutions are compared with the *J*-integral and exact solutions. The results obtained with the use ofmodified shape functions are very accurate and has been listed in the following tables. The results given by Modified Crack Closure Integral[8] are also presented for reference purpose. Displacements at the nodes along the crack-edge are obtained from software and DCT is used to evaluate the SIFs. The examples include both infinite body and finite body problems. Isoparametric graded elements are used to descretize all the geometry. For all examples consistent units are used.

4.1 Edge-Crack Problem

The example involves an edge-cracked plate as shown in the Fig. 4 fixed at the bottom and subjected to far field shear r stress $\tau^\infty = 1$ unit applied at the top. In this example Young's modulus is taken to be $E = 1$ and Poisson's Ratio is taken to be $v = 0.3$. The plate has a length $L = 16$ units and width $W = 7$ units and crack length $a = 0.3$ units. A plane strain condition is assumed.

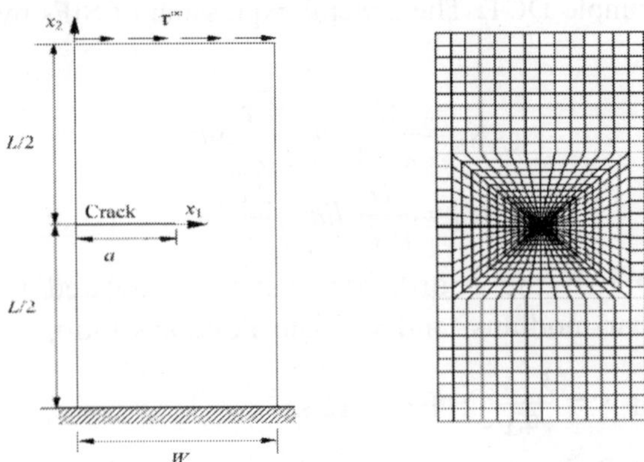

Fig. 4. Edge Cracked Plate: (a) Geometry, loads and boundary conditions, (b) FEM decartelization-2711 nodes, 832 8-noded quad elements, and 48 focused 6-noded QP Elements (ABAQUS & PATRAN).

Table 1. Comparison of SIFs for an edge-cracked plate.

Method	K_I/K_{um}	K_{II}	K_I	K_{II}
		SIFsError Percentage (%) (Absolute)		
SQP-DCT	33.9156	4.3140	0.4411	3.8521
MCCI	33.8451	4.7721	0.6172	6.7645
MQP-DCT	34.0721	4.3765	0.2416	2.3713
J-integral (Reference):	$K_1 = 34.1123,$		$K_{II} = 4.47$	

4.2 Plate with Hole

The example involves a flat-plate with a central hole subjected to far-field normal stress $\sigma = 10000$ units. Young's modulus $E = 10 \times 106$ and Poisson's ratio $\nu = 0.25$, thickness of the specimen is 0.04. The radius of the hole is 0.5. Plane strain condition is assumed. First a plate with a hole at its centre is modeled without a crack and the crack-free analysis is done and the zone of maximum stress is found out. Then a crack is introduced in the zone of high stress and subsequent fracture analysis is undertaken. The crack length introduced is $a = 0.10$.

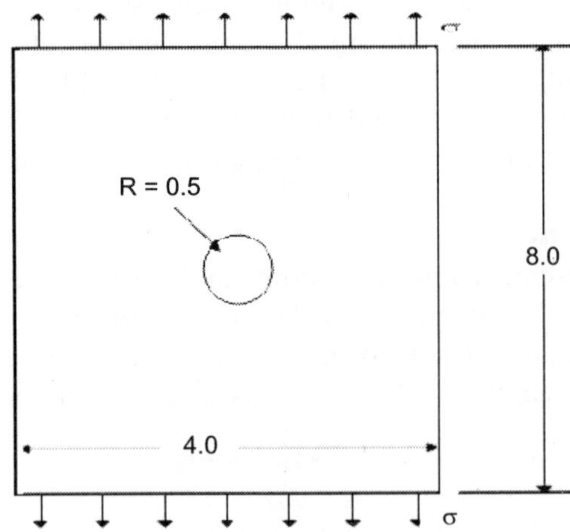

Fig. 5. A flat plate with a central hole.

Table 2. Comparison of SIFs for a flat plate with central hole with the crack initiating from hole.

METHOD	SIFs		Error Percentage (abs)	
	K_I	K_{II}	K_I	K_{II}
SQP-DCT	15531.1321	−4.3535	0.7171	2.0903
MCCI	15296.2178	−2,4981	0.7825	4.1701
MQP-DCT	15345.4155	−4.2103	0.463	1.4309
J-integral (Reference):	K_I = 15416.2171		K_{II} = 4.2691	

4.3 Interior Crack Problem

Mode–I Case ($\theta = 0$)

Infinite plate problem, dimensions of the plate the 2H = 2W = 200 with a a crack length of 2 $a = 0.4$ place centrally. $E = 0.36$ and Poission's taken as $v = 0.3$ $\sigma = 100$. A Place Strain condition is assumed.

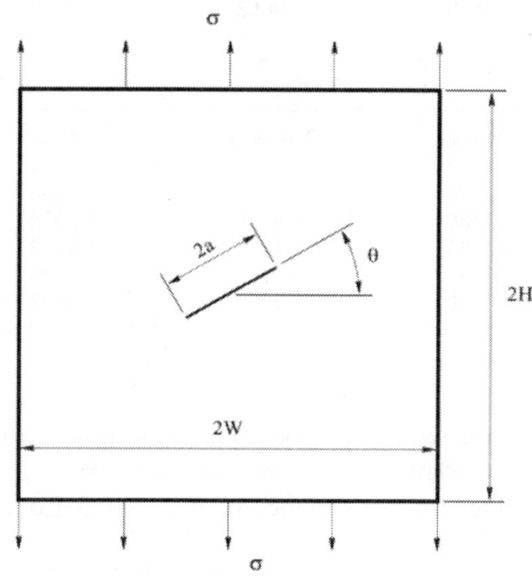

Fig. 6. Single interior crack.

Table 3. Comparison of SIFs for a plate with central crack at $\theta = 0$ or both left and right tips.

METHOD	SIFs (Right Tip)		SIFs (Left Tip)	
	K_I	Error % (abs)	K_I	Error % (abs)
SQP-DCT	81.6143	2.8607	81.5679	2.7851
MCCI	78.6327	0.9580	78.6331	0.9450
MQP-DCT	79.1174	0.2931	79.0711	0.3621
J-integral (Reference):	K_I = 9.3521		K_{II} = 9.2156	

Mode-II Case $(0 < \theta < \pi/2)$

Consider mixed-mode situation where $0 < \theta < \pi/2$. An infinite plate problem with same dimension $2H = 2W = 200$ and crack length is taken as $a/W = 0.1$. A crack length of $2a = 2$ is taken. Young's modulus $E = 0.29e.05$ and Poisson's ratio $\nu = 0.25$ $\sigma = 1$. Values obtained are compared with closed form solutions given by

$$K_I = \sigma\sqrt{\pi a}\, sin^2\alpha$$
$$K_{II} = \sigma\sqrt{\pi a}\, sin\alpha\, cos\alpha \qquad \qquad ...(11)$$

Table 4. Normalized SIFs (w.r.t exact sols.) as a function of θ – RIGHJT TIP & LEFT TIP.

Angle θ (degrees)	S-QPDCT		S-QPDCT	
	K_I/K_{Iexact}	$K_{II}/K_{IIexact}$	K_I/K_{Iexact}	$K_{II}/K_{IIexact}$
15	1.0132	0.9620	1.009	0.9742
30	1.0145	0.9634	1.0093	0.9759
45	1.0162	0.9692	1.0112	0.9816
60	1.0148	0.9660	1.0096	0.9783
75	1.0138	0.9668	1.0087	0.9796

Angle θ (degrees)	S-QPDCT		S-QPDCT	
	K_I/K_{Iexact}	$K_{II}/K_{IIexact}$	K_I/K_{Iexact}	$K_{II}/K_{IIexact}$
15	1.0126	0.9650	1.0096	0.9776
30	1.0130	0.9638	1.0093	0.9762
45	1.0163	0.9692	1.0112	0.9816
60	1.0137	09655	1.0096	0.9779
75	1.0130	0.9663	1.0087	0.9792

4.4 Off-Centre Crack

This is an example of off-centre crack in finite body, crack location is arbitrary and plate is subjected to far field remote tension $\sigma = 1$. Plate Dimensions are $2H = 2W = 2$, $2a = 0.5$, $E = 0.5$, $\nu = 0.3$. The centre of location of the crack is at a distance $e_x = e_y = 0.5$ from the origin.

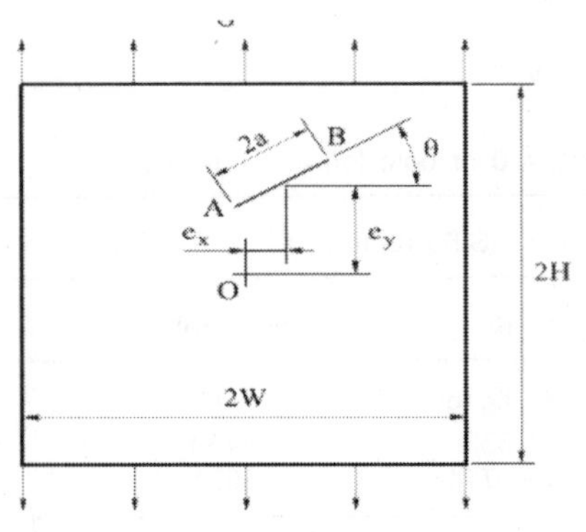

Fig. 7. Off-centre crack.

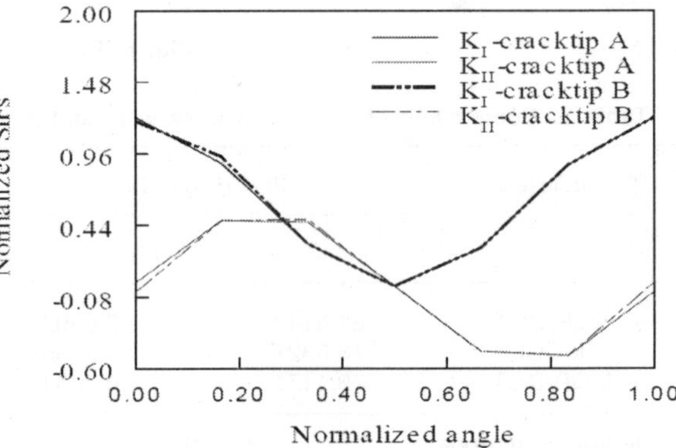

Fig. 8. Normalized SIFs as fuction of angle θ.

Numerical results, normalized by $\sigma\sqrt{\pi a}$ are presented on the Tables 5 and 6 for Mode-I and Mode-II at both crack tips.

Table 7. Normalized SIFs at crack tip A as a function of crack angle θ.

Angle θ	$K_I^A/\sigma\sqrt{\pi a}$			$K_{II}^A/\sigma\sqrt{\pi a}$		
	SQP-DCT	MQP-DCT	J-integral	SQP-DCT	MQP-DCT	J-integral
0	1.2501	1.2249	1.2303	0.0299	0.0262	0.0276
$\pi/6$	0.8937	0.8902	0.892	0.4737	0.4750	0.4887
$\pi/3$	0.3009	0.3025	0.3034	0.4621	0.4622	0.4742
$\pi/2$	0	0	0	0	0	0
$2\pi/3$	0.2781	0.2765	0.2789	−0.4742	−0.4770	−0.4894
$5\pi/6$	0.8846	0.8792	0.8832	−0.4996	−0.5060	−0.5190
π	1.2536	1.2249	1.2319	−0.0446	−0.0412	−0.0431

Table 8. Normalized SIFs at crack tip B as a function of crack angle θ.

Angle θ	$K_I^A/\sigma\sqrt{\pi a}$			$K_{II}^A/\sigma\sqrt{\pi a}$		
	SQP-DCT	MQP-DCT	J-integral	SQP-DCT	MQP-DCT	J-integral
0	1.2345	1.2009	1.2043	−0.0441	−0.0412	−0.0437
$\pi/6$	0.9465	0.9427	0.9446	0.4674	0.4730	0.4851
$\pi/3$	0.3090	0.3077	0.3084	0.4762	0.4822	0.4944
$\pi/2$	0	0	0	0	0	0
$2\pi/3$	0.2787	0.2765	0.2787	−0.4741	−0.4770	−0.4890
$5\pi/6$	0.8841	0.8792	0.8833	−0.4990	−0.5060	−0.5191
π	1.2501	1.2249	1.2345	0.0297	0.0262	−0.0276

CONCLUSION

It is clearly seen that with the usage of new or modified shape functions as has been proposed, the numerical error in the SIFs as predicted by the DCT has reduced as compared to the ones predicted by the standard QP element. Hence the modified quarter point element defined in this paper has been shown to produce highly accurate SIFs.

REFERENCES

1. M.L. Williams, *ASME J. Appl Mech* 19 (1952)526.
2. M.L. Willaims, *ASME J. Appl Mech* 24(1957)109.
3. Raju *et al.*, SIFs, *Theoretical and Applied Fracture Mechanics*, 33(2) (2000) 73.
4. R.D. Henshell, K.G. Shaw. *Int. J. Numer Meth Engng* 9(1975) 495.
5. R.S. Barosum, *Int. J. Numer Meth Engng*, 10 (1976) 25.
6. L.J. Gray, G.H. Paulino, *SIAM J. Appl Math* 58(1998) 428.
7. T.L. Anderson, *Fracture mechanics*, CRC Press LLC (Florida) 1995.
8. E.F. Rybicki, and M.F. Kanninen, *Engineering Fracture Mechanics*, 9(4), (1977), 931-938.
9. J.R. Rice, *ASME J. Appl Mech* 35 (1968)379-86.

41

Simulation of Crack Prediction Using Acoustic Emission and Artificial Neural Network

SANJAY KUMAR SINGH [1], K. SRINIVASAN[2] AND D. CHAKRABORTY[3]

[1]Vikram Sarabhai Space Centre Thiruvananthapuram-695 001, India
[2]Department of Mechanical Engineering, IIT Madras, Chennai-600 036, India
[3]Department of Mechanical Engineering, IIT Guwahati, Guwahati-781 039, India

ABSTRACT

This paper demonstrates the efficacy of artificial neural network in predicting cracks using acoustic emission signals. Acoustic emission (AE) signals have been simulated and filtered as High- and Low- frequency components, based on the fact that cracks would lead to generation of higher frequencies compared to other extraneous sources such as friction. Various combinations of signal metrics such as Maxima, Minima, Mean, and RMS have been used as inputs to train a back propagation neural network. The output of the network is the ratio of the magnitudes of high frequency to low frequency components of the signal which indicates whether the source of AE signal is a crack or not. The neural network seems to hold promise as a potential crack detection tool with acoustic emission signals.

.**Keywords:** Crack; Acoustic Emission; Artificial Neural Network.

1. INTRODUCTION

The use of acoustic emission in structural testing and monitoring is fairly widespread, and routinely used for testing the integrity and life of structures like bridges, pressure vessels, dams and containments, etc. However, precise measures like sources of cracks, evolution of crack formation, growth and propagation still remain challenging problems in this field. Nevertheless, the technique, in conjunction with signal processing tools has proved to be an effective tool for engineers and practicing scientists. Some relevant studies of the past in this direction are discussed below.

The acoustic emission (AE) produced by cracks arises from in-plane sources, which are high frequency extensional and shear waves and very weak low frequency flexure waves. AE produced by impact or friction emanate from out-of plane sources, which are high frequency shear waves, low frequency flexure waves and very weak high frequency extensional waves. Dunegan (1996, 1998) split the AE signals into a low frequency (LF) and high frequency (HF) component and took the peak of the signals from each component, and found the ratio of peak of the high frequency

(HF) component to the peak of low frequency (LF) component. He found that if the source of AE signals are cracks, then ratio of peaks was greater than unity and if the source of AE signals are extraneous noise like friction, impact etc then ratio was less than one. He also found that if the ratio increases with time then it indicates that the crack is in first half of the thickness while if the ratio decreases with time then it indicates that the crack is in the other half thickness of the structure. Hamstad and McColskey (1999) used several AE sensors for detecting and monitoring fatigue crack growth in bridge steels. The array of sensors used arrival time delay to process the crack origin and growth by analyzing several data sets. There have also been theoretical efforts to study crack growth using fracture mechanics (see for example, Lysak (1996)). Neural network has been successfully used as a predictive tool in several fields of engineering. Therefore, it is sufficiently robust to be used in various engineering situations where prediction of a quantity or an event is desired based on a set of inputs for the system to learn. However, in the field of acoustic emission, the use of neural network seems to be very limited. For instance, Walker *et al.* (1997) have used neural networks for using acoustic emission to predict burst pressure in impact damaged composite pressure vessels. Therefore, the objective of this work is to reinforce the applicability of neural networks in crack prediction using simulated AE signals.

2. SIGNAL SIMULATION AND CRACK DETECTION ALGORITHM

AE signals have a very large bandwidth ranging from few kHz to few MHz. Therefore, to generate such complex signals, combination of several signal generators has been used in the MATLAB simulation as shown in Fig. 1. MATLAB's SIMULINK module provides the tools for generating

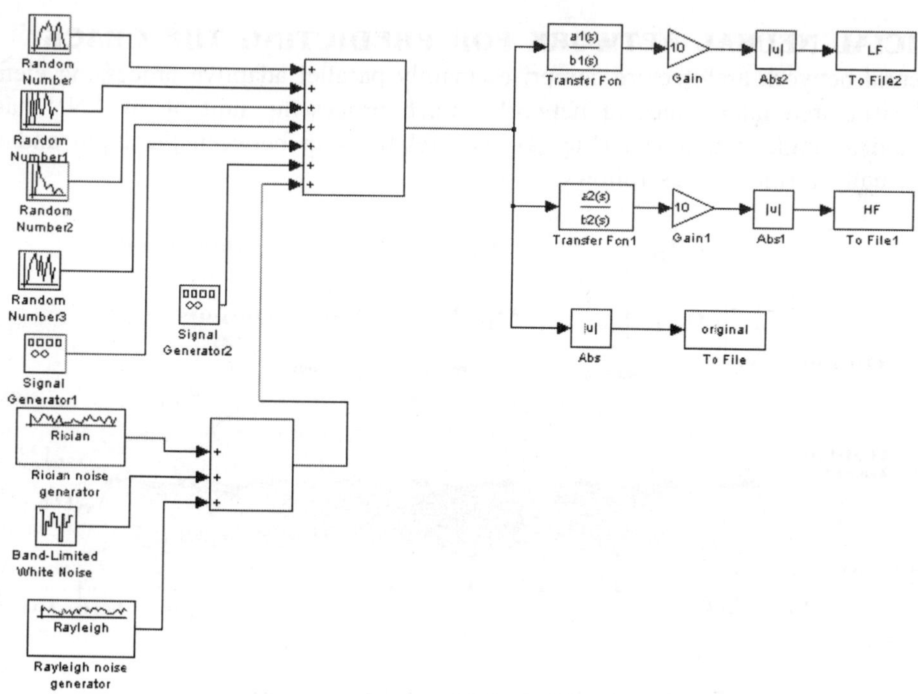

Fig. 1. MATLAB simulation block.

signals of desired features. The random number generator is used to generate random signals, by providing the requisite statistical parameters and sampling time. By varying the sampling time, it is possible to vary the maximum frequency of the random signal. Therefore, in order to generate a signal, which is a combination of both high and low frequency signals, three or four random numbers of different seed and sampling time are combined, to include signals like those generated from friction, impact etc. The noise generators like band-limited white noise generator, Rayleigh noise generator and Rician noise generator are taken and combined to simulate a complex signal similar to that of AE signals.

The characteristics of the signal obtained by MATLAB simulation, like maximum amplitude, mean, range, STD and RMS, are evaluated and all these data are recorded. Then the simulated signal is broken in two parts, one part is passed through high band pass filter (cheby2 filter) (200 kHz-2 MHz) and other part through low band pass filter (20 Hz-200 kHz). Then each filtered signal is passed through an amplifier to avoid loss of signal. The absolute value of the maximum amplitude of the signals is found out for each of the high frequency part (HF) of the signals and low frequency part (LF) of the signals, and the ratio of these amplitudes (HF/LF) is recorded for each case.

The ratio of maximum amplitude of high frequency (HF) and low frequency (LF) signals are used to determine the crack and crack depth in the structure. For example if the ratio is greater than one, then it indicates that there is crack in the specimen while if the ratio is less than one then it indicates that the source of AE signals are extraneous noise like friction, impact etc. The original signal (characterized by its important parameters) and ratio of maximum amplitude of high frequency (HF) and low frequency (LF) signal are used for training a Back Propagation Neural Network.

3. ARTIFICIAL NEURAL NETWORK FOR PREDICTING THE CRACK

Artificial neural network architecture comprises mainly parallel adaptive processing elements with hierarchical structured interconnected networks. Each processing unit of an ANN has multiple input slots and a single output slot (Fig. 2). The relationship between the input and the output signals are usually formulated as follows:

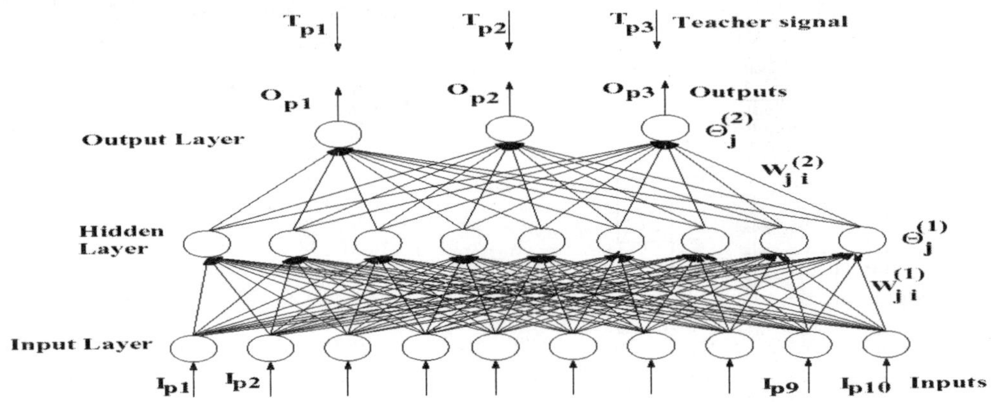

Fig. 2. Three layer back propagation neural network.

$$O_i = f(y_i) = 1/(1 + \exp(y_i)) \qquad \text{...(1)}$$

$$y_i = \sum_{i=1}^{l} w_{ij} I_i - \theta_i \qquad \text{...(2)}$$

where, O_j is the output signal of the *j*th unit, y_j is the potential of the *j*th unit, $f()$ is the activation function that is a sigmoidal function in this case, w_{ji} is the connection weights between *i*th and *j*th units, θ_j is the threshold value of the *j*th unit and l is the number of input signals.

Figure 2 shows a three layer neural network. All the units are formed into multiple layers; i.e., an input layer, a hidden layer, and an output layer. The basic idea of training a neural network is as follows. First, the square error of the pth training pattern Ep is defined as

$$E_p = \sum_{k=1}^{m} \left(T_{pk} - O_{pk}\right)^2 / 2 \qquad \text{...(3)}$$

where, T_{pk} is the teacher signal (desired output) to the kth output unit for p^{th} training pattern, O_{pk} is the output signal to the kth output unit for pth training pattern and m is the number of output units. In the training process, w_{ji} and θ_j are modified repeatedly based on gradient descent method to minimize the above error. This modification proceeds downward. Through such an iterative process, the network attains the ability to promptly output the similar signal to the teacher's one. This training algorithm is called Back Propagation Neural Network (BPNN).

3.1 Characterization of the Simulated Signal

Data obtained from MATLAB simulation like maximum, mean, range, RMS, STD and ratio of amplitude of high frequency (HF) and low frequency (LF) signals are normalized between 0.1 to 0.9 for neural network input-output by using the following equation

$$y = 0.1 + 0.8\frac{x - x_{min}}{x_{max} - x_{min}} \qquad \text{...(4)}$$

where y is in non-dimensional form, x is in dimensional form, x_{max} and x_{min} are maximum and minimum values of a particular data set respectively.

The normalized values of maximum amplitude, mean, range, RMS and STD of complex signals obtained from the combination of different type of signals obtained from MATLAB simulation are arranged in seven different types of input data with one output for artificial neural network. The ratio of maximum amplitude of high frequency and low frequency works as output parameter of artificial neural network and that of the seven types input parameters of the complex signals are given below:

- Maximum amplitude, mean, range, RMS and STD (Type A)
- Maximum amplitude, mean, range and RMS (Type B)
- Maximum amplitude, mean, RMS and STD (Type C)
- Mean, range, RMS and STD (Type D)
- Maximum amplitude, mean and RMS (Type E)
- Mean, range and RMS (Type F)
- Mean, RMS and STD (Type G)

The normalized data sets are shuffled randomly and 70% of the data sets are used for training the network and the rest 30% are used for testing of the network.

3.2 Deciding the Artificial Neural Network Architecture

In the present case a back propagation neural network has been used. The output of the network is the ratio of the magnitude of high frequency to low frequency signal. Therefore the neural network has only one output node. As discussed in the previous section, different combinations of the signal characteristics (i.e. types A through G) have been tried as input to the neural network. This along with different number of hidden nodes (for one hidden layer) has been tried with different values of learning rate (η) and momentum parameter (α). Large number of such runs has been made and for each case training error, testing error as well as the number of iterations required for the convergence of the error has been noted. Table 1 shows some of these results.

Table 1. Training and testing error for different network architectures

Type of Input	Input Nodes	Hidden Nodes	Learning Rate (η)	Momentum Parameter (α)	Number of Iterations	Mean Square Training Error	Mean Square Testing Error
A	5	5	0.3	0.4	11334	0.00525320	0.00607143
A	5	4	0.3	0.4	14772	0.00525320	0.00607427
A	5	6	0.3	0.4	11148	0.00525320	0.00607382
A	5	5	0.5	0.7	46140	0.00525320	0.00525320
B	4	4	0.3	0.4	17109	0.00525320	0.00609608
B	4	3	0.3	0.4	13963	0.00525320	0.00609656
B	4	5	0.3	0.4	12843	0.00525320	0.00609491
B	4	4	0.5	0.7	145832	0.00525320	0.00617165
C	4	4	0.3	0.4	10976	0.00525320	0.00608822
C	4	3	0.3	0.4	12428	0.00525320	0.00606032
C	4	5	0.3	0.4	13199	0.00525320	0.00609297
C	4	4	0.5	0.7	81208	0.00525320	0.00615381
D	4	4	0.3	0.4	203276	0.00525320	0.00601646
D	4	3	0.3	0.4	193286	0.00525320	0.00601817
D	4	5	0.3	0.4	178880	0.00525320	0.00601752
D	4	4	0.5	0.7	158806	0.00525320	0.00600809
E	3	3	0.3	0.4	7652	0.00525320	0.00610824
E	3	2	0.3	0.4	7954	0.00525320	0.00610216
E	3	4	0.3	0.4	5296	0.00525320	0.00610110
E	3	4	0.5	0.7	141981	0.00529658	0.00611890
F	3	3	0.3	0.4	298324	0.00525320	0.00601497
F	3	2	0.3	0.4	330583	0.00525903	0.00602117
F	3	4	0.3	0.4	126191	0.00525320	0.00601768
F	3	4	0.5	0.7	630923	0.00525826	0.00603426
G	3	3	0.3	0.4	102499	0.00525320	0.00599340
G	3	2	0.3	0.4	546096	0.00525538	0.00597659
G	3	4	0.3	0.4	402884	0.00525320	0.00597484
G	3	3	0.5	0.7	438798	0.00528161	0.00599893

4. RESULTS AND DISCUSSION

Table 1 compares the results of training and testing error with number of iterations required in converging to the error. It could be observed from the table that A-type of input patterns is better than others. Keeping in mind the training error, testing error and the number of iterations, among the A-type input pattern results, first one is found to be the best. Therefore, among all the combinations of input-output data set tried here, the optimum network architecture obtained is as follows:

Number of input nodes	:	5
Input parameters	:	maximum amplitude, mean, range, RMS and STD
Number of hidden layer	:	1
Number of hidden nodes	:	5
Learning rate	:	0.3
Momentum parameters	:	0.4
Number of output nodes	:	1
Output parameter	:	ratio of maximum amplitude of high frequency and low frequency

For this network the variation of mean square training and testing errors are shown in Fig. 3. Some times it is possible that the mean square error is small but the actual error is large. Therefore, in order to check the efficacy of the neural network, the actual values of the ratio predicted by the network is compared with the actual values of the ratio obtained from the MATLAB simulation. To enable a proper comparison, the normalized values obtained from the neural network are first converted to the denormalized values using equation (5):

$$ x = x_{min} + (y - 0.1)\frac{x_{max} - x_{min}}{0.8} \qquad \qquad ...(5) $$

where, x is the normalized value and y is the denormalized value.

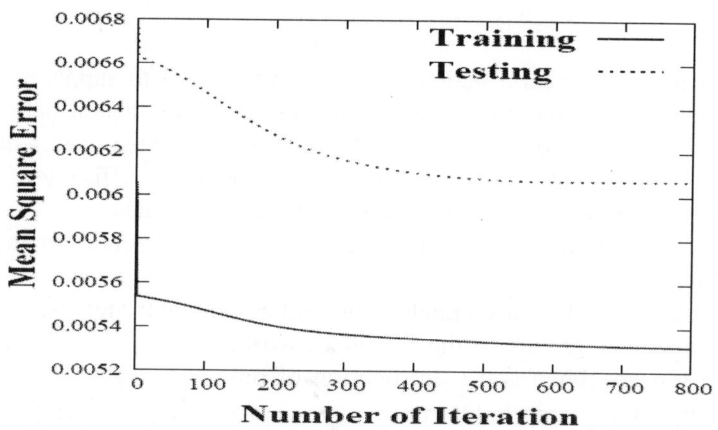

Fig. 3. Variation of mean square training and testing errors.

Figure 4 shows the comparison of the neural network predicted values with those from the MATLAB simulation results. It could be observed that out of the 163 numbers of data tested, most of them are within 20% of the desired values. This shows a very good prediction by the trained neural network.

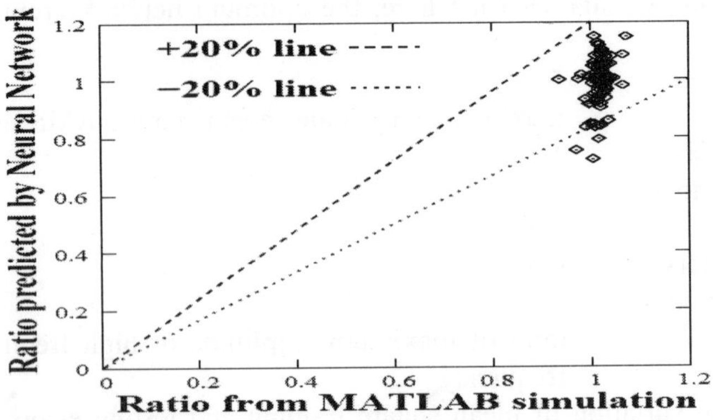

Fig. 4. Comparison of predicted and actual output.

A Hierarchic model for static analysis of laminated composite plate bending has been presented. The numerical results discussed for a wide range of geometry, loading and boundary conditions are in good agreement with the available results in the literature. Also the effects of various parameters such as plate width-to-thickness ratio, fiber orientation angle, number of layers, ratio of thickness of individual lamina, plate aspect ratio and shear correction factor on static deflection are studied in greater detail. Locking phenomenon was found to be suppressed as the element order is upgraded as is evident from the results of the Locking test. Based on these observations, the said p-refinement technique can be recommended for the static analysis of laminated composite plates to predict displacements with adequate accuracy even for very thin plates.

CONCLUSION

The present studies indicate that neural network is a promising tool to supplement AE based crack and fault detection techniques. In the present case, the efficacy of the neural network in conjunction with the AE signal has been studied with simulated data. As a first step in this direction, simulated data has been used for training and testing the network, but the actual efficacy of this approach will be better felt when the network is trained and tested with real life data.

While the present study is of indicative nature, there is a lot of scope for improvement of this paradigm, in the following areas:

Choice of the neural network and compatibility studies with AE signals.

Choice of input parameter groups to the neural network.

Studies on online test specimen and identification of online issues.

In summary, this upcoming area holds room for extensive further investigations.

REFERENCES

1. L. Dunegan Harold, 1996, Use of Plate Wave Analysis in Acoustic Emission Testing to Detect and Measure Crack Growth in Noisy Environments. Proceedings of Structural Materials Technology-An NDT Conference, Feb. 20-23, San Diego California.

2. L. Dunegan Harold, 1998, Modal Analysis of Acoustic Emission Signals, Journal of Acoustic Emission volume 15, 1-4.

3. Hamstad, M.A., McColskey, J.D., 1999, Detectability of slow crack growth in bridge steels by acoustic emission, Materials Evaluation, vol. 57, no. 11, pp. 1165-1174.

4. Lysak, M.V., 1996 Development of the theory of acoustic emission by propagating cracks in terms of fracture mechanics, Engineering Fracture Mechanics, vol. 55, no. 3, pp. 443-452.

5. Walker, J.L., Russell, S.S., Workman, G.L., and Hill, E.V. 1997, Neural network/acostic emission burst pressure prediction for impact damaged composite pressure vessels, Materials Evaluation, vol. 55, no. 8, pp. 903-907.

42

Analysis of Crack Growth Behavior in Metals and Non-Metals Using DCPD Technique

V.D. Palve, R.K. Pandey and P. Mahajan

Department of Applied Mechanics, Indian Institute of Technology, Delhi, New Delhi-110 016, India
email: vikas_palve@yahoo.co.in

ABSTRACT

The direct current potential drop (DCPD) technique has been employed to continually monitor the growth of an edge crack in metals (aluminium and mild steel) and non-metals (polycarbonate).The dimensionless relationship between the potential drop value and the crack length has been obtained and compared by using ASTM standards. Separate equations have been used to calculate the crack length of compact tension (CT) test specimen. The crack length deduced from the DCPD technique is within a difference of 2.5 % of the actual lengths measured from specimen fracture surfaces.

Keywords: Edge crack; Potential drop; crack growth; polycarbonate.

1. INTRODUCTION

Adhesive joints are important in manufacturing sector particularly in the automotive industry where the increasing use of plastics and the need to reduce overall vehicle weight has made adhesive more popular. Despite of these benefits, there is relatively little investigation on the crack growth behavior of adhesively bonded joints [1]. This could be due to difficulties in monitoring crack growth in adhesively bonded joints on continuous basis. In metals and metallic joints such as welds slow crack growth can be measured using the direct current electric potential drop method (DCPD). This method relies on the spreading and oxidation of the advancing crack faces to increase the specimen's electrical resistance [2]. However, this method cannot be used in monitoring crack growth in nonconductive materials and adhesively bonded joints.

The interfacial crack growth measurement has been done using a modified compact tension shear specimen (CTS) involving two aluminum plates bonded by a thin ductile adhesive layer [3]. For this purpose a special crack propagation gauge was installed across the bond line. It was shown that both the fracture load, as well as, the extent of stable crack growth, increases as Mode-II conditions are approached. Similarly the crack growth in adhesively bonded joints subjected to creep loadings has been measured [4]. This method involves supplementing the nonconductive ligament of the joint with a uniform, thin layer of brittle, conductive material such as a thin layer of carbon paint and the changes in its electrical resistance are monitored. It is shown that the crack grows in the middle of the adhesive layer rather than at the interface of the joint. The creep crack growth rate in adhesively bonded joint is related to mode-I energy release rate (G_{Ic}).

The direct current electric potential around a number of penny-shaped cracks was analyzed by Crack-Flow Modification Method (CFMM) in order to investigate the applicability of the Direct Current Electrical Potential Method (DC-EPM) to the detection of damage due to multiple small internal cracks [5].

In our investigation we are incorporating a concept of modified DCPD methodology for monitoring crack growth in nonconductive materials. This method involves supplementing the nonconductive ligament of the joint with the crack gauge, which is used as measuring transducer. Its structure is based on a constantan layer and an electrically insulating backing. The crack gauges are being bonded as a strain gauge onto the specimen. The progress of the crack through the joint also cracks this material and increases the system electrical resistance. Hence there is a potential drop. As a crack propagates the specimen resistance increases in a way that can normally be related to actual values of the crack depth. Thus output of the DCPD equipment is a direct measure of the crack length.

2. DIRECT CURRENT POTENTIAL DROP (DCPD) TEST

DCPD test machine has main display unit, current input and potential drop output cables. The crack growth across the specimen is indicated in terms of potential drop on the main display unit. So it is important to calibrate the DCPD machine output in terms of crack growth.

2.1 Analytical Approach

The relationship between measured potential drop voltage (V) and crack size is given in Eq. (1). It was developed by Hicka and Picard [6] from finite element analysis and was verified through experimental techniques valid for (a/W) ratio as

$$0.24 \leq (a/W) \leq 0.7$$

$$V/V_R = A_0 + A_1(a/W) + A_2(a/W)^2 + A_3(a/W)^3 \qquad ...(1)$$

Here,   V = Measured potential drop voltage (V).

V_R = Reference crack voltage (V).

a = crack size (mm).

W = Specimen Width (mm).

$A_0 = 0.5766, A_1 = 1.9169, A_2 = -1.0712, A_3 = 1.6898$.

OR

In reverse notation, $a/W = B_0 + B_1(V/V_R) + B_2(V/V_R)^2 + B_3(V/V_R)^3$ ...(2)

Here, $B_0 = -0.5051, B_1 = 0.8857, B_2 = -0.1398, B_3 = 0.0002398$

2.2 Test Procedure

First in Eq. (1) the reference voltage V_R is computed for any reference (a/W) ratio and corresponding voltage measurement V. Computing V_R voltage in this way accounts linearly for small changes in applied current. The computed reference voltage can then be used with the second form of the Eq. (2) to determine the crack size for all voltage values V. The compact tension (CT) specimen is shown in Fig. 1 with various lead connections.

W = Specimen Size (80mm).

a = Crack size.

1. Current input leads.
2. Voltage measurement points.
3. Current output leads.

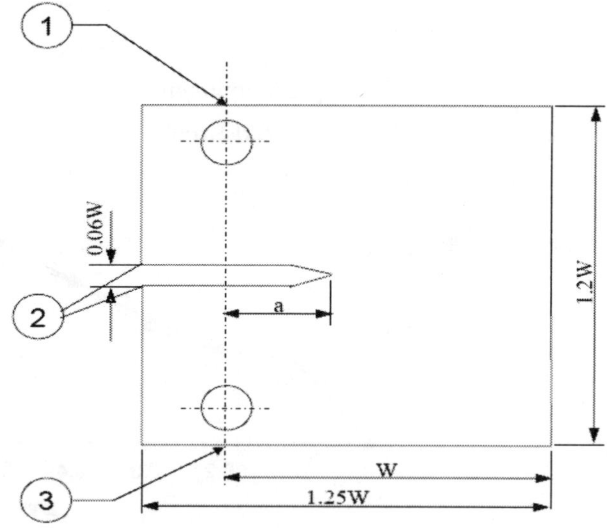

Fig. 1. Geometry and electric potential wire locations.

2.3 Experimental Approach

Experiments were conducted on samples of two different materials is mild steel (M.S.) and Al-alloy of sizes 100 mm (length) × 50 mm (width) and 2 mm (thickness). Both the materials are electrically conducting.

Table 1. DCPD output for metals (Al-M.S.)

S. N.	Crack Length	Potential Drop in µv	
	in mm	Aluminium	Mild Steel
1.	00	376.6	752.6
2.	05	402.1	829.4
3.	10	501.7	965.4
4.	15	651.1	1223.1
5.	20	856.2	1685.0
6.	25	1050.1	1895.1
7.	30	1273.4	2317.2
8.	35	1551.1	2926.3
9.	40	1880.8	3110.9

Sizes of both specimens were kept the same for verifying the results with respect to the changes in the material. An edge crack is made in the specimen and corresponding readings of the potential drop were taken. Then crack length was increased progressively with 5 mm step. Same procedure is repeated till the crack length becomes 40 mm. Table 1 provides the DCPD output for aluminum and mild steel. This output is used in plotting the calibration curves as shown in Fig. 2. The calibration curves for both the material are fairly close upto the (a/W) ratio = 0.7. Then the DCPD output in the dimensionless form plotted against the crack-length is found to be independent of material.

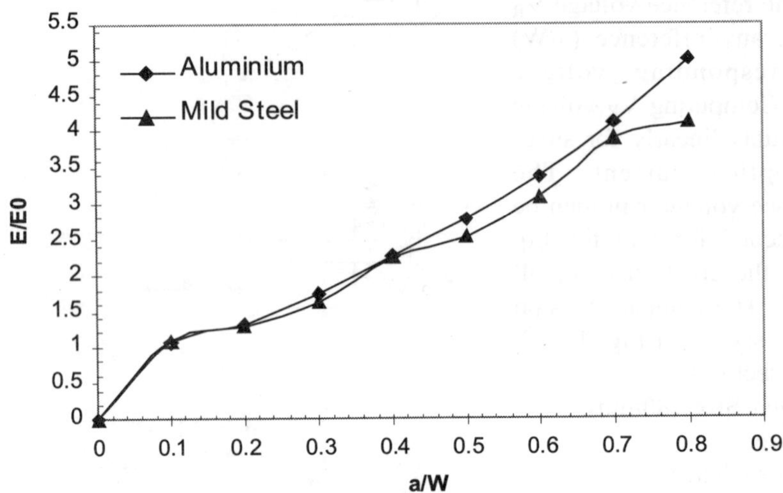

Fig. 2. Potential Drop Ratio against Crack size-width Ratio.

3. DCPD METHOD TO NON-CONDUCTING MATERIALS

DCPD method involves supplementing the nonconductive material with the crack gauge which is used as measuring transducer as shown in Fig. 3. Its structure is based on a constantan layer and an electrically insulating backing. The crack gauges are being bonded as a strain gauge onto the specimen.

The progress of the crack through the specimen also cracks this material and increases the system electrical resistance. A data acquisition system connected to this joint circuit provides continuous monitoring of the crack growth in terms of the potential drop.

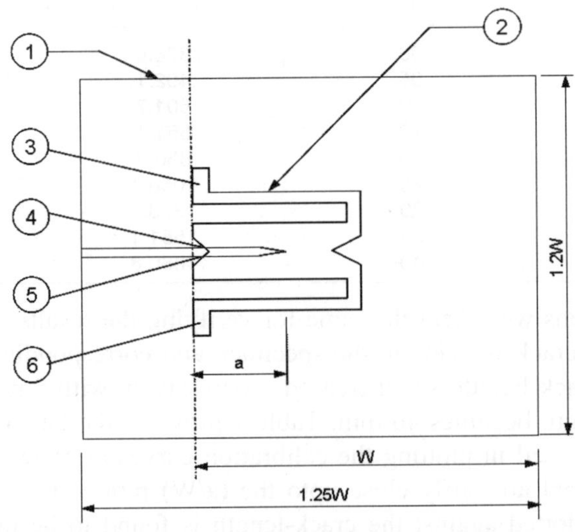

Fig. 3. Geometry and lead connections for DCPD method.

Details of specimen
Size (W) = 80 mm, Crack size = a
1 Polycarbonate
2 Crack gauge
3, 6 Current in and Current out
4, 5 Potential drop.

Table 2. DCPD output corresponding to crack length for Polycarbonate.

S. N.	Crack Length in mm	Potential Drop in mv
1.	00	380.5
2.	02	1056.9
3.	05	1746.4
4.	08	2476.3
5.	11	2988.3
6.	13	4240.2
7.	15	5965.3
8.	17	6281.8
9.	20	6398.1
10.	22	6413.4

Experiments were conducted on non-conducting material (polycarbonate) having size (W) 80 mm bonded with Crack gauge which consists of thin conducting wires made up of constantan (Cu-Ni alloy). An edge crack is made in the specimen and corresponding reading of the potential drop is taken using the DCPD machine. Then the crack size is increased in steps of few millimeters and the corresponding potential drop is measured as given in Table 2. Same procedure is repeated till the length of crack gauge.

4. RESULTS AND DISCUSSION

In DCPD test, experimental results were compared with the analytical procedure mentioned in article 2.1.

4.1 Conducting Materials

The Table 3 shows the comparison between the actual values of crack sizes on the specimen and those calculated from the ASTM standard procedure [7]. It is clear from the readings that as crack size increases, the potential drop across the crack tip increases. The percentage error in the analysis is around 2.5. So the results are much closer and can be accepted.

Table 3. Actual and calculated crack sizes for Aluminium.

Crack size a (mm)	Ratio a/W	Voltage µV	Calculated crack a'(mm)	Error %
26.000	0.325	1230.000	REF	—
34.000	0.425	1452.000	34.950	2.794
40.000	0.500	1610.000	40.680	1.700
46.000	0.575	1809.000	47.060	2.304
52.000	0.650	2024.000	53.250	2.404
56.000	0.700	2224.000	57.400	2.500

4.2 Non-Conducting Material

Fig. 4 shows the results of DCPD technique implied to polycarbonate with the help of crack gauges bonded over the surface. It is clear from the graph that the potential drop cross the crack tip increases with increase in the crack length.

The experiment conducted for the non-conducting materials (polycarbonate) are not satisfactory as the results obtained in Table 4 using the crack gauges. It is giving large errors in the crack size values.

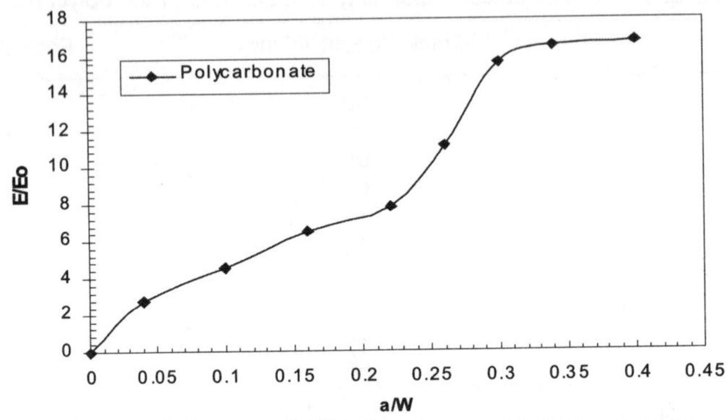

Fig. 4. Potential Drop Ratio against Crack size-width Ratio.

Table 4. Actual and calculated crack sizes for Polycarbonate.

Crack size a (mm)	Ratio a/W	Voltage μV	Calculated crack a'(mm)	Error %
14.000	0.280	6161.600	REF	
20.000	0.400	6175.400	15.500	22.5
25.000	0.500	6182.400	16.400	34.4

CONCLUSION

The objective of experiments to implement the modified DCPD method is fulfilled in case of compact tension aluminium specimen as we obtained desired crack growth from the potential drop. In case of non-conducting material, polycarbonate, potential drop is obtained for surface edge crack using the crack gauge. The results are not appropriate in case of polycarbonate because of limitation of size of the crack gauges.

REFERENCES

1. N. Wang, Cardona D.C., Bowen P., 1997, Finite-element analysis of selectively SiC fibre-reinforced titanium composites (II), International Journal of Fracture 87, 225-40.
2. M.A. Hicks, A.C. Pickard, 1982, A comparison of theoretical and experimental methods of calibrating the electrical potential drop technique for crack length determination, International Journal of Fracture 20, 91-101.
3. Narasimhan R., Madhusudhana K.S., 2002, Experimental and numerical investigations of mixed mode crack growth resistance of a ductile adhesive joint, Engineering Fracture Mechanics 69 ,865-883.
4. Nayeb Hashemi. Swet, A. Vaziri, 2004, New electrical potential method for measuring crack growth in nonconductive materials, Measurement 36, 121-129.
5. N. Tada, Y. Hayashi, T. Kitamura and R. Ohtani, 1997, Analysis on the applicability of direct current electrical potential method to the detection of damage by multiple small internal cracks, International Journal of Fracture 85, 1-9.
6. Hicka and Picard, 2000, Standard Test Method of Measurement of Fatigue Crack Growth Rates, STD, ASTM-E 647-ENGL.

43

Modeling of Damage and Failure Using CDM (Continuum Damage Mechanics) in Plastic Deformation

P.R. ZAGADE[1] AND B.P. GAUTHAM[2]

[1]*Department of Mechanical Engineering, College of Engineering Pune, Pune-411005, India*
email: zagadepramod@yahoo.com
[2]*Tata Research Development and Design Center, Pune-411013, India*
email: bp.gautham@tcs.com

ABSTRACT

Models for CDM based on GTN are incorporated in the large deformation elasto-plastic finite element framework. Model is used for modeling and prediction of life of flat plate subjected to repeated bending leading to plastic loading and unloading. Tensile tests are carried out for determining the material properties related to CDM. Repeated/cyclic bend tests of flat sheet are conducted. Simulation model results are compared with experimental results. These simulation models are used to predict the material response (stress-strain, life of the components etc.) subjected to parameter variations (Thickness, angle of bend etc.) in case of repeated bend test. It is seen that the CDM based models are able to predict the failure point of plates with good accuracy.

1. INTRODUCTION

In many practical applications, failure of metal components subjected to repeated plastic loading takes place after a small and finite number of cycles. Designing with an objective on failure is needed in cases where the component is designed for failure in a preferred and desirable pattern. Besides this, material characterization also involves repeated bending or twisting in plastic range till failure in order to characterize the product's life in use. Hence, it is essential to study and have better understanding of large deformations, large strains and failure phenomenon in the materials under given circumstances. With this understanding and support from computational methods, design optimization of industrial processes becomes efficient. Ductile failure is observed in case of most of the engineering materials (metals) especially steels, aluminum etc. Continuum damage mechanics (CDM) based models are extensively used to describe the behavior of the material in the ductile damage regime leading to failure. Gurson model [1] is most commonly used for modeling ductile failure. It uses void volume fraction as the damage parameter and describes behavior through void nucleation, growth and coalescence.

In the present study, FEA based model is developed for analysis of large strain plasticity with damage based on Gurson model for the axisymmetric and plain strain cases. These are used to study tensile testing of circular rods, flat strips and repeated bending of flat strips. Tensile tests are used for parameter estimation and bending studies are conducted to study the number of bends a plate of particular thickness can withstand. A parametric study of the bend test is done.

2. MODELING OF DUCTILE FRACTURE USING GTN MODEL

Ductile fracture results from initiation, growth and coalescence of cavities. First phase of ductile fracture takes place around the nonmetallic inclusions and second-phase particles. Void initiates at weakest link near the crack tip region. Considering a Gaussian inclusion distribution, an assumption is that the microvoid nucleation rate is mainly controlled by the equivalent plastic strain and defined by the relationship (refer eqn. 1) proposed by Chu and Needleman (1980) [8].

$$\dot{f}_n = \frac{f_N}{S_N \sqrt{2\pi}} \exp\left\{ -\frac{1}{2}\left(\frac{\overline{\varepsilon}_{eq} - \varepsilon_N}{S_N} \right)^2 \right\} \dot{\overline{\varepsilon}}_{eq} \qquad ...(1)$$

Growth of formed cavities takes place because of associated strain rate and stress triaxility. Void growth depends on stress triaxility; this dependence is of exponential nature. The volume growth of the voids is equal to the calculated volume expansion. (refer eqn. 2)

$$\dot{f}_{growth} = (1-f)\varepsilon_{kk}^p \qquad ...(2)$$

Coalescence of neighboring micro voids yields to final material failure. A specific coalescence function f^* replaces the porosity f and also includes the accelerating parameter K.(refer equ. 3,4)

$$\begin{aligned} f^* &= f & if \quad & f \le f_{cr} \\ &= f + K(f - f_{cr}) & if \quad & f > f_{cr} \end{aligned} \qquad ...(3)$$

$$K = \left(\frac{f_u^* - f_{cr}}{f_f - f_{cr}} \right) \qquad ...(4)$$

Model, originally developed by Gurson, approximates yield criteria and flow rules for matrix material being rigid perfectly plastic and obeying von Mises yield criteria. The goal was to describe void growth accounting for plastic dilatancy. It was enhanced by Tvergaard [9] modifying the flow potential function depending on σ_{ij} and state variables. Gurson's yield function [1] (Gurson 1977) improved by Tvergaard (1981) [1, 3, 8, 9] for a ductile porous metal is defined by equation (5):

$$\Phi = \frac{\sigma_{eq}^2}{\sigma_0^2} + 2q_1 f^* \cosh\left(\frac{3q_2 \sigma_m}{2\sigma_0} \right) - \left(1 + q_1^2 f^{*2} \right) \qquad ...(5)$$

In general, evolution of void volume fraction results from both nucleation and growth of existing voids. (Eqn. 6)

$$\dot{f} = \dot{f}_{growth} + \dot{f}_{nucleation} \qquad ...(6)$$

2.1 Validation Using Small Punch Test

Validation of the model is done with a small punch test taken from the literature (Kuna, 2004)

[4, 5]. Test geometry, mesh sizes and material as well as damage parameters are taken same as mentioned in the reference. More details can be found from the work (Kuna, 2004) [4, 5]. The simulation results (LDC) are in good agreement with the experimental measurements. (Refer Fig. 1). This validates the model and we can conclude that the model developed can simulate the ductile behavior.

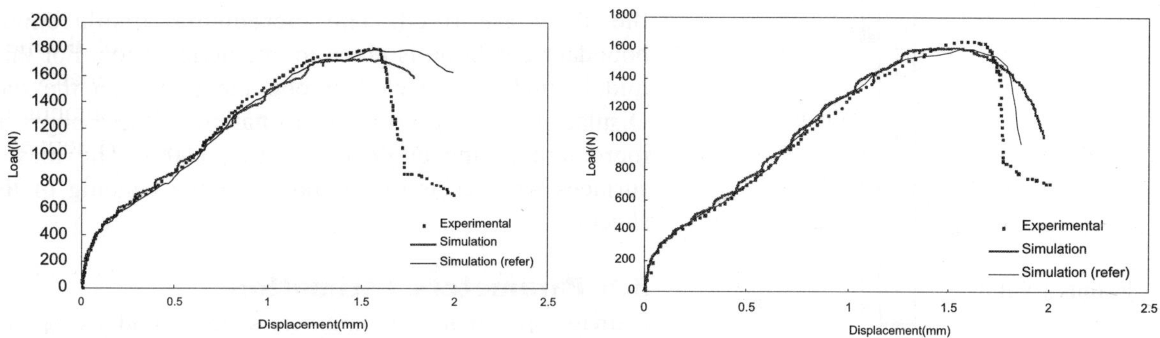

Fig. 1. Validation with small punch test.

3. REPEATED BEND TEST

In many practical applications, failure of metal components subjected to repeated plastic loading takes place after a small and finite number of cycles; this is commonly referred as low cycle fatigue. In the present work, particular case of repeated bending of flat sheets is considered for the fatigue assessment. The life of these flat sheets is expressed in terms of the number of bends it can withstand before crack gets developed when a predetermined amount of bend at the tip of the rod is given. Experiments are carried out on flat plates to find out the life of the plate.

3.1 Repeated Bend Test Experiments

Repeated bend tests are carried on the flat sheets of mild steel. The surface of the sheet is observed after each bend cycle. The initiation and development of the cracks on the surface is noted. Each plate is held tightly in the vice. Then 60^0 bend on each side is applied on the plate. Different plate samples are collected after imposing different number of bend cycles. Then the affected regions i.e. near the clamping are observed. The stretched surface of the plates was observed with a magnification of 8X.(Fig. 2) At 5.25 bend cycles the surface started showing a hair line crack near the master surface.

The instance at which the crack is developed in the plate is not exactly accessible. It can be judged only after it is opened, which may require roughly one additional bend cycle. The plate gets completely fractured in the subsequent 2 or 3 bends.

Fig. 2. Cracks on the plate imposed to repeated bends (2 mm).

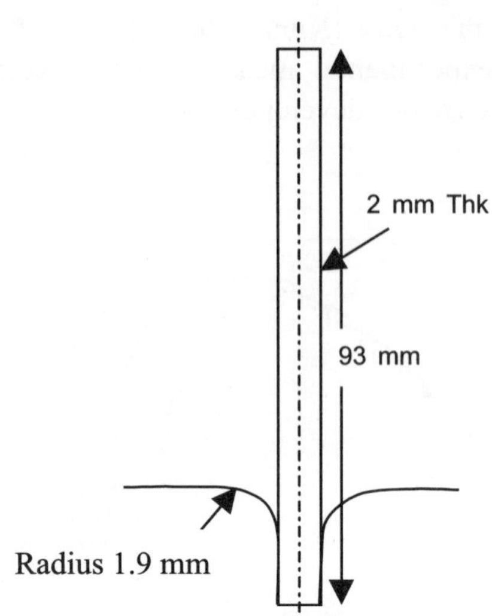

Fig. 3. Geometry of the specimen.

3.2 Simulation of the Repeated Bend Test

As no appreciable strain will be observed in the width direction, plate of infinite width is considered and 2D plain strain model is developed. Figure 3 shows the simulation model for the test. All the bottom nodes of the sheet are fixed. The incremental displacement boundary condition is given to one node on top. For this, middle node is selected among the nodes on the top. Displacement is specified as a rotation of these nodes as increment in the angle of rotation (60°). The master surfaces will act as the supports for the bending of the sheet.

3.3 Parameters Estimation

Damage and material parameters are found by tensile testing of the flat sheets of the material. 2D plain strain code is developed for the simulation of flat plate testing. Plate of larger width (10 times) compared to thickness is used for the tensile testing. For simulations strain in the width direction is considered to be zero. Damage parameters for the flat plate are found as given in Table 1.

Table 1. Damage parameters for flat.

f_0	f_f	q_1	q_2	f_N	S_N	e_N	f_c	K
0.0006	0.338	1.5	1	0.001	0.2	0.4	0.01	2

4. RESULTS AND DISCUSSION

The results of simulation are plotted in terms of different parameters against the number of bends. The failure point of the simulation is considered where the damage in the sheet increases in significant

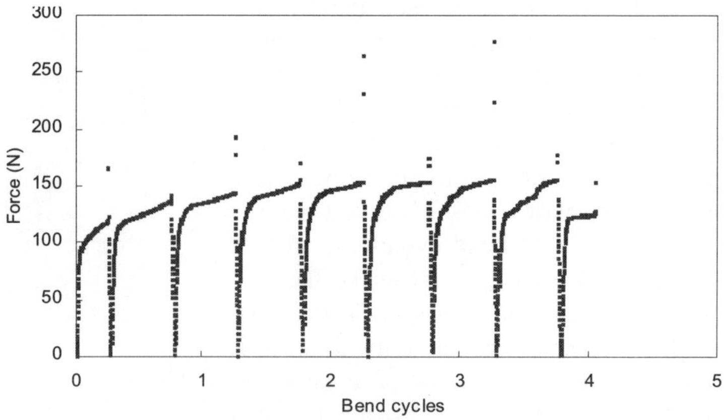

Fig. 4. Force required for bending the sheet.

amount with a very high rate and reaches the critical value. In practice, it can be interpreted as the instant where a point defect is introduced in the material and the sheet will lose its strength in the subsequent one or two bends. Fig. 4 shows the force required for bending of the plate. Due to hardening of the material, an increase in the force required is found as the number of bends increases. This continues till the damage in the material increases drastically. Here the plate loses its stress carrying capacity.

The damage increases with bending (as shown in Fig. 5). The initial rate of increase is not significant with the increase in the number of bends. After certain bends the growth of damage in the most strained region gets highly accelerated. This is point where the crack gets developed. Due to sharp rise in the damage variable material loses its strength. The numerical instabilities in the simulation causes termination of the program execution.

Distribution of damage variable at the time of fracture is shown in the Fig. 5. The zone is highlighted where the crack is likely to initiate.

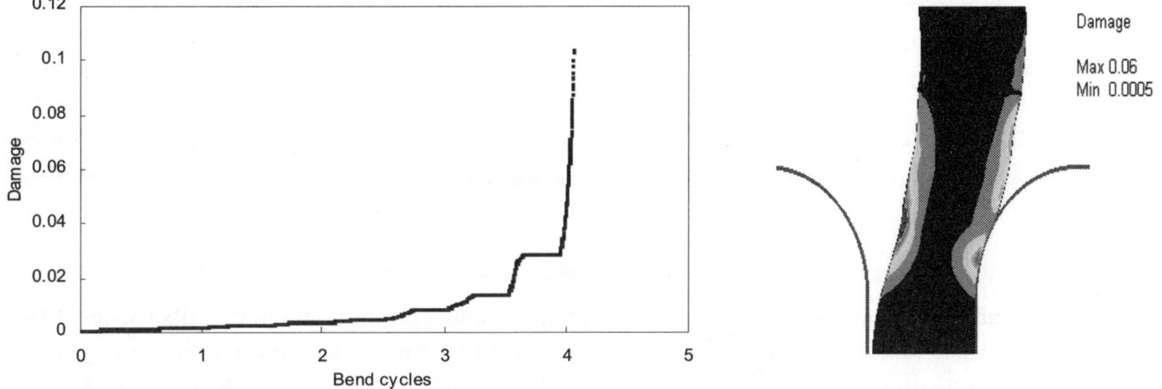

Fig. 5. Damage evolution and distribution at failure.

Results obtained from the simulation of repeated bend test are compared with the results of repeated bend test experiments (Refer Table 2).

Table 2. Comparison of experimental and simulation results (Repeated Bend).

Plate Thickness (mm)	Experimental (Number of bends)	Simulation (Number of bends)
2.0	5.25	4.059
3.5	4.25	3.051

The comparison shows the simulation can predict the crack formation point which is close to the experimental results. Comparison of experiments with the simulation results shows closer agreement. Simulation model predicts the very first opening point of the surface. A better method of monitoring the initiation of crack or simulation of crack formation and growth may provide better comparison.

Life of the plate is dependent on various geometric parameters. Effect of individual parameter variation on the life of the plate is studied. Angle of bend on each side of the plate signifies deformation being imposed in a load cycle. It is observed that as the bend angle increases the number of bends the plate can withstand reduces (Refer Fig. 6). This is obvious as small angle will induce less strain whereas; larger angle inducing more strain fails early.

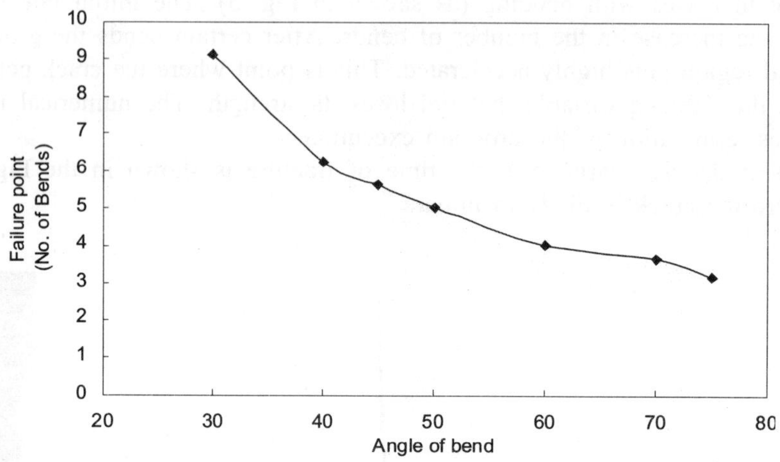

Fig. 6. Effect of changing the angle of bend.

Smaller radius the master surface increases the stress concentration and plate fails earlier. Master surface with larger radius provides a better support avoiding the stress concentration and plate life is more. However after certain radius the improvement in life of plate is not significant (Fig. 7).

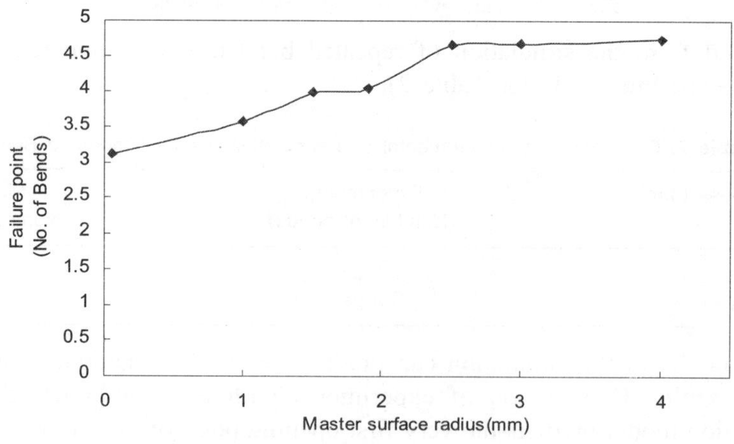

Fig. 7. Effect of changing master surface radius.

The parametric studies help in designing the plates as well as methods of testing as we can identify zones where the effect variation of some of the parameters is low and are ideally suited for tests e.g. MS corner radius should be > 2.6 mm and then the effect is negligible.

CONCLUSION

Modeling of the ductile failure is done using CDM in elastoplastic FE framework. The simulation model developed for modeling repeated bend test can predict the life of the plate. It predicts the point where initiation of micro crack takes place. With each bend damage increases due to increasing plastic strain. At the end sudden rise in the damage variable takes place. Here the material loses all its strength. The force required for bending in this cycle is also less. The crack gets developed in the plate. This simulation model is used for the material characterization and the life prediction. Effect of various parameters on the plate life is studied, the parametric studies have helped identify zones of low influence for certain geometric parameters of test equipment and this helps in design of better testing methods. In repeated bend test experiment, the opening of the crack requires a half or more number of bend cycles after its initiation. Two or three more bend cycles causes the fracture of plate after this point. A sophisticated instrumentation is needed for the testing and observation of the process.

SYMBOLS USED

f_N = initial microvoid volume fraction,

ε_N = mean effective plastic strain of the matrix at incipient nucleation,

S_N = Gaussian standard deviation of the normal distribution of inclusions,

ε_{eq} = equivalent effective plastic strain in the matrix,

f_{cr} = the critical void volume fraction at coalescence onset,

σ_{eq} = von Mises equivalent stress,

σ_m = hydrostatic stress,

σ_0 = yield stress,

q_1 = parameter to account for interactions between voids,

q_2 = parameter to account for void shape.

REFERENCES

1. A.L. Gurson, 1977, Continuum theory of ductile rupture by void nucleation and growth: Part I—Yield criteria and flow rules for porous ductile materials. *J. Engg. Mater. Technol. 99, 2–15.*

2. C.S. Krishnamurthy, 1995, Finite Element Analysis, Tata McGraw Hill Publishing Co. Ltd., New Delhi, 1-127.

3. Jean Lemaitre, Rodrigue Desmorat, 2005, Engineering Damage Mechanics, Springer, New York.

4. Kuna M., Abendroth M., 2006, Identification and Validation of Ductile Damage Parameters by the Small Punch Test, Engineering Fracture Mechanics, 73, Issue 6, 710-725.

5. Kuna M., Abendroth M., 2004, Determination of Ductile Material Properties by Means of the Small Punch Test and Neural Networks, Advanced Engg. Materials, 6(7), 536-540.

6. L.S. Srinath, 1980, Advanced Mechanics of Solids, Tata McGraw Hill Publishing Co. Ltd., First Ed., Tenth Reprint. 1-134.

7. M.A. Crisfield, Non-linear Finite Element Analysis of Solids and Structures, 1, Essentials, John Wiley and Sons,. 1-197.

8. *Needleman, A., Chu, C.C., 1980, Void nucleation effects in biaxially stretched sheets. J. Engng Mater. Technol. 102, 249–256.*

9. *Viggo Tvergaard, 1990, Material failure by void growth to coalescence. Adv. Appl. Mech. 27, 83–151. 23-26.*

44

The Use of the J_k Integral in Thermal Stress Crack Problems of Bi-Material Interface

Ratnesh Khandelwal[1] and J.M. Chandra Kishen[2]

Department of Civil Engineering, Indian Institute of Science, Bangalore-560 012, India
email: [1]ratnesh@civil.iisc.ernet.in, [2]chandrak@civil.iisc.ernet.in

ABSTRACT

A linear bi-material cracked elastic body subjected to thermal loading is analyzed using the J_k integral. This method, used in conjunction with the finite element method is shown to be useful for computing the stress intensity factors for cracks lying between the interface of two dissimilar materials subjected to any type of thermal loading.

1. INTRODUCTION

The study of thermal stress fracture problems due to the local intensification of temperature gradient is of considerable practical importance in the design of turbines, combustion chambers, and nuclear reactors. To determine the stress intensity factor due to thermal effect on homogenous body, Wilson and Yu [1] proposed the modification of the conventional J_1 integral. Shih *et al.* [2] extended the J_1 integral for 3-D thermal problems. Stern [3] using the concept of reciprocal work integral had presented analytical solution for bi-material thermal fracture problem. Yong et al [4] gave an analytical solution for bi-material body subjected to far field heat flux. Sun and Quin [5] recommended the use of MCCI along with displacement ratio method (DR) for better accuracy in SIF. Sun and Ikeda [6] modified Virtual crack extension method and modified crack closed integral method for thermal problem in conjunction with superposition method to get thermal stress intensity factor. Banks-Sill [7] extended conservative M-integral to treat thermal-elastic, bi-material problems.

Khandelwal and Chandra Kishen [8] proposed the analytical expression for J_k integral for bi-material interface problem subjected to mechanical loading and showed that this integral is path independent in a modified sense and is useful in determination of stress intensity factors. J_k integrals are line integrals evaluated along a counterclockwise contour enclosing and shrinking onto the crack tip. Knowles and Sternberg [9] defined J_k as a complex quantity given by

$$J = J_1 - iJ_2 \qquad \qquad ...(1)$$

The quantity J corresponds to the energy release rate for movement of crack edge in any direction. For homogeneous media with traction free crack surface, both component of J, J_1 and J_2 are path indepen-dent. Though J_1 can start anywhere from the lower crack surface and end anywhere along the upper crack surface, the same, does not hold true for J_2 wherein two additional line integrals along the crack surfaces are added up which causes the inclusion of a singular region in the integration. Therefore, direct calculation of J_2 from numerical solution becomes a tedious job.

It will be shown that in the presence of thermal stress, the J_2 line integral over a closed path, which does not enclose singularities, is not equal to zero but is equal to an integral over the area inside the closed path. By combining this area integral with J_k integral, it will be shown that it will behave in the same way as in case of mechanical load and crack tip SIF can be computed for any type of thermal loading.

2. FORMULATIONS

The path independent integral J_k for elastic isothermal homogenous materials in the absence of body force can be expressed as

$$J_k = \int_\Gamma \left(W n_k - \frac{\partial W}{\partial u_{j,i}} u_{j,k} n_i \right) d\Gamma \qquad \ldots(2)$$

where Γ denotes the arbitrary contour which begins and ends at the crack-tip. W is the strain energy density, n_2 is the unit outward normal to the path Γ and u is a displacement vector referred in the cartesian coordinate system. It may be mentioned that the strain energy density W is related to the

stress and strain tensor as $\dfrac{\partial W}{\partial u_{j,i}}$

Applying the above integral to an arbitrary closed integral path Γ, enclosing a portion of the solid can be written as [9]

$$\int_\Gamma \left(W n_k - \frac{\partial W}{\partial u_{j,i}} u_{j,k} n_i \right) d\Gamma = 0 \qquad \ldots(3)$$

However path independent of J_k is not maintained in the presence of body force as well as thermal loadings [10] and for the closed path Γ equation (3) get modified to [10]

$$\int_\Gamma \left(W n_k - \frac{\partial W}{\partial u_{j,i}} u_{j,k} n_i \right) d\Gamma - \int_{A_0} (\overline{F_i} u_{i,k}) dA - \beta \int_{A_0} \left(\frac{1}{2} (\theta \varepsilon_{ii})_{,k} - \varepsilon_{ii} \theta_{,k} \right) dA = 0 \qquad \ldots(4)$$

where A_0 is the area enclosed by the closed path. ε_{ii} is the dilatation and F_i is the prescribed body force and

$$\beta = \frac{E}{(1 - 2\alpha)} \quad \text{for plain strain case}$$

$$\beta = \frac{E}{(1 - \alpha)} \quad \text{for plain stress case}$$

In the absence of body forces equation (4) can be written as,

$$\int_{\Gamma}\left(Wn_k - \frac{\partial W}{\partial u_{j,i}}u_{j,k}n_i\right)d\Gamma - \beta\int_{A_0}\left(\frac{1}{2}(\theta\varepsilon_{ii})_{,k} - \varepsilon_{ii}\theta_{,k}\right)dA = 0 \qquad \ldots(5)$$

Considering two elastic half planes of different materials bounded together as shown in Fig. 1. The plane defined by $x < 0$ and $y = 0$, is unbounded and represents a crack. The material occupying $y > 0$ has Young's modulus E_1, Poisson's ratio v_1 and thermal expansion coefficient α_1 whereas that occupying $y < 0$ has the respective values E_2, v_2 and α_2. Applying separately the above conservation law to the upper and lower closed loops Γ_1 and Γ_2 respectively, which excludes the crack tip as shown in Figure 1.

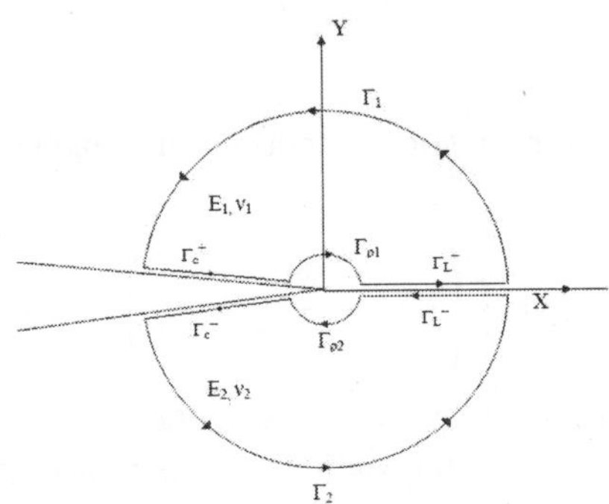

Fig. 1. Contours around the interface crack.

$$\int_{\Gamma^1}\left(Wn_k - \frac{\partial W}{\partial u_{j,i}}u_{j,k}n_i\right)d\Gamma - \beta\int_{A_1}\left(\frac{1}{2}(\theta\varepsilon_{ii})_{,k} - \varepsilon_{ii}\theta_{,k}\right)dA = 0 \qquad \ldots(6)$$

$$\int_{\Gamma^2}\left(Wn_k - \frac{\partial W}{\partial u_{j,i}}u_{j,k}n_i\right)d\Gamma - \beta\int_{A_2}\left(\frac{1}{2}(\theta\varepsilon_{ii})_{,k} - \varepsilon_{ii}\theta_{,k}\right)dA = 0 \qquad \ldots(7)$$

where, $\Gamma^1 = \Gamma_c^+ + \Gamma_1 + \Gamma_{\rho 1} + \Gamma_1^+$ and $\Gamma^1 = \Gamma_c^- + \Gamma_1 + \Gamma_{\rho 2} + \Gamma_1^-$. A_1 and A_2 represent area enclosed within the upper and lower closed loops Γ^1 and Γ^1 respectively. As shown in [8] the above equations can be added and under the condition of traction free crack surface can be reduced to,

$$\int_{\Gamma_\rho}\left(Wn_1 - \frac{\partial W}{\partial u_{j,i}}u_{j,1}n_i\right)d\Gamma = \int_{\Gamma}\left(Wn_1 - \frac{\partial W}{\partial u_{j,i}}u_{j,1}n_i\right)d\Gamma - \beta_1\int_{A_1}\left(\frac{1}{2}(\theta\varepsilon_{ii})_{,1} - \varepsilon_{ii}\theta_{,1}\right)dA - \beta\int_{A_2}\left(\frac{1}{2}(\theta\varepsilon_{ii})_{,1} - \varepsilon_{ii}\theta_{,1}\right)dA$$

$$\qquad \ldots(8)$$

$$\int_{\Gamma_\rho}\left(Wn_2 - \frac{\partial W}{\partial u_{j,i}}u_{j,2}n_i\right)d\Gamma = \int_{\Gamma}\left(Wn_2 - \frac{\partial W}{\partial u_{j,i}}u_{j,2}n_i\right)d\Gamma + \int_{\Gamma_c}[\![W]\!]d\Gamma - \int_{\Gamma_l}\left([\![W]\!] - \sigma_{j2}[\![u_{j,2}]\!]\right)d\Gamma$$

$$-\beta_1\int_{A_1}\left(\frac{1}{2}(\theta\varepsilon_{ii})_{,2} - \varepsilon_{ii}\theta_{,2}\right)dA - \beta\int_{A_2}\left(\frac{1}{2}(\theta\varepsilon_{ii})_{,2} - \varepsilon_{ii}\theta_{,2}\right)dA \qquad \ldots(9)$$

Left hand side of the above equation (8) for the $Lim_{\rho \to 0}$ can be reduced to well known J integral, whereas equation (9) as mentioned in [8] does not converge to some fixed value and is defined for some very small value of ρ lying within the singularity dominated zone.

$$J_1 = \frac{1}{16}\left[\frac{(1+\kappa_1)}{\mu_1} + \frac{(1+\kappa_2)}{\mu_2}\right]\left[(K_1^2 + K_2^2)\right] \qquad ...(10)$$

$$J_{2\rho} = -\frac{1}{32\pi\varepsilon}\left[\frac{(1+\kappa_1)}{\mu_1}(1-e^{-2\pi\varepsilon}) + \frac{(1+\kappa_2)}{\mu_2}(e^{2\pi\varepsilon}-1)\right]\left[(K_1^2 - K_2^2)\sin(2\varepsilon\log\rho) + 2K_1K_2\cos(2\varepsilon\log\rho)\right]$$

$$...(11)$$

where $J_{2\rho}$ can be redefined as

$$J_{2\rho} = \int_\Gamma\left(Wn_2 - \frac{\partial W}{\partial u_{j,i}}u_{j,2}n_i\right)d\Gamma + \int_{\Gamma_c}[\![W]\!]d\Gamma - \int_{\Gamma_I}\left([\![W]\!] - \sigma_{j2}[\![u_{j,2}]\!]\right)d\Gamma$$

$$-\beta_1\int_{A_{1\rho}}\left(\frac{1}{2}(\theta\varepsilon_{ii})_{,2} - \varepsilon_{ii}\theta_{,2}\right)dA - \beta_2\int_{A_{2\rho}}\left(\frac{1}{2}(\theta\varepsilon_{ii})_{,2} - \varepsilon_{ii}\theta_{,2}\right)dA \qquad ...(12)$$

Here for $J_{2\rho}$ all contour segments along the interface start and terminate at a distance ρ away from the crack tip. Therefore, unlike J integral, $J_{2\rho}$ for bi-material interface problem includes extra line integrals as shown in Figure 1 along the material interface besides arbitrary outer contour.

3. VALIDATION

In order to validate the above formulation we have taken the example problem of double edge cracks in jointed dissimilar semi-infinite plates subjected to uniform temperature change of 100°C, solved analytically by Erdogan [11] and computationally by Banks-Sills [7] and Sun and Ikeda [6]. The geometry and material properties of the plane strain problem considered is taken same as that by above authors and is shown in Figure 2 and Table 1.

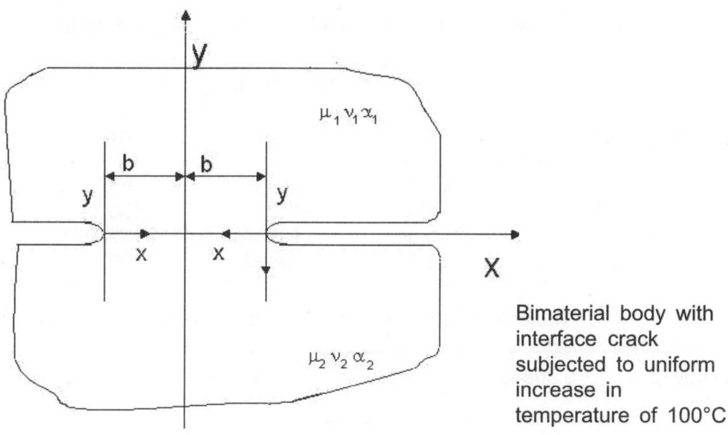

Fig. 2. Bimaterial body subjected to uniform temperature.

Table 1. Material Properties used in the case study.

Properties	Material-1	Material-2
E	1e12	1e11
v	0.3	0.3
α	1.0e $-$ 6	1.0e $-$ 7

Due to the symmetrical nature of the plate along the central vertical line, only the left half of this bi-material semi-infinite plate is modelled using the finite element software ANSYS. The dimension of semi-infinite plate is taken as 200 units by 400 units with jointed part of 1 unit (c = 1). The plate is modelled without any crack tip elements with lowest element size of 5E-04m i.e. 1/100 of crack size. Erdogan [11] presented the exact solution for the SIF as

$$K_1 + iK_2 = \sigma_0(\alpha_2\eta_1 - \alpha_1\eta_2)\Delta T\sqrt{\pi c}(-2\varepsilon + 1)(2c)^{-i\varepsilon} \qquad ...(13)$$

If $K = \sigma_0(\alpha_2\eta_1 - \alpha_1\eta_2)\Delta T\sqrt{\pi c}$ then normalized SIF can be written as

$$K_1^* + iK_2^* = (-2\varepsilon + 1)(2c)^{-i\varepsilon} \qquad ...(14)$$

Eight noded two dimensional elements with a typical element near the crack tip having a size of 5E-04 units are employed. A total of 32,311 elements with 91,022 nodes have been used in the finite element model. The J_1 and $J_{2\rho}$ integrals have been computed by modifying the in-built J integral macro in ANSYS. J_1 integral was obtained using only the outer contour whereas $J_{2\rho}$ integral was computed using the portion of material interface as explained earlier. To demonstrate the path independence of $J_{2\rho}$, three different integration paths are used.

Table 2 shows the results obtained for $J_{2\rho}$ for three different values of ρ and three different integration paths. The analytically obtained values of $J_{2\rho}$ are also shown in the table. It is seen that the error for the range of distances ρ considered is small, thus indicating that $J_{2\rho}$ integral is path independent.

The values of stress intensity factors obtained and compared with Erdogan [11] for the four different distances ρ is shown in Table 3. Results are found to be quite satisfactory within the K-dominance zone.

Table 2. Analytical and Computational value of $J_{2\rho}$ for different values of ρ

ρ	$J_{2\rho}$	Numerical $J_{2\rho}$			Average	% Error
	(Analytical)	Path1	Path2	Path3	$J_{2\rho}$	
6.08e–3	−156.51	−168.48	−164.32	−160.2	−164.33	−4.996
9.80e–3	−126.71	−138.5	−122.61	−133.14	−131.41	−3.700
1.27e-2	−110.41	−105.34	−106.56	−112.76	−108.22	1.1900
1.55e–2	−97.33	−103.78	− 96.56	− 99.12	− 99.82	−2.500

Table 3. Mode I and Mode II SIF's for different values of ρ

ρ	6.08e–3	9.80e–3	1.27e–2	1.55e–2	Average	Erdogan's Solution	% Error
K1	− 0.1963	− 0.2000	− 0.2080	− 0.2025	− 0.2017	− 0.2039	1.067
K2	0.9956	0.9949	0.9932	0.9944	0.9946	0.99065	−0.3960

CONCLUSION

The J_2 integral defined and developed for bi-material interface problem subjected to mechanical loading by Khandelwal and Chandra Kishen [8] has been extended for bi-material body subjected to thermal loading. Integral is shown to be path independent in a modified sense and found to be useful in the computation of the stress intensity factors for bi-material interface problem.

REFERENCES

1. Wilson W.K. and Yu, I.W. The Use of the J-Integral in thermal stress Crack Problem. International journal of fracture mechanics, 15: 377-387, 1979.
3. Shih, C.F., Moran, B. and Nakamura, T. Energy release rate along three dimension crack front in a thermally stress body. International journal of Fracture Mechanics, 30: 79-102, 1986.
4. Stern, M. The numerical calculation of thermally induced stress intensity factor. Journal of Elasticity, 9: 91-95, 1979.
5. Yong K. and Shul, C.W. Determination of stress intensity factors for an interface crack under vertical uniform heat flow. Engineering Fracture Mechanics, 40(6): 1067-1074, 1991.
6. Sun, C.T. and Quin, W. The use of finite extension strain energy release rates in fracture of interfacial cracks. International journal of solids and structure, 34: 2595-2609, 1997.
7. Sun C.T. and Ikeda, T. Stress Intensity factor analysis for an interface crack between dissimilar isotropic materials under thermal stress. International journal of Fracture Mechanics, 111: 229-249, 2001.
8. Leslie Banks-Sills and Orly Dolev. The conservative M-integral for thermal-elastic Problems. International journal of fracture mechanics, 125: 149-170, 2004.
9. Ratnesh Khandelwal and Chandra Kishen, J.M. Complex Variable Method of Computing J_k.Engineering Fracture Mechanics, 73: 1568 - -1580, 2006.
10. J.K. Knowles and E. Sternberg. On a class of Conservation laws in linearized and finite elastostatics. Arch. Rat. Mech. Anal, 44: 187-211, 1972.
11. Chen W.H. and Chen K.T., On the Study of mixed mode thermal fracture using modified J_k integrals. International journal of fracture, 17: R99-R103, 1981
12. Erdogan, F. Stress distribution in Bounded dissimilar materials with cracks. Journal of Applied Mechanics, 32: 403-410, 1965.

45

Boundary Element Analysis of Two-Dimensional Elastic Cracked Body Using Near-Tip Solution

TOSHIO FURUKAWA

Division of Mechanical and System Engineering, Kyoto Institute of Technology, Kyoto-606-8585, Japan
email: furukawa@kit.ac.jp

ABSTRACT

This paper is concerned with a method for the analysis of two-dimensional crack problems using the boundary element method. In this method, a domain to be analyzed is divided into two regions: the near-tip region, which is circular in shape and centered at the crack tip, and the outer region. The displacements and stresses in the near-tip region are presented by the truncated infinite series of the eigenfunction with unknown coefficients. The stress intensity factor is represented by one of the coefficients. The outer region is formulated by the boundary element method. The system of linear algebraic equations with respected to nodal displacements, nodal tractions and the coefficients of the series are constructed by discretized boundary integration in the outer region and the continuity conditions. To extend this method to the crack problem with no symmetry, a domain decomposition method is introduced. The edge cracked rectangular plate under a uniform tension is treated and the calculated dimensionless stress intensity factors are in good accuracy.

Keywords: Elasticity, Crack, Stress intensity factor, Boundary element, Near-tip solution.

1. INTRODUCTION

It is well known that Boundary Element Method is more efficient method comparing with Finite Element Method and Finite Difference Method and widely applied to a lot of problems. By the way, some ideas are necessary in the crack problem to obtain an appropriate solution as well as the other numeric calculation methods because of the stress singularity in the crack-tip. There are a lot of methods to obtain the stress intensity factors, such as, the extrapolation method based on stress or displacement near the crack-tip [1] and the method introducing the singular element [2-5]. The discretization near the crack-tip is required in these methods and severe ideas are required to obtain the solution in good accuracy.

Then, authors proposed the method by which the stress intensity factors could be obtained in good accuracy without discretization near the crack tip [6]. A virtual region including the crack-tip was introduced, and within the virtual region, near-tip infinite series solutions for displacements and stresses that satisfied the governing equations and the boundary conditions on the crack surface were applied and a usual boundary element method was used in other region. At the boundary of the real and virtual regions, the continuity conditions of displacements and stresses were introduced. The stress intensity factors were obtained by the simultaneous equations leaded from the discretized boundary integral equations. To test the accuracy of this method the center slant cracked rectangular plate under a uniform tension with point symmetry has been analyzed. The calculated dimensionless stress intensity factors were in good accuracy. Now, we extend this method to the cracked problem with no symmetry by use of domain decomposition method. To show the effectiveness of this proposed method, we treat the edge cracked rectangular plate with no symmetry and examine several cases.

2. FORMULATIONS

The near-tip solutions of displacements and stresses for two-dimensional elastic body are well known and expressed as follows:

$$\begin{Bmatrix} u_x \\ u_y \end{Bmatrix} = \sum_{n=1}^{\infty} \frac{A_{In}}{2\mu} r^{n/2} \begin{Bmatrix} f_{In}^{(1)}(\theta) \\ f_{In}^{(2)}(\theta) \end{Bmatrix} - \sum_{n=1}^{\infty} \frac{A_{IIn}}{2\mu} r^{n/2} \begin{Bmatrix} f_{IIn}^{(1)}(\theta) \\ f_{IIn}^{(2)}(\theta) \end{Bmatrix} + \begin{Bmatrix} u_{x0} \\ u_{y0} \end{Bmatrix} \qquad \ldots(1)$$

$$\begin{Bmatrix} \sigma_x \\ \sigma_y \\ \tau_{xy} \end{Bmatrix} = \sum_{n=1}^{\infty} A_{In} \frac{n}{2} r^{(n/2)-1} \begin{Bmatrix} g_{In}^{(1)}(\theta) \\ g_{In}^{(2)}(\theta) \\ g_{In}^{(3)}(\theta) \end{Bmatrix} - \sum_{n=1}^{\infty} A_{IIn} \frac{n}{2} r^{(n/2)-1} \begin{Bmatrix} g_{IIn}^{(1)}(\theta) \\ g_{IIn}^{(2)}(\theta) \\ g_{IIn}^{(3)}(\theta) \end{Bmatrix} \qquad \ldots(2)$$

where $f_{In}^{(i)}, f_{IIn}^{(i)}, g_{In}^{(i)},$ and $g_{IIn}^{(i)}$ are functions of the angle θ, A_{In} and A_{IIn} are unknowns, μ is rigidity, and u_x and u_y are rigid body displacements. The stress intensity factors K_I and K_{II} are expressed as

$$K_I = \sqrt{2\pi} A_{I1}, \ K_{II} = -\sqrt{2\pi} A_{II1} \qquad \ldots(3)$$

Now, we consider a cracked body as shown in Fig. 1. A virtual region, which is circular in shape and centered at the crack tip, is introduced. The tractions t_x and t_y at the arbitrary point on this boundary are expressed as follows:

$$\begin{Bmatrix} t_x \\ t_y \end{Bmatrix} = \sum_{n=1}^{\infty} A_{In} \frac{n}{2} a^{(n/2)-1} \begin{Bmatrix} h_{In}^{(1)}(\theta) \\ h_{In}^{(2)}(\theta) \end{Bmatrix} - \sum_{n=1}^{\infty} A_{IIn} \frac{n}{2} a^{(n/2)-1} \begin{Bmatrix} h_{IIn}^{(1)}(\theta) \\ h_{IIn}^{(2)}(\theta) \end{Bmatrix} \qquad \ldots(4)$$

where

$$h_{in}^{(1)}(\theta) = g_{in}^{(1)}(\theta)\cos\theta + g_{in}^{(3)}(\theta)\sin\theta, \quad h_{in}^{(2)}(\theta) = g_{in}^{(2)}(\theta)\sin\theta + g_{in}^{(3)}\cos\theta \qquad \ldots(5)$$

The expression for displacements putted in Eq. (1), and expression (5) for tractions are truncated by first N terms. The final matrix expressions are

$$\{U\} = \begin{Bmatrix} u_x \\ u_y \end{Bmatrix} = \begin{bmatrix} F^{(1)}(a,\theta) \\ F^{(2)}(a,\theta) \end{bmatrix} \{A\} \quad \{T\} = \begin{Bmatrix} t_x \\ t_y \end{Bmatrix} = \begin{bmatrix} H^{(1)}(a,\theta) \\ H^{(2)}(a,\theta) \end{bmatrix} \{A\} \qquad \ldots(6)$$

where the elements of $\{A\}$ are $A_{In} A_{In}$ ($n = 1, 2, \ldots, N$), A_{IIn}, ($n = 1, 2, \ldots, N$) u_x and u_y.

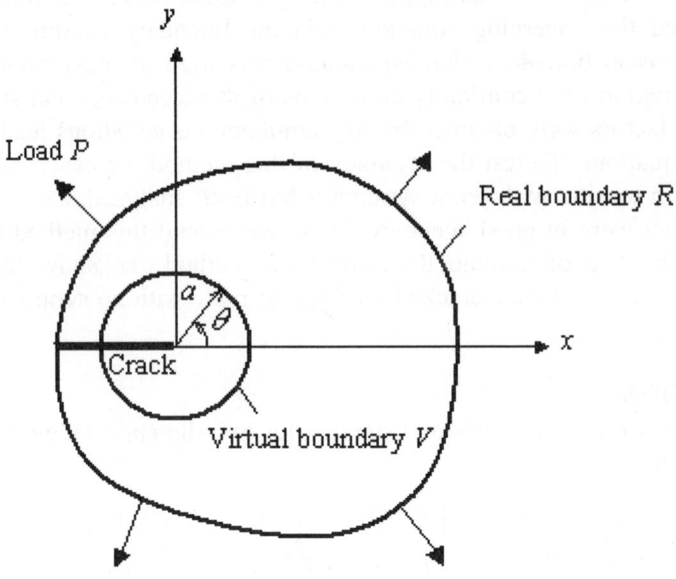

Fig. 1. Cracked body with virtual boundary.

The discretized boundary integral equation for the body excluding the virtual circular region is

$$[\mathrm{H}_R \;\; \mathrm{H}_V]\begin{Bmatrix}\mathrm{U}_R \\ \mathrm{U}_V\end{Bmatrix}=[\mathrm{G}_R \;\; \mathrm{G}_V]\begin{Bmatrix}\mathrm{T}_R \\ \mathrm{T}_V\end{Bmatrix} \qquad ...(7)$$

In this expression, the subscripts R and V denote the real boundary and virtual boundary, respectively. The boundary conditions on this two boundaries are leaded from the continuity conditions for displacements and tractions and are expressed as

$$\mathrm{U}_i^{'} - \mathrm{U}_{Vi} = 0, \;\; \mathrm{T}_i^{'} + \mathrm{T}_{Vi} = 0 \qquad ...(8)$$

where $\mathrm{U}_i^{'}$ and $\mathrm{T}_i^{'}$ are the discretized ith term nodal displacement and traction, respectively.

To treat the problem with no symmetry, we introduce the domain decomposition method. As shown in Fig. 2, we decompose the original configuration to two domains along the crack surface. We combine the near-tip solutions and usual boundary element method for domain I and domain II, respectively. Then we use the combined conditions for displacements and tractions.

$$u(1) = u(2), \;\; t(1) + t(2) = 0 \qquad ...(9)$$

where $u(i)$ and $t(i)$ are the boundary condition of displacements and tractions for domain i.

The final system of linear algebraic equations is

$$[A]\{x\} = \{f\} \qquad ...(10)$$

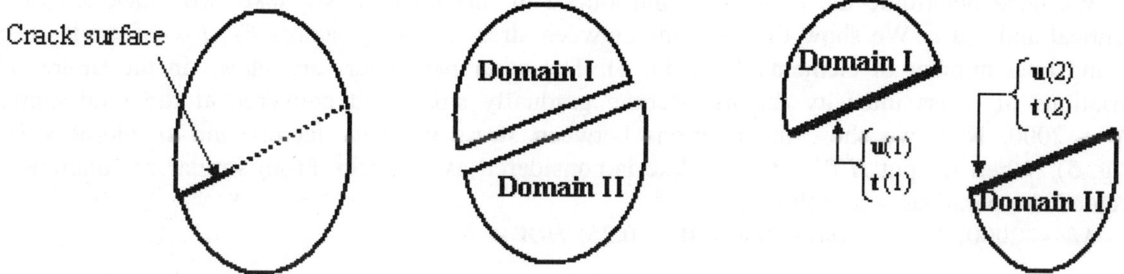

Fig. 2. Outline of domain decomposition.

3. NUMERICAL RESULTS AND DISCUSSION

We consider the edge cracked rectangular plate subjected to uniform tension σ_0 with width W and height H, crack lengths are c_1 and c_2, the distance of these two cracks is d (Fig. 3). We decompose the plate to three domains. The boundary conditions of these domains are shown in the right hand side of this figure. The dimensionless stress intensity factors are defined by

$$F_I = \frac{K_I}{\sigma_0\sqrt{\pi c}} \ , \quad F_{II} = \frac{K_{II}}{\sigma_0\sqrt{\pi c}} \qquad\qquad ...(11)$$

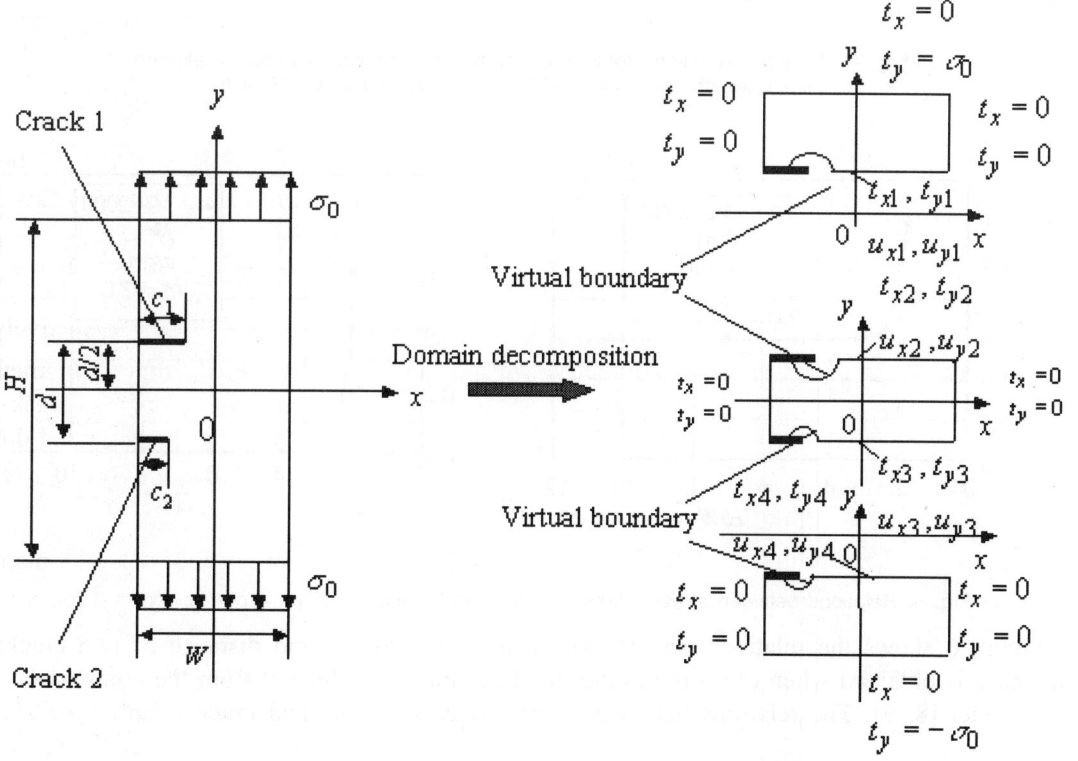

Fig. 3. Edge cracked rectangular plate and boundary conditions.

We need determine the parameters introduced in this method. We take two crack lengths as identical and put c. We show the relations between stress intensity factors F_I, F_{II} of modes I and II, and total number of elements Me (Fig. 4). The other parameters are shown in the figure. The variations of stress intensity factors become gradually small and converge at the total number $Me = 2000$. Next we show the relations between stress intensity factors and slenderness H/W (Fig. 5). When we put $H/W = 6$, the plate is considered as the strip. From similar calculations we decided the parameters as follows.

$Me = 2000$, $N = 15$, $a/c = 0.4$, $c/W = 0.25$, $H/W = 6$

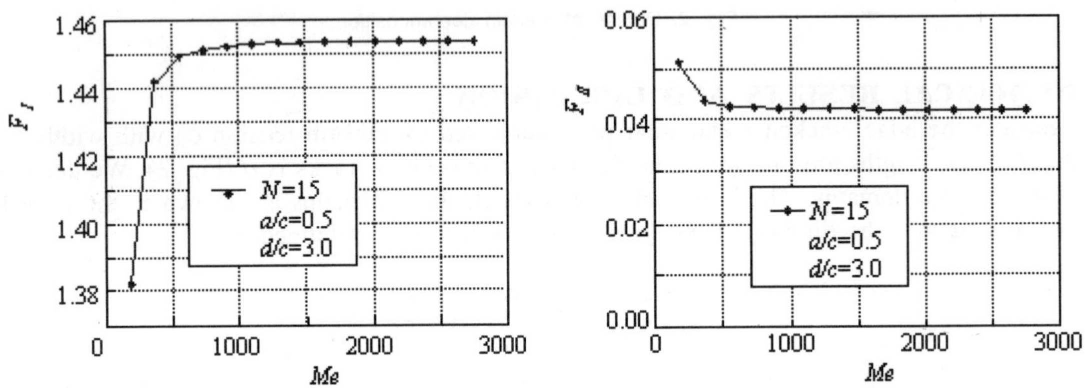

Fig. 4. Relations between stress intensity factors and total number of elements (crack length $c_1 = c_2 = c$; $c/W = 0.25$, slenderness $H/W = 6$).

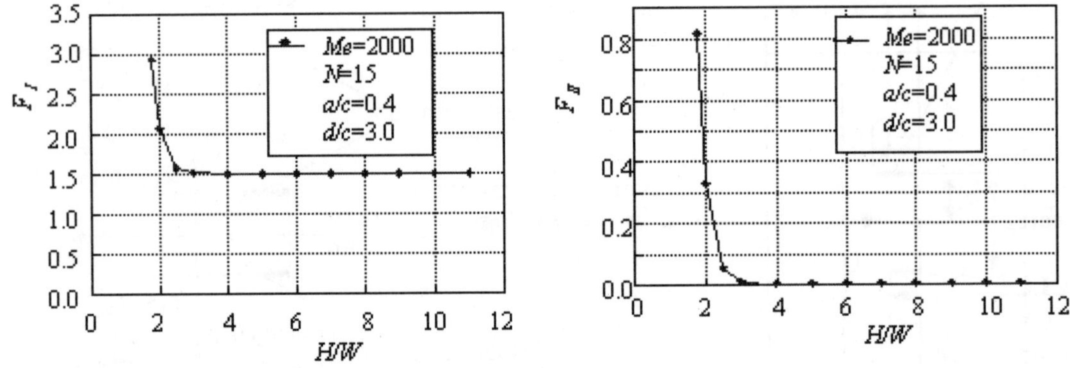

Fig. 5. Relations between stress intensity factors and slenderness ($c_1 = c_2 = c$; $c/W = 0.25$).

Figure 6 shows the relations between stress intensity factors and distance of two cracks. One crack case is obtained when $d/c = 6$ and the good accuracy is achieved from the comparison of one crack model [8, 9]. The relations between stress intensity factors and crack length c_2 is shown in Fig. 7.

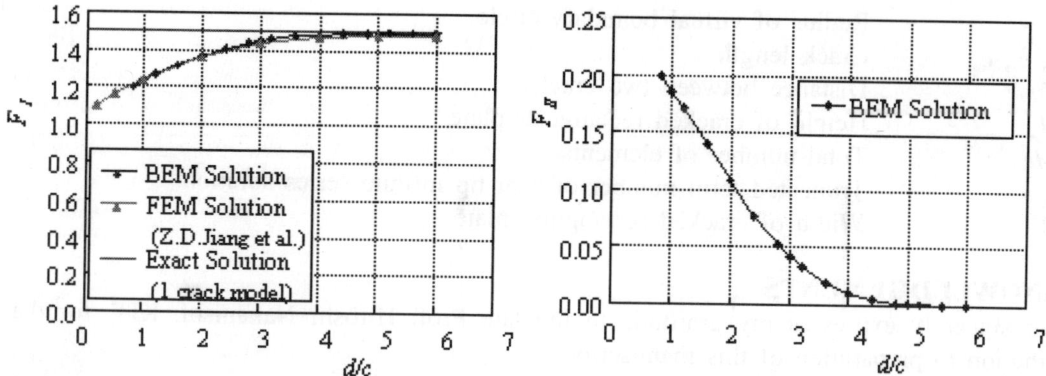

Fig. 6. Relations between stress intensity factors and distance of two cracks ($c_1 = c_2 = c$; $c/W = 0.25$).

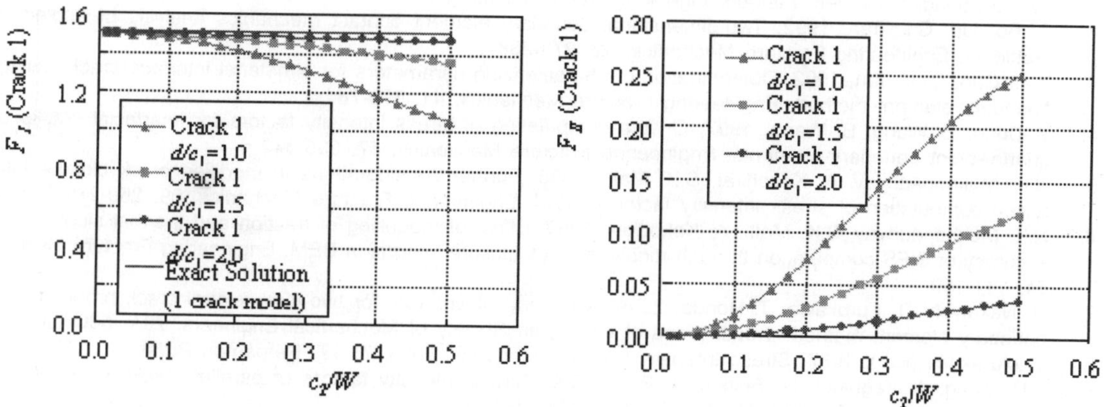

Fig. 7. Relations between stress intensity factors and crack length c_2 ($c_1/W = 0.25$).

When the crack length become small, the stress intensity factor F_I becomes large and coincides the one crack case at the crack length $c_2 = 0$. The stress intensity factor F_{II} becomes small and disappeares at the crack length $c_2 = 0$.

CONCLUSION

A highly accurate and simple solution method related to boundary element method has been presented. The numerical results discussed for two edge cracked rectangular plate with no symmetry are in good agreement with the available results in the literature. Also the effects of various parameters such as plate width-to-height, crack length, radius of virtual boundary circle, truncated term number of near-tip infinite series solution, total number of boundary elements, and distance between two cracks are studied.

NOMENCLATURE

a	Radius of virtual boundary circle
c, c_1, c_2	Crack length
d	Distance between two cracks
H	Height of cracked rectangular plate
Me	Total number of elements
N	Truncated term number of near-tip infinite series solution
W	Width of cracked rectangular plate

ACKNOWLEDGEMENTS

Author sincerely expresses my gratitude to the late Prof. Hiroshi Nakanishi, KIT, for the large contribution to preparation of this manuscript.

REFERENCES

1. For example: M. Tanaka, M. Hamada, Y. Iwata, 1982, Computation of a two-dimensional stress intensity factor by the boundary element method, Ingenieur Archiv. 52, 95-104.
2. S. Bo. Liu, C.L. Tan, 1992, Two-dimensional boundary element contact mechanics analysis of angled crack problems, Engineering Fracture Mechanics, 42, 273-288.
3. Y. L. Gao, C. L. Tan, 1992, Determination of characterizing parameters for bimaterial interface cracks using the boundary element method, Engineering Fracture Mechanics, 41, 779-784.
4. Y. Dong, Z. Wang, B. Wang, 1997, On the computation of stress intensity factors for interfacial cracks using quarter-point boundary elements, Engineering Fracture Mechanics, 57, 335-342.
5. N.K. Mukhopadhyay, A. Kakodkar, S.K. Maiti, 1998, Further considerations in modified crack closure integral based computation of stress intensity factor in BEM, Engineering Fracture Mechanics, 59, 269-279.
6. N.K. Mukhopadhyay, S.K. Maiti, A. Kakodkar, 1999, Effect of modelling of traction and thermal singularities on accuracy of SIFS computation through modified crack closure integral in BEM, Engineering Fracture Mechanics, 64, 141-159.
7. H. Nakanishi, T. Furukawa, T. Honda, 2004, A method of analysis for two-dimensional crack problems by the boundary element method, Transactions of the Japan Society of Mechanical Engineers, 70A, 560-566.
8. Y. Murakami (Ed.), 1987, Stress Intensity Factors Handbook, vol.3, 71-77, Pergamon Press.
9. Z.D. Jiang, A. Zeghloul, G. Bezine, J. Petit, 1990, Stress intensity factors of parallel cracks in a finite width sheet , Engineering Fracture Mechanics, 35, 1073-1079.

46

Prediction of Ductile Fracture for Simple Upsetting Operation Using Genetic Programming (GP)

CHINMAY K. DESAI[1], R.D. SHAH[2] AND A.A. SHAIKH[2]

[1]Department of Mechanical Engineering, C.K. Pithawall College of engineering, Surat-395 007 India
email: chinmay.desai@gmail.com
[2]Department of Mechanical Engineering, Sardar Vallabhbhai National Institute of Technology
(Deemed University), Surat-395 007 India email: aas_svnit@yahoo.com

ABSTRACT

This paper investigates the feasibility of using Genetic Programming (GP) for the modelling of damage evolution during the forging operation and compares it with the prediction accuracy of Artificial Neural Network (ANN). Traditionally predictive models for damage evolution have relied heavily on a large database of isothermal, constant strain rate tests representing the range of processing conditions (strain-rate, temperature, etc.) experienced during forging. Such databases are both expensive to generate and time consuming. In this paper we combine as-forged microstructure measurements with finite element (FEM) predictions of the processing conditions to develop models based on a minimum of mechanical testing. GP is then used to evolutes the non-linear relationship between the input parameters (here the forging conditions) and microstructural outputs (level of damage). Developed model is also tested using completely new dataset to check its capability. Finally prediction capability of this model is compared with that of Artificial Neural Network (ANN).

Keywords: Genetic Programming (GP), Artificial Neural Network (ANN), Finite Element Method (FEM).

1. INTRODUCTION

Workability is measure of the extent of deformation that a material can withstand prior to failure in metal forming processes. In these processes, ductile fracture is the most common reason for failure. An important concern in a metal forming processes is whether the desired deformation can be accomplished without ductile fracture of the work piece. In industrial practice, however the empirical know-how of the designer is decisive for the fracture-free quality of the products, but often requires very costly trial-and-error. Thus, there is a critical need for predicting and preventing fracture, which is a major feature of the forming processes and the products. Ductile fracture is a complicated phenomenon that is dependent upon process parameters such as stress, strain, strain rate friction, and forming temperature as well as upon influential material parameters such as strain hardening and the volume fraction of voids and second phase particles [1].

Owning to this complexity, various criteria have been proposed to evaluate workability [2]. The recent development of the finite-element method for the study of metal forming problems enables fracture initiation to the predicated more precisely using the fracture criteria. Oh *et al.* [3] examined the use of the Cockcroft and Latham criterion and modified McClintock criteria to predict fracture initiation in axisymmetric extrusion and drawing using rigid-plastic finite-element method. Frater and Penza [4] developed a special computer program that used the output of the rigid-plastic finite element analysis as the input and compute the required tensile and compressive strain at the centre of the bulge in the specimen. The results of the study were compared with fracture criteria proposed by Shah and Kuhn [5]. Yoshida and Wanheim [6] attempted to use the model material technique for the surface stress analysis and the prediction of a surface crack. In this paper, Genetic Programming (GP) based methodology for inferenceing between deformed geometry and process parameters are developed. The analysis of the geometrical configuration and the value of ductile fracture are measured by the finite element method. The process parameters are aspect ratio (height/diameter) of the initial billet and height reductions. The training data for GP are coordinates of the free surface and the fracture values for each of the above conditions. The predictions of the GP and experimental results of simple upsetting is compared.

2. BRIEF OVERVIEW OF DUCTILE FRACTURE AND GENETIC PROGRAMMING (GP)

2.1 Ductile Fracture

Cockeroft and Latham developed a ductile fracture criterion which has been successfully applied to various cold forming processes such as extrusion, rolling and upsetting [7–9] for predicting fracture.

Therefore, the Cockcroft and Latham criterion, which is convenient for carrying out experimental and numerical approaches, is used in the present investigation to estimate if and when surface fracture will occur during the deformation process. The formula expressed in Eq. (1) is approximated in incremental from for finite element analysis so that

$$\sum_{i=1}^{n} (\sigma_1 \delta\bar{\varepsilon})_i = C' \ [\text{N mm}^{-2}] \qquad \qquad ...(1)$$

where n is the number of the step in the simulation and $\delta\bar{\varepsilon} = \bar{\varepsilon}_i - \bar{\varepsilon}_{i-1}$ is the incremental effective strain. Equation-1 is calculated numerically for each integration point inside the elements according to the procedure. σ_1 is equal to zero for all compressive stresses. Therefore, the simulation term will be zero, as all compressive stresses do not contribute to the work term. If C' in Eq. 1 reaches the Cockcroft and Latham constant C', obtained from uni-axial tensile test, a fracture should occur in the material at the integration point under consideration. If a fracture is introduced into the numerical simulation the program developed for the analysis of this study stops automatically as no treatment for crack propagation is employed.

2.2 Genetic Programming

GP is probably the most general approach of evolutionary computation methods, which was introduced by J.R.Koza in first half of 1990s [10, 11]. GP deals with structure like chromosome. Evolutionary computation of GP begins with the population of random chromosomes. Each chromosome here is nothing but the mathematical expression represents functional relationship between independent and dependent variables. These chromosomes or tree like structure that can be composed in a recursive manner from a set of function genes $F = \{f_1, f_2, \ldots, fm\}$, where m is the number of function genes. Set of functional genes can be basic arithmetical functions (i.e., operations of addition, subtraction, multiplication and division), power function, natural exponential function, sine function and a set of terminal genes $T = \{a1, a2, \ldots, an\}$, which are in fact independenvariables and floating point numbers. Each individual function f_i from the set F has a certain number of arguments $z(f_i)$. The appropriate number of arguments for function genes from the set F is thus determined with the list $\{z(f_1), z(f_2), \ldots, z(f_m)\}$. The initial population is obtained with the creation of random chromosomes consisting of the available function genes from the set F and the available terminal genes from the set T. The creation of the initial population is a blind random search for solutions in the huge space of possible solutions. The next step is the calculation of adaptation of individuals to the environment (i.e., calculation of fitness for each chromosome). Fitness is a guideline for modifying the structures undergoing adaptation. The initial random population of mathematical expression is further adapted using evolutionary operator like crossover, mutation, and reproduction. In GP, the chromosomes change, in particular, with two genetic operations: reproduction and crossover. The reproduction operation gives a higher probability of selection to more successful organisms. They are copied unchanged into the next generation. The crossover operation ensures the exchange of genetic material between chromosomes. Figure 1 shows the crossover operation of two chromosomes consisting of two chromosomes of several function and terminal genes.

Two new child chromosomes result from two parental chromosomes parental chromosomes parent 1 and parent 2. They are, in fact, mathematical expressions usually written as: (x + y)/ z + xz and x(1–yz). They consist of the set of function genes F= {+, -, *, /} and of the set of terminal genes T = {1, x, y, z}. The appropriate list of arguments for the function genes is {2, 2, 2, 2}. By defining randomly generated crossover points (indicated with dashed lines), two parental crossover fragments (in boldface) are obtained. Child 1 is produced by deleting everything below the crossover point of parent 1 and then inserting the crossover fragment of the parent 2 at the crossover point of the parent 1. Child 2 is produced in a similar manner. Therefore, two child organisms are created: (1-yz)/z + xz and x(x + y). It can be seen that both child organisms include the genetic material from their parents. It is necessary to preserve syntactic structure of the

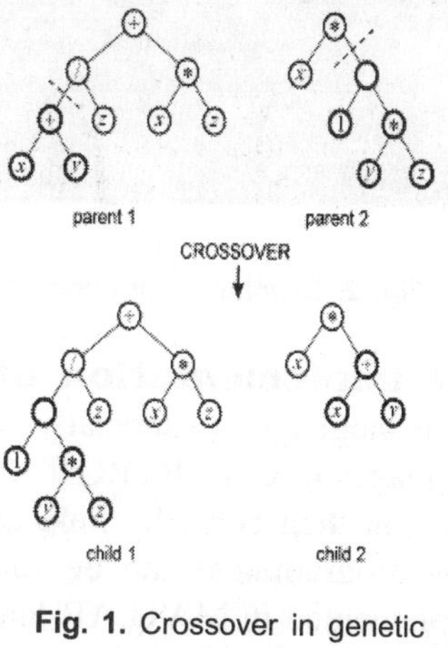

Fig. 1. Crossover in genetic programming.

chromosomes during the crossover operation. The above example of the crossover operation is showed for the chromosomes which are in fact the mathematical expressions. During both the random initialization of population as well as adaptation experimental data set (set of impendent & dependent variables) are used to guide evolutionary process. Evolutionary process terminate when best individual found in the population achieves the prediction accuracy of ± certain percentage on all sample data.

3. NUMERICAL ANALYSES AND GP IMPLEMENTATION

3.1 Finite-Element Simulations and Experiment

The work material used in finite-element analysis and upsetting experiments is commercial aluminium, and its mechanical properties like material constant is 346.49 N mm^{-2}, strain hardening coefficient is 0.173 and the Cockcroft-Latham constant is 22.61 N mm-1. Finite-element analyses are performed with the billets which have three aspect ratios (billet height/diameter), 0.5, 1.0, and 1.5, respectively. The distribution of effective strain is shown in Figs. 2–3 when the ductile fracture occurred. For an aspect ratio of 0.50, the value of the ductile fracture in the centre of the free surface reaches the critical value of ductile facture at a punch stroke of 4.24 mm (Fig. 2) and for an aspect ratio of 1.00, the ductile fracture value occurred in the centre of the free surface at a punch stroke of 8.52 mm (Fig. 3). Thus, the ductile fracture happens in the centre of the free surface for all aspect ratios. To confirm the results of the simulation, upsetting is executed for an aspect ratio of 1.00 for a specimen of 10 mm diameter.

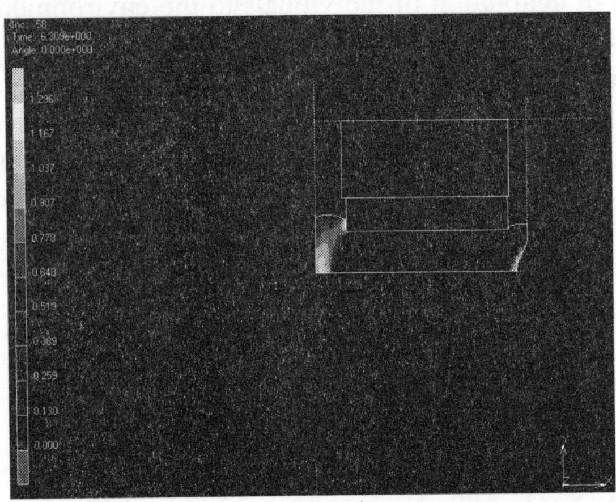

Fig. 2. Distribution of effective strain (h/d = 0.5).   **Fig. 3.** Distribution of effective strain (h/d= 1.00).

3.2 Implementation of GP

LISP language is especially well suited for GP [10, 11]. However, any present day programming languages (e.g. FORTRON, C,C++, JAVA) that can manipulate computer programmes as data and that can then compile, link, and execute the new programmes (or support an interpreter to execute new programmes) can be successfully used for GP [11]. In present work the GP algorithm was implemented in MATLAB language on AMD Athlon™ 64×2 Dual-Core Processor 2400 MHz based work station with system RAM 2GB. MATLAB is general purpose engineering programming environment with in built plotting capability [12].

3.3 Coding of Solution and Fitness Measure

Figure 4 shows the main points of the implemented GP evolutionary algorithm in pseudo code. First, the initial population P (t) of random organisms (i.e., models for prediction of Ductile fracture) consisting of the available function and terminal genes is generated. The organisms are in fact chromosomes of various shapes and size. The variable *t* represents the generation time. In this research, the function genes *F* were: basic arithmetical functions (i.e., operations of addition, subtraction, multiplication, and division), power function, natural exponential function, and sine function.

```
evolutionary algorithm
begin
t ¬ 0
initialize  P(t)
evaluate  P(t)
while (not termination _condition) do
begin
t ® + 1
alter  P(t) by applying genetic operators
evaluate P(t)
end

end
```

Fig. 4. Evolutionary Algorithm in pseudo code.

The latter two function genes have one argument each, whereas the other function genes have two arguments each. Terminal genes *T* were in fact independent variables: Punch stroke ($x1$), aspect ratio ($x2$), coordinates of free surface ($x3$). In order to increase genetic diversity of the organisms the random floating-point numbers from the range [−10, 10] were added to the set of terminals. Therefore, randomly generated chromosome could have the following form:

$$2.391\frac{x_1}{x_2} + 0.289x_1 * x_2 * x_3 - x_1 - 5$$

An average percentage deviation of all sample data for individual organism Δ was introduced as fitness measure. It is defined as:

$$D = \frac{\sum_{i=1}^{n} \Delta i}{n} \qquad ...(4)$$

where *n* is the size of sample data and Δ_i is a percentage deviation of single sample data. The percentage deviation of single sample data, produced by individual organism, is

$$\Delta_i = \frac{|Ei - Gi|}{Ei} \times 100\% \qquad ...(5)$$

Where, E_i and G_i are the actual springback measured and the predicted radial stress calculated by a model, respectively. Only reproduction and crossover were used. It is assumed in this research that the problem is solved successfully if the average percentage deviation Δ of the model is less than 10%.

4. COMPARISON WITH ANN

Back propagation neural network is used to compare the prediction capability of ANN and GP. Both GP and ANN are inspired from nature, and can be used to map non-linear relationship between independent and dependent parameters, both use data sets (pair of input and out put) to develop this relationship. But there is a fundamental difference in working as well as way in which they capture the relationship between input and output. ANN captures the relationship between input and output in the form of weight [13], whereas GP developed the mathematical function which helps to understand (some extent) influence of input parameters over output. Also, generated mathematical model/equation can predict value of dependent parameter (Damage) with reasonable accuracy 3.4% Whereas ANN can predict the same with the accuracy of 10%.

5. RESULTS AND DISCUSSION

Best model for radial stress prediction was found with following evolutionary parameters: population size 1300, maximum number of generations to be run 70, probability of reproduction 0.1, maximum depth for initial random organisms 6, and maximum permissible depth of organisms after crossover 28, both the generative method for the initial random population was ramped half-and-half [8,9]. The probability of crossover as 0.9 and probability of mutation as 0.006.The method of selection for reproduction and crossover was tournament selection with a group size of 8. The models that involve all three independent parameters give the most accurate predictions of the dependent parameter [9]. With the above mentioned genes, the simulated evolution produced the best model for prediction of radial stress presented by,

$$D = -1.87 + x_1 - x_3 * x_2^2 + x_1^2 x_2^2 + \frac{1.26 * x_1 * x_2}{x_2(8.5 - x_2 - 2 * x_1 + x_1 * x_3)} + 2 * x_1 * x_2 * x_1$$

For the best solution, the percentage deviation of the predicted value and sample data for the training and testing data set are shown in Fig. 5. It can be seen that most deviations are considerably smaller than 0.34%.

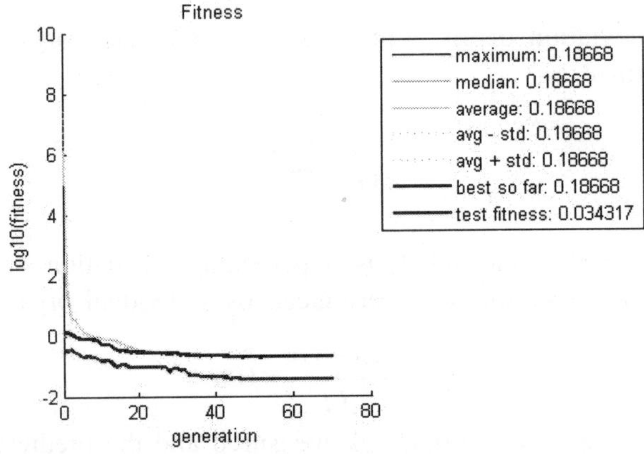

Fig. 5. Fitness vs Generation.

CONCLUSION

The present paper discussed how FEM and AI techniques (ANN, GP) can be best utilized for metal forming process design. For the best solution, the percentage deviation of the predicted value and sample data for the training and testing data set are found 3.4% with GP and 10% with ANN.

REFERENCES

1. G. Dieter, 1998, Evolution of Workability, Metals Handbook, vol 14, pp. 363–372,1998.
2. A. Atkins, 1993, Fracture in forming, Proceedings of the International conference of advanced material and processing technology, 69–77.
3. M.G. Cockeroft and D.J.Latham, 1968, Ductility and the workability of metals, Journal of the Institute of metals, 96, 33–39.
4. F.A. McClintck, 1968, a criterion for ductile fracture by the growth of hole, Journal of applied mechanics, 35, 363-371.
5. M.Oyane, T. Sato, 1980, Criterion of ductile fracture and their application, Journal of Mechanical working technology, 4, 65-81.
6. J.L. Fratar et al., 1986, An empirical formula for workability limits in cold upsetting and bolt heading, journal of applied metal work, 4, 255-261.
7. F.Yoshida and T. wanheim, 1987, The prediction of surface cracking based on the model material technique, Annals CIRP, 36,165-168.
8. S.E.Clift *et al.*, 1990, fracture prediction in plastic deformation, International Journal of Mechanicla Science, 32, 1-17.
9. J.L. frater, B.R.Penza, 1989, predicting fracture in cold upset forging by finite element method, Journal of material shaping technology, 7, 57-62.
10. J.R.Koza, Genetic Programming, the MIT Press, Cambridge, MA, 1992
11. J.R.Koza, Genetic Programming II, The MIT Press, Massachusetts, 1994
12. MATALAB Programming Guide (2005), MATHWORKS.
13. Simon, H. (1998). Neural Networks: Comprehensive foundations, Prentice Hall, U.S.A.

47

A Cohesive Zone Model to Predict Fatigue Crack Growth and Constraint Effects

R. MANIVASAGAM[1], A. BANERJEE[2] AND K. RAMESH[3]

Department of Applied Mechanics, Indian Institute of Technology Madras, Chennai-600 036, India
[1]email: rmanivasagam@iitm.ac.in [2]email: anuban@iitm.ac.in [3]email: kramesh@iitm.ac.in

ABSTRACT

A cohesive law to predict the fatigue crack growth is proposed and analyzed. The law accounts for continuous degradation of the strength due to fatigue by incorporating an evolving damage variable. A three-element mesh which consists of a cohesive zone element embedded between two continuum elements is taken to evaluate the behavior of the proposed law and compared with the existing model. The analysis is carried out for two cases (i) continuum elements having elastic properties (ii) continuum elements having elastic-plastic properties. In addition to this, the effect of constraint near the crack-tip in the modified boundary layer (MBL) formulation subjected to monotonic loading is studied under mode-I plane strain condition. The chosen parameter to quantify the constraint is the second order term of the elastic stress field solution i.e. T-stress.

Keywords: Cohesive zone model, traction separation law or cohesive law, constraint, modified boundary layer formulation.

1. INTRODUCTION

In recent years, cohesive zone modeling (czm) [2, 6] has emerged as a powerful tool for investigating the fracture processes in materials and structures. Cohesive zone represents a narrow band of localized deformation and is idealized as a pair of surfaces on which cohesive traction acts. The cohesive traction T_n is a defined function of the separation displacement, δ_n in the form of traction-separation law (or cohesive law). The exponential cohesive law developed on the basis of atomistic calculation of interfacial separation [4, 5] is used for the present analysis. For a tensile monotonic loading, the law simplifies to

$$T_n = \sigma_{\max} e \exp\left(-\frac{\delta_n}{\delta_0}\right)\frac{\delta_n}{\delta_0} \quad \text{where } e = \exp(1) \qquad ...(1)$$

Following the law [5], the traction increases with increasing separation, till it reaches the peak value of traction, $\sigma_{\max}$ at a characteristic separation, δ_0 as shown in Fig. 1.

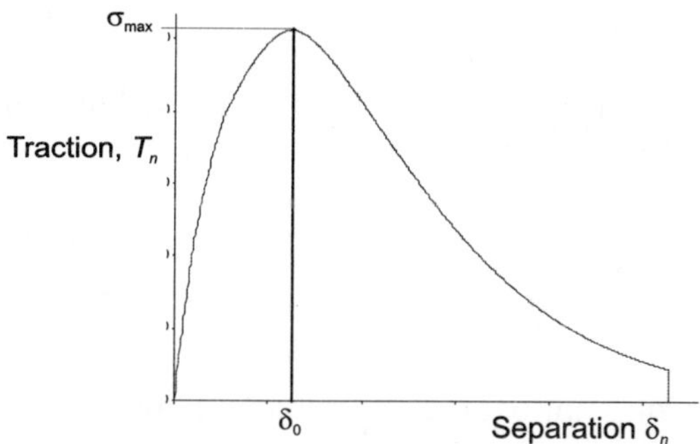

Fig. 1. Exponential traction separation law.

Further increase in separation results in drop in the traction till it vanishes completely, implying the creation of two new traction-free surfaces i.e. crack growth. The energy required for the creation of new surfaces is the area under the traction-separation curve which is given as $\phi_n = \sigma_{max} \delta_0 e$. For extending the cohesive law to model the crack initiation and growth under cyclic loading, the law should incorporate the continuous degradation of the cohesive strength due to fatigue. To incorporate the differences in the crack-tip stress triaxiality (constraint) that exists in structures due to the differences in geometry and loading, a two-parameter description of the stress field is used by including the second order term (T-stress or Q). Under monotonic loading, studies show that the crack configurations with tensile T-stress exhibit high constraint near the crack-tip region whereas geometries with negative T-stress exhibit low constraint.

Currently, studies involving constraint effects under cyclic loading is of interest as constraint affects the fatigue life of stressed members significantly. The focus of this paper is to compare the proposed damage model with an existing model [7, 9] for an idealized three-element case under tension-tension type of cyclic loading with constant amplitude. Also, an attempt has been made to study the effects of constraint under monotonic loading using the modified boundary layer (MBL) formulation for two cases of T-stress namely positive and negative T-stress.

2. CYCLIC DAMAGE MODEL

To incorporate the continuous degradation of the cohesive strength due to cyclic loading, a cyclic damage variable is introduced into the cohesive law. The damage variable D is defined as the effective surface density of micro-defects in the interface [3] and is given as the ratio of damaged cross-sectional area A_D to the initial cross-sectional area A_0, $D = A_D/A_0$. The degradation of cohesive strength due to the accumulation of damage is given as $\sigma_{max}, D = \sigma_{max} (1 - D)$ which is based on the effective stress concept [3]. Correspondingly, the decrease in normal traction value due to damage accumulation is given as $T_{n,D} = T_n(1-D)$.

Based on the continuum damage evolution laws [3], an evolving damage variable D should typically characterize the following requirements (i) damage accumulation starts if a deformation measure, accumulated or current, is greater than a critical magnitude, (ii) there exists an endurance limit which is a stress level below which cyclic loading can proceed infinitely without failure and (iii) the increment of damage is related to the increment of absolute value of damage as weighted by the ratio of current load level relative to strength. Following these requirements, the incremental damage D_c can be given as:

$$\overset{\circ}{D_c} = \frac{\left|\Delta\overset{\circ}{\delta}_n\right|}{\delta_\Sigma}\left[\left(\frac{T_{n,D}}{\sigma_{max}} - \frac{\sigma_f}{\sigma_{max}}\right)\right] H\left(\delta_{n,acc} - \delta_0\right) \text{ and } \overset{\circ}{D_c} \geq 0 \qquad \ldots(2)$$

where $\Delta\overset{\circ}{\delta}_n$ is the separation for an increment, $\delta\Sigma$ is the accumulated cohesive length, σ_f is the endurance limit, H is the Heaviside or unit step function and the accumulated separation $\delta_{n,acc}$ is

given as $\delta_{n,acc} = \int_0^t \left|\Delta\overset{\circ}{\delta}_n\right| dt$. During unloading or reloading, the traction varies linearly with the

separation. In the present model, the stiffness of the unloading/reloading traction-separation equation also evolves due to the accumulation of damage whereas in the existing model, it is treated as constant.

This change in stiffness is also taken into account such that the traction-separation equation during unloading/reloading is given as

$$T_n(t + \Delta t) = \frac{e.\sigma_{max}}{\delta_0}.[1 - D(t)][\delta_n(t + \Delta t) - \delta_n(t)] + T_n(t) \qquad \ldots(3)$$

$D(t)$, $T_n(t)$, $\delta_n(t)$ are damage, traction, separation variables for the increment at time t, $T_n(t + \Delta T)$, $\delta_n(t - \Delta T)$ are traction, separation variables for the increment at time $(t + \Delta t)$, where Δt is the increment time that defines the size of the increment. The total damage D accumulated over the

number of cycles is given as $D = \int \overset{\circ}{D_c} dt$. The damage accumulation model discussed here is

implemented in the user subroutine UEL of the commercial finite element software ABAQUS 6.4 [1].

3. CONSTRAINT EFFECT MODEL

The effect of constraints under monotonic loading is studied numerically using two-parameter modified boundary layer (MBL) formulation. The finite element mesh used to model the formulation has four-noded plane strain elements which represent the continuum part and four-noded cohesive elements which represent the fracture process zone ahead of the crack-tip. The displacements imposed on the boundary follows linear elastic stress field solution which are given as

$$u_x(t) = K_I(t)\sqrt{\frac{r}{2\pi}}\frac{1+v}{E}\cos\frac{\theta}{2}(3 - 4v - \cos\theta) + T(t)\frac{1-v^2}{E}r\cos\theta \qquad \ldots(4)$$

$$u_y(t) = K_I(t)\sqrt{\frac{r}{2\pi}}\frac{1+v}{E}\sin\frac{\theta}{2}(3 - 4v - \cos\theta) - T(t)\frac{v(1+v)}{E}r\sin\theta$$

where $K_1(t)$ is the mode-I stress intensity factor, $T(t)$ is the T-stress, r, θ are the polar coordinates.

The analysis is conducted for two cases of positive and negative T-stress under monotonic loading. The mesh shown in Fig. 2 consists of 2176 number of continuum elements and 27 number of cohesive elements placed ahead of the crack-tip along the crack plane of the model.

Dimensional parameters of the mesh are followed as per [8]. Crack-tip is positioned at the centre of the fine mesh i.e. at $x = 0$, $y = 0$. The length of the fine mesh B is taken as $30\Delta_0$ and the outer radius A is taken as $2000\Delta_0$ where Δ_0 is the length of the square element present at the fine mesh near crack-tip. Elastic-plastic properties are assumed for the continuum elements in both positive and negative T-stress cases.

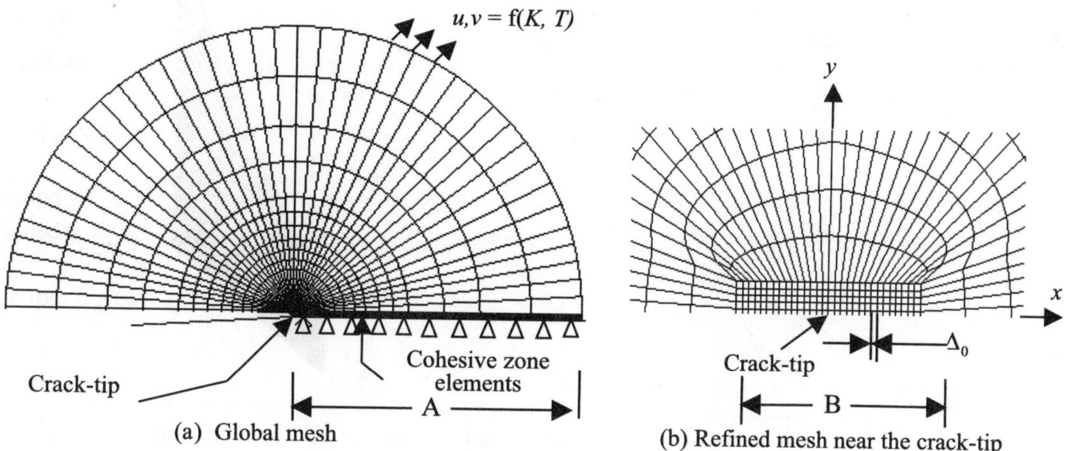

Fig. 2. Finite element mesh for the modified boundary layer formulation.

4. RESULTS AND DISCUSSION

A three elements model with one cohesive zone element embedded between two continuum elements is considered for analyzing the proposed damage model under constant amplitude cyclic loading which is shown in Fig. 3.

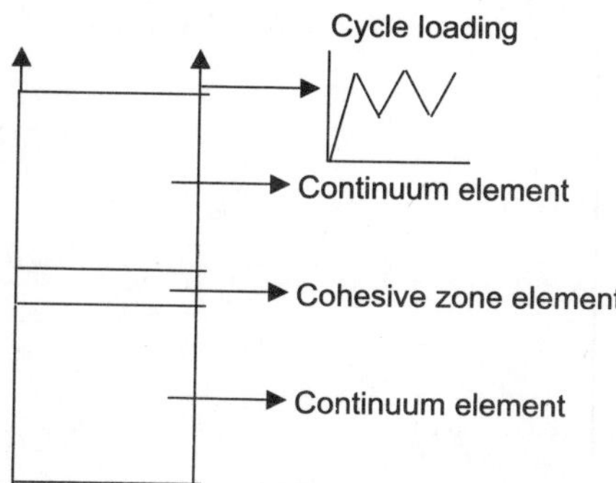

Fig. 3. Three elements mesh subjected to tension-tension type of cyclic loading.

The loading is applied only in the normal direction with tension-tension type of cyclic loading that varies linearly between $6\delta_0$ and $8\delta_0$. Damage accumulation is allowed only in the normal direction, even though the displacement along the tangential direction may occur. Fig. 4a shows the normal traction as a function of normal displacement with the damage starting to accumulate after the characteristic length $\delta_0 = 0.08$ mm. The constant-amplitude cyclic displacement loading applied on the model is shown in Fig. 4b.

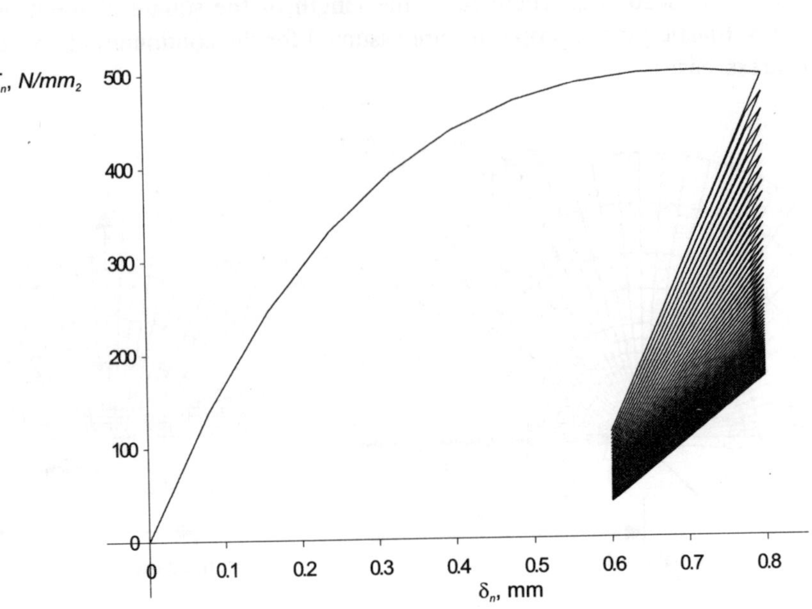

Fig. 4a. Traction separation curve with unloading/reloading effect.

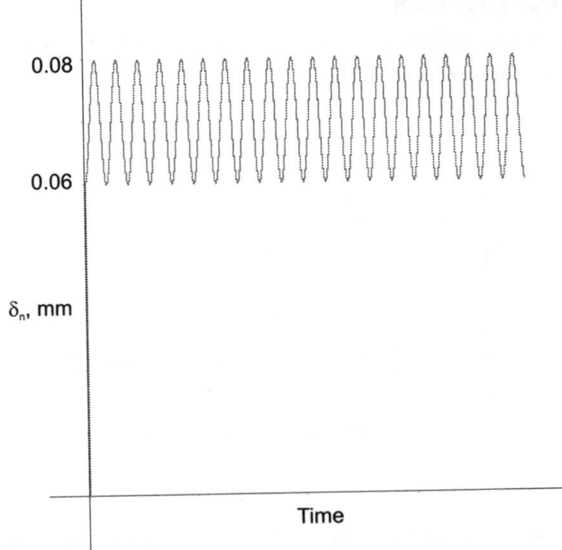

Fig. 4b. Cyclic variation of constant amplitude over a time period.

Accumulation of damage over the increasing no. of cycles for the proposed model is compared with the existing Siegmund's model [7, 9] and is shown in Fig. 5. The continuum elements are assumed to be elastic, with elastic modulus $E = 210000$ N/mm^2, Poisson's ratio $v = 0.3$ and the elements are under mode-I plane strain conditions. The properties of the cohesive zone element are: cohesive strength σ_{max} is 500 N/mm^2, and the characteristic length δ_0 is 0.08 mm.

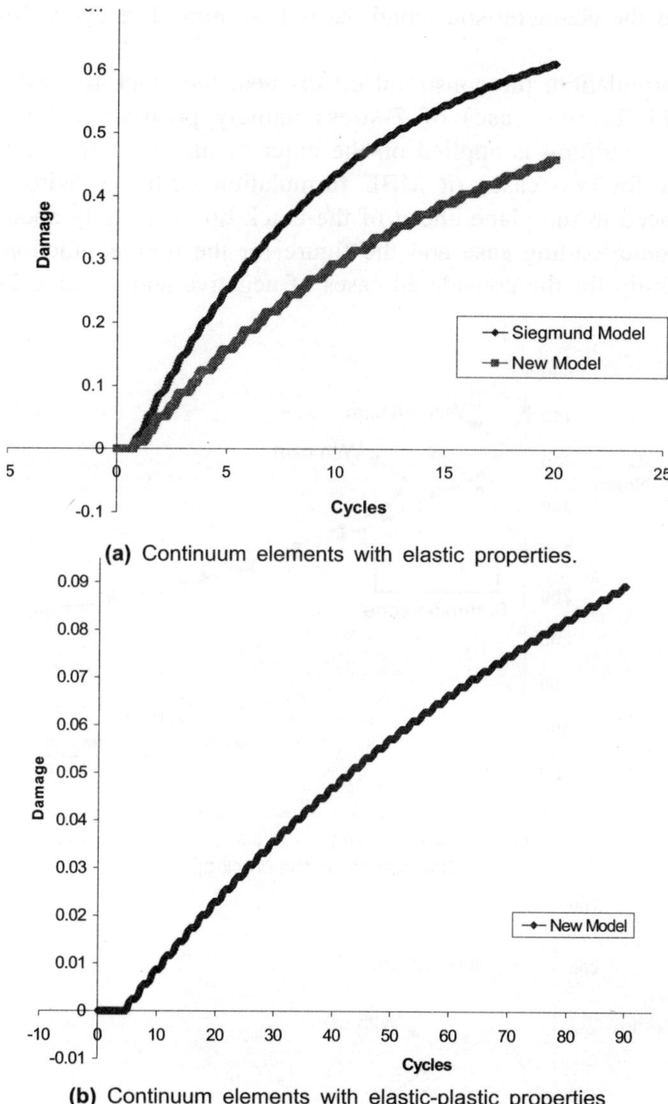

(a) Continuum elements with elastic properties.

(b) Continuum elements with elastic-plastic properties

Fig. 5. Accumulation of damage over no. of cycles.

From Fig. 5, it can be seen that the damage accumulating trend is same in both the cases, whereas the magnitude of the damage accumulated for the same number of cycles varies significantly. In Siegmund's model the traction obtained from TSL is assumed at the start of the unloading and taken as a constant throughout the unloading/reloading path whereas it evolves due to the accumulation of damage in the present model. Accumulation of damage is slower over the increasing number of cycles in the proposed model compared to the Siegmund's model. The damage accumulation as a function of no. of cycles is shown in Fig. 5(b) for the proposed model when the continuum elements have elastic-plastic properties. The elastic-plastic properties are taken as E = 210000 N/mm^2, $v = 0.3$ and yield stress σ_y is $100N/mm^2$ with hardening index as 0.1. The cohesive peak strength

σ_{max} is 100 N/mm^2 and the characteristic length δ_0 is 0.01 mm. The cyclic loading varies between δ_0 and $0.8\delta_0$.

Using the MBL formulation, the constraint effects near the crack-tip under mode-I plane strain condition are analyzed for two cases of T-stress namely positive and negative T-stress. The displacement boundary condition is applied on the outer boundary of the finite element mesh. The analysis is carried out for two cases of MBL formulation with and without the cohesive zone elements which are placed in the plane ahead of the crack tip. The study about the constraint effect is done for the monotonic loading case and the figure for the normal traction as a function of the distance from the crack-tip for the considered cases of negative and positive T-stress respectively is shown in Fig. 6.

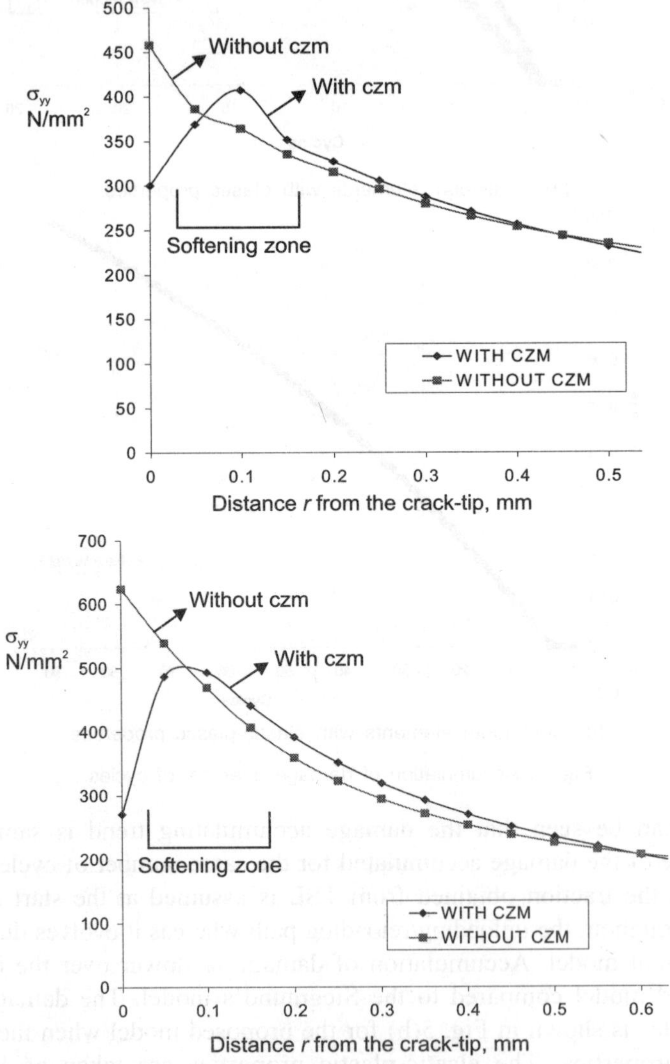

Fig. 6. Variation of normal stresses ahead of the crack-tip with and without cohesive zone elements for negative T-stress and positive T-stress respectively.

In the model without czm elements, the crack-tip opening stress σ_{yy} ahead of the crack-tip is of singular in nature and its magnitude being higher in positive T-stress. In the presence of czm elements, the stresses drop in the region where the normal separation has exceeded δ_0 causing softening. When the separation between cohesive surfaces is maximum for elements nearest to the crack-tip, the stresses developed are lower within the softening zone. For distances beyond the softening zone the presence of czm elements doesn't influence the stress field. The effect of constraint through the parameter T-stress is to lower the cohesive zone stresses when T-stress is negative.

CONCLUSION

To predict the fatigue crack growth, a cohesive law incorporating cyclic damage is proposed. The proposed model is based on the continuous degradation of cohesive strength as well as the elastic modulus. The proposed model shows significantly lower rate of damage accumulation in cohesive elements compared to an existing model, especially under sub critical loading conditions. The effect of constraints near the crack-tip is analyzed using MBL formulation. The presence of cohesive elements result in a zone ahead of the crack tip, which has stresses lowering due to the softening of cohesive elements. The stresses and the extent of softening zone are larger in the case of positive T-stress. Further work will be carried out using the same formulation to study the effect of constraint on the crack growth in cyclic loading.

REFERENCES

1. ABAQUS 6.4 user's manual.
2. De Andres et al, *'Elasto-plastic finite element analysis of three-dimensional fatigue crack growth in aluminium shafts subjected to axial loading'*, International journal of solids and structures, 36 (1999), 2231-2258.
3. Lemaitre, *'A course on damage mechanics'*, Berlin: Springer-Verlag (1996).
4. Needleman A., *'A continuum model for void nucleation by inclusion debonding'*, Journal of Applied Mechanics, 54 (1987), 525-531.
5. Needleman A., *'An analysis of decohesion along an imperfect interface'*, Non-linear fracture mechanics, (1990), 21-40.
6. Roy chowdhury et al, *'A cohesive finite element formulation for modeling fracture and delamination in solids'*, Sadhana, 25 (2000), 561-587.
7. Siegmund T. et al, *'An irreversible cohesive zone model for interface fatigue crack growth simulation'*, Engineering fracture mechanics, 70 (2003), 209-232.
8. Tvergaard V. et al, *'The relation between crack growth resistance and fracture process parameters in elastic-plastic solids'*, Journal of the mechanics and physics of solids, 40 (1992), 1377-1397.
9. Wang B. et al, *'Numerical simulation of constraint effects in fatigue crack growth'*, International journal of fatigue, 27 (2005), 1328-1334.

48

Role of Stiffeners in Resisting Impact Loads on Steel Plates

B.V.Sampathkumar[1], A. Rajaraman[2] and A. MeherPrasad[3]

[1]Student, IITM, Chennai-600 036, India. email: sampath@iitm.ac.in
[2]Visiting Professor, IITM, Chennai-600 036, India. email: arraman_2000@yahoo.com
[3]Professor, IITM, Chennai-600 036, India. email: prasadam@civil.iitm.ernet.in

ABSTRACT

Impact loads are increasingly becoming important with man-made attempts at demolition of structures are on the rise. The focus of the current paper is to assess the role of stiffeners and their location in resisting impact loads. As impact loads are short term high amplitude phenomena, damages occur depending on the impulse and in the first part of the paper this aspect is brought by classifying damages into three types. Later stiffeners are introduced one way and later both ways to focus on their role in dissipating energy to limit damages to specific areas. The paper is concluded with a design recommendation for isolating damages due to impact loads through the proper use of stiffeners.

1. INTRODUCTION

Role of impact on structural response is becoming increasingly a hot topic for research as recent man-made attacks on structures are mostly of short-duration high-amplitude type. These impulse loads are so planned that complete destruction of segments of structural systems results leaving a trail of loss of men, material and life-line utilities. Hence, a revised look at response to impact loads in terms of damage and resistance is needed and the focus of the paper here is on the role of stiffeners in resisting impact loads.

Normal distinction between dynamic and impact loads is made in terms of load-time curves where three types can be identified as (a) impact, (b) impact-dynamic and (c) dynamic depending on amplitude and duration. Hence it is always preferable to mention the time non-dimensionally using the fundamental period of the structure making it structure dependent. But even this sometimes leads to anomalies as in the case of a beam with different boundary conditions. Hence, there is a necessity to perform numerical and computer simulation of the impact dynamics and here the famous finite element technique lends itself as a powerful tool since both domain and mathematical discritisation are possible.

2. STUDIES OF IMPACT AND EXPLOSION

Impact and impact related problems have been an area of considerable research for decades [1, 2] and substantial effort has been invested in order to physically understand and mathematically describe the phenomenon, taking place during ordinance ballistic penetration [3]. So far, considerable progress has been made with experimental investigation of perforation of metal plates, and a large number of studies can be found in the literature [4, 5]. However, due to the complexity and costs related to ballistic experiments it is not optimal to base all impact and explosion related studies on laboratory tests alone. Therefore, a general solution technique is requested as a supplement to high- precision testing in order to reduce the experimental needs to a minimum. One such tool to analyze these types of problems is the use of finite element modelling. Much research has been done on the use of non-linear, large-deformation explicit finite element analysis (FEA). This technique has been used to solve many practical and important applications, which range from simple, small models (a few thousand elements) to complex, large models (hundreds of thousand of elements).

3. SIMULATION OF IMPACT AND EXPLOSION

Most dynamic FEA models for impact and explosion require a 6-step set-up procedure (pre-process)
1. Modelling
2. Material
3. Element property
4. Contact
5. Boundary effects
6. Termination time

3.1 Modelling

Finite element modelling of the spatial domain has to represent the geometry of the structure and to capture the structural response characteristics anticipated in the response. For the problem considered, the structural response involves the penetration of the impactor through the target plate. The contact interaction event now involves more than just the normal face of the impactor. The influence of the sliding contact along the bounding surfaces of the impactor as it penetrates the plate affects the results. As the contact interaction between the impactor and the target plate increases, the importance of the selection of simulation parameters also increases and penetration. The simulation model involves three components:
1. Impactor,
2. Target plate and
3. Stiffeners.

3.2 Impactor

The impactor, shown in the Fig. 1, is modelled using eight node brick elements with a rigid material model and given an initial velocity in the longitudinal direction. The impactor is positioned, which is initially at rest, at the centre of the target plate. The impactor is made of hardened steel with rigid properties. The impactor is cylindrical in shape with blunt end. The mechanical properties of impactor are:

Diameter of impactor:	0.20 m
Length of Impactor:	0.75 m
Mass of the impactor:	182.15 kg

Total numbers of solid elements, roughly of size (0.05 m long, 0.15 m thick and 0.025 m wide) with tapered edges, for impactor are 1080 with contact area of 0.0314 m^2.

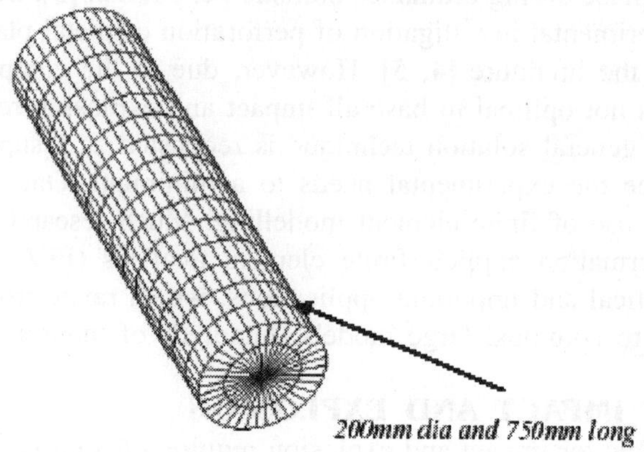

Fig. 1. Impactor model.

3.3 Target Plate

The plate is modelled using 4-noded shell element with elastic plastic strain hardening material. The size of the plate is (2 m × 2 m) with total 900 elements, each having one gauss integration points through the thickness.

3.4 Stiffeners

Flat bar stiffeners are used with 4-noded shell element. Stiffeners are modelled as shown in Fig. 2. Two types of stiffeners are used, viz. one-way and two way, spaced at "s" distance apart. The coincident nodes between plate and the stiffeners are merged so as to have material integrity. The size of the each flat bar stiffener used is (100 mm × 15 mm) with 120 elements each.

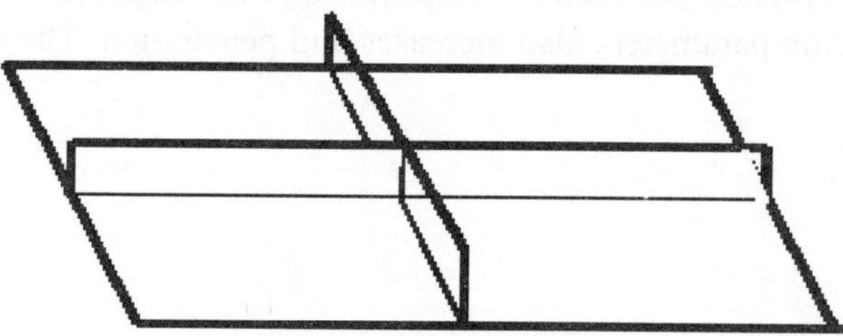

Fig. 2. Stiffener type.

4. STUDIES ON TYPES OF PLATE

Three types of plates are considered with 15 mm thickness.
1. Un-stiffened plate
2. One-way stiffened plate
3. Two-way stiffened plate.

For stiffened plates, stiffeners influence not only the displacement, but also the force transfer path. The contribution of stiffeners depends on the relative stiffness between the stiffener and the plate, the boundary conditions, and loading level. Impact load is mainly transferred from the plate to stiffeners and from stiffeners to the support structures. Although higher stiffness stiffeners reduce the deflection of the panel, they also result in higher shear forces being transferred to the support structures, which may render them vulnerable to buckling. Stiffened panels which are generally used in the structural systems subjected to impact load, ensures that the membrane forces developed due to large displacements do not produce reactions which could cause failure of already highly stressed primary structures.

For the present study flat bar stiffeners of size (100 mm deep by 15mm thick) are adopted. Time history plot for the strain energy is plotted in Fig. 3 and Fig. 4 for 15 and 30 mm plate thickness respectively, for the velocity of impactor equal to 645 m/s. It is found that the energy absorbed by the un-stiffened plate is of lower order than that for the stiffened plate. However, energy absorbed by the one-way and two-way stiffened plate is almost of equal magnitude for the two thickness of the plate considered. Time history plot for the kinetic energy of the plate is presented in Fig. 5. Kinetic energy rises until the end of the impact load and then starts decreasing. The plate absorbs all the input energy as it undergoes deformation and rupture. Fig. 5 also shows the variation of kinetic energy for different types of plates considered.

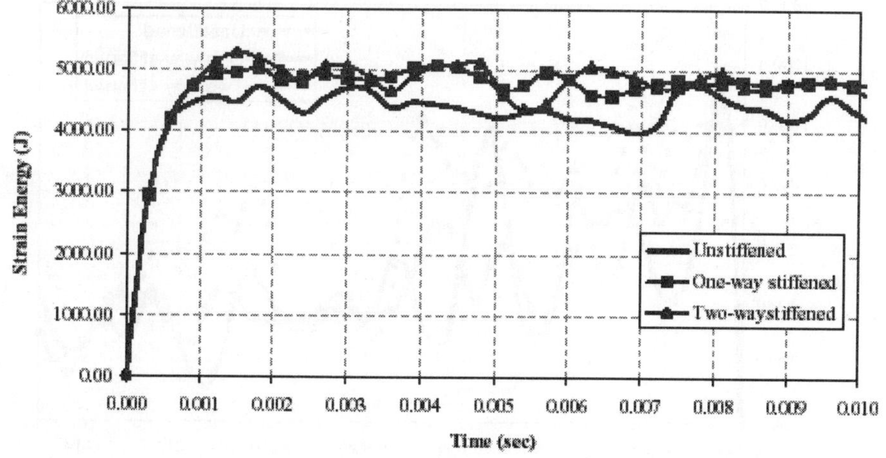

Fig. 3. Variation of SE of different types of plates for V = 645 m/s and d = 15 mm.

A comparison of the deformation for three types of plates considered is given in Fig. 6. The maximum deformation for the un-stiffened plate increases monotonically and reaches a peak value around 8 to 10 millisec. Whereas for the one-way stiffened plate of same size, peak displacement occurs much before this time and is of lower magnitude. The peak displacement is further reduced for the two-way stiffened plate. This represent that the damage becomes more and more localised due to stiffening of the plate.

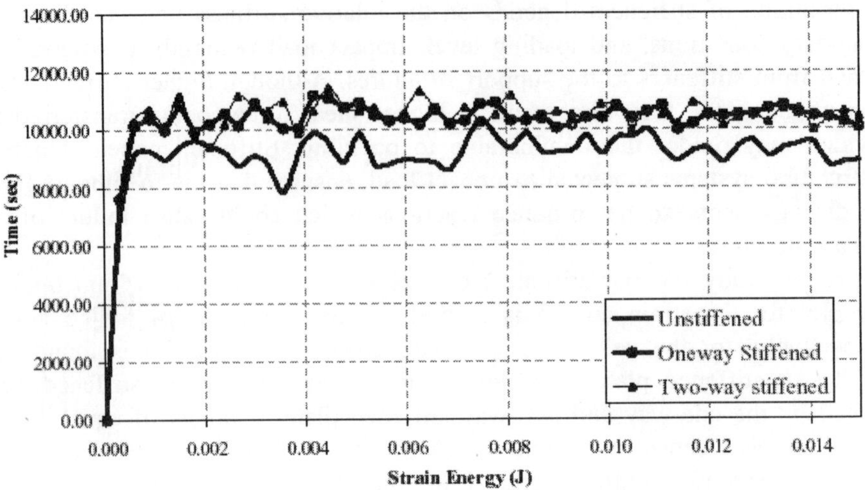

Fig. 4. Variation of SE of plates for V = 645 m/s and d = 30 mm.

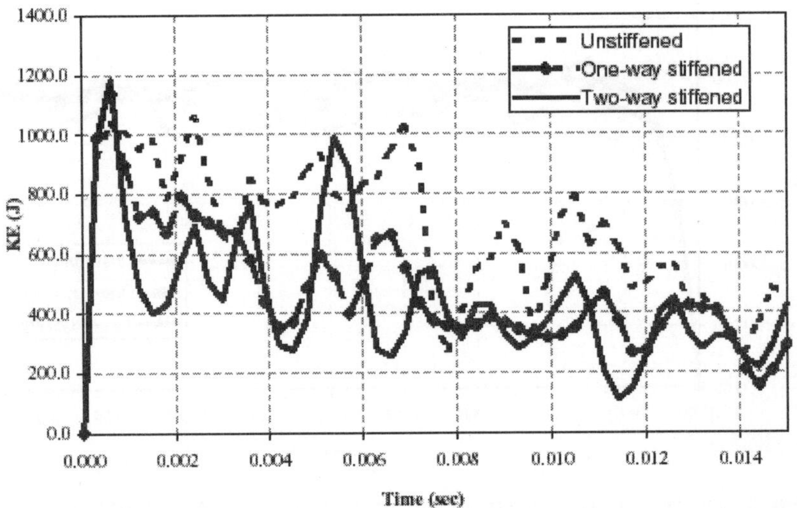

Fig. 5. Variation of KE of plate for V = 645 m/s and d = 15 mm.

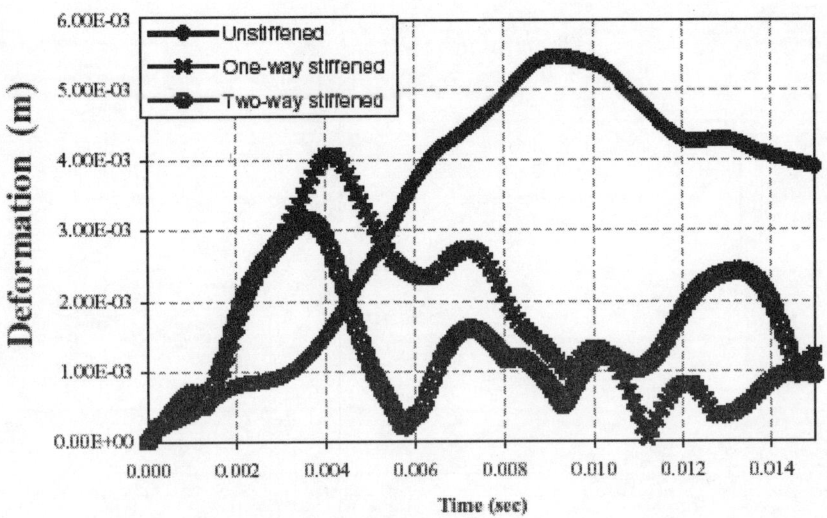

Fig. 6. Plot of deformation vs. time for V = 645 m/s and d = 15 mm.

Figure 7 and Fig. 8 show the Von-Mises stress fringe patterns for one-way stiffened and two-way stiffened plate respectively. It can be noticed that the damage is confined within the stiffener area as against the one-way stiffened plate where the stresses flow along the stiffeners direction.

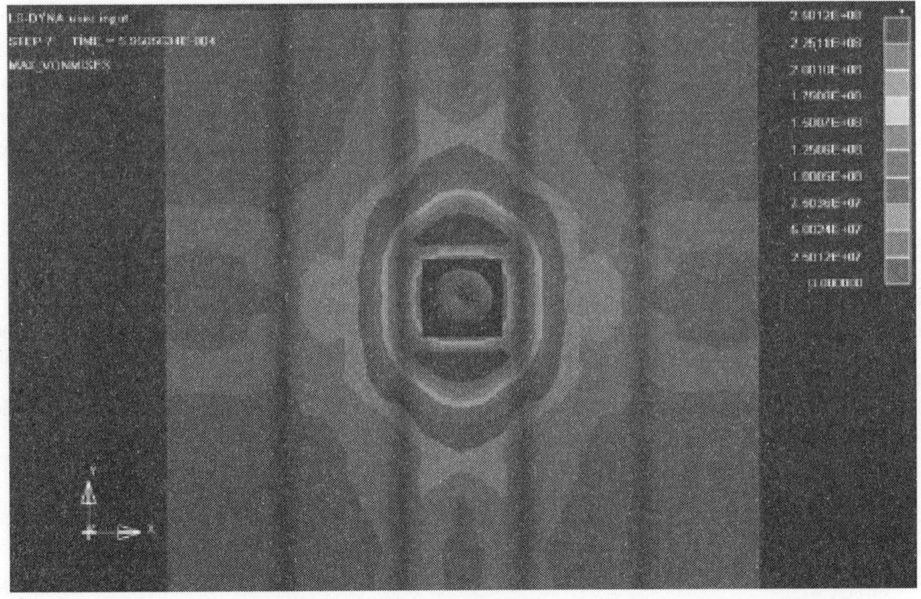

Fig. 7. Von-Mises stress fringe patterns for one-way stiffened plate.

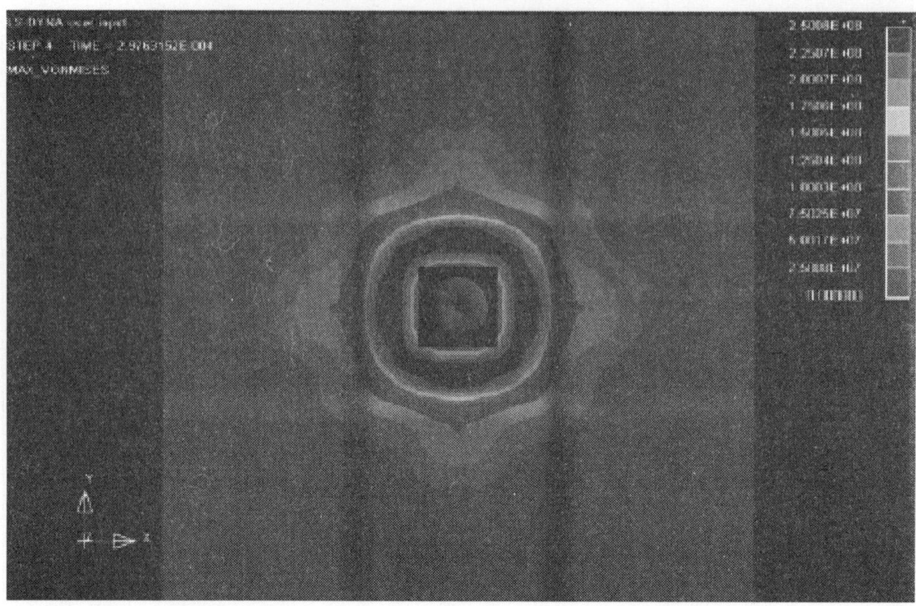

Fig. 8. Von-Mises stress fringe patterns for two-way stiffened plate.

CONCLUSION

Role of impact dynamics on plate configurations is discussed and models are generated to study the effects of a typical impactor on a plate mainly with respect to nonlinear behaviour and type of failure patterns. Three types of failure are identified and the influence of different parameters like velocity of attack, angle of attack, boundary conditions and types of stiffener on the types of failure is indicated. Role of explosion after impact is also modelled to give an idea of the change in failure pattern of the plate. Typical contours of stresses indicate the performance of the plates explicitly giving an idea of the role of impact dynamics, different from conventional dynamics.

REFERENCES

1. Jones, Norman, 1989, Structural Impact, Cambridge University Press, Cambridge.
2. Biggs J.M., 1964, Introduction to Structural Dynamics. McGraw-Hill Book Company, New York.
3. Bangash M.Y.H., 1993, Impact and Explosion: Analysis and Design, Oxford: Blackwell Scientific Publication.
4. Norman F. Knight Jr., 2000, Penetration simulation for uncontained engine debris impact on fuselage-like panels using LS-DYNA. Finite Elements in Analysis and Design, Vol. 36, pp. 99-133.
5. Damodar R. Ambur, 2001, Numerical simulations for high-energy impact of thin plates. Int J. Impact Engg., Vol. 25, pp. 683-702.

49

Nonlinear Analysis of Reinforced Concrete Structures Using Fracture and Damage Mechanics Constitutive Models

RAJESH K. SINGH[1], R.K. SINGH[2] AND T. KANT[3]

[1]Architecture & Civil Engineering Division,
[2]Reactor Safety Division, Bhabha Atomic Research Centre, Trombay, Mumbai-400 085, India
[3]Indian Institute of Technology Bombay, Powai, Mumbai-400 076, India

ABSTRACT

In the present studies fracture and damage mechanics based various numerical models, which have been incorporated in commercial software ADINA, ABAQUS and ANSYS are employed for numerical analysis of a few typical reinforced concrete (RC) composite structures. An effort is made to simulate the experimental routine test based on recommendation of RILEM (1985) for determination of fracture energy of concrete. The crack band model, which is available in ADINA, is used for the numerical simulation of experimental test results. The mesh insensitive numerical results based on fracture energy model are evolved for inelastic analysis. The present numerical study proposes schemes for realistic simulation of RC structures in agreement with observed experimental results. In reinforced concrete, the stress-displacement relationship must also represent the bond action between the concrete and the rebar. This can be accommodated by selecting the value of fracture energy (G_f) in the numerical simulation based on the observed experimental results of notch beam specimens as per RILEM (1985) standard.

1. INTRODUCTION

Plain concrete is not a perfectly brittle material in the Griffith (1924) sense, with an immediate stress drop to zero after the tensile strength has been reached, but some residual load-carrying capacity remains after the crack initiation. The application of fracture mechanics to plain and reinforced concrete has opened up a new field for modelling of phenomena that have often been treated empirically in the past. Cohesive crack model proposed by Hillerborg (1976) and crack band model Bazant *et al.* (1983) with localization limiters are frequently used to study of tension failure of concrete. Many nonlinear fracture mechanics models have been proposed. Bazant *et al.* (1979) recommended the use of the stress intensity factor or the critical strain energy rate as a cracking criterion. An alternative method was developed by Bazant *et al.* (1983) based on the strain softening characteristics as a cracking parameter. The strain softening parameter was adjusted with respect to the element size, so that the total energy under the softening curve represented the fracture energy. Ouyang *et al.* (1994) used an R-curve approach to evaluate the fracture energy required for crack propagation. In their model, concrete is considered as a quasi-brittle material. Kotsovos *et al.* (2004) investigate the causes of size effects in structural-concrete members. Rabczuk *et al.* (2005) describes a two-dimensional approach to model fracture of reinforced concrete structures under (increasing) static loading conditions. Carpinteri et al. (2005) proposed a theoretical model based on fracture mechanics concepts in order to analyze the mechanical damage of ordinary or prestressed reinforced concrete beams. Only a very few published works can be found on the simulation of crack in reinforced concrete using the fictious crack model. Ouyong and Shah (1994) proposed to estimate the fracture energy for reinforced concrete, where in the strain energy, debonding energy and sliding energy on the debonded interface of steel bars and concrete dissipated during cracking are included. The nonlinear fracture mechanics formulation accounts for the presence of fracture zone with a rising fracture resistance curve (R-curve).

2. CONSTITUTIVE MODEL

In this section various fracture energy models for plain concrete is presented with the fictitious crack model and the crack band model, which have been widely used for studying the fracture behaviour of concrete structures. In Fictitious or Cohesive Crack Model (FCM model), the fracture zone is replaced by the cohesive forces acting normal to both the crack surfaces. These surfaces are not fully cracked in the real sense, as they are still able to transfer the tensile stress. They are referred as a fictitious crack boundary. The intensities of these forces are dependent on the crack opening displacement. In this FCM model, the material outside the fracture zone is assumed to behave in linearly elastic manner. The Crack Band Model is the special case of the cohesive crack model. The basic attributes of the crack band model is that the given constitutive relation with strain softening must be associated with a certain width hc of the crack band, which represents a reference width and is treated as a material property. The most important feature of the crack band model is that it can effectively handle the problem of mesh size sensitivity, provided that it is localized within one element. ANSYS provides a dedicated three-dimensional eight-nodded solid isoparametric element, Solid65, to model the nonlinear response of brittle materials based on a constitutive model for the triaxial behavior of concrete after Williams and Warnke (1974). The element includes a smeared crack analogy for cracking in tension zones and a plasticity algorithm to account for the possibility of concrete crushing in compression zones.

The concrete damaged plasticity model uses concepts of isotropic damaged elasticity in combination with isotropic tensile and compressive plasticity to represent the inelastic behavior of concrete. The advantages of this model over concrete smeared crack model is that it can be used for concrete structure subjected to monotonic as well as cyclic loading under the low confining pressure. It consists of the combination of non-associated multi-hardening plasticity and scalar (isotropic) damaged elasticity to describe the irreversible damage that occurs during the fracturing process, and also allows to control of stiffness recovery effects during cyclic load reversals.

3. NUMERICAL ANALYSES AND RESULTS

To overcome the size effect on the value of fracture energy, an effort is made here to determine the fracture energy analytically based on crack band model. Fig. 1 shows a notched beam subject to three-point bending and the material properties used in the analysis are tabulated in Table 1. The typical beam geometry has depth (h) of 100.0 mm, the notch depth (a) of 10.0 mm and the width of beam (b) is 150.0 mm with beam length (l) of 600.0 mm. The two-dimensional 8-noded plane stress element is used for the finite element models of notched beam as shown in Fig. 1 (load control model). The beam was observed to fail in bending in the experiment. The load and displacement control models are used for obtaining the complete failure (Load- displacement) curve as shown in Fig. 2. The post softening curves shows the expected nonlinear behavior of concrete with the fracture energy based damage model.

Table 1. Material properties of concrete.

Material Properties of concrete	
Concrete grade	2
Comp. Strength, f_{ck} *(MPa)*	75.7
Young's Modulus, E_c(MPa)	34300
Fracture Energy, G_f (N/m)	0
Tensile Strength, f_t (MPa)	90.0
	5.3

Table 2. Fracture energy for plain concrete.

Loading condition	Fracture Energy (N/m)
Central point loading	154.0
Loading-unloading	**136.0**
Loading with body force	158.0
Experimental	90.0-140.0 N/m

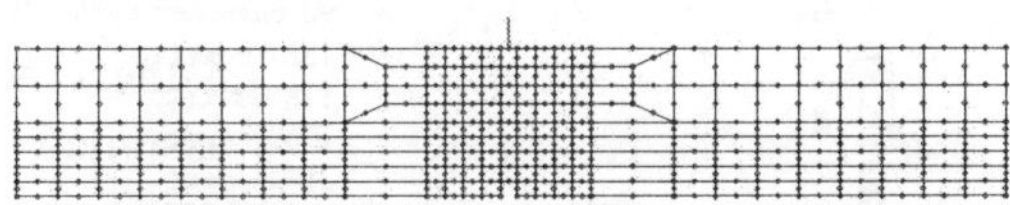

Fig. 1. Finite Element Model of three points bending notched beam (Load control).

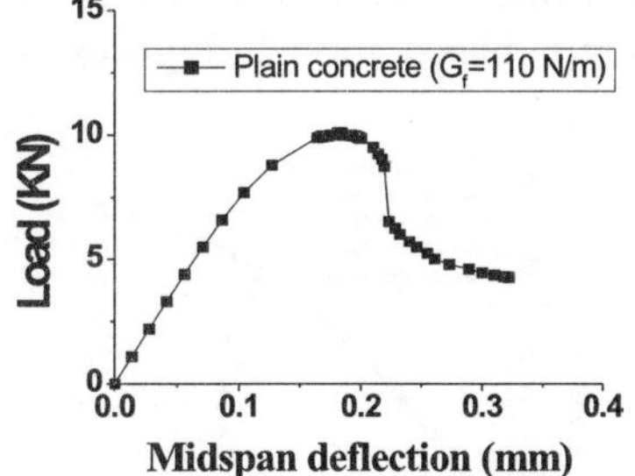

Fig. 2. Load-deflection curve of notched beam using crack band model.

According to RILEM recommendation RILEM fracture energy (G_f) can be calculated as

$$G_f = \frac{A_1 + mg\delta_0}{b(h-a)} \qquad \text{...(1)}$$

From Fig. 2, area under load-deflection curve A_1 can be calculated using trapezoidal numerical integration method. Here

$A_1 = 2.017$ Nm; b = 150.0 mm; h = 100.0 mm; a = 10.0 mm; $\delta_0 = 0.325$ mm; mg = 190.706 N

Putting all the above value of parameter in the above equation 1, fracture energy (G_f) = 154.0 N/m. Bazant *et al.* (1983), introduced the crack band theory in the analysis of a plain concrete panel, which is the one of the simplest types of fictitious cracks models. The final equation derived for determining the strain at which tensile stress is equal to zero (ε_t^f) can be expressed as,

$$\varepsilon_t^f = \frac{2G_f}{b\sigma_t} \qquad \text{...(2)}$$

$$G_f = (2.72 + 0.0214\sigma_t)\frac{\sigma_t^2 D_a}{E_c} \quad \text{Bazant } et\ al.\ (1983) \qquad \text{...(3)}$$

(σ_t and E_c are in psi, D_a is in inch)

The equation 2 can be successfully applied when the finite element mesh size is equal to or smaller than 75 mm. The equation 3 can be used for determination of fracture energy of concrete. The value of fracture energy obtained using above equation can be taken as the first good estimate. The actual value of the fracture energy may be slightly different because the above expression is the result of statistical analysis. For the present problem, the above equation gives a G_f of 147 N/m.

For the analysis of reinforced concrete structures, tension stiffening of concrete is taken into account due to the influence of reinforcement at the vicinity of cracks. The numerical examples presented are based on the three-point bending tested experimentally by Bosco *et al.* (1990). The fracture energy, G_f, is one of the important parameters, which should be carefully determined. For this reason, the various values of G_f in the range of 90.0 to 150.0 N/m were tested in present analysis. It was found that higher value of G_f gave results close to those obtained in the experiment by Bosco *et al.* (1990). Figure 3 shows that load-deflection response for the value of $G_f = 110.0$ N/m seems to be in good agreement with the experimental results. The reinforcement area of 12.7 mm^2 was taken for the above analysis. Parametric studies have been done on the basis of different reinforcement area.

An effort is made here to show the capability of fracture energy model in comparison with stress-strain based model. Figures 4 to 6 present the load-deflection curves obtained with strain

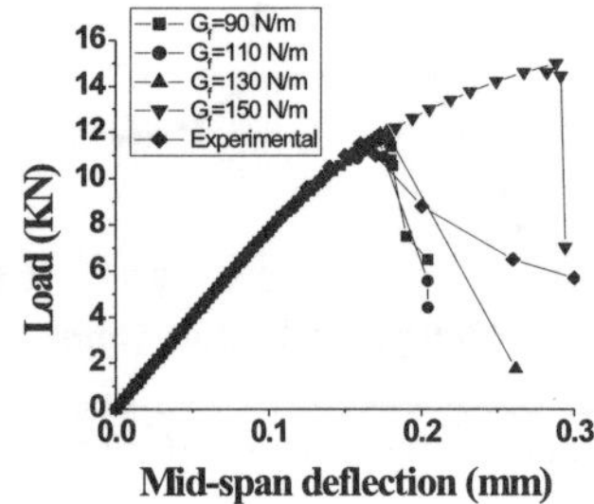

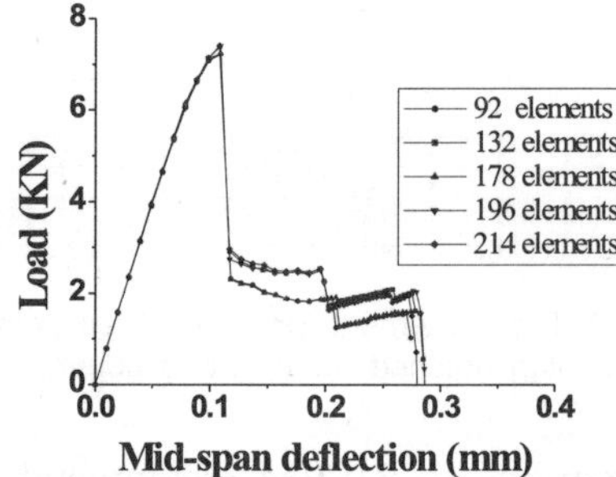

Fig. 3. Load-deflection curve for Different value of fracture energy.

Fig. 4. Load deflection curve showing mesh sensitivity using strain based approach.

based model and the two fracture energy models; namely the cohesive crack model and the crack band model respectively. In Fig. 5, strain based approach shows a sharp peak but it underestimates the maximum load and post-softening response shows oscillations. The crack band fracture energy model traces the peak load (Fig. 6) in true manner in agreement with the experimental value of 10.3 KN and gives the true presentation of softening curve what normally one obtains in experiments. In addition, the convergence for the fracture energy model with mesh refinement is monotonic as noticed in the load-deflection curve in case of fracture energy crack band model compared to the strain based approach. The cohesive crack fracture model computes the peak load in agreement with the experimental results but the post-softening behavior is inferior compared to the crack band model and the mesh sensitivity is noticed for refined grids with 178, 196 and 214 elements Fig. 7 presents the analytical results using elasticity model, stress-strain based plasticity model, cohesive crack fracture energy model with damage plasticity and crack band fracture energy model. In this figure, the stress-strain based plasticity approach shows a sharp peak load of 7.95 KN and it underestimates the maximum load and post-softening response shows oscillations. The crack band

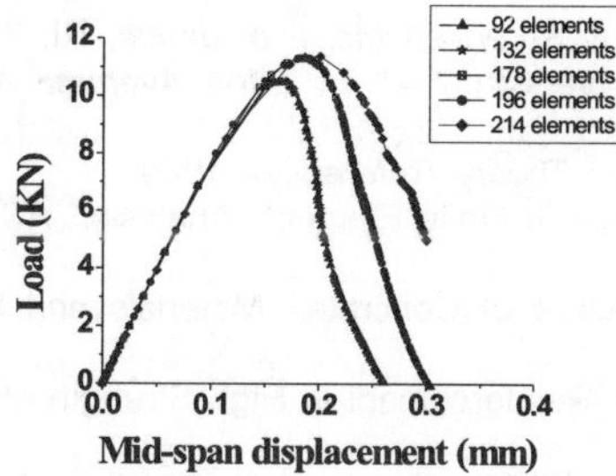

Fig. 5. Load deflection curve showing mesh insensitivity using cohesive crack model approach.

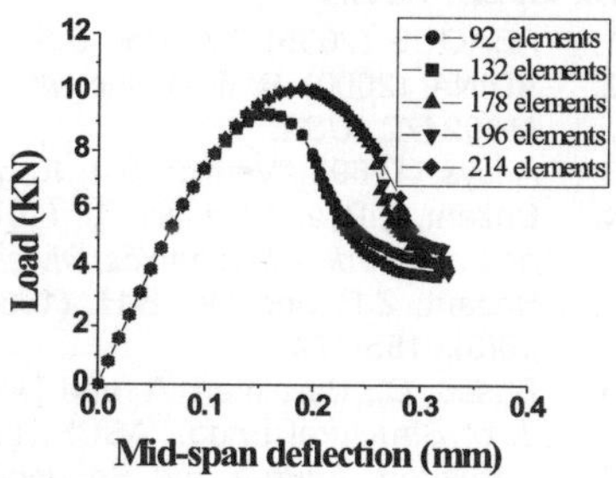

Fig. 6. Load deflection curve showing mesh insensitivity using crack band.

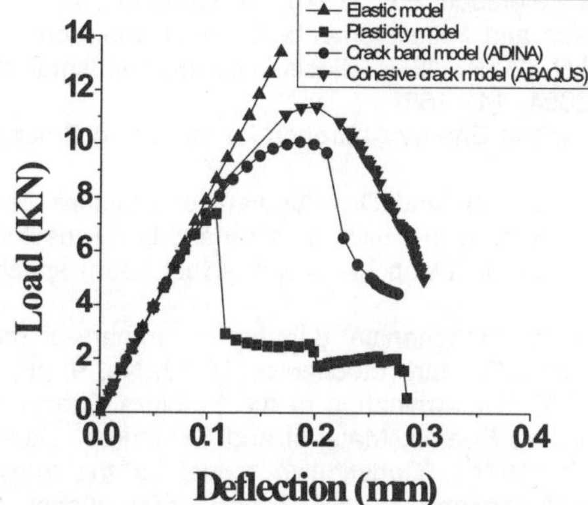

Fig. 7. Comparison of load-deflection curve using strain based approach and fracture energy based approach.

fracture energy model traces the peak load of 10.04 kN accurately in agreement with the experimental value of 10.05 KN and gives the true presentation of softening curve as one obtains in the experiment.

CONCLUSION

The overall structural behaviour predicted by finite element analysis has been represented with the help of load deflection plot at mid-span. Finite element analysis shows good agreement with the experimental results. A numerical model (crack band model with localisation limiters) was used in the present study for the estimation of fracture energy including the effect of body force. The fracture energy of concrete is most sensitive to the presence of reinforcement in the reinforced concrete structures. It was concluded that with an increase in the area of reinforcement, the fracture energy value improves. The present numerical model shows the same response before softening compared to the experimental results but differs slightly in post softening region. It was concluded that fracture energy model traces the ultimate load carrying capacity accurately as observed in the experiment compared to stress-strain based numerical approach. The present study concludes that the fracture energy based concrete model gives mesh insensitive results compared to stress-strain based model.

REFERENCES

1.   ABAQUS (2005). "Version 6.5 User's Manual", *Hibbitt, Karlsson & Sorensen Inc, Providence*, RI, 2005.
2.   ADINA (2000) R & D, Inc. /K. J. Bathe. "The ADINA System, version 7.4," 71 *Elton Avenue, Watermelon, MA02472, USA.*
3.   Ansys (1989). "Version 5.4, Engineering Analysis System", *Ansys Theory References* 1989.
4.   Bazant, Z.P. and Cedolin, L. (1979). "Blunt Crack Band Propagation in Finite Element Analysis," *ASCE Journal of Engineering Mechanics Division*," 106(6), 1287-1306.
5.   Bazant, Z.P. and Oh, B.H. (1983). "Crack Band Theory for Fracture of Concrete," Materials and Structures," 16(3), 155-177.
6.   Bosco, C., Carpinteri, A. and Debernardi, P. G. (1990). "Minimum Reinforcement in High Strength of Concrete", J. of Structural Engg., ASCE, (116), 427-437.
7.   Carpinteri, Andrea and Spagnoli Andrea and Vantadori Sabrina. (2005). "Mechanical damage of ordinary or prestressed reinforced concrete beams under cyclic bending," Engineering Fracture Mechanics, 72(9), 1313-1328.
8.   Griffith, A.A. (1924) "The Theory of rupture," In Proceedings of the first International Conference of Applied Mechanics, pp. 55-63.
9.   Hillerborg, A., Modeer, M. and Petersson, P.E. (1976) "Analysis of crack formation and crack growth in concrete by means of fracture mechanics and finite elements,"Cement and Concrete Res., 6, 773-782.
10. Kotsovos, M.D. and Pavlovi, M.N. (2004). "Size effects in beams with small shear span-to-depth ratios," Computers & Structures, V. 106, No. 6, 2004, 143-156.
11. Ouyang, C. and Shah, S.P. "Fracture Energy Approach for predicting Cracking of Reinforced Concrete Members," ACI J., 91, 1994, pp. 69-78.
12. Petersson, P.E. (1981). "Crack Growth and Development of Fracture Zones in Plain Concrete and   Similar Materials, Report TVBM-1006, Division of Building Materials, Lund Institute of Technology, Lund, Sweden.
13. Leonhardt, F. and Walther, R. (1976). "Deep Leonhardt`s test beam specimen WT3," Deutsher Ausschuss Fur Stahlbeton, Heft 178, Berlin.
14. Rabczuk, T., J. and Eibl, J. (2005). "Mechanical damage of ordinary or prestressed reinforced concrete beams under cyclic bending," Engineering Fracture Mechanics, V. 72, No. 9, pp. 1313-1328.
15. RILEM Recommendations (1985). "Determination of the Fracture Energy of Mortar and Concrete by Means of Three-Point Bend Test on Notched Beams, Material and Structures, 1985, 18 (106), 287-290.
16. William, K.J. and Warnke E.P. (1974). "Constitutive model for the triaxial behaviour of Concrete" Concrete Structures subjected to Triaxial stresses, IABSE Report, International Association of Bridge and Structural Engineers, Zurich, 19, 1974, pp.1-30.

50

Simulation of Materials Damage in the Field of Thermal or Residual Stresses

N.M. Vlasov and I.I. Fedik

Scientific Research Institute Scientific Industrial Association "Luch"
Zheleznodorozhnaya 24, Podolsk Moscow region-142100, Russia
email: iifedik@luch.podolsk.ru

ABSTRACT

Internal stresses occur within a material in the presence of non-uniform deformation. The main types of the internal stresses are the thermal and residual ones and fields of structural defects as well. These stresses have an essential effect on the diffusion processes kinetics. In this case change of the strength material properties takes place. Physical mechanisms of the metal properties changes are rathes various: decreasing surface fracture energy, corrosion cracking and hydrogen embrittlement. These mechanisms are based on interstitial impurities diffusion (for example, hydrogen and oxygen) in the internal stresses field. The kinetics of interstitial atom diffusion is described by an equation of parabolic type under corresponding initial and boundary condition. The purpose of this paper is simulating the material damage as a result of running the diffusion processes. The process of simulating the hydrogen embrittlement of a hollow cylinder with internal stresses is considered as an example. The first invariant of the tensor of internal stresses (thermal or residual) in a hollow cylinder has a logarithmic dependence on the radial coordinate. Such a dependence permits obtaining an exact analytic solution of diffusion kinetics problem in view of internal stresses. If local concentration of the hydrogen atoms exceeds the solubility limit at given temperature, hydrides are formed in some metals (for example, in zirconium). Volume changes of the latter ones lead to forming microcracks on interphase boundaries.

Keywords: Internal stresses, diffusion kinetics, damage simulation.

1. INTRODUCTION

The structural elements of up-to-date equipment are operated under conditions of various physical fields influence. It leads to occurrence of the internal stresses as a result of non-uniform deformation. The basic types of the internal stresses are temperature and residual ones as well as about the structural defects. Alloying elements of the alloy are rather sensitive to the level and character of the corresponding stresses distribution. Diffusion redistribution of the impurity atoms in the stresses field is accompanied by changing the strength characteristics of the alloy. If a local concentration of the impurity atoms exceeds the solubility limit at given temperature, new phases are formed. Considerable volume changes lead to formation of discontinuity flaws on the interphase boundaries. The hydride phases formation within the alloys on the base of zirconium is enough to be mentioned as an illustration. This material is used for claddings of fuel elements of nuclear reactors. Simulation of the material damages taking into account the stresses fields is a rather complicated task. Such complicity is explained by the stresses effect on the diffusion processes kinetics. In general, the internal stresses fields have complex dependence on coordinates. Therefore, essential mathematical difficulties occur when solving a diffusion equation taking into account the stresses. A pleasant exclusion from a general rule is the internal stresses with a logarithmic coordinate dependence. Such dependence allows the exact solution of the diffusion kinetics task to be obtained. In this paper preference is given to just such coordinate dependences of the internal stresses field.

The purpose of this paper is a mathematical simulation of material damages in the internal stresses field with a logarithmic coordinate dependence. Kinetics of the impurity segregations formation in the field of temperature and residual stresses of the cylindrical cladding is considered. The first invariant of these stresses tensor has a logarithmic dependence on the radial coordinate. Kinetics of the new phases growth is used taking into account the internal stresses field. In this case the impurity atoms concentration exceeds the solubility limit at given temperature. The material damages are the results of microcracks formation on the interphase boundaries. The degree of the material damages is defined by volume changes of the new phase in comparison with a matrix material. The material damage is growing while increasing the volume changes of the new phase. Hydrogen embrittlement of zirconium alloy at the expense of hydride phases formation with large volume changes (from 9 to 12%) is considered as an example.

2. PHYSICAL ESSENCE OF THE INTERNAL STRESSES

Self-balanced internal stresses are present in the material without applying an external load. By 'self-balance' we will mean the stresses of a different sign available at a zero value of their integral characteristic. Typical examples of such stresses are temperature and residual stresses as well as ones about the structural defects. Occurrence of the thermal stresses is caused by non-uniform distribution of the temperature field. In general case the temperature stresses have a rather complex dependence on coordinates. Therefore, account of their effect on the diffusion processes kinetics is of essential mathematical difficulties. However, in some cases the thermal stresses logarithmic depend on the coordinates. The thermal stresses in a hollow cylinder are the example of such dependence. It is known that interaction of an impurity atom with the thermal stresses field depends on the first invariant of the stresses tensor. For the thermal stresses in the hollow cylinder this characteristic is determined by the relation [1–3]

$$\sigma_{ll} = \frac{2\alpha\mu\left(1+v\right)\left(T_1 - T_2\right)}{1-v}\left\{\frac{1+2\ln\dfrac{r}{R}}{\ln\dfrac{R}{r_0}} + \frac{2r_0^2}{R^2 - r_0^2}\right\} \qquad \text{...(1)}$$

where α-coefficient of linear expansion, μ-shear module, v-Poisson's ratio, r_0 and R–inner and outer radius of a hollow cylinder, T_1 and T_2—temperatures of inside and outside of a hollow cylinder correspondingly. This relation is obtained for the flat deformation conditions at the boundaries free of the stresses.

Such coordinate dependence is peculiar to the residual stresses in the hollow cylinder. The latter occur in the material when carrying out various technological operations. Let us consider the following version of the residual stresses formation. The cutting edges of the cylinder are moved apart by angle ω, and the missing material is placed there. During this operation the area about outer cylinder surface is under as-pressed condition, and about the inner cylinder surface - as-tensed condition. The first invariant of the residual stresses tensor for the flat deformation conditions also has the logarithmic coordinate dependence [4]

$$\sigma_{ll} = \frac{\omega\mu(1+v)}{2\pi(1-v)}\left\{1+2\ell n\frac{r}{R}+\frac{2\left(\frac{r_0}{R}\right)^2}{1-\left(\frac{r_0}{R}\right)^2}\ell n\frac{r_0}{R}\right\} \qquad ...(2)$$

where ω–angle of the cutting edges opening of the hollow cylinder. The rest notations correspond to the ones adopted before.

3. SEGREGATION OF IMPURITIES IN THE STRESSES FIELD

The alloying elements of the alloy are rather sensitive to the level and character of the internal stresses distribution. The latter ones cause redistribution of the impurity atoms with changing the strength characteristics of the alloy. The typical size of the substitutional impurity is usually more (less) than the corresponding size of a basic material atom. The sizes of interstice (oct- and tetrahedral), as a rule, are less than the substitutional impurities size. Such inconsistency of the typical sizes leads to elastic interaction of the impurity atoms with the internal stresses. The interaction potential is defined by the known relation [5]

$$V = -\frac{\sigma_{ll}}{3}\delta\upsilon \qquad ...(3)$$

where σ_{ll}–first invariant of tensor stresses, $\delta\upsilon$ – change of metal volume at an impurity atom placement. For $\sigma_{ll} > 0$ (tension stresses) and $\delta\upsilon > 0$ (an impurity atom increases a crystal lattice parameter) potential V takes a negative value. It corresponds to attraction of an impurity atom to the tension stresses area and its displacement from the compression stresses area. Relation (3) takes into account only dimensial effect in the energy of the impurity atom connection with the internal stresses field. In other words, inconsistency between the impurity atom sizes and the volume for its placement is accompanied by the elastic interaction of the impurity atom with the stresses fields. The other types of interactions (module, electrostatic, chemical ones) can be easily estimated by renormalization of the constants in relation (3).

The impurity concentration field is determined from the solution of the equation of a parabolic type under the corresponding initial and boundary conditions

$$\frac{1}{D}\frac{\partial C}{\partial t} = \Delta C + \frac{\nabla(C\nabla V)}{kT} \ , \ r_0 < r < R,$$

$$C(r,0) = C_0, \ C(r_0,t) = C_p^1 \ , \ C(R,t) = C_p^2 \qquad \qquad ...(4)$$

where D–coefficient of atoms diffusion, k–Boltzmann constant, T–absolute temperature, C_0– average concentration of impurity atoms, r_0 and R–inner and outer radiuses of the internal stresses action, C_p^1 and C_p^2 –equilibrium concentrations of the impurity atoms on the boundaries of the discussed field. Equation (4) connects physics and mechanics when simulating material damages in the internal stresses field. The complicated coordinate dependence of potential V does not allow us to obtain an analytical solution of task (4) in the class of the known functions and their combinations. This difficulty can be overcome successfully for the logarithmic coordinate dependence of potential V. It is caused by the fact that in this case potential V is a harmonic function, and its gradient is inversely proportional to a radius in the polar coordinate system. Physical meaning of the initial and boundary conditions of task (4) is quite evident. At start time the impurity atoms concentration is equal to an average value. Equilibrium impurities concentration is quickly set on the area boundary and then kept in accordance with potential V on the outer cavity too. From equation (4) one can see that segregation of the impurity atoms is proportional to the gradient of potential V. Therefore, the constants in the relations for $\sigma_{//}$ disappear when differentiating. It simplifies the solution of diffusion kinetics task. Keeping commonality, let us consider the residual stresses in the hollow cylinder. The first invariant of the stresses tensor is determined by relation (2). Taking into account (2) and (3) after simple mathematical conversions we will obtain

$$\frac{1}{D}\frac{\partial C}{\partial t} = \frac{\partial^2 C}{\partial r^2} + \frac{1+\beta}{r}\frac{\partial C}{\partial r}, \ r_0 < r < R,$$

$$C(r,0) = C_0, \ C(r_0,t) = C_p^1 \ , \ C(R,t) = C_p^2 , \qquad \qquad ...(5)$$

$$\beta = \frac{\mu\omega(1+v)\delta\upsilon}{3\pi(1-v)kT}$$

The dimensionless parameter of task β defines the relation between the binding energy of impurity atoms with the field of the residual stresses and heat motion. In case $\beta<<1$, the residual stresses field is slight disturbance of a diffusion flux as a result of the concentration gradient. For $\beta>>1$ the residual stresses field gives the main contribution into the process kinetics. At $\beta\cong1$ diffusion fluxes of the impurity atoms are commensurable at the expense of the concentration gradients and interaction potential. The estimations show that for some systems (for example, Zr-H) value $|\beta| \cong 1$. Taking $\beta = -1$ ($\omega < 0$ for accepted scheme of the residual stresses), we obtain a rather simple variant of task (5)

$$\frac{1}{D}\frac{\partial C}{\partial t} = \frac{\partial^2 C}{\partial r^2}, \ r_0 < r < R,$$

$$C(r,0) = C_0, \ C(r_0,t) = C_p^1 \ , \ C(R,t) = C_p^2 . \qquad \qquad ...(6)$$

An interesting feature of the obtained equation should be noted. It is seen the residual stresses change symmetry of the diffusion equation. Kinetics of the impurity segregations in the hollow cylinder runs according to the law of flat symmetry. The solution of task (6) gives distribution of the impurity atoms concentration taking into account the residual stresses field

$$C - C_0 = \frac{R\left(C_p^1 - C_0\right) - r_0\left(C_p^2 - C_0\right) + r\left(C_p^2 - C_p^1\right)}{R - r_0} +$$

$$+ \frac{2}{\pi} \sum_{n=1}^{\infty} \frac{1}{n}\left[(-1)^n\left(C_p^2 - C_0\right) - \left(C_p' - C_0\right)\right] \times.$$

$$\times \sin\frac{\pi n\left(r - r_0\right)}{R - r_0} \exp\left(-\frac{\pi^2 n^2 Dt}{\left(R - r_0\right)^2}\right) \qquad ...(7)$$

As times goes by, the concentration field of the impurity atoms takes a stationary character in accordance with the interaction potential. In this case redistribution of the impurity atoms takes place when keeping their integral concentration. The strength characteristics of the material are changed and it has a disposition to damages. The limiting link of this process is diffusion of the alloying elements of the alloy in the internal stresses field. If in a local volume of the material the concentration of the impurity atoms exceeds the solubility limit at given temperature, a new phase separations are formed. So for example, in Zr–H system hydrides are formed.

4. NEW PHASES FORMATION IN THE STRESSES FIELD

The new phases are characterized by considerable volume changes in relation to the basic material. The stresses occur on the interphase boundaries, and microcracks are formed. The example of such damage is metal embrittlement when forming hydride phases. The internal stresses also have an effect on kinetics of a new phase growth. Let us consider the residual stresses in a hollow cylinder. Maximal concentration of the impurity atoms occurs on the area boundary, where the new phase formation takes place. Its further growth is realized at the expense of impurity atoms diffusion. The task of defining kinetics of the new phase growth in the hollow cylinder is mathematically formulated as follows

$$\frac{1}{D}\frac{\partial C}{\partial t} = \frac{\partial^2 C}{\partial r^2} + \frac{1 + \beta}{r}\frac{\partial C}{\partial r} \, ,$$

$$C(R_1, t) = C_2, \quad C(r,0) = C_0 \ (r \geq R_0), \quad C(\infty, t) = C_0, \qquad ...(8)$$

$$\left(C_1 - C_2\right)\frac{dR_1}{dt} = D\left\{\left|\left|\frac{dC}{dr}\right| + \left|\frac{C\beta}{r}\right|\right|\right\}_{r=R_1} \, ,$$

where R_0—radius of the a phase centre, R_1 – current radius of a new phase. The rest notations are correspond to the ones accepted before. On the moving interphase boundary the concentration of the impurity atoms is changed in leaps and bounds: $C = C_1$ for the new phase and $C = C_2$ in the surrounding matrix ($C_1 > C_2$, $C_2 < C_0$, where C_0 – average concentration of the impurity atoms). It is supposed that a typical size of the new phase centre is considerably less than the hollow cylinder thickness. Such supposition allows us to consider the new phase growth in an unlimited matrix and obtain the analytical solution of task (8). Change of the new phase radius obeys the law $R_1(t) = \delta\sqrt{Dt}$, where δ –dimensionless parameter of the task. Its value is determined from the mass balance equation on the interphase boundary. For $\beta = -1$ approximately to "stationary interphase boundary" we will obtain a quadratic equation for determining parameter δ

$$\delta^2 - \frac{2\delta}{\sqrt{\pi}}\left|\frac{C_2 - C_0}{C_1 - C_2}\right| - \left|\frac{2C_2}{C_1 - C_2}\right| = 0 \qquad ...(9)$$

If $\beta = 0$, the internal stresses field is not taken into account. For determining parameter d1 of relations $R_1(t) = \delta_1\sqrt{Dt}$ we will obtain a transcendental equation

$$\delta_1 = \frac{2}{\sqrt{\pi}}\left|\frac{C_2 - C_0}{C_1 - C_2}\right|\frac{K_1\left(\delta_1\frac{\sqrt{\pi}}{2}\right)}{K_0\left(\delta_1\frac{\sqrt{\pi}}{2}\right)} \qquad ...(10)$$

where $K_0(x)$ and $K_1(x)$–modified cylindrical functions. Keeping commonality, we take $C_0 = 2 \times 10^{-4}$(at), $C_2 = 10^{-4}$ (at) $C_1 = 3.10^{-4}$ (at). From the solution of equations (9) and (10) we obtain $\delta = 1.3$ and $\delta_1 = 0.8$. The internal stresses field accelerates the diffusion process of the new phase growth. The other values of the boundary conditions change a numerical value of parameters δ and δ_1. The volume changes of the new phase cause the stresses on the interphase boundary. If the stresses level exceeds a critical value, the microcracks formation takes place. The material damage is observed in a macroscopic scale. Simulation of this process also includes kinetics of the new phase growth.

CONCLUSION

The internal stresses have an essential effect on kinetics of diffusion processes in metals and alloys. When changing the concentration of the alloying elements, reducing the strength material characteristics takes place, and probability of damaging under the external load increases. Simulation of the material damage process taking into account the internal stresses field includes a definite sequence of the mathematical operations. These operations define the algorithm of the material damages calculation: definition of the first invariant of the internal stresses tensor, mathematical formulation of the task of impurity atoms diffusion followed by the new phase formation. Kinetics of the impurity atoms segregation obeys the equation of a parabolic type under corresponding initial and boundary conditions. The complicated coordinate dependence of the internal stresses field makes difficulties in obtaining the analytical solution of the diffusion equation in the stresses field. A good exclusion from the general rule is the internal stresses with a logarithmic coordinate dependence. Such dependence allows the analytical solution of the diffusion kinetics task to be obtained. It is caused by the fact that potential V is a harmonic function, and its gradient ∇V is inversely proportional to the radius in the polar coordinate system. The analytical dependences for the impurity atoms concentration taking into account the residual stresses in the hollow cylinder have been obtained. Under the definite conditions some changes in the diffusion equation symmetry have been revealed. Kinetics of the impurity atoms segregation for cylindrical geometry runs according to the flat symmetry law. It increases a rate of the impurity segregation. The results of mathematical simulation of the material damages are of interest for ensuring operational safety of structural elements of modern equipment.

REFERENCES

1.   E. Melan, H. Parkus, 1953, Wärmespannungen infolge stationärer Temperaturfelder, Springer-Verlag, Wien.
2.   H. Parkus, 1959, Instationäre Wärmespannnungen, Springer-Verlag, Wien.
3.   S.P. Timoshenko, G. Gudier, 1979, Theory of Elasticity. Translation from English, Nauka, Moscow.
4.   N.M. Vlasov, I.I. Fedik, 2006, Structural and impurity traps for hydrogen atoms, International Journal of Hydrogen Energy. 31, 265-267.
5.   C. Teodosiu, 1982, Elastic models of crystal defects, Springer, Heidelberg.

51

Mode 3 Spontaneous Crack Propagation Along Functionally Graded Bimaterial Interfaces

D.V. KUBAIR

Computational Dynamic Fracture Mechanics Laboratory, Department of Aerospace Engineering, Institute of Science, Bangalore-560 012, India email: kubair@aero.iisc.ernet.in

ABSTRACT

The effects of spatially varying the material properties on the mode-3 planar crack propagation characteristics are numerically investigated. The spectral scheme that is available for homogeneous materials is modified to account for the asymmetrically varying material properties. Crack propagation along the interface of a functionally graded bimaterial system has been simulated. A parametric study was performed by systematically varying the material inhomogeneity length scale independently in the two half-spaces. Our study indicated that softening type graded materials reduce the resistance to fracture, while a hardening material offers higher fracture resistance with increase in inhomogeneity. Only the transient phase of crack propagation speed was affected by the material property variation, irrespective of whether the material was hardening, softening or an asymmetric type. The crack always reached a quasi-steady-state velocity, which remained unaffected by the material property inhomogeneity.

Keywords: Functionally graded materials, spectral scheme, cohesive zone model, integral equations.

1. INTRODUCTION

Many biological structures such as human bones and shells have material properties that vary smoothly with spatial position in order to efficiently respond to the surrounding mechanical loads (Krassig, 1993; Suresh, 1998). The concept of tailoring the spatial properties in engineering is not new as well, for example case hardened steel components have hardness varying continuously with depth in order to provide better wear resistance. Also, composites both layered as well as particulate, have reinforcing material in either specific directions or at random to increase the specific strength and stiffness of the structure. One draw back of the composite systems is that there exists a distinct interface across the plies or the particle-matrix, which acts as a source of stress concentration and a potential site for delaminations or cracks to initiate. The concept of a functionally graded material tries to resolve the problem of the material property mismatch by continuously varying them as a function of the spatial coordinates. For the purpose of this study, a functionally graded material (FGM) is an inhomogeneous material with a known functional form of the material inhomogeneity. Applications such as thermal barrier coatings (Movchan, 2004), ballistic impact resistance structures (Chin, 1999) and wear resistive coating (Schulz, 2003) are some examples where FGMs are used. Modern day structural designs are based on the concepts of fracture mechanics and a thorough understanding of the fracture response of FGMs is key in designing structures using them. Hence, the primary objective of this study is to understand the fracture response of graded materials. In particular, the effect of material inertia is considered here, that can become either important when the loads applied are time dependent or even when the velocity of crack propagation becomes comparable to the material wave speed.

One of the early works considering the material inhomogeneity was that of Delale and Erdogan (1983), who performed an asymptotic analysis of a stationary tensile crack and extracted the stress-intensity factor. From their analysis they found that the near-tip stress-intensity factor was not affected by the material property inhomogeneity. Eischen (1987) in his asymptotic stress-series (similar to the Williams (1952) expansion for homogeneous materials) confirmed that the most singular term was unaffected by the inhomogeneous material property variation. Parmeswaran and Shukla (2002) have obtained the higher order terms in the asymptotic stress-series expansion and found that the material inhomogeneity affects only the higher order terms. Zhang *et al.* (2003) have studied the effect of material inhomogeneity on the mixed-mode stress-intensity factor ahead of a stationary crack. Due to the complexity of the analysis, only a limited number of studies available in the literature have considered the effect of material inertia in functionally graded materials. Parameswaran and Shukla (1999) have obtained the asymptotic stress and displacement fields for a crack propagating in a graded material. Their analysis assumed exponential and linear variation of the properties in the continuum. Meguid *et al.* (2002) have studied the effect of material property inhomogeneity on the crack speed variation and dynamic stress-intensity factor in a FGM. They solved a singular integral equation in their analysis, which indicated that the material inhomogeneity has no effect on the most singular term. Recently, Wang and Nakamura (2004) have performed rapid crack propagation simulations in FGMs using a finite element method. They have varied the fracture toughness in the direction of the crack propagation by continuously changing the cohesive zone properties. Their study showed that the crack propagation characteristics were affected by the material property inhomogeneity. In the present study, we propose to understand the effects of material inhomogeneity and inertia by performing numerical simulations of rapid crack propagation in functionally graded materials. There are several robust numerical techniques commonly used to simulate rapid crack propagation in solids, this includes, boundary integral method (Israil and Banerjee, 1990) finite differences (Yang and Ravi-Chandar, 1996; Mikumo *et al.*, 1987), finite elements (Wang and Nakamura, 2004; Atluri and Nishioka, 1985; Safjan and Oden, 1993) and cohesive volume finite element (CVFE) (Xu and Needleman, 1994; Geubelle and Baylor, 1998; Camacho and Ortiz, 1996). All of the above-mentioned techniques are computationally intensive and mostly suited for simulations when the crack propagation path is not known apriori. The focus of the present work is to understand the effect of the material property inhomogeneity on the crack propagation characteristics. Hence, we restrict our attention to simulating planar, rectilinear crack propagation. Geubelle and Rice (1995) have developed an efficient numerical tool called the "spectral scheme" that can simulate planar crack propagation. The scheme is versatile and can handle a variety of applied tractions, state- and rate-dependent cohesive. The spectral scheme has been derived for various material systems such as bimaterial interfaces, viscoelastic materials, and orthotropic solids. In this article we extend the spectral formulation to simulate spontaneous crack propagation in functionally graded materials under anti-plane shear loading conditions. The formulation and the details of the numerical implementation are presented in Section 2. In Section 3, we present the results from our parametric study that we have performed using the spectral scheme and illustrate the effect of the material property inhomogeneity on the crack propagation characteristics.

2. FORMULATION

In this section, the elastodynamic relations for an inhomogenous bimaterial system are presented in this section. The interface is assumed to be coincident with the x_2 plane, which is also the weak plane. An interfacial crack of initial length a_0 is allowed to propagate spontaneously on the weak-plane due to the action of an external loading τ_0. In the spectral formulation, the external load can be an arbitrary function of time and spatial position. The materials on the above and below the interface/weak plane are designated by a plus (+) and minus (−), respectively. The rigidity modulus (μ) and density (ρ) are allowed to vary exponentially in the direction perpendicular to the weak plane as

$$\mu\left(x_2^{\pm}\right) = \mu_o^{\pm}\exp\left(\frac{x_2^{\pm}}{L_g^{\pm}}\right),$$

$$\rho\left(x_2^{\pm}\right) = \rho_o^{\pm}\exp\left(\frac{x_2^{\pm}}{L_g^{\pm}}\right), \qquad \qquad \text{...(1)}$$

where $\mu_o^{\pm}$ and $\rho_o^{\pm}$ are the rigidity modulus and density on the weak plane ($x_2 = 0$) for the top (+) and bottom (−) materials, respectively. In the present bimaterial formulation, the material properties are allowed to suffer a discontinuous jump across the weak plane. In Equation (1) $L_g^{\pm}$ corresponds to the natural inhomogeneity length scale of the FGM, which controls how fast or slow the material properties vary along the chosen spatial direction. Unlike in the previous formulations by Kulkarni *et al.* (2006) and Pal *et al.* (2006), the inhomogeneity length scales are also independently varied in the top and bottom halves. The trivial case of a homogeneous material is obtained by assuming the inhomogeneity length scale to be infinity. The inhomogeneity length scale can be either positive or negative in our formulation. In the top half a positive value of the inhomogeneity length scale leads to an inhomogeneous material that progressively becomes more rigid and denser away from the weak plane and is termed as a "hardening" or "strengthening" type functionally graded material. When the inhomogeneity length scale is set to be less than zero (negative) the material properties degenerate away from the weak plane and such a material is termed as a "softening" or ''weakening type functionally graded material. A combination of both a softening material in one half and a hardening material in the other half with different inhomogeneities is possible in the current bimaterial form. The particular case of the unsymmetric (Kulkarni *et al.*, 2006) functionally graded material can be obtained by assuming $\mu_0^{+} = \mu_o^{-}$, $\rho_0^{+} = \rho_o^{-}$ and $L_g^{+} = L_g^{-}$. Similarly a symmetric functionally graded material considered in Pal *et al.* (2006) can be obtained by assuming $\pm L_g^{+} = \mp L_g^{-}$, with identical rigidity modulii and densities for the top and bottom materials. As mentioned earlier, in the present bimaterial formulation we allow the material parameters, namely the rigidity modulii, density and the inhomogeneity length scale to be different in the two halves. As seen from Equation (1), the inhomogeneous variation of the rigidity modulus and densities in the two halves are assumed to be identical mathematical functions. This assumption leads to a homogenous shear wave speed $c_s^{\pm} = \sqrt{\mu_o^{\pm}/\rho_o^{\pm}}$ in each of the half-spaces. We start our spectral formulation from the conservation of linear momentum written in terms of the only non-vanishing out of plane displacements $u_3^{\pm}$ which is given by

$$\frac{\partial^2 u_3^{\pm}}{\partial x_1^2} + \frac{\partial^2 u_3^{\pm}}{\partial x_2^2} + \frac{1}{L_g^{\pm}}\frac{\partial u_3^{\pm}}{\partial x_2} = \frac{1}{\left(c_s^{\pm}\right)^2}\frac{\partial^2 u_3^{\pm}}{\partial t^2}, \qquad \ldots(2)$$

where t is the time coordinate. Equation (2) is a second order hyperbolic equation and will assume the classical form of a scalar wave equation when the inhomogeneity length scale $L_g^{\pm} = \pm\infty$. The out of plane displacements are transformed into the Fourier and Laplace domains as

$$\widehat{\Psi}^{\pm}\left(k, x_2^{\pm}, p\right) = \int\limits_{t=0}^{t=\infty} \int\limits_{x_1=-\infty}^{x_1=\infty} u_3^{\pm}\left(x_1, x_2^{\pm}, t\right) e^{-(ikx_1 + pt)}\, dx_1\, dt, \qquad \ldots(3)$$

where k's are the real and positive spectral mode numbers and p is the complex Laplace transform variable, respectively. Solving the governing ODE in the transformed plane and applying the external tractions $\tau_0(x_1,t)$ we can derive the traction-displacement in the $x_1 - t$ space, which is written as

$$\tau\left(x_1, t\right) = \tau_o\left(x_1, t\right) \mp \frac{\mu_o^{\pm}}{c_s^{\pm}}\dot{u}_3^{\pm}\left(x_1, t\right) \mp \frac{\mu_o^{\pm}}{2L_g^{\pm}}u_3^{\pm}\left(x_1, t\right) + f^{\pm}\left(x_1, t\right), \qquad \ldots(4)$$

where $\dot{u}_3^{\pm}\left(x_1, t\right)$ and $u_3^{\pm}\left(x_1, t\right)$ are the velocities and displacements of the top and bottom half-spaces evaluated on the weak-plane $x_2 = 0$. In the above equation the fourth term $f^{\pm}(x_1, t)$ is the convolution term that corresponds to the tractions related to the history of the displacements. The convolution over the slip history in the spectral domain is written as

$$F^{\pm}\left(k, t\right) = \mp\mu_o^{\pm}k\int\limits_{0}^{t} H^{\pm}\left(kc_s^{\pm}\left(t-t'\right)\right)\psi^{\pm}\left(k, t'\right) kc_s^{\pm} dt', \qquad \ldots(5)$$

where $H^{\pm}(Z)$ is the convolution kernel. The convolution kernel derived here is a well-behaved non-singular function similar to the ones obtained for different types of materials. The elastodynamic relation (4) is solved numerically. A parametric study was performed by systematically varying the inhomogeneity length scale and the results are discussed in the next section.

3. RESULTS AND DISCUSSION

The primary focus of the present study is to understand the effect of varying the material property inhomogeneity independently in the two half-spaces. We performed a parametric study by systematically varying the inhomogeneity length scales $L_g^{\pm}$. We used the spectral scheme that is developed (Section 2) to simulate spontaneous crack propagation along the interface of a bimaterial functionally graded system. In the rest of the discussion in this work we assume that the rigidity modulus on the interface to be μ_0 =1962.6 MPa, density to be ρ_0 = $1230kg - m^{-3}$. The above material properties correspond to an untreated sheet of PMMA. The choice of the above mentioned rigidity modulus and density leads to a homogeneous shear wave speed of c_s =$1263 - m - s^1$. The weak-plane X was assumed to be $16a_0$ and was divided into N = $2048(2^{11})$ equal parts, which was sufficient to achieve spatial convergence. The time step $\Delta t = a_0/(64c_s)$ was used in all our simulations. The interfacial fracture toughness was set to be $G_c = 5KJ - m^{-2}$, which was assumed to have been obtained of a critical crack opening displacement δ_c = $2mm$ and maximum shear strength. A time invariant, uniformly distributed external loading $\tau_0 = \tau_c$ /2 was applied in all our simulations. Several combinations of the inhomogeneity length scale $L_g^{\pm}$ were used in our parametric study. For the purpose of explaining the results four bimaterial combinations are used. They are:

A. Both the top and bottom material is homogeneous with $L_g^{\pm} = \infty$, which is our reference material.

B. Bottom material is a hardening FGM with $- L_g^{-} = B\,\delta_c$ and the top is a homogeneous material with $L_g^{+} = \infty$.

C. Bottom material is again a hardening FGM with $- L_g^{-} = B\,\delta_c$, while the top is a weakening FGM with $L_g^{+} = - C\,\delta_c$.

D. The bottom half is made of a homogeneous material and the top is made of softening FGM with $L_g^{+} = C\,\delta_c$.

We examined the effect of the material property inhomogeneity on the crack sliding displacements by recording snapshots of the displacements while the simulation progressed, which are shown in Fig. 1.

These snapshots are recorded at regular intervals of $0.4a_0/c_s$ apart. The displacements are normalized with the critical crack opening displacement δ_c. The crack sliding displacements in the homogeneous material (A) is symmetric about the weak plane as expected. In the bimaterial combination B the crack sliding is affected by the material inhomogeneity in the bottom half. The material progressively becomes more rigid and denser in this case, which leads to higher fracture resistance and lesser opening of the crack in the bottom half as compared with the homogeneous

case. While in the upper half-space, the crack sliding displacements are of the same order of magnitude as that of the homogeneous material. Due to the material inhomogeneity in the lower half-space alone the crack tip in the bimaterial case B is seen trailing the homogeneous material (A) at all time steps. The crack sliding for the third bimaterial combination also exhibits an effect of the material property inhomogeneity. Unlike the previous two cases, here the material property inhomogeneity is present in both the upper as well as lower half-spaces. The crack sliding in the lower half, where the material is strengthening is lesser than the reference homogeneous material, while in the upper half the crack sliding is larger than in the

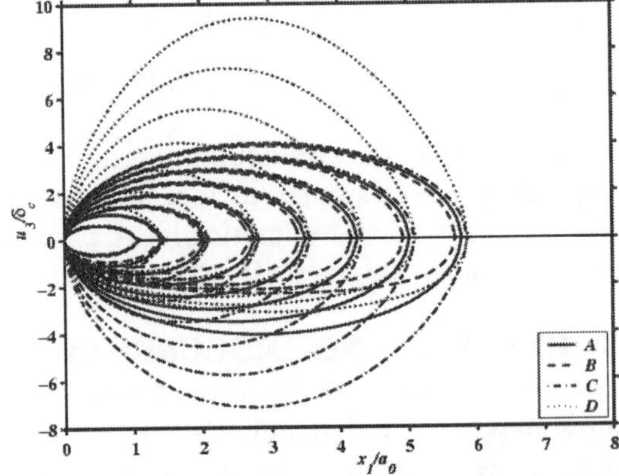

Fig. 1. Crack opening displacements.

homogeneous material. This indicates that a softening material is offering lesser resistance to fracture while the hardening material is offering a larger resistance as seen. The position of the crack tip always leads the homogeneous case as seen. Effectively, the combination of the hardening and softening graded materials considered in case *C* leads to a decrease in the fracture resistance as the crack tip leads, which is similar to the observations of Kulkarni *et al.* (2006) and Pal *et al.* (2006).

From Fig. 1, we see that the crack tip positions are either leading or lagging the homogeneous material, indicating that the crack tip velocities are affected by the inhomogeneous material property variations. The crack tip velocity history shown in Fig. 2 illustrates this. The crack remains stationary until sufficient energy is absorbed in the cohesive failure process. The crack accelerates rapidly once it starts to grow and reaches a quasi-steady-state of propagation in all the bimaterial combinations simulated. The maximum quasi-steady-state velocity of propagation reaches the shear wave speed, which is the limiting speed of propagation under mode 3 loading conditions. The time instant at which the crack begins propagating is altered by the material property inhomogeneity. In comparison with the homogeneous case (A), the homogeneous-hardening FGM system (*B*) delays the start of the crack propagation as more power is absorbed in the cohesive zone. In the bimaterial system of case *C* the initial growth time is advanced. The effect of the material inhomogeneity is to alter the crack propagation velocities. The crack velocity is slower in case *B* while it is faster in case *C*. Only the transient crack propagation seems to be affected while the quasi-steady-state reached after the initial transients have diminished remains unaffected.

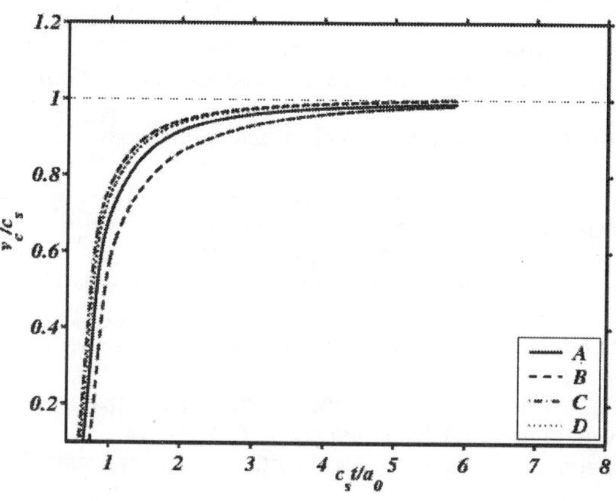

Fig. 2. Crack tip velocity history.

CONCLUSION

We have developed a special form of the spectral scheme to simulate 2D planar crack propagation along the interface of an inhomogeneous bimaterial system. The scheme developed is versatile, similar to its predecessors, in which a variety of traction-separation (cohesive zone) laws or friction laws can be used. The developed spectral scheme was used to perform a parametric study by assuming that the inhomogeneity length scale (L_g) varies independently in either of the two half-spaces. We draw the following conclusions from our parametric study:

1. Combinations of hardening and softening materials in either of the half-spaces tend to increase or decrease the effective fracture resistance on the interface. This is indicated by difference in the crack opening displacements and velocities compared to a homogeneous material.
2. Only the initial accelerating phase of the crack propagation is affected by the material property inhomogeneity. A decrease in the effective fracture resistance on the weak-plane increases the crack tip velocity.
3. The material inhomogeneity affects both the crack opening displacements, cohesive tractions inside the cohesive zone. While the stresses ahead of the cohesive zone tip remain unaffected.

ACKNOWLEDGEMENTS

Kubair acknowledges the support from the Department of Science and Technology through their

FAST young scientist award and the All India Center for Technical Education through their career award for young teachers.

REFERENCES

1. Atluri, S., Nishioka, T., 1985. Numerical studies in dynamic fracture mechanics. International Journal of Fracture 27, 245-261.
2. Camacho, G., Ortiz, M., 1996. Computational modelling of impact damage in brittle materials. International Journal of Solids and Structures 33, 2899-2938.
3. Chin, E., 1999. Army focused research team on functionally graded armor composites. Material science and Engineering A 259, 155-161.
4. Delale, F., Erdogan, F., 1983. The crack problem for nonhomogeneous plane. Journal of Applied Mechanics 50, 609-614.
5. Eischen, J., 1987. Fracture of nonhomogeneous materials. International Journal of Fracture 34, 3-22.
6. Geubelle, P., Baylor, J., 1998. Impact induced delamination of laminated composites:a 2-d simulations. Composite Part B Engineering 29, 589-602.
7. Geubelle, P., Rice, J., 1995. A spectral method for three-dimensional elastodynamic fracture problems. Journal of the Mechanics and Physics of Solids 43, 1791-1824.
8. Israil, A., Banerjee, P., 1990. Advance time domain formulation of bem for two-dimensional transient elastodynamics. International Journal for Numerical Methods in Engineering 29, 1421-1440.
9. Krassig, K., 1993. Cellulose: Structure, accessibility and reactivity. Polymer Monographs 11.
10. Kulkarni, M., Pal, S., Kubair, D. 2006 Mode 3 spontaneous crack propagation in unsymmetric functionally graded materials. International Journal of Solids and Structures. In press.
11. Lambros, J., Santare, M., Sapna, G., 1999. A novel technique for the fabrication of laboratory scale model functionally graded materials. Experimental Mechanics 39 (3), 184-190.
12. Liu, Y., Rizzo, F., 1993. Hypersingular boundary integral equations for radiation and scattering of elastic waves in three dimensions. Computational Methods in Applied Mechanics and Engineering 107, 131-144.
13. Meguid, S., Wang, X., Jiang, L., 2002. On the dynamic propagation of finite crack in functionally graded materials. Engineering Fracture Mechanics 69 (14-16), 1753-1768.
14. Mikumo, T., Hirahara, K., Miyatake, T., 1987. Dynamic rapture process in heterogeneous media. Technophysics 144, 19-36.
15. Movchan, B., Yakovchuk, K., 2004. Graded thermal barrier coatings, deposited by EB-PVD. Surface and Coatings Technology 188-189, 85-92.
16. Pal, S., Kulkarni, M., Kubair, D. 2005 Mode 3 spontaneous crack propagation in symmetric functionally graded materials. International Journal of Solids and Structures. In press.
17. Parameswaran, V., Shukla, A., 1999. Crack tip fields for dynamic fracture in functionally graded materials. Mechanics of Materials 31, 579-596.
18. Parameswaran, V., Shukla, A., 2002. Asymptotic stress fields for stationary cracks along the gradient in functionally graded materials. Transactions of the ASME, Journal of Applied Mechanics 69, 240-243.
19. Safjan, A., Oden, J., 1993. High-order taylor-galerkin and adaptive h-p methods for second order hyperbolic systems application to elastodynamics. Computational Methods in Applied Mechanics and Engineering 103, 187-230.
20. Schulz, U., Peters, M., Bach, F., Tegeder, G., 2003. Graded coatings for thermal, wear and corrosion barriers. Materials Science and Engineering A 362(1-2), 61-80.
21. Suresh, S., Mortensen, A., 1998. Fundamentals of Functionally Graded Materials. Institute of Materials, London.
22. Wang, Z., Nakamura, T., 2004. Simulations of crack propagation in elastic-plastic graded materials. Mechanics of Materials 36, 601.
23. Williams, M., 1952. Stress singularities resulting from various boundary conditions in angular corners of plates in extension. Transactions of the ASME, Journal of Applied Mechanics 19, 526-528.
24. Xu, X., Needleman, A., 1994. Numerical simulations of fast crack growth in brittle solids. Journal of the Mechanics and Physics of Solids 42, 1397-1434.
25. Yang, B., Ravi-Chandar, K., 1996. On the role of the process zone in dynamic fracture. Journal of the Mechanics and Physics of Solids 44(12), 955-1976.
26. Zhang, C., Savaidis, A., Savaidis, G., Zhu, H., 2003. Transient dynamic analysis of a cracked functionally graded material by a BIEM. Computational Material Science 26, 164-174.

52

A Crack Arrest Model for a Cracked Piezoelectric Plate

R.R. BHARGAVA[1] AND NAMITA SAXENA[2]

[1]Department of Mathematics, Indian Institute of Technology, Roorkee-247 667, Uttaranchal, India
email: rajrbfma@iitr.ernet.in,
[2]Department of Mathematics, Indian Institute of Technology, Roorkee-247 667, Uttaranchal, India
email: mitandma@iitr.ernet.in

ABSTRACT

A crack arrest model is proposed for a poled piezoelectric internally cracked ceramic plate. The plate is subjected to uniform anti-plane shear and uniform in-plane electric displacement, acting in the direction of poling of the plate, at infinite boundary. This causes the yielding of the crack both mechanically and electrically. Consequently a yield zone and saturation zone is formed ahead of each tip of the crack. To arrest the crack from further opening these developed zones are subjected to closing forces. The rims of yield zone is subjected to linearly varying yield point shear stress while rims of saturation zone is subjected to in-plane saturation limit electrical displacement. It is assumed that the developed saturation zone length is bigger than that of yield zone. The problem is solved using Fourier transform. Closed form expressions are derived for finding out the length of yield and saturation zones. Crack opening displacement and crack opening potential drop has been calculated for the rims of the crack. Energy release rate expression has also been obtained.

Keywords: Piezoelectric plate, Yield zone, Saturation zone, COD, COP.

1. INTRODUCTION

Smart materials have by now proved their utility in the modern technological apparatus. As a natural consequence, for understanding different aspects of these materials many researchers have taken up the investigations and one such aspect in fracture mechanics of these ceramics. Work in this direction started in big way appropriate boundary condition for a crack in a dielectric medium is investigated in [1]. They also formulated the stress and electric field boundary condition around such mode-III crack in a piezoelectric material. The fracture mechanics problem for a mode-III crack in a piezoelectric material was studied in [2]. In Ref. [3] a displacement based formulation is used to solve the crack problem for nonhomogeneous material under anti-plane shear loading. For a functionally graded material the effects of material gradient on transient dynamic mode III SIF's have been investigated in [4]. The problem of an anti-plane crack situated at the interface between two bonded dissimilar piezoelectric layer is considered under the permeable crack assumption in [5]. Ref. [6] deals with a rectangular piezoelectric ceramic containing an anti-plane shear crack at an arbitrary position. An anti-plane crack problem for a functionally graded material is performed in [7]. A Galerkin method is applied for the numerical solution of the hyper-singular traction BIE. The electrical nonlinear behavior of an anti-plane shear crack in a piezoelectric ceramic layer constrained between two orthotropic layers is examined in [8] using an electric strip saturation model with in the frame work of linear elasticity and nonlinear electroelasticity. More recently a crack arrest model is proposed in [9] for a cracked piezoelectric ceramic plate subjected to anti-plane shear and in-plane electric displacement at infinite boundary, while the crack is arrested by applying sliding shear stress and in-plane cohesive electric displacements on the zones developed.

2. MATHEMATICAL FORMULATION

In x, y, and z coordinate system the out-of-plane displacement component u_i and in-plane electric field components E_i (i = x, y, z) may be defined as

$$u_x = u_y = 0, \quad u_z = w\ (x,\ y) \qquad \qquad ...(1)$$

$$E_x = E_x\ (x,\ y),\ E_y = E_y\ (x,\ y),\ E_z = 0 \qquad \qquad ...(2)$$

Shear stress components σ_{xz}, σ_{yz} and electric displacements potential D_x and D_y are given by

$$\sigma_{xz} = c_{44}\frac{\partial w}{\partial x} + e_{15}\frac{\partial \phi}{\partial x}$$

$$\sigma_{yz} = c_{44}\frac{\partial w}{\partial y} + e_{15}\frac{\partial \phi}{\partial y} \qquad \qquad ...(3)$$

$$D_x = e_{15}\frac{\partial w}{\partial x} - \varepsilon_{11}\frac{\partial \phi}{\partial x}$$

$$D_y = e_{33}\frac{\partial w}{\partial y} - \varepsilon_{11}\frac{\partial \phi}{\partial y} \qquad \qquad ...(4)$$

$$E_i = -\ \phi_j\ (i = x,\ y) \qquad \qquad ...(5)$$

where ϕ is the electric potential and comma denotes the partial derivative of the function with respect to argument following it. Governing continuity equations for stresses and electric displacement are given by

$$c_{44}\nabla^2 w + e_{15}\nabla^2 \phi = 0$$

$$c_{44}\nabla^2 w - \varepsilon_{11}\nabla^2 \phi = 0 \qquad \qquad ...(6)$$

where c_{44} denotes elastic constant, e_{15} denotes piezoelectric constant; e_{11} is a dielectric constant and ∇^2 denotes two-dimensional Laplacian operator.

Introducing vector notation for convenient compact writing as

$$\overline{t}\,(x,y) = \left\{\sigma_{yz}\ D_y\right\}^T \qquad \qquad ...(7)$$

$$\overline{u}\,(x,y) = \left\{w\ \phi\right\}^T \qquad \qquad ...(8)$$

where a bar over the function denote that the quantity is vector.

Taking Fourier transform of Eq. (6) substituting Eq. (7 and 8),

$$\frac{\partial^2 \tilde{\bar{u}}}{\partial q^2} + \tilde{\bar{u}} = 0 \qquad \text{...(9)}$$

where $\sim$ denotes Fourier transform of the quantity; $q = -isy$; s is take on Fourier transform variable. The general solution of Eq. (9) may be written as

$$\tilde{\bar{u}}^+(s,y) = \begin{cases} e^{sy}\overline{C}_1 & s < 0 \\ e^{-sy}\overline{F}_1 & s > 0 \end{cases} \qquad \text{...(10)}$$

$$\tilde{\bar{u}}^-(s,y) = \begin{cases} e^{sy}\overline{C}_2 & s > 0 \\ e^{-sy}\overline{F}_2 & s < 0 \end{cases}$$

superscripts $+$ and $-$ denote the value of the function as approached from $y > 0$ and $y < 0$ planes, respectively.

Then the traction continuity condition on $|x| < \infty$, $y = 0$ may be written as

$$\overline{k}\,(\tilde{\bar{u}}^+ + \tilde{\bar{u}}^-) = 0 \qquad \text{...(11)}$$

where

$$\overline{k} = \begin{cases} -s\overline{B} & s < 0 \\ +s\overline{B} & s > 0 \end{cases} ; \; B = \begin{pmatrix} M & e \\ e & -\varepsilon \end{pmatrix} \qquad \text{...(12)}$$

The jump in displacement and electric potential along $y = 0$ is defined as

$$\Delta \overline{u}(x) = \overline{u}^+(x,0) - \overline{u}^-(x,0) \qquad \text{...(13)}$$

For the crack problem (i) $\Delta \overline{u}(x) = 0$ for $|x| > a$ and (ii) single-valudeness of displacements and electric potential require that

$$\int_{-a}^{a} \overline{f}(x)\,dx = 0 \qquad \text{...(14)}$$

Dislocation and dipole density vector are given by

$$\overline{f}(x) = \{f_1(x)\; f_2(x)\}^T = \frac{d}{dx}\Delta \overline{u}(x) \qquad \text{...(15)}$$

Taking Fourier transform of Eq. (3b) and Eq. (4b), one gets

$$\overline{t}(s,q) = -iB\frac{\partial \overline{u}(s,y)}{\partial y} \qquad \text{...(16)}$$

where and $\overline{C}_1$ and $\overline{F}_1$ are two column constant vector.

Solution of Eq. (16) is written in the form

$$\overline{t}(s,y) = \begin{cases} -is\,e^{sy}\,\overline{B}_1\overline{C}_1 & s < 0 \\ is\,e^{-sy}\,\overline{B}\,\overline{F}_1 & s > 0 \end{cases} \qquad \text{...(17)}$$

Therefore, on y = 0,

$$-\tilde{\bar{t}}^{+}(s,0)=\bar{k}^{+}\tilde{\bar{u}}^{+}(s,0)$$

$$\tilde{\bar{t}}^{-}(s,0)=\bar{k}^{-}\tilde{\bar{u}}^{-}(s,0) \qquad \qquad \dots(18)$$

Taking the inverse Fourier transform of Eq. (17), we get

$$\bar{t}_2(x)=\frac{1}{\pi}\bar{H}\int_{-a}^{a}\frac{\bar{f}(t)}{t-x}dt = \bar{t}^{0}(x)=\left\{t_1^{0}(x),\, t_2^{0}(x)\right\}^{T} \qquad \dots(19)$$

3. THE PROBLEM AND ITS MATHEMATICAL MODEL

An unbounded piezoelectric ceramic plate occupies the xoy plane. The plate is poled along *oy*-direction. A hairline internal crack, L, of length 2a lies on ox axis and occupies the interval [– a, a]. Uniform constant anti-plane shear τ_∞ and uniform constant in-plane displacement D_∞ applied at infinite boundary of the plate opens the crack rims. The crack yields both mechanically and electrically. Consequently a yield zone and saturation zone protrude at each tip of the crack. Under small scale yielding these zones may be assumed to lie along the ox axis. It is assumed that the developed saturation zone is bigger in length as compared to that of yield zone. At the tip x = a the yield zone, Y_1, and saturation zone, S_1, occupy the intervals [a, b] and [a, l] respectively. And at the tip x = – a the yield zone, Y_2, and the saturation zone, S_2, occupy the intervals [– a, – b] a n d [– a, – l] respectively. To arrest the crack from further growing the rims of the yield zones are subject to shear stress $\sigma_{xy}=\frac{t}{b}\tau_s$ and rims of the saturation zone is subjected to in-plane electrical displacement $D_y = D_s$, where denotes the saturation point electric displacement.

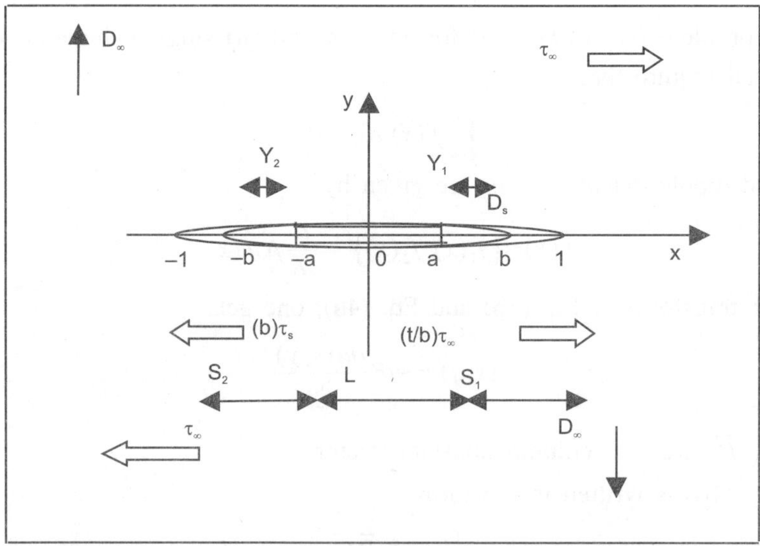

Fig. 1. Schematic representation of configuration of problem.

A mathematical model for the above problem is obtained assuming an infinite piezoelectric plate weakened by a crack of length $[-1, 1]$ where (a<b<l) subjected to following boundary conditions

$$(a)\ t_1^0 = \begin{cases} -\tau_\infty & \text{for } |x| < a \\ -\tau_\infty + \dfrac{t}{b}\tau_s & \text{for } a < |x| < b \end{cases}$$

$$(b)\ t_2^0 = \begin{cases} -D_\infty & \text{for } |x| < a \\ -D_\infty + D_s & \text{for } a < |x| < l \end{cases}$$

(c) mechanical potential $f_1(x) = 0 \quad if |x| > b$

(d) electric potential $f_2(x) = 0 \quad if |x| > l$.

4. SOLUTION OF THE PROBLEM

The solution of the above problem is obtained by solving its mathematical model, the solution of which may be written using Eq. (19) as

$$\frac{1}{\pi}\int_{-b}^{b} \frac{f_1(t)}{t-x} dt = G_{1j}\, t_j^0(x),\ \text{for } |x| < b \tag{20}$$

$$\frac{1}{\pi}\int_{-l}^{l} \frac{H_{2j} f_j(t)}{t-x} dt = t_2^0(x),\ \text{for } |x| < l \text{ and } j = 1, 2 \tag{21}$$

where

$$G = \left[G_{ij}\right] = -\frac{1}{2}\begin{pmatrix} c_{44} & e_{15} \\ e_{15} & -\varepsilon_{11} \end{pmatrix}^{-1} = H^{-1} = \left[H_{ij}\right]^{-1} \tag{22}$$

Since $f_1(x)$ is nonsingular at x = b the Eq. (20) has the solution only if

$$\int_{-b}^{b} \frac{G_{1j}\, t_j(t)}{\sqrt{b^2 - t^2}} dt = 0$$

Integrating it, a transcendental equation is obtained to determine the length of the yield zone from

$$b = a \sec\left\{\frac{\pi}{2}\left(\frac{G_{11}\tau_\infty + G_{12} D_\infty}{G_{12} D_s}\right)\right\} \tag{23}$$

Employing this condition to integrate the Eq. (20) one obtains the dislocation potential $f_1(x)$ as

$$f_1(x) = \frac{G_{12} D_s}{\pi}\{w(x,a,b) - w(-x,a,b)\} + \frac{x\,\tau_s}{\pi}\{w(x,a,b) - w(-x,a,b)\}$$

$$+ \frac{\tau_s}{b}\left(\frac{G_{11}\tau_\infty + G_{12} D_\infty}{G_{12} D_s}\right)\sqrt{b^2 - x^2} \tag{24}$$

where

$$w(x,a,b) = \cos h^{-1}\left\{\frac{b^2 - a^2}{(a-x)b} + \frac{a}{b}\right\}$$

Similarly Eq. (21) has the solution if

$$\int_{-l}^{l} \frac{t_2^0(t)}{\sqrt{l^2 - t^2}}\, dt = 0$$

Solution of this gives the expression to determine the saturation zone length from

$$l = a\sec\left(\frac{\pi}{2}\frac{D_\infty}{D_s}\right) \qquad \qquad ...(25)$$

with help of Eq. (25) may finally write for potential $f_2(x)$ as

$$f_2(x) = \frac{D_s}{\pi H_{22}}\{w(x,a,l) - w(-x,a,l)\} - \frac{H_{21}\,G_{12}}{H_{22}\,\pi}\, f_1(x) \qquad \qquad ...(26)$$

5. CRACK OPENING DISPLACEMENT AND CRACK OPENING POTENTIAL DROP

The crack opening displacement $\Delta w(x)$ of the rims of the crack is obtained from

$$\Delta w(x) = -\int_{-b}^{x} f_1(s)\, ds$$

$$= \frac{G_{12}D_s}{\pi}\Big[\{(x-a)w(x,a,b) + (a+x)w(-x,a,b)\}\Big] + \frac{\tau_s}{b\pi}\left[\frac{x}{2}\{(x-a)w(x,a,b) + (a+x)w(-x,a,b)\}\right]$$

$$+ \frac{\tau_s}{b}\left(\frac{G_{11}\tau_\infty + G_{12}D_\infty}{G_{12}D_s}\right)\times\left[-\frac{x}{2}\sqrt{b^2 - x^2} + \frac{b^2}{2}\cos^{-1}\left(\frac{x}{b}\right) - \pi\right]$$

$$...(27)$$

The crack opening displacement $\Delta w(a)$ of interest tip of the crack is obtained from Eq. (27)

as

$$\Delta w(a) = \left(\frac{2aG_{12}D_s}{\pi} + \frac{a\tau_s}{b\pi}\right)w(-a,a,b) - \frac{a\tau_s}{b}\sqrt{b^2 - a^2} + \frac{\tau_s}{b}\left(\frac{G_{11}\tau_\infty + G_{12}D_\infty}{G_{12}D_s}\right)\times$$

$$\left(\frac{b^2\pi}{4}\left(\frac{G_{11}\tau_\infty + G_{12}D_\infty}{G_{12}D_s}\right) - \pi\right) \qquad \qquad ...(28)$$

Similarly, crack opening potential drop $\Delta\phi(x)$ at the rims of the crack is obtained from

$$\Delta\phi(x) = -\int_{-l}^{x} f_2(s)\, ds = \frac{D_s}{\pi H_{22}}\left[(a-x)w(x,a,l) + (a+x)w(-x,a,l) - \frac{H_{21}}{H_{22}}\Delta w(x)\right] \qquad ...(29)$$

And COP of interest at the tip of $x = a$ of crack similarly obtained from Eq. (29) and may finally written as

$$\Delta\phi(a) = \frac{2aD_s}{\pi H_{22}}w(-a,a,l) - \frac{H_{21}}{H_{22}}\Delta w(a) \qquad \qquad ...(30)$$

6. ENERGY RELEASE RATE

Energy release rate has been calculated at the crack tip $x = a$ using crack closure integral

$$J_a = \lim_{\delta a \to 0} \frac{1}{2\,\delta a}\left[\frac{\tau_s}{b}\left\{r\int_0^{\delta a}\Delta w(\delta a - r)\,dr - \int_0^{\delta a}\int_0^{\delta a}\Delta w(\delta a - r)\,dr\right\}\right] + D_s\,\Delta\phi(a)$$

where δa denotes the yield zone length. Evaluating the above integral using Eq. (27) one gets
   Substituting the value of $\Delta\phi(a)$ from Eq. (30) the energy release rate is obtained.

ACKNOWLEDGMENTS

Authors are grateful to Prof. R.D. Bhargava (Senior Professor (retd.), Indian Institute of Technology, Mumbai, India) for his suggestions and encouragement during the course of this work.

REFERENCES

1. T.Y. Zhang, J.E. Hack, 1992, Mode-III cracks in piezoelectric material, J. Appl. Phys. 71, 5865-5870.
2. T.Y. Zhang, P. Tong, 1996, Fracture Mechanic for a mode III crack in a piezoelectric material, Int. J. Solids Struct. 33, 343-359.
3. Y.S. Chen, G. H. Paulino, A. C. Fannjiang, 2001, The crack problems for nonhomogenous materials under anti-plane shear loading-a displacement based formulation, Int. J. Solids Struct. 38, 2989-3005.
4. Ch. Zhang, J. Sladek, V. Sladek, 2003, Effects of material gradients on transient dynamic mode-III SIF's in a FGM, Int. J. Solids Struct, 40, 5251-5270.
5. X-F Li, G.J. Tong, 2003, Anti-plane interface crack between two bonded dissimilar piezoelectric layers, European J. Mech. A/Solids. 22, 231-242.
6. X-F Li, K.Y. Lee, 2004, Electroelastic behavior of a rectangular piezoelectric ceramic with an anti-plane shear crack at arbitrary position, European J. Mech. A/ Solids, 23, 645-653.
7. Ch. Zhang, J. Sladek, V. Sladek, 2005, Anti-plane crack analysis of a functionally graded material by a BIEM, Comp. Mat. Sci., 32, 611-619.
8. S.F. Kwon, J.H. Shin, 2006, An electrically saturated anti-plane shear crack in a piezoelectric layer constrained between two elastic layer, Int. J. Mech. Sci., 48, 707-716.
9. R.R. Bhargava, N. Saxena, 2006, Crack arrest model for a piezoelectric plate-a generalised Dugdale model, Sadhna, 31, 213-226.

53

Delamination Damage Prediciton in a Single Bolt Single Lap Composite Bolted Joint

P. Ramesh Babu[1] and B. Pradhan[2]

[1]Research Scholar, Department of Mechanical Engineering, IIT, Kharagpur-721 302, India
email: prbmechou@yahoo.com
[2]Professor, Department of Mechanical Engineering, IIT, Kharagpur-721 302, India
email: bpradhan@iitkgp.ernet.in

ABSTRACT

This paper deals with the three-dimensional finite element analysis of a single-bolt, single lap composite bolted joint. The main focus of this work is to predict the delamination damage in the laminated FRP composite bolted joint. The effect of bolt-hole clearance on stress distribution has also been studied. The FE model has been validated with the available literature of McCarthy et al.[8] taking identical specimen geometry and material properties. The surface strains obtained by the finite element model are in good agreement with their experimental results. The delamination initiation has been evaluated by considering delamination onset criterion proposed by Hashin and Roten [10].

Keywords: Bolt-clearance, Delamination Damage, FRP composites, Mechanically Fastened Joint.

1. INTRODUCTION

The use of Fiber Reinforced Plastic (FRP) composite materials in many applications more specific to aeronautical and space has generated the need for developing reliable models to predict the strength of critical aircraft components such as the mechanical fastened composite joints. The bolt joint loaded laminates develop local failure or exhibit local damage such as matrix crack, fiber failure, fiber-matrix shear-outs and delaminations. The ability to predict initiation and growth of such damage is essential for assessing the performance of the joint and design them with safety. A large part of research that has been done on mechanically fastened joints has been concerned with the experimental determination of the influence of geometric factors on the joint strength [1–5]. Most of the researchers have resorted to finite element technique because of the complex contact conditions existing at the interface of the bolt and the hole boundary. With the recent increases in computing power, three-dimensional finite element modeling of composite bolted joints has become feasible and such analyses are being performed by [6,7] to study the stress distribution around the bolt loaded hole. A 3D finite element analysis to study the effects of bolt-hole clearance in composite bolted joints has been reported by [8,9]. In this paper a 3D FE model has been developed to predict the radial and tangential stress distributions around the loaded hole of the laminated composite plate for snug-fit and clearance-fit cases. The focus of the work is to study the damage evolution by prediction of damage initiation location. The delamination initiation has been evaluated by considering delamination onset criterion proposed by Hashin and Roten [10].

2. PROBLEM DESCRIPTION

The specimen geometry along with its dimensions is shown in Fig. 1. A quasi-isotropic lay-up with stacking sequence [45/0/–45/90]5S and a zero-layer dominated lay-up with stacking sequence [(45/02/–45/90)3 /45/02/–45/0]S are considered for the study. Layered 46 element of ANSYS has been used to build the laminates and Solid 45 has been used to model the bolt, nut and washers. Surface-to surface contact elements are employed to simulate the contact conditions.

Material properties for Laminate: E_{11} = 140 GPa, E_{22} = E_{33} = 10GPa, v_{12} = v_{13} = 0.3, v_{23} = 0.5, G12 = G_{13} = 5.2 GPa and G_{23} = 3.9 GPa. Laminate Material Strength data: S^{T}_{11} = 2200 MPa, S^{C}_{11} = 1600 MPa, S^{T}_{22} = 70 MPa, S^{C}_{22} = 250 MPa, S^{T}_{33} = 50 MPa, S^{C}_{33} = 300 MPa, S_{12} = S_{31} = 120 MPa and S$_{23}$ = 50 MPa.

Material properties for Bolt: E = 110 GPa, v = 0.29; Washer and Nut: E = 210GPa, v = 0.3.

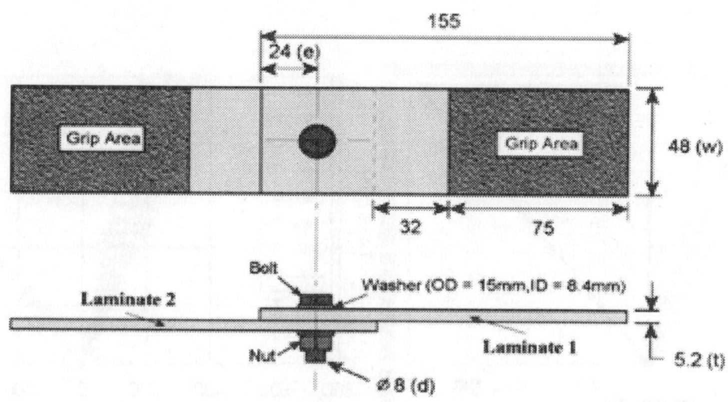

Fig. 1. Specimen geometry.

2.1. Finite Element Model with Boundary Conditions

The fixity and loading boundary conditions used in the analysis are shown in Fig. 2. Contact element is provided at the interface of two laminates, bolt-hole interface, at the interface between top washer-bolt head-top laminate and at the interface between bottom washer-nut-bottom laminate in order to simulate the real situation.

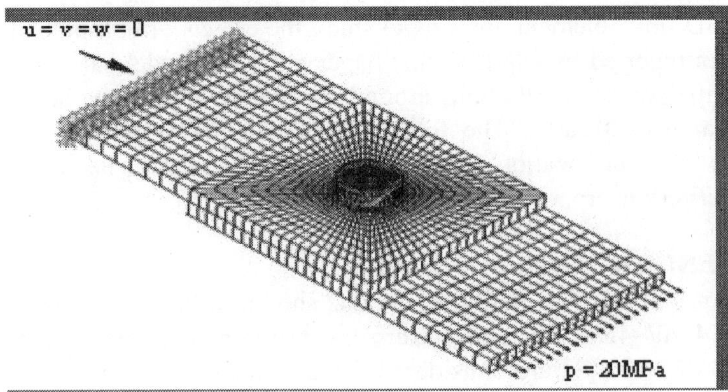

Fig. 2. Finite element model with boundary conditions.

3. MODEL VALIDATION

The model was validated by comparing the experimentally obtained surface strains available in literature [9] with the results obtained from FE analysis. For model validation only the neat-fit type of joint has been considered. Figure 3 shows the experimental and FE results for the quasi-isotropic, snug-fit joint.

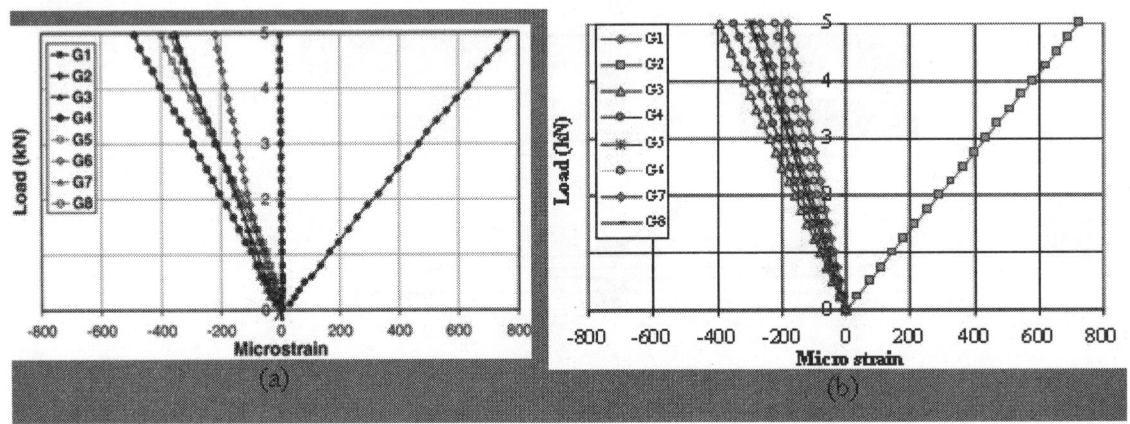

Fig. 3. (a) Experimental Strain Gauge Readings (b) Finite element results.

4. RESULTS AND DISCUSSION

The stresses obtained from the finite element analysis around the periphery of the hole for each layer are depicted in Figs. 4 to 7. All the measurements are taken at different angles starting from the bearing plane. Angles measured clockwise are taken as positive and that measured in anti clockwise are taken as negative.

4.1 Quasi-Isotropic Laminates

4.1.1 Radial Stress Distribution

The radial stress distribution for the snug-fit and clearance-fit quasi-isotropic joint is shown in Fig. 4. The highest radial stress occurs in layer no. 2, which is the second ply from the shear plane and is orientated at 0° with the loading direction in both the cases except the magnitude of stresses are higher in clearance-fit case. Interestingly, quite high stress levels exist in the clearance-fit joint all the way through the thickness, whereas in the neat-fit joint the stresses tend to drop off to very low levels as the free surface of the laminate (i.e. ply no. 40) is approached.

As can be seen, all the 0° plies are most highly stressed at the 0° location (the bearing plane), all the +45° plies are most highly stressed near the +45° location for snug-fit case where as for the clearance-fit case the +45° and –45° plies do not peak at their stiffest locations, but at an angle of approximately +30° and –30° respectively. This is due to the contact pressure being applied over a reduced contact angle.

The 90° plies experience their highest stresses at an angle of 70° which varies through the thickness; the angle is less than 90° since the contact angle is less than 180° where as for the case of clearance-fit the 90° plies experience very low levels of radial stress at all locations around the hole boundary with low peaks occurring at the 0° location.

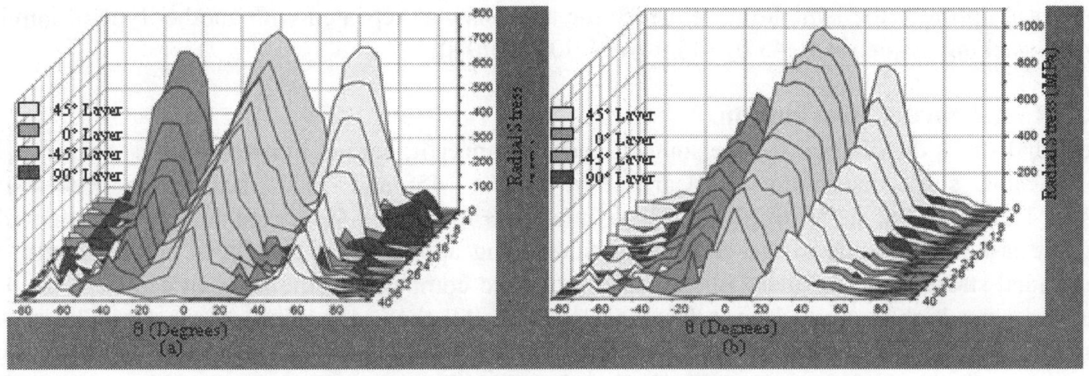

Fig. 4. Radial stress distributions in each layer of the quasi-isotropic joint for (a) Snug-fit (b) Clearance-fit.

4.1.2 Tangential Stress Distribution

The tangential stresses around the hole boundary for the snug-fit and clearance-fit joint are shown in Fig. 5. The tangential stress in each layer is positive whereas at the back of the hole ($\theta = \pm180°$) the stresses developed are negative in both the cases and are pronounced in clearance-fit case. These negative stresses are due to ovalisation of the hole. In case of snug-fit joint the highest tangential stresses occur in all the 0° plies near the 90° location (net-section plane). The highest tangential stresses in the –45° and +45oplies occur near the +45° location and the –45° locations because of their stiffness in the tangential direction at this point. Unlikely the tangential stresses in the case of clearance-fit are negative in a region bounded by the ±30° location in all plies which are more predominant in ±45° layers. This is due to the changes in the deformed shape of the hole due to clearance. The peak tangential stress has also increased slightly and unlike the snug-fit case, occurs in the +45° and − 45° plies, not in the 0° plies. Finally, the stresses in the 90° plies are highest at the 0° location, once again due to their high stiffness in the tangential direction at this location.

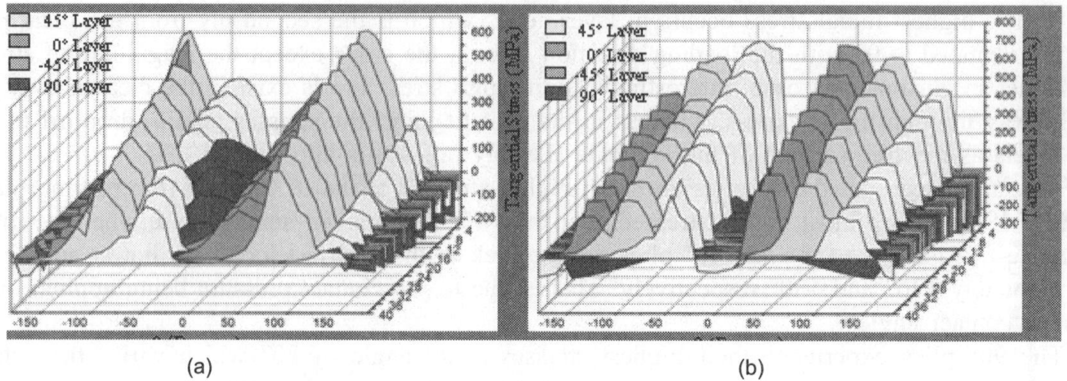

(a)   (b)

Fig. 5. Tangential Stress Distributions in Each Layer of the Quasi-isotropic Join for (a) sung-fit (b) Clearance-fit.

4.2 Zero-Layer Dominated Laminate

In order to optimize the laminating sequence the analysis was repeated with another type of laminate having stacking sequence ($[(45/0_2/–45/90)_3 /45/0_2/–45/0]s$).

4.2.1 Radial Stress Distribution

The radial stress distribution for the snug-fit and clearance-fit, zero-dominated joint is shown in Fig. 6. The radial stress distribution in clearance-fit case at +45° and − 45° plies do not peak at their stiffest locations, but at an angle of approximately +40° and − 40° respectively as compared to snug-fit case. This is due to the contact pressure being applied over a reduced contact angle. The peak radial stress value for these plies is also increased compared to the snug-fit case. All the +45° plies are most highly stressed near the +45° location. All the − 45° plies are most highly stressed near the − 45° location. The 90° plies experience their highest stresses at an angle of 70° which varies through the thickness; the angle is less than 90° since the contact angle is less than 180°.

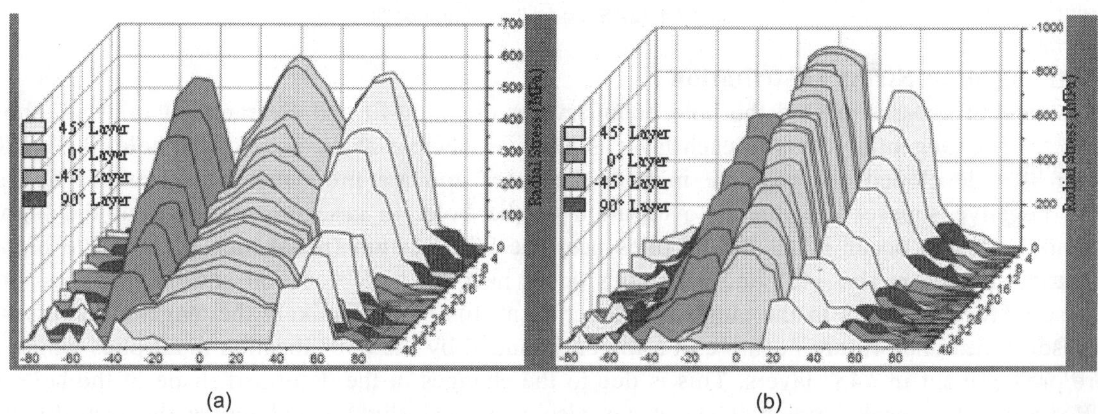

(a)   (b)

Fig. 6. Radial stress distributions in each layer of the zero-layer dominated laminate joint for (a) sung-fit (b) Clearance-fit.

4.2.2 Tangential Stress Distribution

The tangential stresses around the hole boundary for the snug-fit and clearance-fit zero-dominated joint are shown in Fig. 7. The tangential stress in each layer is positive whereas at the back of the hole (θ = ±180°)the stresses developed are negative as shown in Fig.7. These negative stresses are due to ovalisation of the hole as mentioned above. The highest tangential stresses occur in all the 0° plies near the 90° location (net-section plane). The highest tangential stresses in the –45° plies occur near the +45° location and for +45° layers it occurs at –45° locations. Thus, for the snug-fit quasi-isotropic joint, the highest tangential stresses occur in each ply at locations along the hole boundary where they are stiffest in the tangential direction. Unlike the quasi-isotropic clearance fit joints the maximum tangential stresses are developed in 0° layers rather in ±45° layers.

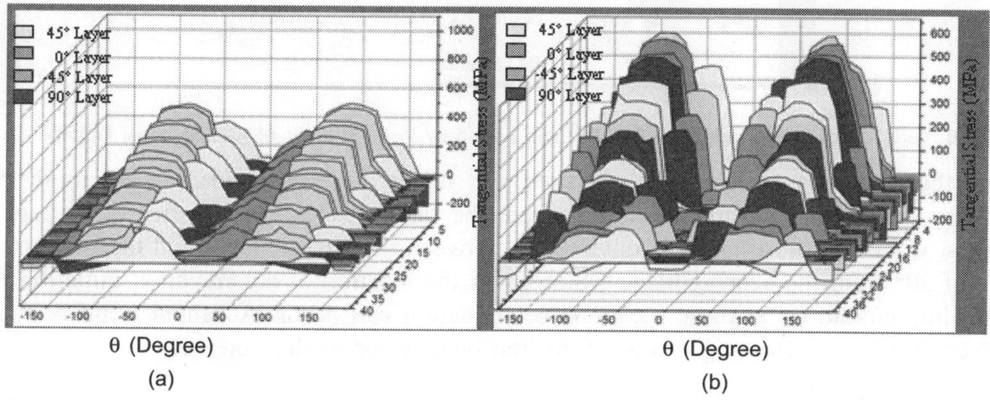

θ (Degree)

(a)　　　　　　　　　　　　　　θ (Degree)

(b)

Fig. 7. Tangential stress distributions in each layer of the zero-layer dominated laminate joint for (a) sang-fit (b) Clearance-fit.

5. PREDICTION OF DELAMINATION INITIATION LOCATION

Failure analysis of laminates with bolted joints is very complex due to the large number of failure mechanisms in composite materials. As a result the existing methods use numerous, mostly empirical failure criteria. Failure analysis was performed by implementing the commonly used Hashin-Rotem polynomial failure criteria[10]. These criteria were selected because they can distinguish between different failures and may be applied with ease in a finite element formulation. As per Hashin-Rotem failure criteria, delamination of a laminate will take place if the following conditions are satisfied:

$$\text{Delamination in tension, for } (\sigma_z \geq)e^2 = \left(\frac{\sigma_z}{Z_T}\right)^2 + \left(\frac{\sigma_{xz}}{S_{xz}}\right)^2 + \left(\frac{\sigma_{yz}}{S_{yz}}\right)^2 \geq 1$$

$$\text{Delamination in compression, for } (\sigma_z \leq 0) = \left(\frac{\sigma_z}{Z_C}\right)^2 + \left(\frac{\sigma_{xz}}{S_{xz}}\right)^2 + \left(\frac{\sigma_{yz}}{S_{yz}}\right)^2 \geq 1$$

where, e = failure index, σ_z = Stress in through-the-thickness direction, σ_{xz} = Shear stress in the plane of through thickness and fiber direction, σ_{yz} = Shear stress in the plane transverse and through thickness plane. The out of plane stress values obtained from the analysis is used in Hashin-Rotem criterion to predict the delamination location and plotted in Fig. 8.

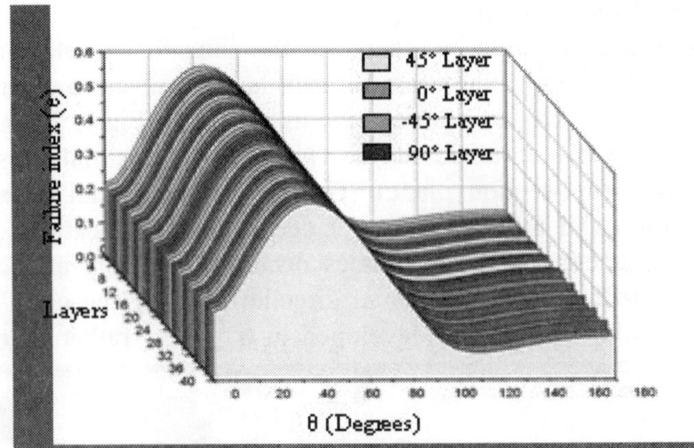

Fig. 8. Variation of failure index around the hole boundary in each layer of neat-fit quasi isotropic joint.

From the above figure it can be observed that under the present loading condition (5kN) no delamination occurs in the laminate as no where the failure index "e" satisfies the failure criteria. The highest value of failure index is obtained in between the first 0° layer and the first 90° layer at an angle of 30° from the bearing plane which gives the location of delamination initiation i.e under higher loading conditions at these locations delamination can occur. A similar trend is obtained in other layers but in a decreasing manner from the inner layer to the outer layer.

CONCLUSION

1. Nonlinear contact finite element analysis is a powerful technique to simulate the deformation characteristics of mechanically fastened joints.
2. It is observed that significant amounts of bending of the laminates occurred (termed "secondary bending"), and the external surface of the joint is in compression, despite the tensile loading applied to the joint.
3. Double-curvature of the surface occurs as a typical deformation mode.
4. There is increase in radial as well as tangential stress as we go from snug-fit to clearance-fit for both quasi-isotropic as well as zero-dominated joints.
5. The maximum tangential stress in case of clearance fit quasi-isotropic joint occurs is ±45° plies whereas it occurs in 0° plies in case of neat-fit zero-degree layer dominated joint.
6. There is decrease in radial as well as tangential stress as we go from quasi-isotropic joint to zero-dominated joints for both types of clearances.

REFERENCES

1. Theo de Jong, 1977, Stresses around pin-loaded holes in elastically orthotropic or isotropic plates, J of Composite Materials, 11:313–331.
2. T.A. Collings, 1977, The strength of bolted joints in multi-directional CFRP laminates, Composites, 8:43-54.
3. L.J. Hart-Smith, 1978, Mechanically fastened joints for advanced composites-phenomelogical considerations and simple analysis, Duglas Paper, 6748:1–32.
4. S.R. Soni, 1981, Failure analysis of composite laminates with a fastener hole, ASTM STP 749, 145-64.
5. L.I. Erikson, 1986, Contact stresses in bolted joints of composite laminates, Composite Structures, 6 (1-3):57-75.

6.   F.L. Matthews, 1987, Joining fiber-reinforced plastics. Elsevier Applied Science, London and New York.

7.   T.Ireman, 1998, Three-dimensional stress analysis of bolted composite single-lap joints, Composite Struct., Vol. 43, 195-216.

8.   M.A. McCarthy, C.T. McCarthy, 2005, Three-dimensional finite element analysis of single-bolt single-lap composite bolted joints: part II- Effect of bolt hole clearance, Composite Structures, 71(2):159–175.

9.   M.A. McCarthy, C.T. McCarthy, V.P. Lawlor, and W.F.Stanly, 2005, Three-dimensional finite element analysis of single-bolt single-lap composite bolted joints: part I– Model development and validation, Composite Structures, 71(2):140-158.

10.   Z. Hashin, and Z. Rotem, 1973, A fatigue failure criterion for fiber reinforced material, J.Compos. mater, 7:448-464.

54

Continuum Damage Mechanics Based Response Prediction in Laminated Composites Under Static Damage

P.M. MOHITE

Department of Aerospace Engineering, Indian Institute of Technology Kanpur-208 016, India
email: mohite@iitk.ac.in

ABSTRACT

In the present paper continuum damage based meso-model proposed by [1, 2] is used to predict the response of initially damaged laminated composites. The model is based on three foundations of meso-model, internal variable approach and method of local state. The present model is refined by replacing the interface as a very thin layer of resin. Further, this layer uses higher order approximation in transverse direction for displacement field over linear variation used in original model. This model has been implemented in the generalized layerwise finite element model developed by author and his co-researcher [3]. It is seen that the fibre failuer mode is dominant comapared to other failure modes in giving significant difference in transverse displacement signature profiling whereas the delamination mode has weak effect. The in-plane strains also show similar behavior. Here, the results are presented for different shapes and sizes of damage under the transverse load only.

Keywords: Continuum damage meso-model, Static damage, Transverse load, Layerwise finite element model.

1. INTRODUCTION

Unidirectional fiber-reinforced composites are widely used in aerospace applications. They are used for the critical structural applications. These structures are subjected to complex aerodynamic and service loads and exhibit progressive failure and hence deterioration of elastic moduli. The mechanics by which failure occurs are numerous. These can be broadly categorized as ply level failure mechanisms like brittle fibre failure, fibre matrix debonding, and matrix cracking; and delamination. A exhaustive literature is available for the initiation of damage in the laminates.

The continuum based damage meso-model is capable of predicting both initiation and growth of damage in these laminates. Further, it is also capable of predicting delamination initiation and its growth. The concept was first introduced by Kachanov [4] for creep damage in metal. This concept is taken by Ladevèze and his co-workers and extended to laminated composites. In the following section, the details of the meso-model are presented for the sake of completeness.

2. MESO-MODEL FOR LAMINATED COMPOSITES

The concept is based on three foundations as

1. *Meso-scale*: In this approach, the laminate is treated as set of alternate homogeneous ply and a very thin interlaminar interface. The approach of meso-model enables the handling of ply level damage and interfacial damage (delamination). The interface is a two dimensional entity, which transfers displacements and forces from one ply to another ply.

2. *Internal variable Approach*: The effect of damage on mechanical behaviour through degradation of the material's elastic moduli and

3. *The method of local state*: It relates damage to thermodynamic forces associated with strain energy.

2.1 Elementary Layer Damage Meso-Model

The form of the damage meso-model of the layer involved in this study is based on the assumption that damage within the layer is governed explicitly by the in-plane stress components, and also that layer damage mechanisms solely affect the in-plane stiffness components. It is assumed that all forms of damage provoked by inter-laminar stresses (i.e. delamination) are completely taken into account by the interface. The damaged strain energy density in the layer is thus written as:

$$E_D = \frac{1}{2(1-d_{11})}\left[\frac{\sigma_{11}^2}{E_1^o} - \left(\frac{v_{12}^o}{E_1^o} + \frac{v_{21}^o}{E_2^o}\right)\sigma_{11}\sigma_{22} - \left(\frac{v_{31}^o}{E_3^o} + \frac{v_{13}^o}{E_1^o}\right)\sigma_{11}\sigma_{33} - \left(\frac{v_{32}^o}{E_3^o} + \frac{v_{23}^o}{E_2^o}\right)\sigma_{22}\sigma_{33}\right]$$

$$+ \frac{1}{2}\left[\frac{<\sigma_{22}>_+^2}{E_2^o(1-d_{22})} + \frac{<\sigma_{22}>_-^2}{E_2^o} + \frac{\sigma_{33}^2}{E_3^o} + \frac{\sigma_{12}^2}{G_{12}^o(1-d_{12})} + \frac{\sigma_{13}^2}{G_{13}^o} + \frac{\sigma_{23}^2}{G_o^{23}}\right] \qquad ...(1)$$

where the subscripts 1, 2, and 3 refer to the principal material coordinate system and, when attached to a damage variable d, indicate the orientation of its associated stress component (i.e. damage modes). For example, the variable $d22$ corresponds to the damage process produced by the normal stress component σ_{22}, which, as defined by the meso-model, is attributed to matrix micro-cracking and tensile failure of the fiber/matrix interface. The bracketed notation $<>_{+or-}$ indicates which term is non-zero depending upon the sign of the normal stress components. It is also important to point out that there is no sign dependence with respect to the contribution of the shear stress σ_{12}. For the damaged strain energy density described in Eq. (1) the thermodynamic force associated with tensile normal stresses is:

$$T_{22} = \frac{\partial <<E_D>>}{\partial d_{12}} = \frac{1}{2(1-d_{22})^2}\left\langle\left\langle\frac{<\sigma_{22}>_+^2}{E_2^o}\right\rangle\right\rangle \qquad ...(2)$$

and the thermodynamic force associated with in-plane shear stresses is:

$$Y_{12} = \frac{\partial <<E_D>>}{\partial d_{12}} = \frac{1}{2(1-d_{12})^2}\left\langle\left\langle\frac{\sigma_{12}^2}{G_{12}^o}\right\rangle\right\rangle \qquad ...(3)$$

where the symbol $\langle\langle\rangle\rangle$ indicates that the value is averaged through the thickness of the layer. This represents the key assumption of the meso-model, that the damaged state is uniform through the thickness of each meso-constituent. With respect to fiber damage, tensile fiber rupture is determined via a maximum strain criterion:

$$d_{11} = 0 \quad \text{if} \quad \varepsilon_{11} < \varepsilon_{11}^{R} \quad \text{else} \quad d_{11} = 1.0 \qquad \text{...(4)}$$

where ε_{11}^{R} corresponds to the experimentally determined strain at tensile fiber failure.

In order to prevent 'healing' of the material, $\underline{Y}_{22}$ and $\underline{Y}_{12}$ introduce the maximum value of the associated thermodynamics forces in a given load history at time t such that:

$$\underline{Y}_{22}(t) = \max_{\tau \leq t}(Y_{22}(\tau)), \quad \underline{Y}_{12}(t) = \max_{\tau \leq t}(Y_{12}(\tau)) \qquad \text{...(5)}$$

The idea that the damage from tension and shear mechanisms are linearly related is then introduced, as follows, via a single thermodynamic force $\underline{\hat{Y}}$:

$$\underline{\hat{Y}} = (\underline{Y}_{12}(t) + b\underline{Y}_{22}(t)) \qquad \text{...(6)}$$

where the material constant b defines the level of coupling between the two effects, a good approximation for which has been shown to be:

$$b = \frac{E_2^o}{G_{12}^o} \qquad \text{...(7)}$$

The values of both internal damage variables of the layer are then defined through:

$$d_{22} = \omega_{22}(\underline{\hat{Y}}) = \frac{\langle\langle\sqrt{\underline{\hat{Y}}} - \sqrt{Y_{22}^o}\rangle\rangle_+}{\sqrt{Y_{22}^c}} \quad \text{if} \quad \omega_{22}(\underline{\hat{Y}}) < 1.0, \quad \text{else} \quad d_{22} = 1.0 \qquad \text{...(8a)}$$

$$d_{12} = \omega_{12}(\underline{\hat{Y}}) = \frac{\langle\langle\sqrt{\underline{\hat{Y}}} - \sqrt{Y_{12}^o}\rangle\rangle_+}{\sqrt{Y_{12}^c}} \quad \text{if} \quad \omega_{12}(\underline{\hat{Y}}) < 1.0, \quad \text{else} \quad d_{12} = 1.0 \qquad \text{...(8b)}$$

where Y_{22}^o and Y_{22}^c are experimentally identified thresholds for the onset and ultimate failure through damage from tensile normal stress σ_{22}. Y_{12}^o and Y_{12}^c are the corresponding values for shear damage.

A further constraint is then applied to the evolution of tensile damage to model the brittle behaviour of tensile failure of the fiber/matrix interface:

$$d_{22} = \omega_{22}(\underline{\hat{Y}}) \quad \text{if} \quad \omega_{22}(\underline{\hat{Y}}) < 1.0 \quad \text{and} \quad \underline{\hat{Y}} < Y_r, \quad \text{else} \quad d_{22} = 1.0 \qquad \text{...(9)}$$

where Y_r is an experimentally-determined threshold which corresponds to rupture of the fiber/matrix interface in tension, and is given as:

$$Y_r = \frac{\langle\sigma_{22}\rangle_+^{2^F}}{2E_2^o(1-d_{22}^F)^2} \qquad \text{...(10)}$$

where the superscript F indicates a value at failure.

Finally, the concept that failure in both modes is obtained when the threshold of one damage mechanism is reached is introduced through the following constraints on the damage evolution laws:

$$d_{22} = \omega_{22}(\hat{\underline{Y}}) \quad \text{if} \quad \omega_{22}(\hat{\underline{Y}}) < 1.0 \quad \text{and} \quad \omega_{12}(\hat{\underline{Y}}) < 1.0, \quad \text{else} \quad d_{22} = 1.0 \qquad \text{...(11a)}$$

$$d_{12} = \omega_{12}(\hat{\underline{Y}}) \quad \text{if} \quad \omega_{12}(\hat{\underline{Y}}) < 1.0; \quad \hat{\underline{Y}} < Y_r \quad \text{and} \quad \omega_{22}(\hat{\underline{Y}}) < 1.0, \quad \text{else} \quad d_{12} = 1.0 \qquad \text{...(11b)}$$

2.2 Damage Meso-Model of the Interface

In a form analogous to that of the layer, the damaged strain energy (per unit area) of the interface can be expressed as follows:

$$E_D^I = \frac{1}{2}\left[\frac{\langle \sigma_{33} \rangle_-^2}{E_{33}} + \frac{\langle \sigma_{33} \rangle_+^2}{E_{33}(1-d_I)} + \frac{\sigma_{32}^2}{G_{32}(1-d_{II})} + \frac{\sigma_{31}^2}{G_{31}(1-d_{III})} \right] \qquad \text{...(12)}$$

As with the layer definition, the contribution to the damaged strain energy from normal stress components is separated into tensile and compressive terms in order to distinguish between the mechanical effects of a crack opening versus that of a crack closing. Differentiation of Eq. (12) with respect to each damage variable then yields the following associated thermodynamic forces for each of the three considered delamination modes:

$$Y_I = \frac{1}{2}\frac{\langle \sigma_{33} \rangle_+^2}{E_{33}(1-d_I)^2}, \quad Y_{II} = \frac{1}{2}\frac{\sigma_{32}^2}{G_{32}(1-d_{II})^2}, \quad Y_{III} = \frac{1}{2}\frac{\sigma_{31}^2}{G_{31}(1-d_{III})^2} \qquad \text{...(13)}$$

Depending on the assumptions made regarding damage in the interface (i.e. the amount of coupling between the three delamination modes, whether or not a given mode exhibits brittle or ductile failure), a variety of damage evolution laws may be considered. The damage evolution law of the interfacial meso-model involved in this study uses the assumption that delamination growth is governed by a so-called "equivalent damage force" which couples the thermodynamic forces from all three delamination modes into a single quantity. Approaches based upon this idea are generally referred to as isotropic damage evolution laws. This coupled thermodynamic force has the form:

$$Y_e = [Y_I^\alpha + (\lambda_1 Y_{II})^\alpha + (\lambda_2 Y_{III})^\alpha]^{\frac{1}{\alpha}} \qquad \text{...(14)}$$

$$\underline{Y}_e(t) = \max_{\tau \le t}(Y_e(\tau))$$

where λ_1 and λ_2 are coupling parameters and is a material parameter linked to classical fracture mechanics. It is important to note that in Eq. (14) the constraint applied is the same as that used in the damage meso-model of the layer to prevent 'healing' of the material as in Eq. (8). The other key assumption for the isotropic damage evolution laws is that, because of its small thickness, the stiffness of the interface in all three directions is reduced to zero when the threshold of one mode is reached. The three damage variables can then all be linked to a single material function $\omega(\underline{Y}_e)$ which relates the coupled thermodynamic force to a single set of damage thresholds:

$$\omega(\underline{Y}_e) = \left[\frac{n}{n+1} \frac{\langle \underline{Y}_e - Y_e^o \rangle_+}{Y_e^c - Y_e^o} \right]^n \qquad \text{...(15)}$$

$$d_I = d_{II} = d_{III} = \omega(\underline{Y}_e) \quad \text{if} \quad \omega(\underline{Y}_e) < 1.0, \quad \text{or} \quad d_I = d_{II} = d_{III} = 1.0 \text{ otherwise}$$

where Y_e^o and Y_e^c are analogous to the experimentally determined parameters used in the meso-model of the layer, and n is a material constant related to the type (brittle, $n > 1$, or ductile, $n < 1$) of failure of the interface.

Remark: In the above model the interface is treated as a very thin orthotropic layer of the same material as that of the ply, with effective orientation as the bisector of the angle between the adjacent layer fibre directions [1]. In the present study the interface is modeled as a very thin layer of homogeneous matrix material.

Remark: Damage in ply is denoted by three scalars d_{11}, d_{22} and d_{12} as (d_{11}, d_{22}, d_{12}).

3. RESULTS AND DISCUSSION

In this section the effect of static damage on the maximum transverse deflection and in-plane strain components is studied, for different sizes, shapes and intensities of damage, under transverse loading. The material used is high modulus M55J/M18 graphite/epoxy. Here, a [0/90/0] laminate of dimension 64×64 mm under transverse uniform load of intensity 0.005N/mm^2 is studied. The thickness of each ply is 0.1 mm and that of interface is 0.01 mm. The plate is simple supported along all edges. The load chosen is such that it does not initiate damage at other places. Even if damage is initiated, its size and magnitude will be negligible as compared to the initial damage taken.

3.1 Effect of Size of Static Damage on the Response

The plate has centrally located damage of (1) Size 1: 10.7×10.7 mm and (2) Size 2: 5.35×5.35 mm. The transverse deflection and in-plane strain components along lines passing through the centre of the laminate parallel to x axis at the top of laminate are computed. The strain components are non-dimensionalized by the maximum strain value of strain when the laminate is undamaged.

The transverse displacement and in-plane strains components for two cases of existing damage, i.e. damage in bottom ply of intensity (0.1, 0.98, 0.1) and damage in top ply of intensity (0.98, 0.1, 0.1), are shown in Fig. 1 and Fig. 2. In this the transverse displacement is also compared with displacement obtained in the absence of damage.

3.2 Effect of Shape of Static Damage on the Response

Here, damage of square size 1 and of circular shape with same area as of square size 1 are considered. The transverse displacement and in-plane strain components are shown in Fig. 3 and Fig. 4.

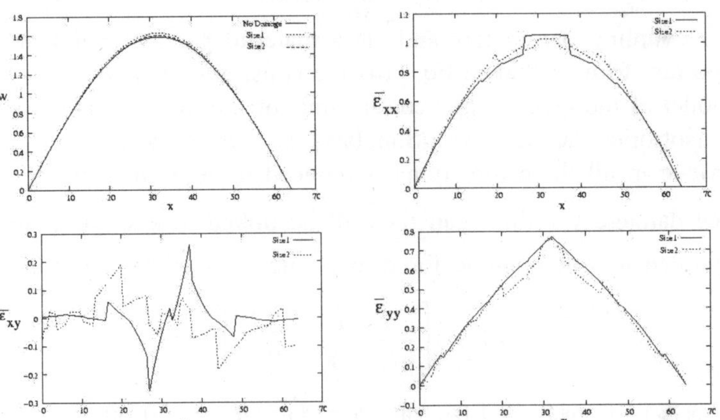

Fig. 1. Effect of size of damage. (Damage in bottom ply of intensity (0.1, 0.98, 0.1)).

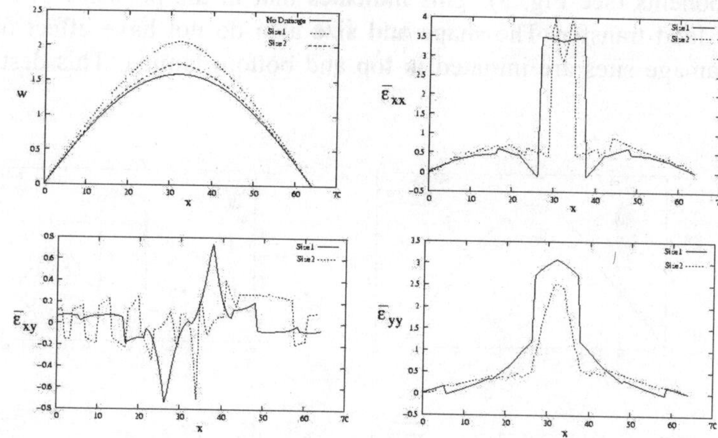

Fig. 2. Effect of size of damage. (Damage in top ply of intensity (0.98,0.1,0.1)).

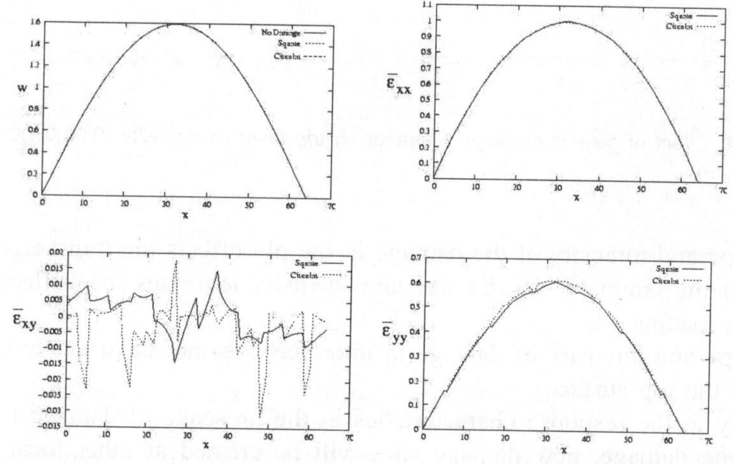

Fig. 3. Effect of shape of damage. (Delamination damage in bottom interface).

From the figures it can be observed that:

1. As the size of damage increases the transverse deflection at the centre of the laminate (where damage is located) increases. This is because the damage reduces the stiffness. Also, the strain components show similar behavior.

2. The fiber failure mode is more dominant than fibre matrix interface debonding and matrix cracking mode. This can be seen through the change in transverse displacement in Fig. 1 and Fig. 2.

3. The change is transverse displacement for square and circular shape is not so significant. The difference for the fiber failure mode is because the circular shape cuts more fibres than square shape of same area thus, reducing more stiffness with circular shape. The strain component becomes more smoothened (see Fig. 4).

4. The delamination damage mode does not have much effect on transverse displacement and

strain components (see Fig. 3). This indicates that in the presence of delamination still there is effective load transfer. The shape and size also do not have effect on the response.

5. The new damage sites are initiated in top and bottom lamina. This destroys the symmetry of profiles.

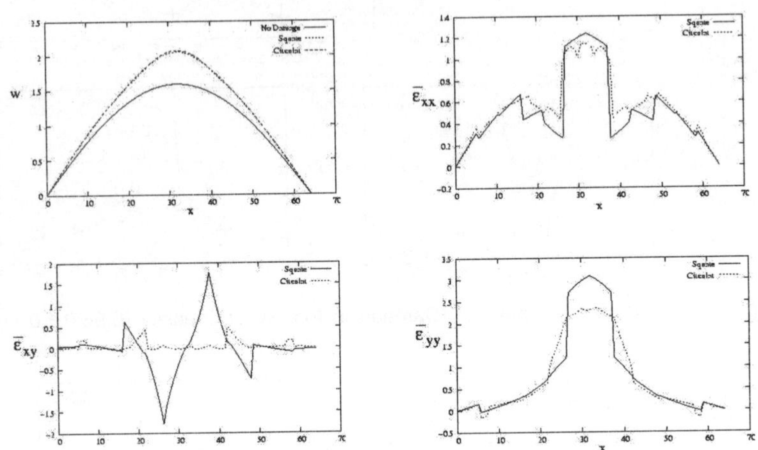

Fig. 4. Effect of size of damage. (Damage in top layer of intensity (0.98,0.1,0.1)).

CONCLUSION

1. The size, shape and intensity of the damage in the ply effects the transverse deflection on the top surface of the laminate. As the size and intensity increases the deflection also increases for transverse loading.
2. The size, shape and intensity of damage in interface does not have much effect on transverse deflection on the top surface.
3. The symmetry in the response characteristics in the presence of damage is lost.
4. Due to existing damage, new damage sites will be created at other locations when a static transverse load is applied.

REFERENCES

1. O. Allix, P. Ladevèze, 1992, Interlaminar interface modelling for the prediction of delamination, Composite Structures, 22, 235-242.
2. E.A. Philips, C.T. Herakovich, L.L. Graham, 2001, Damage development in composites with large stress gradients, Composite Science and Technology, 61, 2169-2182.
3. P.M. Mohite, C.S. Upadhyay, 2006, Accurate computation of critical local quantities in composite laminated plates under transverse loading, Computers and Structures, 84, 657-675.
4. L.M. Kachanov, 1958, Time of the rupture process creep conditions. Izv. Akad. Nauk SSR Otd Tekh Nauk, 8, 26-31.

55

3D Layerwise Stress Analysis and Damage Prediction in Adhesively Bonded Single Lap Joints of Laminated FRP Composites

S.K. PANIGRAHI[1] AND B. PRADHAN[2]

[1]Research Scholar, Department of Mechanical Engineering, IIT Kharagpur-721 302, India
email: skpanigrahi1@yahoo.co.in
[2]Professor, Department of Mechanical Engineering, IIT Kharagpur-721 302, India
email: bpradhan@mech.iitkgp.ernet.in

ABSTRACT

This paper deals with three-dimensional layerwise Finite Element Stress Analysis and Damage Prediction of adhesively bonded Single Lap Joints (SLJ) with laminated FRP composite adherends. The classical theories of Volkersen's, Goland and Reissener's solutions for the SLJs, which neglect adherend shear deformations, forms the basis of the present analysis. Layerwise non-linear finite element analysis is performed in order to study the stress concentration effect due to joint material heterogeneity and discontinuities of loading and geometry. Also, some important three-dimensional issues due to free edge effect, anti-elastic effect and the coupling effect of bending-twisting-stretching are discussed in detail. The effect of different boundary conditions and material anisotropy on the stress and strength of the joint have been assessed thoroughly. Using the point wise stress obtained from the layerwise stress analysis, the failure indices are calculated based on the mechanics of materials approach. Depending on the value of failure indices, the possible location of the damage initiation in the SLJ is identified. It is observed that the free edge of the joint is most critical location for the initiation of damages.

Keywords: FRP Composites, Layerwise Stress Distribution, SLJ, Failure index.

1. INTRODUCTION

Adhesive bonded joints with laminated FRP composites play a significant role in many structural applications especially in space, aircraft and automobile industries. However, designing the composite bonded joints is quite complex, because their performance is limited by the characteristics of the composite laminate adherends and bi-material interface which usually have low interlaminar strengths. The interlaminar stresses induced in the vicinity of the bondline leading edges can cause delamination/

debonding damages in the joint. As such, the cohesive and adhesive failures as in Fig. 1 are the major threats to the adhesive bonded joint performances. Thus, accurate 3D analysis is essential for understanding the joint stress fields, damage initiation and its propagation in practical applications.

The damage study of adhesively bonded SLJ problem is typically approached in one of the two ways; with finite element analysis or through analytical modeling. Accurate analyses of adhesively bonded joint using Finite Element Method are the most demanding tasks [1, 2] and there is a specific need for the analysis that can provide accurate and better results. Many researchers [3-5] investigated the adhesively bonded composite joints using finite element methods. The literature shows that standard finite element codes can analyze the adhesively bonded joints with or without damages under arbitrary loading conditions. In design and sizing, many different joint configurations must be analyzed quickly, and each finite element model can take hours or even days to properly carry out the pre and post-processing. Second, because the stress gradients for bonded joints are very steep especially at the reentrant corner, the accuracy of the method can be highly dependent on mesh refinement. On the contrary, the analytical methods of solution for bonded joint are limited due to many complications like geometry, configuration, eccentric loading path and discontinuous materials etc. The various boundary conditions and material anisotropy also add to the complicacy of damage analyses of bonded joints.

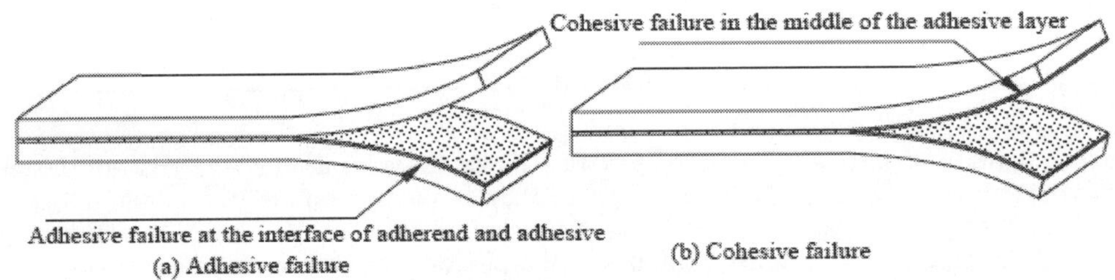

Fig. 1. Adhesive bonded joint failure modes.

Some researchers [4, 5] developed analytical solutions for lap-shear joints with orthotropic adherends using many assumptions for the simplication of the problems which are far away from the reality. These analytical methods mainly focus on obtaining the adhesive stresses with many assumptions and specifically ignoring the out of plane stresses at the interface of the joint, which are known to be the key contributors to failure of the joint. Unfortunately, in almost all the work published so far, only two stress components (shear stress and transverse peel stress) have been considered. However, there is very little work in the literature regarding the complete tri-axial stress state, damage prediction and its propagation in regard to the SLJ with varying material anisotropy and boundary conditions. The present study emphasizes the complete evaluation of the tri-axial stress field and in particular the out-of-plane stresses which play the dominant role of damage initiation and propagation in the adhesive layer.

This paper presents an efficient analysis of bonded joints using FEM consisting of layered brick elements for the adhesive and adherends. The effects of different boundary conditions discussed in below section have been considered. The analyses are confined to the overlap portion only, which is considered to be the most critical region for the bonded joint analysis with constant adhesive layer thickness. The objectives are:

(a) to discuss the three-dimensional issues for the SLJ specimen due to eccentricity of the loading path with a special importance to the out-of-plane stresses with different boundary conditions.

(b) to identify the location of the damage initiation in the joint based on mechanics of material approach.

(c) to discuss the effect of various FRP composite materials viz. Gl/E, B/E and Gr/E on the cohesive failure of adhesively bonded joints.

Based on the observations, suitable design recommendations have been suggested.

2. SLJ SPECIMEN GEOMETRY AND MATERIAL CONSTANTS

The geometry, configuration and loading condition of the specimen analyzed is shown in Fig. 2. Two $[0/90]_s$ laminated FRP composite plates are used as adherends. The laminate considered is graphite/epoxy (Gr/E), boron/epoxy (B/E) and glass/epoxy (Gl/E) and epoxy is used as the adhesive layer. A three-dimensional non-linear finite element analysis is performed using the FE package ANSYS to calculate the out of plane stress distributions in the adhesive layer of the overlap region. Most of the papers in the literature on adhesively bonded joints do not bring out the fact that the deformation and stresses in a joint depend on the type of boundary conditions used. To investigate the influence of boundary conditions on the elastic stress distribution in a single lap joint, two different types of boundary conditions with reference to Fig. 2 are used. The detailed descriptions of the boundary conditions and material properties used are given below:

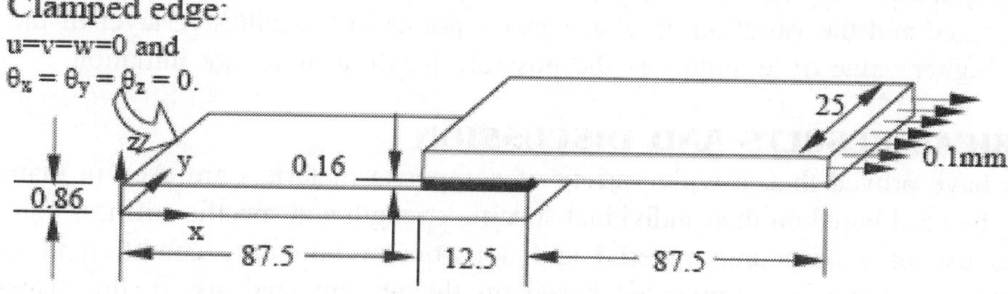

Fig. 2. SLJ test specimen (All dimensions are in mm).

Material Properties

(a) Gr/E – $E_x = 181$ GPa, $E_y = E_z = 10.3$ GPa, $G_{xy} = G_{xz} = 7.17$ GPa, $G_{yz} = 4$ GPa, $\upsilon_{xy} = \upsilon_{yz} = 0.28$, $\upsilon_{xz} = 0.3$

(b) B/E – $E_x = 207$ GPa, $E_y = E_z = 18.63$ GPa, $G_{xy} = G_{xz} = 4.5$ GPa, $G_{yz} = 3.45$ GPa, $\upsilon_{xy} = \upsilon_{yz} = 0.27$, $\upsilon_{xz} = 0.35$

(c) Gl/E – $E_x = 38.6$ GPa, $E_y = E_z = 8.27$ GPa, $G_{xy} = G_{xz} = 4.14$ GPa, $G_{yz} = 4$ GPa, $\upsilon_{xy} = \upsilon_{yz} = 0.25$, $\upsilon_{xz} = 0.27$

(d) Epoxy – $E = 2.4$ GPa, $\upsilon = 0.38$.

Boundary conditions: (Fig. 2)

Type 1 – $u = v = w = 0$, at $x = 0$ and $u = 0.1$, $w = 0$, at $x = 187.5$

Type 2 – $u = v = w = 0$, at $x = 0$ and $w = 0$, at $0 < x \leq 87.5$ and $100 < x \leq 187.5$ (u, v and w are displacements in x, y and z directions, respectively)

3. FINITE ELEMENT MODELLING AND FAILURE ANALYSIS OF SLJ

In the finite element modelling of the single lap joint test specimen, one solid 46 and two solid 45 elements of ANSYS are used to model the adherend and adhesive layer in the thickness direction. To capture the stress concentration effect due to joint material heterogeneity and discontinuities of loading and geometry, optimum meshing scheme is used. In order to avoid the burden of computational effort and shape failure, an element size of 0.25 mm × 0.25 mm is used to model both the adherend and adhesive layer of the joint overlap region. A 3D geometric non-linear finite element analysis is performed in order to capture the large deflection. Various boundary conditions discussed earlier are imposed in the analysis. An in-plane uniform extension of 0.1 mm is applied at the free end in 20 equal steps. The out-of-plane stress distributions over the adhesive layer are obtained from the present analysis in order to study the cohesive failure of the joint. The adhesive layer is considered to be isotropic. The failure in the adhesive layer is known as cohesive failure. The location of initiation of such failures can be predicted in terms of failure index 'e' and may be determined by using a quadratic failure criterion postulated by Raghava et al. [6] given by,

$$(\sigma_1 - \sigma_2)^2 + (\sigma_2 - \sigma_3)^2 + (\sigma_3 - \sigma_1)^2 + 2(|Y_c| - Y_t)(\sigma_1 + \sigma_2 + \sigma_3) = 2|Y_c|Y_t e \qquad ...(1)$$

where, σ_1, σ_2 and σ_3 are principal stresses causing yield and $Y_c = 84.5$ MPa and $Y_t = 65$ MPA are the absolute values of uniaxial compressive and tensile yield strengths of the adhesive material, respectively. It should be noted that, the above paraboloidal yield criterion reduces to more familiar von Mises cylindrical criterion when $Y_c = Y_t$. Thus, using the Eq. (1), the value of failure index 'e' can be determined and the variation of 'e' at various points in the adhesive layer of the joint can be plotted. The higher value of 'e' indicates the possible location of failure initiation.

4. NUMERICAL RESULTS AND DISCUSSION

Experiments have proved that, a wide variety of composite materials are used in many application according to the need based on their individual specific strength and specific stiffness values. However, the judicious use of a particular material and a particular boundary condition in an adhesively bonded single lap joint is recommended based on the present analysis. In this paper, the out-of plane normal and shear stress distribution in the adhesive layer is consodered only, since they are the key factors responsible for the cohesive failure which occurs in the adhesive layer only.

Figures 3 and 4 depict the variation of out-of-plane normal stress (σ_{zz}) known to be peel stress distribution with different boundary conditions (type 1 and type 2 as explained earlier) for various

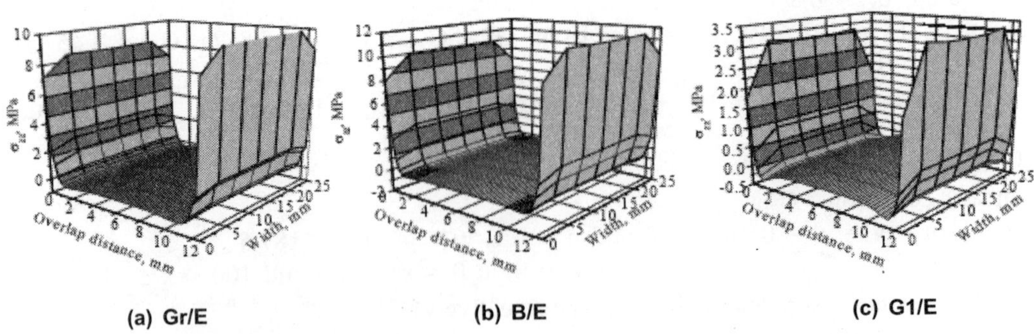

(a) Gr/E   (b) B/E   (c) G1/E

Fig. 3. Peel stress (σ_{zz}) distribution in the adhesive layer for the boundary condition of type 1.

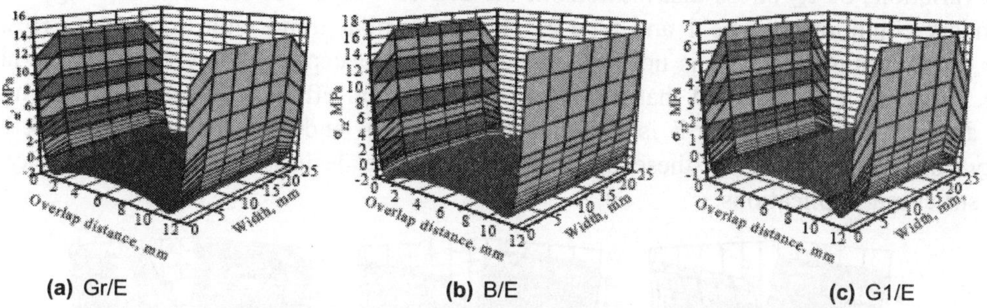

(a) Gr/E   (b) B/E   (c) G1/E

Fig. 4. Peel stress (σ_{zz}) distribution in the adhesive layer for the boundary condition of type 2.

orthotropic materials viz., Gr/E, B/E and Gl/E, respectively. It is observed that the maximum value of σ_{zz} is higher with boundary condition type 2. This increase is about 70% more for Gr/E and B/E, Whereas for G1E it is 200% higher. Further, it may be noted that, the maximum value occurs away from the edge as has been oberved by Adam *et al.* [1, 2]. This observation emphasizes why a 2D analysis is not sufficient for bonded joint analysis. Except at the free edge the σ_{zz} value is almost zero for the central portion of the overlap region of the joint. It indicates that there is stress concentration effect and this effect increases towards the central portion of the joint along the free edge due to many obvious factors such as material discontinuities, loading eccentricity and geometry discontinuities, etc. Thus, the peel stress is highly dependent on the material and type of boundary conditions.

The out-of-plane shear stress (τ_{yz}) distribution is shown in Figs. 5 and 6 for boundary condition of type 1 and type 2, respectively. The τ_{yz} values are insignificant except at the free edge corners of the joint, this shows the influence of τ_{yz} on joint performance and so on the failure are not much irrespective of the material and boundary conditions.

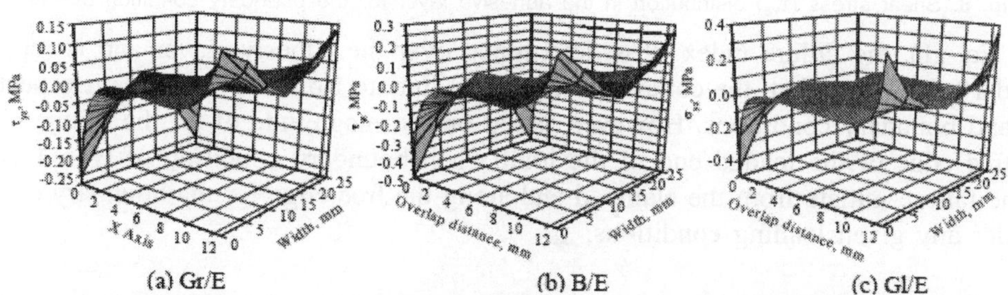

(a) Gr/E   (b) B/E   (c) Gl/E

Fig. 5. Shear stress (τ_{yz}) distribution in the adhesive layer for the boundary condition of type 1.

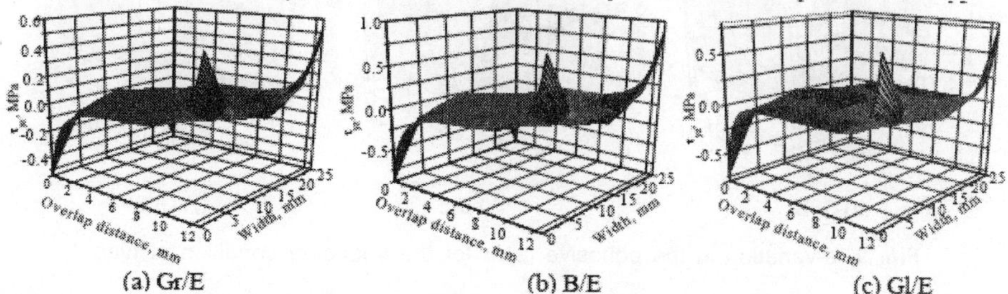

(a) Gr/E   (b) B/E   (c) Gl/E

Fig. 6. Shear (σ_{yz}) distribution in the adhesive layer for the boundary condition of type 2.

The variations of τ_{xz} stress distributions in the adhesive layer are illustrated in Figs. 7 and 8 for the boundary condition of type 1 and type 2. There is similar trend with that of σ_{zz}. However, the increase of maximum value τ_{xz} is not as significant as σ_{zz} except for the Gr/E which is about 30% increase. Further, it is observed that the stress concentration effect is insignificant for most of the overlap area for Gl/E, whereas it is not the case for Gr/E and B/E materials irrespective of the boundary conditions. Based on these observations, suitable design recommendation may be made for any structural applications.

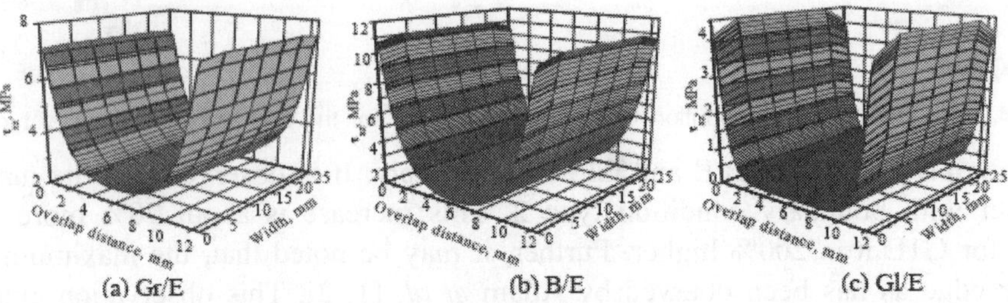

Fig. 7. Shear stress (τ_{yz}) distribution in the adhesive layer for the boundary condition of type 1.

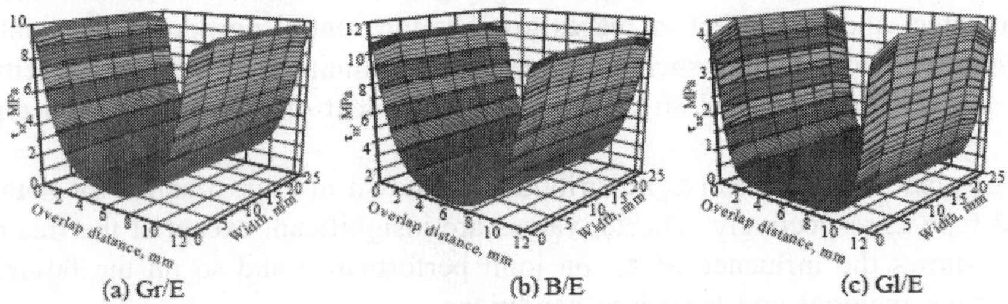

Fig. 8. Shear stress (τ_{yz}) distribution in the adhesive layer for the boundary condition of type 2.

Using Eq. (1), the failure index 'e' is determined over the adhesive layer and is represented in Figs. 9 and 10. It is seen that, the cohesive failure will initiate from the free ends irrespective of the materials and boundary conditions. However, based on the magnitude of 'e', the failure will initiate from the free edge of the loaded end of the joint when boundary condition of type 1 is imposed, whereas the failure initiate from the clamped end along the free edge for the boundary conditionn of type 2 under any given loading conditions.

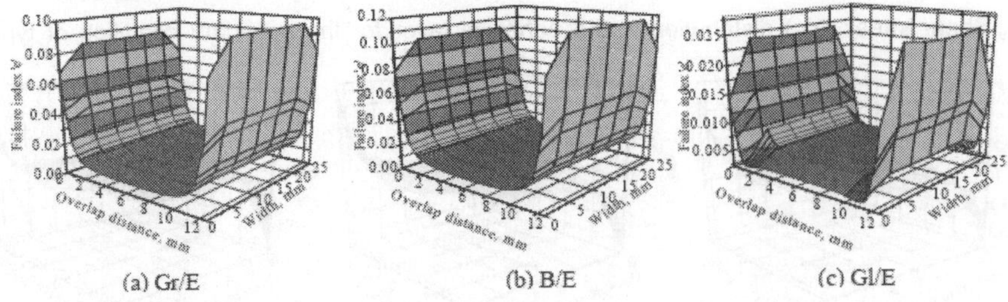

Fig. 9. e-variation in the adhesive layer for the boundary condition of type 1.

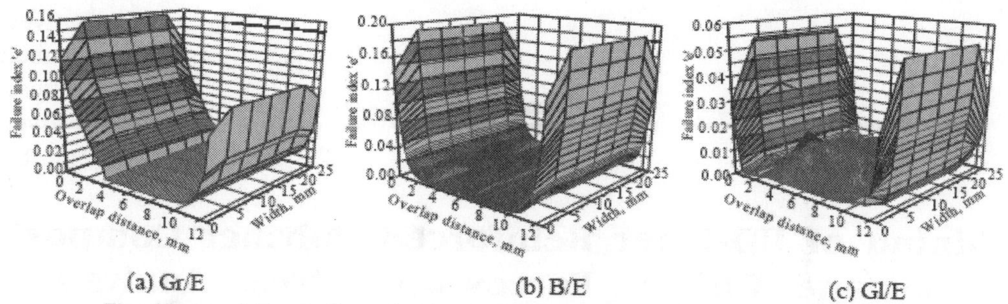

(a) Gr/E (b) B/E (c) Gl/E

Fig. 10. e-variation in the adhesive layer for the boundary condition of type 2.

CONCLUSION

Non-linear finite element analysis of the adhesively bonded single lap joint (SLJ) is performed. The effects of various orthotropic material and boundary conditions on the out-of-plane stress distribution over the adhesive layer are studied and presented. The location of the cohesive failure of the SLJ is predicted using the strength of material approach. Load carrying capacity of the joint are also predicted. Based on the present study, the following conclusions are made:

- Three-dimensional effects are significant for the out-of-plane shear stress component (τ_{yz}). The other two components, σ_{zz} and τ_{yz} are not pronounced irrespective of materials and boundary conditions. The σ_{zz} and τ_{yz} stress components are maximum not at the free edge but away from the free edge towards the central portion of the joint. The same trend is also observed in the earlier works [7, 8].
- The stress distributions and initiation of cohesive failure are largely dependent upon the type of materials and boundary conditions.
- For conservative design of the SLJ, the boundary condition of type 2 should be imposed.
- The cohesive failure will initiate from the loaded end along the free edge of the adhesive layer for the boundary condition of type 1, whereas the failure will initiate from the clamped end when boundary condition of type 2 is imposed.
- The failure index 'e' for Glass/Epoxy is lowest and for Boron/Epoxy it is highest.

REFERENCES

1. A.D. Crocombe, R.D. Adams, 1981, An Effective Stress/Strain Concept in Mechanical Characterization of Structural Adhesive Bonding, Jou. of Adhesion, 13, 141-155.
2. J.A. Harris, R.D. Adams, 1984, Strength Prediction of Bonded Single Lap Joints by Non-linear Finite Element Methods, Int. Jou. of Adhesion and Adhesives, 4, 65-78.
3. M.Y.Tsai, D.W. Oplinger, J.Morton, 1998, Improved Theoretical Solutions for Adhesive Lap Joints, Int. Jou. of Solids Structures, 35, 1163-1185.
4. L.Tong, G.P. Steven, 1999, Analysis and Design of Structural Bonded Joints, Kluwer Academic Publishers.
5. W.C. Carpenter, 1991, A Comparison of Numerous Lap Joint Theories for Adhesively Bonded Joints, Jou. of Adhesion, 35, 55-73.
6. R.S. Raghava, R.M. Cadell, G.S. Yeh, 1973, The Macroscopic Yield Behaviour of Polymers, Jou. of Material Science, 8, 225-232.
7. S.K. Panigrahi, B. Pradhan, 2006, Three-dimensional Failure Analysis and Damage Propagation Behaviour of Adhesively Bonded Single Lap Joints in Laminated FRP Composites, Jou. of Reinforced Plastics and Composites (In Press).
8. M.Y. Tsai, J.Morton, 1994, Three-dimensional Deformations in a Single Lap Joint, Jou. of Strain Analysis, 29, 137-145.

56

Machining of UD-Fiber Reinforced Polymer Composites: Damage, Cutting Forces and Stress Analysis

G. VENU GOPALA RAO, PUNEET MAHAJAN AND NARESH BHATNAGAR

Indian Institute of Technology Delhi, Hauz Khas-110016, New Delhi, India.
email: mahajan@am.iitd.ernet.in

ABSTRACT

The spectrum of present study was simulation of machining response of unidirectional carbon fiber reinforced polymer (UD-CFRP) composites using finite element methods (FEM). A new methodology was developed based on the contact pressure and frictional shear between the cutting tool and work material. This methodology has been used to predict the machining response for various fiber orientation and tool geometry. Although the maximum normal contact pressures did not vary much with fiber orientation (1350 MPa, 1435 MPa, and 1484 MPa for 0.1 mm, 0.15 mm and 0.2 mm depth of cut respectively) the normal force increases with fiber orientation and was largest for 90°. The maximum contact frictional shear stress and shear force increases with fiber orientation. The fiber failure in 90° orientation starts with crushing at the tool-fiber interface, subsequently tensile stresses due to bending at the fiber matrix interface become large enough and fiber failure starts from that interface leading to a combination of crushing and bending which causes the fiber failure.

Keywords: Damage, Cohesive Element, Composites, Contact Pressure, Machining.

1. INTRODUCTION

Carbon Fiber Reinforced Polymer (CFRP) composites are widely used in various applications such as aircraft parts, automobiles, sports goods and other industries. Most of the CFRP products are made to final component level; however, post-production removal of excess material by means of machining is often carried out to meet dimensional requirements and assembly needs. Machining of CFRP products is difficult due to their material discontinuity, inhomogeneity and anisotropic nature. Compared to the machining of metals, studies on machining of composites are few. An experimental work on cutting of UD-CFRP composites was presented by Koplev *et al.* [1]. The machining characteristics were considered for the parallel and perpendicular fiber direction only and results were presented for chip size, cutting forces variations with tool geometry. Bhatnagar *et al.* [2]

ascribed fiber breakage due to axial tension as the cutting mechanism. A review of the machining of composite materials was presented by Gordon *et al.* [3]. It did included different kinds of machining applications as used for different composites and also summarized various experimental, analytical and numerical works. However, very little literature is available regarding the numerical analysis of machining of UD-FRP composites as indicated by Arola *et al.* [4, 5]. The numerical models cited earlier used an Equivalent Homogeneous Material (EHM) for modeling of orthogonal machining operation and this was probably the prime source of deviation between the experimental and numerical results, especially the thrust force.

The objective of present study was to simulate the cutting forces during the machining of UD-CFRP composites using FEM where the fiber and matrix are separately modeled rather than as an EHM and also the interfacial debonding, matrix damage and fiber failure are taken into account while calculating the cutting forces. Both experimental and numerical study is performed to determine the cutting forces (F_h, F_v) and their variation with depth of cut (t), tool rake angle (γ) and fiber orientation (θ) varying between 15° and 90°.

2. EXPERIMENTAL PROCEDURE AND THEIR OBSERVATIONS

The test specimens for machining are prepared by lay-up procedure with desired fiber orientation (15°, 30°, 45°, 60°, 75° and 90°). The fiber orientations are defined in clockwise with reference to the cutting direction as shown in Figure 1. The cutting tool is made of Tungsten Carbide material with different rake angle (5°, 10° and 15°) at a constant edge radius (50 μm) and relief angle (6°). The machining of specimens was carried out on a CNC machine. The signals of the cutting forces were picked up and recorded @ 100 Hz by a Piezoelectric Dynamometer. Figure 2. shows a typical plot between the unit cutting force and time for (a)θ = 90°, (b)θ = 60° @ γ = 15° and t = 0.2 mm.

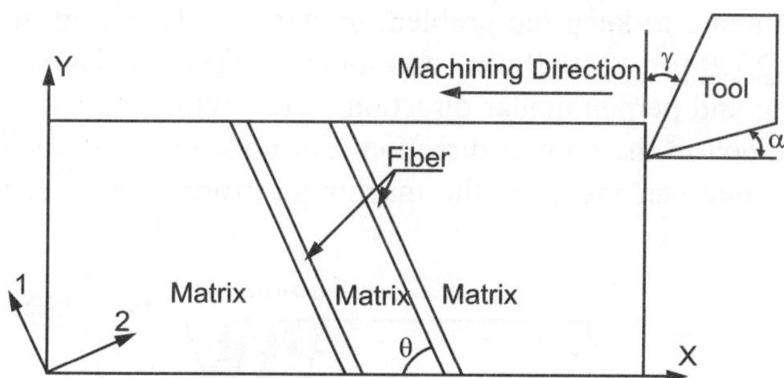

Fig. 1. Notation for the fiber orientation with respect to the cutting direction
(x-y: geometry axis; 1-2: material axis).

The frequency of variation of cutting forces is an inherent characteristic of any continuous fiber reinforced composites machining due to repeated fiber, matrix failure to form chips. For comparison with simulation results, the results on an average of the cutting forces were calculated.

3. IMPLEMENTATION OF TWO-PHASE MICROMECHANICAL FINITE ELEMENT MODEL

Number of process and tool geometry parameters plays an important role in machining of composites. Besides, non-homogeneous and anisotropic nature of the material and machining direction with respect to fiber orientation are additional variables. Therefore, it is quite exhaustive, time consuming

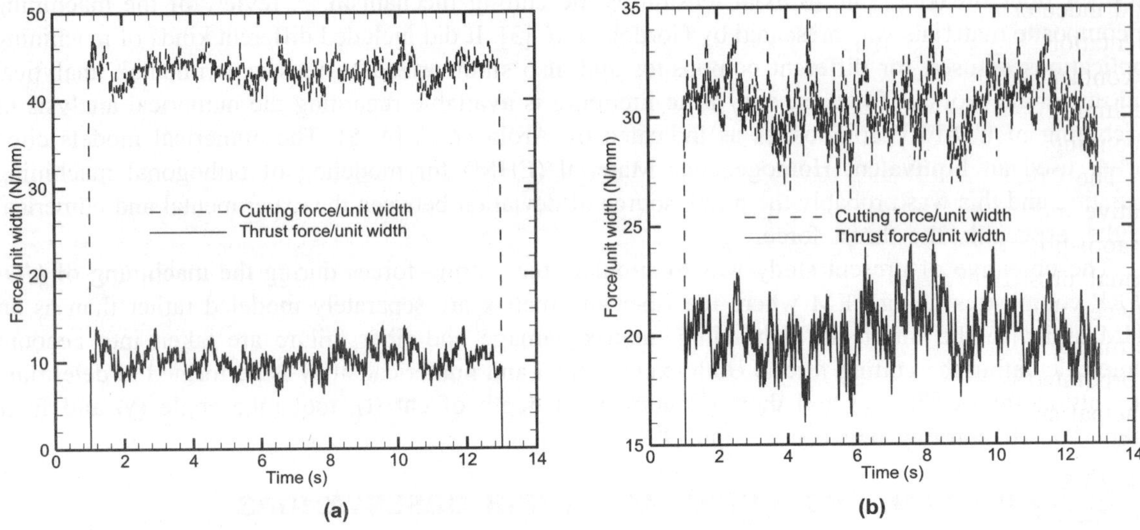

(a)

(b)

Fig. 2. Variation of cutting forces with time (a) θ = 90°, (b) 60° @ γ = 15° and t = 0.2 mm.

and un-economical to determine the desired machining response by experiments alone. FEM is widely used to predict the machining response of metals and the same technique has lately been extended to composites. Commercially available finite element code ABAQUS [6] with plane strain and quasi static option was used for the present study. The material was modeled at micro scale, as a layered material considering fibers and matrix as separate phases rather than as an EHM. The mechanical properties of the present material system used for FE simulation are taken from literature [7-9]. In the numerical model, to keep the problem tractable, only region of the work material close to chip formation zone (2000 μm × 1000 μm) was modeled. The displacements of the bottom of the workpiece in both cutting and perpendicular direction were restrained. The displacements of extreme left side were also restrained in the cutting direction. For understanding the origin and magnitude of cutting and thrust forces one can visualize the machining problem as a contact problem between a

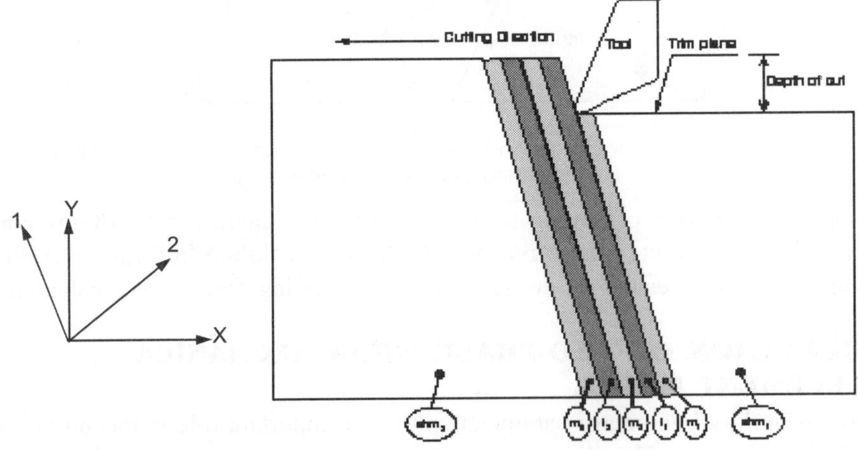

Fig. 3. Schematic view of FE model @ θ = 75°, γ =10° and t = 0.1 mm.

rigid indenter with an edge radius and a layered media as shown in Figure 3 for a case of 75° fiber orientation. In consists of two fibers, three layer of matrix and two EHM. The tool is assumed to be in contact with the first fiber after the first interface of the matrix has been removed by earlier machining. As the tool advances towards the workpiece, there is a possibility of fiber-matrix debonding, matrix cracking and fiber breaking.

The Coulomb friction law option is applied to the contact pair between the fiber and tool and relative motion (slip) between tool-fiber occurs at the contact point when fiber shear stress along the tool-fiber interface 'τ' is more than or equal to the critical friction stress 'μp' where 'p' is the normal pressure at the same point. For the present micro mechanics, the coefficient of friction between the tool-fiber interfaces was taken as a constant 0.3. The damage initiation and evolution in the matrix was modeled based on the yield and ultimate strength of the material respectively. It was incorporated through user subroutine. The failure of the fiber was considered based on the maximum principal stress. The interface failure was taken into account by cohesive zone model.

3.1 Fracture Mechanics Approach for Fiber-Matrix Interface Failure

The interface between the fiber and matrix is simulated by using Cohesive Zone Model (CZM). It is a fracture mechanics approach to study the interfacial effects of either dissimilar material or in the same material when these are initially bonded together as indicated by Xu *et al.* [10]. In the present study, zero thickness cohesive surface elements are introduced between boundaries of elements, at the interface in a normal finite element mesh and these are characterized by a stress-opening displacement potential function (φ). The potential function, used for these elements, allows for both tangential as well as normal separation. The potential function $\phi(\overline{\Delta})$ for the case where the works of separation in normal direction (ϕ_n) and tangential direction (ϕ_t) are equal is written as:

$$\phi(\Delta) = \phi_n - \phi_n \exp\left(-\frac{\Delta_n}{\delta_n}\right)\left\{\left[1+\frac{\Delta_n}{\delta_n}\right]\exp\left(-\frac{\Delta_t^2}{\delta_t^2}\right)\right\} \quad ...(1)$$

Here, δ_n and δ_t are the normal and tangential interface characteristic lengths. The interfacial traction vector $(\overline{T})$ is determined from potential function (φ), using the relation

$$\overline{T} = \frac{\partial \phi}{\partial \overline{\Delta}} \quad ...(2)$$

where, $(\overline{\Delta})$ is a displacement jump vector across the surface. As the interfacial surface separates, the magnitude of the traction at first increases to a maximum and then approaches zero. All through the FE simulation, the solution convergence is very sensitive to the size of cohesive elements at the interface and a number of FE models with different element sizes were tested before finally deciding the element size of 1 μm × 1 μm at the interface. Four noded quadrilateral elements were used throughout the FE model. The cohesive surface elements were implemented numerically through user subroutine.

4. RESULTS AND DISCUSSION

4.1 Contact Pressure and Frictional Shear Stresses

The contact pressure and frictional shear stress between the first fiber and the cutting tool for the given tool displacements resulting in the breaking of the fiber are calculated by FE simulation for a range of fiber orientations, depths of cut and rake angles. The initial contact point along the fiber from the cutting plane varies with 'θ' and 'r' of the tool is given by length (L = r[Tan(θ/2)]). The contact pressures and frictional shear contributes to the chip release from work material and sum of their components normal to and along the cutting direction provide the thrust and the cutting forces respectively. The maximum normal contact pressure did not vary much with fiber orientation and was found constant at 1350 MPa. In the contact region, the fibers experience compression in 'x' and 'y' directions. Carbon fibers have a lower compression modulus (as compared to tension) and this has been considered in the calculations. The contact lengths, for various fiber orientations are found different it decreases from 13 μm to 10 μm as the fiber orientation decreases from 90° to 15°.

4.2 Validation of Cutting and Thrust Forces

The experimental observation of cutting force variations with different fiber orientation, rake angle and depth of cut is studied during the machining of UD-CFRP composites. The cutting force mainly depends on 'θ' and 't' but is less affected by 'γ' for this set of process parameters and material system. The principle cutting force increases significantly with increasing fiber orientation for all the depths of cut. Cutting forces (F_h and F_v) predicted by FE simulations matched well with the experimental results as shown in Figure 4 (a, b), although they tend to be slightly higher than the experimental values for fiber orientations greater than 45°. As the cutting tool was modeled as a rigid as well as deformable body, the magnitude of the cutting forces is not varying with type of

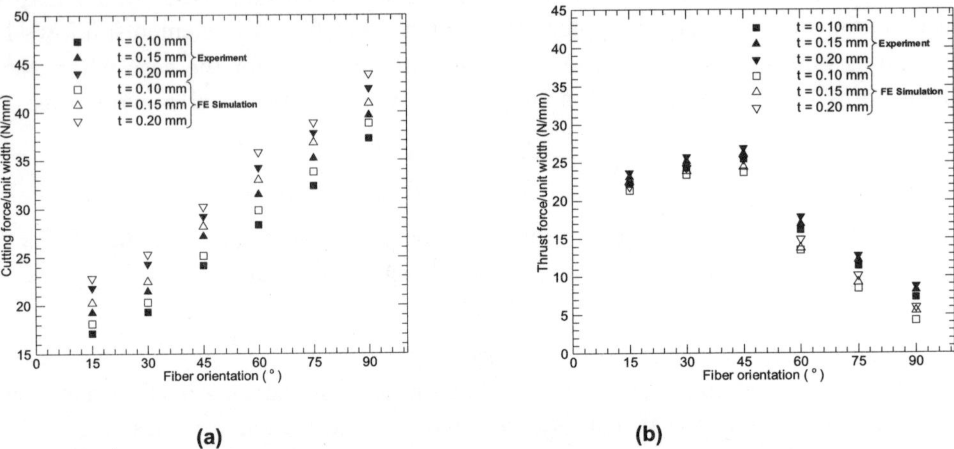

Fig. 4. Validation of cutting forces from the FE Simulation and Experimental observations, (a) Cutting force/unit width, (b) Thrust force/unit width at γ = 10°.

tool and also the distribution of stress is almost the same. For the case of 45° and 75° fiber orientation; the contours of maximum principal stress shown in Figure 5 (a, b). Here, the figures include both the rigid and deformable tool analysis. The variation of these stress contours is insignificant with fiber orientation and also the cutting forces are not varying with type of the cutting tool. This is due to the cutting tool having higher stiffness than the carbon fiber. However, it is advisable to model the cutting tool as a rigid body as it reduces the computational time. The stress distribution at interface between the fiber and cutting tool was compressive and rear side of the first fiber was tensile in nature which contributes to fracture the fiber from the rear side.

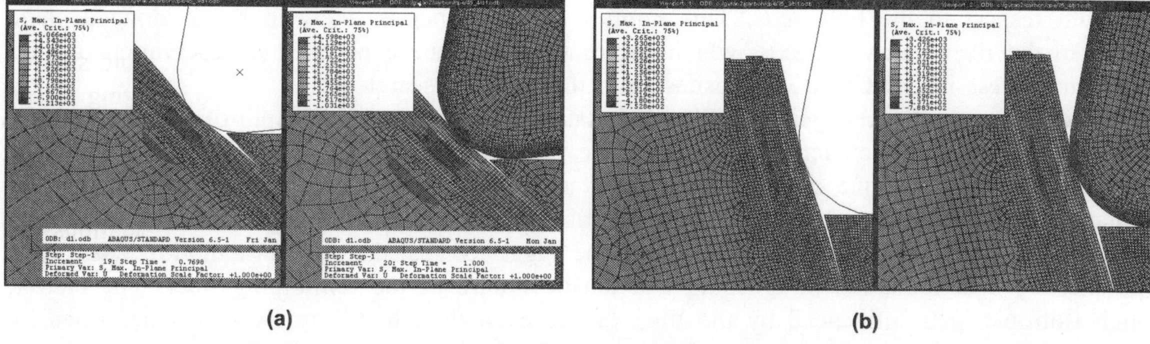

Fig. 5. Contours show the maximum principal stress for the case of (a) 45° and (b) 75° fiber orientations, it includes both the rigid and deformable tool analysis.

The variation of magnitude and direction of resultant forces from the experiment with respect to the fiber orientation and depth of cut are also studied. The magnitude of the resultant force is increasing with fiber orientation and depth of cut. The direction of resultant cutting force is not varying much with depth of cut, but it is observed decreasing with fiber orientation. This is attributed to higher fiber orientation (90°), the direction of resultant force is approximately 10° with respect to the cutting direction and the fiber is almost failing perpendicular to its axis. For the lower fiber orientation (15°), the direction of resultant force is approximately 50° with respect to the cutting direction. This could be one of the possible reasons for the fiber breakage to be observed inclined to its axis as compared to 90° fiber orientation, where the fiber breakage is perpendicular to the axis.

4.3 Mechanism of Chip Formation Process

The maximum compressive principal stress in the fiber was 1130 MPa and 780 MPa in 45° and 75° fiber orientation respectively are observed on the front side of the fiber and on the rear side of the fiber the maximum tensile principal stresses 5000 MPa and 3426 MPa in 45° and 75°, respectively. These maximum principal stress distributions in the fiber are approximately 20 μm below the point where tool touches the fiber. But below this region these stresses in the fiber become compressive in nature. Similar trends are observed for other fiber orientations also. The stress distribution provides an indication as to where the fracture could initiate. However, it is very difficult to determine the actual crack path through the fiber. In absence of any experimental data on fiber fracture initiation and propagation during indentation it is presumed that fiber failure occurs due to maximum compressive principal stress falling below the fiber compressive strength and due to maximum principal stress exceeding the tensile strength of the fiber. The first failure corresponds to fiber crushing, whereas, the second failure corresponds to a tensile crack. As the tool moves towards the fiber, the induced maximum tensile principal stresses on the fiber (rear face) are high enough to cause failure of the fiber. The crack on the rear face of the fiber grows forward but may experience shielding due to the compressive stresses in the contact region. At the same time the maximum principal stresses in the fiber (front face) is also tensile just away from and on both sides of the contact region. However, these stresses are less than the failure strength of the fiber. It can therefore be concluded that a combination of crushing and bending causes the fiber failure. These fiber failures subsequently lead to the formation of a blocky chip.

The direction of the maximum principal stresses in the failed elements varies between $-15°$ to $+20°$ with respect to the fiber axis. This implies that the fiber is likely to break along a plane parallel to the fiber transverse axis, which agrees well with the experimental observation of chip formation process in UD-CFRP composites from the earlier research [2].

The chip length also changes with fiber orientation even for the same depth of cut; this may be due to change of initial contact point between the fiber and the cutting tool edge. For the 50 µm edge radius cutting tool, the initial contact point is 50 µm above the cutting plane for 90° fiber orientations, whereas it is only 6.58 µm for 15° fiber orientation. This initial contact point is also varying with other fiber orientation. Thus, it can be suggested that the chip formation mechanism is dominated by a combination of crushing and tensile failure of the reinforcing fiber. The matrix, though isotropic, gets influenced by the fiber failure even though it may try to get deformed by shear mode. This phenomenon is true for all the depth of cut and rake angle that were used during the present study.

CONCLUSION

This paper developed a new methodology based on the contact pressure and frictional shear between the cutting tool and work material during the orthogonal machining of UD-CFRP composites. This methodology has been used to predict the machining response for a range of fiber orientation and tool geometry. These observations provide a better understanding of the origin of cutting forces. It provides a good agreement with the experimental cutting and thrust forces for the material system investigated. This study shows the possible location of failure in the fiber, damage initiation and evolution in matrix material and interfacial debonding between the fiber and matrix. The induced contact pressure varies mainly with fiber orientation and depth of cut and is less effected rake angle. Both bending and crushing are predominant mode of fiber fracture for the higher fiber orientation whereas bending was the predominant mode for the lower fiber orientation. The chip formation mechanism is dominated by a combination of crushing and tensile failure of the reinforcing fiber.

REFERENCES

1. A. Koplev, A. Lystrup, and T. Vorm, 1983, The cutting process, chips and cutting forces in machining CFRP, Composites, 14(4): 371-376.
2. N. Bhatnagar, N. Ramakrishnan, N.K. Naik, and R. Komanduri, 1995, On the machining of fiber reinforced plastic (FRP) composite laminates, International Journal of Machine Tools and Manufactures, 35(5): 701-716.
3. S. Gordon, and M.T. Hillery, 2003, A review of the cutting of composite materials, Proc. Instn Mech. Engrs Journal of Materials: Design and Applications, 217(Part L): 35-45.
4. D. Arola, and M. Ramulu, 1997, Orthogonal cutting of fiber-reinforced composites: A finite element analysis, International Journal of Mechanical Science, 39(5): 597-613.
5. D. Arola, M.B. Sultan, and M. Ramulu, 2002, Finite element modeling of edge trimming fiber reinforced plastics, Transactions of ASME Journal of Manufacturing Science and Engineering, 124: 32-41.
6. K.A.S. Hibbit, 2005, Theory and User manuals Version 6.5, ABAQUS Inc USA.
7. Thomas, H., Bodo, F., Masaki, H., Shojiro, O., and Karl, S., 2005, Microscopic yielding of CF/epoxy composites and the effect on the formation of thermal residual stresses, Composites Science and Technology, 65: 1626-1635.
8. V.V. Kozey, Hao Jiang, V.R. Mehta, and S. Kumar, 1995, Compressive behavior of materials: Part-II.High performance fibers, Journal of Materials Res., 10(4): 1044-1061.
9. N. Oya, D.J. Johnson, 2001, Longitudinal compressive behavior and microstructure of PAN-based carbon fibers, Carbon, 39, 635-645.
10. X.P. Xu, and A. Needleman, 1994, Numerical Simulations of fast crack growth in brittle solids, Journal of Mechanical Physics of Solids, 42(9): 1397-1434.

57

Assessment of Damage Mode in Composite Material by S-Link Clustering

S. SAMANTA[1], S. GOEL[1] AND G. GOEL[2]

[1]Department of Mechanical Engineering, NERIST (Deemed University), Nirjuli, A.P., India.
email: suta_sama@yahoo.co.in
[2]Department of Civil Engineering, NERIST (Deemed University), Nirjuli, A.P., India.
email: goel2001@email.com

ABSTRACT

In this work, an automated ultrasonic data acquisition system was developed by integrating the PCUS11 ultrasonic board, the QUT99 data acquisition software and a stepper motor controlled C-Scan tank. Longitudinal transducers are employed to perform C-Scan on composite specimens in normal beam through transmission modes. Immersion tests are performed in which QUT99 data acquisition software stored the digitised A scan data in the controlling computer. Ultrasonic C-Scan was performed on flawed glass-epoxy specimen. A Windows compatible software has been developed in the Visual basic platform for automated C-Scan image generation from acquired ultrasonic data by employing S-Link clustering analysis. C-Scan image generated by the developed software and S-Link clustering method, clearly identify the impact damage. However, the resin rich zone could not be detected.

Keywords: S-Link, C-Scan, PCUS11 board.

1. INTRODUCTION

During the past 40 years, materials design has shifted emphasis to pursue lightweight, environment friendliness, low cost, quality and performance. Due to this, applications of composites are growing rapidly in industries. However, the growth in usage also demands for reliable inspection procedures to assess the material's integrity during service life.

Non-destructive testing and evaluation through generation of C-Scan images is a well-known practice. A 'C-Scan' is essentially a plot depicting the variation of certain ultrasonic feature or feature set in two dimensions with respect to the probe locations on the surface of the examined specimen Thus, such techniques are particularly useful for characterization of planar defects, i.e. the defects that are parallel to the plane of scanning. The researchers worldwide have realized the potential of the research endeavour in this field and have starting putting in substantial inputs. The last decade has seen tremendous developments in this field.

Henneke[1] documented a comprehensive review article on ultrasonic NDE of advanced composites. Preuss and Clark [2] used the time of flight of ultrasonic C-Scanning for the detection, sizing and characterization of defects in carbon-fibre composite components. They developed a low cost ultrasonic scanning system, capable of determining the layer-by-layer structure of composite with impact damage and presenting the data in a convenient form. Udaya B. Haiabe and Reynold Franklin [3] developed a conventional method of ultrasonic testing utilizing time domain measurement that is seldom capable of detecting micro-cracks. Hence, there is need to conduct amplitude measurement and frequency domain analysis to enhance the sensitivity of the ultrasonic technique. They presented a methodology to optimise several factors to produce repeatable signals with good signal to noise ratio. Datta *et al.* [4] performed ultrasonic C-Scan in normal beam pulse echo and through transmission mode on laminates with impact and inclusion type damages. Frequency domain features were used, individually or in combination, for generation of C-Scan images. Hierarchical clustering using ward's algorithm was implemented for classification of data set.

In this present investigation an effort has been made to generate a C-Scan image for glass epoxy composite specimen with known flaw. With the help of S-Link clustering technique the scan data set is automatically classified into desired number of groups. The automated grouping helps to generate images in a systematic manner and does not require any prior information regarding the nature and distribution of the data.

2. EXPERIMENTAL SETUP

The present setup developed in-house, is meant for automated immersion scanning of composite laminates in normal beam through transmission mode. The facility is comprised an immersion tank of acrylic glass and a mounting frame furnished with two lead screws in mutually perpendicular directions. Two stepper motors drive the lead screws and a common nut, moving linearly due to their rotation and holding the probe holding device. The transducers fitted in the probe holding device, can move along two mutually perpendicular directions in precise steps and are capable of scanning any predefined two-dimension region. The transducers are connected to an ultrasonic board that acts as the pulser, receiver and digitizer of the ultrasonic waveform. The present ultrasonic board is PCUS11 [5], which can digitise signals with a sampling rate of up to 100 MHz. The board seamlessly interacts with manufacturer supplied software that has the capability to condition, gate and zooming of the digitised signal. The composite laminate is kept immersed in water and is held strictly parallel to the plane of the movement of the transducer(s). The minimum linear movement possible for the transducer is 0.025 mm. Necessary setting to the effect may be used as input to the stepper motor controller via a wired remote control. In the present investigation, immersion type ultrasonic probes with a central frequency of 2 MHz have been used. The scan steps were kept at 2 mm along both directions and sampling rate was maintained at 40 MHz for all measurement.

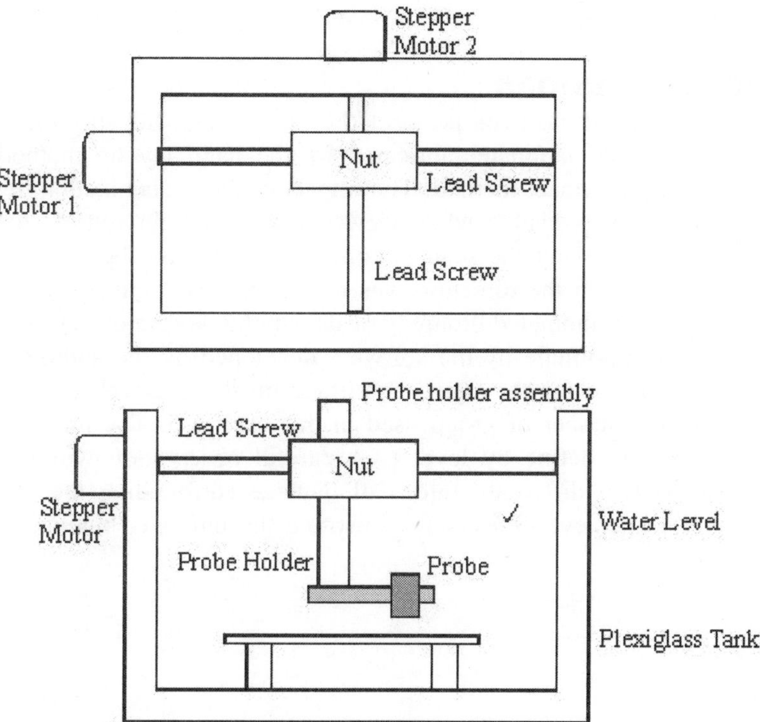

Fig. 1. Schematic Arrangement of the C-Scan Set-Up.

3. CLUSTER ANALYSIS

It has a variety of goals. All relate to grouping or segmenting a collection of objects into subsets of clusters, such that those within each cluster are more closely related to one another than objects assigned to different clusters. Central to all of the goals of cluster analysis is the motion of the degree of similarity (or dissimilarity) between the individual objects being clustered. There are numerous clustering techniques available to researchers such as:

    i.   Single linkage clustering method (S-Link)
   ii.   Complete linkage clustering method (C-Link)
  iii.   Un-weighted pair group method using arithmetic averages (UPGMAA)
  iv.   Ward's minimum variance clustering method.

   In the present investigation the S-Link methodology has been adopted for data segmentation. In this technique the defining feature is taken as the distance between the closet pair of objects, where only pairs consisting of one object from each group are considered.

4. DEVELOPMENT OF THE SOFTWARE

In the present investigation an effort has been made to develop an Ultrasonic C-Scan image generation software for automatic generation of C-Scan images for a set of data. The software has been developed in the Visual Basic platform and can work stand alone. The major advantage of the software is that it is menu driven, interactive and works seamlessly with the data files generated by PCUS 11 board and QUT data acquisition system.

5. RESULTS AND DISCUSSIONS

Immersion type ultrasonic C-Scan has been performed on two composite specimens. Specimen#1 is a 32 ply glass epoxy composite laminate fabricated by the hand lay up method and a damage created in it by the drop weight impact method. The impact creates delamination defects in the plies and the core region of the flaw is visible when viewed against a light source. A square region is identified around the defect zone and the scan is conducted in normal beam through transmission mode. Scanning resolution in both the direction was maintained as 2 mm.

In this study an analysis for automated grouping of the amplitude data objects was done pertaining to the single attribute signal amplitude by the software developed by the authors.

In (Fig. 2) the C-Scan image for the signal amplitude on the backwell echo, generated by the S–Link technique, is shown. Number of group used in this case is 5. The image clearly visualizes the core region of the damage define by level-1. A careful observation will reveal that level-2, level-3 are able to extracts two different regions of damage surrounding the core defect region. Pixels belonging to level-4 and level-5 seem to constitute the unflawed region, unaffected by the impact.

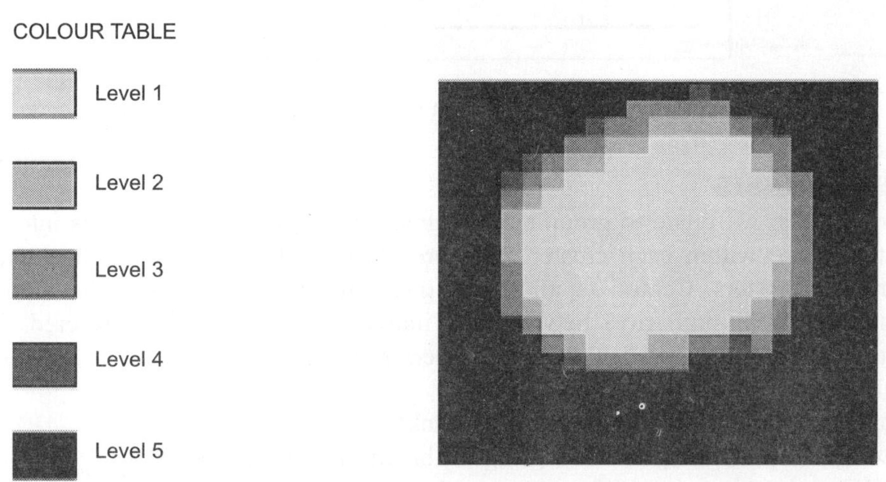

Fig. 2. Ultrasonic C-scan Image of Composite Specimen #1 using S-Link.

Specimen#2 is a 32-ply glass epoxy composite laminate that contains a resin-rich zone. The size of the resin-rich zone is 20 mm by 20 mm and it was created artificially during fabrication of the laminate. The scanned area is a 60 mm by 46 mm rectangle. 492 sampled data has been used in digitizing the waveform. The signal amplitude was extracted from each point scanned and were subjected to clustering by S-Link techniques. The image generated by the S-Link clustering method is shown in (Fig. 3). The image is not in conformity with the scanned domain as specimen. The data points belonging different clusters are found to remain scattered throughout the image and hence are not able to bring out any specific pattern to represent the flawed region. The matter can be correlated to the physical behaviour of the acoustic wave with the composite and the resin-rich zone. It has been found that from the attenuation point of view the normal composite and the resin-rich zone are not much different.

Fig. 3. Ultrasonic C-Scan Image of Composite Specimen # 2 using S-Link.

CONCLUSION

Based on the results and discussions, following conclusion can be made:

1. S-Link hierarchical algorithmic technique is found to be powerful tool for grouping object.
2. S-Link criterion is found to be a quicker yet reasonably effective procedure of grouping compared to the ward's criterion (used by Datta et. al. [4]) in classifying signal amplitude data in finding impact damages. However, the resin-rich zone could not be detected by the signal amplitude based C-Scan image.

ACKNOWLEDGEMENTS

Authors gratefully acknowledge Dr. D. Datta, Bengal Engineering and Science University, Shibpur (W.B.) for providing the facilities to perform the experimental work.

REFERENCES

1. E.G. Henneke II, 1990, Ultrasonic non-destructive evaluation of advanced composites, non-destructive testing fibre reinforced plastics composites, edited by John Summercales, 2, 55-159.
2. T.E. Preuss and G. Clark, 1998, Use of time of flight C-Scanning for assessment of impact damage, composites, 19(2), 145-148.
3. Udaya B Haiabe and Reynold Franklin (March 2001): Fatigue Crack detection in metallic members using ultrasonic Rayleigh waves with time and frequency analysis, Materials Evaluation, 59 (3), 424-431.
4. D. Datta, C.V. Venkatesh and N.N. Kishore, 1995, Application of multidimensional cluster analysis in identification of defects in fibre composites, Non-destructive testing and evaluation, 12, 197-210.
5. PCUS 11 ultrasonic P/R Board Manual, DOC # EBD003-1, Fraunhoffer Institute for non-destructive testing, Saarbrucken, Germany.

58

Hygrothermal Effects on the Initiation and Propagation of Damage in Composite Plates

Anup Ghosh and P.K. Sinha

Department of Aerospace Engineering Indian Institute of Technology, Kharagpur-721302, India
email: anup@aero.iitkgp.ernet.in

ABSTRACT

A finite element analysis procedure is developed to investigate the initiation and propagation of damage in laminated composite plates in hygrothermal environments. A Continuum Damage Mechanics (CDM) model assuming the effective stress concept is introduced with a fourth order damage tensor to model the damage of a lamina. A nine noded isoparametric plate element based on the first order shear deformation theory is used to develop the finite element analysis procedure. The initiation and progress of damage at elevated moisture concentration and temperature in simply-supported square plates due to low velocity impact are studied.

1. INTRODUCTION

Damage in composites develops continuously and grows by various mechanisms at the micro-scale (as coalescence of micro-cracks and voids, fibre debonding, matrix cracking, delamination, etc.). The continuum damage mechanics approach provides a viable framework for the description of distributed damages including material stiffness degradation as well as various forms of defects including initiation, growth and coalescence of micro-cracks and voids. An overall assessment of damage during low velocity impact provides useful data for structural design.

2. PHENOMENOLOGICAL MODEL

Considering the composite structural system as a whole continuum, it is assumed that the model will represent various types of damage mechanisms such as void growth, coalescence in the matrix and fibre fracture, debonding, delamination, etc. In this methodology no distinction is made on the type of damages, as they are all reflected through the fourth-rank overall damage effect tensor M'_{ijkl}.

Following the above considerations and utilizing an overall damage effect tensor M'_{ijkl} for a whole composite structural system, the overall effective stress tensor $\bar{\sigma}$ is given by

$$\bar{\sigma}_{ij} = M_{ijkl}\sigma_{kl} \qquad \qquad ...(1)$$

where, σ_{kl} is the Cauchy stress tensor in the damaged state.

3. CONSTITUTIVE RELATION OF A UNIDIRECTIONAL LAMINA

A unidirectional fibre-reinforced, thin lamina, as shown in Figure 1, is considered to evaluate the elements of the effective constitutive matrix $\overline{E}_{ijkl}$. For a complete representation of damage in a lamina the stress tensor and damage tensors are σ_{ij} and ϕ_{ij}, respectively.

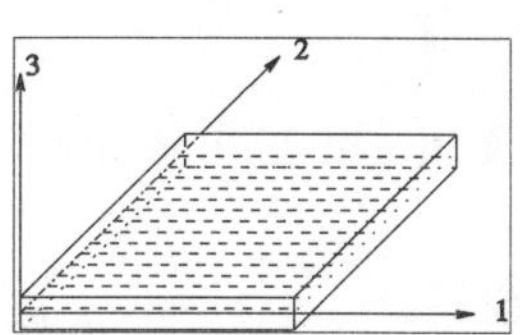

Fig. 1. Unidirectional lamina.

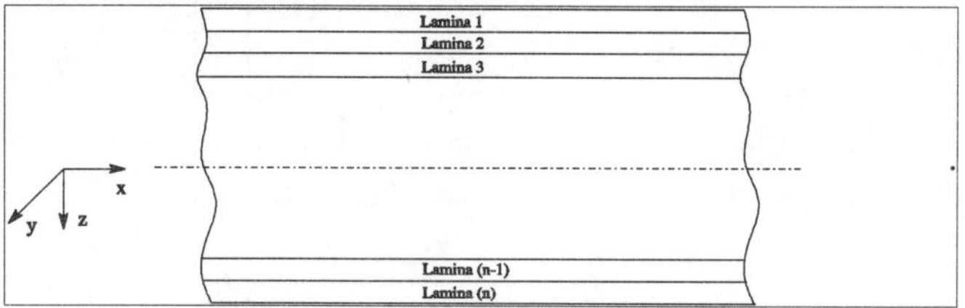

Fig. 2. Stacking sequence of a laminate.

The transformed stresses $\overline{\sigma}_{ij}$ equation (1) give rise to an asymmetric effective stress tensor. Therefore, after symmetrization and rearrangement the relation between effective stress and the damaged state stress becomes (with reduced notation):

$$\begin{Bmatrix} \overline{\sigma}_{11} \\ \overline{\sigma}_{22} \\ \overline{\sigma}_{23} \\ \overline{\sigma}_{13} \\ \overline{\sigma}_{12} \end{Bmatrix} = \begin{bmatrix} M'_{11} & M'_{12} & M'_{13} & M'_{14} & M'_{15} \\ M'_{21} & M'_{22} & M'_{23} & M'_{24} & M'_{25} \\ M'_{31} & M'_{32} & M'_{33} & M'_{34} & M'_{35} \\ M'_{41} & M'_{42} & M'_{43} & M'_{44} & M'_{45} \\ M'_{51} & M'_{52} & M'_{53} & M'_{54} & M'_{55} \end{bmatrix} \begin{Bmatrix} \sigma_{11} \\ \sigma_{22} \\ \sigma_{23} \\ \sigma_{13} \\ \sigma_{12} \end{Bmatrix} \qquad ...(2)$$

Now substituting the values of M'_{ij} in equation 3, derived on the basis of hypothesis of elastic energy equivalence,

$$\overline{E}_{ijkl} = M'^{-1}_{pqkl} E_{rspq} M'^{-T}_{rsij} \qquad ...(3)$$

and considering only principal damage variables, the relations between the engineering constants of damaged state and undamaged state, are derived as follows:

$$\overline{E}_{11} = E_{11}(1-\phi_{11})^2, \quad \overline{E}_{22} = E_{22}(1-\phi_{22})^2, \quad \overline{\nu}_{12} = \nu_{12}\frac{(1-\phi_{11})}{(1-\phi_{22})}$$

$$\overline{G}_{12} = 4 \times G_{12} \times \left[\frac{(1-\phi_{11})(1-\phi_{22})}{(1-\phi_{11})+(1-\phi_{22})}\right]^2, \quad \overline{G}_{23} = 4 \times G_{23} \times \left[\frac{(1-\phi_{22})(1-\phi_{33})}{(1-\phi_{22})+(1-\phi_{33})}\right]^2, \qquad ...(4)$$

$$\overline{G}_{13} = 4 \times G_{13} \times \left[\frac{(1-\phi_{11})(1-\phi_{33})}{(1-\phi_{11})+(1-\phi_{33})}\right]^2$$

For the unidirectional lamina (Figure 1) the stress strain relation in an undamaged state is

$$\begin{Bmatrix} \sigma_{11} \\ \sigma_{22} \\ \sigma_{12} \end{Bmatrix} = \begin{bmatrix} Q_{11} & Q_{12} & 0 \\ Q_{12} & Q_{22} & 0 \\ 0 & 0 & Q_{66} \end{bmatrix} \begin{Bmatrix} \varepsilon_{11} \\ \varepsilon_{22} \\ \varepsilon_{12} \end{Bmatrix} \qquad \text{...(5a)}$$

and,
$$\begin{Bmatrix} \sigma_{23} \\ \sigma_{13} \end{Bmatrix} = \begin{bmatrix} Q_{44} & 0 \\ 0 & Q_{55} \end{bmatrix} \begin{Bmatrix} \varepsilon_{23} \\ \varepsilon_{13} \end{Bmatrix} \qquad \text{...(5b)}$$

Here, Q_{ij} are the elements of on-axis constitutive matrix for a unidirectional lamina and these are defined in terms of engineering constants as follows:

$$Q_{11} = \frac{E_{11}}{1 - v_{12}v_{21}}, \qquad Q_{22} = \frac{E_{22}}{1 - v_{12}v_{21}}, \qquad Q_{12} = \frac{v_{12}E_{22}}{1 - v_{12}v_{21}} \qquad \text{...(6)}$$

$$Q_{66} = G_{12}, \; Q_{44} = G_{23}, \text{ and } Q_{55} = G_{13}$$

The material behavior of the damaged composite can be represented by replacing the above engineering constants defined in equation (6) with the effective ones defined in equation (4). The stresses in any lamina k can be expressed in terms of the laminate mid-surface strains, curvatures, shear rotations and hygrothermal strains of the lamina as:

$$\begin{Bmatrix} \sigma_{xx} \\ \sigma_{yy} \\ \tau_{xy} \end{Bmatrix}_k = \left[Q_{ij} \right]_k \left\{ \{\varepsilon^0\} + z\{\kappa\} \right\} - \left[Q_{ij} \right]_k \{e\}_k \qquad \text{...(7)}$$

where, i, j = 1,2,6 and {e} is the hygrothermal strain components

$$\begin{Bmatrix} \tau_{xz} \\ \tau_{yz} \end{Bmatrix}_k = \left[Q_{ij} \right]_k \begin{Bmatrix} \gamma_{xz} \\ \gamma_{yz} \end{Bmatrix} \qquad \text{...(8)}$$

where, i, j = 4, 5

The evolution of damage and its growth are assumed to depend on the local state of stresses, applied forces, prescribed boundary conditions and the damage state, if any. A convenient way of determining ϕ_i is to utilize the damage law postulated by Matzenmiller [1] for a unidirectional composite lamina. It is given as:

$$\phi = 1 - exp\left(-\frac{1}{me}\left(\frac{\varepsilon}{\varepsilon_f} \right)^m \right) \qquad \text{...(9)}$$

where the ε_f is the failure strain, and the current state of strain, ε at the initiation of failure is determined using the Tsai-Wu strength criteria.

4. RESULTS AND DISCUSSION

Verification of Results

Hygrothermal

Developed FEM code is verified with a set of hygrothermal loads on a cross-ply laminate.

Material properties at different temperatures and at different level of moisture concentrations are tabulated in Tables 1 and 2. Temperature loading from 0 K to 425 K are considered for cross-ply (0/90/0/90), simply-supported laminates. Deflections and moments at the specified points (Fig. 3) are listed in Table 3. Stress resultants due to the variation of moisture concentration are shown in Table 4. In case of rise of temperature, present FEM results matches well with closed form solutions [2]. Stress resultants at different level of moisture concentrations agrees with the results published in open literature [3].

Table 1. Elastic moduli of Graphite/Epoxy unidirectional laminate at different temperature.
$G_{13} = G_{12}$, $G_{23} = 0.5G_{12}$, $\nu_{12} = 0.3$, $\alpha_1 = -0.3 \times 10^{-6}$/K and $\alpha_2 = 2.8 \times 10^{-6}$/K.

Elastic moduli (GPa)	Temperature T (K)					
	300	325	350	375	400	425
E_1	130	130	130	130	130	130
E_2	9.5	8.5	8.0	7.5	7.0	6.75
G_{12}	6.0	6.0	5.5	5.0	4.75	4.5

Table 2. Elastic moduli of Graphite/Epoxy unidirectional laminate at different moisture concentration.
$G_{13} = G_{12}$, $G_{23} = 0.5G_{12}$, $\nu_{12} = 0.3$, $\beta_1 = 0$ and $\beta_2 = 0.44$.

Elastic moduli (GPa)	Moisture concentration C (%)						
	0	0.25	0.5	0.75	1.0	1.25	1.5
E_1	130	130	130	130	130	130	130
E_2	9.5	9.25	9.0	8.75	8.5	8.5	8.5
G_{12}	6.0	6.0	6.0	6.0	6.0	6.0	6.0

Table 3. Deflections and moments of a (0/90/0/90) simply-supported laminated composite plate (Figure 3) at temperature T = 400 K.

Bending Characteristics	Solution	Point			
		A	B	C	D
w (mm)	Closed form [2]	0.0000	0.0085	0.0267	0.0337
	Present FEM	0	0.01	0.03	0.03
M_x (N-mm)	Closed form [2]	−2.753	−2.518	−1.869	−0.966
	Present FEM	−2.770	−2.529	−1.866	−0.971
M_y (N-mm)	Closed form [2]	2.753	2.752	2.657	2.237
	Present FEM	2.770	2.756	2.641	2.238

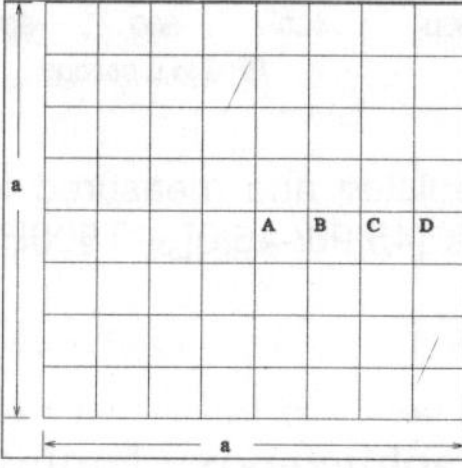

Fig. 3. Details of discretisation and points of measurement of deflections and moments.

Table 4. Stress resultant at different moisture concentrations of a (0/90/0/90)
clamped laminated composite plate (Figure 3).

Moment (N mm)	Solution	Moisture concentration C (%)						
		0	0.25	0.50	0.75	1.00	1.25	1.50
M_x	Sai Ram and Sinha[3]	0	− 0.448	− 0.872	−1.271	−1.646	− 2.057	−2.469
	Present FEM	0	− 0.447	− 0.872	−1.269	−1.644	− 2.055	− 2.466
M_y	Sai Ram and Sinha[3]	0	0.448	0.872	1.271	1.646	2.057	2.469
	Present FEM	0	0.447	0.870	1.269	1.644	2.055	2.466

5. DAMAGE PRODUCED BY IMPACT

The numerical simulation is carried out for a simply supported rectangular plate of dimension 127 mm × 76.2 mm × 4.65 mm under an impact of a steel impactor of mass 314g and initial velocity of 14.6 m/s. The properties of the lamina are furnished in Table 5. Figure 4 shows a comparison of the time-force history calculated by the present finite element method and the experimental data [4].The results are found to compare well in this case also.

Table 5. Stiffness and strength properties of T800H/3900-2 CFRP laminate.

Elastic Parameters		Strength parameters	
Longitudinal modulus, E_{11}	152.4 GPa	Longitudinal tensile, X_t	2089 MPa
Transverse modulus, E_{22}	9.2 GPa	Longitudinal compressive, X_c	1482 MPa
In-plane shear modulus, G_{12}	4.3 GPa	Transverse tensile, Y_t	79 MPa
Poisson's ratio, V_{12}	0.35	Transverse compressive, Y_c	231 MPa
Mass density	1550 kg/m^3	In-plane shear, S	133 MPa

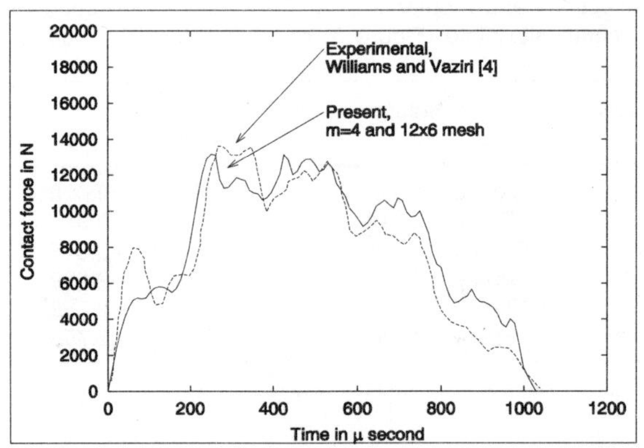

Fig. 4. Comparison of the predicted and measured force-time history for an impact (v = 14.5m/s, M = 314g) on a [45/90/-45/0]$_{3S}$ T800H/3900-2 (Table 5) CFRP plate.

6. CASE STUDIES

In the present investigation T300/934 graphite/epoxy laminates [5] of dimension 200 mm × 200 mm × 4 mm with the laminate configuration [±45]$_4$ are considered. Some of the important material properties are tabulated in Tables 6 and 7 and other material properties are as follows:

Table 6. Stiffness properties of T300/934 CFRP laminate under different temperature and moisture content.

	Temperature in K			Moisture content (%)	
	300	350	400	0.25	0.50
E_{11} (GPa)	124	124	124	124	124
E_{22} (GPa)	11.7	9.8514	8.6229	11.3929	11.0858
G_{12} (GPa)	5.5	5.0419	4.3560	5.5	5.5
G_{23} (GPa)	4.137	3.7924	3.2765	4.137	4.137

Table 7. Percentage reduction in transverse, in-plane and interlaminar limit stresses with moisture for the T300/934 composite.

	Moisture content (%)		
	0.25	0.50	0.75
X_{11}, X_{22}	7.60	15.17	16.33
X_{13}, X_{23}	3.13	7.00	11.60
X_{12}	4.47	7.57	9.60

$$G_{23} = 0.5\,G_{12}, \quad \nu_{12} = \nu_{13} = 0.3, \quad \alpha_1 = -\,0.3 \times 10^{-6}/\text{K}, \quad \alpha_2 = 2.81 \times 10^{-6}/\text{K},$$

$$\beta_1 = 0.0, \quad \beta_2 = 0.44, \quad \beta_3 = 0.44, \quad X_{11}^{t} = 1586 \text{ MPa},$$

$$X_{11}^{c} = 1517 \text{ MPa}, \; X_{22}^{t} = 51.71 \text{ MPa}, \quad X_{22}^{c} = 275.8 \text{ MPa}, \quad X_{12} = 117 \text{ MPa}$$

A steel impactor hits the plate with three different velocities, 40 m/s, 70 m/s and 100 m/s. These impacts are carried out in different hygrothermal environments. Two different moisture contents (C = 0.25% and 0.5%) and temperatures (T = 350K and 400K) are considered individually.

7. VARIATION OF MOISTURE CONCENTRATION

Propagation of damage is observed in the laminates due to the increase of impact velocity as well as due to increase of moisture concentration and temperature. In case of impact velocity of 40 m/s damage is observed in lamina 6, 7 and 8 only (Figure 5) whereas it increases to lamina 1 and lamina 2, 3 under the velocity of 70 m/s (Figure 6) and 100 m/s (Figure 7), respectively. An effect of increased moisture concentration is also observed in all the three categories of impacts. At velocity 40 m/s damage propagates in the bottom layer (lamina 8, Figure 5) and at 100 m/s it propagates in the transverse direction (lamina 2 and 3, Figure 7) due to the increase of moisture concentration.

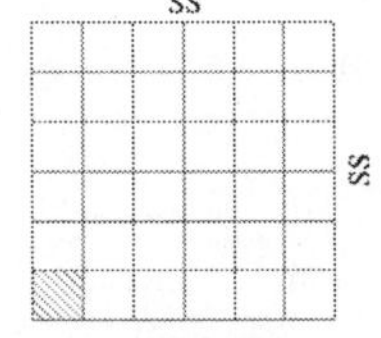

(a) C = 0% and 0.25%, lamina 6, 7 and 8.

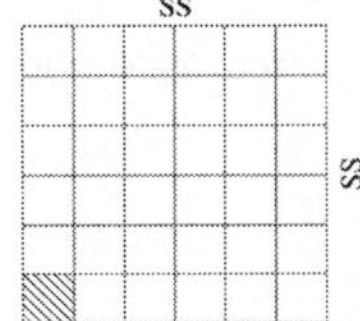

(b) C = 0.5%, lamina 6 and 7.

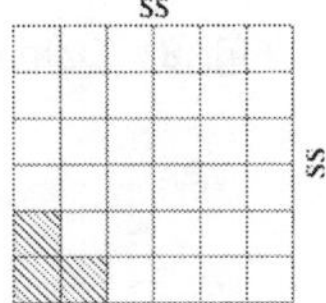

(c) C = 0.5%, lamina 8.

Fig. 5. Damage in [±45]₄ simply-supported plate under impact velocity of 40 m/s, T = 300K.

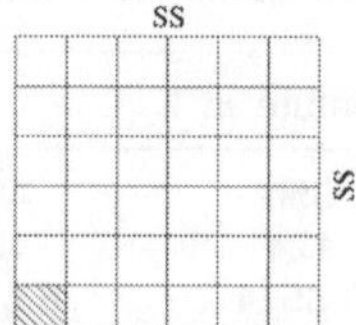

(a) C = 0% and 0.25%, lamina 6, 7 and 8.

Fig. 6. Damage in [±45]₄ simply-supported plate under impact velocity of 70 m/s, C = 0%, 0.25%, 0.50% and T = 300K.

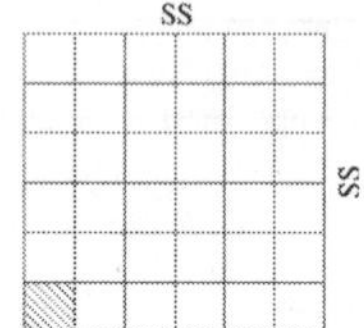

(a) C = 0%, lamina 1, 6, 7 and 8.

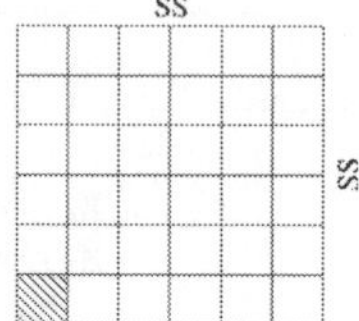

(b) C = 0.25% and 0.5%, lamina 1, 2, 3, 6, 7 and 8.

Fig. 7. Damage in [±45]₄ simply-supported plate under impact velocity of 100 m/s, T = 300K.

8. VARIATION OF TEMPERATURE

In the present investigation impacts are considered at different temperatures. It is observed that propagation of damage is less in case of impacts at elevated temperature. In case of impact velocity 40 m/s damage is observed in three bottom laminae (lamina 6, 7 and 8, Figure 8a) at a temperature of 300K whereas it only in two laminae (lamina 7 and 8, Figure 8b) at 350K and 400K. Similar phenomenon is also observed for the impacts of velocity 70 m/s (Figure 9) whereas no such influence is observed for a velocity of 100 m/s (Figure 10).

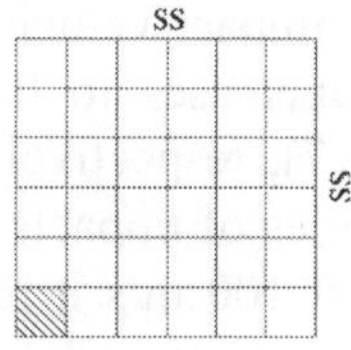

(a) T = 300K, lamina 6, 7 and 8.

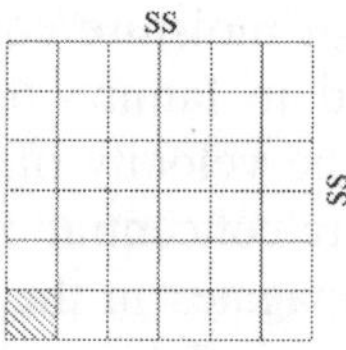

(b) T = 350K and 400K, lamina 7 and 8.

Fig. 8. Damage in [±45]₄ simply-supported plate under impact velocity of 40 m/s, C = 0%.

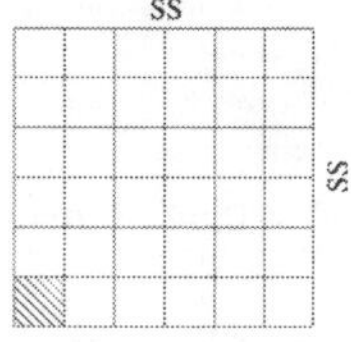

(a) T = 300K lamina 1, 6, 7 and 8.

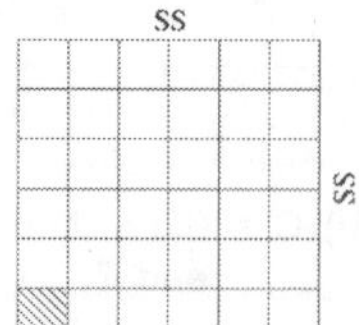

(b) T = 350K lamina 1, 7 and 8.

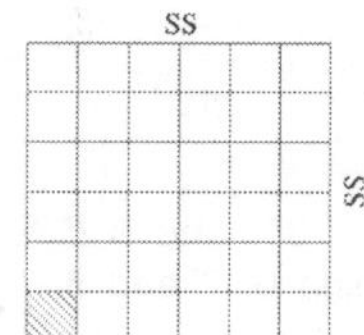

(c) T = 400K lamina 7 and 8.

Fig. 9. Damage in [±45]₄ simply-supported plate under impact velocity of 70 m/s, C = 0%.

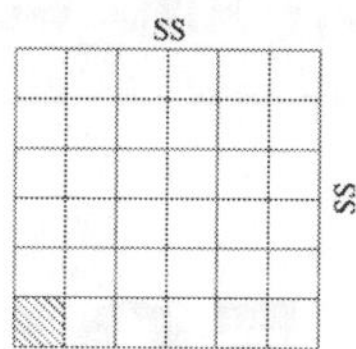

(a) lamina 1, 6, 7 and 8.

Fig. 10. Damage in [±45]₄ simply-supported plate under impact velocity of 100 m/s,
C = 0%, T = 300K, T = 350K and T = 400K.

CONCLUSION

A finite element analysis procedure simulating the damage condition in laminated composite plates is developed using a continuum damage mechanics approach. The present numerical results compare well with the experimental results. The initiation and progress of damage in laminated composite plates under different impact velocities along with different levels of moisture content and elevated temperature levels are investigated. It is interesting to note the variation of lamina damages through the laminate thickness under different impact velocities and hygrothermal loading conditions. The increase of velocity of impactor not only increases the contact force on the plate but also enhances the damaged area in different laminae of the plates. It is also observed that damaged area increases due to the increase of moisture content in the plates. It is noticed that the effect of increase of temperature on the initiation and progress of damage is dependent on the velocity of the impactor. The present study also indicates that the damage analysis based on the present CDM approach can be an effective way of predicting the extent of damages caused due to impact on laminated composite plates.

REFERENCES

1. A. Matzenmiller, J. Lubliner, and R.L. Taylor, "A constitutive model for anisotropic damage in fiber-composites," *Mechanics of Material*, vol. 20, pp. 125-152, 1995.
2. C. H. Wu and T.R. Tauchert, "Thermoelastic analysis of laminated plates," *Journal of Thermal Stresses*, vol. 3, pp. 365-378, 1980.
3. K.S. Sai Ram and P.K. Sinha, "Hygrothermal effects on the bending characteristics of laminated composite plates," *Computers & Structures*, vol. 40, no. 4, pp. 1009-1015, 1991.
4. K.V. Williams and R. Vaziri, "Application of a damage mechanics model for predicting the impact respnse of composite materials," *Computers & Structures*, vol. 79, pp. 997-1011, 2001.
5. A.R. Rispler, G.P. Steven, and L. Tong, "Failure analysis of composite t-joints including inserts," *Journal of Reinforced Plastics and Composites*, vol. 16, no. 18, pp. 1642-1658, 1997.

59

Performance Analysis of Controllers in Active Vibration Control of Beam

S.M. KHOT, A. BAGWE AND S. IYER

Department of Mechanical Engineering, Fr. C. Rodrigues Institute of Technology Vashi,
Navi Mumbai-400703 email: smkhot66@yahoo.co.in

ABSTRACT

For the present study of active vibration control a simply supported beam is selected. The convergence study for finalizing the mesh size is performed and modal analysis of beam structure is carried. The patches are located at the region of maximum strain for better control effect. The active vibration concept is demonstrated through simulation study in ANSYS© for various controllers like proportional, proportional derivative (PD) and proportional integral derivative (PID). The controller design is carried out through simulation study in MATLAB© for a predetermined settling time and estimated controller gains are used for active vibration control. The study shows that PID controller is better in terms of energy requirement and overall response as compared to other controllers.

Keywords: Piezo-patches; Proportional; Proportional-Differential; Proportional-Integral-Differential.

1. INTRODUCTION

The flexible beams are common structural elements employed in many engineering applications and are subject to a wide variety of excitations, including acoustic excitations. It is well-known that significant vibration of beams can have adverse consequences for the overall fatigue life and therefore, it is necessary to address the problem of control of vibration. This undesired vibration can be reduced or eliminated by using active vibration control. Piezoelectric patches bonded on the structure act as sensor to monitor and as actuator to control the response of a structure. The output of piezo sensors is processed by a controller and fed to piezo actuators in terms of voltage. Application of voltage to an actuator introduces the force on the base structure, which is proportional to the displacement measured by the sensors. In recent times, discrete piezo patches have emerged as preferred elements that can sense and cancel elastic vibrations by generating control strains. However, the effectiveness of such vibration control depends on many factors e.g., number of patches, locations, controller design, etc.

In literatures many researchers have studied active vibration control by using piezoelectric material and notable among these are works of Crawley et al [1], Devasia *et al*. [2], Yong Li *et al*. [3] and so on. These methods make use of [1] location of high average strain of first mode of flexible beams, [2] optimal sizing and shaping algorithms, [3] optimal location with corresponding optimal feedback gains etc. to achieve the vibration control objective. Xu *et al*. [4] and followed by Kargulle *et al*. [5] used the commercial finite element packages ANSYS© for their study. The control design carried out [4] by using MATLAB© control system tool box and [5] reported using ANSYS© itself for active vibration control. These authors demonstrated the active vibration control of cantilever type of beam structures. The present study plans to extend the Kargulle *et al*. [5] method to simply supported beams with various controllers and attempt is made to compare performance of controllers.

2. MODELLING AND ANALYSIS

The application of the finite element modelling techniques in the smart structures has grown significantly during the last decade, resulting in availability of piezoelectric elements in commercial finite element codes such as ANSYS©. In the present study, the finite element analysis based method is used for modal as well as transient analysis. In the context ANSYS© is used to model the simply supported beam along with patches. The simply supported beam with piezoelectric patches is modelled by Solid45 elements for the metal part and Solid5 elements for the piezoelectric part of the structure. Initially, only the beam is modelled by using Solid45 elements and modal analysis is performed.

2.1. Modal Analysis of Base Structure

A frequency convergence study of simply supported beam without piezoelectric patch is carried out to arrive at the appropriate mesh size for the analysis. Table 1 provides results of such convergence study.

Table 1. Frequency Convergence study

Mode Shape	Theoretical frequency (Hz)	ANSYS© frequency (Hz)		
		(30*2*1)	(60*4*1)	(120*8*1)
1.	7.151	7.1952	7.1565	7.1527
2.	28.603	29.153	28.692	28.633
3.	64.357	66.839	64.807	64.507

It can be seen from Table 1 that the theoretical frequencies obtained by Euler-Bernoulli's thin beam analysis are closer to the mesh size of 120×8×1. As difference between frequencies for 60 × 4 × 1 and 120 × 8 × 1 mesh size is within 0.1%, therefore, by taking into consideration of computational time required, mesh size selected for the present analysis is 60 × 4 × 1. The modal analysis of only base structure is performed and resulting mode shape diagrams are studied carefully for identifying maximum strain regions and nodal points for locating the piezoelectric patches for effective control of vibration.

2.2. Modal Analysis of Base Structure with Piezo Patches

The dimensions as well as material properties of base structure and actuator, which are used for present study, are given in Table 2.

Table 2. Base structure and actuator properties

Part	Dimensions	Density	Elastic Modulus	PR	Material
Beam	500 × 25.4 × 0.8	2800	68e9	0.32	Aluminum
Actuator	72 × 25.4 × 0.61	7500	—	—	PZT 5H

Modal analysis of integrated structure is performed to include the effect of inertia and stiffness of piezoelectric patches. This also helps in identifying, whether the patches located based on modal analysis of base structure alone are appropriate or not and to determine the time step for transient analysis based on changed natural frequencies due to addition of piezoelectric patches.

2.3. Active Vibration Control

Transient analysis in ANSYS© software is used to demonstrate the active vibration concept for simply supported beam. Vibration generating force (Fc), which is taken unity at $t = \Delta t$ and zero for subsequent time steps. Actuating voltage (V_a) is zero at $t = \Delta t$. The coefficients of Rayleigh damping (α and ß) are defined as $\alpha = \beta = 0.001$. The time step size is chosen as, $1/20 f_n$, where f_n is natural frequency of the mode to be excited. Strain is calculated at the selected sensor location, normally opposite to the actuator, and is multiplied by sensor amplification factor (K_s) and then is subtracted from zero. The zero value is taken as reference value to control the vibration. The difference between the input reference and sensor signal is called as error signal. The error value is multiplied by control and power amplification factor (K_c and K_v) to determine V_a at a time step. The macro is written for closed loop analysis for $t > \Delta t$ for controlling first mode with single patch for proportional controller and is given below:

```
*do, t, 3*dt, ts, dt
*get, u1, node, nr, u, x
*get, u2, node, nr1, u, x
err = 0-Ks*(u2 – u1)/dx
Va = Kc* Kv*err
d, nv, volt, Va
time, t
solve
*enddo
```

The value of K_s and K_v are taken appropriately based on the simulation study in MATLAB©. The values of Kc (control gain) are evaluated for various controllers by using MATLAB© for same settling time for the sake of comparison. The maximum voltage per mm of piezoelectric material is taken as 235 V [5]. This limitation of voltage to be applied to piezo patches is taken into consideration while finalizing K_c value. The modelling approach is validated by similarly modelling the cantilever beam structure of same dimensions as of reference [5] and the results are observed in close agreement.

2.4 Simulation in MATLAB©

The controller design is carried out in MATLAB© by considering single degree of freedom model of beam. The standard transfer function of single degree of freedom system is used and the parameters of beam system such as mass, stiffness and structural damping are estimated for the present analysis. The simulation study by using control system tool box is conducted for achieving settling time as one second for all the controllers and required control gains are estimated.

3. RESULTS AND DISCUSSION

The active vibration control simulation study is performed through transient analysis in ANSYS© by using control gain estimated as discussed in earlier section.

3.1 Proportional Controller

The control gain for proportional controller is obtained from simulation study in MATLAB© to achieve desired response. The control gain (K_c) 6.2 with sensor and actuator gains 1000 each are selected for transient analysis in ANSYS©. In this process controller action is activated at different time intervals i.e., before and after the first peak of transient response and performance is studied. The transient response for proportional controller is given in Fig. 1 and shows that, during the first overshoot, the controlled displacement increases a little than the peak of the uncontrolled displacement and in the next subsequent cycles the amplitude decreases for the case in which control action initiated before the first peak. In the second case when control action is initiated after the peak, controlled and uncontrolled displacements are same and reduce gradually. The maximum actuator voltage required is 163V which is well within the maximum limit.

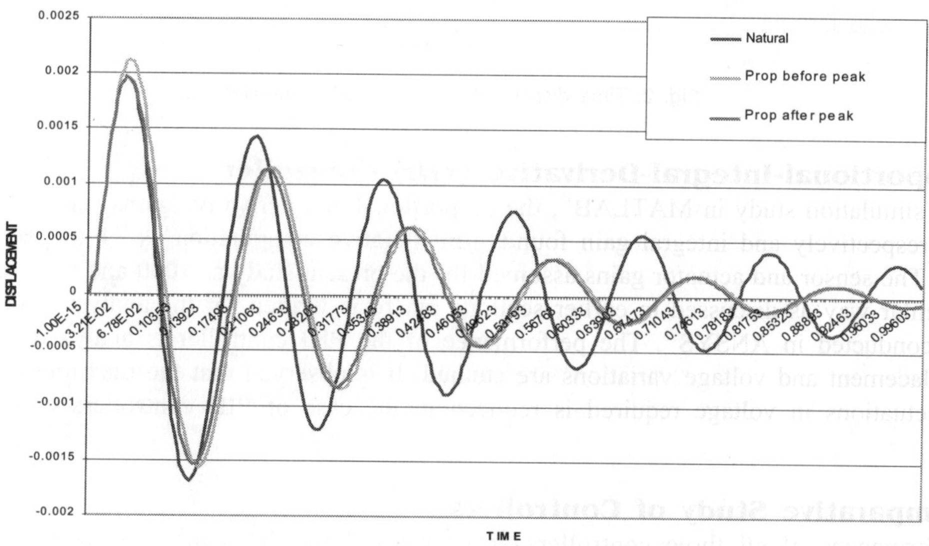

Fig. 1. Time displacement plots for Proportioal controller.

3.2 Proportional-Derivative (PD) Controller

The proportional and differential gains are calculated by time response analysis of a second order system and checked in MATLAB© for desired response. The estimated proportional gain and derivative gain are –1 and – 0.1315, respectively. The sensor gain and actuator gain selected for the present study is 1000 and 600, respectively. The only one change is to be made in macro for calculating Va as per the controller basic equation. The PD controller performance is studied for positive as well as negative proportional gain by maintaining differential gain as negative. Time displacement plot is given in Fig. 2 and shows that, positive and negative proportional gain with same differential gain as estimated gives almost same response.

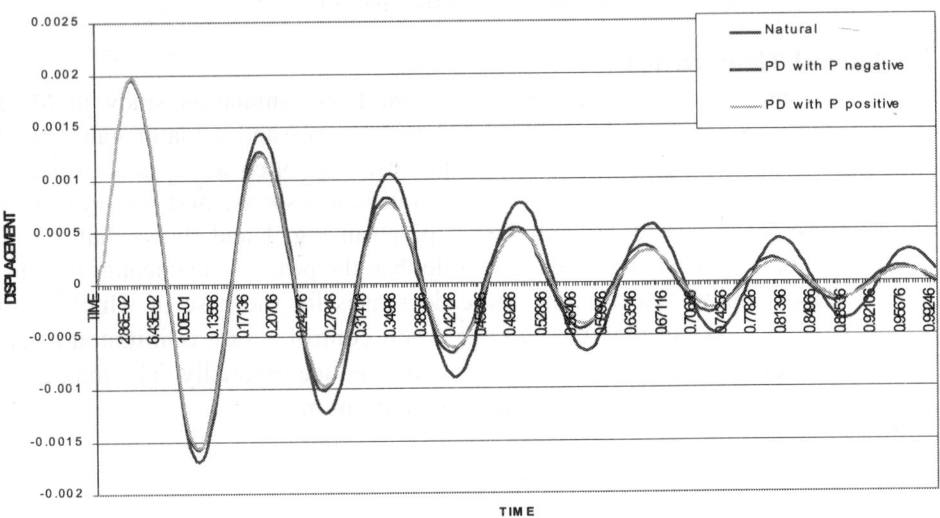

Fig. 2. Time displacement plot for PD controller.

3.3 Proportional-Integral-Derivative (PID) Controller

From the simulation study in MATLAB©, the proportional and derivative gains selected are –1 and –0.1315, respectively and integral gain found from iterative computation as –1 to get the desired response. The sensor and actuator gains assumed for the present study as 1000 and 650, respectively. In the similar way as discussed in earlier sections macro is written for the analysis and simulation study is conducted in ANSYS©. The performance of the PID controller is analyzed and plots of time displacement and voltage variations are studied. It is observed that the maximum voltage and initial fluctuations in voltage required is reduced in the case of PID controllers as compared to others.

3.4 Comparative Study of Controllers

The performances of all three controllers are compared for the same settling time. The time displacement plot and voltage variation plots of all controllers are overlapped and given in Figs. 3 and 4, respectively. It can be seen from Fig. 3 that the settling time is same in all controllers and damping effect is better for proportional and proportional derivative controller than PID controller. The peaks in PD and PID are found to shift to the left while that of the proportional shifts to the right. Fig. 4 shows that the voltage required by proportional controller is more as compared to other controller for the same settling time. The voltage required for PID controller is little less as compared to proportional derivative controller at the beginning of controller action and remains almost same in later part of control action. Therefore, from this discussion it can be concluded that PID controller requires the least energy input as compared to other controllers for achieving the same settling time.

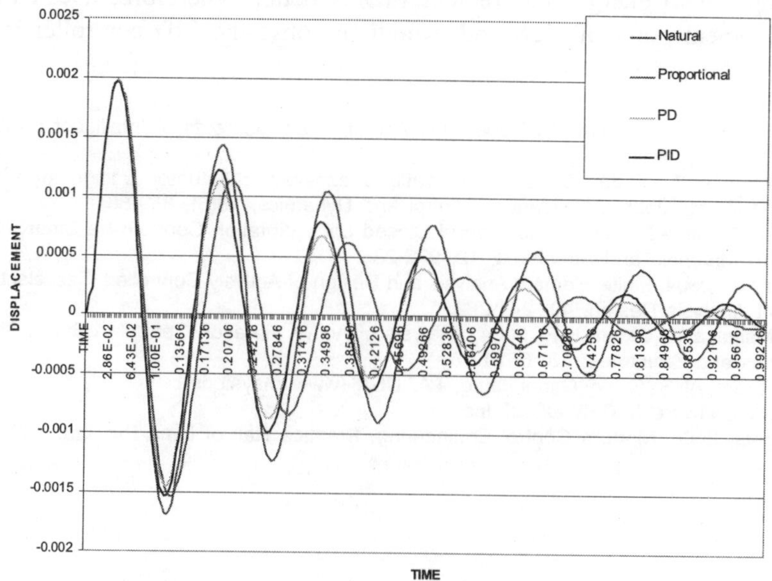

Fig. 3. Time displacement plot for all controllers.

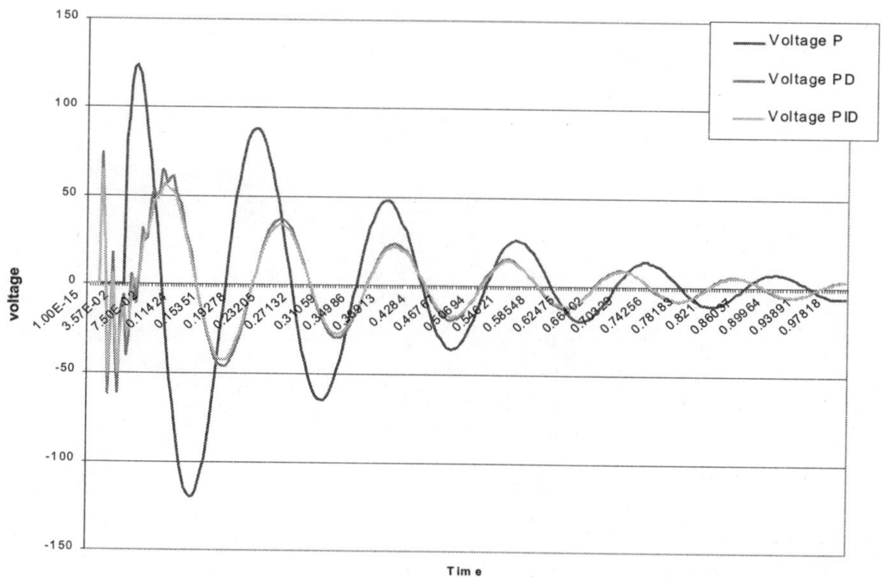

Fig. 4. Voltage variation plot for all controllers.

CONCLUSION

The controllers are designed through simulation study in MATLAB and controller gains are estimated to achieve predetermined settling time. These control gains are used in simulation study of active vibration control for simply supported beam in ANSYS. The transient response and voltage distribution plots are compared and found that amount of damping for PD and proportional are better but as compared to energy requirement PID is better. Therefore, it can be concluded that from energy requirement point of view and overall response the PID controller is better.

REFERENCES

1. Crawley E.F., Luis J., 1987, Use of Piezoelectric Actuators as Elements of Intelligent Structure, AIAA Journal, 1373-1385
2. Devasia S., Meressi T., Paden B., Bayo E., 1993, Piezoelectric Actuator Design for Vibration Suspension: Placement and Sizing, Journal Guidance, Control and Dynamics, 16(5), 859-864.
3. Young H.L., 2003, Finite Element Simulation of Closed Loop Vibration Control of a Smart Plate under Transient Loading, Smart Materials and Structures, 12, 272-286.
4. Xu S. X, Koko T.S., 2004, Finite Element Analysis and Design of Actively Controlled Piezoelectric Smart Structures, Finite Element Analysis Design, 40, 241-262.
5. Kargulle H., Malgaca L., Oktem H.F., 2004, Analysis of Active Vibration Control in Smart Structures by ANSYS, Smart Materials and Structures, 13, 661-667.
6. ANSYS© Software ANSYS Inc. Canonsburg, PA, USA (www.ansys.com)
7. MATLAB© 6.1v Software MATHWORKS Inc.
8. Katsuhiko Ogata, 2000, Modern Control Engineering, Prentice Hall of India Pvt. Ltd.

60

An Investigation of Effect of System Parameters on Natural Frequency of Beams by Using Statistical Methods

S.M. Khot, C.M. Choudhari, Nitesh P. Yelve and Jitendra K. Sardar

Faculty of Department of Mechanical Engineering,
Fr. C. Rodrigues Institute of Technology, Vashi, Navi Mumbai-400 703
email: smkhot66@yahoo.co.in, c.choudhari@rediffmail.com,
niteshpy@yahoo.co.in, jitendrasardar@yahoo.com

ABSTRACT

In order to enhance the performance of mechanical systems, dynamic analysis plays an important role at design stage, which helps in adjusting natural frequencies to avoid resonance. The system parameters plays critical role in controlling natural frequencies. Taguchi's concept of orthogonal array has been used for studying natural frequencies of cantilever and simply supported beams under simultaneous variation of parameters like material, length and thickness. The results obtained under the combination of these parameters are analyzed by using two statistical methods viz. ANOVA (Analysis of Variance) and Factor Plots, to identify the critical parameters affecting the natural frequencies and their contribution.

Keywords: Natural frequency, ANOVA, Factor Plot, System Parameters.

1. INTRODUCTION

The dynamic analysis of a mechanical system plays an important role at design stage for better performance. In order to achieve this, normally, the natural frequencies are adjusted in such a way that resonance is kept away from the operating range. The natural frequencies of the system can be adjusted by controlling the system parameters, so that resonance condition can be kept away from the operating range. The traditional 'vary one factor at a time' experiments are intrinsically incapable of uncovering interaction among factors [1]. Designers generally are curious to understand the design space of the parameters, their interactions and its effect on the performance in the pre and post design process. Statistically designed experiments provide cost-effective solution to such needs. Taguchi's concept of orthogonal arrays is one of the best available such statistical techniques to design the experiments being conducted to understand the performance of the system under 'simultaneous variation' of parameters. This concept is extensively used in mechanical product design and optimizations, but specific areas like applications of these methods to understand the dynamic behaviour of mechanical systems and components have remained more or less unexplored. Hence, an attempt is made to apply Taguchi's concept of orthogonal array in designing experiments to determine the natural frequencies of cantilever beams and simply supported beams under the simultaneous variation of three different parameters.

In the present analysis this technique is used for designing experiments to determine the natural frequency of cantilever and simply supported beams for first mode shape under the simultaneous variation of three different parameters, namely, material, length and thickness. These parameters are identified based on the basic equation of natural frequency; whereas width is not considered as it gets cancelled. Cantilever and simply supported beams are used in many mechanical applications, out of which bridges, supporting structures for machines which are continuously subjected to vibration, are major areas of their applications. The natural frequencies of the system can be adjusted by controlling the system parameters like material, length and thickness so that resonance condition can be kept away from the operating range. And if the most significant factor out of these three system parameters is known special attention can be given to that particular parameter so that system can be kept away from resonance. Keeping this in view, the experimentation is designed by Taguchi's concept of orthogonal array and the obtained results are then analyzed by ANOVA and Factor Plots.

2. FEM ANALYSIS BY ANSYS©

The cantilever and simply supported beams are analyzed by using ANSYS© software and their natural frequencies are determined for first mode shape. The element type selected for the analysis is "beam 3" and the appropriate mesh sizes of beams are selected through convergence study. The natural frequencies of cantilever beams obtained by FEM analysis in ANSYS© software are compared with the experimental and analytical results [2] and found to be in close agreement; which validates the methodology of analysis by ANSYS©. FEM analysis in ANSYS© software is done for cantilever and simply supported beams of different materials, lengths and thicknesses; and natural frequencies are determined for different combinations, for first mode shape and are given in Table 3.

3. DESIGN OF EXPERIMENTS BY TAGUCHI'S METHOD

Taguchi's method of Orthogonal Array is used here to design the experiments for finding out the natural frequencies (f) of cantilever and simply supported beams under the simultaneous variation of three different system parameters namely, material, length and thickness. The natural frequencies are found for different combinations of system parameters according to Taguchi's Orthogonal Array. The experimentation details are given in the Table 1.

Table 1. Factors and their levels for experiment for cantilever and simply supported beams

Control Factors	Levels		
	1	2	3
1. Material (m)	Steel	Aluminium	Copper
2. Length (l)	500 mm	625 mm	781.25 mm
3. Thickness (t)	1 mm	1.25 mm	1.56 mm

For determining required minimum number of experiments to be conducted, the degrees of freedom are required to be calculated. These are determined by using relations given below and shown in Table 2. Hence, it is required to conduct at least 19 experiments to be able to estimate the effect of each factor and the desired interaction. For three factors at three levels L–27 is the available Taguchi's Orthogonal Array, which permits thirteen factors at three levels. Hence, L–27 is chosen as Taguchi's Orthogonal Array for conducting experiments with three factors at three levels.

Degree of freedom of a factor A = (Number of levels of factor A) – 1 and

Degree of freedom for interaction A × B = (Number of levels of factor A × 1) × (Number of levels of factor B – 1)

Table 2. Factors and their DOF

Control Factors	Degrees of Freedom	Control Factors	Degrees of Freedom
Material (m)	3 – 1 = 2	m × t	2 × 2 = 4
Length (l)	3 – 1 = 2	l × t	2 × 2 = 4
Thickness (t)	3 – 1 = 2	Overall Mean	1
m × l	2 × 2 = 4	Total DOF	19

The Linear graph [3] gives the columns in which factors can be assigned. According to the Linear graph of L–27, as shown in Fig. 1, the factor 'm' is assigned to column number 1,' 'l' is assigned to column number 2 and 't' is assigned to column number 5. The columns 3, 4, 6, 7, 8 and 9 take interactions into consideration, whereas the columns 10, 11, 12 and 13 are vacant because number of available columns in L–27 is more than the number of control factors. With the help of Orthogonal Array and Liner graph shown in Figure 1, the experimental scheme is designed and results are obtained by finite element analysis using ANSYS©, which are given in the Table 3.

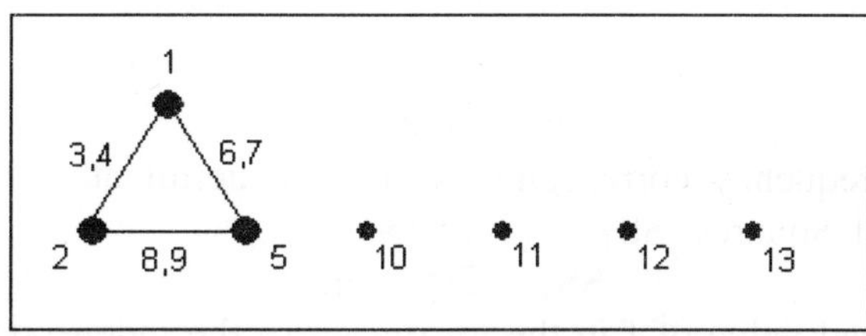

Fig. 1. Linear Graph of L-27.

It is seen from the Table 3 that, for experiment number one, the material of the test specimen is steel; length is 500 mm and thickness is 1.00 mm. The natural frequencies of cantilever and simply supported beams for first mode shape, corresponding to different combinations of material, length and thickness are obtained, which are also shown in the last two columns of Table 3 respectively.

4. STATISTICAL ANALYSIS BY ANOVA

The results obtained from the FEM analysis in ANSYS© software i.e., natural frequencies are analyzed by using the statistical tool ANOVA [3], for getting the relative importance/effect of the individual factor and interactions between the factors on the natural frequency of cantilever and simply supported beams for first mode shape. The analysis by using ANOVA is performed manually. The Grand

Table 3. Experimental Scheme

Exp. No.	1 m	2 l (mm)	3	4	5 t mm)	6	7	8	9	f (Hz) Cant. Beam	f (Hz) Sim. Supp. Beam
1.	Steel	500	-	-	1.00	-	-	-	-	3.3614	9.381
2.	Steel	500	-	-	1.25	-	-	-	-	4.1997	11.727
3.	Steel	500	-	-	1.56	-	-	-	-	5.2477	14.658
4.	Steel	625	-	-	1.00	-	-	-	-	2.1489	6.004
5.	Steel	625	-	-	1.25	-	-	-	-	2.6851	7.505
6.	Steel	625	-	-	1.56	-	-	-	-	3.3553	9.3814
7.	Steel	781.25	-	-	1.00	-	-	-	-	1.3741	3.8426
8.	Steel	781.25	-	-	1.25	-	-	-	-	1.7171	4.8032
9.	Steel	781.25	-	-	1.56	-	-	-	-	2.4157	6.0041
10.	Aluminium	500	-	-	1.00	-	-	-	-	3.328	9.2773
11.	Aluminium	500	-	-	1.25	-	-	-	-	4.1576	11.597
12.	Aluminium	500	-	-	1.56	-	-	-	-	5.9147	14.496
13.	Aluminium	625	-	-	1.00	-	-	-	-	2.1271	5.9375
14.	Aluminium	625	-	-	1.25	-	-	-	-	2.6576	7.4219
15.	Aluminium	625	-	-	1.56	-	-	-	-	3.3208	9.2774
16.	Aluminium	781.25	-	-	1.00	-	-	-	-	1.3622	3.8
17.	Aluminium	781.25	-	-	1.25	-	-	-	-	1.7031	4.75
18.	Aluminium	781.25	-	-	1.56	-	-	-	-	2.1261	5.9376
19.	Copper	500	-	-	1.00	-	-	-	-	2.6006	7.2435
20.	Copper	500	-	-	1.25	-	-	-	-	3.2487	9.0544
21.	Copper	500	-	-	1.56	-	-	-	-	4.0588	11.318
22.	Copper	625	-	-	1.00	-	-	-	-	1.6619	4.6358
23.	Copper	625	-	-	1.25	-	-	-	-	2.0763	5.7948
24.	Copper	625	-	-	1.56	-	-	-	-	2.5943	7.2436
25.	Copper	781.25	-	-	1.00	-	-	-	-	1.0636	2.9669
26.	Copper	781.25	-	-	1.25	-	-	-	-	1.3294	3.7087
27.	Copper	781.25	-	-	1.56	-	-	-	-	1.6603	4.6359

Mean, m' is given by,

$$m' = (\Sigma f_i) / n \qquad \qquad ...(1)$$

where, f_i is the natural frequency corresponding to ith experiment and n is the total number of experiments. Total Sum of Squares, SS_T is given by,

$$SS_T = \Sigma(f_{ij} - m')^2 \qquad \qquad ...(2)$$

Where, f_{ij} is the obtained value of f in the jth run when the independent factor was set at its ith level

4.1 Effect of Individual Parameters on Natural Frequency

The effect of a factor at a particular level is defined as the deviation it causes from the overall mean. The effect of material factor (m) is evaluated here, which is at level one for first nine experiments. The average values of natural frequencies for these experiments, which is denoted by Mm_1, is given by,

$$Mm_1 = (f_1 + f_2 + f_3 + f_4 + f_5 + f_6 + f_7 + f_8 + f_9) / 9 \qquad \qquad ...(3)$$

where, f_1 denotes the natural frequency for experiment number one and so on.

Similarly, the average values of natural frequencies at level two and level three i.e., for next two sets of experiments (Mm_2 and Mm_3) are determined. Since, there are nine values at each level, Sum of Squares due to material factor, $(SS)_m$ with two degrees of freedom is given as,

$$(SS)_m = [(Mm_1 - m')^2 + (Mm_2 - m')^2 + (Mm_3 - m')^2] - 9 \qquad ...(4)$$

Similarly, $(SS)_l$ for length factor and $(SS)_t$ for thickness factor are determined.

4.2 Effect of Interactions of Parameters on Natural Frequency

Since, there are three values at each level for the interactions between material and length, the Sum of Squares for it, $(SS)_{mxl}$ with four degrees of freedom is given by,

$$(SS)_{mxl} = [\Sigma(f_{ij} + m' - f_j - f_i)^2] \times 3 \qquad ...(5)$$

where, f_{ij} is the obtained value of f in the jth run when the independent factor was set at its ith level and f_i and f_j are the average natural frequencies which are shown in the Table 4. Similarly, $(SS)mxt$ for the interaction between material and thickness and $(SS)_{lxt}$ for the interaction between length and thickness are determined.

Table 4. Average table for interaction – m×l

For Cantilever Beams					For Simply Supported Beams			
Length (l) (mm)	Material (m)			f_j	Steel	Aluminium	Copper	f_j
	Steel	Aluminium	Copper					
500	4.2472	4.2001	3.2793	3.9089	11.922	11.7901	9.2053	10.9725
625	2.7182	2.6880	2.0988	2.5017	7.6301	7.5456	5.8914	7.0224
781.25	1.7393	1.7203	1.3432	1.6000	4.8833	4.8292	3.7705	4.4943
f_i	2.9061	2.8695	2.2404		8.1451	8.055	6.2891	

5. STATISTICAL ANALYSIS BY FACTOR PLOTS

The results obtained from the FEM analysis in ANSYS© software i.e., natural frequencies are also analyzed by using the Factor Plots method, to identify the significant parameter affecting the natural frequency of cantilever and simply supported beams for first mode shape., out of material, thickness and frequency. For the parameters namely, material, thickness and frequency the average natural frequencies are calculated at three different levels (as shown in Equation 3 for material factor). Then the graphs are plotted based on these values. If the graph corresponding to a parameter in the Factor Plot has larger slope then that parameter is said to have comparable effect on the natural frequency.

6. RESULTS AND DISCUSSIONS

The results obtained from ANOVA and Factor Plot analysis are discussed here.

6.1 ANOVA Results

The results generated after conducting the manual ANOVA for natural frequencies of cantilever and simply supported beams obtained by FEM analysis in ANSYS©, are given in the Table 5, which is also called ANOVA Table. From Table 5, it can be seen that length contributes 70.85%, which is higher in total variation of natural frequency of cantilever and simply supported beams for first mode shape; hence length is the critical parameter affecting the natural frequency of cantilever and simply supported beams. The next contributing parameter is thickness and followed by material.

Table 5. Manual ANOVA Results

Factors	DOF	Cantilever Beam		Simply Supported Beam	
		SS_T	%Contribution	SS_T	%Contribution
Material	2	02.5015	07.28	19.7147	7.28
Length	**2**	**24.3543**	**70.85**	**191.8813**	**70.84**
Thickness	2	06.3124	18.36	49.7494	18.37
Material × Length	4	00.3167	00.92	2.4932	0.92
Material × Thickness	4	00.0821	00.24	0.6464	0.24
Length x Thickness	4	00.7988	02.32	6.2910	2.32
Residual	1	00.0102	00.03	0.08176	0.03
Total	19	34.3760	100.00	270.8577	100.00

6.2 Factor Plot Results

The resulting Factor Plot is shown in the Fig. 2 and Fig. 3. In the Factor Plots, shown in the Fig. 2 and Fig. 3, slope of the graph corresponding to the length is larger, hence length is the critical parameter affecting the natural frequencies of cantilever and simply supported beams.

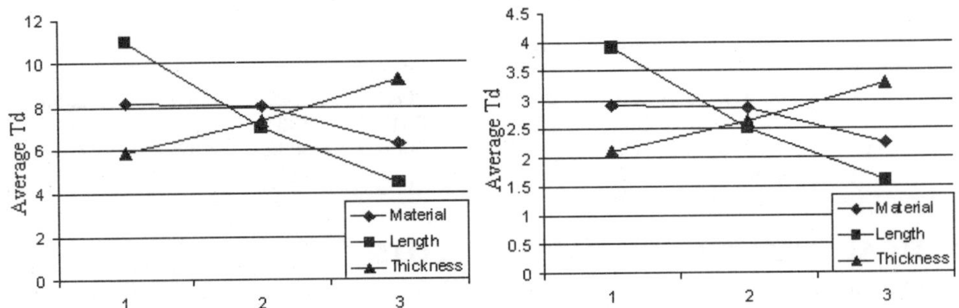

Fig. 2. Factor Plot for Cantilever Beam.   Fig. 3. Factor Plot for Simply Supported Beam.

CONCLUSION

The application of Design of Experiments is attempted to find out the significant parameter influencing the natural frequencies of cantilever and simply supported beams based on Taguchi's Orthogonal Array L–27. According to the scheme obtained by Taguchi's method of Orthogonal Array, FEM analysis in ANSYS© software is conducted on cantilever and simply supported beams, by changing the parameters namely, material, its length and thickness, at three levels. From the results obtained by ANOVA and Factor Plots, it is observed that length contributes higher percentage in total variation of natural frequency of both cantilever as well as simply supported beams for first mode shape; hence, length is the critical parameter affecting the natural frequency of beams. It can be also seen that the contributions of each parameters of cantilever and simply supported beams under the study are same even though total sum of square values are different. This in fact is consistent with the basic theory, where the constant values in basic equation change according to the boundary conditions.

The statistical methods as a part of analysis are found to give a good insight on keeping away the resonance of cantilever beams by adjusting the system parameter according to its contribution in total variation of the results. Thus, the parameter to be given importance while selecting the beam structures for engineering applications are identified by making use of Statistical methods. This approach can be used further to identify the critical parameters and their contribution in overall system parameter effect on natural frequency of various standard machine elements, which in turn helps in adjusting the critical system parameters to control the natural frequency of the system.

REFERENCES

1. Phadke, M.S., 1989, Quality Engineering using Robust Design, Prentice Hall, Angle Wood Cliffd, New Jersey.
2. S.M. Khot, Jitendra K. Sardar, Nitesh P. Yelve, 2005, Analytical, Experimental and FEM (ANSYS©) Analysis of Cantilever Beams for First Three Modes, Proceedings of 1st National Conference on Recent Developments in Mechanical Engineering (YANTRA 2005), S.V.E.R.I.'s College of Engineering, Pandharpur, Dec. 23-24, 2005, pp 32–36.
3. Park, S.H., Robust Design and Analysis for Quality Engineering, Chapman and Hall, London.
4. Phillip, J.R., Taguchi Technique for Quality Engineering, Tata McGraw Hill Publication.
5. Bagchi, T.P., 1993, Taguchi Method Explained Practical Steps to Robust Design, Prentice Hall of India, New Delhi.
6. Malik, A.K., Principles of Vibration Control, East West Press Pvt. Ltd.

61

Vibration Control of Shape Memory Alloy Actuated Structure Using Fast Output Sampling Feedback Control

K. Dhanalakshmi, C. Chandravathanam, M. Umapathy and B. Vasuki

Department of Instrumentation and Control Engineering, National Institute of Technology, Tiruchirappalli-620 015, Tamil Nadu, India.
email: dhanlak@nitt.edu, vathana_c@yahoo.com, umapathy@nitt.edu, bvas@nitt.edu

ABSTRACT

This paper describes the application of Shape Memory Alloy wires as control actuators to minimize structural vibration using fast output sampling feedback controllers. The proposed control scheme uses Pulse Width Modulation (PWM) excitation to the SMA wires while it provides continuous control to the structure. A Linear dynamic model of the structure is obtained using online Recursive Least Square (RLS) parameter estimation. A digital control system that consists of simulink modeling software and dSPACE 1104 controller board has been used for identification and control. The effectiveness of the controller is shown through simulation by exciting the structure at its first mode resonance.

Keywords: Smart structure, Shape Memory Alloy, Vibration Control, System Identification, Fast Output Sampling Feedback.

1. INTRODUCTION

The control of vibration in flexible structures of engineering systems like, in space, aircraft, industrial applications and automotives is a formidable challenge for a control system designer. The application of induced strain actuators in smart structure technologies for vibration control has gained increased attention in the last few years [1]. Actuators based on Shape Memory Effect (SME) of metallic alloys and Piezoelectric Effect (PZE) of some ceramic materials are much promising. Design simplicity for control mechanism, high possibility of miniaturization and low power consumption are the salient properties of the Shape Memory Alloy (SMA) actuators when compared to Piezo-electric materials [2]. SMA actuators can be attached to desired points on the flexible structure in order to generate translational forces for controlling deflections of the structure. Accordingly, this class of actuators has been proven to be successful in vibration control of flexible structures as seen from references sited in [3-5]. System identification is an established modeling tool [6, 7] in engineering and is suitable to identify the dynamics of smart structures. While controller design can be implemented using state and output feedback, this paper proposes one of the piecewise output feedback control scheme-the Fast Output Sampling Feedback control algorithm [8-10], since the output feedback needs only the measurement of system output unlike the state feedback which requires the knowledge of the states or a state estimator. It holds the features of Static Output Feedback and computes piecewise constant output feedback gain for the system. The design of FOS controller for the vibration suppression of flexible structures has been reported in [11]. In this paper, a new control scheme which uses FOS feedback control is proposed. This control scheme applies the control to the structure continuously and allows the actuators to heat and cool alternatively.

2. THE EXPERIMENTAL SCHEME

A flexible aluminum beam with clamped end as shown in Figure 1 is considered in this paper. Two piezoceramic patches are surface bonded at a distance of 30mm from the fixed end of the beam. The patch bonded on the bottom surface acts as a sensor and the one on the top surface is an actuator through which an excitation input can be applied to the structure. The dimensions and properties of the beam and piezoceramic sensor/actuator are listed in Table 1.

Table 1. Dimensions and properties of the Aluminum beam and Piezoceramic actuator

Aluminum beam		Piezoceramic sensor/actuator	
Length (m)	0.480	Length (m)	0.0765
Width (m)	0.01322	Width (m)	0.0127
Thickness (m)	0.00124	Thickness (m)	0.005
Young's modulus (GPa)	71	Young's modulus (GPa)	47.62
Density (kg/m^3)	2700	Density (kg/m^3)	7500
First Natural Frequency (Hz)	6.26	Piezoelectric strain constant (mV^{-1})	-247×10^{-12}
		Piezoelectric stress constant (VmN^{-1})	-9×10^{-3}

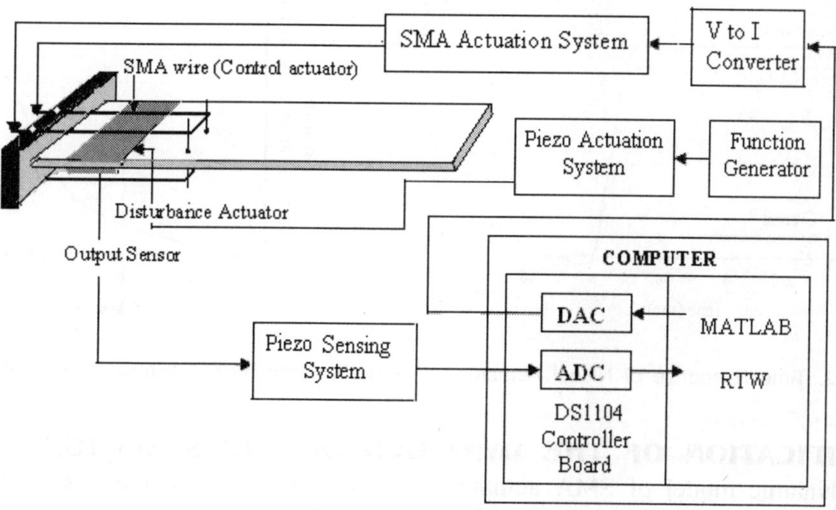

Fig. 1. Schematic of Experimental Arrangement.

The system is excited by a sinusoidal signal through the function generator. Two externally attached SMA wires, one above and the other below the beam are used as control actuators. Each wire is attached from the fixed end and run in parallel to the beam in the form of a rectangular loop of length 0.1m and width 0.01m through two vertical pins, thus connected in a serial manner. The SMA wires are energized through current sources designed using Pulse Width Modulation (PWM) circuit. The dimensions and properties of the Nitinol SMA wire are listed in Table 2.

Table 2. Dimensions and properties of (Nitinol) SMA wire Actuator

Length (m)	0.24	Low Temperature Yield Strength (MPa)	70
Diameter (m)	0.00037	Ultimate Tensile Strength (MPa)	700
Maximum A_s Temperature (°C)	90		
Maximum A_f Temperature (°C)	110		
Maximum M_s Temperature (°C)	85		
Maximum M_f Temperature (°C)	70	Density (g/c.c.)	6.5
Maximum one-way Strain (%)	6 to 8	Resistivity (micro-ohm-cm)	80–89
Hysteresis (°C)	15	Heat Capacity (J/Kg-°K)	837
High Temperature Yield Strength (MPa)	415	Thermal Conductivity (J/m-sec-°K)	18

3. DYNAMIC CHARACTERISTICS OF SMA ACTUATOR

Nitinol (NiTi) the most commonly used Shape memory alloy exhibit change in length and thereby are able to apply force when subjected to a thermal process while it is under a fixed load. An SMA actuator employs the more reliable one-way shape memory effect and can do work as it returns to its cool state. Experiments were conducted to determine the dynamic characteristics of the SMA wires used in this study by applying step current from the PWM circuit. The time response and force-current response obtained experimentally are given in Figures 2 and 3. It is seen that the Nitinol wire exhibits faster heating (actuation) and comparatively a slow response during cooling.

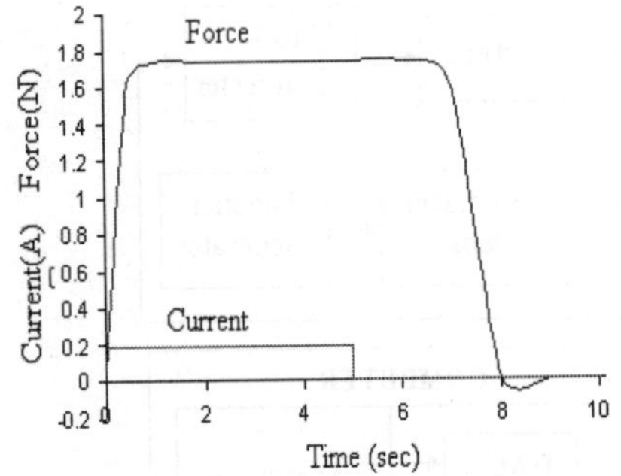

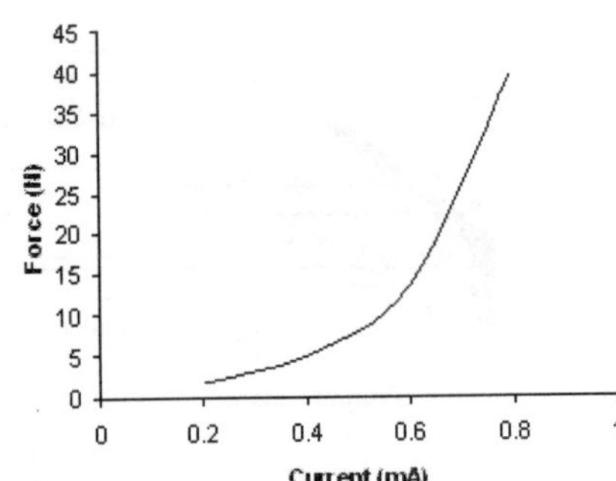

Fig. 2. Time Response of Nitinol Actuator.   **Fig. 3.** Force-current relation of the SMA wire.

4. IDENTIFICATION OF THE DYNAMICS OF THE STRUCTURE

The linear dynamic model of SMA actuated cantilever beam is obtained using online Recursive Least Square (RLS) parameter estimation. The unknown parameters of the smart structure dynamics are estimated using online identification method, since it is proven to be more universal and feasible than analytical and numerical models for the present system. In addition, the RLS method based on

Auto-Regressive (ARX) model is used for linear system identification, which is easy to implement and has fast parameter convergence. Most multi-input single-output plants can be considered by the auto-regressive model. The ARX model for the system shown in Figure 1 is given as,

$$\hat{y}(k) + a_1 y(k-1) + \ldots + a_{n_a} y(k-n_a) = b_{11} u_1(k-1) + \ldots + b_{1n_{b1}} u_1(k-n_{b1})$$

$$+ b_{21} u_2(k-1) + \ldots + b_{2n_{b2}} u_2(k-n_{b2}) + c_1 r(k-1) + \ldots + c_{n_c} r(k-n_c) + e(k) \qquad \ldots(1)$$

where, u(k) is the input signal, y(k) is the piezo sensor output, e(k) is the disturbance input and n_a, n_b, n_c and n_d determine the model order. The vector of unknown parameter to be estimated is thus

$$\theta = \begin{bmatrix} a_1 & a_2 & \ldots a_{n_a} & b_{11} & b_{12} \ldots & b_{1n_{b1}} & b_{21} & b_{22} & \ldots b_{2n_{b2}} & e_1 & e_2 & \ldots e_{n_e} \end{bmatrix}^T \qquad \ldots(2)$$

φ can be calculated based on measured signals and known variables and it is given by

$$\varphi(k) = (-y(k-1), -y(k-2), \ldots, -y(k-n_a), u1(k-1), u1(k-2), \ldots, u1(k-n_{b_1}),$$

$$u2(k-1), u2(k-2), \ldots, u2(k-nb_2), d(k-1), d(k-2), \ldots, d(k-n_d))^T. \qquad \ldots(3)$$

The estimated model output is

$$\hat{y}(k) = \varphi^T(k)\hat{\theta}(k-1) \qquad \ldots(4)$$

In recursive identification methods, the parameter estimates are computed recursively in time. The recursive form is given by

$$\hat{\theta}(k) = \hat{\theta}(k-1) + P(k)\varphi(k)\varepsilon(k) \qquad \ldots(5)$$

where, P(k) is the covariance matrix which is updated using

$$P(k) = P(k-1)\left(1 - \frac{\varphi(k)\varphi^T(k)P(k-1)}{1 + \varphi^T(k)P(k-1)\varphi(k)}\right) \qquad \ldots(6)$$

ε (k) is the prediction error and is given by

$$\varepsilon(k) = y(k) - \hat{y}(k) \qquad \ldots(7)$$

The initial values of $\hat{\theta}(k)$ and P(k) are chosen to be $\hat{\theta}(0) = 0$ and $P(0) = \alpha I_z$ where $\alpha = 10^4$. The natural frequency of the structure is measured experimentally as 6.2Hz. To identify the parameters online, the structure is excited by a sinusoidal signal with first natural frequency and a square wave signal as an input to the control actuators. Control actuator 1 is excited for a period of 5 sec and for the next 5 sec control actuator 2 is actuated by the same square wave signal. The sampling time is chosen to provide five measurements per cycle (sampling time 0.01s). The excitation signal, input signal and sensor output are given to MATLAB/simulink through ADC port of dSPACE 1104 system. The RLS algorithm is implemented by writing a C-file S-function used in MATLAB/simulink. The algorithm is run dynamically until all the parameter values settle down to a final steady value. The continuous state space model derived from the identified second order ARX model parameters is given as

$$\dot{\mathbf{x}} = \mathbf{A}\mathbf{x} + \mathbf{b}_i u + \mathbf{e}r; i=1,2; \; y = \mathbf{c}^T\mathbf{x} \qquad \qquad \text{...(8)}$$

where,

$$\mathbf{A} = \begin{bmatrix} 91.6326 & -99.4428 \\ 105.9147 & -97.9378 \end{bmatrix}, \mathbf{b}_1 = \begin{bmatrix} 0.1692 \\ 1.0647 \end{bmatrix}, \mathbf{b}_2 = \begin{bmatrix} -0.1128 \\ 0.5671 \end{bmatrix}, \mathbf{e} = \begin{bmatrix} -0.0721 \\ 0.5798 \end{bmatrix}, \; \mathbf{c}^T = \begin{bmatrix} 0 & 1 \end{bmatrix}.$$

5. REVIEW OF FAST OUTPUT SAMPLING FEEDBACK CONTROL

A fast output sampling feedback control law shown by Werner and Furta [8–10] is explained briefly in this section. Consider a linearized continuous - time state space model.

$$\dot{\mathbf{x}} = \mathbf{A}\mathbf{x} + \mathbf{B}\mathbf{u}$$
$$y = \mathbf{c}\mathbf{x} \qquad \qquad \text{...(9)}$$

where, $\mathbf{x} \in \Re^n$, $\mathbf{u} \in \Re^m$, $\mathbf{A} \in \Re^{n \times n}$, $\mathbf{B} \in \Re^{n \times m}$, $\mathbf{c} \in \Re^{p \times n}$ and $\mathbf{A}$, $\mathbf{B}$, $\mathbf{c}$ are constant matrices. It is assumed that $(\mathbf{A}, \mathbf{B}, \mathbf{c})$ is controllable and observable. A discrete time invariant linear system is obtained by sampling the system equation (9) at a sampling interval τ during which the control signal μ is held constant. Now we assume that the output measurements are available from the system equation (9) at time instants t $= l\Delta$, where $l = 0,1,2,.....,$ where the sampling interval τ is divided into N subintervals with $\Delta = \tau/N$ and N is equal to or greater than the observability index (v) of the discrete system obtained by sampling the system in equation (9) at the rate $1/\Delta$. The control signal $v(t)$, which is applied during the interval, $k\tau \leq t < (k + 1)\tau$ is then generated according to

$$u(t) = \begin{bmatrix} L_0 & L_1 & . & . & . & . & L_{N-1} \end{bmatrix} \begin{bmatrix} y(k\tau - \tau) \\ y(k\tau - \tau + \Delta) \\ . \\ . \\ . \\ . \\ y(k\tau - \Delta) \end{bmatrix} = \mathbf{L}\mathbf{y}_k, k\tau \leq t < (k+1)\tau \qquad \text{...(10)}$$

where, the matrix block L_j represents output feedback gains.

6. CONTROLLER DESIGN

In this paper active vibration control of cantilever beam through electrical heating and air cooling of SMA wires is examined using Fast Output Sampling (FOS) feedback control. Two fast output sampling feedback controllers are designed to reduce the amplitude of vibration of a cantilever beam at resonance. The control input generated by controller 1 is applied to actuator 1 which is attached on the top surface of the beam. Similarly the control input generated by controller 2 is applied to actuator 2 which is fixed to the bottom of the beam. The controllers are designed as per the design considerations presented in Section 5. Stabilizing state feedback gains for the system with actuator 1 and actuator 2 are

$\mathbf{F}_1$ = [75.5895 –78.0760]; $\mathbf{F}_2$ = [105.5421 – 103.0376]

The fast output sampling feedback gains are obtained using LMI approach [8].

$\mathbf{L}_1$ = [356.0564 – 227.3787 – 403.4308 – 176.1149 448.1070]

$\mathbf{L}_2$ = [147.5960 – 106.8211 – 180.4845 – 75.2033 206.1283]

To assess the performance of the controller, a sinusoidal excitation at first natural frequency with amplitude of 5V peak to peak is applied as disturbance signal through the disturbance actuator. The output of the controller 1 is applied for τ seconds (5 seconds) to the actuator 1 then the control is switched over to controller 2 which applies its output to actuator 2 for the next 5 seconds. The repetition of switching over at equal intervals between the controllers allows the actuators to actuate and cool alternatively. This approach allows applying control continuously to the system and the simulation model of the same is shown in Fig. 4. Simulations are carried out using MATLAB/ SIMULINK. To show the effectiveness of the proposed control scheme simulations were also carried out using single actuator. The disturbance applied, open loop and closed loop responses obtained with single and double actuators are shown in Figs. 5 and 6, respectively.

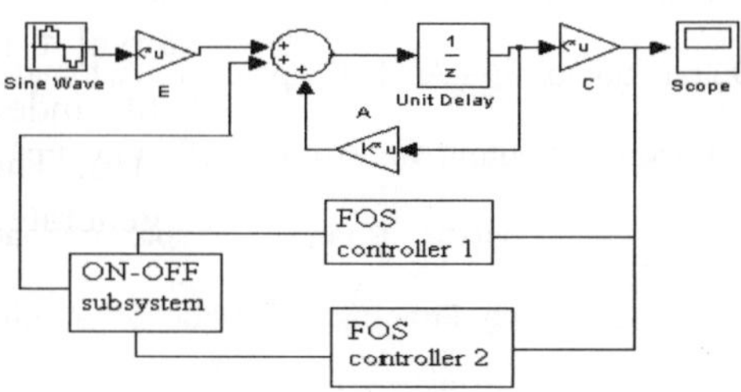

Fig. 4. Simulink model of the structure actuator.

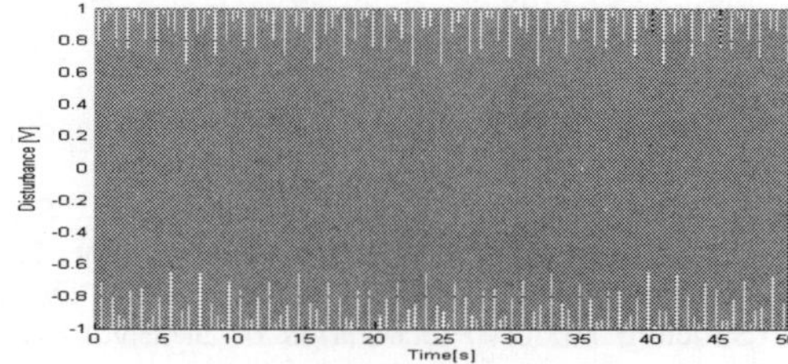

Fig. 5. Disturbance signal applied to PZT with FOS controllers.

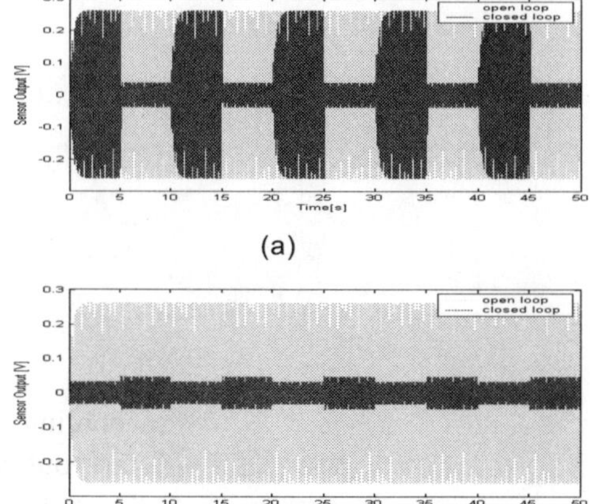

(a)

(c)

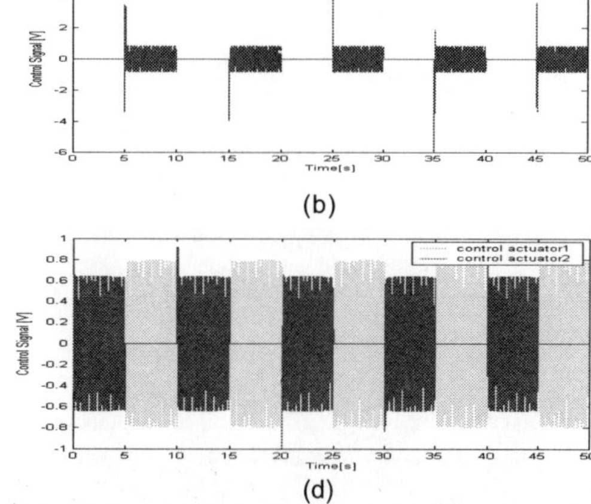

(b)

(d)

Fig. 6. (a) Open loop and closed loop responses of the single SMA actuated beam.
    (b) Control input to the single actuator system.
    (c) Open loop and closed loop responses for the two actuators system.
    (d) Control input to the two actuators system.

CONCLUSION

The model of SMA actuated and piezoelectric sensed cantilever beam structure has been identified using online ARX RLS system identification approach. A unique control scheme which uses two fast output sampling feedback controllers was designed. Simulation results show that the closed loop response obtained with this control scheme employing two actuators exhibits substantial reduction in the amplitude of flexural vibration at its first mode resonance, in comparison with the response obtained with single actuator.

REFERENCES

1. A Srinivasan and D McFarland, 2001, Smart Structure: Analysis and Design, Cambridge University Press.
2. T Waram, 1993, Actuator design using Shape Memory Alloys, 2nd edition, Chap. 2,3,6.
3. A. Baz, K.Imam and J.McCoy, 1990, Active Vibration Control of Flexible Beams using Shape Memory Actuators, Journal of Sound and Vibration, 140(3), 437-456.
4. Seung-Bok Choi and Chae-Cheon Cheong, 1996, Vibration control of a Flexible Beam using Shape Memory Alloy Actuators, Journal of Guidance, Control and Dynamics, 19(5), 1178-1180.
5. J O Salichs, C Shakeri, M N Noori and H Davoodi, 1998, An Experimental Study on the Effect of Nitinol's Memory in Vibration Control of a Cantilever Beam, Engineering Mechanics: A Force for the 21st Century, Proceedings of the 12th Engineering Mechanics Conference, La Jolla, California, 1549-1552
6. Lennart Ljung, 1999, System Identification theory for the user, Prentice Hall PTR, New Jersey.
7. D G Robertson and J H Lee, 2002, On the use of constraints in least squares estimation and control, Automatica, (38), 1113-1123.
8. Herbert Werner, 1998, Multimodel Robust Control by Fast Output Sampling - An LMI Approach, Automatica, 34(12), 1625-1630
9. H Werner and K Furuta, 1995, Simultaneous Stabilisation based on output measurements, Kybernetika, 31(4), 395-411.
10. H Werner and K Furuta, Simultaneous Stabilisation by piecewise constant Periodic Output feedback, Control Theory and Advanced Technology, 10(4), 1995, 1763-1775.
11. M. Umapathy, B. Bandopadhyay, 2002 "Design of Fast Output Sampling Feedback Control for a Smart Structure Model" *Proceedings of SPIE*,vol.4693, pp 222-233.

62

Vertical Dynamic Analysis of a Typical Indian Rail Road Vehicle

HEMANTHA KUMAR[1] AND C. SUJATHA[2]

[1]Research Scholar, [2]Professor Machine Design Section, Department of Mechanical Engineering, Indian Institute of Technology Madras, Chennai-600 036, India.
email: hemantha76@yahoo.co.in, sujatha@iitm.ac.in

ABSTRACT

In this paper numerical simulation of the vertical dynamic behavior of a railway vehicle is presented. A typical Indian railway vehicle of the AC/EMU/T (Alternating Current /Electrical Multiple Unit / Trailer) type running on broad gauge track and modeled as a 17 degree of freedom (dof) rigid body has been used for the analysis. Linear governing equations of motion of the vehicle have been solved and natural frequencies determined. Dynamic response studies were carried out in the frequency domain for the vehicle moving on a straight track, with power spectral densities (PSD) of track vertical profile irregularities and cross level given as input. The excitation forces due to these inputs were introduced through Hertzian springs connected to the vehicle/track interface. The ride dynamic performance of the vehicle has been evaluated in terms of Sperling's ride indices for both ride comfort and ride quality at different speeds.

Keywords: Rail vehicle dynamics, Sperling's ride index, Ride comfort.

1. INTRODUCTION

The dynamic performance of a railroad vehicle as related to safety is evaluated in terms of specific performance indices. The quantitative measurement of ride quality is one of such performance indices. Ride quality is interpreted as the capability of the railroad vehicle suspension to maintain the motion within the range of human comfort and Sperling's ride index (W_z) which is a measure of ride quality and ride comfort is used by Indian Railways for this. Tanifuji [1] described an analytical study on the bending vibration of a bogie vehicle for evaluation of ride quality. He assumed that the car body behaves like a uniform beam supported at the two bogie pivots. Gangadharan [2] conducted studies on the vertical dynamics of Indian railroad vehicles using rigid body as well as finite element models without considering the Hertzian stiffness at the wheel track interface. A report of The Office of Research and Experiments [3] describes PSD functions of various track profiles obtained from four different railways. The objective of the present work is to find the Sperling's ride indices for both ride comfort and ride quality for different operating speeds.

2. MATHEMATICAL MODELLING

A 17 dof rigid body model (Fig. 1) has been used to study the vertical dynamic behavior of an AC/EMU/T (Alternating Current / Electrical Multiple Unit / Trailer) coach used in suburban mass rapid transport in Indian Railways. Vertical translation and rotation about x and y axes (roll and pitch) were the dof considered for the car body and two bogies. The four wheel and axle sets were allowed two dof each (vertical translation and roll and pitch), accounting for a total of 17 dof. Determination of the eigenvalues is the first step towards understanding the dynamic behaviour of any system. The equations of motion of the vehicle are given below:

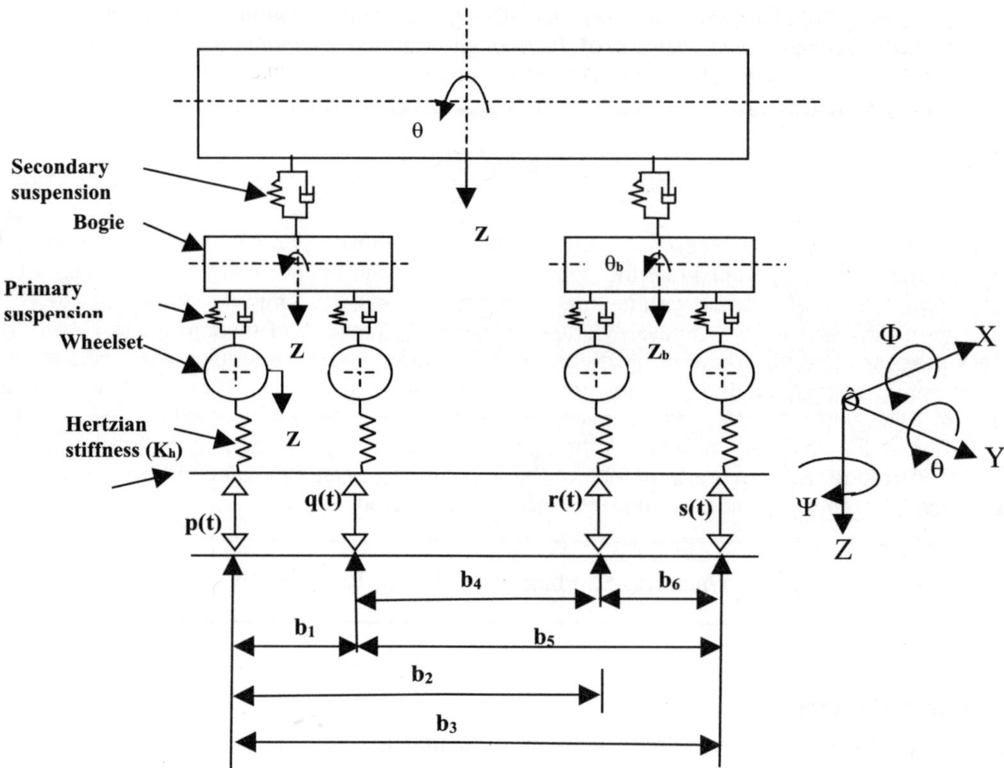

Fig. 1. Vertical dynamic model of a railway vehicle.

2.1 Bogie Bounce

$$m_b \ddot{z}_{b1} + f_{sb11} + f_{sb21} + f_{pb11} + f_{pb21} + f_{pb31} + f_{pb41} = 0$$

$$f_{sb11} = c_s \left[\left(\dot{z}_{b1} + l_s \dot{\phi}_{b1} \right) - \left(\dot{z}_c - l_c \dot{\theta}_c + l_s \dot{\phi}_c \right) \right] + k_s \left[\left(z_{b1} + l_s \dot{\phi}_{b1} \right) - \left(z_c - l_c \dot{\theta}_c + l_s \dot{\phi}_c \right) \right]$$

$$f_{sb21} = c_s\left[\left(\dot{z}_{b1} - l_s\dot{\phi}_{b1}\right) - \left(\dot{z}_c - l_c\dot{\theta}_c - l_s\dot{\phi}_c\right)\right] + k_s\left[\left(z_{b1} - l_s\phi_{b1}\right) - \left(z_c - l_c\theta_c - l_s\phi_c\right)\right]$$

$$f_{pb11} = c_p\left[\left(\dot{z}_{b1} - l_b\dot{\theta}_{b1} + l_p\dot{\phi}_{b1}\right) - \left(\dot{z}_{w1} + l_g\dot{\phi}_{w1}\right)\right] + k_p\left[\left(z_{b1} - l_b\theta_{b1} + l_p\phi_{b1}\right) - \left(z_{w1} + l_g\phi_{w1}\right)\right]$$

$$f_{pb21} = c_p\left[\left(\dot{z}_{b1} - l_b\dot{\theta}_{b1} - l_p\dot{\phi}_{b1}\right) - \left(\dot{z}_{w1} - l_g\dot{\phi}_{w1}\right)\right] + k_p\left[\left(z_{b1} - l_b\theta_{b1} - l_p\phi_{b1}\right) - \left(z_{w1} - l_g\phi_{w1}\right)\right]$$

$$f_{pb31} = c_p\left[\left(\dot{z}_{b1} + l_b\dot{\theta}_{b1} + l_p\dot{\phi}_{b1}\right) - \left(\dot{z}_{w2} + l_g\dot{\phi}_{w2}\right)\right] + k_p\left[\left(z_{b1} + l_b\theta_{b1} + l_p\phi_{b1}\right) - \left(z_{w2} + l_g\phi_{w2}\right)\right]$$

$$f_{pb41} = c_p\left[\left(\dot{z}_{b1} + l_b\dot{\theta}_{b1} - l_p\dot{\phi}_{b1}\right) - \left(\dot{z}_{w2} - l_g\dot{\phi}_{w2}\right)\right] + k_p\left[\left(z_{b1} + l_b\theta_{b1} - l_p\phi_{b1}\right) - \left(z_{w2} - l_g\phi_{w2}\right)\right]$$

$$m_b\ddot{z}_{b2} + f_{sb12} + f_{sb22} + f_{pb12} + f_{pb22} + f_{pb32} + f_{pb42} = 0$$

$$f_{sb12} = c_s\left[\left(\dot{z}_{b2} + l_s\dot{\phi}_{b2}\right) - \left(\dot{z}_c + l_c\dot{\theta}_c + l_s\dot{\phi}_c\right)\right] + k_s\left[\left(z_{b2} + l_s\phi_{b2}\right) - \left(z_c + l_c\theta_c + l_s\phi_c\right)\right]$$

$$f_{sb22} = c_s\left[\left(\dot{z}_{b2} - l_s\dot{\phi}_{b2}\right) - \left(\dot{z}_c + l_c\dot{\theta}_c - l_s\dot{\phi}_c\right)\right] + k_s\left[\left(z_{b2} - l_s\phi_{b2}\right) - \left(z_c + l_c\theta_c - l_s\phi_c\right)\right]$$

$$f_{pb12} = c_p\left[\left(\dot{z}_{b2} - l_b\dot{\theta}_{b2} + l_p\dot{\phi}_{b2}\right) - \left(\dot{z}_{w3} + l_g\dot{\phi}_{w3}\right)\right] + k_p\left[\left(z_{b2} - l_b\theta_{b2} + l_p\phi_{b2}\right) - \left(z_{w3} + l_g\phi_{w3}\right)\right]$$

$$f_{pb22} = c_p\left[\left(\dot{z}_{b2} - l_b\dot{\theta}_{b2} - l_p\dot{\phi}_{b2}\right) - \left(\dot{z}_{w3} + l_g\dot{\phi}_{w3}\right)\right] + k_p\left[\left(z_{b2} - l_b\theta_{b2} - l_p\phi_{b2}\right) - \left(z_{w3} + l_g\phi_{w3}\right)\right]$$

$$f_{pb32} = c_p\left[\left(\dot{z}_{b2} + l_b\dot{\theta}_{b2} + l_p\dot{\phi}_{b2}\right) - \left(\dot{z}_{w4} + l_g\dot{\phi}_{w4}\right)\right] + k_p\left[\left(z_{b2} + l_b\theta_{b2} + l_p\phi_{b2}\right) - \left(z_{w4} + l_g\phi_{w4}\right)\right]$$

$$f_{pb42} = c_p\left[\left(\dot{z}_{b2} + l_b\dot{\theta}_{b2} - l_p\dot{\phi}_{b2}\right) - \left(\dot{z}_{w4} - l_g\dot{\phi}_{w4}\right)\right] + k_p\left[\left(z_{b2} + l_b\theta_{b2} - l_p\phi_{b2}\right) - \left(z_{w4} - l_g\phi_{w4}\right)\right] \quad \ldots(1)$$

2.2 Car Body Bounce

$$m_c\ddot{z}_c + f_{sc1} + f_{sc2} + f_{sc3} + f_{sc4} = 0$$

$$f_{sc1} = c_s\left[\left(\dot{z}_c - l_c\dot{\theta}_c + l_s\dot{\phi}_c\right) - \left(\dot{z}_{b1} + l_s\dot{\phi}_{b1}\right)\right] + k_s\left[\left(z_c - l_c\theta_c + l_s\phi_c\right) - \left(z_{b1} + l_s\phi_{b1}\right)\right]$$

$$f_{sc1} = c_s\left[\left(\dot{z}_c - l_c\dot{\theta}_c - l_s\dot{\phi}_c\right) - \left(\dot{z}_{b1} - l_s\dot{\phi}_{b1}\right)\right] + k_s\left[\left(z_c - l_c\theta_c - l_s\phi_c\right) - \left(z_{b1} - l_s\phi_{b1}\right)\right]$$

$$f_{sc1} = c_s\left[\left(\dot{z}_c + l_c\dot{\theta}_c + l_s\dot{\phi}_c\right) - \left(\dot{z}_{b2} + l_s\dot{\phi}_{b2}\right)\right] + k_s\left[\left(z_c + l_c\theta_c + l_s\phi_c\right) - \left(z_{b2} + l_s\phi_{b2}\right)\right] \qquad \ldots(2)$$

$$f_{sc1} = c_s\left[\left(\dot{z}_c + l_c\dot{\theta}_c - l_s\dot{\phi}_c\right) - \left(\dot{z}_{b2} - l_s\dot{\phi}_{b2}\right)\right] + k_s\left[\left(z_c + l_c\theta_c - l_s\phi_c\right) - \left(z_{b2} - l_s\phi_{b2}\right)\right]$$

2.3 Wheel Bounce

$$m_w\ddot{z}_{w1} + f_{pw11} + f_{pw21} + f_{pw31} + f_{pw41} = 0$$

$$f_{pw11} = c_p\left[\left(\dot{z}_{w1} + l_g\dot{\phi}_{w1}\right) - \left(\dot{z}_{b1} - l_b\dot{\theta}_{b1} + l_p\dot{\phi}_{b1}\right)\right] + k_p\left[\left(z_{w1} + l_g\phi_{w1}\right) - \left(z_{b1} - l_b\theta_{b1} + l_p\phi_{b1}\right)\right]$$

$$f_{pw21} = c_p\left[\left(\dot{z}_{w1} - l_g\dot{\phi}_{w1}\right) - \left(\dot{z}_{b1} - l_b\dot{\theta}_{b1} - l_p\dot{\phi}_{b1}\right)\right] + k_p\left[\left(z_{w1} - l_g\phi_{w1}\right) - \left(z_{b1} - l_b\theta_{b1} - l_p\phi_{b1}\right)\right]$$

$$f_{pw31} = c_h\left[\dot{z}_{w1} + l_g\dot{\phi}_{w1}\right] + k_h\left[z_{w1} + l_g\phi_{w1}\right]$$

$$f_{pw41} = c_h\left[\dot{z}_{w1} - l_g\dot{\phi}_{w1}\right] + k_h\left[z_{w1} - l_g\phi_{w1}\right]$$

$$m_w\ddot{z}_{w2} + f_{pw12} + f_{pw22} + f_{pw32} + f_{pw42} = 0$$

$$f_{pw12} = c_p\left[\left(\dot{z}_{w2} + l_g\dot{\phi}_{w2}\right) - \left(\dot{z}_{b1} - l_b\dot{\theta}_{b1} + l_p\dot{\phi}_{b1}\right)\right] + k_p\left[\left(z_{w2} + l_g\phi_{w2}\right) - \left(\dot{z}_{b1} - l_b\theta_{b1} + l_p\phi_{b1}\right)\right]$$

$$f_{pw22} = c_p\left[\left(\dot{z}_{w2} - l_g\dot{\phi}_{w2}\right) - \left(\dot{z}_{b1} + l_b\dot{\theta}_{b1} - l_p\dot{\phi}_{b1}\right)\right] + k_p\left[\left(z_{w2} - l_g\phi_{w2}\right) - \left(z_{b1} + l_b\theta_{b1} - l_p\phi_{b1}\right)\right]$$

$$f_{pw32} = c_h\left[\dot{z}_{w2} + l_g\dot{\phi}_{w2}\right] + k_h\left[z_{w2} + l_g\phi_{w2}\right]$$

$$f_{pw42} = c_h\left[\dot{z}_{w2} - l_g\dot{\phi}_{w2}\right] + k_h\left[z_{w2} - l_g\phi_{w2}\right]$$

$$m_w\ddot{z}_{w3} + f_{pw13} + f_{pw23} + f_{pw33} + f_{pw43} = 0$$

$$f_{pw13} = c_p\left[\left(\dot{z}_{w3} + l_g\dot{\phi}_{w3}\right) - \left(\dot{z}_{b2} - l_b\dot{\theta}_{b2} + l_p\dot{\phi}_{b2}\right)\right] + k_p\left[\left(z_{w3} + l_g\phi_{w3}\right) - \left(z_{b2} - l_b\theta_{b2} + l_p\phi_{b2}\right)\right]$$

$$f_{pw23} = c_p\left[\left(\dot{z}_{w3} - l_g\dot{\phi}_{w3}\right) - \left(\dot{z}_{b2} - l_b\dot{\theta}_{b2} - l_p\dot{\phi}_{b2}\right)\right] + k_p\left[\left(z_{w3} - l_g\phi_{w3}\right) - \left(z_{b2} - l_b\theta_{b2} - l_p\phi_{b2}\right)\right]$$

$$f_{pw33} = c_h\left[\dot{z}_{w3} + l_g\dot{\phi}_{w3}\right] + k_h\left[z_{w3} + l_g\phi_{w3}\right]$$

$$f_{pw43} = c_h\left[\dot{z}_{w3} - l_g\dot{\phi}_{w3}\right] + k_h\left[z_{w3} - l_g\phi_{w3}\right]$$

$$m_w\ddot{z}_{w4} + f_{pw14} + f_{pw24} + f_{pw34} + f_{pw44} = 0$$

$$f_{pw14} = c_p\left[\left(\dot{z}_{w4} + l_g\dot{\phi}_{w4}\right) - \left(\dot{z}_{b2} + l_b\dot{\theta}_{b2} + l_p\dot{\phi}_{b2}\right)\right] + k_p\left[\left(z_{w4} + l_g\phi_{w4}\right) - \left(z_{b2} + l_b\theta_{b2} + l_p\phi_{b2}\right)\right]$$

$$f_{pw24} = c_p\left[\left(\dot{z}_{w4} - l_g\dot{\phi}_{w4}\right) - \left(\dot{z}_{b2} + l_b\dot{\theta}_{b2} - l_p\dot{\phi}_{b2}\right)\right] + k_p\left[\left(z_{w4} - l_g\phi_{w4}\right) - \left(z_{b2} + l_b\theta_{b2} - l_p\phi_{b2}\right)\right]$$

$$f_{pw34} = c_h\left[\dot{z}_{w4} + l_g\dot{\phi}_{w4}\right] + k_h\left[z_{w4} + l_g\phi_{w4}\right]$$

$$f_{pw44} = c_h\left[\dot{z}_{w4} - l_g\dot{\phi}_{w4}\right] + k_h\left[z_{w4} - l_g\phi_{w4}\right] \qquad ...(3)$$

2.4 Car Body Pitch

$$J_{cy}\ddot{\theta}_c - f_{sc1}l_C - f_{sc2}l_C + f_{sc3}l_C + f_{sc4}l_C = 0 \qquad ...(4)$$

2.5 Bogie Pitch

$$J_{by}\ddot{\theta}_{b1} - f_{pb11}l_b - f_{pb21}l_b + f_{pb31}l_b + f_{pb41}l_b = 0$$

$$J_{by}\ddot{\theta}_{b2} - f_{pb12}l_b - f_{pb22}l_b + f_{pb32}l_b + f_{pb42}l_b = 0 \qquad ...(5)$$

2.6 Car Body Roll

$$J_{cx}\ddot{\phi}_c + f_{sc1}l_s - f_{sc2}l_s + f_{sc3}l_s - f_{sc4}l_s = 0 \qquad \qquad ...(6)$$

2.7 Bogie Roll

$$J_{bx}\ddot{\phi}_{b1} + f_{pb11}l_p - f_{pb21}l_p + f_{pb31}l_p - f_{pb41}l_p + f_{sb11}l_s - f_{sb21}l_s = 0$$

$$J_{bx}\ddot{\phi}_{b2} + f_{pb12}l_p - f_{pb22}l_p + f_{pb32}l_p - f_{pb42}l_p + f_{sb12}l_s - f_{sb22}l_s = 0 \qquad \qquad ...(7)$$

2.8 Wheel and Axle Roll

$$J_{wx}\ddot{\phi}_{w1} + f_{pw31}l_g - f_{pw41}l_g + f_{pw21}l_p - f_{pw11}l_p = 0$$

$$J_{wx}\ddot{\phi}_{w2} + f_{pw32}l_g - f_{pw42}l_g + f_{pw22}l_p - f_{pw12}l_p = 0$$

$$J_{wx}\ddot{\phi}_{w3} + f_{pw33}l_g - f_{pw43}l_g + f_{pw23}l_p - f_{pw13}l_p = 0$$

$$J_{wx}\ddot{\phi}_{w4} + f_{pw34}l_g - f_{pw44}l_g + f_{pw24}l_p - f_{pw14}l_p = 0 \qquad \qquad ...(8)$$

3. DYNAMIC RESPONSE CALCULATIONS

Vertical profile and cross level are the two main irregularities of the track considered for vertical dynamic studies. These irregularities are random in nature and are therefore described by their power spectral densities (PSDs) in the frequency domain. For the present study, the data on Indian railway tracks were obtained from the work of Iyengar and Jaiswal [4]. Figs. 2 and 3 show track irregularity PSDs for vertical profile and cross level for a vehicle speed of 30 kmph.

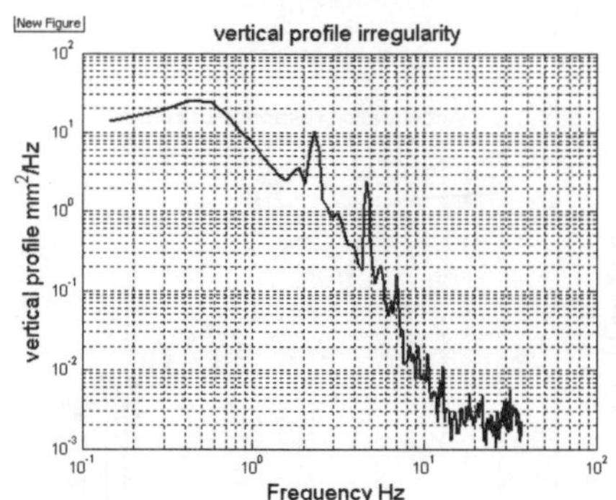

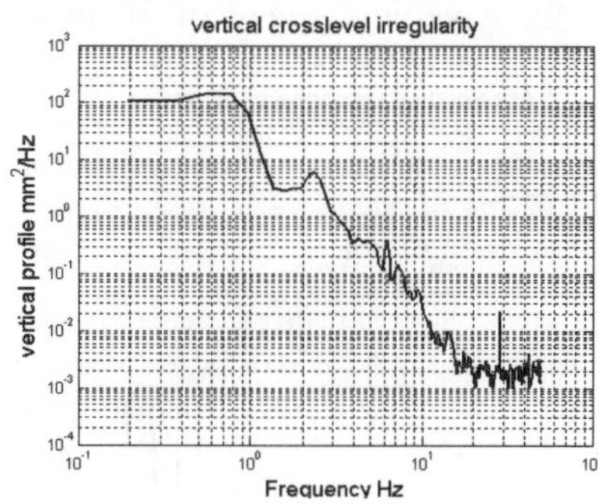

Fig. 2. Track vertical profile irregularity.   **Fig. 3.** Track cross level irregularity.

The 17 dof model can be treated as a system with eight random disturbances due to track irregularities at each of the eight rail wheel contact points. If the input from the left rail is considered to be completely correlated with that of the right rail, then the system can be simplified to a case of four random loadings p(t), q(t), r(t) and s(t) at the wheel rail contact points [1, 4]. It is further assumed that the input is space correlated between the successive wheels on each rail; hence q(t), r(t) and s(t) reproduce p(t) after time lags of τ_1, τ_2 and τ_3 corresponding to wheel bases b_1 (2.896 m), b_2 (14.63 m) and b_3 (17.526 m) as shown in Fig. 1. Therefore, $S_p(f) = S_q(f) = S_r(f)$

= $S_s(f)$ = PSD of p(t), q(t), r(t) and s(t), respectively. From the theory of random vibration, Sx(f), the PSD of response x(t) can be calculated knowing the receptances and PSDs of the inputs from equation (10).

$$|S_x(f)| = \left\{ |\alpha_{xp}|^2 + |\alpha_{xq}|^2 + |\alpha_{xr}|^2 + |\alpha_{xs}|^2 + 2|\alpha_{xp}||\alpha_{xq}|\cos\phi_1 + 2|\alpha_{xr}||\alpha_{xp}|\cos\phi_2 + 2|\alpha_{xs}||\alpha_{xp}|\cos\phi_3 + 2|\alpha_{xq}||\alpha_{xr}|\cos\phi_4 + 2|\alpha_{xq}||\alpha_{xs}|\cos\phi_5 + 2|\alpha_{xr}||\alpha_{xs}|\cos\phi_6 \right\} S_p(f) \qquad \ldots(9)$$

Here α_{xp}, α_{xq}, α_{xr} and α_{xs} are the receptances for harmonic excitation. Phase angles ϕ can be written in terms of the wheel bases and vehicle velocity as

$$\phi_1 = 2\pi f b_1/v, \ \phi_2 = 2\pi f b_2/v, \ \phi_3 = 2\pi f b_3/v, \ \phi_4 = 2\pi f b_4/v, \ \phi_5 = 2\pi f b_5/v, \ \phi_6 = 2\pi f b_6/v \ \ldots(10)$$

The excitation force due to each input is introduced through a Hertzian contact spring at the wheel/rail contact area (Fig. 1). Though this is non-linear and varies with the contact force applied, it has been linearised and an average value of $1.5*10^9$ N/m is used in this analysis.

4. RESULTS AND DISCUSSIONS

Natural frequencies of the above vertical dynamics model are shown in Table 1.

Table 1. Natural frequencies of vehicle (Hz)

1	0.0224	5	4.7458	9	8.5801	13	159.2292	17	214.6872
2	0.0351	6	6.6785	10	159.2292	14	214.6872		
3	0.3427	7	6.6785	11	159.2292	15	214.6872		
4	4.7458	8	8.5801	12	159.2292	16	214.6872		

Figure 4 and 5 show the vertical acceleration response PSDs for vertical and cross level inputs respectively, for a vehicle speed of 30 m/s.

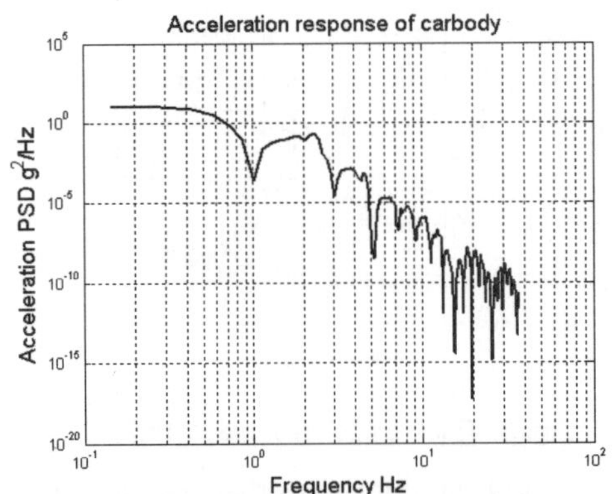

Fig. 4. Car body acceleration for vertical input.

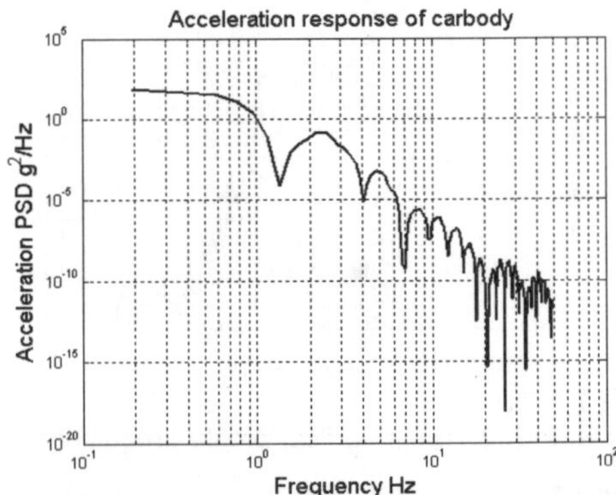

Fig. 5. Car body acceleration for cross level input.

4.1 Sperling's Ride Index (W_z)

The ride index (W_z) was introduced by Sperling, in order to evaluate ride quality and ride comfort of a railroad vehicle (ORE Report, C116/8, 1978).

$$Wz = \left(a^2 B^2\right)^{1/6.67} \qquad \qquad ...(11)$$

where, 'a' is the amplitude of acceleration in cm/s^2 and 'B' is the acceleration weighting factor, which is given by

$$B = 1.14\left[\frac{\left[\left(1-0.056\,f^2\right)^2 + \left(0.0645\right)^2\left(3.35\,f^2\right)\right]}{\left[\left(1-0.252\,f^2\right)^2 + \left(1.547\,f - 0.00444\,f^3\right)^2\right]\left(1+3.55\,f^2\right)}\right]^{1/2} \qquad ...(12)$$

Here, 'f' is the frequency in Hz. The weighting factor B for ride comfort in the vertical direction is given by

$$B_s = 0.588\left[\frac{1.911f^2 + \left(0.25f^2\right)^2}{\left(1-0.277f^2\right)^2 + \left(1.563f - 0.0368f^3\right)^2}\right]^{1/2} \qquad ...(13)$$

The vehicle body vibration is not at a single frequency. The ride index calculation has to be done for the entire spectrum. The W_z ride factor is determined for each individual frequency from the equations mentioned above and the total W_z factor is calculated as

$$Wz_{total} = \left(Wz_1^{10} + Wz_2^{10} + Wz_3^{10} + \cdots + Wz_n^{10}\right)^{1/10} \qquad ...(14)$$

Sperling's indices for the vehicle under study at speeds of 30 to 100 m/s are shown in Table 2.

On comparison with the standard values set by ORE, the values of ride quality are found to be 'Good' and the values of ride comfort are found to be in the **'Clearly noticeable'** region.

Table 2. Ride indices for the different operating speeds(m/s)

Speed	Vertical profile irregularity		Cross level irregularity	
	Ride quality	Ride comfort	Ride quality	Ride comfort
30	1.3633	1.4482	1.4277	1.5219
40	1.3211	1.4082	1.3837	1.4835
50	1.2889	1.3754	1.3503	1.4513
60	1.2624	1.3476	1.3234	1.4237
70	1.2393	1.3232	1.3005	1.3996
80	1.2192	1.3011	1.2810	1.3777
90	1.2011	1.2809	1.2635	1.3574
100	1.1849	1.2621	1.2474	1.3384

CONCLUSION

Vertical dynamic analysis has been carried out for a typical Indian railway vehicle of the AC/EMU/ T (Alternating Current /Electrical Multiple Unit /Trailer) type running on broad gauge track. The natural frequencies have been computed using a 17 dof model. Using random vibration theory, the vertical acceleration response at car cg has been calculated in the frequency domain with PSDs of the random track vertical profile and cross level irregularities given as input. Sperling's ride indices have been found out for the above vehicle. On comparison with the standards set by ORE, the values are found well within satisfactory limit.

NOMENCLATURE AND VEHICLE DATA

m_c Mass of car body (33700 kg)

m_w Mass of wheel set (1500 kg)

l_g Semi gauge length (0.864 m)

l_p Half of primary spring spacing – lateral (1.127 m)

h_c Height of car cg from secondary spring centre (2.4285 m)

J_{cy} Pitch moment of inertia of car body ($7.67*10^5$ kg–m^2)

J_{cx} Roll moment of inertia of car body ($5.24*10^4$ kg-m^2)

J_{bx} Roll moment of inertia of bogie ($2.02*10^3$ kg-m^2)

J_{wx} Roll moment of inertia of wheel set ($7.13*10^2$ kg-m^2)

J_{cz} Yaw moment of inertia of car body ($7.36*10^5$ kg-m^2)

J_{bz} Yaw moment of inertia of bogie ($3.56*10^3$ kg-m^2)

c_{sz} Secondary damping in vertical direction (22000 Ns/m)

k_h Hertzian stiffness of wheel and track ($1.5*10^9$ N/m)

V Linear velocity of wheel

m_b Mass of bogie (3150 kg)

l_b Semi wheel base of bogie (1.448 m)

l_c Half of bogie centre pin spacing (7.315 m)

l_s Half of secondary spring spacing – lateral (0.794 m)

h_b Height of bogie cg from secondary spring centre (0.093 m)

J_{by} Pitch moment of inertia of bogie ($2.02*10^3$ kg–m^2)

J_{wz} Yaw moment of inertia of wheel set ($7.13*10^2$ kg-m^2)

k_{pz} Primary stiffness in vertical direction ($0.7*10^6$ N/m)

k_{sz} Secondary stiffness in vertical direction ($0.41*10^3$ N/m)

c_{pz} Primary damping in vertical direction ($5.88*10^3$ Ns/m)

z_c, z_b, z_w Vertical displacement of car, bogie and wheel set

ϕ_c, ϕ_b, ϕ_w Roll of car body, bogie and wheel set

ψ_c, ψ_b, ψ_w Yaw of car body, bogie and wheel set

REFERENCES

1. K. Tanifuji, 1991, An analysis of the body bending vibration of a bogie vehicle for an evaluation of the ride quality with deflated air spring. Journal of Rail and Rapid Transport, 205, 35-42.

2. K.V. Gangadharan, 2001, Analytical and Experimental Studies on the Dynamics of Railroad Vehicles, Ph.D. Dissertation, IIT Madras, Chennai, India.

3. ORE Report C116/RP 1-9 /EC, 1971-1978, Interaction between Vehicle and Track, ORE, Utrecht.

4. R.N. Iyengar and O.R. Jaiswal, 1995, Random field modelling of railway track irregularities, Journal of Transportation Engineering, July/ August, 303-308.

5. Hemantha Kumar, B. Suresh Kumar and C. Sujatha, 2005, Lateral dynamic analysis of a typical Indian railroad vehicle, Proc. 12th NACCOM-2005, Guwahati, 208-214.

63

Stochastic Finite Element Forced Vibration Analysis with Reliability Based Adaptive Mesh Refinements

M. Manjuprasad[1] and C.S. Manohar[2]

[1]Structures Division, National Aerospace Laboratories, Bangalore-560017, India
email: manjuprasad_m@yahoo.com
[2]Department of Civil Engineering, Indian Institute of Science, Bangalore-560 012, India
email: manohar@civil.iisc.ernet.in

ABSTRACT

The problem of developing an adaptive random field mesh refinement technique for stochastic finite element reliability analysis of structures under forced vibrations is considered in this paper. Spatially varying stochastic system parameters (such as Young's modulus) are modeled as non-Gaussian random fields with prescribed marginal distribution and autocorrelation function in conjunction with Nataf's models. Expansion optimum linear estimation method is used for random field discretisation. The performance function is defined in terms of peak displacement at a given node in the given time duration. Sensitivity measures in terms of gradients of the performance function with appropriate transformations are proposed and used as refinement indicators for carrying out adaptive random field mesh refinements. Reliability index based error indicator is proposed and used for assessing the percentage error in the estimation of notional failure probability. Adaptive random field mesh refinement is carried out using hierarchically graded mesh obtained through bisection of elements. The efficacy of the technique developed is illustrated by a numerical example.

Keywords: Stochastic Finite Element, Random Field Discretisation, Structural Dynamics, Structural Reliability.

1. INTRODUCTION

In general, structural system parameters and loads vary stochastically over space and/or time. Consequently, the response of the structure is also stochastic which may lead to failure with respect to one or more limit states used in design. Stochastic Finite Element Methods (SFEMs) can be effectively used to estimate the stochastic response and reliability of structures with system parameters and loads represented as random fields/processes. The predictions based on mathematical models are inflicted with errors from various sources. The difference between approximate (computable) and exact (non-computable) solution leads to the traditionally well-known error associated with mesh resolution called the discretisation error. Considerable research has been carried out on adaptive mesh refinements, particularly in the context of deterministic finite elements (for example see, [1]). However, literature available on adaptive methods for stochastic finite element analysis is limited. In this paper, a new adaptive algorithm is presented for carrying out stochastic finite element reliability analysis of structures under forced vibrations. It aims at improving the reliability index/failure probability estimates obtained through adaptive refinement of random field (RF) mesh.

2. REFINEMENT INDICATOR

The authors have earlier shown that with appropriate transformations the importance measures for ranking the random variables offer a powerful means to identify the spatial regions in the finite element model where refinement of random field mesh is required [2, 3]. In the present paper the importance measure for performance functions with multiple regions of comparable importance is used to rank the discretised random variables in the order of their relative importance, and, also, to investigate if such ranking could be used for adaptive mesh refinement in stochastic finite element reliability analysis of structures under forced vibrations. If R distinct points are identified in the n-dimensional standard normal space, the Global Importance Measure (GIM) of X_i is computed using the expression,

$$\gamma_i = \sum_{l=1}^{R} w_l \lambda_l^{i^2} \Big/ \sum_{i=1}^{n} \sum_{l=1}^{R} w_l \lambda_l^{i^2} \qquad ...(1)$$

The parameter λ_l^i is defined as

$$\lambda_l^i = \sum_{j=1}^{R} w_l \left\{ \left(\frac{\partial g}{\partial X_i} \right)^i \right\}^2 \qquad ...(2)$$

Here, g is the performance function and w_l are weighting functions. Since, the algorithm is carried out in the standard normal space, the direction cosines for the identified design points $\left(\frac{\partial g}{\partial U} \right)$ are obtained in the standard normal U-space. The corresponding direction cosines in the real X-space are expressed as

$$\left(\frac{\partial g}{\partial X} \right) = J \left(\frac{\partial g}{\partial U} \right) \qquad ...(3)$$

To give a greater weightage to the points closer to the origin in the standard normal U-space, the weights wl are defined in terms of probability content associated with each point in the U-space, and are expressed as

$$w_l = \Phi(-\beta_l) \Big/ \sum_{j=1}^{R} \Phi(-\beta_j) \qquad ...(4)$$

where, βl is the Hasofer-Lind reliability index of the lth point in the U-space.

The global importance measures with respect to element random variables in standard normal space (spatial) are computed using Eq. (1)–Eq. (4) and used as refinement indicators for identifying regions of importance for carrying out adaptive refinement of random field mesh. They are analogous to the element error indicators based on energy error norm used in conventional deterministic adaptive finite element methods.

3. ERROR INDICATOR

In the present method we propose to use the reliability index based norm for estimation of error. Let β^* represent the reliability index computed from a finest random field/finite element mesh and β represent the reliability index computed from a specified mesh. The relative error percentage $\eta\beta$ can therefore be defined as,

$$\eta_\beta(\%) = \frac{\left|\beta - \beta^*\right|}{\left|\beta^*\right|} \times 100 \qquad ...(5)$$

The relative error percentage $\eta\beta$ computed during the adaptive stochastic finite element reliability analysis is used as the error indicator for refinement of random field mesh. The step-by-step algorithm for adaptive stochastic finite element reliability analysis of structures is discussed in the work by Manjuprasad [4].

4. RANDOM FIELD DISCRETISATION

Spatially varying uncertain system parameters and load parameters are modelled as random fields. Discretisation of continuous-parameter random field required for stochastic finite element analysis is carried out using Expansion Optimum linear Estimation (EOLE) method [5] in conjunction with multivariate distribution models based on Nataf's transformations [6]. For the numerical studies carried out in this paper, the following differentiable correlation function is considered for modelling the isotropic, homogeneous random fields:

$$\rho(x, x') = exp\left(-\frac{\left|x - x'\right|^2}{\lambda^2}\right) \qquad ...(6)$$

Here, x refers to the vector of random field nodal coordinates in real space and λ refers to the correlation length which is the measure of fluctuation of random field.

5. RELIABILITY ANALYSIS

Structural reliability analysis is carried out using first order reliability method (FORM) with improvements. Rackwitz-Fiessler method is used to solve the optimization problem [7]. Search for multiple design points of a reliability problem is carried out using a simple heuristic method [8]. An approach using the product of conditional marginals (PCM) method [9] is used to get an approximate value (and not the bounds) of the multinormal integral required to solve the system reliability problem.

Adaptive stochastic finite element reliability analysis is carried out using suitable formulations to compute the performance function and the gradients of the performance function with respect to the required element random variables. Let $X = \left\{X_j\right\}_{j=1}^{N}$ be a vector of N random field variable vectors X_j (in original space) obtained after discretisation of N independent random fields $X_j(x)$. For reliability analysis of linear forced vibration problem, the performance function is defined in terms of peak displacement at a given node $\hat{x}$ in given time duration, which can be expressed as,

$$g = u^* - \max_{0 \leq t \leq T} |u(\hat{x}, t, \Psi)| \qquad \ldots(7)$$

The gradient of the performance function with respect to any given discretised random field variable is obtained through chain rule of differentiation. Details of formulations developed and used for reliability analysis of forced vibration problem including gradients of structural matrices with respect to parameters of interest are given in the work by Manjuprasad [4].

6. NUMERICAL EXAMPLE

The adaptive stochastic finite element method for forced vibration problem proposed in this paper is demonstrated by considering the example of cantilever beam of span 600 mm subjected to random dynamic excitation. The problem involves multiple random field models for spatial variations in structural properties. The example structure considered is shown in Fig. 1.

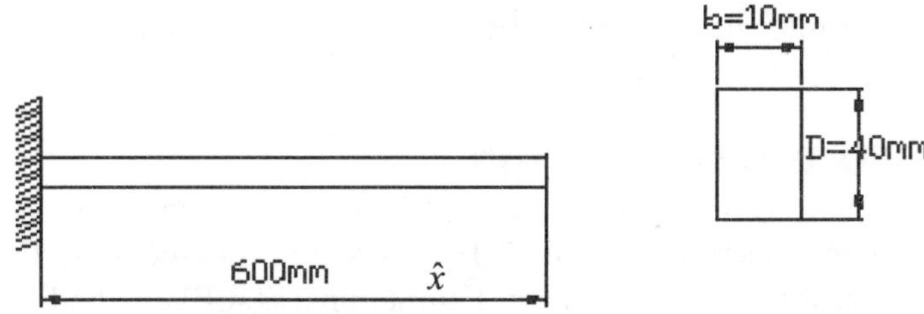

Fig. 1. Geometry of cantilever beam with tip point load.

The load $f(t)$ due to a tip concentrated load is modeled as a zero mean stationary Gaussian random process with a PSD given by band limited white noise with values of variance $=10^4$ kg^2, lower cut-off frequency $= 0$ rad/s and upper cut-off frequency $= 1000$ rad/s. The beam structure is modeled using Euler-Bernoulli beam elements. Viscous damping with a proportional damping model is considered for the beam. The elastic modulus and mass density of the beam are modeled as mutually dependent homogeneous non-Gaussian random fields with a lognormal first order pdf and auto-covariance function of the form in Eq. 6. Mean value of elastic modulus is taken as 2.1E5 MPa and a coefficient of variation 0.20 is assumed. Mean value mass density is taken as 7.8E-6 kg/mm^3 and a coefficient of variation 0.20 considered. The Poisson's ratio is taken as 0.3 and damping ratio of first and second modes are assumed to be 0.02 and 0.01, respectively. Furthermore, the applied load is taken to be independent of the random fields for elastic modulus $E(x)$ and mass density $\rho(x)$.

The performance function is defined in terms of the maximum value of the tip displacement measured over specified time duration after the beam has reached a steady state. In implementing the reliability analysis procedure, the load $f(t)$ is replaced by a Fourier series with random coefficients, with number of terms equal to 10. The set of basic random variables includes the coefficients present in the Fourier expansion for the load in addition to the set of basic random variables emerging during the discretisation of $E(x)$ and $\rho(x)$. It is important to note here that, in addition to choosing the details of displacement field and random field meshes, one needs to select the time step of integration as well. We adopt a direct integration approach for solving this problem using Newmark's method with parameters with $\delta = 0.5$, $\gamma = 0.25$, (corresponding to average acceleration method).

Here, we introduce the notation (N_1, N_2, N_3) to denote number of displacement field elements by N_1, number of random field elements in $E(x)$ by N_2 and number of random field elements in $\rho(x)$ by N_3.

We define β^* using a (10, 10, 10) uniform mesh with time step $\Delta t = 0.05$s. This result along with results for (10, 5, 5), (10, 5, 8) and (10, 10, 10) uniform meshes with a coarser time step of integration $\Delta t = 0.1$s are shown in Table 1.

Table 1. Refinement of RF mesh for mass density of cantilever beam subjected to stationary random excitation due to tip point load

N_1	Δts	N_2	N_3	β	$\eta_\beta(\%) = \dfrac{\|\beta - \beta^*\|}{\|\beta^*\|} \times 100$	Type of RF Mesh
10	0.10	5	5	2.0392	1.3211	Uniform
		5	8	2.0396	1.3017	
		10	10	2.0402	1.2727	
	0.05	10	10	2.0665(ß*)	0.0000	
	0.10	5	8	2.0610	0.2661	Adaptive

By holding Δt at 0.1s we now adaptively refine the random field meshes using GIM derived from (10, 5, 5) uniform mesh with $\Delta t = 0.1$s. Figs. 2 (a) and 2 (b) show the GIM plots for the random fields representing elastic modulus and mass density. A comparison of the figures shows that the GIM values associated with $E(x)$ are negligibly small as compared with GIM for $\rho(x)$. The spatial variation of GIM also shows notably different trends but this difference itself is of marginal interest given the very low values of GIM associated with $E(x)$. The results on reliability index obtained using adaptively refined mesh with $\Delta t = 0.1$s and (10, 5, 8) mesh is shown in Table 1. It should be noted that the adaptive refinement here is restricted only to the mass density field (see Fig. 3 for details of refined mesh for $\rho(x)$. The mesh for elastic modulus field is uniform in nature. Also, it needs to be emphasized that Δt is held uniform at a coarser value of 0.1s. A comparison of results in Table 1 reveals that the reliability index obtained using adaptive random field mesh refinement assuming a coarser Δt, matches well with the results on reliability index with a finer time step size ($\Delta t = 0.05$s) and a finer displacement field and random field meshes (10, 10, 10 uniform mesh). This points towards usefulness of mesh refinement procedures discussed here.

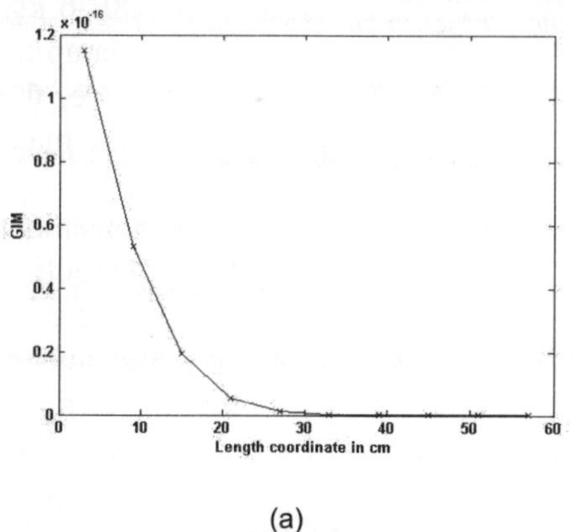

(a)

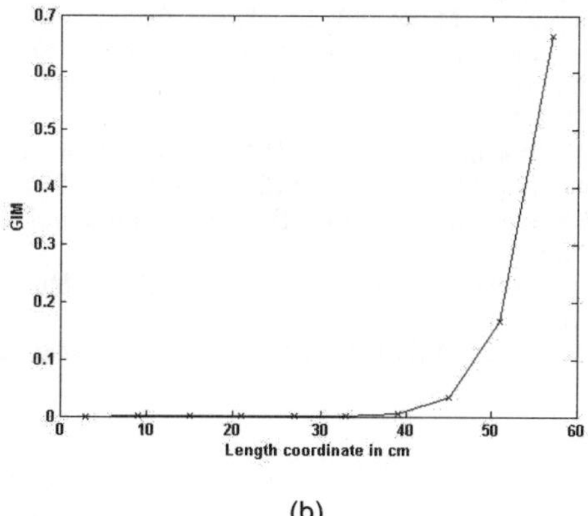

(b)

Fig. 2. (a) GIM plot for elastic modulus with (10, 5, 5) uniform mesh and time step = 0.1s
(b) GIM plot for mass density with (10, 5, 5) uniform mesh and time step = 0.1s.

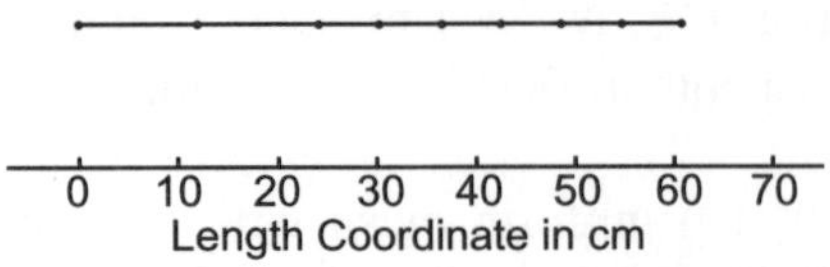

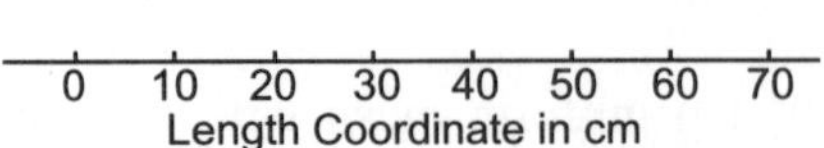

0 10 20 30 40 50 60 70
Length Coordinate in cm

Fig. 3. Adaptive 8-element RF mesh for mass density based on GIM plot for (10, 5, 5) uniform mesh; time step = 0.1s.

CONCLUSION

The problem of developing an adaptive technique for stochastic finite element reliability analysis of structures under forced vibrations is considered in this paper. Sensitivity measures in terms of gradients of the performance function with appropriate transformations are proposed and used as refinement indicators for carrying out adaptive random field mesh refinements. Reliability index based refinement indicator is proposed and used for assessing the percentage error in the estimation of notional failure probability. The efficacy of the technique developed is illustrated by an example problem of forced vibration of a cantilever beam. From numerical studies carried out it is found that, for a given correlation length, good improvement in the estimation of β is attained through adaptive refinement of random field mesh using the proposed refinement indicators. The proposed adaptive technique will be useful to carryout reliability based safety assessment of structures.

ACKNOWLEDGEMENTS

The first author wishes to thank Dr. S. Viswanath, Head, Structures Division, NAL, Bangalore for the kind support and encouragement provided to publish this paper. The author also wishes to thank Dr. A.R. Upadhya, Director, NAL, Bangalore for the kind permission granted to publish this paper.

REFERENCES

1. Ainsworth, M. and Oden, J.T. (2000). *A posteriori error estimation in finite element analysis.* John Wiley and Sons, New York.
2. Manjuprasad, M. and Manohar, C.S. (2005). Adaptive random field mesh refinement in SFEM based reliability analysis. *Proceeding, National symposium on structural dynamics, random vibrations and earthquake engineering,* July 21–22, 2005, Bangalore, India.
3. Manjuprasad, M. and Manohar, C.S. (2005). Stochastic finite element free vibration analysis with reliability based adaptive mesh refinements. *Proceedings, International Conference on Computational & Experimental Engineering and Sciences (ICCES–2005), December 1-6, 2005, Chennai, India.*
4. Manjuprasad, M. (2005). *Stochastic finite element analysis and safety assessment of structures under random excitations.* Ph.D. Thesis, Indian Institute of Science, Bangalore, India.
5. Li, C.C. and Der Kiureghian, A. (1993). Optimal discretisation of random fields. Journal of Engineering Mechanics (ASCE), 119(6): 1136-1154.
6. Der Kiureghian, A. and Liu, P. L. (1986). Structural reliability under incomplete probability information. Journal of Engineering Mechanics (ASCE), 112(1): 85–104.
7. Rackwitz, R., and Fiessler, B. (1978). Structural reliability under combined random load sequences. Computers and Structures, 9:489–494.
8. Der Kiureghian, A. and Dakesian, T. (1998). Multiple design points in first and second order reliability, Structural Safety, 20: 37–49.
9. Pandey, M.D. (1998). An effective approximation to evaluate multinormal integrals. Structural Safety, 20:51–67.

64

Flutter Control Using Resistively Shunted Piezoceramics

S.B. KANDAGAL[1] **AND KARTIK VENKATRAMAN**[2]

Department of Aerospace Engineering, Indian Institute of Science, Bangalore-560 012
[1]*email: sbk@aero.iisc.ernet.in* [2]*email: Kartik@aero.iisc.ernet.in*

ABSTRACT

This paper investigates the feasibility of flutter control using resistively shunted piezoceramics. 2-D airfoil model is elastically restrained in heave and pitch by a set of leaf springs modeled as Bernoulli-Euler beams. The piezoceramic transducers are bonded to the leaf springs. Unsteady aerodynamic theory is used to solve the flutter determinant incorporating the additive damping terms due to resistive shunting. Tuned dampers realized by resistive shunted piezoceramics in heave and pitch springs could enhance flutter speed in the range 25-50% based on additive damping of 10% achieved due to resistive shunting of piezoceramics in either pitch spring, heave spring or both.

Keywords: Flutter control, PZT, resistive shunting, optimal thickness ratio.

1. INTRODUCTION

Swept-back wings of aircraft are prone to dynamic aeroelastic instabilities such as flutter. Since most modern combat aircraft have swept-back wings, active flutter control therapies have become attractive. Several options exist for implementing the aeroelastic control schemes. Aeroelastic control may be effected by altering the aerodynamic forces acting on the wing through articulation of leading or trailing edge control surfaces, or, as in recent proof-of-concept investigations, using strain actuators such as piezoceramics that directly deform the structure through electromechanical coupling [1]. Horikawa and Dowell [2] performed simple feedback control simulations on a two degree-of-freedom (dof) airfoil system and studied the effectiveness of these control laws in changing the flutter speed. Stability analysis with these feedback controls in place was also performed. Control realization was assumed to be through trailing-edge actuator. Lazarus et al.[3] considered a four-dof airfoil system with trailing and leading edge controls. Besides articulated control force realization, strain actuators, such as piezoceramics and magnetostrictives, were also considered.

The present work considers passive control of a two degree-of-freedom model of aeroelastic flutter of a wing section using resistively shunted piezoceramics. The piezoceramic materials are transformers that convert mechanical energy to electrical energy and vice versa. When these piezoceramics are bonded to a structure, the mechanical strain energy generated in piezoceramics is converted to electrical voltage across the poling direction of the piezoceramic device. This voltage or electrical energy is dissipated or shunted to another frequency bandwidth using electrical networks connected to the terminals of piezoceramics, which acts as passive vibration control [4] tool.

Resistively shunted piezoceramic transducers enhance the damping characteristics of the host structure. Optimal thickness and length ratios of the PZT and host-beam structure [5] were used in application of damping enhancement and hence to increase the flutter speed of the wing section. Note that flutter is a dynamic instability caused due to a combination of coupling of bending and torsional oscillations, as well as aerodynamic damping. The aerodynamic damping is a function of the airspeed, and increases with airspeed. The net damping in the aeroelastic system is an algebraic sum of the structural and aerodynamic damping. Therefore, as the airspeed increases, the negative aerodynamic damping will increase, and the net damping will bifurcate from a positive damping to a negative damping system. The flutter speed then is that airspeed where this net damping changes from positive to negative. Thus, the additive damping due to resistive shunting of a piezoceramic transducer will increase the flutter speed.

As already mentioned the 2-D airfoil model is elastically restrained in heave and pitch by a set of leaf springs. These leaf springs can be modeled as Bernoulli-Euler beams. The piezoceramic transducers are bonded to the leaf springs. The location of the piezoceramics is shown in Fig. 1.

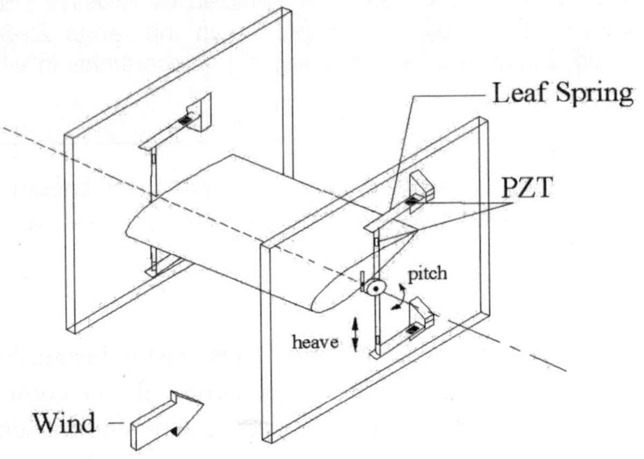

Fig. 1. 2-D airfoil model with PZT in leaf springs.

2. FORMULATION

The airfoil selected has section geometry parameters as $a_h = -0.2$, $x_\alpha = 0.2$, $\mu = 10$, $r_\alpha = 0.45$, $t/2b = 0.12$, wing parameters as $\bar{\omega}_h = 0.46$, $L/b = 7.74$ and steady aerodynamic coefficients as $C_L = 2\pi$, $C_M = 1.885$. The flutter determinant incorporating the additive damping terms due to resistive shunting is represented in the governing equation of motion for a 2-D airfoil as

$$\begin{pmatrix} \mu(1-\overline{\omega}_h^2 X(1+ig_h))+L_h & \mu x_\alpha + L_\alpha - L_h(0.5+a_h) \\ \mu x_\alpha + 0.5 - L_h(0.5+a_h) & \mu r_\alpha^2[1-X(1+ig_h)]-0.5(0.5+a_h)+ \\ & M_\alpha - L_\alpha(0.5+a_h)+L_h(0.5+a_h)^2 \end{pmatrix} \begin{pmatrix} \overline{h} \\ \alpha \end{pmatrix} = \begin{pmatrix} 0 \\ 0 \end{pmatrix}$$

where, g_h and g_α are the additive damping terms in heave and pitch, respectively. Flutter determinant is solved for various cases involving heave and pitch springs (realized through aluminium beams) in association with PZT patches.

3. RESULTS AND DISCUSSIONS

The effect of additive structural damping due to piezoceramic resistive shunting on the flutter speed is studied. Figure 2 shows the variation of the net structural damping with airspeed.

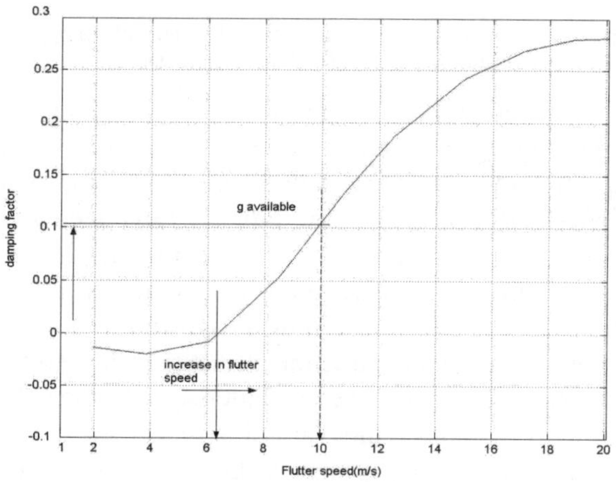

Fig. 2. Effect of damping factor on flutter speed.

Note that the flutter condition is when the net structural damping factor g is zero. Below the flutter speed, the net damping is negative. This is so, because we have assumed the motion to be of the form $\exp(i\dfrac{\omega_\alpha}{\omega}(1+ig)t)$. This figure also indicates that there is 66% increase in flutter speed if the structural damping is augmented by 10%. Now we investigate the effect of resistive shunting of piezoceramic transducers on flutter speed for the two degree-of-freedom model of Fig. 1. There are three cases of interest here. Resistive shunting of the piezoceramic transducers bonded to the leaf springs elastically restraining the heave motion, those bonded to the leaf springs elastically restraining the pitch motion, and resistive shunting of the combination. Figure 3 shows the variation of flutter speed with added damping due to resistive shunting of PZT in the pitch direction.

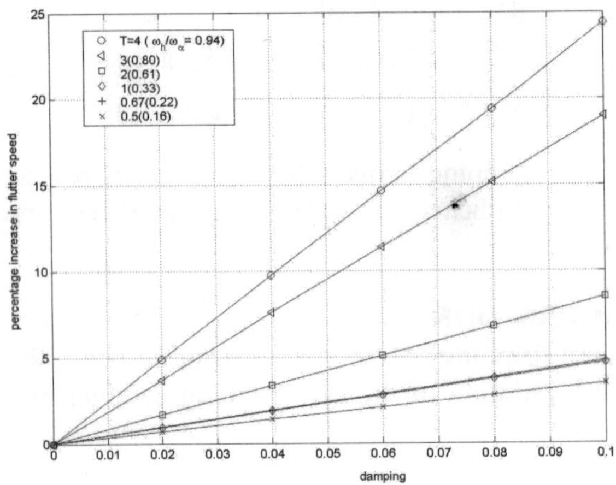

Fig. 3. Percentage increase in flutter speed as a function of damping factor due to resistive shunting of PZTs in pitch springs.

The different curves are for different thickness ratios of the leaf spring relative to the piezoceramic transducer bonded to it. Corresponding to different thickness ratios, the uncoupled natural frequencies in heave and pitch, ω_h / ω_α, also change. Their values are also listed together with the thickness ratios. Note that as the thickness ratio increases, the ratio ω_h / ω_α decreases. There is a consequent decrease in the flutter speeds for higher thickness ratios. However, a truer picture of variation in flutter speed with resistive shunting of the PZT transducer emerges by observing the plot of the percentage increase in flutter speed with added damping as shown in Fig. 3. Here, for higher thickness ratios, and consequently lower values of ω_h / ω_α, the percentage increase in flutter speed is more. For a thickness ratio of 4, there is 25% increase in flutter speed when the added damping due to resistive shunting is 10%. Figure 4 shows the variation of flutter speed for the case when the piezoceramic transducers bonded to the leaf springs in heave are resistively shunted.

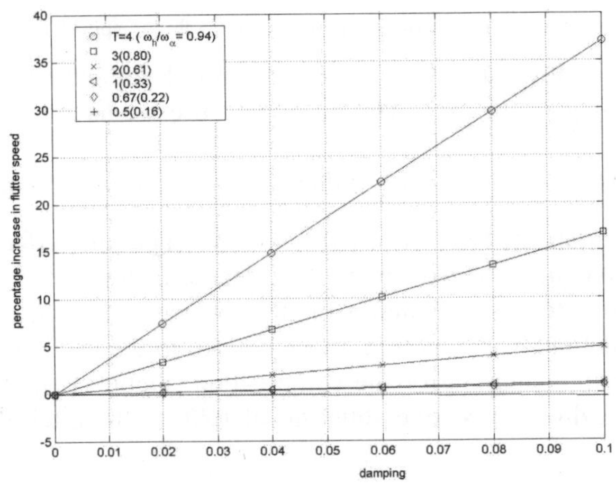

Fig. 4. Percentage increase in flutter speed as a function of damping factor due to resistive shunting of PZT in heave direction.

The trends are similar to that of additive damping in the pitching direction. Note however, that the percentage increase in flutter speed is greater than in the pitching case. For an additive damping factor of 10%, the increase in flutter speed is as high as 37% for a thickness ratio of 4. The effect on flutter speed of combined resistive shunting of the piezoceramic transducers in the heave and pitching directions, $g_h = g_\alpha$, is shown in Fig. 5. Besides the fact that the increase in flutter speed with increase in damping factor shows the same trend as in the earlier two cases, the percentage increase in flutter speed is much higher. For ω_h / ω_α, there is 52% increase in flutter speed at a combined damping factor value of 0.1.

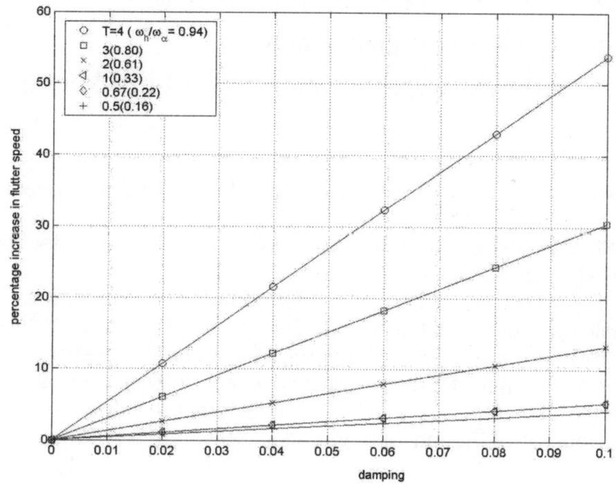

Fig. 5. Percentage increase in flutter speed as a function of damping factor due to resistive shunting of PZT in both heave and pitch directions.

CONCLUSION

Resistive shunting of piezoceramic transducers bonded to lifting surfaces can be used as a passive flutter control technique to increase the flutter speed. In this case, the piezoceramic transducers bonded to the leaf springs that provided elastic restraint in the heave and pitch directions are considered. Resistive shunting of the PZT transducers bonded to the heave and pitch leaf springs separately and together were studied. Tuned dampers realized by resistively shunted piezoceramics in heave and pitch springs could enhance the flutter speed in the range of 25-50% based on additive damping of 10% achieved due to resistive shunting of piezoceramic in either pitch spring, heave spring or both.

NOMENCLATURE

$C(k)$	Theodorsen's function
C_L and C_M	lift and moment coefficients (acting at the elastic axis)
$\bar{q} = U_\alpha^2 / \pi \mu r_\alpha^2$	normalised dynamic presure
r_α	dimensionless radius of gyration
u_h	gain X (force generated by PZT in heave)/ K_h b

u_α gain X (moment generated by PZT in pitch)/K_h b^2

$U_\alpha = U/ \omega_\alpha b$ normalized airspeed

x_α normalized (with respect to semi-chord) distance from the elastic axis to the c.g.

a   position of elastic axis with respect to mid-chord (normalized with respect to semi-chord)

x_s normalized (with respect to semi-chord) distance from the elastic axis to the sensor

$X = (\omega_\alpha/\omega)^2$

ω_h uncoupled natural frequency in heave (rad/s)

ω_α uncoupled natural frequency in pitch (rad/s)

$\bar{\omega}_h = \omega_h/\omega_\alpha$ bending to torsional frequency ratio

T   thickness ratio (leaf spring thickness/PZT thickness)

REFERENCES

1. Crawley, E.F. 1994, Intelligent structures for aerospace: A technology overview and assessment, AIAA Journal, 32(8), 1689-1699.
2. Horikawa, H. and Dowell, E.H.,1979, An elementary explanation of the flutter mechanism with active feedback controls, Journal of Aircraft, 16(4), 225-232.
3. Lazarus, K.B, Crawley, E.F., and Lin, C.Y., 1995, Fundamental mechanics of aeroelastic control with control surface and strain actuation, Journal of Guidance, Control and Dynamics, 18(1), 10-17.
4. Hagood, N.W., and Flotow, A. Von, 1991, Damping of structural vibrations with piezoelectric materials and passive electrical networks, Journal of Sound and Vibrations, 146(2), 243-268,
5. Kandagal, S.B. and Kartik Venkatraman, 2004, Form factors for vibration control of beams using resistively shunted piezoceramics, Journal of Sound and Vibration, 274, 1123-1133.

65

Dynamics of SDOF Oscillators with Bilinear Stiffness and Damping under Sinusoidal Loading

Umesh Kumar Pandey[1] and Gurmail S. Benipal[2]

[1]Research Scholar, [2]Assistant Professor, Department of Civil Engineering, Institute of Technology Delhi, Hauz Khas, New Delhi-110016, India
[1]email: gurmail@civil.iitd.ernet.in [2]email: ukp82002@gmail.com

ABSTRACT

Because of its theoretical significance and practical applications, there is tremendous interest in the non-linear dynamical systems theory. Theoretical investigations have focused on such non-linear SDOF systems as Duffing and van-der-Pol oscillators. The dependence of the stiffness co-efficient and the damping co-efficient, respectively upon vibration amplitude renders these systems non-linear. Thompson and coworkers have modeled the compliant offshore structures as bilinear dynamical system with stiffness co-efficient depending upon the sense of the vibration amplitude. The same model has been employed by the authors to simulate the behavior of SDOF cracked concrete structures under service loads. In this paper, SDOF bilinear dynamical systems with stiffness co-efficient as well as damping co-efficient determined by the sense of the vibration amplitude has been investigated. Such general bilinear dynamical oscillators under sinusoidal loading have been shown to exhibit regular and irregular sub-harmonics, strong dependence upon initial conditions, bilinearity ratio and frequency ratio. The importance of the general bilinear oscillator studied here for the non-linear dynamical systems theory as well as the potential practical applications of the results obtained have been discussed.

Keywords: Bilinear stiffness, Bilinear damping, Sub-harmonics, Jump phenomenon, Phase plots.

1. INTRODUCTION

The theory of linear damped multi-degree-of-freedom (MDOF) mechanical oscillators is quiet well established since the second half of nineteenth century. In contrast, the theory of non-linear oscillators deals mainly with single-degree-of-freedom (SDOF) systems. Even then, their dynamic response turns out to be quite complex. In such systems, either the elastic force or the damping force is non-linear function of the displacement and/or velocity of the vibrating mass. Or equivalently, the stiffness or the damping co-efficient depends upon the system response. For example, in Duffing and van der-Pol oscillators, the stiffness and damping co-efficients, respectively depend upon the vibration amplitude. In contrast, in the bilinear oscillators, the stiffness and damping co-efficients do not vary with magnitude but experience sudden change with change of sense of vibration amplitude [4].

About two decades ago, Thompson, Bokaian, Ghaffari and Elvey investigated [1-4] the dynamic response of a bilinear oscillator. Compliant offshore structures with different stiffness values for positive and negative amplitude constitutes an example of such oscillators. These systems have been shown to exhibit dependence of steady state response on the initial conditions, both regular and irregular sub-harmonic resonances, bifurcations and chaotic motions even under sinusoidal forcing function. Cracked reinforced concrete structures under service loads modeled as such bilinear dynamical systems have been shown by Pandey and Benipal [8] to exhibit similar response. It has been argued that the two values of stiffness imply two values of critical damping, the mass of the vibrating system being constant. Assumption of the constant damping co-efficient implies different values of damping ratio depending upon the sense of the amplitude. Depending upon the values of these damping ratios, such bilinear dynamical systems have also been shown, in certain circumstances, to be simultaneously overdamped and underdamped for positive and negative amplitudes, respectively.

It is well known that mechanical systems are not characterized by their damping co-efficients but by the damping ratios, For example, steel and concrete structures are known to have damping ratios equal to about two and five percentage, respectively. For such systems, different values of stiffness will imply different values of damping co-efficients for positive and negative amplitudes. Dynamic behavior of such SDOF oscillators with bilinear stiffness and damping and under sinusoidal forcing function has been investigated in this paper. For appreciating the distinction between the bilinear oscillators investigated (Model A) by Thompson, *et al.* [1-2], Thompson and Elvey [3], Pandey and Benipal [8] with the one being investigated in this paper, their governing equations are presented below:

Model A
$$m\ddot{y} + c\dot{y} + (a_1 - b_1 \operatorname{sgn} y)y = F_0 \sin \omega_f t \qquad \qquad ...(1)$$

Model B
$$m\ddot{y} + (a_2 - b_2 \operatorname{sgn} y)\dot{y} + (a_1 - b_1 \operatorname{sgn} y)y = F_0 \sin \omega_f t \qquad \qquad ...(2)$$

where, $y > 0$ $\operatorname{sgn} y = 1$; $k_1 = a_1 - b_1$; $c_1 = a_2 - b_2$;

$y < 0$ $\operatorname{sgn} y = -1$; $k_2 = a_1 + b_1$; $c_2 = a_2 + b_2$;

and the constants a_1, b_1, a_2 and b_2 are positive.

The objective of the present paper is to investigate the dynamical behavior of SDOF system with bilinear stiffness and damping co-efficients under sinusoidal loading (Model B). The concepts and techniques of non-linear dynamical systems theory have been employed for the determination and presentation of the dynamic response. The computed results obtained for this general bilinear oscillator have been discussed and the theoretical significance and practical relevance of the study have been delineated.

2. GENERAL BILINEAR OSCILLATOR

The methodology used here is similar to the earlier paper by Pandey and Benipal [8] but with the difference that depending on the positive or negative excursion the appropriate value of damping ratios has been calculated so that the damping ratio of the actual system remains constant and the damping co-efficient varies depending upon the sign of displacement. The general equation of motion as in Eq. (2) for the general bilinear oscillator (Model B) under sinusoidal forcing having mass m is recast as

$$m\ddot{y} + c_i \dot{y} + k_i y = F = F_0 \sin \omega_f t \qquad \qquad ...(3)$$

Depending upon the sense of the amplitude, the actual stiffness k of a bilinear system assumes different values k_1 and k_2 with the corresponding natural time periods T_1 and T_2. The time period T for the undamped free vibration for the full cycle and the equivalent stiffness K are obtained as

$$T = \frac{1}{2}(T_1 + T_2) \qquad K = \frac{4k_1 k_2}{(\sqrt{k_1} + \sqrt{k_2})^2} \qquad \qquad ...(4)$$

In terms of the critical damping $c_{c_i} = 2m\omega_i$, the bilinear circular frequency ω and the equivalent damping ratio ξ can be expressed as follows.

$$\omega = \frac{2\pi}{T} = \sqrt{(K/m)} \qquad \xi = c_i / c_{C_i} = c_i / 2m\omega_i \qquad \qquad ...(5)$$

Using the definitions, frequency ratio $\eta = \omega_f/\omega$, non-dimensionalised amplitude $Y = y/(F_0/K)$, $\tau = \omega_f t$ and bilinearity ratio $\beta = \frac{k_2}{k_1}$, the above governing differential equation can be expressed in the non-dimensional form as

$$\ddot{Y} + 2\frac{\xi_i}{\eta}\dot{Y} + K_i Y = \frac{1}{\eta^2}\sin \tau \quad i = 1, 2 \qquad \qquad ...(6)$$

$$K_1 = (1 + \sqrt{\beta})^2 / 4\beta\eta^2 \qquad K_2 = (1 + \sqrt{\beta})^2 / 4\eta^2$$

where, K_i is either K_1 or K_2 and ξ_i is either ξ_1 or ξ_2 depending upon the sense (positive or negative) of the parameter Y. With the intention of maintaining the same damping ratio ξ, equal to 0.05 in the present case, with different damping co-efficients C_i, the following definitions for ξ_1 and ξ_2 are adopted:

$$\xi_1 = \xi(1 + \sqrt{\beta})/2\sqrt{\beta}; \qquad \xi_2 = \xi(1 + \sqrt{\beta})/2 \qquad \qquad ...(7)$$

The general solution is obtained by superimposing the particular integral and the complimentary solution as

$$Y = A\sin \tau + B\cos \tau + e^{-\xi_i \tau/\eta}(C\sin \omega_D \tau + D\cos \omega_D \tau) \qquad \qquad ...(8)$$

The constants A, B, C and D are obtained from the initial conditions as given below.

$$A = \frac{K_i - 1}{\eta^2[(K_i - 1)^2 + (2\xi_i/\eta)^2]}$$

$$B = \frac{-2\xi_i/\eta}{\eta^2[(K_i - 1)^2 + (2\xi_i/\eta)^2]}$$

$$C = e^{\xi_i \tau/\eta}[Y_0 - A\sin \tau_0 - B\cos \tau_0]\sin \omega_D \tau_0$$

$$+\frac{1}{\omega_D}[\dot{Y}_0 + \frac{\xi_i}{\eta}Y_0 + (B - \frac{\xi_i}{\eta}A)\sin \tau_0 - (A + \frac{\xi_i}{\eta}B)\cos \tau_0]\cos \omega_D \tau_0$$

$$D = e^{\xi_i \tau/\eta}[Y_0 - A\sin \tau_0 - B\cos \tau_0]\cos \omega_D \tau_0 + \frac{1}{\omega_D}[\dot{Y}_0 + \frac{\xi_i}{\eta}Y_0 + (B - \frac{\xi_i}{\eta}A)$$

$$\sin \tau_0 - (A + \frac{\xi_i}{\eta}B)\cos \tau_0]\sin \omega_D \tau_0$$

$$\omega_D = \sqrt{[K_i - (\xi_i/\eta)^2]} \qquad \qquad \text{...(9)}$$

Similarly, the expression for velocity and acceleration are obtained as

$$\dot{Y} = A\cos\tau - B\sin\tau + Ce^{-\xi_i\tau/\eta}(\omega_D\cos\omega_D\tau - (\xi_i/\eta)\sin\omega_D\tau)$$

$$-De^{-\xi_i\tau/\eta}(\omega_D\sin\omega_D\tau + (\xi_i/\eta)\cos\omega_D\tau) \qquad \text{...(10)}$$

$$\ddot{Y} = -A\sin\tau - B\cos\tau + Ce^{-\xi_i\tau/\eta}(-\omega_D^2\sin\omega_D\tau - 2(\xi_i/\eta)\omega_D\cos\omega_D\tau + (\xi_i/\eta)^2\omega_D\sin\omega_D\tau)$$

$$-De^{-\xi_i\tau/\eta}(\omega_D^2\cos\omega_D\tau - 2(\xi_i/\eta)\omega_D\sin\omega_D\tau - (\xi_i/\eta)^2\omega_D\cos\omega_D\tau) \qquad \text{...(11)}$$

3. NUMERICAL RESULTS AND DISCUSSIONS

In the following computational study, the damping ratio and the bilinearity ratio have been assumed to be equal to five percent and eight, respectively. The initial displacement and velocity have both been taken to be equal to zero. The absolute magnitude of the peak force does not have any effect on the non-dimensional dynamic response of the system.

The steady state forced vibration response is attained asymptotically when the initial transient response gets damped out with the passage of time. Practically steady state waveforms showing the temporal variation of non-dimensionalised displacement, velocity, acceleration, elastic force, damping force and inertial force have been plotted in Fig. 1 for the frequency ratio of unity. It can be observed that the vibration amplitude during the positive excursion exceeds that during the negative excursion for bilinearity ratio—the ratio of negative to positive flexural rigidity—higher than unity. The variation of displacement, velocity and elastic force is observed to be smooth whereas the waveforms for damping force, inertial force and acceleration exhibit jump phenomena in the form of sudden changes in their magnitude. This is because of the fact that the value of the damping co-efficient changes suddenly as and when the displacement experiences change in sense. This happens at the peak velocities resulting in a sudden change in the damping force. The applied force being sinusoidal, the sudden change in damping force introduces a corresponding sudden change of opposite sense in the inertial force. Since the mass remains invariant during vibrations, the waveform for acceleration also exhibits jump phenomenon at the same instant.

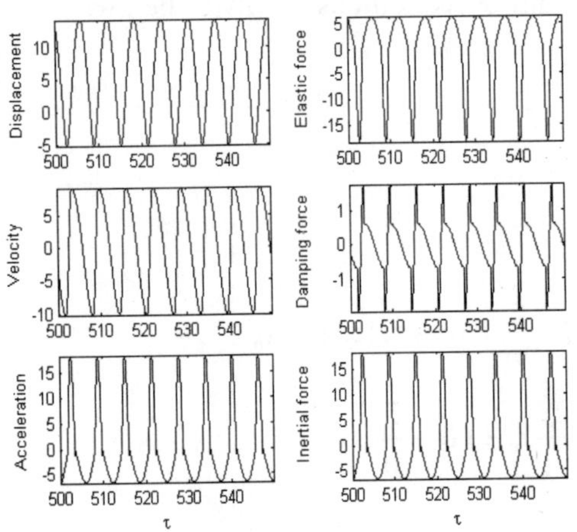

Fig. 1. Response and forces vs τ.

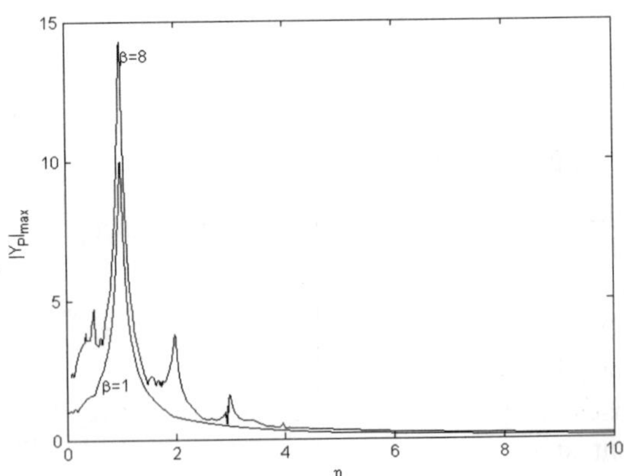

Fig. 2. Resonance response for τ = 565.48.

As is the common practice, maximum absolute peak vibration amplitude has been chosen as a measure of the dynamic response in the frequency domain plot shown in Fig. 2 for the unilinear system ($\beta = 1$) and for the bilinear system ($\beta = 8$). It can be observed that the unilinear system exhibits the expected smooth resonance response with the single peak at its natural frequency and the gradually diminishing response at higher frequency ratios. In contrast, the resonance response curves for the bilinear system exhibit regular and irregular sub-harmonics apart from the fundamental response. The peak resonance response corresponding to frequency ratio of unity is called fundamental response and is of order unity. The regular sub-harmonic resonance peaks are observed approximately at frequency ratios 1.99, 2.99, 3.97, etc., and they pertain to resonance response of various orders ($n = 2, 3, 4$, etc.). The irregular sub-harmonics at frequency ratios 0.5, 0.35, 0.10, etc., all have the same order of unity ($n = 1$). As the order is defined by the ratio of steady state response frequency to the forcing frequency, the bilinear systems have been shown to execute steady state vibrations at frequencies other than the forcing frequency as well.

The phase plots projections have been plotted for the general bilinear system in Fig. 3 for frequency ratios equal to 0.10, 0.50, 1.0, 1.99, 2.99 and 3.97, respectively. It can be observed that the phase plot for the fundamental response is composed of two half ellipses of different lengths of major and minor axes. In contrast, more complicated phase plots are obtained for higher order resonance responses. For example, the third and fourth order response respectively exhibits one and three crossover points, the trajectories in the second order response, like the fundamental one, lack such crossover points. The phase plots for the irregular sub-harmonic resonances confirm the presence of one large excursion succeeded by a number of small amplitude damped vibrations.

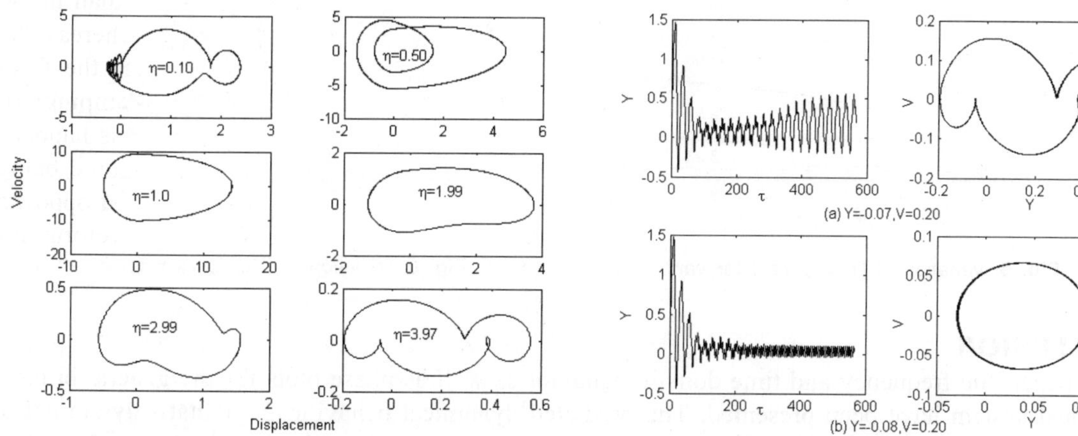

Fig. 3. Phase plots for various values of η.   **Fig. 4.** Sensitivity to initial conditions at $\eta = 3.97$.

Like other non-linear systems, the bilinear mechanical system investigated in this paper also exhibits strong sensitivity to initial conditions. As shown in Fig. 4, two identical systems under the action of identical forcing function ($h = 3.97$) but with different initial conditions ($Y = -0.07$, $V = 0.20$) and ($Y = -0.08$, $V = 0.20$) experience steady state vibrations with different peak amplitudes as shown in their waveforms. The corresponding steady state phase plots reveal that their resonance responses are of order four and one respectively. It can be observed that the waveforms from adjacent starts diverge with time to such an extent that the corresponding steady state responses lack any correlation.

The variation of the maximum absolute peak amplitude for resonance response of various orders ($n = 1, 2, 3, 4$) with bilinearity ratio has been plotted in Figure 5. It can be observed that, for $\beta > 1$, the peak response increases with increase in bilinearity ratio. In contrast, for $\beta < 1$, the absolute peak response decreases with increase in bilinearity ratio. This difference is because of the fact that, in this range of bilinearity ratio, the system is stiffer for the positive excursions of the vibrating mass and the negative peak amplitude are more dominant. Also, a sudden change in the absolute peak resonance response of order three is observed at about the bilinearity ratio equal to 6.3-6.4 with infinitesimally small change in the bilinearity ratio. Such a discontinuity also exists in the frequency domain response plotted in Fig. 2 in the range 2.91-2.96 of frequency ratio. This fact can be confirmed from the blown up image of the relevant portion presented in Figure 6. Thus, like other typical non-linear dynamical systems, the general bilinear oscillator investigated here also exhibits lack of continuous dependence on the frequency ratio (Fig. 2, Fig. 6), initial conditions (Fig. 4) and bilinearity ratio (Fig. 5). The long-term dynamic response of such chaotic non-linear systems cannot be predicted with any certainty.

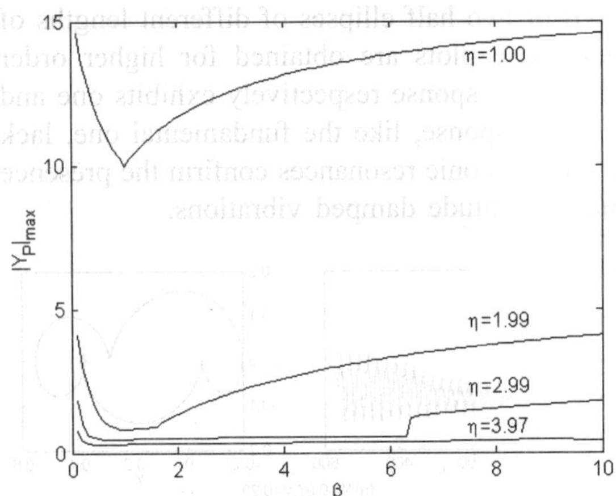

Fig. 5. Variation of $|Y_P|_{max}$ vs β for various η.

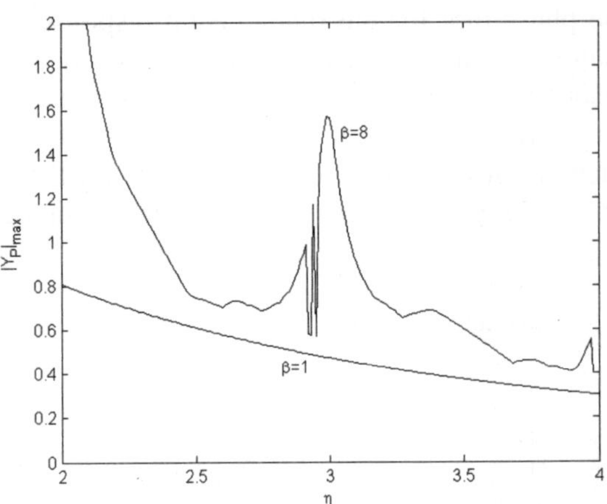

Fig. 6. Variation of $|Y_P|_{max}$ vs η for $\beta = 1, 8$.

CONCLUSION

In this paper, the frequency and time domain behavior as well as phase plots for the general bilinear dynamical system have been presented. The predicted dynamical behavior is qualitatively similar to that for the dynamical system with constant damping investigated earlier by Pandey and Benipal [8]. Both of these models of bilinear dynamical systems predict existence of regular and irregular sub-harmonic resonance response as well as lack of continuous dependence of the system response on the system parameters, initial conditions and forcing frequency. However, the present system does not exhibit simultaneous overdamping/underdamping in the same vibration cycle as exhibited by Model-A. The distinguishing characteristic of the dynamical system studied here is the jump phenomena in its damping force, inertial force and acceleration response.

The dynamic response of the systems is sought to be controlled by designing the system frequency quite different from the forcing or operating frequency. The presence of regular as well as irregular sub-harmonics in bilinear system response renders this vibration control strategy. This fact is of tremendous importance in the seismic design of such bilinear systems as compliant offshore structures, concrete structures, etc. This absence of continuous dependence of system response on its parameters implies the lack of robustness of the system. This property results in non-unique system behavior for real structures where neither the system parameters like bilinearity ratio nor the initial conditions nor the forcing frequency are definitely known. This fact is significant for dynamic analysis, design, vibration control, stability, health monitoring and system identification of these structures.

Of course, SDOF dynamical systems have earlier been employed to model oscillators with clearances [5-7]. The special case pertaining to no clearance or gap assigns two different values to stiffness and damping co-efficients depending upon the sense of the amplitude resulting in two unrelated damping ratios. Obviously, the formulation reduces to Model A or Model B of the present investigation, respectively when the damping co-efficients or damping ratios are assigned constant value independent of the sense of the amplitude. It should be mentioned here that, in these investigations, focusing on the general model, the Model B has not been studied in any detail as a special case.

REFERENCES

1. J.M.T. Thompson, A.R. Bokaian, and R. Ghaffari, 1983, Sub-harmonic and Chaotic Motions of a Bilinear Oscillator, IMA Journal of Applied Mathematics, Vol-31, 207-234.
2. J.M.T. Thompson, A.R. Bokaian, and R. Ghaffari, 1984a, Sub-harmonic and Chaotic Motions of Compliant Offshore Structures and Articulated Mooring Towers, ASME J. Energy Resources Tech., 106, 191-198.
3. J.M.T. Thompson and J.S.N Elvey, 1984b, Elimination of Sub-harmonic Resonances of Compliant Marine Structures, International Journal of Mechanical Science., 26, 419-425.
4. J.M.T. Thompson and H.B. Stewart, 1986, Non-linear Dynamics and Chaos, John Wiley and Sons, New York
5. J.N.Schulman, 1983 Chaos in piecewise linear system Physical Review A Vol-28, 477-479.
6. M. Z. Hossain, K. Mizutani and H. Sawai, 2002, "chaos and multiple periods in an unsymmetrical spring and damping system with clearance" Journal of Sound and Vibration, Vol-250, 229-245.
7. S.Natsiavas, 1990, On the Dynamics of Oscillators with Bilinear Damping and Stiffness, International Journal of Non-linear Mechanics, Vol-25, 535-554.
8. Umesh Kumar Pandey and Gurmail S. Benipal, 2006, Bilinear Dynamics of SDOF Concrete Structures under Sinusoidal Loading, Advances in Structural Engineering, 9, 393-407.

66

Optimal Control of Semi-Active Vehicle Suspension with Sky-Hook Control

R.S. PRABAKAR[1], S. NARAYANAN[2] AND C. SUJATHA[3]

[1]Research Scholar, [2,3]Professor Machine Design Section, Department of Mechanical Engineering,
Indian Institute of Technology Madras-600036 India
email: rsprabakar@yahoo.com, narayans@iitm.ac.in, sujatha@iitm.ac.in

ABSTRACT

Active control using H_∞ controller and semi-active control with sky-hook damper to control vehicular vibration due to random road undulations are investigated. A two degree of freedom (d.o.f.) quarter car vehicle model with sprung and unsprung masses and with passive springs is considered. The vehicle is assumed to be moving at constant velocity and the road roughness is modeled as a spatially homogeneous random process, being the output of second order shaping filter to white noise which takes into account the effect of rolling contact of the tyre with the road. The H_∞ control is effected by minimizing the performance index which is a weighted sum of sprung mass mean square acceleration, suspension stroke, road holding and control force. The parameters of the sky-hook damper are optimized so as to match the performance of the H_∞ controller.

Keywords: H_∞ control; Vehicle response; Sky-hook control.

1. INTRODUCTION

Vehicle suspension system is responsible for transmitting all the forces from the road to the body. The suspension has to be of variable and conflicting characteristics for better directional control, ride comfort and safety. For instance a soft suspension will lead to better ride comfort while stiffer spring is required for better road holding characteristics. A passive suspension will not be able to satisfy these conflicting requirements in which case an active or semi-active suspension may be preferred. An active suspension system has the capability to adjust itself continuously to changing road conditions by means of actuators [1]. Due to the cost, increased complexity, large power supply and difficulty in control hardware and implementation of active suspension systems, semi-active suspension systems which combine the advantages of both passive and active suspension systems have been developed. Karnopp *et al.* [2] have proposed the semi-active dissipative suspension in which an actuator is replaced by a continually adjustable damper. Hac and Youn [3] have dealt with the synthesis of an optimal preview controller for a semi-active dissipative suspension system for a two d.o.f. vehicle model. The behavior of the semi-active suspension can be portrayed using different control algorithms like sky-hook, groundhook and hybrid control. Sammier *et al.* [4] have used H_∞ and sky-hook control approaches for a quarter car model of a vehicle traveling on a step bump and compared the performances with that of a passive system. In the present paper, the stationary response of a two dof model of a vehicle traversing a rough road is controlled using H_∞ control as well as by the use of semi-active sky-hook damper. The random road is modeled as a spatially homogeneous random process being the output of a second order shaping filter to white noise [5] on which the vehicle is assumed to move with constant velocity. The parameters of the sky-hook damper are obtained in an optimal way by equating the active control force generated using $H\infty$ controller to that obtained by the sky-hook damper.

2. MATHEMATICAL MODEL OF VEHICLE

The vehicle modeled as a linear two dof quarter car system with H_∞ and sky-hook control are shown in Figs. 1 and 2, respectively. M and m are the sprung and unsprung masses respectively, c is the damping coefficient, U is the control force, k and k_t are the primary suspension stiffness and tyre stiffness, respectively. Y_1 and Y_2 are the displacements of sprung and unsprung masses, h is the random road excitation, α_s is the sky-hook constant which represents the different proportions in which the damping capacity of the primary suspension and the sky-hook damper is divided and c_s is the sky-hook damping coefficient.

The equations of motion of the vehicle with $H\infty$ control are given by

$$M\ddot{Y}_1 + k(Y_1 - Y_2) - U = 0 \qquad \qquad ...(1)$$

$$m\ddot{Y}_2 + k(Y_2 - Y_1) + k_t(Y_2 - h) + U = 0 \qquad \qquad ...(2)$$

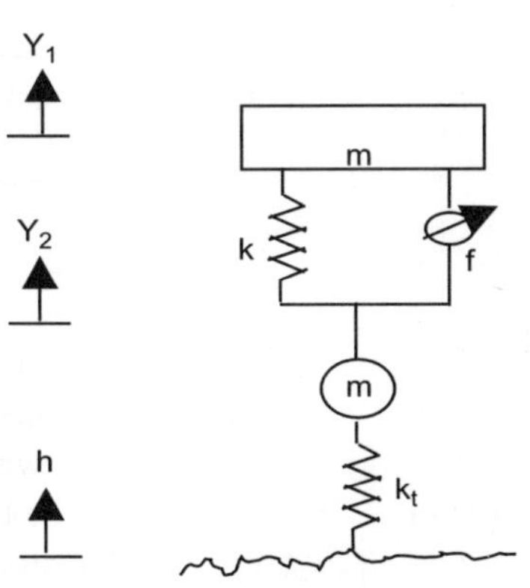

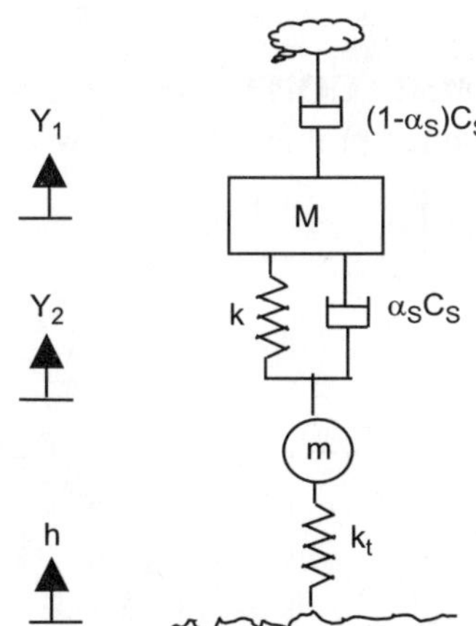

Fig. 1. Quarter car model with $H\infty$ control.　　　　**Fig. 2.** Quarter car model with sky-hook control.

3. ROAD PROFILE MODEL

The road input which is modeled as the output of a second order filter to white noise is given by

$$h''(s) + (\alpha + \beta)h'(s) + \alpha\beta h(s) = \Gamma W(s) \qquad \ldots(3)$$

where α and β are the cutoff wave numbers of the road profile filter and rolling contact filter respectively. $\Gamma = \beta\sigma(2\alpha)^{1/2}$ is the strength of white noise, σ^2 is the variance of the road profile. $W(s)$ is the spatial white noise with $E[W(s_2)W^T(s_1)] = Q\,\delta(s_2 - s_1)$, where Q is the unit matrix, $\delta(.)$ is the Dirac delta function. s is the traverse along the road and the number of primes indicates the order of differentiation with respect to s. The procedure given in reference [5] of covariance equivalent form of white noise from space domain to time domain is adopted to convert Eq. (3) into the time domain and to represent it in the state space form as

$$\dot{h}_1(t) = \dot{s}(t)\,h_2(t); \quad \dot{h}_2(t) = -\dot{s}(t)\alpha\beta\,h_1(t) - \dot{s}(t)\,(\alpha + \beta)h_2(t) + \sigma\beta\sqrt{2\alpha\dot{s}(t)}\;W(t) \qquad \ldots(4)$$

Defining the state variables $X_1 = Y_1$, $X_2 = \dot{Y}_1$, $X_3 = Y_2$, $X_4 = \dot{Y}_2$, $X_5 = h_1$, $X_6 = h_2$, the generalized state space form is given by

$$\dot{X}(t) = F_X X(t) + G_X U(t) + D_X h(t) \text{ and } \dot{h}(t) = F_d h(t) + D_d W(t) \qquad \ldots(5)$$

where,

$$F_x = \begin{bmatrix} 0 & 1 & 0 & 0 \\ -\dfrac{k}{M} & 0 & \dfrac{k}{M} & 0 \\ 0 & 0 & 0 & 1 \\ \dfrac{k}{m} & 0 & -\dfrac{(k+k_t)}{m} & 0 \end{bmatrix}; \; G_x = \begin{bmatrix} 0 \\ \dfrac{1}{M} \\ 0 \\ -\dfrac{1}{m} \end{bmatrix}; \; D_x = \begin{bmatrix} 0 & 0 \\ 0 & 0 \\ 0 & 0 \\ \dfrac{k_t}{m} & 0 \end{bmatrix} \qquad \ldots(6)$$

$$F_d = \begin{bmatrix} 0 & \dot{s} \\ -\dot{s}\alpha\beta & -\dot{s}(\alpha + \beta) \end{bmatrix}; \; D_d = \begin{bmatrix} 0 \\ \sigma\beta\sqrt{2\alpha\dot{s}} \end{bmatrix} \qquad \ldots(7)$$

4. OPTIMAL CONTROL PROBLEM

The H_∞ control problem for the two dof vehicle model can be written in the augmented state space form $X_a^T = \begin{bmatrix} X^T & h^T \end{bmatrix}$ as

$$\dot{X}_a(t) = F X_a(t) + G U(t) + D W(t) \qquad \ldots(8)$$

$$Z(t) = C_{yu} X_a(t) + D_{yu} U(t) \qquad \ldots(9)$$

where,

$$F = \begin{bmatrix} F_x & D_x \\ 0 & F_d \end{bmatrix}; \; G = \begin{bmatrix} G_x \\ 0 \end{bmatrix}; \; D = \begin{bmatrix} 0 \\ D_d \end{bmatrix} \qquad \ldots(10)$$

$$C_{yu} = \begin{bmatrix} -\sqrt{\rho_1}\,\dfrac{k}{M} & 0 & \sqrt{\rho_1}\,\dfrac{k}{M} & 0 & 0 & 0 \\ \sqrt{\rho_2} & 0 & -\sqrt{\rho_2} & 0 & 0 & 0 \\ 0 & 0 & \sqrt{\rho_3} & 0 & -\sqrt{\rho_3} & 0 \\ 0 & 0 & 0 & 0 & 0 & 0 \end{bmatrix}; \; D_{yu} = \begin{bmatrix} \dfrac{\sqrt{\rho_1}}{M} \\ 0 \\ 0 \\ \sqrt{\rho_4} \end{bmatrix} \qquad \ldots(11)$$

where, ρ_1, ρ_2, ρ_3 and ρ_4 are the weighting constants for ride comfort, suspension stroke, road holding and control force, respectively.

The objective function for the stated problem represented in the standard form [6] is

$$J_\gamma = \int_0^T E\left[X^T(t)AX(t) + 2X^T(t)NU(t) + U^T(t)\,BU(t) - \gamma^2 W^T(t)W(t) \right]dt \qquad ...(12)$$

where, $A = C^T_{yu}\,C_{yu}$; $N = C^T_{yu}\,D_{yu}$; $B = D^T_{yu}\,D_{yu}$, γ is the performance bound found out using bisection algorithm [6]. The admissible control law subjected to performance criterion is given by $U(t) = - C(t)X_a(t)$, where $C(t) = B^{-1}(N^T + G^T S(t)$ and $S(t)$ is a symmetric matrix and is the solution of the algebraic Riccati equation.

$$S(t)(F - GB^{-1}N^T) + (F - GB^{-1}N^T)^T S(t) - S(t)(GB^{-1}G^T - \gamma^{-2}DD^T)\,S(t) + (A - NB^{-1}N^T) = 0 \qquad ...(13)$$

The response of the system is found out by using zero-lag covariance matrix of the augmented state vector and is the solution of Lyapunov equation given by

$$P(t)(F(t) - G(t)C(t)) + (F(t) - G(t)C(t))^T P(t) + D(t)QD^T(t) = 0 \qquad ...(14)$$

5. SKY-HOOK CONTROL

The equations of motion of the vehicle with sky-hook damper are given by

$$M\ddot{Y}_1 + k(Y_1 - Y_2) + \alpha_S c_S(\dot{Y}_1 - \dot{Y}_2) + (1-\alpha_S)c_S\dot{Y}_1 = 0 \qquad ...(15)$$

$$m\ddot{Y}_2 + k(Y_2 - Y_1) + k_t(Y_2 - h) + \alpha_S c_S(\dot{Y}_2 - \dot{Y}_1) = 0 \qquad ...(16)$$

Using Eqs. (15) and (16), the generalized state space form is formulated as

$$\dot{X}_a(t) = FX(t) + DW(t) \qquad ...(17)$$

The system matrices F and D are taken from Eq. (10) with minor changes in matrix F_x .i.e., F_x (2, 2), F_x (2, 4), F_x (4, 2), F_x (4, 4) are replaced by $-c_S/M$, $\alpha_S c_S/M$, $\alpha_S c_S/m$ and $-\alpha_S c_S/m$ respectively. The system response is described by the zero lag covariance matrix approach and is obtained by solving the Lyapunov equation

$$P(t)\,F(t) + F(t)^T P(t) + D(t)QD^T(t) = 0 \qquad ...(18)$$

6. EVALUATION OF OPTIMAL SKY-HOOK PARAMETERS

The mean square values of the control force obtained using H_∞ controller is equated with the mean square control force generated using sky-hook controller having the unknown parameters α_S and c_S as in reference [7].

$$c_S = \sqrt{\dfrac{E[U^2]}{E[(\dot{Y}_1 - \alpha_S\dot{Y}_2)^2]}} \qquad ...(19)$$

For each velocity the mean square control force based on H_∞ control is obtained and is substituted in Eq. (19) from which the c_S values for different values of α_S varied in the range 0 to 1 in steps of 0.01 are obtained. Thus, for each velocity we get a combination of α_S and c_S values for which the mean square control force for H_∞ control and sky-hook control are equal. For these combinations of α_S and c_S the sky-hook control scheme is applied and the corresponding rms sprung mass acceleration is calculated. The rms sprung mass acceleration thus obtained for different velocities as a function of α_s is plotted in Fig. 3. The rms sprung mass acceleration obtained by H_∞ control which is constant for a particular velocity (independent of α_s) is also plotted in the same figure by a dotted line. The intersecting points of the H_∞ control and sky-hook control plot shown by the symbols ( ) are obtained by using Lagrangian interpolation. Then the acceleration response using newly found sky-hook parameters are found out for different velocities. The acceleration curve for sky-hook control which is closest to that of H_∞ control (with minimum area in between the two curves) decides the optimal sky-hook parameters. This way of finding the optimal sky-hook parameters is termed as 'acceleration based procedure'. The same procedure is repeated with rms suspension stroke instead of sprung mass acceleration to get the optimal values of the sky-hook parameters using Fig. 4. This approach is termed as 'stroke based procedure'.

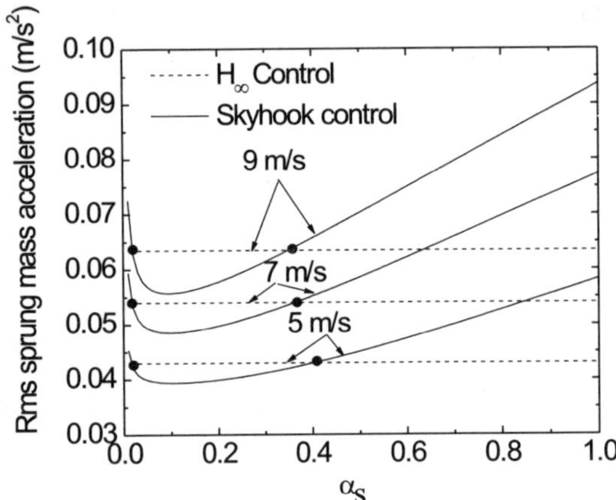

Fig. 3. Rms sprung mass acceleration vs. α_s for different velocities.

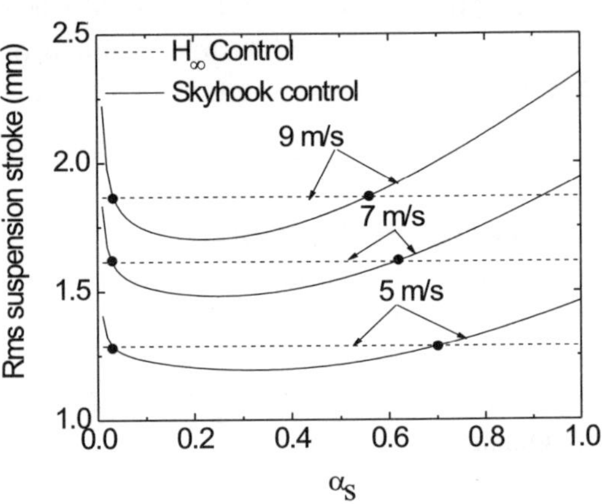

Fig. 4. Rms suspension stroke vs. α_s for different velocities.

7. RESULTS AND DISCUSSION

The active control of the stationary response of a quarter car model using $H\infty$ control and semi-active suspension with sky-hook control is implemented. The numerical data and weighting functions [5] used for analyzing the vehicle model are M = 1000 kg, m = 100 kg, k = 36 kN/m, k_t = 360 kN/m, α = 0.15 rad/m, β = 2, σ^2 = 9 × 10^{-6} m^2, ρ_1 = 1, ρ_2 =10^3, ρ_3 = 10^5 and ρ_4 = 10^{-5}. Figures 5 and 6 show the rms acceleration of the sprung mass and suspension stroke respectively as a function of velocity when the optimization of the sky-hook parameters is as per the acceleration based procedure. The optimal value of sky-hook parameters in this case are α_s = 0.3487 and c_s = 3273.5 Ns/m. It is clear that acceleration responses due to H_∞ control and sky-hook control are almost the same and better than those for the passive system. From Fig. 6 it is clear that the performance of the vehicle suspension in terms of the suspension stroke in the H_∞ control case is worse than that of passive system while the performance with the optimal sky-hook damper parameters is almost as good as the case for the passive suspension system. With respect to the road holding characteristic depicted in Fig. 7, it is seen that the passive system performs better than both the active and semi-active systems. This may be due to the fact that the mean square acceleration may be weighted, respectively higher in the performance index.

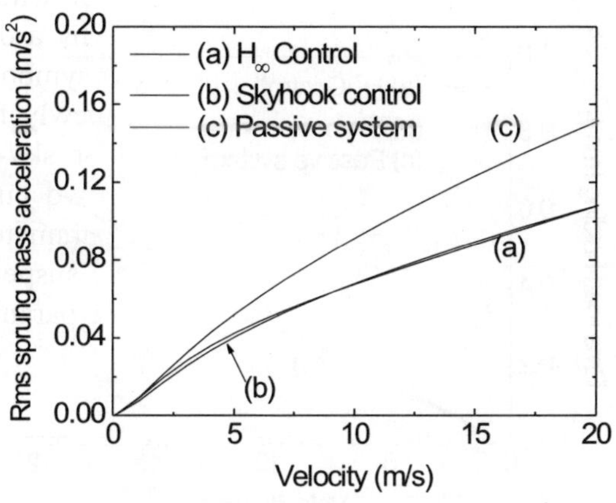

Fig. 5. Variation of sprung mass acceleration-
acceleration based procedure.

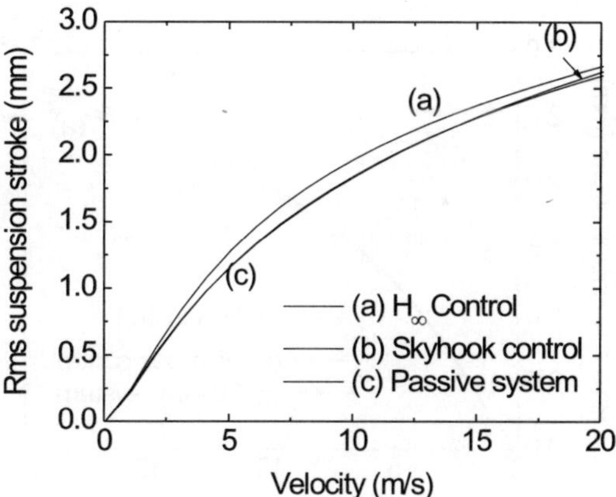

Fig. 6. Variation of suspension stroke-
acceleration based procedure.

The rms values of the sprung mass acceleration, suspension stroke and road holding, for the three cases of H_∞ control, sky-hook damper and passive systems are shown in Figs. 8-10 respectively. In this case the optimum sky-hook parameters are obtained by the 'stroke based response' and they are $\alpha_s = 0.5043$, $c_s = 2829.5$ Ns/m. In this case the performance of the semi-active system with sky-hook damper is worse than the fully active system with H_∞ controller with respect to acceleration response, though the active and semi-active systems perform better than the passive system. It is clear from Fig. 9 that the suspension stroke response for sky-hook control case is almost the same as H∞ controller and better than passive system. This is because that the optimal parameters of sky-hook control is found out by using suspension stroke (refer Fig. 4). The road holding performance depicted in Fig. 10 shows that the passive system performs better than the system with H_∞ controller and sky-hook control.

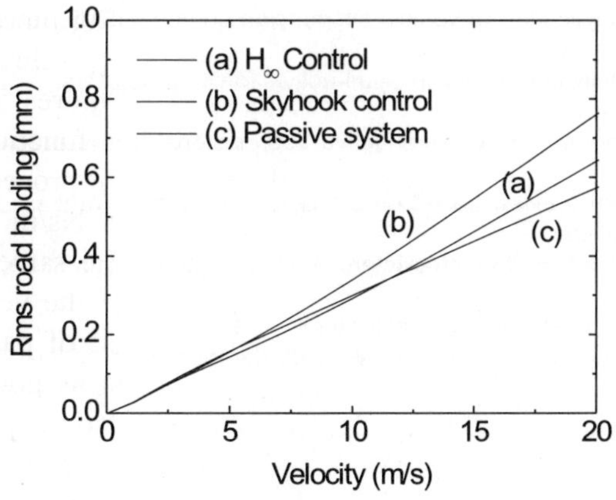

Fig. 7. Variation of road holding—
acceleration based procedure.

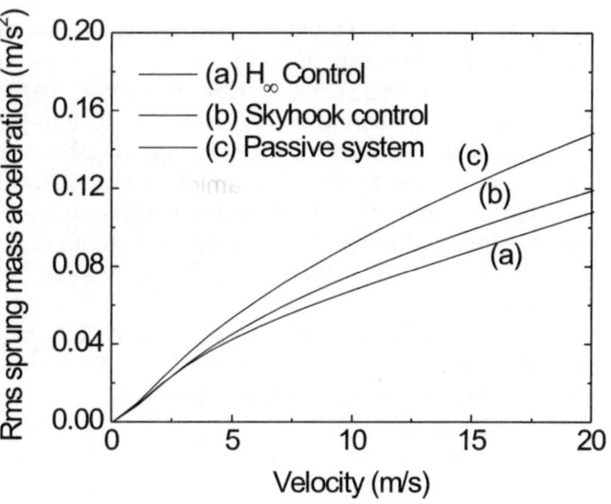

Fig. 8. Variation of sprung mass acceleration—
stroke based procedure.

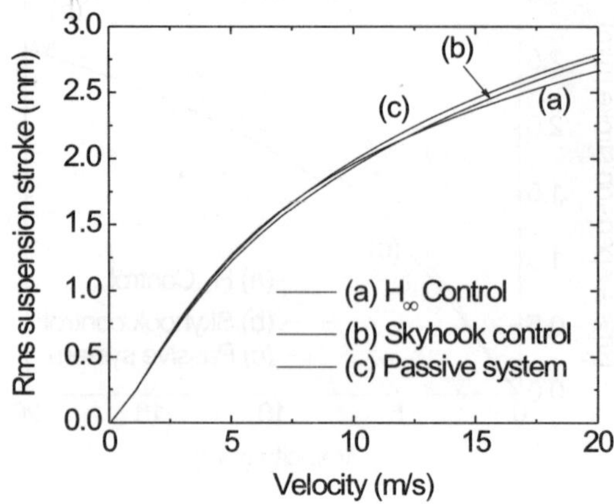

Fig. 9. Variation of suspension stroke—stroke based procedure

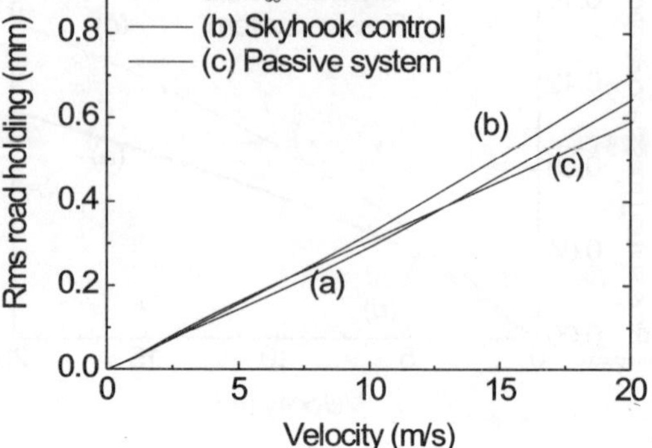

Fig. 10. Variation of road holding—stroke based procedure

CONCLUSION

The stationary response of a two dof vehicle model with H_∞ control and sky-hook control, traversing a rough road is considered. The sky-hook damper parameters are optimized based on acceleration and stroke response by using the control force generated from the H_∞ control. The rms sprung mass acceleration response for the sky-hook control using the acceleration based procedure is better than stroke based procedure. There is not much difference in suspension stroke between passive and sky-hook controlled systems, both with acceleration based response and stroke based response. The road holding response for passive system is better than sky-hook controlled system for both acceleration and stroke based response.

REFERENCES

1. E.K. Bender, 1968, Optimum linear preview control with application to vehicle suspension control, ASME Journal of Basic Engineering, 90, 213-221.
2. D.C. Karnopp, M.J. Crosby, R.A. Harwood, 1974, Vibration control using semi-active force generators, ASME Journal of Engineering for Industry, 96(2), 619-626.
3. D. Sammier, S. Oliver, L. Dugard, 2003, Skyhook and H∞ control of semi-active suspensions: Some practical aspects, Vehicle System Dynamics, 39(4), 279-308.
4. A. Hac, I. Youn, 1992, Optimal semi-active suspension with preview based on a quarter car model, Transactions of the ASME Journal of Vibration and Acoustics, 114, 84-92.
5. R.F. Harrison, J.K. Hammond, 1985, A systems approach to the characterization of rough ground, 99(3), 437-447.
6. K. Zhou, J.C. Doyle, 1998, Essentials of Robust Control, Prentice-Hall, California.
7. L.V.V. Gopala Rao, S. Narayanan, 2005, Semi-active vehicle suspension with skyhook damper, Proc. of the 14th ISME International Conference, Delhi.

67

H Infinity Vibration Controller for Smart Cantilever Beam

S. Soundaravalli[1], M. Umapathy[2], D. Ezhilarasi[3] and J. Balasubramani[4]

Department of Instrumentation and Control Engineering, National Institute of Technology,
Tiruchirappalli-620015, India. email: umapathy@nitt.edu

ABSTRACT

This paper deals with the vibration control of piezoelectric actuated cantilever beam using H_∞ - controller. The controller is designed to control the first vibration mode and suppress the possible instability due to neglected higher modes. The performance of the controller is demonstrated through simulation and experimental results.

Keywords: Smart structure, H_∞ Control.

1. INTRODUCTION

Vibration is a natural phenomenon that may occur in all dynamic systems. Vibration can be detrimental to structural performance and stability and so it is important to find a means of suppressing structural vibrations. Bailey and Hubbard [1] initiated the research on the application of the smart structures in active vibration control. In recent years, piezoelectric materials have been increasingly used as both sensors and actuators for vibration control of cantilever beams and plates, leading to the development of smart structures [2]. The utilization of discrete piezoelectric actuators has been shown to be a viable concept for vibration suppression of one dimensional structure by Crawley and de Luis [3]. Generally, huge flexible structures are distributed parameter systems and to design a feedback controller for such systems using linear control theory, it is necessary to obtain a suitable reduced order model by neglecting higher modes frequencies. The neglected dynamics may induce some severe vibrations on the structure and ultimately damage it. The H_∞ -control is an active control approach that takes care of the spill over instability due to the neglected dynamics [4-6]. In this paper an H-infinity controller is designed to suppress the first vibration mode of a cantilever beam using piezoelectric patches as sensor/actuator. The system uncertainties such as neglected vibration modes (higher order modes) are taken care of at the time of design process. The paper is organized as follows: In Section 2 description of experimental set up and the structure model is presented. H_∞ Based dynamic output feedback controller design and experimental results are presented in Section 3. Conclusions are drawn in the section 4.

2. EXPERIMENTAL SETUP AND STRUCTURE MODEL

A flexible aluminum beam with clamped end as shown in Figure 1 is considered in this paper. Two piezoceramic patches are surface bonded at a distance of 10 mm from the fixed end of the beam. The patch bonded on the bottom surface acts as a sensor and the one on the top surface acts as an actuator. To apply an excitation to the structure another piezoceramic patch is bonded on the top surface at a distance of 153 mm from the fixed end. The dimensions and properties of the beam and piezoceramic patches are given in Table 1.

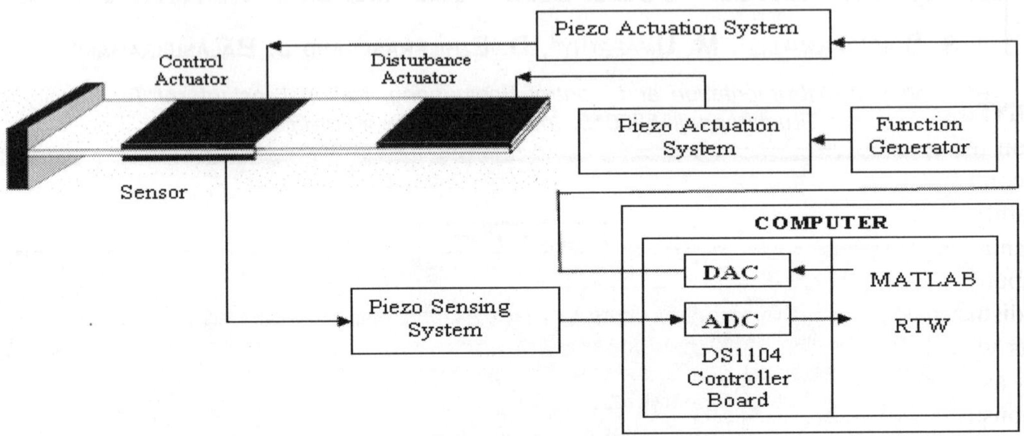

Fig. 1. Schematic diagram of the experimental setup.

Table 1. Properties and dimensions of the aluminum beam and piezoelectric sensor/actuator

Aluminum beam		Beam 1	Piezoceramic sensor/actuator		
Length (m)	l	0.23	Length (m)	l_p	0.0765
Width (m)	b	0.0127	Width (m)	b	0.0127
Thickness (m)	t_b	0.0023	Thickness (m)	t_a	0.005
Young's modulus (GPa)	E_b	71	Young's modulus (GPa)	E_p	47.62
Density (kg/m^3)	ρ_b	2700	Density (kg/m^3)	ρ_p	7500
First natural frequency (Hz)	F1	31.7	Piezoelectric strain constant (mV^{-1})	d_{31}	-247×10^{-12}
Third natural frequency (Hz)	F2	200	Piezoelectric stress constant (Vm N^{-1})	g_{31}	-9×10^{-3}

A flexible aluminum beam with clamped end as shown in Figure 1 is considered in this paper. Two piezoceramic patches are surface bonded at a distance of 10 mm from the fixed end of the beam for structure1, 15 mm from the fixed end of the beam for structure 2 and 20 mm from the fixed end of the beam for structure 3. The patch bonded on the bottom surface acts as a sensor and the one on the top surface acts as an actuator. To apply an excitation input to the structure another piezoceramic patch is bonded on the top surface at a distance of 67 mm from the actuator. The dimensions and properties of the beams and piezoceramic patches are given in Table 1.

The model of the system is identified using recursive least square method [7]. The state space model of the system is

$$\dot{x} = \mathbf{A}x + \mathbf{B}u + \mathbf{E}w$$

$$y = \mathbf{c}^T x$$

where,

$$\mathbf{A} = \begin{bmatrix} -83.0583 & 218.2890 \\ -204.9014 & 76.7292 \end{bmatrix}; \quad \mathbf{B} = \begin{bmatrix} -1.4349 \\ -1.708 \end{bmatrix};$$

$$\mathbf{E} = \begin{bmatrix} -0.2359 \\ -0.0477 \end{bmatrix}; \quad \mathbf{c}^T = \begin{bmatrix} 1 & 0 \end{bmatrix}$$

3. CONTROLLER DESIGN AND EXPERIMENTAL RESULTS

Based on the model obtained in Section 2, H_∞–controller is designed for the smart beam. The goal of the controller is to attenuate the vibrations of the smart beam at its first flexural frequency and gain stabilizes the unmodeled high frequency modes. In H_∞–control design framework, the objective is to minimize the H_∞ norm of the weighted transfer functions from the input disturbance signals to the output error signals. The uncertainties in the plant model can be put in such a form that some of the disturbances and error signals correspond to the channels through which the nominal model interacts with a norm bounded uncertainty block Δ. This generates the set of plants in which the true plant is assumed to exist.

Robust control refers to the control of uncertain plants with unknown dynamics subjected to unknown disturbance signals. The problem is to design a fixed controller, which guarantees acceptable performance norms in the presence of plant and input uncertainty. H_∞–optimal control design technique, directly addresses the problem of robustness by deriving controllers which maintain system response, and error signals to within the prescribed tolerances, in spite of the presence of noise in the system. The objective is to design a dynamic output feedback controller to control the first vibration mode of a cantilever beam and suppress the possible instability due to neglected higher modes. For controller design weighing functions are selected based on mixed sensitivity approach. The weight W_1 determines the bandwidth for vibration suppression. This requires a low pass filter with cut off frequency of 33Hz to provide vibration suppression. The weight W_3 is chosen as a high pass filter. It reflects the increased uncertainty of the model to represent the true dynamics of the system at high frequencies. Weight W_2 determines the shape of the controller and takes into account the additive uncertainty of the model. Since, the controller must not be responsive to higher order un-modeled modes, the amplitude of the controller must 'roll-off' at higher frequencies. Following several iterations of the design process, the weighing functions obtained are

$$W_1 = \frac{3.794 \times 10^4}{s^2 + 389.6\,s + 3.794 \times 10^4}; \quad W_2 = 0.0001; \quad W_3 = \frac{0.0001 s^2}{s^2 + 392.1 s + 3.843 \times 10^4}$$

The controller is designed and then implemented using a digital control system that consists of SIMULINK modeling software and a dSPACE DS 1104 controller board . Bilinear transformation using Tustin's approximation is carried out to transform the continuous system to discrete one with a sampling period of 0.01 second. The controller transfer function is

$$C(z) = \frac{570.5z^6 - 495.9z^5 - 321.8z^4 - 347.2z^3 + 11.79z^2 - 0.1303z + 0.0004746}{z^6 + 0.7376z^5 - 0.07626z^4 + 0.002535z^3 - 3.828 \times 10^5\,z^2 + 2.727 \times 10^7 \times z - 7.463 \times 10^{10}}$$

The open loop and closed loop response obtained experimentally are given in Figure 2 and Figure 3, respectively. The frequency response of the same is given in the Figure 4.

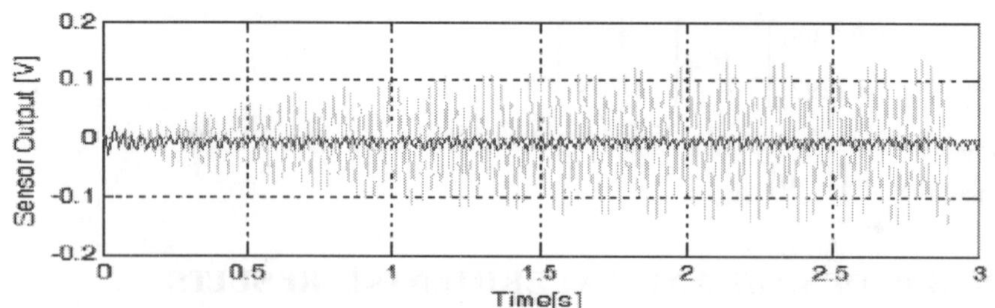

Fig. 2. Open and Closed Loop response when excited with the first mode frequency.

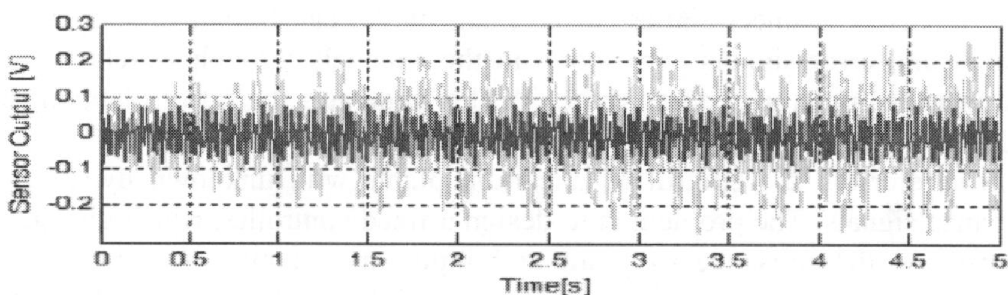

Fig. 3. Open and Closed Loop response when excited with the first mode and third mode frequencies.

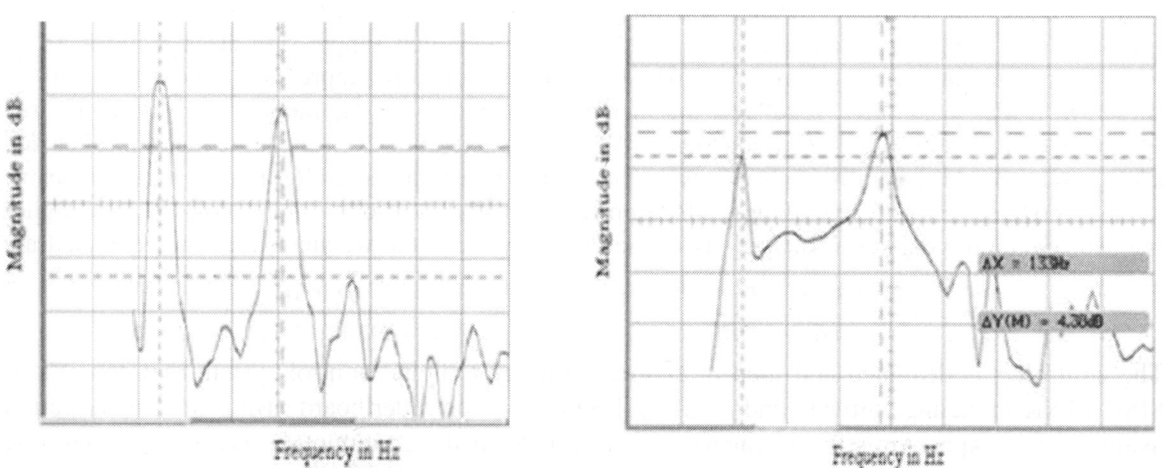

Fig. 4. Frequency response when excited with the first and third mode frequencies (a) open loop, (b) closed loop.

CONCLUSION

An experimentally identified model of the smart beam was utilized in the design of the H_∞ controller. The designed controller was found to suppress the sinusoidally excited vibrations of the smart cantilever beam due to its first flexural mode. It was also shown that the designed controller guarantees the robust performance of the system in the presence of uncertainties due to neglected higher flexural mode vibrations.

REFERENCES

1. Bailey, T. Hubbart, J.E. 1985, Distributed Piezoelectric-Polymer Active Vibration Control of a Cantilever Beam. Journal of Guidance, Control and Dynamics, Vol. 8, No. 5, pp 605–611.
2. Inderjit Chopra, 2002, Review of state of art of smart structures and integrated systems, AIAA Journal, 40(11), pp 2145-2187.
3. E.F. Crawley, J. Louis, 1989, Use of Piezoelectric Actuators as Elements of Intelligent Structures. AIAA Journal, 25, pp 1373-1385.
4. Indra Narayan Kar, Kazuto Seto and Fumio Doi, 2000, Multimode Vibration Control of a Flexible Structure Using H infinity based Robust Control, IEEE/ASME transactions on Mechatronics, Vol 5, and No.1
5. Mario Sznaier, Juanyu Bu, 1998," Mixed l1/ H Infinity Control of MIMO Systems via Convex Optimization", IEEE Transactions on Automatic control, Vol. 43, No. 9.
6. Masami Saeki, 1990, "Search Methods for a Constant Scaling Matrix Attaining Minimum H Infinity Norm", Proceedings of the 29th Conference on Decision and Control Honolulu, Hawaii.
7. Lennart Ljung, 1999, System identification theory for the user, Prentice Hall PTR, New Jersey.

68

Vibration Analysis of Rotating Tapered Thin-Walled Composite Cantilever Beams Including Gyroscopic Coupling

D.N. VADIRAJA AND A.D. SAHASRABUDHE

Department of Mechanical Engineering, Indian Institute of Technology Guwahati, India
email: vadiraja@iitg.ernet.in, ads@iitg.ernet.in

ABSTRACT

In the present work, dynamic modelling method is used to study free vibration characteristics of a tapered rotating cantilever Euler-Bernoulli beam. The mathematical model is derived for arbitrary beam cross-section. Non-classical effects generally exhibited by thin-walled composite material beams such as anisotropy, heterogeneity, warping are included in the mathematical model. The governing system of equations are derived from Hamilton's principle and solution is obtained by extended Galerkin's method. Free vibration characteristics of tapered box beam configuration are presented. Coriolis effect is included during numerical calculation.

Keywords: Thin-Walled Composite Beams, Coriolis Effect, Euler-Bernoulli Beams, Taper.

1. INTRODUCTION

Rotating blades are flexible structures, which are often idealized as cantilever beams. Coriolis forces will appear on a beam whenever there is a radial lengthening or shortening of the beam about the rotational axis as a result of beam bending. The vibration characteristics of these structures should be identified accurately so that one can properly design or control the system.

Non-classical effects in the behaviour of thin walled composite beams such as bending-shear coupling, restrained torsional warping, secondary warping, anisotropy, heterogeneity are considered by various researchers [Refs. 1-3]. A rotating beam differs from a non-rotating beam in terms of coupling of elastic deformations. The importance of the coupling is illustrated in Refs. [4-6]. Suresh and Nagaraj [4] and Chandiramani *et al*. [6] presented HSDT for thin walled composite beams with arbitrary cross-section. Song and Librescu [5] and Chandiramani *et al*. [6] used geometrically non-linear modelling methods to include centrifugal stiffening. However, this approach involves cumbersome formulation procedure and most importantly, gyroscopic coupling cannot be included in numerical analysis, as linearization is possible only after discarding gyroscopic coupling. To overcome this problem, Kane *et al.* [7], Yoo and Shin [8] and Yoo *et al.* [9] used a new method for solid beams called dynamic modeling method. This modelling method employs a non-Cartesian deformation variable in addition to two Cartesian deformation variables.

The purpose of the present study is to extend the dynamic modelling method of Refs. [7-9] to tapered thin-walled composite cantilever beams. In this paper, coupled linear equations of motion describing axial, bending and rotational motions, including gyroscopic coupling and centrifugal stiffening are solved using extended Galerkin's method. Dynamic modelling method is used to study free vibration characteristics of a rotating tapered cantilever Euler-Bernoulli beam. Numerical solution including gyroscopic coupling are illustrated for a box beam configuration.

2. FORMULATION

A tapered beam of length L and hub radius R_0, rotating with constant angular velocity is considered. The Cartesian inertial frame of reference (X, Y, Z) has its origin at the center of the hub. The beam coordinate system (x, y, z) is located at offset R_o from the origin O. Further, (i, j, k) and (I, J, K) represent unit vectors in the (x, y, z) and (X, Y, Z) coordinate systems respectively (Figs. 1(a) and 1(c)). In addition to above, a local coordinate system (s, n, z) associated with the beam is also considered which is shown in Fig. 1(c). Taper parameter in the beam is taken as $\sigma = c_{L=0}/C_{L=L} = b_{L=0}/b_{L-L}$. Further, instead of z, a non-Cartesian variable $\hat{s}$ denoting axial stretch is used in the present study. Due to this transformation, the centrifugal stiffening effect can be captured. The beam rotates about Y-axis with constant angular velocity $\Omega(= \Omega J = \Omega j)$. u_0, v_0 and $\hat{s}$ represent displacements along (x, y, z) axes respectively and Φ represents rotation about Z-axis. The displacements $\hat{s}$ and w_0 are related by [8],

$$\hat{s} = w_0 + \frac{1}{2}\int_0^z \left[\left(\frac{\partial u}{\partial \sigma}\right)^2 + \left(\frac{\partial v}{\partial \sigma}\right)^2\right] d\sigma \qquad \text{...(1)}$$

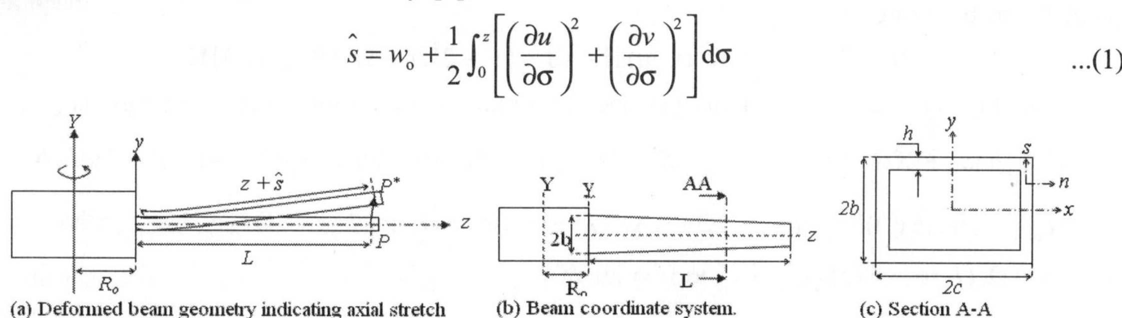

(a) Deformed beam geometry indicating axial stretch   (b) Beam coordinate system.   (c) Section A-A

Fig. 1. Beam configuration and coordinate system.

2.1 Kinematics

The assumptions considered while deriving equations of motion of thin walled beam are [5, 10]: The axial displacement $\hat{s}$ is only due to the shear deformation; No in-plane cross-section deformation, that is ε_{xx}, ε_{yy} and ε_{xy} in-plane strains are neglected; The rate of twist is span wise dependent; The shell forces and moment resultants corresponding to γ_{ss} are negligible; γ_{nn} is negligible compared to γ_{zz} and γ_{sz}; The wall thickness and material properties are invariant along s and n directions. Linear displacements u, v and w representing lag, flap and extensional motion respectively, are obtained for Euler-Bernoulli beam as [10],

$$u = u_0 + z\phi; \quad v = w_0 - x\phi; \quad w = w_0 - [\bar{y} + nm]v_0' - [\bar{x} + nl]u_0' - [F_w + na]\phi' \qquad \text{...(2a-c)}$$

From Eqs. (1) and (2c), the axial displacement component in arc length stretch coordinate system is,

$$w = \hat{s} - [\bar{y} + nm]v_o' - [\bar{x} + nl]u_o' - [F_w + na]\phi' - \frac{1}{2}\int_0^z \left[\left(\frac{\partial u}{\partial \sigma}\right)^2 + \left(\frac{\partial v}{\partial \sigma}\right)^2 \right] d\sigma \qquad \text{...(3)}$$

2.2 Equations of Motion

The variational equation of the Hamilton's principle for the 3-D elasticity theory is

$$\int_{t_1}^{t_2}(\delta V - \delta T)dt = 0; \quad \text{where, } \delta V = \int_{t_2}^{t_2}\int_\tau \sigma_{ij}\delta\varepsilon_{ij}\, d\tau\, dt; \quad \delta T = -\int_{t_2}^{t_2}\int_\tau \rho(\ddot{\mathbf{R}}_i . \delta\mathbf{R}_i)\, d\tau\, dt; \quad \text{...(4a-c)}$$

Using assumptions, the variational strain energy is written as

$$\delta V = \int_{t_1}^{t_2} \int_\tau [\sigma_{zz}\delta\varepsilon_{zz} + \tau_{sz}\delta\gamma_{sz}]\, d\tau\, dt \qquad \text{...(5)}$$

Neglecting the non-linearity arising due to axial displacement and substituting strain field (Appendix) and stress field (Appendix) in Eq. (5), the variational strain energy reduces from a 3-D elasticity problem to a 1-D problem.

$$\delta V = \int_{t_1}^{t_2} \int_0^L [a_{11}\hat{s}'\delta\hat{s}' + a_{17}\phi'\delta\hat{s}' + a_{22}u_o''\delta u_o'' + a_{33}v_o''\delta v_o'' + a_{77}\phi'\delta\phi' + a_{17}\hat{s}'\delta\phi' + a_{66}\phi''\delta\phi'']\, d\tau\, dt \qquad \text{...(6)}$$

The global stiffness coefficients in the above equation match with the one derived in Ref. [6]. The position vector relative to the fixed origin O of a point on the deformed beam is obtained as, $\mathbf{R}=R_o\mathbf{k} + x\mathbf{i} + y\mathbf{j} + z\mathbf{k} + u\mathbf{i} + v\mathbf{j} + w\mathbf{k}$. Differentiating twice with respect to time, the acceleration of point P can be written as,

$$\ddot{\mathbf{R}}=\left(\ddot{u} + 2\Omega\dot{w} - \Omega^2(x+u)\right)\mathbf{I} + \ddot{v}\mathbf{J} + \left(\ddot{w} - 2\Omega\dot{u} - \Omega^2(R_o+z+w)\right)\mathbf{K} \qquad \text{...(7)}$$

From Eqs. (1), (2), (3), (4c) and (7) one can obtain the variational kinetic energy as

$$\delta T = \int_{t_2}^{t_2}\int_\tau [b_1(\ddot{u}_o + 2\Omega\dot{\hat{s}} - \Omega^2 u_o)\delta u_o + (I_{yy}\ddot{u}_o' - \Omega^2(I_{yy}u_o' - R(z)b_1 u_o'))\delta u_o' + b_1\ddot{v}_o\delta v_o' + (I_{xx}\ddot{v}_o' - 2\Omega I_{xx}\dot{\phi}$$

$$-\Omega^2(I_{xx}v_o' - R(z)b_1 v_o'))\delta v_o' + b_1(\ddot{\hat{s}} - 2\Omega\dot{u}_o - \Omega^2(R_o+z+\hat{s}))\delta\hat{s} + (I_{xx}\ddot{\phi} + I_{yy}\ddot{\phi} + 2\Omega I_{xx}\dot{v}_o' - \Omega^2 I_{xx}\phi)\delta\phi \qquad \text{...(8)}$$

$$+(I_{ww}\ddot{\phi}' - \Omega^2(I_{ww}\phi' - R(z)b_1(I_{xx} + I_{yy})\phi'))\delta\phi']dz\, dt$$

where, I_{xx}, I_{yy} and I_{ww}, are inertia quantities (Appendix) and $R(z) = (R_o(L-z) + \dfrac{L^2 - z^2}{2})$.

Neglecting body forces and surface traction and substituting δT and δV in Hamilton's principle, equation of motion yields,

$$\delta u_o: a_{22}u_o^{IV} + b_1(\ddot{u}_o + 2\Omega\dot{\hat{s}} - \Omega^2 u_o) - (I_{yy}\ddot{u}_o'' - \Omega^2(I_{yy}u_o'' - R(z)b_1 u_o'')) = 0$$

$$\delta v_o: a_{33}v_o^{IV} + b_1\ddot{v}_o - (I_{xx}\ddot{v}_o'' - 2\Omega I_{xx}\dot{\phi}' - \Omega^2(I_{xx}v_o'' - R(z)b_1 v_o'')) = 0$$

$$\delta\hat{s} : -a_{11}\hat{s}'' - a_{17}\phi'' + b_1(\ddot{\hat{s}} - 2\Omega\dot{u}_o - \Omega^2(R_o+z+\hat{s})) = 0$$

$$\delta\phi : -a_{77}\phi'' - a_{17}\hat{s}'' + a_{66}\phi^{IV} + I_{xx}\ddot{\phi} + I_{yy}\ddot{\phi} + 2\Omega I_{xx}\dot{v}_o' - \Omega^2 I_{xx}\phi - (I_{ww}\ddot{\phi}'' - \Omega^2(I_{ww} - R(z)b_1(I_{xx} + I_{yy}))\phi'') \quad \text{...(9a-d)}$$

Geometric boundary conditions at z = 0 are $u_o = u_o' = v_o = v_o' = \hat{s} = \phi_o = \phi_o' = 0$ and natural boundary conditions at z = L are

$$\delta u_o: \; -a_{22}u_o''' - (I_{yy}\ddot{u}_o' - \Omega^2(I_{yy}u_o' - R(z)b_1 u_o')) = 0; \quad \delta u_o': a_{22}u_o'' = 0$$

$$\delta v_o: \; -a_{33}v_o''' + I_{xx}\ddot{v}_o' - 2\Omega I_{xx}\dot{\phi} - \Omega^2(I_{xx}v_o' - R(z)b_1 v_o')) = 0; \quad \delta v_o: a_{33}v_o'' = 0$$

$$\delta\hat{s}: a_{11}\hat{s}' + a_{17}\phi' = 0; \quad \delta\phi: a_{77}\phi' + a_{17}\hat{s}' - a_{66}\phi''' + I_{ww}\ddot{\phi}' - \Omega^2(I_{ww} - R(z)b_1(I_{xx}+I_{yy}))\phi') = 0; \quad \delta\phi: a_{66}\phi'' = 0$$

From the equation of motion it can be observed that there exists Coriolis coupling between lag-extension (Eqs. 9a and 9c) and flap-twist (Eqs. 9b and 9d) motion. Extension-twist (Eqs. 9c and 9d) equations are coupled with a_{17} quantity. Hence, these four equations are to be solved simultaneously for coupled analysis.

3. FREE VIBRATION ANALYSIS

In the present study, extended Galerkin's method is used to obtain approximate solution. In this approach discretization is carried out within the Hamilton's equation. The natural boundary conditions are taken back into variational equation by carrying reverse integration in space. Hence $(u_o, v_o, \hat{s}, \phi) = (\varphi_1 q_1, \varphi_2 q_2, \varphi_3 q_3, \varphi_4 q_4)$. For a conservative gyroscopic system, the eigen value problem is written as $AX = \lambda X$, where, $\omega = \sqrt{\lambda}$, $A = \begin{pmatrix} 0 & I \\ -M^{-1}K & -M^{-1}G \end{pmatrix}$, M, K and G are the mass, stiffness and gyroscopic matrices, respectively (see Appendix).

4. NUMERICAL EXAMPLE

The beam considered is of graphite-epoxy [5] material with properties $E_1 = 206.8$ GPa; $E_2 = 5.17$ GPa; $G_{12} = G_{13} = 3.1$ GPa; $G_{23} = 2.5511$ GPa; $\nu_{12} = 0.25$; $\rho = 1528.15$ kg/m^3, with beam geometric configuration (Fig. 1) L = 2.023 m; $R_o = 0.2023$ m; h = 0.01016 m; c = 0.127 m; $b = 0.0254$ m. The trial functions selected to satisfy geometric boundary conditions at the beam root (z = 0) are

$$(\varphi_1, \varphi_2, \varphi_3, \varphi_4) = ([z^2 \; z^3 \; z^4....], [z^2 \; z^3 \; z^4....], [z^1 \; z^2 \; z^3....], [z^2 \; z^3 \; z^4....]).$$

Figure 2. shows the first three normalized eigen modes in lagging and flapping motion, for a 30° ply angle, rotation speed 300 rad/s and taper parameter σ is 1 and 4 with gyroscopic coupling. In first lagging mode, the beam gets straightened for the model without taper (σ = 1), whereas, in second and third the beam gets straightened for the model with taper (σ = 4). Similar trend can be observed in flapping motion also.

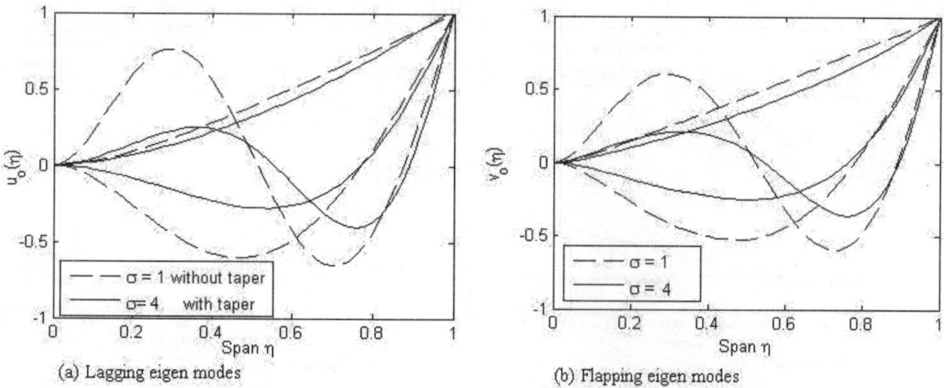

Fig. 2. Variation of the first three lagging and flapping mode shapes in the coupled lagging-flapping motion for two taper parameters (σ = 1 and 4).

Figure 3 shows coupled lag-flap natural frequencies for various rotational speeds and taper parameters for a 30° ply angle. It can be observed that first natural frequency increases monotonically with increase in taper in beam. In other words increase in taper stiffens the beam. From Figs. 3(a) and 3(b), it is observed that the first mode changes from flap to lag and lag to flap respectively, as the rotational speed is increased.

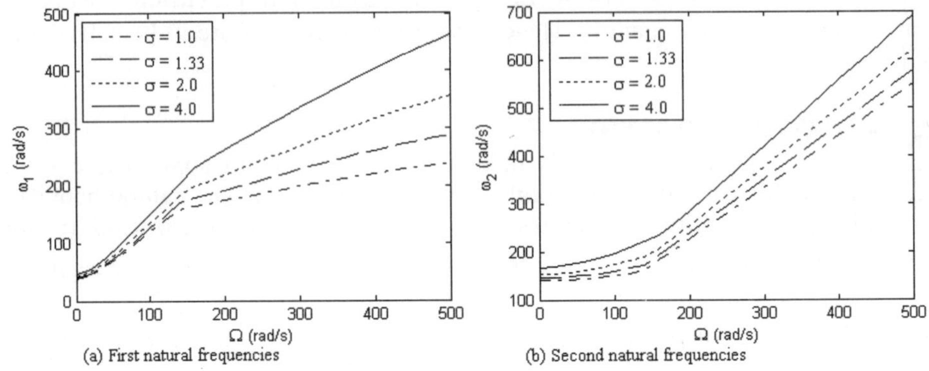

(a) First natural frequencies

(b) Second natural frequencies

Fig. 3. First and second coupled lagging-flapping natural frequencies for various rotational speeds and taper parameters.

Figure 4 shows the first three gyroscopically coupled and uncoupled (solid lines and dotted lines respectively) natural frequencies, for a 30° ply angle, various rotational speeds and for two taper parameters ($\sigma = 1$ and 4). From Fig. 4(a), it is observed that instability occurs for the considered beam configuration at rotational speed $\approx$ 1500 rad/s, which means the dominant first natural frequency, becomes zero. When natural frequency becomes zero, beam will buckle [8]. Hence, buckling speed for the beam configuration considered is $\approx$ 1500 rad/s. Whereas, for beam with taper the buckling occurs at rotational speed $\approx$ 2025 rad/s (see Fig. 4(b)). Therefore, taper in the beam increases the buckling speed of the beam. Moreover, first lag natural frequency with gyroscopic coupling and first extension natural frequency without gyroscopic coupling buckle at the same rotational speed $\approx$ 1500 rad/s. The similar trend was observed for solid isotropic beam by [8]. Hence, the analysis performed neglecting gyroscopic coupling leads to erroneous results.

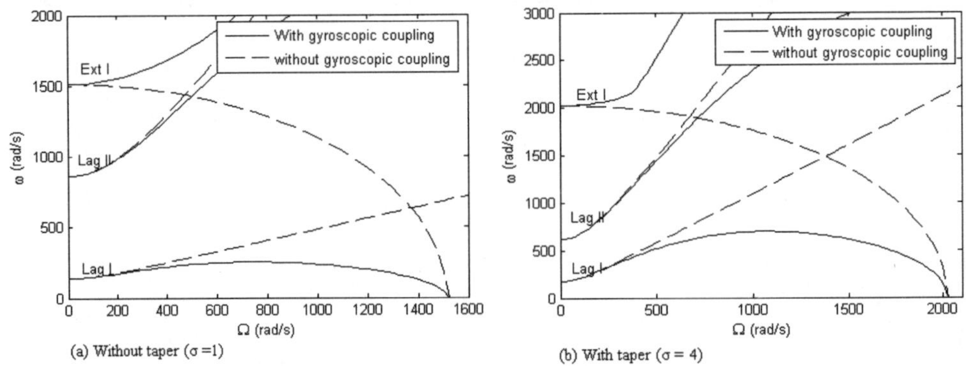

(a) Without taper ($\sigma = 1$)

(b) With taper ($\sigma = 4$)

Fig. 4. First three natural frequencies for a 30° ply angle, various rotational speeds and taper paramaters $\sigma = 1$, 4.

CONCLUSION

In the present study, equations of motion for rotating cantilever Euler-Bernoulli unshearable beams have been derived using hybrid deformation variables, which employ stretch deformation instead of the conventional axial deformation. Gyroscopic effect is included in the numerical example. It is observed that gyroscopic effect is having significant effect on natural frequencies. Discarding gyroscopic coupling leads to erroneous results. Taper in the beam increases the buckling speed of the beam, hence tapered beam is more suitable for higher rotational speeds. Mathematical model is derived for arbitrary configuration, which can be easily extended to other configurations like elliptical cross-section, airfoil cross-section.

REFERENCES

1. A.D. Stemple, S.W. Lee, 1988, Finite-element model for composite beams with arbitrary cross-sectional warping, American Institute of Aeronautics and Astronautics Journal. 26(12), 1512-1520.
2. W.R. Lawrence, R.A. Ali, H.H. Dewey, 1990, Nonclassical behavior of thin-walled composite beams with closed cross-sections, Journal of American Helicopter Society. 35, 42-50.
3. L. Jun, S. Rongying, H. Hongxing, J. Xianding, 2004, Bending torsional coupled dynamic response of axially loaded composite Timoshenko thin walled beam with closed cross-section, Journal of Composite Structures. 64, 23-35.
4. J.K. Suresh, V.T. Nagaraj, 1996, Higher-order shear deformation theory for thin-walled composite beams, Journal of Aircraft. 33(5), 978-986.
5. O. Song, L. Librescu, 1997, Structural modelling and free vibration analysis of rotating composite thin-walled beams, Journal of American Helicopter Society. 42(4), 358-369.
6. N.K. Chandiramani, L. Librescu, C.D. Shete, 2002, On the free-vibration of rotating composite beams using a higher-order shear formulation, Aerospace Science and Technology. 6, 545-561.
7. T.R. Kane, R.R. Ryan, A.K. Banerjee, 1987, Dynamics of beams attached to a moving base, Journal of Guidance, Control, and Dynamics. 10(2), 139-151.
8. H.H. Yoo, S.H. Shin, 1988, Vibration analysis of rotating cantilever beams, Journal of Sound and Vibration. 212(5), 807-828.
9. H.H. Yoo, H.L. Seung, H.S. Sang, 2005, Flapwise bending vibration analysis of rotating multilayered composite beams, Journal of Sound and Vibration. 286(4-5), 745-761.
10. F. Nishino, A. Hasegawa, 1979, Thin-walled elastic members, Journal of the Faculty of Engineering. The University of Tokyo. XXXV(2), 109-190.

APPENDICES

Stress strain relation in (s, n, z) coordinate system and strain field are

$$\begin{bmatrix} \sigma_{ss} \\ \sigma_{zz} \\ \sigma_{nn} \\ \tau_{sz} \end{bmatrix} = \begin{pmatrix} \overline{Q}_{11} & \cdots & \overline{Q}_{14} \\ \vdots & \ddots & \vdots \\ \overline{Q}_{14} & \cdots & \overline{Q}_{44} \end{pmatrix} \begin{bmatrix} \varepsilon_{ss} \\ \varepsilon_{zz} \\ \varepsilon_{nn} \\ \gamma_{sz} \end{bmatrix};$$

$$\varepsilon_{xx} = u_{,x}; \quad \varepsilon_{yy} = v_{,y}; \quad \gamma_{sz} = \psi\phi';$$

$$\varepsilon_{zz} = \hat{s} - [\overline{y} + nm]v_o'' - [\overline{x} + nl]u_o'' - [F_w + na]\phi''$$

$$-\int_0^z \left[\left(\frac{\partial u}{\partial \sigma}\right)' \left(\frac{\partial u}{\partial \sigma}\right) + \left(\frac{\partial v}{\partial \sigma}\right)' \left(\frac{\partial v}{\partial \sigma}\right) \right] d\sigma$$

Inertia Quantities are

$$(M_o, M_2) = \int_{-h/2}^{h/2} \rho(1, n^2)\,dn; \quad (\overline{I}_{xx}, \overline{I}_{yy}, \overline{I}_\mu, \overline{I}_{mm}, I_{ff}, I_{aa}) = \oint (\overline{y}^2, \overline{x}^2, m^2, l^2, F_w^2, a^2)\,ds; \quad b_1 = \oint M_o\,ds;$$

$$I_{xx} = M_o\overline{I}_{xx} + M_2\overline{I}_\mu; \quad I_{yy} = M_o\overline{I}_{yy} + M_2\overline{I}_{mm}; \quad I_{ww} = M_o I_{ff} + M_2 I_{aa};$$

Mass, stiffness and gyroscopic matrices are,

$$
M = \int_0^L \begin{bmatrix} M_{11} & 0 & 0 & 0 \\ 0 & M_{22} & 0 & 0 \\ 0 & 0 & M_{33} & 0 \\ 0 & 0 & 0 & M_{44} \end{bmatrix};\;
K = \int_0^L \begin{bmatrix} K_{11} & 0 & 0 & 0 \\ 0 & K_{22} & 0 & 0 \\ 0 & 0 & K_{33} & K_{34} \\ 0 & 0 & K_{34}^T & K_{44} \end{bmatrix}
G = \int_0^L \begin{bmatrix} 0 & 0 & G_{13} & 0 \\ 0 & 0 & 0 & G_{24} \\ -G_{13}^T & 0 & 0 & 0 \\ 0 & -G_{24}^T & 0 & 0 \end{bmatrix}
$$

where, $M_{11} = b_1\phi_1\phi_1^T + I_{yy}\phi_1'\phi_1'^{T}$; $M_{22} = b_1\phi_2\phi_2^T + I_{xx}\phi_2'\phi_2'^{T}$; $M_{33} = b_1\phi_3\phi_3^T$; $M_{44} = (I_{xx}+I_{yy})\phi_4\phi_4^T + I_{ww}\phi_4'\phi_4'^{T}$;

$K_{11} = a_{22}\phi_1''\phi_1''^{T} - \Omega^2(b_1\phi_1\phi_1^T + (I_{yy} - R(z)b_1)\phi_1'\phi_1'^{T})$; $K_{22} = a_{33}\phi_2''\phi_2''^{T} - \Omega^2(I_{xx} - R(z)b_1)\phi_2'\phi_2'^{T}$;

$K_{33} = a_{11}\phi_3'\phi_3'^{T} - \Omega^2 b_1\phi_3'\phi_3'^{T}$; $K_{34} = a_{17}\phi_3'\phi_4'^{T}$; $K_{44} = a_{77}\phi_4''\phi_4''^{T} + a_{66}\phi_4'\phi_4'^{T} - \Omega^2(I_{xx}\phi_4\phi_4^T + (I_{ww} - R(z)b_1)\phi_4'\phi_4'^{T})$;

$G_{13} = -2\Omega b_1\phi_1\phi_3^T$; $G_{24} = 2\Omega I_{xx}\phi_2'\phi_4'^{T}$;

69

Thermally Induced Vibrations of Beams: A Finite Element Analysis

PRAVIN MALIK AND RAVIKIRAN KADOLI

Department of Mechanical Engineering, National Institute of Technology Karnataka, Surathkal, Srinivasnagar-575 025, India. email: pravin_malik@yahoo.com, rkkadoli@rediffmail.com

ABSTRACT

This paper discusses the numerical results on thermally induced vibrations of a beam with boundary conditions like simple support, clamped-clamped, clamped-simple supported and clamped free subjected to a step heat input. A finite element formulation for an Euler-Bernoulli beam subjected to thermal load is presented. The temperature distribution across the cross-section of the beam subjected to heat source on one surface and insulated on the opposite surface is obtained which produces a time dependent thermal moment on the beam structure. The effect of L/h ratio of the beam on the dynamic displacement and thermal moment is examined. Thermally induced vibrations of beam in presence of axial compressive load are also investigated.

Keywords: Dynamic Displacement, Dynamic Thermal Moment, Finite Element.

1. INTRODUCTION

Structural components subjected to thermal conditions like, high temperature, high heating rate are common in the fields of aerospace, nuclear, castings, forging, radiant burners, heat exchangers, artillery barrels, piping, etc. The heating of the structure may also be due to sudden exposure to very large amount of heat which is typical to launching of rocket, spacecraft structural elements subjected to radiant solar heat, nuclear reactor components, heat treatment of cast and forged components, etc. Pioneering works of Boley [1], Boley and Barber [2] and Boley [3] on thermally induced vibrations of beam structures provides a detailed interrelation of time dependent temperature variation on the structural vibrations. Similarly, of late Blandino and Thornton [4] have examined thermally induced vibrations in tube-like structures.

Analytical solutions for the thermally induced vibrations in simply supported beam structures subjected to step heat input has been presented by Boley [1,3], Boley and Barber[2]. Manolis and Beskos [5] have attempted the same topic through the method of Laplace Transform. In the present article, finite element analysis of structural transients due to the thermal transients are carried out for beams with step heat input on one surface and insulated and convection on the opposite surface. Displacement response of the beam is obtained using Newmark's method. Further, the effect of length to thickness ratio, structural boundary conditions and presence of axial compressive load is also examined.

2. EQUATION OF MOTION OF BEAM SUBJECTED TO EXTERNAL HEAT SOURCE

The curvature of a uniform beam under the simultaneous action of external forces and heat input for Euler-Bernoulli beams is

$$M + M_T = EI\left(\frac{\partial^2 v}{\partial x^2}\right) \qquad \ldots(1)$$

where, M is the bending moment produced by the applied forces, M_T is the thermal moment, v is the transverse deflection in the y direction, E is the Young's modulus and I is the moment of inertia of beam cross-section. For the beam subjected to heat flux on one side and insulated on the other side the thermal moment acts as a forcing function which is given as

$$M_T = \int_A E\,\alpha\,\Delta T\, y\, dA = b\int_y E\,\alpha\,\Delta T\, y\, dy = E\,\alpha\,b\Delta T\left(\frac{y_{i+1}^2 - y_i^2}{2}\right) \qquad \ldots(2)$$

where, ΔT is the change in temperature, α is the coefficient of thermal expansion and A is the cross-sectional area, b is the width of the beam and y_i indicates thickness at i^{th} layer measured along the y axis. The thermal moment is calculated at uniform intervals across the thickness from the top to bottom surfaces of the beam and it is summed up in order to get the total thermal moment across the section. The thermal moment along the length is assumed to be constant as there is no temperature variation along the length of the beam i.e., it is independent of x hence,

$$\frac{\partial^2 T}{\partial x^2} = 0 \quad \text{and} \quad M_T = M_T(t) \qquad \ldots(3)$$

Considering the inertial forces, the governing equation of motion for a beam in the transverse direction in presence of thermal moment is given by [1],

$$\frac{\partial^2 M_T}{\partial x^2} = \frac{\partial^2}{\partial x^2}\left(EI\left(\frac{\partial^2 v}{\partial x^2}\right)\right) + \rho A\left(\frac{\partial^2 v}{\partial t^2}\right) \qquad \ldots(4)$$

where, ρ is the mass density. The following non-dimensional parameters are defined: The non-dimensional time τ is given as

$$\tau = \frac{\kappa t}{h^2} \qquad \ldots(5)$$

where, $\kappa = {k}/{\rho c_p}$ is thermal diffusivity, k is thermal conductivity, c_p is the specific heat and h is the total thickness of beam. The non-dimensional displacement V is given as

$$V = \frac{\pi^4 k v}{192 Q \alpha L^2} \qquad ...(6)$$

where, Q is the heat flux in W/m² and L is the length of the beam. The non dimensional thermal moment m_T is given as

$$m_T(\tau) = \frac{\pi^4 k M_T}{192 E I Q \alpha} \qquad ...(7)$$

The following additional non-dimensional parameters are also introduced

$$B = \frac{h}{L\sqrt{\kappa}} \left(\frac{EI}{\rho A} \right)^{1/4} = \frac{h}{L} \sqrt{\frac{c \sqrt{I/A}}{\kappa}} = \sqrt{\frac{h}{L} \frac{hc}{\kappa} \frac{1}{S}} \qquad ...(8)$$

where, S is the slenderness ratio, $S = \dfrac{L}{\sqrt{I/A}}$ and $c = \sqrt{E/\rho}$ is the velocity of propagation of longitudinal waves. For particular value of B, the thickness of the beam can be calculated as

$$h = \sqrt[6]{\frac{12 \times B^4 L^4 \kappa^2}{c^2}} \qquad ...(9)$$

The parameter B is the square root of the ratio of the characteristic time h^2/k of heat transfer problem to characteristic time $\left(\dfrac{\rho A L^4}{EI} \right)^{1/2}$ of the vibration problem (or proportional to the natural period of vibration). Thus B is large for beams with low diffusivity, low density and high bending rigidity; it is low if beam is slender or dense. Hence, the governing equation of motion, Eq.(4) in non-dimensional notations is

$$B^4 \left(\frac{\partial^4 v}{\partial x^4} \right) + \left(\frac{\partial^2 v}{\partial t^2} \right) = 0 \qquad ...(10)$$

The boundary and initial conditions for the problem are as follows:

$$v(0,t) = v(L,t) = v(x,0) = \dot{v}(x,0) = 0$$
$$v''(0,t) = v''(L,t) = m_T \qquad ...(11)$$

2.1 Determination of Temperature Distribution Across the Beam Cross-Section Using FEM

The finite element equation for temperature evaluation across beam thickness when the beam is exposed to sudden heating on one side and insulated on other side is as follows:

$$\frac{kA}{l}\begin{bmatrix} 1 & -1 \\ -1 & 1 \end{bmatrix}\begin{Bmatrix} T_1 \\ T_2 \end{Bmatrix} + \frac{\rho c_p Al}{6}\begin{bmatrix} 2 & 1 \\ 1 & 2 \end{bmatrix}\begin{Bmatrix} \dot{T}_1 \\ \dot{T}_2 \end{Bmatrix} = \dot{q}A\begin{Bmatrix} 1 \\ 1 \end{Bmatrix} \qquad \text{...(12)}$$

and for the beam exposed to sudden heating on one side and uniform convection on the other surface of the beam is:

$$\left\{ \frac{kA}{l}\begin{bmatrix} 1 & -1 \\ -1 & 1 \end{bmatrix} + \begin{bmatrix} 0 & 0 \\ 0 & h_c A \end{bmatrix} \right\}\begin{Bmatrix} T_1 \\ T_2 \end{Bmatrix} + \frac{\rho c_p Al}{6}\begin{bmatrix} 2 & 1 \\ 1 & 2 \end{bmatrix}\begin{Bmatrix} \dot{T}_1 \\ \dot{T}_2 \end{Bmatrix} = \dot{q}A\begin{Bmatrix} 1 \\ 1 \end{Bmatrix} + h_c A T_\infty \begin{Bmatrix} 0 \\ 1 \end{Bmatrix} \qquad \text{...(13)}$$

In Eq(13) the second matrix on LHS and second vector on RHS are contribution from convection and will be taken into consideration only for last element, h_c is convective heat transfer coefficient, T_1 and T_2 are the nodal temperatures and T_∞ is the ambient temperature. The global finite element equation for time dependent temperature distribution has the following form:

$$\mathbf{K}_{comb}\mathbf{T} + \mathbf{K}_{cap}\dot{\mathbf{T}} = \overline{\mathbf{F}}_Q \qquad \text{...(14)}$$

$\mathbf{K}_{comb}$= elemental conduction and/or convection matrix, $\mathbf{K}_{cap}$ = elemental capacitance matrix and $\overline{\mathbf{F}}_Q$ = force vector. Eq.(14) must be solved for the variation of temperature in space and time domain to obtain the temperature distribution across the thickness of the beam. The close form solution for temperature distribution T, across the thickness of the beam, as given by Boley [1] or Carslaw and Jaeger [6] is

$$T = \left(\frac{hQ}{k} \right)\left[\tau + \frac{1}{2}\left(\frac{y}{h} + \frac{1}{2} \right)^2 - \frac{1}{6} - \frac{2}{\pi^2}\sum_{j=1}^{\infty} \frac{(-1)^j\, e^{-j^2\pi^2\tau}}{j^2}\cos j\pi\left(\frac{y}{h} + \frac{1}{2} \right) \right] \qquad \text{...(15)}$$

provided that the beam is initially at $T = 0$.

2.2 Beam Finite Element Formulation

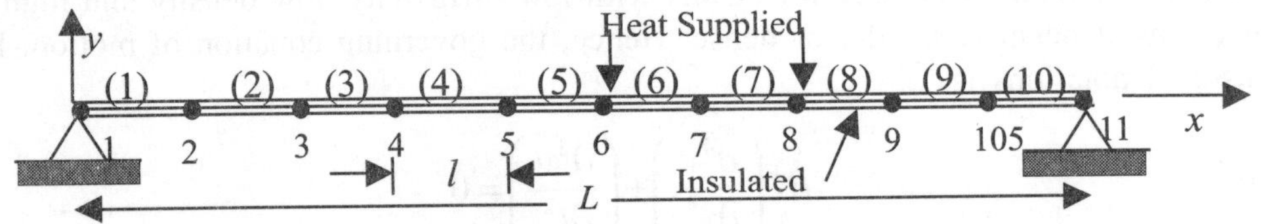

Fig. 1. Finite element idealization of beam for structural analysis.

The weak form of the governing equation Eq. (4) is as follows:

$$EI\left(N^T \frac{\partial^3 v}{\partial x^3}\bigg|_0^l - \frac{\partial N^T}{\partial x}\frac{\partial^2 v}{\partial x^2}\bigg|_0^l + \int_0^l \frac{\partial^2 v}{\partial x^2}\frac{\partial^2 N^T}{\partial x^2}\,dx \right) + \rho A\left(N^T \frac{\partial v}{\partial t}\bigg|_0^l - \int_0^l \frac{\partial v}{\partial t}\frac{\partial N^T}{\partial x}\,dx \right) -$$

$$N^T \frac{\partial M_T}{\partial x}\bigg|_0^l + \int_0^l \frac{\partial M_T}{\partial x}\frac{\partial N^T}{\partial x}\,dx = 0 \qquad \text{...(16)}$$

where, the first term refers to shear force, the second term refers to moment, the third term will give the stiffness matrix, the fourth term will be zero as per boundary conditions, the fifth term will give the mass matrix, the sixth term will give the shear force and the last term will be zero as there is no change in thermal moment along the length of the beam. Hermite shape functions N are used to develop the various finite element matrices. After obtaining the time dependent temperature distribution across the beam thickness, force vector $\{F_T\}$ is evaluated which will contain the thermal moment M_T only. Subsequently, static equation $[K]\{v\} = \{F_T\}$ is solved. The displacement thus obtained at time t is termed as the static displacement v_{st}. Newmark's method is used to solve the second order equation of motion involving the time dependent forcing function

$$[M]\{\ddot{v}\}+[K]\{v\}=\{F_T\} \qquad \qquad ...(17)$$

The displacement obtained by solving the above equation is termed as dynamic displacement v_{dyn}. From the dynamic displacement vector, the displacements for the central element of the beam are extracted to calculate the thermal moment at the centre of the beam and is termed as dynamic thermal moment:

$$\{M_{TD}\}^e = [K]^e \{v_{dyn}\}^e \qquad \qquad ...(18)$$

where, superscript e refers to elemental solution. Hence, the dynamic thermal moment at the centre of the beam is given as,

$$M_{T\,dyn} = M_{TD}^{node} \pm M_{T\,st} \qquad \qquad ...(19)$$

where, $M_{T\,st} = M_T$.

The equation of motion of a beam subjected to axial compressive load and thermal moment is

$$EI\left(\frac{\partial^4 v}{\partial x^4}\right)+P\left(\frac{\partial^2 v}{\partial x^2}\right)+\rho A\left(\frac{\partial^2 v}{\partial t^2}\right)-\frac{\partial^2 M_T}{\partial x^2}=0 \qquad \qquad ...(20)$$

3. NUMERICAL RESULTS AND DISCUSSION

A FORTRAN computer program is written based on the above formulation to study the dynamic response of beams with different support conditions.

3.1 Validation—Simply Supported Beam Subjected to Step Heating

Studies are carried out for the simply supported beam (Fig. 2) subjected to heat source on one surface and insulated on the opposite surface by using the following thermal structural data for aluminium beam: $b = 1\text{m}$, $L = 0.254\text{m}$, $k = 201.87\text{W/mK}$, $\alpha = 22.0 \times 10^{-6}/°\text{C}$, $\rho = 2700\text{kg/m}^3$, $c_p = 869.38\text{J/kg}°\text{C}$, $Q = 1.63\times10^6\text{W/m}^2$, $E = 73.5\times10^9\text{Pa}$, and $G = 26.0\times10^9\text{Pa}$ and other data are mentioned in the Table 1.

Figure 3 shows the temperature distribution obtained across the thickness of the beam for L/h ratio of 88 and it is found that they closely tally with the close form solution (Eq.15) referred from Boley [1]. The results like dynamic mid-span deflection and dynamic mid-span thermal moment for a simply supported beam subjected to step heat input are validated for the various values of B as stated in Boley [1] and Manolis and Beskos [5].

Table 1. Geometric and time data for validation problem.

B	L/h	h (m)	t (sec)	No. of iterations
0	25400	0.000010	0.000002	200
1	165	0.001544	0.04	400
∞	10	0.025400	10.0	400

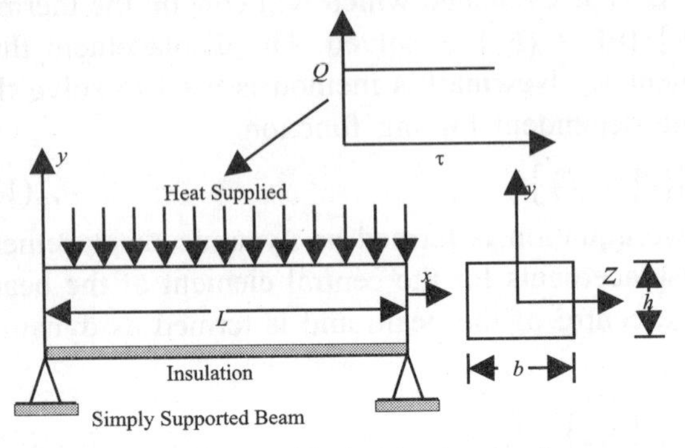

Fig. 2. Beam subjected to step heating.

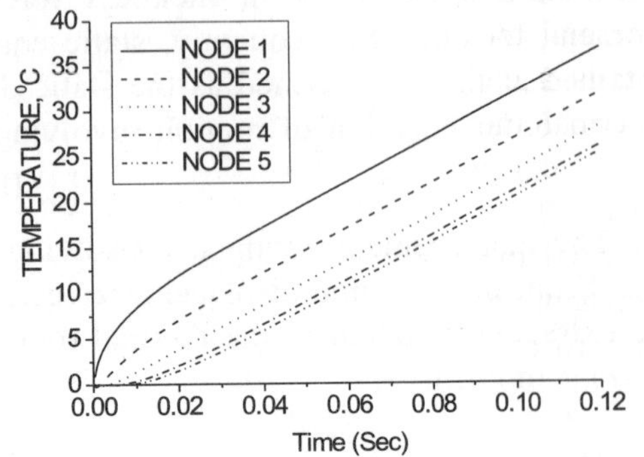

Fig. 3. Temperature distribution across the thickness of the beam for L/h = 88.

3.2 Effect of L/h Ratio on Thermally Induced Vibrations of Beams Subjected to Step Heating

Figure 4 shows the static and dynamic mid-span deflection and Figure 5 shows the static and dynamic mid-span thermal moment for a simply supported beam of *L/h* ratio of 88 which corresponds to $B = 2.5$. It can be seen from Figure 4 that the static displacement increase initially from time $t = 0$ and as time progresses attains steady state. It is also seen that the dynamic displacement oscillates about the static response. Figure 5 shows that the static thermal moment first increases with time and then attains the steady state but the dynamic thermal moment shows the oscillatory trend about zero. These results for $B=2.5$ when compared with results for $B = 1$ (results illustrated in Boley [1]) show that when the beam thickness increase the amplitude of vibration decreases and frequency increases. Apart from that the temperature induced oscillations occur away from the mean position.

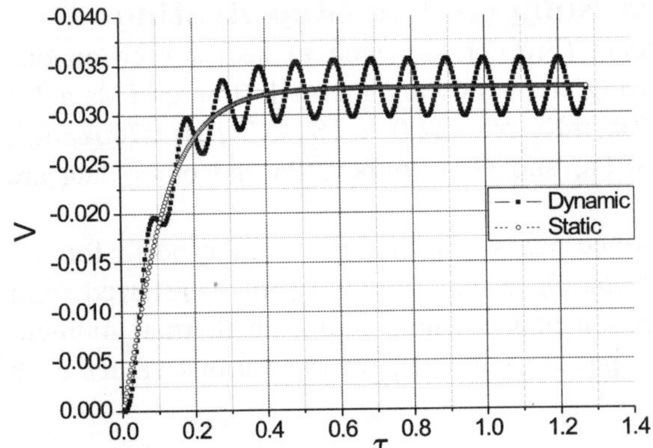

Fig. 4. Static & dynamic mid-span deflection of simply supported beam for L/h = 88.

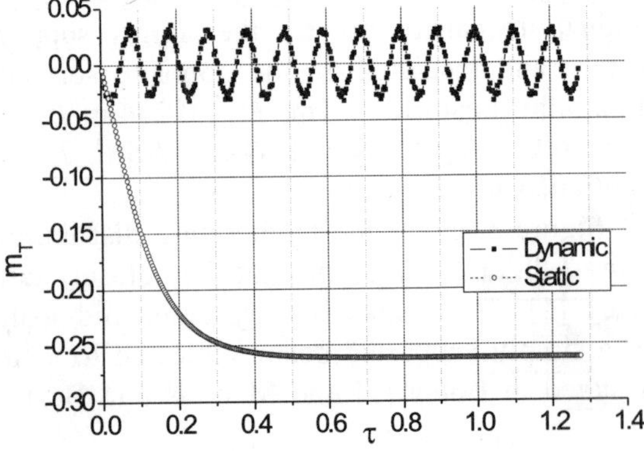

Fig. 5. Static & dynamic mid-span thermal moment of simply supported beam for L/h = 88.

Figure 6 shows the non-dimensional static and dynamic mid-span deflection of cantilever beam for L/h ratio of 88 or $B = 2.5$. Results were also obtained for clamped simply supported beam. The response of the static and dynamic displacement obtained for clamped simply supported and clamped free are similar, but the displacements calculated at the mid-span for clamped simply supported beam are less than that for simply supported beam. The displacement of the cantilever beam at the free end is quite large as compared to other two. The plots are also obtained for static and dynamic thermal moment of clamped simply supported and cantilever beam. The plots revealed that there exists a constant negative value of static thermal moment in all the three cases but the dynamic thermal moment calculated at the mid-span for clamped simply supported beam oscillates about some constant negative value rather than oscillating about zero as in case of simply supported beam. This is because one end is fixed/clamped and the thermal moment at this end is zero but the thermal moment present at the simply supported end puts the beam into oscillatory state about a constant value other than zero. The dynamic thermal moment for the cantilever beam at the free end is zero as shown in Figure 7.

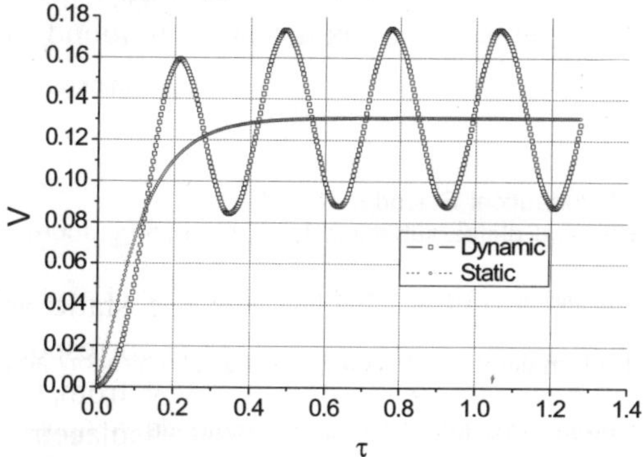

Fig. 6. Non-dimensional static and dynamic mid-span deflection of cantilever beam for L/h = 88.

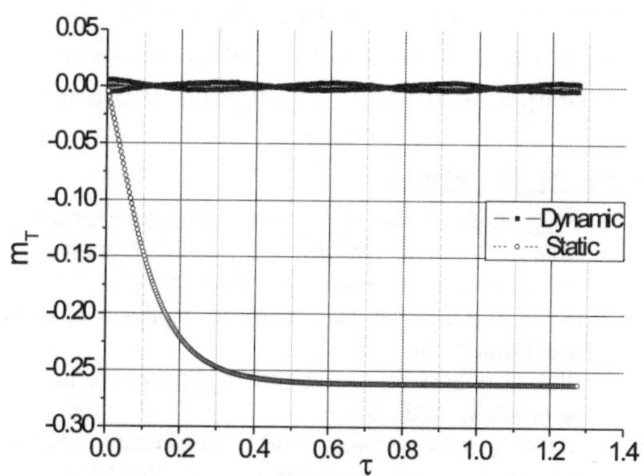

Fig. 7. Non-dimensional static and dynamic mid-thermal moment of cantilever beam for L/h = 88.

3.3 Beam Subjected to Axial Compressive Load and Sudden Heating

Studies are carried out for the simply supported beam with axial compressive load as shown in Fig. 8. The dynamic displacement of simply supported beam for $B = 1$ and various values of axial load (or the ratio of load applied to the critical buckling load of the beam, n) are illustrated in Fig. 9. The plots obtained are same as given by Manolis and Beskos [5].

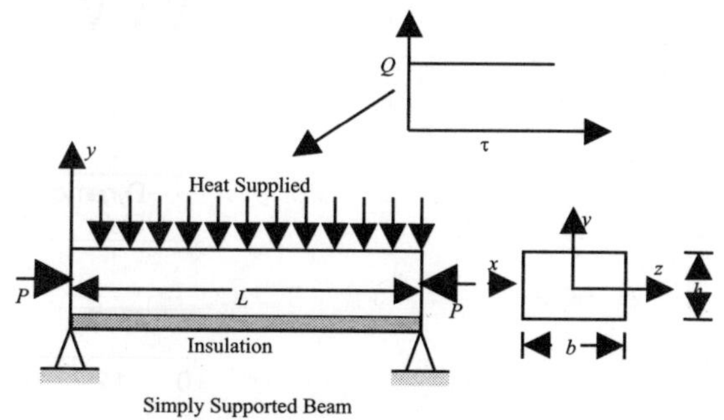

Fig. 8. Beam subjected to axial compressive load and sudden heating.

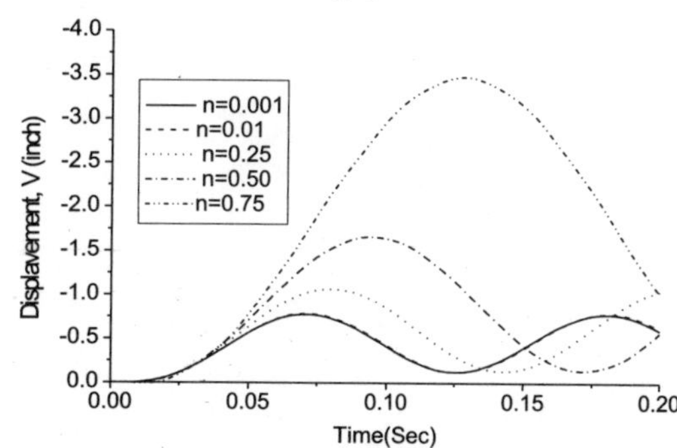

Fig. 9. Dynamic mid-span deflection of simply supported beam (B = 1) for various values of axial load.

3.4 Effect of Convection

Studies are also carried out on the simply supported beam subjected to sudden heating kept under ambient condition i.e., one side subjected to convective heat transfer coefficient of $20W/m^2K$. Since, the beam is exposed to very high heat flux on other side, the convective currents are not that effective in avoiding the thermal vibrations of beam and hence the vibration amplitude is same as that of beam subjected to step heating on one side and insulated on other side as shown in Fig. 4 and Fig. 5. Further studies were also carried out for higher values of convective heat transfer coefficient and it was found that the static part of the displacement response is vary rapid during the initial stages for the small time duration and later it increases continuously with small gradient.

CONCLUSION

Thermally induced vibrations in beams subjected to step heat input on one surface and the opposite surface being insulated or exposed to convective heat loss, with different end conditions and L/h ratio were analyzed using finite element method. The dynamic displacements and dynamic moments have different trends depending on the boundary conditions. The thickness of the beam influences the magnitude of dynamic displacement and its frequency. The presence of axial load was found to increase the amplitude of thermal vibrations.

REFERENCES

1. Boley B.A., 1956, Thermally induced vibrations of beams, J. Aeronautical Science, 23, 179–181.
2. Boley B.A., and Barber A.D., 1957, Dynamic response of beams and plates subjected to rapid heating, Journal of Applied Mechanics, 24, 413-416.
3. Boley B.A., 1972, Approximate analysis of thermally induced vibrations of beams and plates, Journal of Applied Mechanics, 39, 212-216.
4. Blandino and Thornton, 2001, Thermally induced vibrations of internally heated beam, Journal of Vibration and Acoustics, 123, 67-75
5. Manolis and Beskos, 1981, Thermally induced vibrations of beam structures, Computer methods in Applied Mechanics and Engineering, 21, 337-355.
6. Carslaw H.S. and Jaeger J.C., 1959, Conduction of Heat in Solids, Second Edition, Clarendon Press, Oxford.

70

Vibrations of FGM Adhesively Bonded Joints Using Hybrid Finite Element Analysis

S.C. PRADHAN

Department of Aerospace Engineering Indian Institute of Technology Kharagpur-721 302 West Bengal, India. email: scp@aero.iitkgp.ernet.in

ABSTRACT

In the present work, hybrid finite element is developed and employed to study the stress analysis of adhesively bonded joints. Functionally graded materials (FGM) are considered for the adherends and adhesive of the bonded joints. The developed hybrid finite element is able to predict accurately the stresses at the joint interfaces as compared to conventional displacement finite element formulation. Adhesive material is modeled with (a) isotropic material, (b) bi-material and (c) graded adhesive material properties. It is observed that employment of a graded adhesive material has resulted in a stronger bonded structure as compared to bi-material and isotropic material of the adhesive. These results agree well with those reported in the literature. Stress distributions results obtained from displacement and hybrid finite element methods are compared. Developed hybrid finite element could predict more accurately the stress distribution and vibration response of the bonded joint.

Keywords: Bonded joint, Hybrid finite element, functionally graded materials.

1. INTRODUCTION

Joining metallic and composite structural components with adhesively bonded joints has become a common practice in the technologically advanced aerospace and automotive sectors [1-4]. The initial characterization of the problems and analysis difficulties associated with an adhesively bonded joint was the classical shear-lag analysis of a single-lap joint by Volkersen [1]. Volkersen identified the incremental deformation of the adherends, but failed to incorporate bending of the adherends that leads to an overall rotation of the joint. This important physical behavior of a single lap joint was identified in the classical works of Goland and Reissner [2]. Numerous research studies have been conducted since these classical formulations and much of the early work is found in the excellent reviews by Sneddon [3], Erdogan and Ratwani [4]. The research performed by Erdogan and Ratwani [4] developed one of the early analytical solutions for a stepped-lap joint configuration, assumed the adherends to be in a state of plane stress, and provided for orthotropic adherend properties. Notable advancements for adhesively bonded joints with composite adherends were made by Hart-Smith [5-6], Wah [7] and Renton and Vinson [8]. The more recent works by Bigwood and Crocombe [9], Srinivas [10] and Tsai and Morton [11] have made significant advances in the analysis of adhesively bonded joints with composite adherends. Bigwood and Crocombe [9] developed a general joint overlap methodology for evaluating isotropic, adhesively bonded joints with inelastic adhesive behavior and subjected to combined loading. Tsai and Morton [11] evaluated the three-dimensional strain field present in a tension-loaded, single-lap composite joint. A very good compilation of various aspects of adhesively bonded joint, analysis, design and testing are reported by Adam and Wake [12] and Lees [13].

Pradhan [14-16] reported effect of material non-linearity on the bond strength. Pradhan *et al.* carried out finite element analysis of the joint with fracture mechanics approach. Pires, Quintino *et al.* [17] studied the performance of a bi-adhesive bonded joint. Recently, Fitton and Broughton [18] showed that variable modulus in bonding material could reduce stress concentrations, increasing joint strength and reduce experimental scatter and change the mode of failure. Also they concluded that in order to obtain higher strengths in this case a higher shear strength adhesive must be used in the centre of the joint.

Although in the literature various aspects of the bonded joints are discussed. But there is little development of tools to predict accurately the stress distribution in the bonded joint. So, in the present work hybrid finite element is developed and applied to adhesively bonded joints.

2. FORMULATION

The assumed stress hybrid element [19-20] is developed. Accordingly we use the complementary energy principle and an internal stress field that satisfies the differential equation of equilibrium a priori. For a linearly elastic complementary strain energy per unit volume is

$$U_0^* = \frac{1}{2}\{\sigma\}^T [E]^{-1} \{\sigma\} \qquad \text{...(1)}$$

within the element, the assumed stress field is

$$\{\sigma\} = [P]\{\beta\} \qquad \text{...(2)}$$

where, $\{\beta\}$ contains parameters β_i that are to be determined. $[P]$ is a function of the coordinates whose form is such that different equations of equilibrium are satisfied. Complementary strain energy in an element of volume V is

$$U^* = \int U_0^* dV = \frac{1}{2}\{\beta\}^T [H]\{\beta\} \qquad \text{...(3)}$$

where,
$$[H] = \frac{1}{2}[P]^T [E]^{-1} [P] dV$$

Let Φ represent tractions on element boundary S, obtained by evaluating on the boundary. Let boundary displacements $\{u_b\}$ be interpolated from element nodal d.o.f. $\{d\}$.

These relations are symbolized as

$$[\Phi] = [R]\{\beta\} \text{ and } \{u_b\} = [L]\{d\} \qquad \text{...(4)}$$

Total complementary energy in the element is U^* minus work done by tractions $\{\Phi\}$ in moving through displacements $\{u_b\}$; i.e.

$$\Pi_C = U^* - \int \{\Phi\}^T \{u_b\} dS \qquad \ldots(5)$$

or

$$\Pi_C = \frac{1}{2}\{\beta\}^T [H]\{\beta\} - \{\beta\}^T [G]\{d\} \qquad \ldots(6)$$

where,

$$[G] = \int [R]^T [L] dS \qquad \ldots(7)$$

To make Π_C stationary with respect to small changes in stress, one can write

$$\frac{\partial \Pi_C}{\partial \beta_i} = 0, \text{ for } i = 1, 2, \ldots n \text{ or } \left\{ \frac{\partial \Pi_C}{\partial \beta} \right\} = \{0\} \qquad \ldots(8)$$

from which

$$[H]\{\beta\} = [G]\{d\} \text{ or } \{\beta\} = [H]^{-1}[G]\{d\} \qquad \ldots(9)$$

we have asked for the stress field within an element when boundary displacements $\{u_b\}$ are defined by a chosen interpolation $\{u_b\} = [L]\{d\}$ from nodal displacement d.o.f. $\{d\}$ and answered by determining parameters β_i that define a best-fit stress field containing in the approximation $\{\sigma\} = [P]\{\beta\}$. Substitution of (β) from (9) into (3) yields

$$U^* = \frac{1}{2}\{d\}^T [k]\{d\} \text{ where } [K] = [G]^T [H]^{-1}[G] \qquad \ldots(10)$$

we have identified $[k]$ as a stiffness matrix because the form of U^* matches the form of strain energy U in an element. Because we deal here with a linearly elastic material strain energy and the complementary strain energy are equal, $U = U^*$. After assembling the global stiffness matrix $[k]$ from the element stiffness matrix $[k]$ and solving the global system $[K]\{D\} = \{R\}$ for $\{D\}$, nodal displacement d.o.f. $\{d\}$ are known for each element. Element stresses are computed from eqns (2) and (9).

$$\{\sigma\} = [P][H]^{-1}[G]\{d\} \qquad \ldots(11)$$

3. RESULTS AND DISCUSSIONS

Based on the hybrid finite element analysis code is developed and stress analysis is carried out. A rectangular plate under uniform tensile stress is considered to validate the hybrid finite element code. Aluminum material with $E = 70$GPa and Poisson's ratio 0.33 is considered for the validation problem. The uniform consistent displacements in the applied stress direction is shown in Figure 1. It is interesting to note that axial stresses and strains are same all through the plate except near the edges. From the Figure 1 and the uniform stress distribution in the rectangular plate validates the computer code of hybrid elements.

A typical adhesively bonded joint is shown in Figure 2. Various geometrical dimensions and material properties of the adherends and adhesives are listed in Table 1. Load in the axial direction is applied. Employing developed hybrid finite element code, stress analyses of the bonded joint are carried out. Results are shown in Figure 3. Displacement in the axial direction and the finite element mesh are shown in Figure 3. It is observed that the hybrid finite element method is able to predict more accurately the stresses in the interface.

Adhesive material is modeled with (a) isotropic material, (b) bi-material and (c) graded adhesive material properties. Material properties of the adhesive is graded along the overlap length (Fig. 2). It is observed that employment of a graded adhesive material has resulted in a stronger bonded structure as compared to bi-material and isotropic material of the adhesive. Similar observations were made by Pires *et al.* [17]. Vibration response of the FGM joint is also predicted accurately by this tool. From the present analysis it is observed that employing hybrid finite element results in more accurate stress distributions at the joint interfaces. Further employment of graded material properties in the adhesive results in stronger joint. This agrees with the results reported by Pires *et al.* [17].

Table 1. Various lap joint parameters (Fig. 1).

Parameter	Value
Adherend length (l_1, l_2)	75 mm
Overlap length (l)	25 mm
Adherend thickness (t_1, t_2)	2 mm
Modulus of adherend (E_1, E_2) Al	68.9 GPa
Poisson's ratio of adherends (γ_1, γ_2)	0.33
Adhesive thickness (t_s)	0.2 mm
Modulus of adhesive (Es)	3.0 Gpa
Poisson's ratio of adhesive (γ_s)	0.41

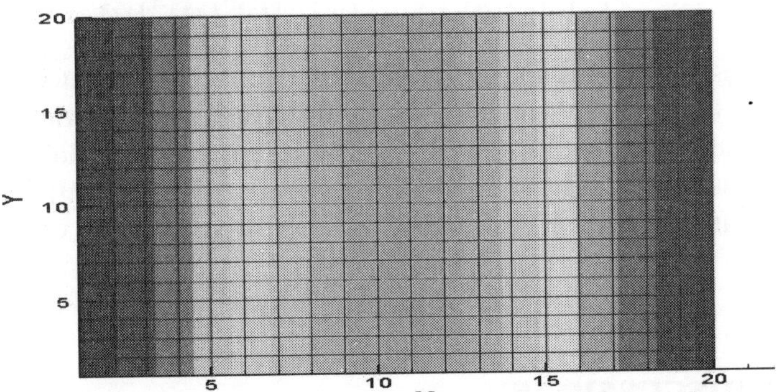

Fig. 1. Schematic of displacements in a rectangular plate.

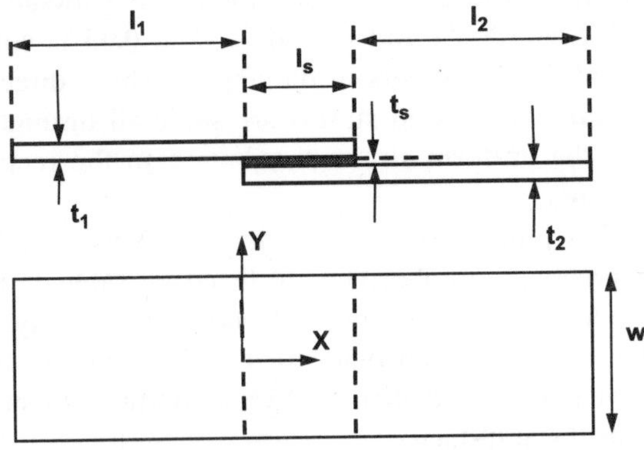

Fig. 2. Schematic of a bonded lap joint.

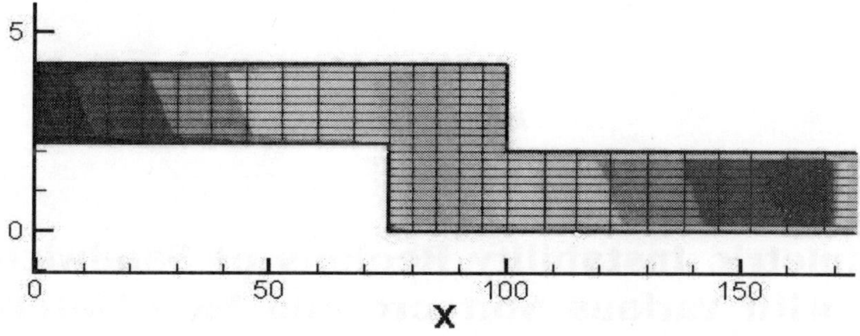

Fig. 3. Finite element mesh and displacements of bonded lap joint.

ACKNOWLEDGEMENTS

The financial support of this research work by AR & DB Structures Panel IIT/SRIC/ /2005-2007/ ABJ grant is gratefully acknowledged.

REFERENCES

1. Volkersen, V.O., "Die Nietkraftverteilung in zugbeanspruchten Nietverbindungen mit konstanten Laschenquerschnitten," Luftfahrforschung, Vol. 15, 1938, pp. 41–47.
2. Goland, M., and Reissner, E., "The Stresses in Cemented Joints," J. of Applied Mechanics, Vol. 66, March 1944, pp. A–18–27.
3. Sneddon, I.S., "Chapter IX - The Distribution of Stress in Adhesive Joints," Adhesion, Eley, D.D., Ed., Oxford University Press, 1961, pp 207-253.
4. Erdogan, F. and Ratwani, M., "Stress distribution in bonded joints," J. of Composite Materials, Vol. 5, 1971, pp. 378-393.
5. Hart-Smith, L.J., "Non-classical adhesive-bonded joints in practical aerospace construction," NASA CR- 112238, January 1973.
6. Hart-Smith, L.J., "Adhesive bonding of aircraft primary structures," SAE Aerospace Congress and Exposition, Los Angeles Convention Center, October 1980, pp. 1-15.
7. Wah T., "Stress distribution in a bonded anisotropic lap joint," J. of Engng. Mater. and Tech., 1973, pp. 174-181.
8. Renton, W.J. and Vinson, J.R., "The analysis and design of composite material bonded joints under static and fatigue loading," AFOSR Report 73-1627, 1973.
9. Srinivas, S., "Analysis of bonded joints," NASA TN D-7855, April 1975.
10. Bigwood, D.A. and Crocombe, A.D., "Non-linear adhesive bonded joint design analyses," Int. J. Adhesion and Adhesives, Vol. 10, No. 1, 1990, pp. 31-41.
11. Tsai, M.Y. and Morton, J., "Three-dimensional deformation in a single-lap joint," J. of Strain Analysis, Vol. 29, No. 1, 1994, pp. 137–145.
12. Adams R.D., Comyn J, Wake W.C.(1997), Structural adhesive joints in engineering, 2nd ed. London, UK: Chapman & Hall..
13. Lees WA. (1977) in: Allen KW, editor. Stresses in bonded joints, Adhesion-12. UK: Applied Science Publishers; pp 141-58.
14. Pradhan, S.C., Non-linear finite lement analysis of adhesively bonded joint: A fracture mechanics approach, Ph.D. thesis, IIT Kanpur, 1994.
15. Pradhan, S.C., Lam K.Y. and Tay, T.E., Determination of fracture parameters of laminated thermoplastic composite materials: a finite element approach, International Journal of Adhesion And Adhesives, 2000, 20(5) 395-401
16. Pradhan, S.C., Loy C.T., Lam K.Y. and Reddy J.N., Vibration characteristics of functionally graded cylindrical shells under various boundary conditions Applied Acoustics, 2000, 61 (1), 111–129.
17. Pires, I., Quintino, L., Durodola, J.F. & Beevers, A., Performance of bi-adhesive bonded aluminum lap joints, International Journal of Adhesion & Adhesives, 2003, 23, 215-223
18. Fitton, M.D. and Broughton, J.G., Variable modulus adhesives: an approach to optimized joint performance, International Journal of Adhesion & Adhesives 2005, 25, 329–336.
19. Pian, T H H, "Derivation of element stiffness matrices by assumed stress functions" AIAA Journal, 1964, Vol 2, No 7, pp 1333-1336.
20. Cook R D, Malkus, D S, Plesha M E and Witt R J, Concepts and applications of finite element analysis, John Wiley, & Sons Inc, 2003, 4th edition.

71

Parametric Instability Regions of Sandwich Beams with Various Softcore and Skin Materials Using Higher Order Theory

N. Mahendra, K.C. Sahu and S.K. Dwivedy

Mechanical Engineering Department, Indian Institute of Technology, Guwahati-781 039, India
email: dwivedy@iitg.ernet.in

ABSTRACT

In this work parametric instability of a three layered unsymmetrical sandwich beam with various skin and soft-core materials subjected to a periodic axial load is considered for simply supported boundary condition. Three different types of soft-core materials viz., H45, H80 and H250, and two different skin materials viz. aluminum and steel are considered in this analysis. Euler Bernoulli beam theory for the skins and a two-dimensional elasticity theory for the core are taken for finding the governing equation of motion, which is reduced to that of Mathieu-Hill equation with complex coefficients. Using modified Hsu method the parametric instability regions are determined.

1. INTRODUCTION

It is well known that vibration can be very much reduced by using a layer of viscoelastic material sandwiched between two elastic metallic layers. Sandwich structures are being used in many applications like airplanes, bridges, ships, space vehicles etc. A typical construction of sandwich beam consists of two skins made of metal or laminated composite and a core. The core is usually made of viscoelastic materials. Now days foam like soft-core materials are used in sandwich beams for weight reduction and flexibility in manufacturing. In such structures, the transverse flexibility of the core affects the overall behavior, as the displacement and stress in the core become nonlinear.

Research work on sandwich beams mostly started with the work of Kerwin [1] and now many researchers are worked on various problems on the free vibration analysis of sandwich structures. There are basically three approaches to analyse the sandwich beams, namely, classical theory, super position theory, and higher order theory [2]. Most of the researchers have worked on classical theory or anti-plane concept in which core is assumed to be incompressible in vertical direction. With the use of foam like materials for core, this anti-plane concept cannot work, as it is flexible in vertical direction.

Frostig and Baruch [2], Sokolinsky *et al.* [3] and Frostig and Thomsen [4] studied the free vibration of softcore sandwich beams using higher order approach. In these works the skins are considered to follow Euler-Bernoulli beam theory and core to follow a two-dimensional elasticity theory for the transverse displacements. Kolster and Wennhage [5] studied the effect of density and frequency on core loss factor of foam material and they have shown that sandwich beams with lower and higher density softcore have higher core-loss factor in comparison to medium density core material.

Many times sandwich structures are subjected to time varying loading where the equation of motion reduces to that of a parametrically excited system. Some researchers (e.g. Ray and Kar [6, 7]) have studied parametric excitation of sandwich beams, but they have considered the core to be incompressible in vertical direction. In present paper, using higher order theory, the equation of motion for unsymmetrical sandwich beam with flexible core subjected to time varying axial loading is carried out and the parametric instability regions are determined for six different combinations of skin and core materials.

2. FORMULATION OF THE PROBLEM

Figure 1 shows a simply supported, three-layered soft-cored unsymmetrical sandwich beam made of metallic skins (steel and aluminum) and foam like core having length L and width b. The top, core and bottom layer thickness are d_1, c and d_b, respectively. The upper and lower skins (face layer) of the beam are of the same or deferent elastic materials and the core is of visco-elastic material. The skins are subjected to an axial periodic load $P(t) = P_0 + P_1 \cos\omega t$, ω being the frequency of the applied load, t being the time and P_0, P_1, are the amplitudes of static and dynamic loads respectively.

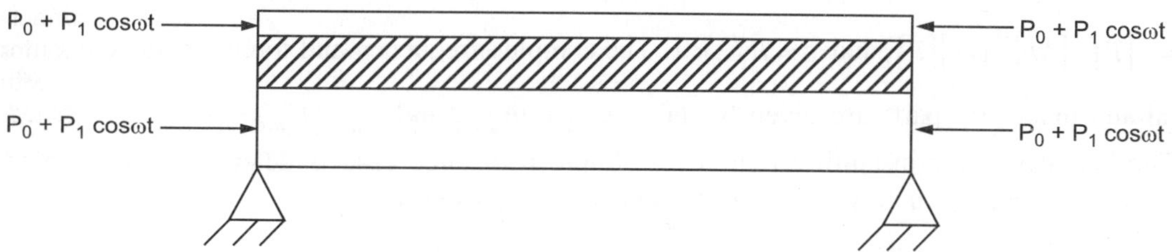

Fig. 1. Schematic diagram of an unsymmetrical sandwich beam.

Following the assumptions similar to that of Frostig and Baruch [8], using extended Hamilton's principle and generalized Galerkin's method the following equation of motion is obtained.

$$[M]\{\ddot{f}\}+[K]\{f\}-\bar{P_1}\cos\bar{\omega}t[F]\{f\}=\{\phi\}, \qquad \dots(1)$$

where

$$[M]=\begin{bmatrix}[M_{11}] & [M_{12}] & [M_{13}] & [M_{14}]\\ [M_{21}] & [M_{22}] & [M_{23}] & [M_{24}]\\ [M_{31}] & [M_{32}] & [M_{33}] & [M_{34}]\\ [M_{41}] & [M_{42}] & [M_{43}] & [M_{44}]\end{bmatrix},\quad [K_1]=\begin{bmatrix}[K_{11}] & [K_{12}] & [K_{13}] & [K_{14}]\\ [K_{21}] & [K_{22}] & [K_{23}] & [K_{24}]\\ [K_{31}] & [K_{32}] & [K_{33}] & [K_{34}]\\ [K_{41}] & [K_{42}] & [K_{43}] & [K_{44}]\end{bmatrix},$$

$$[F]=\begin{bmatrix}[F_{11}] & [\phi] & [\phi] & [\phi]\\ [\phi] & [F_{22}] & [\phi] & [\phi]\\ [\phi] & [\phi] & [\phi] & [\phi]\\ [\phi] & [\phi] & [\phi] & [\phi]\end{bmatrix} \qquad \dots(2)$$

The elements of the above matrices are defined in appendix. The following nondimensional parameters are used in this derivation.

$$t_0 = (ml^4/E)^{1/2}; \quad \bar{t} = (t/t_0); \quad \bar{x} = (x/L); \quad \bar{u} = (u/L); \quad \bar{w} = (w/L); \quad \bar{\omega} = \omega t_0; \quad \bar{P}_0 = P_0 L^2 \big/ (2E_q I_q),$$

$$\bar{P}_1 = P_1 L^2 \big/ (2E_q I_q), \quad E = (E_t I_t + E_b I_b), \quad g = G_c \big/ \left(E_t (c/d_t)(L/d_t)^2 + E_b (c/d_b)(L/d_b)^2 \right),$$

$$\bar{u}_q = (u_{oq}/L), \quad \bar{w}_q = (w_{oq}/L), \quad \bar{m}_q = (m_q/m), \quad \bar{m}_c = (m_c/m), \quad \phi_q = E_q A_q L^2 \big/ E.$$

In the above expressions $\bar{w}_q$, $\bar{u}_q$, $(q = t, b, c)$ represent the vertical and horizontal displacements. Subscript t, b, c are for the top bottom and core layer, respectively. E_q and m_q stand for the Young's modulus and mass per unit length of the qth layer and m is the mass per unit length of the beam.

Equation (1) is a set of coupled Mathieu Hill equations with complex coefficients. In the absence of external forcing these equations are reduced to that of Frostig and Baruch [1], where only free vibration analysis is carried out. The equations are also similar to that obtained by Ray and Kar [6] but the coefficients are different. Taking $[L]$ as normalized modal matrix of $[M]^{-1} [K]$, and using linear transformation $\{f\} = [L]\{U\}$ in equation (1), one may obtain the following equation.

$$\ddot{U}_n + (\omega_n^*)^2 U_n + 2\varepsilon \cos \bar{\omega} \bar{t} \sum_{p=1}^{4N} b_{np}^* U_p = 0 \quad n = 1\ldots 4N \qquad \ldots(3)$$

Here $\left(\omega_n^*\right)^2$ are the distinct eigenvalues of $[M]^{-1}[K]$ and b_{np}^* are the elements of $[B] = -[T]^{-1}[M]^{-1}[H][T]$, and $\varepsilon = \bar{P}_1/2 < 1$. The complex frequency and forcing parameters in terms of real and imaginary parts are given by $\omega_q^* = \omega_{q,R} + j\omega_{q,I}$, and $b_{qp}^* = b_{qp,R}^* + jb_{qp,I}$.   ...(4)

The boundaries for instability region for simple resonance case is determined using modified Hsu method discussed in Ray and Kar [5] which is given below.

$$\left| (\bar{\omega}/2) - \omega_{\mu,R} \right| < \frac{1}{4}\chi_\mu, \text{ where } \chi_\mu = \left[\frac{4\varepsilon^2 \left(b^2{}_{\mu\mu,R} + b^2{}_{\mu\mu,I} \right)}{\omega^2{}_{\mu,R}} - 16\omega^2{}_{\mu,I} \right]^{1/2} \qquad \ldots(5)$$

A code is developed in MATLAB, to determine the instability regions of the sandwich beams using the expressions given in the appendix and equations 2, 4 and 5. The obtained results are discussed in the following section.

3. NUMERICAL RESULTS AND DISCUSSION

In this work, six different sandwich beams are taken considering aluminum and steel as two different skin materials and H45, H80 and H250 as three different core materials. The properties of these materials are given in Table 1.

Table 1. Material properties of the skin and core materials used for the sandwich beams.

Material Properties	Al	Steel	H45	H80	H250
$E(N/m^2)$	67.5×10^9	2.1×10^{11}	4.2×10^7	8.0×10^7	3×10^8
$G(N/m^2)$	25.39×10^9	8.1×10^7	1.8×10^7	3.1×10^7	10.8×10^7
ν	0.33	0.3	0.32	0.32	0.32
Density (Kg/m^3)	2800	7900	48	80	250

The following dimensions of the sandwich beam are considered. Length L = 230 mm, breadth b = 10 mm, thickness of core c = 4 mm, top skin d_t = 2 mm, and bottom skin d_b = 10 mm. Using equation (5), the principal parametric instability regions of first three modes of the simply supported sandwich beam are determined and shown in Figures 2-7. In these Figures the regions bounded by the curves are unstable and regions outside the curves are stable.

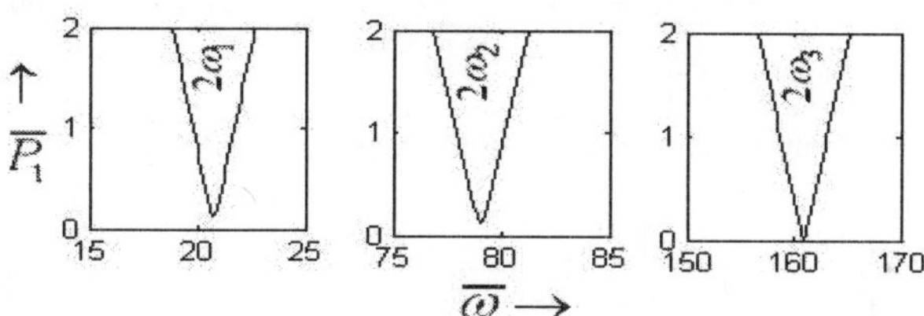

Fig. 2. Parametric instability region of steel-H45-steel sandwich beam, η = 0.1 and g = 0.018.

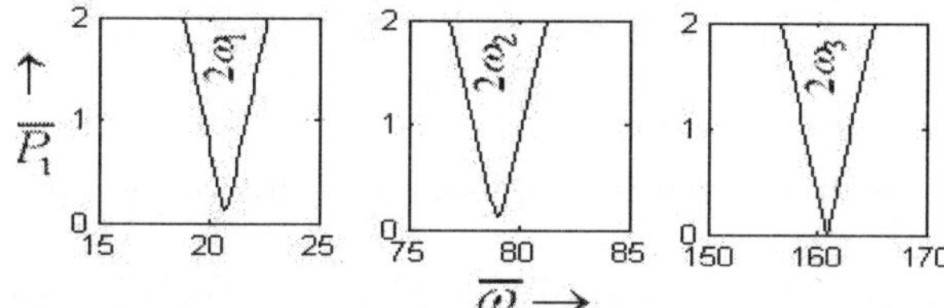

Fig. 3. Parametric instability region of Al-H45-Al sandwich beam, η = 0.05 and g = 0.056.

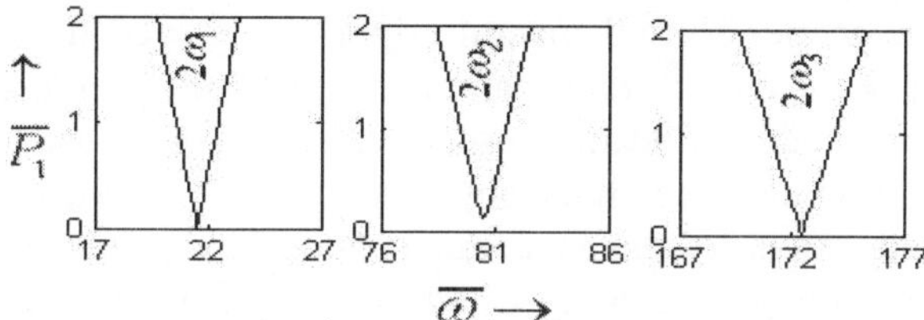

Fig. 4. Parametric instability region of steel-H80-steel sandwich beam, η = 0.05 and g = 0.031.

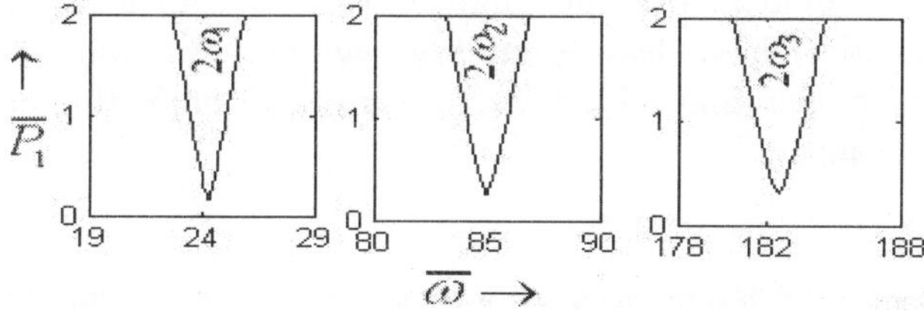

Fig. 5. Parametric instability region of Al-H80-Al sandwich beam, η = 0.05 and g = 0.0964.

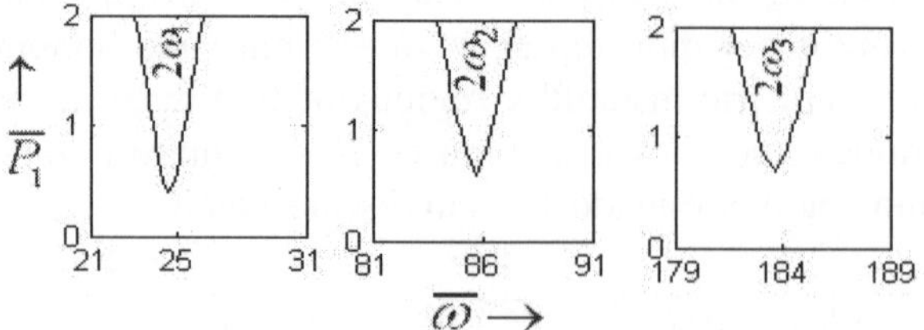

Fig. 6. Parametric instability region of steel-H250-steel sandwich beam, η = 0.1 and g = 0.108.

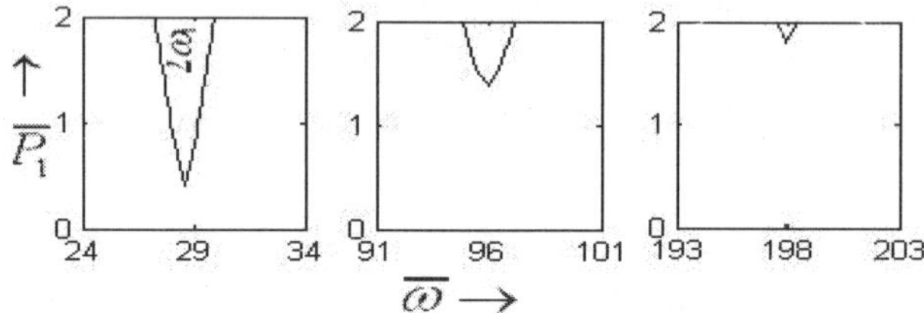

Fig. 7. Parametric instability region of Al-H250-Al sandwich beam, η = 0.1 and g = 0.3359.

Figures 2-3 show the parametric instability regions of steel-H45-steel and aluminium-H45-aluminium sandwich beams, respectively. In the later case the instability region is shifting towards right and lifted upwards thus making system more stable. Figures 4-5 show parametric instability regions of steel-H80-steel and aluminium-H80-aluminium sandwich beams respectively. In Fig. 5 instability regions for three modes shifted towards right and moved up thus resulting more stable region than in Fig. 4. Figures 6 and 7 show parametric instability regions of steel-H250-steel and aluminium-H250-aluminium sandwich beams, respectively. In Fig. 7 instability regions for the three modes are similar to the previous case. It is observed that the stability regions for the sandwich beams with H80 core materials is lowest in compared to H45 and H250, as H80 belongs to the medium grade (H80-H200) softcore materials which have coreloss factor very much less than low (H30-H60) and higher grade (H250-H300) soft core materials (Kolster and Wennhage [5]).

CONCLUSION

From the above results it is evident that the sandwich beams with increase in density of the core (high grade soft-core material) have less instability regions. Aluminum-H250—aluminum is the most stable and has a wider operating region in comparison to other five combinations of skin and core materials sandwich beams.

REFERENCES

1. E.M. Kerwin, 1959, Damping of flexural waves by a constrained viscoelastic layer, journal of acoustic society of America, 31, 952-962.
2. Y. Frostig, and M. Baruch, 1994, Free vibrations of sandwich beams with a transversely flexible core: A higher order approach, Journal of Sound and Vibration, 176(2), 195-208.
3. V.S Sokolinsky, H.F. Bremen, J.A Lavoie and S.R Nutt Analytical and experimental study of free vibration response of soft-core sandwich beams, Journal of Sandwich Structures and Materials, 6, (3), 239-261, 2004.

4. Y. Frostig and O.T Thomsen, High-order free vibration of sandwich panels with a flexible core. International Journal of Solids and Structures, 41, 1697-1724, 2004.
5. H. Kolsters and P. Wennhage, 2000, Density and frequency dependence of loss factor of foam core materials, International Conference on sandwich construction, Switzerland, 5-7 Sept. 2000, 23 pp. 2000.
6. K. Ray, and R.C. Kar, 1996, Parametric instability of multi-layered sandwich beams, Journal of Sound and Vibration, 193(3), 631-644.
7. K. Ray and R.C. Kar, 1995, parametric instability of a sandwich beams under various boundary conditions, Journal of computers and structures, (55), 857-870.
8. Y Frostig, and M Baruch, 1990, Bending of sandwich beams with transversely flexible core, American Institute of Aeronautics and Astronautics Journal, 27, 523-531.
9. A.H. Nayfeh, and D.T. Mook, 1979, Non-linear Oscillations, Wiley Interscience, UK.
10. K.C. Sahu, 2006, Parametric instability regions of a sandwich beam with a soft and magnetorheological elastomer core using higher order theory. M.tech thesis, Department of Mechanical Engineering, IIT Guwahati.

APPENDIX

$$(M_{11})_{ij} = (\bar{m}_t + m_c/3)\left(\int_0^1 w_i w_j d\bar{x}\right) + \left\{(\bar{m}_c/12)(d_t/L)^2\right\}\left(\int_0^1 w_i' w_j' d\bar{x}\right)$$

$$+ \left\{(\bar{m}_c/576)(d_t/c)(1+d_t/c)(c/L)^4(\xi_c/\phi_c)\right\}\left(\int_0^1 w_i'' w_j'' d\bar{x}\right)$$

$$(M_{12})_{ij} = (\bar{m}_c/6)\left(\int_0^1 w_i w_j d\bar{x}\right) - \left\{(\bar{m}_c/24)(d_t d_b/l^2)\right\}\left(\int_0^1 w_i' w_j' d\bar{x}\right)$$

$$+ \left\{(\bar{m}_c/576)(d_b/c)(1+d_t/c)(c/L)^4(\xi_c/\phi_c)\right\}\left(\int_0^1 w_i'' w_j'' d\bar{x}\right)$$

$$(M_{13})_{ij} = \left\{(\bar{m}_t + \bar{m}_c/6)(1/48)(1+d_t/c)(c/L)^3(\xi_c/\phi_c)\right\}\left(\int_0^1 w_i'' u_j' d\bar{x}\right) + \left\{(m_c/6)(d_t/L)\right\}\left(\int_0^1 w_i' u_j d\bar{x}\right)$$

$$(M_{14})_{ij} = -\left\{(1/48)(\bar{m}_b + \bar{m}_c/6)(1+d_t/c)(L)^3(\xi_c/\phi_c)\right\}\left(\int_0^1 w_i'' w_j' d\bar{x}\right)$$

$$+ \left\{(\bar{m}_c/12)(L)\right\}\left(\int_0^1 w_i' u_j d\bar{x}\right)$$

$$(M_{21})_{ij} = (\bar{m}_c/6)\left(\int_0^1 w_i w_j d\bar{x}\right) - \left\{(\bar{m}_c/24)(d_t d_b/L^2)\right\}\left(\int_0^1 w_i' w_j' d\bar{x}\right)$$

$$+ \left\{(\bar{m}_c/576)(d_t/c)(1+d_b/c)(c/L)^4(\xi_c/\phi_c)\right\}\left(\int_0^1 w_i'' w_j'' d\bar{x}\right)$$

$$\left(M_{22}\right)_{ij} = \left(\bar{m}_b + \bar{m}_c/3\right)\left(\int_0^1 w_i w_j\,d\bar{x}\right) + \left\{\left(\bar{m}_c/12\right)(c/L)^2\right\}\left(\int_0^1 w_i' w_j'\,d\bar{x}\right)$$

$$+ \left\{\left(\bar{m}_c/576\right)\left(d_b/c\right)\left(1 + d_b/c\right)(c/L)^4\left(\xi_c/\phi_c\right)\right\}\left(\int_0^1 w_i'' w_j''\,d\bar{x}\right)$$

$$\left(M_{23}\right)_{ij} = \left\{(1/48)\left(\bar{m}_t + \bar{m}_c/6\right)\left(1 + d_b/c\right)(c/L)^3\left(\xi_c/\phi_c\right)\right\}\left(\int_0^1 w_i'' u_j'\,d\bar{x}\right) - \left\{\left(\bar{m}_c/12\right)(c/L)\right\}\left(\int_0^1 w_i' u_j\,d\bar{x}\right)$$

$$\left(M_{24}\right)_{ij} = -\left\{(1/48)\left(\bar{m}_b + \bar{m}_c/6\right)\left(1 + d_b/c\right)(c/L)^3\left(\xi_c/\phi_c\right)\right\}\left(\int_0^1 w_i'' u_j'\,d\bar{x}\right) - \left(\bar{m}_c/6\right)\left(d_b/L\right)\left(\int_0^1 w_i' u_j\,d\bar{x}\right)$$

$$\left(M_{31}\right)_{ij} = \left\{\left(\bar{m}_c/288\right)\left(d_t/c\right)(c/L)^3\left(\xi_c/\phi_c\right)\right\}\left(\int_0^1 u_i'' w_j'\,d\bar{x}\right) + \left\{\left(\bar{m}_c/6\right)\left(d_t/L\right)\right\}\left(\int_0^1 u_i' w_j\,d\bar{x}\right)$$

$$\left(M_{32}\right)_{ij} = \left\{\left(\bar{m}_c/12\right)\left(d_b/L\right)\right\}\left(\int_0^1 u_i' w_j\,d\bar{x}\right) - \left\{\left(\bar{m}_c/288\right)\left(d_b/c\right)(c/L)^3\left(\xi_c/\phi_c\right)\right\}\left(\int_0^1 u_i'' w_j'\,d\bar{x}\right)$$

$$\left(M_{33}\right)_{ij} = -\left\{(1/24)\left(\bar{m}_t + \bar{m}_c/6\right)(c/L)^2\left(\xi_c/\phi_c\right)\right\}\left(\int_0^1 u_i' u_j'\,d\bar{x}\right) - \left(\bar{m}_t + \bar{m}_c/3\right)\left(\int_0^1 u_i u_j\,d\bar{x}\right)$$

$$\left(M_{34}\right)_{ij} = \left\{\left(-\bar{m}_c/6\right)\right\}\left(\int_0^1 u_i u_j\,d\bar{x}\right) + \left\{(1/24)\left(\bar{m}_b + \bar{m}_c/6\right)(c/L)^2\left(\xi_c/\phi_c\right)\right\}\left(\int_0^1 u_i' u_j'\,d\bar{x}\right)$$

$$\left(M_{41}\right)_{ij} = -\left\{\left(\bar{m}_c/12\right)\left(d_t/L\right)\right\}\left(\int_0^1 u_i' w_j\,d\bar{x}\right) + \left\{\left(\bar{m}_c/288\right)\left(d_t/c\right)(c/L)^3\left(\xi_c/\phi_c\right)\right\}\left(\int_0^1 u_i'' w_j'\,d\bar{x}\right)$$

$$\left(M_{42}\right)_{ij} = \left\{\left(\bar{m}_c/6\right)\left(d_b/L\right)\right\}\left(\int_0^1 u_i' w_j\,d\bar{x}\right) + \left\{\left(\bar{m}_c/288\right)\left(d_b/c\right)(c/L)^3\left(\xi_c/\phi_c\right)\right\}\left(\int_0^1 u_i' w_j\,d\bar{x}\right)$$

$$\left(M_{43}\right)_{ij} = \left\{\left(-\bar{m}_c/6\right)\right\}\left(\int_0^1 u_i u_j\,d\bar{x}\right) + \left\{(1/24)\left(\bar{m}_t + \bar{m}_c/6\right)(c/L)^2\left(\xi_c/\phi_c\right)\right\}\left(\int_0^1 u_i' u_j'\,d\bar{x}\right)$$

$$\left(M_{44}\right)_{ij} = -\left\{\left(\bar{m}_b + \bar{m}_c/3\right)\right\}\left(\int_0^1 u_i u_j\,d\bar{x}\right) - \left\{(1/24)\left(\bar{m}_b + \bar{m}_c/6\right)(c/L)^2\left(\xi_c/\phi_c\right)\right\}\left(\int_0^1 u_i' u_j'\,d\bar{x}\right)$$

$$(K_{11})_{ij} = \left\{ (1/4)(1+d_t/c)^2 \, \xi_c \right\} \left(\int_0^1 w_i' w_j' \, d\overline{x} \right) + \left\{ (1/12)(d_t/L)^2 \, \phi_c \right\} \left(\int_0^1 w_i'' w_j'' \, d\overline{x} \right)$$

$$+ \left\{ \phi_c (L/c)^2 \right\} \left(\int_0^1 w_i w_j \, d\overline{x} \right)$$

$$(K_{12})_{ij} = -\left\{ \phi_c (L/c)^2 \right\} \left(\int_0^1 w_i w_j \, d\overline{x} \right) + \left\{ (1/4)(1+d_t/c)(1+d_b/c)\xi_c \right\} \left(\int_0^1 w_i' w_j' \, d\overline{x} \right)$$

$$- \left\{ (1/48)(c/L)^3 (1+d_t/c)(\xi_c/\phi_c)\phi_b \right\} \left(\int_0^1 w_i''' u_j'' \, d\overline{x} \right)$$

$$(K_{13})_{ij} = -\left\{ (1/2)(L/c)(1+d_t/c)\xi_c \right\} \left(\int_0^1 w_i' u_j \, d\overline{x} \right) - \left\{ (1/48)(c/L)^3 (1+d_t/c)(\xi_c/\phi_c)\phi_t \right\} \left(\int_0^1 w_i''' u_j'' \, d\overline{x} \right)$$

$$(K_{14})_{ij} = \left\{ (1/2)(L/c)(1+d_t/c)\xi_c \right\} \left(\int_0^1 w_i' u_j \, d\overline{x} \right) + \left\{ (1/48)(c/L)^3 (1+d_t/c)(\xi_c/\phi_c)\phi_b \right\} \left(\int_0^1 w_i''' u_j'' \, d\overline{x} \right)$$

$$(K_{21})_{ij} = -\left\{ \phi_c (L/c)^2 \right\} \left(\int_0^1 w_i w_j \, d\overline{x} \right) + \left\{ (1/4)(1+d_t/c)(1+d_b/c)\xi_c \right\} \left(\int_0^1 w_i' w_j' \, d\overline{x} \right)$$

$$(K_{22})_{ij} = \left\{ (1/4)(1+d_b/c)^2 \, \xi_c \right\} \left(\int_0^1 w_i' w_j' \, d\overline{x} \right) + \left\{ (1/12)(d_b/L)^2 \, \phi_c \right\} \left(\int_0^1 w_i'' w_j'' \, d\overline{x} \right)$$

$$+ \left\{ \phi_c (L/c)^2 \right\} \left(\int_0^1 w_i w_j \, d\overline{x} \right)$$

$$(K_{23})_{ij} = \left\{ (-1/2)(L/c)(1+d_b/c)\xi_c \right\} \left(\int_0^1 w_i' u_j \, d\overline{x} \right) - \left\{ (1/48)(c/L)^3 (1+d_b/c)(\xi_c/\phi_c)\phi_t \right\} \left(\int_0^1 w_i''' u_j'' \, d\overline{x} \right)$$

$$(K_{24})_{ij} = \left\{ (1/2)(L/c)(1+d_b/c)\xi_c \right\} \left(\int_0^1 w_i' u_j \, d\overline{x} \right) + \left\{ (1/48)(c/L)^3 (1+d_b/c)(\xi_c/\phi_c)\phi_b \right\} \left(\int_0^1 w_i''' u_j'' \, d\overline{x} \right)$$

$$(K_{31})_{ij} = \left\{ (-1/2)(L/c)(1+d_t/c)\xi_c \right\} \left(\int_0^1 u_i' w_j \, d\overline{x} \right), \quad (K_{32})_{ij} = \left\{ (-1/2)(L/c)(1+d_b/c)\xi_c \right\} \left(\int_0^1 u_i' w_j \, d\overline{x} \right)$$

$$(K_{33})_{ij} = (-\phi_t)\left(\int_0^1 u_i' u_j' d\bar{x}\right) - \left\{(L/c)^2 \xi_c\right\}\left(\int_0^1 u_i u_j d\bar{x}\right) - \left\{(1/24)(c/L)^2 (\xi_c/\phi_c)\phi_t\right\}\left(\int_0^1 u_i'' u_j'' d\bar{x}\right)$$

$$(K_{34})_{ij} = \left\{(L/c)^2 \xi_c\right\}\left(\int_0^1 u_i u_j d\bar{x}\right) + \left\{(1/24)(c/L)^2 (\xi_c/\phi_c)\phi_b\right\}\left(\int_0^1 u_i'' u_j'' d\bar{x}\right)$$

$$(K_{41})_{ij} = \left\{(1/2)(L/c)(1+d_t/c)\xi_c\right\}\left(\int_0^1 u_i' w_j d\bar{x}\right)$$

$$(K_{42})_{ij} = \left\{(1/2)(L/c)(1+d_b/c)\xi_c\right\}\left(\int_0^1 u_i' w_j d\bar{x}\right),$$

$$(K_{43})_{ij} = \left\{(L/c)^2 \xi_c\right\}\left(\int_0^1 u_i u_j d\bar{x}\right) + \left\{(1/24)(c/L)^2 (\xi_c/\phi_c)\phi_t\right\}\left(\int_0^1 u_i'' u_j'' d\bar{x}\right)$$

$$(K_{44})_{ij} = (-\phi_b)\left(\int_0^1 u_i' u_j' d\bar{x}\right) - \left\{(L/c)^2 \xi_c\right\}\left(\int_0^1 u_i u_j d\bar{x}\right) - \left\{(1/24)(c/L)^2 (\xi_c/\phi_c)\phi_b\right\}\left(\int_0^1 u_i'' u_j'' d\bar{x}\right)$$

$$(H)_{11} = \int_0^1 w_i' w_j' d\bar{x}$$

In the above relations, $()' = \partial()/\partial\bar{x}$, and the sub matrices, which are not covered by the above elements, should be treated as null.

72

Free Vibration Analysis of Cylindrical Shells Partly Supported on the Edges

S. Kandasamy and A.V. Singh

Department of Mechanical and Materials Engineering, The University of Western Ontario, London, Ontario, Canada, N6A 5B8 email: avsingh@eng.uwo.ca

ABSTRACT

In this study the free vibration analysis of open circular cylindrical shells partly supported on the straight edges is presented. The geometry of the shell is defined by the subtended angle, the length, and the shell thickness. The method of solution implemented here is a modified version of the Rayleigh-Ritz method. A set of displacement grid points on the middle surface are introduced first and each grid point has five degrees of freedom, viz., three translational components along the cylindrical coordinates and two rotational components of the normal to the middle surface. The number of grid points for the analysis depends upon the orders of the polynomials chosen for the displacement components and as a consequence considerably high-order polynomials are used for this purpose. The constitutive equations are derived from the three-dimensional strain-displacement relations in the cylindrical coordinate system for the first order shear deformable cylindrical shell. To validate the method, a convergence study is carried out for the case in which the shell is assumed to be partly supported on the straight edges. Further, the numerical results obtained from the present method are compared with the ones obtained from a commercially available finite element computer code.

Keywords: Cylindrical shells, free vibration, p-type, partly support.

1. INTRODUCTION

The shell vibration research was reported as early as 1888 by Lord Rayleigh and A.E.H. Love, but a significant amount of efforts by other researchers began in 1930s and thereafter. Leissa [1] published a monograph on the vibrations of shells compiling the previous works done on this topic. Subsequently, some other review articles were published in the literature by others. Qatu [2, 3] published two review articles in 2002 and covered the period of 1989-2000.

The present work deals with the free vibration analysis of open cylindrical panel using a variational method and is essentially the extension of the previous work of the authors [4]. The formulation is based of the Reissner-Naghdi shell theory and the effects of the rotary inertia and shear deformation is retained. Due to the increasing application of the open shell panels in aerospace industries, cylindrical shell supported only on a segment of the edge is considered in this investigation. A modified version of the Rayleigh-Ritz method is used over the entire shell surface. The admissible displacement fields are deduced in terms of the displacement components at a pre-selected point, called the displacement node, and it is this form that allows the application of the boundary conditions at discrete points of a continuous system. For the numerical results, first the order of the polynomials for the displacements is established from the convergence study and then additional results from the present method are generated and compared favorably with the ones from a commercially available finite element code. A brief discussion on the frequencies and mode shapes is also presented.

2. METHOD OF ANALYSIS

Thin to moderately thick cylindrical shells are considered in this study using the first order shear deformation theory. The middle surface of the cylindrical shell of thickness h is considered the reference and defined by the cylindrical coordinates (x, θ, z) as shown in Fig. 1, wherein the physical parameters of the shell, viz., length L, radius R and subtended θ_0 angle are also depicted. Let, u', v' and w' be the displacements at an arbitrary location (x, θ, z) in the shell. From the first order shear deformation theory of shells, these three displacement components can be expressed in the following form.

$$u'(x,\theta,z) = u(\theta,x) + \zeta\beta_1(\theta,x)$$

$$v'(x,\theta,z) = v(\theta,x)(1+\zeta) + \zeta\beta_2(\theta,x) \qquad \qquad ...(1)$$

$$w'(x,\theta,z) = w(\theta,x)$$

In the above equation, u, v and w denote the displacement components in the axial (x), circumferential (θ), and radial (z) directions at the reference surface of the shell and the symbols β_1 and β_2 correspond to the components of rotation of the normal to the middle surface of the shell in axial and circumferential directions respectively. In Eq. (1), β_1 and β_2 are made to have the

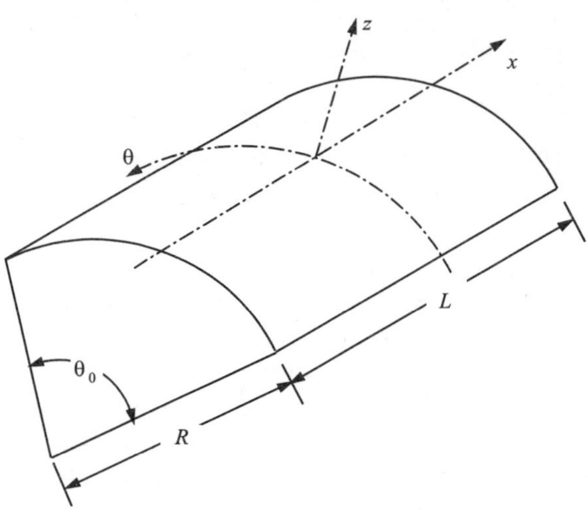

Fig. 1. Circular cylindrical shell.

same unit as a displacement component with the help of R and dimensionless parameter $\zeta = z/R$ is introduced such that $\zeta \ll 1.0$.

The strain-displacement relation for the circular shell in cylindrical coordinate system (Saada [5]) is given as follows:

$$\varepsilon'_{xx} = \frac{\partial u'}{\partial x}$$

$$\varepsilon'_{\theta x} = \frac{1}{R}\left(\frac{1}{1+\zeta}\right)\frac{\partial u'}{\partial \theta} + \frac{\partial v'}{\partial x}$$

$$\varepsilon'_{\theta z} = \frac{\partial v'}{\partial z} + \frac{1}{R}\left(\frac{1}{1+\zeta}\right)\left(\frac{\partial w'}{\partial \theta} - v'\right) \qquad \ldots(2)$$

$$\varepsilon'_{\theta\theta} = \frac{1}{R}\left(\frac{1}{1+\zeta}\right)\left(\frac{\partial v'}{\partial \theta} + w'\right)$$

$$\varepsilon'_{xz} = \frac{\partial u'}{\partial z} + \frac{\partial w'}{\partial x}$$

In the vector-matrix notation Eq. (2) can be written as

$$\{\sigma'\} = [E']\,\{\varepsilon'\} \qquad \ldots(3)$$

where $\{\sigma'\}^T = \{\sigma'_{xx} \quad \sigma'_{\theta\theta} \quad \tau'_{\theta x} \quad \tau'_{xz} \quad \tau'_{\theta z}\}$, $\{\varepsilon'\} = \{\varepsilon'_{xx} \quad \varepsilon'_{\theta\theta} \quad \varepsilon'_{\theta x} \quad \varepsilon'_{xz} \quad \varepsilon'_{\theta z}\}$ and $[E']$ = a fifth order matrix composed of the modulus of elasticity E and the Poisson's ratio v.

The natural coordinate system (ζ, η) is introduced to define the geometry of the shell by prescribing x and θ coordinates (x_j, θ_j for $j = 1, 2, 3,$ and 4 of the four corner points as shown in Fig. 2(a). The (x, θ) coordinates of an arbitrary point can be interpolated by the following [6].

$$\theta(\xi,\eta) = \sum_{j=1}^{4} N_j(\xi,\eta)\theta_j \quad \text{and} \quad x(\xi,\eta) = \sum_{j=1} 4N_j(\xi,\eta)x_j \qquad \ldots(4)$$

In the above, $N_j(\xi,\eta)$ corresponds to the *shape functions* of the jth geometric point. It should be noted that this form of representation of the geometry may not be necessary for the case of rectangular cylindrical panel. However, this kind of geometrical representation can be very well used in the analysis of skewed open cylindrical shells reported earlier by the authors of this paper [4].

The next step is to define the displacement fields in terms of ξ and η for u, v, w, β_1, β_2. To achieve this, a new set of grid points with different identities, termed as the displacement nodes on the reference surface is chosen. These points may fall on the boundary including the four corners coinciding with the geometric nodes as shown in Fig. 2(b) as well as on other locations inside the boundary. The expressions for these displacement fields are of the same form as that for the coordinate points except that these fields are created with relatively high order interpolating polynomials. For example, the displacement component u is expressed as

$$u = \sum_{j=1}^{P} u_j N_j(\xi,\eta) \qquad \ldots(5)$$

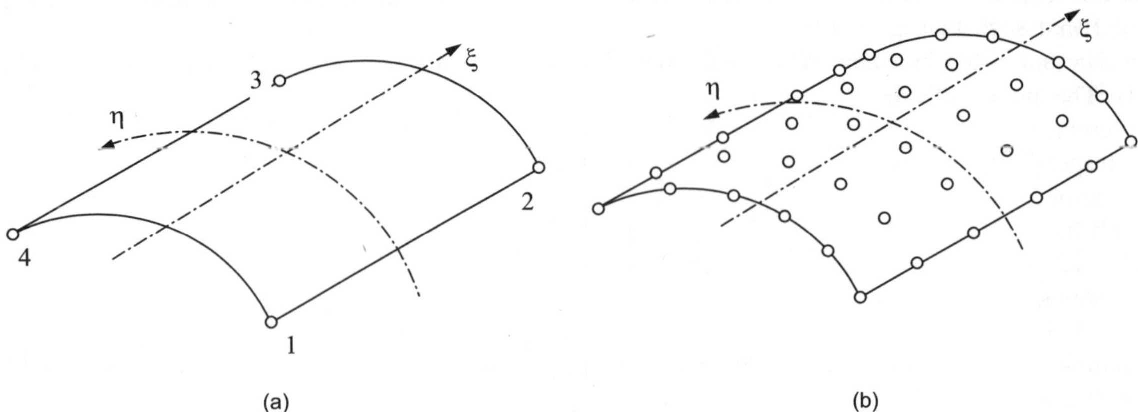

Fig. 2. Cylindrical shell with nodes (a) geometric nodes and (b) displacement nodes.

where $p = (p_1 + 1) \times (p_2 + 1)$, order of the polynomial in ξ, and $p_2 =$ order of the polynomial in η. Also, $N_j\ (\xi, \eta)$ corresponds to the *displacement shape function* and u_j is component u at the jth displacement node.

The strain energy expression can be written as

$$U = \frac{1}{2} \iiint_{x\,\theta\,z} \{\varepsilon'\}^T \{\sigma'\}\, dV \qquad \qquad ...(6)$$

Similarly, the kinetic energy of an isotropic cylindrical shell is given by

$$T = \frac{1}{2} \int_v \rho \left\{ \left(\frac{\partial u'}{\partial t}\right)^2 + \left(\frac{\partial v'}{\partial t}\right) + \left(\frac{\partial w'}{\partial t}\right)^2 \right\} dV \qquad \qquad ...(7)$$

where $\rho =$ mass density, $V =$ volume of the shell and $t =$ time. Using eqs. (1) through (5) and integrating over the thickness of the shell, the following expressions can be obtained.

$$U = \frac{1}{2} \{q\}^T [K] \{q\} \quad \text{and} \quad T = \frac{1}{2} \{\dot{q}\}^T [M]\{\dot{q}\} \qquad \qquad ...(8)$$

where $\{q\}^T = \{u_j\ v_j\ w_j\ \beta_{1j}\ \beta_{2j}\}$, $j = 1 \ldots p$. The "*over-dot*" represents the time derivative of the spatial coordinates. $[K]$ and $[M]$ are the well known stiffness and mass matrices from which differential equation for the free vibration analysis is obtained using the Hamiltonian Principle [4]. The eigenvalue equation is then solved iteratively and non-dimensional frequency parameter $\Omega = \omega R \sqrt{(\rho/E)}$ for the first few modes and corresponding mode shapes are calculated and presented in the following.

3. NUMERICAL RESULTS AND DISCUSSIONS

A numerical method based on the Raleigh-Ritz method has been described briefly in this paper. The modification is made in the displacement fields in such a way that supports at discrete points can be accommodated conveniently. To validate the numerical method in the case of partly supported boundary condition, the cylindrical shell is assumed to be supported on a segment of the straight edges, i.e. the middle third of the straight edge is subject to the condition of $u = v = w = 0$, where as the remaining portion of the straight edges and the curved edges are kept free. The parameters used in the calculation include: $L/R = 1.0$, $h/R = 0.05$, subtended angle $\theta_0 = 30°$ and Poisson's ratio = 0.3.

The numerical results start with the convergence study of the non-dimensional frequencies, wherein, the three different set of displacement grid points/nodes 6×6, 9×9 and 12×12 in ξ and η respectively are used. Table 1 shows the values of the frequencies for the first six modes of vibration. The frequency values are also compared in Table 2 with the results from a finite element analysis software I-DEAS, in which, both linear and parabolic elements are used. In I-DEAS 12×12 elements are used and the middle four elements on the straight edges are subjected to point supported boundary condition. Finally, the mode shapes for the first six frequencies are presented in Fig. 3. In these mode shapes, the symmetry and asymmetry about x- and θ-axes are seen. For example, in the fundamental mode, vibration is symmetric about the x-axis and asymmetric about the θ-axis. Similarly, in the second mode, the shell is vibrating symmetrically about the two axes.

Table 1. Convergence of non-dimensional frequency ($\Omega = \omega R\sqrt{\rho/E}$) for the open circular cylindrical shell partly supported on the straight edges. Parameters: $L/R = 1.0$, $h/R = 0.05$, $v = 0.3$, and $\theta_0 = 30°$.

terms/mode	1	2	3	4	5	6
6	0.29041	0.29715	0.55688	0.65443	1.07962	1.39557
9	0.27423	0.33651	0.59874	0.65942	1.07177	1.35032
12	0.31265	0.33311	0.62642	0.66598	1.05659	1.35518

Table 2. Comparison of non-dimensional frequency ($\Omega = \omega R\sqrt{\rho/E}$) for the open circular cylindrical shell partly supported on the straight edges. Parameters: $L/R = 1.0$, $h/R = 0.05$, $v = 0.3$, and $\theta_0 = 30°$.

Mode	P type 12 terms	I-DEAS 12 Linear Elements	I-DEAS 12 Parabolic Elements
1	0.31265	0.29719	0.29217
2	0.33311	0.33364	0.32987
3	0.62642	0.61261	0.64088
4	0.66598	0.61387	0.64080
5	1.05659	1.04929	1.04929
6	1.35518	1.30690	1.33204

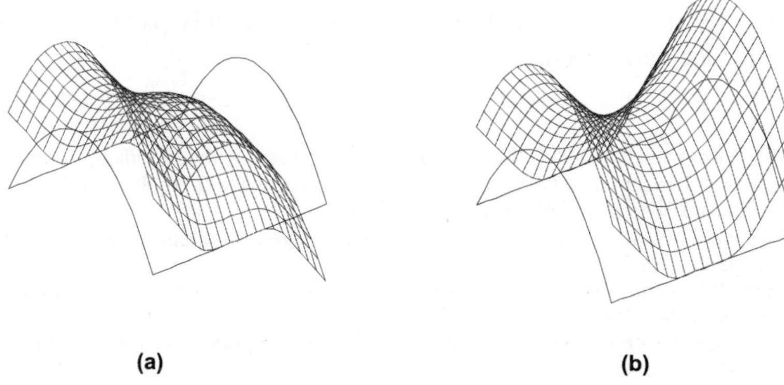

(a)   (b)

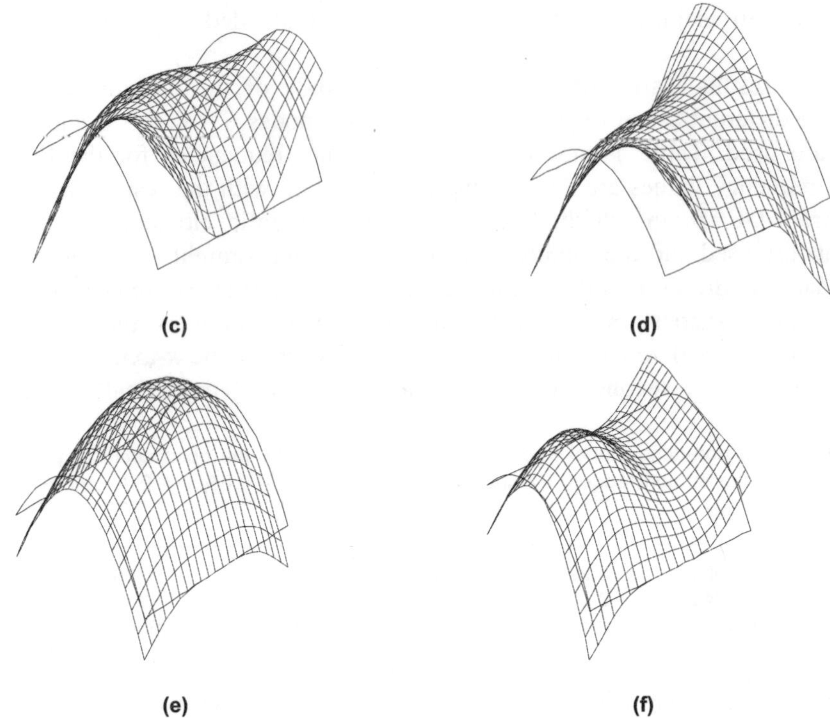

Fig. 3. Mode shapes of cylindrical shell partly supported on the straight edges. (a) mode 1, $\Omega = 0.31625$ (b) mode 2, $\Omega = 0.33311$ (c) mode 3, $\Omega = 0.62642$ (d) mode 4, $\Omega = 0.66598$ (e) mode 5, $\Omega = 1.05659$ (f) mode 6, $\Omega = 0.35578$.

CONCLUSION

A computational scheme to deal with the free vibration of open cylindrical panel supported only on part of an edge or the edges has been presented in this paper. The formulation is based on the first order shear deformable shell theory which includes both the rotary inertia and transverse shear deformation. One numerical example with the support on the middle third of each of the straight edges are presented and discussed. However, segmental support on the curved edges and any combination of the straight and curved edges can be dealt with conveniently. Generally, numerical results, though all that has been calculated are not included in this paper, converge properly and compare very well with the finite element results.

REFERENCES

1. A.W. Leissa, 1973, Vibration of Shells, NASA SP-288, Government Printing Office, Washington, DC.
2. M.S. Qatu, 2002, Recent research advances in the dynamic behavior of shells: 1989-2000, Part 1: laminated composite shells, Applied Mechanics Reviews, 55, 325-350.
3. M.S. Qatu, 2002, Recent research advances in the dynamic behavior of shells: 1989-2000, Part 2: homogeneous shells, Applied Mechanics Reviews, 55, 415-434.
4. S. Kandasamy, and A.V. Singh, 2006, Free vibration analysis of skewed open circular cylindrical shells, Journal of Sound and Vibration, 290, 1111-1118.
5. A.S. Saada, 1993, Elasticity: Theory and Applications, 2nd Edition, Krieger Publishing Company, Florida, Chap.6.
6. W. Jr. Weaver, and P.R. Johnston, 1984, Finite Elements for Structural Analysis, Prentice Hall Inc., Englewood Cliffs, New Jersey.

73

Parametric Instability of a Cantilever Beam with Magnetic Field and Axial Load

Barun Pratiher and Sontosha Kumar Dwivedy

Mechanical Engineering Department, Indian Institute of Technology Guwahati, Guwahati-781 039, India
email: barun@iitg.ernet.in and dwivedy@iitg.ernet.in

ABSTRACT

In this work parametric instability regions of a cantilever beam with and without tip mass subjected to magnetic field, and time varying axial load are investigated. Using extended Hamilton's Principle the equation of motion is derived. The governing equation is reduced to that of damped Hill equation by using the Galarkin principle with assumed shape function. Method of multiple scales is used to study the system instability. Effect of magnetic filed, damping factor, axial force and mass ratio on instability regions are investigated.

Keywords: Hardfacing, MMAW process, ferritic stainless steel, martensitic stainless steel.

1. INTRODUCTION

Nowadays, for many industrial application studies of dynamic instability of structures is very essential. Parametric instability is an important dynamic behavior for foundations of rotating machinery, bridge and many other structural applications. Beam is the most fundamental part of any structural member and it may be subjected to many different types of loading depending on applications. A few literatures related to parametric excitation of beam elements are cited here. Wu [1] studied the instability regions of a simply supported beam subjected magnetic field and thermal loading. He used the incremental harmonic balance (IHB) method to study the stability of this system. Hyun and Yoo [2] investigated the stability of a cantilever beam subjected to axial excitation using both first and second order approximate solutions of method of multiple scales (MMS). Chung *et al.* [3] carried out the dynamic stability of a rotating cantilever beam subjected to parametric excitation. Chen and Yang [4] studied the transverse stability of simply supported and clamed-clamped beams subjected the harmonically axial excitation by using MMS. Liu and Chang [5] determined the interactive behaviors of a magneto-elastic beam in magnetic field, axial loads and external forced. They observed that the effect of magnetic field reduce both deflection and natural frequency simultaneously of the system. Liu *et al* [6] studied dynamic behavior of a simply supported beam subjected to an external axial periodic force in a periodic transverse magnetic field.

From the literature, it is observed that a few works has been carried out for finding the instability regions of a cantilever beam with magnetic field and time varying axial loading. So, in the present work, an attempt has been made to study the dynamic instability of a cantilever beam with and without tip mass subjected to a magnetic filed and time varying axial load. The equation of motion of such system is developed and Galerkin's principle is used to reduce the governing equation to a damped Hill equation. Boundary regions of instability are determined by applying the method of multiple scales. Effect of magnetic filed, damping factor, axial force and mass ratio on instability regions are observed. This work is an extension of the work by Liu and Chang [6] where they have carried out only the free vibration analysis of a similar system supported by additional springs.

2. EQUATION OF MOTION

Figure 1 shows cantilever beams with (Fig. 1a) and without tip mass (Fig. 1b) subjected to axial force $P = P_0 + P_1 \cos\Omega_1 t$ and transverse magnetic field $B_0 = B_m \cos\Omega_2 t$. Here L, d, and h are the length, width, and depth of the beam.

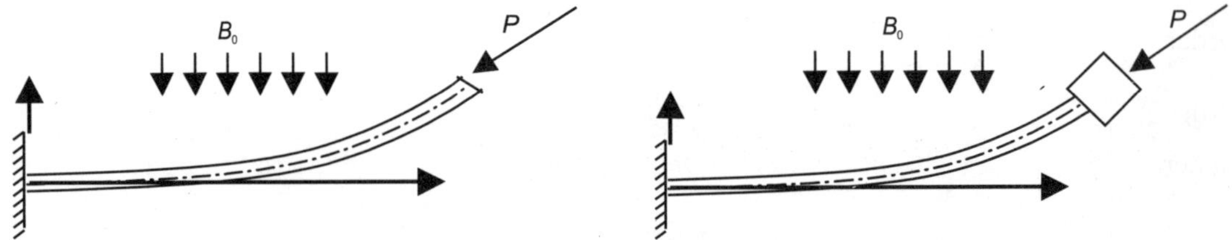

Fig. 1. (a) Cantilever beam, **(b)** Cantilever beam with tip mass.

The equation of motion of this system is derived by using extended Hamilton's principle and then reduced to its temporal form by using Galerkin's Method. The resulting temporal equation of motion can be given by

$$\ddot{u} + 2\varepsilon\mu\dot{u} + u + \varepsilon[\alpha_1\cos\Omega_1\tau - \alpha_2\cos 2\Omega_2\tau]u = 0 \qquad \ldots(1)$$

where u is the transverse displacement of the system, μ is damping factor, $\psi(\bar{x})$ is the assume mode shape of the cantilever beams, B_m is the induced magnetic field, μ_0 is the permeability of the vacuum, μ_r is the relative permeability, χ_m is the magnetic susceptibility. The expression for some other terms natural frequency ω_n, amplitude of external axial force α_1 and amplitude of magnetic force α_2 are given by

$$\omega_n = \sqrt{\left(EI\int_0^1\psi''''(\bar{x})\psi(\bar{x})d\bar{x} - \left(P_0 - \frac{\chi_m}{2\mu_r\mu_0}B_m^2 hd\right)\int_0^1\psi''(\bar{x})\psi(\bar{x})d\bar{x}\right)}$$

$$\alpha_1 = \frac{P_1\int_0^1\psi''(\bar{x})\psi(\bar{x})d\bar{x}}{m\omega_n^2\int_0^1\psi^2(\bar{x})d\bar{x}} \quad \text{and} \quad \alpha_2 = \frac{\chi_m B_m^2 hd\int_0^1\psi''(\bar{x})\psi(\bar{x})d\bar{x}}{2\mu_r\mu_0 m\omega_n^2\int_0^1\psi^2(\bar{x})d\bar{x}}$$

Equation (1) is similar to that of temporal equation (18) of ref. [5] neglecting the external excitation and stiffness terms. But in ref [5] only the free vibration analysis was carried out by solving the temporal equation numerically.

Here, method of multiple scales is used to study the stability of the system. By representing u in terms of different times scale as

$$u(\tau;\varepsilon) = u_0(T_0,T_1) + \varepsilon u_1(T_0,T_1) + \cdots \qquad \qquad ...(2)$$

where, $T_0 = \tau$, $T_1 = \varepsilon\tau$ and transformation of first and second time derivatives are given by

$$\frac{d}{d\tau} = D_0 + \varepsilon D_1 + \cdots,$$

$$\frac{d^2}{d\tau^2} = D_0^2 + 2\varepsilon D_0 D_1 + \cdots \qquad \qquad ...(3)$$

where $D_0 = \dfrac{\partial}{\partial T_0}$ and $D_1 = \dfrac{\partial}{\partial T_1}$. Substituting equations (2), (3) and (4) into equation (1) and equating coefficient of like powers of ε, yields the following equations.

Order ε^0 :
$$D_0^{\,2} u_0 + u_0 = 0 \qquad \qquad ...(4)$$

Order ε^1 :
$$D_0^{\,2} u_1 + u_1 = -2D_0 D_1 u_0 - 2\mu D_0 u_0 - \alpha_1 u_0 \cos(\Omega_1\tau) + \alpha_2 \cos(2\Omega_2\tau) \qquad ...(5)$$

General solutions of equation (5) can be expressed as

$$u_0 = A(T)\exp(iT_0) + \bar{A}(T)\exp(-iT_0) \qquad \qquad ...(6)$$

Substituting equations (6) into equation (5) leads to

$$D_0^{\,2} u_1 + u_1 = -2iA'\exp(iT_0) - 2i\mu A\exp(iT_0) - \frac{\alpha_1}{2}\left[A\exp i(1+\Omega_1)T_0 + \bar{A}\exp i(\Omega_1-1)T_0\right]$$

$$+\frac{\alpha_2}{2}\left[A\exp i(1+2\Omega_2)T_0 + \bar{A}\exp i(2\Omega_2-1)T_0\right] + cc \qquad \qquad ...(7)$$

One may see that any solution of equation (8) will contain secular or small divisor terms when $\Omega_2 \approx 2$, $\Omega_1 \approx 1$ independently or both $\Omega_2 \approx 2$ $\Omega_1 \approx 2$ simultaneously. In the following sections, for finding stability diagrams three cases (i.e. (i) $\Omega_1 \approx 2$ and Ω_2 is away from 1, (ii) $\Omega_2 \approx 1$ and Ω_1 and is away from 2, and (iii) $\Omega_2 \approx 2$ and $\Omega_1 \approx 1$) are discussed separately.

2.1. Case 1: $\Omega_2 \approx 2$ and Ω_2 is away from 1 (i.e. Subharmonic Resonance)

For this case using the detuning parameter, one may write

$$\Omega = \Omega_1 = 2 + 2\varepsilon\sigma, \quad \sigma = O(1) \qquad \qquad ...(8)$$

Substituting equations (8) in equation (7) and eliminating the secular or small divisor terms yields

$$2iA' + 2i\mu A + \frac{\alpha_1}{2}\bar{A}\exp i(2\sigma T_1) = 0 \qquad \qquad ...(9)$$

Substituting A in the form of $(B_r + iB_i)\exp(i\sigma T_1)$ into equation (9) and separating the real and imaginary parts yields

$$2\frac{dB_r}{dT_1} + 2(\sigma + \mu)B_r + \frac{\alpha_1}{2}B_i = 0 \qquad \text{...(10)}$$

$$2\frac{dB_i}{dT_1} + 2(\sigma + \mu)B_i + \frac{\alpha_1}{2}B_r = 0 \qquad \text{...(11)}$$

Equations (10) and (11) admit a solution in the form $(B_r, B_i) = (b_r, b_i)$ ex (γT_1) with constant b_r and b_i. Using this expression in equations (10) and (11) yields

$$2\gamma b_r + 2(\sigma + \mu)b_r + \frac{\alpha_1}{2}b_i = 0 \qquad \text{...(12)}$$

$$2\gamma b_i + 2(\sigma + \mu)b_i + \frac{\alpha_1}{2}b_r = 0 \qquad \text{...(13)}$$

For finding the steady state instability regions, the eigen values of the Jacobian matrix formed from equations (12) and (13) are determined. From these eigenvalues one may note that the transient curves of the first order expansion when $\Omega_1 \approx 2$ may be given by

$$\Omega = \Omega_1 = 2 + 2\varepsilon\left[-\mu \pm \sqrt{\frac{\alpha_1^2}{16}}\right]. \qquad \text{...(14)}$$

2.2. Case 2: $\Omega_2 \approx 1$ and Ω_1 is away from 2 (i.e. Simple Resonance)

Following the method similar to that describes in section 2.1, for the simple resonance case $\Omega_2 \approx 1$ and Ω_1 is away from 2, the transition curves emanating from $\Omega_2 = 1$ are given by

$$\Omega = \Omega_2 = 1 + \varepsilon\left[-\mu \pm \sqrt{\frac{\alpha_2^2}{16}}\right]. \qquad \text{...(15)}$$

2.3. Case 3: $\Omega_1 \approx 2$ and $\Omega_2 \approx 1$ (Subharmonic and Simple Resonance Simultaneously)

Following the same method describes in section 2.1 and 2.2, for subharmonic and simple resonance simultaneously $\Omega_1 \approx 2$ and $\Omega_2 \approx 2$, the transition curves when $(\Omega_1 = 2\Omega_2)$ near 2 may be written by

$$\Omega_1 = 2\Omega_2 = 2 + 2\varepsilon\left[-\mu \pm \sqrt{\frac{(\alpha_1 - \alpha_2)^2}{16}}\right].$$

3. NUMERICAL RESULTS AND DISCUSSIONS

In all simulations, a metallic beam with length $L = 0.336$ m, width $d = 0.001$ m, depth $h = 0.005$ m, Young's Modulus $E = 1.94 \times 10^{11}$ N/m^2 and mass of the beam $m = 0.04$ kg are considered. The permeability of the vacuum, $\mu_0 = 1.26\text{E-06Hm}^{-1}$ and relative permeability $\mu_r = 3.0\text{E03 Hm}^{-1}$ are taken for all numerical calculations. In section 3.1 the results for the case $\Omega_1 \approx 2$ and Ω_2 away from 1, in section 3.2 the results for the case $\Omega_2 \approx 1$ and Ω_1 is away from 2 and in section 3.3 the results for case $\Omega_1 \approx 2$ and $\Omega_2 \approx 1$ are presented. In this analysis, the key parameters that determine the regions of instability of the systems are magnetic field B_0, damping constant C_d, axial periodic load P_1, and mass ratio m_r. Two sets of instability regions viz., transition curves for axial load ($P_1 \sim \Omega$) and transition curves for magnetic field ($B_0^2 \sim \Omega$) are plotted for a number of system parameters. It may be noted that the region bounded by the transition curves is stable and outside the region is unstable.

3.1. Subharmonic Resonance ($\Omega_1 \approx 2$ and Ω_2 away from 1)

In this section, the effect of damping ratio, magnetic field, and mass ratio on the parametric instability regions are investigated when the system is excited by an axial force with nondimensional frequency $\Omega_1 \approx 2$ and a magnetic field with nondimensional frequency Ω_2 away from 1. Here the instability regions are plotted using equation (14).

From Fig. 2(a), it is observed that by increasing damping for a particular value of magnetic filed, the instability curves shift towards left indicating that the system will start vibrating at a lower frequency for higher value of damping. From Fig. 2(a) and 2(b), it is observed that with increase in magnetic field the regions of instability increases. Though for lower value of damping and amplitude of axial forcing, magnetic field strength has little influence on the instability regions, for higher values of damping, the instability regions starts at a lower frequency. As shown in figure Fig. 3(a), one may observe that for a cantilever beam with tip mass, for a particular value of damping with increase in mass ratio, the instability region shift towards right. From Fig. 3(b), it may be noted that with increase in mass ratio, for higher value of damping, the instability region shifts towards right indicating the instability regions starts at a higher frequency.

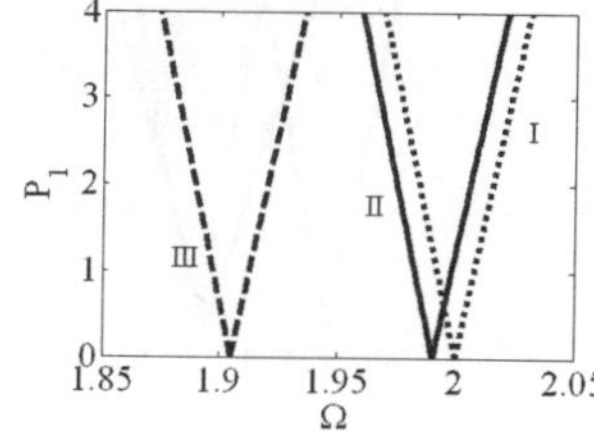
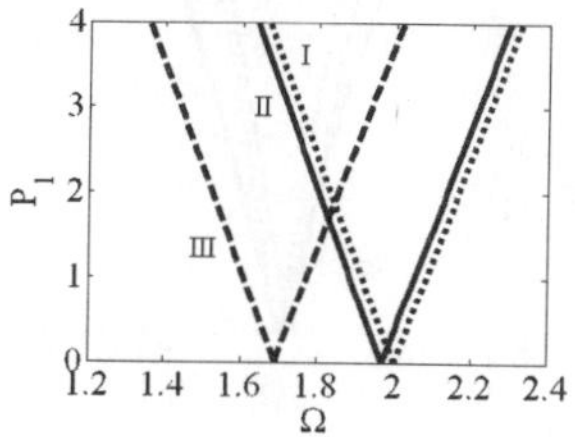

(a)   (b)

Fig. 2. Effect of damping and magnetic field on the parametric instability regions $P_1 \sim \Omega$ for a cantilever beam without tip mass with $P_0 = 1$. (I) $C_d = 0.0$ Ns/m, (II) $C_d = 0.1$ Ns/m, and (III) $C_{d-} = 1.0$ Ns/m; (a) $B_m = 0.3$, (b) for $B_m = 2$.

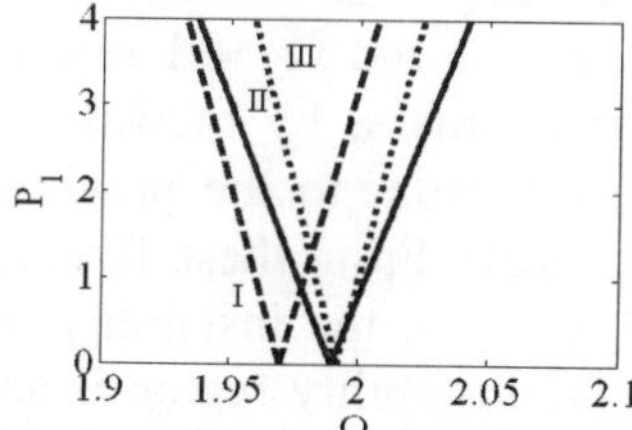
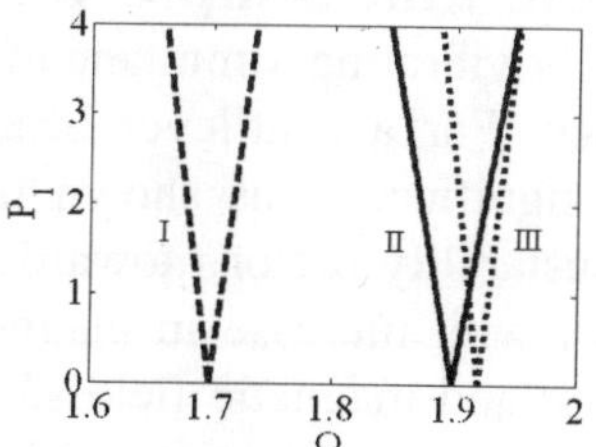

Fig. 3. Effect of damping and tip mass on the parametric instability regions $P_1 \sim \Omega$ for a cantilever beam with tip mass with $B_m = 0.3$, $P_0 = 1$. (I) $m_r = 0.625$, (II) $m_r = 1$, and (III) $m_r = 2$; (a) $C_d = 0.1$ Ns/m, (b) for $C_d = 1.0$ Ns/m.

3.2. Simple Resonance ($\Omega_1 \approx 1$ and Ω_1 is away from 2)

In this section, effect of damping ratio, axial load, and mass ratio on the parametric instability regions with magnetic field are investigated for $\Omega_1 \approx 1$ and Ω_1 is away from 2. In Figure 4(a and b), it is clearly shown that larger value of damping will shift the instability region towards left. Also it is observed that increase in the static amplitude of axial force (P_0) reduces the instability region and the amplitude of dynamic loading has no effect on the instability regions which is clearly evident from equation (15).

The effect tip mass is studied in Fig. 5 and it is observed that increase in mass ratio decreases the instability regions. Also in this case increase in damping shifts the instability region towards left.

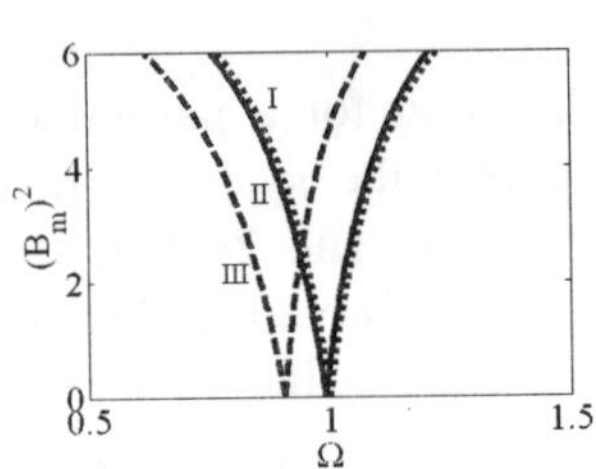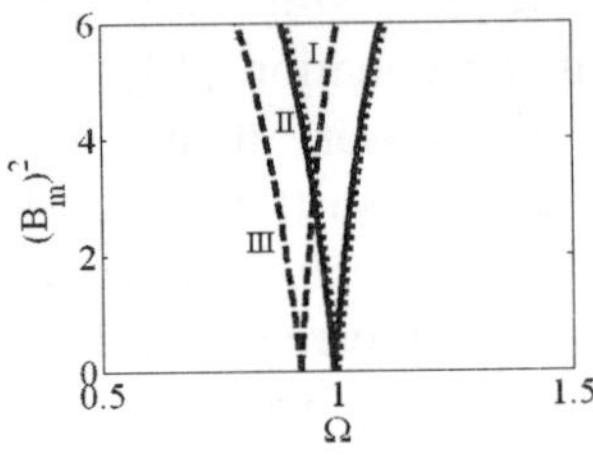

Fig. 4. Effect of the damping and amplitude of static forcing P_0 on the parametric instability regions $B_0^2 \sim \Omega$ for a cantilever beam without tip mass with. (I) $C_d = 0.0$, (II) $C_d = 0.1$Ns/m, and (III) $C_d = 1.0$ Ns/m (a) = 1 and (b) = 10.

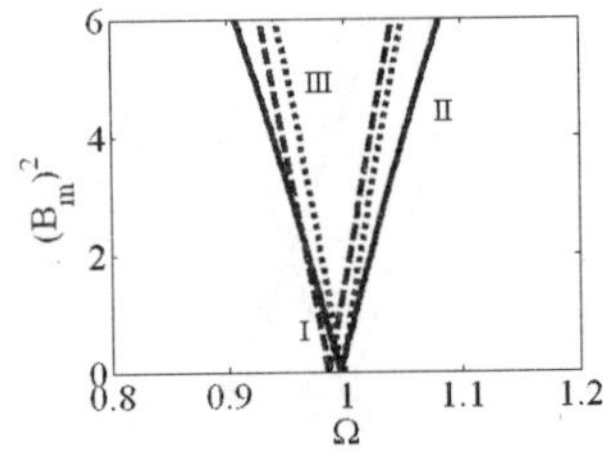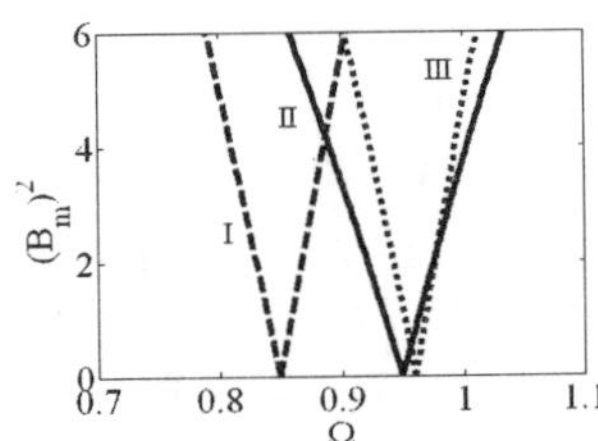

Fig. 5. Effect of tip mass and damping on parametric instability regions $B_0^2 \sim \Omega$ for cantilever beam with tip mass for $P_0 = 1$ (I) $m_r = 0.625$, (II) $m_r = 1$, and (III) $m_r = 2$. (a) $C_d = 0.1$ Ns/m, (b) for $C_d = 1.0$ Ns/m.

3.3. Subharmonic and Simple Resonance ($\Omega_1 \approx 2$ and $\Omega_2 \approx 1$)

In this case, system is vibrating simultaneously with $\Omega_1 \approx 2$ and $\Omega_2 \approx 1$ resulting in subharmonic and simple resonance. For a cantilever beam without tip mass, by increasing P_1, the instability region is increased significantly as shown in Fig. 6(a). Similar to the previous cases increase in damping shifts the instability regions towards left (Fig. 6(b)). From these figures one may note that for same value of P_1 with increase in magnetic field strength, the instability region decreases.

Effect of tip mass and magnetic field strength on the instability region is shown in Fig. 7. It is found that with increase in mass ratio instability region shift towards right and with increase in B_m the stability of the system improves as one has higher operating range in this case.

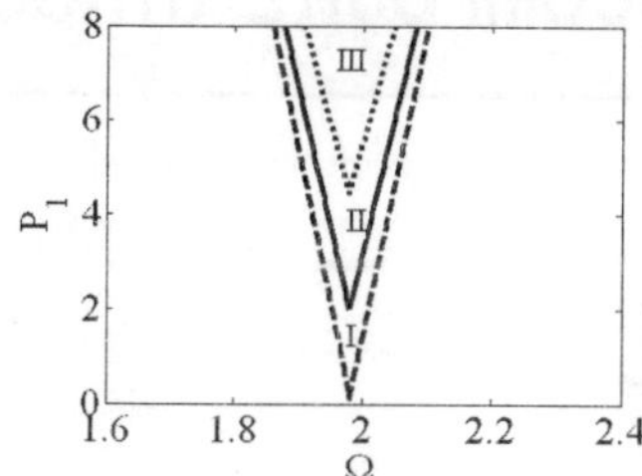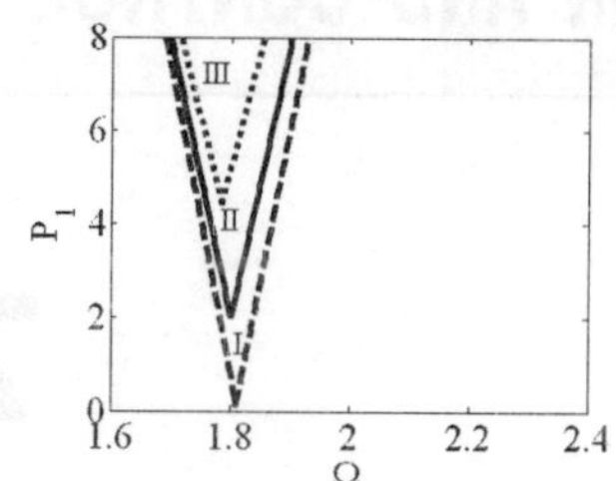

Fig. 6. Effect of the magnetic field on the instability regions $P_1 \sim \Omega$ for cantilever beam without tip mass with $P_0 = 1$. (I) $B_m = 0.0$, (II) $B_m = 1.0$, and (III) $B_m = 2.0$, (a) $C_d = 0.1$ Ns/m, (b) for $C_d = 1.0$ Ns/m.

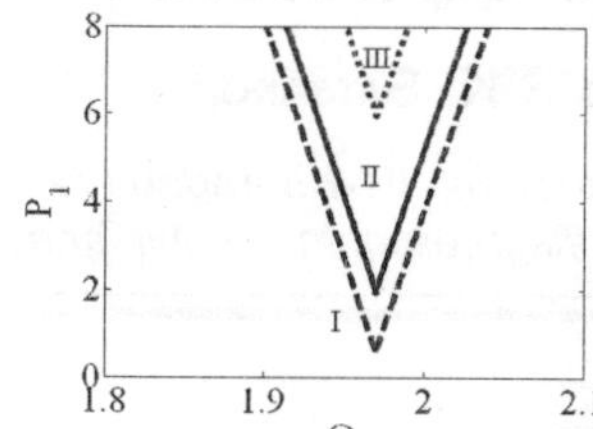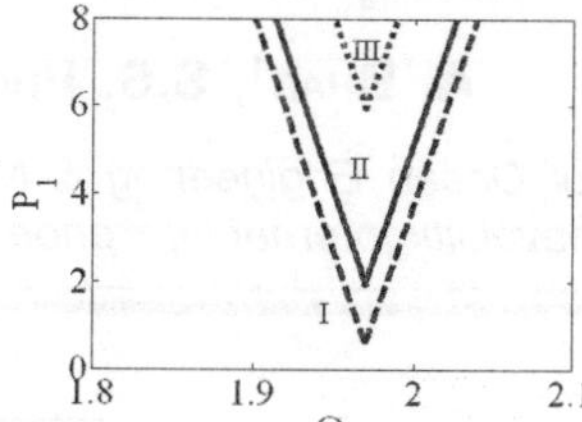

Fig. 7. Effect of the magnetic field on the instability regions $P_1 \sim \Omega$ for cantilever beam with tip mass with $P_0 = 1$. (I) $B_m = 0.0$, (II) $B_m = 1.0$, and (III) $B_m = 2.0$ (a) $m_r = 0.625$, (b) for $m_r = 1.0$.

CONCLUSION

In this work the stability boundaries of the cantilever beam with and without tip mass with transverse magnetic field and axial periodic load are studied. The temporal equation of motion of the system is derived and the instability regions are determined by using method of multiple scales for three different resonance conditions. It is observed that the instability regions and hence the vibrations of the system can be actively controlled by using the magnetic field. In all the cases it is found that increase in mass ratio decreases the instability regions and increase in damping shifts the parametric instability regions towards left. Also it is observed that the change in magnetic field has more influence in the decrease of the instability regions than that of the change in the amplitude of external axial load. Using the developed instability regions, a designer will be able to select the proper system parameters to avoid vibration in the system.

REFERENCE

1. G.Y. Wu, 2005, The analysis of dynamic instability and vibration motions of a pinned beam with transverse magnetic fields and thermal loads, Journal of Sound and Vibration 284, 343-360.
2. S.H. Hyun, H.H. Yoo, 1999, Dynamic modeling and stability analysis of axially oscillating cantilever beams, Journal of Sound and Vibration. 228, 543-558.
3. J. Chung, D. Jubg, H.H. Yoo, 2004, Stability analysis for the flap-wise motions of a cantilever beam with rotary oscillation, Journal of Sound and Vibration. 273, 1047-1062.
4. L.Q. Chen, X.D. Yang, 2005, Stability in parametric resonance of axially moving viscoelastic beams with time-dependent speed, Journal of Sound and Vibration. 284, 879-891.
5. M.F. Liu, T.P. Chang, 2005, Vibration analysis of a magneto-elastic beam with general boundary conditions subjected to axial load and external force, Journal of Sound and Vibration. 288, 399-411.
6. Q.S. Lu, C.W.S. To, K.L. Huang, 1995, Dynamic stability and bifurcation of an alternating load and Magnetic field excited magneto-elastic beam, Journal of Sound and Vibration. 181, 873-891.
7. A.H. Nayfeh, D.T. Mook, 1995, Nonlinear Oscillations, John Willey & Sons, Inc, New York.

74

Vibrational Behavior of Laminated Composite Stiffened Structures

A. BHAR[1], S.S. PHOENIX[2] AND S.K. SATSANGI[3]

Department of Ocean Engineering & Naval Architecture, IIT-Kharagpur-721302, India
email: [1] anindya@naval.iitkgp.ernet.in, [2] phoenix@naval.iitkgp.ernet.in, [3] subir@naval.iitkgp.ernet.in

ABSTRACT

Free vibration analysis of laminated composite stiffened plates with various lamination and stiffening schemes have been carried out using the finite element method. An eight-noded quadratic isoparametric plate element in combination with a three-noded quadratic isoparametric beam element has been used for formulating the stiffened plate element. An improved method of forming the consistent mass matrix of the stiffener element has been used. The model is capable of accommodating stiffener placed anywhere within the plate giving great freedom for mesh generation. Results have been compared with those available in the existing literature.

Keywords: Composite, Stiffened-plate, Free vibration, Finite elements.

1. INTRODUCTION

The recent trend in finite element analysis of stiffened plated structures involves use of stiffened plate bending elements. Based on the quadrilateral plate bending element presented by Rock and Hinton [1] a stiffened plate-bending element for the static analysis of isotropic stiffened plates was introduced by Mukhopadhyay and Satsangi [2]. But that model was capable of accommodating stiffeners only in the plate reference axis directions. A review work of dynamic analyses of stiffened isotropic plates was presented by Mukherjee *et al*. [3]. Later Chattopadhyay *et al*. [4] presented free vibration analysis of eccentrically stiffened composite plates. They considered the lumped mass matrix scheme for both the plate and the stiffeners. Ray *et al*. [5] analyzed laminated plates with closed section hat stiffeners. Though the formulation presented by Ray [6] is capable of accommodating arbitrary stiffener placement, but no such numerical problem has been presented for free vibration analysis.

In present study an eight-noded isoparametric plate-bending element, which is based on Reissner-

Mindlin thick plate theory, has been used in combination with a three-noded isoparametric beam element to form the stiffened plate element. An improved method of formulating consistent mass matrix for the stiffeners as well as the plate has been used. A reduced integration scheme has been employed in the computation of stiffness matrix to avoid spurious shear locking. The versatility of the formulation is due to its ability to handle any generalized pattern of stiffener placement inside the plate.

2. MATHEMATICAL FORMULATION

The equation of equilibrium to be solved for free vibration of an elastic system undergoing small displacement is given by

$$[M]\{\ddot{\delta}\}+[K]\{\delta\}=\{0\} \qquad \qquad ...(1)$$

where $[M]$ and $[K]$ are the overall mass and stiffness matrices of the system, $\{\ddot{\delta}\}$ and $\{\delta\}$ are the acceleration and displacement vectors respectively. This eigenvalue problem has been solved using the simultaneous iteration algorithm presented by Corr and Jennings [7].

2.1 Plate Element Formulation

Cross-section of a typical laminate is shown in Fig. 1. Z_k and Z_{k+1} are the distances of the bottom and top surface of the kth layer from the reference plane. The reference plane of the whole laminated stiffened plate system is assumed to be the middle plane of the plate. Five degrees of freedom namely three translations u, v, w and two rotations θ_x and θ_y have been considered per node. The plate element in physical coordinate system X-Y and in natural coordinate system ξ-η is shown in Fig. 2.

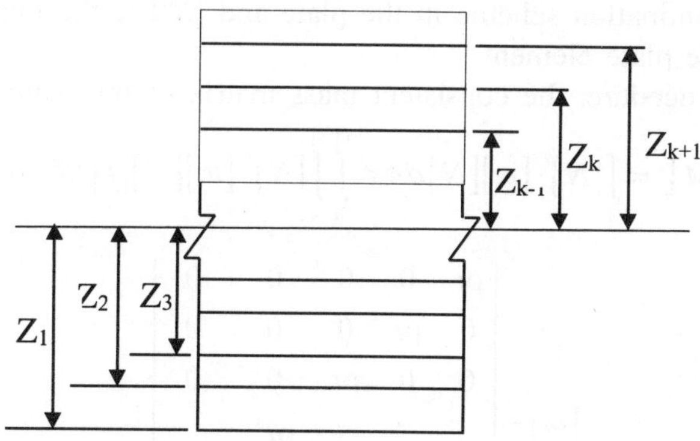

Fig. 1. A typical laminate configuration.

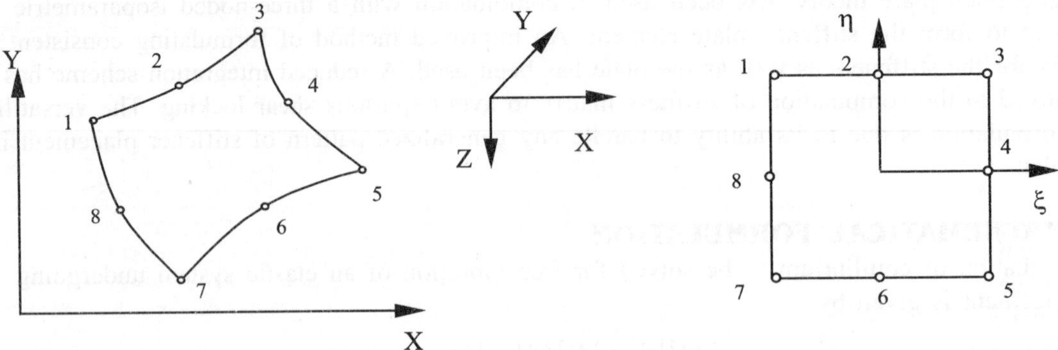

Fig. 2. Isoparametric quadratic plate element **(a)** Physical co-ordinate; **(b)** Isoparametric co-ordinate.

The displacement field at any point within the plate element has been taken according to linear theory in terms of reference plane displacements as

$$U(x,y,z) = u(x,y) - z\theta_x(x,y) \qquad \text{...(2)}$$

$$V(x,y,z) = v(x,y) - z\theta_y(x,y) \qquad \text{...(3)}$$

$$W(x,y,z) = w(x,y) \qquad \text{...(4)}$$

The stiffness matrix of the plate element is given by

$$\left[k^0\right]_e = \int_A \left[B^0\right]^T [D]\left[B^0\right] dA = \int_{-1}^{+1}\int_{-1}^{+1} \left[B^0\right]^T [D]\left[B^0\right]|J|\,d\xi\,d\eta \qquad \text{...(5)}$$

where $|J|$ is the determinant of the jacobian matrix, $[D]$ is the rigidity matrix of the plate laminate which takes care of the lamination scheme in the plate and $[B^0]$ is the linear portion of the strain-displacement matrix of the plate element.

Following standard procedure, the consistent mass matrix of the plate element is found to be

$$[M]_e = \int_A [N]^T [m][N]\,dA = \int_{-1}^{+1}\int_{-1}^{+1}[N]^T [m][N]|J|\,d\xi\,d\eta \qquad \text{...(6)}$$

where

$$[m] = \begin{bmatrix} \rho t & 0 & 0 & 0 & 0 \\ 0 & \rho t & 0 & 0 & 0 \\ 0 & 0 & \rho t & 0 & 0 \\ 0 & 0 & 0 & \dfrac{\rho t^3}{12} & 0 \\ 0 & 0 & 0 & 0 & \dfrac{\rho t^3}{12} \end{bmatrix} \qquad \text{...(7)}$$

and

$$[N] = \sum_{r=1}^{8} N_r\left[I_5\right] \qquad \text{...(8)}$$

where ρ is the mass-density of the plate material, t is the thickness of the plate, Nr is the rth shape function of the plate element, $[I_5]$ is a 5×5 identity matrix.

2.2 Stiffener Element Formulation

The stiffener element is formulated as a three noded beam element (Fig. 3). The displacement field at any point of the stiffener placed along X' direction is represented by

$$\begin{Bmatrix} U' \\ V' \\ W' \end{Bmatrix} = \begin{Bmatrix} u' - z'\theta_{x'} \\ -z'\theta_{y'} \\ w' + y'\theta_{y'} \end{Bmatrix} \qquad ...(9)$$

Such that u', v', w' and $\theta_{x'}$, $\theta_{y'}$ are the displacement and rotation components respectively at the reference plane and y', z' the coordinate variables at any point of stiffener, all with respect to stiffener axis system.

The stiffness matrix, equivalent to plate element stiffness matrix, of an arbitrarily placed stiffener element within a plate element is expressed by

$$[k_s]_e = [T]^T [\Lambda]^T [k_s][\Lambda][T] \qquad ...(10)$$

where

$$[\Lambda] = \begin{bmatrix} [\bar{\Lambda}] & [0] & [0] \\ [0] & [\bar{\Lambda}] & [0] \\ [0] & [0] & [\bar{\Lambda}] \end{bmatrix} \qquad ...(11)$$

such that

$$[\bar{\Lambda}] = \begin{bmatrix} c & s & 0 & 0 & 0 \\ -s & c & 0 & 0 & 0 \\ 0 & 0 & 1 & 0 & 0 \\ 0 & 0 & 0 & c & s \\ 0 & 0 & 0 & -s & c \end{bmatrix} \qquad ...(12)$$

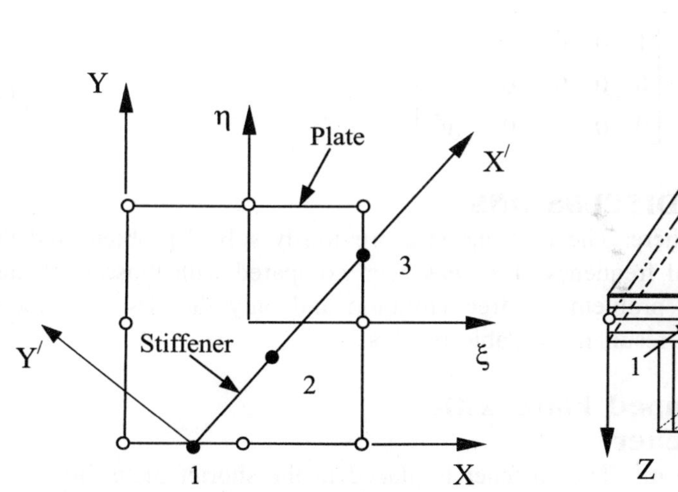

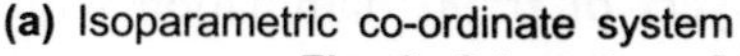

(a) Isoparametric co-ordinate system   **(b)** Original physical system

Fig. 3. Orientation of the stiffener element inside the plate element.

$$c = cos\alpha \quad s = sin\alpha$$

α being the orientation angle of the stiffener element inside the plate element, taken positive in counterclockwise sense from positive X-axis to positive X'-axis direction. The [T] matrix is given by

$$[T] = \sum_{i=1}^{3} \sum_{r=1}^{8} \begin{bmatrix} N_{ir} & 0 & 0 & 0 & 0 \\ 0 & N_{ir} & 0 & 0 & 0 \\ 0 & 0 & N_{ir} & 0 & 0 \\ 0 & 0 & 0 & N_{ir} & 0 \\ 0 & 0 & 0 & 0 & N_{ir} \end{bmatrix} \qquad ...(13)$$

where N_{ir} is the value of the rth plate element interpolation function evaluated at the location of the ith node of stiffener element. This matrix transforms the stiffener element variables to plate element nodal variables. The $[k_s]$ matrix in equation (10) is given by

$$[k_s] = \int_{x'} [B_s]^T [D_s][B_s] dx' \qquad ...(14)$$

which represents the stiffness matrix of the stiffener independent of its position in the plate element.

The plate equivalent element consistent mass matrix due to an arbitrarily placed stiffener is given by

$$[M_s]_e = \int_{x'} [N]^T [\bar{\Lambda}]^T [m_s][\bar{\Lambda}][N] dx' \qquad ...(15)$$

with

$$[m_s] = \int_{y'} \int_{z'} [G]^T [\rho]_s [G] dy' dz' \qquad ...(16)$$

such that

$$[\rho]_s = \rho_s [I_3] \qquad ...(17)$$

where $[I_3]$ is a 3×3 identity matrix, ρ_s is the mass-density of the stiffener material. The [G] matrix in (16) is given by

$$[G] = \begin{bmatrix} 1 & 0 & 0 & -z' & 0 \\ 0 & 0 & 0 & 0 & -z' \\ 0 & 0 & 1 & 0 & y' \end{bmatrix} \qquad ...(18)$$

3. NUMERICAL RESULTS AND DISCUSSIONS

Two numerical problems are presented here. The first one is a previously solved problem and the results obtained, in terms of fundamental frequency (Hz), has been compared with those published in literature. The second one is a new problem for free vibration and only the first four mode shapes are presented since no published result is available on this.

3.1 Stiffened Composite Clamped Plate with Central Rectangular Stiffener

The plate is of dimension (0.375×0.3) m^2. The stiffener is placed in the shorter plate dimension direction. The plate and the stiffener both consist of two layers of 0°/90° lamination with vertical lamination in the stiffener. The whole plate has been analyzed using an 8 × 8 mesh division. The material properties taken for both plate and stiffener are as follows:

$E_x = E_y = 7$ GPa, $G_{xy} = G_{xz} = G_{yz} = 2.6$ GPa, $\nu_{xy} = 0.345$, $\rho = 1504.2$ kg/m^3

The results obtained, in terms of fundamental frequency (Hz) for various system parameters, in the present study along with those available in the literature are indicated in Table 1. It shows that the natural frequency of the composite stiffened plate increases with the increase in the cross sectional area of the stiffener upto a certain limit though the exact position of this critical limit has slightly varied in different studies.

Table 1. Fundamental frequency (Hz) of the composite stiffened clamped plate

Plate thickness (m)	Stiffener Cross-section (b × d) (m)	Fundamental frequency (Hz)			
		Ray C. [6]	Attaf et. al. [8]	Chattopadhyay [9]	Present
0.002400	0 × 0	84.750	84.22	84.30	84.7789
0.002394	0.001 × 0.002	85.643	87.15	85.23	85.5260
0.002378	0.002 × 0.004	92.982	97.62	93.14	92.8831
0.002352	0.003 × 0.006	111.252	115.54	112.38	111.0326
0.002314	0.004 × 0.008	134.973	135.59	136.29	132.9922
0.002208	0.006 × 0.012	146.532	147.74	146.01	151.0791
0.002140	0.007 × 0.014	147.264	148.11	146.61	148.1755
0.002085	0.008 × 0.016	147.743	146.78	146.08	142.8022
0.001866	0.010 × 0.020	144.731	138.54	139.78	129.5617

3.2 Arbitrarily-Stiffened Isotropic Square Clamped Plate

An isotropic stiffened square plate with three stiffeners being oriented in directions as shown in Fig. 4 has been analyzed for free vibration. This same problem was previously solved for static analysis by Ray [6]. The stiffeners are oriented at an angle of +45° with the X-axis. The material properties taken for both plate and stiffener are:

Modulus of elasticity $E = 2 \times 10^{11}$ N/m^2 and Poisson's ratio $\nu = 0.3$

As free vibration results of plates with arbitrarily oriented stiffener is not available in published literature, in the present case only the first four mode shapes obtained is presented in Fig. 5.

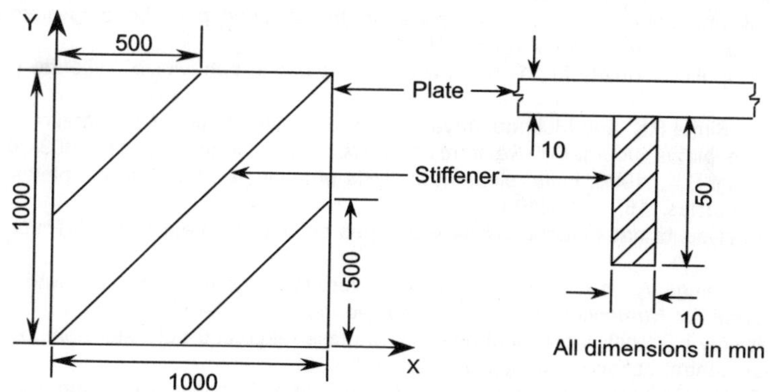

Fig. 4. Isotropic clamped plate with inclined stiffeners.

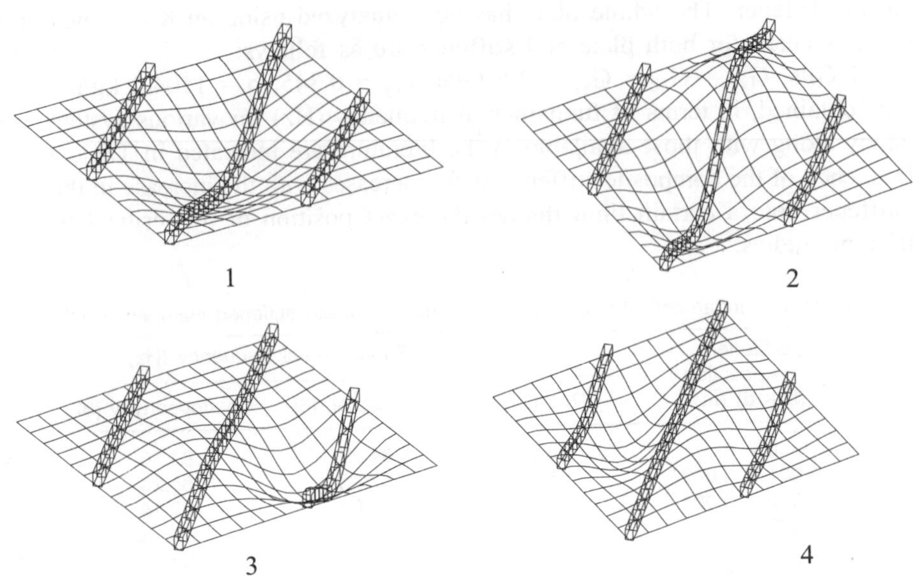

Fig. 5. Mode shapes for the Arbitrarily-Stiffened clamped isotropic plate.

CONCLUSION

Free vibration analysis of stiffened composite plates using the present formulation gives quite satisfactory results keeping the method an economical one. The formulation can easily be extended to accommodate for various types of stiffener cross-sectional shapes and plate planform. However, example on dynamic analysis of arbitrarily stiffened composite plates is still scanty in the literature and needs further attention.

REFERENCES

1. Rock T.A. and Hinton E., 1976, A finite element method for the free vibration of plates allowing for transverse shear deformation, *Comput. Struct.* 6, 37-44.
2. Mukhopadhyay M. and Satsangi S.K., 1984, Isoparametric stiffened plate bending element for the analysis of ships' structures, *Trans. R. Inst. Naval Arch.* 125, 144-151.
3. Mukherjee A. and Mukhopadhyay M., 1986, A review of dynamic behavior of stiffened plates, *Shock Vibr. Dig.* 18, 3-8.
4. Chattopadhyay B., Sinha P.K. and Mukhopadhyay M., 1992, Finite element free vibration analysis of eccentrically stiffened composite plates, *Journal of Reinforced Plastics and Composites*, 11, 1003-1034.
5. Ray C. and Satsangi S.K., 1996, Finite element analysis of laminated hat-stiffened plates, *Journal of Reinforced Plastics and Composites,* 15, 1174-1193.
6. Ray C. (1998), Analysis of hat-stiffened composite plates by finite element method, *Ph.D Thesis*, IIT-Kharagpur-721302, INDIA.
7. Corr R.B. and Jennings A., 1976, A simultaneous iteration algorithm for symmetric eigenvalue problems, *International Journal for Numerical Methods in Engineering*, 10, 647-663.
8. Attaf B. and Hollaway L., 1990, Vibrational analysis of glass reinforced polyester composite plates reinforced by a minimum mass central stiffener, *Composites* 21, 425-430.
9. Chattopadhyay B. (1993), Linear and nonlinear static and dynamic analysis of composite stiffened plates by the finite element method, *Ph.D Thesis*, IIT-Kharagpur-721302, India.

75

Effect of Stacking Sequence on Damping Loss Factors of Laminated Composite Beams

H. Ravisankar[1] and P. Bangarubabu[2]

[1]Department of Mechanical Engineering, National Institute of Technology Warangal-506 004, India
email: ravi_hota@rediffmail.com
[2]Department of Mechanical Engineering, National Institute of Technology Warangal-506 004, India
email: bangaru@nitw.ernet.in

ABSTRACT

Damping behavior of various laminated composite beams with different stacking sequences is analytically analyzed. Theoretical analysis is carried out to find loss factors in x, y, xy directions. The analysis is carried out based on Ni-Adams theory Models. In this work damping loss factors are analyzed for 8, 12 ply symmetric laminated beams. Results showed that the stacking sequence has considerable effect on damping loss factors. Further It is also analyzed the effect of fiber Orientation in each lamina on damping loss factors.

Keywords: Damping factors; Composite Beam; Stacking sequence.

1. INTRODUCTION

Passive damping is a useful parameter in controlling Vibrations. Material damping is one of the parameters that can control vibrations in structures. The material damping of laminated composite beam is highly dependent on the fiber direction. Many of the researchers are working on characterization and evaluation of damping loss factors of laminated composites. The prediction of damping factors in fiber reinforced composites is based on micromechanical and macromechanical models. R.D. Adams [1], considered the unidirectional composite beams and evaluated the effect of fiber orientation of laminated composites on dynamic properties. C.T. Sun *et al.* [2] introduced complex modulii for predicting the damping in laminated composites by using classical lamination theory. Roger. M. Crane and Gillespie [3] characterized damping loss factors experimentally for unidirectional fiber composites. A.S. Hadi and Ashton [4] evaluated the damping loss factors and damping properties both experimentally and analytically and worked on zero degree beams. It is found that damping factors are independent of frequency at lower range of frequencies. Yim *et al.* [5] studied the damping of 0° laminated composite beams with viscoelastic layers and found the effect of thickness of viscoelastic layers on damping properties. Recently, Zhang *et al.* [6] studied the damping characteristics of laminated composites with integrated viscoelastic layers by using commercial package ANSYS7.0. Berthlot [7] analyzed laminated beams and plates by using Ritz method and found modal loss factors at various modes. From previous research, it appears that damping factors are evaluated for unidirectional composites. In the present work, theoretical analysis is carried out for finding the damping factors of laminated composites and their dependence on different stacking sequences. It is also analyzed the effect of fiber orientation in each lamina on damping loss factors.

2. ANALYTICAL MODEL

For evaluating loss factors Ni-Adams model is considered and equations are formulated by using classical lamination theory. Neglecting the transverse shear stress in the composite laminated beam the loss factors are evaluated for various stacking sequences. The effect of the fiber orientation in each layer on over all damping factor of the laminate is also modeled.

For unidirectional composites the constitutive relations for each layer can be given as

$$\begin{Bmatrix} \sigma_{xx} \\ \sigma_{yy} \\ \sigma_{xy} \end{Bmatrix} = \begin{pmatrix} Q_{xx} & Q_{xy} & 0 \\ Q_{yx} & Q_{yy} & 0 \\ 0 & 0 & Q_{ss} \end{pmatrix} \begin{Bmatrix} \varepsilon_{xx} \\ \varepsilon_{yy} \\ \varepsilon_{xy} \end{Bmatrix} \text{ or } \{\sigma_{xy}\} = [Q]_{xy}\{\varepsilon_{xy}\} \qquad ...(1)$$

$[Q]_{xy}$ is on axis stiffness matrix.

If loading is not coincident with fiber orientation, (x-y direction) strain need to be transformed to material coordinate system (loading axis, 1–2 direction).

$\{\sigma_{12}\} = [T]_s \{\sigma_{xy}\}$ where $[T]\sigma$ is stress transformation matrix whose elements are functions of Directional cosines

$\{\sigma_{12}\} = [Q]_{12} \{\varepsilon_{xy}\}$ where $[Q]_{12} = [T]_{12} [Q]_{xy} [T]_\varepsilon$.

$[Q]_{12}$ is referred as off axis stiffness matrix, the expressions for elements in the matrix is given in ref. [11].

2.1 Flexure of Thin Laminate

Considering the Kirchoff's theory, the stress in each lamina can be expressed as

$$\begin{Bmatrix} \sigma_1 \\ \sigma_2 \\ \sigma_{12} \end{Bmatrix} = [Q]_{12} \begin{Bmatrix} \varepsilon_1^0 \\ \varepsilon_{22}^0 \\ \varepsilon_{12}^0 \end{Bmatrix} + [Q]_{12} \; z \begin{Bmatrix} k_1 \\ k_2 \\ k_{12} \end{Bmatrix} \qquad ...(2)$$

where

$\varepsilon_1, \varepsilon_2, \varepsilon_{12},$ are the strains in 1, 2, 12 directions respectively.

$\varepsilon_1^0, \varepsilon_2^0, \varepsilon_{12}^0$ are the mid plane strains and independent of thickness coordinate z.

k_1, k_2, k_{12} are the plate curvatures and equal to $-\dfrac{\partial^2 w_0}{\partial x^2}, -\dfrac{\partial^2 w_0}{\partial y^2}, -\dfrac{\partial^2 w_0}{\partial x \partial y}$

w is the displacement in z direction.

The stress in a laminate vary from lamina to lamina as $[Q]_{12}$ vary from lamina to lamina. Hence the resultant forces and moments can be evaluated by interating stress through thickness.

$$\begin{Bmatrix} N_1 \\ N_2 \\ N_{12} \end{Bmatrix} = \begin{Bmatrix} \int_{-h/2}^{h/2} \sigma_1 dz \\ \int_{-h/2}^{h/2} \sigma_2 dz \\ \int_{-h/2}^{h/2} \sigma_{12} dz \end{Bmatrix} \text{ and } \begin{Bmatrix} M_1 \\ M_2 \\ M_{12} \end{Bmatrix} = \begin{Bmatrix} \int_{-h/2}^{h/2} \sigma_1 z dz \\ \int_{-h/2}^{h/2} \sigma_2 z dz \\ \int_{-h/2}^{h/2} \sigma_{12} z dz \end{Bmatrix} \quad ...(3) \ \& \ (4)$$

Substituting the equation (3) into equation (4) and (5)

$$\begin{Bmatrix} N_1 \\ N_2 \\ N_{12} \end{Bmatrix} = \begin{pmatrix} A_{11} & A_{12} & A_{13} \\ A_{21} & A_{22} & A_{23} \\ A_{31} & A_{32} & A_{33} \end{pmatrix} \begin{Bmatrix} \varepsilon_1^0 \\ \varepsilon_{22}^0 \\ \varepsilon_{12}^0 \end{Bmatrix} + \begin{pmatrix} B_{11} & B_{12} & B_{13} \\ B_{21} & B_{22} & B_{23} \\ B_{31} & B_{32} & B_{33} \end{pmatrix} \begin{Bmatrix} k_1 \\ k_2 \\ k_{12} \end{Bmatrix} \quad ...(5)$$

$$\begin{Bmatrix} M_1 \\ M_2 \\ M_{12} \end{Bmatrix} = \begin{pmatrix} B_{11} & B_{12} & B_{13} \\ B_{21} & B_{22} & B_{23} \\ B_{31} & B_{32} & B_{33} \end{pmatrix} \begin{Bmatrix} \varepsilon_1^0 \\ \varepsilon_{22}^0 \\ \varepsilon_{12}^0 \end{Bmatrix} + \begin{pmatrix} D_{11} & D_{12} & D_{13} \\ D_{21} & D_{22} & D_{23} \\ D_{31} & D_{32} & D_{33} \end{pmatrix} \begin{Bmatrix} k_1 \\ k_2 \\ k_{12} \end{Bmatrix} \quad ...(6)$$

where

$$A_{ij} = \sum_{k=1}^{n} [Q_{ij}]^k (h_k - h_{k-1}), \quad [Q_{ij}]^k \text{ is off axis stiffness matrix for kth layer.}$$

A_{ij} is extension-to-extension coupling.

$$B_{ij} = \frac{1}{2} \sum_{k=1}^{n} [Q_{ij}]^k (h_k^2 - h_{k-1}^2), \quad B_{ij} \text{ bending extensional coupling}$$

$$D_{ij} = \frac{1}{3} \sum_{k=1}^{n} [Q_{ij}]^k (h_k^3 - h_{k-1}^3). \quad D_{ij} \text{ is bending bending coupling.}$$

2.2 Damping Prediction Of Laminated Beams

Adams and Bacon [1] damping criterion states that Energy dissipation in thin unidirectional lamina is sum of the separable energy dissipations due to σ_x, σ_y, σ_{xy}.

The specific damping capacity (SDC) $\Psi = \dfrac{\Delta U_x}{U} + \dfrac{\Delta U_y}{U} + \dfrac{\Delta U_{xy}}{U}$ $\quad ...(7)$

where ΔU_x is energy dissipation in x direction, ΔU_y is energy dissipation in y direction and ΔU_{xy} is energy dissipation in shear direction. U is total strain energy stored in lamina

$$\Delta U_x = \frac{1}{2} \iiint \Psi_L [T]_\varepsilon^{-1} \varepsilon_1 [T]_\sigma^{-1} \sigma_1 dv \text{ is } \Psi_L \text{ loss factor in X direction.} \quad ...(8)$$

where $[T]_\varepsilon^{-1}, [T]_\sigma^{-1}$ are the inverse transformation matrices for strain and stress to transfer from 1 and 2 directions to X and Y directions.

In the present analysis, free flexure LAMINATED COMPOSITE BEAM is considered in which moment M_1 alone is applied about the axis 1. In free flexure beam the moments and forces in other directions are zero, and hence the inverse transformation matrix for stress is $\cos^2\theta$.

In absence of axial forces the mid plane strains $\begin{Bmatrix} \varepsilon_1^0 \\ \varepsilon_{22}^0 \\ \varepsilon_{12}^0 \end{Bmatrix} = 0$, and total strains

$$\begin{Bmatrix} \varepsilon_1 \\ \varepsilon_2 \\ \varepsilon_{12} \end{Bmatrix} = z \begin{Bmatrix} k_1 \\ k_2 \\ k_{12} \end{Bmatrix} \text{ and } \begin{pmatrix} A_{11} & A_{12} & A_{13} \\ A_{21} & A_{22} & A_{23} \\ A_{31} & A_{32} & A_{33} \end{pmatrix} = 0. \text{ For symmetric laminate } \begin{pmatrix} B_{11} & B_{12} & B_{13} \\ B_{21} & B_{22} & B_{23} \\ B_{31} & B_{32} & B_{33} \end{pmatrix} = 0$$

$$M_1 = D_{11}K_1 + D_{12}K_2 + D_{13}K_3 \text{ and } \begin{Bmatrix} k_1 \\ k_2 \\ k_3 \end{Bmatrix} = \left[D_{ij}\right]^{-1} M_1 \qquad \qquad ...(9) \ \& \ (10)$$

Substituting the equation (4) and (10) in equation (8),

$$\Delta U_x = \left[\frac{1}{2}\right] 2 \int_0^{1/2} M_1^2 dx \, 2 \int_0^{h/2} \left[T_{ij}\right]_\varepsilon^{-1} \left[D_{ij}\right]^{-1} \left[Q\right]_{12} \left[D_{ij}\right]^{-1} m z^2 dz \qquad ...(11)$$

Total strain energy stored in the beam

$$U = \int_0^{1/2} M_1 k_1 dx = \left[D_{ij}\right]^{-1} \int_0^{1/2} M_1^2 dx \qquad \qquad ...(12)$$

$$\Psi_x = \frac{\Delta U_x}{U}, \ \Psi_x = \frac{2}{\left[D_{ij}\right]^{-1}} \int_0^{h/2} \left[T_{ij}\right]_\varepsilon^{-1} \left[D_{ij}\right]^{-1} \left[Q\right]_{12} \left[D_{ij}\right]^{-1} m^2 z^2 dz$$

If all layers of equal thickness, $h = Nh_0$ where N is total number of layers and h_0 is the thickness of each layer and integrating and substituting strain transformation matrix

$$\Psi_x = \frac{8\Psi_L}{N^3 \left[D_{11}^*\right]^{-1}} \sum_{k=1}^{N/2} m^2 \left[m^2 \left[D_{11}^*\right]^{-1} + n^2 \left[D_{12}^*\right]^{-1} + mn \left[D_{16}^*\right]^{-1} \right]$$

$$\left[\left[D_{11}^*\right]^{-1} Q_{11}^k + \left[D_{12}^*\right]^{-1} Q_{12}^k + \left[D_{16}^*\right]^{-1} Q_{16}^k \right] \left[k^3 - (k-1)^3 \right] \qquad ...(13)$$

Similarly,

$$\Psi_y = \frac{8\Psi_T}{N^3 \left[D_{11}^*\right]^{-1}} \sum_{k=1}^{N/2} n^2 \left[n^2 \left[D_{11}^*\right]^{-1} + m^2 \left[D_{12}^*\right]^{-1} - mn \left[D_{16}^*\right]^{-1} \right]$$

$$\left[\left[D_{11}^{*}\right]^{-1}Q_{11}^{k}+\left[D_{12}^{*}\right]^{-1}Q_{12}^{k}+\left[D_{16}^{*}\right]^{-1}Q_{16}^{k}\right]\left[k^{3}-(k-1)^{3}\right] \qquad ...(14)$$

$$\Psi_{xy}=\frac{8\Psi_{LT}}{N^{3}\left[D_{11}^{*}\right]^{-1}}\sum_{k=1}^{N/2}mn\left[2mn\left[D_{11}^{*}\right]^{-1}-2mn\left[D_{12}^{*}\right]^{-1}-\left(m^{2}-n^{2}\right)\left[D_{16}^{*}\right]^{-1}\right]$$

$$\left[\left[D_{11}^{*}\right]^{-1}Q_{11}^{k}+\left[D_{12}^{*}\right]^{-1}Q_{12}^{k}+\left[D_{16}^{*}\right]^{-1}Q_{16}^{k}\right]\left[k^{3}-(k-1)^{3}\right] \qquad ...(15)$$

Total loss factor $\Psi = \Psi_x + \Psi_y + \Psi_{xy}$ $\qquad\qquad ...(16)$

For numerical simulation two composite materials are considered for case study. In the present analysis loss factors are evaluated for 8 ply and 12 ply laminates for both carbon fiber and glass fiber laminates with different sequences and found that stacking sequence influences loss factors considerably.

The effect of the fiber orientation in each lamina on loss factors is also evaluated.

3. RESULTS AND DISCUSSION

The overall damping factors of 913C and 913G composites [1] for different stacking sequence are given in Table 1 to 5.

Table 1. Material Properties [1]

Composite	E_1 Mpa]	E_t Mpa]	G_{lt} Mpa]	v	Ψ_L %	Ψ_T %	Ψ_{LT} %
FIBERDUX9 13C	12.5	8.4	4.8	0.3	0.74	7.3	6.6
FIBERDUX9 13G	41.5	12.5	5.0	0.2	1.61	6.7	7.3

Table 2. Fiberedux 913C 8 ply laminate

STACKING SEQUENCE	Ψ_X	Ψ_Y	Ψ_{XY}	Ψ
[90/90/90/90]₂	0	6.7	0	6.7
[0/30/45/60]₂	0.17045	2.6818	2.65512	5.50738
[60/45/30/0]₂	0.656316	0.100997	0.655059	1.41237

Table 3. Fibredux 913C, 12 layered

STACKING SEQUENCE	Ψ_X	Ψ_Y	Ψ_{XY}	Ψ
[0/45/45/45/60/60]₂	0.061821	3.66742	2.73288	6.46211
[0/0/45/45/45/60]₂	0.216739	2.45856	2.44363	5.11898

Table 4. Fibredux 913G, 8 layered

STACKING SEQUENCE	Ψ_X	Ψ_Y	Ψ_{XY}	Ψ
[90/90/90/90]₂	0	6.7	0	6.7
[0/30/45/60]₂	0	2.4236	3.54249	6.212
[60/45/30/0]₂	1.3943	0.181021	1.2027	2.62072

Table 5. Fibredux 913G, 12 layered

STACKING SEQUENCE	Ψ_X	Ψ_Y	Ψ_{XY}	Ψ
[0/45/45/45/60/60]$_2$	0.09578	2.959	3.64164	6.69651
[0/0/45/45/45/60]$_2$	0.261472	2.23518	3.6791	6.17575

From the analysis, it is observed from Table 1 that the stacking sequence affects the damping loss factors of laminated composites considerably.

The effect of the fiber orientations in each layer on over all damping loss factors are numerically simulated and shown in Figures 1–7. Fig. 1 shows the effect of fiber orientation in the layer at the line of symmetry on the damping of $\theta/0/0/0)_2$ composite (θ represents the fiber orientation is varied from 0 to 90^0 and keeping remaining same). It is observed that the layers at line of symmetry will not affect the damping properties considerably and almost negligible. Fig. 2 shows the effect of the fiber orientations in second layer from line of symmetry. A little effect is observed on the damping properties. It is observed from Fig. 3 and Fig. 4, the influence of the orientations in 3rd and 4th layers from line of symmetry is considerably high on damping loss factors. Similar trend is observed for 12 layer symmetric composite from Fig. 5 and Fig. 6. It is observed from Fig. 7 that the effect of fiber orientation in any layer on overall loss factors of (45/ 45/ 45/ 45)$_2$ laminate (when keeping orientations in other layers are 45°) is not considerable.

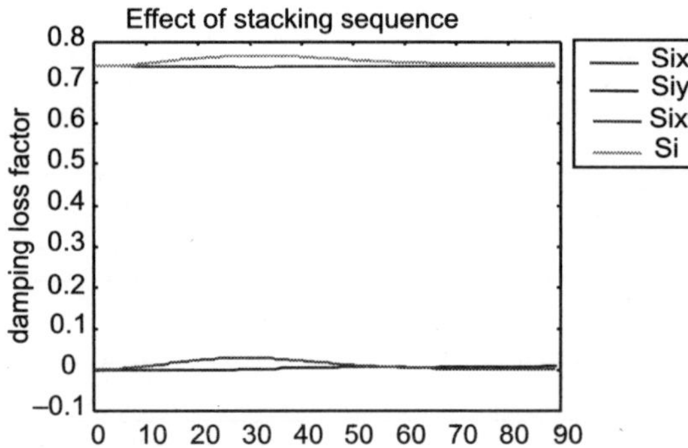

Fig. 1. Efect of fiber orientation in layer at line of symmetry of (0000) 2 laminate.

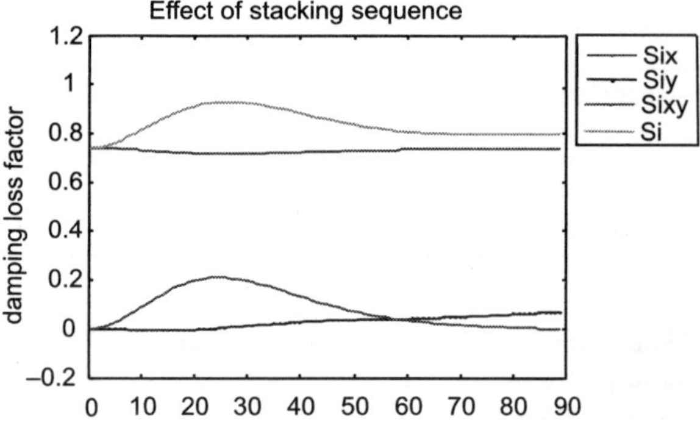

Fig. 2. Efect of fiber orientation in 2nd layer from line of symmetry of (0000) 2 laminate.

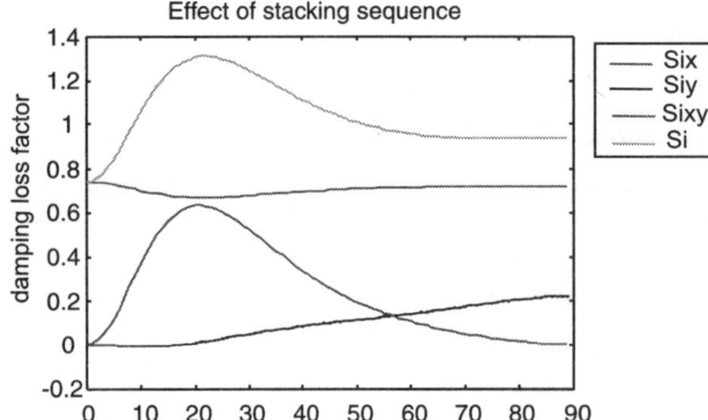

Fig. 3. Efect of fiber orientation in 3rd layer from line of symmetry of (0000) 2 laminate.

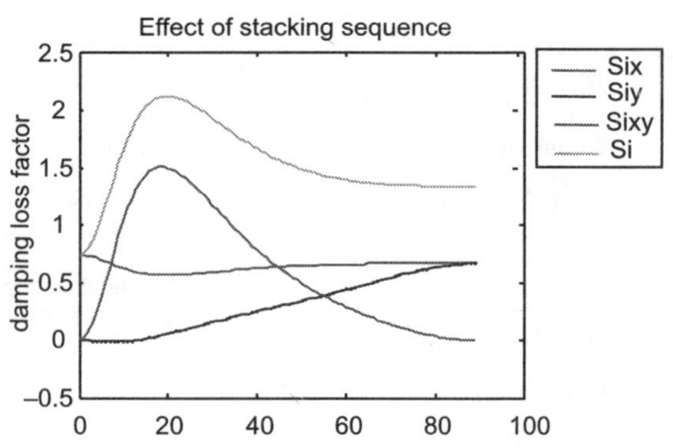

Fig. 4. Efect of fiber orientation in top layer of (0000) 2 laminate.

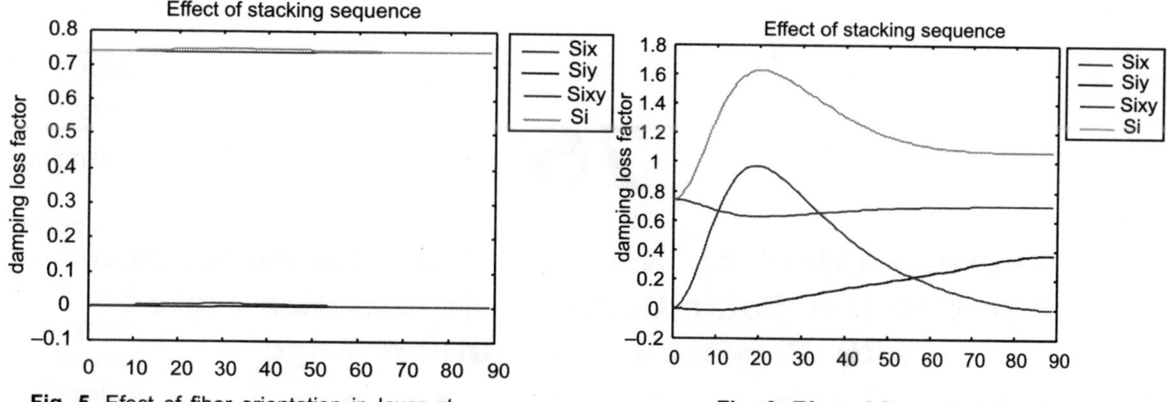

Fig. 5. Efect of fiber orientation in layer at line of symmetry of (0000) 2 laminate.

Fig. 6. Efect of fiber orientation in top layers of (0000) 2 laminate.

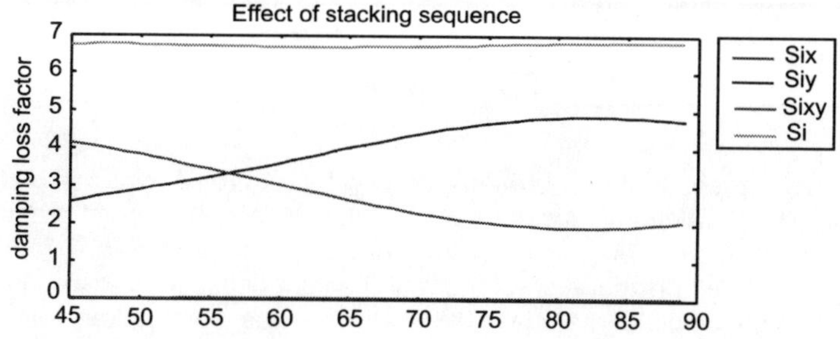

Fig. 7. Efect of fiber orientation in top layers of (45 45 45 45) 2 laminate.

CONCLUSION

The damping is highly depending on the stacking sequence for both glass and carbon fiber composites. The layers away from the line of symmetry plays major roll in defining the damping of the material. The layers at line of symmetry have little effect on damping factors of the composites. Maximum damping is observing when the orientations in extreme layers are in the range 20° to 60°.

REFERENCES

1. R.D. Adams & D.C.G. Bacon, J. Comp. Materials, V7. Oct 1973, p402.
2. C.T. Sun et all, J. Material Science, 1987, p1005-1012.
3. Roger. M. Crane &. Gillespie, Composite Science and Technology, V40. 1991, p335-375.
4. A. Hadi & J.N. Aston, Composite Structures,V34, 1996, p 381–385.
5. Jong Hee yim *et al.*, Composite Structures, V60, 2003,p 367-374.
6. Shao huy zhang *et al.* Composite Structures, Article in press.
7. Berthelot, Composite Structures, Article in press.
8. Ohta, Composite Structures, V.57, 2002, p169–175.
9. J.N. Reddy, ASME Journal, 1984, V51, p 745.
10. Chandra. R. *et al.*, Composite Structures, V46, 1999, p 41–51.
11. M.W. Hyer, Stress analysis of fiber reinforced composite materials, McGraw-hill, 1997.

76

Thermal Effect on Axisymmetric Vibrations of Non-Uniform Polar Orthotropic Circular Plates with Elastically Restrained Edge

U.S. GUPTA, ROSHAN LAL AND SEEMA SHARMA

Department of Mathematics, Indian Institute of Technology Roorkee, Roorkee-247 667, India

ABSTRACT

In the present paper, axisymmetric vibrations of polar orthotropic circular plates of non-linear thickness variation with restrained elastic edge subjected to constant thermal gradient are discussed on the basis of classical plate theory. Ritz method has been employed to obtain the approximate solution of the problem. The consideration of thermal gradient causes non-homogeneity i.e. variation in mechanical properties of plate material. This variation has been taken into consideration by assuming that Young's modulus of the plate varies linearly with the radius vector. The first three natural frequencies have been obtained for different values of flexibility conditions and for classical edge conditions: clamped, simply supported(SS) and free. The effect of edge conditions and that of orthotropy, thermal gradient and thickness variation on the natural frequencies has been investigated for the first three modes of vibration. Normalized displacements for specified plate have been drawn for all the plates. The results for linear thickness variation (LTV) as well as parabolic thickness variation (PTV) have been obtained as special cases. Comparison studies have been carried out which establish the accuracy of present method.

Keywords: Axisymmetric vibration; Polar orthotropic; Thermal gradient; Elastically restrained edge.

1. INTRODUCTION

The analysis of the vibration of plates with elastically restrained edge is an important problem in aeronautical and naval structural engineering. In aircraft structures, the individual plates are connected to the other plates or stiffeners at their boundaries and thus have elastic restraint at their edges [1-2]. Thermally induced vibrations of non-uniform polar orthotropic circular plates are of great interest in air-craft, machine design and also in nuclear, astronautical and chemical engineering. In presence of thermal gradient, the elastic coefficients become functions of space variable, causing non-homogeneity in the material. Most of the engineering materials are found to have a linear relationship between the modulus of elasticity and the temperature [3-4]. Due to the increasing use of modern materials in structural components, it is important to study their vibrational behaviour in presence of thermal gradient. Studies dealing with the effect of thermal gradient on vibration of isotropic/orthotropic plates of various geometries with uniform/non-uniform thickness have been carried out by a number of researchers and are reported in references [5-8], to mention a few. However, no work has been done to study the effect of thermal gradient on vibration of polar orthotropic circular plate of quadratically varying thickness with elastically restrained edges.

In present paper, Ritz method has been used to analyse the effect of constant thermal gradient on the vibration of polar orthotropic circular plate of quadratically varying thickness with elastically restrained edge conditions, where basis functions based upon the static deflection for isotropic plates have been used. Convergence and comparison studies verify the accuracy of the present method.

2. ANALYSIS

Consider a thin circular plate of radius a, thickness $h(r)$, elastically restrained against translation and rotation by springs of stiffness k and k_ϕ, referred to cylindrical polar coordinates (r, θ, z) where the axis of the plate is taken as the line $r = 0$ and its middle surface as the plane $z = 0$. Let the plate be subjected to a steady one-dimensional temperature distribution T.

The maximum kinetic energy of the plate is given by

$$T_{\max} = \frac{1}{2}\rho\omega^2 \int_0^a \int_0^{2\pi} hW^2 r\,d\theta\,dr \qquad \qquad ...(1)$$

where W is the transverse deflection, ρ the mass density and ω the frequency in rad/sec.

The maximum strain energy of the plate is given by

$$U_{\max} = \frac{1}{2}\int_0^a \int_0^{2\pi}\left[D_r\left\{\left(\frac{\partial^2 W}{\partial r^2}\right)^2 + 2\upsilon_\theta \frac{\partial^2 W}{\partial r^2}\left(\frac{1}{r}\frac{\partial W}{\partial r}\right)\right\} + D_\theta\left(\frac{1}{r}\frac{\partial W}{\partial r}\right)^2\right] r\,d\theta\,dr$$

$$+ \frac{1}{2}ak_\phi \int_0^{2\pi}\left(\frac{\partial W(a,\theta)}{\partial r}\right)^2 d\theta + \frac{1}{2}ak\int_0^{2\pi} W^2(a,\theta)d\theta \qquad ...(2)$$

where k and $1/k_\phi$ are the translational and rotational flexibility of the springs and

$$D_r = \frac{E_r h^3}{12(1-\upsilon_r\upsilon_\theta)}, \quad D_\theta = \frac{E_\theta h^3}{12(1-\upsilon_r\upsilon_\theta)} \text{ are flexural rigidities of the plate.}$$

3. METHOD OF SOLUTION: RITZ METHOD

Ritz method requires that the functional

$$J(W) = U_{\max} - T_{\max}$$

$$= \frac{1}{2}\int_0^a \int_0^{2\pi}\left[D_r\left\{\left(\frac{\partial^2 W}{\partial r^2}\right)^2 + 2\upsilon_\theta \frac{\partial^2 W}{\partial r^2}\left(\frac{1}{r}\frac{\partial W}{\partial r}\right)\right\} + D_\theta\left(\frac{1}{r}\frac{\partial W}{\partial r}\right)^2\right] r\,d\theta\,dr$$

$$+ \frac{1}{2}ak_\phi \int_0^{2\pi}\left(\frac{\partial W(a,\theta)}{\partial r}\right)^2 d\theta + \frac{1}{2}ak\int_0^{2\pi} W^2(a,\theta)d\theta - \frac{1}{2}\rho\omega^2 \int_0^a \int_0^{2\pi} hW^2 r\,d\theta\,dr \qquad ...(3)$$

be minimized.

Introducing the non-dimensional variables $\overline{W}=W/a$, $R=r/a$, we consider the thickness variation as $h=h_0\left(1+\alpha R+\beta R^2\right)$ and temperature distribution given by

$$T=T_0\left(1-R\right) \qquad \qquad ...(4)$$

where h_0 is the thickness of plate at center, T is the temperature excess above the reference temperature at any point R, T_0 is the temperature excess at the centre $R = 0$ above the reference temperature at any point on the boundary of the plate. For most engineering materials, the temperature dependence of the modulus of elasticity is given by a relation of the type

$$E_r\left(T\right)=E_1\left(1-\gamma T\right) \text{ and } E_\theta\left(T\right)=E_2\left(1-\gamma T\right). \qquad \qquad ...(5)$$

Using relation (4), relations (5) reduce to

$$E_r\left(T\right)=E_1\left(1-\zeta\left(1-R\right)\right) \text{ and } E_\theta\left(T\right)=E_2\left(1-\zeta\left(1-R\right)\right) \qquad \qquad ...(6)$$

Assuming the deflection function as

$$\overline{W}=\sum_{i=0}^{m}A_iF_i\left(R\right)=\sum_{i=0}^{m}A_i\left(1+\alpha_iR^4+\beta_iR^2\right)R^{2i} \qquad \qquad ...(7)$$

where, A_i are undetermined coefficients, α_i, β_i are unknown constants to be determined from boundary conditions

$$K_\phi\frac{d\overline{W}\left(1\right)}{dR}=-\left(1+\alpha+\beta\right)^3\left[\frac{d^2\overline{W}}{dR^2}+\upsilon_\theta\left(\frac{1}{R}\frac{d\overline{W}}{dR}\right)\right]_{R=1} \qquad \qquad ...(8)$$

$$K\overline{W}\left(1\right)=\left(1+\alpha+\beta\right)^3\left[\frac{d}{dR}\left(\frac{d^2\overline{W}}{dR^2}+\frac{1}{R}\frac{d\overline{W}}{dR}\right)\right]_{R=1} \qquad \qquad ...(9)$$

Using non-dimensional variables $\overline{W}$ and R along with the relations (6) and (7), the functional $J(W)$ given by Eq. (3) becomes

$$J\left(\overline{W}\right)=\frac{D_{r_0}}{2}\left[\int_0^1\int_0^{2\pi}\left[\left(1+\alpha R+\beta R^2\right)^3\left(\zeta_1+\zeta R\right)\left\{\left(\frac{\partial^2\overline{W}}{\partial R^2}\right)^2+\frac{2\upsilon_\theta}{R}\frac{\partial^2\overline{W}}{\partial R^2}\frac{\partial\overline{W}}{\partial R}\right.\right.\right.$$

$$\left.\left.\left.+p^2\left(\frac{1}{R}\frac{\partial\overline{W}}{\partial R}\right)^2\right\}\right]Rd\theta\,dR+K\int_0^{2\pi}\overline{W}^2\left(1\right)d\theta+K_\phi\int_0^{2\pi}\left(\frac{\partial\overline{W}\left(1\right)}{\partial R}\right)^2d\theta\right.$$

$$\left.-\Omega^2\int_0^1\int_0^{2\pi}\left(1+\alpha R+\beta R^2\right)\overline{W}^2Rd\theta\,dR\right] \qquad \qquad ...(10)$$

where $D_{r_0}=\dfrac{E_1h_0^3}{12\left(1-\upsilon_r\upsilon_\theta\right)}$, $\zeta_1=1-\zeta$, $p^2=\dfrac{E_\theta}{E_r}$, $K=\dfrac{a^3k}{D_{r_0}}$, $K_\phi=\dfrac{ak_\phi}{D_{r_0}}$, $\Omega^2=\dfrac{a^4\omega^2\rho h_0}{D_{r_0}}$.

The minimization of the functional $J\left(\overline{W}\right)$ given by Eq. (10) requires

$$\frac{\partial J\left(\overline{W}\right)}{\partial A_i} = 0 \, , \; i = 0, \, 1, \, 2, \ldots .m. \tag{11}$$

This leads to a system of homogeneous equations in A_i, $i = 0, \, 1, \ldots, \, m$, whose non-trivial solution leads to the frequency equation.

$$\left| A - \Omega^2 B \right| = 0 \, , \tag{12}$$

where, $A = [a_{ij}]$ and $B = [b_{ij}]$ are square matrices of order (m +1) given by

$$a_{ij} = \int_0^1 \left(1 + \alpha R + \beta R^2\right)^3 \left(\zeta_1 + \zeta R\right)\left[F_i'' F_j'' + \frac{\upsilon_\theta}{R}\left(F_i'' F_j' + F_j'' F_i'\right) + \frac{p^2}{R^2} F_i' F_j' \right] R dR$$
$$+ K_\phi F_i'(1) F_j'(1) + K F_i(1) F_j(1) \tag{13}$$

and

$$b_{ij} = \int_0^1 \left(1 + \alpha R + \beta R^2\right) F_i F_j R dR \tag{14}$$

for $i = 0, \, 1, \ldots, \, m$; $j = 0, \, 1, \, \ldots, \, m$.

4. NUMERICAL RESULTS AND DISCUSSIONS

The frequency equation (12) has been solved to obtain first three natural frequencies for various values of plate parameters such as rigidity ratio p(= 0.5, 0.75, 1.0, 2.0, 3.0, 4.0, 5.0), thermal gradient ζ(= 0.0(0.1)0.5), taper parameters α(= − 0.5(0.1)0.5); β(= − 0.5(0.1)0.5) such that $\alpha + \beta > - 1.0$, flexibility parameters K(= 0, 10, 100, $10^{20} \cong \infty$); K_ϕ(= 0, 10, 100, $10^{20} \cong \infty$) and $\upsilon_\theta = 0.3$. All natural frequencies obtained from Ritz method are upper bounds of the exact ones and therefore convergence should be monotonic from above as the number of terms of admissible functions increases. The convergence graph for clamped plate is shown in Fig. 1 for $\zeta = 0.5$, $\alpha = - 0.3$, $\beta = - 0.2$, p = 5.0. It is observed that 11 terms of admissible function give first three frequency parameters at least accurate to four significant digits.

The numerical results are presented in Figures (2-5). Figure 2 shows the plots for frequency parameter Ω versus thermal gradient ζ for different values of rigidity ratio p(= 0.5, 1.0, 2.0) for clamped, simply supported and free plate for first mode of vibration. It is observed that frequency parameter Ω decreases with increasing values of thermal gradient ζ. It can be seen that the effect of orthotropy decreases in the order of plates free, simply supported and clamped. Figures 3 and 4 depict the variation of frequency parameter Ω with taper parameters a and ß for clamped, simply supported and free plate for first mode of vibration. It is found that the frequency increases with increasing values of α as well as β. The rate of increase of Ω with α as well as β for free plate is higher than that for simply supported plate and less than that for clamped plate. Also, the frequency for linearly as well as parabolically tapered plate is smaller than that for quadratically tapered plate for same values of taper parameters. The behavior of Ω with respect to above parameters for plates vibrating in second and third mode is same except that the rate of change is higher as compared to first mode (Figures not given for sake of brevity).

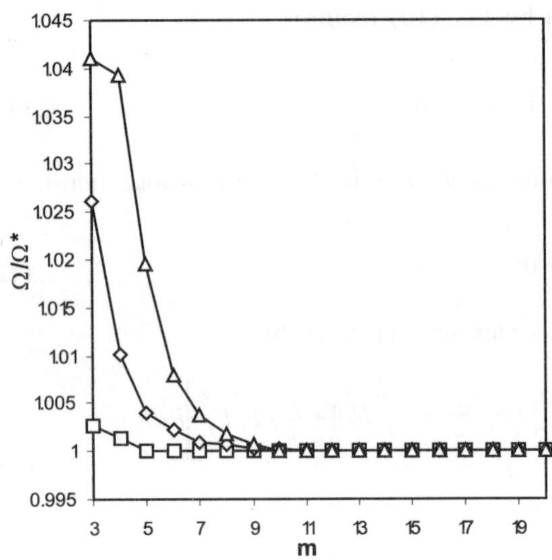

Fig. 1. Normalized frequency parameter for p = 5.0, ξ = 0.5, α = − 0.3, β = − 0.2 for clamped plate. –□–, first; –◇– , second; − Δ − , third mode. Ω^* − the frequency using 20 terms.

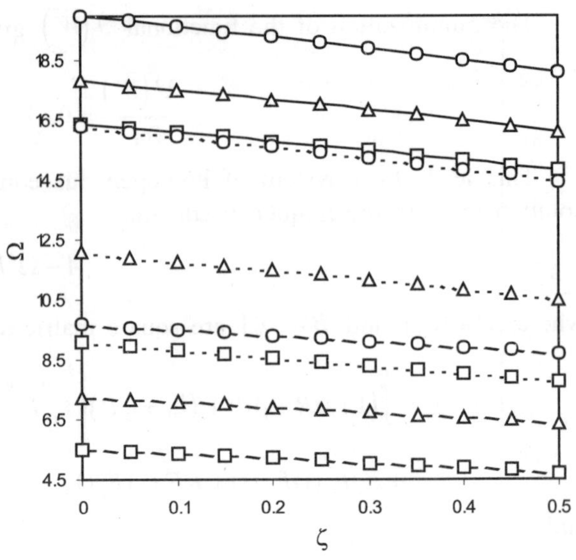

Fig. 2. Frequency parameter of plates vibrating in first mode for α = 0.5, β = 0.5. —, clamped; – – –, SS; ------, free. □, p = 0.5; Δ, p = 1.0; O, p = 2.0.

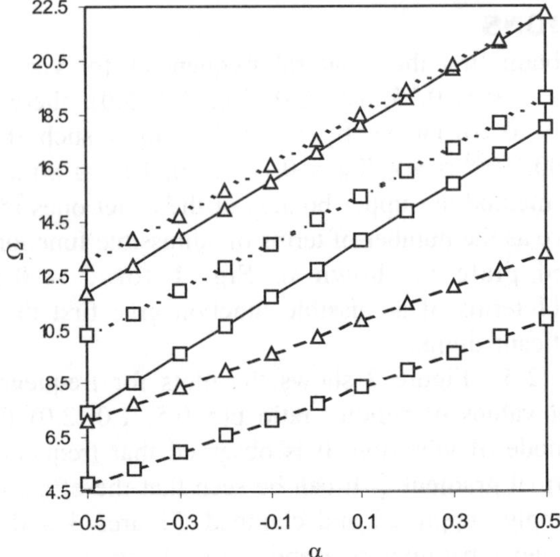

Fig. 3. Frequency parameter of plates vibrating in first mode for ζ = 0.5, p = 5.0. —, clamped; – – –, SS; ------, free. □, β = 0.0; Δ, β = 0.5.

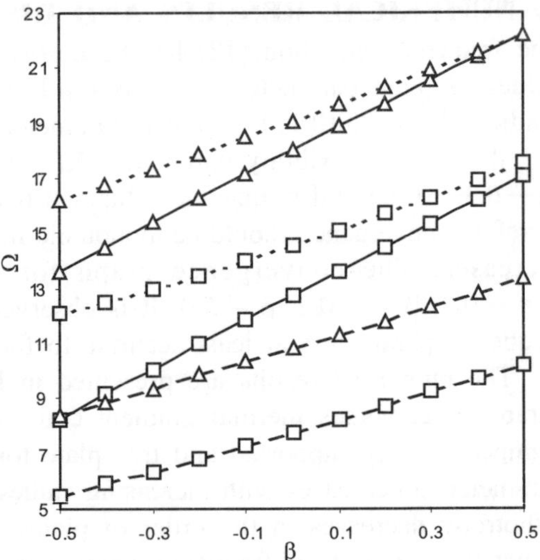

Fig. 4. Frequency parameter of plates vibrating in first mode for ? = 0.5, p = 5.0. —, clamped; – – –, SS; ------, free. □, α = 0.0; Δ, α = 0.5.

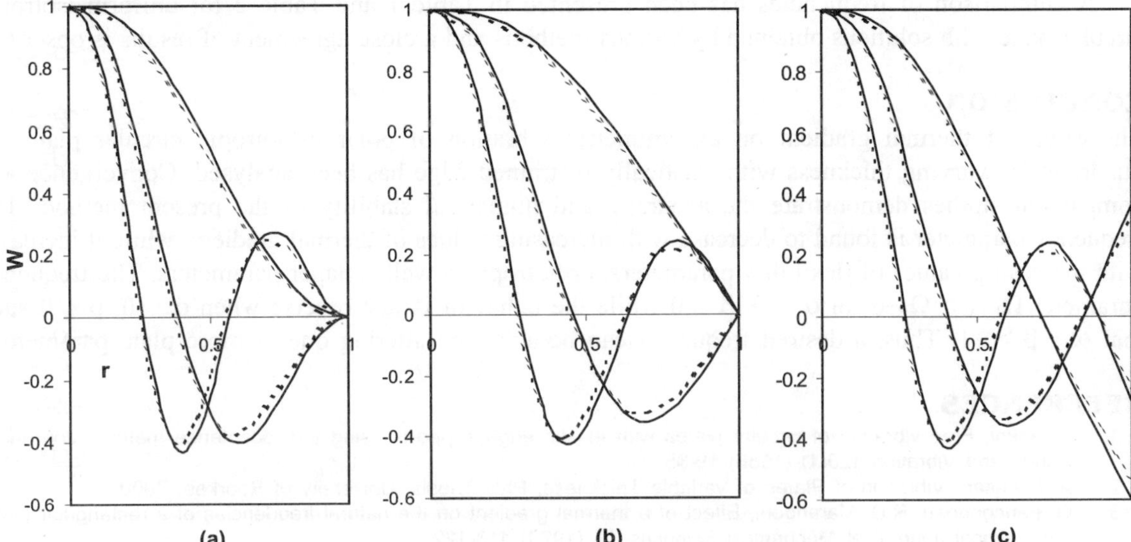

Fig. 5. Normalized displacement for **(a)** clamped **(b)** simply supported and **(c)** free plates for $\alpha = 0.5$, $\beta = 0.5$, $p = 5.0$. ——————, $\zeta = 0.0$; - - - - - - - -, $\zeta = 0.5$.

Figure 5(a, b, c) shows the plots for normalized displacements for first three modes of vibration for clamped, simply supported and free plates respectively. It is observed that effect of thermal gradient decreases the radii of nodal circles.

Table 1. Comparison of frequency parameter O for uniform isotropic circular plate

Method	Kϕ / Mode	K = 0				K = ∞			
		0	10	100	∞	0	10	100	∞
a		9.0030	13.5130	14.5390	14.6820	4.9350	8.7520	10.0190	10.2160
b	I	9.0031	13.5129	14.5388	14.6820	4.9351	9.7519	10.0192	10.2158
present		9.0031	13.5129	14.5388	14.6820	4.9351	8.7519	10.0192	10.2158
a		38.4430	45.7640	48.7460	49.2180	29.7200	35.2180	39.0290	39.7710
b	II	38.4432	45.7643	48.7457	49.2159	29.7200	35.2190	39.0288	39.7711
present		38.4432	45.7643	48.7457	49.2185	29.7200	35.2190	39.0288	39.7711
a		87.7490	97.0450	102.5180	103.5000	74.1560	80.6850	87.4880	89.1030
b	III	87.7502	97.0428	102.5204	103.4995	74.1560	80.6869	87.4900	89.1041
present		87.7502	97.0428	102.5204	103.4995	74.1561	80.6870	87.4901	89.1041

[a]values obtained by Azimi [1].   [b]values obtained by Ansari [2].

Table 2. Comparison of frequency parameter O for uniform isotropic circular plate

Method	Clamped			S-S		
F.E.M.	10.2159	39.7766	89.1708	4.9352	29.7222	74.1938
receptence	10.2160	39.7710	89.1030	4.9350	29.7200	74.1560
Ritz	10.2158	39.7711	89.1041	4.9351	29.7200	74.1560
Exact	10.2158	39.7711	89.1041	4.9352	29.7200	74.1561
present	10.2158	39.7711	89.1041	4.9351	29.7200	74.1561

[a]values obtained by Pardoen [10].   [b]values obtained by Azimi [1].
[c]values obtained by Ansari [2].   [d]values obtained by Leissa [9].

A comparison of frequencies has been presented in Table 1 and Table 2 for uniform isotropic circular plate with solutions obtained by various methods and a close agreement of results is observed.

CONCLUSION

The effect of thermal gradient on axisymmetric vibration of polar orthotropic circular plates of quadratically varying thickness with elastically restrained edge has been analysed. Convergence and comparison studies demonstrate the accuracy and numerical stability of the present method. The frequency parameter is found to decrease with increasing values of thermal gradient, while it increases with increasing values of flexibility parameters, orthotropy as well as taper parameters. The frequency parameter $\Omega_{LTV} > \Omega_{PTV}$ for $\alpha > 0$, $\beta > 0$ while the behavior is just reverse when $\alpha < 0$, $\beta < 0$ such that $\alpha + \beta > -1$. Thus, a desired frequency can be achieved altering one or more plate parameters.

REFERENCES

1. S. Azimi, Free vibration of circular plates with elastic edge supports using the receptence method, Journal of Sound and Vibration 120(1) (1988) 19-35.
2. A.H. Ansari, Vibration of Plates of Variable Thickness, PhD Thesis, University of Roorkee, 2000.
3. G. Fauconneau, R.D. Marangoni, Effect of a thermal gradient on the natural frequencies of a rectangular plate, International Journal of Mechanical Sciences 12 (1970) 113-122.
4. W. Nowacki, Thermoelasticity, Pergamon Press, New York, 1962.
5. N. Ganesan, M.S. Dhotarad, Influence of a thermal gradient on the natural frequencies of tapered orthotropic plates, Journal of Sound and Vibration 66(4) (1979) 621-625.
6. J.S. Tomar, A.K. Gupta, Thermal effect on axisymmetric vibration of an orthotropic circular plate of variable thickness, AIAA Journal 22(7) (1984a) 1015-1017.
7. D.G. Gorman, Thermal gradient effects upon the vibrations of certain composite circular plates, Part I, Journal of Sound and Vibration 101 (1985a) 325-336.
8. H.N. Arafat, A.H. Nayfeh, W. Faris, Natural frequencies of heated annular and circular plates, International Journal of Solids & Structures 41 (2004) 3031-3051.
9. A.W. Leissa, Vibration of Plates, NASA SP-160, Washington-DC, 1969.
10. G.C. Pardoen, Asymmetric vibration and stability of circular plates, Computers and Structures 9 (1978) 89-95.

77

Stability and Vibration Behavior of Composite Shell Panels

J. GIRISH[1] AND L.S. RAMACHANDRA[2]

[1]Assistant Professor, Department of Civil Engineering, Bapatla Engineering College, Bapatla-522 101, India. email: girish_iitkgp@yahoo.co.in
[2]Associate Professor, Department of Civil Engineering, Indian Institute of Technology, Kharagpur-721 302, India. email: lsr@civil.iitkgp.ernet.in

ABSTRACT

The work presented in this paper describes the buckling, postbuckling and small amplitude vibration behavior about the prebuckling and postbuckling equilibrium states of laminated composite shallow shell panels, subjected to mechanical compressive edge loads. The structural model used in the present study is based on a higher-order transverse shear deformation theory of shallow shells that includes the von Kármán-type geometric nonlinearities and initial geometric imperfections. The solutions to the governing nonlinear partial differential equations are sought using the multi-term Galerkin's method. A convergence study has been carried out by taking various numbers of terms in the displacement field approximation. The equilibrium configurations for a given shell panel are obtained using a Newton-Raphson procedure in conjunction with arc-length procedure. The small amplitude free vibration frequencies in prebuckling and in postbuckling range of a shell panel are obtained solving the associated linear eigenvalue problem.

1. INTRODUCTION

The buckling and postbuckling behavior of laminated panels (plate/shell) are investigated extensively and reported in the literature. Librescu *et al.* [1–3] presented the analytical results of simply supported single-layer flat and curved panels made from transversely isotropic materials. The authors solved the nonlinear boundary-value problem using Airy's stress function and one-term Galerkin approximation. The research on the postbuckling of plates and shells reported in the literature (Girish and Ramachandra [4, 5]) shows that the multi-term Galerkin method gives better results than the one-term solution. The work presented in this paper focuses on the postbuckling and postbuckled vibrations of laminated composite shell panels using multi-term Galerkin's method for a wide range of loading and structural parameters.

Adopting Galerkin's procedure, the governing nonlinear partial differential equations are converted into a set of nonlinear algebraic equations in the case of postbuckling analysis and nonlinear ordinary differential equations in the case of free vibration analysis. The critical buckling load is obtained from the solution of linear eigenvalue problem. The postbuckled equilibrium paths are obtained by solving the nonlinear algebraic equations, using the Newton-Raphson iterative procedure. In the case of free vibration analysis of buckled panel, the solution of differential equations is assumed to be the sum of time dependent solution (vibration amplitude) and time independent solution (postbuckling deflection). The vibration amplitude is considered small compared to the postbuckled deflection and hence higher order time dependent terms are neglected. The free vibration frequencies of a postbuckled panel about the static equilibrium state are obtained by solving the eigenvalue problem for different postbuckled deflections. Numerical results are presented for symmetric [0/90/0] and antisymmetric [0/90] cross-ply, composite shell panels under simply supported boundary conditions.

2. FORMULATION

Consider a doubly curved shell on rectangular planform of constant thickness h composed of a finite number of orthotropic layers of uniform thickness. The coordinate system is such that the surface coordinates (x, y) of the orthogonal coordinate system (x, y, z) are located on the middle surface of the laminate, and are coincident with the lines of principal curvature; the z coordinate is normal to the middle plane; R_x and R_y are the principal radii of curvature of the middle surface. The displacement fields used in the present study are (Reddy and Liu [6]):

$$u = \left(1 + z/R_x\right)u^o + z\varphi_1 + z^3(4/3h^2)[-\varphi_1 - w_{,x}^o]$$

$$v = \left(1 + z/R_y\right)v^o + z\varphi_2 + z^3(4/3h^2)[-\varphi_2 - w_{,y}^o] \qquad \text{...(1)}$$

$$w = w^o$$

Here u, v, w are displacement components respectively along x, y, z directions; u^o, v^o, w^o are the displacements of a generic point on the mid-plane; and φ_1 and φ_2 are the rotations of the cross-sections perpendicular to the x and y axes, respectively.

In the present theory the governing equations of shallow shell are established using Love's first order geometric approximation (neglecting z/R_x and z/R_y in comparison with unity) and Reissner's shallow shell simplifications [Kraus (7)], which are identical to Donnell's assumptions for cylindrical shells [i.e. (i) the transverse shearing force makes a negligible contribution to the equilibrium of forces in the circumferential direction and (ii) neglecting the tangential displacements and their derivatives for the midsurface changes in curvature and twist] The above displacement fields can be rearranged as:

$$u = \left(1 + z/R_x\right)u^o - zw_{,x}^o + f(z)\phi_1 \quad v = (1 + z/R_y)v^o - zw_{,y}^o + f(z)\phi_2 \qquad \text{...(2)}$$

where $\phi_1 = \varphi_1 + w_{,x}^o$; $\phi_2 = \varphi_2 + w_{,y}^o$; $f(z) = z[1 - (4/3)(z/h)^2]$

The introduction of $f(z)$ in the displacement field reduces certain higher-order moment and transverse shear force resultants. These are due to the particular form of the proposed displacement field. It is apparent that the unknown functions ϕ_1 ($\varphi_1 + w_{,x}^o$) and ($\varphi_2 + w_{,y}^o$) represent the action of transverse shear strains on the shell middle surface. This action, by means of the first partial derivatives of ϕ_1 and ϕ_2, gives rise to some additional changes of curvature, ϕ_1 and ϕ_2, and twist ($\phi_{1,x} + \phi_{1,y}$), of the shell middle surface.

The stress resultants can be defined as:

$$\left(\left(\begin{array}{c} N_x \\ N_y \\ N_{xy} \end{array} \right), \left(\begin{array}{c} M_x \\ M_y \\ M_{xy} \end{array} \right), \left(\begin{array}{c} P_x \\ P_y \\ P_{xy} \end{array} \right) \right) = \int_{-h/2}^{h/2} \left(\begin{array}{c} \sigma_x \\ \sigma_y \\ \sigma_{xy} \end{array} \right) \left(1, \; z, \; f(z) \right) \; dz \qquad \text{...(3)}$$

and $\qquad (V_{xz}, V_{yz}) = \int_{-h/2}^{h/2} (\sigma_{xz}, \sigma_{yz}) f'(z) dz \qquad$ where $\; f'(z) = \dfrac{d}{dz} f(z) \qquad$ ...(4)

where N_x, N_y, N_{xy}, and M_x, M_y, M_{xy} are the force and moment resultants; P_x, P_y, P_{xy} are additional moment resultants due to additional changes of curvature, $\phi_{1,x} \; \phi_{2,y} \; (\phi_{2,x} + \phi_{1,y})$; V_{xz}, V_{yz} are transverse shear force resultants.

The equations of motion of the shell derived from the Hamilton's principle can be written as:

$$N_{x,x} + N_{xy,y} = 0$$
$$N_{xy,x} + N_{y,y} = 0$$
$$M_{x,xx} + 2M_{xy,xy} + M_{y,yy} - (N_x / R_x) - (N_y / R_y) +$$
$$N_x w,_{xx} + 2N_{xy} w,_{xy} + N_y, w,_{yy} + q = \rho h w_{,tt}^o \qquad \text{...(5)}$$
$$P_{x,x} + P_{xy,y} - V_{xz} = 0$$
$$P_{xy,x} + P_{y,y} - V_{yz} = 0$$

where $(\;)_{,x}$ denotes partial differentiation with respect to x; q is the distributed transverse load; ρ is the mass per unit area of the shell. Expressing the stress resultants in terms of displacements, the governing equilibrium equations are obtained in displacement variables (Girish and Ramachandra [5]).

3. SOLUTION PROCEDURE

3.1 Nonlinear Static Response

Let a and b denote lengths of a cylindrical shell panel ($R_x = \infty$, $R_y = R$) along x and y directions respectively. It is assumed that the cylindrical shell considered is subjected to the following set of simply supported boundary conditions:

$$N_x = v^o = w^o = P_x = \phi_2 = M_x = 0 \; at \; x = 0, a$$
$$u^o = N_y = w^o = \phi_1 = P_y = M_y = 0 \; at \; y = 0, b \qquad \text{...(6)}$$

The displacement fields appropriate to simply supported boundary conditions are represented as:

$$u = \sum_{m=1}^{i} \sum_{n=1}^{j} U_{mn} \cos\left(\frac{m\pi x}{a}\right) \sin\left(\frac{n\pi y}{b}\right); \quad \phi_1 = \sum_{m=1}^{i} \sum_{n=1}^{j} \alpha_{mn} \cos\left(\frac{m\pi x}{a}\right) \sin\left(\frac{n\pi y}{b}\right)$$

$$v = \sum_{m=1}^{i} \sum_{n=1}^{j} V_{mn} \sin\left(\frac{m\pi x}{a}\right) \cos\left(\frac{n\pi y}{b}\right); \quad \phi_2 = \sum_{m=1}^{i} \sum_{n=1}^{j} \beta_{mn} \sin\left(\frac{m\pi x}{a}\right) \cos\left(\frac{n\pi y}{b}\right) \qquad \text{...(7)}$$

$$w = \sum_{m=1}^{i} \sum_{n=1}^{j} W_{mn} \sin\left(\frac{m\pi x}{a}\right) \sin\left(\frac{n\pi y}{b}\right)$$

where i and j denote the number of modes/terms associated in x and y directions in the multi-term Galerkin's method. Applying Galerkin procedure, one obtains a system of nonlinear algebraic equations in constant coefficients U_{mn}, V_{mn}, W_{mn}, α_{mn} and β_{mn}. The critical buckling loads are obtained from the solution of the linear eigenvalue problem. Using the Newton-Raphson method in conjunction with the Riks approach; the system of nonlinear algebraic equations is solved for deflections. The nonlinear algebraic equations based on multi-term Galerkin procedure are not presented for the sake of brevity.

3.2 Small Amplitude Vibration About a Static Equilibrium State

For the free vibration analysis about a static equilibrium state, the unknown modal amplitudes are assumed to be the sum of time-independent and time-dependent solution, which may be written as:

$$\{w\} = \{w_s\} + \{w_t\} \qquad ...(8)$$

where $\{w_s\}$ is the static deflection and $\{w_t\}$ is the dynamic deflection about a static equilibrium state. In the case of small amplitude vibration $\{wt\}^2 << \{w_s\}$, hence higher order time dependent terms are neglected. Substituting the displacement field (Eq. 7) into the governing partial differential equations and adopting the Galerkin's technique, we obtain a set of ordinary differential equations. The equation of motion may be written as:

$$\left([M]\{\ddot{w}_t\} + [K] + [N]\right)\{w_t\} = 0 \qquad ...(9)$$

where $[M]$ is the mass matrix; $[K]$ is the linear elastic stiffness; $[N]$ is the nonlinear stiffness due to large deformation. The nonlinear stiffness matrix $[N]$ is a function of only time-independent amplitude (large deformation). For the free vibration analysis at frequency 'ω', the following one-harmonic approximation is assumed to solve the ordinary differential equations (9):

$$\{w_t\} = \{\tilde{w}_t\}\sin \omega t \qquad ...(10)$$

In the analysis, the tangential and rotary inertia terms are neglected and consequently the in-plane displacements become a function of $\sin^2\omega t$. Now condensing the equations of motion, the standard eigenvalue problem is obtained. By substituting the converged static deflection values obtained from the nonlinear static analysis, the free vibration frequencies are obtained by solving the linear eigenvalue problem.

4. RESULTS AND DISCUSSION

Numerical results are presented for simply supported, cross-ply cylindrical shell panels ($R_x = \infty$, $R_y = R$). The following lamina material properties are used (Reddy and Liu [6]) in the analysis:

$$E_1 = 25E_2, \quad G_{12} = G_{13} = 0.5E_2, \quad G_{23} = 0.2E_2, \quad v_{12} = 0.25, \quad \alpha_2/\alpha_1 = 3, \quad a/b = 1$$

Figure 1 shows the postbuckled equilibrium paths of a three-layered [0/90/0] symmetric cross-ply cylindrical shell panel ($R_y/a = 10$, $a/h = 10$) under uni-axial edge compression (N^*). The equilibrium paths are traced by taking 1-term ($m = n = 1$), 3-term (($m = n = 1$), ($m = 1$, $n = 3$), ($m = 3$, $n = 1$)) and 4-term (($m = n = 1$), ($m = 1$, $n = 3$), ($m = 3$, $n = 1$), ($m = n = 3$)) in the displacement fields (see Eq. (7)). It can be seen that the 3-term and 4-term solution compare well.

However, the discrepancy between the 1-term and 3-term results is noticeable. At a load level $N^* = 2$, the difference in the central deflections as calculated by considering 1-term and 3-term solutions is 12.11%, whereas at $N^* = 3$, the difference is 22.45%, whereas the difference between 3-term and 4-term solution solutions is less than 1%. Thus, three terms are sufficient to obtain accurate results for the problem under consideration.

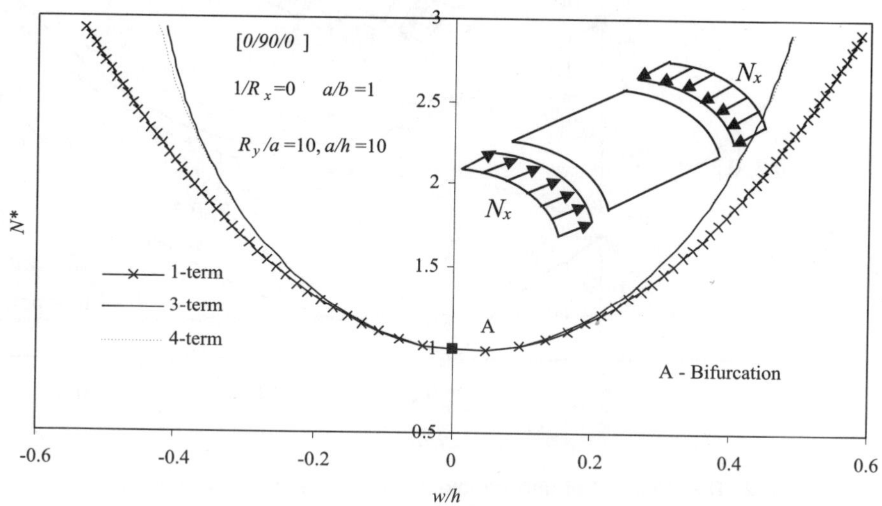

Fig. 1. Comparison of 1-term, 3-term and 4-term solutions for the postbuckling behavior of a composite cylindrical shell panel [0/90/0] under uni-axial edge compression.

Figure 2 shows the influence of uniform lateral pressure and bi-axial edge load on the postbuckling behavior of symmetric cross-ply shell panel. The load ratio No (N_y/N_x) is the ratio of compressive (positive) or tensile (negative) edge load in the y-direction ($\pm N_y$) and compressive edge load in the x-direction (N_x) respectively. In this study, the lateral pressure is kept constant whereas the edge load is varied. It can be observed that the initial ($N^* = 0$) positive (inward) and negative (outward) deflections of the panel are due to positive and negative lateral pressure respectively. For the case $q^* = 0.2$, No = 0.2, the results shows that as soon as mechanical edge load is applied, the positive deflection decreases and with the increase of mechanical edge load the displacements transit to negative deflection. Whereas, for $q^* = 0.2$, No = -0.2, the initial positive deflection increases monotonically with the increase of edge load, the shell panel shows a stiffening behavior. In the case of $q^* = -0.2$, No = -0.2, the initial negative deflection transit to positive deflection. However, for $q^* = -0.2$, No = -0.1, the shell panel exhibit a limit point instability response. For the case $q^* = -0.2$, No = 0.2, the initial negative deflection increases monotonically with the increase of edge load. The frequency results presented in Fig. 3 are corresponding to the static equilibrium paths shown in Fig. 2. The decrease in the fundamental frequency is due to flattening of deformed surface.

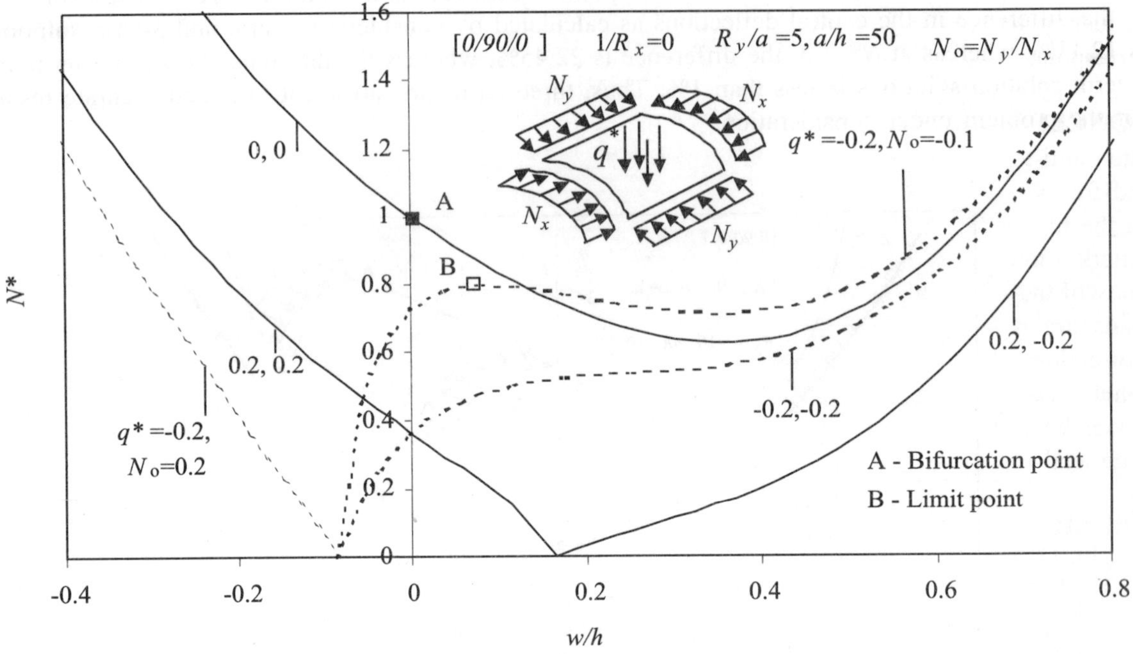

Fig. 2. The influence of uniform lateral pressure and bi-axial edge load
on postbuckling behavior of cross-ply cylindrical shell panel.

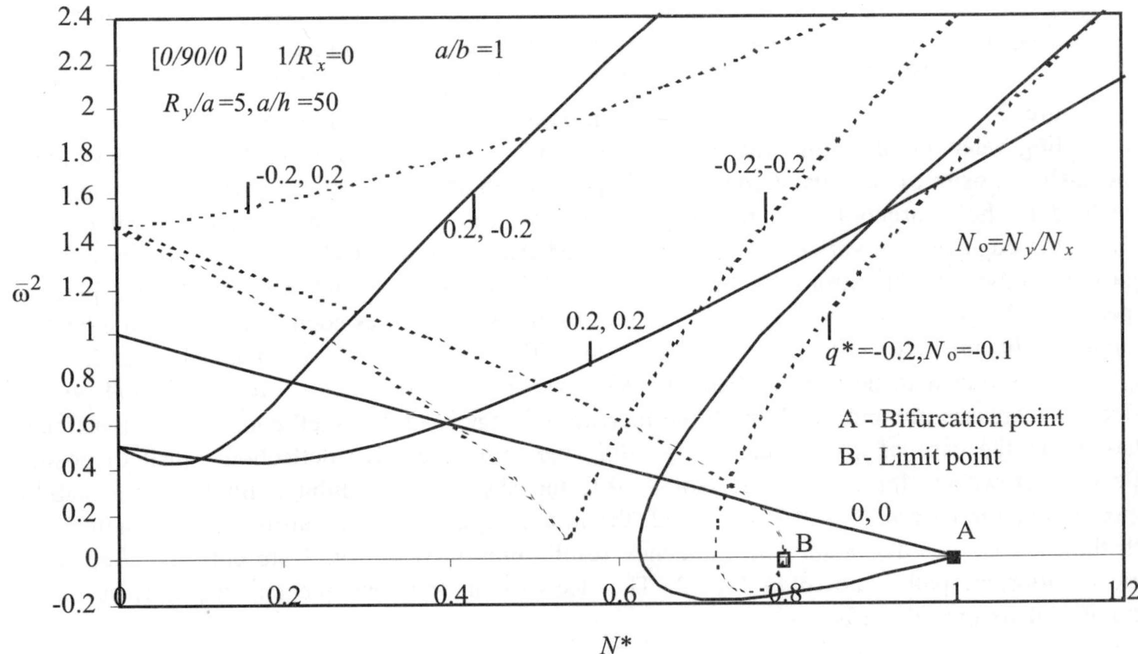

Fig. 3. The influence of uniform lateral pressure and bi-axial edge load
on fundamental frequencies of cross-ply cylindrical shell panel.

After flattening, the deformed surface begins to develop curvature as a result the fundamental frequency increases with the increase of bending stiffness. The bifurcation point shown in Fig. corresponds to critical buckling load of a uni-axially compressed shell.

CONCLUSION

The multi-term Galerkin method is used to obtain analytical solutions for the nonlinear static response and free vibration behavior of simply supported cross-ply laminated cylindrical shell panels, based on the higher-order transverse shear deformation theory. It is observed from results that multi-term Galerkin method gives better results than the single-term solution. Numerical results of a parametric study of the nonlinear static response and vibration behavior of curved panels subjected to mechanical loads are presented. When the cylindrical panel is subjected to uni-axial compression in the x-direction, the postbuckled equilibrium path is asymmetric about the bifurcation point and the panel is sensitive to the magnitude and direction of the lateral load and edge load in the y-direction. It can be concluded that, by suitably adjusting the geometric parameters and loading conditions beneficial results can be obtained.

REFERENCES

1. L. Librescu, W. Lin, M.P. Nemeth, and J.H. Starnes Jr, 1996, Frequency-load interaction of geometrically imperfect curved panels subject to heating, AIAA Journal, 34(1), 166-177.
2. L. Librescu, W. Lin, M.P. Nemeth, and J.H. Starnes Jr, 1996, Vibration of geometrically imperfect panels subjected to thermal and mechanical loads, Journal of Spacecraft and Rockets, 33(2), 285-291.
3. L. Librescu, and W. Lin, 1997, Vibration of thermomechanically loaded flat and curved panels taking into account geometric imperfections and tangential edge restraints, Int. Journal of Solids and Structures, 34(17), 2161-2181.
4. J. Girish, and L.S. Ramachandra, 2005, Thermomechanical postbuckling analysis of symmetric and antisymmetric composite plates with imperfections, Composite Structures, 67(4), 453-460.
5. J. Girish, and L.S. Ramachandra, 2006, Thermomechanical postbuckling analysis of cross-ply laminated cylindrical shell panels, Journal of Engineering Mechanics, 132(2), 133-140.
6. J.N. Reddy, and C.F. Liu, 1985, A higher-order shear deformation theory of laminated elastic shells, Int. J. Engineering Science, 23(3), 319-330.
7. H. Kraus, Thin Elastic Shells, John Wiley and Sons, Inc., New York.
8. A. Khdeir, M.D. Rajab, and J.N. Reddy, 1992, Thermal effects on the response of cross-ply laminated shallow shells, Int. Journal of Solids and Structures, 29, 653-667.

78

Second Order Statistics of Buckling of Laminated Composite Plate Supported on Elastic Foundation with Uncertain Foundation Stiffness Parameters

ACHCHHE LAL[1], B.N. SINGH[2] AND RAKESH KUMAR[3]

[1]Research Scholar, Department of Applied Mechanics, MNNIT Allahabad-211 004, India.
[2]Assistant Professor, Department of Aerospace Engineering, IIT Kharagpur-721302, India.
email: bnsingh@aero.iitkgp.ernet.in.
[3]Sr. Lecturer, Department of Applied Mechanics, MNNIT Allahabad-211 004, India.

ABSTRACT

The present work investigates the effect of randomness of foundation stiffness parameters on the initial buckling of laminated composite plates which are resting on elastic foundation and subjected to uniform in-plane edge compression. The scatter in the buckling load is due to variation in random foundation stiffness parameters. The plate foundation interaction is taken into account in the analysis through two-parameter Pasternak model. The uncertain foundation stiffness parameters are modelled as independent random variables. Higher order shear deformation theory has been used to model the displacement field. A C° finite element method in conjunction with a mean cantered first order perturbation technique (FOPT) is used to evaluate the stochastic characteristics of buckling load. Typical numerical results are presented for laminate having all edges clamped to show the influence of variation in foundation stiffness parameters, plate side to thickness ratio and plate aspect ratio on the second-order statistics of buckling loads. The results have been validated with independent Monte Carlo simulation.

Keywords: Hardfacing, MMAW process, ferritic stainless steel, martensitic stainless steel.

1. INTRODUCTION

Plates on elastic foundation have been extensively used in structural and applied mechanics field, such as aerospace, structure operational activities of large transport air crafts on runways, foundation of deep wells. Many studies have been carried out to investigate the buckling response of laminated composite plate resting on elastic foundation on the basis of deterministic analysis [1]. However, limited literature is available on such type structures with random material properties [2-3]. To the best of author's knowledge, no work dealing with buckling analysis of laminated composite plate resting on elastic foundation based on higher order shear deformation theory with random foundation stiffness parameters has been reported in the literature.

2. FORMULATION

Consider a rectangular laminated composite plate of length a, width b, and thickness h, which consist of N number of orthotropic layers (Fig. 1). All orthotropic layers of the composite plate are of uniform thickness. The mid plane of the plate considered the reference plane. The thickness coordinates z of the top and bottom surfaces of any (k^{th}) layer are denoted by $h^{(k+1)}$ and h^k, respectively. The fibers of kth layer are oriented at an angle θ_k to the x-axis. The plate is assumed to attach to the foundation so that no separation takes place in the process of deformation. The load displacement relation between the plate and the supporting foundation

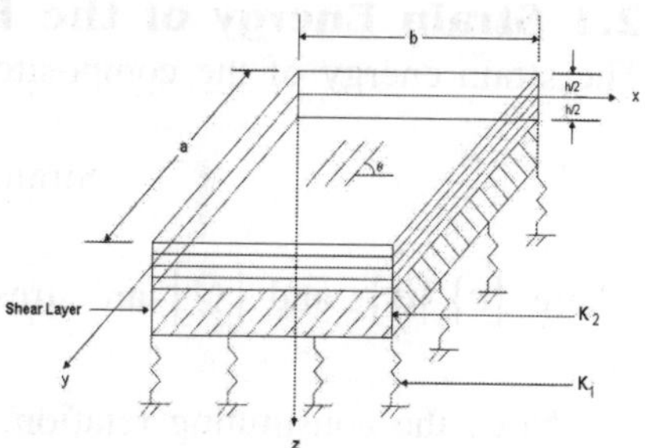

Fig. 1.

fallows the two- parameter Pasternak model [1]. In the Pasternak type foundation modeling, influence of the shear effect on the foundation is inserted into the formulation by a shear layer K_2 beside the vertical spring element K_1.

The load-displacement relation for this model is as follows:

$$P = K_1 w - K_2 \nabla^2 w ; \quad \text{where,} \quad \nabla^2 = \frac{\partial}{\partial^2 x} + \frac{\partial}{\partial^2 y} \qquad ...(1)$$

where K_1 is Winkler foundation stiffness, P is density of reaction forces of foundation, and K_2 is Pasternak foundation stiffness. As a special case when shear layer is neglected ($K_2 = 0$) in the formulation, the Pasternak model converge to Winkler model.

Displacement field can be assumed as [4]:

$$\bar{u} = u + z\psi_x - z^3 4/3h^2(\psi_x + \partial w/\partial x) = u + f_1(z)\psi_x + f_2(z)\partial w/\partial x$$

$$\bar{v} = v + z\psi_y - z^3 4/3h^2(\psi_y + \partial w/\partial y) = v + f_1(z)\psi_y + f_2(z)\partial w/\partial y; \quad \bar{w} = w \qquad ...(2)$$

where, $\bar{u}$ $\bar{v}$ and $\bar{w}$ denote the displacements of a point along the (x, y, z) coordinates and u, v, and w are corresponding displacements of a point on the mid plane. ψ_x and ψ_y are the rotations of normal to the mid plane about the y-axis and x-axis respectively.

A C^o continuous finite element model having seven degrees of freedom (DOF) have been used for the buckling analysis of laminated composite plates together with HSDT. A C^o finite element formulation by assuming the slopes as a separate degree of freedom for the plate analysis has been used [3].

The strain displacements relations are obtained by using small deformation theory [4]

$$\varepsilon_{xx} = \varepsilon_1 = \partial\bar{u}/\partial x = \varepsilon_1^0 + z(k_1^0 + z^2 k_1^2); \; \varepsilon_{yy} = \varepsilon_2 = \partial\bar{v}/\partial x = \varepsilon_2^0 + z(k_2^0 + z^2 k_2^2)$$

$$\gamma_{xy} = \varepsilon_6 = \partial\bar{u}/\partial y + \partial\bar{v}/\partial x = \varepsilon_6^0 + z(k_6^0 + z^2 k_6^2); \; \gamma_{yz} = \varepsilon_4 = \partial\bar{v}/\partial z + \partial\bar{w}/\partial y = \varepsilon_4^0 + z^2 k_4^2$$

$$\gamma_{xz} = \varepsilon_5 \; \partial\bar{u}/\partial z + \partial\bar{w}/\partial x = \varepsilon_5^0 + z^2 k_5^2 \qquad ...(3)$$

The linear stress-strain relationship for an orthotropic layer is given by Reddy [4].

$$\{\sigma\} = \left[\overline{Q}\right]\{\varepsilon\} \qquad\qquad \text{...(4)}$$

2.1 Strain Energy of the Plate

The strain energy of the composite plate can be written as.

$$\text{Strainl energy } (U_I) \;=\; \frac{1}{2}\iiint_V \{\varepsilon\}\{\sigma\}\,dV \qquad\qquad \text{...(5a)}$$

where $\{\varepsilon\}\,\{\sigma\}$; and $\left[\overline{Q}\right]$ are stress vector, strain vector and reduced stiffness matrix, respectively [1].

Now, the constituting relation, the strain energy can be written as

$$U_1 = \frac{1}{2}\int_A^{-T} \varepsilon D \overline{e}\, dA\, dV; \qquad\qquad \text{...(5b)}$$

where $\{\overline{\varepsilon}\} = \left(\varepsilon_1^0 \;\; \varepsilon_2^0 \;\; \varepsilon_6^0 \;\; k_1^0 \;\; k_2^0 \;\; k_6^0 \;\; k_1^2 \;\; k_2^2 \;\; k_6^2 \;\; \varepsilon_4^0 \;\; \varepsilon_5^0 \;\; k_4^2 \;\; k_5^2 \right)^T$;

$$D = \sum_{k=1}^{NL}\int_{k-1}^{z\,k} [T]^T\left[\overline{Q}\right][T]\,dz = \begin{bmatrix} [A_1] & [B] & [E] & 0 & 0 \\ [B] & [C_1] & [F_1] & 0 & 0 \\ [E] & [F_1] & [H] & 0 & 0 \\ 0 & 0 & 0 & [A_2] & [C_2] \\ 0 & 0 & 0 & [C_2] & [F_2] \end{bmatrix} \text{ with }$$

$$\left(A_{1_{ij}},\, B_{ij},\, C_{1_{ij}},\, E_{ij},\, F_{1_{ij}},\, H_{ij}\right) = \sum_{k=1}^{NL}\int_{k-1}^{z\,k} \overline{Q}_{ij}^{(k)}\left(1, z, z^2, z^3, z^4, z^6\right)dz \text{ for } i, j = 1,\, 2,\, 6$$

$$\left(A_{2_{ij}},\, C_{2_{ij}},\, F_{2_{ij}}\right) = \sum_{k=1}^{NL}\int_{k-1}^{z\,k} \overline{Q}_{ij}^{(k)}\left(1, z^2, z^4\right)dz \text{ for } i, j = 4,\, 5$$

After summing over total number of element NE, strain energy of vibrating plate

$$U_1 = \sum_{e=1}^{NE} U^{(e)} = \frac{1}{2}\sum_{e=1}^{NE}\left\{\varepsilon\right\}^{(e)^T}[D]\{\varepsilon\}^{(e)} = \sum_{e=1}^{NE}\{q\}^{(e)^T}[K]\{q\}^{(e)} \qquad\qquad \text{...(6)}$$

2.2 Strain Energy due to Foundation

The strain energy of the laminate due to foundation can be expressed as

$$U_2 = \frac{1}{2}\int_A \begin{Bmatrix} w \\ \dfrac{\partial w}{\partial x} \\ \dfrac{\partial w}{\partial y} \end{Bmatrix}^T \begin{bmatrix} K_1 & 0 & 0 \\ 0 & K_2 & 0 \\ 0 & 0 & K_2 \end{bmatrix} \begin{Bmatrix} w \\ \dfrac{\partial w}{\partial x} \\ \dfrac{\partial w}{\partial y} \end{Bmatrix}^T dA;$$

Summing over total number of element NE, potential energy of vibrating plate

$$U_2 = \sum_{e=1}^{NE} U^{(e)} = \frac{1}{2} \sum_{e=1}^{NE} \{B_g\}^{(e)^T} [D_f] \{B_g\}^{(e)} = \sum_{e=1}^{NE} \{q\}^{(e)^T} [K_f] \{q\}^{(e)} \qquad ...(7)$$

2.3 External Work Done

The work done by the in plane forces in producing out of plane displacement 'w' in the domain of small displacement is,

$$W = \frac{1}{2} \left\{ \begin{array}{c} \dfrac{\partial w}{\partial x} \\ \dfrac{\partial w}{\partial y} \end{array} \right\}^T \left[\begin{array}{cc} N_x & N_{xy} \\ N_{xy} & N_y \end{array} \right] \left\{ \begin{array}{c} \dfrac{\partial w}{\partial x} \\ \dfrac{\partial w}{\partial y} \end{array} \right\} dA \qquad ...(8)$$

where, N_x, N_y and N_{xy} are in plane forces.

Summing over total number of element NE, total work done is given by

$$W = \sum_{e=1}^{NE} W^e = \{q\}^{e^T} \lambda^* [k^*] \{q\}^{(e)} \qquad ...(9)$$

3. GOVERNING EQUATION

The governing equation for buckling analysis can be derived using the principle of total potential Energy (TPE). This gives

$$\delta (U + V) = 0 \qquad ...(10)$$

where potential energy $V = -W$

Substituting Eqs. (6), (7) and (9) in Eq. (10), ones obtain as:

$$\{[K_{sf}] - \lambda [K_g]\} \{q^*\} = 0 \quad \text{with } K_{sf} = K_s + K_f \qquad ...(11)$$

where, $\{q^*\} = \sum_{e=1}^{NE} \{q\}^{(e)}$; $[K] = \sum_{e=1}^{NE} [K]^{(e)}$, $[K_f] = \sum_{e=1}^{NE} [K_f]^{(e)}$, $[K_g] = \sum_{e=1}^{NE} [K_g]^{(e)}$, and

also $\{q^*\}, [K], [K_f], [K_g]$ and λ are defined as a global displacement vector, foundation stiffness matrix, global geometric stiffness matrix and eigenvalue respectively.

Equation (11) is random in nature, being dependent on the system material properties. A mean centered first order perturbation technique in conjunction with C^0 finite element method has been used to obtain the solution of governing equations.

4. SOLUTION APPROACH: PERTURBATION TECHNIQUE

We consider a class of problems where the random variation is very small as compared to the mean part of random system properties. Further, it is quit logical to assume that the dispersion in the derived quantities like $[K_f]$, and λ also small to their mean values. In the present analysis, since foundation stiffness parameters are totally dependent on the material properties of the supporting elastic medium which possess random fluctuations. So, randomness in both K_1 and K_2 is taken into consideration.

In general, any arbitrary random variable b_i can be represented as the sum of the mean variable and a zero mean random variable, denoted by superscripts 'd' and 'r', respectively,

$$b_i = b_i^d + b_i^r \qquad \qquad ...(12)$$

Similarly, all of the quantity in Eq. (11) can be broken into

$$\left[K_{fi} \right] = \left[K_{fi}^d \right] + \left[K_{fi}^r \right]; \quad \lambda_i = \lambda_i^d + \lambda_i^r \quad \text{and} \quad \{q_i\} = \{q_i^d\} + \{q_i^r\} \qquad ...(13)$$

where $i = 1, 2,......p$.

Substituting Eq. (13) in Eq. (11) and collecting same order of the magnitude term and keeping only up to first order terms, we obtain

$$\left[K_{fi}^d \right]\{q_i^d\} = \lambda_i^d \{q_i^d\} \qquad \qquad ...(14)$$

$$\left[K_{fi}^d \right]\{q_i^r\} + \left[K_{fi}^r \right]\{q_i^d\} = \lambda_i^r \{q_i^d\} + \lambda_i^d \{q_i^r\} \qquad ...(15)$$

Because Eq. (14), is the deterministic equation relating to the mean values, the mean eigen value and corresponding mean eigen vectors can be determined by convention eigen solution procedures. Using orthogonality properties the Eq. (15) may be decoupled and the same using Taylor series expansion, the following can obtain as [3].

$$Var(\lambda_i) = \sum_{j=1}^{p} \sum_{k=1}^{p} \frac{\partial \lambda_i^{d}}{\partial b_j} \frac{\partial \lambda_i^{d'}}{\partial b_k} Cov\left(b_j^r, b_k^r\right) \qquad ...(16)$$

$$Var\{q_i^l\} = \sum_{j=1}^{p} \sum_{k=1}^{p} \frac{\partial \{q_j^d\}}{\partial b_j} \frac{\partial \{q_k^d\}}{\partial b_k} Cov\left(b_j^r, b_k^r\right) \qquad ...(17)$$

where $Cov\left(b_j^r, b_k^r\right)$ is the cross variance between b_j^r and b_k^r. The standard deviation (SD) is obtained by the square root of the variance.

5. RESULTS AND DISCUSSION

A mean centered first order perturbation technique (FOPT) in conjunction with C° finite element method has been used to obtain the solution of governing equations. Based on convergence (4 × 4) mesh has been used for computation of the results. Random variations of foundation stiffness parameter are incorporated into the prediction of the normalized buckling load of the laminate having side-to-thickness ratio of 10 and 5 respectively.

The following mean values of the material properties are used for the computation.

$$E_{11} = 40E_{22}, \quad G_{12} = G_{13} = 0.6E_{22}, \quad G_{23} = 0.5E_{22}, \quad v_{12} = v_{21} = 0.25, \quad \rho = 1$$

The following non-dimensionalysed buckling load and foundation parameters have been used in this study

$$\lambda_{cr} = \overline{N} \, b^2 / \left(E_{22} \, h^3 \right); \quad k_1 = K_1 b^2 / E_{22} \, h^3; \quad k_2 = K_2 b^2 / E_{22} \, h^3$$

For uni-axial compression, $N_x = 1$; $N_{xy} = 0$; $N_y = 0$ and for bi-axial compression $N_x = 1$; $N_{xy} = 0$; $N_y = 1$ where λ_{cr}, N_x, N_y, N_{xy}, k_1 and k_2 are the dimensionalyzed mean buckling load, applied reference loads in x and y direction, applied reference loads in $x - y$ direction, non-dimensional Winkler foundation stiffness and non-dimensional Pasternak foundation stiffness respectively.

The outlined approach has been validated by comparing the results with independent MCS. Fig. 2 shows the comparison of results obtained using proposed outlined approach with independent MCS for all edges clamped square 4-layer anti-symmetric cross-ply plate resting on elastic foundation. In this case only foundation stiffness parameters k_1 and k_2 are taken random. The results are in good agreement.

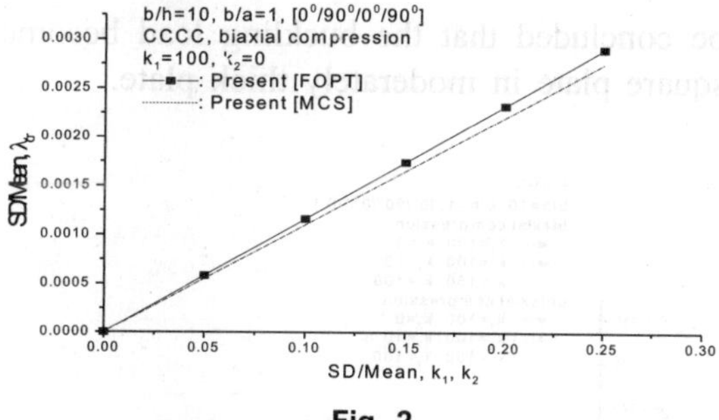

Fig. 2.

Table 1 shows the results of non-dimensionalyzed mean buckling load for anti-symmetric cross ply [0°/90°/0°/90°] with all edges clamped plate having side to thickness ratio a/h=10 and 5. It is observed that foundation stiffness parameters and side to thickness ratio have a significantly affect on the buckling load of the foundation stiffness. It can be noticed that, increase in the foundation stiffness parameters in Pasternak foundation, the buckling load increases more significantly.

Table 1. Results of non-dimnesionalyzed mean buckling load, λ_{cr} of all edges clamped biaxial compression anti-symmetric [0°/90°/0°/90°] laminates $N_x = 1$; $N_{xy} = 0$; $N_y = 1$, a/h = 10, 5)

b/h	b/a	Non-dimensionalyzed buckling load		
		$k_1 = 100, k_2 = 0$	$k_1 = 100, k_2 = 10$	$k_1 = 150, k_2 = 100$
10	1	28.9609	38.9609	130.579
	2	39.50052	49.5052	139.9601
5	1	11.7101	21.7101	112.1881
2		12.5505	22.5505	112.7

Figure 3 examines the influence of scattering in foundation stiffness parameters k_1 and k_2 for 4-layer anti-symmetric [0°/90°/0°/90°] cross-ply square laminate under uniaxial and biaxial compression with b/h = 10. The material properties of the constituent materials kept constant. In such a case, the non-dimensionalyzed buckling load becomes more sensitive to the variation in k_1 and k_2 as their mean values increase for all cases. It is found that the presence of biaxial compression increases the sensitivity of buckling load of the laminate.

The influence of scattering in foundation stiffness parameters k_1 and k_2 for 4-layer anti-symmetric [0°/90°/0°/90°] cross-ply square biaxial compression laminated plate with b/h = 10 and 5 is shown in Fig. 4. The material properties of the constituent materials kept constant. The buckling load becomes more sensitive for moderately thick plate as compare to thin plate.

Figure 5 examines the influence of scattering in foundation stiffness parameters k_1 and k_2 for 4-layer anti-symmetric [0°/90°/90°/0°] cross-ply biaxial compressed laminate with b/h = 10 and b/a = 1 and 2. The material properties of the constituent materials kept constant. It can be noticed that the buckling load becomes more sensitive for rectangular plate as compare to square plate in thin plate.

The influence of scattering in foundation stiffness parameters k_1 and k_2 for 4-layer anti-symmetric [0°/90°/90°/0°] cross-ply biaxial compression laminate with b/h = 5 and b/a = 1 and 2 is shown in Fig. 6. The material properties of the constituent materials kept constant. It can be concluded that the buckling load becomes more sensitive for rectangular plate as compare to square plate in moderately thick plate.

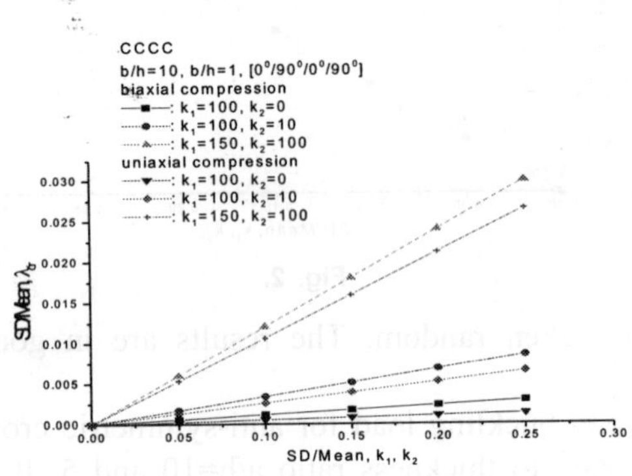

Fig. 3.

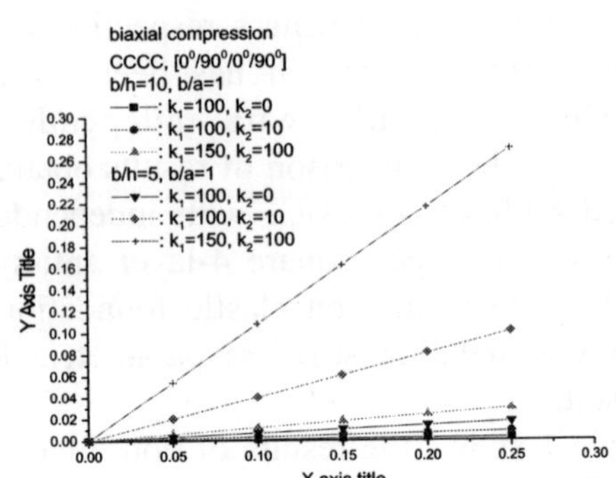

Fig. 4.

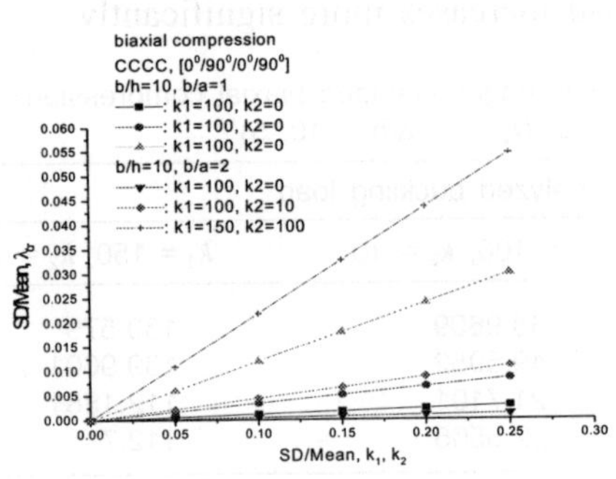

Fig. 5.

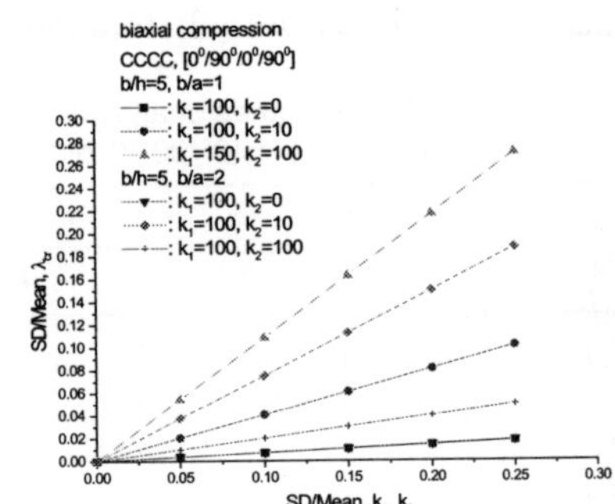

Fig. 6.

CONCLUSION

Based on the limited study conducted, the following conclusions can be drawn:

1. The biaxial compression plate is more sensitive to uniaxial compression plate.
2. The buckling load dispersion increase as the mean foundation stiffness parameters increases.
3. The scattering of buckling load increases, as the thickness of the laminate increases.
4. As the aspect ratio increases the dispersion in buckling load increases.

REFERENCES

1. H.S. Shen, J.J. Zheng, and X.L. Huang, 2003, Dynamic response of shear deformable plates under thermo-mechanical loading and resting on elastic foundations, Journal of Composite Structure, 60, 57-66.
2. Z. Zhang, and S. Chen, 1990, The standard deviation of the eigen solutions for random multi degree freedom system, 39(6), 603-607.
3. B.N. Singh, N.G.R. Iyengar, and D. Yadav, 2002, A C° finite element investigation for buckling of shear deformable laminated composite plates with random material properties, International Journal of Structural Engineering Mechanics, 13(1), 53-74.
4. J.N. Reddy, 1984, A simple higher-order theory for laminated composite plates, ASME Journal of Applied Mechanics, 51, 745-752.

79

Free Vibration Response of Laminated Composite Plate Resting on Elastic Foundation with Uncertain Foundation Stiffness Parameters

ACHCHHE LAL[1], B.N. SINGH[2] AND RAKESH KUMAR[3]

[1]Research Scholar, Department of Applied Mechanics, MNNIT Allahabad-211 004, India
[2]Assistant Professor, Department of Aerospace Engineering, IIT Kharagpur-721 302, India
email: bnsingh@areo.iitkgp.ernet.in
[3]Sr. Lecturer, Department of Applied Mechanics, MNNIT Allahabad-211 004, India.

ABSTRACT

The present work considers free vibration study of laminated composite plates resting on elastic foundation with the objective of determining the second order statistics (mean values and standard deviation) of fundamental frequency. In modeling, the foundation stiffness parameters are considered as independent random variables. Higher order shear deformation theory (HSDT) has been used to model the displacement field. A $C°$ finite element method (FEM) in conjunction with mean-centered first order perturbation technique (FOPT) is employed to determine the second-order statistic of the fundamental frequency of laminated composite plate having all edges clamped. The typical numerical results showing the effect of foundation parameters, plate side to thickness ratio and plate aspect ratio on the fundamental frequency and its dispersion with respect to various random foundation stiffness parameters are presented. The results have been compared with independent Monte Carlo simulation (MCS).

Keywords: FEM; Fundamental frequency; foundation parameters.

1. INTRODUCTION

Composite laminated structures supported on elastic foundations are being increasingly used in a great variety of engineering applications. The exact modeling of the elastic foundation is one of the areas of the research. It is not possible to model the foundation in exact sense. To represent the characteristic behavior of practical foundation, the two-parameter Pasternak is considered to be most effective and very useful mainly from a design perspective. However, this advantage may be offset by dispersion in foundation stiffness parameters as the dispersion in stiffness parameters is inherent in nature due to inaccurate modeling. For accurate prediction of behavior, suitable modeling of the foundation parameters in stochastic sense is essential for sensitive applications.

Considerable research has been done to characterize the free vibration response of laminated composite plate resting on elastic foundation. Much of the work in free vibration response is based on deterministic analysis [1]. Limited literature is available on analysis of composite structures with random material properties [2]. To the best of authors' knowledge, no work dealing with free vibration behaviour of laminated composite plates resting on an elastic foundation using higher order shear deformation theory (HSDT) with random foundation stiffness parameter has been reported in the literature.

2. FORMULATION

Figure 1 shows geometry of the plate resting on elastic foundation. The plate has been modeled using higher shear deformation theory. The displacement fields, strain displacement relations based on small deformation, constitutive relations, etc. can be seen in Ref. [3]. The load-displacement relation for elastic foundation based on two parameter Pasternak model, are as fallows:

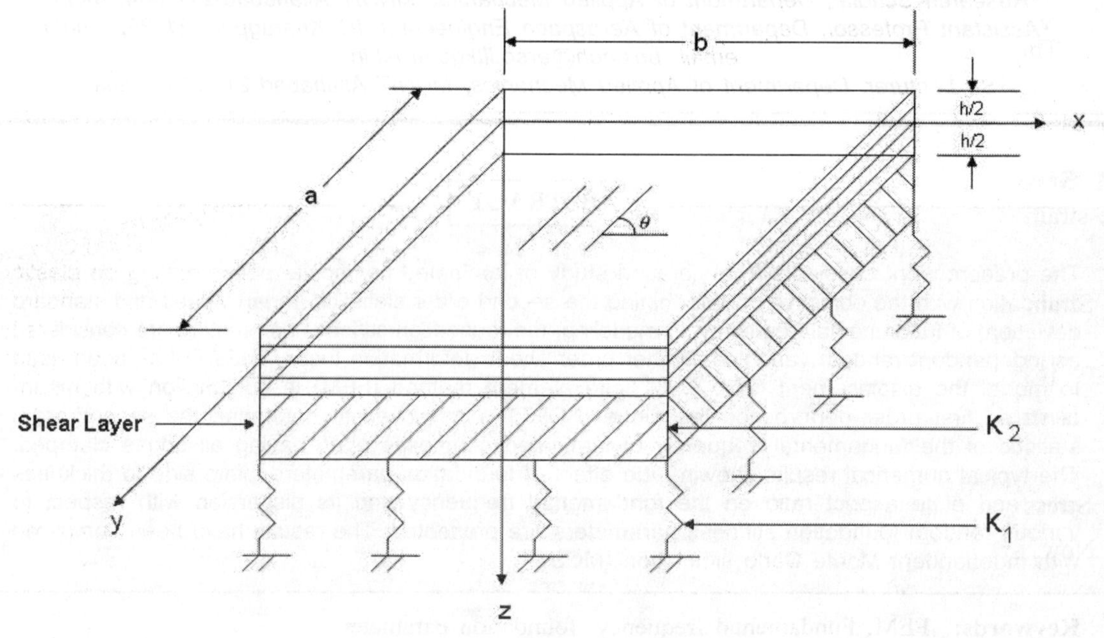

Fig. 1.

$$P = K_1 w - K_2 \nabla^2 w; \quad \nabla^2 = \frac{\partial}{\partial^2 x} + \frac{\partial}{\partial^2 y} \qquad \qquad ...(1)$$

where, K_1 is Winkler foundation stiffness, P is density of reaction forces of foundation, and K_2 is Pasternak foundation stiffness. As a special case when shear layer is neglected ($K_2 = 0$) in the formulation, the Pasternak model converge to Winkler model [1].Displacement field can be expressed as [3]:

$$\bar{u} = u + z\psi_x - z^3 4/3h^2(\psi_x + \partial w/\partial x) = u + f_1(z)\psi_x + f_2(z)\partial w/\partial x$$

$$\bar{v} = v + z\psi_y - z^3 4/3h^2(\psi_y + \partial w/\partial y) = v + f_1(z)\psi_y + f_2(z)\partial w/\partial y; \quad \bar{w} = w \qquad ...(2)$$

where, $\bar{u}$ $\bar{v}$ and $\bar{w}$ denote the displacements of a point along the (x, y, z) coordinates and u, v, and w are corresponding displacements of a point on the mid plane. Ψ_x and Ψ_y are the rotations of normal to the mid plane about the y-axis and x-axis respectively.

A C^o continuous finite element model, developed by Shankara and Iyenger [4], having seven degrees of freedom (DOF) have been used for the free vibration analysis, together with a higher order shear deformation theory (HSDT). A C^o finite element formulation by assuming the slopes as a separate degree of freedom for the plate analysis has been used. Though this increases the nodal degrees of freedom, the choice of element is made simpler.

The strain displacements relations are obtained by using small deformation theory [3]

$$\varepsilon_{xx} = \varepsilon_1 = \partial\bar{u}/\partial x = \varepsilon_1^0 + z(k_1^0 + z^2 k_1^2); \; \varepsilon_{yy} = \varepsilon_2 = \partial\bar{v}/\partial x = \varepsilon_2^0 + z(k_2^0 + z^2 k_2^2)$$

$$\gamma_{xy} = \varepsilon_6 = \partial\bar{u}/\partial y + \partial\bar{v}/\partial x = \varepsilon_6^0 + z(k_6^0 + z^2 k_6^2); \; \gamma_{yz} = \varepsilon_4 = \partial\bar{v}/\partial z + \partial\bar{w}/\partial y = \varepsilon_4^0 + z^2 k_4^2$$

$$\gamma_{xz} = \varepsilon_5 = \partial\bar{u}/\partial z + \partial\bar{w}/\partial x = \varepsilon_5^0 + z^2 k_5^2 \qquad \text{...(3)}$$

The linear stress-strain relationship for an orthotropic layer can be expressed as [3],

$$\{\sigma\} = [\bar{Q}]\{\varepsilon\} \qquad \text{...(4)}$$

2.1 Strain Energy of the Plate

The strain energy of the composite plate is equal to the potential energy. Hence, it can be expressed as,

Strain energy $\qquad (U_1) = \dfrac{1}{2}\iiint_V \{\varepsilon\}\{\sigma\}dV \qquad \text{...(5)}$

where $\{\varepsilon\}$, $\{\sigma\}$ and $[\bar{Q}]$ are stress vector, strain vector and reduced stiffness matrix, respectively and is stress vector, and it can be expressed as:

$$\{\varepsilon\} = \{\varepsilon_1 \; \varepsilon_2 \; \varepsilon_6 \; \varepsilon_4 \; \varepsilon_5\}^T ; \; \{\sigma\} = \{\sigma_1 \, \sigma_2 \, \sigma_6 \, \sigma_4 \sigma_5\}^T$$

and Stress vector $\{\sigma\} = [Q]\{\varepsilon\}$

Now, strain energy can be expressed as:

$$U_1 = \frac{1}{2}\int_V \{\varepsilon\}\{\sigma\}dV$$

$$\{\bar{\varepsilon}\} = \left(\varepsilon_1^0 \quad \varepsilon_2^0 \quad \varepsilon_6^0 \quad k_1^0 \quad k_2^0 \quad k_6^0 \quad k_1^2 \quad k_2^2 \quad k_6^2 \quad \varepsilon_4^0 \quad \varepsilon_5^0 \quad k_4^2 \quad k_5^2\right)^T$$

Potential energy becomes $U_1 = \dfrac{1}{2}\int_A \bar{\varepsilon}^T D\bar{\varepsilon}\,dAdV$

where, $\qquad D = \displaystyle\sum_{k=1}^{NL} \int_{k=1}^{z\,k} [T]^T[\bar{Q}][T]dz = \begin{bmatrix} [A_1] & [B] & [E] & 0 & 0 \\ [B] & [C_1] & [F_1] & 0 & 0 \\ [E] & [F_1] & [H] & 0 & 0 \\ 0 & 0 & 0 & [A_2] & [C_2] \\ 0 & 0 & 0 & [C_2] & [F_2] \end{bmatrix}$

$$\left(A_{1_{ij}}, B_{ij}, C_{1_{ij}}, E_{ij}, F_{1_{ij}}, H_{ij} \right) = \sum_{k=1}^{NL} \int_{z_{k-1}}^{z_k} \overline{Q}_{ij}^{(k)} \left(1, z, z^2, z^3, z^4, z^6 \right) dz \quad \text{for } i, j = 1, 2, 6$$

$$\left(A_{2_{ij}}, C_{2_{ij}}, F_{2_{ij}} \right) = \sum_{k=1}^{NL} \int_{z_{k-1}}^{z_k} \overline{Q}_{ij}^{(k)} \left(1, z^2, z^4 \right) dz \quad \text{for } i, j = 4, 5$$

Summing over total number of element NE, potential energy of vibrating plate

$$U_1 = \sum_{e=1}^{NE} U_1^{(e)} = \frac{1}{2} \sum_{e=1}^{NE} \{\varepsilon\}^{(e)^T} [D]\{\varepsilon\}^{(e)} = \sum_{e=1}^{NE} \{q\}^{(e)^T} [K]\{q\}^{(e)} \qquad \text{...(6)}$$

2.2 The Kinetic Energy of the Laminate

The kinetic energy of laminate is expressed as:

$$T = \frac{1}{2} \int_V \rho \{\dot{u}\}^T \{\dot{u}\}\, dV \qquad \text{...(7a)}$$

For *NL* number of layers of composite plate $T = \dfrac{1}{2} \left(\int_A \sum_{k=1}^{NL} \int_{Z_{k-1}}^{Z_k} \rho^{(k)} \{\dot{u}\}^T \{\dot{u}\}\, dz \right) dA$

$$T = \frac{1}{2} \int_A \{\dot{q}\}^T [m]\{\dot{q}\}\, dA \quad \text{where,} \quad [m] = \sum_{k=1}^{NL} \int_{Z_{k-1}}^{Z_k} \rho \left[\overline{N}\right]^T \left[\overline{N}\right] dz$$

Elemental kinetic energy is given as $T^{(e)} = \dfrac{1}{2} \int_{A^{(e)}} \left\{\dot{q}^{(e)}\right\}^{(T)} [m]\left\{\dot{q}^{(e)}\right\} dA$

Summing over total number of element *NE*, Kinetic energy of vibrating plate

$$T = \sum_{e=1}^{NE} T^{(e)} = \frac{1}{2} \sum_{e=1}^{NE} \left\{\dot{q}^{(e)}\right\}^T [M]\left\{\dot{q}^{(e)}\right\} \qquad \text{...(7b)}$$

2.3 The Strain Energy due to Foundation

Strain energy due to foundation can be expressed as

$$U_2 = \frac{1}{2} \int_A \begin{Bmatrix} w \\ \dfrac{\partial w}{\partial x} \\ \dfrac{\partial w}{\partial y} \end{Bmatrix}^T \begin{bmatrix} K_1 & 0 & 0 \\ 0 & K_2 & 0 \\ 0 & 0 & K_2 \end{bmatrix} \begin{Bmatrix} w \\ \dfrac{\partial w}{\partial x} \\ \dfrac{\partial w}{\partial y} \end{Bmatrix} dA \ ; \qquad \text{...(8a)}$$

Summing over total number of element *NE*, potential energy of vibrating plate

$$U_2 = \sum_{e=1}^{NE} U_2^{(e)} = \frac{1}{2} \sum_{e=1}^{NE} \{B_g\}^{(e)^T} [D_f]\{B_g\}^{(e)} = \sum_{e=1}^{NE} \{q\}^{(e)^T} \left[K_f\right]\{q\}^{(e)} \qquad \text{...(8b)}$$

2.4 Governing Equation of Motion

The governing equation of the motion is obtained using Hamilton's variational principle,

$$\delta \int_{t_1}^{t_2} (U - T) dt = 0 \qquad \text{...(9)}$$

Substituting, Eqs. (6), (7b) and (8b) in Eq. (9), ones obtain as:

$$[M]\{\ddot{q}\} + \left[K_{sf}\right]\{q\} = 0 \quad \left\{\left[K_{sf}\right] - \lambda [M]\right\}\{q\} = 0 \qquad \text{...(10)}$$

where,

$$\left[K_{sf}\right] = [K] + \left[K_f\right] \quad \{q\} = \sum_{e=1}^{NE}\{q\}^{(e)}, [K] = \sum_{e=1}^{NE}[K]^{(e)}, \left[K_f\right] = \sum_{e=1}^{NE}\left[K_f\right]^{(e)}, [M] = \sum_{e=1}^{NE}[M]^{(e)}$$

Also $\left\{q^*\right\}$, $[K]$, $\left[K_f\right]$, $[M]$, $\lambda(=\omega^2)$ and ω are define as a global displacement vector, foundation stiffness matrix, global mass stiffness matrix ,eigenvalue and fundamental frequency respectively.

Equation (10) is random in nature, being dependent on the system material properties. A mean centered first order perturbation technique in conjunction with C^0 finite element method has been used to obtain the solution of governing equations.

3. SOLUTION APPROACH: PERTURBATION TECHNIQUE

We consider a class of problems where the random variation is very small as compared to the mean part of random system properties. Further, it is quit logical to assume that the dispersion in the derived quantities like $[K_f]$, and λ also small to their mean values. In the present analysis, since foundation stiffness parameters are totally dependent on the material properties of the supporting elastic medium which possess random fluctuations. So, randomness in both k_1 and k_2 is taken into consideration.

In general, any arbitrary random variable b_i can be represented as the sum of the mean variable and a zero mean random variable, denoted by superscripts 'd' and 'r', respectively,

$$b_i = b_i^d + b_i^r \qquad \text{...(11)}$$

Similarly, all of the quantity in Eq. (10) can be broken into

$$\left[K_{fi}\right] = \left[K_{fi}^d\right] + \left[K_{fi}^r\right] \; ; \; \lambda_i^r = 2\omega_i^d \omega_i^r + \omega_i^{r^2} \text{ and } \{q_i\} = \{q_i^d\} + \{q_i^r\}; \qquad \text{...(12)}$$

where $i = 1, 2,……p$.

Substituting Eq. (12) in Eq. (10) and collecting same order of the magnitude term and keeping only up to first order terms, we obtain

$$\left[K_{fi}^d\right]\{q_i^d\} = \lambda_i^d\{q_i^d\} \qquad \text{...(13)}$$

$$\left[K_{fi}^d\right]\{q_i^r\} + \left[K_{fi}^r\right]\{q_i^d\} = \lambda_i^r\{q_i^d\} + \lambda_i^d\{q_i^r\} \qquad \text{...(14)}$$

Because Eq. (13), is the deterministic equation relating to the mean values, the mean eigenvalues and corresponding mean eigenvectors can be determined by convention eigen solution procedures. Using orthogonality properties the Eq. (14) may be decoupled and using Taylor series expansion, the following can obtained as [2]:

$$Var\left(\lambda_i\right)=\sum_{j=1}^{p}\sum_{k=1}^{p}\frac{\partial\lambda_i^d}{\partial b_j}\frac{\partial\lambda_i^d}{\partial b_k}Cov\left(b_j^r,b_k^r\right)\ ,\qquad\qquad ...(15)$$

$$Var\left\{q_i^l\right\}=\sum_{j=1}^{p}\sum_{k=1}^{p}\frac{\partial\left\{q_j^d\right\}}{\partial b_j}\frac{\partial\left\{q_k^d\right\}}{\partial b_k}Cov\left(b_j^r,b_k^r\right)\ ,\qquad\qquad ...(16)$$

where $Cov\left(b_j^r,b_k^r\right)$ is the cross variance between b_j^r and b_k^r. The standard deviation (SD) is obtained by the square root of the variance.

4. RESULTS AND DISCUSSION

A typical results are presented for all edges clamped square plate for a/h = 5 and 10 with various stacking sequences. Based on convergence study conducted for the fundamental frequency, a (4 × 4) mesh with eight nodded serendipity element is used for computation of the results in this study.

The following mean values of the material properties are used for the computation.

$$E_{11}=40E_{22},\ \ G_{12}=G_{13}=0.6E_{22},\ \ G_{23}=0.5E_{22},\ \ v_{12}=v_{21}=0.25,\ \ \rho=1.$$

The following non-dimensionalyzed mean natural frequency and foundation parameters k_1 and k_2 have been used in this study as $\bar{\omega}=(\omega a^2\sqrt{\rho/E_{22}})/h$, $k_1=K_1b^4/E_{22}h^3$ and $k_2=K_2b^2/E_{22}h^3$

where ω,ρ,a,b,h,k_1 and k_2 are non-dimensionalized mean natural frequency, density of foundation, length of plate, width of the plate, thickness of the plate, non-dimensionalized Winkler foundation stiffness parameter and non-dimensional Pasternak foundation respectively.

The outlined approach has been validated by comparing the results with independent MCS. Fig. 2 shows a comparison of results obtained using proposed outlined approach with independent MCS for an all edges clamped square [0°/90°/90°/0°] plate resting on Winkler elastic foundation. For MCS, based on convergence, 10000 samples have been used to obtain second order statistics. The results are in good agreement.

Table 1 shows the results of non-dimensionalized mean fundamental frequency for symmetric cross-ply [0°/90°/90°/0°] square plate with all edges clamped plate having side to thickness ratio

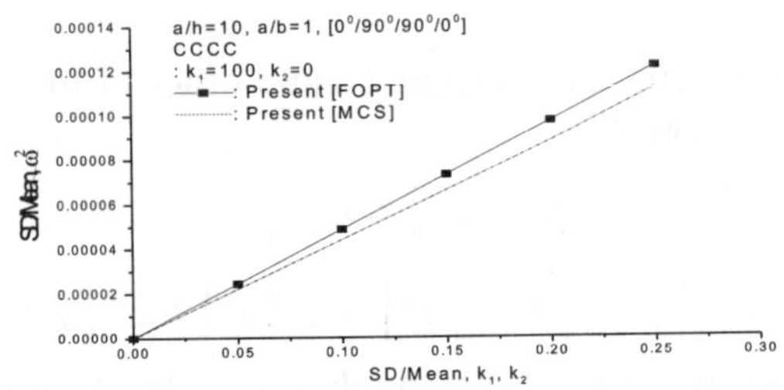

Fig. 2. Validation study.

Table 1. Non-dimensionalized mean fundamental frequency for a [0°/90°/90°/0°] laminated composite plate with all edges clamped

a/h	a/b	Non-dimensionalized fundamental frequency		
		$k_1 = 100, k_2 = 0$	$k_1 = 100, k_2 = 10$	$k_1 = 150, k_2 = 100$
10	1	24.6254	28.5453	51.6101
	2	51.6996	59.6264	107.9455
5	1	16.5818	21.7842	47.8829
	2	45.8167	53.9951	103.9996

a/h = 10 and 5. It is observed that increase in the foundation stiffness parameters in Pasternak foundation, the fundamental frequency increases more significantly. The results have also been generated for various cases. However, these are not presented here. The results show that uncertainty in foundation stiffness has significant impact on response of free vibration of the plate.

Figure 3 shows the influence of scattering in foundation stiffness parameters k_1 and k_2 for 4-layer symmetric $[0^0/90^0/90^0/0^0]$ cross-ply square laminate with a/h = 10. The material properties of the constituent materials kept constant. In such a case, the nondimensionalysed fundamental frequency becomes more sensitive to the variation in k_1 and k_2 as their mean values increases for all cases.

The influence of scattering in foundation stiffness parameters k_1 and k_2 for 4-layer symmetric $[0^0/90^0/90^0/0^0]$ cross-ply square laminate with a/h = 5 is shown in Fig. 4. The material properties of the constituent materials kept constant. The fundamental frequency becomes more sensitive for moderately thick plate as compare to thin plate.

Figure 5 plots the influence of scattering in foundation stiffness parameters k_1 and k_2 for 4-layers symmetric [0°/90°/90°/0°] cross-ply laminate with a/h = 10 and a/b = 2. The material properties

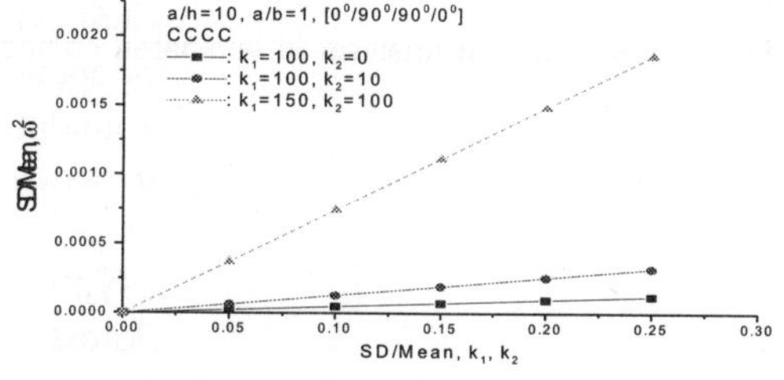

Fig. 3.

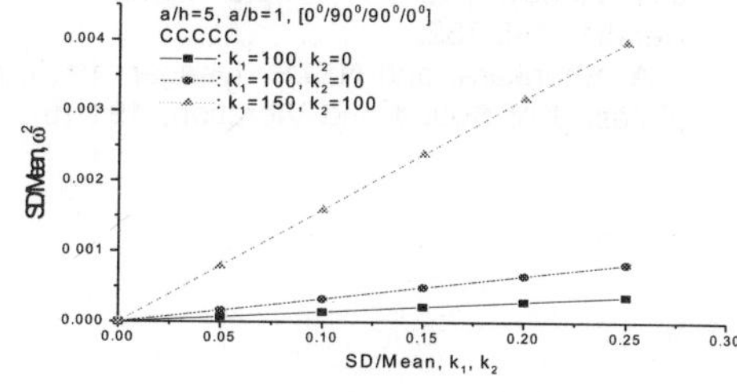

Fig. 4.

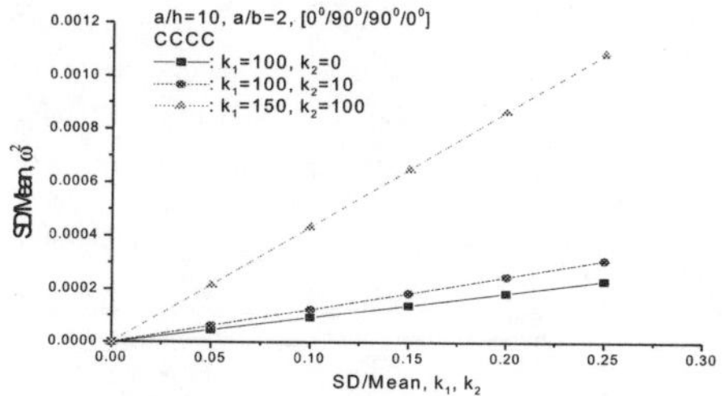

Fig. 5.

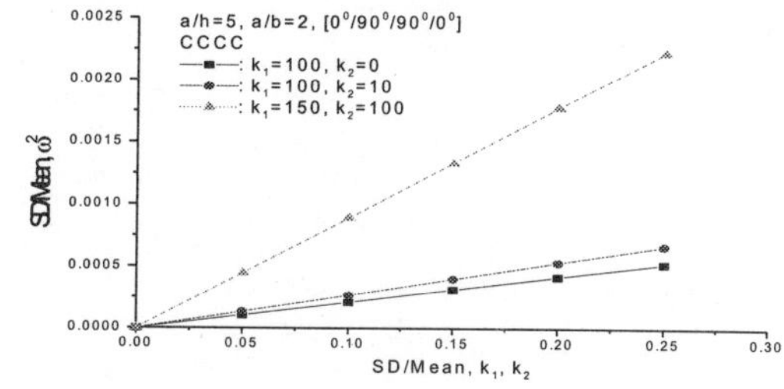

Fig. 6.

of the constituent materials kept constant. In such a case, the nondimensionalysed fundamental frequency becomes more sensitive to the variation in k_1 and k_2 as their mean values increases. The fundamental frequency becomes more sensitive for rectangular plate as compare to square plate.

The influence of scattering in foundation stiffness parameters k_1 and k_2 for 4-layer symmetric [0°/90°/90°/0°] cross-ply laminate with a/h = 5 and a/b = 2 is shown in Fig. 6. The material properties of the constituent materials kept constant. It can also be concluded from Figs 4 and 6 that the fundamental frequency becomes more sensitive for rectangular plate as compare to square plate.

CONCLUSION

The following conclusions are noted for the limited study.

1. The dispersion in the square of non-dimensionalyzed fundamental frequency (ω^2) which is linear with variation in random parameters involved is greatly influenced by foundation parameters, side to thickness ratio and plate aspect ratio.
2. The fundamental frequency response of the laminated plate increases with as the mean foundation stiffness parameters increase.
3. As the thickness of composite plate increases, scattering of square of fundamental frequency square (ω^2) increases.
4. The scattering of square of fundamental frequency square (ω^2) increases as the aspect ratio of the plate increases.

REFERENCES

1. H.S. Shen, J.J. Zheng, and X.L. Huang, 2003, Dynamic response of shear deformable plates under thermo-mechanical loading and resting on elastic foundations, Journal of Composite. Structure, 60, 57-66.
2. B.N. Singh, D. Yadav, and N.G.R. Iyengar, 2001, Natural frequencies of composite plates with random material properties using higher-order shear deformation theory, International J of Mechanical Sciences, 43, 2193-2214.
3. J.N. Reddy, 1984, A simple higher-order theory for laminated composite plates, J of Applied Mechanics, Dec-51, 745-752.
4. C.A. Shankara, and N.G.R. Iyengar, 1996, A $C°$ element for the free vibration analysis of laminated composite plates, J of Sound and Vibration, 191 (5), 721-738.

80

Higher Order Refined Computational Model for the Free Vibration Analysis of Antisymmetric Angle-Ply Laminated Plates

K. Swaminathan[1] and S.S. Patil[2]

[1]Department of Civil Engineering, National Institute of Technology Karnataka-575 025, India.
email: swami@vasnet.co.in
[2]Department of Civil Engineering, National Institute of Technology Karnataka-575 025, India.
email: spatilss@yahoo.com

ABSTRACT

Analytical formulations and solutions to the natural frequency analysis of simply supported antisymmetric angle ply composite and sandwich plates hitherto not reported in the literature based on a higher order refined computational model with twelve degrees of freedom already reported in the literature are presented. The theoretical model presented herein incorporates laminate deformations which account for the effects of transverse shear deformation, transverse normal strain/stress and a nonlinear variation of in-plane displacements with respect to the thickness coordinate thus modeling the warping of transverse cross sections more accurately and eliminating the need for shear correction coefficients. In addition, another higher order computational model with five degrees of freedom already available in the literature is also considered for comparison. The equations of motion are obtained using Hamilton's principle. Solutions are obtained in closed form using Navier's technique and by solving the eigenvalue equation. Accuracy of the theoretical formulations and the solution method is first ascertained by comparing the results with that already available in the literature. After establishing the accuracy of the solutions, numerical results with real properties using above two computational models are presented for the free vibration analysis of multilayer antisymmetric angle ply composite and sandwich plates, which will serve as a benchmark for future investigations.

Keywords: Free vibration; Higher-order theory; Shear deformation; Angle-ply plates; Analytical solutions.

1. INTRODUCTION

Laminated composite plates are being increasingly used in the aeronautical and aerospace industry as well as in other fields of modern technology. To use them efficiently a good understanding of their structural and dynamical behaviour is needed. The Classical Laminate Plate Theory [1] which ignores the effect of transverse shear deformation becomes inadequate for the analysis of multilayer composites. The First Order Shear Deformation Theories (FSDTs) based on Reissner [2] and Mindlin [3] assume linear in-plane stresses and displacements, respectively, through the laminate thickness. Since FSDTs account for layerwise constant states of transverse shear stress, shear correction coefficients are needed to rectify the unrealistic variation of the shear strain/stress through the thickness and which ultimately define the shear strain energy. In order to overcome the limitations of FSDTs, higher order shear deformation theories (HSDTs) that involve higher order terms in Taylor's expansions of the displacement in the thickness co-ordinate were developed. Hildebrand *et al.* [4] were the first to introduce this approach to derive improved theories of plates and shells. Using the higher-order theory of Reddy [5] free vibration analysis of isotropic, orthotropic and laminated plates was carried out by Reddy and Phan [6]. Kant [7] was the first to derive the complete set of variationally consistent governing equations for the flexure of a symmetrically laminated composite plate incorporating both distortion of transverse normals and effects of transverse normal stress/ strain by utilizing the complete three-dimensional generalized Hooke's law and presented results for isotropic plate only. Later finite element solutions by Mallikarjuna [8], Mallikarjuna and Kant [9], Kant and Mallikajuna [10,11] and analytical solutions for the isotropic, orthotropic and cross ply multilayer plates by Kant and Swaminathan [12,13] were presented using a set of higher order refined theories already reported in the literature by Pandya and Kant [14-18] and Kant and Manjunatha [19] for the free vibration analysis. In this paper, analytical formulations developed and solutions obtained for the first time is presented for the free vibration analysis of antisymmetric angle ply laminated composite and sandwich plates using a higher order refined computational model with twelve degrees of freedom. Solutions obtained using this model is also compared with the results of another model considered in the present investigation. Correctness of the solutions is first established and then benchmark results with real properties using both the models are presented for the antisymmetric angle ply composite and sandwich plates.

2. DISPLACEMENT MODELS

For the purpose of evaluation, the following higher order displacement model is considered. Analytical formulations and solutions for the free vibration analysis of anti-symmetric angle ply laminated composite and sandwich plates are obtained for the first time in this investigation and results obtained using this model is referred to as present in all the tables.

Model-1 [19]

$$u(x,y,z,t) = u_o(x,y,t) + z\theta_x(x,y,t) + z^2 u_o^*(x,y,t) + z^3\theta_x^*(x,y,t)$$

$$v(x,y,z,t) = v_o(x,y,t) + z\theta_y(x,y,t) + z^2 v_o^*(x,y,t) + z^3\theta_y^*(x,y,t)$$

$$w(x,y,z,t) = w_o(x,y,t) + z\theta_z(x,y,t) + z^2 w_o^*(x,y,t) + z^3\theta_z^*(x,y,t) \qquad ...(1)$$

In addition to the above, another higher order theory shown below as Model-2 (Reddy, 1984) already reported in the literature are also considered for the evaluation purpose. Analytical formulations developed and numerical results generated independently using this model are also being presented here with a view to have all the results obtained using both the models on a common platform.

Model-2 [5]

$$u(x,y,z,t) = u_o(x,y,t) + z\left[\theta_x(x,y,t) - \frac{4}{3}\left(\frac{z}{h}\right)^2\left\{\theta_x(x,y,t) + \frac{\partial w_o}{\partial x}\right\}\right]$$

$$v(x,y,z,t) = v_o(x,y,t) + z\left[\theta_y(x,y,t) - \frac{4}{3}(\frac{z}{h})^2\left\{\theta_y(x,y,t) + \frac{\partial w_o}{\partial y}\right\}\right]$$

$$w(x,y,z,t) = w_o(x,y,t) \qquad\qquad ...(2)$$

where the terms u, v and w are the displacements of a general point (x, y, z) in the laminate domain in x, y and z directions respectively. The parameters are the in-plane displacements and is the transverse displacement of a point (x, y) on the middle plane. The functions θ_x, θ_y are rotations of the normal to the middle plane about y and x axes respectively. The parameters u_o^*, v_o^*, w_o^*, θ_x^*, θ_y^*, θ_z^* and θ_z are the higher-order terms in the Taylor's series expansion and they represent higher-order transverse cross sectional deformation modes.

3. NUMERICAL RESULTS AND DISCUSSIONS

For all the problems a simply supported plate with SS-2 boundary conditions is considered for the analysis. Results are obtained in closed-form using *Navier's* solution technique for the above geometry and the accuracy of the solution is established by comparing the results with the solutions wherever available in the literature. The non-dimensionalized natural frequencies computed for two, four and eight layer antisymmetric angle-ply laminate with layers of equal thickness are given in Table 1 and Table 2. The orthotropic material properties of individual layers in all the above laminates considered are E_1/E_2 = open, $E_2 = E_3$, $G_{12} = G_{13} = 0.6E_2$, $G_{23} = 0.5E_2$, $\upsilon_{12} = \upsilon_{13} = \upsilon_{23} = 0.25$. The variation of natural frequencies with respect to side-to-thickness ratio a/h is presented in Table 1. The natural frequencies obtained using the present theory are compared with Reddy's theory. In the case of thick plates (a/h ratios 2, 4 and 10) there is a considerable difference exists between the results computed using Model-1 and Model-2. The variation of natural frequencies with respect to side-to-thickness ratio a/h for different E_1/E_2 ratio is presented in Table 2. For a thick plate with a/h ratio equal to 2 and E_1/E_2 ratio equal to 3 and 10, the percentage difference in values predicted by Model-1 are 0.13% and 3.51% lower as compared to Model-2. At higher range of E_1/E_2 ratio equal to 20-40, the percentage difference in values obtained using Model-2 is very much higher. For a thick plate with a/h ratio equal to 2 and E_1/E_2 ratio equal to 20, 30 and 40, the percentage difference in values predicted by Model-1 are 6.08%, 7.99% and 9.70% lower as compared to

Table 1. Non-dimensionalized fundamental frequencies $\bar{\omega} = (\omega b^2/h)\sqrt{\rho/E_2}$ for

a simply supported anti-symmetric angle-ply square laminated plate with E_1/E_2 = 40

Lamination and number of layers	Theory	a / h					
		2	4	10	20	50	100
$(45°/-45°)_1$	Model–1 (Present)	5.332	8.842	12.911	14.170	14.601	14.667
	Model–2[a]	6.283	9.759	13.263	14.246	14.572	14.621
$(45°/-45°)_2$	Model–1 (Present)	5.567	10.073	17.877	21.623	23.195	23.449
	Model–2[b]	6.106	10.650	18.322	21.806	23.223	23.450
$(45°/-45°)_4$	Model–1 (Present)	5.923	10.747	19.126	23.183	24.896	25.174
	Model–2[a]	6.283	10.991	19.266	23.239	24.905	25.174

[a] Results using this theory are computed independently and are found to be the same as reported in the reference [6].
[b] Results using this theory are computed independently for the first time.

Table 2. Non-dimensionalized fundamental frequencies for a simply supported four layered anti-symmetric angle-ply (45°/ –45° / 45°/ –45°) square laminated plate.

E_1/E_2	Theory	a/h					
		2	4	10	20	50	100
3	Model-1 (Present)	4.649	6.459	7.634	7.872	7.944	7.954
	Model-2[b]	4.655	6.455	7.627	7.865	7.937	7.947
10	Model-1 (Present)	5.206	8.345	11.412	12.229	12.495	12.535
	Model-2[b]	5.389	8.512	11.467	12.238	12.487	12.524
20	Model-1 (Present)	5.414	9.331	14.473	16.257	16.895	16.993
	Model-2[b]	5.743	9.685	14.661	16.315	16.896	16.985
30	Model-1 (Present)	5.508	9.797	16.454	19.232	20.313	20.484
	Model-2[b]	5.948	10.278	16.775	19.349	20.328	20.481
40	Model-1 (Present)	5.567	10.073	17.877	21.623	23.195	23.449
	Model-2[b]	6.107	10.651	18.322	21.806	23.224	23.451

Model-2. The difference between the models tends to reduce for thin and relatively thin plates. The percentage difference in values in above theories increases with the increase in the degree of anisotropy. As the number of layer increases, the percentage difference in values obtained using the above two theories decreases significantly.

The variation of fundamental frequency with respect to the various parameter like the side-to-thickness ratio (a/h), thickness of the core to thickness of the flange (t_c/t_f) and the aspect ratio (a/b) of a five layer sandwich plate with antisymmetric angle-ply face sheets are given in Table 3 to Table 5. The following material properties are used for face sheets and the core [12]:

Face sheets (Graphite-Epoxy T300/934)

$E_1 = 19 \times 10^6$ psi (131 GPa) $E_2 = 1.5 \times 10^6$ psi (10.34 Gpa)

$E_2 = E_3$ $G_{12} = 1 \times 10^6$ psi (6.895 GPa)

$G_{13} = 0.90 \times 10^6$ psi (6.205 GPa) $G_{23} = 1 \times 10^6$ psi (6.895 GPa)

$\upsilon_{12} = 0.22$, $\upsilon_{13} = 0.22$, $\upsilon_{23} = 0.49$

$\rho = 0.057$ lb/inch^3 (1627 kg/m^3).

Core properties (Isotropic)

$E_1 = E_2 = E_3 = 2G = 1000$ psi (6.90 $\times$ 10^{-3} GPa)

$G_{12} = G_{13} = G_{23} = 500$ psi (3.45 $\times$ 10^{-3} GPa)

$\upsilon_{12} = \upsilon_{13} = \upsilon_{23} = 0$

$\rho = 0.3403 \times 10^{-2}$ lb/inch^3 (97 kg/m^3).

The results clearly show that for all the parameters considered, there is a considerable difference exists between the results computed using the Model-1 and Model-2. The Reddy's theory very much overestimates the frequency values. From the results of natural frequencies shown in Table 3 to Table 5, it can be concluded that the effect of transverse shear moduli of stiff layers is more pronounced in thick laminates than for thin laminates.

Table 3. Non-dimensionalized fundamental frequencies $\bar{\omega} = (\omega b^2 / h)\sqrt{(\rho/E_2)_f}$ for a simply supported anti-symmetric ($45°/ - 45°/$ core $/ 45°/ - 45°$) sandwich plate with $a/b = 1$ and $t_c / t_f = 10$

Theory	a / h							
	2	4	10	20	30	40	50	100
Model-1 (Present)	1.281	2.191	5.065	9.274	12.439	14.669	16.206	19.309
Model-2[b]	1.693	3.217	7.489	12.696	15.721	17.441	18.460	20.135

Table 4. Non-dimensionalized fundamental frequencies $\bar{\omega} = (\omega b^2 / h)\sqrt{(\rho/E_2)_f}$ for a simply supported anti-symmetric (45o $/ -45°$ / core $/ 45°/ - 45°$) sandwich plate with $a/b = 1$ and $a/h = 10$

Theory	t_c/t_f						
	4	10	20	30	40	50	100
Model-1 (Present)	9.819	5.065	3.215	2.890	2.861	2.897	3.085
Model-2[b]	12.051	7.489	4.539	3.572	3.229	3.108	3.109

Table 5. Non-dimensionalized fundamental frequencies $\bar{\omega} = (\omega b^2 / h)\sqrt{(\rho/E_2)_f}$ for a simply supported anti-symmetric (45° $/ - 45°$ / core $/ 45°/- 45°$) sandwich plate with $a/h = 10$ and $t_c / t_f = 10$

Theory	a / b						
	0.5	1.0	1.5	2.0	2.5	3.0	5.0
Model-1 (Present)	15.451	5.065	2.910	2.056	1.605	1.327	0.818
Model-2[b]	22.220	7.489	4.331	3.063	2.388	1.966	1.173

CONCLUSION

Analytical formulations and solutions to the natural frequency analysis of simply supported antisymmetric angle ply composite and sandwich plates hitherto not reported in the literature based on a higher order refined theory which takes in to account the effects of both transverse shear and transverse normal deformations are presented. The accuracy of the present computational model with twelve degrees of freedom in comparison to other higher order model with five degrees of freedom considered in the present investigation in predicting natural frequency has been established. After ascertaining the accuracy, new results for multilayered sandwich plates with antisymmetric angle ply face sheets are presented which will serve as a benchmark for future investigations.

REFERENCES

1. E. Reissner, and Y. Stavsky, 1961, Bending and stretching of certain types of heterogeneous aelotropic elastic plates, ASME Journal of Applied Mechanics, 28, 402-408.
2. E. Reissner, 1945, The effect of transverse shear deformation on the bending of elastic plates, ASME Journal of Applied Mechanics, 12, 69–77.
3. R.D. Mindlin, 1951, Influence of rotary and shear on flexural motions of isotropic, elastic plates, ASME Journal of Applied Mechanics, 18, 31-38.
4. F.B. Hildebrand, E. Reissner, and G.B. Thomas, 1949, Note on the foundations of the theory of small displacements of orthotropic shells, NACA TN–1833.
5. J.N. Reddy, 1984, A simple higher order theory for laminated composite plates, ASME Journal of Applied Mechanics, 51, 745–752.

6. J.N. Reddy, and N.D. Phan, 1985, Stability and vibration of isotropic, orthotropic and laminated plates according plates according to a higher order shear deformation theory, Journal of Sound and Vibration, 98(2), 157-170.

7. T. Kant, 1982, Numerical analysis of thick plates, Computer Methods in Applied Mechanics and Engineering, 31, 1-18.

8. Mallikarjuna, 1988, Refined Theories with C° Finite Elements for Free Vibration and Transient Dynamics of Anisotropic Composite and Sandwich Plates, Ph.D. thesis, Indian Institute of Technology Bombay, Mumbai, India.

9. Mallikarjuna, and T. Kant, 1989, Free vibration of symmetrically laminated plates using a higher order theory with finite element technique, International Journal for Numerical Methods in Engineering, 28, 1875-1889.

10. T. Kant, and Mallikarjuna, 1989, A higher order theory for free vibration of unsymmetrically laminated composite and sandwich plates – finite element evaluation, Computers and Structures, 32, 1125-1132.

11. T. Kant, and Mallikarjuna, 1989, Vibration of unsymmetrically laminated plates analysed by using a higher order theory with a C° finite element formulation, Journal of Sound and Vibration, 134, 1-16.

12. T. Kant, and K. Swaminathan, 2001, Analytical solutions for free vibration of laminated composite and sandwich plates based on a higher order refined theory, Composite Structures, 53(1), 73-85.

13. T. Kant, and K. Swaminathan, 2001, Free vibration of isotropic, orthotropic and multilayer plates based on higher order refined theories, Journal of Sound and Vibration, 241(2), 319-327.

14. B.N. Pandya, and T. Kant, 1987, A consistent refined theory for flexure of a symmetric laminate, Mechanics Research Communications, 14, 107-113.

15. B.N. Pandya, and T. Kant, 1988, Higher order shear deformable theories for flexure of sandwich plates – finite element evaluations, International Journal of Solids and Structures, 24(12), 1267-1286.

16. B.N. Pandya, and T. Kant, 1988, Flexure analysis of laminated composites using refined higher order C° plate bending elements, Computer Methods in Applied Mechanics and Engineering, 66, 173-198.

17. B.N. Pandya, and T. Kant, 1988, A refined higher order generally orthotropic C° plate bending element, Computers and Structures, 28, 119-133.

18. B.N. Pandya, and T. Kant, 1988, Finite element stress analysis of laminated composite plates using higher order displacement model, Composite Science and Technology, 32, 137-155.

19. T. Kant, and B.S. Manjunatha, 1988, An unsymmetric FRC laminate C° finite element model with 12 degrees of freedom per node, Engineering Computation, 5(3), 300-308.

81

Optimisation of Stiffener Arrangement of Composite Hypar Shell with respect to Fundamental Frequency

Sarmila Sahoo[1] and Dipankar Chakravorty[2]

[1]Department of Civil Engineering, Jadavpur University, Kolkata-700 032, India
email: sarmila_ju@yahoo.com
[2]Department of Civil Engineering, Jadavpur University, Kolkata-700 032, India
email: dc_ju@yahoo.com

ABSTRACT

An eight noded shell element combined with a three noded beam element is used to study the free vibration behaviour of laminated composite stiffened hypar shells. A variety of problems are solved taking simply supported boundary conditions. Free vibration response of stiffened composite hypar shells is studied with respect to fundamental frequency by varying the number and depth of stiffeners. The results are further analysed to suggest guidelines to select optimum stiffener arrangement.

Keywords: Hypar shell, stiffened, free vibration, finite element analysis, laminated composite.

1. INTRODUCTION

Skewed hypar shells (Fig. 1) can offer a number of civil engineering advantages particularly when the material is laminated composite with high strength to weight ratio. The hypar shells may be stiffened to have enhanced rigidity when subjected to point loads or provided with cutouts for some service requirements. Sinha and Mukhopadhyay [1] echoed this fact in their review paper. The initial studies about vibrations of stiffened shell panels were about cylindrical shells. Recently Nayak and Bandyopadhyay [2,3] carried out free vibration studies of isotropic stiffened shell panels in details including stiffened hypar shells. Free and forced vibrations of unstiffened composite hypar shell was reported by Chakravorty *et al.* [4]. In a recent paper Sahoo and Chakravorty [5] presented results of static analysis of stiffened composite hypar shells.

An overall look at the volume of literature that has accumulated till date dealing with stiffened shell panels reflect the fact that vibration of stiffened composite skewed hypar shells have not received due attention by researchers. This, no doubt, defines a wide area of research and the present paper attempts towards suggesting guidelines for optimization of stiffener

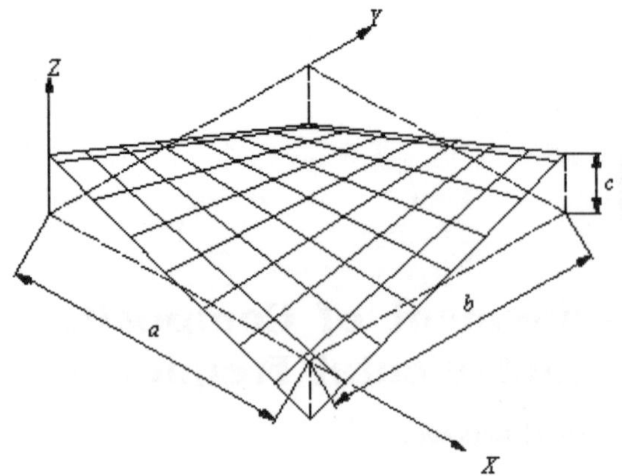

Fig. 1. Surface of a skewed hypar shell.

Surface equation: $z = \dfrac{4c}{ab}\left(x - \dfrac{a}{2}\right)\left(y - \dfrac{b}{2}\right)$

arrangement to minimize the material consumption of composite hypar shells with respect to fundamental frequency.

2. MATHEMATICAL FORMULATION

The stiffened shell element is formulated by combining an eight noded shell bending element and three noded curved beam element and the systematic development of the stiffness and mass matrices are reported elsewhere [6]. The same paper reports about the validation of the formulation through solution of benchmark problems.

3. NUMERICAL EXAMPLES

Fundamental frequencies are obtained for symmetric angle ply laminate (+45°/–45°/–45°/+45°) for different biaxial arrangement of stiffeners and stiffener depth. The individual lamina properties are assumed to be as $E_{11} = 25E_{22}$, $G_{12} = G_{13} = 0.5E_{22}$, $G_{23} = 0.2E_{22}$, $v_{12} = v_{21} = 0.25$. However, in all the cases the fibers in the stiffeners are considered to be arranged in a single layer along the length. In all the cases only the converged results are presented.

4. RESULTS AND DISCUSSIONS

45°/–45°/–45°/+45° stacking order is taken up to investigate the effect of number of stiffeners, which may vary from 1 to 10 in either or both of the directions, on the fundamental frequency. The results furnished in Table 1 reveal that for any given value of n_x (0 to 8) when n_y increases, the nondimensional fundamental frequency ($\overline{\omega}$) always increases upto $n_y = 8$. However, when n_y is varied beyond 8 for given value of $n_x = 9$ or 10, $\overline{\omega}$ sometimes shows a marginal change and even may show a decreasing tendency. Addition of stiffeners to a bare shell surface increases both its stiffness and mass. When increase of stiffness is more significant than mass addition, the fundamental frequency tends to increase. But if the mass contribution prevails over the stiffness contribution of the stiffeners the shell may suffer from a resulting flexibility and $\overline{\omega}$ may decrease. This explains the above observation. For the same reason when n_y is fixed between 0 to 5 or at 8 and n_x is increased, $\overline{\omega}$ increases monotonically but when n_y is fixed at 6, 7, 9 or 10, the variation of $\overline{\omega}$ with n_x is not always monotonic.

As evident from Table 1, for simply supported shells for almost all given values of n_x, $\overline{\omega}$ tends to attain a saturation when n_y exceeds 5 and a shell with $n_y = 5$ (for any given value of n_x except 7) has $\overline{\omega}$ equal to 90% or more than that of a shell with $n_y = 10$. For $n_x = 7$ above value is marginally less than 90% (89.63%). Hence it may be concluded that for any given value if n_x, providing more than 5 y-stiffeners is of no practical use. In fact, in some cases ($n_x = 0$ or 1) a shell with 4 y-stiffeners can attain a frequency more than 90% of that with $n_y = 10$. Interestingly for any given value of n_y also, 5 x-stiffeners may be regarded practically to be as efficient than 10 x-stiffeners according to the criterion mentioned above. If one looks along the diagonal of Table 1 it becomes evident that among shells with $n_x = n_y = n$(say), a shell with $n = 6$ has $\bar{\omega}$ more than 90% of that with $n = 10$ ($0.9 \times 18.87 = 16.98$). Other stiffener combinations where $n_x \neq n_y$ may yield $\bar{\omega}$ more than 16.98 provided the number of stiffeners along either of the directions is not less than 4.

The variations of $\bar{\omega}$ with d_{st}/h ratio are presented in Fig. 2 with $n_x = n_y = n$ varying between 1 to 6. The graphs show that for any given value of n when the dst/h ratio is increased the ratio of frequencies of stiffened and bare shells ($\bar{\omega}_{st}/\bar{\omega}_b = k$) initially increases but then attains a saturation. For $n = 6$ such saturation values of k is 2.8.

Table 1. Non-dimensional fundamental frequency of simply supported laminated composite hypar shell of 45°/ –45°/ – 45°/ + 45° lamination and different combinations of stiffener arrangements

$n_y \rightarrow$ $n_x \downarrow$	0	1	2	3	4	5	6	7	8	9	10
0	8.97	9.12	9.30	9.50	9.85	10.06	10.40	10.52	10.68	10.74	10.78
1	9.09	10.55	10.84	11.07	11.49	11.74	12.18	12.33	12.57	12.64	12.72
2	9.26	10.82	11.75	12.14	12.62	12.91	13.45	13.62	13.92	13.98	14.09
3	9.45	11.05	12.13	12.95	13.61	13.96	14.55	14.74	15.06	15.12	15.26
4	9.80	11.45	12.61	13.60	14.58	15.08	15.66	15.84	16.10	16.14	16.28
5	10.00	11.71	12.90	13.95	15.06	15.87	16.61	16.78	17.03	17.05	17.20
6	10.34	12.14	13.42	14.52	15.63	16.57	17.43	17.66	17.74	17.74	17.86
7	10.45	12.28	13.59	14.71	15.80	16.74	17.60	18.28	18.52	18.43	18.68
8	10.61	12.52	13.87	15.02	16.05	16.97	17.68	18.45	18.55	18.60	18.70
9	10.67	12.57	13.93	15.07	16.09	16.99	17.67	18.36	18.59	19.05	18.88
10	10.71	12.65	14.04	15.21	16.22	17.14	17.80	18.51	18.64	18.86	18.87

$a/b = 1$, $a/h = 100$, $c/a = 0.2$; $E_{11} = 25E_{22}$, $G_{12} = G_{13} = 0.5E_{22}$, $G_{23} = 0.2E_{22}$, $n_{12} = n_2 = 0.25$, $b_{st}/h = 1$, $d_{st}/h = 2$. Each stiffener has a single lamina with fibers along its length.

Table 2 is derived from Fig. 2 and furnishes the values of r (which is the ratio of weights of stiffened and bare shells) for different values of n and k. The values of d_{st}/h ratios are given in parentheses. If a particular value of k is unattainable for a given value of n the corresponding place of the table is left blank. It is found from the Table 2 that for any given value of k, there are different options of r and the minimum value of r among these is obviously the most economical solution provided the

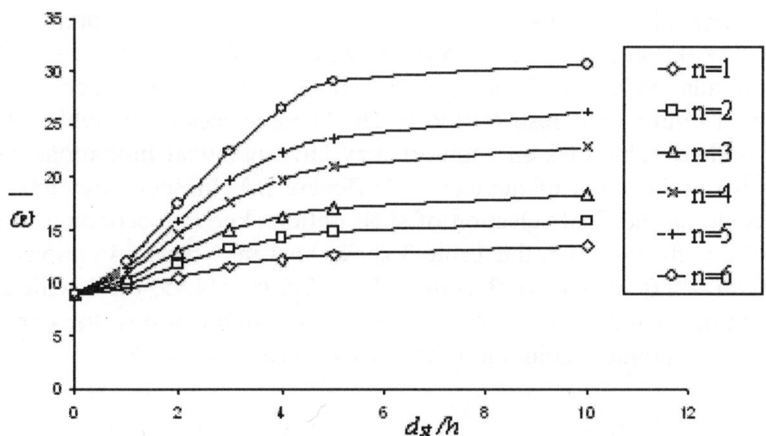

Fig. 2. Variation of nondimensional fundamental frequency with stiffener depth to shell thickness ratio for orthogonally stiffened composite hypar shell.

corresponding value of d_{st}/h ratio is not unacceptable due to other considerations such as aesthetics and headroom. Interestingly, for any given value of k, although d_{st}/h decreases monotonically with

n variation of r with n is quite arbitrary. In some cases a large value of d_{st}/h ratio combines with a small value of n, to give economy (low value of r) and in some other situations a combination of a greater number of biaxial but shallow stiffeners seems to be most acceptable economically. Hence, this table is expected to be very useful to practicing engineers to select a particular set of values of n and d_{st}/h to achieve a given value of k considering all the aspects of economy and other architectural and functional criteria.

Table 2. Weight ratios (W_{st}/W_b = r) of stiffened to bare shell for different values of frequency ratios ($k = \bar{\omega}_{st}/\bar{\omega}_b$) of stiffened to bare shell and $n_x = n_y = n$ (Ref. Fig. 2)

	Simply supported (+45°/ − 45°/ − 45°/ + 45°)					
$k \to$	1.3	1.6	1.9	2.2	2.5	2.8
$n_x = n_y = n$ $\downarrow$						
1	1.0645 (3.24)	–	–	–	–	–
2	1.0768 (1.94)	1.1584 (4.0)	–	–	–	–
3	1.0904 (1.53)	1.1531 (2.59)	1.2660 (4.5)	–	–	–
4	1.0972 (1.24)	1.1474 (1.88)	1.2078 (2.65)	1.3183 (4.06)	1.7542 (9.62)	–
5.	1.1034 (1.06)	1.1580 (1.62)	1.2126 (2.18)	1.2925 (3.00)	1.4076 (4.18)	1.8375 (8.59)
6	1.1059 (0.91)	1.1641 (1.41)	1.2118 (1.82)	1.2875 (2.47)	1.3562 (3.06)	1.4214 (3.62)

The values in the parentheses are d_{st}/h ratios.

Table 3 presents the values of the number of biaxial stiffeners ($n_x = n_y = n$) in each direction for different values of the d_{st}/h ratio for which a particular value of k is just exceeded with the corresponding values of r in parentheses. This table is prepared taking practical upper limits of $d_{st}/h = 4.0$ and $n = 6$. For example when $d_{st}/h = 3.5$ and $n = 3$ the value of k just exceeds 1.6. A particular value of d_{st}/h ratio may not be sufficient to achieve a particular value of k since the upper limit of n is restricted to 6. The blank spaces in the table refer to such cases. In order to attain a desired value of k an engineer may have practical limitations either in terms of headroom (d_{st}/h ratio) or in terms of number of stiffeners (n). In such situations Table 3 will be of great help to decide on the final selection of shell options keeping economy in consideration. If one has to attain $k^3 1.6$ with d_{st}/h 3.5, the Table 3 shows that they have 14 options: 1 with $d_{st}/h = 1.5(n = 6)$, 2 with $d_{st}/h = 2.0$ ($n = 4, 6$), 3 with $d_{st}/h = 2.5$ ($n = 4, 5, 6$), 4 with $d_{st}/h = 3.5$ ($n = 3, 4, 5, 6$) and 4 with $d_{st}/h = 3.5$ ($n = 3, 4, 5, 6$). Of these combinations, the one with $d_{st}/h = 2.0$ and $n = 4$ is the most economical solution as it has the least value of r.

Table 3. Values of $n_x = n_y = n$ for different values of frequency ratio of stiffened to bare shell and $(k = \overline{\omega}_{st} / \overline{\omega}_b)$ d_{st}/h ratio.

	Simply supported Lamination: $+45°/-45°/-45°/+45°$					
$k \rightarrow$ $d_{st}/h \downarrow$	1.3	1.6	1.9	2.2	2.5	2.8
1.0	5 (1.0975)	–	–	–	–	–
1.5	3 (1.0887)	6 (1.1746)	–	–	–	–
2.0	2 (1.0792)	4 (1.1568)	6 (1.2328)	–	–	–
2.5	2 (1.0990)	4 (1.1960)	5 (1.2438)	6 (1.2910)	–	–
3.0	1 (1.0597)	3 (1.1773)	4 (1.2352)	5 (1.2925)	6 (1.3492)	–
3.5	1 (1.0697)	3 (1.2955)	4 (1.2744)	5 (1.3413)	6 (1.4074)	
4.0	1 (1.0796)	2 (1.1584)	4 (1.3136)	4 (1.3136)	5 (1.3900)	6 (1.4660)

Values in the parentheses indicate weight ratios (r) of stiffened shell to bare shell.

CONCLUSION

The following conclusions are drawn from the present study.

1. With the increase of the number of stiffeners, either uniaxial or biaxial, the fundamental frequency increases but reaches a saturation whereafter there is no appreciable increase in frequency with the provision of additional stiffeners.

2. For any given number of stiffeners in one of the directions, there is no point in providing more than five stiffeners for simply supported shells along the other direction. Among shells with equal number of x and y stiffeners there is no appreciable increase in frequency when the number of stiffeners in either direction exceeds 5.

3. The present study proves that there are two ways of increasing the fundamental frequency of a bare shell by stiffening, either by increasing the stiffener depth keeping the number of stiffeners fixed or by increasing the number of stiffeners with a given stiffener depth. Table 2 combines these possibilities with the ratios by which the weight of bare shell increases on stiffening and will help a practicing engineer to choose his optimum solution considering both economy and other practical limitations such as aesthetics and headroom. In many practical situations an engineer has to restrict the stiffener depth and also the number of stiffeners in either direction. Table 3 is prepared taking practical upper limits of $d_{st}/h = 4$ and $n_x = n_y = 6$ and will help a practicing engineer to decide on the number and depth of stiffeners which will economically achieve a given increase of the fundamental frequency of a bare shell.

NOMENCLATURE

a, b, c	length, width and rise of shell
b_{sx}, b_{sy}	width of x and y stiffener
b_{st}	width of stiffener
d_{sx}, d_{sy}	depth of x and y stiffener
d_{st}	depth of stiffener
D	flexural rigidity
E_{11}, E_{22}	elastic moduli
G_{12}, G_{13}, G_{23}	shear moduli of a lamina
np	number of plies in a laminate
n_x, n_y	number of stiffeners in x and y directions
z_k	distance of bottom of kth ply from mid-surface of a laminate
v_{12}, v_{21}	Poisson's ratios
ρ	density of material
ω	natural frequency

ACKNOWLEDGEMENTS

The first author gratefully acknowledges the financial assistance of CSIR (India) through the Senior Research Fellowship vide grant no. 9/96 (412) 2003-EMR-I.

REFERENCES

1. G. Sinha, and M. Mukhopadhyay, 1995, Static and dynamic analysis of stiffened shells – A review. Proc. Indian Natn. Sci. Acad. 61A (3 & 4), 195-219.
2. A.N. Nayak, and J.N. Bandyopadhyay, 2002, Free vibration analysis and design aids of stiffened conoidal shells, Journal of Engineering Mechanics. 128(4), 419-427.
3. A.N. Nayak, and J.N. Bandyopadhyay, 2002, On the free vibration of stiffened shallow shells. Journal of Sound and Vibration. 255(2), 2002b, 357-382.
4. D. Chakravorty, J.N. Bandyopadhyay, and P. K. Sinha, 1998, Applications of FEM on free and forced vibrations of laminated shells, ASCE Journal of Engineering Mechanics. 124 (1), 1-8.
5. S. Sahoo, and D. Chakravorty, 2006, Deflections, forces and moments of composite hypar shell roofs under concentrated load, Journal of Strain Analysis for Engineering Design. 41(1), 81-97.
6. S. Sahoo, and D. Chakravorty, 2005, Free vibration of laminated composite stiffened hypar shell roofs by finite element, RTDMT, Coimbatore, 17-18 March, E-Proc., Session S15, Paper 5, 1-11.

82

Factorial Analysis of a Magneto-Rheological Damper Used for Semi-Active Vibration Control

SHIVARAM A.C.[1] AND K.V. GANGADHARAN[2]

[1]M.Tech Student, ac.shivaram@gmail.com, [2]Assistant Professor, kvganga@nitk.ac.in
Department of Mechanical Engineering, National Institute of Technology Karnataka,
Surathkal, Mangalore-575025, India.

ABSTRACT

Various factors like vibration frequency, amplitude, shearing gap, volume fraction and magneto-motive force influence the total damping coefficient of an MR damper. In this work, initially an MR damper is tested using two-level factorial design to identify the most significant factors and there interactions. Then, using response surface method, a statistical model is proposed, which can predict damping coefficient in the considered range of factors. A response surface of damping coefficient with all other factors varying is plotted, which could be of assistance in choosing optimal design parameters.

Keywords: Magneto-Rheological damper, Factorial design, Response surface method.

1. INTRODUCTION

A Magneto-Rheological (MR) damper is similar to any other damper with piston and cylinder arrangement; however, the only difference being type of fluid used. An MR damper uses MR fluid, whose viscosity may be varied from a light oil level up to semi-solid state by applying magnetic field [1]. Such a wide range of controllable viscosity has opened the doors for a variety of applications like brakes, clutches and dampers [2, 3]. MR dampers are mainly being tried out in the area of structural damping as well as automotive applications. Machine vibration control could be one more interesting application area. For effective use of this controllability offered by MR dampers, an efficient controller should be designed which can accommodate all the non-linearities inherent in the system. Such a controller demands a detailed mathematical model [4, 5], which clearly defines the system behavior in the entire working range. One may find few articles in literature related to modeling of MR dampers; however, none are exhaustive covering entire spectrum of structural, automotive and machine vibration control applications. Another approach is to do away with the idea of model based control system design, and work on adaptive algorithms based on experimental and run-time system response data [6]. Objective of this work is to generate and communicate such an experimental data, which could be of use in designing an efficient MR Damper and a control system for it.

2. EXPERIMENTATION

2.1 Approach

Damping coefficient (C) of an MR damper may be split into three parts, namely $C_{Friction}$, $C_{Viscous}$, and $C_{Magnetization}$ as shown in Fig. 1, Eq. (1) and Eq. (2). A designer would wish to minimize $C_{Friction}$ to the least value possible, and maximize the controllable component $C_{Magnetization}$. $C_{Viscous}$ togetherwith $C_{Friction}$ would decide the damping available in MR off state (zero current state).

$$C_{MR\text{-}Damper} = C_{Friction} + C_{Viscous} + C_{Magnetization} \qquad ....(1)$$

$$C_{MR\text{-}Damper} = \frac{4F}{\pi\omega X} + \mu_o\left[\frac{3\pi D_o L}{4g^3}\left(D_o^2 - D_i^2 + 2D_o g\right)\right] + C_{Magnetization} \qquad ...(2)$$

$C_{Friction}$ is a function of friction force (F), which is mainly due to oil seal, circular frequency (w) and amplitude of vibration (X). $C_{Viscous}$ is determined by MR off state viscosity (μ_o), piston geometry like length (1), diameter (D_0), piston rod diameter (D_i), shearing gap (g) and viscosity of the carrier fluid used.

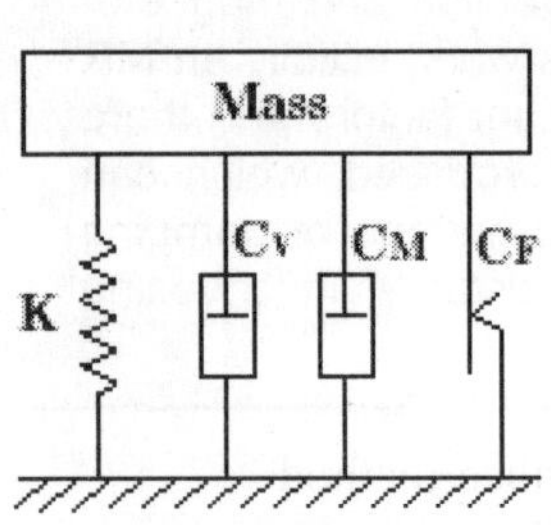

Fig. 1. A spring mass damper system.

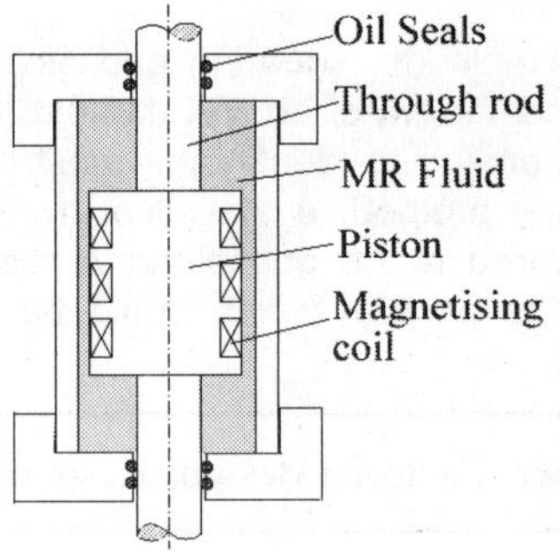

Fig. 2. Piston cylinder arrangement.

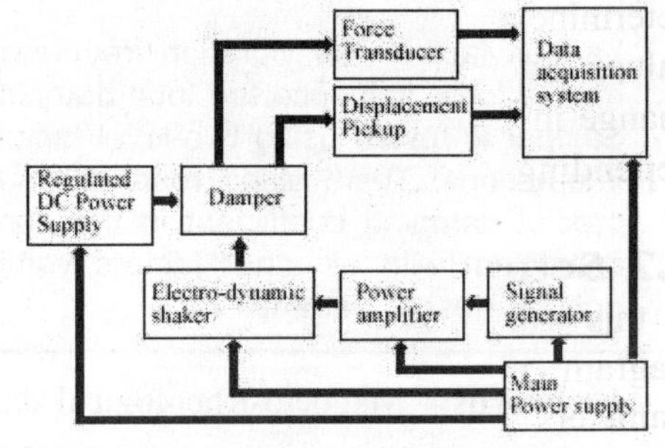

Fig. 3. Experimental Setup.

MR off state viscosity is again dependent on volume fraction of iron powder in dispersant medium and anti-settling agents used. C Magnetization is believed to be influenced by factors like magnetizing current, number of turns in the magnetizing coil, volume fraction, and average particle size of iron powder, shearing gap, type of anti-settling agents used, vibration frequency and amplitude. Also, there could be interaction between few of the above factors, which could result in a more desirable output, than by changing only one factor at a time.

In present work, factors like vibration frequency, amplitude, shearing gap, volume fraction of iron powder and Magneto-motive force MMF (Ampere-turns) are considered for analysis. Initially all factors are varied in two levels (high, low) and response variable the damping coefficient (C-Ns/m) is determined at each combination level. Experimentation is fully randomized for three factors namely, frequency, amplitude and MMF. Three samples (replicates) are taken at each combination level. Other two factors are varied in steps as shown in Table 1. Above experimental design is historically known as 2n factorial design [8]. Aim of these tests is to screen through all the factors, and identify the critical factors which have a lot of influence on the response variable.

Table 1. Factor levels for 2^5 factorial design.

Factors	Low	High
A. Volume fraction of iron powder	10%	40%
B. Shearing gap	1.5 mm	2.5 mm
C. Magneto-motive force (MMF)	0 Ampere-turns	120 Ampere-turns
D. Frequency	0.2 Hz	8 Hz
E. Stroke	1 mm	3 mm

Table 2. Factor levels for Response surface method

Factors	Low	Mid	High
A. Volume fraction	10%	25%	40%
B. Shearing gap	1.5 mm	2 mm	2.5 mm
C. Magneto-motive force (MMF)	0 Ampere-turns	60 Ampere-turns	120 Ampere-turns
D. Frequency	4 Hz	6 Hz	8 Hz
E. Stroke	3 mm	3 mm	3 mm

Once critical factors have been identified, those factors are varied in three levels as shown in Table 2 and response variable determined. Face centered central composite design [8] is used as a guide to acquire and analyze the experimental data in this stage. Based on these experimental data, a response surface is plotted, which fits a best polynomial. Aim of this set of experiments, is to determine the response surface of damping coefficient as all other factors varies over a range of values. Such a surface would give a rough estimate about how the response variable changes with change in various factors. Also, it will be an aid to choose optimal design and operating conditions, depending on the region of operation.

2.2 Setup

In this work a single tube, through rod type damper [9] is used for experimentation. A schematic diagram of damper and experimental setup is shown in Fig. 2 and Fig. 3. Pistons of different diameters are used to achieve different shearing gaps. MR fluid of Carbonyl iron powder dispersed in silicone oil was prepared in-house at different % volume fractions of iron powder. No additives were used during fluid preparation. Current through the piston coils was varied so as to achieve 0 to 120 ampere-turns of MMF per coil. Entire damper assembly was mounted on an electro-dynamic shaker which in turn is driven by a power amplifier and signal generator. A data acquisition card (National Instrument's 4472 dynamic signal analyzer) was used to acquire force data (HBM force-transducer), and displacement data (Micro-epsilon non-contact laser pickup). All this was managed with the help of programs written on LabVIEW platform. Data thus obtained was analyzed in Design-Expert, a statistical software tool from Stat-Ease, Inc.

3. RESULTS AND DISCUSSION

Summary of individual as well as interaction effects of all the factors are listed in Table 3. From the table it can be observed that frequency, MMF and interaction of frequency and stroke are the major contributing factors. Same may be observed in Fig. 4 the half-normal probability plot of the effects. Fig. 5 shows normal probability plot of residuals.

Table 3. Factor effect estimates

Factor	Sum of squares	% Contribution
A. Volume fraction	0.085	0.69
B. Shearing gap	0.049	0.4
C. MMF	0.71	5.81
D. Frequency	9.13	74.26
E. Stroke	0.16	1.32
DE	0.63	5.14
ACD	0.27	2.21
CD	0.27	2.2
BCD	0.19	1.54
CDE	0.18	1.45
AB	0.1	0.84
AD	0.062	0.5

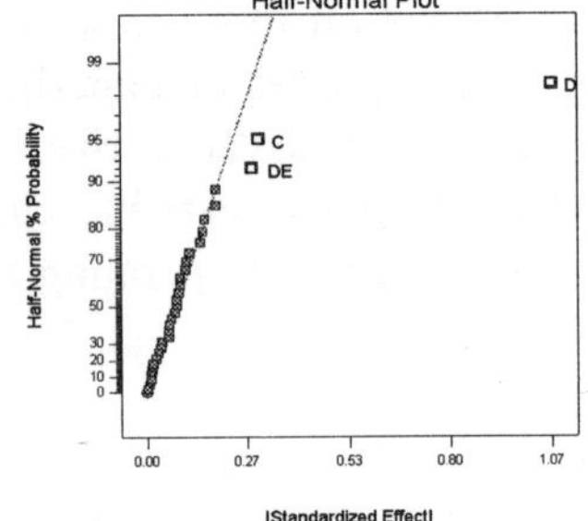

Fig. 4. Half-Normal probability plot of effects.

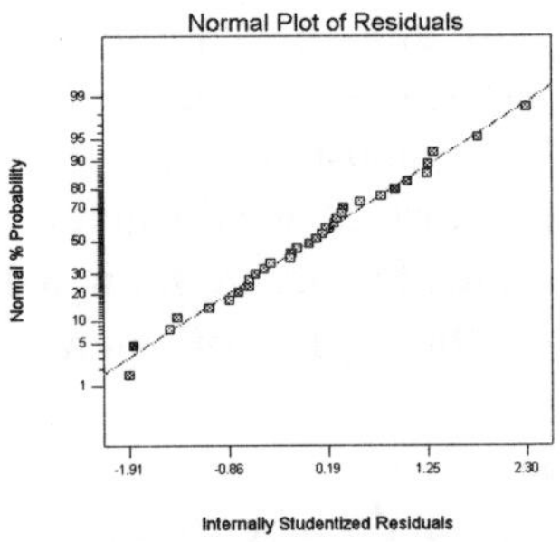

Fig. 5. Normal probability plot of residuals.

Figures 6 to 9 show the interaction effects of various factors. It is observed that lower the frequency and larger the stroke, higher will be the damping coefficient. However, it is predominantly

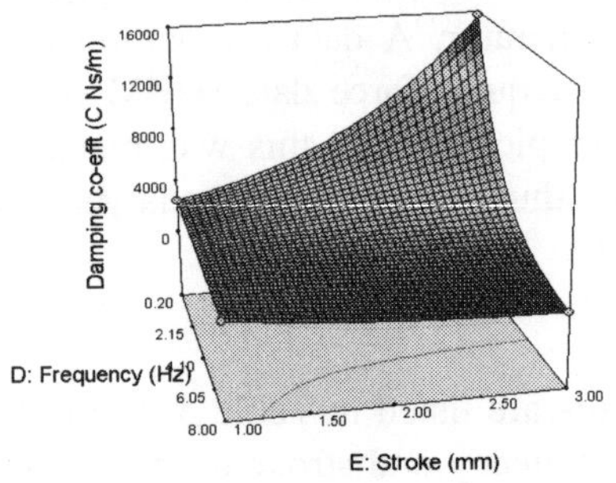

Fig. 6. MR on state with 120 Ampere-turns.

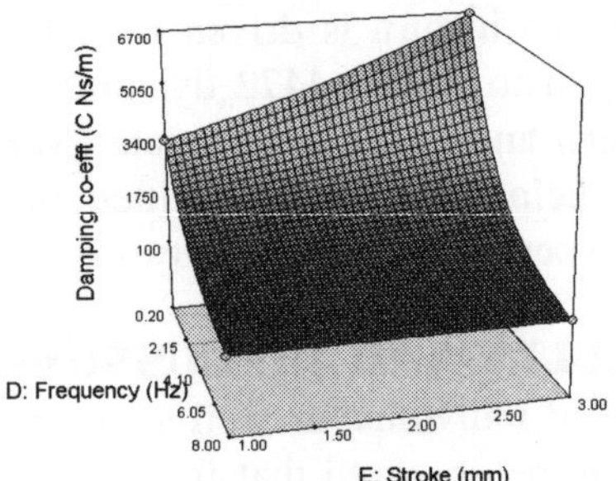

Fig. 7. MR off state.

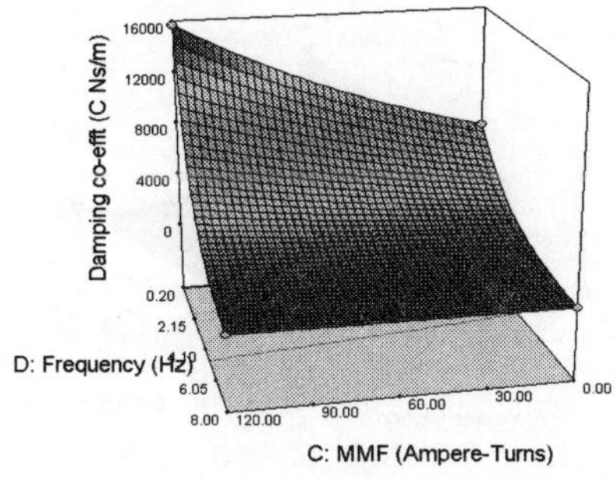

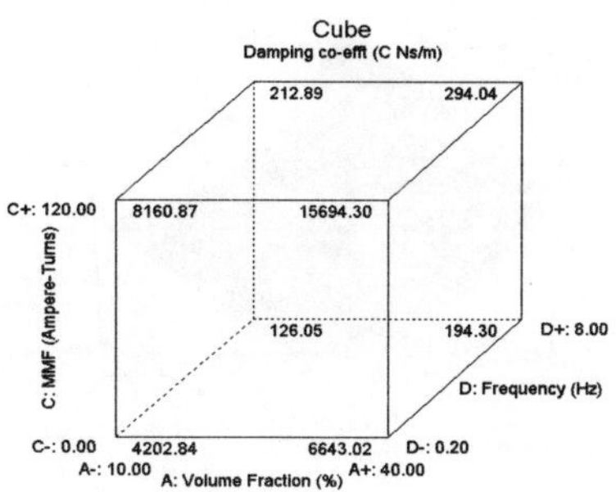

Fig. 8. 40% volume fraction, 2.5 mm gap, Stroke 3 mm.

Fig. 9. 2.5 mm gap, Stroke 3 mm.

friction damping at lower frequencies as observed from force-displacement loops [9] shown at Figs. 12 to 21. Also, it is observed that stroke has a minimal effect on the $C_{Magnetization}$. Hence, this factor will be held constant at 3 mm in response surface tests. Other factors are varied in three levels as summarized earlier in the Table 2. It may be observed from Table 4 that a quadratic surface would be the best fit surface, and a cubic surface would be aliased. Such response surface are shown in Figs. 10 and 11.

Table 4. Summary of lack of fit tests for the response surface method's data

Source	Sum of Squares	df	Mean Square	F Value	p-value Prob > F	
Linear	529.777308	20	26.48887	211.8316	< 0.0001	
2FI	408.2275449	14	29.15911	233.1855	< 0.0001	
Quadratic	58.60704645	10	5.860705	46.86808	< 0.0001	Suggested
Cubic	28.28281025	2	14.14141	113.0889	< 0.0001	Aliased
Pure Error	4.626732928	37	0.125047			

Eq. (3) predicts the damping coefficient variation in terms of actual factors A,B,C,D,E as explained in Table 2. Power transformation is used to have a manageable ratio between minimum and maximum response values.

4. DAMPING COEFFICIENT PREDICTION EQUATION IN TERMS OF ACTUAL FACTORS

$$\text{Damping Coefficient}^{0.37} = 51.02651 - 0.34099A - 1.00528B + 0.092546C - 12.81731D - 0.018560AB$$
$$+9.54696 \times 10^{-4} AC - 0.022931AD - 0.018673BC + 0.48354BD - 7.23506 \times 10^{-3} CD + 0.011660A^2$$
$$-0.62759B^2 + 5.32536 \times 10^{-5} C^2 + 0.92046D^2$$

$$...(3)$$

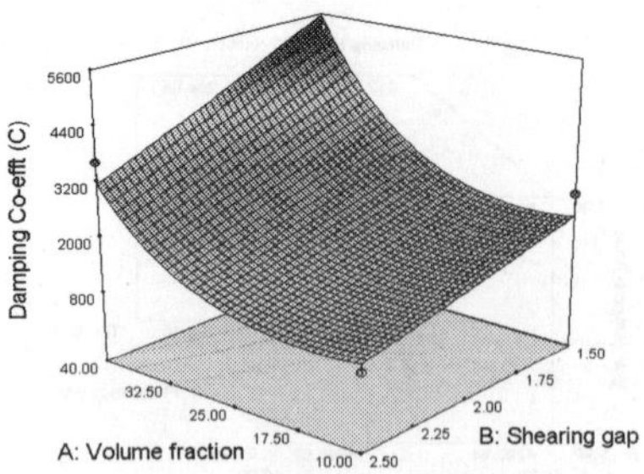

Fig. 10. 4 Hz, 120 Ampere-turns.

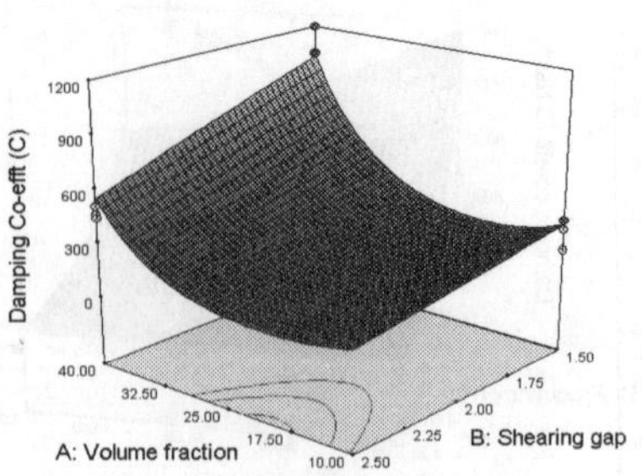

Fig. 11. 8 Hz, 120 Ampere-turns.

Figures 12 to 21 show force versus displacement and force versus velocity curves at various conditions as indicated in the title. Stroke is maintained constant at 3 mm in all the cases.

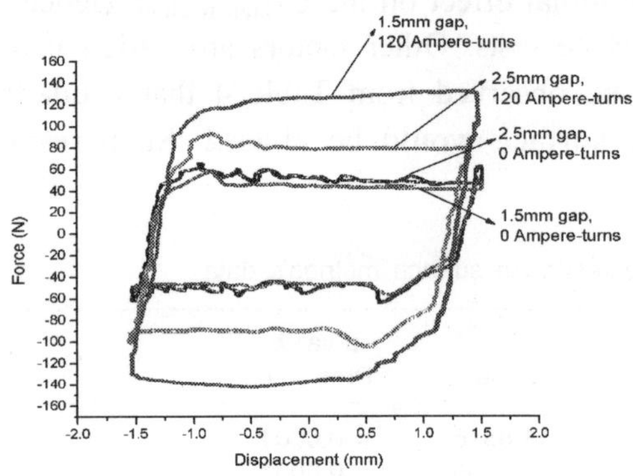

Fig. 12. 10% volume fraction, 0.2 Hz.

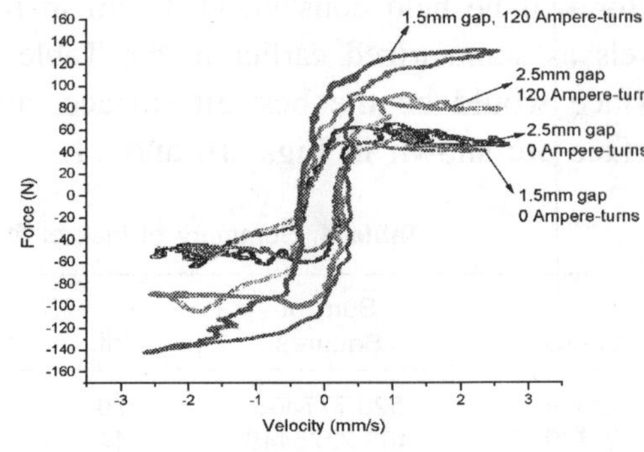

Fig. 13. 10% volume fraction, 0.2 Hz.

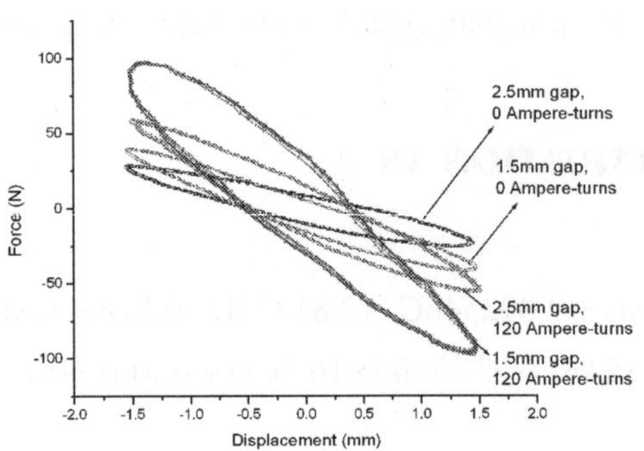

Fig. 14. 10% volume fraction, 8 Hz.

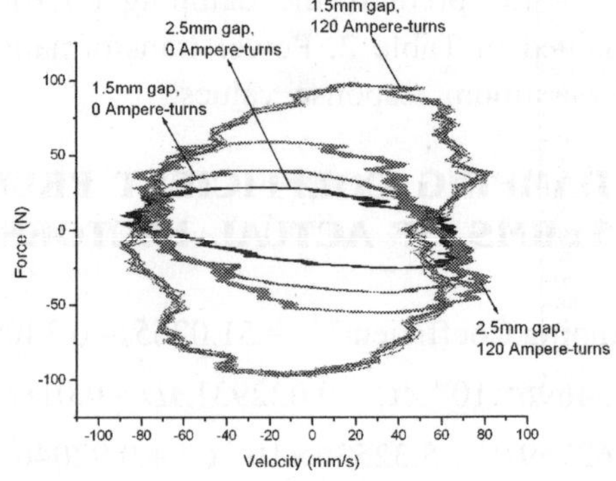

Fig. 15. 10% volume fraction, 8 Hz.

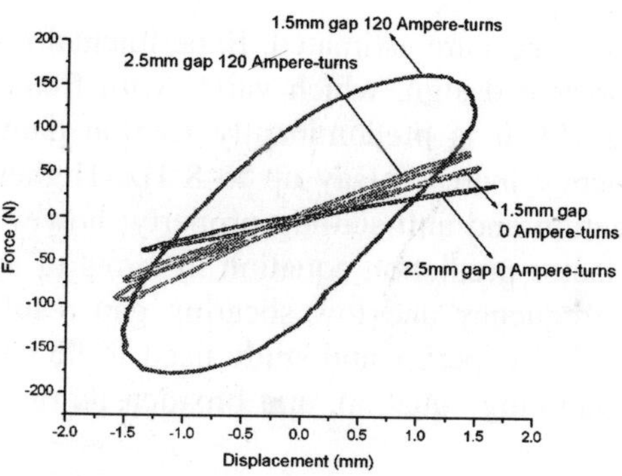

Fig. 16. 25% volume fraction, 6 Hz.

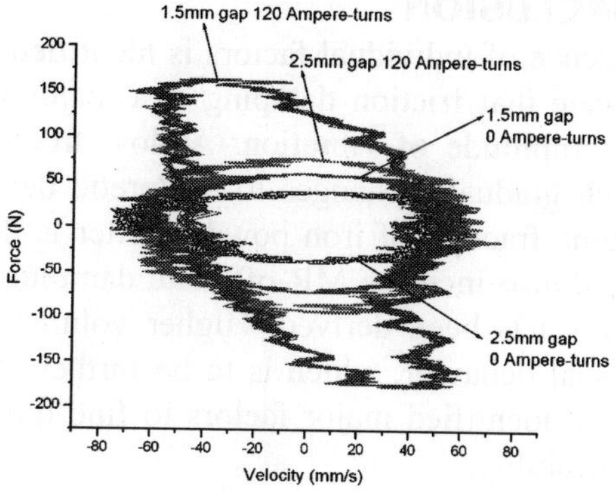

Fig. 17. 25% volume fraction, 6 Hz.

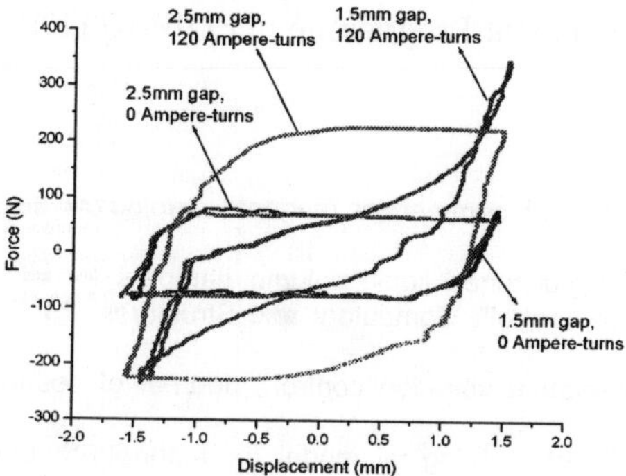

Fig. 18. 40% volume fraction, 0.2 Hz.

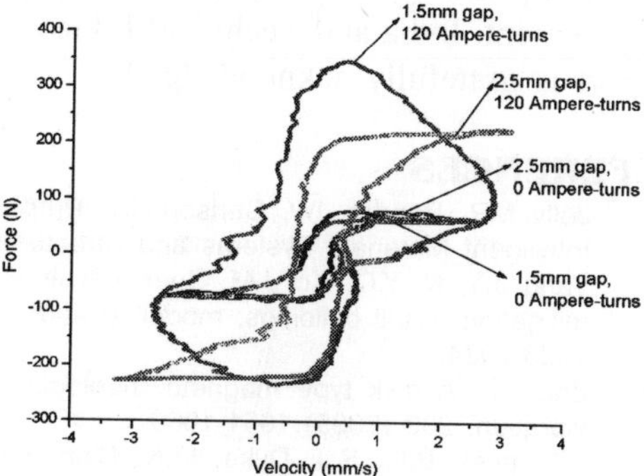

Fig. 19. 40% volume fraction, 0.2 Hz.

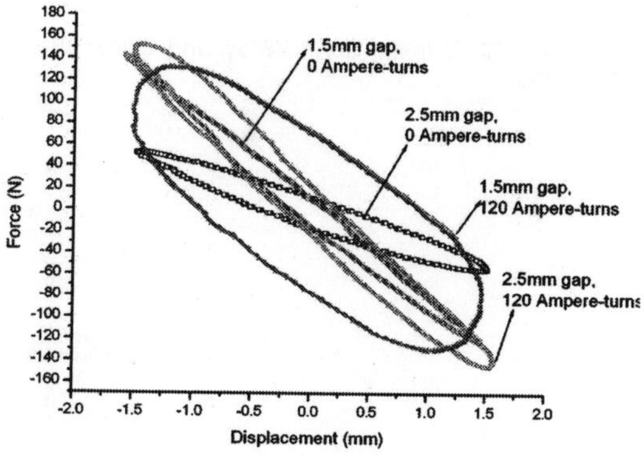

Fig. 20. 40% volume fraction, 8 Hz.

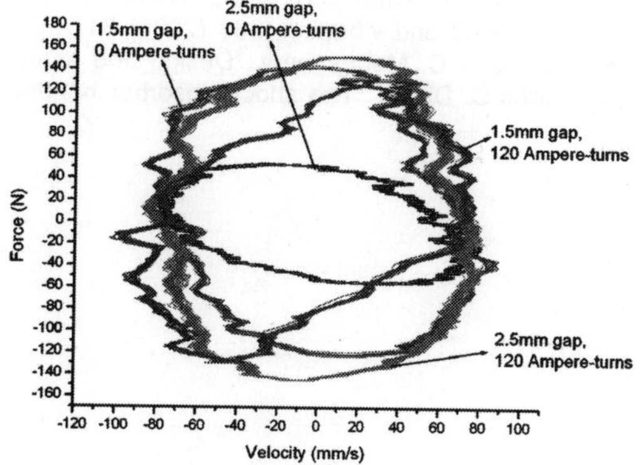

Fig. 21. 40% volume fraction, 8 Hz.

CONCLUSION

Influence of individual factors is identified and interaction effects are estimated. Experimental results indicate that friction damping is a major factor in the present design, which varies with frequency and amplitude of vibration. At low frequencies say 0.2 Hz, it is predominantly friction damping which gradually changes to hysteretic damping as frequency increases say up to 8 Hz. Higher the volume fraction of iron powder better is the C magnetization and anti-settling property; however, it would also increase MR-off state damping. A damping factor prediction equation in terms of actual factors has been derived. Higher volume fraction, low frequency and low shearing gap results in unusual behavior, which is to be further explored. A detailed experimental study need to be carried out on identified major factors to fine tune the derived damping equation, and broaden its range of applicability.

ACKNOWLEDGEMENTS

The work presented in this paper was supported under R&D grant from Ministry of Human Resource Development, India and Technical Education Quality Improvement Programme (TEQIP-NITK). This support is gratefully acknowledged.

REFERENCES

1. Jolly MR, Bender JW, Carlson JD. Properties and applications of commercial magneto-rheological fluids. J Intelligent Materials Systems and Structures 1999; 10:5-13.

2. Yang J.Y, Ni Y.Q, Ko J.M, Spensers Jr B.F, "Magneto-rheological tuned liquid column dampers for vibration mitigation of tall buildings: modeling and analysis of open loop control", Computers and Structures, 83 (2005), 2023-2034.

3. Zhu. C, "A disk type magneto rheological damper for rotor system vibration control", Journal of sound and vibration, 283 (2005) 1051-1069.

4. Spencer, B.F., S.J. Dyke, M.K. Sain and J.D. Carlson, "Phenomenological model for magneto-rheological dampers", Journal of Engineering Mechanics, 230-238, March 1997.

5. Tsang .H.H, Su .R.K.L, Chandler .A.M, "Simplified inverse dynamics model for MR fluid dampers", Engineering Structures 28 (2006) 327-341.

6. Song X, Ahmadian M, Southward S, Miller .L.R, "An adaptive semi-active algorithm for magneto-rheological suspension systems", Journal of vibration and Acoustics, Vol 127, 493-503, October 2005.

7. Hong. S.R, Choi. S.B, *et al.*, "Non-dimensional analysis and design of a magneto-rheological damper", Journal of sound and vibration, 288 (2005) 847-863.

8. Douglas C. Montgomery, "Design and analysis of experiments", 5th Edition, John Wiley and Sons Inc, 2001.

9. John.C. Dixon, "The shock absorber handbook", SAE Inc, 1999.

83

Axisymmetric Vibrations of Variable Thickness Moderately Thick Non-Homogeneous Polar Orthotropic Annular Plates

ROSHAN LAL AND SHUCHITA SHARMA

Department of Mathematics, I.I.T. Roorkee-247 667, India email: rlatmfma@iitr.ernet.in

ABSTRACT

Free vibration analysis of non-homogeneous annular plates of linearly varying thickness possessing polar orthotropy has been presented using the first order shear deformation plate theory of Mindlin. Hamilton's energy principle has been used to derive the coupled differential equations governing the motion of such plates. The consideration of polar orthotropy together with thickness variation and non-homogeneity of the plate material further complicates these resulting equations. The frequency equations for three different boundary conditions have been obtained employing Chebyshev collocation technique, which has minimax property. The effect of various structural parameters such as radii ratio, thickness variation, non-homogeneity and density along with the shear deformation and rotatory inertia effects has been studied on the vibrational characteristics of the plate for the first three modes of vibration. Mode shapes are computed for a specified plate. A comparison of frequencies with the corresponding values obtained from classical plate theory has been presented.

Keywords: Non-homogeneous; polar orthotropic; variable thickness; Mindlin annular plates.

1. INTRODUCTION

Nowadays, engineers are able to tailor advanced materials by combining two or more materials, which are lighter, stiffer and stronger than the previously used conventional materials. These materials have a wide range of operating temperatures besides high damping and resistance to corrosion. The use of such high technology composite materials in the design of plate type structural components has necessitated to study the vibrational behaviour of anisotropic plates. Variable thickness plates fabricated out of modern composites are not only reduced in size and weight but also meet the desirability of high strength. Furthermore, their use in various technological situations such as in space shuttle and high speed aircraft, structures under high-temperature environments, etc. demand that non-homogeneity of the material should be taken into account to predict their dynamic behaviour with a fair amount of accuracy.

As the plates used in actual practice have appreciable thickness, it is important to include the effects of transverse shear and rotatory inertia, neglected in classical plate theory. In the present work, both these effects have been considered to study the axisymmetric vibrations of non-homogeneous polar orthotropic annular plates of variable thickness. A more general model for the non-homogeneity of the plate material has been proposed and used, from which most of the earlier proposed models in the literature [1] can be regarded as particular cases. An approximate solution of the coupled differential equations governing the motion of such plates has been obtained by using Chebyshev polynomials. A fibre-reinforced plastic (glass-epoxy) has been taken as an example of a polar orthotropic material.

2. FORMULATION

According to Mindlin's shear theory, the differential equations which govern the axisymmetric motion of annular plates of inner and outer peripheral radii b and a respectively, with thickness h(r), referred to a cylindrical polar coordinate system (r, θ, z) are

$$M_{r,r} + (M_r - M_\theta)/r - Q_r - (\rho h^3/12)\,\psi_{r,tt} = 0\,, \qquad Q_r/r + Q_{r,r} - \rho\,h\,w_{,tt} = 0 \qquad ...(1,2)$$

where, t is the time, ρ is the mass density per unit volume, w is the transverse deflection, Ψ_r is the angle of rotation in the rz – plane and M_r, M_θ and Q_r are the moment and shear resultants per unit length.

For the case of axially symmetric displacements and orthotropic plate material

$$M_r = D_r(\psi_{r,r} + (\upsilon_\theta/r)\psi_r)\,, \quad M_\theta = D_\theta(\psi_r/r + \upsilon_r\psi_{r,r})\,, \quad Q_r = k_s G_{r\theta} h\,(\psi_r + w_{,r}) \qquad ...(3)$$

where, $(D_r, D_\theta) = (E_r, E_\theta)h^3(r)/12(1 - \upsilon_r\upsilon_\theta)$ are the flexural rigidities, and $E_r, E_\theta,\ \upsilon_r, \upsilon_\theta$ and $G_{r\theta}$ are the elastic constants, with $\upsilon_r D_\theta = D_r\upsilon_\theta$, $k_s(= \pi^2/12)$ an averaging shear coefficient and a comma followed by a suffix, represents the partial differentiation with respect to that variable.

For elastically non-homogeneous material, we assume that the Young's moduli E_r , E_θ and ρ are the functions of r only and the shear modulus is taken as $\sqrt{E_r E_\theta}/2\,(1 + \sqrt{\upsilon_r\upsilon_\theta})$ following Lekhnitskii [2], and now substitution of the relations (3) into equations (1) and (2), together with

$$w(r,t) = \bar{w}(r)e^{i\,\omega t} \text{ and } \psi_r(r,t) = \psi(r)e^{i\,\omega t} \text{ (for harmonic motion) leads to the equations}$$

$$\phi - k_2 12 k_s G_{r\theta}((1 - \upsilon_r\upsilon_\theta)/E_r)h\,(\psi + \bar{w}^{\,i}) = 0\,, \quad \{r\phi\}^i + 12\rho\,\omega^2((1 - \upsilon_r\upsilon_\theta)/E_r)rh\,\bar{w} = 0, \qquad ...(4,\,5)$$

$$\phi = h^3(\psi^{i\,i} + \psi^i/r - (E_\theta/E_r)\psi/r^2)$$

$$+\{h^3 E_r^{\,i}/E_r + 3h^2 h^i\}(\psi^i + \upsilon_\theta\,\psi/r) + k_1\rho\omega^2((1 - \upsilon_r\upsilon_\theta)/E_r)h^3\psi,$$

ω being the radian frequency, k_1, and k_2 are tracers to identify rotatory inertia and transverse shear terms and dashes denote the differentiation with respect to r. Elimination of $\bar{w}$ from equations (4) and (5) leads to an uncoupled differential equation in Ψ. The resulting equation for linear variation in thickness such that $H = h_0\,(1 - \alpha x)$, and with exponential variation for non-homogeneity of the plate as $E_r = E_1 e^{\mu x}, E_\theta = E_2 e^{\mu x}, \rho = \rho_0 e^{\beta x}$, is reduced to a non-dimensional fourth order linear homogeneous differential equation with variable coefficients, given by

$$A_4\psi^{iv} + A_3\psi^{iii} + A_2\psi^{ii} + A_1\psi^{i} + A_0\psi = 0 \qquad \text{...(6)}$$

where, A_i's, $i = 0,1,...,4$ are given in Appendix A.

Here dashes denotes the differentiation with respect to x, α is the taper parameter, E_1, E_2 are the elastic constants at $x = 0$, and $x = r/a$, $W = \overline{w}/a$, and $H = h/a$ are the dimensionless variables. Also W can be expressed in terms of Ψ and its derivatives.

$$W(x) = -(h_0^2/12\Omega^2)e^{(\beta-\mu)x}(1-\alpha x)^2(B_3\psi^{iii} + B_2\psi^{ii} + B_1\psi^{i} + B_0\psi), \qquad \text{...(7)}$$

$$B_3 = 1, \quad B_2 = 2(1+\mu x)/x, \quad B_1 = \left\{-p + \mu^2 x^2 + (2+\upsilon_\theta)\mu x + k_1\Omega^2 x^2 e^{(\beta-\mu)x}\right\}/x^2,$$

$$B_0 = \left\{(p + \upsilon_\theta\mu^2 x^2) - p\mu x + k_1\Omega^2 x^2 e^{(\beta-\mu)x}(1+\beta x)\right\}/x^3.$$

Equation (6) together with the boundary conditions at the inner and outer edges constitutes a well defined two point boundary value problem in the range $(\varepsilon, 1)$, $\varepsilon = b/a$, which is solved by Chebyshev collocation technique. Accordingly, the range $\varepsilon \le \times \le 1$ is transformed to the applicability range of the technique, $-1 \le y \le 1$, by introducing a new independent variable $y \equiv \{2\times - (1+\varepsilon)\}/(1 - \varepsilon)$. Equations (6) and (7) now become

$$P_4\psi^{iv} + P_3\psi^{iii} + P_2\psi^{ii} + P_1\psi^{i} + P_0\psi = 0, \qquad \text{...(8)}$$

$$W(y) = -(h_0^2/12\Omega^2)e^{(\beta-\mu)x}(1-\alpha x)^2\{W_3\psi^{iii} + W_2\psi^{ii} + W_1\psi^{i} + W_0\psi\}, \qquad \text{...(9)}$$

where, $P_i = \eta^i A_i$, $W_i = \eta^i B_i$, $\eta = 2/(1-\varepsilon)$, $i = 0,1,2,3,4$. Following [1, 3], one can assume

$$\psi^{iv} \equiv \frac{d^4\psi}{dy^4} = \sum_{k=0}^{m-5} c_{k+5} T_k, \text{ and its successive integrations lead to}$$

$$\Psi = c_1 + c_2 T_1 + c_3 T_1^1 + c_4 T_1^2 + \sum_{k=0}^{m-5} c_{k+5} T_k^4.$$

where, $c_j (j = 1, 2, ..., m)$ are unknown constants, T_j $(j = 0, 1, 2, ..., m-5)$ are the Chebyshev polynomials and T_k^j represents the jth integral of T_k. Substitution of Ψ and its derivatives in equation (8) and satisfaction of this resultant equation at (m-4) collocation points given by

$$y_k = \cos\{(2k+1)\pi/2(m-4)\}, \quad k = 0, 1, 2, ... m-5,$$

provides a set of (m-4) equations, which can be written in matrix form as

$$[N][C^*] = [0], \qquad \text{...(10)}$$

where, N and C^* are matrices of order (m-4) x m and m × 1, respectively.

A set of four more equations is obtained for (i) C-C, clamped at both the inner and outer edges; (ii) C-S, clamped at the inner and simply supported at the outer edge; (iii) C-F, clamped at the inner and free at the outer edge; boundary conditions, by satisfying the relations: $W = \Psi = 0$ and $W = \eta(d\Psi/dy) + (\upsilon_\theta/x)\Psi = 0$, and $\eta(d\Psi/dy) + (\upsilon_\theta/x)\Psi = +\eta(d\Psi/dy) = 0$, at a clamped, simply supported and free edge, respectively. Thus, a set of m homogeneous equations for C-C, C-S and C-F plates can be written

$$\text{as } \left[\begin{array}{c} N \\ \hline N^{CC} \end{array}\right]\left[C^*\right]=[0]\ ,\quad \left[\begin{array}{c} N \\ \hline N^{CS} \end{array}\right]\left[C^*\right]=[0],\quad \left[\begin{array}{c} N \\ \hline N^{CF} \end{array}\right]\left[C^*\right]=[0]\ ,\quad \dots(11, 12, 13)$$

For a nontrivial solution of equations (11-13), the frequency determinants must vanish and hence

$$\left|\begin{array}{c} N \\ \hline N^{CC} \end{array}\right|=0\ ,\quad \left|\begin{array}{c} N \\ \hline N^{CS} \end{array}\right|=0\ ,\quad \left|\begin{array}{c} N \\ \hline N^{CF} \end{array}\right|=0\ ,\qquad \dots(14, 15, 16)$$

respectively.

3. NUMERICAL RESULTS AND DISCUSSION

The frequency equations (14-16) provide the values of the frequency parameter Ω for various values of the plate parameters. In the work reported here, the first three modes of vibration have been computed for non-homogeneity parameter μ (= $-0.5(0.2)$, 0.9, 0.0, 1.0), density parameter β(= $-0.5(0.2)$, 0.9, 0.0, 1.0), taper parameter α(= $-0.5(0.2)$, 0.5), and thickness parameter h_0(= 0.01, 0.03, 0.05, 0.1, 0.15, 0.2) for two values of radii ratio ε (=0.3, 0.5) by Mindlin plate theory (MPT; Ω_s) and classical plate theory (CPT; Ω_s). For CPT, the governing equation of motion is obtained by putting $\psi = \partial w / \partial r$ and tracer $k_1 = 0$ in equations (4) and (5). The numerical values of elastic constants used for the plate material (i.e., glass epoxy) are taken from Cheng and He [6].

$$E_2 / E_1 = 0.3333,\ G / E_1 = 0.1666,\ \upsilon_\theta = 0.0833,\ \upsilon_r = 0.2500$$

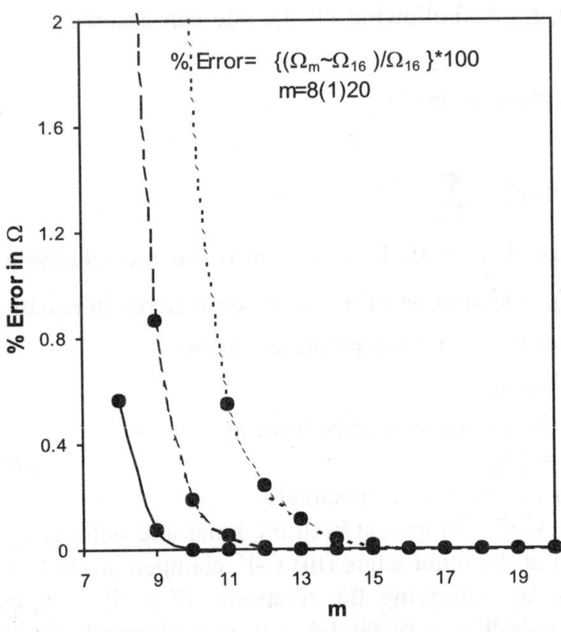

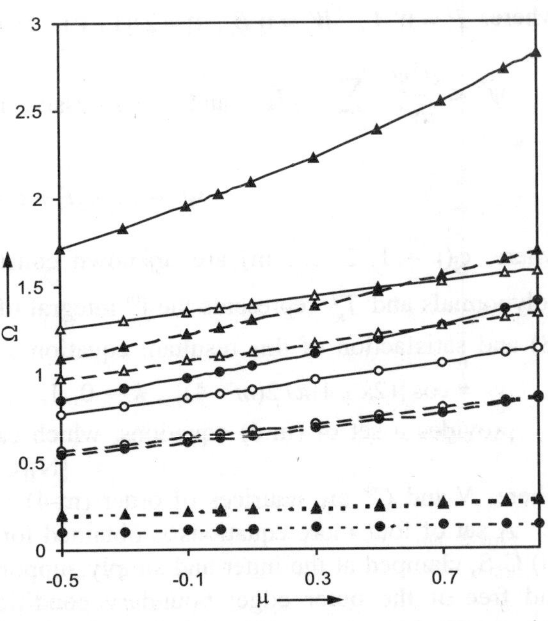

Fig. 1. Convergence graph for C-C plate for $h_0 = 0.2$, for $\mu = 0.5$, $\alpha = -0.5$, $\beta = -0.5$ and $\varepsilon = 0.5$. —, first mode; - - -, second mode; ......, third mode.

Fig. 2. Natural frequencies for the first mode of vibration for $\beta = -0.5$, $\alpha = -0.5$ and $\varepsilon = 0.5$. —, C-C; - - -, C-S; ......, C-F. : ● $h_0 = 0.05$; ▲ : $h_0 = 0.1$. 0, $\triangle$: shear theory ; ●, ▲ : classical theory.

Figure 1 shows the convergence of the solution with the number of collocation points. Here, m = 16 has been taken in all the computations for the accuracy of four decimals. Figures (2-4) show the behaviour of Ω with non-homogeneity parameter μ, density parameter β and taper parameter α for the first mode of vibration, respectively. It is found that Ω increases with the increasing values of μ and α and decreases with increasing values of β for all the three plates. For second and third modes, the behaviour of frequency parameter with these plates parameters remains same except that the rate of increase/decrease of Ω with μ, α and β is higher as compared to the first mode (Figs. are not given here). The effect of shear theory is more pronounced for negative values of taper parameter $\alpha(<0)$ as compared to the positive values i.e., $\alpha(>0)$. The effect of thickness parameter h_0 on the frequency parameter Ω_C and Ω_S for $\varepsilon = 0.3, 0.5$ has been shown in Figs. 5(a, b, c). The frequency parameter increases with the increase in h_0 whatever be the other plate parameters. This effect increases with the increase in number of modes and decreases in the order of boundary conditions C-C, C-S, C-F.

Modes shapes are shown in Figs. 6(a, b, c). The nodal circles are seen to shift towards the inner edge as the plate becomes more and more stiff towards the outer edge. In Table 1, a comparative percentage change in the values of frequency parameter due to the inclusion of rotatory inertia and transverse shear for homogeneous to non-homogeneous plates has been presented. It is noticed that the non-homogeneity parameter μ plays a major role for this change.

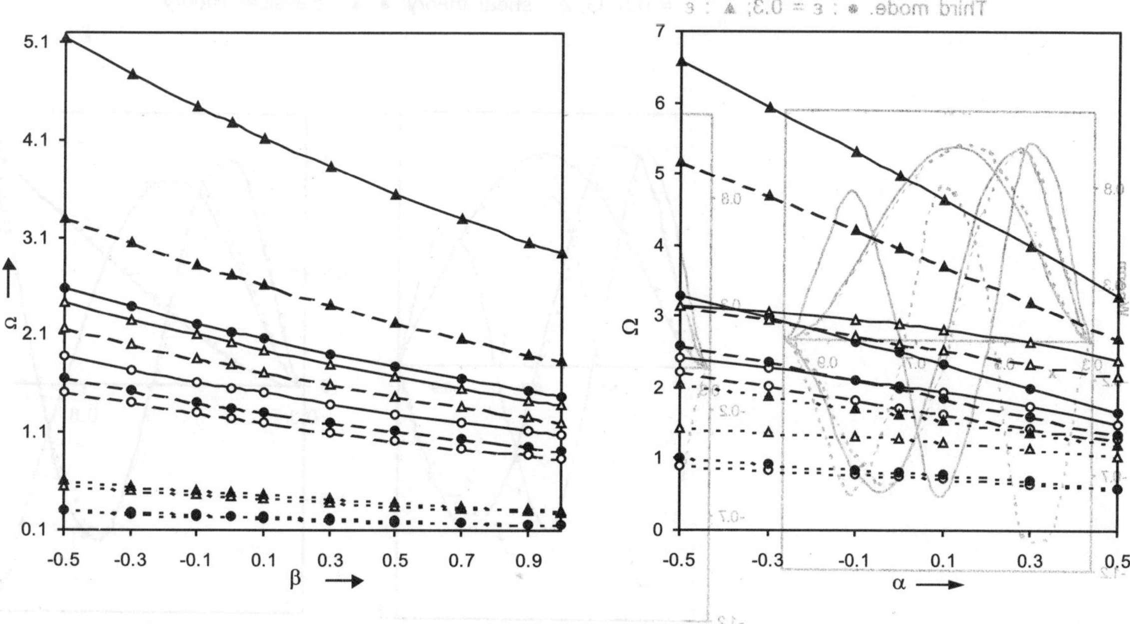

Fig. 3. Natural frequencies for first mode of vibration for $\alpha = -0.5$, $\mu = 0.5$ and $\varepsilon = 0.5$. ———, C-C; - - - ,C-S; ....., C-F. $\bullet$: $h_0 = 0.05$; $\blacktriangle$: $h_0 = 0.1$. O, $\triangle$: shear theory ; $\bullet$, $\blacktriangle$: classical theory.

Fig. 4. Natural frequencies for first mode of vibration for $\beta = -0.5$, $\mu = 0.5$ and $\varepsilon = 0.5$. ———, C-C; - - -, C-S; ....C-F. $\bullet$: $h_0 = 0.05$; $\blacktriangle$: $h_0 = 0.1$. 0, $\triangle$: shear theory; $\bullet$, $\blacktriangle$: classical theory.

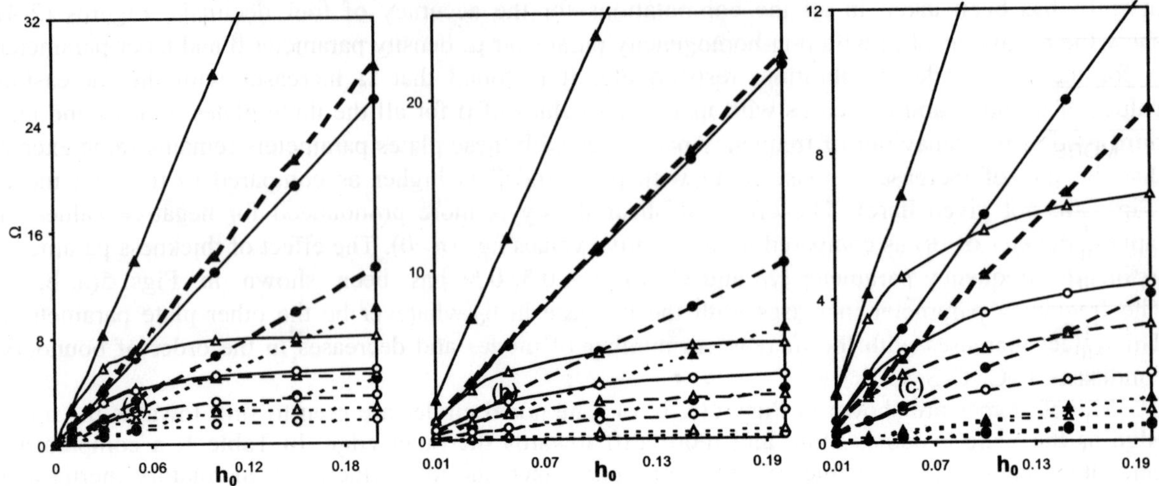

Fig. 5. Natural frequencies for the first three modes of vibration for $\mu = 0.5$, $\alpha = -0.5$ and $\beta = -0.5$, (a) C-C; (b) C-S; (c) C-F. ....., First mode; $---$, Second mode; ———, Third mode. $\bullet$: $\varepsilon = 0.3$; $\blacktriangle$: $\varepsilon = 0.5$. O, $\triangle$: shear theory; $\bullet$, $\blacktriangle$: classical Theory.

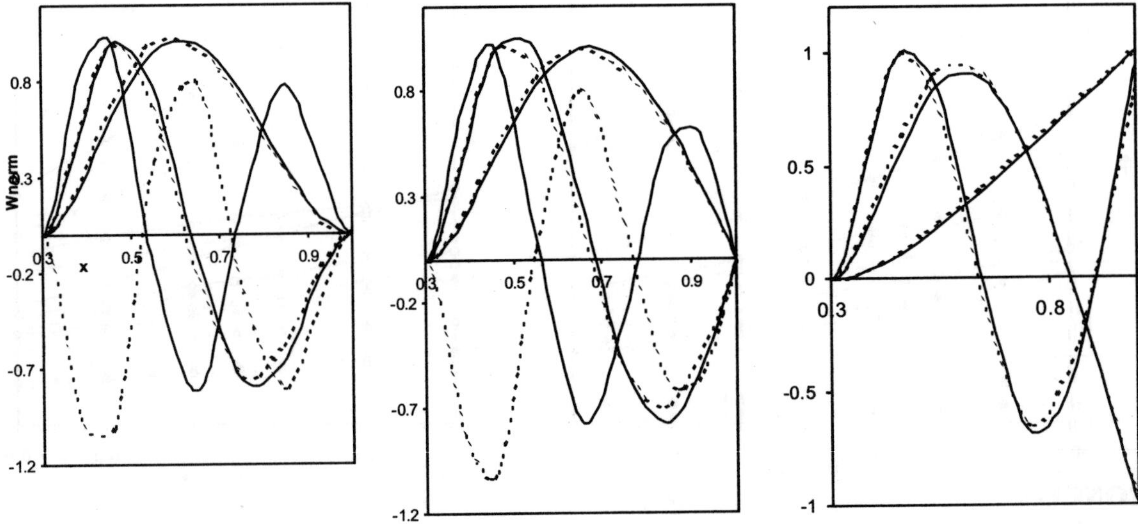

Fig. 6. Normalized displacements for the first three modes of vibration; (a) C-C; (b) C-S; (c) C-F. For $\alpha = -0.5$, $\beta = -0.5$, $\varepsilon = 0.3$ and $h_0 = 0.1$. ———, $\mu = -0.5$; ---------, $\mu = 0.5$.

Table 1. Percentage change in the frequencies $\{(\Omega_C - \Omega_S)/\Omega_S\}^* 100$ for homogeneous/non-homogeneous plates for $h_0 = 0.2$, $\propto = -0.5$.

MODE	β / μ	$\varepsilon = 0.3$			$\varepsilon = 0.5$		
		−0.5	0	1	−0.5	0	1
		C-C					
I		107.6553	107.3973	107.6057	175.1137	174.9832	175.3139
II	−0.5	184.8708	185.2495	187.1063	288.0011	288.2814	288.5668
III		251.7527	252.0286	253.7422	382.9225	382.9832	384.0070
I		135.4373	135.5131	136.5889	224.4093	224.5530	225.5836
II	0	224.8996	225.7084	228.5000	354.3606	354.8708	356.6892
III		303.9512	304.7019	307.5940	472.1844	472.6676	474.7342
I		211.2810	212.5828	216.5818	360.2544	361.3823	364.7168
II	1	328.5952	330.6120	336.1563	536.6987	538.1637	542.2323
III		440.7175	443.1375	449.9869	715.9542	717.7738	722.9922
		C-S					
I		58.8768	59.0912	59.9839	104.2712	104.6615	105.8757
II	−0.5	123.2344	122.9234	122.9874	198.4985	198.5604	199.1076
III		40.7793	116.0992	212.4054	303.6171	303.7981	139.1836
I		69.9637	70.0650	70.7495	125.9631	126.2498	127.2803
II	0	145.5223	145.4649	146.1792	241.6733	242.0388	243.3784
III		63.2123	147.7114	149.4719	374.6783	375.2861	258.7563
I		99.7428	99.6362	99.9616	182.3379	182.5085	183.3094
II	1	207.3558	208.2211	211.2680	363.2603	364.5753	368.2667
III		164.8824	230.9137	235.4451	5687278	408.0485	411.4456
		C-F					
I		10.3886	9.7142	8.4809	21.4586	20.5525	18.8485
II	−0.5	79.9378	77.3880	72.6411	124.9135	121.7581	115.4502
III		128.7963	124.4365	111.2049	101.7727	101.3682	101.0183
I		13.4564	12.7150	11.3750	29.2838	28.2796	26.3857
II	0	95.9862	93.1804	87.9014	151.9813	148.3829	141.1486
III		155.2816	150.7277	135.5085	144.5386	144.3214	144.3173
I		21.8488	20.9460	19.2767	51.6159	50.3094	47.8919
II	1	134.7541	131.5261	125.4422	221.1097	216.8134	208.1091
III		226.7887	222.6590	210.1736	253.6250	253.7659	254.5799

CONCLUSION

The effect of transverse shear and rotatory inertia together with non-homogeneity of the plate material on the natural frequencies of Mindlin annular plates has been studied. It is observed that the effect of non-homogeneity brings significant percentage change in the frequencies. The difference $(\Omega_C - \Omega_S)$ is found to increase with the increasing values of h_0 and the number of modes. It is also seen that transverse shear deformation accounts almost the entire discrepancy. Thus, the effect of rotatory inertia and transverse shear cannot be neglected while predicting the dynamic behaviour of moderately thick ($h_0 > 0.1$) plates. A similar inference was obtained by Mindlin and Deresiewicz [5] for homogeneous isotropic and Gupta and Lal [6] for homogeneous polar orthotropic plates of variable thickness.

REFERENCES

1. R. Lal and S. Sharma, 2004, Axisymmetric vibrations of non-homogeneous polar orthotropic annular plates of variable thickness, Journal of sound and vibration. 272, 245-265.
2. Lekhnitskii, S., 1968 Anisotropic Plates, Gordon and Breach, New York.
3. L. Fox and Parker, 1968, Chebyshev Polynomials in Numerical Analysis, Oxford University Press, Oxford.
4. S. Cheng, and F.B. He, 1984, Theory of orthotropic and composite cylindrical shells, accurate and simple fourth-order governing equations, Jl. of Applied Mechanics, 51, 736-744
5. H. Deresiewicz, and R.D. Mindlin, 1955, Axially symmetric flexural vibrations of a circular disk, Jl. of Applied Mechanics, 22, 86-88
6. Gupta, U.S. and Lal, R., 1985, Axisymmetric vibrations of polar orthotropic Mindlin annular plates of variable thickness, Jl. of Sound and Vibration. 98, 565-573.

APPENDIX A

$$A_4 = 1, \qquad A_3 = \left(2 + 3\mu x - \beta x - 8\alpha x/(1 - \alpha x)\right)/x,$$

$$A_2 = \left\{ \begin{array}{l} -(2 + p) + (4 + \upsilon_\theta)\mu x - 2\beta x + \mu x(3\mu x - 2\beta x) + \Omega^2 x^2 \left(k_1 + k_2/k_0\right)e^{(\beta-\mu)x} \\ -\{\alpha x/(1 - \alpha x)\}\left(10 + 3\upsilon_\theta + 16\mu x - 6\beta x\right) + 12\alpha^2 x^2/(1 - \alpha x)^2 \end{array} \right\}/x^2,$$

$$A_1 = \left\{ \begin{array}{l} 3p - \mu x\left(2 + p + \upsilon_\theta(1 - \mu x)\right) - (\beta - \mu)x\left[-p + \mu x\left(2 + \upsilon_\theta + \mu x\right)\right] + \Omega^2 x^2 e^{(\beta-\mu)x}(k_1(1 + \beta x)) \\ +k_2(1 + \mu x)/k_0) - \{\alpha x/(1 - \alpha x)\}\{-6 - 5p - 3\upsilon_\theta + 2\mu^2 x^2 + 4\mu x(1 + 2\upsilon_\theta) - 3(\beta - \mu)x(2 + \\ \upsilon_\theta + 2\mu x) + \Omega^2 x^2 e^{(\beta-\mu)x}\left(5k_1 + 3k_2/k_0\right) + 3\alpha^2 x^2\left(2 + 3\upsilon_\theta + 4\mu x - 2\beta x\right)/(1 - \alpha x)^2 \end{array} \right\}/x^3$$

$$A_0 = \left\{ \begin{array}{l} p\left(-3 + 2\mu x\right) - \upsilon_\theta \mu^2 x^2 + (\beta - \mu)x\left(p(-1 + \mu x) - \upsilon_\theta \mu^2 x^2\right) + \Omega^2 x^2 e^{(\beta-\mu)x}(-k_1 + k_2(-p + \upsilon_\theta \mu x + \\ k_1\Omega^2 x^2 e^{(\beta-\mu)x})/k_0) - \{\alpha x/(1 - \alpha x)\}\{8p - 2\mu x(p + 3\upsilon_\theta - \upsilon_\theta \mu x) - (\beta - \mu)x(6\upsilon_\theta \mu x - 3p) \\ +\Omega^2 x^2 e^{(\beta-\mu)x}(2k_1(1 + \beta x) + 3\upsilon_\theta k_2/k_0)\} \\ +\{3\alpha^2 x^2\left(-p - 2\upsilon_\theta(1 + \beta x - 2\mu x) + k_1\Omega^2 x^2 e^{(\beta-\mu)x}\right) - 12\Omega^2 x^4 e^{(\beta-\mu)x}/h_0^2\}/(1 - \alpha x)^2 \end{array} \right\}/x^4$$

$$k_0 = k_s G\left(1 - \upsilon_r \upsilon_\theta\right)/E_1, \quad \Omega^2 = \rho_0 a^2 \omega^2\left(1 - \upsilon_r \upsilon_\theta\right)/E_1, \quad G = \sqrt{E_1 E_2}/2\left(1 + \sqrt{\upsilon_r \upsilon_\theta}\right),$$

$$p = E_2/E_1, \quad h_0 = H\big|_{x=0}.$$

84

Static and Free Vibration Analysis of Laminated Conoids Using a Higher-Order Theory

S. PRADYUMNA[1] AND J.N. BANDYOPADHYAY[2]

Department of Civil Engineering, IIT Kharagpur-721 302, West Bengal, India
email: [1] pradyumna@civil.iitkgp.ernet.in, [2] jnb@civil.iitkgp.ernet.in

ABSTRACT

A C^0 finite element formulation using a higher-order shear deformation theory is developed and used to analyze static and dynamic behavior of laminated composite conoids. The element consists of nine degrees-of-freedom per node with higher-order terms in the Taylor's series expansion which represents the higher-order transverse cross-sectional deformation modes. The formulation includes Sanders' approximation for doubly curved shells considering the effects of rotary inertia and transverse shear. A realistic parabolic distribution of transverse shear strains through the shell thickness is assumed and the use of shear correction factor is avoided. The accuracy of the formulation is validated by comparing the results with those available in the existing literature.

1. INTRODUCTION

Shell structures are widely used in all industrial applications; especially those related to automobile, marine, nuclear, civil, aerospace and petrochemical engineering. In civil engineering construction, conoid shells are commonly used as roofing units to cover large column-free areas. Conoids (Fig. 1) provide ease of fabrication and allow sunlight to come in. They are most suitable when they are truncated i.e., when lower rise is provided at one end.

Study of static and dynamic behavior of laminated composite shells has gained much interest of many researchers from past few decades. Obtaining closed form solutions for such problems are complex; therefore, to efficiently and conveniently solve the problem, the finite element method is widely used. Many of the classical theories developed for thin elastic shells are based on the Love-Kirchhoff assumptions, in which it is considered that normals to the mid-plane before deformation remain straight and normal to the plane after deformation. It underpredicts deflections and overpredicts natural frequencies and buckling loads. These deficiencies are mostly due to the neglect of transverse shear strains. The errors are even higher for structures made of advanced composites, whose elastic modulus to shear modulus ratios are very large.

Yang [1] developed a higher-order shell element with two principal radii, orthogonal to each other and one twist radius. The displacement functions u, v and w are composed of products of one-dimensional Hermite interpolation formulae. Reddy and Liu [2] modified Sanders' theory to develop a higher-order shear deformation theory of laminated elastic shells, which accounts for tangential stress free boundary conditions. They also presented Navier-type solutions for bending and free vibration problems. Shu and Sun [3] developed an improved higher-order theory for laminated composite plates satisfying the stress continuity across each layer interface and also including the influence of different materials and ply-up patterns on the displacement field. Liew and Lim [4] proposed a higher-order theory by considering the Lamé parameter $(1 + z/R_x)$ and $(1 + z/R_y)$ for the transverse strains, which were neglected by Reddy and Liu [2]. This theory accounts for cubic distribution (non-even terms) of the transverse shear strains through the shell thickness in contrast with the parabolic shear distribution (even-terms) of Reddy

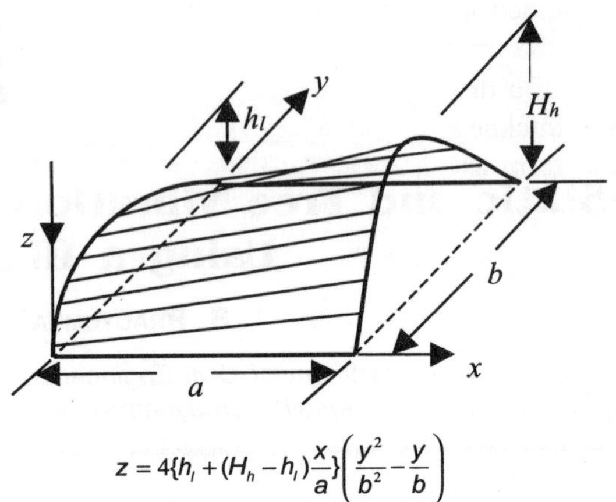

$$z = 4\{h_l + (H_h - h_l)\frac{x}{a}\}\left(\frac{y^2}{b^2} - \frac{y}{b}\right)$$

Fig. 1. Conoid.

and Liu [2]. Kant and Khare [5] presented a higher-order facet quadrilateral composite shell element. Bhimaraddi [6], Mallikarjuna and Kant [7], Cho et al. [8] are among the others to develop higher-order shear deformable shell theory. It is observed that except the theory of Yang [1], remaining higher-order theories do not account for twist curvature $(1/R_{xy})$, which is very much essential while analyzing shell forms like conoid shells.

Static and dynamic analyses of shell panels were carried out by many researchers in the past few decades. Choi [9], with modified isoparametric element, analyzed the conoidal shells by adding four extra non-conforming displacement modes to transverse displacement. Ghosh and Bandyopadhyay [10, 11] studied bending behavior of conoidal shells using the finite element method and Galerkin method. Chakravorty et al. [12,13] carried out finite element dynamic analysis of doubly curved shells which include hyperbolic paraboloids, hypars and conoids. Stavridis [14], by analytical treatment of Marguerre equations with inertia terms added, studied free vibration of elliptical, hyperbolic paraboloid, hypar, conoid and soap bubble isotropic shells. The effects of presence of the stiffeners and the cutouts on these types of shells were studied by Nayak and Bandyopadhyay [15,16]. Higher-order theories were widely used for free vibration analysis of laminated composite shells by many researchers [2, 4, 6].

To the best of the authors' knowledge, however, literature is not available related to the application of higher-order theory for studying the static and dynamic behavior of laminated composite shells with the combination of all three radii of curvature. Therefore, in the present analysis, static and free vibration behavior of laminated composite anticlastic shells are studied employing a Higher-order shear deformation theory (HSDT), developed by Kant and Khare [5] and extending it to the shells with all three radii of curvature.

2. THEORY AND FORMULATION

Let us consider a laminated shell element made of a finite number of uniformly thick orthotropic layers (Figs. 2 and 3), oriented arbitrarily with respect to the shell co-ordinates (x, y, z). The co-ordinate system (x, y, z) is chosen such that the plane $x - y$ at $z = 0$ coincides with the mid-plane. In order to approximate the three-dimensional elasticity problem to a two-dimensional one, the displacement components $u(x, y, z)$, $v(x, y, z)$ and $w(x, y, z)$ at any point in the shell space are expanded in Taylor's series in terms of the thickness co-ordinates. The elasticity solution indicates that the transverse shear stresses vary parabolically through the element thickness. This requires the use of a displacement field in which the in-plane displacements are expanded as cubic functions of the thickness co-ordinate. The displacement fields, which satisfy the above criteria are assumed in the form as given by Kant and Khare [7]

$$u(x, y, z) = u_0(x, y) + z\theta_y + z^2 u_0^*(x, y) + z^3 \theta_y^*(x, y)$$

$$v(x, y, z) = v_0(x, y) - z\theta_x + z^2 v_0^*(x, y) - z^3 \theta_x^*(x, y) \qquad ...(1)$$

$$w(x, y, z) = w_0$$

where, u, v and w are the displacements of a general point (x,y,z) in an element of the laminate along x, y and z directions, respectively. The parameters u_0, v_0, w_0, θ_x and θ_y are the displacements and rotations of the middle plane, while u_0^*, v_0^*, θ_x^* and θ_y^* are the higher-order displacement parameters defined at the mid-plane.

The linear strain-displacement relations according to Sanders' approximation are,

$$\varepsilon_x = \frac{\partial u}{\partial x} + \frac{w}{R_x}, \ \varepsilon_y = \frac{\partial v}{\partial y} + \frac{w}{R_y}, \ \gamma_{xy} = \frac{\partial v}{\partial x} + \frac{\partial u}{\partial y} + \frac{2w}{R_{xy}}$$

$$\gamma_{xz} = \frac{\partial u}{\partial z} + \frac{\partial w}{\partial x} - C_1\frac{u}{R_x} - C_1\frac{v}{R_{xy}}, \ \gamma_{yz} = \frac{\partial v}{\partial z} + \frac{\partial w}{\partial y} - C_1\frac{v}{R_y} - C_1\frac{u}{R_{xy}} \qquad ...(2)$$

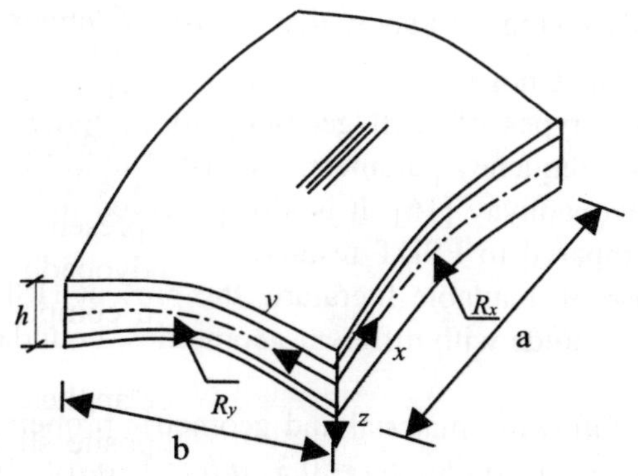

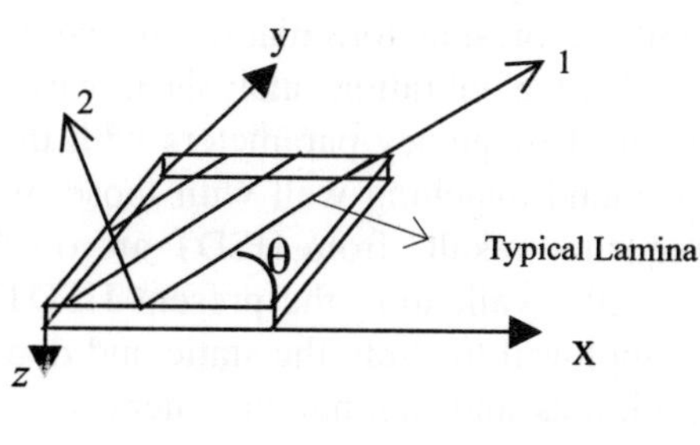

Fig. 2. Laminated composite doubly curved shell element.   **Fig. 3.** Lamina reference axis and fibre orientation.

3. RESULTS AND DISCUSSIONS

A computer program is developed based on the above formulation. A parallel program is developed based on the first order shear deformation theory (FSDT) in order to compare the results with those of HSDT. The shell-forms mainly considered here are hyperbolic, paraboloid, hypar and conoid shells.

The following two boundary conditions are used in the present analysis:

(i) Simply supported boundary (S-S-S-S) having SS1 for isotropic and cross-ply laminates and SS2 for angle-ply laminates;

$$\text{SS1: } v_0 = w_0 = \theta_y = v_0^* = \theta_y^* = 0, \text{ at } x = 0, a; \quad u_0 = w_0 = \theta_x = u_0^* = \theta_x^* = 0, \text{ at } y = 0, b.$$

$$\text{SS2: } u_0 = w_0 = \theta_y = u_0^* = \theta_y^* = 0, \text{ at } x = 0, a \text{ ; and } \quad v_0 = w_0 = \theta_x = v_0^* = \theta_x^* = 0, \text{ at } y = 0, b.$$

(ii) Clamped boundary (C-C-C-C): $u_0 = v_0 = w_0 = \theta_x = \theta_y = u_0^* = v_0^* = \theta_x^* = \theta_y^* = 0$, at $x = 0, a$ and $y = 0, b$.

Non-dimensional length parameters are designated by $\bar{x} = x/a$ and $\bar{y} = y/b$.

Non-dimensional center deflection parameter $\hat{w} = (wh^3 E_2 / w_z a^4) \times 1000$.

Non-dimensional frequency parameter $\hat{\omega} = \omega a^2 / h \sqrt{(\rho / E_2)}$

Unless otherwise specified the elastic properties of the structure is taken as $E_1 / E_2 = 25$;

$G_{23} = 0.2 E_2$ $G_{12} = G_{13} = 0.5 E_2$.

In order to validate the present formulation, the following problems are taken up from the existing literature.

1. A conoid shell with $a =$ 2.521m, $b =$ 1.828m, $H_h =$ 0.457m, $h_l =$ 0.228m, thickness $=$ 12.7mm, E $=$ 38.843kN/mm², $v =$ 0.15, with simply supported (SS1) boundary condition and subjected to a uniformly distributed pressure of 2.8734×10^{-3} N/mm², which was earlier solved by Choi [9]. The variations of transverse displacement (w) along $\bar{x} = 0.5$ and $\bar{y} = 0.5$ to 1.0 (Fig. 4a) and also transverse displacement (w) along $\bar{x} = 0$ to 1.0 and $\bar{y} = 0$ (Fig. 4b) show that the results obtained from the present formulation are comparable with those of Choi [9].

2. Free vibration analysis is carried out for different types of shell geometry to compare the obtained frequency parameters with the available results. Frequency parameters w, listed in Table 1, are found matching well with those of Nayak and Bandyopadhyay [16]. It is also observed that the frequency results from HSDT are on the lower side compared to FSDT results.

After validating the present HSDT results with those of available literature, the present HSDT is employed to study the static and dynamic behavior of conoids with different geometries, boundary conditions and lamination schemes.

Firstly, a full conoid ($h_l = 0$) is considered with the following material and geometric properties $E_1 = 138$ GPa, $E_2 = 8.96$ GPa, $G_{12} = G_{13} = 7.1$ GPa, $G_{23} = 6.21$GPa, $v_{12} = 0.3$ $a/b = 1$, $a/h = 100$ and $a/ H_h = 2.5$. Four lamination schemes (0/90, 0/90/0, 0/90/0/90 and 0/90/90/0) are employed to study behavior of the conoid subjected to uniformly distributed load. Figure 5 showing the variation of non-dimensional displacement $\bar{w}$ along $\bar{x} = 0.5$ and $\bar{y} = 0$ to 1 reveals that conoid with 0/90 lamination scheme is undergoing the maximum central displacement. It is also seen that for symmetric laminate (0/90/0 and 0/90/90/0), value of $\bar{w}$ decreases towards the centre showing two distinct points having maximum value of $\bar{w}$. However, antisymmetric laminates have the maximum value of $\bar{w}$ at the center.

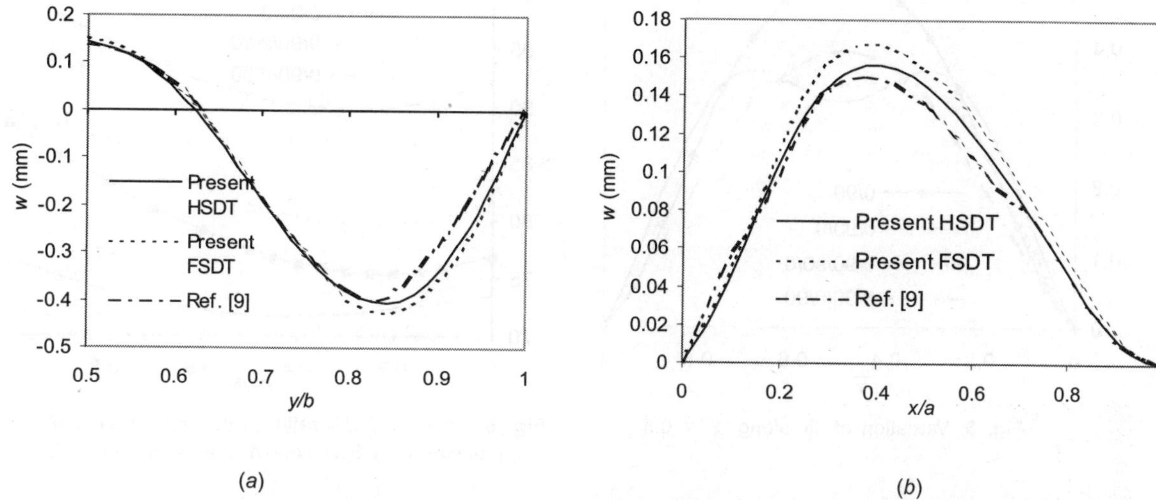

(a)   (b)

Fig. 4. Variation of transverse displacement (w) along (a) $\bar{x}$ = 0.5 and (b) $\bar{y} = 0$.

Figure 6 shows the effect of h_l/H_h ratio on the frequency parameter $(\hat{\omega})$ for a simply supported (SS2) conoid shell of $a/h = 100$; $a/b = 1$ for four different lamination schemes with $a/H_h = 5$. The material properties are $E_1/E_2 = 25$; $G_{23} = 0.2E_2$; $G_{12} = G_{13} = 0.5E_2$. The values of $\hat{\omega}$ are increasing with the increase of h_l/H_h. This increase of $\hat{\omega}$ has two distinct zones in most of the stacking sequences; initially with steep and almost constant slope upto h_l/H_h around 0.3 and then with much reduced slopes. Critical study of Fig. 6 further reveals marginal reduction of ω in the second zone for some of the orientations. Figure 7 shows the effect of h_l/H_h ratio on the frequency parameter $(\hat{\omega})$ for a clamped conoid shell for $a/H_h = 5$. The nature of variation of $\hat{\omega}$ is somewhat different for the clamped conoids (Fig. 7) for all four types of lamination scheme, where there is a tendency of reduction with the increase of h_l/H_h initially, followed by the increase of $\hat{\omega}$ at the later stage.

Table 1. Non-dimensional frequencies $\hat{\omega}$ of simply supported cross-ply $[0°/90°]_4$ laminated composite shells

Shell Type	Ref. [16]	Present FSDT	Present HSDT
Elliptic paraboloid	47.384	47.380	47.341
Hyperbolic paraboloid	14.743	14.742	14.596
Hypar	52.002	52.014	51.291
Conoid	81.097	80.981	80.918

Note: $a/b = 1$, $a/h = 100$, for elliptic paraboloid, $h/R_x = h/R_y = 1/300$; for hyperbolic paraboloid, $h/R_x = -h/R_y = 1/300$; for hypar, $c/a = 0.2$; for conoid, $a/H_h = 2.5$, $h_l/H_h = 0.25$;

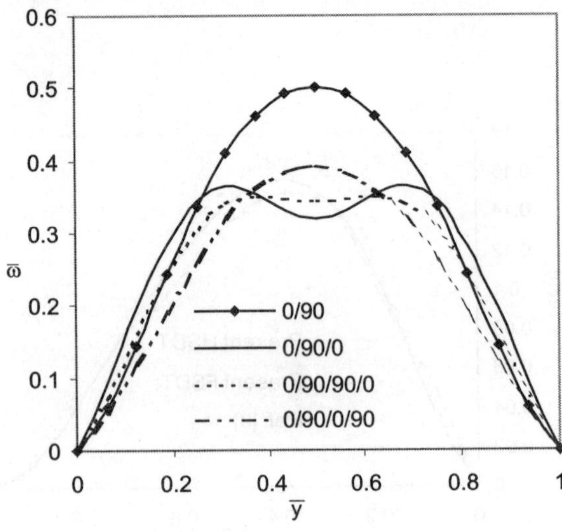

Fig. 5. Vartiation of $\bar{\omega}$ along $\bar{x}$ = 0.5.

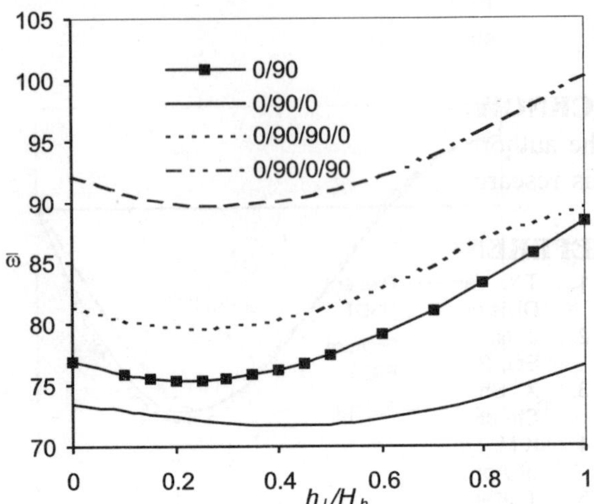

Fig. 6. Effect of h_l/H_h ratio on the non-dimensional frequencies of a SS2 conoid shell with a/H_h = 5.

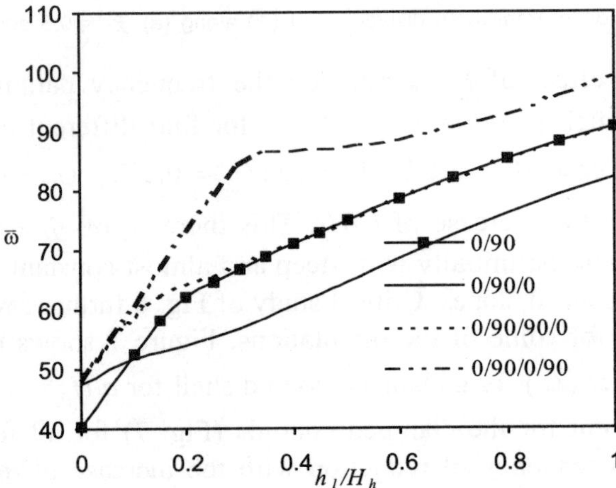

Fig. 7. Effect of h_l/H_h ratio on the non-dimensional frequencies of a clamped conoid shell with a/H_h = 5.

CONCLUSION

Static and free vibration analyses of laminated anticlastic shells are carried out using a higher-order theory taking into account all the three radii of curvature. From the present study, the following conclusions are made:

1. A comparative study of the results of conoids employing present formulation with those of available literature shows that the present higher-order formulation gives fairly good results.
2. Authors' results of free vibration of conoids show the superiority of the clamped conoids

having comparatively lower range. Simply supported conoids, however, show an increasing frequency initially with steep slope upto h_l/H_h around 0.3 and then with much reduced slopes.

ACKNOWLEDGEMENTS

The authors are thankful to the AICTE, New Delhi, for providing financial assistance to carry out this research work under the National Doctoral Fellowship Scheme.

REFERENCES

1. T.Y. Yang, 1973, High order rectangular shallow shell finite element. Journal of the Engineering Mechanics Division. 99, EM1, 157-181.
2. J. N. Reddy, C. F. Liu, 1985, A higher-order shear deformation theory of laminated elastic shells. Int. J. Engng. Sci. 23, 3, 319-330.
3. X. Shu, L. Sun, 1994, An improved simple higher-order theory for laminated composite plates. Computers & Structures. 50, 2, 231-236.
4. K.M. Liew, C.W. Lim, 1996, A higher-order theory for vibration of doubly curved shallow shells. ASME, Journal of Applied Mechanics. 63, 587-593.
5. T. Kant, R.K. Khare, 1997, A higher-order facet quadrilateral composite shell element, Int. J. for Numerical Methods in Engng. 40, 4477-4499.
6. Bhimaraddi, 1984, A higher order theory for free vibration analysis of circular cylindrical shells. Int. J. Solids Structures. 20, 623-630.
7. Mallikarjuna, T. Kant, 1992, A general fibre-reinforced composite shell element based on a refined shear deformation theory. Computers & Structures. 42, 3, 381-388.
8. M. Cho, K. Kim, M. Kim, 1996, Efficient higher-order shell theory for laminated composites. Composite Structures. 34, 197-212.
9. C.K. Choi, 1984, A conoidal shell analysis by modified isoparametric element, Computers & Structures, 18, 5, 921-924.
10. B. Ghosh, J.N. Bandyopadhyay, 1989, Bending analysis of conoidal shells using curved quadratic isoparametric element, Computers & Structures. 33, 4, 717-728.
11. B. Ghosh, J.N. Bandyopadhyay, 1990, Approximate bending analysis of conoidal shells using the Galerkin method, Computers & Structures, 36, 5, 801-805.
12. D. Chakravorty, J.N. Bandyopadhyay, P.K. Sinha, 1996, Finite element free vibration analysis of doubly curved laminated composite shells. Journal of Sound and Vibration, 191, 4, 491-504.
13. D. Chakravorty, J.N. Bandyopadhyay, P.K. Sinha, 1998, Application of FEM on free and forced vibration of laminated shells. ASCE, Journal of Engineering Mechanics, 124, 1, 1-8.
14. L.T. Stavridis LT, 1998, Dynamic analysis of shallow shells of rectangular base. Journal of Sound and Vibration, 218, 5, 861-882.
15. A.N. Nayak, J.N. Bandyopadhyay, 2002, Free vibration analysis and design aids of stiffened conoid shells. ASCE, J. Eng. Mech. 124, 4, 419-427.
16. A.N. Nayak, J.N. Bandyopadhyay, 2005, Free vibration analysis of laminated stiffened shells. ASCE, J. Eng. Mech. 131, 1, 100-105.

85

Quintic Splines in the Study of Buckling and Vibration of Non-Homogeneous Orthotropic Rectangular Plates with Variable Thickness

ROSHAN LAL[1] AND DHANPATI[2]

[1]*Professor, Department of Mathematics, IIT Roorkee, Roorkee-247667.*
[2]*Research Scholar, Department of Mathematics, IIT Roorkee, Roorkee-247667.*

ABSTRACT

In this paper transverse vibrations of non-homogeneous orthotropic rectangular plates of variable thickness having two opposite edges ($y = 0$ and b) simply supported and these are subjected to constants in-plane force have been analyzed on the basis of classical plate theory. The other two edges ($x = 0$ and a) may be clamped, simply supported or free. For non-homogeneity of the plate material, Young's modulii and density are assumed to vary exponentially along one direction. Assuming the transverse displacement (w) to vary as, $\sin(p\pi y/b)$, the governing partial differential equation of motion is reduced to an ordinary differential equation in x with variable coefficients. Applying boundary conditions at $x = 0$ and a, frequency equations for three different combinations of boundary conditions have been obtained employing quintic splines interpolation technique. The effect of in-plane force parameter together with orthotropy, non-homogeneity, aspect ratio and thickness variation on the natural frequencies of vibration is illustrated for the first three modes of vibration. Transverse displacements are presented for specified plates. Critical buckling loads in compression for various values of plate parameters have been computed. A comparison of results has been presented.

Keywords: Non-homogeneous; rectangular orthotropic; variable thickness; quintic splines.

1. INTRODUCTION

The study of buckling and flexural vibrations of plates of various geometries is of importance in connection with various engineering applications. In the recent past, there has been increasingly great interest in high strength materials for structural components used in mechanical, aerospace, ocean engineering, electronic and optical equipments. Plate type structural components of varying thickness fabricated out of modern composite materials (such as glass epoxy, boron epoxy, kevalar and graphite etc.) are lighter, stiffer and stronger than any other material used earlier and helps the designer to reduce the weight and size of the structure. Under normal working conditions, these plates may be subjected to in-plane stressing arising from hydrostatic, centrifugal and thermal [1, 2] stresses. In many practical situations their use under high-temperature environmental conditions demand that non-homogeneity of the material should be taken into account. Keeping in view, the work dealing with earlier proposed models [3], a more general model for the non-homogeneity of the plate material has been proposed and used in the study of transverse vibrations of non-homogeneous rectangular orthotropic plates of exponentially varying thickness when the two opposite edges of the plate are simply supported and these are subjected to constant in-plane axial force. The analysis is based upon classical plate theory and differential equation governing the motion of such plates has been solved by the quintic splines interpolation technique for three different combinations of clamped, simply supported and free boundary conditions at the other two edges.

2. FORMULATION

Consider a rectangular orthotropic plate of length a, breadth b, thickness $h = h(x, y)$, density ρ, and referred to a system of rectangular cartesian coordinates (x, y, z). The middle surface being $z = 0$ and the origin is at one of the corners of the plate. The x–and y–axes are taken along the principle directions of orthotropy, the axis of z is perpendicular to the xy– plane and a constant in-plane axial force N_y per unit length is applied along the edges $y = 0$ and b as shown in Fig. 1. The governing differential equation of motion for such plates, is given by

$$D_x \frac{\partial^4 w}{\partial x^4} + D_y \frac{\partial^4 w}{\partial y^4} + 2H \frac{\partial^4 w}{\partial x^2 \partial y^2} + 2\frac{\partial H}{\partial x}\frac{\partial^3 w}{\partial x \partial y^2} + 2\frac{\partial H}{\partial y}\frac{\partial^3 w}{\partial y \partial x^2} + 2\frac{\partial D_x}{\partial x}\frac{\partial^3 w}{\partial x^3} + 2\frac{\partial D_y}{\partial y}\frac{\partial^3 w}{\partial y^3} +$$

$$\frac{\partial^2 D_x}{\partial x^2}\frac{\partial^2 w}{\partial x^2} + \frac{\partial^2 D_y}{\partial y^2}\frac{\partial^2 w}{\partial y^2} + \frac{\partial^2 D_1}{\partial y^2}\frac{\partial^2 w}{\partial x^2} + \frac{\partial^2 D_1}{\partial x^2}\frac{\partial^2 w}{\partial y^2} + 4\frac{\partial^2 D_{xy}}{\partial x \partial y}\frac{\partial^2 w}{\partial y \partial x} + \rho h \frac{\partial^2 w}{\partial t^2} - N_y \frac{\partial^2 w}{\partial y^2} = 0, \qquad \text{...(1)}$$

where, $\qquad D_x = E_x^* h^3 / 12, \quad D_y = E_y^* h^3 / 12, \quad D_{xy} = G_{xy} h^3 / 12, \quad D_1 = E^* h^3 / 12,$

$$H = D_1 + 2D_{xy}, \; E^* = v_y E_x^* = v_x E_y^* \; (E_x^*, E_y^*) = (E_x, E_y)/(1 - v_x v_y),$$

$w(x, y, t)$ is the transverse deflection, t is the time, ρ is the mass density and E_x, E_y, v_x, v_y and G_{xy} are material constants in proper directions defined by an orthotropic stress-strain law.

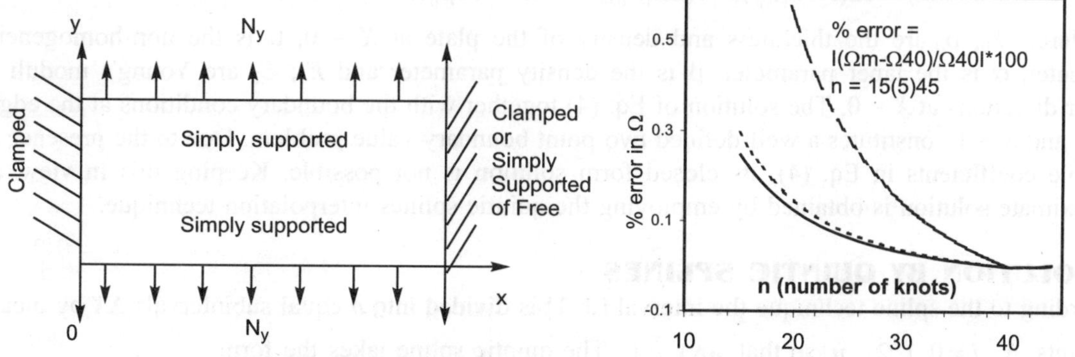

Fig. 1. Rectangular plate loaded by constant force N_y along edges $y = 0$ and b.

Fig. 2 Percentage error in Ω for C-C plate a/b = 1, α = 0.5, μ = 0.5, β = – 0.5. ——, first mode; ....., second mode;. - - -, third mode.

Let us assume that the two opposite edges $y = 0$ and $y = b$ are simply supported and the thickness of the plate varies in the x-direction only i.e., $h = h(x)$. For a harmonic solution, the deflection w (Lévy approach) is assumed to be

$$w(x, y, t) = \bar{w}(x)\sin(p\pi y/b)e^{i\omega t} \qquad \qquad ...(2)$$

where, p is a positive integer and ω is the frequency in radians.

Further, for elastically non-homogeneous material, we assume that the Young's moduli E_x, E_y and density ρ are the functions of space variable x only and following Leikhnitskii [4], the shear modulus G_{xy} is taken as $\sqrt{E_x E_y}/2(1+\sqrt{v_x v_y})$. Introducing the non-dimensional variables $X = x/a$, $Y = y/b$

$$\bar{h} = h/a, \quad W = \bar{w}/a \quad \text{Eq. (1) reduces to}$$

$$E_x \bar{h}^3 W^{iv} + [2(\bar{h}^3 E_x' + 3\bar{h}^2 \bar{h}' E_x)]W''' + [(6\bar{h}\bar{h}'^2 + 3\bar{h}^2 h'')E_x + 6\bar{h}^2 \bar{h}' E_x' + \bar{h}^3 E_x''$$

$$-2\lambda^2 \bar{H}12\,(1 - v_x v_y)]W'' - [2\lambda^2 \{3(v_y E_x + 2(1 - v_x v_y)G_{xy})\bar{h}^2 \bar{h}' + (v_y E_x' + 2(1 - v_x v_y)G_{xy}')\bar{h}^3\}]W'$$

$$+[\lambda^4 E_y \bar{h}^3 - \lambda^2 v_y \{E_x'' \bar{h}^3 + 6\bar{h}^2 \bar{h}' E_x' + (6\bar{h}\bar{h}'^2 + 3\bar{h}^2 \bar{h}'')E_x\} - 12\,(1 - v_x v_y)(\rho\bar{h}a^2\omega^2 + \lambda^2 \bar{N}_y)]W = 0, \qquad ...(3)$$

where, $\lambda^2 = \pi^2 p^2 a^2/b^2$ and primes denote differentiation with respect to X.

For exponential variation in thickness, i.e., $\bar{h} = h_0 e^{\alpha X}$ and non-homogeneity of the plate material in

X direction as $E_x = E_1 e^{\mu X}$, $E_y = E_2 e^{\mu X}$, $\rho = \rho_0 e^{\beta X}$, Eq. (3) now reduces to

$$A_0 W^{iv} + A_1 W''' + A_2 W'' + A_3 W' + A_4 W = 0, \qquad \qquad ...(4)$$

$$A_0 = 1, \quad A_1 = 2(\mu + 3\alpha), \quad A_2 = (\mu + 3\alpha)^2 - 2\sqrt{E_2/E_1}\,\lambda^2, \quad A_3 = -2\lambda^2(\mu + 3\alpha)\sqrt{E_2/E_1},$$

$$A_4 = \lambda^4 E_2/E_1 - \lambda^2 v_y(\mu + 3\alpha)^2 + \bar{N}_y e^{(\mu+3\alpha)x} - \Omega^2 e^{(-\mu-2\alpha+\beta)}$$

$$\bar{N}_y = 12(1 - v_x v_y)N_y/(aE_1 h_0^3), \qquad \Omega^2 = 12\rho_0(1 - v_x v_y)a^2\omega^2/E_1 h_0^2.$$

Here, h_0, ρ_0 are the thickness and density of the plate at $X = 0$, μ is the non-homogeneity parameter, α is the taper parameter, β is the density parameter and E_1, E_2 are Young's moduli in proper directions at $X = 0$. The solution of Eq. (4) together with the boundary conditions at the edges $X = 0$ and $X = 1$ constitutes a well-defined two-point boundary value problem. Due to the presence of variable coefficients in Eq. (4), its closed form solution is not possible. Keeping this in view, an approximate solution is obtained by employing the quintic splines interpolation technique.

3. SOLUTION BY QUINTIC SPLINES

According to the spline technique the interval [0, 1] is divided into n equal subintervals ΔX by means of points $X_i, i = 0, 1, 2 \ldots n$ so that $n\Delta X = 1$. The quintic spline takes the form

$$W(X) = a_0 + \sum_{j=1}^{4} a_j (X - X_0)^j + \sum_{i=0}^{n-1} b_i (X - X_i)_+^5, \qquad \ldots(5)$$

where,
$$(X - X_i)_+ = \begin{cases} 0 & if \quad X \leq X_i \\ X - X_i & if \quad X > X_i \end{cases},$$

and, $a_0, \ldots, a_4, b_0, \ldots, b_{n-1}$ are (n + 5) unknown constants.

Substitution for W(X) and its derivatives into Eq. (4) and satisfying the resulting equation at (n + 1) knots, $X_i, i = 0, 1, 2 \ldots n$, one obtains a set of (n + 1) homogeneous equations having (n + 5) unknowns $a_i, i = 0(1)4, \quad b_j, j = 0, 1 \ldots (n-1)$, which can be represented by the matrix equation

$$[A]\{B\} = \{0\}, \qquad \ldots(6)$$

where, A is a matrix of order (n + 1) × (n + 5) and, $\{B\}$ and $\{0\}$ are column vectors of order(n + 5) × 1.

4. BOUNDARY CONDITIONS AND FREQUENCY EQUATIONS

A set of four more equations is obtained for three sets of boundary conditions, namely, C-C, C-S, C-F have been considered in which the first symbol represents the condition at the edge $X = 0$ and second symbol at the edge $X = 1$ and C, S, F stand for clamped, simply supported and free edge, respectively. The relations that should be satisfied at clamped, simply supported and free edges are

$$W = dW / dX = 0; \quad W = (d^2W / dX^2) - (E^* / E_x^*)\lambda^2 W = 0; \quad and$$

$$d^2W / dX^2 - (E^* / E_x^*)\lambda^2 w = d\{E_x \overline{h}^3 (d^2 w / dx^2 - \lambda^2 v_y W)\} / dx - 4\lambda^2 (1 - v_x v_y) G_{xy} \overline{h}^3 dW / dX = 0,$$

respectively. These equations together with field the field Eqs. (6) give a complete set of (n + 5) equations in (n + 5). For C-C, C-S C-F plates these can be written in matrix form as

$$\begin{bmatrix} A \\ B^{CC} \end{bmatrix}\{B\} = \{0\}, \quad \begin{bmatrix} A \\ B^{CS} \end{bmatrix}\{B\} = \{0\}, \quad \begin{bmatrix} A \\ B^{CS} \end{bmatrix}\{B\} = \{0\}, \qquad \ldots(7, 8, 9)$$

For a non-trivial solution of Eqs. (7-9), the frequency determinant must vanish and hence,

$$\begin{vmatrix} A \\ B^{CC} \end{vmatrix} = 0, \quad \begin{vmatrix} A \\ B^{CS} \end{vmatrix} = 0, \quad \begin{vmatrix} A \\ B^{CF} \end{vmatrix} = 0, \qquad \ldots(10, 11, 12)$$

respectively.

5. NUMERICAL RESULTS AND DISCUSSIONS

In the work reported here, the frequency Eqs. (10-12) have been solved numerically to compute the values of frequency parameter Ω for the first three modes of vibration for different values of in-plane force parameter $\overline{N}_y = -50, -30, 0, 30, 50$, non-homogeneity parameter $\mu = -0.5(0.2)0.9, 0.0, 1.0$, density parameter $\beta = -0.5(0.2)0.9, 0.0, 1.0$, taper parameter $\alpha = -0.5(0.2)0.9, 0.0, 1.0$, and aspect ratio a/b = 0.5(0.5)2.0 for $p = 1$ and $h_0 = 0.1$. The values of elastic constants for the plate material

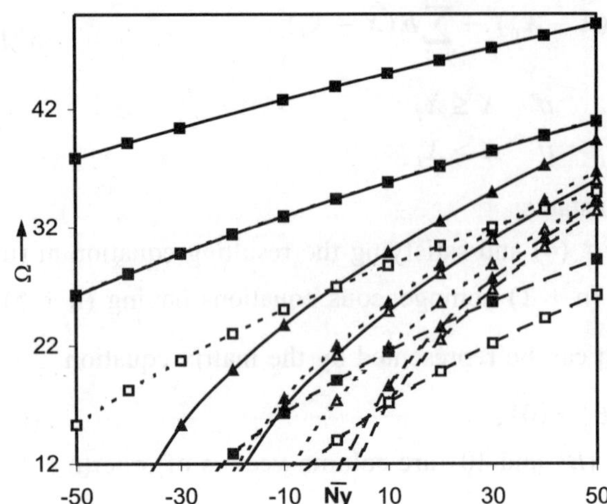

Fig. 3. Natural frequencies for the first mode
of vibration for

$\beta = -0.5$, a/b = 1; —— C-C, ------ C-S – – – – C-F;

Δ, $\mu = -0.5$, $\alpha = -0.5$; ▲, $\mu = 0.5$, $\alpha = -0.5$;

□, ($\mu = -0.5$, $\alpha = 0.5$; ■, ($\mu = 0.5$, $\alpha = 0.5$.

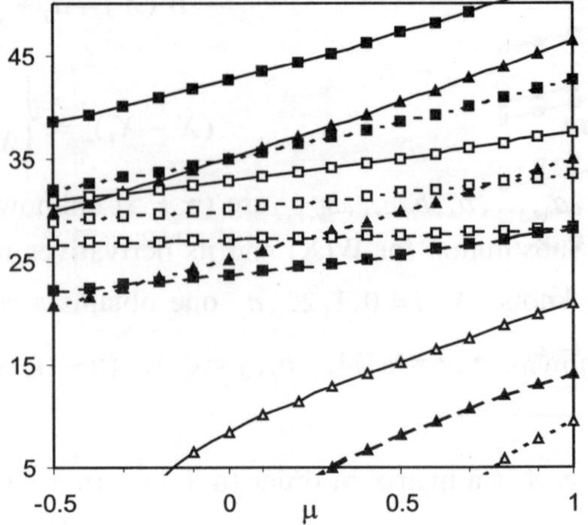

Fig. 4. Natural frequencies for the first **mode**
of vibration for

$\beta = -0.5$, a/b = 1; —— C-C, ------ C-S – – – – C-F;

Δ, $\alpha = -0.5$, $\overline{N}y = -30$; ▲, $\alpha = 0.5$, $\overline{N}y = -30$;

□, $\alpha = 0.5$, $\overline{N}y = 30$; ■, ($\alpha = 0.5$, $\overline{N}y = 30$.

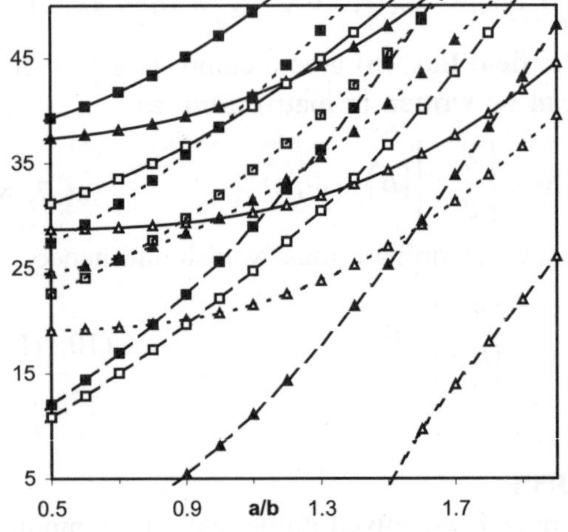

Fig. 5. Natual Frequencies for the first
mode of vibration

for $\beta = -0.5$, $\alpha = 0.5$; —— C-C, ------C-S – – – – C-F;

Δ, μ, $= -0.5$, $\overline{N}y = -30$; ▲, $\mu = 0.5$, $\overline{N}y = -30$;

□, $\mu = -0.5$, $\overline{N}y = 30$, ■, $\mu = 0.5$, $= 30$,

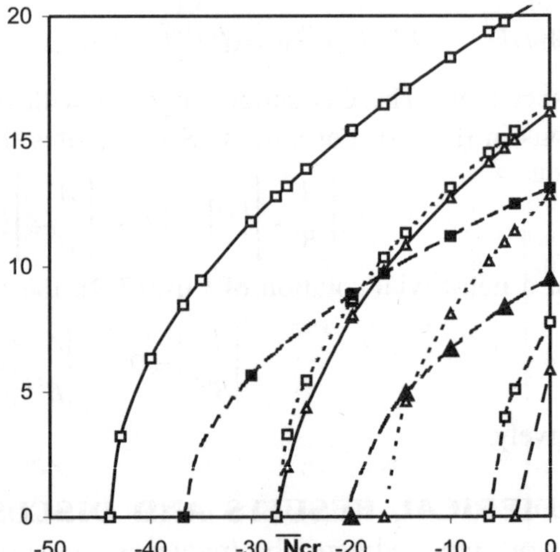

Fig. 6. Buckling load for the first
mode of vibration

for $\beta = -0.5$, a/b = 1; —— C-C, ------C-S – – – – C-F;

Δ, $\alpha = -0.5$, m $= -0.5$; ▲, $\alpha = 0.5$, $\mu = -0.5$;

□, $\alpha = -0.5$, $\mu = 0.5$; ■, $\alpha = 0.5$, $\mu = 0.5$.

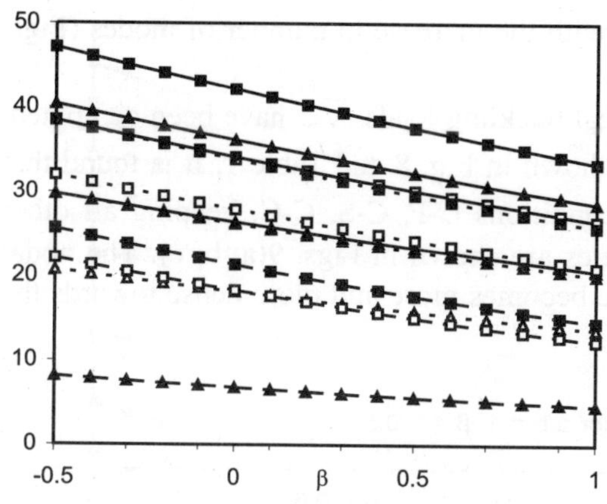

Fig. 7. Natural frequencies for the first
mode of vibration

for $\alpha = 0.5$, a/b = 1; —— C-C, ------ C-S – – – – C-F;

Δ, $\mu = -0.5$, $\overline{N}_y = -30$; ▲, $\mu = 0.5$, $\overline{N}_y = -30$;

□, $\mu = -0.5$, $\overline{N}_y = 30$, ■, $\mu = 0.5$, $\overline{N}_y = 30$,

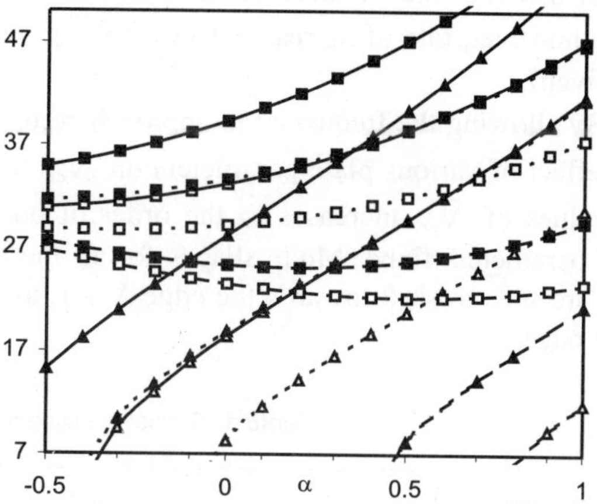

Fig. 8. Natural frequencies for
the first mode of vibration

for $\beta = -0.5$, a/b = 1; —— C-C, ------ C-S C-F;

Δ, $\alpha = -0.5$, $= -30$; ▲, $\alpha = 0.5$, $\overline{N}_y = -30$;

□, $\mu = 0.5$, $\overline{N}_y = 30$; ■, $\mu = 0.5$, $\overline{N}_y = 30$.

'ORTHO1' are taken as $E_1 = 1 \times 10^{10}$, $E_2 = 1 \times 10^9$ from [5]. Fig. 2 shows the convergence of the solution with the number of knots. Here $n = 40$ has been taken in all the computations for the accuracy of four decimal places. Figs. (3-7) show the behaviour of Ω with the in-plane force parameter $\overline{N}_y$, non-homogeneity parameter μ, density parameter β, taper parameter α and aspect ratio a/b for the first mode of vibration, respectively. It is found that Ω increases with the increasing values of $\overline{N}_y$, μ, α, a/b while decreases with increasing values of β for all the three plates. For second and third modes, the behaviour of frequency parameter with these parameters remains almost the same

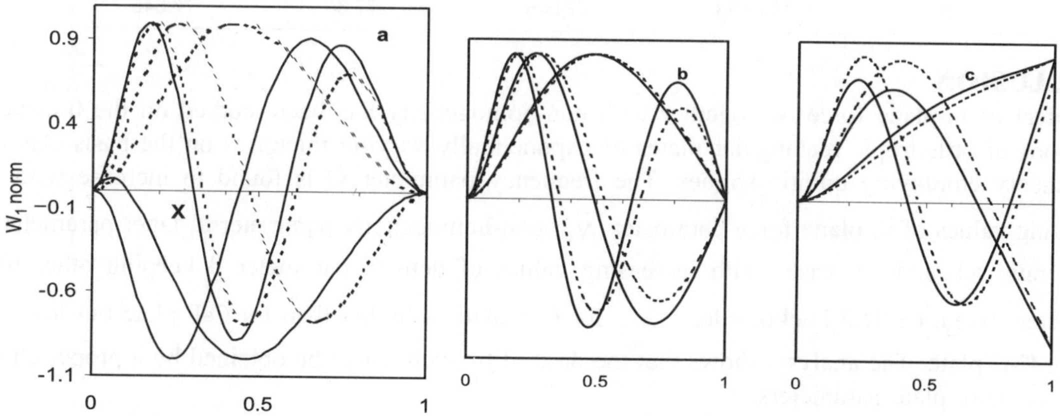

Fig. 9. Normalized displacements for the first three modes of vibration; (a) C-C; (b) C-S; (c) C-F.
for $\overline{N}_y = 30$, $\alpha = 0.5$, $\mu = 0.5$, a/b = 1; ——, $\beta = -0.5$; -------, $\beta = 0.5$.

except that the rate of increase/decrease of Ω with μ, α, a/b and β is higher as compared to first mode however, rate of increase of Ω with $\overline{N}_y$ decreases with the increase in number of modes (Figs. not given).

By allowing the frequency to approach zero, the critical buckling loads $\overline{N}_{cr}$ have been computed. The effect of various plate parameters on $\overline{N}_{cr}$ has been shown in Fig. 8 and Table 1. It is found that the values of $\overline{N}_{cr}$ increases in the order of boundary conditions C-F, C-S, C-C, keeping all other plate parameters fixed. Mode shapes for all the three plates are shown in Figs. 9(a, b, c). The nodal lines are seen to shift towards the edge $X = 1$, as the plate becomes more and more dense towards the outer edge.

Table 1. Critical buckling loads $\overline{N}_{cr}$ for a/b = 1, β = − 0.5.

Boundary condition	μ\mode	$\alpha = -0.5$ I	II	III	$\alpha = 0.5$ I	II	III
C-C	− 0.5	− 26.4208	− 163.245	− 581.121	− 119.767	− 746.755	− 2670.13
	0.5	− 44.0598	− 274.716	− 982.287	− 195.225	− 1206.23	− 4293.93
C-S	− 0.5	− 16.6556	− 116.155	− 447.589	− 73.7054	− 517.587	− 2021.51
	0.5	− 27.5062	− 193.857	− 752.688	− 120.048	− 828.307	− 3227.7
C-F	− 0.5	− 3.4758	− 34.4023	− 179.241	− 19.8469	− 156.934	− 803.886
	0.5	− 6.0952	− 57.2872	− 299.823	− 36.791	− 260.15	− 1291.99

Table 2. Comparison of non-dimensional critical buckling load $\overline{N}_{cr}$ for isotropic plates.

a/b	present	[6]	[7]	[8
0.4	93.209	93.247	93.2	93.247
0.5	75.887	75.91	75.9	75.91
0.6	69.604	69.632	69.6	69.632
0.7	69.072	69.095	69.1	69.095
0.8	72.067	72.084	71.9	72.084
0.9	77.533	77.545	77.3	77.545

CONCLUSION

The effect of in-plane force N_y together with non-homogeneity has been studied on the transverse vibrations of orthotropic rectangular plates of exponentially varying thickness on the basis classical plate theory employing quintic splines. The frequency parameter Ω is found to increase with the increasing values of in-plane force parameter $\overline{N}_y$, non-homogeneity parameter μ, taper parameter α, aspect ratio a/b and decreases with increasing values of density parameter β keeping other plate parameters fixed. Critical buckling load $\overline{N}_{cr}$ for C-S plate is higher than for C-F plate but less than that for C-C plate. The analysis shows that the desired frequency can be obtained by a proper choice of one or more plate parameters.

ACKNOWLEDGEMENTS

Author[2] is grateful to Council of Scientific & Industrial Research India for providing the junior research fellowship.

REFERENCES

1. A.W. Leissa, 1982, Advances and trends in plate buckling research, Research in Structural and Solid Mechanics. 441, 1-20.
2. D.G. Gorman, 1983, Vibration of thermally stressed polar orthotropic annular plates, Earthquake Engineering and Structural Dynamics, 11, 843-855.
3. R. Lal and S. Sharma, 2004, Axisymmetric vibrations of non-homogeneous polar orthotropic annular plates of variable thickness, Journal of Sound and Vibration. 272, 245-265.
4. S.G. Lekhnitskii, 1968, Anisotropic plates, Translated by S.W. Tsai and T. Cheron, Gordon and Breach, New York.
5. M.E. Biancolini, C. Brutti, and L. Reccia, 2005, Approximate solution for free vibration of thin orthotropic rectangular plates, Journal of Sound and Vibration. 288, 321-344.
6. X. Wang, L.Gan and Y. Wang, 2006, A differential quadrature analysis of vibration and buckling of an SS-C-SS-C rectangular plate loaded by linearly varying in-plane stress, In Press.
7. S. Timoshenko, and J. Gere, 1963, Theory of elastic stability, second edition, McGraw-Hill Book Company, Inc. New York.
8. A.W. Leissa and J.H. Kang, 2002, Exact solution for vibration and buckling of an SS-C-SS-C rectangular plate loaded by linearly varying in plane stresses, International Journal of Mechanical Sciences. 44, 1925-1945.

86

Linear Vibration Analysis of Composite Rectangular Plates: A State Space Approach

V. Anjani Kumar[1], Ramesh Pandey[1], K.K. Shukla[1] and Jin H. Huang[2]

[1]Applied Mechanics Deptt., M.N.N.I.T. Allahabad-211 004, India
[b]Mechanical Engg. Deptt., Feng Chia University, Taichung, Taiwan-40724
email: kkshukla@mnnit.ac.in

ABSTRACT

The aim of the present work is to obtain an analytical solution for the free vibration of a simply supported composite rectangular plate. The formulation is based on state space approach utilizing three-dimensional equations of elasticity. The formulation is remarkably simple and deals with only six variables in place of nine independent variables. Employing double Fourier series, the eigen value problem is solved. The natural frequency and the associated mode shapes of the composite plate are obtained. Numerical results for the natural frequency for different mode shapes of the plate with various plate parameters are obtained. The contours of transverse displacement and stresses are plotted.

Keywords: Vibration, State Space, Composite Plate, Analytical.

1. INTRODUCTION

Composites are one of the most basic structural elements and have been increasingly applied in many engineering fields such as aviation, automobiles, marine, submarine vehicles, and aerospace structures. This is mainly due to the fact that these components have great advantages of promising high stiffness and high strength to weight ratios and high rigidity over the conventional and traditional structures. Dynamic analyses of the composite plates have been the interest of many researchers. A semi-analytical method is developed for free vibration analysis of cross-ply laminated composite plates by Chen and Lue [1]. Ray [2] developed Zeroth order shear deformation theory for dynamic analysis and applied for both thin and thick laminated composite plates with high accuracy. Tarn [3] presented the state space formalism for rectilinear anisotropic elasticity. Sheng and Ye [4] used finite element method based on state space approach for the analysis of laminate composite plates. The aim of the present work is to obtain an exact analytical solution of a simply supported cross-ply laminated plate based on three-dimensional elasticity theory, utilizing state space formalism.

The results obtained could serve as a benchmark solution to asses other approximate methodologies or as a basis for establishing simplified laminated composites plate theories.

2. PROBLEM FORMULATION

A laminated composite rectangular plate with dimensions a, b and h, consisting of N linearly elastic orthotropic lamina is considered. The constitutive relations for the laminated composite plate are given by;

$$\sigma_{ij} = C_{ijkl} u_{k,l} \qquad \qquad ...(1)$$

where, C_{ijkl} denotes the elastic constant, σ_{ij} is the component of stress, u_k is the displacement component and $u_{k,l} = \dfrac{\partial u_k}{\partial x_l} + \dfrac{\partial u_l}{\partial x_k}$ is the component of stain.

In the absence of body force the field equations of equilibrium are;

$$\sigma_{ij,j} = \rho \partial_{,tt} u_i \qquad \qquad ...(2)$$

The stress displacement relations are expressed as;

$$
\begin{bmatrix} \sigma_{11} \\ \sigma_{22} \\ \sigma_{33} \\ \sigma_{12} \\ \sigma_{23} \\ \sigma_{13} \end{bmatrix}
=
\begin{bmatrix}
C_{11} & C_{12} & C_{13} & 0 & 0 & 0 \\
C_{12} & C_{22} & C_{23} & 0 & 0 & 0 \\
C_{12} & C_{23} & C_{33} & 0 & 0 & 0 \\
0 & 0 & 0 & C_{44} & 0 & 0 \\
0 & 0 & 0 & 0 & C_{55} & 0 \\
0 & 0 & 0 & 0 & 0 & C_{66}
\end{bmatrix}
\cdot
\begin{bmatrix}
\dfrac{\partial u_1}{\partial x_1} \\[2mm]
\dfrac{\partial u_2}{\partial x_2} \\[2mm]
\dfrac{\partial u_3}{\partial x_3} \\[2mm]
\dfrac{\partial u_1}{\partial x_2} + \dfrac{\partial u_2}{\partial x_1} \\[2mm]
\dfrac{\partial u_2}{\partial x_3} + \dfrac{\partial u_3}{\partial x_2} \\[2mm]
\dfrac{\partial u_1}{\partial x_3} + \dfrac{\partial u_3}{\partial x_1}
\end{bmatrix}
\qquad ...(3)
$$

By eliminating the stress (σ_{11}, σ_{22}, σ_{12}) and differential operators $\dfrac{\partial}{\partial x_1}$ and $\dfrac{\partial}{\partial x_2}$ in equations (1) and (2), followed by collecting the displacement u_i and the stress σ_{3i} ($i = 1 - 3$), the state equations can be compactly written in the matrix form as

$$\frac{\partial}{\partial x_3} \begin{bmatrix} \Pi \\ \Gamma \end{bmatrix} = \begin{bmatrix} 0 & A \\ B & 0 \end{bmatrix} \cdot \begin{bmatrix} \Pi \\ \Gamma \end{bmatrix} \qquad \qquad ...(4)$$

and,
$$\begin{bmatrix} \sigma_{11} \\ \sigma_{22} \\ \sigma_{12} \end{bmatrix} = H.\Pi \qquad \qquad ...(5)$$

where, the state variables are $\Pi = \begin{bmatrix} u_1 \\ u_2 \\ \sigma_{33} \end{bmatrix}$ and $\Gamma = \begin{bmatrix} \sigma_{13} \\ \sigma_{23} \\ u_3 \end{bmatrix}$

The operator matrices A, B and H consists of $A = \begin{bmatrix} \dfrac{1}{C_{66}} & 0 & -\partial_1 \\ 0 & \dfrac{1}{C_{55}} & -\partial_2 \\ -\partial_1 & -\partial_2 & \rho\partial_{tt} \end{bmatrix}$, $H = \begin{bmatrix} k_1\partial_1 & k_2\partial_2 & k_3 \\ k_4\partial_1 & k_5\partial_2 & k_6 \\ k_7\partial_2 & k_8\partial_1 & 0 \end{bmatrix}$,

$$B = \begin{bmatrix} \rho\partial_{tt} - k_1\partial_{11} - k_7\partial_{22} & -(k_2 + k_8)\partial_{12} & -k_3\partial_1 \\ -(k_4 + k_7)\partial_{12} & \rho\partial_{tt} - k_8\partial_{11} - k_5\partial_{22} & -k_6\partial_2 \\ -k_3\partial_1 & -k_6\partial_2 & k_9 \end{bmatrix}$$

where,

$$k_1 = C_{11} - \frac{C_{13}^2}{C_{33}}, \quad k_2 = C_{12} - \frac{C_{13}C_{23}}{C_{33}}, \quad k_3 = \frac{C_{13}}{C_{33}}, \quad k_4 = C_{12} - \frac{C_{13}C_{23}}{C_{33}}, \quad k_5 = C_{22} - \frac{C_{23}^2}{C_{33}}, \quad k_6 = \frac{C_{23}}{C_{33}},$$

$$k_7 = C_{44}, \quad k_8 = C_{44}, \quad k_9 = \frac{1}{C_{33}}$$

Boundary conditions on all four edges (simply supported)
$\sigma_{11} = 0$ & $u_2 = u_3 = 0$ at x_1 and $x_1 = a$
$\sigma_{22} = 0$ & $u_1 = u_3 = 0$ at $x_2 = 0$ and $x_2 = b$
Boundary conditions at the bottom surface ($x_3 = 0$)

$$\sigma_{13} = X^-(x_1, x_2), \quad \sigma_{23} = Y^-(x_1, x_2), \quad \sigma_{33} = Z^-(x_1, x_2)$$

Boundary conditions at the top surface ($x_3 = h$)

$$\sigma_{13} = X^+(x_1, x_2), \quad \sigma_{23} = Y^+(x_1, x_2), \quad \sigma_{33} = Z^+(x_1, x_2)$$

3. SOLUTION PROCEDURE

The state variables that satisfy the boundary conditions is assumed as

$$\Pi = \begin{bmatrix} u_1 \\ u_2 \\ \sigma_{33} \end{bmatrix} = \sum_{m=1}^{\infty}\sum_{n=1}^{\infty} \begin{bmatrix} U_{mn}(x_3)\cos\left(\dfrac{m\pi x_1}{a}\right)\sin\left(\dfrac{n\pi x_2}{b}\right)e^{(i\omega_{mn}t)} \\ V_{mn}(x_3)\sin\left(\dfrac{m\pi x_1}{a}\right)\cos\left(\dfrac{n\pi x_2}{b}\right)e^{(i\omega_{mn}t)} \\ Z_{mn}(x_3)\sin\left(\dfrac{m\pi x_1}{a}\right)\sin\left(\dfrac{n\pi x_2}{b}\right)e^{(i\omega_{mn}t)} \end{bmatrix} \qquad ...(6)$$

$$\Gamma = \begin{bmatrix} \sigma_{13} \\ \sigma_{23} \\ u_3 \end{bmatrix} = \sum_{m=1}^{\infty} \sum_{n=1}^{\infty} \begin{bmatrix} X_{mn}(x_3)\cos\left(\dfrac{m\pi x_1}{a}\right)\sin\left(\dfrac{n\pi x_2}{b}\right)e^{\left(i\omega_{mn}t\right)} \\ Y_{mn}(x_3)\sin\left(\dfrac{m\pi x_1}{a}\right)\cos\left(\dfrac{n\pi x_2}{b}\right)e^{\left(i\omega_{mn}t\right)} \\ W_{mn}(x_3)\sin\left(\dfrac{m\pi x_1}{a}\right)\sin\left(\dfrac{n\pi x_2}{b}\right)e^{\left(i\omega_{mn}t\right)} \end{bmatrix} \qquad ...(7)$$

where, ω_{mn} are natural frequencies. For specified loading conditions, the unknown Fourier coefficients U_{mn}, V_{mn}, W_{mn}, X_{mn}, Y_{mn}, and Z_{mn} in the preceding equations can be determined by solving equation (4).

Substituting equations (6) and (7) into equation (4), the following matrix equation is obtained.

$$\frac{\partial M(x_3)}{\partial x_3} = K_{mn}M(x_3) \qquad ...(8)$$

where, $M(x_3) = \begin{bmatrix} U_{mn}(x_3) & V_{mn}(x_3) & Z_{mn}(x_3) & X_{mn}(x_3) & Y_{mn}(x_3) & W_{mn}(x_3) \end{bmatrix}^{T}$ and $K_{mn} = \begin{bmatrix} 0 & K_3 \\ K_4 & 0 \end{bmatrix}$ with K_3, K_4 being given as

$$K_3 = \begin{bmatrix} \dfrac{1}{C_{66}} & 0 & \dfrac{-m\pi}{a} \\ 0 & \dfrac{1}{C_{55}} & \dfrac{-n\pi}{b} \\ \dfrac{m\pi}{a} & \dfrac{n\pi}{b} & -\rho\omega_{mn}^2 \end{bmatrix},$$

$$K_4 = \begin{bmatrix} -\rho\omega_{mn}^2 + k_1\left(\dfrac{m\pi}{a}\right)^2 + k_7\left(\dfrac{n\pi}{b}\right)^2 & (k_2 + k_8)\dfrac{m\pi}{a}\dfrac{n\pi}{b} & -k_3\dfrac{m\pi}{a} \\ (k_4 + k_7)\dfrac{m\pi}{a}\dfrac{n\pi}{b} & -\rho\omega_{mn}^2 + k_8\left(\dfrac{m\pi}{a}\right)^2 + k_5\left(\dfrac{n\pi}{b}\right)^2 & -k_6\dfrac{n\pi}{b} \\ k_3\dfrac{m\pi}{a} & k_6\dfrac{n\pi}{b} & k_9 \end{bmatrix}$$

The solutions to the equation (8) is

$$M(x_3) = \left[A_{mn}^{k}\right]M_{k-1}(x_3) \qquad ...(9)$$

where, $A_{mn}^{k}(x_3) = \exp\left(K_{mn}\cdot\left(x_3 - (x_3)_{k-1}\right)\right)$, which can be further expanded as a matrix polynomial as follows:

$$A_{mn}^{k}\left(x_{3}\right)=\sum_{i=0}^{\infty}\frac{\left(K_{mn}\cdot\left(x_{3}-\left(x_{3}\right)_{k-1}\right)\right)^{i}}{i!}$$, which is a converging series for all real and complex

numbers. The convergence for the present solution is obtained for $i = 5$. From equation (9) it can be written as

$$M_{k}\left(x_{3}\right)=\left[A_{mn}^{k}\right]M_{k-1}\left(x_{3}\right) \qquad ...(10)$$

where, $A_{mn}^{k}\left(x_{3}\right)=\exp\left(K_{mn}.h_{k}\right)$

Similarly,
$$M_{k+1}\left(x_{3}\right)=\left[A_{mn}^{k+1}\right]M_{k}\left(x_{3}\right) \qquad ...(11)$$

The above two equations (10) and (11) result in;

$$M_{k+1}\left(x_{3}\right)=\left[A_{mn}^{k+1}\right].\left[A_{mn}^{k}\right].M_{k-1}\left(x_{3}\right) \qquad ...(12)$$

From equation (12), the following simplification is obtained;

$$M\left(h\right)=T_{mn}\left(h\right)M\left(0\right) \qquad ...(13)$$

or,
$$\begin{bmatrix} U_{mn}\left(h\right) \\ V_{mn}\left(h\right) \\ Z_{mn}\left(h\right) \\ X_{mn}\left(h\right) \\ Y_{mn}\left(h\right) \\ W_{mn}\left(h\right) \end{bmatrix} = \begin{bmatrix} T_{11}\left(h\right) & T_{12}\left(h\right) & T_{13}\left(h\right) & T_{14}\left(h\right) & T_{15}\left(h\right) & T_{16}\left(h\right) \\ T_{21}\left(h\right) & T_{22}\left(h\right) & T_{23}\left(h\right) & T_{24}\left(h\right) & T_{25}\left(h\right) & T_{26}\left(h\right) \\ T_{31}\left(h\right) & T_{32}\left(h\right) & T_{33}\left(h\right) & T_{34}\left(h\right) & T_{35}\left(h\right) & T_{36}\left(h\right) \\ T_{41}\left(h\right) & T_{42}\left(h\right) & T_{43}\left(h\right) & T_{44}\left(h\right) & T_{45}\left(h\right) & T_{46}\left(h\right) \\ T_{51}\left(h\right) & T_{52}\left(h\right) & T_{53}\left(h\right) & T_{54}\left(h\right) & T_{55}\left(h\right) & T_{56}\left(h\right) \\ T_{61}\left(h\right) & T_{62}\left(h\right) & T_{63}\left(h\right) & T_{64}\left(h\right) & T_{65}\left(h\right) & T_{66}\left(h\right) \end{bmatrix} \begin{bmatrix} U_{mn}\left(0\right) \\ V_{mn}\left(0\right) \\ Z_{mn}\left(0\right) \\ X_{mn}\left(0\right) \\ Y_{mn}\left(0\right) \\ W_{mn}\left(0\right) \end{bmatrix}$$

where, $T_{mn}=\prod_{k=N}^{1}A_{mn}^{k}$ is called the global transfer matrix which can be easily calculated by the built-in functions in Matlab or Mathematica. The natural frequency of the plate is evaluated assuming the following boundary conditions, that stress on the top and bottom surfaces are traction free.

$$X_{mn}(0) = Y_{mn}(0) = Z_{mn}(0) = 0 \text{ and } X_{mn}(h) = Y_{mn}(h) = Z_{mn}(h) = 0$$

Substituting these in equation (13) at $x_3 = h$, it is obtained as;

$$\begin{bmatrix} U_{mn}\left(h\right) \\ V_{mn}\left(h\right) \\ 0 \\ 0 \\ 0 \\ W_{mn}\left(h\right) \end{bmatrix} = \begin{bmatrix} T_{11}\left(h\right) & T_{12}\left(h\right) & T_{13}\left(h\right) & T_{14}\left(h\right) & T_{15}\left(h\right) & T_{16}\left(h\right) \\ T_{21}\left(h\right) & T_{22}\left(h\right) & T_{23}\left(h\right) & T_{24}\left(h\right) & T_{25}\left(h\right) & T_{26}\left(h\right) \\ T_{31}\left(h\right) & T_{32}\left(h\right) & T_{33}\left(h\right) & T_{34}\left(h\right) & T_{35}\left(h\right) & T_{36}\left(h\right) \\ T_{41}\left(h\right) & T_{42}\left(h\right) & T_{43}\left(h\right) & T_{44}\left(h\right) & T_{45}\left(h\right) & T_{46}\left(h\right) \\ T_{51}\left(h\right) & T_{52}\left(h\right) & T_{53}\left(h\right) & T_{54}\left(h\right) & T_{55}\left(h\right) & T_{56}\left(h\right) \\ T_{61}\left(h\right) & T_{62}\left(h\right) & T_{63}\left(h\right) & T_{64}\left(h\right) & T_{65}\left(h\right) & T_{66}\left(h\right) \end{bmatrix} . \begin{bmatrix} U_{mn}\left(0\right) \\ V_{mn}\left(0\right) \\ 0 \\ 0 \\ 0 \\ W_{mn}\left(0\right) \end{bmatrix} \text{ or}$$

$$\begin{bmatrix} T_{31}\left(h\right) & T_{32}\left(h\right) & T_{36}\left(h\right) \\ T_{41}\left(h\right) & T_{42}\left(h\right) & T_{46}\left(h\right) \\ T_{51}\left(h\right) & T_{52}\left(h\right) & T_{56}\left(h\right) \end{bmatrix} . \begin{bmatrix} U_{mn}\left(0\right) \\ V_{mn}\left(0\right) \\ W_{mn}\left(0\right) \end{bmatrix} = \begin{bmatrix} 0 \\ 0 \\ 0 \end{bmatrix}$$

$$...(14)$$

The determinant of the transfer matrix is solved to obtain the natural frequency of the rectangular plate with simply supported boundary conditions.

4. NUMERICAL RESULTS

Numerical results for non-dimensional frequency parameter of a simply supported cross-ply laminated composite plates are obtained and compared with the available results. The comparisons are shown in Tables 1-2. The results show very good agreement with the 3D elastic solution due to Chen and Lue [1].

Table 1. Comparison of fundamental frequency parameter $\omega_{mn}h\sqrt{(\rho/E_2)}(b/h)^2$ for the simply supported square laminated plate. ($E_1 = 40E_2$, $E_2 = E_3$, $G_{12} = 0.5\,E_2$, $G_{23} = G_{13} = 0.6\,E_2$, $v_{12} = v_{23} = v_{13} = 0.25$)

0/90	Theory	b/h = 2	5	20	50	100
	Present	4.798	8.308	10.873	11.101	11.134
	3D[1]	4.953	8.527	11.036	11.263	11.297
	% Error	3.12	2.56	1.47	1.43	1.44
0/90/90/0	Present	5.162	10.436	17.394	18.421	18.573
	3D[1]	5.315	10.682	17.636	18.669	18.835
	% Error	2.87	2.3	1.37	1.33	1.39

Table 2. Comparison of frequency parameter $\omega_{mn}h\sqrt{(\rho/E_2)}(b/h)^2$ for different modes for simply supported (0/90) cross-ply square laminates. (b/h = 10, E_1 = 173.01, E_2 = 173.01, E_3 = 5.17, G_{12} = 9.38, G_{23} = 3.24, G_{13} = 8.27, •12 = 0.036, •23 = 0.25, •13 = 0.171)

Mode (m, n)	Method of Solution	Non Dimensional Frequency	% Error
1, 1	3D[1]	0.06027	1. 31
	Present Method	0.05948	
1, 2/2, 1	3D[1]	0.14548	1.27
	Present Method	0.14363	
2, 2	3D[1]	0.02229	1. 35
	Present Method	0.02199	

Figures 1-4, show the lateral displacement variation for (i) m = n = 1 (ii) m = 1, n = 2 (iii) m = 2, n = 1 and (iv) m = n = 2, respectively assuming $W_{mn}(x_3)$ equal to unity for simply supported square cross-ply laminated plates.

Figures 5-8, show the variation in the stress σ_{11} for (i) m = n = 1 (ii) m = 1, n = 2 (iii) m = 2, n = 1 and (iv) m = n = 2, respectively assuming the Fourier coefficients $U_{mn}(x_3)$, $V_{mn}(x_3)$, $Z_{mn}(x_3)$ as unity.

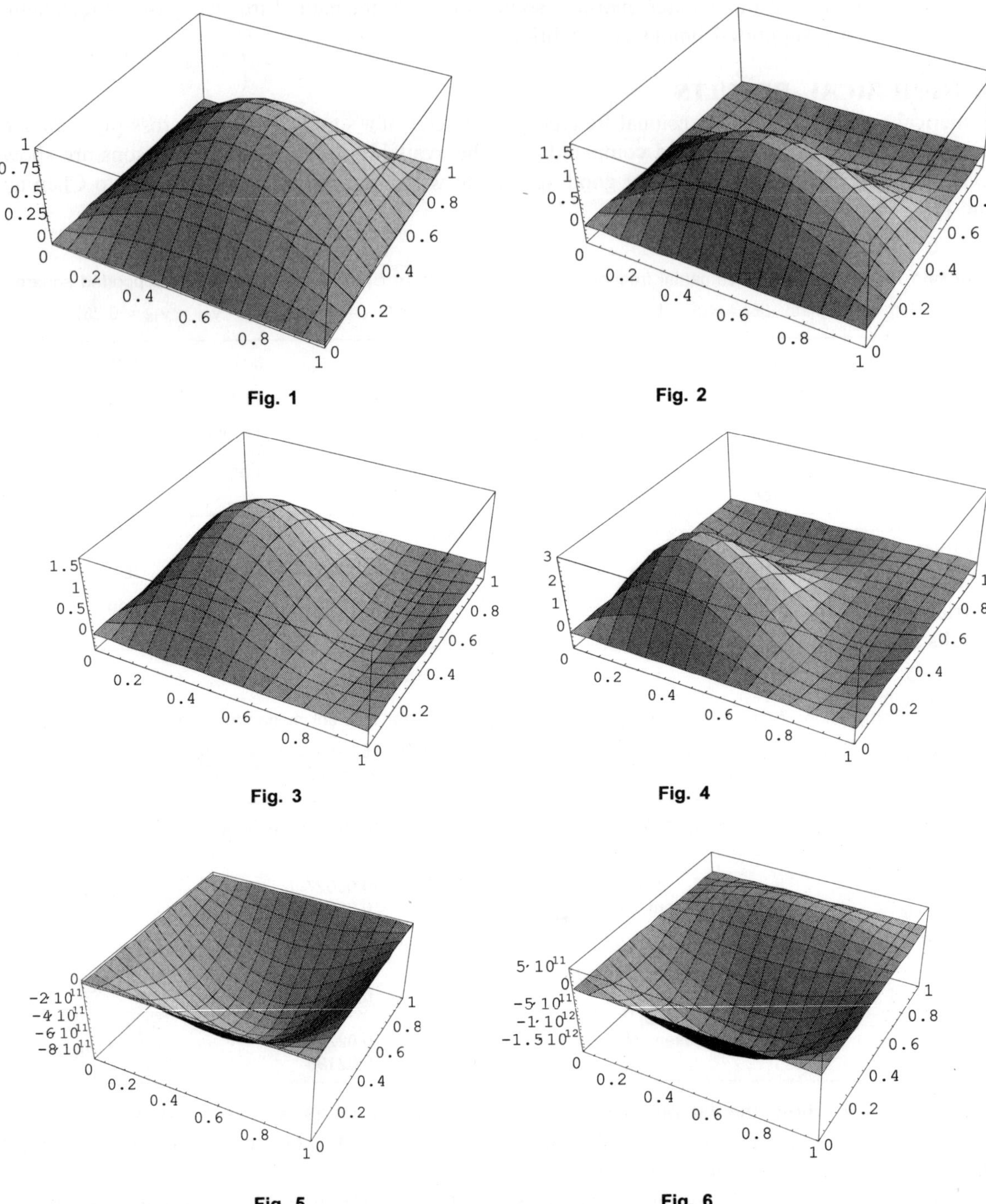

Fig. 1

Fig. 2

Fig. 3

Fig. 4

Fig. 5

Fig. 6

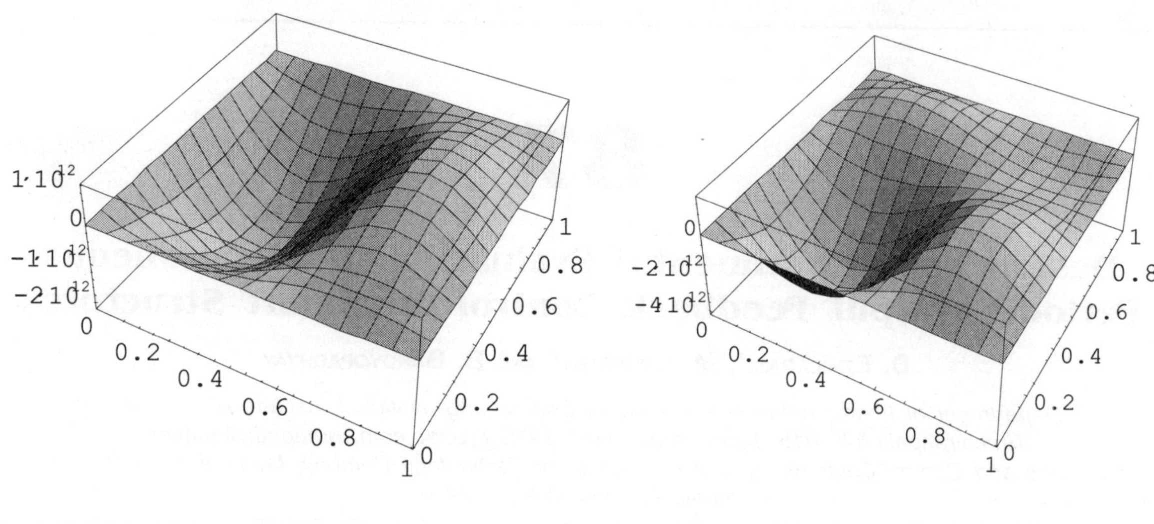

Fig. 7   **Fig. 8**

CONCLUSION

The free vibration analysis of cross-ply laminated composite plates with simply supported edges is carried out by solving the state equations. The present state space method is validated with available results and very good agreements among the results are obtained with 3D elasticity solution of cross-ply laminated composite plate. The present method provides an effective alternative solution for thick laminated composites.

REFERENCES

1. W.Q. Chen, and C.F. Lue, 2005, 3D free vibration analysis of cross-ply laminated plates with one pair of opposite edges simply supported, *Journal of Composite Structures*, 69, 77-87.
2. M.C. Ray, 2003, Zeroth-Order shear deformation theory for laminated composite plates, *ASME Journal of Applied Mechanics*, 70, 374-380.
3. Jiann-Quo Tarn, 2002, A state space formalism for anisotropic elasticity, Part I: Rectilinear Anisotropy, *International Journal of Solids and Structures*, 39, 5143-5155.
4. H.Y. Sheng, and J.Q. Ye, 2002, A state space finite element for laminated composite plates, *Computational Methods in Applied Mechanics and Engineering*, 191, 4259-4276.

87

Design and Experimental Evaluation of Simultaneous Periodic Output Feedback Control for Smart Structures

D. EZHILARASI[1], M. UMAPATHY[2] AND B. BANDYOPADHYAY[3]

[1,2]Department of Instrumentation and Control Engineering, National Institute of Technology, Tiruchirappalli-620 015, India email: ezhil_1979@yahoo.com, umapathy@nitt.edu
[3]Systems and Control Engineering, Indian Institute of Technology Bombay, Mumbai-400 076, India email: bijnan@ee.iitb.ernet.in

ABSTRACT

This paper presents the design and experimental implementation of simultaneous periodic output feedback controller to minimize structural vibration using collocated piezoelectric actuators and sensors. The linear dynamic model of piezoelectric bonded cantilever beams is obtained using online recursive least square parameter estimation. A digital control system that consists of simulink modeling software and dSPACE 1104 controller board is used for identification and control. The performance of simultaneous controller is evaluated by considering a cantilever beam structure of three different lengths. The effectiveness of the controller is demonstrated experimentally by exciting the structures at resonance.

Keywords: Smart structure, Simultaneous control, Periodic output feedback, System identification.

1. INTRODUCTION

Vibration control of flexible structures has been a major research topic over the past few decades. In recent years, a great number of research results have been produced in active structural vibration control using piezoelectric materials as distributed sensors and actuators [1-4]. Distributed piezoelectric materials experimentally have proven to be practical in sensing and controlling the vibrations of flexible structures [5-7]. Smart structures represent an interesting challenge for system identification methods to identify the dynamics of smart structures. System identification is an established modeling tool in engineering and numerous successful applications have been reported. [8-10].

The problem of simultaneously stabilizing a whole family of plants has received considerable attention for many years. One way of approaching the simultaneous stabilization problem with incomplete state information is to use observer based control laws, i.e., dynamic compensators. The problem with observer based controller is that the state feedback and state estimation cannot be separated in face of the uncertainty represented by a whole family of systems. Assuming that a simultaneously stabilizing state feedback gain has been found, it is possible to use convex programming algorithm to search for a simultaneously stabilizing full order observer gain, but this search is dependent on the state feedback gain previously obtained. If no stabilizing observer for this state feedback exists, nothing can be said because there may exist stabilizing observer for different feedback gains. In order to search directly for the compensator parameters, the problem can be transformed into an equivalent static output feedback problem. The algorithm given in [11], to solve this problem does not guarantee its convergence in general. H. Werner and K. Furuta [12, 13] have used the convex programming algorithm proposed in [14] and shown that existence of a simultaneously stabilizing state feedback gain for a family of systems generically implies the existence of a simultaneously stabilizing periodic output feedback gain. As evident from previous studies, numerous researchers have designed controllers for vibration control of cantilever beam using smart materials as sensors/actuators. Since, the control effort required is different if there is a small change in the dimension or property of the structure. Not much research has been made to device a common output feedback controller that can suppress the selected structural vibration modes for the structures with parameter variations. If such a controller exists one would not need to change the controller at all in smart structure. In this paper a single simultaneous periodic output feedback controller is designed and experimentally implemented to suppress the first vibration mode of a smart cantilever beam with three different lengths. The authors believe that the implementation of simultaneous periodic output feedback control for a smart structure in real time is first of its kind.

This paper is organized as follows: Section 2 gives the experimental set up used for identification and control. A recursive identification method used to identify the structures is presented in Section 3. Design of simultaneous periodic output feedback control is given in Section 4. In Section 5 experimental implementation of simultaneous periodic output sampling feedback control is presented and conclusions are drawn in Section 6.

2. EXPERIMENTAL SET-UP

A flexible aluminum beam with clamped end as shown in Figure 1 is considered in this paper. Two piezoceramic patches are surface bonded at a distance of 10 mm from the fixed end of the beam for structure 1, 15 mm from the fixed end of the beam for structure 2 and 20 mm from the fixed end of the beam for structure 3. To demonstrate simultaneous control, three different structural models are considered by introducing small variation in length (parameter variation) of the beam. The patch bonded on the bottom surface acts as a sensor and the one on the top surface acts as an actuator. To apply an excitation input to the structure another piezoceramic patch is bonded on the top surface at a distance of 67 mm from the actuator. The dimensions and properties of the beams and piezoceramic patches are given in Table 1.

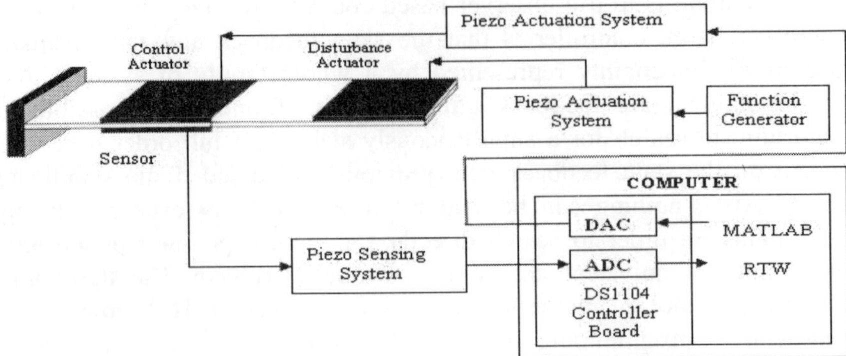

Fig. 1. Schematic diagram of the experimental set up.

Table 1. Properties and dimensions of the Aluminum beams and piezoceramic sensor/actuator.

Aluminum beam		Beam1	Beam2	Beam3	piezoceramic sensor/actuator		
Length (m)	l	0.23	0.235	0.24	Length (m)	l_p	0.0765
Width (m)	b	0.0127	0.0127	0.0127	Width (m)	b	0.0127
Thickness (m)	t_b	0.0023	0.0023	0.0023	Thickness (m)	t_a	0.005
Young's modulus(GPa)	E_b	71	71	71	Young's modulus (GPa)	E_p	47.62
Density (kg/m^3)	ρ_b	2700	2700	2700	Density (kg/m^3)	ρ_p	7500
First natural frequency (Hz)	f	31.7	28.8	27.8	Piezoelectric strain constant (m V^{-1})	d_{31}	-247×10^{-12}
					Piezoelectric stress constant (V m N^{-1})	g_{31}	-9×10^{-3}

The sensor output is given to the piezo sensing system which consists of high quality charge to voltage converting signal conditioning amplifier with variable gain. The conditioned piezosensor signal is given as analog input to dSPACE1104 controller board. The control algorithm is developed using simulink software and implemented in real time on dSPACE 1104 system using RTW and dSPACE real time interface tools. The simulink software is used to build control block diagrams and real time workshop is used to generate C code from the simulink model. The C code is then converted to target specific code by real time interface and target language compiler supported by dSPACE1104. This code is then deployed on to the rapid prototype hardware system to run hardware in-the-loop simulation. The control signal generated from simulink is interfaced to piezo actuation system through analog input/output unit of dSPACE 1104 system. The piezo actuation system drives the actuator and the excitation signal is applied from simulink environment through a DAC port of dSPACE system.

3. IDENTIFICATION OF SMART STRUCTURE DYNAMICS

The unknown parameters of the smart structure dynamics are estimated using online identification method, which is proven to be more universal and feasible than analytical and numerical models for the present system. The Recursive Least Square (RLS) method based on ARX model is used for linear system identification, which is easy to implement and has fast parameter convergence. The ARX model for the system shown in figure 1 is given as,

$$\hat{y}(k) + a_1 y(k-1) + \ldots + a_{n_a} y(k-n_a) = b_1 u(k-1) + \ldots + b_{n_b} u(k-n_b) + c_1 r(k-1) + \ldots$$

$$+ c_{n_c} r(k-n_c) + e(k). \qquad \ldots(1)$$

Where, u(k) is the input signal, r(k) is the excitation signal, y(k) the piezo sensor output, e(k) is white noise and n_a, n_b and n_c determine the model order.

The unknown parameter and data vector is thus

$$\theta = (a_1, a_2, \ldots, a_{n_a}, \quad b_1, b_2, \ldots, b_{n_b}, \quad c_1, c_2, \ldots, c_{n_c})^T \qquad \ldots(2)$$

$$\varphi(k) = (-y(k-1), -y(k-2), \ldots, -y(k-n_a), \quad u(k-1), u(k-2), \ldots, u(k-n_b),$$

$$r(k-1), r(k-2), \ldots, r(k-n_c))^T. \qquad \ldots(3)$$

The estimated model output is

$$\hat{y}(k) = \varphi^T(k)\hat{\theta}(k-1) \qquad \ldots(4)$$

For the RLS algorithm to update the parameters at each sample time, it is necessary to define an error. The model prediction error, $e(k)$ is the key variable in the RLS algorithm and is defined as

$$\varepsilon(k) = y(k) - \hat{y}(k) \qquad \ldots(5)$$

The $\varepsilon(k)$ is used to update the parameter estimate as

$$\hat{\theta}(k) = \hat{\theta}(k-1) + P(k)\varphi(k)\varepsilon(k) \qquad \ldots(6)$$

where, the covariance matrix $P(k)$ is updated using

$$P(k) = P(k-1)\left(1 - \frac{\varphi(k)\varphi^T(k)P(k-1)}{1 + \varphi^T(k)P(k-1)\varphi(k)}\right) \qquad \ldots(7)$$

The initial values of $\hat{\theta}(k)$ and P(k) are chosen to be $\hat{\theta}(0) = 0$ and $P(0) = \alpha I_Z$ where $\alpha = 10^4$. The natural frequency of the three beam structures are measured experimentally as 31.7Hz, 28.8Hz and 27.8 Hz, respectively. To identify the parameters in online, the structures are excited by a sinusoidal signal with their first natural frequency and a square wave signal as an input to the control actuator. The sampling time is chosen to provide approximately four measurements per cycle (sampling frequency 120Hz). The excitation signal, input signal and sensor output are given to MATLAB/simulink through ADC port of dSPACE 1104 system. The RLS algorithm is implemented by writing a C-file S-function used in MATLAB/simulink. The continuous state space model derived from the identified second order ARX model is given as

$$\dot{x} = Ax + bu + er; \quad y = c^T x \qquad \ldots(8)$$

where,

$$A = \begin{bmatrix} -83.0583 & 218.2890 \\ -204.9014 & 76.7292 \end{bmatrix}, \quad b = \begin{bmatrix} -1.4349 \\ -1.708 \end{bmatrix}, \quad e = \begin{bmatrix} -0.2359 \\ -0.0477 \end{bmatrix}, \quad c^T = \begin{bmatrix} 1 & 0 \end{bmatrix}. \quad \text{(Beam1)}$$

$$A = \begin{bmatrix} -47.5764 & 192.9894 \\ -179.6632 & 40.4213 \end{bmatrix}, \quad b = \begin{bmatrix} -1.9431 \\ -2.6129 \end{bmatrix}, \quad e = \begin{bmatrix} -0.2821 \\ -0.0039 \end{bmatrix}, \quad c^T = \begin{bmatrix} 1 & 0 \end{bmatrix}. \quad \text{(Beam2)}$$

$$A = \begin{bmatrix} -33.3311 & 183.5194 \\ -169.2012 & 25.2079 \end{bmatrix}, \quad b = \begin{bmatrix} -2.1026 \\ -2.2612 \end{bmatrix}, \quad e = \begin{bmatrix} -0.2920 \\ -0.0060 \end{bmatrix}, \quad c^T = \begin{bmatrix} 1 & 0 \end{bmatrix}. \quad \text{(Beam3)}$$

4. DESIGN OF PERIODIC OUTPUT FEEDBACK CONTROLLER

The simultaneous periodic output feedback controller [12, 13] is designed to reduce the amplitude of vibration of a cantilever beams at their first mode resonance. Let $(\Phi_{\tau,i},\Gamma_{\tau,i},\mathbf{c}_i^T),(\Phi_i,\Gamma_i,\mathbf{c}_i^T)$ be the systems obtained by sampling the system $(\mathbf{A}_i,\mathbf{b}_i,\mathbf{c}_i^T)$ at rate $1/\tau$ and $1/\Delta$, respectively. Here, sampling time $\tau = 0.01$ ms, $\Delta = \dfrac{\tau}{N} = 0.00125$ ms where N is equal to or greater than the controllability index of $(\Phi_{\tau,i},\Gamma_{\tau,i})$. For N = 8, $(\tilde{\Phi},\tilde{\Gamma})$ is controllable. A stabilizing output injection gain $\mathbf{G}$ is designed for the system $(\Phi_{\tau,i},\Gamma_{\tau,i},\mathbf{c}_i^T)$, $i = 1, 2, 3$ such that the eigen values of $(\Phi_{\tau,i}+\mathbf{G}_i\mathbf{c}_i^T)$ lie inside the unit circle. The output injection gains obtained are:

$\mathbf{G_1}$ = [0.6318 0.8677], $\mathbf{G_2}$ = [0.3926 0.8591], $\mathbf{G_3}$ = [0.2745 0.8503]

For the system $(\Phi_{\tau,i},\Gamma_{\tau,i},\mathbf{c}_i^T)$ the control signal is generated according to

$$u(t) = K_1\, y_i(k\tau),\ k\tau + l\Delta \le t < k\tau + (l+1)\Delta,\ K_{l+N} = K_l, \qquad \text{...(9)}$$

For l = 0, 1, ... N-1.

where,
$$\mathbf{K} = \begin{bmatrix} K_0 & K_1 & \cdots & K_{N-1}\end{bmatrix}^T \qquad \text{...(10)}$$

The simultaneous periodic output feedback gain $\mathbf{K}$ is obtained by solving $\Gamma_i\mathbf{K}=\mathbf{G}_i$ with the following performance index weight matrces $\mathbf{R}=[1],\quad \tilde{\mathbf{Q}}=2000\,[\,\mathbf{I}\,]_{6\times 6},\tilde{\mathbf{P}}=3000\,[\,\mathbf{I}\,]_{6\times 6}.$ where,

$$\Gamma_i = \begin{bmatrix} \Phi_i^{N-1}\Gamma_i & \Phi_i^{N-2}\Gamma_i & \Phi_i^{N-3}\Gamma_i & \cdots & \Gamma_i \end{bmatrix}$$

$$\mathbf{K} = \begin{bmatrix} 32.2526 & 22.4895 & 13.0922 & 4.4134 & -3.3057 & -9.9169 & -15.4373 & -19.5808\end{bmatrix}^T \ \text{...(11)}$$

5. EXPERIMENTAL EVALUATIONS

In the experimental set up to measure and control the vibration of cantilever beam is shown in Figure 2. The simultaneous periodic output feedback controller designed in section 4 is used to suppress the vibration of beam 1, beam 2 and beam 3. To assess the performance of the control system, a sinusoidal excitation at first natural frequency of each beam with amplitude of 10V (p-p) is applied. As required by periodic output feedback control, the sensor output was sampled at 0.01 sec (τ) through ADC port of dSPACE and MATLAB/simulink. The control signal is updated and applied to the control actuator at the sampling interval of 0.00125 sec ($\Delta = \tau/N$) through DAC port of dSPACE system. The controller is implemented by developing a real time simulink model using MATLAB RTW in simulink. The open loop response, closed loop response with simultaneous periodic output feedback control and control signal acquired from dSPACE control desk are shown in Figures 3, 4 and 5.

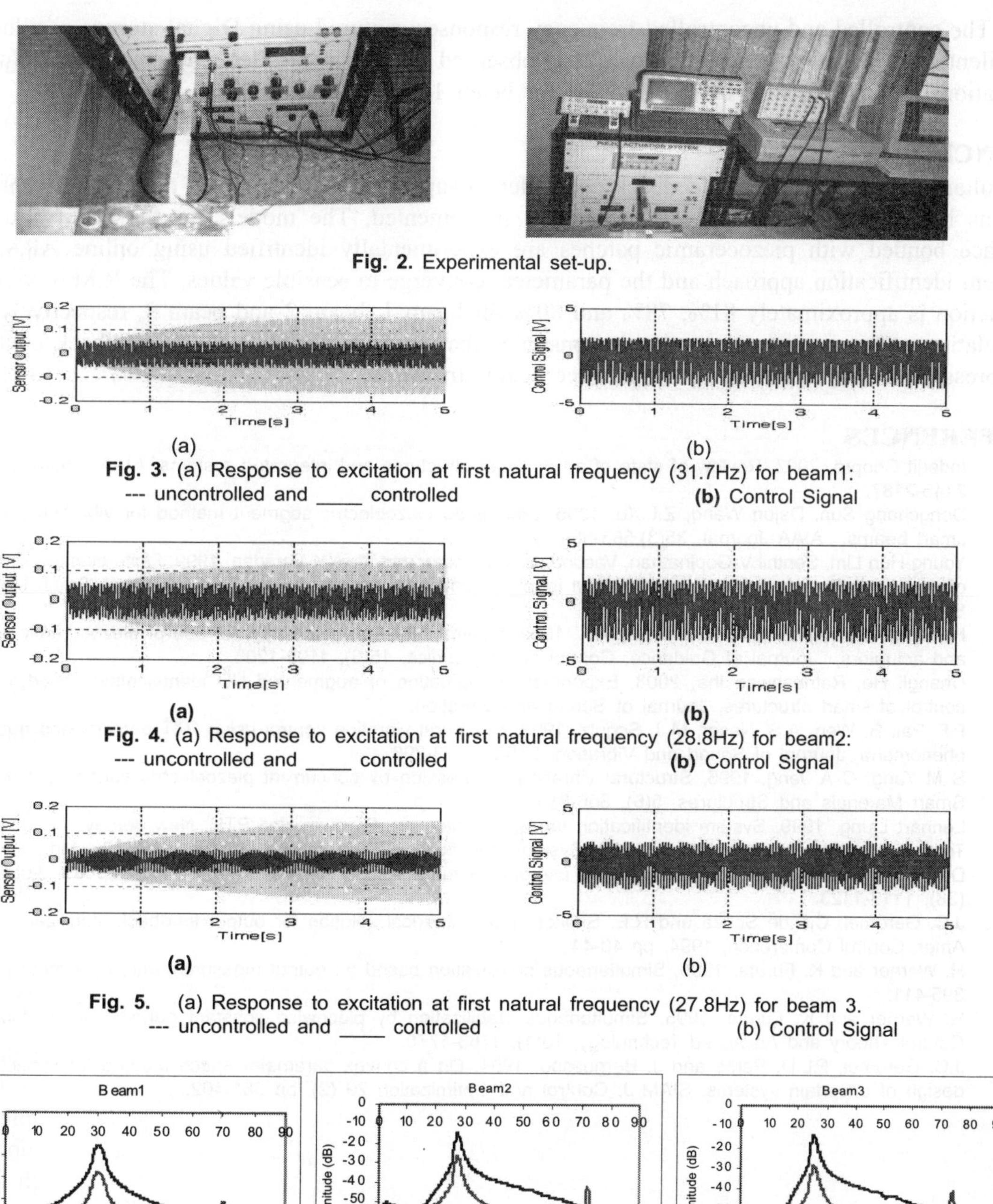

Fig. 2. Experimental set-up.

(a)

(b)

Fig. 3. (a) Response to excitation at first natural frequency (31.7Hz) for beam1:
--- uncontrolled and ____ controlled **(b)** Control Signal

(a)

(b)

Fig. 4. (a) Response to excitation at first natural frequency (28.8Hz) for beam2:
--- uncontrolled and ____ controlled **(b)** Control Signal

(a)

(b)

Fig. 5. (a) Response to excitation at first natural frequency (27.8Hz) for beam 3.
--- uncontrolled and ____ controlled (b) Control Signal

(a)

(b)

(c)

Fig. 6. Response to excitation at first natural frequency for beam1(31.7Hz),
beam2(28.8Hz) and beam3(27.8Hz): _ Uncontrolled and ____ controlled.

The controlled and uncontrolled frequency responses captured using Digital storage oscilloscope (Agilent 54621A) is shown in figure 6. It is observed that the controller reduces the magnitude of vibration by 12.8 dB, 11.5 dB, and 15 dB for beam 1, beam 2 and beam 3, respectively.

CONCLUSION

Simultaneous periodic output feedback controller to suppress the vibration of three smart cantilever beams has been designed and experimentally implemented. The models for the beam structures surface bonded with piezoceramic patches are experimentally identified using online ARX RLS system identification approach and the parameters converge to sensible values. The R.M.S. vibration reduction is approximately 81%, 78% and 80% for beam 1, beam 2 and beam 3, respectively. The simulation and experimental results demonstrates that a single periodic output feedback controller suppresses the first vibration mode of three beam structures.

REFERENCES

1. Inderjit Chopra, 2002, Review of state of art of smart structures and integrated systems, AIAA Journal, 40(11), 2145-2187.
2. Dongchang Sun, Dajun Wang, Z.L.Xu, 1996, Distributed piezoelectric segment method for vibration control of smart beams, AIAA Journal, 35(3),583-584.
3. Young-Hun Lim, Senthil V Gopinathan, Vasundara V Varadan and Vijay K Varadan, 1999, Finite element simulation of smart structures using an optimal output feedback controller for vibration and noise control, Smart Materials, Structures, (8), 324-337.
4. Hanagud, S.,Obal, M.W., and Callise, A.J., 1992, Optimal vibration control by the use of piezoceramic sensors and actuators, Journal of Guidance, Control and Dynamics, 15(5), 1199-1206.
5. Chengli He, Ratneshwar Jha, 2003, Experimental evaluation of augmented UD identification based vibration control of smart structures, Journal of Sound and Vibration.
6. P.F. Pai, B. Wen, A.S. Naser, M.J. Scgulz, 1998, Structural vibration control using PZT patches and non-linear phenomena, Journal of Sound and Vibration, 215(2), 273-296.
7. S M Yang, C A Jeng, 1996, Structural vibration suppression by concurrent piezoelectric sensor and actuator, Smart Materials and Structures, 5(6), 806-813.
8. Lennart Ljung, 1999, System identification theory for the user, Prentice Hall PTR, New Jersey.
9. Torsten Soderstrom, Petre Stoica, 1989, System identification, Prentice Hall International (UK) Ltd.
10. D.G. Robertson and J.H. Lee, 2002, On the use of constraints in least squares estimation and control, Automatica, (38), 1113-1123.
11. J.C. Geromel, C.C.de Souza and R.E. Skelton, LMI numerical solution for output feedback stabilization, Proc. Amer. Control Conference, 1994, pp 40-44
12. H. Werner and K. Furuta, 1995, Simultaneous stabilization based on output measurements, Kybernetika, 31(4), 395-411.
13. H. Werner and K. Furuta, 1995, Simultaneous stabilization by piecewise constant periodic output feedback, Control Theory and Advanced Technology, 10(4), 1763-1775.
14. J.C. Geromel, P.L.D. Peres and J. Bernussou, 1991, On a convex parameter space method for linear control design of uncertain systems, SIAM J. Control and Optimization 29 (2), pp 381-402.

88

Modeling of Thermal Stresses on Cantilever Magneto-Electro-Elastic Strip Using ANSYS

A. KUMARAVEL, N. GANESAN AND RAJU SETHURAMAN

*Department of Mechanical Engineering,
Indian Institute of Technology Madras-636 006, India.
email: kumaravel@iitm.ac.in, nganesan@iitm.ac.in, sethu@iitm.ac.in*

ABSTRACT

In this paper, the commercial finite element code ANSYS has been employed to analyze the thermal stresses on magneto-electro-elastic cantilever strip under thermal environment. ANSYS contains analysis of piezoelectric material using coupled field elements under ANSYS package. It does not contain explicitly analysis of piezomagnetic materials. In this work, an innovative approach has been made use of to analyze layered piezoelectric/piezomagnetic cantilever strip under thermal environment using the input listings used for piezoelectric analysis. The influences of piezoelectric coupling and magneto elastic coupling on thermal displacement, electric, magnetic potential and thermal stresses for $BaTiO_3$ /$CoFe_2O_4$/$BaTiO_3$ and $CoFe_2O_4$/ $BaTiO_3$/$CoFe_2O_4$ layered cases are studied.

Keywords: Magneto-electro-elastic; thermal environment; composite; ANSYS.

1. INTRODUCTION

Nowadays, composite materials are applied to various areas of science and engineering applications. More recently, the piezoelectric and/or piezomagnetic material is a potential candidate in the design of smart systems and structures due to their ability of converting the magnetic, electric, mechanical and thermal energies from one to another. These composites have excellent piezoelectric and piezomagnetic properties, which leads to use in applications such as magnetic field probes, electronic packaging, sensors and actuators etc. [1]. Sedaghati et al. [2] studied the optimization of inchworm-type linear piezoelectric actuator using ANSYS finite element software. Sunar *et al.* [3] derived the linear constitutive equations of thermopiezomagnetism with the aid of a thermodynamic potential and variational approach of obtaining general coupled field equations for thermopiezomagnetic composites. Ootao et al. [4] has investigated the behavior of simply supported multilayered magneto-electro-thermo elastic strip due to non-uniform heat supply using Laplace and finite sine transformations. Kumaravel *et al.* [5] studied the thermal displacement on electro-magneto-elastic beam subjected to non-uniform temperature rise with simply supported boundary condition. The main aim is to study the influences of piezoelectric coupling and magneto-elastic coupling on thermal displacement, electric potential, magnetic potential and transverse stresses on magneto-electro-elastic solid across thickness direction using ANSYS.

2. FINITE ELEMENT FORMULATION

2.1 Thermal Analysis

To evaluate the thermal stresses of the layered magneto-electro-elastic cantilever strip, the prerequisite is the evaluation of temperature distribution with in the domain. In the present analysis, the temperature distribution is evaluated using the ANSYS commercial finite element code by solving steady state two-dimensional Fourier heat conduction equation. Figure 1 shows the geometrical model along with thermal boundary conditions. The strip is subjected to temperature T_t at the top and maintained the room temperature T_b at bottom surface. Left and right side are insulated.

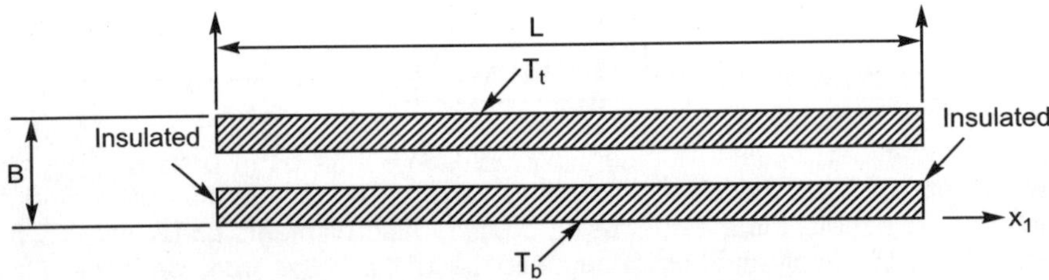

Fig. 1. Geometrical model of layered magneto-electro-elastic strip along with thermal boundary conditions.

2.2 Magneto-Electro-Elastic Problem Subjected to Thermal Loading

In fixed Cartesian coordinate system x_i ($i = 1, 2, 3$), the governing equations for linearly magneto-electro-elastic three-dimensional solid can be written as,

$$\sigma_{ij} = c_{ijkl}S_{kl} - e_{kij}E_k - q_{kij}H_k - \beta_{ij}\theta$$

$$D_i = e_{ikl}S_{kl} + \eta_{ik}E_k + m_{ik}H_k$$

$$B_i = q_{ikl}S_{kl} + m_{ik}E_k + \mu_{ik}H_k \qquad \qquad \text{...(1)}$$

$$S_{ij} = \frac{1}{2}(u_{i,j} + u_{j,i}), \quad E_i = -\phi_{,i}, \quad H_i = -\psi_{,i} \qquad \text{...(2)}$$

$$\sigma_{ij,j} = 0, \quad D_{i,i} = 0, \quad B_{i,i} = 0 \qquad \text{...(3)}$$

Where, comma denotes the partial differentiation. σ_{ij}, D_i and B_i are the components of stress, electric displacement and magnetic induction respectively. u_i, ϕ and Ψ denote the thermal displacement, the electric potential and the magnetic potential, respectively. c_{ijkl}, η_{ik} and μ_{ik} are the elastic, dielectric and magnetic permeability coefficients respectively. e_{kij}, q_{kij}, m_{ik} and β_{ij} are the piezoelectric, piezomagnetic, magneto-electric and stress temperature coefficients respectively. S_{ij}, E_k, H_k and θ are strain, the electric field and magnetic field and temperature change, respectively. In the present analysis, the coupled three-dimensional constitutive Eq. (1) for magneto-electro-elastic solid in $x_1 - x_2$ plane are assumed to be isotropic. The non-zero components of material constants of Eq. (1) for transversely isotropic magneto-electro-elastic solid can be written in matrix form as

$$[c] = \begin{bmatrix} c_{11} & c_{12} & c_{13} & 0 & 0 & 0 \\ & c_{11} & c_{23} & 0 & 0 & 0 \\ & & c_{33} & 0 & 0 & 0 \\ & & & c_{44} & 0 & 0 \\ & Sym & & & c_{44} & 0 \\ & & & & & c_{66} \end{bmatrix} \quad [e] = \begin{bmatrix} 0 & 0 & e_{31} \\ 0 & 0 & e_{31} \\ 0 & 0 & e_{33} \\ 0 & e_{15} & 0 \\ e_{15} & 0 & 0 \\ 0 & 0 & 0 \end{bmatrix} \quad [q] = \begin{bmatrix} 0 & 0 & q_{31} \\ 0 & 0 & q_{31} \\ 0 & 0 & q_{33} \\ 0 & q_{15} & 0 \\ q_{15} & 0 & 0 \\ 0 & 0 & 0 \end{bmatrix}$$

$$[\eta] = \begin{bmatrix} \eta_{11} & 0 & 0 \\ 0 & \eta_{11} & 0 \\ 0 & 0 & \eta_{33} \end{bmatrix} \quad [\mu] = \begin{bmatrix} \mu_{11} & 0 & 0 \\ 0 & \mu_{11} & 0 \\ 0 & 0 & \mu_{33} \end{bmatrix} \quad [m] = \begin{bmatrix} m_{11} & 0 & 0 \\ 0 & m_{11} & 0 \\ 0 & 0 & m_{33} \end{bmatrix} \quad \{\beta\} = \begin{Bmatrix} \beta_{11} \\ \beta_{11} \\ \beta_{33} \\ 0 \\ 0 \\ 0 \end{Bmatrix} \quad ...(4)$$

The different material constants for magneto-electro-elastic solid are given in Table.1. For plane stress problems, the stress components $\sigma_{22} = \sigma_{23} = \sigma_{12} = 0$, electric displacement $D_2 = 0$ and magnetic induction $B_2 = 0$. The coefficients $c_{ik}, \eta_{ik}, \mu_{ik}, e_{ki}, q_{ki}$ and m_{ik} are replaced by reduced coefficients $\bar{c}_{ik}, \bar{\eta}_{ik}, \bar{\mu}_{ik}, \bar{e}_{ki}, \bar{q}_{ki}$ and $\bar{m}_{ik}$, respectively. The reduced coefficients are derived in terms of material constants based on plane stress conditions [4]. The magneto-electro-elastic strip under thermal environment is discretized using four nodded plane stress element having three nodal degrees of freedom viz. thermal displacement in x_1, x_3 directions, electric potentials. It can be represented by suitable shape functions such as,

$$u_i = [N_u]\{u\}; \phi = [N_\phi]\{\phi\}; \psi = [N_\psi]\{\psi\} \quad ...(6)$$

Where, $\{u\}$, $\{u_1 \; u_3\}^T$, u_1 and u_3 are displacements in x_1 and x_3 directions, respectively. After evaluating the thermal displacements, the electric potential ϕ and magnetic potential Ψ can be derived at each nodal points using the following equations [6],

$$\phi = [K_{\phi\phi}]^{-1}[K_{u\phi}]^T \{u\}; \quad \psi = [K_{\psi\psi}]^{-1}[K_{u\psi}]^T \{u\} \quad ...(7)$$

The thermal stresses are evaluated using the constitutive Equation (2) shows the material properties used in the present analysis.

Table. 1. Material properties of $BaTiO_3$—$CoFe_2O_4$ [3]

	c_{11}	c_{12}	$c_{13} = c_{23}$	c_{33}	$c_{44} = c_{55}$	c_{66}	$\eta_{11} = \eta_{22}$	η_{33}	$\lambda_{11} = \lambda_{33}$
$BaTiO_3$	166.0	77.0	78.0	162.0	43.0	44.5	11.2	12.6	2.5
$CoFe_2O_4$	286.0	173.0	170.5	269.5	45.3	56.5	0.08	0.093	3.2

	e_{31}	e_{33}	e_{15}	$q_{31} = q_{32}$	q_{33}	q_{15}	$\mu_{11} = \mu_{22}$	μ_{33}	$\alpha_{11} = \alpha_{22}$	α_{33}	$m_{11} = m_{33}$
$BaTiO_3$	−4.4	18.6	11.6	0.0	0.0	0.0	5.0	10.0	15.7	6.4	0
$CoFe_2O_4$	0.0	0.0	0.0	583.0	699.7	550.0	−590.0	157.0	10.0	10.0	0

Units: c-GPa, $\eta - 10^{-9}$ C/V-m, λ − W/mK, e − C/m², q − N/A-m, $\mu - 10^{-6}$ − Ns/VC, − $\alpha - 10^{-6}$ 1/K, m − Ns/VC,

3. RESULTS AND DISCUSSIONS

For the present study, Barium Titanate ($BaTiO_3$) and Cobalt iron oxide ($CoFe_2O_4$) are used as piezoelectric and magnetostrictive materials. The sandwich structures is made of the combination of piezoelectric and magnetostrictive materials, which is in the form of two stacking sequences as $BaTiO_3$/ $CoFe_2O_4$/ $BaTiO_3$ (B/F/B) and $CoFe_2O_4$/ $BaTiO_3$ / $CoFe_2O_4$ (F/B/F). The length (L) and thickness (B) of the strip is used for the analysis are 0.06 m and 0.01 m, respectively. First, the nodal temperature is evaluated using thermal analysis separately. The magneto-electro-elastic cantilever strip is discretized using 540 elements. Plane55 two-dimensional thermal solid element used for performing the thermal analysis along with thermal boundary conditions as shown in Fig. 1. The temperature of top (T_t) and bottom (T_b) surface is assumed as 60°C and 27°C, respectively. Figure 2 shows the variation of temperature distribution across thickness direction.

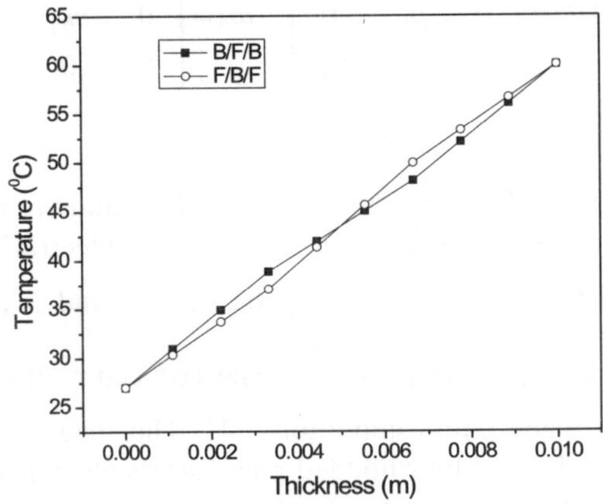

Fig. 2. Temperature distribution across thickness direction for B/F/B and F/B/F stacking sequences.

For structural analysis, cantilever magneto-electro-elastic strip modeled using ANSYS8.0 software package. ANSYS contains analysis of piezoelectric material and it does not contain explicitly analysis of piezomagnetic material. For piezoelectric analysis PLANE13 coupled-field quadrilateral solid element is used with three nodal degrees of freedom viz. displacement in x_1, x_3 directions and electric potential ϕ. The displacement, electric potential and magnetic potential are constrained in the fixed edge. The piezoelectric model requires the dielectric matrix, piezoelectric matrix and elastic coefficient matrix to be specified as material properties. In order to evaluate the magnetic potential Ψ, the piezomagnetic model requires the magnetic permeability matrix, piezomagnetic matrix and elastic coefficient matrix to be specified as material properties. This can be achieved by replacing the dielectric matrix, piezoelectric matrix and elastic coefficient matrix respectively in piezoelectric code. The present studies make use of the approach that the magnetic potential can be viewed similar to electric potential.

For sandwich structures, the analysis can be carried out two times, first run to evaluate the electric potential ϕ and second run to evaluate the magnetic potential Ψ by changing the corresponding material properties. The stress components can be evaluated by adding the stress components due to purely elastic constants, due to piezoelectric coupling and due to piezomagnetic coupling.

Figure 3(a) illustrates the displacement variation along the length at the middle of B/F/B and F/B/F layered magneto-electro-elastic beam. It is clear that the deflection at the tip is more in case of B/F/B stacking sequence than F/B/F case. It is due to the stiffness of F/B/F layered beam is higher. Figure 3(b) shows the variation of displacement across thickness direction. It is felt that the piezoelectric coupling and magneto elastic coupling effect on B/F/B stacking sequence. It shows the displacement of magneto-electro-elastic cantilever beam is higher as compared to elastic case. In the case of F/B/F layered beam there is no significant variation between magneto-electro-elastic and elastic case. Figures 4(a) and (b) show the variation of electric and magnetic potential across thickness direction at the free end of the cantilever beam. Figure 4(a), it is shown that the electric potential is higher for B/F/B stacking sequence as compared to F/B/F case. It is due to piezoelectric constants for magnetostrictive materials are zero. Figure 4(b) shows the magnetic potential for F/B/F layered

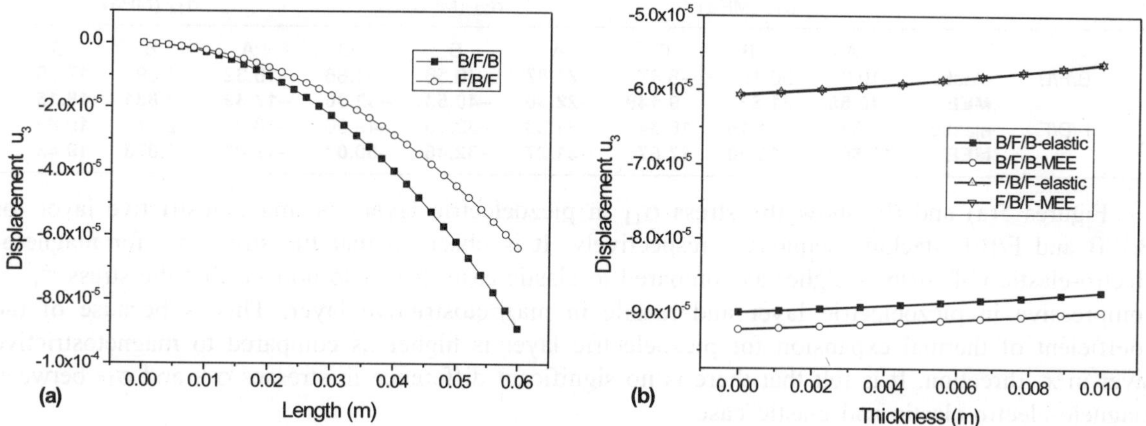

Fig. 3. Displacement variation in B/F/B and F/B/F layered cantilever beam (a) along the length (b) across thickness direction at the free end (Note: MEE - magneto-electro-elastic).

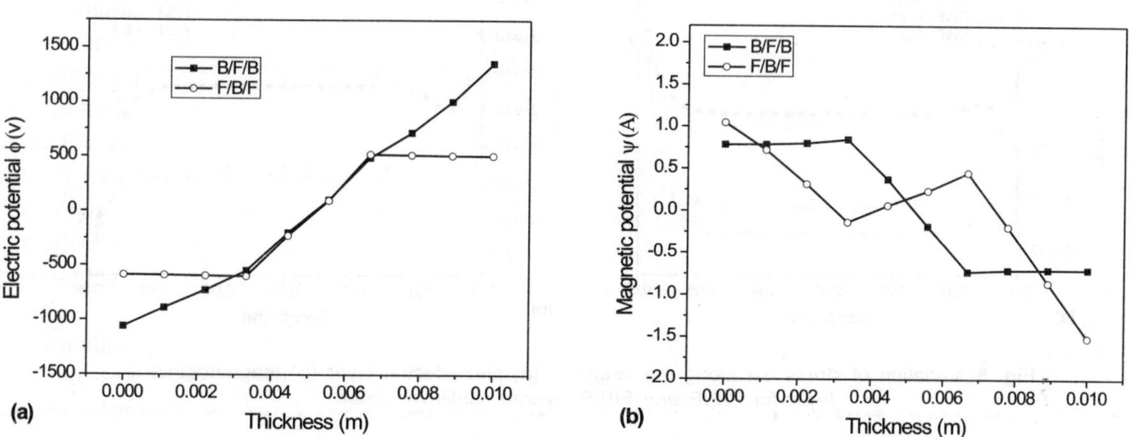

Fig. 4. Variation of (a) electric potential (b) magnetic potential across thickness direction at the free end (L = 0.06 m) for B/F/B and F/B/F stacking sequence.

beam is higher as compared to B/F/B case due to the piezomagnetic constants for piezoelectric layer are zero. Table 2. shows the stress histories of three reference points for B/F/B and F/B/F stacking sequences. It is observed that the stresses for magneto-electro-elastic beam are higher as compared to elastic case. It is noticed that the stress σ_{11} is compressive in piezoelectric layer and tensile in magnetostrictive layer. Stress σ_{33} are compressive due to the thermal expansion is restricted in x_3 direction at the clamped edge. Stress σ_{13} is compressive in bottom fiber and tensile in top fiber.

Table 2. Stress near the clamped edge for a layered magneto-electro-elastic cantilever beam 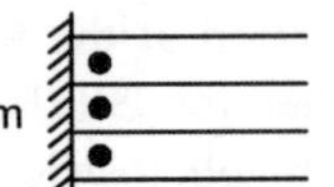

		σ_{11} (MPa)			σ_{33} (MPa)			σ_{13} (MPa)		
		A	B	C	A	B	C	A	B	C
B/F/B	elastic	−10.07	30.19	−8.87	−21.87	−40.56	−31.86	−16.32	−1.59	17.36
	MEE	**−10.55**	**31.62**	**−9.149**	**−22.30**	**−40.53**	**−32.30**	**−17.30**	**−1.831**	**18.15**
F/B/F	elastic	12.63	−14.15	16.34	−31.33	−32.10	−49.96	−18.27	−2.87	19.82
	MEE	**13.66**	**−15.40**	**17.67**	**−31.27**	**−32.46**	**−50.07**	**−17.62**	**−3.073**	**19.48**

Figures 5(a) and (b) show the stress σ_{11} in piezoelectric layer and magnetostrictive layer for B/F/B and F/B/F stacking sequences respectively. It is observed that the stress σ_{11} for magneto-electro-elastic C-F strip is higher as compared to elastic case. It is also noticed that the stress σ_{11} is compressive in piezoelectric layer and tensile in magnetostrictive layer. This is because of the coefficient of thermal expansion for piezoelectric layer is higher as compared to magnetostrictive layer in x_1 direction. It is felt that there is no significant difference in stresses σ_{33} and σ_{13} between magneto-electro-elastic and elastic case.

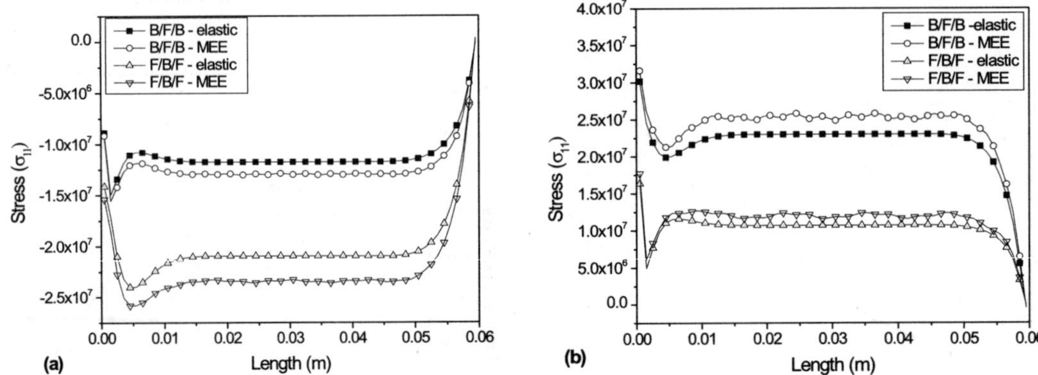

Fig. 5. Variation of stress σ_{11} along the length in (a) piezoelectric layer (b) magnetostrictive layer for B/F/B and F/B/F layered cantilever beam.

CONCLUSION

The magneto-electro-elastic layered cantilever strip under thermal environment is investigated using ANSYS commercial finite element software, we conclude the following:

1. The transverse displacement is higher for magneto-electro-elastic cantilever strip as compared to elastic case due to piezoelectric coupling and magneto-elastic coupling. It is noticed that during the analysis the transverse displacement is higher for B/F/B layered strip as compared to F/B/F layered strip.

2. The electric potential ϕ is nearly constant in piezomagnetic layer and linearly varying in the piezoelectric layer. The electric potential is higher in B/F/B stacking sequence as compared to F/B/F stacking sequence. The variation of magnetic potential Ψ is vice versa.

3. The axial normal stress σ_{11} is compressive in piezoelectric layer and tensile in magnetostrictive layer. Magneto-electro-elastic σ axial normal stress σ_{11} is higher as compared to elastic case. It is felt that the stacking sequences which affects the stress distribution.

REFERENCES

1. Nan C W, 1994, Magnetoelectric effect in piezoelectric and piezomagnetic phases, Phys Rev B, 50, 6082-6088.
2. Jian Li, Ramin Sedaghati, Javad Dargahi, David Waechter, 2005, Design and development of a new piezoelectric linear Inchworm actuator, Mechatronics, 15, 651-681.
3. Sunar. M, Ahmed Z. Al-Garni, M. H. Ali and R. Kahraman, 2002, Finite Element modeling of thermopiezomagnetic smart structures, AIAA Journal, 40, 1846-1851.
4. Yoshihiro Ootao and Yoshinobu Tanigawa, 2005, Transient analysis of multilayered and magneto-electro-thermo elastic strip due to nonuniform heat supply, Composite Structures, 68, 471-479.
5. Kumaravel A, Ganesan N and Raju Sethuraman, 2005, Studies on control of thermal deflections of magneto-electro-elastic smart beam under thermal environment, Proceedings of ISSS 2005 International Conference on Smart Materials Structures and Systems, July 28-30, PS 99-106.
6. ANSYS theory manual, 1999, Coupled field analysis guide, ANSYS Inc.

89

Geometrically Nonlinear Analysis of Composite Structures with Integrated Piezo-fiber Reinforced Composite Actuator

J. Shivakumar[1] and M.C. Ray[2]

[1]Research Scholar, Department of Mechanical Engineering, Indian Institute of Technology Kharagpur-721 302, India email: jskumar@mech.iitkgp.ernet.in
[2]Assistant Professor, Department of Mechanical Engineering Indian Institute of Technology Kharagpur-721 302, India

ABSTRACT

The nonlinear static analysis of simply supported angle-ply substrate plates integrated with a layer of piezoelectric fiber reinforced composite material is presented. The Von Ka'rma'n type non-linear strain displacement relations and first order shear deformation theory are used to formulate the variational model of this electromechanical coupled problem. Subsequently, Galerkin procedure is employed to derive the non-linear algebraic governing equations which are solved by the use of Newton-Raphson method. The results suggest the potential use of **PFRC** material for distributed control of nonlinear deformations of smart composite structures. Particular emphasis has been placed on investigating the effect of variation of piezoelectric fiber orientation on the actuating capability of the **PFRC** layer for counteracting the non-linear deformations of the smart composite plates.

Keywords: Piezoelectric; Fiber-reinforced; Composites; Nonlinear; Smart structures.

1. INTRODUCTION

The increasing use of piezoelectric material as distributed sensors and/or actuators, to achieve active control of high performing lightweight flexible smart structures is attributed to their inherent properties of direct and converse piezoelectric effects [1-3]. In applications such as space and aircraft structures which demand lightweight and thin flexible wall may undergo large deformation. Linear analyses may lead to inaccurate and inadmissible results under certain conditions and loadings. When the transverse deflections of a plate are not small compared to its thickness, the interaction between the membrane stresses and the curvatures must be considered. This type of interaction results in the stretching of the median surface, which in turn leads to nonlinear terms in the strain-displacement relations [4]. In this paper, analysis of control of nonlinear deformations of smart laminated angle-ply composite substrate plates using Piezoelectric Fiber Reinforced Composites has been carried out. The analytical model is a rectangular laminate composed of fiber-reinforced laminate and piezoelectric layer integrated on it. The Von Ka`rma`n strains are introduced to treat non-linear deformation. The governing equations are solved using the Galerkin Method.

2. FORMULATIONS

Figure 1 illustrates a simply supported rectangular laminated substrate plate made of number of N orthotropic layers. The length, width and thickness of plate are denoted by a, b and h, respectively. The top surface of the plate is integrated with a layer of the piezoelectric fiber reinforced composite (**PFRC**) material, acting as a distributed actuator of the plate. The thickness co-ordinate z of the top and bottom surfaces of any layer is denoted by h_{k+1} and h_k, respectively with k denoting the layer number of the layer. The fibers of the **PFRC** layer are aligned with the length of the substrate plate. When the **PFRC** layer is activated with applied electric field, the overall laminated plate acts as a smart composite plate.

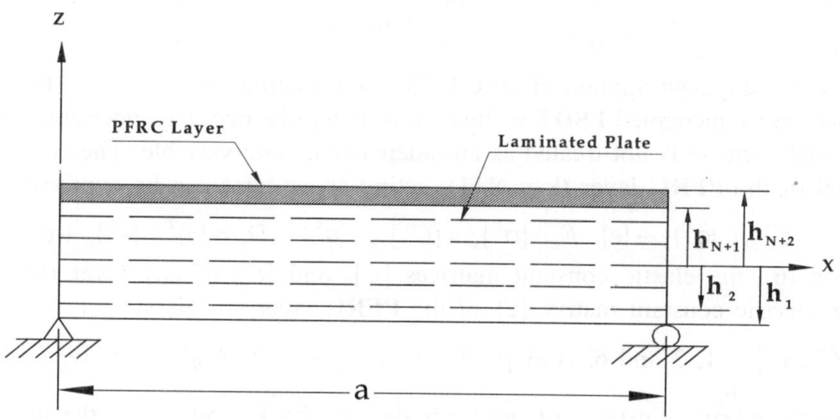

Fig. 1. Laminated Composite Plate integrated with a PFRC Layer.

The thickness of the actuator layer is denoted by h_p. The overall plate is considered to be subjected to an uniformly distributed load P(x, y) on its top surface Z = h/2+h_p and the materials of the plate are assumed to be linearly elastic. The first order shear deformation theory (FSDT) is used to describe the kinematics of deformation of the plate. The axial displacements u and v along the x and y direction, respectively and the transverse displacement, w at any point in the overall plate are given by

$$u = u_o\ (x,y) + z\ \theta_x(x,y),\quad v = v_o\ (x,y) + z\ \theta_y(x,y)\ \text{ and }\ w = w_o\ (x,y) \qquad ...(1)$$

where, u_o, v_o, w_o are the displacements at any point on the reference plane along x, y and z directions, respectively; θ_x and θ_y are the rotation of the transverse normal to the reference plane about y and x, axes respectively. The state of strain at any point in the overall plate is considered by the following two strain vectors: $\{\epsilon\}_b = [\epsilon_x\ \ \epsilon_y\ \ \epsilon_{xy}]^T$ and $\{\epsilon\}_s = [\epsilon_{xz}\ \ \epsilon_{yz}]^T$ $\qquad ...(2)$

where, $\in_x$ and $\in_y$ are the normal strains along x and y directions, respectively, $\in_{xy}$ is the in-plane shear strain, $\in_{xz}$ and $\in_{yz}$ are the transverse normal strains. Similarly, the state of stress at any point in the overall plate is described by the following two stress vectors:

$$\{\sigma\}_b = [\sigma_x \quad \sigma_y \quad \sigma_{xy}]^T \text{ and } \{\sigma\}_s = [\sigma_{xz} \quad \sigma_{yz}]^T \qquad ...(3)$$

Considering Von Ka`rma`n type geometric non-linearity and using the strain-displacement relations, the strain vectors can be expressed as

$$\{\in\}_b = [\in_{xo} \quad \in_{yo} \quad \in_{xyo}]^T + z \ [\kappa_x \quad \kappa_y \quad \kappa_{xy}]^T, \text{ and } \{\in\}_s = \left[\frac{\partial w}{\partial x} + \theta_x \quad \frac{\partial w}{\partial x} + \theta_x \right]^T \qquad ...(4)$$

in which the membrane strains $\in_{x0}$, $\in_{y0}$ and $\in_{xy0}$ and the flexural strains κ_x, κ_y and κ_{xy} are given by

$$\in_{xo} = \frac{\partial u_o}{\partial x} + \frac{1}{2}\left(\frac{\partial w_o}{\partial x}\right)^2, \ \in_{yo} = \frac{\partial v_o}{\partial y} + \frac{1}{2}\left(\frac{\partial w_o}{\partial y}\right)^2, \ \in_{xyo} = \frac{\partial u_o}{\partial y} + \frac{\partial v_o}{\partial x} + \left(\frac{\partial w_o}{\partial x}\right)\left(\frac{\partial w_o}{\partial y}\right),$$

$$\kappa_x = \frac{\partial \theta_x}{\partial x}, \ \kappa_y = \frac{\partial \theta_y}{\partial y} \text{ and } \kappa_{xy} = \frac{\partial \theta_x}{\partial y} + \frac{\partial \theta_y}{\partial x} \qquad ...(5)$$

In the present study, the applied electric field is considered to act only in the z-direction and analysis is based on an uncoupled **FSDT** without considering the two-way electromechanical coupling, in which potential variable is not treated as an independent state variable. The constitutive relations for the material of the **PFRC** layer (k = N+1), acting as actuator can be expressed as

$$\{\sigma^k\}_b = [C^k]_b \ \{\in^k\}_b - \{e\} \ E_z \ \{\sigma^k\}_s = [C^k] \ _s \ \{\in^k\}_s \ D_z = \{e\}^T \ \{\in^k\}_b + \varepsilon_{33} \qquad E_z \qquad ...(6)$$

In Equation (6), the elastic constant matrices $[C]_b$ and $[C]_s$ of any layer (k^{th}) of the overall plate, the piezoelectric constant matrix $\{e\}$ of the **PFRC** layer are given by

$$[C^k]_b = C_{ij}^k, \text{ i, j = 1, 2 and 6. And } [C^k]_s = C_{ij}^k, \text{ i, j = 4, 5. And } \{e\} = [e_{31} \quad e_{32} \quad e_{36}]^T \ ...(7)$$

where, C_{ij}^k are the elastic constants of the layer denoted by k, and e_{ij} are the piezoelectric stress coefficient with respect to the reference coordinate system. Also, in Eq. (6) E_z and D_z are the electric field and electric displacement, respectively along the z-direction and ε_{33} is a dielectric constant.

The total potential energy of the laminated plate coupled with the piezoelectric layer can be written as

$$T_p = \frac{1}{2} \sum_{K=1}^{N+1} \int_{h_K}^{h_{K+1}} \int_0^a \int_0^b [\{\in^K\}_b^T \{\sigma^K\}_b + \{\in^K\}_s^T \{\sigma^K\}_s] dxdydz - \frac{1}{2}$$

$$\int_{h_{N+1}}^{h_{N+2}} \int_0^a \int_0^b E_z D_z dxdydz - \int_0^a \int_0^b PW \ \Big|_{z=h_{N+2}} dxdy \qquad ...(8)$$

The mathematical form of the principle of minimum potential energy which yields the governing equation of the problem as well as the associated boundary conditions is given by

$$\delta T_P = 0 \qquad ...(9)$$

where, δ denotes the operator for first variation.

On substitution of Eqs. (4)-(6) into Eq. (8) and then using Eq. (9), the following variational statement of the problem which would yield the governing differential equation of the overall plate can be obtained:

$$\int_0^a \int_0^b (L_1\delta u_o + L_2\delta v_o + L_3\delta w_o + L_4\delta\theta_x + L_5\ \delta\theta_y)\ dx\ dy - \int_0^a N_{xy}\ \delta u_o \Big|_0^b dx - \int_0^b N_{xy}\ \delta v_o \Big|_0^a dy -$$

$$\int_0^b [M_x - \frac{1}{2}e_{31}\ (h_{N+2}^2 - h_{N+1}^2)]\ E_z\,\delta\theta_x \Big|_0^a dy - \int_0^a [M_y - \frac{1}{2}e_{32}\ (h_{N+2}^2 - h_{N+1}^2)]E_z\,\delta\theta_y \Big|_0^b dx = 0 \qquad ...(10)$$

in which $L_1 = \dfrac{\partial N_x}{\partial x} + \dfrac{\partial N_{xy}}{\partial y}$, $L_2 = \dfrac{\partial N_{xy}}{\partial x} + \dfrac{\partial N_y}{\partial y}$,

$$L_3 = \frac{\partial Q_x}{\partial x} + \frac{\partial Q_y}{\partial y} + N_x \frac{\partial^2 w_o}{\partial x^2} + \frac{\partial N_x}{\partial x}\frac{\partial w_o}{\partial x} + N_y \frac{\partial^2 w_o}{\partial y^2} + \frac{\partial N_y}{\partial y}\frac{\partial w_o}{\partial y} + \frac{\partial N_{xy}}{\partial y}\frac{\partial w_o}{\partial x} +$$

$$2N_{xy}\frac{\partial^2 w_o}{\partial x\partial y} + \frac{\partial N_{xy}}{\partial x}\frac{\partial w_o}{\partial y} + P - h_p\ (e_{31}\frac{\partial^2 w_o}{\partial x^2} + e_{32}\frac{\partial^2 w_o}{\partial y^2} + 2\ e_{36}\frac{\partial^2 w_o}{\partial x\ \partial y})\ E_z,$$

$$L_4 = \frac{\partial M_x}{\partial x} + \frac{\partial M_{xy}}{\partial y} - Q_x,\quad L_5 = \frac{\partial M_{xy}}{\partial x} + \frac{\partial M_y}{\partial y} - Q_y \qquad ...(11)$$

The stress resultants N_x, N_y, N_{xy} and the moment resultants M_x, M_y, M_{xy} appearing in equation (11) are given by

$$\begin{Bmatrix} N_x \\ N_y \\ N_{xy} \\ M_x \\ M_y \\ M_{xy} \end{Bmatrix} = \begin{bmatrix} A_{11} & A_{12} & A_{16} & B_{11} & B_{12} & B_{16} \\ A_{12} & A_{22} & A_{26} & B_{12} & B_{22} & B_{26} \\ A_{16} & A_{26} & A_{66} & B_{16} & B_{26} & B_{66} \\ B_{11} & B_{12} & B_{16} & D_{11} & D_{12} & D_{16} \\ B_{12} & B_{22} & B_{26} & D_{12} & D_{22} & D_{26} \\ B_{16} & B_{26} & B_{66} & D_{16} & D_{26} & D_{66} \end{bmatrix} \begin{Bmatrix} \in_{xo} \\ \in_{yo} \\ \in_{xyo} \\ \kappa_x \\ \kappa_y \\ \kappa_{xy} \end{Bmatrix}, \quad [Q_x \quad Q_y] = K\begin{bmatrix} A_{55} & A_{45} \\ A_{45} & A_{44} \end{bmatrix}\begin{Bmatrix} \dfrac{\partial W_o}{\partial X} + \theta_x \\ \dfrac{\partial W_o}{\partial Y} + \theta_y \end{Bmatrix}$$

$$...(12)$$

wherein, $A_{ij} = \displaystyle\sum_{K=1}^{N+1} \int_{h_k}^{h_{k+1}} C_{ij}^k\ dz$; $\quad B_{ij} = \displaystyle\sum_{K=1}^{N+1} \int_{h_k}^{h_{k+1}} z\ C_{ij}^k\ dz$; $\quad D_{ij} = \displaystyle\sum_{K=1}^{N+1} \int_{h_k}^{h_{k+1}} z^2\ C_{ij}^k\ dz$

and K is the shear correction factor and its value is taken as 5/6. The variational principle also yields the following homogeneous form of simply supported boundary conditions associated with the above variational statement for the overall plate:

$$u_o = w_o = \theta_y = N_{xy} = M_x = 0 \ \text{ at } x = 0 \text{ and } a \ \& \ v_o = w_o = \theta_x = N_{xy} = M_y = 0 \ \text{ at } y = 0 \text{ and } b ...(13)$$

For a particular mode of deformation, the displacement functions which satisfy the boundary conditions given by (13) are chosen as

$$u_o = U_{mn} \sin\alpha x \cos\beta y, \quad v_o = V_{mn} \cos\alpha x \sin\beta y, \quad w_o = W_{mn} \sin\alpha x \sin\beta y$$

$$\theta_x = \theta_{xmn} \cos\alpha x \sin\beta y \quad \text{and} \quad \theta_y = \theta_{ymn} \sin\alpha x \cos\beta y \qquad \text{...(14)}$$

in which U_{mn}, V_{mn}, W_{mn}, θ_{xmn} and θ_{ymn} are the unknown constants to be determined and $\alpha = \dfrac{m\pi}{a}$, $\beta = \dfrac{n\pi}{b}$ with m, n being the mode numbers. Substituting the displacement functions given by (14) into Eq. (10), the following Galerkin governing equilibrium equations of the overall plate are obtained.

$$\int_0^a \int_0^b L_1 \sin\alpha x \cos\beta y \, dxdy - \int_0^a N_{xy} \sin\alpha x \cos\beta y \Big|_0^b dx = 0;$$

$$\int_0^a \int_0^b L_2 \cos\alpha x \sin\beta y \, dx \, dy - \int_0^b N_{xy} \cos\alpha x \sin\beta y \Big|_0^a dy = 0 \int_0^a \int_0^b (L_3 \sin\alpha x \cos\beta y) dxdy = 0;$$

$$\int_0^a \int_0^b L_4 \cos\alpha x \sin\beta y \, dxdy - \int_0^b (M_x - \frac{1}{2}e_{31}(h_{N+2}^2 - h_{N+1}^2) \ E_z)\cos\alpha x \sin\beta y \Big|_0^a dy = 0$$

$$\int_0^a \int_0^b L_5 \sin\alpha x \cos\beta y \, dxdy - \int_0^a (M_y - \frac{1}{2}e_{32}(h_{N+2}^2 - h_{N+1}^2) \ E_z) \ \sin\alpha x \cos\beta y \Big|_0^b dx = 0 \qquad \text{...(15)}$$

Finally substituting Eqs. (11) and (14) into Eq. (15) and carrying out the explicit integrations with respect to the space coordinates, the following system of nonlinear algebraic equations are obtained:

$$R_1 U_{mn} + R_2 V_{mn} + R_3 \theta_{xmn} + R_4 \theta_{ymn} + R_5 W_{mn}^2 = 0$$

$$R_2 U_{mn} + R_6 V_{mn} + R_4 \theta_{xmn} + R_7 \theta_{ymn} + R_8 W_{mn}^2 = 0$$

$$R_{11} W_{mn} + R_{12} \theta_{xmn} + R_{13} \theta_{ymn} + R_9 U_{mn} W_{mn} + R_{10} V_{mn} W_{mn} + R_{14} \theta_{xmn} W_{mn} +$$
$$R_{15} \theta_{ymn} W_{mn} + R_{16} W_{mn}^3 + R_{17} = 0;$$

$$R_3 U_{mn} + R_4 V_{mn} + R_{12} W_{mn} + R_{18} \theta_{xmn} + R_{19} \theta_{ymn} + R_{20} W_{mn}^2 + R_{21} = 0$$

$$R_4 U_{mn} + R_7 V_{mn} + R_{13} W_{mn} + R_{19} \theta_{xmn} + R_{22} \theta_{ymn} + R_{23} W_{mn}^2 + R_{24} = 0 \qquad \text{...(16)}$$

The various coefficients are given as,

$$R_1 = (A_{11}\alpha^2 + A_{66}\beta^2), \quad R_2 = (A_{12} + A_{66}) \ \alpha\beta, \quad R_3 = 2 \ B_{16} \ \alpha\beta, \quad R_4 = (B_{16}\alpha^2 + B_{66}\beta^2)$$

$$R_5 = [A_{26}\beta^3 \ (32/(9\pi^2)) \ -A_{26}\alpha^2(16/(3\pi b))], \quad R_6 = (A_{22}\beta^2 + A_{66}\alpha^2), \quad R_7 = 2 \ B_{26}\alpha\beta$$

$$R_8 = [A_{16}\alpha^3 \ (32/(9\pi^2)) \ -A_{16}\alpha^2(16/(3\pi a))], \quad R_9 = -A_{26}\beta^3(32/9\pi^2)$$

$$R_{10} = -A_{16}\alpha^3(32/(9\pi^2)), \quad R_{11} = [A_{55}\alpha^2 + A_{44}\beta^2 - (e_{31}\alpha^2 + e_{32}\beta^2)h_p E_z], \quad R_{12} = KA_{55}\alpha,$$

$$R_{13} = KA_{44}\beta, \quad R_{14} = -[B_{11}\alpha^3 + (B_{12} - B_{66})\alpha\beta^2] \ (32/(9\pi^2)),$$

$$R_{15} = -[(B_{12} - B_{66}) \ \alpha^2\beta - 3 \ B_{22} \ \beta^3] \ (32/(9\pi^2))$$

$$R_{16} = [9(A_{11}\alpha^4 + A_{22}\beta^4) + 4\ (A_{12} + A_{66})\alpha^2\beta^2]\ (1/64)$$

$$R_{17} = -\frac{16}{\pi^2}P\ ,\ \ R_{18} = (D_{11}\alpha^2 + D_{66}\beta^2 + K\ A_{55})\ ,\ \ R_{19} = (D_{12} + D_{66})\ \alpha\beta$$

$$R_{20} = -[\{(B_{12} - B_{66})\alpha\beta^2 - 2\ B_{11}\alpha^3\}\ (16/(9\pi^2))\ + B_{11}\alpha^2(16/(3\pi a))]$$

$$R_{21} = -8\ e_{31}E_z(h_{N+2}^2 - h_{N+1}^2)\ V/(\pi a h_p)\ ,\ \ R_{22} = (D_{22}\beta^2 + D_{66}\alpha^2 + K\ A_{44})$$

$$R_{23} = -[\{(B_{12} - B_{66})\alpha^2\beta - 2B_{22}\beta^3\}(16/(9\pi^2)) + B_{22}\beta^2(16/(3\pi b))]$$

$$R_{24} = -8\ e_{32}E_z(h_{N+2}^2 - h_{N+1}^2)\ V/(\pi b h_p))$$

3. NUMERICAL RESULTS AND DISCUSSIONS

The set of Eqs.(16)-(20) are solved using Newton-Raphson method to evaluate the numerical results. Angle-ply substrate laminates of equal thickness are considered for evaluation of the numerical results. The following set of material properties for the layers of the substrates are considered for presenting the numerical results: (E1/E2=25, G12/E2=0.5, G23/E2=0.2, v12=0.25). The thickness of laminate is considered as 2 mm. The piezoelectric fibers and the matrix of the **PFRC** layer are made of PZT5H and epoxy, respectively. Considering 40% fiber volume fraction, the effective elastic and piezoelectric coefficients of the **PFRC** layer are obtained by using the micromechanics model derived by Mallik and Ray [5] and are given by

$$C_{11} = 32.6\ GPa,\ C_{12} = 4.3\ GPa,\ C_{22} = 7.2\ GPa,\ C_{44} = 1.05\ GPa,\ C_{55} = C_{66} = 1.29\ GPa,$$
$$e_{31} = -6.76\ C/m^2,\ \ e_{32} = -0.076\ C/m^2.$$

The applied load is uniformly distributed; the predominant mode of deflection is the first mode. Assuming the values of m = 1, n = 1 in the definition of α and β the numerical results are evaluated with and without applying the applied voltages to the **PFRC** layer for different values of aspect ratios, S (= a/h) of the laminated substrates. The following non-dimensional quantities are used for presenting the results.

$Q = (q_0.S_4/E_T)$, $S = (a/h)$, $\overline{\sigma}_x = (\sigma x\ .\ S^2/E_2)$ Next, considering the thickness of the **PFRC** layer as 250 µm, the results are evaluated for the substrate plates (a/h = 200) with and without applying the electrical potential distribution on the actuator surface. Because the thickness of the **PFRC** layer is very low, the variation of the electric potential function across the thickness may be considered as linear.

Figure 2 illustrates the responses of the plate when the **PFRC** layer is active ($V \neq 0$) or passive (**V = 0**). It may be observed from these figures that the active **PFRC** layer counteracts the nonlinear deformation of the substrate plate caused by the applied mechanical load. Hence, the activated **PFRC** layer can act as distributed actuator for controlling nonlinear vibrations of smart structures. Since, the applied voltage in the **PFRC** layer remains same while the applied mechanical load varies, the magnitude of actuation by the **PFRC** layer activated with a particular voltage decreases with the increase in the applied load. Similar behavior was also noticed for the four layered anti-symmetric cross-ply (0°/90°/0°/90°) plate [7]. The important aspect of developing this analytical model is to investigate the effect of variation of fiber orientation ψ in the **PFRC** layer on its capability of actuating the laminated substrate plates. Fig. 3 illustrates this effect of variation of the piezoelectric fiber orientation ψ in the **PFRC** layer on the center deflection of the laminated composite substrate plate for which S = a/h = 200. The mechanical load is considered to be in the linear range i.e., say Q = 20. For this, a prescribed uniformly distributed electric potential say, -300 voltage is considered on the exposed surface of the **PFRC** layer such that the resultant deflection under the combined action of mechanical and electrical load becomes opposite to that caused by the uniformly distributed mechanical load alone acting vertically upward only. It may be observed from Fig. 3 that for angle-ply substrates the control authority of the **PFRC** layer becomes maximum when ψ = -45° or 45° according as the fiber orientation in the top layer of the substrate plate being integrated with the **PFRC** layer with fiber orientation at 45° or –45°, respectively. This result is found to be matching with the result for the analysis of four-layered angle-ply substrates [6]. Figure 4 shows actuating capability of **PFRC** layer with the piezoelectric fiber angle considering the values of the applied mechanical load which cause nonlinear deformations of the substrate laminates. The distribution of in-plane normal stress across the thickness of thin (a/h=200) angle-ply (-45°/45°/-45°/45°) substrate is illustrated in the Fig. 5.

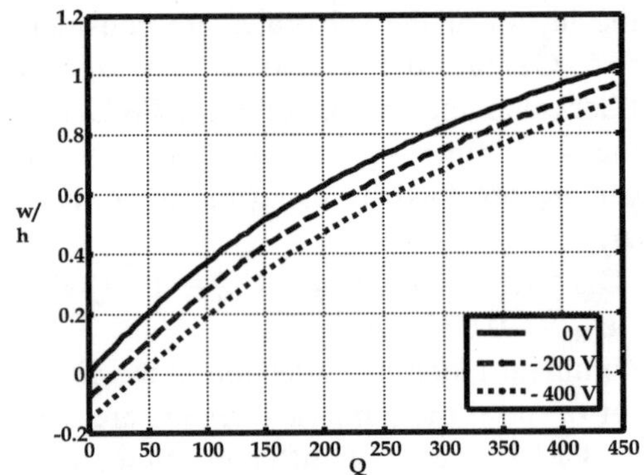

Fig. 2. Variation of center deflection of angle-ply (–45°/45°/–45°/45°) substrate plate with the applied mechanical load.

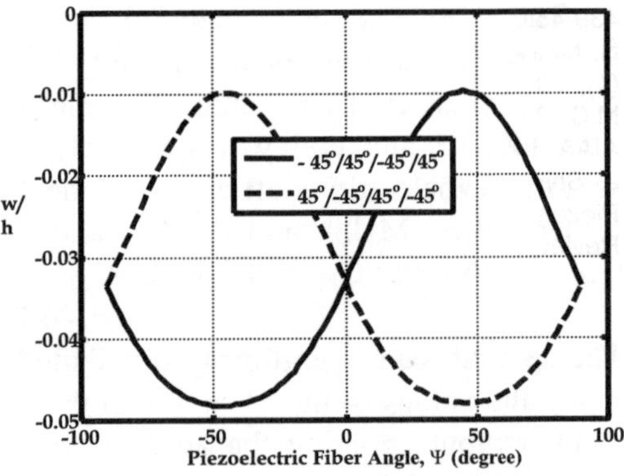

Fig. 3. Variation of center deflection of angle-ply substrate plates with the piezoelectric fiber angle in the PFRC layer (Q = 20, V = –300 V).

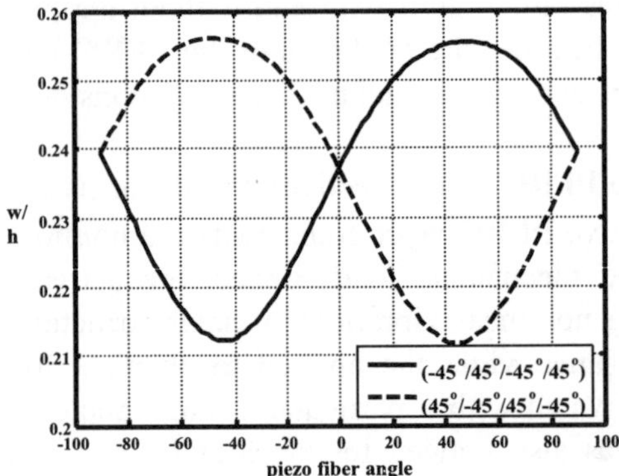

Fig. 4. Variation of center deflection of angle-ply Substrate plate with the piezoelectric fiber angle in the PFRC layer (Q = 100, V = –300 V).

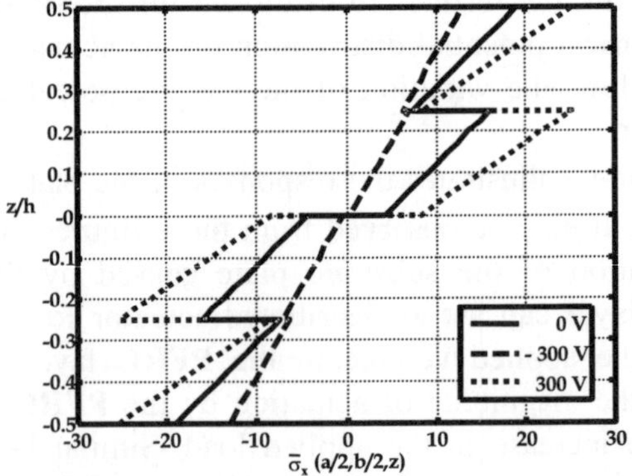

Fig. 5. Distribution of normal stress (σ_x) across the thickness of angle-ply (-45°/45°/-45°/45°) substrate plate (Q = 100).

CONCLUSION

A semi-analytical model of a laminated smart composite plate has been investigated. Results are presented for angle-ply substrate plates. The results indicate that the distributed actuator made of **PFRC** material significantly encounters the nonlinear deformation of laminated composite plates. Hence, this distributed actuator can be used for controlling nonlinear vibrations of smart laminated composite structures. The analysis revealed that the fiber orientation in the PFRC layer plays significant role in counteracting the nonlinear deformations of angle-ply substrate plates.

REFERENCES

1. T. Bailey, J.E. Hubbard, 1985, Distributed Piezoelectric Polymer Active Vibration Control of a Cantilever Beam, Journal of Guidance, Control and dynamics. 8, 605-611.
2. E.F. Crawley, J.D. Luis, 1987, Use of Piezoelectric Actuators as Elements of Intelligent Structures, AIAA Journal, 25, 1373-1385.
3. A. Baz, S. Poh, Performance of an active control system with piezoelectric actuators, Journal of Sound and Vibration, 126, 327-343.
4. C.Y. Chia, 1988, geometrically nonlinear behavior of composite plates: A review, Applied Mechanics Review, 41, 439-451.
5. N. Mallik, M.C. Ray, 2003, Effective Coefficients of Piezoelectric Fiber Reinforced Composites, AIAA Journal, 41 (4), 704-710.
6. M.C. Ray, N. Mallik, 2005, Performance of smart damping treatment using piezoelectric fiber reinforced composites, AIAA Journal, 43, 184-193.
7. J. Shivakumar, M.C. Ray, 2006, Nonlinear Analysis of Smart Composite Plates Integrated with a Distributed Piezoelectric Fiber Reinforced Composite Actuator, Mechanics and Advanced Material Structures Journal. (Under Review).

90

Experimental Evaluation of Different Active Vibration Control Algorithms for a Smart Structure

SUDEEP U.[1], A.S.N. PRASAD[2] AND C. SUJATHA[3]

[1]Department of Applied Mechanics. email: sudeepnss@yahoo.co.in.
[2]Department of Ocean Engineering,
[3]Department of Mechanical Engineering, Indian Institute of Technology Madras, Chennai-600 036, India.

ABSTRACT

This paper deals with the experimental evaluation of active vibration control strategies for a smart beam using PZT actuators and sensors. Following an initial computational and experimental modal analysis to identify the natural frequencies of the system, different control strategies such as Feed forward, Negative velocity feedback and Linear Quadratic Gaussian (LQG) control have been attempted. Feed forward and Negative velocity feedback control are implemented using LabVIEW 7.1, which is a graphic programming language that can be used with a personal computer (PC) based data acquisition system. LQG control is implemented using the DSP control box developed by MIDE technologies. An aluminium beam (280×262 ×2 mm) in the cantilever mode with 3 PZT patches (50 × 25 × 0.9 mm) bonded as sensor, actuator and exciter is used as the test structure.

Keywords: Active vibration control, PZT, Smart structure.

1. INTRODUCTION

Vibration often leads to undesired noise and causes operating problems in many engineering applications, especially in space structures. It is usually the low frequency modes that need to be controlled because of the lower damping associated with them and the resulting slow decay of these vibrations. This necessitates active control of vibrations, apart from the commonly employed passive control methods. Recent developments in smart materials such as Lead Zirconium Titanate (PZT), Magneto Rheological (MR) fluids, etc. have helped to achieve active control [1].

Piezoelectric materials are crystalline materials which become electrically polarized when subjected to mechanical forces. Tension and compression generate voltages of opposite polarity and in proportion to the applied force. The converse of this relationship is true: if this voltage generating crystal is exposed to an electric field, it lengthens or shortens according to the polarity of the field and in proportion to the strength of the field. These behaviors were named as the *piezoelectric*

effect and the ***inverse piezoelectric effect*** respectively [2]. This can be suitably used for vibration control.

$$\{T\} = [c]\{S\} - [e]\{E\} \qquad \text{...(1)}$$
$$\{D\} = [e]^{T}\{S\} + [\varepsilon]\{E\} \qquad \text{...(2)}$$

where, $\{T\}$ = stress vector $\{D\}$ = electric flux density vector $\{S\}$ = strain vector
  $\{E\}$ = electrical field vector $[c]$ = elasticity matrix $[e]$ = piezoelectric matrix
  $[\varepsilon]$ = dielectric matrix.

2. EIGENVALUE ANALYSIS OF STRUCTURE

An aluminium beam (280 × 262 × 2 mm) in the cantilever mode with 3 PZT patches (50 × 25 × 0.9 mm) bonded as sensor, actuator and exciter was analyzed. The beam was discretized into 2500 elements, each element having 8 nodes. Figure 1 shows the finite element model of the beam developed using ANSYS. The PZT elements can be seen near the left fixed end. As piezoelectrics involve interaction between electrical and mechanical fields, they can be modeled using coupled field elements in ANSYS [3]. This analysis takes into account the interaction (coupling) between two or more disciplines (fields). In the present work, SOLID5 which is a 3-D Coupled field solid element was used to model the piezoelectrics. To find out the natural frequencies of the beam, modal analysis was done using the general equation of motion of the system

$$[M]\{\ddot{u}\} + [K]\{u\} = 0 \qquad \text{...(3)}$$

where, [M] is the mass matrix, [K] is the stiffness matrix and [u] is the displacement vector

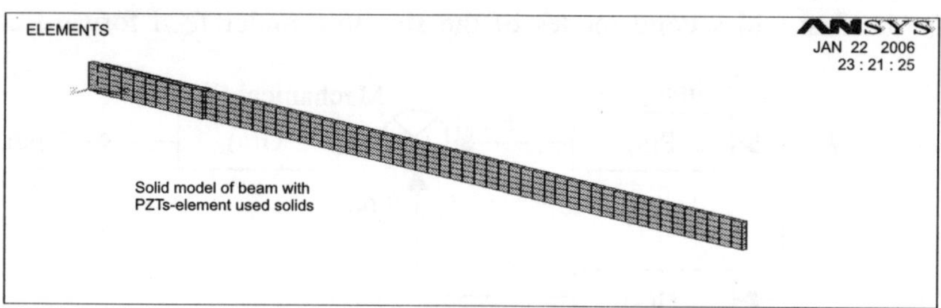

Fig. 1. Finite Element Model of Beam.

3. EXPERIMENTAL MODAL ANALYSIS

In order to validate the analytical results, natural frequencies of the structure were found experimentally. The beam was excited using one PZT patch and the response was sensed using another. Excitation was given using a periodic chirp signal with a frequency range 10 to 800 Hz. Table 1 shows a comparison of the results obtained using experimental and computational modal analysis.

Table 1. Results of Experimental and Computational Modal Analysis

Sl. No.	Natural Frequencies (Hz)		% Error	Damping Ratio (Experiment)
	Experiment	ANSYS		
1	21.5	23.0	6.5	0.0073
2	123.0	124.6	1.3	0.033
3	315.5	323.5	2.5	0.093
4	620.0	615.0	-0.8	0.018

Large difference in natural frequency values between experiment and ANSYS for the first frequency may be due to experimental error in ensuring required boundary condition.

4. ACTIVE VIBRATION CONTROL

The idea of active vibration control is to generate a secondary disturbance, which destructively interferes with the primary disturbance causing vibration. This is shown in Fig. 2.

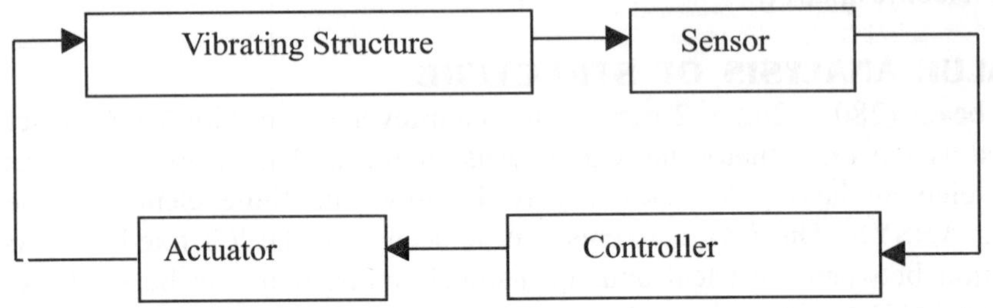

Fig. 2. Components of Active Vibration Control System.

4.1 Feed Forward Control

This method relies on the availability of a reference signal correlated to the primary disturbance as shown in Fig. 3. The controller generates a secondary signal which is out of phase with the primary such that the overall response Y(s) is reduced. There is no guarantee in this method that the global response is reduced at locations other than the sensor locations. Figures 4 and 5 show amplitude vs. time plots for the first and second modes of the structure under feed forward control.

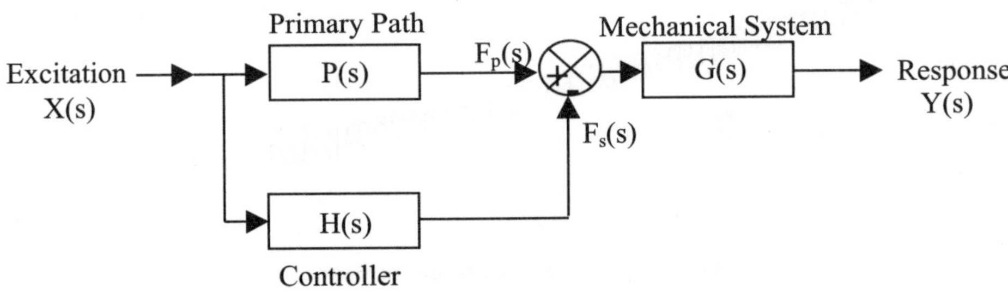

Fig. 3. Block Diagram of a Feedforward Control System.

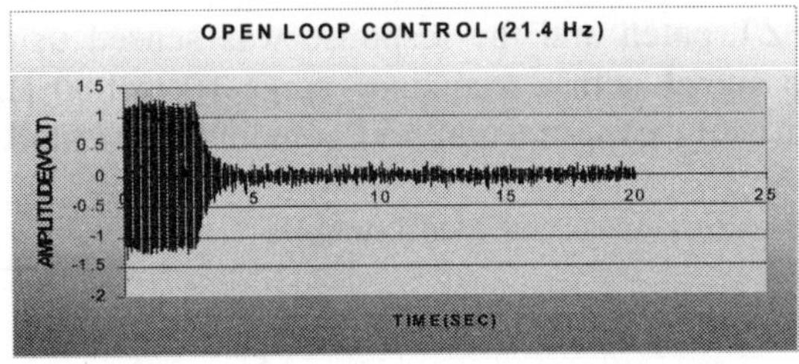

Fig. 4. Amplitude vs. Time (first mode).

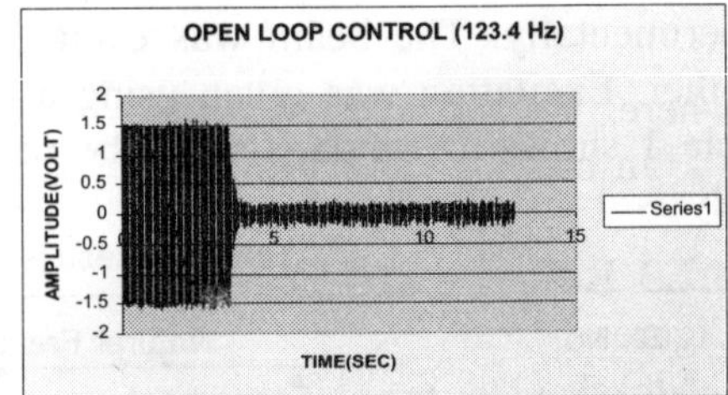

Fig. 5. Amplitude vs. Time (second mode).

Reductions of 23 dB and 23.6 dB are observed for the first mode (21.4 Hz) and the second mode (123.4 Hz), respectively. For both modes, it is observed that the amplitude of vibration drops to almost the noise floor level after closing the loop.

4.2 Feedback Control

This method shown in Fig. 6 is suited for vibration control of large structures and also when a reference signal correlated to the primary disturbance is not available. The output R(s) of the system is detected by a sensor which is manipulated by the controller so as to minimize the overall response. The design problem consists of finding the appropriate compensator H(s) such that the closed loop system is stable and behaves appropriately [4].

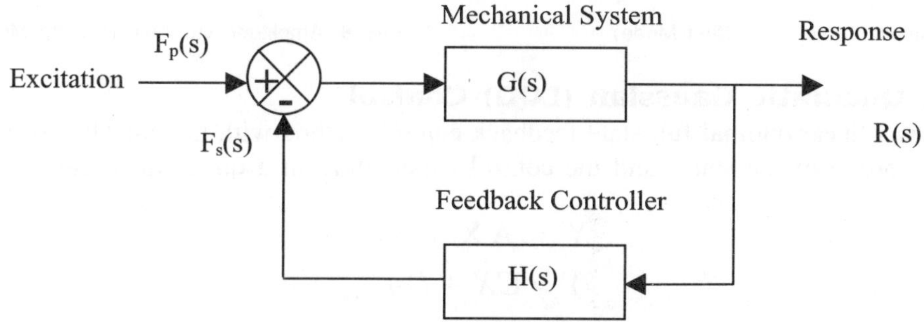

Fig. 6. Block Diagram of a Feedback Control System.

4.2.1 Spillover

When an attempt is made to control one particular mode or a particular set of modes of the structure, it is noticed that the control energy excites other frequencies of the beam also, which are known as residual frequencies. As a result, the system gets excited at those residual modes. This concept is referred to as "spillover" in vibration control. Spillover is averted by passing the control signal through a bandpass filter that allows only the frequencies in the range of interest to pass. A fourth order Butterworth bandpass filter was used in the experiments to reduce spillover effects.

4.2.2 Negative Velocity Feedback Algorithm

Theoretically the control output (velocity feedback) can be given as

$$V_a = -G \int V_s \, dt \qquad \ldots(4)$$

where, V_a = actuator voltage, V_s = sensor voltage and G = gain

In order to implement the control algorithm, a subroutine was developed in LabVIEW.

4.2.3 Results and Discussion

Figures 7 and 8 show amplitude vs. time plots for the first two modes of the structure under feedback control. The reduction in vibration amplitude for the first mode is 10 *dB* or (68.4%) and that for the second mode is 13.74 *dB* or (83%.)[5].

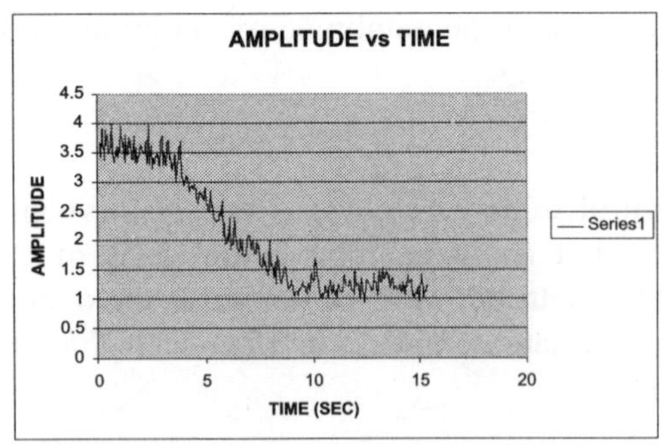

Fig. 7. Amplitude vs. Time (first Mode).

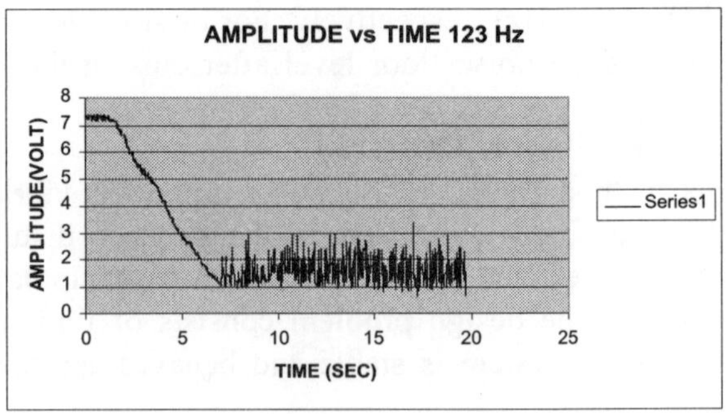

Fig. 8. Amplitude vs. Time (second Mode).

4.3 Linear Quadratic Gaussian (LQG) Control

LQG Control is a linear optimal full state feedback control method with the objective of minimizing the impulse response of the states and the control expenditure in a quadratic sense.

$$\dot{X} = A X + B u \qquad \qquad ...(5)$$

$$Y = CX + Du \qquad \qquad ...(6)$$

where,

X = state vector
A = system matrix
B = input matrix

Y = sensor output
C = output matrix
D = feed through matrix

In LQG Control, Quadratic Performance index J is minimized

$$J = \int_0^\infty (X^T Q X + u^T R u)\,dt \qquad \qquad ...(7)$$

where, Q = state weighting matrix
R = input weighting matrix

LQG control has been implemented with the help of DSP box along with the software packages DynaMod and ControlForge developed by MIDE Technologies. The experimental frequency response function of the beam is acquired using DSP box and imported into DynaMod which is a system identification software that fits a mathematical model from measurement by specifying a suitable model order based on the required number of modes. The model order is then refined using a Logarithmic Least Squares (LLS) tuning algorithm to achieve a lower model order fitting the actual data as closely as possible. The output of DynaMod represents the model in state space form. ControlForge is then used to design the LQG controller for the state space model obtained from Dynamod. DSP box can work as a stand alone unit to perform active control of vibrations. The flow chart and experimental set-up for LQG Control design and implementation are shown in Figs. 9 and 10.

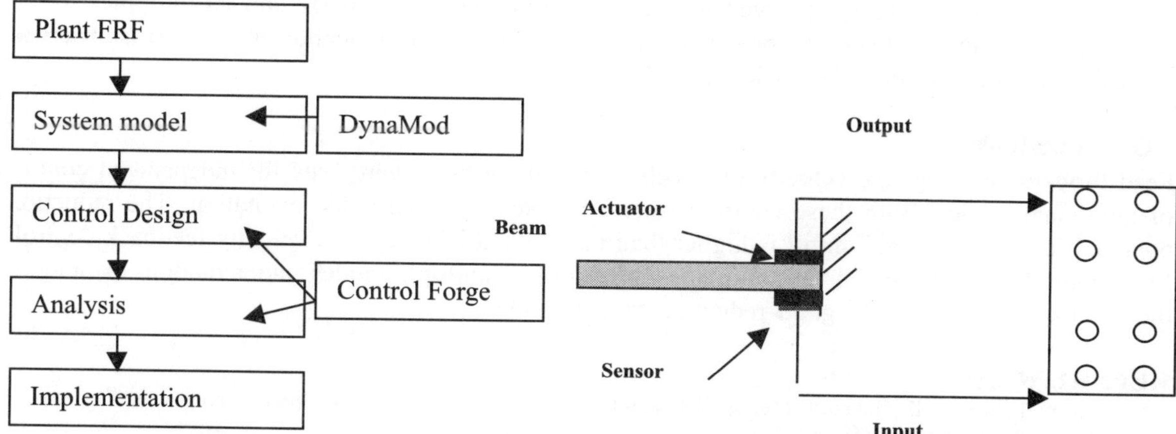

Fig. 9. Block diagram of LQG controller.

Fig. 10. Experimental setup for control.

4.3.1 Results and Discussions

With random excitation, the acceleration response power spectral density (PSD) plots for uncontrolled and controlled responses are shown in Fig. 11 and Fig. 12 shows the acceleration response PSD zoomed around the first mode.

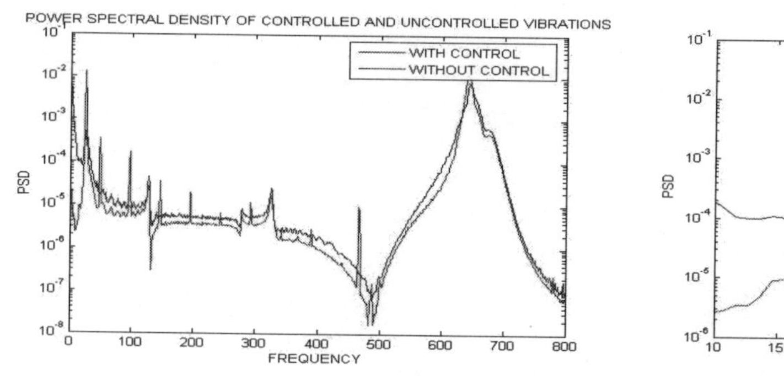

Fig. 11. PSD of controlled and uncontrolled responses.

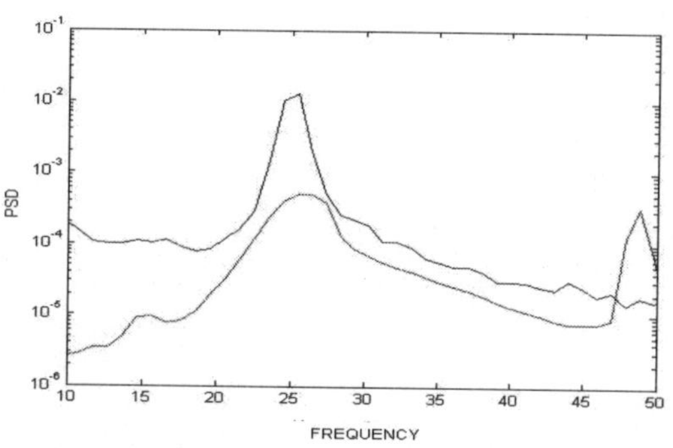

Fig. 12. PSD zoomed around first mode.

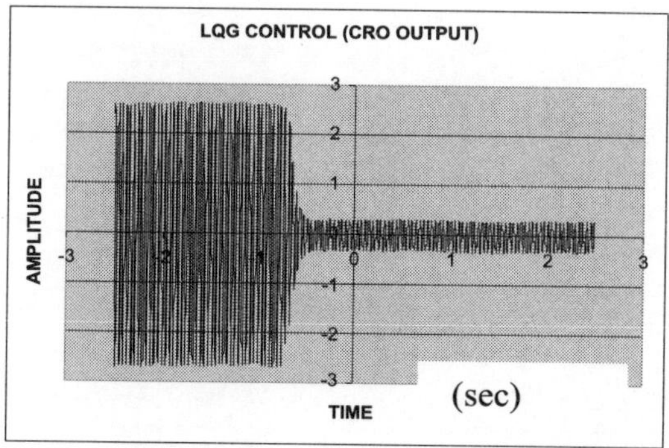

Fig. 13. Time response before and after control.

A reduction of 32 dB is observed for the first mode. Peaks at 50 Hz and its multiples in the controlled spectrum are due to the power line frequency. An overall reduction of 22.7 dB is observed from the time domain plot shown in Fig. 13.

CONCLUSION

Feed forward and Negative velocity feedback control algorithms were used for independent control of individual modes. Both these controls were attempted with harmonic excitation. The reduction obtained for feed forward control is higher than that obtained in negative velocity feedback control. LQG control is implemented for simultaneous control of multiple modes under random excitation; the results show reasonably good reduction in amplitude.

REFERENCES

1. Bailey T., and J. E. Hubbard, (1985) Distributed piezoelectric-polymer active vibration control of a cantilever beam, *AIAA Journal*, 8(5), 605-611.
2. Batta R. G., and A. Chen, (1991) An experimental system for the study of active vibration control—development and modeling, *IEEE Transactions*. 432-435.
3. Fuller C. R., Elliott, S. J., and Nelson, P. A, (1996) *Active Control of Vibration,* Academic Press, San Diego, CA, USA.
4. Chen S. H., Z. D. Wang, and X. H. Liu, (1997) Active vibration control and suppression for intelligent structures, *Journal of Sound and Vibration*, 200(2), 167-177.
5. Sudeep U. *Active vibration control studies on smart structures* M.Tech. Thesis IITM, Chennai 2006.

91

Simulation of Flapping Wing Motion Using a PZT-PVA-PAA Smart Hybrid Structure

S. Kamle[1], R.S. Jeyevijeyan[2] and A. Gupta[3]

Department of Aerospace Engineering, Indian Institute of Technology Kanpur, Kanpur, India.
email: [1] kamle@iitk.ac.in, [2] jeyvijeyan@yahoo.co.in, [3] alokg@iitk.ac.in

ABSTRACT

This paper presents the analysis of a smart hybrid structure made using PZT and PVA-PAA polymer gel for simulating the flapping motion of a bird. In the first phase, the electro-mechanical response of the swollen hydrogels was measured by placing the gel between a pair of electrodes. This enabled the study of extent and response of the bending with varying electric field as well as the nature of the electrolyte-solution. Next, the mechanical properties (such as tensile strength and modulus of elasticity) of the polymer gel were determined using a specially designed experimental setup. In the second phase, a flapping wing based adaptive structure using polymer gel actuators was designed and fabricated. When a potential difference is applied across the polymer gel, the gel undergoes large displacements. This is in contrast to a piezo actuator which produces large forces while undergoing small displacement. The hybrid structure was designed in such a way that the piezo-composite structure could be used to provide the same functionality as the arm bones of a bird. Similarly, the polymer gel actuators were used to provide functionalities of wing muscles and feathers. This arrangement was designed to resemble the feather arrangement in bird's wing. PWM inverter modules were used for inputting voltages to the hybrid structure. One set of modules was used for giving voltage inputs to piezo-composites while the other set of modules was used for giving inputs to the polymer gel actuators. Synchronization between these modules was achieved using an external reference source. The PVA-PAA-PZT system was found to simulate the bird's wing motions effectively.

Keywords: ¹Smart, PZT, Polymer Gel, Flapping, Composite.

1. INTRODUCTION

Recently, research interest towards developing unmanned aerial vehicles that can fly like birds has gained momentum. To fly like birds, micro air vehicles should follow the flight principles of the birds, especially their wing movement. With the aim to understand the flight principles of the birds, lot of research work is currently underway [1], [2]. Therefore, we can evaluate and adopt their findings for engineering design of flapping wings. Reproducing the wing movements of the birds artificially is a challenging process. The bird mimicking flapping devices must be designed such that they can create twisting motion in the course of up and downstroke.

Currently, pneumatic and motor-driven actuators are widely used in the aerospace application to produce wing movement, but they are not plausible for actuation of small or micro-scale flapping robots due to their large payload and system-complexity. Artificial muscle-type actuators must be alternative actuators suitable for this purpose. We have suggested engineering designs that can mimic the bird wing motion based on biological observations. [3]

The variation of twist with respect to time for a given voltage that an artificial wing made of PVA-PAA-PZT can undergo is determined by carrying out the bending test. Further, the amount of energy stored by the gel muscle when loaded up to its yield point is found out by carrying out the tensile test. This data is useful to determine the load carrying capacity of the gel muscle during flight.

2. BENDING OF PVA-PAA GELS

The electro-active behaviors of the gels were studied in detail. The polymer hydrogels bend towards the positive electrode when the gel strips are placed mid way between the electrodes, which are 22 mm apart and an electric field is applied. It was absorbed that the gel facing the negative electrode gains weight and swells and this leads to the bending of the gel towards the positive electrode. Bending is actually the deformation with swelling.

Following figure shows the bending angle of PVA-PAA gel in three different concentrations of $NaHCo_3$ (0.01M, 0.05 M and 0.005 M), at varying voltages as a function of time. These graphs indicate that gels respond almost linearly with the electric field. Both the bending angle and bending speed increased with the increase in voltage across the gel at fixed concentration of the electrolyte. The rate of bending was fast at higher voltages than that at lower voltages. From Fig. 1 is clear that at 15 V maximum and fastest bending is achieved for the PVA-PAA gel. The deformation of the polymer hydro-gel under an electric field is due to the voltage induced motion of the ions inside and outside the gel resulting in change in the osmotic pressure. As the voltage is increased the drift of the ions are enhanced and it leads to faster bending.

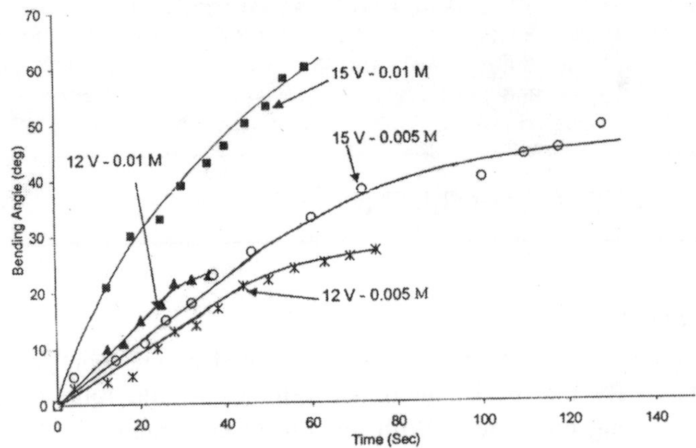

Fig. 1. Bending of PVA-PAA gel system in different molarities of sodium bicarbonate at different voltages.

3. TENSILE TEST

The tensile behavior of a polymer is probably the most fundamental mechanical property used in the evaluation of polymers. The stress-strain data is employed to determine the yield strength, ultimate tensile strength and elongation properties as well as various moduli which are important in designing with polymer gels. Tensile experiment is carried out in a specially designed mini load frame capable of applying loads of the order of few grams.

The stress-strain diagram obtained can also be used to measure the ability of a polymer to absorb energy. Fig. 2 shows the stress-strain curve for PVA-PAA. Here, polymer is loaded elastically up to the stress σmax; hence, it stores elastic energy per unit volume. From Fig. 2 it is evident that the yield strength of PVA-PAA gel is 0.09 MPa. Further the stress-strain plot is nonlinear and a Neo hookean hyper-elastic model can fit the uniaxial stress-strain data.

Equation 1 [4] is the neo-hookean hyper elastic equation used to represent the non-linear stress-strain relationship for the polymer gels shown in the Fig. 1.

$$\sigma = \frac{E}{3}[\lambda - \frac{1}{\lambda^2}]$$

...(1)

Here, σ is the linear stress in the material, E is the young's modulus of the material and λ is the stretch. Young's modulus of the polymer gel is determined by the plot shown in the Fig. 3. Sigma is plotted along the y axis and $[\lambda - 1/\lambda^2]$ is plotted along x axis. The slope of the line gives the value of E/3, from which E is calculated to be 0.555 MPa.

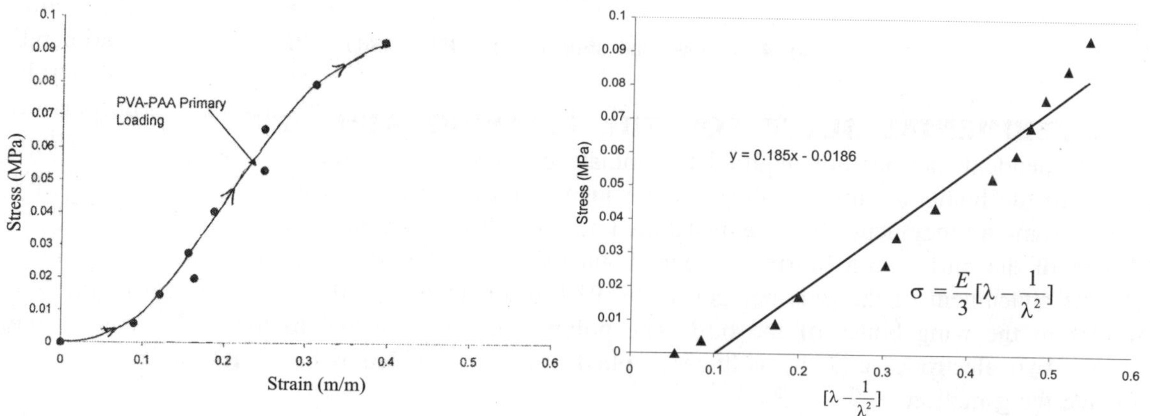

Fig. 2. Stress-strain curve for PVA-PAA gel system.   **Fig. 3.** Young Modulus for PVA-PAA gel system.

4. PVA-PAA-PZT SYSTEM TO SIMULATE BIRD'S WING MOTION

Bird's wing has an aerofoil shape. As the bird moves forward the air flows over the wing and produces sufficient forces for the bird to stay afloat and move forward. Initially let us consider that the bird's wing moves down. During its downward motion sufficient lift and thrust are generated for the bird to move in the forward direction. During upward stroke of the bird the wing angle has to change to avoid negative lift generation, hence, the bird twists its wing through an angle to ensure that it gets positive lift.

Therefore, both up-down motion of the wing and the twist of the wing are essential for the bird to fly. Fig. 4 shows the bone structure of the bird which makes it possible for the bird to have up-down motion and twisting motion simultaneously. At the base of the humerus portion of the bone structure the wings join with the body. This joint allows the bird's wing to have up and down movement. The elbow joint after the humerus allows for twisting of the front portion of the wing. These two motions are independent of each other and are controlled by the bird according to the requirement.

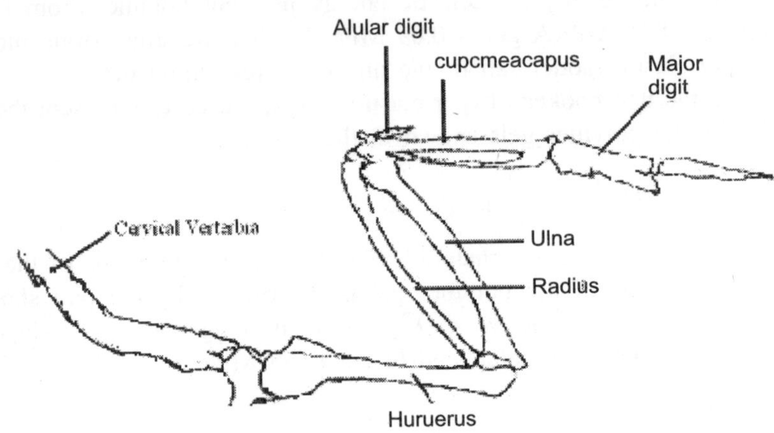

Fig. 4. Typical bone structure of a Bird's wing.

5. EXPERIMENTAL SETUP FOR THE FLAPPING WING BASED SYSTEM

Two independent motions are required to simulate bird's wing movement as discussed above. The motion of the humerus bone of bird's wing which is up-down as shown in Fig. 4 is achieved by a PZT beam arrangement. Voltage is applied to the PZT beam to move it up and down. The twisting of the bird's wing (during up and down motion) is simulated by applying voltage to the gel. The attachment of the polymer gel to the PZT beam is similar to the attachment of the wing muscles to the wing bones of the bird. The polymer gel so attached to the PZT beam is hung between two electrodes and the voltages applied are synchronized with each other to effectively simulate wing motion.

Figure 5 shows the schematic diagram of the PZT-PVA-PAA setup used to simulate wing motion. Electrodes are placed in the beaker containing 0.01M $NaHCO_3$ solution. PVA-PAA gel is hung in between the electrodes. A varying voltage is applied to the gel at some initial frequency which causes gel to twist in the z axis. The twist of the gel and the movement of the beam are controlled by varying individual frequencies such that they synchronize with each other effectively. It is found that at a frequency of 0.1 HZ synchronization of the shaker movement with the gel twisting motion is achieved.

A cantilever beam of size 225 mm × 25 mm × 0.4 mm was used in the setup. Two Piezo-electric crystals were put on both sides of the beam at a distance of 10 mm from the fixed end. One accelerometer was put at a distance of 200 mm from the fixed end of the beam. The free end of the beam was used to move a connecting rod which was further connected to the gel.

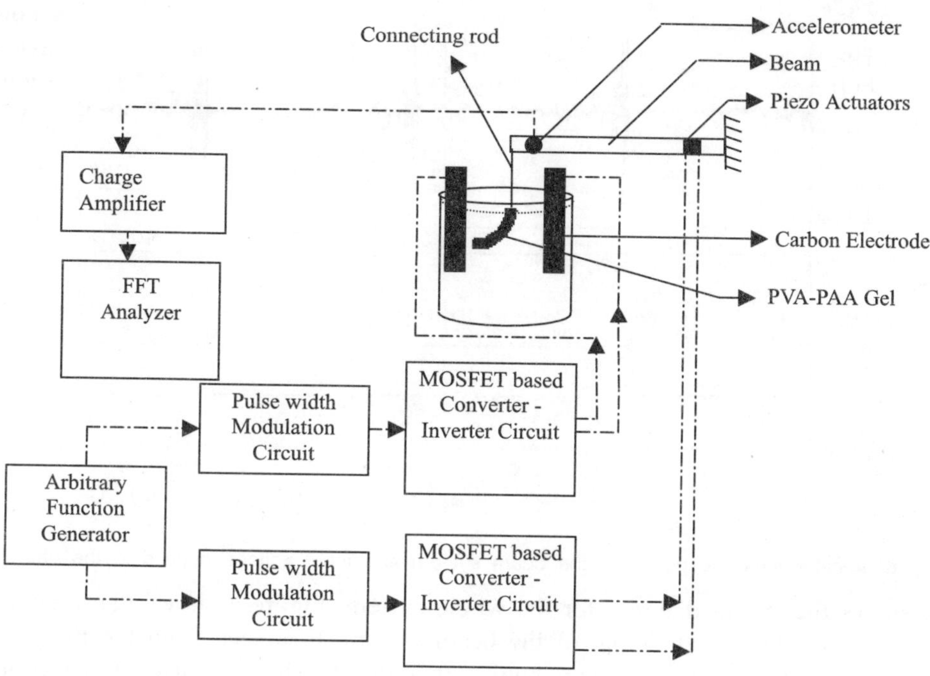

Fig. 5. PVA-PAA-PZT gel system.

Synchronization pulses were generated using an arbitrary function generator. This pulse was used as a reference source in two different circuits. The first circuit was used to provide excitation to the PVA-PAA gel and the other circuit was used to provide excitation to the piezo crystals on the beam. The advantage of using two independent circuits was that the gel system could be excited and controlled independently of the beam system. In a flapping wing system the motion of the humerus bone is independent of the motion of the digits (fingers) but the motions are synchronized. Typically, the feathers are light weight structures which undergo large oscillations while the humerus bone provides the motion of the overall wing structure. Additionally, a bird would be executing different types of flapping motions during take-off, landing and cruise situations. To be able to simulate all types of motions, it is necessary to have two independent and synchronized motions.

The first circuit consisted of a PWM circuit followed by a MOSFET based converter-inverter circuit. The output of this circuit was fed to the gel system. Similarly, the second circuit consisted of a PWM circuit followed by a MOSFET based converter-inverter circuit whose output was fed into the beam system. The typical voltage applied to the gel system was 60 V while a voltage of 200 V was applied to the piezo actuated beam system. A spectrum analyzer AD-3524 was used to display the response of the beam and gel system. GPIB bus was used to transfer the spectrum analyzer data to the PXI system.

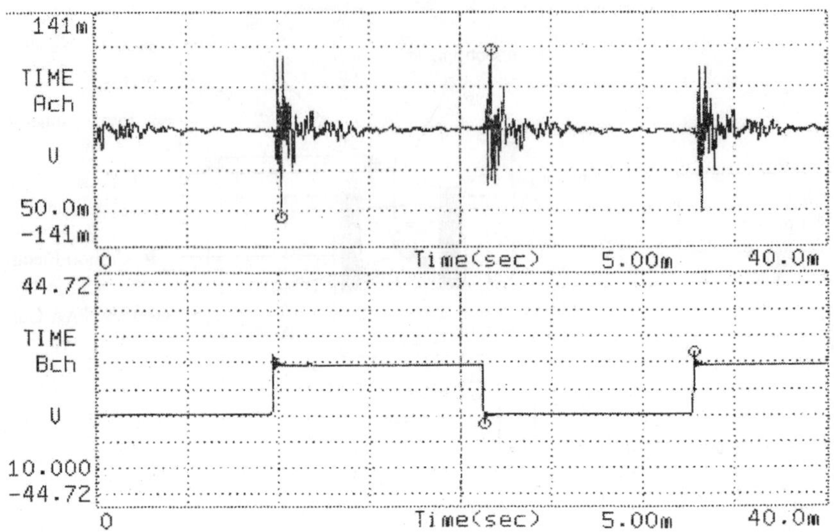

Fig. 6. Accelerometer response of the beam system and input voltage applied to the gel.

Figure 5 shows the circuit diagram for the whole system. Figure 6 shows the input voltage of the gel system and accelerometer output of the beam system. It is seen from the Figure 6 that as the voltage is applied to the piezo, beam begins to deform. The response of the beam shows damped oscillations. It can also be seen that the deformation of the beam and the input applied to the gel are synchronized.

CONCLUSION

Two different actuator systems, namely, PVA-PAA gel system and composite beam with PZT were successfully integrated to form the hybrid structure. Synchronization of the gel motion with the oscillatory motion of the beam was achieved. Mechanical characterization and electrical property studies were conducted for the hydrogel system for usage in flapping wing application. The gel was able to produce twisting motions as seen in birds wings with ease, which otherwise would require complex mechanical systems to produce the same motion. Hence, the gels potential for usage in flapping wing was established.

REFERENCES

1. Sibley, D.A., *The Sibley Guide to Bird Life & Behavior*, Chanticleer Press, 2001.
2. Whittow, G.C., *Avian Physiology*, Academic Press, 2000.
3. Bar-Cohen, Y., *Electroactive Polymer (EAP) Actuators as Artificial Muscles*, SPIE Press, 2001.
4. Shahinpoor M and Kim J., 2004, "Ionic polymer metal composites:III.Modeling and simulation as biomimetic sensors, actuators, transducers, and artificial muscles", *Smart Mater. Struct.*, 13, 1362-88.

92

Vertically Reinforced 1-3 Piezoelectric Composites for Active Constrained Layer Damping (ACLD) Treatment of Composite Plates

A.K. PRADHAN[1] AND M.C. RAY[2]

[1]Research Scholar, Department of Mechanical Engineering, IIT Kharagpur-721 302, India
email: arun@mech.iitkgp.ernet.in
[2]Assistant Professor, Department of Mechanical Engineering, IIT Kharagpur-721 302, India

ABSTRACT

This study deals with the use and performance of vertically reinforced 1-3 piezoelectric composite material as a constraining layer for active constrained layer damping (**ACLD**) treatment. A finite element model has been developed for analyzing the active constrained layer damping of laminated symmetric and anti-symmetric cross-ply and angle-ply composite plates integrated with the patches of such **ACLD** treatment. Both in-plane and out-of-plane actuation of the constraining layer of the **ACLD** treatment have been utilized for deriving the finite element model. The analysis revealed that the vertical actuation dominates over the in-plane actuation. Particular emphasis has been placed on investigating the performance of the patches when the orientation angle of the piezoelectric fibers of the constraining layer is varied in two mutually orthogonal vertical planes. The analysis also revealed that the vertically reinforced 1-3 piezoelectric composites, which are in general used as the distributed sensors, can also be potentially used as distributed actuators of high performance light-weight smart structures.

Keywords: Active Constrained Layer Damping (ACLD), Piezocomposite, Smart Structures.

1. INTRODUCTION

The inherent properties of direct and converse piezoelectric effects possessed by piezoelectric materials make them suitable as distributed sensors and actuators, respectively. The flexible structures when integrated with layers or patches of these materials acting as distributed sensors and actuators are known as "Smart Structures" [1-4]. Among the various piezoelectric composites proposed by the researchers till today, laminae of vertically reinforced 1-3 piezoelectric composites are commercially available (Piezocomposites, Materials Systems Inc., 543 Great Road, Littleton, MA 01460) and are being effectively used as sensors. The constructional feature of a lamina made of this vertically reinforced 1-3 piezoelectric composite is shown in Fig. 1 and Fig. 2.

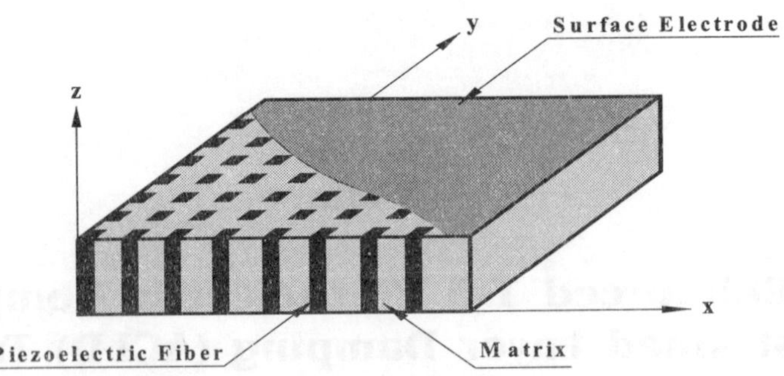

Fig. 1. Schematic representation of a lamina of vertically reinforced 1-3 piezoelectric composite material.

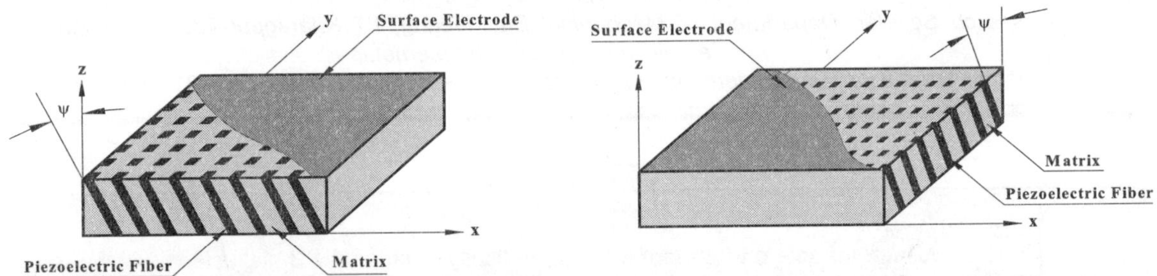

Fig. 2. Schematic diagrams of the laminae of vertically reinforced 1-3 piezocomposites in which the piezoelectric fibers are either coplanar with the vertical **xz** or **yz** plane while the fiber orientation angle with respect to the **z** axis is **ψ** .

2. FINITE ELEMENT MODELING

Figure 3 illustrates a simply supported laminated composite plate composed of **N** number of orthotropic layers. The top surface of the composite plate is integrated with the rectangular patches of **ACLD** treatment. The constraining layer of the patches of **ACLD** treatment [5] is made of the vertically reinforced 1-3 piezoelectric composite material. The thickness of the substrate laminated plate, piezoelectric composite layer and the viscoelastic layer are denoted by **h**, $\mathbf{h_p}$ and $\mathbf{h_v}$, respectively. Denoting by **k** (**k** = 1, 2, 3... N + 2), the layer number of any layer of the overall plate, the thickness coordinates **z** of the top and bottom surface of any (**k**[th]) layer are represented by $\mathbf{h_{k+1}}$ and $\mathbf{h_k}$, respectively. In Fig. 4, the kinematics of axial deformations of the overall plate based on the first order shear deformation theory has been illustrated. The displacements at any point in any layer are given by

$$u(x,y,z,t) = u_0(x,y,t) + (z - \langle z - h/2 \rangle)\theta_x(x,y,t) + (\langle z - h/2 \rangle - \langle z - h_{N+2} \rangle)\phi_x(x,y,t)$$
$$+ \langle z - h_{N+2} \rangle \gamma_x(x,y,t) \qquad \ldots (1)$$

$$v(x,y,z,t) = v_0(x,y,t) + (z - \langle z - h/2 \rangle)\theta_y(x,y,t) + (\langle z - h/2 \rangle - \langle z - h_{N+2} \rangle)\phi_y(x,y,t)$$
$$+ \langle z - h_{N+2} \rangle \gamma_y(x,y,t) \qquad \ldots (2)$$

$$w(x,y,z,t) = w_0(x,y,t) + (z - \langle z - h/2 \rangle)\theta_z(x,y,t) + (\langle z - h/2 \rangle - \langle z - h_{N+2} \rangle)\phi_z(x,y,t)$$
$$+ \langle z - h_{N+2} \rangle \gamma_z(x,y,t) \qquad \ldots (3)$$

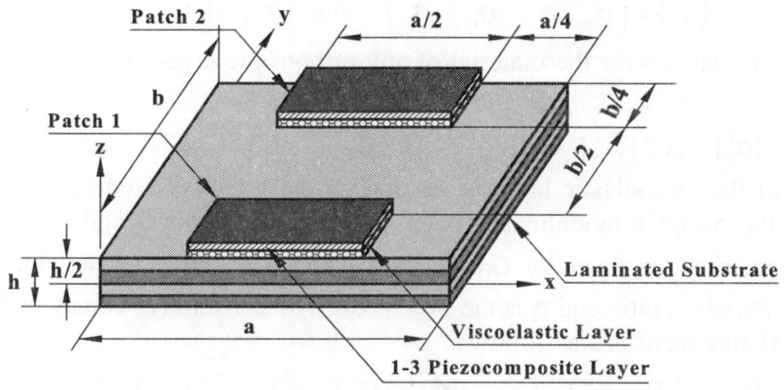

Fig. 3. Schematic representation of a laminated composite plate integrated with the patches of ACLD treatment composed of vertically reinforced 1-3 piezoelectric composite constraining layer.

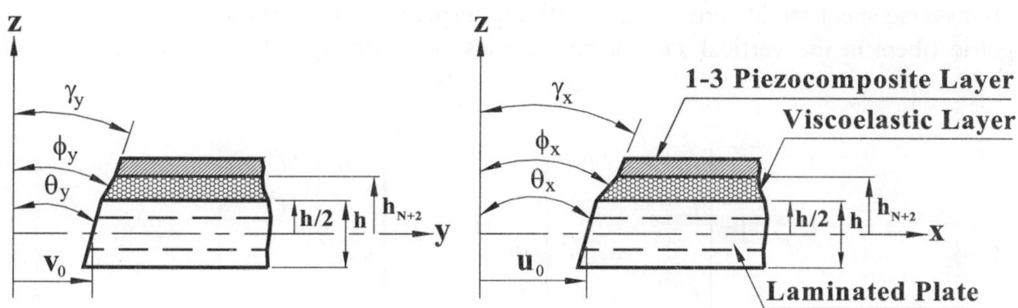

Fig. 4. Kinematics of deformation.

The generalized displacement variables are grouped into the two following vectors:

$$\{d_t\} = [u_0 \quad v_0 \quad w_0]^T \text{ and } \{d_r\} = [\theta_x \quad \theta_y \quad \theta_z \quad \phi_x \quad \phi_y \quad \phi_z \quad \gamma_x \quad \gamma_y \quad \gamma_z]^T$$

The state of strain at any point in the overall plate is:

$$\{\in_b\} = [\in_x \quad \in_y \quad \in_{xy} \quad \in_z]^T \text{ and } \{\in_s\} = [\in_{xz} \quad \in_{yz}]^T.$$

The state of in-plane and transverse normal strains at any point in the substrate composite plate, viscoelastic layer and the active constraining layer, respectively, can be expressed as

$$\{\in_b\}_c = \{\in_{bt}\} + [Z_1]\{\in_{br}\}, \quad \{\in_b\}_v = \{\in_{bt}\} + [Z_2]\{\in_{br}\} \text{ and } \{\in_b\}_p = \{\in_{bt}\} + [Z_3]\{\in_{br}\} \qquad ...(4)$$

Similarly, defining the state of transverse shear strains at any point in the substrate composite plate, viscoelastic layer and the active constraining layer, respectively, can be expressed as

$$\{\in_s\}_c = \{\in_{st}\} + [Z_4]\{\in_{sr}\}, \quad \{\in_s\}_v = \{\in_{st}\} + [Z_5]\{\in_{sr}\} \text{ and } \{\in_s\}_p = \{\in_{st}\} + [Z_6]\{\in_{sr}\} \qquad ...(5)$$

The state of stresses at any point in the overall plate is described by the following stress vectors:

$$\{\sigma_b\}=\begin{bmatrix}\sigma_x & \sigma_y & \sigma_{xy} & \sigma_z\end{bmatrix}^T \text{ and } \{\sigma_s\}=\begin{bmatrix}\sigma_{xz} & \sigma_{yx}\end{bmatrix}^T \qquad ...(6)$$

The constitutive relations for the material of any orthotropic layer of the substrate plate are given by

$$\{\sigma_b^k\}=[C_b^k]\{\epsilon_b^k\} \text{ and } \{\sigma_s^k\}=[C_s^k]\{\epsilon_s^k\};\ (k = 1, 2, 3, \ldots, N) \qquad ...(7)$$

The material of the viscoelastic layer is assumed to be linearly viscoelastic and isotropic and is modeled by using the complex modulus approach. The shear modulus **G** and the Young's modulus **E** of the viscoelastic material are given by $G = G'(1+i\eta)$ and $E = 2G(1+\upsilon)$, in which G' is the storage modulus, υ is the Poisson's ratio and η is the loss factor. The constitutive equations for the constraining layer of the **ACLD** treatment are as follows:

$$\{\sigma_b^k\}=[\bar{C}_b^k]\{\epsilon_b^k\}+[\bar{C}_{bs}^k]\{\epsilon_s^k\}-\{e_b\}E_z,\ \ \{\sigma_s^k\}=[\bar{C}_{bs}^k]^T\{\epsilon_b^k\}+[\bar{C}_s^k]\{\epsilon_s^k\}-\{e_s\}E_z,\ \ k = N+2$$

and,
$$D_z =[e_b]^T\{\epsilon_b^k\}+[e_s]^T\{\epsilon_s^k\}+\bar{\epsilon}_{33}E_z \qquad ...(8)$$

Here, $\mathbf{E_z}$ and $\mathbf{D_z}$ represent the electric field and displacement along the **z**-direction, respectively and $\bar{\epsilon}_{33}$ is the dielectric constant. It may be noted from the above form of the constitutive relations that the transverse shear strains are coupled with the in-plane normal strains due to the orientation of piezoelectric fibers in the vertical **xz** - or **yz** - planes. The corresponding coupling elastic constant matrices $[\bar{C}_{bs}^{N+2}]$

are given by
$$[\bar{C}_{bs}^{N+2}]=\begin{bmatrix}\bar{C}_{15}^{N+2} & 0 \\ \bar{C}_{25}^{N+2} & 0 \\ 0 & \bar{C}_{46}^{N+2} \\ \bar{C}_{35}^{N+2} & 0\end{bmatrix} \text{ or } [\bar{C}_{bs}^{N+2}]=\begin{bmatrix}0 & \bar{C}_{14}^{N+2} \\ & \bar{C}_{24}^{N+2} \\ \bar{C}_{56}^{N+2} & \\ & \bar{C}_{34}^{N+2}\end{bmatrix} \qquad ...(9)$$

according as the piezoelectric fibers are coplanar with the vertical **xz** - or **yz** - plane. Also, the piezoelectric constant matrices $\{e_b\}$ and $\{e_s\}$ appearing in Eq. (8) contain the following transformed effective piezoelectric coefficients of the 1-3 piezoelectric composite:

$$\{e_b\}=[\bar{e}_{31} \quad \bar{e}_{32} \quad \bar{e}_{36} \quad \bar{e}_{33}]^T \text{ and } \{e_s\}=[\bar{e}_{35} \quad \bar{e}_{34}]^T \qquad ...(10)$$

The total potential energy $\mathbf{T_p}$ and the kinetic energy $\mathbf{T_k}$ of the overall plate/**ACLD** system are given by [6, 7];

$$T_p =\frac{1}{2}\left[\sum_{k=1}^{N+2}\int_\Omega \left(\{\epsilon_b^k\}^T\{\sigma_b^k\}+\{\epsilon_s^k\}^T\{\sigma_s^k\}\right)d\Omega -\frac{1}{2}\int_\Omega D_z E_z d\Omega\right]-\int_A\{d\}^T\{f\}\,dA \qquad ...(11)$$

and,
$$T_k =\frac{1}{2}\sum_{k=1}^{N+2}\int_\Omega \rho^k (\dot{u}^2 + \dot{v}^2 + \dot{w}^2)d\Omega \qquad ...(12)$$

in which ρ^k is the mass density of the $\mathbf{k}^{th}$ layer, $\{\mathbf{f}\}$ is the externally applied surface traction acting over a surface area **A** and Ω represents the volume of the concerned layer. The overall plate is discretized by eight noded isoparametric quadrilateral elements. Total potential energy T_p^e and the kinetic energy T_k^e of a typical element augmented with the **ACLD** treatment can be expressed as

$$T_p^e = \frac{1}{2}\Big[\{d_t^e\}^T[K_{tt}^e]\{d_t^e\} + \{d_t^e\}^T[K_{tr}^e]\{d_r^e\} + \{d_r^e\}^T[K_{tr}^e]^T\{d_t^e\} + \{d_r^e\}^T[K_{rr}^e]\{d_r^e\}$$

$$-2\{d_t^e\}^T\{F_{tp}^e\}V - 2\{d_r^e\}^T\{F_{rp}^e\}V - 2\{d_t^e\}^T\{F^e\}\Big] - \varepsilon_{33}\frac{V^2}{h_p} \qquad ...(13)$$

and,

$$T_k^e = \frac{1}{2}\int_0^{a_e}\int_0^{b_e} \bar{m}\{\dot{d}_t^e\}^T[N]^T[N]\{\dot{d}_t^e\}\,dx\,dy \qquad ...(14)$$

Applying the virtual work principle;

$$[M^e]\{\ddot{d}_t^e\} + [K_{tt}^e]\{d_t^e\} + [K_{tr}^e]\{d_r^e\} = \{F_{tp}^e\}V + \{F^e\} \qquad ...(15)$$

and,

$$[K_{rt}^e]\{d_t^e\} + [K_{rr}^e]\{d_r^e\} = \{F_{rp}^e\}V \qquad ...(16)$$

The elemental equations of motion are assembled to obtain the open loop global equation of motion as

$$[M]\{\ddot{X}\} + [K_{tt}]\{X\} + [K_{tr}]\{X_r\} = \sum_{j=1}^{q}\{F_{tp}^j\}V^j + \{F\} \qquad ...(17)$$

and,

$$[K_{rt}]\{X\} + [K_{rr}]\{X_r\} = \sum_{j=1}^{q}\{F_{rp}^j\}V^j \qquad ...(18)$$

where, **[M]** is the global mass matrix, **[K_{tt}]**, **[K_{tr}]** and **[K_{rr}]** are the global stiffness matrices, **{F_{tp}}**, **{F_{rp}}** are the global electro-elastic coupling vectors, **{X}** and **{X_r}** are the global nodal generalized displacement vectors, **{F}** is the global nodal force vector, **q** is the number of patches and **V^j** is the voltage applied to the **j**-th patch.

3. CLOSED LOOP MODEL

Thus, the control voltage for each patch can be expressed in terms of the derivatives of the global nodal degrees of freedom as follows:

$$V^j = -k_d^j\dot{w} = -k_d^j[U_t^j]\{\dot{X}\} - k_d^j(h/2)[U_r^j]\{\dot{X}_r\} \qquad ...(19)$$

where, k_d^j is the control gain for the jth patch, $[U_t^j]$ and $[U_r^j]$ are the unit vectors for expressing the transverse velocity of the point concerned. The final equations of motion governing the closed loop dynamics of the overall plate/**ACLD** system can be obtained as follows:

$$[M]\{\ddot{X}\} + [K_{tt}]\{X\} + [K_{tr}]\{X_r\} + \sum_{j=1}^{m}k_d^j\{F_{tp}^j\}[U_t^j]\{\dot{X}\} + \sum_{j=1}^{m}k_d^j(h/2)\{F_{tp}^j\}[U_r^j]\{\dot{X}_r\} = \{F\} \qquad ...(20)$$

and,

$$[K_{rt}]\{X\} + [K_{rr}]\{X_r\} + \sum_{j=1}^{m}k_d^j\{F_{rp}^j\}[U_t^j]\{\dot{X}\} + \sum_{j=1}^{m}k_d^j(h/2)\{F_{rp}^j\}[U_r^j]\{\dot{X}_r\} = 0 \qquad ...(21)$$

4. RESULTS AND DISCUSSION

PZT-5H/spur composite with 60% fiber volume fraction has been considered for the material of the constraining layer of the **ACLD** treatment. The effective material properties of this vertically reinforced piezoelectric composite computed by standard micromechanics procedure are:

$$C_{12} = 6.18\text{GPa}, \quad C_{13} = 6.05\text{GPa}, \quad \mathbf{C_{33} = 35.44\text{GPa}}, \quad C_{23} = C_{13}, \quad C_{44} = 1.58\text{GPa}, \quad C_{66} = 1.54\text{GPa},$$

$$C_{55} = C_{44}, \quad e_{31} = -0.1902\text{C}/\text{m}^2, \quad e_{32} = e_{31}, \quad e_{33} = 18.4107\text{C}/\text{m}^2, \quad e_{24} = 0.004\text{C}/\text{m}^2 \text{ and } e_{15} = e_{24}.$$

The material of the orthotropic layers of the substrate plates is a graphite/epoxy composite and its material properties are $\mathbf{E_L}$ = 172GPa, $\mathbf{E_L/E_T}$ = 25, $\mathbf{G_{LT}}$ = 0.5$\mathbf{E_T}$ and $\boldsymbol{\upsilon_{LT}}$ = 0.25. The aspect ratio (**a / h**) and thickness of the square substrate plates are considered as 100 and 0.003m. The complex shear modulus, Poisson ratio and the density of the constrained viscoelastic layer are 20(1 + i) **MNm**$^{-2}$, 0.49 and 1140 **kg m**$^{-3}$. The thicknesses of the viscoelastic layer and the piezoelectric layer are 50.8 **mμ** and 250 **μm**. The plates are harmonically excited by a force of 1 N applied to a point (a/2, a/ 4, h/2) on the substrate plates. The control voltage supplied to the first patch is negatively proportional to the velocity of the point (a/2, a/4, h/2,) and that supplied to the other patch is negatively proportional to the velocity of the point (a/2, 3a/4, h/2).

Figure 5 illustrates the frequency response functions for the transverse displacement of a point (a/2, a/4, h/2) of a simply supported symmetric cross-ply (0°/90°/0°) when the fiber orientation angle (ψ) in the constraining layer is 0°. This figure displays that the constraining layer made of vertically reinforced 1-3 piezoelectric composite material being studied here significantly attenuates the amplitude of vibrations, enhancing the damping characteristics of the laminated composite plates over the passive damping (uncontrolled). Similar effects are observed for antisymmetric cross-ply (0°/90°/0°/90°) and anti-symmetric angle-ply (– 45°/45°/– 45°/45°) plate, respectively when the fiber orientation angle (ψ) in the constraining layer is 0°. The maximum value of the control voltage required to compute the controlled response have been found to be quite low (around 45 V). From

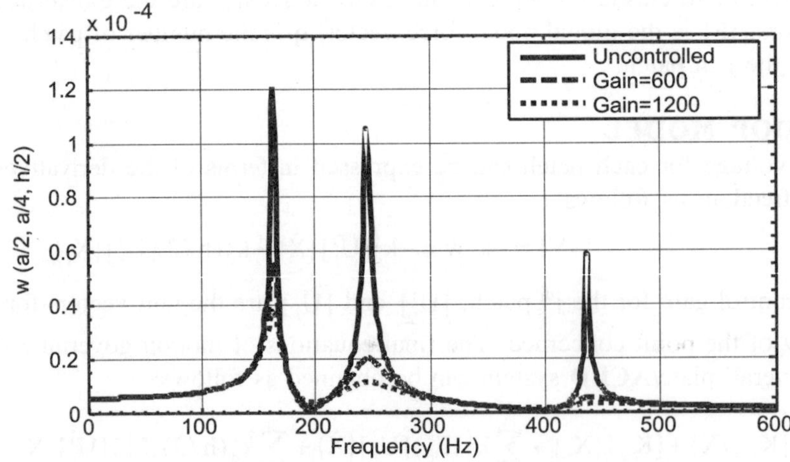

Fig. 5. Frequency response functions for the transverse displacement w (a/2,a/4,h/2) of a simply supported symmetric cross-ply (0°/90°/0°) plate (a/h = 100, ψ = 0°).

Fig. 5, active control responses of the symmetric cross-ply plate for a particular value of gain (1200) are plotted in Fig. 6 with and without considering the values of e_{33} and e_{31}. It is observed that the contribution of the piezoelectric coefficient e_{33} of the constraining vertically reinforced 1-3 piezoelectric composite layer on the attenuating capability of the **ACLD** treatment is significantly larger than that of the piezoelectric coefficient e_{31} of the constraining layer for controlling the modes.

Figure 7 illustrates that the attenuating capabilities of the patches become maximum when the orientation angle (Ψ) of the fibers in the yz plane is 0°. Similar effect is also obtained when the fiber orientation in the constraining layer of the ACLD treatment is varied in the vertical xz plane.

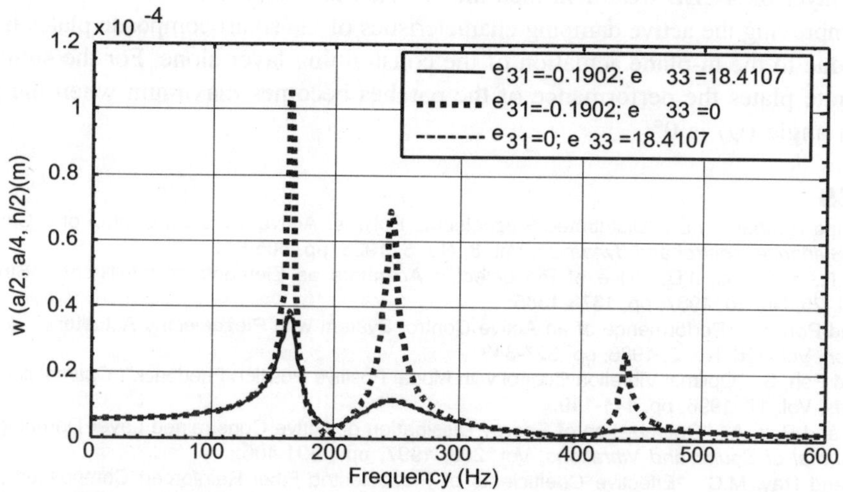

Fig. 6. Frequency response functions showing the contribution of e_{33} on the controlled responses for the transverse displacement w (a/2,a/4,h/2) of the symmetric cross-ply $(0^0/90^0/0^0)$ plate($\psi = 0^0$).

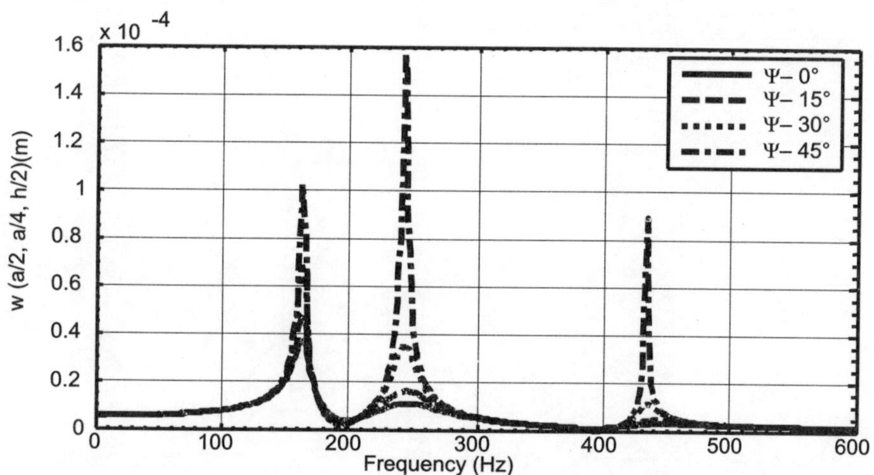

Fig. 7. Effect of variation of fiber orientation (ψ) on the performance of the patches for controlling a simply supported symmetric cross-ply (0°/90°/0°) plate when the piezoelectric fibers of the constraining layer are coplanar with the **yz** plane (Gain = 1200, a/h = 100).

CONCLUSION

A finite element model has been developed to describe the dynamics of the plates integrated with the patches of **ACLD** treatment and the constraining layer of the **ACLD** treatment is considered to be composed of the vertically reinforced 1-3 piezoelectric composite material. The frequency responses of the symmetric cross-ply, anti-symmetric cross-ply and angle-ply composite plates indicate that the active constraining layer of the **ACLD** treatment made of vertically reinforced piezoelectric composite material significantly enhances the damping characteristics of the plates over the passive damping. The analysis revealed that if the vertically reinforced 1-3 piezoelectric composite material is used for the constraining layer of **ACLD** treatment then the contribution of vertical actuation of the constraining layer alone for improving the active damping characteristics of the smart composite plates is significantly larger than that due to the in-plane actuation of the constraining layer alone. For the simply supported laminated substrate plates the performance of the patches becomes maximum when the piezoelectric fiber orientation angle (ψ) is $0°$.

REFERENCES

1. Bailey, T., and Hubbard, J.E., "Distributed Piezoelectric Polymer Active Vibration Control of a Cantilever Beam", *Journal of Guidance, control and dynamics*, Vol. 8, No. 5, 1985, pp. 605-611.
2. Crawley, E.F., and Luis, J.D., "Use of Piezoelectric Actuators as Elements of Intelligent Structures", *AIAA Journal*, Vol. 25, No. 10, 1987, pp. 1373-1385.
3. Baz, A., and Poh, S., "Performance of an Active Control System with Piezoelectric Actuators", *Journal of Sound and Vibration*, Vol. 126, No. 2, 1988, pp. 327-343.
4. Baz, A., and Poh, S., "Optimal Vibration Control with Modal Positive Position Feedback", *Optimal control applications and methods*, Vol. 17, 1996, pp. 141-149.
5. Ray, M.C., and Baz, A., "Optimization of Energy Dissipation of Active Constrained Layer Damping Treatments of Plates", *Journal of Sound and Vibrations*, Vol. 208, 1997, pp. 391-406.
6. Mallik, N. and Ray, M.C., "Effective Coefficients of Piezoelectric Fiber-Reinforced Composites", *AIAA Journal*, Vol. 41, No. 4, 2003, pp. 704-710.
7. Ray, M.C. and Mallik, N, "Active Control of Laminated Composite Plates using Piezoelectric Fiber Reinforced Composite Layer", *Smart Materials and Structures*, 2004, Vol. 13, No. 1, 2004, pp. 146-152.

93

3D Layered Finite Element for Modeling and Control of Smart Structure

T. Roy, M. Mastan Pasha and D. Chakraborty

Department of Mechanical Engineering, Indian Institute of Technology Guwahati-781 039, India email: tarapada@iitg.ernet.in, chakra@iitg.ernet.in

ABSTRACT

The present article deals with the three-dimensional layered finite element formulation and optimal vibration control of smart FRP composite structure using linear quadratic regulator (LQR) under sinusoidal as well as transverse impact loading. An eight-noded layered three-dimensional element has been formulated for analysis of smart laminates. Optimal vibration control performance of beams/plates with piezoelectric sensors and actuators layers has been analyzed. It has been observed from the present study that vibration control of smart FRP composite structure could be achieved using LQR schemes under sinusoidal as well as contact impact loading.

Keywords: Smart FRP composite; 3D layered element; LQR; vibration control.

1. INTRODUCTION

The increasing interest for the lightweight structures with superior structural performance has lead to the development of smart or intelligent structures. In these structures, the load bearing substrates are, in general made of composite materials for their higher specific strength and stiffness. These structures are integrated with distributed piezoelectric materials that act as sensors and actuators. The distributed piezoelectric sensor layer monitors the structural shape deformation and the distributed actuator layer controls the deflection. This activeness of the structure is achieved by using smart materials. The basic feature of smart FRP laminates lies in its ability to actively control the vibration of structures. The layers/patches of smart materials act as sensors and actuators, which affect the controlling feature of the smart structures. Some of the important works in this direction are presented in the following paragraph.

Allik et al. [1] presented two finite element models in which isoparametric hexahedral and tetrahedral elements were applied in piezoceramic transducer design. Structural identification and control of plate model with distributed piezoelectric sensors/actuators studied by Tzou and Tseng [2] and presented thin piezoelectric hexahedron finite element with three internal degrees of freedom. Ha et al. [3] have developed an eight-noded three-dimensional composite brick element and studied the response of laminated composite containing distributed piezoelectric ceramics subjected to both mechanical and electrical loadings. K. Y. Sze and L. Q. Yao [4] developed a finite element models for comprehensive modeling of smart structures with segmented piezoelectric sensing and actuating patches. These include an eight-node solid-shell element for modeling homogeneous and laminated host structures as well as an eight-noded solid-shell and four-node piezoelectric membrane elements for modeling surface bonded piezoelectric sensing and actuating patches. Piefort et al. [5-6] developed the theory of piezolaminated shells and they also presented a general finite element formulation for piezoelectrically coupled systems. S. Narayanan and V. Balamurugan [7] developed a finite element modeling of laminated structures with distributed piezoelectric sensor and actuator layers and control electronics and also presented that LQR approach is found to be more effective in vibration control with lesser peak voltages applied in the piezoelectric actuator layers as in this case the control gains are obtained by minimizing a performance index.

The aim of the present work is to develop a 3D layered solid element for finite element analysis of smart composite laminate and using the response obtained from the finite element analysis to study the vibration control of such smart structures using linear quadratic regulator (LQR) control strategy under sinusoidal as well as transverse impact loading.

2. FINITE ELEMENT MODELING

A layered eight-noded solid element (Fig. 1) has been used for finite element modeling of the laminated plate.

The shape functions of the element are:

$$N_i = \frac{1}{8}(1+rr_i)(1+ss_i)(1+tt_i)$$

$$i = 1,2,3,...8$$

...(1)

To adequately simulate the flexural response, extra shape functions have been introduced which are given by,

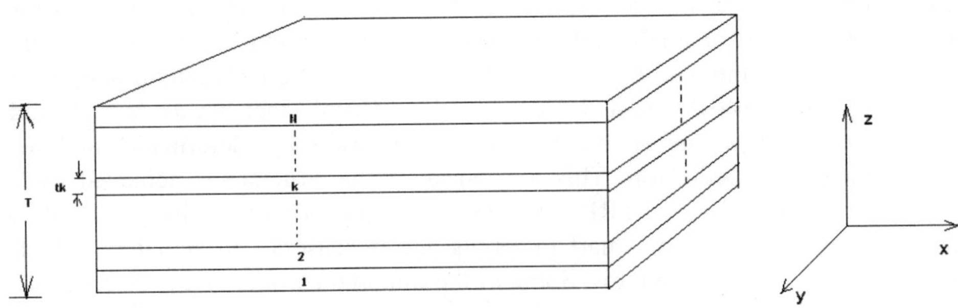

Fig. 1. Layered 3D element with N layers.

$$P_1 = (1 - r^2) \quad P_2 = (1 - s^2) \quad P_3 = (1 - t^2) \qquad ...(2)$$

The displacement variation for an eight-noded solid element with incompatible modes is expressed as,

$$\{d\} = \begin{Bmatrix} u \\ v \\ w \end{Bmatrix} = \sum_{i=1}^{8} N_i \begin{Bmatrix} u_i \\ v_i \\ w \end{Bmatrix} + [P]\{\Psi\} \qquad ...(3)$$

Where,

$$[P] = \begin{bmatrix} P_1 & P_2 & P_3 & 0 & 0 & 0 & 0 & 0 & 0 \\ 0 & 0 & 0 & P_1 & P_2 & P_3 & 0 & 0 & 0 \\ 0 & 0 & 0 & 0 & 0 & 0 & P_1 & P_2 & P_3 \end{bmatrix}$$

$$[\Psi]^T = \begin{bmatrix} \Psi_1 & \Psi_2 & \Psi_3 & \Psi_4 & \Psi_5 & \Psi_6 & \Psi_7 & \Psi_8 & \Psi_9 \end{bmatrix}$$

$\Psi_1, \Psi_2 ... \Psi_9$ are constants.

The electromechanical constitutive equations for linear material behavior are:

$$\{T\} = [C]\{S\} - [e]\{E\} \qquad ...(4)$$

$$\{D\} = [e]^T \{S\} + [\varepsilon]\{E\} \qquad ...(5)$$

Above equations are the usual constitutive equations for structural and electrical fields, respectively, except for the coupling terms involving the piezoelectric matrix $[e]$. Here, $\{T\}$ is the stress vector, $\{D\}$ is the electric flux density vector, $\{S\}$ is the strain vector, $\{E\}$ is the electric field vector, $[c]$ is the constitutive matrix and $[e]$ is the dielectric matrix. Establishing nodal solution variables and element shape functions over an element domain, which approximates the solution, finite element discretization has been performed.

$$\{u_c\} = \left[N^u \right]^T \{u\} \qquad ...(6)$$

$$\phi_c = \left\{ N^\phi \right\}^T \{\phi\} \qquad ...(7)$$

After the application of the variational principle and finite element discretization, the coupled finite element matrix equation derived for a one-element model is:

$$\left(\begin{bmatrix} [M] & [0] \\ [0] & [0] \end{bmatrix} \right) \begin{Bmatrix} \{\ddot{U}\} \\ \{\ddot{\phi}\} \end{Bmatrix} + \left(\begin{bmatrix} [K] & \left[K^z\right] \\ \left[K^z\right] & \left[K^d\right] \end{bmatrix} \right) \begin{Bmatrix} \{U\} \\ \{\phi\} \end{Bmatrix} = \begin{Bmatrix} \{F\} \\ \{G\} \end{Bmatrix} \qquad ...(8)$$

Structural mass:
$$[M] = \int_{vol} \rho \left[N^u \right] \left[N^u \right]^T d(vol) \qquad ...(9)$$

Structural stiffness:
$$[K] = \int_{vol} \left[B^u \right]^T [C][B_u] d(vol) \qquad ...(10)$$

Dielectric conductivity:
$$\left[K^d \right] = -\int_{vol} [B_v]^T [\varepsilon][B_u] d(vol) \qquad ...(11)$$

Piezoelectric coupling matrix:

$$\left[K^z\right] = \int_{vol} \left[B_u\right]^T \left[e\right]\left[B_v\right] d\left(vol\right) \qquad ...(12)$$

$\{F\}$ = Vector of nodal forces, surface forces, and body forces, $\{G\}$ = Applied nodal charge vector

Considering a laminate made up of N layers with a total thickness of T, the above equation is transformed as below,

Structural stiffness:
$$[K] = \frac{2}{T} \int_{-1}^{1} \int_{-1}^{1} \sum_{k=1}^{N} \frac{t_k - t_{k-1}}{2} \int_{-1}^{1} G(r,s,t)\,drdsdt \qquad ...(13)$$

Dielectric conductivity:
$$\left[K^d\right] = \frac{2}{T} \int_{-1}^{1} \int_{-1}^{1} \sum_{k=1}^{N} \frac{t_k - t_{k-1}}{2} \int_{-1}^{1} G_1(r,s,t)\,drdsdt \qquad ...(14)$$

Piezoelectric coupling matrix:
$$\left[K^z\right] = \frac{2}{T} \int_{-1}^{1} \int_{-1}^{1} \sum_{k=1}^{N} \frac{t_k - t_{k-1}}{2} \int_{-1}^{1} G_2(r,s,t)\,drdsdt \qquad ...(15)$$

where,
$$G(r, s, t) = \left[B_u\right]^T \left[C\right] \left[B_u\right] |J|,$$

$$G_1(r, s, t) = -\left[B_\phi\right]^T \left[\varepsilon\right] \left[B_\phi\right] |J| \text{ , and}$$

$$G_2(r, s, t) = \left[B_u\right]^T \left[e\right] \left[B_\phi\right] |J|$$

Numerical integration for stiffness matrix has been evaluated using Gauss quadrature ($2 \times 2 \times 2$) scheme.

After applying the boundary conditions (both structural and electrical) and assembling all elemental matrices, the global matrices were derived, the global system of equations are given as:

$$\begin{pmatrix} [M] & [0] \\ [0] & [0] \end{pmatrix} \begin{Bmatrix} \{\ddot{U}\} \\ \{\ddot{\phi}\} \end{Bmatrix} + \begin{pmatrix} [K_{uu}] & [K_{u\phi}] \\ [K_{\phi u}] & [K_{\phi\phi}] \end{pmatrix} \begin{Bmatrix} \{U\} \\ \{\phi\} \end{Bmatrix} = \begin{Bmatrix} \{F\} \\ \{G\} \end{Bmatrix} \qquad ...(16)$$

where, $[K_{uu}]$, $[K_{u\phi}]$, $[K_{\phi\phi}]$ are global structural, global piezoelectric, global capacitance matrices, respectively.

3. LINEAR QUADRATIC REGULATOR (LQR) OPTIMAL CONTROL

The dynamic equation of the system in the state space representation given by:

$$\begin{Bmatrix} \dot{x} \\ \ddot{x} \end{Bmatrix} = \begin{pmatrix} 0 & I \\ -\omega^2 & 0 \end{pmatrix} \begin{Bmatrix} x \\ \dot{x} \end{Bmatrix} - \begin{bmatrix} 0 \\ \mu^{-1}[\psi]^T[K_{u\phi}] \end{bmatrix} \{\phi\} \qquad ...(17)$$

It is in the form of
$$\dot{x}(t) = Ax(t) + B\phi \qquad ...(18)$$

where, $[\psi]$, $[\omega]$ and $[\mu]$ are the mode shape, eigen frequencies and modal mass, respectively.

Linear quadratic regulator (LQR) optimal control theory is used to determine the control gains. In this, the feedback control system is designed to minimize a cost function or a performance index, which is proportional to the required measure of the system's response. The cost function used in this case is given by

$$J = \tfrac{1}{2} \int_{t_0}^{t_f} (X^T Q X + \phi^T R \phi)\,dt \qquad \ldots(19)$$

Where, $[Q]$ and $[R]$ are the semi-positive-definite and positive-definite weighting matrices on the outputs and control inputs, respectively. Physical meaning is that larger (relatively) elements in $[Q]$ mean that demand more vibration suppression ability from the controller. The purpose of the second term $[R]$ is to account for the effort being expended by the control system, so that small reductions in the output response are not obtained at the expense of physically unreasonable actuator input levels.

Assuming full state feedback, the control law is given by

$$\{\phi_a\} = -[R]^{-1}[B]^T[K]\{X\} = -[G_c]\{X\} \qquad \ldots(20)$$

Where, $[G_c]$ is the control gain.

$[K]$ can be obtain by solution of Riccati equation given by

$$([\dot{K}] + [K][A] + [A]^T[\lambda] + [K][B][R]^{-1}[B]^T[K] - [Q])X = 0 \qquad \ldots(21)$$

4. RESULTS AND DISCUSSIONS

A computer code has been developed in C for dynamic analysis of smart FRP composites using 3D layered finite element and MATLAB has been used for vibration control of the structure using LQR. After validating the FE code, the optimal vibration control was studied for two cases. Material properties used in the both the cases are listed in Tables 1 to 3 and the ply orientation of the laminate has been taken as $[p(/0/45/-45/90)_{2s}/p]$. Total number of layers is 18 out of which two 'p' layers are piezoelectric layers (one is actuator and the other is sensor). Calculations are performed using $6 \times 6 \times 4$ uniform finite element mesh with time step of $1\mu s$. In the first case, the smart FRP laminated cantilever beam with dimensions 0.1 m $\times$ 0 .05 m $\times$ 0.01135 m subjected to sinusoidal load of magnitude 150 N with frequency 600 rad/s, applied at tip, for 0.5 second and the uncontrolled

Table 1.

Material	Modulus of Elasticity (all in GPa)			Major Poisson's ratio	Other Poisson's ratio	Density Kg/m³
	E_1	E_2	G_{12}	ν_{12}	ν_{23}	ρ
Graphite/Epoxy	181.00	10.3	7.17	0.28	0.28	1600
PZT	82.05	4.140	2.780	0.25	0.25	7500

Table 2.

Material	Dielectric matrix (in 10^{-9} F/m)			
	ε_{11}	ε_{22}	ε_{33}	$\varepsilon_{12} = \varepsilon_{13} = \varepsilon_{21} = \varepsilon_{23} = \varepsilon_{31} = \varepsilon_{32}$
PZT	7.124	7.124	5.841	0

Table 3.

Material	Piezoelectric matrix (in C/ m^2)					
	e_{13}	e_{23}	e_{33}	e_{52}	e_{61}	$e_{11} = e_{12} = e_{21} = e_{22} = e_{31} = e_{32} = e_{41} = e_{42} = e_{43} = e_{51} = e_{53} = e_{62} = e_{63}$
PZT	−4.1	−4.1	14.1	10.5	10.5	0

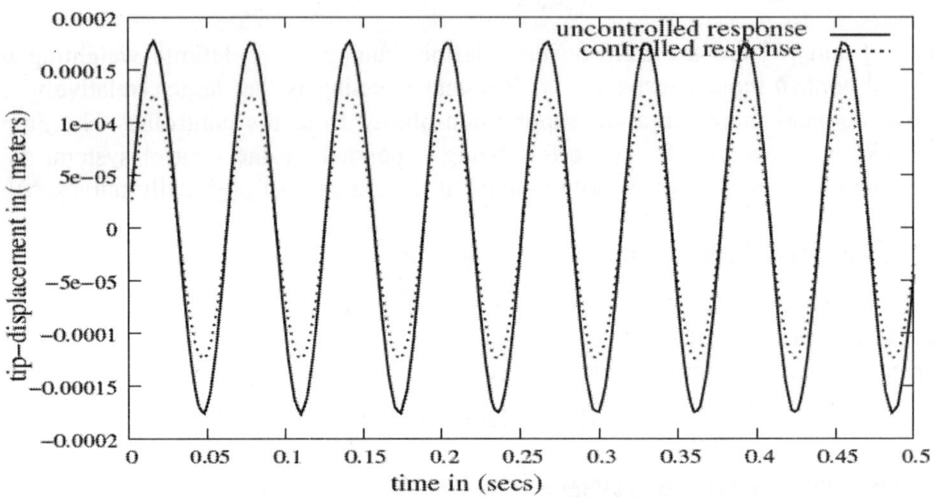

Fig. 2. Response of cantilever beam subjected to sinusoidal load.

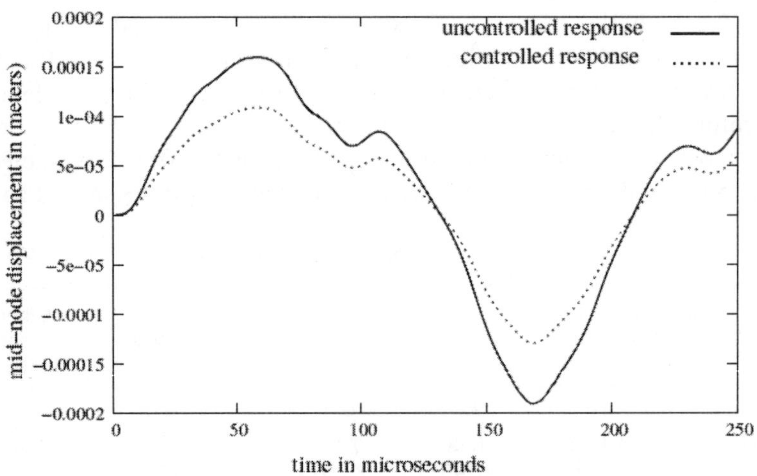

Fig. 3. Response of plate for impactor velocity 12 m/s.

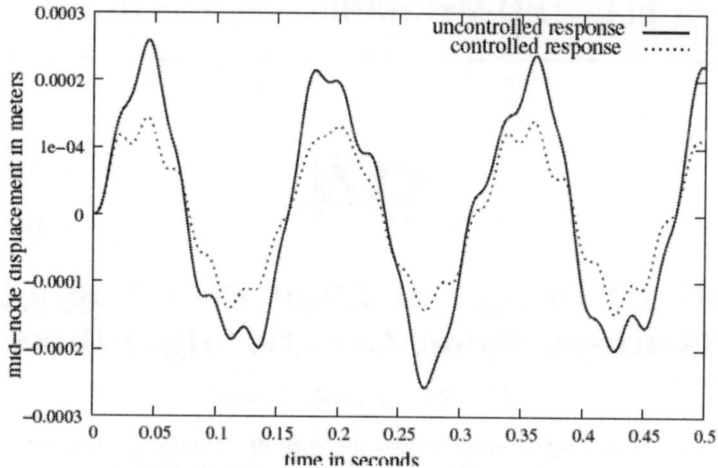

Fig. 4. Response of plate for sinusoidal load.

and control response presented as shown in Fig. 2. In the second case, the smart FRP laminated plate with dimensions 0.070 m × 0.070 m × 0.0035 m is impacted at the center by an aluminum sphere of diameter of 0.0127 m and the plate is assumed to be clamped along four edges. The uncontrolled and controlled response is presented for impactor velocity 12 m/s as shown Fig. 3. Fig. 4 shows the uncontrolled and controlled displacement of the plate under sinusoidal loading.

CONCLUSION

This article presents the three-dimensional layered finite element formulation and optimal vibration control of smart FRP composite structure using linear quadratic regulator (LQR). And also the responses of piezoelectric laminated composite plate under harmonic excitation as well as transverse impact considering Hertzian contact theory have been analyzed. It is observed that using LQR control strategy the maximum deflection is reduced by 29% in case of cantilever beam subjected to sinusoidal load and in case of plate subjected to impact loading maximum deflection is reduced by 31%. It can be concluded that LQR control scheme could be effectively used in vibration control of smart FRP structures both in the case of impact loading as well as sinusoidal loading.

REFERENCES

1. H. Allik, and T.J. Hughes, 1979, Finite Element Method for Piezoelectric Vibration, Intl. J. of Numerical Methods in Engineering, 2, 151-168.
2. H.S Tzou and C.I. Tseng, 1990, Distributed Piezoelectric Sensor/Actuator Design for Dynamic Measurement/ Control of Distributed Parameter Systems: A Finite Element Approach, J. of Sound and Vibration, 138 (1), 17-34.
3. S. K. Ha, C. Keilers and F.K. Chang, 1992, Finite Element Analysis of Composite Structures Containing Distributed Piezoceramic Sensors and Actuators, AIAA 30 (3), 772-780.
4. K.Y. Sze and L.Q. Yao, 2000, Modeling Smart Structures with Segmented Piezoelectric Sensors and Actuators, Journal of Sound and Vibration 235 (3), 495-520.
5. V. Piefort, N. Loix and A. Premont, 1998, Modeling of Piezolaminated composite shells for vibration control, Active Structure Laboratory, Universite Libre de Bruxelles.
6. V. Piefort and A. Premont, 2001, Finite element modeling of piezoelectric structures, Active Structure Laboratory, Universite Libre de Bruxelles.
7. S. Narayanan and V. Balamurugan, 2003, Finite element modeling of piezolaminated smart structures for active vibration control with distributed sensors and actuators, Journal of Sound and Vibration 262, 529-562.

94

Optimal Design of Eight Pole Magnetic Bearings Using Genetic Algorithms

J.S. RAO[1] AND R. TIWARI[2]

Department of Mechanical Engineering, Indian Institute of Technology Guwahati-781 039, India
email: [1] jsrao@iitg.ernet.in [2] rtiwari@iitg.ernet.in

ABSTRACT

Optimal design of radial magnetic bearings has been carried out by using Single Objective Genetic Algorithms (SOGAs). Two objective functions, namely, the powerloss and the weight are considered for the minimization. The load required to be supported, the maximum space available, the maximum current density that can be supplied in the coil, the maximum flux density that is allowed in the stator iron are the constraints considered. The geometries of optimized magnetic bearings are compared in tabular form for different cases. It has been observed that the two objective functions are mutually conflictive and a weighted sum approach result is also presented.

Keywords: Design optimization; Genetic Algorithms; Magnetic Bearings.

1. INTRODUCTION

Active magnetic bearing (AMB) systems are complex interdisciplinary systems which involve mechanical, electrical, electronic, and control disciplines making the AMB design and analysis complex. Systematic design approaches have not been still available for the design of AMB which makes it only possible by ones expertise. The gradient-based deterministic design of radial magnetic bearings can be performed with the help of simple closed form formulas of power losses and weight in terms of design variables [1]. However, the complexity of the problem increases drastically for the case of integrated design with control systems [2], for which case it may be very difficult and some times not possible to obtain a closed formula by the Lagrange's or any other method. Moreover, when the analysis has to be carried out by numerical schemes such as finite element methods [3], the numerical optimization of the design integrated with analysis has to be performed. The difficulty increases further for the multi-objective optimization with objective functions such as powerloss, weight, cost, etc. GAs have the ability to handle large number of objective functions and constraints of any complexity conveniently without posing problem of the local optimum and no presumptions of nature of the search space [4]. The authors have applied GAs to the optimal design of thrust magnetic bearings for the minimization of powerloss and load to weight ratio [5].

In the present paper the design optimization of radial active magnetic bearings has been performed using GAs as an initial attempt of solving more complex problems as mentioned above. Geometrical details are given in section 2, followed by the mathematical formulation of the objective functions and associated constraints in the next section. Two objective functions are considered, namely, the minimization of powerloss and minimization of weight. A weighted sum approach is also presented. Implementation algorithm is given in section 4, followed by the sections of results and conclusions.

2. GEOMETRY

bsThe geometry of a typical eight pole radial magnetic bearing is shown in Figure 1. Bias magnets are made of permanent magnetic materials like neodymium iron boron [6] and are designed to support the minimum load acting on the system. Thus, the control current flowing in coils is required to compensate the additional change in the position and the load [5]. Different dimensional parameters of the bearing are the radius of the shaft r_s, the thickness of the laminated disk t_d, the radius of the laminated disk r_d, the length of the air gap l_g, the thickness of bias magnets l_m, the width of the pole c, the width of the bearing or the axial length of the pole b, the thickness of the coil t_c, the width of the pole-coil unit c_o, the axial length of the bearing-coil unit b_o, the height of the pole h, the height of the coil h_c, the back wall thickness of the bearing t_b, and the outer radius of the bearing r_0. In the present paper the design vector is chosen as $[b,c,h,h_c,t_c,l_m]$. Input variables are $[r_d,l_g,F,B_r,B_{sat},\alpha_{max},\alpha_{min},r_{omax},w_{bmax}]$ and dependent parameters are $[c_o,b_o,r_o]$. From Figure 1 it can be observed that c_o,b_o,r_d and r_o can be expressed as

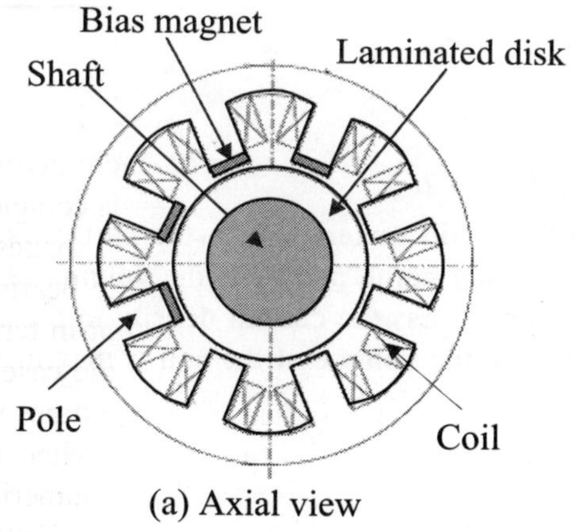

(a) Axial view

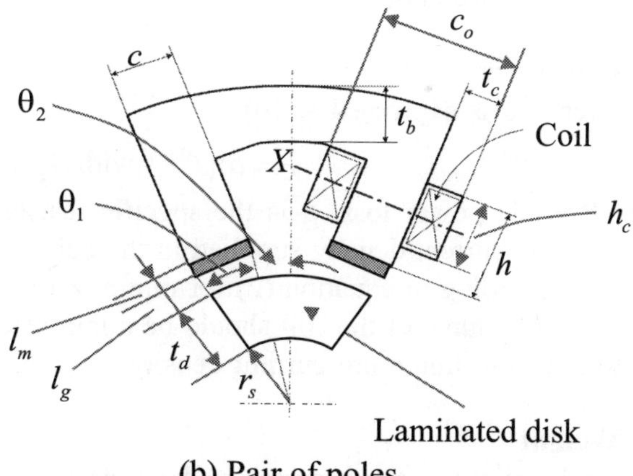

(b) Pair of poles

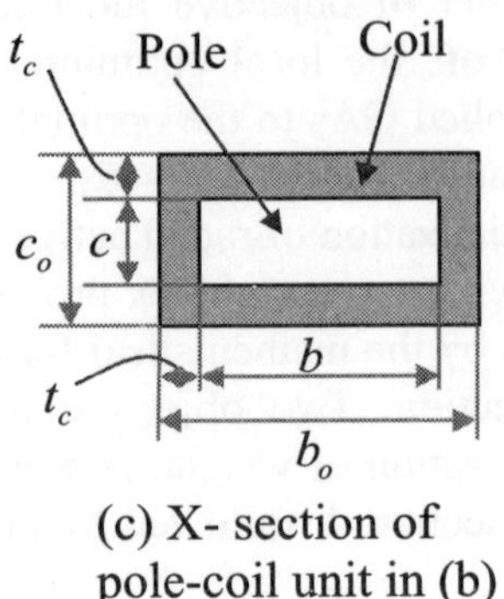

(c) X- section of
pole-coil unit in (b)

Fig. 1. Radial magnetic bearing geometry.

$$c_o = c + 2t_c; \qquad b_o = b + 2t_c; \qquad r_d = r_s + t_d; \qquad r_o = r_d + l_g + l_m + h + t_b \qquad ...(1)$$

and t_b can be shown to be equal to c by assuming that the area of the flux flow A_g is constant [3].

3. MATHEMATICAL MODEL

The optimization problem, in general, consists of objective function(s) which should be optimized and constraints which should be satisfied by the design vector. In this section expressions are presented for the design of the radial magnetic bearing with bias magnets. Expressions for the case of no bias magnets can be obtained by eliminating terms involving parameters of bias magnets [5].

3.1 Objective Functions

Powerloss represents the running cost of magnetic bearing while the weight represents the carrying or transportation cost and the initial cost. With the weight densities of the materials of the magnetic bearing fixed, the overall weight also represents the space occupied by the bearing. Thus, both the powerloss and weight get important roles in the selection of magnetic bearing design. In the present work three objective functions, namely, the powerloss, the weight, and their weighted sum are considered for single objective optimization.

3.1.1 Powerloss

The power loss is expressed as [6]

$$P = \rho \eta J^2 V_c \text{ with } V_c = 16 t_c h_c (b + c + 2t_c) \qquad ...(2)$$

where, P is the power loss, ρ is the specific resistance of the coil material, η is the coil packing factor, J is the current density supplied in the coil at the operating point, and V_c is the volume of all the coils. According to equation (2) for minimization of the powerloss the current density in the coil as well as the volume of the coil should be minimized, however, the volume of the coil is maximized so as to maintain minimum current density.

3.1.2 Weight

The overall weight of the bearing can be expressed as [5]

$$W = W_c + W_s + W_m \qquad ...(3)$$

where, W is the weight of components of magnets. Subscripts: c, s and m represent the coil, the stator, and the magnet (bias or permanent), respectively. Maximization of the load to weight ratio decreases the volume of the bearing while increasing the load that can be supported. With reference to Figure 1 the components of the weight of the bearing can be expressed as

$$W_c = 16\gamma_c t_c h_c (b + c + 2t_c); \quad W_s = \gamma_s \left(8bch + 4bcl_m + \pi bt_b (2r_o - t_b)\right); \quad W_m = 4\gamma_m bcl_m \quad ...(4)$$

where, γ is the weight density of the corresponding material. Powerloss represents the running cost while weight represents the carrying cost.

3.1.3 Normalized Weighted Sum

The normalized weighted sum is defined as,

$$\sigma = \varepsilon_1 \left|(P - P_{\min})/(P_{\max} - P_{\min})\right| + \varepsilon_2 \left|(W - W_{\min})/(W_{\max} - W_{\min})\right| \quad ...(5)$$

where, σ is the normalized weighted sum, ε_1 and ε_2, ($\varepsilon_1 = \varepsilon_2$), are weights of objective functions considered $P_{\min}$ and $W_{\min}$ are minimum feasible values and, $P_{\max}$ and $W_{\max}$ are maximum feasible values of the individual objective functions, P and W are the feasible values of objective functions of the member considered for calculating the weighted sum.

3.2 Constraints

The major design requirements and constraints for the problem are explained in this section. The bearing is required to support a designed load. The temperature cannot be more than that of insulation of the coil can sustain which depends on the capacity of the cooling system being used. Thus, the current density supplied in the coil is limited [8]. The magnetic flux density flowing in the stator iron cannot go beyond the saturation value [2]. The space occupied by the bearing is limited [5]. These constraints are summarized in the following subsections.

3.2.1 Load to be supported

All the loads acting on the bearing can be resolved into two normal components in x and y directions. If F_x and F_y are two components acting in two directions. The maximum and minimum loads in x and y directions can be calculated [2, 5] as

$$F_{x\max} = F_x(1 + \kappa_x); \quad F_{x\min} = F_x(1 - \kappa_x); \quad F_{y\max} = F_y(1 + \kappa_y); \quad F_{y\min} = F_y(1 - \kappa_y) \quad ...(6)$$

where, $F_{x\max}$ and $F_{x\min}$ are the maximum and minimum forces acting, K_x is the load variation factor in x-direction and suffix y represents the corresponding quantities in y-direction. In the present work the loads acting in both the directions are assumed to be equal. The bearing should be able to support the designed load (s) [2] i.e.,

$$(B^2_{\max,\min}/\mu_0)A_g \cos\theta = F_{\max,\min} \quad ...(7)$$

where, $F_{\max}$ and $F_{\min}$ are the maximum and minimum loads acting, μ_0 is the permeability of vacuum, A_g is the area of the air gap at poles, $\theta = 0.5\theta_1 + \theta_2 = 22.5° = 22.5°$. θ_1 and θ_2 are the pole face angles as shown in Figure 1.

3.2.2 Maximum Current Density Allowed in the Coil

The current density supplied in the coil should not be more than the maximum current density that the coil material can sustain [7, 8]. The flux density is expressed in terms of magneto-motences as [9]

$$B_{\max,\min} = 0.5\mu_0 \left(K_i n i_{\max,\min} + 2B_r l_m/\mu_0\right)/\left(K_a l_{g\max,\min} + K_f l_m\right) \quad ...(8)$$

where, B_r is the remanence flux of bias magnets, K_i is the coil mmf loss factor, K_a is the actuator loss factor, K_f is the flux leakage factor, n is the number of turns, l_{gmax} $(= l_g + x_{max})$ is the maximum gap allowed, lgmin $(= l_g - x_{max})$ is the minimum gap allowed and x_{max} is the maximum displacement of the rotor from the operating position which can be expressed as a percentage of the operating gap l_g. ni_{max} and ni_{mix} are respectively, the maximum and minimum magneto-motences. l_m can be determined from equation (8) using bias flux and zero magneto-motence [6]. ni_{max} and ni_{min} can be determined by using maximum and minimum fluxes, respectively and l_m in equation (8). Magneto-motences can also be expressed as [6]

$$ni_{\max} = \eta J_{\max} A_c \quad \text{and} \quad ni_{\min} = \eta J_{\min} A_c \qquad \text{with} \quad A_c = t_c h_c \qquad \text{...(9)}$$

where, A_c is the cross-sectional area of the coil perpendicular to the flow of the current, η is the coil packing factor, J_{max} and J_{min} are the maximum and minimum current densities to be supplied in the coil, respectively. From equations (8) and (9), we get

$$J_{\max, \min} = \left(2B_{\max,\min}\left(K_a l_{g\max, \min} + K_f l_m\right) - 2B_r l_m\right)\Big/\left(\mu_0 K_i \eta A_c\right) \text{ with } J_{ub} \geq \max\left(\left|J_{\max}\right|, \left|J_{\min}\right|\right) \text{ ...(10)}$$

where, J_{ub} is the upper bound of the current density that the coil can sustain.

3.2.3 Maximum Flux Density Allowed in the Stator Iron

The flux density in the stator iron should not be more than the saturation flux density B_{sat}, beyond which, there won't be any response of the actuator with the increase in the current. For a linear range of operation [2], the actuator can be designed to have the flux density between two linear range limits.

$$\alpha_{\max} B_{sat} \geq B_{\max} \quad \text{and} \quad B_{\min} \geq \alpha_{\min} B_{sat} \qquad \text{...(11)}$$

where, α is the iron saturation factor [5].

3.2.4 Maximum Space Occupied by the Bearing

The space available for the whole bearing is specified [2] i.e.

$$r_{o\max} \geq r_o \quad \text{and} \quad b_{o\max} \geq \left(b + 2t_c\right) \qquad \text{...(12)}$$

where, $r_{o\,max}$ is the maximum bearing outer diameter, and $b_{o\,max}$ is the maximum width of the bearing-coil unit or axial length.

3.2.5 Maximum Space Available for Pole-Coil Unit

The maximum space occupied by each pole-coil unit cannot be more than the available space for each pole-coil unit, $c_{o\,max}$ and is expressed as (Figure 1)

$$c_{o\max} \geq c + 2t_c \quad \text{with} \quad c_{o\max} = 2\left(r_d + l_g + l_m + h - h_c\right)\tan\theta \qquad \text{...(13)}$$

Equations (10) to (13) are inequality constraints while equation (7) is an equality constraint.

4. RESULTS AND DISCUSSION

The non-dominated sorting genetic algorithm-II (NSGA-II) developed by Kanpur Genetic Algorithm Laboratory (KanGAL), IIT Kanpur [10] is implemented to the present single objective optimisation problem. For the optimisation, SBX (simulated binary crossover) parameters namely, the probability of the crossover p_c, the probability of the mutation p_m, the crossover distribution index η_c, and the mutation distribution index η_m are assumed to be 0.8, $1/p$, 5, and 10, respectively [11] where, p is the number of real variables which is equal to 5 in the present case. A population of 500 is taken with 100 generations and 5 runs. The input data assumed for the design is tabulated in Table 1 and the geometries of optimized radial magnetic bearings are given in Table 2. The convergence plots of the objective functions namely, the minimum powerloss and minimum weight versus generation number are shown in Figure 2 for the case of using bias magnets. Similar plots are observed for other cases also.

Table 1. Input parameters assumed for the radial magnetic bearing design.

Parameter	Value	Parameter	Value
Disc radius of the bearing, r_s (mm)	25.00	Iron saturation factor, α_{min}	0.0
Operating air gap, l_g (mm)	4.00	Iron saturation factor, α_{max}	0.8
Operating load, F (N)	50.00	Coil mmf loss factor, K_i	1.394
Variation in gap	±5%	Actuator loss factor, K_a	1.072
Variation in load	±10%	Flux leakage factor, K_f	0.840
Saturation flux density, B_{sat} (T)	1.00	Specific gravity of stator iron, γ_s (g/cm^3)	7.77
Remanence flux density, B_r (T)	1.20	Specific gravity of copper, γ_c (g/cm^3)	8.91
Packing factor, η	0.85	Specific gravity of Ne-Fe-B, γ_m (g/cm^3)	7.50

Table 2. Optimized bearing geometries.

*Objective Functions	Design variables							Volumes			Performance parameters at the operating point			
	b	c	h	h_c	t_c	l_m	r_o	V_c	V_s	V_p	J	B	P	W
	(mm)	(mm)	(mm)	(mm)	(mm)	(mm)	(mm)	(cm^3)	(cm^3)	(cm^3)	(A/mm^2)	(T)	(kW)	(kg)
1	25.39	9.10	49.23	33.73	12.75	–	85.33	412.95	208.29	-	3.63	0.542	11.57	5.30
2	16.21	8.75	18.50	18.31	7.65	2.06	56.32	90.30	68.54	1.17	1.99	0.692	0.76	1.35
3	75.06	16.44	56.55	33.60	12.46	–	100.0	780.57	1270.4	-	1.61	0.235	4.32	16.82
4	76.59	16.93	55.63	34.37	11.70	0.43	100.0	752.72	1325.1	2.25	0.17	0.229	0.05	17.02
5	30.21	15.10	57.84	41.55	10.36	–	99.95	455.17	476.31	-	2.58	0.386	6.44	7.75
6	21.57	14.94	47.01	41.73	6.33	1.06	90.02	207.98	289.69	1.37	0.60	0.459	0.16	4.11

*1→ Minimum weight without bias magnets
3 → Minimum powerloss without bias magnets
5 → Minimum normalized weighted sum without bias magnets

2 → Minimum weight with bias magnets
4 → Minimum powerloss with bias magnets
6 → Minimum normalized weighted sum with bias magnets

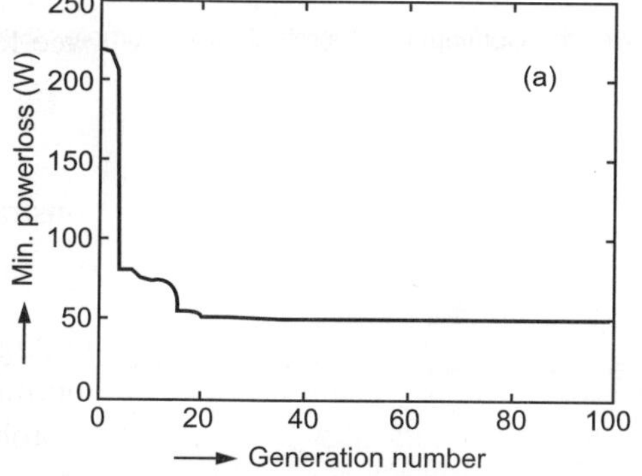

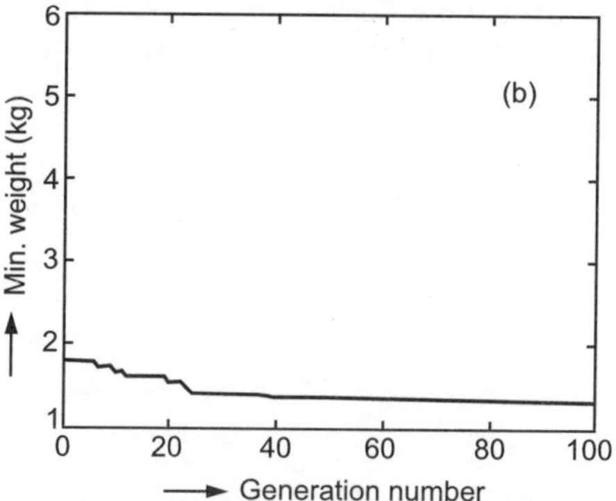

Fig. 2. The convergence of the objective functions for the case of with bias magnets at load 50N versus the generation number (a) for the minimum powerloss (b) for the minimum weight.

It is observed that the weight and powerloss are mutually conflicting quantities. A weighted sum optimized design with equal weights is also presented. A 21.26% of increase in the weight resulted in a 70.76% of decrease in the powerloss in case of without bias magnets and a 17.61% of increase in the weight resulted in an 84.50% decrease of the powerloss in the case of using bias magnets. The satisfactory results GAs ensured the implementation to the problems of more complexity such as integration with dynamics, control and analysis.

CONCLUSION

In the present work an optimal design of eight pole radial magnetic bearings using GAs has been carried out. Mathematical models of objective functions and associated constraints have been formulated and discussed. Two types of bearings i.e., without and with bias magnets have been taken into consideration. Two objectives have been considered, namely, minimizing the powerloss and minimizing the weight and their combined effect is studied by weighted sum approach. It is observed that the weight and powerloss are mutually conflictive. The methodology can be extended to the optimal design with complexity including integration with the control and by using the multi-objective optimization.

REFERENCES

1. S.L. Chen, C.T. Hsu, 2002, Optimal Design of A Three Pole Active Magnetic Bearing, IEEE Transactions on Magnetics, 38(5), 3458-3466.
2. H. Chang, S.C. Chung, 2002, Integrated Design of Radial Active Magnetic Bearing Systems using Genetic Algorithms, Mechatronics, 12(1), pp. 19-36.
3. A. Kenny, A.B. Palazzolo, 2003, Single Plane Radial Magnetic Bearings Biased with Poles Containing Permanent Magnets, Journal of Mechanical Design, 125, 178-185.
4. D.E. Goldberg, 1989, Genetic Algorithms in Search, Optimization, and Machine Learning, Addison-Wesley.
5. J.S. Rao, R. Tiwari, 2006, Design Optimization of Thrust Magnetic Bearings Using Genetic Algorithms, 7th IFToMM-Conference on Rotor Dynamics, Vienna, Austria, 25-28 Sept.
6. V.D. Bloodgood Jr., N.J. Groom, C.P. Britcher, 2000, Further development of an optimal design approach applied to axial magnetic bearings, NASA-2000-7ismb-vdb.
7. A. Chiba, T. Fukao, O. Ichikawa, M. Oshima, M. Takemoto, D.G. Dorrell, 2005, Magnetic Bearings & Bearingless Drives, Newnes, Elsevier.
8. G. Schweitzer, H. Bleuler, A. Traxler, 2003, Active Magnetic Bearings: Basics, Properties and Applications of Active Magnetic Bearings, Authors Reprint: Zürich.
9. N.J. Groom, V.D. Bloodgood Jr., 2000, A Comparison of Analytical and Experimental Data for a Magnetic Actuator, NASA-2000-tm210328.
10. K. Deb, A. Pratap, S. Agarwal, T. Meyarivan, 2002, A fast and elitist multiobjective genetic Algorithm: NSGA-II, IEEE Trans. Evol. Comput., 6, pp. 182-197.
11. K. Deb, R.B. Agarwal, 1995, Simulated Binary Crossover for Continuous Search Space, Complex Systems, 9, pp. 115-148.

95

Magnetic Fluid Based Squeeze Film Between Infinitely Long Rectangular Plates

G.M. Deheri[1], H.C. Patel[2] and Rakesh M. Patel[3]

[1] Department of Mathematics, Sardar Patel University, Vallabh Vidyanagar, Anand, Gujarat-388 120, India. email: gmdeheri@rediffmail.com
[2] Gujarat Council on Science and Technology, 7th Floor, M. S. Building, Sector-11, Gandhinagar, Gujarat-382 011, India. email: prof_himanshu@rediffmail.com
[3] Department of Mathematics, Gujarat Arts and Science College, Ahmedabad, Gujarat-380 006, India. email: jrmpatel@rediffmail.com

ABSTRACT

Efforts have been directed to study and analyze the performance of a magnetic fluid based squeeze film between infinitely long rectangular plates. A magnetic fluid is used as the lubricant and the external magnetic field is oblique to the lower plate. The associated Reynolds' equation is solved with suitable boundary conditions to get the pressure distribution, which is then used to obtain the expression for load carrying capacity leading to the calculation of the response time. The results are presented graphically as well as in tabular form. This investigation tends to suggest that the bearing system registers a slightly enhanced performance as compared to that of a bearing system working with a conventional lubricant. The graphical and the tabular representations make it clear that the load carrying capacity and response time increase with increasing magnetization parameter. Further, it is revealed that the performance of the bearing system gets strongly influenced by the aspect ratio. In addition, the present study indicates that although the combined negative influence of the porosity and the aspect ratio dominates the positive effect induced by the magnetization parameter, there is a scope for improving the performance of the bearing system by choosing a suitable combination of the magnetization parameter and the aspect ratio.

Keywords: Magnetic fluid, rectangular plates, Reynolds' equation, load carrying capacity.

1. INTRODUCTION

The classical theory of squeeze film between plane parallel surfaces was discussed by Archibald [1]. Wu [2-4] analyzed the squeeze film performance of a porous bearing for mainly two types of geometries; annular and rectangular. Subsequently, Prakash and Vij [5] investigated the behavior of squeeze film between porous plates of various shapes using Morgan-Cameron approximation. This article also made it clear that the aspect ratio played a crucial role on the performance of this type of bearing system. In fact, it has been established in this paper that increasing values of the aspect ratio caused decreased load carrying capacity. All these above studies considered conventional lubricants.

Oil based or other lubricant fluid based magnetic fluid can be used as a lubricant. The advantage of magnetic fluid as a lubricant over the conventional ones is that the former can be retained at the desired location by an external magnetic field. The magnetic fluid is prepared by suspending fine magnetic grains coated with surfactants and dispersing it in non-conducting and magnetically passive solvents such as kerosene, hydrocarbons and fluorocarbons. When magnetic field is applied to the magnetic fluid each particle experiences a force. Hence, with the proper application of magnetic field the magnetic fluid can be made to adhere to any desired surface.

Use of magnetic fluid as a lubricant modifying the performance of the bearing has been very well recognized. Verma [6] and Agrawal [7] investigated the application of magnetic fluid as a lubricant. While Verma [6] considered tangential slip velocity at the porous matrix-lubricant interface, Agrawal [7] considered no slip condition. Bhat and Deheri [8,9] analyzed the performance of a squeeze film between annular disks and curved circular plates lubricated with a magnetic fluid and found that its performance with the magnetic fluid as a lubricant was better than with a conventional lubricant. Patel and Deheri [10] studied the behavior of squeeze film formed by a magnetic fluid between curved annular plates. Recently, Patel and Deheri [11] discussed the performance of a magnetic fluid based squeeze film between rough annular plates.

All these above studies establish that the performance of the bearing system gets enhanced by the presence of a magnetic fluid lubricant. Hence, it was deemed proper to investigate and analyze the behavior of a magnetic fluid based squeeze film between infinitely long porous rectangular plates.

2. ANALYSIS

The configuration of the bearing system is shown in the Fig. 1.

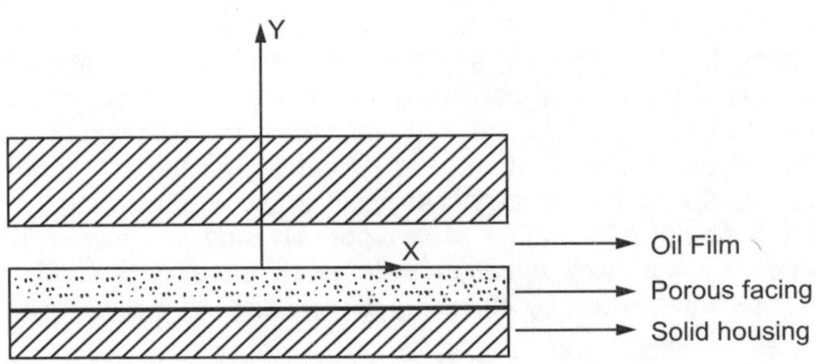

Fig. 1. Geometry and coordinates of the problem.

In the analysis the assumptions of a conventional lubrication theory are retained. Assuming axially symmetric flow of the magnetic fluid between the rectangular plates under an oblique magnetic field $\overline{H}$ whose magnitude H is a function of z vanishing at $\pm \frac{b}{2}$, the modified Reynolds' equation governing the film pressure p is obtained as (c.f.[5, 9]).

$$\frac{d^2}{dz^2}\left[p - 0.5\mu_0\,\overline{\mu}\,H^2\right] = \frac{12\mu\,\dot{h}}{h^3 + 12\phi H} \qquad \ldots(1)$$

where,

$$H^2 = \left(\frac{b}{2} - z\right)\left(\frac{b}{2} + z\right)$$

(c.f. [8, 10]). Integration of Eq. (1) with the concerned boundary conditions $p(\pm b/2) = 0$ leads to the pressure distribution:

$$p = 0.5\mu_0\,\bar{\mu}\,H^2 + \frac{6\mu\dot{h}}{h^3 + 12\phi H}\left[z^2 - \frac{b^2}{4}\right]$$

Now the pressure distribution in dimensionless form is given by

$$\bar{p} = \frac{1}{\beta}\left[\frac{1}{4} - \bar{z}^2\right]\left\{\frac{\mu^*}{2} + \frac{6}{1 + 12\psi}\right\} \qquad \ldots(2)$$

The load carrying capacity w can be expressed in dimensionless form as

$$\bar{w} = \frac{1}{\beta}\left[\frac{\mu^*}{12} + \frac{1}{1 + 12\psi}\right] \qquad \ldots(3)$$

wherein,

$$w = a \int_{-b/2}^{b/2} p(z)\,dz$$

The time Δt taken by the upper plate to reach a film thickness h_1 at t_1 starting from an initial film thickness h_0 at t_0 is described by

$$\Delta t = -\frac{\mu\bar{w}\,a^2 b^2}{w}\int_{h_0}^{h_1}\frac{dh}{h^3 + 12\phi H}$$

which then in non-dimensional form is given by

$$\Delta\bar{t} = \frac{\bar{w}}{6\alpha^2}\left[\ln\left\{\frac{(1+\alpha)^2(\bar{h}^2 - \bar{h}\alpha + \alpha^2)}{(\bar{h} + \alpha)^2(1 - \alpha + \alpha^2)}\right\} + 2\sqrt{3}\,\tan^{-1}\left\{\frac{\sqrt{3}\alpha(1 - \bar{h})}{2\alpha^2 - (1 + \bar{h})\alpha + 2\bar{h}}\right\}\right] \qquad \ldots(4)$$

3. RESULTS AND DISCUSSIONS

Expressions for dimensionless pressure, load carrying capacity and response time for the present case are given in Eq. (2), (3) and (4), respectively. Those for the corresponding non-magnetic case (c.f. [5]) can be obtained by setting $\mu^* = 0$ in these expressions. It is clear from the Eq. (3) that the load carrying capacity increases by $\dfrac{\mu^*}{12\beta}$.

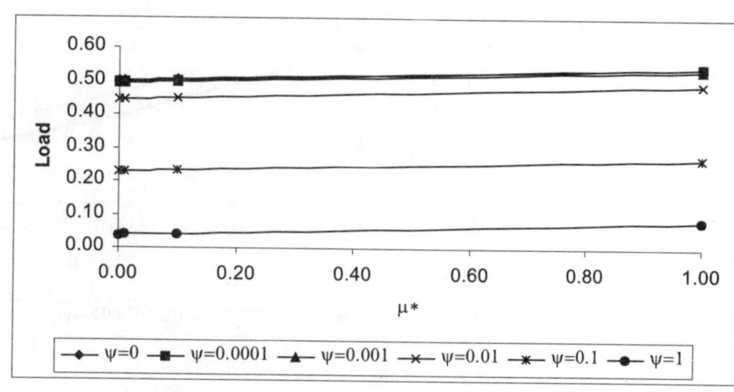

Fig. 2. Variation of load carrying capacity with respect μ^* for $\beta = 2$.

Figures 2 and 3 present the variation of load carrying capacity with respect μ^* and ψ for a suitable value of the aspect ratio β. We have the distribution of load carrying capacity with respect to magnetization parameter and the aspect ratio in Figs. 4 and 5.

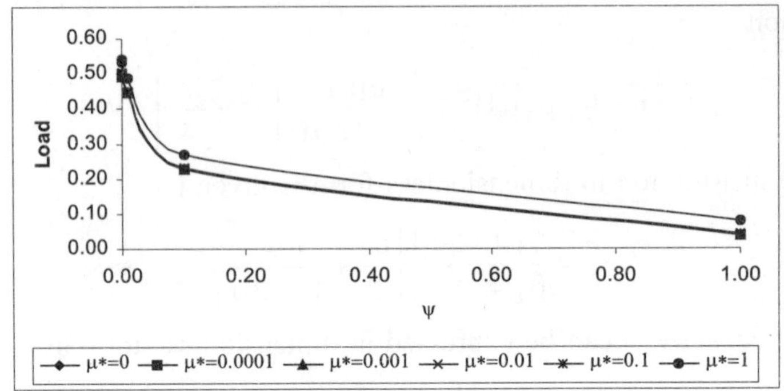

Fig. 3. Variation of load carrying capacity with respect ψ for $\beta = 2$.

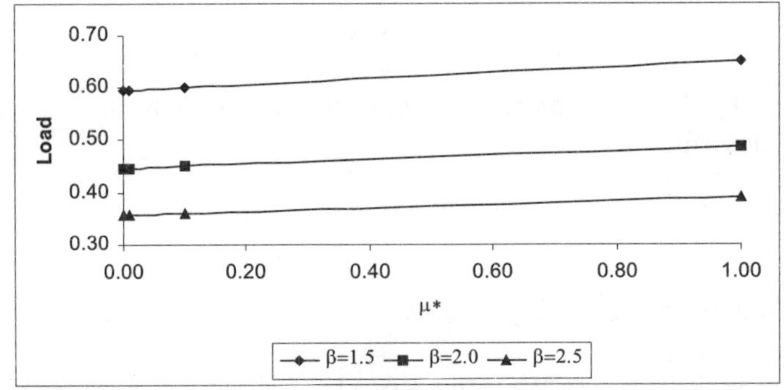

Fig. 4. Variation of load carrying capacity with respect μ^* for $\psi = 0.01$.

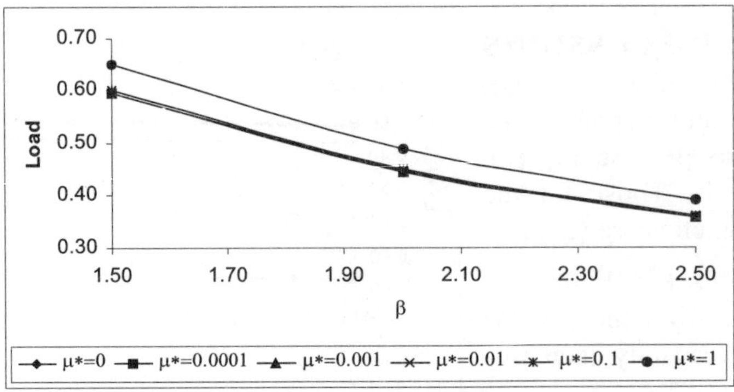

Fig. 5. Variation of load carrying capacity with respect β for $\psi = 0.01$.

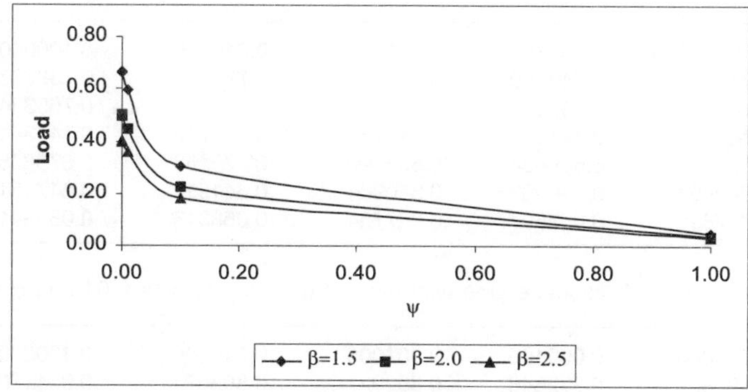

Fig. 6. Variation of load carrying capacity with respect ψ for $\mu^* = 0.01$.

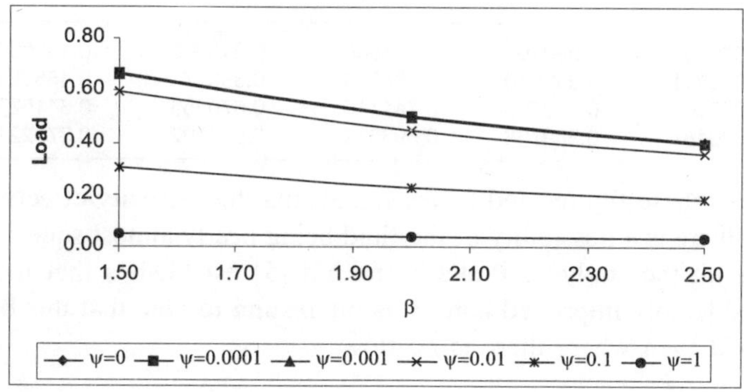

Fig. 7. Variation of load carrying capacity with respect β for $\mu^* = 0.01$.

Figures 6 and 7 give the variation of load carrying capacity with respect to the porosity parameter and the aspect ratio. These figures tend to suggest that the performance of the bearing system improves significantly for a suitable value of the aspect ratio β. In fact, increasing values of the magnetization parameter leads to the increasing values of load carrying capacity. Further, the aspect ratio and the porosity give rise to the decreasing values of load carrying capacity wherein the effect of the aspect ratio is more sharp.

Next the variation of response time with respect to the magnetization parameter μ^*, porosity ψ and the aspect ratio β is presented in Tables 1-3. From these Tables it can be easily observed that the response time more or less follows the trends of the load carrying capacity. However, the time taken in reducing a prescribed film thickness is greater as compared to the corresponding situation described in Prakash and Vij [5].

Table 1. Variation of response time with respect μ^* and ψ for $\beta = 2$ and $\alpha = 0.001$

ψ / μ^*	0.000000	0.000100	0.001000	0.010000	0.100000	1.000000
0.000000	0.749983	0.749990	0.750046	0.750608	0.756233	0.812482
0.000100	0.749084	0.749091	0.749147	0.749709	0.755334	0.811583
0.001000	0.741090	0.741097	0.741153	0.741715	0.747340	0.803589
0.010000	0.669628	0.669634	0.669690	0.670253	0.675878	0.732127
0.100000	0.340902	0.340908	0.340964	0.341527	0.347151	0.403400
1.000000	0.057691	0.057697	0.057754	0.058316	0.063941	0.120190

Table 2. Variation of response time with respect μ^* and β for $\psi = 0.01$ and $\alpha = 0.001$

β / μ^*	0.000000	0.000100	0.001000	0.010000	0.100000	1.000000
1.500000	0.892837	0.892846	0.892921	0.893671	0.901170	0.976169
2.000000	0.669628	0.669634	0.669690	0.670253	0.675878	0.732127
2.500000	0.535702	0.535707	0.535752	0.536202	0.540702	0.585701

Table 3. Variation of response time with respect ψ and β for $\mu^* = 0.01$ and $\alpha = 0.001$

β / ψ	0.000000	0.000100	0.001000	0.010000	0.100000	1.000000
1.500000	1.000811	0.999613	0.988954	0.893671	0.455369	0.077755
2.000000	0.750608	0.749709	0.741715	0.670253	0.341527	0.058316
2.500000	0.600487	0.599768	0.593372	0.536202	0.273221	0.046653

A closed scrutiny of these figures and tables signals that the porosity effects may be neglected up to certain extent while there is a possibility of the fluid being nearly almost squeezed out. A comparison of this investigation with the study of Prakash and Vij [5] establishes that the performance of the bearing system is considerably improved here. It is interesting to note that this bearing with magnetic fluid can support a load even where there is no flow.

CONCLUSION

The present study indicates that although the combined negative influence of the porosity and the aspect ratio dominates the positive effect induced by the magnetization parameter, there is considerable scope for improving the performance of the bearing system by choosing a suitable combination of the magnetization parameter and the aspect ratio.

NOMENCLATURE

μ_0	Permeability of the free space	N/A^2
μ	fluid viscosity	Pa.s
$\bar{\mu}$	Magnetic susceptibility	m^3/kg
μ^*	$= -\dfrac{\mu_0 \bar{\mu} h^3}{\mu \dot{h}}$ Magnetization parameter	
ϕ	Permeability of the porous matrix	
ψ	$= \dfrac{\phi H}{h^3}$	

$$\alpha = \left(12\,\psi_0\right)^{\tfrac{1}{3}}$$

$$\bar{h} = \frac{h_1}{h_0}$$

$$\bar{z} = \frac{z}{b}$$

$$\beta = \frac{a}{b}$$

$$\psi_0 = \frac{\phi H}{h_0^3}$$

REFERENCES

1. Archibald, F.R., 1956, Load capacity and time height relations for squeeze films, Journal of Basic Engineering, Series D, Vol. 78, p.p. 231-245.
2. Wu, H., 1970, Squeeze film behavior for porous annular disks, Journal of Lubrication Technology, Vol. 92, p.p. 593-596.
3. Wu, H., 1972a, An analysis of the squeeze film between porous rectangular plates, Journal of Lubrication Technology, Vol. 94, p.p. 64-68.
4. Wu, H., 1972b, Effect of velocity slip on the squeeze film between porous rectangular plates, Wear, Vol. 20, p.p. 67-71.
5. Prakash, J. and Vij, S. K., 1973, Load capacity and time height relations for squeeze films between porous plates, Wear, Vol. 24, p.p. 309-322.
6. Verma, P.D.S., 1986, Magnetic fluid based squeeze films, International Journal of Engineering Sciences, Vol. 24(3), p.p. 395-401.
7. Agrawal, V.K., 1986, Magnetic fluid based porous inclined slider bearing, Wear, Vol. 107, p.p. 133-139.
8. Bhat, M.V. and Deheri, G.M., 1991, Squeeze film behavior in porous annular disks lubricated with magnetic fluid, Wear, Vol. 151 (), p.p. 123-128.
9. Bhat, M.V. and Deheri, G.M., 1993, Magnetic fluid based squeeze film in curved porous circular disks, Journal of Magnetism and Magnetic Material, Vol. 127, p.p. 159-162.
10. Patel, R. M. and Deheri, G.M., 2002, On the behavior of squeeze film formed by a magnetic fluid between curved annular plates, Indian Journal of Mathematics, Vol. 44, p.p. 353-359.
11. Patel, R.M. and Deheri, G.M., 2004, Magnetic fluid based squeeze film behavior between annular plates and surface roughness effect, Proceedings of AIMETA International Tribology Conference, Perma, Italy, p.p. 631-638.

96

Dynamic Analysis of Orthotropic Clamped-Free Cylindrical Shells with Constrained Electro-Rheological Fluid Damping

SHARNAPPA, N. GANESAN AND RAJU SETHURAMAN

Department of Mechanical Engineering, Indian Institute of Technology Madras, Chennai-600 036, India
email: me05d003@iitm.ac.in

ABSTRACT

This paper deals with the dynamic characteristics of orthotropic cylindrical sandwich structure with electro-rheological (ER) fluid core under clamped free condition. The contribution of composite and electro-rheological damping to the total damping of the structure is studied. There are two types of electro-rheological fluid cores are used in the present study. The damping variation of cylindrical shell for different length to radius (L/R) ratio, core to facing thickness ratio (t_c/t_f) are carried out. The variation of damping properties of ER fluid with electric field also investigated.

Keywords: Composite structures, Electro-rheological fluid, Cylindrical shell, Damping.

1. INTRODUCTION

The study of the dynamic characteristics of a composite cylindrical shell structures has a direct and important bearing on the structural problems of missile and rockets. Composite materials are now widely used in space vehicles because of their lightweight and high specific strength. The use of sandwich structures with core material in shells of revolution is an effective way of damping the vibration in the structures subjected to dynamic loading. Most commonly studied sandwich composite cylindrical shells are made up of sandwich structures with a viscoelastic core.

In recent past a lot of research has been done on the sandwich shell structures with viscoelastic core. Ramasamy and Ganesan [1] have studied the vibration and damping analysis of fluid filled cylindrical shells with viscoelastic damping by using finite element method. Okazaki *et al.* [2] have investigated the damping properties of two layered cylindrical shells with unconstrained viscoelastic layer. They derived the fundamental equation for symmetric and non-symmetric vibration using Flugge's theory, and effects of shear deformation and thickness-wise extensional deformation considered. Wilkins *et al.* [3] have studied the free vibration analysis of orthotropic conical shell with viscoelastic core sandwich structures. Yeh and Chen [4] studied the dynamic stability of a sandwich plate with a constrained layer and electro-rheological fluid core. They extended the study on dynamic stability analysis of a rectangular orthotropic sandwich plate with an electro-rheological fluid core [5]. There are no studies in the literature available on the composite cylindrical shell with electro-rheological fluid core sandwich structures. The present study deals with dynamic analysis of cantilever cylindrical shell with electro-rheological fluid core.

2. FINITE ELEMENT FORMULATION

Semi-analytical method is used to calculate the frequency and loss factor of cylindrical shell. The finite element formulation is used in the present study is same as given by Ramasamy and Ganesan [1]. The three-noded beam element is used along the length direction with seven degree of freedom per node and Fourier series is assumed in circumferential direction. The model is made up of sandwich structure, which consists of graphite epoxy composite facings with electro-rheological fluid core as shown in Fig. 1. Two types of electro-rheological (ER) fluid cores are used in the present study. The properties of ER fluid and graphite epoxy composite materials are given in the Table 1 [4,6]. There are three different damping cases are considered in order to know the effect of each damping on the structure such as 1) only composite damping 2) only electro-rheological fluid damping and 3) total damping, for different core to facing (t_c/t_f) thickness ratios.

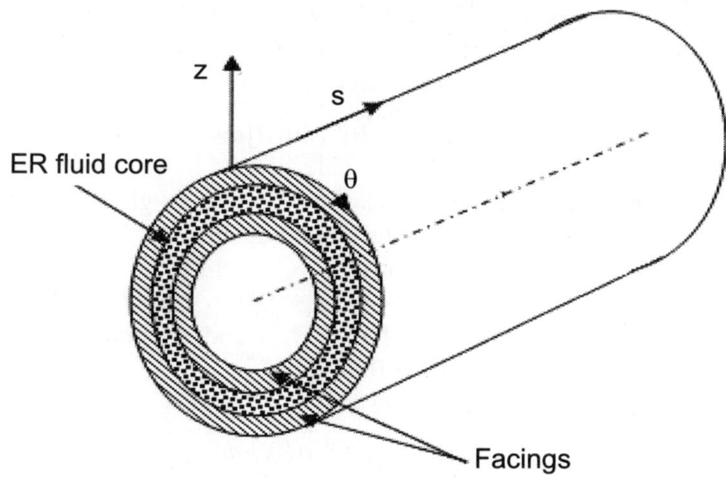

Fig. 1. Sandwich cylindrical shell with electro-rheological fluid core and orthotropic facings.

Table 1. Properties of ER core and composite facings

Shear modulus (G)	ER fluid core		Graphite-epoxy composite facings
N/m^2	Type 1 (ER1)	Type 2 (ER2)	E_{LL} = 137 GPa; E_{TT} = 8.96 GPa;
Real part (G_R)	$\approx 15000\,E_*^2$	$\approx 50000\,E_*^2$	G_{LT} = 7.1 GPa; ρ = 1600 Kg/m^3;
Imaginary part (G_I)	≈ 6900	$\approx 2600\,E_*^2 + 1700$	η_{LT} = 0.0045; η_{LT} = 0.0422;
			η_{LT} = 0.075; v_{LT} = 0.3;

where, E_* is the electric field in kV/mm

The displacement field used in the present analysis is proposed by Wilkins *et al.* [3] The strain displacement relations for core, inner and outer facings referred in the present study are given by Ramasamy and Ganesan [1]. In the present analysis three-noded element with seven degrees of freedom per node is used. The vector of displacement per element u_e is expressed as

$$\{u_e\} = \{u_{0,1}\ v_{0,1}\ w_{0,1}\ \psi_{s,1}\ \psi_{\theta,1}\ \phi_{s,1}\ \phi_{\theta,1}\ \ldots\ldots\ldots\phi_{s,3}\ \phi_{\theta,3}\}$$ where the subscripts 1,2 and 3 denote the node number.

The strain vectors can be represented as

$$\{\varepsilon\} = \{\varepsilon_{ss}\ \varepsilon_{\theta\theta}\ \gamma_{s\theta}\ \gamma_{\theta z}\ \gamma_{sz}\}^T = [B]\{u_e\} \qquad \ldots(1)$$

Here, [B] is the derivative of shape function matrix.
The element stiffness matrix $[K_e]$ is expressed as

$$[K_e] = \int_v [B]^T [D][B]\,dv \qquad \ldots(2)$$

where, $[D] = [T]^T [Q][T]$ is the transformed elasticity matrix in global coordinates
   $[Q]$ is the elasticity matrix in the material coordinates.
   $[T]$ is the standard transformation matrix
$$[K_e] = [K]_R + [K]_I$$

The stiffness matrix $[K_e]$ consists of real part $[K]_R$ and imaginary part $[K]_I$ due to complex material properties of facings, ER1 and ER2 fluid core.

The element mass matrix $[M_e]$ is given by

$$[M_e] = \int_v [N]^T \rho [N]\,dv \qquad \ldots(3)$$

where ρ is density of the structure.

Evaluation of Frequency and Loss factor

The following eigen value problem has to be solved to get the natural frequencies.
$$[K]_R - \omega^2 [M] = 0 \qquad \ldots(4)$$
where, ω is the natural frequency

The composite and ER fluid loss factor for n^{th} mode can be calculated using modal strain energy method.

$$\eta_n = \frac{\phi_n^T [K]_I \phi_n}{\phi_n^T [K]_R \phi_n} \qquad \qquad ...(5)$$

where, $[K]_R$ and $[K]_I$ are the real and imaginary parts of the stiffness matrices $[K]$, respectively and ϕ_n is n^{th} mode eigenvector.

3. RESULTS AND DISCUSSION

The frequency and damping analysis of sandwich cylindrical shell by using two types of electro-rheological (ER1 and ER2) core materials are investigated. The parametric study carried out for different core to facing (t_c/t_f) thickness and length to radius (L/R) ratio in clamped free (C-F) boundary condition. The results are plotted for first axial mode of a cylindrical shell.

Figure 2 shows the loss factor of ER fluids (type 1 and type 2) with 20 circumferential modes for different electric field. In case of ER1, loss factor decreases with increasing the electric field it is observed clearly at 10th circumferential mode shown in Fig. 2(a). In case of ER2, loss factor increase with increasing the electric field up to 1.5 kV/mm and reaches a maximum value then onwards decreases with increasing the electric field as shown in Fig. 2 (b). The variation of loss factor with electric field is same as given by Yeh and Chen [4].

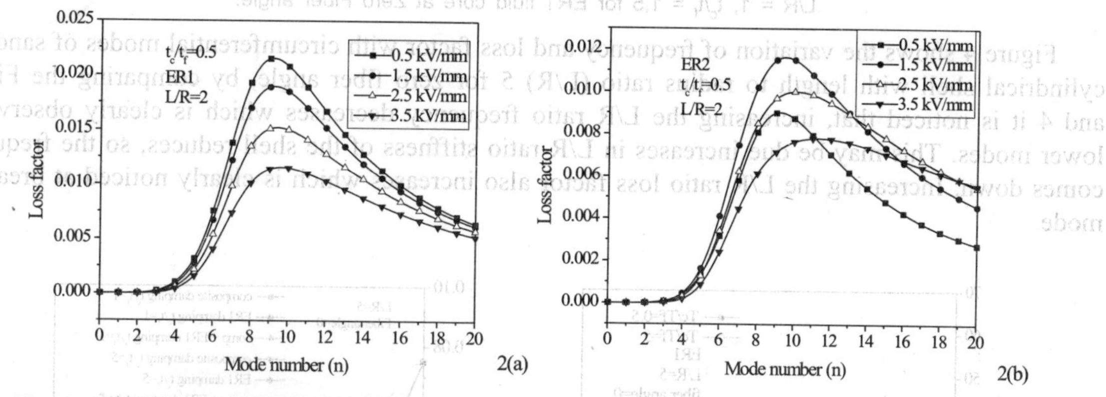

Fig. 2. Variation of loss factor with mode number (n) for different voltages in case of ER1 and ER2.

Figure 3 shows the variation of frequency and loss factor of ER1 fluid core at L/R ratio 1 for different core to facing thickness ratio (t_c/t_f) at zero fiber orientation. Frequency decreases with increasing the mode number and reaches a minimum value then increases with increase in mode number. Increasing the core to facing thickness ratio frequency decreases at all the circumferential modes. This effect may be due to increases in core to facing ratio the mass of the structure increases so the frequency decreases. Increasing the t_c/t_f ratio the frequency variation is observed more at lower and higher modes compared to intermediate modes (n = 8 to 14) this effect may happens because in the lower mode the membrane effect is more and at higher modes the bending effect is more.

From the Fig. 3 it is noticed that increasing the t_c/t_f ratio, ER fluid damping increases and also composite-damping increases, so the net result is increase in the total damping, which is observed clearly at higher circumferential modes. ER1 fluid damping is not having much influence at lower circumferential modes, as the mode number increases (from n = 6) the ER fluid damping increases.

The composite damping is higher at lower modes. This variation may be due to membrane effect is more at lower modes and bending effect is more at higher modes. In case of composite sandwich structure, the composite damping is more compared to ER fluid damping at all circumferential modes. So, ER damping is not having much influence on the dynamic properties of the structure.

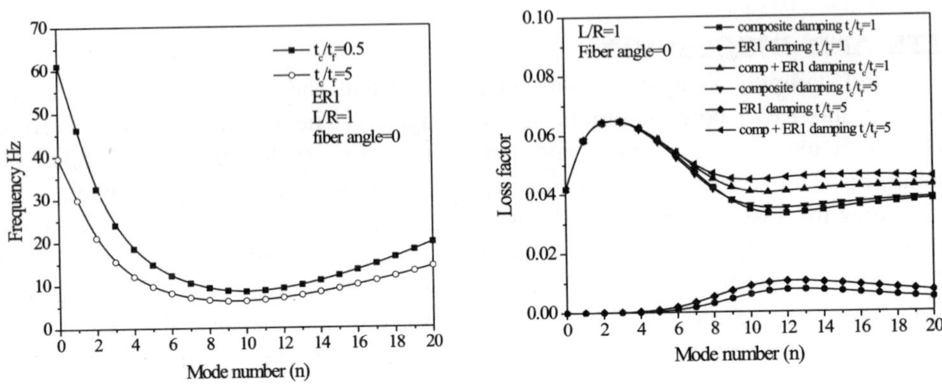

Fig. 3. Variation of Frequency and Loss factor with mode number
L/R = 1, t_c/t_f = 1,5 for ER1 fluid core at Zero Fiber angle.

Figure 4 shows the variation of frequency and loss factor with circumferential modes of sandwich cylindrical shell with length to radius ratio (L/R) 5 for zero fiber angle, by comparing the Figs. 3 and 4 it is noticed that, increasing the L/R ratio frequency decreases which is clearly observed at lower modes. This may be due increases in L/R ratio stiffness of the shell reduces, so the frequency comes down. Increasing the L/R ratio loss factor also increases which is clearly noticed at breathing mode.

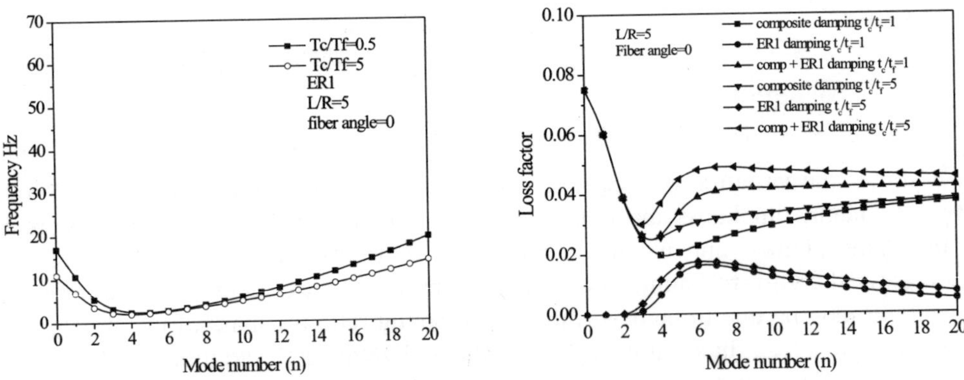

Fig. 4. Variation of Frequency and Loss factor with mode number
L/R = 5, t_c/t_f = 1, 5 for ER1 fluid core at Zero Fiber angle.

Figures 5 and 6 show the variation of frequency and loss factor at 90° fiber orientation for length to radius ratio (t_c/t_f) 1 and 5, respectively. Frequency decreases with increasing the mode number and reaches a minimum value then increases with increasing the mode number. Composite damping is more compared to ER1 fluid damping at all modes. There is a drastic fall in the composite loss factor at lower modes, so the total damping also follows the same trend of curves. Comparing the Figures 5 and 6 increasing the L/R ratio frequency decreases and loss factor increases which is clearly observed at lower modes particularly at breathing mode. Comparing the Figs. 4 and 6 the composite damping is more at zero fiber angle at higher modes. From the Figs. 3 and 5 it is noticed that the frequency at zero fiber angle is low compared to 90° fiber orientation at lower and higher modes.

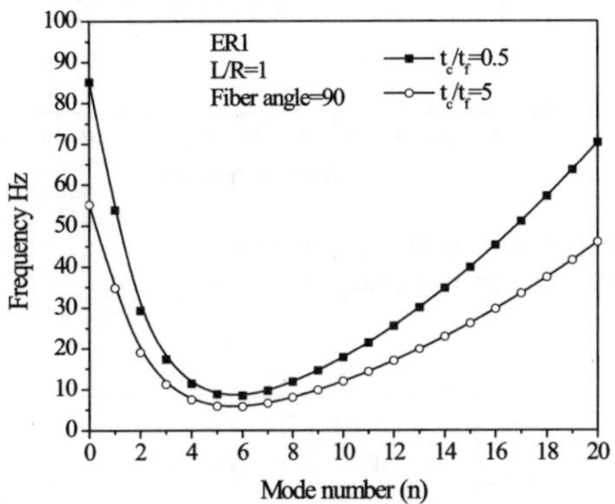
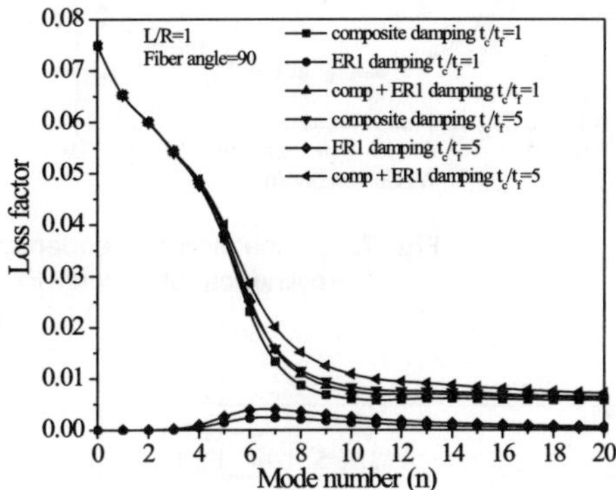

Fig. 5. Variation of frequency and loss factor with mode number
L/R=1,t_c/t_f = 1,5 for ER1 fluid core at 90 Fiber angle

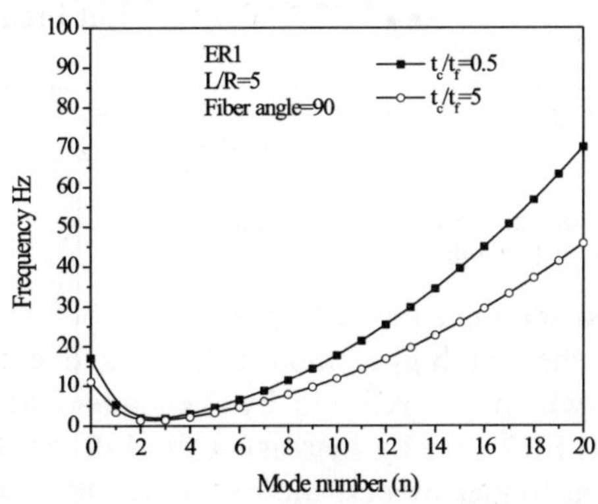
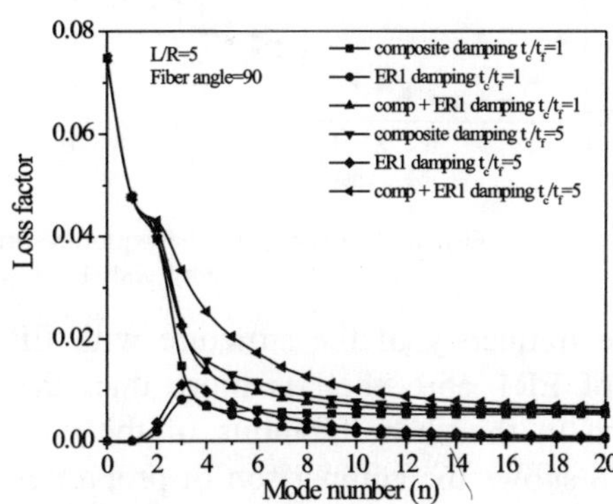

Fig. 6. Variation of frequency and loss factor with mode number
L/R=5, t_c/t_f = 1,5 for ER1 fluid core at 90 fiber angle

Figures 7 and 8 show the variation of frequency and loss factor with circumferential modes for ER1, ER2 fluid cores at L/R ratio 1 and 5, respectively. The comparison of properties with ER1 and ER2 core shell is carried out at zero fiber orientation at t_c/t_f ratio 0.5. Figure 7 shows that the frequency decreases with increasing the circumferential modes and reaches a minimum value then increases with increase in mode number.

loss factor at lower modes, so the total damping also follows the same trend of curves. Comparing the Figures 5 and 6 increasing the L/R ratio frequency decreases and loss factor increases which is clearly observed at lower modes, particularly at breathing mode. Comparing the Figs. 4 and 6 the composite damping is more at zero fiber angle at higher modes. From the Figs. 3 and 5 it is noticed that the frequency at zero fiber angle is low compared to 90° fiber orientation at lower and higher modes.

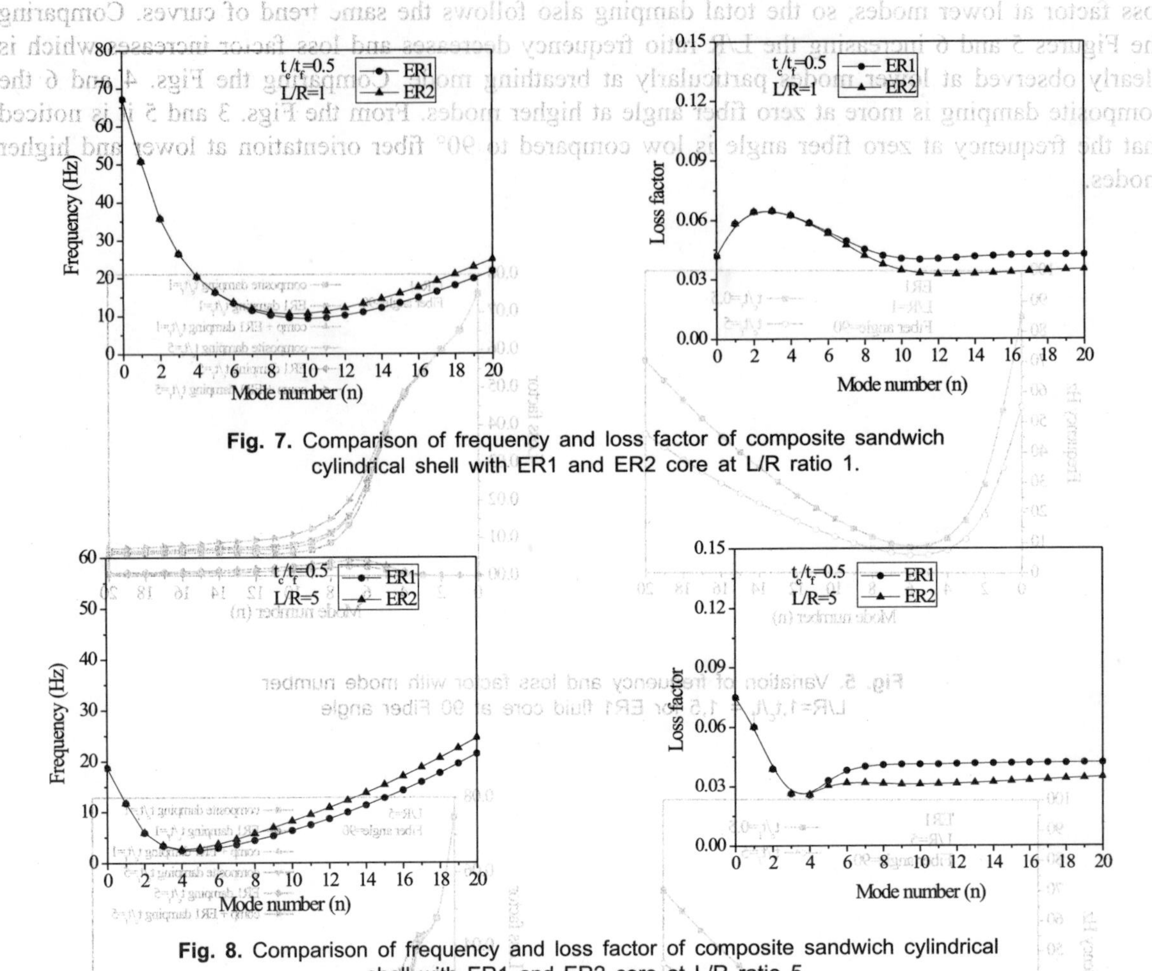

Fig. 7. Comparison of frequency and loss factor of composite sandwich cylindrical shell with ER1 and ER2 core at L/R ratio 1.

Fig. 8. Comparison of frequency and loss factor of composite sandwich cylindrical shell with ER1 and ER2 core at L/R ratio 5.

The frequency of the structure with ER1 core is lower than the ER2 core shell, but the loss factor of ER1 core shell is more than the ER2 core shell at higher modes. This is due to the variation in the shear modulus of the core materials, which are referred by Yeh and Chen [4]. Figure 8 shows the comparison of properties of ER1 and ER2 core by increasing the L/R ratio to 5. In this case also the effect of core material is observed at higher modes, the reason for this may be at higher modes the bending effect is more predominant, which may have more effects on frequency and loss factor of the shell with different core material.

CONCLUSION

The dynamic analysis of orthotropic sandwich cylindrical shell with electro-rheological fluid core in clamped free condition is investigated, the following conclusions are arrived with the study.

1. Increasing the electric field ER1 fluid damping decreases, ER1 fluid damping is more compared to ER2 fluid damping.

2. Increasing the core to facing thickness ratio frequency decreases and loss factor increases.

3. Composite damping is more compared to ER fluid damping at all modes and also composite damping is more at zero fiber orientation compared to 90^0 fiber orientation.

4. Frequency of ER1 fluid core is lower than ER2 fluid core shell but loss factor of ER1 is more than the ER2 core shell at higher modes.

REFERENCES

1. R. Ramasamy, and N. Ganesan, 1999, Vibration and damping analysis of fluid filled orthotropic cylindrical shells with constrained viscoelastic damping, Computers and Structures, 70, 363-376.
2. A. Okazaki, A. Tatemichi, and S. Mirza, 1994, Damping properties of two layered cylindrical shells with an unconstrained viscoelastic layer, Journal of sound and vibration, 176, (2), 145-161.
3. D.J. Wilkins, C.W. Bert, and D.M. Egle, 1970, Free vibration of orthotropic sandwich conical shells with various boundary conditions, Journal of sound and vibration, 13, 211-28.
4. J.Y. Yeh, and L.W. Chen, 2005, Dynamic stability of a sandwich plate with a constrained layer and electro-rheological fluid core, Journal of sound and vibration, 285, 637-652.
5. J.Y. Yeh, and L.W. Chen, 2006, Dynamic stability analysis of a rectangular orthotropic sandwich plate with an electro-rheological fluid core, Composite Structures, 72, 33-41.
6. J.S. Chang, and J.W. Shyong, 1994, Thermally induced vibration of laminated circular cylindrical shell panels, Composite Science and Technology, 51, 419-427.

97

A Computational Model for Thermal Evolution and Sintering Kinetics in Selective Laser Sintering of Titanium Powder

S. Dilip Srinivas, B. Vignesh and E. Raghu

Department of Production Engineering, PSG College of Technology, Coimbatore-641 004, Tamil Nadu India email: dilip441@yahoo.co.in

ABSTRACT

In the present work, a computational model of selective laser sintering of titanium powder is developed. An Nd: YAG laser beam of power 2 W, scan speed 1 mm/s and spot radius 0.025 mm is focused on the bed. The beam operates in continuous wave mode and has a wavelength 1.06 µm. The thermal conductivity of the powder is identified as the most critical property in modeling the process. The results give an insight of the mechanism of heat transfer in SLS through the peak temperatures and the depth of penetration. The results clearly indicate the error that is introduced in the results when non-linearity in thermal conductivity is not addressed. The results obtained are in good agreement with the experimental results reported by researchers.

Keywords: Laser; Titanium; Sintering; Modeling.

1. INTRODUCTION

Selective laser sintering makes it possible to create fully functional parts directly from metals. Powders are sintered under the influence of a laser beam and there is a localized melting of powders. The part obtained after the process is usually porous and not strong enough for functional use. An infiltration process usually accompanies the process to produce functionally rigid parts. The mechanical properties of the component are strongly influenced by the temperature distribution, which is a function of spatial coordinates and time. The complexity of SLS process is so high that assumptions are needed to model the same as reported by previous researchers. Previous studies [1-3] have made assumptions to simplify the problem to a two-dimensional model. Bugeda *et al*. [4] have assumed that the thermal conductivity is directly proportional to solid fraction. The model by Kolossov et al. [5] incorporates thermal evolution through temperature dependent properties. The model, however, assumes a constant thermal conductivity at later stages due to high computational demand [5].

Tolochko *et al.* [6] have modeled the process by considering neck growth and sintering kinetics. They have also reported that the thermal conduction phenomenon in powder beds is due to contact thermal conduction between powder particles and radiation between powder particles. Gusarov *et al.* [7] have described in detail the phenomenon of contact thermal conduction of powder beds in SLS. They have cited contact thermal conductivity for different lattice structures and variation of contact thermal conductivity with relative neck size. The radiation component of the thermal conductivity in a bed of spheres is well addressed by Sih [8].

Titanium has good corrosion resistance and strength at high temperatures. As it is compatible with human system, it is used for medical purposes. Hence, the selection of rapid prototyping for making prototypes with complex profiles. This paper deals with the modeling of selective laser sintering using computational fluid dynamics code FLUENT. In this work, a three dimensional thermal model of selective laser sintering of titanium powder has been developed. An initial solid fraction of 45 % is considered. Bulk thermal properties of titanium such as thermal conductivity and specific heat are modeled as temperature dependent properties.

2. MODELING DETAILS

2.1 Model and Mesh

The model consists of a titanium powder bed of dimensions 5 × 5 × 2 mm. A moving laser source of spot radius 0.025 mm is located over the surface of the bed of titanium powder. The model is shown in Fig. 1.

In this study, the analysis of heat transfer in SLS process is described by the transient heat conduction equation and is given by Eq. (1).

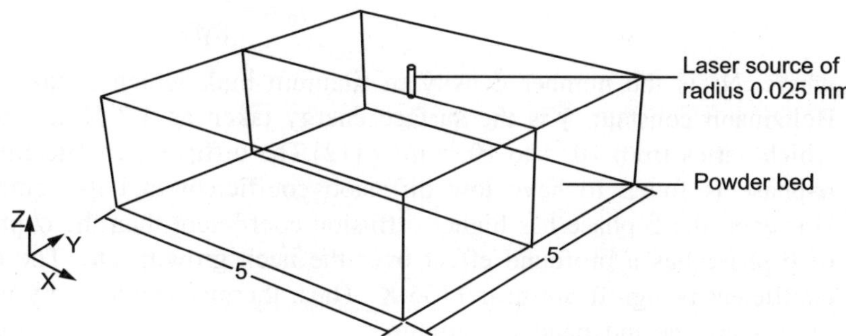

Fig. 1. Model used for computation.

$$\rho Cp\,\frac{\partial T}{\partial t} = \nabla^2(kT) \qquad \qquad ...(1)$$

where, ρ is the density of the titanium bed, Cp is the specific heat capacity of the titanium bed, k is the thermal conductivity, T is the temperature and t is time.

2.2 Material Properties

2.2.1 Thermal Conductivity

The principal mechanisms of heat transfer in selective laser sintering are contact thermal conduction between powder particles and inter-particle radiation [6]. The general relation determining the contact thermal conduction k_c is given by Eq. (2) [7].

$$k_c = k_{norm}\,k_{bulk}x \qquad \qquad ...(2)$$

where, k_{norm} is the normalized contact conductivity, k_{bulk} is the bulk thermal conductivity and x is the relative neck size. The normalized contact conductivity is reported for different lattice structures.

Titanium exists as α-phase at room temperatures. The α-phase is reported to exist as hexagonal closed packing (HCP) structure [9]. The value of normalized thermal conductivity for HCP is reported to be 0.433 [7]. However, after 1156 K, α-β transformation takes place [6]. The β-phase is associated with body centered cubic (BCC) structure and the normalized thermal conductivity for BCC is reported to be 1.732 [7]. Thus, the normalized thermal conductivity is found to vary from 0.433 to 1.732 during the process.

The bulk thermal conductivity k_{bulk} is also temperature dependent. The variation of thermal conductivity of the titanium bulk with temperature is taken from a handbook [10]. The growth of the relative neck size during the process is given by the Kuczynski's theory of volume diffusion and is given by Eq. (3) [11].

$$\frac{dx}{dt} = \frac{1}{t_0 x^4} \qquad \qquad ...(3)$$

where, t_0 is the characteristic time. The analytical relation governing the value of t_0 is given by Eq. (4) [11] as

$$t_0 = \frac{N_0 KTR^3}{8\gamma D_s} \qquad \qquad ...(4)$$

where, N_0 is the number density of titanium bulk which is taken as 5.667×10^{28} /m^3, K is the Boltzmann constant, γ is the surface energy taken as 1.7 N/m, and D_s is the diffusion coefficient which varies from 10^{-11} to 10^{-18} m^2/s [12]. The diffusion coefficient is a variant of temperature. The α-phase is found to have low diffusion coefficient and its contribution to neck growth is low. However, the β-phase has higher diffusion coefficient than the α-phase. Consequently, the diffusion of β-phase has a profound effect over the neck growth rate. The change in the value of diffusion coefficient is significant near 1156 K. Thus, thermal conductivity is a property that is dependent on more than one independent variable.

2.2.2 Radiative Thermal Conductivity

The thermal conductivity of the powder bed due to radiation among the powder particles is given by Eq. (5) [8] as

$$k_r = 4F\sigma T^3 D \qquad \qquad ...(5)$$

where, F is the geometric view factor, which is taken to be as 0.333 [8], σ is the Stefan-Boltzmann constant, T is the temperature and D is the diameter of the powder particle, which is taken to be 63 μm [6]. The contribution of radiative thermal conductivity is highly significant at higher temperatures and its contribution during earlier stages of sintering is very low as cited by Tolochko *et al.* [6].

2.2.3 Melting of Titanium Powder

The melting point of titanium is 1944 K [13]. Above this temperature, the powder particles join as a melt pool. Thus, in this model the thermal conductivity of the melted powder is taken to be equal to bulk thermal conductivity.

Thus, the effective thermal conductivity of the powder during the process is split into two levels and is given by Eq. (6) as

$$k = \begin{Bmatrix} k_r + k_c & \text{if } T < 1944\,K \\ 22 & \text{if } T >= 1944\,K \end{Bmatrix} \qquad \qquad ...(6)$$

2.2.4 Specific Heat

The specific heat is modeled as a temperature dependent function. The temperature dependence of specific heat is taken from a handbook [10].

3. BOUNDARY CONDITIONS

The boundary condition at surfaces of the bed except the bottom surface of the bed is given by Eq. (7)

$$-k\frac{\partial T}{\partial z} = \varepsilon\sigma(T^4 - T^4_{sur}) + h(T - T_{sur}) \qquad (7)$$

where, ε is the external emissivity of the bed, T_{sur} is the ambient temperature, T is temperature at surface of the bed, h is the convective heat transfer coefficient and σ is Stefan-Boltzmann constant. It is also assumed that the bottom of bed does not exchange heat with the environment. Thus, the appropriate boundary condition at the bottom face of the bed is given by Eq. (8)

$$k\frac{\partial T}{\partial z}\Big|_{z=0} = 0 \qquad (8)$$

The convective heat transfer coefficient is calculated from Nusselt Number for free convection over a flat plate. The emissivity of the bed is assumed dependent the initial solid fraction and is weighted based on area occupied by holes and solid and is given by Eq. (9)

$$\varepsilon = A_h\varepsilon_h + (1 - A_h)\varepsilon_s \qquad (9)$$

where, A_h is the area fraction of surface occupied by holes and, ε_s is the emissivity of solid particle and ε_h is the emissivity of the air-filled hole.

4. RESULTS AND DISCUSSIONS

4.1 Non-Linear Thermal Conductivity

Figure 2 shows variation of temperature along X-axis after a laser displacement of 0.65 mm. The peak temperature is found to be 3100 K. Good spread of temperature distribution along X-axis is also observed.

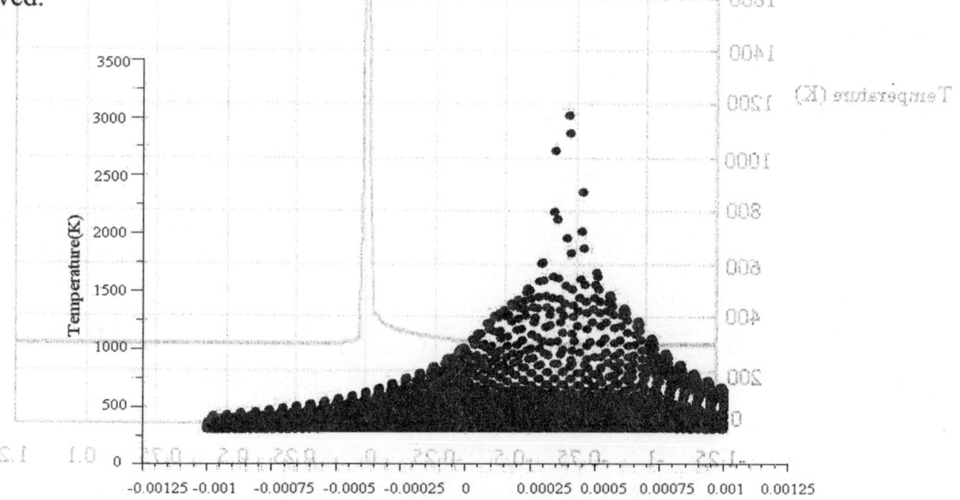

Fig. 2. Non-linear thermal conductivity-Variation of temperature along X-axis after laser scan of 0.65 mm.

The penetration depth is high because the model allows a comparatively steep increase in contact thermal conductivity at temperatures near the α-β transition temperature (1156 K). As the diffusion coefficient of β-phase is higher than α-phase, the neck size grows rapidly near this temperature resulting in steep increase in contact conductivity. Since there is no phase change, this increase in conductivity can be compensated only by heat flux for energy continuity. Therefore, a good penetration of temperature is observed. The reported experimental value of peak temperature is 3023 K [5]. The difference of 2.5 % may be attributed to the simplifying assumptions made in the model. The predicted peak temperatures and the resulting temperature profiles of the present model are in good agreement with the experimental values reported by Kolossov *et al.* [5]. The depth of penetration along X-axis and the resulting temperature profiles are also in good agreement with the experimental values reported by Kolossov *et al.* [5].

4.2 Temperature Dependent Thermal Conductivity

Figure 3 shows variation of temperature at centre of the laser source along X-axis after a laser displacement of 0.65 mm. The temperature plot shows us the prohibitive error that results when the model incorporates temperature dependent thermal conductivity.

The error in peak temperature when compared with the experimental value is now as much as 32.68 %. Thus, an assumption of temperature dependence of thermal conductivity is not valid because of high error in peak temperature. In addition, the depth of penetration is confined to a small area below the source. This is because the thermal conductivity only increases gradually with temperature. As a result, there is no driving mechanism such as high temperature gradient to facilitate high flux dispersion. Thus, the depth of penetration is confined to a limited area.

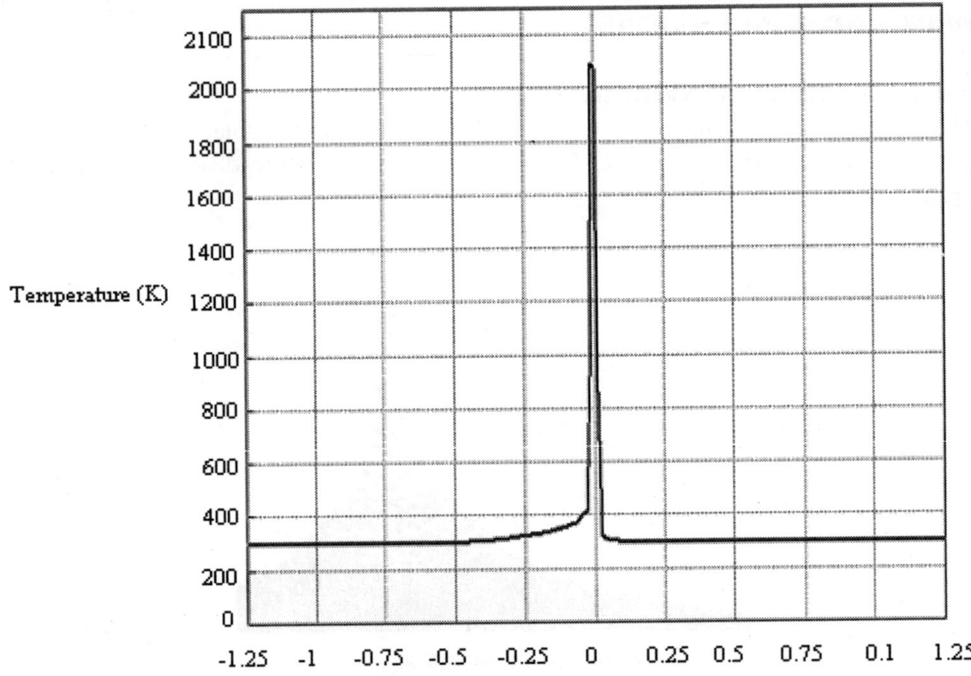

Fig. 3. Temperature dependent thermal conductivity-Variation of temperature along X-axis after laser scan of 0.65 mm.

CONCLUSION

Modeling and simulation of selective laser sintering of titanium powder is presented. Irradiation of bed of titanium powder of 2 mm thickness with an Nd: YAG laser of 1.06 μm wavelength, 0.025 mm spot radius and 2 W power is simulated. The non-linear treatment of thermal conductivity is found to be essential. The peak temperature is found to be around 3050 K. The error in peak temperature is predicted to be 32.68 % when non-linear treatment is not considered. In addition, the depth of penetration is also affected by this assumption. Thus, an error in thermal conductivity manifests itself as errors in peak temperatures and depth of penetration.

REFERENCES

1. M. Matsumoto, M. Shiomi, K. Osakada, and F. Abe, 2001, Finite element analysis of single layer forming on metallic powder bed in rapid prototyping by selective laser processing, International Journal of Machine Tools and Manufacture, 42, 61-67.
2. J.C. Nelson, S. Xue, J.W. Barlow, J.J. Beaman, H.L. Marcus, and D.L. Bourell, 1993, Model of the selective laser sintering of bisphenol-a polycarbonate, Industrial engineering and Chemical Research, 32, 2305-2317.
3. N.K.Vail, B.Balasubramanian, J.W Barlow, and H.L Marcus, 1996, A thermal model of polymer degradation during selective laser sintering of polymer coated ceramic powder, Rapid Prototyping Journal, 2, 24-40.
4. G.Bugeda, M.Cervera, and G.Lombera, 1999, Numerical prediction of temperature and density distributions in selective laser sintering process, Rapid prototyping Journal, 5, 21-26.
5. S.Kolossov, E.Boillat, R.Glardon, P.Fischer, and M.Locher, 2004, 3D FE simulation for temperature evolution in the selective laser sintering process, International Journal of Machine Tools and Manufacture, 44,117-123.
6. N.K.Tolochko, M.K.Arshinov, A.V.Gusarov, V.I. Titov, T.Laoui, and L.Froyen, 2003, Mechanisms of selective laser sintering and heat transfer in Ti powder, Rapid Prototyping Journal, 9,314-326.
7. A.V. Gusarov, T.Laoui, L.Froyen, and V.I. Titov, 2003, Contact thermal conductivity of a powder bed in selective laser sintering, International Journal of Heat and Mass transfer, 46, 1103-1109.
8. S.S. Sih, 1996, The thermal and optical properties of powders in selective laser sintering, Ph.D. thesis, University of Texas at Austin, Austin.
9. P.Fischer, H.Leber, V. Romano, H.P Weber, N.P Karapatis, C.Andre, and R. Glardon, 2004, Microstructure of near-infrared pulsed laser sintered titanium samples, Applied Physics A, 78, 1219-1227.
10. Y.S. Touloukian, and C.Y. Ho, 1970, Thermophysical Properties of Matter, The TPRC Data series, New York-Washington.
11. G.C. Kuczynski, 1949, Self-diffusion in sintering of metallic particles, Journal of Metals, 1(2), 169-178.
12. I.S. Grigoriev, and E.Z. Meilikhov, 1991, Physical Quantities Handbook, Energoatomizdat, Moscow.
13. M.J. Donachie, 2000, Titanium-A technical guide, ASM International, Ohio.

98

Harmonic and Transient Electromagnetic Analysis for an Eddy Current Sensor

T. KATHIRVEL[1], BRUCE MAXFIELD[2], C.V. KRISHNAMURTHY[2] AND KRISHNAN BALASUBRAMANIAM[1*]

[2] Centre for Non-destructive Evaluation

[2] Department of Mechanical Engineering, IIT Madras, Chennai-600 036, India
email: kathir.thiyagarajan@iitm.ac.in, balas@iitm.ac.in

ABSTRACT

Harmonic and transient electromagnetic analyses are based on the magnetic vector potential (MVP) formulation. 2D axisymmetric modeling of eddy-current Non-destructive Evaluation (NDE) is carried out analytically for analyzing the eddy-current probes by harmonic and transient excitation. MVP formulation uses the Maxwell's equation as the basis for the magnetic field analysis. In harmonic magnetic analysis, boundary conditions and applied loads vary sinusoidal with respect to time (i.e., single frequency excitation) and transient magnetic analysis involves variation in boundary conditions and loading as a function of time (i.e., pulsed excitation containing various frequencies). Since most commercial probes use ferrite cores, the model also calculates the induced magnetization in the ferrite core which acts as secondary source for the electromagnetic field where the primary source is the externally applied excitation current to the coil [1]. Hence, the model developed can be used for both air cored coils as well as ferrite cored coils. The variation of conductivity, frequency and lift off parameters are studied in the model for both harmonic and transient excitation and the results are discussed.

Keywords: Non-destructive Evaluation, Eddy-current testing, Harmonic excitation and Transient excitation.

1. INTRODUCTION

Mathematical models are used to stimulate the eddy-current phenomenon and its application in non-destructive testing. Model based quantitative eddy-current testing has evolved gradually with improvements in computing power. A focus on accurate modeling had led to a thorough understanding of eddy-current testing. Modeling is performed by solving the Maxwell's equations and the solution can be expressed either analytically or numerically [2]. In this paper volume integral equation in addition with method of moments (for discretizing the ferrite cores in the eddy-current probes) are used to obtain a semi analytical solution. Other approaches such as finite element, boundary element and finite volume method results in a numerical solution. Integral equation approaches are most efficient because a large proportion of the computation is treated analytically by using Green's functions which automatically account for many boundary and interface conditions. This mathematical model developed can be used for coil design, test frequency selection and interpretation of test data. The model computes the induced eddy- current distribution in the specimen as well as the resulting impedance change in the coil. Calculation and visualization of impedance plane loci are used for comparison with actual test signals.

2. MATHEMATICAL FORMULATION

The mathematical model describing the eddy-current problem in the conductors at low frequencies is given by the quasi-static Maxwell equations. The model efficiently solves for the distribution of induced eddy-currents in the specimen using a volume integral equation. In volume integral equation, the electric field at a point is calculated by summing the effects of the sources at all other points including the field points. In summing these effects the sources must be weighted by a function that depends upon the distance between the source and field points. This function is called Green function. The Green's functions doesn't formulate for ferrite core regions, hence, the integral equation for the magnetization of the ferrite is discretized into regular grids as shown in the Fig. 1. The discrete equations are formed using the method of moments and solved by biconjugate gradient methods.

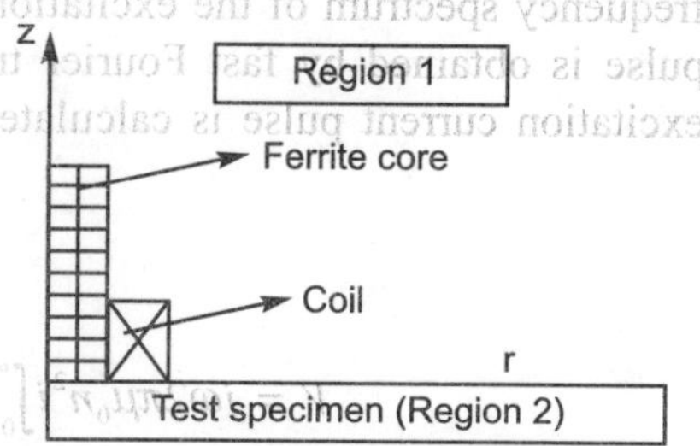

Fig. 1. 2D axisymmetric EC Model (Coil with Ferrite core).

2.1. Harmonic Excitation

A model of two-dimensional axisymmetric coils with air core or ferrite core in the presence of a conducting half space is developed. The half space is accounted for by computing the appropriate Green's functions. The problem consists of a cylindrical ferrite core, excited by a coaxial coil and in the presence of conducting half-space (Fig. 1). Since, most commercial probes use ferrite cores, the model calculates the induced magnetization in the ferrite core which acts as secondary source for the electromagnetic field where the primary source is the externally applied excitation current to the coil. Hence, the induced magnetization in the core depends on the excitation current applied to the coil. The magnetization or induced magnetic field of the ferrite core is calculated by introducing equivalent Amperian currents within the core which is taken into account while deriving the volume integral equation. The volume integral equation is transformed by method of moments into a vector matrix equation which is then solved using biconjugate gradient methods. The solution for MVP is determined in terms of Green's function by using Green's vector identity and scalar triple product. The boundary conditions for obtaining the solutions are continuity of MVP and its tangential components across the various regions. Green's function is the response to a current filament of

radius r' located on the plane $z = z'$. The Green's function notation $G_{ij}\left(r,z;r',z'\right)$ denotes that the

field point $\left(r,z\right)$ to be calculated lies in the region i and the source point $\left(r',z'\right)$ lies in the region j. The integral equation for the induced magnetization in the ferrite derived from the curl of MVP is discretized using method of moments and a vector matrix equation is formed which is solved using biconjugate gradient method. The impedance of the coil is derived from the MVP consisting of two terms viz. the EMF induced into the coil in the absence of core i.e., self-EMF of the coil and the second term is the additional voltage induced into the coil due to induced magnetization within the core.

2.2. Transient Excitation

A rectangular or square pulse is used as a current excitation in the pulsed eddy-current (PEC) technique [3]. The superposition of delta function coil is considered for the modeling of the rectangular cross section coil in harmonic analysis. For transient analysis different frequencies are superimposed to obtain the response of a PEC system. The transient response of the PEC system is computed by applying the inverse Fourier transform (IFFT) to the product of voltage frequency spectrum and frequency spectrum of the excitation current pulse. The frequency spectrum of the excitation current pulse is obtained by fast Fourier transform (FFT). The voltage response at each frequency of the excitation current pulse is calculated to obtain the voltage frequency spectrum.

$$Z = \frac{V}{i}$$

$$V = j\omega 2\pi\mu_0 n^2 i \int_0^\infty \left[\frac{RF_1^2(z_1,z_2) + F_3(z_1,z_2,z_1,z_2)}{2\alpha_0} \right] \frac{I^2(r_1 l, r_2 l)}{l^3} dl$$

$$+ \qquad\qquad\qquad\qquad\qquad ...(1)$$

$$j\omega 2\pi \left(1 - \frac{\mu_0}{\mu}\right) n \sum_{j=1}^{N_c} \frac{\left(b_j^{(r)} F_j^{(r)} + b_j^{(z)} F_j^{(z)}\right)}{\mu_0 n i}$$

where, ω-Angular frequency, μ_0-Permeability of free space, (Henry/m),
μ-Permeability of ferrite, (Henry/m), ε_0-Permittivity of free space (Farads/m),
N_c-Total number of cells discretized in r and z direction, n – No. of turns in the coil

i-Excitation current to the coil, $R = \left(\dfrac{\alpha_0 - \alpha_1}{\alpha_0 + \alpha_1}\right)$, $\alpha_0 = \sqrt{l^2 - k_0^2}$, $k_0^2 = \omega^2 \mu_0 \varepsilon_0$, $\alpha_1 = \sqrt{l^2 - k_1^2}$,

$k_1^2 = \omega^2 \mu_0 \varepsilon_0 - j\omega\mu_0\sigma$, σ-Conductivity of half space, (Siemens/m),

$I(r_1 l, r_2 l) = \int_{r_1 l}^{r_2 l} x J_1(x) dx$ $J_1(x)$, -First order Bessel function of first kind,

$F_1(z_1,z_2) = \int_{z_1}^{z_2} \exp(-\alpha_0 x) dx$, $F_3(z_1,z_2,c,d) = \int_{z_1}^{z_2} dx \int_c^d \exp\left(-\alpha_0 |y - y'|\right) dy$ $b_j^{(r)}$ and $b_j^{(z)}$ are the r and

z components of the induced magnetic field in the ferrite core.

$F_j^{(r)}$ and $F_j^{(z)}$ are the r and z components of the forcing function vectors required for the vector

matrix equation.

The model also calculates the impedance of the coil for both harmonic and transient excitation proposed in [4] were the differential equations for MVP are derived using variable separation method from the Maxwell's equations and the 'closed-form' solutions are obtained in the form of integrals of first order Bessel functions resulting in MVP, from which the other electromagnetic quantities such as radial and tangential magnetic flux densities, induced eddy- currents and, induced voltages are derived and visualized. The constant in the MVP solution are determined by enforcing the boundary conditions viz. vector potential and its tangential components are continuous across the boundaries. The base of the coil is at a height l_1 above the half-space surface. The coil parameters of importance are number of turns (n), inner and outer radii r_1 and r_2, and coil length ($l_2 - l_1$). The conductivity of first layer is σ_1 and for the second layer σ_2. Both the layers can be of either magnetic or non magnetic accordingly the equation for $C(\alpha)$ changes. The thickness of the first layer is denoted by c. Eq. (1) and Eq. (2) for the coil impedance are solved numerically using Gauss Legendre polynomials for 80 points [5].

The impedance for the rectangular cross-section coil above the conducting half space is given by

$$Z = K \int_0^\infty \frac{1}{\alpha^5} I^2\left(r_2, r_1\right)\left[2L + \frac{1}{\alpha}\left\{A(\alpha) + B(\alpha)C(\alpha)\right\}\right]d\alpha \qquad \qquad \text{...(2)}$$

The air impedance is derived from above equation where σ_i are zero which leads $C(\alpha)$ to zero

$$[\alpha_i = \alpha] \quad Z_{air} = K \int_0^\infty \frac{1}{\alpha^5} I^2\left(r_2, r_1\right)\left[2L + \frac{1}{\alpha}A(\alpha)\right]d\alpha$$

where, $L = l_2 - l_1$, $\quad K = \dfrac{j\omega\pi\mu n^2}{L^2\left(r_2 - r_1\right)^2}$, $\quad I\left(r_2, r_1\right) = \int_{\alpha r_1}^{\alpha r_2} x J_1(x)dx$, $\quad A(\alpha) = \left(2\exp[-\alpha L] - 2\right)$,

$$B(\alpha) = \left(\exp(-\alpha l_2) - \exp(-\alpha l_1)\right)^2$$

$$C(\alpha) = \frac{\left((\alpha + \alpha_1)(\alpha_1 - \alpha_2) + (\alpha - \alpha_1)(\alpha_1 + \alpha_2)\exp(2\alpha_1 c)\right)}{\left((\alpha - \alpha_1)(\alpha_1 - \alpha_2) + (\alpha + \alpha_1)(\alpha_2 + \alpha_1)\exp(2\alpha_1 c)\right)} \quad \text{for a two layer conducting half space.}$$

$$\alpha_i = \sqrt{\alpha^2 + j\omega\mu\sigma_i}$$

3. NUMERICAL RESULTS

3.1. Harmonic Excitation

Figure 2 is the impedance plot for varying the frequencies and lift-off (distance between the surface of the conducting material to the bottom surface of the coil) computed by Eq. (1) for air cored coil. The impedance is normalized, using the inductive reactance of the coil in air as the normalizing factor. Fig. 2 can demonstrate the optimum frequency for a specific test. The optimum frequency produces the best phase difference between the loci of two parameters. The conducting half-space is aluminum ($\sigma = 35.4$ MS/m) and the solid curve represents the locus produced by varying the excitation frequency. The dashed lines are the lift-off curves and represent the impedance variation with coil lift-off. Since, the conductivity and frequency appears as a product in Eq. (1) and Eq. (2) the same curve is produced for constant excitation frequency and a varying conductivity as shown in the Fig. 3 without normalization. The computation of coil impedance (Eq. 1) based on Green's function formulation is validated by computing the same using the Dodd and Deeds solution (Eq. 2). Figure 4 shows the impedance display for the same parameters except a cylindrical ferrite core is introduced into the coil. Due to the presence of ferrite core, there is a change in the resistive and inductive reactive components of the coil impedance. The resistance and inductive reactance for the ferrite core are lesser than for the air cored coil for an each point in the impedance display. Figure 5 shows the difference between the ferrite cored and air cored coil for Aluminum half-space at various frequencies at a lift-off of 1mm. The parameters used in the computation for harmonic excitation are inner radius of the coil ($r_1 = 2$ mm), outer radius of coil ($r_2 = 4$ mm), coil width ($L = 1$ mm), number of turns in the coil is 50, relative permeability ($\mu_r = 1$ for air and $\mu_r = 1000$ for ferrite), conductivity of half spaces are 35.4MS-m^{-1} (Aluminum) (Fig. 2), 0.1276 MS-m^{-1} (Graphite), 0.58 MS-m^{-1} (Titanium), 1.45 MS-m^{-1} (Stainless steel), 6.38 MS-m^{-1} (Bronze), 20.3 MS-m^{-1} (Aluminum) and 60.09 MS-m^{-1} (Copper). Lift off (l_1) varied are 0 mm, 0.25 mm, 0.5 mm & 1mm. Frequencies used for excitation of the coil are 100 Hz, 316 Hz, 1 kHz, 3.16 kHz, 10 kHz, 31.6 kHz, and 100 kHz.

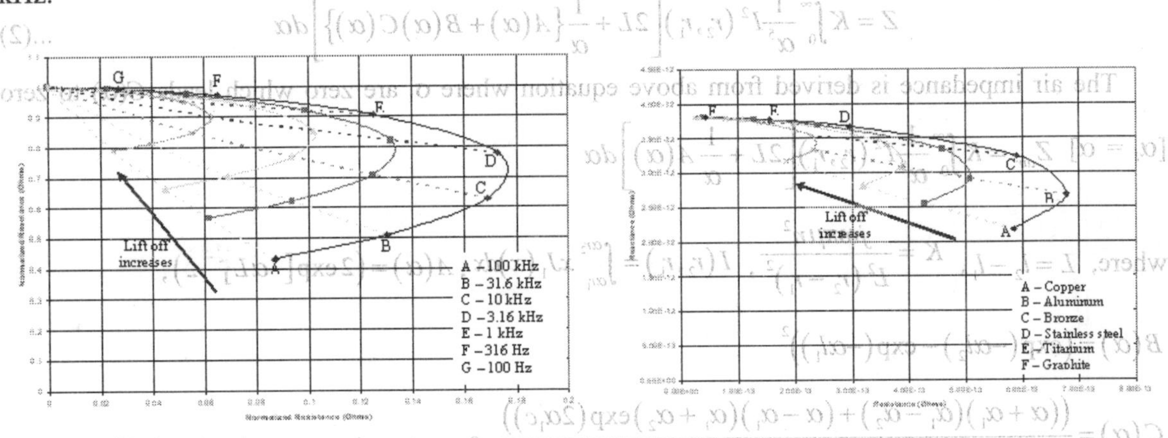

Fig. 2. Air cored coil Impedance display for various frequencies and lift off by Eq. (1)

Fig. 3. Air cored coil Impedance display for frequency (10 kHz), various conductivities and lift-off

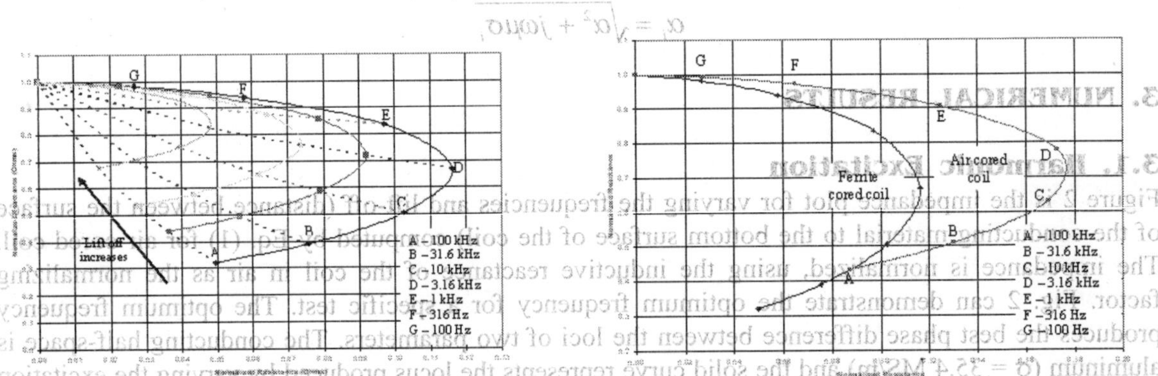

Fig. 4. Ferrite cored coil Impedance display for various frequencies and lift off.

Fig. 5. Impedance display for Ferrite cored (pink) and air cored coil (dark blue) for various frequencies and lift off of 1 mm.

3.2. Transient Excitation

PEC response for different coating thickness is obtained using the model [6-8]. The coil impedance given in Eq.2 is used for calculating the transient response of PEC system to an input square wave excitation as shown in the Fig. 6. The current difference in the frequency domain $\Delta I(\omega)$, is obtained by multiplying admittance difference (ΔY) by the input voltage spectrum $V(\omega)$ and inverse Fourier transform of $\Delta I(\omega)$ is taken to obtain the transient current response. The impedance for the coating layer is subtracted from the impedance of the half-space which results in the transient response. As the thickness of the coating increases, peak height of the transient response increases as shown in the Fig.7. The parameters used in the computation for transient excitation are inner radius of the coil (r1 = 2.75 mm), outer radius of the coil (r2 = 5.64 mm), coil width (L = 2.65 mm), number of turns in the coil is 638, relative permeability ($\mu_r = 1$ for air), thickness of the coating layer are 0.1 mm, 0.2 mm, 0.4 mm and 1 mm. Conductivity of coating layer $\sigma_1 = 2.03$ MS-m^{-1} (Titanium), conductivity of half space $\sigma_2 = 23.2$ MS-m^{-1} (Aluminum) and lift off (l_1) of 0.33 mm.

$$\Delta Y = \frac{1}{Z_L} - \frac{1}{Z_{HSP}}, \quad \Delta I(\omega) = \Delta Y(\omega)V(\omega), \quad \Delta i(t) = IFT\left[\left(\Delta I(\omega)\right)\right] \quad \ldots(3)$$

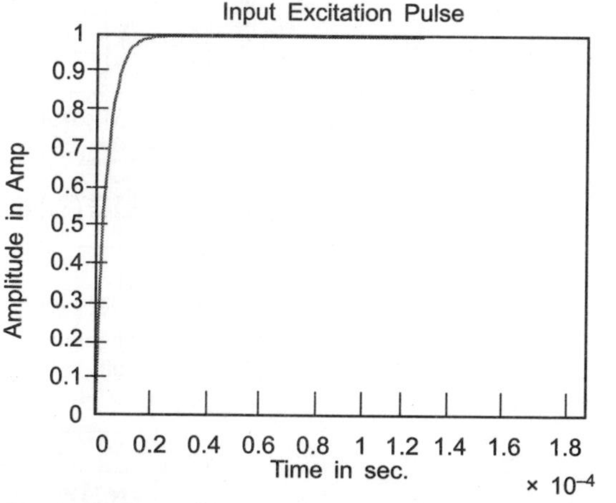

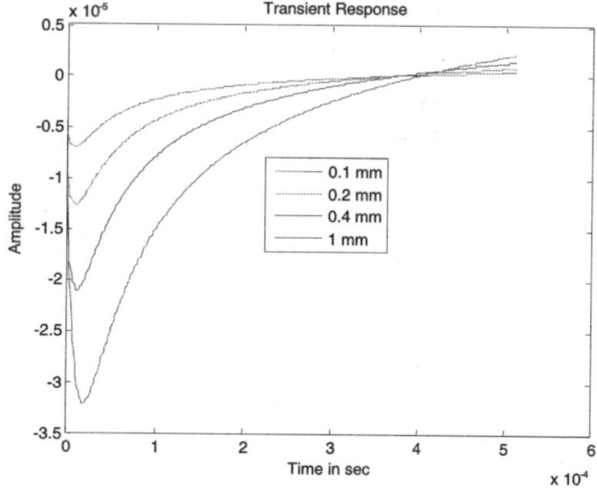

Fig. 6. Input excitation current to the coil for 512 µsec.   **Fig. 7.** Transient response for different coating thicknesses.

CONCLUSION

A two-dimensional axisymmetric model with air core or ferrite core coil in the presence of a conducting half space is developed [9-10]. In case of harmonic analyses, the variation of the coil impedance is shown by varying the conductivity of the half-space, lift-off parameters and various frequencies. The analytical expressions derived for the MVP and other electromagnetic quantities such as induced eddy-currents, magnetic flux densities can be computed and visualized. The coil impedances computed using Green's functions formulation fully agrees with the values computed from the Dodd and Deeds solution. In transient analysis, the response is obtained from the calculation of coil impedance given in Eq. (2) for different coating thickness layers. The results for transient excitation are to be validated.

REFERENCES

1. H.A. Sabbagh, 1987, "A model of eddy-current probes with ferrite cores", IEEE Transactions on Magnetics, 23(2), 1888-1904.
2. S.S. Udpa, 2004, Non-destructive Testing Handbook-Electromagnetic Testing, Third Edition, 5, ASNT, Chap-4.
3. J. Bowler, 1997, "Pulsed eddy-current response to a conducting half-space", IEEE Transactions on Magnetics, 33(3), 2258-2264.
4. C.V. Dodd and W.E. Deeds, 1968, "Analytical solutions to eddy-current probe-coil problems", Journal of Applied Physics, 39(6), 2829-2838.
5. M. Abramowitz and I. Stegun, 1972,"Handbook of Matematical Functions with Formulas, Graphs, and Mathematical Tables" Dover Publications, Newyork, Chap-25, 916-919.
6. C.C. Tai, J.H. Rose and J.C. Moulder, 1996, "Thickness and conductivity of metallic layers from pulsed eddy-current measurements", Review of Scientific Instruments, 67(11), 3965-3971.
7. H.C. Yang and C.C. Tai, 2002, "Pulsed eddy-current measurement of a conducting coating on a magnetic metal plate", Measurement Science and Technology, 13, 1259-1265.
8. H.C. Yang, 2003, "The Design and Applications of Eddy-Current Non-destructive Inspection System", Ph.D. thesis, National Cheng Kung University, Taiwan.
9. J.C. Moulder, E. Uzal, J.H. Rose, 1992, "Thickness and conductivity of metallic layers from eddy current measurements", Review of Scientific Instruments, 63(6), 3455-3465.
10. H.L.Libby, 1971, "Introduction to Electromagnetic Non-destructive Test Methods", Wiley-Interscience, New York.

99

Analysis of Electro-Mechanical Deformation

S.N. KHADERI AND S. BASU

Department of Mechanical Engineering, Indian Institute of Technology Kanpur-208 016, India.
email: syed@iitk.ac.in, sbasu@iitk.ac.in

ABSTRACT

When a deformable body is placed under the influence of an electric field it experiences a system of stress within it. To equilibrate these stresses the body deforms. Generally, these stresses are small and generally neglected. Problems where the influence of electrostatic forces cannot be neglected are miniature switches used in MEMS, soft structures like thin films and cases like the ultra high voltage insulation where high electrical fields are involved. The current work involves developing a general purpose Finite Element Code to analyze such deformations. As a case study the deformation of a cylinder and a sphere in an uniform electric field are analysed.

Keywords: Electromechanics, FEM.

1. INTRODUCTION

In this paper on deformation of elastic dielectrics Toupin [1] using virtual work principle derived the constitutive relations for the mechanical stresses and electrical field. Non-linear equations for thermoelectroelasticity were given by Tiersten [2]. Both these formulations do not have the flexibility to incorporate different form of the electrical body force and electrostatic stress. In this work the electromechanical formulation given by McMeeking and Landis [3] is used. Here no specific form of the body force or the electrostatic stress is assumed. Depending on the material in question, the electrostatic stress and the body force will take specific form.

The deformation analysis is performed using Finite element method. Both material and geometric non-linearities are incorporated in the analysis. The crossed triangle quadrilateral design of elements is used to circumvent problem arising out of incompressibility.

2. FORMULATION

2.1 Electrostatic Governing Equation

The Maxwell's equations form the governing equations for the electrostatic problem, which when combined results in a Poisson's equation

$$\nabla \cdot \nabla \phi = \rho / \varepsilon \,, \qquad \qquad ...(1)$$

where, ϕ is the electrostatic potential (Volts), ρ is the space charge density (coloumb/m^3) and ε is the dielectric permittivity. All the equations are solved in the reference configuration and equation (1) in the reference configuration becomes

$$\nabla_0 \cdot F^{-1} \cdot F^{-T} \cdot \nabla_0 = -\frac{\rho}{\varepsilon} \qquad \qquad ...(2)$$

where, F is the deformation gradient and ∇_0 is the gradient operator in the reference configuration. Equation (2) is solved for the potentials and then the electric field in the domain of analysis.

2.2 Mechanical Governing Equation

In this work a total Lagrangian formulation of the mechanical boundary value problem is employed. The virtual work rate equation, which is an alternative form of representing equilibrium of a body, in reference configuration, is expressed as

$$\int_V S^{ij} \delta \dot{\eta}_{ij} \, dV + \int_S T^i \delta v_i \, dS = 0 \qquad \qquad ...(3)$$

where, S^{ij} are the components of the Piola-Kirchoff stress tensor, η_{ij} are the components of Green-Lagrange strain tensor, T^i is the traction applied in the reference configuration and v_i is the velocity.

2.3 Mechanical Constitutive Relations

Rate tangent formulation of Pierce et.al. [4] is employed in this work for the mechanical constitutive relations. The Jaumann rate of mechanical Kirchoff's stress is

$$\overset{\triangledown}{\tau}_{mech} = L^{\tan} : D \qquad \qquad ...(4)$$

where, D is the rate of deformation tensor and $L^{\tan}$ is the tangent modulus.

2.4 Electrical Constitutive Relation

The formulation of electro-mechanial boundary value problem by McMeeking and Landis [3] is employed in this work. For isotropic conservative material the electrostatic stress is given by

$$\tau_{es} = \varepsilon \left(EE - \frac{1}{2} E^2 I \right)$$

The sum of mechanical stress and electrical stress $\tau = \tau_{es} + \tau_{mech}$ obeys the usual laws of equilibrium [3].

3. A CYLINDER IN A UNIFORM ELECTRICAL FIELD

A dielectric cylinder in a uniform electric field is analysed for deformation and stresses. The problem is defined as follows: If a dielectric cylinder(material 1) of permittivity ε_1, immersed in an infinite medium(material 2) permittivity of ε_2, is exposed to a uniform electric field E_0, what will be the distribution of stresses induced inside the cylinder?

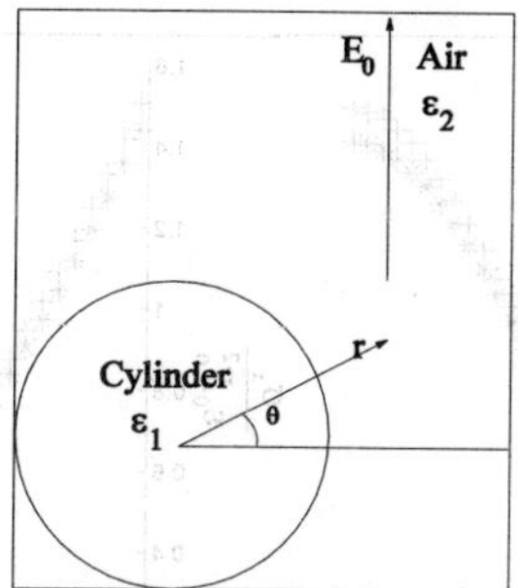

Fig. 1. Dielectric sphere in an uniform electric field.

The analytical elastic solutions for the stresses and displacements (obtained by choosing a suitable Airy's stress function) in the cylinder are

$$\sigma_r = \frac{11k}{2} - \frac{9k}{2}\cos 2\theta,$$

$$\sigma_\theta = \frac{11k}{2} - 9k\left(\frac{1}{2} + \frac{r^2}{a^2}\right)\cos 2\theta,$$

$$\sigma_{r\theta} = \frac{9k}{2}\left(1 - \frac{r^2}{a^2}\right)\sin 2\theta,$$

$$2\mu u_r = \frac{11k}{2}r(1-2v) - \frac{9k}{2}r\cos 2\theta + \frac{3kv}{a^2}r^3\cos 2\theta,$$

$$2\mu u_t = \frac{9k}{2}r\sin 2\theta - (6-4v)\frac{3k}{4a^2}r^3\sin 2\theta,$$

$$...(5)$$

where, μ is the modulus of rigidity, u_r and u_θ are the displacements in the radial and tangential direction.

3.1 Results

In this section, we use the analytical solution obtained in the previous section to benchmark our FE code. To this end, a linear elastic cylinder is modeled. Only one half of the cylinder with appropriate symmetry conditions at x = 0 needs to be analysed. Plane strain conditions are assumed. The material outside the cylinder is assumed to be air having an elastic modulus 10 times lower than the cylinder. The constant ambient field is applied by assuming rigid capacitor plates at $y = \pm H$. Then $\phi(H) = \phi_0$ and $\phi(-H) = -\phi_0$ leading to $E_0 = \phi/h$. The height and width of the domain have to be large enough so that the uniform ambient field conditions are faithfully reproduced far away from the cylinder.

The variation of stresses in the cylinder are compared with the analytical forms (equations 1.7) plotted in Fig. 2(a) and Fig. 2(b). Fig. 2(a) shows the variation of σ_{xx} with θ along the circumference r = a. The parabolic variation of σ_{xx} along $\theta = \pi/2$ is also shown in Fig. 4(b).

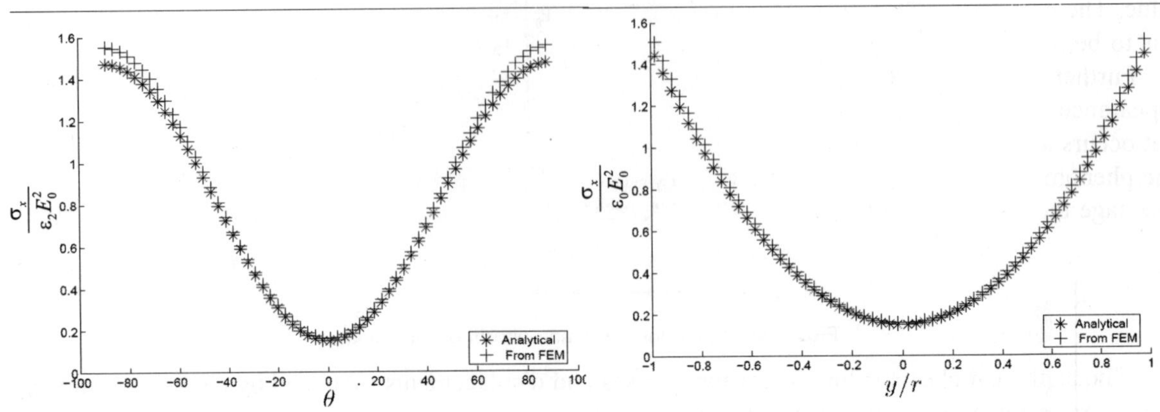

(a) Variation of σ_{xx} with θ at $r = 2$.   **(b)** Variation of σ_{xx} with r at $\theta = 90$.

Fig. 2. Stress variation in cylinder.

The analytical value of σ_{xx} is found by appropriately transforming the stresses in equation (5) to Cartesian coordinate system. From Fig. 2(a) and 2(b) it is clear that the mismatch between the analytical and numerical results are highest around $\theta = 90°$. This is due to the inadequate resolution of highly non-uniform electric field in this region.

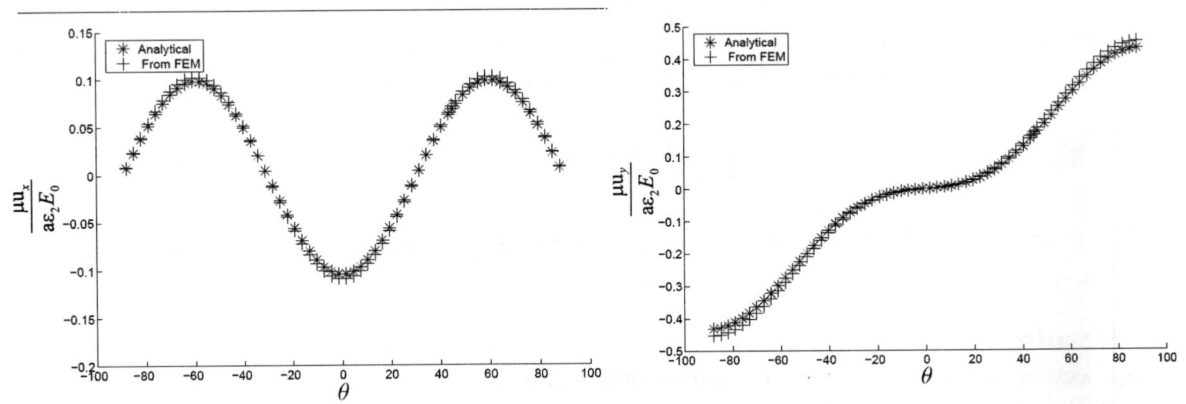

Fig. 3. Variation of u_x along circumference.   **Fig. 4.** Variation of u_y along circumference.

Finite element solution of the u_x and u_y components of the displacement are compared in Fig. 3 and Fig. 4 against the analytical solution from equations (5). The deformation of the interface is well captured by the finite element solution.

4. SPHERE IN AN UNIFORM ELECTRIC FIELD

To see what happens when an elasto-plastic sphere is placed in an uniform electric field, we do a fully non-linear analysis (geometric and material non-linearities). At low fields the stress pattern inside the sphere is that the σ_{yy} is constant along y axis (not shown). As the applied field is increased (Fig. 4(a),

4(b)), the tractions at the poles of the sphere increase and the stress at the poles increase to a high value. The core of the sphere as well as the poles of the sphere enter into the plastic regime. However, it is to be noted that the plastic strain in the core is much higher than the plastic strains at the poles.

Further application of fields (Fig. 4(c), 4(d)) induce large plastic strains at the poles. The sudden appearance of this unusually high plastic activity at the poles is likely to be an interfacial instability that occurs at the sphere-air interface. This leads to the interface getting corrugated close to the poles. The phenomenon is unstable and the corrugation pattern is not symmetric at the two poles (Fig. 5). At this stage the simulations have to be stopped as the stiffness matrix does not remain positive definite.

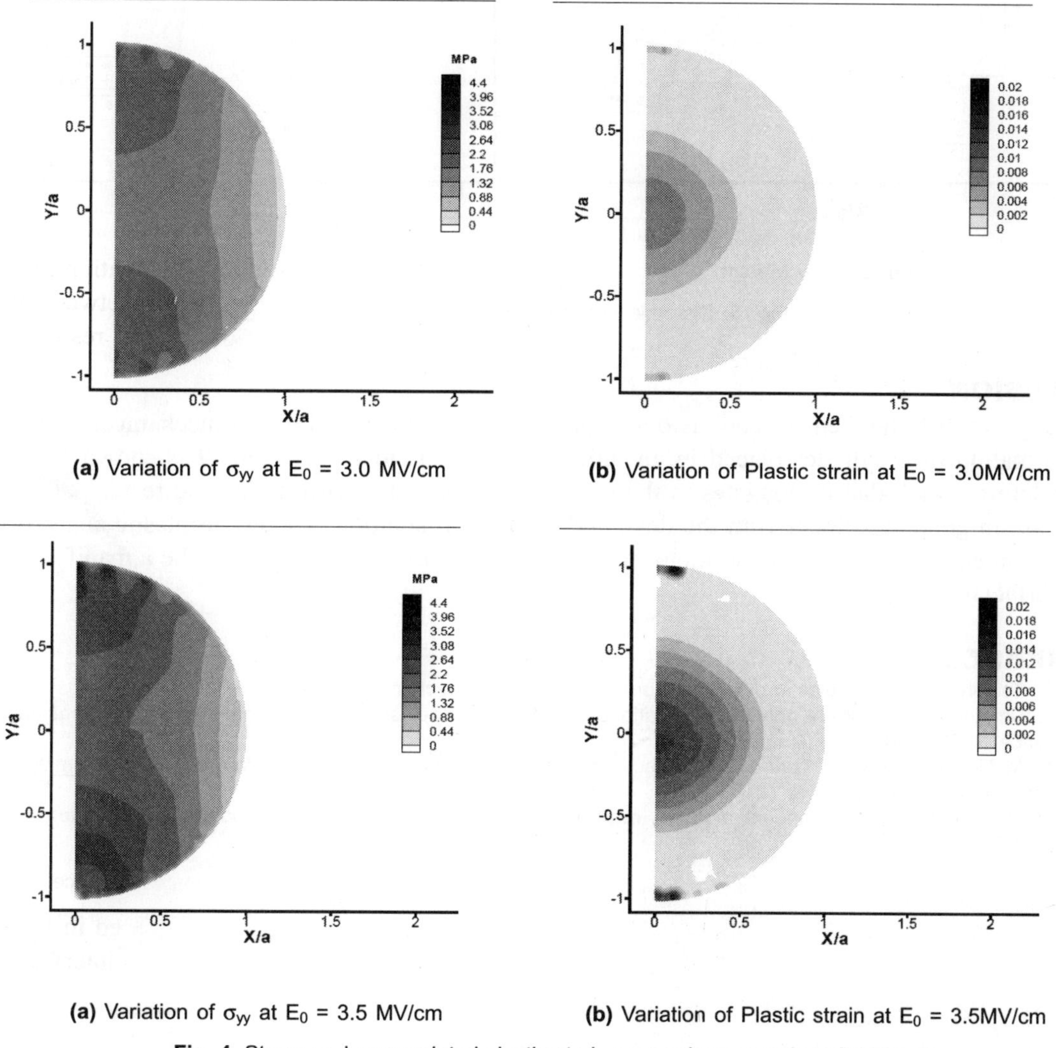

(a) Variation of σ_{yy} at E_0 = 3.0 MV/cm

(b) Variation of Plastic strain at E_0 = 3.0MV/cm

(a) Variation of σ_{yy} at E_0 = 3.5 MV/cm

(b) Variation of Plastic strain at E_0 = 3.5MV/cm

Fig. 4. Stress and accumulated plastic strain comparison at various field levels.

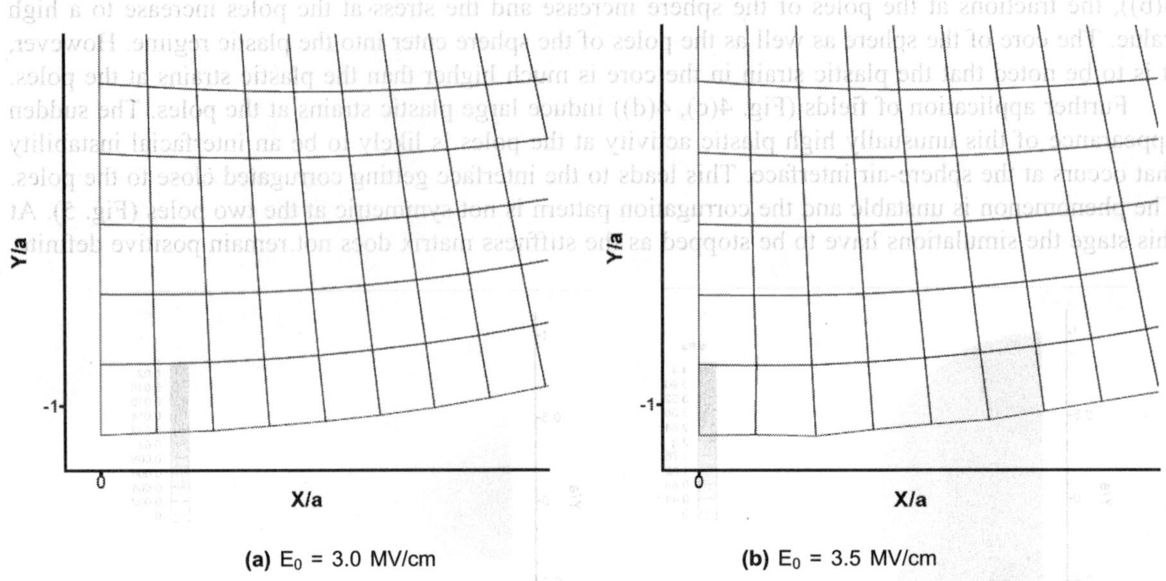

(a) E_0 = 3.0 MV/cm   **(b)** E_0 = 3.5 MV/cm

Fig. 5. Distorsion of mesh in plastic regime.

CONCLUSION

A general purpose finite element code is developed to solve the coupled electromechanical problem. The deformation of a cylinder placed in an uniform electric field is analysed. It is shown that the cylinder deforms such that it elongates in the direction of applied electric field. The results from the FE code are in good agreement with the developed analytical solution. The elasto-plastic analysis of a sphere in an electric field shows enormous plastic straining at the poles. This could be a manifestation of an interfacial instability.

REFERENCES

1. R.A. Toupin, 1956, The elastic dielectric, Journal of Rational Mechanics and Analysis, 5, 849-916.
2. H.F. Tiersten, 1971, On the non-linear equations of thermoelectroelasticity. International Journal of Engineering Science, 9, 587-604.
3. R.M. McMeeking and C. M. Landis, 2005, Electrostatic forces and stored energy for deformable dielectric materials, Journal of Applied Mechanics, 72, 581-590.
4. D. Pierce *et al.*, 1984, A tangent modulus method for rate dependent solids, Computers and Structures, 18, 875-887.

100

Control of a Rotating Beam with Tip Rotor Using a Coupled Electro-Mechanical Approach

N.K. Chandiramani[1] and L.I. Librescu[2]

[1]*Department of Civil Engineering, Indian Institute of Technology
Bombay Mumbai 400076, India. email: naresh@civil.iitb.ac.in*
[2]*Department of Engineering Science and Mechanics, Virginia Tech.
Blacksburg, VA 24061, USA. email: librescu@vt.edu*

ABSTRACT

Active control of a thin-walled rotating beam with pretwist, double-taper, and a tip rotor, is considered using the higher-order shear deformation theory (HSDT). The beam comprises an orthotropic host with surface-embedded transversely isotropic (PZT-4) sensor-actuator pairs. Span- and thickness- wise variation is considered for the electric field applied to actuators. This yields a coupled electro-mechanical system, wherein displacement variables are coupled via the electric field. Optimal LQR control with state feedback is used to obtain the control input (charge density applied to actuators). Parametric studies involving ply-angle, rotation speeds of beam and rotor, pretwist, taper, rotor mass, and saturation constraint on actuator voltage, are performed. The present model yields an order-of-magnitude reduction in settling time and control voltage/power, and lower response, vis-a-vis the decoupled approach.

Keywords: Smart composite structures, Shear deformation, Optimal control.

1. INTRODUCTION

Fiber-reinforced composites with embedded piezoelectric elements provide a synthesis of passive and active control. Design of optimal controllers—yielding reduced settling time and control energy—require an accurate plant model incorporating shearability, satisfaction of traction free boundary conditions (BCs), warping restraint, etc. Kim and White [1] analyzed a non-rotating thick-walled beam using a cubic variation of axial displacements to satify the traction free BCs. Eigenvibration analyses for rotating blades were done by Jung et. al. [2] within the First-Order Shear Deformation Theory (FSDT), and by Chandiramani *et al.* [3] considering a composite blade and the HSDT. Optimal control of flexural vibration in a composite plate was studied by Ray [4] using output feedback and coupled charge-mechanical equations. Herein, a HSDT model for pretwisted, composite blades [3] is extended to include a tip rotor, double-taper, and spanwise distributed PZT-4 sensor-actuator pairs. The coupled electromechanical system is solved to obtain the optimal control input.

2. FORMULATION

Consider a straight, pretwisted, doubly-tapered, single celled box beam with tip rotor (mass m_R, spin speed $\bar{\Omega}\mathbf{k}_R$) mounted on a rigid hub (radius R_0, rotation speed $\Omega\mathbf{J} = \Omega j$) (Fig. 1). The beam-fixed coordinate system (x, y, z) originates at the beam root, (s, n, z) is a local (surface) coordinate system, and (x^p, y^p, z^p) are local coordinates along the cross-section principal axes. The beam-fixed and rotor-fixed bases are $(\mathbf{i}, \mathbf{j}, \mathbf{k})$ and $(\mathbf{i}_R, \mathbf{j}_R, \mathbf{k}_R)$, respectively. The coordinate relations are $\bar{x}[s,z] = \bar{x}^p[s]\cos\beta - \bar{y}^p[s]\sin\beta$; $\bar{y}[s,z] = \bar{x}^p[s]\sin\beta + \bar{y}^p[s]\cos\beta$; $z = z^p$ $z = z^p$ where the overbar denotes mid-surface (n = 0) quantities and $\beta[z] = \gamma + \beta_0 z/L + \beta_1 (z/L)^2$ is the quadratically varying pretwist.

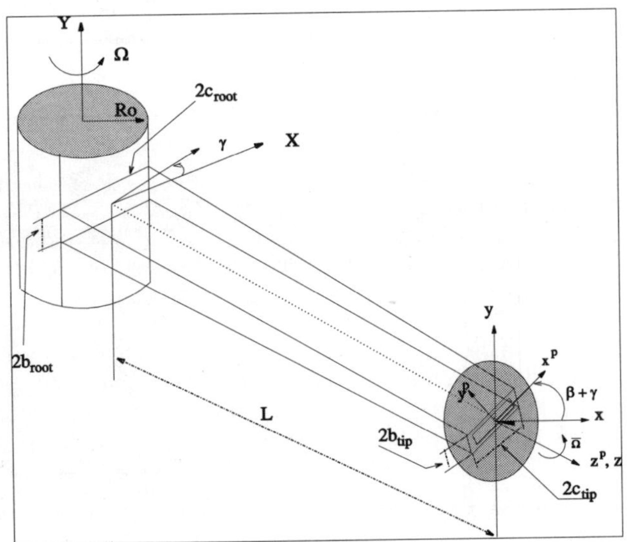

Fig. 1. Tapered beam with tip rotor.

Kinematics: The assumptions include a spanwise dependent twist rate, quadratic variation of transverse shear strains through wall thickness, secondary warping, and no in-plane cross-section distortion. Imposing the traction free BC's yields the transverse shear strain distributions $\gamma_{xz} = (1 - 4n^2/h^2)\bar{\gamma}_{xz}[z;t]$ and $\gamma_{yz} = (1 - 4n^2/h^2)\bar{\gamma}_{yz}[z;t]$ (HSDT). The lag ($u[x, y, z; t]$), flap ($v[x, y, z; t]$), and extensional ($w[x, y, z; t]$) displacements are obtained in terms of the corresponding displacements $u_0[z; t]$, $v_0[z; t]$, $w_0[z; t]$ of a reference point $0[0, 0]$ on the cross-section, the twist $\phi[z; t]$, and the rotations $\theta_x[z; t]$ and $\theta_y[z; t]$ about x and y axes, respectively [3].

Piezopatch distribution: The PZT-4 sensor and actuator patches are embedded on the bottom and top face, respectively, with their surface of isotropy being parallel to the mid-surface of the pretwisted beam (Fig. 2). The electric potential distribution is considered as $\Psi[s, n, z; t] = n\Psi_0[z; t]$, yielding

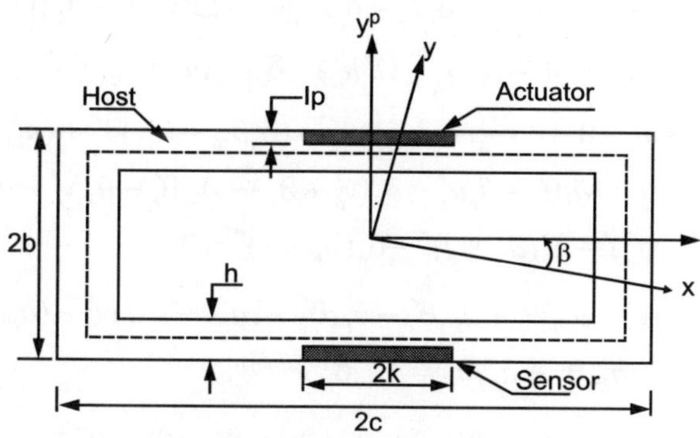

Fig. 2. Cross-sectional view.

the electric field $E_1 = -\partial\psi/\partial s = 0$, $E_2 = -\partial\psi/\partial z = -n\psi'_o$, $E_3 = -\partial\psi/\partial n = -\psi_o[z;t]$.

Stress Field: The stress-strain relation for a constituent beam-wall layer is,

$$\begin{Bmatrix}\sigma_{ss}\\\sigma_{zz}\\\sigma_{sz}\end{Bmatrix} = \begin{bmatrix}\bar{Q}_{11} & \bar{Q}_{12} & \bar{Q}_{16}\\\bar{Q}_{12} & \bar{Q}_{22} & \bar{Q}_{26}\\\bar{Q}_{16} & \bar{Q}_{26} & \bar{Q}_{66}\end{bmatrix}\begin{Bmatrix}\varepsilon_{ss}\\\varepsilon_{zz}\\\gamma_{sz}\end{Bmatrix} - \begin{Bmatrix}\bar{\varsigma}_{31}E_3\\\bar{\varsigma}_{31}E_3\\0\end{Bmatrix}; \quad \begin{Bmatrix}\tau_{nz}\\\tau_{ns}\end{Bmatrix} = \begin{bmatrix}\bar{Q}_{44} & \bar{Q}_{45}\\\bar{Q}_{45} & \bar{Q}_{55}\end{bmatrix}\begin{Bmatrix}\gamma_{nz}\\\gamma_{ns}\end{Bmatrix} - \begin{Bmatrix}\bar{\varsigma}_{15}E_2\\0\end{Bmatrix} \quad ...(1)$$

where, $\bar{Q}_{ij}$ and $\bar{\varsigma}_{ij}$ are transformed reduced stiffnesses and reduced piezoelectric coefficients of the composite beam (details omitted for brevity). Introducing the strain field and integrating over the cross-section, these 3-D constitutive equations are reduced to a 1-D dependency, thus yielding the beam forces and moments.

Governing System: Hamiltons principle for the beam with rotor reads,

$$\int_{t_0}^{t_1}\left[\int_{\tau}\left(\delta\tilde{H}+\rho\ddot{\mathbf{R}}\cdot\delta\mathbf{R}\right)d\tau + \int_{\tau_R}\left(\rho_R\ddot{\mathbf{R}}_R\cdot\delta\mathbf{R}_R\right)d\tau_R - \int_{\zeta}\left(\bar{\sigma}_k\,\delta U_k - \bar{\sigma}\,\delta\psi\right)d\zeta\right]dt = 0 \quad ...(2)$$

Here, $\bar{\sigma}_k$, U_k, $\bar{\sigma}$, Ψ, $\tilde{H}\,(=0.5[\sigma_{zz}\varepsilon_{zz}+\tau_{sz}\gamma_{sz}+\tau_{nz}\gamma_{nz}]-\varsigma_{31}E_3\varepsilon_{zz}-0.5[\xi_{11}E_1^2+\xi_{22}E_2^2+\bar{\xi}_{33}E_3^2])$,

$\mathbf{R}\,(=[x+u]\mathbf{i}+[y+v]\mathbf{j}+[R_0+z+w]\mathbf{k})$, $\mathbf{R}_R\,(=u_o\mathbf{i}+v_o\mathbf{j}+(L+w_o+R_0)\mathbf{k}+x\mathbf{i}_R+y\mathbf{j}_R+(r_m+z)\mathbf{k}_R)$, and ξ_{ij} denote surface traction vector, displacement vector, applied surface charge density, electric potential, electric enthalpy, position vector for deformed beam, position vector for rotor, and dielectric constants (details omitted), respectively. A vertical (y-directed) line load, $p_y[z;t]$ is assumed. The resulting seven electromechanical equations of motion (EOM) and BC's exhibit a 1-D dependency on the spanwise (z) coordinate, and are in terms of displacement field variables u_0, v_0, θ_x, θ_y, w_0, ϕ, and electric potential Ψ_0. Since the beam is directed radially outward from the hub, Coriolis effects due to beam rotation (Ω) are negligible. The Circumferentially Uniform Stiffness (CUS) ply angle configuration is considered [3]. This yields a linearized and coupled system governing the motion (bending - transverse shear, i.e., flap-lag, and extension-twist) and the electric potential. The coupling occurs via Ψ_0. When a spatially constant electric potential is assumed, flap-lag and extension-twist motions decouple in contrast to the present case. Representative (u_0 and Ψ_0) EOM's are:

$$\delta u_o : \{[\delta_h I_1 + \delta_e I_7](\ddot{u}_o' - \Omega^2 u_o')\}' + \{[\delta_h I_5 + \delta_e I_9](\ddot{v}_o' - \Omega^2 v_o')\}' + \delta_h\{[I_3 - I_1](\ddot{\theta}_y - \Omega^2 \theta_y)\}' -$$

$$\delta_h\{[I_6 - I_5](\ddot{\theta}_x - \Omega^2 \theta_x)\}' - b_1(\ddot{u}_o - \Omega^2 u_o) - \delta_e(a_{22}u_o'' + a_{23}v_o'')'' + (b_1\Omega^2 Ru_o')' +$$

$$\delta_t\{(a_{44} + \delta_h\tilde{a}_1)(u_o' - \theta_y) + (a_{43} + \delta_h\tilde{a}_2)\theta_x' + (\delta_h\tilde{a}_{30} - a_{42})\theta_y' + (a_{45} + \delta_h\tilde{a}_{31})(v_o' + \theta_x)\}' +$$

$$\delta_h\{[\tilde{a}_3 v_o'' + \tilde{a}_{32}u_o'']' + [\tilde{a}_5\theta_y' - \tilde{a}_6 u_o'' - \tilde{a}_4(v_o' + \theta_x) - \tilde{a}_{34}\theta_x' - \tilde{a}_{35}v_o'' - \tilde{a}_{32}(u_o' - \theta_y)]''\} +$$

$$\delta_t(\tilde{a}_{P2}\psi_o - \tilde{a}_{P13}\psi_o')' - \delta_e(\tilde{a}_{P9}\psi_o)'' - \delta_h(\tilde{a}_{P8}\psi_o)'' = 0 \qquad \ldots(3)$$

$$\delta\psi_o : L_{P22}\psi_o'' - L_{P33}\psi_o + \tilde{a}_{P1}w_o' + \tilde{a}_{P4}\theta_x' - \tilde{a}_{P5}\theta_y' - (\tilde{a}_{P7} - \tilde{a}_{P6})v_o'' - (\tilde{a}_{P9} + \tilde{a}_{P8})u_o'' + \tilde{a}_{P10}\phi' -$$

$$\tilde{a}_{P2}u_o' - \tilde{a}_{P3}v_o' + \tilde{a}_{P4}\theta_x + \tilde{a}_{P2}\theta_y = a_{P10}\bar{\sigma} \qquad \ldots(4)$$

The BC's at the clamped end ($z = 0$) are $u_o = v_o = \theta_y = \theta_x = u_o' = v_o' = w_o = \phi = \phi' = \psi_o = 0$. Here ($\tilde{a}_{ij}[z]$, $a_{ij}[z]$) are global stiffnesses, the former including pretwist and HSDT effects, ($I_1, \cdots, I_9, I_{xx}^p, I_{yy}^p, I_{\omega\omega}^p, b_1, \hat{I}_P$) are structural and mass quantities, ($\tilde{a}_{Pi}, a_{Pi}$) are global piezoelectric coefficients, and $R[z]$ contains the centrifugal stiffening effect (details omitted).

The surface charge density on actuators due to applied voltage is $\bar{\sigma}[z;t] = \varphi_7^T[z]\hat{\sigma}[t]$, where $\hat{\sigma}[t]$ is the control input to be determined via LQR control. Considering $(u_0, v_0, w_0, \theta_x, \theta_y, \phi, \Psi_0) = (\varphi_1^T[z]\mathbf{q}_1[t], \cdots, \varphi_7^T[z]\mathbf{q}_7[t])$, using the extended Galerkin method [3], and eliminating the electrical degree of freedom ψ_o via Eq. (4), the discretized system $\mathbf{M\ddot{q}} + \mathbf{G\dot{q}} + \mathbf{Kq} = \mathbf{Q}[t] - \mathbf{F}\hat{\sigma}[t]$ results from the displacement governing equations. Here $\mathbf{q} = \{\mathbf{q}_1^T | \cdots | \mathbf{q}_6^T\}^T$. The quantities $\mathbf{G}|_{6N\times6N}, \mathbf{Q}|_{6N\times1}, \mathbf{F}|_{6N\times N}$, and $\hat{\sigma}|_{N\times1}$ represent the gyroscopic matrix due to rotor, external forcing, piezoelectrically induced forcing coefficients, and time dependent charge density vector applied on actuators, respectively.

3. OPTIMAL CONTROL

Sensor Output: Applying Gauss' law on the exposed surface of sensors, the total charge generated is $\tilde{q}_P = \int_z \oint_s D_3 S_s f_z \, ds \, dz \,|_{n=h/2}$. Since, no voltage is applied to sensors and $\varepsilon_{ss} = 0$, the n-component of electric displacement is $D_3 = \varsigma_{31}\varepsilon_{zz}$. Introducing the strain $\varepsilon_{zz}(= w')$ and performing the spatial discretization yields $\tilde{q}_P[t] = \mathbf{Cq}$. Here, $\mathbf{C} = [\mathbf{C}_1 | \cdots | \mathbf{C}_6]$, with $\mathbf{C}_i, i = 1, \cdots, 6$, being N-dimensional row vectors (details omitted). The sensor patches are treated as capacitors with capacitance $C_P = \varsigma_{33}A_P / t_P$, where A_P is the patch surface area. Hence, the voltage applied on actuators is $u[t] = 1/C_P \int_0^L \int_0^{2\kappa} \varphi_7^T\hat{\sigma}[t]A_s f_z \, ds \, dz$. Here S_s, A_s denote s-wise distribution of sensors and actuators, respectively, and f_z is their spanwise distribution. The current from sensors is $I[t] = \dot{\tilde{q}}_P$ and power required is $P = uI$.

LQR Control: Using $\mathbf{x} = \{\mathbf{q}^T \mid \dot{\mathbf{q}}^T\}^T$, the state space represenatation of the system is

$$\dot{\mathbf{x}} = \mathbf{A}\mathbf{x} + \mathbf{B}\mathbf{Q} + \mathbf{W}\hat{\sigma}; \qquad \mathbf{A} = \begin{bmatrix} \mathbf{0} & \mathbf{I} \\ -\mathbf{M}^{-1}\mathbf{K} & \mathbf{0} \end{bmatrix}; \qquad \mathbf{B} = \begin{bmatrix} \mathbf{0} \\ \mathbf{M}^{-1} \end{bmatrix}; \qquad \mathbf{W} = \begin{bmatrix} \mathbf{0} \\ -\mathbf{M}^{-1}\mathbf{F} \end{bmatrix} \qquad ...(5)$$

One seeks the control input $\hat{\sigma}[t]$ that minimizes the cost index $J_a = \int_{t_0}^{t_f} (\mathbf{x}^T \mathbf{Z}\mathbf{x} + \hat{\sigma}^T \mathbf{R}\hat{\sigma})dt$, where $\mathbf{R}$ is the positive definite control weighting matrix chosen as $\mathbf{R} = \eta \mathbf{F}^T \mathbf{K}^{-1} \mathbf{F}$, $\mathbf{Z}$ is the positive semi-definite state weighting matrix representing mechanical energy, i.e., $\mathbf{Z}\mathbf{x} = [\alpha \mathbf{q}^T \mathbf{K}^T \mid \mu \dot{\mathbf{q}}^T \mathbf{M}^T]^T$, and α, μ, η are suitably chosen weights. The cost minimization yields $\hat{\sigma} = -\mathbf{G}\mathbf{X}$ for the optimal control input with $\mathbf{G} = \mathbf{R}^{-1}\mathbf{W}^T\mathbf{P}$, where $\mathbf{P}$ is the solution of $\mathbf{A}^T\mathbf{P} + \mathbf{P}\mathbf{A} - \mathbf{P}\mathbf{W}\mathbf{R}^{-1}\mathbf{W}^T\mathbf{P} + \mathbf{Z} = 0$ (the Algebraic Riccati Equation (ARE)). The ARE is solved using the stable eigenvectors of the Hamiltonian matrix of the LQR system. Hence, the actuator voltage for optimal control is $u[t] = -\hat{\mathbf{G}}\mathbf{x}$, where $\hat{\mathbf{G}} = 1/C_P \int_0^L \int_0^{2\kappa} \varphi_7^T \mathbf{R}^{-1}\mathbf{W}^T \mathbf{P} A_s f_z \, ds dz$. In order to avoid saturation of the piezoactuators, the voltage is limited to $\mathrm{sgn}[u]V_{\max}$ whenever $|u| \geq V_{\max}$, where $V_{\max}$ is the actuator saturation voltage.

4. RESULTS AND DISCUSSIONS

A single ply Graphite-Epoxy host structure is considered. The properties for the host and piezopatches (PZT-4) are taken from [3]. The data used is $R_0 = 0.2032$ m, $L = 2.032$ m, $b = 0.0254$, $c = 0.127$ m, $h = 0.0127$ m, $t_P = 0.0127$ m, $\kappa = 0.0381$ m, taper ratio $\sigma = c_{\text{tip}} / c_{\text{root}} = 0.25$, $m_R = m = 1$ kg. The trial functions satisfying BC's at the root are $\varphi_1 = \varphi_2 = \varphi_6 = \{z^2 \ z^3 \ z^4 \ ...\}^T$, $\varphi_3 = \varphi_4 = \varphi_5 = \varphi_7 = \{z \ z^2 \ z^3 \ ...\}^T$. The default case is for HSDT and a linearly pretwisted beam, with $\Omega = 100$ rad/s, $p_m = 875.63$ Nm^{-1}, and piezopatch pair extending over the span. The non-dimensional controlled tip response $v_0 = v_0 [L; t] / L$ control voltage u, and power P are plotted.

Figure 3 shows the effect of taper on the eigenfrequencies for pretwist $\beta_0 = 15°$, ply-angle $\theta = 45°$, beam speed $\Omega = 200$ rad/s, and rotor speed $\bar{\Omega} = 200$ rad/s. There is a rapid decrease in the eigenfrequencies in the range $0 \leq \sigma \leq 0.05$. For $\sigma \geq 0.2$ the taper has a negligible effect on the first two frequencies, and the third one shows a stiffening effect when the taper is reduced. The gyroscopic softening effect appears in Fig. 4 where the rotor mass causes a reduction in eigenfrequencies (especially the first and third ones). Figure 5 shows the comparison of the three formulations for linear and quadratically varying pretwist with $\theta = 90°$, $\beta_1 = 45°$, $\beta_2 = 45°$, $\beta_0 = 90°$, $\bar{\Omega} = 250$ rad/s. The HSDT formulation yields the lowest coupled natural frequencies, thus providing conservative data for use in attaining non-resonant passive as well as active control designs. This emphasizes the importance of considering variations in transverse shear across the beam wall.

A comparison of the present control scheme (electric field varying along span) with the one considered in [3] (uniform electric field) is done for the rotorless beam with $\beta_0 = 30°$. The present scheme yields an order-of-magnitude reduction in settling time, control voltage, and power required, as well as lower response, as evident from Figs. 8-9 when compared to Figs. 6-7 (reproduced from [3]). When considering a saturation constraint on the actuator voltage, the peak power requirement is reduced fivefold as seen in Fig. 10. Gyroscopic forces due to the the tip rotor appear to have a pronounced qualitative effect on the response when considering stuctural tailoring along with active control. In contrast with the rotorless system (Fig. 8), when the tip rotor is present the response attenuation is greater for a smaller ply-angle beam (Fig. 11). Figure 12 shows the effect of taper on tip response, for $\beta_0 = 30°$, $\theta = 30°$, $\Omega = 400$ rad/s, $\bar{\Omega} = 40$ rad/s, and step forcing. As expected, the response increases with taper, i.e., a uniform cross-section beam has the lowest response, due to its bending rigidity being uniformly higher over the span. Due to centrifugal stiffening arising from increased rotor mass, the response gets attenuated as shown in Fig. 13. When comparing the untwisted, linearly pretwisted, and parabolically pretwisted beam, the response is intermediate for parabolic pretwist and lowest for linear pretwist as shown in Fig. 14. However, the control voltage and power is approximately the same for linear and parabolic pretwist.

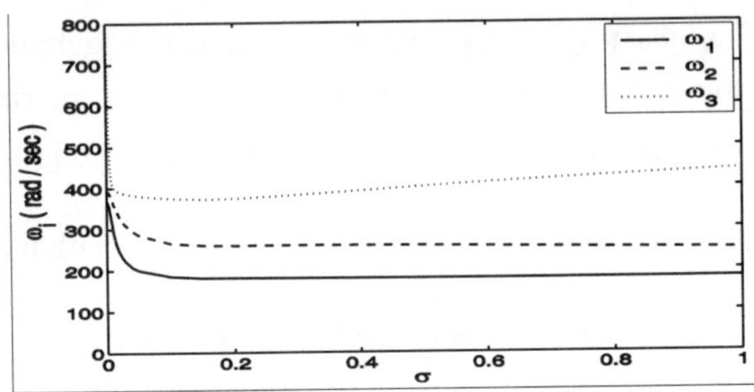

Fig. 3. Effect of taper on eigenfrequencies.

Fig. 4. Effect of tip mass on eigenfrequencies.

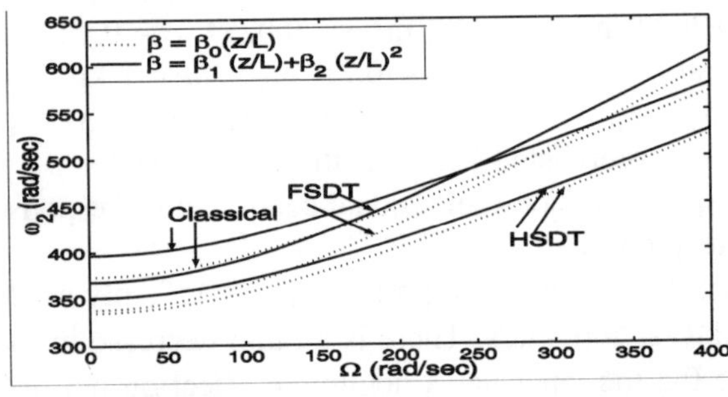

Fig. 5. Comparison of three theories.

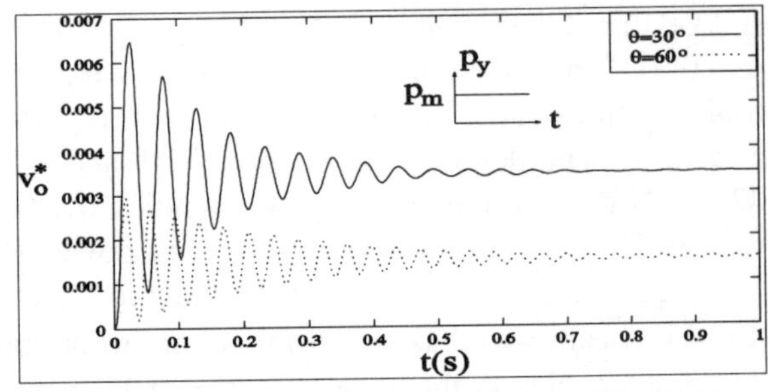

Fig. 6. Response for uniform electric field[1].

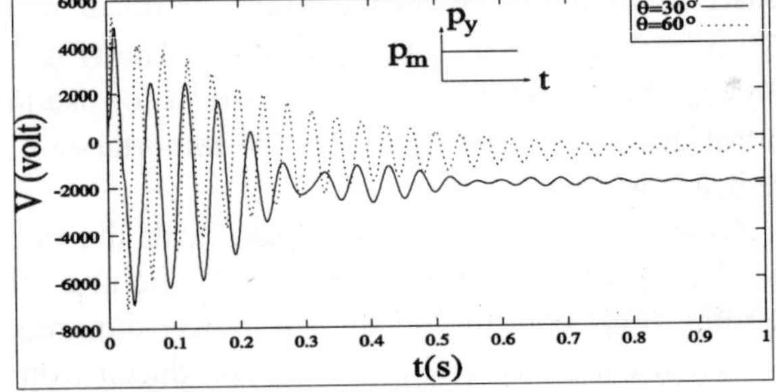

Fig. 7. Voltage for uniform electric field [1].

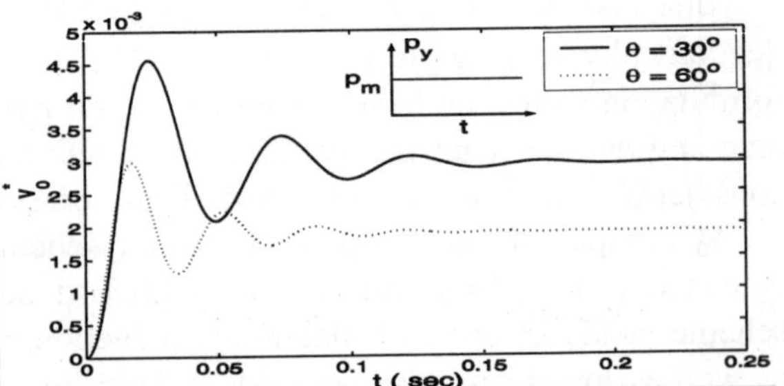

Fig. 8. Response—Nonuniform electric field.

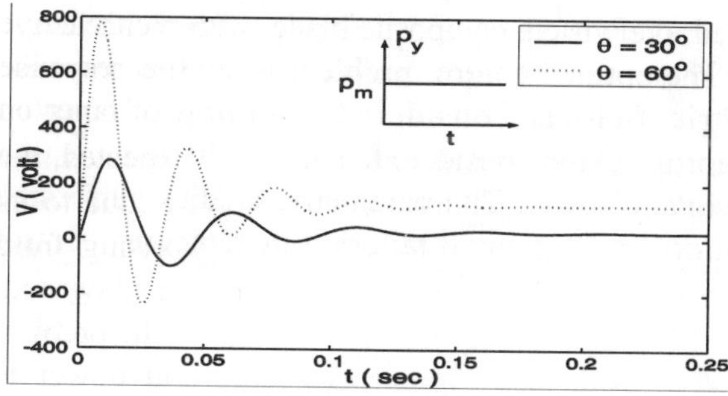

Fig. 9. Voltage—Nonuniform electric field.

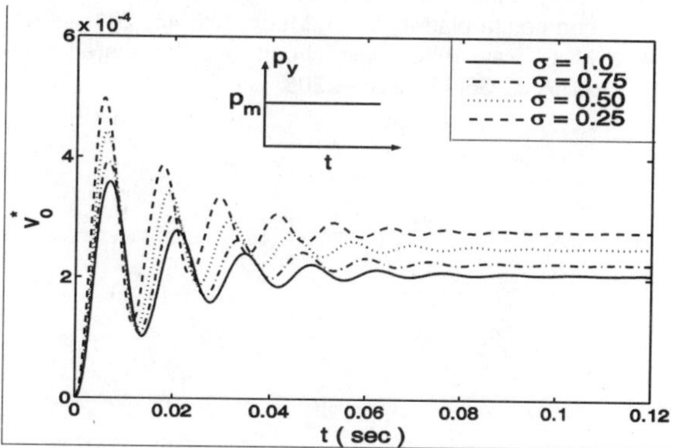

Fig. 10. Power—Effect of voltage constraint.

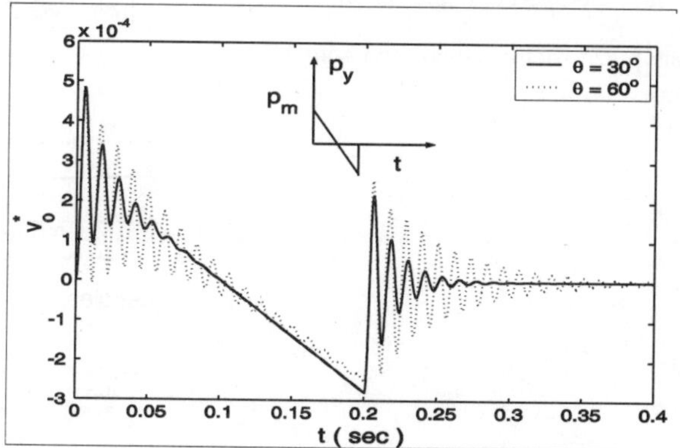

Fig. 11. Response due to sonic boom.

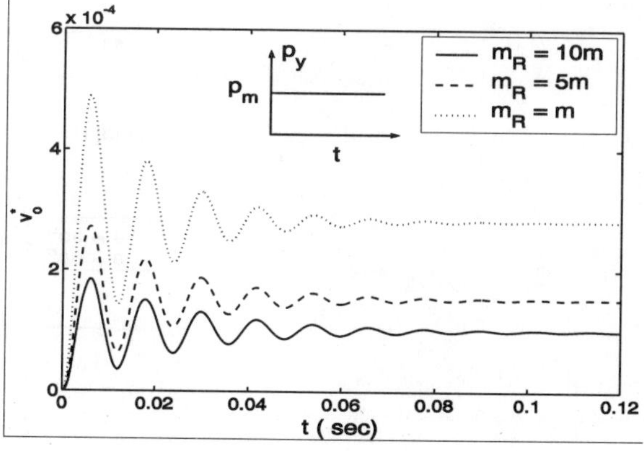

Fig. 12. Effect of taper ratio on response.

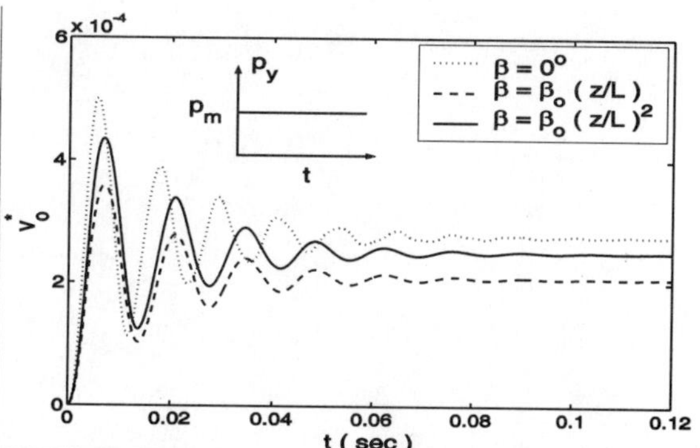

Fig. 13. Effect of tip mass on response.

Fig. 14. Effect of pretwist on response.

CONCLUSION

A HSDT structural model for a rotating, doubly-tapered, pretwisted, composite blade, with piezoelectric sensors-actuator pairs, and a tip rotor is developed. The optimal control problem is studied for wide range of excitations. A spanwise varying electric field is considered, yielding a coupled electromechanical system, as opposed to when a uniform field is considered. This results in increased attenuation, and reduced settling time and control voltage/power. The parametric studies performed underscore the importance of synthesizing active control and structural tailoring in achieving control effective designs.

REFERENCES

1. C. Kim, S. R. White, 1997, Thick-walled composite beam theory including 3-D elastic effects and torsional warping Int. J. Solids & Structures, 34(31-32), 4237-4259.
2. S. N. Jung, V. T. Nagaraj, and I. Chopra, 2001, Refined structural dynamics model for composite rotor blades, AIAA J., 39(2), 339-348.
3. N. K. Chandiramani, C. D. Shete, and L. I. Librescu, 2003, Vibration of higher-order-shearable pretwisted rotating composite blades, Int. J. Mech. Sci, 45, 2017-2041.
4. M. C. Ray, 1998, Optimal control of laminated plate with piezoelectric sensor and actuator layers, AIAA J., 36(12), 2204-2208.

101

Finite Element Analysis of Elastic-Plastic Contact of Fractal Surfaces

N. GHOSH[1] AND P. SAHOO[2]

[1]Department of Mechanical Engineering, Jadavpur University, Kolkata-700032, India
email: niloyghosh2001@yahoo.co.in
[2]Department of Mechanical Engineering, Jadavpur University, Kolkata-700032, India
email: psahoo@mech.jdvu.ac.in

ABSTRACT

In the present study, finite element methods are used to study non-adhesive, frictionless contact between a rigid plane and an elastic-plastic solid with self-affine fractal surfaces. Three-dimensional rough surfaces are generated using a modified two-variable Weierstrass-Mandelbrot function with given fractal parameters. Parametric studies are done to establish general relations between contact properties and key material and surface parameters with the help of commercial FEM package ANSYS 8.0. Non-dimensional contact area and displacement are plotted against non-dimensional load for varying tangent modulus to illustrate the elastic-plastic contact behaviour of fractal surfaces.

Keywords: Elastic-plastic contact, Finite Element, Self-affine.

1. INTRODUCTION

Contact analysis of rough surfaces plays a central role in studies of adhesion, friction, lubrication and wear. The adhesive and frictional forces between two solids arise from regions where their atoms are close enough to interact. The forces of friction and adhesion between two surfaces are determined by the interactions between atoms at their interface. Most surfaces are rough enough that atoms are only close to interact strongly in areas where peaks or asperities on opposing surfaces overlap. This real area of contact A is often much smaller than the projected area A_0 of the surfaces. Bowden and Tabor [1] have examined the plastic limit where the local pressure is large enough to flatten asperities. The mean pressure in the contacts is W/A and in the simplest model this has a constant value that is proportional to the hardness. The resulting linear relation between W and A is often given as an explanation for the linear rise of friction with the load [10]. A number of models, treating asperities as an elasto-plastic material, have been developed based on Greenwood and Williamson assumptions [3]. Relevant examples of elasto-plastic statistical models are found in Chang *et al.* [4], Zhao *et al.* [5]. However, this approach ignores some or all of the correlation between asperities, implying that asperities are far apart. This assumption may prove especially limiting in the context of plastic deformation where large and closely spaced contact clusters may develop.

Finite element analysis is a robust approach for addressing some of these shortcomings. Liu et al. [6] developed an isothermal elasto-plastic asperity model, which is solved with the increment form of a simplex-type algorithm for the contact of a cylinder with a rigid plane. The result of a finite-element simulation obtained by Faulkner and Arnell [7] were fitted in power-law type for the expression of normal load and shear forces between two hemispherical asperities. These finite element models did not provide a method to determine the onset of the fully plastic deformation regime. Finite element analysis of fractal surface contact is rarely reported. Hyun *et al.* [8] adopted a fully three-dimensional finite element analysis for elastic contact between rough surfaces with a range of self-affine fractal scaling behaviour. Pie *et al.* [9] extend the results established for elastic contact of Hyun *et al.* [8] to the case of elasto-plastic asperities. Considering a wide range of self-affine surface topographies and varying the ratio of the yield stress to the Young's modulus and the rate of strain hardening in the constitutive law. However, these approaches are cumbersome, time consuming and complicated. The present study aims at analysing the elastic- plastic contact of self-affine fractal surfaces using commercial software ANSYS 8.0. Three-dimensional rough surfaces are generated using a modified two-variable Weierstrass-Mandelbrot function [10] with given fractal parameters and used in commercial finite element package ANSYS 8.0 for contact analysis. The contact morphology including contact area, contact pressure and contact displacement is analysed for varying material properties.

2. SURFACE MODELING

Rough surface topographies have been traditionally quantified in terms of the height variance, slope and curvature of surface summits. However, a realistic multi-scale roughness description can only be done using scale-independent fractal parameters. A three-dimensional fractal surface topography can be generated using a modified (truncated) two-variable Weierstrass-Mandelbrot function that can be written as [10]

$$z(x,y) = L\left(\frac{G}{L}\right)^{(D-2)} \left(\frac{\ln\gamma}{M}\right)^{1/2} \sum_{m=1}^{M} \sum_{n=0}^{n_{max}} \gamma^{(D-3)n} \times$$

$$\left\{\cos\phi_{m,n} - \cos\left[\frac{2\pi\gamma^n (x^2 + y^2)^{1/2}}{L} \times \cos\left(\tan^{-1}\left(\frac{y}{x}\right) - \frac{\pi m}{M}\right) + \phi_{m,n}\right]\right\} \qquad ...(1)$$

where, L is the sample length, G is the fractal roughness, D is the fractal dimension $(2 < D < 3)$, γ $(\gamma > 1)$ is a scaling parameter, M is the number of superposed ridges used to construct the surfaces, n is a frequency index, with $n_{max} = \text{int}[\log(L/Ls)/\log\gamma]$ representing the upper limit of n, where L_s is the cut-off length, and $\phi_{m,n}$ is a random phase. The scaling parameter γ controls the density of frequencies in the surface profile. Based on surface flatness and frequency distribution density considerations, $\gamma = 1.5$. For a truncated series (i.e., starting at $n = 0$ rather than $n = -\infty$ (Eq. (1)), the scaling property is approximate, i.e., scaling is satisfied only to within a small additive term. Thus, the surface function given by Eq. (1) possesses a scale-invariant (fractal) behaviour only within a finite range of length scales, outside of which, the surface topography can be represented by a deterministic function. In practice, the smallest length corresponds to the instrument resolution and the upper length to the length of the profile. Because frequencies outside the range determined by the lower and upper wavelengths do not contribute to the observed profile, self-similarity is satisfied at all scales only approximately. The fractal roughness G is a height scaling parameter independent of frequency. The magnitude of the fractal dimension D determines the contribution of high and low frequency components in the surface function $z(x,y)$. Thus, high values of D indicate that high-frequency components are more dominant than low-frequency components in the surface topography profile. The surface height function given by Eq. (1) is continuous, non-differentiable, scale-invariant (in the range determined by the upper and lower wavelengths used in the truncated series), and self-affine (asymptotically self-affine). The latter implies that as the surface is repeatedly magnified, more and more surface features appear and the magnified image shows a close resemblance to that of the original surface obtained at a different scale. These properties make the function given by Eq. (1) suitable for constructing surfaces possessing topographies closely resembling the actual surfaces from which the fractal parameters D and G were determined experimentally. Figure 1 shows a sample Weierstrass-Mandelbrot (W–M) surface produced in MATLAB 7.0 using D = 2.3, G = 1.36×10^{-11} m, L = 9×10^{-7} m, L_s = 1.5×10^{-7} m, M =10 and γ = 1.5 (height scale is in m).

3. FINITE ELEMENT MODELING

The finite element analysis was carried out by using the static analysis of ANSYS 8.0 which calculates the contact area after the application of the load. MATLAB 7.0 is used to generate the z(x, y) values from Eq. (1) as per the supplied x and y values. The length of sample is set to be 9×10^{-7} m with $M = 10$ and $L_s = 1.5 \times 10^{-7}$ m. The points so generated, is imported to ANSYS 8.0 as key points. A surface is created by joining the key points. The surface so created is made solid in ANSYS. The upper surface of the solid is identified as CONTACT surface. A rigid surface is made to just touch the contact surface from the top. The rigid top surface is set as TARGET surface.

The elastic solid body is modeled using 3-D solid element SOLID45, which is defined by eight nodes having three degrees of freedom at each node: translations in the nodal x, y, and z directions. The element SOLID45 has plasticity, creep, swelling, stress stiffening, large deflection, and large strain capabilities. TARGE170 is used to represent the 3-D "target" surface for the associated contact element (CONTA173). The contact elements themselves overlay the solid elements describing the boundary of a deformable body and are potentially in contact with the target surface, defined by TARGE170. This target surface is discretized by a set of target segment elements (TARGE170) and is paired with its associated contact surface via a shared real constant set. One can impose any translational or rotational displacement, temperature, voltage, and magnetic potential on the target segment element. One can also impose forces and moments on target elements. CONTA173 is used to represent contact and sliding between 3-D "target" surfaces (TARGE170) and a deformable surface, defined by this element. The element is applicable to 3-D structural and coupled field contact analyses. This element is located on the surfaces of 3-D solid.

In order to validate the model, mesh convergence must be satisfied. The mesh density was iteratively increased until the contact force and contact area differed by less than 1% between iterations. Depending on the fractal parameters and expected region of contact, the number of elements in resulting mesh varies. The resulting mesh consists of at least 28,848 SOLID45 elements, 577 TARGE170 elements and 3200 CONTA173 elements. Figure 2 shows a typical geometry of a finite element mesh in an elastic solid with a self-affine surface with D = 2.4 and G = 1.36×10^{-10} m. In order to restrict any

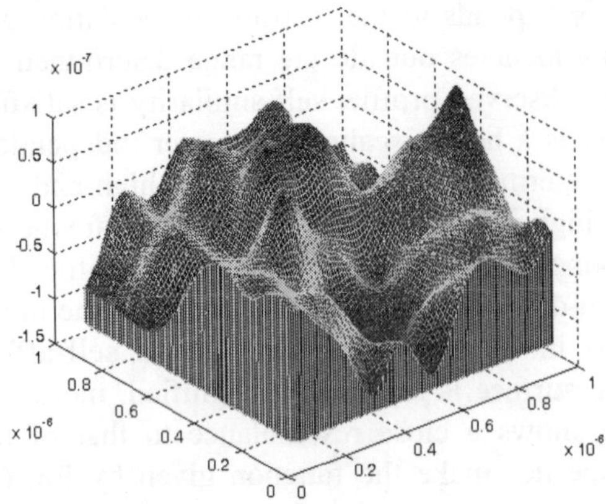

Fig. 1. A sample Weierstrass-Mandelbrot surface.

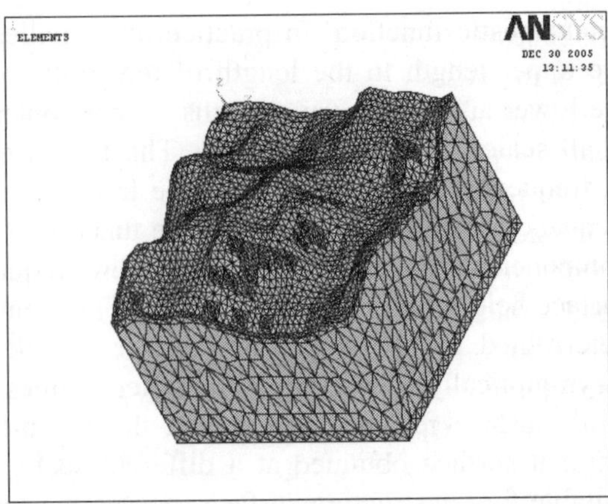

Fig. 2. Geometry of finite element mesh in elastic solid.

movement of the base of the solid, the nodes lying in the xy plane at $z = 0$ were rigidly constrained from moving in the z direction. The rigid plane is allowed to move in the z-direction only. Each analysis consisted of single load step with the maximum and minimum numbers of sub-steps being set to 100 and 10, respectively. These values were chosen in order to establish a small initial sub-step. Within each sub-step a maximum of 30 equilibrium iterations was allowed. To avoid element distortion, ANSYS uses bisection method to increase the number of load sub-steps, so that load can be applied at a slower rate. The contact with the rigid plane is realized using surface-to-surface contact elements that uses Augmented Lagrangian method. In our approach, a force is applied to the rigid plane causing it to move incrementally downward and a consecutive node is assumed to come into contact when its distance from the plane becomes zero. To obtain a converged solution in a reasonable period of time, the displacement tolerance was set to 0.001. The computations were performed on a Xeon processor. The run times varied according to the number of displacement steps involved and the machine used. However, on average each set of results took approximately 3 h to complete. In addition to mesh convergence, the present contact model is used to analyse the elastic contact of a sphere with a rigid plane and that compares well with the Hertz elastic solution. The contact force differs from the Hertzian solution by no more than 2% and the contact radius differs by a maximum of 10%. The favourable comparison of the results illustrates the suitability of the finite element model for the present analysis involving only global variables, such as contact pressure and contact area and the correctness of the assumed boundary conditions.

4. COMPARISON PROBLEMS

4.1 Comparison for Load Area Relationship

The effectiveness of the proposed model is evaluated using the data and results given in Kucharski *et al.* [11]. They performed measurement of contact load, area and approach. In their work, steel specimens are employed with Young's modulus 200 GPa, Tangent modulus 20GPa, Poisson's ratio 0.3 and tensile yield strength 400 MPa. The three-dimensional profilometry of sand-blasted surface (E60) are provided there, but no data is available regarding fractal dimension and fractal roughness parameter. From the three-dimensional profilometric data, average values of fractal dimension and roughness parameter are chosen for comparative study in the present case as: $D = 2.3$, $G = 1.36 \times 10^{-11}$ m. The equivalent Young's modulus E' is evaluated as $E/(1-v^2)$ which is equal to 219.78 GPa and Hardness H is taken as 1.12 GPa (2.8 times yield strength) for analytical solution. The same material property is used for the finite element solution in ANSYS. Figure 3 shows the comparison of the present model with the experimental results of Kucharski *et al.* [11] and analytical solution Yan and Komvopoulos [10] for non-dimensional contact area variation with non-dimensional contact load. Clearly the present model presents favourable agreement between the predicted contact area/ load values to that obtained by measurements. However, the analytical solution differs significantly from both the ANSYS and experimental results. This is probably due to the fact that analytical model ignores the asperity interaction effect.

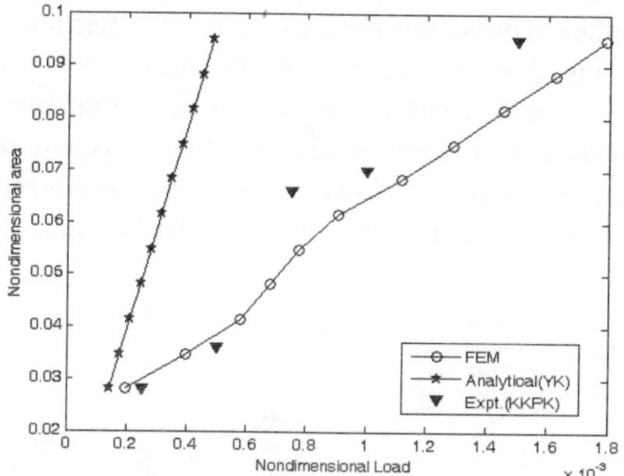

Fig. 3. Comparison between FEM (ANSYS) results, analytical results (YK Model [10]) and experimental results (KKPK[11]).

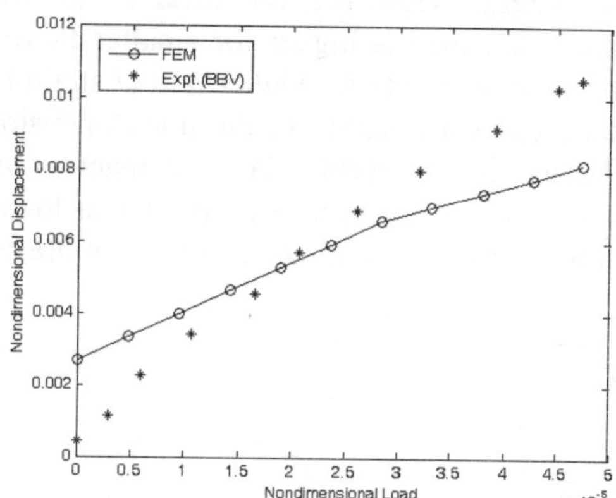

Fig. 4. Comparison between ANSYS solution and experimental results of BBV [12].

4.2 Comparison for Load-Deflection Relationship

Buzio *et al.* [12] investigated the load deflection behaviour between atomic force microscope (AFM) probes and self-affine fractal carbon films. Thus, they presented the experimental result for multiple asperity contact between nanostructured carbon and AFM probe. For nanostructured carbon, Young's modulus is 0.8 GPa, Hardness is 45 MPa, Tangent modulus is 0 and Poisson's ratio is 0.3. The AFM probe can be considered rigid. Thus for this system, equivalent Young's modulus ($E' = E / 1- v^2$) is 0.88 GPa and yield strength is taken as H/2.8 = 16.07 MPa. In Buzio et al. [12], AFM imaging provides fractal dimension $D = 2.3$ and fractal roughness $G = 1.55 \times 10^{-10}$ m along with the nominal contact area $A_0 = 4.8 \times 10^{-12}$ m^2. The same material and fractal properties are used for the ANSYS solution. Fig. 4 depicts the comparison between the experimental result of BBV model and ANSYS solution. It may clearly be seen that there is favourable agreement between the two.

5. NUMERICAL RESULTS AND DISCUSSION

Contact simulations are performed for different rough surfaces with varying fractal dimension and fractal roughness. The deformable solid is considered to be elastic with elastic modulus 200 GPa, yield stress 250 MPa and Poisson's ratio 0.3. For each of the numerical analyses performed, the contact load, contact area and displacement of the rigid plane are recorded and normalized as: Non-dimensional load: $P/(A_0E^*)$, Non-dimensional contact area: A_r/A_0, Non-dimensional displacement:

d/L. Here, P is the contact load, A_0 is the nominal contact area (= L^2, L being the length of sample), E* is the composite elastic modulus [= E/ $(1- v^2)$], A_r is the real contact area and d is the displacement of the rigid plane due to applied load. Fig. 3 shows the contact morphology at varying applied load for a surface with D = 2.4, G = 1.36 × 10^{-12} m. In order to consider the elastic-plastic contact of fractal surfaces, the applied load is now so chosen that the maximum contact pressure crosses the yield strength of the material. The analysis is carried out by keeping the D and G values constant and varying the value of tangent modulus (Et). Fig. 6 shows the plots of the non-dimensional displacement as function of the non-dimensional load for fractal parameter D = 2.4 and G = 1.36e-11m for different values of the ratio of E_t and E. It is seen that load-displacement relation for elastic-plastic contact is nearly linear for higher E_t values while it is highly non-linear at E_t = 0, i.e., for elastic-perfectly plastic material. However, for linear strain hardening material (with non-zero E_t), the displacement at a particular load is higher for smaller E_t values. In other words, the load capacity at a particular displacement is higher for higher rate of strain hardening. Fig. 5 shows the plot of the non-dimensional contact area as a function of the non-dimensional load for various values of E_t. It is seen that load-area behaviour for elastic-plastic contact is linear for different E_t values except at E_t = 0 where, bilinear nature is observed. At a particular load, contact area is more for lower E_t. In other words, for the same contact area, load capacity is more for higher rate of strain hardening (i.e., higher E_t).

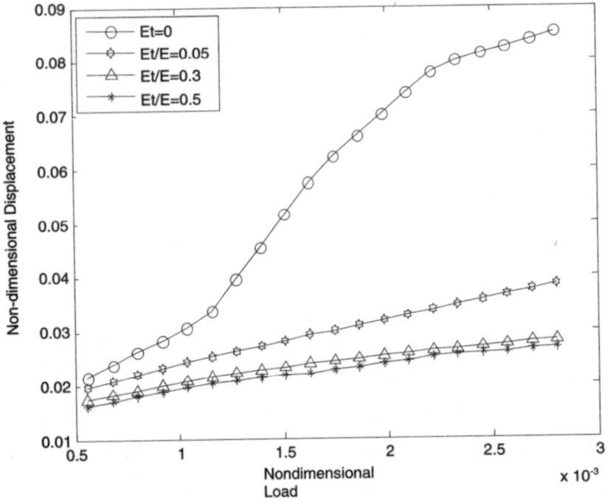

Fig. 5. Non-dimensional displacement against non-dimensional load: effect of varying E_t/E.

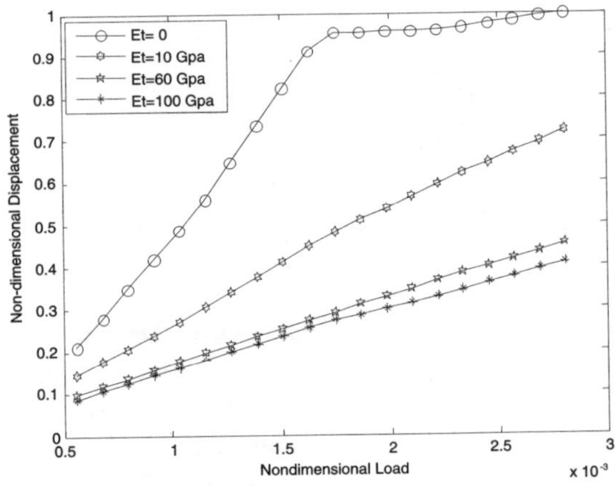

Fig. 6. Non-dimensional area against non-dimensional load: effect of varying E_t, keeping D = 2.4 and G = 1.36e-11 m.

CONCLUSION

Fractal surfaces are simulated and the finite element contact analysis of the same are performed using commercial FEM software ANSYS 8.0 in the present study. The favourable comparison of the results from the present modeling with Hertz solutions illustrates the suitability of the finite element model for the present analysis involving only global variables, such as contact pressure and contact area and the correctness of the assumed boundary conditions. Elastic-plastic contacts are considered for fractal surfaces with a fixed fractal dimension D and fractal roughness parameter G. In the elastic-plastic regime the load capacity at a particular displacement is found to be higher for higher rate of strain hardening. Also at a particular load contact area is found to be more for lower rate of strain hardening.

REFERENCES

1. F. P. Bowden, and D. Tabor, 1964, The Friction and Lubrication of Solids, Oxford University Press, Oxford.
2. K. L. Johnson, 1985, Contact Mechanics, Cambridge University Press, New York.
3. J. A. Greenwood and J. P. B. Williamson, 1966, Contact of nominally flat surfaces, Proceedings of the Royal Society of London, A 295, 300-319.
4. W. R. Chang, I. Etsion, and D. B. Bogy, 1987, An elastic-plastic model for the contact of rough surfaces, Journal of Tribology, 109, 257-263.
5. G. Y. Zhou, M. C. Leu, and D. Blackmore, 1993, Fractal geometry model for wear prediction, Wear, 170, 1-14.
6. G. Liu, J. Zhu, and Q. J. Wang, 2001, Elastic-plastic contact of rough surfaces, Tribol. Trans., 44, 437-443.
7. A. Faulkner, and R. D. Arnell, 2000, The Development of a finite element model to simulate the sliding interaction between two, three- dimension, elasto-plastic, hemispherical asperities, Wear, 242, 114-122.
8. S. Hyun, L. Pei, J. F. Molinari, and M. Robbins, 2004, Finite element analysis of contact between elastic self-affine surfaces, Phy. Rev. E , 70, 026117 (1-12).
9. L. Pei, S. Hyun, J. F. Molinari, and M. Robbins, 2005, Finite element modeling of elasto-plastic contact between rough surfaces, Journal of the Mechanics and Physics of Solids, 53, 2385-2409.
10. W. Yan, and K. Komvopoulos, 1998, Contact analysis of elastic-plastic fractal surfaces, J. Appl. Phys., 84, 3617-3624.
11. K. Kucharski, T. Klimczak, A. Polijaniuk, and J. Kaczmarek, 1994, Finite element model for the contact of rough surfaces, Wear, 177, 1-13.
12. R. Buzio, C. Boragno, and U. Valbusa, 2003, Contact mechanics and friction of fractal surfaces probed by atomic force microscopy, Wear, 254, 917-923.

102

Modelling of Deformation Behaviour of Municipal Solid Waste

V. KRASE[1] AND D. DINKLER[2]

[1]Institute for Structural Analysis, Technical University Braunschweig, Germany
email: v.krase@tu-bs.de
[2]Institute for Structural Analysis, Technical University Braunschweig, Germany
email: d.dinkler@tu-bs.de

ABSTRACT

A constitutive model for the deformation behaviour of municipal solid waste including a split of the material into two different solid phases within the frame of the mixture theory is presented in this contribution. The first phase is composed of grainy particles, while the second phase includes more fibrous constituents with a great impact on the composite strength. Non-linear stress-strain relation and time depending deformation behaviour for both phases is described within the theory of plasticity for large displacement and large inelastic strains. The constitutive model is implemented into a Finite-Element code for plain strain condition. A numerical case study shows the capability of the developed tool.

Keywords: Finite element, large strain plasticity, composite material.

1. INTRODUCTION

Municipal solid waste is a quite unique and complex material. In a microscopic sense the refuse consists of single particles exhibiting very different shapes and sizes. For a macroscopic description parameters such as density, porosity or strength vary in wide ranges. Biological, chemical and physical processes change the composition and alter the macro-mechanical parameters in time. Municipal solid waste undergoes large inelastic and time depending settlements under load. Experimental investigations carried out in geotechnical laboratories in the 1990s by Jessberger and Kockel (1993) proof the large influence of fibrous particles on the shear strength.

2. FORMULATIONS

The key idea in the formulation of a constitutive model is to consider municipal solid waste as a composite material (Fig. 1 left). The more grainy particles form the basic matrix in which fibrous constituents, consisting mainly of wood, plastic, metal and paper, are embedded. The fibres are orientated horizontally due to the emplacement process and act as additional reinforcement. On a macroscopic scale the microstructure is homogenised in the representative elementary volume. Using the concept of superimposed continua both matrices occupy every material point at the same time. The assumption that both matrices undergo the same displacement field leads to equivalence of total strains.

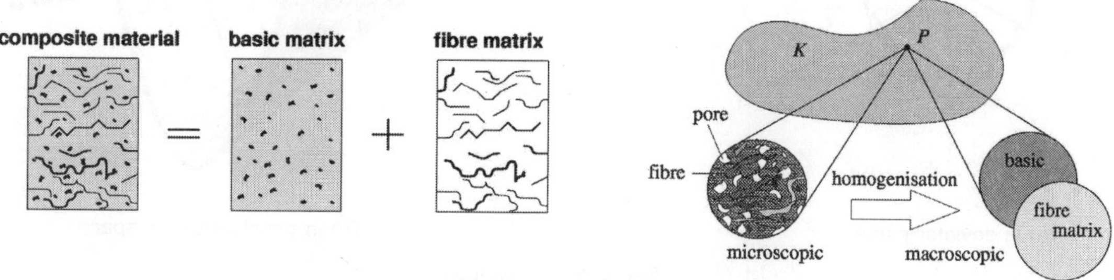

Fig. 1. Composite material and homogenisation.

2.1 Material Model of Basic Matrix

The non-linear stress-strain behaviour of the basic matrix is modelled in the framework of plasticity theory. The Kirchhoff-stresses $\mathbf{T}_B$ are linked to the elastic Almansi-strains $\mathbf{A}_{B,el}$ by generalised Hookean law with the Lamè constants λ and μ due to the fact that only small reversible deformations occur

$$\mathbf{T}_B = \lambda \, \mathrm{tr}(\mathbf{A}_{B,\,el})\mathbf{1} + 2\,\mu\,\mathbf{A}_{B,\,el} \qquad \ldots(1)$$

The yield envelope (Ehlers 1995) limits the stresses using invariants J_1, J_{2D}, J_{3D} of the stress tensor and its deviator

$$f\left(\mathbf{T}_B, \varepsilon_{B,Vin}\right) = \sqrt{J_{2D}\left(1 + \gamma\frac{J_{3D}}{J_{2D}^{3/2}}\right)^m} + \alpha_F\,J_1 - \kappa \leq 0 \qquad \ldots(2)$$

Fig. 2 compares the applied yield envelope with well-known Drucker-Prager and Mohr-Coulomb criterion of soil mechanics. The factor under the root includes the Lode-angle by the ratio of second and third invariant of the stress deviator and considers different behaviour of granular materials in compression and extension. The parameters α_F and κ are related to classical shear parameters angle of internal friction ϕ and cohesion c

$$\alpha_F = \frac{\sin\left[\varphi\left(\varepsilon_{B,Vin}\right)\right]}{3}, \qquad \kappa = c\left(\varepsilon_{B,Vin}\right)\cos\left[\varphi\left(\varepsilon_{B,Vin}\right)\right] \qquad \ldots(3)$$

If the material densifies, the shear strength increases, too. Thus, the shear parameters are formulated dependent on the inelastic volume strain $\varepsilon_{B,Vin}$ expressed by the invariants I_1, I_2 and I_3 of the inelastic Almansi strain tensor $\mathbf{A}_{B,in}$

$$\varepsilon_{B,Vin} = 1 - \sqrt{1 - 2\,I_1 + 4\,I_2 - 8\,I_3} \qquad \ldots(4)$$

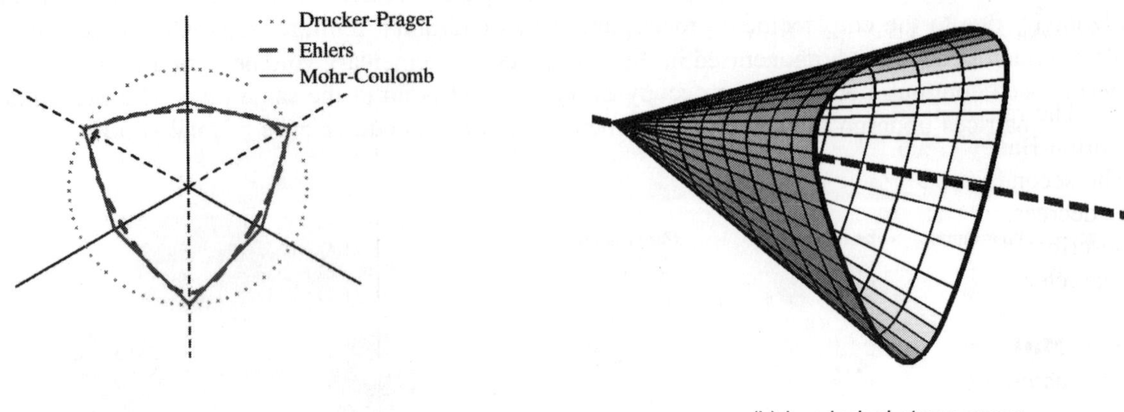

Fig. 2. Yield envelope.

The following functions are introduced in the model approach

$$\varphi\left(\varepsilon_{B,Vin}\right)=\frac{\mathbf{a}_{\varphi}\left(-\varepsilon_{B,Vin}\right)}{\mathbf{b}_{\varphi}+\left(-\varepsilon_{B,Vin}\right)}, \qquad c\left(\varepsilon_{B,Vin}\right)=\frac{\mathbf{a}_{c}\left(-\varepsilon_{B,Vin}\right)}{\mathbf{b}_{c}+\left(-\varepsilon_{B,Vin}\right)} \qquad \ldots(5)$$

Fitting to data of laboratory tri-axial or direct shear tests allows the parameter identification. The parameters γ and m can be estimated knowing the value of the angle of internal friction and the cohesion by comparison with the Mohr-Coulomb criterion.

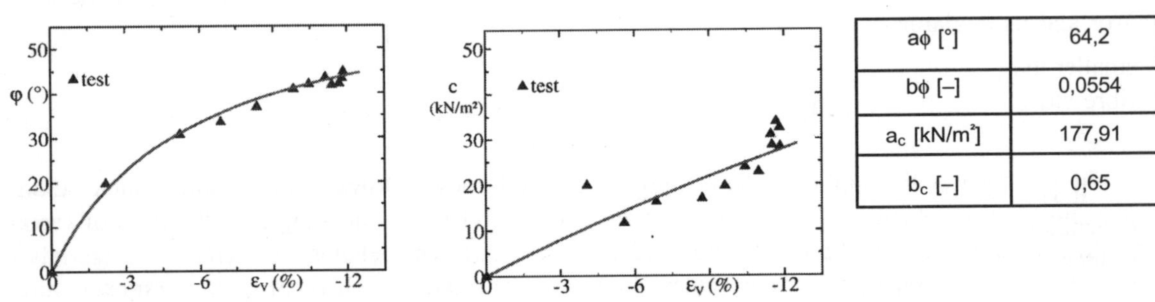

$a\phi$ [°]	64,2
$b\phi$ [–]	0,0554
a_c [kN/m²]	177,91
b_c [–]	0,65

Fig. 3. Fitting angle of internal friction and cohesion to laboratory tests.

The rate of inelastic strains and the flow rule is determined by introducing a plastic potential

$$q\left(\mathbf{T}_B,\varepsilon_{B,Vin}\right)=\sqrt{J_{2D}}+\alpha_q\,J_1, \qquad \alpha_q=\frac{\sin\left[\psi\left(\varepsilon_{B,Vin}\right)\right]}{3} \qquad \ldots(6)$$

The angle of dilatancy ψ determines the amount of inelastic volume strains produced during plastic flow and is defined similar to the angel of internal friction.

Municipal solid waste undergoes time depending inelastic deformation under compression. A creep law describes the visco-plastic behaviour. The covariant Oldroyd rate of the inelastic strain tensor is isotropically connected with the compaction rate $\dot{\varepsilon}_{V,c}$

$$\overset{\Delta}{\mathbf{A}}_{B,in} = \dot{\varepsilon}_{V,c}\, \mathbf{1}, \ \dot{\varepsilon}_{V,c} = \left[\mathbf{B}\left(\frac{|\mathbf{J}_1|}{\sigma_0}\right)^n\right]\cdot\left[\ln\left(\frac{\varepsilon_{V,E}}{\varepsilon_{V,E}-\varepsilon_{V,0}+\varepsilon_{B,Vin}}\right)\right], \ \varepsilon_{V,E} = \frac{b_V\,\varepsilon_{V,0}+|\mathbf{J}_1|^{av}\,\varepsilon_{V,max}}{b_V+|\mathbf{J}_1|^{av}} \qquad ...(7)$$

The rate depends on the stress state of the basic matrix and on the state of compaction itself. A Norton-Bailey-creep law using the first invariant of the stress tensor $\mathbf{J}_1$ includes the stress dependency. The second function governs the dependency on the inelastic volume strain. In denser states the rate is decreasing and has to be zero in the limit state of maximal density. The model parameter $\varepsilon_{V,E}$ describes the final value of compaction in an applied stress state, while the parameters $\varepsilon_{V,max}$ and $\varepsilon_{V,0}$ characterise the maximal capacity of densification and the initial state.

2.2 Material Model of Fibre Matrix

The material model for the fibre matrix derives from tension tests developed by Kölsch (1996). The test material is filled into a large box and vertically loaded. After consolidation of the material the box is separated into two parts and the tension load is applied in one horizontal direction. Fig. 4 (left) pictures a typical stress-strain curve obtained in this laboratory experiment. The tension force carried by the fibres is increasing with the vertical load due to the increase of the anchoring effect. Plotting the maximal tension force against the vertical load a linear relation is obtained (Fig. 4 right). In a critical load point the fibre specific strength is reached and the tension force cannot be increased by further vertical loading.

A transversal-isotropic elasticity relation between Kirchhoff-stresses $\mathbf{T}_F$ and elastic strains $\mathbf{A}_{F,el}$ of the fibre matrix considers the anisotropy induced by the emplacement process

$$\mathbf{T}_F = \overset{4}{\mathbf{C}}_F : \mathbf{A}_{F,el} \qquad ...(8)$$

Only few fibres are orthogonal to the fibre plane. Thus, the stiffness in this direction is much smaller than in the plane directions. The stresses are limited by a Rankine criterion. All three principal fibre stresses $\mathbf{t}_{F,ii}$ have to fulfil the yield criterion

$$\sigma_D \le t_{F,ii} \le \sigma_Z\left(\mathbf{A}_{F,in}, \mathbf{T}_B\right). \qquad ...(9)$$

It is assumed that the fibres cannot bear any compression stress. The bearable tension stress depends on the stress state of the basic matrix and the current state of inelastic deformation. The proposal for the yield stress in tension zone is given by

$$\sigma_Z\left(\mathbf{A}_{F,in}, \mathbf{T}_B\right) = z_0 + \mathbf{z}_m\left(\mathbf{T}_B\right)\cdot h_1\left(\mathbf{A}_{F,in}\right) - \left(z_0 + \mathbf{z}_m\left(\mathbf{T}_B\right) - z_R\right)\cdot h_2\left(\mathbf{A}_{F,in}\right) \qquad ...(10)$$

The parameter z_0 characterises the load independent part of the tension strength, the function $z_m(\mathbf{T}_B)$ includes the bilinear dependency on surrounding basic matrix stresses, while the parameter z_r describes the residual strength. Strain hardening and softening is governed by the function h_1 and h_2

$$h_1 = 1 - a^{-b\,\varpi_{ii}}, \qquad h_2 = 1 - c^{-d\,(\varpi_{ii})^e}, \qquad \varpi_{ii} = A_{F,in}\left(t_{F,ii}\right). \qquad ...(11)$$

The parameter ϖ_{ii} is the component of the inelastic strain tensor in the direction of the investigated principal stress. The flow rule is associated to the yield condition. Fig. 4 left shows a good agreement of simulation with the proposed model and test data.

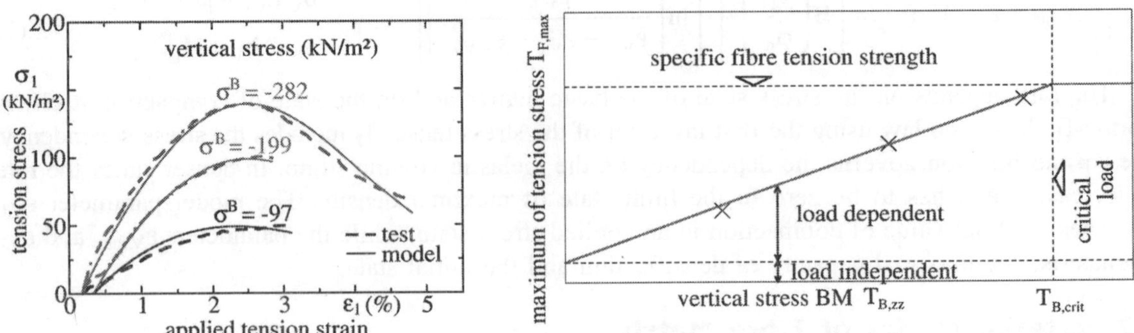

Fig. 4. Tension test and load depending tension strength.

2.3 Plasticity in Large Deformation Theory and Numerical Aspect

Municipal solid waste undergoes large irreversible deformations. The basic concept of plasticity in large deformation theory is the multiplicative split of the deformation gradient $\mathbf{F} = \text{Grad }(\mathbf{u})$ into an elastic $\mathbf{F}_{el}$ and an inelastic part $\mathbf{F}_{in}$ (Lee, 1969)

$$\mathbf{F} = \mathbf{F}_{el} \cdot \mathbf{F}_{in} = \mathbf{V}_{el} \cdot \mathbf{R}_{el} \cdot \mathbf{R}_{in} \cdot \mathbf{U}_{in} \approx \mathbf{R}_{el} \cdot \mathbf{R}_{in} \cdot \mathbf{U}_{in} \,. \qquad \ldots(12)$$

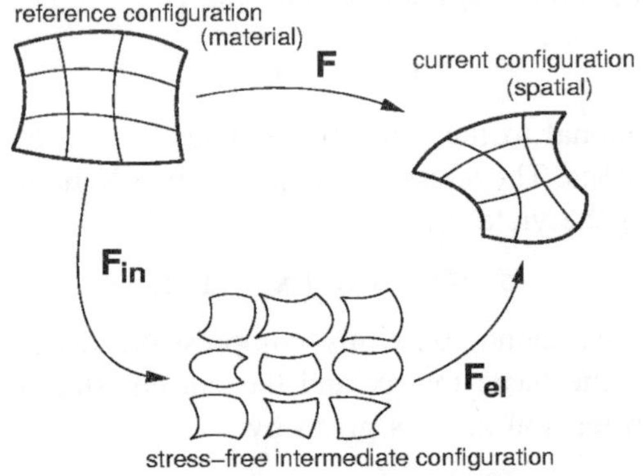

Fig. 5. Decomposition of deformation gradient, different configurations.

A stress-free intermediate configuration arises, where purely elastic and inelastic deformation measures can be established (Fig. 5). Therefore, the intermediate configuration is the natural basis for the constitutive equations. From the polar decomposition of the elastic deformation gradients $\mathbf{F}_{el}$ into the elastic left stretch tensor $\mathbf{V}_{el}$ and the orthogonal elastic rotation $\mathbf{R}_{el}$ and of the inelastic deformation gradient $\mathbf{F}_{in}$ into the inelastic right stretch tensor $\mathbf{V}_{in}$ and the inelastic orthogonal tensor $\mathbf{R}_{in}$ follows the non-uniqueness of the intermediate configuration. Assuming small reversible deformation as observed in laboratory experiments of municipal solid waste the elastic right stretch tensor $\mathbf{V}_{el}$ is approximately equal to the identity tensor $\mathbf{1}$. Thus, current and intermediate configuration are distinguished by a pure rotational deformation. Due to the principle of material objectivity the material equations are transformed from the intermediate into the current configuration applying stress and strain measures defined in the latter.

The developed material model is implemented in a non-commercial Finite-Element code of the Institute for Structural Analysis, TU Braunschweig, for plane strain condition using standard nine-node-displacement-elements and 3×3 Gaussian integration scheme. Neglecting inertial terms, the balance laws of mass, linear momentum and moment of momentum are formulated in the weak form via the Principle of Virtual Work at the reference configuration

$$\int_{V_0} S^t : \delta E^t \, dV_0 - \int_{V_0} \rho_0 \, b_0^t \cdot \delta u \, dV_0 - \int_{A_0} t_0^t \cdot \delta u \, dA_0 = 0. \qquad ...(13)$$

Thus, the second Piola-Kirchhoff-stress tensor $S = F^{-1} \cdot T \cdot F^{-T}$ is introduced by the pullback of the Kirchhoff-stress tensor used in the material equations in chapter 2.1 and 2.2. The virtual Green-strains can be computed from the virtual displacement field δ_u and the deformation gradient $\mathbf{F}$. Volume loads result from the density ρ_0 of the refuse and the vector of gravity $\mathbf{b}_0$. Prescribed forces $\mathbf{t}_0$ additionally act on the Von-Neumann part of the boundary of the domain, while displacements $\mathbf{u}_0$ are defined on Dirchlet part. The discrete non-linear equation system is solved by the Newton-Raphson-method including a consistent linearisation with respect to the unknown node displacements $\mathbf{u}$. The total stresses in equation (13) are the sum of the volume-averaged stresses of basic and fibre matrix.

The integration of the constitutive equations is carried out by the predictor-corrector scheme. In a spontaneous elastic step of the global incremental step n + 1 the trial stresses $\mathbf{T_{tr}}^{n+1}$ are determined after transformation of the inelastic strains $\mathbf{A_{in}}^n$ from the last converged increment n into the new current configuration with the incremental deformation gradient $\mathbf{F_u}$ by the elastic-stress-strain relations

$$T_{tr}^{n+1} = \overset{4}{C} : \left(A^{n+1} - F_u^{-T} \cdot A_{in}^n \cdot F_u^{-1} \right). \qquad ...(14)$$

If the trial stress state is violating the yield conditions or produces time depending compaction the corrector step has to be carried out. Within the corrector step the current configuration is fixed. Thus, the co- and contravariant Oldroyd rates equal the material time derivatives. The time integration is done by the implicit Eulerian backward scheme. The algorithm is similar to the Radial-Return-Method of small strain plasticity (Simo and Hughes 1998). Due to the decoupling of the material equations of the basic matrix from the equations of the fibre matrix the analysis can be carried out separately for both matrices.

3. NUMERICAL RESULT AND DISCUSSIONS

The time depending settlement behaviour of an artificial 15 m high landfill with an inclination ration 1:1 is simulated for one year. Geometric and boundary conditions are pictured in Fig. 6a. The time-settlement diagram in Fig. 6b shows the vertical displacements of two selected points. Initial settlements of about 0.75 m are created due to the dead load after emplacement of the material within few hours. The increase of stiffness in the beginning is caused by the densification of the material.

Then time depending compaction produces further settlements. The total settlements at the end of the simulation period of one year are about 1.75 m. The contour plots in Fig. 7 show the change of the inelastic volume strain and characterise zones of high deformations. The lower layers have the essential part on the total settlements due to the high-pressure states there. It is a well-known fact that the long time behaviour of landfills over decades is mainly governed by biodegradation of organic matter. The expansion of the presented model to consider this phenomenon is an important part of the ongoing research work.

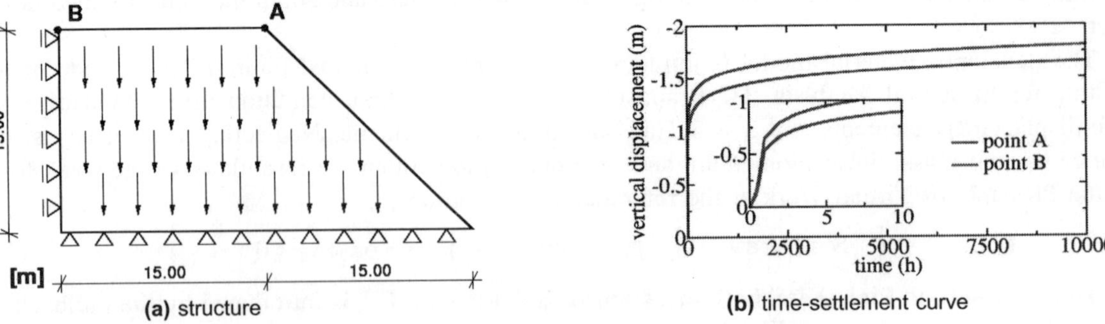

(a) structure

(b) time-settlement curve

Fig. 6. Landfill under compaction - sketch and time-settlement curve.

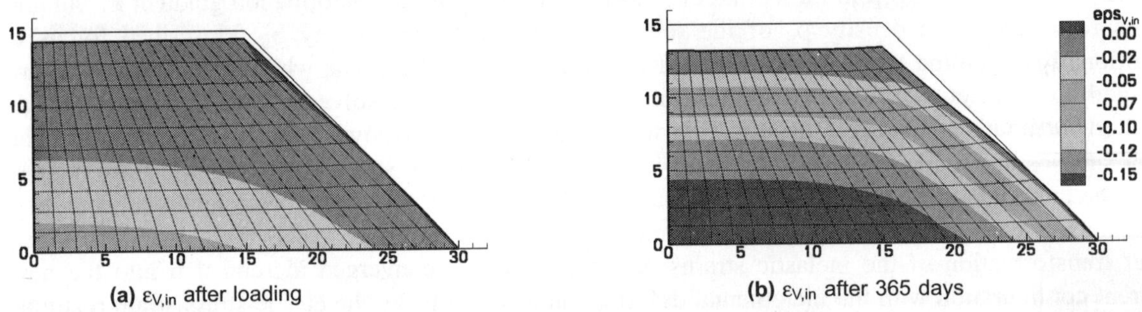

(a) $\varepsilon_{V,in}$ after loading

(b) $\varepsilon_{V,in}$ after 365 days

Fig. 7. Landfill under compaction-inelastic volume strain.

CONCLUSION

A material model for municipal solid waste is presented. Splitting the material into two solid phases allows the detailed description of different phenomena. The first phase is the basic matrix including all grainy particles. The constitutive model for this matrix includes as well elastic-plastic behaviour as a creep model. Fibrous constituents form the second phase and act as additional reinforcement and increase the strength of the composite material enormously. The model parameters of the constitutive equations are determined by fitting on laboratory tests. The material model is implemented in a Finite-Element code. A case study shows the capacity of the model.

ACKNOWLEDGEMENTS

This work is carried out in the Special Collaborative Research Centre 477 "Life Cycle assessment of structures via innovative monitoring" at TU Braunschweig, Germany. The Deutsche Forschungs-gemeineschaft (DFG) grants financial support.

REFERENCES

1. H.L. Jessberger, R. Kockel, 1993, Determination and assessment of mechanical properties of waste materials, Proc. 4th Int. Landfill Symp., Santa Margherita di Pula, CISA, Cagliari, Italy, pp. 1883-1392.
2. W. Ehlers, 1995, A single-surface yield function for geomaterials, Arch. Appl. Mech. 65, pp. 246-259.
3. F. Kölsch, 1996, Der Einfluss der Faserbestandteile auf die Scherfestigkeit von Siedlungsabfall (The influence of fibrous constituents on the shear strength of municipal solid waste), Ph.D.-thesis, Leichtweiss-Institut für Wasserbau, TU Braunschweig, Germany.
4. J.C. Simo, T.J.R. Hughes 1998, Computational Inelasticity, Springer, New York, Berlin.
5. E.H. Lee, 1969, Elastic-plastic deformation at finite strains, Journal of applied mechanics, 36, pp. 1-6.

103

Comprehensive Nonlinear Analysis of Dynamics of Multi-Body Systems Involving Anisotropic Elastic Members: Validations

P. HEMARAJU AND D. HARURSAMPATH

Nonlinear Multifunctional Composites Analysis and Design Lab (NMCAD Lab), Department of Aerospace Engineering, Indian Institute of Science, Bangalore, Bangalore-560 012, Karnataka, India email: raju@aero.iisc.ernet.in, dinesh@aero.iisc.erent.in

ABSTRACT

This work is concerned with the dynamic analysis of nonlinear multi-body systems involving elastic beams made of laminated, anisotropic composite materials *using object-oriented framework*. The deformations and rotations in the individual members are allowed to be large resulting in a geometrically nonlinear analysis. However, the strains within each elastic body are assumed to remain small resulting in a materially linear analysis. The elastic elements are modeled in an asymptotically correct manner using the finite element method. This is accomplished using the concepts of decomposition of the rotation tensor and the Variational Asymptotic Method (VAM) to accurately and efficiently model dimensionally reducible composite structures. In this work, unconditionally stable time-integration schemes presenting high-frequency numerical dissipation are used to solve the ensuing governing equations.

All three sub-tasks of this work (viz. nonlinear cross-sectional analysis, nonlinear beam analysis and nonlinear flexible multi-body system analysis) are to be accomplished on a single platform based on Object-Oriented Programming (OOP).

Keywords: Geometrically nonlinear; materially linear; asymptotically correct; VAM; unconditionally stable.

1. INTRODUCTION

This paper is concerned with the dynamic analysis of flexible, nonlinear multi-body systems [1-3], i.e. a collection of bodies in arbitrary motion with respect to each other while each body is undergoing large displacements [4] and rotations with respect to a body attached frame of reference. The focus is on problems where the strains within each elastic body remain small [5, 6].

The elastic bodies are modeled using the finite element method. The use of beam and flexible joint elements will be demonstrated for multi-body systems. The location of each node is represented by its Cartesian coordinates in an inertial frame, and the rotation of the cross-section at each node is represented by a finite rotation tensor [7] expressed in the same inertial frame. The kinematic constraints among the various bodies are enforced via the Lagrange multiplier technique. Although this approach does not involve the minimum set of coordinates, it allows a modular development of finite elements for the enforcement of the kinematic constraints. The formulation of universal and revolute joints are also implemented.

The equations of motion resulting from the modeling of multi-body systems with the above methodology present distinguishing features: they are stiff, nonlinear, differential-algebraic equations [2]. The stiffness of the system stems from the presence of high frequencies in the elastic members, but also from the infinite frequencies associated with the kinematic constraints. Indeed, no mass is associated with the Lagrange multipliers giving rise to algebraic equations coupled to the other equations of the system, which are differential in nature.

2. FORMULATION OF BEAM ELEMENTS MADE OF ANISOTROPIC MATERIALS

2.1. Nonlinear Analysis of the Beam Cross-Section

The purpose of the nonlinear beam cross-sectional analysis is to determine the elements of the matrix C^* and the recovering relations. The nonlinear beam cross-sectional stiffness matrix C^* relates the cross-sectional stress f^* resultants to the generalized strain measures, a 1-D constitutive law. Recovering relations provide the relationship between the strain tensor components and the generalized strain measures e^* [8].

In the present notation, the 1-D constitutive law can be written as

$$f^* = C^* \, e^* \qquad \qquad ...(1)$$

where , $f^* = [f_1 \ \ f_2 \ \ f_3 \ \ m_1 \ \ m_2 \ \ m_3]^\mathrm{T}$ with f_1 as the axial force, f_2 and f_3 as transverse shear forces, m_1 as the twisting moment, m_2 and m_3 as the bending moments; $e^* = [\gamma_{11} \ \ 2\gamma_{12} \ \ 2\gamma_{13} \ \ \kappa_1 \ \ \kappa_2 \ \ \kappa_3]^\mathrm{T}$ with γ_{11} as the axial stretching measure of the beam, $2\gamma 12$ and $2\gamma 13$ as transverse shear measures, κ_1 as the elastic twist per unit length, κ_2 and κ_3 as elastic components of the curvature.

2.1.1. Strain Energy of the Pretwisted Strips

For developing a rigorous non-classical non-linear asymptotic theory for beams which could be considered as pretwisted arbitrary assemblies of thin rectangular members. A thin pretwisted composite strip is modeled as a beam, including those non-linear effects that need to be considered, by virtue of the special geometry, to establish asymptotic correctness. This is achieved by deriving its strain energy function in terms of 1-D quantities only. A geometrically non-linear shell theory is derived [9] and the VAM [5, 6] is used as a tool to carry out the dimensional reduction from 2-D [10-12] to 1-D [4].

Consider the strip to be a 2-D elastic body [13-15], its strain energy density (i.e., energy per unit middle surface area) is given by

$$U_{2D} = \frac{1}{2} \begin{Bmatrix} \varepsilon_{11} \\ \varepsilon_{22} \\ 2\varepsilon_{12} \\ \rho_{11} \\ \rho_{22} \\ 2\rho_{12} \end{Bmatrix}^{T} \begin{bmatrix} A & B \\ B & D \end{bmatrix} \begin{Bmatrix} \varepsilon_{11} \\ \varepsilon_{22} \\ 2\varepsilon_{12} \\ \rho_{11} \\ \rho_{22} \\ 2\rho_{12} \end{Bmatrix}, \qquad \qquad ...(2)$$

where, $\varepsilon_{\alpha\beta}$ and $\rho_{\alpha\beta}$ are the 2-D strain measures; and A, D and B are the membrane, bending and coupling (3×3) stiffness matrices, respectively.

The beam strain energy density (i.e., energy per unit length of the strip) is given by $U_{1D} = \langle U_{2D} \rangle$

and the notation $\langle \bullet \rangle \equiv \int_{-b/2}^{b/2} (\bullet) \ dx_2$. The strain energy is given by

$$U = \int_{0}^{L} U_{1D} \left(\gamma_{11}, \ \kappa_1, \ \kappa_2, \ \kappa_3 \right) \ dx_1 \qquad \qquad ...(3)$$

where L is the wavelength of deformation along the strip. The width and thickness of the strip are denoted by b and h respectively. Now we can integrate U_{2D} with respect to across the width and obtain 1-D strain energy density as

$$U_{1D} = \frac{1}{2}\varepsilon_l^{T} \ [S_l]\varepsilon_l + \varepsilon_l^{T} [S_{ln}]\varepsilon_n + \frac{1}{2}\varepsilon_n^{T} [S_n]\varepsilon_n, \qquad \qquad ...(4)$$

where, the linear and non-linear 1-D strain vectors, ε_l and ε_n, respectively, are defined as follows:

$$\varepsilon_l = \{ \gamma_{11} \quad \kappa_1 \quad \kappa_2 \quad \kappa_3 \}^{T},$$

$$\varepsilon_n = \{ \ \kappa_1^2 \quad \kappa_2^2 \quad \kappa_2\gamma_{11} \quad \kappa_2\kappa_3 \quad \kappa_2\kappa_1 \}^{T}, \qquad \qquad ...(5)$$

and the matrices $[S_l]$, $[S_{ln}]$ and $[S_n]$ can be thought of as partitions of a 9×9 matrix $[S]$.

Substitute ε_l, ε_n, $[S_l]$, $[S_{ln}]$ and $[S_n]$ defined [9] in U_{1D} functional and take partial derivatives with respect to γ_{11}, κ_1, κ_2, and κ_3 gives the 4×4 nonlinear stiffness matrix from the nonlinear analysis of the beam cross-section. This can be written as:

$$\left[\frac{\partial U_{1D}}{\partial \gamma_{11}} \quad \frac{\partial U_{1D}}{\partial \kappa_1} \quad \frac{\partial U_{1D}}{\partial \kappa_2} \quad \frac{\partial U_{1D}}{\partial \kappa_3} \right]^{T} = \left[C^* \right] = \left[\gamma_{11} \quad \kappa_1 \quad \kappa_2 \quad \kappa_3 \right]^{T} = \left[f_1 \quad m_1 \quad m_2 \quad m_3 \right]^{T} \ ...(6)$$

The stiffness matrix in Eq. (6) is called classical or Euler-Bernoulli beam stiffness matrix obtained from the Euler-Bernoulli model (also called classical beam model) using the first approximation. This can treat extension, torsion and bending in two directions of composite beams.

The classical model works pretty well for slender beams, which does not have thin-walled open sections and undergoes with large wavelength. However, refined theories are required for better prediction when the beam is fat, or has thin walled open sections, or vibrate in short-wavelength modes. A generalized Timoshenko beam model is used to treat transverse shear deformation, which becomes significant for fat beams vibrating in higher frequencies. This can be written as:

$$\begin{bmatrix} f_1 & f_2 & f_3 & m_1 & m_2 & m_3 \end{bmatrix}^T = \begin{bmatrix} C^* \end{bmatrix} \begin{bmatrix} \gamma_{11} & 2\gamma_{12} & 2\gamma_{13} & \kappa_1 & \kappa_2 & \kappa_3 \end{bmatrix}^T \qquad ...(7)$$

Once the 1-D constitutive models are obtained, one can use them as inputs for all kinds of global beam analyses [16] because it is derived using an intrinsic formulation, which is geometrically exact. Use the initial guess values $\{\gamma_{11}\ \kappa_1\ \kappa_2\ \kappa_3\}_0$ for $\{\gamma_{11}\ \kappa_1\ \kappa_2\ \kappa_3\}$ and calculate the 6´6 nonlinear stiffness matrix C^*. This is input along with some other data to the 1-D code to perform beam analysis. Repeat the process till $\{\gamma_{11}\ \kappa_1\ \kappa_2\ \kappa_3\}$, and $\{\gamma_{11}\ \kappa_1\ \kappa_2\ \kappa_3\}_{k+1}$ converges (where, k = 0, 1, 2, ...).

3. NUMERICAL RESULTS AND DISCUSSIONS

3.1. Extension-Twist Coupling in Pretwisted Antisymmetric Laminates

The theoretical development described above has been implemented using object-oriented framework. The code can handle arbitrary cross-sectional geometries of initially curved and twisted composite beams.

A well-documented test case is provided with both experimental and theoretical results by Armanios *et al.* [13]. They considered two specimens with the same material properties, given in Table 1 and a pretwist of $-5°$. The lay up for specimen 1 is $[20_2 / -70_4 / 20_2 / -20_2 / 70_4 / -20_2]^T$ while for specimen 2 it is $[30_2 / -60_4 / 30_2 / -30_2 / 60_4 / -30_2]^T$. As the cross section departs from a strip configuration, the trapeze effect, stems from non-linearity of the cross-sectional analysis, becomes less and less important compared to the overall torsional rigidity. Its variation as the cross-section gradually changes from a thin strip to a square configuration is represented in Fig. (1, 2) and as a function of the aspect ratio (h/b) of the cross-section.

Table 1. Geometrical and material characteristics for the extension-twist coupling experiments.

L $= 254$ mm	2b $= 25.4$ mm	h $= 1.168$ mm	
$E_{11} = 135.6$ GPa	$E_{22} = E_{33} = 9.9$ GPa	$G_{12} = G_{13} = 4.2$ GPa	$G_{23} = 2.3$ GPa
$v_{12} = v_{13} = 0.3$	$v_{23} = 0.5$		

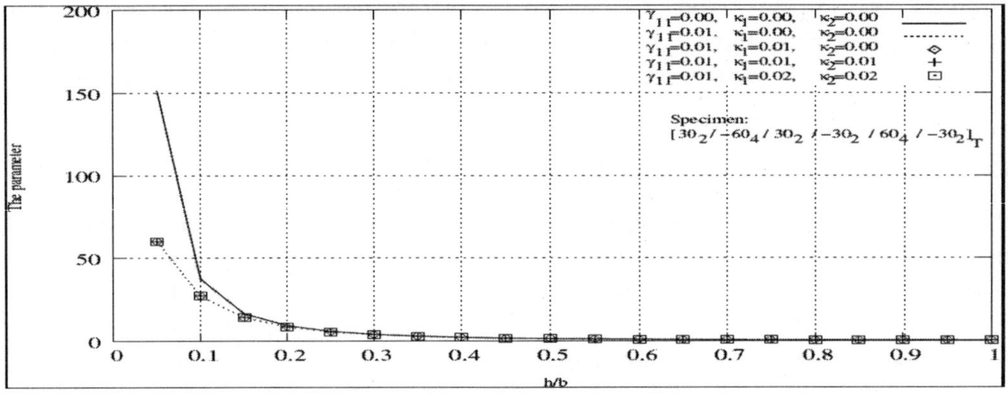

Fig. 1. Trapeze effect variation with the aspect ratio (specimen: 1).

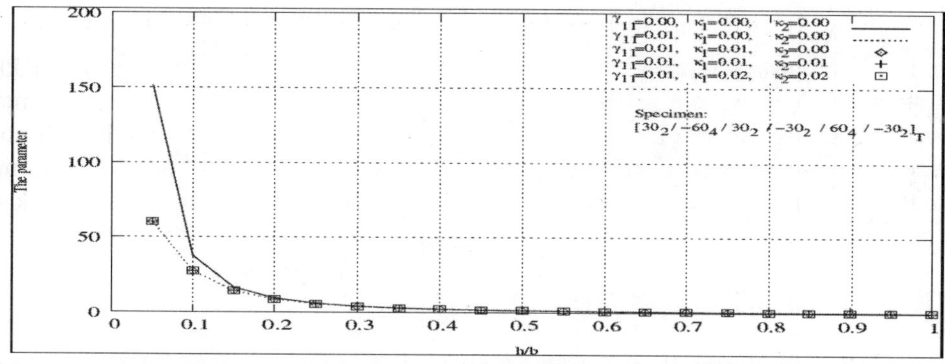

Fig. 2. Trapeze effect variation with the aspect ratio (specimen: 2).

3.2. Example: Nonlinear Spring-Mass System

Consider the nonlinear spring mass oscillator defined by a kinetic energy $K = \frac{1}{2} m \, \dot{u}^2$, a strain energy $V = \frac{1}{2} k \, \varepsilon^2$ and a strain $\varepsilon = u^2$. For this example $m = k = 1.0$. Though the trapezoidal rule scheme is unconditionally stable for linear system, there is no guarantee of stability when applied to nonlinear systems. Figures 3-6 show the response of the system for initial conditions $u_0 = 1.0$, $\dot{u}_0 = 0.0$ predicted by the energy preserving and energy decaying schemes (EPS & EDS). As expected, the total energy of the system is exactly preserved in Fig. 4 and decays in Fig. 6. The results of a convergence study shown in Fig. 7 indicate second order accuracy for the trapeze rule and the energy preserving scheme and third order accuracy for the energy decaying scheme.

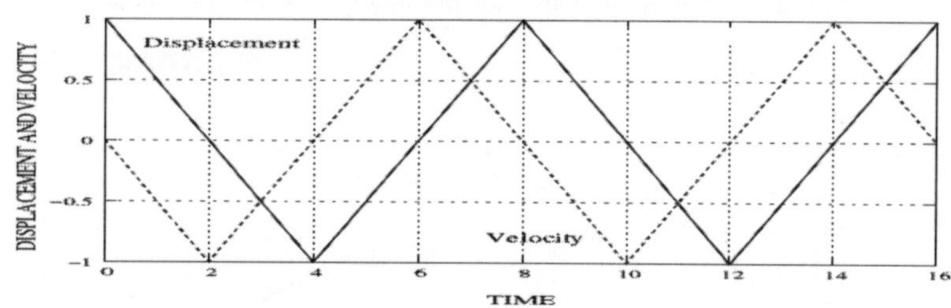

Fig. 3. Displacement response for the energy preserving scheme.

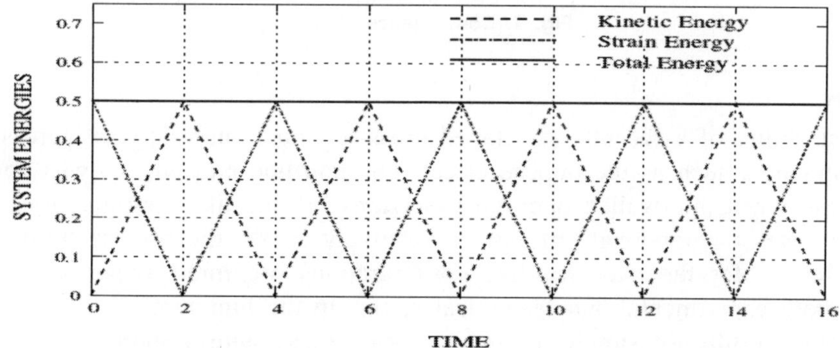

Fig. 4. Energy response for the energy preserving scheme.

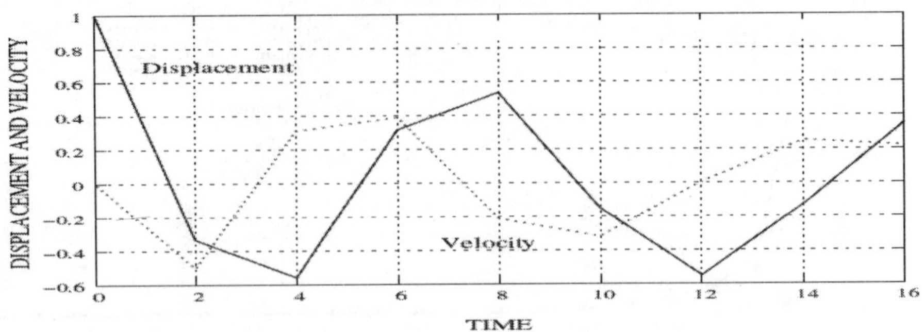

Fig. 5. Displacement response for the energy decaying scheme.

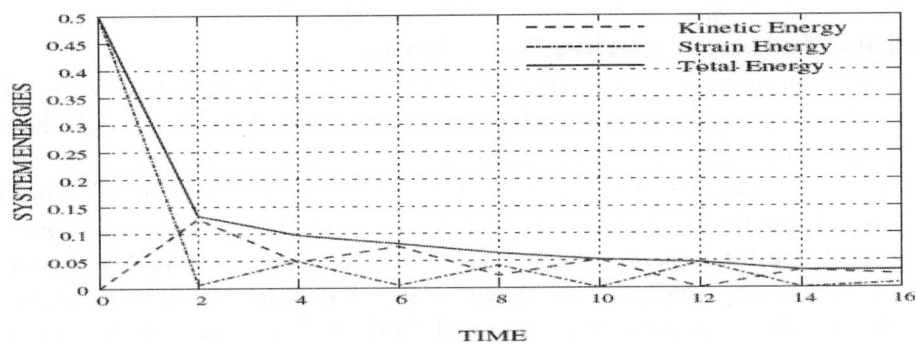

Fig. 6. Energy response for the energy decaying scheme.

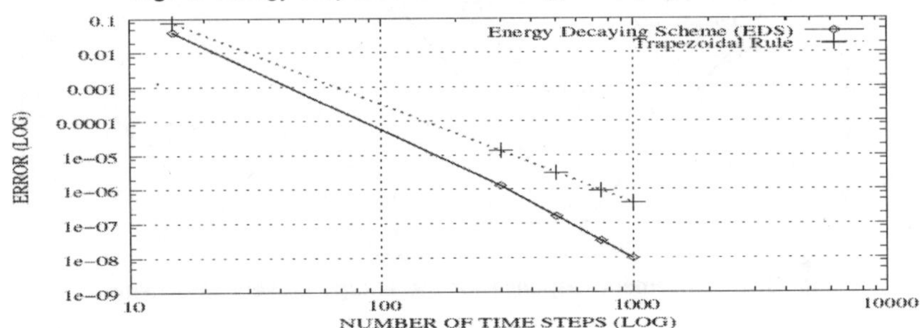

Fig. 7. Convergence study.

CONCLUSION

The modeling capability of a recently developed cross-sectional analysis was increased to include non-linear phenomenon, such as the trapeze effect. The non-linear extension-twist coupling occurs especially the case when thin-walled open cross-sections or strip-like configurations are involved. Since, for these cases the torsional stiffness is relatively small, the trapeze terms may become dominant. The results for the two strip-like configurations presented above were in very good agreement with both experimental and theoretical results in the literature.

Also, the above results for simple non-linear spring mass system shows that though the EPS perform well, their lack of high frequency numerical dissipation can be a problem. The selection of a smaller time step does not necessarily help the convergence process, as a smaller time step allows even higher frequency oscillations to be present in the response. This could prove to be a real limitation of EPS when applied to more and more complex models. For such models, the use of integration schemes presenting high frequency numerical dissipation become increasingly desirable. It appears that the development of EDS is desirable. This is particularly important when dealing with problems presenting a complex dynamic response such as constrained flexible multi-body problems.

REFERENCES

1. O. A. Bauchau and N.K. Kang, 1993, A multi-body formulation for helicopter structural dynamics analysis, Journal of the American Helicopter Society. 38, 3-14.
2. O. A. Bauchau, 1998, Computational Schemes for Flexible, Nonlinear Multi-body Systems, Multi-body System Dynamics. 2, 169-225.
3. O. A. Bauchau, D. H. Hodges, 1999, Analysis of Nonlinear Multi-body Systems with Elastic Couplings, Multi-body System Dynamics. 4, 168-188.
4. M. Borri and T. Merlini, 1986, A large displacement formulation for anisotropic beam analysis, Meccanica. 21, 30-37.
5. V. L. Berdichevsky, 1979, Variational-asymptotic method of constructing a theory of shells, PMM. 43, 664-687.
6. V. L. Berdichevsky, 1982, On the energy of an elastic rod, PMM. 45, 518-529.
7. D. A. Danielson and D. H. Hodges, 1987, Nonlinear Beam Kinematics by Decomposition of the Rotation Tensor, Journal of Applied Mechanics. 54, 258-262.
8. C. E. S. Cesnik, D. H. Hodges, B. Popescu and D. Harursampath, 1996, Composite beams cross-sectional modeling including obliqueness and trapeze effects. In Proceedings of 37th Structural Dynamics and Materials Conference, 1384-1397, Salt Lake City, Utah, April 15-17, AIAA Paper 96-1469.
9. D. Harursampath, 1998, Non-classical Non-linear Effects in Thin-walled Composite Beams. Ph.D.Thesis, School of Aerospace Engineering, Georgia Institute of Technology, Atlanta, Georgia, U.S.A.
10. C. E. S. Cesnik and D. H. Hodges, 1997, VABS: a new concept for composite rotor blade cross-sectional modeling, Journal of the American Helicopter Society. 42, 27-38.
11. B. Popescu and D. H. Hodges, 1999, Asymptotic treatment of the trapeze effect in finite element cross-sectional analysis of composite beams , International Journal of Non-linear Mechanics, 34, 709-721.
12. Y. Wenbin, 2002, Variational Asymptotic Modeling of Composite Dimensionally Reducible Structures. Ph.D.Thesis, School of Aerospace Engineering, Georgia Institute of Technology, Atlanta, Georgia, U.S.A.
13. E. A. Armanios, A. Makeev and D. H. Hodges, 1996, Finite Displacement Analysis of Laminated Composite Strips with Extension-Twist Coupling, Journal of Aerospace Engineering. 9, 80-91.
14. H. Hodges, D. Harursampath, V. V. Volovoi and C. E. S. Cesnik, 1999, Non-classical effects in non-linear analysis of pretwisted anisotropic strips, International Journal of Non-Linear Mechanics. 34, 259-277.
15. D. H. Hodges, 1990, A review of composite rotor blade modeling, AIAA Journal. 28, 561-565.
16. Y. Wenbin, An Integrated Approach for Efficient High-fidelity Analysis of Composite Structures, AIAA-2005-1911, 46th AIAA/ASME/ASCE/AHS/ASC Structures, Structural Dynamics and Material Conference, Austin, Texas, Apr. 18-21, 2005.
17. T. R. Kane and D. A. Levinson, Dynamics: Theory and Applications, McGraw-Hill, Inc. New York, 1985.

104

Geometrically Nonlinear Free Flexural Vibrations of Composite Cylindrical Shells with Cutouts

NAMITA NANDA AND J.N. BANDYOPADHYAY

Department of Civil Engineering, IIT Kharagpur-72130 West Bengal, India
email: namita@civil.iitkgp.ernet.in, jnb@civil.iitkgp.ernet.in

ABSTRACT

The large amplitude free flexural vibration of laminated composite shallow cylindrical shells in the presence of cutouts is investigated. Nonlinear strains of Von Karman type are incorporated into the first-order shear deformation theory. The finite element model using an eight-noded C^0 continuity, isoparametric quadrilateral element with five degrees of freedom per node is used to study the dynamic behavior. The non linear eigenvalue problem is solved by using the direct iteration method. The validity of the present approach is established by comparing the results obtained by the present study with those available in the literature. The effects of cutout sizes, radii of curvature, lamination schemes and thickness ratios on nonlinear vibration behavior of cylindrical shell are investigated.

Keywords: Cutouts; Finite element analysis; Free vibration; Large amplitude.

1. INTRODUCTION

The increasing application of fiber-reinforced composite materials in spacecraft, high speed aircraft, naval vessels, etc. has resulted in a renewed interest in the vibration of composite plates and shells at large amplitude. Cutouts are inevitable in structures. Cutouts are made to lighten the structure, for venting, to provide accessibility to other parts of the structure and for altering the resonant frequency. Effects of cutouts are likely to be quite considerable when the structure is undergoing large oscillation. It is, therefore, important to predict the natural frequencies of these structural members accurately. Extensive literature reviews on nonlinear vibrations of plates and shells have been reported in Refs [1-4]. Various finite element methods are used for nonlinear vibration of plates in Refs [5-9]. Large amplitude flexural vibration of layered composite plates with cutouts are investigated in Refs [10, 11] employing the finite element method.

Nowinski [12] investigated the large amplitude vibration of thin closed circular cylindrical shells and observed the hard spring behavior. Chia and Chia [13] studied large amplitude flexural vibration of an antisymmetrically laminated angle ply shallow spherical shell with rectangular plan form employing Galerkin's procedure. Sathyamoorthy [14, 15] investigated the effects of large amplitude on the free flexural vibrations of isotropic and orthotropic spherical shells, respectively. He obtained solutions to the system of thick shell equations by means of Galerkin's method and the numerical Runge-Kutta procedure. Ganapathi and Varadan [16] studied nonlinear free vibration of laminated anisotropic circular cylindrical shells using finite element method. Shin [17] investigated large amplitude vibration behavior of symmetrically laminated moderately thick doubly curved shallow open shells with simply supported sides by applying Galerkin's approximation. To the best of authors' knowledge, literature on large amplitude vibration of laminated composite shells with cutouts is not available. Therefore, the present work is aimed at investigating large amplitude free vibration of laminated composite cylindrical shells with cutouts by employing the finite element method. The effects of cutout sizes, radii of curvature, lamination schemes and thickness ratios on nonlinear vibration behavior of cylindrical shell are investigated.

2. ELEMENT DESCRIPTION AND FORMULATION

An eight-noded isoparametric element with five degrees of freedom viz., u, v, w, θ and θ_y at each node is used. The displacement vector at any point on the mid surface is given by

$$w = \sum_{i=1}^{8} N_i(x, y)w_i \qquad \qquad ...(1)$$

where, w_i and N_i are the displacement vector and the interpolating function, respectively, associated with the node i. In this paper, numerical results are presented for laminated cylindrical shells with rectangular plan form dimensions a and b and rectangular cutouts of dimensions c and d in the c and y directions, respectively (Fig. 1). Quarter shell model is used for biaxial symmetry problems. The strain displacement equations of the shear deformable theory of doubly curved shells as given by Reddy [18] are:

$$\varepsilon_x' = \varepsilon_x + z\kappa_x, \qquad \varepsilon_y' = \varepsilon_y + z\kappa_y \qquad \qquad ...(2)$$

$$\gamma_{xy}' = \gamma_{xy} + z\kappa_{xy} \qquad \gamma_{xz}' = \gamma_{xz} \qquad \gamma_{yz}' = \gamma_{yz}$$

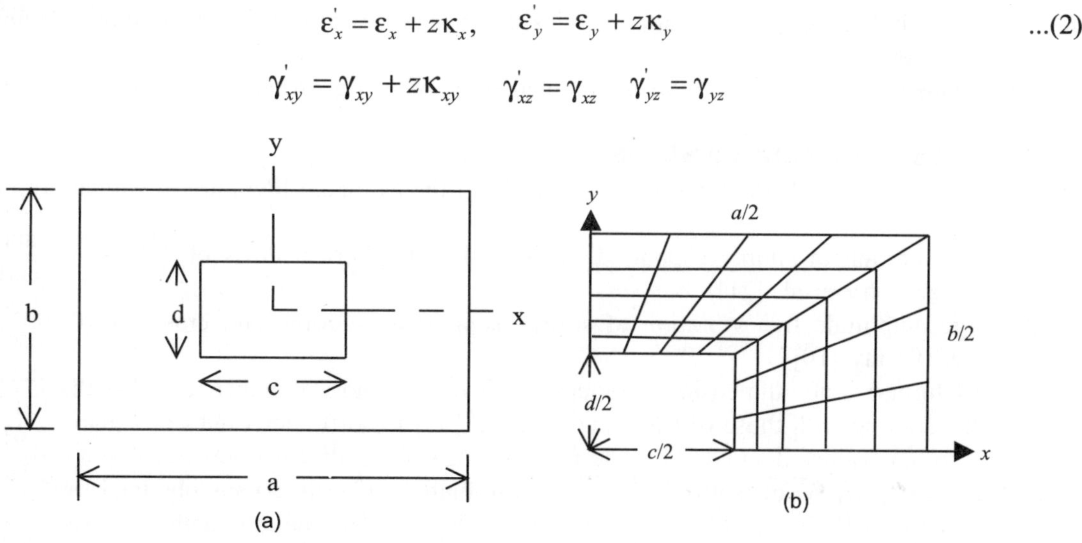

Fig. 1. (a) Geometry of shell with cutout in plan,   **(b)** Finite element mesh of a quadrant.

The strains and curvatures of the middle surface are related to the displacements by

$$\varepsilon_x = \frac{\partial u}{\partial x} + \frac{w}{R_x} + \frac{1}{2}\left(\frac{\partial w}{\partial x}\right)^2 \quad \varepsilon_y = \frac{\partial v}{\partial y} + \frac{w}{R_y} + \frac{1}{2}\left(\frac{\partial w}{\partial y}\right)^2 \quad \gamma_{xy} = \frac{\partial u}{\partial y} + \frac{\partial v}{\partial x} + \frac{2w}{R_{xy}} + \left(\frac{\partial w}{\partial x}\right)\left(\frac{\partial w}{\partial y}\right)$$

$$\gamma_{xz} = \theta_x + \frac{\partial w}{\partial x} - \frac{u}{R_x} - \frac{v}{R_{xy}}$$

$$\gamma_{yz} = \theta_y + \frac{\partial w}{\partial y} - \frac{v}{R_y} - \frac{u}{R_{xy}}$$

$$\kappa_x = \frac{\partial \theta_x}{\partial x} \quad \kappa_y = \frac{\partial \theta_y}{\partial y} \quad \kappa_{xy} = \frac{\partial \theta_x}{\partial y} + \frac{\partial \theta_y}{\partial x} + C_0\left(\frac{\partial v}{\partial x} - \frac{\partial u}{\partial y}\right) \qquad \ldots(3)$$

where, R_x, R_y and R_{xy} (here $R_x = R$ and $R_y = R_{xy} = \infty$) are the usual radii of curvatures.

Term $C_0 \left(= \frac{1}{2}\left(\frac{1}{R_y} - \frac{1}{R_x}\right)\right)$ is a result of Sanders' theory [19] which accounts for the condition of zero strain for rigid body motion.

The free vibration analysis involves determination of natural frequencies from the condition

$$([K] - w^2[M])\{\delta\} = 0 \qquad \ldots(4)$$

where, $[K]$ is the secant stiffness matrix and can be expressed (Wood and Schrefler [20])) as

$$[K] = [K_L] + \frac{1}{2}[N_1] + \frac{1}{3}[N_2] \qquad \ldots(5)$$

$[K_L]$ is the linear stiffness matrix which includes the shear terms also, $[N_1]$ and $[N_2]$ are the nonlinear stiffness matrices.

The frequency amplitude relation is obtained by solving Eq. (4) iteratively.

3. RESULTS AND DISCUSSIONS

The following two sets of problems are taken up from the available literature to validate the present formulation:

1. Free vibration of simply supported cylindrical and spherical cross ply $(0°/90°)_4$ shells with cutouts (Chakravotry et al. [21]).

2. Large amplitude free vibration of simply supported isotropic and cross-ply $(0°/90°)$ plates with cutouts (Reddy [10]).

Table 1 furnishes non-dimensional frequencies (w_l) of cylindrical and spherical shells as obtained by the authors along with those of Chakravorty et al. [21] for different values of square cutout sizes c/a from 0 to 0.5. Tables 2 and 3 show the ratio of linear to nonlinear frequency (ω_l/ω_{nl}) for various amplitude ratios (w_{max}/h) of isotropic and cross-ply plates, respectively, as obtained by the authors along with those of Reddy [10]. Good agreement is observed between the authors' results and those of earlier investigators.

Table 1. Non dimensional fundamental frequencies (ω_l) for shells with concentric cutouts

Shell type	c/a	Authors	Chakravotry et at. [21]
Cylindrical	0.0	26.9872	26.990
	0.1	27.0393	27.042
	0.2	27.2981	27.291
	0.3	27.9642	27.913
	0.4	28.8021	28.711
	0.5	29.2354	29.472
Spherical	0.0	47.3322	47.109
	0.1	47.6728	47.524
	0.2	48.9788	48.823
	0.3	51.1672	50.925
	0.4	54.0276	53.789
	0.5	58.5524	57.524

Note: Lamination = $(0°/90°)_4$, $\omega_l = \omega a^2 (\rho/E_{22}h^2)^{1/2}$, $a/b = 1$, $h/a = 0.01$, $c/d = 1$; For cylindrical, $R_x/h = \infty$, $R_y/h = 300$; for spherical, $R_x/h = R_y/h = 300$.

Table 2. Linear to nonlinear frequency ratio (ω_l/ω_{nl}) of isotropic plate with square cutout

c/a	$a/h = 10$		$a/h = 20$	
	Authors	Reddy [10]	Authors	Reddy [10]
0.2	0.6655	0.6613	0.6707	0.6691
0.5	0.7920	0.7502	0.7548	0.7924
0.6	0.8320	0.8082	0.8355	0.8111

(Note: $w_{max}/h = 1$, | $\upsilon = 0.3$|, $a/b = 1$, $c/d = 1$)

Table 3. Linear to nonlinear frequency ratio (ω_l/ω_{nl}) of cross ply ($0°/90°$) laminate with square cutout

w_{max}/h	Authors	Reddy [10]
0.2	0.9560	0.9625
0.4	0.8553	0.8696

(Note: $E_{11} = 40E_{22}$, $G_{12} = G_{13} = G_{23} = 0.5E_{22}$, $\upsilon_{12} = 0.25$, $a/b = 1$, $a/h = 1000$, $c/d = 1$, $c/a = 0.2$)

After validating the formulation by comparing the authors' results with those of respective authors, the present approach is employed to study the large amplitude vibration of cylindrical shells with lamination scheme, and varying cutout sizes, radii of curvature and side to thickness ratios. The material properties are as follows: $E_{11} = 25E_{22}$, $G_{12} = G_{13} = 0.5E_{22}$, $G_{23} = 0.2 E_{22}$ and $\upsilon_{12} = 0.25$. The non-dimensional natural frequency parameter is $\omega_l = \omega a^2(\rho/E_{22}h^2)^{1/2}$.

To demonstrate the effects of cutout, cutout sizes are varied from 0 to 0.5 using R/h as 300 and h/a as 0.01. The results presented in Fig. 2 reveal that for all cutout sizes (c/a), there is consistent increase of frequency ratio (ω_{nl}/ω_l) with the increase of amplitude ratio (w_{max}/h), indicating that the nonlinear vibration behavior is of hardening type. Further, it is also observed that there is decrease in the hardening effect with the increase of cutout size. Figure 3 shows the effect of R/h on large amplitude vibration of cylindrical shell with c/a as 0.2. Here R/h is varied from 300 to 2000 keeping h/a as 0.01.

Hard spring behavior is observed from the increasing trend of frequency ratio with the increase of amplitude ratio in Fig 3. Moreover, such consistent increase of frequency ratio at any particular amplitude is also observed for R/h from 300 to 2000 indicating that there is positive stiffening effect of the shell. From Fig. 4, showing the effect of lamination angle on nonlinear behavior of cylindrical shell with c/a as 0.2, it is observed that $(0°/90°)$ lamination scheme produces higher nonlinearity than other orientations discussed here. From Fig. 5, showing the effect of thickness ratio on nonlinear behavior of cylindrical shell with c/a as 0.2, it is observed that for any thickness ratio (ω_{nl}/ω_t) from 0.01 to 0.1, there is consistent increase of frequency ratio h/a with the increase of amplitude ratio (w_{max}/h), indicating that the nonlinear vibration behavior is of hardening type. For $h/a = 0/001$, the frequency ratios are reduced at higher amplitudes.

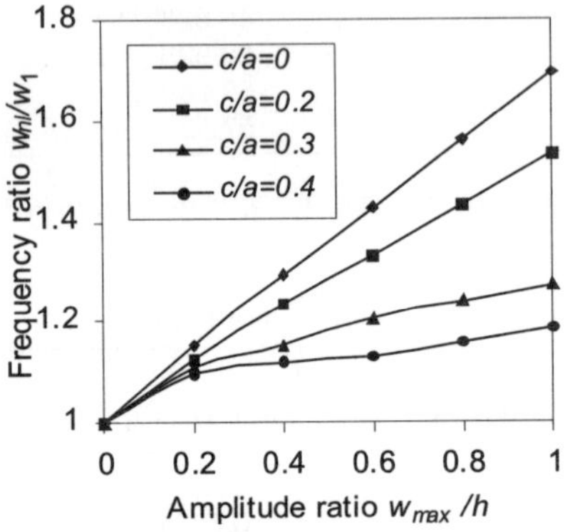

Fig. 2. Effect of cutout size on the frequency ratio of cylindrical shell (h/a = 0.01, R/h = 300, lamination = $(0°/90°)$).

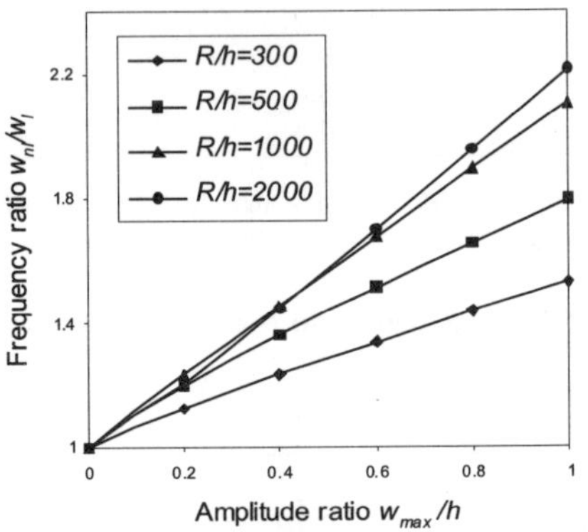

Fig. 3. Variation of frequency ratio with amplitude ratio for various radii of curvature (c/a = 0.2, h/a = 0.01, $(0°/90°)$).

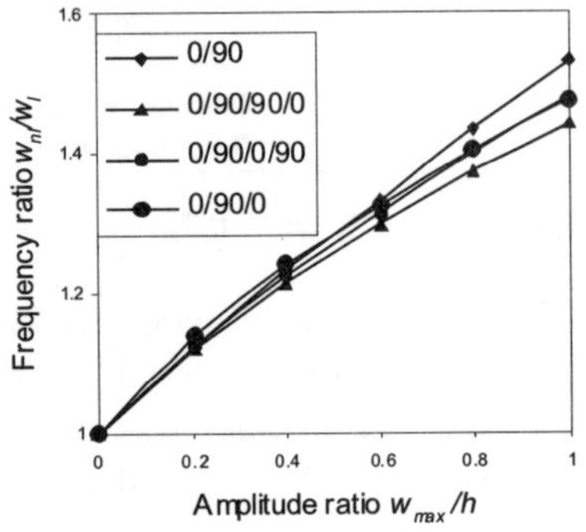

Fig. 4. Effect of lamination angle on the frequency ratio of cylindrical shell (h/a = 0.01, c/a = 0.2, R/h = 300).

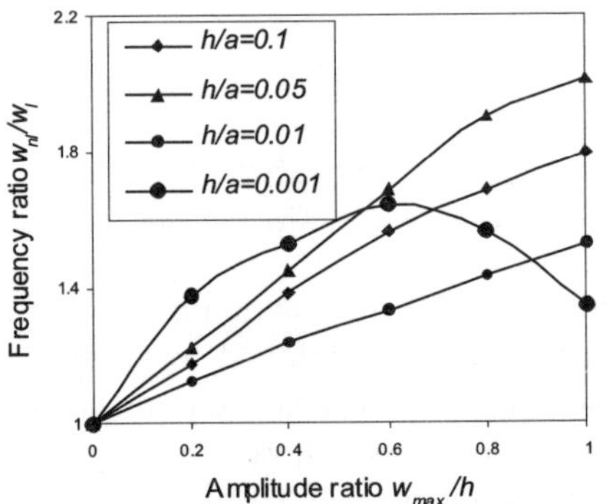

Fig. 5. Variation of frequency ratio with amplitude ratio for various radii of curvature (c/a = 0.2, R/h = 100, $(0°/90°)$).

CONCLUSION

The following are the conclusions of the present study

    1. The finite element formulation as developed here gives comparable results for plates and shells for the problems solved earlier and presented in Tables 1-3.

2. The cylindrical shell shows hard spring behavior for all values of c/a, R/h and h/a parameters which is identical to those of plates as obtained by Reddy [10] and Sivakumar *et al.* [11].
3. There is decrease in the hardening effect with the increase of cutout size.
4. For shallow shells hardening effect is more.
5. For all the lamination angle considered, nonlinear effect is more for two layer cross-ply shell and less for four layer symmetric cross-ply shell. Hardening type nonlinearity increases with the increase in thickness ratio except for certain higher amplitudes when h/a is 0.001.

REFERENCES

1. C. Y. Chia, 1980, Nonlinear analysis of plates, McGraw Hill, New York, USA.
2. M. Sathyamoorthy, 1987, Nonlinear vibration analysis of plates: a review and survey of current developments, Applied mechanics reviews. 40, 1553-1561.
3. C. Y. Chia, 1988, Geometrically nonlinear behavior of composite plates: a review, Applied mechanics reviews. 41, 439-451.
4. K. A. Alhazza, A. A. Alhazza, 2004, A review of the vibrations of plates and shells, The shock and vibration digest. 36, 377-395.
5. K. Kanaka Raju, E. Hinton, 1980, Nonlinear vibrations of thick plates using Mindlin plate elements, International Journal of Numerical Methods in Engineering. 16, 247-257.
6. J. N. Reddy, W. C. Chao, 1981, Large deflection and large amplitude free vibrations of laminated composite material plates, Computers & Structures. 13, 341-347.
7. J. N. Reddy, W. C. Chao, 1982, Nonlinear oscillations of laminated, anisotropic, rectangular plates, Journal of applied mechanics, 49, 396-402.
8. M Ganapathi, T. K. Varadan, B. S. Sarma, 1991, Nonlinear flexural vibrations of laminated orthotropic plates, Computers & Structures. 3, 685-688.
9. T. Kant, J. R. Kommineni, 1994, Large amplitude free vibration analysis of cross-ply composite and sandwich laminates with a refined theory and C0 finite elements, Computers & Structures. 50, 123-134.
10. J. N. Reddy, 1982, Large amplitude flexural vibration of layered composite plates with cutouts, Journal of sound and vibration. 83, 1-10.
11. K. Sivakumar, N. G. R. Iyengar, K. Deb, 1999, Free vibration of laminated composite plates with cutout, Journal of sound and vibration. 221, 443-470.
12. J. L. Nowinski, 1963, Nonlinear transverse vibrations of orthotropic cylindrical shells, AIAA journal. 3, 617-620.
13. C. Y. Chia, D. S. Chia, 1992, Nonlinear vibration of moderately thick anti-symmetric angle-ply shallow spherical shell, Computers & Structures, 44, 797-805.
14. M. Sathyamoorthy, 1994, Vibrations of moderately thick shallow spherical shells at large amplitudes, Journal of sound and vibration. 172, 63-70.
15. M. Sathyamoorthy, 1995, Nonlinear vibration of moderately thick orthotropic shallow spherical shells, Computers & Structures, 57, 59-65.
16. M. Ganapathi, T. K. Varadan, 1995, Nonlinear free flexural vibrations of laminated circular cylindrical shells, Composite Structures. 30, 33-49.
17. D. K. Shin, 1997, Large amplitude free vibration behavior of doubly curved shallow open shells with simply supported edges, Computers & Structures. 62, 35-49.
18. J. N. Reddy, 1984, Exact solutions of moderately thick laminated shells, ASCE Journal of Engineering Mechanics, 110, 794-809.
19. J. L. Sanders, 1959, An improved first approximation theory for thin shells, NASA TR-24.
20. R. D. Wood, B. Schrefler, 1978, Geometrically nonlinear analysis: A correlation of finite element notations, International Journal of Numerical Methods in Engineering. 12, 635-642.
21. D. Chakravorty, P. K. Sinha, J. N. Bandyopadhyay, 1998, Applications of FEM on free and forced vibration of laminated shells, ASCE Journal of Engineering Mechanics 124, 1-8.

105

Asymptotically Accurate Nonlinear Analysis of Inflatable Structures

B. Ramesh Gupta[1] and Dinesh Kumar Harur Sampath[2]

[1]Department of Aerospace Engineering, Institute of Science, Bangalore-560012, India
email: ramesh@aero.iisc.ernet.in
[2]Department of Aerospace Engineering, Indian Institute Science, Bangalore-560012, India
email: dinesh@aero.iisc.ernet.in

ABSTRACT

The focus of this work is on the development of an intrinsic formulation of an asymptotically correct theory for inflatable structures. The problem is both geometrically and materially nonlinear. The geometric nonlinearity is handled by allowing for finite deformations and generalized warping [4] while the material nonlinearity is incorporated through a recently validated hyperelastic material model [3]. The development, based on the Variational Asymptotic Method (VAM) [1], begins with three-dimensional nonlinear elasticity and mathematically splits the analysis into a one-dimensional through-the-thickness analysis and a two-dimensional membrane analysis [2].

The through-the-thickness analysis provides constitutive relation between the generalized, two dimensional strain and stress tensors for the membrane analysis and a set of recovery relations to approximately express the three-dimensional displacement, strain and stress fields in terms of two-dimensional variables determined from solving the equations of the membrane analysis. Numerical examples are presented to compare with existing analytical or semi-analytical or finite element solutions. Results based on this model will be demonstrated for specific inflatable structures.

Keywords: Shell; Variational Asymptotic Method; hyperelastic model; 3-D stress-strain recovery.

1. INTRODUCTION

Inflatable structures are membrane structures or any ultra-low-mass hybrid structure with extensive use of membranes. Inflatable structures have special properties like lightweight, low deflated volume, and high strength-to-mass ratio. These are suitable for large space applications such as large solar array, solar antennas, solar sails and optical mirrors, etc. inflatable structures, can potentially revolutionize the designs and applications of large space structures. These structures are analyzed for both geometric and material nonlinearity. The geometric nonlinearity is handled by allowing for finite deformations and generalized warping [4] through Green strain, while the material nonlinearity is incorporated through a recently validated hyperelastic material model.

An asymptotically correct shell theory for modeling/analysis of inflatable structures can be developed by taking the advantage of small parameters associated with geometry and stiffness (material).

Shells are dimensionally reducible structures; this reduction is carried out using the variational asymptotic method. Thus, 3-D problem splits into a nonlinear, one-dimensional (1-D), through-the-thickness analysis and a nonlinear, two-dimensional (2-D), shell analysis.

Through thickness analysis (1-D) provides a 2-D nonlinear constitutive law for the shell equations and a set of recovery relations that can be used to express the 3-D field variables (displacements, strains and stresses) through the thickness in terms of 2-D shell variables calculated in the shell analysis (2– D).

Validation of present theory is carried out for the two test cases. First, Henky's problem, inflation of circular membrane. Comparisons are made with existing analytical, semi-analytical and finite element solutions. The second validation is related to dynamic analysis of rectangular plate. Comparisons are made with finite element solutions.

Further present method will be applied to specific deployable structure. The ease of handling such a large space based structures with combined nonlinearity can be demonstrated.

2. INTRINSIC FORMULATIONS

A shell is a 3-D body with a relatively small thickness h and a smooth reference surface usually chosen to be the mid surface. The geometry of the reference surface can be mathematically represented by a set of arbitrary curvilinear coordinates, x_i. (here and throughout the formulation, Greek indices assume values 1 and 2 while Latin indices assume 1, 2, and 3. Dummy indices are summed over their range except where explicitly indicated). Where, x_α are inplane coordinates and x_3 is normal coordinate. Lines of curvatures are assumed to be the curvilinear coordinates to simplify the formulation. Let b_i, B_i denote the unit vectors along x_i for the undeformed and deformed configurations respectively.

One can then describe the position of any material point in the undeformed configuration and deformed configuration by its position vectors $\hat{r}$ and $\hat{R}$, respectively from a fixed point O, such that

$$\hat{r}(x_1, x_2, x_3) = r(x_1, x_2) + x_3 b_3(x_1, x_2) \qquad \text{...(1)}$$

$$\hat{R}(x_1, x_2, x_3) = R(x_1, x_2) + x_3 B_3(x_1, x_2) + w_i(x_1, x_2, x_3) B_i(x_1, x_2) \qquad \text{...(2)}$$

and covariant and contravariant vectors in undeformed state are given by , $g_i = \dfrac{\partial \hat{r}}{\partial x_i}$,

$$g^i(x_l) = \frac{1}{2\sqrt{g}} e_{ijk} g_j \times g_k \text{ where, } g = \det(g_i \cdot g_j) \qquad \text{...(3)}$$

i.e.,
$$g_1 = A_1(1 + x_3 k_{11}) b_1 , g_2 = A_2(1 + x_3 k_{22}) b_2 , g_3 = b_3$$
and

$$g^1 = \frac{b_1}{A_1(1 + x_3 k_{11})} , \quad g^2 = \frac{b_2}{A_2(1 + x_3 k_{22})} , g^3 = b_3 \qquad \text{...(4)}$$

Similarly, covariant base vectors for deformed configuration are given by . $G_i = \dfrac{\partial \hat{R}}{\partial x_i}$ ...(5)

The relation between B_i and b_i can be specified by an arbitrarily large rotation specified in terms of the matrix of direction cosines $C(x_1, x_2)$ so that

$$B_i = C_{ij} b_j, \quad C_{ij} = B_i b_j \qquad \text{...(6)}$$

2.1 Green Strain

In the present scheme all possible deformations (large displacements and rotations) are allowed and corresponding Green strain components are

$$E_{ij} = \frac{1}{2}\left(F^T_{\ ik} F_{kj} - I_{ij}\right) \qquad \text{...(7)}$$

where, F_{ij} are components of deformation gradient tensor, given by $F_{ij} = B_i . G_k g^k . b_j$ ...(8)

2.2 Strain Energy Functions

Two hyperelastic material constitutive models are used for numerical purposes:

1. The Saint-Venant Kirchhoff model, for which the strain energy is

$$\psi = \frac{\lambda}{2}(trE)^2 + \mu E : E \qquad \text{...(9)}$$

   Where, λ and m are material parameters and

2. The compressible isotropic model described by a strain energy of neo-Hookean type

$$\psi = \frac{\mu}{2}(trC - 3) - \mu \ln J + \frac{\lambda}{2}(\ln J)^2 \qquad \text{...(10)}$$

where, $C = (F^T F) J = \det (C)$

2.3 Potential due to Load

Load contribution from top and bottom surfaces and body loads on the normal are

$$PL = \tau^T w^+ + \beta^T w^- + \langle \phi^T w \rangle \qquad \text{...(11)}$$

where $(\ )^+ = (\ )^{h/2}, (\ )^- = (\ )^{-h/2}$ and $\langle \ \rangle$ = integration through the thickness.

   Total potential energy: . $\Pi = \psi - PL$ ...(12)

First. Variation of the functional w.r.t to warping functions subjected to constraints is $\delta \Pi = 0$.

Till this point its an alternative to 3-D elasticity formulation. If we attempt to solve this problem directly, we will meet the same difficulty as solving any full 3-D elasticity problem. Fortunately, as shown below, VAM can be used to calculate the 3-D unknown warping functions asymptotically.

3. DIMENSIONAL REDUCTION

Now, to rigorously reduce the original 3-D problem to a 2-D shell problem, one must attempt to reproduce the energy stored in the 3-D structure in a 2-D formulation. This dimensional reduction can only be done approximately, and one way to do it is by taking advantage of the smallness of h/l, h/r and ε. As mentioned before, although the reduced models based on ad hoc kinematic assumptions regularly appear in the literature, there is no basis whatsoever to justify such assumptions. Rather, in this work, the VAM will be used to mathematically perform a dimensional reduction of the 3-D problem to a series of 2-D models. To proceed by this method, one has to assess and keep track of the orders of all the quantities in the formulation.

Where, Δ is the order of the maximum strain in the shell and μ is the order of the material constants (all of which are assumed to be of the same order). VAM requires one to find the leading terms of the functional according to the different orders. The total potential energy consists of quadratic expressions involving the warping and the generalized strains. In addition there are terms that involve the loading along with interaction terms between the warping and the both of the other types of quantities. Since, only the warping is varied, one needs the leading terms that involve warping only and the leading terms that involve the warping and other quantities (i.e., the generalized strain and loading).

For the first approximation, these leading terms are denoted by Π_0

Stationary points, for the unknown 3-D warping functions are found by minimizing the functional subject to constraint, $\langle w_i(x_1, x_2, x_3) \rangle = 0$

The first approximation of the energy functional consist of terms not higher than $(h/R)^0, (h/l)^0$ which coincides with classical laminated shell theories. However, we do not use ad hoc kinematic assumptions such as the Kirchhoff assumption to obtain this result. The result is also the same as the first approximation of plate theory because we have not yet included the geometrical correction due to the initial curvatures.

The characteristics of the first approximation may be summarized as: (1) normal line elements of the undeformed shell remain straight and normal to the deformed shell average surface; (2) since, warping component w_3 for each layer is a quadratic polynomial, the normal strain is not zero; (3) the transverse normal stress is zero; (4) the transverse shear stresses are zero; and (5) the transverse energy is zero

Let us obtain the correction coming from h/R first to include the effect of initial curvatures of the structure. Usually to find the refinement, one needs to calculate the refined warping functions based on the next approximation. However, since we are only interested in obtaining energy asymptotically correct up to the order of h/R which is sufficient for most of the engineering applications, it is unnecessary to calculate the refined warping with respect to h/R which makes no contribution to the energy up to the order of h/R. At this stage we will postpone the consideration of the load contribution to the step of h/l correction if there are any. A strain energy that is asymptotically correct through the order of h/R can be expressed as . Present theory is implemented using a symbolic manipulator MathematicaTM. Due to space constraints analytical expressions are not presented.

4. NONLINEAR SHELL THEORY

From above, we have found an asymptotically accurate shell model. It is the purpose of this section to develop a geometrically exact nonlinear shell theory consistent with the obtained generalized 2-D nonlinear constitutive law to obtain 2-D results (displacements, strains and stresses). The formulation of the 2-D nonlinear shell theory will still based on the choice of lines of curvatures as the curvilinear coordinates. After developing a set of the compatibility equations between 2-D strain measures, we derive the 2-D kinematics relating 2-D strain measures and the deformation of the reference surface. Then we express the variations of intrinsic strain measures in terms of virtual displacements, which are independent of the displacement or rotational variables chosen. Finally, the principle of virtual work is used to derive the intrinsic equilibrium equations and the associated boundary conditions. The compatibility equations, kinematics and equilibrium equations along with the constitutive law comprise a complete set of governing equations for a 2-D nonlinear shell theory.

5. RECOVERY RELATIONS

From the above, we have obtained an asymptotically accurate model for composite shells which is as close as possible to the total energy. One can use this model to carry out various analyses (for example, static, dynamic and aeroelastic) for composite shells. In many applications, however, while it is necessary to accurately calculate the 2-D displacement field of composite shells, this is not sufficient. Ultimately, the fidelity of a reduced-order model such as this depends on how well it can predict the 3-D results in the original 3-D structure. Hence, recovery relations should be provided to complete the theory so that the results can be compared with those of the original 3-D model. By recovery relations, then, we mean expressions for 3-D displacement, strain, and stress fields in terms of 2-D quantities and x_3. For an energy that is asymptotically correct through the second approximation, we can recover the 3-D displacement, strain, and stress fields only through the first approximation in a strict sense of asymptotical correctness.

Using Eqs. (1), (2), and (6), one can recover the 3-D displacement field through the first order as

$$U_i = u_i + x_3 C_{3i} + C_{ij} w_j$$

One can use the 3-D constitutive law to obtain 3-D stresses . However, according to the present approach, the transverse normal stress σ_{33} is a second-order quantity and not available in the first approximation. Despite that it is usually much smaller than other components; σ_{33} is a very critical component for some failures such as debonding or delamination. In order to obtain a reasonable recovery for the transverse normal stress, VAM needs to be applied one more time to find the warping functions of the order of $(h/l)^2$. Using the same procedure listed in previous section, the second-order warping function w_3 can be obtained.

6. NUMERICAL RESULTS AND DISCUSSIONS

Two test cases are chosen for validation of present theory.

First, Central deflection of Circular plate of Nonlinear-elastic, isotropic material model.

- Neo-Hookean strain energy function (hyperelastic material model).
- Fixed boundary condition along the periphery.
- Incremental uniform pressure on the surface.
- Geometry and material properties: radius a = 0.1375 m, equivalent Young's modulus E* = 600kPam, Poisson's ratio v = 0.3.

Analysis is carried out for low pressure range from 0-1000kPa and results are compared well with existing literature [3] Figure (1).

Second, Time domain analysis of cantilever rectangular plate subjected to time varying load at tip.

- Linear elastic isotropic material model.
- Step load, 1 kN is applied from 0–0.1 sec and observations are taken up to 0.5 sec.
- Geometry and material properties: length = 0.45 m, width = 0.3, thickness = 0.01, Young's modulus = 210e9, Poisson's ratio = 0.3125.

Results are compared in terms of transverse deflection at the center of the plate with ANSYS commercial FEM software. Good agreement results are observed (Figure 2).

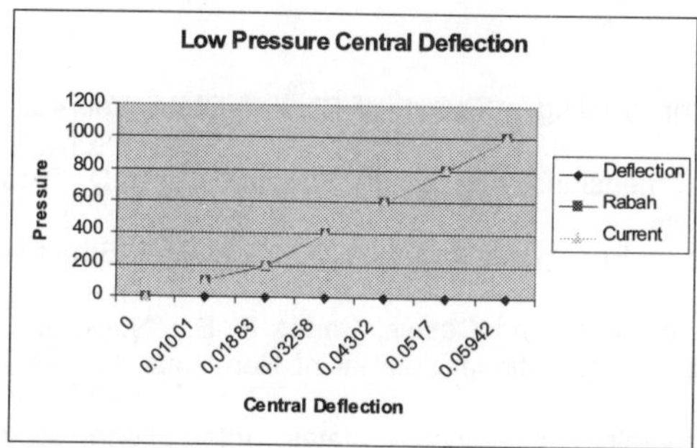

Fig. 1. Central Deflection Vs the pressure, neo-Hookean potential, low pressure range.

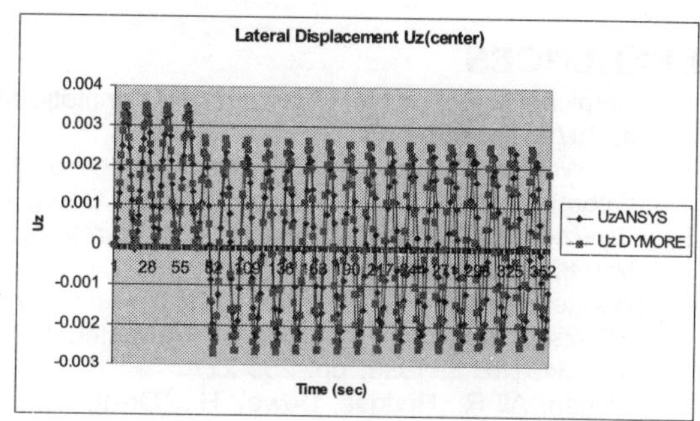

Fig. 2. Tip deflection.

CONCLUSION

An asymptotically accurate theory has been developed. Validation is carried out for isotropic, static nonlinear material model as well as linear elastic dynamic analysis, and present results give good agreement with existing literature. Further, validation is taking place for dynamic analysis of membrane structures made of composite materials as well as 3-D results in the form of recovery relations, and finally present theory will be applied for specific inflatable structures.

NOMENCLATURE

U_i, u_i	3-D displacement components, 2-D (shell) displacement components
$\hat{r}$, $\hat{R}$	Position vector undeformed configuration, deformed configuration
x_i	Coordinates
b_i, B_i	Base vectors
g_i, g^i	Covariant, contra-variant vectors
G_i	Covariant vectors
F	Deformation gradient tensor
C	Direction cosine matrix
E	Green strain
Ψ	Strain energy per unit volume

PL	Potential due to load
Π	Total potential energy
h, R, l	Thickness, radius, length
λ, μ	Material constants
τ, β, ϕ	Top surface, bottom surface load, body load
w	Warping
σ_{ij}	Stress components

ACKNOWLEDGEMENTS

I gratefully acknowledge many fruitful discussions with Prof. Dinesh Kumar Harursampath, IISc, Bangalore, for the preparation of this manuscript.

REFERENCES

1. Berdichevsky, Victor L., "Variational-Asymptotic Method of Constructing a Theory of Shells," PMM, Vol. 43, No 4, 1979, pp. 664-687.
2. Yu, Wenbin., "Variational Asymptotic Modeling of Composite Dimensionally Reducible Structures," Ph. D. Thesis, Georgia Institute of Technology, Atlanta, Georgia, U.S.A., 1998.
3. Bouzidi, R., "Numerical Solution of Hyperelastic Membranes by Energy Minimization," Computers & Structures, Vol. 82, 2004, pp. 1961-1969.
4. Hodges, Dewey H., Harursampath, Dinesh Kumar, Volovoi, Vitali V., and Cesnik, Carlos E. S., "Non-classical Effects in Non-linear Analysis of Pretwisted Anisotropic Strips," International Journal of Non-linear Mechanics, Vol. 34, No. 2, 1999, pp. 259-277.
5. Atilgan, Ali R., Hodges, Dewey H., "On the strain energy of laminated composite plates," International Journal of Solids & Structures, Vol. 29, No. 20, 1992, pp. 2527-2543.
6. Yu, W., Hodges, D.H., Volovoi, V.V., 2002. "Asymptotic construction of Reissner-like models for composite plates with accurate strain recovery," International Journal of Solids and Structures, Vol. 39, pp. 5185-5203.
7. Jenkins, Christopher H.M., "Gossamer spacecraft: Membrane and inflatable structures technology for space application," Vol. 191, Progress in astronautics and aeronautics, 2001.
8. Salama, M., Kuo, C. P., Lou, M., "Simulation of Deployment Dynamics of Inflatable Structures," AIAA Journal, Vol. 38, No. 12, 2000, pp. 2277- 2283.
9. Wang, John. T., Johson, A. R., "Deployment simulation of ultra-lightweight inflatable structures," AIAA-2002-1261.
10. Lou, M.C, Fang, H., "Analytical characterization of space inflatable structures-an overview," AIAA-99-1272.
11. Fichter WB. Some solutions for the large deflections of uniformly loaded circular membranes. NASA Technical Paper 3658-NASA Langley Research Center; Hampton, VA; 1997.
12. Reddy, J. N., Mechanics of Laminated Composite Plates: Theory and Analysis. CRC Press, Boca Raton, Florida, 1997.
13. Wempner, G., Mechanics of Solids with Applications to Thin Bodies. Sijthoff & Noordhoff, 1981.
14. Bonet, Javier., Wood, Richard D., Nonlinear continuum mechanics for finite element analysis. Cambridge University Press, 1997.
15. Malvern, Lawrence E., Introduction to the mechanics of a continuous medium. Prentice Hall, Inc., 1969.
16. Forray, Marvin J., Variational calculus in science and engineering. Mc-Grahill Book Company, 1968.

106

Vibration Control of Nonlinear Beams Using Optimal Dynamic Inversion with Discrete Actuators

SK. FARUQUE ALI[1] AND RADHAKANT PADHI[2]

[1]Department of Civil Engineering, Indian Institute of Science, Bangalore-560012, India.
email: skfali@civil.iisc.ernet.in
[2]Department of Aerospace Engineering, Indian Institute of Science, Bangalore-560012, India.
email: padhi@aero.iisc.ernet.in

ABSTRACT

Most of the structural elements, like beams, cables, etc. are distributed parameter systems. The usual approach of modal representation is not an accurate one for those structures. Therefore, one has to control excessive vibrations of these structures using the system partial differential equation (PDE) for better representation of the reality. In this paper, we propose to use a recently developed optimal dynamic inversion technique to design such a set of controllers for this purpose. We assume that the control force to the structure is applied through finite number of actuators, which are distributed along the spatial domain and located at pre-defined locations. The formulation has better practical significance, both because it leads to closed form solution of the controller (hence, avoids computational issues) as well as because a set of discrete actuators along the spatial domain can be implemented with relative ease as compared to a continuous actuator.

Keywords: Dynamic Inversion; Optimal Dynamic Inversion; Nonlinear Beam; Vibration Control.

1. INTRODUCTION

Beams, cables, plates and shells are important structural elements and are modeled as distributed parameter systems (DPS). Structural engineers have to deal with them very often. They are often subjected to excessive vibration under various dynamic loads, like wind, earthquake, etc. Control is necessary to suppress the excessive structural vibrations, since natural damping of these structures are usually small. Control of bridge and building structures using active, passive, semi-active and hybrid means are studied by many authors, review can be found in [1, 2]. They are designed based on approximate-then-design philosophy. Where the system with infinite degree of freedoms is reduced to finite degrees of freedom discretized in space known as full-order-systems (FOS) using a set of orthogonal basis functions (Galerkin projection). This FOS results in a large number of degree of freedom and consists of high-frequency dynamics which requires large computational time for controller simulation. Therefore, a reduced-order-system (ROS) model is developed using Guyan reduction, static and dynamic condensation, or eigen mode reduction. These methods are very much problem specific. Therefore, when ROS-based control design is synthesized and applied to a real structure, inevitable errors like control and observer spillover and possible instability are introduced [2]. To minimize these errors one has to design a controller based on system PDE.

Long span bridges, highway bridges, chimney, etc, are designed as beams in civil engineering field. All previous studies of controlling beam either have taken a linear model or used lumped mass approximation. Yang [3] studied the problem of discrete control of a continuous Euler-Bernoulli beam in frequency domain. It has been shown in [4] that the modeling of beams as lumped mass and providing control force based on the model may lead to instability under control force. Therefore, practical application of control engineering lies in controlling beams as distributed parameter system directly from the system PDE. Recently, Padhi [5] developed optimal dynamic inversion technique to control distributed parameter system.

This paper controls nonlinear Euler-Bernoulli beam vibration, in time domain, using the philosophy proposed by one of the authors of the paper in [5]. The control design formulation assumes a constant control action provided by the actuators in the region of its contact. Therefore, the method has more practical significance as one can implement a set of discrete controllers with relative ease. Moreover, the formulation leads to a closed form solution for the control variable and hence, can be implemented easily in real time. Numerical results with random initial condition show that beam vibration can be effectively bought to zero with only few actuators.

2. NONLINEAR BEAM MODEL

The problem of nonlinear free vibrations of a simply supported elastic Euler-Bernoulli beam is revisited and controlled using a set of discrete actuators. The equations of motion with finite deformation and small rotation is obtained after application of Hamilton's Principle as in [6] and is given as

$$m\ddot{x}+c\dot{x}+EI\,x''''-\frac{EA}{2l}\left[x''\int_0^l\{x'\}^2\,dy\right]=f(y,t) \qquad ...(1)$$

where, $(')$ and $(\cdot)$ represents derivatives with respect to spatial variable y and time t. x is displacements along transverse direction. m is the mass per unit length of the beam. c, E, A, I are the damping co-efficient, elastic modulus, area of cross-section and moment of inertia of the beam section, respectively. $f(y, t)$ represents external force applied to the beam.

In the formulation the nonlinearities above cubic order are truncated and longitudinal inertia is assumed negligible in comparison to transverse inertia. Nonlinear strain displacement is used to derive the equation of motion which couples axial and transverse stiffness. The axial stiffness term makes the system nonlinear,

When we substitute $f(y, t)$ with the control force $(\bar{u}_n)$, the Eq.(1) becomes

$$m\ddot{x}+c\dot{x}+EI\,x''''-\frac{EA}{2l}\left[x''\int_0^l\{x'\}^2\,dy\right]=\frac{1}{m}\sum_{n=1}^{N}\bar{u}_n\,F(y,y_n,w_n) \qquad ...(2)$$

where, function $F(y, y_n, w_n)$ is defined as:

$$F(y, y_n, w_n) \triangleq \begin{bmatrix} 1 & \forall\ y_n - \dfrac{w_n}{2} \le y \le y_n + \dfrac{w_n}{2} \\ 0 & Otherwise \end{bmatrix} \qquad ...(3)$$

The boundary conditions are given as

$$x(0,t) = 0; \quad x(l,t) = 0; \quad x''(0,t) = 0; \quad x''(l,t) = 0; \qquad ...(4)$$

where, t represent time in seconds. As simply supported beam is assumed the displacement and moments at the supports are zero.

A set of discrete controllers $\bar{u}_n$ (see Fig. 1) that are located at y_n ($n = 1, ..., N$) locations of the beam are assumed. Width of the controller at y_n is considered to be w_n.

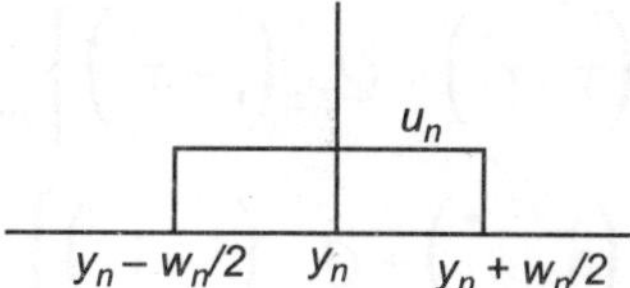

Fig. 1. Control action at y_n, $(\bar{u}_n)$.

Here the following assumptions are made:

- In the interval $\left[y_n - w_n/2,\ y_n + w_n/2\right]$, the controller $\bar{u}_n(y,t)$ is assumed to have a constant magnitude. Outside this interval, $\bar{u}_n = 0$. Note that the interval w_n may or may not be small. This property is obtained by using the function $F(y, y_n, w_n)$ as defined in Eq. (3).
- There is no overlapping of the controller located at y_n with its neighboring controllers.
- No controller is placed exactly at the boundary. This eliminates the situations where control action enters the system dynamics through boundary actions.

The dynamical equation of motion for system Eq. (2) can now be written as follows

$$\ddot{x} = -\left(\frac{c}{m}x + \frac{EI}{m} - \frac{EA}{2ml}\left[x''\int_0^l \{x'\}^2\, dy\right]\right) + \frac{1}{m}\sum_{n=1}^{N} \bar{u}_n F(y, y_n, w_n) \qquad ...(5)$$

with appropriate boundary given in Eq. (4).

3. OPTIMAL DYNAMIC INVERSION CONTROLLER SYNTHESIS

The goal of the proposed controller is to ensure that the state variables go to target values, *i.e.*, $x(y,t) \to x^*(y,t)$ and $\dot{x}(y,t) \to \dot{x}^*(y,t)$ as $t \to \infty$ for all $y \in [0, L]$. In our beam problem we have taken the target states to be zeros.

First, we define an output (an integral error) term as follows

$$z(t) = \frac{1}{2}\int_0^l \left\{[\Delta x\ \ \Delta\dot{x}]\, Q \begin{bmatrix} \Delta x \\ \Delta\dot{x} \end{bmatrix}\right\} dy \qquad ...(6)$$

where, $\Delta x = (x - x^*)$, $\Delta \dot{x} = (\dot{x} - \dot{x}^*)$, and $Q > 0$ is weighing matrix. Note that when $z(t) \to 0$, $x(y,t) \to x^*(y,t)$ and $\dot{x}(y,t) \to \dot{x}^*(y,t)$ everywhere in $y \in [0, 1]$. The motivation for taking an integral error term is to make sure that the system reaches its goal in an integral sense. Therefore, the L_2 norm of the difference of the desired and actual state should approach zero. Next, the principle of dynamic inversion [7] is used to design a controller such that the following first-order equation is satisfied

$$\dot{z} + k\, z = 0 \qquad \qquad ...(7)$$

where, $k > 0$ serves as a gain; an appropriate value of k has to be chosen by the control designer. For a better physical interpretation, one may choose it as $k = (1/\tau)$, where $\tau > 0$ serves as a "time constant" for the error $z(t)$ to decay. Using Eqs. (5) and (6) in Eq. (7) and then carrying out the necessary algebra we get

$$\left[\int_{y_1 - \frac{w_1}{2}}^{y_1 + \frac{w_1}{2}} \left\{ Q_{1,2}\left(x - x^*\right) + Q_{2,2}\left(\dot{x} - \dot{x}^*\right) \right\} \frac{1}{m}\, dy \right] \bar{u}_1 + \cdots +$$

$$\left[\int_{y_1 - \frac{w_n}{2}}^{y_1 + \frac{w_n}{2}} \left\{ Q_{1,2}\left(x - x^*\right) + Q_{2,2}\left(\dot{x} - \dot{x}^*\right) \right\} \frac{1}{m}\, dy \right] \bar{u}_N = \gamma \qquad ...(8)$$

where

$$\gamma = -\int_0^l \left[\begin{array}{c} \left(Q_{1,1}\,\Delta x \Delta \dot{x} + Q_{1,2}(\Delta \dot{x})^2 \right) + \left\{ \left(Q_{2,1}\,\Delta x + Q_{2,2}\,\Delta \dot{x}\right) \left(-\frac{c}{m}x - \frac{EI}{m} + \frac{EA}{2ml}\left[x'' \int_0^l \{x'\}^2\, dy \right] \right) \right\} + \cdots \\[2mm] \cdots \frac{k}{2}\left(\left\{ [\Delta x \;\; \Delta \dot{x}]\, Q \begin{bmatrix} \Delta x \\ \Delta \dot{x} \end{bmatrix} \right\} \right) \end{array} \right]\, dy\, F$$

or convenience, we define

$$I_n \triangleq \int_{y_n - \frac{w_n}{2}}^{y_n + \frac{w_n}{2}} \left\{ Q_{1,2}\,\Delta x + Q_{2,2}\,\Delta \dot{x} \right\} \frac{1}{m}\, d y, \qquad \qquad n = 1, ..., N \qquad ...(9)$$

Then from Eqs. (8) and (9), we can write

$$I_1 \bar{u}_1 + \quad \cdots \quad + I_N \bar{u}_N = \gamma \qquad \qquad ...(10)$$

Equation (10) will eventually guarantee that $z(t) \to 0$ as $t \to \infty$.

However, note that Eq. (10) is a single equation with M variables u_n, $n = 1, ..., N$ and hence we have infinitely many solutions. We are interested in obtaining an optimal solution for the problem, so that the controller input to the system remains optimal to the integral error approach zero. To obtain a optimal solution, we aim to obtain a solution that will not only satisfy Eq. (10), but at the same time will also minimize the following cost function

$$J = \frac{1}{2}\left(r_1\, w_1\, \bar{u}_1^2 + \quad \cdots \quad + r_N\, w_N\, \bar{u}_N^2 \right) \qquad ...(11)$$

i.e., we wish to minimize the cost function in Eq. (11), subjected to the constraint in Eq. (10). An implication choosing this cost function is that we wish to obtain the solution that will lead to a minimum control effort. In Eq. (11), choosing appropriate values for $r_1, ..., r_N > 0$ gives a control designer the flexibility of putting relative importance of the control magnitude at different spatial locations y_n.

The constrained optimization problem was solved and following expression for the controller was found (we omit the details for brevity).

$$\bar{u}_n = \frac{I_n \gamma}{r_n w_n \sum_{n=1}^{N} I_n^2 / (r_n w_n)} \quad , \quad n = 1,\ldots,N \qquad \ldots(12)$$

As a special case, we consider $r_1 = \cdots = r_N$ (i.e., equal important to all controllers) and $w_1 = \cdots = w_n$ (i.e. widths of all controllers are same), we have

$$\bar{u}_n = \frac{I_n \gamma}{\|I\|_2^2} \qquad \ldots(13)$$

where $I \triangleq \begin{bmatrix} I_1 & \cdots & I_N \end{bmatrix}^T$.

3.1 Singularity in Control Solution and Revised Goal

From Eqs. (12) and (13), it is clear that when $\|I\|_2^2 \to 0$ (i.e., $I_1,\ldots,I_N \to 0$) and $\gamma \not\to 0$, we have the problem of control singularity in the sense that $\bar{u}_n \to \infty$. Note that if the number of controllers N is large, probably the occurrence of such a singularity is a rare possibility, since all of $I_1,\ldots,I_N \to 0$ simultaneously is a strong condition. Nonetheless such a case may arise in transition. More important, this issue of control singularity will always arise when $x(y,t) \to x^*(y,t)$, $\forall y \in [0,L]$. Hence, whenever such a case arises, to avoid the issue of control singularity, we propose to redefine the goal as follows.

Let us define a new error vector as $e_n \triangleq [\Delta x_n]$ where e_n is the nth element of the error vector E and Δx_n is $x_n - x^*_n$. Next, design a controller such that $E \to 0$ as $t \to \infty$, i.e., the aim is to guarantee that the values of the state variable at the node points (y_n, $n = 1,\ldots,N$) track their corresponding desired values. To accomplish this goal, select a positive definite gain matrix K and $\mathbb{C}$ such that:

$$\ddot{E} + \mathbb{C}\dot{E} + K E = 0 \qquad \ldots(14)$$

One-way of selecting such a gain matrix K and $\mathbb{C}$ is to choose it as diagonal matrices. In such a case, the nth channel of Eq. (14) can be written as

$$\ddot{e} + c_n \dot{e}_n + k_n e_n = 0 \qquad \ldots(15)$$

Expanding the expressions for e_n, $\dot{e}_n$ and $\ddot{e}_n$, then solving for $\bar{u}_n$ ($n = 1,\ldots,N$), we obtain

$$\bar{u}_n = -m\left(-\frac{C}{m}\dot{x}_n - \frac{EI}{m}x_n + c_n \Delta \dot{x}_n + k_n \Delta x_n + \ddot{x}^*_n \right) \qquad \ldots(16)$$

where,
$$x_n \triangleq x(y_n,t), \quad \dot{x}_n \triangleq \dot{x}(y_n,t)$$

Combining the results in Eqs.(12) and (16), we finally write the control solution as

$$\bar{u}_n = \begin{cases} -m\left(-\dfrac{c}{m}\dot{x}_n - \dfrac{EI}{m}x_n + c_n\,\Delta\dot{x}_n + k_n\,\Delta x_n + \ddot{x}_n^*\right), & \text{if } \|I\|_2 < tol \\[2ex] \dfrac{I_n\,\gamma}{r_n w_n \sum\limits_{n=1}^{N} I_n^2 /(r_n w_n)}, & \text{otherwise} \end{cases} \qquad \text{...(17)}$$

where, *tol* represents a tolerance value. An appropriate value for it can be fixed by the control designer.

4. NUMERICAL RESULTS

For numerical simulation we choose a simply supported beam with $m = 19.6250\ kg/m$, $c = 0.01\%$, $EI = 1.04 \times 10^5\ N - m^2$, $EA = 5 \times 10^8\ N$ and $l = 10$ m. We choose a random initial curve [5] for the beam and simulated both uncontrolled and controlled dynamics. For controlled case we have taken control gain $k = 1$; $Q = \left[\{2\ \ 0.19\}; \{0.19\ \ 0.02\}\right]$; and the tolerance value $tol = 0.005$. After switching, the control gain $K = diag(k_1 \ ... \ k_N)$ was used and we selected $k_n = 1$, $c_n = 0.2$ for $n = 1$, ..., N. Furthermore, $r_1 = ... = r_N$ and $w_1 = ... = w_N = 0.4$ m are taken. With this assumption, the simplified expression for the controller could be used when $\|I\|_2 > tol$.

Figure 2 shows the uncontrolled displacement plot of the beam. The displacement magnitude of the uncontrolled simulation decays due to the presence of 0.01% damping in the system. Control simulations were run for 3 equally spaced control actuators ($N = 3$) for about 10s. Figures 3 and 4, show the displacement profile and the control input to the beam. Figure (3) shows that the displacement amplitude of the controlled vibration decreases sharply and approaches the goal. The input control amplitude (shown in Fig. 4) to the system is within 0.8kN force. Corresponding plots for displacement and control input with 6 actuators are shown in Figs. 5 and 6.

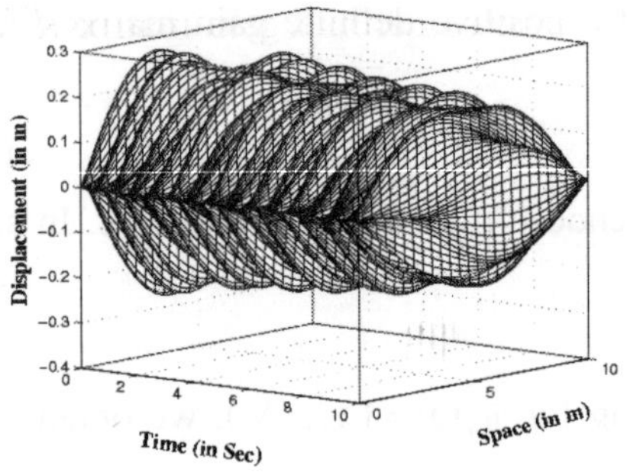

Fig. 2. Uncontrolled displacement.

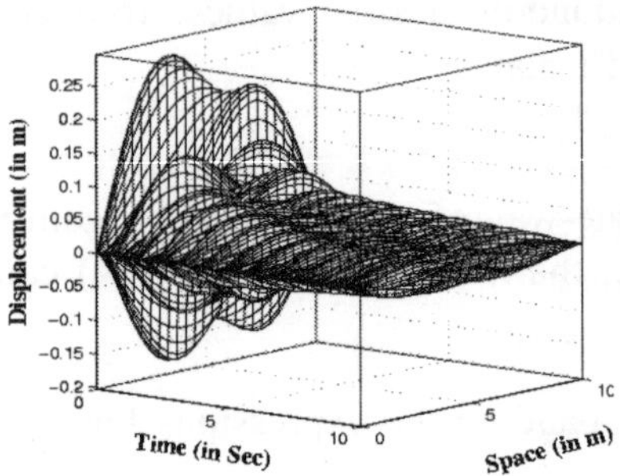

Fig. 3. Controlled displacement.

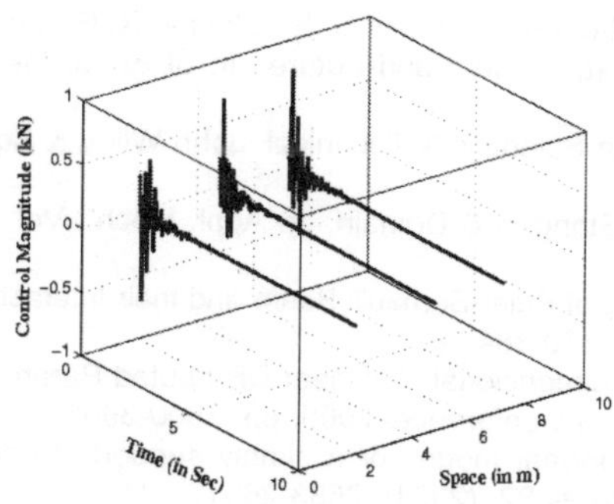

Fig. 4. Control input.

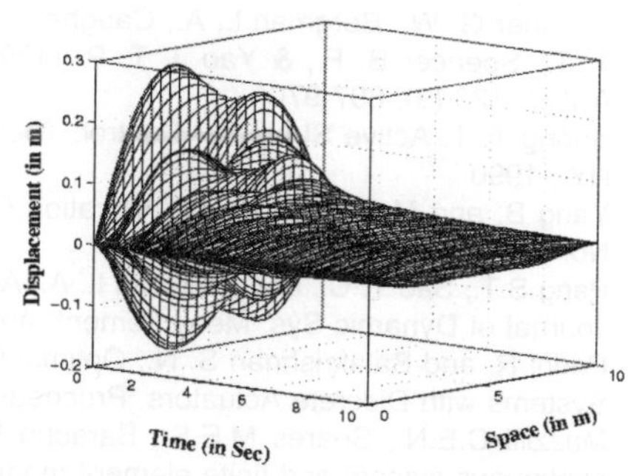

Fig. 5. Controlled displacement (six actuators).

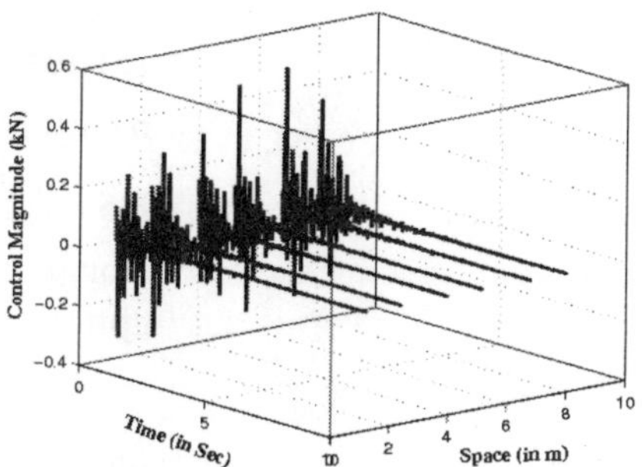

Fig. 6. Control input (six actuators).

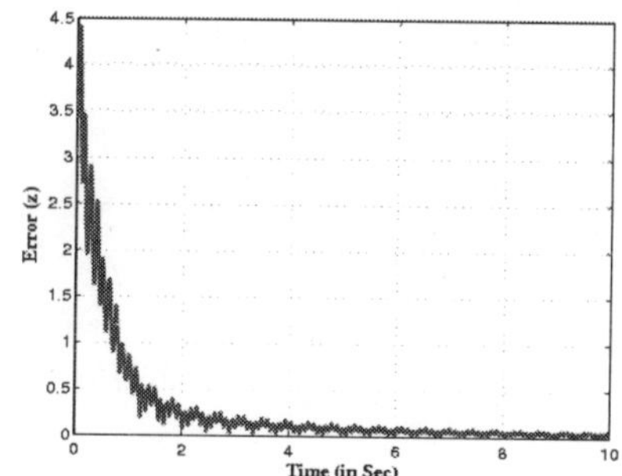

Fig. 7. Error function (z).

It is evident from Figs. 3-6, that increase in number of actuators provides better control with less control input. For uncontrolled system, displacement amplitude was seen to be 0.25 *m* even after *10* seconds whereas; the controlled vibration amplitudes are well below 0.1 *m* in almost about 2 seconds. The error functional, $z(t)$ (see Eq. (6)) plot (shown in Fig. 7) shows that our control goal is almost reached within 10s of vibration.

CONCLUSION

A set of discrete actuators, has been successfully used to control nonlinear beam vibration with simply supported support condition using the system PDE directly. The control goal has been reached by minimizing the error between actual and desired state (in the L_2 norm sense) using recently developed optimal dynamic inversion approach. The formulation presented here has good practical significance, since one can implement a set of discrete actuators easily (as compared to a continuous actuator). All computations necessary can be carried out online since we obtain a *closed form state feedback solution*.

ACKNOWLEDGEMENTS

The authors are very much thankful to Prof. Ananth Ramaswamy, Department of Civil Engineering, Indian Institute of Science, Bangalore, India for the productive intellectual discussions within connection with the example problem.

REFERENCES

1. Housner G. W., Bergman L. A., Caughey T. K., Chassiakos A. G., Claus R. O., Masri S. F., Skelton R. E., Soong T. T., Spencer B. F., & Yao J. T. P. (1997). Structural Control: Past, Present and Future, Jr. of Engg. Mech., ASCE, 123 (9), 897-970.
2. Soong T. T. Active Structural Control: Theory and Practice, Longman Scientific & Technical, John Wiley & Sons, NY., 1990
3. Yang B. and Mote C.D., Active Vibration Control of Axially Moving String in S Domain, Jr. Appl. Mech., Vol. 58, No. 1, 1991, 189-196.
4. Pang S.T., Sao T. C., and Bergman L. A., Active and Passive Damping of Euler-Bernaulli Beam and their Interaction, Journal of Dynamic Sys. Measurement, and Control, Vol. 115, 1993, 379-384.
5. Padhi R. and Balakrishnan S. N., Optimal Control of a Class of One-dimensional Nonlinear Distributed Parameter Systems with Discrete Actuators, Proceedings of the American Control Conference, 2005, pp. 3900-3905.
6. Mazzilli C.E.N., Soares M.E.S., Baracho Neto O.G.P., Non-linear normal modes of a simply supported beam: continuous system and finite element models, Computers and Structures 82, (2004), 2683-2691
7. Slotine, J-J. E. and Li, W., Applied Nonlinear Control, Prentice Hall, 1991.

107

Linear and Geometrically Nonlinear Dynamic Response of Plate and Shell Structures by DKT Element

ASOKENDU SAMANTA

Research and Coordination Division Indian Register of Shipping, Powai, Mumbai-400 072, India
email: a.samanta@irclass.org

ABSTRACT

DKT (Discrete Kirchhoff Triangle) element being very much efficient has been used effectively for the analysis of various plate/shell structures. But mostly all the previously published papers involving this element are related to linear static, buckling, vibration and nonlinear static analyses. In this paper a step forward is taken and nonlinear dynamic response analysis of the plates/shells by DKT element has been done for the first time. Variations of deflections and stresses with time at various points of the structure have been plotted for different types of plate/shell structures. The results show the efficacy of the DKT element in the nonlinear dynamic response analysis of these structures.

Keywords: Dynamic response, Finite element method, Nonlinear, Plate/Shell.

1. INTRODUCTION

For the effective treatment of a broad class of practical problems involving the nonlinear analysis of structures, simple and efficient plate/shell elements are taken as prerequisite. In an assessment of plate-bending elements with 3 degrees of freedom at the corner nodes only, Batoz et al. [1] have concluded that the DKT (Discrete Kirchhoff Triangle) and HSM (Hybrid Stress Model) elements are the most efficient, cost-effective and reliable elements of their classes. Batoz [2] has also shown that the convergence properties of the DKT element do not deteriorate with increase in the element aspect ratio which is not so for the HSM element. Being triangular in shape, the geometrical approximation to the curved shell can also be done effectively with this element. Thus, the DKT element becomes quite attractive to users over other 9 degrees of freedom plate bending triangular elements.

Originally this element was proposed by Stricklin *et al.* [3]. Later on some more work has been done on this element by Batoz, *et al.* [1], Batoz [2], Jayachandrabose et al.[4], Meek and Tan [5],

Dhatt *et al.* [6], Talaslidis and Sous [7], Samanta and Mukhopadhyay [8, 9, 10, 11, 12], Satish Kumar and Mukhopadhyay [13] and Samanta [14].

In all the above papers, mostly the analyses have been restricted to linear static, buckling, vibration or nonlinear static problem. In this paper the nonlinear dynamic formulation has been done with DKT element and the nonlinear time history analysis of plate/shell structures has been performed to show the efficiency of the element in the dynamic response analysis.

2. MATHEMATICAL FORMULATION

The governing nonlinear differential equation of motion for an un-damped structural system can be expressed as

$$\{\Psi\} = \int [K_s]\{\delta(t)\} + \int [M]\{\ddot{\delta}(t)\} - \{F(t)\} \qquad ...(1)$$

where, $[K_s]$ is the global secant stiffness matrix, $[M]$ global mass matrix, $\{\delta(t)\}$ is the displacement at time t, $\{\ddot{\delta}(t)\}$ is the acceleration at time t and $\{F(t)\}$ is the external time dependent load vector. The governing un-damped differential equations of motion have been solved by Newmark's method for direct time integration.

The basic flat shell element is a combination of the DKT (Stricklin et al. 3) plate bending element and Allman's [15] plane stress triangular element. The element is a three noded triangle. The degrees of freedom at each node are u, v, w, θ_x, θ_y and θ_z at three corner points of the element. The derivation of linear stiffness matrix has been given in details in Samanta and Mukhopadhyay [8].

The nonlinear strain of the shell element can be written as

$$\{\varepsilon\} = \{\varepsilon_0\} + \{\varepsilon_L\} \qquad ...(2)$$

The subscript, 0 and L are used to indicate the linear and nonlinear part, respectively.

The nonlinear strain vector $\{\varepsilon_L\}$ can be represented as

$$\{\varepsilon_L\} = \frac{1}{2} \begin{bmatrix} \dfrac{\partial w}{\partial x} & 0 \\ 0 & \dfrac{\partial w}{\partial y} \\ \dfrac{\partial w}{\partial y} & \dfrac{\partial w}{\partial x} \\ 0 & 0 \\ 0 & 0 \\ 0 & 0 \end{bmatrix} \begin{Bmatrix} \dfrac{\partial w}{\partial x} \\ \dfrac{\partial w}{\partial y} \end{Bmatrix} = \frac{1}{2}[A]\,\{\theta\} \qquad ...(3)$$

Here, both $[A]$ and $\{\theta\}$ are dependent on $\{\delta\}$.

The tangential stiffness and secant matrix of the shell element is given by

$$[K_T] = [K_0] + [N_1] + [N_2] \qquad ...(4)$$

$$[K_s] = [K_0] + \frac{1}{2}[N_1] + \frac{1}{3}[N_2] \qquad ...(5)$$

where, $[K_0]$ is the linear stiffness matrix, $[N_1]$ is the nonlinear stiffness matrix linearly dependent upon $\{\delta\}$, $[N_2]$ is the nonlinear stiffness matrix quadratically dependent upon $\{\delta\}$.

3. RESULTS AND DISCUSSION

The following examples have been presented for the dynamic analysis of plates/shells.

Example 1: Square Glass Window Panel

Response of a square glass window panel under uniform pressure has been determined and the results are compared with the previously published results. The glass panel is 2.44 m × 2.44 m × 6.35 mm (8ft × 8ft × 0.25 in) with a specific weight of 24.74 KN/m^3 (157.5 Ib/ft^3), $E = 6.895 \times 10^{10}$ N/m^2, $v = 0.23$ and subjected to uniform pressure of 47.9 N/m2 (1psf). Boundary conditions have been considered as simply supported with immovable in-plane degrees of freedom.

The linear and nonlinear responses of this panel were determined previously by Bayles et al. [16] and Ali and Al-Noury [17]. Both have solved the problem by finite difference method. Taking advantage of the double symmetry, a 4´4 mesh in quarter panel has been taken in the present analysis. Time step has been chosen as 1 milli-second for linear analysis and 0.5 milli-second for nonlinear analysis.

Fig. 1 shows the linear and nonlinear displacement time history curve of the center point of the panel. A good agreement has been found with the previous results of Bayles et al. [16] and Ali and Al-Noury [17].

It has been clearly observed from Fig. 1 that the nonlinear effect is predominant. The maximum center point transverse displacement, 8.846 mm, which is obtained from linear analysis is reduced to 5.834 mm due to nonlinear effect. The period of oscillation in nonlinear case is also shorter. The stress history both for linear and nonlinear at the center point of the plate is also plotted in Fig. 2. The comparison has been made with general purpose software package ANSYS. A 10 × 10 mesh division in full plate with shell93 element has been taken in ANSYS for analysis. Shell93 is an eight noded element with six degrees of freedom per node. These degrees of freedom are three displacements (u, v, w) and three rotations (θ_x, θ_y, θ_z). The agreement is excellent. This shows clearly the efficiency of the present programming code.

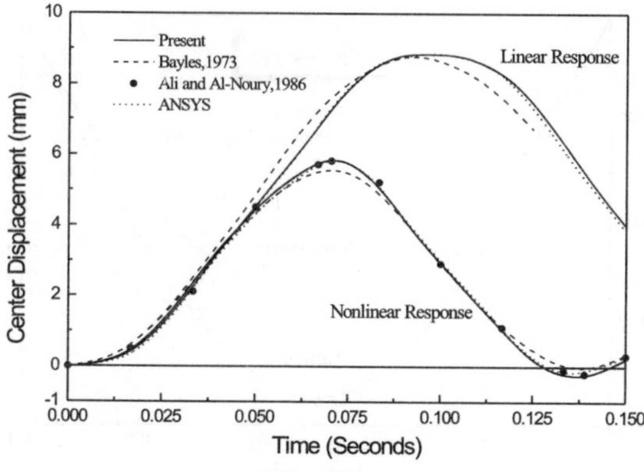

Fig. 1. Displacement time history of a glass panel.

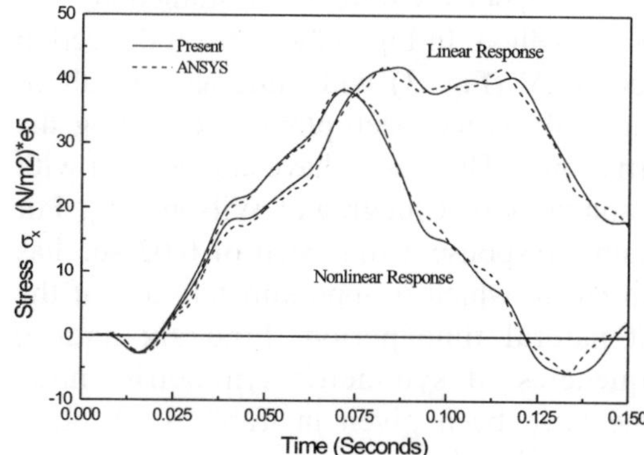

Fig. 2. Variation of central stress with time history of a glass panel.

Example 2: Diaphragm Supported Cylindrical Shell Panel

A cylindrical shell panel (shown in Fig. 3) subjected to a vertical uniformly distributed half sinusoidal wave impulsive loading with a peak intensity of 90 psf (4310.77 N/m^2) (Fig. 4) is studied.

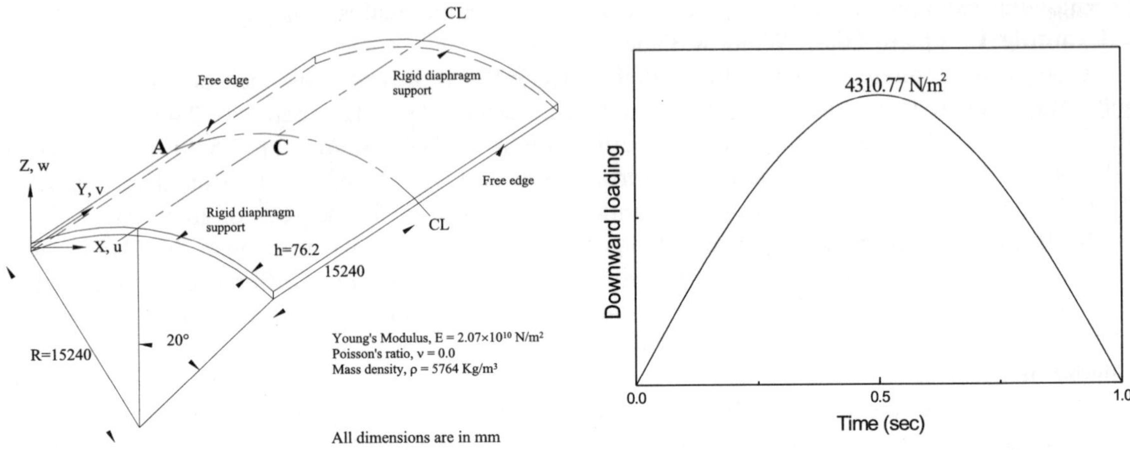

Fig. 3. Cylindrical shell - dynamic analysis.

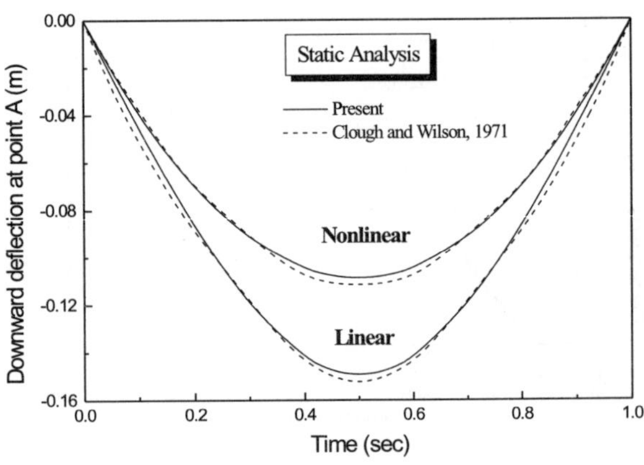

Fig. 4. Dynamic loading.

The same problem was previously solved by Clough and Wilson [18] and by Tsai and Palazotto [19] using boundary element and finite element, respectively. The two straight edges of the shell panel are assumed to be free and the curved edges are supported on rigid diaphragms. The diaphragms are rigid against the in-plane degrees of freedom (u, w, θ_y) whereas, it is free for out-of-plane degrees of freedom (v, θ_x, q_z). The materials were taken to be purely elastic with Young's modulus, = 3 × 10^6 psi (2.07 × 1010 N/m^2), Poisson ratio, v = 0.0 and weight density = 90 psf (5764.176 kg/ m^2). Due to symmetry, only a quadrant of the panel is analyzed. A 4 × 4 and an 8 × 8 mesh division have been taken by the present method.

The linear and nonlinear static and dynamic responses have been studied by the present method. In Fig. 5 the vertical deflection of point 'A' (Fig. 3) with time as a result of static analysis has been plotted neglecting the inertia force. The results have agreed well with the solutions of Clough and Wilson [18]. For dynamic response a time step of 0.02 sec has been taken, which is approximately 3% of the first natural time period. First six natural frequencies of symmetric-symmetric mode shape have been given in Table 1. A close agreement has been found with the previous results of Clough and Wilson [18].

The linear and nonlinear dynamic response

Fig. 5. Static analysis of cylindrical shell.

of vertical displacement at point 'A' have been plotted along with the results of Clough and Wilson [18] in Fig. 6 and Fig. 7. These figures also show the closeness of the results with the available results of Clough and Wilson [18].

Table 1. Frequencies (rad/sec) of first six symmetric- symmetric mode of a diaphragm ended cylindrical shell

Mesh*	Source	Mode Number					
		1	2	3	4	5	6
2 × 2	Present	9.02	24.35	24.71	30.73	61.36	96.20
	Clough♣	8.61	22.52	-	-	-	-
4 × 4	Present	9.77	24.05	33.64	41.98	64.33	66.62
	Clough♣	9.64	24.49	33.88	42.81	64.20	68.95
6 × 6	Present	9.80	23.99	33.96	44.73	68.03	69.53
	Clough♣	9.77	24.24	34.17	44.99	68.99	70.31
8 × 8	Present	9.81	23.98	33.86	45.78	69.22	69.81
	Clough♣	9.78	24.09	34.08	45.95	69.73	70.93

* Mesh indicates in quarter shell ; ♣ Clough and Wilson (1971)

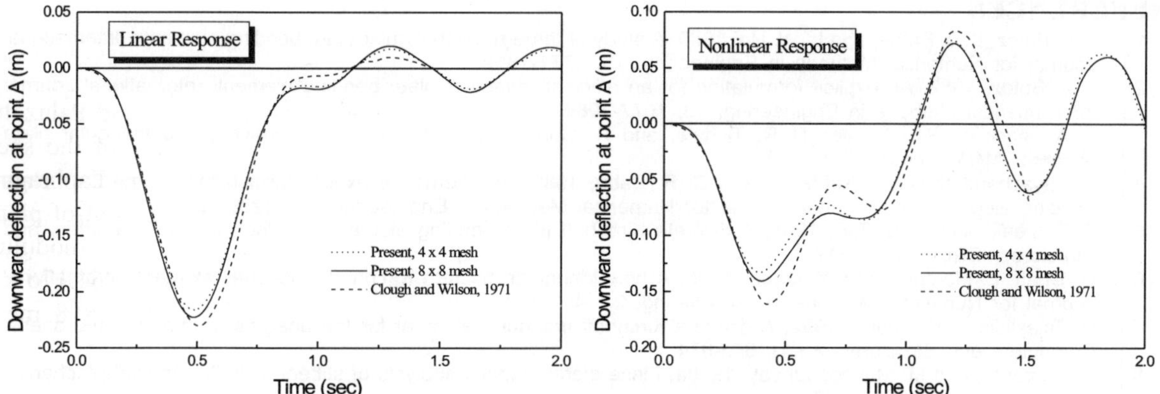

Fig. 6. Linear dynamic response at point A.

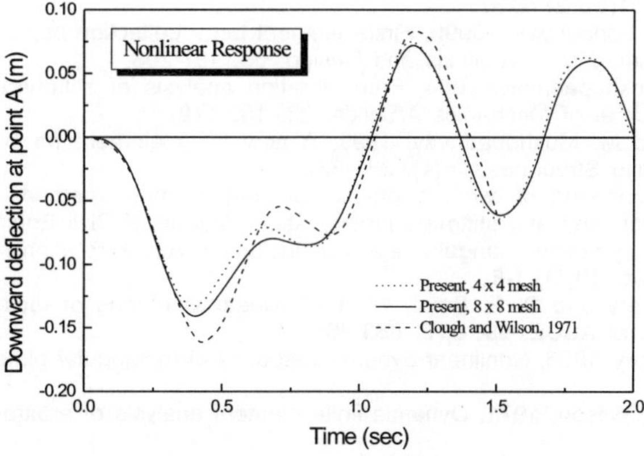

Fig. 7. Nonlinear dynamic response at point A.

CONCLUSION

The effectiveness of the DKT element for static, buckling and free vibration analyses has been shown previously by different authors. In this paper the effectiveness of this element in linear and nonlinear dynamic response analyses of plate/shell structures has been shown. The nonlinear dynamic response analysis with this element has been done for the first time. The element formulation has been presented elaborately. Several examples have been solved with various types of loading such as uniform pressure with triangular time variation, step load, half sinusoidal load. In example 1 of glass panel, stress history along with the displacement history has been shown. It can be helpful for the future researchers. In this example the predominance of nonlinearity has been shown over linear response. In Example 2 of shell panel, a complete study (free vibration, linear and nonlinear static and dynamic response) has been done. From all these results it can be concluded that the element is very much efficient for the dynamic response analysis of the plate/shell structures.

Notations		Abbreviation	
$[K_S]$	overall secant stiffness matrix	DKT	Discrete Kirchhoff Triangle
$[K_T]$	overall tangent stiffness matrix	FEM	Finite Element Method
$[M]$	overall mass matrix	HSM	Hybrid Stress Model

REFERENCES

1. J. L. Batoz, K. J. Bathe, and L. W. Ho, 1980, A study of three-node triangular plate bending elements, International Journal for Numerical Methods in Engineering, 15, 1771-1812.
2. J. L. Batoz, 1982, An explicit formulation for an efficient triangular plate bending element, International Journal for Numerical Methods in Engineering, 18, 1077-1089.
3. J. A. Stricklin, W.E. Haisler, P. R. Tisdale, and R. Gunderson, 1969, A rapidly convergence triangular plate element, AIAA Journal, 7, 180-181.
4. C. Jayachandrabose, J. Kirkhope, and C. R. Babu, 1985, An alternative explicit formulation for the DKT plate bending element, International Journal for Numerical Methods in Engineering, 21, 1289-1293.
5. J. L. Meek, and H. S. Tan, 1985, A discrete Kirchhoff plate bending element with loof nodes, Computers and Structures, 21(6), 1197-1212.
6. G. Dhatt, L. Marcotte, and Y. Matte, 1986, A new triangular discrete Kirchhoff plate/shell element, International Journal for Numerical Methods in Engineering, 23, 453-470.
7. D. Talaslidis, and I. Jons, 1992, A discrete Kirchhoff triangular element for the analysis of thin stiffened shells, Computers and Structures, 43(4), 663-674.
8. A. Samanta, and M. Mukhopadhyay, 1998a, Finite element static analysis of stiffened shells, Applied Mechanics and Engineering, 3(1), 55-87.
9. A. Samanta, and M. Mukhopadhyay, 1998b, Tension-induced conveyor belts: A finite element free vibration study, Journal of Vibration and Control, 4, 673-696.
10. A. Samanta, and M. Mukhopadhyay, 1999a, Finite element static and dynamic analyses of folded plates, Engineering Structures, 21(3), 277-287.
11. A. Samanta, and M. Mukhopadhyay, 1999b, Finite element large deflection static analysis of shallow and deep stiffened shells, Finite Elements in Analysis and Design, 33, 187-208.
12. A. Samanta, and M. Mukhopadhyay, 2004, Free vibration analysis of stiffened shells by the finite element technique, European Journal of Mechanics A/Solids, 23, 159-179.
13. Y. V. Satish Kumar, and M. Mukhopadhyay, 1999, A new finite element for buckling analysis of laminated stiffened plates, Composite Structures, 46(4), 321-331.
14. A. Samanta, 2000, Comparison of two approaches of finite element methods for the large amplitude free vibration analysis of unstiffened and stiffened plates, Asian Journal of Civil Engineering, 1(2), 27-36.
15. D.J. Allman, 1984, A compatible triangular element including vertex rotations for plane elasticity analysis, Computers and Structures, 19(1), 1-8.
16. D. J. Bayles, R. L. Lowery and D. E. Boyd, 1973, Nonlinear vibrations of rectangular plates, Journal of the Structural Division, Proc. of ASCE, 99, ST5, 853-864.
17. S. A. Ali, and S. I. Al-Noury, 1986, Nonlinear dynamic response of rectangular plates, Computers and Structures, 22(3), 433-437.
18. R. W. Clough, and E. L. Wilson, 1971, Dynamic finite element analysis of arbitrary thin shells, Computers and Structures, 1, 33-56.
19. C. T. Tsai, and A. N. Palazotto, 1991, On the finite element analysis of non-linear vibration for cylindrical shells with high-order shear deformation theory, International Journal of Non-linear Mechanics, 26(3/4), 379-388.

108

Nonlinear Dynamic Analyses of Deep Water Marine Risers

RIZWAN A. KHAN AND SUHAIL AHMAD

Department of Applied Mechanics, Indian Institute of Technology, Delhi, New Delhi-110016
email: rizwan_iitd@yahoo.co.in

ABSTRACT

The dynamic response of marine risers to both regular and random waves is obtained in the time domain using finite element solver ABAQUS/Aqua. For random sea states, time histories of sea surface elevation, water particle kinematics and vessel top motion are simulated by a harmonic superposition technique. Results are presented which illustrate the effects of nonlinearities, long term drift oscillation and instantaneous motion of the vessel and current velocity on the bending stresses in marine risers. The bending stress response time histories are obtained and presented in terms of power spectra showing the contribution of various harmonics which is significant because of a non-linear system.

Keywords: Marine riser, random waves, vessel motion, dynamic response.

1. INTRODUCTION

A marine riser is a major component of offshore drilling and production systems that are part of either fixed or floating offshore platforms. It is a slender pipe that serves as a conduit between the platform and the sub sea well head. Marine riser is excited by ocean currents, waves and the vessel motion. These excitations produce significant dynamic stresses in the risers, the natural frequencies of which fall within the range of most excitation frequencies. Generally, deep risers have natural frequencies that lie within the dominant frequencies of most of the frequently occurring sea states and, consequently have a large dynamic response. Generally, three types of riser dynamic analysis are specified in the literature:

1. deterministic steady-state or frequency domain analysis;
2. deterministic time history analysis; and
3. non-deterministic random vibration analysis

A good numbers of research papers appeared in various journals and conference proceedings on various aspects of dynamic analysis of marine risers. Ahmad and Datta [1] studied the dynamic response of marine risers to both regular and random waves in the time domain using a time marching numerical integration scheme to solve the equation of motion. Atadan et al.[2], Yousun Li [3] studied the force dynamics of the riser system, connected to a floating platform and conveying fluid, in the presence of ocean waves and ocean currents. Huyse *et al.* [4] have presented a static three-dimensional analytical method for drilling risers experiencing large displacements and slip at the top joint. Vessel motion has not been taken into account. Chatjigeorgiou and Mavrakos [5] studied the non-linear dynamic response in the transverse direction of vertical marine risers/tensioned cable legs subjected to parametric excitation at the top of the structure. The dynamic model contains both elastic and bending effects. Kaewunruen *et al.* [6] employed finite element method to analyze the nonlinear free vibrations of marine risers/pipes conveying internal fluids based on the energy approach.

In the present study, a nonlinear dynamic analysis of marine riser Figure (1), under regular and random waves has been investigated. The responses are obtained using finite element solver ABAQUS/Aqua in the time domain. Riser is modeled as tensioned beam with three degrees of freedom at each node. The equation of motion has been solved considering the nonlinearities due to large deformation and time-wise variation of the submergence, buoyancy, added mass and resultant hydrodynamic loading. For random waves, sea states have been represented by the DNV version of the Pierson-Moskowitz (PM) sea spectrum which is defined by the two parameters, H_s (1/3rd significant wave height) and T_z (average time period). To adequately simulate the vessel motion, a model which includes terms representing mean vessel offset from the well bore, long term drift- i.e., modeling the response of the positioning system (mooring or dynamic)- and the instantaneous response of vessel motion to irregular waves has been adopted.

Following parametric studies on the behavior of marine riser due to dynamic excitation have been undertaken to investigate their effect on riser response:

- Long-term periodic motion;
- Instantaneous vessel motion produced by random seas;
- Current together with wave.

2. MATHEMATICAL FORMULATION

The main features of the model and the assumptions are as follows:

1. A riser is idealized as a tensioned beam, single line, two-dimensional structure undergoing motion in the plane of fluid loading caused by waves and currents in the same or opposite directions. For analysis, the riser is discretized into a number of finite beam elements.
2. The bottom end of the riser is hinged and assumed to be restrained in the horizontal and vertical directions; top end of the riser is supported on a roller which allows horizontal restraint but allows movement in the vertical direction.
3. It is assumed that no buoyancy tanks are provided along the riser length.

2.1 Equation of Motion

With the above assumptions for the analysis of the riser system figure (1), the equation of motion for the resultant multi-degree-of freedom system is given as

$$[M]\{\ddot{x}\}+[C]\{\dot{x}\}+[K]\{x\}=\{F(t)\} \qquad ...(1)$$

where, $[M]$ is the consistent mass matrix, $[C]$ is the damping matrix, $[K]$ is the system stiffness matrix and $\{F(t)\}$ is the time- dependent random hydrodynamic loading $\{\ddot{x}\}, \{\dot{x}\}, \{x\}$ are the vectors of structural acceleration, velocity and displacement respectively.

The stiffness matrix $[K]$ is made of both elastic and geometric stiffness matrices corresponding to the degrees of freedom.

The consistent mass matrix $[M]$ for the complete riser is made up from the assembly of the element mass matrices in global coordinates. The elemental mass matrix consists of two parts, one due to the added mass effect. The latter is considered for the submerged of the riser up to still water level (SWL).

2.2 Treatment of Dynamic Loading

The sources of dynamic loading considered in this study consist of hydrodynamic loading due to random wave and current. The hydrodynamic loads $\{F(t)\}$ has two components viz. drag force and inertia force which depend on the velocity and acceleration w.r.t the riser as well as the velocity and accelerations of the water particle and currents. The hydrodynamic load per unit length on the riser at any instant of time is determined by using Morison's equation:

$$f_i = \frac{1}{2}\rho_w C_D D \left(\dot{U}_i - \dot{x}_i + V_{c_i}\right)\left|\dot{U}_i - \dot{x}_i + V_{c_i}\right| + \rho_w \frac{\pi}{4}D^2 C_M \ddot{U}_i - m \cdot \ddot{x}_v \qquad ...(2)$$

where m is the mass of the riser per unit length and is given by

$$m = \frac{\pi}{4}\left(D^2 - D_i^2\right)\rho_s + \frac{\pi}{4}D_i^2\rho_o + \frac{\pi}{4}D^2\rho_w\left(C_M - 1\right) \qquad ...(3)$$

in which $\dot{U}_i$, $\ddot{U}_i$ are the water particle velocity and acceleration, $\dot{x}_i$ is the velocity of the riser at point 'i' for any instant of time and $\ddot{x}_v$ is the acceleration of the vessel imparted to the top of the riser at the same time. Other variables are defined in Table 1.

2.3 Simulation of Random Waves

The sea surface is assumed to be Gaussian ergodic process and the surface elevation is assumed to be a super position of infinite small harmonic waves having randomly distributed phases. The spectral density of the sea surface elevation is represented by the DNV version of the Pierson-Moskowitz (PM) sea spectrum, which is given by

$$S_{\eta\eta}(f) = \frac{H_s^2 T_z}{8\pi^2}(T_z f)^{-5} . \exp\left[-\frac{1}{\pi}(T_z f)^{-4}\right] \qquad ...(4)$$

where, f is the frequency in cycles/s, H_s is the significant wave height in m, T_z is the zero up crossing period in s, and $S_{\eta\eta}(f)$ is the PM (single-sided) sea surface elevation spectrum. The random waves and the corresponding water particle kinematics are simulated by wave superposition techniques from their respective spectra.

Table 1. Marine Riser Specifications

Riser Tension	=	3×10^6N
Water depth	=	500 m
Riser length	=	500 m
Outer diameter of riser	=	0.406 m
Inner diameter of riser	=	0.374 m
Effective outer diameter	=	0.66 m
Mass density of steel	=	7840 kg/m^3
Mass density of water	=	1025 kg/m^3
Coefficient of drag, Cd	=	1.1
Coefficient of Inertia, Cm	=	2.5
Current velocity,Vc	=	2 m/sec
Modulus of elasticity	=	2.1×10^{11} N/m^2

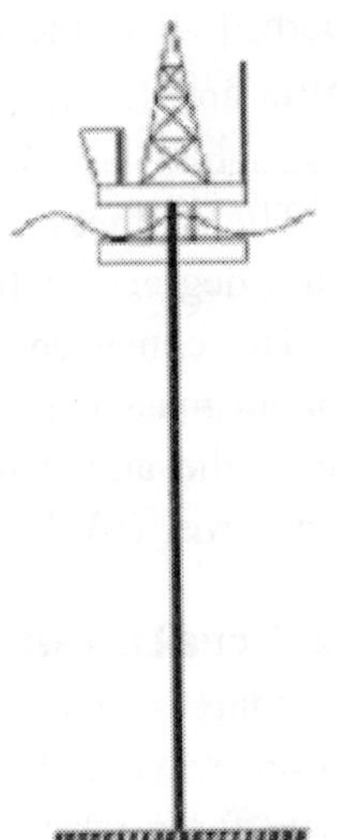

Fig. 1. Marine Riser.

3. NUMERICAL STUDY

For the numerical study, a marine riser of length 500 m signifying long riser is considered. Riser properties are given in Table 1. The applied tension at the top of the riser is considered as variable. The first five natural frequencies of the riser for a specified level of applied tension is shown in Table 2.

Table 2. Natural frequencies of Marine Riser

Riser length (m)	Top tension (kN)	Natural Frequencies (cycles/s)				
		Mode1	Mode2	Mode3	Mode4	Mode5
500	3000	0.077	0.156	0.235	0.316	0.398

Dynamic response is obtained for both regular and random waves. For irregular waves, twelve different sea states has been selected (Table 3). These combinations are chosen in order to study the behavior of risers under extreme resonating and non-resonating dynamic excitations. For random waves the, sea states are represented by the DNV version of the PM spectrum which is defined by two parameters, H_s (1/3rd significant wave height) and T_z (average time period). Combinations of H_s and T_z used in the numerical examples are sea state 1&2. These sea states adequately cover the

Table 3. Simulated Sea states

Sea States	Significant wave height H_s(m)	Zero crossing period T_z(s)	Wind velocity U (m/s)
S1	17.15	13.26	24.38
S2	15.65	12.66	23.29
S3	14.15	12.04	22.15
S4	12.65	11.39	20.94
S5	11.15	10.69	19.66
S6	9.65	9.94	18.29
S7	8.15	9.14	16.81
S8	6.65	8.26	15.18
S9	5.15	7.26	13.36
S10	3.65	6.12	11.25
S11	2.15	4.69	8.63
S12	0.65	2.58	4.75

condition of significant dynamic excitation. Two types of vessel motion are considered: long-term drift oscillation and the instantaneous response of the vessel to random waves. The long term drift oscillation is modeled by a harmonic motion whose amplitude is taken as 3m and the period is varied. The phase angle between drift motion and the wave is taken as zero.

3.1 Effect of Vessel Motion on Riser Response

Long-term drift motion and instantaneous motion due to random waves are investigated. Long-term drift motion is assumed to be harmonic. For regular waves, the combined response is obtained by assuming that the vessel motion has the same period as that of the wave and is in phase with the fluctuation of sea surface elevation. In figure 2, the envelopes of maximum bending stresses obtained by the 17.15m/13.26 s (regular) wave motion and the combined motion (same wave and vessel motion of 3m/13.26 s) are presented. Since the riser's fundamental period s almost 15 sec a near resonating condition is caused in the dynamic response. The significance of the vessel motion on the riser response is well depicted in the figure. Not only does the stresses at the top of the riser becomes maximum, but considerable stresses are also produced in the middle of the riser.

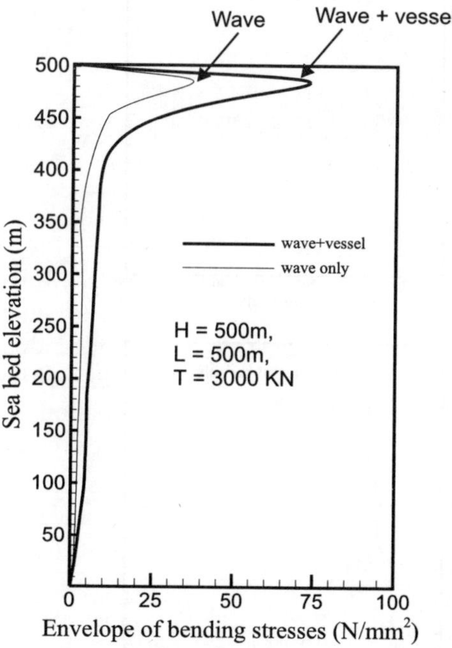

Fig. 2. Effect of vessel motion on response of long riser (sea state 1).

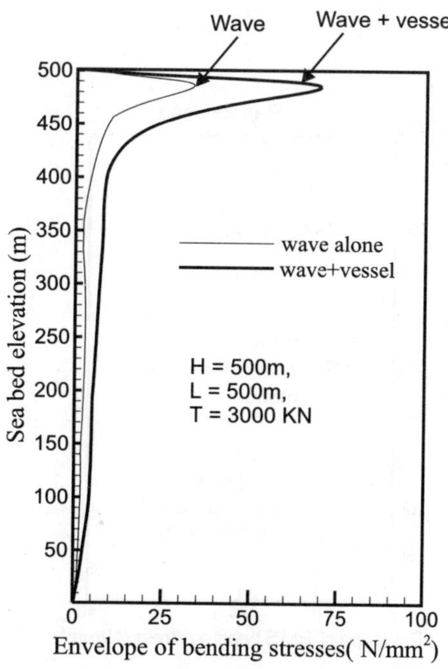

Fig. 3. Effect of vessel motion on response of long riser (sea state 2).

It is clear from the figure that the long term drift motion of the vessel produces stresses far in excess of those produced by waves alone. In the present case, the maximum response produced by combined effect of wave and vessel is about three times that produced by the wave alone. In figure 3 the same results are shown for the 500 m riser under the 15.65m /12.66 s regular wave and 3m/ 12.66s vessel motion. In this case the, the main contribution to the response is from the first two modes of the structure. The results show the same trend. The ratio of the maximum stress produced by the combined effect of wave and vessel to that of wave alone is about 2.1.

3.2 Effect of Current on Dynamic Response

The water particle velocity is increased by the current velocity non-varying with time. The drag force is accordingly increased nonlinearly. The current velocity does significantly influence the dynamic response. The steady drift is enhanced due to the presence of current velocity introducing a pseudo static effect. The change in the dynamic response induced by current velocity greatly depends upon the ratios of wave period to the structure's periods. When current velocity is considered in the computation of hydrodynamic forces, hydrodynamic loading gives harmonics at both even and odd multiples of wave frequency, and also at zero frequency. If the ratios are such that one of the first few frequencies of the structure nearly coincides with any of the even harmonic components of the loading, the effect of current on dynamic response is severely magnified. In figure 4a comparison is made between the responses obtained by considering forces produced by wave alone, wave and current, and wave, current and vessel motion. Current is found to increase the response considerably. This probably due to the resonating condition described above.

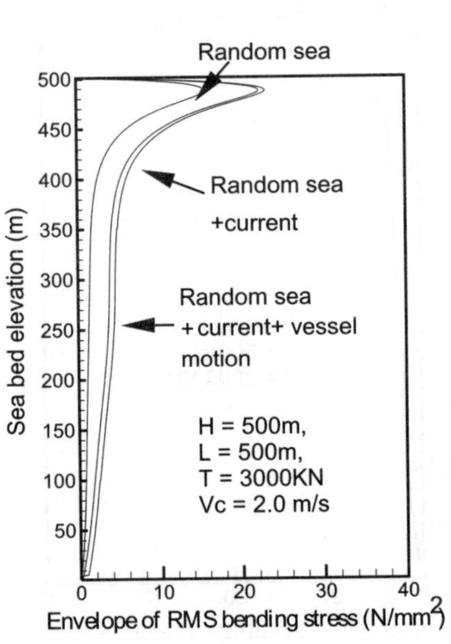

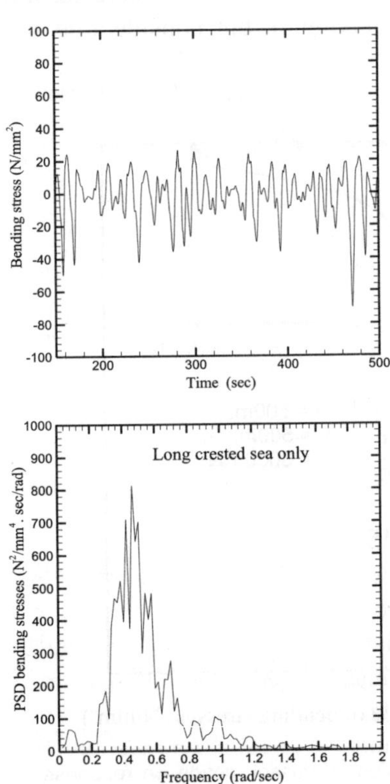

Fig 4. Effect of current together with vessel motion on dynamic response of riser for random sea (sea state 1).

Fig 5. Time history and PSD of the bending stresses (sea state 1).

The effect of current, together with the long crested sea on riser bending stress is shown in Figure 5 and 6. It has been observed that the simple addition of current of 2 m/sec velocity(Vc) to the long crested random sea results in increase of bending stresses twice as compared to long

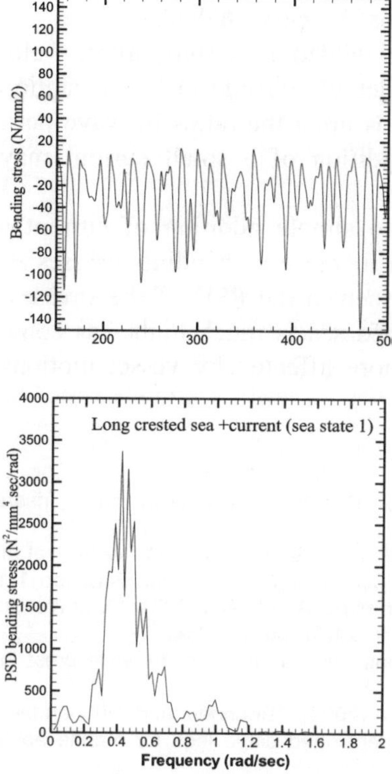

Fig 6. Time history and PSD of the bending stresses (sea state 1)

crested random sea alone. It can also be seen in the power spectral density (PSD) of the stresses as the maximum energy content of the current plus long crested sea is much higher as compared to long crested sea alone.

CONCLUSION

Dynamic response of marine risers to both regular and random waves has been investigated. In the dynamic response analysis, the effects of the relative velocity squared drag term, the long-term drift oscillation and the current velocity were all considered. The response was obtained in time domain using finite element package ABAQUS/AQUA. For the random sea states, the time histories of sea surface elevation, the water particle kinematics and the top vessel motions were simulated by a harmonic superposition technique. The following results can be drawn from the numerical study.

(1) Long-term drift motion of the vessel has profound effect on the envelope of maximum stress in the riser. Combined vessel and wave motion results in increase in stresses near the top and in middle of the riser. Moreover, the combined effect may produce stresses far in excess of those produced by waves alone.

(2) When periodic long-term vessel drift and wave motion are considered together, the maximum response may occur when the structure's frequencies are away from the wave frequency.

(3) The ratio of the maximum stress produced by the combined effect of wave and vessel to that of wave alone is about 2.1 for the case studied.

(4) When the current velocity is added to water-particle velocity for the computation of the hydrodynamic loading, considerable change in the dynamic response of the riser may result. The amount of change depends upon the ratios of wave period to the structure's periods. For the resonating conditions, addition of a small current may lead to a large change in the dynamic response.

(5) It has been observed that the simple addition of current of 2 m/sec velocity to the long crested random sea results in increase of bending stresses twice as compared to long crested random sea alone. It is also seen in the PSD of the stresses as the maximum energy content of the current plus long crested sea is much higher as compared to long crested sea alone.

(6) In general, long risers are more affected by vessel motions (long-term drift) than by wave forces.

REFERENCES

1. Ahmad, S. and Datta, T. K.(1989). "Dynamic response of marine risers," *Engineering Structures*, Vol. 11, pp 179-188.

2. Atadan, A. S., Calisal, S. M., Modi, V. J., Guo, Y. (1997). "Analytical and numerical analysis of a marine riser connected to a floating platform," *Ocean Engineering*, Vol. 24(2) pp 111-131.

3. Luc Huyse, Mansa C. Singh and Marc A. Maes (1997). "A static drilling riser model using free boundary conditions," *Ocean Engineering*, Vol. 24(5), pp 431-444.

4. Yousun Li (1999). "Dynamic response of marine risers to wave crest loads," *Offshore Technology Conference* (OTC), pp 1-6.

5. Chatjigeorgiou, K., Mavrakos, S. A.(2002). "Bounded and Unbounded transverse response of parametrically excited vertical marine risers and tensioned cable legs for marine applications".*Applied Ocean Research*, Vol. 24, Pp 341-354.

6. Kaewunruen, S., Chiravatchradrej, J. and Chucheepsakul, S.(2005). "Nonlinear free vibration of marine risers/ pipes transporting fluid," *Ocean Engineering*, Vol. 32, pp 417-440.

109

Geometric Nonlinear Analysis of Piezoelectric Laminated Cylindrical Shells

C.K. KUNDU, D.K. MAITI AND P.K. SINHA

Department of Aerospace Engineering, Indian Institute of Technology Kharagpur-769008, India
email: ckkundu@gmail.com dkmaiti@aero.iitkgp.ernet.in

ABSTRACT

It is necessary to study the electromechanical post buckling behaviour of doubly curved shells (Kraus [1]) under transverse and electrical loads. In the present paper geometric nonlinear bending analysis of laminated doubly curved shells is presented using finite element method. A nine noded isoparametric (Ziekiewicz and Taylor [2]) nonlinear doubly curved shell element is developed. The nonlinear strain-displacement relations (Kundu and Sinha [3]) are expressed in curvilinear coordinates system. The piezoelectric material is used in the form of layers or patches embedded and/or surface bonded on laminated composite shells (Crawley and Luis [4]). The static postbuckling behaviour of piezoelectric laminated shells can be controlled by the use of piezoelectric layers. The principle of virtual work forms the basis to derive the nonlinear finite element equations. To solve the nonlinear finite element equations an incremental iterative technique based on the Arc-length method (Crisfield [5]) is employed. The present results are found to compare well with those available in the literature (Varelis and Saravanos [6]). The nonlinear responses of cylindrical shells are analyzed and the results are discussed.

Keywords: Finite element; Geometric nonlinear; Piezoelectric; Laminated; Composite; Shells.

1. INTRODUCTION

Laminated composite shells are extensively used as structural components in aerospace and other engineering sectors due to their various superior properties. The shells have the capability to carry higher loads due to the coupling between membrane and flexure terms. Piezoelectric laminated structures enhance the use of shells due to the adaptive capabilities. Some of these shells experience large bending deformations due to transverse load. Similarly, piezoelectric laminated shells may undergo large bending deformations (Pai et al. [7]). The composite material properties can be improved by variation of the fiber orientation and stacking sequence of the layer of the composite laminate (Jones [8]), thereby enhancing the flexibility in designing any structural component by the designer.

Piezoelectrics are the most popular smart materials. It is known that piezoelectric material have the properties to generate electric charge under mechanical load or deformation and in reverse, applying electric fields to the piezoelectric material resulting mechanical strains and stresses. When a piezoelectric layer is integrated into the composite laminated shells the adaptive capabilities of the shells are enhanced (Mukherjee and Chaudhuri [9]). Integration of a piezoelectric layer into the composite laminated shells enhances the adaptive capabilities of the shells. The most widely used piezoceramics, such as lead zirconatetitanate (PZT), are in the form of thin sheets that can be readily attached or embedded in the composite structures or stacked to form discrete piezo stack actuators. They undergo deformation (strain) when an electric field is applied across them (converse effect), and conversely produce voltage when strain is applied (direct effect). The piezoelectric effect is a phenomena resulting from a coupling between the electric and mechanical properties of a material.

2. NONLINEAR FINITE ELEMENT FORMULATION

2.1 Piezoelastic Constitutive Relations

Consider a general thin piezoelectric laminated shell (Fig. 1) of thickness t composed of an arbitrary number of thin unidirectional lamina layers, including the piezoelectric layers. Each lamina may have different thickness as well as arbitrary fiber orientation. The piezoelectric material is used in the form of layers or patches embedded and/or surface bonded on laminated composite shells. In a piezoelectric material, the interaction between the mechanical and electrical fields is defined by [4, 10]

$$\{\sigma\}=\left[\bar{Q}\right]\{\varepsilon\}-\left[\bar{e}\right]\{E\}$$
$$\{D\}=\left[\bar{e}\right]^{T}\{\varepsilon\}+\left[\bar{p}\right]\{E\} \qquad \qquad ...(1)$$

where , σ, ε , D and E are the elastic stress vector, elastic strain vector, electric displacement vector and electric field vector, respectively. $\bar{e}$ and $\bar{p}$ are the piezoelectric stress coefficient matrix and dielectric coefficient matrix, respectively.

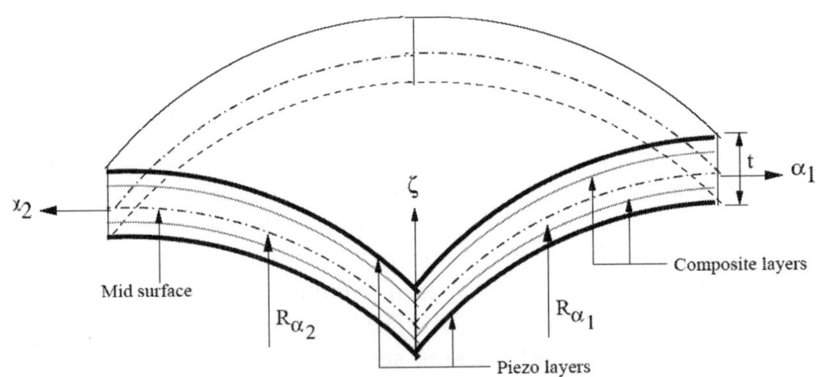

Fig. 1. A doubly curved piezoelectric laminated shell panel.

2.2 Displacement Models

The displacements at any point can be described as

$$u\left(\alpha_1,\alpha_2,\zeta\right)=u^0(\alpha_1,\alpha_2)+\zeta\theta_1$$
$$v(\alpha_1,\alpha_2,\zeta)=v^0(\alpha_1,\alpha_2)+\zeta\theta_1 \qquad\qquad ...(2)$$
$$w(\alpha_1,\alpha_2,\zeta)=w^0(\alpha_1,\alpha_2)$$

where, u_1^0, u_2^0 and w^0 are the displacements on the reference surface along $\propto_1$, $\propto_2$ and ξ and directions, respectively; and θ_1 and θ_2 are the rotations.

2.3 Strain-Displacement Relations

The strains at any point are expressed as follows [3]:

$$\varepsilon_{11}=\varepsilon_1^0+\frac{1}{2}\left[\left(\varepsilon_1^0\right)^2+\left(\varepsilon_3^0\right)^2+\left(\varepsilon_5^0\right)^2\right]+\zeta\kappa_1$$

$$\varepsilon_{22}=\varepsilon_2^0+\frac{1}{2}\left[\left(\varepsilon_2^0\right)^2+\left(\varepsilon_4^0\right)^2+\left(\varepsilon_6^0\right)^2\right]+\zeta\kappa_2$$

$$\gamma_{12}=\left(\varepsilon_3^0+\varepsilon_4^0\right)+\left[\varepsilon_1^0\varepsilon_4^0+\varepsilon_2^0\varepsilon_3^0+\varepsilon_5^0\varepsilon_6^0\right]+\zeta(\kappa_3+\kappa_4) \qquad\qquad ...(3)$$

$$\gamma_{13}=\varepsilon_5^0+\theta_1$$

$$\gamma_{23}=\varepsilon_6^0+\theta_2$$

where, ε_1^0, ε_2^0, ε_3^0, ε_4^0, ε_5^0, ε_6^0 correspond to the mid-surface strains, and κ_1, κ_2, κ_3, κ_4 correspond to the curvatures. The strain and curvature terms of Eqn. (3) are expressed in terms of the field variables as

$$\varepsilon_1^0=\frac{1}{A_1}\frac{\partial u_1^0}{\partial\alpha_1}+\frac{u_2^0}{A_1A_2}\frac{\partial A_1}{\partial\alpha_2}+\frac{w^0}{R_1},\qquad \varepsilon_2^0=\frac{1}{A_2}\frac{\partial u_2^0}{\partial\alpha_2}+\frac{u_1^0}{A_1A_2}\frac{\partial A_2}{\partial\alpha_1}+\frac{w^0}{R_2}$$

$$\varepsilon_3^0=\frac{1}{A_1}\frac{\partial u_2^0}{\partial\alpha_1}-\frac{u_1^0}{A_1A_2}\frac{\partial A_1}{\partial\alpha_2},\qquad \varepsilon_4^0=\frac{1}{A_2}\frac{\partial u_1^0}{\partial\alpha_2}-\frac{u_2^0}{A_1A_2}\frac{\partial A_2}{\partial\alpha_1}$$

$$\varepsilon_5^0=\frac{1}{A_1}\frac{\partial w^0}{\partial\alpha_1}-\frac{u_1^0}{R_1},\qquad \varepsilon_6^0=\frac{1}{A_2}\frac{\partial w^0}{\partial\alpha_2}-\frac{u_2^0}{R_2} \qquad\qquad ...(4)$$

$$\kappa_1=\frac{1}{A_1}\frac{\partial\theta_1}{\partial\alpha_1}+\frac{\theta_2}{A_1A_2}\frac{\partial A_1}{\partial\alpha_2},\qquad \kappa_2=\frac{1}{A_2}\frac{\partial\theta_2}{\partial\alpha_2}+\frac{\theta_1}{A_1A_2}\frac{\partial A_2}{\partial\alpha_1}$$

$$\kappa_3=\frac{1}{A_1}\frac{\partial\theta_2}{\partial\alpha_1}-\frac{\theta_1}{A_1A_2}\frac{\partial A_1}{\partial\alpha_2},\qquad \kappa_4=\frac{1}{A_2}\frac{\partial\theta_1}{\partial\alpha_2}-\frac{\theta_2}{A_1A_2}\frac{\partial A_2}{\partial\alpha_1}$$

3. NUMERICAL RESULTS AND DISCUSSIONS

The post-buckling response of an isotropic cylindrical panel ($a=b=508$ m, $R_1=2544.53$ mm, $h=12.7$ mm) is considered, having additionally two piezoelectric continuous layers attached on the upper and lower surfaces of the substrate. The thickness of each piezoelectric layer is 0.1 mm. The material properties of the piezoelectric laminated cylindrical shell panel are as follows [6]:

E = 3100 N/mm^2, v = 0.3

PZT 5:

$E_{11} = E_{22}$ = 62 GPa, $G_{12} = G_{13}$ = 23.6 GPa, G_{23} = 18 GPa, v_{12} = 0.31,

$d_{31} = d_{32}$ = -200×10^{-12} m/V, $d_{24} = d_{15}$ = 670×10^{-12}, $p_{11} = p_{22} = p_{33}$ = 2992×10^{-12} farad/m

The mechanical point load is applied on the center of the cylindrical panel in combination with bipolar electric fields imposed on the actuators (+ve Volts on the outer electrode and the inner electrode is connected to earth). The nonlinear load-deflection curves, for various applied electric volt are shown in Fig. 2. When positive voltage is applied, the shell buckle under higher load value compared to zero electric voltage. Conversely, for negative voltage the shell buckle at a lower load compared to that of zero voltage. It is noted here that the negative voltage indicates an early buckling. The present results and the published results (Varelis and Saravanos [6]) are in very good agreement.

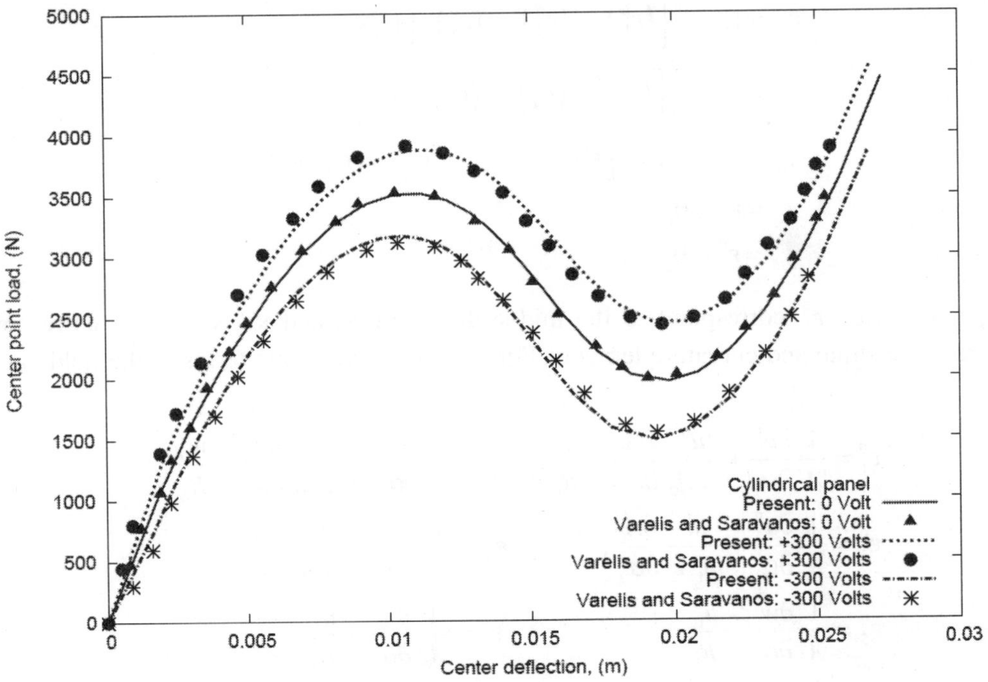

Fig. 2. Nonlinear load-deflection curves of smart cylindrical shells.

3.1 Post-Buckling Analysis of Smart Cylindrical Shells

The curved edges of the cylindrical shell panels are assumed to be free and the straight edges are hinged. The two opposite curved edges of the spherical panel are assumed to be free and the other opposite curved edges are hinged. The deflections are presented at the center of the panels. The panels ($a = b$ = 500 mm) are subjected to a central transverse point load. The total thickness of the each panel is considered as 6.25 mm, and the thickness of each upper and lower piezoelectric layer is assumed as 0.1 mm. For a cylindrical shell, R_1 = 2500 mm and $R_2 = \infty$. The material property used for the analysis of graphite-epoxy shell panels are as follows:

E_{11} = 181.0 GPa, E_{22} = 10.3 GPa, G_{12} = G_{13} = 7.17 GPa, G_{23} = 3.58 GPa, u_{12} = 0.28

The post-buckling behaviour of $[0°/ \pm 45°/90°]_s$ laminated cylindrical panel is analyzed with two piezoelectric layers, one is on the top surface and another is on the bottom surface of the panel. The outer electrode is connected to $+ve$ Volts and the inner electrode is connected to earth. The material properties of piezoelectric layer PZT-5 are assumed. Along with the action of a concentrated vertical load, electric voltage is applied to the piezoelectric layers. The transverse center deflection versus the applied concentrated load for various applied electric voltage values are shown in Fig. 3.

It is observed that the curve corresponding to +300 Volts undergoes snap-through at a higher load values compared to that of 0.0 Volt. The shell undergoes snap-through at a lower values when negative voltage of −300 Volts is applied to the piezoelectric layers. The transverse point load induces compressive mechanical stress and the nonlinear deflection increases till limit point load is reached. At the limit point the stiffness matrix becomes negative. All the strain energy stored as potential energy during the loading process converts to kinetic energy. So, the smart shell jumps from one equilibrium state to another state which is dynamic in nature and the smart shell undergoes snap though. When the applied +600 Volts electric voltage is increased, the limit point load is observed to be also increased.

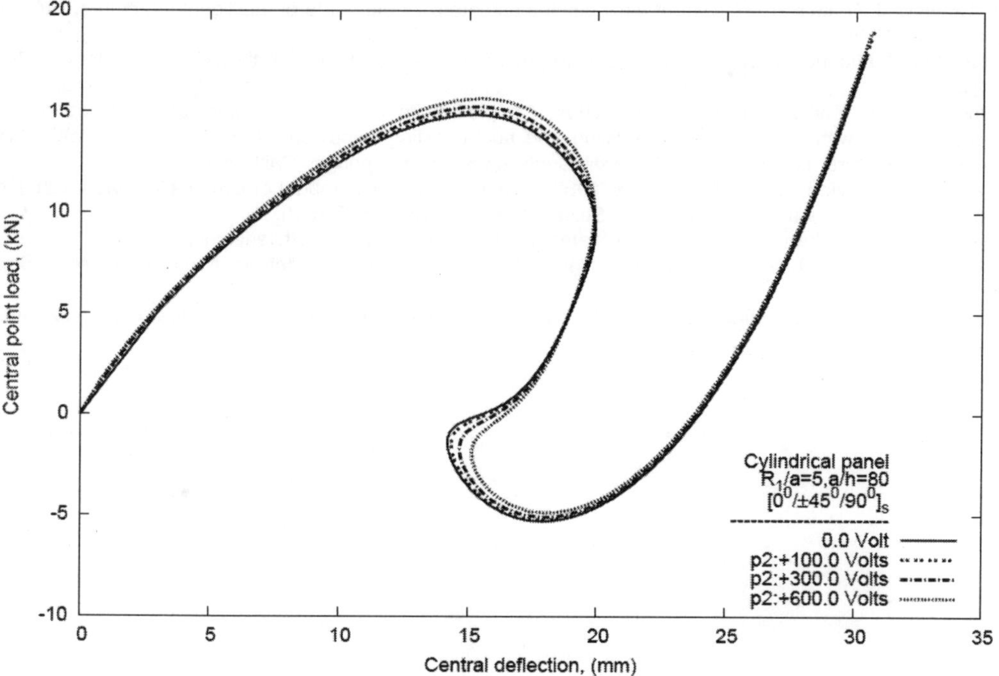

Fig. 3. Nonlinear load-deflection curves of laminated smart cylindrical shell panels for different electric potentials.

CONCLUSION

The geometrically nonlinear analysis of doubly curved piezoelectric laminated shells is carried out using finite element method. A nine-noded isoparametric geometrically nonlinear coupled shell element is developed. The coupled nonlinear responses of smart cylindrical shells undergoing large displacement (snap through and snap back) are predicted. An incremental iterative technique based on arc-length method is implemented in order to capture and predict the snap-through responses of laminated smart shell panels under transverse center point load values. Bipolar electric potential is applied to the piezoelectric layers. The positive electric voltage is connected to the outer electrodes of the piezoelectric surface and the other surface electrodes are connected to earth. The linear variation of potential is assumed across the transverse direction of the piezoelectric layer. The buckling load increases for positive electric potential applied. Thin smart shells having been integrated piezoelectric actuators undergo large deformation resulting in snap-through and snap-back phenomenon from one equilibrium state to another. The present results are found to compare well with those available in the open literature. The nonlinear responses of $[0°/ \pm 45°/90°]_s$ laminated cylindrical shells with piezoelectric layer are analyzed and the numerical results are presented.

REFERENCES

1. H. Kraus, 1967, Thin elastic shells, John Wiley and Sons, Inc: New York.
2. O. C. Ziekiewicz, R. L. Taylor, 1991, The Finite Element Method, Vols, I, II, McGraw Hill: New York.
3. C .K. Kundu, P. K. Sinha, 2005, Post-buckling analysis of laminated composite shells. Composite Structures (in press).
4. E. F. Crawley, J. D. Luis, 1987, Use of piezoelectric actuators as elements of intelligent structures, AIAA Journal. 25(10), 1373-1385.
5. Crisfield MA. A fast incremental iterative solution procedure that handles snap-through. Computers \& Structures 1981;13:55-62.
6. D. Varelis, D. A. Saravanos, 19-22 April 2004, Coupled finite element for nonlinear laminated piezoelectric composite shells with applications to buckling and post-buckling behaviour. 45th AIAA/ASME/ASCE/AHS/ASC Structures, Structural Dynamics and Material Conference: Palm Springs, California.
7. P. F. Pia, A. H. Nayfeh, K. Oh, D. T. Mook, 1993, A refined non-linear model of composite plates with integrated piezoelectric actuators and sensors, Int J Solids Struct, 30(12), 1603-1630.
8. R. Jones, 1999, Mechanics of Composite Materials, Taylor and Francis: Philadelphia.
9. A. Mukherjee, A. S. Chaudhuri, 2002, Piezolaminated beams with large deformations. Int J Solids Struct. 39, 4567-4582.
10. J. N. Reddy, 1999, On laminated composite plates with integrated sensors and actuators, Engg. Struct. 21, 568-593.

110

Intelligent Chaos Controller Using Recurrent Neural Networks and Genetic Algorithms

J. Krishnaiah[1], C.S. Kumar[2] and M.A. Faruqi[3]

[1]Formerly Scholar at Deptt. of Mech. Engg., IIT Kharagpur-721302, India
email: j.krishnaiah@gmail.com
[2]Robotics and Intelligent Systems Lab, Deptt. of Mech Engg., IIT Kharagpur-721302, India
email: kumar@mech.iitkgp.ernet.in
[3]Formerly Professor at Deptt. of Mech. Engg., IIT, Kharagpur-721302, India
email: aslamfaruqi@yahoo.com

ABSTRACT

We present a Chaos Controller which works based on the modelling capability of Recurrent Neural Networks(RNN) and global searching capability of Genetic Algorithms (GA). In reality many physical and non-physical systems are very difficult to represent using a mathematical form; even if mathematical models exist, it would be a difficult task to make a controller which works in real-time. Moreover, if the considered system is chaotic in nature there are very few methods to tackle the real-world systems for controlling task. On contrary there are large number of Chaos Control techniques when the considered system is/has a mathematical model. Based on the fundamental idea of these techniques, i.e., a small perturbation at appropriate time is enough to control such chaotic systems; the present method uses the global search capability of GA to find a best perturbation to the control parameter at each step by using RNN model of considered system as an objective function. This approach was realized only after demonstration of the plausibility of developing a RNN model that could be used to construct a Bifurcation Diagram and estimate the Largest Lyapunov Exponents of a given system from its observed data in the recent research works.

Keywords: Chaos Controller; Recurrent Neural Networks; Genetic Algorithms.

1. INTRODUCTION

The recognition that the occurrence of chaos is widespread in nature, and it even forms the basis of intelligence in brain and provides control mechanism for heart and other systems. And also its prevalence in engineering and scientific systems has been fueling research in this area. Many books [1-3] have been written focusing on different aspects of the chaotic systems. A recent literature review provided by Andrieveskii and Fradkov [4, 5] showing the diversity in chaos related research.

They have also noted that using chaos as a basis for designing engineering systems is growing which can be seen from its use in anti-control of systems, coding and encryption in communication systems, etc.

Literature shows that the study of chaos can be divided into two broad categories. One that deals with engineering applications, where the existence of chaos and its control methods in practical problems, and the other that deals with the scientific applications where the focus is on control theory, methods of discovering new regularities in physical systems, etc.

Most of the chaos controlling methods work on the fact that the relationship at fixed point on a Poincaré section is linear [6]. In these methods instead of a deliberate action to control the chaos in the system, wait till the system reaches the vicinity of the fixed point then the controlling action is applied in the form of a small one-time or continuous perturbation to the accessible control parameter. In reality waiting for the system to reach the vicinity of the fixed point is an indefinite time-taking process. In some cases of chaos control methods a very heavy computing may be necessary to find the magnitude of the perturbation and the point at which the perturbation is required to apply. Moreover, knowing the exact mathematical models for real-world systems is also a difficult proposition, and also the real-world systems are prone to noise and observational errors. At the same time the long-term predictability remains a challenge due to its extreme sensitivity to initial conditions.

Studies of real-world systems have to additionally contend with the prevalence of noise in systems, its role in changing the initial conditions of the system and its capability to cause bifurcations [7]. The problem gets further complicated with time scale of the phenomena to be the controlled, if the computations for control actions have to be done in real-time. There are three main methods of controlling chaos that have been established [8]:

1. Open loop method based as periodic excitation of the system,
2. OGY method (Poincaré map linearization), and
3. Method of time delayed feedback (Pyragas method).

The open loop method has been in use for a long time specially in systems with high frequency operations, to alter their behaviour with small changes in inputs. Well developed procedures are available today.

It is the OGY [6] method, which made the control of chaos possible in large variety of continuous and discrete systems. The method aims at stabilizing the periodic orbits of a system at an unstable fixed point. Today numerous examples [9-11] are available of its successful application even in presence of noise [11, 12]. The original method requires waiting for suitable orbits to occur before stabilization is attempted. However, it has been modified by Weeks and Bergees [13] where suitable impulses to the system are given in efforts towards stabilizing the orbits without waiting. Neural network modules could be trained to predict the impulses to be given. But this effort cannot be undertaken where the basic model of the system and the trained neural networks are not available. This limitation could be especially true for systems which are observed only through input/output data.

However, yet an important way of controlling a chaotic process may be through the modification of the attractor of the process, leading to the transformation of chaotic oscillations into periodic ones or vice versa [8]. It is relevant to real-world chemical and metallurgical process control fields and also to the fields of biology and medicine. As an example it can be easily seen that in chemical process sometimes it may be necessary to increase the chaoticity to increase mixing in the reactor etc. This approach has also been utilized in efforts to control certain biological systems as well.

Pyragas'[14] method has been used successfully for stabilizing chaotic systems without a complete model of the system, through suitably derived time delayed feedbacks to the system [15]. However, formal analysis for understanding the process and the conditions of stability are still ongoing research activities.

The basic problem in identifying the presence of chaos in real-world system has been addressed through identification procedure using observed data [16, 17]. In [16] dynamic Neural Networks has been developed for the identification and stabilization of chaos. De Feo [17] has developed a method of chaos identification in irregular periodic data, using ARMA techniques along with Genetic algorithms. De Feo has also observed a phenomenon of 'self emergence' of chaos indicating that the characteristics of chaotic system appear quite early in the identification procedures.

From the above literature review it can be seen that when a real-world chaos (with changes in initial conditions) is observable only through its input/output data, techniques for using this multivariate noisy data, and developing control strategies for the bifurcation control or stabilization or attractor modification have not been sufficiently explored.

2. REVIEW ON TECHNIQUES

2.1 Recurrent Neural Networks

Neural Networks with Feed-forward architecture are capable of modeling only static, nonlinear relationships between the input space to output space of a given nonlinear system. These networks are known as Feed-forward Neural Networks (FNN). In reality most of the physical and non-physical systems are dynamic in nature in addition to being nonlinear i.e., time dependent nature as well as parameter dependent. It has been found that the Neural Networks with Recurrent architectures are suitable for the dynamic and non-linear modeling of systems.

After the discovery of Backpropagation algorithms for Feed-forward neural networks, many researchers explored the possibility of extending the Backpropagation algorithms to the cases of networks with recurrent connections. Pineda [18], Almeida [19], and Rohwer and Forrest [20] have independently pointed out that Backpropagation can be extended to arbitrary networks as long as they converge to stable states. In the present case RNN is based on the idea of Narendra architecture [21].

2.2 Chaos Control

There are several approaches evolved after the introduction of famous OGY method of chaos control. Till the introduction of OGY method it was assumed that the chaos cannot be controlled and one has to avoid the region of chaos and take safer route and keep in non chaotic region. After understating the nature of chaos, it has been shown that the chaos is a co-existence state of many periodicities which are unstable in nature. In the OGY method utilising the richness of the very property has been demonstrated by controlling to a unstable periodic orbit. Meaning, it is possible to switch to any periodic orbit once the system is in chaotic mode. This gives a wide opportunity of easily changing the mode with less amount of energy. And also, sometimes operating the system in non-chaotic region would be an expensive when comparing to chaotic region. Hence, controlling and stabilizing the system in chaotic region is very important.

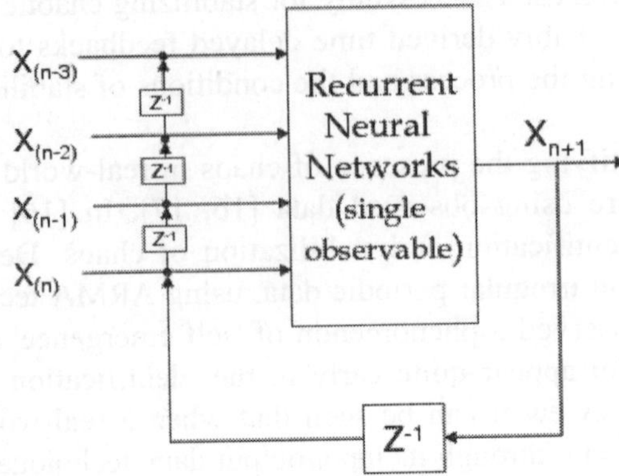

Fig. 1. Recurrent Neural Networks used for modelling Hénon map attractor proposed by J. C. Principe.

The chaos control techniques can be broadly classified into two: one is feedback based chaos control and other is non-feedback chaos control.

2.3 Application of RNN to Chaos Control

While RNNs have been extensively used for modelling complex non-linear dynamical systems their evolution towards representing chaotic systems has been reported only in few publications.

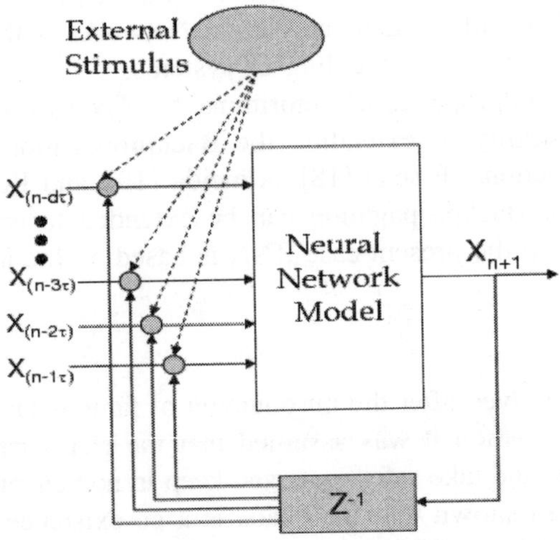

Fig. 2. Recurrent Neural Networks used for modelling and controlling oflkeda map attractor by A.Jones.

J.C.Principe [22] proposed a method shown in Figure 1 where it is possible to model a chaotic system and be able to predict time-series for a larger number of steps for a given initial condition. The main limitation of this model is that it considers only one set of accessible control parameters, which may not be enough to model and control real-world systems.

Another variation of the above approach has been used by Jones et.al. [23] to model more complex systems for control by providing additional information at feedback input points as shown in Figure. 2.

Jones et.al. [24] have explored the possibility of synchronising chaotic systems, using recurrent neural networks, trained to low values of mean squared error of training. The possibilities of using these for real-world system were not explored.

From the literature it can thus be seen that recurrent neural networks have been used extensively with various architectures for one step or several steps prediction from a given starting point. Their capability of being used as a model for predicting larger number of steps in time to represent evolution of chaotic systems has not been studied. Further the need for modelling real-world systems with multiple inputs and outputs, and the role of noise has not been addressed to.

2.4 Genetic Algorithms

In 1970s John Holland at University of Michigan formally introduced Genetic algorithms. The continuous reduction of price/performance of computational systems has made them attractive for some types of optimization. They are less susceptible to getting 'stuck' at local optima than gradient search methods. But they tend to be computationally expensive in some cases where objective function is heavy. As by now genetic algorithms are common detail treatment is not provided and only discussed on the application of GA in Chaos Control.

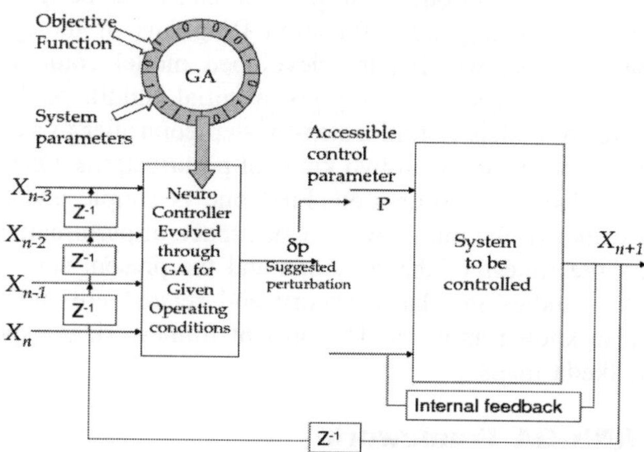

Fig. 3. Framework used by Weeks and Burgess to stabilise unstable periodic orbit of the Logistic and Hénon maps. The controller was evolved for each attractor through Genetic Algorithms.

There has been some work to implement GA to search controller to control chaotic systems. Among these important is Weeks' and Burgess [13]. Weeks and Burgess proposed a method for controlling chaos where the controller is a Neural Network model, getting feedback from the plant itself. This Neuro-controller is evolved through a Genetic Algorithm based search, as the one shown in Figure 3. This Neuro-controller gives a perturbation to the control parameters based on last three time states of the system. This method is also basically devoted to control of chaotic system through orbit stabilisation procedures [6].

3. INTELLIGENT CHAOS CONTROLLER

The present framework is inspired by the OGY method. The creative thought that has been implemented in the OGY method and Weeks' GA based Neuro-controller are the origin to the present way of controlling chaos through a RNN model and GA based approach. In OGY method one has to wait to apply the control action till the system reach the vicinity of the fixed point or the periodic orbit that being sought to achieve i.e., period one, period two, period four, etc. In the present approach the control action starts immediately.

With the advent of plausibility of Constructing Bifurcation Diagram based on the model developed using observed data shown in [25], Estimation of Largest Lyapunov Exponents and control architecture suggested in [26], a RNN-GA based method to drive the system towards the fixed point is implemented for Hénon and Ikeda maps. The advantage of the present method is that, it is not required to know the fixed point to be able to reach the fixed point. In this it is not required to wait for the system to reach the vicinity of the fixed point of the system, instead this RNN-GA based controlling is a very deliberate way of perturbing the system to drive it towards a fixed point. That is, control action starts immediately from the point of the method activation. In GA-Chaos controller, the GA searches the perturbation required to control parameters that minimizes the difference between the present state and the next state.

The present approach is targeted to control real-world chaotic systems. It is known that the real-world systems are difficult to represent through any mathematical equations. They can be represented through the observed multivariate data. A newly developed methodology that could be verified the trueness of the developed model was introduced in [26]. In that, it has been shown that the developed model could be used for Constructing the Bifurcation Diagrams of the system for various input conditions. Further, it has been shown that the developed model could be used for calculating maximum Lyapunov Exponents of system for any given initial conditions. This possibility gave the strength to use those developed models to find a step-by-step controller to stabilize the given system.

The present approach has been tried with two kinds of perturbations. One is uniform perturbation and second is differential (adaptive) perturbation. First one is simple and one range of space for perturbation. Second one works on the multiple ranges of space of perturbation. And the perturbation range is decided based on the fitness of the solutions and the present state of the system.

In general most of the studies in Chaos Theory and related are first experimented on well known discrete systems also known as maps. The present framework is experimented on two such maps namely Hénon and Ikeda maps.

3.1 How to Build RNN-GA Framework?

The main targeted application of the present framework is to build an intelligent Chaos Controller for Industrial Processes/Systems. Therefore the procedure has been developed keeping in mind that how these processes operated and observed for the data collection. Moreover they are prone to noise and available short (short because some process may take one hr to one day for a single set of observation, therefore it may take lot of time to have large set of data).

Once the data for a given process is available, using the generic architecture of RNN described in [26], model is built. After analysing the model for its Bifurcation Diagram, which represents the behaviour of the given system in-terms of time and as well as control parameters.

Using a standard GA for searching the perturbation required to be given the present set of control parameters of the considered system is build. The RNN model in hand will be used as objective function in the GA search.

3.2 Control of H\'enon Map

H\'enon map is well-known discrete system, which represents a two-dimensional chaotic system. Most of the ideas in the chaos theory have been first verified and evolved for general chaotic systems. The H\'enon map can be represented by two equations with two constants called control parameters or bifurcation parameters. Figure 4 shows the application of intelligent chaos controller on controlling Hénon map.

4. CONTROLLING IKEDA MAP

Ikeda map is another well-known discrete chaotic system of dimension three. It models a laser system in which a string of light pulses is phase-shifted because of the presence of a non-linear dielectric. Using the developed RNN-GA based approach the Ikeda map has been tried to stabilize on to a fixed point, though the fixed point is not known. Given some initial conditions, GA suggests amount of perturbation to be given so that the objective function is minimised. In the present test case it took less than 100 iterations to minimize the variation. For the complete stabilization the GA took more that 300 iteration. Figure 5 shows the application of intelligent chaos controller to control Ikeda map.

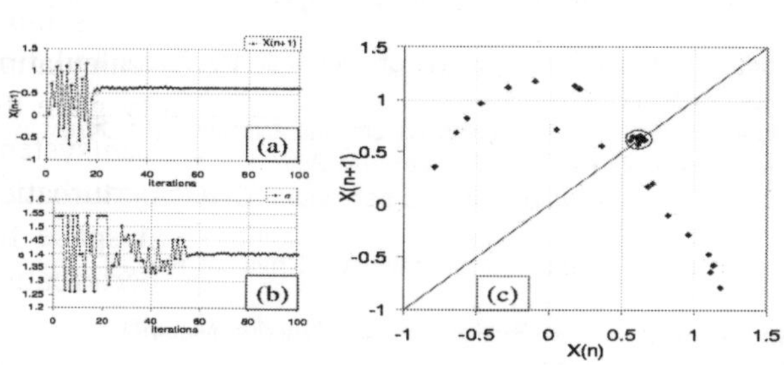

Fig. 4. Controlling chaos in the H´enon map using RNN-GA based approach.

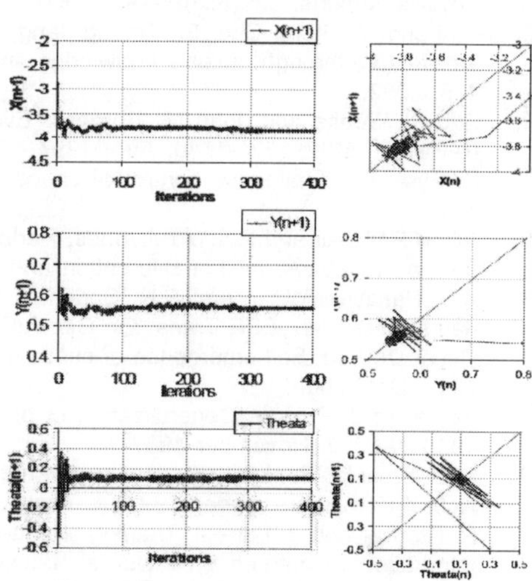

Fig. 5. Controlling chaos in the Ikeda map using RNN-GA based approach.

CONCLUSION

The experimentation with these two maps gave encouraging results for the chaos control approach through RNN-GA. It has been observed its simplicity and fastness in stabilizing the system within the chaotic regions, without actually knowing the fixed point. It has been generalized for data driven systems using Recurrent Neural Networks models.

Further, this shows the general applicability of the framework described in [26]. This framework may be applied to any process or system, which has complex nonlinear dynamic behaviour.

ACKNOWLEDGEMENTS

The authors are thankful to SERC-DST, Government of India which provided grants as per Sanction no: SR/FTP/ET-83/2001 under its scheme of fast track proposal for young scientists-2001.

REFERENCES

1. Ali H. Nayfeh and Balakumar Balachandran. Applied Nonlinear Dynamics: Analytical, Computational and Experimental Methods. John Willey and Sons, New York, 1995.
2. Edward Ott. Chaos in Dynamical Systems. Cambridge University Press, Maryland, USA, 1993.
3. Robert C.Hilborn. Chaos and Nonlinear Dynamics: An Introduction for Scientists and Engineers. Oxford University Press, New York, 2nd edition,2000.
4. A.L.Fradkov and R.J.Evans. Control of chaos: Survey 1997-2000. In Preprints of 15th IFAC World Congress on Automatic Control. Plenary papers, Survey papers, Milestones, pages 143-154, Barcelona, July 2002.
5. B.R.Andrievskii and A.L.Fradkov. Control of chaos: Methods and applications. 2.applications. Automation and Remote control, 65(4):505- 533, November 2003.
6. E. Ott, C. Grebogi, and J. A. Yorke. Controlling of chaos. Phys. Rev. Lett., 64:1192-1196, 1990.
7. Klaus R. S., Bifurcations of the randomly perturbed logistic map-Numerical Study and Visualizations-. URL:http://www.iew.unizh.ch/home/klaus/logistic/intro.html.
8. B.R.Andrievskii and A.L.Fradkov. Control of chaos: Methods and applications. 1 methods. Automation and Remote Control, 64(5):673-700, October 2003.
9. W.L.Ditto, S.N.Rauseo, and M.L.Spano. Experimental control of chaos. Phys. Rev. Lett., 65(26):3211-3214, Dec. 1990.
10. Devid J. Christini and James J. Collins. Controlling neuronal chaos using chaos control. URL: http://arxiv.org/abs/chao-dyn/9503003.
11. S. Boccaletti, C. Grebogi, Y.C. Lal, H. Mancini, and D. Maza. The control of chaos: Theory and applications. Physics Reports, 329(3):103-197, 2000.
12. Po-Feng, Ji-Zheng Chu, Shi-Shang Jang, and Shyan-Shu Shieh. Developing a robust model predictive control architecture through regional knowledge analysis of artificial neural networks. Journal of Process Control, 13:423-435, 2002.
13. Eric R. Weeks and John M. Burgess. Evolving artificial neural networks to control chaotic systems. Physical Review E, 56(2):1531-1540, Aug. 1997.
14. K. Pyragas. Continuous control of chaos by self-controlling feedback. Physics Letters A, pages 421-428, 170 1992.
15. Alban P.M.Tsui and Antonia J.Jones. Periodic response to external stimulation of a chaotic neural network with delayed feedback. Int.J.of Bifurcation and Chaos, 9(4):713-722, 1999.
16. A.S. Panzyak, W.Yu, and E.N. Sanchez. Identification and control of unknown chaotic systesm via dynamical neural networks. IEEE Trans. On Circuits and Systems, 46(12):1491-1495, Dec. 1999.
17. Oscar De Feo. Self-emergence of chaos in the identification of irregular periodic behaviour. CHAOS, 13(4):1205-1215, Dec. 2003.
18. Fernando J. Pineda. Generalization of backpropagation to recurrent neural networks. Physical Review Letters, 19:2229-2232, November 1987.
19. L.B. Almeida. Backpropagation in perceptrons with feedback. In R. Eckmiller and Von der Malsburg, editors, Neural Computers, pages 199-208, New York, 1988. Springer-Verlag.
20. R. Rohwer and B. Forrest. Training time-dependence in neural networks (vol. 2). proceedings of the first. In IEEE International Conference on Neural Networks, volume 2, pages 701-708, San Diego, CA, 1987.
21. Narendra K.S. and Parthasarathi K., 1990, Identification and Control of Dynamical Systems Using Neural Networks. IEEE Transactions on Neural Networks, Vol. 1, pp. 4-27.
22. Jyh-Ming Kuo, J.C.Principe, and Bert de Vries. Prediction of chaotic time-series using recurrent neural networks. IEEE Workshop Neural Networks for Signal Processing, pp. 436-443.
23. Ana Guedes de Oliveira, Alban P. Tsui, and Antonia J. Jones. Using a neural network to calculate the sensitivity vectors in synchronization of chaotic maps. In In Proceedings 1997 International Symposium on Nonlinear Theory and its Applications (NOLTA'97), volume 1, pages 46- 49, Honolulu, U.S.A., 1997. Research Society of Nonlinear Theory and its Applications, IEICE.
24. Antonia J. Jones, A.P.M Tsui, and Ana G. Oliveira. Neural models of arbitrary chaotic systems: construction and the role of time delayed feedback in control and synchronization. Complexity International, 9, 2002. URL: http://www.complexity.org.au/vol09/tsui01/.
25. Krishnaiah, J., Kumar, C.S., and Faruqi, M.A. (2002), "Constructing Bifurcation Diagram for a chaotic time-series data through a recurrent neural network model," Proceedings of the 9th International Conference on Neural Information Processing, 2002, pp. 2354- 2358.
26. Krishnaiah, J. Kumar, C.S., and Faruqi, M.A. (2006) Modelling and control of chaotic processes through their Bifurcation Diagrams generated with the help of Recurrent Neural Network models: Part 1 simulation studies Journal of Process Control, 16(1):53-66, January 2006.

111

Analysis of Cable Net Structures

PRASAD JOSHI

*Final year ME (Structures) Student of Maharashtra Institute of Technology, Pune.
Anant Smruti, D-9, S.No. 127/1A, Ambegaon Road, Behind Bharati Vidyapeeth,
Katraj, Pune-411 046, Maharashtra, India.*

ABSTRACT

Cable net structures are the structures formed by cable nets having different configurations, in which main load carrying elements are cables. These structures usually have special erection methods as well as structural systems depending on their load carrying capacity. The initial shape of cable net depends on the internal force distribution, which cannot be described by simple geometrical models. A key step in the design of these structures is determination of their stable geometrical configuration, known as form finding.

This paper describes the form finding method that includes the principle arrangements for the cables to develop a stable configuration. This method is based on tension truss concept, in which the main load carrying elements are the cables connected with pin jointed cable network by forming a net of triangular elements. The main advantage of this method is that we can achieve greater surface accuracy and stability of the network by increasing number of triangles without increasing number of supports.

In this paper an attempt is made to describe a suitable form finding method to generate a stable configuration for a reflector surface of an antenna use to transmit the electromagnetic signals for space applications, which is formed by cable net. The problem statement was given by Institute of Armament Technology, (IAT) Pune and an attempt has been made to solve this problem using Multivariable Newton-Raphson Technique, which is an iterative method use to solve non-linear algebraic equations involving more than single variables and a software package known as MATLAB, use for high performance numerical computations and visualisation.

Keywords: Cable net structures, form finding, tension truss, tensegrity structures.

1. INTRODUCTION

Cable net structures are the structures formed by cable nets having different configurations, in which the main load carrying elements are cables. The cables are flexible tensile members and are made up of high strength steel wires and silicon-coated glass fibre membranes. The main advantage of these tensile members over compression members is that, they can be as light as the tensile strength permits. Cable net structures have many applications. They are used to form different shaped roofs as well as they had been used to develop paraboloidal surfaces of reflector antennas for space applications. This is because cable net structures represent construction with minimum amount of materials. With the help of special arrangement of cables and struts we can achieve stable configuration with accuracy in the shape of these structures.

However, for tensile structures because of the negligible flexural stiffness of cables, the initial configuration of these structures must be stressed and their behaviour for the external loads should be performed before the analysis. After the initial reference configuration is determined, the structural members are described by matrices and force vectors. The single cable is modelled by one or several other elements depending on the sag to span ratio. The final step in the analysis process is to define the external loads applied on the structures.

Initially the analysis of cable net structures was based on tensegrity principle, which was invented by R B Fuller. The word tensegrity is a contraction of tensile integrity and was successfully used to analyse and design the reflector antennas for space applications. A tensegrity system is an internally prestressed, free standing and pin jointed network in which cables or tendons are tensioned against the systems of bars and struts. To explain the mechanical principle of tensegrity structures, Tibert G [6] elaborates a balloon analogy. If the enclosed air is at higher pressure than the surrounding air it pushes outwards the skin of the balloon. If the air inside the balloon is increased, the stresses in the skin become greater and balloon will harder to deform. In tensegrity structure the struts plays a role of air and cables are behave like a skin of balloon. Increasing the forces in the elements, bearing capacity of the structure increases strength and load bearing capacity of the structure by achieving the accuracy in the shape.

2. AIM AND SCOPE

The aim of the present work was to study mechanical aspects of cable supported structures falls under the category of the reflector antennas for space applications. Formulation of a suitable geometry using cable net and also to decide principal arrangement for the cables connected with pin joints for a reflector antenna using the form finding method used by Tibert G [5]. Further, this geometry was analysed using multivariable Newton-Raphson Technique to solve the non-linear algebraic equations generated for cables to find displacements of joints and develop a paraboloidal surface of a pin jointed cable network subjected to external loading.

3. ANALYSIS OF REFLECTOR ANTENNA USING MULTIVARIABLE NEWTON-RAPHSON METHOD WITH MATLAB PROGRAMMING

A major step in analysis and design of cable net structures is the determination of their equilibrium configuration, known as form finding. The next step in the analysis is to develop equilibrium equations for the cables meeting in different joints to find the joint displacements subjected to external loads. This section will discuss in detail about the formulation of three-dimensional cable net system using method suggested by Tibert G [6] and further about the Newton-Raphson technique and solution of non-linear algebraic equations using MATLAB [2], which is a software package used for high performance numerical calculations involved in solving such equations.

3.1 Analysis of Reflector Antenna

Reflector antennas for space applications are used to transmit the electromagnetic signals between geostatic stations and the spaceships or satellites. In these antennas surface accuracy is an important factor to reflect the waves of the electromagnetic signals in the right directions. A problem statement was given by Institute of Armament Technology (IAT), Pune to generate a surface for a reflector antenna using a cable net and analyse it by using a suitable method which involve iterations and capable to solve the non-linear algebraic equations containing more than one variable. For deciding a stable configuration of an antenna the form finding method based on tension truss concept [3] was used. In this concept the reflector surface is divided into a net of triangular elements and flexible cables are used for the sides of the triangles. All the cables are made to take tensions subjected to the external loading; used to generate a parabolic surface for reflection. The main advantage of tension truss antenna is that we can increase the surface accuracy by decreasing the sides of triangles without increasing number of supports.

The shape of the paraboloidal surface was determined by the lengths of the cables and not by the prestress distribution. The number of joints and number of cables required to generate a cable net surface for the antenna were found by the expressions suggested by Tibert G [5] for general tension truss by deciding six bays i.e., six sided polygon and three divisions i.e., three triangular subdivisions. Hence, the number of joints 'j' and number of cables 'b' were found by the equations 1 and 2 as:

$$j = 1 + 6\,\frac{3(1+3)}{2} = 37 \text{ nos.} \qquad \qquad \text{...(1)}$$

$$b = 6\,\frac{3(1+3\times3)}{2} = 90 \text{ nos.} \qquad \qquad \text{...(2)}$$

After finding the number of joints and number of cables the final shape for the reflector antenna was decided which is as shown in the Figure 1.

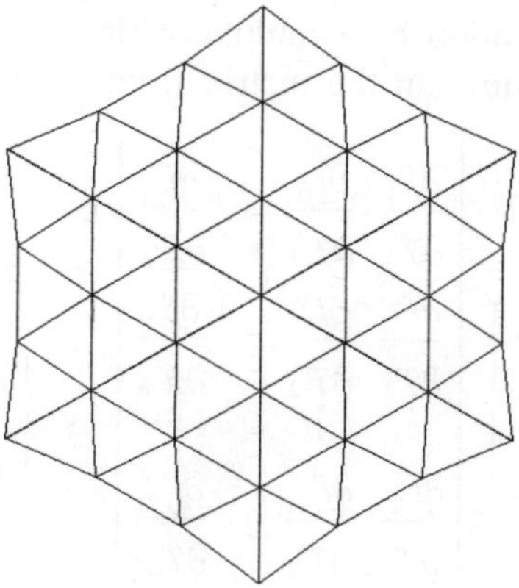

Fig. 1. Geometry for a reflector Antenna.

This geometry was then tested for symmetrical loading with the maximum load at the central node. Because of the symmetry in the configuration it was decided to analyse a single bay of the above cable net system as shown in Figure 2.

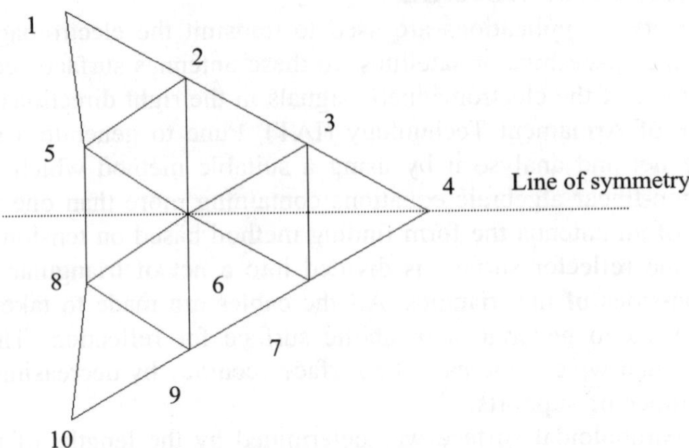

Fig. 2. Geometry of an antenna considered for the analysis.

This is a three-dimensional cable net system in x, y and z planes. The known things about this system were the lengths of all the cables meeting in joints 2, 3, 4, 5, 6, 7, 8 and 9 and the forces applied at these joints. Our aim was to find the co-ordinates of these joints in x, y and z directions and the tensions in all the cables meeting at these joints after application of the loads.

This system was then analysed using Newton-Raphson Technique and also a programme was prepared in MATLAB to solve the equilibrium equations for cables meeting at these joints. There are twenty one equations obtained, written in the form of non-linear algebraic equations, which were involved twenty one different variables like T_1, T_2, T_3, T_4, T_5, T_6, T_7, T_8, T_9, T_{10}, x_2, y_2, z_2, x_3, y_3, z_3, y_4, x_5, y_5, z_5 and x_6, assuming that there is only vertical movement at joint '4' and horizontal movement at joint '6'. All the twenty one equations were defined by separate functions from f_1 to f_{21} and multivariable Newton-Raphson Technique was used to solve the set of these twenty one equations. After application of a multivariable Taylor series of expansion to all these equations and putting all the equations in the matrix form we obtain:

$$\begin{Bmatrix} f_1 \\ f_2 \\ \vdots \\ f_{21} \end{Bmatrix} \begin{bmatrix} \dfrac{\partial f_1}{\partial T_1} & \dfrac{\partial f_1}{\partial T_2} & \cdots & \dfrac{\partial f_1}{\partial X_6} \\ \dfrac{\partial f_2}{\partial T_1} & \dfrac{\partial f_2}{\partial T_2} & \cdots & \dfrac{\partial f_2}{\partial X_6} \\ \vdots & \vdots & & \vdots \\ \dfrac{\partial f_{21}}{\partial T_1} & \dfrac{\partial f_{21}}{\partial T_2} & \cdots & \dfrac{\partial f_{21}}{\partial T_{21}} \end{bmatrix} \times \begin{Bmatrix} T_1 \\ T_2 \\ \vdots \\ X_6 \end{Bmatrix} \qquad \ldots(3)$$

The above equation 3 can be written as:

$$\{F\} = \{F\} \times \{X\} \qquad \ldots(4)$$

where, $\{A\}$ is a jacobian matrix containing jacobian elements obtained by partial derivatives of the twenty one equations with respect to the twenty one variables, hence it is a matrix of constants of the order 21×21. The initial values in the matrices were calculated by considering some initial guess for the variables involved in the algebraic equations. We obtain the exact values for the variables involved in the matrix $\{X\}$ by premultiplying both the sides of eq. 4 with the inverse of the matrix $\{A\}$ and by using the relation:

$$\{X\}^{(k+1)} = \{X\}^{(k)} - \{A^{(k)}\}^{-1} \times \{F^k\} \qquad \ldots(5)$$

The major step in the analysis using multivariable Newton-Raphson technique is that to calculate inverse of the matrix $\{A\}$ once in every iteration and reuse it further to obtain new values for the variables involved in the equations using the relation 5. For this a programme was prepared in MATLAB's language to carry out iterations until we get the converging solution. Initially we have to give some values for the variables T_1, T_2, T_3, T_4, T_5, T_6, T_7, T_8, T_9, T_{10}, x_2, y_2, z_2, x_3, y_3, z_3, y_4, x_5, y_5, z_5, and x_6 as an initial guess and also we have to put all the known values like lengths of all cables and the forces applied at the joints 2, 3, 4, 5 and 6 to calculate the initial values for matrices $\{F\}$, $\{A\}$ and $\{X\}$ for the eq. 3. This is our first approximation. After knowing the initial values we have to calculate the inverse of matrix $\{A\}$ and use it to obtain new values of the variables for next iteration by substituting in the eq. 5. A loop command was used to repeat this procedure and finally we get the improved values for the variables that converge to the exact values. All the twenty one equations were defined by separate functions from f_1 to f_{21}, to obtain the values for matrix $\{F\}$ and the new values for variables obtained after every iteration was indicated by 'Y'. We get the output in the sequence like, first the set of new values of variables 'Y' and following to this the new values for the functions, 'F' based on 'Y'.

3.2 The Final Values of the Variables Obtained by Solving the Matrices and Using the Relation 5 for the Reflector Antenna Shown in Figure 1 are as Follows:

Variables	Values	
T_1	2.7800	kN
T_2	3.1700	kN
T_3	1.4000	kN
T_4	0.9500	kN
T_5	3.2400	kN
T_6	3.6000	kN
T_7	1.0700	kN
T_8	1.6000	kN
T_9	3.4500	kN
T_{10}	2.5000	kN
X_2	0.8700	m
Y_2	0.4500	m
Z_2	$-$ 1.0000	m
X_3	1.7400	m
Y_3	0.7100	m
Z_3	$-$ 0.5000	m
Y_4	0.8000	m
X_5	0.1200	m
Y_5	0.1000	m
Z_5	$-$ 0.5000	m
X_6	0.8700	m

These values are obtained after the twenty iterations using the Newton-Raphson technique. The final parabolic shape of the geometry of the reflector antenna using the above values for the given set of loading is shown in the Figure 3.

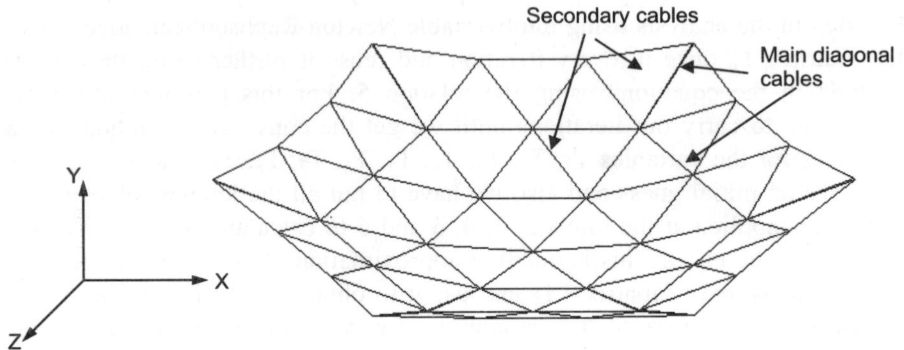

Fig. 3. Final shape of a Reflector Antenna after the Analysis.

This project is having a practical application for the defence operations; hence the Institute of Armament Technology provided us the approximate scaled dimensions for the analysis of the problem given by them because the IAT not want to expose the original dimensions. The requirement of IAT was only to suggest them a suitable form finding method as well as the method of analysis that also includes some programming part, so that they may apply it for their main project. According to this, the analysis has been carried out using the initial guess for the programming in MATLAB based on the dimensions provided by IAT and we obtain the final shape for the reflector antenna after the application of given set of loading with maximum sag of 0.8 m at the central node for 6 m diameter as shown in Figure 3.

This sag may be adjusted by applying telescopic struts at each node inside the geometry based on tensegrity principle, according to the requirement of the accuracy in the reflection of the electromagnetic signals as well as we can increase the surface accuracy of the antenna by decreasing the size of the triangles without increasing the number of supports.

4. COMMENTS ON ANALYSIS

1. The form finding method used to form the geometry of the surface for the reflector antenna was based on the concept of tension truss [3].
2. In this concept the reflector surface is divided into a net of triangular elements. Major advantage of this concept is that we can use lightweight flexible members for the sides of the triangles and all the cables are made to take tensions.
3. Using the expressions derived by Tibert G we can achieve a stable system of cable net by knowing the number of joints and the cables.
4. The main advantage of the tension truss concept used in the analysis over the other types is that by decreasing the size of the triangles without increasing number of supports we can achieve accuracy and geometrical stability in the surface of reflector antennas.
5. The analysis of these types of structures using the standard software packages like Staad-Pro and ANSYS is quite difficult task because we are not able to create a geometrical model of

the structural elements i.e., model of flexible cables in these software packages, hence we have to use the analysis method based on solution of equilibrium equations, which are in the form of non-linear algebraic equations.

6. The Newton-Raphson technique used to solve linear and non-linear algebraic equations was proved to be a very useful technique as it gives converging solution for obtaining exact values of variables involved in the equations.

5. COMMENTS ON THE ANALYSIS RESULTS

1. The Newton-Raphson technique used for analysis is an iterative method and the major step in analysis is to carry out inverse of the jacobian matrix once in every iteration.

2. The accurate results are obtained by preparing a programme in a software package known as MATLAB used for high performance numerical computations and visualisation, which provides excellent tools for solution of linear and non-linear algebraic equations.

REFERENCES

1. Subramanian N, "Principles of Space Structures", Second Edition, Wheelers Publications, New Delhi.
2. Rudra Pratap, "Getting Started with MATLAB", Oxford University Press Publications, New York, 2003.
3. Miura K and Miyazaki Y, "Concept of the Tension Truss Antenna", AIAA Journal, 28 (6): 1098 – 1104, 1990.
4. S Pellegrino and Tibert G, "Deployable Tensegrity Structures for Space Applications", Royal Institute of Technology, Department of Structural Engineering, Stockholm, Sweden.
5. Tibert A G and S Pellegrino, "Form Finding Methods for Tensegrity Structures", 42nd AIAA/ASME/ASCE/AHS/ASC Structures, Structural Dynamics and Materials Conference, Norfolk VA, 14 - 17 May 2001.
6. Tibert G, "Optimal Design of Tension Truss Antennas", 44th AIAA/ASME/ASCE/AHS/ASC Structures, Structural Dynamics and Materials Conference, 7 – 10 April 2003.

112

Numerical and Digital Photoelastic Study of a Spline Shaft

K. ASHOKAN AND K. RAMESH

Department of Applied Mechanics, IIT Madras, Chennai-600 036, India.
email: ukashokan@iitm.ac.in, kramesh@iitm.ac.in

ABSTRACT

Digital photoelasticity is an experimental method for determining stresses in 2D and 3D models. Stereolithography, one of the Rapid Prototyping (RP) techniques is being widely used to produce 3D photoelastic models. The stereolithography models can be stress frozen and then mechanically sliced for evaluating the stress field experimentally. The Computer Aided Design (CAD) model used for preparing Stereolithography model can be used for Finite Element (FE) analysis and the FE results are post processed to plot simulated photoelastic fringe contours of slices cut from the model. Such an approach helps in developing an appropriate slicing plan so that experimental analysis is cost effective. In this paper, finite element analysis of circular shaft under torsional load is carried out using commercial software package ABAQUS V6.4. Simulated photoelastic contours are obtained from FE results for different slicing plans and this knowledge of circular shaft analysis has been extended to analyze spline shaft numerically.

Keywords: Digital Photoelasticity, rapid prototyping, stereolithography, spline shaft, finite element.

1. INTRODUCTION

Digital photoelasticity is a whole field technique, which provides the information of principal stress difference (isochromatics) and the orientation of principal stress direction (isoclinics) in the form of fringes based on intensity processing [1]. Photoelasticity is the only optical experimental technique available to study the stresses interior to the model. However, model making for 3D photoelastic analysis is difficult. The recent advances in Rapid Prototyping have made it possible to produce 3D prototypes of very complicated shapes directly from CAD models [1]. Among the various RP techniques available, stereolithography has found wide acceptance by photoelasticians because of its ability to provide models that are photoelastically sensitive [2, 3]. These 3D stereolithography models can be stress frozen and then mechanically sliced for evaluating the stress field experimentally. Stress freezing is a process in which the model under load is subjected to a heat treatment process and at the end of it when the loads are removed, the model retains the stress field introduced due to the applied load. Appropriate slices of finite thickness are cut mechanically and then analysed by using the principles of 2D photoelasticity. The slices cut from the model will retain the same stress distribution, as it was part of the 3D model, provided the slicing operation is carried out carefully so that no machining stresses are introduced. One of the problems in experiments is the selection of appropriate load for stress freezing and the slicing plan for complex models. To overcome this, the same CAD model used for preparing stereolithography model is used for FE analysis. The FE results are post processed to plot simulated photoelastic fringe contours for slices cut from the model.

Splined connections are widely used as a coupling mechanism in rotating machinery. The primary function of a spline shaft is to transfer torque from one rotating member to another. Splines allow loads to be distributed over several teeth. The application of the splined connection for the transmission of torque is well documented in the literature [4, 5]. In this paper, finite element analysis of a circular shaft under torsional load is initially carried out. The FE analysis is done by using ABAQUS V6.4 and the FE results are post processed to plot photoelastic fringe contours for different slicing plans. Based on this, numerical analysis of spline shaft is done to obtain photoelastic fringe contours for different slices cut from it.

2. FINITE ELEMENT ANALYSIS OF CIRCULAR SHAFT

A circular shaft of diameter 40 mm and length 200 mm is modelled (Fig. 1a) and the FE analysis is done by using ABAQUS V6.4. The model is discretised with 8 noded hexahedral elements. The Boundary Conditions (BC) and the loading are applied as shown in Fig. 1b. One end of the shaft is fixed and at the other end, torsional load is applied tangential to the circumferential surface of the shaft. The material properties used for numerical simulation are $E = 3290$ MPa, $v = 0.4$. The details of nodal co-ordinates, nodal connectivity and stress components are obtained from ABAQUS software for post processing of the results to obtain fringe contours.

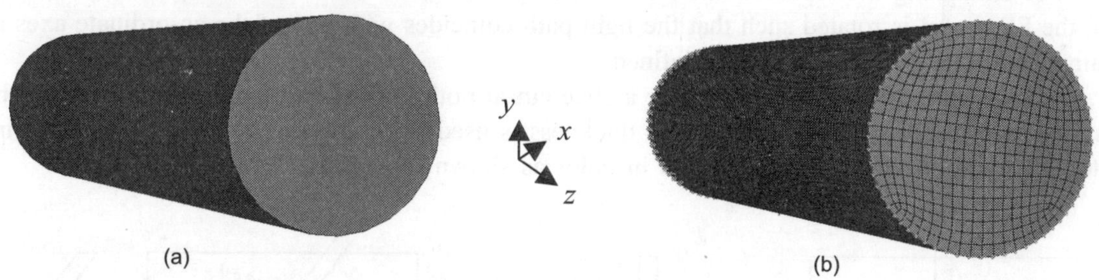

Fig. 1. (a) Circular shaft model (b) Finite element mesh with BCs.

3. PLOTTING OF FRINGE CONTOURS

A fringe plotting software named 'femfrn3D' developed in-house in VC++ environment is used for plotting photoelastic fringe contours using the results obtained from FE analysis. To simplify the implementation of the scanning scheme, a twenty noded element is degenerated to a eight noded element by renaming some nodes with the same global number as shown in Fig. 2. For interpolation, the shape functions of twenty noded elements are directly used [6]. The fringe contours are plotted in grey scale and in color.

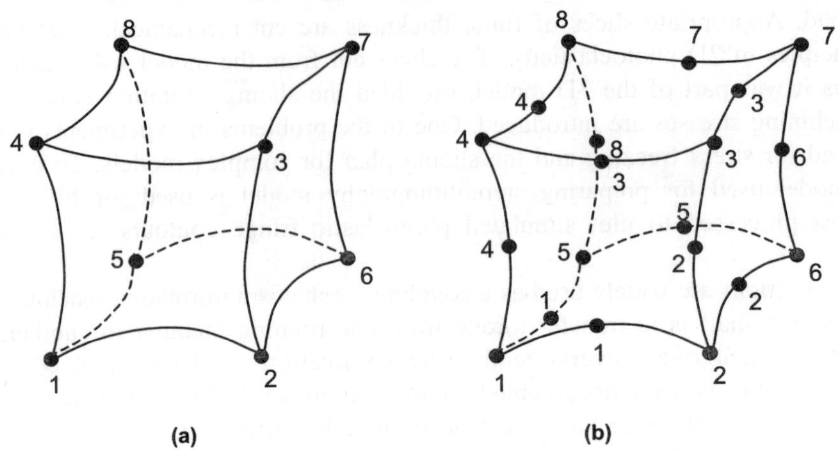

Fig. 2. (a) Representation of 8 noded element in global co-ordinates (b) Representation of
8 noded element as degenerated elements of 20 noded element.

To plot isochromatics of slices, a pair of cutting planes is defined by specifying a point lying on the
plane along with its normal. The thickness of the slice is determined by the normal distance between
the cutting planes. The element faces is scanned in the local coordinate system as though it is a two
dimensional element using appropriate scanning intervals. The values of secondary principal stress
difference at each scanning points are calculated from the nodal values using shape functions
themselves as interpolation functions [7, 8]. For a given data point, the points lying on the cutting
planes, and the points that lie on all the element faces between the cutting planes are considered for
total fringe order evaluation. For ease of computations, one of the coordinates axes is considered as
the viewing direction. For normal incidence fringe plotting, this may be straightforward in most
cases. For simulating the isochromatics in inclined plane, rather than considering an inclined light
path, the FE model is rotated such that the light path coincides with one of the co-ordinate axes and
a pair of arbitrarily cutting planes is defined.

A pair of cutting planes is defined for a slice cut at both normal and inclined at 45^0 to the shaft
axis as shown in Fig. 3. A slice of 4 mm thickness is used for numerical computations. The fringe
contours are simulated in grey scale and in color as shown in Fig. 4.

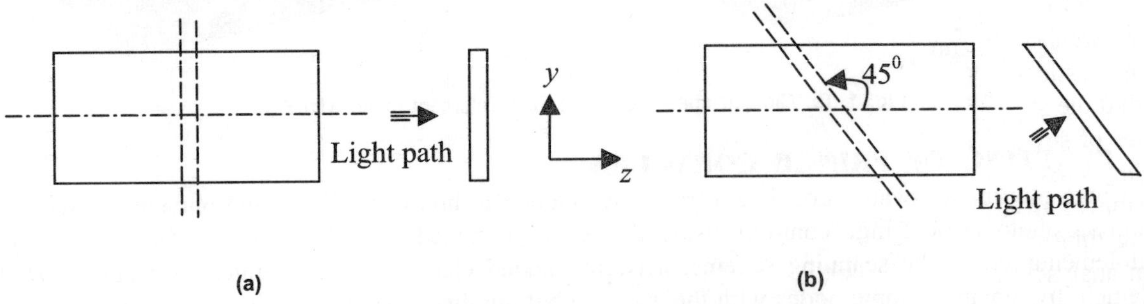

Fig. 3. (a) Slice cut normal to z axis (c) Slice taken at an inclined plane of 45^0 to shaft axis.

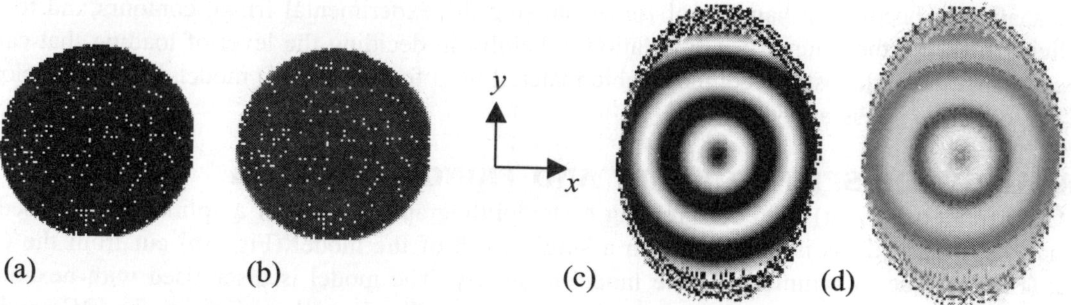

Fig. 4. Numerically simulated fringe contours for the slice cut normal to the slice axis in (a) grey scale (b) color. Numerically simulated fringe contours for slice cut 45^0 to shaft axis in (c) grey scale (d) color.

Only stress components perpendicular to the light path cause formation of photoelastic contours. For a circular shaft under torsion, the stress tensor at any point is

$$\begin{bmatrix} 0 & 0 & 0 \\ 0 & 0 & \tau_{\theta z} \\ 0 & \tau_{\theta z} & 0 \end{bmatrix}$$

...(1)

Thus in normal incidence, no fringes should be seen which is the case in Fig. 4a. However, noise appears as dots, which could be attributed to the approximations involved in numerical modeling. On the other hand, while viewing a slice at 45^0, the shear stress variation cause the formation of fringes as concentric circles [9]. The stress tensor corresponding to 45^0 inclined plane is

$$\begin{bmatrix} 0 & 0 & 0 \\ 0 & -\tau_{\theta z} & 0 \\ 0 & 0 & \tau_{\theta z} \end{bmatrix}$$

...(2)

The numerically simulated fringe pattern for this case is concentric circles thus, validating the numerical code for fringe simulation. The geometric features of the fringes are captured well in numerical simulation.

4. STATES OF STRESS ACTING ON A SPLINE SHAFT

In a circular shaft subjected to pure torsion at any given point, only shear stress is present. The photoelastic analysis can be done on the circular shaft by using thin slices cut at 45^0 to the shaft axis to find the shear stress. But in a spline shaft subjected to torsion, at any given point two superimposed states of stress are present [4]. The first is the state of bending stress acting on the teeth in the plane of any cross-section cut normal to the spline axis. The second is the state of shear stress due to torsion of the spline shaft about its axis. The photoelastic analysis can be done on the spline by using thin slices cut normal and at 45^0 to the spline axis. The slice cut normal to the slice axis is then analysed in normal incidence to get bending stress in the plane of the slice. For obtaining the torsional shear stress, the slices cut at an angle of 45^0 is used in normal incidence of light beam. Experimentally one does not know *a priori* how the photoelastic fringe contours of slices from complex problems will appear. Numerical simulations of the fringe contours of the slices cut from spline shaft thus becomes a handy tool for visualising the experimental fringe contours and to plan the slicing plan. Further, numerical simulation is helpful in deciding the level of loading that can be applied so that the high cost Stereolithographic material used for making 3D models can be judiciously used for effective stress analysis.

5. ANALYSIS OF SPLINED SHAFT AND FRINGE PLOTTING

The CAD model (Fig. 5a) used for creating a stereolithography model of a spline shaft is used for FE analysis. The analysis is performed on a single tooth of the model (Fig. 5b) cut from the CAD model (Fig. 5a) due to symmetry in the model geometry. The model is discretised with hexahedral elements (8 noded linear elements) as shown in Fig. 5c. The Boundary Conditions (BC) and the loading used in the analysis are similar to the one to be used in actual photoelastic model. The material properties used for numerical simulation are same as that of the circular shaft. The female part is fixed and the torsional load is applied tangentially on the circumferential surface of the male part of the spline shaft as shown in Fig. 5c. The FE results are post processed to plot simulated fringe contours. A pair of cutting planes is defined for the slice cut at both normal and inclined at 45^0 to the spline axis as was done for the circular shaft. A slice of 3 mm thickness is used. The fringe contours are simulated in grey scale and in color as shown in Fig. 6.

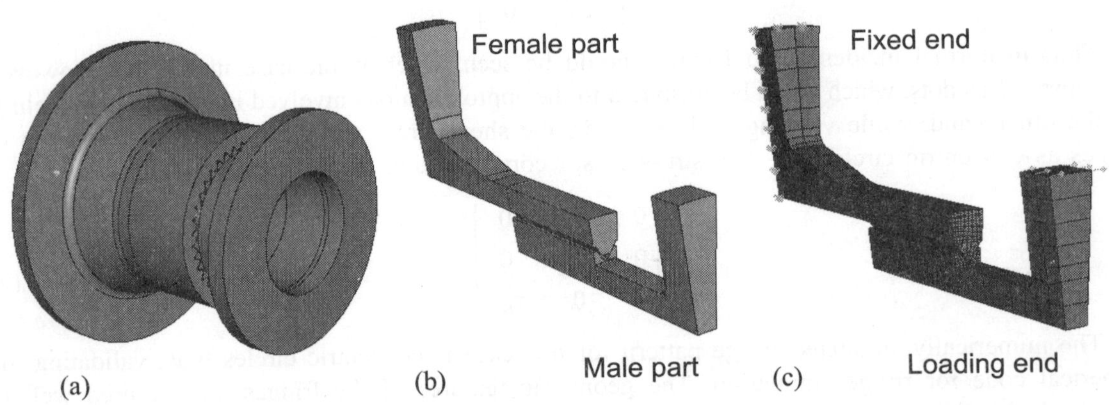

Fig. 5. (a) CAD model of spline shaft (b) Single teeth of the spline shaft model (c) Finite element mesh with BC's

6. RESULTS AND DISCUSSION

Figure 6 shows the numerical fringe contours obtained for the slice cut at normal and at 45^0 inclined plane in grey scale and in color. The typical behaviour of fringes due to torsional load is captured in the numerically simulated fringe contours. One can identify the location of critical stress zone in the slices cut from the spline shaft using numerically simulated fringe contours. Figure 6 shows zone of high stresses occurring in the teeth due to bending stresses where the torque is being transferred from one member to the other. Figure 7 shows the high stresses due to shear stresses is occurring in the spline tooth root fillet. The results obtained from FE analysis are thus very much helpful in visualizing the fringe contours and to identify the stress concentration zone in the slices cut from the spline shaft experimentally.

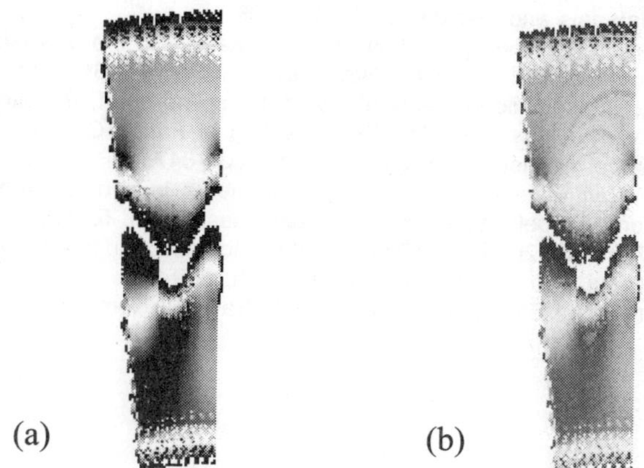

(a)   (b)

Fig. 6. Numerically simulated fringe contours for the slice cut normal to the slice axis in (a) grey scale (b) color

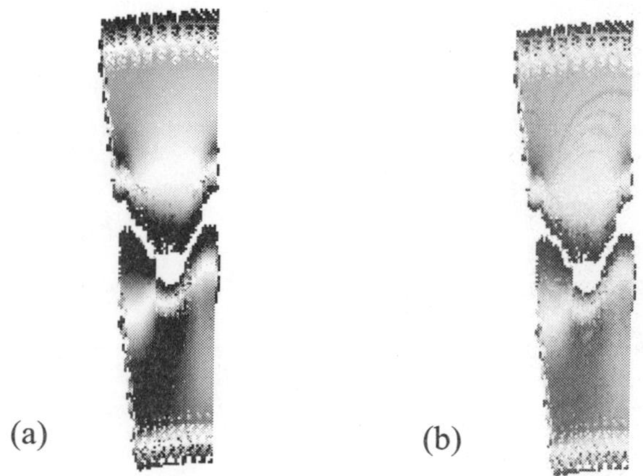

(a)   (b)

Fig. 7. Numerically simulated fringe contours for the slice cut at 45° to the spline axis in (a) grey scale (b) color

CONCLUSION

In order to evolve an optimal slicing plan for effective photoelastic analysis of stereolithography built spline shaft, FE analysis is done on CAD model of spline shaft and the FE results obtained are post processed to plot simulated photoelastic fringe contours by using femfrn3D software developed in VC++. This methodology reduces the number of experiments to be carried out on spline shaft and gives the idea of how photoelastic fringe contours will appear experimentally. Numerically simulated photoelastic fringe contours for the slices cut at normal and at 45° inclined planes to the spline axis are shown.

REFERENCES

1. K. Ramesh, 2000, Digital Photoelasticity: Advanced Techniques and Applications, Springer-Verlag, Berlin, Germany.
2. J.D. Curtis, S.D. Hanna, E.A. Patterson, Taroni M, 2005, On the use of stereolithography for the manufacture of photoelastic models, Experimental Mechanics. 42(2),148-162.

3. D.E. Karalekas, A. Ajelopoulos, 2005, On the use of stereolithography built photoelastic models for stress analysis investigations, Materials and Design. 27(2), 100-106.

4. D.G. Salyards, H.J. Macke, 1990, The application of photoelasticity to the analysis of shaft splines, Proc. SEM Spring conference on Experimental Mechanics, New Mexico, USA, 4th -6th June, 117-124.

5. Hiroyuki Yoshitake, 1962, Photoelastic stress analysis of the spline shaft, Bulletin of JSME. 5(17), 195-201.

6. K. Ramesh, A.K. Yadav, Vijay A. Pankhawalla, 1995, Plotting of Fringe Contours from Finite Element Results, Communications in Numerical Methods in Engineering. 11, 839-847.

7. P.R.D. Karthick Babu, K. Ramesh, 2006, Plot of finite elements in evolving the slicing plan for photoelastic analysis of three-dimensional models, Experimental Techniques. 30(3), 52-56.

8. P.R.D. Karthick Babu, K. Ramesh, 2006, Development of Photoelastic fringe plotting scheme from 3D FE results, Communications in Numerical Methods in Engineering. 22(7), 809-821.

9. L.S. Srinath, 1984, Experimental Stress Analysis, Tata McGraw-Hill Company, New Delhi.

113

A Study on the Elasto-Plastic Behavior of a Rotating Solid Disk Having Variable Thickness

S. Bhowmick[1], D. Das[2] and K.N. Saha[3]

Department of Mechanical Engineering, Jadavpur University, Kolkata-700 032, India
email: [1] shubh.ju@gmail.com, [2] debu235@yahoo.co.in, [3] kashinathsaha@gmail.com

ABSTRACT

A computational model based on Von-Mises yield criterion and linear strain hardening rule is developed to study the elasto-plastic behaviour of rotating solid disks having uniform and variable thickness and a numerical solution of the stress and deformation states of the disk is reported. The problem is formulated through a variational method, where the radial displacement field is taken as unknown variable. Assuming a series solution and using Galerkin's principle, the solution of the governing partial differential equation is obtained. The rotational speed is high enough to cause yielding of the disk and the stress distribution is estimated for various rotational speeds.

Keywords: Elasto-plastic, Variational method, Von-Mises criterion, Plastic front, Limit angular speed.

1. INTRODUCTION

Due to widespread applications, the analysis of rotating disk behavior has been of greatest interest to many researchers. The study has become much more feasible in the past few decades due to the intensified application of numerical methods as well as advent of computational machines. The plane stress state in a rotating disk of constant thickness of elastic plastic material with linear strain hardening behaviour has been studied in [1-3]. The analyses were carried out using Tresca's yield condition and its associated flow rule and initiation of a plastic core at the axis of the disk was reported. The plastic core was shown to consist of two plastic adjacent regions governed by different forms of yield condition. The work of [1-3] was extended to rotating annular disks with variable thickness in [4-7]. In [4] the effect of radial density gradient on elastic-plastic stresses and radial displacement of a rotating disk with variable thickness under the assumption of Tresca's yield condition and its associated flow rule have been studied. In [5] the study of linearly hardening rotating solid disk with variable thickness for fully plastic state has been carried out. The thickness was considered to vary hyperbolically along the radius. Exponential variation of thickness was first discussed in [6] although it could not present a solution satisfying all boundary and continuity conditions. In [7] analysis of rotating disks with power function thickness profile has been carried out. Elasto-plastic deformation of rotating solid and annular disks of uniform thickness made of elastic-perfectly plastic material and comparison of the solutions obtained from the two different failure criterion have been reported in [8]. The study of elasto-plastic deformations of disks with different parameter values of parabolic thickness functions, representing a wide range of non uniform cross-sectional profiles has been carried out in [9] where closed form solutions in terms of hypergeometric functions, by performing displacement based formulation was reported. In [10] an analytical solution of a solid convex disk with exponentially varying thickness has been obtained. A similar work was carried out for power function thickness variation in [11]. However, in [12] elasto-plastic deformation behaviour of variable thickness solid disks having concave profiles and its difference from that of uniform thickness disk was reported. Due to non-linearity involved with the application of Von Mises criterion, the analysis demands for a numerical solution. But the advantage in using Von Mises criterion is that unlike Tresca's Criterion, here a single formulation takes care of the whole plastic region. In [13] inelastic stress state of linear hardening solid disks with exponential thickness variation using both Tresca's and Von Mises criterion has been studied. In a recent paper [14] analytical solution for rotating disks with elliptical thickness variation and made of linearly hardening material using Tresca's criterion and its associated flow rule has been presented. In [15] numerical schemes based on perturbation method and power series solution method to investigate elastic-plastic deformations of rotating disks with uniform thickness have been presented. This work was further extended in [16] by using Runge-Kutta numerical procedure, to compute elastic-plastic stresses in rotating annular disks of variable thickness and variable density. Recently, the application of variational method, proposed in [17] has yielded a generalized approach to study the behaviour of rotating disks of variable thickness in the elastic regime.

The present work employs variational principle to study the elasto-plastic behaviour of rotating disks of variable thickness using Von Mises yield criterion and linear strain hardening material behaviour. The validations of the present model with the existing works have been presented. Also the result showing the advancement of the plastic front with angular speed has been reported.

2. MATHEMATICAL FORMULATION

A disk of exponentially varying thickness subjected to centrifugal loading is shown in Fig. 1. The mathematical model is based on the assumptions that the material of the disk is homogeneous, isotropic and follows a linear strain hardening flow rule (Fig. 2). Whenever the Von Mises stress at a particular radial location of a rotating disk reaches the uni-axial yield stress value, the plastic front initiates at that location and the corresponding rotational speed is termed as elastic limit angular speed (Ω_1). The determination of Ω_1 has been carried out in [17] by the variational principle. The present work is carried out to obtain the solution of a rotating disk above elastic limit angular speed.

Variational principle states that $\delta(U+V)=0$ $\qquad\qquad$ ...(1)

where, $U = U_e + U_p$ i.e., total strain energy, U consists of an elastic (U_e) and a plastic (U_p) part and V is the potential of the external forces. The interface between the outer elastic and the inner post elastic region is demarcated by the radius $r = x_c$.

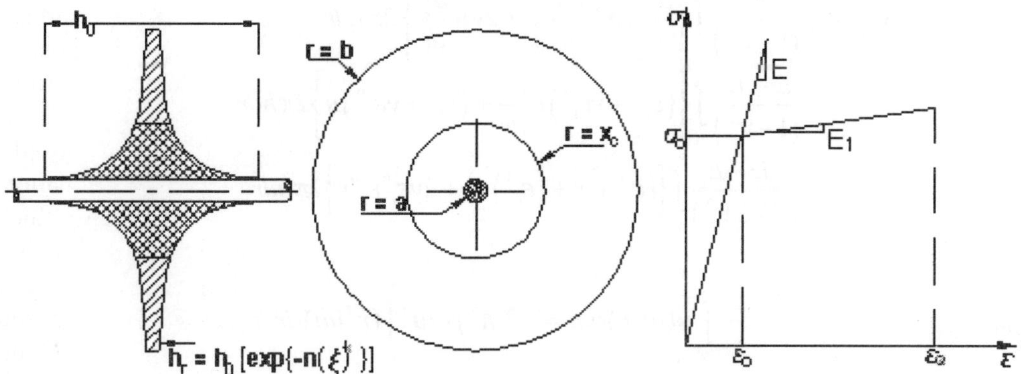

Fig. 1. An annular disk having varying thickness.   **Fig. 2.** Linear strain hardening material.

Elastic part of strain energy,
$$U_e = \frac{\pi E}{1-\mu^2} \int_{x_c}^{b} \left\{ \frac{u^2}{r} + 2\,\mu\,u\left(\frac{du}{dr}\right) + r\left(\frac{du}{dr}\right)^2 \right\} h\,dr \qquad ...(2)$$

and post elastic part of strain energy,

$$
\begin{aligned}
U_p &= \int (area\,under\,radial\,'\sigma - \varepsilon'\,curve + area\,under\,tangential\,'\sigma - \varepsilon'\,curve)dv \\
&= \int (dU_r + dU_t)dv
\end{aligned}
\qquad ...(3)
$$

dU_r and dU_t are given by, (considering $\varepsilon_r^{\,p} = \dfrac{du}{dr} - \varepsilon_r^{\,0}$ and $\varepsilon_t^{\,p} = \dfrac{u}{r} - \varepsilon_t^{\,0}$)

$$
\begin{aligned}
dU_r =\; & \frac{1}{2}\frac{E_1}{(1-v^2)}\left[\left(\frac{du}{dr}\right)^2 + v\frac{du}{dr}\frac{u}{r}\right] + \frac{E-E_1}{(1-v^2)}\left[\varepsilon_r^{\,0} + v\varepsilon_t^{\,0}\right]\frac{du}{dr} \\
& + \frac{1}{2}\frac{E_1 - E}{(1-v^2)}\left[(\varepsilon_r^{\,0})^2 + v\varepsilon_r^{\,0}\varepsilon_t^{\,0}\right] + \frac{1}{2}\frac{E_1}{(1-v^2)}\left[\frac{du}{dr}\varepsilon_t^{\,0} - \frac{u}{r}\varepsilon_r^{\,0}\right]v \quad \text{and}
\end{aligned}
\qquad ...(4)
$$

$$
\begin{aligned}
dU_t =\; & \frac{1}{2}\frac{E_1}{(1-v^2)}\left[\left(\frac{u}{r}\right)^2 + v\frac{du}{dr}\frac{u}{r}\right] + \frac{E-E_1}{(1-v^2)}\left[\varepsilon_t^{\,0} + v\varepsilon_r^{\,0}\right]\frac{u}{r} \\
& + \frac{1}{2}\frac{E_1 - E}{(1-v^2)}\left[(\varepsilon_t^{\,0})^2 + v\varepsilon_r^{\,0}\varepsilon_t^{\,0}\right] + \frac{1}{2}\frac{E_1}{(1-v^2)}\left[\frac{u}{r}\varepsilon_r^{\,0} - \frac{du}{dr}\varepsilon_t^{\,0}\right]v
\end{aligned}
\qquad ...(5)
$$

So, substituting Eq. (4) and Eq. (5) in Eq. (3), we have

$$U_p = \frac{1}{2}\frac{E_1}{\left(1-v^2\right)}\int_a^{x_c}\left\{\frac{u^2}{r}+r\left(\frac{du}{dr}\right)^2+2vu\frac{du}{dr}\right\}2\pi h dr$$

$$+\frac{E-E_1}{\left(1-v^2\right)}\int_a^{x_c}\left\{\left(\varepsilon_r^{\,0}+v\varepsilon_t^{\,0}\right)r\frac{du}{dr}+\left(\varepsilon_t^{\,0}+v\varepsilon_r^{\,0}\right)u\right\}2\pi h dr$$

$$+\frac{1}{2}\frac{E-E_1}{\left(1-v^2\right)}\int_a^{x_c}\left\{\left(\varepsilon_r^{\,0}\right)^2 r+\left(\varepsilon_t^{\,0}\right)^2 r+2v\varepsilon_t^{\,0}\varepsilon_r^{\,0}r\right\}2\pi h dr \qquad \text{...(6)}$$

Again,
$$V=-\int_{Vol} u(\omega^2 r)dm=-2\,\pi\,\rho\,\omega^2\int_a^b (r^2 uh)dr \qquad \text{...(7)}$$

The normalization of the radial coordinate (r) is carried out with three parameters (Δ, Δ_1 and Δ_2) and three normalized coordinates (ξ, ξ_1 and ξ_2), where

$$\Delta=b-a \ \text{ and } \ \xi=(r-a)/\Delta, \ \Delta_1=x_c-a \ \text{ and } \ \xi_1=(r-a)/\Delta_1, \ \Delta_2=b-x_c$$

and,
$$\xi_2=(r-x_c)/\Delta_2, \qquad \text{...(8)}$$

Substituting the expressions of U_e, U_p and V from Eq. (2), Eq. (6) and Eq. (7), respectively in Eq. (1) and using the normalization given by Eq. (8), the governing equilibrium equation becomes,

$$\frac{E}{1-\mu^2}\int_0^1\left\{\frac{u\delta u}{\left(\Delta_2\xi_2+x_c\right)}+\frac{\mu}{\Delta_2}\left[u\delta\left(\frac{du}{d\xi_2}\right)+\frac{du}{d\xi_2}\delta u\right]+\frac{\left(\Delta_2\xi_2+x_c\right)}{\left(\Delta_2\right)^2}\frac{du}{d\xi_2}\delta\left(\frac{du}{d\xi_2}\right)\right\}h\Delta_2 d\xi_2$$

$$+\frac{E_1}{\left(1-v^2\right)}\int_0^1\left[\left\{\frac{\left(\Delta_1\xi_1+a\right)}{\left(\Delta_1\right)^2}\frac{du}{d\xi_1}\delta\left(\frac{du}{d\xi_1}\right)+\frac{u\delta u}{\left(\Delta_1\xi_1+a\right)}\right\}+\frac{v}{\Delta_1}\left\{\frac{du}{d\xi_1}\delta u+u\delta\left(\frac{du}{d\xi_1}\right)\right\}\right]h\Delta_1 d\xi_1$$

$$+\frac{E-E_1}{\left(1-v^2\right)}\int_0^1\left[\varepsilon_r^{\,0}\left(v\delta u+\frac{\left(\Delta_1\xi_1+a\right)}{\Delta_1}\delta\left(\frac{du}{d\xi_1}\right)\right)+\varepsilon_t^{\,0}\left(\delta u+v\frac{\left(\Delta_1\xi_1+a\right)}{\Delta_1}\delta\left(\frac{du}{d\xi_1}\right)\right)\right]h\Delta_1 d\xi_1 \qquad \text{...(9)}$$

$$-\rho\,\omega^2\int_0^1\left\{\left(\Delta\xi+a\right)^2 h\delta(u)\right\}\Delta d\xi=0$$

The global displacement function $u(\xi)$ in Eq. (9) is approximated by $u(\xi)\cong\sum c_i\phi_i$, i=1, 2,..., n, where ϕ_i is the set of orthogonal functions developed through Gram-Schmidt scheme. The necessary starting function to generate the higher order orthogonal functions is selected by satisfying the relevant boundary conditions ($\sigma_r\big|_{(a)}=0$ and $\sigma_r\big|_{(b)}=0$), of a rotating disk in elastic regime. For the purpose of computation, displacement functions in the elastic and post-elastic regions are expressed as $u(\xi_1)\cong\sum c_i\phi_i^{\,p}$ and $u(\xi_2)\cong\sum c_i\phi_i^{\,e}$, respectively. Substituting these assumed displacement functions and replacing operator δ by $\partial/\partial c_j$, Eq. (9) is obtained in matrix form as follows:

$$\frac{E}{1-\mu^2}\sum_{j=1}^{n}\sum_{i=1}^{n}c_i\int_0^1\left\{\frac{\phi_i^e\phi_j^e}{(\Delta_2\xi_2+x_c)}+\frac{\mu}{\Delta_2}(\phi_i^{e'}\phi_j^e+\phi_i^e\phi_j^{e'})+\frac{(\Delta_2\xi_2+x_c)}{(\Delta_2)^2}\phi_i^{e'}\phi_j^{e'}\right\}h\Delta_2 d\xi_2$$

$$+\frac{E_1}{(1-v^2)}\sum_{j=1}^{n}\sum_{i=1}^{n}c_i\int_0^1\left[\left\{\frac{(\Delta_1\xi_1+a)}{(\Delta_1)^2}\phi_i^{p'}\phi_j^{p'}\right\}+\left\{\frac{\phi_i^p\phi_j^p}{(\Delta_1\xi_1+a)}\right\}+\frac{v}{\Delta_1}\left\{\phi_i^{p'}\phi_j^p+\phi_i^p\phi_j^{p'}\right\}\right]h\Delta_1 d\xi_1 \qquad \ldots(10)$$

$$=\sum_{j=1}^{n}\left[\rho\omega^2\int_0^1\left\{(\Delta\xi+a)^2\phi_j\right\}h\Delta d\xi-\frac{E-E_1}{(1-v^2)}\right.$$

$$\int_0^1\left\{\varepsilon_r^0\left(v\phi_j^p+\frac{(\Delta_1\xi_1+a)}{\Delta_1}\phi_j^{p'}\right)+\varepsilon_t^0\left(\phi_j^p+v\frac{(\Delta_1\xi_1+a)}{\Delta_1}\phi_j^{p'}\right)\right\}h\Delta_1 d\xi_1\left.\right]$$

In Eq. (10), ( )' indicates differentiation with respect to normalised coordinates.

3. SOLUTION ALGORITHM

The governing equation can be expressed in matrix form as, $[K]\{c\}=\{R\}-\{R'\}$ and the required solution of unknown coefficients $\{c\}$ is obtained numerically by using an iterative scheme. To obtain the values of ε_r^0 and ε_t^0, to generate the right hand side of Eq. (10) for a particular load step above elastic limit angular speed, the ratio of σ_t and σ_r in each radial coordinate of the complete field is assumed to be same to that of previous load step. As the plastic front originates from r = a (for the disk profiles taken for analysis) at the elastic limit angular speed, with each subsequent increase in the speed above the elastic limit angular speed, plastic front proceeds towards the outer radius. For each load step, the location of plastic front is given a small increment starting from its exact location solved for the previous load step and attainment of Von Mises stress at the plastic front location equal to the value of unidirectional yield stress gives the required solution for that load step.

4. NUMERICAL RESULTS

The validation of this work is carried out with [15] which gives a solution methodology for the elasto-plastic analysis of rotating disk for a non-linearly strain hardening material using Von Mises yield criterion. An equivalent linear strain hardening stress-strain relation and the solution using that relation by finite element method are also given in [15]. The normalized (by yield stress) radial and tangential stresses for a solid disk with uniform thickness obtained by the present scheme using the equivalent linear strain hardening stress-strain relation mentioned in [15] is compared with corresponding finite element solution and the graphical comparisons for ω = 1620 rad/sec are given in Fig. 3 and Fig. 4. The comparison is carried out by using E = 207 GPa, E_1 =33.63 GPa, ρ = 7850 kg/m³, σ_y = 232.97 MPa, v = 0.3 and b = 0.2 m.

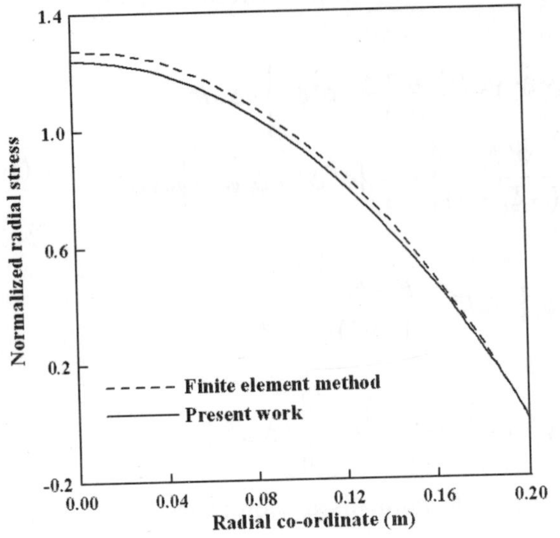

Fig. 3. Comparison of radial stress

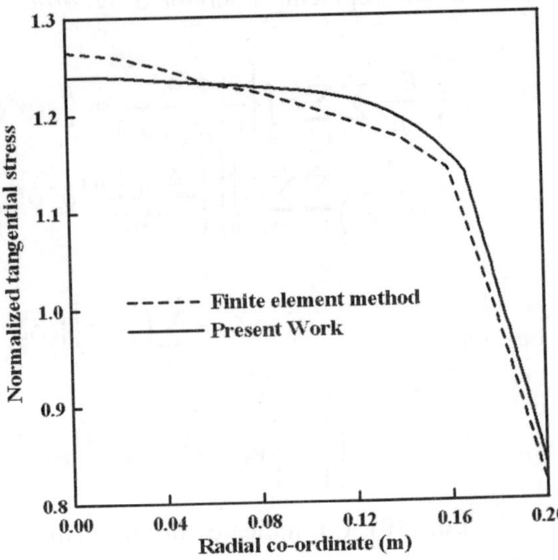

Fig. 4. Comparison of tangential stress

The fully plastic speed in non-dimensional form, Ω_2 (speed at which the whole disk attains the plastic stress state) has been calculated for a solid disk having uniform thickness and following linear strain hardening material behavior using Von Mises yield criterion in [13], taking hardening parameter $H = 0.5$ and $\nu = 1/3$. Using these parameters, the comparison of the fully plastic speed obtained by

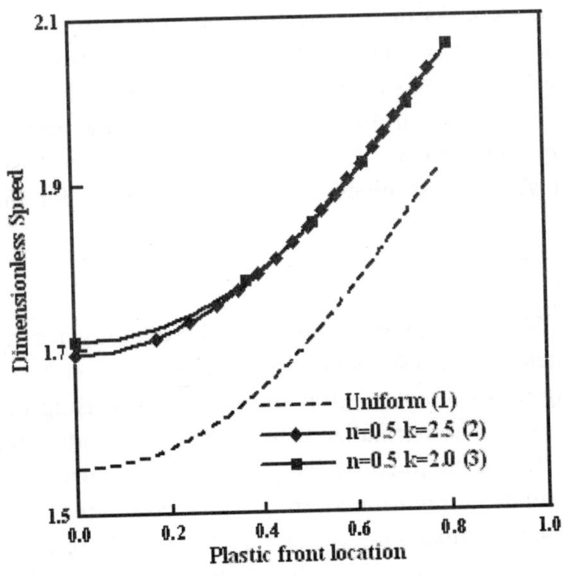

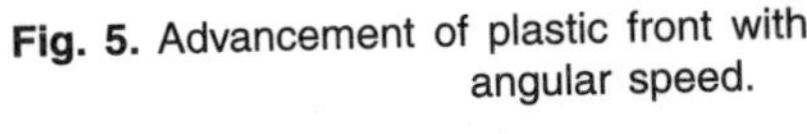

Fig. 5. Advancement of plastic front with angular speed.

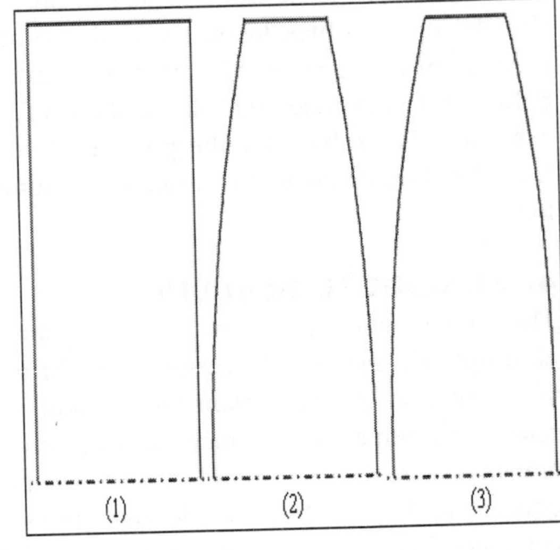

Fig. 6. Disk profiles corresponding to the cases of Fig. 3.

the present method and by [13] is given in Table 1. The speed Ω_2 is represented in non-dimensional form by using the relation $\Omega = wb \sqrt{(\rho/\sigma_y)}$. It may be pointed out that the ratio Ω_2/Ω_1 gives the limit design factor for rotating disks. This factor is very important to the designers because it gives an indication of the reserve in hand, once the yielding has started.

Table 1. Comparison of fully plastic speed

	Present work	By [13]
Ω_2	2.16997	2.11747

The variation of the advancement of the plastic front (initiated at the root) with angular speed has been carried out and shown in Fig. 5 for a solid disk with uniform thickness and for solid disks with exponential variation of thickness with $n = 0.5$, $k = 2.0$ and $n = 0.5$, $k = 2.5$. The graphical variations of the disk profiles have been shown in Fig. 6 in which the exponential variation of thickness is given by $h(\xi) = h_0 \exp[-n(\xi)^k]$. The analysis is carried out for $E = 210$ GPa, $E_1 = 70$ GPa, $\rho = 7850$ kg/m^3, $\sigma_y = 350$ MPa, $\nu = 0.3$, and $b = 1.0$ m.

CONCLUSION

The present work gives an approximate solution in the elasto-plastic region of a solid rotating disk of varying thickness assuming linear strain hardening material behavior following Von Mises yield criterion. The results obtained by the present methodology have been validated and it showed a close conformity with the existing results of similar problem. Also a plot showing the advancement of the plastic front with increase in rotational speed has been presented. The present work has all the potential to determine the limit design factors, and essential tool for the rotating disk design, in loaded and unloaded condition.

NOMENCLATURE

a, b	Inner and outer radii of the disk respectively
c_i	The vector of unknown coefficients
h_0, h	Thickness at the root and thickness at any radius r of the disk
n, k	Parameters controlling the thickness variation of disk
r, t	Subscripts, correspond to radial and tangential directions
u	Displacement field of the disk
x_c	Radial location of yield front
U	Strain energy of the disk
V	Potential energy of the disk due to rotation
ρ, ν, E, E_1	Density, Poisson's ratio, elasticity modulus and tangent modulus of the disk material
H	Hardening parameter as mentioned in [13]
ω, Ω	Angular speed and dimensionless angular speed of the disk
Ω_1	Elastic limit angular speed
Ω_2	Plastic limit angular speed
$\varepsilon_r, \varepsilon_t$	Strains in radial and tangential direction
$\varepsilon_r^0, \varepsilon_t^0$	Strains in radial and tangential direction in yield condition

σ_r, σ_t	Radial and tangential stresses
σ_y	Yield stress of the disk material
ϕ_i	The set of orthogonal polynomials used as coordinate functions
ξ	Normalized radial co-ordinate
ξ_1, ξ_2	Normalized radial co-ordinate in plastic and elastic region
e, p	Superscripts, correspond to elastic and plastic state of stress

REFERENCES

1. U. Gamer, 1984, Elastic-plastic deformation of the rotating solid disk, Ing.-Arch., 54, 345-354.
2. U. Gamer, 1984, The rotating solid disk in the fully plastic state, Forsch. Ing. Wes., 50, 137-140.
3. U. Gamer, 1985, Stress distribution in the rotating elastic-plastic disk, ZAMM, 65, T136-T137.
4. U. Güven, 1992, Elastic-plastic stresses in rotating annular disk of variable thickness and variable density, Int. J. Mech. Sci., 34(2), 133-138.
5. U. Güven, 1994, The fully plastic rotating disk of variable thickness, ZAMM, 74, 61-65.
6. U. Güven, 1995, On the applicability of Tresca's yield condition to the linear hardening rotating solid disk of variable thickness, ZAMM, 75, 397-398.
7. U. Güven, 1995, Tresca's yield condition and the linear hardening rotating solid disk of variable thickness, ZAMM, 75, 805-806.
8. D.A. Rees, 1999, Elastic-plastic stresses in rotating disks by Von Mises and Tresca, ZAMM, 79, 281-288.
9. A.N. Eraslan, 2003, Elastic-plastic deformations of rotating variable thickness annular disks with free, pressurized and radially constrained boundary conditions, Int. J. Mech. Sci., 45, 643-667.
10. A.N. Eraslan and Y. Orcan, 2002, Elastic-plastic deformation of a rotating solid disk of exponentially varying thickness, Mech. Mat., 34, 423-432.
11. Y. Orcan and A.N. Eraslan, 2002, Elastic-plastic stresses in linearly hardening rotating solid disks of variable thickness, Mech. Res. Commun., 29, 269-281.
12. A.N. Eraslan and Y. Orcan, 2002, On the rotating elastic-plastic solid disks of variable thickness having concave profiles, Int. J. Mech. Sci., 44, 1445-1466.
13. A.N. Eraslan, 2002, Inelastic deformations of rotating variable thickness solid disks by Tresca and Von Mises criteria, Int. J. Comp. Engrg. Sci., 3, 89-101.
14. A.N. Eraslan, 2005, Stress distributions in elastic-plastic rotating disks with elliptical thickness profiles using Tresca and Von Mises criteria, ZAMM, 85, 252-266.
15. L.H. You and J.J. Zhang, 1999, Elastic-plastic stresses in a rotating solid disk, Int. J. Mech. Sci., 41, 269-282.
16. L.H. You, Y.Y. Tang, J.J. Zhang and C.Y. Zheng, 2000, Numerical analysis of elastic-plastic rotating disks with arbitrary variable thickness and density, Int. J. Solids Struct., 37, 7809-7820.
17. S. Bhowmick, G. Pohit, D. Misra and K.N. Saha, 2004, Design of High Speed Impellers, Proc. Int. Con-HERP, IIT, Roorkee, 229-241.

114

MLS Based Response Surface Method for Reliability Analysis

RAJIB CHOWDHURY, B.N. RAO AND A. MEHER PRASAD

Structural Engineering Division Department of Civil Engineering,
Indian Institute of Technology Madras-600 036, India email: rajib@iitm.ac.in

ABSTRACT

This paper presents a novel response surface method for predicting failure probability of structural or mechanical systems subjected to random loads, material properties and geometry. The method involves multivariate function decomposition technique to approximate the original implicit performance function into explicit performance function. Moving least square approach has been employed to generate the response surface. Failure probability is calculated by response surface using Monte Carlo simulation. Result of a ten bar truss structure with uncertain area of bars is presented here. A comparison of result obtained from different methods indicates that the present method predicts failure probability with sufficient accuracy with least number of function evaluations.

Keywords: Response surface method; Moving least square; Decomposition technique; Failure probability.

1. INTRODUCTION

The basic purpose of structural reliability analysis is to predict the stochastic behaviour of structure under uncertain characteristics of random variables, such as loadings, material properties and the geometry. The performance of a structure or any other system is governed by the state function of basic random variables. This limit state of basic random variables $\mathbf{X}$ is defined by $g(\mathbf{X})$. It divides the design space into two regions namely, if $g(\mathbf{X}) < 0$, then it is termed as failure region, otherwise it is termed as safe region.

The probability of failure is expressed as the integral of joint probability density function $f_x(\mathbf{x})$ over the failure domain. Let there are N number of random variables denoted by $\mathbf{X} = \{X_1, X_2, X_3, ..., X_N\}^T \in \Re^N$, the failure probability is expressed by

$$P_f = \int_{g(\mathbf{X})<0} f_{\mathbf{X}}(\mathbf{x})d\mathbf{x}$$

$$...(1)$$

For most of the practical problems, the exact evaluation of this integral, either analytically or computationally, is not possible because N is large, $f_X(\mathbf{x})$ is generally non-Gaussian, and $g(\mathbf{X})$ is highly non-linear function of $\mathbf{X}$.

The most common approach to solve the failure probability in Eq. (1) involves first- and second-order reliability methods (FORM/SORM) [1, 2]. These methods are based on linear (FORM) or quadratic (SORM) approximation of the performance function at design point. Generally, FORM/SORM yields sufficiently accurate result for engineering purpose, provided that performance function at design point is close to being linear or quadratic and no multiple design points exist.

The performance function for large and complex structural/mechanical system may not be explicitly defined. FORM/SORM concept is not convenient to apply directly to solve the problem with implicit performance function. Monte Carlo simulation (MCS) seems to be a suitable surrogate. The advantages of MCS are obvious, but an intrinsic disadvantage of it is the formidable computational effort for problems involving high reliability or the problems that requires a considerable amount of computation in each sampling cycle. Therefore, some approximate methods need to develop to achieve the computational efficiency for evaluating Eq. (1) without compromising the accuracy of result. Within the context of this paper, a new interpolation scheme is suggested which enables fairly accurate representation of the structural behaviour by a response surface. This response surface approach utilizes elementary statistical information of the basic random variables. Thus, the response surface generated through the present scheme is independent of the type of distribution or correlation among the basic variables.

2. FORMULATION

2.1 Multivariate Function Decomposition

Any continuous, differentiable multivariate function can be decomposed into a set of lower variate functions either by product decomposition [3] or additive decomposition [4]. In this present study, an elegant function decomposition method was used, based on additive decomposition.

Consider a continuous, differentiable, real-valued function $g(\mathbf{x})$. This function depends on random variables $\mathbf{x} = \{x_1, x_2, x_3, ..., x_N\}^T \in \mathfrak{R}^N$. The transformed performance function $g(\mathbf{u}) = 0$ maps original performance function into standard Gaussian space ($\mathbf{u}$ space).

Suppose that $g(\mathbf{u})$ has a convergent Taylor series expansion at mean $\mu = (\mu_1, \mu_2, \mu_3, ..., \mu_N\}^T$ and can be expressed by

$$g(u) = g(\mu) + \sum_{j=1}^{\infty} \frac{1}{j!} \sum_{i=1}^{N} \frac{d^j g}{du_i^j}(\mu)(u_i - \mu_i)^j + \mathbb{R}_2 \qquad ...(2)$$

where, the remainder term $\mathbb{R}_2$, denotes all terms with dimension two and higher.

Now consider a uni-dimensional approximation of $g(\mathbf{u})$, denoted by

$$\hat{g}(\mathbf{u}) \equiv \hat{g}(u_1, u_2, ..., u_N) = \sum_{i=1}^{N} \left[g(\mu_1, ...\mu_{i-1}, u_i, u_{i+1}, ..., \mu_N) - g(\mu) \right] + g(\mu) \qquad ...(3)$$

where, each term in the summation is a function of only one variable and can be subsequently expanded in a Taylor series at $\mathbf{u} = \mu$, yielding

$$\hat{g}(u) = \hat{g}(\mu) + \sum_{j=1}^{\infty} \frac{1}{j!} \sum_{i=1}^{N} \frac{d^j g}{du_i^j}(\mu)(u_i - \mu_i)^j \qquad \qquad ...(4)$$

Comparison of Eq. (2) and Eq. (4) indicates that the uni-dimensional approximation leads to the residual error $g(\mathbf{u}) - \hat{g}(\mathbf{u}) = \mathbb{R}_2$, which includes contributions from terms of two and higher order dimension. For sufficiently smooth $g(\mathbf{u})$ with convergent Taylor series, the coefficients associated with higher dimensional terms are usually much smaller than that with one-dimensional terms. As such, higher dimensional terms contribute less to the function, and therefore, can be neglected. Nevertheless, Eq. (3) includes all higher order univariate terms, as compared with FORM and SORM, which only retain linear and quadratic terms, respectively. Hence, $\hat{g}(\mathbf{u})$ estimates $g(\mathbf{u})$ more accurately than FORM/SORM. Furthermore, Eq. (3) represents exactly the same function as $g(\mathbf{u})$ when, $g(\mathbf{u}) = \Sigma g_i(u_i)$ i.e., when $g(\mathbf{u})$, can be additively decomposed into functions $g_i(u_i)$ of single variable.

It should be noted here that, this uni-dimensional approximation in Eq. (3) cannot be viewed as first- or second-order Taylor series expansions nor does it limit the non-linearity of $g(\mathbf{u})$. According to Eq. (4), all higher order univariate terms of $g(\mathbf{u})$ are included in the $\hat{g}(\mathbf{u})$. In fact, the univariate component function $g_i(u_i)$ can be highly non-linear and therefore should provide in general higher-order representation of a performance function than those by FORM/SORM. Furthermore, the approximations contain contributions from all variables.

Finally, the decomposition method depends on the selected reference point. Improper or careless selection of the reference point can spoil the approximation. The work by Xu and Rahman [4] indicates that the mean point of random input is a good candidate for defining the reference point. In this present work also, mean value of random variables were taken as reference point.

2.2 Response Surface Construction

Consider the uni-dimensional component function $g_i(u_i) \equiv g\left(\mu_1, ..., \mu_{i-1}, u_i, \mu_{i+1}, ..., \mu_N\right)$ in Eq. (3). If for $u_i = u_i^j, n$ function values

$$g_i\left(u_i^j\right) = g\left(\mu_1, ..., \mu_{i-1}, u_i, \mu_{i+1}, ..., \mu_N\right); \quad j = 1, 2, ..., n \qquad ...(5)$$

are given, the function value for arbitrary u_i can be obtained by moving least square (MLS) interpolation [5] as

$$g_i\left(u_i\right) = \sum_{j=1}^{n} \phi_j\left(u_i\right) g_i\left(u_i^j\right) \qquad \qquad ...(6)$$

where, the interpolation function $\phi_j\left(u_i\right)$ can be obtained from MLS interpolation scheme. The derivation of MLS interpolation function is explained in the next section.

By using Eq. (6), arbitrarily many values of $g_i\left(u_i\right)$ can be generated if n function values are given. The same procedure is repeated for all univariate functions, i.e., $g_i(u_i), i = 1, 2, ..., N$.

Therefore, the total cost of function evaluation entails a maximum of $(n-1) \times N + 1$.

2.3 Moving Least Square Method

Consider a function, $g(\mathbf{u})$ over a domain, $\Omega \subseteq \Re^K$, where $K = 1, 2,$ or 3. Let $\Omega_X \subseteq \Omega$ denote a sub-domain describing the neighbourhood of a point, $\mathbf{u} \in \Re^K$ located in Ω. According to the MLS, the approximation, $g^h(\mathbf{u})$ of $g(\mathbf{u})$ is

$$g^h(\mathbf{u}) = \sum_{i=1}^{m} p_i(\mathbf{u}) a_i(\mathbf{u}) = \mathbf{p}^T(\mathbf{u}) \mathbf{a}(\mathbf{u}) \qquad ...(7)$$

where $\mathbf{p}^T(\mathbf{u}) = \{p_1(\mathbf{u}), p_2(\mathbf{u}), \cdots, p_m(\mathbf{u})\}$ is a vector of complete basis functions of order m and

$\mathbf{a}(\mathbf{u}) = \{a_1(\mathbf{u}), a_2(\mathbf{u}), ..., a_m(\mathbf{u})\}$ is a vector of unknown parameters that depend on u. The basis

functions should satisfy the following properties: (1) $p_1(u) = 1$, (2) $p_i(\mathbf{u}) \in C^s(\Omega)\, i = 1, 2, ..., m$,

where $C^s(\Omega)$ is a set of functions that have continuous derivatives up to order s on Ω, and (3) $p_i(u)$, $i = 1, 2, ..., m$ constitute a linearly independent set. For example, in one-dimension ($K = 1$) with u-coordinate

$$\mathbf{p}^T(\mathbf{u}) = \{1, u\}, \; m = 2 \qquad ...(8)$$

representing linear basis functions.

   In Eq. (7), the coefficient vector, $a(\mathbf{u})$ is determined by minimizing a weighted discrete L_2 norm, defined as

$$J(\mathbf{u}) \;=\; \sum_{I=1}^{n} w_I(\mathbf{u}) \left[\mathbf{p}^T(\mathbf{u}_I) \mathbf{a}(\mathbf{u}) - d_I \right]^2 = \left[\mathbf{Pa}(\mathbf{u}) - \mathbf{d} \right]^T \mathbf{W} \left[\mathbf{Pa}(\mathbf{u}) - \mathbf{d} \right] \qquad ...(9)$$

where, $\mathbf{u}_I$ denotes the coordinates of sampling points I, $d^T = \{d_1, d_2, ...d_n\}$ with d_I representing the sampling parameter (not the sample values of $g^h(\mathbf{u})$) for sampling point I, $\mathbf{W} = \text{diag}\left[w_1(\mathbf{u}), w_2(\mathbf{u}), \cdots, w_N(\mathbf{u}) \right]$ with $w_I(\mathbf{u})$ denoting the weight function associated with sampling point I such that $w_I(\mathbf{u}) \geq 0$ for all $\mathbf{u}$ in the support Ω_u of $w_I(\mathbf{u})$ and zero otherwise, n is the number of sampling points in Ω_u for which $w_I(\mathbf{u}) > 0$, and

$$\mathbf{P} = \left[\mathbf{p}^T(\mathbf{u}_1) \mathbf{p}^T(\mathbf{u}_2) ... \mathbf{p}^T(\mathbf{u}_n) \right]^T \in L\left(\Re^n \times \Re^m \right) \qquad ...(10)$$

   The stationarity $J(\mathbf{u})$ of with respect to a $(\mathbf{u})$ yields

$$\mathbf{A}(\mathbf{u}) \mathbf{a}(\mathbf{u}) = \mathbf{C}(\mathbf{u}) \mathbf{d} \qquad ...(11)$$

where,

$$\mathbf{A}(\mathbf{u}) = \sum_{I=1}^{n} w_I(\mathbf{u}) \mathbf{p}(\mathbf{u}_I) \mathbf{p}^T(\mathbf{u}_I) = \mathbf{P}^T \mathbf{W} \mathbf{P} \qquad ...(12)$$

$$\mathbf{C(u)} = \left[w_1(\mathbf{u})\mathbf{p(u_1)}, \cdots, w_n(\mathbf{u})\mathbf{p(u_n)}\right] = \mathbf{P}^T\mathbf{W} \qquad \qquad ...(13)$$

Solving $a(x)$ from Eq. (11) and then substituting it in Eq. (7) gives

$$g^h(\mathbf{u}) = \sum_{I=1}^{n} \phi_I(\mathbf{u})d_I = \mathbf{\Phi}^T(\mathbf{u})\mathbf{d} \qquad \qquad ...(14)$$

where,

$$\mathbf{\Phi(u)} = \left\{\Phi_1(\mathbf{u}), \Phi_2(\mathbf{u}), \cdots \Phi_n(\mathbf{u})\right\} = \mathbf{p}^T(\mathbf{u})\mathbf{A}^{-1}(\mathbf{u})\mathbf{C(u)} \qquad ...(15)$$

This interpolation function can be used in Eq. (6) to generate arbitrarily many values of $g_i(u_i)$.

3. MONTE CARLO SIMULATION

MCS was employed to the proposed response surface for estimating the failure probability.

$$P_F = \frac{1}{N_S}\sum_{i=1}^{N_S} I\left[\hat{g}\left(\mathbf{u}^{(i)}\right) < 0\right], \qquad \qquad ...(16)$$

where, $\mathbf{u}^{(i)}$ is the ith realization of $\mathbf{U}$, N_s is the sample size, and $I[.]$ is an indicator function such that $I = 1$ if $\mathbf{u}^{(i)}$ is in failure set and zero otherwise. Since, the proposed method leads to an explicit limit state function of original implicit performance function, the embedded MCS can be conducted for any sample size.

4. NUMERICAL EXAMPLE

For the proposed method, $n(= 3, 5, 7,$ or $9)$ uniformly distributed points

$$\mu_i - (n-1)\sigma_i/2, \mu_i - (n-3)\sigma_i/2, ..., \mu_i, ..., \mu_i + (n-3)\sigma_i/2, \mu_i + (n-1)\sigma_i/2$$ were deployed at the

coordinate, leading to $(n-1)\times N + 1$ function.

When comparing computational efforts by various methods, the number of original performance functions evaluations is chosen as the primary comparison tool in this paper. For the direct MCS, the number of original function evaluations is the same as the sample size. While evaluating the original function through direct MCS, CPU time demands high computational effort, because it involves actual finite-element analysis repeated times. However, in the proposed method the MCS embedded in decomposition method are conducted using their response surface approximations. Here, although the same sample size as in direct MCS is considered, but the number of original function evaluation is very less. Hence, the computational effort expressed in terms of function evaluations alone should be carefully interpreted for problems involving explicit functions.

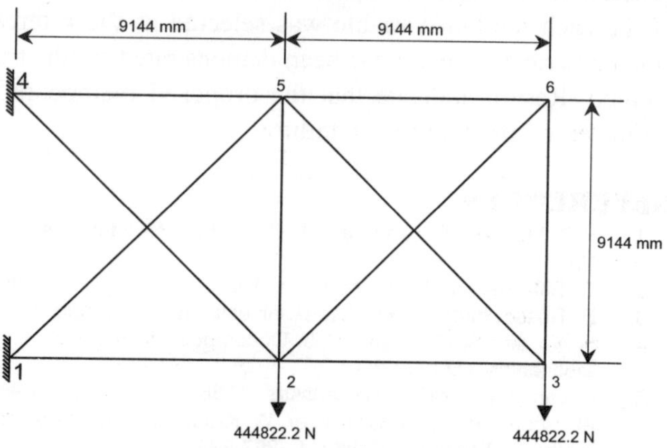

Fig. 1. A 10-bar truss structure.

A 10-bar, linear-elastic, truss structure, shown in Fig. 1, was studied in this example to examine the accuracy and efficiency of the proposed reliability method. The Young's modulus of the material is 68947.57 N/mm^2. Two concentrated forces of 444822.2 N are applied at nodes 2 and 3, as shown in Fig. 1. The cross-sectional area $X_i, i = 1, 2, ..., 10$ for each bar follows truncated normal distribution clipped at $x_i = 0$ and has mean $\mu = 1,612.9\,\text{mm}^2$ and standard deviation $\sigma = 322.58\,\text{mm}^2$. According to the loading condition, the maximum displacement $\left[u_3 \left(X_1, X_2, ..., X_{10} \right) \right]$ occurs at node 3, where a permissible displacement is limited to 457.2 mm^2. Hence, the performance function is

$$g(\mathbf{X}) = 18 - u_3 \left(X_1, X_2, ..., X_{10} \right) \qquad \qquad ...(16)$$

Table 1 compares the prediction of failure probability using proposed method, FORM, SORM due to Hohenbichler *et al.* (1987) and Monte Carlo simulation. In this example a value of $n = 7$ was selected.

Table 1. Failure probability of 10-bar truss structure

Method	Failure probability	No. of function evaluation[a]	Relative Error
Proposed method	0.1357	61[b]	2.65%
FORM	0.0862	127	38.16%
SORM (Hohenbichler et al., 1987)[c]	0.1524	506	−9.33%
Direct Monte Carlo simulation[d]	0.1394	1,000,000	—

[a]Total number of times the original performances is calculated. [b]$(n - 1) \times N + 1 = (7 - 1) \times 10 + 1 = 61$ [c]See Ref. [6] [d]See Ref. [4]

CONCLUSION

A new response surface method has been proposed for reliability of structural/mechanical system, where FORM/SORM cannot be used conveniently. This method employs additive decomposition method to approximate the implicit function and moving least square method for response surface generation, in order to express implicit performance function into an explicit one. The mean value of the each random variable was selected as the reference point for uni-dimensional approximation. One numerical example has been demonstrated to illustrate the efficiency and accuracy of the proposed method. Results indicate that the proposed method provides accurate and computationally efficient estimates of probability of failure.

REFERENCES

1. H.O. Madsen, S. Krenk and N.C. Lind, 1986, Methods of Structural Safety, Prentice-Hall, Inc., Englewood Cliffs, NJ.
2. O. Ditlevsen and H.O. Madsen, 1996, Structural Reliability Methods, John Wiley & Sons Ltd., Chichester.
3. E. Rosenblueth, 1981, Two-point estimate in probabilities. Applied Mathematical Modelling. 5, 329-335.
4. H. Xu and S. Rahman, 2005, Decomposition methods for structural reliability analysis. Probabilistic Engineering Mechanics. 20, 239-250.
5. P. Lancaster and K. Salkauskas, 1986, Curve and surface fitting; An introduction. London: Academic Press.
6. M. Hohenbichler, S. Gollwitzer, W. Kruse and R. Rackwitz, 1987, New light on first- and second-order reliability methods. Structural Safety. 4, 267-284.

115

Forced Vibration of Stiffened Conoidal Shell Roofs

S.S. Hota[1] AND D. Chakravorty[2]

Department of Civil Engineering, Jadavpur University, Kolkata-700 032, India
email: [1] sasankhota@yahoo.co.in
email: [2] chakravortydipankar@yahoo.co.in

ABSTRACT

A finite element code is developed for analyzing the forced vibration characteristics of stiffened conoidal shells by combining an eight noded curved shell element with a three noded curved beam element. The code is validated by solving problems available in the literature and comparing the present results with established ones. A number of problems are further solved where the number of stiffeners and the nature of the transient loads are varied for simply supported and clamped edges of full and truncated conoids. The effects of these parametric variations on shell actions, important from design point of view are studied and discussed.

Keywords: Stiffened conoidal shell, finite element analysis, forced vibration.

1. INTRODUCTION

Singly ruled doubly curved conoidal shells are preferred as roofing units in places requiring large column free areas such as airport lobbies, automobile manufacturing and repairing yards, food processing, medicine industries and so on. For additional stability and stiffness the conoids often need to be stiffened Research on conoids started as early as 1964 with Hadid [1] analyzing the static aspects of clamped conoids using a variational method and integral equation technique. The finite element method came up later and was adopted by Choi [2], Ghosh and Bandyopadhyay [3, 4] and Das and Bandyopadhyay [5] for static analysis. Chakravorty *et al.* [6, 7, 8] studied free and forced vibrations of composite unstiffened conoids. The stiffened conoidal shell received attention due to Nayak and Bandyopadhyay [9, 10, 11] where free vibration aspects were studied only. Excellent review papers dealing with isotropic and composite shells, both stiffened and unstiffened, were published by Sinha and Mukhopadhyay [12] and later by Qatu [13, 14]. Information about forced vibration of stiffened conoids is found to be missing in the papers referred by them.

Hence, it is found that though conoidal shells occupied a place of importance in the mind of researchers from the seventh decade of the last century, stiffened conoidal shells were taken up only by Nayak and Bandyopadhyay for free vibration. The only report on forced motions of conoidal shells was due to Chakravorty *et al.* [8] dealing with unstiffened conoids only. Hence in the present study the authors intend to study transient response of stiffened conoids (Fig. 1) with different boundary conditions and stiffening schemes with three different types of load-time histories.

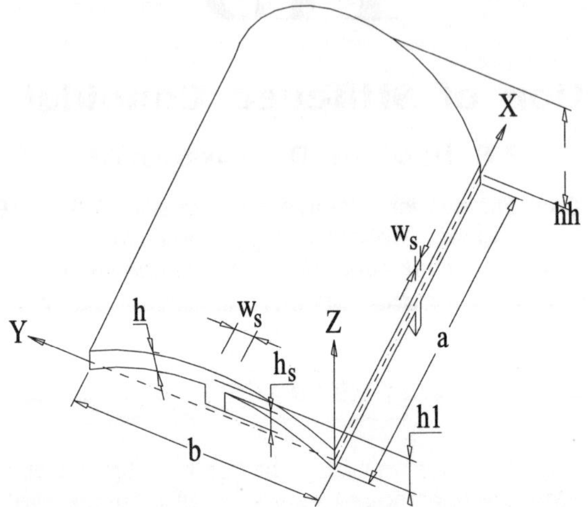

Fig. 1. Conoidal shell with eccentric stiffeners.

2. MATHEMATICAL FORMULATION

An eight noded curved element is used to model the shell surface together with a three noded curved beam element to model the stiffeners. The shell element has five degrees of freedom per node except the inplane rotation and the stiffener element has four degrees of freedom per node including axial and transverse displacements, flexural and torsional rotations. The shell formulation is reported in reference [4].

The constitutive relation of the x-stiffener can be expressed as below.

$$\{F_{sx}\} = [D_{sx}] \ \{\varepsilon_{sx}\} \qquad \qquad ...(1)$$

The force vector $\{F_{sx}\} = \{N_{sxx}, \ M_{sxx}, \ T_{sxx}, \ Q_{sxx}\}^T$ where

$$z = 4\left[hl + (hh - hl)\frac{x}{a}\right]\left[\frac{y}{b} - \frac{y^2}{b^2}\right]$$

$N_{sxx}, \ M_{sxx}, \ T_{sxx}, \ Q_{sxx}$ are the usual force, moment, torsion and shear resultants of x-stiffener element. The elasticity matrix of x-stiffener element is given as follows:

$$[D_{sx}] = \begin{bmatrix} EA & & & \\ ES & EI & symmetric & \\ 0 & 0 & GJ & \\ 0 & 0 & 0 & \dfrac{GA}{1.5} \end{bmatrix} \qquad ...(2)$$

where, A, S, I and J are cross-sectional area, first moment of area, moment of inertia about the reference axis and equivalent polar moment of inertia of the isotropic x-stiffener, respectively.

$$\{\varepsilon_{sx}\}=\{\varepsilon_{sxx}\ ,\ k_{sxx}\ ,\ k_{sxxy}\ ,\ \gamma_{sxz}\}^{T}\ =\left\{\frac{\partial u_{sx}}{\partial x}\ ,\ \frac{\partial \alpha_{sx}}{\partial x}\ ,\ \frac{\partial \beta_{sx}}{\partial x}\ ,\ \alpha_{sx}+\frac{\partial w_{sx}}{\partial x}\right\}^{T} \qquad ...(3)$$

For the y-direction stiffeners similar relationship exists but due to the curvature of the y-stiffener the shell displacements are to be multiplied by the following transformation matrix to obtain the stiffener displacements.

$$[T]=\begin{bmatrix} 1+\dfrac{e}{R_{Y}} & symmetric & & \\ 0 & 1 & & \\ 0 & 0 & 1 & \\ 0 & 0 & 0 & 1 \end{bmatrix}$$

The element matrices of the stiffened shell are sum of the corresponding matrices of the shell and the stiffeners. The element matrices are assembled to get the global matrices on which the boundary conditions are imposed Newmark's direct time integration scheme of constant average acceleration [15] is employed to solve the dynamic equation of motion. Dynamic stresses are obtained in each time step (Δt) from the global displacements.

3. PROBLEMS AND DISCUSSIONS

Two benchmark problems are solved to validate the correctness of the stiffened shell formulation and the approach to solve the forced vibration problems. The results from literature [10, 16] along with the authors' are presented in Table 1 and Fig. 2, respectively. After validation forced vibrations of simply supported and clamped conoids with a set of biaxial stiffeners (1, 3 and 5 along each plan direction and equally spaced) are studied for 2 seconds and the dynamic magnification factors (DMF) are obtained under three different transient loading conditions, such as:

Load case I: Uniformly distributed load of infinite duration

Load case II: Uniformly distributed load of finite duration (1sec)

Load case III: Uniformly distributed half sine load of finite duration (1sec)

A preliminary bending analysis for stiffened full conoid (Table 2) and that for stiffened truncated conoid (Table 3) are done to obtain the values and locations of critical transverse deflection, compressive inplane force and bending moment which govern the shell depth. The DMF values for these shell actions are presented in Table 4 and Table 5, respectively.

The different geometric and material properties are furnished along with the tables. The results reflect that DMF value for load case III are comparatively less where the sinusoidal time variation of the load equals in magnitude the corresponding static load only for an instant. Moreover these values are somewhat non-responsive to any variation of boundary condition and stiffening schemes.

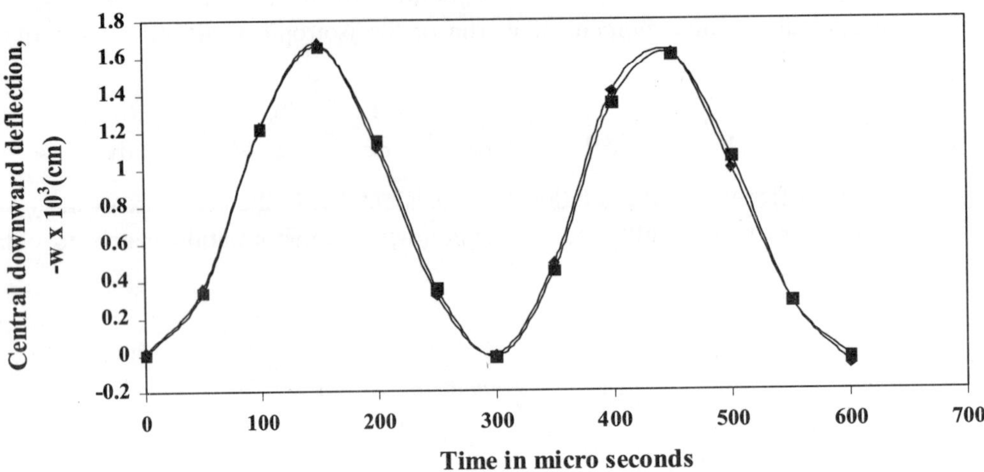

Fig. 2. Dynamic response of a simply supported square plate subjected to uniform pulse loading [a = b = 25cm, h = 5cm, ρ = 8 ×10^{-6} N s^2/cm^4, v = 0.25, E = 2.1 × 10^6 N /cm^2, q_0 = 10 N/cm^2, Δt = 5μ s. ——◆—— Reddy, ——■—— Present.]

Table 1. Fundamental frequencies (rad/sec) of clamped conoidal shell with central stiffeners.

Stiffener position	Stiffener along x-direction		Stiffener along y-direction		Stiffener along both x and y directions	
	Nayak and Bandyopadhyay [9]	Present	Nayak and Bandyopadhyay [9]	Present	Nayak and Bandyopadhyay [9]	Present
Concentric	17.28	17.31	20.83	20.76	20.90	20.84
Eccentric at top	17.83	17.73	21.57	21.45	29.39	29.32
Eccentric at bottom	17.55	17.52	22.31	22.24	22.92	22.87

a = 50m, b = 50m, h = 0.2m, hh = 10m, hl = 2.5m, E = 25.4910 × 10^9N/m^2, υ = 0.15, ρ = 2500kg/m^3, w_s = 0.3m, h_s = 1m.

Table 2. Static shell actions of stiffened full conoid with simply supported (SS) and clamped (CL) edges under uniformly distributed load.

Shell actions	$n_x = n_y = 1$		$n_x = n_y = 3$		$n_x = n_y = 5$	
	SS	CL	SS	CL	SS	CL
$\overline{-w}$	-0.13×10^{-2}	-0.38×10^{-3}	-0.40×10^{-3}	-0.21×10^{-3}	-0.29×10^{-3}	-0.16×10^{-3}
	(.21,.25)	(.25,.33)	(.25,.50)	(.33,.42)	(.25,.46)	(.33,.50)
$\overline{-N_y}$	-0.16×10^1	-0.13×10^1	-0.11×10^1	-0.10×10^1	-0.97	-0.89
	(.17,.42)	(.33,.08)	(.33,.20)	(.50,.50)	(.33,.50)	(.42,.42)
$\overline{-M_x}$	-0.81×10^{-2}	-0.25×10^{-2}	-0.29×10^{-2}	-0.15×10^{-2}	-0.19×10^{-2}	-0.13×10^{-2}
	(.08,.17)	(.17,.25)	(.08,.08)	(.17,.42)	(.08,.25)	(.25,.42)

a/b = 1, a/w_s = 50, a/h = 100, a/hh = 5, h_s/h = 3, υ = 0.3.

Table 3. Static shell actions of stiffened truncated conoid with simply supported (SS) and clamped (CL) edges under uniformly distributed load.

Shell actions	$n_x = n_y = 1$		$n_x = n_y = 3$		$n_x = n_y = 5$	
	SS	CL	SS	CL	SS	CL
$\overline{-w}$	-0.73×10^{-3}	-0.23×10^{-3}	-0.19×10^{-3}	-0.13×10^{-3}	-0.15×10^{-3}	-0.99×10^{-4}
	(.17,.21)	(.25,.29)	(.33,.42)	(.33,.42)	(.25,.42)	(.29,.42)
$\overline{-N_y}$	-0.18×10^{1}	-0.12×10^{1}	-0.87	-0.89	-0.79	-0.76
	(.17,.33)	(.25,.33)	(.17,.42)	(.33,.42)	(.25,.42)	(.33,.50)
$\overline{-M_x}$	-0.58×10^{-2}	-0.17×10^{-2}	-0.28×10^{-2}	-0.11×10^{-2}	-0.17×10^{-2}	-0.11×10^{-2}
	(.08,.17)	(.17,.25)	(.08,.08)	(.08,.08)	(.08,.08)	(.08,.08)

$a/b = 1$, $a/w_s = 50$, $a/h = 100$, $a/hh = 5$, $a/hl = 20$, $h_s/h = 3$, $\upsilon = 0.3$.

Table 4. Dynamic magnification factors of critical transverse deflection $(-w)$ and critical governing force $(-N_y)$ and moment $(-M_x)$ of full conoid.

Load case	Shell actions	$n_x = n_y = 1$		$n_x = n_y = 3$		$n_x = n_y = 5$	
		SS	CL	SS	CL	SS	CL
I	$\overline{-w}$	1.71	1.52	1.69	1.53	1.65	1.52
	$\overline{-N_y}$	1.74	1.44	1.66	1.55	1.70	1.54
	$\overline{-M_x}$	1.71	1.61	1.34	1.60	1.35	1.51
II	$\overline{-w}$	1.71	1.46	1.64	1.53	1.65	1.52
	$\overline{-N_y}$	1.74	1.44	1.63	1.54	1.70	1.54
	$\overline{-M_x}$	1.71	1.59	1.34	1.52	1.35	1.51
III	$\overline{-w}$	1.02	1.00	1.03	1.02	1.04	1.03
	$\overline{-N_y}$	1.03	1.03	1.04	1.02	1.04	1.04
	$\overline{-M_x}$	1.03	1.01	1.01	1.03	1.03	1.03

$a/b = 1$, $a/w_s = 50$, $a/h = 100$, $a/hh = 5$, $h_s/h = 3$, $\upsilon = 0.3$.

Table 5. Dynamic magnification factors of critical transverse deflection $(-w)$ and critical governing force $(-N_y)$ and moment $(-M_x)$ of truncated conoid.

Load case	Shell actions	$n_x = n_y = 1$		$n_x = n_y = 3$		$n_x = n_y = 5$	
		SS	CL	SS	CL	SS	CL
I	w	1.65	1.37	1.51	1.45	1.52	1.43
	-Ny	1.61	1.39	1.61	1.42	1.57	1.52
	-Mx	1.62	1.69	1.28	1.14	1.18	1.09
II	w	1.65	1.37	1.51	1.45	1.47	1.42
	-Ny	1.61	1.36	1.59	1.42	1.53	1.50
	-Mx	1.62	1.51	1.28	1.14	1.18	1.09
II	w	0.99	0.96	1.01	0.99	1.02	1.01
	-Ny	1.01	1.00	1.01	1.03	1.03	1.02
	-Mx	1.01	1.00	1.01	0.99	1.00	0.99

$a/b = 1$, $a/w_s = 50$, $a/h = 100$, $a/hh = 5$, $a/hl = 20$, $h_s/h = 3$, $\upsilon = 0.3$.

It is also observed that the DMF values for load case I and II in most of the cases are equal except for a few places where the values for load case I are marginally higher. The results show that the DMF values monotonically decrease with increase in the number of stiffeners. It is also observed that magnification of shell actions under transient load is more for simply supported boundary condition where the number of support constraints is relatively less than clamped shells. The DMF values for governing inplane force and moment may be more than those for transverse deflection. This establishes the fact that dynamic analysis should involve study of all shell actions and not merely transverse deflection.

CONCLUSION

The mathematical model developed here works well for solving stiffened conoidal shells under forced vibration. This may be concluded by observing close agreement of results obtained by the present approach and those reported in the literature. It is inferred from the study that the DMF values for any particular shell configuration and transient load case are not necessarily maximum for deflections and hence applying a DMF value of deflection to get the dynamic values of other shell actions cannot be recommended. To get the magnitudes of individual shell actions a detail study is needed. DMF values for sinusoidal load are the least of the three load cases considered here. DMF values for sinusoidal load are non-responsive to parametric changes DMF values decrease monotonically with increase in the number of stiffeners. DMF values are more for simply supported boundary conditions than for clamped ones. DMF values for full conoids are always greater than those for truncated conoids for the shell actions considered here.

NOMENCLATURE

a, b	length and width of shell in plan
e	eccentricities of both x and y direction stiffeners with respect to shell midsurface
E, G	Young's and shear moduli, respectively
h, h_s	shell and stiffener thicknesses, respectively
hl, hh	lower and higher heights of truncated conoid
$\overline{M_x}$	non-dimensional moment resultants per unit length $[= (M_x)/q_0a^2]$
n_x, n_y	number of stiffeners in x and y directions, respectively
$\overline{N_y}$	non-dimensional inplane force resultants per unit length $[= (N_y)/q_0a]$
q_0	magnitude of distributed static load and peak of transient distributed load.
Q_x, Q_y	transverse shear resultants of shell per unit length
$u, w,$	displacements along x and z axes, respectively
w_s	width of stiffener
$\overline{w}$	non-dimensional transverse deflection $[= wEh^3/(q_0a^4)]$
α, β	rotations vectorially along y and x axes, respectively

$\varepsilon_x, \varepsilon_y$	normal strains
$\gamma_{xy}, \gamma_{xz}, \gamma_{yz}$	shear strains
υ	Poisson's ratio
ξ, η	local natural coordinates of an element
ρ	mass density of shell and stiffeners
Δt	time step

REFERENCES

1. H.A. Hadid, 1964. An Analytical and Experimental Investigation into the Bending Theory of Elastic Conoidal Shells, Ph. D. dissertation, University of Southampton.
2. C.K. Choi, 1984,A conoidal shell analysis by modified isoparametric element, Computers & Structures, 18 (5), 921-924.
3. B. Ghosh and J.N. Bandyopadhyay, 1989, Bending analysis of conoidal shell using curved quadratic isoparametric element, Computers & Structures, 33 (3) , 717-728.
4. B. Ghosh and J.N. Bandyopadhyay, 1994, Bending analysis of conoidal shells with cutouts, Computer & Structures, 33 (1), 9-18.
5. A.K. Das and J.N. Bandyopadhyay, 1993, Theoretical and experimental studies on conoidal shells, Computers & Structures, 49 (3), 531-536.
6. D. Chakravorty, J.N. Bandyopadhyay and P.K. Sinha, 1995, Free vibration analysis of point-supported laminated composite doubly curved shells - A finite element approach, Computers & Structures, 54 (2), 191-198.
7. D. Chakravorty, J.N. Bandyopadhyay and P.K. Sinha, 1996, Finite element free vibration analysis of doubly curved laminated composite shells, Journal of Sound and Vibration, 191 (4) 491-504.
8. D. Chakravorty, J.N. Bandyopadhyay and P.K. Sinha, Application of FEM on free and forced vibration of laminated shells, 1998, ACSE Journal of Engineering Mechanics, 124 (1), 1-8.
9. A.N. Nayak and J.N. Bandyopadhyay, 2002, Free vibration and design aids of stiffened conoidal shells, Journal of Engineering Mechanics, 128 (4), 419- 427.
10. A.N. Nayak and J.N. Bandyopadhyay, 2002, On the free vibration of stiffened shallow shells, Journal of Sound & Vibration, 255 (2), 357- 382.
11. A.N. Nayak and J.N. Bandyopadhyay, Free vibration analysis of laminated stiffened shells, Journal of Engineering Mechanics, 131 (1) (2005) 100-105.
12. G. Sinha and M. Mukhopadhyay, (1995), Static and dynamic analysis of stiffened shells - A review, Proceedings of Indian National Science Academy, 61A (3 & 4), 195-219.
13. M.S. Qatu, (2002), Recent research advances in the dynamic behavior of shells:1989-2000, Part 1: Laminated composite shells, Appl. Mech. Rev., 55 (4), 325-350.
14. M.S. Qatu, (2002), Recent research advances in the dynamic behavior of shells:1989-2000, Part 2: Homogeneous shells, Appl. Mech. Rev., 55 (5), 415-434.
15. K.J. Bathe, (2002), Finite Element Procedure, Prentice-Hall of India Private Limited, New Delhi.
16. J.N. Reddy, 1982, Dynamic (transient) analysis of layered anisotropic composite material plates, Int. J. Num. Meth. Engg., 19, 237-255.

116

Flexural Analysis of an Asymmetrically Loaded Circular Top Plate with Eccentric Oblong Penetration

RAKESH KUMAR MOURYA, NILAYENDRA CHAKRABORTY, B.K. SREEDHAR,
P. KALYANASUNDARAM AND G. VAIDYANATHAN

Sodium Technology Group, Fast Reactor Technology Group
Indira Gandhi Centre for Atomic Research, Kalpakkam-603 102, Tamil Nadu, India.
email: bksd@igcar.gov.in

ABSTRACT

A vessel top cover plate, having an eccentrically located oval opening is analyzed to study the deflection and bending pattern. The plate covers the test vessel in the sodium test facility and supports the fuel handling machine of the Prototype Fast Breeder Reactor (PFBR). The plate is 2490 mm diameter and 95 mm thick and the oblong penetration is of 360 mm width and 572.5 mm center-to-center distance. The weight of the fuel handling machine is 250 kN. The deflection and bending pattern of the plate is critical because it will affect the position of the handling head of the machine which is ~ 9000 mm below the level of the plate. As code/standard formulae are not available for similar openings, an approximate calculation was first done by replacing the oblong opening with a circular opening. Detailed analysis was later carried out by FEM using 8-noded shell element.

Keywords: Circular plate, oblong opening, asymmetrically loading.

1. INTRODUCTION

The Prototype Fast Breeder Reactor (PFBR), under construction at Kalpakkam, Tamil Nadu, India, is a pool type fast breeder reactor of 500 MWe capacity. It uses liquid sodium as the primary and secondary coolants from neutronic and thermal considerations. The temperature of the primary sodium is 547°C at the exit of the core. The choice of fuel in fast reactor is governed by factors like feasibility of breeding of plutonium, closing of fuel cycle, achievement of high power density and high burn up, ease of fabrication and fuel cycle economics. Mixed oxide of Pu and U is selected as the fuel of PFBR.

Refueling is done after 180 effective full power days. The reactor is shut down and the sodium temperature is 200°C during refueling. The fuel handling machine, located on the small rotating plug in the roof slab of the reactor, weighs about 250 kN and is 24 m high. The machine is supported at its middle and a portion of it is in the cover gas space below the roof slab while the rest of it is above the roof slab.

The fuel handling machine, called Transfer Arm, will be tested in sodium before it is installed in the reactor. A test vessel has been constructed and erected for this testing. The vessel, of SS 316 LN, is 13150 mm tall and 2104 mm in outer diameter with a Y shaped limb at 3954 elevation (Fig. 1). The vessel has a ring flange that is bolted to the top cover plate. The top cover plate has an eccentrically located oblong opening through which the Transfer Arm is introduced into the vessel. (Fig. 1)

2. DESCRIPTION OF THE TOP COVER PLATE

The top cover plate (Fig. 2) is 2490 mm in diameter and 100 mm thick. It has an oblong shaped opening of 360 mm wide and 572.5 mm centre to centre distance for Transfer Arm mounting. The total weight of TA will act on the periphery of the opening. The maximum sodium temperature inside the test vessel is 550°C. The top cover plate has 49 nos. of thermal baffles mounted at its bottom to increase the resistance to heat flow and ensure that the cover plate temperature is below 120°C. The oblong opening is eccentrically located with respect to the plate center by 133.75 mm in north and 142.5 mm in west directions, respectively.

The top cover plate and the vessel ring flange are made of carbon steel equivalent to ASTM A516 Grade 65 steel. This material is chosen for its better weldability, through thickness ductility, better resistance to lamellar tearing of weld and good compatibility with sodium vapor.

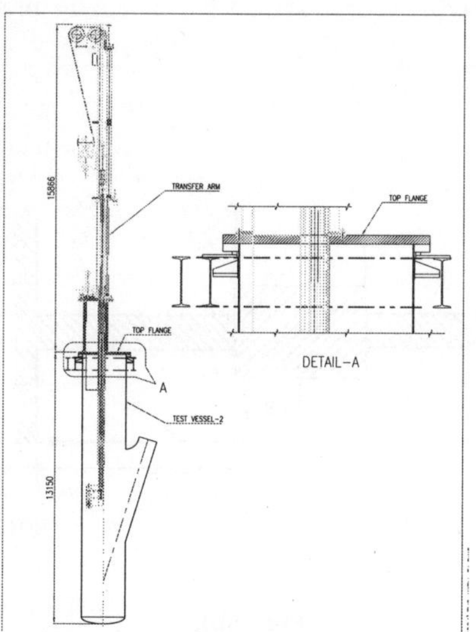

Fig. 1. Transfer arm in test vessel-2.

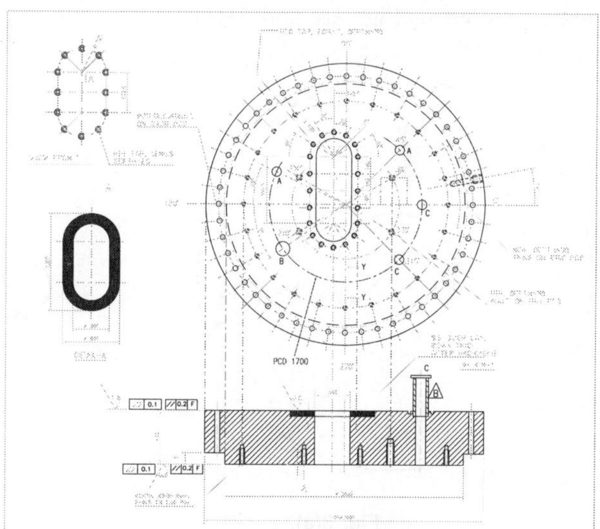

Fig. 2. Top cover plate of test vessel-2.

3. ANALYTICAL APPROACH

The minimum thickness of flat cover plate and blind flange can be calculated by ASME Section VIII Division 1, UG-34. General rules of reinforcement required for openings in flat cover plates are given in paragraph UG-39; here requirements of large circular, centrally located openings are discussed (UG-39(c)(1). There is no clear guideline for oblong shaped off centered positioned opening *("UG-39(c) (3) when large opening is any other shape than that describe in (c) (3) above, there are no specific rules given.'')*. In UG-39(d) the increase in thickness to provide the necessary reinforcement in flat cover plates with a single opening with diameter that does not exceed one half of the head diameter is mentioned.

The top cover plate thickness, calculated as per UG-39(d) (1) replacing C by 2C (where C = a factor depending upon method of attachment and shell dimensions) is found to be 84 mm. A plate of 100 mm thickness is selected from availability consideration. [1 & 2]

Since analytical formulas are not available for these types of openings, the analysis is first done using the formula available for circular openings by approximating the elliptical opening with a circular opening. Since the flatness of the plate is an important consideration during operation of the Transfer Arm, the problem is then analyzed using standard FEM software.

3.1 Conventional Method Using Basic Equations of Plate

The top cover plate is fastened to the vessel ring flange by a set of 20 mm bolts at a PCD of 2438 mm. The flange of the Transfer Arm is bolted to the top cover plate by means of 20 nos. of 30 mm bolts in the periphery of the oval opening. Hence, both boundaries are considered fixed, i.e., all degrees of freedoms are constrained, for the calculations. The load of the machine is considered as uniformly distributed on the periphery of the oval opening. The self-weight of the top cover plate is included in the analysis.

For the analytical calculation the opening is assumed as circular and located at the center of the plate [shown in Fig. 3A and 3B].

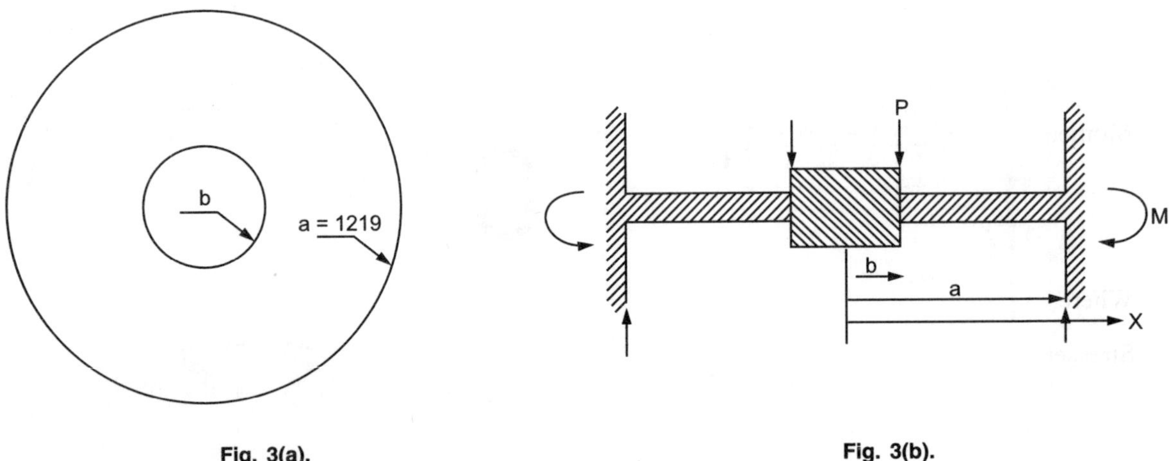

Fig. 3(a).   Fig. 3(b).

The diameter of the circular hole is calculated with the following approximations:
  i. Equivalent circle in area: The circular opening has a radius whose area is same as the area of actual oblong opening.
  ii. Equivalent circle with circumference: The circular opening has a radius whose circumference is same as the area of actual oblong opening.

iii. Major diameter circle: The circular opening whose diameter is same as the major diameter of actual oblong opening.

iv. Minor diameter circle: The circular opening whose diameter is same as the minor diameter of actual oblong opening.

Using the following basic equations for symmetrically loaded circular plate of uniform thickness with a circular central hole [3 & 4], the deflection and stress are calculated.

Slop from radial axis,
$$\phi = -\frac{Px}{8\pi D}(2\ln x - 1) + \frac{C_1 x}{2} + \frac{C_2}{x} \qquad \ldots(1)$$

Deflection,
$$w = \frac{Px^2}{8\pi D}(\ln x - 1) - \frac{C_1 x^2}{4} - C_2 \ln x + C_3 \qquad \ldots(2)$$

Where;
$$D = \frac{EI}{\left(1-\mu^2\right)} = \frac{Eh^3}{12\left(1-\mu^2\right)}$$

Assuming $\mu = 0.3$

Applying the boundary conditions

$$\phi = 0 \text{ at } x = a, \quad \phi = 0 \text{ at } x = b \text{ and } w = 0 \text{ at } x = a$$

Solving the above equations (1, 2) for the given boundary conditions, we get the value of constants as,

$$C_1 = \frac{P}{4\pi D\left(a^2 - b^2\right)}\left\{2\left(a^2 \ln a - b^2 \ln b\right) - \left(a^2 - b^2\right)\right\},$$

$$C_2 = -\frac{Pa^2 b^2 \ln\frac{a}{b}}{4\pi D\left(a^2 - b^2\right)} \text{ and } C_3 = -\frac{Pa^2}{8\pi D}(\ln a - 1) + \frac{C_1 a^2}{4} + C_2 \ln a$$

Moments are given as

$$M_1 = D\left(\frac{d\phi}{dx} + \mu\frac{\phi}{x}\right) \text{ and } M_2 = D\left(\frac{\phi}{x} + \mu\frac{d\phi}{dx}\right)$$

Where;
$$\frac{d\phi}{dx} = -\frac{P}{8\pi D}(2\ln x + 1) + \frac{C_1}{2} - \frac{C_2}{x^2}, \text{ and } \frac{\phi}{x} = -\frac{P}{8\pi D}(2\ln x - 1) + \frac{C_1}{2} + \frac{C_2}{x^2}$$

Stresses are given as,

$$\sigma_1 = \frac{6M_1}{h^2} \text{ and } \sigma_2 = \frac{6M_2}{h^2}$$

3.2 Finite Element Analysis

Since the deflection of the plate during operation of the Transfer Arm governs the proper functioning of the machine it was decided to analyze the problem using standard FEM software and confirm that the deflections are within limits.

3.2.1 Modeling

The top cover plate of the vessel is modeled using standard FEM software. The 8-noded shell element is used (SHELL 93). This element has six degrees of freedom at each node, viz. translations in the nodal x, y, and z directions and rotations about the nodal x, y, and z-axes. The deformation shapes are quadratic in both in-plane directions. The element has plasticity, stress stiffening, large deflection, and large strain capabilities [5]. As the plate is not axi-symmetric full modeling of the plate is required. The mesh is refined in the periphery of the oblong opening but fairly coarse size is used in the regions where the stress variation is expected to be small (Figures 4A & 4B). The resulting model consists of 2640 elements and 8276 nodes.

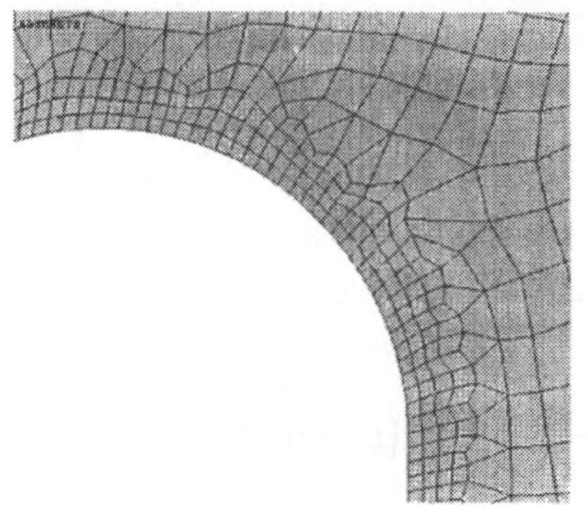

Fig. 4(a).

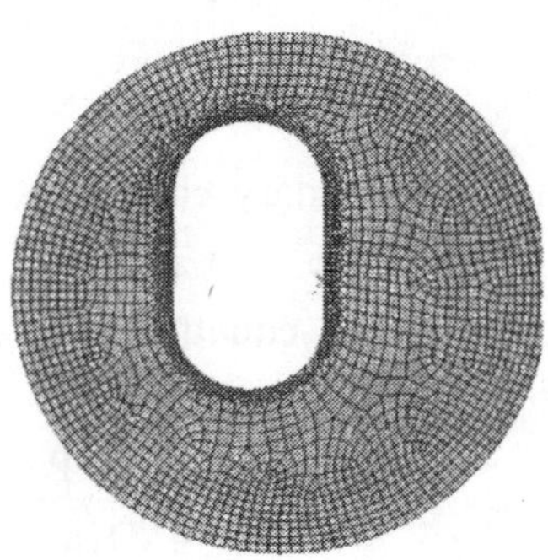

Fig. 4(b).

3.2.2 Boundary Conditions and Loading

All nodes at a diameter 2438 mm, which is the PCD of the bolts connecting the top cover plate and the vessel ring flange are restrained from all degrees of freedom i.e., fixed/clamped condition. All the nodes on the PCD connecting the Transfer Arm flange to the periphery of the oval opening in the top cover plate are restrained from rotation about any axis. Total weight of the fuel handling machine 250 kN is distributed over all nodes on the oval shaped PCD. Properties of Carbon steel material equivalent to ASTM A 516 Grade 65 steel has been used (at 120°C).

3.2.3 Results

Approximation: The results obtained from analytical calculation and finite element analysis for the various approximations show negligible variation. The results are compared in Table 1. The closeness in the results is a validation of the type of element and the mesh density used.

Table 1. Maximum stress and deflection

Different approximations	Deflection w (μm)		Von Mises Stress (MPa)	
	Analytical	FEM	Analytical	FEM
Case 1: The circular opening has a radius whose circumference is same as the area of actual oblong opening.	81.49	87.48	12.16	12.24
Case 2: The circular opening has a radius whose circumference is same as the area of actual oblong opening.	69.59	75.14	11.01	11.18
Case 3: The circular opening whose diameter is same as the major diameter of actual oblong opening.	40.291	44.63	8.31	8.38
Case 4: The circular opening whose diameter is same as the minor diameter of actual oblong opening.	154.13	162.60	18.40	18.52

Actual case: The actual case of the eccentrically located oval opening was analysed using FEM software with the following boundary conditions :

Uniformly distributed load of 97.7 N/mm along the periphery of the oval opening

All dof fixed at 2438 mm PCD

Rotation = 0 on the periphery of oval opening

The maximum Von Mises stress is **27 MPa**, which is well below the allowable stress value (118 MPa) of the material at 120°C. The deflection in the radial direction and Von Mises stresses of the plate are shown in the Figs. 5 &6. Figure 5A shows that deflection is maximum (0.115 mm) on the right side of the opening and that is because of greater span length on this side. Figure 5B indicates the variation of deflection along the section AA (shown in Fig. 5A).

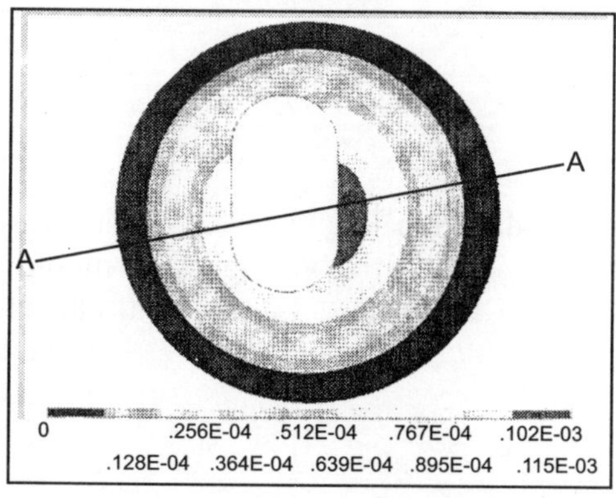

Fig. 5(a). Deflection of top flange.

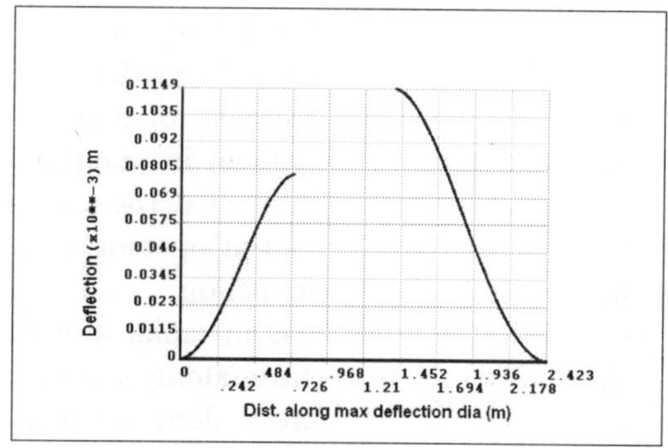

Fig. 5(b). Deflection of top flange.
(across-section AA, mentioned in Fig. 5A).

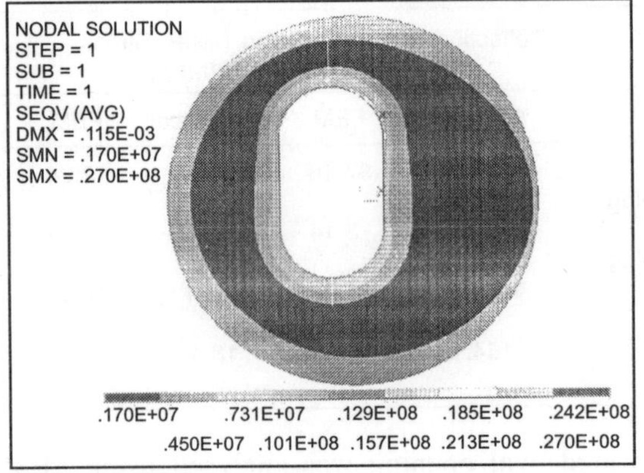

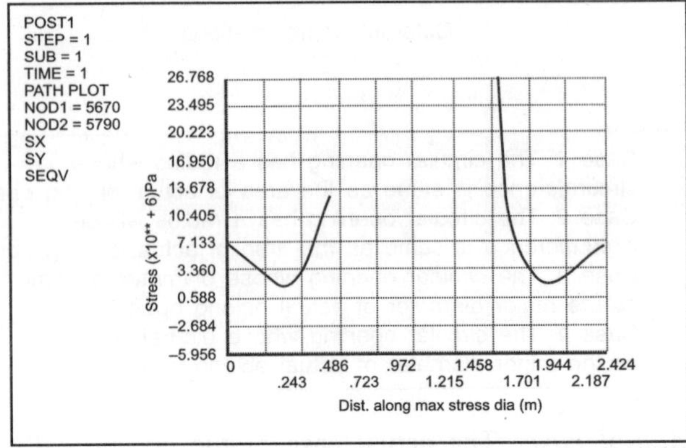

Fig. 6(a). Von mises Stresses for top flange.   **Fig. 6(b).** Stresses for top flange.

CONCLUSION

The results of the analysis using analytical formulae, with suitable approximation, and standard FEM software shows that the stresses and deflection in the top cover plate are within limits.

NOMENCLATURE

a	radius of flange
b	radius of opening
C_1, C_2 & C_3	constants
D	flexural rigidity
E	Young's modulus of elasticity
h	thickness of plate
M_1	moment in radial direction
M_2	moment perpendicular to diametrical direction
P	load at opening edge
w	deflection
x	axis in radial direction
σ_1	stress along cylindrical section
σ_2	stress along diametrical section
Φ	slope from radial axis
μ	Poisson's ratio

REFERENCES

1. American Society of Mechanical Engineers, New York - Boiler and Pressure Vessel Code (2004), Section VIII, Division 1.
2. American Society of Mechanical Engineers, New York - Boiler and Pressure Vessel Code (2004), Section 2D, for material properties.
3. Timoshenko and Woinowsky-krieger, 'Theory of Plates and shell', Second edition.
4. John. F. Harvey, P.E., "Theory and Design of Pressure Vessels", First edition.
5. Michel A. Potter *et al.*, 'Comparison of limit load, linear and non-linear FE analysis of atypical vessel nozzle' Pressure Vessel and Piping Design Analysis-2001, ASME2001, Volume-430.

117

Analysis and Design of Metallic Plates for Biaxial Loading Using FEM

S. Sambamurty, R. Velmurugan, R. Palaninathan and T. Sekar

Indian Institute of Technology, Madras, Chennai-600 036, Tamil Nadu, India
email: rvel@iitm.ac.in

ABSTRACT

To evaluate the mechanical behavior of metals under high temperature and pressure applications, it is necessary to develop experimental techniques to realize multi axial testing under large deformation. One technique for investigating material behavior under such complex stress states is to use in-plane biaxial testing. For experimental investigations concerning the mechanical behavior under biaxial stress state of metals, mostly cruciform flat specimens are used. By means of empirical methods, different specimen geometries have been proposed earlier. In the present paper, the design of cruciform specimen under in-plane biaxial loading is carried out by using the finite element technique. The basic constraints of the cruciform specimen design are analysed to get optimal size i.e., the maximum stress in the specimen should occur within the gage area and the distribution of stress within the gage area should be relatively uniform.

Keywords: In-plane biaxial tension test, material properties, cruciform specimen design, finite element method.

1. INTRODUCTION

In pressure vessels, more than one stress will be induced under service conditions. In case of cylindrical pressure vessels, the circumferential stresses are double the longitudinal stresses whereas in the case of spherical pressure vessels, the longitudinal stresses and the circumferential stresses are equal. Generally, pressure vessels are designed on the basis of uniaxial properties of the material whereas in practical, these are subjected to multi axial loading.

The uniaxial tension test is obviously the simplest way of characterizing the material behavior. The specimen under uniaxial tension extends and fails under well-defined loading conditions and tests are performed on dog-bone shaped test specimens. These specimens are predefined and standard type. But we cannot predict the exact mechanical behavior of materials under severe service conditions, by using the standard uniaxial tension test. To predict the accurate stress flow in the material, it is necessary to realize the exact boundary and loading conditions in the material.

Earlier, several experimental approaches have been developed to predict the material stress strain flow under combined loading. One method is by using the thin walled tube specimens under either combined tension and torsion [1, 2], or combined tension and internal pressure [3]. Other experimental techniques include the hardness test [4], hydraulic bulge test and biaxial stretching of a cross-shaped plate specimen. For predicting the inelastic behavior of metal, a tube specimen is not applicable. The properties of the metal will change when the sheet metal is converted into tube form. The hardness test is only applicable as a qualitative evaluation tool. Because of the non uniform stress and strain distribution, the hydraulic bulge test is not recommendable [5]. Thus, the realistic experimental approach in studying the material behavior is through a direct biaxial loading system.

2. BIAXIAL TENSION TEST

An experimental method for precisely measuring the stress-strain distribution and predicting the material behavior under certain specialized service conditions has been developed using in-plane biaxial testing. The basic requirement for an ideal biaxial testing is that the stress and strain distribution in the gage area of the specimen (region of uniform thickness, where the strain measurements is carried out) should be homogeneous; it should be possible to determine properly the developed stress state i.e., the values of the stresses and the orientation of the principal stress directions; and the initial yielding should occur in the central gage area of the specimen.

The most realistic experimental method to create a known homogeneous in-plane biaxial stress state in metals is the direct biaxial test as developed on a flat cruciform specimen [6]. Using this approach, cruciform specimens fabricated from plate material are gripped at four locations and loaded along two orthogonal axes. Servo hydraulic loading systems are used in this application.

The cross-shaped specimen of uniform thickness by Shimada [7] induces only a relatively small area of homogeneous stresses and observed that the stress concentrations are high near the corner fillets. Other investigations propose cruciform specimens with a thickness-reduced test section and slotted arms [8].These modifications improve the homogeneity of the stress distribution in the thinned gage section and avoids stress concentrations in other regions.

It is observed in literature that all proposed specimen have been obtained empirically, by means of photo elastic tests [9]. In the present paper analysis is carried out on cruciform specimen by using the finite element technique. Verification is carried out on the basic constraints of the cruciform specimen design i.e., the maximum stress in the specimen has to occur within the gage area and the distribution of stress within the gage area should be relatively uniform. Many trails have been carried out by changing the size, thickness of the gage area and the slots to get the optimum dimensions of the cruciform specimen.

3. BIAXIAL SPECIMEN DESIGN

The primary step in the design process is to select overall specimen dimensions. Consideration is given here to a number of test system features and characteristics including load capacity of specimen grips; size envelope available within the load frame for specimen installation and gripping; gripping arrangement and size restrictions imposed by the specimen grips and specimen gage area needed to meet strain measurement and other instrumentation needs.

A major complication in using the flat cruciform specimen is that it induces coupling between the two loading directions. One method of minimizing this effect is to incorporate fairly complicated arrangement of slots [10]. These slots provide the transversal flexibility to the arms and thus facilitate to transfer the load to central region.

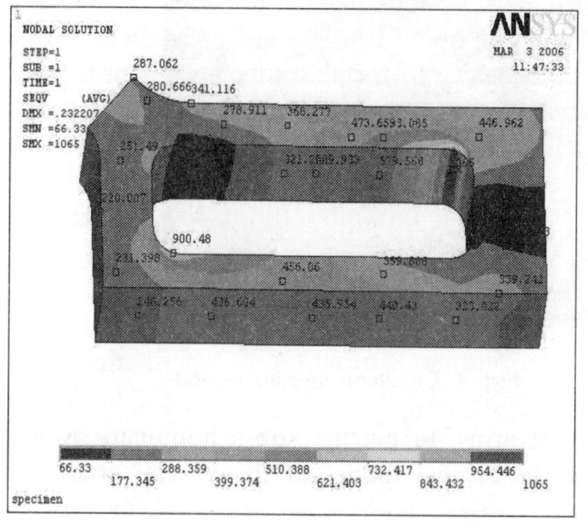

Fig. 1. Stress distribution in slots.

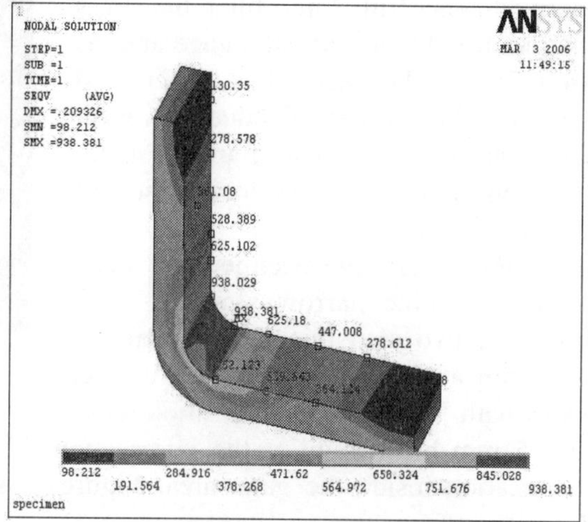

Fig. 2. Stress distribution at junction of loading arms.

One drawback in using slots is that regions of high stress concentrations are developed at the slot corners in close proximity to gage area as shown in Fig. 1. This problem has been solved by providing thickness transition from gripping area to gage area. Analyze the specimens with single step thickness transition from 6 mm to 3 mm and other with the two step thickness transition from 16 mm to 2 mm with 4.8 mm intermediate step. From the stress distribution shown in Fig. 3, it is clear that in case of single step thickness transition specimen, the gage area stresses are less than half of the stress concentrations developing at the intersection of loading arms, whereas in case of two step thickness transition the variation in stress distribution from gage area to rest of the gage area is less, but still the stress concentration is dominating. This raises the possibility that failure may be initiated outside the gage area. If the thickness transition ratio is 10, then the stress distribution in the gage area is more uniform and maximum stress develops in the gage area.

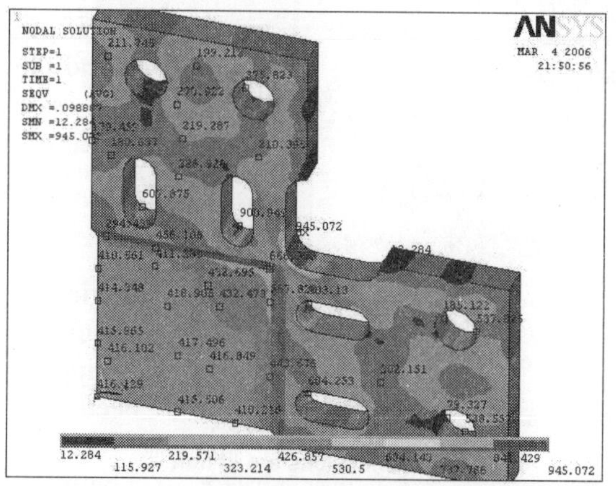

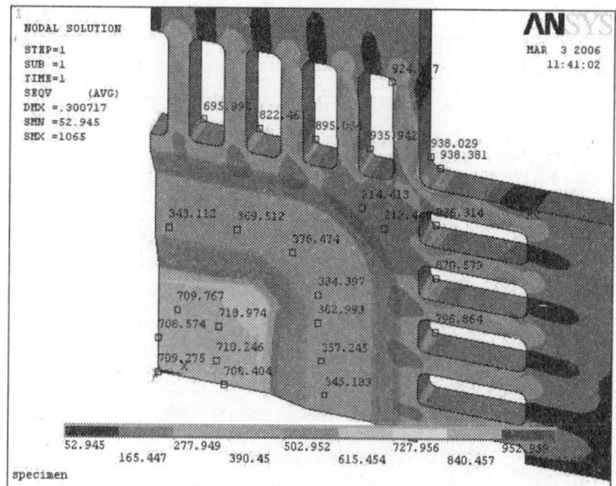

Fig. 3. Stress distribution in specimens with single and two step thickness transitions.

At the intersection of the loading arms the 2 mm blend radii is provided to eliminate the sharp corners. With 2 mm fillet, the stress concentration is higher than the gage area as shown in Fig. 2. The fillet radius is increased from 2 mm to higher optimal value to prevent the failure at the intersection of arms. Finally at 3.175 mm blend radii, the localization of stresses is eliminated.

Thus the cruciform specimen has been designed with nine narrow slots on each loading arm, two step thickness transitions from 16 mm at gripping area to 1.6 mm at gage area with 'ten' thickness transition ratio,

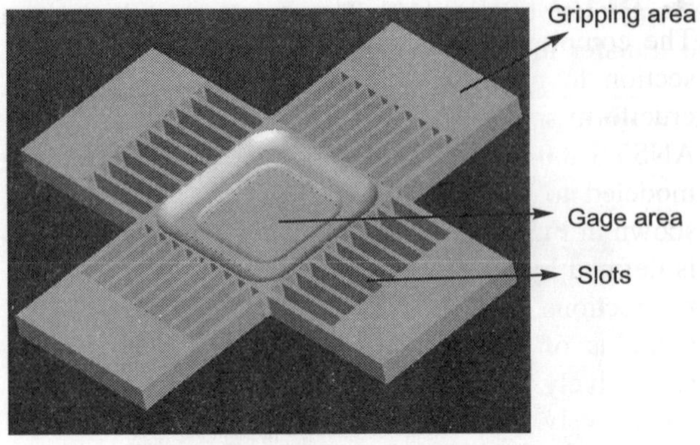

Fig. 4. Cruciform specimen model.

and 3.175 mm blend radii at the intersection of loading arms, to get the stress homogeneity and failure location inside the gage area. Figure 4. show the typical model of the biaxial cruciform specimen.

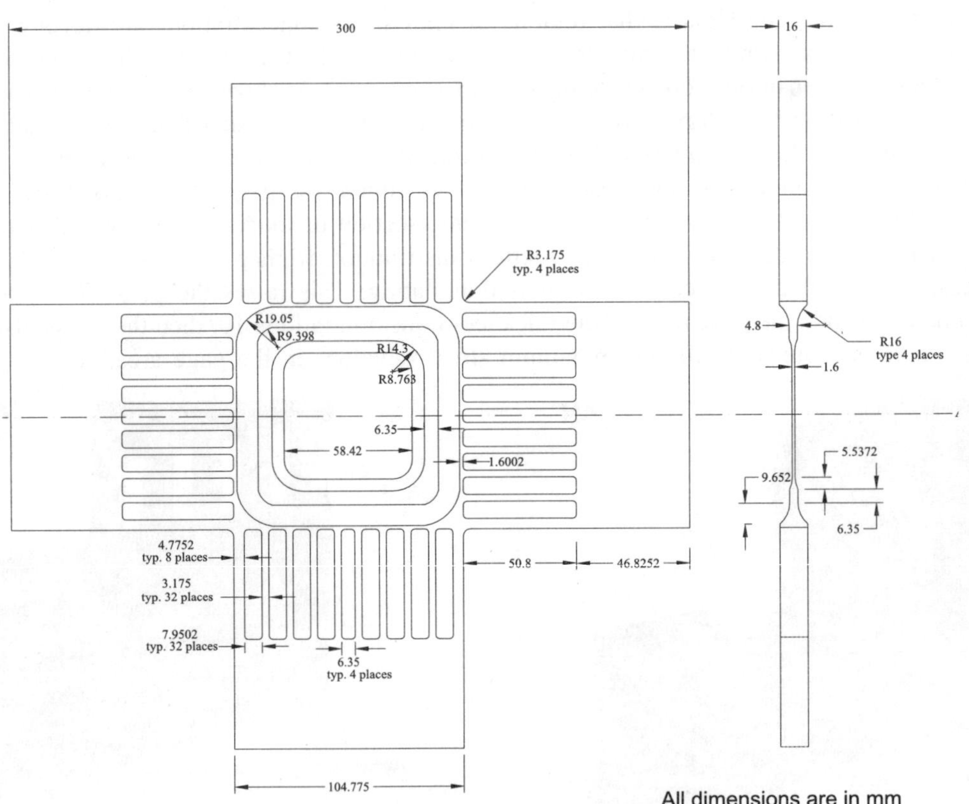

Fig. 5. Biaxial cruciform specimen geometry.

4. BIAXIAL SPECIMEN ANALYSIS

The complex design features are included in the cruciform specimen as explained in the previous section to meet the biaxial test requirements. To verify the basic requirements of the proposed cruciform specimen design, finite element analysis is carried out by using the finite element package ANSYS 8.0. Because of the material and geometric symmetry, only one eighth of the specimen is modeled to minimize the computational time. The finite element meshing of specimen, which is shown in Fig. 6, comprises three-dimensional 20-node structural solid element SOLID95. The element is defined by 20 nodes having three degrees of freedom per node: translations in the nodal x, y, and z directions. The Ti-6Al -4V Grade 5 annealed material is considered for specimen analysis. The modulus of elasticity and poison's ratio of the titanium alloy material are 114 GPa and 0.342 respectively. The yield strength and ultimate strength of the material are 880 MPa and 950 MPa respectively. The symmetry boundary conditions about X, Y, and Z-axes are imposed in the finite element model. 146.33 MPa pressure is applied on the loading area of the specimen along two perpendicular directions.

5. RESULTS AND DISCUSSION

Numerical results on stress distribution from finite element analysis are plotted in Fig. 7. The maximum von mises stress in the gage area is 945.484 MPa whereas in the outside of the gage area is 923.509 MPa. The ratio of the maximum stresses from the gage area to the outside of gage area is observed to be 1.024. The stress distribution show that the maximum von mises stress is developed more closely to a line with 45° inclination to the direction of maximum principal stress.

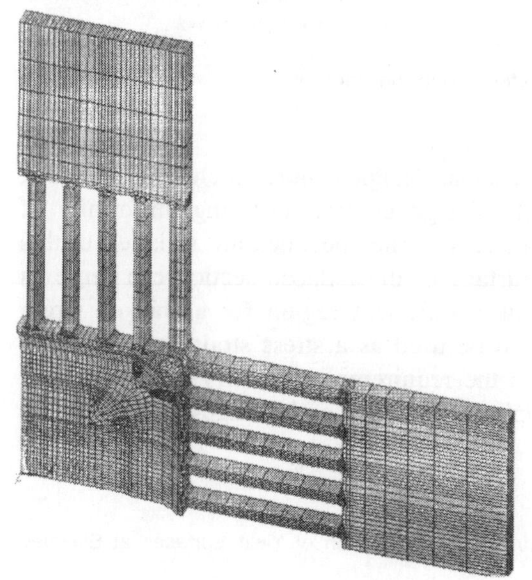

Fig. 6. Finite element meshing.

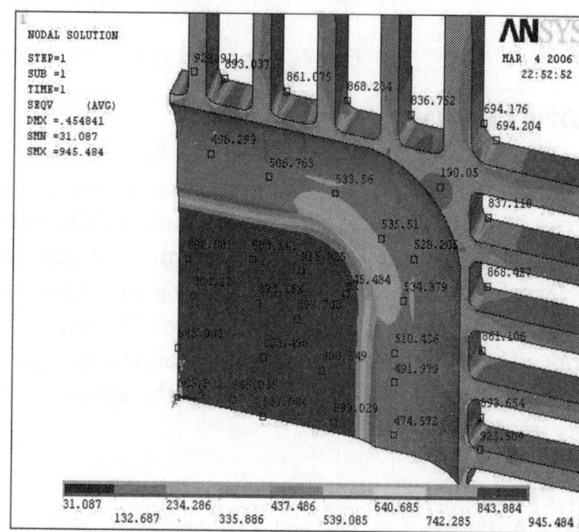

Fig. 7. Von Mises stress distribution of the specimen.

The maximum von mises stress in the outside of the gage area is observed at the corner of central slot along the horizontal direction. The von mises stresses in the corners of the slots from central slot to outer slots in the horizontal direction are observed as 923.5, 892.6, 861.1, 868.4, 837.1 and 694.2 MPa, respectively, which are smaller than that in the gage area.

Figure 8 shows the vonmises stress and strain distribution in the gage area against the horizontal distance from center of gage. The deviation observed in the von mises stresses in the test region is found to be 0.28% only in the 20mm span from the center of the specimen. Thus, the homogeneity of stress distribution in the 20mm by 20mm gage area of the cruciform specimen is observed to be in precise limits to predict the stress strain distribution of the test material.

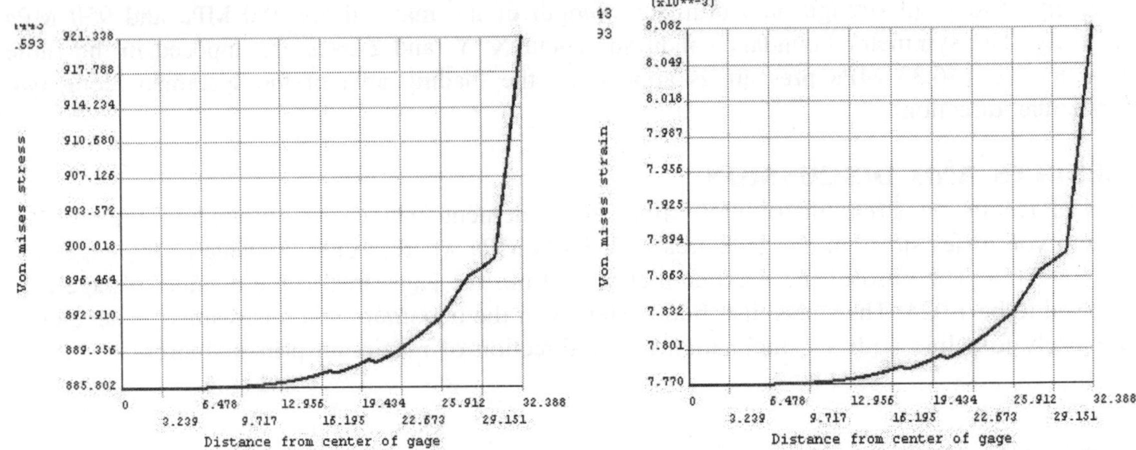

Fig. 8. Von Mises stress and strain distribution along the gage area.

CONCLUSION

The biaxial cruciform specimen has been designed with special design features including the slots, thickness transition and incorporation of blend radii. The design criteria regarding uniformity of stress in the gage area and the location of the maximum stress in the specimen are satisfied in this design. The central region of 20 mm × 20 mm on the surface of the reduced section can serve as the working area. This working area provides a more than sufficient region for a stacked strain rosette with an active gage section of 20 mm × 20 mm to be used as a stress strain sensor in this work. Therefore, this cruciform specimen design satisfies the requirements to carry out the biaxial tensile testing. The configuration of the specimen is complex, presenting a challenge to fabricate the specimen.

REFERENCES

1. A. Phillips, C.S. Liu, and J.W. Justusson, 1972, An Experimental Investigation of Yield Surfaces at Elevated Temperatures, Acta Mech., 14, 119.
2. M.J. Michno, JR., and W.N. Findley, 1973, Experiments to Determine Small Offset Yield Surfaces for 304L Stainless Steel Under Combined Tension and Torsion, Acta Mech., 18, 163.
3. S.S. Hecker, 1971, Yield Surfaces in Prestrained Aluminum and Copper, Metal. Trans., 2, 2077.
4. D. Lee, F.S. Jabara, and W.A. Backofen, 1967, Knoop-Hardness Yield Loci for 2 Titanium Alloys, Trans. TMS-AIME, 239, 1476.

5. J. Woodthorpe, and R. Pearce, 1970, The Anomalous Behavior of Aluminum Sheet Under Balanced Biaxial Tension, Int. J. Mech. Sci., 12, 341.

6. E. Shiratori, and K. Ikegami, 1967, A New Biaxial Testing Machine With Flat Specimen, Bulletin of the Tokyo Institute of Technology, 82, 105-118.

7. H. Shimada, K. Shimizu, M. Obata, K. Chikugo, and M. Chiba, 1976, A New Biaxial Testing Machine for the Flat Specimen and a Fundamental Study on the shape of the Specimen, Technology Reports, Tohoku University, 41, 351-369.

8. D.A. Kelly, 1976, Problems in Creep Testing Under Biaxial Stress Systems, J. Strain Analysis, 11(1), 1-6.

9. R. Hill, 1990, Constitutive Modelling Of Orthotropic Plasticity In Sheet Metals, J.Mech.Phys.Solids, 38, 405.

10. J.R. Ellis, and A. Abul-Aziz, 2003, Specimen Designs for Testing Advanced Aeropropulsion Materials Under In-Plane Biaxial Loading, NASA/TM -2003-212090.

118

ABI Simulation for the Material Property Estimation

Kamal Sharma[1], Vivek Bhasin, R.K. Singh, K.K. Vaze and A.K. Ghosh

Reactor Safety Division, BARC Trombay, Mumbai-400 085, India.
[1] email: hello_kamal@yahoo.com

ABSTRACT

A combined mechanical property evaluation methodology with ABI (Automated Ball Indentation) simulation and Artificial Neural Network (ANN) analysis is evolved to evaluate the mechanical properties for SA 333 grade steel. The experimental load deflection data is converted into meaningful mechanical properties for this material. An ANN database is generated with the help of contact type finite element analysis by numerically simulating the ABI process for various magnitudes of yield strength (σ_{yp}) (200 MPa–400 MPa) with a range of strain hardening exponent (n) (0.1-0.5) and strength coefficient (K) (200 MPa-600 MPa). For the present problem, a ball indenter of 2 mm diameter having Young's Modulus approximately 100 times more than the test piece is used to minimize the error due to indenter deformation. Test piece dimension is kept large enough in comparison to the indenter configuration in the simulation to minimize the deflection at the outer edge of the test piece.

Further, this database after the neural network training; is used to analyze measured material properties of different test pieces. The ANN predictions are reconfirmed with contact type finite element analysis for an arbitrary selected test sample. The methodology evolved in this work can be extended to predict material properties for any irradiated nuclear material in the service. Extensions of the ABI tests and the associated database analysis could lead to evaluation of the indentation energy to fracture needed for the structural integrity assessment of aged components.

Keywords: Automated Ball Indentation, ANN, Finite Element Simulation, Irradiated Nuclear Material, Miniature Specimen Testing.

1. INTRODUCTION

Nuclear reactor components and power piping are generally subjected to various forms of thermal cycling. As a result, the mechanical properties of the materials of the components get degraded. It is therefore of prime importance, that the altered mechanical properties of the degraded materials be known for life assessment of the components of the nuclear and thermal power plants. So determination of mechanical properties of materials by using non-conventional techniques has been an active area of research for a long time. Among some non-destructive methods for determining mechanical properties of materials, a semi-destructive type of testing, called *Automated Ball Indentation* (ABI) has been developed. Figure 1 is a schematic representation of the indentation profile in an ABI test. It is one of the most promising techniques to evaluate mechanical properties of materials, as it requires small size of test materials. For the validity of this material property evaluation procedure with ABI, the tests reported here have been initially carried out for standard materials with known mechanical properties. The *Automated Ball Indentation* technique is capable of extracting degraded mechanical behavior and properties of thermally aged or irradiated materials from very small specimens. The objective of miniaturized specimen testing technology is to enable the characterization of mechanical behavior while using a greatly reduced minimum volume of material. In most of the cases extracting specimen from components, for conducting conventional tests for evaluation of properties of the material, is neither possible nor permissible. The significance of this technology is obvious to the nuclear industry where neutron irradiation space is limited and irradiation cost scales up with specimen volume. In addition, the substantial advantages resulting from the application of this technology in non-nuclear industries are now beginning to be realized. For this evaluation the specimen undergoes multiple indentations by a spherical ball indenter. Furthermore, this method can be used to characterize weldments and associated Heat Affected Zone (HAZ), it also avoids the need to fabricate test specimen, and it is relatively rapid.

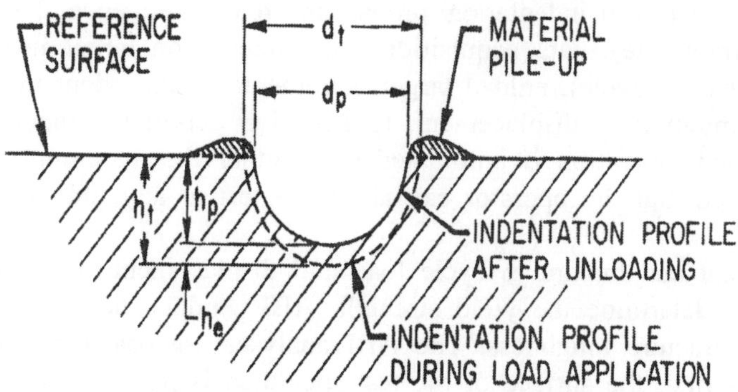

Fig. 1. Automatic Ball Indentation Process.

2. REVIEW OF EARLIER WORK

A few research groups have published a series of investigations on ball indentation technique to evaluate mechanical properties. For evaluating mechanical properties many theories and models have been developed and some of the mechanical properties such as yield strength, ultimate tensile strength, strength coefficient, true stress-true strain curves and strain hardening coefficient are evaluated with these theories. It was Mayer [1] who first developed a relationship between the mean pressure and indentation diameter to evaluate the yield strength of materials. Tabor [2] gave an empirical relationship to find the representative strain of materials while indentation is done through a hard spherical ball.

However, Tabor's relation holds very close to the test observation when the indentation process becomes fully plastic. Haggag *et al.* [3] did extensive work and developed an automated ball indentation test set up for determining flow properties directly from the test around a small volume of material. They used various base materials and found excellent agreement with standard ASTM uniaxial tensile and fracture toughness data. The location dependence of the mechanical properties was successfully measured by Murthy *et al.* [4]. Gradients in mechanical and fracture properties of SA 533B steel welds were studied using ball indentation technique. Haggag and Nansted [5] also described a simple technique for estimating the fracture toughness by coupling the measured flow properties with a modified but empirically correlated critical fracture strain model. This technique predicted fracture toughness values that differed by less than 11% from measured values for both A 515 grade 70 carbon steel and A 533-grade B class 1 pressure vessel steels, respectively. Mathew *et al.* [6] studied the effects of low temperature aging (673K) up to 18 months on the mechanical and fracture properties of cast CF-8 stainless steel in the range of 173-423K. These studies have been carried out using non-destructive ball indentation tests. A theoretical model is proposed to estimate fracture toughness of ferritic steel in the transition region from ball indentation test data by Byun, kin, Hang [7]. The key concept of the model is that the indentation energy to a critical load is related to fracture energy of material.

3. AUTOMATIC BALL INDENTATION

The ABI test is based on multiple indentation cycles on a polished metallic surface with a spherical indenter. Each cycle consists of indentation, unload and reload sequence. The values of indentation penetration speed (strain rate), data acquisition rate, indentation target incremental displacement (penetration depth for each cycle), unload target incremental load, indentation maximum final load and indentation maximum final displacement, (penetration depth) are input before the test starts. Once the test is started, operation is automated until either the maximum load or the maximum displacement is reached, but the operator can abort the test at any time if system malfunction is detected.

These displacement and maximum cycle load and displacement values from each indentation sequence are used to determine the yield strength, ABI produce derived true stress/ true plastic strain curve and the fracture toughness. The ABI analyses are based primarily on elasticity and plasticity theories. At first, the surface of the test specimen is polished with a polishing tool. The testing head is then properly secured to the test specimen such that the load cell is perpendicular to the polished surface of the region where testing is to be performed. The specimen undergoes multiple indentations by the spherical indenter. With increasing indentation loads, an increasing volume of material is forced to flow under multi-axial compression caused by the indenter. The material has less constraint at the surface around the indentation in the ABI test. In each cycle, a loading-holding-unloading-reloading sequence is maintained. The indentation load and indentation diameter are then converted into true stress-strain data of material being tested. Multiple cycles for the above loading and the load increases approximately linearly with penetration depth. The linear increase is the consequence of two non-linear but opposing processes occurring simultaneously; the non-linear increase in the applied load with penetration depth because of the spherical geometry of the indenter is offset by the power-law work hardening behaviour of the test material. The total and plastic indentation depths (h_t and h_p) and applied indentation load (P_1) are then measured for first cycle. Typically, the load is increased for each succeeding cycle and the values of P, h_t, h_p, are measured for each cycle. Several load cycles are conducted at the same indentation location. For determining a full true stress-true plastic strain curve at this location, this cyclic loading can be continued; however, the maximum total indentation depth (h_t) should not exceed one half of diameter D of indenter. However, in the present experimental setup the unloading is not possible and elastic spring back is neglected in determining the plastic strain values (Fig. 2).

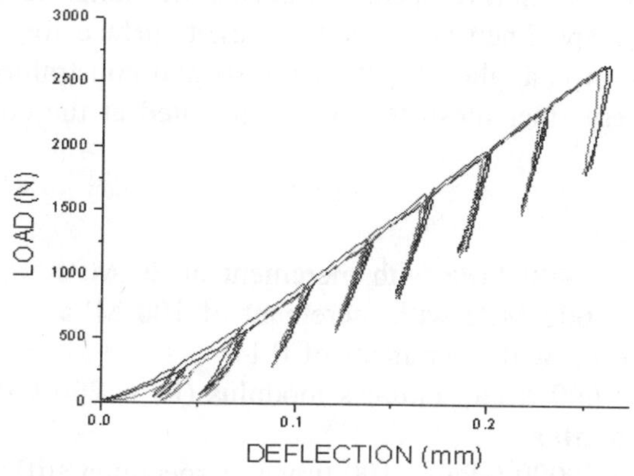

Fig. 2. Variation of indentation load with penetration depth.

4. NUMERICAL MODEL AND ANALYSIS PROCEDURE

In the present work due to spherical nature of indenter and circular test specimen, fully axisymmetric 2D model is created. The finite element model of the ball indenter and specimen is shown in the Fig. 3 and the deformed model enlarged at contact area due to the loading is shown in Fig 4. After

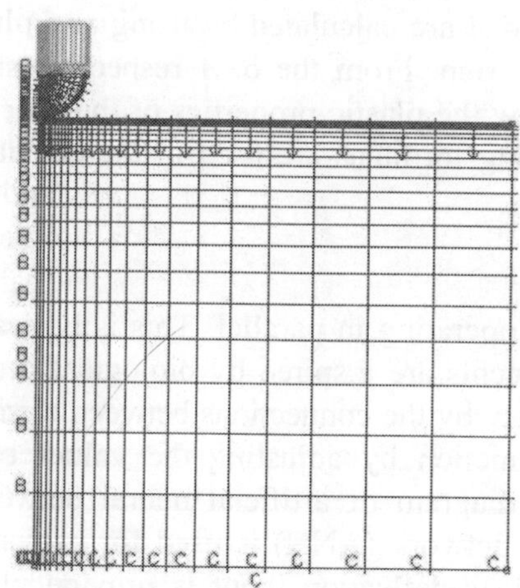

Fig. 3. FE model of Indenter and Test Specimen.

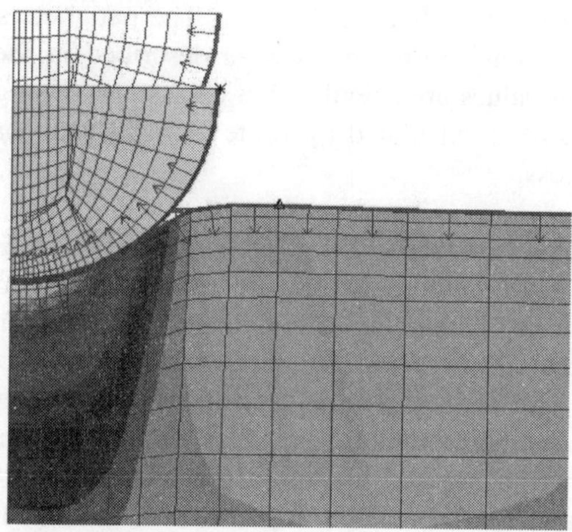

Fig. 4. Deformed FE model at the contact area.

validation plane 42 and contact 48 elements of ANSYS are chosen. For the present analysis, SA 333 material properties are used. The indenter has been simulated with Young's modulus 100 times them the test specimen. This helps in simulating the hardened ball stiffness. The test specimen is modeled with 10 mm × 10 mm dimensions. The indenter in this analysis has 1mm radius. The indenter and specimen have been modeled in two sections (areas), respectively. The lower section (area) of indenter and upper section of test specimen consist of contact 48 elements. The upper section of indenter and lower section of specimen consist of planer 42 elements. Indenter is defined as contactor surface and the specimen is defined as target surface for the contact pair. Large strain and large displacement option is applied for the analysis and constrained function contact algorithm is used for the solution. Very finer mesh has been generated at the contact element location.

4.1 Material Data

Material data for specimen

σ_{yp} = 200 to 400 MPa with increment of 20 MPa

K = 200 to 600 MPa with increment of 100 MPa

n = 0.1 to 0.5 with increment of 0.1

Ultimate limit (σ_{uts}) = 600 MPa; Young's modulus (E) = 200 GPa; ν(Poisson Ratio) = 0.25

Material data for stiff indenter

Young's modulus (E) = 20000 GPa (~ 100 times of specimen stiffness); ν(Poisson Ratio) = 0.0.

4.2. Analysis

Analysis of indenter and specimen FE model has been carried out for different values of σ_{yp}, K and *n* and for each case the input material data is varied and load is incremented to simulate the indentation process. For the analysis purpose the power law is taken into consideration to generate the stress-strain data for the different combinations of σ_{yp}, K and *n* and the generated database is used as material property input for the analysis. The non-linear analysis is carried out for the case with varying steps of pressure from 3 MPa to 300 MPa. After each analysis load-deflection curves are obtained and at each point the respective stress-strain values have been calculated using different parameters such as δ, A, d_p, ε_p. The parameters such as δ and A are calculated by using an in-house code with iterative process as described in the following section. From the δ, A respective stress-strain values are calculated at each point. Using the power law the plastic properties of the specimen have been calculated by finite element analysis. The following formulas are used in the calculation process.

5. ARTIFICIAL NEURAL NETWORKS

Artificial neural networks are composed of simple elements operating in parallel. This is basically a network or interconnections of artificial neurons. These elements are inspired by biological nervous system. As in nature, the network function is determined largely by the connections between elements. We can train a neural network to perform a particular function by adjusting the values of the connections (weights) between elements. Generalized line diagram of artificial neural network is shown in Figure 6. In the present work the artificial neural network (ANN) is used for performing logical function on its input such as load-deflection. The load-deflection input is provided to the ANN for the range of K, n, and σ_{yp} values. This data has been used to train the ANN. The Neural Network used in this case for the purpose of data inversion is an example of Multi layer Perceptron

(MLP) network with back propagation algorithm. It consists of 6 layers before the output stage, 5 layers consisting of the intermediate neurons also known as hidden layers. The first layer is an input stage.

The command TRAINLM is used for training the network while the command PURELIN is the transfer function used to activate the neurons. Arbitrarily some load-deflection plots are chosen from the database and are input to the trained network for cross checking the network accuracy. In the subsequent step experimental data collection is carried out. After that from the ANN the respective properties for the given input will be obtained.

6. RESULTS AND DISCUSSIONS

Overall methodology used for the material property evaluation is shown in Figure 5. For the different combinations of n, K and yield strength, stress-strain data was generated, which was used for the input in the FE analysis. After the analysis, load-deflection graphs were plotted for different cases with varying strain hardening exponent, strength coefficient and yield strength. Typical plots are

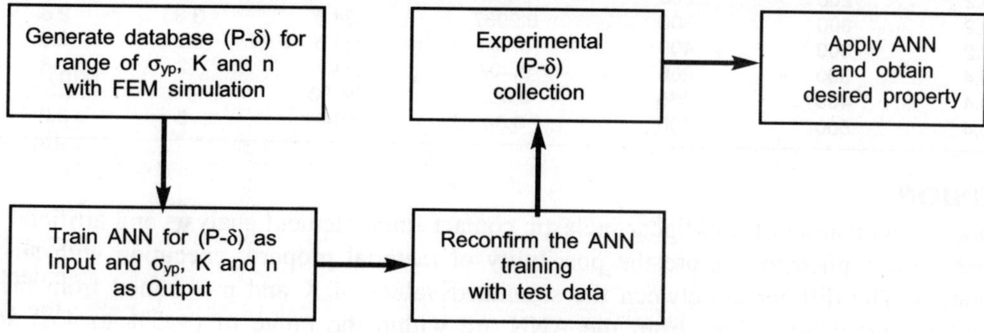

Fig. 5. Methodology for the Evaluation.

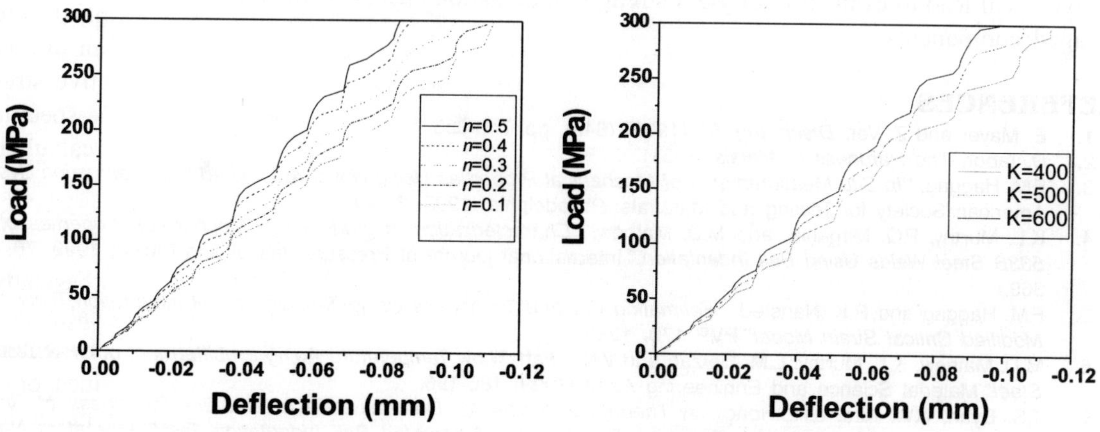

Fig. 6. Load-deflection for different n (strain hardening coefficient) values Yield 300 MPa and K 400 MPa.

Fig. 7. Load-deflection for different K (Strength coefficient) values Yield 300 MPa and n = 0.2.

shown in Figures 6 to 7. These plots data are used for the analysis with the help of ANN according to flow chart description (Figure 5). The load-deflection input is provided to the ANN for the range of K, n and σ_{yp} values. This data has been used to train the ANN. This database after network training is used to analyze measured material property on different material test pieces. The ANN predictions are reconfirmed with contact type finite element analysis. Table 1 shows that for some arbitrary chosen combination of strength coefficient and yield strength for strain hardening exponent of 0.2 and 0.4, it gives maximum error up to 4.5% in strain hardening exponent and 3.2% in strength coefficient.

Table 1. Comparison between simulated value and analyses value for n = 0.2 & 0.4

N	Simulated Value		Analytical value		%Error in n	%Error in K
	K (MPa)	σ (MPa)	n	K (MPa)		
0.2	200	240	0.193	196.5	3.50	1.75
0.2	200	280	0.205	196.8	2.5	1.6
0.2	300	300	0.2067	294.8	3.35	2.6
0.2	300	400	0.194	291.6	3.00	2.8
0.4	300	260	0.408	291.7	2	2.8
0.4	400	360	0.382	394.93	4.5	1.37
0.4	500	400	0.38	484	5	3.2

CONCLUSION

In this work a combination of non-linear inelastic contact finite element analysis and artificial neural network has been applied to explore the possibility of material property execution with automatic ball indentation. The difference between the simulated values of K and n obtained from the FEM analysis and the predicted values from the ANN are within the range of ($\sim$ 3% to 4%) and the agreement is very good. The methodology evolved in this work can be extended to predict material properties for any irradiated nuclear material. Further extensions of the ABI tests and their database analysis will lead to evaluation of yield strength, indentation energy to fracture and structural integrity of aged components.

REFERENCES

1. E. Mayer and Z. Ver, *Dtsch. Ing.* 52 (1908) (645), pp. 740-835.
2. D. Tabor, *The Hardness of Metals* (1951).
3. F.M. Haggag, *"In Situ Measurements of Mechanical Properties Using Novel Automated Ball Indentation System"* American Society for Testing and Materials, Philadelphia, 1993, 27-44.
4. K.L. Murthy, P.Q. Mirgania, and M.D. Mathew, *"Characterization of gradients in Mechanical Properties of SA-533B Steel Welds Using Ball Indentation,"* International journal of Pressure Vessel and Piping, 1999, 76, 361-369.
5. F.M. Haggag and R.K. Nansted, *"Estimating Fracture Toughness Using Tension or Ball Indentation Tests and a Modified Critical Strain Model"* PVP 170, 1989.
6. M.D. Mathew, K.L. Murthy, L.M. Lietzan and V.N. Shah, *"Low Temperature Aging Embitterment of CF-8 Stainless Steel"* Material Science and Engineering A269 (1999) 186-196.
7. T.S. Byan, J.W. Kim, J.H. Hong, *"A Theoretical Model for Determination of Fracture Toughness of Reactor Pressure Vessel Steels in the Transition Region from Automated Ball Indentation Test,"* Journal of Nuclear Materials, 1998, 252, 187-194.
8. Y.H. Lee, W.J. Ji and D. Kwon, *"Stress Measurement of SS400 Steel Beam Using the Continuous Indentation Technique"* Experimental Mechanics, February 2004, Volume 44.

119

Finite Element Modeling and Stress Analysis of Forward Curved Radial Tipped Centrifugal Fan

MS. KINNARI SHAH[1], N.N. VIBHAKAR[2] AND S.A. CHANNIWALA[3]

[1] Lecturer, Sarvajanic College of Engineering and Technology, Surat, email: shahkh_25@ yahoo.co.in
[2] Assistant Professor, Sarvajanic College of Engineering and Technology, Surat, India
email: v_nitin1@rediffmail.com
[3] Prof. and Head of the Department, Sardar Vallabhbhai National Institute of Technology,
Ichhanath, Surat-395007, Gujarat, India. email:sac@svrec.ernet.in

ABSTRACT

The design of turbomachinery has been practiced in the last half of the previous century with increasing degree of sophistication. This trend of development is not complete because design of any turbomachine is interdisciplinary process involving aerodynamics, thermodynamics, fluid dynamics, stress analysis, vibration analysis, the selection of materials, and the requirements for manufacturing. So, products have not reached a level of full maturity. Today there is considerable room for further development of most turbomachinery products when looked at with the perspective of the broadest possible design world and when it comes to development at the design stage then various types of analysis of every component are performed on computer before the first prototype is built. Among these the major one end the most frequently used in the manufacturing of any mechanical part is stress analysis.

The present work concentrates on to generate the finite element model of forward curved radial tipped impeller which behaves mathematically like structure model and carry out stress analysis using I-deas master series 9i (high end software). The model is solved to suggest the optimum thickness with which it would have been fabricated as per developed unified design and to gain advantages like reduction in kinetic energy to rotate the impeller and saving power consumption due to reduced weight and finally reduction in cost.

Keywords: Forward curved radial tipped centrifugal fan, unified design, finite element analysis.

1. INTRODUCTION

Finite element method provides the solution domain into simply shaped regions or elements. An approximate solution for partial difference equation can be developed for each of these elements. The total solution is then generated by linking together or assembly the individual solutions taking care to ensure continuity at the inter-element boundaries. Thus, the partial difference equation is satisfied in piecewise fashion.

The finite element method is endowed with three basic features that account for its superiority over other competing methods:

Geometrically complex domain of the problem is represented as a collection of geometrically simple sub domains called 'finite elements'.

Over each finite element the approximation, functions are derived using the basic idea that any continuous function can be represented by a linear combination of algebraic polynomials.

Algebraic relations among the undetermined coefficient (i.e., nodal values) are obtained by satisfying the governing equation often in a weighted integral sense over each element.

1.1 Selection of Software for Modeling and Stress Analysis

The accurate result of the software analysis depend upon the effectiveness and simplicity of analysis software together with its speed and simple user interface. There for among many options, I-deas master series 9i is selected for modeling and stress analysis of forward curved radial tipped centrifugal fan which is having variety of modules for different tasks.

1.2 Finite Element Model

For finite element modeling the simulation module of program is used. Finite element programme divides the element into a grid of 'element', which form a model of the real structure.

A finite model is complete idealization of the entire structural problem, including the node locations, the elements, physical and materials properties, loads and boundary condition. The model is supposed to be defined differently for different types of analysis: static structural loads, dynamics or thermal analysis. The goal of finite element model is not to make a model look like the structure. The purpose of finite element modeling is to make a model that behaves mathematically like the structure modeled, not necessarily one looks like the real structure.

2. FINITE ELEMENT MODELING AND STRESS ANALYSIS

It consists of three steps which are

1. Pre-processing
2. Solution
3. Post-processing.

2.1 Pre-Processing

This includes the entire process of developing the geometry of a finite element model, entering physical and material properties, describing the boundary conditions and loads and checking the model. For variable thickness of the impeller, first made model the 'surface model' along with 'split surface' which is shown in Figure 1 and Figure 2.

2.1.1 Entering the Material Properties of Forward Curved Radial Tipped Impeller

Poisson's ratio = 0.3,
Modulus of elasticity $E = 20,680$ kg/mm^2,
Density = 7,850 kg/mm^2,
Shear modulus $G = 7953.84$ kg/mm^2

2.1.2 Meshing

Nodes and elements are generated by one of the two methods, mapped mesh or free mesh. Here, the free mesh is used for meshing because of more flexibility in defining mesh areas, it will automatically created by a algorithm, which tries to minimize element distortion and can easily have internal holes which is shown in Figure 3.

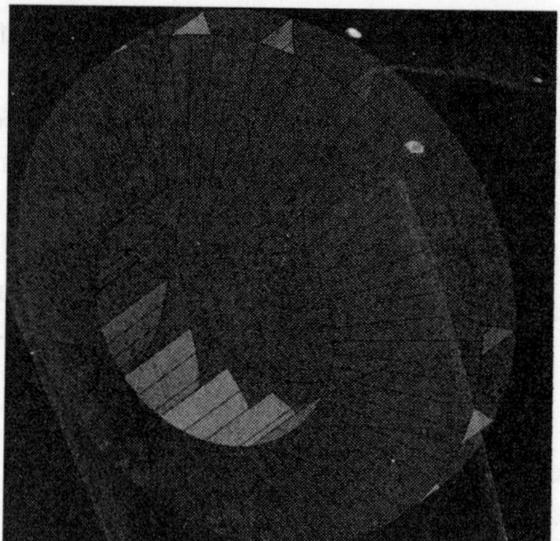

Fig. 1. Finite Element Model of Forward Curved Radial Tipped Centrifugal Fan.

Fig. 2. Pre-processing Involves Spit surfaces.

Fig 3. Meshing.

2.1.3 Boundary Condition

The boundary condition is applied to build analysis cases containing loads and restraint of the model .in finite element analysis, the model is considered to be in equilibrium. So, the loads and moments should be such that the condition of $\Sigma F = 0$ and $\Sigma M = 0$ are satisfied.

Structural Loads: Structural loads can be nodal forces or pressure on face or edge of an element. A nodal force has six values, for three forces and three moments.

Restraints: Restraints are used to restrain the model to ground. Restraints also have six values at nodes, three translations and three rotations. In this case we gave following restraints.

Translation ----- X constant, Y constant, Z constant

Rotation--------- X free, Y free, Z free

2.2 Solution

The solution phase can be performed in the model solution task of the simulation application, or an external finite element analysis program. Model solution can solve linear and non-linear static, dynamics, buckling, conduction heat transfer and potential flow analysis problems.

The model solution task in the simulation applications is where the finite element model is solved. In this case, the solution set is created for obtaining displacements and stress as the output data. Then model is solved using the 'solve command'. After the solution displacements will be stored in one data set and stresses will be another.

2.3 Post-Processing

From the type of material, the yield stress was known.

Now by Von Mises theory the induced stress should be less than 0.66

Permissible yield stress $\sigma x = 0.577$ (yield stress) $= 248.26$ N/mm^2

These involve plotting deflections and stresses and comparing these results with failure criteria on the design such as maximum deflection allowed, the material static, fatigue strengths, etc. The post-processing task of the simulation application provides tools to display and interpret the results after the solution is finished. Result of variable thickness impeller with different angular velocity can also be brought in from external finite element analysis for post-processing which is shown in Figure 4.

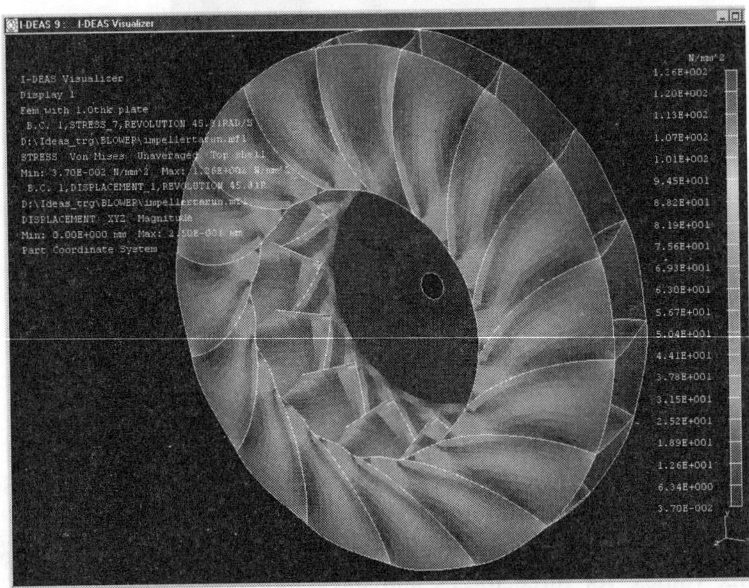

Fig. 4. Stress Analysis of Forward Curved Radial Tipped Impeller with Thickness of Material 1 mm.

From these analysis, the maximum value of stresses obtained forward curved radial tipped impeller with different angular velocities for various thickness of material like 2 mm, 1.5 mm, 1.0 mm explained in result and discussion and values were compared with above limiting value of stresses.

3. RESULT AND DISCUSSION

The results in Table 1 invariably imply that at 2.03 mm the stresses created are too less. Hence, it will need to carry out the same analysis for lesser thick material. Gradually, moved to lesser thickness and stress analysis of impeller having material thickness of 1.5 mm. For this case, maximum and minimum value of the stresses are given in Table 2 which is too much less.

Table 1. Stresses obtained for Forward Curved Radial Tipped Impeller With Thickness of Material

ω rad/sec	Maximum value of stress N/mm²	Minimum value of stress N/mm²	Maximum displacement (mm)
36.65	57.6	0.0471	0.0581
45.81	90.0	0.0737	0.0908
54.97	130	0.1060	0.131

Table 2. Stresses obtained for Forward Curved Radial Tipped Impeller with Thickness of Material 1.5 mm.

ω rad/sec	Maximum value of stress N/mm²	Minimum value of stress N/mm²	Maximum displacement (mm)
36.65	67.2	0.0466	0.0395
45.81	105	0.072	0.140
54.97	151	0.105	0.201

Further, we carried out the stress analysis of the impeller of material having still less thickness and until 1 mm of thickness the stress which we obtained were much below the yield value which is clearly represented in Table 3.

Table 3. Stresses obtained for Forward Curved Radial Tipped Impeller with Thickness of Material 1.00 mm.

ω rad/sec	Maximum value of stress N/mm²	Minimum value of stress N/mm²	Maximum displacement (mm)
36.65	80.7	0.0237	0.160
45.81	126	0.037	0.250
54.97	181	0.0532	0.360

Here, stresses for 54.97 rad/sec are quiet high.

CONCLUSION

Looking to these results, for forward curved radial tipped impeller with various thicknesses and taking into consideration various other criterions like wear, life, etc., it was inferred that to suggest thickness of material plate less than 1 mm would be feasible and practically appropriate. Thus, it was concluded that the optimum thickness of impeller should be taken as 1 mm which will lead to advantages like:

- Reduction in required kinetic energy to rotate the impeller.
- Saving in power consumption.
- Reduction in cost.

ACKNOWLEDGEMENTS

Authors are very much thankful to Prof. Dipak Sharma and project students of C.K.P.C.E.T, Surat.

REFERENCES

1. Stepanoff, A.J, 1955, "Turbo Blowers", John Wiley & Sons.
2. Andre Kovats, 1964, "Design and Performance of Centrifugal and Axial Flow Pumps and Compressors", Pergamon Press.
3. Vibhakar N. N., 1998, "Experimental Investigations on Radial Tipped Centrifugal Blower", M.E. Dissertation Report submitted to South Gujarat University, Surat, India.
4. Church, A.H, 1966, "Centrifugal Pumps and Blowers", John Wiley & Sons.
5. Osborne, W.C, 1966, "Fans", Pergamon Press.
6. Earl Logan. Jr., "Hand Book of Turbo Machinery"
7. Chandraputla ,T.R. 'Introduction to finite elements in engineering.
8. A project report year 2001-2002 on 'Comparative Assessment & stress analysis of forward and backward curved radial tipped centrifugal blower.' undergraduate student, C.K.P.ET, Surat.
9. Ms. Kinnari Shah, 2002, "Unified Design and Comparative Performance Evaluation of Forward and Backward Curved Radial Tipped Centrifugal Blower/Fan", M.E. Dissertation Report submitted to South Gujarat University.

120

Dynamics of Helmet and Head Impact: Finite Element Modeling

Praveen Kumar Pinnoji and Puneet Mahajan

Department of Applied Mechanics, Indian Institute of Technology Delhi-110 016, India.

ABSTRACT

In the event of accident, motorcycle helmet protects the rider by absorbing impact energy and reducing the loading imparted to the head. A non-linear explicit finite element code, LS-DYNA™, is used to perform the dynamic analysis of helmeted-head impact to see the mechanical behaviour of the helmet under impact. Alternative designs of helmet are investigated to improve the ventilation in helmets. The effect of ventilation on the performance of helmet has been studied as these ventilation openings may reduce the impact protection and structural integrity of the helmet. The presence of grooves enhanced remarkably the airflow velocities around the top of the head. Pressure and stresses in the brain were investigated and were found not to change significantly due to the presence of grooves in the helmet.

Keywords: Helmets; Head model; Ventilation; Finite Elements; Dynamics.

1. INTRODUCTION

In South Asia, excessive sweating and resulting discomfort due to hot and humid weather conditions discourages motorcycle riders from using helmets unless it is mandatory by law. The space between the head and helmet is small and air velocities in this gap are also low; as a result the sweat is unable to evaporate making the driver uncomfortable. To enhance the evaporation, alternative designs of the helmet with 11 mm wide grooves in the foam were considered. In first ventilation design, the helmet had only one groove in the central plane with 7 mm deep. In second design, the helmet had three parallel grooves of (each separated by 28 mm) 7 mm deep. In third design, the helmet had three parallel grooves cut through the foam thickness but not the shell. In the fourth design a 10 mm hole was provided in the shell of the third design. Front and side impact simulations for impact velocities upto 10 m.s^{-1} were performed to see if the presence of grooves has a detrimental effect on the dynamic performance of the helmet. In dynamic studies, a deformable human head model was used. Head Injury Criterion (HIC), Von Mises stresses and Pressures were obtained in all these cases.

2. DYNAMICS OF MOTORCYCLE HELMET

Dynamic performance of the helmet with and without ventilation was studied using explicit Finite Element code LS-DYNA^{TM}. In the past the FE analysis of drop test of helmet used a rigid head and results were reported in the form of Head Injury Criterion (HIC) values and accelerations of the head form. Lately, FE is also commonly used to model the head [3-5]. Computed Tomography (CT) scanning and Magnetic Resonance Imaging (MRI) have been used to generate geometrical data in digital format for the head models. Once the geometry is available the material modeling of brain tissue is done next. Generally, the models do not contain all the details of the head and are much simpler than the actual head. One such 3D finite element model of human head developed by Willinger *et al.* [5] having skin, skull, CSF and brain is used here. The material properties of the various parts of the head were assumed to be homogeneous, isotropic and linearly elastic, except for the brain, which was assumed as viscoelastic in nature.

Initially, the frontal impact simulations were carried out at a velocity of 6.3 m.s^{-1} with FE head model and a rigid sphere of 48 mm radius. The pressure responses at coup and contra-coup are compared well with available literature. Horgan and Gilchrist [3] and Zong *et al.* [8] have also constructed three-dimensional Finite element models of the human head. The former used it for simulating the pedestrian accidents whereas the latter authors use a SI (Structural Intensity) approach to study the power flow distribution inside head in frontal, rear and side impacts. The results using human head models are presented in the form of pressures and stresses in the brain although a clear relation between stresses and brain injury are still to be fully established.

2.1 Shell, Foam and Head-Material Properties

A motorcycle helmet has two major parts, namely, the outer shell and the energy absorbing liner, which we call as 'foam'. The energy absorbing liner is made of expanded polystyrene or EPS and outer shells are made from composite material, like fiberglass, carbon fiber and Kevlar, or a molded thermoplastic like ABS or polycarbonate. The outer shell resists the penetration of any foreign object and distributes the impact load on a wider area thus, increasing the foam liner energy absorbing capacity. Material Model 3 (*MAT_PLASTIC_KINEMATIC) available in LS-DYNA^{TM} is used for outer shell in finite element analysis with following properties of Acrylo-Butadiene Styrene (ABS) (Table 1).

Table 1. Material properties for Helmet shell

Part	Density (kg.m^{-3})	Elastic Modulus (N.m^{-2})	Yield stress (N.m^{-2})	Poisson's ratio
Shell	2000	1.7×10^9	34.3×10^6	0.3

The liner considered here was manufactured from expanded polystyrene (EPS). Figure 1 shows the stress-strain behavior of EPS foam. The foam depicts linear elasticity at low stresses followed by a collapse plateau, truncated by a regime of densification in which the stress rises steeply. The longer the plateau region more is the energy absorbed. Densification in the foam starts at 65% strain and stresses rise sharply after that.

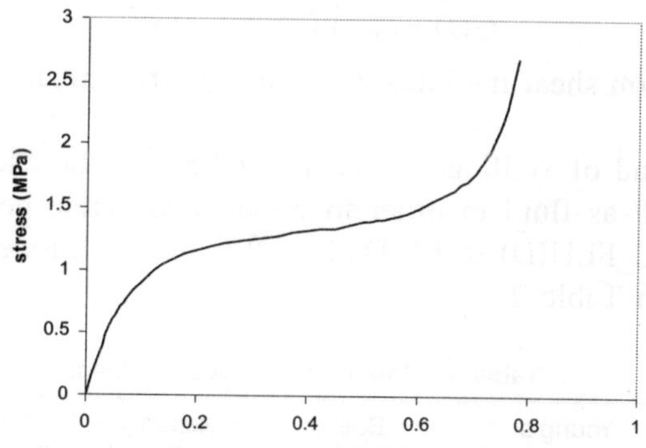

Fig. 1. Stress-Strain relationship for 44 kg.m^{-3} density of expanded polystyrene under quasi-static loading

Material Model 63 (*MAT_CRUSHABLE_FOAM) available in LS-DYNA™ was used for foam material. The model transforms the stresses into the principal stress space where the yield function is defined. If the principal stresses exceed the yield stress they are scaled back to the yield surface and transformed back to the original stress space. The yield surface and its evolution are defined by the equations:

Yield surface description:

$$f_t = |\sigma_i| - Y = 0$$

Hardening formulation:

$$Y = Y^0 + H(e_v)$$
$$Y_t = Y_t^0$$

where,

Y is the yield stress, Y° is initial compressive yield stress, Y_t is tensile cut off stress and σ_i is the principal stresses and H is strain hardening.

Here e_v is the volumetric strain defined by natural logarithm of relative volume. An associative flow rule is assumed and the plastic strains are derived from

Flow of plastic strains:

$$\dot{\varepsilon}_{ij}^p = \dot{\lambda} \frac{\partial F}{\partial \sigma_{ij}}$$

The flow surface is same as the yield surface.

In LS-DYNA™ the data for stress versus volumetric strain for the liner are given in tabular form and it fits the above equations to this curve. The stress-strain curves, for three different densities of liner, for uniaxial loading are taken from Yettram [6].

The various layers of the head like other biological materials do not follow the constitutive relations for common engineering materials and are generally non-homogeneous, anisotropic, non-linear and viscoelastic. However, for modeling purposes here, they are assumed as homogeneous, isotropic and linearly elastic, except for the brain, which is assumed as viscoelastic in nature. The shear characteristics of viscoelastic behaviour of the brain are expressed by

$$G(t) = G_\infty + \left(G_0 - G_\infty\right)e^{-\beta t}$$

Here, G_∞ is the long-term shear modulus, G_0 is the short-term shear modulus and β is the decay factor.

In the FE model of head of Willinger *et al.* [5], CSF was modeled as solid. Here, it has been modified and modeled CSF as fluid in order to obtain reasonable head impact response. Material Model 1 (*MAT_ELASTIC_FLUID) of LS-DYNA™ is used to model fluid in impact analysis and the properties are shown in Table 2.

Table 2. Material properties of Head.

Part	Density (kg.m^{-3})	Young's Modulus (N.m^{-2})	Bulk Modulus (N.m^{-2})	Poisson's ratio	G0 (N.m^{-2})	GI (N.m^{-2})	β (s^{-1})
Skin	1200	16.7×10^6	-	0.42	-	-	-
Cranium	1800	15.0×10^9	-	0.21	-	-	-
Face	3000	5.0×10^9	-	0.21	-	-	-
Tentorium	1140	31.5×10^6	-	0.23	-	-	-
Falx	1140	31.5×10^6	-	0.23	-	-	-
CSF	1040	2.19×10^7	2.19×10^7	0.49	-	-	-
Brain	2000	-	1.125×10^9	-	49.0×10^{-9}	16.7×10^{-9}	145

3. FINITE ELEMENT MODEL OF HELMET-HEAD IMPACT

The FE model of head provided by Willinger *et al.* [5] was combined with a commercially available helmet to determine the reduction in force experienced by the head and intracranial pressures during impact due to the presence of helmet. The foam or liner in helmet was of variable thickness with 28mm on front portion, 32 mm on side and 40 mm on top. Outer ABS shell was of 3 mm thickness. For Finite element simulations, motorcycle helmet model required input data like geometry, initial and boundary conditions, interface conditions and material properties. Figure 2 shows FE models of the head with a helmet undergoing frontal impact and side impact with a flat rigid surface.

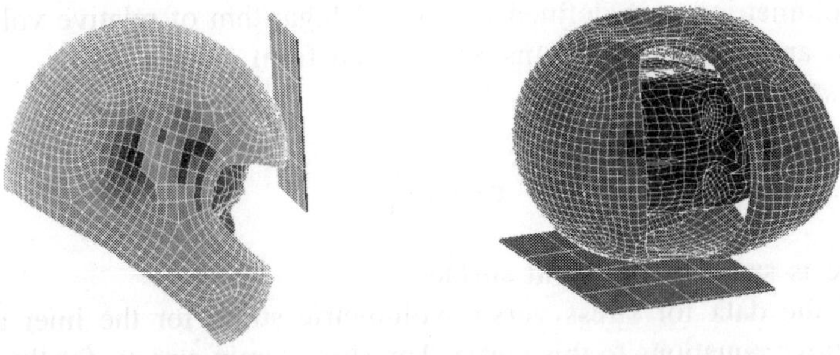

Fig. 2. Finite Element model of helmet-head in front and side impacts.

Fig. 3 shows the contact forces between helmet and head for different velocities during the frontal impact. The maximum contact force for helmet without ventilation at 10 m.s^{-1} velocity is 13000N; at 8m.s^{-1} velocity is 8000N and at 6 m.s^{-1} velocity it is 5700N. The maximum force was reached at 5ms in all these cases. The foam in the impact region was completely bottomed out at 10 m.s^{-1} velocity. At 8 m.s^{-1} velocity, compression of foam was 80%; whereas with 6 m.s^{-1} velocity it was 60% approx. Figure 4 shows the contact forces experienced by the head at impact velocity of 6m.s^{-1} in helmet with single groove 7 mm deep, helmet with three grooves 7 mm deep and helmet with three grooves going through the thickness of the foam.

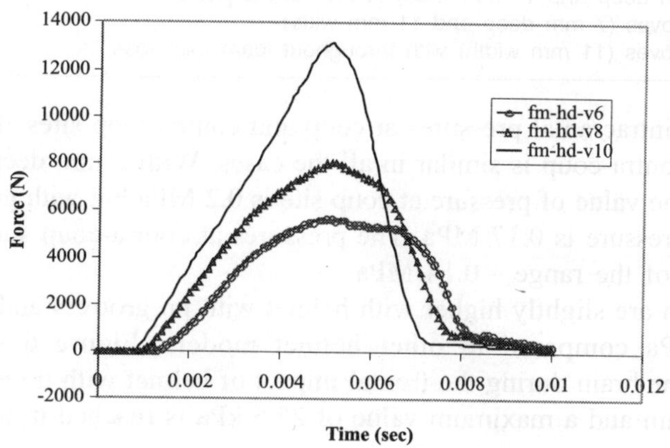

Fig. 3. Forces between helmet (no grooves) and head in front impact at 6, 8 and 10 m.s^{-1} velocity

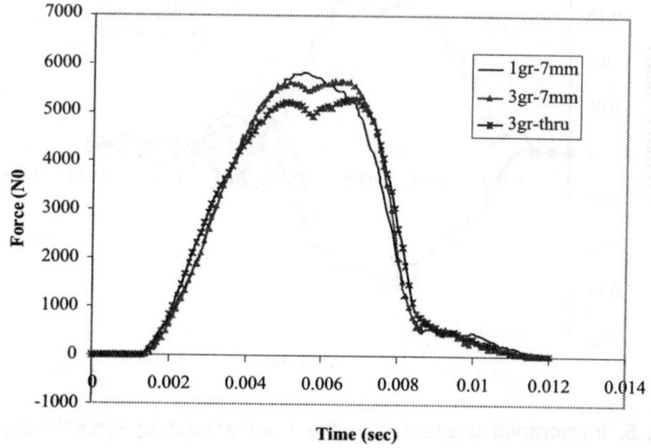

Fig. 4. Forces between helmet foam and head in front impact at 6 m.s^{-1} velocity for different groove models

The contact force does not change appreciably when grooves do not run through the full thickness of the foam. In the helmet with three grooves of throughout the foam thickness the maximum contact force was lower at 5190N. This reduction is similar to what one observes with lower density foam except that no bottoming out of the foam occurs in the present case. The helmet impact duration is almost same in all these cases and is around 6.5 ms. Similar results were obtained when the deformable head was replaced by a rigid head and HIC values, as shown in Table 3, were calculated. The HIC value for the helmet with three parallel grooves throughout the foam thickness was the lowest at 848.

Table 3. Head Injury Criterion (HIC) with different ventilation models

Helmet ventilation model	HIC
No grooves	977
One groove (7 mm deep and 11 mm wide) in the central plane	999.2
Three parallel grooves (7 mm deep and 11 mm wide)	975.5
Three parallel grooves (11 mm width) with throughout foam thickness	848

Figure 5 compares intracranial pressures at coup and contra-coup sites. The trend of intracranial pressures at coup and contra-coup is similar in all the cases. With 7 mm deep single groove or three grooves in the helmet, the value of pressure at coup site is 0.2 MPa but with grooves going throughout thickness of foam the pressure is 0.17 MPa. The pressures at contra-coup site are almost same in all the cases and they are of the range − 0.14 MPa.

Stresses in the brain are slightly higher with helmet with no grooves and Von Mises stresses are approximately 30.6 kPa compared to other helmet models. Figure 6 shows the contours of Von Mises stresses in the brain during the frontal impact of helmet with three grooves going through full thickness of the foam and a maximum value of 27.5 kPa is reached in the brain stem at 7.5 ms.

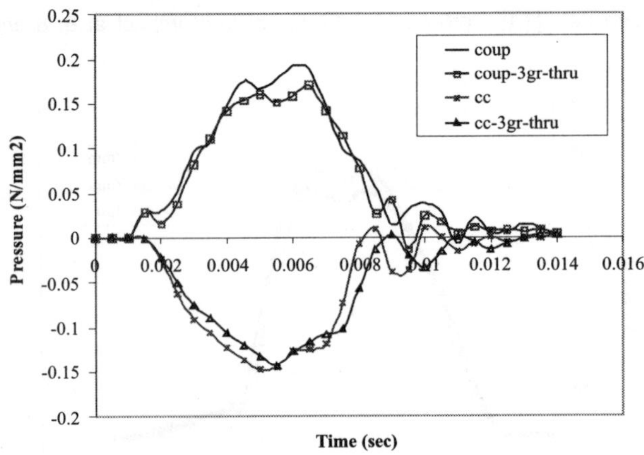

Fig. 5. Intracranial pressures in the front impact at 6m.s^{-1} velocity

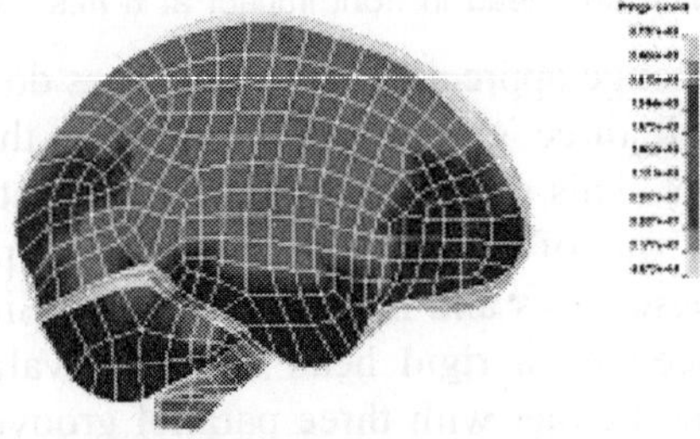

Fig. 6. Von Mises stresses in brain with three groover goind throughout the foam thickness in helmet at time = 7.5 m.s

Figure 7 shows the contact forces between head and foam for different helmet models for side impact at 6 m.s^{-1}. The behavior was similar to one seen in front impact. The contact force experienced by the head for a helmet with grooves through the thickness was 7600N as compared to 8000N experienced with other designs. The impact duration in side impact was same for all the cases and is 7 m.s.

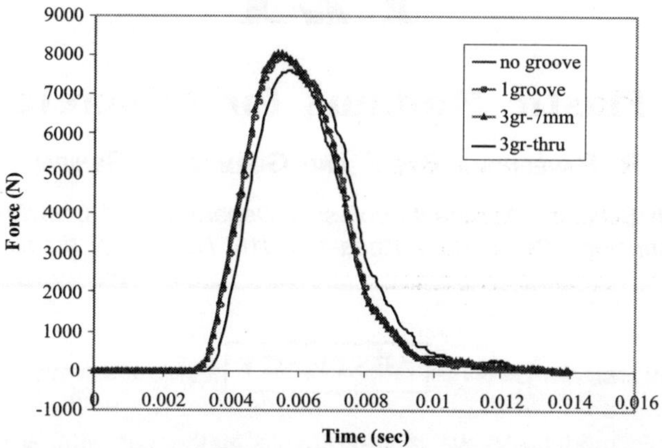

Fig. 7. Forces between helmet foam and head in side impact at 6 m.s^{-1} velocity for different groove models.

CONCLUSION

Dynamic analysis of helmet-head impact had been carried out for various ventilated helmet models and compared with helmet of no ventilation. The helmet with three parallel grooves running through the thickness of foam gave lower contact forces, HIC value and stresses in the brain when compared to other ventilation designs. The frontal and side impact results indicate that provision of grooves in the foam is not detrimental to the dynamic performance of the helmet.

REFERENCES

1. Gilchrist and Mills, 1994, Impact deformation of ABS and GRP motorcycle helmet shells, *Plastics, Rubber and Composites Processing and Applications,* 21, 141-150.
2. Hallquist J.O.,1994, LS-DYNA3D Theoretical Manual, LSTC Report 1018.
3. Horgan T.J. and Gilchrist M.D, 2004, Influence of FE model variability in predicting brain motion and intracranial pressure changes in head impact simulations, *Int. J. of Crashworthiness*, 9, 401-418.
4. Ruan et al, 1994, Dynamic response of the human head to impact by three-dimensional finite element analysis, *Journal of Biomechanics,* Transactions of ASME, 116, 44-50.
5. Willinger et al, 2000, Finite element modeling of skull fractures caused by direct impact, *Int. Journal of Crashworthiness,* 5, 249-258.
6. Yettram and et al,1994, Materials for motorcycle crash helmets—A finite element parametric study, *Plastics, Rubber and Composites Processing and Applications,* 22, 215-221.
7. Zhang Jun et al., 1998, Constitutive modeling of polymeric foam material subjected to dynamic crash loading, *Int. J. Impact Engineering,* 21, 369-386.
8. Zong Z., Lee H.P., Lu C., 2006, A three-dimensional human head finite element model and power flow in a human head subjected to impact loading, *J. of Biomechanics,* 39,284-92.

121

Plastic Modulus for Concrete

R. RAVEENDRA BABU[1] AND GURMAIL S. BENIPAL[2]

[1]Research Scholar, [2]Assistant Professor Department of Civil Engineering,
Indian Institute of Technology Delhi, Hauz Khas-110 016, New Delhi email: gurmail@civil.iitd.ac.in

ABSTRACT

The concept of plastic modulus plays a central role in the constitutive modelling of classical hardening elasto-plastic behaviour of solids. The objective of the present paper is to derive analytical expressions for the plastic modulus for different choices of hardening parameter for hardening elasto-plastic concrete in terms of loading function and empirical data from its mechanical response under the applied state of uniaxial compressive stress.

Keywords: Plastic modulus, Hardening function, Hardening parameter, Effective stress, Concrete elasto-plasticity.

1. INTRODUCTION

The concept of plastic modulus plays a central role in the constitutive modelling of classical hardening elasto-plastic behaviour of solids. In the current literature [5, 1] on concrete, plastic modulus is defined as the rate of change of effective stress with effective plastic strain which is chosen as the hardening parameter. The plastic modulus is assumed to be universally valid for all stress/strain histories and is obtained by using the empirical data from the simplest test like uniaxial compressive stress test. In that context, the effective stress is defined as the equivalent uniaxial stress for any general multiaxial state of stress. In order to obtain empirical effective stress-effective plastic strain relation from the above test, the effective plastic strain is additionally assumed to be given by the axial compressive plastic strain. In view of the definition of effective plastic strain, this assumption is unjustified and at best superfluous. A survey of the recent literature reveals that the concept of effective stress is no longer in vogue and various other parameters like plastic work, volumetric plastic strain, etc., are also being employed as hardening parameters. While plastic work can be defined only in terms of the axial plastic strain, the effective plastic strain and the volumetric plastic strain require knowledge about the lateral plastic strains as well. This fact has rarely been acknowledged in the available literature. Also, realistic uniaxial stress-strain curves have not been used in general. For example, while concrete under uniaxial compressive stress exhibits peak stress at a definite strain value, the stress-strain relations like Ramberg-Osgood model have been employed which show monotonic increase in stress with axial strain.

The objective of the present paper is to derive analytical expressions for the plastic modulus for different choices of hardening parameter for hardening elasto-plastic concrete in terms of loading function and empirical data from its mechanical response under the applied state of uniaxial compressive stress. For illustrative purpose, Drucker-Prager loading function with isotropic hardening rule have been assumed. The scope of the paper is limited to proposing a realistic expression for the plastic modulus and the prediction of the material response for different load histories is excluded from its scope. The work done in the paper has been discussed and its implications for the hardening elasto-plasticity theory of concrete have been delineated.

2. PLASTIC MODULUS

The plastic modulus H_p is defined as the rate of increase of hardening function κ with the chosen hardening parameter p and is assumed to be valid for all stress/strain histories. The plastic modulus is determined from some simple test such as uniaxial compressive stress test. The relation between hardening function κ and the absolute value of the uniaxial compressive stress σ is obtained from the loading function. The stress-strain curves in this test are used for obtaining the relation between and the chosen hardening parameter p. Then, the expression for the plastic modulus will take different forms based on the chosen hardening parameter and can be expressed as

$$H_p = \frac{d\kappa}{dp} = \frac{d\kappa}{d\sigma}\frac{d\sigma}{dp} \qquad \qquad ...(1)$$

A loading function is defined in terms of the instantaneous stress components and a hardening function representing the load history. For illustrative purpose, Drucker-Prager loading function with isotropic hardening rule has been assumed. Mathematically, this loading function can be expressed as

$$f = \alpha I_1 + \sqrt{J_2} - k = 0 \qquad \qquad ...(2)$$

where, α is a material constant and κ is the hardening function monotonically increasing with plastic flow.

From the loading function, the following relation between κ and α can be formulated for the uniaxial compressive stress case

$$\kappa = \frac{1}{3}\left(\sqrt{3} - \alpha\right)\sigma \qquad \qquad ...(3)$$

On differentiating the above expression, one obtains

$$\frac{d\kappa}{d\sigma} = \frac{(\sqrt{3} - \alpha)}{3} \qquad \qquad ...(4)$$

It has been observed [5] that the peak axial stress is reached at an axial strain of ε_0 of 0.002 and at lateral strain $\overline{\varepsilon}_0$ of 0.00075. In this paper, the following well-known realistic axial stress σ-axial strain ε relation called Madrid parabola [4] for the uniaxial compressive stress test has been used:

$$\sigma = f_c' \left[\frac{2\varepsilon}{\varepsilon_0} - \left(\frac{\varepsilon}{\varepsilon_0} \right)^2 \right] \qquad \qquad ...(5)$$

$$\frac{\varepsilon}{\varepsilon_0} = 1 \pm \sqrt{1 - \frac{\sigma}{f_c'}} \qquad \qquad ...(6)$$

The above equation can be rewritten in terms of minor principal stress and strain as

$$\frac{\varepsilon_{33}}{\varepsilon_0} = -1 + \sqrt{1 + \frac{\sigma_{33}}{f_c'}} \qquad \qquad ...(7)$$

At the elastic limit, $\sigma_{33} = -0.3 f_c'$, the equation (7) gives $\varepsilon_{33} = -0.163 \varepsilon_0$. Thus, the Young's

modulus of elasticity is determined as $E_0 = \dfrac{\sigma_{33}}{\varepsilon_{33}} = 1.8405 \dfrac{f_c'}{\varepsilon_0}$.

In elastic range,

$$\frac{\varepsilon_{33}^e}{\varepsilon_0} = 0.543 \frac{\sigma_{33}}{f_c'} \qquad \qquad ...(8)$$

In the elasto-plastic range

$$\frac{\varepsilon_{33}^p}{\varepsilon_0} = \frac{\varepsilon_{33}}{\varepsilon_0} - \frac{\varepsilon_{33}^e}{\varepsilon_0} \qquad \qquad ...(9)$$

Thus,

$$\frac{\varepsilon_{33}^p}{\varepsilon_0} = -\left(1 + 0.543 \frac{\sigma_{33}}{f_c'} \right) + \sqrt{1 + \frac{\sigma_{33}}{f_c'}} \qquad \qquad ...(10)$$

$$d\varepsilon_{33}^p = \frac{0.001086}{f_c'} \left[-1 + \frac{0.9208103}{\sqrt{1 + \dfrac{\sigma_{33}}{f_c'}}} \right] d\sigma_{33} \qquad \qquad ...(11)$$

Assuming the parabolic axial stress-lateral strain curve, the following relations are obtained.

$$\frac{\varepsilon_{11}}{\varepsilon_0} = \frac{\varepsilon_{22}}{\varepsilon_0} = 0.68612 \left[1 - \sqrt{1 + 0.79072 \frac{\sigma_{33}}{f_c'}} \right] \qquad \qquad ...(12)$$

Using the above procedure, one obtains

$$d\varepsilon_{11}^p = d\varepsilon_{22}^p = \frac{0.00022}{f_c'} \left[\frac{0.9366865}{\sqrt{1 + 0.79072 \dfrac{\sigma_{33}}{f_c'}}} \right] d\sigma_{33} \qquad \qquad ...(13)$$

Using the well-known definition, the expression for effective plastic strain increment $d\varepsilon_p$ for the uniaxial compressive stress test can be expressed as

$$d\varepsilon_p = \sqrt{\left(d\varepsilon_{11}^p\right)^2 + \left(d\varepsilon_{22}^p\right)^2 + \left(d\varepsilon_{33}^p\right)^2} \qquad ...(14)$$

Using the expressions (11) and (14), the following useful expression is derived in terms of the absolute value of the axial compressive stress σ by replacing σ_{33} by σ.

$$\frac{d\sigma}{dp} = \frac{d\sigma_{33}}{d\varepsilon_p} = \frac{f_c'}{\sqrt{1.2737\times10^{-6} + \dfrac{8.27827\times10^{-8}}{1-0.79072\dfrac{\sigma}{f_c'}} + \dfrac{0.999999\times10^{-6}}{1-\dfrac{\sigma}{f_c'}} - \dfrac{17.675\times10^{-8}}{\sqrt{1-0.79072\dfrac{\sigma}{f_c'}}} - \dfrac{2.172\times10^{-6}}{\sqrt{1-\dfrac{\sigma}{f_c'}}}}}$$

$$...(15)$$

The expression for incremental plastic work dW_p for the uniaxial compressive stress turns out to be

$$dW_p = \sigma_{33}d\varepsilon_{33} \qquad ...(16)$$

$$\frac{d\sigma}{dp} = \frac{-d\sigma_{33}}{dW_p} = \frac{1}{\sigma\varepsilon_0\left(\dfrac{-0.543}{f_c'} + \dfrac{1}{2f_c'}\dfrac{1}{\sqrt{1+\dfrac{\sigma}{f_c'}}}\right)} \qquad ...(17)$$

Substituting Equations (4) and (15) in Equation (1), the expression for plastic modulus for the case of plastic strain as hardening parameter can be obtained.

$$H_p = \frac{(\sqrt{3}-\alpha)f_c'}{3\sqrt{1.2737\times10^{-6} + \dfrac{8.27827\times10^{-8}}{1-0.79072\dfrac{\sigma}{f_c'}} + \dfrac{0.999999\times10^{-6}}{1-\dfrac{\sigma}{f_c'}} - \dfrac{17.675\times10^{-8}}{\sqrt{1-0.79072\dfrac{\sigma}{f_c'}}} - \dfrac{2.172\times10^{-6}}{\sqrt{1-\dfrac{\sigma}{f_c'}}}}} \qquad ...(18)$$

Substituting Equations (4) and (17) in Equation (1), the expression for plastic modulus for the case of work as hardening parameter can be obtained.

$$H_p = \frac{(\alpha-\sqrt{3})}{3\sigma_{33}\varepsilon_0\left(\dfrac{-0.543}{f_c'} + \dfrac{1}{2f_c'}\dfrac{1}{\sqrt{1+\dfrac{\sigma_{33}}{f_c'}}}\right)} \qquad ..(19)$$

3. DISCUSSION

A realistic uniaxial stress-strain curve is an essential thing for getting a proper plastic modulus and hence a realistic material model. But, realistic uniaxial stress-strain curves have not been used in general. For example, while concrete under uniaxial compressive stress exhibits peak stress at a definite strain value, the stress-strain relations like Exponential hardening model, Ramberg-Osgood model have been employed which show monotonic increase in stress with axial strain. These idealized stress-stain models are far from the actual representation of concrete behaviour as shown in Fig. 1. Therefore, the constitutive model should use an actual stress-stain curve instead of an idealized one for better prediction of its behaviour. Also, in the current literature, it is assumed without any justification that the axial strain in the uniaxial compressive stress test measures the effective plastic strain.

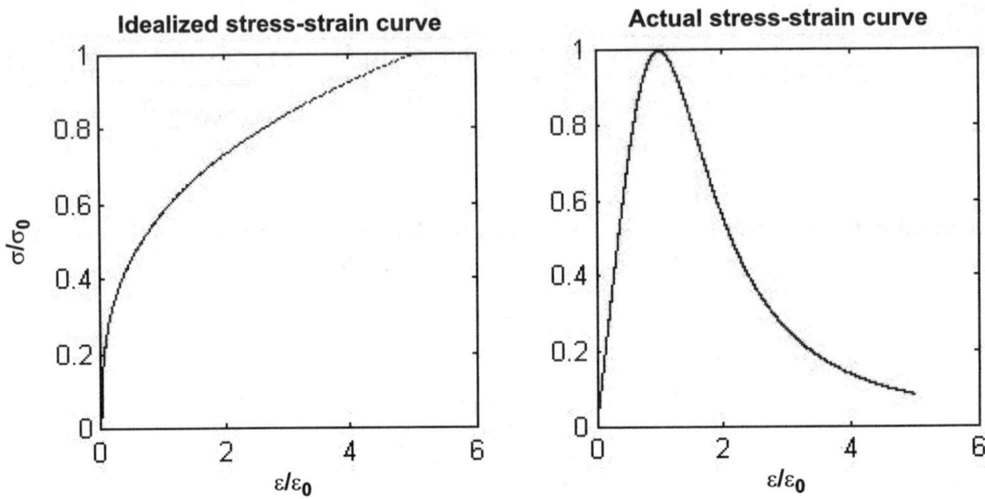

Fig. 1. Idealized (Ramberg-Osgood) and actual stress-strain curves.

Here, a simple relation in the form of Madrid parabola proposed for axial stress-axial strain relation (Figure 1) has been assumed to be valid for lateral strains as well. Using these expressions, the equations depicting the variation of the axial as well as lateral plastic strains with axial stress have been obtained and the incremental relations between the chosen hardening parameter and the axial stress have been derived. Defining the plastic modulus as the rate of change of hardening function with hardening parameter, analytical expressions for the same have been derived for different choices of hardening parameters like effective plastic strain and plastic work. Here, plastic modulus is obtained as the product of the rate of change of hardening function with compressive stress and of the compressive stress with the hardening parameter. This latter rate of change of stress with hardening parameter, and so the plastic modulus, vanishes as the peak stress is reached. This observation is confirmed in Figure 2 showing the variation of hardening function with the chosen hardening parameters as well as the variation of plastic modulus with the hardening function. These κ-p and H_p-κ relations are valid for all load histories. The load history determines the value of κ which in turn determines the σ. Also, the absolute value of the axial compressive stress σ can be interpreted to play the same role as the effective stress in the existing literature.

As the plastic modulus appears in the denominator of the expression for the plastic compliance tensor coefficients, infinitely large plastic strain increments are introduced by the applied stress increments at that stage of load history. Infinitely large compliance coefficients imply vanishing elasticity tensor coefficients. This implies that the tangent elasto-plasticity tensor may become singular or may have some vanishing eigen values. In this paper, such a state of affairs is interpreted as

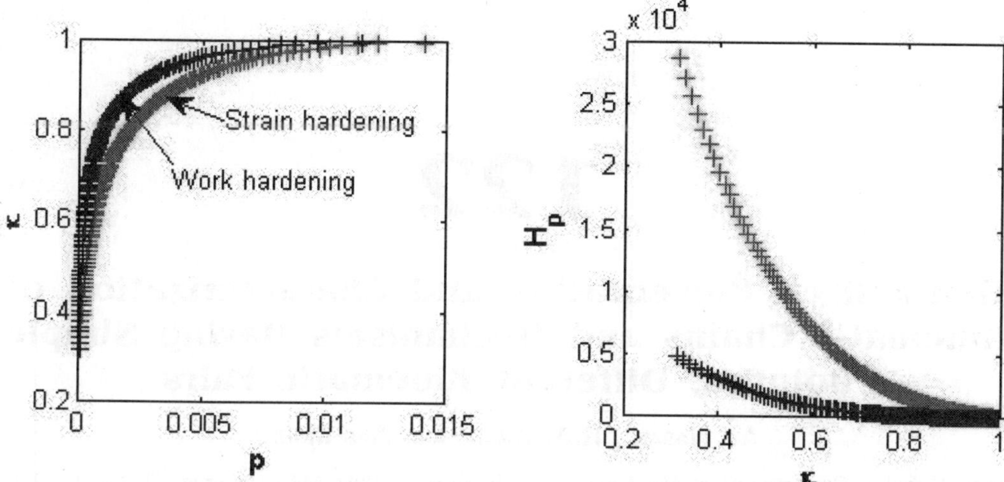

Fig. 2. Plastic modulus for strain and work hardening cases.

constituting elasto-plastic failure. The concept of elasto-plastic failure is motivated by Coon and Evans' [6] concept of hypoelastic failure and has recently been proposed by Babu and Benipal [2]. In this paper, Drucker-Prager loading function has been used for illustrative purpose. The same approach has been followed for constitutive modelling of concrete elasto-plasticity using appropriate loading function [3].

CONCLUSION

The plastic modulus plays an important role in the elasto-plasticity theory. In this paper, the determination of plastic modulus for concrete is based on realistic stress-strain curves and without the use of superfluous assumptions. The vanishing of the plastic modulus has been interpreted as elasto-plastic failure of concrete.

REFERENCES

1. R.R. Babu, Gurmail S. Benipal., and A.K. Singh, 2005, Constitutive modelling of concrete: an overview, Asian. J. of Civil Engrg., 6(4), 211-246.
2. R.R. Babu and Gurmail S. Benipal, 2005, Elasto-plastic failure of concrete, In Proc. Indian Society of Theoretical and Applied Mechanics, 66-74.
3. R.R. Babu, Gurmail S. Benipal and A.K. Singh, 2006, Plasticity-based constitutive model for concrete in stress space, Latin American J. of Solids and Structures. (Submitted)
4. M. Cervera and E. Hinton, 1986, Non-linear analysis of reinforced concrete plates and shells using a three-dimensional model, In E. Hinton and R. Owen, editors, Comp. Modeling of RC struct., 327-370, Pineridge Press, UK.
5. W.F. Chen, 1994, Constitutive Equations for Engineering Materials, Vol. 2: Plasticity and Modelling, Elsevier Publications.
6. M.D. Coon and R.J. Evans, 1972, Incremental Constitutive laws and their associated failure criteria with application of plain concrete, Int. J. of solids structures, 8, 1169-1183.

122

Generalized Presentation and Characterization of Kinematic Chains and Mechanisms Having Simple Jointed, Different Kinematic Pairs

ALI HASAN[1], R.A. KHAN[2] AND AAS MOHD[3]

[1] Research Scholar, Department of Mechanical Engineering, Jamia Millia Islamia, Delhi-110 025, India.
email: alihasan786@rediffmail.com
[2] Prof. and Head, Department of Mechanical Engineering, Jamia Millia Islamia, Delhi-110 025, India.
email: rasheed_jmi@hotmail.com
[3] Lecturer, Department of Mechanical Engineering, Jamia Millia Islamia, Delhi-110 025, India.
email: am200647@rediffmail.com

ABSTRACT

This paper presents a systematic method to determine a set of identification number for a given mechanisms, having simple jointed different kinematic pairs. A physical connectivity matrix [PCM] has been defined and developed leading to a mathematical equation (connectivity function) and an identification set which is an invariant of a given mechanism. The method is based on theoretic approach, sum of absolute polynomial coefficient values [PCMΣ] and maximum value of polynomial coefficient [PCMmax] of [PCM] matrix. Each kinematic pair has been assigned a specific numeric value keeping in view of the type of kinematic pair.

Keywords: [PCM] matrix, Kinematic pair, Simple joints, Polynomial coefficient.

1. INTRODUCTION

Over the past several years much work has been reported in the literature on the structural analysis and synthesis of mechanisms [1-17]. Motives behind these studies range from the desire for an orderly classification system, to studies of mechanism degree of freedom to the hope of identifying the mechanisms. Methods for the recognition and identification of a given mechanism's kinematic structure can be divided into two main categories, graphical methods based on the visual inspection of various forms of simplified systematic diagrams and numerical methods [13-15, 17] many of which are based on the theory of graphs. However, even with the current numerical methods no efficient generalized method is known which will characterize the given mechanisms by inspection.

In this paper the authors present a generalized methods to characterize any given mechanisms having simple jointed various kinematic pairs.

2. REPRESENTATION OF KINEMATIC PAIRS

The two links are either connected by lower pair or higher pair. To distinguish them in physical connectivity matrix [PCM], the lower pairs and higher pairs are represented by numeric '1' and '2' respectively. Lower pairs and higher pairs are further distinguished on the basis of the respective pairs with the number followed by decimal. These are explained in [PCM] matrix.

3. PHYSICAL CONNECTIVITY MATRIX [PCM]

One possible symbolism for the connectivity of a kinematic mechanism is the link-link form of kinematic pair. Once the links of the mechanism have been numbered from 1 to n, [PCM] is defined as a square matrix of order n, each row and each column representing the link with the corresponding number. The elements of [PCM] are then entered either with zero or the type of kinematic pair [2], depending on the absence or presence of a direct kinematic connection (pair) between the links corresponding to that row and column.

Let us define a [PCM] matrix of the kinematic chain and mechanism as

[PCM] = $\{P_{ij}\}$,

Where $P_{ii} = 0$, if i^{th} link is not fixed and $P_{ii} = 1$, if i^{th} link is   fixed,

P_{ij} = type of joint or type of kinematic pair between i^{th} link and j^{th} link, when i ≠ j.

Let R = 1.1, P = 1.2, C = 1.3, SL = 1.4, F = 1.5, G = 1.6, HP = 21 and HL = 2.2.

4. CHARACTERISTIC POLYNOMIALS

A mat lab software has been used to determine the characteristic polynomials coefficients of a given kinematic mechanisms. A new identification scheme for mechanisms can be based on these polynomials by listing the coefficient of the corresponding polynomial in order. For example, a six bar Stephenson's chain or mechanism having 6 revolute joint and 1 prismatic pair would be identified as the 1/0/-8.7/0/14.0118/-7.0277/0 chain or mechanism. Further this kinematic chain/mechanism may be represented by a mathematical equation as $x^6 - 8.7x^4 + 14.0118x^2 - 7.0277x$. This would provide a highly selective identification for each possible kinematic chain and mechanism having same or different kinematic pairs.

5. ILLUSTRATIVE EXAMPLES

Example 1: Consider a 4 bar slider crank mechanism, n = 4, n_2 = 4, j = 4 (6-R and 1P), F = 1 and [PCM] is C1 (Fig. 1).

Example 2: 3 cam and follower mechanism, n = 3, n_2 = 3, j = 2 (2-R), h = 1(1-HL), F = 1 and [PCM] is C2 (Fig. 2)

Example 3: 6 bar slider crank mechanism, n = 6, n_2 = 4, n_3 = 2, j = 7 (6-R and 1-P), F = 1 and[PCM] is C3 (Fig. 3).

Example 4: 6 bar slider crank mechanism, n = 6, n_2 = 4, n_3 = 2, j = 7(6-R and 1P), F = 1 and [PCM] is C4 (Fig. 4).

Example 5: 10 bar slider crank mechanism, n = 10, n_2 = 7, n_3 = 2, n_4 = 1, j = 12(10-R and 2-P), and [PCM] is C5, F = 3 (Fig. 5).

Example 6: 7 bar slider crank mechanism, $n = 7$, $n_3 = 2$, $n_2 = 5$, $j = 7$(6-R and 1-P), $h = 1$(1-HP), $F = 3$ and [PCM] is C7 (Fig. 7).

$$C1 = \begin{pmatrix} 0 & R & 0 & P \\ R & 0 & R & 0 \\ 0 & R & 0 & R \\ P & 0 & R & 0 \end{pmatrix} = \begin{pmatrix} 0 & 1.1 & 0 & 1.2 \\ 1.1 & 0 & 1.1 & 0 \\ 0 & 1.1 & 0 & 1.1 \\ 1.2 & 0 & 1.1 & 0 \end{pmatrix}$$

$$C2 = \begin{pmatrix} 0 & 1.1 & 1.2 \\ 1.1 & 0 & 2.2 \\ 1.2 & 2.2 & 0 \end{pmatrix} \quad C3 = \begin{pmatrix} 0 & 0 & 1.1 & 1.1 & 0 & 1.2 \\ 0 & 0 & 1.1 & 1.1 & 1.1 & 0 \\ 1.1 & 1.1 & 0 & 0 & 0 & 0 \\ 1.1 & 1.1 & 0 & 0 & 0 & 0 \\ 0 & 1.1 & 0 & 0 & 0 & 1.1 \\ 1.2 & 0 & 0 & 0 & 1.1 & 0 \end{pmatrix}$$

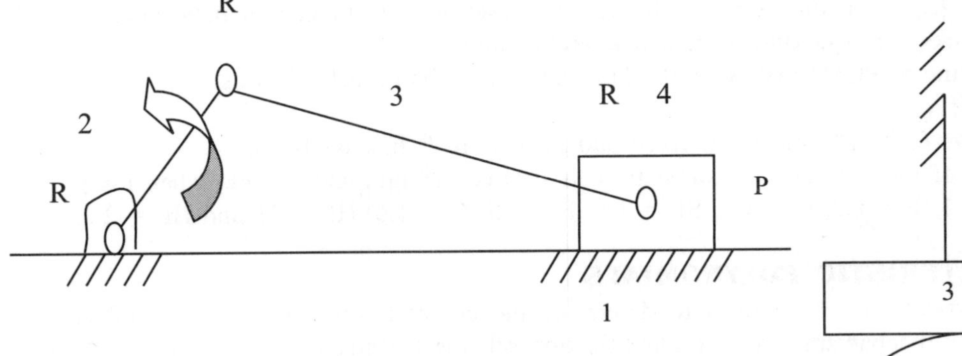

Fig. 1. 4 Bar slider crank mechanism.

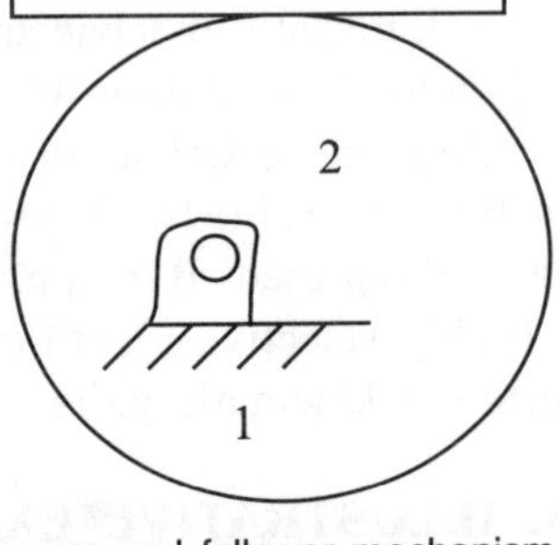

Fig. 2. 3 Bar cam and follower mechanism.

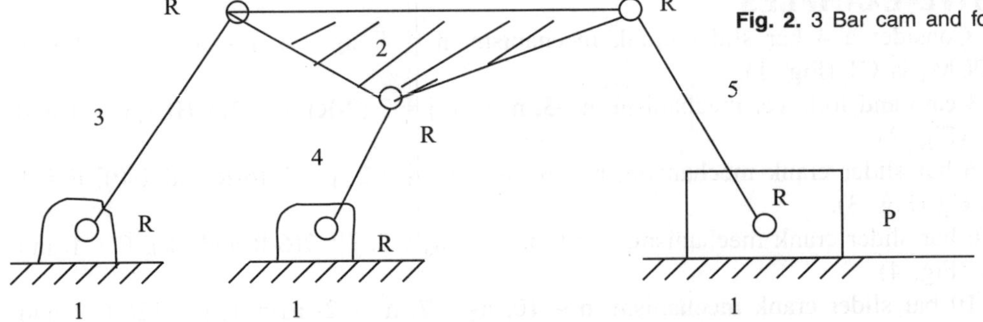

Fig. 3. 6 Bar slider crank mechanism.

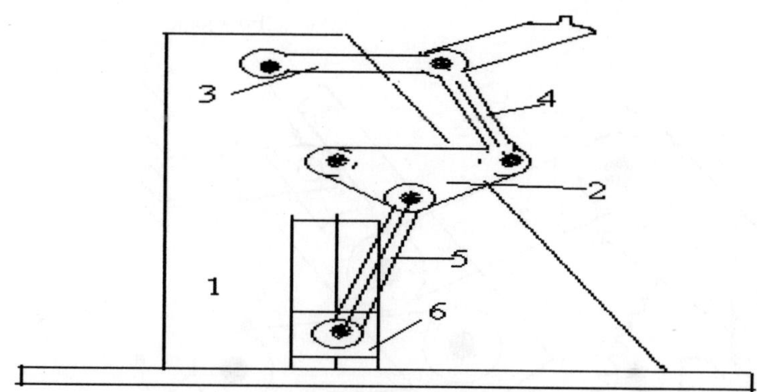

Fig. 4. 6 Bar slider crank mechanism.

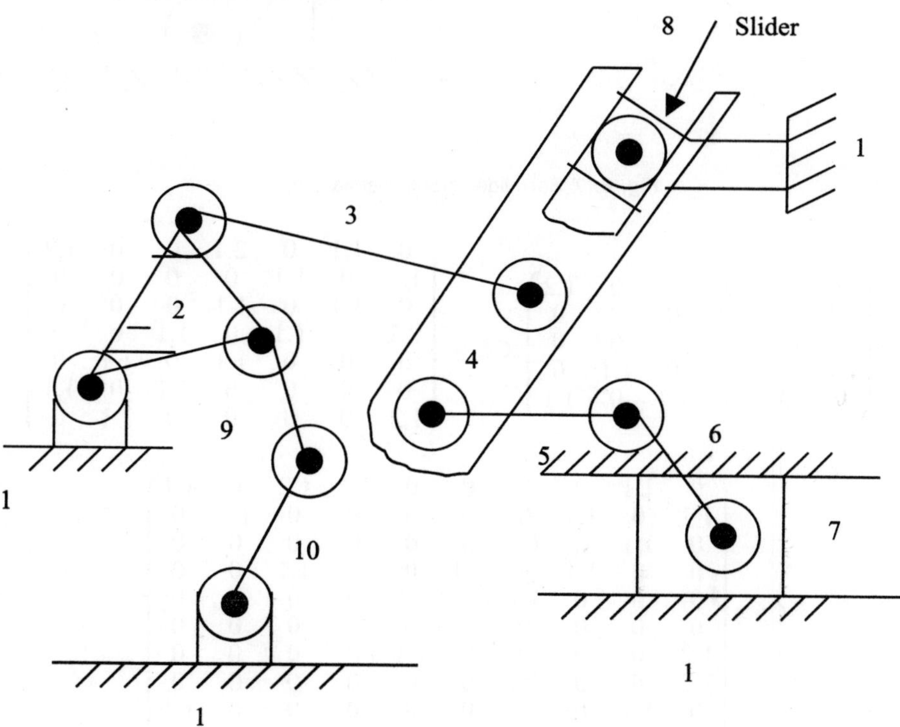

Fig. 5. 10 Bar slider crank mechanism.

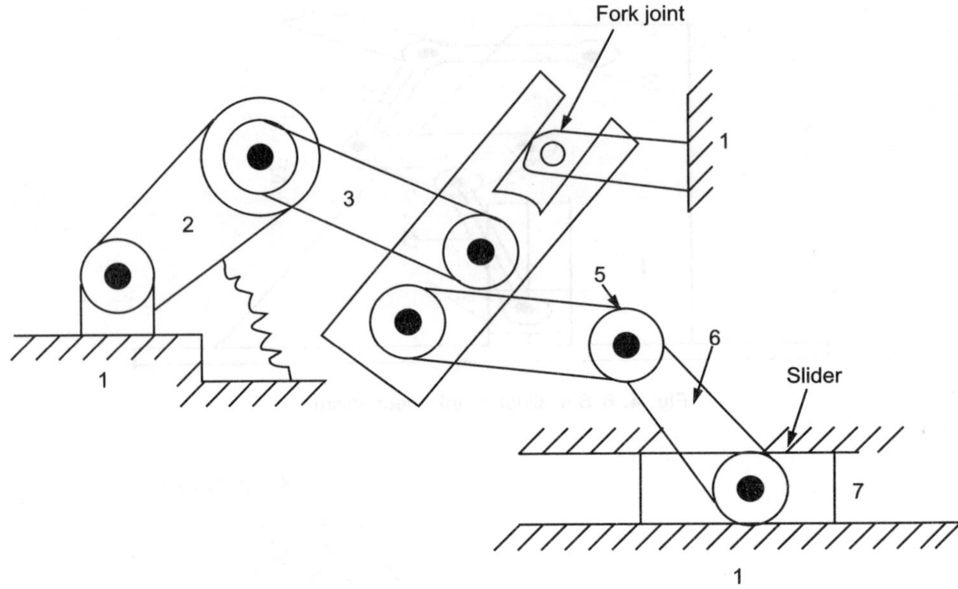

Fig. 6. 7 Bar slider crank mechanism.

$$
C4\begin{pmatrix}
0 & 1.1 & 0 & 1.1 & 0 & 1.2 \\
1.1 & 0 & 1.1 & 0 & 0 & 0 \\
0 & 1.1 & 0 & 1.1 & 0 & 0 \\
1.1 & 0 & 1.1 & 0 & 1.1 & 0 \\
0 & 0 & 0 & 1.1 & 0 & 1.1
\end{pmatrix}
\quad
C6 = \begin{pmatrix}
0 & 1.1 & 0 & 2.1 & 0 & 0 & 1.2 \\
1.1 & 0 & 1.1 & 0 & 0 & 0 & 0 \\
0 & 1.1 & 0 & 1.1 & 0 & 0 & 0 \\
2.1 & 0 & 1.1 & 0 & 1.1 & 0 & 0 \\
0 & 0 & 0 & 1.1 & 0 & 1.1 & 0 \\
0 & 0 & 0 & 0 & 1.1 & 0 & 1.1 \\
1.2 & 0 & 0 & 0 & 0 & 1.1 & 0
\end{pmatrix}
$$

$$
C5 = \begin{pmatrix}
0 & 1.1 & 0 & 0 & 0 & 0 & 1.2 & 1.2 & 0 & 1.1 \\
1.1 & 0 & 1.1 & 0 & 0 & 0 & 0 & 0 & 1.1 & 0 \\
0 & 1.1 & 0 & 1.1 & 0 & 0 & 0 & 0 & 0 & 0 \\
0 & 0 & 1.1 & 0 & 1.1 & 0 & 0 & 1.1 & 0 & 0 \\
0 & 0 & 0 & 1.1 & 0 & 1.1 & 0 & 0 & 0 & 0 \\
0 & 0 & 0 & 0 & 1.1 & 0 & 1.1 & 0 & 0 & 0 \\
1.2 & 0 & 0 & 0 & 0 & 1.1 & 0 & 0 & 0 & 0 \\
1.2 & 0 & 0 & 1.1 & 0 & 0 & 0 & 0 & 0 & 0 \\
0 & 1.1 & 0 & 0 & 0 & 0 & 0 & 0 & 0 & 1.1 \\
1.1 & 0 & 0 & 0 & 0 & 0 & 0 & 0 & 1.1 & 0
\end{pmatrix}
$$

6. RESULT AND DISCUSSION

The characteristic polynomials for Figs. 1, 2, 3, 4, 5 and 6 are $x^4 - 5.07x^2 + 0.0121$, $x^3 - 7.49x - 5.808$, $x^6 - 8.7x^4 + 14.0118x^2 - 7.0277x$, $x^6 - 8.7x^4 + 10.817x^2 - 2.1083$, $x^{10} - 14.98x^8 + 69.7807 x^6 - 3.5138x^5 - 120.2758x^4 + 4.2517x^3 + 74.0867x^2 - 12.3471$ and $x^7 - 13.11x^5 + 32.0287 x^3 - 6.7082x^2 - 12.7962x + 3.8652$, respectively.

We can characterize mechanisms given in Figs.1, 2, 3, 4, 5 and 6 as:
1./0/-5.07/0/0.0121, 1/0/-7.49/-5.8080, 1/0/-8.7/0/14.0118/-7.0277/0,1/0/-8.7/0/10.8174/0/-2.1083, 1/0/-14.98/0/69.7807/-3.5138/-120.2758/ 4.2517/74.0867/0/-12.3471, and 1/0/-13.11/0/32.0287/-6.7082/ -12.7962/ 3.8652, respectively.

On the basis of result obtained, we can conclude that many of the coefficients can be given physical significance and may be found directly by inspection itself. According to the results, the leading coefficient is always unity [2]. Also because diagonal elements of the [PCM] must be zero, i.e., no kinematic pair can connect a link to itself, the second coefficient must be zero. The third coefficient is always the negative. The fourth coefficient is always zero for the mechanisms. The fifth coefficient is always positive. The sixth and seventh coefficients are always negative and the eighth coefficient will be positive.

CONCLUSION

The authors do not claim that this paper presents an exhaustive study of this new method of characterization and structural analysis of kinematic chains and mechanisms having simple jointed, different kinematic pairs. There is much work which is yet to be done before a final, "best" form is chosen. However, it is hoped that this paper presents a new concept on which a new characterization system for mechanisms can be based. Such a new characterization system would be extremely selective and would minimize, if not completely eliminate the possibility of duplicate characterization for structurally different mechanisms.

NOTATION

R: revolute lower pairs., P: prismatic lower pairs., C: cylinder lower pairs., SL: screw lower pairs.

F: planer lower pairs., G: spheric lower pairs., HP: point contact higher pairs., HL: line contact higher pairs.

[PCM]:   physical connectivity matrices.

[PCMΣ]:   sum of absolute polynomial coefficient values of [PCM].

[PCMmax]: maximum value of polynomial coefficient of [PCM].

REFERENCES

1. Preben W. Jensen, Classical and Modern Mechanism for Engineers and Inventors, *Marcel Dekker*, 1998, Inc. New York.
2. Uicher J.J. and Raicu A.,1985, A method for identification and recognition of equivalence of kinematic chains, *Mech. Mach. Theory*, 10, 375-383.
3. Rao. A.C., 1988, Kinematic chains, Isomorphism, Inversions and Type of Freedom using the concept of hamming distance, *Indian J. of Tech.*, 26, 105-109.
4. Mruthyunjaya T.S. and Raghavan M.R., 1984, Computer Aided Analysis of the structural synthesis of kinematic chains, *Mech. Mach. Theory*, 19, (3) 357-368.
5. Nageswara Rao C. and A.C. Rao, 1996, Selection of best frame, input and output links for function generators modeled as probabilistic systems, *Mech. Mach. Theory*, 31, 973-983.
6. Aas Mohammad and Agarwal V.P., 1999, Identification and Isomorphism of kinematic chains and mechanisms, proceedings of the 11th *ISME conference*, IIT Delhi, New Delhi.
7. A.G. Ambekar and Agrawal V.P.,1987, Identification of kinematic Generator using Min. Codes, *Mech. Mach. Theory*, 22, (5) 463-471.
8. Ambekar AG. And Agarwal V.P., 1985, Identification and classification of kinematic chains using identification code, The Fourth International Symposium on Linkage and Computer aided design methods,, Bucharest, Romania, 1, 3 545-552.

9. Agarwal V.P., Yadav J.N and Pratap C.R.,1996, Mechanism of Kinematic chain and the degree of structural similarity based on the concept of link path code, *Mech. Mach. Theory*, 31(7) 865-871.

10. Aas Mohd., 1998, Kinematic chains structural modeling and correlation with machine performance, M.Tech. Dissertation, IIT Delhi.

11. Deo. Narsingh, 2002, Graph Theory with applications to Engineering and Computer Science, PHI, Delhi, India

12. Crossley F.R.E., 1964, A contribution to Grubler's theory in the number synthesis of plane mechanisms, Trans.ASME, J.Engg.Ind, 86B (I), 1-8.

13. Dobrjanskyj L. and Freudenstein F.,1967, Some applications of graph theory to the structural synthesis of mechanisms, Trans.ASME, J.Engg.Ind, 89B (I), 153-158.

14. Buchsbaum F. and Freudenstein F., 1970, Synthesis of kinematic structure of geared kinematic chains and other mechanisms, *J.Mechanism*, 5(3), 357-392.

15. Manolescu N.I.,1973, A method based on Baranov trusses and using graph theory to find the set of planer jointed kinematic chains and mechanisms, *Mechanism and Machine Theory* ,8(I),3-22.

16. Verho A., 1973, An extension of the concept of the group, *Mechanism and Machine Theory*, 8(2)249-256.

17. Raicu A., 1974, Matrices associated with kinematic chains from 3 to 5 members, *Mechanism and Machine Theory*, (1)123-130.

123

Finite Element Simulation of Nosing of Metallic Circular Tubes

P.K. Gupta[1] AND R.N. Khapre[2]

[1]Civil Engineering Department, Indian Institute of Technology, Roorkee-333 031, UA, India.
email: spramod_3@yahoo.com
[2]Department of Civil Engineering, RKNEC, Nagpur-440 013, Maharashtra, India
email: raju_khapre@yahoo.co.in

ABSTRACT

In this paper, finite element simulation of nosing process is presented by performing axial compression of circular tube against circular die surface. A code is developed using the finite element method and employed for the attempted study. A case study is presented where circular tube is compressed between a flat die and circular profile die. Material of the deforming tube has been modeled rigid-viscoplastic. Deformed shapes at various stages of compression process are plotted and presented. Load-compression and energy-compression relationships are obtained and presented.

Keywords: Finite element method; Tube nosing; Large deformation.

1. INTRODUCTION

Metallic tubular sections are commonly used as energy absorbing devices [1,2] to absorb energy released during crashes of light fixed wing and rotary wing aircrafts. These energy absorbing devices are classified based on three types of energy absorption mechanisms namely, material deformation, material extrusion and friction. When metallic tubes are used as energy absorbing devices, the mechanism of energy absorption is plastic deformation of tubes. Generally, rectangular and circular cross-sections are used for such purposes aligning in axial as well as lateral directions [3-7]. The tubes can be flattened, made to turn inside out, made to expand, made to contract, made to change in cross-sectional shape, made to fold or spilt and curl up. Collapse mechanism of these tubular elements would be different due to the different loading directions. Circular tubes are collapsed in concertina and multiple barreling modes of collapse [4] and externally inverted by sliding it over a curved surface [5] when axially loaded. On the other hand, flattening of circular and rectangular tubes occurs when they are compressed laterally [6,7].

2. BACKGROUND

To study the plastic deformation mechanism of such tubular elements finite element method could be used effectively [3]. Commercial finite element softwares like FORGE2 [3-6] and ANSYS [7] are generally employed for such purposes. These softwares generally work on PC's and hence in some cases analysis becomes very time-consuming. In order to reduce the computational time, supercomputers could be used successfully. One such effort was made by Gupta and Mishra [8] by developing a finite element software FEMLD for analysis of large deformation problems on supercomputer PARAM 10000. Matrix Inversion Method parallel solver [9] was used in their software to obtain quick solution of system of linear equations generated during FEA.

This software was further improved by simulating flat die surfaces. Using this modified software, Khapre and Gupta [10] simulated flattening process of rectangular tube specimens. They also carried out axial compression of circular tube specimen between two flat dies [11]. In both the cases [10, 11], the obtained results were compared with the similar results obtained from commercial software FORGE2 and with the results available in literature. It was observed that the results match fairly well. The software FEMLD was further improved by modeling curved die surface and an attempt has been made to simulate problem of external inversion of circular tube specimen [12].

The objective of this paper is to simulate the process of nosing of circular tube specimen. In the nosing process circular tube gets invert on a circular die of higher radius as compared to the inversion process. In order to simulate such process, circular die surface is modeled in the code FEMLD developed earlier. Using this code, axial compression of circular tube specimen against circular die is simulated. The deformation process of this tube is studied by plotting deformed shapes at various stages of deformation. Load-compression and absorbed energy-compression relationships are also studied and presented.

3. FINITE ELEMENT FORMULATION

Finite element formulation given by Kobayashi *et al.* [13] is quite useful for metal forming analysis. In this formulation, variational method is used for the derivation of the basic equations for finite element analysis.

$$\pi = \int_v E(\dot{\varepsilon}_{ij})dV - \int_{S_F} F_i u_i dS \qquad \qquad ...(1)$$

The stiffness equation is as follows

$$\frac{\partial \pi}{\partial v_i} = \sum_j \left(\frac{\partial \pi}{\partial v_i} \right)_{(j)} = 0 \qquad \qquad ...(2)$$

which can be rewritten in the form of

$$\mathbf{K} \, \Delta \mathbf{v} = \mathbf{f} \qquad \qquad ...(3)$$

where,

$$\mathbf{K} = \frac{\partial^2 \pi}{\partial v_i v_j} = \int_v \frac{\overline{\sigma}}{\dot{\overline{\varepsilon}}} P_{ij} dV + \int_v \left(\frac{1}{\dot{\overline{\varepsilon}}} \frac{\partial \overline{\sigma}}{\partial \dot{\overline{\varepsilon}}} - \frac{\overline{\sigma}}{\dot{\overline{\varepsilon}}^2} \right) \frac{1}{\dot{\overline{\varepsilon}}} P_{ik} v_k v_m P_{mj} dV + \int_v KC_j C_i dV \qquad ...(4)$$

and,

$$\mathbf{f} = \frac{\partial \pi}{\partial v_i} = \int_v \frac{\overline{\sigma}}{\dot{\overline{\varepsilon}}} P_{ij} v_j dV + \int_v KC_j v_j C_i dV - \int_{sf} F_j N_{ji} dS \qquad \qquad ...(5)$$

The term K is introduced to keep the volumetric strain nearer to zero. The limiting strain rate is also introduced to incorporate the incompressibility. The material behavior is expressed by following expression:

$$\bar{\sigma} = K_{ot}\left(1 + a\bar{\varepsilon}\right)\dot{\bar{\varepsilon}}^{m} \qquad \qquad ...(6)$$

4. CONTACT TREATMENT

The contact between internal surface of the tube and circular die is modeled as frictionless throughout the deformation process. Each node that is in contact with circular die has been modeled to roll over the circular path of radius R provided by circular die. Consider any typical node $P^{(i)}$, which is in contact with die before the ith iteration (see Fig. 1). During the ith iteration as the friction between die and tube interface is ignored, so the incremental displacement of this node will be v.Δt Hence, this node $P^{(i)}$ will travel a distance of v.Δt along the circular path of radius R and will reach to a new location $P^{(i+1)}$ at the end of the ith iteration. The new coordinates of node $P^{(i+1)}$ has been computed as

$$x^{(i+1)} = x^{(i)} - dx, \quad y^{(i+1)} = y^{(i)} - dy \qquad \qquad ...(7)$$

where, dx and dy have been computed as

$$dx = \cos(\theta) - \cos(\theta + d\theta), \quad dy = \sin(\theta + d\theta) - \sin(\theta) \qquad \qquad ...(8)$$

Angles θ and $d\theta$ have been found by using trigonometry as the coordinates of node $P^{(i)}$ and point O are known.

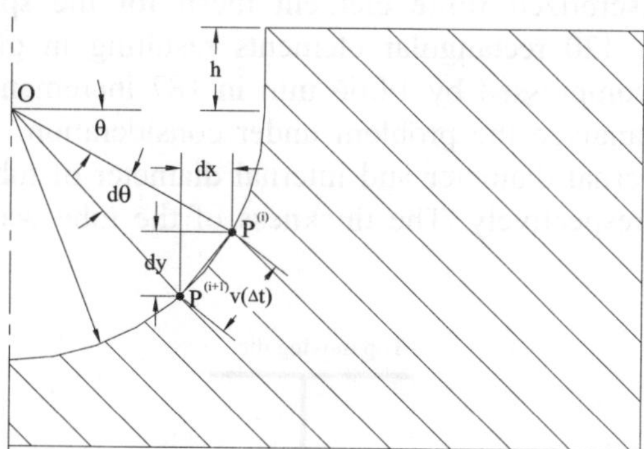

Fig. 1. A node in contact with circular die

5. FINITE ELEMENT PROCEDURE

The entire deformation process is sub-divided into several numbers of increments. Stepwise solution for one increment is in the following sequence:

1. Assume guess nodal velocities on all nodes of the discretized problem domain.
2. Generation of stiffness matrix and residual force vector for each element. Assemble them to get stiffness equation for the entire problem domain.
3. Solve the stiffness equation to get the solution velocities.

4. Check the solution velocities with guess velocities through convergence criteria based on two error norms namely, velocity error norm $\|\Delta v\|/\|v\|$ and force error norm $\|\partial \pi/\partial v\|$ (where

$$\|v\| = \left(v^T v\right)^{1/2} \text{ and } v \text{ is the velocity at nodal points})$$

5. If convergence criteria's are not satisfied then follow the Direct iteration method or Newton Raphson method till the convergence is achieved.
6. Once the solution gets converge, updated geometry of problem domain is generated. This is followed by strain rate, strain and stress evaluation, which ultimately gives the complete solution of one increment.

6. CASE STUDY

Aluminium circular tube specimen AL6060 adopted by Rosa *et al.* [14] to investigate experimentally and theoretically process of internal inversion of thin walled tubes was considered in the present study. The tube specimen has external diameter (d_0), thickness (t) and initial height (h_i) of 40 mm, 1 mm and 70 mm, respectively. Height of straight part of die (h) is considered as 15 mm. The material property of the deforming tube is expressed by the expression $\bar{\sigma} = k\bar{\varepsilon}^m$, where numerical values of k and m were taken as 298.3 MPa and 0.09, respectively. The radius of curvature of the circular die is taken as 20 mm. The friction between the deforming material and rigid die interface is ignored. This tube is compressed between rigid dies with stationary bottom die and moving top die with the downward velocity of 1.17 mm/s.

Figure 2 shows the discretized finite element mesh for the specimen tube AL6060. Mesh consists of 564 nodes with 420 rectangular elements resulting in global stiffness matrix of size 1128 × 1128. This tube is compressed by 14.66 mm in 187 increments. It was observed that 1584 iterations were required to analyze the problem under consideration. After 14.66 mm compression it was observed that the external diameter and internal diameter of tube at the nose was reduced to 33.13 mm and 31.00 mm, respectively. The thickness of the tube was increased to 1.14 mm after 14.66 mm compression.

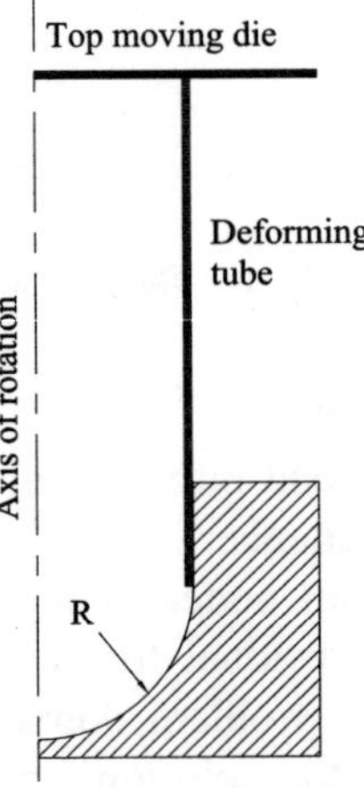

Fig. 2. Discretized finite element mesh for aluminum circular tube.

Figure 3 shows the deformed shapes at various stages of compression. It can be observed that as the deformation progresses, the diameter of tube reduces. Figures 4(a) and 4(b) show the load-compression and energy compression curve, respectively. It can be observed from Fig. 4(a) that load is zero initially and starts increasing after 2.03 mm compression (point A). Further, load increases smoothly with the compression. At 6.35 mm compression, load increases suddenly (point B). After point B, increase in load is marginal and continuous (points C and D). Deformed shapes shown in Fig. 3 (a), (b), (c) and (d) refers to the compression at point A, B, C and D as shown in Fig. 4(a) respectively. Figure 4(b) shows the energy-compression curve. It can be seen that increase in energy is continuous with compression. Sudden change in the slope of energy compression curve can be seen at 6.35 mm compression.

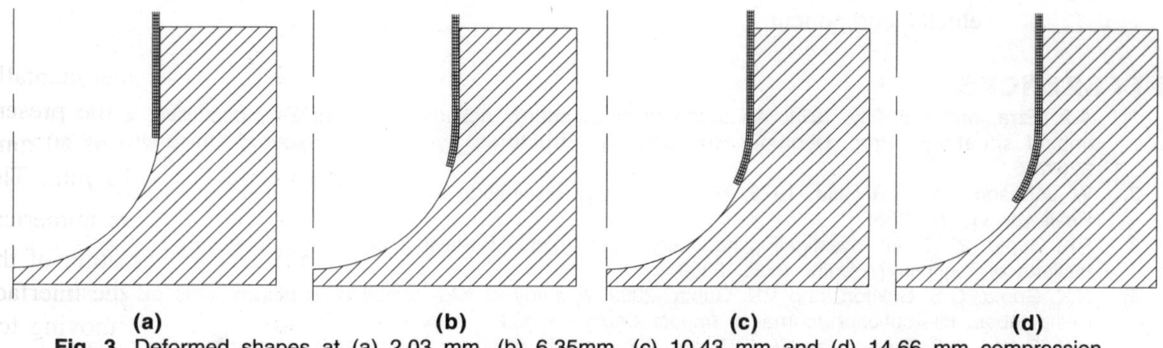

(a)   (b)   (c)   (d)

Fig. 3. Deformed shapes at (a) 2.03 mm, (b) 6.35mm, (c) 10.43 mm and (d) 14.66 mm compression

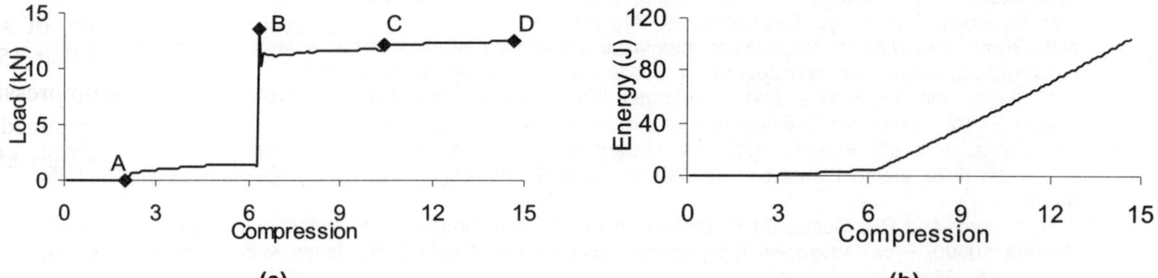

(a)   (b)

Fig. 4. (a) Load-compression and (b) energy-compression relationships for tube.

CONCLUSION

This paper presents a finite element modeling of circular tube between flat and curved surface dies. Computer code FEMLD is modified by incorporating the modeling of curved die surface. With the help of this modified code, a case study problem of tube nosing is presented. In this case study problem, aluminium circular tube is compressed against circular die of different radii. Deformed shapes at various stages of deformations are obtained. Load compression and energy compression relationships were also obtained, studied and discussed. It can be concluded that the developed code is capable to model and simulate the large deformation compression process of metallic tubes.

NOMENCLATURE

$E(\dot{\varepsilon}_{ij})$	work function
Δv	velocity correction term
F_i	surface tractions
$\dot{\bar{\varepsilon}}$	effective strain-rate
$\mathbf{f}$	force vector
$\bar{\varepsilon}$	effective strain
$\mathbf{K}$	stiffness matrix
$\bar{\sigma}$	effective stress
v	velocity component

REFERENCES

1. A.A. Ezra, and R.J. Fay, 1971, An assessment of energy absorbing devices for prospective use in aircraft impact situations, Proc. Dynamic Response of Structures Symposium, Stanford University, California, 225-246.
2. W. Johnson, and S.R Reid, 1978, Metallic energy dissipating systems, ASCE Journal Applied Mechanics Reviews. 31, 277-288.
3. P.K. Gupta, 2000, An Investigation into Large Deformation Behaviour of Metallic Tubes, Ph.D. Thesis, Indian Institute of Technology, Delhi, India, 2000.
4. N.K. Gupta, G.S. Sekhon, and P.K. Gupta, 2002, A study of fold formation in axisymmetric axial collapse of round tubes, International Journal of Impact Engineering. 27, 87-117.
5. N.K. Gupta, G.S. Sekhon, and P.K. Gupta, 2005, Study of lateral compression of round metallic tubes, Thin-Walled Structures. 43, 895-922.
6. N.K. Gupta, G.S. Sekhon, and P.K. Gupta, 2001, A study of lateral collapse of square and rectangular metallic tubes, Thin-Walled Structures. 39, 745-772.
7. N.K. Gupta, and Nagesh, 2004, Experimental and numerical studies of the collapse of thin tubes under axial compression, Latin American Journal of Solids and Structures. 1, 233-360.
8. P.K. Gupta, and J.P. Mishra, 2004, Computer Simulation of Large Deformations Process, Research Project Report PI-10, Center for Development of Advanced Computing, Pune.
9. P.K. Gupta, and R.N. Khapre, 2005, An efficient parallel solver using matrix inversion method for linear and non-linear finite element analysis, Journal of Institution of Engineers, Civil Engineering Division, India. 86, 44-48.
10. R.N. Khapre, and P.K. Gupta, 2006, Simulation of lateral compression of rectangular tubes on supercomputer PARAM 10000, Proc. Advances in Mechanical Engineering Conference, Jamia Millia Islamia University, New Delhi, 155-165.
11. R.N. Khapre, and P.K. Gupta, 2006, Simulation of axial compression of round metallic tube on supercomputer PARAM 10000, Proc. 14th National Seminar on Aerospace Structures Conference, Visvesvaraya National Institute of Technology, Nagpur, 447-453.
12. P.K. Gupta and R.N. Khapre, 2006, Finite element modeling of circular tube subjected to axial compression between flat and curved surface dies, Proc. National Conference on Rapid Prototyping Applications Conference, Visvesvaraya National Institute of Technology, Nagpur, 47-55.
13. S. Kobayashi, S.I. Oh, and T. Altan, 1989, Metal Forming and The Finite-Element Method, Oxford University Press, New York, Chap. 1-6.
14. P.A. Rosa, J.M.C. Rodrigues, and P.A.F. Matrins, 2004, Internal inversion of thin walled tubes using a die: experimental and theoretical investigation, International Journal of Machine Tool & Manufacture. 44, 775-784.

124

Finite Element Post-Buckling Analysis of Plates with Initial Imperfections

SUHAS C. VHANMANE[1], ABDUL HAMID SHEIKH[2] AND ASOKENDU SAMANTA[1,2]

[1] Research and Coordination Division, Indian Register of Shipping, Mumbai-400 072, India
[2] Department of Ocean Engineering and Naval Architecture, Indian Institute of Technology, Kharagpur-721 302, India

ABSTRACT

The paper presents a finite element formulation of rectangular plates to evaluate the postbuckling behavior under constant edge displacement loading. Eight noded isoparametric quadratic plate-bending element and Mindlin's hypothesis are used to formulate the plate. The analysis is based on displacement incremental Newton-Raphson method. A computer program is developed to examine the post-buckling behavior of the rectangular plates with initial imperfection which may occur due to ill manufacturing process. Initial imperfections are simulated by assuming some pre defined deformed shapes. The results obtained by the present method are compared with the available literatures whenever available and may serve as useful resources for the future researchers.

Keywords: Finite element method; Initial imperfection; Newton-Raphson method; Post-buckling.

1. INTRODUCTION

When the structure is subjected to inplane loading, it does not collapse immediately reaching the critical load or buckling load. It sustains additional load beyond buckling. The post-buckling behavior is the reserve capacity, which exists after the initial buckling due to the development of membrane strains. Large deformations of an elastic body produce stretching effect and thus lead to curvature-displacement nonlinearity, making the post-buckling behavior a geometric nonlinear problem.

Murray and Wilson [1] studied post-buckling behavior of simply supported thin square plates using updated Lagrangian approach. The approximate incremental stiffness was determined by combining a geometric stiffness with the stiffness of the unstressed, but displaced, structure. Post-buckling behavior of rectangular plates with small initial curvature, loaded in edge compression was solved numerically by Yamaki [2]. The procedure was based on Marguerre's fundamental equations of initially deflected thin plates. Pica and Wood [3] used the geometric nonlinear Mindlin shallow shell formulation for the 9-noded Lagrangian element to demonstrate post-buckling behavior of plates subject to direct inplane loading which was the extension of the work of Pica *et al.* [4].

The main purpose of the present investigation is to study the post-buckling behavior of the rectangular plates with initial imperfection by finite element method. The formulation of large deflection analysis is based on von Karman's nonlinear plate equations where the bending deformation follows Mindlin's hypothesis. The initial deflection is incorporated to simulate imperfection of the plate flatness. The isoparametric quadratic plate-bending element, (Satsangi and Mukhopadhyay, 5) with five degrees of freedom per node is used to study the post-buckling behavior of plates. Further, the element being isoparametric, transverse shear deformation of the plate is incorporated. The nonlinear equations are solved using displacement incremental Newton-Raphson technique. A Fortran program is developed and several numerical examples are solved involving various boundary conditions and different initial imperfections.

2. MATHEMATICAL FORMULATION

For a continuum undergoing small or large displacements, the equation of equilibrium of the external and the internal forces is expressed by the following virtual work equation.

$$\{d\delta\}^T\{\psi\} = \int_A \{d\varepsilon\}^T\{\sigma\}dA - \{d\delta\}^T\{P\} = \{0\} \qquad ...(1)$$

in which $\{d\delta\}$ is the variation of the nodal displacement vector of discrete continuum, (ψ) is the sum of the internal and the external generalized forces and $\{P\}$ is the generalized nodal force vector.

An isoparametric quadratic 8-noded plate-bending element with five degrees of freedom per node $\left(u,v,w,\theta_x,\theta_y\right)$ has been used in the present analysis. The details for this element have been given by Satsangi and Mukhopadhyay [5].

The generalized strain of the plate in the simplest form can be written as

$$\{\varepsilon\} = \{\varepsilon_0\} + \{\varepsilon_L\} + \{\varepsilon_I\} \qquad ...(2)$$

The subscript *0* and *L* are used to indicate the linear and nonlinear parts, respectively. The subscript I is used to indicate the strain vector component due to initial imperfection. The complete generalized strain vector can be written as

$$\{\varepsilon\} = \left([B_0] + \frac{1}{2}[B_L] + [B_I]\right)\{\delta\} \qquad ...(3)$$

The variation of the generalized strain vector with respect to $\{d\delta\}$ can be written as

$$\{d\varepsilon\} = \{d\varepsilon_0\} + \{d\varepsilon_L\} + \{d\varepsilon_I\} \qquad ...(4)$$

where, the variation or increment of any parameter is indicated by d of the parameter. The complete variation of the generalized strain vector can be expressed as

$$\{d\varepsilon\} = ([B_0] + [B_L] + [B_I])\{d\delta\} = [B]\{d\delta\} \qquad ...(5)$$

An incremental equation is required to solve the Eq. (1) and it can be expressed as

$$\{d\psi\} = [K_T]\{d\delta\} \qquad ...(6)$$

where, $[K_T]$ = Tangent stiffness matrix and is obtained by differentiating residual vector $\psi(\delta)$ with respect to $\{\delta\}$. In the simplest form the tangent stiffness matrix can be written as (the detailed procedure is given in Pica, [3] and Pica *et al.* (4).

$$[K_T] = [K_0] + [K_L] + [S] \qquad \ldots(7)$$

where, $[K_0]$ is the linear stiffness matrix for plate, $[K_L]$ is the initial displacement matrix, and $[S]$ is the initial stress matrix

During the fabricating process or later in the service, the plate usually inherits initial curvatures and it becomes imperfect. The initial deflections are considered in the order of the magnitude of the plate thickness ($W_\delta = 0.1t, 0.2t, etc.$). Double Fourier series are used to define the initial deflection for various shapes (Fig. 1).

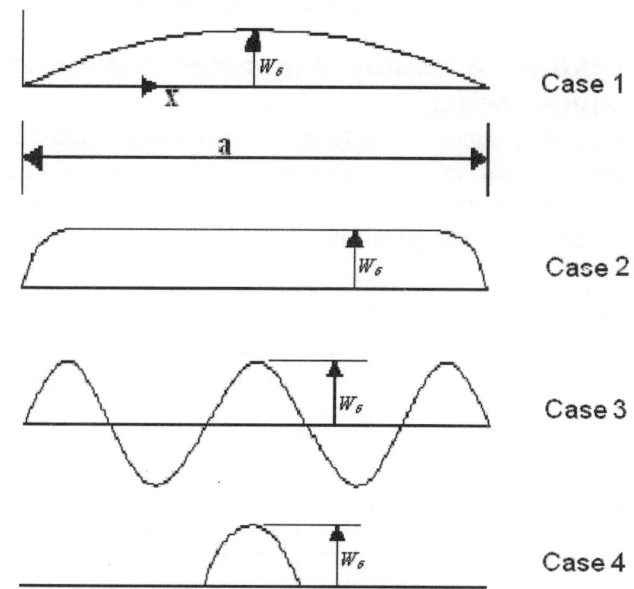

Fig. 1. Different types of initial imperfections in terms of deflection.

Case 1

$$w_i = W_\delta \sin\left(\frac{\pi x}{a}\right)\sin\left(\frac{\pi y}{b}\right)$$

Case 2

$$w_i = W_\delta \sin\left(\frac{5\pi x}{a}\right)\sin\left(\frac{\pi y}{b}\right) \quad (x/a < 0.1, x/a > 0.9),$$

$$w_i = W_\delta \qquad\qquad (0.1 \le x/a \le 0.9)$$

Case 3

$$w_i = W_\delta \sin\left(\frac{5\pi x}{a}\right)\sin\left(\frac{\pi y}{b}\right)$$

Case 4

$$w_i = W_\delta \sin\left(\frac{5\pi x}{a}\right)\sin\left(\frac{\pi y}{b}\right) \qquad (0.4 \geq x/a \leq 0.6),$$

$$w_i = 0.0 \qquad\qquad\qquad (0.4 > x/a < 0.6)$$

Case 1 is chosen in the present study, as this is the common imperfection form in the plate. The formulations of nonlinear equations for initially deflected plate are developed and incremental displacement method (Batoz and Dhatt [6]) is used to solve the plate problems.

3. NUMERICAL EXAMPLES

Following numerical examples are presented in this section to support the theoretical derivation.

Example 1. Post-Buckling of Simply Supported Square Plate Loaded in Edge Compression

Post-buckling of simply supported square plate (Fig. 2) with small initial curvature, loaded in edge compression was investigated numerically by Yamaki [2], using series solution, for various edge conditions. Pica and Wood [3] used nine noded total Lagrangian element to solve the same problem. The boundary condition of the plate are given as

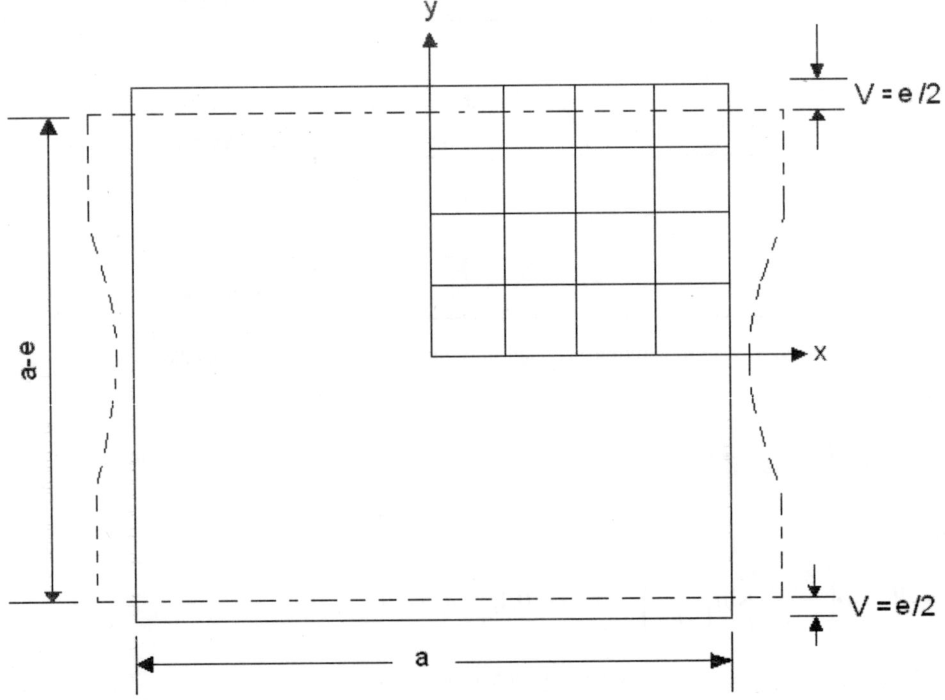

Fig. 2. Simply supported square plate.

$x = \pm a/2$; u, v, θ_x = free; $w = \theta_y = 0$

$y = \pm a/2$; u, θ_y = free; v constant; $w = \theta_x = 0$

The initial imperfection in this case is taken as, $\quad w_i = 0.1t \cos\dfrac{\pi x}{a}\cos\dfrac{\pi y}{a}$

The plate geometry (shown in the Fig. 2) and the material properties are
$a = 400$ mm ; $t = 1.5$ mm ; $E = 60 \times 10^6$ N/mm^2; $\mu = 1/3$; Edge shortening, $\Delta = ea/t^2$;
Load, $\lambda = \sigma_y a^2 / \pi^2 E t^2$; where, σ_y is the average edge compressive stress in the y-direction.

Maximum deflection at the center of the plate, $W = w_s(0,0)/t$; Effective width, $R = \sigma_y / E(e/a)$

'Effective width' is the ratio of the actual load carried by the plate to the load the plate would have carried if the stress had been uniform, and equal to the Young's modulus times the average edge strain.

Displacement increment Newton-Raphson iterative solution technique has been used in this problem with 8×8 mesh division in the quarter plate. The results are compared with the available references in the graphs shown in Fig. 3 to Fig. 5 and the agreement is very good. Figs. 3 and 4 show that even after critical load ($\lambda > 1$), the failing strength of a thin plate can exceed the buckling stress. Fig. 5 shows that the effective width ratio (R) decreases with increase of edge shortening.

Convergence study of the same plate is also done and Fig. 6 shows the graph of maximum deflection (W) against number of elements used in quarter plate. It is observed that the maximum deflection is stabilized after 50 numbers of elements in the quarter plate.

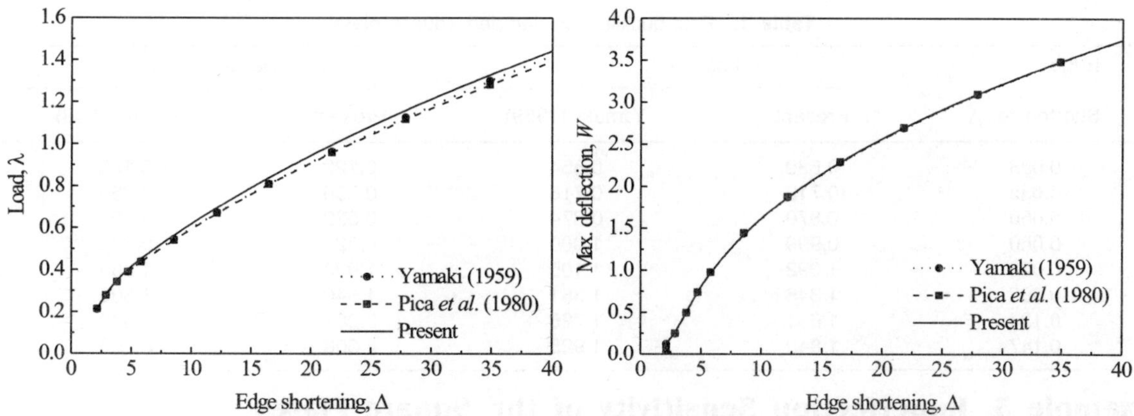

Fig. 3. Load, λ vs edge shortening, Δ for a simply supported square plate.

Fig. 4. Maximum deflection, W vs Edge shortening, Δ.

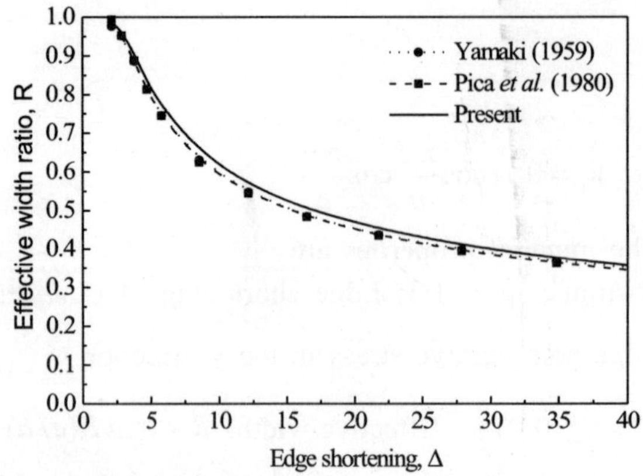

Fig. 5 Effective width ratio, *R* vs edge shortening, Δ for a simply supported square plate.

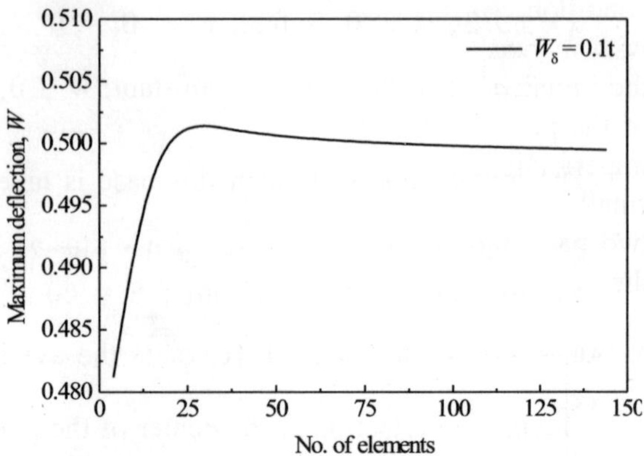

Fig. 6 Convergence study of simply supported square plate.

Example 2. Post-Buckling of Clamped Square Plate

The same plate is analyzed for clamped boundary condition. The boundary conditions are given below.

$$x = \pm a/2 \;\; ; u, v = \text{free}; w = \theta_x = \theta_y = 0$$

$$y = \pm a/2 \;\; ; u = \text{free} \; v = \text{constant}; w = \theta_x = \theta_y = 0$$

The initial imperfection in this case is taken as, $w_i = 0.1t\sin^2\dfrac{\pi x}{a}\sin^2\dfrac{\pi y}{a}$

The results are compared with Yamaki (1959) in Table 1 and the agreement is found good.

Table 1. Post-buckling of clamped square plate

Edge	Load, λ		Max. Deflection, W	
Shortening, Δ	Present	Yamaki (1959)	Present	Yamaki (1959)
0.028	0.539	0.554	0.122	0.125
0.038	0.712	0.718	0.256	0.250
0.050	0.870	0.874	0.522	0.500
0.060	0.999	1.000	0.821	0.775
0.068	1.092	1.105	1.033	1.000
0.092	1.346	1.381	1.540	1.500
0.123	1.631	1.726	2.007	2.000
0.147	1.843	1.925	2.306	2.250

Example 3. Imperfection Sensitivity of the Square Plates

The plate as shown in the Fig. 2 is analyzed for varying initial imperfection. The W_δ value is taken as 0.1t, 0.5t, 1.0t, 2.0t. The initial imperfection has been defined by Fourier series,

$w_i = W_\delta \cos\dfrac{\pi x}{a}\cos\dfrac{\pi y}{a}$. Fig. 7 shows the behavior of the square plate for various initial imperfection values for simply supported boundary condition while Fig. 8 shows results for clamped boundary condition. In the post-buckling region (beyond point of inflection), the deflection (w_s) decreases with increase in imperfection for a particular edge shortening. The inflection point of the curve for the initially deflected plate corresponds almost to the critical load. But, the total deflection ($w_s + w_i$) of the plate with higher initial imperfection is always large than that of the plate with lesser initial imperfection. For plate with less initial imperfection (W_δ/t = 0.1), the initial deflection is very small as compared to the plate dimensions. For this plate the defection curve (Fig. 8) consists of two parts joining almost at the buckling load. The first part, which is straight line, corresponds to the simple compression and the second part represents the effect of initial deflection.

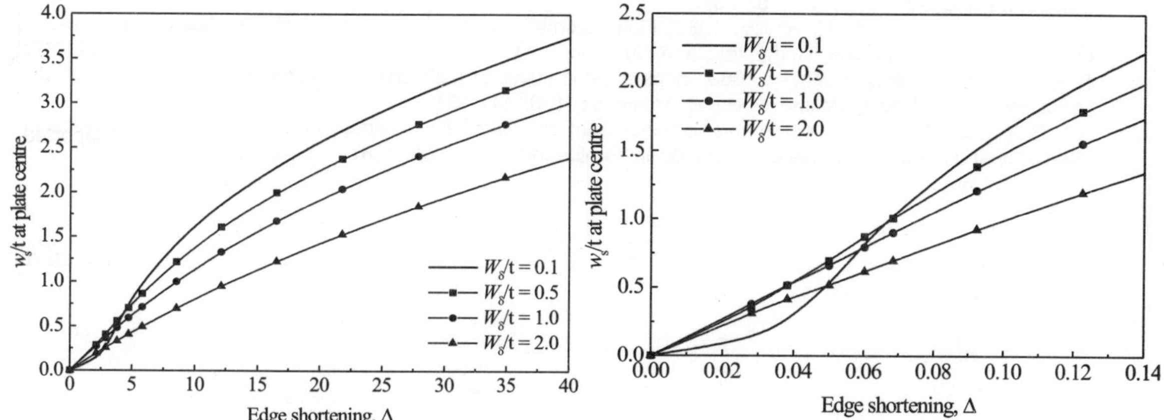

Fig. 7. Imperfection sensitivity of the simply supported square plate

Fig. 8. Imperfection sensitivity of the clamped square plate

CONCLUSION

The post-buckling behavior of the plates with different boundary conditions and with various initial imperfections has been analyzed with the help of eight-noded isoparametric plate finite element. Geometric nonlinearity has been taken into consideration in formulating plate stiffness matrix, whereas material nonlinearity has been ignored. Following points can be summarized from the above-mentioned analysis.

i)   In the post-buckling region (beyond point of inflection), the deflection (w_s) decreases with increase in imperfection for a particular edge shortening.

ii) The failing strength of a thin plate can exceed appreciably the buckling stress in the elastic region.

iii) Effective width decreases with the increase of edge shortening for a particular initial imperfection.

NOTATIONS

$[K_G]$	geometric stiffness matrix
$[K_S]$	secant stiffness matrix
$[K_T]$	tangent stiffness matrix
$[S]$	initial stress matrix
w_i	deflection due to initial imperfection

w_s deflection due to load

$\{\varepsilon_o)$ first component of linear strain vector

$\{\varepsilon_1)$ second component of linear strain vector due to initial imperfection

REFERENCES

1. D. W. Murray, E. L. Wilson, 1969, Finite element large deflection analysis of plates, Journal of Engineering Mechanics Division, Proceedings of ASCE, EM 1,143-165.
2. N. Yamaki, 1959, Post-buckling behavior of rectangular plates with small initial curvature loaded in edge compression, J. Applied Mechanics, ASME, 26, 407-414.
3. A. Pica, R. D. Wood, 1980, Post-buckling behaviour of plates and shells using a Mindlin shallow shell formulation, Computers and Structures, 12, 759-768.
4. A. Pica, R. D. Wood, and E. Hinton, 1980, Finite element analysis of geometrically nonlinear plate behaviour using a Mindlin formulation, Computers and Structures, 11, 203-215.
5. S. K. Satsangi, M. Mukhopadhyay, 1983, Isoparametric stiffened plate bending element for the analysis of ships' structures, Trans. Royal Institute of Naval Architects, 126, 141-151.
6. J. L. Batoz, G. Dhatt, 1979, Incremental displacement algorithms for nonlinear problems, short communications, International Journal for Numerical Methods in Engineering, 14, 1262-1267.

125

Wavelet Basis Finite Element Method for Solution of Engineering Problems

K. Gopi Krishna[1] and Manish Shrikhande[2]

[1] Research Scholar, [2] Assistant Professor Department of Earthquake Engineering Indian Institute of Technology Roorkee, Roorkee, Uttaranchal, India.
email: mshrifeq@iitr.ernet.in

ABSTRACT

Finite element method is the most versatile tool for numerical solution of engineering problems, described by partial differential equations. The conventional formulation of finite element method uses polynomial shape functions and hence belongs to the class of single scale methods. There are several problems which exhibit multi-scale behaviour, for example, problems with geometric singularity (e.g., holes), material singularity (e.g., boundary layers), or detailed features in the loads, etc. The use of single scale methods for solution of such problems requires the use of very fine mesh and accordingly increases the problem size significantly. The use of a multi-scale method, on the other hand, facilitates an efficient way of improving the accuracy of computations without increasing the computational effort significantly. The use of wavelets as basis functions in finite element formulation holds some promise due to their compact support, localization and multi-resolution properties. In this article, we discuss various critical issues related to the use of wavelet basis in finite element method for solution of engineering problems. The formulation of a wavelet-basis finite element system is demonstrated for a one-dimensional problem. A brief description of pertinent concepts on wavelets and their properties is also presented.

Keywords: Wavelet basis, finite element, function approximation, Galerkin method, weighted residual, hierarchical formulation.

1. INTRODUCTION

Wavelet based representations are receiving increased attention in recent years in various disciplines like signal processing, numerical analysis, etc. because of their intrinsic capability of representing data in various levels of resolution. Furthermore, multi-resolution analysis using orthogonal, compactly supported wavelets, which are localized in space, helps in analyzing the local variations of the problem at various levels of resolution. This feature helps in development of hierarchical solutions that facilitate adaptivity. This means that an initial rough estimate of solution can be obtained based on coarse description of the problem, which then can be refined adaptively by adding details as the investigation progresses, with minimal computational effort and without significantly increasing the problem size. Hence, finite element formulation using these bases appear to be ideal in analyzing problems having local variations or discontinuities, having strong oscillations, or sharp spikes, etc. (like the ones that arise in various engineering problems namely stress concentrations, transient problems, etc.). The use of wavelet basis in finite element method is to provide an improvement in the approximation of the solution compared to traditional finite element method because of excellent approximation properties of wavelets [7].The approximation can be improved by increasing the level of resolution and also by increasing the order of wavelet used for approximation. In this article, Daubechies compactly supported wavelets have been considered as basis functions for demonstrating the finite element formulation for the governing partial differential equation (PDE) and to discuss various difficulties that arise in computing the approximate solution. Different approaches to address these difficulties are discussed.

2. WAVELET THEORY

A wavelet is a spatially localized function having finite energy. The term localized refers to compact support of the function, which means that outside some interval the amplitude of wavelet vanishes or decays exponentially. The most popular among the classical wavelet constructions are the orthonormal, compactly supported wavelets developed by Daubechies [7]. In this section we briefly review the properties of Daubechies' wavelets, as they are well-suited for efficient representation of localized features.

To define Daubechies' wavelets, let us consider the two functions $\phi(x)$ and $\varphi(x)$ which satisfy the scaling or refinement relations given by

$$\phi(x) = \sum_{k=0}^{L-1} h_k \phi(2x - k) \qquad \qquad ...(1)$$

$$\varphi(x) = \sum_{k=0}^{L-1} (-1)^k h_{L-k} \phi(2x - k) \qquad \qquad ...(2)$$

where, $\{h_k\}$ are set of filter coefficients that categorize the specific wavelet basis, and L is the support or order of the wavelet. The function $\phi(x)$ is called as scaling function and the associated wavelet function is defined as $\varphi(x)$, k is the translation parameter of scaling and wavelet functions.

The scaling function $\phi(x)$ is normalized to have unit area, i.e., $\int_{-\infty}^{\infty} \phi(x)dx = 1$, which implies

$$\sum_k h_k = 2 \qquad \qquad ...(3)$$

The translates of the scaling function $\phi(x)$ are required to be orthonormal, i.e.,

$$\int \phi(x - k)\phi(x - m) = \delta_{k,m} \qquad \qquad ...(4)$$

where, $\delta_{k,\,m}$ is the Kronecker delta.

From the scaling relation defined in Eq. (1)., the above condition implies that

$$\sum_{k=0}^{L-1} h_k h_{k-2m} = \delta_{0,m} \qquad \qquad ...(5)$$

The set of filter coefficients $\{h_k\}$ satisfying above conditions are chosen so that dilations (defined by parameter j) and translations (defined by parameter k) of the wavelet function $\varphi_k^j(x)$ form a complete orthonormal basis of L^2 space (a vector space wherein the energy norm is defined, a Hilbert space). In other words, $\varphi_k^j(x)$ will satisfy

$$\delta_{k,l}\delta_{j,m} = \int_{-\infty}^{\infty} \varphi_k^j(x)\varphi_l^m(x)dx \qquad \qquad ...(6)$$

Moreover, the wavelet function $\varphi(x)$ has M vanishing moments, which means that polynomials having terms of the form $1, x, x^2, ..., x^{M-1}$ can be expressed as linear combinations of $\varphi(x)$ as given in Eqs. (7) and (8). In other words, this condition—known as approximation condition or smoothness condition—means that scaling function is capable of approximating a polynomial of degree M. A Daubechies' wavelet of length L is usually denoted as D_L and is related to number of vanishing moments M by $2M = L$.

$$\int_{-\infty}^{\infty} \varphi(x)x^m dx = 0 \quad \forall \ m = 0, 1, 2, ...M\text{-}1 \qquad \qquad ...(7)$$

In terms of filter coefficients the above condition is stated as

$$\sum_k (-1)^k k^m h_k = 0 \qquad \qquad ...(8)$$

The spaces spanned by scaling function $\phi_m^j(x)$ and wavelet function $\varphi_m^j(x)$ over the parameter m, at a fixed resolution j are denoted as V_j and W_j and are defined as

$$V_j = \{2^{j/2}\phi(2^j x - m); m = ..., -1, 0, 1, ...\}, \qquad \qquad ...(9)$$

$$W_j = \{2^{j/2}\varphi(2^j x - m); m = ..., -1, 0, 1, ...\} \qquad \qquad ...(10)$$

where, W_j is known as an orthogonal complement of V_j in $V_j + 1$. Multi-resolution analysis is defined as a nested sequence of subspaces V_j, which satisfy the following properties:

$$V_{j+1} = V_j \oplus W_j \qquad \qquad ...(11)$$

$$V_0 \subset V_1 \subset ...... \subset V_{j+1,} \qquad \qquad ...(12)$$

$$\bigcap_{j\in Z} V_j = \{0\} , \ \bigcup_{j\in Z} V_j = L^2(R) \qquad \qquad ...(13)$$

$$V_{j+1} = V_0 \oplus W_0 \oplus W_1 \oplus ...... \oplus W_j \qquad \qquad ...(14)$$

where, Z is a set of non-negative integers.

To solve the governing differential equations, the function $f(x)$ and its derivatives are expanded in terms of basis functions. The wavelet expansion of the function in L^2 space is of the form

$$f(x) = \sum_{l\in Z} c_{0l}\phi_{0l}(x) + \sum_{j=0}^{\infty}\sum_{k\in Z} c_{jk}\varphi_{jk}(x) \qquad \qquad ...(15)$$

If this approximation is truncated at level j then the approximation is given by

$$\tilde{f}(x) = \sum_{l \in Z} c_{jl} \phi_{jl}(x) \qquad ...(16)$$

In case of derivatives of the function $f(x)$, Daubechies [7] showed that there exists a factor $\lambda > 0$ such that a wavelet of length N has $\lambda \, (N/2 - 1)$ continuous derivatives and for small value of N, $\lambda \geq 0.55$. More details regarding properties of compactly supported wavelets may be found in references [6, 7].

3. WAVELET-GALERKIN METHOD

Galerkin method is one of the best known methods for numerical solution of PDE, because of its simplicity in formulation. In this section the formulation of Galerkin method using wavelet basis functions is demonstrated by means of following differential equation:

$$u_{,xx} + \beta^2 u = f \qquad ...(17)$$

with associated boundary conditions. This equation generally arises in engineering problems for studying the instability analysis of structural members (beams and columns), wave propagation problems, etc. The solution is approximated by means of wavelet basis functions as given by

$$u(x) = \sum_{k} u_k \phi(X - k) \quad \text{where } X = 2^j x \qquad ...(18)$$

Similarly,

$$f(x) = \sum_{l} f_l \phi(X - l) \qquad ...(19)$$

Substituting the approximation in Eq. (17) and taking inner product with $\phi(X - m)$ on both sides, we get an equation of the form $\boldsymbol{Tu} = \boldsymbol{f}$. The matrix '$\boldsymbol{T}$' is a circulant matrix having entries of the form

$$\Omega_{k,j}^{d1,d2} = \int_{-\infty}^{\infty} \phi_k^{d1} \phi_l^{d2} dx \qquad ...(20)$$

where, $\Omega_{k,j}^{d1,d2}$ are known as connection coefficients and indices $d1$, $d2$ denoting the derivative order of wavelet function $\phi(x)$. The connection coefficients depend on the differential operator. The linear system of equations $\boldsymbol{Tu} = f$ can be solved for the unknown u after incorporating specified boundary conditions.

4. CRITICAL ISSUES

Some critical issues for successful formulation of wavelet basis finite element method are discussed below.

4.1 Evaluation of Connection Coefficients

The wavelet Galerkin approximation relies heavily on reliable evaluation of connection coefficients in the form of Eq. (20), where the integrands are products of wavelet bases, their translates and derivatives. Compactly supported wavelets have a finite number of filter coefficients, however, the derivatives of these wavelets, if exist, are highly oscillatory in nature. Hence, precise evaluation of these integrals by standard numerical quadrature methods is difficult and unstable. To circumvent this problem Latto *et al.* [1], devised an exact method, known as the connection coefficient method, for computing these coefficients by exploiting the scaling relation and moment condition of scaling function to reduce the calculation to an eigenvector problem. This method is designed for unbounded domains and when used for problems defined on a finite interval, the connection coefficients computed by this method are found to be inaccurate near the boundary and have to be corrected. There are two approaches to solve this problem on bounded intervals. One, suggested by Romine *et al.* [3], is to compute the connection coefficients on an interval (called by them as proper connection coefficients) at scale j = 0 with the limit of integration being 0 to L-1 and the ones at higher resolutions are derived from connection coefficients computed at j = 0. Other suggested by Lu *et al.* [4], was to artificially extend the interval and use the connection coefficients as suggested by Latto *et al.* The connection coefficients to be evaluated are not problem dependent, hence it can be computed *a priori* and can be used whereever necessary in order to save computational time.

4.2 Incorporation of Boundary Conditions

The excellent properties attributed to Daubechies' wavelets are based on the assumption that the computational domain is unbounded (i.e., the PDE is periodic in the computational domain) and do not carry over well to bounded domains. Hence approaches that assume periodicity of PDE complicate the treatment of boundary conditions in a finite domain. For instance, the connection coefficients computed using the method developed by Latto *et al.* [1], do not provide correct inner product at the endpoints of bounded interval, thus making implementation of boundary conditions difficult [3]. Amaratunga *et al.* [5] suggested a polynomial extrapolation technique to address this problem. In this method the scaling functions are classified as regular scaling functions (which lie entirely within the boundaries), irregular scaling functions (which lie partially inside the intervals of boundaries) and exterior scaling functions (which lie outside the interval). To deal with the difficulty of treating irregular scaling function, they predicted the coefficients of exterior scaling function by using the polynomial extrapolation method and proposed a modified form of wavelet Galerkin equations for irregular scaling functions. Lu *et al.* [4] suggested the idea of fictitious boundary approach in which Galerkin method can be applied directly. The original boundary of the problem is embedded in a boundary of larger size called as fictitious boundary. The fictitious boundary is assumed so that original boundaries remain unaffected by the inaccuracies introduced by connection coefficients that were computed for unbounded domains by Latto *et al.* [1]. In this method, the unknown function is expanded by scaling functions and fictitious boundary is assumed to handle the difficulty of treating boundary conditions and an additional condition is considered so that solution incorporates specified boundary conditions on the original or actual boundary. This method can be adapted to all classes of boundary value problems [4].

4.3 Condition Number of System and Connection Coefficient Matrices

The potential for numerical instabilities in the solution of simultaneous equations ($T\,u = f$) obtained by application of wavelet Galerkin method to the governing PDE at a chosen resolution is determined by means of the condition number of the system matrix (T) at that resolution. This is defined for a positive definite matrix T as the ratio of largest to smallest eigen value; the lower the condition number the more reliable the numerical result would be

$$\kappa([T]) = \frac{\lambda_n}{\lambda_1} \quad \text{where} \quad \sigma([T]) = \{\lambda_1 \leq \dots \leq \lambda_n\}. \qquad \dots(21)$$

The condition number increases with increase in resolution for the matrices that arise in computation of connection coefficients (as shown in Fig. 1), and also for the overall system matrix which is circulant in nature and influenced by connection coefficients. The loss of accuracy in evaluation of connection coefficients also affects the accuracy of solution. The condition number for higher order wavelets is consistently lower than lower order wavelets as shown in Fig. 1. Condition number also depends on order of derivatives and increases with increase in order of derivatives.

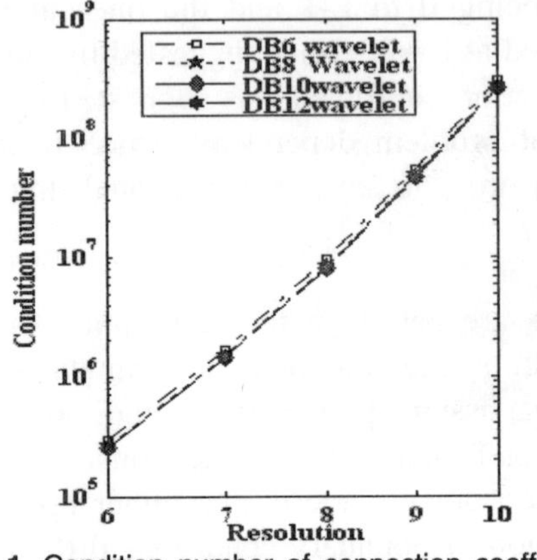

Fig. 1. Condition number of connection coefficients.

Fig. 2. Convergence rate with resolution.

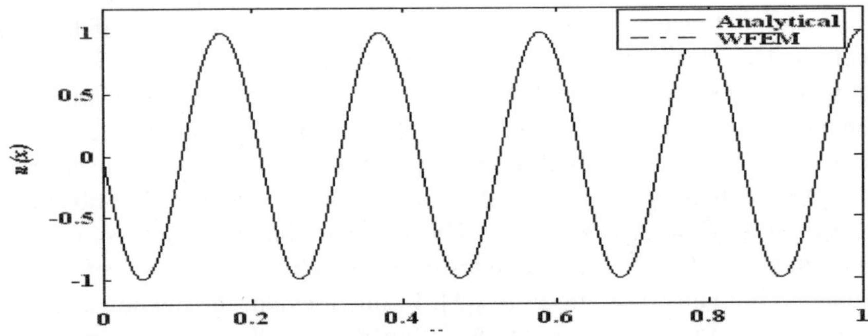

Fig. 3. Wavelet Finite element solution using D 10 wavelets.

5. RESULTS

To demonstrate the viability of wavelet basis finite element formulation described in this article, a numerical problem has been solved in 1D for the differential equation given in Eq. (17) on a bounded domain [0, 1]. Dirichlet boundary conditions are imposed at the end points as $u(0) = 0$ and $u(1) = 1$. Daubechies' wavelets have been used as basis functions and the boundary conditions are incorporated using fictitious boundary approach defined by Lu *et al.* [4]. Here for simplicity in demonstrating the solution in wavelet approximation space β^2 is taken as 891 and results are shown in Fig. 3. The error is defined as

$$\left\| error \right\|_{L^2} = \left\| \frac{u_{WFEM} - u_{Analytical}}{u_{Analytical}} \right\|_{L^2} \qquad \qquad ...(22)$$

CONCLUSION

Finite element formulation using wavelet basis has been demonstrated as shown in Fig. 3, and for evaluating the numerical solution as shown in Fig. 3, the connection coefficients are computed using the Matlab® code developed based on the algorithm of Restrepo *et al.* [2]. It should be noted that adequate care has to be taken for computing connection coefficients as it would in turn dictate the incorporation of boundary conditions. In case of approach suggested by Amaratunga *et al.* [5] the method gives good results but simplicity of Galerkin method is lost in case of engineering problems, as one should resort to obtain the modified form of Galerkin system of equations. In case fictitious boundary approach, Galerkin method can be used directly and original boundary condition is changed to be an additional constraint to get solution on the original boundary. However, these methods address the issues indirectly and there is a need to work on extending these methods, so as to resolve the boundary conditions in a natural way as suggested by Romaine *et al.* [3]. Also special attention has to be given regarding increase in condition number of the matrices that evolves from Galerkin discretization and also for computing connection coefficients with respect to resolution,as it would also affect the convergence rate as shown in Fig. 2. This preliminary investigation suggests that wavelet based finite element method shows some promise in obtaining the solution for engineering problems described by PDE and are more suitable for analyzing localized features.

REFERENCES

1.   A. Latto, H.L. Resnikoff, and E. Tannenbaum,1999,The evaluation of connection coefficients of compactly supported wavelets, Aware, Inc.Cambridge, MA 02142 U.S.A.
     URL: http://www2.appmath.com:8080/site/con3_5/con3_5.html (Last Accessed:30/06/06)
2.   J.M. Restrepo G.K. Leaf, Inner Product Computations Using Periodised Daubechies Wavelets, Mathematics Department, University of California, Los Angeles, CA 90095 U.S.A.
     URL: http://www-fp.mcs.anl.gov/division/publications/abstracts/abstracts94.htm(Last-Accessed 30/06/06)
3.   C.H. Romine, and B.W. Peyton, 1997, Computing Connection Coefficients of Compactly Supported Wavelets on Bounded Intervals, Contract No. DE-AC05-96OR22464, Mathematical Sciences Section, Oak Ridge National laboratory, Oak Ridge, TN 37831-6367 U.S.A.
     URL: http://citeseer.ist.psu.edu/romine97computing.html (Last Accessed:30/06/06)
4.   D. Lu, T. Ohyoshi, and L. Zhu, 1996, Treatment of boundary conditions in the application of wavelet galerkin method to a SH wave problem, Akita University, Japan.
     URL: http://citeseer.ist.psu.edu/84953.html   (Last Accessed: 30/06/06)
5.   K. Amaratunga , J.R. William, S. Qian, and J. Weiss, 1994, Wavelet Galerkin solutions for one dimensional partial differential equations, Int. J. Numer. Methods Eng., 37, 2703-2716.
6.   G. Strang, 1989, Wavelets and dilation equations: A brief introduction, SIAM review, 31, 614-627.
7.   I. Daubechies, 1988, Orthonormal bases of compactly supported wavelets, Journal of Commun. Pure Appl. Math., 41, 909-996.

126

The Development of a Portable Hybrid System Simulation Model

S.P. SINGH[1] AND S. ASHOK[2]

[1]Department of Mechanical Engineering, NIT Calicut-673 601, India. email: [1] m050141em@nitc.ac.in
email: [2] ashoks@nitc.ac.in

ABSTRACT

Embedded generation has been described as a "paradigm shift" in the way in which electricity is produced with the focus of power production shifting away from large centralized generation plants to production of heat and power close to the point of use. This paper describes recent work to develop and test a simple mathematical model of a Portable Hybrid System and its integration within the various technical domains of a building simulation tool. Specifically, the paper will describe: a) the concept of the portable hybrid system; b) the development of the mathematical model; c) analysis using Computational fluid dynamics and d) the integration of the model into a building simulation tool. This paper will show the CFD analysis of the model in which Navier stroke equation is being solved by generating the grid and programming, from where the flow and the power available at rotor is calculated. Also the rotor blades (HAWT) are designed and tested in the gambit software by analyzing the pressure drop across the blades. The purpose of providing solar panel with this model is to increase the efficiency of solar cells. Experiments in the lab show considerable change in efficiency of silicon solar cells. The same was verified by developing the CuO solar cell in the lab and checking their efficiency with cooling. A better model of the Portable Hybrid System (PHS) is also discussed which can exploit the wind in two directions. This paper also investigates the development of a portable hybrid system (PHS) that could be introduced into an urban environment and address the problems associated with the large scale devices by being integrated into the building fabric and generating the electricity onsite.

Keywords: Ducted wind turbine; Mathematical model; Building simulation.

1. INTRODUCTION

As we move into the 21st century, technological innovation is changing the means by which heat and power can be delivered to the built environment. New ''micro-gird'' type technologies offer the potential of supplying heat and power locally from 'clean' and energy-efficient-type technologies.

Examples of these technologies include micro-CHP using Stirling engines, photovoltaic (PV) and fuel cells [2]. To assess the effectiveness of these devices and also to assess the impact of their diffusion into the built environment it is necessary to develop models to simulate their performance in a realistic operational context. Building simulation offers a means to do this and can reveal important performance characteristics of heat and power output and their compatibility with the loads that they are designed to serve [1].

The ducted wind turbine (DWT) is an emerging micro-grid technology. The ducted wind turbine overcomes many of the problems associated with the use of conventional wind turbines in an urban environment, which are hampered by high levels of turbulence in the air stream, and are also constrained by concerns over visual impact, noise and public safety. In contrast DWT units are purposely designed for attachment to building and are both robust and unobtrusive. This paper also describes the integration of a simple DWT model [5] {Grant *et al.* 2002} into a building simulation tool.

2. THE DWT MODEL

The model described here has been derived from a design that attaches to the roof edge of rectangular section buildings (Fig. 1) making use of the pressure differentials that are naturally created by the action of the wind.

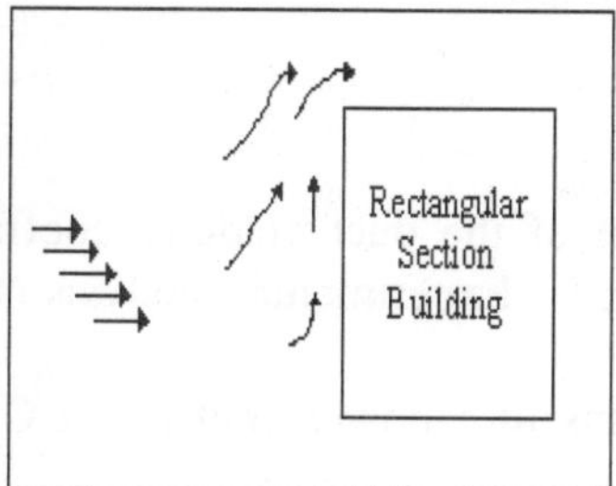

Fig. 1. Building.

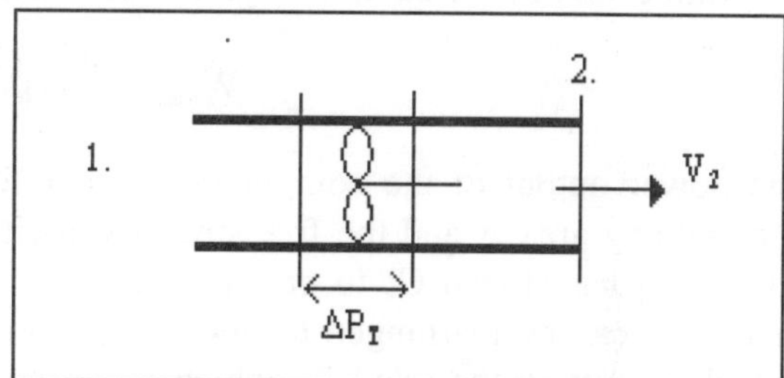

Fig. 2. Simple Duct with Turbine.

2.1 Calculations

To illustrate the potential for power production from device a simple, validated mathematical model (Fig. 2) can be constructed for 1-D flow. When turbine is introduced into a duct it creates a pressure drop δP_T.

2.2 Mathematical Model

To illustrate the potential for power production from such a device, a simple, validated mathematical model has been constructed for one dimensional flow (Grant *et al.* 2002)[2,3]. This is shown in (Fig. 4). Analysis of this model shows that the power extracted from the air stream is

$$W_T = \rho A U_2 \frac{\{\delta U_2^2 - U_\infty^2\}}{2} - \frac{U_\infty^2}{2Cv^2} \qquad ...(1)$$

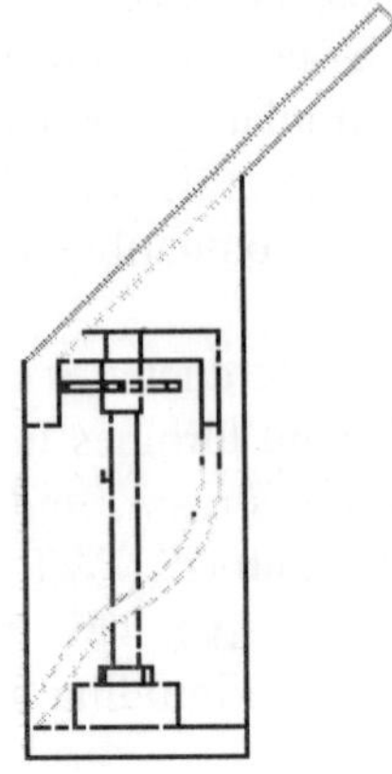

Fig. 3. DWT.

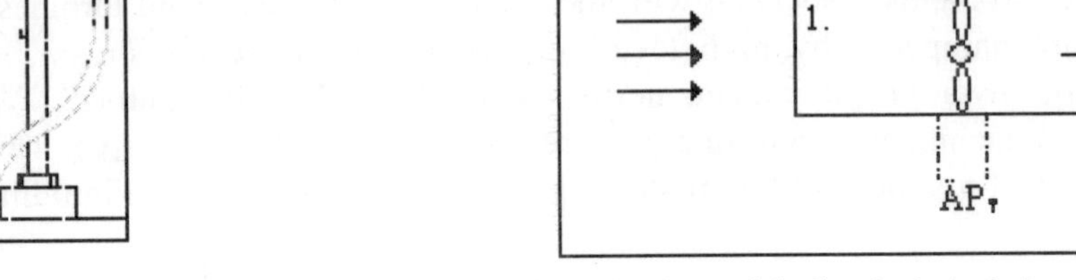

Fig 4. A simple model of a ducted wind energy conversion device.

Differentiating Equation 1 w.r.t. U_2 gives the maximum power condition:

$$U_2 = Cv\sqrt{\frac{8}{3}}U_? \qquad ...(2)$$

Substituting this into Equation 1 gives the maximum power output:

$$W_{T\,max} = \frac{C_v}{3^{3/2}}\rho A\delta^{3/2}U_\infty^3 \qquad ...(3)$$

The power output of the component is, therefore, a function of the duct velocity coefficient (C_V), the opening area A and the free stream velocity of the wind, U. Experimental analysis (Grant *et al.* 2002) [3] has shown C_v to be close to 1.0.

Equation 3 can be rearranged to give the power output in terms of a power coefficient C_p and the available power in the wind P_w which is defined as:

$$P_w = \frac{1}{2}AU_\infty^3\rho \qquad ...(4)$$

By inspection of Equation 3 the power coefficient is:

$$C_p = \frac{2}{3^{3/2}}C_v\delta^{3/2} \qquad ...(5)$$

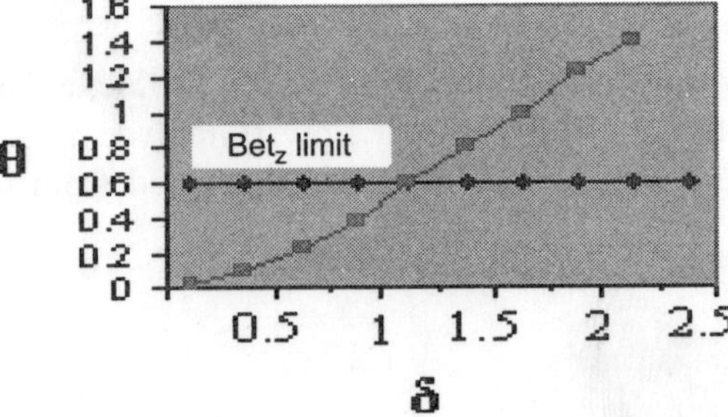

Fig. 5. Variation of C_p with δ.

And the power extracted from the air stream is:

$$W_{Tmax} \; = \; C_p \; P_w \qquad \text{...(6)}$$

Figure 5 shows the variation in C_p with the pressure coefficient differential across the turbine. In conventional wind turbines the power coefficient peaks around 0.593 (the Betz limit). However, it is clear from Figure 5 that the power coefficients considerably greater than the Betz limit can be obtained for ducted wind turbines if losses in the duct are kept to a reasonable minimum.

The reason for this is that the achievable pressure differential across the DWT, a × P, is far higher than that achievable with a conventional wind turbine. This is due to the fact that for the DWT the a × P is being created by the air flow over the building, whereas in a conventional turbine the a × P is caused by the air flow through the blades themselves [4,5].

3. RESULTS AND INTEGRATION WITH BUILDING SIMULATION

Equation 3 forms the basis of the DWT model that has been integrated into the ESP-r building simulation tool as a building-integrated renewable energy component. Examples of other renewable components that already exist in the tool include solar collectors and photovoltaic. These components utilize boundary condition data provided by ESP-r (solar radiation, temperature, etc.) to calculate their heat and/or power output. They are fully integrated with the rest of the building model. For example, the PV model is defined as a component of the building fabric, which interacts with both the thermal and electrical domains of an ESP-r model (Kelly *et al.*, 2001) [6]. In a similar fashion, to calculate its power output, the DWT model uses the wind velocity and direction, and ESP-r calculation of inlet and outlet surface pressure coefficients δ_i and δ_o, respectively.

$$\Delta P = \frac{1}{2}\delta_i \rho U_\infty^2 - \frac{1}{2}\delta_0 \rho U_\infty^2 \qquad \text{...(7)}$$

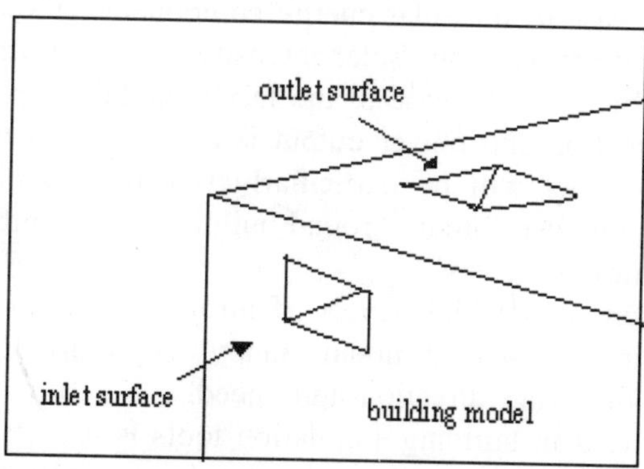

Fig. 6. Integration of DWT model into a building model

These pressure coefficients and their differential (used in equations 1, 2, 3 and 5) are derived from the pressure difference, a × P, (based on stagnation pressures) across the inlet and outlet surfaces of the DWT. The differential pressure coefficient is therefore:

$$\delta = \delta_i - \delta_0 \qquad \text{...(8)}$$

It follows that the pressure difference across the DWT can be found from:

$$\Delta P = \underline{0.5} \; \delta \; \rho \; U_\infty^2 \qquad \qquad \text{...(9)}$$

ESP-r can calculate values of δ_i and δ_o and supply them to the DWT model so that can be calculated. To achieve this DWT component is linked to two surfaces on the ESP-r building model (Fig. 6). At any time t during a simulation ESP-r's calculated values of δ_i and δ_o for these surfaces, and the values of the free stream wind velocity, U_w, held in the simulation climate file are fed into the DWT model. This then calculates the instantaneous electrical power output (W).

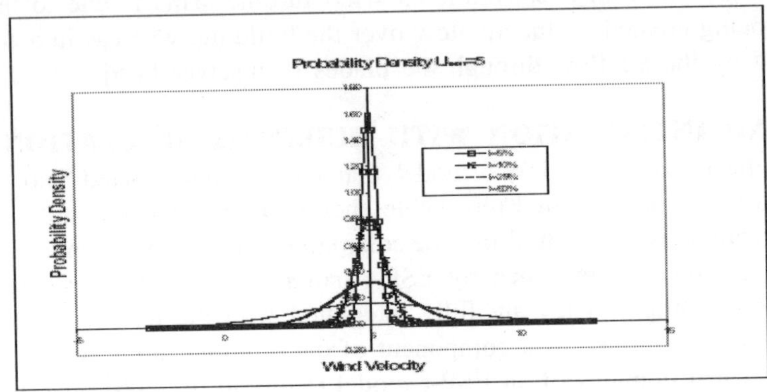

Fig. 7. Result with ESP-r model

4. CLIMATE DATA

The output of all ESP-r's renewable energy components is heavily influenced by the climate data that provides the boundary conditions for their solution. ESP-r simulations generally use hourly averaged climate data. In the case of the solar energy components (PV and solar thermal collectors) the output is largely a linear function of the solar intensity falling on the component and so the use of averaged data is not a problem. In the case of the DWT model the power output is influenced by the local wind speed and direction and power output is a function of the cube of the incident free stream wind speed. So the model will be particularly sensitive to higher speeds, which have a disproportionate impact on the power output. Around buildings in the urban environment such speeds frequently occur in short duration ''gusts''.

Gusting is a manifestation so the high levels of turbulence, e.g. over 30% turbulent intensity (Feranec *et al.*, 2001), found in the so-called 'urban canopy'. High turbulence levels lead to significant, high frequency changes in both wind direction and speed.

As mentioned, the data used in building simulation tools is usually time-averaged, hourly data, often collected at rural weather stations, where the local microclimate and characteristics of the atmospheric boundary layer are often very different to that found in urban areas. This is problematic for the DWT model. Firstly, the use of time averaged data will filter out effects such as wind gusting, which will impact on the power output calculations of the model. Secondly, the wind speed data is often recorded data different height to that which DWT's may be located on a building. Consequently, direct use of ''raw'' climate data as read from simulation climate file may give misleading results as in the potential for power output from the DWT device.

Various techniques have been developed to overcome these problems. The effect of height differences between the measurement point and the location of a simulation component on the wind velocity is often addressed by adjusting the measured data according to as assumed wind velocity profile. An example is the power law profile

$$\frac{U_1}{U_{10}} = Kz_1 \qquad \qquad ...(10)$$

However, profiles must be applied with caution, particularly in urban environments, where their use is often inappropriate as wind patterns within the urban canopy may be dominated by neighboring buildings. Note that this statistical approach does not give a picture of the short-term temporal variation in power output as would be provided by the use of high- frequency monitored wind data with the model. The use of such data would be necessary if the model was used in power quality analysis [7].

To calculate the distribution of wind speeds and the wind directions about their mean values, the two readings [U∞, θ] are recast into two component velocities.

$$\overline{u} = U_\infty \sin\theta, \overline{v} = U_\infty \cos\theta \qquad \qquad ...(11a,b)$$

The instantaneous values of these two components is given by

$$u = \overline{u} + u', v = \overline{v} + v' \qquad \qquad ...(12ab)$$

Where the barred values are the components of mean wind speed and the desired components are the fluctuating velocities. The instantaneous values of the two components are assumed to follow a normal distribution about the mean speed which can be described using the following standard probability density function.

$$f(u) = \frac{1}{\sigma_u \sqrt{2\pi}} \exp\left[-\frac{1}{2}\left(\frac{u - \overline{u}}{\sigma_u}\right)^2 \right] \qquad \qquad ...(13)$$

The standard deviation can be expressed in terms of the turbulent intensity (i) of the air flow. Turbulent flow is given by-

$$I = \frac{\sqrt{\overline{u'^2} + \overline{v'^2} + \overline{z'^2}}}{3U_\infty} \qquad \qquad ...(14)$$

Now in turbulent flow the root mean squared value of fluctuations in all directions are assumed to be identical so that the turbulent intensity can be expressed as:

$$I = \frac{\sqrt{\overline{u'^2}}}{U_\infty} = \frac{\sqrt{\dfrac{1}{n}\sum_{i=1}^{n}(u - \overline{u})}}{U_\infty} \cong \frac{\sigma_u}{U_\infty} \qquad \qquad ...(15)$$

The actual probability of wind speeds occurring over a range [a, b] is given by:

$$p = \int_{u=a}^{u=b} (u)du \qquad \qquad ...(16)$$

CONCLUSION

The effect of changing the turbulent intensity of the air flow in Equation 14 is shown in figure 5. Increasing I increases the spread of velocities and reduces the height of the probability density curve at the mean wind speed. Equation 11 is embedded in the DWT model. At each simulation time step the model makes n samples and calculates the probability density and then probabilities of a range of u and v values, and the combinatorial probabilities of [u, v] pairs. A 3-D distribution of the time duration for a range of wind speeds and directions can then be developed for each time step based on the turbulent intensity [9]. The time duration of each [u, v] pair is given by:

$$t_{u,v} = p(u, v)^* \Delta t$$

NOMENCLATURE

a, K:	coefficients for calculation of wind speed from a profile, A :duct area (m^2),
Cv:	duct velocity coefficient,
I:	Turbulence intensity %,
PW:	Power contained in the wind (W),
$P(u, v)$:	probability of velocity components [u, v] occurring in a simulation time step (%),
n:	number of samples in a time step, t, Ät time, time step length (s),
u,v:	velocity components (m/s),
$\bar{u}, \bar{v}$:	Average velocity components (m/s), v u fluctuating velocity components (m/s), z, z_l velocity component, height,
u:	general velocity component (m/s),
U,U_l:	free stream wind speed, wind speed at height 1 (m/s),
U_{10} :	measured wind speed (m/s),
W_T :	Work extracted from turbine (W),
a:	pressure coefficient differential.

ACKNOWLEDGEMENTS

I am grateful to my project guide Dr. S. Ashok and Prof. Jayaraj, NIT Calicut for their help in completion of the same and the preparation of this manuscript.

REFERENCES

1. Feranec et al 2001-"Wind load on buildings and structures in groups".
2. Kelly N J, Clarke J A, 2001-"Integrating power flow modeling with building   simulation".
3. Born F J, Johnstone C M 2002-"Development of simulation based-decision support tool for renewable energy integration and demand-supply matching".
4. Macdonald I J A, French P, Gillan S, Glover C, Tatton D, Devlin J and Mann R 2000-"The Deployment of photovoltaic components within the lighthouse building".
5. Grant, A D, Dannecker, R K and Nicolson-2002, "Deployment of building integrated wind turbines".
6. Webster G W -1979, "Devices for utilizing the power of the power of the wind".
7. Morgan T.R., Marshall R.H., Brinkworth B. J.(1997),'Ares'-A refined Simulation Program for the Sizing and Optimization of Autonomous Hybrid Energy Systems, Solar Energy, Vol 59, No 4-6, pp 205-215.
8. www.nrel.gov, National Renewable Energy Laboratory.
9. Dannekar R, 2001, "Wind energy in the built envoirnment : An experimental and numerical investigation of a building integrated wind turbines".

127

Fuzzy Logic Based Simulation Approach for Military Aerospace Demand Forecasting

Samir Chabra, S.G. Deshmukh and O.P. Gandhi

Indian Institute of Technology Delhi, India

ABSTRACT

This paper proposes a fuzzy logic based approach to facilitate the forecasting of spares, which is a complex problem in the domain of military aerospace maintenance. Intermittent demands, lack of information sharing and inflexible procurement processes in military aerospace sector makes spares forecasting even more difficult. Military aerospace after sales support poses certain unique challenges and accounts for a significant portion of revenue in the industry. It is important to predict correct support as non-availability of spare parts can adversely affect weapon system availability. As limited studies have been carried out on demand forecasting using adaptive techniques in military aerospace, this paper aims to stimulate thought process of the academicians and practitioners in this area.

Keywords: Fuzzy logic, Military aerospace, Demand forecasting, Simulation.

1. INTRODUCTION

In military aerospace domain, supply chain for repairable commodities begins with the demand forecast, procurement based on inventory policy, manufacture, and distribution of a part; continues with its delivery to a source of repair; and ends with the distribution of the now serviceable asset to retail accounts and maintenance customers in order to return weapon systems to mission capable status. In this context, accuracy of demand forecasting is crucial to the success of the military aerospace supply chain. With the use of high-end technology in military aerospace, the need for spare parts to optimize the utilization of equipment is becoming paramount. Sound spare parts management improves weapon system availability and helps in optimizing after sales spending. Spares provisioning is a complex problem and requires an accurate analysis of all conditions and factors that affect the selection of appropriate spares provisioning models. Various techniques employed in spare parts provisioning have been discussed in many literatures [3, 15, 16, and 18]. Most of these papers deal with the repairable systems and spares inventory management [1, 20, and 23]. Queuing theory approach to determine the spare parts stock on hand to ensure a specified availability of the system has been also used extensively [11, 14]. These models have been further extended to incorporate the inventory management aspect of maintenance [12, 13, and 21]. Also, use of deterministic, stochastic, economic and simulation models is common in supply chain design [2].

This works suggests fuzzy logic based approach towards demand forecasting. Section 1 enumerates peculiarities of military aerospace segment from provisioning viewpoint. In Section 2, existing practices in demand forecasting are briefly reviewed. A fuzzy logic based approach in demand forecasting in military aerospace provisioning environment is presented in Section 3. Finally, Section 4 concludes the work.

2. PECULIARITIES OF MILITARY AEROSPACE PROVISIONING

Military Aerospace industry offers certain unique challenges from provisioning viewpoint as enumerated below:

(a) The defense sector is not as open to public influence as other aspects of public sector and there is close monitoring of defense companies by government [24].

(b) There is a regular dependence on OEMs, as complete design and repair capabilities are not transferred to even major buyers, even under license production and Transfer of Technology arrangements, for commercial, strategic and technological reasons [4].

(c) Characteristics of airborne equipment with an emphasis on factors like weight, speed, strength and performance lead to quality and price difference, when compared to ground based equipments. This is mainly due to stiff screening/ environmental testing requirements for airworthy material qualification.

(d) Procurement is more of joint military-industry partnership and there is a strong dependence on government sales [10]. This implies usually a longer order processing time and also, at times, has restrictive influence in adopting procurement strategies purely on financial considerations. There is tendency of end user to keep higher levels of safety stocks thereby increasing inventory carrying costs.

(e) The product volumes are not very high like automobile or retail industry and the demands are also cyclic [6]. Therefore, competitive manufacturing and vendor development strategies prevalent in automobile or retail industry may not always be applicable here.

(f) The product cost and lifespan is considerably high. Typically, a combat aircraft has a lifespan of around 30 years. Therefore, after sales support accounts for a significant portion of revenue in the industry.

(g) System reliability and maintainability are key drivers in determining spares requirements and life cycle cost.

(h) The product support is also governed by tight resupply window [5], particularly during operations. For instance, if a spare part required for a combat aircraft operating from an aircraft carrier in a sea is to be drop shipped, the delivery options are limited.

(i) Demand-Supply lead-time is generally high due to a combination of various factors. In some cases, OEMs have military as well as commercial aircraft parts production facilities multiplexed for better commercial viability and there may be order queue for made to order parts [17].

(j) Managing obsolescence throws up many challenges. Earlier, an OEM would offer end of life sales to its customers before closing down a product line. Now, there is an effort towards transferring design and technology to a customer selected vendor for sustained production of spare parts, which are under obsolescence [5]. However, this is restricted to smaller parts, which do not require intensive capital investments for setting up production facilities.

(k) Accurate spare parts forecasting is a major area of concern not only for customers and main suppliers but even more for tier-I and tier-II suppliers. The main reasons are application of antiquated techniques in demand forecasting by end customers and lack of inter-organization information sharing due to security restrictions.

(l) Most of the countries permit military hardware/ software sales through designated state agencies, mainly for monitoring purpose. This brings in intermediatories between a customer and an OEM, reducing visibility and increasing bullwhip effect.

3. DEMAND FORECASTING IN MILITARY AEROSPACE

The problem of demand forecasting in a military aerospace maintenance environment is characterized by stochastic intermittent demands, stochastic lead times, multi-item and multi-level inventories and certain intangibles like threat perception, geopolitical dimensions, which have a direct bearing on utilization rate. The current methods of demand forecasting in military aerospace maintenance are based on regulations and somewhat expert knowledge but are largely arbitrary. The forecasts are largely governed by past consumption patterns, stock positions and future equipment usage envisaged. Overstocking of critical long lead-time components is done to reduce backorders. Although there is dependence on historical data, forecasting decisions are significantly influenced by judgment. Many current solution methods for demand forecasting and determining optimal stocking quantities are based on the assumption that parameters are known deterministically. In some cases sensitivity analysis has been performed on inventory models in stochastic environments [15, 9]. Spare parts provisioning models based on Poisson process and renewal theory are quite popular [8]. The Poisson process can be used whenever the failure rate is constant. This means that each failure mode and other factors, which influence the demand, should follow the exponential distribution [7].

Simulation modeling paradigm can create simplified representation of a complex military aerospace demand forecasting system, which may be difficult to mathematically tract using analytical approach. Simulation modeling not only helps in observing system behavior and statistics overtime but also allows controlling time. Therefore, we can quickly observe demand forecasts over long time horizons. Simulation approach may be used to test certain hypothesis about how or why certain phenomenons occur in system. For example, hypothesis could be that system reliability decreases in ageing aircraft. In military context, 'what-if' analysis of unfamiliar situations like wartime spares requirement may be evaluated though simulation-modeling approach.

4. FUZZY LOGIC BASED DEMAND FORECASTING

It is well-known that fuzzy systems have the ability to make use of knowledge expressed in the form of linguistic rules without completely resorting to the precise mathematical models [19]. In recent years, researchers have also attempted to apply it on inventory control problems [22]. The basic configuration of the fuzzy logic system is shown in Fig. 1.

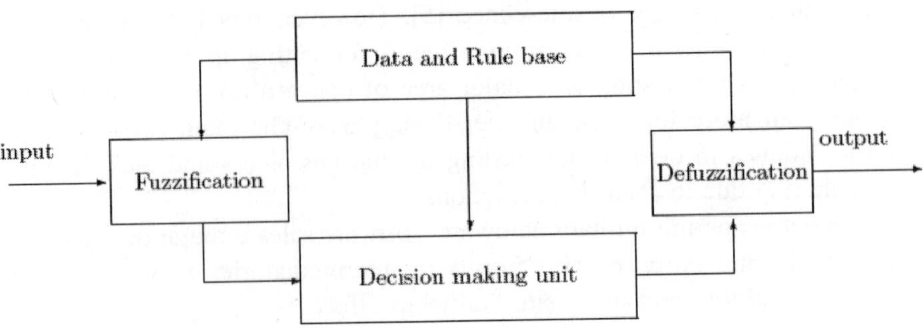

Fig. 1. General structure of a fuzzy logic simulation system.

It is composed of four functional blocks. A rule base contains a number of fuzzy if-then rules and database defines the membership functions of the fuzzy sets used in the fuzzy rules. The decision-making unit performs the inference operations on the rules. A fuzzification interface transforms the crisp inputs into degrees of match with linguistic value. A defuzzification interface transforms the fuzzy results of the inference into a crisp output. The fuzzy rule base consists of a collection of fuzzy if-then rules expressed as the form if 'a' is 'A' then 'b' is 'B', where 'a' and 'b' denote linguistic variables, 'A' and 'B' represent linguistic values that are characterized by membership functions. The input and output variables of a fuzzy system are called linguistic variables as they take linguistic values (e.g., large, small, very large, very small, etc.). The linguistic sets are described by their membership functions. The rule-based presentation of a fuzzy logic system does not include any dynamics, and the computational structure of a fuzzy logic-control consisting of fuzzification, inference, and defuzzification is highly nonlinear. A fuzzy logic system is a kind of gain-scheduling systems, which can cover a wide range of operating regions. This specific feature is appropriate for military aerospace demand forecasting since these demands are characterized by many variables, some of which are difficult to determine accurately.

Main factors, which influence demand patterns of an item required in military aerospace MRO, are past consumption, procurement lead time, utilization rate, reliability, quantum of back orders, fill rates envisaged, assigned life of the component, residual useful life, etc. Precise values of some of these input variables may not be easily available. In such cases, fuzzy logic based approach provides inherent robustness in system modeling. Also, relationship between these variables may be difficult to express through a set of equations or as empirical relations. Fuzzy logic approach will enable use of expert knowledge in military aerospace domain to be translated in terms of linguistic rules. To present the concept of fuzzy logic based demand forecasting, four input variables viz. Past Consumption (C), Procurement Lead Time (LT), Weekly Utilization Rate (UR) and Reliability Index (MTBF) have been considered that influence the output variable viz. Demand (D). Each of these variables, let us assume, can take three linguistic values viz. Low (L), Medium (M) and High (H). A rule base framed is shown in Table 1. Membership functions may be assigned values based on Fuzzy Statistics method and Mamdani Implication may be used along with Centroid defuzzification. Precise shapes of membership function curves are not so important as approximate placement of the curves on universe of discourse, the number of curves used and the overlapping [19].

Table 1. Fuzzy rule base

Rule	Antecedents				Consequence
	C	LT	UR	MTBF	R
1	H	H	H	L	H
2	L	H	H	H	L
3	M	H	M	M	M
4	H	L	M	M	M
5	M	M	L	H	L
6	H	M	L	L	H
7	L	L	M	-	L
8	H	L	H	M	M
9	M	H	L	H	M
10	L	H	H	H	M

CONCLUSION

In this paper, fuzzy logic based demand forecasting for military aerospace maintenance requirements is proposed. Demand forecasting in a complex and multi echelon environment like military aerospace requires consideration of a host of stochastic as well as certain intangible variables. It is imperative to capture appropriate system details and dependencies in the model for accurate results. Availability of past data is expected in the proposed approach to validate the model. The output of the proposed approach is essentially an estimate and may not be as accurate as an output of a deterministic model solved analytically. The fusion of knowledge based techniques like fuzzy logic, neural networks and genetic algorithm may be used to improve the proposed concept. Neural networks may be used to impart learning ability to the fuzzy logic based demand forecasting system and genetic algorithms may be used to optimize the parameters of the membership functions. Further, the concept may be extended to area of prediction and analysis of system reliability, which is one of the main drivers in determining after sales spares requirement. The analysis of prevailing approaches in demand forecasting in military aerospace domain indicate that the fuzzy logic based system may provide better weapon system availability, reduced operating and maintenance costs.

REFERENCES

1. K.P. Aronis, I. Magou, R. Dekker and G. Tagaras. Inventory control of spare parts using a Bayesian approach: A case study, Econometric Institute Report, No. EI-9950/A. Erasmus University, Rotterdam, 1999.
2. B. M. Beamon. Supply chain design and analysis: Models and methods. International Journal of Production Economics, 55:281-294, 1998.
3. A. Chelbi and D. Ait-Kadi. Spare provisioning strategy for preventively replaced systems subjected to random failure. International Journal of Production Economics, 74(1):183-189, 2001.
4. I. Fan, S. Russell, and R. Lunn. Supplier knowledge exchange in aerospace product engineering. European Business Review, 72(1):14-17, 2000.
5. M. T. Farris, M. Wittmann, and R. Hasty. Aftermarket support and the supply chain: Exemplars and implications from aerospace industry. International Journal of Physical Distribution and Logistics Management, 35(1):6-19, 2005.
6. A. A. Ghobbar and C. H. Friend. Sources of intermittent demand for aircraft spare parts within airline operations. Journal of Air Transport Management, 8.
7. B. Ghodrati and U. Kumar. Operating environment-based spare parts forecasting and logistics - A case study. International Journal of Logistics: Research and Applications, pages 490-499, 2004.
8. B. Ghodrati and U. Kumar. Reliability and operating environment based spare parts estimation approach. Journal of Quality Maintenance Engineering, 11(2):169-184, 2005.
9. B. C. Giri and K. S. Chaudhuri. Heuristic models for deteriorating items with shortages and time-varying demand and costs. International Journal of Systems Science, 28(2):153-159, 1997.

10. G. Graham and G. Hardaker. Defence sector procurement and supply chain relationships. European Business Review, 3(3):142-148, 1998.

11. S. C. Graves. A multi-echelon inventory model for a repairable item with one-for-one replacement. Management Science, 31(10):1247-1256, 1985.

12. D. Gross, D. R. Miller, and R. M. Soland. On some common interests among reliability, inventory and queuing. IEEE Transactions on Reliability, 34(3):204-208, 1985.

13. F. Hall and A. J. Clark. ACIM: Availability Centered Inventory Model. Proceeding of the Annual Reliability and Maintainability Symposium, IEEE, New York, NY, pp. 247-52, 1987.

14. J. Huiskonen. Maintenance spare parts logistics: Special characteristics and strategic choices. International Journal of Production Economics, 71(3):125-133, 2001.

15. W. J. Kennedy, J. W. Patterson, L. D. Fredendall. An overview of recent literature on spare parts inventories. International Journal of Production Economics, 76(2):201-215, 2002.

16. J. W. Langford. Logistics: Principles and Applications. McGraw-Hill, New York, 1995.

17. D. Manchester and B. H. Kleiner. What defense companies have done to be more commercially successful. Aircraft Engineering and Aerospace Technology, 69(3):282-286, 1997.

18. D. K. Orsburn. Spares Management Handbook. McGraw-Hill, New York, 1991.

19. T. J. Ross. Fuzzy Logic with Engineering Applications. McGraw-Hill, Singapore, 1997.

20. R. Sarker and A. Haque. Optimization of maintenance and spares provisioning policy using simulation. Applied Mathematical Modeling, 24(10):751-760, 2000.

21. C. C. Sherbrooke. Optimal Inventory Modeling of Systems. John Wiley, New York, 1992.

22. C. S. Shieh. Non-linear rule based controller for missile terminal guidance. IEE Proc. on Control Theory Applications, 150(1):45-48, 2003.

23. C. H. Smith and M. K. Schaefer. Optimal inventories for repairable redundant systems with aging components. Journal of Operation Management, 5(3):339-349, 1985.

24. P. R. J. Trim. Public-private partnerships and the defense industry. European Business Review, 13(4):227-233, 2001.

128

Application of Response Spectrum in Principal Direction for Analysis of Unsymmetric Building

R.K. Ingle[1] AND P.P. Bangosavi[2]

[1]Professor, Department of Applied Mechanics, Visvesvaraya National Institute of Technology, Nagpur, South Ambazari Road, Nagpur-440 011 email: rkingle@vnitnagpur.ac.in
[2]Research Student

ABSTRACT

Unsymmetrical buildings are generally analysis using any one axis of co-ordinate system parallel to one side of building or whichever is simple. IS 1893[1] suggest application of response spectrum in principal direction of the buildings. In view of non-application of the earthquake excitation in principal direction, 100-30-30 rules are suggested. In this paper, an attempt is made to analyse two unsymmetrical buildings with co-ordinate system parallel side of building in one case and parallel to principal direction in another case.

Keywords: Dynamics, Principal direction, Unsymmetrical buildings.

1. INTRODUCTION

Generally seismic forces are applied unidirectional i.e., X and Y-direction separately for the symmetrical building or for unsymmetrical building taking 100% in considered direction and 30% in another direction, if excitation is not applied along principal direction. In this an attempt has been made to apply the seismic excitation in the principal direction and compared the results. It may be noted that the number of load combinations reduces drastically in case the earthquake force is applied along the principal direction in comparison to the application conventionally.

Principal axes of a building are generally two mutually perpendicular horizontal directions in plan of a building along which the geometry of the building is oriented. Principal directions are calculated by the conventional method. Equations 1 and 2 give equations to calculate excitation angle show the maximum and minimum moment of inertia of building and also the excitation angle. This excitation angle is calculated only in the unsymmetrical building.

$$I_{max/min} = \frac{I_x + I_y}{2} \pm \sqrt{\left(\frac{I_x + I_y}{2}\right)^2 + I_{xy}}$$

$$...(1)$$

$$\tan 2\theta = \frac{-2I_{xy}}{I_x - I_y} \qquad \qquad ...(2)$$

Sometimes it may be difficult to obtain principal direction of building plan using above equations. The principal direction may change if there is vertical irregularity in the building. Approximately and also appropriately the mode shape can be used for defining the principal directions of the building as a whole. In case of building it can be seen that major mass gets excited in first fundamental mode and hence the first mode direction can be considered as major principal axis of the building and second mode shape will give minor principal direction.

2. UNSYMMETRICAL BUILDING

Two unsymmetrical buildings are considered for the study, one with shear wall and another as framed structure with column and beams only. The buildings are briefly described below.

2.1 Unsymmetrical Building with Shear Wall

The G + 9 residential reinforced concrete building as shown in Figure 1 is used to study. Building is unsymmetrical about both axes. The analysis and design is considered for the gravitational load

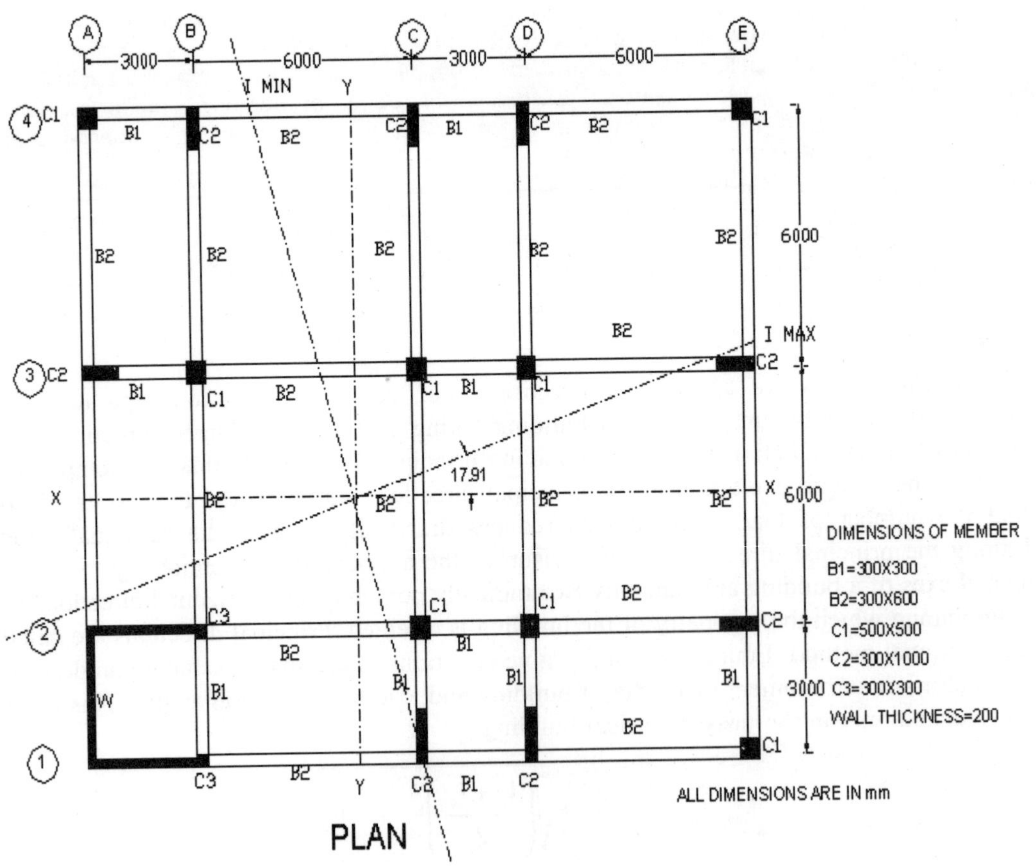

Fig. 1. Plan of (G + 9) unsymmetrical Building.

coming on the frame of the building and seismological load for zone III. Hard type of soil condition is assumed with columns supporting on isolated footing. Importance factor of building is taken as 1 and response reduction factor is taken as 5 considering moment resisting frame with shear wall. For calculation of base shear, seismic weight as the sum of dead load and 0.25 times the live load is used.

In this the channel shaped shear wall with boundary elements (C3) is used. Model is analyzed by using response spectrum analysis in SAP2000.

2.2 L-Shape Building

The G + 4, L-shaped building with specified dimensions, plan as shown in Fig. 2 is taken to study. In this model only bare frame is modeled.

In this building, the moment of Inertia about both axis are same, so the principal direction of the building is 45°. Analysis is done using the SAP2000 and the first three modes are shown in Fig. 3(a-c). It can be seen that the first mode indicates the major principal direction whereas the second mode shape indicate the minor principal axis.

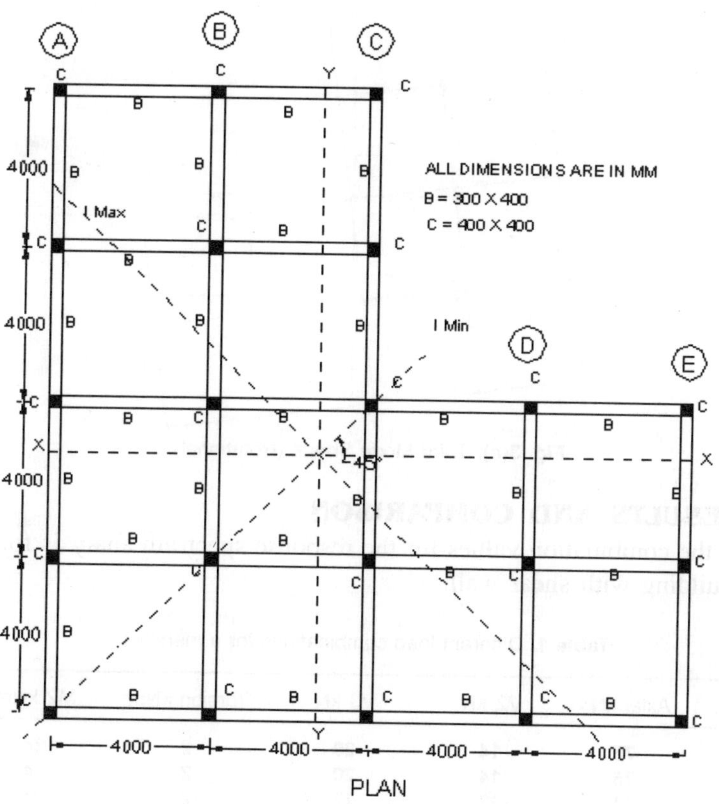

Fig. 2. Plan of (G+4) L-shape building.

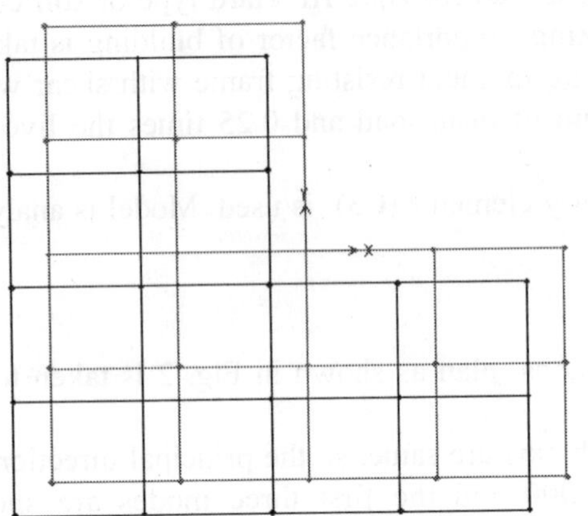

Fig. 3(a). First Mode Shape (Diagonal).

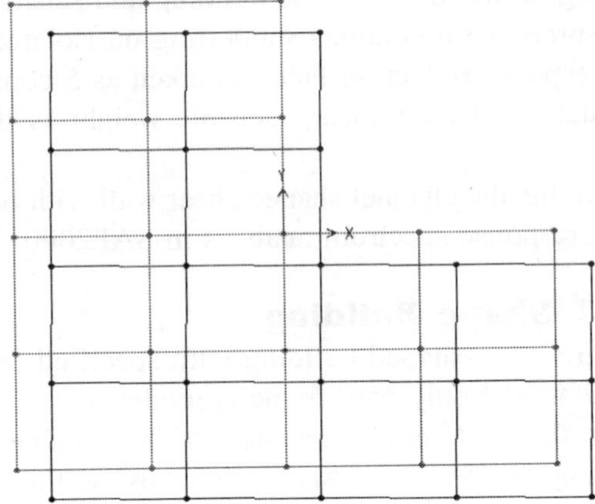

Fig. 3(b). Second Mode Shape (Diagonal).

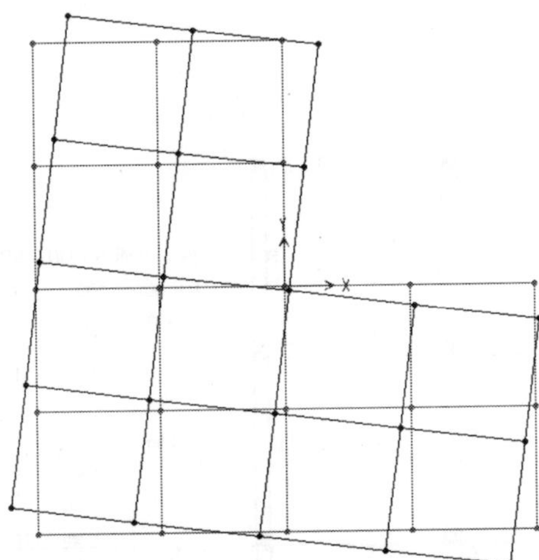

Fig. 3(c). Third Mode Shape (Rotational).

3. ANALYSIS RESULTS AND COMPARISON

Tables 1 and 2 give the combination values for the response spectrum analysis for different members in unsymmetrical building with shear wall.

Table 1. Different load combinations for a member.

Combinations	Axial kN	V2 kN	V3 kN	Torsion kNm	M2 kNm	M3 kNm
Max Principal	**75**	**14**	**20**	**2**	**24**	**23**
SRSS	**75**	**14**	**20**	**2**	**24**	**22**
100 + 30 + 30	74	13	17	2	21	21
100 − 30 + 30	46	8	15	0	10	12
100 − 30 − 30	29	8	15	1	10	12
100 + 30 − 30	57	13	17	1	21	21

Contd..

−100 + 30 + 30	−29	−8	−15	2	−10	−12
−100 − 30 + 30	−57	−13	−17	−1	−21	−21
−100 − 30 − 30	−74	−13	−17	−2	−21	−21
−100 + 30 − 30	−46	−8	−15	−1	−10	−12

Table 2. Different load combinations for shell element.

Combinations	F11 KN	F22 kN	F12 kN	M11 kNm	M22 kNm	M12 kNm
Max Principal	**21**	**107**	**52**	**3**	**16**	**2**
SRSS	**21**	**107**	**52**	**3**	**16**	**2**
100 + 30 + 30	19	96	51	3	15	1
100 − 30 + 30	10	48	31	2	8	1
100 − 30 − 30	8	42	29	2	8	1
100 + 30 − 30	18	91	49	3	15	1
−100 + 30 + 30	−8	−42	−29	−2	−8	−1
−100 − 30 + 30	−18	−91	−49	−3	−15	−1
−100 − 30 − 30	−19	−96	−51	−3	−15	−1
−100 + 30 − 30	−10	−48	−31	−2	−8	−1

Table 3 gives the combination values for the response spectrum analysis for a member in L-shaped unsymmetrical building.

Table 3. Different load combinations for a typical member.

Combinations	Axial kN	V2 kN	V3 kN	Torsion kNm	M2 kNm	M3 kNm
Max Principal	**33**	**7**	**7**	**0**	**12**	**12**
SRSS	**33**	**7**	**7**	**0**	**12**	**12**
100 + 30 + 30	31	6	6	0	11	11
100 − 30 + 30	17	3	3	0	6	6
100 − 30 − 30	16	3	3	0	6	6
100 + 30 − 30	30	6	6	0	11	11
−100 + 30 + 30	−16	−3	−3	0	−6	−6
−100 − 30 + 30	−30	−6	−6	0	−11	−11
−100 − 30 − 30	−31	−6	−6	0	−11	−11
−100 + 30 − 30	−17	−3	−3	0	−6	−6

The above results show that applying the forces in principal direction or using SRSS method can perform the seismic analysis. This will reduce the number of load combinations to be considered in the 100-30-30 rules.

CONCLUSION

Following are some conclusions drawn from the above study:

i. In SRSS combination and earthquake loads applying in principal direction gives conservative results with respect to the 100/30/30 combination.

ii. Applying response spectrum along principal direction and the SRSS combination results are same.

iii. The mode shapes can be used in evaluating the principal direction of the building.

iv. The number of load combinations in SRSS reduces as compared to 100/30/30 combination.

NOMENCLATURE

Axial	Axial force
V2	Shear in local 2 direction
V3	Shear in local 3 direction
M2	Moment about local 2 axis
M3	Moment about local 3 axis
F11, F22, F12	Normal and shear Stress Components in shell element
M11, M22, M12	Moment Components in shell element

REFERENCE

1. IS: 1893 (part1): 2002, "*Criteria for Earthquake Resistant Design of Structure*", Bureau of Indian standard, New Delhi.

129

Scale-Dependent Adhesive Friction Based on Plastic Asperity Concept

S. Mohamed Ali[1] and Prasanta Sahoo[2]

[1]Department of Mechanical Engineering, Pondicherry Engineering College, Pondicherry-605 014, India.
email: smdalipec@hotmail.com
[2]Department of Mechanical Engineering, Jadavpur University, Kolkata-700 032, India.
email: psahoo@mech.jdvu.ac.in

ABSTRACT

The paper describes an analysis of adhesive friction between rough surfaces with small-scale surface asperities using an elastic-plastic model of contact deformation based on fictitious plastic asperity concept and the single asperity friction model of Hurtado and Kim is integrated with the Johnson-Kendall-Roberts adhesion model to consider the adhesive friction behavior of rough surfaces.The model considers a realistic asperity deformation, i.e., simultaneous occurrence of elastic and plastic behaviours for an asperity. Results show that coefficient of friction depends strongly on applied load, surface roughness and elastic adhesion index. Comparison with previous elastic-plastic model that was based on elastic-then-plastic assumption is made showing significant differences.

Keywords: Scale-dependence; Adhesive friction; Plastic asperity.

1. INTRODUCTION

Adhesion at the contact of rough solids has been studied analytically in great detail by Chang *et al.* [1] [CEB model] and Roy Chowdhury and Ghosh [2] [RG model] using Greenwood and Williamson's rough surface model [3] (Fig. 1).The RG model used the JKR [4] adhesion model. In RG model, it is considered that some contacting asperities remain purely elastic and other purely plastic and the model overlook a wide intermediate range of interest where elastic-plastic contact prevails. In CEB model an attempt was made to bridge this gap. However, the simplifying assumption introduces a discontinuity in the contact load at the transition from elastic to elastic-plastic contact. Further modifications based on mathematical, rather than physical, considerations to smooth the discontinuity in the CEB model were proposed but failed to provide a solution to the basic problem of lacking accuracy in the elastic-plastic contact regime. Such accurate solution calls for the use of a Finite Element Method (FEM).

Recently, Kogut and Etsion [5] presented an accurate elastic-plastic finite element solution for the contact of a deformable sphere pressed by a rigid flat [KE model]. Their solution provides convenient dimensionless expressions for contact load and area covering a large range of interference from yielding inception to fully plastic contact of the sphere. The more accurate FEM solution [5] provides a solution to the basic problem in the intermediate elastic plastic contact regime.

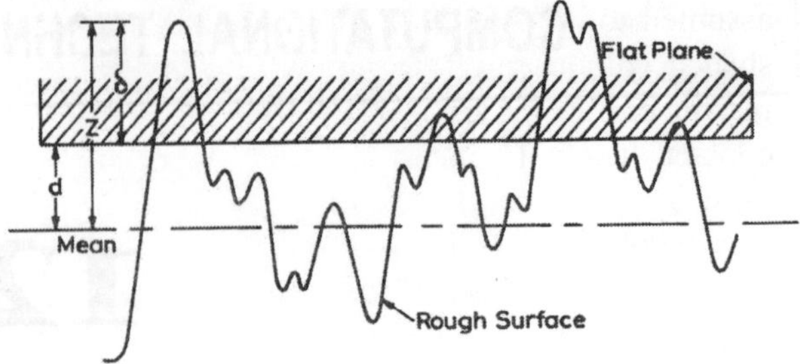

Fig. 1. Contact between a rough deformable surface and a rigid flat plane

The present work considers a new elastic plastic model (termed as PAC hereafter) for adhesive contact analysis of JKR contacts following the plastic asperity concept by Abdo and Farhang [6] that provides the more realistic picture of plastic deformation of rough surfaces through the definition of fictitious plastic asperities that are assumed to be embedded at a critical depth within the actual surface asperities. The present work also analyses adhesive friction for contacting rough surfaces using the single asperity friction model of Hutrado and Kim [7-8] [HK] and the JKR model of adhesion. The present study includes two new considerations; incorporation of a more realistic elastic-plastic contact model and the JKR adhesive contact situations.

2. PLASTIC ASPERITY CONCEPT

In a recent pioneering work, Abdo and Farhang [6] presented the plastic asperity concept for modelling the elastic-plastic contact of rough surfaces. The salient feature of their approach is outlined here in brief in order to set a scene for the present analysis. Considering the contact between one single asperity on rough surface and a rigid plane, the behaviour of the asperity is initially elastic. As the load is increased the elastic behaviour continues to describe the deformation until a critical interference is reached. At this critical load and beyond, the asperity deforms as a purely plastic body. Hence, for every asperity there are two types of interactions. The first is the elastic contact between the plane and the asperity. If the interference (δ) exceeds the critical interference (δ_c), then the interaction also includes plastic contact. Abdo and Farhang [6] considered elastic-plastic contact through the introduction of a fictitious asperity that can only deform plastically. As

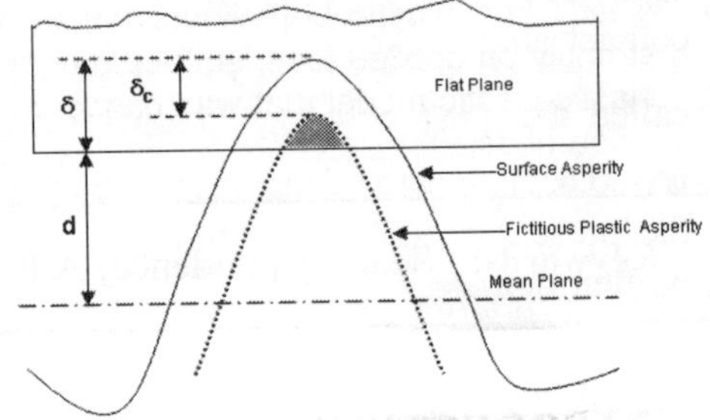

Fig. 2. Elastic-plastic interaction of an asperity with a rigid flat plane

shown in Fig. 2, the critical interference (δ_c) is used to define the plastic asperity. The surface of this plastic asperity is obtained by displacement of every point on the physical asperity by δ_c along the direction normal to the original surface asperity. Using mathematical mapping of a point on the physical asperity to a corresponding point on the plastic asperity, Abdo and Farhang [6] derived the expression for the radius of curvature for such a fictitious plastic asperity (Rp) as

$$R_P = R - \delta_c \qquad \qquad ...(1)$$

where, **R** is the radius of curvature for the original surface asperity. Thus, plastic asperities are assumed to be embedded at a critical depth of δ_c within the actual surface asperities. In Fig. 2, the shaded volume representing the interference of the plastic asperities and the plane contribute to the plastic portion of contact whereas the remaining volume of interference contributes to the elastic contact. It may be noted that with the plastic contact pressure there exists an elastic quantity, which must be subtracted to obtain the net elastic contribution. Therefore, if P be the contact load, then it may be obtained by appropriately accounting for the aforementioned interactions. Thus,

$$P = P_{e1} - P_{e2} + P_{p2} \qquad \ldots(2)$$

where, suffix 1, 2 represent actual surface asperity and fictitious plastic asperity, respectively and e, p represent elastic and plastic interactions, respectively.

3. FRICTION ANALYSIS

In analyzing the adhesive friction it is necessary to consider both the normal load and the friction force in the presence of surface forces. In the present model, it is considered that the surfaces contain asperities that deform elastically and within these asperities are embedded the fictitious plastic asperities that can deform only plastically. Following the works of Mukherjee *et al.* [9] for contact between a smooth sphere and a flat in the presence of surface forces [EP model], the total load on all asperities in contact per unit area for the present case can be written following Eq.(2)

$$P_1 = N\int_d^\infty P_e\varphi(z)dz - N\int_{d+\delta_{c1}}^\infty P_e\varphi(z)dz + N\int_{d+\delta_c}^\infty P_p\varphi(z)dz \qquad \ldots(3)$$

where, the first two integrals represent the net elastic contribution to applied load and the third integral represents the plastic contribution to applied load. In Eq.(3) P_e and P_p are the expressions for load on an elastically and plastically deformed asperities respectively [9]. It may be noted here that the plastic asperity peaks are considered as being farther away from the plane by δ_c. Thus the mean plane of plastic asperity is considered as being δ_c below the mean plane of the surface asperities and hence the limits of integration need to be shifted by δ_c for plastic asperities as presented in Eq.(3). Now in any contact analysis of rough surfaces, the nature of contact is indicated by the value of the parameter called the plasticity index, $\psi = (E*/H)\sqrt{\sigma/R}$ defined by Greenwood and Williamson [3]. For $\psi < 0.6$, the contact is predominantly elastic and for $\psi > 1.0$, the contact is predominantly plastic. In the present elastic-plastic analysis, ψ is used as one of the governing parameters. Eq.(3) may be written non-dimensionally in terms of elastic adhesion index θ and plasticity index ψ as

$$\overline{P}_1 = \int_{\Delta_0}^\infty \left(\Delta^{3/2} - \frac{4.34}{\theta^{1/2}}\Delta^{3/4}\right)\overline{\phi}(\Delta)d\Delta - (1 - \Delta_c\alpha^2)^{1/2}\int_{\Delta_{c1}}^\infty \left(\Delta^{3/2} - (1 - \Delta_c\alpha^2)^{1/4}\frac{4.34}{\theta^{1/2}}\Delta^{3/4}\right)\overline{\phi}(\Delta)d\Delta$$

$$+ (1 - \Delta_c\alpha^2)^{1/2}\int_{\Delta_c}^\infty \left(\frac{4.75}{\psi}\Delta - \frac{6.28}{\theta}\right)\overline{\phi}(\Delta)d\Delta \qquad \ldots(4)$$

where, $\overline{P}_1 = P_1/KNR^{1/2}\sigma^{3/2}$, $\Delta = \delta/\sigma$, $\Delta_c = \delta_c/\sigma$, $\Delta_{c1} = \delta_{c1}/\sigma$, $\alpha = \sqrt{\sigma/R}$ and $\overline{\phi}(\Delta)$ is the normalized asperity height distribution function. For Gaussian surfaces $\overline{\phi}(\Delta) = (1/\sqrt{2\pi})e^{-(h+\Delta)^2/2}$.

Here, Δ_0 is the non-dimensional apparent displacement corresponding to actual displacement $\delta = 0$.

With the above substitutions we may get the limits of Eq.(4) from following equations which are written using the elastic adhesion index (θ) and plasticity index (ψ) [9], $\Delta_{c1}^{3/4} - (8.65/\psi)\Delta_{c1}^{1/4} - (4.34/\theta^{1/2}) \geq 0$, $\Delta_c = \Delta_{c1} - (2.89/\theta^{1/2})\Delta_{c1}^{1/4}$ and Δ_0 may be obtained from this by substituting $\Delta_c = 0$ with $\Delta_{c1} = \Delta_0$ and written as $\Delta_0 = (4.125/\theta^{2/3})$. In the expression for non-dimensional load in Eq.(4), the first part of the right hand side of the equation refers to elastic load on all the asperities and the third part refers to plastic load on all asperities that undergo deformation beyond the critical interference (Δ_c) while the second part refers to elastic load of plastic asperities. Now following the works of Ali and Sahoo [10], which is based on the **HK** single asperity frictional slip model extended to elastic-plastic regime and plastic regime in addition to elastic regime [11], the corresponding friction force for the present case may be written following Eq.(2) as

$$T = N\int_d^\infty T_e\phi(z)dz - N\int_{d+\delta_{c1}}^\infty T_e\phi(z)dz + N\int_{d+\delta_c}^\infty T_p\phi(z)dz \qquad \text{...(5)}$$

with the same non-dimensional scheme as

$$\overline{T} = \frac{3(1-\upsilon)}{8\alpha\beta^2}\left[2\pi\overline{\tau}_{f1}\int_0^{\overline{a}_1} \overline{a}^2\overline{\phi}(\Delta)d\Delta + 2\pi10^B\int_{\overline{a}_1}^{\overline{a}_2} \overline{a}^{M+2}\overline{\phi}(\Delta)d\Delta + 2\pi\overline{\tau}_{f2}\int_{\overline{a}_2}^\infty \overline{a}^2\overline{\phi}(\Delta)d\Delta\right] \qquad \text{...(6)}$$

where, $\overline{\tau}_{f1} = 1/43$, $\overline{\tau}_{f2} = \overline{\tau}_{f1}/30$, $\overline{a}_1 = 28$, $\overline{a}_2 = 8\times10^4$, $\overline{T} = T/KNR^{1/2}\sigma^{3/2}$, $\alpha = \sqrt{\sigma/R}$, $\beta = (\sigma R)^{1/2}/b$ and υ is Poisson's ratio. Now, for evaluation of the total friction force using Eq. (5) non-dimensional contact radius ($\overline{a} = a/b$) needs to be expressed in terms of Δ. For the three integrals in Eq.(5), the relations between $\overline{a}$ and Δ are expressed, respectively, as $\overline{a} = \beta\Delta^{1/2}$, $\overline{a} = (1-\alpha^2\Delta_c)^{1/2}\beta\Delta^{1/2}$ and $\overline{a} = \sqrt{2}(1-\alpha^2\Delta_c)^{1/2}\beta\Delta^{1/2}$. While evaluating $\overline{T}$, the integrands in Eq.(5) has to be replaced by the right hand side of the Eq.(6) and the respective relation between $\overline{a}$ and Δ needs to be used and proper limit has to be supplied based on the dimensionless contact radius ($\overline{a}$). i.e., whether it falls on the concurrent slip region or single dislocation assisted slip region or multiple dislocations co-operated slip region, depending on the nature of deformation of the asperity. The coefficient of friction is then evaluated as $\mu = \overline{T}/\overline{P}_1$.

4. RESULTS AND DISCUSSION

Equations established in the previous section are evaluated numerically. The load-separation behaviour and frictional characteristics for Gaussian surfaces are investigated for typical combinations of θ and ψ. Limiting value of θ for JKR contacts is usually quoted as 10, beyond which effect of adhesion becomes insignificant due to surface roughness effect. In adhesive contact analysis of rough surfaces using JKR model, elastic adhesion index θ and plastic adhesion index λ are generally used as parameters. In the present analysis of adhesive contact for rough surfaces, θ and ψ are used as parameters in order to consider the effect of adhesion and contact condition simultaneously. However, θ, λ and ψ may be related as $\psi = 0.65(\theta^{1/2}/\lambda^{1/4})$ and thus for any combination of θ and ψ, the corresponding λ may be evaluated. Limiting value of θ for JKR contacts is usually quoted as 10, beyond which effect of adhesion becomes insignificant due to surface roughness effect. In the present analysis, typical values of θ between 2 and 15 are considered in order to analyse the effect of adhesion. The value of the plasticity index, ψ determines the nature of contact. For $\psi < 0.6$, the contact is predominantly elastic and for $\psi > 1$, the contact is predominantly plastic, while for $0.6 < \psi < 1.0$, the contact is elastic-plastic. Thus ψ values are considered in the range of 0.5 to 2 in order to consider the whole range of deformation from predominantly elastic to predominantly plastic including elastic-plastic. The non-dimensional mean separation ($h = d/\sigma$) is considered between +1 to –4. At larger separations, $h > 1$ the number of asperities in contact is very small, as evidenced from the magnitude of external load. On the other hand minimum value of h has been taken as –4 to avoid any bulk deformation and for $h < -4$, the present asperity based contact model may not yield proper results. In the present model, a dimensionless number α appears in the non-dimensional load expression. This may be termed as roughness parameter and its practical value varies in the range of 0.005 to 0.05 for smaller-scale systems. However, in the present analysis it is found that the variation of α has insignificant effect on load value and for all loading characteristics presented here α is taken as 0.01. Another dimensionless number, β used in frictional force equation has the practical values for micro nano contact situations are in the order of 100 to 1000.

Figure 3 shows the plots of non-dimensional applied load against mean separation for typical combinations of θ and ψ. It also shows the comparison of present PAC model and the earlier EP model for the same parametric combinations in the same figure.

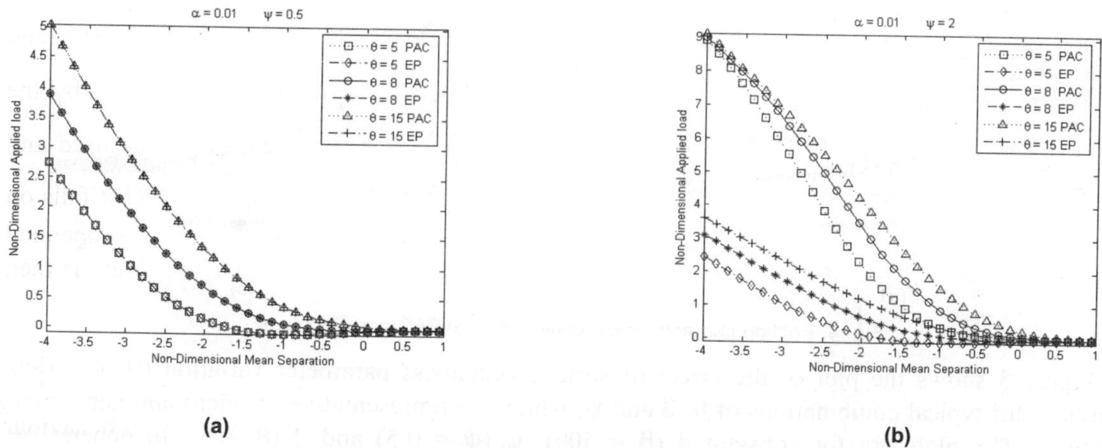

Fig. 3. Loading characteristics: Comparison of PAC and EP model.

At low plasticity index ($\psi = 0.5$), as can be seen in Fig. 3(a), the present analysis predict the same loading force as that for the EP model for all mean separation. Thus, for predominantly elastic contact situation, both the models give same result. This is clearly because the present formulation differs from the EP model with respect to plastic behaviour modelling. When there is no plastic behaviour, the two models should yield the same result. At $\psi = 0.5$, the contact is predominantly elastic and hence the two models predict the same loading force. For elastic-plastic contact situation, i.e., for $\psi = 0.9$, the loading force for the present model differs significantly particularly at low (negative) mean separation, the same has been omitted for brevity here. For predominantly plastic contact situation, i.e., for $\psi = 2$, as can be seen in Fig. 3(b), the difference in loading force from the two models is still higher. Thus, for predominantly elastic contact situation, both the models give same result. But for elastic-plastic and predominantly plastic contact conditions, EP model underestimates the loading force particularly at low mean separations.

Figure 4 shows the plots of comparison of friction behavior between the EP model and the present PAC model in elastic and plastic regimes for different values of ψ and typical combinations of α, β and for varying θ. In general, the trends are same and the load dependence of friction coefficient is always present. PAC model predicts higher coefficient of friction than the EP model in predominantly elastic ($\psi = 0.5$) and elastic-plastic situations. The difference is relatively very high at low load situations and high adhesive situations, i.e., at lower values of elastic adhesion index. In predominantly plastic regime ($\psi = 2$) the prediction of friction coefficient from the two models are comparable. At higher θ, the EP model predicts comparatively higher coefficient of friction. This is due to fact that the EP model predicts lower applied load which in turn predicts higher friction coefficient.

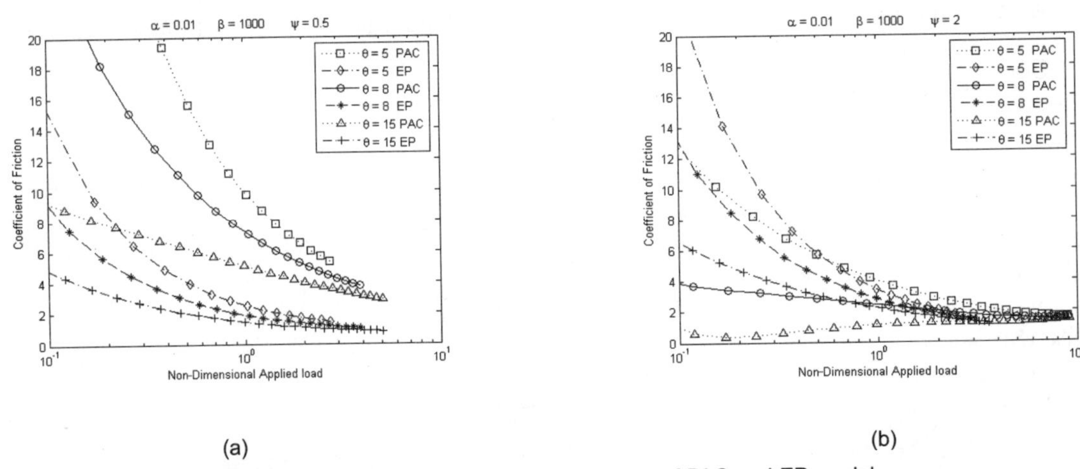

Fig. 4. Friction characteristics: comparison of PAC and EP model.

Figure 5 shows the plot of the effect of surface roughness parameter variation (α) on friction behaviour for typical combinations of θ, β and ψ, which are representative of micro and nano contact situations. The plots are for constant β ($\beta = 500$), ψ ($\psi = 0.5$) and θ ($\theta = 5$). In general, with increase in the surface roughness parameter α, the friction coefficient decreases significantly. This effect is more pronounced at low θ, i.e., for adhesive contact. The same trend is observed for varying θ values. Figure 6 shows the plot of the effect of friction regime parameter variation (β) on friction

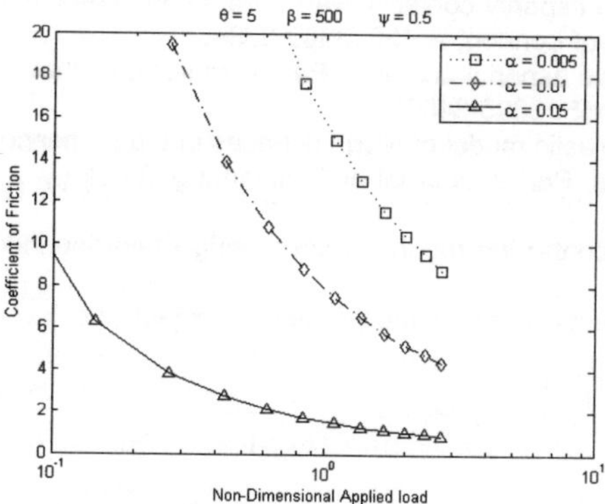

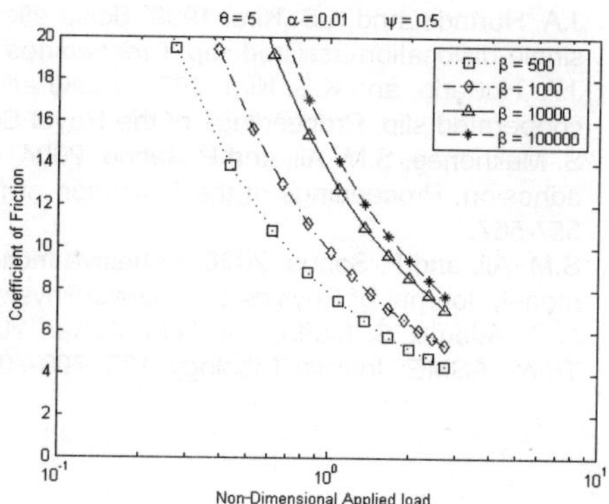

Fig. 5. Friction characteristics of PAC model for varying surface roughness parameter.

Fig. 6. Friction characteristics of PAC model for varying friction regime parameter.

behaviour for typical combinations of θ, α and Ψ. For low values of elastic adhesion index and predominantly elastic contact condition, the coefficient of friction is higher at lower applied load and decreases with increase in applied load. At a particular normal load the coefficient of friction is more for higher β. The trend of variation of friction coefficient with β values remains the same for different θ values though the magnitude of coefficient of friction decreases for higher θ values.

CONCLUSION

The adhesive friction is analyzed using the plastic asperity concept considering the simultaneous occurrence of elastic and plastic behavior within the same asperity. Thus, it represents more realistic situation than the previous elastic-plastic model, which can be characterized as elastic-then-plastic contact. The HK frictional slip model is incorporated to study the frictional behavior. Non-dimensional parameters like surface roughness parameter and friction regime parameters in addition to well established elastic adhesion index and plasticity index are used to consider different conditions that arise as a result of varying load, surface and material parameters. The coefficient of friction in general decreases with increasing applied load. Friction coefficient is found to be strongly dependent on surface roughness parameter and elastic adhesion index. The friction coefficient is very high for low roughness and high adhesion. Comparative study with the earlier model, i.e., elastic-plastic model shows substantial difference particularly at low load and high adhesive situations.

REFERENCES

1. W.R. Chang, I. Etsion, and D.B. Bogy, 1988, Adhesion Model for Metallic Rough Surfaces, Trans. ASME: Journal of Tribology, 110, 50-56.
2. S.K. Roy Chowdhury, and P. Ghosh, 1994, Adhesion and adhesional friction at the contact between solids, Wear, 174, 9-19.
3. J.A. Greenwood, and J.P.B. Williamson, 1966, Contact of nominally flat surfaces, Proceedings of the Royal Society of London, A 295, 300-319.
4. K.L. Johnson, K. Kendall, and A.D. Roberts, 1971, Surface energy and the contact of elastic solids, Proceedings of the Royal Society of London, A 324, 301-313.
5. L. Kogut, and I. Etsion, 2002, Elastic-plastic contact analysis of a sphere and a rigid flat, Trans ASME: Journal of Applied Mechanics, 69, 657-662.

6.   J. Abdo, and K. Farhang, 2005, Elastic-Plastic model for rough surfaces based on plastic asperity concept, International Journal of Non-linear Mechanics, 40 (4), pp. 495-506.

7.   J.A. Hurtado, and K.S. Kim, 1999, Scale effects in friction of single asperity contacts: Part I; from concurrent slip to single-dislocation-assisted slip, Proceedings of the Royal Society of London, A 455, 3363-3384.

8.   J.A. Hurtado, and K.S. Kim, 1999, Scale effects in friction in single asperity contacts: Part II; multiple-dislocation-cooperated slip, Proceedings of the Royal Society of London, A 455, 3385-3400.

9.   S. Mukherjee, S.M. Ali, and P. Sahoo, 2004, An improved elastic-plastic model of rough surfaces in the presence of adhesion, Proceedings of the Institution of Mechanical Engineers, Part J: Journal of Engineering Tribology, 218, 557-567.

10.   S.M. Ali, and P. Sahoo, 2006, Adhesive friction for elastic-plastic contacting rough surfaces using scale-dependent model, Journal of Physics D: Applied Physics, 39, 721-729.

11.   G.G. Adams, S. Muftu, and N.M. Azhar, 2003, A scale-dependent model for multi-asperity contact and friction, Trans. ASME: Journal Tribology, 125, 700-708.

130

Triangular Mesh Generation over an Arbitrary Two-Dimensional Domain by Hybrid Advancing Front Technique

S. Sai Babu[1] and G. Saravana Kumar[2]

[1]Department of Mechanical Engineering, Indian Institute of Technology Guwahati-781 039, India. email: sanka.s@iitg.ernet.in
[2]Department of Mechanical Engineering, Indian Institute of Technology, Guwahati-781 039, India. email: gskumar@iitg.ernet.in

ABSTRACT

The paper presents a hybrid advancing front technique (HAFT) for triangular mesh generation over arbitrary two-dimensional domains. The boundary of the domain to be meshed is modeled using non-uniform rational b-spline curves (NURBS). First, the boundary of the domain is discretized into field points from the parametric model. The boundary mesh thus constructed is then advanced by field points defined following a hybrid technique consisting of a *pseudo* implicit boundary offset and conventional advancing front technique (AFT), along with the connectivity. Mesh generation for many case studies have been done to show that the quality of the mesh obtained by HAFT is better in comparison with that by conventional AFT.

Keywords: Mesh generation; Advancing front technique; NURBS offsetting; Delaunay triangulation.

1. INTRODUCTION

In recent years the Finite Element Method and Boundary Element Method have become integral components in the design/synthesis process in virtually all engineering sciences. Domain descritisation plays an important role in finite element analysis because the convergence criteria will mostly depends on it. Most of mesh generation tasks involve a great deal of effort to generate numerical data, such as node numbers, nodal point coordinates, parameter values and element connectivity. This work could be very time-consuming and error prone if done manually, especially when a large number of elements or very complex geometry is encountered. So algorithms for the automatic mesh generation gained the popularity. The last decade has seen a considerable amount of effort devoted to automatic mesh generation [1-9], resulting in two very powerful and by now mature techniques: the Advancing

Front and the Delaunay Triangulation. The *quality* of the mesh is related to the shape and the size of the mesh elements and significantly affects the accuracy of the numerical simulation. In the present paper, we present an unstructured mesh generation method that yields good quality mesh using a hybrid technique consisting of a *pseudo* implicit boundary offset and advancing front. In a conventional AFT, the selection of field points involves checking various validity and *quality* criteria involving computationally expensive procedures. If the geometry of the domain could be defined implicitly using distance function representation, all field points at a specified distance from the domain boundary can be obtained using the distance function. This aids in advancing the front with lesser computations and yielding meshes with high quality. In order to derive this advantage of implicit geometric definition, we create field points by *pseudo* implicit offset of the domain boundary.

2. PREVIOU-781 039S WORK

In recent years a number of algorithms have been developed for the creation of elements (nodes as well as their connectivity) inside the geometry for mesh generation. There are two approaches for the node placement during the mesh generation. The first one is incrementally inserting the nodes one by one into the existing mesh and the second one is placing all the nodes in the geometry first and then connecting them into a mesh. Seven major mesh generation approaches and a scheme for classifying mesh generation methods is proposed by Ho-Le [1]. Most popular meshing algorithms in the literature include advancing front technique [2-5] and Delaunay triangulation methods [6]-[7]. El-Hamalawi [8] proposed a mesh generation technique combining advancing front and Delaunay triangulation techniques. Mesh generation methods like bubble meshing method [9], Monte Carlo method [10], *dist mesh* [11] and offsetting method [12] place the nodes into optimal locations before connecting them into a mesh.

3. DOMAIN DESCRIPTION USING NURBS AND BOUNDARY DESCRETISATION

NURBS curves have been used to describe the boundary of the domain to be meshed. The formulation of NURBS renders it the ability to represent simple analytical curves like line, conics, etc. to complex free from shape. Consider a NURBS curve represented by $P(t)$ defined for t $\in$ [0 1] and $k+n+1$ non uniform knot vectors in the domain. It is controlled by the $n+1$ vertices of the control polygon B_i and h_i, the weights of kth order b-spline basis functions $N_{i,k}(t)$ defined or each vertex. The boundary of the domain may be composed of number of NURBS curves. Both for the AFT and the proposed HAFT, these curves are to be sub-divided into number of linear segments of equal length as specified by the element size. Exact methods are computationally expensive and thus an approximate method is used. It involves first the computation of chord distance between two sufficiently close points say P_j and P_{j+1} of a NURBS curve $P(t)$ and summing such distances till the required element size say ρ is reached. The approximation tends to an exact procedure if one takes more number of segments. In this process, many times the actual size of the last element of the curve is not equal to the desired element size within acceptable error. This is corrected by removing the last element and equally distributing it's length to other elements as an acceptable error and the summation procedure is repeated.

4. CONVENTIONAL ADVANCING FRONT TECHNIQUE

This method consists of the following steps for the mesh generation:

1. The domain to be meshed is modeled using boundary curves defined by NURBS.
2. The boundary is discretized into linear segments to form the initial front Γ (Fig. (1a)).
3. An active edge AB is selected from the front Γ. Generate an element by inserting a new node N or by taking one of the nodes in the front Γ as shown in Fig. (1b).
4. Update the front Γ and it advances further into the domain as shown in Fig. (1c).
5. The node and element generation is successive and the steps 3 and 4 are repeated until no edge is left available for the element generation as shown in Fig. (1d).

Element is formed by insertion of a new node N provided the following conditions are satisfied:

➢ edges formed say $\overline{AN}$ and $\overline{BN}$ should not intersect with any of the edges in the advancing front Γ

➢ triangle ABN does not contain any existing node in the mesh

➢ N lies interior to Γ

For the element formation with an existing node in the advancing front Γ, a valid node set $\Lambda(\overline{AB},\Gamma)$ with respect to segment $\overline{AB}$ is constructed. A valid node set is defined as the collection of all the nodes in Γ, such that the above three conditions are satisfied with respect to $\overline{AB}$. Either one of the above mentioned methods is used depending on the quality of the resulting element. Two quality measures as defined below are used. The deviation of the actual element size (S) from the desired value (ρ) as well as shape quality is considered to define the coefficient (μ) as given in Eq. (1). The shape quality α of a triangular element ABC is also defined in Eq. (1). The ideal (equilateral) triangle will attain a maximum value of 1 as per this equation.

$$\mu = \frac{S}{\rho}\left(2 - \frac{S}{\rho}\right)\alpha \quad \text{where} \quad \alpha(ABC) = 2\sqrt{3}\,\frac{\|AB \times AC\|}{\|AB\|^2 + \|BC\|^2 + \|AC\|^2} \qquad \ldots(1)$$

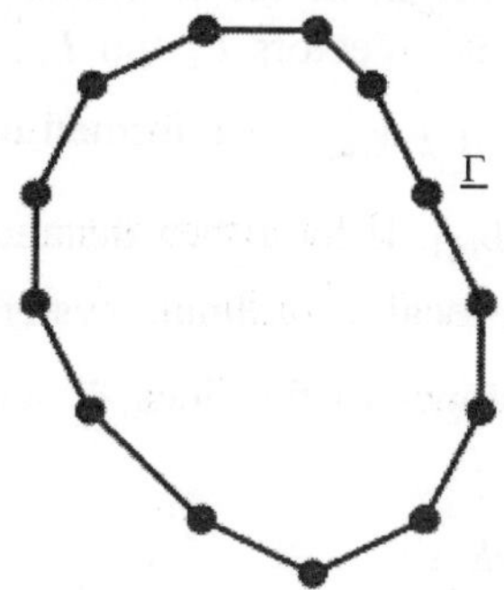

(a) Initialization of generation front.

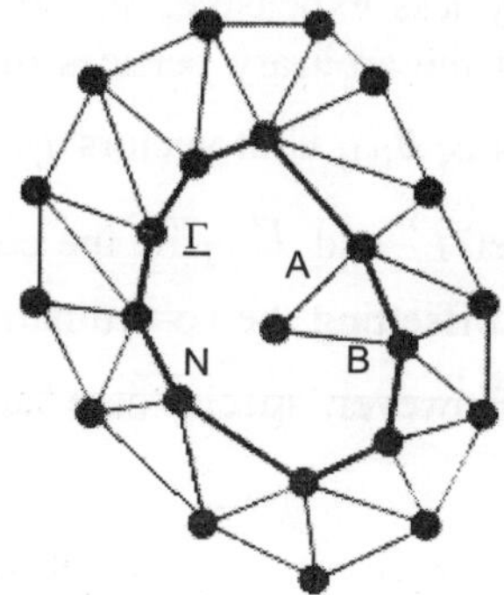

(b) Generation of new front.

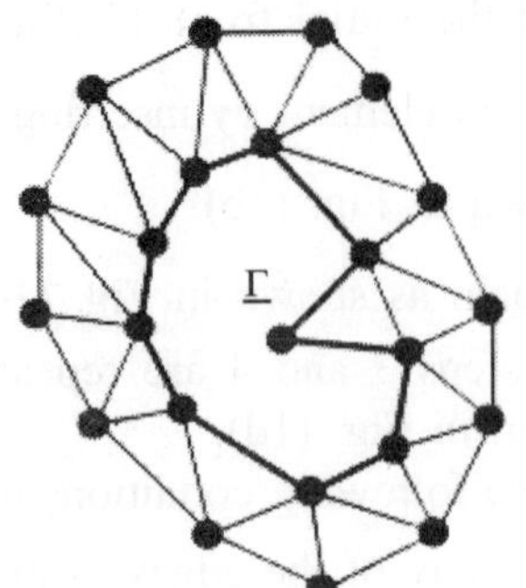
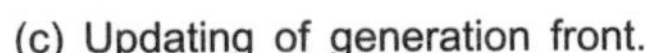

(c) Updating of generation front.

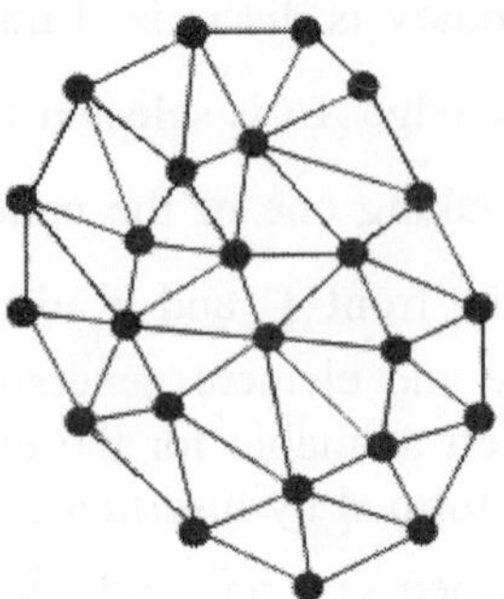

(d) Complete mesh generation.

Fig. 1. Triangular mesh generation by convertional AFT.

5. HYBRID ADVANCING FRONT TECHNIQUE

The proposed hybrid advancing front technique (HAFT) consists of the following steps:

1. The domain to be meshed is modeled using NURBS.
2. *Pseudo* implicit boundary offset by desired element size S to generate interior nodes.
3. Repeat step 2 iteratively till the offset curve self intersects.
4. Node generation on the boundary and the offset NURBS curves.
5. The nodes generated are triangulated by Delaunay triangulation.
6. Discretize the remaining region of the domain into triangular elements by using conventional advancing front technique.

The complete procedure is detailed further and illustrated using an example shown in Fig. (3a-c).

5.1 Offsetting of NURBS Curve

The *pseudo* implicit boundary offset involves offsetting the boundary defined by a NURBS curve inward by offsetting its control polygon. Though this is an approximate method, but it is computationally less expensive. Without loss of generality, the origin of the temporary coordinate system is set at the arbitrary vertices of the control polygon say Bi. Vectors L_i and L_{i+1} are formed by $B_i-1\ B_i$ and $B_i\ B_{i+1}$. Unit vectors $\vec{l}_i = x_i\vec{i} + y_i\vec{j}\,\vec{t}_i$ and $\vec{l}_{i+1} = x_{i+1}\vec{i} + y_{i+1}\vec{j}$ are formed using vectors L_i and L_{i+1}. Let L'_i and L'_{i+1} be the equidistant lines of L_i and L_{i+1}. If the offset distance is R, then for the inward offsetting the co-ordinates (x, y) of offset node in local co-ordinate system is derived as in Eq. (2). However, special care has to be taken when the slopes of the lines L'_i and L'_{i+1} are same.

$$x = \frac{(x_{i+1} - x_i)R}{\begin{vmatrix} x_i & y_i \\ x_{i+1} & y_{i+1} \end{vmatrix}}\; ;\;\; y = \frac{(y_{i+1} - y_i)R}{\begin{vmatrix} x_i & y_i \\ x_{i+1} & y_{i+1} \end{vmatrix}} \qquad\qquad ...(2)$$

5.2 Termination Criteria for the Offsetting Process

The offsetting method has to be terminated just before the occurrence of self intersection of the offset curve, which will arise while offsetting the curves beyond a particular point. This self intersection may happen in two ways as shown in Fig. (2). These problems are identified by checking the intersection of the edges of the control polygon with any other edges of the same as well as other control polygon.

(a) Local intersection.   (b) Global intersection.

Fig. 2 Self intersection of offset curves.

5.3 Node Insertion and Delaunay Triangulation

Node insertion is done by dividing all the curves (the boundary as well as the offset curves) into linear segments by introducing the nodes according to the element size as shown in Fig. (3a). Mid points of the nodes on the successive curves are taken to get good quality triangular elements. The nodes are triangulated by the popular Delaunay triangulation technique. However, since the Delaunay triangulation always results in a mesh representing a convex hull and also only for a part of the domain nodes have been generated, unwanted triangles have to be eliminated. This elimination is done by determining *in center* of each triangle and removing the triangular element if the *in center* lies outside the external curve or lies inside the innermost offset curve using *ray tracing* algorithm. Figure (3b) shows the meshed domain (the triangular elements removed in the non convex portions and also inside the interior most offset curve are shown in grey). The remaining area of the domain i.e., interior to the innermost offset curve is meshed into triangular elements using conventional AFT. The complete triangular mesh for the example meshing is shown in Fig. (3c).

6. RESULTS

To compare the overall shape quality of the triangular mesh generated, a parameter $\bar{\alpha}$ defined in Eq. (3) called geometric mean α-quality is used. Where NT is number of triangular elements in the mesh and α_i is the shape quality of ith triangular element. Mesh generation for many case studies were done both by conventional AFT and by HAFT and only few cases are shown as example. Figures (3c) and (3d) show one example mesh obtained by the two methods. The grey level of each triangular mesh corresponds to it's α quality (darker $\Rightarrow$ poorer). Table 1 enumerates the quality details for the mesh obtained by conventional AFT and HAFT for this example. Two more case studies are illustrated in Fig. (4). Figure (4b) illustrates the need for better offsetting method for HAFT to obtain quality elements in regions of high curvature.

$$\bar{\alpha} = \left(\prod_{i=1}^{NT} \alpha_i \right)^{\frac{1}{NT}} \qquad \qquad \dots(3)$$

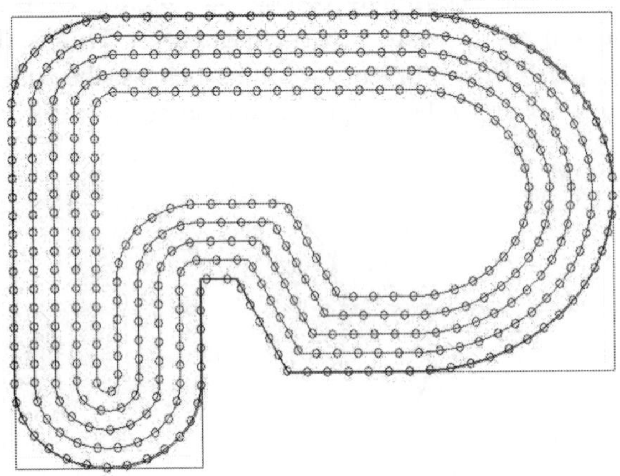

(a) The boundary NURBS curve, control polygon and nodes inserted by offsetting.

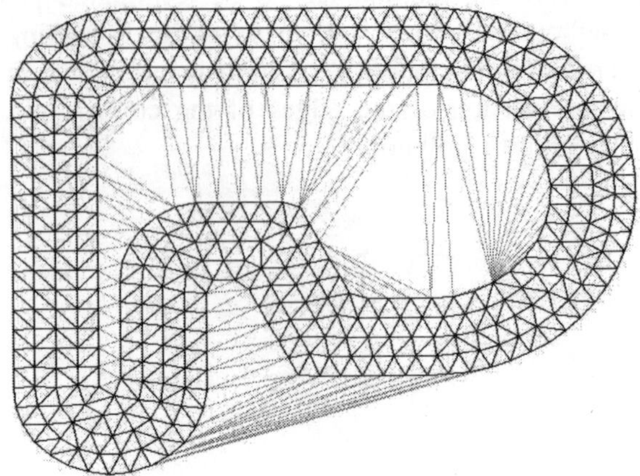

(b) Delaunay triangulation and removal of unwanted elements.

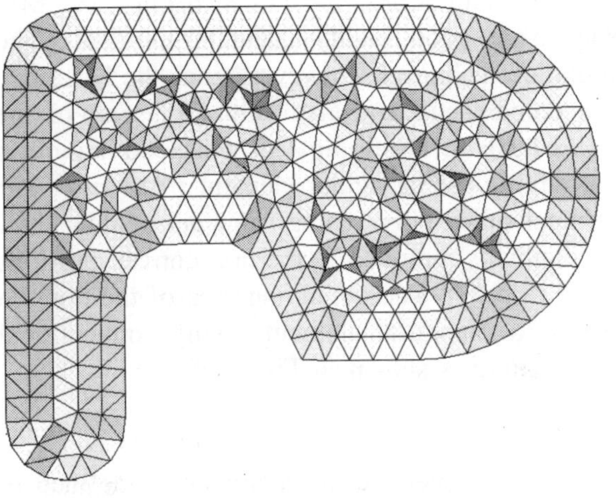

(c) Complete mesh by HAFT.

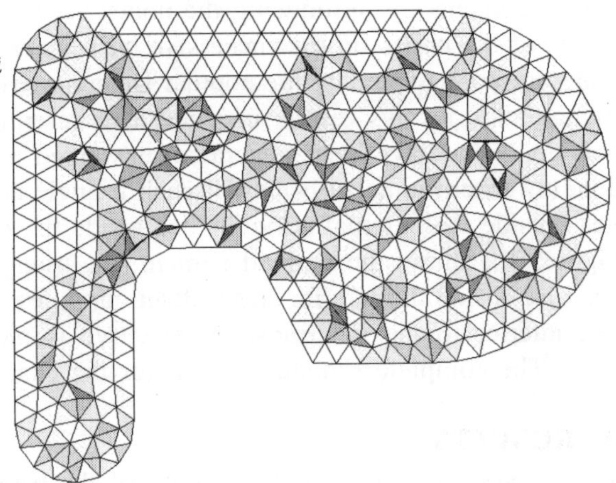

(d) Complete mesh by conventional AFT.

Fig. 3. Triangular mesh generation by HAFT and conventional AFT.

Table 1. Quality details of the triangular mesh generated by conventional AFT and hybrid AFT

S.No.	Description of quality measure	AFT	HAFT
1	Number of triangular elements	900	829
2	Number of nodes	495	634
3	Geometric mean α-quality, $\bar{\alpha}$	0.9271	0.9344
4	Elements with aspect ratio more than 0.7	83.2 %	86.1%

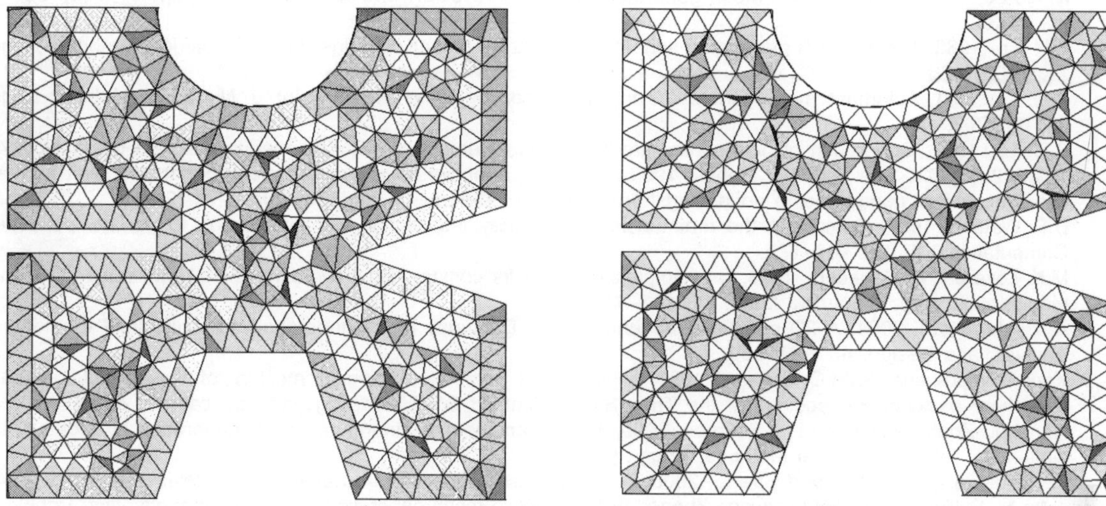

(a) Mesh by HAFT and conventional AFT for a cross-section of a machine block.

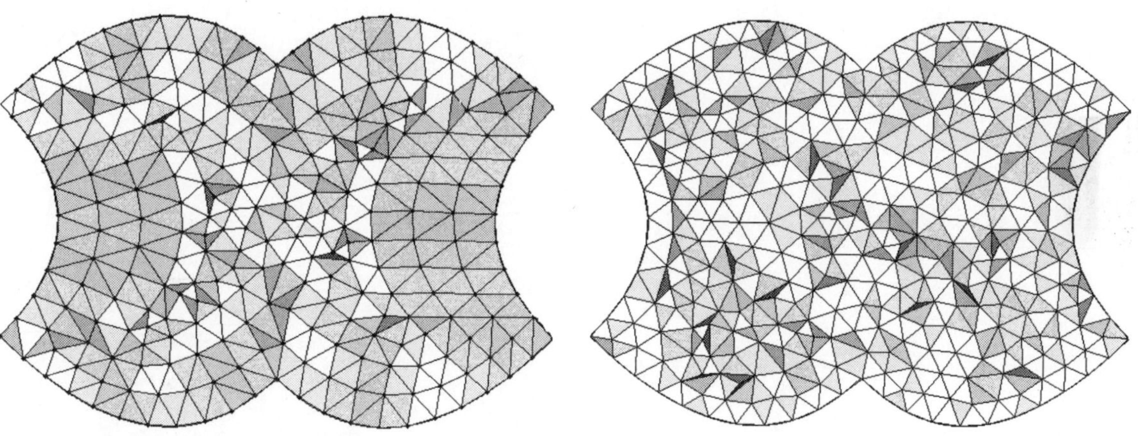

(b) Mesh by HAFT and conventional AFT for an arbitrary domain.

Fig. 4. Some more examples of triangular mesh generation by HAFT and conventional AFT.

CONCLUSION

A hybrid advancing front technique using *pseudo* implicit boundary offset and advancing front for two-dimensional domains has been presented. The quality and uniformity of the mesh obtained by hybrid advancing front method is better than the advancing front technique. This improvement is significant if the node insertion by *pseudo* implicit boundary offset is done in predominant regions of the domain, though this is limited by boundaries having high curvature changes.

REFERENCES

1. K. Ho-Le, 1988, Finite element mesh generation methods: A review and classification, Computer Aided Design, 20, 27-38.
2. S.H. Lo, 1985, A new mesh generation scheme for arbitrary planar domains, Int. J. Numerical Methods. Engg., 21, 1403-1426.
3. S.H. Lo, 1991, Automatic mesh generation and adaptation by using contours. Int. J. Numerical Methods Engg., 31, 689-707.
4. S.H. Lo, 1992, Generation of high-quality gradation finite element mesh, Engineering Fracture Mechanics, 41, 191-202. [5] C.K. Lee and R.E. Hobbs, 1999, Automatic adaptive finite element mesh generation over arbitrary two-dimensional domain using advancing front technique, Computers and Structures, 71, 9-34.
6. D.F. Watson, 1981, Computing the n-dimensional Delaunay tessellation with application to Voronoi polytopes, Computational J., 24, 167–172.
7. M.K. Loze and R. Saunders, 1993, Two simple algorithms for constructing a two-dimensional constrained Delaunay triangulation. App. Numerical Math, 11, 403–418.
8. A. El-Hamalawi, 2004, A 2D combined advancing front-Delaunay mesh generation scheme Finite Elements in Analysis and Design, 40, 967–989.
9. Kenji Shimada and David C. Gossard, 1995, Bubble mesh: automated triangular meshing of non-manifold geometry by sphere packing, Proceedings of the Third Symposium on Solid Modeling and Applications, 409-419.
10. Hanzhou Zhang and Andrei V. Smirnov, 2005, Node placement for triangular mesh generation by Monte Carlo simulation, Int. J. Numerical Methods. Engg., 0, 1-16.
11. Per-Olof Persson and Gilbert Strang, 2000, A simple mesh generator in matlab, SIAM Review, 46(2), 329-345.
12. John M. Sullivan Jr, James Qingyang Zhang, 1997, Adaptive mesh generation using a normal offsetting technique, Finite Elements in Analysis and Design, 25, 275-295.

131

A Software for Computer Assisted Learning of Finite Element Method

V.S. Purani[1], D.P. Parmar[2] AND S.C. Patodi[3]

[1]Assistant Professor of Applied Mechanics, Government Engineering College,
Distt. Gandhinagar, Chandkheda, Gujarat, India. email: vinaypurani@yahoo.com
[2]Structural Engineer, Reliance Industries Limited, Motikhavadi, Jamnagar, Gujarat, India
[3]Professor and Head of Applied Mechanics Department, Faculty of Tech. & Engg.
The M.S. University of Baroda, Vadodara, Gujarat, India. email: scpatodimsu@yahoo.com

ABSTRACT

A number of packages are available for the analysis of different types of structures under different types of loading based on the well-known Finite Element Method (FEM). Using these packages now it is very easy to analyze any complex structure within few minutes and get the output in desired form. However, most of these packages are black box type and do not give any idea about the procedure followed to obtain the results. The main aim of the present work is to develop an educational purpose software which performs numerical computations in front of the user so as to facilitate Computer Assisted Learning (CAL) of FEM. An attempt is made to create an interactive CAL software with the options for the analysis of bar using 2 or 3 noded axial element, beam using 4 DOF beam element and plane stress/strain problems using Constraint Strain Triangular element by blending FEM and contemporary Microsoft Foundation Classes (MFC) on the Windows platform. To facilitate menu driven input and graphical output, three modules i.e., Pre-processor, Main Processor and Post-processor modules are developed. In order to include the interactive features in pre- and post-processor, variety of classes and functions of developer studio are used. A sound file management system is evolved to interact with input and output parameters. An efficient implementation of 96 source and header files with 53 classes is managed in Object Oriented Program (OOP) using Visual C++. The forte of this software lies in quick grasp of the structural analysis fundamentals by the novice user at the solution stage itself, which significantly contribute in the domain of generation of academic tutorials, on-line help and interactive parametric studies of selected structures through FEM.

Keywords: Computer assisted learning, Finite element method, Pre- and post-processors, Visual C++.

1. INTRODUCTION

The finite element method is one of the most popular numerical techniques used for obtaining an approximate solution of complex problems in various fields of engineering [1]. The generality of its application coupled with the easy availability of high speed personal computers, has put finite element method in wide use. There has been significant advancement in the use of computer graphics for pre- and post-processing at the input and output stages of finite element programs. Also, the recent developments in software engineering like the use of problem oriented language, database management and graphics user interface has lead to the development of large number of packages for finite element analysis. Although most of the commercially available packages are quite attractive, they are black box type and do not give any idea about the procedure followed to obtain the results. Blind use of such packages provides sometimes quite erroneous results and, therefore, a basic understanding of the procedure followed in these packages is a must.

Keeping this in view, a software is developed in the present work for computer assisted learning of finite element method. The software which is named as *fempro*, is developed to display all the calculations required for the analysis on the computer screen so as to help the user in learning the basic concepts and complete procedure of the finite element analysis. As it is not possible to include different types of elements in a tutor type software, only few basic one- and two- dimensional elements are included in the present software.

2. GENERAL DESCRIPTION OF FEM

The basic philosophy of finite element method is to replace the continuum having infinite number of unknowns by a mathematical model which has a finite number of unknowns at certain selected discrete points. For solving a problem using FEM, a continuum is discretized and idealized by using a mathematical model which is an assembly of subdivisions or discrete elements. These elements are assumed to be interconnected only at the joints called nodes. Simple function, such as polynomials, are selected in terms of unknown displacements at the nodes to approximate the variation of the actual displacement over each finite element. The external loading is also transformed into equivalent forces applied at the nodes. Next, the behavior of each element independently and later as an assembly of these elements is obtained by relating their response to that of the nodes in such a way that the basic conditions such as the equation of equilibrium, the compatibility equations and the material constitutive relationship are satisfied at each node.

The equations, which are obtained in the form of force-displacement relationship, are then modified to incorporate the given boundary conditions and then these modified equations are solved to obtain displacements at the nodes which are the basic unknowns in the finite element method. Finally, strains and stresses in different elements are obtained, at desired points, using strain- displacement and stress-strain relationship, respectively.

3. DEVELOPMENT OF *FEMPRO*

It may be mentioned that the FEM involves extensive computations, mostly repetitive in nature. Hence, the method is suited for computer programming on digital computers. Further, to obtain the analysis solution in the format of explanation given in the textbook with explicit presentation of intermediate steps and visual graphical feedback, some special features are necessary to be programmed [2]. From the computational point of view the basic steps in finite element analysis can be well organized in the following three basic components.

3.1 Pre-Processor

As a first step of the analysis the given structure is to be dicretized into finite elements. This step requires knowledge of the physical behavior of the structure to decide on the type of analysis and elements to be used to arrive at the finite element model. After discretization, nodes are numbered keeping in view the minimum bandwidth requirements. Graphics based pre-processor automatically generates the mesh and numbers the nodes and elements. The present version of the software is limited to use of 2 or 3-noded axial element, 2-noded beam element and plane stress/strain triangular element.

Further, the pre-processor of *Fempro* software developed here includes features such as flexibility, default values, transparency, error checking, editable, visual feedback, etc. An interactive menu driven and dialogue based input device is developed and series of programs are linked interactively using OOP methodology for efficient use of options of pull-down menus, dialogue and toolbar for geometry, discretization, element and node numbering, loading, etc. The data being assigned is mainly through mouse or alternatively via keyboard. The pre-processor can be activated either through toolbar, pull down menu or by the defining the short cut. Selection of the type of structure, material properties, element properties, boundary conditions, loading, etc. can be defined interactively through different available options. Some typical screen shots of pre-processor are shown in Figs. 1 to 3.

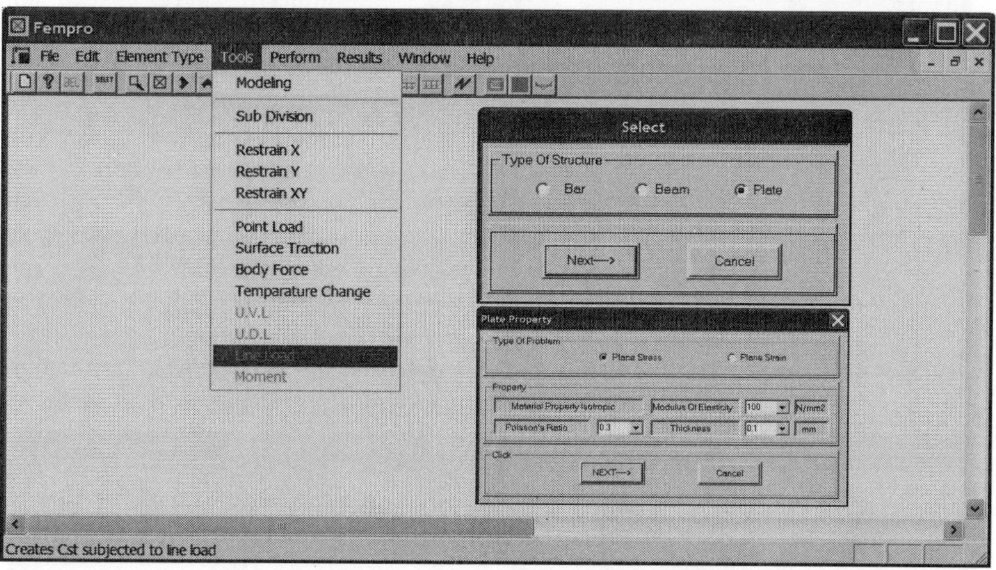

Fig. 1. Screen shot of pre-processor of *Fempro*.

3.2 Main Processor

The processor module involves mathematical computations that require large arithmetic operations. The strain-displacement linking matrix, material rigidity matrix, element stiffness matrix and nodal load vector are computed for each element and then using direct stiffness method element properties are assembled to generate overall equilibrium equations. After applying boundary conditions, the

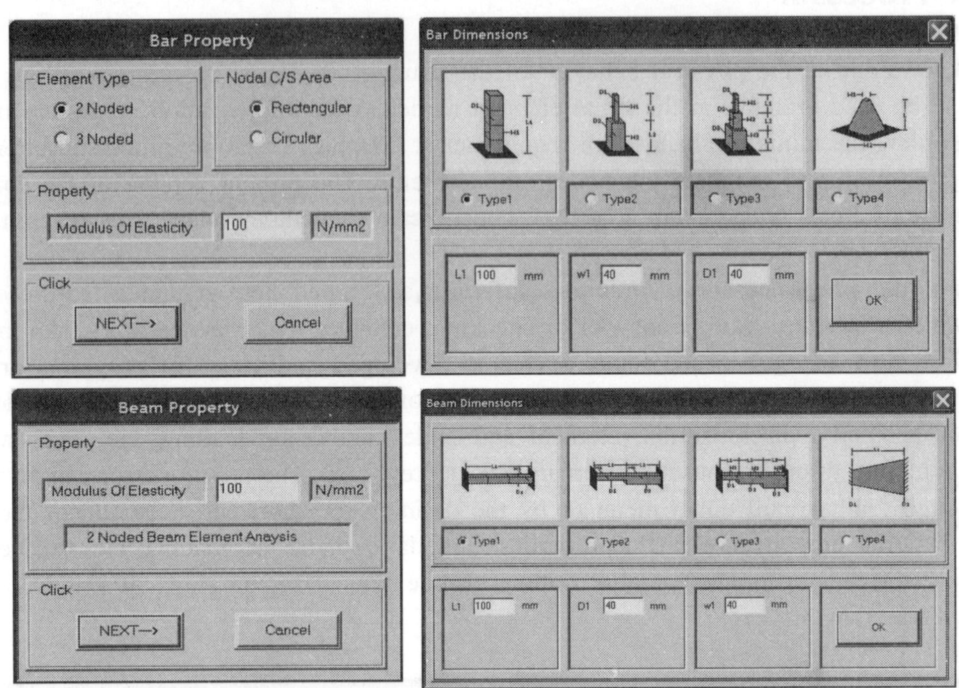

Fig. 2. Pre-processor input forms for bar and beam structures.

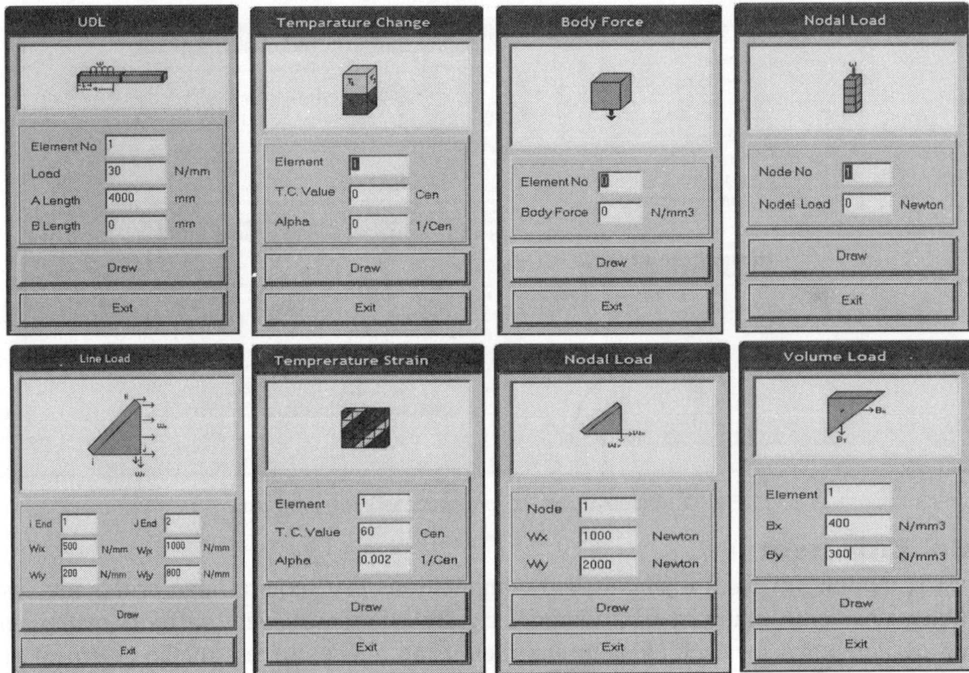

Fig. 3. Different loading options of *Fempro.*

equations are solved using skyline technique in this module to obtain nodal displacements. Finally, strains and stresses are computed at user defined locations. Provision is made in the analysis/processor module to display all these steps on the computer screen to give clear cut idea about the analysis procedure.

3.3 Post-Processor

This module produces the results in graphical as well as text output form. In graphics form it produces the stress contours [3] as well as deflected shape. The text output displays the detailed or summary type of the solution for the selected problem as per the users choice. In detailed output option, user can view a well formatted complete solution in step by step stages. Text output report can be obtained by selecting "Report" menu from the "Results" pull-down menu while graphics output can be obtained by selecting "Maps" menu and "Deformed Shapes" menu from the "Results" pull-down menu. Alternatively, a toolbar option is also designed which can be used to activate the desired option for the output of the post-processor.

4. ILLUSTRATIVE EXAMPLE

In this section, a square plate shown in Fig. 4 having dimensions of 100 mm × 100 mm × 10 mm thick is analyzed using *Fempro* package for plane stress condition. Modulus of elasticity is assumed as 1 N/mm^2 and Poisson's ratio is considered as 0.3.

Input to the *Fempro* is interactively assigned. Analysis condition is selected as plane stress. Fig. 5 shows dimension and discretization related input. Here plate is discretized into 100 CST elements. The complete discretized plate model with boundary condition and loading is shown in Fig. 6.

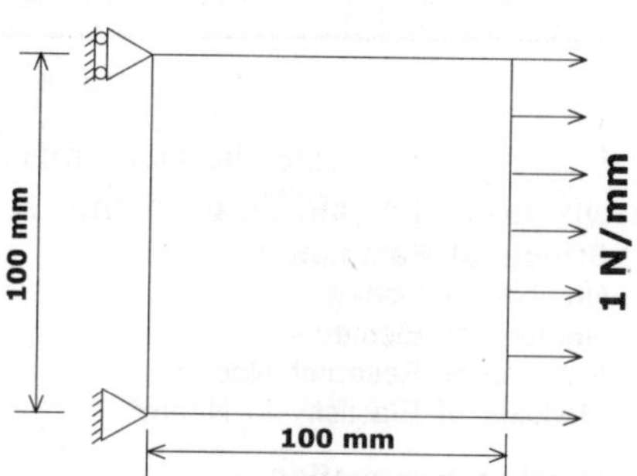

Fig. 4. Plate structure for the case study.

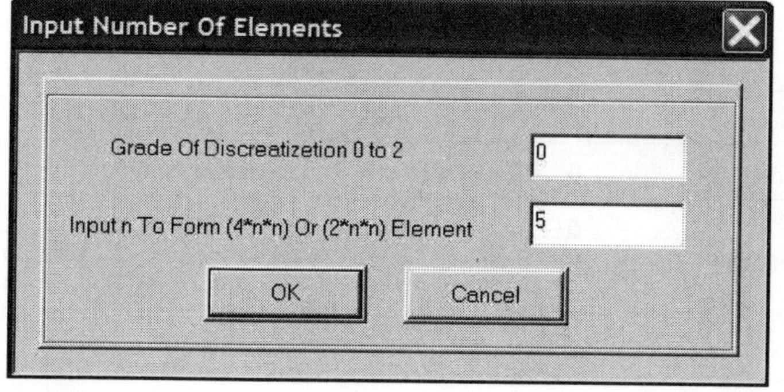

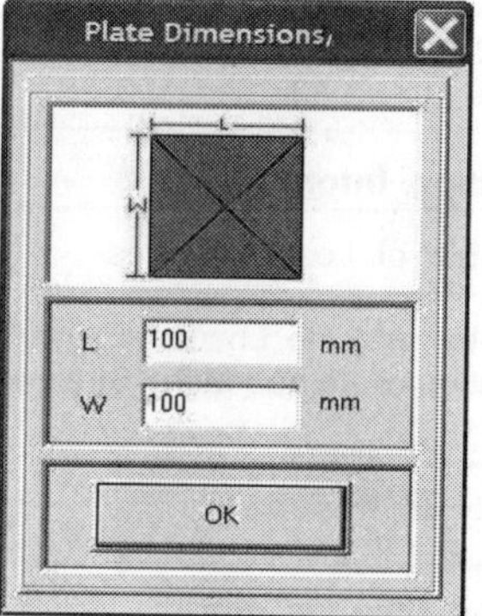

Fig. 5. Input for the plate geometry.

After successful modeling of the square plate, analysis module of the *Fempro* is activated. User may select buttons either from toolbar or pull-down menu to invoke the analysis program of the respective structure generated in the pre-processor. For the present case, the plate structure is analysed for plane stress condition and the detailed output is generated.

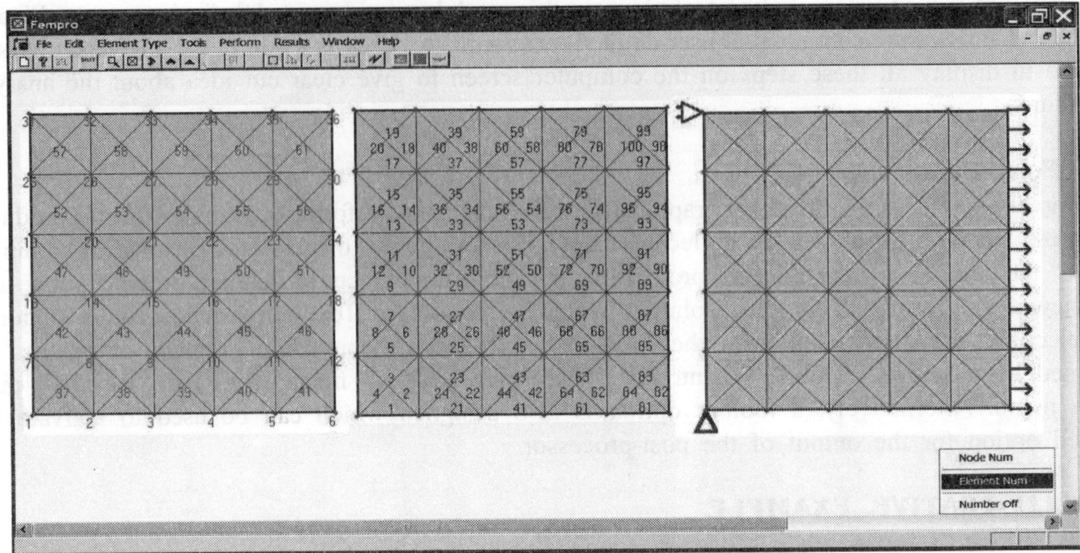

Fig. 6. *Fempro* modeling for plate example.

User may generate the customized text output report by selecting Typical, Compact or Custom options from the pull-down menu. A part of the generated text output report is given below.

Structural Parameters

Number of Nodes	**61**	Number of Degree of Freedom	**119**
Number of Elements	**100**	Number of Nodal Restraints	**3**
Number of Restraint Nodes	**2**	Number of Terms In Sn	**4278**
Modulus of Elasticity in N/mm^2	**1**	Poisson's Ratio	**0.3**

Member Information

Node	X mm	Y mm	Element	I	J	K	RESTRAINED NODE LIST		
							NODE	R1	R2
1	0	100	1	2	1	37	1	1	1
2	20	100	2	8	2	37	31	1	0
3	40	100	3	8	37	7			

Loading Information

Number of Loaded Nodes		0		Line Load N/mm		

		I	J	BL1	BL2	BL3	BL4
Number of Line Loaded Elements	5						
Number of Temperature Strained Element	0	6	12	1	1	0	0
		12	18	1	1	0	0

Analysis Results

Nodal Disp. in mm			Element Stresses N/mm^2					Support Reactions N		
Node	D1	D2	Ele.	Sx	Sy	Sxy	Sz	Node	Ar1	Ar2
1	0	0	1	0.515	−0.095	−0.069	0	1	−49.99	−4.766e-005
2	10.863	9.063	2	0.247	0.028	−0.031	0	31	−49.99	0
3	17.039	10.033	3	0.010	0.012	−0.099	0			
4	20.644	9.579	4	0.277	−0.113	−0.138	0			

In the post-processor's graphical output option, stress distribution and deflected shape are depicted as shown in Fig. 7.

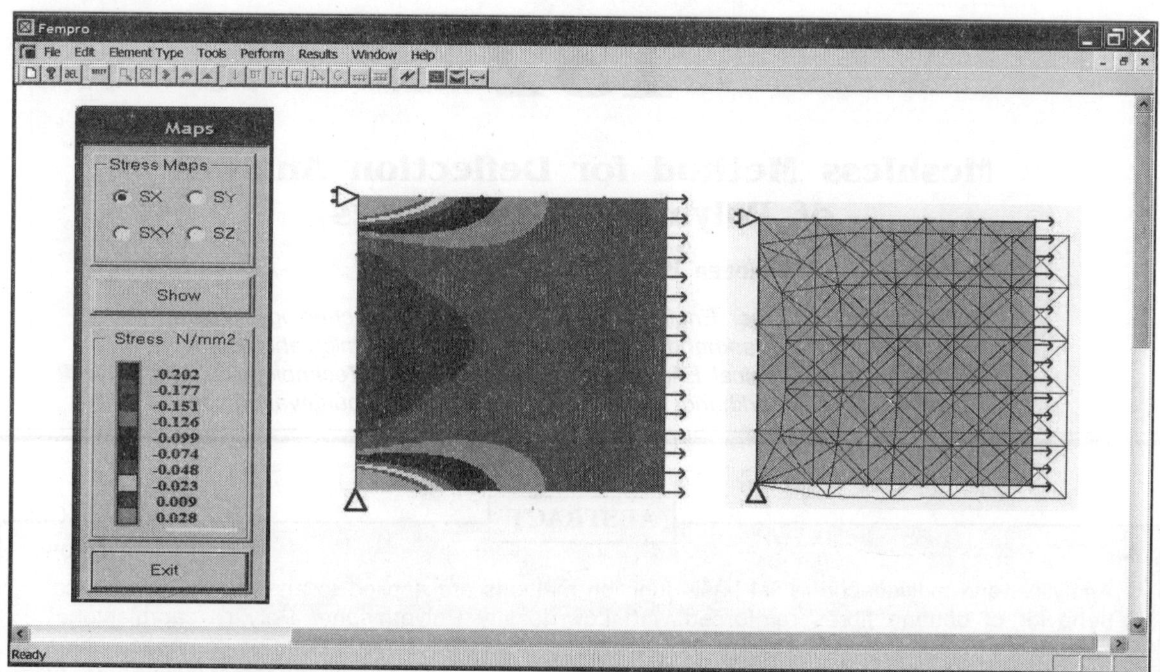

Fig. 7. Screen shot from Post-processor.

CONCLUSION

User friendliness, interactive input facility, display of all intermediate steps of analysis with suitable animation on screens, graphical presentation of structure and results and generation of the analysis report in desired format are some of the salient features of the software developed in Visual C++ on most widely used Windows environment. Because of its glass box type of features, it definitely puts the learner of the finite element method in the comfortable position.

Graphical view of deformed shape and stress distribution through stress contours adds to the ease and speed of data interpretation.

The developed software may prove very helpful as an assistant to the faculty teaching FEM in explaining the procedure and in checking intermediate steps of analysis.

REFERENCES

1. R. J. Melosh, 1990, Structural Engineering Analysis by Finite Elements, Prentice Hall International Publication.
2. P. Paultre, P. Leger and Prolux, 1990, Computer Graphics for Computer Assisted Learning of Structural Analysis, Computer and Structures, Vol. 36, No. 36.
3. T. C. Gopalkrishan and M. Korttom, 1993, An Algorithm for Contouring and Interpolation of Data using Bilinear Finite Elements, Finite Element in Analysis and Design Journal, V-14, 37-54.

132

Meshless Method for Deflection Analysis of Polymeric Composites

Sandeep Kumar[1] and R.K. Misra[2]

[1]Department of Polymer Engineering, Birla Institute of Technology, Mesra, Ranchi-835215, Jharkhand, India. email: sandeepasom@yahoo.com
[2]Department of Mechanical Engineering, Birla Institute of Technology, Mesra, Ranchi-835215, Jharkhand, India. email: mishrark_kanpur@yahoo.com

ABSTRACT

Analytical and multiquadric radial basis function methods are applied to analyze the mechanical behavior of banana fibres reinforced with Low-density Polyethylene/ Poly ($\in$-caprolactone) composites. The paper also reports the fabrication and mechanical properties of banana fibres reinforced thermoplastic composites determined experimentally. Mechanical properties of composites were determined in longitudinal, transverse as well as random direction.

Keywords: Multiquadric radial basis function; Banana fibres; hydrophobicity; mechanical properties.

1. INTRODUCTION

Polymeric Composites structures are used extensively in various applications such as missiles, aircraft, submarines and automobiles due to its high strength low weight ratio. Its vast applications in industry have emphasized the development of new efficient numerical technique for the analysis of composites. Mostly, finite element method is used for the analysis of composites. It consumes more time in mesh generation than the execution. Recently, as an alternative of finite element method, a number of mesh less methods are developed, which avoid time-consuming mesh generation. These methods are element free Galerkin method (EFG) [1], partition of unity method [2], generalized finite difference method [3], wavelet Galerkin method [4], the method of finite sphere [5], meshless local Petrov-Galerkin (MLPG) method [6], smooth particle hydrodynamics [7], etc. In 1990, Kansa developed radial basis function (RBF) for the analysis of partial differential equation [8]. It has drawn the attention of many researchers [9-14]. Franke [15] has ranked multiquadric radial basis function (MQRBF) as the best interpolation method based on its accuracy, visual aspect, and sensitivity to parameters, execution time, storage requirements and ease of implementation. Here multiquadric radial function is developed for static analysis. In addition, fabrication and mechanical properties of banana fibers reinforced low-density polyethylene/ polycaprolacton composites have also been mentioned.

Studies on coir and jute as reinforcement of thermoset resins (such as Unsaturated Polyester resins) have been carried out in our laboratories [16-18]. The presence of surface impurities and the large amount of hydroxyl groups on these lignocellulosic fibres makes them less attractive for reinforcement for thermoplastics. In order to make the natural fibres more compatible with thermoplastics, their surface modifications have been carried out. For example, hemp, sisal, jute and kapok fibres have been modified by alkalization using sodium hydroxide [19]. Jute fibre has been graft copolymerised with vinyl monomers such as methyl methacrylate, ethyl acrylate, styrene, vinyl acetate, acrylonitrile and acrylamide in the presence of different redox initiator systems [20, 21]. Except for acrylamide, the other entire reagents were useful in reducing moisture regain. Their use for reinforcement of polyester resins [22] and resol type phenol formaldehyde resins has been investigated in the past.

2. GOVERNING EQUATIONS

The coordinate system, geometry and loading are shown in Fig. 1. Neglecting the in-plane and rotary inertia. Equation of thin plate composite is expressed as

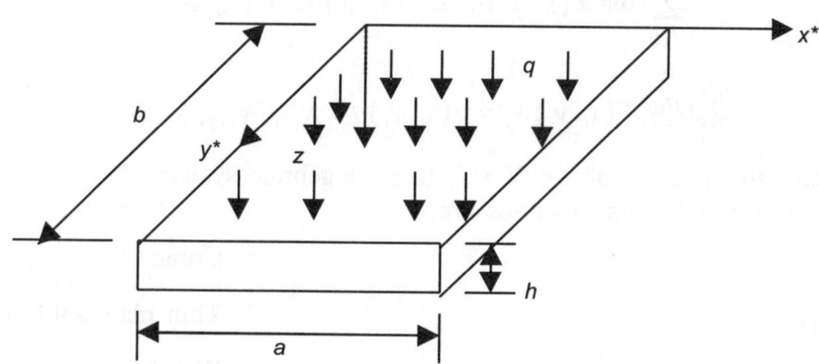

Fig. 1. Geometry of the composite plate.

$$D_{11}w_{xxxx} + 2(D_{12} + 2*D_{66})w_{xxyy} + D_{22}w_{yyyy} = q(x,y) \qquad ...(1)$$

The boundary conditions considered are as follows:
(1) For all four edges simply supported

$$x = 0, a \quad w = 0, \ w_{xx} = 0 \qquad ...(2)$$

$$y = 0, b \quad w = 0, w_{yy} = 0 \qquad ...(3)$$

3. RADIAL BASIS FUNCTION METHOD

Consider a general differential equation

$$Aw = f(x,y) \text{ in } \Omega \qquad ...(4)$$

$$Bw = g(x, y) \text{ on } \partial\Omega \qquad \ldots(5)$$

where A is a linear differential operator and B is a linear boundary operator imposed as boundary conditions. Ω is domain and $\partial\Omega$ is boundary. Let us denote $\{P_i = (x_i, y_i)\}_{i=1}^{N}$ to be N collocation points in Ω of which $\{(x_i, y_i)\}_{i=1}^{N_I}$ are interior points; $\{(x_i, y_i)\}_{i=N_1+1}^{N}$ are boundary points. In Kansa's method, it is assumed that the approximate solution for problem (4) can be expressed as

$$w(x, y) = \sum_{j=1}^{N} w_j \varphi_j(x, y) \qquad \ldots(6)$$

$$\varphi_j = \sqrt{(x - x_j)^2 + (y - y_j)^2 + c^2} = \sqrt{r_j^2 + c^2} \qquad \ldots(7)$$

where $\{w_j\}_{j=1}^{N}$ are the unknown coefficients to be determined, and $\varphi_j(x_j, y_j) = \varphi\|P - P_j\|$ is a RBF.

Here $r = \|P - P_j\|$ is the Euclidean norm between points $P = (x, y)$ and $P_j = (x_j, y_j)$. By substituting eq. (6) into eq. (4) and eq. (5), we have

$$\sum_{j=1}^{N} (A\varphi_j)\,(x_i, y_i)\,w_j = f(x_i, y_i), i = 1, 2 \ldots\ldots N_I \qquad \ldots(8)$$

$$\sum_{j=1}^{N} (B\varphi_j)\,(x_i, y_i)\,w_j = g(x_i, y_i)\ i = N_{I+1}, N_{I+2}, \ldots\ldots N \qquad \ldots(9)$$

Hence, we have to solve the above $N \times N$ linear algebraic system
Most widely used radial basis functions are:

$\varphi(r) = r^3$ Cubic

$\varphi(r) = r^2 \log(r)$ Thin plate splines

$\varphi(r) = (1 - r)^m_+ \, p(r)$ Wend land functions

$\varphi(r) = e^{-(cr)^2}$ Gaussian

$\varphi(r) = \sqrt{c^2 + r^2}$ Multiquadrics

$\varphi(r) = (c^2 + r^2)^{-1/2}$ Inverse multiquadrics

4. MULTIQUADRIC METHOD FOR GOVERNING DIFFERENTIAL EQUATION

Substituting radial basis function in equation (1), we get

$$(\sum_{j=1}^{N} w_j \frac{\partial^4}{\partial x^4}\varphi_j + 2\lambda^2\eta\sum_{j=1}^{N} w_j \frac{\partial^4}{\partial x^2 y^2}\varphi_j + \sum_{j=1}^{n} \lambda^4\psi w_j \frac{\partial^4}{\partial y^4}\varphi_j) + \sum_{j=1}^{N} w_j \frac{\partial^2}{\partial t^2}\varphi_j - C_v w_j \frac{\partial}{\partial t}\varphi_j - Q = 0 \qquad \ldots(10)$$

4.1 Boundary Conditions

(a) For Simple supported edge

$$x = 0, a \qquad \sum_{j=1}^{N} w_j \varphi_j = 0 \qquad \qquad \dots(11)$$

$$y = 0, b \qquad \sum_{j=1}^{N} w_j \varphi_j = 0 \qquad \qquad \dots(12)$$

$$x = 0, a \qquad \sum_{j=1}^{N} w_j \frac{\partial^2}{\partial x^2} \varphi_j = 0 \qquad \qquad \dots(13)$$

$$y = 0, b \qquad \sum_{j=1}^{N} w_j \frac{\partial^2}{\partial y^2} \varphi_j = 0 \qquad \qquad \dots(14)$$

5. CASE STUDIES

The material used for the ply is banana fibres reinforced LDPE/PCL Composites.

Properties are: $E_x = 18 \dfrac{GN}{m^2}, E_y = 8 \dfrac{GN}{m^2}, \nu_{xy} = 0.36$

6. EXPERIMENTAL

The banana fibres bundles were obtained from Regional Research Laboratory (CSIR) Jorhat. Sodium hydroxide, toluene, petroleum ether, TDI, sebacoyl chloride and LDPE (16MA400) (obtained from Indian Petrochemicals Corporation Limited, Vadodara) were used. PCL Tone TM polymer, from Union Carbide Company, USA was used as such.

6.1 Chemical Treatment of Banana Fibers

In order to separate the individual fibres, organic solvents such as acetone, toluene, and petroleum ether were used. Better separation of fibres was achieved by using toluene followed by washing with petroleum ether. Fibres were treated with 5% or 10% aq-NaOH solution, TDI and sebacoyl chloride at room temperature for 4 h.

6.2 Fabrication of Composites

A Klockner Windsor single screw extruder having L/D ratio of 21:1 and screw diameter of 30 mm was used for blending HDPE and PCL. The banana fibres (10% alkali treated) were separated and dried in oven for 3 hours at 80°C for further processing; these fibers are used for composite fabrication due to their better mechanical properties. The sample designation and mass of the constituent materials used for composites fabrication are given in Table 1. The matrix (LDPE: PCL, 70:30) has been designated as M, while ML and MT has designated the longitudinal and transverse composites respectively.

Table 1. Composition of banana fiber reinforced composites.

Sample designation	LDPE (g)	PCL (g)	Banana Fiber (g)
M	140	60	0
ML	140	60	10(longitudinal)
MT	140	60	10(transverse)

6.3 Mechanical Tests

Tensile testing was done on a Zwick tensile tester, (Model Z010) at ambient temperature on compression molded test specimens according to the ASTM D-638 standard. A load cell of 10KN capacity used. An average of 5 specimens were used for the various samples. The following conditions were employed: gauge length = 5 cm, cross-head speed = 50 mm/min.

7. RESULTS AND DISCUSSION

7.1 Mechanical Properties of Composites

Fiber reinforced LDPE/PCL (70:30) random composite was fabricated using untreated and 10% alkali treated banana fibres. The results of tensile strength, tensile modulus and elongation at break of composites are given in Table 2.

Table 2. Mechanical properties of composites.

Sample Designation	Tensile strength (MPa)	Tensile Modulus (GPa)	Elongation at break (%)
M	21.4	11.4	4.50
ML	80.8	18.0	7.23
MT	32.9	8.0	6.78
MR	28.1	12.6	6.24

7.2 Analysis of Composites

The static results of composite plates in longitudinal, transverse and random directions are shown in tables 3 to 5.

Table 3. Deflection of simple supported longitudinal fiber composite plate with variation of aspect ratio.

a (mm)	b (mm)	a/b	H (mm)	q (N/m^2)	Analytical method [20] w_{max} (mm) $\times 10^{-2}$	Radial basis method w_{max} (mm) $\times 10^{-2}$
300	300	1.0	10	1000	3.677	3.641
300	200	1.5	10	1000	1.638	1.657
300	150	2.0	10	1000	0.739	0.770
300	120	2.5	10	1000	0.362	0.357
300	100	3.0	10	1000	0.193	0.195
300	50	6.0	10	1000	0.014	0.014

Table 4. Deflection of simple supported transverse fiber composite plates with variation of aspect ratio.

a (mm)	B (mm)	a/b	H (mm)	Q (N/m^2)	Analytical method [20] w_{max} (mm) $\times 10^{-2}$	Radial basis w_{max} (mm) $\times 10^{-2}$
300	300	1.0	10	1000	3.669	3.732
300	200	1.5	10	1000	1.143	1.149

contd.

300	150	2.0	10	1000	0.429	0.436
300	120	2.5	10	1000	0.190	0.192
300	100	3.0	10	1000	0.096	0.095
300	50	6.0	10	1000	0.006	0.006

Table 5. Simply Supported Random fiber composites.

b (mm)	a (mm)	a/b	H (mm)	Q (N/m^2)	Analytical method [20] w_{max} (mm) × 10^{-2}	Radial basis method w_{max} (mm) × 10^{-2}
200	200	1.00	10	1000	0.24157	0.24285
400	200	0.50	10	1000	0.60270	0.60269
600	200	0.33	10	1000	0.72768	0.72760
800	200	0.25	10	1000	0.77469	0.77464

There was a large decrease (99.61%) in deflection of longitudinal and transverse composites by increasing the aspect ratio from 1.0 to 6.0. Same results were found in the case of random composites by varying the aspect ratio from 0.25 to 1.00. This may be attributed to the increase in concentration of fiber, which in turn increasing the strength of the composite.

CONCLUSION

Alkali treatments of banana fibers give better mechanical properties. Multiquadric radial basis function has been applied for static analysis of banana fibers reinforced low-density polyethylene/polycaprolacton composites because of many advantages, which are following:

- No connectivity is needed because the discretization process is based on the employment of a set of randomly distributed boundary and internal nodes.
- After performing the analysis on classical laminated plates, it is observed that this method is very suitable for computer implementation.
- Numerical examples have shown the accuracy and good convergence of the results that prove the effectiveness of the approach.

NOTATIONS

a, b	Dimension of plates
$D_{11}, D_{22}, D_{12}, D_{66}$	Flexural rigidity of plates
E_x, E_y	Young's modulus in x, y direction respectively
h	Thickness of plates
q	Transverse load
λ	Aspect ratio (a/b)
v	Poisson's ratio
w	Displacement in z direction

REFERENCES

1. T. Belystcho, and L.GU, 1994, Element free Galerkin methods, International Journal of Numerical Methods in Engg, 37, 229-256.
2. I. Babuska and J.M. Melenk, 1997, The partition of unity method. International Journal of Numerical Methods in Engineering 40,727-758.
3. T. Liszka, 1984, An interpolation method for irregular net of nodes. International Journal of Numerical Methods in Engineering, 20, 1599-1612.

4. S. Qian, and J. 1993, Weiss, Wavelet and the numerical solution of partial differential equations. Journal of Computational Physics, 106,155-168.

5. S. De, and K.J. Bathe, The method of finite spheres, Computational Mechanics, 2000, 25, 329-341.

6. Atluri, S.N. and Zhu, T. 1998, A new meshless local Petrov-Galerkin approach in computational mechanics Computational Mechanics, 22, 117-187.

7. J.J. Monaghan, and A. Annu, 1992, Smoothed particle hydrodynamics Rev. Astron.Astrophys, , 30, 543-553.

8. E.J. Kansa, 1990, Multiquadrics-a scattered data approximation scheme with applications to computational fluid dynamics, i: Surface approximations and partial derivative estimates, Comput. Math. Appl. 19 (8/9) 127-139.

9. Y.C. Hon, M.Lu, W.M. Xue, and Y.M. Zhu, 1997, Multiquadric method for the numerical solution of by phasic mixture model, Appl.Math. Comput. 88 153-171.

10. Y.C. Hon, K.F. Cheung, X.Z. Mao, and E.J. Kansa, 1999, A multiquadric solution for the shallow water equation, ASCE J. Hydraul. Engrg. 125 (5) 524-538.

11. J.G. Wang, G.R. Liu, and P. Lin, 2002, Numerical analysis of Biot's consolidation process by radial point interpolation method, Int. J. Solids Struct. 39 (6) 1557-1569.

12. G.R.Liu, and Y.T. GU, 2001, A local radial point interpolation method for free vibration analyses of 2-d solids, J. Sound Vibration. 246 (1) 29-38.

13. G.R. Liu, and J.G. Wang, 2002, A point interpolation meshless method based on radial basis functions, Int. J. Numer. Methods Engrg. 54, 1623-1639.

14. J.G. Wang, and G.R. Liu, 2002, On the optimal shape parameters of radial basis functions used for 2-d meshless methods, Comput.Methods Appl. Mech. Engrg. 191 2611-2631.

15. R.Franke, 1982, Scattered data interpolation: tests of some methods. Math. Comput., 48 181-197.

16. D.S. Varma, M. Varma and I.K. Varma, 1985, Studies on Jute Fibre reinforced thermoplastic composite, J. Reinforced Plast. Compos. 4, 419-429.

17. D.S. Varma, M. Varma and I. K. Varma, 1984, Studies on Jute Fibre reinforced thermoplastic composite, Text. Res. J. 54, 827-839.

18. K. Varma, S.R. Ananthakrishnan and S. Krishnanmoorthy, 1989, Studies on Coir Fibre reinforced thermoplastic composite, Text. Res. J. 59, 368-381.

19. L.Y. Mwaikambo and M.P. Ansell, 2001, Surface modifications natural fibres by alkalization J. Appl. Polym. Sci. 84, 2222-2238.

20. Ashton, J.E.; Whitney, J.M.: Theory of Laminated plates, Technomic Publishing Co., Inc, (1970).

21. S.R. Ananthakrishnan, Modification of Jute Fibres for application in Fibre reinforced Composites, Ph.D. Thesis (1988), IIT Delhi.

22. D.S. Varma and S. Murthy, 1989, Studies on natural Fibre reinforced thermoplastic composite, Text. Res. J. 60, 269-278.

23. N.A. Elshinawy and S.F. Elkalyoubi, 1985, studies on reinforcement of polyester resins, J. Appl. Polym. Sci. 30, 2171-2183

133

Cost Effective Distributed Computing in Structural Engineering

P.V. PATEL AND S.C. PATODI

Department Applied Mechanics Department, Faculty of Technology and Engineering,
The M.S. University of Baroda-390 001, India
email: parpat1@yahoo.com, scpatodimsu@yahoo.com

ABSTRACT

Recent advances in hardware and software have made parallel processing a viable and attractive approach for the solution of computationally intensive problems. The various issues related with parallel processing are hardware, software, algorithm and performance evaluation. From the literature review, it is found that majority of applications have been dealt either using supercomputers or using network of workstations and message passing libraries. As structural engineers are not exposed to these kinds of tools they find it difficult to use parallel and / or distributed computing for their application. The main objective of the present paper, therefore, is to present not only a cost effective alternative but to evolve a methodology by which any user can develop distributed computing application very easily. WebDedip (Web enabled Development Environment for Distributed Image Processing), which has been developed using JAVA technology, simplifies the distributed application development using Local Area Network (LAN). In the present work a brief account is given of various distributed data processing applications dealt by the authors such as static, dynamic, linear and nonlinear analyses of large structures in addition to some special applications such as training of artificial neural network and optimization using genetic algorithm. From various implementation of WebDedip environment a computational efficiency of about 60% to 80% is observed.

Keywords: Distributed Computing; WebDedip; Finite element; Data processing applications.

1. INTRODUCTION

The literature review indicates that there is increasing interest among researchers in use of available parallel computers and development of algorithms for various structural engineering applications. Adeli and Kamal [1] implemented analysis of large size frame structures on supercomputers. Adeli and Kumar [2] presented distributed implementation of finite element analysis over network of workstation using Parallel Virtual Machine (PVM). Kumar and Adeli [3] discussed distributed algorithm for minimum weight design of large structure using Genetic Algorithm and implemented the same over network of workstations using PVM. Topping et al. [4] described a parallel processing implementation for neural computing and its application to finite element mesh decomposition. Appa Rao [5] discussed algorithm for parallel structural analysis and suggested use of network of computers as economical alternative of parallel processing. Annamalai and Krishnamoorthy [6] presented parallel implementation of Adaptive Finite Element Analysis on a cluster of workstation and a library PAVE (Parallel Virtual Environment) was developed to take care of all the essential needs of message passing/communication. Sziveri and Topping [7] implemented transient dynamic finite element analysis over MIMD supercomputer using Message Passing Interface (MPI). Umesha and Venuraju [8] presented parallel version of Computer Adaptive Language (CAL) to analyse large structural mechanics problems on message passing systems. Sotelino [9] discussed various algorithms for parallel implementation of structural engineering applications. Rao et al. [10] described the parallel overlapped domain decomposition algorithm for nonlinear dynamic finite element analysis and implemented using MPI and PARAM10000 MIMD parallel computer. Kant et al. [11] implemented thermo-mechanical analysis of FRP structures on distributed memory high performance parallel machine PARAM10000. Gupta and Khapre [12] presented analysis of stresses developed in anchorage zone in prestressed post-tensioned concrete beam using finite element method on supercomputer PARAM10000. Most of the researchers have used supercomputer or network of workstations along with message passing libraries for parallel or distributed structural engineering applications. This requires some theoretical background of parallel processing and networking.

In the present research work, an account of various highly computational intensive structural engineering applications implemented on LAN using WebDedip environment is given. WebDedip simplifies development of distributed applications by eliminating use of message passing functions from program [13]. In use of WebDedip, the user has to simply visualize the parallel or distributed computing potential of application and accordingly divide it into number of small tasks. These tasks are performed on various computers connected through network and communication between them is done through intermediate files using File Transfer Protocol (FTP).

Various structural engineering applications like static, dynamic and nonlinear analysis of large structures, finite element analysis of laminated composites, optimization of structures and neural computing, etc. demands large computation power. In this paper distributed implementation of structural engineering application over LAN is discussed along with algorithm used and efficiency observed. In this implementation network of Pentium-IV computers having 256 MB RAM running on WINDOWS operating system at 2.6 GHz is used.

2. DISTRIBUTED PROCESSING ENVIRONMENT: WEBDEDIP

WebDedip has three-tier architecture: GUI, DedipServer and Agents as shown in Fig. 1. The GUI is the web enabled graphical user interface to make the entire user interaction truly system independent. It supports various Java Applets for application configuration, application building, application operation initiation, application process monitoring and session controlling. WebDedip provides a full-fledged operation environment for executing the user application and as it is browser based, it enables the user to operate from any node over the network. The user initiates the interaction by visiting a predefined site using a standard web browser. The standard web server loads the required GUI on the web browser.

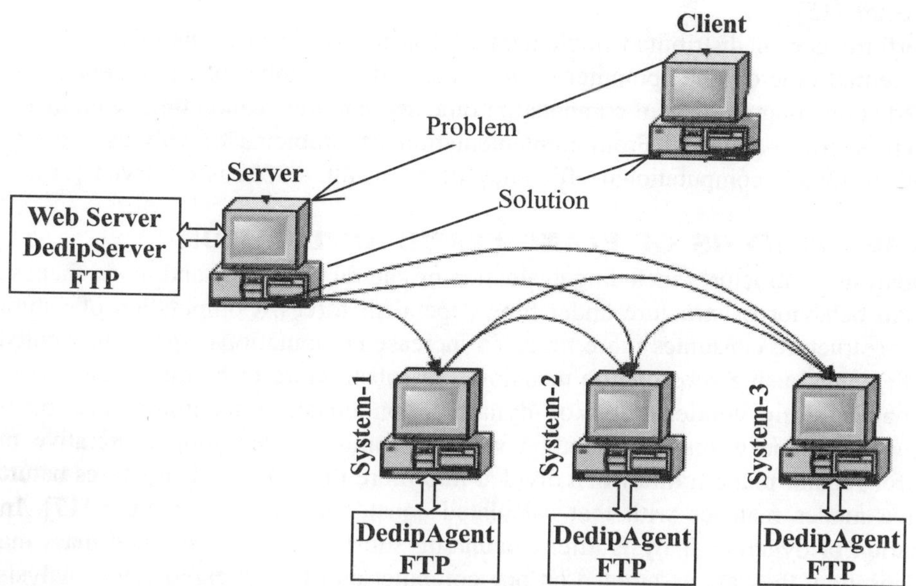

Fig. 1. WebDedip environment.

It has back-end DedipServer running on the website, which based on request of client, reads the application configuration information from the configuration file and initiates the execution of first process in the interdependency chart. It informs the agent on the target node to start the execution of the process and the agent sends the status information back to the DedipServer when process is completed. It finds out the dependent process on the successful completion of the process and initiates the execution of each such process. The required files are transferred from one node to another using FTP.

3. STATIC FINITE ELEMENT ANALYSIS APPLICATION

To implement static finite element over distributed computing WebDedip environment the substructure technique is used [14]. In this technique generation of substructure stiffness matrix and load vector corresponding to boundary nodes for each substructure is carried out in parallel on different computers. After assembly of substructure stiffness matrices and load vectors, displacements corresponding to boundary nodes are calculated. Subsequently, displacements of internal nodes and stresses for each substructure are calculated on different computers in parallel. In this process data is communicated four times. In first communication data of substructure is sent to slave computers from master computer. In second communication slave computers send substructure stiffness matrix and load vector to master computer for assembly. After evaluation of interface DOF, in third communication master computer sends displacements corresponding to interface node to respective slave computers. Finally in fourth communication slave computer sends displacements of nodes and element stresses of assigned substructure to master computer for getting results. It is observed that most of the time is consumed in generation of substructure stiffness matrix which is based on condensation of internal DOF from equilibrium equations of substructure. As internal nodes are numbered first and all interface nodes are numbered last, the bandwidth of stiffness matrix generated from equilibrium equation increases which in turn increases the time required for derivation of substructure stiffness matrix and load vector [15].

The performance of distributed implementation is measured in terms of speedup, which is the ratio of sequential time on one computer to time required by number of computers. As time required in distributed computing consists of computation time and communication time, with increased number of computers speedup reduces. From implementation of problems of various sizes (DOF ranges from 17000 to 43000) computational efficiency of about 80 – 90% is observed [16].

4. DYNAMIC ANALYSIS OF PLANE FRAME APPLICATION

Dynamic analysis of structures including evaluation of natural frequency and mode shapes is important to understand behavior of structure under time dependent forces. Computation of natural frequency for large size structure consumes more time. To increase computational speed distributed processing can be implemented using substructure technique. In substructure technique to generate substructure stiffness matrix static condensation or dynamic condensation technique can be used. Static condensation technique is one step method while dynamic condensation is iterative method. It is found that for regular plane frame when divided into more than three substructures natural frequency of first three modes matches with that calculated considering whole structure [17]. In distributed implementation of dynamic analysis after calculating substructure stiffness and mass matrices using static condensation, they are assembled on one computer and finally eigen value analysis of reduced system gives natural frequency of structure. An example of 300 storied plane frame (DOF 9933) [18] is implemented over distributed computing WebDedip environment considering 2 to 5 substructures and computational efficiency of 60 – 80% is observed.

5. GEOMETRIC NONLINEAR ANALYSIS OF PLATES APPLICATION

As nonlinear analysis is iterative in nature for larger size problem it is highly time consuming and so distributed computing can be beneficial. To implement geometric nonlinear finite element analysis of plates over distributed computing environment two approaches are considered. In first approach for entire structure unbalanced load vector, geometric stiffness matrix and initial stress stiffness matrix are calculated in parallel on different computers while solution and check for convergence are carried out sequentially. In second approach the substructure technique is used in which tangent stiffness matrix and unbalanced load vector of each substructure are calculated in parallel and finally solution of reduced system and check for convergence are done sequentially. Both the approaches are implemented using LAN and found that first approach is suitable for small size problem while second approach is more suitable for large size problem [19]. Based on solution of a problem of clamped square plate subjected to uniformly distributed load (20165 DOF) using second approach an efficiency of about 60 – 80% is observed.

6. FINITE ELEMENT ANALYSIS OF LAMINATED COMPOSITE APPLICATION

Due to higher stiffness to weight ratio and strength to weight ratio, laminated composite materials are widely used in aerospace, military and civil engineering structures. Finite element analysis of large size laminated composite plates using higher order shear deformation theory requires larger computation time. To improve computational efficiency distributed computing is adopted and two alternatives are explored for implementation. In first alternative stiffness matrix and load vector for each lamina are calculated in parallel considering entire structure and solution of equations is carried out sequentially. Finally, stresses in each lamina are calculated in parallel. In this approach as the entire structure is considered, the size of stiffness matrix to be transferred from one computer to other is larger which consume more time in communication. So, it is less efficient for larger problems. In second alternative substructure technique is used. Stiffness matrix and load vector for each substructure of laminate are calculated in parallel and calculation of displacements of boundary nodes is carried out on one computer. Finally, displacements of internal nodes and stresses in each lamina of each substructure are calculated in parallel. For the distributed finite element analysis of simply supported laminate subjected to sinusoidal loading (29766 DOF) a computational efficiency of about 60 – 80% is observed [20].

7. OPTIMIZATION USING GENETIC ALGORITHM APPLICATION

Structural optimization using Genetic Algorithm (GA) which is based on the principle of survival of the fittest includes evaluation of objective functions for the individuals of population and based on their fitness generation of new population. Generally, evaluation of objective function requires structural analysis. So for large size structure, having large population size, it consumes lot of computation time. To improve computational speed using GA, evaluation of objective functions of subpopulation can be distributed to various computers. In the present work, distributed GA is developed by distributing subpopulation to different computers for objective function evaluation and genetic operations like reproduction, crossover, mutation are carried out on one computer. Small size problems of weight optimization of plane and space truss indicated computational efficiency of 40 – 50% [21]. But it is expected to give more efficiency for larger size problem.

8. DISTRIBUTED TRAINING OF NEURAL NETWORK APPLICATION

Artificial Neural Networks (ANN) based on learning function of brain is very useful in solving complex problems where significant number of training sets are available for mapping the relationship between input and output. Various useful properties of ANN are self-organization, generalization, fault tolerance and massive parallelism. As brain learns from examples, ANN is trained by presenting it the set of input-output patterns. ANN consists of input, hidden and output layers and training of ANN is carried out by adjusting initially randomized weights which is computational intensive part due to iterative nature. To improve computational efficiency, training of ANN is implemented over network of computers. Counter Propagation Neural Network (CPNN) algorithm is used because of fix topology and less inter-processor communication requirement during training. An example of training of neural network for design of bi-axially loaded column is considered. More than 7400 training set are used for training. Each training pattern consists of four inputs and one output. When training is implemented over 2 to 5 computers connected through LAN an efficiency of about 60 –80% is observed [22].

CONCLUSION

- Local Area Network and WebDedip can be used to implement structural engineering applications over distributed computing environment without any additional resources. WebDedip, which is developed using JAVA, simplifies distributed application development.
- Ideal speedup equal to number of computers is not possible because sometime is spent in dividing the problem, in data transfer for synchronization of computation, etc. The efficiency

of computation depends on type of problem, hardware, software and algorithm used for the solution of problem.

- Total time spent for solution of problem on distributed computing includes computation time and communication time. Higher computational efficiency can be achieved by minimizing the communication time. In finite element analysis the communication time can be minimized by keeping number of interface nodes of substructure as minimum.

- Several structural engineering applications programmed in C++ like static, dynamic, linear and nonlinear analysis, optimization using genetic algorithm and training of neural network indicated about 60 – 80% computational efficiency for the problems considered when implemented over different number of computers. But higher computational efficiency is expected for still larger size problem.

REFERENCES

1. H. Adeli and O. Kamal, 1992, Concurrent Analysis of Large Structures - I Algorithms, Computers and Structures, 42, 413-424.
2. H. Adeli and S. Kumar, 1995, Distributed Finite Element Analysis on Network of Workstations – Algorithms, Journal of Structural Engineering, ASCE, 121, 1448-1455.
3. S. Kumar and H. Adeli, 1995, Minimum Weight Design of Large Structures on a Network of Workstations, Microcomputers in Civil Engineering, 10, 423-432.
4. B. H. V. Topping, A. I. Khan and A. Bahreininejad, 1997, Parallel Training of Neural Networks for Finite Element Mesh Decomposition, Computers and Structures, 63, 696-707.
5. T. V. S. R. Appa Rao, 1999, Some Recent Advances in Computational Methods for Structural Analysis, Proc. International Conference on Structural Engineering, Advances in Structural Engineering, S. K. Kaushik, ed., Ghaziabad, 673-689.
6. V. Annamalai and C. S. Krishnamoorthy, 1999, Adaptive Finite Element Analysis on Parallel and Distributed Environment, Parallel Computing, 25, 1413-1434.
7. J. Sziveri and B. H. V. Topping, 2000, Transient Dynamic Nonlinear Analysis using MIMD Computer Architectures, Journal of Computing in Civil Engineering, ASCE, 14, 79-91.
8. P. K. Umesha and M. T. Venuraju, 2000, A Parallel Computer Adaptive Language for Structural Analysis on Message Passing Systems, Journal of Structural Engineering, 27, 183-194.
9. E. D. Sotelino, 2003, Parallel Processing Techniques in Structural Engineering Applications, Journal of Structural Engineering, ASCE, 129, 1698-1706.
10. A. Ram Mohan Rao, T. V. S. R. Appa Rao and B. Dattaguru, 2003, New Parallel Overlapped Domain Decomposition for Nonlinear Dynamic Finite Element Method, Computers and Structures, 80, 2441-2454.
11. T. Kant, M. S. Shah and B.B. Mahanta, 2003, Thermo-Mechanical Analysis of Fibre Reinforced Composite Plates and Shells on Supercomputers, Proc. Structural Engineering Convention-2003, S. K. Bhattacharya, ed., IIT-Kharagpur, 715-723.
12. P. K. Gupta and R. N. Khapre, 2003, Finite Element Analysis of Anchorage Zone using Supercomputer PARAM10000, Proc. Structural Engineering Convention-2003, S. K. Bhattacharya, ed., IIT Kharagpur, 465-474.
13. A. K. Agrawal, S. C. Patodi, P. V. Patel and Bhatt H. S., 2000, Web based Meta-computing using DEDIP : Application Development Perspectives, Proc. 26th Annual Convention and Exhibition of IEEE India council, IEEE Bombay section.
14. S. C. Patodi, P. V. Patel and H. S. Bhatt, 2001, Distributed Finite Element Analysis using WebDedip Environment, Proc. Structural Engineering Convention SEC-2001, An International Meet, S. K. Kaushik, ed., Indian Institute of Technology Roorkee, 628-635.
15. P. V. Patel and S. C. Patodi, 2004, LAN Based Distributed Computing in Finite Element Analysis, Proc. 49th Congress of Indian Society of Theoretical and Applied Mechanics, NIT Rourkela.
16. P. V. Patel, V. P. Saxena and S. C. Patodi, 2005, Substructure based Distributed Finite Element Analysis of Plates, Proc. International Conference on Recent Advances in Concrete and Construction Technology, M. Lakshmipathy, ed., SRM Engineering College, Chennai, 379-390.
17. P. V. Patel and S. C. Patodi, 2004, Application of Substructure Technique in Dynamic Analysis of Plane Frames, Proc. National Conference on Structural Engineering and Mechanics (SEM-04), P. K. Gupta et al., eds., Birla Institute of Technology and Science, Pilani, 239-244.

18. P. V. Patel and S. C. Patodi, 2004, Application of Distributed Processing in Dynamic Analysis of Plane Frames, Proc. International Congress on Computational Mechanics and Simulation (ICCMS-04), N. G. R. Iyengar et al., eds., Indian Institute of Technology Kanpur, 372-379.

19. P. V. Patel and S. C. Patodi, 2005, Distributed Geometric Nonlinear Finite Element Analysis of Plates, Proc. Structural Engineering Convention, SEC 2005–An International Meet, J. M. Chandrakishan et al., eds., Indian Institute of Science, Bangalore.

20. P. V. Patel and S. C. Patodi, Distributed Finite Element Analysis of Laminated Composites, Journal of Institution of Engineers (I) – Civil Engineering Division (Under Communication).

21. P. V. Patel and S. C. Patodi, 2005, Personal Computers Based Distributed Genetic Algorithm for Optimization Problems, Journal of Computer Society of India, 59, 59-66.

22. P. V. Patel, V. S. Purani and S. C. Patodi, 2004, Distributed Training of Neural Network for Design of Column, Proc. National Seminar of Emerging Trends in Soft Computing based Artificial Intelligence, K. R. Chowdhary, ed., M. B. M. Engineering College, Jodhpur.

134

An Inverse Finite Element Procedure for the Prediction of Constitutive Properties of Aluminum Alloy

G. Partheepan[1], D.K. Sehgal[2] and R.K. Pandey[2]

[1]Research Associate [2]Professor, [1]Department of Civil Engineering
[2]Department of Applied Mechanics, Indian Institute of Technology Delhi, New Delhi-110 016, India
email: partheepan@gmail.com, profsehgal@yahoo.com, rkp@am.iitd.ac.in

ABSTRACT

The objective of this paper is to present a method for determining elastic modulus, yield strength and the true stress-true strain diagram of an aluminum alloy in a virtually non-destructive manner. Standard test methods for predicting mechanical properties require the removal of large material samples from the in-service component, which is impractical. To circumvent this situation, a new dumb-bell shaped miniature specimen has been designed and fabricated from the material of which the properties are to be determined. Also the test fixtures were developed to perform tension test on this proposed miniature specimen in the Zwick testing machine. An inverse finite element algorithm is proposed to find the material properties of the alloy using the load-elongation diagram obtained as miniature test output. The predicted results corroborate well with the experimental results. The proposed miniature specimen has an additional advantage in finite element modelling with respect to computational time and memory space.

Keywords: Inverse finite element; Miniature test; Tensile properties; Load-elongation.

1. INTRODUCTION

Inverse problems could be described as problems where the answer is known, but not the question. In other words, the results or consequences are known, but not the cause. This unknown cause can be found out using the inverse finite element procedure. The inverse simulation approach is of practical value for a number of reasons and has attracted particular attention in the solution of nonlinear problems where interest is focused upon the control action needed to achieve a particular form of output response [1].

Manahan *et al.* [2] have demonstrated the use of large strain finite element analysis to measure material uniaxial tensile stress-strain behaviour from the small punch test. The approach consists of matching the observed small punch test load-displacement curve to a curve from a database of curves via a series of finite element analysis of the small punch test. Similarly, an inverse methodology to determine the elastic-plastic stress-strain curve of material in the form of constitutive law having unknown parameters was attempted from the small punch load-displacement curves [3].

Researchers also performed the finite element simulation of ball indentation test to estimate the flow properties based on the measurement of load (P) vs. penetration (h) curve [4, 5]. Although the overall trend of predicted true stress and strain is consistent with the input constitutive law, the constitutive behaviour at low strains is not always well estimated by the ball indentation method by analyzing P-h data.

For the present study a dumb-bell shaped miniature specimen is designed and used for predicting the properties of any unknown material using the experimental setup described elsewhere [6]. The developed inverse finite element technique is used to determine the elastic modulus (E), yield strength (σ_y) and true stress-true strain diagram of the aluminum alloy. The computed values are compared against the uniaxial tensile test results. This may further be used to predict tensile and fracture properties of the material.

2. EXPERIMENTAL WORK

The aluminum alloy (AR66) used in the present work has the chemical composition as given in Table 1.

Table 1. The chemical compositions (%by wt) of aluminum alloy.

Material	Zn	Al	Cu	Zr
AR66 alloy	6.30	89.70	1.55	0.14

Figure 1 shows the configuration of the developed miniature specimen for the present study. The advantage of this miniature specimen is that it requires minimum number of machining operations. The specimens were mechanically polished and a final finish was given using 600 grit emery paper. The thickness of the specimen was maintained at 0.50 mm with an accuracy of 1%. Then the specimens were marked with gage length.

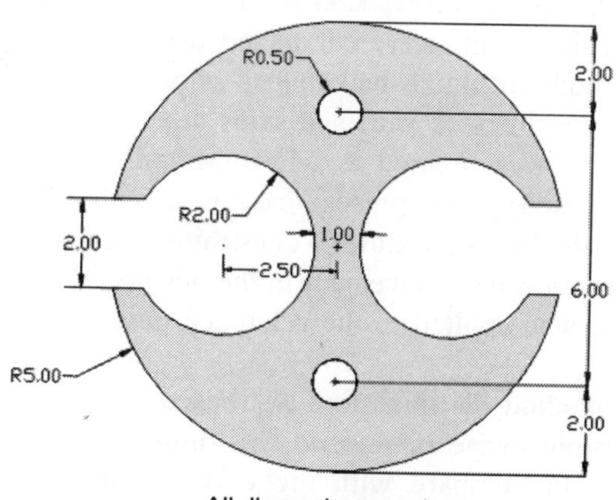

All dimensions are in mm

Fig. 1. Configuration of the miniature dumb-bell specimen.

A special test fixture was used to hold the miniature specimen in the test machine as the small size of specimen does not permit the movable cross head to come very close to the top cross head. The complete assembly consisting of fixture along with the specimen holders, test specimen and loading pins is shown in Fig. 2. The complete assembly is installed in the Zwick testing machine having a load cell of 5kN capacity and testware named *testXpert®* interfaced with it. Then the fixture attached to the specimen holder is detached smoothly. An extensometer is attached with the testing machine for the measurement of elongation of the miniature specimen during the test. The sensor arm of the extensometer is placed just near the miniature specimen. During the miniature test the load-elongation diagram was recorded. The typical load-elongation diagram as obtained from the test is shown in Fig. 3.

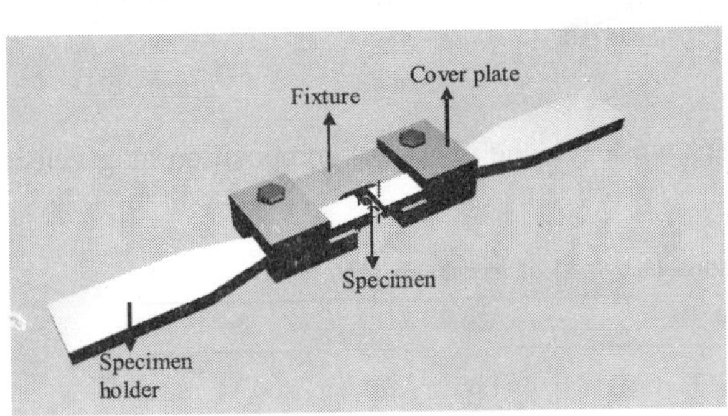

Fig. 2. Assembly of fixture attaching specimen holders and the specimen

Fig. 3. Comparison of load-elongation curve obtained from the miniature test and the inverse finite element for the specimen from the aluminum alloy

3. INVERSE FINITE ELEMENT SIMULATION

The finite element modeling calculations were conducted using the finite element code, ABAQUS. The 2-dimensional analytical model of dumb-bell shaped miniature specimen used for the simulation is shown in Fig. 4. The test specimen is modeled with eight-noded quadratic quadrilateral plane stress elements. The loading pins were treated as 2-D rigid bodies with a low friction coefficient of 0.01. The boundary condition used in the present case is as follows: The loading pin in the top fixture is fixed, and the one in the bottom fixture is constrained against translation in the X direction. However, it was allowed to experience displacement in the negative Y direction as in the experimental situation. A concentrated force was applied to the reference point of the bottom pin in the negative Y direction.

In inverse finite element method the miniature test based experimental load-elongation curve is given as an input in a linear piecewise manner to the finite element software. The experimental load-elongation curve is used to compare with the curve produced from inverse finite element procedure. The algorithm used for the determination of properties of materials using miniature specimen test along with the inverse finite element procedure can be explained in stepwise manner as follows.

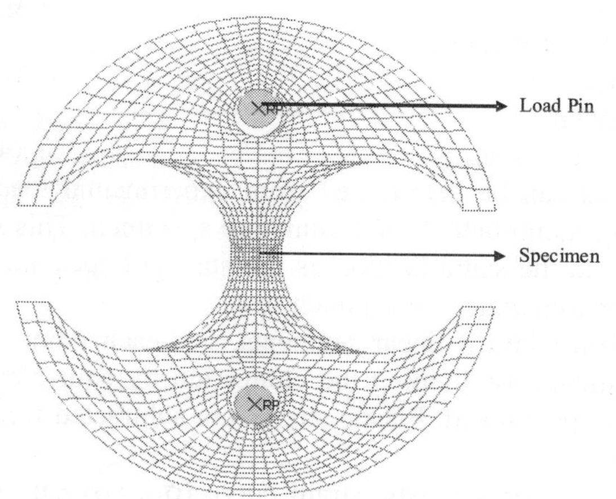

Fig. 4. Finite element model showing the mesh.

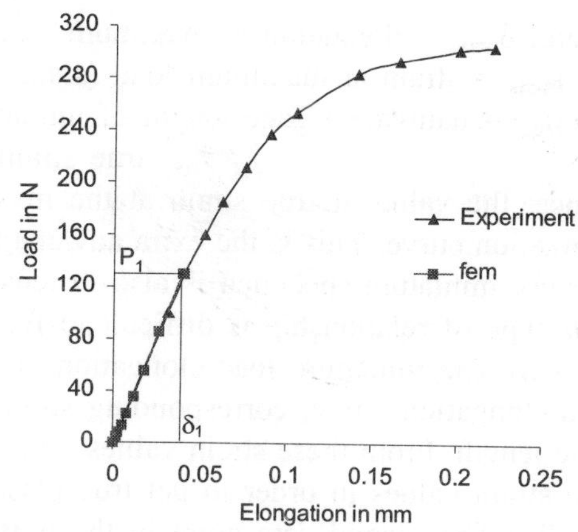

Fig. 5. Comparison of load-elongation curve for the determination of elastic modulus.

(i) The first iteration of inverse finite element analysis was carried out with an initial Young's modulus value E_{as}. In the present study E_{as} is taken as 70 GPa. The inputs to the inverse finite element analysis are E_{as} as well as load P_1 and elongation δ_1(for comparison purpose only), where the latter corresponds to the load point P_1 on the miniature load-elongation (P-δ) curve where the linear portion ends or the non-linear portion starts.

The inverse finite element analysis is carried out with the above parameters as inputs and after the elastic analysis is completed, the elongation of the specimen (δ_p) is noted. Then the value of δ_p is compared with δ_1(elongation corresponding to load P_1 in the miniature test load-elongation curve) as follows:

$$|\delta_p - \delta_1| \leq \alpha \qquad \qquad ...(1)$$

where, α has a very small value (tolerance) set by the user. In the present study this tolerance is set to 0.001mm. If the above condition is not satisfied, then the value of E_{as} is changed as follows:

$$(E_{as})_{m+1} = \left[1 + \left(\frac{\delta_p - \delta_1}{\delta_1}\right)\beta\right]_m (E_{as})_m \qquad \qquad ...(2)$$

where, β is the convergence control parameter ($0 < \beta \leq 1$), $(E_{as})_{m+1}$ is the value of E at $(m + 1)^{th}$ iteration, $(E_{as})_m$ is the value of E at m^{th} iteration. Then the analysis is carried out till the above condition is satisfied. At this iteration, the E_{as} value is the Young's modulus (E) of the material. Figure 5 shows the comparison between the load-elongation curve obtained from the FE analysis and the miniature test. A good agreement is noticed between the load-elongation curves obtained from both the methods up to the load point P_1. The maximum Von Mises stress present in the specimen at load P_1 is the yield strength of the material (σ_{t1}). The corresponding true plastic strain ε_{t1} is zero. Hence, we have determined the first point (σ_{t1}, ε_{t1}) on the true stress-true plastic strain curve.

(ii) Additional points in the true stress-true plastic strain curve are obtained by considering the miniature test load-elongation curve. It was observed that the elongation at the maximum load in the miniature test and strain at the maximum load in the standard uniaxial test bears a fixed ratio and this ratio is termed as equivalent gage length of miniature specimen.

$$\delta_{max} = l_{eq}\, \varepsilon_{max} \qquad \text{...(3)}$$

where, δ_{max} = elongation at maximum load (miniature specimen)

ε_{max} = strain at maximum load (standard uniaxial test)

l_{eq} = equivalent gage length of miniature specimen

$$\text{true strain } (\varepsilon_t)_{max} = \ln(1 + \varepsilon_{max}) \qquad \text{...(4)}$$

Hence, the value of true strain at the maximum load can be determined from experimental load-elongation curve. This is the extra advantage of using dumb-bell shaped miniature specimen. This is because miniature specimen is also in tension mode in the same fashion as the standard specimen. This type of relationship is difficult to formulate for miniature shear punch tests.

(iii) The miniature load-elongation curve is divided into n linear segments. For each point of load-elongation curve, corresponding strain is determined by dividing the elongation by equivalent gage length. From these strain values, true strains are determined. Elastic strains are subtracted from true strain values in order to get true plastic strains.

(iv) The second data point in the uniaxial true stress-true plastic strain curve (σ_{t2}, ε_{t2}) can be obtained by performing the finite element analysis by giving the inputs as load P_2, elastic modulus E, the yield stress σ_{t1} and corresponding true plastic strain ε_{t1} which is zero, the true plastic strain ε_{t2} and corresponding assumed value of σ_{t2}. The assumed value of σ_{t2} may be taken a little higher than the σ_{t1} to start iterations. Now the elongation δ_p given by finite element analysis is compared against δ_2 which is experimental value of elongation at load P_2. If $\left|\delta_p - \delta_2\right| \geq \alpha$ then σ_{t2} is modified as follows.

$$\left(\sigma_{t2}\right)_{m+1} = \left[1 + \left(\frac{\delta_p - \delta 2}{\delta_2}\right)\beta\right]_m \left(\sigma_{t2}\right)_m \qquad \text{...(5)}$$

Where, $(\sigma_{t2})_{m+1}$ = true stress at load P_2 in $(m+1)^{\text{th}}$ iteration

$(\sigma_{t2})_m$ = true stress at load P_2 in m^{th} iteration

β = Convergence control parameter $(0 < \beta \leq 1)$

Iterations may be stopped when $\left|\delta_p - \delta_2\right| \leq \alpha$. At the final iteration (σ_{t2}, ε_{t2}) is the required second data point in the uniaxial true stress-true plastic strain curve.

(v) Similarly, the third data point in the true stress-true strain curve is obtained by considering the load P_3, elastic modulus E, the previous points on the true stress-true strain curve i.e., (σ_{t1}, ε_{t1}), (σ_{t2}, ε_{t2}), the true strain (ε_{t3}) as obtained from the miniature test and the assumed value of stress σ_{t3}. Now the inverse FE analysis is carried out by iterating on σ_{t3} till the complete match is obtained between the load-elongation curves of the miniature test and the inverse FE. At the final iteration the assumed stress value σ_{t3} and the true strain ε_{t3} is the required third data point on the uniaxial true stress-true strain curve. Likewise rest of the data points on the true stress-true strain curve are obtained by proceeding in the same manner.

(vi) The material properties of the unknown material are determined in terms of Young's modulus, yield stress and data points on uniaxial true stress-true plastic strain curve. Once the material properties are known, the direct finite element analysis can simulate miniature specimen test in much better way and can do in-depth analysis of miniature specimen test.

4. RESULTS

The inverse finite element algorithm developed in the present study for the analysis of dumb-bell shaped miniature specimen under in-plane tensile load was implemented with the help of the ABAQUS code. The miniature test load-elongation curves in combination with the inverse finite element method may provide best estimates of conventionally measured tensile properties for different materials. Figure 3 shows that the load-elongation curves as obtained from the inverse finite element procedure are matching completely against the one obtained from the miniature test on the specimen from the AR66 alloy for a unique combination of constitutive parameters.

Table 2 shows the comparison of elastic modulus value obtained from the inverse finite element model and from the standard test. The starting value of Young's modulus is taken as 70 GPa.

Table 2. Comparison of elastic modulus (E) from standard test and inverse FE for AR66 alloy

Method	(E) in GPa
Uniaxial test	70.65
Inverse FEM	71.00

The developed procedure requires repeated finite element analysis as different data points on the true stress-true strain curve are obtained by considering the different segments of miniature test load-elongation curve. The procedure employed here is the one that consists simply of matching the load-elongation curve obtained from the inverse finite element analysis to the experimental curve upto the segment under consideration. When a complete match is found between the load-elongation curves of the miniature test and the one from the inverse finite element method, it gives the constitutive behaviour of the unknown material. Figure 6 shows the comparison of true stress-true strain curve obtained from the standard tensile test with that obtained by using inverse finite element procedure for the miniature specimen from used in the present investigation. It is observed from the figure that the results are encouraging and the predicted true stress-true strain diagram corroborates well with the standard test curve.

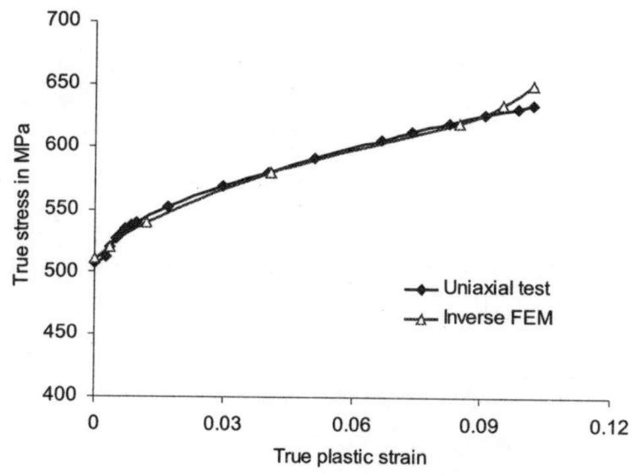

Fig. 6. Comparison of true stress-true plastic strain by inverse FE and uniaxial test for the AR66 alloy.

CONCLUSION

The determination of constitutive tensile behavior of any unknown material is made possible using the inverse finite element procedure as outlined above. The elastic modulus of unknown material is determined with the help of the initial slope of the miniature specimen's load-elongation curve and repeated inverse finite element analysis. The results from the inverse finite element are comparing well with the standard test results. The approach seems to have potential to predict the mechanical properties of the material in the in-service structures, which could be used in remaining life estimation.

The proposed dumb-bell shaped miniature specimen and the designed loading geometry in the present investigation have some special advantages in finite element modeling. The two dimensional finite element modeling is sufficient and so it can be used in any basic finite element software. The computation time and memory space required for the analysis of this miniature test is considerably less as compared to the conventional small punch test where a 3D finite element modeling is necessitated. Typically the computational time for a single run of dumb-bell shaped miniature specimen takes around 240-300 seconds whereas for the 3D small punch test simulation it takes 2000-2500 seconds.

REFERENCES

1. D.J. Murray-Smith, 2000, The inverse simulation approach: A focused review of methods and applications, Journal of Mathematics and Computer in Simulation, 53, 239-247.
2. M.P. Manahan, A.E. Browning, A.S. Argon and O.K. Harling, 1986, Miniaturized Disk Bend Test Techniques-Development and Application, ASTM STP 888, 17-49.
3. J.L. Bucaille, S. Stauss, E. Felder and J. Michler, 2003, Determination of plastic properties of metals by instrumented indentation using different sharp indenters, Acta Materialia, 51, 1663-1678.
4. M.Y. He, G.R. Odette, G.E. Lucas and Schroeter B, 2002, A finite element study relating load-penetration curves and indentation topography to yield and post-yield constitutive behavior, ASTM STP 1418, 306-320.
5. B. Taljat, T. Zacharia and F. Kosel, 1998, New analytical procedure to determine stress-strain curve from spherical indentation data, International Journal of Solids and Structures, 35(33), 4411-4426
6. G. Partheepan, D.K. Sehgal, R.K. Pandey, 2005, Design and Usage of a Simple Miniature Specimen Test Setup for the Evaluation of Mechanical Properties, International Journal of Microstructure and Material Properties, 1(1), 38-50.

135

Performance of the Superconvergent Patch Recovery Technique in Quadratic Triangular Element Meshes

Satya Narayan Chouksey[1] and K.S.R.K. Murthy[2]

[1]Engineers India Limited, New Delhi
[2]Department of Mechanical Engineering, Indian Institute of Technology Guwahati, India
email: ksrkm@iitg.ernet.in

ABSTRACT

Earlier and recent investigations have indicated that the Gaussian integration points viz. {(1/2, 1/2, 0) – Barlow points} and {(1/6, 1/6, 2/3) – Best-fit points} are possible candidates for optimal stress sampling points in quadratic triangular elements. In the present investigation a thorough numerical study has been carried out on performance of the SPR technique with the above sets of sampling points in 2D domains containing stress concentration zones. The results show that except at a few points in the domain, average finite element stresses are more accurate compared to the recovered nodal stresses from the SPR technique. As far as quadratic triangular elements are concerned it is desirable to use simple averaging process of nodal stresses than to use the SPR technique with the currently available sampling points.

Keywords: Stress recovery, SPR, optimal points, best-fit points, error estimation

1. INTRODUCTION

Commonly used 2D elements in displacement finite element procedures are of the C_0 type, and therefore do not display interelement continuity of stress fields. For this reason the element stresses are in general not continuous across element boundaries. Post-processing methods are usually employed in order to estimate more accurate values of stresses (derivatives) from finite element solutions. Such techniques are commonly known as "stress smoothing or stress recovery techniques".

The recovered stresses from various smoothing techniques are expected to be more accurate and continuous than the finite element solutions. More recently the difference between recovered stresses from various smoothing techniques and raw FEA results are employed to estimate the discretization error in a given finite element mesh and subsequent adaptive procedures [1, 2]. It has been observed that the quality and the reliability of the error estimator greatly depends on the accuracy of the recovered stresses and therefore on effectiveness of the smoothing techniques.

As a result of various applications of the recovered stresses, several recovery procedures have been proposed [1, 3-5]. Amongst, the superconvergent patch recovery (SPR) method introduced by Zienkiewicz and Zhu [1] is the most widely employed recovery technique in the linear elasticity problems. In the SPR technique, the nodal values of stresses are recovered using a least- squares fit of the elemental stresses evaluated at certain points known as *sampling points*. These are the locations within the element at which finite element stresses are more accurate than nodal values. Thus, the SPR technique depends upon the precise location of these sampling points, and hence, the accurate location of the sampling points is very important for the success of the SPR.

Zienkiewicz and Zhu [1], Strang and Fix [6], Barlow [7] and have suggested that the midpoints of the edges may be optimal points for the quadratic triangular elements. As a result of no mathematical proof, it appears that considerable lack of agreement on the optimal sampling points of the quadratic triangular element. This viewpoint is further supported by recent studies on optimal sampling points in triangular elements by Rajendran and Liew [8]. They have shown that well-known *Barlow points* {1/2, 1/2, 0} do not exist for all components of strains. Moreover, they have also shown that the Gaussian quadrature points {1/6, 1/6, 2/3} can also be candidates for optimal stress sampling points (*Best-fit points*) and shown that the SPR technique with best-fit sampling points recovers more accurate nodal stresses than Barlow points. Evidently, uncertainty in selection of optimal sampling points can lead to inaccurate extraction of nodal stresses by the SPR technique. In view of importance of this technique in error estimation and adaptivity, the present work aims at thorough investigation of performance of the original superconvergent patch recovery technique with Barlow and best-fit points as optimal sampling points on quadratic triangular element meshes.

2. FORMULATION

In the present investigation, the original superconvergent patch recovery (SPR) technique [1] has

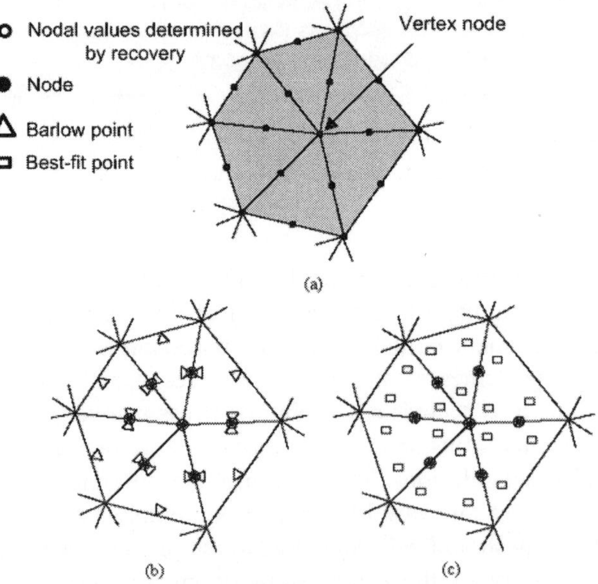

Fig. 1. A node patch and associated optimal points

been considered. In this method the finite element mesh of the domain is divided into many overlapping node patches. Such a patch represents a union of elements that are connected to a node (vertex node). A typical patch is shown in Fig.1. Barlow points and best fit points are shown in Fig. 1(b) and (c) respectively.

In this recovery process, the nodal superconvergent stresses $\{\overline{\sigma_*}\}$ belong to a polynomial expansion σ_*^P of the same complete polynomial order p as that present in the basis functions and which is assumed valid over a node patch. Thus, σ_*^P within a node patch is given by

$$\sigma_*^P = \{P\}\{a\} \qquad \qquad ...(1)$$

where, $\{P\}$ contains the complete second order polynomial terms (for T6 element) and $\{a\}$ is a set of unknown parameters corresponding to the chosen stress component σ_*^P. Thus, σ_*^P in a node patch of the T6 elements can be written as

$$\sigma_*^P = \{1 \quad x \quad y \quad xy \quad x^2 \quad y^2\}\{a_1 \quad a_2 \quad a_3 \quad a_4 \quad a_5 \quad a_6\}^T \qquad ...(2)$$

In order to determine unknown coefficients $\{a\}$ that define the superconvergent stress σ_*^P within a node patch, a discrete least square fit has been employed. This fit is made over a set of superconvergent sampling points existing in the chosen patch in order to minimize the function.

$$F(a) = \sum_{i=1}^{n}\left(\sigma(x_i,y_i) - \{P(x_i,y_i)\}\{a\}\right)^2 \qquad ...(3)$$

where, $F(a)$ are the co-ordinates of a group of total number of sampling points 'n'. The suggested sampling points for the T6 element are mid-side nodes. However, it is important to recognize that such a suggestion is based on the numerical experimentation by Zienkiewicz and Zhu [1] rather than the rigorous mathematical analysis. The total sampling points n in a chosen patch with m number of elements connected to the vertex node is given by

$$n = mk \qquad ...(4)$$

where, k is the number of sampling points available in an element of a total m elements. Thus, for the node patch shown in Fig. 1(b) total number of sampling points are 18. It can be shown that minimization condition of $F(a)$ implies that $\{a\}$ satisfies

$$\{a\} = [A]^{-1}\{b\} \qquad ...(5)$$

where $$[A] = \sum_{i=1}^{n}\{P(x_i,y_i)\}^T\{P(x_i,y_i)\} \quad \text{and} \quad \{b\} = \sum_{i=1}^{n}\{P(x_i,y_i)\}^T \sigma(x_i,y_i) \qquad ...(6)$$

Once the parameters $\{a\}$ are determined, the recovered nodal values of σ_* are simply calculated by inserting appropriate co-ordinates into the Eq. (2). The entire procedure can be repeated for all components of the stress or the formulation can be adjusted so that the unknown coefficients $\{a\}$ for each component of the stress can be found in a single run.

Table 1. Notation for SPR with different optimal points

Notation	Sampling points	Figure	Description
SPR1	(1/2, 1/2, 0) - *Barlow points* Gaussian quadrature scheme 1		Superconvergent Patch Recovery technique with Barlow points or mid-side nodes as optimal sampling points
SPR2	(1/6, 1/6, 2/3) - Best-fit points Gaussian quadrature scheme 2		Superconvergent Patch Recovery technique with Best-fit points as optimal sampling points
FEM			Average finite element stresses
Exact			Exact solution

3. NUMERICAL RESULTS AND DISCUSSION

The following notation (Table 1) has been used while illustrating the results of the present investigation. The problem to be studied here is the L-shaped domain as shown in Fig. 2. It is loaded by the tractions (Eq. (7)) corresponding to the following exact stress fields [9, 10]

$$\left.\begin{aligned}
\sigma_{11} &= A\lambda r^{\lambda-1}\left\{\left[2-\phi(\lambda+1)\right]\cos(\lambda-1)\theta-(\lambda-1)\cos(\lambda-3)\theta\right\} \\
\sigma_{22} &= A\lambda r^{\lambda-1}\left\{\left[2+\phi(\lambda+1)\right]\cos(\lambda-1)\theta+(\lambda-1)\cos(\lambda-3)\theta\right\} \\
\sigma_{12} &= A\lambda r^{\lambda-1}\left[(\lambda-1)\sin(\lambda-3)\theta+\phi(\lambda+1)\sin(\lambda-1)\theta\right]
\end{aligned}\right\} \qquad \text{...(7)}$$

where, A is a constant similar to the mode-I SIF. The following parameters have been employed so that the solution satisfies the equations of equilibrium and the stress free boundary conditions on the re-entrant edges AB and AF.

$$\text{Where,} \qquad \left.\begin{aligned}
A &= 1.0 \\
\lambda &= 0.544483737 \\
\phi &= 0.543075579
\end{aligned}\right\} \qquad \text{...(8)}$$

The re-entrance corner point 'A' in Fig. 2 is the singular point at which stresses are infinite. Thus, this point has also been considered for accuracy examination. The other non-singular points at which accuracy is tested are shown in Fig. 2 with solid circles. Three levels of uniform meshes have been employed for the accuracy observations and these are shown in Fig. 3.

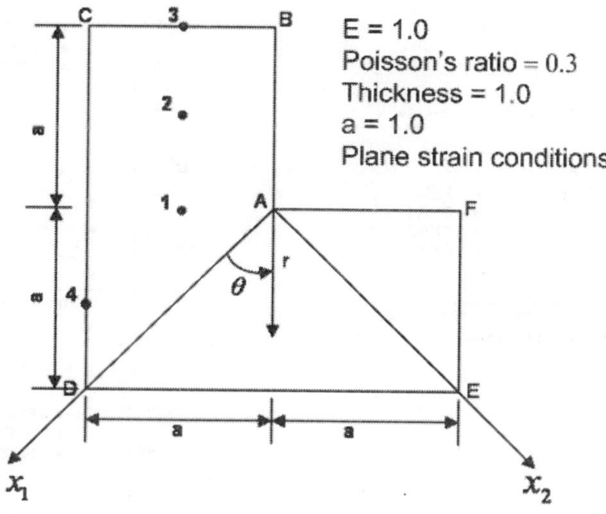

Fig. 2. L-shaped domain

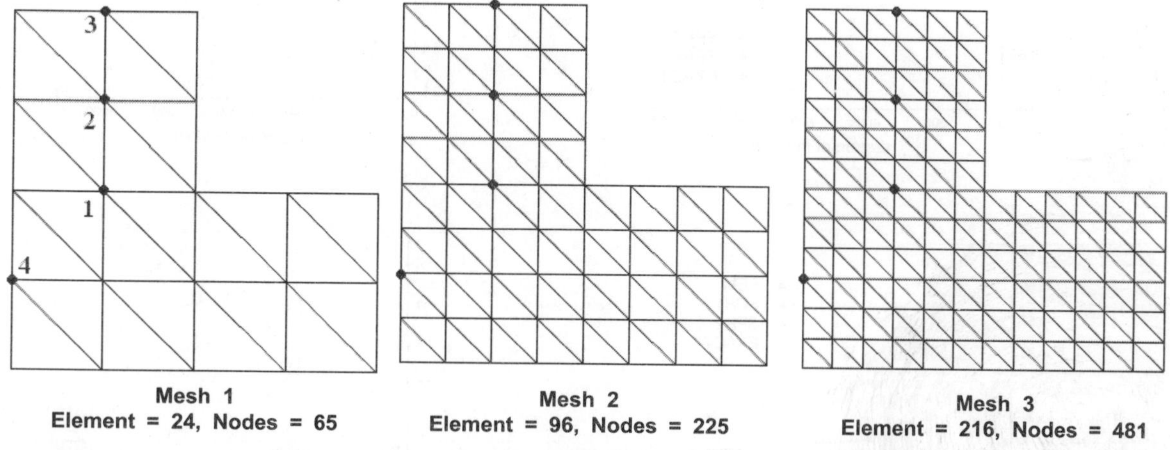

Fig. 3. Mapped meshes used for accuracy observations

Figure 4(a) shows the convergence of the recovered stresses at the non-singular point 1 with respect to the nodal degree of freedom. Figure 4(b) shows the convergence of recovered stresses at the singular point 'A' at which the exact stresses are infinite. Results indicate that, similar to the other examples no monotonic increase in accuracy of the recovered stresses by the SPR1 and SPR2 at the selected points for all components of the stresses is observed, as the meshes are refined. Moreover, similar and unacceptable performance has been exhibited by the SPR1 and SPR2 in comparison with the average finite element stresses. Again no superior performance of the SPR2 method is observed over the SPR1 and averaging process of the finite element solutions. This is contrary to the observations of Rajendran and Liew [8]. However, for certain components of the stresses and at some locations these methods (SPR1 and SPR2) have shown superior accuracy compared to the average finite element stresses (For example, σ_y and τ_{xy} at the point 1). Unfortunately such performance is not exhibited at all points and in all the components of the stresses.

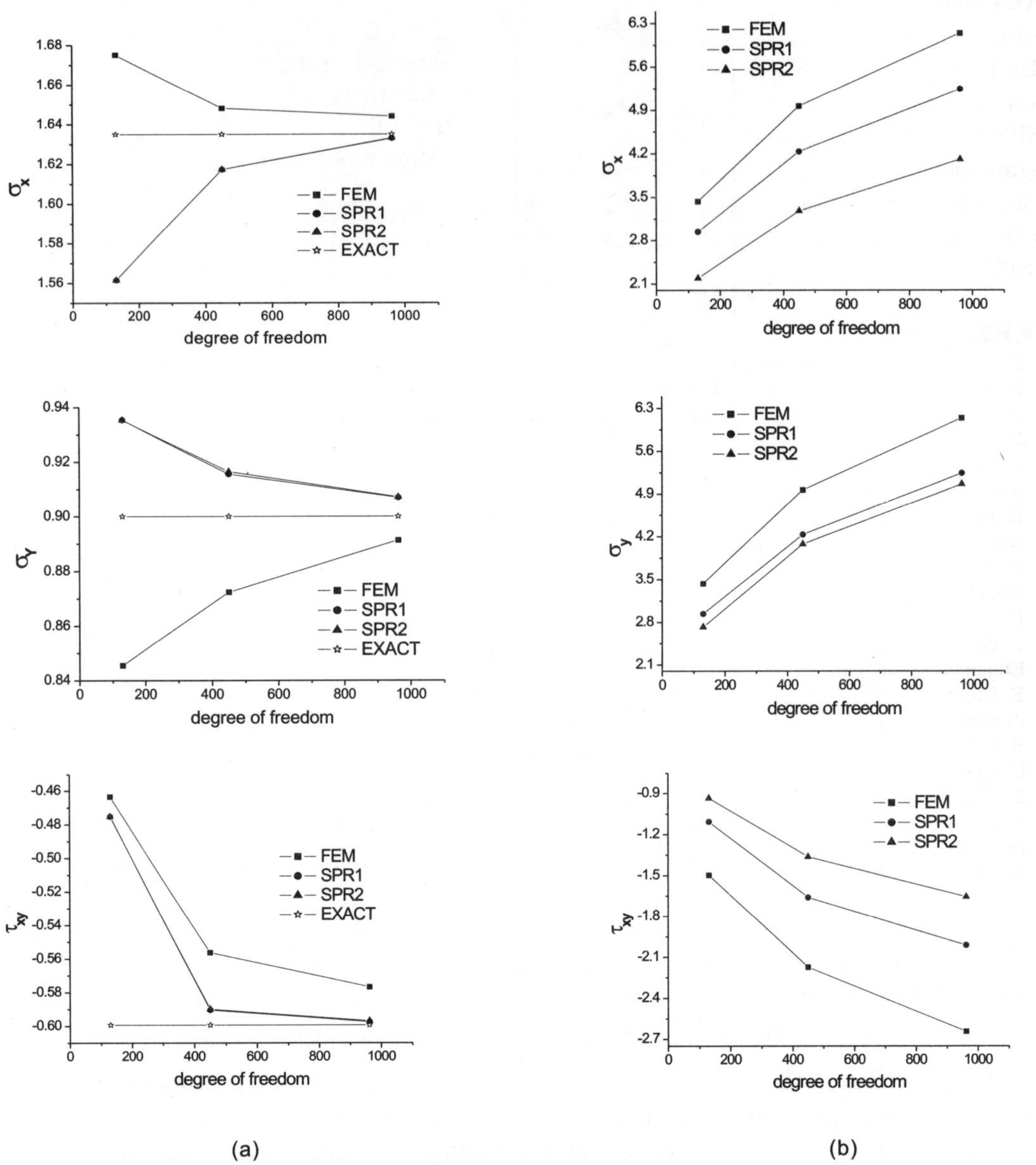

Fig. 4. Convergence of the recovered stresses (a) at the point 1 and (b) re-entrant corner A.

Poor accuracy of the results can be observed from the SPR1 and SPR2 as compared to the average finite element stresses at the singular point. Moreover, the performance of SPR2 is inferior to that of SPR1. This is again contrary to the observations of Rajendran and Liew [8]. This result coupled with others clear indicates that average finite element stresses more reliable, consistent and more accurate compared to results of the SPR1 and SPR2.

CONCLUSION

A thorough comparison has been made between superconvergent patch recovery (with the Barlow and Best fit points as the sampling points) and standard averaging process of the finite element stresses. The performance of both the SPR1 (Barlow points as sampling points in the SPR technique) and SPR2 (Best-fit points as sampling points in the SPR technique) are inferior compared with the averaging process of finite element solutions. Only at few locations of a given domain, both the methods have shown satisfactory recovery than the entire domain. In contrast to observations made by Rajendran and Liew [8], no superior and consistent performance of the SPR2 technique than SPR1 has been observed. Instead similar performance has been observed in many cases.

REFERENCES

1. O.C. Zienkiewicz, and J.Z. Zhu, 1992, The superconvergent patch recovery and a posterior error estimates. Part 1: The recovery technique, International Journal for Numerical Methods in Engineering, 33, 1331-1364.
2. O.C. Zienkiewicz, and J.Z. Zhu, 1992, The superconvergent patch recovery and a posteriori error estimates. Part 2: Error estimates and adaptivity, International Journal for Numerical Methods in Engineering, 33, 1331-1364.
3. E. Hinton, and J.S. Campbell, 1974, Local and global smoothing of discontinuous finite element functions using a least squares method, International Journal for Numerical Methods in Engineering, 8, 461-480.
4. B. Boroomand, and O.C. Zienkiewicz, 1997, Recovery by equilibrium in patches (REP), International Journal for Numerical Methods in Engineering, 40, 137-164.
5. J. Aalto, and H. Isoherranen, 1997, An element by element recovery method with built in equations, Computers and Structures, 64, 177-196.
6. G. Strang, and G.J. Fix, 1973, An Analysis of the Finite Element Method, Prentice-Hall, Englewood Cliffs, NJ.
7. J. Barlow, 1976, Optimal stress location in finite element method, International Journal for Numerical Methods in Engineering, 10, 243-251.
8. S. Rajendran, and K.M. Liew, 2003, Optimal stress sampling points of plane triangular elements for patch recovery of nodal stress, International Journal for Numerical Methods in Engineering, 58, 579-607.
9. B.A. Szabo, I. Babuska, and B.K. Chayapathy, 1989, Stress computations for nearly incompressible materials by the p-version of the finite element method, International Journal for Numerical Methods in Engineering, 28, 2175-2190.
10. C.K. Lee, and R.E. Hobbs, 1997, On using different finite elements with an automatic adaptive refinement procedure for the solution of 2-D stress analysis problems, International Journal for Numerical Methods in Engineering, 40, 4547-4576.

136

Modeling and Performance Analysis of a Solar Absorption System with Parabolic Trough Collector for 24 Hours

M. MAZLOUMI, M. NAGHASHZADEGAN AND K. JAVAHERDEH

Department of Mechanical Engineering, Guilan University, Rasht, Guilan-3756, Iran
email: Naghash@guilan.ac.ir

ABSTRACT

Ahwaz is one of the sweltering cities in Iran where enormous energies is being consumed to cool residential spaces in the year, so the cooling absorption systems are the most suitable for this region. A computer program has been written to simulate a solar absorption cooling system to supply the cooling load of a typical house during 24 hours of the design days when the cooling load peak is about 17.5kw (5 refrigeration tons) that occur in July. The generator energy is provided by parabolic trough collector with direct water and thermal storage tank which has been insulated. The results shown that the collector mass flow rate has negligible effect on minimum required collector area which its mean value is about 104m^2, but the storage tank volume depends on the collector mass flow rate.

Keywords: Solar air-conditioning, parabolic trough, simulation.

1. INTRODUCTION

The solar lithium bromide (LiBr)-water absorption air-conditioning system has been become attractive recently, because maximum cooling load occur when more solar radiations are available. Wide researches have been performed to simulate solar absorption systems [1-3].

Ahwaz city lies to the south of Iran where latitude is equal to 31.3°. Its residential places need to cool for about six months of the year that the daily average of global solar radiation is almost 24MJ/m^2; therefore, solar air-conditioning systems are suitable in this region. A simulation computer program has been written to design a solar single effect lithium bromide-water absorption system for a given residential place. Using mean meteorological data has been obtained by processing of 20 years hourly measurements. Solar energy is gained by parabolic trough collector with horizontal axis lies to north-south direction. Water is pumped to collector directly and the storage tank has been insulated to remove heat losses. Thermodynamic model has been used to simulate absorption cycle.

2. SYSTEM SIMULATION

The system performance depends on the weather data, therefore the hourly direct solar radiation, ambient temperature and wind velocity have been used to simulate the solar absorption system. The cooling load variations for 24 hours of design days are shown in Fig. 1.

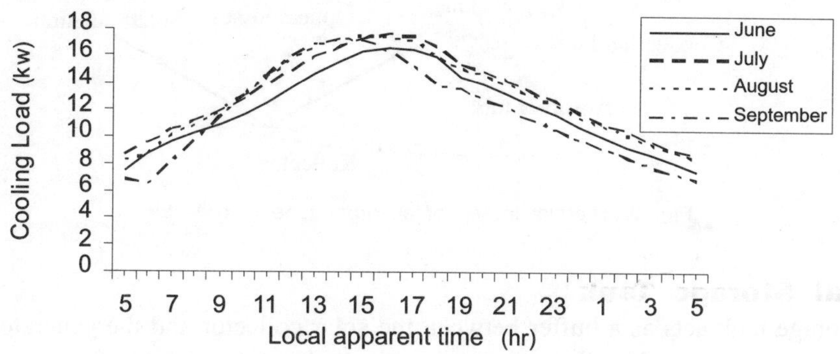

Fig. 1. Cooling load variations of the design days.

The given solar absorption cooling system consists of a parabolic trough collector, a hot water storage tank and a single effect lithium bromide-water absorption chiller which is shown in Fig. 2.

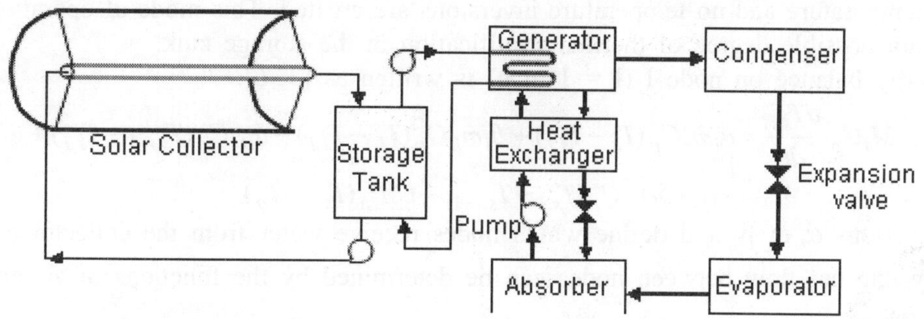

Fig. 2. Schematic diagram of solar absorption cooling system.

2.1 Solar Collector

The parabolic trough collector with north-south horizontal focal axis is the most conventional. This collector is a type of medium concentration collectors which is rotated about an axis parallel to its focal axis and adjusted continuously so that the solar direct radiations make minimum angle of incidence with the aperture plane of collector at all sunshine times of a day. The uniform solar heat flux on absorber tube of collector is assumed and its useful heat gain rate is obtained by the following [4]:

$$Q_u = F_r A_c [S - \frac{U_L}{C}(T_{fi} - T_a)] \qquad ...(1)$$

Figure 3 shows the thermal losses which are considered to thermal analysis of collector [5].

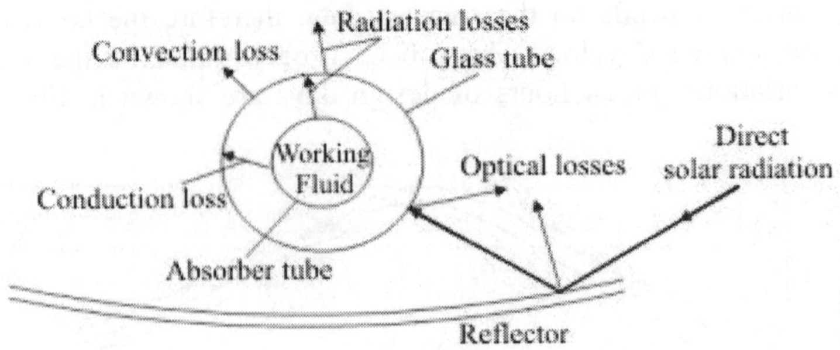

Fig. 3. Thermal losses of absorber tube of collector.

2.2 Thermal Storage Tank

The thermal storage tank acts as a buffer between the solar collector and the generator of absorption cycle. It stores solar energy when the amount of collected energy is more than the requirement of application and discharges it when the collected amount is inadequate. The thermally stratified situation has been considered to simulate thermal storage tank. In this simulation, the storage tank is divided into n fully mixed volume (nodes) and modeled by multinode one dimensional model, also the variable inlet positions are assumed, i.e., the flows enter the nodes that are closest in density or temperature and no temperature inversions are created. This mode of operation preserves the maximum possible degree of thermal stratification in the storage tank.

An energy balance on node I (I = 1 to n) is written as [3, 6]:

$$M_i C_p \frac{dT_{s,i}}{dt} = \alpha_i \dot{m}_c C_p (T_c - T_{s,i}) + \beta_i \dot{m}_L C_p (T_L - T_{s,i}) + \delta_i \gamma_i C_p (T_{s,i-1} - T_{s,i}) +$$

$$(1 - \delta_i)\gamma_i C_p (T_{s,i} - T_{s,i+1}) - UA_{s,i}(T_{s,i} - T_a) \qquad \ldots(2)$$

The functions α_i of β_i and define which nodes receive water from the collector and the load respectively, the net flow between nodes can be determined by the functions of δ_i and γ_i.

$$\alpha_i = \begin{cases} 1 & \text{if } i = 1 \text{ and } T_c > T_{s,i} \\ 1 & \text{if } T_{s,i-1} \geq T_c > T_{s,i} \\ 0 & \text{otherwise} \end{cases} \qquad \beta_i = \begin{cases} 1 & \text{if } i = n \text{ and } T_L < T_{s,n} \\ 1 & \text{if } T_{s,i} \geq T_L > T_{s,i+1} \\ 0 & \text{otherwise} \end{cases}$$

$$\delta_i = \begin{cases} 1 & \text{if } \gamma_i > 0 \\ 0 & \text{if } \gamma_i \leq 0 \end{cases} \qquad \gamma_i = \begin{cases} -\dot{m}_L \sum\limits_{j=2}^{n} \beta_j & \text{If } i = 1 \\ \dot{m}_c \sum\limits_{j=1}^{n-1} \alpha_j & \text{If } i = n \\ \dot{m}_c \sum\limits_{j=1}^{i-1} \alpha_j - \dot{m}_L \sum\limits_{j=i+1}^{n} \beta_j & \text{If } i = 2,\ldots,n-1 \end{cases}$$

The cylindrical thermal storage tank has been insulated all round with glass wool.

2.3 Absorption Cycle

The lithium bromide-water solution is the most conventional for solar absorption air-conditioning applications. The main components of absorption cycle are the generator, condenser, evaporator and absorber as fig. 2. In the generator, solution is heated to vaporize the refrigerant and separate from the solution, the superheat vapor of refrigerant is condensed to deliver heat to cooling water, the condense liquid flows through an expansion valve to the evaporator where refrigerant is evaporated to produce cooling effect, the refrigerant vapor then goes to the absorber where it is absorbed by high concentration solution coming from the generator and reject heat to the cooling water. Finally, the low concentration solution is pumped through heat exchanger to the generator; the solution is preheated in heat exchanger. This cycle operates at low pressure to evaporate water in low temperatures. The performance of absorption cycle can be simulated well by the thermodynamic model [1, 7]. The basic assumptions of thermodynamic model of absorption cycle are as follows:

1. There are steady state conditions.
2. Pressure drops and heat losses in components and tubes are negligible.
3. Saturated solution leaves the absorber and generator.
4. There is saturated refrigerant at condenser and evaporator outlets.
5. Pressure drops and heat losses in components and tubes are negligible.
6. Expansion valves are adiabatic.

The mass and energy balances are performed to analyze absorption cycle at its each component. The equilibrium temperature and enthalpy of lithium bromide-water solution can be obtained as:

$$T_{sol} = T_{ref} \sum_{i=0}^{3} a_i x^i + \sum_{i=0}^{3} b_i x^i \qquad \qquad ...(3)$$

$$h_{sol} = \sum_{i=0}^{4} c_i x^i + T_{sol} \sum_{i=0}^{4} d_i x^i + T_{sol}^2 \sum_{i=0}^{4} e_i x^i \qquad \qquad ...(4)$$

The constant coefficients of $a_i, b_i, ... e_i$ were presented by ASHRAE handbook [8].

The actual coefficient of performance of absorption cooling cycle is obtained by:

$$Cop = \frac{\dot{Q}_e}{\dot{Q}_g} \qquad \qquad ...(5)$$

2.4 Simulation Assumptions

Since performance of system depends on the weather data, it is necessary to assume the variable temperatures of generator and condenser, so it has been assumed that the generator temperature is 5°C lower than the temperature of the upper part of the storage tank and the condenser temperature is 10°C higher than the wet bulb temperature of ambient. The absorption cycle has been simulated based on the evaporator temperature of 6°C, the solution pump flow rate of 0.053Liter/sec and heat exchanger efficiency of 50%. The generator of absorption cycle operates by delivery water temperatures 65-95°C [1], and water isn't saturated in collector. Initial temperature of storage tank is about 61°C that is resulted from the storage water temperature at the end of each day operation. The specifications of parabolic trough collector are presented in Table 1.

Table 1. Parabolic trough collector specifications.

Aperture Width	2m
Overall optical efficiency	73%
Absorber tube inner diameter	0.035m
Absorber tube outer diameter	0.042m
Evacuated glass tube diameter	0.07m
Absorber tube emissivity	0.25
Glass tube emissivity	0.95

3. RESULTS AND DISCUSSION

The solar absorption cooling system has been simulated in design days. The results shown the largest required collector area occurred in design day of July. Figure 4 shows the effect of collector mass flow rate on minimum required collector area and optimum storage tank volume for operation from sunrise to sunset of the design day in July and it is presented for 24 hours operation as Fig. 5.

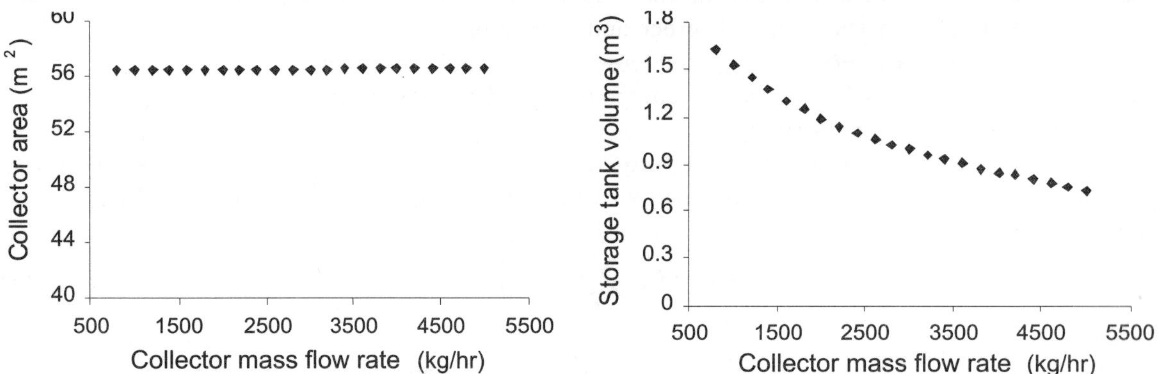

Fig. 4. The effect of collector mass flow rate on minimum required collector area and optimum storage tank volume for operation from sunrise to sunset of the design day in July.

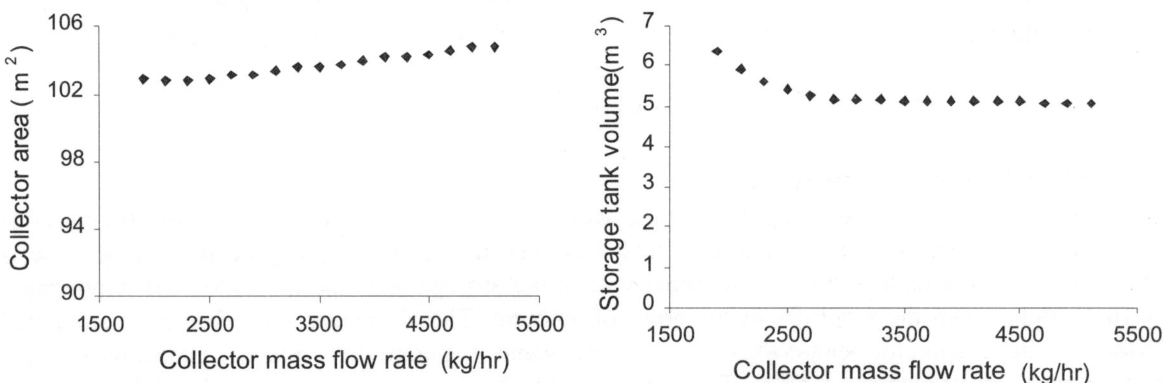

Fig. 5. The effect of collector mass flow rate on minimum required collector area and optimum storage tank volume for 24 hours operation in the design day of July.

The results showed the collector mass flow rate has negligible effect on the minimum required collector area, because the collector heat losses are low and they have been considerably reduced by low operation temperature of collector which has good agreement with the results of Odeh, et al .[5]. Reducing the collector mass flow rate lowers the collector heat removal factor but also tends to increase storage tank thermal stratification, which may improve the overall system performance [6].

The low collector mass flow rates have significant effect on the optimum storage tank volume because the water temperature has been limited.

Figure 6 shows the cop variations of absorption cooling cycle. Its values are within the given values of conventional lithium bromide-water absorption systems by Duffie and Beckman [9].

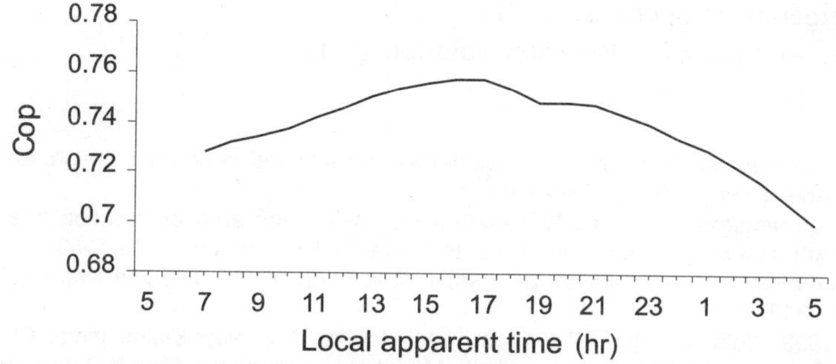

Fig. 6. Cop variations for 24 hours operation in the design day of July.

CONCLUSION

The generator energy of a single effect lithium bromide-water absorption cooling system can be supplied by solar parabolic trough collector with direct water. The collector mass flow rate has negligible effect on minimum required collector area. In Ahwaz climate, the minimum required collector area is about 56.5m^2 to supply the cooling loads of a given residential place from sunrise to sunset and it is about 104m^2 for 24 hours operation when the cooling load peak is about 17.5kw. The storage tank volume depends on collector mass flow rate.

NOMENCLATURE

Q_u	Useful heat gain rate of collector (W)
F_r	Heat removal factor
A_c	Effective aperture area of collector (m^2)
S	Absorbed solar flux (W/m^2)
U_L	Overall heat loss coefficient (W/m^2.°C)
C	Concentration ratio
T_{fi}	Inlet fluid temperature of collector (°C)
T_a	Ambient temperature (°C)
M_i	Fluid mass of node I (kg)
C_p	Specific heat capacity of fluid (KJ/kg.°C)
$T_{s,j}$	Fluid temperature of node I (°C)
dt	Time step
m_c	Collector mass flow rate (kg/hr)
T_c	Outlet fluid temperature of collector (°C)
$\dot{m}_L$	Load mass flow rate (kg/hr)

T_L	Fluid temperature coming from the load (°C)
$UA_{s,i}$	Heat loss coefficient-area product of node I (KJ/hr.°C)
T_{sol}	Temperature of LiBr-water solution (°C)
h_{sol}	Enthalpy of LiBr-water solution (KJ/kg)
Cop	Coefficient of performance of absorption cooling cycle
$\dot{Q}_e$	Heat transfer rate of evaporator (KW)
$\dot{Q}_g$	Heat transfer rate of generator (KW)
T_{ref}	Refrigerant temperature (°C)
X	Concentration of LiBr-water solution (%)

REFERENCES

1. N.K. Ghaddar, M. Shihab and F. Bdeir, 1997, Modeling and simulation of solar absorption system performance in Beirut, Renewable Energy, 10(4), 539-558.
2. F. Assilzadeh, S.A. Kalogirou, Y. Ali and K.Sopian, 2005, Simulation and optimization of a LiBr solar absorption cooling system with evacuated tube collectors, Renewable Energy, 30, 1143-1159.
3. Z.F. Li and K. Sumathy, 2001, Simulation of a solar absorption air conditioning system, Energy Conversion & Management, 42, 313-327.
4. S.P. Sukhatme, 1993, Solar Energy, 9th Edition, Tata McGraw-Hill, New Delhi, India, Chap. 6.
5. S.D. Odeh, G.L. Morrison and M. Behnia, 1998, Modeling of parabolic trough direct steam generation solar collectors, Solar Energy, 62(6), 395-406.
6. E.M. Kleinbach, W.A. Beckman and S.A. Klein, 1993, Performance study of one-dimensional models for stratified thermal storage tanks, Solar Energy, 50(2), 155-166.
7. G.A. Florides, S.A. Kalogirou, S.A. Tassou and L.C. Wrobel, 2003, Design and construction of a LiBr-Water absorption machine, Energy Conversion & Management, 44, 2483-2508.
8. ASHRAE, 1997, Handbook of fundamentals, Atlanta, USA.
9. J.A. Duffie and W.A. Beckman, 1991, Solar engineering of thermal processes, John Wiley & Sons, New York, USA, Chap. 15.

137

Object Oriented Modeling of Rotor Shaft with Internal Damping Using Bondgraph

V. Rastogi[1], A. Mukherjee[2], A. Dasgupta[2] and A.K. Samantaray[2]

Department of Mechanical Engineering
[1]Sant Longowal Institute of Engineering and Technology, Longowal-148106, Punjab, India
email: rastogivikas@yahoo.com
[2]Indian Institute of Technology, Kharagpur-721302, India
email: amalendu@mech.iitkgp.ernet.in, anir@mech.iitkgp.ernet.in, samantaray@mech.iitkgp.ernet.in

ABSTRACT

This paper presents the dynamic behaviour of a hollow rotor shaft with internal damping driven through a dissipative coupling. The coupling in the system is absolutely flexible in transverse and bending but torsionally rigid. The bondgraph technique is used as a modeling tool and bond graph model of the system is created using object-oriented capsules. Simulation results show interesting phenomenon of limiting behaviour of rotor shaft with internal damping beyond the threshold speed of instability. It is further shown that the entrainment of whirling speed at different natural frequency of the rotor shaft primarily depends on the ratio of external and internal damping.

Keywords: Rotor Shaft, internal damping, dissipative coupling, bondgraph, natural modes of vibration.

1. INTRODUCTION

The role of damping in rotor dynamics has been well known for more than half century [1, 2]. All damping which can be associated with the non-rotating parts of the machine has the usual stabilizing role as in structural dynamics, but damping of rotating elements brings the element of instability above the critical speed. The reason behind this is that internal damping of rotor shaft couples rotational motion and vibration causing energy to be transferred from the former to latter. This phenomenon is generated occurring every time energy dissipation takes place in a rotating system and this internal damping of rotor causes instability above the critical speed. The researchers and technocrats have long been aware of the destabilization of the rotor systems due to internal damping. Gunter [3] work included an evaluation of internal viscous damping. This contribution demonstrated the destabilizing effect of internal viscous damping at speeds beyond the first forms of internal damping and noted that internal viscous damping, which is destabilizing at speeds beyond the first critical speed whereas hysteretic damping is destabilizing at all speeds.

The effects of internal damping were studied by Dimarogonas [4], who had presented a finite element description of rotor systems. The effects of internal damping were studied by Lazen [5] and Crandall [6]. The incorporation of internal damping mechanism into linear finite element simulation was achieved by Zorzi and Nelson [7]. Mazzei *et al.* [8, 9] in their papers included the external and internal damping in a model of a rotating flexible shaft driven through a universal joint. All these analyses are based on either finite element method or the other. Limited studies are reported in literature to analyze higher natural mode. Moreover, no such studies have been attempted to model the rotor shaft with internal damping through bondgraph modeling [10]. Bondgraphs may be used as a modeling tool to obtain the object oriented reusable capsules of model. The present work shows modeling of a rotor shaft, with internal and external damping, resided in the spinning hub and driven by a constant speed source through dissipative coupling. Simulation study reveals interesting phenomena of entrainment of rotor speed at different natural frequencies, existence of limiting orbit of the rotor.

2. OBJECT ORIENTED MODELING OF ROTOR SHAFT

The physical system of a hollow flexible rotor shaft with external and internal damping driven through dissipative coupling is shown in Fig. 1. The boundary conditions of the rotor shaft are taken as pin-pin. The coupling in the system is absolutely flexible in transverse and bending but torsionally rigid. The torsional vibrations are not considered in the modeling. The analysis for these object-oriented capsules may be given in brief in the following sub-sections.

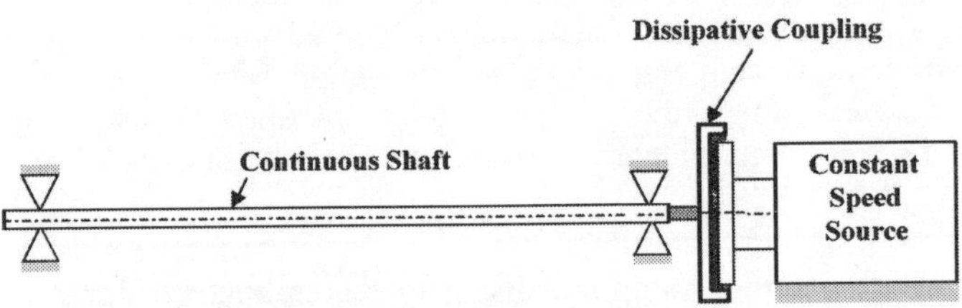

Fig. 1. Schematic diagram of a continuous shaft with internal and external damping driven by a constant speed source through dissipative coupling.

2.1 Modeling of Velocity Transformation

There are two kinds of damping effects, which may be modeled in the following way:

1. The damping force, which acts on the shaft, reflecting the external damping, can be modeled to be proportional to the transitional motion of the centre of the shaft with respect to the stationary system of coordinates.

2. Forces due to the internal damping can be modeled as proportional to the rate of change of curvature of the shaft as it can be seen from the rotating system of coordinates. So in the whirling of the shaft, the material damping is shown in a frame rotating with the shaft. Therefore, the transverse velocities of the shaft are to be obtained in the rotating frame to compute the material damping force. Velocities in the rotating frame can be obtained through coordinate transformation fixed to rotating frame.

The fixed frame coordinate system is shown in Fig. 2, where subscripts "f" and "r" stand for the fixed and the rotating coordinate frame, respectively and given as

$$\begin{bmatrix} X_r \\ Y_r \end{bmatrix} = \begin{bmatrix} \cos\alpha & \sin\alpha \\ -\sin\alpha & \cos\alpha \end{bmatrix} \begin{Bmatrix} X_f \\ Y_f \end{Bmatrix}, \qquad \text{...(1)}$$

where, α is the angular rotation of the shaft. Differentiating Eq. (1) with respect to time, one obtains

$$\begin{bmatrix} \dot{X}_r \\ \dot{Y}_r \end{bmatrix} = \begin{bmatrix} \cos\alpha & \sin\alpha \\ -\sin\alpha & \cos\alpha \end{bmatrix} \begin{Bmatrix} \dot{X}_f \\ \dot{Y}_f \end{Bmatrix} - \omega \begin{bmatrix} \sin\alpha & -\cos\alpha \\ \cos\alpha & \sin\alpha \end{bmatrix} \begin{Bmatrix} X_f \\ Y_f \end{Bmatrix}, \qquad \text{...(2)}$$

and $\alpha = \omega$

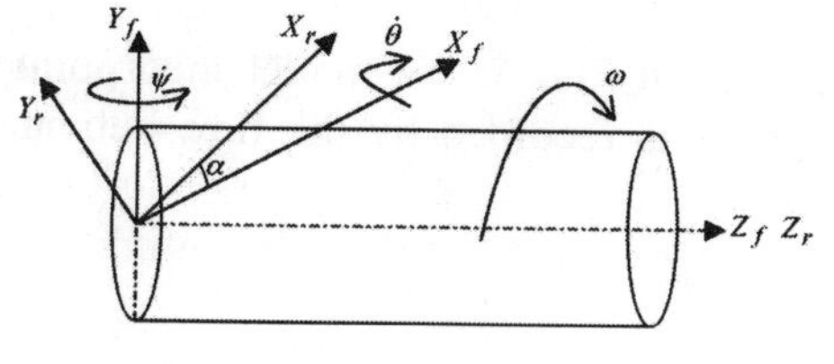

Fig. 2. Fixed coordinate frame of the shaft.

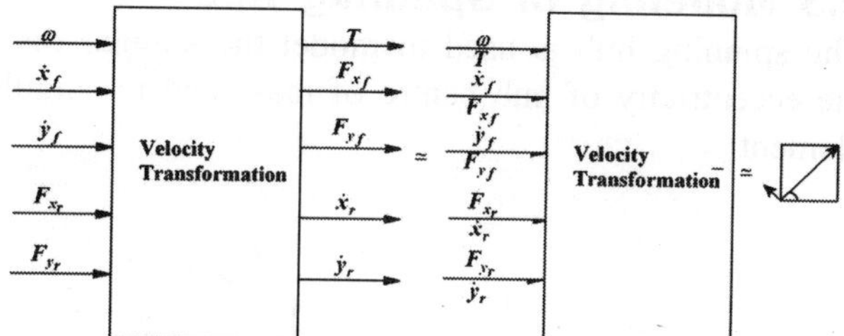

Fig. 3. Fixed rotating frame velocity transform in block diagram, word-bondgraph and capsule representation.

2.2 Modeling of Spinning Shaft

The shaft model is based on the Rayleigh beam model, where rotary inertia of the beam is included. Shaft elements with shear forces and moments acting on it can be modeled where the stiffness of the shaft elements relates the generalized Newtonian forces to generalized displacements at the ends of the element. The stiffness matrix can be modeled as 4-port compliance fields storing energy due to the four generalized displacement. The two compliance fields model the bending of the shaft in XZ and YZ directions, respectively. Mathematically, compliance field [10] in XZ directions may be represented by

$$\begin{Bmatrix} V_{x_l} \\ V_{x_r} \\ M_{x_l} \\ M_{x_r} \end{Bmatrix} = \frac{EI}{l^3} \begin{bmatrix} 12 & 6l & -12 & 6l \\ 6l & 4l^2 & -6l & 2l^2 \\ -12 & -6l & 12 & -6l \\ 6l & 2l^2 & -6l & 4l^2 \end{bmatrix} \begin{Bmatrix} \dot{x}_{f_l} \\ \dot{x}_{f_r} \\ \psi_{f_l} \\ \psi_{f_r} \end{Bmatrix} \qquad \text{...(3)}$$

where, subscripts l and r represent the left and right side of the model. Compliance field in YZ direction may also be represented likewise. The fixed to rotating frame velocity transform capsule has been used four times to transform $\dot{X}_f, \dot{Y}_f, \psi_f$ and $\dot{\theta}_f$ to rotating frame to model internal damping of the shaft in the rotating frame through the R-fields, which bring out the tensorial nature of the rotating internal damping. The left interface vector glue port has four scalar components namely, the two translational and two rotational velocities in fixed frame. The right interface vector glue port has also the same attributes. The top interface glue ports have spinning speed port. The object oriented capsule representation may be shown through word-bondgraphs as in Fig. 4.

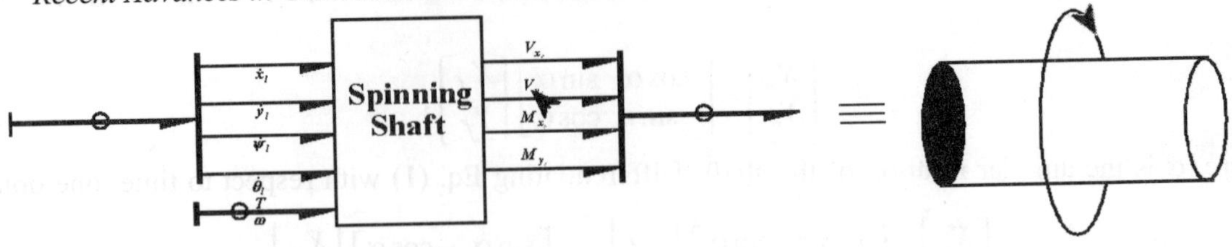

Fig. 4. Spinning shaft word bondgraph and capsule representation.

2.3 Modeling of Spinning Hub

The spinning hub is used to model the annular disks. As shown in Fig. 5, the model incorporates the eccentricity of hub centre of mass and its angular orientation is recorded by the flow-activated element.

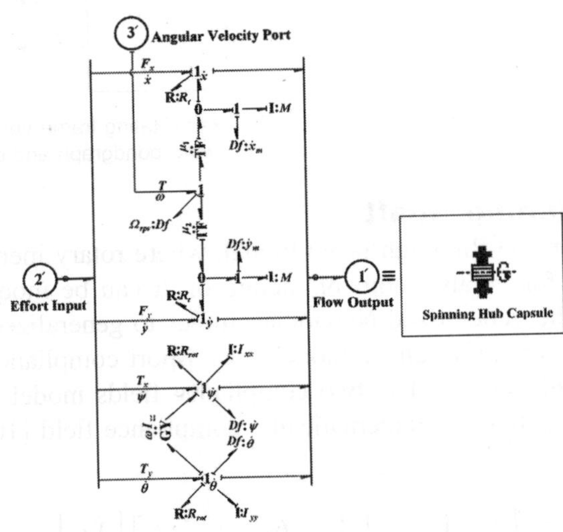

Fig. 5. Spinning hub word bondgraph and capsule representation.

An angular speed port is taken separately to model the spinning speed of the hub. Inertial element attached to X and Y direction models the masses (shaft mass + hub mass) whereas in rotating frame, inertial elements represent the moment of inertia of the hub for rotation motions about diametrical axes, respectively, which can be defined as

$$I_{xx} = I_{yy} = \frac{\rho\pi}{8}\left(L_{prev} + L_{next}\right)\left(R_o^4 - R_i^4\right) + MI_{hub} \qquad \text{...(4)}$$

The expressions of the mass and the rotary inertia contribution from the interfacing shaft elements are added to those of the hub. Flow activated elements record the displacement of the hub centre of mass with zero initial condition of the corresponding state vectors give the displacement of the hub axis. Resistive elements in X and Y directions represent the translational damping (R_t) whereas in rotating frame, it is rotational damping (R_{rot}) about diametrical axes. The gyroscopic forces are also included by the gyrator element in the hub model, whose modulus may be given as

$$\omega I_{zz} = \left\{ \frac{\rho\pi}{4}\left(L_{prev} + L_{next}\right)\left(R_o^4 - R_i^4\right) + I_{p_{hub}} \right\}\Omega. \qquad \ldots(5)$$

The modulus of transformer may be expressed as

$$\mu_1 = -\varepsilon\sin\left(\Omega t + \phi\right), \qquad \ldots(6a)$$

$$\mu_2 = \varepsilon\cos\left(\Omega t + \phi\right), \qquad \ldots(6b)$$

where, ϕ is the angle of equivalent single unbalance mass measured from the vertical axis and ε is the eccentricity. There are three glue ports where one is for spinning speed port; the other two are of order four for interfacing with the shaft elements on either side.

3. INTEGRATED MODEL OF ROTOR SHAFT SYSTEM

The integrated model of rotor shaft with dissipative coupling is a vectorized bondgraph model interfacing different capsules, respectively by their respective icons. The shaft is modeled with ten reticules (shaft elements) with nine elements of the spinning hub in between them. The significance of using more reticules improves the modelling accuracy. The dissipative coupling is modeled by the resistive element (R_{cp}) and shaft is rotated by a constant speed Ω. Inertial element (I_p) represents the total polar moment of inertia and given as

$$I_P = \frac{\rho\pi}{2}\left(R_0^2 - R_i^2\right)\times\text{Total length of than shaft} \qquad \ldots(7)$$

The pin-pin end conditions are modeled as shown in Fig. 6, where source of flow (SF) is put to zero and inertial element may be expressed as

$$I_d = \frac{\rho\pi}{8}L_{elem}\left(R_o^4 - R_i^4\right), \quad \text{where } L_{elem} = \frac{Length\,of\,the\,shaft}{No.of\,elements} \qquad \ldots(8)$$

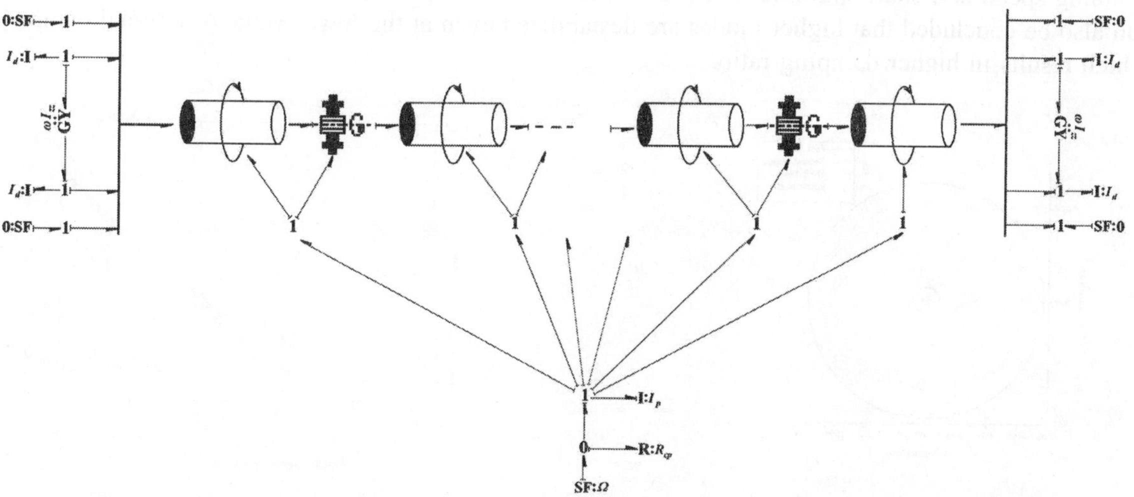

Fig. 6. Bondgraph model of integrated system of rotor shaft through dissipative couplingh using capsules.

The other conditions, which are considered in the integrated model includes the uniform circular cross-section, modulus of elasticity, material damping coefficient and material density. In present analysis mass of hub, moment of inertia and polar moment of inertia of the hub are assumed to be zero.

4. NUMERICAL SIMULATION AND DISCUSSIONS

Simulations of the rotor shaft used the bondgraph simulation tool, SYMBOLS-2000 [11].For integration of state equations; a Runge-Kutta fourth order method was adopted. Values of the system parameters for the simulations are presented in Table 1.

Table 1. Simulation Parameters.

L_{beam}	Length of shaft	5 m
N_{elem}	Number of elements	10
E	Modulus of elasticity	104.5 e9 N/m^2
R_i	Internal Diameter	0.01 m
R_o	External Diameter	0.02 m
ρ	Material density	4420 kg/m^3
R_{cp}	Dissipative coupling coefficient	0.001
Ω	Excitation frequency	16Hz, 40Hz, 100Hz, 160Hz For mode 1,2,3,4, respectively
μ_i	Internal Damping Coefficient	2.0 e-4 Ns/m
μ_{ex}	External Damping Coefficient	3.00e-5, 6.00e-4, 3.00e-3, 1.00e-2 Ns/m For mode 1,2,3, 4, respectively

Figure 7 shows that the trajectories of the rotating shaft reach the limiting orbit. This is due to the loading of the source and the angular speed of the shaft gets entrained at 42.113 rad/sec, the first threshold speed of instability, $\dot{\theta}_{1^{st} \text{mod} e}$. If the external damping is further increased as per Table 1, one obtains the other whirl speed of the rotor shaft at various frequencies as shown in Fig. 7.

The various plots shown in Fig. 8 exhibit that damping ratio is increasing with non-dimensional spinning speed and shaft spinning speed are entrained at various natural modes of the rotor shaft. It can also be concluded that higher modes are destabilized even at the lower value of internal damping, which results in higher damping ratio.

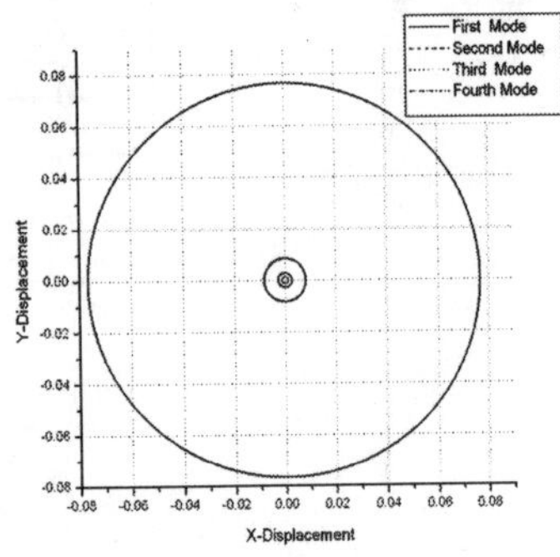

Fig. 7. Limiting orbit of shaft at different natural mode.

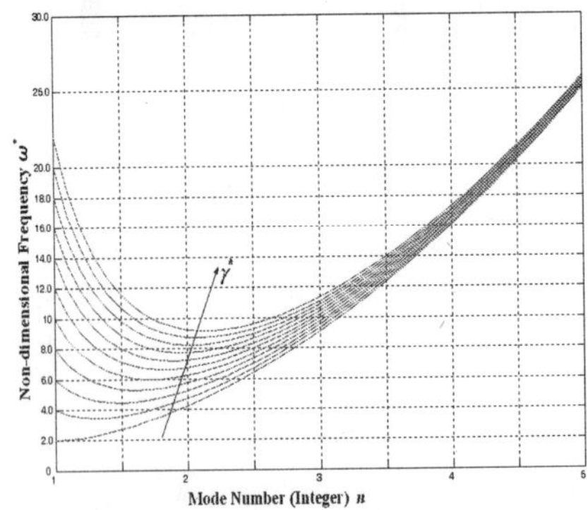

Fig. 8. Entrainment of 2nd mode of rotor shaft with frequency range (ω^* = 0 – 30).

Through the simulation data, the following empirical relation may be found out in Ref. [12] as

Shaft spinning speed = Natural frequencies (1 + Damping ratio)*

CONCLUSION

A bondgraph model of a rotor shaft with internal damping driven through dissipative coupling was constructed that exhibited the object-oriented reusable capsules, used to model the system. The internal damping of the shaft is modeled in the rotating directions whereas external damping is modeled in the fixed directions. It is clear from the work that bondgraph modeling of this system effectively incorporates the internal damping in the model and provides an added dimension in studying tubomachinery stability. The study has confirmed that internal damping is a destabilizing influence on rotor systems, when rotated beyond first critical speed. Simulation results of the system has revealed the entrainment of the whirl speed of the shaft at different natural frequencies and it has been observed that whirl speed of the shaft mainly depends on the damping ratio and natural frequencies of the rotor shaft.

REFERENCES

1. Den Hartog, J.P,, 1934, Mechanical Vibrations, McGraw-Hill, New York.
2. Dimentberg, M., 1961, Flexural Vibrations of Rotating Shafts, Butter worth, London, England.
3. Gunter, E.J., 1972 Rotor-Bearing Stability, Proceeding of the First Turbomachinery Symposium, pp.119-141.
4. Dimarogonas, A.D., 1975, A general method for stability of rotating shafts, Ingenieur-Archiv, 44, pp. 9-20.
5. Lazen, B.J., 1968, Damping of materials and members in structural mechanics, Pargamon Press.
6. Crandall, S.H., 1970, The role of damping in vibration theory, Journal of sound and Vibration, 11, pp. 3-18
7. Zorzi, E.S., and Nelson, H.D., 1977, Finite Element Simulation of Rotor Bearing Systems with Internal damping, Journal of Engineering Power, ASME trans., 99, pp.71-76.
8. Mazzei Jr., A.J., Argento, A., Scoot, R.A., 1999, Dynamic stability of a rotating shaft driven through a universal joint, Journal of Sound and Vibration, 222(1), pp. 19-47.
9. Mazzei Jr., A.J., Scoot, R.A., 2003, Effects of internal viscous damping on the stability of a rotating shaft driven through a universal joint, Journal of sound and vibration, 265, pp. 863-885.
10. Mukherjee, A. and Karmakar, R., 2000, Modelling and Simulation of Engineering System through Bond Graph, Narosa publishing House, New Delhi, reprinted by CRC press for North America and by Alpha Science For Europe
11. Mukherjee, A. and Samantaray, A.K., 2000, SYMBOLS-2000 User's Manual, High-Tech Consultants, STEP Indian Institute of Technology, Kharagpur, India.
12. Rastogi Vikas, 2005, Extension of Lagrangian-Hamiltonian Mechanics, Study of Symmetries and invariants, Ph. D Thesis, Indian institute of Technology, India.

138

An *r-h* Adaptive Strategy Based on Material Forces and Error Assessment for Analysis of Reissner-Mindlin Plates

A. RAJAGOPAL[1] AND S.M. SIVAKUMAR[2]

[1]Research Scholar, Department of Civil Engineering,
Indian Institute of Technology Madras-600 036, India. email: kalya76@yahoo.com
[2]Professor, Department of Applied Mechanics and Civil Engineering,
Indian Institute of Technology Madras-600 036, India. email: mssiva@iitm.ac.in

ABSTRACT

A new *r-h* adaptive strategy is proposed and formulated for Reissner Mindlin Plates. The *r*-adaption is based on configurational driving force that arises out of imbalance in material force equilibrium. The *h*-refinement is based on node patch based super convergent error estimators enhanced by strain energy density ratios. An efficient polak reberie conjugate gradient algorithm has been used for node relocation in r-adaption procedure.

Keywords: Configurational Forces; Mesh adaption; Polak Reberie algorithm; *r-h*-refinement.

1. INTRODUCTION

In present work a combined *r-h* adaptive refinement strategy has been implemented for Reissner-Mindlin plate bending problems. The *r*-adaption is based on configurational force method together with modified algorithms for node relocation. The *h*-refinement is based on node patch based super-convergent error estimator enhanced by strain energy density ratios. Configurational or material driving forces are defined as the conjugate forces to the nodal motion with respect to the potential energy [1], and these forces vanish when the potential energy is a minimum. Further, the minimization of the potential energy reduces error norm due to the orthogonality of error with respect to solution space for linear problems.

2. MATERIAL FORCES FOR *r*-ADAPTION OF REISSNER MINDLIN PLATES

In this section the variational form of material force balance is derived to determine the configurational forces that occur at the nodes. The element formulations are based on Mindlin's plate theory, which includes the effects of transverse shear deformation. The finite element formulations are described by Cook *et al.* [2]. The Lagrangian of the plate is equal to the negative sum of the strain energy density W and the potential of external forces V.

$$L = -(W + V) = L\left(x_k, \theta_i, \theta_{i,j}, w, w_{,i}\right) \qquad ...(1)$$

The equilibrium equations of the plate in terms of stress resultants is given by

$$M_{ij,j} - Q_i = 0 \qquad ...(2)$$

$$Q_{i,j} + p = 0$$

Now considering the gradient of the Lagrangian, we can write

$$grad\,(W + V) = (W + V)_{,k}$$

$$= M_{ji}\theta_{i,jk} + Q_i(\theta_{i,k} + w_{,ik}) - pw_{,k} + \frac{\partial(W + V)}{\partial x_k} \qquad ...(3)$$

The explicit derivative of $W + V$ with respect to the independent variables $x_k\,(k = 1, 2)$ accounts for material inhomogeneties produced, for example, by stiffness distribution $D(x_k)$ and $C_s(x_k)$ smoothly or discontinuously varying with x_k (defects), and for physical inhomogeneties, produced by $p(x_k)$ Integration by parts yields:

$$(W + V)_{,k} = \left(M_{ji}\theta_{i,k}\right)_{,j} - \left(M_{ji,j} - Q_i\right)\theta_{i,k}$$

$$+ \left(Q_i w_{,k}\right)_{,j} - \left(Q_{j,j} + p\right)w_{,k} + \frac{\partial(W + V)}{\partial x_k} \qquad ...(4)$$

Using the equilibrium equations from (6.26) and $(W + V)_{,k} = (W + V)_{,j}\delta_{jk}$ we can write:

$$\left[(W + V)\delta_{jk} - M_{ji}\theta_{i,k} - Q_j w_{,k}\right]_{,j} = \frac{\partial(W + V)}{\partial x_k} \qquad ...(5)$$

The left hand side of equation (6.29) is equal to the Eshelby tensor of plate theory and can be written as

$$C_{jk} = (W + V)\delta_{jk} - M_{ji}\theta_{i,k} - Q_j w_{,k} \qquad ...(6)$$

In the absence of inhomogeneties, the Eshelby tensor given in equation (6.30) is divergence free and can be written as

$$C_{jk,j} = 0 \qquad ...(7)$$

In a finite element context, it is required to compute the discrete configurational forces arising out of discretization. In the absence of body forces the weighted residual form of the balance law equation using a vectorial test function η and integrating over the domain Ω_0 is given by

$$\int_{\Omega_0} C_{ij,j}\eta_i \, d\Omega_0 = 0 \qquad ...(8)$$

The weak form in the absence of body forces obtained by integrating by parts can be written as

$$-\int_{\Omega_0} C_{ij}\eta_{i,j}\,d\Omega_0 + \int_{\Gamma} C_{ij}\,n_j\,\eta_i\,d\Gamma + \int_{\Gamma_e\,\alpha\Gamma} C_{ij}\,n_j\,\eta_i\,d\Gamma = 0 \qquad \text{...(9)}$$

For a homogeneous body in continuous case, linear momentum balance implies material force balance and thus the divergence of the energy momentum tensor is zero for a homogeneous body without body forces. However, in the discrete case nodal force balance does not imply nodal material force balance due to the presence of nodes and hence element interface. The FE approximation introduces spurious material forces originates from the fact that the approximation is not smooth with respect to strains (and stresses). Therefore, in the interior of the domain the numerical value of the nodal material force do not vanish. Thus, in a discretized form considering the material force equilibrium the non-vanishing of the divergence of energy momentum tensor at the inter element boundaries is taken as an error indicator[3]. This is used to check the discrete solution obtained through finite element analysis.. The standard polak reberie algorithm [4] is used for node relocation.

2.1 *h*-Refinement Strategy

The error is a sum of error in energy from moments and shears and is given by:

$$\|E\|^2 = \|E_M\|^2 + \|E_S\|^2 \qquad \text{...(10)}$$

where,

$$\|E_M\|^2 = \int_{\Omega}\left[\hat{M} - M_{FE}\right]^T D_b^{-1}\left[\hat{M} - M_{FE}\right]d\Omega \qquad \text{...(11)}$$

$$\|E_S\|^2 = \int_{\Omega}\left[\hat{S} - S_{FE}\right]^T D_b^{-1}\left[\hat{S} - S_{FE}\right]d\Omega \qquad \text{...(12)}$$

The quantities under curl in the above equations are obtained form a node patch based super convergent estimator. The adaptive refinement strategy is based on strain energy density ratios and is given by:

$$\frac{h_n}{h_0} = \overline{\eta}\left[\frac{D_g}{D_e}\right]\left(\frac{\sum_{e=1}^{M}\|U_{est}\|_e^2 A_e}{A}\right)\left(\frac{1}{\|E\|_e}\right) \qquad \text{...(13)}$$

3. NUMERICAL RESULTS AND DISCUSSIONS: COMBINED *R-H* STRATEGY

A simply supported isotropic square plate of dimensions 2 m × 2 m as shown in figure is considered. The plate is analyzed using RM elements. The material properties of the plate are Youngs modulus E = 100Gpa and Poisson's ratio ν = 0.30. Reduced and full integration schemes are considered for shear terms. A symmetric quadrant of plate is considered. The initial mesh is as shown in Figure (1(b)). Mesh adaption by node relocation is performed and the final *r*-adapted mesh is shown in Figure (1(c)). There is a substantial decrease in potential energy with node relocation iterations as seen in Figure (1(d)). The order of integration used for shear terms greatly influences the flexibility of the system as is seen from Figure (1(d)). The potential energy decreases when shear terms are selectively integrated.

The combined refinement strategy is performed in cycles. The *h*-adaptive mesh is shown in Figure 2(a). The *r-h* adaptive mesh is shown in figure 2(b). From the convergence characteristics as shown in figure 2(c), it seen that we get a better estimation of error and convergence.

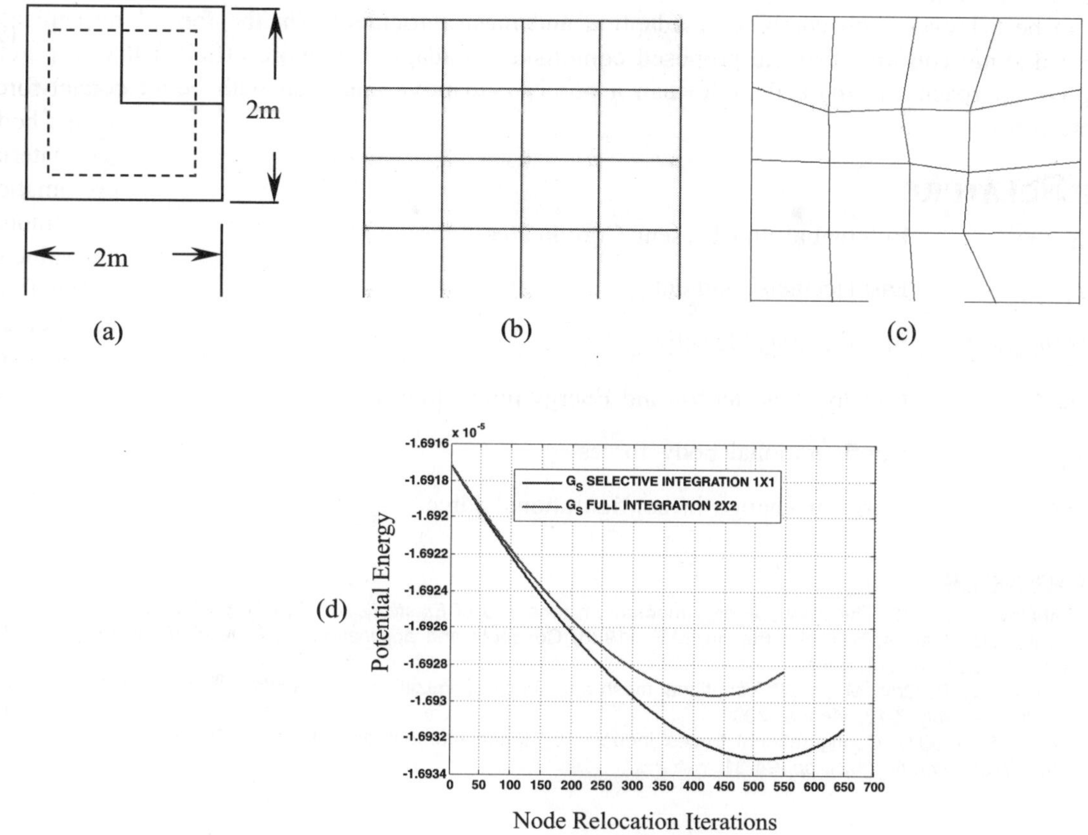

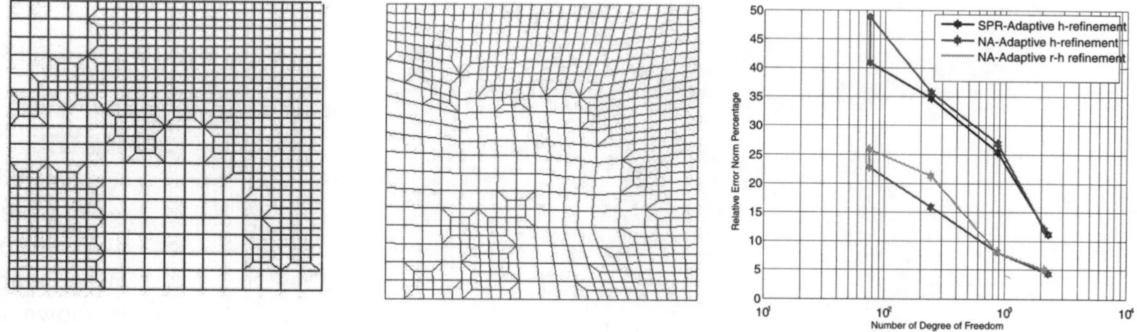

Fig. 1. *r*-adaption for quarter plate

Fig. 2. (a) h-adaptive meshes (b) *r-h* adaptive meshes (c) Convergence characteristics.

CONCLUSION

In this study a new *r-h* adaptive scheme has been proposed and formulated. The *r*-adaption is based on a modified configurational force methods. It is seen that a combined r-h strategy with *r*-adaption followed by *h*-adaption in cyclic sequence performs well than a single cycle of *r*- followed by *h*-adaption. A further reduction in the potential energy and the relative error norm of the system is found to be achieved with combined *r*-adaption and mesh enrichment (in the form *h*-refinement). Numerical study confirms that the proposed combined *r-h* adaption is more efficient than a purely *h*-adaptive approach and more flexible than a purely *r*-adaptive approach with better convergence characteristics.

NOMENCLATURE

X_A, x_a	Referential and Present Coordinates
F_{iA}	Displacement gradient
$W(u_{i,j}, x_k)$	Strain energy density
σ_{ij}, C_{ij}	Cauchy stress tensor and Energy momentum tensor
g_k	Configurational body forces
$\|\Delta G_i\|$	Norm of change in configurational force

REFERENCES

1. Eshelby, J.D. 1975: The elastic energy-momentum tensor. *J of Elasticity, vol.* 5(3-4), pp. 321-335.
2. Cook, R.D., Malkus, S.D. and Plesha, M.E. (1988) Concepts and applications of finite element analysis, Third edition, John Wiley & Sons.
3. Thoutireddy, P.; Ortiz, M. (2002): Variational mesh and shape optimization. *Fifth World Congress on Computational Mechanics, July 7-12*, Vienna. 2002.
4. Braun, M. (1997): Configurational forces induced by Finite element discretization. *Proc. Estonian Acad.Sci. Phys. Math. vol.* 46 (1/2), pp. 24-31.

139

Time Domain Parametric Identification of Plate Bending Rigidity Coefficients Using Substructural Approach

SANDESH S.[1] AND K. SHANKAR[2]

[1]Research Scholar, [2]Assistant Professor, Machine Design Section,
Department of Mechanical Engineering, Indian Institute of Technology Madras, India-781 039
email: [1] sandesh@iitm.ac.in, [2] skris@iitm.ac.in

ABSTRACT

In this article substructure identification method is extended for identification of plate bending rigidities in time domain. Unknown bending rigidity coefficients are identified by Genetic Algorithm (GA) by updating the finite element model of the plate using vibration data. Numerical simulation studies are performed for isotropic steel plate and symmetric carbon/epoxy laminate.

Keywords: Substructural identification; Genetic algorithm; Bending rigidity coefficients.

1. INTRODUCTION

Structural identification is based on the modification of the structural model matrices such as mass, stiffness and damping to reproduce as closely as possible the measured static or dynamic response data. These methods solve for the updated matrices (or perturbation to the nominal model that produces the updated matrices) by forming a constrained optimization problem based on the structural equations of the motion and the measured data. The methods use a common basic set of equations and the differences in the various algorithms can be classified as:

- Objective function
- Constraints placed on the problem
- Numerical scheme used to implement the optimization

It can be used for structural health monitoring and damage assessment in a non-destructive way by tracking changes in relevant structural parameters. The structural identification methods can be classified under various categories e.g., frequency domain and time domain methods, parametric and non-parametric models [1]. The main categorization is perhaps by means of the frequency and time domain in which structural identification is carried out.

The method of substructural identification is first proposed by Koh *et al.* [2]. M. Grediac [3,4] determined the bending rigidities of the anisotropic plates by modal analysis. Dunn SA [5] demonstrated how genetic algorithm can be used with experimentally measured frequency response data to identify finite element model of the structure from dynamic analysis. Many researchers used Genetic Algorithm (GA) for model updating technique to identify the unknown elastic constants of the plate using measured/simulated eigenvalues and eigenvectors [6,7]. Koh and Shankar [8-9] formulated the substructure identification method without interface measurement in frequency domain. T. Lauwagie *et al.* [10] determined the in plane elastic properties of laminates by Resonalyser procedure.

In the literature it is found that all identification techniques used to identify elastic constants of plate's, use eigenvalues and eigenvectors (natural frequencies and mode shapes) of the complete plate. From practical point of view natural frequencies can easily be measured with an accelerometer. However, the mode shapes cannot be measured easily with commercially available experimental devices. In the present work a substructural identification approach [11], is extended to identify the bending rigidities of the plate. Finite element model of the plate is formulated using Classical Plate Theory and model is updated using Genetic Algorithm to reproduce as closely as possible the measured dynamic response data.

2. EQUATION OF MOTION FOR PLATES

The partial differential equation governing the transverse vibration of a symmetrically laminated plate is

$$\rho h \frac{\partial^2 w}{\partial t^2} + D_{11}\frac{\partial^4 w}{\partial x^4} + 4D_{16}\frac{\partial^4 w}{\partial x^3 \partial x} + 2(D_{12}$$

$$+D_{66})\frac{\partial^4 w}{\partial x^2 \partial x^2} + 4D_{26}\frac{\partial^4 w}{\partial x^2 \partial y^2} + D_{22}\frac{\partial^4 w}{\partial x \partial y^3} + D_{22}\frac{\partial^4 w}{\partial y^4} = f(x,y,t) \qquad ...(1)$$

where, ρ and h are the density and thickness of the plates, respectively. The transient analysis of the plate is done by finite element method. The four-noded plate bending element is used to model the plate. The global mass matrix ([M]) and stiffness matrix ([K]) are assembled using elemental mass matrix ([m]) and stiffness matrix ([k]).

$$[k]=\int_V [B]^T[D][B]dV \quad [m]=\int_V [N]^T \rho [N]dV \qquad ...(2)$$

where, D, B and N are bending rigidity matrix, strain-displacement matrix and shape function matrix. The differential equations of motion of transverse vibration is given by

$$[M]\ddot{\vec{w}}(t)+[C]\dot{\vec{w}}+[K]\vec{w}(t)=\vec{F}(t)$$

It is important to note that stiffness matrix depends upon the elastic properties that are to be identified, i.e.,

$$K=K(D_{11},D_{22},...,D_{66}) \qquad ...(3)$$

where, D_{ij} are the unknown plate bending rigidity coefficients. This stiffness matrix is updated using GA till error between computed and experimental vibration data is minimized. In GA finite element model is updated using Darwin's theory of natural selection and survival of the fittest.

2.1 Substructure Formulation

The equations of motion of the structure are given by Newton second law motion as:

$$
\begin{bmatrix}
M_{uu} & M_{uf} & & & \\
M_{fu} & M_{ff} & M_{fr} & & \\
 & M_{rf} & M_{rr} & M_{rg} & \\
 & & M_{gr} & M_{gg} & M_{gd} \\
 & & & M_{dg} & M_{dd}
\end{bmatrix}
\begin{Bmatrix} \ddot{u}_u \\ \ddot{u}_f \\ \ddot{u}_r \\ \ddot{u}_g \\ \ddot{u}_d \end{Bmatrix}
+
\begin{bmatrix}
C_{uu} & C_{uf} & & & \\
C_{fu} & C_{ff} & C_{fr} & & \\
 & C_{rf} & C_{rr} & C_{rg} & \\
 & & C_{gr} & C_{gg} & C_{gd} \\
 & & & C_{dg} & C_{dd}
\end{bmatrix}
\begin{Bmatrix} \dot{u}_u \\ \dot{u}_f \\ \dot{u}_r \\ \dot{u}_g \\ \dot{u}_d \end{Bmatrix}
+
$$

$$
\begin{bmatrix}
K_{uu} & K_{uf} & & & \\
K_{fu} & K_{ff} & K_{fr} & & \\
 & K_{rf} & K_{rr} & K_{rg} & \\
 & & K_{gr} & K_{gg} & K_{gd} \\
 & & & K_{dg} & K_{dd}
\end{bmatrix}
\begin{Bmatrix} u_u \\ u_f \\ u_r \\ u_g \\ u_d \end{Bmatrix}
=
\begin{Bmatrix} P_u \\ P_f \\ P_r \\ P_g \\ P_d \end{Bmatrix}
\qquad \dots(4)
$$

where, subscript 'r' denotes internal DOFs of the substructure concerned, subscripts 'f' and 'g' represent the interface DOFs , 'u' and 'd' represent DOFs of the remaining structure as shown in Fig. 2. For the substructure considered, the equations of motion may be extracted from the above system of equations.

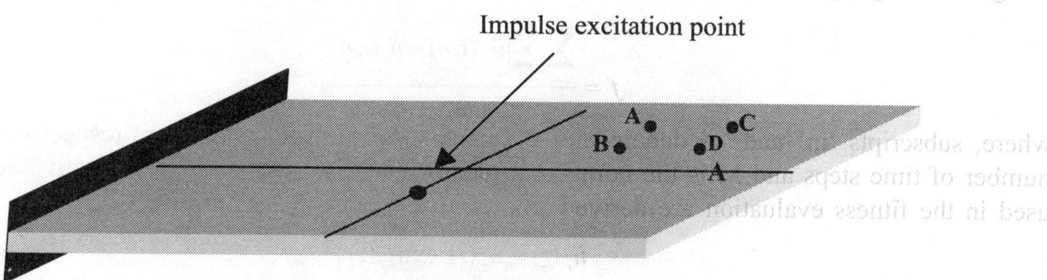

Fig. 1 Cantilever plate.

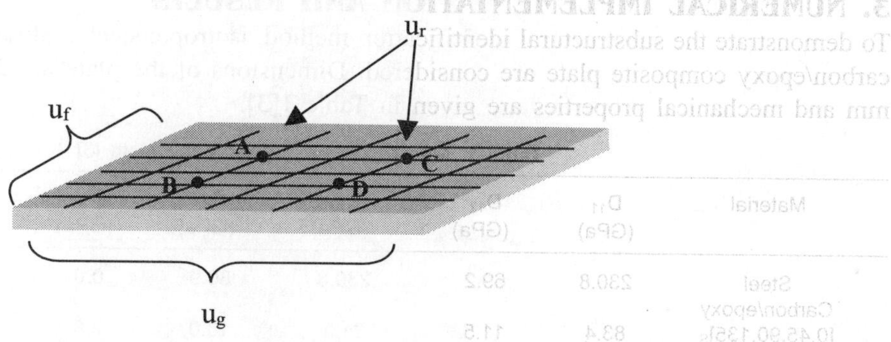

Fig. 2. Substructure of cantilever plate.

$$\begin{bmatrix} M_{rf} & M_{rr} & M_{rg} \end{bmatrix} \begin{Bmatrix} \ddot{u}_f \\ \ddot{u}_r \\ \ddot{u}_g \end{Bmatrix} + \begin{bmatrix} C_{rf} & C_{rr} & C_{rg} \end{bmatrix} \begin{Bmatrix} \dot{u}_f \\ \dot{u}_r \\ \dot{u}_g \end{Bmatrix} + \begin{bmatrix} K_{rf} & K_{rr} & K_{rg} \end{bmatrix} \begin{Bmatrix} u_u \\ u_r \\ u_g \end{Bmatrix} = \{P_r\} \qquad ...(5)$$

The interaction effect at interface are treated as 'input', the above equation can be rearranged as

$$M_{rr}\ddot{u}_r(t) + C_{rr}\dot{u}_r(t) + K_{rr}u_r(t) = P_r - M_{rj}\ddot{u}_j(t) - C_{rj}\dot{u}_j(t) - K_{rj}u_j(t) \qquad ...(6)$$

The subscript 'j' denotes all interface DOFs (i.e., f and g included). The displacements for internal DOFs are split into quasi-static displacement [12] (u_r^s) and relative displacement (u_r^*), i.e.,

$$u_r(t) = u_r^s(t) + u_r^*(t) \qquad ...(7)$$

Quasi-static displacements are obtained by solving Eq(6) while ignoring the applied force, inertia effect and damping effect. Hence,

$$K_{rr}u_r^s = -K_{rj}u_j$$

or $\qquad\qquad\qquad\qquad\qquad\qquad\qquad\qquad\qquad\qquad\qquad\qquad$ ...(8)

$$u_r^s = -K_{rr}^{-1}K_{rj}u_j = Ru_j$$

$$M_{rr}\ddot{u}_r^*(t) + C_{rr}\dot{u}_r^*(t) + K_{rr}u_r^*(t) = P_r - \left(M_{rj} + M_{rr}R\right)\ddot{u}_j(t) - \left(C_{rj} + C_{rr}R\right)\dot{u}_j(t) \qquad ...(9)$$

where, 'R' is called the influence coefficient matrix which relates internal DOFs under the quasi-static condition. The response of the internal DOFs are obtained using GA and fitness value, evaluated using following equation

$$f = \frac{\displaystyle\sum_{i=1}^{M}\sum_{n=1}^{L}\left|\ddot{u}_m(i,n) - \ddot{u}_e(i,n)\right|^2}{ML} \qquad ...(10)$$

where, subscripts 'm' and 'e' denote measurement and estimated quantities, respectively, L is the number of time steps and M is the number of measurement sensors used. The relative accelerations used in the fitness evaluation are derived from Eqs. (7) and (8) as follows:

$$\ddot{u}_r^*(t) = \ddot{u}_r(t) - R\ddot{u}_j(t) \qquad ...(11)$$

3. NUMERICAL IMPLEMENTATION AND RESULTS

To demonstrate the substructural identification method, isotropic steel cantilever plate and symmetric carbon/epoxy composite plate are considered. Dimensions of the plate are 200 mm x 200 mm x 1 mm and mechanical properties are given in Table 1[3].

Table 1. Mechanical properties of the plate [3]

Material	D_{11} (GPa)	D_{12} (GPa)	D_{22} (GPa)	D_{66} (GPa)	D_{16} (GPa)	D_{26} (GPa)	Density (kg/m^3)
Steel	230.8	69.2	230.8	80.9	0.0	0.0	7800.0
Carbon/epoxy [0,45,90,135]$_s$	83.4	11.5	31.3	13.0	7.8	7.8	1510.0

A four-noded rectangular plate element with 3 degrees of freedom per node (one transverse and two rotations) is considered for finite element mesh (12×12 mesh). The size of the stiffness matrix of the complete plate is 507×507, Fig. 1, and that of the substructure as shown in Fig. 2 is 147×147. Finite element analysis is done using MATLAB. Impulse responses of the plate are simulated by solving the equations of motion (Eq. 6) using Newmark's constant acceleration method. Impulse response is computed with constant time step of 0.001s. Accelerations at the points A, B C and D are assumed to be available for the purpose of identification (Fig. 2). To simulate experimental results 5% and 10% noise is added to the numerically obtained results. The quarter plate containing the free end is considered as substructure. Since, input force is outside the substructure considered, time history of excitation is not required for identification. So external force term P_r vanishes in Eq. 6 and results in 'output only identification'. The responses at the interface nodes are assumed to be known for identification purpose. The GA parameters used are: crossover rate = 60%, mutation rate = 1%, population size = 1000 and generations = 25. The initial population is randomly generated within the specified parameter range: no good initial guess is needed. The GA search range of each unknown is taken as ±50% of the corresponding exact values given in Table 1. The mass of the plate is assumed to be known in all following examples.

3.1 Example 1: Complete Structure

The complete model of the cantilever plate is considered for identification of bending rigidity coefficients and damping constants. Measurements of input force time histories are required for identification in the force vector of Eq. (4). The complete finite element model of the plate is updated by GA and acceleration responses at points A, B, C and D are used in the fitness function evaluation (Eq. 11). Numerical simulation study has been made for isotropic steel plate with known damping and unknown damping for without noise. For known damping case, the commonly adopted proportional damping matrix is assumed: $C = \alpha[M] + \beta[K]$ where, α and β are damping constants that can be related to damping ratio of any two selected modes. A damping ratio of 1% is assumed for all modes. Results are tabulated in the Table 2.

Table 2. Percentage error in identified values of the isotropic steel plate

	Isotropic steel plate (0% Noise)	
Bending rigidities	Known Damping ($<10^{-6}$)	Unknown Damping
D_{11}	0.0	0.007
D_{66}	0.0	0.008
D_{12}	0.0	0.010
α	–	0.142
β	–	0.032

3.2 Example 2: Substructure with Known Damping

The damping constants α and β are assumed as in example 1. Since, the force is applied outside the substructure considered, no information is required about the force. The identified values D_{ij} are shown in Table 3 for both the plates. The uniqueness is numerically illustrated by ideal case of zero noise, where by exact values (within the numerical error $<10^{-6}\%$) are obtained for all unknown parameters. When the noise is present, uniqueness and robustness are issues for all system identification methods. In this respect, 5% and 10% noise is added to the numerically obtained responses. All identified results are presented in Tables 3 and 4 (examples 2 and 3) are average results over 5 runs. The noise pattern was freshly generated for each run to avoid any bias that might results from using the same for all of the 5 runs.

Table 3. Percentage error in identified values of the isotropic steel plate

Bending rigidities	Isotropic steel plate			Carbon/epoxy composite plate		
	Noise			Noise		
	0%($<10^{-6}$)	5%	10%	0%	5%	10%
D_{11}	0.0	0.72	1.22	0.014	0.12	5.27
D_{22}	0.0	0.72	1.27	0.009	1.35	8.65
D_{66}	0.0	0.87	2.08	0.113	4.86	3.28
D_{12}	0.0	0.72	5.85	0.036	0.97	2.37
D_{16}	–	–	–	0.979	4.62	7.87
D_{26}	–	–	–	0.727	4.14	9.31

3.3 Example 3: Substructure with Unknown Damping

To test the uniqueness of the substructure method, the damping is assumed to be unknown, there by increasing the difficulty of identification. In the substructure equation of motion (Eq. 9) the damping matrix 'C' which appears on both sides of the equation is updated along with stiffness matrix till error between estimated and measured responses reduces. Table 4 gives the error in identification of bending rigidity coefficients 'D_{ij}' and damping constants of the steel and carbon/epoxy composite plate. Since, number of unknown parameters is more and also because of unknown damping, error in identified value are comparatively more than Example 2.

Table 4. Percentage error in identified values of the isotropic steel plate

Bending rigidities		Isotropic steel plate			Carbon/epoxy composite plate		
		Noise			Noise		
		0%	5%	10%	0%	5%	10%
D_{11}		0.017	0.65	0.05	0.001	1.45	6.72
D_{22}		0.017	0.65	0.05	0.004	1.60	8.56
D_{66}		0.869	3.06	11.80	0.151	5.40	2.53
D_{12}		0.056	0.10	0.61	0.047	1.14	2.87
D_{16}		–	–	–	2.015	8.52	11.32
D_{26}		–	–	–	2.308	11.46	15.13
Damping	α	3.372	9.50	13.97	4.299	8.67	17.28
Constants	β	0.062	0.72	7.61	0.033	3.58	8.20

CONCLUSION

In the present study numerical simulation of substructure identification method has been done for parametric identification of bending rigidities and damping constants of the isotropic and composites plates. The Genetic Algorithm is used for identification because of its robustness and ease of implementation. The finite element model is updated till the error between measured and estimated acceleration responses is reduced. For complete structure identification measurement of input time history is required. In general, measurements of input time histories are very difficult. In substructural identification methods, the effect of the input force is expressed in terms of the responses at the interfaces within the whole structure. Then the identification can be performed without measuring the actual input force to the whole structure. By substructure identification method, substantial saving in computational time about by 80% is noticed in examples 2 and 3 as compared to complete structure (example 1). The effect noise is considered on identification of unknown parameters. The maximum error of 5.85% and 9.31% in identified values is noticed in example 2, 13.97% and 17.28% in example 3, for steel and composite plate respectively with 10% noise. This can be further reduced by optimizing the GA parameters such as population size, number of generation and mutation rate.

REFERENCES

1. Koh C.G. and See L.M., 1999, Technique in the identification and uncertainty estimation of parameters, Structural Dynamics System Computational Techniques and optimization, International Series in Engineering Technology and Applied Sciences, Netherlands, pp. 241-273, 1999.
2. Koh C.G., See L.M., and Balendra T., 1991, Estimation of structural parameters in time domain: A substructure approach. Earthquake Engineering and Structural Dynamics, 20, 787-801.
3. Grediac M. Paris P.A., 1996, Direct identification of elastic constants of anisotropic plates by modal analysis: theoretical and numerical aspects, Journal of Sound and Vibration, 195(3), 401-415.
4. Grediac M., Vautrin A., 1990, A new method for determination of bending rigidities of thin anisotropic plates, Journal of Applied Mechanics, 57, 964-968.
5. Dunn S.A., 1998, The use of genetic algorithm and stochastic hills climbing in dynamic finite element model identification, Computers and Structures, 66 (4), 489-497.
6. Chakaraborthy A., Mukhopadhyay M., 2000, Estimation of elastic parameters of stiffened composite plate, AIAA Journal, 38(9), 1716-1724.
7. Cunha J., Cogan S., Berthos C., 1999, Application of genetic algorithm for the identification of elastic constants of composite materials from dynamic tests, International Journal for Numerical Methods in Engineering, 45, 891-900.
8. Koh C.G. and Shankar K., 2003, Substructural identification methods without interface measurement, Journal of Engineering Mechanics, 129(7), 769-776.
9. Koh C.G. and Shankar K., 2003, Stiffness identification by a substructural approach in frequency domain, International Journal of Structural Stability and Dynamics, 3(2), 267-281.
10. Lauwagie T., Sol H., Heylen W., Roebben G., 2004, Determination of the in-plane elastic properties of the different layers of laminated plates by means of vibration testing and model updating, Journal Sound and Vibration, 274, 529-546.
11. Koh C.G., Hong B., and Liaw C.Y., 20003, Substructural and progressive structural identification methods, Engineering Structures, 25, 1551-1563.
12. Ray W. Clough and Joseph Penzien, 1993, Dynamic of Structures, 2nd Edition, McGraw-Hill, New York.

140

A Numerical Integration Technique for Elastic Analysis of Thick Orthotropic Axisymmetric Circular Cylinders of Finite Length

Tarun Kant and Payal Desai

Department of Civil Engineering Indian Institute of Technology Bombay, Powai, Mumbai-400 076, India
email: tkant@civil.iitb.ac.in

ABSTRACT

Theory of three-dimensional (3D) elasticity is utilized here for boundary value problems (BVPs) of finite orthotropic cylinders. A simply supported orthotropic cylinder under axisymmetric mechanical load is first considered as a two-dimensional (2D) generalized plane strain problem of elasticity. The 2D problem is reduced to one dimensional (1D) one by assuming an analytical solution in longitudinal direction in terms of Fourier series expansion which satisfies the simply (diaphragm) supported boundary conditions completely. Fundamental (basic) dependent variables are chosen in the radial direction of the cylinder. The resulting simultaneous first order ordinary differential equations (ODEs) are integrated through an effective numerical integration technique. The numerical solutions for orthotropic cylinders are first validated for their accuracy with available 1D solution for infinitely long cylinder.

Keywords: Elasticity theory, Numerical integration, Composites, Boundary value problems.

1. INTRODUCTION

Orthotropic thick cylinders are widely used to configure aircraft, nuclear pressure vessels, piping and civil engineering structures. Accurate analysis of deformations and stresses induced by mechanical loading is certainly needed for such an important structural element. Klosner and Kempner [1] obtained the stresses and displacements by the use of 3D elasticity theory and several shell theories for the problem of a long isotropic circular cylinder subjected to an uniform, external, circumferential, radial line load. Iyengar and Yogananda [2] presented an elasticity solution by using a Love function approach for the analysis of a semi-infinite circular cylindrical shell subjected to a concentrated uniform, circumferential, radial line load at the end. Misovec and Kempner [3] obtained an approximate solution to the Navier equations of the 3D elasticity for an axisymmetric orthotropic circular cylinder subjected to internal and external pressure, axial loads, and closely spaced periodic radial loads. Chandrashekhara and Kumar [4] obtained an exact solution for a thick, transversely isotropic, simply supported, circular cylindrical shell subjected to axisymmetric load using a transfer matrix approach.

In this paper, governing differential equations from exact theory of 2D elasticity equations, which govern the behaviour of finite length circular orthotropic cylinder in a state of axisymmetric plane strain under external pressure loading which is a function of both radial and axial coordinates, are taken. By assuming a global analytical solution in the longitudinal direction satisfying the two end boundary conditions exactly, the 2D problem is reduced to a 1D problem in the radial direction. The equations are reformulated to enable application of an efficient and accurate numerical integration technique for the solution of the BVP of a cylinder.

2. FORMULATION

Basic governing equations of an axisymmetric cylinder [5] in cylindrical coordinates are (Fig.1).

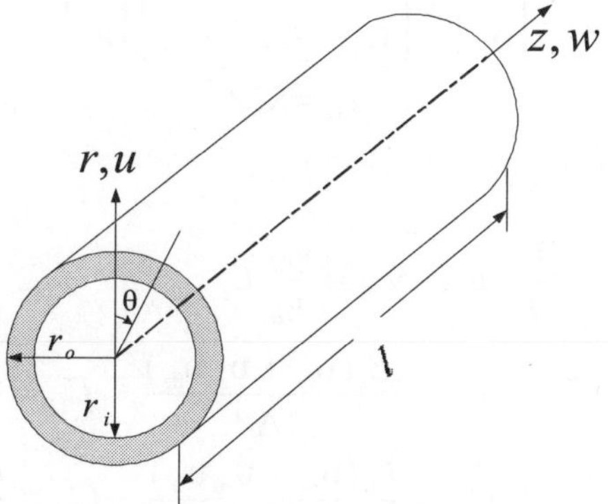

Fig. 1. Coordinate system and geometry of cylinder

Equilibrium equations

$$\frac{\partial \sigma_r}{\partial r} + \frac{\partial \tau_{zr}}{\partial z} + \frac{\sigma_r - \sigma_\theta}{r} = 0$$

$$\frac{\partial \tau_{zr}}{\partial r} + \frac{\partial \sigma_z}{\partial z} + \frac{\tau_{zr}}{r} = 0 \qquad \qquad ...(1a)$$

Strain displacement relations

$$\varepsilon_r = \frac{\partial u}{\partial r}, \varepsilon_\theta = \frac{u}{r}, \varepsilon_z = \frac{\partial w}{\partial z}$$

$$\gamma_{rz} = \frac{\partial w}{\partial r} + \frac{\partial u}{\partial z} \qquad \qquad ...(1b)$$

Stress-strain relations for cylindrically orthotropic materials

$$\varepsilon_r = \frac{\sigma_r}{E_r} - \nu_{\theta r}\frac{\sigma_\theta}{E_\theta} - \nu_{zr}\frac{\sigma_z}{E_z}$$

$$\varepsilon_\theta = -\nu_{r\theta}\frac{\sigma_r}{E_r} + \frac{\sigma_\theta}{E_\theta} - \nu_{z\theta}\frac{\sigma_z}{E_z}$$

$$\varepsilon_z = -\nu_{rz}\frac{\sigma_r}{E_r} - \nu_{\theta z}\frac{\sigma_\theta}{E_\theta} + \frac{\sigma_z}{E_z}$$

$$\gamma_{rz} = \frac{\tau_{rz}}{G_{rz}} \qquad \qquad \text{...(1c)}$$

These equations can also be written as,

$$\begin{Bmatrix} \sigma_r \\ \sigma_\theta \\ \sigma_z \end{Bmatrix} = \begin{bmatrix} C_{11} & C_{12} & C_{13} \\ C_{21} & C_{22} & C_{23} \\ C_{31} & C_{32} & C_{33} \end{bmatrix} \begin{Bmatrix} \varepsilon_r \\ \varepsilon_\theta \\ \varepsilon_z \end{Bmatrix} \qquad \qquad \text{...(1d)}$$

$$\tau_{rz} = G\gamma_{rz}$$

where,

$$\nu_{r\theta} = \frac{\nu_{\theta r}}{E_\theta}E_r, \quad \nu_{rz} = \frac{\nu_{zr}}{E_z}E_r, \quad \nu_{z\theta} = \frac{\nu_{\theta z}}{E_\theta}E_z$$

$$C_{11} = \frac{E_r(1-\upsilon_{\theta z}\upsilon_{z\theta})}{\Delta}, \quad C_{12} = \frac{E_r(\upsilon_{\theta r}+\upsilon_{zr}\upsilon_{\theta z})}{\Delta}, \quad C_{13} = \frac{E_r(\upsilon_{zr}+\upsilon_{\theta r}\upsilon_{z\theta})}{\Delta}$$

$$C_{22} = \frac{E_\theta(1-\upsilon_{rz}\upsilon_{zr})}{\Delta}, \quad C_{32} = \frac{E_\theta(\upsilon_{z\theta}+\upsilon_{r\theta}\upsilon_{zr})}{\Delta} \quad C_{33} = \frac{E_z(1-\upsilon_{r\theta}\upsilon_{\theta r})}{\Delta}$$

where, $\Delta = (1-\nu_{r\theta}\nu_{\theta r} - \nu_{\theta z}\nu_{z\theta} - \nu_{zr}\nu_{rz} - 2\nu_{\theta r}\nu_{z\theta}\nu_{rz})$

$$C_{21} = C_{12}, \quad C_{23} = C_{32}, \quad C_{31} = C_{13}$$

and boundary conditions in the longitudinal and radial directions are,

$$\text{at } z=0,1 \quad u=0; \ \sigma_z = 0 \text{ at } r=r_i; \ \sigma_r = \tau_{zr} = 0, \ r=r_o; \ \sigma_r = p(r,z), \ \tau_{zr} = 0 \qquad \text{...(2)}$$

and $p(r,z) = p_0\sin\dfrac{\pi z}{l}$ is sinusoidal load.

in which l is the length, r_i is the inner radius, r_o is the outer radius of a hollow cylinder and p_0 is the maximum intensity of pressure.

Radial direction r is chosen to be a preferred independent coordinate. Four fundamental dependent variables, viz., displacements, u and w and corresponding stresses, σ_r and τ_{rz} are chosen in the radial direction. Circumferential stress σ_θ and axial stress σ_z are treated here as auxiliary variables [6, 7] since these are found to be dependent on the fundamental variables. A set of four first order partial differential equations in independent coordinate r which involves only fundamental variables is obtained through algebraic manipulation of Eqs. (1a)-(1c). These are,

$$\frac{\partial u}{\partial r} = \frac{\sigma_r}{C_{11}} - \frac{C_{12}}{C_{11}}\frac{u}{r} + \frac{C_{13}}{C_{11}}\frac{\partial w}{\partial z}$$

$$\frac{\partial w}{\partial r} = \frac{\tau_{rz}}{G} - \frac{\partial u}{\partial z}$$

$$\frac{\partial \sigma_r}{\partial r} = -\frac{\partial \tau_{rz}}{\partial z} - \frac{\sigma_r}{r}\left[1 - \frac{C_{21}}{C_{11}}\right] + \left[C_{22} - \frac{C_{21}C_{12}}{C_{11}}\right]\frac{u}{r^2} + \left[C_{23} - \frac{C_{21}C_{13}}{C_{11}}\right]\frac{1}{r}\frac{\partial w}{\partial z}$$

$$\frac{\partial \tau_{rz}}{\partial r} = -\frac{\tau_{rz}}{r} - \frac{C_{31}}{C_{11}}\frac{\partial \sigma_r}{\partial z} + \frac{C_{31}C_{12}}{C_{11}}\frac{1}{r}\frac{\partial u}{\partial z} + \frac{C_{31}C_{13}}{C_{11}}\frac{\partial^2 w}{\partial z^2} - \frac{C_{32}}{r}\frac{\partial u}{\partial z} - C_{33}\frac{\partial^2 w}{\partial z^2} \qquad ...(3)$$

Variations of the four fundamental dependent variables which completely satisfy the boundary conditions of diaphragm supports at z = 0, l can then be assumed as,

$$\sigma_r(r,z) = \sigma(r)\sin\frac{\pi z}{l}, \qquad \tau_{rz}(r,z) = \tau(r)\cos\frac{\pi z}{l}$$

$$u(r,z) = U(r)\sin\frac{\pi z}{l}, \qquad w(r,z) = W(r)\cos\frac{\pi z}{l} \qquad ...(4)$$

Substitution of Eq.(4) in Eq.(3) and simplification resulting from orthogonality conditions of trigonometric functions leads to the following four simultaneous ordinary differential equations involving only fundamental variables. These are,

$$U'(r) = \frac{\sigma(r)}{C_{11}} - \frac{C_{12}}{C_{11}}\frac{U(r)}{r} - \frac{C_{13}}{C_{11}}\left[\frac{\pi}{l}\right]W(r)$$

$$W'(r) = \frac{\tau(r)}{G} - \left(\frac{\pi}{l}\right)U(r)$$

$$\sigma'(r) = \left(\frac{\pi}{l}\right)\tau_{rz}(r) - \left[1 - \frac{C_{21}}{C_{11}}\right]\frac{\sigma(r)}{r} + \left[C_{22} - \frac{C_{21}C_{12}}{C_{11}}\right]\frac{U(r)}{r^2} - \left(\frac{\pi}{l}\right)\left[C_{23} - \frac{C_{21}C_{13}}{C_{11}}\right]\frac{W(r)}{r}$$

$$\tau'(r) = -\frac{\tau(r)}{r} - \left(\frac{\pi}{l}\right)\frac{C_{31}}{C_{11}}\sigma(r) + \left[\frac{C_{31}C_{12}}{C_{11}} - C_{32}\right]\left(\frac{\pi}{l}\right)\frac{U(r)}{r} + \left(\frac{\pi}{l}\right)^2\left[C_{33} - \frac{C_{31}C_{13}}{C_{11}}\right]W(r) \qquad ...(5)$$

The above system of first order simultaneous ordinary differential equations together with the appropriate boundary conditions at the inner and outer edge of the cylinder forms a two-point BVP. However, a BVP in ODEs cannot be numerically integrated as only a half of the dependent variables (two) are known at the initial edge. It becomes necessary to transform the problem into a set of initial value problems (IVPs). For this, procedure is used which is given in Ref. [7]. Runge-Kutta-Gill fourth order algorithm is used for the numerical integration of the IVPs.

3. NUMERICAL RESULTS AND DISCUSSIONS

A hollow cylinder is analyzed by taking three h/R ratios of 1/5, 1/20 and 1/50 which cover thick, moderately thick and thin hollow cylinders. Maximum intensity of pressure p0 is taken as 1000

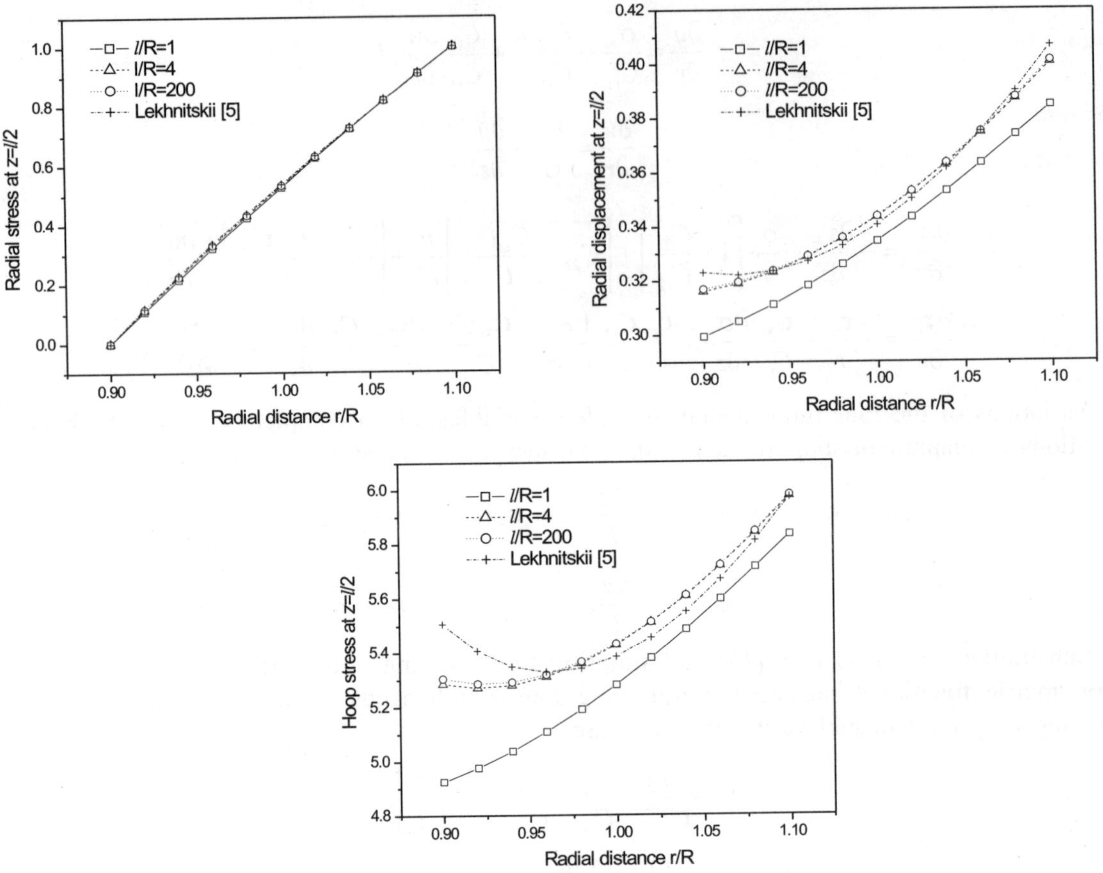

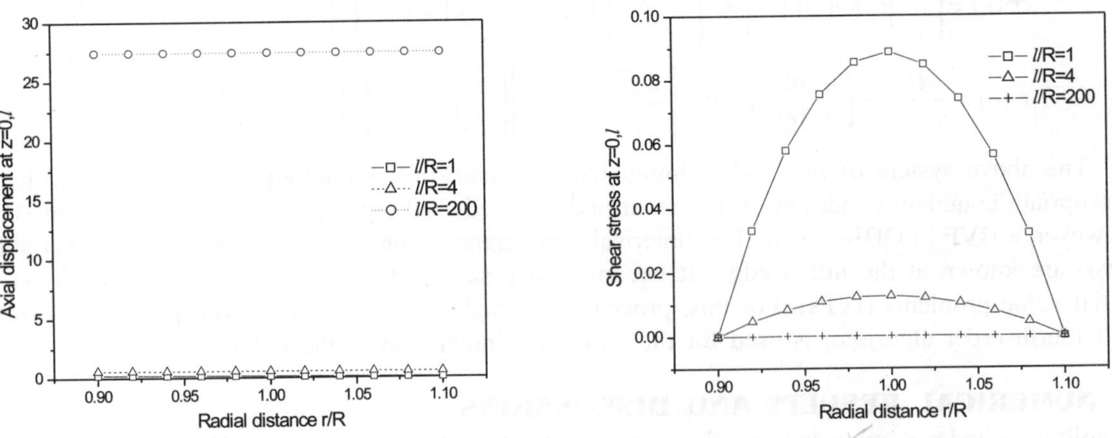

Fig. 2. Distribution of radial stress $\overline{\sigma}_r$, radial displacement $\overline{u}$ and hoop stress $\overline{\sigma}_\theta$ through thickness with h/R = 1/5 in an orthotropic cylinder.

Fig. 3. Distribution of shear stress $\overline{w}$ and axial displacement $\overline{\tau}_{rz}$ through thickness for h/R = 1/5 in an orthotropic cylinder.

KN/m^2. Variations of fundamental quantities through thickness and axial quantities are shown in Figs. 2 and 3 for thick cylinder case. l/R ratios of 1, 4 and 200 are taken to cover both short and infinitely long (plane strain) cylinders.

Material properties taken from Kollar and Springer [8] for Graphite-epoxy material are (fibres are oriented in circumferential direction)

$$E_r = 9.65 \times 10^6, \ E_\theta = 148 \times 10^6, E_z = 9.65 \times 10^6, G_{zr} = 3.0156 \times 10^6$$
$$v_{\theta r} = 0.3, v_{zr} = 0.6, v_{\theta z} = 0.3$$

Non-dimensionalized parameters are chosen as follows for mechanical loading. R is the mean radius of cylinder

$$\bar{r} = \frac{r}{R}, \ \ R = \frac{1}{2}(r_0 + r_i), \ \ (\bar{u}, \bar{w}) = \frac{E_r}{p_0 R}(u, w), \ \ (\overline{\sigma_r}, \overline{\sigma_\theta}, \overline{\sigma_{z,}\tau_{rz}}) = \frac{1}{p_0}(\sigma_r, \sigma_\theta, \sigma_{z,}\tau_{rz})$$

Numerical results are shown in Tables 1 and 2 for thick orthotropic cylinder.Results are compared with the plane strain elasticity solution given in Lekhnitskii [5] for infinitely long cylinder. It is clearly seen that for a long cylinder the results are very close to the plane strain solution for both thick and thin cases.

Table 1. Non-dimensional radial stress and radial displacement through thickness for a diaphragm supported elastic orthotropic cylinder under external pressure with h/R = 1/5.

r	$\overline{\sigma_r}$	Present (z = l/2)	Lekhnitskii [5]	$\bar{u}$	Present (z = l/2)		Lekhnitskii [5]
	l/R = 1	l/R = 4	l/R = 200		l/R = 1	l/R = 4	l/R = 200
0.9	0	0	0	0	0.2995	0.3162	0.317	0.3231
0.94	0.2158	0.2246	0.2251	0.2304	0.3113	0.3232	0.3237	0.3236
0.98	0.4247	0.4326	0.4331	0.4386	0.326	0.3355	0.3358	0.3328
1.02	0.6259	0.6289	0.6292	0.6327	0.3434	0.3527	0.353	0.3501
1.04	0.7233	0.7239	0.724	0.7263	0.353	0.363	0.3633	0.3615
1.08	0.9106	0.9091	0.9091	0.9095	0.3737	0.3868	0.3874	0.3898
1.1	1	1	1	1	0.3845	0.4003	0.4011	0.4066

Table 2. Non-dimensional hoop stress through thickness for a diaphragm supported elastic orthotropic cylinder under external pressure with h/R = 1/5.

r	$\overline{\sigma_r}$	Present (z=l/2)		Lekhnitskii [5]
	l/R = 1	l/R = 4	l/R = 200	
0.9	4.9253	5.2846	5.3043	5.5062
0.94	5.0373	5.2798	5.2907	5.3485
0.98	5.1895	5.3597	5.3653	5.3405
1.02	5.3782	5.5093	5.5122	5.4538
1.04	5.484	5.6064	5.6089	5.5494
1.08	5.7122	5.8394	5.8428	5.8086
1.1	5.8315	5.9733	5.978	5.9693

CONCLUSION

It is clearly seen from numerical results that for long cylinder the results are very close to the plane strain solution given in Lekhnitskii [5]. Numerical integration technique adopted here is found to be very effective and accurate.

NOMENCLATURE

$\sigma_r, \sigma_\theta, \sigma_z$	Normal stress components parallel to r, θ, and z axis
τ_{zr}	Shearing stress in cylindrical coordinate
h	Thickness of the cylinder
R	Mean radius
p	sinusoidal external pressure
[C]	Stiffness matrix for orthotropic materials

ACKNOWLEDGEMENTS

Partial support of USIF project 95IU001 is gratefully acknowledged.

REFERENCES

1. J.M. Klosner, and J. Kempner, 1963, Comparison of elasticity and shell-theory solutions, AIAA Journal, 1 (3), 627-630.
2. K.T.S.R. Iyengar, and C. V. Yogananda, 1966, Comparison of elasticity and shell theory solutions for long circular cylindrical shells, AIAA Journal 4(12), 2090-2096.
3. A.P. Misovec, and J. Kempner, 1970, Approximate elasticity solution for orthotropic cylinder under hydrostatic pressure and band loads, ASME Journal of Applied Mechanics, 37(1), 101-108.
4. K. Chandrashekhara, and B.S. Kumar, 1993, Static analysis of a thick laminated circular cylindrical shell subjected to axisymmetric load, Composite Structures, 23, 1-9.
5. S.G. Lekhnitskii, 1968, Anisotropic plates, Gordon and Breach Science, New York.
6. T. Kant, 1976, Thick Shells of Revolution-Some Studies. PhD Thesis, Deptt. of Civil Engg., Indian Institute of Technology Bombay.
7. T. Kant, and C.K. Ramesh, 1981, Numerical integration of linear boundary value problems in solid mechanics by segmentation method, International Journal for Numerical Methods in Engineering, 17, 1233-1256.
8. L. P. Kollar, and G.S. Springer, 2003, Mechanics of Composite Structures, Cambridge University Press, New York.

141

Mindlin-Reissner Theory Based New Nine-Node Lagrangian Plate Bending Finite Element Using Integrated Force Method

H.R. Dhananjaya[1], P.C. Pandey[2], J. Nagabhushanam[3] and T.C. Bimal Mohan[4]

[1]Reader, Department of Civil Engineering Manipal Institute of Technology, Manipal-576 104, India
email: djaya_hr@yahoo.com, djaya.hr@manipal.edu
[2]Professor, Department of Civil Engineering, Indian Institute of Science, Bangalore-560 012, India
email: pcpandey@civil.iisc.ernet.in
[3]Emeritus Professor, Department of Aerospace Engineering, Indian Institute of Science,
Bangalore-560 012, India email: naga@aero.iisc.ernet.in
[4]Post Graduate Student, Department of Civil Engineering Manipal Institute of Technology,
Manipal-576 104, India

ABSTRACT

The Integrated Force Method (IFM) has been developed in recent years for analyzing the civil, mechanical and aerospace engineering structures. In this method all independent or internal forces are treated as unknown variables which are calculated by simultaneously imposing equations of equilibrium and compatibility conditions. In this paper the formulation of a new nine-node quadrilateral Lagrangian plate bending element for IFM is presented. The Mindlin-Reissner theory is employed in the formulation which accounts for the shear deformation. This element considers three degrees of freedom namely a transverse displacement and two rotations at each node. The performance of this element is studied by analyzing plate bending benchmark problems and results are compared with those of displacement based finite elements along with the exact solutions. The performance of the element is quite excellent in both thin and moderately thick plate bending situations.

Keywords: Integrated Force Method; displacement fields; stress-resultant fields; equilibrium matrix; flexibility matrix.

1. INTRODUCTION

A new formulation, termed the Integrated Force Method, has been developed in recent years to analyze problems in structural mechanics [1, 2]. It is a new formulation for computerizing the classical force method of analysis. In the Integrated Force Method all independent forces, not just the redundants, are treated as unknown quantities that can be calculated by simultaneously imposing both equilibrium and compatibility conditions. Procedures have been developed for generating compatibility conditions that yield sparse and banded matrices and can be easily adapted to computer automation [3]. The initial applications of the Integrated Force Method to static analysis, vibration analysis, and optimization of trusses have shown that the Integrated Force Method has certain advantages over the displacement method, both in accuracy and computer efficiency [4-8].

This paper presents the formulation and testing of a new proposed nine-node quadrilateral plate bending element (MQP9) to analyze the plate bending problems using IFM. Mindlin-Reissner theory has been used in the formulation which accounts for the shear deformation. This element considers three degrees of freedom namely a transverse displacement and two rotations at each node. Suitable displacement and stress-resultants fields are chosen over the element and element equilibrium and flexibility matrices are developed. The shear correction factor [9] has been considered in the formulation. Standard plate bending benchmark problems are considered to study the performance of the proposed element. The results are compared with those of displacement based 8-node quadrilateral elements. The results show that the proposed element can be successfully used to analyze the plate bending problems accounting shear deformation.

2. FORMULATION OF ELEMENT EQUILIBRIUM AND FLEXIBILITY MATRICES

These elements are based on the Mindlin-Reissner theory where a line that is straight and normal to mid-surface of the undeformed plate remain straight but not necessarily normal to the mid surface of the deformed plate. This leads to the following definition of the displacement components u,v,w in the x,y,z Cartesian coordinates system (Fig. 1).

$$u = -z\theta_x(x,y) \qquad v = -z\theta_y(x,y) \qquad w = w(x,y) \qquad \qquad ...(1)$$

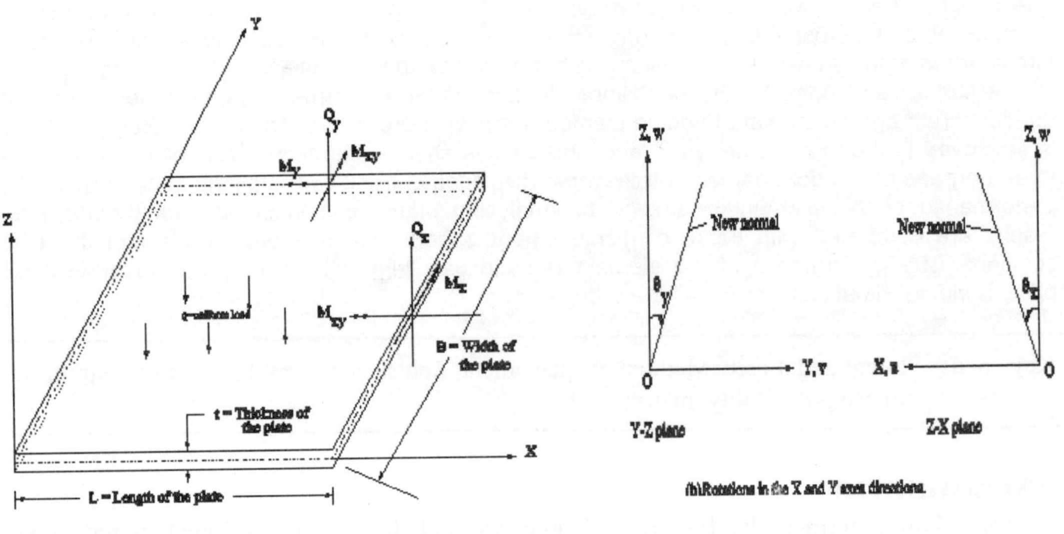

Fig. 1: (a) Plate with stress resultants. (b) Rotations in the X and Y directions.

where,

x, y are coordinates in the reference mid-surface

z is the coordinate through the thickness t with $-t/2 < z < t/2$

w is the transverse (lateral) displacement

θ_x and θ_y represents the rotations of the normal in x-z and y-z planes, respectively.

Engineering strains for Mindlin-Reissner theory can be written as

$$\{\varepsilon\} = -z\{k_1\} \qquad \ldots(2)$$

where,

$$\{\varepsilon\} = \begin{bmatrix} \varepsilon_x & \varepsilon_y & \gamma_{xy} & \gamma_{yz} & \gamma_{xz} \end{bmatrix}^T$$

$$\{k_1\} = \begin{bmatrix} \dfrac{\partial\theta_x}{\partial x} & \dfrac{\partial\theta_y}{\partial y} & \dfrac{\partial\theta_x}{\partial y}+\dfrac{\partial\theta_y}{\partial x} & \dfrac{\theta_y-\dfrac{\partial w}{\partial y}}{z} & \dfrac{\theta_x-\dfrac{\partial w}{\partial x}}{z} \end{bmatrix}^T$$

The stress-strain relation for an isotropic two-dimensional plate material is given by

$$\{\sigma\} = [C_{con}]\{\varepsilon\} \qquad \ldots(3)$$

$$\{\sigma\} = \text{Stress Components} = \begin{bmatrix} \sigma_x & \sigma_y & \tau_{xy} & \tau_{yz} & \tau_{xz} \end{bmatrix}^T$$

where,

$$\{\varepsilon\} = \text{Strain Components} = \begin{bmatrix} \varepsilon_x & \varepsilon_y & \gamma_{xy} & \gamma_{yz} & \gamma_{xz} \end{bmatrix}^T$$

$$[C_{con}] = \text{Constitutive matrix}$$

$$= \frac{E}{(1-v^2)} \begin{bmatrix} 1 & v & 0 & 0 & 0 \\ v & 1 & 0 & 0 & 0 \\ 0 & 0 & \dfrac{(1-v)}{2} & 0 & 0 \\ 0 & 0 & 0 & \dfrac{(1-v)}{2} & 0 \\ 0 & 0 & 0 & 0 & \dfrac{(1-v)}{2} \end{bmatrix}$$

$E = $ Young's modulus

$v = $ Poisson's ratio

The stress-resultants $\{M\}$ can be written as

$$\{M\}^T = \int_{-t/2}^{t/2} \begin{bmatrix} \sigma_x z & \sigma_y z & \tau_{xy} z & \tau_{yz} & \tau_{xz} \end{bmatrix} \qquad \ldots(4)$$

where,

$$\{M\}^T = \begin{bmatrix} M_x & M_y & M_{xy} & Q_y & Q_x \end{bmatrix}$$

$$\{\sigma_r\}^T = \begin{bmatrix} \sigma_x & \sigma_y & \tau_{xy} & \dfrac{\tau_{yz}}{z} & \dfrac{\tau_{xz}}{z} \end{bmatrix}$$

Equations (2) to Eq. (4) yield the moment curvature relation:

$$\{M\} = [C_1]\{k\} \qquad \qquad ...(5)$$

where,

$$\{k\}^T = \begin{bmatrix} \dfrac{\partial \theta_x}{\partial x} & \dfrac{\partial \theta_y}{\partial y} & \dfrac{\partial \theta_x}{\partial y} + \dfrac{\partial \theta_y}{\partial x} & \theta_y - \dfrac{\partial w}{\partial y} & \theta_x - \dfrac{\partial w}{\partial x} \end{bmatrix}$$

= vector of curvatures

$[C_1]$ = matrix relating stress-resultants to curvatures.

From the Eq. (2), curvature-moment relation becomes

$$\{k\} = [C_1]^{-1}\{M\} = [H]\{M\} \qquad \qquad ...(6)$$

where, [H] is the matrix relating curvatures to stress-resultants and it can be written with Reissner's shear correction factor of 5/6 as:

$$[H] = \dfrac{1}{D_1} \begin{bmatrix} 1 & -v & 0 & 0 & 0 \\ -v & 1 & 0 & 0 & 0 \\ 0 & 0 & 2(1+v) & 0 & 0 \\ 0 & 0 & 0 & \dfrac{t^2(1+v)}{5} & 0 \\ 0 & 0 & 0 & 0 & \dfrac{t^2(1+v)}{5} \end{bmatrix} \qquad \qquad ...(7)$$

where,

$D_1 = Et^3/12$ $\qquad\qquad$ t = thickness of the plate;

The strain energy U_p of the plate in general is given by

$$U_p = \iint \dfrac{1}{2}\{k\}^T \{M\} dxdy \qquad \qquad ...(8)$$

For a discrete plate bending element the {M} and {k} can be expressed in terms of assumed stress-resultant and displacement fields, respectively in the matrix form as

$$\{M\} = [\psi]\{F_e\} \qquad \qquad ...(9)$$

$$\{k\} = [D_{op}][\phi_1]\{\alpha\} = [D_{op}][\phi]\{X_e\} \qquad \qquad ...(10)$$

where,

[Ψ] = matrix of polynomial terms for stress-resultant fields

$\{F_e\}$ = vector of force components of the discrete element

$[\phi_1]$ = matrix of polynomial terms for displacement fields

$[\phi] = [\phi_1][A]^{-1}$

$[A]$ = matrix formed by substituting the coordinates of the element nodes into the polynomial of displacement fields

$\{\alpha\}$ = coefficients of the displacement field polynomial

$\{X_e\}$ = vector of displacements of the discrete element

$$[D_{op}] = \text{Differential operator matrix} = \begin{bmatrix} 0 & \dfrac{\partial}{\partial x} & 0 \\[2mm] 0 & 0 & \dfrac{\partial}{\partial y} \\[2mm] 0 & \dfrac{\partial}{\partial y} & \dfrac{\partial}{\partial x} \\[2mm] \dfrac{\partial}{\partial y} & 0 & 1 \\[2mm] \dfrac{\partial}{\partial x} & 1 & 0 \end{bmatrix}$$

Substituting Eq. (9) and Eq. (10) into the Eq. (8), the strain energy for the discrete element can be expressed as

$$U_p = \frac{1}{2}\{X_e\}^T [B_e]\{F_e\} \qquad \ldots(11)$$

where, $[B_e]$ represents the element equilibrium matrix and is given by

$$[B_e] = \iint [\phi]^T \left[D_{op} \right]^T [\psi]\, dxdy \qquad \ldots(12)$$

Using the Eq. (7) the complementary strain energy of the element is written as

$$U_c = \iint \frac{1}{2}\{M\}^T [H]\{M\}\, dxdy = \frac{1}{2}\{F_e\}^T [G_e]\{F_e\} \qquad \ldots(13)$$

where, $[G_e]$ represents the element flexibility matrix and is given by

$$[G_e] = \iint [\psi]^T [H][\psi]\, dxdy \qquad \ldots(14)$$

The Eq. (12) and Eq. (14) can be used to obtain element equilibrium and flexibility matrices ($[B_e]$ and $[G_e]$) respectively. These element equilibrium matrix $[B_e]$ and element flexibility matrix $[G_e]$ of all elements are assembled to obtain the global equilibrium matrix $[B]$ and global flexibility matrix $[G]$ of the structure and they are used to set up the IFM governing equation to analyze the structure by IFM.

2.1 Displacement and Stress-resultant Fields for MQP9

MQP9 is a nine-node quadrilateral plate bending element developed using IFM. Independent description of assumed displacement and stress-resultant fields are required for the development of element equilibrium and flexibility matrices. The assumed polynomials for displacement fields should satisfy the convergence requirements. Three degrees of freedom namely, the transverse displacement (w) and two rotations (θ_x and θ_y) are considered at each node of this element. The Eq. (15) shows the assumed independent displacement fields for the transverse displacement (w) and two rotations (θ_x and θ_y).The Eq. (16) shows the assumed polynomials with the unknown generalized force parameters for stress-resultant fields of this element.

The assumed displacement fields for w, θ_x and θ_y in terms of generalized displacement parameters: α_1, $\alpha_2 \ldots \alpha_{27}$ are

$$w = \alpha_1 + \alpha_2 x + \alpha_3 y + \alpha_4 x^2 + \alpha_5 xy + \alpha_6 y^2 + \alpha_7 x^2 y + \alpha_8 xy^2 + \alpha_9 x^2 y^2$$

$$\theta_x = \alpha_{10} + \alpha_{11} x + \alpha_{12} y + \alpha_{13} x^2 + \alpha_{14} xy + \alpha_{15} y^2 + \alpha_{16} x^2 y + \alpha_{17} xy^2 + \alpha_{18} x^2 y^2 \quad \ldots(15)$$

$$\theta_y = \alpha_{19} + \alpha_{20} x + \alpha_{21} y + \alpha_{22} x^2 + \alpha_{23} xy + \alpha_{24} y^2 + \alpha_{25} x^2 y + \alpha_{26} xy^2 + \alpha_{27} x^2 y^2$$

and the assumed stress-resultant fields in terms of generalized independent forces F_1, $F_2 \ldots F_{24}$ are

$$M_x = F_1 + F_2 x + F_3 y + F_4 x^2 + F_5 xy + F_6 y^2 + F_7 x^2 y + F_8 xy^2 + F_9 x^2 y^2$$

$$M_y = F_{10} + F_{11} x + F_{12} y + F_{13} x^2 + F_{14} xy + F_{15} y^2 + F_{16} x^2 y + F_{17} xy^2 + F_{18} x^2 y^2 \quad \ldots(16)$$

$$M_{xy} = F_{19} + F_{20} x + F_{21} y + F_{22} x^2 + F_{23} xy + F_{24} y^2$$

By substituting Eq. (7), Eq. (15) and Eq. (16) into the Eq. (12) and Eq. (14), the element equilibrium and flexibility matrices are obtained for this plate bending element.

3. NUMERICAL TESTS AND DISCUSSIONS

To study the performance of the proposed element MQP9, the following standard benchmark plate bending problems are considered.

1. A square thin plate (t/L = 0.01) with simply supported/clamped boundary conditions subjected to uniform load/central point load. The parameters of the problem are: size of the plate = 100 × 100, t = 1, E = 10000000, n = 0.3, q = 10 or 400.[10]
2. A rectangular thin plate (aspect ratio = 2 or 3) with simply supported/clamped boundary conditions subjected to uniform load. The parameters of the problem are: size of the plate = 200 × 100 or 300 × 100, t = 1, E = 10000000, v = 0.3, q = 10.
3. A square thick plate (t/L = 0.1) with simply supported/clamped boundary conditions subjected to uniform load/central point load. The parameters of the problem are: size of the plate = 100 × 100, t = 10, E = 200000, v = 0.3, q = 10 or 400.
4. The Morley's plate problem-The parameters of the problems are: L = 100, B = 100, t = 1, E = 1092000.0, v = 0.3 and q = 1, θ = 30^0, w =0 on all boundaries.[11]
5. The Razzaque's plate problem-The parameters of the problems are: L = 100, B=100, t = 1, E = 1092000.0, v = 0.3 and q = 1, inclination of the plate θ = 60^0.[12]
6. A skew cantilever plate (θ = 30°) subjected to uniform load over the whole plate. The parameters of the problems are: L = 100, B=100, t = 1, E = 100, v = 0.3 and q = 1.

Results of the proposed element are compared with those of displacement based eight-node quadrilateral plate bending elements (QH1, QH2, QH3, QH4) available in literature [10], IFM based 8-node quadratic plate bending element MQP8 and those computed using similar elements available in commercial software package NISA[13]. The results are also compared with the exact solutions given references [14] and [15] for thin plates and moderately thick plates, respectively. All the above example (test) problems are analyzed using the proposed element and the results are reported in the reference [16]. A few results of them are presented in this paper.

Normalized central deflections for simply supported thin square plate with uniform load are summarized in Table 1. It shows that the performance of the proposed element is better compared to the other elements. Table 2 shows the normalized central deflections for a simply supported thin rectangular plate (aspect ratio=2) with uniform load. The performance of the proposed element is better compared to all the elements.

Table 1. Normalized central deflections for simply supported thin square plate with uniform load (t/L = 0.01)

Elements	QH1	QH2	QH3	QH4	MQP8	MQP9
1 × 1	0.955	0.890	0.900	0.835	0.954	0.981
2 × 2	1.000	0.985	0.990	0.980	1.000	1.000
3 × 3	1.000	0.990	0.990	0.990	1.000	1.000
4 × 4	1.000	0.990	0.990	0.990	1.000	1.000

Table 2. Normalized central deflections for simply supported thin rectangular plate (aspect ratio = 2) with uniform load

Elements	QH1	QH2	QH3	QH4	MQP8	MQP9
1 × 1	0.975	0.870	0.910	0.855	0.960	0.975
2 × 2	1.000	0.990	1.010	0.980	1.000	0.999
3 × 3	1.000	1.000	1.000	0.990	1.000	1.002
4 × 4	1.000	1.000	1.000	0.995	1.000	1.000

CONCLUSION

A new nine-node quadrilateral plate bending element (MQP9) has been proposed using the Integrated Force Method to analyze the plate bending problems. Mindlin-Reissner theory has been employed in the formulation which accounts for the shear deformation. Various standard plate bending benchmark problems are analyzed using this new proposed element via Integrated Force Method. The proposed element (MQP9) has yielded in general excellent results in all the example problems considered. The proposed element MQP9 is free from spurious energy modes and it does not lock under thin plate bending situations. Hence, the same element can be used to analyze both thin and moderately thick plate bending problems.

REFERENCES

1. S.N. Patnaik, 1973, An integrated force method for discrete analysis, International Journal for Numerical method and engineering, 41, 237-251.
2. S.N. Patnaik, 1986, The variational energy formulation for the integrated force method, AIAA Journal, 24(1), 129-134.
3. J. Nagabhusanam and S.N. Patnaik, 1990, General purpose program to generate compatibility matrix for the integrated force method, AIAA Journal, 28(10), 1838-1842.
4. S.N. Patnaik, and S. Yadagiri, 1982, Frequency analysis of structures by integrated force method, Journal of Sound and Vibration, 83, 93-109.
5. S.N. Patnaik, I. Kaljevic and D.A. Hopkins, 1996, Development of finite elements for two-dimensional structural analysis using the integrated force method, NASA Technical Memorandum 4655, NASA, Lewis Research Center, Cleveland, Ohio.
6. N.R.B. Krishnam Raju, and J. Nagabhusanam, 2000, Non-linear structural analysis using integrated force method, Sadhana, 25(4), 353-365.

7. H.R. Dhananjaya, P.C. Pandey and J. Nagabhushanam, 2005, Mindlin-Reissner theory based new 4-node quadrilateral plate bending element for integrated force method, Journal of Structural Engineering, 32(3), 179-194.

8. I. Kaljevic, S.N. Patnaik, and D.A. Hopkins, 1996, Three-dimensional structural analysis by integrated force methods, Computers and Structures, 58(5), 869-886.

9. E. Reissner, 1945, The effect of transverse shear deformation on bending of plates, Journal of Applied Mechanics, 12, A69 - A77.

10. R.L. Spilker, 1982, Invariant 8-node hybrid stress elements for thin and moderately thick plates, International Journal for Numerical methods and engineering, 18, 1153-1178.

11. L.S.D. Morley, 1963, Skew plates and structures, Pergaman press, Oxford.

12. A. Razzaque, 1973, Program for triangular plate bending element with derivative smoothing, International Journal for Numerical Methods and Engineering, 6, 333-345.

13. NISA Software and manual, (Version 9.3).

14. S.P. Timoshenko and S. W. Krieger, 1959, Theory of plates and shells, Second Edition, McGraw Hill international editions.

15. Jane Liu, H.R. Riggs and Alexander Tessler, 2000, A four node shear-deformable shell element developed via explicit Kirchhoff constraints, International journal for numerical methods and engineering, 49, 1065-1086.

16. T.C. Bimal Mohan, 2006, Mindlin-Reissner Theory Based New Nine-Node Lagrangian Plate Bending Finite Element Using Integrated Force Method, M. Tech Thesis, Department of Civil Engineering, Manipal Institute of Technology, Manipal, India.

142

A General Modelling Structure for Dynamic Engine Simulation

D.G. THOMBARE AND S.K. VERMA

Mechanical Engineering Department, Dr. Babasaheb Ambedkar Technological University, Lonere, Dist. Raigad, Maharashtra-402 103, India email: dgtrit@yahoo.co.in

ABSTRACT

The subject of this paper is presentation of a consistent and generalized approach in building engine system models. This approach guaranties the flexibility to modification and addition of modern systems. It also solves the causality problems arising from insertion of new components to an existing model. The method is implemented successfully in the development of a dynamic model of a Stirling cycle engine. The engine selected here for dynamic simulation is a gamma type, single acting twin cylinder reciprocating engine. This paper presents simulation results of generalized modular engine structure, developed using MATLAB/simulink software. Use of generalized block simplifies the overall structure of the model and allows user to modify the model. This flexibility is important so that the user can experiment with new concepts and can see effect on the model. The model in this paper goes behind others in the literature; this is certainly not an exhaustive list of important dynamic engine models but provides the reader with a cross section of the major models being used. The model is written using the Simulation X software and the overall modular structure of the model is characteristic of and an extension of what was been developed.

Keywords: Engine model, Stirling engine, generalized modular structure.

1. INTRODUCTION

From last two decades the analytical models of engine systems have taken an increasingly important role in the engine design and development process. This change has come due to competitive need to shorten the development time and reduce the cost of the development process. The engine designer and manufacturer can get advantage in being more responsive to the product and gaining the advantage of faster and closer results of the analysis of the product. Another strong influence that the computer simulation has ability to model, analyse, compare, remodel and again analyse till better results. Also it reduces cost of prototype development as the final design has gone through number of simulation stages and optimized for best results which reduces failure chances of prototype developed.

The exponential increase in the computational power of computers and continues decrease in cost have influenced to use them for modeling and simulation. Computer based simulations with developing simulation code were expensive a decade ago which have became common today because of availability of softwares. An example is vehicle crash simulation and the effect of that is reducing cost, improving safety and decreasing developing time. Computer modelling and simulation does not replace experimental methods because they each have different attributes and disadvantages. The key to successful implementation and use of computer models is to appreciate their strengths and imitations and to use both experimentations and simulations in a manner that utilize these strengths. At this point of time the development of engine system models are becoming very important issues in the design, development and performance evaluation process of the engine.

Engine systems contain wide range of multi domain physical modelling challenges as described by C.D. McCartan [1]. Developing a dynamic model of Stirling engine is very challenging because of the broad range of components that comprises the engine, as well as the range of engineering disciplines that are brought to bear the engine design. The task is further complicated because the goals of such a model can very immensely depending on its use, and in a high quality dynamic model these goals will drive the overall structure and fidelity of the model. Quality and usefulness of the dynamic model depend upon fidelity of model. Table 1 gives a list of physical processes and modeling domain that could be considered in the modeling of Stirling engine system depending upon particular analysis and desired level of details.

Table 1. Physical engine processes and modelling domain for an engine system.

Physical process	Modelling domain
Piston and crankshaft motion	M
Flow of working fluid in the passages	F, T
Flow of working fluid through regenerator	F, T, Th
In cylinder fluid motion	F, T
Heat transfer between working fluid and metal surfaces of heater wall	Th
Heat transfer between working fluid and metal surfaces of cooler,	Th
Heat transfer between working fluid and material regenerator	Th
Frictional effect between meeting surfaces of engine	M, Th
Coolant flow	F, Th
C-Chemical, F-Fluid, M-Mechanical, T-Thermodynamic, Th-Thermal	

Due to wide variety of physical processes and modelling domains along with inherent interactions, it is imperative to have a descriptive language that is capable of modelling across the physical domains. Simulation X with its high level, declarative formulation of physical modelling in an ideal component level modelling: Every engine component will be modelled as an object, using mainly physical techniques. Simulation X with its high level, declarative formulation for physical modeling is an ideal language for multi domain system simulation. The simulation X standard mechanical, rotational, multi-body and thermal libraries contain the connector definitions, interfaces and the basic models that provide framework for modelling engine system.

Methodologies of Modelling and Simulation

To build a mathematical model for a system, it is necessary to define boundaries, inputs and outputs. The model is build up from cause effect relationships between inputs and outputs at certain boundary condition. For the Stirling engine inputs are design parameters such as engine geometry, operational parameters such as phase angle and other parameters such as heat transfer coefficients, heat capacity and so on. The outputs are both the thermal and mechanical performance parameters. A schematic diagram of a simple model for illustration is shown in Fig. 1. For simplification and avoiding errors in computation of engine performance a top-down approach is selected for modelling and simulation.

Generalised Block Structure

An engine system is modelled as a simple thermal system as a level I model. The system is configured with minimum inputs and outputs as a level I model.

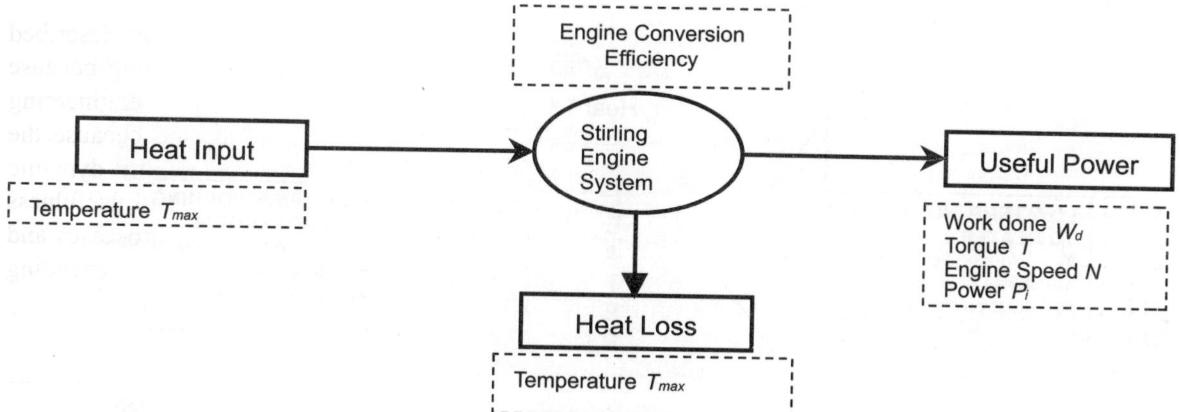

Fig. 1. An engine system model: Generalised block structure, Level I.

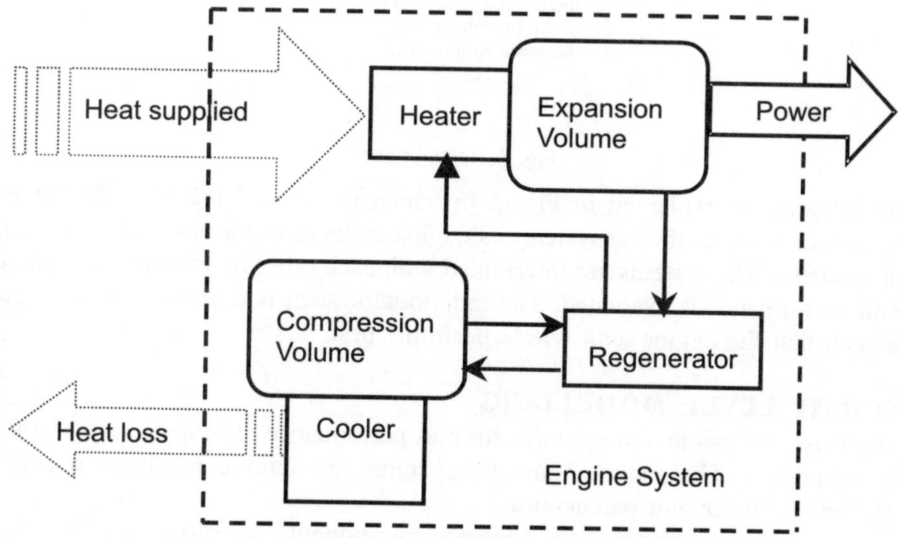

Fig. 2. Stirling engine system model, Level II.

In transient dynamic system models of the system components must be interconnected in a way that each state/variable is determined from solving exactly one equation at each integration step and avoiding algebraic loops or causal mismatches. In most cases simplified models represent system components. The simplicity of the model can help to resolve the causal mismatch by rearranging the equations describing the components so that the variables determination fits in the overall system causality

Mechanical modelling of engine system includes a combination of 1D and multidimensional dynamics. A 1D approach is useful in engine modeling because 1D framework can include either kinematics relationship or dynamic behavior. Thermodynamic modelling is a crucial part of engine system modelling. A several control volumes are formulated for which fundamental equation of energy and mass conservation is used.

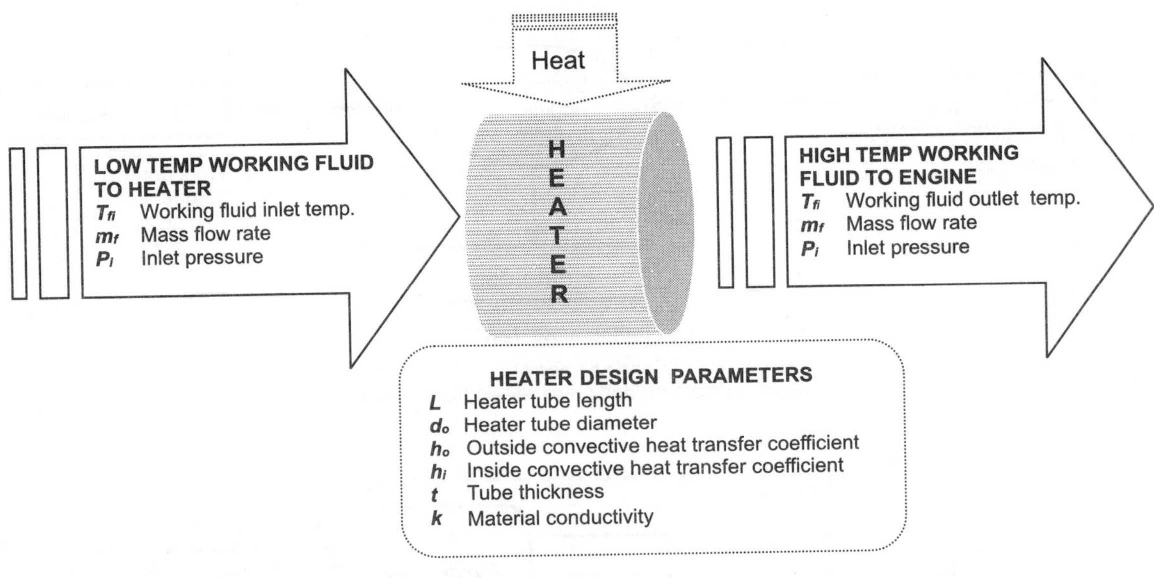

Fig. 3. Heater model.

The level II model is explained in Fig. 2 in which the level I model is further elaborated to system level. This model uses the subsystems as engine cylinder crank mechanism, heater, cooler and regenerator as systems. The systems are interlinked with each other by mechanical, fluid or heat flow and the overall performance is evaluated. The individual system is designed and configured for best performance such that the engine as a whole performs best.

2. COMPONENT LEVEL MODELLING

For making a precise analysis and to optimize for best performance the engine subsystems/components are modeled separately. The main engine subsystems are engine mechanism (cylinder piston arrangement), heater, cooler and regenerator.

These components are modeled as separate components as shown in Fig. 3, Fig. 4 and Fig. 5.

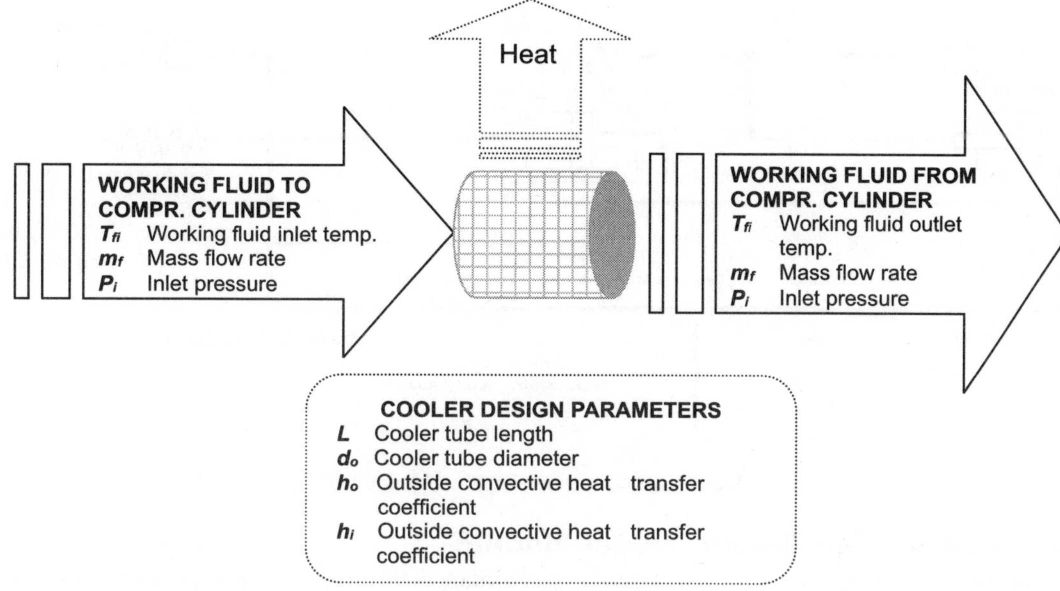

Fig. 4. Cooler model.

The generalize block inputs and outputs variables have consistent schemes for all subsystems, which allows flexibility in modification of overall system. With best performance as a individual subsystem the interfacing allowed to model the engine as a whole system. The implementation of the generalised block to build a complete engine system model can be considered as a straightforward task.

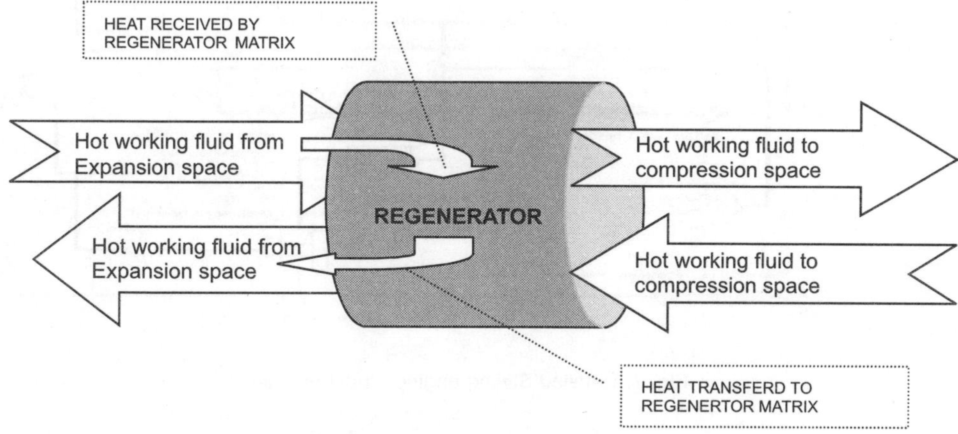

Fig. 5. Regenerator model.

2.1 Stirling Engine Simplified Model

The Fig. 6 gives a simplified model of Stirling engine based on the layout of Fig. 2. The simplified model gives flexibility to change design parameters with least possible iteration for optimization.

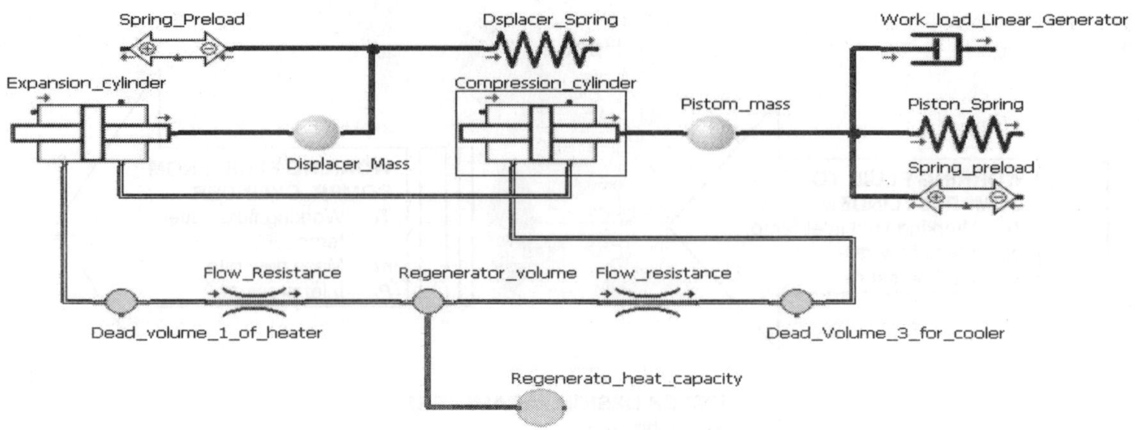

Fig. 6. Stirling engine simplified model.

2.2 Detailed Stirling Engine Model Module

After evaluation of the engine for best suitable performance the model is further expanded to the detailed model as shown in Fig. 6.

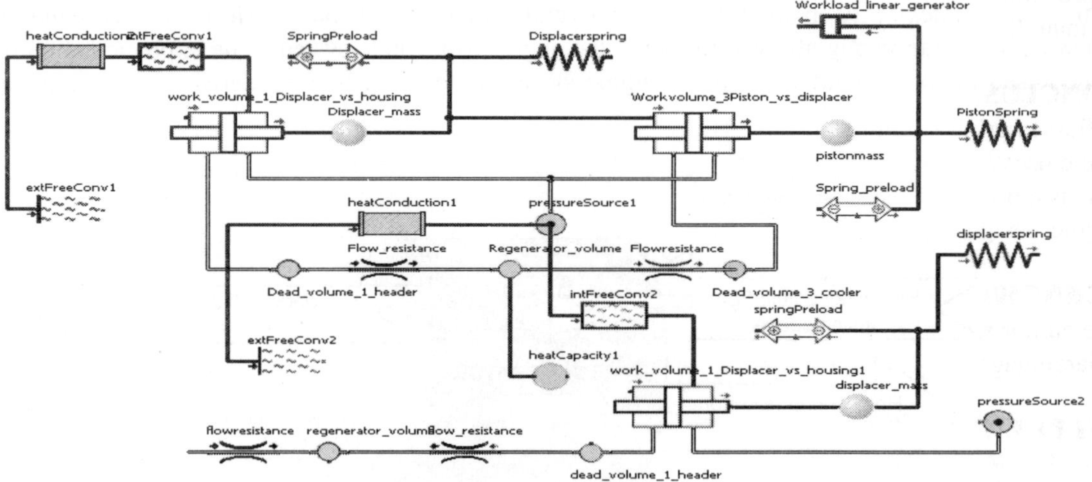

Fig. 7. Detailed Stirling engine model module.

3. DISCUSSIONS

Some of the sample performance results are shown in Figs. 7 and 8. The araiation in expansion cylinder variation for the engine and regenerator heat flow is shown as a sample result. The results of simulation gives a greater insight in to the component and system for further advancement where the component design can be modified further.

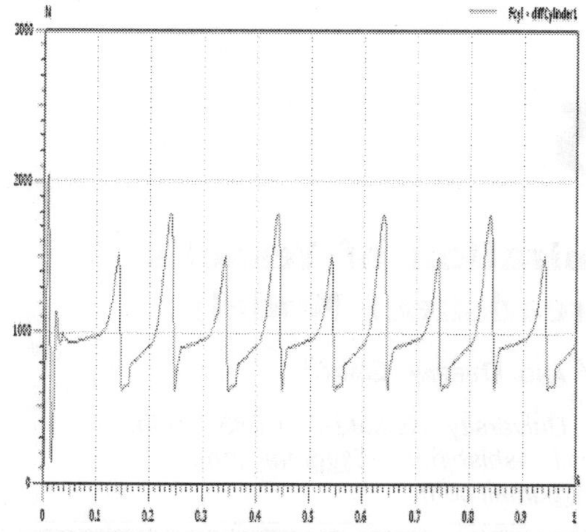

Fig. 7. Expansion cylinder pressure variation.

Fig. 8. Regenerator heat flow.

4. FUTURE PLANS

The dynamic engine model takes physical inputs in the form of key dimensions and masses. An improvement will be made in the framework of the model after investigations. The final model will be manufactured as per design and tested for performance evaluation.

CONCLUSION

A framework for an end to end dynamic model of Stirling cycle engine is presented in detail. The frame uses advantages to modify, alter and rearrange the engine system for best suitable performance. This is a powerful engine modeling approach that has many implementations in future applications for engine design and optimization.

ACKNOWLEDGEMENTS

The author would like to acknowledge the support provided by Rajarambapu Institute of Technology, Rajaramnagar in developing engine system model by using Simulation X, simulation software.

REFERENCES

1. C.D. McCartan, P.T. McEntee, P. Fleck, G.P. Blair, DF. O. Mackey, 2002, Computer simulation of the performance of a 1.9 liter direct injection diesel engine, SAE Transactions on Engines, 2002-01-0070, 334-342.
2. Y.H. Zweiri, J.F. Whidborne and L.D. Seneviratne, 1999, Dynamic simulation of a single cylinder diesel engine including dynamometer modelling and friction, Proc. of Institution of Mechanical engineers, Volume 213, part D, 391-415.
3. B.J. Huang, H.Y. Chen, 2000, Modelling of integral type Stirling refrigerator using system dynamic approach, International journal of refrigeration, Volume 23, 632-641.
4. John Batteh, Michael Tiller and Charles Newman, 2003, Simulation of engine systems in Modelica, Proceedings of 3rd International Conference on System Modeling, 342-358.
5. Z.S. Filipi, D.N. Assanis, 2001, A non-linear, transient, single-cylinder diesel engine simulation for prediction of instantaneous engine speed and torque, 2001, Journal of engineering for gas turbine and power, Volume 123, 951-959.
6. S.Lu., 1999, Dynamic modelling and simulation of power plant systems, Proc. of Institution of Mechanical engineers, Volume 213, part A, 7-22.

143

"Opti-Marine-Ware" (Optimization of Vessel's Parameters Through Spreadsheet Model)

ABHIJIT DE[1], ASHISH KUMAR[2] AND DEEPAK GARG[3]

Department of Marine Engineering, Jadavpur University, Kolkata-700 088, India.
[1] email: abhijitde549128@gmail.com, [2] email: ashishjha5615@gmail.com,
[3] email: deepakgarg5625@gmail.com

ABSTRACT

The objective of this paper is to describe and evaluate a scheme of engineering-economic analysis for determining optimum ship's main dimensions and power requirement at basic design stage. An optimization designs the problem and is arranged into five main parts, namely, *Input, Equation, Constraint, Output and Objective Function*. The constraints, which are the considerations to be fulfilled, become the director of this process and a **minimum and a maximum value are set on each constraint** so as to give the working area of the optimization. The *outputs* (decision variables) are optimized in favor of **minimizing** *the objective function*. Microsoft Excel-Premium Solver Platform (a spreadsheet modeling tool is utilized to model the optimization problem. This paper is commenced by the description of the general optimization problems, and is followed by the model construction of the optimization. A case study on the determination of ship's main dimensions and its power requirement is performed with the main objective to minimize the Economic Cost of Transport (ECT). After simulating the model and verifying the results, it is observed that the spreadsheet model yields considerably comparable results with the main dimensions and power requirement data of the real operated ships (tanker). It is also experienced that this kind of optimization process needs no exhaustive efforts in producing programming codes, if the problem and the optimization model have been well defined.

Keywords: Optimization, design, Ship power requirement.

1. INTRODUCTION

The problems in designing ship and marine machinery appear due to numerous considerations that must be taken into account. These conditions increase the capital cost and the complexity of the design option. Therefore, ship's design and its selected machinery must guarantee that the ship and its machinery will operate with low level of failure, safely and efficiently, with high level of availability and will deliver an optimum rate of return on the capital being employed.

Thorp and Armstrong [3] utilized a comprehensive method to select the machinery arrangement for a Panamax-size bulk carrier of 70.000 DWT. Their economic assessment was only focused on two alternatives of slow speed diesel installation and medium speed diesel installation. Some parameters that were included in their study are also taken in our study. *One of the major differences with their study is that our study takes the problem since the basic design process which allows the optimization process determines the ship's main dimension and its machinery characteristics within the given constraints.*

This paper proposes an alternative method for optimizing marine designs, particularly in determining ship's main dimension and its power requirement at basic design stage. Spreadsheet modeling is utilized and non-linear programming (hereafter NLP) can express our problem. *The Generalized-reduced gradient (hereafter GRG) method can work in conjunction with the NLP problems.* Basic diagrammatic concepts of the optimization process and a case study are also given comprehensively.

2. BASIC DESIGN OPTIMIZATION PROCESS FOR TANKER WITH SPECIFIED THROUGHPUT

2.1 Problem Statement

At the basic design stage, it is required to design a numbers of series ships (tanker) delivering contract of a certain throughput., which have optimum main dimension and optimum specified power. *Economic Cost of Transport* is utilized as the objective of the optimization problems. Port characteristics require such constraints, as the ship must not exceed 200-m in length and 11-m in draught.

The conceptual problem is shown in Figure 1. Some economic data are employed during the optimization process, as shown in Table 1.

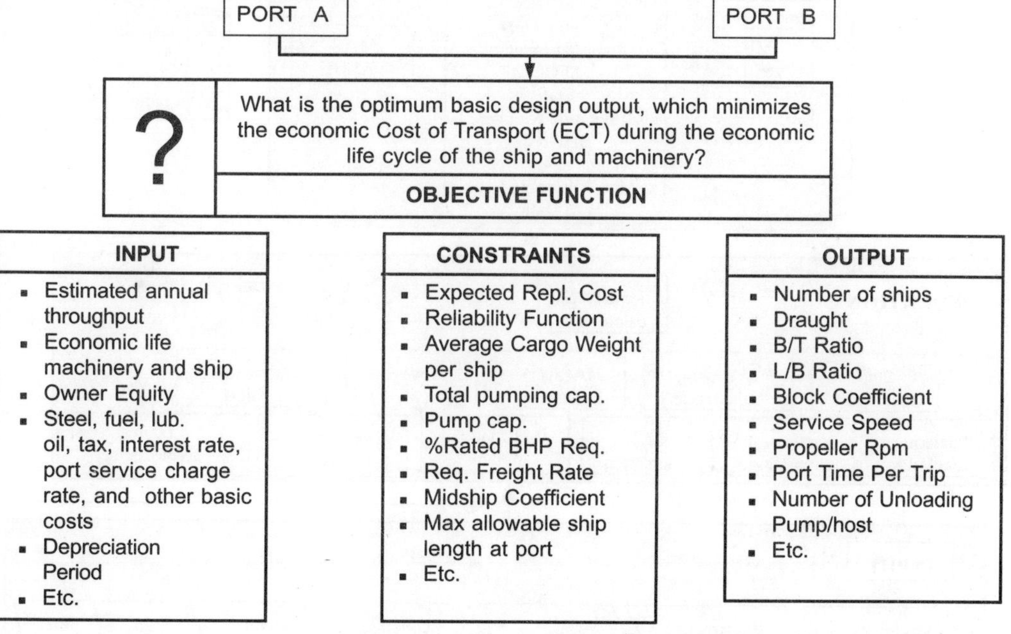

Fig. 1. Problem statement.

Table 1. Economic data input* (Source Ref [9]).

Economic life of machiner	Years	20.00
Loan repayment period	Years	20.00
Interest rate	%	0.10
Rate of return on equity	%	0.12
Economic life of ship	Years	20.00
Ship depreciation period	Years	15.00
Machinery depreciation period	Years	15.00
Tax rate	%	0.30
Annual inflation rate	%	0.01
Average fuel price (HFO/DO)	US$/lb.	0.08
Average crew cost per month	US$/month	1,250.00
Average LO price (ME/GE)	US$/ton	750.00
Steel cost	US$/ton	493.70

2.2 Model Structure

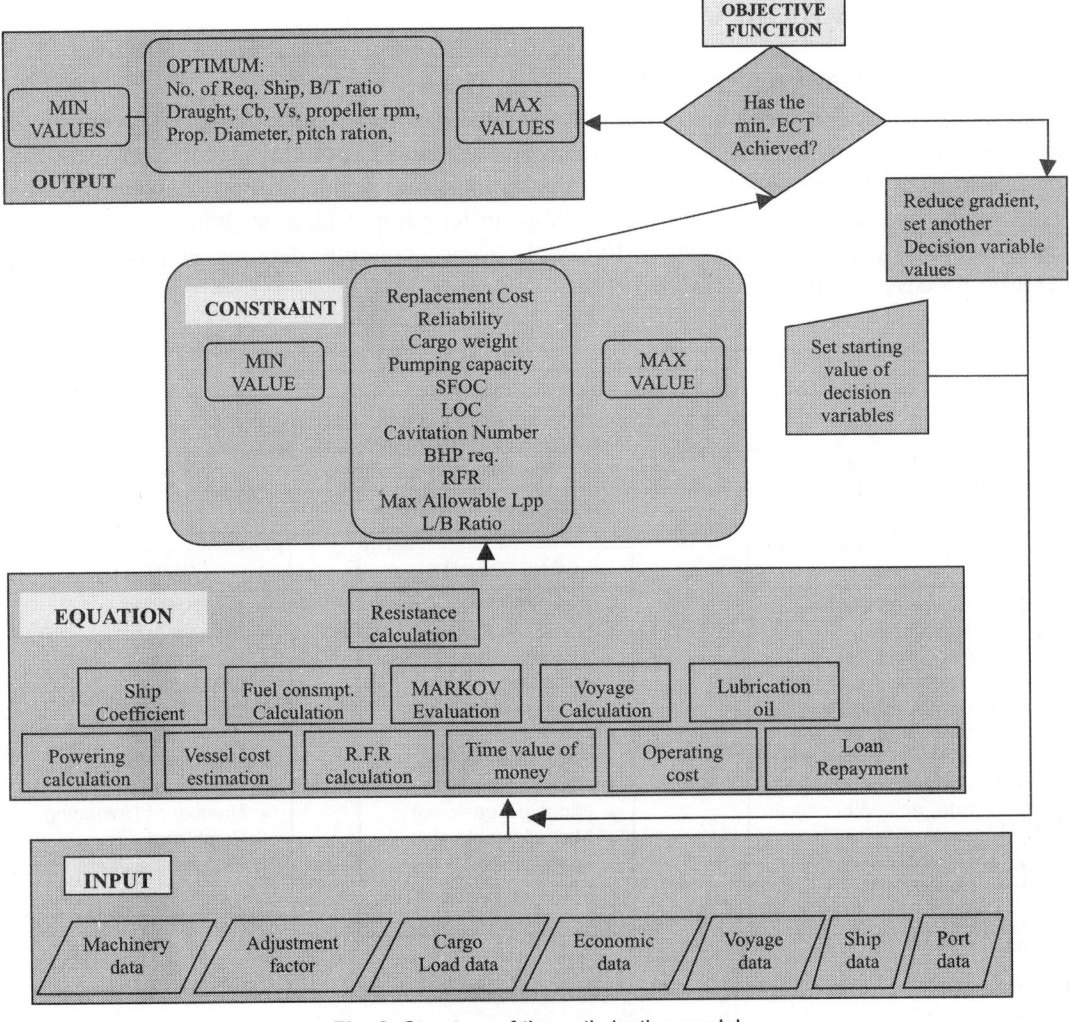

Fig. 2. Structure of the optimization model.

In favor of making the optimization problem easier, the INPUT folder and the EQUATION folder are grouped into several directories. In this particular optimization, the INPUT folder covers: the ship data, machinery data, reliability data,. Each directory represents collection of parameters that are used in the calculation process.

The EQUATION folder consists of several directories such as the ship coefficient, machinery, reliability, loading and unloading, fuel, operating cost and the economic considerations. The CONSTRAINT folder comprises of the expected replacement cost, reliability index, unloading pump capacity, specific fuel oil consumption (hereafter SFOC) for Maine Engine (ME) and and the maximum allowable ship length in port. The OUTPUT folder yields the optimum preventive maintenance interval, block coefficient, optimum design draught, optimum, B/T ratio, and the number of ships. These values are sought with the main objective to minimize the ECT of the ship. ECT, the objective for this particular optimization problem is composed by several variables, namely the required freight rate (hereafter RFR), the inventory cost of cargo and the annual tons of cargo carried (ATC) [4].

Table 2. Optimization statement.

	Find				
	X_1	Min value	≤ Time (t) independent variable	≤	Max. value
	X_2	Min value	≤ Number of ships	≤	Max. value
	X_3	Min value	≤ Draught	≤	Max. value
	X_4	Min value	≤ B/t ratio	≤	Max. value
	X_5	Min value	≤ Block coefficient	≤	Max. value
	X_6	Min value	≤ Service speed	≤	Max. value
	X_7	Min value	≤ Propeller rpm	≤	Max. value
	X_8	Min value	≤ Diameter propeller	≤	Max. value
	X_9	Min value	≤ Pitch ratio	≤	Max. value
	X_{10}	Min value	≤ Time required for preventive replacement	≤	Max. value
	X_{11}	Min value	≤ Port time per trip (loading)	≤	Max. value
	X_{12}	Min value	≤ Number of unloading pump/host	≤	Max. value

Which minimizes: Economic Cost of Transport (ECT) (f(X))

RFR	Total cost Annual port cost *f* (unit cost, grt, voyage per year, no. of operated ship)
	Annual insurance cost *f* (voyage per year, weight of cargo, unit insurance, no. of ship)
	Annual overhead cost *f* (constant, no. of ship)
	Annual crew cost *f* (unit of crew cost, no. of crew, no. of ship)
	Annual expected replacement cost *f* (reliability, no. of ship)
	Annual m/r cost *f* (reliability, no. of ship)
	Annual dry docking expenses *f* (constant, no. of ship)
	Annual administration cost *f* (constant, no. of ship)
	Annual operating cost *f* (lo cost, do cost, hfo cost, etc)
Owner equity	Constant
Throughput	Given
Cargo cost unit	Constant
Number of voyage	Operating day it(docking days, unscheduled maintenance days, time at port)
	Turn round time
Interest rate	*Constant*
Subject to	

$g_1(X)$ Min value $\leq$ Exptd. replacement cost, f (Reliability index, Cost of fail. rep, Cost of Prev. rep)	$\leq$ Max. value
$g_2(X)$ Min value $\leq$ Reliability function, f (failure distribution parameters)	$\leq$ Max. value
$g_3(X)$ Min value $\leq$ Ave. cargo wt./ship, f (throughput,No. of ship, voy./ year,Load factor)	$\leq$ Max. value
$g_4(X)$ Min value $\leq$ Total pumping capacity, f (Pump capacity, No. of req. pump)	$\leq$ Max. value
$g_5(X)$ Min value $\leq$ Pump capacity f (Cargo weight, Port time, Cargo density)	$\leq$ Max. value
$g_6(X)$ Min value $\leq$ SFOC for full load ME f (DHP, engine rpm)	$\leq$ Max. value
$g_7(X)$ Min value $\leq$ SFOC for full load GE f (DHP, diesel generator rpm)	$\leq$ Max. value
$g_8(X)$ Min value $\leq$ Cavitation no f (THP,Projected Blade Area (Ap),dyn. press. at tip radius)	$\leq$ Max. value
$g_9(X)$ Min value $\leq$ Local cavitation no. f (press. at the screw centerline, dyn.press. at tip radius)	$\leq$ Max. value
$g_{10}(X)$ Min value $\leq$ %Rated BHP requirement f (min resulted SFOC at feasible region)	$\leq$ Max. value
$g_{11}(X)$ Min value $\leq$ Required freight rate f (Ann.Vessel cost, total opr.Cost, throughput)	$\leq$ Max. value
$g_{12}(X)$ Min value $\leq$ Midship coefficient f (Displacement, Breadth, Draught, Lpp)	$\leq$ Max. value
$g_{13}(X)$ Min value $\leq$ L/B ratio f (Lpp/Breadth)	$\leq$ Max. value
$g_{14}(X)$ Min value $\leq$ Max allowable ship length at port f (Vol.Displ, Breadth, Draught, Block coef.)	$\leq$ Max. value
$g_{15}(X)$ Min value $\leq$ Length of water line (LWL) f (LOA)	$\leq$ Max. value
$g_{16}(X)$ Min value $\leq$ Length between perpendicular (LPP) f (LWL)	$\leq$ Max. value

The optimum value of RFR itself depends on the annual capital recovery of the vessel cost, the annual operating cost, and the annual throughput [5]. The sequence of this design process indicates strict relationship among each design consideration.

For instance, it might be not a simple work to relate the optimum number of shore connection, which must be fitted on a tanker with the resulted RFR or outcomes of the loan repayment scheme. However, it is believed that those variables somehow interconnect and affect each other. Hence, the basic nature of ships and its machinery design optimization process would lie on the ability of the engineers to accommodate all of the design considerations and to provide adequate flexibility in altering the decision variables, while fulfilling the main objective of the optimization process.

Figure 2 shows the general structure of this optimization problem. The optimization process is commenced by setting the initial value of the decision variables. Using relevant basic parameters located in the INPUT folder, all basic calculations are executed in the EQUATION folder. The results are then exported to the CONSTRAINT folder to calculate all constraints accordingly the optimization problem can be mapped as shown in Table 2. The objective is to minimize f (X), which is the ECT while determining the optimum value of X1 to X12 subject to constraint g1 (X) to g16 (X). The basic ship design and ship resistance formulae are mainly taken form references [6, 7, 8, 9] and the economic parameters and major assumptions related to cost calculation from references [4, 10].

2.3 Results Verification

To verify the performance of this optimization program, comparison on BHP, DWT, and T (draught) has been made on several tanker data (obtained from different shipping companies). The comparisons are shown in Figures 3-4. Generally, it is observed that the results of the simulation very closely conforms to the real data.

At some points the optimization result drastically shift to a new point. This is caused by any adjustment made to the optimization program, which is different from that of the previous one. For instance, if the throughput is less than 300.000 ton, then we could set the maximum cargo carrying capacity of the constraint at the value of 25.000 ton. Once we increase the throughput, the optimization cannot produce optimum results, until we increase the maximum value of the cargo carrying capacity.

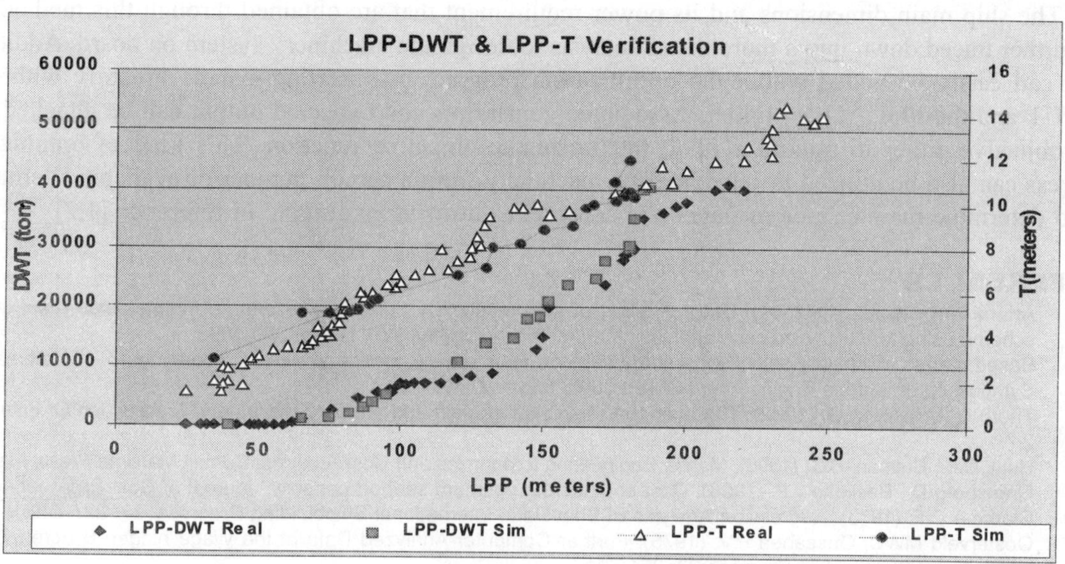

Fig. 3. LPP-DWT and LPP-T Verification.

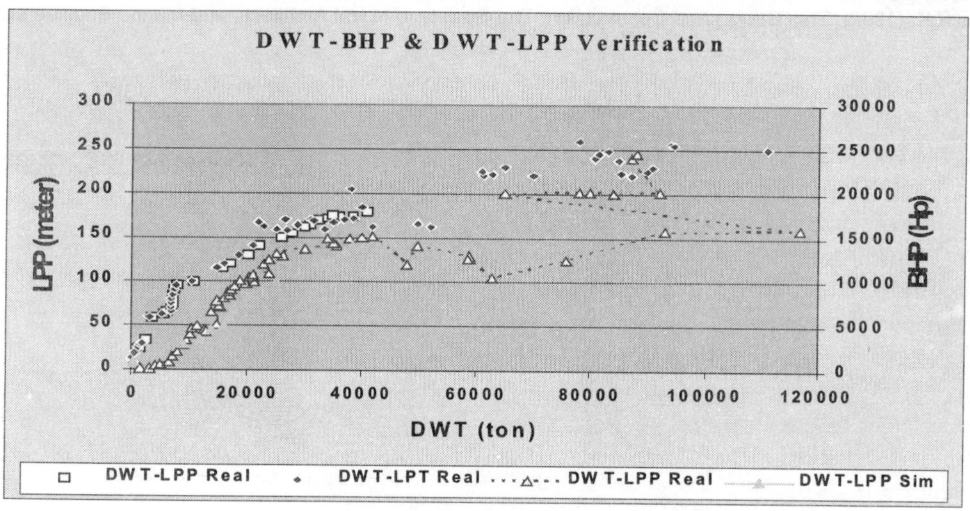

Fig. 4. DWT-BHP and DWT-LPP Verification.

CONCLUSION

For basic design stage or feasibility study purposes, this method could be employed before commencing any further design stage. The case study presented here shows how this optimization program can effectively and precisely consistent with the real ship's design. Furthermore, it is also observed that the difficulties of using PSP to solve optimization problems do not manifest in the PSP construction viewpoint. Instead, is via the way to express every optimization problem in mathematical expressions, which can be executed by the spreadsheet.

The ship main dimensions and its power requirement that are obtained through this method, can be further traced down into a more detail analysis to design the machinery system on board. Additional task can easily be added within the optimization program by inserting a new directory within the INPUT and the EQUATION folder. Associated constraints and expected output can be attached with the objective either to minimize or to maximize the objective function. This kind of optimization process can also be utilized to select marine machinery from a certain number of available alternatives or to determine maintenance management scheme, as utilized by authors in reference [1,2].

REFERENCES

1. Artana K.B., Ishida K. (2001). Determination of ship machinery performance and its maintenance management scheme using MARKOV process analysis. Marine Technology IV, WIT Press: 379-389.
2. Based marine machinery selection: a study case on main engine cooling system. Proceedings: Sixth International Symposium on Marine Engineering (ISME 2000), Tokyo. 2: 791-796.
3. Thorp I., Armstrong G. (1982). The economic selection of main and auxiliary machinery. Transaction Of ImarE. 95: 2-7.
4. Hunt, C.E., Butman, B.S. (1995). Marine Engineering Economics and Cost Analysis. Cornell Maritime Press, Maryland
5. Gransberg D., Basilotto J.P. (1998). Cost engineering optimum seaport capacity. Journal of Cost Engineering. 4.
6. Clarke A.C.F. (1975). Regression Analysis of Ship Data. International Shipbuilding Progress. 22: 227-249.
7. Oosterveld MWC, Oossanen P.V. (1975). Further Computer-Analyzed Data of the Wage ningen B-Screw Series. International Shipbuilding Progress.22: 251-261.
8. Harvald Sv. A.A. (1983). Resistance and Propulsion of Ships, John Wiley & Sons.
9. SNAME (1967). Principle of Naval Architecture. The Society of Naval Architecture and Marine Engineers, New York.
10. Kiss R.K. (1992). Ship Design and Construction. The Society of Naval Architects and Marine Engineers, New York.

144

Application of Finite Element Method and Artificial Neural Network to the Design Tooling for Tube Nosing Operation

CHINMAY K. DESAI[1], R.D. SHAH[2] AND A.A. SHAIKH[2]

[1]Department of Mechanical Engineering, C.K. Pithawall College of Engineering, Surat-395 007, India.
email: chinmay.desai@gmail.com
[2]Department of Mechanical Engineering, Sardar Vallabhbahi National Institute of Technology (Deemed University), Surat-395007 India email: aas_svnit@yahoo.com

ABSTRACT

The present investigation adapts the finite element method (FEM) and the artificial neural network (ANN) to plan the tube nosing of work-hardened materials to yield the optimal die designed. The process parameters considered herein are initial tube thickness (t_0), initial tube diameter (d_0), initial tube length (L), semi-die angle, friction factor, punch speed, fillet radius of the die, and shoulder length of the die (CL), work hardening coefficient (n). There effect on critical nosing ratio as well as final shape of the tube is simulated using explicit FEM. This data set is then used to obtain revere mapping model for the design of die. For a design engineer there are several situations in which die design has to carry out. Keeping this in mind the reverse mapping model is determine by considering six different design conditions and according to that combinations of input and output is varied.

Keywords: Artificial neural network, Finite element method, Tube nosing, Rigid-plastic theory.

1. INTRODUCTION

A reducer is often used in pipe fitting industry to join two pipes, with different diameters. The reducer is usually manufactured by tube nosing with a conical die [1]. In this processes, a circular short tube is first pushed into a conical die with a suitable travelling punch, then ejected out of the finished tube. Finished tube is then subjected to trimming operation to remove rough and uneven ends of the tube. The noising process is similar to the sinking or drawing process of metal tubes, although for the noising process a short tube is pushed into a conical die with suitable travelling punch, whereas a long tube is pulled through out a conical die for the drawing process.

Guo et al. [2] dealt with tube sinking using an upper bound approach. Ruminski *et al.* [3] analysed the effect of die shape on the distribution of mechanical properties and strain field in the tube sinking process. Park *et al.* [4] proposed a backward-tracing method to the determination of optimal preform for shell nosing. Zhu [5] proposed a preform design method by forward-deformation FEM simulation plus geometrical modification of the initial shape for shell nosing. Reid and Harrigan [6] analysed transient effects in the quasi-static and dynamic internal inversion and nosing of metal tubes. Dai and Wang [7] analysed the shear stress distribution in the drawing of a thin wall tube with a conical die. However, most studies of tube nosing, by laterally constraining the tube specimens before they enter the die the onset of buckling is inhibited. Nevertheless, in the industrial reducer production, the tube nosing process is only by partially laterally constraining the tube specimens in order to determine the critical nosing ratio of the tube nosing in order to prevent the specimens bulking during process design.

In the past decade advancement in computer hardware and software technology offers the possibility of designing the entire manufacturing process on computer using Finite Element Method (FEM) without being performing it physically [8-10]. Even though very powerful FEM is costly in terms of computational time, particularly non-linear analysis of forming processes. On the other hand process design by numerical simulation or by experimentation on prototype developed is a trial and error procedure, which is costly in both the terms of time and money where the objective is to find a cost-effective process to manufacture a defect free product by varying process parameters. Recently trend has been set to use new techniques such as Artificial Neural Network (ANN), Fuzzy logic, and Genetic Algorithms along with the FEM to reduce trial and error to find the condition of optimal design parameters rapidly for designing various metal forming processes [11].

In this paper, cold nosing process of partially laterally constrained metal tubes with a conical die is simulated using the ANSYS/LS-DYNA finite element software. Also, few experiments are carried out with Aluminium alloy 1050-O tube billets at room temperature. This data set is then used to obtain revere mapping model for the design of conical die with ANN. For a design engineer there are several situations in which die design has to carry out. Keeping this in mind the reverse mapping model is determine by considering six different design conditions and according to that combinations of input and output is varied.

2. SIMULATION MODELLING

The schematic diagram of the tube nosing process is shown in Fig. 1. In this process, a die shoulder length (CL in Fig. 1) is set for partially laterally constrained tubes, and a punch moves downwards toward the tube billet whilst the conical die is stationary, then a circular tube billet is formed into a reducer after the punch travelling a suitable distance.

A commercial FE code ANSYS/LS-DYNA Version 10.0 [12], with explicit FE formulation is used to analyse the plastic deformation of the tube nosing from a circular tube billet. During the

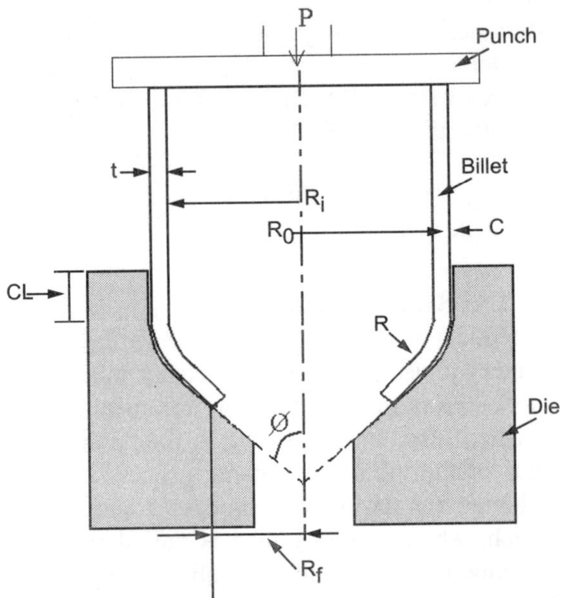

Fig. 1. Variation of maximum relative error in displacement (L$_2$ norm).

analyses, the die and the axial punch are assumed rigid. The form of flow stress equation for the billet material, is $\overline{\sigma} = A\overline{\varepsilon}^n$, where $\overline{\sigma}$, A, $\overline{\varepsilon}$ and n present effective stress, strength co-efficient, effective strain and strain hardening exponent respectively, and the Coulomb coefficient of friction is assumed at the interface between the workpiece and die. All of the analyses reported are for tubes with an outer diameter of mm. The radial clearance between the die shoulder and the tube (c in Fig. 1) was set 0.025 mm. Others forming process parameters used in the FE-simulation on the tube nosing are summarized in Table 1. The FE-model of the tube nosing simulated in ANSYS/LS-DYNA is shown in Fig. 2. The 3D solid (solid 168) elements are used in the FE model for the tube billet in axisymmetric simulation. Total 550 elements and 610 nodes were meshed.

Table 1. Maximum reduction of response for different inner radius of baffle.

1	Initial tube thickness (t_o)	2.8, 3.2, 3.4, 4.5 mm
2	Initial tube diameter (d_o)	88.9 mm
3	Initial tube length (L)	90, 135, 180 mm
4	Semi-die angle (°)	30° , 45° , 60°
5	Punch speed (mm/s)	0.00, 0.06, 0.15
6	Fillet radius of the die (R)	5 mm/s
7	Shoulder length of the die (CL)	30 mm

2.1 Explicit FEM Formulation

The Finite Element Method (FEM) is a very common and accurate way to solve boundary-value problems in mechanics of continua [13]. In the case of biological tissue, we have to deal with large deformations and also with anisotropic, inhomogeneous, and non-linear materials. Furthermore, as organs and surgical instruments interact, numerical contact problems arise. Nonetheless, provided an adequate formulation is chosen, even in these cases FEM is a very powerful tool because it is physically based and, therefore, can be adapted for these highly non-linear problems. Within the FEM a body is subdivided by a finite number of well defined elements (such as hexahedrons, tetrahedrons, quadrilaterals). Displacements and positions in an element are interpolated from discrete nodal values. For every element, the partial differential equations governing the motion of material points of a continuum can be formulated, resulting in the following discrete system of differential equations (Eq. 1):

$$M\ddot{u} + C\dot{u} + K\delta u = f - r \qquad \qquad ...(1)$$

where

M is the mass matrix
C is the damping matrix
K is the incremental stiffness matrix
u is the vector of the nodal displacements
f are the external node forces, and
r are the internal node forces
All these matrices and vectors may be time dependent.

One approach is to solve equation (1) in a quasi-static manner [14]. In this case the dynamic part of the equation is neglected ($du/dt = d^2u/d^2t = 0$); and the solution is reached iteratively. Within each iteration a huge set of linear algebraic equations must be solved. In problems with large deformation and contact interactions, the iteration often does not converge and a stable simulation over several minutes is nearly impossible to achieve. Otherwise, the dynamic equation has to be integrated with respect to time. The time integration of the equation can, in principle, be performed using implicit or explicit integration schemes. Using the implicit method, the solution has to be calculated iteratively at every discrete time step, like in the quasi-static problem. On the other hand, the explicit time integration can be performed without iteration and without solving a system of linear algebraic equations. This integration scheme is only conditionally stable; that is, only very small time steps lead to a stable solution. An estimation for the critical time step is made by $dt_{max} = dL / c$, where c is the maximal wave propagation speed in the medium and dL is the smallest element length. This equation leads to 10,000 time steps per second ($dt = 100\ \mu s$). These short time steps increase the computational effort but lead to a very stable contact formulation. The disadvantage of this method (that many steps have to be taken) is not so significant because each step is much less time-consuming than in an implicit algorithm.

3. EXPERIMENTAL DETAILS

An Experimental set-up was developed for the tube nosing process. The set-up was in a 100-ton UTM (see Fig. 2).

Fig. 2.

The die and the tube billet dimensions of the experiments are listed in Table 2. The SDK11 (JIS) die steel was chosen to manufacture the nosing punch and the die.

Table 2. Maximum reduction of response for different inner radius of baffle.

1.	Initial tube thickness (t_o)	2.8, 3.2, 3.4, 4.5 mm
2.	Initial tube diameter (d_o)	88.9 mm
3.	Initial tube length (L)	90, 135, 180 mm
4.	Semi-die angle (°)	30°, 45°, 60°
5.	Punch speed (mm/s)	0.00, 0.06, 0.15
6.	Fillet radius of the die (R)	5 mm/s
7.	Shoulder length of the die (CL)	30 mm

While the Aluminium alloy 1050-O was chosen as the working material. In order to obtain stress-strain curve, a tension test was performed at room temperature on a material testing system, with specimens having 25 mm in width, 50 mm in length and 2.75 mm in thickness which is cut from tube, the stress-strain relationship obtained from curve fitting is given by,

The tube billets were lubricated with machine oil. Experiments for the tube nosing process were carried out in the 100-ton UTM machine at room temperature. The practical nosing dies and the formed tubes as well as the tube billets are shown in Figs. 3 and 4, respectively.

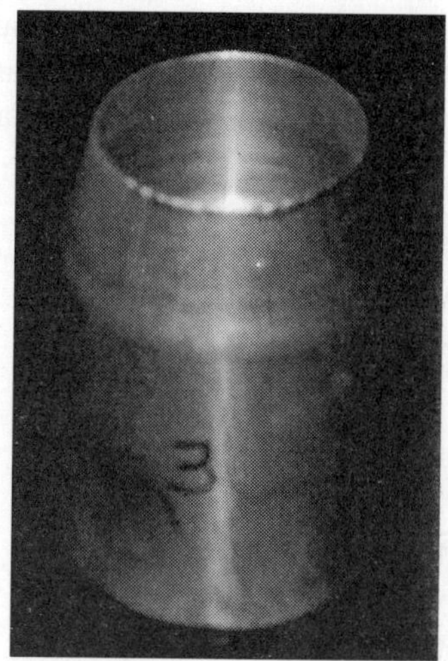

Fig. 3. Variation of maximum relative error in displacement (L2 norm).

4. ARTIFICIAL NEURAL NETWORK THEORY AND IMPLEMENTATION

Back propagation neural network (BPNN) has been used in the present work. Basic structure of back propagation neural network having input, hidden and output layers is shown in Fig. 1. Input layer receives information from the external sources and passes this information to the network for processing. Hidden layer receives information from the input layer, and does all the information processing, and output layer receives processed information from the network, and sends the results out to an external receptor. The input signals are modified by interconnection weight, known as weight factor w_{ij}, which represents the interconnection of ith node of the first layer to j^{th} node of the second layer. The sum of modified signals (total activation) is then modified by a sigmoidal transfer function. Batch mode type of supervised learning has been used in the present case, where, all input-output pattern sets are presented to the neural network one by one, and then adjusted using average gradient information. During training, the calculated output is compared with the target output and the mean square error is calculated. If the mean square error is more than a prescribed limiting value error, it is back propagated, i.e. from output to input then weights are further modified till the error is within a prescribed limit. Mean square error E, is calculated by the Eq. (1) as:

$$E = \frac{1}{2}\left(\sum_{p=1}^{p}\sum_{k=1}^{n}\left(d_k^p - c_k^p\right)^2 \right)$$

...(2)

where p is the number of pattern, n is the number of node in output layer, d_k^p is the desired output of kth node of pth pattern and c_k^p is the calculated output of kth node of pth pattern.

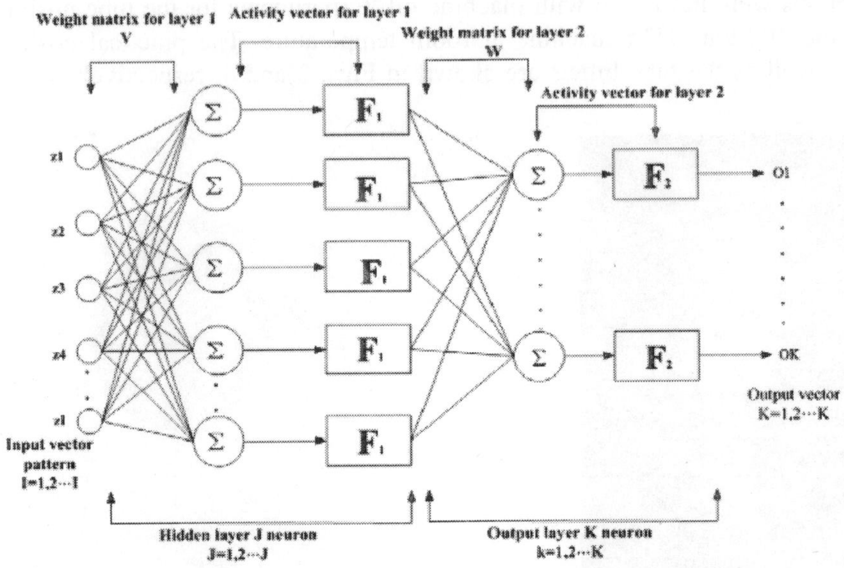

Fig. 4. Variation of maximum relative error in displacement (L$_2$ norm).

5. RESULTS AND DISCUSSION

The process parameters considered herein are initial tube thickness (t_0), initial tube diameter (d_0), initial tube length (L), semi-die angle, friction factor, punch speed, fillet radius of the die, and shoulder length of the die (CL), work hardening coefficient (n) and there effect on critical nosing ratio as well as final shape of the tube is simulated using explicit FEM. This data set is then used to obtain revere mapping model of ANN for the design of die. For a design engineer there are several situations in which die design has to carry out. Keeping this in mind the reverse mapping model is determine by considering six different design conditions and according to that combinations of input and output is varied.

These cases are as follows:

Case-1: Training ANN by considering flange shape as input and rest of the parameters as output.

Case-2: Training ANN by considering flange shape and fillet radius of die as input and rest of the parameters as output.

Case-3: Training ANN by considering flange shape and shoulder length as input and rest of the parameters as output.

Case-4: Training ANN by considering flange shape and friction factor as input and rest of the parameters as output.

Case-5: Training ANN by considering flange shape, friction factor and work hardening coefficient as input and rest of the parameters as output.

Case-6:   Training ANN by considering flange shape, friction factor, work hardening coefficient, shoulder length of the die as input and rest of the parameters as output.

For each of the case mentioned above 5 hidden nodes are found sufficient. Learning rate 0.3 is found optimum whereas momentum, factor found to be 0.09. With each of the hidden nodes sigmoid activation function used except out put node. With output nodes linear activation function is used. For different parameter combinations, the calculated error rates of the flange shape are below 0.0004. This value is acceptable in precision metal-forming processing. The simulation of the input parameter combinations in the above six cases reveals that the error rate of the flange shape is not critical, but the position of maximum error in each case is different. There fore during the design process engineers can use different input and output parameter combinations to meet their individual requirements.

CONCLUSION

In this paper DEM and ANN is integrated for the design of tube nosing process. Six different conditions or design cases are reverse mapped by generating input and output data pairs by using FEM. With all six cases learning error of ANN is found to be 0.0004 which is acceptable in precision forming.

REFERENCES

1. Guo N.C., Lou Z.J., Cui G., 1993, Analysis of tube sinking by an upper bound approach, Journal of Applied Mechanics. 12, 69-77.
2. Ruminski M., Lusksza J., Kusiak J. et al., 1998m Analysis of the effect of die shape on the distribution of mechanical properties and strain field in the tube sinking process, J. Mater Process Technol, 80-81: 683-689.
3. Park jj, Rebelo N., Kobayashi S, 1994, A new approach to preform design in metal forming with FEM. Int. J. Mach Tool Design Res, 27:71-79.
4. Zhu J., 1997, A new approach to preform design in shell nosing, J. Mater Process Technol, 63:640-644.
5. Reid S.R., Harrigan J.J., 1998, Transient effects in the qasi static and dynamic internal inversion and nosing of metal tube, Int J. Mech Sci, 40:263-280.
6. Dai K., Wang Z.R., 2000, A graphical description of shear stress distribution in the drawing of a thin wall tube with conical die, J. Mater Process Technol, 102:174-178.
7. Harrijan J.J. et al, 1995, Internal inversion and nosing of laterally constrained metal tubes, J. Mater Process technol, 134:343-394.
8. B.S. King, N.S. Kim et al, 1990, Computer aided perform design in forging of an airfoil section blade, International journal of Machine tools and Manufacture, Vol.32, 43-52.
9. N.S. KIM and S. Kobayshi, 1990, Perform design in H-shaped cross sectional axi-symmetric forging by the finite element method, International journal of machine tools and manufacture, 243-268.
10. D.H. KIM, D.J. KIM, et al, 1998, Process design of multi stage drawing using ANN, Journal of Korean society for technology of plasticity,127-138.
11. Dixt U.S., Chandra S., 1996, A neural snetwork based methodology for the prediction of roll force and roll torque in fuzzy form for cold flat rolling process, International Journal of Advance Manufacturing Technology, Vol.23, pp. 883-889.
12. ANSYS Corporation, 2002, ANSYS/LS-DYNA. Version 6.0, ANSYS Corporation, Southpointe.
13. O.C. Ziekiewicz, R.L. Taylor, 1991, The Finite Element Method, Vols, I, II, McGraw Hill: New York.
14. D.S. Malkus, T.J.R. Hughes, 1978, Mixed finite element methods-reduced integration techniques: a unification of concepts, Computer Methods in Applied Mechanics and Engineering. 15, 63-81.
15. S.P. Timoshenko, S. Winowsky-Krieger, 1959, Theory of Plates and Shells, 2nd Edition, McGraw-Hill, New York.

145

Finite Element Simulation of Metal Forming Process: With Spring Back Effects

H.S. Suresh AND K. Badari Narayana[1]

[1]Tata Consultancy Services Ltd., Salarpuria G.R. Tech Park White Field Road, Bangalore-560 066
[1] email: bn.kantheti@tcs.com

ABSTRACT

High strength materials used for various industrial applications in sheet and wire forms have formability problems as their ductility is limited; leading to large amount of dimensional mismatch in the form of spring back. The objective of the present work is to understand the spring back aspects in a typical staple manufacturing process and to perform a spring back analysis in ABAQUS/Standard. The various steps involved in the process are illustrated with an example of staple formation of a titanium wire using ABAQUS [1].

Keywords: Metal staple forming, Finite elements, Explicit, Implicit, Spring back and ABAQUS.

1. INTRODUCTION

Metal forming is the most commonly used manufacturing process. Sheet metal forming/stamping aspects are extensively used in aerospace, automotive and allied manufacturing industry. It is important to have good dimensional tolerances and repeatability in the components. A major concern in these processes is spring back. For this process to be cost effective, accurate predictions must be made of its formability. A typical forming process is affected by excessive downtime due to operational and maintenance issues. Often the produced components are out of specifications, increasing their rejection rate.

Spring back relates to the change in shape between the fully loaded and unloaded configurations the material encounters during a forming operation. This results in the formed component being out of tolerance and can create major problems in the assembly. Bending type of forming operations have been used widely in sheet metal forming industries to produce structural stamping parts such as braces, brackets, supports, hinges, angles, frames, channel and other sheet metal parts. The effect of the various process and material parameters on spring-back and spring-in are studied by a number of investigators [2-7]. Largely spring back has been correlated with an increase in normal anisotropy and decrease in strain hardening exponent [8] of the material. It is shown [9] that only advanced models like combined isotropic-kinematic reliably predict the spring back behavior.

Finite element is a powerful numerical technique that has been applied in the past many years to a wide range of engineering problems [10]. Although much FE analysis is used to verify the structural integrity of designs, more recently this method has been used to model fabrication processes. When modeling fabrication processes that involve deformation, such as sheet metal forming, the formation process must be evaluated in terms of stresses, strain states and contact issues of the bodies under consideration.

The objective of the present work is to understand the spring back aspects in a typical staple manufacturing process and to perform a spring back analysis in ABAQUS/Standard. The first stage of forming process is simulated in ABAQUS/Explicit; in the 2nd stage, the spring back analysis is performed in ABAQUS/Standard. The import capability is used to transfer the ABAQUS/Explicit results to ABAQUS/Standard.

2. STAPLE FORMATION

A typical staple forming process consists of a blank which is gripped between the forming head and the supporting guide, (Fig. 1) and is pushed down between two rollers. The wheels force the blank to take the shape of the head and the staple is made.

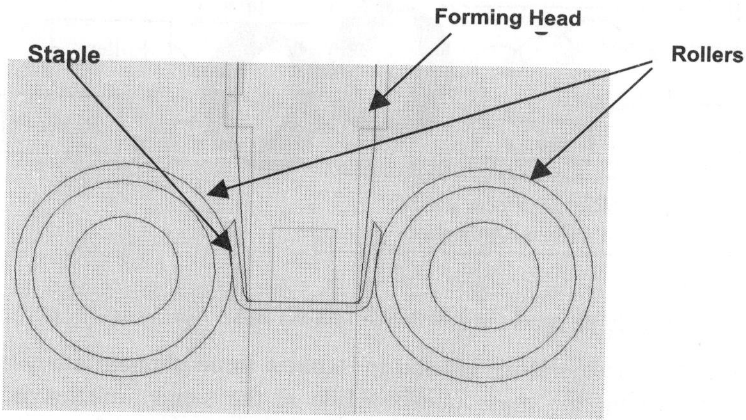

Fig. 1. Schematic of Staple Forming Process.

3. FINITE ELEMENT MODEL

Element Selection: For the metal forming simulation any of the three types of elements [11] namely, the membrane, shell and continuum element are considered. Membrane Computational efficiency and better convergence in contact analysis, but can not consider bending effect. Shell elements can capture the combination of stretching and bending, there by these elements give more number of degrees of freedom to capture accurate stress distribution. But, it takes a substantial amount of computational time and space. Where as the continuum are used where fully 3-D theory is needed to describe the deformation process. They can handle through thickness compressive straining resulting a large system of equations to be solved [12]. The present Finite Element (FE) Model representing the staple forming process is shown in Fig. 2. As stated earlier ABAQUS explicit and standard modules are used for the simulation. The head, rollers, gripper and the head backing surface are all assumed to be rigid (modeled with rigid elements, R3D4 in ABAQUS). The wire is modeled with hex elements (R3D8R in ABAQUS). SPRINGA in ABAQUS elements are used to model the springs. General contact algorithm with explicit solver is used to solve the first stage of the problem.

Material Data: High strain rates require high strain rate material models. Abaqus supports a large variety of material models. For most metal a popular choice is to use piece-wise linear plasticity in terms of true stress and strain description, the same is used in the current simulation.

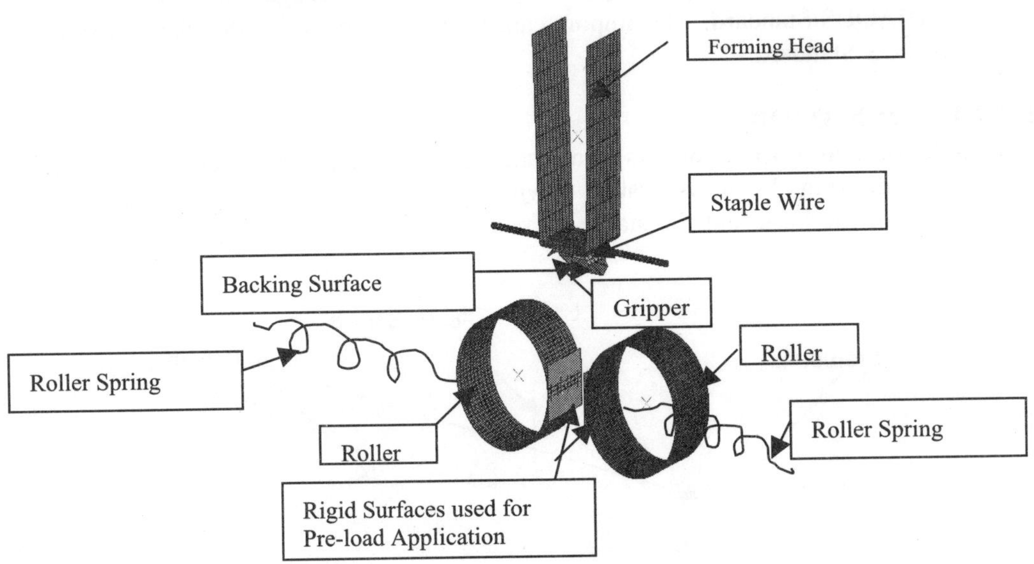

Fig. 2. FE Model of the Staple forming Head Assembly.

Mass Scaling: Variable mass scaling is used in explicit finite element analyses to increase the critical time step by modifying the mass matrix while at the same time leaving the rigid body translational behavior unaffected. Variable mass scaling of the wire is employed for speeding up the analysis. However, care is taken to keep the kinetic energy of the overall system as low as possible (less than 10%) compared to the total internal energy of the system. Certain elements may have hourglass modes that affect their behavior. An indication of whether hour glassing has any effect on the solution is the indicated through the artificial energy, variable ALLAE. The artificial energy should always be much less than the internal energy (generally less than 0.5%). Figure 3 shows the various energy contents of the system during the down ward motion of the forming head.

Spring back: Spring back refers to the shape discrepancy between the fully loaded and unloaded configurations. The stress-strain plot shown in Fig. 5 illustrating the spring back phenomenon. Upon unloading during a metal forming process there is elastic recovery, which is the release of the elastic strains and the redistribution of the residual stresses through the thickness of wire, causing spring back. Spring back causes changes in shape and dimensions that can create major problems in the assembly; hence spring back prediction is an important issue in metal forming process.

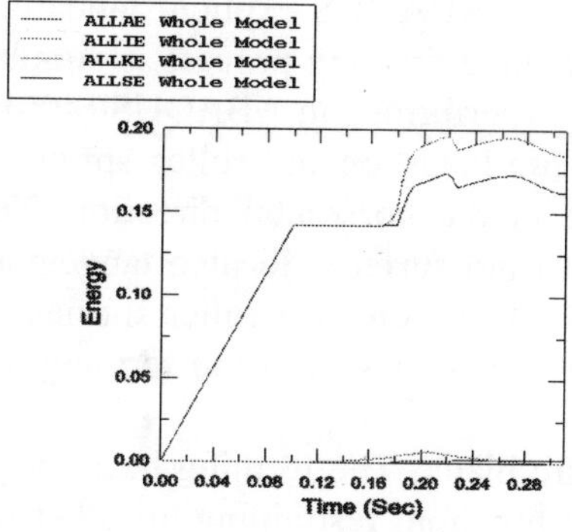

Fig. 3. Energy History Plot for Staple Simulation.

Fig. 4. Typical Shape of the Formed Staple from FEA.

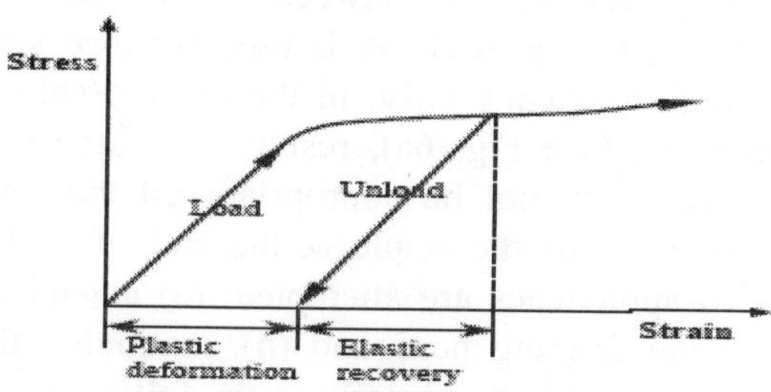

Fig. 5. Spring Phenomena.

Many factors could affect spring back in the process, such as variation in the mechanical properties of the material, sheet thickness, tooling geometric parameters and lubricant condition. A typical shape of the formed staple from present simulation is shown in Fig. 4.

4. SOLUTION PROCEDURES

There are two solution procedures generally used for the global equilibrium equation of metal forming. These are dynamic explicit and static implicit. Investigations have shown that explicit method is well suited for solving large sheet forming models, and implicit codes handle the spring back calculation very efficiently. Hence, recently researchers have adopted a new method of coupling the implicit and explicit methods to solve complex sheet metal forming processes to save design effort and production time. Narasimhan et al. [13] have used ANSYS/LS-DYNA explicit coupled with ANSYS implicit for spring back simulation. Finn et al. [14] combined the commercial codes LS-DYNA3D and NIKE3D for prediction of spring back in automotive body panels. Taylor et al. [15] discussed the numerical solution of sheet-metal forming applications using the ABAQUS general purpose implicit and explicit finite-element modules. In the present work also both explicit modules (for initial staple forming) and standard module (for predicting spring back) of ABAQUS are used.

5. RESULTS AND DISCUSSION

FE simulation: The simulation program used in this investigation is ABAQUS explicit/standard. As stated earlier the analysis is performed in two stages. In the first stage forming process is simulated using ABAQUS/Explicit; in the 2nd stage the spring back analysis is performed in ABAQUS/Standard.

Load Step1: In this step the springs are pre-loaded. The pre-loads on the roller springs are applied as follows. The spring free ends are allowed to move in the horizontal direction. These spring ends are displaced by a specified distance (of 2") against rigid surfaces located tangential to the rollers. This displacement produces a required Pre-load (of 1.24 lbs) on each roller springs. The gripper is also moved vertically upwards to allow the gripping of the wire against the forming head. The forming head and the two rollers are completely restrained.

Load Step 2: In this step, the staple is formed out of the wire blank. The forming head, gripper and head back surface are allowed to move only in the vertical direction, restraining all other dofs. The free ends of the springs are completely restrained (retaining the pre-load displacement). The gripper and the forming head back surface are also moved with the give velocity profile so that all the surfaces of the forming head and the gripper move together gripping the wire. General contact algorithm is used to define contact interactions between various parts. ABAQUS Explicit solver is used to solve the problem. An event time of .1 sec is used for load step1 and .21 sec for the load step2. It is observed that the high-frequency noise in the explicit solution caused primarily by the oscillations of the blank's free end (see Fig. 6a), results in unstable staple tip condition, in this condition any measurement taken may not be appropriate till the tip of staple reaches a stable condition. Typical deformation shape of the staple at the end of 0.31 sec is shown in Fig.7. To arrive at a stable condition two approaches are attempted: (a) extending the explicit solution time beyond down ward motion of the forming head and (b) continuing the analysis with ABAQUS/ standard. These two approaches are mentioned briefly in the following sections.

Extending the ABAQUS/explicit analysis: The staple FEA analysis is restarted from step 2 (forming head down stroke) and continued beyond .31 second. The analysis is continued for an additional duration of 0.2 second. As can be seen, (Fig. 6) there is continuation of tip oscillations till the end of analysis. However, reduction in the amplitude of oscillations can be observed. A swing of 0.0027 inch is observed at end of analysis (.51 sec). This translates to 0.0054" for the full model.

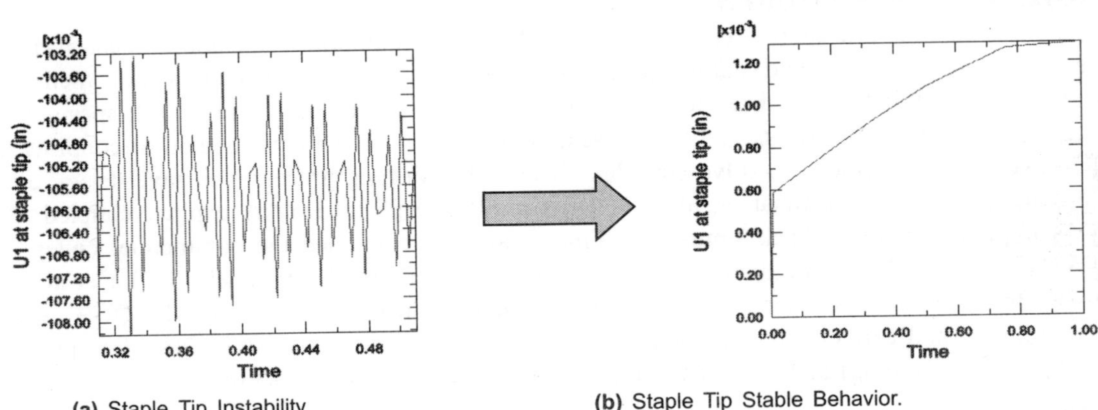

(a) Staple Tip Instability   (b) Staple Tip Stable Behavior.

Fig. 6. Displacement responses of the staple tip during the ABAQUS explicit and standard analysis.

Analysis with ABAQUS/Standard: In the manufacturing process the part is removed after the forming has been completed and the material is free to spring back into an unconstrained state. To understand the final shape after this physical effect, we perform a spring back analysis in ABAQUS/ Standard. The analysis is continued beyond 0.31 sec in ABAQUS/STANDARD. For this purpose, only the staple model/results information are imported from the previous ABAQUS/Explicit run to ABAQUS/STANDARD using *IMPORT command in ABAQUS. The BCs are applied to the staple model and the analysis is carried out in ABAQUS/STANDARD without any loading. Finally, the spring back of the material is performed in ABAQUS/Standard. Staple deformed shape shown in Fig.7. (at the end of .31 sec) imported to ABAQUS/Standard. X = 0 (Symmetry BC), Y = Z = 0 at two points in the plane (to arrest rigid motion). When the current state of a deformed body in an explicit dynamic analysis is imported into a static analysis, the model will not initially be in static equilibrium. The initial out-of-balance forces are applied to the deformed body in dynamic equilibrium to achieve static equilibrium. During the process of removing the out-of-balance forces, the body will deform further and a redistribution of internal forces will occur, resulting in a new stress state. (This is essentially what occurs during "spring back," when a formed product is removed from the work tools.)

Figure 6b shows the additional deformation of the staple tip from the ABAQUS/standard run. As can be seen from the figure, the staple tip deforms further under initial out-of-balance forces and reaches equilibrium position. An additional deformation of 0.00129 inch is observed at the end of

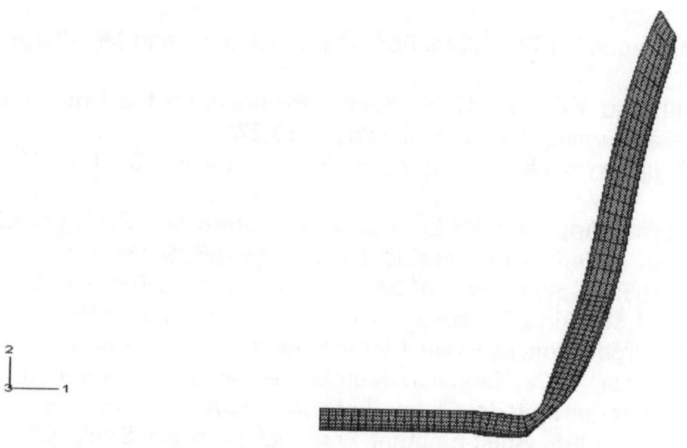

Fig. 7. Staple deformed shape at the end of explicit analysis.

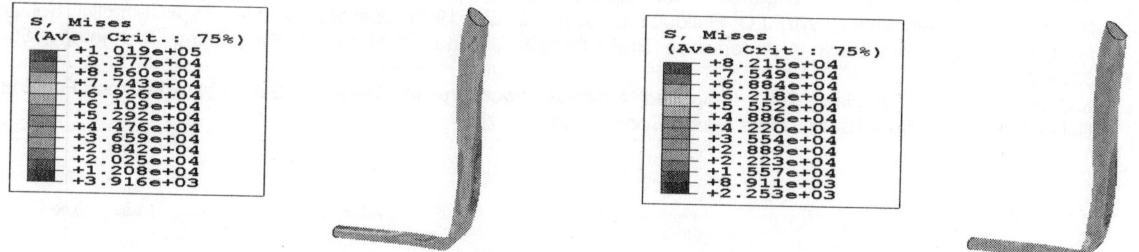

Fig. 8. Von Mises Stress Distribution in the staple at the end of explicit.

Fig. 9. Von Mises Stress Distribution in the staple at the end of standard run.

analysis. For the full model, this translates to 0.00258 inch. Hence at location 1 in the tip deformation the staple will be: value at the end of explicit run + .00258 inch = 0.14578 inch. Same analogy is extended for measurement of final dimension at location 2 of the staple. Typical Von mises stress plots at the end of explicit and standard runs are given below (Figs. 8 and 9).

CONCLUSION

ABAQUS explicit/implicit has shown a right tendency in simulating the staple forming process. The dynamic explicit procedure is more time efficient, while the implicit procedure comes out with better results. The simulation in ABAQUS given a good predication of the spring back behaviour as indicated by the comparison of simulated and experimentally estimated staple features for the 2 locations considered; (with in 8.5%) with measured value being higher than the current FEA results. Varying the friction coefficient between 0.05 and 0.1 has no effect on the final results.

REFERENCES

1. ABAQUS Analysis User's Manual, 2001, Ver 6.4.
2. Crisbon Delfina Joseph, 2003, Experimental measurement and finite element simulation of spring back in stamping aluminum alloy sheets for auto-body panel application, A Ph.D. Thesis, Department of Mechanical Engineering, Mississippi.
3. Z.T. Zhang and D. Lee, 1995, Effect of Process Variables and Material Properties on the spring back Behavior of 2D-Draw Bending Parts, Automotive stamping technology, 11-18.
4. H. Livatyali and T. Altan, 2001, Prediction and Elimination of Spring back in Straight Flanging Using Computer Aided Design Methods, Part 1: Experimental Investigation, Journal of Materials Processing Technology, 117, 262-268.
5. W.F. Hosford and R.M. Caddell, 1993, Metal Forming: Mechanics and Metallurgy, 2nd Ed. New York, NY: Prentice-Hall.
6. R.G. Davies, C.C. Chu, and Y.C. Liu, 1985, Recent Progress on the Understanding of Spring back, Computer Modeling of Sheet Metal Forming Process: Theory, 259-271.
7. Sen Jiang M S, 1997, spring back investigations, A Ph.D. thesis, Dept. of Mechanical Engineering, Ohio state Univ.
8. D.K. Leu, 1997, A Simplified Approach for Evaluating Bendability and Spring back in Plastic Bending of Anisotropic Sheet Metals, Journal of Materials Processing Technology, 66, 9-17.
9. W. Kubli, J. Reissner, 1995, Optimization of Sheet-Metal Forming Process Using the Special-Purpose Program AUTOFORM". Journal of Materials Processing, Technology, 50, 292-305.
10. J.N. Reddy, 1993, An Introduction to Finite Element Method, Mc Graw Hill, Inc., 2nd Ed.
11. Hoon Huh, Tae Hoon Choi, 1999, Modified Membrane Finite Element Formulation for Sheet Metal Forming Analysis of Planar Anisotropic Materials," *International Journal of Mechanical Sciences,* 42, 1623-1643.
12. M. Kawka, A. Makinouchi, 1995, Shell-Element Formulation in the Static Explicit FEM Code for the Simulation of Sheet Stamping". *Journal of Materials Processing Technology,* 50, 105-115.
13. Narkeeran Narasimhan, Michael Lovell, 1999, Predicting Spring back in Sheet Metal Forming: An Explicit to Implicit Sequential Solution Procedure," Finite Elements in Analysis and Design, 33, 29-42.
14. M.J. Finn, P.C. Galbraith, L. Wu, J.O. Hallquist, L. Lum, T.L. Lin, 1995, Use of a Coupled Explicit-Implicit Solver for Calculating Spring-Back in Automotive Body Panels, Journal of Materials Processing Technology, 50, 395-409.
15. K.C. Ho, J. Lin, and T.A. Dean, 2001, Integrated Numerical Procedures for Simulating Spring back in Creep Forming Using ABAQUS, ABAQUS UK Users Group Conference.

146

Simulation of Inclusion Removal Process in a Multi-Strand Billet Caster Tundish

P. Srinivasa Rao, Pradeep K. Jha and Anupam Dewan

Department of Mechanical Engineering Indian Institute of Technology, Guwahati
Guwahati-781039, Assam, India

ABSTRACT

The Reynolds averaged Navier-Stokes equations have been solved to obtain a steady state three dimensional velocity field inside the multi-strand billet caster tundish using the standard K-ϵ model of turbulence. These steady flow fields so obtained are then used to predict the removal of alumina inclusions from molten steel by solving the inclusion transport equation which includes the effect of height and position of the flow modifiers in the tundish and size of the inclusion particles. Inclusion particles that touch the bottom and sidewalls are assumed to be reflected back and at the top surface they are assumed to be trapped. It has been found that with the increase in the height of the dams, the inclusion removal tendency increases whereas the shifting of the dams towards the outlets decreases the inclusion removal. An increase in the inclusion removal has been found with increase in the inclusion particle size.

Keywords: Continuous casting tundish, inclusion removal, particle size, dams.

1. INTRODUCTION

Inclusions in steel have detrimental effects on its mechanical properties. Non-metallic inclusions mainly come from sources like de-oxidation products, re-oxidation products, slag entrapment, exogenous inclusions and many chemical reactions [1-2]. Tundish is the last device before the entry of the molten steel into the continuous casting mold. Inclusions not being removed from the tundish will enter the mold, thereby reducing the quality of the slabs/billets being produced. The inclusion removal mechanism in a tundish mainly depends upon the flow behaviour of steel inside the tundish, which primarily is dependent upon different parameters like configuration of the tundish, use of flow modifiers to alter the flow inside the tundish, shape and size of the inclusion particles.

A number of literatures on the inclusion removal mechanism are reported [3-5]. Most of them discuss about the inclusion removal mechanism in single strand slab caster tundish. Study of multi-strand billet caster tundish has not been carried out so far. The presence of outlets more than one makes the study of multi-strand tundish more important as the inclusions passing through each outlet will determine the quality of the billets coming through each strand. In the present study, a multi-strand (six strand) industrial size delta shape tundish has been considered for the study of inclusion removal mechanism. The effect of flow modifiers, especially dams of different combinations of height and different positions have been used to see its effect on the inclusion removal through each outlet. Effect of size of the inclusion particle has also been investigated.

2. PHYSICAL DESCRIPTION OF THE PROBLEM

The geometry of the multi strand tundish used in present study is shown in Figure 1a. Half of the tundish is shown because of the symmetry about the inlet plane. The depth of the tundish is 572

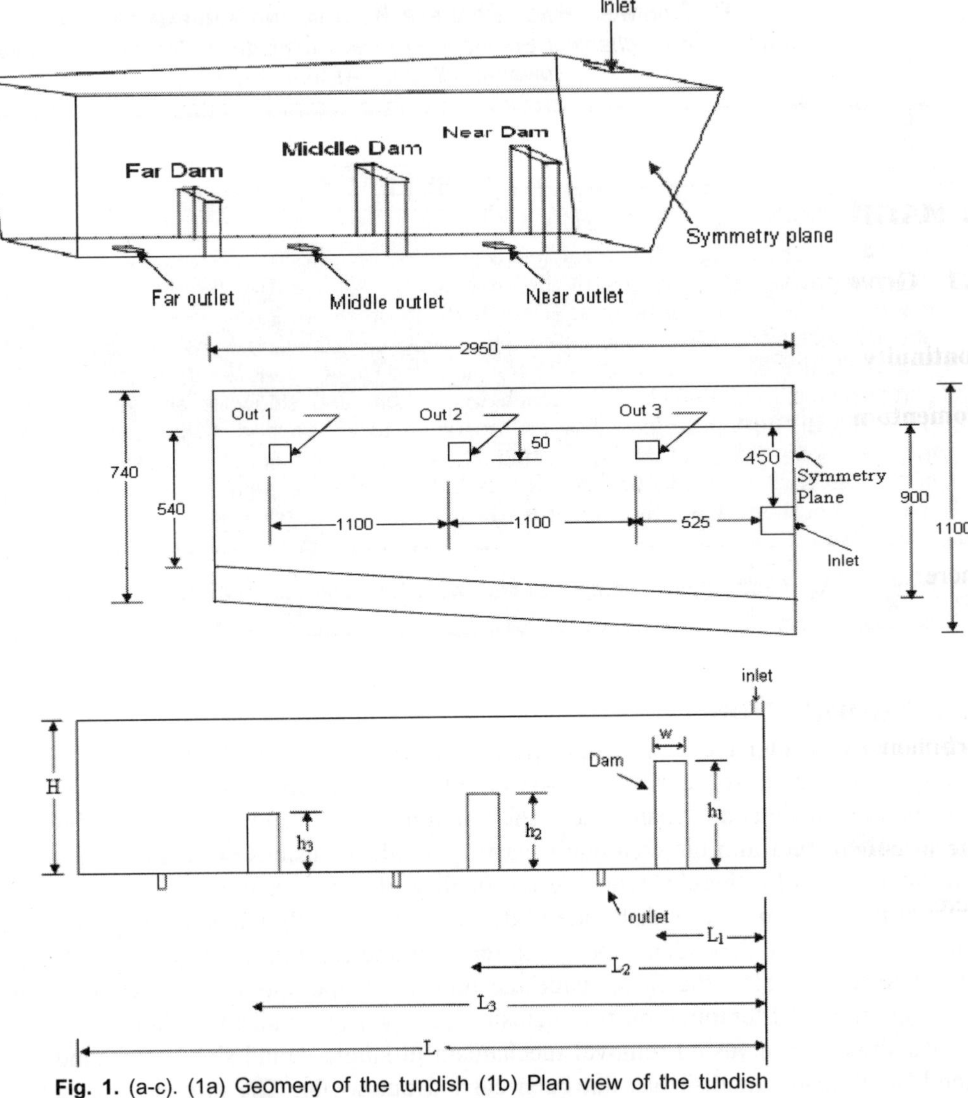

Fig. 1. (a-c). (1a) Geomery of the tundish (1b) Plan view of the tundish
(1c) Description of the physical parmeters.

mm with the size of the inlet as 25 mm × 50 mm and all other dimensions are shown in a plan view in Fig. 1 b. The outlets are placed 50 mm away from the wall. Fig. 1 c shows the different parameters like the height and the distance of the dams which have been varied in the present study. The different value of these parameters are presented in Table 1.

Table 1. The variation of dam heights and positions in the present study.

Heights of Dams				Distance of the dams from the inlet			Case	
Height	h_1	h_2	h_3	Positions	L_1	L_2	L_3	
H1	200	150	100	a	250	950	1950	H1a	H1a,H 1b, H1c
				b	300	1100	2150	H1b	
				c	350	1250	2350	H1c	
H2	250	200	150	----Do----				H2a, H2b, H2c	
H3	300	250	200					H3a, H3b, H3c	

3. MATHEMATICAL FORMULATIONS

3.1 Governing Equations

Continuity equation

$$\frac{\partial u_i}{\partial x_i} = 0 \qquad \qquad \qquad \dots(1)$$

Momentum equation

$$\frac{\partial}{\partial x_i}\rho u_i u_j = -\frac{\partial P}{\partial x_i} + \frac{\partial}{\partial x_i}\left(\mu_{eff}\left(\frac{\partial u_i}{\partial x_j} + \frac{\partial u_j}{\partial x_i}\right)\right) \qquad \dots(2)$$

where

$$\mu_{eff} = \mu_0 + \mu_t = \mu_0 + \rho C_\mu \frac{k^2}{\varepsilon}$$

Turbulent kinetic energy

$$U_j \frac{\partial k}{\partial x_j} = D_k + P - \varepsilon \qquad \qquad \dots(3)$$

Rate of dissipation of k

$$U_j \frac{\partial \varepsilon}{\partial x_j} = D_\varepsilon + C_1 P \frac{\varepsilon}{k} - C_2 \frac{\varepsilon^2}{k} \qquad \dots(4)$$

where,

$$\overline{u_i u_j} = \frac{2}{3} k \delta_{ij} - \nu_t \left(\frac{\partial U_i}{\partial x_j} + \frac{\partial U_j}{\partial x_i}\right)$$

$$\nu_t = 0.09 \frac{k^2}{\varepsilon}, \quad \mu_{eff} = \rho \nu_t + \mu \quad D_\phi = \frac{\partial}{\partial x_j}\left[\left(\mu + \frac{\mu_t}{\sigma_\phi}\right)\frac{\partial \phi}{\partial x_j}\right]$$

$$P = -\overline{u_i u_j} \frac{\partial U_i}{\partial x_j}$$

Steady state velocity field is obtained by solving equation (1) to (4) using finite volume method with the use of CFD software FLUENT.

3.2 Inclusion Removal Model with Trajectory Calculation

Inclusion trajectories were calculated using a Lagrangian particle tracking method, which solves a transport equation for each inclusion as it travels through the previously calculated steady state flow field. The mean local inclusion velocity components, u_p needed to obtain the particle path, are obtained from the following force balance, which includes drag and buoyancy forces relative to the steel.

$$\frac{du_p}{dt} = F_D \left(u - u_p \right) + \frac{g_x \left(\rho_p - \rho \right)}{\rho_p} + F_x \qquad \ldots(5)$$

$F_D \, (u - u_p)$ is the drag force per unit mass of the particle.

$$F_D = \frac{18\mu}{\rho_p d_p^{\,2}} \frac{C_D R_e}{24} \qquad \ldots(6)$$

Here, u is the fluid phase velocity, u_p the particle velocity, μ is the molecular viscosity of the fluid, ρ is the fluid density, ρ_p is the density of the particle, and d_p is the particle diameter. Re is the relative Reynolds number, which is defined as

$$R_e = \frac{\rho d_p \left| u_p - u \right|}{\mu} \qquad \ldots(7)$$

3.3 Assumptions and Boundary Conditions

Flow is assumed to be steady and incompressible, air and gas entrainment by incoming metal stream is neglected, the surface of tundish is considered to be perfectly flat and slag depth was considered to be insignificant, the generation of inclusions in the tundish due to any erosion of refractory or any chemical reactions like deoxidation etc. was ignored.

The 3-D domain of this tundish is divided into about 54000 cells, making a finer mesh in the zone of the incoming and outgoing liquid jet in order to get high resolution in these complex zones and also to visualize in more details the effects of velocity and turbulence gradients.

'All velocities are set to zero at wall (no slip condition). A log-law wall function is employed to deduce K,ε, wall shear stress, and all velocity components parallel to the boundary at the first computational grid point adjacent to this wall. The velocity components normal to planes of symmetry are set equal to zero. The upper top surface is assumed as a free surface. At the inlet, the velocity is 1.4 m/sec. Turbulence intensity at inlet is specified as 2%.

For inclusion trajectory removal cases, inclusions touching the sidewall are assumed to reflect. This assumption should be investigated with further work, as inclusions might alternatively stick to walls. Inclusions that are reaching the top surface are assumed to be trapped and at the outlet they are assumed to be escaped.

4. RESULTS AND DISCUSSIONS

Variation of percentage removal of inclusions through each outlet and overall removal with varying dam heights at the different outlet positions respectively, are presented with the help of Fig.2(a-c). It is clear that with the use of dams before the outlets, there is a sharp increase in the inclusion removal from the tundish. It can also be seen that as the height of the dams are increased, the percentage removal of the inclusions increases. This is true for the dams placed at any of the three positions. This is because the inclusions get the tendency to flow towards the top surface because of the placement of the dams. The tendency of the particles to flow through the outlet is obstructed due to the presence of the dams before the outlets. Hence, the inclusion removal increases, as the inclusions having the tendency to go upward are trapped at the top free surface. It can be seen from Fig. 2a to Fig. 2c that for constant height of dams, say H1, the inclusion removal is seen to be

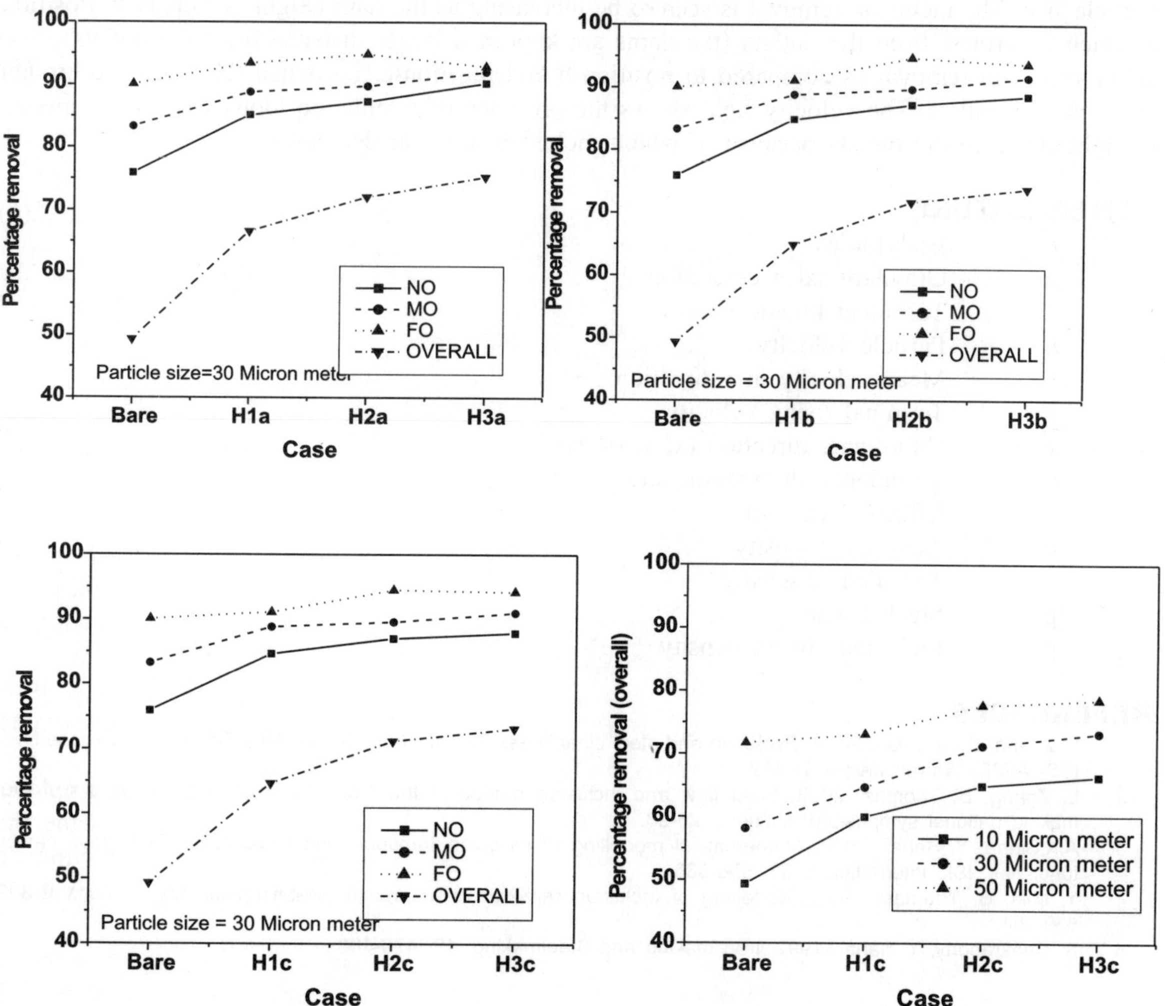

Fig. 2. (a–d). (a–c) Variation of percentage inclusion removal from each outlet and overall removal with increase in dam height at three different postitions (d) Variation of overall removal with varying particle size.

higher when the dams are at position-a than that at position-b or position-c. This suggests that if thedams are shifted towards the outlets at position-b or position-c, the inclusion removal tendency decreases. Fig. 2d shows the variation of overall percentage removal of inclusions with changing height of dams for different inclusion cluster sizes. The graph shows when the dams are at Position-c(other two positions have not been shown to save the space). It is seen that as the size of the inclusion particle increases, the inclusion removal also increases. This is true for the dams of all the three height combinations i.e. H1, H2 and H3. The reason may be attributed to the increase in the terminal rise velocity of the inclusion particles with increase in the particle diameter.

CONCLUSION

Numerical investigation of inclusion removal process in a multi-strand billet caster tundish has been carried out. The parameters which has been varied are dam height, dam positions and the inclusion particle size. The inclusion removal is seen to be increasing as the dam height is increased. Position-a which is farthest from the outlets (the dams are kept at a larger distance from the outlets) gives better inclusion removal as compared to position-b and position-c, i.e. when the dams are brought closer to the outlets. The velocity field shows the presence of circulation loops when the dams are brought closer to the outlets because of which inclusion removal decreases.

NEMENCLATURE

F	Bodyforces
g	Gravitational acceleration
K	Turbulent kinetic energy
u_p	Particle velocity
u_i	Mean velocity in i direction
V_s	Terminal rising velocity
x_i, x_j	Coordinate direction (x, y, or z)
ε	Turbulence dissipation rate
μ_{eff}	Effective viscosity
μ_0	Aaminar viscosity
μ_1	Turbulent viscosity
ρ	Steel density
ρ_c	Inclusion cluster density

REFERENCES

1. L. Zhang, B. Thomas, 2002, Evolution and steel cleanliness-Review, 85th Steelmaking Conference Proceedings, ISS- AIME, Warren dale. 431-452.
2. L. Zhang, B. Thomas, 2003, Fluid flow and inclusion motion in the Continuous casting strand, XXIV steel making national symposium, Mexico. 26-28.
3. A.K.Sinha, Y. Sahai, 1993, Mathematical modelling of inclusion transport and removal in Continuous casting tundishes, ISIJ international. 33, 556-566.
4. Y. Miki, B. Thomas, 1999, Modelling of inclusion removal in a tundish, Metallurgical Transactions B.30B, 639-654.
5. S. Chakraborty, Y. Sahai, 1992, Iron making and Steelmaking. 19, 479-488.

147

Simulation of Thermoelastic Instability in Disc Brakes Through Finite Element Programming

R. Vasudevan[1] and M.M. Mayuram[2]

[1]Research Scholar, [2]Professor Department of Mechanical Engineering
Indian Institute of Technology, Madras, Chennai-600036 India

ABSTRACT

A transient analysis of the thermoelastic instability of disc brakes with frictional heat generation is analyzed using the finite element. The numerical simulation of the thermoelastic behavior is obtained for the short or emergency brake application by solving the heat and the elastic problem as the coupled one. The temperature distribution on the friction surfaces and along the thickness of the disc and the pads are computed. The computational results are presented for the distribution of the temperature and also the contact pressure at both friction surfaces.

Keywords: Disc brakes, Sliding, Thermo Elastic Instability, Contact pressure, FEM.

1. INTRODUCTION

Brakes are one of the most important safety devices in automobiles. Development of brakes has constantly focused on increasing braking power and reliability. A significant factor affecting braking performance is the temperature rise in the brake components. Heat dissipation in disc brakes are much more effective than drum brakes and are increasingly being used in new class of vehicles. Disc brakes are exposed to large thermal stresses during routine and extraordinary thermal stresses during hard braking. A temperature of the order 900^0C at the disc surface can generate within a fraction of a second under such conditions. Such high temperatures and the resulting thermal stresses that accompany them produce a number of disadvantageous effects. The main effects are thermal shock that generates surface cracks, large amounts of plastic deformation in the brake rotor, thermally induced judder and friction induced squeal in braking. The major difficulty of the thermal problem is to determine how much is the heat generated and how it is distributed in the two components namely the disc and the pad, during the transient state of the contact. Therefore it is important to predict the temperature rise of a given brake system and assess its thermal performance in the early design stage.

The contact pressure and the temperature on the friction surfaces are greatly influenced by the frictional heating, thermal deformation and elastic contact in sliding contact systems. If the sliding speed is excessively high, the coupled thermal and mechanical behaviors can be unstable which leads to localized regions of high temperature contact, called "hot spots" on the sliding interface [1, 2].This phenomenon is known as Thermoelastic Instability (TEI). TEI can affect the tribological performance of the sliding components by causing changes in frictional and material properties which lead to increased power consumption, excess vibration, heat checking (formation of radial cracks caused by high thermal stresses), and many other problems. [3].

Sliding system such as the brakes and clutches experience intense period of operation with varying sliding speed. So, a numerical method to simulate the behavior of the coupled transient thermoelastic contact problem is the most realistic approach. The first numerical simulation was obtained by Keddedy and Ling [4] for aircraft type multi-disc systems. The thermo-mechanical phenomena of each disc are assumed to be symmetric about the disc's midplane. The analysis of temperatures and stresses of a steel multi disc wet clutch was performed by Zagrozki [5]. Here it was assumed that the distribution of normal pressures on friction surfaces is uniform and has no relation with that of temperatures. The numerical computation under this assumption gives rise to the under estimation of thermal stresses in the discs. The shortcomings were overcome by the Zagrodzki [6] who performed a transient thermoelastic analysis for axissymmetric multidisc clutches and brakes without making the above said assumption. A two dimensional thermoelastic contact problem of a stationary layer between two sliding layers with frictionally excited thermoelastic instability was investigated by Zagrodzki [7]. A three dimensional analytical model was developed by Gao and Lin [8] in which the brake pressure is assumed to be distributed uniformly over the contact area of the in-board and out-board surfaces. A finite element model for an axisymetric geometry was developed by Choi and Lee [9] to investigated the TEI during the dragging process.

The consideration that the contact conditions and the frictional heat flux transfer are independent of the angular direction θ, may lead to pseudo thermal elastic distortion on the surface of the disc. Also the consideration of axisymmetric problem leads to the improper assumption since even though the geometry is axisymmetric, the contact load conditions and the thermal boundary conditions are non-axisymmetric. The assumptions of uniform brake pressure on both the contact surfaces of the disc and the two pads and the symmetry of the temperature distribution with respect to the central plane of the disc lead to the inaccurate temperature distribution at the disc surface.

In the present study, the finite element code has been developed using MATLAB for solving the three dimensional unsteady state heat conduction equations in cylindrical coordinate system with the convection boundary conditions for the non-axisymmetric disc. This can be achieved by solving the coupled heat and elastic finite elements. The single stop barking condition is considered in the numerical computation of the transient thermoelastic behavior of disc brakes.

2. FORMULATION OF THE PROBLEM

The present study aims to analyze a transient thermoelastic contact problem of the disc brakes with frictional heat generation using the finite element method. The numerical simulation for the thermoelastic behavior of disc brake is to be obtained for the single stop condition. It involves two types of problem.

1. Thermal problem, 2. Elastic problem.

2.1 THERMAL PROBLEM

This module involves finding the temperature distribution and the heat flux distribution in the disc at various locations. It can be found by solving the unsteady state heat conduction equation in the cylindrical coordinates with the appropriate boundary conditions.

The unsteady state heat conduction equation is given as

$$\frac{1}{r}\frac{\partial}{\partial r}\left[rk_r\frac{\partial T}{\partial r}\right]+\frac{1}{r^2}\frac{\partial}{\partial\theta}\left[k_\theta\frac{\partial T}{\partial\theta}\right]+\frac{\partial}{\partial Z}\left[k_z\frac{\partial T}{\partial Z}\right]=\rho c\frac{\partial T}{\partial t} \qquad ...(1)$$

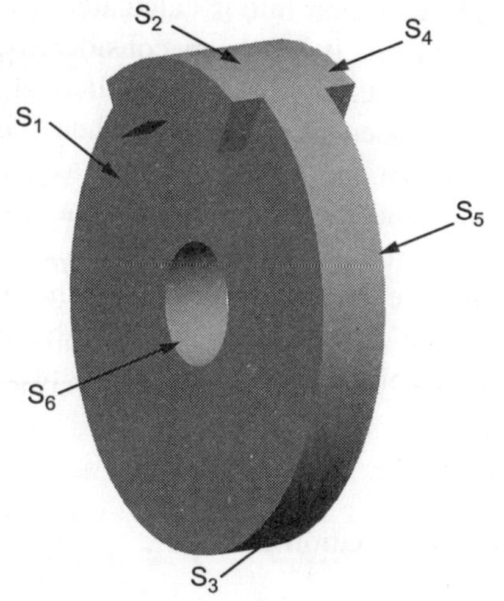

Fig. 1. Disc brake and pad model.

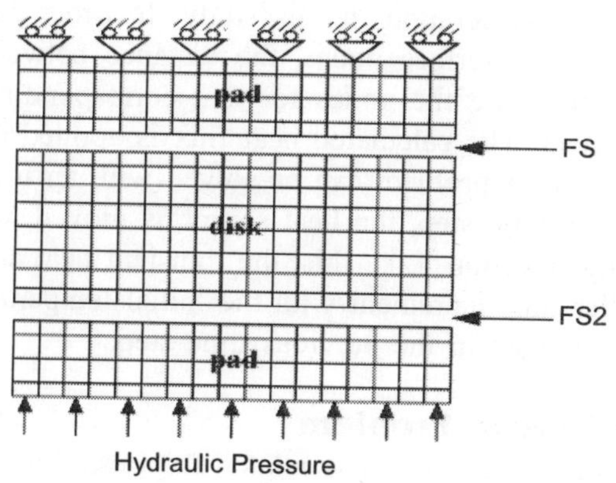

Fig. 2. Elastic Finite Element Model of disc and pad.

Initial conditions: $T(r, Z, \theta, 0) = T\alpha$ on S_1...(2)

Initial temperature = 20° C.

After applying the Galerkin method, the element matrices

$$[C]\dot{T}+[K]T=\{F\} \qquad ...(3)$$

Where

$$[c]=\int_V \rho c N_i N_j dv \qquad ...(4)$$

$$[k]=\int_V\left\{k_r\frac{\partial N_i}{\partial r}\frac{\partial N_j}{\partial r}+\frac{k_\theta}{r^2}\frac{\partial N_i}{\partial\theta}\frac{\partial N_j}{\partial\theta}+k_z\frac{\partial N_i}{\partial Z}\frac{\partial N_j}{\partial Z}\right\}dV+\int_{S_1}N_iN_jh_1ds_1+\int_{S_3}N_iN_jh_3ds_3+\int_{S_5}N_iN_jh_5ds_5 \qquad ...(5)$$

$$\{f\}=\int_{S_1}N_jh_1T_\alpha ds_1+\int_{S_2}\gamma\mu p\omega rN_j ds_2+\int_{S_3}N_jh_3T_\alpha ds_3+\int_{S_4}\gamma\mu p\omega rN_jh_4 ds_4+\int_{S_5}N_jh_5T_\alpha ds_5 \qquad ...(6)$$

Where N_i and N_j the shape functions which are defined for the iso parametric eight nodded brick element in the cylindrical coordinates.

By using the backward difference technique for the derivative term

$$\dot{T} = \frac{T^{t+\Delta t} - T^t}{\Delta t} \qquad \qquad ...(7)$$

and substituting in (3), we have

$$\{[C] + \Delta t[K]\}\{T\}^{t+\Delta t} = \Delta t\{F\} + [C]\{T\}^t \qquad ...(8)$$

To solve the equation (8), the transient finite element technique is used. During the simulation, the heat fluxes are assigned to the elements within the contact zone at every time step. The location of the contact zone at each time step is to be determined by the relative position of the disc and the pad from the calculation of the deceleration rate of the disc. Deceleration rate is calculated using the time taken to stop the vehicle and the initial velocity of the wheel. It has been considered as the uniform deceleration in this study. In order to make sure that the heat flux is applied at every contact or interface nodes, the distance between the adjacent nodes is calculated. Then the time taken to rotate the nodes for that corresponding distance between the nodes is taken as the time increment. The calculated heat flux is applied to the FE model and the three dimensional transient heat transfer problem can be solved with an immovable heat source at that particular time step. At the next time step, the heat source is moved to the neighboring elements according to the relative sliding direction. Like this, the transient heat conduction problem is repeatedly solved by using the smaller time increment with the initial temperature distribution corresponds to the temperature that was obtained in the previous time step.

2.2 Elastic Problem

The mechanical strain is related to stress through a constitutive equation as

$$\{\sigma\} = [D]\{\in^{me}\} \qquad ...(9)$$

where [D] is the material property matrix.

The total strain, the sum of the mechanical and thermal strains, is given by

$$\{\in\} = [\in^{me}] = \{\in^{th}\} \qquad ...(10)$$

where superscripts me and *th* denote mechanical and thermal strains, respectively.

Eq (9) becomes

$$\{\sigma\} = [D]\{\{\in\} - \{\in^{th}\} \qquad ...(11)$$

where $\{\sigma\} = \{\sigma_r\ \sigma\theta\ \theta z\ \sigma_{r\theta}\ \sigma_{\theta z}\ \sigma_{zr}\}$, $\{\in\} = \{\in_r\ \in_\theta\ \in_z\ \gamma_{r\theta}\ \gamma_{\theta z}\ \gamma_{zr}\}$

For an isotropic material, temperature change results in a body expansion or shrinkage but no distortion. In other words the temperature change affects the normal strains but no shear strains. Thus the thermal strain vector is expressed as

$$\{\in^{th}\} = \{\alpha\Delta T\ \alpha\Delta T\ \alpha\Delta T\ 0\ 0\ 0\}$$

in which α is the coefficient of thermal expansion and ΔT indicates the temperature change.

The total strain is expressed in terms of the nodal displacements as

$$\{\in\} = [B]\{d\} \qquad ...(12)$$

in which [B] is the kinematics matrix.

Substituting (12) into (11), we have $\{\sigma\} = [D][B]\{d\} - [D]\{\in^{th}\}$ $\qquad ...(13)$

Applying the weighted residual technique to eq.(13) will result in the following equation.

$$[K]\{d\} = \{F^{th}\} + \{F^{me}\} \qquad ...(14)$$

where the element stiffness matrix for elasticity is given as

$$\left[K^e\right] = \int_{\Omega^e} [B]^T [D][B] d\Omega \qquad \text{...(15)}$$

$\{F^{th}\}$ and $\{F^{me}\}$ are the thermal and mechanical force vectors which are denoted as

$$\{F^{Th}\} = \int_{\Omega} N^T N d\Omega$$

$$\{F^{me}\} = \int_{s} N^T N dS$$

The elastic problem is solved by using the constitutive equation. During the numerical modeling, special attention is required on satisfying the continuity of normal displacements on the contact surface and the overlap conditions. For satisfying these conditions, the following conditions of displacements and stresses are imposed on each pair of nodes on the interface.

$$w_i = w_j \text{ when } P > 0; \ w_i \neq w_j \text{ otherwise} \qquad \text{....(16)}$$
$$\sigma_{zj} - \sigma_{zj} \text{ when } P > 0; \ \sigma_{zj} \neq -\sigma_{zj} \text{ otherwise.} \qquad \text{...(17)}$$

The following constraint conditions of temperature and heat flux are imposed on each pair of nodes on the interface

$$T_i = T_j \text{ when } P > 0; \ T_i \neq T_j \text{ otherwise} \qquad \text{...(18)}$$
$$q = \mu p \omega r \text{ when } P > 0; \ q^* = 0 \text{ otherwise} \qquad \text{...(19)}$$

3. RESULTS AND DISCUSSION: VALIDATION OF THE FINITE ELEMENT CODE

To validate the finite element code, a comparison of the transient results with the steady state solution of thermoelastic behaviors was performed for a typical the operation condition. If the transient solution of this operation condition converges to the steady state as time elapses, it can be regarded as the validation of the applied transient scheme. Both the transient and the steady state solution was performed and the results were found to get converged.

Figure 3 shows the temperature variation of the disc surface in angular direction at various radius at $t = 4.3$ sec. It can be concluded that the temperature varies in the angular direction in a

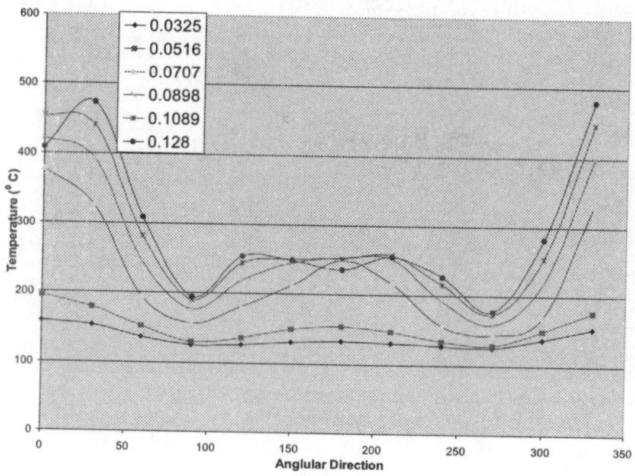

Fig. 3. Temperature variation along angular direction of disc at $t = 4.3$ sec.

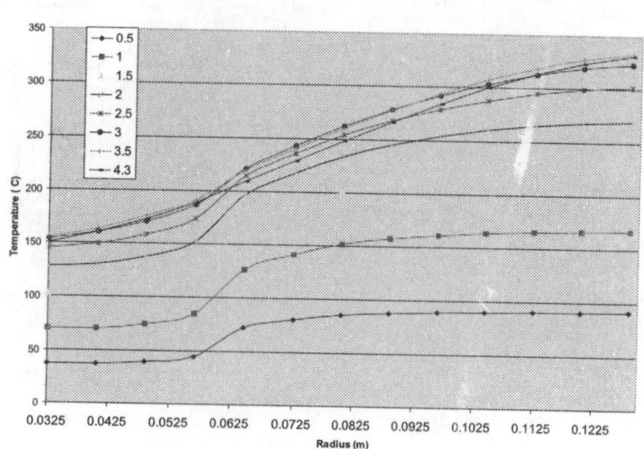

Fig. 4. Temperature variation along radius at surface 2 at various time.

significant manner. It also shows that the negligence of the variation of temperature in the angular direction reported in the literature produces incorrect temperature values.

The Figs. 4 and 5 show the temperature variation along the radius at different time at the surfaces 1 and 2 respectively. The temperature produced at the surface 2 is high compared to the surface 1 since the surface 2 is constrained in the Y direction.

The Fig. 6 shows the temperature variation of the disc along the thickness at different radius at $t = 4.3$ sec. It can be seen that the temperature goes on decreasing along the thickness but it starts increasing after reaching the middle layer of the disc. This can be justified as the heat is supplied at both the surfaces of the disc, the heat flux enters into the disc on both the surfaces and reaches an equilibrium values at the mid of the disc.

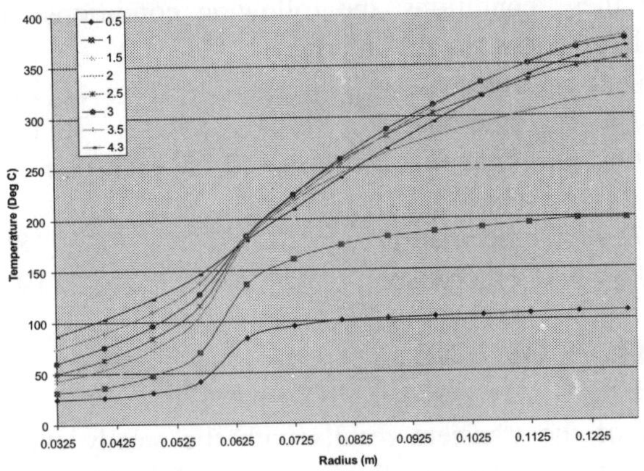

Fig. 5. Temperature variation along radius at surfaces 1 at various time.

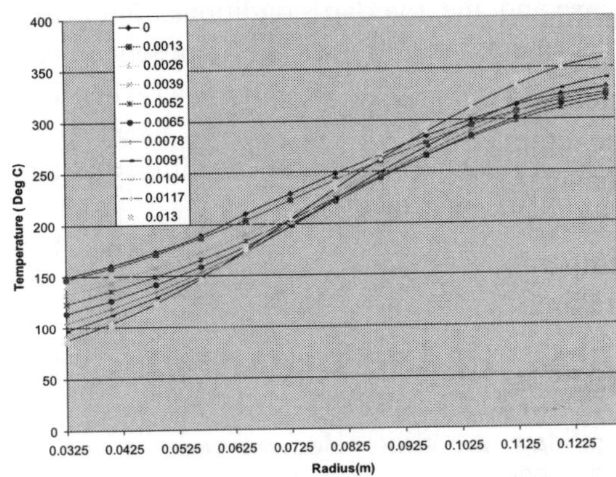

Fig. 6. Temperature variation along thickness at at different radius at time $t = 4.3$ sec.

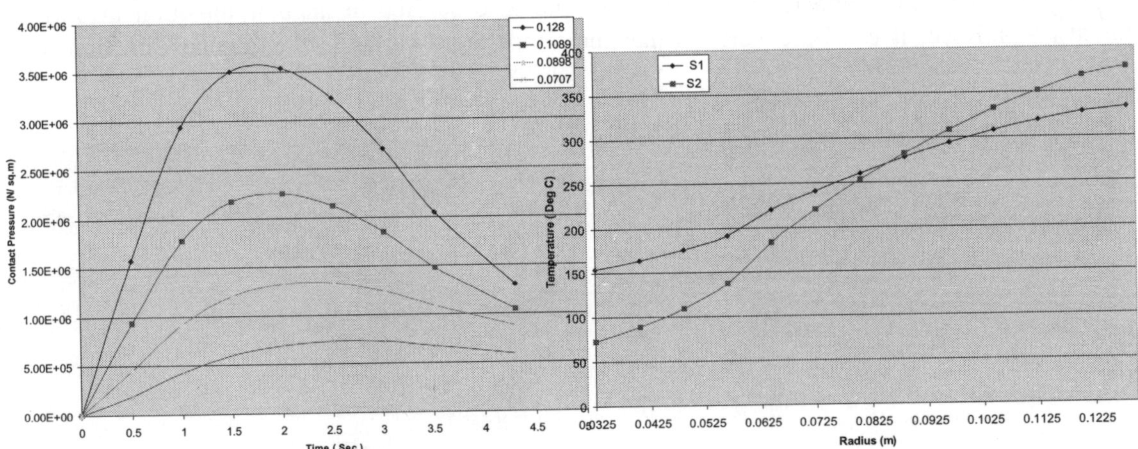

Fig. 7. Contact Pressure variation at different. radius at various times.

Fig. 8. Temperature variation at surface 1 and 2 along different radii.

The Fig. 7 shows the contact pressure variation at the disc surface at various times along the radii. It can be seen that the contact pressure starts increasing unto 1.6 sec after which it starts decreasing. The major cause of the phenomena is that the contact conditions on the friction surfaces are changed to satisfy the new equilibrium conditions due to the rise in temperature.

Figure 8 shows the temperature variation of the disc surfaces 1 and 2 along the radius at time $t = 4.3$ sec. It can be seen that the temperature variation may be uniform at the earlier stage but it also changes during the braking process. This can be justified as even though the pressure distribution on the friction surface is uniform in the early stages of braking application, it becomes non-uniform in the radial direction as the braking progress.

CONCLUSION

In this study, the transient therrmoelastic analysis of disc brakes in single brake application has been performed. The finite element method is applied to thermoelastic contact problem with frictional heat generation. To obtain the simulation of the thermoelastic behavior in disc brakes, the coupled heat conduction and elastic equations are solved with contact conditions. The consideration of the axisymmetric disc brake model is also proved to be ineffective. The contact pressure variation at the disc surface has also been analyzed and the effect of that on the temperature also analyzed. The present study can provide a useful design tool and improve the braking performance of brake system.

REFERENCES

1. W. Krietlow, F. Schrodter, H. Matthai, Vibration and hum of disc brakes under load, SAE 850079, 1985.
2. A.E. Anderson, R.A.Knapp, Hot spotting in automotive friction systems, Wear 135 (1990) 319-337.
3. Effect of temperature on thermoelastic instability in thin discs, Journal of tribology, 124 (2002), 429-437.
4. F.E. Kennedy, F.F. Ling, A thermal, ASME J. Lubrication Technology. 97 (1974) 497-507.
5. P. Zagrodzki, Numerical analysis of temperature fields and thermal stresses in the friction discs of a multidisc wet clutch, Wear 101(1985) 255-271.
6. P. Zagrodzki, Analysis of thermo mechanical phenomena in mutidisc clutches and brakes, Wear 140 (1990) 291-308.
7. P. Zagrodzki, K.B. Lam, E. Al Bahkali, J.R. Barber, Non-linear transient behavior of a sliding system with frictionally excited thermoelastic instability, ASME Journal of Tribology. 123 (2001) 699-708.
8. Transient temperature field analysis of a brake in a non-axisymmetric three dimensional model, Journal of materials processing technology 129 (2002) 513-517.
9. J.H. Choi, I. Lee, Transient thermoelastic analysis of disk brakes in frictional contact, Journal of Thermal Stresses 26 (2003) 223-244.

148

Choice of Scaling Parameter in Meshless Local Petrov-Galerkin (MLPG) Method

P. PAL AND S.K. BHATTACHARYYA

Department of Civil Engineering, IIT, Kharagpur-721302, India.

ABSTRACT

The focus of the present paper is on the development of a scaling parameter for Gaussian type weight function for the solution of linear potential problems using MLPG method is developed. A truly meshless method, based on the local symmetric weak form (LSWF) and the moving least squares approximation, is presented for solving potential problems with higher accuracy. The essential boundary conditions in the present formulation are imposed by a penalty method. All integrals are easily evaluated over regularly shaped domains (in general, circle in 2-D problems) and their boundaries. A few numerical examples are presented herein. The present meshless method based on the LSWF is found to be simple, efficient, and attractive with a great potential in engineering applications.

Keywords: Meshless method; local Petrov-Galerkin method; MLS approximation; scaling parameter.

1. INTRODUCTION

Finite element method has been widely used for the solution of problems in solid and fluid mechanics. It is well established that the construction of finite element meshes is usually an involved and time-consuming task. The employment of conforming meshes for the solution of fluid problems is not quite suitable for design purposes, as the spatial grid restrict changes of the parameters of the geometric model such that it is difficult or even impossible to change the shape of the model without remeshing. In order to overcome these problems, the focus of research is directed on the development of alternative techniques, which do not have an explicit boundary between the pre-process and the main process. In particular, the meshless methods, and a number of techniques have been recently proposed by different researchers. Various meshless methods belonging to this family are: smooth particle hydrodynamics [9], element-free Galerkin (EFG) method [4], reproducing kernel particle method [7], the partition of unity method [3] and meshless local Petrov-Galerkin (MLPG) method [2]. In one of such approaches, the meshless Petrov-Galerkin (MLPG) method, proposed by

Atluri and his coworkers based on moving least square (MLS) approximations, has attracted a lot of attention of the researchers. Cho, J.Y. and Atluri, S.N. [6] have analyzed the problems of shear flexible beams by the MLPG method based on a locking-free weak formulation. Ma, Q. W. [8] has presented meshless local Petrov-Galerkin (MLPG) method for nonlinear water wave problems. In the present work, a true meshless formulation based on the local symmetric weak form (LSWF) is proposed to solve the linear potential problems. Computer codes are developed for two-dimensional linear potential problems.

2. MOVING LEAST-SQUARES APPROXIMATION SCHEME

Based on MLS approximations, the shape function $\Phi_i(x)$ corresponding to node i may be written as

$$\Phi_i(x) = \sum_{j=1}^{m} P_j(x)\left(A^{-1}(x)B(x)\right)_{ji} = P^T A^{-1} B \qquad \text{...(1)}$$

Where P is the complete monomial basis. A and B are the matrices detemined by that of the basis functions and of the weight functions. Gaussian type weight function with compact support is considered in the present work. The Gaussian weight function corresponding to node i may be written as

$$w_i(x) = \begin{cases} \dfrac{\exp\left[-\left(d_i/c_i\right)^{2k}\right] - \exp\left[-\left(d_i/c_i\right)^{2k}\right]}{1 - \exp\left[-\left(r_i/c_i\right)^{2k}\right]} & 0 \le d_i \le r_i \\[2em] 0 & d_i \ge r_i \end{cases} \qquad \text{...(2)}$$

where $d = \|x - x_i\|$ is the distance from node x_i to point x. c_i is a constant which control the shape of the weight function. r_i is the size of the support for the weight function. In the present work, k is chosen as 1 for C° continuity over the entire domain Ω.

A necessary condition for a well defined MLS approximation is that at least m weight functions are non-zero for each sample point $x \in \Omega$. The smoothness of the shape functions is determined by that of the basis functions and of the weight functions. To construct a well defined shape function, the number of nodes (n) influencing the concerned point must be greater than m i.e. $n \ge 3$ for linear polynomial basis and $n \ge 6$ for quadratic polynomial basis and so on. For a two-dimensional problem, the radius of a local sub-domain Ω_x of each internal node (radius of the compact support of weight function) which is influenced by n number of nodes over the problem domain Ω may be written as

$$r_i = s_i\, h_i \qquad \text{...(3)}$$

Where s_i = scaling parameter and h_i = distance between two consecutive nodes in the directions of x, y and z. if the distance between two consecutive nodes into the global domain Ω in the direction of X is dx and in the directions of Y and Z are dy and dz then h_i should be equal to the value of dz if $dx > dy > dz$. The value of scaling parameter s_i should be greater than the value of dx/dz. Scaling parameter s_i may be chosen in such a way that the surface plot of the value of shape functions at every node of the problem domain is smooth enough. The appropriate value of c_i may be calculated by minimizing the undulation of the surface plot of the altitude of the shape functions. The magnitude of c_i is chosen in such a way that the size (radius) of each local sub-domain Ωx should be big enough such that the union of all local subdomains covers as much as possible of the global domain.

3. LOCAL SYMMETRIC WEAK FORM (LSWF)

The present approach is fully general in solving general boundary value problems. The linear Poison's equation [2] is solved to demonstrate the present formulation. A generalized local weak form of the Poison's equation and applying boundary conditions for all nodes over a local sub-domain Ω_s and a local boundary Γ_{su} and Γ_{sq} leads to the following discretized system of linear equations

$$[K]\{\bar{u}\} = \{f\} \qquad \qquad ...(4)$$

where, the "stiffness" matrix [K] and the "load" vector {f} are defined by

$$K_{ij} = \int_{\Omega} \Phi_{j,k}(x) v_{,k}(x,x_I) d\Omega + \alpha \int_{\Gamma_{su}} \Phi_j(x) v(x,x_I) d\Gamma - \int_{\Gamma_{su}} \Phi_{j,n} v(x,x_I) d\Gamma \qquad ...(5)$$

and

$$f_i = \int_{\Gamma_{sq}} \bar{q}(x) v(x,x_I) d\Gamma + \alpha \int_{\Gamma u} \bar{u}(x) v(x,x_I) d\Gamma - \int_{\Omega_s} w(x) v(x,x_I) d\Omega \qquad ...(6)$$

In which Φ_j is the shape function from the MLS approximation, and $(\ )_{,n} = \partial(\ )/\partial n$. where $\bar{u}$ and $\bar{q}$ are the prescribed potential and normal flux on the essential boundary Γ_u and on the flux boundary Γ_q.

4. NUMERICAL EXAMPLES

A computer code is developed based on the algorithms described in the previous sections. A few numerical examples are solved and some parametric case studies are presented to validate the developed computer code and to further establish the applicability of the algorithm in potential problems.

4.1 Example 1.

A cantilever beam of length L subjected to a parabolic traction at the free end as shown in Fig. 1 is considered. The beam has a depth D and is assumed to be in a state of plane stress. The present MLPG formulation is used to study the characteristics of the cantilever beam the exact solution of which is given in literature [10].

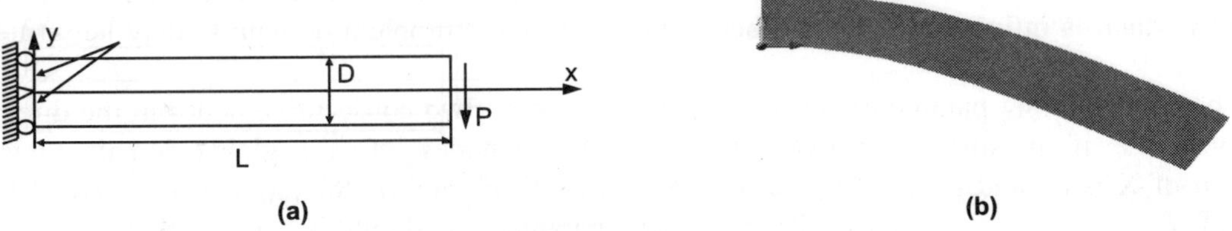

Fig. 1. (a) A cantilever beam (b) Deflected shape.

The geometric and the material properties for the problem are selected from literature [4]. Regular meshes of 27 (9 × 3), 85 (17 × 5), 175 (25 × 7) and 637 (49 × 13) nodes are used in the directions of length and depth of the cantilever beam. In all cases, $r_i = 2h$ to $3h$ and $c_i = r_i / 4$ are used, with h being the mesh size. In the computation, 9 Gauss points are used on each section of boundary and 5 × 8 points are used in each local domain for numerical quadratures. Figure 2 shows the error level in the computation of displacements for the cantilever beam for different scaling parameters. Based on this study, a scaling parameter value of 2.5 is chosen for the present solution. Table 1 compares the present results for the vertical displacement at the free end of the beam with the analytical solution [4]. Present results are in excellent agreement with the analytical ones.

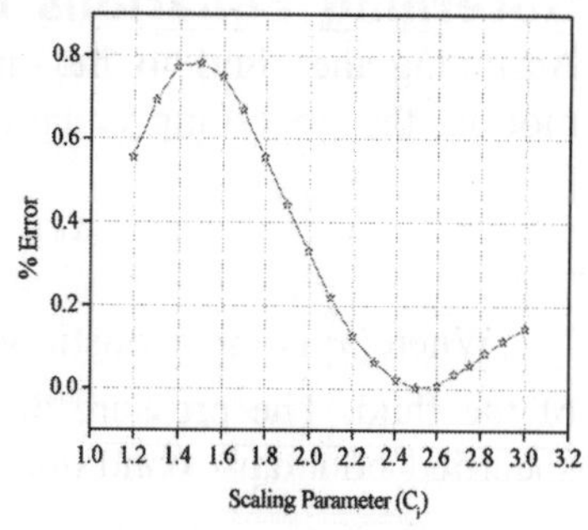

Fig. 2. Relative errors with scaling parameter for beam problem.

Table 1. Comparison of vertical displacement at end of beam.

No. of Nodes	Analytical [4]	Present
27		− 8.6235900307 E-003
85		− 8.7736632953 E-003
175	− 8.9000000000 E-003	− 8.8770269781 E-003
637		− 8.8940370953 E-003

4.2 Example 2.

Figure 3 shows the problem geometry of a typical 2-D prismatic partially liquid filled rigid container under external excitation. The base of the tank is assumed to be rigid and the container fluid is assumed to be compressible and inviscid. The effect of the static pressure is not considered in the analysis. The container is rectangular with uniform thickness and its material is homogeneous, isotropic and linearly elastic in nature.

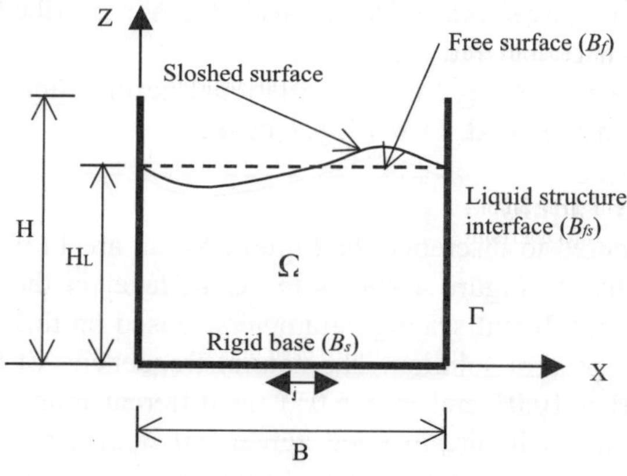

Fig. 3. Model used for analysis.

Governing Equations for the Fluid Domain

Assuming the fluid to be linearly compressible and inviscid with a small amplitude irrotational motion, the governing equation may be given by the Helmholtz equation.

$$\nabla^2 p(x,z,t) = \frac{1}{c^2}\ddot{p}(x,z,t)a \qquad \text{...(7)}$$

Where, c is the acoustic wave velocity in the liquid, k is the bulk modulus and ρ_f is the density of the fluid. The pressure distribution in the fluid domain is obtained by solving Eq. 4.1 with the specified boundary conditions.

Boundary Conditions

At the bottom of the tank

$$\frac{\partial p}{\partial n}(x,0,t) = 0 \quad \text{on } B_s \qquad \text{...(8)}$$

At the liquid-structure interface

$$\frac{\partial p}{\partial n}(0,z) = \frac{\partial p}{\partial n}(L,z) = -\rho_f a_0 \omega^2 \sin\omega t \quad \text{on} \quad B_{fs} \qquad \text{...(9)}$$

Where, a_0 is the exciting amplitude, ω is the exciting frequency of the external excitation and ρ_f is the density of fluid.

At the liquid free surface

The actual free surface is non-linear in nature. However, a linear approximation of the free surface may be made without much error when the amplitude of the surface wave is very small in comparison to the depth.

$$\frac{\partial p}{\partial n}(x,H,t) = -\frac{1}{g}\frac{\partial^2 p}{\partial t^2} \quad \text{on } B_f \qquad \text{...(10)}$$

The developed computer code is used to evaluate the slosh frequencies, sloshing displacements and the hydrodynamic pressure over the container wall for a prescribed excitation. The material properties of the fluid domain considered are

Mass density of fluid, $\rho_f = 1000$ kg / m^3, Gravitational acceleration, g = 9.81 m/sec^2

Acoustic wave velocity in the fluid, C = 1438.7 m/sec

4.2.1 Slosh Frequencies of Liquid

The numbers of nodes considered to discretize the liquid domain are 11 × 7 and 13 × 9 in the width and depth directions, respectively. Figure 4 shows the error level in the computation of the mode shape of sloshing of liquid for different scaling parameters. Based on this study, a scaling parameter value of 2.2 is chosen for the present solution. The natural frequencies of liquid in a tank, computed for H/B ratio equal to 0.5 (B = 10.0^m and H = 5.0^m) for different modes are presented in Table 2. It is observed that the present results are in good agreement with the natural frequencies obtained from the analytical expression reported in NASA monograph [1] and the antisymmetric mode frequencies of liquid given by Choun *et al.* [5].

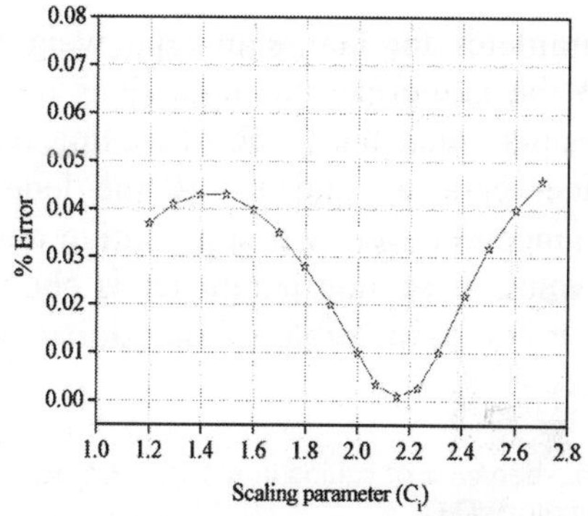

Fig. 4. Relative errors with scaling parameter for potential flow problem.

Table 2. Slosh frequencies, f_m (Hz) of liquid in a rectangular container.

Mode No.	Abramson [1]	Choun et al. [5]	Present 11×7	Present 13×9
1	0.2676	0.2675	0.2712	0.2681
2	0.3944	–	0.4021	0.3962
3	0.4834	0.4838	0.4962	0.4873
4	0.5588	–	0.5855	0.5674

4.2.2 Slosh Response–Sinusoidal Lateral Base Excitation

The dynamic pressure of fluid within a rectangular container (H = 1.0^m and B = 1.0^m) is evaluated for sinusoidal ground acceleration ($a_0 \sin \omega t$) with the boundary conditions as discussed in 4.2. The range of frequency considered is such that the lower frequencies are close to the sloshing frequency of the free surface. Typical time response of the free surface wave is shown in Fig. 5 for the situation when the amplitude and frequency of the base acceleration excitation are 0.02^m and 5.5 rad/sec, respectively and the scaling parameter is considered as 2.2. The response shown in the Figure 5(a) matches closely with that reported by Washizu *et al.*[11]. Figure 5(b) shows the free surface profile at both the walls of rectangular tank.

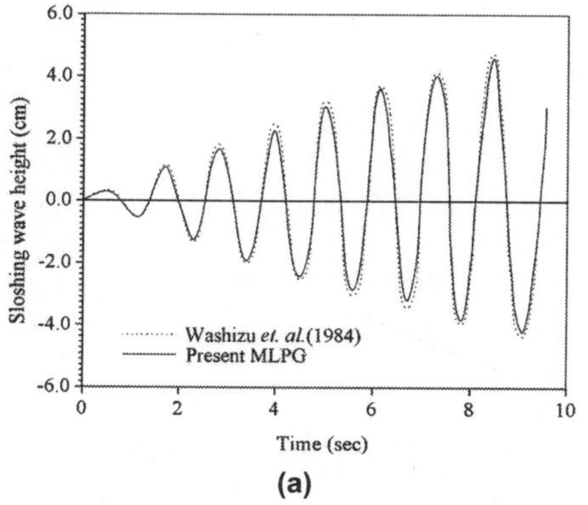

(a)

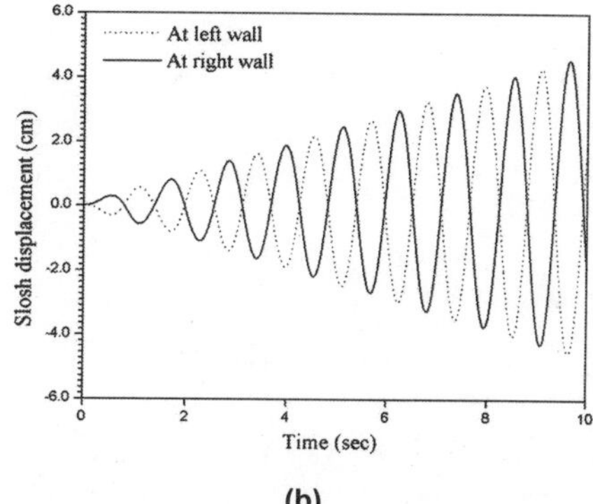

(b)

Fig. 5. Slosh displacements on the free surface of a rectangular tank (a) at right wall (b) at both walls.

CONCLUSION

The development of a scaling parameter for Gaussian type weight function in MLPG method is developed in the present paper. In the numerical examples it is observed that the computed results match closely with the reported results when the value of scaling parameter is greater than equal to $2.0 \times (L_x / d_x) / (L_y / d_y)$ or vice-versa. L_x and L_y are the length in the direction of x and y respectively. dx and dy are the distance between two consecutive nodes in the direction of X and Y. For more satisfactory results, the value of scaling parameter is chosen in such a way that the union of all local subdomains covers as much as possible of the global domain.

REFERENCES

1. H.N. Abramson, 1966, The dynamic behavior of liquids in moving containers, NASA SP-106, National Aeronautics and Space Administration, Washington, D.C.
2. S.N. Atluri, T. Zhu, 1998, A new Meshless Local Petrov-Galerkin (MLPG) approach in computational mechanics, Journal of Computational Mechanics. 22, 117-127.
3. I. Babuska, M. Melenk, 1997, The partition of unity method, International Journal for Numerical Methods in Engineering. 40, 727-758.
4. T. Belytschko, Y. Y. Lu, L. Gu, 1994, Element-free Galerkin methods, International Journal for Numerical Methods in Engineering. 37, 229-256.
5. Y.S. Choun, C.B. Yun, 1999, Sloshing Analysis of Rectangular Tanks with a Submerged Structure by using Small-amplitude Water wave Theory, Earthquake Engineering and Structural Dynamics. 28, 763-783.
6. J.Y. Cho, S.N. Atluri, 2001, Analysis of shear flexible beams, using the meshless local Petrov-Galerkin method based on a locking-free formulation, Journal of Engineering Computations. 18(1/2), 215-240.
7. W.K. Liu, Y. Chen, C.T. Chang, T. Belytschko, 1996, Advances in multiple scale kernel particle methods, Journal of Computational Mechanics. 18(73), 73-111.
8. Q.W. Ma, 2005, Meshless local Petrov-Galerkin method for two-dimensional non-linear water wave problems, Journal of Computational Physics. 205, 611-625.
9. J.J. Monaghan, 1988, An introduction to SPH, Journal of Computational Physics Communications. 48, 89-96.
10. S.P. Timoshenko, J.N. Goodier, 1951, Theory of Elasticity, McGraw-Hill Book Company, Inc., New York.
11. K. Washizu, T. Nakayama, M. Ikegawa, Y. Tanaka, T. Adachi, 1984, Some finite element techniques for the analysis of non-linear sloshing problems, Journal of Finite Elements in Fluids. 5, 357-376.

149

Configuration Optimization of Trusses Using Genetic Algorithm with Fuzzy Constraints

N.K. Solanki[1] and S.C. Patodi[2]

[1]Research Scholar, [2]Professor, Applied Mechanics Department, Faculty of Technology and Engineering M.S. University of Baroda, Vadodara, Gujarat-390 001, India.
email: nksolankimsu@yahoo.com, scpatodimsu@yahoo.com

ABSTRACT

Configuration of pin joined structure significantly affects the weight of the structure. Hence in optimum design of such structures, selection of configuration is very important. In the traditional Genetic algorithm (GA) optimum solution is obtained by imposing hard constraints. In the actual design methods constraint evaluation involves many sources of imprecision and approximation. Some sort of relaxation of constraints is therefore recommended. A genetic evolution based search technique with soft (fuzzy) constraints is explored for configuration optimization of trusses in this paper. A user friendly computer program based on the GA-Fuzzy approach is developed and used in minimum weight design of truss structures. To improve the performance of the genetic algorithm various design constraints such as member stress, joint displacement and buckling stress are slightly relaxed using Fuzzy Logic. The numerical examples, solved using the approach, are presented here to show the improvement in the search process due to involvement of fuzzy logic.

Keywords: Genetic Algorithm, Fuzzy Logic, Constraints, Hybrid Approach, Trusses.

1. INTRODUCTION

The area of design optimization, especially structural optimization, has been and continues to be an active area of research. There are several reasons for the interest, including the need to handle wider class of problems, to include realistic definitions of design variables, to find techniques to locate the global optimum solution and to improve the efficiency of the numerical procedure.

GAs are very effective at finding optimal solution to a variety of problems. This innovative technique performs especially well when solving complicated real world problems because it does not impose similar limitations as those of the traditional optimization methods. It finds an optimal solution by generating population of solution strings randomly and improving the solutions in succeeding generations using operations that mimic those of natural evolution such as reproduction, crossover and mutation etc. GAs have been successfully used in the optimal design of trusses by Rajan [1], Soh and Yang [2, 3], Rajeev and Krishnamoorthy [4] and many other research workers.

In traditional optimization algorithm the constraints are satisfied strictly due to which it can miss true optimum solution within the confine of practical and realistic approximation. In real life design problems, design methods and constraint evaluation involves fuzziness and imprecision. Also during the design state, the decision making in the shape design of structures is mainly based on the conceptual understanding. Unfortunately such information is usually vaguely defined by the experts. Therefore consideration of imprecise and vague information becomes an important aspect in the structural optimization of trusses. Moreover, if some organisms (candidate solutions) are thrown out because they violate one or more constraints during the early optimization process, then the search may miss the potentially global optimum solution. In other words, by treating the constraints as fuzzy constraints the chance of obtaining global optimum can be increased.

Fuzzy Logic is based on fuzzy set theory and provides an easy way of dealing with complex problems containing vagueness and imprecision. Several papers have focused on structural optimization in fuzzy environment. Soh and Yang [2] presented fuzzy rule-based system to control GA based search technique in structural shape optimization. Sarma and Adeli [5] presented a fuzzy augmented lagrangian GA for optimization of steel structures taking in to account the fuzziness in the constraints.

The structural optimization problem for truss structures can usually be described using three different types of design variables: (1) sizing variables, (2) geometric variables and (3) topological variables. In the present article the hybrid GA-fuzzy approach has been used for configuration and size optimization of plane truss structures. In this approach the fuzzy optimization problem is solved by dividing it in to three known fuzzy problems and solving each of them by using pure GA. The stress, displacement and buckling constraints of truss structure are fuzzified by using linear membership function due to its simplicity. A computer program has been developed in Visual Basic 6.0 due to its extraordinary graphic features and wide varieties of inbuilt functions and procedures. Numerical examples solved by the program are presented and the results are compared with those available in literature to confirm the validity of the developed program.

2. CONFIGURATION OPTIMIZATION PROBLEM FORMULATION

Configuration optimization of truss structures involves simultaneously arriving at optimum values for the nodal coordinates R and member cross-sectional areas A that minimize the structural weight. The general form of truss configuration optimization can be described as

$$\text{Minimize, structural weight } W(x) = \rho \sum_{i=1}^{n} A_i L_i , \qquad \qquad ...(1)$$

$$\text{Subject to, } g_j^l \leq g_j \leq g_j^u, \quad j = 1, 2, ..., q, \qquad \qquad ...(2)$$

where ρ is the material density, L_i and A_i are length and cross sectional area of ith member and g_j^l and g_j^u are the lower and upper bounds on the inequality constrained function g_j. The upper and lower bounds on the constraint functions of Eq. (2) include the following: (i) nodal co-ordinates $(R_i^l \leq R_i \leq R_i^u, i = 1,...,m)$, (ii) member cross-sectional areas $(A_i^l \leq A_i \leq A_i^u, i = 1,...,n)$, (iii) member stresses $(\sigma_i^l \leq \sigma_i \leq \sigma_i^u, i = 1,...,n)$, (iv) nodal displacements $(\delta_i^l \leq \delta_i \leq \delta_i^u, i = 1,...,m)$ and (v) member buckling stresses $(\sigma_i^b \leq \sigma_i \leq 0, i = 1,...,n)$.

In this optimization problem, all the crisp constraints are converted in the fuzzy constraints and fuzzy optimization problem thus formulated is solved by GA. Various aspects of GA based mathematical model for configuration optimization problem described in Eq. (1) and Eq. (2), are outlined as under.

2.1 Design Variables

As mentioned earlier, x and y co-ordinates of movable nodes (nodes whose co-ordinates are allowed to change during the search process), and member cross-sectional areas are the design variables in simultaneous size and configuration optimization.

2.2 Objective Function

Weight of the truss under consideration is taken as the objective function to be minimized. The objective function is calculated by adding weights of the truss members only as formulated in Eq. (1) without considering weight of fasteners and gusset plates.

2.3 Constraints and Penalty Functions

Three main constraints required to be imposed as per design codes are stress constraints, displacement constraints and buckling constraints. The constraint functions are evaluated based on degree of constraint violation as discussed in the succeeding subsections.

(a) Stress constraint: If σ_j and σ_{ja} are calculated and permissible stresses for j^{th} member the constraint function is,

$$g(x) = \max(\sigma_j/\sigma_{ja} - 1, 0). \tag{3}$$

(b) Displacement constraint: For u_i and u_{ia} as calculated and permissible displacement of ith joint respectively, the displacement constraint function can be written as,

$$g(x) = \max(u_i/u_{ia} - 1, 0). \tag{4}$$

(c) Buckling constraint: In this constraint the calculated stress in the jth truss member is not allowed to exceed the permissible buckling stress (σ_{jb}). The constraint function is given by,

$$g(x) = \max(\sigma_j/\sigma_{jb} - 1, 0). \tag{5}$$

If C is the summation of all such constraint functions for a candidate solution violating constraints, it is penalized using the penalty function given as:

$$P(x) = (1 + K.C), \tag{6}$$

where K is penalty parameter which is selected judiciously. In the present study K is kept low in the initial generations and is gradually increased to large values in the subsequent generations using the equation: $K = K_{initial} \{ 1 + 0.2 (n_g - 1) \}$, where n_g is the generation number. A value of 10 has been found suitable for $K_{initial}$. The penalty function value obtained in Eq. (6) is multiplied with objective function of Eq. (1) to get penalized objective function $Op(x)$.

2.4 Fitness Function

To avoid negative fitness values following fitness function has been used in the present work.

$$f(x) = 1/(1 + O_p(x)) \tag{7}$$

3. HYBRID APPROACH FOR FUZZY OPTIMIZATION

The hybrid approach employed here divides the fuzzy optimization problem in three non-fuzzy optimization problems. Each of these problems is then solved by pure GA. In the initial step the optimum solution is obtained by GA where constraints are considered crisp and are handled by penalty function method. In the next step constraints are relaxed by allowing small violation d_j of constraints during the search process and the optimum solution is obtained using GA. For first two GA runs, weight of the truss is used to find fitness value as given Eq. (7). The third step involves the calculation of membership functions for objective function and constraints. Linear membership function has been used due to simplicity and found to give good results. An additional variable λ is entered in the existing design variables in third GA run. It represents the membership grade and takes any value between 0 and 1. Membership grades for objective function and design constraints are used to find penalty parameters in the last GA run. In this step, λ is directly taken as fitness function to be maximized. Optimum value of λ and corresponding design variables give the final optimum solution.

Initial population for first and second GA runs is generated randomly. The information gathered during first two runs is used in the final run. The first one third of the initial population is taken from the last generation of the first GA run and second one third is adopted from the second GA run. The remaining string length, which corresponds to the additional variable λ is made up by generating random numbers. The last one third of the population is generated entirely using random number generation. This technique leads to faster convergence of GA.

4. ILLUSTRATIVE EXAMPLES

Two illustrative examples have been solved to demonstrate the efficiency of suggested hybrid GA-Fuzzy approach for shape optimization of structures. GA is used with single-point crossover and constant rate mutation operator. Two selection schemes-Roulette wheel and Tournament are tried along with elitism operator.

4.1 Cantilever Plane Truss Example

The 18-bar cantilever plane truss, shown in Fig. 1, is one of the most popular truss design optimization problems. Due to its simple configuration, this structure has been used as a benchmark problem to verify the efficiency of various optimization methods [2, 3, 4, 6]. The material density is 2768 kg/m^3 (0.1 lb/in^3) and the modulus of elasticity is 68947.6 MPa (10,000 ksi). The cross-sectional areas of the members are categorized in to four groups: (1) $A_{G1} = A_3 = A_7 = A_{11} = A_{15} = A_{18}$, (2) $A_{G2} = A_1 = A_5 = A_9 = A_{13} = A_{17}$, (3) $A_{G3} = A_4 = A_8 = A_{12} = A_{16}$ and (4) $A_{G4} = A_2 = A_6 = A_{10} = A_{14}$. The single loading condition considered in this study is a set of vertical loads, P = 88.96 kN (20 kips), acting on the upper nodal points of the truss, as depicted in Fig 1. The lower nodes 3, 5, 7 and 9 are allowed to move in any direction in the x-y plane. Thus, there are a total of 12 independent design variables that included four sizing and eight co-ordinate variables. Population size and maximum number of generations used for the search in the present study are 30 and 50 respectively. Cross-over and mutation probabilities considered are 0.91 and 0.05 respectively. Optimum solution is obtained after ten runs with different seed values for random number generation.

The purpose of the study is to design such a configuration for the truss that produces a minimum weight with all the constraints satisfied to a desired level. In this problem, the allowable tensile and compressive stresses are ± 137.89 Mpa (20 ksi). The Euler buckling compressive stress limit used for the buckling constraint is computed as–$4AE/L^2$.

Table 1 lists the best solution vector and weight form hybrid GA-Fuzzy approach and also the results reported using other optimization methods [2-4, 6]. The optimal configuration is displayed in Fig. 2.

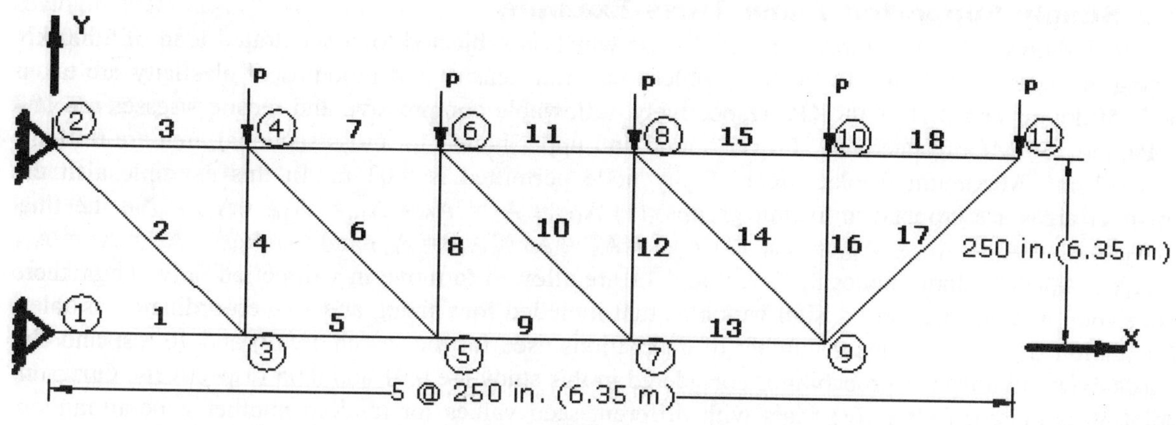

Fig. 1. 18-bar plane truss-initial configuration.

Table 1. Optimal results of 18-bar plane truss (Comparative study).

Ref. No.	Optimal sectional areas (in²)						Optimal co-ordinates R (in)						W (lb)
	A_{G1}	A_{G2}	A_{G3}	A_{G4}	X_3	Y_3	X_5	Y_5	X_7	Y_7	X_9	Y_9	
2	12.59	17.91	5.50	3.55	200.9	32.0	410.0	97.0	640.3	147.8	909.8	184.5	4531.9
3	12.33	17.97	5.60	3.66	202.1	30.9	413.9	102.0	643.3	149.2	907.2	184.2	4520.0
4	12.50	16.25	8.00	4.00	184.4	23.4	385.4	72.5	610.6	118.2	891.9	145.3	4616.8
6	12.65	17.22	6.17	3.55	195.3	30.6	402.1	90.5	630.3	136.3	903.1	174.3	4515.6
Present	12.57	17.92	5.52	3.35	202.7	35.4	414.5	102.3	640.8	145.2	913.2	184.6	4508.4

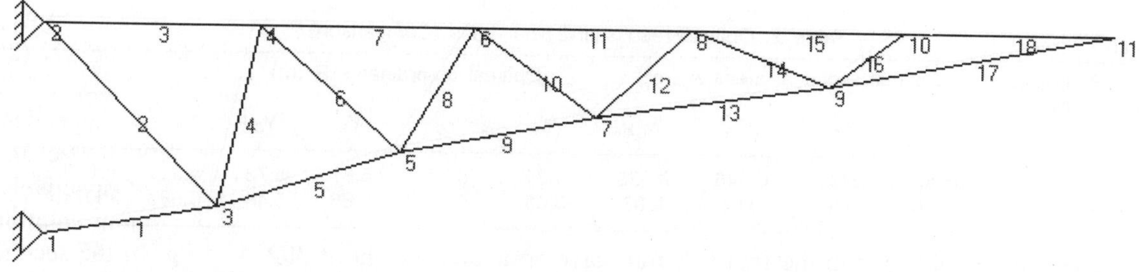

Fig. 2. Optimum configuration for 18-bar plane truss example.

The optimum solution obtained in the first GA run where, crisp constraints are imposed, gave an optimum structure weight of 2064.2 kg (4553.35 lb). In the second GA run, when constraints are relaxed by 15%, the optimum weight obtained is 1946.4 kg (4291 lb). In the last GA run optimum value of λ and corresponding weight of the structure obtained are 0.75 and 2045.0 kg (4508.4 lb) respectively.

4.2 Simply Supported Plane Truss Example

Figure 3 shows a 21-bar simply supported truss which is subjected to concentrated load of 3000 kN acting at nodes 2, 4, 6, 8, 10. In this problem material density and modulus of elasticity are taken as 7850 kg/m^3 and 2.1E + 08 KPa respectively. Allowable compressive and tensile stresses are 104 MPa and 130 MPa respectively. Lower bound and upper bound for cross sectional area are 0.01 m^2 and 0.1 m^2. Maximum displacement of any node permitted is 0.03 m. In this example also the member areas are grouped in to four groups: (1) $A_{G1} = A_2 = A_6 = A_{10} = A_{14} = A_{18} = A_{21}$, (2) $A_{G2} = A_1 = A_4 = A_8 = A_{12} = A_{16} = A_{20}$, (3) $A_{G3} = A_5 = A_9 = A_{13} = A_{17}$ and (4) $A_{G4} = A_3 = A_7 = A_{11} = A_{15} = A_{19}$. The upper nodes 3, 5, 7, 9 and 11 are allowed to move in y direction only. Thus, there are a total of 9 independent design variables that included four sizing and five co-ordinate variables. Population size and maximum number of generations used for the search are 30 and 40 respectively. Cross-over and mutation probabilities considered in this study are 0.91 and 0.05 respectively. Optimum solution is obtained after five trials with different seed values for random number generation.

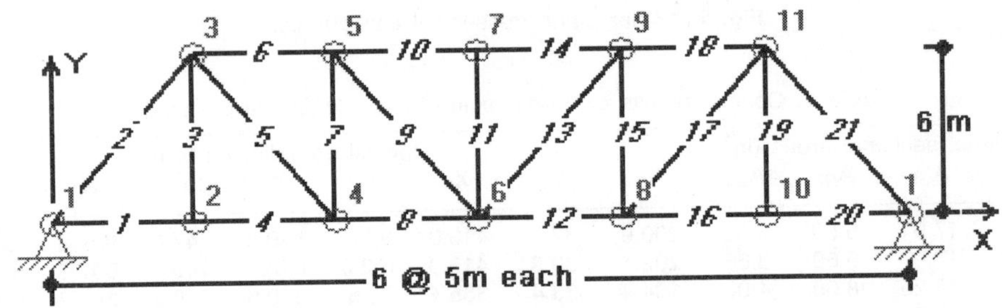

Fig. 3. 21-Bar plane truss-initial configuration

Table 2 lists the best solution vector form hybrid GA-Fuzzy approach and also the results obtained using pure GA [10]. The optimal configuration obtained is depicted in Fig. 4.

Table 2. Optimal results of 21-bar truss (Comparison)

Reference No.	Optimal sectional areas A (m^2)				Optimal co-ordinates R (m)					W (kg)
	A_{G1}	A_{G2}	A_{G3}	A_{G4}	Y_3	Y_5	Y_7	Y_9	Y_{11}	
1	0.082	0.022	0.046	0.028	4.31	5.73	5.87	5.73	4.31	39014.3
Present work	0.096	0.01	0.01	0.03	3.68	5.44	5.86	5.44	3.68	35074.4

The solution obtained in the first GA run, gave structural weight of 39234.45 kg. In the second GA run, when constraints are relaxed by 10%, the optimum weight obtained is 32760.4 kg. In the last GA run, optimum value of λ and corresponding weight of the structure are 0.58 and 35074.4 kg respectively.

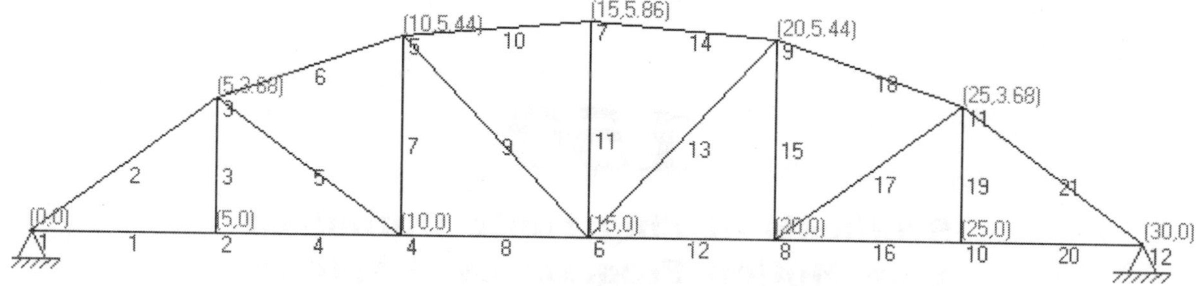

Fig. 4. 21-Bar plane truss-optimal configuration.

CONCLUSION

The developed program can handle cantilever and simply supported plane trusses with different support and loading conditions. The graphical features of the program provide facility to easily define support conditions, loading conditions and joint movability. Tournament selection scheme is found to yield better results. It has been clearly demonstrated that the hybrid approach selected here, not only effectively takes in to account the fuzziness and imprecision involved in the analysis and design, but also increases the probability of getting optimum solution.

In the first example, the weight of the truss obtained in the second run is found to be 5.76% lower than that obtained in first run due to relaxation in the constraints. The solution obtained by the combined GA-Fuzzy approach is @1% cheaper than the pure GA. The optimum solution obtained through hybrid approach is found to violate the constraints by very little amount (i.e. 0.2% to 1.45%), which is acceptable due to fuzziness involved in the analysis and design. The solution obtained through GA-Fuzzy approach is better than four other algorithms reported in the literature as indicated in Table 1.

In the second example, the optimum weight obtained in the second run is 16.5% lower than that obtained in first run. The solution obtained by the combined GA-Fuzzy approach is 10.6% cheaper than pure GA. Only two constraints are found to be violated by only 1.38 % and 0.19% in the optimum solution. Moreover optimum solution obtained through proposed GA-Fuzzy approach is significantly cheaper than that obtained by pure GA.

REFERENCES

1. S.D. Rajan, 1995, Sizing, Shape and Topology Design Optimization of Trusses Using Genetic Algorithm, Journal of Structural Engineering, ASCE, 121 (10), 1480-1487.
2. C.K. Soh, J. Yang, 1996, Fuzzy Controlled Genetic Algorithm Search for Shape Optimization, Journal of Computing in Civil Engineering, ASCE, 10 (2), 143-150.
3. J.P. Yang, C.K. Soh, 1997, Structural Optimization by Genetic Algorithm with Tournament Selection, Journal of Computing in Civil Engineering, ASCE, 11 (3), 195-200.
4. S. Rajeev, C.S. Krishnamoorthy, 1997, Genetic Algorithm-based methodologies for Design Optimization of Trusses, Journal of Structural Engineering, ASCE, 123 (3), 350-358.
5. K.C. Sarma, H. Adeli, 2000, Fuzzy Genetic Algorithm for Optimization of Steel Structures, Journal of Structural Engineering, ASCE, 126 (5), 596-604.
6. S.L. Kang, W.G. Zong, 2005, A new Meta-heuristic Algorithm for Continuous Engineering Optimization: Harmony Search Theory and Practice, Computer Methods in Applied Mechanics and Engineering, Elsevier, 194, 3902-3933.

150

Synthesis of Ping Finite Optimized Cam Motion Program by B-Splines

R. Mishra[1] and T.K. Naskar[2]

[1]M.E. Student, [2]Reader, Mech. Engineering Department, Jadavpur University, Kolkata-700032, India, email: rbm_mis@yahoo.com, tknaskar@mech.jdvu.ac.in

ABSTRACT

In this paper ping finite cam motion program is designed instead of jerk finite program. B-spline is used as cam displacement function. Control points (CP), characterized by angular positions of intermediate knots, are introduced in B-splines to minimize jerk. Study is done how the CPs affect the peak value of jerk (J_P). Values of jerk at end position (J_e) and ping at end position (P_e) also influence the value of the J_P. Here AP and either J_e or P_e are varied independently or simultaneously to optimize the J_P.

Keywords: B-splines, CP, jerk finite, ping finite.

1. INTRODUCTION

Essential objective of synthesis of cam motion program is to optimize the values of the acceleration and jerk of the follower. The synthesis of cam displacement function is guided by the fundamental principles [1], which state that:

1. A displacement function must be continuous through the first and second derivatives (i.e., velocity and acceleration) across the entire cycle;
2. The jerk function must be finite across the entire interval.

The principle ensures third order continuity (function plus two derivatives, i.e., velocity and acceleration) of a cam function at all boundaries with jerk function giving finite values across the interval. Here this function is called *jerk finite* function.

For jerk finite function acceleration curve is continuous with low peak value and discontinuity in jerk function. If it is ping finite function both acceleration and jerk curves will be continuous with higher peak values of acceleration but lower peak values of jerk [2]. It was concluded in [2] that if noise is no factor in a cam-drive jerk finite function is acceptable. However, if the concern is for smooth operation and low noise, i.e., precision work, the jerk should be continuous. The fundamental principles were revised to get jerk continuous in the following way:

1. A displacement function should have fourth order continuity (the function plus three derivatives) across the entire cycle;
2. The ping function must be finite across the entire interval.

Hence, a function that satisfies the above-modified principle is called *ping finite function*. In [2] the ping finite function was designed by polynomials. Here it is designed by B-splines. Many works are done in the field of kinematics and computer-aided design by manipulating the knots in mathematical splines [3] and B-Splines [4-7]. In this work 6-order and 8-order B-splines are taken for constructing cam displacement functions. Polynomials are combined piecewise at knots. CPs were introduced in natural splines [8] for optimization of acceleration and jerk of cam follower. Here CPs are introduced in B-splines for the same objective.

2. B-SPLINES AS CAM DISPLACEMENT FUNCTION

We can write B-spline as [1] $g_{(m,t,\theta)} = [0, \quad \theta < t \ (\theta - t)^{m-1} \ t <= \theta]$ ...(1)

This is a spline of order m with a single knot at t and a jump discontinuity in the $(m-1)$ derivative. These single knot splines are called the truncated power basis (TPB). For a sequence of knots of $(t_1, t_2, \ldots t_n)$ a displacement function can be expressed as a general polynomial spline of order m of the form,

$$S(\theta) = a_1 g_{(m,t_1,\theta)} + a_2 g_{(m,t_2,\theta)} + \ldots + a_n g_{(m,t_n,\theta)} \qquad ...(2)$$

3. SYNTHESIS OF A SINGLE DWELL CAM FUNCTION BY 6TH ORDER TPB: JERK FINITE: A CASE STUDY

Specifications for the synthesis of a single dwell cam function (Table 1): Rise h unit in $\pi/2$ (term *unit* is omitted in future), fall h in $\pi/2$, dwell at zero displacement for π and constant angular velocity of the camshaft. There will be a knot at $\pi/2$ with displacement of h and two end knots at 0 and π with finite jerk values. Two intermediate knots are assumed for which the values of the APs are γ and 3γ. These two knots are CPs – the values of the APs of which control the velocity, acceleration and jerk. Also J_e, which equals to $\pm$ x at two end points, imposes control over the A_P and J_P.

Here there are nine boundary conditions, which form nine equations. However, we can bring down the total number of equations. We note that according to Table 1, every $g_{(m-t,\theta)}$, with t larger than or equal to zero, will have its value and the value of the first and second derivatives equal to 0 at 0. So if we use knots that are at 0 or to the right of 0, the first three equations will be dropped and the number of equations will be reduced to 6. Therefore, there will be three knots at 0. Now the convention [1] for using multiple knots at a single point is: for a general polynomial spline of order m, with knot at t, the TPB function is $g_{(m,t,\theta)}$ for the first occurrence of the knot; for the second occurrence, $g_{(m-1,t,\theta)}$ and for the j_{th} occurrence $g_{(m-j+1,t,t\theta)}$.

According to the above convention, if $\gamma = \pi/4$, for a six-order spline the displacement function is,

$$S(\theta) = a_1 g_{(4,0,\theta)} + a_2 g_{(5,0,\theta)} + a_3 g_{(6,0,\theta)} + a_4 g_{(6,\frac{\pi}{4},\theta)} + a_5 g_{(6,\frac{\pi}{2},\theta)} + a_6 g_{(6,\frac{3\pi}{4},\theta)} \qquad ...(3)$$

The Eq. (3) is differentiated successively for three times and the boundary conditions in

Table 1 are used to obtain 6 equations. The 6 unknown coefficients $a_1 \ldots a_6$ are determined by solving the equations. The S, V, A, J curves of this TPB splines are drawn (Fig. 1). The figure shows that for $\gamma = \pi/4$ and $J_e = \pm 14$, which are taken as initial choice (IC) values, AP = 2.22 and $J_P = 14$.

3.1. Minimization of JP–Jerk Finite

Attempt is made to minimize the J_P by manipulating the CPs. In this process each of the two parameters–the AP and the J_e - are varied keeping the other one fixed, and lastly both are varied. First, AP is varied and the resultant maximum jerk, J_P vs. AP curve is shown in Fig. 2. Secondly J_e is varied and the J_P vs. J_e is shown in Fig. 3. Lastly both AP and J_e are varied. Figures 4 and 5 show the acceleration and jerk curve obtained by optimization of J_P. Table 2 shows the result of the jerk minimization program for jerk finite cam function. It proves that the minimization program adopted here can lower the values of the J_P considerably. In Table 2 A_P and J_P denote the values that are obtained on the basis of the inputs of A_p and J_e.

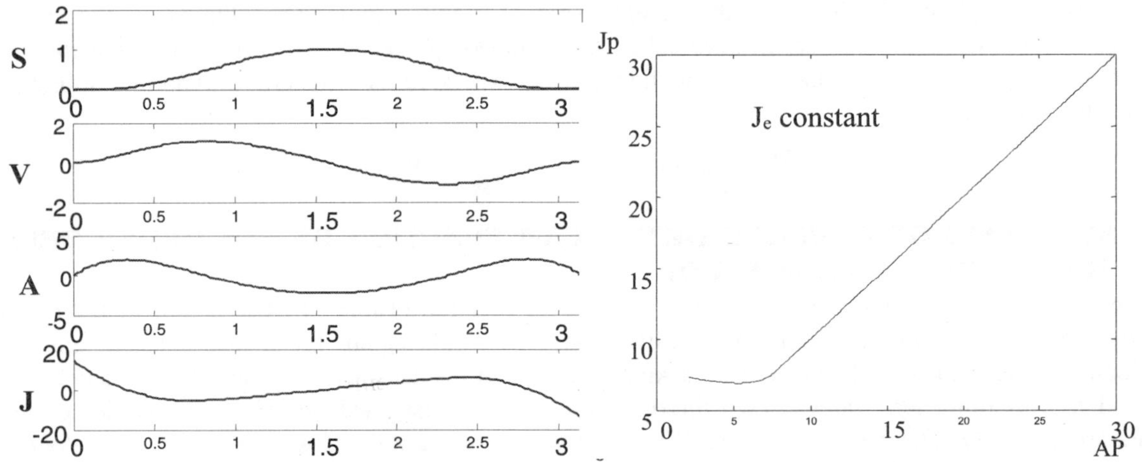

Fig. 1. S-V-J Curves: Jerk Finite.

Fig. 2. Jp-AP.

Table 1.

Angular Position of knot (radians)	0	γ	$\pi/2$	3γ	π
	AP			AP	
S	0	–	h	-	0
V	0	–	–	-	0
A	0	–	–	-	0
J	x	–	–	-	−x

4. SYNTHESIS OF A SINGLE DWELL CAM FUNCTION BY 8TH ORDER TPB: PING FINITE–A CASE STUDY

In order to accommodate ping we assume 8th order B-spline. Specifications for the synthesis of a ping finite single dwell cam function are described in Table 3. The displacement function is,

$$S(\theta) = a_1 g_{(5,0,\theta)} + a_2 g_{(6,0,\theta)} + a_3 g_{(7,0,\theta)} + a_4 g_{(8,0,\theta)} + a_5 g_{(8,\frac{\pi}{60},\theta)} + a_6 g_{(8,\frac{\pi}{2},\theta)} + a_7 g_{(8,\frac{59\pi}{60},\theta)} \qquad \ldots(4)$$

The S, V, A, J, P curves of this TPB splines are shown in Fig. 6. For AP, i.e. $\gamma = \pi/60$ and $P_e = \pm 14$ (IC values), $A_P = 3.461$ and $J_e = 8.8143$.

Table 2. Results of minimization of J_P: Jerk finite.

	AP	J_e	A_P	J_P
IC	$\pi/4$	14	2.22	14
AP varying, J_e constant	$0.0161\,\pi$	14	3.123	6.9045
J_e varying, AP constant	$\pi/4$	7.1	2.36	7.1842
Both AP and J_e varying	$\pi/11$	5	2.89	6.4392

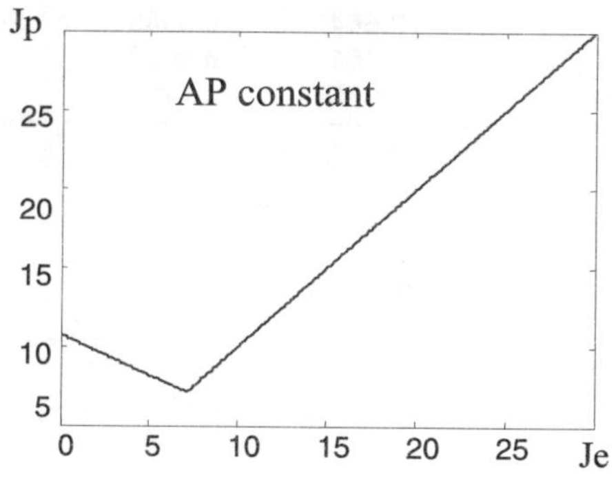

Fig. 3. Jp-AP

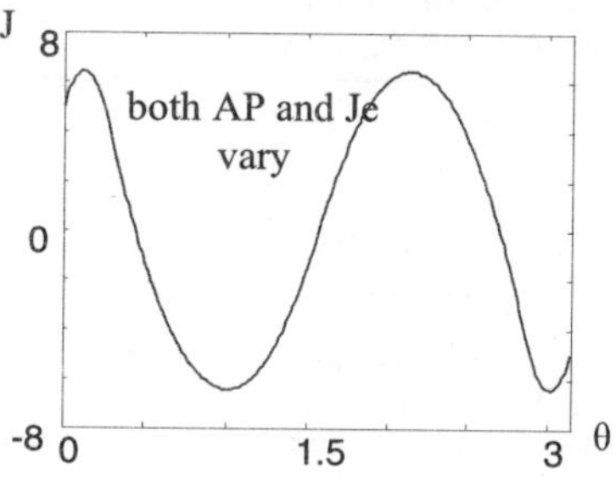

Fig. 4

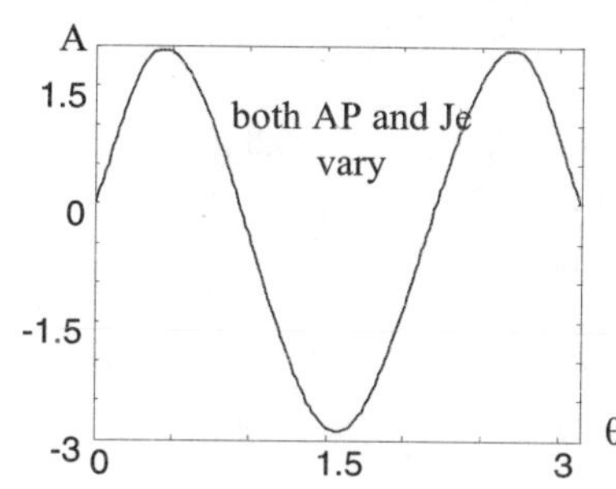

Fig. 5

Fig. 4. and 5. J-θ and corresponding A-θ (both AP and Je vary)

Table 3

Angular Position of knot	AP			AP	
	0	γ	$\pi/2$	3γ	π
S	0	-	h	-	0
V	0	-	-	-	0
A	0	-	-	-	0
J	0	-	-	-	0
P	x	-	-	-	$-x$

4.1. Minimization of JP–Ping Finite

The same procedure is adopted like jerk finite to minimize jerk. The J_P vs. AP and the J_P vs. P_e curves are shown in Fig. 7 and Fig. 8. Figures 9 and 10 show the jerk and acceleration curves obtained on optimizing J_P. Table 4 shows the results of the jerk minimization program for ping finite. It again proves that the minimization program adopted here lowers the values of the J_P.

Table 4. Results of minimization of J_P: ping finite.

	AP	P_e	A_P	J_P
IC	0.0166 π	14	3.461	8.8143
AP varies, P_e const.	0.4763 π	14	3.664	9.5943
P_e varies, AP const.	0.0166 π	0	3.461	8.8052
Both AP and P_e vary	0.01 π	0	3.4387	8.7155

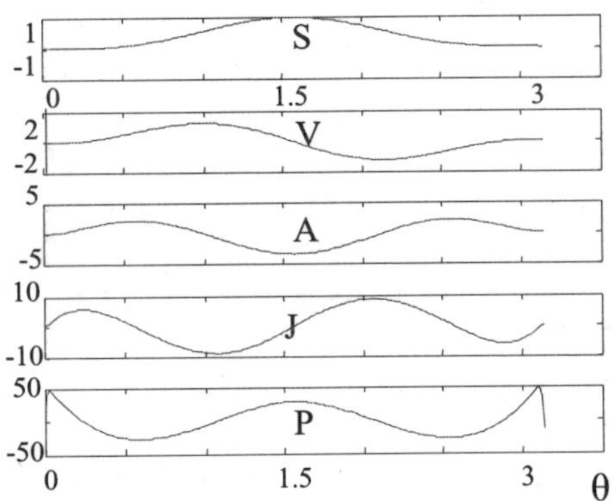

Fig. 6. S-V-A-J-P Curves: Ping Finite

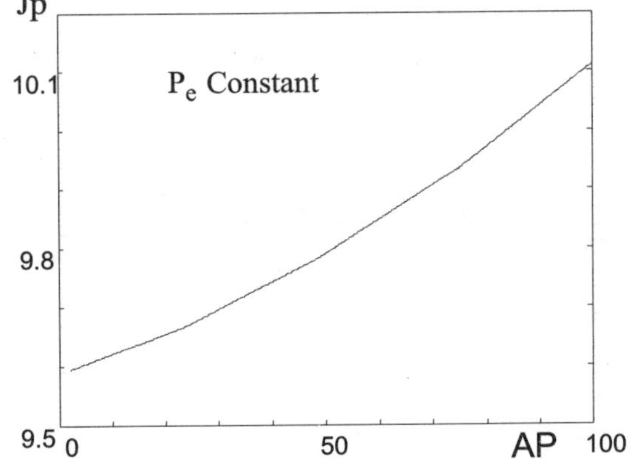

Fig. 7. J_P-AP

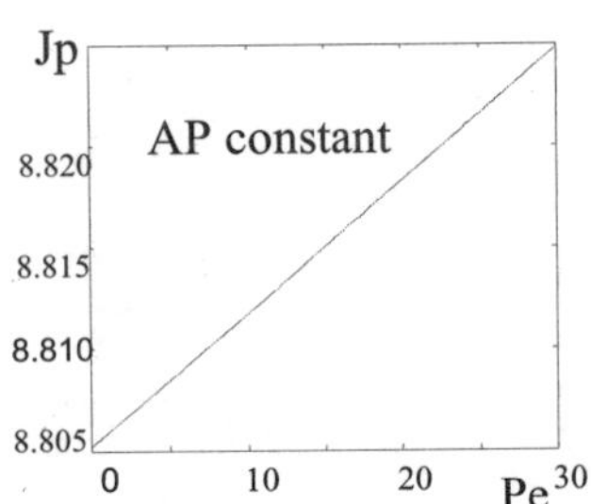

Fig. 8. J_p–P_e

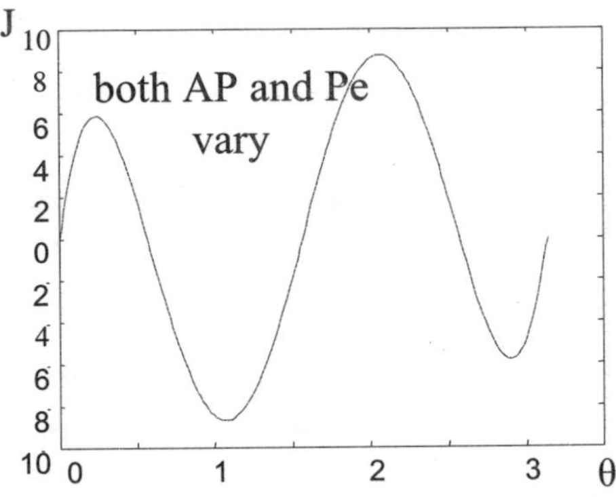

Fig. 9.

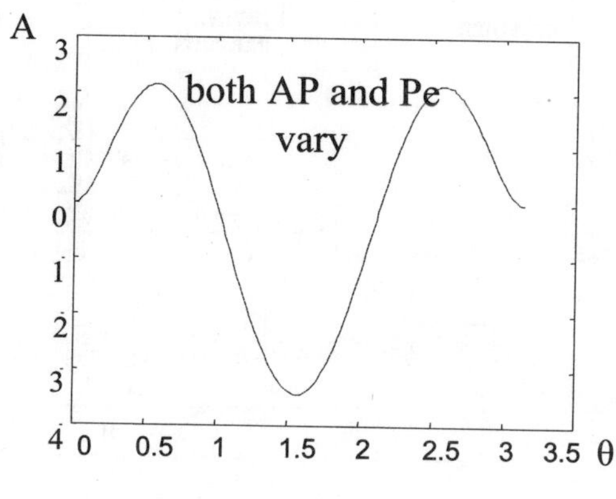

Fig. 10.

Fig. 9. and 10. J-θ and corresponding A-θ.

5. EXPERIMENTAL ANALYSIS

Experiment is done to find out the displacement, velocity, acceleration and jerk of the follower for validation. For this purpose cams are manufactured with the optimized profile based on theoretical displacement curve. Cam profile is drawn (Fig. 11) by a method of analytical design of cam with an offset roller follower [9]. Two materials, stainless steel and EN-31 are used for manufacturing both ping finite and jerk finite cams. Test is done at 600-900 rpm with steps of 50. The data is stored by computer aided data acquisition system (ADC-216, Pico). Figure 12 shows the test rig. The displacement (S) of the cam is obtained in voltage. Values of S in mm are computed by multiplying the output voltage by a conversion factor 1.99. The S-V-A-J curves are drawn as trend line by MS excel and are shown for jerk finite and ping finite at different rpm with EN31 as cam material in Fig. 13 and Fig. 14 respectively.

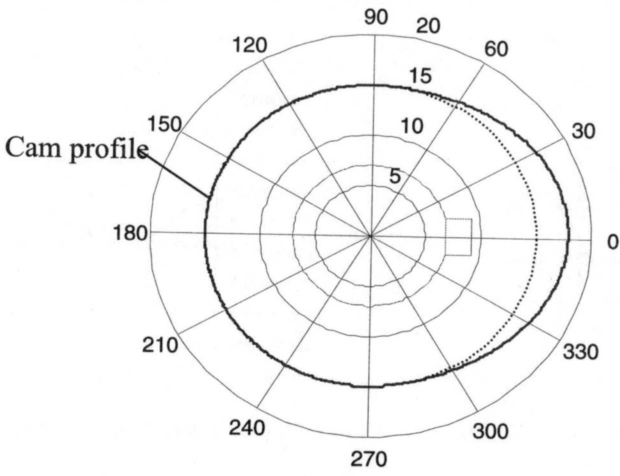

Fig. 11. Optimized cam profile.

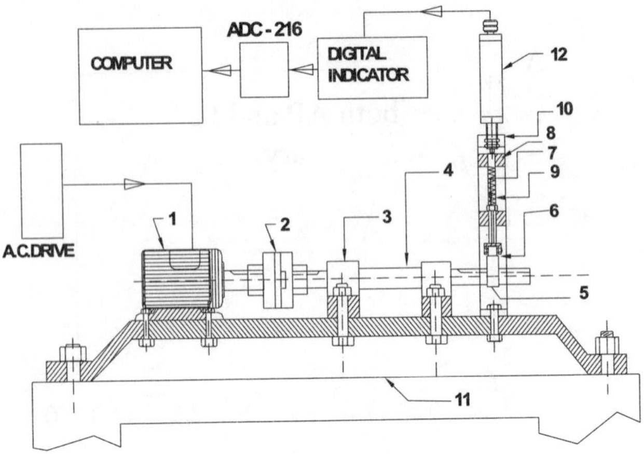

Fig. 12. Test Rig: 1. Motor, 2. Coupling, 3. Plummer block, 4.Shaft, 5. Cam, 6. Roller follower, 7. Spring, 8. Guide, 9. Follower rod, 10. Follower stand, 11. Base, 12. LVDT.

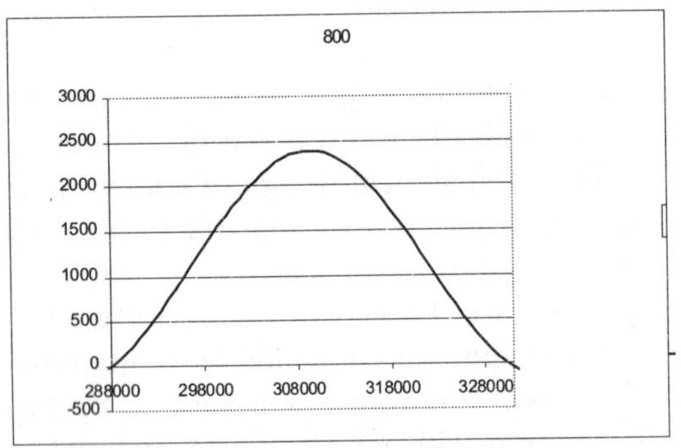

Fig. 13 (i). S: jerk finite, 800 rpm, EN31.

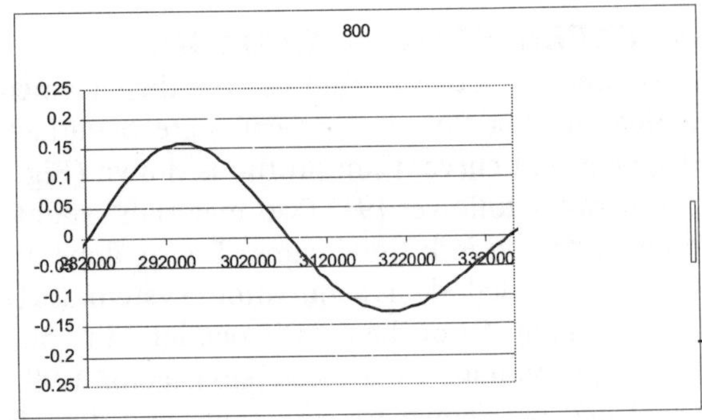

Fig. 13 (ii). V: jerk finite, 800 rpm, EN31.

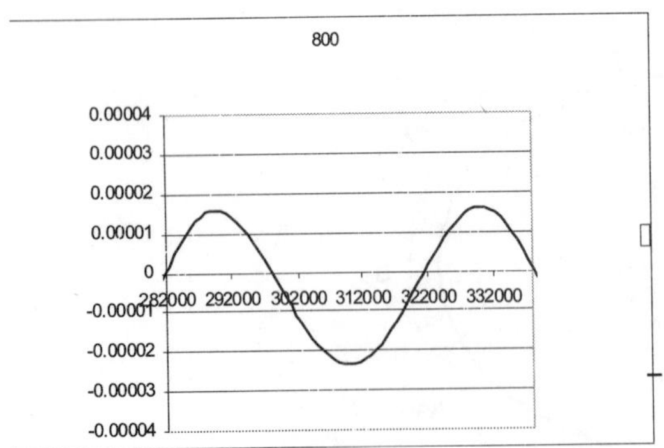

Fig. 13 (iii). A: jerk finite, 800 rpm, EN31.

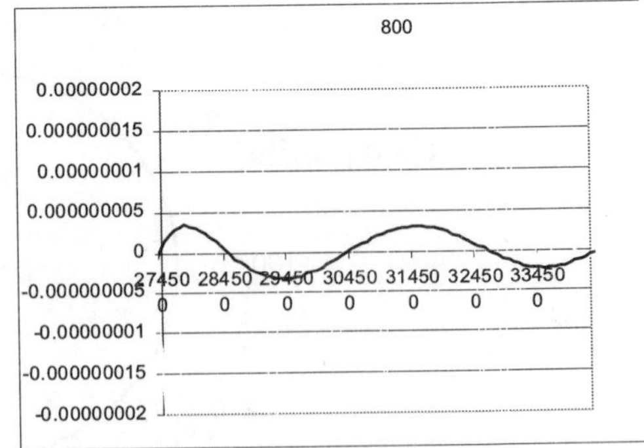

Fig. 13 (iv). J: jerk finite, 800 rpm, EN31.

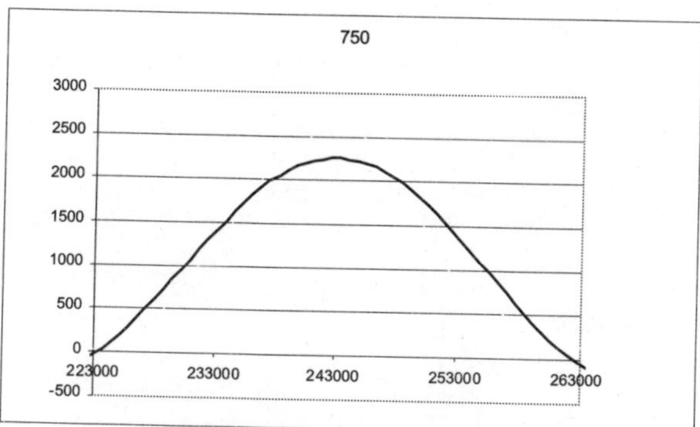

Fig. 14 (i). S: ping finite, 750 rpm, EN31.

Fig. 14 (ii). V: ping finite, 750 rpm, EN31.

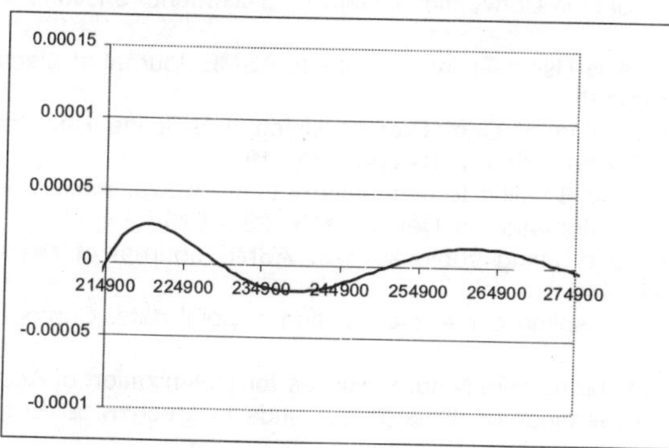

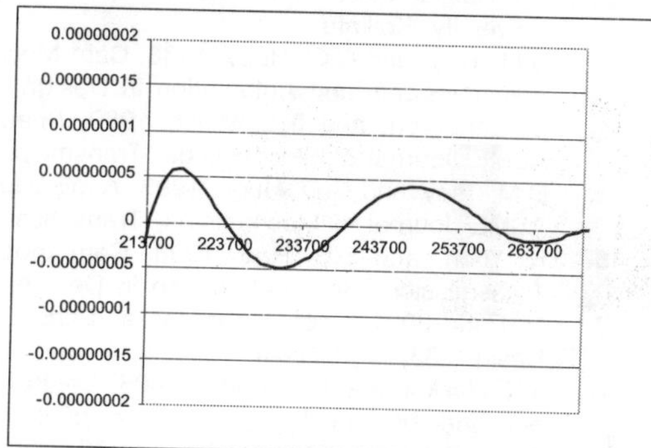

Fig. 14 (iii). A: ping finite, 750 rpm, EN31.   Fig. 14 (iv). J: ping finite, 750 rpm, EN31.

CONCLUSION

Ping finite function is designed with B-splines. Introduction of CPs in B-splines minimize the peak values of the acceleration and jerk. It only requires suitable manipulation of the CP parameter. Running optimization program does appropriate manipulation of the CP parameter. Comparing the Tables 2 and 4 we see that when both the AP and J_e or P_e are varied, the value of J_P becomes the lowest. Also we see that J_P value of ping finite function is not much higher than the J_P value of jerk finite function. So for the sake of smooth and noiseless operation of a cam-drive slightly higher J_P values may be acceptable.

The displacement, velocity, acceleration and jerk curves, which are obtained experimentally for jerk finite and ping finite are mostly similar in nature to the theoretical curves.

NOMENCLATURE

S = Cam displacement function
V = Follower velocity
A = Follower acceleration
J = Follower jerk
P = Follower ping

θ = Cam rotation angle in radian

h = Displacement at mid position

J_e = Jerk at end position

P_e = Ping at end position

A_P = Peak value of acceleration

J_P = Peak value of jerk

P_P = Peak value of ping

AP = Angular position of intermediate knots

$a_1 \ldots a_8$ = B-spline co-efficient

REFERENCES

1. L. Robert, and P.E. Norton, 2002, Cam Design and Manufacturing Handbook, Industrial Press, NY.
2. T.K. Naskar, 2005, Ph.D. Theses, Towards Synthesis of Non-Conventional Cam Displacement Functions, Jadavpur University, Kolkata.
3. D.M. Tsay, and C.O. Huey, 1988, Cam Motion Synthesis Using Spline Functions, ASME Journal of Mechanisms, Transmissions, and Automation in Design, 110, 161-165.
4. E. Sandgren, and R.L. West, 1989, Shape Optimization of Cam Profiles Using a B-Spline Representation, ASME Journal of Mechanisms, Transmissions, and Automation in Design, 111, 195.
5. D.M. Tsay, and C.O. Huey, 1993, Application of rational B-spline to synthesis of cam follower motion programs, ASME Journal of Mechanisms, Transmissions, and Automation in Design, 115, 621- 626.
6. K. Yoon, and S.S. Rao, 1993, Cam motion synthesis using cubic splines, ASME Journal of Mechanisms, Transmissions, and Automation in Design, 115, 441-446.
7. H. Park, 2001, Choosing nodes and knots in closed B-spline curve interpolation to point data, Computer Aided Design, 33(13), 967-74.
8. T.K. Naskar, and M. Mondal, 2005, Control Point Manipulation in Natural Splines for Optimization of Acceleration and Jerk of Cam Followers, Proc. of the All India Seminar on Emerging Trends in Mechanical Engineering, Allied Publishers, New Delhi, 141-46.
9. S. Doughty, 1987, Mechanics of Machines, John Wiley and Sons, Inc. NY, 120-156.